Circle

$(x-h)^2+(y-k)^2=r^2$

Parabola

$y=\frac{1}{4p}(x-h)^2+k$

Ellipse

$\frac{(x-h)^2}{a^2}+\frac{(y-k)^2}{b^2}=1$

Hyperbola

$\frac{(x-h)^2}{a^2}-\frac{(y-k)^2}{b^2}=1$

$\frac{(y-k)^2}{a^2}-\frac{(x-h)^2}{b^2}=1$

Arithmetic Sequence

General term: $a_n=a_1+(n-1)d$

Sum of first n terms: $S_n=\frac{n}{2}(a_1+a_n)$

$S_n=\frac{n}{2}[2a_1+(n-1)d]$

Geometric Sequence

General term: $a_n=a_1r^{n-1}$

Sum of first n terms: $S_n=\frac{a_1(1-r^n)}{1-r}$

Sum of terms of infinite series: $S=\frac{a_1}{1-r}$

Combinations

${}_nC_k=\binom{n}{k}=\frac{n!}{(n-k)!k!}$

Permutations

${}_nP_k=n(n-1)(n-2)\ldots(n-(k-1))$

Binomial Expansion Formu

$(x+y)^n=\sum_{i=0}^{n}\binom{n}{i}x^{n-i}y^i$

Sum of Constants Propert

$\sum_{i=1}^{n}k=nk$

Constant Factor Property

$\sum_{i=1}^{n}[k\cdot f(i)]=k\cdot\sum_{i=1}^{n}f(i)$

Sum of Terms Property

$\sum_{i=1}^{n}[f(i)+g(i)]=\sum_{i=1}^{n}f(i)+\sum_{i=1}^{n}g(i)$

Sum of Integer Series

$\sum_{i=1}^{n}i=\frac{n(n+1)}{2}$

Sum of Squares of Integers Series

$\sum_{i=1}^{n}i^2=\frac{n(n+1)(2n+1)}{6}$

Sum of Cubes of Integers Series

$\sum_{i=1}^{n}i^3=\left(\frac{n(n+1)}{2}\right)^2$

Probability of an Event A

$P(A)=\frac{n(A)}{n(S)}$

Probability of Mutually Exclusive Events

$P(A \text{ or } B)=P(A)+P(B)$

General Addition Rule of Probability

$P(A \text{ or } B)=P(A)+P(B)-P(A \text{ and } B)$

Probability of Complementary Events

$P(A')=1-P(A)$

Area and Perimeter of Several Plane Figures

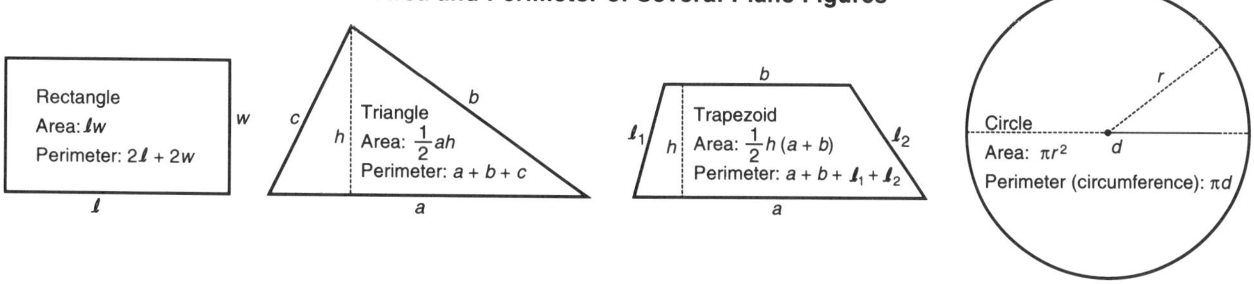

COLLEGE ALGEBRA
AND
TRIGONOMETRY
WITH **APPLICATIONS**

COLLEGE ALGEBRA
AND
TRIGONOMETRY
WITH APPLICATIONS

TERRY H. WESNER
HENRY FORD COMMUNITY COLLEGE

PHILIP H. MAHLER
MIDDLESEX COMMUNITY COLLEGE

WCB Wm. C. Brown Publishers
Dubuque, Iowa • Melbourne, Australia • Oxford, England

Book Team

Editor *Paula-Christy Heighton*
Developmental Editor *Theresa Grutz*
Production Editor *Eugenia M. Collins*
Designer *K. Wayne Harms*
Cover designer *Anna Manhart*
Photo Editor *Carrie Burger*

Wm. C. Brown Publishers
A Division of Wm. C. Brown Communications, Inc.

Vice President and General Manager *Beverly Kolz*
Vice President, Publisher *Earl McPeek*
Vice President, Director of Sales and Marketing *Virginia S. Moffat*
Marketing Manager *Julie Joyce Keck*
Advertising Manager *Janelle Keeffer*
Director of Production *Colleen A. Yonda*
Publishing Services Manager *Karen J. Slaght*
Permissions/Records Manager *Connie Allendorf*

Wm. C. Brown Communications, Inc.

President and Chief Executive Officer *G. Franklin Lewis*
Corporate Senior Vice President, President of WCB Manufacturing *Roger Meyer*
Corporate Senior Vice President and Chief Financial Officer *Robert Chesterman*

Cover photo © Bill Brooks/Masterfile

Copyedited by Patricia Steele

A Times Mirror Company

Library of Congress Catalog Card Number: 93–71067

ISBN 0–697–11655–7

Printed in the United States of America by Wm. C. Brown Communications, Inc., 2460 Kerper Boulevard, Dubuque, IA 52001

10 9 8 7 6 5 4 3 2 1

To Thomas Stewart Wesner

Dad

To Margot

Phil

Contents

Chapter 5 The Trigonometric Functions

Chapter 6 Radian Measure, Properties of the Trigonometric and Inverse Trigonometric Functions

Chapter 7 Trigonometric Equations

Chapter 8 Additional Topics in Trigonometry

Chapter 9 Exponential and Logarithmic Functions

CHAPTER 10

Systems of Linear Equations and Inequalities

CHAPTER 11

The Conic Sections

CHAPTER 12

Topics in Discrete Mathematics

Preface

Intent

This book is designed to prepare college students for the mathematics they need in the social sciences, computer science, business and economics, and physical sciences up to the precalculus level. It is also intended to serve a course that has as its objective an introduction to or review of what are currently called "precalculus" topics. In addition, some of those topics that are amplified in modern discrete mathematics and finite mathematics courses are introduced.

Assumptions

It is assumed that students have completed an intermediate algebra course, and therefore have been introduced to solving equations, factoring, radicals and graphing linear equations. It is also assumed students own a scientific calculator and are familiar with the basic keys for arithmetic computation. Keystrokes for a typical scientific calculator are presented where they go beyond the basic arithmetic operations.

Graphing calculators/computers

The text acknowledges the growing availability of graphing calculators. These are not yet so available or even accepted that we can assume their use throughout the text. Thus, the material, as presented, does not *require* these devices, and the directions for the examples and problem sets are suitable for situations with or without a graphing device.

The text specifically indicates how to use these devices in places where graphing calculators would be appropriate. Explicit references to graphing calculators are indicated by the graphic symbol which marks this section of the preface. Examples are presented based on the TEXAS INSTRUMENTS TI-81 graphing calculator. The Computer-Aided Mathematics section introduces the basic principles involved in using a graphing calculator, employing the TI-81 as an example.

Certain exercises require extensive (nongraphing) calculator use. These are noted with the symbol .

Several exercises invite the student to write a program for a programmable calculator or computer. These are indicated by the symbol that marks this paragraph. In addition, manuals on using graphing, programmable calculators, which are specific to this text, are available from the publisher. The manuals are calculator specific and available for the most prevalent calculators.

Content

Chapter 1, Fundamentals of Algebra, should be largely review. It is the experience of most teachers that this material needs to be explicitly covered, but at a rapid pace. Students who are not prepared for this treatment are in the wrong course, and will not finish the course in any case. It is better for such students to find out early and change to a more appropriate course rather than struggle to delay the inevitable.

Chapter 2, Equations and Inequalities, also has a large percentage of review content. The instructor should note that completing the square is not introduced here, and the quadratic formula is therefore not derived at this point. The formula is adequately justified, even from a theoretical viewpoint, as an exercise. Quadratic equations are not solved by completing the square, consistent with actual practice. Completing the square is delayed until chapter 3 and coverage of the equations of circles, where it is really necessary for the first time. The part of section 2–4 that treats nonlinear inequalities, as well as section 2–5 on equations and inequalities involving absolute value, will be new for many students.

Chapter 3, Relations, Functions, and Analytic Geometry, has several objectives. The first is to introduce functions. The second is to acquaint students with the graphs of straight lines, second-degree equations, and certain ''standard'' relations. The third objective is to introduce the use of linear translations and symmetry to graph appropriate relations. The fourth objective is to present the idea that analytic geometry and plane Euclidean geometry are parallel concepts, with the former modeling the latter.

Functions are introduced as sets of ordered pairs, not as an abstract function machine or rule of correspondence. The transition from functions as sets of ordered pairs to an algebraic viewpoint is smooth and natural. The viewpoint of functions as sets of ordered pairs expedites teaching (and understanding) some of the harder concepts of functions, such as the one-to-one property and inverse functions, in both this and future chapters.

Chapter 4, Polynomial and Rational Functions, and the Algebra of Functions, introduces the more important features of polynomial and rational functions as well as the ''algebra'' of functions. The interrelationship between zeros of a function, factors of a function, and x-intercepts of a graph is stressed. The section on the decomposition of rational functions (partial fractions) could be skipped entirely without impact on the following chapters.

Chapter 5, The Trigonometric Functions, begins with the trigonometry of right triangles and degree measure. This has the advantage of introducing substantive applications and building on things students probably already know. Analytic trigonometry, applied to any angle in standard position, is introduced next. This chapter also includes an early introduction to trigonometric equations, both identity and conditional.

Chapter 6, Radian Measure, Properties of the Trigonometric and Inverse Trigonometric Functions, first introduces radian measure for angles and arc length. Graphs and properties (i.e., periodicity) of the trigonometric functions are treated next. Period and phase shift are not treated separately but are integrated into one step. It is our experience that the graphing of quite complicated functions is easily and quickly taught and learned by the method presented here. The chapter finishes with the inverse trigonometric functions. Those types of applications necessary for the calculus are stressed.

Chapter 7, Trigonometric Equations, is a full treatment of both conditional equations and identities. It builds on the experience from chapter 5.

Chapter 8, Additional Topics in Trigonometry, covers the solution of oblique triangles using the laws of sines and cosines. It then treats vectors, complex numbers in polar form, and polar coordinates.

Chapter 9, Exponential and Logarithmic Functions, is a modern treatment of these functions. That is, only a passing reference is made to tables and to applications in calculation. Instead, the use of these functions to describe and model phenomena in the sciences and business world is stressed.

Chapter 10, Systems of Linear Equations and Inequalities, introduces these topics. The concept of elimination for solving linear systems is stressed. This is the method of choice in all applied areas. Solution by substitution is not covered in this chapter; it is covered in section 3–2 (linear systems) and chapter 11 (nonlinear systems of equations). The fundamentals of matrices and determinants, matrix algebra, and Cramer's rule are also covered; these are also important for more advanced work in this area and in courses that many students will take in other disciplines. Solving systems of equations via matrix equations is covered. There are now calculators that perform matrix arithmetic (the TI-81 is used to illustrate), and thus solving systems of equations in this way has become more important than in the past. Linear programming is introduced in this chapter for what it is, an important justification for studying systems of linear inequalities, and a topic that would be referred to in business and management courses.

Chapter 11, The Conic Sections, provides a solid introduction to the conic sections and includes a treatment of nonlinear systems of equations and the substitution method of solution of systems of equations.

Chapter 12, Topics in Discrete Mathematics, begins with the important topics of sequences and series. Section 12–3, after introducing the binomial theorem, provides more experience with manipulation of sigma notation. This is important in future courses in mathematics and economics and is quite important for computer science majors, but may be omitted without impact on future sections. Finite induction follows. In addition to higher mathematics, this topic is used in computer science in the analysis of algorithms. An introduction to combinatorics and probability is next. The final section is directed at computer science majors, and includes material that has become popular only in the last decade.

Appendixes

Appendix A is the development of several lengthy formulas, which may or may not be used by the instructor. Its inclusion within the text does not promote reading the text by the student or a better understanding of the material. Appendix B gives the answers to odd-numbered exercise problems and to all chapter review and chapter test problems. Solutions are provided for trial exercise and skill and review problems. Appendix C provides graphical material that students may want to reproduce and have handy as an aid in studying.

Features

Clarity

- Exposition—An attempt has been made to write the exposition of the material in clear, logically sequenced, understandable prose. Where the exposition builds on material from an earlier section, that section is referenced.
- Structure—The structure of each section of a chapter is designed to provide both easy reading and clear examples of the skills that students are expected to master in that section. The mix of prose and examples is designed to flow smoothly and make explicit those skills that are most important.
- Mastery points—At the end of each section is a list, called *Mastery points,* of skills from that section that students are expected to master. The testing package to support this text is based on the mastery points.

Problem sets

- Problem solutions—The answers to all problems, except even-numbered exercise problems, appear at the end of the text. The complete solutions to selected problems, indicated by boxing the problem number, also appear at the end of the text.
- Core problems—Students with a strong background in a particular topic may profit by choosing to work just those problems whose numbers are in color. In those parts of the exercises where there are sequences of similar problems, these are a subset of the problems that are sufficient to exercise the skills required for that group of problems.
- Review problems—Each problem set ends with skill and review problems. These either review old material or prepare for the next section. The solutions to these problems appear in appendix B.
- Progressive difficulty—The exercises progress from straightforward application of the material covered in the exposition to problem solving via more difficult application problems and then to problems that require some ingenuity and creativity. Those problems requiring exceptional ingenuity or amount of work are marked with the symbol . Students who confront the complete range of problems will have a good introduction to the way in which this material is bent, shaped, and modified to serve a variety of disciplines.

Review

The skill and review problems review the primary skills from previous sections and chapters. These problem sets reinforce recognition of the type of problem and appropriate solution procedures, not to provide drill of skills. These problem sets are kept short so that their presence is an aid to progress, not an impediment. Every student is most concerned with the current material, and a long review set will not be used. The solutions to these problems also appear at the end of the text.

Closure, the last link

Each chapter terminates with a *chapter summary,* which represents the highlights of that chapter, a *chapter review,* which presents review problems from that chapter, keyed to sections, and a *chapter test,* designed to help the student practice the material as it might appear on a test, out of the context of each section. The chapter test may well be longer than what would be confronted in class—its objective is to provide material out of the context of being surrounded by similar problems. In the exercise sets, there are inevitably many clues to the method of solution, including nearby problems and temporal and physical proximity to explanations. The chapter test is an aid to make that last link in learning—recognition of problem type, with attending method of solution. The answers to all chapter review and chapter test problems appear at the end of the text.

Applications

As the title of the text indicates we have tried to provide a cross section of applications, chiefly in the problem sets. These are not intended to be obtrusive on the coverage of the mathematics, but to motivate students with a variety of interests and goals by showing that this material is used in many disciplines.

Certain applications, such as mixture problems, are considered integral to a mathematics course of this type. These are fully treated. Note that applications problems that are beyond the typical syllabus can always be done without mastering the problem domain (physics, business, finance, etc.) of the application.

Mathematics in culture

We have attempted to provide an historical aspect to the material by frequent but unobtrusive references to relevant names and dates, as well as an occasional problem taken from ancient and non-Western cultures. It is hoped that students will come to appreciate that mathematics has a historical and cultural side as well as its ''applied'' side.

Mathematics ability is not in the genes

Many other industrialized cultures understand better than ours that *anyone* can do mathematics. Americans tend to feel that difficulties in learning mathematics indicate lack of ability and that mathematics should therefore be avoided by an individual encountering problems. Other cultures react by expecting the individual to work harder, and anyone who reads the newspaper has seen reports that other cultures are correct—anyone can do mathematics.

We tend to apply this misunderstanding mostly, although not exclusively, to gender. The authors have tried to be totally nonsupportive of this misconception by using a ''gender-free'' exposition throughout the text, including problem sets. Except where discussing a specific individual, the gender-oriented pronouns are scrupulously avoided.

We have also tried to avoid cultural bias. For example we do not assume students understand simple interest, perimeter and area, or even the makeup of a standard deck of playing cards.

The electronic age is here

It is assumed throughout the text that the student always has access to a scientific or business calculator, with, as a minimum, the trigonometric and logarithmic functions. No special treatment is given to calculators–the National Council of Teachers of Mathematics' (NCTM) *Standards* advises that they be a part of every student's tool kit from the fifth grade on. We do acknowledge the modern orientation toward viewing numbers in decimal form by frequently providing the decimal approximation of answers where appropriate, such as for intercepts of graphs and answers to applications.

Pedagogical highlights—to the teacher

We want this to be a text that works. It does this if a student, with the correct level of preparation, who is willing to work, masters the material at a reasonable level of competence, without stress caused by the text itself. We have attempted to remove those things that a text can do to introduce unwarranted stress. Some identifiable characteristics for doing this follow.

Exposition

Most students do not rely much on the exposition of the material in a textbook. This is the single largest duty of the instructor. However, when a student is motivated, or has missed a class, or has not understood the instructor, the student will hopefully turn to the text. We have been careful to make the exposition clear and logical, without making undue assumptions about prior knowledge. We have worked hard to provide a logical flow of the material. When a skill previously covered is revisited, a reference to that part of the text is presented.

The predominant goal of a course at this level is skill building. We have been careful to explicitly present a sequence of steps that applies to a particular class of problem whenever possible. These are found labeled and in boxes throughout the text. Please do not confuse this material with ''spoon feeding'' students. Most students profit from explicitly seeing the steps they are performing. To leave these steps out is to teach by example, leaving the student to deduce the steps being performed. The fact that some students, and most mathematics teachers, can learn this way does not justify teaching in this manner.

Examples

For students, this is the heart of the text. In this text the examples are carefully designed and graded to correspond well to the skills being covered and to the problem sets. The examples often provide asides that indicate what is being done at certain steps.

Problem sets

Perhaps the largest source of discouragement in a mathematics course is attempting to work a homework assignment without success. We have designed the problem sets to avoid this. Problems that emphasize skills are similar to the examples, and they are carefully graded. The solutions to a representative subset of the problems are included in appendix B. These problems are indicated in the problem sets by having the problem number in a box.

In addition, *all* of the problems in this text were solved by the authors themselves. This is the best way to ensure that the problem sets are useful and commensurate with the exposition of the material in the text. Also, many of the chapters were explicitly classroom tested in prepublication form, and all of the chapters have profited from the experience we have gained in writing previous textbooks.

Success on tests

There are several levels of understanding that must be present for students to be successful on a mathematics test, and therefore to succeed in a mathematics course. This text is designed to support students and teachers in achieving this understanding. The lowest level of understanding is pure skill. Given a certain class of problems, can students apply the procedures that solve these problems. This level of understanding is arrived at by drill. As in any good textbook the exercises in the problem sets are predominantly of this nature. These types of problems parallel specific examples in the chapter. Students require examples to learn skills. It is the responsibility of the text to provide them, in a clear, explicit manner.

The next level of understanding required is the ability to apply skills to problems where the sequence of procedures used is not clearly defined. The later problems in the exercise sets are of this nature.

The next level, which is insufficiently recognized as essential, is the ability to look at a problem, out of the context of the text, and determine which set of procedures should be applied. The student who attempts to ''solve'' the expression $x^2 - 4$, even given the instructions that often accompany a problem, has not been able to classify this problem as a factoring problem and not an equation-solving problem. This level of understanding is supported here in two ways. First, the chapter review problems are temporally separate from the coverage of the material. These problems are keyed by section, however, and there are usually several exercises that apply to each problem classification. Thus, there are still some clues as to methodology that the student will not have on a test. The chapter test is designed to provide problems in a context containing as few clues as encountered on an exam. *This makes the chapter test an integral part of the learning process*—it is where the student obtains the ability to classify problems by type, and to recall the appropriate procedures for solving.

Critical thinking

The final level of understanding (in the implied taxonomy begun under "Success on Tests") is critical thinking or problem solving. The NCTM *Standards* ask for more of this in our mathematics education system. The implications for a mathematics course at this level are still fuzzy. Problem solving presupposes a minimal level of skill, and the primary function of this text is skill building. It would be misleading to say that this text presents problem solving in the sense of presenting "real-world" applications in which the problem domain must first be mastered. This type of material requires considerable time to cover in any meaningful way, and we have never met a mathematics teacher who felt that there was enough time in a typical curriculum to even sufficiently cover basic skills. Also, this type of material can be very disconcerting for students who do not already have a knowledge of the problem domain in the application: physics, finance, chemistry, etc.

The problem sets do contain problems, toward the end of the problem set, that involve problem solving and critical thinking, but in the context of material covered in this text. Some of these problems require real synthesis of what has been covered, and we have tried to make these problems interesting but achievable.

We believe that if a student can work, or even seriously engage, these problems, that student will be able to apply the skills learned in this text to other problem domains—in a physics, finance, or biology course—where the required subject background is obtained.

Maintaining interest

Most students feel that mathematics is a plot designed to make education painful. (If you don't believe this, ask them.) To feel good about mathematics, students need to know that they will use it in whatever career path they choose.

We have tried to show, unobtrusively, that mathematics is used across all disciplines, both by explicit statements in the text and by problems selected from many disciplines. Also we have tried to show that there is a long history to mathematics, and that it has been important across time and across cultures.

We have also tried to not stifle interest! This comes from poor exposition, unpleasant surprises in homework sets (i.e., inappropriate problems), and culturally inappropriate material (assumptions about background, sexist writing). The features of this text (above) discusses these issues.

Graphing calculators

The increased use of graphing calculators is a positive development for mathematics. It can free the student from the necessity of "plotting" points—always a tiresome and error prone process. Graphing calculators also provide new ways to visualize solutions to certain classes of problems. This text welcomes these devices for those who have access to them. The teacher will see that the text can be easily used to allow the free use of graphing calculators. Instructions in problem sets demand that such important things as asymptotes and intercepts be stated. This requires an active participation by students in completing a graph whose basic shape has been determined by a pattern in silicon crystals, while removing some drudgery.

We have tried to create a happy medium with regard to this new technology. We think we can throw out the bath water and keep the baby.

Bibliography

The following books and articles have been particularly inspirational in writing this text. This is in addition to the books and articles cited within the text itself.

Computer-Aided Mathematics

Introduction

The increasingly widespread availability of electronic computing and graphing devices is causing the mathematics community to rethink every aspect of mathematics education. The computer age has led to two developments that have a contradictory character. First, mathematics is used more than ever in all areas of human knowledge, and is therefore more important than ever. Second, computers can do a lot of the mathematics that formerly had to be done by hand.

Unfortunately, as the technology develops, different people have access to different levels of computing power. This has put many mathematics teachers in a quandary about what to teach, what to stress, and what, in terms of technology, to allow.

In this book we are taking a middle road. We think that graphing calculators should be used whenever they are available. We present the material in a calculator-independent fashion, for those who do not yet have access to graphing calculators, but we also show where and how a graphing calculator can be used.

The graphing calculator

As of this writing, there are a half-dozen graphing calculators on the market. The proliferation of new models and features is guaranteed. We show many examples based on the Texas Instruments TI-81 graphing calculator throughout the text. It is popular, easy to use, and similar to other brands in its use. We must assume the student will learn the specifics about a calculator from its manual.

In addition to numeric calculations, all graphing calculators have a few graphing capabilities that we use extensively:

- setting the ''range'' for the screen,
- graphing an equation in which y is described in terms of x,
- tracing and zooming, and
- finding an x- or y-intercept.

We describe the first three capabilities in this introduction, and illustrate how they are accomplished with a TI-81 graphing calculator. Finding an x- or y-intercept is discussed in section 3–1.

All students at this level have had at least an introduction to the x-y coordinate system. This topic is more formally developed in section 3–1—here we present the bare essentials very informally, by example, which is enough to describe using the graphing calculator in the first few chapters.

Set the range for the screen

We graph using the x-y rectangular coordinate system. Recall that an **ordered pair** is a pair of numbers listed in parentheses, separated by a comma. In the ordered pair (x,y) x is called the *first component* and y is called the *second component;* $(5,-3)$, $(9,3)$, $(4,\frac{2}{3})$ are examples of ordered pairs. The graphing system we use is formed by sets of vertical and horizontal lines; one vertical line is called the y-axis, and one horizontal line is called the x-axis. The geometric plane (flat surface) that contains this system of lines is called the *coordinate plane.* See figure 1.

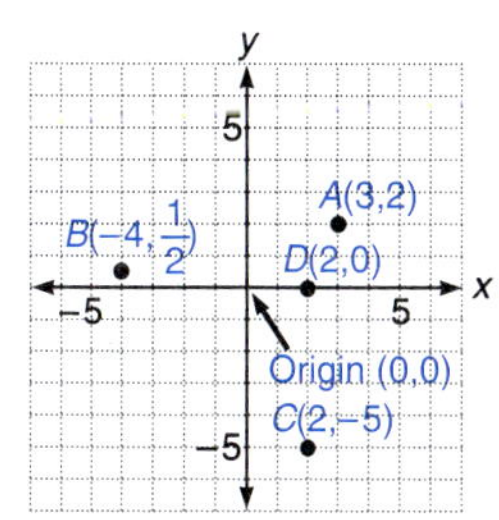

Figure 1

The **graph** of an ordered pair is the geometric point in the coordinate plane located by moving left or right, as appropriate, according to the first component of the ordered pair, and vertically a number of units corresponding to the second component of the ordered pair.

The graphs of the points $A(3,2)$, $B(-4,\frac{1}{2})$, $C(2,-5)$, and $D(2,0)$ are shown in the figure. The first and second elements of the ordered pair associated with a geometric point in the coordinate plane are called its **coordinates.**

Graphing calculators have a way to describe which part of the coordinate plane will be displayed. It is called setting the RANGE. Using the [RANGE] key shows a display similar to that in table 1. The Xmin and Xmax values refer to the range of x values that will be displayed. The Ymin and Ymax values refer to the range of y values that will be displayed. The Xscl and Yscl values refer to the tick marks that will appear on the screen. The Xres refers to the number of x values that will be calculated. It should be left at 1. Throughout the text we show the Xmin, Xmax, Ymin, and Ymax values, in this order, in a box labeled RANGE. For the values shown above we would write [RANGE −10,10,−10,10]. Unless otherwise stated we assume Xscl = Yscl = 1.

Range
Xmin=−10
Xmax=10
Xscl=1
Ymin=−10
Ymax=10
Yscl=1
Xres=1

Table 1

By entering numeric values and using the [ENTER] key to move down the list, the values in the RANGE can be changed. Note that to obtain a negative number the [(−)] (change sign) key is used, not the [−] (subtract) key.

Figure 2 shows the screen appearance for various settings of Xmax, Xmin, Ymax, Ymin. Xscl and Yscl are 1 except where labeled Yscl=3 and Xscl=2. After setting these values with the [RANGE] key, use the [GRAPH] key to show the screen. Using the [CLEAR] button readies the calculator for numeric calculations again. The settings in part (a) of figure 2 are the "standard" settings, obtained by selecting [ZOOM] 6.

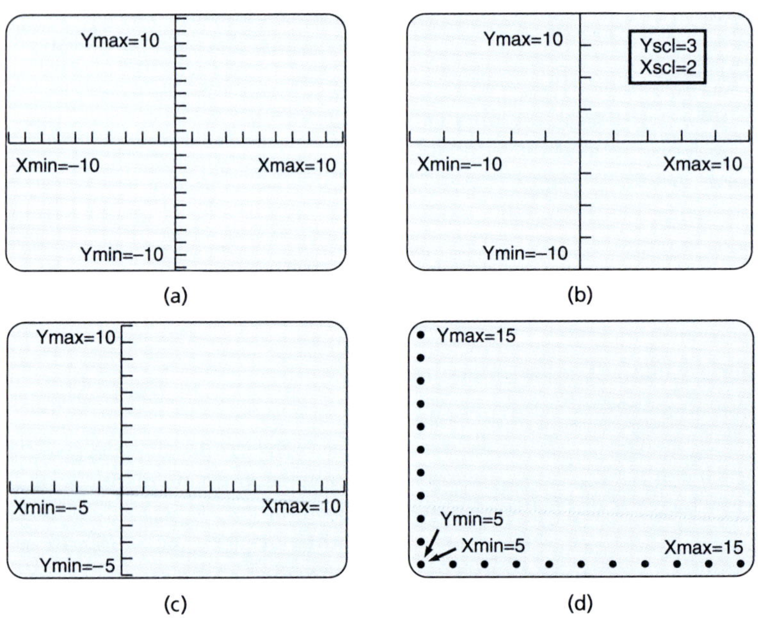

Figure 2

Observe that the distance between units is not the same on the screen. The calculator automatically makes horizontal units 1.5 times as long as vertical units. To have horizontal and vertical distances the same, use the [ZOOM] function, where option 5 says SQUARE. This makes the screen use the same scale for distance vertically and horizontally by changing the values of Xmin and Xmax.

Graph an equation in which y is described in terms of x

If an equation describes values for a variable y in terms of a variable x, the graphing calculator can be used to view the graph of the equation.

Example A

Graph each equation.

1. Graph $y = 2x - 3$.

This could be done without a graphing calculator with practically no knowledge of graphing by a table of values, by letting x take on many values, such as $-3, -2, -1, 0, 1, 2, 3$, etc., and computing y for each one. In fact, this table is shown here. The y values are computed by computing $2x - 3$ for the given x value. Each pair of values for x and y represents an ordered pair (x,y) (we always write the x value first). If we plot enough of these values in a coordinate system we start to see a picture emerge. In this case it is a straight line.

x	$y\ (2x-3)$
−3	−9
−2	−7
−1	−5
0	−3
1	−1
2	1
3	3

Of course, the point of this section is to have the calculator automatically calculate the x and y values and plot them. Assuming the described standard RANGE settings, proceed as follows to obtain the graph:

[Y=]	Allows us to enter up to four equations
[2]	
[X|T]	The variable x
[−]	
[3]	The display looks like
[GRAPH]	

:Y_1=2X−3
:Y_2=
:Y_3=
:Y_4=

Note If there are any equations already entered for Y_1 use the [CLEAR] key before entering the equation. If there are any extra equations entered for Y_2, Y_3, or Y_4, move down with the down arrow key, ▼ to that equation and use the [CLEAR] key to clear that entry. The figure shows what the display will look like.

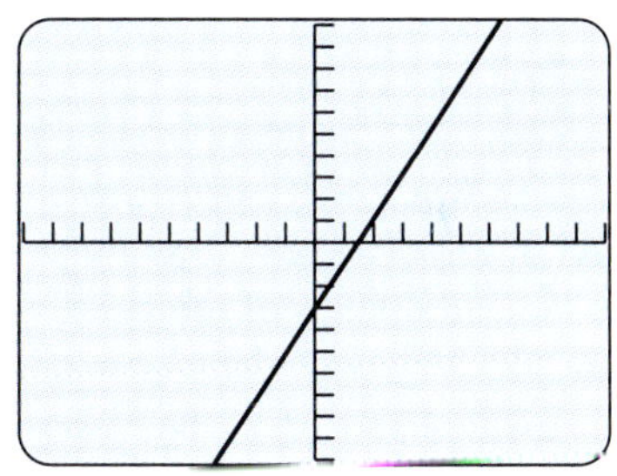

2. Graph $y = x^3 - 4x^2 + x + 1$.

The following steps would produce a graph similar to that shown in the figure.

Steps	Explanation
Enter the x and y-axis limits.	
[RANGE]	
[(−)] 2 [ENTER]	Xmin becomes −2.
5 [ENTER]	Xmax becomes 5.
1 [ENTER]	Xscl becomes 1.
[(−)] 6 [ENTER]	Ymin becomes −6.
4 [ENTER]	Ymax becomes 4.
1 [ENTER]	Yscl becomes 1.

Enter the function into Y_1.

Y= X|T MATH 3 − 4 X|T x^2 +
X|T + 1 GRAPH

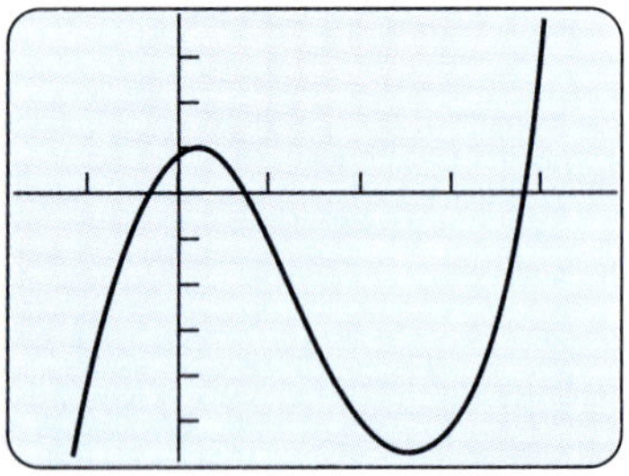

RANGE −2,5,−6,4 ■

Tracing and zooming

A calculator's trace capability displays the x- and y-coordinates for a particular point. A zoom capability allows us to expand a graph around some particular point. We illustrate this here to find approximate values for the coordinates of the point where two straight lines cross.

Example B

Graph the two straight lines $y = 2x - 3$ and $y = -\frac{1}{3}x + 2$. Then estimate the coordinates of the point where these two lines cross.

To graph both lines using standard RANGE settings, proceed as follows:

Y = CLEAR 2 X|T − 3
ENTER CLEAR
((−) 1 ÷ 3) X|T + 2
ZOOM 6 Standard settings

The graph shown appears.

Using the trace feature we can position the box around the point of intersection of these two lines: select TRACE and then use the ▶ and ◀ keys to move the blinking box as close to the point of intersection as possible. The display shows x is about 2.2 and y is about 1.4.

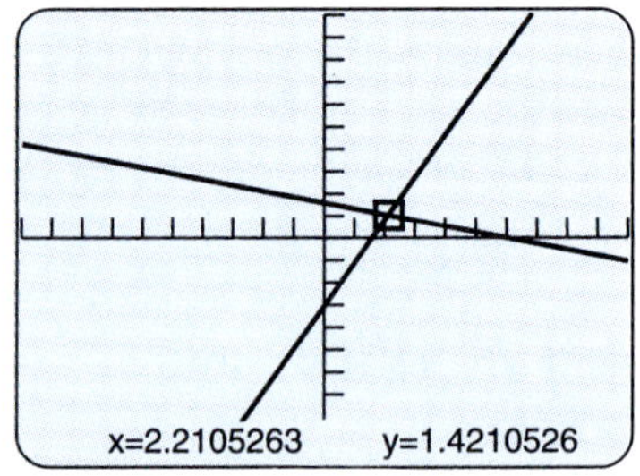

Using ZOOM 2 ENTER (zoom in) expands the graph around the point selected using the trace feature. It produces a new graph. Tracing shows that the coordinate of the point where the lines cross is about (2.16,1.32).

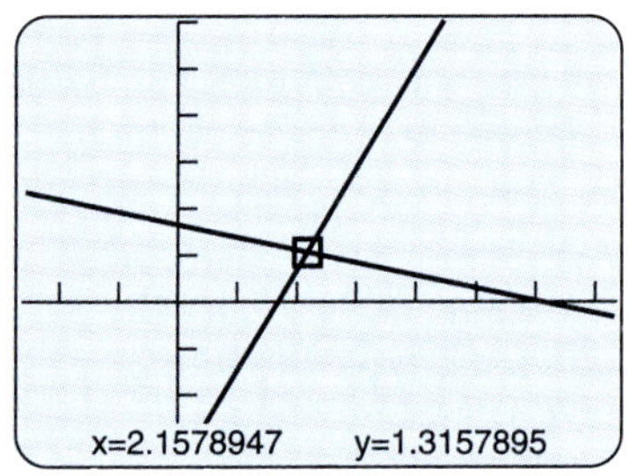

Repeatedly zooming and tracing will show that $x \approx 2.14$ and $y \approx 1.28$. Using methods shown in section 3–2 we could show that x is *exactly* $2\frac{1}{7}$ and y is exactly $1\frac{2}{7}$. ■

At this point you should have some idea about how to

- set the RANGE for a graph,
- graph an equation in which y is expressed in terms of x, and
- use trace and zoom to expand a particular part of a graph.

These capabilities and others are shown throughout the text, wherever they are appropriate.

C H A P T E R 1

Fundamentals of Algebra

This chapter reviews the basics of algebra. Much more can be said about everything in this chapter, but we have focused on what is essential for the rest of this book and for the mathematics most students encounter in other college courses.

1–1 *Basic properties of the real number system*

Suppose the postage (in cents) for a certain category of mail is as follows:

Maximum weight (in ounces)	2	3	5	10
Postage (in cents)	15	20	30	40

Above 10 ounces, the rate is 3.5 cents per ounce.

If we wanted to write a computer program to compute the postage for a given item we would have to be able to describe this situation in a mathematical way. A way of doing this is presented in this section.

The set of real numbers

The terminology of sets is often used to describe the *real number system. This is the number system we will use most of the time in this text.*

Set
A set is a collection of things. These things are called elements of the set.

The **natural numbers** is the set $N = \{1, 2, 3, . . .\}$, which is read "N is the set of numbers one, two, three, etc." Sets are often indicated symbolically using braces "{" and "}". The three dots . . . , called an **ellipsis,** represents the repetition of a pattern.

We use the symbols[1] ε, which is read "is an element of," and ∉, which is read "is not an element of." Thus for example $3 \in N$ (3 is an element of N) but $0 \notin N$ (zero is not an element of N).

The set of **whole numbers** is $W = \{0, 1, 2, 3, \ldots\}$ and the set of **integers** is $J = \{\ldots, -3, -2, -1, 0, 1, 2, 3, \ldots\}$. Every element of W is also an element of J. We say that W is a **subset** of J.

We specified the sets above by listing some of their elements, but to be able to express certain sets of numbers we need the concept of a *variable* and of *set-builder notation.* Generally a **variable** is represented by a lowercase letter, such as x or y. *A variable is a symbol that represents an unspecified element of a set that contains two or more elements.* This set is called the **replacement set.** For example,

$$\{x \mid x \in N \text{ and } x < 6\}$$

is verbalized "the set of all elements x such that x is a natural number less than six." (Remember, $<$ means "less than.") This is equivalent to the set $\{1, 2, 3, 4, 5\}$. Here x is a variable since it may represent any one of the values 1, 2, 3, 4, 5.

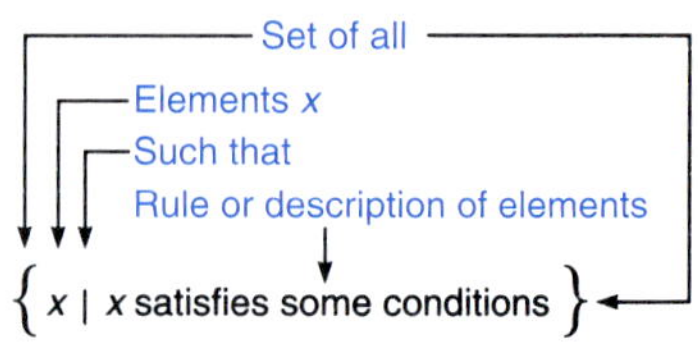

Figure 1–1

In general **set-builder notation** has the pattern shown in figure 1–1.

■ *Example 1–1 A*

Describe each set by listing the elements of the set.

1. $\{x \mid x \in W \text{ and } x > 3\}$
$= \{4, 5, 6, \ldots\}$ — The first element in W greater than 3 is 4

2. $\left\{\dfrac{a}{a+1} \,\middle|\, a \in \{2, 5, 8\}\right\}$ — Replace a in $\dfrac{a}{a+1}$ by 2, 5, 8

$$= \left\{\frac{2}{2+1}, \frac{5}{5+1}, \frac{8}{8+1}\right\} = \left\{\frac{2}{3}, \frac{5}{6}, \frac{8}{9}\right\}$$ ■

The set of **rational numbers** is $Q = \left\{\dfrac{p}{q} \,\middle|\, p, q \in J, q \neq 0\right\}$. That is, the rational numbers are written as quotients, $\dfrac{p}{q}$, of integers, where the denominator, q, may not be zero.

Every rational number has a decimal form, found by dividing the denominator into the numerator (most conveniently with a calculator). *It can be proven that the decimal form of any rational number either terminates or repeats.*

■ *Example 1–1 B*

Write the decimal form of each rational number.

1. $\frac{5}{8} = 0.625$ — Terminating decimal

2. $\frac{2}{7} = 0.285714\overline{285714}$ — Repeating decimal ■

[1] ε is the Greek letter epsilon.

Thus we could also say that

$$Q = \{x \mid \text{The decimal representation of } x \text{ terminates or repeats}\}$$

This way of describing the rational numbers gives us the tool to define the irrational numbers. The **irrational numbers** are denoted by *H,* where

$$H = \{x \mid \text{The decimal representation of } x \text{ does not terminate or repeat}\}$$

It can be shown that an irrational number cannot be expressed as a quotient of two integers (i.e., a rational number); in fact ''irrational'' means ''not rational.''

Examples of irrational numbers are $\sqrt{5}$ (square root of 5), $-\sqrt[3]{100}$ (opposite of the cube root of 100), and π (pi).[2] The decimal values of these numbers can be approximated using a scientific calculator as shown.

Irrational number	Calculator approximation	Typical calculator keystrokes
$\sqrt{5}$	2.236067977	5 [$\sqrt{x}$] TI 81: [2nd] [x^2] 5 [ENTER]
$-\sqrt[3]{100}$	−4.641588833	100 [$\sqrt[x]{y}$] 3 [=] [+/−] TI 81: [(−)] [MATH] 4 100 [ENTER]
π	3.141592654	[π] TI 81: [2nd] [∧]

The set of **real numbers** is $R = \{x \mid x \in Q \ \textbf{ or } \ x \in H\}$. That is, the real numbers are all numbers that can be represented by repeating, terminating, or nonrepeating and nonterminating decimals.

Note In this text, whenever a specific replacement set for a variable is not indicated it is understood that the replacement set is *R*.

The real number line

The **number line** (figure 1–2) helps visualize the set of real numbers. We assume that for any point on a line there is a real number, and that for every real number there is a point on the line. The number associated with a point is called the **coordinate** of that point, and the point is called the **graph** of the number. Numbers to the right of 0 are called **positive,** while those to the left are **negative.** The graph of the value zero is called the **origin.** Figure 1–2 shows the graphs for the coordinate values $-2\frac{1}{3}$, $-\sqrt{2}$, 2, 2.5, and π.

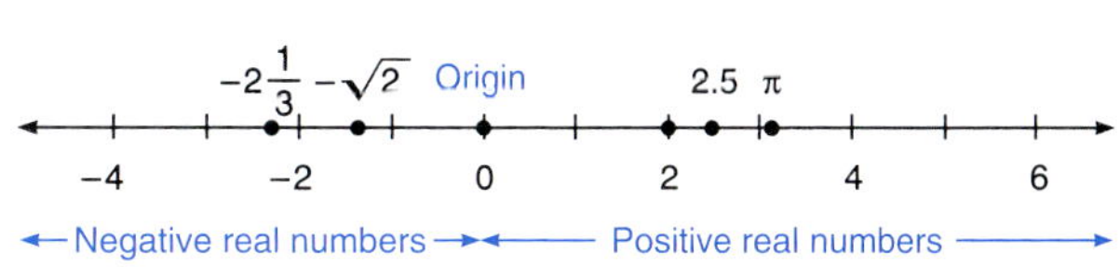

Figure 1–2

[2]The use of the Greek letter pi to represent the number it represents was introduced in 1706 in a book by the Englishman William Jones.

Properties of the real numbers

There are certain rules, called axioms, that we assume all real numbers obey. In the following axioms we assume that there are two operations called addition and multiplication. The variables a, b, and c represent any real numbers; $a + b$ represents their sum, a real number; and ab or $a \cdot b$ represents their product, a real number.[3]

Axiom	For addition	For multiplication
Commutative	$a + b = b + a$	$ab = ba$
Associative	$a + (b + c) = (a + b) + c$	$a(bc) = (ab)c$
Identity	There is a unique number 0 such that $a + 0 = a$.	There is a unique number 1 such that $a(1) = a$.
Inverse	For every a there is a value $-a$ such that $a + (-a) = 0$.	For every a except 0 there is a value $\frac{1}{a}$ such that $a\left(\frac{1}{a}\right) = 1$.
Distributive	$a(b + c) = ab + ac$	

Note 0 is called the additive identity, 1 is the multiplicative identity, $-a$ is the additive inverse of a, and $\frac{1}{a}$ is the multiplicative inverse of a.

We make the following definitions using the properties above.

Subtraction: $a - b = a + (-b)$

Division: $a \div b = a\left(\frac{1}{b}\right)$

Factor: In the product ab, a and b are called factors.

Term: In the sum $a + b$, a and b are called terms.

Like terms: Terms with identical variable factors. Like terms can be combined by combining the numerical factors. For example, $2a + 3a = 5a$, and $-3xy^2 + xy^2 = -2xy^2$.

Note $3xy$ and $3xy^2$ are not like terms and cannot be combined into one term using addition or subtraction.

Expression: A meaningful collection of variables, constants, grouping symbols, and symbols of operation. For example, $x + 3(5 - y)$ is an expression.

Natural number exponents: If $n \in N$ then a^n means n factors of a. For example, x^4 means $x \cdot x \cdot x \cdot x$.

Order of operations: Expressions with mixed operations should be performed in the following order:

1. Operations within symbols of grouping (parentheses, fraction bars, radical symbols)
2. Indicated powers and roots

[3]The juxtaposition of symbols, as in xy, to indicate multiplication is due to René Descartes (1637).

3. Multiplications and divisions from left to right
4. Additions and subtractions from left to right

Fraction: A fraction is an expression of the form $\frac{a}{b}$, $b \neq 0$.

Operations for fractions: (assuming $a, b, c, d \in R$ and $c, d \neq 0$)

$$\frac{a}{c} + \frac{b}{c} = \frac{a+b}{c} \qquad \frac{a}{c} + \frac{b}{d} = \frac{ad+bc}{cd}$$

$$\frac{a}{c} - \frac{b}{c} = \frac{a-b}{c} \qquad \frac{a}{c} - \frac{b}{d} = \frac{ad-bc}{cd}$$

$$\frac{a}{c} \cdot \frac{b}{d} = \frac{ab}{cd} \qquad \frac{a}{c} \div \frac{b}{d} = \frac{a}{c} \cdot \frac{d}{b},\ b \neq 0$$

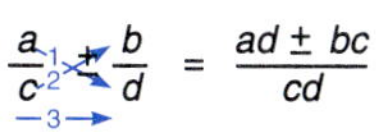

Figure 1–3

The patterns for the addition and subtraction of fractions with unlike denominators can be viewed as three multiplications, indicated in figure 1–3 by arrows (1, 2, and 3) in the order in which the products are formed. The symbol ± is read "plus or minus."

■ *Example 1–1 C*

Perform the indicated operations.

1. $5x^2 \cdot 2x^3$
$= 10(x \cdot x)(x \cdot x \cdot x) = 10x^5$

2. $-4x^2y + 6x^2y - 2xy$
$= 2x^2y - 2xy$

3. $\frac{2x}{y} \cdot \frac{3x}{4y} = \frac{\cancel{2}^1 x}{y} \cdot \frac{3x}{\cancel{4}^2 y}$ Divide out common factors
$= \frac{3x^2}{2y^2}$

4. $\frac{2x}{y} + \frac{3x}{4} = \frac{(2x)(4) + (y)(3x)}{4y}$ Use $\frac{2x}{y} \times \frac{3x}{4}$
$= \frac{8x + 3xy}{4y}$ ■

An expression like -3^2 often causes confusion. It is *not* the same as the expression $(-3)^2$.

$$-3^2 \quad \text{means} \quad -(3^2) \quad \text{which is } -(3 \cdot 3) = -9$$

That is, *square the value 3 first,* then take the opposite of the result.

$$(-3)^2 \quad \text{means} \quad (-3)(-3) = 9.$$

That is, *change the sign of 3 first,* then square the result. For example,

$$15 - 3^2 = 15 - 9 = 6.$$

On a calculator,

-3^2 is calculated as 3 [x^2] [+/−]

$(-3)^2$ is calculated as 3 [+/−] [x^2]

Order

The set of real numbers is ordered. That is, for any two distinct numbers, one is greater than the other. If value a is greater than value b we write $a > b$. The symbol $>$ is read "**is greater than.**" Similarly, $a < b$ means that the value a **is less than** the value b. These symbols of inequality are called **strict inequalities.** The symbols $\geq$ and $\leq$ are read "**is greater than or equal to**" and "**is less than or equal to.**" These are called **weak inequalities.**[4]

The fact that the real numbers are ordered is called the law of trichotomy.

> **Law of trichotomy**
>
> For any real numbers a and b, exactly one of the following is true:
>
> $$a > b, \qquad a = b, \qquad a < b$$

We can determine which of the three possibilities is true in one of two ways. The easiest is to observe that $a > b$ if and only if a is to the right of b on the number line. The second method is with the following formal definition of inequality.

> **Definition of $a > b$**
>
> If $a - b$ is a positive value, then $a > b$.

Intervals

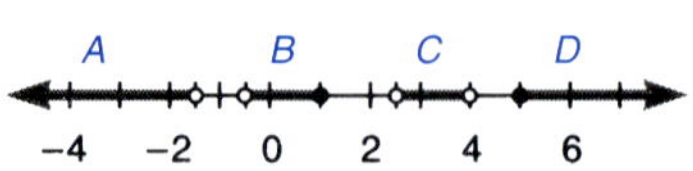

Figure 1–4

Most of the time we simply rely on our mental picture of the number line to determine which of two values is greater. We also use the concept of the number line, as well as the terminology of set-builder notation, to talk about **intervals** on the number line. Consider figure 1–4. The figure shows four intervals, A, B, C, and D. A is an interval that has no lower limit; the open circle shows that it also does not include the point at $-1\frac{1}{2}$. In B the solid circle is used to emphasize that the graph of 1 is included in the interval. C represents all values between 2.5 and 4, but specifically excludes 2.5 and 4 themselves. D is an interval with no upper limit. These intervals can be described using either set-builder notation or **interval notation.** These descriptions are shown in figure 1–5. Observe in the figure that the symbols "(" and ")" are associated with strict inequalities, and that "[" and "]" are associated with weak inequalities. We also use the symbol ∞ (infinity) to indicate the concept of no upper or lower limit. We use the symbols "$(-\infty$" to indicate an interval with no lower limit, and the symbols "$+\infty)$" to indicate that an interval has no upper limit.

The notation $2\frac{1}{2} < x < 4$ (interval C in figure 1–5) is read "x is greater than $2\frac{1}{2}$ *and* x is less than 4." This notation is called a **compound inequality.**

[4]The symbols $>$ and $<$ are attributed to Thomas Harriot in his *Artis analyticae praxis,* 1631. The symbols $\geq$ and $\leq$ were used by the Frenchman P. Bougher in 1734.

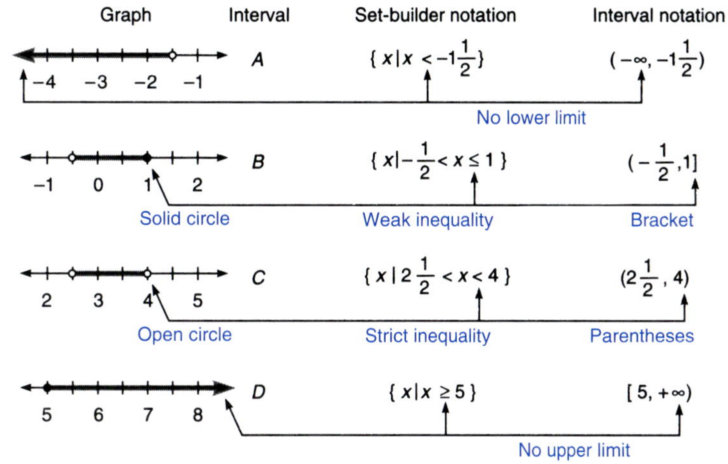

Figure 1–5

> **Compound inequality: $a < x < b$**
> $a < x < b$ means that x is greater than a and x is less than b.
>
> ***Concept***
> x is between a and b.

A similar statement applies (as in B in figure 1–5) when the symbol $\leq$ is used. Compound inequalities correspond to intervals on the number line.

■ ***Example 1–1 D***

Graph the interval, then describe the interval in set-builder notation if given in interval notation, and describe in interval notation if given in set-builder notation.

1. $\{x \mid -3 < x \leq 5\}$
Interval notation: $(-3, 5]$

Graph: 

2. $[-2, \sqrt{10})$ $\sqrt{10} \approx 3.2$ (calculator)
Set-builder notation: $\{x \mid -2 \leq x < \sqrt{10}\}$

Graph: 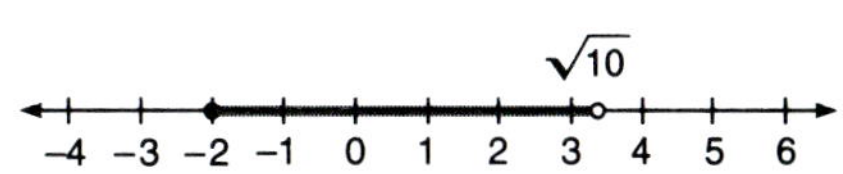

3. $(-\infty, 5]$
Set-builder notation: $\{x \mid x \leq 5\}$

Graph: 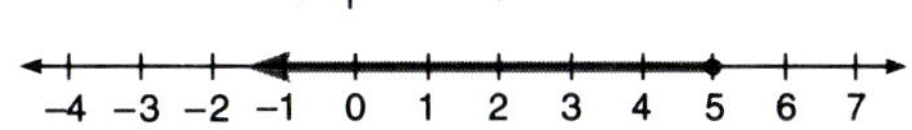

■

■ **Example 1–1 E**

Describe each interval shown in the graph in both set-builder and interval notation.

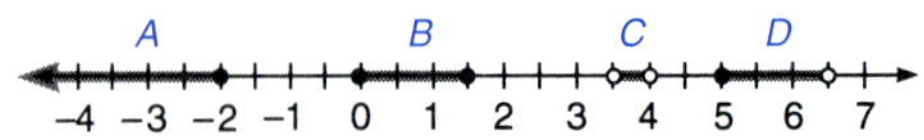

	Set	Set-builder notation	Interval notation
1.	A	$\{x \mid x \leq -2\}$	$(-\infty, -2]$
2.	B	$\{x \mid 0 \leq x \leq 1\frac{1}{2}\}$	$[0, 1\frac{1}{2}]$
3.	C	$\{x \mid 3\frac{1}{2} < x < 4\}$	$(3\frac{1}{2}, 4)$
4.	D	$\{x \mid 5 \leq x < 6\frac{1}{2}\}$	$[5, 6\frac{1}{2})$

■

Absolute value

The absolute value of a real number measures the undirected distance that the number is from the origin. Note that an **undirected distance** is nonnegative (positive or zero). The symbol $|x|$ is read "the absolute value of x."[5] The formal definition of absolute value uses the fact that the opposite of a negative number is positive.

> **Absolute value**
>
> $$|x| = \begin{cases} x \text{ if } x \text{ is positive or zero} \\ -x \text{ if } x \text{ is negative} \end{cases}$$
>
> ***Concept***
>
> If a number is nonnegative, its absolute value is the number itself. If a number is negative, the absolute value of that number is its opposite.

Note The symbol $-x$ does not necessarily represent a negative number. It symbolically states the opposite of the value that x represents.

■ **Example 1–1 F**

Write the following without absolute value by using the definition of absolute value.

1. $\left|-\frac{\sqrt{2}}{2}\right| = -\left(-\frac{\sqrt{2}}{2}\right) = \frac{\sqrt{2}}{2}$ $\quad -\frac{\sqrt{2}}{2} < 0;\ |a| = -a \text{ if } a < 0$

2. $|3 - \sqrt{10}| = -(3 - \sqrt{10}) = -3 + \sqrt{10} = \sqrt{10} - 3$ $\quad 3 - \sqrt{10} < 0$

3. $|a| = a$ if $a \geq 0$ or $-a$ if $a < 0$. ■

[5]The symbol $|x|$ was introduced by the German mathematician Karl Weierstrass in 1841.

Properties of absolute value

There are certain properties that can be proved valid concerning absolute value for any real numbers a, b. The proof would rely on the previously stated axioms for the real numbers as well as the definition of absolute value stated above. These properties are:

[1] $|a| \geq 0$ [2] $|-a| = |a|$

[3] $|a| \cdot |b| = |ab|$ [4] $\dfrac{|a|}{|b|} = \left|\dfrac{a}{b}\right|$

[5] $|a - b| = |b - a|$

■ ***Example 1–1 G***

Use the preceding properties and the definition of absolute value to simplify and remove the absolute value symbol from the following expressions.

1. $|2a|$

$= |2| \cdot |a|$ $|a| \cdot |b| = |ab|$

$= 2|a|$ $|2| = 2$

$= 2a$ if $a \geq 0$ or $-2a$ if $a < 0$ Definition of $|a|$

2. $|x^2y|$

$= |x^2| \cdot |y|$ $|a| \cdot |b| = |ab|$

$= x^2|y|$ $|x^2| = x^2$ since $x^2 \geq 0$

$= x^2y$ if $y \geq 0$ or $-x^2y$ if $y < 0$ Definition of $|y|$ ■

Mastery points

Can you

- List the elements of a set when the set is described in set-builder notation?
- Find the decimal form of any rational number, and describe it as terminating or repeating?
- Combine simple algebraic expressions?
- Recognize and describe intervals in set-builder, interval, and graph form?
- Find the absolute value of expressions?
- Use properties of absolute value to simplify expressions?

Exercise 1–1

Describe each set by listing the elements of the set.

1. $\{x \mid 3 < x < 12 \text{ and } x \in W\}$

2. $\{x \mid x \in W \text{ and } x \notin N\}$

3. $\{x \mid x \text{ is odd and } x \in N \text{ and } x < 21\}$

4. $\left\{x \mid x \text{ is represented by a digit in the decimal expansion of } \dfrac{178}{185}\right\}$

5. $\{3x \mid x \in \{-2, -1, 0, 1, 2, 3, 4\}\}$

6. $\left\{\dfrac{3x}{x+1} \,\middle|\, x \in \{1, 2, 3, 4, 5\}\right\}$

Give the decimal form of each rational number. Describe the form as terminating or repeating, as appropriate.

7. $\frac{2}{5}$ **8.** $\frac{2}{3}$ **9.** $\frac{3}{13}$ **10.** $\frac{1}{18}$

Simplify the given algebraic expressions.

11. $-5(2-8)^2 - 12(11-3)$

12. $(-8^2 - 3) - 5(2-5)^2 + (1-4)(4-1)$

13. $\frac{5}{8} - \frac{3}{5} + \frac{3}{4}$

14. $\frac{1}{2} - \frac{1}{4} + \frac{2}{3}$

15. $\frac{8}{15} \div (\frac{5}{9} - \frac{3}{10})$

16. $(\frac{3}{7} - \frac{7}{12}) \div (\frac{3}{7} + \frac{7}{12})$

17. $\dfrac{-3[5 - 2(9-12) + 4] - (8-2)(2-8)}{5(3-7)^2 - (3-7)^3}$

18. $\dfrac{-5(8) - (-2)(4)^2}{3^2 - 5^2} + \dfrac{18}{6 - 4(-4)}$

19. $\dfrac{x}{a} - \dfrac{y}{b}$

20. $\dfrac{2a-3}{3a} + \dfrac{3b+2}{2b}$

21. $\dfrac{x-y}{4x} - \dfrac{2x+y}{3y}$

22. $\dfrac{4a}{b} - \dfrac{2a}{3c} + \dfrac{1}{2}$

23. $(7x^2)(-2x^5)$

24. $(-3xy)(2xy^2z)$

25. $\dfrac{3x}{2y} \div \dfrac{5y}{2x} \cdot \dfrac{x}{5y}$

26. $\dfrac{5a}{3b} \cdot \dfrac{a}{2b} \div \dfrac{2b}{a}$

27. $\left(\dfrac{a}{2b} + \dfrac{b}{a}\right) \cdot \dfrac{3b}{5a}$

28. $\dfrac{3y^6}{4a^2} - \dfrac{2y^2}{4a^2}$

Graph the interval and describe in interval notation.

29. $\{x \mid -2 < x < 8\}$ **30.** $\{x \mid 0 < x \le 10\}$ **31.** $\{x \mid -8 \le x < 0\}$

32. $\{z \mid -3\frac{1}{4} \le z < -2\frac{3}{4}\}$ **33.** $\{y \mid -\sqrt{2} < y \le \pi\}$ **34.** $\{x \mid -\frac{1}{3} < x < \frac{2}{3}\}$

35. $\{x \mid x < 4\}$ **36.** $\{y \mid y \le -1\}$ **37.** $\{x \mid x \ge -2\}$

Graph the interval and describe in set-builder notation.

38. $[-2, 5]$ **39.** $[-5, -1]$ **40.** $[-\frac{2}{3}, -\frac{1}{3})$

41. $(-\infty, 1]$ **42.** $(1.8, +\infty)$ **43.** $\left[-\dfrac{\pi}{2}, \dfrac{3\pi}{2}\right)$

Describe each interval shown in the graph in both set-builder and interval notation.

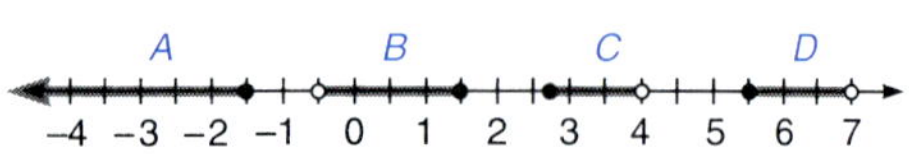

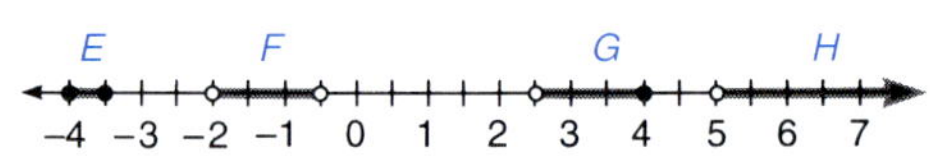

44. *A* **45.** *B* **46.** *C* **47.** *D* **48.** *E* **49.** *F* **50.** *G* **51.** *H*

Write the expression without the absolute value symbol.

52. $-|8\frac{1}{2}|$ **53.** $|-4|$ **54.** $|-3\frac{1}{3}|$ **55.** $-|2|$

56. $|-\sqrt{3}|$ **57.** $|-\sqrt{10} - 3|$ **58.** $-|\sqrt{10} - 6|$ **59.** $|2\frac{1}{4} - 4|$

60. $|6 - \sqrt{10}|$ **61.** $|(-5)^2|$ **62.** $-|-\sqrt{2}|$ **63.** $-|\sqrt{2} - 3|$

64. $|x^6|$ **65.** $|2x^4|$ **66.** $\left|\dfrac{y^4}{3}\right|$ **67.** $\left|\dfrac{x^2y^6}{z^8}\right|$

Use the properties for absolute value and the definition of absolute value to simplify the following expressions and rewrite without the absolute value symbol.

68. $|3a|$ **69.** $-|-5x^2|$ **70.** $\left|\dfrac{3x^2}{2y}\right|$ **71.** $\left|\dfrac{5x}{2y^2}\right|$

72. $|5x - 10y|$ **73.** $|(x-2)^2(x+1)|$ **74.** $\dfrac{|x|}{x}$ **75.** $\dfrac{x^2}{|x|}$

76. Put the following fractions in order, from smallest to largest: $\frac{3}{8}, \frac{5}{17}, \frac{21}{38}, \frac{22}{39}, \frac{15}{42}, \frac{13}{40}$. *Hint:* Use the decimal form of the number.

77. Compute the average of the following temperatures: -8, -6, 5, 2, 4. To compute the average one adds up the values and divides by the number of values.

78. Suppose the postage (in cents) for a certain category of mail is as follows:

Maximum weight (in ounces)	2	3	5	10
Postage (in cents)	15	20	30	40

Above 10 ounces, the rate is 3.5 cents per ounce. For example, any piece of mail weighing greater than 2 ounces and less than or equal to 3 ounces will cost 20 cents to mail.

a. Graph the weights as intervals on the number line.
b. Describe the intervals using interval notation.
c. Compute the postage rate in cents per ounce for each interval, to the nearest 0.1 cents.

For example, for the first interval it is $\frac{15 \text{ cents}}{2 \text{ ounces}} = 7.5$ cents per ounce.

79. Some calculators have a key labeled $\boxed{a^{b}/_{c}}$. This computes fraction expressions in exact form. For example $\frac{1}{3} + \frac{1}{3}$ results in $\frac{2}{3}$, not 0.6666666. If your calculator has such a key use it to compute the result of problems 13 through 16.

Skill and review

1. Simplify $2x^2 \cdot 3x^3$.
2. Simplify $2n + (-2n)$.
3. Simplify $8 - (-3)$.
4. $2{,}000{,}000{,}000 =$ **a.** $2 \cdot 10^{10}$ **b.** $2 \cdot 10^9$ **c.** 2^{10} **d.** 2^9
5. $0.00003 =$ **a.** $\frac{3}{1{,}000}$ **b.** $\frac{3}{10{,}000}$ **c.** $\frac{3}{100{,}000}$ **d.** $\frac{3}{1{,}000{,}000}$
6. Calculate $-3[2(4[\frac{1}{2}(2-3)+2]-1)+7]+4$.
7. Multiply $2a(2a - 2b - ac)$.
8. Multiply $(2a - c)(3a + 2c)$.

1–2 Integer exponents and polynomials

> The amount of solar radiation reaching the surface of the earth is about 3.9 million exajoules a year. (An exajoule is one billion joules of energy.) The combustion of a ton of oil releases about 45.5 joules of energy. How many tons of oil would have to be burned to equal the total amount of solar radiation reaching the surface of the earth in 1 hour? The annual consumption of global energy is about 350 exajoules. How many tons of oil would have to be burned to yield this amount of energy?

Questions like this are very important to those trying to find safe, alternate, renewable fuels for society. The concept of exponents is very helpful in dealing with this kind of problem, and exponents is one of the subjects of this section.

Integer exponents

We defined natural number exponents in section 1–1. For example, x^4 means $x \cdot x \cdot x \cdot x$. It has proved useful to extend the definition of exponents to include a definition for an exponent of zero and a definition for negative exponents.[6]

[6]It was understood by 1553, by Michael Stifel, that a quantity with an exponent zero had the value one. The superscript notation for positive exponents was introduced by Descartes in 1637. Newton introduced our modern notation for negative (and fractional) exponents in a letter in 1676. He wrote "Since algebraists write a^2, a^3, a^4, etc. for *aa, aaa, aaaa,* etc., so I write . . . a^{-1}, a^{-2}, a^{-3}, etc. for $\frac{1}{a}, \frac{1}{aa}, \frac{1}{aaa}$, etc."

Zero exponent

$$a^0 = 1 \text{ if } a \neq 0$$

Negative integer exponents

If $n \in N$, then

$$a^{-n} = \frac{1}{a^n} \text{ if } a \neq 0$$

Example 1–2 A

Simplify each expression; assume no variable expression represents zero.

1. $5^0ab^0 = (1)a(1) = a$
2. $(a - 3b)^0 = 1$
3. $5^{-2} = \frac{1}{5^2} = \frac{1}{25}$
4. $(x + y)^{-1} = \frac{1}{x + y}$ ■

There are several properties that apply to expressions with integer exponents.

Properties of integer exponents

If $a, b \in R$ and $m, n \in J$, then

[1] $a^ma^n = a^{m+n}$ [2] $\frac{a^m}{a^n} = a^{m-n}, a \neq 0$

[3] $(ab)^m = a^mb^m$ [4] $\left(\frac{a}{b}\right)^m = \frac{a^m}{b^m}, b \neq 0$

[5] $(a^m)^n = a^{mn}$

Example 1–2 B

Simplify each expression. Leave the answer with only positive exponents. Assume no variable represents zero.

1. $(5x^3y^4)(-2x^2y^2) = -10x^{3+2}y^{4+2}$ $\quad a^ma^n = a^{m+n}$
   $= -10x^5y^6$
2. $\frac{12x^8y^5}{3x^3y^{-1}} = \frac{12}{3}x^{8-3}y^{5-(-1)}$ $\quad \frac{a^m}{a^n} = a^{m-n}$
   $= 4x^5y^6$
3. $(a^3)^4(b^4)^{-1} = a^{12}b^{-4}$ $\quad (a^m)^n = a^{mn}$
   $= \frac{a^{12}}{b^4}$
4. $\left(\frac{3x^2}{y^5}\right)^3 = \frac{(3x^2)^3}{(y^5)^3}$ $\quad \left(\frac{a}{b}\right)^m = \frac{a^m}{b^m}$
   $= \frac{27x^6}{y^{15}}$ $\quad (ab)^m = a^mb^m$
5. $2x^{-2}y + x^2y^{-1} = \frac{2y}{x^2} + \frac{x^2}{y}$
   $= \frac{2y^2 + x^4}{x^2y}$

6. $\dfrac{x^n y^{2m}}{x^2 y^m} = x^{(n-2)}y^{(2m-m)}$ $\qquad \dfrac{a^m}{a^n} = a^{m-n}$

$= x^{n-2}y^m$ ■

Scientific notation

Very large and very small numbers appear in most of the physical and social sciences. For example, the mass of a hydrogen atom is 0.000 000 000 000 000 000 000 001 67 gram, and there are over 5,000,000,000 (5 billion) people on this planet. The English language probably permits at least 8,000,000,000,000 three-word sentences.

If we observe that 1,000,000 (one million) is 10^6, and $\dfrac{1}{1{,}000{,}000}$ (one-millionth) is 10^{-6}, we see that integer exponents might prove useful in expressing these quantities.

We can convert any decimal number into what is called **scientific notation.** We define this form of a number Y to be

$$Y = a \times 10^n$$

where $1 \leq |a| < 10$. The steps to put Y into this form are as follows.

> **To put a number *Y* into scientific notation**
>
> 1. Move the decimal point to the position immediately following the first nonzero digit in Y.
> 2. The absolute value of n is the number of decimal places that the decimal point was moved.
> 3. If the decimal point is moved *left,* n is *positive.*
>   If the decimal point is moved *right,* n is *negative.*
>   If the decimal point is not moved, n is 0.

To put a number back into decimal form we reverse these steps.

To understand why these steps are correct consider the value 53,000:

$$\begin{aligned} 53{,}000 &= 53{,}000 \times 1 \\ &= 53{,}000 \times (10^{-4} \times 10^4) \\ &= (53{,}000 \times 10^{-4}) \times 10^4 \\ &= 5.3 \times 10^4 \end{aligned}$$

■ ***Example 1–2 C***

Convert each number into scientific notation.

1. $3{,}500{,}000\,000 = 3.\underbrace{500\,000\,000} \times 10^9 = 3.5 \times 10^9$

2. $-0.000\,000\,000\,000\,805\,4 = -0.\underbrace{000\,000\,000\,000\,8}.054 \times 10^{-13}$

$= -8.054 \times 10^{-13}$ ■

■ Example 1–2 D

Convert each number into a decimal number.

1. -2.8903×10^{12}
   $= -2{,}890{,}300{,}000{,}000$ Move decimal 12 places to the right

2. 2.8903×10^{-10}
   $= 0.000\ 000\ 000\ 289\ 03$ Move decimal 10 places to the left ■

Polynomials

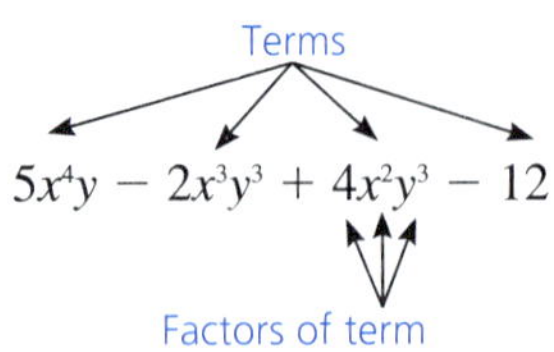

Recall from section 1–1 that a **term** is part of a sum and a **factor** is a part of a product. The constant factor of a term is called its **numerical coefficient** (-2 in $-2x^3y^2$). An **algebraic expression** is any meaningful collection of variables, constants, grouping symbols, and symbols of operations. A special kind of expression is a polynomial.

> **Polynomial**
> A **polynomial** is an expression containing one or more terms in which each factor is either a constant or a variable with a natural number exponent.

A **monomial** is a polynomial of one term, a **binomial** is a polynomial of two terms, and a **trinomial** is a polynomial of three terms.[7]

A polynomial is an expression in which the variables are not found in radicals or denominators, and have only natural number exponents. The idea is that a polynomial is made up of constants and variables using only the operations of addition, subtraction, and multiplication—as a result, a polynomial is always defined. Observe that $\frac{1}{x}$ uses division. Thus it is not a polynomial, and is not defined for the replacement value zero.

■ Example 1–2 E

The following illustrate monomials, binomials, and trinomials.

1. Monomials: 5, $3x^2$, $2x^3y^5$, $-\sqrt{2}x$, $\frac{1}{2}x$ (note that $\sqrt{2}$ and $\frac{1}{2}$ are constants)
2. Binomials: $5x^2 + 2$, $-4a + 3b$, $(3x^4 - 2y^5) - z^4$
3. Trinomials: $2x^5 - 3x^2 + 5$, $3x^2 - x + 12$, $(a + 2b)^2 - 3(a + 2b) + 7$ ■

The following expressions are not polynomials: $5\sqrt{x}$, $\frac{5}{x}$, and $5x^{-3}$. They include operations other than addition, subtraction, and multiplication.

[7]Mono, bi, tri, and poly are prefixes taken from the Greek language. They mean "one," "two," "three," and "many," respectively.

We often categorize polynomials in one variable by their degree.

> **Degree**
> The **degree of a term** of a polynomial in one variable is the exponent of the variable factor. The degree of a constant term is zero. The **degree of a polynomial** is the degree of its term of highest degree.

For example, the degree of $4x^5$ is 5, the degree of $2x^5 - 3x^2$ is 5, and the degree of $3x^4 + 2x^3 - 11x^2 - 5$ is 4.

Substitution of value

It is important to remember that an expression represents a numerical value. Something like $a + b$ makes sense only if a and b represent actual real numbers. To find the value of an expression, given a value for each variable in an expression, use what we will call **substitution of value.** This means to *replace each variable in the expression by the value associated with that value.* Then perform the indicated arithmetic.

■ ***Example 1–2 F***

Find the value that each expression represents when $a = -5$, $b = 4$, $c = -\frac{1}{2}$, and $d = 6$.

	Problem	Solution
1.	$5a^2b$	$5(-5)^2(4) = 5(25)(4) = 500$
2.	$(2a - \frac{1}{2}b)(6c + d)$	$[2(-5) - \frac{1}{2}(4)][6(-\frac{1}{2}) + 6]$ $= [-10 - 2][-3 + 6] = -36$

■

Polynomials are a very important part of mathematics. There is an ''arithmetic'' of polynomials. We will review how to add, subtract, multiply, and divide polynomials.

Multiplication of polynomials

Multiplication of monomials proceeds using the properties for exponents covered earlier in this section. Basically, *multiply coefficients and add exponents.* For example,

$$(4x^2y)(-3x^5y^3) = -12x^7y^4$$

Multiplication of general expressions uses the distributive axiom. Recall that this states that

$$a(b + c) = ab + ac$$

■ ***Example 1–2 G***

Multiply.

1. $-3a^2b(5a^2 - 3ab)$

$$= -3a^2b(5a^2 - 3ab)$$
$$= -3a^2b(5a^2) + 3a^2b(3ab) = -15a^4b + 9a^3b^2$$

2. $(5a - 3b)(2x + 3y)$

In this case we apply the distributive axiom to multiply each term of $(5a - 3b)$ by $(2x + 3y)$. This amounts to multiplying each term in the second factor, $2x$ and $3y$, by each term in the first factor, $5a$ and $-3b$. This is illustrated as

Multiply by $5a$

$$(5a - 3b)(2x + 3y) = (5a - 3b)(2x + 3y)$$

Multiply by $-3b$

$$= 5a(2x + 3y) - 3b(2x + 3y)$$
$$= 10ax + 15ay - 6bx - 9by$$

3. $(2a - 3b - 5c)(4x - 7y + z)$

In this case each term in the second factor is multiplied by each term in the first factor; this is a total of nine products.

$$(2a - 3b - 5c)(4x - 7y + z)$$

$$2a\,(4x - 7y + z) - 3b\,(4x - 7y + z) - 5c\,(4x - 7y + z)$$
$$= 8ax - 14ay + 2az - 12bx + 21by - 3bz - 20cx + 35cy - 5cz$$ ■

Addition and subtraction of polynomials

Addition and subtraction is equivalent to combining like terms. Basically, *add and subtract coefficients—do not change exponents.* Recall from section 1–1 that like terms are terms with identical variable factors. The distributive axiom gives the method for combining like terms. For example, by the distributive axiom,

$$5a^2b + 8a^2b = (5 + 8)a^2b$$
$$= 13a^2b$$

We also need to observe that $-(a + b) = -a - b$. That is, *when a grouping symbol is preceded by a negative sign, the grouping symbol may be removed if we change the sign of every term within the grouping symbol.* This can be viewed as multiplication by -1; that is, $-(a + b) = (-1)(a + b)$, which is $-a - b$.

■ ***Example 1–2 H***

Perform the indicated operations.

1. $(5x^2 - 3x + 4) - (2x^2 - 6x + 9) + (6 - 3x - x^2)$

$= 5x^2 - 3x + 4 - 2x^2 + 6x - 9 + 6 - 3x - x^2$ Rewrite without parentheses

$= 5x^2 - 2x^2 - x^2 - 3x + 6x - 3x + 4 - 9 + 6$ Terms reordered for clarity

$= 2x^2 + 1$ Combine like terms

2. $(2x - 3)(8x + 3)$

$= 16x^2 + 6x - 24x - 9$ Multiply

$= 16x^2 - 18x - 9$ Combine like terms ■

Division of polynomials

Division by a monomial uses the rules for exponents as well as the division of numbers: $\dfrac{6x^6}{3x^3} = 2x^3$. That is, basically, *divide coefficients and subtract exponents.* If the numerator has more than one term we *rewrite the quotient using separate fractions:* $\dfrac{8x^4 - 6x^2 + 12x^2y}{2x^2}$ should be rewritten as $\dfrac{8x^4}{2x^2} - \dfrac{6x^2}{2x^2} + \dfrac{12x^2y}{2x^2}$. Simplifying each fraction produces the result

$$4x^2 - 3 + 6y$$

To divide by a polynomial of more than one term we use the **long division algorithm.** An algorithm is a precise procedure for doing something. The long division algorithm repeats the same three steps over and over until the degree of the remainder is less than the degree of the divisor:

1. Divide the first term of the divisor into the first term of the dividend.
2. Multiply the divisor by the result.
3. Subtract and bring down remaining terms.

It is practically the same as the long division used in arithmetic. This algorithm is illustrated in parts 3 and 4 of the following example.

■ ***Example 1–2 I***

Divide.

	Problem	**Solution**
1.	$\dfrac{24x^6y^3}{36x^2y^3}$	$\dfrac{24}{36}\cdot\dfrac{x^6}{x^2}\cdot\dfrac{y^3}{y^3} = \dfrac{2}{3}x^4(1) = \frac{2}{3}x^4$
2.	$\dfrac{24x^6y^3 - 20x^2y^8 + 12x^2y^2}{12x^2y^2}$	$= \dfrac{24x^6y^3}{12x^2y^2} - \dfrac{20x^2y^8}{12x^2y^2} + \dfrac{12x^2y^2}{12x^2y^2}$ $= 2x^4y - \frac{5}{3}y^6 + 1$

3. $\dfrac{4x^3 - 2x^2 - 8}{2x + 1}$.

Division by an expression of more than one term; use long division. $2x + 1$ is the divisor and $4x^3 - 2x^2 - 8$ is the dividend. We insert a term $0x$ for clarity.

$$\begin{array}{r@{\,}l}
 & 2x^2 - 2x + 1 \\
2x + 1 \,\big)\!\! & \overline{4x^3 - 2x^2 + 0x - 8} \\
 & \underline{-(4x^3 + 2x^2)} \\
 & \quad -4x^2 + 0x - 8 \\
 & \quad \underline{-(-4x^2 - 2x)} \\
 & \qquad\qquad 2x - 8 \\
 & \qquad\qquad \underline{-(2x + 1)} \\
 & \qquad\qquad\quad -9
\end{array}$$

Thus, $(4x^3 - 2x^2 - 8) \div (2x + 1) = 2x^2 - 2x + 1$ with a remainder of -9. We would usually write the result as

$$\frac{4x^3 - 2x^2 - 8}{2x + 1} = 2x^2 - 2x + 1 - \frac{9}{2x + 1}.$$

To check, we would compute $(2x + 1)(2x^2 - 2x + 1) - 9$, and make sure the result is $4x^3 - 2x^2 - 8$.

4. $\dfrac{2x^4 + x^3 - 3x^2 + 3}{x^2 - 2x + 1}$. The divisor has more than one term so use long division.

$$\begin{array}{r}
2x^2 + 5x + 5 \\
x^2 - 2x + 1 \overline{\smash{)}\ 2x^4 + x^3 - 3x^2 + 0x + 3} \\
\underline{-(2x^4 - 4x^3 + 2x^2) } \\
5x^3 - 5x^2 + 0x + 3 \\
\underline{-(5x^3 - 10x^2 + 5x) } \\
5x^2 - 5x + 3 \\
\underline{-(5x^2 - 10x + 5)} \\
5x - 2
\end{array}$$

Thus, $\dfrac{2x^4 + x^3 - 3x^2 + 3}{x^2 - 2x + 1} = 2x^2 + 5x + 5 + \dfrac{5x - 2}{x^2 - 2x + 1}$. ■

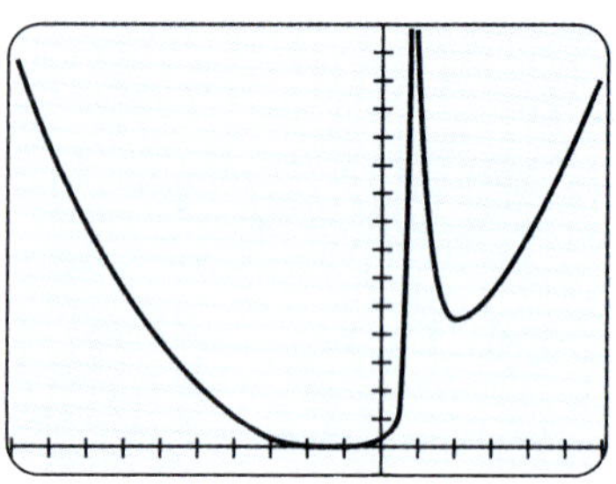

Figure 1–6

When using the long division algorithm it is important to arrange the terms of the divisor and dividend in decreasing order of degree. The algorithm stops when the degree of the remainder is less than that of the divisor.

Note A graphing calculator can actually be used to help verify that the last calculation was correct. The graph of $Y_1 = (2x^4 + x^3 - 3x^2 + 3)/(x^2 - 2x + 1)$ is shown in figure 1–6 (using x-values from -10 to 6, and y-values from -2 to 150). The graph of $Y_2 = 2x^2 + 5x + 5 + ((5x - 2)/(x^2 - 2x + 1))$ is identical to it. Thus, graphically, $Y_1 = Y_2$ for any value of x. This would indicate our result was correct.

Subscripted variables

Subscripted variables are used in many situations. For example, if one measured the temperature of something at two different times these temperatures might be called t_1 and t_2 (read "t sub 1 and t sub 2"). The 1 and 2 are subscripts, in this case indicating the first and second temperature. *Variables with different subscripts are not like terms,* and we never perform any arithmetic operations on subscripts. Thus, for example, $t_1 + 4t_1 + t_2$ can combine into $5t_1 + t_2$, but that is all. Similarly,

$$\begin{aligned}(t_2 - t_1)(2t_2 + t_1) &= 2t_2^2 + t_1t_2 - 2t_1t_2 - t_1^2 \\ &= 2t_2^2 - t_1t_2 - t_1^2\end{aligned}$$

Mastery points

Can you

- Simplify expressions involving integer exponents?
- Convert numbers between scientific notation and decimal notation?
- Determine whether an expression is a monomial, binomial, or trinomial, or not a polynomial at all?
- Evaluate expressions when given values for the variables, using substitution of value?
- Perform the indicated operations of addition, subtraction, multiplication, and division to combine and transform expressions?

Exercise 1–2

Use the properties of exponents to simplify the following expressions.

1. $2x^5 \cdot x^2 \cdot x^4$
2. $-2x^4 \cdot (2^2x^3)$
3. $(-2^2)(2^3)$
4. $(-2)^2(-3^2)$
5. $(3a^5b)(2a^2b^2)$
6. $(-4x^5yz^2)^2(xy^3z^2)$
7. $2x^{-2}(2^6x^3)$
8. $3a^{-1}b^4(3^{-1}a^2b)$
9. $3x^4y^{-3}$
10. $-5x^{-2}y$
11. $(2x^3y^5)^3$
12. $(-3a^2bc^3)^4$
13. $(-3^2a^2b^{-3})^2$
14. $(2x^{-3}y^2)^3$
15. $\dfrac{1}{x^2y^{-3}}$
16. $\dfrac{2}{2^{-1}x^{-5}}$
17. $(-3)^{-3}$
18. $\dfrac{1}{2^{-2}}$
19. $(-2x^{-2})^2$
20. $(2^{-3}x^3y)^2$
21. $(2x^4y)(-3x^3y^{-2})$
22. $(3^2x^{-5}y)(-2^2x^5y^{-4})$
23. $(6x^4)^0$
24. $\dfrac{3}{x^{-2}(y^4)^5}$
25. $\dfrac{3^2x^{-4}y}{3xy^{-4}}$
26. $\dfrac{2x^{-2}y^0z^3}{-2^{-3}x^2y^2z^{-1}}$
27. $\left(\dfrac{9a^5b^4}{18ab^{11}}\right)^{-3}$
28. $\left(\dfrac{6x^5y^2}{2x^2y^{-2}}\right)^2$
29. $\left(\dfrac{-2x^2y}{5x^{-3}y^3}\right)^3$
30. $\left(\dfrac{3x^2y^{-2}}{6x^{-3}y}\right)^0$
31. $\left(\dfrac{3^3a^{-4}bc^6}{3^4ab^{-2}c^{-2}}\right)^2$
32. $\left(\dfrac{4a^{-3}}{12a^{-5}}\right)^2$
33. $\left(\dfrac{3x^2}{12x^4}\right)\left(\dfrac{16x^{-1}}{2x}\right)$
34. $\left(\dfrac{a^2b^3c^0}{a^3b^{-4}c^2}\right)\left(\dfrac{a^4c}{a^2b}\right)$
35. $\left(\dfrac{x^2y}{x^2y^2}\right)^2\left(\dfrac{3x^0y^{-2}}{4xy}\right)^2$
36. $\left(\dfrac{a^5bc^2}{2abc}\right)^3\left(\dfrac{2^2a^3b^2c}{ab^{-2}c^3}\right)^{-1}$
37. $x^{2n-3}x^{2n+3}$
38. $a^nb^{2n}(a^2b^{-n})$
39. $\dfrac{x^{n-2}}{x^{n-3}}$
40. $\dfrac{2^{-3}a^{-n}}{2ab^{-2}}$
41. $\left(\dfrac{x^ny^{2-n}}{x^{-n}y^{n-2}}\right)^4$
42. $\left(\dfrac{x^{2n}}{x^{n+1}}\right)^{-3}$

Convert each number into scientific notation.

43. 3,650,000,000,000,000
44. 2,003,000,000,000
45. −19,002,000,000,000
46. 0.000 000 000 203
47. −0.000 000 000 000 029 2
48. −0.000 000 5
49. 0.000 000 000 003 502
50. −21,500,000,000,000

Convert each number given in scientific notation into a decimal number.

51. 2.502×10^{13}
52. 4.31×10^{-8}
53. -1.384×10^{-10}
54. -5.11×10^7
55. 9.23×10^6
56. 3.002×10^{12}

57. The mass of an electron is

0.000 000 000 000 000 000 000 000 000 91 gram.

Put this value into scientific notation.

58. The half-life of lead[8] 204 is

14,000,000,000,000,000,000 years.

Put this value into scientific notation.

Categorize each expression as a monomial, binomial, trinomial, polynomial (if more than three terms), or not a polynomial. Also state the degree of those expressions that are polynomials.

59. $5x^2 - 3x - 2$
60. $2x - 1$
61. $4x^3 - 3x^2 - x + 3$
62. $3x^5 - 2x^4 + x^2 - 3x - 7$
63. $\sqrt{3}\,x^4 - \frac{1}{2}x^6 + 4$
64. $3(x + 1) + (3x - 2) + 9$
65. $3\sqrt{x} - 4x - 2$
66. $9x^2$

In the following problems find the value that each expression represents, assuming that $a = -5$, $b = 4$, $c = -\frac{1}{2}$, $d = 6$.

67. $3a^3 - 5a^2 + 11a - 2$
68. $(a - 2b)(3c + d)$
69. $\frac{1}{3}a - 2cd^2 + \sqrt{b}$
70. $(3a^2 - 2a + 1)(a - 1)^2$
71. $2c^3 - 5c + bc$
72. $-2(4[\frac{1}{2}(-b + 3) + 2] - 1) + 7$

Perform the indicated operations.

73. $(5x^2 - 3x - 2) - (3x^2 + x - 8)$
74. $(-3x + 4x^2 + 11) + (4 - 2x - 5x^2)$
75. $(2a - 3b - c) + (5a - 4b + 2c) - a$
76. $-(a - 4b) + (2b - 3a) - (5a - 5b)$
77. $(3x^2y - 2xy^2 - xy) + (2xy^2 - 5x^2y + 3xy)$
78. $(5ab^2 - 2a^2b^2 + 3a) - (a^2b^2 - 2ab^2 + 3a^2)$
79. $3x - [x - y - (7x - 3y)]$
80. $7x - [4x + 3y + (x - 2y)]$
81. $[-(3a - b) - (2a + 3b)] - [(a - 6b) - (3b - 10a)]$
82. $-(a - 4) + [3a - (2a + 5) - 4]$
83. $2x^2(5x^3 - 2x^2 + 7)$
84. $-5xy^2(2x^2 - 3xy - y^2)$
85. $-2a^2b(5a^2b + 3a^2b^2 - 2ab^3)$
86. $\frac{1}{2}ab^4(4a^4b - 8a^3b^2 + 2a^2b^3 - 10)$
87. $(5a - 3)(5a + 3)$
88. $(2x + 4y)(x - 2y)$
89. $(3x + y)(5x - y)$
90. $(a - 3b)(2a + 7b)$
91. $(2a - b)(a - 2b + c)$
92. $(x + 3y)(2x - y - 4)$
93. $(2x^2 - 3x + 1)(5x^2 - 2x + 7)$
94. $(a^2 + a - 3)(2a^2 - 6a + 4)$
95. $(5b^2 - 2b + 3)(b^2 + b - 3)$
96. $(4x^2 - 5x - 3)(x^2 + 2x + 1)$
97. $(x + 2y)(x - 3y)(x + y)$
98. $(2a - b)(a + b)(a - b)$
99. $(3a + 2b)(3a - b)(a + 2b)$
100. $(x + 4)(2x - 3)(2x + 5)$
101. $(2a - 3)(3a - 2b + c)$
102. $(3x - 2)^2(x + 5)$
103. $(x - 2y)(x + 2y)^2$
104. $(a + 2b)(a - b)^2$
105. $(2x + 5)^3$
106. $(a - 2b)^3$

Perform the indicated divisions.

107. $\dfrac{12x^5y^2}{18x^2y}$
108. $\dfrac{24a^8b^3c}{8a^4bc}$
109. $\dfrac{6a^6b^2 - 8a^4b^4 + 12a^4b^6}{2a^4b^2}$
110. $\dfrac{20x^2y^3 + 5xy^5 - 10xy^3}{10xy^3}$
111. $\dfrac{10x^4y^2z^2 + 15x^3y^2z - 20x^2y^4}{15x^2y^2}$
112. $\dfrac{-30a^9b^3 - 12a^6b^3 + 18a^3b^3}{12a^3b^3}$
113. $\dfrac{x^2 - 3x + 2}{x - 1}$
114. $\dfrac{y^3 - 1}{y - 1}$
115. $\dfrac{x^4 - 3x^3 + 8}{x + 2}$
116. $\dfrac{2z^5 - 3z^3 + z^2 - z - 1}{z + 1}$
117. $\dfrac{6x^3 + x^2 - 10x + 5}{2x + 3}$
118. $\dfrac{2x^4 - 3x + 5}{x - 3}$
119. $\dfrac{4x^3 - x^2 + 3x + 4}{x - 2}$
120. $\dfrac{2x^2 - 11x + 3}{x - 8}$
121. $\dfrac{4x^3 - x^2 + 5}{x^2 - x + 1}$
122. $\dfrac{2x^4 - x^3 + 2x - 2}{x^2 + 3}$
123. $\dfrac{3x^4 - 2x^2 - x + 1}{x^2 - 3}$
124. $\dfrac{x^3 - x^2 + 2x - 5}{x^2 - 2x - 1}$

[8]The time necessary for half of the material present to decay radioactively.

125. Perform the indicated operations on the subscripted variables.

a. $(3t_1 + 4t_2) - (t_1 + 6t_2 - 3t_3)$
b. $(3t_1 + 4t_2)(t_1 + t_2 - 3t_3)$
c. $(-3x_1^2x_2^4)(2x_1x_2)^3$

126. Divide.

a. $(2x^3 - 3x^2 - x + 4) \div (2x + 3)$
b. $(2x^3 - 3x^2 - x + 4) \div (2x^2 + 3x - 1)$
c. $(2x^3 - 3x^2 - x + 4) \div (2x^3 + 3x^2 - 3x + 2)$

127. The equations

$$(a^2 + b^2)(c^2 + d^2) = (ac + bd)^2 + (ad - bc)^2$$
$$= (ac - bd)^2 + (ad + bc)^2$$

played important roles in medieval algebra and, still, in trigonometry, an important part of mathematics. Show that all three expressions are equivalent.

128. Srinivasa Ramanujan (1897–1920) was a mathematical genius from India at the beginning of this century. Among numerous incredible accomplishments at the highest levels of mathematics, he also developed and used the following formula:

$$(a + 1)(b + 1)(c + 1) + (a - 1)(b - 1)(c - 1)$$
$$= 2(a + b + c + abc)$$

Compute the left member to show that it is the same as the right member.

129. In a certain class, four one-hour tests are given and a final exam. The final exam counts 40% of the course grade, so each test counts 15%. Under these conditions the course grade G is described by the formula $G = 0.15(T_1 + T_2 + T_3 + T_4) + 0.4E$, where T_1 represents the grade of the first test, etc., and E is the grade on the final exam. Compute a student's final exam grade, to the nearest 0.1, if the student's test grades were 68, 78, 82, and 74, and the final exam grade was 81.

130. The perimeter of a geometric object is the distance around the object. Write and simplify an expression for the perimeter of each of the objects shown in the figure.

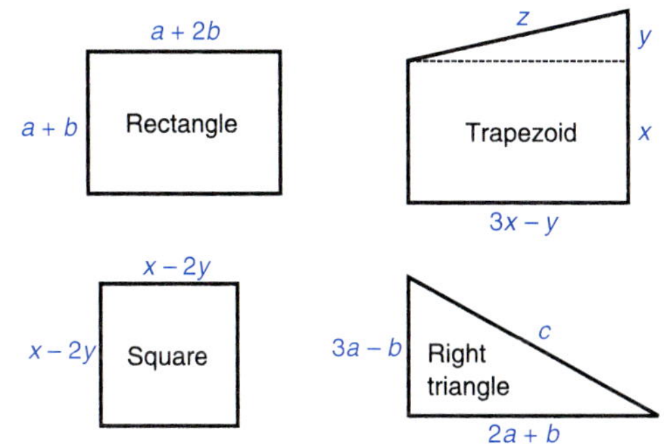

131. The area of a square or rectangle is the product of its two dimensions. Find an expression for the area of the square and rectangle shown in the figure with problem 130.

132. The area of a triangle is $\frac{1}{2}bh$ (see the figure), where b means base and h means height. Find and simplify an expression for the area of the right triangle in the figure with problem 130.

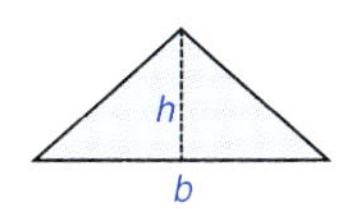

133. Find and simplify an expression for the area of the trapezoid shown in the figure with problem 130. *Hint:* The figure can be decomposed into a rectangle and a triangle.

134. In his book *The Schillinger Method of Musical Composition,* Schillinger shows how to apply algebraic concepts to the study of music.[9] One application is rhythmic continuity. Since $(A + B)^2 = A^2 + AB + BA + B^2$, if we let $A = 1$ and $B = 2$ we obtain $9 = 3^2 = (1 + 2)^2 = 1 + 2 + 2 + 4 = 9$, and if we let $A = 2$ and $B = 1$ we obtain $9 = 3^2 = (2 + 1)^2 = 4 + 2 + 2 + 1$. These are two sequences of attacks, which total 9 beats.

a. Use this idea to generate two sequences of attacks that total 16 beats.
b. Use the expansion of $(A + B + C)^2$ with values for A, B, and C to generate two sequences that total 36 beats.

[9]The authors acknowledge former student and professional musician, Gary Leach, for bringing this to their attention.

135. The Rhind Mathematical Papyrus is an Egyptian work on mathematics. It dates to the sixteenth century B.C., and contains material from the nineteenth century B.C. It contains 84 problems, including tables for manipulations of fractions.

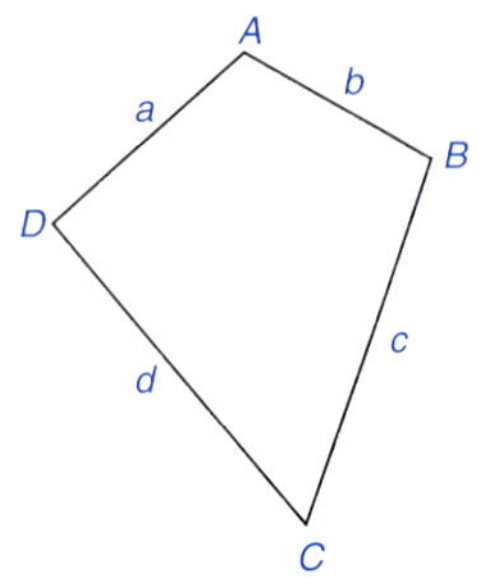

Problems 51–53 of the Papyrus include the following formula for finding the area of a four-sided figure like $ABCD$ in the figure: $\frac{1}{2}(a + c) \cdot \frac{1}{2}(b + d)$. (Observe that $\frac{1}{2}(a + c)$ is the average length of the two sides a and c; the same is true for $\frac{1}{2}(b + d)$.) Show that this formula is equivalent to

$$\frac{1}{4}(ab + ad + bc + cd)$$

By the way, this formula is inaccurate. Except for rectangles, it gives an answer that is too large. It was used for the purpose of taxing land. There is not always an economic incentive to get the correct answer.

136. One early notation for algebra put the coefficient first and the exponent last, as we do today.[10] Exponents were not written above the line, however. Thus, a^2 was $a2$, and $5a^2b$ was $5a2b$. Using this notation, simplify the following expressions. Write the result in the early notation.

a. $(3a2b3)(2ab2)$ **b.** $(ab3)(3ab)$
c. $(aaa)(3abb)$ **d.** $(5a5)(2b2)$
e. $\dfrac{9a9}{3a3}$

137. Most computers take longer to multiply two values than to add or subtract them. Suppose a certain computer takes five units of time for a multiplication but only one for an addition/subtraction.

Now consider the equivalent expressions $a(b + c) = ab + ac$. The expression $a(b + c)$ represents one addition followed by one multiplication. This would take $1 + 5 = 6$ units of time to calculate. The expression $ab + ac$ has two multiplications followed by one addition. This would take 11 units of time. Thus, given a choice, the expression $a(b + c)$ should be used when writing programs for this computer.

Analyze each of the following expressions for this computer by performing the indicated operations and simplifying, then comparing the number of units of time each expression would take. The number of units are shown for the given expression.

a. $(a + b)(c + d) : 7$ **b.** $(2a + 3b)(a - 2b) : 22$
c. $a(b + c - d) : 7$ **d.** $(a + b)(c + d + e) : 8$
e. $(a + b)(c + d)(e + f) : 13$

Any scientific calculator will accept numbers in scientific notation. In fact this is the only way to enter a very large or small number. Calculators usually have a key marked [EE], for enter exponent, or [EXP], for exponent. For example, to perform a calculation such as

$$(3.8 \times 10^8)(-2.5 \times 10^{-12})$$

one would enter a sequence like

3.8 [EE] 8 [×] 2.5 [+/−] [EE] 12 [+/−] [=]

TI-81: 3.8 [EE] 8 [×] [(−)] 2.5 [EE] [(−)] 12 [ENTER]

Example

Calculate (3,500,000,000,000,000)(51,000,000,000). Leave the result in scientific notation.

$$\begin{aligned}(3{,}500{,}000{,}000{,}000{,}000)&(51{,}000{,}000{,}000)\\ &= (3.5 \times 10^{15})(5.1 \times 10^{10})\\ &= (3.5)(5.1)(10^{15} \times 10^{10})\\ &= 17.85 \times 10^{25}\\ &= 1.785 \times 10^{26}\end{aligned}$$

On a calculator the steps would be 3.5 [EE] 15 [×] 5.1 [EE] 10 [=]. On the TI-81 use [ENTER] for [=]. ■

Compute the values in problems 138–141 using a calculator.

138. (31,000,000,000,000)(5,300,000,000,000,000)
139. (5,000,000,000,000) ÷ 0.000 000 000 25
140. $(39{,}100{,}000{,}000)^3$
141. $\sqrt{4{,}000{,}000{,}000{,}000{,}000{,}000}$

[10]Pierre Hérigone, a French mathematician, advocated this in his book *Cursus mathematicus,* which appeared in 1634.

142. The amount of solar radiation reaching the surface of the earth is about 3.9 million exajoules a year. An exajoule is one billion joules of energy. The combustion of a ton of oil releases about 45.5 joules of energy.

a. Compute how many tons of oil would have to be burned to equal the total amount of solar radiation reaching the surface of the earth in one hour.

b. The annual consumption of global energy is about 350 exajoules. How many tons of oil would have to be burned to yield this amount of energy?

Skill and review

1. Write 360 as a product of prime integers. Prime integers are the positive integers greater than one that are not divisible by any other positive integer except one. They are 2, 3, 5, 7, 11, 13, 17, etc.
2. $3x^5y^4 - 12x^3y^3 + 6x^2y^3 = 3x^2y^3(\_\_\_\_ - \_\_\_\_ + \_\_\_\_)$
3. $x^2 - 16 = (x - 4)(\_\_\_\_ + \_\_\_\_)$
4. $x^2 + 6x + 8 = (x + 2)(\_\_\_\_ + \_\_\_\_)$
5. $x^2 + 6x - 16 = (x - 2)(\_\_\_\_ + \_\_\_\_)$
6. $6x^2 - 7x - 3 = (2x - 3)(\_\_\_\_ + \_\_\_\_)$
7. $(x - 1)(x^2 + x + 1) = \_\_\_\_$

1–3 Factoring

A computer program must calculate the quantity $A^2 - B^2$ for thousands of different values of A and B. Also, the computer can store only the first six digits of any number. Since $A^2 - B^2$ is the same as $(A - B)(A + B)$, would one form be better than the other to use in the program?

As a matter of fact, the form $(A - B)(A + B)$ is better. It requires only one multiplication, whereas $A^2 - B^2$ is $AA - BB$, and requires two. Multiplication is much slower than addition on many computers. (See problem 137 in section 1–2, also.) The form $(A - B)(A + B)$ is a factored form of $A^2 - B^2$.

Also, the factored form can be more accurate than the unfactored form. Suppose the computer can store only two more digits after the leftmost nonzero digit of a number[11] (in reality they can store about five more such digits, but the idea is the same in either case). Let A be 3.00 and B be 2.99. Then consider the two calculations, where all intermediate results cannot contain more than two digits after the leftmost nonzero digit.

$(A - B)(A + B)$	$A^2 - B^2$	
$(3.00 - 2.99)(3.00 + 2.99)$	$3.00^2 - 2.99^2$	
$0.0100(5.99)$	$9.00 - 8.9401$	
0.0599	$9.00 - 8.94$	Round 8.9401 to three digits
	0.0600	

Thus the first computation, using the factored form, gives the accurate value 0.0599, whereas the second gives 0.0600. This error is only one part in one thousand, but can be important in some situations.

[11]In science this is called three significant digits.

This example illustrates one of the ways in which factoring of algebraic expressions can be useful. As we will see in later sections, another important use is in simplifying algebraic fractions. In this section we will investigate several ways in which algebraic expressions can be ''factored.'' Recall that a factor is part of an indicated product—to factor an expression means to write it as a product.

The greatest common factor

This method of factoring applies to expressions with two or more terms and *should always be tried first.* The greatest common factor (GCF) is the greatest expression that divides into each term of the expression. The variables in the GCF are those that appear in every term. The exponent on a variable in the GCF is the *smallest exponent of that variable* found in the terms.

■ ***Example 1–3 A***

Factor.

1. $6x^5y^2 + 12x^3y^4 - 48x^5y^2z^3$ — The GCF is $6x^3y^2$

$= \underline{6x^3y^2} \cdot x^2 + \underline{6x^3y^2} \cdot 2y^2 - \underline{6x^3y^2} \cdot 8x^2z^3$

$= 6x^3y^2(x^2 + 2y^2 - 8x^2z^3)$ — This expression is a product

2. $-20x^6y^3z + 12x^2y^4 - 4x^2y^3$

When the leading coefficient is negative, we usually factor out the negative of the GCF; the GCF is $4x^2y^3$, so we will factor out $-4x^2y^3$.

$= (-4x^2y^3)(5x^4z) + (-4x^2y^3)(-3y) + (-4x^2y^3)(1)$

$= -4x^2y^3(5x^4z - 3y + 1)$

3. $3a(x - 4) + 2b(x - 4) - x + 4$

$= 3a(x - 4) + 2b(x - 4) - (x - 4)$ — Group the last two terms

$= (x - 4)(3a + 2b - 1)$ — Factor out the GCF $(x - 4)$ ■

Factoring by grouping

Some expressions with four or more terms can be factored as illustrated in the following example.

■ ***Example 1–3 B***

Factor $2am^2 - bn^2 - bm^2 + 2an^2$.

There are no common factors in the first two or second two terms. Therefore we try a different arrangement.

$= 2am^2 - bm^2 + 2an^2 - bn^2$ — Rearrange the order of the terms

$= m^2(2a - b) + n^2(2a - b)$ — Factor the common factor from the first two and second two terms

$= (2a - b)(m^2 + n^2)$ — Factor out the common factor $(2a - b)$ ■

Quadratic trinomials

Many three-termed expressions are quadratic trinomials.

Quadratic trinomial

A quadratic trinomial is a polynomial of the form

$$ax^2 + bx + c$$

We will concern ourselves with the case where $a, b, c \in J$.

There are several ways to factor quadratic trinomials,[12] but we will illustrate factoring by inspection. This method relies on the fact that if a quadratic trinomial factors, it factors into a product of two binomials.

■ ***Example 1–3 C***

Factor.

1. $9x^2 - 21x - 8$

There are only two possibilities for the first term in the binomial factors, since we need to form $9x^2$ in the first term:

$$(9x + \cdots)(x + \cdots) \quad \text{or} \quad (3x + \cdots)(3x + \cdots)$$

To obtain -8 in the third term, the missing terms in our binomial are either 1, -8 or -1, 8, or 2, -4 or -2, 4. By trying these possibilities and then multiplying we obtain the correct factors.

$$\begin{array}{lcl} (9x + 1)(x - 8) & \text{or} & (3x + 1)(3x - 8) \\ (9x - 1)(x + 8) & \text{or} & (3x - 1)(3x + 8) \\ (9x + 8)(x - 1) & \text{or} & (3x + 8)(3x - 1) \\ (9x - 8)(x + 1) & \text{or} & (3x - 8)(3x + 1) \\ (9x + 2)(x - 4) & \text{or} & (3x + 2)(3x - 4) \\ & \text{etc.} & \end{array}$$

The product $(3x - 8)(3x + 1)$ is $9x^2 - 21x - 8$, so these are the factors.

2. $x^2 + 3x - 28$

$(x + \cdots)(x + \cdots)$, with missing factors of 1, -28 or -1, 28 or 4, -7 or -4, 7. Trial and error produces the result $(x - 4)(x + 7)$. ■

The difference of two squares

Difference of two squares

An expression of the form $a^2 - b^2$ is a difference of two squares.

[12] In particular the algorithmic method presented in *Intermediate Algebra with Applications* by Terry H. Wesner and Harry L. Nustad, by the same publisher as this text. This method does not depend upon trial and error.

It is easy to verify that

$$a^2 - b^2 = (a - b)(a + b)$$

Thus, a difference of two squares can always be factored into two binomials that are identical except for the sign of the second term. Binomials with this property are called **conjugates.**

Note that *the exponents of any variable that is a perfect square is even.* To see why, observe that $(x^n)^2$ describes any variable being squared, and $(x^n)^2 = x^{2n}$, and $2n$ is an even number.

■ ***Example 1–3 D***

Factor the following.

1. $x^4 - 81$

$= (x^2)^2 - 9^2$ $\quad (x^2)^2 = x^4,\ 9^2 = 81$

$= (x^2 - 9)(x^2 + 9)$

$= (x - 3)(x + 3)(x^2 + 9)$ $\quad x^2 - 9$ is a difference of two squares

Note *A sum of two squares,* such as $x^2 + 9$, *does not factor* using real numbers.

2. $x^6 - 4$

$= (x^3)^2 - 2^2$ $\quad x^6 = (x^3)^2$

$= (x^3 - 2)(x^3 + 2)$ ■

The difference and sum of two cubes

Difference and sum of two cubes

An expression of the form $a^3 - b^3$ is a difference of two cubes.
An expression of the form $a^3 + b^3$ is a sum of two cubes.

Each of these two types of expressions can be written as the product of a binomial and trinomial in the following manner:

$$a^3 - b^3 = (a - b)(a^2 + ab + b^2)$$
$$a^3 + b^3 = (a + b)(a^2 - ab + b^2)$$

The following illustrates how to remember these two patterns. It is illustrated for the difference of two cubes.

$a^3 - b^3$ = binomial · trinomial

The binomial is the difference of the two factors that were cubed: $(a - b)$

The trinomial can be formed from the binomial in the following way:

1. Square the first term of the binomial: $(a - b)(a^2$
2. Use the opposite of the sign in the binomial: $(a - b)(a^2 +$
3. Multiply the values of the two terms in the binomial (disregard the minus sign): $(a - b)(a^2 + ab$
4. Square the second term of the binomial: $(a - b)(a^2 + ab + b^2)$

The same method is valid for the sum of two cubes. See part 2 of example 1–3 E. Also, note that *to find the factor that was cubed* (i.e., a and b), *divide any exponents by 3.* It is also useful to know the cubes of the first few natural numbers:

$$1^3 = 1,\ 2^3 = 8,\ 3^3 = 27,\ 4^3 = 64,\ 5^3 = 125$$

■ ***Example 1–3 E***

Factor the following.

1. $x^3 - 27$

Form the binomial:	$x - 3$	$x^3 - 27 = x^3 - 3^3$
Now form the trinomial:		
Square the first term:	$(x - 3)(x^2$	
Change the sign:	$(x - 3)(x^2 +$	
Multiply the two terms:	$(x - 3)(x^2 + 3x$	
Square the last term:	$(x - 3)(x^2 + 3x + 9)$	

Thus, $x^3 - 27 = (x - 3)(x^2 + 3x + 9)$.

2. $x^6 + 64y^3$

Binomial:	$(x^2 + 4y)$	$x^6 + 64y^3 = (x^2)^3 + (4y)^3$
Square the first term:	$(x^2 + 4y)(x^4$	
Change the sign:	$(x^2 + 4y)(x^4 -$	
Multiply the two terms:	$(x^2 + 4y)(x^4 - 4x^2y$	
Square the last term:	$(x^2 + 4y)(x^4 - 4x^2y + 16y^2)$	

Thus, $x^6 + 64y^3 = (x^2 + 4y)(x^4 - 4x^2y + 16y^2)$. ■

It is a good idea to review when each factoring method is used.

Factoring techniques

When factoring, remember the following:

- Whenever possible, factor out any common factor (i.e., the GCF).
- When there are two terms, think about a difference of two squares or sum/difference of two cubes.
- When there are an even number of terms, and more than two terms, think about grouping.
- When there are three terms, think about a quadratic trinomial.

Substitution for expression

The following two expressions are both quadratic trinomials of the form $y^2 - 2y - 3$; in one the variable y is replaced by $(z + 1)$, and in the other by $(2x - 1)$.

$$y^2 - 2y - 3$$
$$(z + 1)^2 - 2(z + 1) - 3$$
$$(2x - 1)^2 - 2(2x - 1) - 3$$

Since they are of the same form they factor in a similar manner. The method of **substitution for expression** can help factor expressions like the last one. To use this form of substitution, *replace any expression that appears more than once with some variable.* After factoring *replace this variable with the original expression.* This method is a form of the general concept of substitution, a very useful theme that we will see many times throughout this text. (Substitution of value, section 1–2, was another instance of this concept.)

■ Example 1–3 F

Factor using substitution.

1. $(x - 4)^2 + x - 4 - 6$

$(x - 4)^2 + (x - 4) - 6$	
$z^2 + z - 6$	Replace $(x - 4)$ by z
$(z - 2)(z + 3)$	Factor
$[(x - 4) - 2][(x - 4) + 3]$	Replace z by $(x - 4)$
$(x - 6)(x - 1)$	Simplify each factor

2. $(2x - 3)^3 - 27$

$(2x - 3)^3 - 27$	
$z^3 - 27$	Replace $(2x - 3)$ by z
$(z - 3)(z^2 + 3z + 9)$	Factor
$[(2x - 3) - 3][(2x - 3)^2 + 3(2x - 3) + 9]$	Replace z by $(2x - 3)$
$(2x - 6)(4x^2 - 6x + 9)$	Simplify each factor
$2(x - 3)(4x^2 - 6x + 9)$	Common factor in $(2x - 6)$ ■

Several methods of factoring must be used to factor some expressions.

■ Example 1–3 G

Factor.

$a^2(4 - x^2) + 8a(4 - x^2) + 16(4 - x^2)$

$= (4 - x^2)(a^2 + 8a + 16)$	Factor out the common factor $(4 - x^2)$
$= (2 - x)(2 + x)(a + 4)^2$	$4 - x^2$ is a difference of two squares and $a^2 + 8a + 16$ is a quadratic trinomial ■

Mastery points

Can you

- Factor using
  1. greatest common factor?
  2. grouping?
  3. inspection (quadratic trinomials)?
  4. difference of two squares?
  5. sum and difference of two cubes?
- Factor using substitution for expression?

Exercise 1–3

Factor the following expressions using the greatest common factor.

1. $12x^2 - 9xy - 18$
2. $30a^3b^2 - 10ab + 60ab^2$
3. $-20a^4b^2 + 60a^3b - 24a^2b^2$
4. $-40x^4y^3 + 16x^3y^4 + 20x^3y^3$
5. $6x(a - b) + 5y(a - b)$
6. $2a(x + 3) - b(x + 3)$
7. $5a(2x - y) - 2x + y$
8. $3x(4a + 3b) - 4a - 3b$
9. $2m(n + 5) - n - 5 - p(n + 5)$
10. $m - 2n - 5a(m - 2n) + 2b(m - 2n)$

Factor the following expressions using grouping.

11. $ac + ad - 2bc - 2bd$
12. $2ax + 6bx - ay - 3by$
13. $5ax - 3by + 15bx - ay$
14. $3ac - 2bd - 6ad + bc$

Factor the following expressions by inspection.

15. $6x^2 + 13x + 6$
16. $4x^2 - 11x + 6$
17. $x^2 + 7xy + 12y^2$
18. $6x^2 - 17xy + 5y^2$
19. $6a^2 + 13ab - 5b^2$
20. $3x^2 + 4xy - 4y^2$
21. $x^2 - 18x + 32$
22. $y^2 - 4yz - 12z^2$

Factor the following expressions using the difference of two squares.

23. $9x^2 - 25$
24. $2a^4 - 2c^4$
25. $x^4 - 16y^4$
26. $y^4 - 81$

Factor the following expressions using the sum/difference of two cubes.

27. $27x^3 - 1$
28. $a^3 + 8$
29. $8a^3 + 125$
30. $x^3y^3 - 27z^3$
31. $a^9 - b^9$
32. $x^3y^6 + 8z^9$

Factor the following expressions using substitution.

33. $(y - 2)^2 + 5(y - 2) - 36$
34. $(x + 3)^2 + 8(x + 3) + 12$
35. $4(m - n)^2 - 28(m - n) - 32$
36. $(a + b)^2 - a - b - 12$
37. $(2x - y)^2 - (x + y)^2$
38. $4(a + b)^2 - (a - b)^2$
39. $9(3x + 1)^2 - (x - 3)^2$
40. $(a - 2b)^2 - (a + 2b)^2$

Factor the following expressions using several methods.

41. $16x^{12} + 2x^3$
42. $x^2(x^2 - 9) + 2x(x^2 - 9) - 15(x^2 - 9)$
43. $2x^2(a^2 - b^2) + x(a^2 - b^2) - 3a^2 + 3b^2$
44. $x^2(x^2 - 25) - 25(x^2 - 25)$

Factor the following expressions.

45. $m^2 - 49$
46. $81 - x^2$
47. $x^2 + 6x + 5$
48. $a^2 + 11a + 10$
49. $7a^2 + 36a + 5$
50. $3x^2 + 13x + 4$
51. $2a^2 + 15a + 18$
52. $5b^2 + 16b + 12$
53. $a^2b^2 + 2ab - 8$
54. $x^2y^2 - 5xy - 14$
55. $27a^3 + b^3$
56. $x^3 + 64y^3$
57. $25x^2(3x + y) + 5x(3x + y)$
58. $36x^2(2a - b) - 12x(2a - b)$
59. $10x^2 - 20xy + 10y^2$
60. $6a^2 - 24ab - 48b^2$
61. $4m^2 - 16n^2$
62. $9x^2 - 36y^2$
63. $(a - b)^2 - (2x + y)^2$
64. $(3a + b)^2 - (x - y)^2$
65. $3x^6 - 81y^3$
66. $32a^4 - 4ab^9$
67. $12x^3y^2 - 30x^2y^3 + 18xy^4$
68. $9x^5y - 6x^3y^3 + 3x^2y^2$
69. $4x^2 - 36y^2$
70. $36 - a^2b^4$
71. $3ax - 2by - bx + 6ay$
72. $6am - 3an + 4bm - 2bn$
73. $27a^9 - b^3c^3$
74. $x^{12} - y^3z^6$
75. $5a^2 - 32a - 21$
76. $7a^2 + 16a - 15$
77. $a^4 - 5a^2 + 4$
78. $x^4 - 37x^2 + 36$
79. $4a^2 - 4ab - 15b^2$
80. $6x^2 + 7xy - 3y^2$
81. $y^4 - 16$
82. $a^4 - 81$
83. $4a^2 + 10a + 4$
84. $6x^2 + 18x - 60$
85. $(x + y)^2 - 8(x + y) - 9$
86. $(a - 2b)^2 + 7(a - 2b) + 10$
87. $6a^2 + 7a - 5$
88. $4x^2 + 17x - 15$
89. $4ab(x + 3y) - 8a^2b^2(x + 3y)$
90. $3x^2y(m - 4n) + 15xy^2(m - 4n)$
91. $4a^2 - 20ab + 25b^2$
92. $9m^2 - 30mn + 25n^2$
93. $80x^5 - 5x$
94. $3b^5 - 48b$
95. $3a^5b - 18a^3b^3 + 27ab^5$
96. $3x^3y^3 + 6x^2y^4 + 3xy^5$
97. $9a^2 - (x + 5y)^2$
98. $7x(a^2 - 4b^2) + 14(a^2 - 4b^2)$
99. $3x^2 + 8x - 91$
100. $3x^2 - 32x - 91$
101. $24x^5y^9 + 81x^2z^6$

102. $a^8b^4 + 27a^2b^7$

103. $a^2(9 - x^2) - 6a(9 - x^2) - 9(x^2 - 9)$

104. $x^2(16 - b^2) - 10x(b^2 - 16) + 25(16 - b^2)$

105. The area of a circle is $A = \pi r^2$, where r is the radius. The area of an annular ring with inner radius r_1 and outer radius r_2 would therefore be $\pi r_2^2 - \pi r_1^2$. Factor this expression. (If a computer program were being written to compute the area of many annular rings, the factored form would be far more efficient and, under certain circumstances, more accurate. See the discussion at the beginning of this section for the reasons why this is true.)

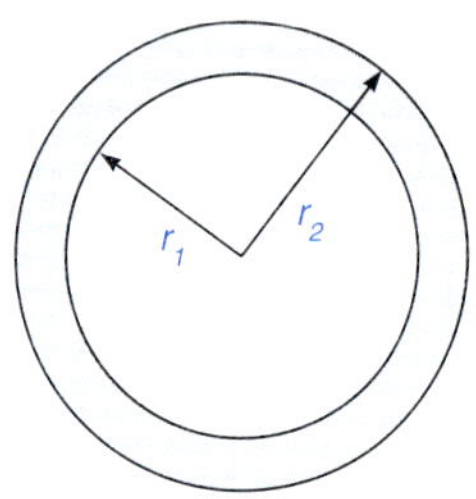

106. A freely falling body near the surface of the earth falls a distance $s = \frac{1}{2}gt^2$ in time t. The difference of two such distances measured for two different times would therefore be $\frac{1}{2}gt_2^2 - \frac{1}{2}gt_1^2$. Factor this expression.

107. Multiplication and factoring of expressions sometimes have a geometric interpretation. For example, since the area of a rectangle is the product of its two dimensions, the square shown represents $(x + y)^2 = x^2 + 2xy + y^2$. Construct a rectangle that would represent $2x^2 + 7xy + 3y^2 = (2x + y)(x + 3y)$.

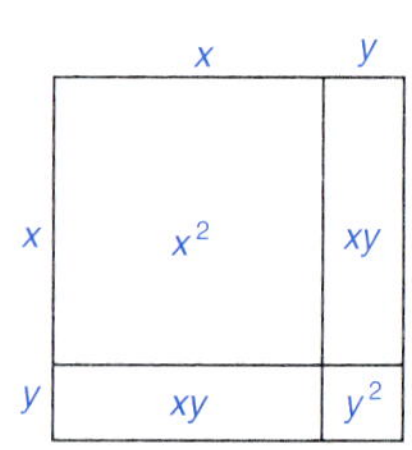

108. (See problem 107.) Construct a square that represents $(x - y)^2 = x^2 - 2xy + y^2$.

109. Factor $x^6 - 1$ two different ways. First, as a difference of two squares and second as a difference of two cubes. Draw a conclusion about the expression $x^4 + x^2 + 1$.

110. If the top of a pyramid with a square base is cut off parallel to the base, the resulting figure is called a *frustum* (the solid shown in the figure).

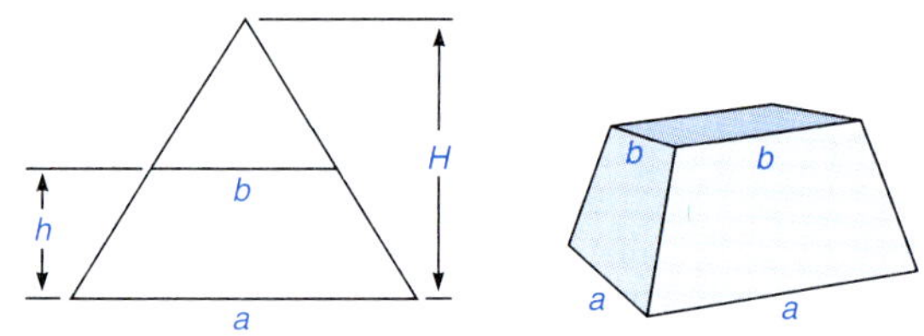

The volume of a frustum can be determined in the following manner.

The volume of a pyramid is $\frac{1}{3}hA$, where h is the height of the pyramid and A is the area of its base. This can be used to find the volume of a frustum: find the volume of the entire pyramid and subtract the volume at the top, which is missing. This missing volume is also a pyramid. Use this idea to show that the volume of a frustum with square base of length a, with height h, and with top horizontal dimension b (see the figure) is $\frac{1}{3}h(a^2 + ab + b^2)$. Use the following hints.

The volume of the entire pyramid shown is $\frac{1}{3}a^2H$. The volume of the top, missing, pyramid is $\frac{1}{3}b^2(H - h)$. The required volume V is the difference between these two values, or[13]

$$V = \frac{1}{3}[a^2H - b^2(H - h)].$$

Also, the geometric property of proportion between similar figures allows us to conclude that $\dfrac{a}{H} = \dfrac{b}{H - h}$. Use this to solve for H, and substitute the resulting expression for H in the formula for V. This will leave the formula for V using only the variables a, b, and h.

[13]A problem that correctly finds the volume of the frustum of a pyramid is found on an Egyptian papyrus in the Moscow Museum of Fine Arts. The papyrus is at least 3,600 years old.

111. Write a calculator or computer program that will compute the greatest common factor of two integers. The greatest common factor of two integers is the largest integer that evenly divides into both integers. One method of finding the greatest common factor of two integers is "Euclid's method." It is described as follows:

[1] Let x,y be two integers, $x > y$.
[2] Let d be the integer result of x/y.
[3] Let $r = x - dy$.
[4] If $r = 0$, the greatest common factor is y, and the process stops.
[5] Otherwise, replace x by y, and replace y by r.
[6] Proceed to the second step.

112. Write a calculator or computer program that will compute the integer coefficients for the factors of quadratic trinomials. For example, for the trinomial $3x^2 + 10x - 8$, the program would provide the coefficients 1, 4, 3, and -2, which means that $3x^2 + 10x - 8 = (x + 4)(3x - 2)$. This is a *hard* problem. Some hints for one solution follow.

Assume you are given $Ax^2 + Bx + C$. The object is to find a, b, c, and d so that $(ax + b)(cx + d)$. First, find two integers m and n such that $mn = AC$ and $m + n = B$. Under these conditions, $Ax^2 + Bx + C$ can be rewritten $Ax^2 + mx + nx + C$. Then a is the greatest common factor of A and m, and b is the greatest common factor of n and C. Also, the sign of b is the same as the sign of n. Further, c is A/a, and d is C/b. See problem 111 for a discussion of finding the greatest common factor of two integers.

Skill and review

1. For what value(s) of x is $2x - 5$ equal to 0?
2. For what value(s) of x is $x^2 + 4$ negative or zero?
3. $-\frac{-1}{2}$ is the same as **a.** $\frac{1}{-2}$ **b.** $-\frac{1}{2}$ **c.** $\frac{1}{2}$ **d.** 2
4. Reduce $\frac{3x^3}{6x^6}$.
5. Multiply $\frac{3}{4} \cdot \frac{12}{5}$.
6. Add $\frac{2}{3} + \frac{1}{4}$.
7. Simplify $(2x - 3)(x + 1) - (x - 2)(x - 1)$.
8. Divide $\frac{5}{8} \div 2$.

1–4 Rational expressions

If one printer can print x pages per hour and another can print $x + 2$ pages per hour, then the combined rate for these printers is $\frac{1}{x} + \frac{1}{x+2}$. Combine this expression.

This problem illustrates the fact that many physical situations are represented by expressions in which a variable appears in the denominator. In this section we examine expressions that are fractions in which the numerator and denominator are polynomials. This is reflected in the following definition.

Rational expression

A rational expression is an expression of the form $\frac{P}{Q}$, where P and Q are polynomials and $Q \neq 0$.

The following are rational expressions: $\frac{2x}{3}$, $\frac{3x - 2}{x^2 - 5x - 9}$, and $\frac{x - 2}{x + 8}$.

Reducing rational expressions

Just as the fraction $\frac{2}{8}$ can be reduced to $\frac{1}{4}$, so many rational expressions can also be reduced. This is done by using the fundamental principle of rational expressions.

Fundamental principle of rational expressions

If P, Q, and R are polynomials, $Q \neq 0$, and $R \neq 0$, then $\frac{P \cdot R}{Q \cdot R} = \frac{P}{Q}$.

This principle states that factors that are common to the numerator and denominator of a rational expression (R) may be eliminated. When all common factors except 1 or -1 are eliminated, we say the rational expression is *reduced to its lowest terms.*

We need to comment about the signs of rational expressions. A useful principle is that the following are equivalent:

$$\frac{-a}{b} = \frac{a}{-b} = -\frac{a}{b}$$

■ ***Example 1–4 A***

Reduce each rational expression to its lowest terms.

1. $\frac{6x^5y}{8x^3y} = \frac{3x^2}{4}$ — Reduce by GCF $2x^3y$

2. $\frac{2x - 6}{2x^2 - x - 15} = \frac{2(x - 3)}{(x - 3)(2x + 5)}$ — Factor

$= \frac{2}{2x + 5}$ — Reduce by GCF $(x - 3)$

3. $-\frac{3 - x}{x^2 - 9} = -\frac{-(x - 3)}{(x - 3)(x + 3)}$ — $3 - x = -(x - 3)$

$= -\frac{-1}{x + 3} = \frac{1}{x + 3}$ — $-\frac{-a}{b} = \frac{a}{b}$ ■

Just as we saw with polynomials (section 1–2), there is an arithmetic of rational expressions. We will see how to perform addition, subtraction, multiplication, and division with rational expressions.

Multiplication and division of rational expressions

Multiplication and division of rational expressions are done in the same way as with fractions in arithmetic.

> **Multiplication of rational expressions**
>
> If P, Q, R, and S are polynomials, $Q \neq 0$, and $S \neq 0$, then
>
> $$\frac{P}{Q} \cdot \frac{R}{S} = \frac{PR}{QS}$$
>
> **Division of rational expressions**
>
> If P, Q, R, and S are polynomials, $Q \neq 0$, $R \neq 0$, $S \neq 0$, then
>
> $$\frac{P}{Q} \div \frac{R}{S} = \frac{P}{Q} \cdot \frac{S}{R}$$

The terms $\frac{R}{S}$ and $\frac{S}{R}$ are called **reciprocals** of each other. To divide by an expression is therefore equivalent to multiplying by the reciprocal of the divisor.

■ ***Example 1–4 B***

Find the indicated products or quotients. Assume all denominators represent nonzero numbers.

1. $\dfrac{3x - 1}{x - 4} \cdot \dfrac{x - 4}{3x^2 + 14x - 5}$

$$= \frac{(3x - 1)(x - 4)}{(x - 4)(3x - 1)(x + 5)}$$ Factor; $\frac{P}{Q} \cdot \frac{R}{S} = \frac{PR}{QS}$

$$= \frac{1}{x + 5}$$ Reduce by $(x - 4)$, $(3x - 1)$

2. $\dfrac{x^2 + 2x}{5x} \div (2x^2 + x - 6)$

$$= \frac{x^2 + 2x}{5x} \cdot \frac{1}{2x^2 + x - 6}$$ $\frac{P}{Q} \div \frac{R}{S} = \frac{P}{Q} \cdot \frac{S}{R}$; $a = \frac{a}{1}$

$$= \frac{x^2 + 2x}{5x(2x^2 + x - 6)} = \frac{x(x + 2)}{5x(2x - 3)(x + 2)} = \frac{1}{5(2x - 3)}$$ ■

Addition and subtraction of rational expressions

Two properties govern addition and subtraction of rational expressions.

> **Addition/subtraction of rational expressions**
>
> If P, Q, R, and S are polynomials, $Q \neq 0$, and $S \neq 0$, then
>
> [1] $$\frac{P}{Q} + \frac{R}{Q} = \frac{P + R}{Q} \text{ and } \frac{P}{Q} - \frac{R}{Q} = \frac{P - R}{Q}$$
>
> ***Concept***
>
> If two rational expressions have the same denominator they may be added or subtracted by adding or subtracting the numerators only and retaining the denominator.
>
> [2] $$\frac{P}{Q} + \frac{R}{S} = \frac{PS + QR}{QS} \text{ and } \frac{P}{Q} - \frac{R}{S} = \frac{PS - QR}{QS}$$
>
> ***Concept***
>
> If two rational expressions have different denominators they may be added or subtracted by "cross multiplying" (see the following discussion).

Property [1] is the definition of addition and subtraction of rational expressions, and is used when adding or subtracting rational expressions with the same denominator. Property [2], sometimes called "cross multiplying," is a short cut for adding/subtracting rational expressions with different denominators. (We saw the same principle in section 1–1 for real numbers.) It avoids having to find the least common denominator. Property [2] can be remembered as a sequence of three multiplications, which are shown graphically as $\frac{P}{Q} \overset{1 \quad 2}{\underset{3 \;\to}{\times}} \frac{R}{S}$. Observe that the first product should be PS, not QR; this is necessary in the case of subtraction. A demonstration that property [2] is valid is left as an exercise.

■ ***Example 1–4 C***

Add or subtract.

1. $$\frac{3x - 1}{x - 3} - \frac{5x + 4}{x - 3} = \frac{(3x - 1) - (5x + 4)}{x - 3} \qquad \frac{P}{Q} - \frac{R}{Q} = \frac{P - R}{Q}$$

$$= \frac{-2x - 5}{x - 3} = -\frac{2x + 5}{x - 3}$$

Note A common error is to omit the parentheses in expressions where subtraction is involved, as in $\frac{3x - 1 - 5x + 4}{x - 3}$. This error produces the wrong result.

2. $\dfrac{3x}{2a} + \dfrac{5y}{3b} = \dfrac{3x(3b) + 2a(5y)}{2a(3b)}$

$= \dfrac{9bx + 10ay}{6ab}$

$\dfrac{P}{Q} + \dfrac{R}{S} = \dfrac{PS + QR}{QS}$
(cross multiply)

3. $\dfrac{3x - 1}{x - 2} - \dfrac{5x + 4}{x - 3}$

$= \dfrac{(3x - 1)(x - 3) - (x - 2)(5x + 4)}{(x - 2)(x - 3)}$

$\dfrac{P}{Q} - \dfrac{R}{S} = \dfrac{PS - QR}{QS}$
(cross multiply)

$= \dfrac{-2x^2 - 4x + 11}{x^2 - 5x + 6}$ ■

The properties above serve well when the least common denominator of the rational expressions is their product. However, in other cases it pays to find the least common denominator first and convert each expression to one having this as its denominator. The **least common denominator** (LCD) of two or more rational expressions is the smallest expression into which each denominator will divide. For example, for the fractions $\frac{5}{12}$, $\frac{3}{20}$, and $\frac{7}{30}$, the LCD is 60, since 12, 20, and 30 all divide evenly into 60, but they do not divide into any smaller number. The following rules describe how to find the LCD of two or more rational expressions when we cannot determine this by inspection.

> **To find the LCD of two or more rational expressions:**
> - Factor each denominator completely. The resulting factors are called **prime factors.**
> - Write the product of each prime factor.
> - Apply the greatest exponent of each factor found in the previous step.

Note that when an integer is factored completely, the factors are called **prime numbers.** These are the natural numbers that are divisible (evenly) only by one and itself. The first few prime numbers are 2, 3, 5, 7, 11, 13, 17, 19, 23, 29, 31, ... There are an infinite number of prime numbers.

■ *Example 1–4 D*

Add or subtract.

1. $\dfrac{3}{2x^2} - \dfrac{5}{6x} + \dfrac{5}{4xy}$

To find the LCD we first factor each denominator completely:

$$\left.\begin{aligned} 2x^2 &= 2x^2 \\ 6x &= 2 \cdot 3x \\ 4xy &= 2^2xy \end{aligned}\right\} \text{The prime factors are 2, 3, } x\text{, and } y.$$

The highest exponent of the factor 2 is 2; for 3 it is 1; for x it is 2; for y the highest exponent is 1.

$2 \cdot 3xy$	Form the product of the prime factors
$2^2 \cdot 3x^2y$	Apply the largest exponent of each prime factor
$12x^2y$	Simplify

Now multiply both the numerator and denominator of each fraction by the factor that makes that denominator the LCD:

$$\frac{3}{2x^2} \cdot \frac{6y}{6y} - \frac{5}{6x} \cdot \frac{2xy}{2xy} + \frac{5}{4xy} \cdot \frac{3x}{3x}$$
$$= \frac{18y}{12x^2y} - \frac{10xy}{12x^2y} + \frac{15x}{12x^2y}$$
$$= \frac{18y - 10xy + 15x}{12x^2y}$$

2. $\dfrac{2x}{x+2} + \dfrac{4}{x^2-4} - \dfrac{2}{x^2-4x+4}$

$$= \frac{2x}{x+2} + \frac{4}{(x+2)(x-2)} - \frac{2}{(x-2)^2}$$

The LCD is $(x + 2)(x - 2)^2$.

$$= \frac{2x}{x+2} \cdot \frac{(x-2)^2}{(x-2)^2} + \frac{4}{(x+2)(x-2)} \cdot \frac{x-2}{x-2} - \frac{2}{(x-2)^2} \cdot \frac{x+2}{x+2}$$
$$= \frac{2x^3 - 8x^2 + 8x}{(x+2)(x-2)^2} + \frac{4x-8}{(x+2)(x-2)^2} - \frac{2x+4}{(x+2)(x-2)^2}$$
$$= \frac{2x^3 - 8x^2 + 8x + (4x-8) - (2x+4)}{(x+2)(x-2)^2}$$
$$= \frac{2x^3 - 8x^2 + 8x + 4x - 8 - 2x - 4}{(x+2)(x-2)^2}$$
$$= \frac{2x^3 - 8x^2 + 10x - 12}{x^3 - 2x^2 - 4x + 8}$$

Complex rational expressions

A complex rational expression is one in which the numerator or denominator is itself a rational expression. Such expressions can be simplified in either of two ways.

■ ***Example 1–4 E***

Simplify the complex rational expression $\dfrac{\dfrac{6}{a^2} - 4}{6 - \dfrac{2}{a}}$.

Method 1: Perform the indicated division.

$$\frac{\dfrac{6}{a^2} - 4}{6 - \dfrac{2}{a}}$$

$$= \left(\frac{6}{a^2} - \frac{4}{1}\right) \div \left(\frac{6}{1} - \frac{2}{a}\right) \qquad 4 = \frac{4}{1};\ 6 = \frac{6}{1}$$

$$= \frac{6(1) - 4a^2}{1(a^2)} \div \frac{6a - 2(1)}{1(a)} \qquad \frac{a}{b} - \frac{c}{d} = \frac{ad - bc}{bd}$$

$$= \frac{6 - 4a^2}{a^2} \cdot \frac{a}{6a - 2} \qquad \frac{a}{b} \div \frac{c}{d} = \frac{a}{b} \cdot \frac{d}{c}$$

$$= \frac{a(6 - 4a^2)}{a^2(6a - 2)}$$ Multiply numerators and denominators

$$= \frac{2a(3 - 2a^2)}{2a^2(3a - 1)}$$ Factor

$$= \frac{3 - 2a^2}{a(3a - 1)}$$

Method 2: Multiply the numerator and denominator by the LCD of $\dfrac{6}{a^2}$ and $\dfrac{2}{a}$, which is a^2.

$$\frac{\left(\dfrac{6}{a^2} - 4\right) \cdot a^2}{\left(6 - \dfrac{2}{a}\right) \cdot a^2}$$

$$= \frac{\dfrac{6}{a^2} \cdot \dfrac{a^2}{1} - 4 \cdot a^2}{6 \cdot a^2 - \dfrac{2}{a} \cdot \dfrac{a^2}{1}}$$ Distributive axiom

$$= \frac{6 - 4a^2}{6a^2 - 2a}$$

$$= \frac{2(3 - 2a^2)}{2a(3a - 1)}$$

$$= \frac{3 - 2a^2}{a(3a - 1)}$$ ■

Mastery points

Can you

- Reduce rational expressions to lowest terms?
- Perform addition, subtraction, multiplication, and division involving rational expressions?
- Simplify complex rational expressions?

Exercise 1–4

Reduce each rational expression to lowest terms.

1. $\frac{24p^5q^7}{18pq^3}$
2. $\frac{-27mn^3}{36m^4n}$
3. $\frac{4a+12}{3a+9}$
4. $\frac{-36x}{42x^3+24x}$
5. $\frac{a^2-9}{4a+12}$
6. $\frac{6a+3b}{4a^2-b^2}$
7. $\frac{64-49p^2}{7p-8}$
8. $\frac{m^2-4m-12}{m^2-m-6}$
9. $\frac{6a^3-6b^3}{a^2-b^2}$
10. $\frac{2a^2-3a+1}{2a^2+a-1}$
11. $\frac{8-2a}{a^3-64}$
12. $\frac{6x^2+17x+7}{12x^2+13x-35}$
13. $\frac{3a^2+16a-12}{6-7a-3a^2}$
14. $\frac{x^2+3x-10}{x^3-8}$

Perform the indicated additions and subtractions.

15. $\frac{3x}{2y}-\frac{2y}{5x}$
16. $\frac{a-2b}{3a}+\frac{5a-b}{2b}$
17. $\frac{x}{x+1}-\frac{3-x}{x-1}$
18. $\frac{2x-1}{x+3}-\frac{5x-1}{x+3}$
19. $\frac{3x-5}{x-4}+\frac{3x+2}{4-x}$
20. $\frac{3}{4}-\frac{2x}{3x+5}$
21. $\frac{3}{5}-\frac{2}{a}+\frac{9}{2b}$
22. $\frac{x-1}{x}-\frac{x-1}{x^2+x+1}$

Perform the indicated operations.

23. $\frac{12b}{7a}\cdot\frac{28a^3}{4b^3}$
24. $\frac{a-6}{6a+18}(a+3)$
25. $\frac{2}{x-3}+\frac{5}{x^2-3x}$
26. $\frac{5x}{x^2-4}-\frac{3}{x-2}$
27. $\frac{-6a}{a^2-a-6}-\frac{7a}{a^2+7a+10}$
28. $\frac{12}{5y-10}+\frac{7}{2y+4}$
29. $\frac{2a-1}{2a-3}+\frac{4-a}{3-2a}$
30. $\frac{x+2}{x-8}-\frac{x-6}{8-x}$
31. $\frac{4a+8}{3a-12}\cdot\frac{a-2}{5a+10}$
32. $\frac{4x^2-49}{8x^3+27}\cdot\frac{4x^2+12x+9}{2x^2-13x+21}$
33. $(x^2-2x-3)\div\frac{4x-4}{x^2-1}$
34. $\frac{r+4}{r^2-1}\div\frac{r^2-16}{r+1}$
35. $\frac{3y}{5y-10}+\frac{19}{2y+4}$
36. $\frac{10}{4a-6}-\frac{13}{3a+9}$
37. $\frac{3y}{y^2+5y+6}-\frac{5}{4-y^2}$
38. $\frac{3}{6b^2-4bc}-\frac{4}{6c^2-9bc}$
39. $\frac{m^2-9}{3m+4}\cdot\frac{9m^2-16}{m^2+6m+9}$
40. $\frac{a^2+2a+1}{1-4a}\cdot\frac{16a^2-1}{a^2-1}$
41. $\frac{x^2-25}{2x+10}\div(x^2-10x+25)$
42. $(4x^2-9)\div\frac{4x+6}{x+3}$

Simplify each complex rational expression.

43. $\dfrac{\frac{1}{3}+\frac{1}{2}}{\frac{2}{3}-\frac{1}{6}}$

44. $\dfrac{3-\frac{1}{5}}{5+\frac{3}{10}}$

45. $\dfrac{1+\dfrac{1}{a}}{2-\dfrac{3}{ab}}$

46. $\dfrac{\dfrac{2}{a}-\dfrac{3}{b}}{\dfrac{5}{ab}+\dfrac{3}{2a}}$

47. $\dfrac{\dfrac{3x-4}{8}}{\dfrac{2x+1}{10}}$

48. $\dfrac{\dfrac{2}{3a+2}}{\dfrac{4}{3a-2}}$

49. $\dfrac{\frac{2}{3}-\frac{1}{2}}{\frac{2}{3}+\frac{1}{2}}$

50. $\dfrac{2-\frac{3}{5}}{5+\frac{1}{5}}$

51. $\dfrac{\dfrac{m^2-n^2}{n}}{\dfrac{1}{n}-\dfrac{1}{m}}$

52. $\dfrac{y-\dfrac{x}{y}}{x-\dfrac{y}{x}}$

53. $\dfrac{5-\dfrac{3}{x+2}}{3+\dfrac{2}{x+2}}$

54. $\dfrac{\dfrac{6}{x^2-x}-2}{\dfrac{3}{x-1}+2}$

55. $\dfrac{\dfrac{1}{x-1}-\dfrac{2}{x+1}}{\dfrac{6}{x+1}-\dfrac{3}{x-1}}$

56. $\dfrac{\dfrac{x}{x-4}+\dfrac{3}{x}}{5-\dfrac{9}{x^2-4x}}$

57. $\dfrac{\dfrac{a}{a-b}+\dfrac{a}{a+b}}{\dfrac{3}{a^2-b^2}}$

58. $\dfrac{\dfrac{2}{x}-\dfrac{3}{x+y}}{\dfrac{2}{x}+\dfrac{3}{x+y}}$

59. In electricity theory the following expression arises: $\dfrac{\dfrac{V_1}{R_1}+\dfrac{V_2}{R_2}+\dfrac{V_3}{R_3}}{\dfrac{1}{R_1}+\dfrac{1}{R_2}+\dfrac{1}{R_3}}$. Simplify this complex fraction.

60. Simplify this complex fraction from electricity theory: $\dfrac{\dfrac{V_1}{R_1}+\dfrac{V_2}{R_2}}{\dfrac{1}{R_1}+\dfrac{1}{R_2}}$.

61. When making a round trip whose one-way distance is d, the average rate (speed) traveled, r, is $\dfrac{2d}{\dfrac{d}{r_1}+\dfrac{d}{r_2}}$, where r_1 and r_2 are the average rates for each direction of the trip. Simplify this expression.

62. If one printer can print x pages per hour and another can print $x + 2$ pages per hour, then the combined rate for these printers is $\dfrac{1}{x}+\dfrac{1}{x+2}$. Combine this expression.

63. If x is the speed of a boat in still water (in knots), and $x + 3$ is the speed of the boat in a 3 knot current, then the difference in times it will take to cover one knot under these two different conditions is $\dfrac{1}{x}-\dfrac{1}{x+3}$. Combine this expression.

64. If two investments have a rate of return of r_1 and r_2, with $r_1 > r_2$, then the difference in time it would take each investment to produce an amount of interest I on a principal P is $\dfrac{I}{Pr_2}-\dfrac{I}{Pr_1}$. Combine these expressions.

65. The following were presented as properties of addition and subtraction of rational expressions:

$$\frac{P}{Q}+\frac{R}{S}=\frac{PS+QR}{QS} \text{ and } \frac{P}{Q}-\frac{R}{S}=\frac{PS-QR}{QS}.$$

Prove that these are true by computing

$$\frac{P}{Q}+\frac{R}{S} \text{ and } \frac{P}{Q}-\frac{R}{S},$$

first obtaining a common denominator.

66. In studying the reliability of mechanical systems, MTBF means mean (average) time between failures. If a system is made up of a series of three devices with MTBF of $MTBF_1$, $MTBF_2$, and $MTBF_3$, then the MTBF for the system, $MTBF_S$, is

$$MTBF_S = \left(\frac{1}{MTBF_1}+\frac{1}{MTBF_2}+\frac{1}{MTBF_3}\right)^{-1}.$$

Rewrite so the three rational expressions are combined and the negative exponent is removed.

67. See the previous problem for terminology. A certain computer contains three main sections: CPU (central processor unit), power supply, and disk memory. Suppose the CPU has a MTBF of 1,000 hours, the power supply has a MTBF of 3,000 hours, and the disk memory has a MTBF of 1,800 hours. Since the failure of any of these parts causes the failure of the computer, the series formula of problem 66 applies for finding the MTBF of the computer. Find this value to the nearest hour for this computer.

68. In pulmonary function testing in medicine, pulmonary compliance is the volume change per unit of pressure change for the lungs (C_L), the thorax (C_T), or the lungs-thorax system (C_{LT}). The compliance for all three is recorded in liters per centimeter of water. These values are related by $\frac{1}{C_L} + \frac{1}{C_T} = \frac{1}{C_{LT}}$. Find an expression for C_{LT}.

69. Write a program for a computer or programmable calculator that will determine if a given natural number is prime. The program either prints out the first prime factor of the number or the word "Prime."

Since no even number after two is prime (why?), and even numbers are easy to recognize (how?) it is only necessary to test odd natural numbers. The easiest method is to simply divide the number by all odd integers beginning with three. It is only necessary to go as far as the square root of the number.

By way of example, to determine whether 323 is prime we could divide by 3, 5, 7, 9, . . . , 17, since $\sqrt{323} \approx 17.97$.

Skill and review

1. Compute **a.** $\sqrt{25}$ **b.** $\sqrt{100}$ **c.** $\sqrt{400}$

2. Compute **a.** $\sqrt[3]{8}$ **b.** $\sqrt[3]{64}$ **c.** $\sqrt[6]{64}$

3. Compute **a.** $\sqrt{4 \cdot 9}$ **b.** $\sqrt{4} \cdot \sqrt{9}$

4. Multiply $2x^2y^3(2^3x^2y)$.

5. Find the smallest integer that is greater than or equal to the given integer and is divisible by 4.

Example: 10 The answer is 12 because 12 is divisible by 4 but 10 and 11 are not.

Example: 37 The answer is 40 because 40 is divisible by 4 but 37, 38, and 39 are not.

a. 7 **b.** 8 **c.** 13 **d.** 44 **e.** 54

6. Consider $81a^4b^8 = 3a^2b^5(3^xa^yb^z)$. Find x, y, and z.

1–5 Radicals

The advertisement for an apartment says it has an enormous square living room with over 900 square feet. What is the length and width of the room?

The area of a square is the square of its length. For this living room the length must be about 30 feet, because 30^2 is 900. We could say that $\sqrt{900}$ (the square root of 900) is 30.

Definition of a radical

Consider the symbol $\sqrt{9}$. It means "the square root of 9"; it is "the number that, when squared, gives 9." This number is 3. (Note that -3 squared also gives 9, however.) Similarly, $\sqrt[3]{8}$ means "the cube root of 8"; this is 2, since $2^3 = 8$. We will generalize this idea in the following definition.[14]

[14]The $\sqrt{\ }$ symbol, now called the radix symbol, was introduced by Christoff Rudolff in 1526. The more general notation $\sqrt[n]{\ }$ was used in 1891 by Giuseppe Peano.

> **Principal *n*th root of *a***
>
> Let $a, b \varepsilon R$, a and b have the same sign (or are both zero), $n \varepsilon N$. Then $\sqrt[n]{a} = b$ means $b^n = a$. b is called the principal nth root of a.
>
> ***Concept***
>
> b is the number whose nth power is a; if $a > 0$ then $b > 0$.

Note We require b and a to have the same sign so that b is positive when n is even; this makes $\sqrt{9}$ be 3 and not -3, for example.

We read $\sqrt[n]{a}$ "the principal nth root of a" although we often omit the word "principal" for convenience. $\sqrt[n]{a}$ is called a **radical,** n is the **index,** and a is the **radicand.** When the index is 2 it is omitted; this is called the square root. When the index is 3 this is called the cube root. The rest are verbalized as the fourth root, fifth root, etc.

■ ***Example 1–5 A***

Find the indicated root.

1. $\sqrt[3]{8} = 2$ — $2^3 = 8$
2. $\sqrt[4]{625} = 5$ — $5^4 = 625$
3. $\sqrt[3]{-27} = -3$ — $(-3)^3 = -27$
4. $-\sqrt{25} = -5$ — $-\sqrt{25} = -(\sqrt{25}) = -5$
5. $\sqrt{-25}$ Not a real number ■

There is no real value b such that $b^2 = -25$ (since any real number squared produces a nonnegative value). Thus, $\sqrt{-25}$ is not a real number. For the same reason, *when the radicand is negative and the index is even, the radical does not represent a real number.* We will see how to deal with these cases when we examine the system of complex numbers in section 1–7.

The nth root of most real numbers is an irrational number and can only be approximated. This can be done with a calculator, which will be illustrated in the next section, after discussing rational exponents.

Simplifying radicals

We will state what it means to simplify a radical later in this section, but first we examine specific cases where radicals can be written more simply.

Whenever the index of a radical is equal to the exponent of the radicand the following property can be applied.

> ***n*th root of *n*th power property**
>
> $$\sqrt[n]{a^n} = \begin{cases} |a| & \text{if } n \text{ is even} \\ a & \text{if } n \text{ is odd} \end{cases}$$
>
> In particular $\sqrt{x^2} = |x|$.

Note If $a \geq 0$ then $\sqrt[n]{a^n} = a$ whether n is even or odd, since in this case $|a| = a$.

Thus, $\sqrt[3]{(-2)^3} = -2$ — n is odd so $\sqrt[n]{a^n} = a$

but $\sqrt{(-2)^2} = 2$ — n is even so $\sqrt[n]{a^n} = |a|$

This property can also be used to simplify a radical like $\sqrt[3]{8x^3y^6}$. The radicand can be viewed as a cube by rewriting as $\sqrt[3]{(2xy^2)^3}$. The nth root of nth power property then tells us that this is $2xy^2$.

■ ***Example 1–5 B***

Simplify. Assume the variables are nonnegative in 1 through 3.

1. $\sqrt{5^6} = \sqrt{(5^3)^2} = 5^3 = 125$

2. $\sqrt{81a^8b^2c^4} = \sqrt{9^2a^8b^2c^4}$ — Rewrite 81 as 9^2

$= \sqrt{(9a^4bc^2)^2} = 9a^4bc^2$

3. $\sqrt[5]{32a^5b^{10}c^{20}} = \sqrt[5]{2^5a^5b^{10}c^{20}}$

$= \sqrt[5]{(2ab^2c^4)^5} = 2ab^2c^4$

In the following examples, variables may represent negative values. Thus, we use $\sqrt{x^2} = |x|$.

4. $\sqrt{4x^2y^4} = \sqrt{(2xy^2)^2} = |2xy^2| = 2y^2|x|$

5. $\sqrt{x^2 - 4x + 4} = \sqrt{(x-2)^2} = |x - 2|$

6. $\sqrt{16a^4b^8} = \sqrt{(4a^2b^4)^2} = |4a^2b^4|$

$= 4a^2b^4$ — $a^2 \geq 0$, $b^4 \geq 0$ ■

Multiplication and division of radicals, and more on simplification

The procedure illustrated above applies when the index of the radical divides evenly into the exponent of each factor of the radicand. However, the index of the radical may not divide evenly into each exponent of the radicand; in this case, the following property can be used. It applies *when the exponent of any factor of the radicand is greater than or equal to the index.*

> **Product property of radicals**
> If $a \geq 0$, $b \geq 0$, and $n \in N$, then $\sqrt[n]{ab} = \sqrt[n]{a}\sqrt[n]{b}$.

The product property tells us how to multiply radicals as well as how to simplify certain radicals.

■ ***Example 1–5 C***

Simplify. Assume all variables are nonnegative.

1. $\sqrt{8a^5b^8} = \sqrt{2^3a^5b^8}$

Observe that the index 2 does not divide into all of the exponents; we can use the product rule for radicals as follows.

$= \sqrt{2^2 \cdot 2 \cdot a^4ab^8}$ — $2^3 = 2^2 \cdot 2$; $a^5 = a^4 \cdot a$

$= \sqrt{2^2a^4b^8}\sqrt{2a}$ — Product rule for radicals

We have factored the radical into two radicals; in the first, each exponent is divisible by the index, 2, and in the second, each exponent is less than the index.

$$= \sqrt{(2a^2b^4)^2}\sqrt{2a}$$
$$= 2a^2b^4\sqrt{2a} \qquad n\text{th root of } n\text{th power}$$

2. $\sqrt{32} = \sqrt{2^5} = \sqrt{2^4}\sqrt{2} = 2^2\sqrt{2} = 4\sqrt{2}$

3. $\sqrt[3]{54x^7y^2} = \sqrt[3]{2 \cdot 3^3x^6 \cdot xy^2}$ $\qquad 54 = 2 \cdot 27 = 2 \cdot 3^3;\ x^7 = x^6 \cdot x$

$$= \sqrt[3]{3^3x^6}\sqrt[3]{2xy^2} = 3x^2\sqrt[3]{2xy^2}$$

4. $\sqrt[4]{96a^5b^8c^{15}} = \sqrt[4]{2^5 \cdot 3a^5b^8c^{15}} = \sqrt[4]{2^4a^4b^8c^{12}}\sqrt[4]{2 \cdot 3ac^3} = 2ab^2c^3\sqrt[4]{6ac^3}$

Perform the indicated multiplications, then simplify. Assume all variables are nonnegative.

5. $(2\sqrt{3})(5\sqrt{27})$

$$= (2 \cdot 5)(\sqrt{3})(\sqrt{27}) \qquad (ab)(cd) = (ac)(bd)$$
$$= 10\sqrt{3 \cdot 27} \qquad \text{Product rule for radicals}$$
$$= 10\sqrt{81} = 10 \cdot 9 = 90$$

6. $\sqrt[3]{4x^2y^4}\sqrt[3]{2xy}$

$$\sqrt[3]{(4x^2y^4)(2xy)} = \sqrt[3]{8x^3y^5} = 2xy\sqrt[3]{y^2}$$ ■

The following property can help simplify radicals involving rational expressions as well as tell how to perform some division operations with radicals.

Quotient property of radicals

If $a \geq 0$, $b > 0$, and $n \in N$, then $\sqrt[n]{\dfrac{a}{b}} = \dfrac{\sqrt[n]{a}}{\sqrt[n]{b}}$.

The quotient property will be illustrated in the following example, along with the following property.

Index/exponent common factor property

Given $\sqrt[n]{a^m}$ and $a \geq 0$. If m and n have a common factor, this factor can be divided from m and from n.

Concept

If the index and every exponent of a radicand can be reduced by some common factor, this reduction is valid.

For example, given $\sqrt[6]{x^4y^2}$, $x \geq 0$, $y \geq 0$. Since the value 2 is a common factor for every exponent and the index, we can rewrite this radical by dividing every exponent and the index by 2, obtaining $\sqrt[3]{x^2y}$.

Before illustrating these properties we will now make clear what we mean by "simplifying" a radical.

A radical is simplified if

1. The exponent of all factors of the radicand is less than the index.
2. There is no radical in a denominator.
3. There is no fraction under the radical.
4. The index of the radical and the power of each factor in the radicand have no common factor other than 1.

■ Example 1–5 D

Simplify. Assume all variables are nonnegative and do not represent division by zero.

1. $\sqrt{\dfrac{16a^4b^5c}{25d^8}} = \dfrac{\sqrt{2^4a^4b^5c}}{\sqrt{5^2d^8}}$ — We have a fraction in a radical

$= \dfrac{2^2a^2b^2\sqrt{bc}}{5d^4} = \dfrac{4a^2b^2\sqrt{bc}}{5d^4}$ — Quotient rule for radicals

2. $\dfrac{6x}{\sqrt[3]{4x}}$ — We have a radical in a denominator

In this case we eliminate a radical in a denominator by multiplying it (and the numerator) by a factor that will make all of the exponents multiples of the index (in this case 3).

$= \dfrac{6x\sqrt[3]{2x^2}}{\sqrt[3]{2^2x}\sqrt[3]{2x^2}}$ — We want each exponent under the radical in the denominator to be a multiple of 3

$= \dfrac{6x\sqrt[3]{2x^2}}{\sqrt[3]{8x^3}} = \dfrac{6x\sqrt[3]{2x^2}}{2x} = 3\sqrt[3]{2x^2}$

3. $\dfrac{6}{\sqrt{3}}$ — For square roots multiply the numerator and denominator by the radical in the denominator

$\dfrac{6\cdot\sqrt{3}}{\sqrt{3}\cdot\sqrt{3}} = \dfrac{6\sqrt{3}}{\sqrt{3^2}} = \dfrac{6\sqrt{3}}{3} = 2\sqrt{3}$

4. $\dfrac{6}{\sqrt[3]{3}}$ $\quad \dfrac{6}{\sqrt[3]{3}}\cdot\dfrac{\sqrt[3]{3^2}}{\sqrt[3]{3^2}} = \dfrac{6\sqrt[3]{3^2}}{\sqrt[3]{3^3}} = \dfrac{6\sqrt[3]{9}}{3} = 2\sqrt[3]{9}$

5. $\sqrt[5]{\dfrac{3a^6}{64b^5c^4}} = \dfrac{\sqrt[5]{3a^6}}{\sqrt[5]{2^6b^5c^4}}$ — Quotient rule for radicals

$= \dfrac{a\sqrt[5]{3a}}{2b\sqrt[5]{2c^4}}$ — Simplify the radicals in the numerator and denominator

$= \dfrac{a\sqrt[5]{3a}\sqrt[5]{2^4c}}{2b\sqrt[5]{2c^4}\sqrt[5]{2^4c}}$ — $2\cdot 2^4 = 2^5$; $c^4\cdot c = c^5$

$= \dfrac{a\sqrt[5]{2^4(3)ac}}{2b\sqrt[5]{2^5c^5}} = \dfrac{a\sqrt[5]{48ac}}{2b(2c)} = \dfrac{a\sqrt[5]{48ac}}{4bc}$

6. $\sqrt[4]{ab} \cdot \sqrt[4]{ab^3}$

$= \sqrt[4]{a^2b^4}$ — Multiply

$= \sqrt{ab^2}$ — Since the index (4) and each exponent are divisible by 2, reduce them all by dividing each by 2. Recall that we do not usually show the index 2 in a square root. ■

Addition and subtraction with radicals—combining like terms

The usual rules for manipulating expressions, along with the properties illustrated above, provide the means of simplifying expressions involving radicals. The product and quotient properties of radicals tell us how to multiply and divide with radicals. *We add and subtract radicals by combining like terms,* as with any algebraic expression. For example,

$$5\sqrt{6} + 2\sqrt{6} = 7\sqrt{6}$$

just as

$$5a + 2a = 7a$$

■ ***Example 1–5 E***

Perform the indicated operations. Assume all variables are nonnegative.

1. $7\sqrt{2} + 3\sqrt{8} - \sqrt{32}$

$= 7\sqrt{2} + 3(2\sqrt{2}) - 4\sqrt{2}$ — $\sqrt{8} = \sqrt{4}\sqrt{2} = 2\sqrt{2}$ and

$= 7\sqrt{2} + 6\sqrt{2} - 4\sqrt{2}$ — $\sqrt{32} = \sqrt{16}\sqrt{2} = 4\sqrt{2}$

$= 9\sqrt{2}$ — Combine like terms

2. $\sqrt[3]{81x^4y} - x\sqrt[3]{24xy}$

$= 3x\sqrt[3]{3xy} - 2x\sqrt[3]{3xy}$ — Simplify each radical

$= x\sqrt[3]{3xy}$ — Combine like terms

3. $5\sqrt{2}(3\sqrt{2} - 6\sqrt{3})$

$= 5\sqrt{2} \cdot 3\sqrt{2} - 5\sqrt{2} \cdot 6\sqrt{3}$ — $a(b - c) = ab - ac$

$= 15(2) - 30\sqrt{6}$ — $\sqrt{2}\sqrt{2} - 2$, $\sqrt{2}\sqrt{3} = \sqrt{6}$

$= 30 - 30\sqrt{6}$

4. $(\sqrt{2x} - \sqrt{6})(\sqrt{2x} + 3\sqrt{6})$

$= 2x + 3\sqrt{12x} - \sqrt{12x} - 3(6)$ — $(a - b)(c + d) = ac + ad - bc - bd$

$= 2x + 2\sqrt{12x} - 18$

$= 2x + 4\sqrt{3x} - 18$ ■

Many expressions happen to have two terms in the denominator in which one or both expressions involve square roots. *These radicals can be eliminated from the denominator by multiplying the numerator and denominator by the conjugate of the denominator.* This relies on the fact that $(a - b)(a + b) = a^2 - b^2$; $a - b$ and $a + b$ are conjugates (section 1–3), and, if a or b is a square root, then neither a^2 nor b^2 will contain a square root. This process of rationalizing the denominators is illustrated in the following examples.

■ Example 1–5 F

Rationalize the denominators. Assume all variables are nonnegative and no denominator equals zero.

1. $\dfrac{8}{\sqrt{11} - 3}$

The conjugate of the denominator is $\sqrt{11} + 3$

$$= \frac{8}{\sqrt{11} - 3} \cdot \frac{\sqrt{11} + 3}{\sqrt{11} + 3}$$

$$= \frac{8(\sqrt{11} + 3)}{(\sqrt{11})^2 - 3^2}$$

$(a - b)(a + b) = a^2 - b^2$

$$= \frac{8(\sqrt{11} + 3)}{11 - 9} = \frac{8(\sqrt{11} + 3)}{2}$$

$$= 4(\sqrt{11} + 3)$$

2. $\dfrac{\sqrt{ab} + \sqrt{b}}{\sqrt{a} + \sqrt{b}}$

The conjugate of the denominator is $\sqrt{a} - \sqrt{b}$

$$= \frac{\sqrt{ab} + \sqrt{b}}{\sqrt{a} + \sqrt{b}} \cdot \frac{\sqrt{a} - \sqrt{b}}{\sqrt{a} - \sqrt{b}}$$

$$= \frac{\sqrt{a^2b} - \sqrt{ab^2} + \sqrt{ab} - b}{(\sqrt{a})^2 - (\sqrt{b})^2}$$

$$= \frac{a\sqrt{b} - b\sqrt{a} + \sqrt{ab} - b}{a - b}$$ ■

In advanced applications we often come across expressions like those shown in the following example.

■ Example 1–5 G

Perform the indicated operations and simplify.

1. $\dfrac{\sqrt{3}}{2} \cdot \dfrac{1}{\sqrt{5}} - \left(-\dfrac{1}{2}\right)\left(\dfrac{3}{\sqrt{5}}\right)$

$$= \frac{\sqrt{3}}{2\sqrt{5}} + \frac{3}{2\sqrt{5}} = \frac{\sqrt{3} + 3}{2\sqrt{5}} = \frac{\sqrt{3} + 3}{2\sqrt{5}} \cdot \frac{\sqrt{5}}{\sqrt{5}} = \frac{\sqrt{15} + 3\sqrt{5}}{10}$$

2. $\sqrt{\dfrac{1 - \dfrac{\sqrt{3}}{2}}{2}}$

$$= \sqrt{\frac{1}{2}\left(1 - \frac{\sqrt{3}}{2}\right)} = \sqrt{\frac{1}{2}\left(\frac{2}{2} - \frac{\sqrt{3}}{2}\right)} = \sqrt{\frac{2 - \sqrt{3}}{4}} = \frac{\sqrt{2 - \sqrt{3}}}{2}$$ ■

Mastery points

Can you

- Find exact values of indicated roots?
- Simplify radicals using the nth root of nth power property?
- Simplify radicals using the product and quotient properties?
- Perform algebraic operations on expressions which contain radicals?
- Rationalize denominators?

Exercise 1–5

Find the indicated root.

1. $\sqrt{289}$ 2. $\sqrt{625}$ 3. $\sqrt[5]{32}$ 4. $\sqrt[4]{81}$ 5. $\sqrt[3]{-125}$ 6. $\sqrt[3]{125}$

Simplify the following radical expressions. Do not assume that variables represent nonnegative real numbers.

7. $\sqrt{4x^2}$ 8. $\sqrt{9x^4}$ 9. $\sqrt{25x^6y^8}$ 10. $\sqrt{\dfrac{1}{4a^4}}$

11. $\sqrt{\dfrac{16x^6}{9y^{10}}}$ 12. $\sqrt{4x^6(x-3)^4}$ 13. $\sqrt{x^4-6x^2+9}$ 14. $\sqrt{\dfrac{x^2}{x^2+12x+36}}$

Simplify the following radical expressions. Assume that all variables represent nonnegative real numbers.

15. $\sqrt[3]{40}$ 16. $\sqrt[4]{32}$ 17. $\sqrt{200}$ 18. $\sqrt[3]{8{,}000}$

19. $\sqrt[5]{64a^7b^5}$ 20. $\sqrt{50x^6y^7z^2}$ 21. $\sqrt{200x^2y^9z^{12}}$ 22. $\sqrt[3]{52a^4b^5c^6}$

23. $\sqrt[4]{16x^3y^6}$ 24. $\sqrt[4]{625a^6b^9c^4}$ 25. $\sqrt[3]{a^4}\sqrt[3]{a}\sqrt[3]{a^7}$ 26. $\sqrt[4]{4a^2b}\sqrt[4]{4a^2b^6}$

27. $\sqrt[3]{25a^2b^4c}\sqrt[3]{25a^2b^5c^2}$ 28. $\sqrt[3]{12x^3y^2z^5}\sqrt[3]{12x^2yz^2}$ 29. $\dfrac{8}{\sqrt{2}}$ 30. $\dfrac{9}{4\sqrt{3}}$

31. $\dfrac{12a}{\sqrt{50a}}$ 32. $\dfrac{5x}{\sqrt{20x}}$ 33. $\sqrt{\dfrac{8}{27}}$ 34. $\sqrt{\dfrac{7}{18}}$

35. $\sqrt[3]{\dfrac{8}{9}}$ 36. $\sqrt[3]{\dfrac{3}{20}}$ 37. $\sqrt{\dfrac{12x^5y^6}{5z}}$ 38. $\sqrt{\dfrac{x^{14}y^7}{27z^3w}}$

39. $\sqrt[5]{\dfrac{64x^{14}y^6}{x^2yz^4w^8}}$ 40. $\sqrt[3]{\dfrac{a^5b}{4ab^3c}}$ 41. $\sqrt[3]{\dfrac{16a^5}{b^7c^2}}$ 42. $\sqrt[4]{\dfrac{x^4y^9}{8z^2}}$

43. $\dfrac{\sqrt[4]{16x^6y^5}}{\sqrt[4]{27x^4y^3z^6}}$ 44. $\dfrac{\sqrt[3]{5x^7y^2}}{\sqrt[3]{30xy^7z^3}}$ 45. $\dfrac{\sqrt[4]{8x^5y^7}}{\sqrt[4]{50x^3y^9}}$ 46. $\dfrac{\sqrt[3]{5a^2b^6}}{\sqrt[3]{10a}}$

Simplify the following radical expressions. Assume that all variables represent nonnegative real numbers.

47. $4\sqrt{2}-8\sqrt{2}+\sqrt{8}$ 48. $\sqrt{5}-6\sqrt{5}+\sqrt{45}$ 49. $5\sqrt{3}+7\sqrt{12}-\sqrt{75}$

50. $2\sqrt{8}-\sqrt{50}+3\sqrt{2}$ 51. $2\sqrt{48}-3\sqrt{27}$ 52. $5\sqrt{2x^3}-2x\sqrt{8x}$

53. $\sqrt[3]{16}+5\sqrt[3]{24}$ 54. $2b\sqrt[4]{32b}-\sqrt[4]{162b^5}$ 55. $-2\sqrt{36a^2b}+4\sqrt{25a^2b}$

56. $3\sqrt{18xy^2}-2y\sqrt{2x}$ 57. $2\sqrt{3a}(\sqrt{3a}-2\sqrt{27a})$ 58. $3\sqrt{x}(2\sqrt{xy}-\sqrt{x})$

59. $\sqrt[3]{4x}(\sqrt[3]{2x^2}+\sqrt[3]{4x}-\sqrt[3]{x^2})$ 60. $(\sqrt{2}-3)(\sqrt{2}+3)$ 61. $(3-4\sqrt{x})(4-2\sqrt{x})$

62. $(\sqrt{8a}-\sqrt{2})(\sqrt{2a}-\sqrt{8})$ 63. $(5\sqrt{3}-2\sqrt{6})(\sqrt{3}+\sqrt{12})$ 64. $\sqrt{2xy}(\sqrt{2xy}-\sqrt{2x}+\sqrt{6y})$

65. $(2\sqrt{2x^2}-\sqrt{4x})^2$ 66. $(\sqrt{5}-\sqrt{10})^2$

Rationalize the denominators. Assume that all variables represent nonnegative real numbers.

67. $\dfrac{\sqrt{5}}{\sqrt{5}-\sqrt{2}}$

68. $\dfrac{\sqrt{6}-\sqrt{2}}{\sqrt{6}+\sqrt{2}}$

69. $\dfrac{a+\sqrt{b}}{a-\sqrt{b}}$

70. $\dfrac{\sqrt{6}}{\sqrt{6}-12}$

71. $\dfrac{\sqrt{2x}}{\sqrt{6x}+\sqrt{2}}$

72. $\dfrac{2a}{\sqrt{a}+\sqrt{ab}}$

Perform the indicated operations and simplify.

73. $\dfrac{1}{3}\cdot\dfrac{\sqrt{7}}{5}-\dfrac{\sqrt{2}}{3}\cdot\dfrac{\sqrt{2}}{5}$

74. $\dfrac{1}{\sqrt{5}}\left(-\dfrac{\sqrt{5}}{4}\right)+\dfrac{3}{\sqrt{5}}\left(-\dfrac{1}{4}\right)$

75. $\dfrac{2}{\sqrt{5}}\cdot\dfrac{1}{\sqrt{3}}-\dfrac{3}{\sqrt{5}}\cdot\dfrac{2}{\sqrt{3}}$

76. $-\dfrac{5}{2\sqrt{2}}\left(-\dfrac{1}{\sqrt{2}}\right)-\dfrac{1}{2}\left(\dfrac{\sqrt{5}}{3}\right)$

77. $\sqrt{\dfrac{3-\dfrac{1}{\sqrt{2}}}{2}}$

78. $\sqrt{\dfrac{4+\dfrac{\sqrt{3}}{2}}{2}}$

79. $\sqrt{\dfrac{4-\dfrac{\sqrt{5}}{8}}{2}}$

80. $\sqrt{\dfrac{1-\dfrac{\sqrt{3}}{\sqrt{5}}}{5}}$

Simplify the following expressions. Assume all variables are nonnegative.

81. $\sqrt[4]{a^6}$

82. $\sqrt[6]{x^4y^2}$

83. $\sqrt[10]{32x^5y^{10}}$

84. $\sqrt[8]{a^8b^{12}c^{16}}$

85. If $x \geq 0$ and $y \geq 0$, then $x-y$ can be factored into $(\sqrt{x}-\sqrt{y})(\sqrt{x}+\sqrt{y})$ by viewing it as the difference of two squares. Similarly, viewed as the difference of two cubes, $x-y=(\sqrt[3]{x}-\sqrt[3]{y})Q$, where Q represents a quadratic expression in the variables $\sqrt{x}$ and $\sqrt{y}$.

a. Determine what expression is represented by Q.

b. Factor $x+y$ using a similar strategy with the factorization of a sum of two cubes.

c. Factor $8x-y$, first viewing it as a difference of two cubes, and then viewing it as a difference of two squares.

86. Use the factorization for a^3+b^3 as a guide to a way to rationalize the denominator of the fraction $\dfrac{3xy}{\sqrt[3]{x}+\sqrt[3]{y}}$. That is, use the fact that

$$a+b=(\sqrt[3]{a})^3+(\sqrt[3]{b})^3=(\sqrt[3]{a}+\sqrt[3]{b})(\;.\;.\;.\;).$$

87. Using problem 86 as a hint, rationalize the denominator of the fraction $\dfrac{\sqrt[3]{x}}{\sqrt[3]{2x^2}-\sqrt[3]{3x}}$.

88. Using problems 86 and 87 as a hint, rationalize $\dfrac{3}{\sqrt[4]{x}-\sqrt[4]{y}}$. Assume $x>0$ and $y>0$.

89. Show that $\sqrt{\dfrac{2-\sqrt{3}}{2+\sqrt{3}}}=\dfrac{1-\dfrac{1}{\sqrt{3}}}{1+\dfrac{1}{\sqrt{3}}}$.

90. About 3,000 years ago the Babylonians developed the following method for computing an approximate value for a square root. Today it is known as Newton's method.

Let $x=\sqrt{a}$ be the desired value. Let a_1 be a first approximation to $\sqrt{a}$. Let $b_1=\dfrac{a}{a_1}$. If $a_1=b_1$, then since $a_1b_1=a$, $a_1=\sqrt{a}$. Of course this is unlikely to happen. If a_1 is too small then b_1 is too large, and vice versa. This means that their average will be a better approximation to the root. Thus, let $a_2=\dfrac{a_1+b_1}{2}$ and $b_2=\dfrac{a}{a_2}$. Again, let $a_3=\dfrac{a_2+b_2}{2}$ and $b_3=\dfrac{a}{a_3}$ etc. This method can be continued indefinitely, until a_i is sufficiently accurate.

With a calculator, use this method to compute $\sqrt{19}$ to 4-digit accuracy using only the operations of addition, subtraction, multiplication, and division. Since $4<\sqrt{19}<5$, use 4.5 for a_1, and note how many iterations are necessary to achieve the desired accuracy. $\sqrt{19}\approx 4.359$.

91. There is a property of radicals that allows us to rewrite $\sqrt{\sqrt[3]{5}}$ using only one radical.

a. Guess what this one radical might be, then try to demonstrate that your guess is correct.

b. Deduce what this property of radicals is, in general (always assuming variables to be nonnegative for simplicity).

Skill and review

1. Add $\frac{1}{3} + \frac{3}{4}$.
2. Multiply $\frac{1}{3} \cdot \frac{3}{4}$.
3. Simplify $3a^{-2}$.
4. Simplify $\sqrt{9a^4b^6}$, $a,b \geq 0$.
5. Compute $\frac{1}{\sqrt[3]{-8}}$.
6. Simplify $\frac{8a^8}{2a^2}$.
7. Simplify $\frac{-2a^{-2}}{8a^8}$.

1–6 Rational exponents

A formula which relates the leg diameter D_l necessary to support a body with body length L_b for large vertebrates is $D_l = cL_b^{1.5}$, where c is a constant depending on the vertebrate in question. Suppose body length increases by a factor of 4. What needs to happen to leg diameter to support the body?

The expression that is the right member of the equation above contains the exponent 1.5, which is not an integer. In this section we learn how to interpret and use exponents that are not simply integers.

Meaning of integer exponents

Recall that we have defined the meaning of expressions with integer exponents.

If $n \in N$, then $x^n = n$ factors of x.
If $x \neq 0$, then $x^0 = 1$.
If $n \in N$ and $n > 0$, then $x^{-n} = \frac{1}{x^n}$.

Thus, $9^2 = 9 \cdot 9 = 81$, $9^0 = 1$, and $9^{-2} = \frac{1}{9^2} = \frac{1}{81}$.

We also noted that the following properties apply to integer exponents, assuming that $a,b \in R$ and $m,n \in J$ and that no variable represents zero where division by zero would be indicated.

[1] $a^m a^n = a^{m+n}$ [2] $\frac{a^m}{a^n} = a^{m-n}$ [3] $(ab)^m = a^m b^m$

[4] $\left(\frac{a}{b}\right)^m = \frac{a^m}{b^m}$ [5] $(a^m)^n = a^{mn}$

Definition of rational exponents

It is natural to ask what might be the value of $9^{\frac{1}{2}}$. If we define this expression involving a rational exponent to have some meaning, we want to do this in a fashion that does not violate the properties of exponents above.

Since $(\sqrt{9})^2 = 9$, and we would expect $\left(9^{\frac{1}{2}}\right)^2 = 9^{\frac{1}{2}\cdot 2} = 9^1 = 9$, it is natural to define $9^{\frac{1}{2}}$ to mean $\sqrt{9}$. Thus, $9^{\frac{1}{2}} = \sqrt{9} = 3$. This reasoning leads to the following definition of an exponent of the form $\frac{1}{n}$, $n \in N$, $n > 1$.

Definition of $a^{\frac{1}{n}}$

If $a \in R$, $n \in N$, $n > 1$, and $\sqrt[n]{a} \in R$, then

$$a^{\frac{1}{n}} = \sqrt[n]{a}.$$

This definition is extended to rational exponents in which the numerator is not one[15] in the following definition.

Definition of $a^{\frac{m}{n}}$

If $m,n \in N$ and $\sqrt[n]{a} \in R$, then

$$a^{\frac{m}{n}} = \left(a^{\frac{1}{n}}\right)^m.$$

The following property can also be shown to be true.

Equivalence of $\left(a^{\frac{1}{n}}\right)^m$ and $(a^m)^{\frac{1}{n}}$ for $a \geq 0$

If $a \geq 0$ and $m,n \in N$, then

$$a^{\frac{m}{n}} = \left(a^{\frac{1}{n}}\right)^m = (a^m)^{\frac{1}{n}}.$$

This last property is important because it allows us to rewrite expressions in whichever way is easier for us. For example, we can rewrite $9^{\frac{3}{2}}$ as $\left(9^{\frac{1}{2}}\right)^3$ (numerator of fraction outside the parentheses), which becomes $3^3 = 27$. We can also write something like $(\sqrt{7})^{\frac{2}{3}}$ as $((\sqrt{7})^2)^{\frac{1}{3}}$ (numerator of fraction inside the parentheses), which becomes $7^{\frac{1}{3}}$ or $\sqrt[3]{7}$.

[15] a^x where x is a negative integer or fraction was introduced in concept by John Wallis in 1656, but Newton introduced our modern notation in 1676.

We need one more definition before we are finished.

Definition of $a^{-\frac{m}{n}}$

If $a \neq 0$, $m,n \in N$, and $a^{\frac{m}{n}} \in R$, then

$$a^{-\frac{m}{n}} = \frac{1}{a^{\frac{m}{n}}}.$$

■ ***Example 1–6 A***

Rewrite each expression in terms of radicals. Simplify if possible. Assume all variables represent nonnegative values.

1. $(16x)^{\frac{1}{4}} = \sqrt[4]{16x}$ — Definition: $a^{\frac{1}{4}} = \sqrt[4]{a}$

$= \sqrt[4]{16}\sqrt[4]{x} = 2\sqrt[4]{x}$ — Simplify the radical

2. $8^{\frac{2}{3}} = \left(8^{\frac{1}{3}}\right)^2$ — Definition: $a^{\frac{2}{3}} = \left(a^{\frac{1}{3}}\right)^2$

$= (\sqrt[3]{8})^2$ — Definition: $a^{\frac{1}{3}} = \sqrt[3]{a}$

$= 2^2 = 4$

3. $(32x^{10})^{-\frac{3}{5}} = \dfrac{1}{(32x^{10})^{\frac{3}{5}}}$ — Definition of $a^{-\frac{m}{n}}$

$= \dfrac{1}{(\sqrt[5]{32x^{10}})^3} = \dfrac{1}{(2x^2)^3} = \dfrac{1}{8x^6}$ ■

If $a < 0$, $a^{\frac{m}{n}}$ is not defined whenever n is even, since in this case $\sqrt[n]{a}$ is not defined. Rational exponents that are not relatively prime may be reduced when both expressions are real numbers. By way of example, $(-8)^{\frac{5}{3}}$ is defined and has the value -32, but $(-8)^{\frac{10}{6}}$ is not defined since $\sqrt[6]{-8}$ is not real. It is true that *rational exponents may always be reduced when the base is nonnegative.*

Simplifying expressions with rational exponents

It can be shown that if the values of all bases are nonnegative then the properties of exponents [1]–[5] stated at the beginning of this section can be applied to rational exponents. Thus, for example, to multiply we add exponents, etc. This is illustrated in the following example.

■ ***Example 1–6 B***

Simplify using the properties of exponents [1]–[5] as well as the definitions and theorems of this section. Assume all variables represent nonnegative values.

1. $9^{\frac{3}{4}} \cdot 9^{\frac{3}{4}} = 9^{\frac{3}{4}+\frac{3}{4}} = 9^{\frac{6}{4}} = 9^{\frac{3}{2}} = (\sqrt{9})^3 = 3^3 = 27$

2. $\left(3x^{\frac{1}{2}}\right)\left(5x^{\frac{2}{3}}y\right) = 15x^{\frac{1}{2}+\frac{2}{3}}y = 15x^{\frac{7}{6}}y$ — $a^m a^n = a^{m+n}$

3. $\left(27x^{\frac{1}{2}}y^4\right)^{\frac{1}{3}} = 27^{\frac{1}{3}}x^{\frac{1}{2}\cdot\frac{1}{3}}y^{4\cdot\frac{1}{3}} = 3x^{\frac{1}{6}}y^{\frac{4}{3}}$ — $(ab)^m = a^m b^m$

4. $\dfrac{x^{-\frac{1}{2}}y^{\frac{2}{3}}}{4^{-\frac{1}{2}}x^2y^{-\frac{1}{3}}} = \dfrac{1}{4^{-\frac{1}{2}}} \cdot \dfrac{x^{-\frac{1}{2}}}{x^2} \cdot \dfrac{y^{\frac{2}{3}}}{y^{-\frac{1}{3}}}$ — Separate for clarity

$= 4^{\frac{1}{2}}x^{\left(-\frac{1}{2}-2\right)}y^{\frac{2}{3}-\left(-\frac{1}{3}\right)}$ — Definition of $a^{-\frac{m}{n}}$; $\frac{a^m}{a^n} = a^{m-n}$

$= 2x^{-\frac{5}{2}}y^1 = \dfrac{2y}{x^{\frac{5}{2}}}$ ■

Approximate values for rational exponents and radicals

As stated in the previous section most radicals are irrational. We can only obtain decimal approximations to these values. Many scientific/engineering calculators have a key marked $\boxed{\sqrt[x]{y}}$ designed to approximate roots. If a calculator does not have this key it probably has a key marked $\boxed{x^{1/y}}$, which is used the same way, since $\sqrt[y]{x} = x^{1/y}$ for $x \geq 0$. The Texas Instruments TI-81 has neither key. The value of $\sqrt[x]{y}$ is obtained by the sequence y $\boxed{\wedge}$ x $\boxed{x^{-1}}$ $\boxed{\text{ENTER}}$.

Some calculators do not have a $\boxed{\sqrt[x]{y}}$ or $\boxed{x^{1/y}}$ key. In this case the sequence $\boxed{x^y}$ "index" $\boxed{1/x}$ must be used,[16] which has the same effect as the $\boxed{x^{1/y}}$ key. This is illustrated in parts 3, 4, 5 of the next example.

The $\boxed{x^y}$ key is also the key to computing expressions directly expressed with rational exponents. The following examples illustrate various ways calculators compute numeric values of expressions with rational exponents.

■ ***Example 1–6 C***

Compute the following values to six digits of accuracy.

1. $5^{\frac{3}{4}}$ — 5 $\boxed{x^y}$ $\boxed{(}$ 3 $\boxed{\div}$ 4 $\boxed{)}$ $\boxed{=}$

$= 3.34370$ — TI-81: 5 $\boxed{\wedge}$ $\boxed{(}$ 3 $\boxed{\div}$ 4 $\boxed{)}$ $\boxed{\text{ENTER}}$

[16]Nonalgebraic calculators (Hewlett-Packard™, for example) use "index" $\boxed{1/x}$ $\boxed{y^x}$.

2. $(-8.2)^{-\frac{2}{3}}$

Calculators will not accept a negative base with the $\boxed{x^y}$ key. In this case it is necessary to determine the sign of the answer ourselves. To see the sign of this result, rewrite in radical form:

$$(-8.2)^{-\frac{2}{3}} = \left((-8.2)^{-\frac{1}{3}}\right)^2 = \left(\frac{1}{\sqrt[3]{-8.2}}\right)^2.$$

$\sqrt[3]{-8.2}$ is negative, but when squared the result is positive. Thus, we actually compute $8.2^{-\frac{2}{3}}$, since it is the same value:

$= 0.245918$ $\quad 8.2\ \boxed{x^y}\ \boxed{(}\ 2\ \boxed{\div}\ 3\ \boxed{)}\ \boxed{+/-}\ \boxed{=}$

TI-81: $8.2\ \boxed{\wedge}\ \boxed{(}\ \boxed{(-)}\ 2\ \boxed{\div}\ 3\ \boxed{)}\ \boxed{\text{ENTER}}$

3. $\sqrt[3]{21.6}$ $\quad 21.6\ \boxed{\sqrt[x]{y}}\ 3\ \boxed{=}$ or $21.6\ \boxed{x^y}\ 3\ \boxed{1/x}\ \boxed{=}$

$= 2.78495$ $\quad$ TI-81: $\boxed{\text{MATH}}\ 4\ 21.6\ \boxed{\text{ENTER}}$

4. $\sqrt[5]{2003^2}$ $\quad 2003\ \boxed{x^2}\ \boxed{\sqrt[x]{y}}\ 5\ \boxed{=}$ or $2003\ \boxed{x^2}\ \boxed{x^y}\ 5\ \boxed{1/x}\ \boxed{=}$

$= 20.9253$ $\quad$ TI-81: $2003\ \boxed{x^2}\ \boxed{\wedge}\ 5\ \boxed{x^{-1}}\ \boxed{\text{ENTER}}$ ■

Mastery points

Can you

- Rewrite expressions with rational exponents in terms of radicals?
- Simplify expressions with rational exponents using the properties of exponents?
- Evaluate indicated numeric roots?
- Compute approximate values to numeric expressions involving radicals and rational exponents?

Exercise 1–6

Rewrite each expression in terms of radicals. Simplify if possible. Assume all variables represent nonnegative values.

1. $64^{\frac{1}{4}}$ **2.** $16^{\frac{3}{4}}$ **3.** $8^{-\frac{2}{3}}$ **4.** $100^{-\frac{1}{2}}$ **5.** $(-8)^{-\frac{2}{3}}$

6. $81^{-\frac{3}{4}}$ **7.** $(16x)^{\frac{1}{2}}$ **8.** $(50x^3)^{\frac{1}{2}}$ **9.** $(81x^3)^{\frac{1}{4}}$ **10.** $(27x^6)^{\frac{1}{4}}$

11. $(32x^6y^7)^{\frac{1}{4}}$ **12.** $100^{\frac{1}{4}}$ **13.** $(8x^3)^{\frac{1}{2}}$

Simplify.

14. $5^{\frac{3}{2}} \cdot 5^{\frac{1}{2}}$ **15.** $25^{\frac{1}{4}} \cdot 25^{\frac{1}{4}}$ **16.** $2b^{\frac{3}{4}}\left(3b^{\frac{1}{4}}\right)$ **17.** $2a^{\frac{1}{2}}\left(4a^{\frac{1}{4}}\right)$

18. $\left(a^{\frac{1}{3}}\right)^3$ **19.** $\left(b^{\frac{3}{4}}\right)^{\frac{8}{3}}$ **20.** $\left(x^{\frac{3}{4}}y^{\frac{3}{5}}\right)^{\frac{8}{3}}$ **21.** $\left(4x^{\frac{4}{7}}y^{\frac{6}{5}}z\right)^{\frac{1}{2}}$

22. $\left(\frac{2}{3}x^{\frac{2}{3}}y^{-\frac{1}{3}}\right)\left(3x^{\frac{1}{3}}y^{\frac{4}{3}}\right)$

23. $\left(\frac{1}{4}x^{\frac{1}{4}}y^{\frac{1}{2}}\right)\left(\frac{1}{2}x^{\frac{1}{4}}y^{\frac{1}{4}}\right)$

24. $\dfrac{x^{\frac{3}{4}}}{x^{\frac{1}{4}}}$

25. $\dfrac{ab^{\frac{1}{3}}}{a^2b^{\frac{5}{6}}}$

26. $\dfrac{x^{\frac{2}{3}}}{x}$

27. $\dfrac{a}{a^{\frac{1}{4}}}$

28. $\dfrac{4a^{\frac{1}{4}}b^{\frac{3}{2}}}{8a^{\frac{3}{4}}b^{\frac{1}{2}}}$

29. $\dfrac{6a^{\frac{1}{2}}}{2a^{\frac{1}{4}}}$

30. $\dfrac{4x^{\frac{-3}{4}}}{x^{\frac{1}{4}}y^{\frac{3}{4}}}$

31. $\dfrac{x^{\frac{5}{8}}y^{\frac{2}{3}}}{x^{\frac{3}{8}}y^{\frac{-2}{3}}}$

32. $\left(\dfrac{x^{\frac{-3}{5}}y^{\frac{3}{4}}}{z^{\frac{3}{10}}}\right)^{20}$

33. $\left(\dfrac{x^{\frac{4}{3}}y^{\frac{8}{5}}}{z^{-8}}\right)^{\frac{1}{8}}$

34. $\left(x^{\frac{a}{b}}y^{\frac{a}{c}}\right)^{\frac{bc}{a}}$

35. $\left(a^{\frac{m}{n}}b^{\frac{2}{n}}\right)^{2n}$

36. $\dfrac{3^{\frac{1}{n}}}{3^{\frac{3}{2n}}}$

37. $\left(\dfrac{2^{\frac{m+n}{m}}x^{\frac{n}{m}}}{2^{\frac{n}{m}}x^{\frac{n}{m}}y^{\frac{-n}{m}}}\right)^{m}$

Compute the following values to four decimal places of accuracy using your calculator.

38. $28^{\frac{1}{3}}$

39. $1{,}345^{\frac{3}{5}}$

40. $(-19)^{\frac{2}{3}}$

41. $(-200)^{\frac{1}{5}}$

42. $\sqrt[4]{96}$

43. $\sqrt[3]{-900}$

44. $\sqrt[6]{19^8}$

45. $\sqrt[4]{37.5^3}$

46. $(\sqrt[4]{5})^{\frac{2}{3}}$

47. $(\sqrt[5]{19})^{\frac{3}{5}}$

48. $(\sqrt[4]{97.1})^{\frac{5}{3}}$

49. $(\sqrt[3]{-500})^{\frac{4}{3}}$

50. $\sqrt[4]{19.5}$

51. $\sqrt[3]{0.125}$

52. $\sqrt[5]{0.825}$

53. $\sqrt[3]{18^7}$

54. $\sqrt[3]{0.5^4}$

55. $(\sqrt[5]{603.25})^4$

56. $(\sqrt[3]{31.5})^5$

57. A formula that relates the leg diameter D_l necessary to support a body with body length L_b for large vertebrates is $D_l = cL_b^{1.5}$, where c is a constant depending on the vertebrate in question. Suppose body length increases by a factor of 4. What needs to happen to leg diameter to support the body? *Hint:* Replace L_b by $4L_b$.

58. The volume of a sphere is $V = \frac{4}{3}\pi r^3$, where r is the radius. Thus, $r = \sqrt[3]{\dfrac{3V}{4\pi}}$. Since the cross section of a sphere is a circle, the area of the cross section at its widest part is $A = \pi r^2$, or in terms of the volume of the sphere, $A = \pi\left(\dfrac{3V}{4\pi}\right)^{2/3}$. Show that this can be transformed into $A = \dfrac{\sqrt[3]{\pi}}{4}(6V)^{2/3}$.

59. The formula $P = A\left[\dfrac{i}{1 - (1 + i)^{-N}}\right]$ gives the monthly payment on a fixed rate mortgage, where A is the amount borrowed, i is the monthly interest rate (yearly rate ÷ 12), and N is the total number of monthly payments (number of years × 12). By way of example, if \$50,000 is borrowed at 9% yearly interest for 30 years, then $N = 30 \times 12 = 360$ and $i = 0.09/12 = 0.0075$, so the monthly payment is

$$P = 50{,}000\left(\frac{0.0075}{1 - 1.0075^{-360}}\right) = \$402.31.$$

Calculate the monthly payment on a loan of \$45,000 at 10% interest for 30 years.

60. (Refer to problem 59.) The formulas $B_n = A\left[\dfrac{i(1 + i)^{n-1}}{(1 + i)^N - 1}\right]$ and $I_n = P[1 - (1 + i)^{n-1-N}]$ are related to monthly mortgage payments. If n is the nth monthly payment, then B_n is the amount of principal being paid that month, and I_n is the amount of interest being paid that month. N, A, and i are defined in problem 59. Find B_n and I_n for the first month of the loan in problem 59 (\$45,000, 10%, 30 years).

61. The following expression gives the value of what is called the second Fibonacci number (discussed further in chapter 12); compute this value:

$$\frac{1}{\sqrt{5}}\left[\left(\frac{1+\sqrt{5}}{2}\right)^2 - \left(\frac{1-\sqrt{5}}{2}\right)^2\right].$$

62. How can you find the fourth root of a number on a calculator using only the square root key? The eighth root?

63. Compute (with a calculator) the value of the expression: $\sqrt[3]{2+\sqrt{5}} + \sqrt[3]{2-\sqrt{5}}$.

64. Evaluate the expression $\dfrac{\dfrac{1}{\sqrt{3}} + \sqrt{\dfrac{2-\sqrt{3}}{2+\sqrt{3}}}}{1 - \dfrac{1}{\sqrt{3}}\sqrt{\dfrac{2-\sqrt{3}}{2+\sqrt{3}}}}$ with a calculator.

Skill and review

1. Select the correct statement(s) about $\sqrt{-4}$.

a. not real **b.** $= -2$ **c.** $= 2^{-1}$ **d.** $= \dfrac{1}{\sqrt{4}}$

2. Compute $(5i - 3)(2i + 7)$.

3. Compute **a.** $(-1)^5$ **b.** $(-1)^{20}$ **c.** $(-1)^{47}$

4. Compute $(\sqrt{2} - \sqrt{3})(3\sqrt{2} + \sqrt{6})$.

5. Rationalize the denominator of $\dfrac{2\sqrt{3}}{\sqrt{6}+\sqrt{2}}$.

1–7 Complex numbers

> If two impedances Z_1 and Z_2 are connected in parallel in an electronic circuit, the total impedance, T, is related by the statement $T = \dfrac{Z_1 + Z_2}{Z_1 Z_2}$.

Here Z_1 and Z_2 are complex numbers. Numbers of this type have proved very useful in electronics theory. We study complex numbers in this section.

Definitions

For the last thousand years it was recognized that there are no real number solutions to the equation $x^2 = -4$. Nevertheless, such equations are often encountered in mathematics and the physical sciences (particularly electronics theory in physics).

Solutions to such equations would involve numbers like $\sqrt{-4}$, which cannot be a real number. This is because if $x = \sqrt{-4}$, then $x^2 = -4$. However $x^2 \geq 0$ for all real numbers. Thus, x could not be real.

The problem was solved several hundred years ago by extending the real number system in the way described by the definitions below. The result was the set of **complex numbers,** C. It is based on the idea of assuming that there is a square root of negative one. We call it the imaginary unit.

> **Imaginary unit i**
> There exists a number i such that $i = \sqrt{-1}$.

The imaginary unit[17] i was first used in 1777 by the famous mathematician Leonhard Euler. Engineers use the letter j for the same value. The imaginary unit is not part of the real number system.[18]

> **Square root of a negative real number**
> If b is a positive real number, then $\sqrt{-b} = i\sqrt{b}$.

This definition now gives us a way to write, for example, $\sqrt{-4}$:

$$\sqrt{-4} = i\sqrt{4} = 2i$$

Complex numbers are defined based on a notation suggested by W. R. Hamilton in 1837.

> **Complex numbers**
> The set of complex numbers = C =
> $$\{a + bi \mid a, b \in R, \text{ and } i \text{ is the imaginary unit}\}$$
> ***Concept***
> A complex number is a binomial; the first term is a real number and the second is the product of a real coefficient and the imaginary unit.

In $a + bi$, a is called the **real part** and b is called the **imaginary part.** For example, $3 - 8i$ is a complex number with real part 3 and imaginary part -8. An expression of the form $a + bi$ is said to be the **standard form** for a complex number. A complex number like $0 - 3i$ might be simply written as $-3i$, but it is nevertheless a complex number. We can also consider an expression such as 4 to represent $4 + 0i$, and thus view it as a complex number also. Equality of complex numbers is defined as follows.

> **Equality of complex numbers**
> Two complex numbers $a + bi$ and $c + di$ are equal if and only if $a = c$ and $b = d$.
> ***Concept***
> Complex numbers are equal when their real parts and imaginary parts are each the same.

[17]H. Cardan was the earliest mathematician to seriously consider imaginary numbers. He demonstrated calculations with the square roots of negative numbers in his work *Ars magna* (Nurenberg, 1545).

[18]"I met a man recently who told me that, so far from believing in the square root of minus one, he did not even believe in minus one. This is at any rate a consistent attitude" (E. C. Titchmarsh).

Operations with complex numbers

Since $i = \sqrt{-1}$, $i^2 = -1$. Addition, subtraction, multiplication, and division of complex numbers are defined so that *we can use the algebra of the real number system, with the provision that i^2 be replaced by -1 wherever it appears.* The statement of the formal rules for addition, subtraction, multiplication, and division of complex numbers are left as exercises. (They are not necessary to perform computations by hand—they would be necessary if we wished to program a computer to perform these operations.[19])

■ Example 1–7 A

Perform the operations shown on the complex numbers.

1. $(5 - 3i) - (12 + 3i)$

$= 5 - 12 - 3i - 3i$ Remove grouping symbols

$= -7 - 6i$ Combine like terms

2. $(-2i)(8 - 3i)$

$= -16i + 6i^2$ Multiply as with real-valued expressions

$= -16i + 6(-1)$ $i^2 = -1$

$= -6 - 16i$ Rewrite so real part is first

3. $(5 - 3i)(12 + 3i)$ $(a - b)(c + d) = ac + ad - bc - bd$

$= 60 + 15i - 36i - 9i^2$

$= 60 - 21i - 9(-1)$ $i^2 = -1$

$= 69 - 21i$ ■

Division is handled in a special way. The value $a - bi$ is called the **complex conjugate** of $a + bi$, and vice versa. *To divide by a complex number multiply by a fraction in which both the numerator and denominator are the complex conjugate of the divisor.* This works because, as will be seen below, the product of a complex number with its conjugate is a real number.

■ Example 1–7 B

Divide.

1. $\dfrac{2 - 3i}{5 + 2i}$

$= \dfrac{2 - 3i}{5 + 2i} \cdot \dfrac{5 - 2i}{5 - 2i}$ Multiply by the conjugate of the denominator

$= \dfrac{10 - 4i - 15i + 6i^2}{25 - 10i + 10i - 4i^2}$

$= \dfrac{4 - 19i}{29}$ The denominator is now the real number 29

$= \dfrac{4}{29} - \dfrac{19}{29}i$ Put the answer in standard form $a + bi$

[19]The FORTRAN programming language is preprogrammed to recognize and use complex values. Some programmable calculators (TI-85 for example) will perform complex arithmetic.

2. $\dfrac{6-3i}{4i} = \dfrac{6-3i}{4i} \cdot \dfrac{-4i}{-4i}$ The conjugate of $0 + 4i$ is $0 - 4i$, or just $-4i$

$$= \frac{-24i + 12i^2}{-16i^2} = \frac{-12 - 24i}{16} = -\frac{3}{4} - \frac{3}{2}i$$ ■

Note 1. An answer such as $\dfrac{4-19i}{29}$ (part 1 example 1–7 B) is not complete. It should be written in the standard form (a binomial) $\dfrac{4}{29} - \dfrac{19}{29}i$.

2. The conjugate of a complex number such as $0 + bi$ is $0 - bi$, or simply $-bi$ (as in part 2 example 1–7 B).
3. Part 2 of example 1–7 B could be done by simply using $-i$ instead of $-4i$.

It is very important to avoid using the square roots of negative real values in calculations. Convert them to "i-notation" first. This is shown by the computation of $\sqrt{-2}\sqrt{-2}$. If we simply use the rules for radicals, we obtain

$$\sqrt{-2}\sqrt{-2} = \sqrt{(-2)(-2)} = \sqrt{4} = 2$$

whereas if we proceed with i-notation we obtain

$$\sqrt{-2}\sqrt{-2} = i\sqrt{2} \cdot i\sqrt{2} = i^2\sqrt{4} = -2$$

The second value is correct, not the first. The first is wrong because we assume the rules for radicals, such as $\sqrt{a}\sqrt{b} = \sqrt{ab}$, apply when a or b is negative. This is not true.[20]

■ Example 1–7 C

Simplify.

1. $\sqrt{-18}\sqrt{-6}$

$$= i\sqrt{18} \cdot i\sqrt{6}$$ Rewrite in i-notation before any computation

$$= 3i\sqrt{2} \cdot i\sqrt{6} = 3i^2\sqrt{12}$$
$$= -3(2)(\sqrt{3}) = -6\sqrt{3}$$

2. $\dfrac{\sqrt{-3} - 4}{\sqrt{-2} - 1}$

$$= \frac{i\sqrt{3} - 4}{i\sqrt{2} - 1} \cdot \frac{i\sqrt{2} + 1}{i\sqrt{2} + 1}$$ Multiply numerator and denominator by conjugate of the denominator

$$= \frac{i^2\sqrt{6} + i\sqrt{3} - 4i\sqrt{2} - 4}{i^2(2) + i\sqrt{2} - i\sqrt{2} - 1}$$

$$= \frac{-\sqrt{6} - 4 + i\sqrt{3} - 4i\sqrt{2}}{-3}$$ $i^2 = -1$; separate the real and imaginary terms in the numerator

[20]Mistakes of this type were made even by famous mathematicians as late as the eighteenth century.

$$= \frac{-(\sqrt{6} + 4)}{-3} + \frac{i(\sqrt{3} - 4\sqrt{2})}{-3}$$ Break up into real and imaginary parts

$$= \frac{\sqrt{6} + 4}{3} - \frac{\sqrt{3} - 4\sqrt{2}}{3}i$$ ■

The value of i^n, n a whole number, can only have one of four values. Computation shows that

$i^1 = i$
$i^2 = -1$
$i^3 = -i$ $\quad i^3 = i^2 \cdot i = -i$
$i^4 = 1$ $\quad i^4 = (i^2)^2 = (-1)^2$

For values of n above four, the result is always one of the four values obtained above. For powers above four, *all multiples of four can be eliminated,* since they correspond to factors of one (see example 1–7 D).

■ Example 1–7 D

Compute the value of each expression.

Problem **Solution**

1. i^{43} $\quad i^{43} = i^{40} \cdot i^3 = 1 \cdot i^3 = i^3 = i^2 \cdot i$
$= (-1) \cdot i = -i$ $\quad i^2 = -1$

2. Evaluate $5x^7 - 3x^5 + x^4 - 6x^3 + 4$ for $x = i$.

$5i^7 - 3i^5 + i^4 - 6i^3 + 4$
$5i^3 - 3i + 1 - 6i^3 + 4$ $\quad i^7 = i^3$ and $i^5 = i$
$-i^3 - 3i + 5$
$-(-i) - 3i + 5$ $\quad i^3 = -i$
$5 - 2i$ ■

Mastery points

Can you
- Simplify the square root of a negative number?
- Add, subtract, multiply, and divide complex numbers?

Exercise 1–7

Perform the operations shown on the complex numbers.

1. $(-3 + 5i) - (2 - 3i)$
2. $(6 + 2i) + (6 - 12i)$
3. $(5 - 3i) - i + 4(2 + 3i)$
4. $-5(4 + 2i) - (5 + 3i) + 2i$
5. $(-8 + 3i)(2 - 7i)$
6. $2i(6 - 4i)$
7. $(5 - 3i)(5 + 3i)$
8. $(2 + i)(2 - i)$
9. $(5 - 2i)^2$
10. $i(2 + 5i)(\frac{1}{2} + 4i)$
11. $i[(5 - 3i)(-2 + 4i) - (2 - i)^2]$
12. $[(3 - i) - (9 + 2i)][(2 + 3i)(2 - 3i)]$

Divide.

13. $\frac{3 - 4i}{2 + 5i}$ **14.** $\frac{8 + 2i}{4 - i}$ **15.** $\frac{6 + 4i}{6 - 4i}$ **16.** $\frac{5 - i}{i}$ **17.** $\frac{6i}{2 - 3i}$ **18.** $\frac{4 - 3i}{2 - 3i}$

Simplify the following expressions.

19. $\sqrt{-6}\sqrt{-12}$ **20.** $\sqrt{-8}\sqrt{-4}$ **21.** $\sqrt{5}\sqrt{-10}$ **22.** $\sqrt{-12}\sqrt{8}$

23. $(8 - \sqrt{-8}) - (-5 + \sqrt{-50})$ **24.** $(2 + 3\sqrt{-20}) + (\sqrt{-45} - 3\sqrt{-80})$

25. $(3 - \sqrt{-3})(4 + \sqrt{-3})$ **26.** $\sqrt{-10}(\sqrt{-6} - 4\sqrt{-8})$ **27.** $(2 + \sqrt{-6})(3 - \sqrt{-2})$

28. $(3\sqrt{-8} + \sqrt{2})(\sqrt{-2} - 4\sqrt{8})$ **29.** $\frac{4 - \sqrt{-6}}{2 + 3\sqrt{-2}}$ **30.** $\frac{5 + 2\sqrt{-14}}{4 - \sqrt{-6}}$

31. $\frac{3 - 4\sqrt{-8}}{i(2 - 3\sqrt{-2})}$ **32.** $(2 - \sqrt{-2})^4$ **33.** $\frac{\sqrt{-6} + \sqrt{6}}{\sqrt{-2} - \sqrt{2}}$ **34.** $\frac{2\sqrt{30} + \sqrt{-6}}{\sqrt{-12}}$

Compute the value of each expression.

35. i^{10} **36.** i^{21} **37.** i^{15} **38.** i^{19} **39.** i^{-5} **40.** i^{-3} **41.** i^{-15} **42.** i^{-2}

43. Evaluate $5x^9 - 6x^6 + 4x^5 - 2x^3 + 12x^2 - 1$ for $x = i$.

44. Evaluate $(2x^5 - 3x^2)(x^3 + 2x)$ for $x = i$.

In the following problems compute the value of the expression if $Z_1 = 2 - i$, $Z_2 = 9 + 2i$, and $Z_3 = 5 - 3i$.

45. $\frac{Z_1Z_2}{Z_1 - Z_2}$ **46.** $(Z_2 - Z_1)(Z_2 + Z_1)$ **47.** $Z_1^3 - 3Z_1^2 + Z_1$ **48.** $\frac{Z_1 - Z_3}{Z_1 + Z_3}$

49. If two impedances, Z_1 and Z_2, are connected in parallel in an electronic circuit, the total impedance, T, is related by the statement $T = \frac{Z_1 + Z_2}{Z_1Z_2}$. If $Z_1 = 10 - 3i$ and $Z_2 = 20 + i$, find T.

50. The impedance in an electrical circuit is the measure of the total opposition to the flow of an electric current. The impedance, Z, in a series circuit is

$$Z = R + i(X_L - X_C),$$

where R is resistance and X_L (read "X sub L") is inductive reactance and X_C ("X sub C") is capacitive reactance. Find Z (in ohms) if $R = 25$ ohms, $X_L = 15 + 2i$ ohms, and $X_C = 20 + 4i$ ohms.

51. (Refer to problem 50.) Suppose that $Z = 12 - 3i$ ohms, $R = 6$ ohms, and $X_L = 3 - 8i$ ohms. Find X_C.

52. For what values of x is $\sqrt{4 - x} \in C$?

53. For what values of x is $\sqrt{x - 16} \in C$?

54. Show that the product of a complex number and its conjugate is always a real number by forming the product $(a + bi)(a - bi)$ and examining the result.

55. If we add two complex numbers $a + bi$ and $c + di$, we obtain a third complex number, $e + fi$. For addition, $e = a + c$ and $f = b + d$. We can see this in the following derivation:

$$\begin{aligned}(a + bi) + (c + di) &= (a + c) + (bi + di)\\ &= (a + c) + (b + d)i.\end{aligned}$$

In other words,

$$(a + bi) + (c + di) = (a + c) + (b + d)i$$

is the rule for the addition of two complex numbers. Derive the rules for the subtraction, multiplication, and division of two complex numbers.

56. Julia sets are sets of numbers that appear in the study of fractals, an area of mathematics that has become much more interesting with the advent of modern computer graphics. Fractals are simple formulas that can actually generate complex graphics images. Julia sets are generated by calculating the value of the expression $z^2 + c$ repeatedly, for some fixed value of c and some starting value of z. The value calculated is the value used in the next computation of $z^2 + c$. For example, if $c = 1 + i$ and the first value of z (called the "seed value") is $2 - i$, then we proceed as shown.

z	$z^2 + c$	New value of z
$2 - i$	$(2 - i)^2 + (1 + i)$	$4 - 3i$
$4 - 3i$	$(4 - 3i)^2 + (1 + i)$	$-24 - 23i$
$-24 - 23i$	$(-24 - 23i)^2 + (1 + i)$	$48 - 1{,}103i$
	etc.	

This process produces the sequence of values $2 - i$, $4 - 3i$, $-24 - 23i$, ...

Calculate the first four values in the Julia set formed with $c = 2 - i$ and a seed value of $1 + 2i$. (The seed is the first value, so calculate three more values.)

57. See problem 56. Let an initial value of z be $0.5 - 0.2i$, and let c be $0.1 + 0.05i$. Write a program for a programmable calculator or computer to compute the successive values in the Julia set created by these values. Use the program to calculate the first 20 values. Look for a pattern.

Skill and review

1. Simplify $-5[3x - 2(1 - 4x)]$.
2. For what value(s) of x is the statement $x + 5 = 12$ true?
3. For what value(s) of x is the statement $5x = 20$ true?
4. For what value(s) of x is the statement $\frac{x}{6} = 48$ true?
5. Is the statement $3(2 - 3x) = 1 - 10x$ true when x represents -5?
6. For what value(s) of x is the statement $x + x = 2x$ true?
7. The formula $C = \frac{5}{9}(F - 32)$ converts a temperature in degrees Fahrenheit (F) to one in degrees centigrade (C). Convert 72° F to centigrade.
8. Multiply $0.06(1{,}000 - 2x)$.
9. $8\% =$ **a.** 800 **b.** 80 **c.** 0.8 **d.** 0.08
10. Find 8% of 12,000.
11. Add 6% of 4,000 to 10% of 12,000.

Chapter 1 summary

Definitions

- Some important sets
  - N **natural numbers** $= \{1, 2, 3, \ldots\}$
  - W **whole numbers** $= \{0, 1, 2, 3, \ldots\}$
  - J **integers** $= \{\ldots, -3, -2, -1, 0, 1, 2, 3, \ldots\}$
  - Q **rational numbers** $= \left\{\frac{p}{q} \middle| p \in J, q \in N\right\}$

    The decimal form of any rational number either terminates or repeats.
  - H **irrational numbers** $= \{x \mid$ The decimal representation of x does not terminate or repeat$\}$
  - R **real numbers** $= \{x \mid x \in Q$ or $x \in H\}$
  - C **complex numbers** $= \{a + bi \mid a, b \in R$, and i is the imaginary unit$\}$
- **Order of operations** Exponents, operations within symbols of grouping, multiplications and divisions from left to right, additions and subtractions from left to right.
- **Operations for fractions** (assume $a,b,c,d \in R$, and $c,d \neq 0$)

  $$\frac{a}{c} \pm \frac{b}{c} = \frac{a \pm b}{c} \quad \frac{a}{c} \pm \frac{b}{d} = \frac{ad \pm bc}{cd}$$

  $$\frac{a}{c} \cdot \frac{b}{d} = \frac{ab}{cd} \quad \frac{a}{c} \div \frac{b}{d} = \frac{a}{c} \cdot \frac{d}{b}, \ b \neq 0$$
- **Absolute value** $|x| = \begin{cases} x \text{ if } x \text{ is positive or zero} \\ -x \text{ if } x \text{ is negative} \end{cases}$
- **Zero exponent** $a^0 = 1$ if $a \neq 0$
- **Negative exponent** If $n \in R$, then $a^{-n} = \frac{1}{a^n}$ if $a \neq 0$

- **Rational exponent** $a^{\frac{1}{n}} = \sqrt[n]{a}$ if $a \varepsilon R$, $n \varepsilon N$, and $\sqrt[n]{a} \varepsilon R$
  $a^{\frac{m}{n}} = \left(a^{\frac{1}{n}}\right)^m$ if $m,n \varepsilon N$ and $\sqrt[n]{a} \varepsilon R$
- **Prime number** Any natural number greater than one that is divisible only by itself and one. The first primes are
  2, 3, 5, 7, 11, 13, 17, 19, 23, 29, ...
- $\boldsymbol{i = \sqrt{-1}}$; i is the imaginary unit.
- $\boldsymbol{\sqrt{-b}} = i\sqrt{b}$ if $b \varepsilon R$, $b > 0$
- $\boldsymbol{a - bi}$ is the **complex conjugate** of $a + bi$.

Rules

- **Absolute value**

  [1] $|a| \geq 0$ [2] $|-a| = |a|$

  [3] $|a| \cdot |b| = |ab|$ [4] $\dfrac{|a|}{|b|} = \left|\dfrac{a}{b}\right|$

  [5] $|a - b| = |b - a|$
- **Real exponents** If a, b, m, $n \varepsilon R$ and no variable represents zero where division by that variable is indicated, then

  [1] $a^m a^n = a^{m+n}$ [2] $\dfrac{a^m}{a^n} = a^{m-n}$ [3] $(ab)^m = a^m b^m$

  [4] $\left(\dfrac{a}{b}\right)^m = \dfrac{a^m}{b^m}$ [5] $(a^m)^n = a^{mn}$
- **Factoring** The general types of factoring are
  Greatest common factor
  Difference of two squares
  $$a^2 - b^2 = (a - b)(a + b)$$
  Grouping
  Quadratic trinomial
  The difference and sum of two cubes
  $$a^3 - b^3 = (a - b)(a^2 + ab + b^2)$$
  $$a^3 + b^3 = (a + b)(a^2 - ab + b^2)$$
- **Fundamental principle of rational expressions** If P, Q, and R are polynomials, $Q \neq 0$, and $R \neq 0$, then $\dfrac{P}{Q} = \dfrac{P \cdot R}{Q \cdot R}$.
- **Multiplication of rational expressions** If P, Q, R, and S are polynomials, $Q \neq 0$, and $S \neq 0$, then $\dfrac{P}{Q} \cdot \dfrac{R}{S} = \dfrac{PR}{QS}$.
- **Division of rational expressions** If P, Q, R, and S are polynomials, $Q \neq 0$, $R \neq 0$, and $S \neq 0$, then $\dfrac{P}{Q} \div \dfrac{R}{S} = \dfrac{P}{Q} \cdot \dfrac{S}{R}$.
- **Addition/subtraction of rational expressions** If P, Q, R, and S are polynomials, $Q \neq 0$, and $S \neq 0$, then

  [1] $\dfrac{P}{Q} \pm \dfrac{R}{Q} = \dfrac{P \pm R}{Q}$ [2] $\dfrac{P}{Q} \pm \dfrac{R}{S} = \dfrac{PS \pm QR}{QS}$
- **nth root of nth power** $\sqrt[n]{a^n} = \begin{cases} |a| & \text{if } n \text{ is even} \\ a & \text{if } n \text{ is odd} \end{cases}$
- **Square root of a square** If $x \varepsilon R$ then $\sqrt{x^2} = |x|$.
- **Product property of radicals** If $a \geq 0$, $b \geq 0$, and $n \varepsilon N$, then $\sqrt[n]{ab} = \sqrt[n]{a}\sqrt[n]{b}$.
- **Quotient property of radicals** If $a \geq 0$, $b > 0$, and $n \varepsilon N$, then $\sqrt[n]{\dfrac{a}{b}} = \dfrac{\sqrt[n]{a}}{\sqrt[n]{b}}$.
- **Index/exponent common factor property** Given $\sqrt[n]{a^m}$ and $a \geq 0$. If m and n have a common factor, this factor can be divided from m and from n.
- $i^1 = i$, $i^2 = -1$, $i^3 = -i$, $i^4 = 1$; the value of i^n, n a whole number, can only have one of these four values.

Chapter 1 review

[1–1] The following sets are given in set-builder notation; write the sets as a list of the elements.

1. $\{x \mid x > -3 \text{ and } x < 8 \text{ and } x \varepsilon W\}$
2. $\{x \mid x \varepsilon W \text{ and } x \varepsilon N\}$
3. $\left\{\dfrac{x}{x+1} \,\middle|\, x \varepsilon \{1, 2, 3, \ldots, 100\}\right\}$

Give the decimal form of each rational number.

4. $\frac{5}{12}$
5. $\frac{3}{13}$

Simplify the given algebraic expressions.

6. $-\dfrac{5(2-8)^2}{6} - \dfrac{14(3-14)}{8}$
7. $\frac{5}{6} - \frac{3}{8} - \frac{3}{4}$
8. $\dfrac{3-8}{15} \div \dfrac{5}{(4+3^2) - \frac{10}{3}}$
9. $\dfrac{5(5 - 2(9-7) + \frac{4}{5}) - (8-12)(2-8)}{5(3-7)^3 - (3-7)^2}$
10. $\dfrac{3x}{5a} - \dfrac{y}{4b}$
11. $\dfrac{a+3}{4a} + \dfrac{3b+2}{b}$
12. $\dfrac{3x}{2y} \div \dfrac{5y}{2x} \cdot \dfrac{x}{5y}$
13. $\left(\dfrac{2a}{b} + \dfrac{b}{3a}\right) \cdot \left(\dfrac{3b}{5a}\right)$

In problems 14 and 15 an interval is indicated in set-builder notation; give the interval notation and the graph of the set.

14. $\{z \mid -2\frac{1}{2} \le z < -\frac{3}{4}\}$ **15.** $\{y \mid -\frac{1}{4} < y\}$

In problems 16 and 17 an interval is indicated in interval notation; give the set-builder notation and the graph of the set.

16. $(-\infty, -1]$ **17.** $\left[-\frac{\pi}{3}, \pi\right)$

18. Describe the interval in the figure in both set-builder and interval notation.

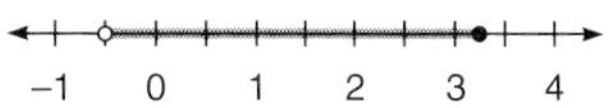

19. Describe the interval in the figure in both set-builder and interval notation.

In problems 20–22 express the value of the expression without the absolute value symbol.

20. $-\left|-\frac{1}{2}\right|$ **21.** $|-\pi - 9|$ **22.** $-|\sqrt{2} - 5|$

In problems 23–26 use the rules for absolute value to rewrite the expression with as few factors as possible in the absolute value operator.

23. $|-5x^2|$ **24.** $\left|\frac{3x^2}{2y^3}\right|$ **25.** $|10x - 5|$

26. $-|(x-2)^2(x+1)|$

[1–2] Use the rules for exponents to simplify the following expressions.

27. $(3x^2)^3y^{-3}$ **28.** $-3xy^{-2}$ **29.** $5^{-2}x^3y^2$

30. $(-3x^5y)(-2x^{-4}y^{-2})$ **31.** $(-3^2xy^{-5})(2^2x^5y^{-1}z^0)$

32. $\dfrac{-3^2x^{-4}y}{-3xy^{-4}}$ **33.** $\left(\dfrac{-2x^{-2}y^{-1}}{8x^{-2}y^{-3}}\right)^3$ **34.** $\left(\dfrac{2a^{-2}}{12a^{-12}}\right)^{-2}$

35. $x^{3-4n}x^{2n-3}$ **36.** $\left(\dfrac{x^{3n}y^{3-n}}{x^{-n}y^{n-3}}\right)^2$

Convert each number into scientific notation.

37. 42,182,000,000,000,000

38. −0.000 000 000 046 05

Convert each number given in scientific notation into a decimal number.

39. 4.052×10^{-7} **40.** -3.409×10^{11}

In problems 41–43 find the value that each expression represents, assuming that $a = 2$ and $b = -6$.

41. $a^3 - 2a^2 + 12a + 3$ **42.** $(6a^2 - 2a + 1)(a-1)^2$

43. $-2b(\frac{1}{3}(4(\frac{1}{2}b(-b+4) - 2) - 7) + 2) + 1$

Multiply.

44. $-3x^3(5x^3 + 7x - \frac{1}{3} - 2x^{-3})$

45. $(a-5)^2(5a+1)$

46. $(x^2 - 5x + 5)(-3x^3 - 2x^2 + 3)$

Perform the indicated divisions.

47. $\dfrac{-12x^6 - 4x^4 + 4x^2}{4x^2}$

48. $\dfrac{30a^8b^4 + 9a^4b^4 + 12a^4b^{12}}{-3a^4b^4}$

49. $\dfrac{y^4 - 1}{y - 1}$

50. $\dfrac{4x^4 - 4x^3 - 5x^2 - 10x - 5}{2x + 1}$

[1–3] Factor the following expressions.

51. $25x^3 - x^5$ **52.** $x^2 + 13x + 36$

53. $8a^3 - 14a^2 + 5a$ **54.** $3a^2b^2 + 2ab - 8$

55. $8a^3b + 125b^4$

56. $45ax^2(x^2 - 1) - 5a(x^2 - 1)$

57. $5x^2 - 51xy + 10y^2$ **58.** $(a-b)^2 - (2x+y)^2$

59. $54x^6 - 2y^3$ **60.** $12x^2 - 48y^2$

61. $-2by + 3ax - bx + 6ay$ **62.** $8a^9 - b^3c^3$

63. $7a^2 - 32a - 21$ **64.** $4x^4 - 24a^2x^2 + 36a^4$

65. $3(x^3 - 1)^2 - 2(x^3 - 1) - 8$

66. $4b(x + 3y) - 16a^2b(x + 3y)$

67. $a^2 - 4(x + 5y)^2$ **68.** $3x^2 - 8x - 91$

69. $3x^5y^9 + 81x^2z^6$

70. $x^2(x^2 - 1) - 4x(x^2 - 1) - 4(1 - x^2)$

[1–4] Reduce each rational expression to lowest terms.

71. $\dfrac{3a^2 - a}{9a^2 - 1}$ **72.** $\dfrac{2y - 3xy}{6x^2 + 5x - 6}$

73. $\dfrac{4x^2 - 1}{8x^3 - 1}$ **74.** $\dfrac{2x(x-2)^2 + (x-2)^2}{4x^4 - 17x^2 + 4}$

Perform the indicated operations.

75. $\dfrac{x - 2y}{5x} + \dfrac{3x - y}{3y}$ **76.** $\dfrac{x - 5}{x - 2} + \dfrac{6 - 2x}{4 - 2x}$

77. $\dfrac{1}{4x} - \dfrac{2x}{4x - 5}$ **78.** $\dfrac{18b^2}{14a^3} \cdot \dfrac{28a^3}{3b^3}$

79. $\dfrac{2x^3 - 4x^2 + 2x}{x^2 + 3x - 4} \cdot \dfrac{x^2 - 16}{4x^2 - x - 3}$

80. $\dfrac{x^2 - x - 6}{x^2 - 2x - 3} \div (x^2 + 4x + 4)$

81. $\dfrac{3x}{x^2 - 5x + 6} - \dfrac{2x - 5}{x - 3}$

82. $\dfrac{2x - 1}{x^2 + 2x - 3} + \dfrac{x + 1}{9 - x^2} - \dfrac{3}{x - 3}$

Simplify each complex rational expression.

83. $\dfrac{\frac{5}{2x} - \frac{3}{y}}{\frac{1}{x} + \frac{2}{y}}$ **84.** $\dfrac{\frac{3a - 2b}{2a} + \frac{5}{3}}{\frac{2a + 3b}{4a} - \frac{5}{6a}}$ **85.** $\dfrac{\frac{3}{a - b} - 2}{5 - \frac{3}{a - b}}$

[1–5] Find the indicated root.

86. $\sqrt[5]{-32}$ **87.** $\sqrt[4]{256}$

Simplify the following radical expressions. Assume that all variables represent nonnegative real numbers.

88. $\sqrt[3]{432}$ **89.** $\sqrt{54x^4y^7z^2}$ **90.** $\sqrt[3]{48a^6b^5c^8}$

91. $(\sqrt[3]{9a^4b^2})^2$ **92.** $\sqrt[4]{25a^2b^4c}\sqrt[4]{25a^2b^5c^2}$

93. $\sqrt[6]{a^4b^6}$ **94.** $\sqrt{128a^{15}b^{10}}$

95. $\dfrac{18a}{\sqrt{108a^3}}$ **96.** $\sqrt[3]{\dfrac{16x^4y^7}{25w^5z^2}}$

Simplify the following radical expressions. Assume that all variables represent nonnegative real numbers.

97. $-\sqrt{72a^2b^3} + 3\sqrt{50a^2b^3}$

98. $\sqrt[4]{48b^3c^6} - \sqrt[4]{243b^3c^6}$

99. $\sqrt{3xy^3}(\sqrt{3xy} - \sqrt{3x} + \sqrt{6y})$

100. $(\sqrt[4]{9x^2} - 3\sqrt[4]{3x^3})^2$

Rationalize the denominators. Assume that all variables represent nonnegative real numbers.

101. $\dfrac{\sqrt{6x} - 5}{\sqrt{6x} + \sqrt{2}}$ **102.** $\dfrac{a + \sqrt{a}}{\sqrt{a^3} + \sqrt{ab}}$

Perform the indicated operations and simplify.

103. $\dfrac{1}{4}\left(\dfrac{3}{2\sqrt{2}}\right) - \dfrac{\sqrt{3}}{4}\left(-\dfrac{1}{\sqrt{2}}\right)$ **104.** $\sqrt{\dfrac{3 - \frac{\sqrt{2}}{3}}{3}}$

[1–6] Rewrite each expression in terms of radicals. Simplify if possible. Assume all variables represent nonnegative values.

105. $(25x^3)^{\frac{1}{2}}$ **106.** $(16x^8)^{\frac{3}{4}}$ **107.** $(81x^6)^{-\frac{1}{4}}$

Simplify. Variables may represent negative values.

108. $\sqrt{8x^6y^9}$ **109.** $\sqrt{\dfrac{x^6}{8y^{10}}}$

110. $\sqrt{16x^6(x - 3)^2}$

Simplify. Assume all variables represent nonnegative values.

111. $\left(x^{\frac{3}{4}}y^{-\frac{1}{4}}\right)\left(x^{\frac{2}{3}}y^{\frac{1}{2}}\right)$ **112.** $\left(8x^{\frac{4}{5}}y^{-\frac{3}{4}}z\right)^{\frac{2}{3}}$

113. $\dfrac{\left(4x^{-\frac{3}{2}}y\right)^{-\frac{1}{2}}}{x^{\frac{1}{2}}y^{\frac{3}{4}}}$ **114.** $\left(\dfrac{64x^{-\frac{3}{5}}y^{-\frac{3}{4}}}{8z^{\frac{3}{4}}}\right)^{-\frac{8}{3}}$

115. $\left(x^{\frac{3a}{b}}y^{\frac{2a}{c}}\right)^{\frac{bc}{6a}}$ **116.** $\left(\dfrac{8x^{\frac{m+n}{m}}}{2x^{\frac{n}{3m}}y^{-\frac{n}{3m}}}\right)^{3m}$

Compute the following values to four places of accuracy.

117. $(-356)^{\frac{2}{3}}$ **118.** $\sqrt[4]{25^8}$

119. $\sqrt[5]{8.2^3}$ **120.** $(\sqrt[4]{200})^{\frac{1}{3}}$

[1–7] Perform the operations shown on the complex numbers.

121. $(-13 + 12i) - (6\frac{1}{2} - 3i)$

122. $(-8 + 3i)(2 - 7i)$

123. $(\frac{2}{3} - 3i)(-3 + \frac{1}{3}i) - (\frac{1}{3} - 3i)^2$

124. $\dfrac{3 - 6i}{2 + 5i}$ **125.** $\dfrac{-3 + 2i}{i(4 - i)}$

Simplify the following expressions.

126. $(1 - \sqrt{-8})(4 + \sqrt{-8})$

127. $\dfrac{6 + 3\sqrt{-12}}{4 - \sqrt{-12}}$ **128.** $(2 - \sqrt{-2})^4$

In problems 129 and 130 compute the value of the expression if $A = 1 - 5i$ and $B = -2 + 2i$.

129. $\dfrac{B - A}{B^2 - A^2}$ **130.** $A(B + A)$

131. If two impedances, A and B, are connected in parallel in an electronic circuit the total impedance, T, is related by the statement $\dfrac{1}{T} = \dfrac{1}{A} + \dfrac{1}{B}$. If $A = 1 - 3i$ and $B = 4 + i$, find T.

132. Evaluate

$$x^8 - 5x^7 + x^6 - 3x^5 + 2x^4 + x^3 + 5x^2 - 11x + 1$$

for $x = i$.

Chapter 1 test

1. Write the set as a list of elements $\left\{\frac{x}{x+2} \,\middle|\, x \in \{1, 2, 3, 4, 5\}\right\}$.

Simplify the given algebraic expressions.

2. $-\frac{2(2-5)^3}{6} - (2^2 - 3^2)$

3. $\frac{5}{4} - \frac{3}{8} - \frac{2}{3}$

4. $\frac{a+3b}{4b} - \frac{3b+2a}{a}$

5. $\left(\frac{2a}{b} + \frac{b}{3a}\right) \cdot \left(\frac{3b}{5a}\right)$

6. Give the interval notation and the graph of the set $\left\{z \,\middle|\, -2\frac{1}{2} < z \le -\frac{3}{4}\right\}$.

7. Give the set-builder notation and the graph of the set $[-3, \infty)$.

8. Describe the interval in both set-builder and interval notation.

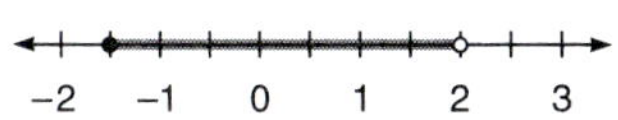

Express the value of the expression without the absolute value symbol.

9. $-|-4|$

10. $|-\pi + 2|$

11. Rewrite the expression with as few factors as possible in the absolute value operator. $|-3x^2y|$

Simplify the following expressions.

12. $(-2x^5y)(-3x^4y^2)$

13. $(-2^2x^{-1}y^5)(2^2x^5y^{-1}z^0)$

14. $\frac{3^2x^4y}{-3xy^{-4}}$

15. $\left(\frac{2a^{-2}}{12a^{-12}}\right)^{-2}$

16. $x^{-3-2n}x^{n+3}$

17. Convert into scientific notation: 205,000,000,000.

18. Convert into a decimal number: 2.13×10^{-4}.

In problems 19 and 20 find the value that each expression represents, assuming that $a = -\frac{2}{3}$, $b = 2$.

19. $(9a^2 - 6a + 1)(a + 1)$

20. $\frac{1}{3}(4(\frac{1}{2}b(-b + 4) - 2) - 7)$

Multiply.

21. $-2x^3(x^3 + \frac{1}{2}x - 3 - 2x^{-3})$

22. $(a - 5)^2(a + 5)^2$

23. Divide: $\frac{2x^3 - x^2 + 4x - 5}{x - 2}$.

Factor the following expressions.

24. $4a^3 - 16a$

25. $9x^2 - 3x - 2$

26. $x^4 - 16$

27. $64x^6 - 1$

28. $(x - 2)^2 + 2(x - 2) - 3$

29. $3ac - 2bd + ad - 6bc$

Reduce each rational expression to lowest terms.

30. $\frac{2x^2 - 2x}{2x^2 - 2}$

31. $\frac{2x^2 + x - 1}{4x^2 - 1}$

Perform the indicated operations.

32. $\frac{x-5}{x-2} - \frac{6+3x}{6-3x}$

33. $\frac{x}{2x+1} - \frac{x-1}{3x}$

34. $\frac{x^2 - 2x + 1}{x^2 - 4} \cdot \frac{x^2 - 4x + 4}{x^2 - 1}$

Simplify each complex rational expression.

35. $\dfrac{\frac{2}{3a} - \frac{3}{b}}{\frac{2}{3ab} + 2}$

36. $\dfrac{3 - \frac{3}{a-b}}{3 + \frac{3}{a-b}}$

Simplify the following radical expressions. Assume that all variables represent nonnegative real numbers.

37. $\sqrt[3]{128}$

38. $\sqrt{50x^4y^3z}$

39. $(\sqrt[3]{16ab^5})^2$

40. $\sqrt[4]{36ab^4c^2}\sqrt[4]{36a^2b^5c^2}$

41. $\frac{12x^3}{\sqrt{24x^5}}$

42. $\sqrt[3]{\frac{8x^4}{y^2z}}$

Simplify the following radical expressions. Assume that all variables represent nonnegative real numbers.

43. $\sqrt{45a^3b} - a\sqrt{20ab} + \sqrt{5a^3b}$

44. $-\sqrt{2x^2y}(\sqrt{2xy} - \sqrt{8x} + \sqrt{6y})$

45. Rationalize the denominator of $\frac{3\sqrt{a} - 3}{\sqrt{6a} + \sqrt{3}}$. Assume that all variables represent nonnegative real numbers.

46. Perform the operations and simplify: $\frac{1}{\sqrt{3}} \cdot \frac{1}{2} - \frac{\sqrt{5}}{\sqrt{3}}\left(-\frac{\sqrt{3}}{2}\right)$.

Rewrite each expression in terms of radicals. Simplify if possible. Assume all variables represent nonnegative values.

47. $(16x^3)^{\frac{1}{3}}$

48. $(16x^{11}y^4)^{-\frac{1}{4}}$

Simplify. Variables may represent negative values.

49. $\sqrt{20x^6y^8}$ **50.** $\sqrt{25x^4(x-3)^2}$

Simplify. Assume all variables represent nonnegative values.

51. $\left(x^{\frac{1}{4}}y^{-\frac{3}{4}}\right)\left(x^{\frac{3}{4}}y^{\frac{1}{2}}\right)$ **52.** $\left(27x^{\frac{3}{4}}y^{-\frac{3}{8}}z\right)^{\frac{4}{3}}$

53. $\dfrac{\left(8x^{\frac{3}{2}}y\right)^{-\frac{1}{3}}}{2x^{-\frac{1}{3}}y^{\frac{2}{3}}}$ **54.** $\left(\dfrac{64x^{-\frac{3}{2}}y^{-\frac{3}{4}}}{8z^{\frac{3}{4}}}\right)^{\frac{8}{3}}$

55. $\left(a^{\frac{m}{n}}b^{\frac{2}{n}}\right)^{\frac{n}{2m}}$

56. Compute $91^{-\frac{2}{3}}$ to four places of accuracy.

Perform the operations shown on the complex numbers.

57. $(-8+3i)^2(2-7i)$ **58.** $\dfrac{3-6i}{2+5i}$

Simplify the following expressions.

59. $(\sqrt{-4}-\sqrt{-9})(3+\sqrt{-12})$

60. $\dfrac{6+\sqrt{-12}}{3+\sqrt{-12}}$

61. If two impedances, Z_1 and Z_2, are connected in parallel in an electronic circuit, the total impedance, T, is related by the statement $\dfrac{1}{T}=\dfrac{1}{Z_1}+\dfrac{1}{Z_2}$. If $Z_1=2+4i$ and $Z_2=1+2i$, find T.

C H A P T E R

2 Equations and Inequalities

2–1 *Linear equations*

A book of the sixth century A.D., called the *Greek Anthology,* contains the following problem: If one pipe fills a cistern in one day, a second in two days, a third in three days, and a fourth in four days, how long will it take all four running together to fill it?

This section introduces the techniques necessary to solve this ancient problem.

Linear equations in one variable

An **equation** is a statement that two expressions are equal. Thus, $5(2) = 10$ and $5(2) = 12$ are both equations.[1] The first is true and the second is false. The equation $5x = 10$ is neither true nor false; such equations are called **conditional equations.** The left side of an equation is the **left member,** and the right side is the **right member.**

In this section we are concerned with **first-degree conditional equations in one variable,** also called **linear equations in one variable.**[2] In such an equation the exponent of the variable is one.

If the variable x in the equation $5x = 10$ is replaced by a number, the result is an equation that is true or false. The equation is true only if x is replaced by 2. Any replacement value of the variable, such as 2 here, for which the equation is true is called a **root** or **solution** of the equation. The set of all solutions to an equation is called the **solution set.** To solve the equation means to find the solution set.

[1]The = symbol was first used by Robert Recorde in *The Whetstone of Witte,* published in 1557. "I will sette as I doe often in woorke use, a paire of paralleles, or Gemowe [twin] lines of one lengthe, thus: ═══, bicause noe 2 thynges, can be moare equalle."

[2]The algebra of ancient Egypt (some 4,000 years ago) was much concerned with linear equations.

If the solution set to an equation is all permissible replacement values of the variable, the equation is an **identity.** The equation $x + x = 2x$ is an identity because the left member equals the right member regardless of the value that x represents. In this case the solution set is R, the set of real numbers. The equation $\frac{2x}{x} = 2$ is true for any value of x except zero, which is not permissible. Thus, this is an identity for all real numbers except zero.

We solve a linear equation by forming a sequence of equivalent equations, until we come to one that is sufficiently simple to solve by inspecting it. Two equations are said to be **equivalent equations** if they have the same solution set. We form these equations by using the following two properties of equality.

Addition property of equality

For any algebraic expressions A, B, and C,

$$\text{if } A = B \text{ then } A + C = B + C$$

Multiplication property of equality

For any algebraic expressions A, B, and C,

$$\text{if } A = B \text{ then } AC = BC$$

One additional property that is useful is called cross multiplication for equations.

Cross multiplication for equations

For any real numbers a, b, c, d, $b \neq 0$ and $d \neq 0$,

$$\text{if } \frac{a}{b} = \frac{c}{d} \text{ then } ad = bc$$

Note The proof of this property is left as an exercise. It can be visualized as if $\frac{a}{b} \times \frac{c}{d}$ then $ad = bc$

The cross-multiplication property states that the products formed across the two diagonals are equal. For example, if $\frac{x+1}{3} = \frac{5}{8}$, then we can conclude that $8(x + 1) = 15$. This property *only* applies when *one* fraction equals *one* other fraction. It does not apply to the equation $\frac{x+1}{3} + \frac{1}{3} = \frac{5}{8}$, for example, because the left side is not simply one fraction.

These properties allow us to perform the usual transformations on equations, as illustrated below. The basic procedure for solving linear equations follows.

> **Solving linear equations**
> - Clear any denominators by multiplying each term by the least common denominator of all the terms (or cross multiply if possible).
> - Perform any indicated multiplications (remove parentheses).
> - Use the addition property of equality so that all terms with the variable are in one member of the equation, and all other terms are in the other member.
> - If necessary factor out the variable from the terms containing it.
> - Divide both members of the equation by the coefficient of the variable.

Observe how these steps are used in examples 2-1 A, 2-1 B, and 2-1 C.

■ ***Example 2–1 A***

Find the solution set.

1. $\dfrac{x-2}{3} = \dfrac{x}{6}$

$6(x-2) = 3x$
$6x - 12 = 3x$
$6x - 3x = 12$
$3x = 12$
$x = 4$
$\{4\}$

2. $\dfrac{5}{6}x - \dfrac{3}{4} = 11 - \dfrac{x}{2}$

$12\left(\dfrac{5}{6}x - \dfrac{3}{4}\right) = 12\left(11 - \dfrac{x}{2}\right)$
$10x - 9 = 132 - 6x$
$16x = 141$
$x = \frac{141}{16} = 8\frac{13}{16}$
$\{8\frac{13}{16}\}$

3. $-2(x-4) + 1 = 3(x+3) - 5x$

$-2x + 8 + 1 = 3x + 9 - 5x$
$0 = 0$

This is a true statement regardless of the value of x, indicating the solution set is R and that the equation is an identity.

R Solution set

4. $5x - 4(x-3) = x + 11$

$x + 12 = x + 11$
$12 = 11$

This statement is a contradiction—that is, it is never true, regardless of the value of x; thus, the solution set is the empty set, indicated by the symbol ϕ.

ϕ Solution set ■

Note A set which contains no elements is called the **empty set,** ϕ.

Parts 3 and 4 of example 2–1 A illustrate *how to recognize an identity and an empty solution set.* In the first case we arrive at a statement that is true independent of the value of x, such as $1 = 1$ or $2x = x + x$, and in the second case we arrive at a contradiction, such as $1 = 0$.

■ ***Example 2–1 B***

Find a four-digit approximation to the solution to the equation $3.5x - 4.1(2x - 3) = 7.04$.

$$3.5x - 8.2x + 12.3 = 7.04$$
$$-4.7x = -5.26$$
$$x = \frac{-5.26}{-4.7} \approx 1.119$$
$$\{1.119\}$$
■

A graphing calculator can be used to verify numeric solutions to linear equations. This is a special case of what is covered in the *Computer-Aided Mathematics* section. For example, part 2 of example 2–1 A can be verified by graphing the two equations $Y_1 = \frac{5}{6}x - \frac{3}{4}$ and $Y_2 = 11 - \frac{x}{2}$ (see figure 2–1). At the point where the two lines cross $Y_1 = Y_2$, so $\frac{5}{6}x - \frac{3}{4} = 11 - \frac{x}{2}$. Therefore, the x-value at this point is the solution to the equation. Use the trace and zoom features to move the cursor to the point at which the lines cross. This will show that these two graphs cross at $x \approx 8.74$, which is close to the exact solution $8\frac{13}{16}$.

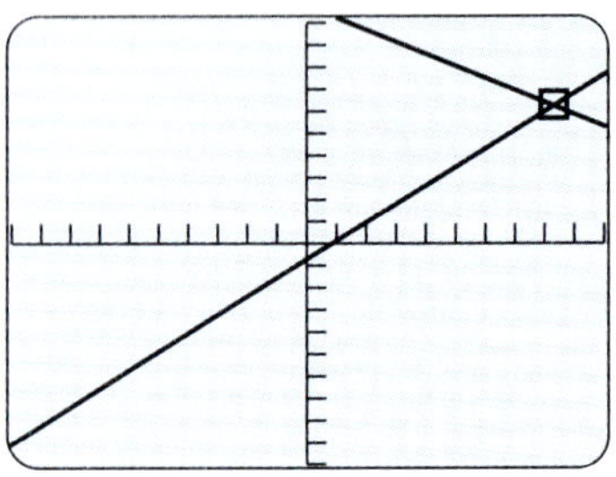

Figure 2–1

Thus, by graphing the right member of an equation and the left member we can find approximate numeric solutions using the trace calculator feature to find the value of x where the two graphs cross. Note that if the lines cross somewhere off the graphing area the RANGE must be adjusted and the lines regraphed. This process can be tedious, and thus this graphing solution method is not too practical.

Unfortunately the graphing calculator cannot be used to verify the solutions to literal equations, which we now examine.

Literal equations and formulas

A **literal equation** is one in which the solution is expressed in terms of nonnumeric symbols (letters). We may use the word **formula** for a literal equation in which the variables apply to some known situation.

To solve a literal equation for a variable means to rewrite it so it expresses that variable in terms of the others, by isolating that variable as the only term in one member of the equation. Literal equations that are linear in the unknown are solved *using the same steps* for solving linear equations *stated above.*

■ ***Example 2–1 C***

Solve the literal equation for the specified variable.

1. The formula for the perimeter of a rectangle is $P = 2\ell + 2w$. Solve for ℓ.

$$P = 2\ell + 2w$$
$$P - 2w = 2\ell$$
$$\frac{P - 2w}{2} = \ell$$
$$\ell = \frac{P - 2w}{2}$$

2. $\dfrac{x + y}{x - y} = z$; solve for x.

$$\frac{x + y}{x - y} = z$$
$$(x - y)\left(\frac{x + y}{x - y}\right) = z(x - y)$$
$$x + y = zx - zy$$
$$x - zx = -zy - y$$
$$x(1 - z) = -y(z + 1)$$
$$\frac{x(1 - z)}{1 - z} = \frac{-y(z + 1)}{1 - z}$$
$$x = \frac{-y(z + 1)}{1 - z}$$

Note The expression for x could also be written $\dfrac{y(z + 1)}{z - 1}$ by multiplying the numerator and denominator by -1. ■

Interest problems

One class of problems for which linear equations are appropriate is **interest problems.** To solve these requires an understanding of the basic principles of simple interest. The formula $I = Prt$ describes the interest I, which is earned by a principal P at a simple yearly interest rate r, in t years.

If $t = 1$ (year), as it will be below, the formula is

$$I = Pr$$

For example, \$850 at 6% yields $I = (850)(0.06) = 51.00$. Thus, \$850 at 6% yearly interest rate produces \$51 interest.

Now suppose \$5,000 was invested in two accounts; \$1,000 at 3% and the remainder at 8%. The total interest would be

$$I = (1{,}000)(0.03) + (5{,}000 - 1{,}000)(0.08) = \$350.00$$

Now suppose that \$5,000 is invested, part at 3% and the rest at 8%, and that the interest earned is \$200. How much is invested at each rate? If x is the amount invested at 3% (like the \$1,000 above) then $5{,}000 - x$ is the rest, invested at 8% (like the $5{,}000 - 1{,}000$ above). Then the equation for $I = \$200$ is

$$200 = 0.03x + 0.08(5{,}000 - x)$$

which can be solved to find x, the amount invested at 3%:

$$\begin{aligned} 200 &= 0.03x + 400 - 0.08x && \text{Multiply } 0.08(5{,}000 - x) \\ -200 &= -0.05x && \text{Combine terms, add } -400 \text{ to both members} \\ \frac{-200}{-0.05} &= x && \text{Divide each member by } -0.05 \\ x &= 4{,}000. \end{aligned}$$

Thus, \$4,000 is invested at 3%, and \$1,000 (\$5,000 − \$4,000) at 8%.

■ ***Example 2–1 D***

A total of \$15,000 is invested, part at 5% and the remainder at 7.5%. If the total interest earned in a year is \$825, how much was invested at each rate?

If x represents the amount invested at 5%, then $15{,}000 - x$ was invested at 7.5%. The amount of interest earned by the 5% investment is $0.05x$, and the amount of interest earned by the 7.5% investment is $0.075(15{,}000 - x)$. These two amounts of interest total \$825; this is expressed by the equation:

$$\text{Interest at 5\%} + \text{Interest at 7.5\%} = \text{Total interest}$$

$$\overbrace{0.05x}^{5\%} + \overbrace{0.075(\underbrace{15{,}000 - x}_{})}^{7.5\%} = 825$$

15,000

$$\begin{aligned} 0.05x + 1125 - 0.075x &= 825 \\ -0.025x &= -300 \\ x = \frac{-300}{-0.025} &= 12{,}000 \end{aligned}$$

Thus, \$12,000 is invested at 5%, and the remainder, \$15,000 − \$12,000 = \$3,000, is invested at 7.5%. ■

Mixture problems

A similar process is used to solve problems about mixtures. It is important to understand the physical setting for such problems. For example, what does it mean to say that a 20-liter container is full of an 8% acid/water solution? It means that 8% of the 20 liters is acid and the rest, 92%, is water. Thus, $(0.08)(20) = 1.6$ liters are acid, and $(0.92)(20) = 18.4$ liters are water. If we could separate the acid and water, we could represent the situation as shown in figure 2–2.

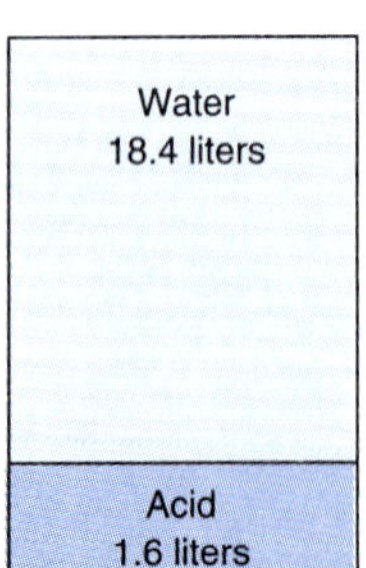

Figure 2–2

The following example illustrates a mixture problem. Just as with interest problems, we have information about several quantities that are percentages of other amounts. We usually obtain an equation by adding these percentages and equating them to some known total. Note that we do not add the *percents*, such as 80% + 35%. We do add *percentages*, such as 80% of one quantity + 35% of another quantity.

■ **Example 2–1 E**

What quantities of an 80% silver alloy and a 35% silver alloy must be mixed to obtain 400 grams of a 50% silver alloy?

A problem like this can often be solved by writing an algebraic statement about the amount of silver present. The total amount of silver in the resulting mixture is 50% of the 400 grams, or $0.50(400) = 200$ grams. Where does this 200 grams of silver come from? It comes from the 80% and 35% alloys. If x is the amount of 35% alloy, then it contains $0.35x$ grams of silver. The amount of 80% alloy must be $400 - x$, since the two alloys must combine to give us 400 grams. The amount of silver in this alloy is thus $0.80(400 - x)$ grams.

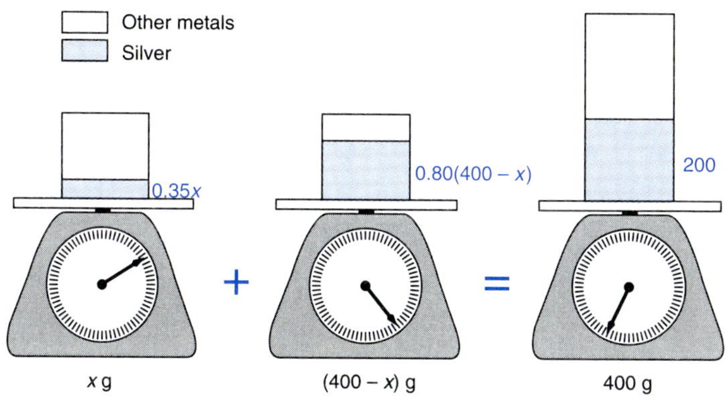

An expression that describes the amount of silver in each of the alloys is $0.35x + 0.80(400 - x) = 200$. This is all illustrated in the figure. We can find x, the amount of the 35% alloy, by solving the equation

$$0.35x + 0.8(400 - x) = 200$$
$$0.35x + 320 - 0.8x = 200$$
$$-0.45x = -120$$
$$x = \frac{-120}{-0.45} \approx 266.7 \text{ grams}$$

Thus, 266.7 grams of the 35% alloy and $400 - 266.7 = 133.3$ grams of the 80% alloy must be mixed to obtain 400 grams of a 50% silver alloy. ■

Rate problems

Another type of problem uses rates; the rate at which a person or machine works, or at which a pipe fills a pool, or at which a computer computes. Of interest to us here are situations in which the rates can be added. This is illustrated in example 2–1 F.

■ **Example 2–1 F**

Solve.

1. One painter can paint a certain size room in 6 hours; the painter's partner requires 10 hours to do a room of the same size. How long would it take them to paint such a room, working together?

   The first painter paints at a rate of $\frac{1}{6}$th of a room per hour, and the second at a rate of $\frac{1}{10}$th of a room per hour. We assume that working together they can paint $\frac{1}{6} + \frac{1}{10} = \frac{4}{15}$ of a room per hour. Now, if we let t be the time required to paint one room, we use the idea that rate · time = one (job) or $\frac{4}{15} \cdot t = 1$.

We solve for t.

$$\tfrac{4}{15}t = 1$$

$$(\tfrac{15}{4})(\tfrac{4}{15})t = (\tfrac{15}{4})1 \qquad \text{Multiply both members by } \tfrac{15}{4}$$

$$t = \tfrac{15}{4} = 3\tfrac{3}{4} \text{ hours}$$

Thus, it would take the painters $3\frac{3}{4}$ hours to paint the room, working together.

Note The rates to complete a job add, not the times needed to complete the job. This is illustrated above.

2. A boat travels at 20 miles per hour (mph) in still water. It travels downstream to a certain destination and back to the starting point in 3 hours. The speed of the current is 5 mph. How far downstream is the destination?

Let x represent this unknown distance downstream (and upstream). Using distance = rate × time, or $d = rt$ and solving for t, $t = \dfrac{d}{r}$, we summarize this information in a table.

	d **miles**	r **miles per hour**	t **hours**
Downstream	x	$20 + 5 = 25$	$\dfrac{x}{25}$
Upstream	x	$20 - 5 = 15$	$\dfrac{x}{15}$

Since total time downstream and back upstream is 3 hours, we know that

$$\text{time downstream} + \text{time upstream} = 3 \text{ hours}$$

$$\frac{x}{25} + \frac{x}{15} = 3$$

$$\frac{75}{1} \cdot \frac{x}{25} + \frac{75}{1} \cdot \frac{x}{15} = 75(3) \qquad \text{Multiply by the LCD, 75}$$

$$3x + 5x = 225$$

$$x = 28\tfrac{1}{8}$$

Since x represents the distance downstream, this distance is $28\frac{1}{8}$ miles. ■

Mastery points

Can you

- Solve linear equations in one variable?
- Solve literal equations in one variable?
- Solve certain word problems by setting up appropriate linear equations in one variable and solving?

Exercise 2–1

Solve the following linear equations[3] by specifying the solution set.

1. $13x = 5 - 3x$
2. $-12x + 4 = 8 - x$
3. $9 - 2x = -5(3 + x)$
4. $9a - 3 = 4(3 - 2a)$
5. $\frac{1}{4}x + 3 = \frac{3}{8}x - 8$
6. $-2(5 - 3x) = x - (3 + 6x)$
7. $\frac{2}{3}x - 3 = 4x + 1$
8. $\frac{2x}{5} = 3 - x$
9. $-5(3x - 2) + x = 0$
10. $\frac{x + 3}{2} = \frac{2x}{5}$
11. $\frac{2 - 3x}{4} = \frac{x}{2}$
12. $\frac{1}{2}x = 19 - 4(5 - 2x)$
13. $6x - 5 = x + 5(x - 1)$
14. $2 - \frac{3}{4}x = 5x$
15. $2[(3x - 1) - (5x + 2) - 4] = 2x$
16. $2(x + 1) = x - (3x - 1)$
17. $\frac{3}{4}(4x - 3) + x = -17$
18. $5[3 - 2(5 - x) + 2] - x = 0$
19. $\frac{1 - x}{4} = \frac{1 + x}{4}$
20. $\frac{x + 3}{4} = \frac{x + 4}{3}$
21. $-4[2x - 3(3x - 2) - (5 - x)] + 6(1 - x) = 0$
22. $4 - 2y = 4 + 2y$
23. $3x - 12 = 12 - 3x$
24. $x + 1 = x + 2$
25. $5(x - 3) = -(15 - 5x)$
26. $\frac{1}{4}x - \frac{1}{3}x = x$

Find approximate answers to the following problems. Round the answer to four digits of accuracy.

27. $9.6x - 2.4(3 - 1.8x) = 10.0$
28. $-19.6x = 4.55(3 - 2x)$
29. $150x - 13.8 = 0.04(1{,}500 - 1{,}417x)$
30. $13.5 + 1.5x - 2.3[9 - 3.25(67 - 4.25x) - 15x] = 0$

Solve the following literal equations for the variable indicated.

31. $V = k + gt$; for t
32. $m = -p(\ell - x)$; for x
33. $2S = 2Vt - gt^2$; for V
34. $R = W - b(2c + b)$; for c
35. $S = \frac{n}{2}[2a + (n - 1)d]$; for a
36. $P = n(P_2 - P_1) - c$; for P_1
37. $d = d_1 + (k - 1)d_2 + (j - d_2)d_3$; for d_2
38. $\frac{x + 2y}{x - 2y} = 4$; for x
39. $\frac{x + 2y}{x - 2y} = 4$; for y
40. $\frac{x + y}{x} = y$; for y
41. $2x - y = 5x + 6y$; for x
42. $3(4x - y) = 2x + y + 6$; for y
43. $V = r^2(a - b)$; for a
44. $\frac{x + y}{3} = \frac{2x - y}{5}$; for y
45. $\frac{x + y}{3} = \frac{2x - y}{5}$; for x
46. $2S = 2Vt - gt^2$; for g
47. $V = \frac{1}{3}\pi h^2(3R - h)$; for R
48. $b(y - 4) = a(x + 3)$; for y
49. $b(y - 4) = a(x + 3)$; for x
50. $T = \frac{R - R_0}{aR_0}$; for R
51. $T = \frac{aR_0}{R - R_0}$; for R_0
52. $A = \frac{1}{2}h(b_1 + b_2)$; for b_1
53. Amdahl's law is an equation used to measure efficiency of parallel algorithms in parallel processors of computers. One form is $\frac{1}{s} = f + \frac{1 - f}{p}$. Solve this equation for f.
54. Solve Amdahl's law (see problem 53) for s.

Solve the following interest problems.

55. A total of \$15,000 is invested, part at 8% and part at 6%. The income from these investments for one year is \$1,100. How much was invested at each rate?
56. A total of \$28,000 is invested, part at 4% and part at 8%. The income from these investments for one year is \$2,000. How much was invested at each rate?

[3] "Someone told me that each equation I included in the book would halve the sales. I therefore resolved not to have any equations at all." Stephen Hawking, talking about his wonderful book *A Brief History of Time*.

57. \$18,000 was invested; part of the investment made a 14% gain, but the rest had a 9% loss. The net gain from the investments was \$680. How much money was invested at each rate?

58. \$25,000 was invested; part of the investment made an 8% gain, but the rest had a 9% loss. The net loss from the investments was \$890. How much was invested at each rate?

59. Two investments were made. The larger investment was at 8%, and the smaller was at 5%. The larger investment earned \$230 more than the smaller investment, and was \$1,000 more than the smaller investment. How much money was invested at each rate?

60. Two investments were made. The larger investment was at 10%, and the smaller was at 6%. The larger investment earned \$410 more than the smaller investment, and was \$500 more than the smaller investment. How much money was invested at each rate?

61. A total of \$18,000 was invested, part at 5% and part at 9%. If the income for one year from the 9% investment was \$100 less than the income from the 5% investment, how much was invested at each rate?

62. \$6,000 has been invested at 5% interest. How much money would have to be invested at 8% so that the interest rate on the two investments would be 6%?

Solve the following mixture problems.

63. A trucking firm has two mixtures of antifreeze; one is 35% alcohol, and the other is 65% alcohol. How much of each must be mixed to obtain 80 gallons of a 50% solution?

64. A company has 2.5 tons of material that is 30% copper. How much material that is 75% copper must be mixed with this to obtain a material that is 50% copper?

65. A company has 3,000 gallons of a 10% pesticide solution. It also can obtain as much of a 4% pesticide solution as it needs. How much of this 4% solution should it mix with the 3,000 gallons of 10% solution so that the result is an 8% solution?

66. A drug firm has on hand 400 lb of a mixture that is 3% sodium. The firm can buy as much of a 0.8% sodium mixture as it needs. It wishes to sell a mixture that is 2% sodium. How much of the 0.8% sodium mixture must it mix with the 400 lb of on-hand material to obtain a 2% mixture?

67. A drug firm has an order for 300 liters of 40% hydrogen peroxide. It only stocks 20% and 55% solutions. How much of each should be mixed to fill the order?

Solve the following rate problems.

68. A printing press takes 35 minutes to print 5,000 flyers; a second takes 50 minutes. (a) Running together, how long would it take to print 5,000 flyers, to the nearest second? (b) 8,000 flyers?

69. A conveyor belt takes 3 hours to move 50 tons of iron ore. A newer belt takes $2\frac{1}{4}$ hours to do the same amount of work. (a) How long (to the nearest minute) would it take both belts running together to move 50 tons of iron ore? (b) 235 tons?

70. One logging crew takes 18 hours to log 2 acres; a second crew takes 14 hours for the same job. How long would it take the crews to log 2 acres working together?

71. A book of the sixth century A.D., called the *Greek Anthology,* contains the following problem: If one pipe fills a cistern in one day, a second in two days, a third in three days, and a fourth in four days, how long will it take all four running together to fill it? Solve this problem.

72. A street sweeper takes 8 hours to sweep 20 miles of street. A new model takes 6 hours to do the same thing. (a) Working together, how long would it take both machines to sweep 20 miles of street? (b) 35 miles of street?

73. An automobile can travel 200 miles in the same time that a truck can travel 150 miles. If the automobile travels at an average rate that is 15 mph faster than the average rate of the truck, find the average rate of each.

74. A boat moves at 16 mph in still water. If the boat travels 20 miles downstream in the same time it takes to travel 14 miles upstream, what is the speed of the current?

75. An airplane can cruise at 300 mph in still air. If the airplane takes the same time to fly 950 miles with the wind as it does to fly 650 miles against the wind, what is the speed of the wind?

76. A boat travels 40 kilometers upstream in the same time that it takes to travel 60 kilometers downstream. If the stream is flowing at 6 km per hour, what is the speed of the boat in still water?

77. It takes one jogger 2 minutes longer to jog a certain distance than it does another. What is this distance, if the faster jogger jogs at 7 mph and the slower at 5 mph?

78. An individual averages 10 miles per hour riding a bicycle to deliver papers. The same individual averages 30 mph to deliver the papers by car. If it takes one-half hour less time by car, how long is the paper route?

79. Prove the cross-multiplication property. It states

For any real numbers $a, b, c, d, b \neq 0$ and $d \neq 0$, if $\frac{a}{b} = \frac{c}{d}$ then $ad = bc$.

(*Hint:* Multiply both members by the least common denominator bd.)

Skill and review

1. Solve for x: $(x - 2)(x + 5) = 0$.
2. Multiply: $(2x - 3)(x + 2)$.
3. Factor: $4x^2 - 16x$.
4. Factor: $4x^2 - 1$.
5. Factor: $6x^2 - 5x - 4$.
6. Simplify: $\sqrt{-20}$.
7. Simplify: $\sqrt{8 - 4(3)(-2)}$.
8. Simplify: $\frac{8 - \sqrt{32}}{4}$.
9. Find the area and perimeter of a rectangle whose length is 8 inches and width is 6 inches.
10. How long will it take a vehicle that is going 45 miles per hour to travel 135 miles?

2–2 Quadratic equations

One printing press takes 3 hours longer than another to print 10,000 newspapers. Running together they produce the 10,000 papers in 8 hours. Find the time required for each to do this job alone.

This section shows the mathematics necessary to deal with this problem and related types of problems.

Recall from section 1–3 that a quadratic expression is an expression of the form $ax^2 + bx + c$.

Quadratic equation

A quadratic equation in one variable is an equation that can be put in the form

$$ax^2 + bx + c = 0$$

$a, b, c \in R$, $a > 0$.

This form is called the **standard form** for a quadratic equation.

Solution by factoring

If the quadratic expression in a quadratic equation in standard form can be factored, the equation can be solved using the zero product property.

> **Zero product property**
> For any algebraic expressions A and B,
> $$AB = 0 \text{ if and only if } A = 0 \text{ or } B = 0$$
> ***Concept***
> A product can be zero if and only if one of its factors is zero.

■ ***Example 2–2 A***

Solve the quadratic equations by factoring.

1. $6p^2 = 3p$
$6p^2 - 3p = 0$
$3p(2p - 1) = 0$
$3p = 0$ or $2p - 1 = 0$
$p = 0$ or $p = \frac{1}{2}$
$\{0, \frac{1}{2}\}$

2. $x^2 + \frac{14}{3}x = \frac{5}{3}$
$x^2 + \frac{14}{3}x - \frac{5}{3} = 0$
$3x^2 + 14x - 5 = 0$
$(3x - 1)(x + 5) = 0$
$3x - 1 = 0$ or $x + 5 = 0$
$x = \frac{1}{3}$ or $x = -5$
$\{-5, \frac{1}{3}\}$ ■

Solution by extracting the roots

When b in $ax^2 + bx + c = 0$ is zero, we have a simpler equation of the form $ax^2 + c = 0$. This can be solved by the method called **extracting the roots,** which uses the fact that

$$\text{if } x^2 = k, \text{ then } x = \sqrt{k} \text{ or } x = -\sqrt{k}$$

This can be abbreviated using the symbol ±, which means "plus or minus," to $x = \pm\sqrt{k}$.

■ ***Example 2–2 B***

Solve $(4x - 2)^2 = 8$ by extracting the roots.

$(4x - 2)^2 = 8$	The member with the variable is a perfect square
$4x - 2 = \pm\sqrt{8}$	Extract the roots
$4x - 2 = \pm 2\sqrt{2}$	$\sqrt{8} = \sqrt{4}\sqrt{2} = 2\sqrt{2}$
$4x = 2 \pm 2\sqrt{2}$	Add 2 to both members
$x = \dfrac{2 \pm 2\sqrt{2}}{4}$	Divide both members by 4
$= \dfrac{2(1 \pm \sqrt{2})}{4}$	See following note
$= \dfrac{1 \pm \sqrt{2}}{2}$	Reduce by 2
$\left\{\dfrac{1 \pm \sqrt{2}}{2}\right\}$ or $\left\{\dfrac{1 + \sqrt{2}}{2}, \dfrac{1 - \sqrt{2}}{2}\right\}$	Solution set

Note "$2 \pm 2\sqrt{2}$" means "$2 + 2\sqrt{2}$ or $2 - 2\sqrt{2}$." Each has a common factor 2 and can be rewritten "$2(1 + \sqrt{2})$ or $2(1 - \sqrt{2})$," which can be abbreviated as "$2(1 \pm \sqrt{2})$." ■

Solution by the quadratic formula

When the methods mentioned above do not apply (or even if they do), a quadratic equation can be solved using the quadratic formula.

The quadratic formula

If $ax^2 + bx + c = 0$ and $a \neq 0$, then

$$x = \frac{-b \pm \sqrt{b^2 - 4ac}}{2a}$$

This formula, developed in the seventeenth century,[4] can be used to solve any quadratic equation. The derivation of the formula is an exercise in section 4–1 after a discussion of a procedure called "completing the square." A proof that the formula gives solutions to the quadratic equation is in the exercises of this section.

The formula is applied by determining the given values of a, b, and c and using substitution of value (section 1–2).

■ ***Example 2–2 C***

Solve $3 + \frac{6}{y^2} = \frac{4}{y}$ using the quadratic formula.

$3y^2 + 6 = 4y$ — Multiply each term by y^2

$3y^2 - 4y + 6 = 0$ — Put in standard form: $a = 3$, $b = -4$, $c = 6$

$y = \frac{-(-4) \pm \sqrt{(-4)^2 - 4(3)(6)}}{2(3)}$ — Substitute into the formula

$= \frac{4 \pm 2i\sqrt{14}}{6}$ — $\sqrt{-56} = i\sqrt{56} = i\sqrt{4 \cdot 14} = 2i\sqrt{14}$

$= \frac{2(2 \pm i\sqrt{14})}{6}$ — Factor 2 from the numerator

$= \frac{2 \pm i\sqrt{14}}{3}$ — Reduce

$= \frac{2}{3} \pm \frac{\sqrt{14}}{3}i$ — Standard form for complex number

$\left\{\frac{2}{3} \pm \frac{\sqrt{14}}{3}i\right\}$ — Solution set ■

[4]Quadratic equations were solved in ancient civilizations, often using geometric constructions, but negative and complex roots were rejected as not being part of the real world. Algebraic formulas for solving the quadratic equation were developed by Rafael Bombelli (ca. 1526–1573).

As can be observed in the examples above, when $b^2 - 4ac < 0$ the solutions of the quadratic equation are complex. The expression $b^2 - 4ac$ is called the **discriminant** of the quadratic expression $ax^2 + bx + c$. The solutions of the quadratic equation can be categorized as follows:

Value of discriminant	Solutions of $ax^2 + bx + c = 0$
$b^2 - 4ac > 0$	Two real solutions
$b^2 - 4ac = 0$	One real solution
$b^2 - 4ac < 0$	Two complex solutions.

As shown in section 2–1, a graphing calculator can be used to verify numeric solutions to equations. For example, the solutions to $2x^2 = 4x - 1$ can be verified by graphing the two equations $Y_1 = 2x^2$ and $Y_2 = 4x - 1$ (see figure 2–3). Using the "trace" feature will show that these two graphs cross at the approximate x-values 0.31 and 1.79. The solutions are $\frac{2 - \sqrt{2}}{2} \approx 0.29$ and $\frac{2 + \sqrt{2}}{2} \approx 1.71$. Using the zoom feature can produce better approximations. Note that this method cannot be used to verify complex roots.

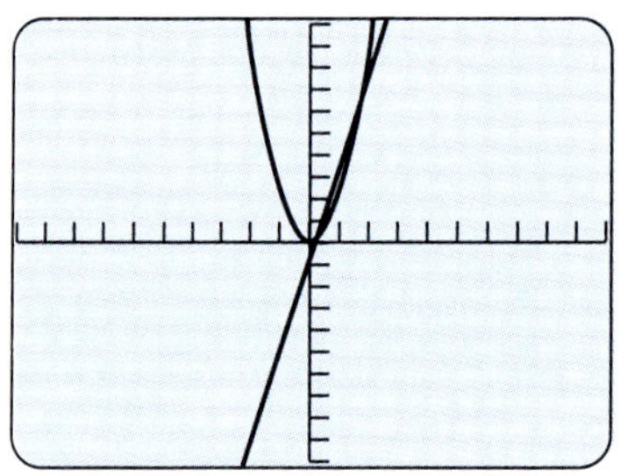

Figure 2–3

General factors of a quadratic expression

In chapter 1 we reviewed factoring quadratic trinomials when the resulting factors have integer coefficients. The quadratic formula also allows us to factor *any* quadratic expression by the following theorem.

> **Factors of a quadratic expression**
> Given $ax^2 + bx + c$, $a \neq 0$, then
> $$ax^2 + bx + c = a\left(x - \frac{-b + \sqrt{b^2 - 4ac}}{2a}\right)\left(x - \frac{-b - \sqrt{b^2 - 4ac}}{2a}\right)$$

■ ***Example 2–2 D***

Factor the quadratic expression $3y^2 - 4y + 6$.

From example 2–2 C the zeros are $\frac{2}{3} \pm \frac{\sqrt{14}}{3}i$, so $3y^2 - 4y + 6$

$$= 3\left[y - \left(\frac{2}{3} + \frac{\sqrt{14}}{3}i\right)\right]\left[y - \left(\frac{2}{3} - \frac{\sqrt{14}}{3}i\right)\right].$$

■

Some geometry

A **right triangle** is a triangle with one right (90°) angle (see figure 2–4). The side opposite that angle is called the **hypotenuse** and is always the longest side of the triangle. If c represents the length of the hypotenuse, and a and b represent the lengths of the remaining two sides, then the following relation holds: $a^2 + b^2 = c^2$. This is the **Pythagorean theorem.**[5]

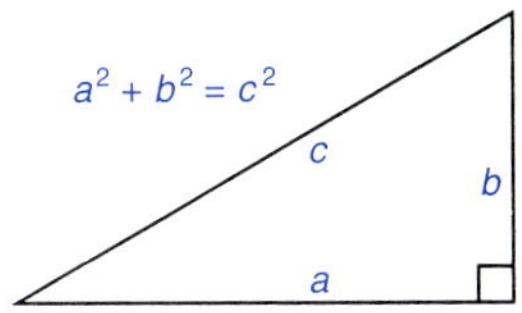

Figure 2–4

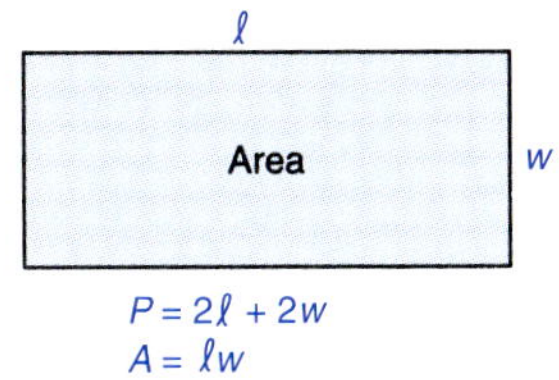

Figure 2–5

A **rectangle** is a four-sided figure with four right angles (see figure 2–5). The lengths of its sides are called the length and width. The **perimeter** is the distance around the figure; the **area** of the figure describes the size of its surface. If ℓ means length, w means width, P means perimeter, and A means area then the following formulas hold for rectangles: $P = 2\ell + 2w$ and $A = \ell w$.

■ *Example 2–2 E*

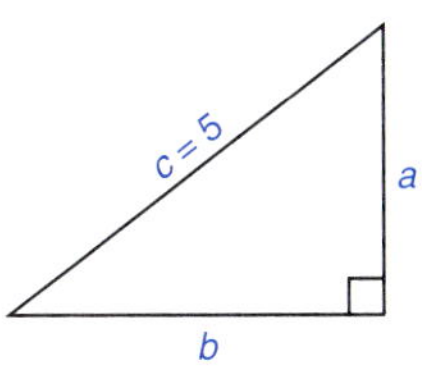

Solve the following problems.

1. In the right triangle in the figure the length c of the hypotenuse is 5 and length b is 1 unit larger than length a. Find a and b.

In this triangle we are told that side b can be described as $a + 1$. We begin with the Pythagorean theorem, which we know to be true for any right triangle.

$a^2 + b^2 = c^2$	Pythagorean theorem
$a^2 + (a + 1)^2 = 5^2$	$b = a + 1$
$a^2 + (a^2 + 2a + 1) = 25$	Perform indicated operations
$2a^2 + 2a - 24 = 0$	Put in standard form
$a^2 + a - 12 = 0$	Divide each member by 2
$(a + 4)(a - 3) = 0$	Factor
$a + 4 = 0$ or $a - 3 = 0$	Zero product property
$a = -4$ or $a = 3$	Solve each linear equation

The value of a cannot be -4 since it represents the length of the side of a triangle. Thus, $a = 3$.

We now find b.

$$b = a + 1$$
$$= 3 + 1 = 4$$

Thus $a = 3$ and $b = 4$.

[5]Named for the Greek mathematician Pythagoras of Samos (ca. 580–500 B.C.), but known to the Mesopotamians 4,000 years ago.

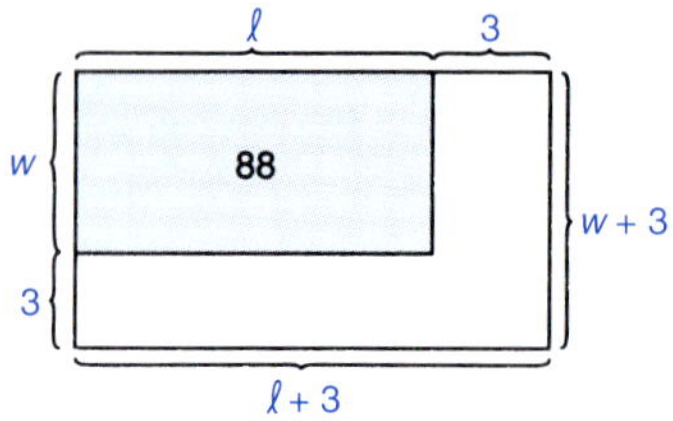

2. The area of a rectangle is 88 in.^{2}. If the length and width are each increased by 3 inches the new area is 154 in.^{2} (see the figure). Find the original length and width, ℓ and w.

$\ell w = 88$	Area of original figure
$w = \dfrac{88}{\ell}$	Solve for w
$(\ell + 3)(w + 3) = 154$	Area of new, larger figure
$(\ell + 3)\left(\dfrac{88}{\ell} + 3\right) = 154$	Replace w by $\dfrac{88}{\ell}$
$88 + 3\ell + \dfrac{264}{\ell} + 9 = 154$	Perform the multiplication
$3\ell + \dfrac{264}{\ell} - 57 = 0$	Add -154 to both members; combine like terms
$3\ell^2 - 57\ell + 264 = 0$	Multiply both members by ℓ
$\ell^2 - 19\ell + 88 = 0$	Divide both members by 3
$(\ell - 8)(\ell - 11) = 0$	Factor the quadratic trinomial
$\ell - 8 = 0$ or $\ell - 11 = 0$	Use the zero product property
$\ell = 8$ or $\ell = 11$	Solve each linear equation

Since $w = \dfrac{88}{\ell}$, we obtain the solutions $w = 11$ when $\ell = 8$, or $w = 8$ when $\ell = 11$. We select the second solution, since we usually think of length as greater than width. (Either solution is valid; they are just different ways to label the same geometric figure.) Thus, the original length is 11 inches and the original width is 8 inches.

3. One printing press takes 2 hours longer than another to print all of one edition of a certain newspaper. Running together they produce the edition in $1\frac{1}{3}$ hours. Find the time required for each to do this job alone.

This type of problem requires the principle that rate × time = part done. In this problem the part done is one edition of the newspaper, represented by 1. Thus we use rate × time = 1. Observe that this means that rate $= \dfrac{1}{\text{time}}$, which we use below.

If it takes x hours for the faster machine to print the edition, then its rate is $\dfrac{1}{x}$ papers per hour $\left(\text{rate} = \dfrac{\text{one job}}{\text{time}}\right)$. The slower machine takes $x + 2$ hours, so its rate is $\dfrac{1}{x + 2}$ papers per hour. Their combined rate is $\dfrac{1}{x} + \dfrac{1}{x + 2}$.

It takes $1\frac{1}{3}$ hours to print the edition when the presses are running together, so the combined rate is $\frac{1}{\text{time}} = \frac{1}{1\frac{1}{3}} = \frac{1}{\frac{4}{3}} = \frac{3}{4}$. Thus, we know that $\frac{1}{x} + \frac{1}{x+2} = \frac{3}{4}$, which we solve for x.

$$\frac{1}{x} + \frac{1}{x+2} = \frac{3}{4}$$

$$\frac{4x(x+2)}{1} \cdot \frac{1}{x} + \frac{4x(x+2)}{1} \cdot \frac{1}{x+2} = \frac{4x(x+2)}{1} \cdot \frac{3}{4}$$

Multiply by the LCD, $4x(x + 2)$

$$4(x+2) + 4x = 3x(x+2)$$

$$8x + 8 = 3x^2 + 6x$$

$$3x^2 - 2x - 8 = 0$$

Perform indicated operations, combine like terms, and put in standard form

$$(3x+4)(x-2) = 0$$

Factor

$$3x + 4 = 0 \text{ or } x - 2 = 0$$

Zero product property

$$x = -\frac{4}{3} \text{ or } x = 2$$

$$x = 2$$

We reject the negative value; it could not describe the running time for the presses

Thus the times for the presses running alone are $x = 2$ and $x + 2 = 4$, or 2 hours and 4 hours.

4. When flying directly into a 20 mph wind it takes an aircraft $1\frac{1}{2}$ hours longer to travel 400 miles than when there is no wind. Find the time required for the aircraft to fly this distance in no wind.

We use the relation rate × time = distance, or $rt = d$. Let x be the desired value—the time for the aircraft to fly 400 miles in no wind. Then $x + 1\frac{1}{2}$ is the time when there is a wind. We can show this information graphically with a table. Since $r = \frac{d}{t}$, the rate for a trip under no wind and wind conditions is $\frac{400}{x}$ and $\frac{400}{x + 1\frac{1}{2}}$.

	$r = \frac{d}{t}$	t	d
Into wind	$\frac{400}{x + 1\frac{1}{2}}$	$x + 1\frac{1}{2}$	400
No wind	$\frac{400}{x}$	x	400

The wind slows down the aircraft by the speed of the wind, so we know that $\dfrac{400}{x} - \dfrac{400}{x + 1\frac{1}{2}} = 20$. We can now solve for x.

$$\frac{20}{x} - \frac{20}{x + \frac{3}{2}} = 1 \qquad \text{Divide each term by 20}$$

$$\frac{20}{x} - \frac{2}{2} \cdot \frac{20}{x + \frac{3}{2}} = 1 \qquad \text{Multiply the second term by } \tfrac{2}{2}$$

$$\frac{20}{x} - \frac{40}{2x + 3} = 1$$

$$\frac{x(2x + 3)}{1} \cdot \frac{20}{x} - \frac{x(2x + 3)}{1} \cdot \frac{40}{2x + 3} = x(2x + 3)(1)$$

Multiply by the LCD, $x(2x + 3)$

$$20(2x + 3) - 40x = x(2x + 3)$$

$$2x^2 + 3x - 60 = 0$$

Perform operations and put in standard form

$$x = \frac{-3 \pm \sqrt{9 - 4(2)(-60)}}{2(2)} = \frac{-3 \pm \sqrt{489}}{4}$$

We choose the positive value for x, obtaining $\dfrac{-3 + \sqrt{489}}{4} \approx 4.8$ hours for the aircraft to fly 400 miles in no wind conditions. ■

The domain of a rational expression

One application of solving linear and quadratic equations is determining when a rational expression is not defined. Recall that section 1–4 introduced rational expressions. All values for which an expression is defined are called its domain.

> **Domain of an expression**
> The domain of an expression is the set of all replacement values of the variable for which the expression is defined.

In the case of rational expressions the denominator presents a special problem. Division by zero is not defined; thus, the domain of a rational expression must exclude all values that cause the denominator to have the value 0. For example, in the expression $\dfrac{2x - 1}{x - 3}$, we cannot permit x to take on the value 3, since this would evaluate to $\dfrac{2(3) - 1}{3 - 3} = \dfrac{5}{0}$, and this expression is not defined. We would express the domain of this expression as $\{x \mid x \neq 3\}$. *To determine which values to exclude we write an equation in which one side is the denominator and the other is zero and solve.* The solutions must be excluded from the domain.

■ ***Example 2–2 F***

Determine the domain of the following rational expressions.

1. $\frac{x-3}{2x+5}$

$2x + 5 = 0$	Set the denominator equal to zero and solve
$2x = -5$	Add -5 to both members (sides) of the equation
$x = -\frac{5}{2}$	Divide both members by 2

Thus the domain is $\{x \mid x \neq -\frac{5}{2}\}$. All real numbers except $-\frac{5}{2}$

Note We do not concern ourselves with zeros of the numerator. This is because division *into* zero is defined.

2. $\frac{x-3}{x^2-3x-10}$

$x^2 - 3x - 10 = 0$	Set the denominator equal to 0; the result is a quadratic equation
$(x-5)(x+2) = 0$	Factor the quadratic expression
$x - 5 = 0$ or $x + 2 = 0$	Set each factor to 0
$x = 5$ or $x = -2$	

Thus the domain is $\{x \mid x \neq 5, x \neq -2\}$. All real numbers except -2 and 5

3. $\frac{3z}{z^2+4}$

We know that $z^2 \geq 0$, so that $z^2 + 4$ cannot ever be zero (it cannot be less than 4). Thus, z can take on any value, and the domain is R (the set of real numbers). ■

Solution using substitution

The expressions in many equations are quadratic, not in a simple variable such as x, but in some more complicated variable expression, such as x^2, x^{-1}, $\sqrt{x}$, or $(x+1)^2$. For example, the following expressions are quadratic in the variable expression shown.

Expression	Variable expression	Quadratic form of the equation
$3x^2 - 2x - 5$	x	$3x^2 - 2x - 5$
$3x^4 - 2x^2 - 5$	x^2	$3(x^2)^2 - 2(x^2) - 5$
$3x^{-2} - 2x^{-1} - 5$	x^{-1}	$3(x^{-1})^2 - 2(x^{-1}) - 5$
$3x - 2\sqrt{x} - 5$	$\sqrt{x}$	$3(\sqrt{x})^2 - 2(\sqrt{x}) - 5$
$3(x+1)^2 - 2(x+1) - 5$	$(x+1)$	$3(x+1)^2 - 2(x+1) - 5$

The technique of substitution for expression (section 1–3) can help solve equations that are quadratic in more complicated expressions like those above.

Solving quadratic equations using substitution for expression

1. Determine the variable expression; this is usually the variable factor with the exponent with the smallest absolute value.
2. Let u represent the variable expression. Calculate u^2.
3. Substitute u and u^2 as appropriate.
4. Solve the quadratic equation for u.
5. Substitute the variable expression for u in the solutions.
6. Solve these equations for the variable.

■ ***Example 2–2 G***

Solve each equation.

1. $3x^{-2} - 2x^{-1} - 5 = 0$ — Step 1: The variable factor with exponent of smallest absolute value is x^{-1}

Step 2: Let $u = x^{-1}$; then $u^2 = x^{-2}$

$3u^2 - 2u - 5 = 0$ — Step 3: Replace u by x^{-1}, x^{-2} by u^2

$u = \dfrac{5}{3}$ or $u = -1$ — Step 4: Solve the quadratic equation for u

$x^{-1} = \dfrac{5}{3}$ or $x^{-1} = -1$ — Step 5: Replace u by x^{-1}

$\dfrac{1}{x} = \dfrac{5}{3}$ or $\dfrac{1}{x} = -1$ — Step 6: Solve; recall that $x^{-1} = \dfrac{1}{x}$

$x = \frac{3}{5}$ or $x = -1$

$\{-1, \frac{3}{5}\}$ — Solution set

2. $3x - 2\sqrt{x} - 5 = 0$ — Let $u = \sqrt{x}$; then $u^2 = x$

$3u^2 - 2u - 5 = 0$ — Replace $\sqrt{x}$ by u

$u = \frac{5}{3}$ or $u = -1$ — Solve the quadratic equation for u

$\sqrt{x} = \frac{5}{3}$ or $\sqrt{x} = -1$ — Replace u by $\sqrt{x}$

$x = \frac{25}{9}$ — Square both members to obtain x

$\sqrt{x} = -1$ does not yield a solution. The square root of any number, real or complex, cannot be a negative real number.

$\{\frac{25}{9}\}$ — Solution set ■

Mastery points

Can you

- Solve certain quadratic equations by factoring?
- Solve equations of the form $ax^2 + c = 0$ by extracting the roots?
- Solve any quadratic equation by using the quadratic formula?
- Solve certain word problems that involve quadratic equations?
- Recognize and solve equations that are of quadratic type using substitution of expression?
- Determine the domain of a rational expression?

Exercise 2–2

Solve the following quadratic equations for x by factoring.

1. $x^2 = 7x + 8$
2. $y^2 - 18y + 81 = 0$
3. $q^2 = 5q$
4. $2x^2 - 3x - 2 = 0$
5. $\frac{x^2}{6} = \frac{x}{3} + \frac{1}{2}$
6. $x - 1 = \frac{x^2}{4}$
7. $3x^2 + 5x - 2 = 0$
8. $6x^2 = 20x$
9. $\frac{x}{2} + \frac{7}{2} = \frac{4}{x}$
10. $(p + 4)(p - 6) = -16$
11. $(3x + 2)(x - 1) = 7 - 7x$
12. $x^2 - 4ax + 3a^2 = 0$
13. $5x^2 - 6y^2 = 7xy$
14. $3x^2 - 13xy + 4y^2 = 0$
15. $12x^2 = 8ax + 15a^2$
16. $12xy - 6y^2 = 6x^2$

Solve the following equations by extracting the roots.

17. $3y^2 = 27$
18. $3y^2 - 72 = 0$
19. $m^2 - 40 = 0$
20. $x^2 = 10$
21. $9x^2 - 40 = 0$
22. $4x^2 = 25$
23. $5x^2 = 32$
24. $7x^2 - 20 = 10$
25. $(x - 3)^2 = 10$
26. $(2x - 1)^2 + 6 = 0$
27. $3(x + 1)^2 = 8$
28. $5(2x + 5)^2 + 8 = 0$
29. $a(bx + c)^2 = d$, assuming $a, d > 0$
30. $(y - a)^2 - b^2 = 0, b > 0$

Solve the following quadratic equations using the quadratic formula.

31. $3y^2 - 5y = 6$
32. $3x - 7 = 5x^2$
33. $2y^2 + \frac{4y}{3} = 5$
34. $3x - \frac{3}{x} = 8$
35. $(z + 1)(z - 1) = 6z + 4$
36. $\frac{1}{x + 2} - x = 5$

Factor the following quadratic expressions.

37. $5x^2 - 8x - 12$
38. $3x^2 + 2x - 9$
39. $2x^2 + 6x - 4$
40. $x^2 - x - 6$

Solve the following problems.

41. In a right triangle the length of the hypotenuse is 20, and one side is four units longer than the other. Find the length of the longer side.

42. In a right triangle one of the sides is five units longer than the other. The length of the hypotenuse is 50. Find the length of the shorter side.

43. A rectangular playground has a length 5 feet less than three times the width, and the distance from one corner to the opposite corner is 100 feet. Find the dimensions of the playground.

44. A rectangle that is 6 centimeters long and 3 centimeters wide has its dimensions increased by an equal amount. The area of this new rectangle is three times that of the old rectangle. What are the dimensions of the new rectangle?

45. The area of a rectangular floor is 1,196 square meters. The width is 3 meters less than half the length. Find the dimensions of the floor.

46. A cuneiform tablet from Mesopotamia, thousands of years old, asks for the solution to the system $xy = 7\frac{1}{2}$, $x + y = 6\frac{1}{2}$. By the first equation $y = \frac{15}{2x}$, so the second equation is $x + \frac{15}{2x} = 6\frac{1}{2}$. Solve this new equation for x, then find the value of y.

47. The demand equation for a certain commodity is given by $D = \frac{2{,}000}{p}$, where D is the demand for the commodity at price p dollars per unit. The supply equation for the commodity is $S = 300p - 400$, where S is the quantity of the commodity that the supplier is willing to supply at p dollars per unit. Find the equilibrium price (where supply equals demand).

48. A manufacturer finds that the total cost C for a solar energy device is expressed by $C = 50x^2 - 24{,}000$, and the total revenue R at a price of \$200 per unit to be $R = 200x$, where x is the number of units sold. What is the break even point (where total cost = total revenue)?

49. One printing press takes 3 hours longer than another to print 10,000 newspapers. Running together they produce the 10,000 papers in 8 hours. Find the time required for each to do this job alone.

50. Two pipes can be used to fill a swimming pool. When both pipes run together it takes 18 hours to fill the pool. When they run separately it takes the smaller pipe 2 hours longer to fill the pool than the larger pipe. Find the time required for the larger pipe to fill the pool alone.

51. When flying directly into a 15 mph wind it takes an aircraft 30 minutes longer to travel 600 miles than when there is no wind. Find the time required for the aircraft to fly this distance in no wind.

52. A certain river boat makes 75-mile trips up and down a river. The river flows at 4 mph. It takes the boat 7 hours longer to make the trip upstream than downstream. Find the time it would take the boat to make the trip in still water, to the nearest 0.1 hours.

Determine the domain of the following rational expressions. State the domain in set-builder notation.

53. $\dfrac{x-2}{2x-3}$ **54.** $\dfrac{x+1}{x}$ **55.** $\dfrac{5z}{2z+4}$ **56.** $\dfrac{2x+3}{x^2-3x}$

57. $\dfrac{3m-1}{3m^2-6m}$ **58.** $\dfrac{4x+1}{x^2+8x+16}$ **59.** $\dfrac{2-3x}{x^2-4x-21}$ **60.** $\dfrac{2x+9}{x^2-9}$

61. $\dfrac{x^2}{x^2+4}$ **62.** $\dfrac{4x^2-2x+1}{x^2+1}$

Solve the following quadratic equations using substitution.

63. $x^4+3x^2-28=0$ **64.** $x^6+3x^3-28=0$ **65.** $(x-3)^2-4(x-3)-9=0$

66. $(m+3)^2+3(m+3)=20$ **67.** $4t-3\sqrt{t}=8$ **68.** $3x-\sqrt{x}=2$

69. $x^{2/3}-10x^{1/3}=-9$ **70.** $3y^{1/3}+5y^{1/6}-8=0$ **71.** $4y^{-4}+4=17y^{-2}$ **72.** $x^{-4}=5x^{-2}-4$

73. Show by direct substitution of each value for x that both $\dfrac{-b+\sqrt{b^2-4ac}}{2a}$ and $\dfrac{-b-\sqrt{b^2-4ac}}{2a}$ are solutions to the equation $ax^2+bx+c=0$ if $a \neq 0$.

74. Show by multiplication that, if $a \neq 0$, then

$$a\left(x-\frac{-b+\sqrt{b^2-4ac}}{2a}\right)\left(x-\frac{-b-\sqrt{b^2-4ac}}{2a}\right)$$
$$= ax^2+bx+c$$

75. Find a complex number $a+bi$ so that $(a+bi)^2 = -5-12i$.

76. Find a complex number $a+bi$ so that $(a+bi)^2 = 7-24i$.

77. Find expressions for a and b (in terms of c and d) so that, for any complex number $c+di$, $(a+bi)^2 = c+di$. Make sure that the values for a and b are real, and not themselves complex.

78. The *Sulvasutras* (800 B.C.) is an ancient Hindu manual on geometry. In it appears the problem of the Great Altar (the Maha-Vedi).[6] The altar is trapezoidal in shape with the dimensions $a = 24$, $b = 30$, and $h = 36$. The area of a trapezoid is $A = h\left(\dfrac{a+b}{2}\right)$, so in this case the area was $36\left(\dfrac{24+30}{2}\right) = 972$ square units.

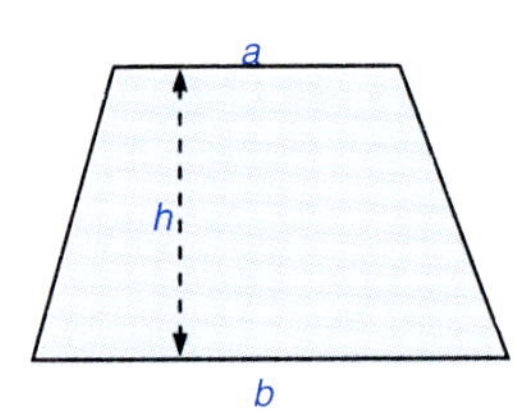

The area of the altar was to be increased by m square units to $972+m$. The dimensions are to be increased proportionately by p units, to $24p$, $30p$, and $36p$.

a. Show that the new area will be $972p^2$ units.

b. Use the fact that $972p^2 = 972+m$ to find the new dimensions of an altar that is 100 square units larger than the original (i.e., m is 100).

[6]From "Hindu Romance with Quadratic Equations" by Dr. Gurcharan Singh Bhalla, *The Amatyc Review,* Fall, 1987.

Skill and review

1. Compute: $(3\sqrt{2x})^2$.
2. Compute: $(3 + \sqrt{2x})^2$.
3. Solve: $\sqrt{x} = 4$.
4. Solve: $\sqrt[3]{x} = 4$.
5. $\sqrt{\frac{2}{3} + 2} =$ **a.** $\frac{2}{3}$ **b.** $\frac{2\sqrt{2}}{3}$ **c.** $\frac{2\sqrt{3}}{3}$ **d.** $\frac{2\sqrt{6}}{3}$

2–3 Equations involving radicals

At an altitude of h feet above level ground the distance d in miles that a person can see an object is given by $d = \sqrt{\frac{3h}{2}}$. How many feet up must a person be to see an object that is 8 miles away?

This section discusses how to deal with the type of equation in this problem.

Equations with at least one term containing a radical expression involving a variable are solved using the following principle.

Property of *n*th power

If $n \in N$ and P and Q are algebraic expressions, then all of the solutions of the equation $P = Q$ are also solutions of $P^n = Q^n$.

This says that we can solve equations by raising both members to the same positive integer power n. It also implies that *we may get solutions that are not solutions to the original equation.* In particular, this can happen when n is even, but not odd. For example,

$x = -2$	$x = -2$	Original equation
$x^2 = (-2)^2$	$x^3 = (-2)^3$	Raise both members to a power
$x^2 = 4$	$x^3 = -8$	New equation

$x^2 = 4$ has two solutions, ± 2, whereas the solution to the original equation is only -2. $x^3 = -8$ has only the single real solution -2.

These extra solutions (such as 2 in $x^2 = 4$, above) are called **extraneous roots.** The best way to detect this situation is to *check all solutions in the original equation when we raised both members of an equation to an even power.* Of course it is always good practice to check any solution, and we shall do so in the following examples.

Raising radical expressions to a power

We will use the fact that $(\sqrt{x})^2 = x$, $(\sqrt[3]{x})^3 = x$, $(\sqrt[4]{x})^4 = x$, and so on quite a bit. If a member of an equation has a factor that is a radical, we will raise both members of the equation to the power that corresponds to the index of the radical. This eliminates the radical.

Example 2–3 A illustrates some of the algebra we will encounter.

■ Example 2–3 A

Perform the indicated operations.

1. $(\sqrt{x-2})^2$ — One term being squared
   $= x - 2$
2. $(\sqrt{x} - 2)^2$ — Square a two-termed expression
   $= (\sqrt{x} - 2)(\sqrt{x} - 2)$ — $(a - b)^2 = (a - b)(a - b)$
   $= \sqrt{x}\sqrt{x} - 2\sqrt{x} - 2\sqrt{x} + 4$ — $(a - b)(a - b) = a^2 - ab - ab + b^2$
   $= x - 4\sqrt{x} + 4$ — Combine like terms
3. $(2\sqrt{3x}\sqrt{x-2})^2$ — Square a one-termed expression
   $= 2^2(\sqrt{3x})^2(\sqrt{x-2})^2$ — $(ab)^n = a^n b^n$
   $= 4(3x)(x - 2)$
   $= 12x^2 - 24x$ ■

Solutions of equations by raising each member to a power

Example 2–3 B illustrates how to solve equations involving radicals by raising both members to a power.

■ Example 2–3 B

Solve the equations.

1. $x - \sqrt{x + 12} = 0$
   $x = \sqrt{x + 12}$ — Put the radical alone in one member of the equation before proceeding
   $x^2 = (\sqrt{x + 12})^2$ — Square both members
   $x^2 = x + 12$
   $x^2 - x - 12 = 0$ — Standard form for a quadratic equation
   $(x - 4)(x + 3) = 0$ — Factor the quadratic trinomial
   $x = 4$ or -3

   Check the solutions: $4 = \sqrt{4 + 12}$
   $4 = \sqrt{16}$ — True; the value 4 checks in the original equation
   $-3 = \sqrt{-3 + 12}$
   $-3 = \sqrt{9}$ — False; the value -3 does not check in the original equation, so it is not a solution

   $\{4\}$ — Solution set
2. $\sqrt[3]{5x - 2} + 3 = 0$
   $\sqrt[3]{5x - 2} = -3$ — Put the radical alone in one member by adding -3 to both members
   $(\sqrt[3]{5x - 2})^3 = (-3)^3$ — Raise both members to the value of the index, 3
   $5x - 2 = -27$
   $x = -5$

   Check: $\sqrt[3]{5(-5) - 2} + 3 = 0$
   $\sqrt[3]{-27} + 3 = 0$
   $-3 + 3 = 0$ — True, so -5 is a solution to the original equation

   $\{-5\}$ — Solution set

3. $\sqrt{3x+1} + \sqrt{x-1} = 6$

It is a good idea to not square with two radicals on one side of an equation. *Rewrite the equation so that one radical is the only term on one side.*

$\sqrt{3x+1} = 6 - \sqrt{x-1}$	A radical is the only term in the left member of the equation
$(\sqrt{3x+1})^2 = (6 - \sqrt{x-1})^2$	Square both members
$3x + 1 = (6 - \sqrt{x-1})(6 - \sqrt{x-1})$	Rewrite the right member as a product, for convenience
$3x + 1 = 36 - 12\sqrt{x-1} + (x-1)$	Multiply
$2x - 34 = -12\sqrt{x-1}$	Make the radical the only term in one member before squaring again
$x - 17 = -6\sqrt{x-1}$	Divide each term by 2
$(x-17)^2 = (-6\sqrt{x-1})^2$	Square both members again
$(x-17)(x-17) = (-6)^2(\sqrt{x-1})^2$	
$x^2 - 34x + 289 = 36(x-1)$	
$x^2 - 70x + 325 = 0$	Standard form
$(x-5)(x-65) = 0$	Factor
$x = 5$ or 65	Possible solutions
$\{5\}$	Solution set; 65 does not check ■

Mastery points

Can you

- Compute powers of certain expressions?
- Solve certain equations involving radicals by raising both members to powers as necessary?

Exercise 2–3

Perform the indicated operations.

1. $(\sqrt{2x+3})^2$ **2.** $(\sqrt{5-x})^2$ **3.** $(\sqrt[3]{4x+3})^3$ **4.** $(\sqrt[3]{x+11})^3$

5. $(\sqrt{x}-5)^2$ **6.** $(\sqrt{3x}-1)^2$ **7.** $(3\sqrt{x}+2)^2$ **8.** $(2\sqrt[3]{3-5x})^3$

9. $(\sqrt{2x}-2)^2$ **10.** $(1-\sqrt{x})^2$ **11.** $(\sqrt{x-1}-2)^2$ **12.** $(x+\sqrt{x+1})^2$

13. $(2\sqrt{x-1})^2$ **14.** $(\sqrt{2x-1}-1)^2$ **15.** $(\sqrt{x+3}+3)^2$ **16.** $(3-\sqrt{x-1})^2$

17. $(1-\sqrt{1-x})^2$ **18.** $(\sqrt{3x-2}-2)^2$

Find the solution set for each equation.

19. $\sqrt{2x} = 8$ **20.** $\sqrt[3]{4x} = -2$ **21.** $\sqrt[3]{\frac{x}{2}} = 3$ **22.** $\sqrt{\frac{2x}{3}} = 5$

23. $\sqrt{3-x} = -2$ **24.** $\sqrt[3]{3-x} = -2$ **25.** $\sqrt[5]{x+3} = -1$ **26.** $\sqrt[4]{x+3} = -1$

27. $\sqrt{2-3x}-5=0$

28. $\sqrt[3]{x+2}+2=0$

29. $\sqrt[3]{2x+3}+5=0$

30. $\sqrt[4]{2y-3}=2$

31. $\sqrt[3]{2-3x}=-4$

32. $\sqrt[5]{2-6x}=-2$

33. $\sqrt[4]{x^2-24x}=3$

34. $\sqrt{\dfrac{2x+1}{2}}-4=0$

35. $\sqrt{w^2-6w}=4$

36. $\sqrt{x^2+6x+9}-6=0$

37. $\sqrt{x^2-5x+2}-4=0$

38. $\sqrt{x^2-x+6}+6=0$

39. $\sqrt{5x+1}-\sqrt{11}=0$

40. $\sqrt{9a+5}=\sqrt{3a-1}$

41. $\sqrt{2p+5}-\sqrt{3p+4}=0$

42. $2\sqrt{2z-1}-\sqrt{3z}=0$

43. $\sqrt{m}\sqrt{m-8}=3$

44. $\sqrt{x}\sqrt{x-2}=2\sqrt{6}$

45. $\sqrt{y}\sqrt{y-5}-6=0$

46. $\sqrt{2m}\sqrt{m+2}-4=0$

47. $\sqrt{u-1}=u-3$

48. $\sqrt{3x+10}-3x=4$

49. $\sqrt{x-2}=x-2$

50. $\sqrt{2a}=4-a$

51. $\sqrt{2n+3}-\sqrt{n-2}=2$

52. $3-\sqrt{y+4}=\sqrt{y+7}$

53. $\sqrt{p+1}=\sqrt{2p+9}-2$

54. $\sqrt{y+4}+\sqrt{y+7}=3$

55. $(2y+3)^{1/2}-(4y-1)^{1/2}=0$

56. $(1-2y)^{1/2}+(y+5)^{1/2}=4$

57. $(4x+2)^{1/2}-(2x)^{1/2}=0$

58. $(x-2)^{1/2}=(5x+1)^{1/2}-3$

Solve the following formulas for the indicated variable.

59. $\sqrt{\dfrac{y^2-x^2}{3}}=y-3$; for x

60. $r=\sqrt{\dfrac{A}{\pi}-R^2}$; for A

61. $D=\sqrt[3]{\dfrac{6A}{\pi}}$; for A

62. $v=\sqrt{\dfrac{2gKE}{W}}$; for W

63. $\sqrt{s-t}=t+3$; for s

64. $4x\sqrt{xy}=3$; for y

65. On wet pavement the velocity V of a car, in miles per hour, which skids to a stop is approximated by $V=2\sqrt{3S}$, where S is the length of the skid marks in feet. How long would the skid marks be if a car was traveling at 60 mph when the brakes were applied?

66. At an altitude of h feet above level ground the distance d in miles that a person can see an object is given by $d=\sqrt{\dfrac{3h}{2}}$. How many feet up must a person be to see an object that is 8 miles away?

67. The volume of a sphere is related to its radius by $r=\sqrt[3]{\dfrac{3V}{4\pi}}$. Solve this for V.

68. If a falling object falls a distance s in t seconds then $t=\sqrt{\dfrac{2s}{g}}$. Solve this for s.

69. Assume a bank account pays a simple yearly interest rate i, compounded every six months. If A is the amount in the account after one year, on a deposit P, then $i=2\left(\sqrt{\dfrac{A}{P}}-1\right)$ is true. Show that $A=P\left(1+\dfrac{i}{2}\right)^2$.

70. Use the formula of problem 69 to find the simple yearly interest rate i if a deposit earns \$324 on a deposit of \$4,500 after one year. (This means that A is \$4,500 + \$324 = \$4,824.)

Skill and review

1. Is the statement $3(2x+1)>x+6$ true for $x=\frac{1}{2}$?

2. Graph the interval $-2<x<3$.

3. Is the statement $\frac{1}{2}<\frac{1}{3}$ true?

4. Is the statement $-\frac{1}{2}<-\frac{1}{3}$ true?

5. Is the statement $\dfrac{x}{x-3}>x$ true for $x=-2$?

6. Which of the following are true statements?

a. $5>2$

b. $5+3>2+3$

c. $5-3>2-3$

d. $3(5)>3(2)$

e. $\dfrac{5}{3}>\dfrac{2}{3}$

f. $(-1)(5)>(-1)(2)$

2–4 *Inequalities in one variable*

The zoning bylaws of Carlisle, Massachusetts, require, among other things, that building lots have a ratio of area A to perimeter P conforming to the relation $\frac{16A}{P^2} > 0.4$.

As in this example, an inequality is a statement that two expressions are related by order; that is, that one is greater or less than another. For example, $5x > 10$ is an inequality that states that $5x$ is greater than 10, or equivalently, $10 < 5x$ (10 is less than $5x$). Recall from section 1–1 that a simple linear inequality can be graphed as an interval.

Any statement involving the symbols $<$, $>$, $\leq$, and $\geq$ is an inequality. Some are linear, and some are nonlinear. We will examine **linear** and **nonlinear inequalities.** Nonlinear inequalities involve expressions in which some of the variables have exponents other than one or are in denominators. Examples of each are

Linear		Nonlinear	
$5x < 4$		$5x^2 < 4$	Variable with exponent
$\frac{x}{3} \geq 9$	Number in denominator	$\frac{3}{x} \geq 9$	Variable in denominator
$2x - 3 > 4x + 9$		$3x^2 - 2x > 9$	
$x - 4 > 0$		$(x - 4)(x + 3) > 0$	Variable with exponent, if we multiply

Linear inequalities

Algebraic methods for solving linear inequalities in one variable are much the same as those for linear equations. They use the following properties.

Addition property of inequality

For any algebraic expressions A, B, and C,

$$\text{if } A < B \text{ then } A + C < B + C$$

Multiplication property of inequality

For any algebraic expressions A, B, and C,

1. if $C > 0$ and $A < B$ then $AC < BC$
2. if $C < 0$ and $A < B$ then $AC > BC$.

Thus we solve linear inequalities exactly the same way as linear equations with one exception: *if we multiply (or divide) both members of an inequality by a negative value we must reverse the direction of the inequality.*

We write the solution sets using interval or set-builder notation and graph the solution sets on the number line using intervals, as shown in section 1–1.

■ *Example 2–4 A*

Find and graph the solution sets.

1. $3(2 - 4x) > 18 - 22x$

$6 - 12x > 18 - 22x$ Expand the left member

$22x - 12x > 18 - 6$ Add $22x$ and -6 to both members

$10x > 12$ Combine like terms

$x > \frac{12}{10}$ or $x > \frac{6}{5}$ Divide both members by 10; reduce

$\{x \mid x > 1\frac{1}{5}\}$ Set-builder notation

$(1\frac{1}{5}, \infty)$ Interval notation

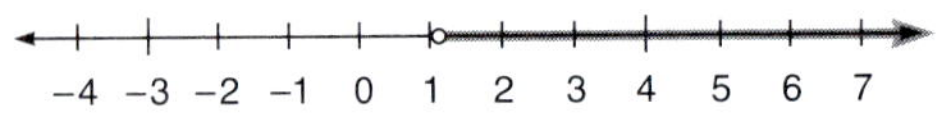

2. $\frac{5(3 - 2x)}{2} \geq 12 - x$ A weak inequality is solved in the same way as a strict one

$5(3 - 2x) \geq 24 - 2x$ Multiply both members by 2

$15 - 10x \geq 24 - 2x$ Multiply in the left member

$-8x \geq 9$ Add $2x$ and -15 to both members

$x \leq -\frac{9}{8}$ Divide each member by -8; by the multiplication principle we must therefore *reverse the direction of the inequality*

$\{x \mid x \leq -1\frac{1}{8}\}$ Set-builder notation

$(-\infty, -1\frac{1}{8}]$ Interval notation

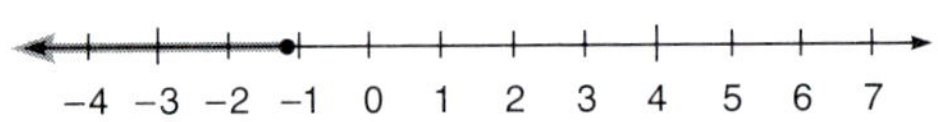

3. $-8 < 2 - 3x \leq 14$

This is a **compound inequality** (section 1–1). It means that

$$-8 < 2 - 3x \textit{ and } 2 - 3x \leq 14$$

The compound inequality can be solved in the same way as simple inequalities if we apply the same rule to all three members at a time. *This process is equivalent to solving both the inequalities it represents.*

$-10 < -3x \leq 12$ Add -2 to each member

$\frac{-10}{-3} > \frac{-3x}{-3} \geq \frac{12}{-3}$ Divide each member by -3; this means we must reverse the direction of the inequalities

$3\frac{1}{3} > x \geq -4$

$-4 \leq x < 3\frac{1}{3}$ We usually write the lesser quantity on the left in a compound inequality

$\{x \mid -4 \leq x < 3\frac{1}{3}\}$

$[-4, 3\frac{1}{3})$ Solution set

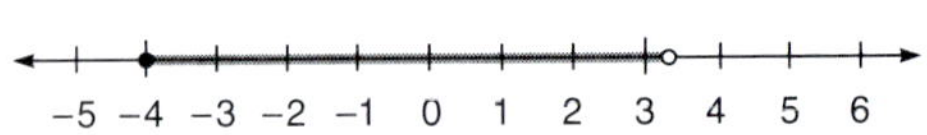

■

Graphical methods can be employed to find approximate solutions to linear inequalities. There are at least two ways to do this. This is illustrated in example 2–4 B, where we do parts 1 and 3 of example 2–4 A one way and part 2 of example 2–4 A a second way.

■ Example 2–4 B

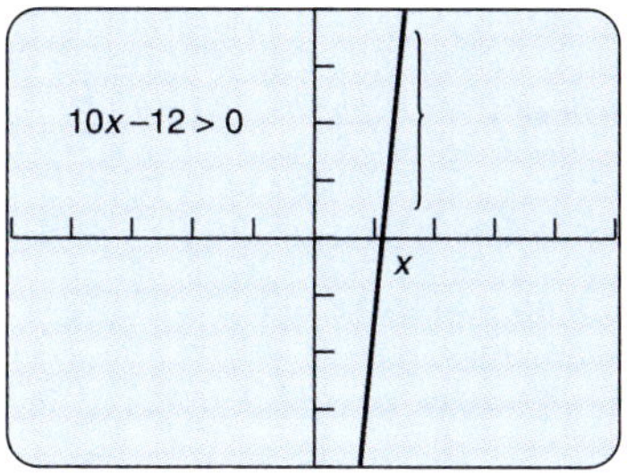

Find the solution sets by graphical methods.

1. Solve the inequality $3(2 - 4x) > 18 - 22x$.

We can simplify as
$$\begin{aligned} 3(2 - 4x) &> 18 - 22x \\ 6 - 12x &> 18 - 22x \\ 6 - 12x - 18 + 22x &> 0 \\ 10x - 12 &> 0 \end{aligned}$$

We graph $y = 10x - 12$; the solution is where the graph is greater than zero.

We can see that the graph is greater than zero when x is to the right of where it crosses the x-axis. The value x can be found by tracing and zooming to be about 1.2. Thus, $10x - 12 > 0$ when $x > 1.2$, and the solution to the original inequality is $x > 1.2$.

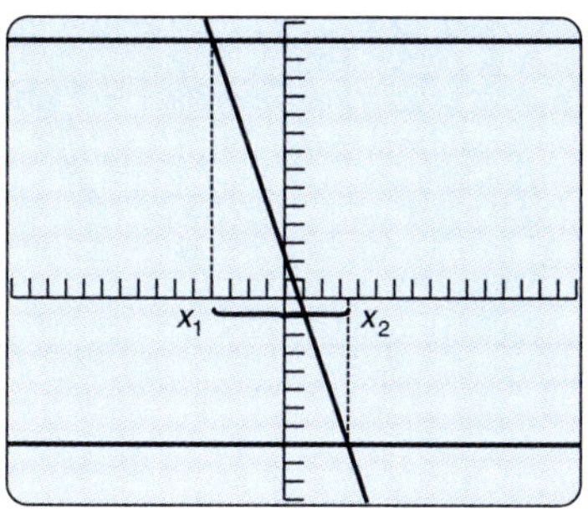

2. Solve the inequality $-8 < 2 - 3x \leq 14$ graphically.

Graph the three lines $Y_1 = -8$, $Y_2 = 2 - 3x$, $Y_3 = 14$. See the graph. Use Ymax=15, Ymin=−10, Xmax=15, Xmin=−15.

We can see that the solution is the values of x between x_1 and x_2, and including x_1 (since we did say $2 - 3x \leq 14$). We can use the trace function on the graph of Y_2. Do this by selecting [TRACE] and using the up and down arrows to move the cursor to the slanted line, which is the graph of Y_2. In this manner we can determine that the value of x_1 is about −4 and the value of x_2 is about 3.3. Algebraically we determined the solution is exactly: $-4 \leq y < 3\frac{1}{3}$ (example 2–4 A).

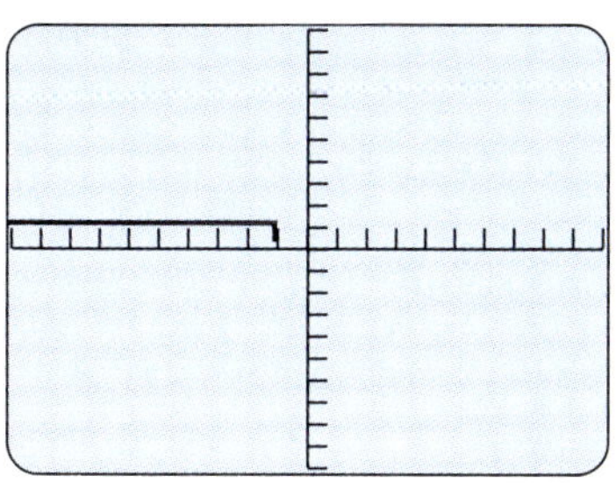

3. $\dfrac{5(3 - 2x)}{2} \geq 12 - x$ (part 2 of example 2–4 A).

Use the following keystrokes:

[Y =]

5 [(] 3 [−] 2 [X | T] [)] [÷] 2

[TEST] ([2nd] [MATH])

4 (≥)

12 [−] [X | T]

[GRAPH]

The TI-81 is programmed to graph the value 1 where a function is TRUE, and 0 where it is FALSE. Thus the graph indicates the expression $\dfrac{5(3 - 2x)}{2} \geq 12 - x$ is true (1) for x slightly less than −1, and is false (0) everywhere else. Zooming could refine the value, obtaining an answer close to the actual value of $-1\frac{1}{8}$. ■

Nonlinear inequalities

To solve nonlinear inequalities algebraically we employ a method called the **critical point/test point method.** We can also solve them, approximately, by graphing. This is shown after we examine the algebraic methods.

We first investigate the critical point/test point method.

The **critical points** of an inequality are

1. the solutions to the corresponding equality and
2. the zeros of any denominators.

Critical points are used to divide the real numbers into intervals in which the inequality is either always true or always false.

To get a feeling for why this all works, consider for example the simple nonlinear inequality $\frac{5x+5}{x+2} > 4$. The value of the expression $\frac{5x+5}{x+2}$ can be divided into three categories relative to 4: greater than 4, equal to 4, or less than 4. Look at the following table of values as x varies from -5 to 6 as shown in table 2–1. The information shown in table 2–1 is also shown graphically on the number line in figure 2–6. Observe that the number line divides into three intervals. In two of the intervals the statement $\frac{5x+5}{x+2} > 4$ is true, and it is false in one. The points -2 and 3 separate the intervals. We call -2 and 3 critical points.

x	$\frac{5x+5}{x+2}$	**Greater than 4**	**Equal to 4**	**Less than 4**
-5	$6\frac{2}{3}$	True		
-4	$7\frac{1}{2}$	True		
-3	10	True		
-2	undefined			
-1	0			True
0	$2\frac{1}{2}$			True
1	$3\frac{1}{3}$			True
2	$3\frac{3}{4}$			True
3	4		True	
4	$4\frac{1}{6}$	True		
5	$4\frac{2}{7}$	True		
6	$4\frac{3}{8}$	True		

Table 2–1

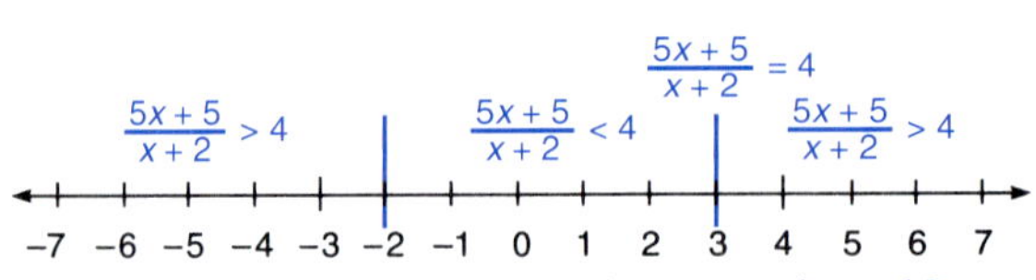

Figure 2–6

In an inequality critical points can occur in two places. The first is where the left and right members are equal. The point $x = 3$ is where the left and right members of the inequality $\frac{5x + 5}{x + 2} > 4$ are equal, or in other words, it is a solution to $\frac{5x + 5}{x + 2} = 4$. The second place that critical points can occur is where a denominator is 0. The denominator of $\frac{5x + 5}{x + 2}$ is 0 at $x = -2$, and this is a critical point.

Note that an expression is undefined wherever a denominator is 0, so such a point cannot be part of a solution set.

To locate where $\frac{5x + 5}{x + 2} > 4$ is true we have to check only one value in each interval, since the inequality is true throughout the entire interval or is false throughout the entire interval. We call such a value a test point. Any value in the interval may serve as a test point. According to table 2–1 $\frac{5x + 5}{x + 2} > 4$ is true for $x < -2$ or $x > 3$.

The critical point/test point method for solving nonlinear inequalities is summarized as follows.

The critical point/test point method for nonlinear inequalities

Step 1: Find critical points.
- a. Change the inequality to an equality and solve.
- b. Set any denominators involving the variable to zero and solve.

Step 2: Find test points.
- a. Use the critical points found in step 1 to mark intervals on the number line.
- b. Choose one test point from each interval. Any point will do.

Step 3: Locate the intervals which form the solution set.
- a. Try the test points in the original inequality.
- b. Note the intervals where the test point makes the original problem true.

Step 4: Include any of the critical points which make the original inequality true in the solution set.
This will only occur when the original inequality was a weak one.

Step 5: Write the solution set (we will use both set-builder and interval notation).

Note Do not attempt to solve a nonlinear inequality such as $\frac{5x + 5}{x + 2} > 4$ by multiplying both members by $x + 2$. We do not know whether $x + 2$ is positive or negative, so we do not know whether we should reverse the $>$ or not.

■ Example 2–4 C

Solve and graph the solutions to the following nonlinear inequalities.

1. Solve the nonlinear inequality $\dfrac{x+3}{x-4} \geq 5$.

Step 1: Find the critical points

a. $\dfrac{x+3}{x-4} = 5$	Change ≥ to = and solve
$x + 3 = 5(x - 4)$	Multiply both sides by $x - 4$
$x + 3 = 5x - 20$	
$23 = 4x$	
$5\frac{3}{4} = x$	Solve for x
b. $x - 4 = 0$	Set the denominator $x - 4$ equal to zero
$x = 4$	Solve for x
Critical points: 4, $5\frac{3}{4}$	At $5\frac{3}{4}$ we also have equality. Since this is a weak inequality, $5\frac{3}{4}$ is part of the solution set

Step 2: Find test points.

Plot the points 4 and $5\frac{3}{4}$ from step 1. These form the three intervals shown in the figure. Choose one point from each interval to be a test point. We choose 0, 5, and 6.

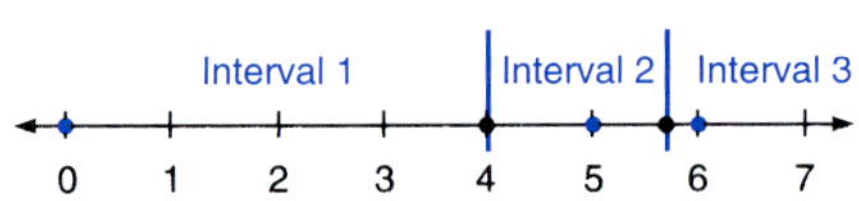

Step 3: Test the original inequality using these test points and note where the original inequality is true.

The original inequality is $\dfrac{x+3}{x-4} \geq 5$. We test it for each value of x, 0, 5, and 6.

	Interval 1	**Interval 2**	**Interval 3**
Test point	0	5	6
	$\dfrac{0+3}{0-4} \geq 5$	$\dfrac{5+3}{5-4} \geq 5$	$\dfrac{6+3}{6-4} \geq 5$
	$-\frac{3}{4} \geq 5$	$\dfrac{8}{1} \geq 5$	$\dfrac{9}{2} \geq 5$
	False	True	False

The original inequality is true in interval 2.

Step 4: From step 1, the inequality is satisfied at $5\frac{3}{4}$, so this is part of the solution set.

Step 5: The solution is those intervals that had a test point that made the original inequality true, as well as any critical points for which the original inequality is true. In this case the solution is interval 2 together with the critical point $5\frac{3}{4}$. This result can be graphed as shown in the figure and described algebraically as

$\{x \mid 4 < x \leq 5\frac{3}{4}\}$ Set-builder notation
$(4, 5\frac{3}{4}]$ Interval notation

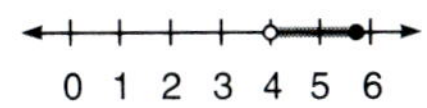

2. Solve the nonlinear inequality $x^3 - x^2 - 9x + 9 \geq 0$.
We solve by using the five steps illustrated above.

$x^3 - x^2 - 9x + 9 = 0$ Change $\geq$ to $=$ and solve
$x^2(x - 1) - 9(x - 1) = 0$ Factor by grouping
$(x - 1)(x^2 - 9) = 0$
$(x - 1)(x - 3)(x + 3) = 0$ Set each factor to 0
$x - 1 = 0$ or $x - 3 = 0$ or $x + 3 = 0$
$x = 1$ or $x = 3$ or $x = -3$

Thus, our critical points are -3, 1, and 3. These points are also part of the solution set since they are solutions to the original weak inequality. These form the four intervals shown in the figure. Choose -4, 0, 2, and 4 for test points.

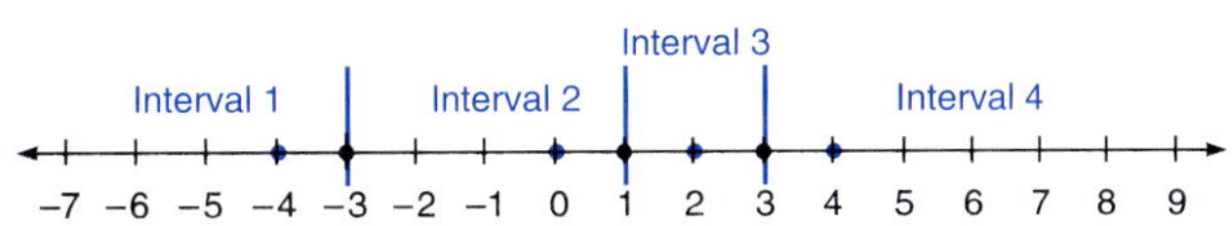

The original inequality is $x^3 - x^2 - 9x + 9 \geq 0$. We compute with the factored form of the left expression $(x - 1)(x - 3)(x + 3)$, since these computations are easier and give the same result.

	Interval 1	**Interval 2**	**Interval 3**	**Interval 4**
Test point	-4	0	2	4
	$(-5)(-7)(-1) \geq 0$	$(-1)(-3)(3) \geq 0$	$1(-1)(5) \geq 0$	$3(1)(7) \geq 0$
	$-35 \geq 0$	$9 \geq 5$	$-5 \geq 0$	$21 \geq 0$
	False	True	False	True

The solution is intervals 2 and 4, together with the critical points. This is written as

$\{x \mid -3 \leq x \leq 1 \text{ or } x \geq 3\}$ Set-builder notation
$[-3, 1]$ or $[3, \infty)$ Interval notation

and is graphed as shown in the figure.

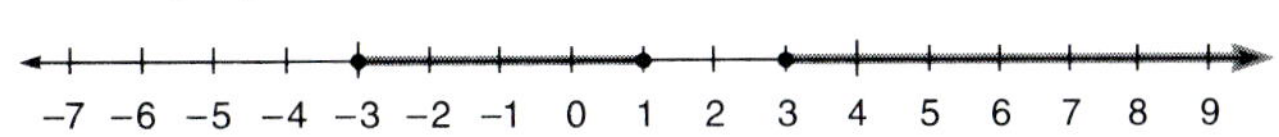

Example 2–4 D shows how to solve nonlinear inequalities by graphing.

■ Example 2–4 D

Solve the following nonlinear inequalities.

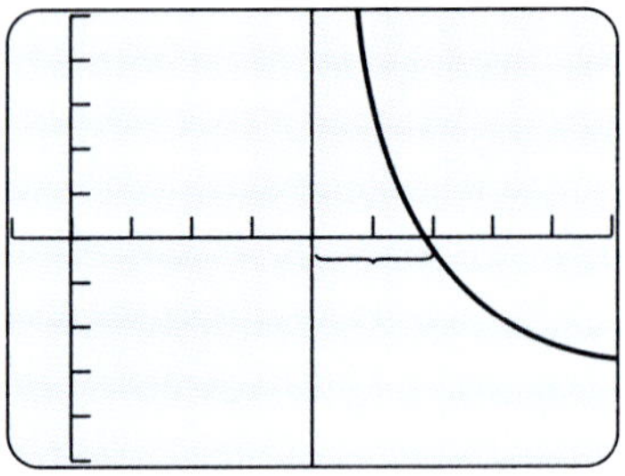

1. $\dfrac{x+3}{x-4} \geq 5$ (This is example 2–4 C, part 1.)

[Y=] [(] [X|T] [+] 3 [)] [÷] [(] [X|T] [−] 4 [)] [−] 5

[RANGE −1,9,−5,5]

It is easier to graph the equivalent expression $\dfrac{x+3}{x-4} - 5 \geq 0$. By suitably using trace and zoom we can see that this is greater than or equal to zero above 4 and less than or equal to 5.7, approximately, or $4 < x \leq 5.7$. (Note that the expression is not defined at $x = 4$.) In example 2–4 C we found the exact answer to be $4 < x \leq 5\frac{3}{4}$.

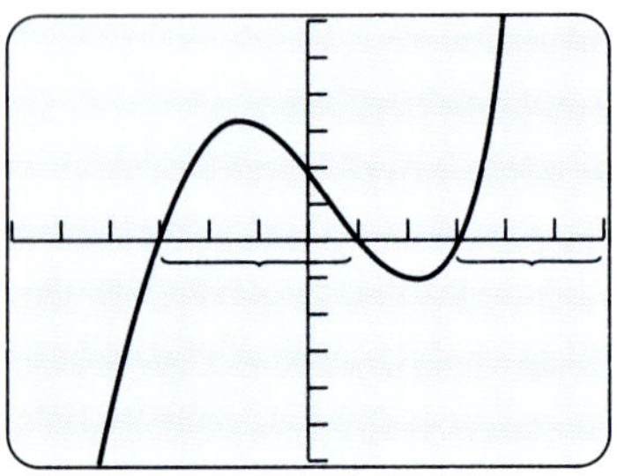

2. $x^3 - x^2 - 9x + 9 \geq 0$ (part 2 from example 2–4 C).
Graph $y = x^3 - x^2 - 9x + 9 \geq 0$.

[Y=] [X|T] [MATH] 3 [−] [X|T] [x^2] [−] 9 [X|T] [+] 9

[RANGE −6,6,−30,30]

YSCL = 5.

The values of x where the graph is greater than or equal to zero can be seen to be $-3 \leq x \leq 1$ or $x \geq 3$. This can be confirmed by using the trace and zoom functions.

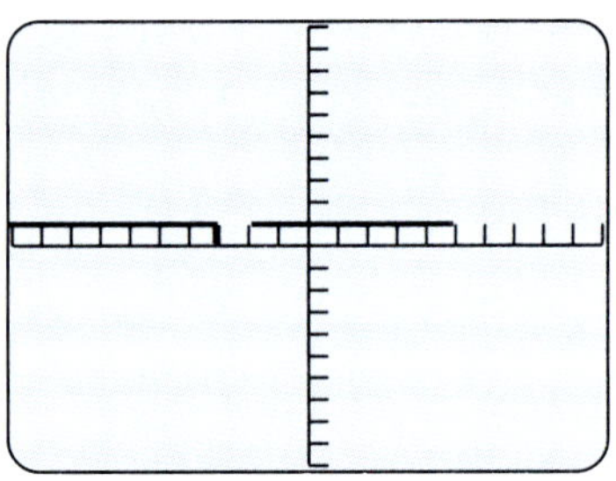

3. Solve $\dfrac{3x}{x-5} - 3 \leq \dfrac{2}{x+2}$.

Here we show a second way to use the graphing calculator to obtain solutions to inequalities. As illustrated in section 2–3 we can use the capability of the calculator to graph the value one where an expression is true, and zero where it is false.

[Y=] [(] 3 [X|T] [÷] [(] [X|T] [−] 5 [)] [)] [−] 3

[TEST] 6 2 [÷] [(] [X|T] [+] 2 [)] [RANGE −10,10,−10,10]

Zooming and tracing could be used to see that the values are −2 and 5, and the third value is close to $-3\frac{1}{13}$, or ≈ -3.077. Algebraic solution shows that the solution is $(-\infty, -3\frac{1}{13})$ or $(-2, 5)$. ■

Several examples of applications of linear and nonlinear inequalities are illustrated in example 2–4 E.

■ *Example 2–4 E*

Applications of inequalities.

1. Find the domain of the expression $\sqrt{2x - 3}$.
We require that the expression under a square root radical be nonnegative to obtain a real value. Thus,

$$2x - 3 \geq 0$$
$$2x \geq 3$$
$$x \geq \tfrac{3}{2}$$

The domain is $\{x \mid x \geq \tfrac{3}{2}\}$.

2. To pass a company's aptitude test a candidate must average 80 or above on three tests. If a certain candidate has had scores of 72 and 81 on the first two tests, what is the minimum score that must be obtained on the third test to have a passing average? If T represents the test score on the third test, we require that $\dfrac{72 + 81 + T}{3} \geq 80$.

$$\frac{72 + 81 + T}{3} \geq 80$$
$$72 + 81 + T \geq 240 \quad \text{Multiply each member by 3}$$
$$T \geq 87 \quad \text{Subtract (72 + 81) from both members}$$

Thus the candidate must score 87 or above on the third test. ■

Mastery points

Can you

- Solve linear inequalities in one variable using the algebraic methods presented?
- Solve nonlinear inequalities in one variable using the critical point/test point method?

Exercise 2–4

Find and graph the solution sets to the following linear inequalities.

1. $5x + 2 > 3x - 8$
2. $12x - 3 < 6x + 6$
3. $3x - 7 \leq x - 4$
4. $9x + 2 \geq 3 - 2x$
5. $2(3x - 3) \geq 9x + 1$
6. $5 - 3(x - 1) < 6x$
7. $9x - 4(2 - x) < 20$
8. $16 - 3(2x - 3) \geq 4x + 6$
9. $-3(x - 2) + 2(x + 1) \geq 5(x - 3)$
10. $2x + 17 < -4(2x - 5) - x + 7$
11. $x < 27 - 9x$
12. $0 \geq 9x - 36$
13. $5x + 18 \geq 0$
14. $4(2 - 3x) + 3 \geq 20$
15. $6x - 12 \leq 3x - 2(x + 6)$
16. $3x - 5 \leq 5x + 2$
17. $9x \geq 4x + 3$
18. $3(3x - 3) \geq 11$
19. $5x - 3(x - 1) \geq 6x$
20. $9 - 3(1 - x) < 20x$
21. $12 + 3(3 - 2x) > 4(x + 6)$
22. $10(x - 2) - 2(x + 1) \geq 5(x - 3)$
23. $17 \leq 3(3x - 5) - (x + 7)$
24. $3(x + 2) < 27 - 9x$
25. $6 - x > 9x - 36$
26. $12 - 6x \leq 3x - 2(x + 5)$

Find and graph the solution set to the following nonlinear inequalities.

27. $x^2 + 5x \leq 24$

28. $\dfrac{x - 2}{x + 1} \leq 0$

29. $(x - 3)(x + 1)(x - 1) \leq 0$

30. $x^2 - 5x - 24 > 0$

31. $q(3q - 5) < 0$

32. $2m^2 - 3m \geq 0$

33. $r(2r - 3)(r + 1) \leq 0$

34. $3w^2 + 5 \geq 16w$

35. $(x^2 - 4)(x - 1) < 0$

36. $(x - 9)(x^2 - 9) \geq 0$

37. $(x^2 - 4x + 4)(x^2 - 4) \leq 0$

38. $(x - 1)^2(x + 2)^3 > 0$

39. $\dfrac{(x + 1)^2}{2x - 3} \geq 0$

40. $\dfrac{3 - x}{3 + x} - 4 < 0$

41. $\dfrac{3}{x} > 0$

42. $\dfrac{4}{x - 2} < 0$

43. $\dfrac{x^2 - 1}{x - 3} \leq 0$

44. $\dfrac{x + 3}{x^2 - 4} \geq 0$

45. $\dfrac{x^2 - 10x + 25}{x^2 - x - 6} \leq 0$

46. $\dfrac{2x}{x + 1} \geq 3$

47. $\dfrac{x - 3}{x + 5} + x \geq 3$

48. $\dfrac{2p - 1}{2p} \leq 4$

49. $\dfrac{x}{x + 1} - \dfrac{2}{x + 3} \leq 1$

50. $\dfrac{x + 2}{x - 3} \geq \dfrac{x + 4}{x - 6}$

Find the domain of the following expressions.

51. $\sqrt{4x - 10}$

52. $\sqrt{3x + 1}$

53. $\sqrt{9 - 2x}$

54. $\sqrt{6 - \dfrac{x}{2}}$

55. $\sqrt{x^2 - 5x - 6}$

56. $\sqrt{x^2 + x - 30}$

57. $\sqrt{4x^2 - 4x - 3}$

58. $\sqrt{3x^2 + 2x - 5}$

59. The final grade in a certain course is the average of four exams. Each exam is scored from 0 to 100 points. A student has received grades of 66, 71, and 84 on the first three exams. The student would like to achieve a final average grade of at least 75.
   a. What is the minimum score that this student must have on the fourth exam?
   b. What is the highest average that this student can achieve for the course?

60. A student in the course described in problem 59 has received a grade of 60 on the first test.
   a. If the student would like to have an average of 75 or above at the end of the course what must be the average of the remaining three exams?
   b. What is the highest average which this student can achieve for the course?

61. The perimeter of a certain rectangle must be less than 100 feet. The length of this rectangle is 30 feet. Find all values of the width that would meet these conditions. The width must be a positive number.

62. The perimeter of a certain rectangle must be between 50 feet and 200 feet. The length of this rectangle is 20 feet. Find the range of values that the width must be to meet these conditions.

63. The perimeter of a square must be greater than 16 inches but less than 84 inches. Find all values of the length of a side that will meet these conditions. Note that, since all four sides of a square have the same length, the perimeter P of a square is $4s$, where s is the length of a side.

64. In a course there are four in-class tests and a final exam. The exam counts 30% of the grade, and the tests are all counted equally. If T = test average, E = exam, and G = final grade, then a relation that expresses this would be $0.7T + 0.3E = G$. A certain student has an average of 78 for the four tests, and would like to get a final average of 80 or above. Write an inequality that describes the final exam grade E necessary for this result.

65. An electronics circuit has two resistances in parallel. The value of one resistance is x (ohms) and the second resistance is 10 ohms greater than this. The total resistance must not exceed 40 ohms. Under these circumstances the relation $\dfrac{1}{x} + \dfrac{1}{x + 10} \geq \dfrac{1}{40}$ will be true. Solve for x.

66. In a certain rectangle the width is 7 less than the length. The area must exceed 30 square units. Find the restrictions on the length that will give this result.

67. A certain printing press can print 1,500 newspapers in a certain time t (in minutes). A second press is always three minutes faster, so it takes $t - 3$ minutes to do the same thing. Running together it is necessary for the presses to produce 3,000 papers per minute. Under these circumstances the relation $\dfrac{1{,}500}{t} + \dfrac{1{,}500}{t - 3} \geq 3{,}000$ holds. Solve this inequality, and then note that $t > 0$ and $t - 3 > 0$ must also be true to find a range of values for t that make sense.

68. The zoning bylaws of a certain town (Carlisle, Mass.) require, among other things, that building lots have a ratio of area A to perimeter P conforming to the relation $\frac{16A}{P^2} > 0.4$.

a. Solve this relation for A.
b. Solve this relation for P.

69. Determine which of the following building lots conform to the zoning bylaw of the previous problem.

	Length	Width		Length	Width
a.	100′	50′	**b.**	200′	50′
c.	300′	50′	**d.**	400′	50′
e.	500′	50′	**f.**	600′	50′

70. Consider how the relation stated in problem 68 applies to rectangular lots. If L is length and W is width for a rectangular lot, then

$$A = LW \text{ and } P = 2(L + W).$$

Let k be the ratio of length to width $\left(\text{i.e., } k = \frac{L}{W}\right)$. Show that k must fall in the interval $4 - \sqrt{15} < k < 4 + \sqrt{15}$.

71. A machine can produce 30 bolts per hour when the cutting tool is new. Then for the first 10 hours of use, it produces two fewer bolts per hour than the previous hour. Under these conditions the total number of bolts produced after x hours is given by the expression $30x - x^2$, $0 \le x \le 10$. To find out how many hours are required for the total production to be 100 bolts or more we solve $30x - x^2 \ge 100$. Solve this inequality and find x to the nearest minute.

72. The critical point/test point method can be used to solve simple linear inequalities. Use it to solve $3x - 5 \le 6x$.

Skill and review

1. Solve $\frac{2x - 3}{4} = 2$.

2. Solve $\frac{2x^2 - 4}{7} = x$.

3. If $|x| = 8$, then
a. $x = 8$ **b.** $x = -8$ **c.** $x = 8$ or $x = -8$
d. $-8 < x < 8$

4. If $|x| < 8$, then
a. $x = 0$ **b.** $-8 < x < 8$ **c.** $x < -8$ or $x > 8$
d. $x < 7$

5. If $|x| > 8$ then
a. $x = 0$ **b.** $-8 < x < 8$ **c.** $x < -8$ or $x > 8$
d. $x < 7$

6. Is the value 3 a solution to $\left|\frac{1 - x}{2}\right| < x$?

7. For what value(s) of x is $|x| = |-x|$ true?

2–5 Equations and inequalities with absolute value

A computer program is used to buy and sell stock. A certain stock is selling for $32\frac{1}{8}$ points (\$32.125 per share). The computer is set to alert someone if the price of the stock changes by more than $1\frac{1}{8}$ points. Thus the prices x that would set off this alert are described by $\left|32\frac{1}{8} - x\right| \ge 1\frac{1}{8}$.

As this problem illustrates, absolute values and inequalities can be used to describe certain applied situations. The combination is most useful where the difference between two quantities must meet some restriction. We can often use the properties described in this section to find the solution set when one or both members of an equation or inequality is the absolute value of an expression involving a variable.

Equations with absolute value

What would $|x| = 4$ mean about x? It can be seen that x must be either 4 or -4. That is, if $|x| = 4$ then $x = 4$ or $x = -4$. We can generalize this into the following principle.

> **Equations with absolute value**
> [1] If $|x| = b$ and $b \geq 0$, then $x = b$ or $x = -b$.

■ ***Example 2–5 A***

Solve the equation $|4x - 2| = 8$.

$$|4x - 2| = 8$$

Rewrite using property [1] then solve each equation

$$4x - 2 = 8 \quad \text{or} \quad 4x - 2 = -8$$
$$4x = 10 \qquad\qquad 4x = -6$$
$$x = \tfrac{5}{2} \qquad\qquad x = -\tfrac{3}{2}$$

$\{-\tfrac{3}{2}, \tfrac{5}{2}\}$ Solution set ■

Inequalities with absolute value

The following two properties describe many situations involving absolute values and inequalities. These can be thought of as rewriting rules, just as property [1] is. These properties allow us to rewrite a statement that involves absolute values as equivalent sets of statements that do not involve absolute value.

> **Inequalities with absolute value**
> [2] If $|x| < b$ and $b > 0$, then $-b < x < b$.
> [3] If $|x| > b$ and $b \geq 0$, then $x > b$ or $x < -b$.

Figure 2–7 illustrates these rewriting rules.

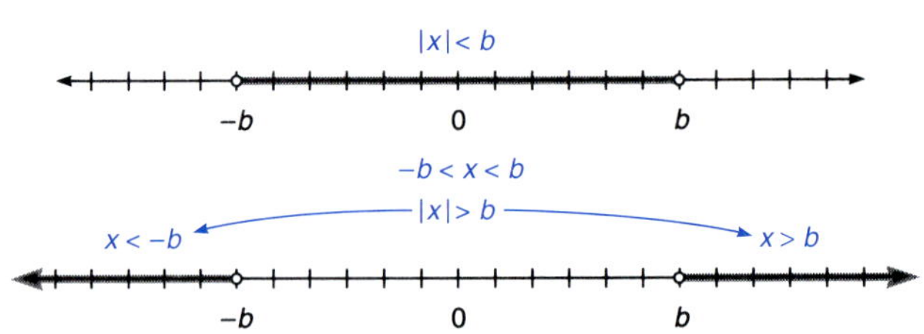

Figure 2–7

Note 1. For property [2], if $b < 0$, then the solution set is the empty set.
2. For property [3], if $b < 0$ then any value of x is a solution, since $|x| \geq 0 > b$, if b is negative.

Similar properties apply for weak inequalities.

Although properties [2] and [3] are always true, they are most useful when the expression represented by x is linear. These cases are illustrated in example 2–5 B.

■ ***Example 2–5 B***

Solve. State the solution in set-builder notation and graph the solution set.

1. $\left|\dfrac{3x-1}{4}\right| > 3$

$\dfrac{3x-1}{4} > 3$ or $\dfrac{3x-1}{4} < -3$		Rewrite using property [3]
$3x - 1 > 12$	$3x - 1 < -12$	Multiply each member by 4
$3x > 13$	$3x < -11$	
$x > 4\frac{1}{3}$	$x < -3\frac{2}{3}$	Solve the linear inequality
$\{x \mid x < -3\frac{2}{3} \text{ or } x > 4\frac{1}{3}\}$		Solution set, set-builder notation

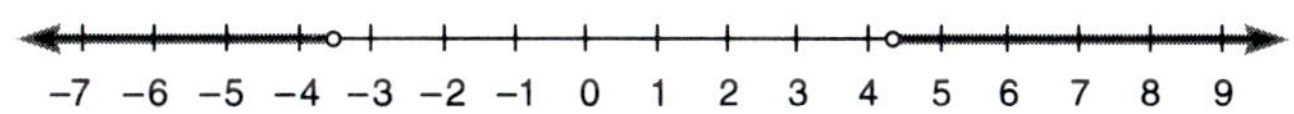

2. $|5 - 2x| \le 8$

$-8 \le 5 - 2x \le 8$	Rewrite using property [2]
$-13 \le -2x \le 3$	Add -5 to each of the three members
$\frac{13}{2} \ge x \ge -\frac{3}{2}$	Divide each member by -2
$-\frac{3}{2} \le x \le \frac{13}{2}$	It is customary to put the smallest member on the left
$\{x \mid -1\frac{1}{2} \le x \le 6\frac{1}{2}\}$	Solution set

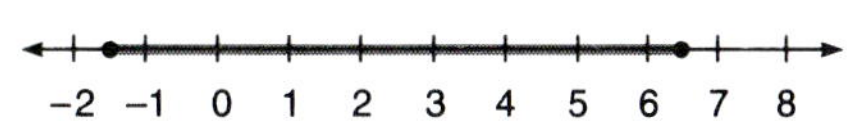

3. $|x - 1| > -1$

Since $|x - 1| \ge 0$ for any value of x, and since $0 > -1$, we can conclude that $|x - 1| > -1$ is true for any value of x. Thus, the solution set is R. The graph is the entire number line. ■

Example 2–5 C shows how to find approximate solutions to these problems using the graphing calculator.

■ ***Example 2–5 C***

Solve $\left|\dfrac{3x-1}{4}\right| > 3$ (part 1 of example 2–4 B).

As illustrated in the last two sections, there are two ways to solve this problem graphically.

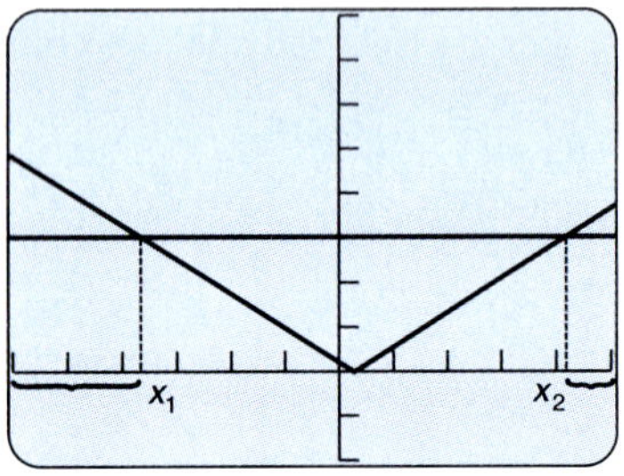

Method 1: Graph the equations $Y_1 = \left|\dfrac{3x-1}{4}\right|$ and $Y_2 = 3$. The graph is shown. The solution is where $Y_1 > Y_2$. This is to the right of x_2 and the left of x_1. We can use trace and zoom to estimate the values as 4.3 and -3.7. Thus the approximate solution is $x > 4.3$ or $x < -3.7$.

[Y=] [ABS] [(] [(] 3 [X|T] [−] 1 [)] [÷] 4 [)] [ENTER] 3
[RANGE −6,6,−2,8]

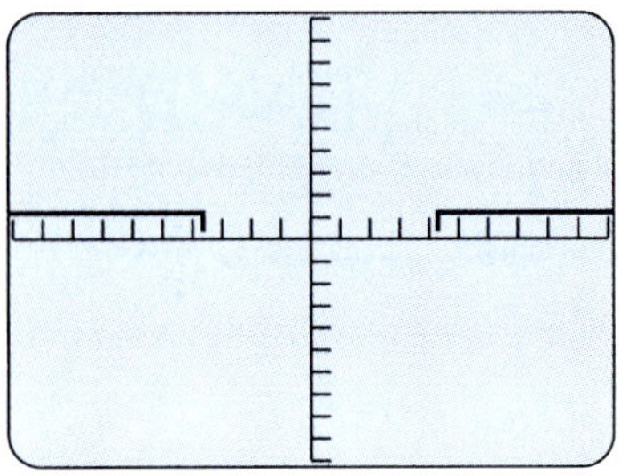

Method 2: Graph Y_1=abs((3X − 1)/4) > 3. (Remember, > is [2nd] [MATH] 3.)

The second graph is shown for standard settings ([ZOOM] 6). ■

Mastery points

Can you

- Solve equations involving absolute value?
- Solve inequalities involving absolute value?

Exercise 2–5

Solve the following equations involving absolute value.

1. $|5x| = 8$
2. $|3x - 1| = 3$
3. $|2x + 5| = 6$
4. $\left|x + \frac{3}{5}\right| = 1$
5. $|3 - 2x| = 5$
6. $|9x + 1| = 7$
7. $|2x + 8| = 0$
8. $\left|\dfrac{3 - 2x}{4}\right| = 0$
9. $\left|\dfrac{3x - 5}{4}\right| = 1$
10. $|x^2 - 2x| = 3$
11. $|2x^2 - 5x| = 9$
12. $\left|\dfrac{x^2 + 6x}{2}\right| = 8$

Solve the following inequalities involving absolute value. State the solution in set-builder notation and graph the solution set.

13. $|3 + 6x| > 4$
14. $|4 - 3x| \le \frac{1}{2}$
15. $\left|\dfrac{x - 4}{3}\right| \le 10$
16. $\left|\dfrac{4x - 1}{5}\right| < 1$
17. $|2 - 5x| < -3$
18. $|2 - 5x| > -3$
19. $|-5(2x + 3)| \ge 8$
20. $\left|\dfrac{2x}{3} - x\right| \le 4$
21. $\left|3x - \dfrac{x}{3}\right| > 3$
22. $|-x + 11| < \frac{3}{8}$
23. $|x + 5| > -2$
24. $\left|3x - \frac{1}{5}\right| < 0$

Solve the following equations and inequalities involving absolute value. State the solution in set-builder notation.

25. $|3x| > 22$
26. $|4x + 3| > 1$
27. $25 < |5 - 2x|$
28. $17 < |4x - 1|$
29. $\left|\dfrac{x - 2}{4}\right| > 9$
30. $\left|\dfrac{3 - 2x}{5}\right| > 8$
31. $|3x| = 22$
32. $|4x + 3| = 1$
33. $|5 - 2x| = 25$
34. $17 = |4x - 1|$
35. $\left|\dfrac{x - 2}{4}\right| = 9$
36. $8 = \left|\dfrac{3 - 2x}{5}\right|$
37. $|3x| < 22$
38. $|4x + 3| < 1$
39. $|5 - 2x| < 25$
40. $17 > |4x - 1|$

41. $\left|\frac{x-2}{4}\right| < 9$ **[42.]** $8 > \left|\frac{3-2x}{5}\right|$ **43.** $|4x| = 5$ **44.** $|3x+1| < 4$

45. $5 > |6x-3|$ **46.** $|9x-17| > 5$ **47.** $|3x| \geq 8$ **48.** $6 \leq |9x+1|$

49. $\left|\frac{2x-3}{4}\right| \leq 17$ **50.** $\frac{1}{2} < \left|\frac{2x+5}{9}\right|$ **51.** $\left|\frac{4x+7}{8}\right| \geq \frac{4}{9}$ **52.** $16 \geq \left|\frac{3x-1}{5}\right|$

53. $2 < |-x|$ **54.** $4 > |11-6x|$ **55.** $2 \leq \left|\frac{x+2}{3}\right|$ **[56.]** $\frac{3}{4} \leq \left|\frac{3x+1}{8}\right|$

57. In the diagram the dimension z is $6\frac{1}{2}$ inches. If the circle has radius $\frac{7}{8}$ inches, then the dimensions x and y are the two solutions to the equation $|6\frac{1}{2} - w| = \frac{7}{8}$. Solve this for w, and thereby find the values of x and y.

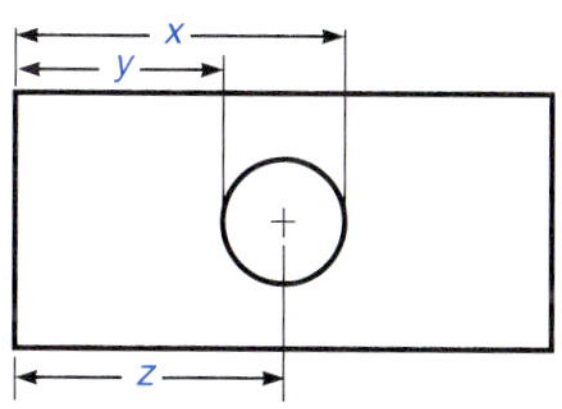

58. A computer program is used to buy and sell stock. A certain stock is selling for $32\frac{1}{8}$ points (\$32.125 per share). The computer is set to alert someone if the price of the stock changes by more than $1\frac{1}{8}$ points. Thus the prices x that would set off this alert are described by $|32\frac{1}{8} - x| \geq 1\frac{1}{8}$. Solve this inequality.

59. The total number of bolts produced after x hours by a certain machine is given by the expression $x^2 - \frac{x}{3}$. A different machine produces x^2 bolts after x hours. A production engineer is interested in knowing when the two machines have produced the same number of bolts, plus or minus 6 bolts. This can be determined by solving $\left|\left(x^2 - \frac{x}{3}\right) - (x^2)\right| \leq 6$. Solve this expression for x. Assume $x \geq 0$.

[60.] Beginning with $(x-y)^2 \geq 0$, show that

$$\frac{x^2+y^2}{2} \geq xy.$$

Skill and review

1. If $x = 1$ and $y = -3$ is the statement $2x - y = 5$ true?
2. If $x = -3$ and $y = -11$ is the statement $2x - y = 5$ true?
3. If $x = 3$ and $y = 1$ is the statement $2x - y = 5$ true?
4. If $x = 0$ and $y = -5$ is the statement $2x - y = 5$ true?
5. If $x = 2$ and $y = -3$, which of the statements is true?
   a. $3x + y = 3$ **b.** $-x + 5y = -17$ **c.** $y + 9 = 3x$
   d. $x = y + 5$
6. If $x = -2$ and $y = 4$, which of the statements is true?
   a. $3x + y = -2$ **b.** $-x + 5y = 18$ **c.** $y + 10 = 3x$
   d. $x = y + 6$
7. Solve $2x + y = 8$ for y.
8. Solve $x - 2y = 4$ for y.

Chapter 2 summary

- **To solve a linear equation** Form a sequence of equivalent equations until one is sufficiently simple to solve by inspecting it. Form these equations by using the following steps:
  1. Clear any denominators by multiplying each term by the least common denominator of all the terms.
  2. Perform indicated multiplications (remove the parentheses).
  3. Use the addition property of equality so that all terms with the variable are in one member of the equation, and all other terms are in the other member.
  4. If necessary factor out the variable from the terms containing it.
  5. Divide both members of the equation by the coefficient of the variable.

- **Quadratic equation in one variable** An equation which can be put in the standard form $ax^2 + bx + c = 0$, $a,b,c \in R$, $a \neq 0$.
- **The quadratic formula and the factors of a quadratic expression** If $ax^2 + bx + c = 0$ and $a \neq 0$, then $x = \dfrac{-b \pm \sqrt{b^2 - 4ac}}{2a}$ and $ax^2 + bx + c$ $= a\left(x - \dfrac{-b + \sqrt{b^2 - 4ac}}{2a}\right)\left(x - \dfrac{-b - \sqrt{b^2 - 4ac}}{2a}\right)$.
- **Solving quadratic equations**

  Put the equation in standard form.

  If the quadratic expression can be factored use the zero product property.

  When b in $ax^2 + bx + c = 0$ is zero use the method called extracting the roots.

  When the methods mentioned above do not apply use the quadratic formula.
- **Pythagorean theorem** In any right triangle where c is the longest side (the hypotenuse, always opposite the right angle), and a and b are the remaining two sides, $a^2 + b^2 = c^2$.
- **Solving radical equations** Raise each member of the equation to the appropriate power. We may get solutions that are not solutions to the original equation.
- **Addition property of inequality** For any algebraic expressions A, B, and C, if $A < B$ then $A + C < B + C$.
- **Multiplication property of inequality** For any algebraic expressions A, B, and C,

  1. if $C > 0$ and $A < B$ then $AC < BC$,
  2. if $C < 0$ and $A < B$ then $AC > BC$.
- **Solving linear inequalities** Solve linear inequalities exactly the same way as linear equations except that if we multiply (or divide) both members of an inequality by a negative value we must reverse the direction of the inequality.
- **Solving nonlinear inequalities** To solve nonlinear inequalities employ a method called the critical point/test point method:

  1. Solve the corresponding equality.
  2. Solve for any zeros of any denominators. Steps 1 and 2 produce critical points.
  3. Use the critical points found in steps 1 and 2 to mark intervals on the number line.
  4. Choose a test point from each interval.
  5. Try the test points in the original problem.
  6. Note the intervals where the test point makes the original problem true. These are part of the solution set.
  7. Include any of the critical points which make the original inequality true in the solution set.
  8. Write the solution set using set-builder notation.
- **Solving equations involving absolute value** Use the property

  [1] If $|x| = b$ and $b \geq 0$, then $x = b$ or $x = -b$.
- **Solving absolute value and inequality problems** Use the properties

  [2] If $|x| < b$ and $b > 0$, then $-b < x < b$.

  [3] If $|x| > b$ and $b \geq 0$, then $x > b$ or $x < -b$.

Chapter 2 review

[2–1] Solve the following linear equations by specifying the solution set.

1. $\frac{3}{5}x - 4 = 2 - \frac{3}{4}x$
2. $\frac{3}{4}(5x - 3) - 3x + 19 = 0$
3. $-2[\frac{1}{2} - 2(5 - x) + 2] - \frac{3}{2}x = 0$
4. $4(x - 2) = -(7 - 4x) - 1$
5. $x - \frac{3}{8}x = \frac{1}{4}x$

Find approximate answers to the following problems. Round the answer to 4 digits of accuracy.

6. $13.5x - 22.3x = 0.03(1{,}200 - 2{,}113x)$
7. $11.4 - 3.5x - \sqrt{2}[9.2 - 1.5(\frac{3}{8}x - 5.3) - x] = 0$

Solve the literal equations in problems 8 through 12 for the variable indicated.

8. $m = -p(Q - x)$; for Q
9. $R = W - k(2c + b)$; for b
10. $P = \dfrac{n}{5}(P_2 - P_1) - c$; for P_1
11. $\dfrac{x + 2y}{3 - 2y} = x$; for x
12. $\dfrac{x + y}{x} = y$; for x
13. A total of \$8,000 is invested, part at 7% and part at 5%. The income from these investments for one year is \$471. How much was invested at each rate?

14. A total of \$15,000 was invested; part of the investment made a 12% gain, but the rest had a 10% loss. The net *loss* from the investments was \$720. How much was invested at each rate?

15. A total of \$5,000 was invested, part at 5% and part at 9%. If the income for one year from the 9% investment was \$44 more than the income from the 5% investment, how much was invested at each rate?

16. A company has a fertilizer that is 10% phosphorous, and a second that is 25% phosphorous. How much of each must be mixed to obtain 2,000 pounds of a mixture that is 15% phosphorous?

17. A company has 15 tons of material that is 40% copper. How much material that is 55% copper must be mixed with this to obtain a material that is 45% copper?

18. One computer printer can print out 12,000 labels in 35 minutes; a second printer takes 1 hour 10 minutes to do the same job. How long would it take the printers, running at the same time, to print out a total of 12,000 labels?

19. A boat moves at 12 mph in still water. If the boat travels 30 miles downstream in the same time it takes to travel 18 miles upstream, what is the speed of the current?

20. A boat travels 50 kilometers upstream in the same time that it takes to travel 80 kilometers downstream. If the stream is flowing at 5 km per hour, what is the speed of the boat in still water?

[2–2] Solve the following quadratic equations for x by factoring.

21. $2x^2 - 7x - 30 = 0$ **22.** $x = \frac{5}{2} + \frac{3}{2x}$

23. $(2x - 5)^2 = 40 - 16x$ **24.** $6a^2x^2 + 7abx = 5b^2$

Solve the following equations by extracting the roots.

25. $27x^2 - 40 = 0$ **26.** $(2x - 3)^2 = 12$

27. $4(x + 1)^2 = 8$

28. $a(bx + c)^2 = d$; assume $a, d > 0$

Solve the following quadratic equations using the quadratic formula.

29. $5y^2 - 15y - 6 = 0$ **30.** $2y^2 - \frac{2y}{3} + 1 = 0$

31. $(z + 4)(z - 1) = 5z + 4$ **32.** $\frac{1}{x + 2} - 6x = 25$

Use the quadratic formula to help factor the following quadratic equations.

33. $3x^2 - 8x - 12$ **34.** $5x^2 + 6x - 4$

35. In a right triangle the length of the hypotenuse is 26, and one side is 14 units longer than the other. Find the length of the longer side.

36. A rectangle that is $8\frac{1}{2}$ centimeters long and 4 centimeters wide has its dimensions increased by adding the same amount. The area of this new rectangle is 90 cm^2. What are the dimensions of the new rectangle?

37. A manufacturer finds that the total cost C for a product is expressed by $C = 250x^2 - 24{,}000$, and the total revenue R at a price of \$200 per unit to be $R = 200x - 1{,}000$, where x is the number of units sold. What is the break even point (where total cost = total revenue) to the nearest unit?

38. Two pipes can be used to empty the water behind a dam. When both pipes run together it takes 40 hours to empty the dam. When they are used separately it takes the smaller pipe 12 hours longer to empty the dam than the larger pipe. Find the time required for the larger pipe to empty the dam alone, to the nearest hour.

39. When flying directly into a 25 mph wind it takes an aircraft 1 hour longer to travel 300 miles than when there is no wind. Find the time required for the aircraft to fly this distance in no wind.

40. Determine the domain of the expression $\frac{3x}{x - 3}$.

41. Determine the domain of the expression $\frac{2x - 5}{x^2 + 8x + 12}$.

Solve the following quadratic equations. You may want to use substitution.

42. $2x^4 - 27x^2 + 81 = 0$

43. $(x - 3)^2 - 8(x - 3) - 20 = 0$

44. $2t + 7\sqrt{t} - 15 = 0$ **45.** $x^{3/2} - 9x^{3/4} + 8 = 0$

46. $y^{-4} - 32 = 4y^{-2}$

[2–3] Find the solution set for each equation.

47. $\sqrt{x(6x + 5)} = 1$ **48.** $\frac{2}{3}\sqrt{9z - 1} - \sqrt{5z} = 0$

49. $\sqrt{5w + 1} = 5w - 19$

50. $\sqrt{3n + 1} - \sqrt{n + 1} = 2$

51. $\sqrt[4]{2y - 3} = 2$

Solve the following formulas for the indicated variable.

52. $\sqrt{(2x - 1)(2k + 1)} = x + k$; for x

53. $r = \sqrt{\dfrac{A}{\pi} - AR^2}$; for A

[2–4] Find and graph the solution sets to the following linear inequalities.

54. $-9x - 4(2x - 3) < 0$

55. $3(x - 1) + 2(3 - 2x) \geq 5(x - 3)$

56. $2x - 12 \leq 3x - \frac{1}{2}(x - 6)$

57. $12 + \frac{5}{2}(3 - \frac{2}{3}x) < -4(x - 6)$

Find and graph the solution set to the following nonlinear inequalities.

58. $r(r - 3)(6r + 12) \geq 0$

59. $w^2 - 1 < \frac{7}{12}w$

60. $(4x^2 - 25)(x^2 - 16) \leq 0$

61. $(x^2 - 6x + 9)(x^2 - 1) \leq 0$

62. $(x - 4)^2(x^2 + 2)^3 > 0$

63. $\dfrac{x - 1}{x - 3} \geq 0$

64. $\dfrac{x - 1}{x - 3} \geq 2$

65. $\dfrac{x + 3}{x^2 - x - 6} < 0$

66. $\dfrac{x - 4}{x + 5} - 2x \geq 1$

67. $\dfrac{3x}{x - 1} - \dfrac{2}{x + 3} < 1$

68. $\dfrac{x - 3}{x - 4} \geq \dfrac{x + 4}{x - 6}$

[2–5] Solve the following equations and inequalities involving absolute value.

69. $\left|\frac{3}{8} - 2x\right| = \frac{3}{4}$

70. $\left|x^2 - 5x\right| = 50$

71. $\left|x^2 + 1\right| = 1$

72. $\left|\dfrac{x - 4}{4}\right| \leq 5$

73. $\left|x^2 - x\right| = 2$

74. $\left|\dfrac{1 - 2x}{5}\right| < 10$

75. $\left|\dfrac{x - 2}{4}\right| > 9$

76. $\left|\dfrac{2 - x}{4}\right| = 9$

77. $5 > \left|-2x - 3\right|$

78. $\left|\dfrac{x + 7}{4}\right| \geq \dfrac{2}{3}$

79. $\left|3x - \frac{4}{5}\right| > 2$

Chapter 2 test

Solve the following linear equations by specifying the solution set.

1. $7x - 4 = 7(4 - x)$

2. $\dfrac{2x - 3}{5} = \dfrac{3 - 4x}{2}$

3. $3x - \dfrac{3}{4}x = \dfrac{1}{4}x + 2$

4. Find an approximate answer to four digits of accuracy $2.4x - 8.7x = 2(7.2 - 0.2x)$.

Solve the literal equations in problems 5 through 7 for the variable indicated.

5. $m = -p(Q - x)$; for x

6. $P = \dfrac{n}{5}(P_2 - P_1) - c$; for P_2

7. $\dfrac{x + 2y}{3 - 2y} = x$; for y

8. A total of $12,000 is invested, part at 9% and part at 5%. The income from these investments for one year totals $720. How much was invested at each rate?

9. A company has 28 tons of alloy that is 30% copper. How much alloy that is 80% copper must be mixed with this to obtain an alloy that is 50% copper?

10. A printing machine can print out 4,000 labels in 20 minutes; a second machine takes 50 minutes to do the same job. How long would it take the printers, running at the same time, to print out a total of 4,000 labels?

11. A boat moves at 10 mph in still water. If the boat travels 20 miles downstream in the same time it takes to travel 15 miles upstream, what is the speed of the current?

Solve the following quadratic equations for x by factoring.

12. $3x^2 + 4x - 15 = 0$

13. $10 + \dfrac{13}{x} = \dfrac{3}{x^2}$

Solve the following equations by extracting the roots.

14. $4x^2 - 50 = 0$

15. $(3x - 3)^2 = 24$

Solve the following quadratic equations using the quadratic formula.

16. $m^2 - 3m - 6 = 0$

17. $(z - 1)(z + 1) = \frac{1}{3}(-6z - 7)$

18. Use the quadratic formula to help factor the quadratic equation $2x^2 - x - 12$.

19. In a right triangle the length of the hypotenuse is 26, and the length of one side is 4 units longer than twice the length of the other side. Find the length of the longer side.

20. A manufacturer finds that the total cost C for a product is expressed by $C = 3x^2 - 8$, and the total revenue R by $R = 3x - 2$, where x is the number of units sold. What is the break even point (where total cost = total revenue) to the nearest unit?

21. Two printers can be used to print an edition of a newspaper. When both printers run together it takes 12 hours to print the edition. When they are used separately it takes the slower press one hour longer to print the edition than the faster press. Find the time required for the faster press to print the edition alone. Round the answer to the nearest 0.1 hour.

22. When flying directly into a 20 mph wind it takes an aircraft one hour longer to travel 400 miles than when there is no wind. Find the speed of the aircraft in no wind.

23. Find the domain of the expression $\dfrac{x-1}{x^2-81}$.

Solve the following quadratic equations using substitution.

24. $4x^4 - 37x^2 + 9 = 0$ **25.** $x^3 - 7x^{3/2} - 8 = 0$

Find the solution set for each equation.

26. $\sqrt{2 + w - w^2} - 1 = w$

27. $\sqrt{5n+5} - \sqrt{13-n} = 2$

28. Solve the formula for b: $r = \sqrt{\dfrac{A}{\pi} - Ab}$.

Find and graph the solution sets to the following linear inequalities.

29. $4x - 3(2x - 3) < x$ **30.** $x - 5 \leq 2x - \frac{1}{2}(6 - x)$

Find and graph the solution set to the following nonlinear inequalities.

31. $(x - 3)(x^2 - 4) > 0$

32. $(x^2 - 4x + 4)(x^2 - 1) \leq 0$

33. $\dfrac{x-10}{x-3} > 2$ **34.** $\dfrac{x+3}{x^2-2x-8} \geq 0$

Solve the following equations and inequalities involving absolute value.

35. $|5 - 2x| = 8$ **36.** $|3x - 2| \leq 10$

37. $\left|\dfrac{2-3x}{4}\right| > 2$ **38.** $|5x - 1| \geq 6$

39. $3 > |2x - 3|$ **40.** $|x^2 + 2x| = 10$

Solve the following equations.

41. $\dfrac{2x-3}{3} + \dfrac{5-x}{4} = 1 - \dfrac{2x+3}{6}$

42. $\dfrac{2x}{x+1} + \dfrac{2(x+2)}{x} = 3$

43. $\dfrac{2}{x+1} + \dfrac{2(x+2)}{x} = 3$ **44.** $\dfrac{1}{x+1} + \dfrac{2}{x+2} = 1$

45. $\dfrac{5}{2x-3} = \dfrac{4}{x+1}$ **46.** $\dfrac{5}{2x-3} = \dfrac{4x}{x+1}$

47. $\dfrac{2x-5}{x+2} = 3$ **48.** $\dfrac{2x-5}{x+2} = 3x$

CHAPTER 3

Relations, Functions, and Analytic Geometry

In this chapter we examine the concepts of relation, function, and analytic geometry. We consider relations and analytic geometry before moving on to functions in section 3–3. Analytic geometry can loosely be described as the representation of geometric figures by equations.

3–1 Points and lines

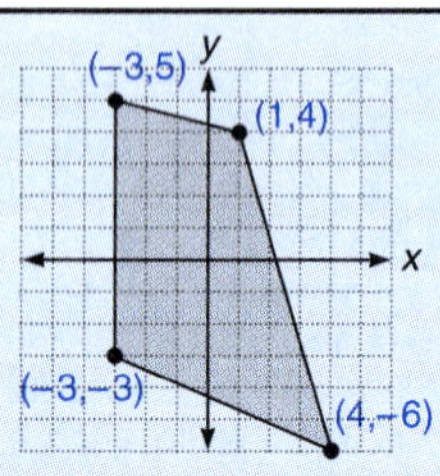

A surveyor has surveyed a piece of property and plotted the measurements shown in the figure (each unit represents 100 feet). The surveyor's client also wants to know the area of the property. How can this be done?

This section introduces how algebra can be used to solve geometric problems, such as that just shown.

Geometry was highly developed in the ancient world; the Greeks had a very sophisticated knowledge of it by 500 B.C. Algebra developed somewhat more slowly. It began in Mesopotamia 4,000 years ago, developed in the Persian and Hindu worlds of this era, and was brought to Europe by the Arabs some 800 years ago. It reached a high degree of development by the sixteenth century. Up to that time, algebra and geometry were two disconnected fields of study. What could be derived geometrically was considered true, in conformance with the reality of the universe; the world of algebra was considered to be a contrivance—useful but artificial.

Algebra has acquired its own credibility in the last three hundred years. Part of the reason for this is **analytic geometry**—the connection between algebra and geometry. We begin looking at this subject with the ordered pair.

> **Ordered pair**
> An **ordered pair** is a pair of numbers listed in parentheses, separated by a comma.
>
> **Equality of ordered pairs**
> Two ordered pairs (x,y) and (a,b) are *equal* if and only if
> $$x = a \quad \text{and} \quad y = b$$

In the ordered pair (x,y), x is called the *first component* and y is called the *second component.*[1] Examples of ordered pairs include $(5,-7)$, $(9,\pi)$, $(\sqrt{3},\frac{2}{3})$. Because of the definition of equality of ordered pairs, the ordered pairs (5,8) and (8,5) are not equal; the order of the values of the components is important. Ordered pairs often have meaning in some application. For example, the ordered pairs (3,9) and (4,16) could represent the lengths of a side of a square and the corresponding area of the square (found by squaring the length of the side).

We speak of sets of ordered pairs so often that we give them a name: relation.

> **Relation**
> A **relation** is a set of ordered pairs.

For example, the set of ordered pairs

$$A = \{(1,2), (2,4), (3,-5), (3,4), (8,-5)\}$$

is a relation.

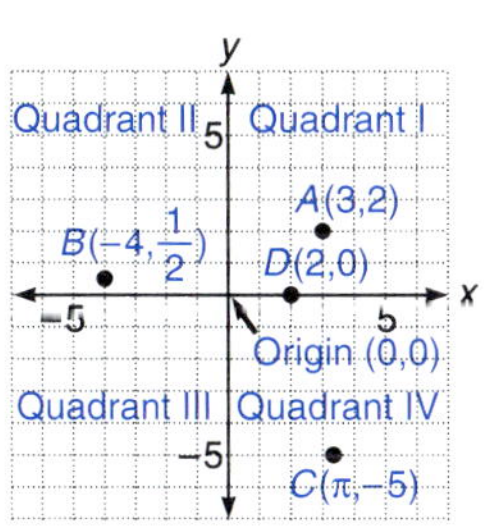

Figure 3–1

The French philosopher-mathematician-soldier René Descartes (1596–1650) developed analytic geometry with his **rectangular,** or Cartesian, **coordinate system.**[2] This system is formed by sets of vertical and horizontal lines; one vertical line is called the y-axis, and one horizontal line is called the x-axis. See figure 3–1.

The x- and y-axes divide the "coordinate plane" into four **quadrants,** labeled I, II, III, and IV. The **graph of an ordered pair** is the geometric point in the coordinate plane located by moving left or right, as appropriate, according to the first component of the ordered pair, and vertically a number of units corresponding to the second component of the ordered pair. The graphs of the points $A(3,2)$, $B(-4,\frac{1}{2})$, $C(\pi,-5)$, and $D(2,0)$ are shown in figure 3–1. The first and second elements of the ordered pair associated with a geometric point in the coordinate plane are called its **coordinates.**

[1]The first component is also called the *abscissa,* and the second component is also called the *ordinate.*

[2]There are other types of coordinate systems; popular ones are polar, log, and log log. Descartes is the rationalist philosopher credited with the statement "Cogito, ergo sum" (I am thinking, therefore I must exist).

Points and lines: definitions

In this section we introduce the concepts basic to analytic geometry. We intentionally define algebraic objects using the same names as geometric objects, such as point and line. We make the definitions so that the algebraic objects have the same properties as the geometric objects. We begin with point.

Point

A point is an ordered pair.

Graph of a point

The graph of a point is the geometric point in the coordinate plane associated with the ordered pair that defines that point. We say we *plot* the point when we mark its graph on a coordinate system.

Graph of a relation

The graph of a relation is the set of graphs of all ordered pairs in the relation.

An equation such as $2x - y = 3$ defines a relation—all those ordered pairs (x,y) that make it true. For example, $(1,-1)$ makes the statement true because, using substitution of value (section 1–2) we find that $2(1) - (-1) = 3$ is true. We say that $(1,-1)$ is a solution to the equation, and belongs to the relation.

In geometry a line is a set of points. Lines must have certain properties, such as that any two points belong to a unique line. Experience has told us how to define a line algebraically with the same properties as a line in geometry.

Straight line

A straight line is the relation described by any equation that can be put in the form

$$ax + by + c = 0$$

with at least one of a or b not zero. The equation $ax + by + c = 0$ is called the **standard form** of a straight line.

To find solutions to an equation involving two variables, such as the equation of a straight line, choose a value (at random) for one of the two variables, x or y, use substitution of value (section 1–2) on this variable, then solve the equation for the other variable.

If we organize these values into a table of x- and y-values, such as the table in example 3–1 A, we say we have created a **table of values** for the equation.

■ ***Example 3–1 A***

Find five solutions to the straight line $3x + 2y = 6$.

Note that this can be put in the form $ax + by + c = 0$ by subtracting 6 from both sides. Two solutions are easy to find; let x be 0, then let y be 0:

$$x = 0\text{: } 3(0) + 2y = 6,\ y = 3, \text{ so } (0,3) \text{ is a solution.}$$
$$y = 0\text{: } 3x + 2(0) = 6,\ x = 2, \text{ so } (2,0) \text{ is a solution.}$$

For more solutions, *it is useful to solve the equation for y:*

$$
\begin{aligned}
3x + 2y &= 6 \\
2y &= -3x + 6 \\
y &= -\tfrac{3}{2}x + \tfrac{6}{2} \\
y &= -\tfrac{3}{2}x + 3
\end{aligned}
$$

Now we can conveniently calculate y for any x. Let us choose x to be -2, 4, and 6. (If we choose even integers for x, the denominator, 2, of the fraction will be reduced, eliminating fractions from the resulting value[3] of y.)

x	y
0	3
2	0
-2	6
4	-3
6	-6

Table 3–1

$x = -2$: $y = -\frac{3}{2}x + 3$, $y = -\frac{3}{2}(-2) + 3 = 6$, so $(-2,6)$ is a solution.
$x = 4$: $y = -\frac{3}{2}x + 3$, $y = -\frac{3}{2}(4) + 3 = -3$, so $(4,-3)$ is a solution.
$x = 6$: $y = -\frac{3}{2}x + 3$, $y = -\frac{3}{2}(6) + 3 = -6$, so $(6,-6)$ is a solution.

Table 3–1 shows the points. This is a convenient way to list ordered pairs. Thus, five solutions to $3x + 2y = 6$ are $(0,3)$, $(2,0)$, $(-2,6)$, $(4,-3)$, and $(6,-6)$. ■

If the graphs of the points in table 3–1 were plotted, along with as many other solutions as we wished, we would find that these points form a straight line. Also, any point that was not a solution would not lie on the line. (The graph is shown in part 1 of example 3–1 B.) With this statement in mind we talk more about graphing straight lines.

Graphs of straight lines

The easiest way to graph a straight line is to locate any two points that lie on the line. It is an axiom of geometry that any two points determine a unique line; this same fact is a matter of definition in analytic geometry. The easiest two points to locate are usually the ***x*- and *y*-intercepts.** These are the points where the straight line crosses the axes. A few tests quickly show that, when a point is on the x-axis the y component is zero, and that when a point is on the y-axis the x component is zero.

We thus obtain the following procedure for locating the intercepts of any graph that is described by an equation.

> **Locating intercepts of any graph described by an equation**
> - To locate the x-intercept, set the y-variable equal to 0 and solve for x.
> - To locate the y-intercept, set the x-variable equal to 0 and solve for y.

Note An intercept is a point (an ordered pair). However, for convenience we often refer to the appropriate component as the intercept. For example, we may say that an x-intercept is $(3,0)$ or just 3.

Plotting the intercepts generally allows us to graph a straight line; however, it is a good idea to plot a third point as a check on our work.

[3]Most people, including mathematicians, avoid fractions whenever possible!

■ *Example 3–1 B*

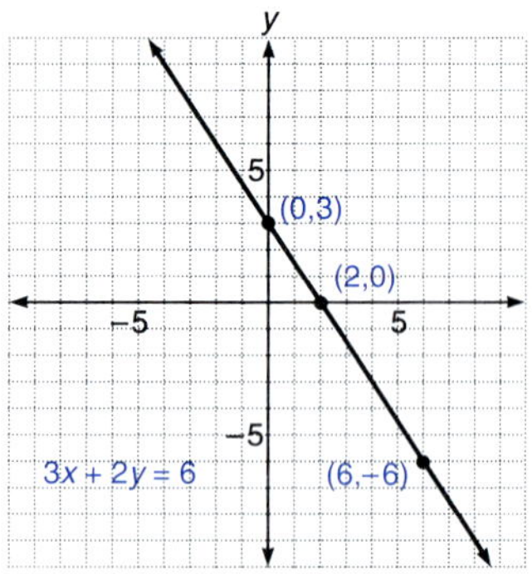

Graph the following straight lines by plotting the intercepts. Graph a third check point also.

1. $3x + 2y = 6$

Let $x = 0$: $2y = 6$

$y = 3$

giving the point (0,3); this is the y-intercept.

Let $y = 0$: $3x = 6$

$x = 2$

giving the point (2,0); this is the x-intercept.

Let $x = 6$: $18 + 2y = 6$

$2y = -12$

$y = -6$

giving the check point $(6,-6)$.

To graph the line we plot the two intercepts and draw the straight line that passes through them. This is shown in the figure.

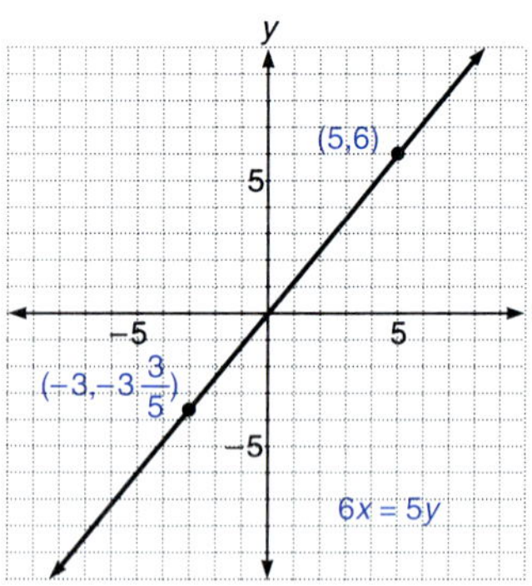

2. $6x = 5y$

Let $x = 0$: $y = 0$, giving (0,0), the origin.

We get the same result, (0,0), when we let $y = 0$. We need two points to graph the line. Thus, let x be something (anything) other than 0. Let $x = 5$.

Let $x = 5$: $30 = 5y$

$y = 6$: (5,6)

Let $x = -3$: $-18 = 5y$

$y = -3\frac{3}{5}$ $\quad (-3,-3\frac{3}{5})$ Check point

3. A graphing calculator can be used to graph a nonvertical straight line. *The equation must first be solved for y.* For example, to graph $3x + 2y = 6$ (part 1 of this example) we must first solve the equation for y: $2y = -3x + 6$

$y = -\frac{3}{2}x + 3$

It is best to set the calculator to SQUARE to see the line as we expect. The steps for graphing this problem would be

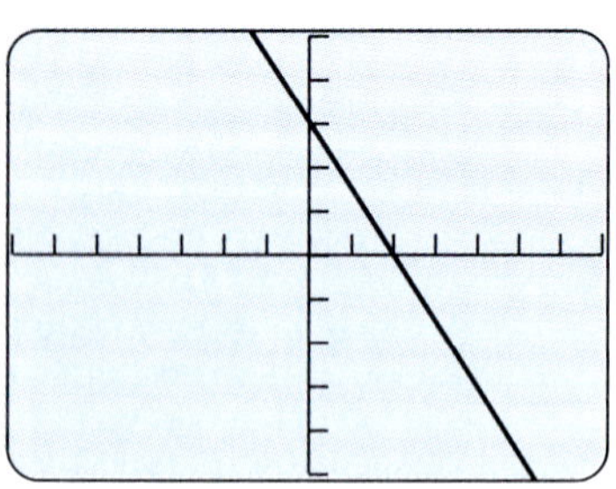

[Y=]

[CLEAR]

[(] [(−)] 3 [÷] 2 [)] [X|T] [+] 3

[ZOOM] 6 Standard RANGE settings

[ZOOM] 5 Square

[ZOOM] 2 [ENTER] Expand the display

[GRAPH] Remove cursor coordinates from screen

The trace and zoom functions can be further used to find or verify the values of the intercepts. ■

In the context of analytic geometry we consider an equation such as $y = -3$ to represent the line $0x + y = -3$. Any point for which the second (y) component is -3 will satisfy this equation, such as $(-3,-3)$, $(1,-3)$, $(10,-3)$, and so on.

■ Example 3–1 C

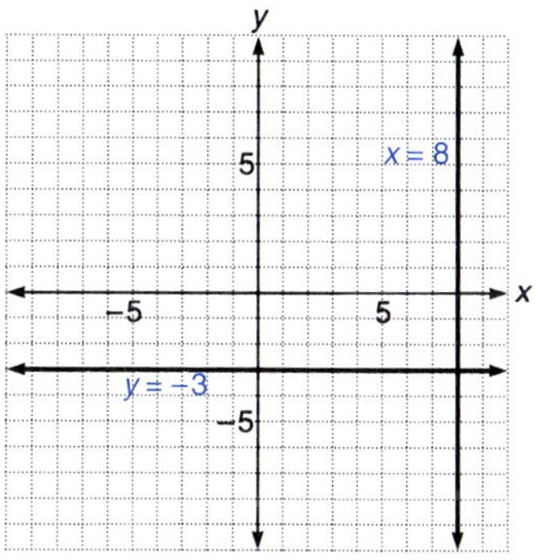

Graph the following straight lines.

1. $y = -3$
   Any point (x,y) for which the second component, y, is -3 will do. This is shown in the figure as a horizontal line.
2. $x = 8$
   Any point for which x is 8, regardless of the value of y, is on this line. This is a vertical line, as shown. ■

Based on example 3–1 C we make the following definitions.

> **Horizontal and vertical lines**
> A line of the form $x = k$ is a vertical line, and a line of the form $y = k$ is a horizontal line.

Although a line is defined as a set of points, for convenience we often speak of an equation as a line. Also, a point (ordered pair) is said to be "*on a line*" (an equation of the form $ax + by + c = 0$) if the ordered pair is a solution to an equation that defines the line.

Note that *two different equations can describe the same line.* This happens whenever the coefficients of one equation are multiples of the other. For example, the following equations all describe the same line:

$$2x - y = 3$$
$$4x - 2y = 6$$
$$-6x + 3y = -9.$$

Many situations can be modeled by using straight lines; that is, a linear equation can be used to approximate some situations in the real world. However, in these situations the values being used are often quite large or quite small. In these cases, we need to mark our vertical and horizontal axes in a scale other than one unit per mark on each axis. We also often use different scales on each axis, and we may or may not use the intercepts to draw the line. Example 3–1 D illustrates this.

Example 3–1 D

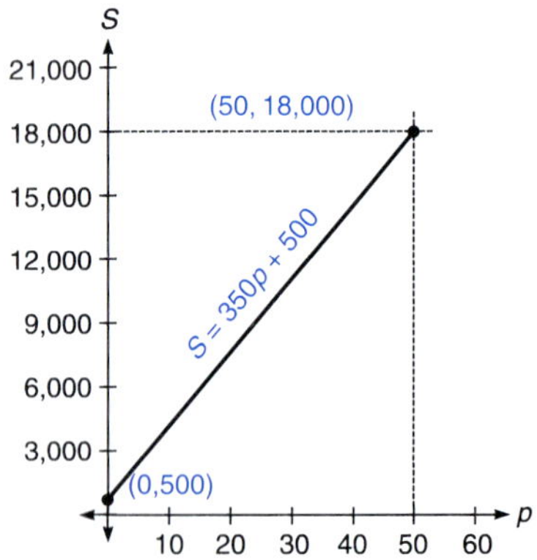

Suppose that the supply curve for a certain commodity is approximated by the equation

$$S = 350p + 500$$

where S is the number of the commodity that will be produced when the price is p dollars per unit, for values $0 \leq p \leq 50$. Graph this equation.

We use an axis system labeled p and S, where p plays the role of x, and S the role of y. Letting p be 0 and then 50, we obtain the ordered pairs (p,S) of (0,500) and (50, 18,000). For this large range of values we mark a scale on the S-axis every 3,000 units, and on the p-axis every 10 units. ■

Midpoint of a line segment

A line is imagined as having no beginning or end. A **line segment** is a portion of a line with both a beginning and end. It is useful to be able to find the midpoint of a line segment. For example, suppose a line segment has terminal points at (1,2) and (6,8) (figure 3–2). Let $P(a,b)$ be the midpoint of this segment. What are its coordinates?

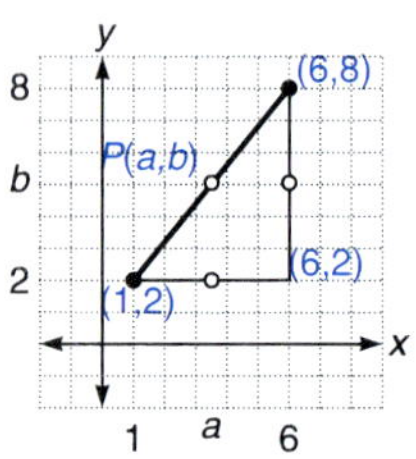

Figure 3–2

In figure 3–2 we can see that a is halfway between 1 and 6, and b is halfway between 2 and 8. The value halfway between two other values is their **average,** which is half of their sum. Thus, $a = \frac{1 + 6}{2} = 3\frac{1}{2}$, and b is $\frac{2 + 8}{2} = 5$. Thus the point P is $(3\frac{1}{2},5)$. This example leads to the following definition.

> **Midpoint of a line segment**
> If $P_1(x_1,y_1)$ and $P_2(x_2,y_2)$ are the end points of a line segment, then M, the midpoint of the line segment is $\left(\frac{x_1 + x_2}{2},\frac{y_1 + y_2}{2}\right)$.

This definition simply states that the x-coordinate of the midpoint is the average of the two given x-coordinates, and the y-coordinate is the average of the two y-coordinates.

Example 3–1 E

Find the midpoint of the line segment with end points $(-3,8)$ and $(4,-2)$.

Choose $(x_1,y_1) = (-3,8)$ and $(x_2,y_2) = (4,-2)$. Thus $x_1 = -3$ and $x_2 = 4$, $y_1 = 8$ and $y_2 = -2$. Using substituton of value (section 1–2) we proceed:

$$M = \left(\frac{x_1 + x_2}{2},\frac{y_1 + y_2}{2}\right) = \left(\frac{-3 + 4}{2},\frac{8 + (-2)}{2}\right) = \left(\frac{1}{2},3\right)$$

Note We could have chosen $(x_1,y_1) = (4,-2)$ and $(x_2,y_2) = (-3,8)$. The result would be the same. ■

Distance between two points

Another important definition in analytical geometry is distance.

> **Distance between two points**
> If $P_1(x_1,y_1)$ and $P_2(x_2,y_2)$ are two different points, the distance between them is called $d(P_1,P_2)$ and is defined as
> $$d(P_1,P_2) = \sqrt{(x_2 - x_1)^2 + (y_2 - y_1)^2}$$

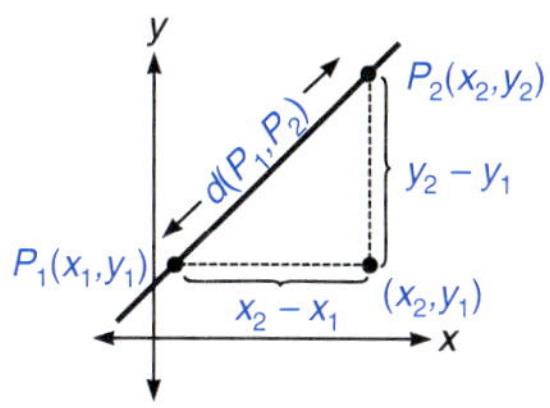

Figure 3–3

This definition is motivated by the Pythagorean theorem (section 2–2) and figure 3–3. Here $d(P_1,P_2)$ represents the distance from point P_1 to P_2. The vertical distance $|y_2 - y_1|$ is the length of one leg of a right triangle, and $|x_2 - x_1|$ is the length of the second leg.[4] According to the Pythagorean theorem,

$$d(P_1,P_2)^2 = |x_2 - x_1|^2 + |y_2 - y_1|^2, \text{ so}$$
$$d(P_1,P_2) = \sqrt{(x_2 - x_1)^2 + (y_2 - y_1)^2}$$

Note that we drop the absolute value symbol since $|a|^2 = (a)^2$; also, $d(P_1,P_2)$ is defined to be nonnegative. To use this formula with known values we use the substitution of value method of section 1–2.

■ Example 3–1 F

1. Find the distance between the two points $(-3,5)$ and $(4,-1)$.

Assume $P_1 = (-3,5)$ and $P_2 = (4,-1)$. Then,

$$\begin{aligned} d(P_1,P_2) &= \sqrt{(x_2 - x_1)^2 + (y_2 - y_1)^2} \\ &= \sqrt{(4 - (-3))^2 + (-1 - 5)^2} \\ &= \sqrt{7^2 + (-6)^2} = \sqrt{49 + 36} = \sqrt{85} \end{aligned}$$

Note The same result will be obtained if we assume $P_1 = (4,-1)$ and $P_2 = (-3,5)$.

2. Describe the set of all points (x,y) that are equidistant from the points $(-2,1)$ and $(4,-1)$.

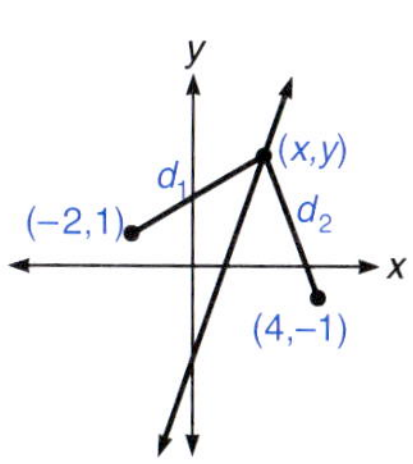

Plotting some points that are equal distances from these two points will show that all these points lie on a straight line as shown in the figure. We need to find its equation. Let (x,y) represent any point that is equidistant from the given points. We know that the two distances d_1 and d_2 are equal. Thus, we proceed as follows.

$$d_1 = \sqrt{(x + 2)^2 + (y - 1)^2}$$ Apply the distance formula using $(-2,1)$ and (x,y)

$$d_2 = \sqrt{(x - 4)^2 + (y + 1)^2}$$ Apply the distance formula using $(4,-1)$ and (x,y)

$$\sqrt{(x + 2)^2 + (y - 1)^2} = \sqrt{(x - 4)^2 + (y + 1)^2}$$

[4]Either $x_2 - x_1$ or $y_2 - y_1$ may be negative; this is not important because these quantities are squared in the distance formula.

Replace d_1 and d_2 by the values shown above. This is the substitution for expression procedure (section 1–3).

$$(x + 2)^2 + (y - 1)^2 = (x - 4)^2 + (y + 1)^2$$

Square both members of the equation

$$x^2 + 4x + 4 + y^2 - 2y + 1 = x^2 - 8x + 16 + y^2 + 2y + 1$$
$$3x - y - 3 = 0$$

Thus the required equation is the straight line $3x - y - 3 = 0$. ■

Mastery points

Can you

- Define relation?
- Recognize ordered pairs and determine when ordered pairs are equal?
- Find solutions to equations that determine straight lines?
- Graph relations that are straight lines?
- Find the midpoint of line segments?
- Find the distance between two points?

Exercise 3–1

Find three points that lie on the following straight lines; that is, find three solutions to the following straight lines.

1. $y = 3x - 8$ **2.** $2x - 5y = 10$ **3.** $3x - 4y = 12$ **4.** $2x - 3 = y$

5. $x = y + 2$ **6.** $2y - x = 19$ **7.** $y = 3x - 2$ **8.** $5x - 2y + 4 = 0$

9. $x = y$ **10.** $x = 5$ 11. $7 - y = 0$ **12.** $x - 5 + y = 0$

13. $\frac{1}{2}x - \frac{1}{3}y = 1$ **14.** $\frac{2x - 1}{4} - y = 0$ **15.** $0.2x - 0.3y = 1$ **16.** $y - 0.4x = 0.2$

Graph each line by plotting the intercepts and one check point.

17. $y = 3x - 8$ **18.** $2x - 5y = 10$ **19.** $3x - 4y = 12$ **20.** $2x - 3 = y$

21. $x = y + 2$ **22.** $2y - x = 19$ **23.** $y = 3x - 2$ **24.** $5x - 2y + 4 = 0$

25. $x = y$ **26.** $x = 5$ 27. $7 - y = 0$ **28.** $x - y + 5 = 0$

29. $\frac{1}{2}x - \frac{1}{3}y = 1$ **30.** $\frac{2x - 1}{4} - y = 0$ **31.** $0.2x - 0.3y = 1$ **32.** $y - 0.4x = 0.2$

33. $\sqrt{3}x - \sqrt{2}y = \sqrt{6}$ **34.** $2\sqrt{5} = x$ **35.** $0.5x - 0.25y = 2$ **36.** $3x - 0.5 = 0.5y$

Solve the following problems.

37. An investment account pays 15% interest on money invested, but deducts \$50 per year service charge. The amount of interest I that will be paid on an amount of money p if that amount does not change over the year is $I = 0.15p - 50$. Graph this equation for values of p from 0 to \$10,000. Use a scale of \$1,000 on the p-axis and \$200 on the I-axis.

38. A savings account pays 10% interest on money invested, but deducts \$20 per year service charge. The amount of interest I that will be paid on an amount of money p if that amount does not change over the year is $I = 0.10p - 20$. Graph this equation for values of p from 0 to \$10,000.

39. The perimeter P of a rectangle is defined by $P = 2\ell + 2w$, where ℓ and w are the length and width of the rectangle. If we solve this for w we obtain the equation $w = -\ell + \frac{P}{2}$. Graph this equation for $P = 20$, $P = 50$, $P = 100$; put all three graphs on the same set of axes for comparison. Use a scale of 10 units on both axes.

40. An aircraft flying with a ground speed of 250 mph is 50 miles from its starting point. From this point in time on, the distance from the starting point is given by $d = 250t + 50$, where t is the time in hours. Graph for $0 \leq t \leq 6$.

41. An automobile moving at 45 mph still has 500 miles to go to get to its destination. The distance from the destination is given by $d = -45t + 500$, where t is the time in hours. Graph this equation for $t \geq 0$ up to the point where the auto reaches its destination.

42. A certain production worker is paid \$50 per week plus \$0.35 for each item produced. If n is the number of items produced, then gross pay p for the worker is given by the equation $p = 0.35n + 50$. Graph this for $0 \leq n \leq 1{,}500$.

43. A person who waits on tables is paid \$50 per week and expects an average tip of \$5 per table waited on. If t is the number of tables serviced, then gross pay p for the worker is given by the equation $p = 5t + 50$. Graph this equation for $0 \leq t \leq 50$.

44. The amount h in feet that a railroad track with incline 0.05 would rise over a distance of d feet might be expressed by $h = 0.05d$. Graph this equation for $0 \leq d \leq 2{,}000$.

Find the coordinates of the midpoint of the line segment determined by the following points.

45. $(1,5), (-3,9)$ **46.** $(-5,2), (6,8)$ **47.** $(1,5), (7,5)$ **48.** $(-4,8), (-4,-20)$
49. $(\frac{1}{2},-4), (3\frac{1}{2},1)$ **50.** $(2,-\frac{1}{3}), (-2,\frac{2}{3})$ **51.** $(\frac{2}{3},3), (4\frac{1}{2},\frac{1}{3})$ **52.** $(-\frac{4}{5},-2), (\frac{9}{10},-1\frac{4}{5})$
53. $(-3,4), (-2,8)$ **54.** $(\sqrt{2},5), (\sqrt{8},9)$ **55.** $(-3,\sqrt{50}), (5,\sqrt{8})$ **56.** $(5m,-3n), (2m,n)$

Find the distance between the following sets of two points.

57. $(-3,4), (2,-1)$ **58.** $(5,2), (-3,-4)$ **59.** $(0,3), (2,-3)$ **60.** $(-2,-1), (3,-1)$
61. $(8,-3), (8,-4)$ **62.** $(7,1), (5,5)$ **63.** $(-3,6), (3,-6)$ **64.** $(0,0), (3,4)$
65. $(0,3), (3,0)$ **66.** $(-10,2), (-10,2)$ **67.** $(\frac{1}{2},3), (2,5)$ **68.** $(-2,\frac{1}{3}), (3,\frac{4}{3})$
69. $(3,\frac{1}{5}), (-1,\frac{3}{5})$ **70.** $(\frac{3}{4},1), (-2,\frac{1}{4})$ **71.** $(-3,8), (3,-8)$ **72.** $(3m,n), (m,-n)$
73. $(2a,-b), (-a,5b)$ **74.** $(\sqrt{2},-3), (\sqrt{8},1)$ **75.** $(4,\sqrt{75}), (2,\sqrt{27})$ **76.** $(-\sqrt{20},2), (\sqrt{45},-2)$

Find the equation that describes all the points that are equal distances from the given points.

77. $(1,2)$ and $(9,8)$ **78.** $(-3,-1)$ and $(4,5)$ **79.** $(-4,1)$ and $(-1,-6)$

80. The ordered pair $(a + 3, b - 12)$ is the same as the ordered pair $(3,-5)$. Find a and b.

81. The ordered pair $(3a - 2, 5b + 7 - a)$ is the same as the ordered pair $(3,-6)$. Find a and b.

82. The ordered pair $(x, x^2 - y)$ is the same as the ordered pair $(2,12)$. Find x and y.

83. Heron's formula for finding the area of a triangle with sides of length a, b, and c is

$$A = \sqrt{s(s - a)(s - b)(s - c)},$$

where $s = \frac{a + b + c}{2}$ is half of the perimeter. Use this formula, along with the definition of distance, to find the area of the triangle with vertices at $(1,2)$, $(1,-2)$, and $(4,-2)$.

84. Use Heron's formula (problem 83) to find the area of the triangle with vertices at $(1,1)$, $(1,-7)$, and $(7,-7)$.

85. "Taxicab geometry" describes a situation in which distance can only be measured parallel to the x- and y-axis, and in which the only points are those with integer coordinates. It is like a taxicab that, to go from one point in a city to another, must stay in the streets. We can define distance to be the length of the *shortest* path from one point to another. Three paths are shown in the figure for measuring the distance between (0,0) and (5,3). Two give the distance 8, and one the distance 14. The distance, using our definition, is 8. (a) Give a definition for computing distance between two points $P_1 = (x_1,y_1)$ and $P_2 = (x_2,y_2)$ using this definition of distance; that is, a formula involving x_1, x_2, y_1, and y_2. (b) What can be said about "taxicab" distance versus the "straight-line" distance we defined in this section?

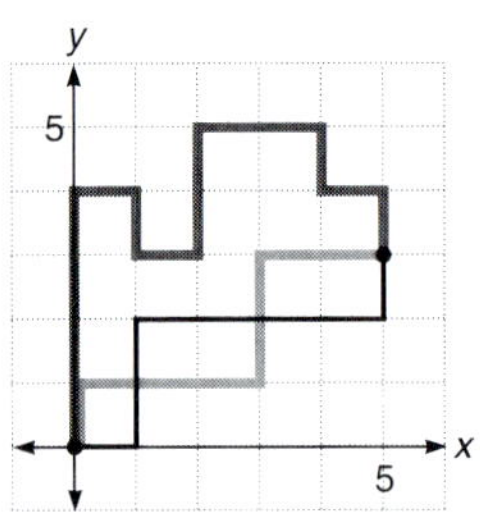

86. In geometry, the distance from the midpoint of a line segment to either end point is the same. We want our analytic definitions of midpoint and distance to reflect this geometric property. Let $P_1(x_1,y_1)$ and $P_2(x_2,y_2)$ represent the end points of an arbitrary line segment. Compute the midpoint, and show that the distance from this point to each end point is equal.

87. A triangle has vertices at (0,0), (6,5), and $(a,0)$, as shown in the figure. The horizontal line shown parallel to the x-axis is above the x-axis by half the height of the triangle. This means that it bisects (meets the midpoint of) both sides of the triangle. Find the value of a.

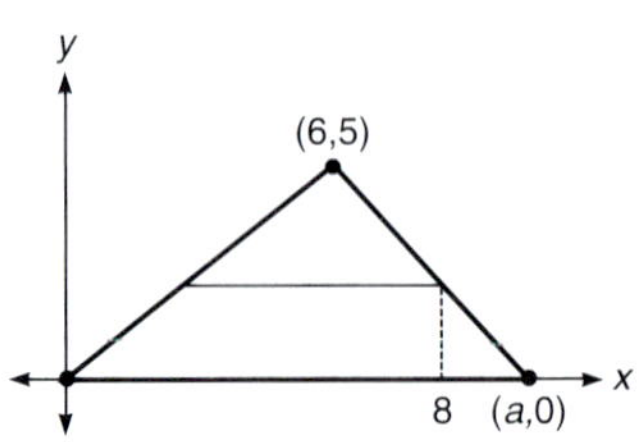

88. The point (3,0) is the midpoint of the line segment from (0,0) to (a,b) in the figure. Also, (5,2) is the midpoint of the line segment from (3,0) to (c,d), and the line passing through (5,2) and (e,f) is parallel to the x-axis, and therefore bisects the line segment from (a,b) to (c,d). Find the values of a, b, c, d, e, and f.

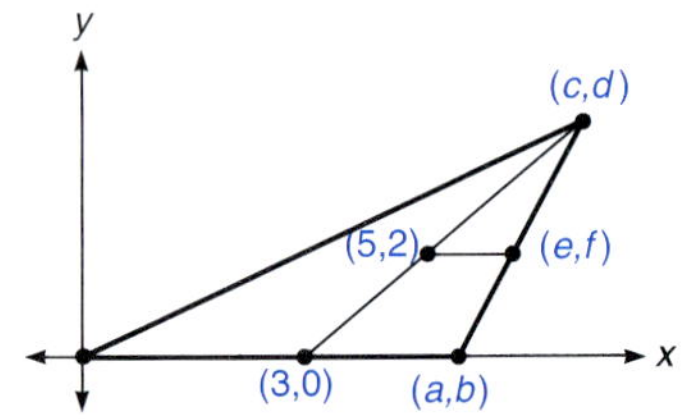

89. Show that if the coefficients of one linear equation are multiples of the coefficients of another linear equation, then the two equations describe the same line. Do this by assuming two lines, $a_1x + b_1y + c_1 = 0$ and $a_2x + b_2y + c_2 = 0$, and a value $k \neq 0$ such that $a_2 = ka_1$, $b_2 = kb_1$, and $c_2 = kc_1$.

90. If the coordinates of the four vertices of a quadrilateral (four-sided figure) are (x_1,y_1), (x_2,y_2), (x_3,y_3), and (x_4,y_4), then the area of the quadrilateral is the absolute value of $\frac{1}{2}\{(x_1y_2 - x_2y_1) + (x_2y_3 - x_3y_2) + (x_3y_4 - x_4y_3) + (x_4y_1 - x_1y_4)\}$. Find the area of the quadrilateral whose vertices are $(-3,5)$, $(1,4)$, $(4,-6)$, and $(-3,-3)$.

91. If the coordinates of the vertices of a five-sided figure are (x_1,y_1), (x_2,y_2), (x_3,y_3), (x_4,y_4), and (x_5,y_5), then its area is the absolute value of

$$\frac{1}{2}\{(x_1y_2 - x_2y_1) + (x_2y_3 - x_3y_2) + (x_3y_4 - x_4y_3) + (x_4y_5 - x_5y_4) + (x_5y_1 - x_1y_5)\}.$$

Find the area of a five-sided figure whose vertices are $(2,4)$, $(5,2)$, $(5,-3)$, $(3,-4)$, and $(-4,-5)$.

Skill and review

1. Compute $\frac{a-b}{c-d}$ if $a = 9$, $b = -3$, $c = -5$, and $d = -1$.
2. Solve $3y - 2x = 5$ for y.
3. Solve $ax + by + c = 0$ for y.
4. If $y = 3x - b$ contains the point $(-2,4)$, find b.
5. Graph the following lines using the same coordinate system:
   a. $y = 2x - 2$ **b.** $y = 2x$ **c.** $y = 2x + 2$
6. Solve the equation $2x^2 - x = 3$.
7. Solve the equation $|2x - 3| = 5$.
8. Simplify $\sqrt{48x^4}$.
9. Calculate $\frac{5}{8} - \frac{1}{4} + \frac{2}{3}$.

3–2 Equations of straight lines

At 8:00 A.M. the outdoor temperature at a certain location was 28° F. At noon the temperature was 59°. Estimate what the temperature was at 10:30 A.M. Estimate at what time the temperature was 40°.

Straight line modeling can be used to deal with problems like this one. In this section we discuss the properties of straight lines that are important for these and other problems.

Slope

Any nonvertical line is said to have a **slope.** For a given line, the slope is defined using two points taken from that line; slope is designed to correspond to the common notion of steepness.

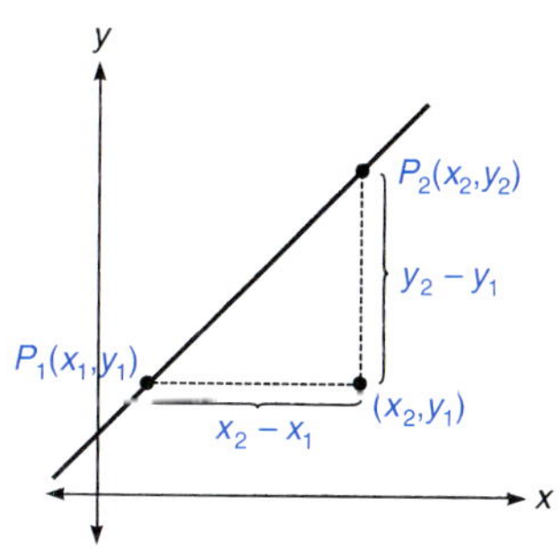

Figure 3–4

Slope

If $P_1 = (x_1,y_1)$ and $P_2 = (x_2,y_2)$ are any two different points on a nonvertical line, then the slope m of the line is given by

$$m = \frac{y_2 - y_1}{x_2 - x_1}$$

Note Slope is not defined for vertical lines because in this case $x_1 = x_2$, so $x_2 - x_1 = 0$, and the quotient is not defined.

A geometric interpretation of slope is that it is a ratio of the vertical distance between two points to the horizontal distance between the same two points. This is shown in figure 3–4.

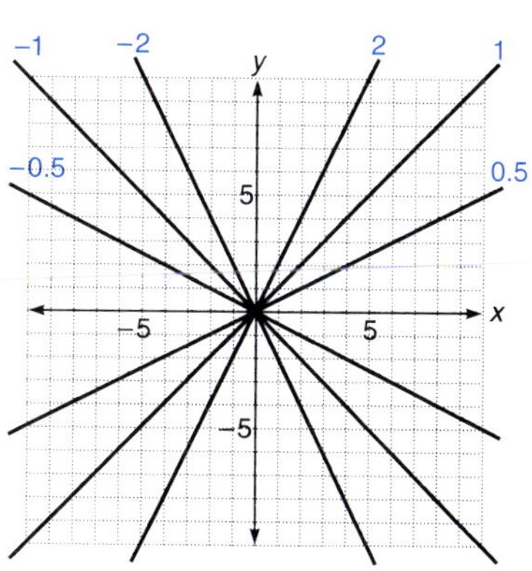

Figure 3–5

Figure 3–5 shows the value of the slope m for some representative lines. The definition of slope has the following properties:

- "Uphill" lines have positive slopes, and "downhill" lines have negative slopes.
- The steeper the line the greater is the absolute value of the slope.
- A horizontal line has slope 0.
- Slope is not defined for vertical lines.

Although the slope of a line is defined in terms of points on the line we will relate the slope of a line to its equation later in this section.

To use the slope formula we use the substitution of value method (section 1–2).

■ *Example 3–2 A*

Use the definition of slope to find the slope of the line that contains the points $(5,-2)$ and $(3,6)$.

Let $P_1 = (x_1,y_1) = (5,-2)$ and $P_2 = (x_2,y_2) = (3,6)$. Thus, $x_1 = 5$, $x_2 = 3$, $y_1 = -2$, $y_2 = 6$. Then $m = \dfrac{y_2 - y_1}{x_2 - x_1} = \dfrac{6 - (-2)}{3 - 5} = -4$. ■

Most scientific calculators can find the slope of a line from two points that lie on the line. See example 3–2 D, where we illustrate finding both the slope and y-intercept from two points.

The slope of a line is independent of the choice of the two points that are used to determine its value—that is, the value of m is always the same for a given line, regardless of the choice of points. This is shown in example 3–2 B.

■ *Example 3–2 B*

Let (a,b) and (c,d) be two different points on a nonvertical line; show that the choice of which point is designated as P_1 and which is designated as P_2 is not important when calculating the value of m.

Let $P_1 = (a,b)$ and $P_2 = (c,d)$. Then $m_1 = \dfrac{y_2 - y_1}{x_2 - x_1} = \dfrac{d - b}{c - a}$.

Now let $P_1 = (c,d)$ and $P_2 = (a,b)$. Then $m_2 = \dfrac{y_2 - y_1}{x_2 - x_1} = \dfrac{b - d}{a - c}$.

However,
$$m_1 = \frac{d - b}{c - a} = \frac{-(b - d)}{-(a - c)} = \frac{b - d}{a - c} = m_2.$$

Thus we arrive at the same value for m in both choices of P_1 and P_2. ■

The slope-intercept form of the equation of a line

As we have seen, the slope of a (nonvertical) line can be found using any two points that lie on the line. However, if we know the equation of a line, we do not have to use this process; we can find the slope by transforming the equation itself.

In particular, if we take a nonvertical straight line $Ax + By + C = 0$, and solve this for y, we obtain the equivalent equation $y = -\dfrac{A}{B}x - \dfrac{C}{B}$. Note that division by B is defined because if the line is not a vertical line then $B \neq 0$. It is customary to replace the value $-\dfrac{A}{B}$ by m, and $-\dfrac{C}{B}$ by b, writing the equation as $y = mx + b$. It can be proven that the value of m is the slope of the line (see the exercises). By letting $x = 0$, we see that the y-intercept is $(0,b)$. Because m is the slope and b is the second element of the y-intercept, this form of the equation is customarily called the **slope-intercept** form of the equation of a straight line. The most important use of this form is in finding the slope of a line when given its equation.

> **Slope-intercept form of a straight line: $y = mx + b$**
> If the equation of a nonvertical straight line is put in the form
> $$y = mx + b$$
> then m is the slope of the line, and b is the y-intercept.
>
> ***Concept***
> To find the slope of a line when given its equation, solve the equation for y. The slope is then the coefficient of x, and the constant term is the y-intercept.

Note The slope-intercept form of a straight line is the most practical form for graphing a line using a graphing calculator or computer. This is because graphing calculators require equations in the form "Y = ", followed by an appropriate expression in the variable x. This was illustrated in example 3–1 B.

Example 3–2 C

Find the slope m of the line $4x - 2y = 6$.

$-2y = -4x + 6$	Add $-4x$ to both members
$y = 2x - 3$	Divide each member by -2
$m = 2$	Slope is the coefficient of x

Finding the equation of a line that contains two points

An important type of problem is finding the equation of a line that contains two given points. The solution to this problem is contained in the following property.

> **Point-slope formula**
> If $P_1 = (x_1,y_1)$ and $P_2 = (x_2,y_2)$ are two different points and $x_1 \neq x_2$, then the equation of the line that contains these points is obtained by the formula
> $$y - y_1 = m(x - x_1)$$
> where m is the slope determined by P_1 and P_2.

Note The same results are obtained from $y - y_2 = m(x - x_2)$.

To see why the point-slope formula is valid, let (x_1,y_1) and (x_2,y_2) be two points such that $x_1 \neq x_2$. Consider the line that contains the points (x_1,y_1) and (x_2,y_2) and let (x,y) be any other point on that line. Since the slope m of a line is the same no matter which two points are chosen, we can calculate it using the points (x_1,y_1) and (x,y): $m = \dfrac{y - y_1}{x - x_1}$, and multiplying both sides by the denominator $x - x_1$ we obtain $y - y_1 = m(x - x_1)$.

This point-slope formula is useful for finding the equation of a line when we know either (1) two points or (2) the slope and one point.

■ Example 3–2 D

Find an equation of the line in each case. Leave the equation in slope-intercept form.

1. A line contains the points $(-1,2)$ and $(4,-5)$. Find the equation of the line.

Let $P_1 = (-1,2)$ and $P_2 = (4,-5)$.

$m = \dfrac{-5-2}{4-(-1)} = -\dfrac{7}{5}$	Use the definition of m with P_1 and P_2
$y - y_1 = m(x - x_1)$	Point-slope formula
$y - 2 = -\frac{7}{5}(x - (-1))$	Substitute the values of x_1, y_1, and m
$y - 2 = -\frac{7}{5}(x + 1)$	$x - (-1) = x + 1$
$5y - 10 = -7(x + 1)$	Multiply each member by 5
$5y - 10 = -7x - 7$	Distribute the -7
$5y = -7x + 3$	Solve for y
$y = -\frac{7}{5}x + \frac{3}{5}$	Slope-intercept form

2. A line has slope -3 and x-intercept at 4. Find its equation.

The x-intercept is $(4,0)$; this can serve as P_1. The slope is $m = -3$.

$y - y_1 = m(x - x_1)$	Point-slope formula
$y - 0 = -3(x - 4)$	Substitute the values of x_1, y_1, and m
$y = -3x + 12$	Slope-intercept form

Most scientific calculators can find the equation of a straight line that passes through two points. This process is called linear regression, and, when restricted to two points, it achieves the same results we just obtained. Part 1 is illustrated for two calculators. Note that on some calculators the slope-intercept form of the equation is described as $y = A + Bx$.

3. Use a calculator to find the slope and y-intercept of the line that passes through the two points $(-1,2)$, $(4,-5)$.

CASIO fx-115N

[MODE] 2	Linear regression mode
[SHIFT] [AC]	Clear constants
1 [+/−] [x_D,y_D] 2 [DATA]	Use [[(---] for [x_D,y_D]
4 [x_D,y_D] 5 [+/−] [DATA]	Use [M+] for [DATA]
[SHIFT] 8	m −1.4 (called B on this calculator)
[SHIFT] 7	b 0.6 (called A on this calculator)

TI-81

[2nd] [MATRX] ▶ ▶ 2 [ENTER]	ClrStat (clear any old data)
[2nd] [MATRX] ▶ ▶ [ENTER]	Edit data
[(−)] 1 [ENTER] 2 [ENTER] 4 [ENTER] [(−)] 5	
[2nd] [MATRX] 2 [ENTER]	LinReg

m is -1.4 (called b on this calculator)
b is 0.6 (called a on this calculator)

A value $r = -1$ is also displayed. It is called the regression coefficient and will always be one in absolute value when two points are used to find the equation of a straight line.

The resulting line is $y = -1.4x + 0.6$.

Note To graph this line on the TI-81 use the LR (linear regression) feature:

[Y=] [CLEAR] [VARS] ► ► 4 [GRAPH] ■

Linear interpolation

One important use for finding the equation of a straight line is called linear interpolation. This is a method for estimating unknown values from known values. To use linear interpolation we find the equation of the straight line that passes through (interpolates) two known pairs of data values. We then use this equation to find an unknown member of a third pair of data values.

We will illustrate linear interpolation using table 3–2, which represents the *wind chill factor,* often mentioned in media weather forecasts. For example, the table shows that if the temperature is 25° F and the wind is blowing at 15 miles per hour, then the wind chill factor is 1° F. That is, to exposed human skin the cooling rate is equivalent to a temperature of 1° F under no wind conditions. (The entries that are darkened in are used in example 3–2 E.)

Wind chill factor

Wind (miles per hour)	Degrees Fahrenheit										
	30°	**25°**	**20°**	15°	10°	5°	0°	−5°	−10°	−15°	−20°
5	27°	21°	16°	12°	7°	1°	−6°	−11°	−15°	−20°	−26°
10	16°	8°	2°	−2°	−9°	−15°	**−22°**	−27°	−34°	−40°	−45°
15	9°	1°	−6°	−11°	−18°	−25°	**−31°**	−38°	−45°	−51°	−58°
20	3°	**−4°**	**−9°**	−17°	−24°	−32°	−40°	−46°	−52°	−60°	−68°
25	0°	−7°	−15°	−22°	−29°	−37°	−45°	−52°	−58°	−67°	−75°
30	−2°	−11°	−18°	−26°	−33°	−41°	−49°	−56°	−63°	−70°	−78°

Table 3–2

Observe that the table would not tell us what to expect at a temperature of, say, 22° F, or at a wind speed of 18.5 mph. The following example illustrates how to estimate values that are not in the table by first creating a straight line connecting the closest known values. This is called linear interpolation.

■ Example 3–2 E

Use table 3–2 and linear interpolation to estimate the wind chill factor (wcf) for a temperature of 22° F at a wind speed of 20 mph.

At 20 mph we have the following (temperature, wcf) ordered pairs: (25, −4) and (20, −9). We want the y value for the ordered pair $(22,y)$. First find the equation of the straight line that passes through (interpolates) the two known points:

$$m = \frac{-4-(-9)}{25-20} = 1$$

$$y = mx + b$$

$$-4 = 1(25) + b, \text{ so } b = -29$$

Thus the equation of the line is $y = x - 29$.

Now find the y associated with $x = 22$: $y = 22 - 29 = -7$. Thus the wind chill factor is −7° F. ■

As shown earlier graphing calculators can find the equation of a straight line when given two points that lie on the line. This also provides a way to solve the problems of example 3–2 E. Example 3–2 F shows how to do example 3–2 E with the same two calculators shown earlier.

■ Example 3–2 F

Estimate the wind chill factor (wcf) for a temperature of 22° F at a wind speed of 20 mph.

As in example 3–2 E, at 20 mph we have the following (temperature,wcf) ordered pairs: (25, −4) and (20, −9). We want the y value for the ordered pair $(22,y)$. (We know from example 3–2 E that y is −7.) We first find the values of m and b, (i.e., the interpolating linear equation) as in example 3–2 D.

[*CASIO fx-115n*]

[MODE] 2 [SHIFT] [AC]

25 [x_D,y_D] 4 [+/−] [DATA] 20 [x_D,y_D] 9 [+/−] [DATA]

22 [$\hat{y}$] −7 — [$\hat{y}$] is under the [—)]] key

[*TI-81*]

[2nd] [MATRX] ▶ ▶ 2 [ENTER] — ClrStat (clear any old data)

[STAT] — [2nd] [MATRX]

Select [DATA] with the right arrow key, then [ENTER].
$x1=25$, $y1=-4$, $x2=20$, $y2=-9$

[STAT] 2 [ENTER] — Select LinReg

[Y=] [CLEAR] [VARS]

Select [LR] with the right arrow key. LR means linear regression.
4 — Regular equation

The equation is stored in Y_1. Do the following:

[QUIT] [2nd] [CLEAR]

22 [STO▸] [X|T] [ENTER]

[Y-VARS] 1 [ENTER] [2nd] [VARS]

Thus, when x is 22, y is -7. ■

Parallel and perpendicular lines, and the intersection of lines

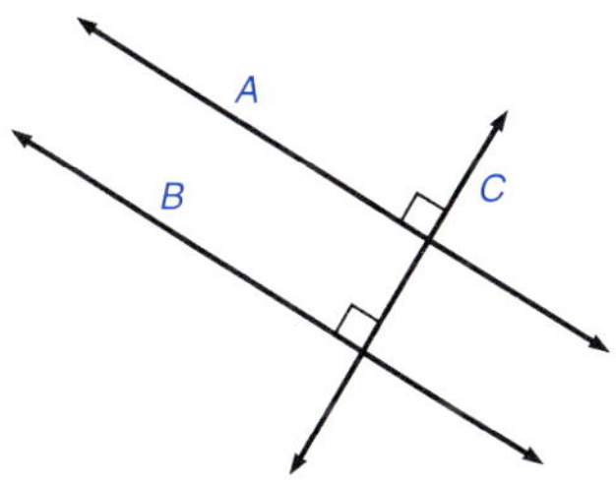

Figure 3–6

Other concepts of geometry that are important are the ideas of parallel lines, perpendicular lines, and the intersection of lines. Recall from geometry that parallel lines in a plane have no points in common. See lines A and B in figure 3–6. Perpendicular lines intersect at one point and form a right angle. In figure 3–6, line C is perpendicular to both A and B. We define these concepts for the lines of analytic geometry as follows.

Parallel lines

If two different lines have the same slope they are said to be parallel. Also, vertical lines are parallel.

Perpendicular lines

If the product of the slopes of two lines is -1 the lines are said to be perpendicular. Also, vertical and horizontal lines are perpendicular.

Intersection of lines

Two different lines are said to intersect if they have one point in common.

These definitions are made to give to the lines of analytic geometry the same properties we expect of parallel and perpendicular lines in geometry. For example, it can be proven that two parallel lines of analytic geometry do not intersect; this is left as an exercise.

Parallel and perpendicular lines

It is worth noting that if the product of two values is -1 then one value is formed from the other by changing its sign and inverting it; the result is called the **negative reciprocal** of the first value. This is seen generally as $\frac{a}{b}\left(-\frac{b}{a}\right) = -1$. Thus, *to find the slope of a line that is perpendicular to a line with a known slope, compute the negative reciprocal of the known slope.* For example, if the known slope is 2, we invert, obtaining $\frac{1}{2}$, and change the sign, obtaining $-\frac{1}{2}$. The values 2 and $-\frac{1}{2}$ are negative reciprocals of each other, and lines having slopes of 2 and $-\frac{1}{2}$ are perpendicular.

■ ***Example 3–2 G***

1. Find the slope-intercept equation of a line that has y-intercept -4 and is parallel to the line $5x - 2y = 3$.

The line we want has y-intercept $(0,-4)$. This can be P_1. To use the point-slope formula we still need the slope m.

$5x - 2y = 3$	We want a line parallel to this line
$y = \frac{5}{2}x - \frac{3}{2}$	Solve for y, to find m
$m = \frac{5}{2}$	m is the coefficient of x

The slope of the line we are looking for must also have slope $\frac{5}{2}$, since the lines are to be parallel. Thus we want a line that passes through the point $P_1 = (0,-4)$ with slope $m = \frac{5}{2}$.

$y - (-4) = \frac{5}{2}(x - 0)$	Substitute into the point-slope formula: $y - y_1 = m(x - x_1)$
$y + 4 = \frac{5}{2}x$	$x - 0 = x$
$y = \frac{5}{2}x - 4$	

2. Find a line that is perpendicular to the line $y = \frac{2}{3}x$ and passes through the point $(-2,4)$.

We will use the point-slope formula with $P_1 = (-2,4)$; we still need the value of m.

$y = \frac{2}{3}x$	We want a line perpendicular to this line, which has slope $\frac{2}{3}$
$m = -\frac{3}{2}$	The negative reciprocal of $\frac{2}{3}$ is $-\frac{3}{2}$

Thus, we want a line with slope $m = -\frac{3}{2}$ and passing through the point $P_1 = (-2,4)$:

$y - 4 = -\frac{3}{2}(x - (-2))$	Substitute into the point-slope formula: $y - y_1 = m(x - x_1)$
$y - 4 = -\frac{3}{2}(x + 2)$	$x - (-2) = x + 2$
$y - 4 = -\frac{3}{2}x - 3$	
$y = -\frac{3}{2}x + 1$	■

Verifying solutions by graphing

One can visually verify that two straight lines are parallel or perpendicular by sketching their graphs. If using a graphing calculator (or computer program), one must be careful. In the case of parallel lines there is no problem. However, when verifying that two lines are perpendicular we must make sure that the scale on the x- and y-axis is the same and that any "scaling factors" for the axes are the same. For example, on the TI-81, " [ZOOM] 5:Square" must be selected.

Point of intersection

If two different lines are not parallel then they intersect at some point. There are many ways to find where two lines intersect; these are treated more extensively later in the chapter "Systems of Linear Equations and Inequalities." Here we use the method of substitution for expression, introduced in section 1–3.

A collection of two equations in two variables is called a **system of two equations in two variables.** An example is the system

$$3x - y = 5$$
$$2x + 3y - 6 = 0$$

To solve such a system means to find a set of ordered pairs that satisfies both equations.

When the equations in a system of two equations represent two different nonparallel straight lines, the solution is the one point where the two lines intersect.

In the following description of using the method of substitution for expression we assume the equations are in terms of the variables x and y, but the method applies regardless of the names of the variables.

> Using the method of substitution for expression to solve a system of two equations in two variables
>
> 1. **Solve one equation for *y*.** The other member is an expression involving constants and x.
> 2. **Substitute for *y* in the other equation.** Use the expression involving constants and x found in step 1.
> 3. **Solve this new equation for *x*.**
> 4. **Substitute the known value of *x* in either of the original equations to find *y*.**

Note a. This method works equally well by first solving one equation for x, and replacing x in the second equation.
b. This method does not apply to parallel lines, since they do not intersect.

■ ***Example 3–2 H***

Use the method of substitution for expression to solve each problem.

1. Find the point at which the lines $3x - y = 5$ and $2x + 3y - 6 = 0$ intersect. Graph both lines to verify the solution.

We apply the method of substitution to the system of two equations in two variables.

Step 1: Solve one equation for y.

$$3x - y = 5$$
$$3x - 5 = y$$

This states that in this equation every y is equivalent to $3x - 5$. At the point where the lines intersect, the y-value is the same for both equations. Therefore we can replace y in the *second* equation by $3x - 5$.

Step 2: In the other equation, substitute the expression $3x - 5$ for y.

$2x + 3y - 6 = 0$ — The equation of the second line

$2x + 3(3x - 5) - 6 = 0$ — Replace y by $3x - 5$

Step 3: Solve for x.

$$11x = 21$$
$$x = \tfrac{21}{11}$$

Step 4: To find the y-value we substitute this value of x into *either* of the given equations. This process is shown here for *both* equations to show that either equation will give the same value for y.

Put the value of x into either equation to find y.

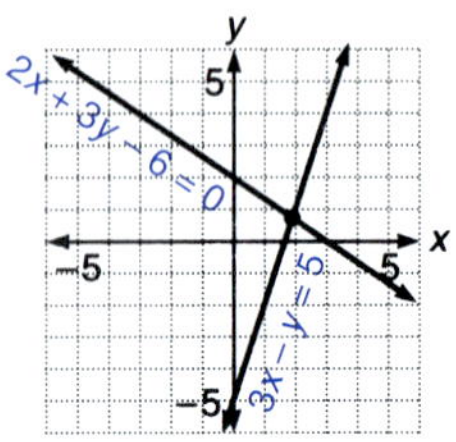

$3x - y = 5$	$2x + 3y - 6 = 0$
$3(\frac{21}{11}) - y = 5$	$2(\frac{21}{11}) + 3y - 6 = 0$
$(\frac{63}{11}) - 5 = y$	$3y = 6 - \frac{42}{11}$
$\frac{63}{11} - \frac{55}{11} = y$	$3y = \frac{66}{11} - \frac{42}{11}$
$\frac{8}{11} = y$	$3y = \frac{24}{11}$
	$\frac{1}{3}(3y) = \frac{1}{3} \cdot \frac{24}{11}$
	$y = \frac{8}{11}$

Thus, either way, $y = \frac{8}{11}$, and the point where the two given lines intersect is $(1\frac{10}{11}, \frac{8}{11})$. The figure shows the graph of each equation and the point of intersection.

The next part of this example shows that the method of substitution can be used to solve systems in which one or both equations is not a straight line. This will occur, for example, if one of the variables has an exponent other than one.

2. Solve the system of equations $y = 2x - 1$ and $y = x^2 - x - 5$.

Step 1: Both equations are already solved for y.

Step 2: $y = 2x - 1$

$x^2 - x - 5 = 2x - 1$ — Substitute $x^2 - x - 5$ for y in the first equation

Step 3: $x^2 - 3x - 4 = 0$ — Collect terms on the left

$(x - 4)(x + 1) = 0$ — Factor the left member

$x - 4 = 0$ or $x + 1 = 0$ — Zero product property

$x = 4$ or -1

Step 4:

When $x = 4$	When $x = -1$
$y = 2x - 1$	$y = 2x - 1$
$y = 2(4) - 1$	$y = 2(-1) - 1$
$y = 7$	$y = -3$

Thus the solutions are $(4,7)$ and $(-1,-3)$. ■

Graphing calculators can be used to approximate solutions to systems of two equations in two variables by graphing each curve (equation) and then using the trace and zoom features to move the cursor to the point of intersection. This topic is revisited in more detail in the chapter "The Conic Sections: Systems of Nonlinear Equations and Inequalities."

Systems of equations appear wherever mathematics is applied to the real world. Thus, the method of substitution is very important. We use it over and over throughout this text.

Mastery points

Can you

- Find the slope of a straight line?
- Given two nonvertical points, find the slope of the straight line that contains them?
- Given two points, find the equation of the straight line that contains them?
- Find the equation of a straight line that meets certain requirements, such as being parallel to another line and passing through a given point?
- Use the method of substitution to solve systems of two equations in two variables?
- Use the method of substitution to find the point of intersection of two different nonparallel lines?

Exercise 3–2

Find the slope of the line that contains the given points.

1. $(-3,2), (5,1)$
2. $(4,9), (-4,-2)$
3. $(5,-1), (8,12)$
4. $(0,-9), (11,5)$
5. $(-3,-5), (-7,-10)$
6. $(2,0), (0,0)$
7. $(7,\frac{7}{12}), (-5,-3)$
8. $(-2,\frac{1}{2}), (1,-\frac{1}{2})$
9. $(4,-3), (4,5)$
10. $(-3,4), (5,4)$
11. $(-6,3), (0,3)$
12. $(\frac{2}{3},-5), (-\frac{1}{3},-2)$
13. $(-\frac{1}{2},-2), (\frac{1}{2},8)$
14. $(\frac{1}{4},\frac{1}{8}), (\frac{3}{4},-\frac{1}{8})$
15. $(\sqrt{2},-5), (3\sqrt{2},-15)$
16. $(\sqrt{3},-1), (2\sqrt{3},4)$
17. $(\sqrt{27},3), (\sqrt{12},9)$
18. $(-2,\sqrt{48}), (1,2\sqrt{75})$
19. $(p+q,q), (p-q,p), q \neq 0$
20. $(m,m+2n), (n,m-2n), m \neq n$
21. Choose any two points that lie on the line $y = -2x + 3$ and show that these two points give a slope of -2 when used in the definition of slope.
22. Choose any two points that lie on the line $y = \frac{5}{8}x - 1$ and show that these two points give a slope of $\frac{5}{8}$ when used in the definition of slope.

Graph each line and find its slope m.

23. $3x - 5y = 15$
24. $2x + y = 6$
25. $y = 6 - 4x$
26. $x - y = 6$
27. $x = 3y - 4$
28. $2x - 6 = 0$
29. $4y = \frac{1}{3}x - 5$
30. $5x - 5y = 1$
31. $3y = -8$
32. $3 + 4x = -2y$
33. $\frac{1}{3}x - \frac{5}{6}y = 2$
34. $x - 11y = 20$
35. $x = -\frac{4}{9}$
36. $9 - y = x$
37. $y - 2 = 0$
38. $x = 1$

Find the slope-intercept equation of the line that contains the given point and has the given slope.

39. $(-3,5), m = -2$
40. $(0,2), m = \frac{1}{3}$
41. $(2\frac{1}{4},\frac{3}{4}), m = -4$
42. $(\frac{1}{2},3), m = 6$
43. $(a,b), m = \dfrac{1}{a}$
44. $(8h,4k), m = \dfrac{1}{2h}$

Find the slope-intercept equation of the line that contains the given points.

45. $(-3,1), (5,4)$
46. $(-4,9), (3,1)$
47. $(6,5), (-2,-2)$
48. $(1,-3), (3,1)$
49. $(15,-10), (18,12)$
50. $(0,-9), (6,-5)$
51. $(-\frac{4}{5},2), (\frac{1}{5},-2)$
52. $(3,\frac{4}{5}), (-1,\frac{4}{15})$
53. $(4,\frac{3}{8}), (12,-\frac{1}{4})$
54. $(\pi,-2\pi), (5\pi,-4\pi)$
55. $(\sqrt{2},-5), (3\sqrt{2},-15)$
56. $(\sqrt{3},-1), (2\sqrt{3},4)$
57. $(m,m+2n), (n,m-2n), m \neq n$
58. $(p+q,q), (p-q,p), q \neq 0$

Find the slope-intercept equation of the line in each case.

59. A line with slope 5 and y-intercept at -3.

60. A line with slope -6 and x-intercept at 4.

61. A line with slope -2 that passes through the point $(\frac{1}{2}, -3)$.

62. A line that passes through the origin and has slope $\frac{4}{3}$.

63. A line that passes through the point $(2, -5)$ and has slope 0.

64. A line that passes through the point $(2, -5)$ and has no slope (its slope is undefined).

65. A line that is parallel to the line $5y - 3x = 4$ and passes through the point $(-5, 2)$.

66. A line that is perpendicular to the line $2x + y = 1$ and passes through the point $(-\frac{3}{5}, 4)$.

67. A line that is perpendicular to the line $4y + 5 = 3x$ and passes through the point $(\frac{3}{2}, -\frac{1}{5})$.

68. A line that is parallel to the line $y = 4$ and passes through the point $(2, -3)$.

69. A line with slope 5 and x-intercept at -3.

70. A line with slope -3 and y-intercept at 2.

71. A line that passes through the origin and has slope -1.

72. A line with slope $\frac{1}{2}$ that passes through the point $(-4, 2)$.

73. A line that passes through the point $(-1, 4)$ and has undefined slope.

74. A line that passes through the point $(-2, 3)$ and has slope 0.

75. A line that is perpendicular to the line $x - 2y = 5$ and passes through the point $(4, -10)$.

76. A line that is parallel to the line $3y - 5x = 1$ and passes through the point $(-3, 6)$.

77. A line that is parallel to the line $x = 4$ and passes through the point $(2, -3)$.

78. A line that is perpendicular to the line $3y - 5 = x$ and passes through the point $(6, -1)$.

Use table 3–2 to answer problems 79–81.

79. Compute the wind chill factor to the nearest 0.1° when the temperature and wind speed are (a) 20° and 22 mph; (b) −5° and 14 mph.

80. Compute the wind chill factor to the nearest 0.1° when the temperature and wind speed are (a) 22° and 20 mph; (b) −4° and 30 mph.

81. Find the wind chill factor when the temperature is −11.5° and the wind speed is 18.5 mph. (You will have to interpolate with respect to *both* the wind speed and the temperature. *Hint:* Interpolate with respect to one factor, temperature or wind speed, at a time.)

82. In Plymouth the population in 1965 was 18,517. In 1980 the population was 29,112. Use linear interpolation to approximate the population in (a) 1970, and (b) 1975.

83. In Canton the per capita average income in 1960 was \$12,875. In 1980 it was \$22,565. Use linear interpolation to approximate the average per capita income in (a) 1968 and (b) 1977, to the nearest dollar.

84. At 8:00 A.M. the outdoor temperature at a certain location was 28° F. At noon the temperature was 59°. Use linear interpolation to estimate (a) what the temperature was at 10:30 A.M. (b) at what time the temperature was 40°.

85. Point C in the figure is at $(-8, 4)$. Point A is at $(-5, 8)$. It is fairly easy to verify that the distance AC is 5. We want to locate the point B so that it is 12 units from point C and on a line that is perpendicular to the line that contains A and C. In this case the distance from A to B will be 13 (since $5^2 + 12^2 = 13^2$). Find B. (*Hint:* One way to do this is to find the equation of the line that contains B and C, by finding the equation of the line through A and C. Then use the distance formula, point C, and the equation of this line.)

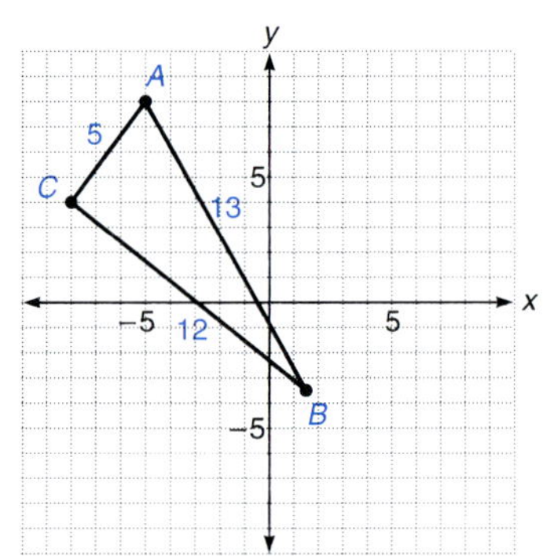

86. Show that any two points on the line $y = 5x - 2$ will produce a slope of 5. Do this by letting $P_1(a,b)$ and $P_2(c,d)$ be any two points on the line and noting that $b = 5a - 2$ and $d = 5c - 2$. Then apply the definition of slope.

87. Show that any two distinct points that lie on the line $y = 3x - 4$ will give a value of 3 for the slope when used in the definition of m (slope). See the suggestion for problem 86.

88. Show that any two points on the line $y = 7x - 6$ will produce a slope of 7.

89. Show that any two points on the line $y = \frac{1}{3}x + 2$ will produce a slope of $\frac{1}{3}$.

90. Prove that in the equation $y = ax + b$ a is the slope of the line. *Hint:* Let $P_1(x_1,y_1)$ and $P_2(x_2,y_2)$ be two different points on the line. Then $y_1 = ax_1 + b$, and $y_2 = ax_2 + b$. Use the two points and this information with the definition of slope.

91. Prove that if $P_1 = (x_1,y_1)$ and $P_2 = (x_2,y_2)$ are two different points and $x_1 \neq x_2$, then a line that contains these points is obtained by the formula $y - y_1 = m(x - x_1)$, where $m = \dfrac{y_2 - y_1}{x_2 - x_1}$. (Do this by showing that P_1 and P_2 are each solutions to this equation and that therefore this line contains these points.)

92. Prove that parallel lines do not intersect. Do this by assuming two lines, $y = mx + b_1$ and $y = mx + b_2$, where $b_1 \neq b_2$, and assume there is a point (x_1,y_1) that lies on both lines.

Find the point at which each of the two lines intersect.

93. $2x + y = 4$
$x - y = 6$

94. $5y - x = 1$
$x = y$

95. $y = x + 6$
$3y + x = 5$

96. $-x + 2y = 2$
$y = 3x + 1$

97. $\frac{1}{2}x - 3y = 1$
$\frac{2}{3}y = x + 4$

98. $\frac{1}{4}y = 2x + 7$
$x - y = 4$

99. The lines $y = -3x + 15$ and $y = \frac{1}{3}x + 2$ are perpendicular since their slopes are negative reciprocals. The figure shows a triangle formed by these two lines and the y-axis. Its vertices are at the points (0,2), (0,15), and (h,k), as shown in the figure. The triangle should be a right triangle. Show that $a^2 + b^2 = c^2$, which will prove that the triangle is a right triangle.

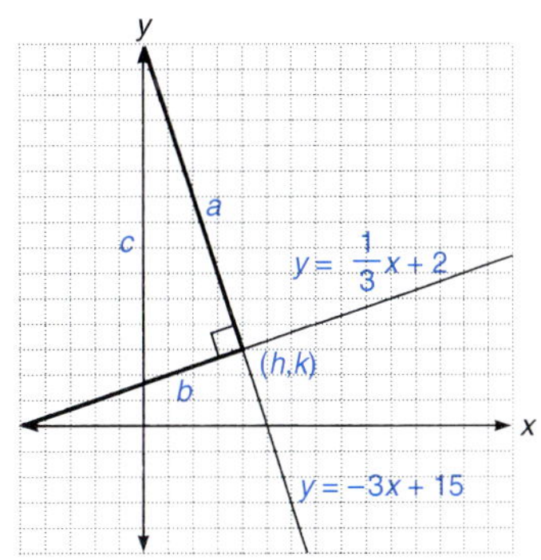

Solve the system of two equations in two variables.

100. $2A - B = 1$
$A + B = 2$

101. $-A + B = 10$
$A + B = 4$

102. $3A + 2B = 5$
$A - B = -1$

103. $2A - 2B = 5$
$A + 3B = 0$

104. $y = x^2 - 23$
$y = 2x - 8$

105. $y = x^2 + 24$
$y = 10x + 3$

106. $y = x^2 + 9x - 73$
$y = 6x - 3$

107. $y = 2x^2 - 6x + \frac{3}{2}$
$y = x - \frac{1}{2}$

108. $y = 4x^2 + 6x - 1$
$y = -2x + 4$

109. $y = x^2 - 3x - 5$
$y = 2x + 1$

110. $y = x^2 + 2x - 6$
$y = -2x - 1$

111. $y = -x^2 + 7$
$y = 2x - 1$

112. $y = 3 - x - x^2$
$y = 3x - 2$

113. $y = x^2 + 3x - 70$
$y = x^2 - x - 2$

114. $y = x^2 - 5x - 3$
$y = x^2 + 6x - 8$

115. $y = x^2 - 9$
$y = -x^2 + 3x - 4$

116. $y = x^2 + 3x + 13$
$y = -x^2 - 6x + 9$

Skill and review

1. Evaluate $3x^2 + 2x - 10$ for $x = -5$.
2. Evaluate $3x^2 + 2x - 10$ for $x = c + 1$.
3. Solve $2x^2 - 2x - 5 = 0$.
4. Simplify $\sqrt{\dfrac{2x}{5y^3}}$.
5. Solve $|2x - 6| < 8$.
6. Solve $\dfrac{x - 2}{4} = \dfrac{2x + 1}{3}$.
7. Compute $\left(\dfrac{2}{3} - \dfrac{1}{4}\right) \div 5$.

3–3 Functions

> A formula that relates temperatures in degrees centigrade C and degrees Fahrenheit F is $F = \frac{9}{5}C + 32$. We could say it describes temperature in degrees Fahrenheit as a function of temperature in degrees centigrade. Describe temperature in degrees centigrade as a function of temperature in degrees Fahrenheit.

In this section we investigate the concept of function. It is one of the most important concepts in modern mathematics. It finds application in any situation in which we wish to express the concept that one quantity depends on another, as in the box above. As another example, one's electric bill depends on, and is therefore a function of, the amount of electricity used.

Definition of function

Recall that a relation is a set of ordered pairs. The set of all first components of a relation is called the **domain** of the relation, and the set of all second components is called the **range** of the relation. For example, the set of ordered pairs

$$A = \{(1,2), (2,4), (3,-5), (3,4), (8,-5)\}$$

is a relation. Its domain[5] D is $D = \{1, 2, 3, 8\}$ and its range R is $R = \{-5, 2, 4\}$. Observe that the value 3 appears as the first component of two different ordered pairs. We say that we have a repeated first component. Relations in which no first component repeats are called functions. This is stated in the following definition.

> **Function**
> A function is a relation in which all of the first components of the ordered pairs are different.

■ ***Example 3–3 A***

State the domain and range of each relation and determine which relations are also functions.

1. $\{(3,1), (5,1), (8,10), (9,10)\}$

Domain: $\{3, 5, 8, 9\}$

Range: $\{1, 10\}$

This is a function since all the first components are different.

2. $\{(3,1), (3,5), (8,10)\}$

Domain: $\{3, 8\}$

Range: $\{1, 5, 10\}$

This is not a function since the first components are not all different. The first component 3 repeats. ■

[5]We always assume that duplicate elements in a set are deleted. Thus, we view the set $\{1, 2, 3, 2\}$ as the set $\{1, 2, 3\}$.

Definition of a one-to-one function

In some functions, such as that in part 1 of example 3–3 A, one or more second components repeat. In this case it was the values 1 and 10. Functions in which no second components repeat are called one-to-one functions.

> **One-to-one function**
> A function is one to one if all of the second components of the ordered pairs are different.

■ ***Example 3–3 B***

State whether each function is one to one or is not one to one.

1. $\{(1,3), (3,1), (5,2), (8,10), (9,11)\}$
   This is a one-to-one function since no second component repeats; thus, all the second components are different.
2. $\{(3,1), (4,5), (8,5)\}$
   This function is not one to one since a second component, 5, repeats. ■

These relations and functions are described by listing their elements. Relations and functions are usually described by rules. For example,[6]

$$A = \{(x,y) \mid y = 2x, x \in \{1, 3, 5\}\}$$

describes a relation because it describes a set of ordered pairs. The domain is the set of all first components, or the set of all x's: $\{1, 3, 5\}$. If we calculate each value of y, the second components, we obtain the relation

$$A = \{(1,2), (3,6), (5,10)\}$$

Since all of the second components are different, A is also a one-to-one function. Its range is $\{2, 6, 10\}$. Observe that A was described by defining the domain and a rule that described each range component.

■ ***Example 3–3 C***

List each relation as a set of ordered pairs, state the domain and range, and note whenever the relation is a function. For each function, state whether it is one to one or not.

Domain	Range
x	$3x - 2$
0	-2
1	1
4	10

1. $A = \{(x,y) \mid y = 3x - 2, x \in \{0, 1, 4\}\}$.
   The table shows the computations of the range components.
   We find $A = \{(0,-2), (1,1), (4,10)\}$.
   Domain: $\{0, 1, 4\}$
   Range: $\{-2, 1, 10\}$
   This relation is a one-to-one function.

[6]Read "A is the set of all ordered pairs (x,y) such that each value of y is found by the rule $y = 2x$, and x is 1, 3, or 5."

Domain	Range	
x	$\sqrt{x}$	$-\sqrt{x}$
4	2	-2
9	3	-3
100	10	-10

2. $A = \{(x,y) \mid y = \pm\sqrt{x},\ x \in \{4, 9, 100\}\}$.
The table shows the computations of the range components.
We find $A = \{(4,2), (4,-2), (9,3), (9,-3), (100,10), (100,-10)\}$.
Domain: $\{4, 9, 100\}$
Range: $\{\pm 2, \pm 3, \pm 10\}$
This relation is not a function since there are first components that repeat; in fact, all of the first components repeat once. ■

f(*x*) notation

A special notation was devised for functions by Leonhard Euler in 1734. It is called "$f(x)$" (read "f of x") notation, and *it is fundamental to higher mathematics.* The symbol $f(x)$ represents the range component associated with a given domain component for a given function f. $f(x)$ is defined by an expression involving x. For example, if the function f is defined by

$$f(x) = x - 1,\ x \in \{3, 5, 9\}$$

the notation $f(x) = x - 1$ is a pattern that we use to compute the ordered pairs in f for the first components 3, 5, 9. In this notation,

$f(x) = x - 1$, so — Read "f of x is x minus 1"
$f(3) = 3 - 1 = 2$ — Replace x with 3; read "f of 3 is 2"
$f(5) = 5 - 1 = 4$ — Replace x with 5; read "f of 5 is 4"
$f(9) = 9 - 1 = 8$ — Replace x with 9; read "f of 9 is 8"

then

$$f = \{(3,2), (5,4), (9,8)\}$$

All of the following notations would describe the function just discussed:

$$f(x) = x - 1,\ x \in \{3, 5, 9\}$$
$$f = \{(x,y) \mid y = x - 1,\ x \in \{3, 5, 9\}\}$$
$$f = \{(x,f(x)) \mid f(x) = x - 1,\ x \in \{3, 5, 9\}\}$$
$$f = \{(3,2), (5,4), (9,8)\}$$

■ ***Example 3–3 D***

For each function compute the function's value for -2, $\frac{3}{4}$, and a.

1. $f(x) = \dfrac{x}{x+1}$

$$f(-2) = \frac{-2}{-2+1} = 2$$ Replace x by -2

$$f\left(\tfrac{3}{4}\right) = \frac{\frac{3}{4}}{\frac{3}{4}+1} = \frac{\frac{3}{4}}{\frac{3}{4}+1} \cdot \frac{4}{4} = \frac{3}{3+4} = \frac{3}{7}$$ Replace x by $\frac{3}{4}$

$$f(a) = \frac{a}{a+1}$$ Replace x by a

2. $h(x) = 3x^2 - 2x - 5$

$$h(-2) = 3(-2)^2 - 2(-2) - 5 = 11$$
$$h(\tfrac{3}{4}) = 3(\tfrac{3}{4})^2 - 2(\tfrac{3}{4}) - 5 = -\tfrac{77}{16}$$
$$h(a) = 3a^2 - 2a - 5$$

3. A programmable calculator can be used to compute numeric values of functions that are described with $f(x)$ notation. For example to do part 2 on a TI-81 we enter the function $h(x) = 3x^2 - 2x - 5$ as follows:

a. [Y=] 3 [X|T] [x^2] [−] 2 [X|T] [−] 5

[QUIT] — [2nd] [CLEAR]

To compute, say, $h(\frac{3}{4})$ we proceed as follows:

b. 3 [÷] 4 [STO▶] [X|T] [ENTER] — Put $\frac{3}{4}$ into X

[Y-VARS] 1 [ENTER] — −4.8125 appears in the display

Note that -4.8125 is $-\dfrac{77}{16}$.

To compute h for other values of x we repeat the steps as at b. ■

Implied domain of a function

Unless we are told otherwise we assume that *the domain of a function is the set of all real numbers for which the expression defining the function is defined.* This is called the **implied domain** of the function. The implied domain must exclude real numbers that cause division by zero and even-indexed roots of negative values (i.e., square roots, fourth roots, etc.).

■ ***Example 3–3 E***

Find the domain of each function.

1. $f(x) = \dfrac{x-3}{x^2-4}$

The denominator must not take on the value 0. To find out where this happens we set the denominator equal to 0:

$$x^2 - 4 = 0$$
$$(x-2)(x+2) = 0$$
$$x = \pm 2$$

Thus, $D = \{x \mid x \neq \pm 2\}$.

2. $f(x) = \dfrac{x}{\sqrt{x^2-4}}$

We require that $x^2 - 4 \geq 0$, because of the radical. Also, this expression may not be zero since it is in a denominator.

Thus, we require that $x^2 - 4 > 0$. This is a nonlinear inequality and can be solved by the critical point/test point method illustrated in section 2–4. We find that $x > 2$ or $x < -2$ is the solution. Thus, $D = \{x \mid x > 2 \text{ or } x < -2\}$, which can also be written $\{x \mid |x| > 2\}$.

3. $f(x) = 2x^3 - x^2 + 9$
There are no operations such as radicals or division that could restrict the values of x, so the domain is all real numbers, R. ■

Expressions involving $f(x)$ notation

Expressions may involve $f(x)$ notation within them. If we are given the defining expression for $f(x)$ we can use this to remove the $f(x)$ notation from the expression containing it.

■ ***Example 3–3 F***

1. The expression $\dfrac{f(x+h) - f(x)}{h}$ is called the *difference quotient.* It is very important in the study of calculus.
Evaluate the quotient if $f(x) = x^2 - 3x - 2$.

We need to replace $f(x + h)$ in this quotient by an expression.

$$f(x) = x^2 - 3x - 2$$
$$f(x + h) = (x + h)^2 - 3(x + h) - 2 \qquad \text{Replace } x \text{ by } (x + h)$$

Replace $f(x + h)$ by the expression $(x + h)^2 - 3(x + h) - 2$, and $f(x)$ by $x^2 - 3x - 2$:

$$\frac{f(x+h) - f(x)}{h} = \frac{\overbrace{[(x+h)^2 - 3(x+h) - 2]}^{f(x+h)} - \overbrace{(x^2 - 3x - 2)}^{f(x)}}{h}$$
$$= \frac{h(2x + h - 3)}{h} = 2x + h - 3$$

2. Given $f(x) = x^2 - 3x + 4$ and $g(x) = \sqrt{9 - x}$, compute
   a. $f(g(-6))$ — Read "f of g of -6"
   b. $3f(-2) + 4g(5)$

   a. $g(-6) = \sqrt{9 - (-6)}$
   $= \sqrt{15}$ — Compute $g(-6)$ first, $g(-6) = \sqrt{15}$

   $f(g(-6)) = f(\sqrt{15})$ — Replace $g(-6)$ by $\sqrt{15}$
   $= (\sqrt{15})^2 - 3\sqrt{15} + 4$ — Compute $f(\sqrt{15})$
   $= 19 - 3\sqrt{15}$

   b. $f(-2) = (-2)^2 - 3(-2) + 4 = 14$
   $g(5) = \sqrt{9 - 5} = 2$
   $3f(-2) + 4g(5) = 3(14) + 4(2) = 50$ — Replace $f(2)$ by 14, $g(5)$ by 2 ■

Linear functions and their graphs

Linear functions are functions whose graphs are straight lines. In section 3–4 we will consider the graphs of other types of functions.

> **Linear function**
> A linear function is a function of the form
> $$f(x) = mx + b$$

Recall that a function is a set of ordered pairs in which no first component repeats. The graph of a function f is the graph of the set of ordered pairs (x,y) where $y = f(x)$; in other words, we plot the values $f(x)$ with respect to the y-axis, as a vertical distance. Thus, for example, to graph $f(x) = 3x + 2$ it would be more convenient to rewrite this as $y = 3x + 2$. In this form it is obvious that a linear function's graph is a straight line.

In general, to graph a function f we replace the symbol $f(x)$ by y. The new equation represents a set of points (x,y), which can be plotted.

■ Example 3–3 G

Graph the linear function $f(x) = -4x + 2$.

Replace $f(x)$ by y: $y = -4x + 2$. By setting x and y to zero we obtain intercepts at $x = \frac{1}{2}$ and $y = 2$. The result is the straight line shown in the figure.

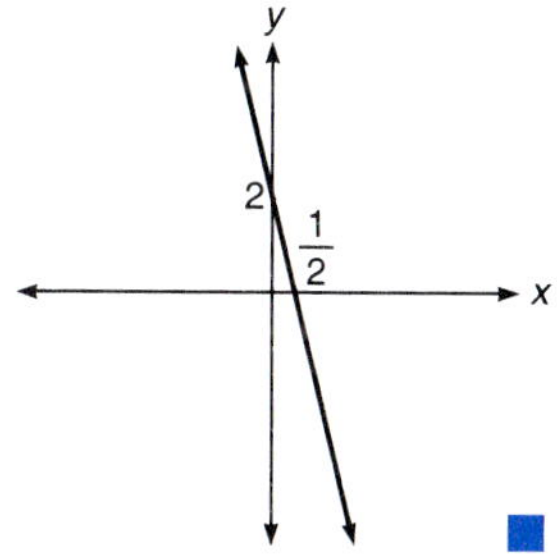

■

■ Example 3–3 H

Finding mathematical descriptions of applied situations is called **mathematical modeling.** The following example illustrates this.

A company found that five salespeople sold \$600,000 worth of its products in a year. It increased its sales force to eight people and found that they sold \$1,400,000 worth of its products. Find a linear function that describes sales s as a function of the number of salespeople p and use it to predict sales for a sales force of nine people.

To make things easier the sales can be described in units of \$100,000. Thus we rewrite 600,000 as 6 and 1,400,000 as 14. We will use these smaller values to develop the function.

For a given value of p we want to calculate s. Use ordered pairs (x,y) to correspond to (p,s). We are given the coordinates of two such ordered pairs: $P_1 = (5,6)$ and $P_2 = (8,14)$.

$m = \dfrac{y_2 - y_1}{x_2 - x_1}$	Definition of slope
$= \dfrac{14 - 6}{8 - 5} = \dfrac{8}{3}$	$x_1 = 5$, $y_1 = 6$, $x_2 = 8$, $y_2 = 14$; the units are dollars per person
$y - y_1 = m(x - x_1)$	Point-slope formula
$y - 6 = \dfrac{8}{3}(x - 5)$	Substitute known values
$3y - 18 = 8(x - 5)$	Clear the denominator
$3y = 8x - 40 + 18$	Remove parentheses; add 18 to both members
$y = \dfrac{8x - 22}{3}$	Divide both members by 3
$y = \frac{8}{3}x - \frac{22}{3}$	Rewrite right member as two terms
$s = \frac{8}{3}p - \frac{22}{3}$	Describe using s and p

Thus, the function is $s = \frac{8}{3}p - \frac{22}{3}$.
To predict sales for nine people we let $p = 9$ and compute s:

$$s = \frac{8}{3}(9) - \frac{22}{3} = 16\frac{2}{3}$$

This is in units of \$100,000 so the actual value is $16\frac{2}{3}(100{,}000)$ = \$1,666,666.67 in sales for nine salespeople. ■

Mastery points

Can you

- Define a function and a one-to-one function?
- State the ordered pairs in a relation that is defined by a rule?
- Determine when a set of ordered pairs is a function, and if so, if it is one to one?
- Determine the implied domain of a function?
- Compute $f(x)$, given an expression that defines $f(x)$, for various values of x?
- Compute expressions that involve $f(x)$?
- Graph linear functions?
- Find expressions for linear functions that can model a given applied situation?

Exercise 3–3

1. Define function.

2. Define one-to-one function.

State the domain and range of each relation. State whether each relation is or is not a function, and if a function, whether or not it is one to one.

3. $\{(5,1), (-3,8), (4,2), (1,5)\}$

4. $\{(-3,0), (-2,4), (-1,3), (0,4), (1,17)\}$

5. $\{(-10,12), (4,13), (2,9), (2,-5)\}$

6. $\{(100,\pi), (3,-\sqrt{2}), (17,\frac{8}{13}), (\pi,\sqrt{2})\}$

List each relation as a set of ordered pairs; note whenever the relation is a function. For each function state whether it is one to one or not. Also, state the domain and range of the relation.

7. $\{(x,y) \mid x + y = 8, x \in \{-2, 3, 5, \frac{3}{4}, 7\}\}$

8. $\{(x,y) \mid y = 3x - 1, x \in \{1, 2, 3, 4\}\}$

9. $\{(x,y) \mid y = \sqrt[3]{x}, x \in \{\pm1, \pm8, \pm27\}\}$

10. $\{(x,y) \mid y = \pm\sqrt{2x - 3}, x \in \{2, 3, 4, 5\}\}$

For each function state the implied domain D of the function, then compute the function's value for the domain components $x = -4, 0, \frac{1}{2}, 7, 3\sqrt{2}$, and $c - 1$ (unless not in the domain of the function; assume $c - 1$ is in the domain of each function).

11. $f(x) = 5x - 3$

12. $g(x) = \dfrac{1 - 2x}{5}$

13. $g(x) = \sqrt{2x - 1}$

14. $h(x) = \sqrt{4 - x}$

15. $f(x) = \dfrac{2x - 1}{x + 3}$

16. $f(x) = \dfrac{1 - x^2}{4x}$

17. $m(x) = 3x^2 - x - 11$

18. $v(x) = 3 - 2x - x^2$

19. $f(x) = \dfrac{\sqrt{x - 1}}{\sqrt{x + 1}}$

20. $g(x) = \dfrac{-4}{\sqrt{x^2 + 5x - 6}}$

21. $h(x) = x^3 - 4$

22. $f(x) = \dfrac{1}{x - 3}$

23. $g(x) = \dfrac{x}{x + 3}$

24. $h(x) = x^3 + x + 3$

Solve the following problems.

25. Compute an expression for $\dfrac{f(x+h)-f(x)}{h}$ if $f(x) = x^2 - 3x - 5$.

26. Compute an expression for $\dfrac{f(x+h)-f(x)}{h}$ if $f(x) = 3x - 4$.

27. If $f(x) = 2x^4 - 3x^2 + 1$ and $g(x) = \sqrt{3x-1}$, compute (a) $f(g(3))$ and (b) $f(g(\frac{2}{3}))$.

28. If $f(x) = \dfrac{x+1}{x}$ and $g(x) = \dfrac{1}{x-1}$, compute (a) $f(g(4))$; (b) $f(g(\frac{2}{5}))$; (c) $g(f(5))$; (d) $g(f(-7))$.

29. If $f(x) = 2x - 5$ and $g(x) = \dfrac{2x}{x+1}$, compute (a) $f(g(1))$; (b) $f(g(3))$; (c) $g(f(0))$; (d) $g(f(\frac{1}{2}))$.

30. If $f(x) = x^2 - x + 1$ and $g(x) = x + 2$, compute (a) $f(g(-2))$; (b) $f(g(5))$; (c) $g(f(\frac{1}{2}))$; (d) $g(f(3))$.

In problems 31–40, let $f(x) = 5x - 1$ and $g(x) = 2x + 2$ and compute the value of the given expression.

31. $f(2) - 3g(1)$

32. $4f(-1) + 2g(3)$

33. $\dfrac{f(1) - g(2)}{f(1) + g(2)}$

34. $\dfrac{4 - 3g(-2)}{f(8)}$

35. $(f(-3))^2 - 3(g(1))^2$

36. $\sqrt{f(3) + g(3)}$

37. $f(x) + 3$

38. $f(x + 3)$

39. $g(x - 2)$

40. $g(x) - 2$

Graph the following linear functions.

41. $f(x) = 2x - 6$

42. $f(x) = \frac{1}{2}x + 2$

43. $g(x) = 3 - 5x$

44. $g(x) = 2 - x$

45. $h(x) = x$

46. $h(x) = 2$

47. $f(x) = 0$

48. $h(x) = -x$

Solve the following problems.

49. An automobile rental company has found that it costs \$500 per year and \$0.34 per mile to own a car. Create a linear function that describes cost of ownership C as a function of miles driven m for one year.

50. The company of the previous problem rents its cars at a rate that averages \$0.40 per mile. Describe the company's income I as a linear function of the number of miles driven m.

51. The company of the previous two problems breaks even on a car when the cost of ownership equals the income from the car. Find the number of miles which a car must be driven for the company to break even on that car.

52. A formula that relates temperatures in degrees centigrade C and degrees Fahrenheit F is $F = \frac{9}{5}C + 32$. Describe temperature in degrees centigrade as a function of temperature in degrees Fahrenheit.

53. The wind at the top of a building was measured to be 25 mph; the wind at the bottom was 8 mph at the same moment. The building is 340 feet high. Assuming the velocity of the wind varied linearly along the height of the building, describe the velocity v as a function of height h above the ground.

54. A giant supermarket found that with five checkout lanes open the average length of a line at peak business times was 6.4 persons, and with eight lanes open the average length was 5.2. Create a linear function that describes the length of a line L as a function of the number of checkout lanes open n, and then use this to predict the average length of a line with ten lanes open.

55. The average weight of a 20-year-old male in a certain population was found to be 150 lb. At the age of 45, the average weight was 194 lb. Create a linear function that models this situation, viewing weight as a function of age, and use it to predict the average weight in the population at the age of 40.

56. The temperature at which a paint blisters was being studied by continuously raising the temperature while recording the temperature and time, and photographing the paint on video tape. At 3:05 P.M. the experiment began, and the temperature was 74° F. At 4:15 P.M. a technician discovered that the paint had blistered, but that the strip recorder recording the temperature had failed. The technician noted that the temperature at that time was 625° F. The video tape showed that the paint had blistered at 4:00 P.M. Assuming that the temperature increased linearly, create a function that describes the temperature T as a function of time t, from 3:05 to 4:15 P.M., then use this function to determine the temperature at which the paint blistered.

57. The surface area of a cylinder is the surface area of its top, bottom, and side. If r is the radius and h is the height of the cylinder, then the surface area A is $A = 2\pi r^2 + 2\pi rh$. If $r + h = 20$, then $r = 20 - h$, and the expression for A can be put in terms of h only. Rewrite the expression for A in terms of h only.

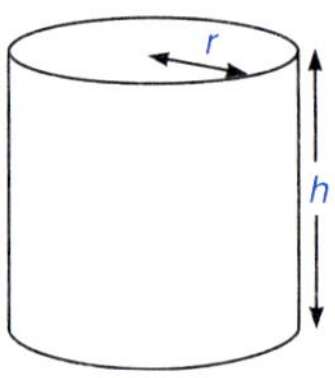

58. Given the conditions in problem 57, describe A as a function of radius r only.

59. A piece of copper of dimensions 40 inches by 30 inches is folded into a tray by cutting squares of side length x from each corner, as shown in the figure. Write the volume V of the resulting tray as a function of x.

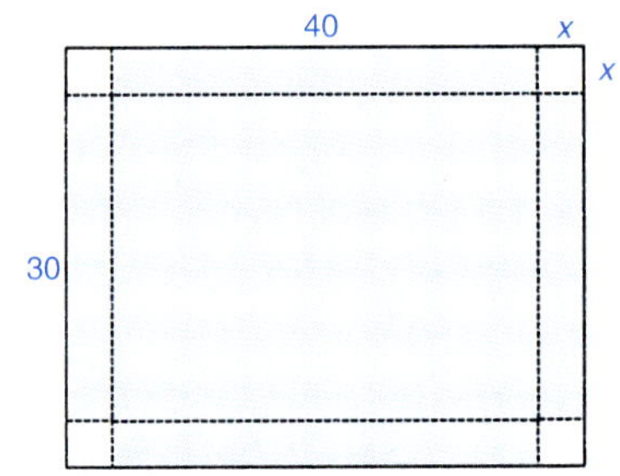

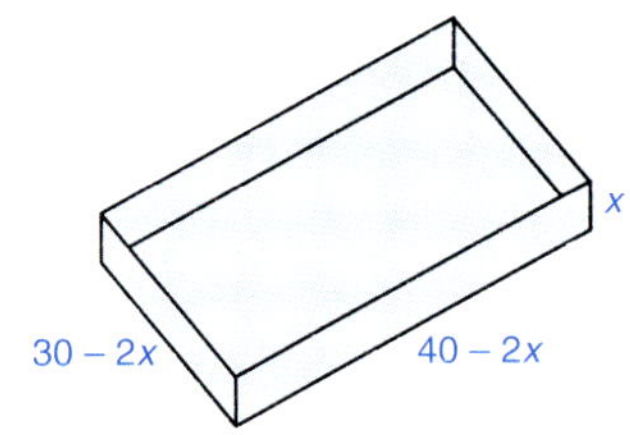

60. Write the surface area of the outside of the box of problem 59 (four sides and bottom) as a function of x.

Skill and review

1. Find the equation of the line that contains the points (1,3) and (−2,4).
2. Find the equation of the line that is parallel to the line $2y - x = 4$ and has y-intercept 3.
3. Factor $8x^3 - 1$.
4. Compute $(3x - 2)^3$.
5. Solve $\dfrac{2x - 1}{3} - \dfrac{x - 1}{2} = 6$.
6. Solve $|x - 3| > 1$.
7. Solve $\dfrac{x + 3}{x - 1} < 1$.

3–4 The graphs of some common functions, and transformations of graphs

A certain worker takes one hour to produce 50 plastic toys on an injection molding machine. If another worker takes x hours to perform the same task, then their combined rate for producing toys is $f(x) = \dfrac{1}{x} + 1$. Graph this function.

In this section we investigate the graphs of some functions that appear often in the study and application of mathematics. We also look at three ways in which a graph can change in a predictable way: vertical translations, horizontal translations, and vertical scaling.

Vertical and horizontal translations

$f(x) = x^2$

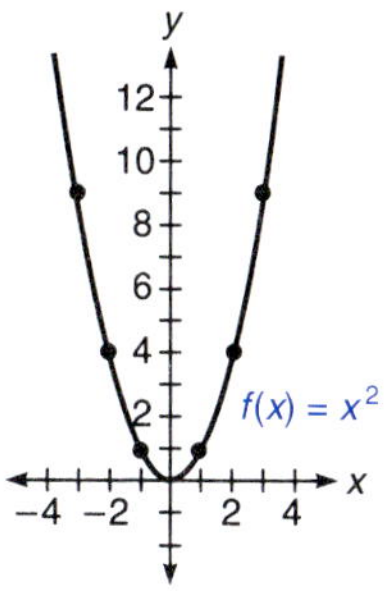

Figure 3–7

We introduce the ideas of translations by examining the function $f(x) = x^2$. The table is a table of values of $y = x^2$, which are plotted, then connected with a smooth curve in figure 3–7.

x	−4	−3	−2	−1	0	1	2	3	4
$y = x^2$	16	9	4	1	0	1	4	9	16

This curve is called a **parabola.** Its low point, at (0,0), is called its **vertex.** The vertical line passing through the vertex is called the **axis of symmetry,** because the graph forms a mirror image about this line. We now consider the graphs of three functions:

$$f(x) = x^2$$
$$f(x) = x^2 + 2$$
$$f(x) = (x + 2)^2$$

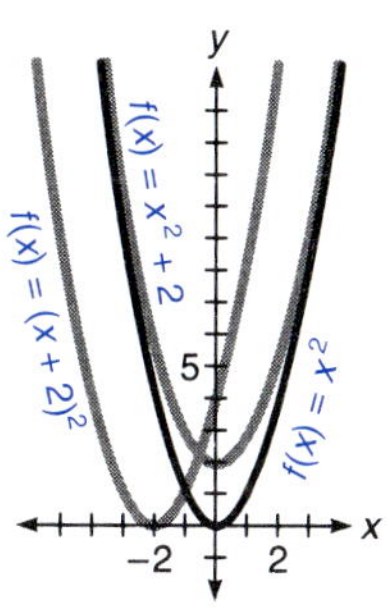

Figure 3–8

We replace $f(x)$ by y in each equation, and graph $y = x^2$, $y = x^2 + 2$, and $y = (x + 2)^2$. The following table shows the computation of a set of values to plot to obtain the graph of each of these functions. The graphs are shown in figure 3–8.

x	−5	−4	−3	−2	−1	0	1	2	3
x^2	25	16	9	4	1	0	1	4	9
$x^2 + 2$	27	18	11	6	3	2	3	6	11
$(x + 2)^2$	9	4	1	0	1	4	9	16	25

The TI-81 calculator can conveniently store up to four functions to be graphed. To create figure 3–8 on this calculator, set Xmin, Xmax, Ymin, and Ymax to the values [−4,4,−2,14], then use the

[Y=] key and enter $Y_1 = X^2$ [X|T] [x^2] [ENTER]

$Y_2 = X^2 + 2$ [X|T] [x^2] [+] 2 [ENTER]

$Y_3 = (X+2)^2$ [(] [X|T] [+] 2 [)] [x^2]

[ZOOM] 5 Square

[GRAPH]

Observe that $f(x) = x^2 + 2$ has the same graph as $f(x) = x^2$, but is shifted up two units. This makes sense because $y = x^2 + 2$ has a value that is two greater than $y = x^2$. We say that $f(x) = x^2 + 2$ is a **vertical translation** of $f(x) = x^2$.

The graph of $f(x) = (x + 2)^2$ is also the same as $f(x) = x^2$, but shifted two units to the left. This is because, to compute $y = (x + 2)^2$ we first add two to x—thus, x can be two units less than the value of x in $y = x^2$ and produce the same value. Two units less, when referring to x-values, is two units to the left. We say that $f(x) = (x + 2)^2$ is a **horizontal translation** of $f(x) = x^2$.

Saying that $f(x) = (x + 2)^2$ is a translation to the left is confusing, since we naturally think of $+2$ as to the right of zero. Thus, we often write $f(x) = (x - (-2))^2$ instead. This will be reflected in the generalization below. One more observation, on notation:

$$\begin{aligned} &\text{If } f(x) = x^2, \\ &\text{then } f(x) + 2 = x^2 + 2, \\ &\text{and } f(x - 2) = (x - 2)^2. \end{aligned}$$

With this notation in mind we can generalize this discussion as follows.

> Given the graph of a function $f(x)$,
>
> **Vertical translation**
>
> The graph of
> $$y = f(x) + c$$
> is the graph of $f(x)$ shifted up if $c > 0$ and down if $c < 0$.
>
> **Horizontal translation**
>
> The graph of
> $$y = f(x - c)$$
> is the graph of $f(x)$ shifted right if $c > 0$ and left if $c < 0$.

Note that to have a horizontal translation of a function f, *every* instance of x in $f(x)$ must be replaced by an expression of the form $x - c$.

Example 3–4 A

In each case, functions g and f are given. Describe the graph of f as a horizontal and/or vertical translation of the graph of g.

1. $g(x) = x^2$
$f(x) = (x - 3)^2$

The graph of f is a horizontal translation of the graph of g, 3 units to the *right*.

2. $g(x) = x^2$
$f(x) = x^2 + 3$

The graph of f is the graph of g shifted *upward* three units.

3. $g(x) = x^5 + x$
$f(x) = (x - 2)^5 + (x - 2) + 3$

The graph of f is the graph of g translated *upward* 3 units and to the *right* 2 units. ■

We can use knowledge about vertical and horizontal translations as an aid in graphing many functions. For example, since we know what the graph of $y = x^2$ looks like we can graph functions whose graphs are vertical and/or horizontal translations of this graph.

Recall from section 3–3 that to graph a function f we replace the symbol $f(x)$ by y. This is not necessary, but is convenient. Recall also that, to find a y-intercept, replace x by zero and solve for y, and to find an x-intercept, replace y by zero and solve for x.

When using a graphing calculator it is useful to predict the appearance of the graph by comparing it to the graph of one of some basic function whose graph is familiar—this is especially useful when setting the RANGE (limits of x and y that will be plotted) on graphing calculators. For each example we show the keystrokes used to enter the function into a graphing calculator. We also show the values of Xmin, Xmax, Ymin, and Ymax. These are shown *in a box, in this order.*

As illustrated in example 3–4 B we try to compare the appearance of the graph of a given function to the graph of a function with whose graph we are familiar.

■ ***Example 3–4 B***

The function $f(x) = (x + 3)^2 + 2$ is a translation of the parabola $y = x^2$. Describe the appearance of the graph compared to the graph of $y = x^2$ and then sketch the graph. Compute all intercepts.

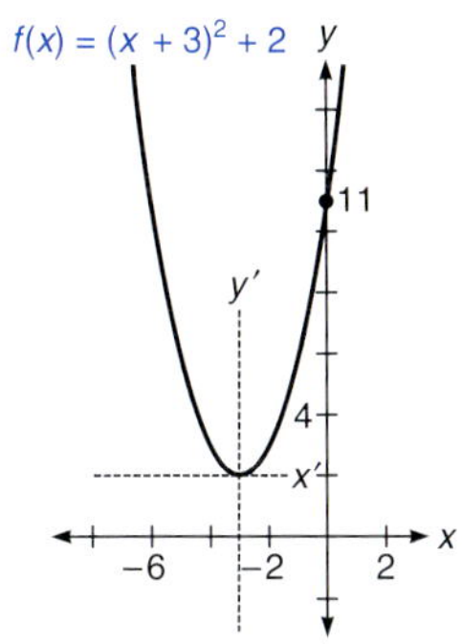

Y= | (| X|T | + | 3
) | x^2 | + | 2
RANGE −7,1,−1,12

This is equivalent to graphing $y = (x - (-3))^2 + 2$. The graph is the same as the graph of $y = x^2$ shifted up 2 units and to the left 3 units. This moves the vertex from (0,0) to (−3,2). We draw the graph f by drawing the graph of $y = x^2$ with vertex at (−3,2), as shown. It is convenient to draw in a second set of axes as shown, labeled x' and y' (read x-prime and y-prime).

y-intercept:

$y = 3^2 + 2$ Set $x = 0$ in $y = (x + 3)^2 + 2$

$y = 11$

x-intercept: We can see from the graph that there are no x-intercepts; algebraically we see this as follows:

$0 = (x + 3)^2 + 2$ Set $y = 0$ in $y = (x + 3)^2 + 2$

$-2 = (x + 3)^2$ This has no real solutions ■

We will continue to use the idea of horizontal and vertical translations as we examine the graphs of some basic functions.

$f(x) = \sqrt{x}$

We want to graph $y = \sqrt{x}$. By plotting points (or using a graphing calculator) we obtain the graph shown in figure 3–9; it is in fact half of a parabola (the parabola $y^2 = x$) with a horizontal axis of symmetry. This function has intercepts at the origin.

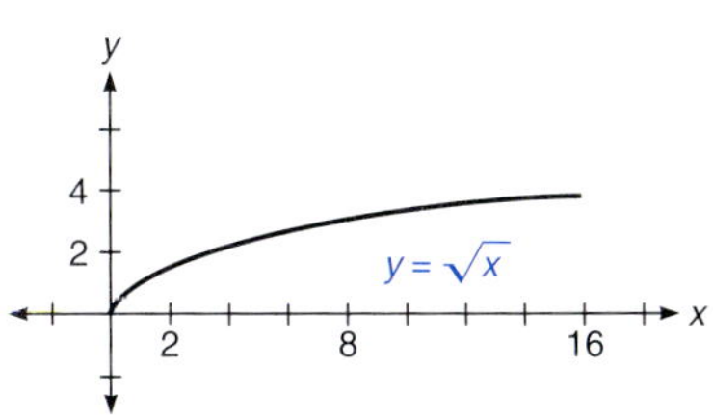

Figure 3–9

Y= | √ | X|T
RANGE −1,17,−1,5

■ Example 3–4 C

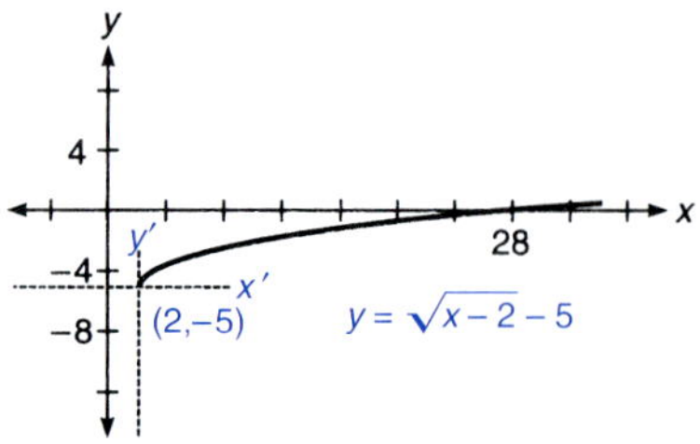

[Y=] [√] [(] [X|T]
[−] 2 [)] [−] 5
[RANGE −1,30,−8,4] Xscl=4

Describe the appearance of the graph of $f(x) = \sqrt{x-2} - 5$ compared to the appearance of the graph of $y = \sqrt{x}$, then sketch the graph. Compute all intercepts.

We rewrite as $y = \sqrt{x-2} - 5$; this is the graph of $y = \sqrt{x}$ but shifted down 5 units and to the right 2 units. We can picture a new axis system x' and y' centered at $(2,-5)$.

y-intercept:

$y = \sqrt{-2} - 5$ — Let $x = 0$ in $y = \sqrt{x-2} - 5$; there is no real solution

x-intercepts:

$0 = \sqrt{x-2} - 5$ — Let $y = 0$ in $y = \sqrt{x-2} - 5$
$5 = \sqrt{x-2}$ — Add 5 to both members
$25 = x - 2$ — Square both members
$27 = x$ — Add 2 to both members; the x-intercept is at 27

We thus sketch the graph of $y = \sqrt{x}$ with "origin" at $(2,-5)$ and x-intercept at $(27,0)$. ■

$f(x) = x^3$

By plotting some points or using a graphing calculator, we obtain the graph of $y = x^3$. See figure 3–10. This function has an intercept at the origin.

[Y=] [X|T] [MATH] 3
[RANGE −2,2,−5,5]

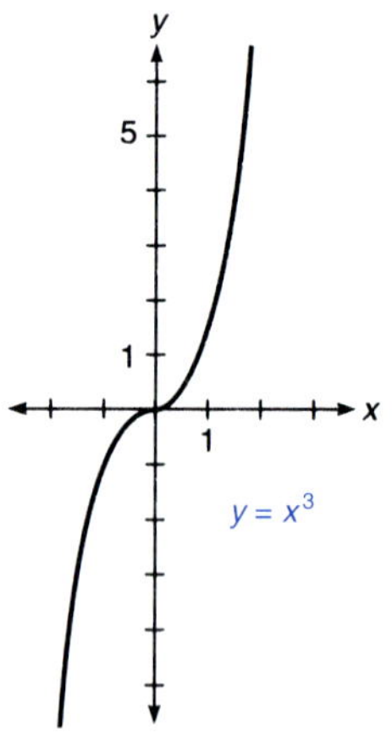

Figure 3–10

■ Example 3–4 D

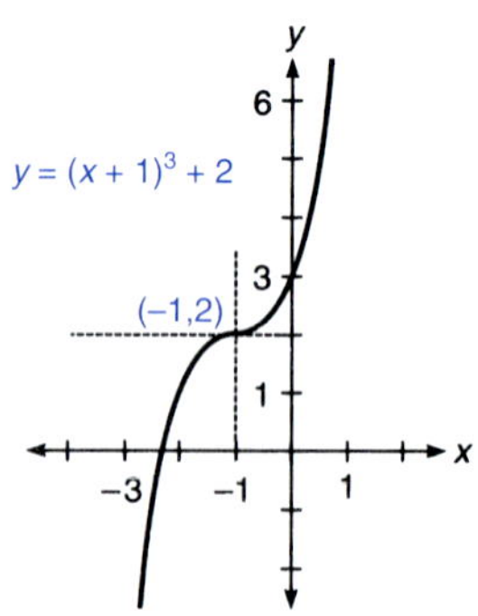

[Y=] [(] [X|T] [+] 1 [)]
[MATH] 3 [+] 2
[RANGE −4,2,−1,7]

Graph the function $f(x) = (x+1)^3 + 2$ after describing its appearance compared to that of $y = x^3$. Compute all intercepts.

We rewrite as $y = (x - (-1))^3 + 2$; the graph of this equation is the graph of $y = x^3$, but shifted up 2 units and to the left 1 unit. We thus sketch the graph of $y = x^3$ but using a x'–y' axis system centered at $(-1,2)$.

y-intercept:

$y = 1^3 + 2 = 3$ — Let $x = 0$ in $y = (x+1)^3 + 2$

x-intercepts:

$0 = (x+1)^3 + 2$ — Let $y = 0$ in $y = (x+1)^3 + 2$
$-2 = (x+1)^3$ — Add -2 to both members
$\sqrt[3]{-2} = x + 1$ — Take cube root of both members
$-1 + \sqrt[3]{-2} = x$ — Add -1 to both members
$x \approx -2.3$ — Approximate value of x-intercept ■

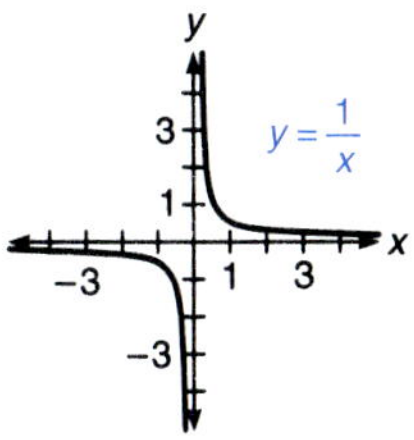

Figure 3–11

$f(x) = \dfrac{1}{x}$

We rewrite as $y = \dfrac{1}{x}$ and plot points or use a graphing calculator. See figure 3–11.

> [Y=] 1 [÷] [X|T]
> [RANGE −4,4,−4,4]

This graph has no intercepts, and is undefined at $x = 0$.

Note [Y=] [X|T] [x^{-1}] is another way to enter this function.

■ *Example 3–4 E*

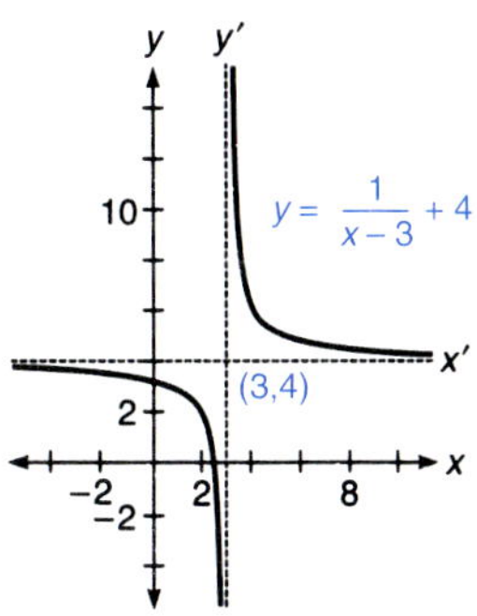

> [Y=] 1 [÷] [(] [X|T]
> [−] 3 [)] [+] 4
> [RANGE −2,8,−2,12]

Graph the function $f(x) = \dfrac{1}{x-3} + 4$ after describing its appearance compared to the graph of $y = \dfrac{1}{x}$, then sketch the graph. Compute all intercepts.

Rewrite as $y = \dfrac{1}{x-3} + 4$; this is the graph of $y = \dfrac{1}{x}$ but shifted up 4 units and to the right 3 units. We sketch the graph of $y = \dfrac{1}{x}$ about an x'–y' axis with origin at (3,4).

y-intercept:

$y = \dfrac{1}{-3} + 4 = 3\frac{2}{3}$ Let $x = 0$ in $y = \dfrac{1}{x-3} + 4$

x-intercept:

$0 = \dfrac{1}{x \quad 3} + 4$ Let $y = 0$ in $y = \dfrac{1}{x-3} + 4$

$-4 = \dfrac{1}{x-3}$ Add −4 to both members

$-4x + 12 = 1$ Multiply both members by $x - 3$

$x = 2\frac{3}{4}$

Note [Y=] [(] [X|T] [−] 3 [)] [x^{-1}] [+] 4 could also be used to enter this function. ■

$f(x) = |x|$

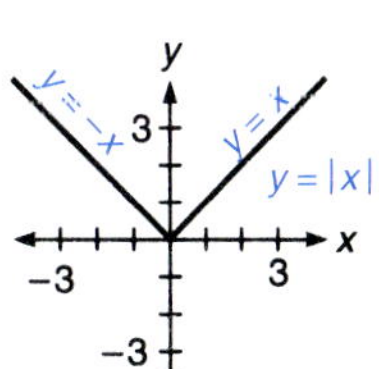

Figure 3–12

Rewrite as $y = |x|$. Note that, for $x \geq 0$ this is the same graph as $y = x$, and for $x < 0$ it is the same as $y = -x$. This is because $|x| = x$ when $x \geq 0$, and $|x| = -x$ when $x < 0$. There are intercepts at the origin. See figure 3–12.

> [Y=] [ABS] [X|T]
> [RANGE −3,3,−3,3]

■ Example 3–4 F

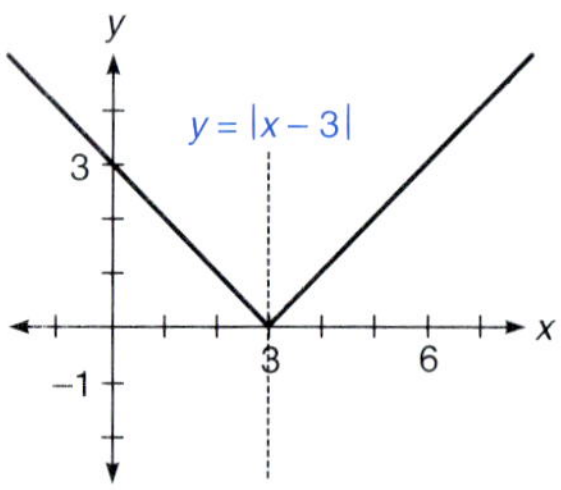

Y= ABS (X|T
– 3)
RANGE –1,7,–1,7

Graph the equation $f(x) = |x - 3|$ and relate its graph to that of $y = |x|$. Compute all intercepts.

Rewrite as $y = |x - 3|$; this is the graph of $y = |x|$ but shifted right 3 units.

y-intercept:

$y = |-3| = 3$ Let $x = 0$ in $y = |x - 3|$

x-intercept:

$0 = |x - 3|$ Let $y = 0$ in $y = |x - 3|$

$0 = x - 3$ If $|a| = 0$, then $a = 0$

$x = 3$ x-intercept ■

Vertical scaling of functions

Some graphs are versions of these graphs that are vertically scaled. We say that a function g is a **vertically scaled** version of a function f if $g(x) = kf(x)$ for some nonzero real number k. Some examples follow.

■ Example 3–4 G

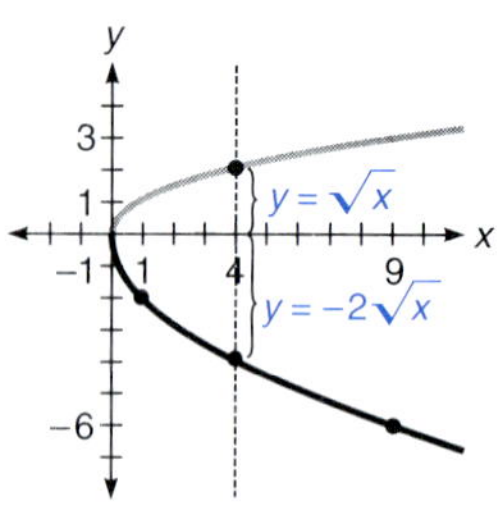

Y= (–) 2 √ X|T
ENTER √ X|T
RANGE –1,9,–6,3

Compare the graph of each function to that of an appropriate basic graph cited previously, then graph the function. Compute all intercepts.

1. $f(x) = -2\sqrt{x}$

The graph of $y = -2\sqrt{x}$ is the same as the graph of $y = \sqrt{x}$ except that each of the y-values ($\sqrt{x}$) is doubled in value *and its sign is changed.* Thus each point in the graph moves twice as far from the x-axis and to the other side. Both graphs are shown in the figure. Note that at $x = 4$ the upper graph is $\sqrt{4} = 2$ while the lower one is $-2(\sqrt{4}) = -4$.

Additional points: $(1,-2)$, $(4,-4)$, $(9,-6)$

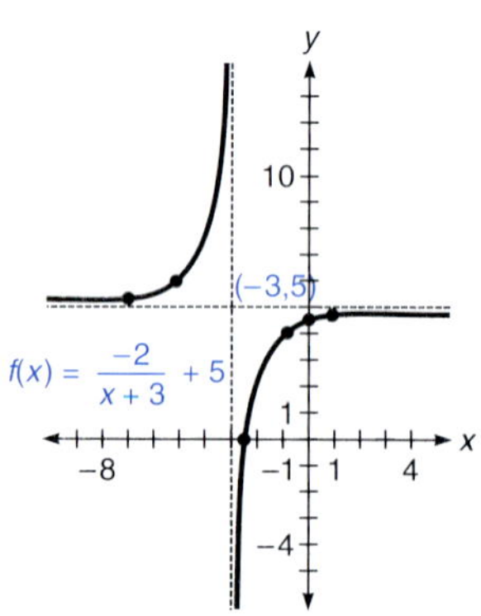

2. $f(x) = \dfrac{-2}{x+3} + 5$

Rewrite as $y = \dfrac{-2}{x-(-3)} + 5$; this is the graph of $y = \dfrac{1}{x}$ but shifted left 3 units, up 5 units, and scaled by -2. The negative scaling factor will "flip over" the graph. The "origin" moves 3 units left and 5 up to $(-3,5)$.

Y= (–) 2 ÷ (
X|T + 3) + 5
RANGE –9,3,–4,12

y-intercept:

$$y = \frac{-2}{3} + 5 = 4\tfrac{1}{3} \qquad \text{Let } x = 0 \text{ in } y = \frac{-2}{x+3} + 5$$

x-intercept:

$$0 = \frac{-2}{x+3} + 5 \qquad \text{Let } y = 0 \text{ in } y = \frac{-2}{x+3} + 5$$

$$\frac{2}{x+3} = 5 \qquad \text{Add } \frac{2}{x+3} \text{ to each member}$$

$$2 = 5x + 15 \qquad \text{Multiply each member by } x + 3$$

$$-2\tfrac{3}{5} = x \qquad y\text{-intercept}$$

Additional points: $(-7,5\frac{1}{2})$, $(-5,6)$, $(-1,4)$, $(1,4\frac{1}{2})$

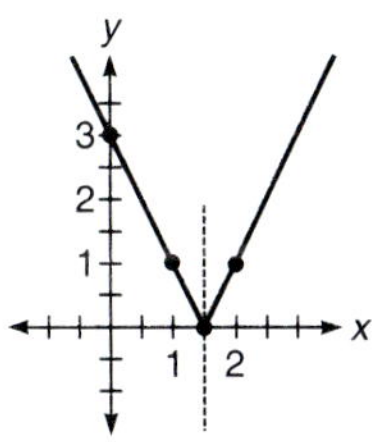

Y=	ABS	(2	X|T
− 3	)		
RANGE −1,3,−1,3			

3. $f(x) = |2x - 3|$

$y = |2x - 3|$

$y = |2(x - \frac{3}{2})|$

$y = 2\,|2x - \frac{3}{2}|$

The graph of this is the same as the graph of $y = |x - \frac{3}{2}|$ but vertically scaled 2 units. The "origin" is moved to $(1\frac{1}{2},0)$.

Intercepts:

$x = 0$: $y = |0 - 3| = 3$

$y = 0$: $0 = |2x - 3|$

$2x - 3 = 0 \qquad$ If $|a| = 0$, then $a = 0$

$x = \frac{3}{2}$

Additional points: $(1,1)$, $(2,1)$ ■

Certain functions appear quite often throughout the study of mathematics, and it is often helpful to make a quick sketch of these functions. Many of these functions have been covered in this section. Thus, you should be familiar with the graphs of $y = x^2$, x^3, $\sqrt{x}$, $\dfrac{1}{x}$, and $|x|$ and be able to sketch them quickly, with or without a graphing calculator.

Mastery points

Can you

- Make a sketch of the functions $y = x^2$, $\sqrt{x}$, x^3, $\dfrac{1}{x}$, and $|x|$?
- Compare the graph of certain functions to the graphs of $y = x^2$, $\sqrt{x}$, x^3, $\dfrac{1}{x}$, and $|x|$?
- Graph equations involving linear translations and vertical scaling of the functions $y = x^2$, $\sqrt{x}$, x^3, $\dfrac{1}{x}$, and $|x|$?

Exercise 3–4

Describe the appearance of the graphs of the following functions compared to the graph of $y = x^2$. Then graph and state the x- and y-intercepts.

1. $f(x) = x^2 - 4$
2. $f(x) = x^2 + 2$
3. $f(x) = x^2 + 3$
4. $f(x) = x^2 - 9$
5. $f(x) = (x - 1)^2$
6. $f(x) = (x - 2)^2$
7. $f(x) = (x + 3)^2$
8. $f(x) = (x + 1)^2$
9. $f(x) = (x + 3)^2 - 3$
10. $f(x) = (x + 1)^2 + 2$
11. $f(x) = (x + 2)^2 + 1$
12. $f(x) = (x - 1)^2 - 3$

Describe the appearance of the graphs of the following functions compared to the graph of $y = \sqrt{x}$. Then graph and state the x- and y-intercepts.

13. $f(x) = \sqrt{x} - 2$
14. $f(x) = \sqrt{x} + 1$
15. $f(x) = \sqrt{x} + 2$
16. $f(x) = \sqrt{x} - 3$
17. $f(x) = \sqrt{x + 1}$
18. $f(x) = \sqrt{x - 2}$
19. $f(x) = \sqrt{x - 3}$
20. $f(x) = \sqrt{x + 4}$
21. $f(x) = \sqrt{x - 3} + 2$
22. $f(x) = \sqrt{x + 2} + 3$
23. $f(x) = \sqrt{x + 5} + 5$
24. $f(x) = \sqrt{x + \frac{1}{2}} - 2$

Describe the appearance of the graphs of the following functions compared to the graph of $y = x^3$. Then graph and state the x- and y-intercepts.

25. $f(x) = (x - 2)^3$
26. $f(x) = (x + 1)^3$
27. $f(x) = x^3 - 8$
28. $f(x) = x^3 + 1$
29. $f(x) = (x + 1)^3 - 2$
30. $f(x) = (x - 2)^3 + 1$
31. $f(x) = (x + 2)^3 + 2$
32. $f(x) = (x + 2)^3 - \frac{27}{64}$

Describe the appearance of the graphs of the following functions compared to the graph of $y = \dfrac{1}{x}$. Then graph and state the x- and y-intercepts.

33. $f(x) = \dfrac{1}{x} + 2$
34. $f(x) = \dfrac{1}{x} - 1$
35. $f(x) = \dfrac{1}{x - 6}$
36. $f(x) = \dfrac{1}{x + 1}$
37. $f(x) = \dfrac{1}{x - 3} - 5$
38. $f(x) = \dfrac{1}{x + 2} + 3$
39. $f(x) = \dfrac{1}{x - 1} + 1$
40. $f(x) = \dfrac{1}{x - 6} - 1$

Describe the appearance of the graphs of the following functions compared to the graph of $y = |x|$. Then graph and state the x- and y-intercepts.

41. $f(x) = |x + 2|$
42. $f(x) = |x| - 2$
43. $f(x) = |x| + 2$
44. $f(x) = |x + 1|$
45. $f(x) = |x - 5| - 4$
46. $f(x) = |x + 3| + 2$
47. $f(x) = |x + 3| + 3$
48. $f(x) = |x - 2| + 1$

Describe the appearance of the graphs of the following functions compared to the graph of $y = x^2, x^3, \sqrt{x}, \dfrac{1}{x}$, and $|x|$. Then graph and state the x- and y-intercepts.

49. $f(x) = 3(x - 1)^2 + 2$
50. $f(x) = 3\sqrt{x - 1} - 1$
51. $f(x) = \frac{1}{2}(x - 2)^3$
52. $f(x) = \dfrac{3}{x - 3} + 1$
53. $f(x) = |3x - 6| - 2$
54. $f(x) = 4(x - 2)^2 - 8$
55. $f(x) = -4\sqrt{x + 3} + 2$
56. $f(x) = -8x^3 + 11$
57. $f(x) = \dfrac{-2}{x + 3} - 4$
58. $f(x) = -|x + 3| + 5$
59. $f(x) = -2(x + 1)^2 + 3$
60. $f(x) = \sqrt{4x - 8} - 3$
61. $f(x) = -\frac{3}{2}(x + 1)^3$
62. $f(x) = \dfrac{-2}{x + 5} - 3$
63. $f(x) = \left|\dfrac{x}{3}\right| + 1$

64. A certain worker takes one hour to produce 50 plastic toys on an injection molding machine. If another worker takes x hours to perform the same task, then their combined rate for producing toys is $f(x) = \dfrac{1}{x} + 1$. Graph this function.

65. The temperature of a certain chemical process is measured with reference to a base temperature of 2° C. What is important is the difference between the temperature and 2°. Under these conditions the function $f(x) = |x - 2|$ describes this difference. Graph this function.

66. The cost of filling a cube of length x ft with concrete, where concrete costs \$1 per square foot, and where \$8 is charged for delivering the concrete, is described by $f(x) = x^3 + 8$. Graph this function.

67. To plate a square piece of metal with chrome a company charges \$5 plus \$0.50 per square inch. If x represents the length of a side of a piece of metal to be chrome plated, then the cost in dollars to plate it is described by $f(x) = \frac{1}{2}x^2 + 5$. Graph this function.

68. A rectangular solid with thickness 1 in. and equal length and width is to be made from x cubic inches of plastic. It is estimated that $1\frac{1}{4}$ in.^{3} will be lost in the construction process. Under these conditions the length (or width) of each side of the solid is described by $f(x) = \sqrt{x - 1\frac{1}{4}}$. Graph this function.

Skill and review

1. Find the distance between the points (1,2) and (6,8).
2. Find the midpoint of the line segment that joins the points (1,2) and (6,8).
3. Find the equation that describes all points equidistant from the two points (1,2) and (6,8).
4. Find where the lines $2y - 3x = 5$ and $x + y = 3$ intersect.
5. Find the equation of a line that is perpendicular to the line $y = -2x + 3$ and passes through the point $(1,-2)$.
6. Solve $x^2 - 4x = 32$.
7. Solve $\dfrac{x}{x + y} = 3$ for x.

3–5 Circles and more properties of graphs

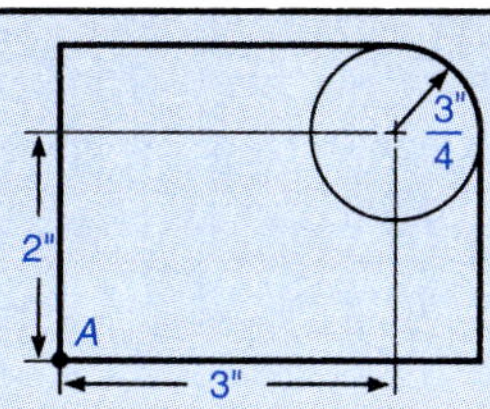

A computer-controlled robot is being programmed to grind the outer edge of the object shown in the figure. For this purpose, the equation of the circle must be known in terms of a coordinate system centered at the point *A*. Find this equation.

This section discusses the equations of circles, and then proceeds to investigate some helpful properties of the graphs of functions.

Circles

In geometry a circle is defined as the set of all points equidistant from a given point (the center). This is what we use for our definition in analytic geometry. If the center of the circle is to be $C = (h,k)$, and the radius is $r > 0$, then we want the circle to be the set of all points that are r units from the point (h,k). Thus, if $P = (x,y)$ is any point on the circle,

$d(P,C) = r$ — The distance from the point to the center is r

$\sqrt{(x - h)^2 + (y - k)^2} = r$ — Distance formula with (x,y) and (h,k)

$(x - h)^2 + (y - k)^2 = r^2$ — Square both members

This leads to the following definition of what is called the standard form for the equation of a circle.

> **Standard form of the equation of a circle**
> A circle is the relation defined by an equation of the form
> $$(x - h)^2 + (y - k)^2 = r^2, \quad r > 0$$
> The center *C* is at (h,k), and r is the radius.

■ ***Example 3–5 A***

Graph the circle $(x - 3)^2 + (y + 2)^2 = 12$.

$$(x - 3)^2 + (y + 2)^2 = 12$$
$$(x - 3)^2 + (y - (-2))^2 = 12 \quad \text{Rewrite in standard form}$$

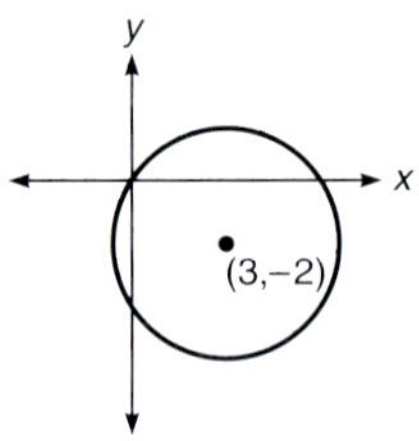

The center of the circle is $(3,-2)$.

$$r^2 = 12$$
$$r = \sqrt{12} = \sqrt{4 \cdot 3} = 2\sqrt{3} \approx 3.5 \quad \text{Radius is about 3.5}$$

With the center and radius we can draw the graph as shown.

Note It is actually easier to graph a circle by hand than to use a graphing calculator, since all that is needed is the center and radius, and it is an easy figure to draw. The chapter "The Conic Sections" does, however, show a method by which circles can be graphed on the calculator. ■

Completing the square

If we are given an equation of the form $x^2 + ax + y^2 + by + c = 0$, such as $x^2 - 4x + y^2 + 2y - 12 = 0$, we can put the equation in the standard form of a circle using a method called **completing the square.** To see how this method works we make the following observation.

If we square the binomial $(x - h)^2$ we obtain $x^2 - 2hx + h^2$. In this trinomial, we can see that the last coefficient, h^2, is the square of half the coefficient, $-2h$, of the middle, linear term. Thus, for example, if we wanted to determine c so that $x^2 - 6x + c$ would be the square of a binomial, then c should be the square of half of -6. Half of -6 is -3, and $(-3)^2 = 9$, so we know that $x^2 - 6x + 9$ is a perfect square. In fact, $x^2 - 6x + 9 = (x - 3)^2$.

This example suggests that to complete the square on an expression of the form $x^2 + bx$, **add the square of half the coefficient of x,** $\frac{b^2}{4}$.

Some examples of completing the square follow.

1. $x^2 - 8x$ — Original expression
-4 — Half of the coefficient of x
16 — Square this value
$x^2 - 8x + 16$ — Add to original expression
$(x - 4)^2$ — Rewrite as a square

2. $x^2 - \frac{3}{5}x$ — Original expression
$-\frac{3}{10}$ — Half of $-\frac{3}{5}$ is $\frac{1}{2}(-\frac{3}{5}) = -\frac{3}{10}$
$\frac{9}{100}$ — Square this value
$x^2 - \frac{3}{5}x + \frac{9}{100}$ — Add to original expression
$(x - \frac{3}{10})^2$ — Rewrite as a square

When the expression of interest appears in an equation we must add the same value, $\frac{b^2}{4}$, to *both* members of the equation. It is only necessary to complete the square on x or y when there is a linear term on that variable. This is illustrated in the following example.

■ **Example 3–5 B**

(a) Transform each equation into the standard form for a circle. (b) State the center and radius of the circle. (c) Graph the circle.

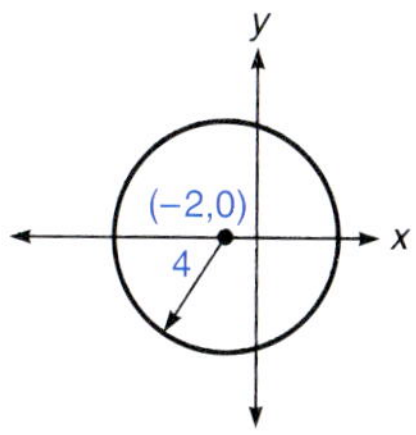

1. $x^2 + 4x + y^2 - 12 = 0$ — We complete the square on the x terms since there is a linear term, $4x$, in this variable

$x^2 + 4x + y^2 = 12$ — Put the constant on the other member of the equation

$x^2 + 4x + 4 + y^2 = 12 + 4$ — Add 4 to $x^2 + 4x$; to maintain the equality add it to the other member of the equation also

$(x + 2)^2 + y^2 = 16$ — $x^2 + 4x + 4 = (x + 2)^2$; $12 + 4 = 16$

$(h,k) = (-2,0)$; $r^2 = 16$ — Taking h, k, and r from the standard form

This is a circle with center at $(-2,0)$ and radius 4.

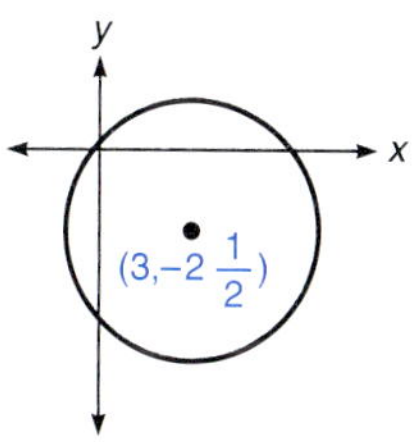

2. $x^2 - 6x + y^2 + 5y = 1$

We first complete the square on x. The constant term is already in the right-hand member of the equation.

$x^2 - 6x + 9 + y^2 + 5y = 1 + 9$ — Half of -6 is -3; $(-3)^2 = 9$

$(x - 3)^2 + y^2 + 5y = 10$ — $x^2 - 6x + 9 = (x - 3)^2$

Now complete the square on y.

$(x - 3)^2 + y^2 + 5y + \frac{25}{4} = 10 + \frac{25}{4}$ — Half of 5 is $\frac{5}{2}$; $(\frac{5}{2})^2 = \frac{25}{4}$

$(x - 3)^2 + (y + \frac{5}{2})^2 = \frac{65}{4}$ — $10 + \frac{25}{4} = \frac{4 \cdot 10}{4} + \frac{25}{4} = \frac{65}{4}$

This is the equation of a circle with center at $(3,-2\frac{1}{2})$ and radius $\sqrt{\frac{65}{4}} = \frac{\sqrt{65}}{2} \approx 4.03$. ■

Finding the equation of a circle from given information

Given the center and radius of a circle we can find the equation of the circle by replacing (h,k) and r in the standard form of the equation of a circle.

■ **Example 3–5 C**

Find the equation of the circle. Leave the equation in standard form.

1. Center at $(2,-4)$, radius $= 3$.

$(x - h)^2 + (y - k)^2 = r^2$ — General equation of a circle

$(x - 2)^2 + (y - (-4))^2 = 3^2$ — Replace (h,k) by $(2,-4)$ and r by 3

$(x - 2)^2 + (y + 4)^2 = 9$ — Equation in standard form

2. Center at $(-3,1)$, passes through the point $(2,4)$.

To find the equation of a circle we need the center and the radius. The center is $(-3,1) = (h,k)$. We can find the radius because this must be the distance between the center $(-3,1)$ and any point on the circle—in this case, one of these points is $(2,4)$. Thus we find this distance between $(-3,1)$ and $(2,4)$, using the distance formula from section 3–1:

$$d = \sqrt{(-3 - 2)^2 + (1 - 4)^2} = \sqrt{34}$$

This same value is the radius. Using $r = \sqrt{34}$ and $(h,k) = (-3,1)$ we create the equation of the circle:

$$(x - (-3))^2 + (y - 1)^2 = (\sqrt{34})^2$$
$$(x + 3)^2 + (y - 1)^2 = 34$$

Graphical analysis of relations for the function and one-to-one properties

The graph of a relation can be used to determine whether that relation is a function, and whether or not a function is one to one. This is done by the vertical line test and the horizontal line test.

> **Vertical line test for a function**
> If no vertical line crosses the graph of a relation in more than one place, the relation is a function.

The vertical line test works for the following reason. Assume a vertical line crosses a graph at more than one point. Since these two points are in a vertical line their first components (the x-values) are equal. Therefore the function must have two points in which the first element repeats, and it is therefore not a function.

> **Horizontal line test for a one-to-one function**
> If no horizontal line crosses the graph of a function in more than one place, the function is one to one.

The horizontal line test works for reasons similar to those for the vertical line test. If a horizontal line crosses a function at two (or more) points, then these are different domain elements (first components) with the same range elements (second component). Therefore the function is not one to one.

■ ***Example 3–5 D***

Tell which relations are functions, and which functions are one to one.

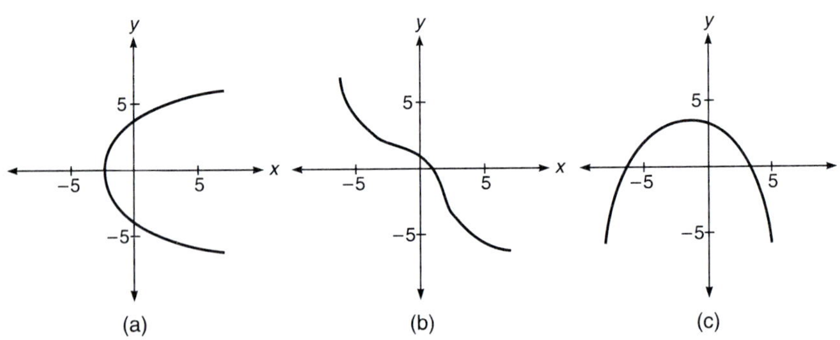

1. Relation (a) is not a function since there are clearly many vertical lines that would intersect the graph in at least two places.
2. Relation (b) is a function since no vertical line will intersect the graph in more than one place. It is also one to one since no horizontal line will intersect the graph in more than one place.
3. Relation (c) is a function by the vertical line test, but not a one-to-one function. ■

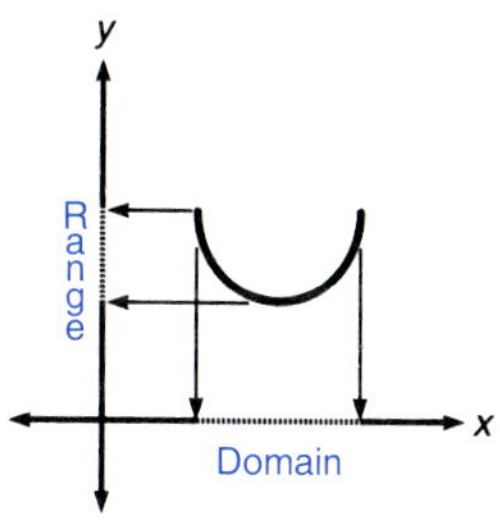

Figure 3–13

The domain and range for a given relation or function can also be determined from its graph. The domain is that portion of the x-axis that lies below or above the graph of the relation. The range is that portion of the y-axis that is to the right or left of the graph of the relation. Figure 3–13 illustrates this for a function that happens to lie in the first quadrant.

The domain and range for the relations of example 3–5 D are shown in figure 3–14. Formally, these intervals on the x- and y-axes are called **projections.** Thus, for example, the domain of the relation shown in (a) is a projection of the relation on to the x-axis, and the range is the projection of the relation on to the y-axis.

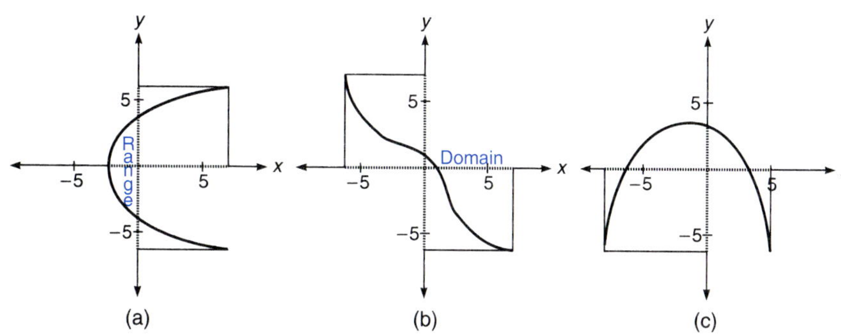

Figure 3–14

Increasing/decreasing property of functions

A function is said to be decreasing if it looks like: 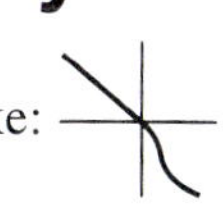 . It is said to be increasing when it looks like  . Most functions exhibit both decreasing and increasing behavior 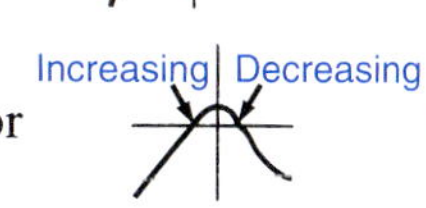

for different parts of their domain.

For this reason it is often necessary to restrict the values of x to a certain interval if we expect to describe the behavior of the function as increasing or decreasing.

To obtain an algebraic description of these properties, observe that if a function is decreasing, the y, or $f(x)$, values decrease (move down in the graph) as the x values increase (move to the right in the graph). This can be stated as follows.

Decreasing function
A function f is decreasing on an interval if for any x_1 and x_2 in the interval $f(x_2) < f(x_1)$ whenever $x_2 > x_1$.

Similar reasoning leads to the definition of an increasing function.

Increasing function
A function f is increasing on that interval if for any x_1 and x_2 in the interval $f(x_2) > f(x_1)$ whenever $x_2 > x_1$.

The following discussion, as well as example 3–5 E, will illustrate this concept.

Functions as tables of values and reading $f(x)$ values from a graph

Table 3–3 shows the office vacancy rates, in percentages, for Phoenix, Arizona, for the years 1978 to 1985. This data can be plotted using the horizontal axis for the year and the vertical axis for the vacancy rate. The resulting points are plotted and connected by straight lines[7] in figure 3–15.

Year	Vacancy rate
1978	10.5
1979	5.3
1980	6.7
1981	8.2
1982	10.1
1983	13.2
1984	19.7
1985	23.1

Table 3–3

Source: *Quantitative Methods for Financial Analysis*, Stephen J. Brown and Mark P. Kritzman, Editors (Homewood, Illinois: Dow Jones-Irwin, 1987).

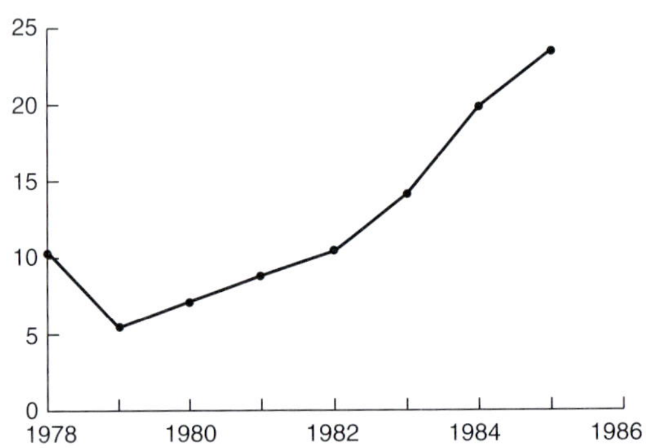

Figure 3–15

[7]The original graph was generated using a spreadsheet computer program. Many spreadsheet programs provide the capability to graph tables of data in various formats. The TI-81 can graph data with its STAT: DRAW: xy Line feature.

Table 3–3 is a function. It is a set of ordered pairs (year, vacancy rate) in which no first element repeats. We could say it is decreasing for the interval [1978, 1979], and increasing for the interval [1979, 1985]. (Of course we do not know whether the actual vacancy rate got even lower sometime in the [1978, 1979] interval. We apply the terms decreasing and increasing to the set of data in table 3–3 only.)

Every issue of any large newspaper presents graphs like that in figure 3–15 without the table of data or mathematical formula from which it was obtained. The ability to interpret functions presented in this way is so important that it is on tests to get into law, medical, or practically any other advanced school, and on many preemployment tests.

■ ***Example 3–5 E***

Solve the following problems based on the figure at the top of page 185. The figure represents how the depth of slush on a runway affects the takeoff distance of a large passenger aircraft. The depth of the slush and its density ρ (rho) is measured and reported to the aircraft's crew. Takeoff distance must be adjusted accordingly.

1. At a slush depth of 19 millimeters (mm) and density of 0.8, find an approximation to the percentage of increase in takeoff distance.

Point *a*, circled in the figure, is about halfway between 20% and 40%. Thus we could estimate a 30% increase in takeoff distance.

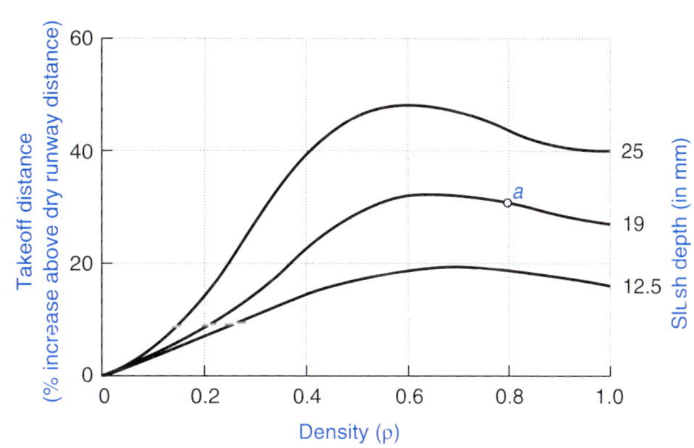

Source: Based on a figure in D. P. Davies, *Handling the Big Jets*, 3rd ed. (London: Civil Aviation Authority, 1975).

2. If the slush thickness is 25 mm, for what densities is the takeoff distance increased 40% or above?

The curve representing a 25-mm thickness is above 40% for densities from 0.4 to 1.0.

3. For a slush depth of 12.5 mm, for what densities is the percentage increasing, and for what densities is it decreasing?

The curve representing the 12.5-mm depth is increasing up to a density of about 0.7 (more than halfway between 0.6 and 0.8). It is decreasing for densities above 0.7. ■

Example 3–5 F also illustrates reading values from the graph of a function.

Example 3–5 F

The figure represents the graph of $y = f(x)$ for some function f. Use it to answer the following questions.

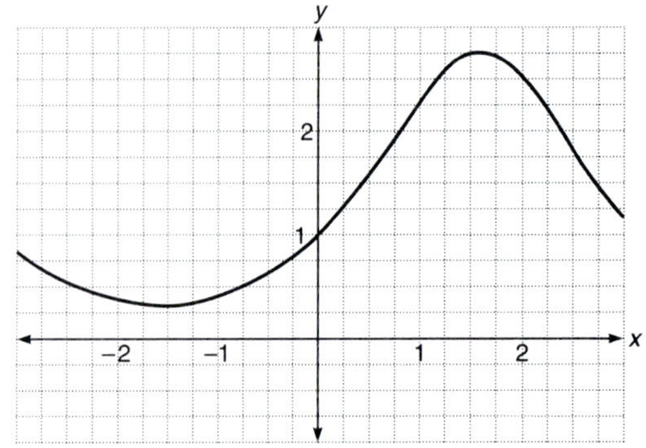

1. Estimate $f(1)$ to the nearest 0.1.

Moving right to $x = 1$ and moving up to where that vertical line meets the graph of f, we estimate a value of 2.3 (slightly above 2.25).

2. For what value of x is $f(x) = 2$?

Examining where the horizontal line that passes through $y = 2$ crosses the graph of f, we see that $x \approx 0.8$ or 2.3.

3. For what values of x is f increasing?

Approximately for x-values between -1.5 and 1.6.

Even/odd functions and symmetry

The graphs of many functions exhibit some form of symmetry. Two forms of symmetry are related to whether a function is even or odd.

> **Even/odd property of functions**
> A function f is **odd** if for all x in its domain $f(-x) = -f(x)$.
> A function f is **even** if for all x in its domain $f(-x) = f(x)$.

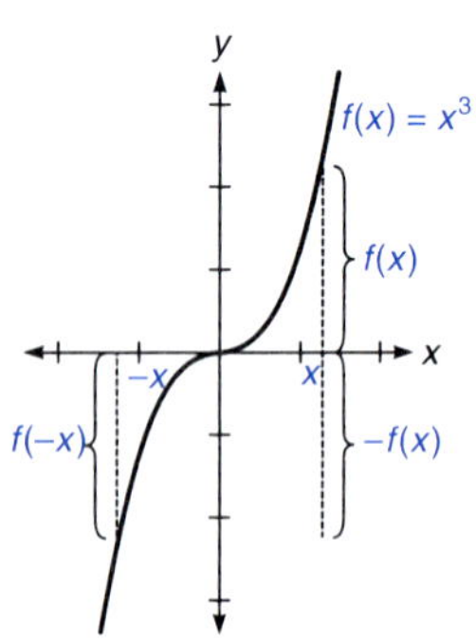

Figure 3–16

The choice of names odd and even is not coincidental. $f(x) = x^2$ is an even function, and its exponent is even. For example, $f(-3)$ is 9, and $f(3)$ is 9. $g(x) = x^3$ is an odd function, and its exponent is odd. Observe that $g(-3) = -g(3)$, since both values are -27.

The graph of an odd function is symmetric about the origin. This is illustrated with the function $f(x) = x^3$ in figure 3–16, where we see graphically that $f(-x) = -f(x)$. *The graph of an even function is symmetric about the* y-axis. We see this in the graph of $f(x) = x^2$ (figure 3–17), where we see that $f(-x) = f(x)$.

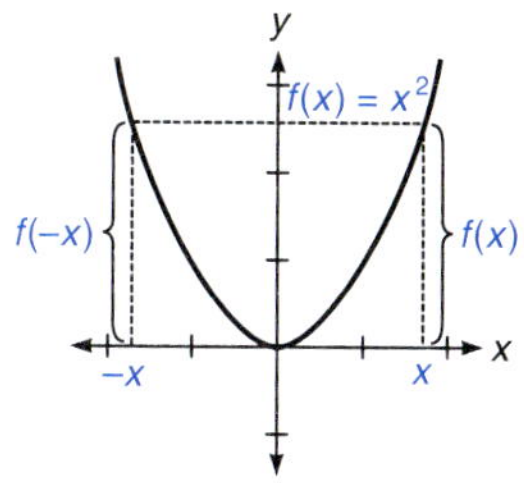

Figure 3–17

When a function is defined by an expression in one variable, say x, the following will tell whether the function is even or odd (or neither).

> **To determine if a function is even or odd**
> Compute expressions for $f(-x)$ and for $-f(x)$ and compare them.
> - If $f(-x) = f(x)$ the function is even.
> - If $f(-x) = -f(x)$ the function is odd.

By examining where the ordered pairs of a function would have to lie for a function to be both even and odd we can establish that the only function that is defined for all real numbers, which is both even and odd, is the function $f(x) = 0$. Its graph is the x-axis. Thus we generally do not have to check to see if a function has both the even and odd properties—*it can only have one of them.*

The algebra we use relies heavily on the property that

$$(-x)^n = \begin{cases} x^n & \text{if } n \text{ is even} \\ -x^n & \text{if } n \text{ is odd} \end{cases}$$

Thus $(-x)^2 = x^2$, $(-x)^3 = -x^3$, $(-x)^4 = x^4$, $(-x)^5 = -x^5$, and so on.

■ *Example 3–5 G*

Test the function for the even/odd property. State which type of symmetry the graph would have based on being even, odd, or neither even nor odd.

If you have access to a graphing calculator graph the function to check your answer.

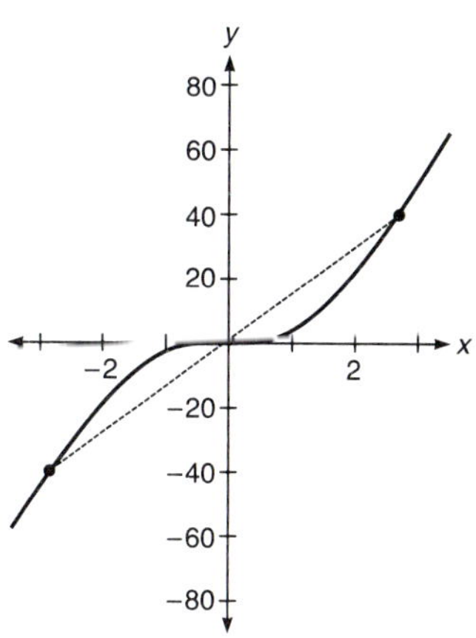

1. $f(x) = 2x^3 - x$

$f(-x) = 2(-x)^3 - (-x) = -2x^3 + x$ — Compute $f(-x)$

$-f(x) = -(2x^3 - x) = -2x^3 + x$ — Compute $-f(x)$

Since $f(-x) = -f(x)$, f is an odd function and the graph has symmetry about the origin. Two points are shown that illustrate the origin symmetry, which illustrates that any point on the graph has a mirror image on a line that passes through the origin.

> [Y=] 2 [X|T] [MATH] 3
> [−] [X|T]
> [RANGE −3,3,−60,60]
> Yscl=10

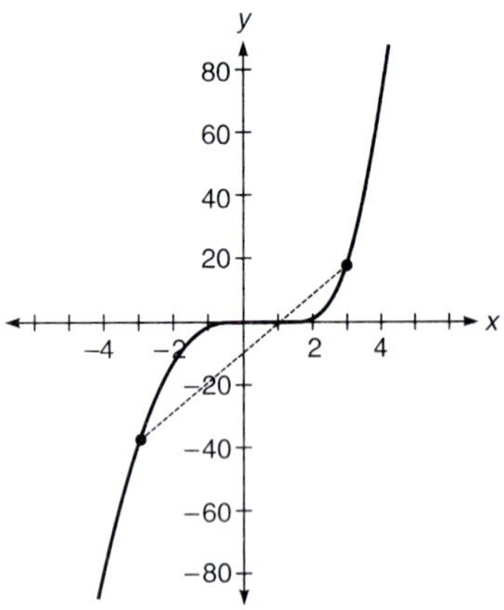

[Y=] [X|T] [MATH] 3
[−] [X|T] [x^2]
[RANGE −5,5,−60,60]
Yscl=10

2. $f(x) = x^3 - x^2$

$$f(-x) = (-x)^3 - (-x)^2 = -x^3 - x^2$$
$$-f(x) = -(x^3 - x^2) = -x^3 + x^2$$

Since $f(-x) \neq f(x)$, f is not even; since $f(-x) \neq -f(x)$, f is not odd. Thus f is neither even nor odd. The graph would not have y-axis or origin symmetry. In fact the graph clearly shows no y-axis symmetry, and the two points shown indicate no origin symmetry. Observe that one point is $(3,f(3))$, and the second is $(-3,f(-3))$. If $f(-3)$ were the same as $-f(3)$, then the line connecting them would pass through the origin. ■

Mastery points

Can you

- Graph a circle when its equation is given in standard form?
- Complete the square to put the equation of a circle in standard form?
- Determine the equation of a circle that has certain properties?
- Apply the vertical line test to the graph of a relation to determine if that relation is a function?
- Apply the horizontal line test to the graph of a function to determine if that function is one to one?
- Test equations for the even/odd properties?

Exercise 3–5

State the center and radius of the circle and graph it.

1. $x^2 + y^2 = 16$
2. $x^2 + (y - 1)^2 = 8$
3. $(x + 2)^2 + y^2 = 20$
4. $x^2 + y^2 = 9$
5. $x^2 + (y - 4)^2 = 9$
6. $(x + 3)^2 + y^2 = 16$
7. $(x - 1)^2 + (y - 4)^2 = 8$
8. $(x + 2)^2 + (y - 5)^2 = 12$
9. $(x + 3)^2 + (y - 2)^2 = 20$
10. $(x - 3)^2 + (y - 5)^2 = 36$

Transform the equation into the standard form for a circle. Then state the center and radius of the circle and graph it.

11. $x^2 + y^2 - 6y = 6$
12. $x^2 + 8x + y^2 = 0$
13. $x^2 + 5x + y^2 = 4$
14. $x^2 + y^2 - 3y = 2$
15. $x^2 - x + y^2 - 4y = 9$
16. $x^2 + 3x + y^2 - 5y = 1$
17. $x^2 - 2x + y^2 + 4y + 5 = 0$
18. $2x^2 - 6x + 2y^2 + 4y = 0$
19. $x^2 + 4x + y^2 + 6 = 0$
20. $x^2 - 10x + y^2 + 2y = 0$
21. $3x^2 + 3y^2 - y - 10 = 0$
22. $2x^2 + 2y^2 = 3$
23. $2x - 5x^2 - 5y - 5y^2 + 3 = 0$
24. $4x^2 - 8x - 12y + 4y^2 = 1$
25. $(2x - 3)^2 + (2y + 1)^2 = 2$
26. $9x^2 + (3y - 2)^2 = 14$

Find the equation of the circle with the given properties. Leave the equation in the form $Ax^2 + Bx + Cy^2 + Dy + E = 0$.

27. radius 2, center at $(-3,2)$
28. radius 5, center at $(4,2)$
29. radius $\sqrt{5}$, center at $(2,3 - \sqrt{2})$
30. radius $\frac{2}{3}$, center at $(\frac{4}{5},-\frac{1}{2})$
31. center at $(0,3)$ and passes through the point $(0,-5)$
32. center at $(-2,4)$ and passes through the point $(-5,4)$
33. center at $(1,-3)$ and passes through the point $(-2,5)$
34. center at $(-2,0)$ and passes through the point $(2,6)$
35. end points of a diameter are at $(-4,2)$ and $(10,8)$
36. end points of a diameter are at $(-3,4)$ and $(7,-6)$

Tell which relation is a function, and which functions are one to one.

37.

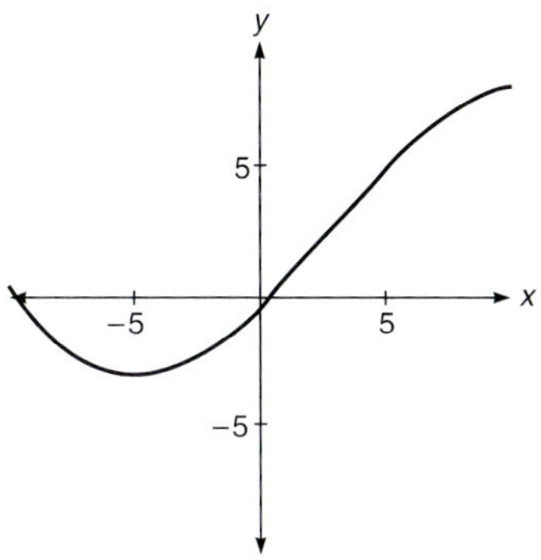

38.

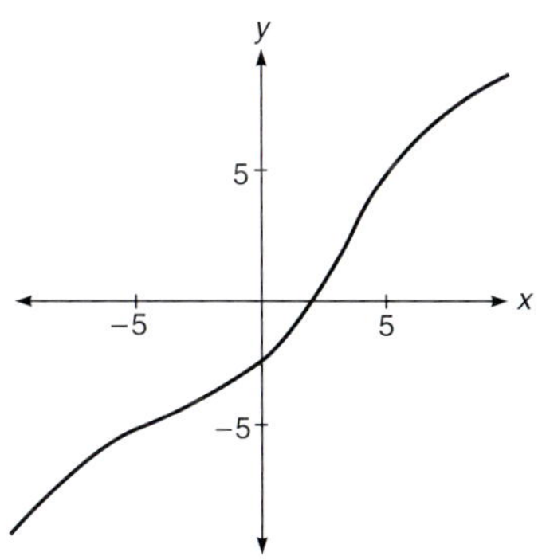

39.

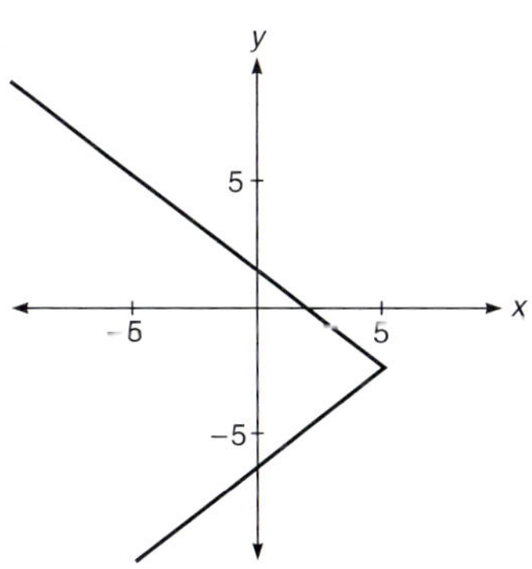

40.

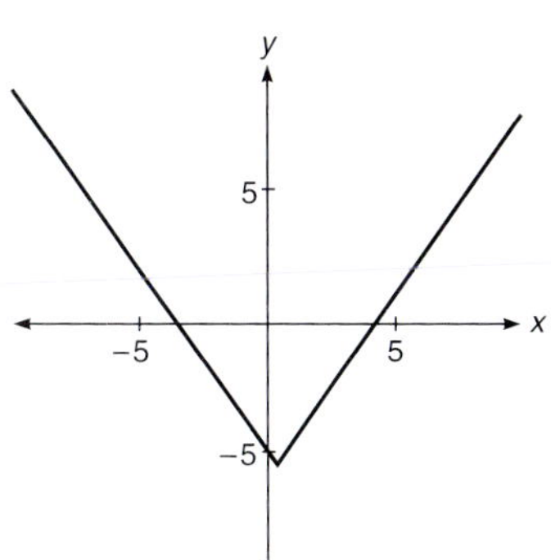

Problems 41 through 44 apply to the figure on takeoff distances.

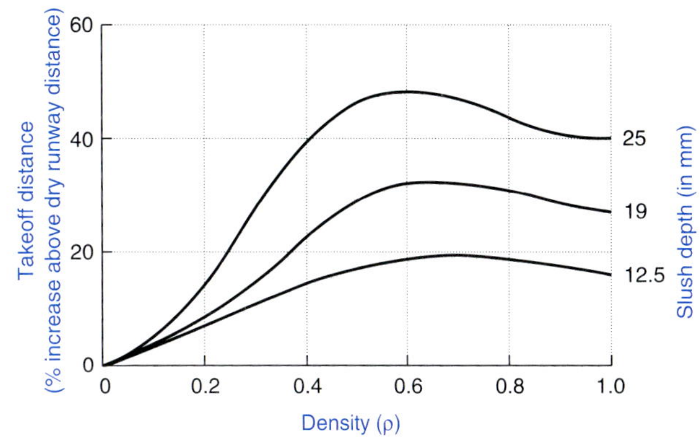

Based upon a figure in D. P. Davies, *Handling the Big Jets,* 3rd ed. (London: Civil Aviation Authority, 1975).

41. At a slush depth of 25 mm and density of 0.4, what is the approximate percentage increase in takeoff distance?

42. At a slush depth of 12.5 mm and density 0.2, what is the approximate percentage increase in takeoff distance?

43. At a slush depth of 19 mm, what does the density have to be for the percentage increase in takeoff distance to be at least 20%?

44. Suppose that the dry runway takeoff distance for the aircraft is 3,400 feet, but there is 25 mm of slush on the runway with a measured density of 0.4. What is the revised takeoff distance, to the nearest 100 feet?

Problems 45 through 50 refer to the figure, which is the graph of $y = f(x)$ for a function f. Estimate all required values to the nearest 0.1.

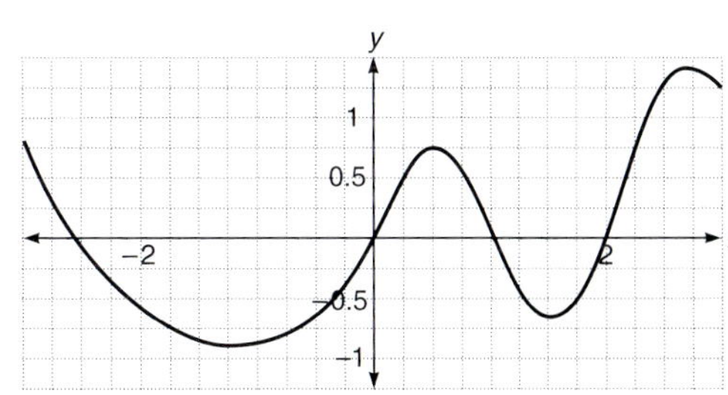

45. What is the value of (a) $f(0.5)$ and (b) $f(-2)$?

46. For what value(s) of x is $f(x) = 0$?

47. For what value(s) of x is $f(x) = -0.5$?

48. For what values of x is f decreasing?

49. For what values of x is f increasing?

50. For what values of x is $f(x) = x$?

Test the function for the even/odd property. State which type of symmetry the graph would have based on being even, odd, or neither even nor odd.

51. $f(x) = \sqrt{9 - x^2}$

52. $g(x) = x$

53. $h(x) = \dfrac{4}{x}$

54. $f(x) = x^5 - 4x^3 - x$

55. $g(x) = x^2 - 4$

56. $g(x) = x^4 - 9$

57. $h(x) = x^3$

58. $f(x) = x^4 - x - 2$

59. $f(x) = x^3 - 5$

60. $h(x) = \dfrac{x^2 - 4}{x^2 + 9}$

61. $g(x) = \dfrac{x}{x^3 - 1}$

62. $f(x) = \dfrac{x^2}{x^2 + 3}$

63. $f(x) = \dfrac{x}{x^2 + 3}$

64. $f(x) = 2 - 3x^2$

65. $f(x) = \sqrt{4 - x^2}$

66. $f(x) = \sqrt{x^2 + 2}$

67. $f(x) = \dfrac{x}{\sqrt{x^2 + 1}}$

68. $f(x) = \dfrac{x^2}{\sqrt{x^2 + 2}}$

69. The circle $(x - 1)^2 + (y + 2)^2 = 13$ passes through the point $(3,-5)$. Find the equation of the line L that passes through this point and is tangent to the circle. See the figure. *Hint:* A radius from the center to the point of tangency is perpendicular to the tangent line.

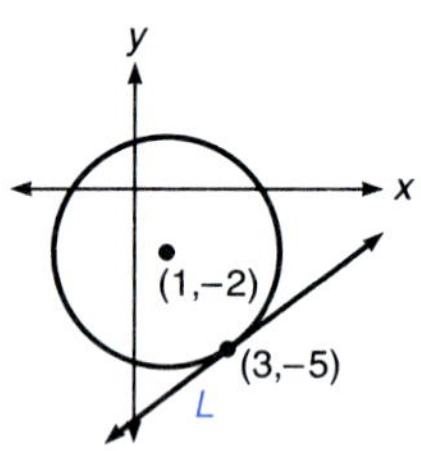

70. The function written as $f(x) = [x]$ is called the *greatest integer function* and $[x]$ is defined as the greatest integer that is less than or equal to x. For example, $[3.8] = 3$, $[0.8] = 0$, $[-0.8] = -1$, $[-3.8] = -4$. Graph this function.

71. A computer-controlled robot is being programmed to grind the outer edge of the object shown in the figure. For this purpose the equation of the circle must be known in terms of a coordinate system centered at the point A. Find this equation.

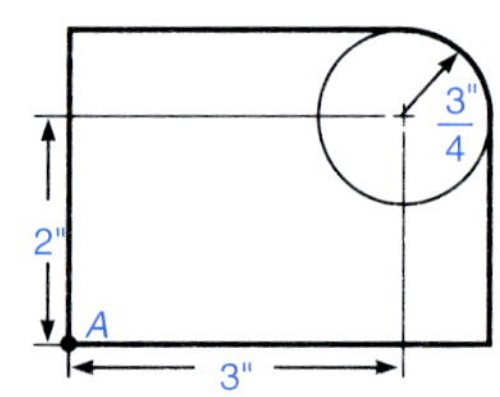

72. If you have a graphing calculator, graph those functions in problems 51 through 68 that you determined had even or odd symmetry, and verify the symmetry by examining the graph.

Skill and review

1. Graph $f(x) = x^2 - 4$.
2. Graph $f(x) = (x - 4)^2$.
3. Graph $f(x) = (x - 4)^2 - 4$.
4. Solve $|2x - 3| = 8$.
5. Factor $x^6 - 64$. (Note: $64 = 2^6$.)
6. Find the equation of the line that passes through the points $(-4,1)$ and $(3,-5)$.

Chapter 3 summary

- **Relation** A set of ordered pairs. The set of all first components is called the **domain** of the relation, and the set of all second components is called the **range** of the relation.
- **Point** An ordered pair.
- **Straight line** The set of solutions (points) to any equation that can be put in the form $ax + by + c = 0$, with at least one of a or b not zero.
- A **horizontal line** is a line of the form $y = k$.
- A **vertical line** is a line of the form $x = k$.

- **Slope of a straight line** If $P_1 = (x_1,y_1)$ and $P_2 = (x_2,y_2)$ are two different points on a nonvertical line then the slope of the line, m, is $m = \dfrac{y_2 - y_1}{x_2 - x_1}$.
- **Slope-intercept form of a straight line** $y = mx + b$, where m is called the **slope** of the line, and b is the **y-intercept.**
- **Point-slope formula of a straight line** If $P_1 = (x_1,y_1)$ and $P_2 = (x_2,y_2)$ are two different points and $x_1 \neq x_2$, then the equation of the line that contains these points is obtained by the formula $y - y_1 = m(x - x_1)$, where m is the slope determined by P_1 and P_2.
- **Parallel lines** Two different lines with the same slope.
- **Perpendicular lines** Two lines, the product of whose slopes is -1.
- **The method of substitution** To solve a system of two equations in two variables, x and y:
  1. Solve one equation for y.
  2. Replace y in the other equation.
  3. Solve this new equation for x.
  4. Use the known value of x in either of the original equations to find y.
- **Distance between two points** If $P_1(x_1,y_1)$ and $P_2(x_2,y_2)$ are two different points, then the distance between them is called $d(P_1,P_2)$ and is defined as $d(P_1,P_2) = \sqrt{(x_2 - x_1)^2 + (y_2 - y_1)^2}$.
- **Midpoint of a line segment** If $P_1(x_1,y_1)$ and $P_2(x_2,y_2)$ are the end points of a line segment, then M, the midpoint of the line segment, is $M\left(\dfrac{x_1 + x_2}{2},\dfrac{y_1 + y_2}{2}\right)$.
- **Function** A relation in which every first component is different.
- **One-to-one function** A function in which every second component of the ordered pairs is different.
- **Implied domain of a function** Unless otherwise stated, the set of all values in R for which all expressions that define the function are defined and real valued. We ordinarily just say domain instead of implied domain.
- **Linear function in one variable** A function of the form $f(x) = mx + b$; its graph is a straight line.
- Given the graph of a function $f(x)$,

  Vertical translation The graph of $y = f(x) + c$ is the graph of $f(x)$ shifted up if $c > 0$ and down if $c < 0$.

  Horizontal translation The graph of $y = f(x - c)$ is the graph of $f(x)$ shifted right if $c > 0$ and left if $c < 0$.
- If $g(x) = kf(x)$ for some functions g and f and a real number k, then g is a **vertically scaled** version of f.
- **Circle** $\{(x,y) \mid (x - h)^2 + (y - k)^2 = r^2, r > 0\}$; (h,k) is the center, and r is the radius.
- **Vertical line test for a function** If no vertical line crosses the graph of a relation in more than one place, the relation is a function.
- **Horizontal line test for a one-to-one function** If no horizontal line crosses the graph of a function in more than one place, the function is one to one.
- **Decreasing function** A function f is decreasing on an interval if for any x_1 and x_2 in the interval $f(x_2) < f(x_1)$ whenever $x_2 > x_1$.
- **Increasing function** A function f is increasing on an interval if for any x_1 and x_2 in the interval $f(x_2) > f(x_1)$ whenever $x_2 > x_1$.
- **Symmetry/even-odd property** A function f is even if for all x in its domain $f(-x) = f(x)$. The graph of an even function is symmetric about the y-axis. A function f is odd if for all x in its domain $f(-x) = -f(x)$. The graph of an odd function is symmetric about the origin.

Chapter 3 review

[3–1] For each problem, list three points which lie on the lines; then graph the line. Compute the x- and y-intercepts.

1. $2y = -5x - 8$ **2.** $\frac{1}{3}x - 3 = 4y$
3. $3x - 2y + 8 = 0$ **4.** $x = -4$
5. $\frac{3}{4}x - \frac{1}{3}y = 1$ **6.** $1.5x - y = 6$

7. An investment account pays 9% interest on money invested, but deducts \$100 per year service charge. The amount of interest I that will be paid on an amount of money p, if that amount does not change over the year, is therefore $I = 0.09p - 100$. Graph this equation for values of p from 0 to \$10,000.

Find the coordinates of the midpoint of the line segment determined by the following points.

8. $(-\frac{1}{3},-2), (4\frac{1}{3},-3)$ **9.** $(-2,\sqrt{8}), (-6,\sqrt{2})$

Find the distance between the given pairs of points.

10. $(-3,-1\frac{1}{2}), (4,2\frac{1}{2})$ **11.** $(\sqrt{8},2), (\sqrt{2},-3)$

[3–2] Graph each line and find its slope, m.

12. $5x - 9y = 15$ **13.** $2y = 6 - 3x$ **14.** $2x = 3y - 4$
15. $y = \frac{4}{9}$ **16.** $x - 8.5 = 0$

Find the slope of the line that contains the given points.

17. $(-3,-2\frac{1}{3}), (5,\frac{2}{3})$ **18.** $(\frac{1}{2},\frac{2}{3}), (\frac{3}{4},\frac{1}{3})$
19. $(\sqrt{3},-1), (2\sqrt{3},4)$

Find the slope-intercept equation of the line that contains the given points.

20. $(-4,9)$, $(2,1)$ **21.** $(2,-3)$, $(3,1)$

Find the slope-intercept equation of the line in each case.

22. A line with slope $-\frac{2}{3}$ and y-intercept at -3.
23. A line with slope -6 that passes through the point $(\frac{1}{2},-4)$.
24. A line that is parallel to the line $2y - 3x = 4$ and passes through the point $(3,2)$.
25. A line that is perpendicular to the line $2y + 5x = 8$ and passes through the point $(4,-\frac{1}{5})$.

Find the point at which each of the two lines intersect.

26. $x + 2y = 4$
$3x - y = 6$

27. $y = 3x + 2$
$-y + x = -5$

28. Find the equation that describes all the points that are equal distances from the points $(-4,3)$ and $(4,6)$.

29. Show that any two distinct points on the line $y = 3x - 4$ will produce a slope of 3.

30. The records of a retail store show that a customer must wait 2 minutes 40 seconds, on average, when 6 checkout lines are open. Its records also show that under similar conditions a customer waits 35 seconds, on average, when 20 checkout lines are open. Use linear interpolation to estimate how long a customer might wait when 10 checkout lines are open. Round to the nearest 5 seconds.

[3–3] In problems 31–34, (a) state the domain and range of the relation and (b) note whenever the relation is a function. For each function state whether it is one to one or not.

31. $\{(1,4), (-3,8), (4,2), (1,5)\}$
32. $\{(-3,4), (-2,5), (-1,3), (0,4), (1,1)\}$
33. $\{(-10,12), (4,13), (2,2), (3,-5)\}$
34. $\{(3,\pi), (3,-\sqrt{2}), (17,\frac{8}{13}), (\pi,\sqrt{2})\}$

List each relation as a set of ordered pairs; note whenever the relation is a function. For each function state whether it is one to one or not.

35. $\{(x,y) \mid x + 3y = 6, x \in \{-3,9,\sqrt{18},\frac{3}{4},\pi\}$
36. $\{(x,y) \mid y = \pm\sqrt{4 - 3x}, x \in \{-3,-2,-1,0,1\}\}$

In problems 37–40, for each function, state the implied domain D and compute the function's value for the domain elements $x = -4, 0, \frac{1}{2}, 3\sqrt{5}, c - 2$ (unless not in the domain of the function). Do not rationalize complicated denominators. Assume $c - 2$ is in the domain of the function.

37. $f(x) = \dfrac{1 - x^2}{4x - 3}$ **38.** $g(x) = \sqrt{12 - 24x}$

39. $v(x) = 3 - 2x - x^2$ **40.** $g(x) = \dfrac{-4}{\sqrt{x^2 + x - 6}}$

41. If $f(x) = x^2 - 5x - 5$, compute an expression for $\dfrac{f(x + h) - f(x)}{h}$.

42. If $f(x) = x^6 - 3x^3 + 1$ and $g(x) = \sqrt[3]{4x - 1}$, compute (a) $f(g(3))$; (b) $f(g(\frac{1}{2}))$; (c) $f\left(g\left(\dfrac{a^3}{4} + \dfrac{1}{4}\right)\right)$.

43. If $f(x) = 2x + 3$ and $g(x) = 1 - 4x$, compute (a) $f(g(-3))$; (b) $f(g(\frac{1}{2}))$; (c) $g\left(f\left(\dfrac{a}{a + b}\right)\right)$.

44. If $f(x) = 3x - 2$ and $g(x) = -2x + 1$, compute the value of the given expression.

a. $\dfrac{3g(2)}{2g(3)}$ **b.** $\dfrac{2f(5) - f(\frac{1}{3})}{2f(-5)}$

45. Graph the linear function $f(x) = 5x - 3$.

46. With a wind speed of 10 miles per hour the wind chill factor makes an actual temperature of 30° F feel like 16° F and an actual temperature of 10° F feel like −9° F. Assuming a wind speed of 10 mph, create a linear function that describes perceived temperature (according to the wind chill factor) as a function of actual temperature. Then use this function to compute perceived temperature when the actual temperature is 14° F. Note that the 10 mph is not part of the function, since we are assuming it to be the same for both cases.

[3–4] Graph each of the following. Compute x- and y-intercepts.

47. $f(x) = (x - 3\frac{1}{2})^2 + 5$ **48.** $f(x) = (x + 1)^2 - 2$
49. $f(x) = \sqrt{x + \frac{5}{2}} - 5$ **50.** $f(x) = \sqrt{x + 4} + 4$
51. $f(x) = (x - 2)^3 + 8$ **52.** $f(x) = x^3 - 8$
53. $f(x) = \dfrac{1}{x - 3} - 5$ **54.** $f(x) = \dfrac{1}{x} + 2$
55. $f(x) = |x + 5| - 5$ **56.** $f(x) = |x + 2| + 3$
57. $f(x) = -3(x - 1)^2 + 4$ **58.** $f(x) = 3\sqrt{x - 8} - 5$
59. $f(x) = 2(x - 2)^3 + 1$ **60.** $f(x) = \dfrac{4}{x + 4} - 1$

[3–5] If necessary, transform the equation into the standard form for a circle. Then state the center and radius of the circle and graph it.

61. $x^2 + (y + 4)^2 = 12$ **62.** $x^2 - 6x + y^2 = 16$
63. $x^2 - 2x + y^2 - 4y = 4$
64. $x^2 - x + y^2 + 3y - 5 = 0$
65. $(2x - 5)^2 + (2y + 3)^2 = 8$

Find the equation of the circle with the given properties.

66. radius $4\sqrt{5}$, center at $(-3\frac{1}{2},2)$
67. radius 3, center at $(1,-3)$
68. center at $(1,3)$ and passes through the point $(-2,-5)$

Tell which relation is a function, and which functions are one to one.

69.

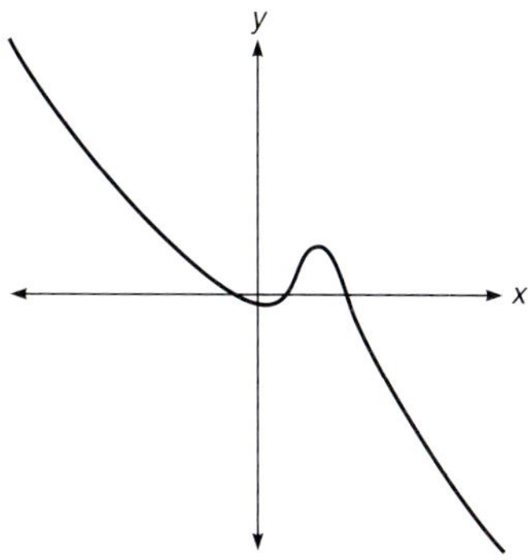

70.

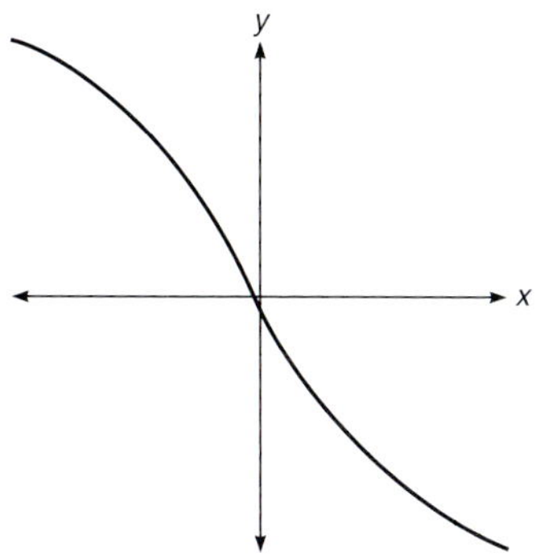

71.

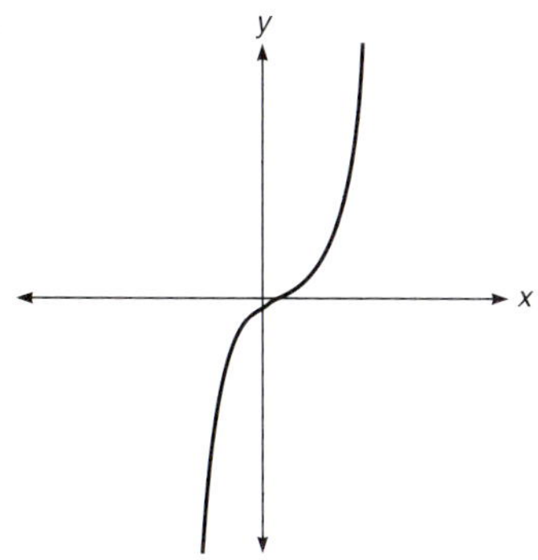

Test the function for the even/odd property. State which type of symmetry the graph would have based on being even, odd, or neither even nor odd.

72. $f(x) = 3x^4 - 5x^2$

73. $f(x) = \dfrac{-x}{x^2 + 1}$

74. $h(x) = \dfrac{x^2}{x^3 - x}$

75. $h(x) = x\sqrt{x^2 - 3}$

76. $f(x) = \dfrac{2x}{x^2 + 3x}$

77. $g(x) = \dfrac{3}{2 - x}$

78. The circle $(x - 2)^2 + (y - 3)^2 = 25$ passes through the point (7,0). Find the equation of the line L that passes through this point and is tangent to the circle.

The chart describes gene drift.[8] It gives the frequency of a gene in two populations that split from one population. Over time, the frequency of a given gene can diverge widely, as the graph shows. Use the chart to answer questions 79 through 82. Since the data is plotted for every five generations, answers should be to the nearest five generations.

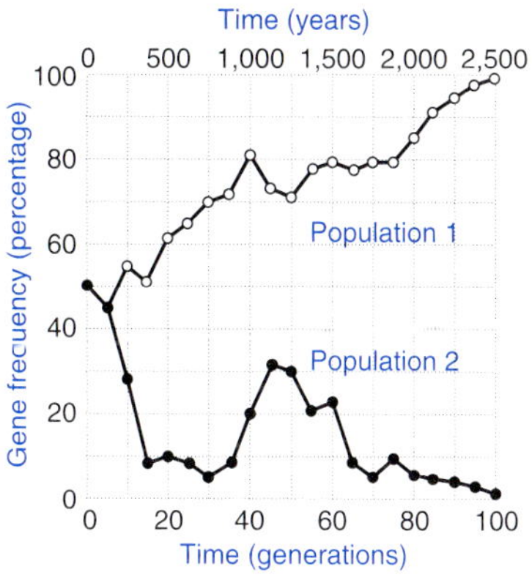

79. In what generation(s) does the frequency of the gene go above 75% in population 1?

80. What is the percentage of frequency of the gene in population 2 in the 65th generation?

81. In what generation does the difference in frequency of the gene in the two populations first differ by more than 70%?

82. In what time intervals is population 2 increasing?

[8]From Luigi Luca Cavalli-Sforza, "Genes, Peoples and Languages," *Scientific American,* November 1991.

Chapter 3 test

For each line, list three points that lie on the line; then graph the line. Compute the x- and y-intercepts.

1. $3y + 5x = 15$

2. $x - 3 = \frac{1}{2}y$

3. $x = 5$

4. $1.2x - 1.5y = 6$

5. Find the coordinates of the midpoint of the line segment determined by the points $(-2,5\frac{1}{2})$, $(3,2\frac{1}{2})$.

6. Find the distance between the points $(2,-5)$, $(-3,-1)$.

Graph each line and find its slope m.

7. $2x - y = 7$ **8.** $5x + 4y = -20$ **9.** $y = -1$

10. Find the slope of the line that contains the points $(-3,-2)$, $(5,-4)$.

11. Find the slope-intercept equation of the line that contains the points $(-2,3)$, $(2,1)$.

12. Find the slope-intercept equation of the line with slope -5 and x-intercept at -3.

13. Find the slope-intercept equation of the line that is perpendicular to the line $y - \frac{1}{5}x = 1$ and passes through the point $(2,-3)$.

14. Find the point at which the two lines intersect: $2x + y = 4$ and $3x - 2y = 6$.

15. Find the equation that describes all the points that are equal distances from the points $(2,3)$ and $(4,6)$.

16. List the relation as a set of ordered pairs; note if the relation is a function and whether it is one to one or not. $\{(x,y) \mid y = \sqrt{2x - 1}, x \in \{1, 2, 3, 4\}\}$

For each function in problems 17–19, state the implied domain and compute the function's value for the domain elements $x = -2, 0, 3$, and $c - 3$ (unless not in the domain of the function); assume $c - 3$ is in the domain.

17. $f(x) = \dfrac{x}{x - 3}$ **18.** $g(x) = \sqrt{6 - 2x}$

19. $v(x) = 3 - 2x - x^2$

20. If $f(x) = 2x^2 - 3$, compute $\dfrac{f(x + h) - f(x)}{h}$.

Graph the function. Compute the intercepts and the vertex of those that are parabolas.

21. $f(x) = (x + 1)^2 - 2$ **22.** $f(x) = \sqrt{x + 4} - 2$

23. $f(x) = (x - 2)^3 + 3$ **24.** $f(x) = \dfrac{1}{x - 3} - 2$

25. $f(x) = |x + 2| + 3$ **26.** $f(x) = \frac{1}{2}(x - 2)^3 + 1$

If necessary, transform the equation into the standard form for a circle. Then state the center and radius and graph it.

27. $(x - 2)^2 + (y + 4)^2 = 16$

28. $x^2 - 6x + y^2 + 3y - 5 = 0$

Find the equation of the circle with the given properties.

29. radius $4\sqrt{5}$, center at $(-3\frac{1}{2}, 2)$

30. center at $(1,3)$ and passes through the point $(-2,-5)$

Questions 31 through 37 refer to the figure.

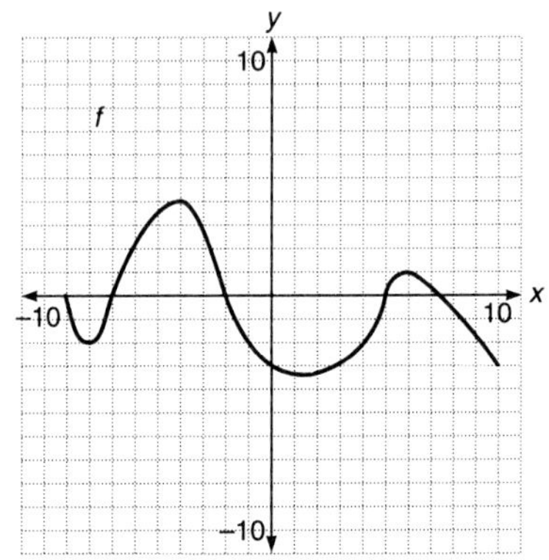

31. Find **a.** $f(-8)$ **b.** $f(-4)$ **c.** $f(0)$ **d.** $f(3)$

32. If $f(x) = 0$, $x \approx$ ____ **33.** If $f(x) = -2$, $x \approx$ ____

34. Where is f increasing? **35.** Where is f decreasing?

36. What is the domain of f? **37.** What is the range of f?

Test the function for the even/odd property. State which type of symmetry the graph would have, based on being even, odd, or neither even nor odd.

38. $f(x) = \dfrac{-x^2}{x^2 + 1}$ **39.** $h(x) = \dfrac{-5x}{x - x^2}$

40. $h(x) = x^2\sqrt{x^2 - 3}$

41. The table shows the approximate boiling point of water for various altitudes in feet. Denver, Colorado, is about 1 mile (5,280 feet) above sea level. (a) Use linear interpolation to estimate the boiling point of water in Denver, to the nearest tenth of a degree. (b) National parks in the mountains near Denver are at altitudes above 12,000 feet. Use linear interpolation to estimate the boiling point of water at a camp site at 12,000 feet, to the nearest tenth of a degree.

Altitude (feet)	Boiling point (° C)
15,430	84.9
10,320	89.8
6,190	93.8
5,510	94.4
5,060	94.9
4,500	95.4
3,950	96.0
3,500	96.4
3,060	96.9
2,400	97.6
2,060	97.9
1,520	98.5
970	99.0
530	99.5
0	100

CHAPTER 4

Polynomial and Rational Functions, and the Algebra of Functions

In this chapter we discuss two classes of functions: polynomial and rational functions. We also discuss the algebra of functions. We begin with functions whose defining expression is quadratic. The graphs of these functions are parabolas.

4–1 *Quadratic functions and functions defined on intervals*

A garden is being laid out as a semicircle attached to a rectangle. The perimeter is to be surrounded by a chain, of which 500 feet is available. What should be the dimension x so that the area of the garden will be maximized, and what is this area?

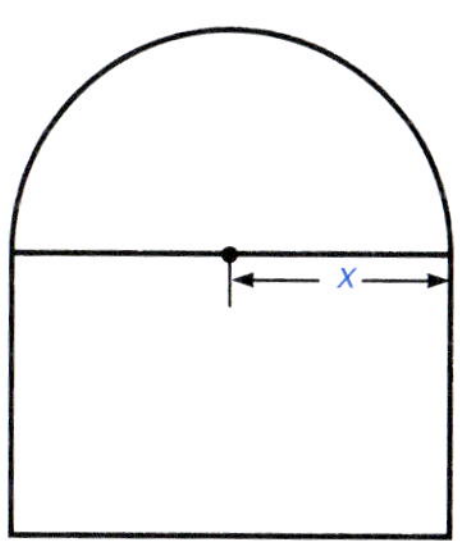

The solution to this problem involves quadratic functions, with which we begin this section.

Quadratic functions—parabolas

If a function is defined by a polynomial of one variable that is quadratic (degree 2) we call that function a quadratic function.

Quadratic function

A quadratic function is a function defined as

$$f(x) = ax^2 + bx + c, \quad a \neq 0$$

We graph such a function by plotting all the points (x,y) where $y = f(x)$. In other words, we first rewrite the function as $y = ax^2 + bx + c$. The graphs of these functions are parabolas. The path of a football thrown through the air and the path of a space vehicle that has exactly enough velocity to escape the attraction of the earth are both parabolic. A function that might be used to maximize the profit of a company in certain circumstances is a quadratic function, so graphing this function produces a parabola.

In section 3–4 we used the idea of horizontal and vertical translations and vertical scaling to graph certain quadratic functions. These quadratic functions were in a form in which we could apply these ideas. We call this form the vertex form for a quadratic function.

Vertex form for a quadratic function

The graph of

$$f(x) = a(x - h)^2 + k, \quad a \neq 0$$

has vertex at (h,k) and is the graph of $y = x^2$, vertically scaled by a units. If $a < 0$ the parabola opens downward, and if $a > 0$ the parabola opens upward.

The vertex is at (h,k) because there is a horizontal translation of h units and a vertical translation of k units.

Figure 4–1 summarizes the features of a parabola that we use in this section: x-intercepts, y-intercept, vertex.

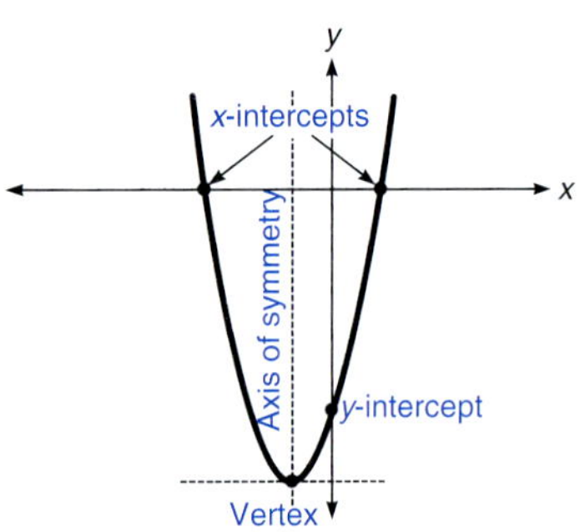

Figure 4–1

Since the vertex of $y = x^2$ is at (0,0), the location of the vertex of a general parabola is determined by the amount of horizontal and vertical translation of the point (0,0). This is illustrated in the examples.

When a quadratic function is not in vertex form, it can be put in that form by completing the square on the variable x. Rewriting the equation in this form allows us to see it as a translated, scaled version of $y = x^2$.

Complete the square as in section 3–5, by taking half the value of the coefficient of the x-term (the linear term) and squaring it. However, in this case, do not add this value to both members of the equation—instead add it and its negative to the same member of the equation. This is equivalent to adding zero, which has no effect on the value of the expression.

When graphing, it is also helpful to note that *the parabola is symmetric about the vertical line that passes through the vertex.* This line is called the **axis of symmetry.**

Example 4–1 A illustrates graphing quadratic functions. They may be graphed by plotting the intercepts and vertex and using the general shape of a parabola, or, of course, by letting a graphing calculator plot points for us. The steps for hand sketching are shown here as well as the steps for a TI-81 graphing calculator, along with a suggested range for the calculator's display. When using a graphing calculator we can graph the function and then use the trace and zoom capabilities to find approximate values of the vertex.

■ *Example 4–1 A*

Graph the parabola after putting the function in vertex form. Compute the vertex and intercepts.

1. $f(x) = x^2 + 2x - 3$

Complete the square on x:

$y = x^2 + 2x + 1 - 1 - 3$ — Half of 2 is 1, and $1^2 = 1$; Add +1 and −1 to the right member

$y = (x + 1)^2 - 4$ — $x^2 + 2x + 1 = (x + 1)^2$

$y = (x - (-1))^2 - 4$ — Vertex form

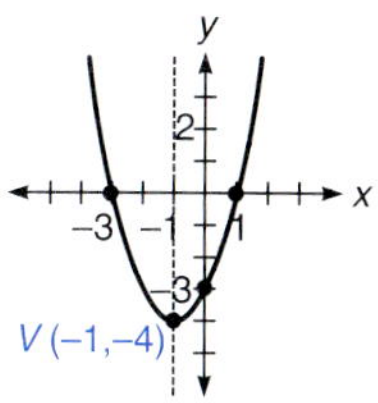

Y= X|T x^2 + 2
X|T − 3
RANGE −4,2,−5,2

Vertex: $(-1,-4)$

y-intercept:

$y = 0^2 + 2(0) - 3$ — Let $x = 0$ in $y = x^2 + 2x - 3$

$y = -3$ — y-intercept

x-intercept:

$0 = x^2 + 2x - 3$ — Let $y = 0$ in $y = x^2 + 2x - 3$

$0 = (x + 3)(x - 1)$ — Factor

$x + 3 = 0$ or $x - 1 = 0$ — Zero product property

$x = -3$ or $x = 1$ — x-intercepts

Since the coefficient of x^2, 1, is positive, the parabola opens upward. To find approximate values of the vertex in this problem with the TI-81 we graph it with the calculator, as shown, and then proceed as follows.

TRACE	The cursor appears on the graph. Move it as close to the vertex as possible.
ZOOM 2	This expands the graph.

Repeat this trace and zoom process several times. Each time we do this we find a more and more accurate value. The most accurate values are found by algebraic processes, such as by completing the square. In practical cases where this must be done for many quadratic equations we could develop a formula that would give exact results. (See problem 58 in the exercises.) Here, however, we are as interested in getting experience with this type of function as we are in obtaining numeric results. When $a \neq 1$ we must first factor a from the expression to complete the square.

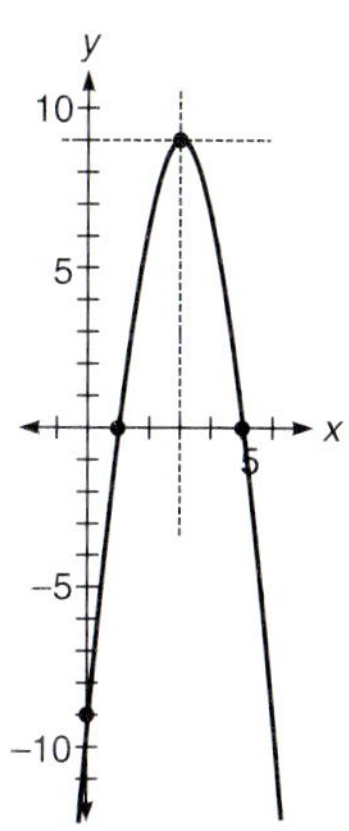

Y= (−) 2 X|T x^2
+ 1 2 X|T − 9
RANGE −1,7,−10,10

2. $f(x) = -2x^2 + 12x - 9$

Complete the square:

$y = -2(x^2 - 6x) - 9$ — Since $a \neq 1$ we factor it from the x terms

$y = -2(x^2 - 6x + 9) + 2(9) - 9$ — We added $-2(9)$, so we must add $+2(9)$

$y = -2(x - 3)^2 + 9$ — Vertex form

Vertex: $(3,9)$

y-intercept

$y = -2(0)^2 + 12(0) - 9 = -9$ — Let $x = 0$ in $-2x^2 + 12x - 9$

x-intercept

$0 = -2x^2 + 12x - 9$ — Let $y = 0$ in $y = -2x^2 + 12x - 9$

$0 = 2x^2 - 12x + 9$

$x = \dfrac{6 \pm 3\sqrt{2}}{2} \approx 0.9, 5.1$ — Quadratic formula

Since $a < 0$ the parabola opens downward. Plot the two x-intercepts, the y-intercept, and the vertex. Draw the parabola that passes through these points. ■

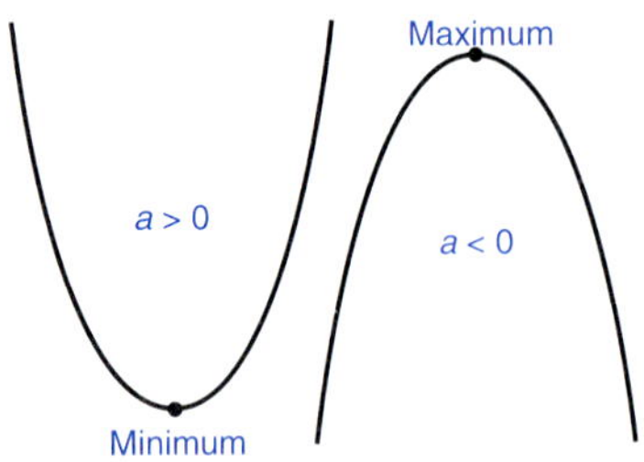

Figure 4–2

Maximum and minimum values of quadratic functions

The vertex (h,k) of a parabola $y = ax^2 + bx + c$ is the lowest point on the graph when $a > 0$, and the highest point when $a < 0$ (figure 4–2). The value of k at these points is said to be a *minimum* or *maximum value* at these points, and h is the value of x at which this minimum or maximum occurs. This has many applications, one of which is illustrated in example 4–1 B.

■ *Example 4–1 B*

A homeowner has 48 feet of fencing to fence off an area behind the home. The home will serve as one boundary. What are the dimensions of the maximum area that can be fenced off, and what is the area?

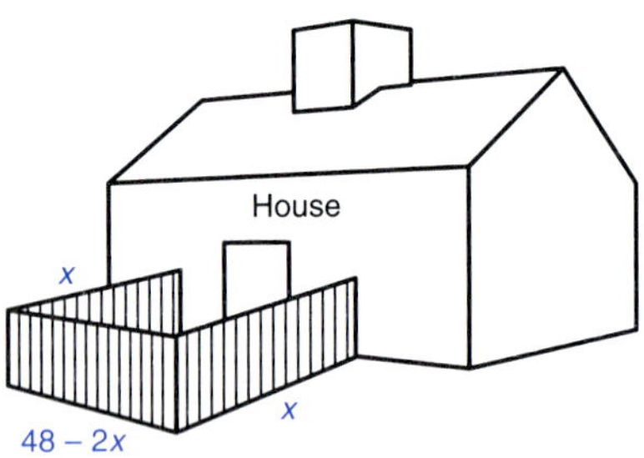

As shown in the figure, let x represent the length of each of the two sides that are perpendicular to the house. Since there is 48 feet of fencing available, there is $48 - 2x$ feet remaining for the third side of the fence. The area of a rectangle is the product of its length and width, so, if A represents area, we have the equation

$$A = x(48 - 2x)$$
$$A = -2x^2 + 48x$$
$$A = -2(x^2 - 24x)$$
$$A = -2(x^2 - 24x + 144) + 2(144) \qquad \text{Complete the square}$$
$$A = -2(x - 12)^2 + 288$$

The vertex is (12,288).

Since the coefficient of x^2 is negative, this is a parabola that opens downward, so that its vertex is at its highest point, and this is therefore the maximum value of A. The x-coordinate of the vertex is 12. We use this value of x to compute A:

$$A = -2x^2 + 48x$$
$$A = -2(12^2) + 48(12)$$
$$A = 288$$

Observe that this is the y-coordinate at the vertex. This value represents the maximum value that A can take on in this situation.

The third side is $48 - 2x = 48 - 24 = 24$ feet. Thus, to obtain the maximum area of 288 ft^2 the dimensions are 12 ft by 24 ft. ■

Functions defined on intervals

Functions are often defined with more than one expression. Each expression applies to a particular interval. Each expression is graphed as usual, but we use only the part of the graph indicated by the inequalities. This is illustrated in example 4–1 C.

Part 2 of example 4–1 C shows how to graph functions defined on intervals using the TI-81 graphing calculator.

■ *Example 4–1 C*

Graph the following functions. State all intercepts.

1. $f(x) = \begin{cases} -2x, & x < -1 \\ x + 3, & x \geq -1 \end{cases}$

For the interval $x < -1$ we graph the straight line $y = -2x$. For the interval $x \geq -1$ we graph the straight line $y = x + 3$, which gives a y-intercept at $y = 3$. The figure shows this process.

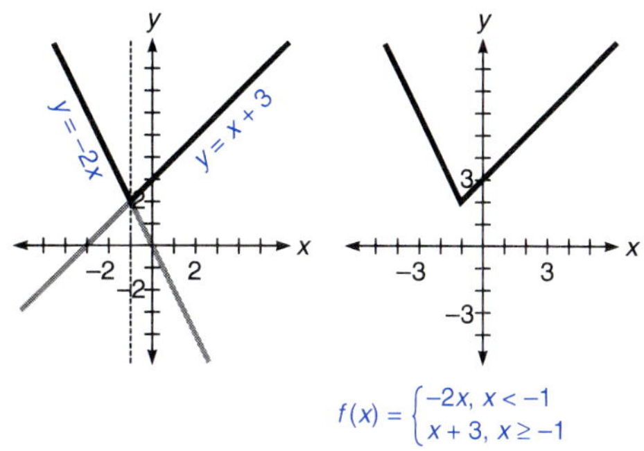

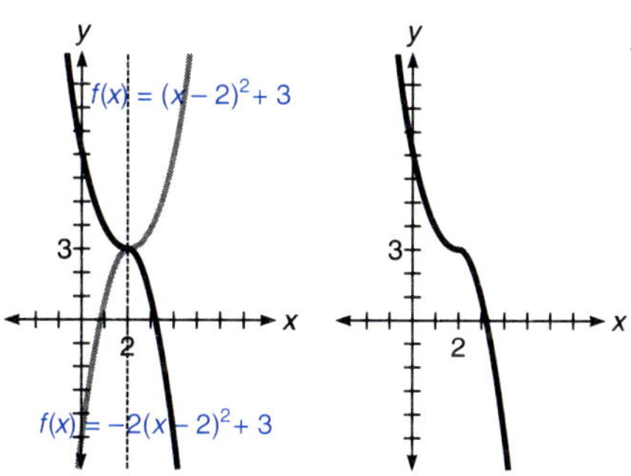

2. $f(x) = \begin{cases} (x - 2)^2 + 3, & x < 2 \\ -2(x - 2)^2 + 3, & x \geq 2 \end{cases}$

Interval: $x < 2$

$y = (x - 2)^2 + 3$; parabola, opens upward, vertex at (2,3), y-intercept at (0,7).

Interval: $x \geq 2$

$y = -2(x - 2)^2 + 3$; parabola, opens downward, vertex at (2,3), y-intercept at (0,−5).

On the TI-81 a "TEST" evaluates to zero if false and one if true. Therefore, to graph this function we enter the following for Y_1 (using the [Y=] key). Use the standard range setting.

$:Y_1 = ((X-2)^2 + 3)(X<2) + (-2(X-2)^2 + 3)(X\geq 2)$

The < and ≥ symbols are found using the TEST window ([2nd] [MATH])

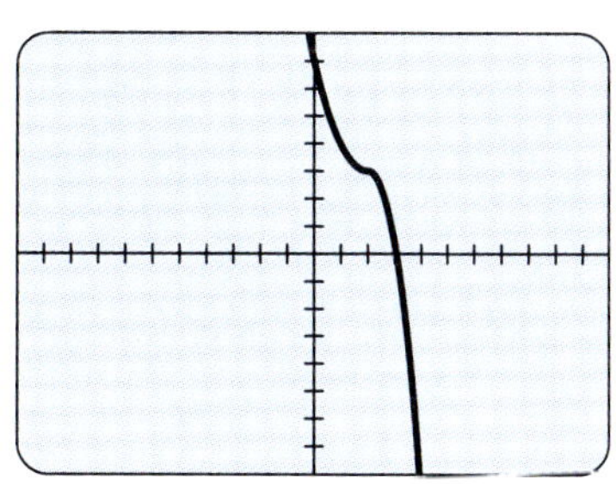

The idea is that if $x < 2$ then Y_1 is equivalent to $((x - 2)^2 + 3)(1) + (-2(x - 2)^2 + 3)(0) = (x - 2)^2 + 3$. Similarly, when $x \geq 2$ Y_1 is equivalent to $-2(x - 2)^2 + 3$. ■

Mastery points

Can you

- Graph quadratic functions using the vertex and intercepts?
- Use the vertex of a parabola to maximize or minimize a value determined by a quadratic equation?
- Graph functions defined on intervals?

Exercise 4–1

Graph the following parabolas. State the intercepts and vertex.

1. $f(x) = (x - 1)^2 + 3$
2. $f(x) = (x + 2)^2 - 1$
3. $f(x) = 2(x + 3)^2 - 4$
4. $f(x) = -2(x - 1)^2 - 6$
5. $f(x) = -(x - 5)^2 - 1$
6. $f(x) = \frac{1}{2}(x + 4)^2 + 8$

Graph the following parabolas, completing the square if necessary. State the intercepts and vertex.

7. $y = x^2 - x - 6$
8. $y = x^2 + 2x - 15$
9. $y = 3x^2$
10. $y = 2x^2 - x$
11. $y = x^2 + 3x$
12. $y = \frac{1}{2}x^2$
13. $y = -x^2 + 3x + 40$
14. $y = 5x - x^2$
15. $y = x^2 - 4$
16. $y = 9 - x^2$
17. $y = 3x^2 + 6x - 2$
18. $y = 3x^2 - 9x + 8$
19. $y = x^2 - 5x - 8$
20. $y = 4 - x - x^2$
21. $y = 2x^2 - 4x - 4$
22. $y = x^2 + 3x - 3$
23. $y = x^2 + 4$
24. $y = x^2 + x - 3$
25. $y = x^2 - x + 5$
26. $y = -2x^2 - 2x + 1$
27. $y = -x^2 - x$
28. $y = 3x^2 + 9x$

Solve the following problems by creating an appropriate second-degree equation and finding the vertex.

29. A homeowner has 260 feet of fencing to fence off a rectangular area behind the home. The home will serve as one boundary (the fence is only necessary for three of the four sides). What are the dimensions of the maximum area that can be fenced off, and what is the area?

30. The homeowner of problem 29 has 300 feet of fence available. What are the dimensions of the maximum area that can be fenced off, and what is the area?

31. What is the area of the largest rectangle that can be created with 260 feet of fence? What are the length and width of this rectangle?

32. A garden is being laid out as a semicircle attached to a rectangle (see the figure). The perimeter is to be surrounded by a chain, of which 500 feet is available. What should be the dimension x so that the area of the garden will be maximized, to the nearest foot? What is this area, to the nearest square foot? Recall that the area of a circle is given by $A = \pi r^2$, and the perimeter (circumference) is $C = 2\pi r$, where r is the radius of the circle (x in the figure).

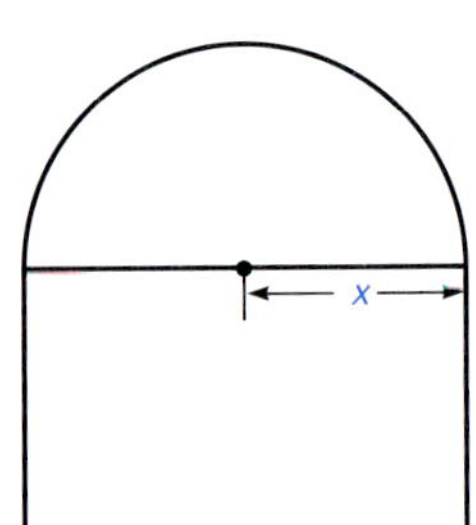

33. If an object is thrown into the air with an initial vertical velocity of v_0 ft/s, then its distance above the ground s, for time t, is given by $s = v_0 t - 16t^2$. Suppose an object is thrown upward with initial velocity 64 ft/s; find how high the object will go (when s is a maximum) and when it will return to the ground (when $s = 0$).

34. An arrow is shot into the air with an initial vertical velocity of 48 ft/s. Find out how high the arrow will go, and when it will return to the ground. See problem 33.

35. Suppose the velocity distribution of natural gas flowing smoothly in a certain pipeline is given by $V = 6x - x^2$, where V is the velocity in meters per second and x is the distance in meters from the inside wall of the pipe. What is the maximum velocity of the gas, and where does this occur?

36. The power output P (in watts), of an automobile alternator that generates 14 volts and has an internal resistance of 0.20 ohms is given by $P = 14I - 0.20I^2$. At what current I (in amperes) does the generator generate maximum power, and what is the maximum power?

37. If a company's profit P (in dollars) for a given week when producing u units of a commodity in that week is $P = -u^2 + 100u - 1{,}000$, how many units must be made to produce the maximum profit, and what is this profit?

38. Find the two numbers whose sum is 100 and whose product is a maximum.

39. Given that the difference between two numbers is 8, what is the minimum product of the two numbers? Also, what are the numbers?

40. Find the minimum product of two numbers whose difference is 12; what are the numbers?

41. For a given perimeter, will a circle or rectangle contain a greater area? To find out, assume a length P for the total perimeter, and maximize the area for a rectangle with perimeter P (as in problem 31), then compute the area of a circle with perimeter (circumference) P.

42. The method of completing the square can be used to derive the quadratic formula. This states:

If $ax^2 + bx + c = 0$ and $a \neq 0$, then

$$x = \frac{-b \pm \sqrt{b^2 - 4ac}}{2a}.$$

Derive the formula by assuming that $ax^2 + bx + c = 0$; then add $-c$ to both sides and complete the square on x.

43. An alternate method of deriving the quadratic formula (see problem 42) was devised by the Hindus. About the year 1025, the Indian Sridhara presented the following method.[1]

- **a.** Multiply each member of $ax^2 + bx + c = 0$ by $4a$.
- **b.** Add b^2 to both members.
- **c.** Subtract $4ac$ from both members.
- **d.** Observe that the left member is now a perfect square, then take the square root of both members.

Use this method to derive the quadratic formula.

[1]From K. R. S. Sastry, "The Quadratic Formula: A Historic Approach," *The Mathematics Teacher*, November 1988.

Graph the following functions defined on intervals.

44. $f(x) = \begin{cases} -2x + 1, & x \geq 1 \\ 3x + 1, & x < -1 \end{cases}$

45. $g(x) = \begin{cases} -\frac{1}{2}x, & x < 0 \\ -3x, & x \geq 0 \end{cases}$

46. $h(x) = \begin{cases} x + 1, & x < -1 \\ \frac{1}{2}x + \frac{1}{2}, & x \geq -1 \end{cases}$

47. $f(x) = \begin{cases} 2x - 1, & x < \frac{1}{2} \\ \frac{2}{3}x - \frac{1}{3}, & x \geq \frac{1}{2} \end{cases}$

48. $g(x) = \begin{cases} x^2 - 4x + 4, & x < 2 \\ -\frac{3}{4}x + \frac{3}{2}, & x \geq 2 \end{cases}$

49. $h(x) = \begin{cases} x^2 - 2x + 4, & x < 1 \\ -2x + 5, & x \geq 1 \end{cases}$

50. $f(x) = \begin{cases} x^2 + 2, & x < 0 \\ -2x^2 + 2, & x \geq 0 \end{cases}$

51. $g(x) = \begin{cases} -2x^2 + 8x - 10, & x < 2 \\ x^2 - 4x + 2, & x \geq 2 \end{cases}$

52. $g(x) = \begin{cases} -x, & x < 0 \\ \sqrt{x}, & x \geq 0 \end{cases}$

53. $g(x) = \begin{cases} x, & x < 0 \\ x^3, & x \geq 0 \end{cases}$

54. $g(x) = \begin{cases} x^3, & x < 0 \\ \sqrt{x}, & x \geq 0 \end{cases}$

55. $g(x) = \begin{cases} x^2, & x < 0 \\ -\sqrt{x}, & x \geq 0 \end{cases}$

56. $g(x) = \begin{cases} x, & x < 1 \\ x^3 + 1, & x \geq 1 \end{cases}$

57. $g(x) = \begin{cases} x - 1, & x < 1 \\ \sqrt{x - 1}, & x \geq 1 \end{cases}$

58. There is a formula for each of the coordinates of the vertex of the general parabola when the equation is given in the form $f(x) = ax^2 + bx + c$. This formula can be found by completing the square using the literal constants a, b, and c instead of numeric values. Find the formula for the x and the y coordinates of the vertex.

Skill and review

1. Factor $3x^2 + x - 10$.
2. Factor $3x^2 + 13x - 10$.
3. Factor $x^4 - 16$.
4. List all the prime divisors of 96.
5. List all the positive integer divisors of 96.
6. If $f(x) = 2x^3 - x^2 - 6x + 20$, find $f(-2)$.
7. Use long division to divide $2x^3 - x^2 - 6x + 20$ by $x^2 + 2$.
8. Graph $f(x) = \sqrt{x - 2} - 3$.
9. Compute $f \circ g(x)$ and $g \circ f(x)$ if $f(x) = x^4 - 6x^2 + 8$ and $g(x) = \sqrt{x + 1}$.

4–2 Polynomial functions and synthetic division

Napthalene ($C_{10}H_8$) is a very stable chemical.[2] To determine its stability one needs to solve the equation $x^{10} - 11x^8 + 41x^6 - 65x^4 + 43x^2 - 9 = 0$. It has 10 solutions. Find them to four decimal places of accuracy.

Solving equations like the one in the section opening problem is very common in science, engineering, business and finance, and anywhere mathematics is applied. Most sections of this book include problems that require solving some type of equation. In this and the next few sections, we investigate this important part of mathematics. The problem concerning napthalene appears in exercise set 4–3. This section presents algebraic methods for solving polynomial equations. Section 4–3 presents ways to graph polynomial functions and ways to solve equations with graphical calculators.

Some terminology

A zero of a function f is a value c such that $f(c) = 0$. If the function is defined by a polynomial, this means the polynomial evaluates to zero when x is replaced by c. In this section, we examine algebraic and graphical methods for finding zeros of functions whose defining rule is a polynomial in one variable with real coefficients. The implied domain of such functions is the set of real numbers. *Although most of what we discuss in this section extends to include the complex number system we restrict our discussion to real numbers.*

We begin with a definition of polynomial functions.

Polynomial function

A polynomial function in one variable is a function of the form

$$f(x) = a_nx^n + a_{n-1}x^{n-1} + \cdots + a_2x^2 + a_1x + a_0,\ a_n \neq 0$$

where $a_0, a_1, a_2, \ldots, a_n$ are constant, real-valued coefficients.

[2]From Professor Jun-Ichi Aihara, "Why Aromatic Compounds Are Stable," *Scientific American,* March 1992.

The coefficient a_n is called the **leading coefficient,** and n is the **degree of the polynomial.** The term a_0 is the **constant term.** When the polynomial is written as above, with exponents in descending order, we say it is written in the **standard form** for a polynomial.

A linear function is a polynomial function of degree 0 or 1, and a quadratic function is a polynomial function of degree 2. A polynomial function of degree 0 is also called a **constant function.** By way of example:

Function	Degree	
$f(x) = 4$	0	Constant function
$f(x) = -3x + 4$	1	Linear function
$f(x) = 5x^2 - 3x + 4$	2	Quadratic function
$f(x) = x^3 + 5x^2 - 3x + 4$	3	Polynomial function of degree 3

An important property of a function is those values of the domain for which $f(x)$ is zero. As noted above these values are called zeros of a function. When we graph these functions in section 4–3, we will see that *a zero of a function corresponds to an x-intercept of the graph of the function.*

> **Zeros of a function**
> If $f(x)$ defines a function, and for some real number c in the domain of f, $f(c) = 0$, then c is called a zero of the function.

Zeros of polynomial functions of degree 0, 1, and 2

We have algebraic methods to find zeros of linear and quadratic polynomial functions (functions of degrees 0, 1, and 2). We *replace $f(x)$ by 0* and solve.[3] Methods for solving linear and quadratic equations were covered in chapter 2.

Zeros of polynomial functions of degree 3 and above

We begin our discussion of algebraic methods for finding zeros of functions by stating several facts, without proof, that can be of help.

> **Maximum number of zeros**
> A polynomial function of degree n has at most n zeros.
>
> ***Concept***
> For a given polynomial function $f(x)$ of degree n there are at most n values $c_1, c_2, \ldots, c_n$ such that $f(c_1) = 0, f(c_2) = 0, \ldots, f(c_n) = 0$.

This tells us that *the graph of a polynomial function of degree n has at most n x-intercepts.*

[3]The Babylonians of 4,000 years ago could deal with all three of these situations.

Another important fact is that it is possible to find all *rational* zeros of polynomial functions.

> **Rational zero theorem**
>
> If $\frac{p}{q}$ is a rational number in lowest terms and is a zero of a polynomial function, then the denominator q is a factor of the leading coefficient, a_n, and the numerator p is a factor of the constant term, a_0.

As a memory aid remember that any rational zero is of the form

$$\frac{\text{Factor of } a_0}{\text{Factor of } a_n}$$

Example 4–2 A illustrates the use of the rational zero theorem.

■ ***Example 4–2 A***

List all possible rational zeros for the polynomial function $f(x) = 6x^4 + 25x^3 - 15x^2 - 25x + 9$.

According to the theorem above, if $f\left(\frac{p}{q}\right) = 0$ then p, the numerator, divides 9 and q, the denominator, divides 6. Thus, we need to check all positive and negative fractions where the numerator is 1, 3, or 9, and the denominator is 1, 2, 3, or 6. We obtain these values as shown.

Numerator is 1: $\pm\frac{1}{1}, \pm\frac{1}{2}, \pm\frac{1}{3}, \pm\frac{1}{6}$

Numerator is 3: $\pm\frac{3}{1}, \pm\frac{3}{2}, \pm\frac{3}{3}, \pm\frac{3}{6}$

Numerator is 9: $\pm\frac{9}{1}, \pm\frac{9}{2}, \pm\frac{9}{3}, \pm\frac{9}{6}$

After reducing, the set of possible rational zeros of f is

$$\pm 1, \pm 3, \pm 9, \pm\tfrac{1}{2}, \pm\tfrac{3}{2}, \pm\tfrac{9}{2}, \pm\tfrac{1}{3}, \pm\tfrac{1}{6}$$ ■

The remainder theorem

To find out which possible rational zeros are actually zeros we will use a method called synthetic division; to understand it we must first consider the following theorem.

> **Remainder theorem**
>
> If f is a nonconstant polynomial function and c is a real number, then the remainder when $x - c$ is divided into $f(x)$ is $f(c)$.

This implies that c is a zero of f *if and only if there is no remainder* when $f(x)$ is divided by $x - c$. We could use long division to test values as potential zeros. Rather than go back to long division, however, we introduce a faster algorithm called **synthetic division.** This algorithm is *equivalent to long division by a linear factor* $x - c$.

Synthetic division

We illustrate synthetic division with the example:

$$f(x) = x^3 - 2x^2 - 5x - 6 \quad \text{and} \quad c = 4$$

Computation would show that $f(4) = 6$. The remainder theorem states that this can also be found by computing $(x^3 - 2x^2 - 5x - 6) \div (x - 4)$.

We construct a table using the coefficients of the polynomial dividend and the value of c at the left.

	1	−2	−5	−6
4				

We proceed with the first column on the left, moving column by column to the right, *performing the same steps on each column:*

1. Add the values in the first two rows; write the result in the third row.
2. Multiply the third row by the value c and put the result in the second row of the next column.

The first column will have only one value initially, so we do not need to perform any addition in that column—we simply bring down the value.

	1	−2	−5	−6
		4		
4	1			

Bring down the 1

Multiply by 4

	1	−2	−5	−6
		4	8	
4	1	2		

Add 4 + (−2)

Multiply the result by 4

	1	−2	−5	−6
		4	8	12
4	1	2	3	

Add 8 + (−5)

Multiply the result by 4

	1	−2	−5	−6
		4	8	12
4	1	2	3	[6]

Add (−6) + 12

Looking at the last table we see the value $f(4) = 6$ appear in the last computation.

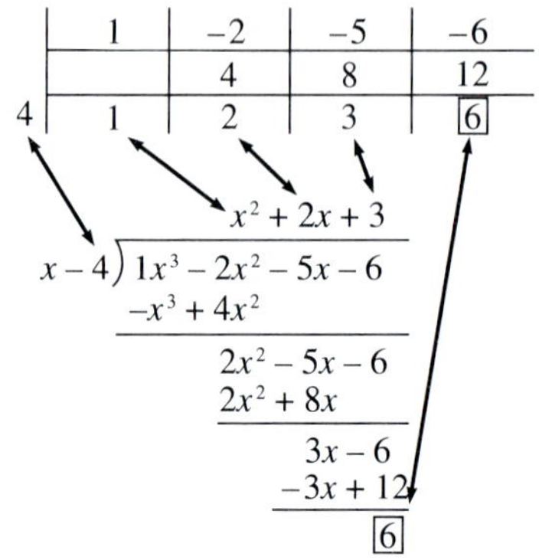

Figure 4–3

Figure 4–3 shows the similarities between long division and synthetic division. (In the long division we changed the signs of the terms that are subtracted.) Observe also that the coefficients of the quotient $x^2 + 2x + 3$ are in the synthetic division table.

Example 4–2 B shows the use of synthetic division to evaluate a function and to perform algebraic division by a linear factor.

■ ***Example 4–2 B***

1. $f(x) = 2x^3 - 4x^2 + 2x - 1$; (a) find $f(-2)$, (b) divide f by $x + 2$.

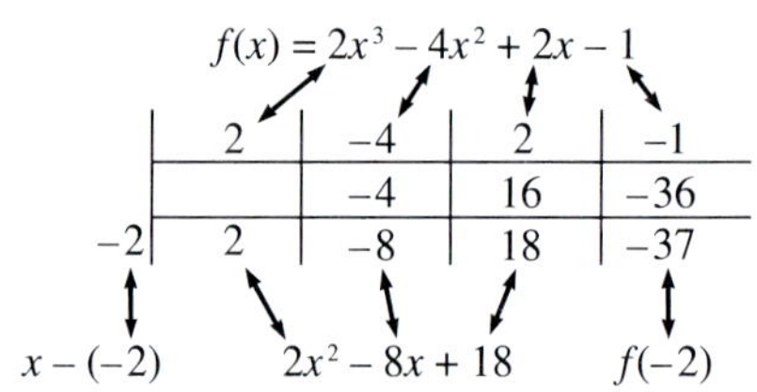

The bottom line of this table gives two results. It tells us that $f(-2)$ is -37, which is the answer to (a). It also tells us what would happen if we divided by $x - c$, or $x - (-2) = x + 2$, which is the answer to (b). It tells us that $\dfrac{2x^3 - 4x^2 + 2x - 1}{x + 2} = 2x^2 - 8x + 18$, with remainder -37. We know that the result is of degree 2 since we have divided a polynomial of degree 3 by a divisor of degree 1, leaving a quotient of degree $3 - 1 = 2$.

2. Divide $6x^2 + x - 2$ by $x - \frac{1}{2}$.

	6	1	−2
		3	2
$\frac{1}{2}$	6	4	0

The first two values of the last line produce a first-degree polynomial (a linear expression), which is $6x + 4$. The 0 is the remainder. Hence we know that $\frac{6x^2 + x - 2}{x - \frac{1}{2}} = 6x + 4$, with no remainder.

This last result means that $x - \frac{1}{2}$ is a factor of $6x^2 + x - 2$. In particular, $6x^2 + x - 2 = (x - \frac{1}{2})(6x + 4)$.

3. Divide $x^5 - 3x^3 + x$ by $x - 2$.

We first rewrite this expression as $x^5 + 0x^4 - 3x^3 + 0x^2 + 1x + 0$ to obtain the coefficients for the table: 1, 0, −3, 0, 1, and 0.

	1	0	−3	0	1	0
		2	4	2	4	10
2	1	2	1	2	5	10

This last line gives the result $\frac{x^5 - 3x^3 + x}{x - 2} = x^4 + 2x^3 + x^2 + 2x + 5$ with remainder 10.

Note This result also means that if we evaluate $x^5 - 3x^3 + x$ for 2 we obtain the value 10. That is, $2^5 - 3(2^3) + 2 = 10$. ■

Zeros and factoring

Whenever we get a remainder of zero using synthetic division with $x - c$, we have found a zero, c, and a factor of $f(x)$, namely $x - c$. Thus, we can use zeros to help factor a polynomial, and we can use factoring to find zeros.

Example 4–2 C illustrates using synthetic division to find linear factors and to thereby factor a polynomial expression.

■ ***Example 4–2 C***

Find all rational zeros of the given function. If possible find all real zeros. Also, factor using rational values.

1. $f(x) = 6x^4 - 5x^3 - 39x^2 - 4x + 12$

Using the rational zero theorem all possible rational zeros of this function can be determined to be $\pm 1, \pm\frac{1}{2}, \pm\frac{1}{3}, \pm\frac{1}{6}, \pm 2, \pm\frac{2}{3}, \pm 3, \pm\frac{3}{2}, \pm 4, \pm\frac{4}{3}, \pm 6$, and ± 12.

We start with 1, then −1, etc. It can be verified that neither 1 nor −1 is a zero. Therefore try $\frac{1}{2}$ next.

	6	−5	−39	−4	12
		3	−1	−20	−12
$\frac{1}{2}$	6	−2	−40	−24	0

Since the remainder is 0, this is a zero of f, and $(x - \frac{1}{2})$ is a factor of $f(x)$

$$f(x) = (x - \tfrac{1}{2})(6x^3 - 2x^2 - 40x - 24) \quad \text{From the last line of the table}$$
$$= 2(x - \tfrac{1}{2})(3x^3 - x^2 - 20x - 12) \quad \text{Common factor of 2}$$

Any remaining zero of f will also be a zero of $3x^3 - x^2 - 20x - 12$, so we attack this problem with synthetic division. All possible rational zeros of this expression are $\pm 1, \pm\frac{1}{3}, \pm 2, \pm\frac{2}{3}, \pm 3, \pm 4, \pm\frac{4}{3}, \pm 6$, and ± 12.

However, if ± 1 were zeros they would have been zeros of f, so we do not check them again. Neither $\frac{1}{3}$ nor $-\frac{1}{3}$ is a zero; neither is 2.

	3	−1	−20	−12
		−6	14	12
−2	3	−7	−6	0

$(x + 2)$ is a factor of $f(x)$

$$f(x) = 2(x - \tfrac{1}{2})(x + 2)(3x^2 - 7x - 6)$$
$$f(x) = 2(x - \tfrac{1}{2})(x + 2)(3x + 2)(x - 3)$$ Factor the quadratic trinomial
$$f(x) = 2(x - \tfrac{1}{2})(x + 2)(3)(x + \tfrac{2}{3})(x - 3)$$ $3x + 2 = 3(x + \frac{2}{3})$
$$f(x) = 6(x - \tfrac{1}{2})(x + 2)(x + \tfrac{2}{3})(x - 3)$$

Factoring $3x + 2$ in this way is not necessary, but it makes all of the linear factors of the form "$x + a$" (that is, leading coefficient one). This factorization has given the four zeros of f: $\frac{1}{2}$, -2, $-\frac{2}{3}$, and 3.

2. $f(x) = x^3 - 3x^2 - 2x + 8$
Possible rational zeros are ± 1, ± 2, ± 4, and ± 8. Checking 2 gives

	1	−3	−2	8
		2	−2	−8
2	1	−1	−4	0

Thus $f(x) = (x - 2)(x^2 - x - 4)$.

The quadratic trinomial $x^2 - x - 4$ does not factor, but its zeros can be found with the quadratic formula. If $0 = x^2 - x - 4$ then $x = \dfrac{1 \pm \sqrt{17}}{2}$.

Thus, the real zeros of f are 2, $\dfrac{1 + \sqrt{17}}{2}$, $\dfrac{1 - \sqrt{17}}{2}$, and

$$f(x) = (x - 2)\left(x - \frac{1 + \sqrt{17}}{2}\right)\left(x - \frac{1 - \sqrt{17}}{2}\right).$$ ■

Note See ways to factor quadratic expressions in section 2–2. Any quadratic expression can be factored using the quadratic formula as shown in that section. Thus, *we never need to use synthetic division on quadratic polynomials.*

Prime factorization of polynomials

In part 2 of example 4–2 C we saw a quadratic polynomial that factored over the real number system. When the zeros of a quadratic expression are complex we say that the expression is **prime over the real number system.** This is determined by the value of the discriminant $b^2 - 4ac$. If $b^2 - 4ac < 0$ then the quadratic is prime (over the real number system) since the formula produces complex zeros. In this case we choose not to factor the quadratic.

The next theorem comes from work done by the famous German mathematician Karl Friedrich Gauss by the year 1799.

> **Prime factorization of polynomials over the real number system**
> Every polynomial of positive degree n with real coefficients can be expressed as the product of a real number, linear factors, and prime quadratic factors.

This theorem states that if a polynomial function has only real coefficients, then it can be factored into a product of a real number and linear and quadratic factors, where the quadratic factors have only complex zeros. It is not always easy to find all these factors, but Gauss proved that it is possible in theory. In this book we examine only selected problems in which we have some chance of success.

Before illustrating Gauss's theorem, we should also discuss the idea of *multiplicity* of a zero. By way of example, if $f(x) = (x - 2)(x + 3)^2(x - \frac{1}{2})^4$, then 2 is a zero of multiplicity one, -3 a zero of multiplicity 2, and $\frac{1}{2}$ a zero of multiplicity 4. This is formalized in the following definition.

> **Multiplicity of zeros**
> If $f(x)$ is defined by a polynomial expression, c is a real number in the domain of f, and $(x - c)^n$ divides $f(x)$, but $(x - c)^{n+1}$ does not divide $f(x)$, then we say that c is a zero of multiplicity n.

Example 4–2 D illustrates Gauss's theorem about the prime factorization of polynomials over the real number system, and the idea of multiplicity of zeros.

Example 4–2 D

Given $f(x) = 4x^5 - 10x^4 + 6x^3 - 4x^2 + 10x - 6$, (a) factor the polynomial completely and (b) list the zeros of the polynomial; note if any zeros have multiplicity greater than one.

$$f(x) = 4x^5 - 10x^4 + 6x^3 - 4x^2 + 10x - 6$$
$$= 2(2x^5 - 5x^4 + 3x^3 - 2x^2 + 5x - 3) \qquad \text{Common factor of 2}$$

We now factor $2x^5 - 5x^4 + 3x^3 - 2x^2 + 5x - 3$.

Possible rational zeros are ± 1, ± 3, $\pm\frac{1}{2}$, and $\pm\frac{3}{2}$.

	2	−5	3	−2	5	−3	
		2	−3	0	−2	3	
1	2	−3	0	−2	3	0	$(x - 1)$ is a factor of $f(x)$

$$f(x) = 2(x - 1)(2x^4 - 3x^3 - 2x + 3)$$

$2x^4 - 3x^3 - 2x + 3$ has the same possible rational zeros: ±1, ±3, $\pm\frac{1}{2}$, and $\pm\frac{3}{2}$, and we could proceed by synthetic division. However, the pattern of the coefficients suggests we can *factor using grouping:*

$$\begin{aligned} 2x^4 - 3x^3 - 2x + 3 &= x^3(2x - 3) - 1(2x - 3) \\ &= (2x - 3)(x^3 - 1) \\ &= (2x - 3)(x - 1)(x^2 + x + 1) \\ &= 2(x - \tfrac{3}{2})(x - 1)(x^2 + x + 1) \end{aligned}$$

$x^3 - 1$ is a difference of two cubes

The zeros of $x^2 + x + 1$ are complex ($b^2 - 4ac < 0$), so we would say that we have completely factored the expression over the real numbers. Thus, $f(x) = 2(x - 1)[2(x - \frac{3}{2})(x - 1)(x^2 + x + 1)]$ and the answer to part (a) is

$$f(x) = 4(x - 1)^2(x - \tfrac{3}{2})(x^2 + x + 1)$$

The answer to part (b) is 1 and $\frac{3}{2}$, with 1 having multiplicity 2. ■

Bounds for real zeros—where to look

It is often possible to avoid testing all of the possible rational zeros that are given by the rational zero theorem. When we apply synthetic division, the last line can give an indication about bounds for real zeros.

Upper and lower bounds for real zeros

- The real number U is an **upper bound** for the real zeros of a polynomial function if U is greater than or equal to all the zeros of the function.
- The real number L is a **lower bound** for the real zeros of a polynomial function if L is less than or equal to all the zeros of the function.

The following theorem can tell us when a value is an upper or lower bound for the real zeros of a polynomial function.

Bounds theorem for real zeros

Let c be a real number and $f(x)$ be a polynomial function with real coefficients and positive leading coefficient; consider all the coefficients in the last line of the synthetic division algorithm as applied to the value c. Then c is

- an *upper bound* if $c \geq 0$ and these coefficients are all positive or zero.
- a *lower bound* if $c \leq 0$ and the signs of the values in the last row alternate, except that zero may be written as $+0$ or -0 when considering sign alternation.

There is no need to check possible rational zeros that are greater than an upper bound or that are less than a lower bound. (Note that this theorem applies to irrational as well as rational zeros.) This is illustrated in example 4–2 E.

■ ***Example 4–2 E***

Find upper and lower bounds for the real zeros of the function $f(x) = x^4 + 3x^3 + 2x^2 - 5x + 12$.

Possible rational zeros are ±1, ±2, ±3, ±4, ±6, and ±12.

	1	3	2	−5	12
		1	4	6	1
1	1	4	6	1	13

The value 1 is an upper bound because all the coefficients in the last row are positive or zero. Thus, there would be no reason to check the values 2, 3, 4, 6, or 12.

	1	3	2	−5	12
		−3	0	−6	33
−3	1	0	2	−11	45

The value −3 is a lower bound because the coefficients in the last row alternate signs if we write the zero in the last row as −0: 1 −0 2 −11 45. Thus, there would be no reason to check the possible zeros −4, −6, or −12. ■

Number of real zeros

The next theorem can be used to obtain the number of possible positive and negative real zeros. It refers to variations in the signs of the coefficients of a polynomial. If a polynomial expression is written in standard form, ignoring terms with coefficients that are 0, a **variation in sign** is said to occur if a succeeding coefficient has a different sign than the one that preceded it (see the example 4–2 F).

> **Descartes' rule of signs**
>
> Let $f(x)$ be a function defined by a polynomial with real coefficients. Then
> - the number of *positive* real zeros is equal to the number of variations in sign in $f(x)$ or is less than this number by a multiple of 2.
> - the number of *negative* real zeros is equal to the number of variations in sign in $f(-x)$ or is less than this number by a multiple of 2.

Descartes' rule of signs is illustrated in the next example.

Example 4–2 F

a. Investigate the real zeros of the given function in terms of the number of possible positive and negative real zeros.

b. List all possible rational zeros.

c. Find all rational zeros.

d. Find irrational zeros when possible; when not possible, find bounds for the irrational zeros.

e. Write the function as a product of linear and prime quadratic factors if practical.

1. $f(x) = x^5 - 8x^4 + 11x^3 + 22x^2 - 26x - 28$

a. $f(x) = x^5 - 8x^4 + 11x^3 + 22x^2 - 26x - 28$

Three changes of sign; 1 or 3 positive zeros

$f(-x) = -x^5 - 8x^4 - 11x^3 + 22x^2 + 26x - 28$

Two changes in sign; 0 or 2 negative zeros

b. The possible rational zeros are ±1, ±2, ±7, ±14, and ±28.

	1	−8	11	22	−26	−28
		−1	9	−20	−2	28
−1	1	−9	20	2	−28	0

$(x + 1)$ is a factor of $f(x)$

$f(x) = (x + 1)(x^4 - 9x^3 + 20x^2 + 2x - 28)$

	1	−9	20	2	−28
		−1	10	−30	28
−1	1	−10	30	−28	0

$(x + 1)$ is a factor of $f(x)$ for the second time; thus $(x + 1)^2$ is a factor of $f(x)$

$f(x) = (x + 1)^2(x^3 - 10x^2 + 30x - 28)$

Checking −1 again shows it is not another zero; this is also true of +1.

	1	−10	30	−28
		2	−16	28
2	1	−8	14	0

$(x - 2)$ is a factor of $f(x)$

$f(x) = (x + 1)^2(x - 2)(x^2 - 8x + 14)$

The zeros of $x^2 - 8x + 14$ are $4 \pm \sqrt{2}$ (using the quadratic formula). Thus,

$$x^2 - 8x + 14 = [x - (4 - \sqrt{2})][x - (4 + \sqrt{2})]$$
$$= (x - 4 + \sqrt{2})(x - 4 - \sqrt{2})$$

c. Rational zeros: −1 (multiplicity 2), 2.

d. Irrational zeros: $4 \pm \sqrt{2}$.

e. $f(x) = (x + 1)^2(x - 2)(x - 4 + \sqrt{2})(x - 4 - \sqrt{2})$.

2. $f(x) = x^4 - 5x^3 + 4x^2 + 3x - 1$

a. $f(x) = x^4 - 5x^3 + 4x^2 + 3x - 1$ Three changes of sign so there are one or three positive real zeros

$f(-x) = x^4 + 5x^3 + 4x^2 - 3x - 1$ One change in sign so there is one negative real zero

b. The only possible rational zeros are ±1.

	1	−5	4	3	−1
		1	−4	0	3
1	1	−4	0	3	2

We see that 1 is not a zero or an upper bound. Since there are no other positive rational zeros possible, there is no sense looking for more positive integer zeros. However, we do need an upper bound.

Looking at the first two columns above shows that nothing less than 5 will serve as an integer upper bound, since anything less than 5 will cause a negative value in the second column and a change of sign. Thus, we proceed directly to the value 5:

	1	−5	4	3	−1
		5	0	20	115
5	1	0	4	23	114

This table shows that 5 is the *least positive* integer upper bound.

We now look for negative zeros/bounds.

	1	−5	4	3	−1
		−1	6	−10	7
−1	1	−6	10	−7	6

The alternation of signs tells us that -1 is a lower bound for real zeros, although it is not a zero itself.

c. There are no rational zeros.

d. There are one or three positive irrational zeros, less than 5, and there is one negative irrational zero, greater than -1. ■

In example 4–2 F, part 2, we would say that 5 is the *least positive integer upper bound* since no positive integer less than 5 is an upper bound. Similarly -1 is the *greatest negative integer lower bound.*

A historical note

We can solve polynomial functions of degree 0, 1, and 2 (linear and quadratic functions) by way of general formulas and methods. We might ask if there are formulas, like the quadratic formula, for the zeros of polynomial functions of degree greater than 2. Such solutions were sought for thousands of years, and solutions, although much more complicated, were discovered for polynomials of degrees 3 and 4 in the sixteenth century.[4] They were published in the book *Ars magna* by the Italian Geronimo Cardano (1501–1576). Cardano noted that the solution to the general cubic equation was due to Niccolo Tartaglia (ca. 1500–1557).

[4]The methods of exact solution of polynomial functions of degree 3 or 4 are presented in David M. Burton, *The History of Mathematics—An Introduction,* 2 ed. (Dubuque, Iowa: Wm. C. Brown Publishers, 1991).

In 1824, Niels Henrik Abel (1802–1829) of Norway published a proof that there was no similar formula for solving the general polynomial equation of degree 5. Évariste Galois (1811–1832) of France proved that *there is no general method for solving general polynomial equations of degree 5 and above.* This ended the quest for such formulas.

Mastery points

Can you

- List all possible rational zeros of a polynomial function?
- Use synthetic division to find all rational zeros of a polynomial function?
- Use synthetic division to divide a polynomial function by a linear function?
- Use synthetic division and the rational zero theorem to factor certain polynomials?
- Use Descartes' rule of signs to determine the number of possible positive and negative real zeros of a polynomial function?
- Use the bounds theorem to determine least positive upper and greatest negative lower integer bounds for the zeros of a polynomial function?

Exercise 4–2

Find all zeros of the following polynomial functions of degree less than three.

1. $f(x) = 7$ **2.** $g(x) = \frac{1}{2}$ **3.** $f(x) = 12x - 8$ **4.** $g(x) = -5x + 10$
5. $h(x) = x + 11$ **6.** $f(x) = x^2 - 4x - 4$ **7.** $g(x) = -x^2 + 3x + 10$ **8.** $h(x) = x^2 - 8$

List all possible rational zeros for a function defined by the following polynomials.

9. $x^4 - 3x^2 + 6$ **10.** $2x^3 - x - 3$ **11.** $3x^5 - x^3 - 4$ **12.** $x^4 - 8$
13. $6x^2 - 5 + 2x^3$ **14.** $2x^5 - x^6 + 7$ **15.** $3x^4 - 2x^2 + x - 9$ **16.** $5 - 2x^2 + 4x^4$
17. $5x^4 - 2x^3 + 3x - 10$ **18.** $6x^4 - 3x^3 + x - 8$ **19.** $8x^3 - 2x^2 + 4$ **20.** $8x^3 - 3x - 6$
21. $2 - 3x^2 + 4x^3$ **22.** $6 - 2x - 3x^2 + 6x^6$ **23.** $10x^4 - 3x^3 + 2x^2 - 4$ **24.** $4x^4 - 3x^3 + 2x^2 - 10$
25. $8x^3 - 8x + 16$ **26.** $4x^4 - 4x^2 + 4$

Use synthetic division to (a) divide each polynomial by the divisor indicated and (b) to evaluate the function at the value indicated.

27. $f(x) = 3x^4 - 2x^3 + x^2 - 5$
a. $x - 4$ **b.** $f(4)$

28. $g(x) = -2x^3 + 4x^2 - 3x + 1$
a. $x + 2$ **b.** $g(-2)$

29. $f(x) = x^3 - 2x^2 + 3x + 1$
a. $x - 1$ **b.** $f(1)$

30. $h(x) = 3x^4 - x^2 - 6$
a. $x + 1$ **b.** $f(-1)$

31. $h(x) = x^5 - 3x^3 - x^2 + 5$
a. $x + 3$ **b.** $h(-3)$

32. $f(x) = 2x^3 - x - 1$
a. $x - 6$ **b.** $f(6)$

33. $f(x) = \frac{1}{2}x^3 - 3x^2 + \frac{3}{4}x - 3$
a. $x - 6$ **b.** $f(6)$

34. $g(x) = 11x^3 - 2x^2 - 8$
a. $x + 2$ **b.** $g(-2)$

Assume each polynomial below defines a function $f(x)$; for each polynomial

a. Use Descartes' rule of signs to find the number of possible real zeros.
b. List all possible rational zeros of each polynomial.
c. Find all rational zeros; state the multiplicity when greater than one.
d. Write the function as a product of linear and prime quadratic factors if possible.
e. State any irrational zeros found in part d. If there are any other possible irrational zeros state the least positive integer upper bound and the greatest negative integer lower bound for these zeros.

35. $x^4 - x^3 - 7x^2 + x + 6$
36. $x^4 - 5x^2 + 4$
37. $4x^3 - 12x^2 + 11x - 3$
38. $x^4 - 6x^3 + 54x - 81$
39. $x^4 - 8x^3 + 30x^2 - 72x + 81$
40. $x^5 - 4x^3 - 8x^2 + 32$
41. $x^6 - 4x^3 - 5$
42. $6x^3 + 7x^2 - x - 2$
43. $3x^4 + 5x^3 - 11x^2 - 15x + 6$
44. $4x^5 - x^3 - 32x^2 + 8$
45. $x^5 - 4x^2 + 2$
46. $x^3 - 3x^2 - 3$
47. $2x^4 + 6x^3 - 2x - 6$
48. $2x^4 - 5x^3 - 8x^2 + 17x - 6$
49. $3x^4 + 2x^3 - 4x^2 - 2x + 1$
50. $2x^4 - x^3 - 10x^2 + 4x + 8$
51. $x^3 + x^2 - 7x - 10$
52. $2x^3 + x^2 + x - 1$
53. $x^5 - 4x^3 + x^2 + 4x + 4$
54. $x^4 - x^2 + 2x - 2$
55. $9x^4 - 82x^2 + 9$
56. $6x^5 - 23x^4 + 25x^3 - 3x^2 - 7x + 2$
57. $4x^5 + 16x^4 + 37x^3 + 43x^2 + 22x + 4$
58. $x^7 + x^6 - 7x^4 - 7x^3 - 8x - 8$

59. The value $\sqrt[4]{3}$ is a zero of the expression $x^4 - 3$. Use synthetic division to divide $x^4 - 3$ by $x - \sqrt[4]{3}$.

60. The value $\sqrt[5]{6}$ is a zero of the expression $x^5 - 6$. Use synthetic division to divide $x^5 - 6$ by $x - \sqrt[5]{6}$.

61. Use the rational zero theorem to find all possible "rational" zeros of the polynomial $acx^3 + (ad - bc - ace)x^2 + (-ade - bd + bce)x + bde$. Assume that a, b, c, d, and e are integers with no common factors.

62. Use synthetic division and the results of problem 61 to find all the rational zeros of the polynomial $acx^3 + (ad - bc - ace)x^2 + (-ade - bd + bce)x + bde$; use the zeros to factor this polynomial.

63. Referring to the definition in the text of a polynomial of one variable of degree n, define a polynomial function of degree 4. That is, apply the definition when n is 4.

Skill and review

In problems 1–4 graph each function; label all intercepts.

1. $f(x) = (x - 2)^2 - 1$
2. $f(x) = x^2 + x - 4$
3. $f(x) = |x - 2| - 3$
4. $f(x) = x^3 - 1$
5. Find all zeros of $f(x) = 2x^5 + 7x^4 + 2x^3 - 11x^2 - 4x + 4$.
6. Solve $|2x - 3| < 9$.

4–3 The graphs of polynomial functions, and finding zeros of functions by graphical methods

In section 4–2 we studied algebraic methods for finding zeros of polynomial functions. In this section we study the graphs of these functions. We will see that zeros of functions have a clear graphical interpretation, and that graphs provide a powerful tool for finding zeros.

Two very useful pieces of information when graphing polynomial functions are the intercepts of the polynomial and its behavior for large values of x.

Intercepts

A *zero of a function is also an x-intercept of its graph.* This is because x-intercepts are found by replacing y by 0 in an equation. Replacing y by 0 in $y = f(x)$ we solve $0 = f(x)$, which gives the zeros of the function f. Similarly, *the y-intercept of a function is the value of the function when x is zero,* or, for a function f, the value of $f(0)$. This is summarized as follows.

> To find the x-intercepts of a function f, solve $f(x) = 0$.
> To find the y-intercept of a function f, compute $f(0)$.

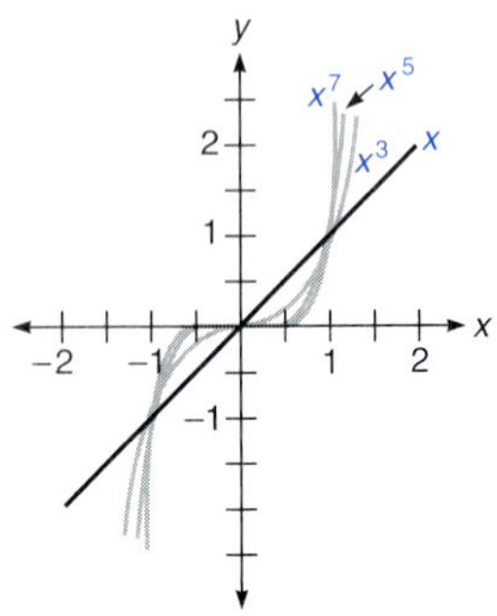

Figure 4–4

Behavior for large values of $|x|$

The graphs of $f(x) = x^n$ for $n = 1, 2, 3, 4, 5, 6, 7, \ldots$ fall into two categories, shown in figure 4–4 for odd powers and figure 4–5 for even powers. When n is odd, the graph is negative for negative values of x. When n is even the graph is always positive.

The graphs are similar in each figure. Considering polynomial functions with positive coefficient, it is an important fact that for large values of $|x|$ all odd degree polynomial functions behave like the functions in figure 4–4 and all even degree polynomials behave like those in figure 4–5. This is shown in figure 4–6.

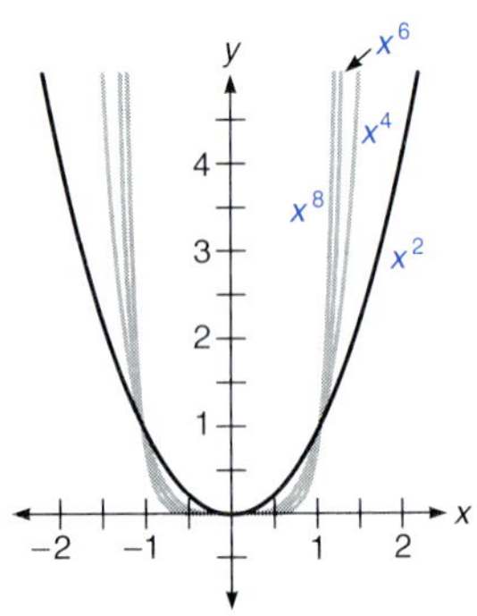

Figure 4–5

To get a feeling for why all polynomials of a certain degree behave in a similar way for large values of $|x|$, consider the values in the following table for the functions $f(x) = x^3$, $f(x) = 2x^3 + x^2$, and $f(x) = x^3 + 5x^2 + 100$.

x	x^3	$x^3 + x^2$	$x^3 + 5x^2 + 100$
−1,000	−1,000,000,000	−999,000,000	−994,999,900
−100	−1,000,000	−990,000	−949,900
−10	−1,000	−900	−400
10	1,000	1,100	1,600
100	1,000,000	1,010,000	1,050,100
1,000	1,000,000,000	1,001,000,000	1,005,000,100

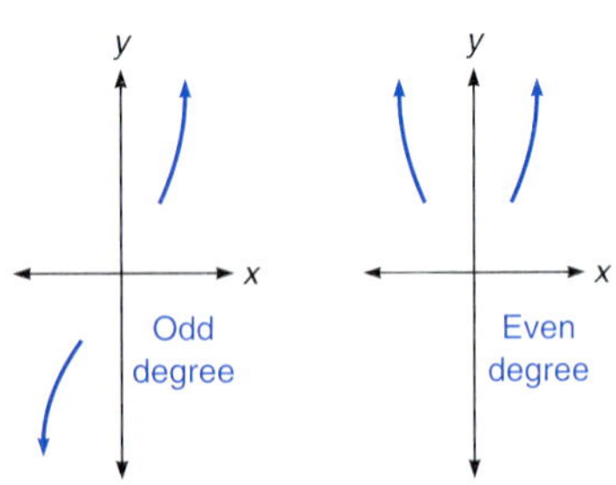

Figure 4–6

Examination shows that as $|x|$ grows the difference in the three functions becomes smaller and smaller, as a percentage of the value of x^3. A graph of these three functions would become practically indistinguishable as $|x|$ gets larger and larger. Thus, for $|x|$ large enough we might just as well work with the simplest of the functions, $f(x) = x^3$.

Figure 4–6 shows the behavior of odd and even degree polynomials (with positive leading coefficient) for large values of $|x|$; this means somewhat to the right and left of the origin. Near the origin, the behavior varies depending on the values of the zeros of the function.

Graphing using intercepts and behavior for large values of $|x|$

To graph polynomial functions using algebraic properties as a guide, we use several pieces of information, including

- Intercepts.
- The fact that for large values of $|x|$ (i.e., to the right and left of any zeros) all polynomial functions of the same degree are similar (figure 4–6).
- Reflection of the graph about the x-axis when the coefficient of the term of highest degree is negative.
- Plotting points other than the intercepts.

This is illustrated in example 4–3 A.

Of course graphing calculators provide a tool that can achieve similar results, often with much less work. The algebraic properties we are studying still provide an understanding of why a graph behaves as it does, and can be valuable in deciding how to set the RANGE, where to zoom, etc. These functions are graphed on a graphing calculator as shown in earlier sections.

■ Example 4–3 A

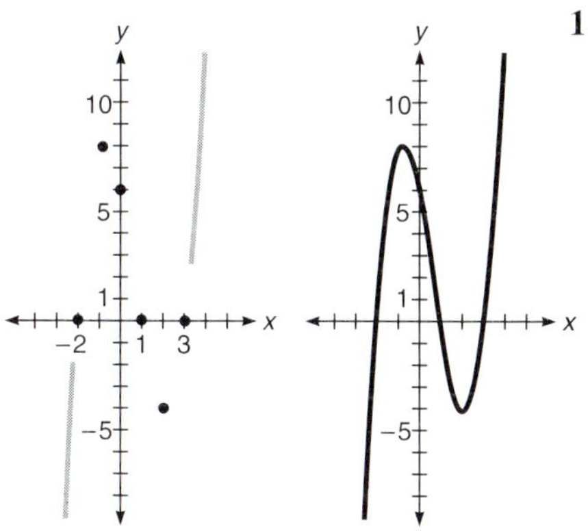

(a) (b)

Y= (X|T − 1)
(X|T + 2) (
X|T − 3)
RANGE −3,4,−7,10

Graph the function. Compute all intercepts.

1. $f(x) = (x - 1)(x + 2)(x - 3)$

Since $f(x) = (x - 1)(x + 2)(x - 3) = x^3 - 2x^2 - 5x + 6$, f is a function of degree 3.

x-intercepts:

$0 = (x - 1)(x + 2)(x - 3)$	Let $f(x) = 0$
$x - 1 = 0$ or $x + 2 = 0$ or $x - 3 = 0$	Zero factor property
$x = 1$ or $x = -2$ or $x = 3$	x-intercepts

y-intercept:

$f(0) = (-1)(2)(-3) = 6$ — $y = f(0)$ is the y-intercept

Additional points:

$f(-1) = (-1 - 1)(-1 + 2)(-1 - 3) = 8$; plot $(-1,8)$
$f(2) = (2 - 1)(2 + 2)(2 - 3) = -4$; plot $(2,-4)$

We show the intercepts, the points $(-1,8)$ and $(2,-4)$, and also the similarity with the curve of $y = x^3$ for x to the right and left of the intercepts in part a of the figure. Connecting these with a smooth curve produces the graph of the function in part b.

2. $f(x) = x^4 - 4x^3 - 6x^2 + 21x + 18$

y-intercept:

$f(0) = 0^4 - 4(0^3) - 6(0^2) + 21(0) + 18 = 18$ The y-intercept is $f(0)$

x-intercepts: To find x-intercepts we must look for possible rational zeros of f. Use the rational zero theorem to establish that $f(x) = (x + 2)(x - 3)(x^2 - 3x - 3)$.

The zeros of the quadratic $(x^2 - 3x - 3)$ are found by the quadratic formula to be $\dfrac{3 \pm \sqrt{21}}{2}$. Thus, the x-intercepts are -2, 3, and $\dfrac{3 \pm \sqrt{21}}{2}$ (approximately 3.8 and -0.8).

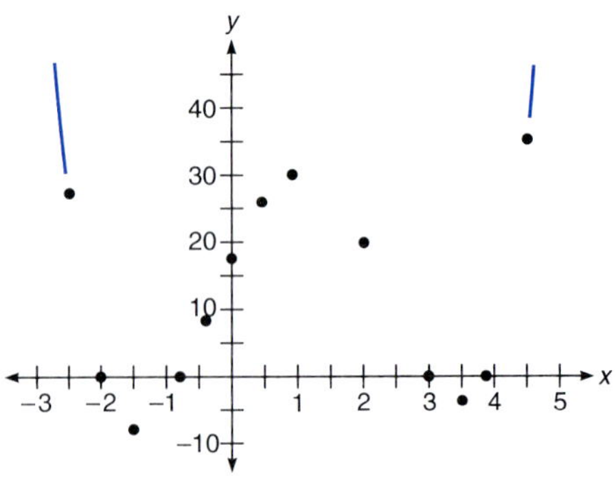

(a)

Additional points: In addition to the x-intercepts and y-intercept we usually need to *choose some additional points to plot* to obtain a graph with approximately the correct proportions. (Unless using a graphing calculator, of course!) As a minimum *choose at least one point between any two intercepts* (-1.5, 1, and 3.5 in the figure), and *one point to the right and one to the left of all intercepts* (-2.5 and 4.5). *Choose additional points when the distance between intercepts is large,* as between -0.8 and 3 in this case.

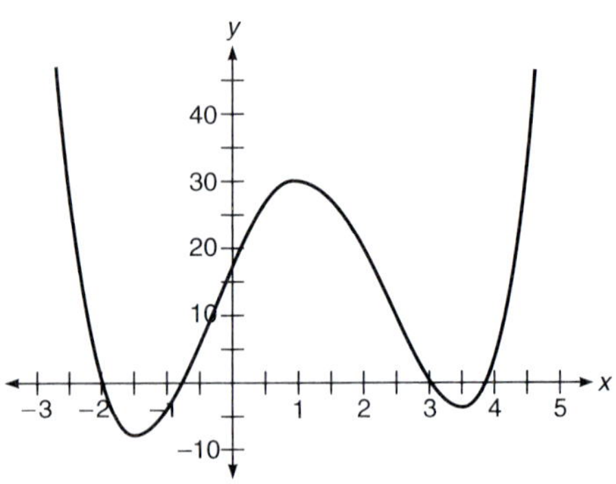

(b)

Synthetic division is a convenient way to compute the function value for these additional points. This is shown for the value -1.5.

	1	−4	−6	21	18
		−1.5	8.25	−3.375	−26.4375
−1.5	1	−5.5	2.25	17.625	−8.4375

$f(-1.5) \approx -8.4$

With the large range of y-values we use a different scale for the x- and y-axes. We plot the intercepts and additional points (part a). Since the degree of the function is even, it behaves as shown in part a of the figure. This information allows us to sketch the smooth curve shown in part b of the figure.

x	−2.5	−1.5	0.5	1	2	3.5	4.5
y	29.5	−8.4	26.6	30	20	−3.4	36.6

Solution using a graphing calculator

For the TI-81 we would enter

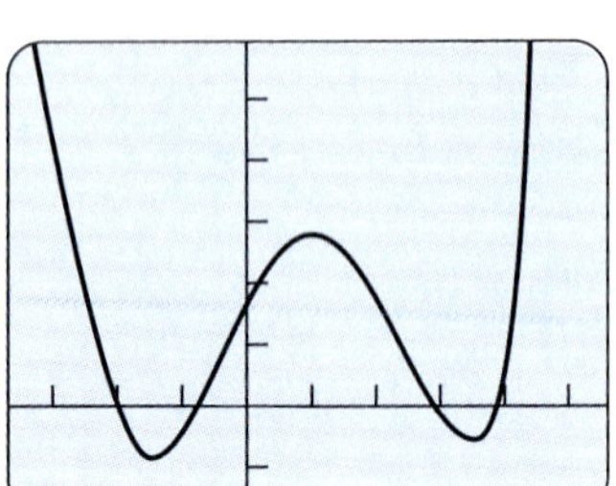

[Y=] [X|T] [^] 4 [−] 4 [X|T] [MATH] 3 [−] 6 [X|T] [x^2] [+] 21 [X|T] [+] 18

[RANGE −3,5,−10,50] Yscl=10

The result would look like that shown in the figure. Approximate values of the zeros can be found by tracing and zooming but Newton's method is better. ■

The TI-81 and Newton's method

There are numeric methods for finding zeros of polynomial functions quickly and to great accuracy by using a programmable calculator and writing a program that searches for a zero of a function. This is useful when the function is well behaved around the zero. For our purposes here we mean that one continuous, smooth line could be used to draw the graph of the function near the zero in question. A good method is called Newton's method. The Texas Instruments TI-81 calculator handbook presents a program called NEWTON that implements this method. Enter the program into the calculator as follows

[PRGM] ▸ 1 Program EDIT Mode

Assumes entering the program as Prgm1. Enter the keys that correspond to N E W T O N. For example, T is over the [4] key.

[ENTER]

Now type in the program as shown. Use [ENTER] after each line.

Program	**Keystroke hints**
:(Xmax − Xmin)/100→D	Xmax is in VARS RNG. Xmin is in VARS RNG. →D is [STO▸] [x^{-1}].
:Lbl 1	Lbl is in PRGM CTL.
:X–Y_1/NDeriv(Y_1,D)→R	Y_1 is in Y-VARS. NDeriv is [MATH] 8. ",D" is [ALPHA] [.] [ALPHA] [x^{-1}]. →R is [STO▸] [×].
:If abs (X–R)≤abs (X/1E10)	If is in [PRGM] CTL. ≤ is in TEST ([2nd] [MATH]). 1E10 is 1 [EE] 10.
:Goto 2	Goto is in [PRGM] CTL.
:R→X	Use [ALPHA] [×] [STO▸] [X|T].
:Goto 1	
:Lbl 2	
:Disp "ROOT"	Use [PRGM] I/O 1 [A-LOCK] [+] R O O T [+].
:Disp R	Use [2nd] [CLEAR] when finished entering the program.

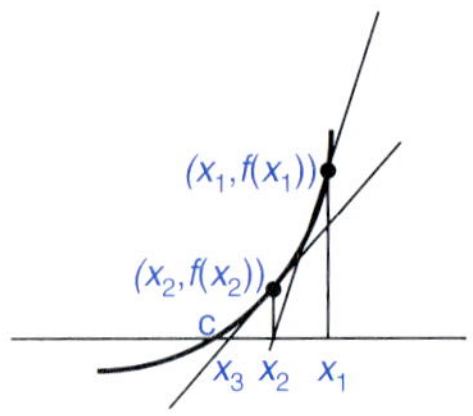

Figure 4–7

Figure 4–7 illustrates how this method gets closer and closer to a root. Assume a function f has a zero at c in the figure. Suppose x_1 is a value of x near c. The program uses the line that is tangent to (i.e., just touches) the function f at the point $(x_1,f(x_1))$ to locate the point x_2, which is closer to c. The program then uses the line that is tangent to the function f at the point $(x_2,f(x_2))$ to locate the point x_3, which is even closer to c. The program repeats this until the difference between the last x-value and the newest x-value is less than a predetermined error value.

The algebraic way in which the program discovers the tangent line at each step is left for a course in calculus. With a little background in this subject, it is not hard to understand.

■ ***Example 4–3 B***

Use the program NEWTON to find the zero near -1 and the zero near 4 in part 2 of example 4–3 A.

Graph the function on the calculator as shown in example 4–3 A.

Use [TRACE] to move the cursor near the zero located near -1.

Then select [PRGM] 1 (assumes the program NEWTON is stored as the first program). The display shows Prgm1 in the display. Use [ENTER] to run the program.

The approximate value of this zero is displayed: $-.7912878475$.

Rerun the program by selecting [GRAPH] and then [TRACE] again, placing the cursor near the zero near 4. Then select [PRGM] 1 [ENTER]. This will show that this zero is approximately 3.791287847. ■

Graphs at zeros of multiplicity greater than one

The multiplicity of a zero affects the graph. When a zero has **even multiplicity** the function does not cross the x-axis at the corresponding intercept but rather just touches the axis. The reason is discussed after example 4–3 C. A function does cross the x-axis at intercepts corresponding to zeros of **odd multiplicity.**

■ ***Example 4–3 C***

Graph $f(x) = -x^3 + 4x^2 - 5x + 2$.

It would be easiest to think of this as $f(x) = (-1)(x^3 - 4x^2 + 5x - 2)$, and first graph $y = x^3 - 4x^2 + 5x - 2$, then flip the graph about the x-axis.

$$y = x^3 - 4x^2 + 5x - 2$$
$$y = (x - 1)^2(x - 2)$$ Use synthetic division to factor

x-intercepts:

$$0 = (x - 1)^2(x - 2)$$ Let $y = 0$ in $y = (x - 1)^2(x - 2)$
$$x = 1 \text{ or } 2$$ 1 has even multiplicity, so the graph just touches but does not cross the x-axis at $x = 1$

y-intercept:

$y = -2$ Let $x = 0$ in $y = x^3 - 4x^2 + 5x - 2$

Additional points:

$y = (1.5 - 1)^2(1.5 - 2) = -0.125$ Let $x = 1.5$ in $y = (x - 1)^2(x - 2)$

We first plot the intercepts and the additional point (1.5,−0.125). We also know the function behaves like that shown in figure 4–4 for odd-degree functions. We also know that the function crosses the x-axis at the intercept at 2 but not at 1 (part a of the figure).

We next draw a smooth curve to represent $y = x^3 - 4x^2 + 5x - 2$ (part b of the figure), and finally, to obtain the finished graph we draw a graph symmetric to this one about the x-axis (part c of the figure).

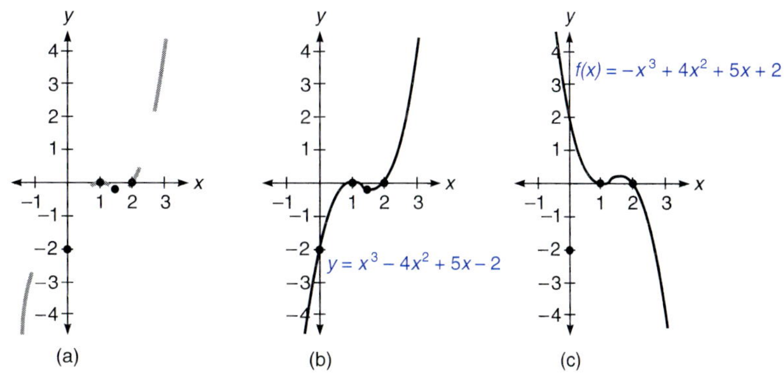

Y=	(−)	X|T	MATH 3	+ 4	X|T	x^2	− 5
X|T	+ 2	RANGE −1,3,−3,4					

■

To see why the function in example 4–3 C does not cross the x-axis at 1, consider the equation $y = (x - 1)^2(x - 2)$. When x is less than 1, $x - 1$ is negative. When x is greater than 1, $x - 1$ is positive. However, in both cases $(x - 1)^2$ is positive. Thus, $(x - 1)^2$ provides a constant influence on the sign of the product $(x - 1)^2(x - 2)$ regardless of the value of x. This would be true whatever the exponent of $(x - 1)$, as long as the exponent is even.

Example 4–3 D further illustrates using the multiplicity of a zero to help graph the function, or when using a graphing calculator, to help understand the behavior of the function.

■ Example 4–3 D

Graph the functions.

1. $f(x) = (x - 2)^3(x - 5)^2$

The only zeros are at 2 (odd multiplicity) and 5 (even multiplicity); the y-intercept is at $f(0) = -200$. Thus the graph does not cross the x-axis at 5, but does cross it at 2. We plot some more points to obtain the graph.

Additional points:

x	0.5	1	1.5	2.5	3	4	5.5	6
y	−68.3	−16	−1.5	0.78	4	8	10.7	64

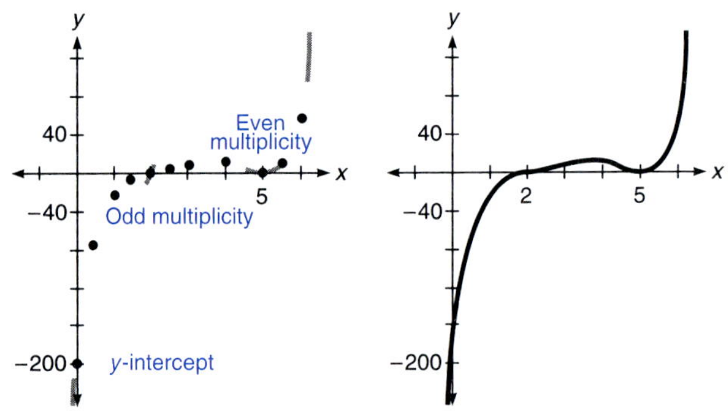

[Y=] [(] [X|T] [−] 2 [)] [MATH] 3 [(] [X|T] [−] 5 [)] [x^2]

[RANGE −0.5,6,−250,250] YSCL=40

2. $f(x) = (x + 3)^2(x^2 + x + 1)$

It can be found, by the quadratic formula, that $x^2 + x + 1$ has only complex zeros. Thus, f has only one x-intercept, at −3. This zero is of even multiplicity so the function just touches the axis at −3.

The function f is a fourth-degree polynomial, so for large values of $|x|$ it behaves as indicated in figure 4–5.

The y-intercept is $f(0) = 9$.

Additional points:

x	−4	−3.5	−2.5	−2	−1.5	−1	−0.5	0.5
y	13	2.4	1.2	3	3.9	4	4.7	21.4

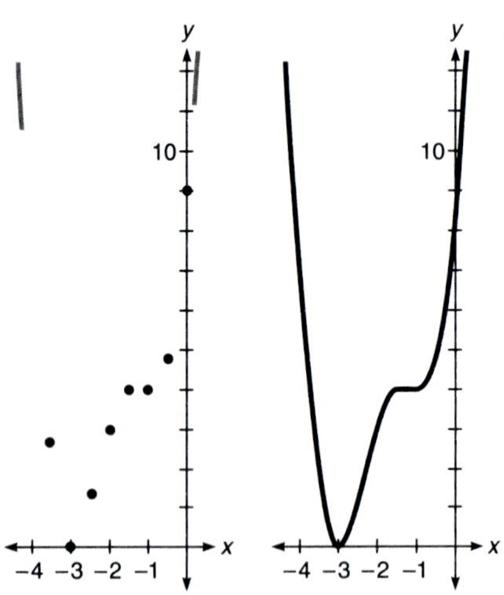

[Y=] [(] [X|T] [+] 3 [)] [x^2] [(] [X|T] [x^2] [+] [X|T] [+] 1 [)]

[RANGE −4,0.5,−0.5,12]

Note The shape of this graph in the area of $x = -1$ can be hard to detect without plotting sufficient points. Observe that the complex roots of $x^2 + x + 1$ are $-\frac{1}{2} \pm \frac{\sqrt{-3}}{2}$. It is a good idea to plot additional points at and near the real part of these values, $-\frac{1}{2}$. ■

Applications

If a function is modeling an applied situation in science, business, and the like, then the graph of that function is a powerful tool for answering many types of questions about the situation being modeled.

■ Example 4–3 E

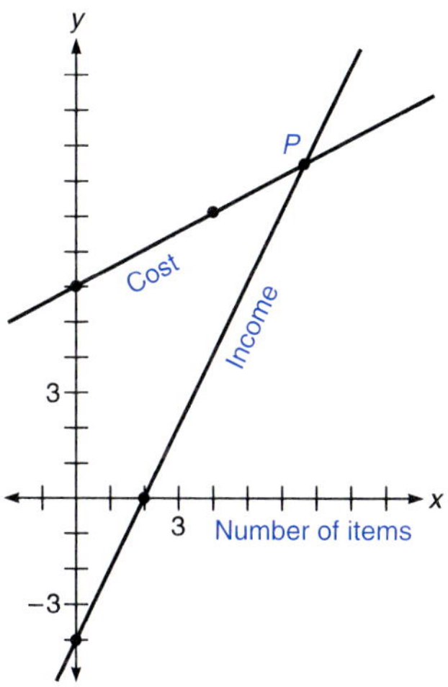

A company has discovered that, with respect to a certain product, its income I for i items is $I(i) = 2i - 4$. Its costs for i items is $C(i) = \frac{1}{2}i + 6$.
(a) Graph both of these functions in the same coordinate system, and
(b) determine the minimum number of items that must be produced to make a profit P.

Since it is easier to think in terms of x and y let us graph the relations

$$y = 2x - 4 \qquad \text{Income}$$
$$y = \tfrac{1}{2}x + 6, \qquad \text{Cost}$$

where x represents the number of items produced.

These are linear functions, so we plot two points for each function, normally the intercepts. For the cost function the x-intercept (-12) is inconvenient, so we plot another point—in this case $x = 4$ so $y = \frac{1}{2}(4) + 6 = 8$.

To make a profit the *income must be greater than cost.* This happens when the number of items, x is to the right of the point P shown in the figure. We find P, the point of intersection, as in section 3–2.

$$y = 2x - 4$$
$$y = \tfrac{1}{2}x + 6$$
$$\text{so } 2x - 4 = \tfrac{1}{2}x + 6$$

Solving for x:

$$2x - 4 = \tfrac{1}{2}x + 6$$
$$4x - 8 = x + 12 \qquad \text{Multiply each member by 2}$$
$$3x = 20$$
$$x = 6\tfrac{2}{3}$$

Thus income exceeds cost when the number of items produced is above $6\frac{2}{3}$.

Of course the graphing calculator can also be used to graph both functions. Then tracing along either function will find the approximate (x,y) coordinate where the lines cross. Zooming can be used to increase the accuracy. ■

Mastery points

Can you

- Graph polynomial functions of degree greater than 2 when that polynomial is factored into a product of linear and quadratic factors?
- Construct polynomial functions of any degree that describe stated conditions, and graph these functions?

Exercise 4–3

Graph the following polynomial functions. Label all intercepts.

1. $f(x) = (x - 2)(x + 1)(x + 3)$
2. $g(x) = 2(x - 4)(x^2 - 4)$
3. $h(x) = (x^2 - 1)(x^2 - 9)$
4. $f(x) = (2x - 1)(x - 1)(x - 2)$
5. $g(x) = (x^2 - 4)(4x^2 - 25)(x + 3)$
6. $h(x) = (x^2 - x - 6)(x^2 - 9)(2x + 1)$
7. $f(x) = (x - 1)^2(x + 1)$
8. $g(x) = (x - 2)^2(x + 3)^2$
9. $h(x) = (x + 2)^3(2x - 3)^2$
10. $f(x) = (2x^2 - 3x - 5)^2$
11. $g(x) = (x - 2)^2(x^2 + 3x + 6)$
12. $h(x) = (x - 2)(x^2 + 2x + 5)$
13. $f(x) = x^4 - x^3 - 7x^2 + x + 6$
14. $h(x) = x^4 - 5x^2 + 4$
15. $g(x) = 4x^3 - 12x^2 + 11x - 3$
16. $f(x) = x^4 - 6x^3 + 54x - 81$
17. $f(x) = x^4 - 8x^3 + 30x^2 - 72x + 81$
18. $h(x) = x^5 - 4x^3 - 8x^2 + 32$
19. $g(x) = 6x^3 + 7x^2 - x - 2$
20. $f(x) = 3x^4 + 5x^3 - 11x^2 - 15x + 6$
21. $h(x) = 4x^5 - x^3 - 32x^2 + 8$
22. $g(x) = x^5 - 4x^3 + 8x^2 - 32$

Solve the following problems.

23. A company's income I from a product is described by $I(x) = 3x - 6$, where x is the number of items sold. Its cost for x items is $C(x) = x + 4$.
    a. Graph both of these functions on the same coordinate system.
    b. Determine the minimum number of items that the company must produce to make a profit.

24. A company's income I from a product is described by $I(x) = 4x - 2$, where x is the number of items sold. Its cost for x items is $C(x) = 2x + 4$.
    a. Graph both of these functions on the same coordinate system.
    b. Determine the minimum number of items that the company must produce to make a profit.

25. The Ajax Car Rental Company rents cars for \$30 per day and \$0.30 per mile. The Zeus Car Rental Company rents cars for \$42 per day with unlimited mileage. Under these conditions, the costs of renting from each company for a day can be described as $A(x) = 0.30x + 30$ and $Z(x) = 42$.
    a. Graph both functions in the same coordinate system.
    b. Determine from the graph under which conditions it is cheaper to rent from each company.

26. A photographer is trying to decide whether it is cheaper to use slide film or print film to shoot prints. Slide film costs \$6 for 36 exposures, including developing, and then it costs \$0.25 to make a print from a slide. Print film costs \$10 for 36 exposures, including developing into prints. Under these conditions the cost for x prints, up to 36, for slides s and for prints p is $s(x) = 0.25x + 6$, and $p(x) = 10$.
    a. Graph both functions s and p in the same coordinate system.
    b. Determine the maximum number of good prints that must be kept out of every 36 to make shooting slides cheaper.

27. A rectangular carpet costs \$1 per square foot. It is made in varying widths, but the length is always 3 feet more than the width. Create and graph a function that describes the cost of such a carpet as a function of width.

28. The same company of problem 27 makes carpet that costs \$1.50 per square foot. For these carpets the length must be 1 foot more than twice the width. Create and graph a function that describes the cost of such a carpet as a function of width.

29. A company makes a decorative box of varying sizes, but the proportions of length, width and height are $x + 6$, $x - 2$, and x, respectively, where x is the height in inches. The material that covers the box costs one cent per square inch. Create a function that describes the cost of covering a box as a function of its height. Graph this function.

30. The company described in problem 29 fills these boxes with a different material. Create a function that describes the volume of a box as a function of its height. (The volume of a rectangular solid is the product of its length, width, and height.) Graph this function.

31. Let us call a function "k-scalable" if for any integer k, $k(f(x)) = f(kx)$ for all x in its domain. For example, a function is 2-scalable if $2f(x) = f(2x)$ for all x in its domain.
    a. Show that the function $f(x) = 5x$ is 2-scalable.
    b. Show that the function $f(x) = -3x$ is k-scalable.
    c. Show that the function $f(x) = x^2 - 2x - 8$ is not 3-scalable.

32. Let us call a function "additive" if for all a and b in its domain, $f(a + b) = f(a) + f(b)$.
    a. Show that the function $f(x) = 5x$ is additive.
    b. Show that the function $f(x) = -3x + 1$ is not additive.
    c. Show that the function $f(x) = x^2 - 2x - 8$ is not additive.

Use graphical methods and the programmable capabilities of graphing calculators to find the zeros of the following functions to five decimal places.

33. $f(x) = x^3 - 2x^2 + 5x - 2$

34. $f(x) = 2x^3 - 2x^2 + 5x - 3$

35. $f(x) = 2x^4 - 2x^2 + 5x - 3$

36. $f(x) = x^5 - 3x^3 + 5x^2 - 3$

37. $f(x) = x^5 + 2x^4 + 5x^2 - 3$

38. $f(x) = x^7 - 5x^5 - 2x^4 + 2x^3 + 5x^2 - 1$

39. Napthalene ($C_{10}H_8$) is a very stable chemical. To determine its stability, one needs to solve the equation $x^{10} - 11x^8 + 41x^6 - 65x^4 + 43x^2 - 9 = 0$. It has ten solutions. Find them to five decimal places of accuracy.

40. The equation $x^{10} - 11x^8 + 41x^6 - 61x^4 + 31x^2 - 3 = 0$ must also be solved to fully solve the problem suggested in the previous problem. This equation also has ten solutions. Find them to five decimal places of accuracy.

Skill and review

1. Graph $f(x) = x^2 + 2x - 1$.

2. Solve $\left|\dfrac{2x - 3}{4}\right| \geq \dfrac{1}{2}$.

3. Solve $x^{-2} - x^{-1} - 12 = 0$.

4. Solve $\sqrt{2x - 2} = x - 5$.

5. Combine $\dfrac{3}{x - 1} - \dfrac{2}{x + 1} + \dfrac{1}{x}$.

6. Simplify $\sqrt[3]{\dfrac{4x^2y^7}{3z^8}}$.

7. Rewrite $|5 - 2\pi|$ without absolute value symbols.

4–4 Rational functions

Assume that gasoline costs \$1.00 per gallon, that a car is driven 15,000 miles per year, and that the car's fuel economy is *x* miles per gallon (mpg). Let *y* represent the annual savings in fuel costs for increasing the mileage by 10 mpg. Then $y = \dfrac{15{,}000}{x} - \dfrac{15{,}000}{x + 10}$. Graph this function.

In this section we investigate some of the properties of functions defined as a quotient of two polynomials. These functions are called rational functions. They may be used to study the rate at which a machine can do a job, or to investigate improving gas economy in an automobile, or to describe resistance or capacitance in an electronics circuit.

We begin by defining this class of functions.

Rational function

A rational function is a function of the form

$$f(x) = \frac{P(x)}{Q(x)}$$

where $P(x)$ and $Q(x)$ are polynomial expressions in the variable x. Unless otherwise stated, the domain of f is all real numbers for which $Q(x) \neq 0$.

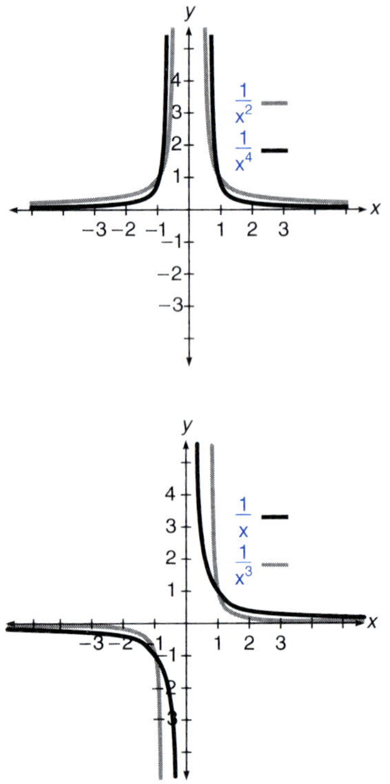

Figure 4–8

Rational functions that are translations of $f(x) = \dfrac{a}{x^n}$

As with polynomial functions (sections 4–2 and 4–3) we are interested in the graphs of rational functions. A basic rational function is $f(x) = \dfrac{1}{x}$. We graphed this function in section 3–4; its graph is shown in figure 4–8, which also shows the graph of $f(x) = \dfrac{1}{x^n}$ for $n = 2, 3, 4$. The graph for $n = 2$ is typical of all even powers, and the graph for $n = 1$ is typical for all odd powers. For the functions in figure 4–8, the y-axis is a **vertical asymptote** and the x-axis is a **horizontal asymptote.** A vertical asymptote is a vertical line that is not part of the graph of the function but that indicates that the function gets larger and larger (or smaller and smaller for negative values) as x gets closer and closer to some given value. A horizontal asymptote is a horizontal line, also not part of the graph of the function, to which the graph of the function gets closer as x gets larger and larger, or smaller and smaller (more and more negative).

We use the following strategy for using algebraic information for graphing rational functions whose graphs are translated and vertically scaled versions of $y = \dfrac{1}{x^n}$. Of course, graphing calculators can be used; this is illustrated for the TI-81.

> To graph functions of the form $f(x) = \dfrac{a}{(x-h)^n} + k$
>
> - There is a vertical asymptote wherever the denominator is zero: at h.
> - The line $y = k$ is a horizontal asymptote.
> - We plot intercepts and a few additional points to determine the basic shape of the graph.

The value a is a vertical scaling factor (section 3–4). It is the primary reason we must plot a few points to obtain an accurate rendition of the graph. The value h represents a horizontal translation, and that of k a vertical translation.

Remember that the y-intercept is at $f(0)$ and the x-intercepts are the solutions to the equation $f(x) = 0$.

■ ***Example 4–4 A***

Describe the function in terms of the graph of $f(x) = \dfrac{1}{x^n}$ for the appropriate value of n. State all intercepts and asymptotes. Then sketch the graph.

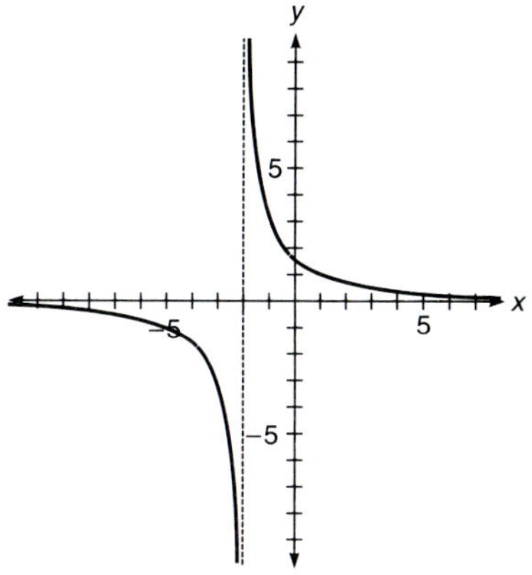

Y= 3 ÷ (X|T + 2)

RANGE −10,6,−8,8

1. $f(x) = \dfrac{3}{x+2}$

$y = \dfrac{3}{x-(-2)}$ Rewrite

This graph is the same as that of $y = \dfrac{1}{x}$ except that it is horizontally translated 2 units to the left and vertically scaled by 3.

y-intercept:

$f(0) = \dfrac{3}{0+2} = 1\dfrac{1}{2}$: $\left(0, 1\dfrac{1}{2}\right)$

x-intercept:

$0 = \dfrac{3}{x+2}$ Solve $f(x) = 0$

$0 = 3$ No x-intercept

Vertical asymptote at $x = -2$ (where the denominator is 0).
Horizontal asymptote at $y = 0$ (the x-axis).

Additional points:

x	-5	-4	-3	-2.5	-1.5	-1	1
y	-1	$-1\frac{1}{2}$	-3	-6	6	3	1

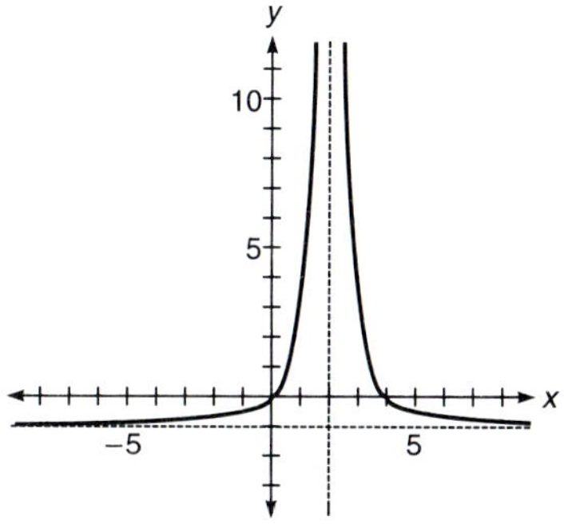

Y= 4 ÷ (X|T − 2) x^2 − 1

RANGE −5,9,−2,10

2. $g(x) = \dfrac{4}{(x-2)^2} - 1$

This is the graph of $y = \dfrac{1}{x^2}$ translated right 2 units, down 1 unit, and vertically scaled by 4.

Vertical asymptote: $x = 2$

Horizontal asymptote: $y = -1$

y-intercept:

$$g(0) = \frac{4}{(0-2)^2} - 1 = 0$$

x-intercepts:

$0 = \dfrac{4}{(x-2)^2} - 1$ $f(x) = 0$

$1 = \dfrac{4}{(x-2)^2}$ Add 1 to each member

$(x-2)^2 = 4$ Multiply each member by $(x-2)^2$

$x - 2 = \pm 2$ Extract the square root of each member

$x = 2 \pm 2$ Add 2 to both members

$x = 0$ or 4

Additional points:

x	-2	-1	1	1.5	2.5	3	5
y	$-\frac{3}{4}$	$-\frac{5}{9}$	3	15	15	3	$-\frac{5}{9}$

■

Rational functions in which the denominator contains more than one linear factor

The denominator of a rational function may have more than one zero. *There will be a vertical asymptote for each zero of the denominator* that is not also a zero of the numerator. We will see what happens at a zero of both the numerator and denominator later in this section.

As long as the degree of the numerator is less than the degree of the denominator the *x-axis will be a horizontal asymptote,* unless there is a vertical translation. In that case the horizontal asymptote is also translated vertically.

These graphs cannot be compared to graphs of $y = \frac{a}{x^n}$. We graph them by noting where they have asymptotes and by plotting a few additional points.

■ Example 4–4 B

Graph the function. State all asymptotes and intercepts.

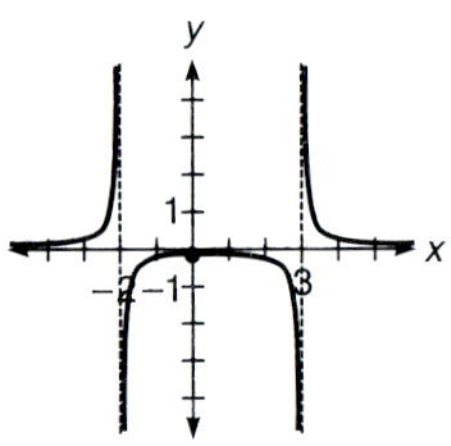

Y= 1 ÷ (X|T x² − X|T − 6)
RANGE −4,5,−3,3

1. $f(x) = \dfrac{1}{x^2 - x - 6}$

   Factoring the denominator gives $f(x) = \dfrac{1}{(x-3)(x+2)}$.

   Vertical asymptotes: $x = 3$ and $x = -2$

   Horizontal asymptote: $y = 0$

   y-intercept: $f(0) = \dfrac{1}{0^2 - 0 - 6} = -\dfrac{1}{6}$

   x-intercepts: $0 = \dfrac{1}{x^2 - x - 6}$ No solution, so no x-intercepts

   Additional points:

x	−3	−2.5	−1.5	1	2	2.5	3.5	4
y	$\frac{1}{6}$	$\frac{4}{11}$	$-\frac{4}{9}$	$-\frac{1}{6}$	$-\frac{1}{4}$	$-\frac{4}{9}$	$\frac{4}{11}$	$\frac{1}{6}$

2. $f(x) = \dfrac{x}{x^2 - 9}$

   First, observe that the degree of the numerator (1) is less than the degree of the denominator (2). Thus the x-axis is a horizontal asymptote.

   $$f(x) = \frac{x}{(x-3)(x+3)}$$ Vertical asymptotes at ± 3

   y-intercept:

   $$f(0) = \frac{0}{0^2 - 9} = 0$$

   x-intercepts:

   $$0 = \frac{x}{x^2 - 9}$$
   $$0 = x$$

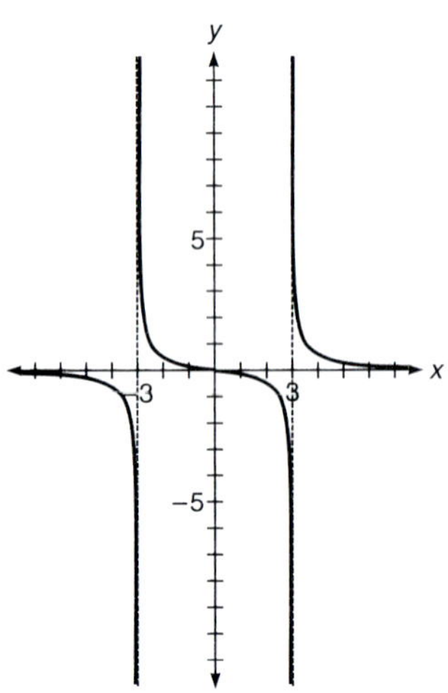

Y= X|T ÷ (X|T x² − 9)
RANGE −8,8,−5,5

Additional points:

x	-5	-4	-2	-1	1	2	4
y	$-\frac{5}{16}$	$-\frac{4}{7}$	$\frac{2}{5}$	$\frac{1}{8}$	$-\frac{1}{8}$	$-\frac{2}{5}$	$\frac{4}{7}$

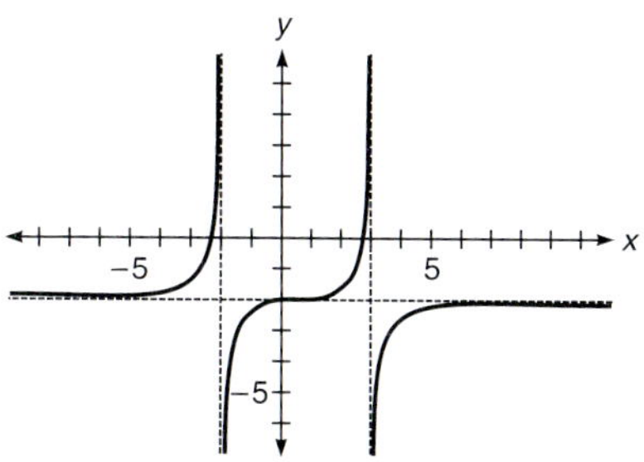

[Y=] [(] [(−)] [X|T]
[+] 1 [)] [÷] [(] [(]
[X|T] [−] 3 [)] [(]
[X|T] [+] 2 [)] [)] [−]
2 [RANGE −7,9,−6,3]

3. $f(x) = \dfrac{-x + 1}{(x - 3)(x + 2)} - 2$

The degree of the numerator is less than the degree of the denominator, so we expect a horizontal asymptote. Since there is a vertical translation of -2 the horizontal asymptote will be at $y = -2$ instead of the x-axis. There are vertical asymptotes at -2 and 3.

y-intercept: $f(0) = \dfrac{-0 + 1}{(0 - 3)(0 + 2)} - 2 = -2\frac{1}{6}$

x-intercepts:

$$0 = \frac{-x + 1}{(x - 3)(x + 2)} - 2$$

$$2 = \frac{-x + 1}{(x - 3)(x + 2)}$$

$$2(x - 3)(x + 2) = -x + 1$$

$$2x^2 - x - 13 = 0$$

$$x = \frac{1 \pm \sqrt{105}}{4} \approx -2.3, 2.8$$

Additional points:

x	-5	-4	-3	-1	1	2	4	5
y	$-1\frac{3}{4}$	$-1\frac{9}{14}$	$-1\frac{1}{3}$	$-2\frac{1}{2}$	-2	$-1\frac{3}{4}$	$-2\frac{1}{2}$	$-2\frac{2}{7}$ ■

Rational functions in which the degree of the numerator is equal to the degree of the denominator

When the degree of the numerator is *equal to* the degree of the denominator, we first divide the denominator into the numerator, using long division (section 1–2). As illustrated below, this produces a horizontal asymptote.

■ ***Example 4–4 C***

Graph the function $f(x) = \dfrac{x^2}{x^2 - 9}$. State all asymptotes and intercepts.

Observe that the degree of the numerator is equal to the degree of the denominator; therefore divide, using long division (see section 1–2):

$$\begin{array}{r|l} & \quad\quad\quad\quad 1 \\ x^2 - 9 & x^2 + 0x + 0 \\ & -(x^2 \quad\quad - 9) \\ \hline & \quad\quad\quad\quad 9 \end{array}$$

Long division

$$f(x) = 1 + \frac{9}{x^2 - 9} \text{ or } \frac{9}{(x - 3)(x + 3)} + 1$$

Vertical asymptotes at ± 3; horizontal asymptote at $x = 1$

y-intercept:

$$f(0) = \frac{0^2}{0^2 - 9} = 0$$

x-intercepts:

$$0 = \frac{x^2}{x^2 - 9}$$
$$0 = x^2$$
$$0 = x$$

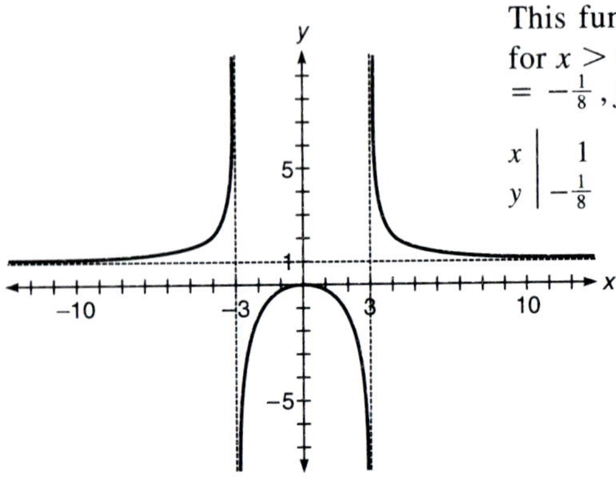

Additional points:

This function has *y*-axis symmetry (see section 3–5). We calculate *y* values for $x > 0$ and use these to fill in the rest of the table. For example, since $f(1) = -\frac{1}{8}$, $f(-1)$ is also $-\frac{1}{8}$.

x	1	2	2.5	3.5	4	5	−1	−2	−2.5	−3.5	−4	−5
y	$-\frac{1}{8}$	$-\frac{4}{5}$	$-2\frac{3}{11}$	$3\frac{10}{13}$	$2\frac{2}{7}$	$1\frac{9}{16}$	$-\frac{1}{8}$	$-\frac{4}{5}$	$-2\frac{3}{11}$	$3\frac{10}{13}$	$2\frac{2}{7}$	$1\frac{9}{16}$

[Y=] [X|T] [x^2] [÷] [(]
[X|T] [x^2] [−] 9 [)]
[RANGE −10,10,−5,7]

■

Rational functions in which the degree of the numerator is greater than the degree of the denominator

When the degree of the numerator is *greater than* the degree of the denominator, we also first divide. In this case, we do not get a horizontal asymptote. If the difference in the degrees is 1 we get a **slant asymptote.**

If the difference in degrees is greater than 1 we get a function of degree 2 or above which the given function approaches when the absolute value of x is large. These are called nonlinear asymptotes. We will restrict ourselves to cases where we get a linear, slant asymptote.

■ ***Example 4–4 D***

Graph the rational function. State all asymptotes and intercepts.

$$f(x) = \frac{x^2 + 3x + 2}{x - 2}$$

$$f(x) = \frac{12}{x - 2} + x + 5 \qquad \text{Use long division}$$

The line $y = x + 5$ is a slant asymptote. There is a vertical asymptote at $x = 2$.

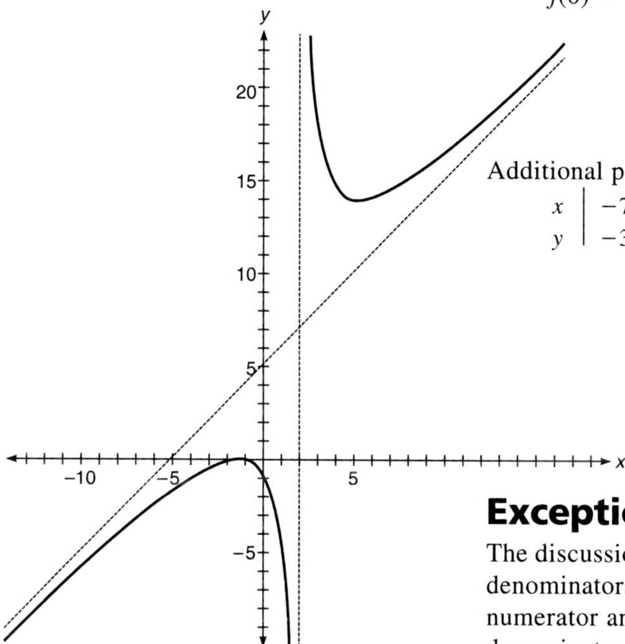

y-intercept:

$$f(0) = -1$$

x-intercepts:

$$0 = \frac{x^2 + 3x + 2}{x - 2}$$
$$0 = x^2 + 3x + 2$$
$$0 = (x + 2)(x + 1)$$
$$x = -2 \text{ or } -1$$

Additional points:

x	−7	−5	−3	−1.5	1	1.5	2.5	3	5	7
y	−3.3	−1.7	−0.4	0.07	−6	−17.5	31.5	20	14	14.4

[Y=] [(] [X|T] [x^2] [+] 3 [X|T]
+ 2 [)] [÷] [(] [X|T] [−] 2 [)]
[RANGE −12,12,−10,25]

Exceptional cases

The discussion so far has focused on those rational functions with zeros in the denominators. Two other situations are worth noting. The first is where the numerator and denominator have a common factor. The second is where the denominator is never zero. These two cases are illustrated in example 4–4 E.

■ *Example 4–4 E*

Graph the function.

1. $f(x) = \dfrac{x^2 + x - 2}{x^2 + 2x - 3}$

Factoring gives $f(x) = \dfrac{(x - 1)(x + 2)}{(x - 1)(x + 3)}$, which is not defined at $x = 1$ and $x = -3$, and otherwise reduces to $\dfrac{x + 2}{x + 3}$. Thus, the function f can be described as $f(x) = \dfrac{x + 2}{x + 3}$, $x \neq 1$.

This function falls into the category of the degree of the numerator equal to the degree of the denominator, so we graph it as illustrated in example 4–4 C.

To show that the function is not defined at 1, we show a hole in the graph for $x = 1$. When this function is graphed on a graphing calculator the hole at $x = 1$ will not necessarily be visible. This is because of the

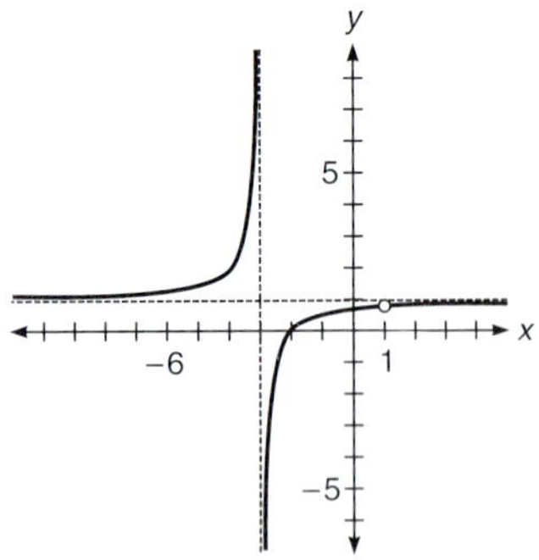

limitations of technology. However, it can cause problems. To illustrate this enter the function into Y_1 and try to evaluate $f(x)$ at $x = 1$. Do this as follows:

1 [STO▶] [X|T] [Y-VARS] [ENTER]

This sequence produces an error, because it attempts to compute Y_1 at $x = 1$, which causes an attempt to divide by zero.

[Y=] [(] [X|T] [x^2] [+] [X|T] [−] 2 [)] [÷] [(] [X|T] [x^2] [+] 2 [X|T] [−] 3 [)]

2. $f(x) = \dfrac{5x - 1}{x^2 + 9}$

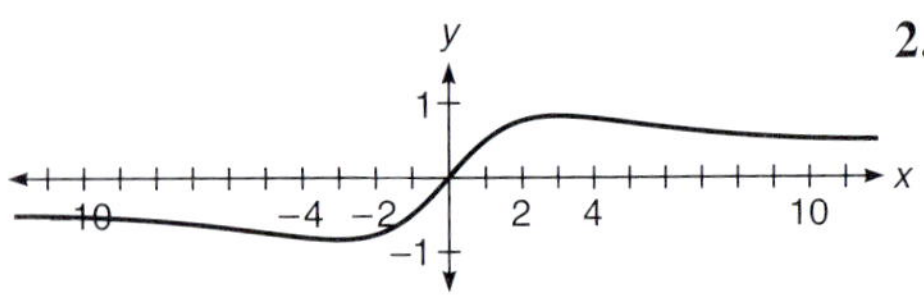

[Y=] [(] 5 [X|T] [−] 1 [)] [÷] [(] [X|T] [x^2] [+] 9 [)]

[RANGE −12,12,−2,2]

The denominator of this function is never zero, and so there are no vertical asymptotes. *The degree of the numerator is less than the degree of the denominator* and there is no vertical translation *so the x-axis is a horizontal asymptote.*

y-intercept: $y = -\frac{1}{9}$

x-intercept: $x = \frac{1}{5}$

Additional points:

x	−10	−5	−4	−2	−1	1	2	4	5	10
y	−0.47	−0.76	−0.84	−0.85	−0.6	0.4	0.69	0.76	0.7	0.45

Mastery points

Can you

- Graph a rational function of the form $\dfrac{a}{(x-h)^n} + k$, and compare its graph to the graph of $y = \dfrac{1}{x^n}$ for a suitable value of n?
- Graph a general rational function when the degree of the numerator is less than the degree of the denominator by analyzing its vertical and horizontal asymptotes and point plotting?
- Graph a rational function when the degree of the numerator is greater than or equal to the degree of the denominator by first doing long division?

Exercise 4–4

Describe the function in terms of the graph of $f(x) = \frac{1}{x^n}$ for an appropriate value of n. Then graph the function. State all intercepts and asymptotes.

1. $f(x) = \frac{3}{x - 2}$
2. $g(x) = \frac{1}{x + 3}$
3. $h(x) = \frac{-2}{x + 4}$
4. $f(x) = \frac{2}{x - 1}$
5. $f(x) = \frac{3}{(x - 2)^2}$
6. $f(x) = \frac{2}{(x + 3)^3}$
7. $g(x) = \frac{1}{(x + \frac{1}{2})^4}$
8. $g(x) = \frac{4}{(x - 1)^2}$
9. $h(x) = \frac{3}{(x - 2)^3}$
10. $h(x) = \frac{-4}{(x + 5)^4}$
11. $f(x) = \frac{-4}{(x - 2)^2}$
12. $g(x) = \frac{-2}{x^3}$
13. $h(x) = \frac{1}{x - 1} + 2$
14. $g(x) = \frac{1}{(x + 1)^2} - 2$
15. $f(x) = \frac{1}{(x - 3)^2} + 2$
16. $g(x) = \frac{1}{x + 3} + 3$

Graph the following rational functions. State all intercepts and asymptotes.

17. $f(x) = \frac{3}{x^2 - 3x - 18}$
18. $h(x) = \frac{1}{x^2 - x - 20}$
19. $g(x) = \frac{-4}{(x - 2)(x + 4)}$
20. $f(x) = \frac{-1}{x^2 - 1}$
21. $h(x) = \frac{2x - 1}{x^2 - 4}$
22. $g(x) = \frac{x}{x^2 - 9}$
23. $h(x) = \frac{2x}{x^2 - 4x - 5}$
24. $g(x) = \frac{-3x + 1}{x^2 - x - 12}$
25. $f(x) = \frac{-2x - 3}{x^2 - 4x}$
26. $f(x) = \frac{x}{x^2 - 6x + 5}$
27. $g(x) = \frac{3x^2 - 1}{(x - 2)(x^2 - 9)}$
28. $h(x) = \frac{x - 3}{(x - 1)(x^2 - 2x - 8)}$
29. $h(x) = \frac{3x - 1}{(x^2 - 2x)(x + 1)}$
30. $g(x) = \frac{2x + 3}{x^2 - 16}$
31. $h(x) = \frac{2}{(x - 1)(x + 3)} + 2$
32. $f(x) = \frac{x}{(x + 2)(x - 1)} - 2$
33. $g(x) = \frac{x}{x^2 - 2x - 15} - 4$
34. $h(x) = \frac{1}{x^2 + 5x + 4} + 1$

Graph the following rational functions. State all intercepts and asymptotes.

35. $f(x) = \frac{x}{x + 1}$
36. $g(x) = \frac{2x}{x - 2}$
37. $h(x) = \frac{2x}{x - 3}$
38. $f(x) = \frac{-x}{x - 1}$
39. $f(x) = \frac{x^2 - 3}{x^2 - 4x - 5}$
40. $g(x) = \frac{x^2 - x + 1}{x^2 + 3x - 4}$
41. $h(x) = \frac{-3x^2 + 2x - 1}{x^2 - 4}$
42. $f(x) = \frac{4x^2 - 1}{x^2 - 1}$

Graph the following rational functions. State all intercepts and asymptotes.

43. $f(x) = \frac{\frac{1}{2}x^2}{x - 1}$
44. $g(x) = \frac{x^3}{x^2 - 1}$
45. $h(x) = \frac{x^3}{x^2 - 2x + 1}$
46. $f(x) = \frac{x^2}{x + 2}$
47. $h(x) = \frac{x^3 - x}{x^2 - 4}$
48. $h(x) = \frac{3x^3 - x^2 - 2}{x^2 - x - 6}$
49. $g(x) = \frac{x^3 + 8}{x^2 + 3x - 4}$
50. $f(x) = \frac{3x^4 - x^3}{(x - 1)(x^2 - 4)}$
51. $g(x) = \frac{2x^4 - x^2}{(x + 2)(x^2 - 1)}$
52. $f(x) = \frac{-x^3 - x^2 + 11x - 9}{x^2 + 2x - 8}$

Graph the following rational functions. State all intercepts and asymptotes.

53. $f(x) = \dfrac{x^2 - 1}{x^2 + 2x - 3}$ **54.** $f(x) = \dfrac{x^2 - x - 6}{x - 3}$ **55.** $g(x) = \dfrac{x - 2}{x^3 - x^2 - 4x + 4}$

56. $h(x) = \dfrac{x^3 - 2x^2 - 25x + 50}{(x - 5)(x^2 - 7x + 10)}$ **57.** $h(x) = \dfrac{x^3 + 4x^2 + 3x + 12}{x^2 + 3}$ **58.** $f(x) = \dfrac{x + 1}{2x^2 + x - 1}$

59. $f(x) = \dfrac{8x - 2}{x^2 + 1}$ **60.** $g(x) = \dfrac{2x + 1}{x^4 + 1}$ **61.** $g(x) = \dfrac{x^2 - 4}{x^2 + 4}$ **62.** $f(x) = \dfrac{x^3 - 1}{x^2 + 1}$

63. Iron ore is being moved 120 feet by a conveyor belt that travels at r feet per minute, and then an additional 160 feet by another belt that is 5 feet per minute slower than the first. Since distance equals the product of rate and time, or $d = rt$, then $t = \dfrac{d}{r}$. Thus the total time T taken to move the ore, as a function of the rate r of the first belt, is $T(r) = \dfrac{120}{r} + \dfrac{160}{r - 5}$, or $T(r) = \dfrac{280r - 600}{r(r - 5)} = 40\left(\dfrac{7r - 15}{r(r - 5)}\right)$. Graph the relation $y = \dfrac{7x - 15}{x(x - 5)}$.

64. It takes a certain machine m minutes to print a certain number of pages. Because of a fixed warm-up time, a second machine always takes $m + 2$ minutes to print the same number of pages. Under these conditions the rate at which these machines print pages when running together is $\dfrac{1}{m} + \dfrac{1}{m + 2} = \dfrac{2m + 2}{m(m + 2)}$. Graph the relation $y = \dfrac{2x + 2}{x(x + 2)}$.

65. In an electronics circuit with two resistances, x and y in parallel (see the figure), and where the total resistance is to be 20 ohms, the relation $\dfrac{1}{x} + \dfrac{1}{y} = \dfrac{1}{20}$ will be true.
(a) Solve this relation for y as a function of x (note that $x > 0$ and $y > 0$ are reasonable assumptions to make).
(b) Graph this function.

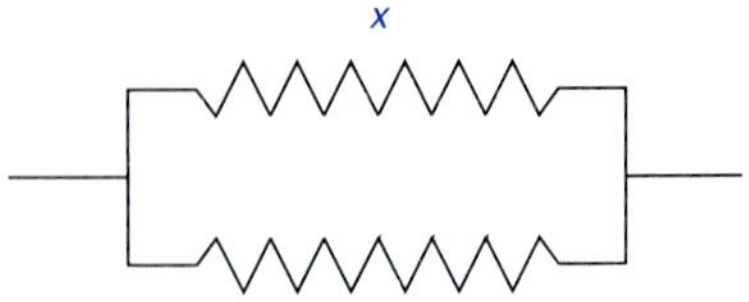

66. If x represents the value, in ohms, of a certain resistor in an electronics circuit, and a second resistor in parallel with it is 10 ohms more, then the reciprocal of their combined resistance y is given by $y = \dfrac{1}{x} + \dfrac{1}{x + 10}$, or $y = \dfrac{2x + 10}{x^2 + 10x}$. Graph this relation.

67. In problem 66 the relation described the reciprocal of the combined resistance. The actual resistance would be described by the relation $y = \dfrac{x^2 + 10x}{2x + 10}$. Graph this relation.

68. Assume that gasoline costs \$1.00 per gallon, that a car is driven 15,000 miles per year, and that the car's fuel economy is x miles per gallon (mpg). Let y represent the annual savings in fuel costs for increasing the mileage by 10 mpg. Then $y = \dfrac{15{,}000}{x} - \dfrac{15{,}000}{x + 10}$. Combine the right member into one term, then graph this function. Use a scale of 5,000 per unit on the y-axis.[5]

69. Use the formula of problem 68 to compute the annual savings obtained by increasing the mileage of a car by 10 mpg if the car now gets (a) 10 mpg, (b) 20 mpg, and (c) 30 mpg.

[5]This problem presented by Floyd Vest, North Texas State University, in *Consortium,* by COMAP, Arlington, Mass., Summer 1990.

Skill and review

If $f(x) = 2x - 3$ and $g(x) = x^2 + 2x + 3$, compute:

1. $f(5)$ **2.** $g(-4)$ **3.** $f(g(2))$ **4.** $g(f(-1))$

5. Solve $x = 2y + 7$ for y.

6. Solve $x = \dfrac{1}{y - 2}$ for y.

7. Graph $f(x) = 2x^4 - x^3 - 14x^2 + 19x - 6$.

8. Solve $|4 - 3x| = 16$.

4–5 Composition and inverse of functions

A company charges \$0.50 per cubic foot for a plastic it makes. Thus, if x is the number of cubic feet of this plastic, the price paid in dollars is $P(x) = \frac{1}{2}x$. The volume of a cube is $V(x) = x^3$, where x is the length of one of its dimensions. Let C be the cost function that will give the cost of a cube of this plastic when the length of a side is x feet. Compute an expression for $C(x)$.

The process used to compute the expression for C in the preceding problem is called composing functions. It is one of the operations that can be performed on functions that we study in this section.

The basic operations for functions

The concept of a function is so important in advanced mathematics that an algebra for functions has developed. This algebra is a system for performing computations in which the elements are functions, not numbers. The operations of addition/subtraction and multiplication/division are easy to define for functions.

Definition of addition, subtraction, multiplication, and division of functions

Let f and g be functions. Then, for every element x in the domain of both functions

$$(f + g)(x) = f(x) + g(x)$$
$$(f - g)(x) = f(x) - g(x)$$
$$(f \cdot g)(x) = f(x) \cdot g(x)$$
$$(f/g)(x) = \frac{f(x)}{g(x)}, \text{ if } g(x) \neq 0$$

Note that "$f + g$," "$f - g$," etc. are the names of new functions.

■ ***Example 4–5 A***

Find expressions for $f + g$, $f - g$, $f \cdot g$, and f/g for the given functions.

1. $f(x) = x^2 - 3$, $g(x) = x$

$(f + g)(x) = f(x) + g(x)$ — Definition of $(f + g)(x)$ is $f(x) + g(x)$
$= (x^2 - 3) + x$ — Replace $f(x)$ by $x^2 - 3$, and $g(x)$ by x
$= x^2 + x - 3$ — Simplify

$$(f - g)(x) = f(x) - g(x) = (x^2 - 3) - x = x^2 - x - 3$$
$$(f \cdot g)(x) = f(x) \cdot g(x) = (x^2 - 3)(x) = x^3 - 3x$$
$$(f/g)(x) = \frac{f(x)}{g(x)} = \frac{x^2 - 3}{x} \text{ if } x \neq 0$$

2. $f(x) = \sqrt{x-3}$, $g(x) = \sqrt{x+1}$

$$(f+g)(x) = f(x) + g(x) = \sqrt{x-3} + \sqrt{x+1}$$
$$(f-g)(x) = f(x) - g(x) = \sqrt{x-3} - \sqrt{x+1}$$
$$(f \cdot g)(x) = f(x) \cdot g(x) = \sqrt{x-3} \cdot \sqrt{x+1} = \sqrt{x^2 - 2x - 3}$$
$$(f/g)(x) = \frac{f(x)}{g(x)} = \frac{\sqrt{x-3}}{\sqrt{x+1}}$$

■

Composition of functions

There is an operation defined for functions, called **composition** of functions, which does not have a direct analogy with arithmetic, as do the operations defined above. It gives us a new way to look at functions.

Consider this function: $f(x) = \sqrt{x^2 - x - 3}$. How would one compute $f(4)$? First, we would evaluate $x^2 - x - 3$ for $x = 4$. This is 9. We would then compute $\sqrt{9}$, giving 3. Thus, $f(4) = 3$. We have viewed it as a two-stage operation. We can formalize this idea as composition of functions.

> **Composition of two functions *f* and *g***
> Let f and g be functions, $x \in$ domain of g, $g(x) \in$ domain of f. Then
> $$(f \circ g)(x) = f(g(x))$$

Note $(f \circ g)$ is read "f composed with g."

First we focus on how this operation is used. It is important to look again at $f(x)$ notation. Consider as an example $f(x) = 2x^2 - x + 1$. Whatever replaces x in $f(x)$ also replaces x in $2x^2 - 3x + 1$. Consider the following sequence of examples using this function.

$f(x)$	$= 2x^2$	$- 3x$	$+ 1$
$f(1)$	$= 2(1)^2$	$- 3(1)$	$+ 1$
$f(a)$	$= 2a^2$	$- 3a$	$+ 1$
$f(c+2)$	$= 2(c+2)^2$	$- 3(c+2)$	$+ 1$
$f(x^4-2)$	$= 2(x^4-2)^2$	$- 3(x^4-2)$	$+ 1$
$f(g(x))$	$= 2(g(x))^2$	$- 3(g(x))$	$+ 1$

The statement $f(x) = 2x^2 - x + 1$ can be viewed as a pattern; any meaningful mathematical expression can replace the x in the entire statement.

■ ***Example 4–5 B***

Compute an expression for $(f \circ g)(x)$ and $(g \circ f)(x)$ for the given functions.

1. $f(x) = x^2 - 3x + 2$; $g(x) = 2x - 5$

$(f \circ g)(x) = f(g(x))$	Definition of $(f \circ g)(x)$ is $f(g(x))$
$= [g(x)]^2 - 3[g(x)] + 2$	Replace x by $g(x)$ in $x^2 - 3x + 2$
$= [2x-5]^2 - 3[2x-5] + 2$	Replace $g(x)$ by $2x - 5$
$= 4x^2 - 26x + 42$	Simplify

Thus, $(f \circ g)(x) = 4x^2 - 26x + 42$.

$(g \circ f)(x) = g(f(x))$	Definition of $(g \circ f)(x)$ is $g(f(x))$
$= 2(f(x)) - 5$	Replace x by $f(x)$ in $2x - 5$
$= 2(x^2 - 3x + 2) - 5$	Replace $f(x)$ by $x^2 - 3x + 2$
$= 2x^2 - 6x - 1$	Simplify

Thus, $(g \circ f)(x) = 2x^2 - 6x - 1$.

2. $f(x) = \sqrt{x^2 - 3}$ and $g(x) = \dfrac{1}{x}$

$(f \circ g)(x) = f(g(x))$	
$= \sqrt{[g(x)]^2 - 3}$	$f(x) = \sqrt{x^2 - 3}$, so $f(g(x)) = \sqrt{[g(x)]^2 - 3}$
$= \sqrt{\left(\dfrac{1}{x}\right)^2 - 3}$	Replace $g(x)$ by $\dfrac{1}{x}$
$= \sqrt{\dfrac{1 - 3x^2}{x^2}}$	$\left(\dfrac{1}{x}\right)^2 - 3 = \dfrac{1}{x^2} - \dfrac{3x^2}{x^2} = \dfrac{1 - 3x^2}{x^2}$
$(g \circ f)(x) = g(f(x))$	
$= \dfrac{1}{f(x)}$	$g(x) = \dfrac{1}{x}$ so $g(f(x)) = \dfrac{1}{f(x)}$
$= \dfrac{1}{\sqrt{x^2 - 3}}$.	Replace $f(x)$ by $\sqrt{x^2 - 3}$ ■

The domain of $f \circ g$ is the set of all x in the domain of g such that $g(x)$ is in the domain of f. This subset of the domain of g can be difficult to find, and we will not pursue this problem in this text except to illustrate the difficulty as follows.

Consider the functions $f(x) = \sqrt{12 - x^2}$, and $g(x) = \sqrt{x - 4}$. Then

$$(f \circ g)(x) = f(g(x)) = \sqrt{12 - (g(x))^2} = \sqrt{12 - (\sqrt{x - 4})^2} = \sqrt{16 - x}$$

The implied domain of $(f \circ g)(x) = \sqrt{16 - x}$ is $x \le 16$. This includes the value 0, for example. However, since $g(0)$ is not defined, $f(g(0))$ is not defined, so 0 is not in the domain of $f \circ g$. In complicated situations the expression for $f \circ g$ cannot be relied on to determine its domain.

Inverses of functions

Consider $f(x) = 2x + 3$, and $g(x) = \dfrac{x - 3}{2}$ (see figure 4–9). By computation we could determine the following facts.

$$f(1) = 5 \text{ and } g(5) = 1$$
$$f(2) = 7 \text{ and } g(7) = 2$$
$$f(-5) = -7 \text{ and } g(-7) = -5$$

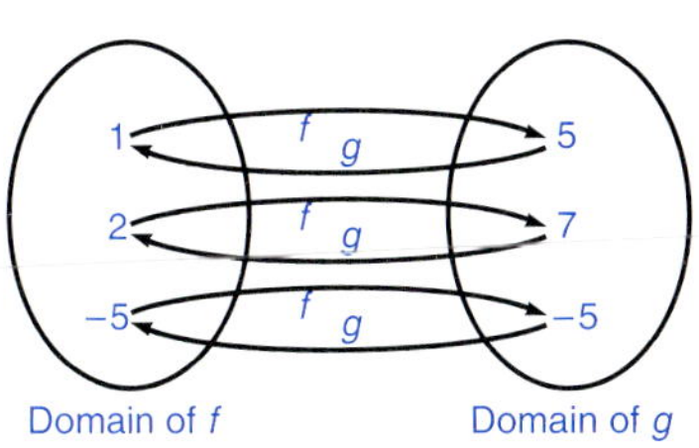

Figure 4–9

Whatever value z we try, f sends z to some value z', and g sends z' back to z (see figure 4–10). In fact we can prove this; let z represent any real number.

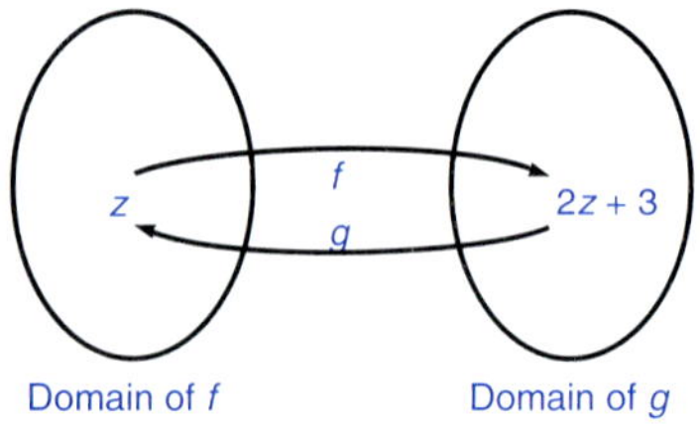

Figure 4–10

Then,

$$f(z) = 2z + 3 \text{ and}$$
$$g(f(z)) = g(2z + 3) = \frac{(2z + 3) - 3}{2} = z$$

Also

$$g(z) = \frac{z - 3}{2} \text{ and}$$
$$f(g(z)) = f\left(\frac{z - 3}{2}\right) = 2\left(\frac{z - 3}{2}\right) + 3 = z$$

When two functions f and g act this way we say they are inverse functions.

> **Function inverse**
> If $(f \circ g)(x) = x$ and $(g \circ f)(x) = x$ for all x in the domains of functions f and g, then f and g are said to be inverse functions. The inverse of a function f is symbolized as f^{-1}.

Thus, if $f(x) = 2x + 3$ we can say that $f^{-1}(x) = \frac{x - 3}{2}$. Note that the superscript -1, when applied to the name of a function, is not an exponent; it does not indicate division, as it does if applied as an exponent of a real valued expression. Thus, $f^{-1}(x)$ **does not mean** $\frac{1}{f(x)}$.

> **To show that two functions f and g are inverses of each other**
> Show that [1] $(f \circ g)(x) = x$ and [2] $(g \circ f)(x) = x$

It is possible for one and not the other of equations [1] and [2] to be true. For example, take $f(x) = x^2$ and $g(x) = \sqrt{x}$. $(f \circ g)(x) = (\sqrt{x})^2 = x$, but $(g \circ f)(x) = \sqrt{x^2} = |x|$, not x.

■ ***Example 4–5 C***

Show that f and g are the inverse functions of each other.

1. $f(x) = \frac{1}{3}x - 1$
$g(x) = 3x + 3$
$(f \circ g)(x) = f(g(x)) = \frac{1}{3}(3x + 3) - 1 = x$
$(g \circ f)(x) = g(f(x)) = 3(\frac{1}{3}x - 1) + 3 = x$

2. $f(x) = \sqrt{x}$
$g(x) = x^2, x \geq 0$
$(f \circ g)(x) = f(g(x)) = \sqrt{x^2} = |x|$, but since $x \geq 0$ for g, $|x| = x$.
$(g \circ f)(x) = g(f(x)) = (\sqrt{x})^2 = x$ ■

The following two theorems are quite useful in working with functions and their inverses.

> **Ordered pairs reverse in f^{-1}**
> If an ordered pair (a,b) is in a function f, and if f has an inverse function f^{-1}, then (b,a) is an ordered pair in f^{-1}.

To see this, note that $f(a) = b$, so that $(a,b) \varepsilon f$, and $f^{-1}(b) = f^{-1}(f(a)) = a$, so that $(b,a) \varepsilon f^{-1}$.

> **Only one-to-one functions have inverse functions**
> A function f has an inverse function f^{-1}, if and only if it is one to one.

The following explains why only one-to-one functions have inverse functions.

If a function f is not one to one, then there are at least two elements in it in which the ordered pairs have the same second component. Let these points be (x_1,b) and (x_2,b). Now if f^{-1} exists, then (b,x_1) and (b,x_2) are elements of that function. However, these points have a first element that repeats, making f^{-1} a relation, but not a function.

Similarly, if f is a one-to-one function, then no second element repeats. Therefore, the relation created when we reverse all the ordered pairs is a function, since there is no repetition of first elements, and this function meets the definition of the inverse of f. Thus we determine that the above theorem is true.

The graph of a function's inverse

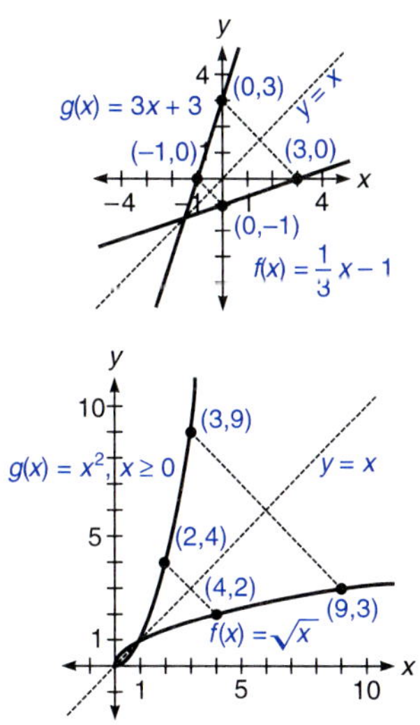

Figure 4–11

The fact that the ordered pairs reverse in the inverse function of a function means that *the graph of f^{-1} is a reflection of the graph of f about the line $y = x$.* By way of example, observe the graphs of the functions in the last example. These are shown in figure 4–11. To draw a graph that is symmetric about the line $y = x$ to a given graph, we draw lines perpendicular to the line $y = x$, as shown, and plot points at equal distances from this line, but on the other side of this line. Since the ordered pairs of f all reverse in f^{-1} *the domain of f is the range of f^{-1}, and the range of f is the domain of f^{-1}.*

Finding an expression for the inverse of a function

The fact that the ordered pairs reverse in a function's inverse provides a method that can be used to find the inverse of a function.

> **To find $f^{-1}(x)$ for a one-to-one function f**
> - Replace $f(x)$ by y.
> - Replace each x by y and y by x (this is "reversing" the ordered pairs of f).
> - Solve for y.
> - Replace y by $f^{-1}(x)$.

This method is useful when the resulting expression can be solved for y. It is illustrated in example 4–5 D.

■ Example 4–5 D

Find f^{-1} for each function f. Also, graph f and f^{-1}.

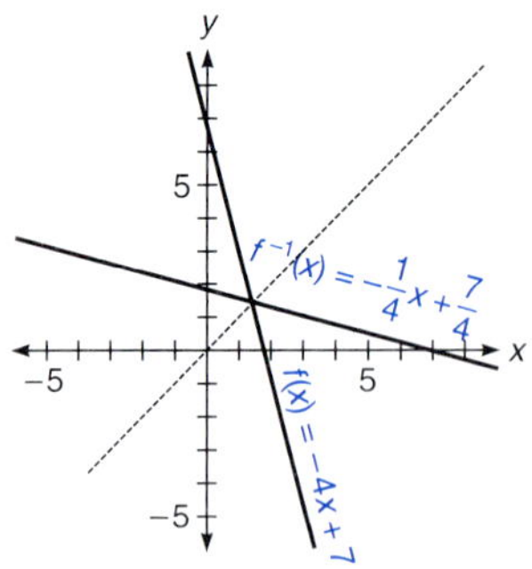

1. $f(x) = -4x + 7$

First note that the graph of this function is a straight line that passes the horizontal line test (see section 3–5), and thus is one to one. It thus has an inverse function.

$y = -4x + 7$	Replace $f(x)$ by y (this relation describes f)
$x = -4y + 7$	Replace x by y and y by x (this relation describes f^{-1})
$4y = -x + 7$	Solve for y
$y = -\frac{1}{4}x + \frac{7}{4}$	Divide each member by 4
$f^{-1}(x) = -\frac{1}{4}x + \frac{7}{4}$	Rewrite with $y = f^{-1}(x)$

Thus, $f^{-1}(x) = -\frac{1}{4}x + \frac{7}{4}$.

The graphs of f and f^{-1} are both straight lines that are graphed by plotting their intercepts.

Intercepts for f: $(0,7)$, $(\frac{7}{4},0)$

Intercepts for f^{-1}: $(7,0)$, $(0,\frac{7}{4})$

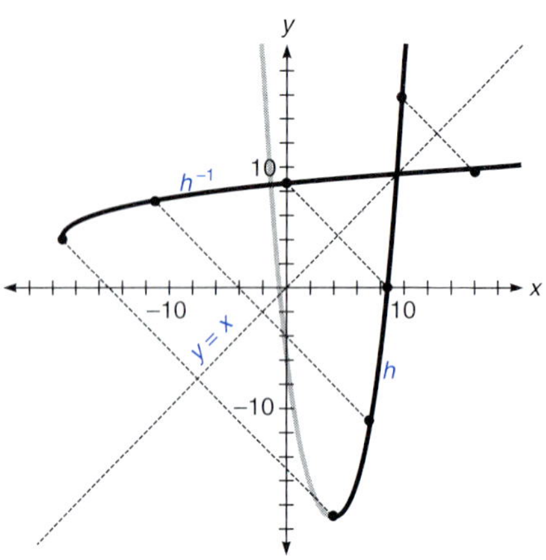

2. $h(x) = x^2 - 8x - 3$, $x \geq 4$

The graph of h is a parabola. We graph it by completing the square.

$$y = x^2 - 8x + 16 - 16 - 3$$
$$y = (x - 4)^2 - 19$$

Vertex at $(4,-19)$.

y-intercept:

$$h(0) = 0^2 - 8(0) - 3 = -3$$

x-intercepts:

$$0 = (x - 4)^2 - 19$$
$$(x - 4)^2 = 19$$
$$x - 4 = \pm\sqrt{19}$$
$$x = 4 \pm \sqrt{19} \approx -0.4,\ 8.4$$

With $x \geq 4$ we can see by the horizontal line test that h is one to one.

We draw the graph of h^{-1} from the graph of h, by reflecting points in h about the line $y = x$.

We proceed to find an expression for h^{-1}.

$y = x^2 - 8x - 3$ and $x \geq 4$	$y = h(x)$
$x = y^2 - 8y - 3$ and $y \geq 4$	Interchange x and y
$0 = y^2 - 8y + (-x - 3)$ and $y \geq 0$	To solve for y use the quadratic formula; this requires that the equation be set to 0

$$y = \frac{-(-8) \pm \sqrt{(-8)^2 - 4(1)(-x - 3)}}{2(1)}$$
$$= \frac{8 \pm \sqrt{64 + 4(x + 3)}}{2}$$
$$= 4 \pm \sqrt{x + 19} \text{ and } y \geq 4$$

$y = 4 + \sqrt{x + 19}$ since $y \geq 4$, and $4 - \sqrt{x + 19}$ is less than or equal to 4

Thus, $h^{-1}(x) = 4 + \sqrt{x + 19}$. ■

The graphing calculator can help verify that we have found the correct expression for the inverse of a function. We graph the original function and its inverse in the same graph. If one is the mirror image of the other across the line $y = x$, then we have correctly found the inverse.

Mastery points

Can you

- Find expressions for $f + g$, $f - g$, $f \cdot g$, and f/g when given expressions that define functions f and g?
- Compute an expression for $(f \circ g)(x)$ and $(g \circ f)(x)$ when given expressions that define functions f and g?
- Demonstrate that two functions are inverses of each other?
- Find f^{-1} when given a function f, and graph both functions?

Exercise 4–5

Find expressions for $f + g$, $f - g$, $f \cdot g$, f/g, and $(f \circ g)(x)$ and $(g \circ f)(x)$ for the given functions f and g.

1. $f(x) = 3x - 5$; $g(x) = -2x + 8$

2. $f(x) = 2x + 3$; $g(x) = \frac{1}{2}x - 3$

3. $f(x) = x + 4$; $g(x) = \sqrt{x - 4}$

4. $f(x) = x^2 - 3$; $g(x) = \sqrt{8 - x}$

5. $f(x) = \dfrac{x - 3}{2x}$; $g(x) = \dfrac{x}{x - 1}$

6. $f(x) = x^3 - 3x^2 + x - 4$; $g(x) = x^2 - 1$

7. $f(x) = x^4 - x^2 + 3$; $g(x) = \sqrt{\dfrac{x}{x + 1}}$

8. $f(x) = -x$; $g(x) = x$

9. $f(x) = x$; $g(x) = 3$

10. $f(x) = 2$; $g(x) = 3$

11. $f(x) = \sqrt[3]{x - 5}$; $g(x) = x^3 + 5$

12. $f(x) = \dfrac{x}{x^2 - 2x - 15}$; $g(x) = \dfrac{1}{x}$

(a) Show that the following functions f and g are inverses of each other. Assume the domains as indicated are correct.
(b) Graph each function and its inverse in the same coordinate system.

13. $f(x) = 2x - 7$; $g(x) = \frac{1}{2}x + 3\frac{1}{2}$

14. $f(x) = -\frac{1}{3}x + \frac{1}{2}$; $g(x) = -3x + \frac{3}{2}$

15. $f(x) = \frac{1}{3}x + \frac{8}{3}$; $g(x) = 3x - 8$

16. $f(x) = x - 1$; $g(x) = x + 1$

17. $f(x) = x^2 - 9$, $x \geq 0$; $g(x) = \sqrt{x + 9}$

18. $f(x) = \sqrt{4 - 2x}$; $g(x) = 2 - \frac{1}{2}x^2$, $x \geq 0$

19. $f(x) = x^3$; $g(x) = \sqrt[3]{x}$

20. $f(x) = x^3 - 3$; $g(x) = \sqrt[3]{x + 3}$

21. $f(x) = x^2 - 2x + 3$, $x \geq 1$; $g(x) = \sqrt{x - 2} + 1$

22. $f(x) = \sqrt{x + 9} - 2$; $g(x) = x^2 + 4x - 5$, $x \geq -2$

23. $f(x) = \dfrac{2x}{x - 3}$; $g(x) = \dfrac{3x}{x - 2}$

24. $f(x) = \dfrac{x - 3}{x - 2}$; $g(x) = 2 - \dfrac{1}{x - 1}$

Find the inverse function of the given function. All the functions are one to one.

25. $f(x) = 4x - 5$

26. $g(x) = \dfrac{3x - 2}{4}$

27. $h(x) = 12 - \frac{5}{2}x$

28. $f(x) = \dfrac{x}{2}$

29. $g(x) = x^2 - 9$, $x \geq 0$

30. $h(x) = x^2 + 3$, $x \geq 0$

31. $f(x) = \sqrt{9 - x^2}$, $x \geq 0$

32. $g(x) = \sqrt{2x^2 - 16}$, $x \geq 0$

33. $h(x) = \sqrt{x - 4}$

34. $f(x) = \sqrt{5 - x}$

35. $g(x) = 2x^3 - 9$

36. $h(x) = x^3 + 20$

37. $f(x) = \sqrt[3]{4x - 5}$

38. $g(x) = \sqrt[3]{1 - x} + 3$

39. $f(x) = \dfrac{3 - 5x}{4x}$

40. $g(x) = \dfrac{x}{3 - x}$

41. $g(x) = \dfrac{x}{x + 1}$

42. $h(x) = \dfrac{2x - 3}{x}$

43. $h(x) = \dfrac{x - 1}{x + 1}$

44. $f(x) = \dfrac{1 - 2x}{x - 3}$

45. $h(x) = x^2 - 2x - 9,\ x \geq 1$

46. $f(x) = x^2 - 8x + 1,\ x \geq 4$

47. $g(x) = 2x^2 + 3x - 2,\ x \geq -\frac{3}{4}$

48. $h(x) = 3x^2 - 6x - 1,\ x \geq 1$

49. A company charges \$0.50 per cubic foot for a plastic it makes. Thus, if x is the number of cubic feet of this plastic the price paid in dollars is $P(x) = \frac{1}{2}x$. The volume of a cube is $V(x) = x^3$, where x is the length of one of its dimensions. Thus, if x is in feet, then the cost function $C = P \circ V$ will give the cost of a cube of this plastic when the length of a side is x feet. Compute an expression for $C(x)$.

50. The volume of a sphere is $V = \frac{4}{3}\pi^3$, where r is the radius of the cube. Using the price function from problem 49 compute an expression for $C(x)$, the cost of a sphere of radius r composed of the same plastic.

51. A railroad car is accelerating slowly so that its forward velocity after t seconds, in feet per second, is $R(t) = \frac{1}{4}t$. A person in the car is walking in the opposite direction so that the person's velocity relative to the car is $V_r(t) = 2$ feet per second. Under these circumstances the person's velocity relative to the earth is $V_e = R - V_r$. Compute an expression for $V_e(t)$.

52. The velocity, relative to the earth, of an aircraft flying with the wind is $V_e = V_a + V_w$, where V_a is the velocity of the aircraft relative to the air and V_w is the velocity of the wind. Find an expression for the velocity of the aircraft after t seconds, $V_e(t)$, when $V_w(t) = 25$ (miles per hour) and $V_a(t) = 80 + \frac{1}{10}t$ (miles per hour).

53. A stunt person runs off a platform horizontally at 5 feet per second. The falling person describes a parabola. A movie director wants the fall to be in front of a curtain. The area of the curtain is important to the director because its cost is a function of its area. Under these conditions the stunt person's horizontal distance traveled after t seconds is $d_h(t) = 5t$ feet, and the vertical distance fallen is $d_v(t) = 16t^2$ feet. The area of the curtain is therefore $A = d_h d_v$. Find an expression for $A(t)$. Note that the units for A is square feet.

54. The cost of the material for the curtain of problem 53 in dollars for x square feet is $C(x) = 0.5x$. Under these circumstances the cost to the director for the curtain for a fall of t seconds is $C \circ A$. (a) Find an expression for $(C \circ A)(t)$ and (b) use this expression to predict the cost of the curtain for a 3-second fall.

55. The area of a rectangle with width 4 and length $x + 4$, $x \geq 0$, is $A(x) = 4(x + 4)$. Find the inverse of A, which would give the value of x for a given area.

56. A falling object with no initial vertical velocity falls a distance $d(t) = 16t^2$ feet in t seconds, $t \geq 0$. Find the inverse of this function, which would give the time necessary to fall a distance d.

57. In an electronic circuit in which two resistances are in parallel, and the value of the resistances are 20 ohms and x ohms, the total resistance is $R(x) = \dfrac{20x}{20 + x}$. Find the inverse of this function, which would give the value of x required for a total resistance R.

58. $C(t) = \frac{5}{9}(t - 32)$ gives the centigrade temperature for a given temperature t in degrees Fahrenheit. Find the inverse of this function, which would find the Fahrenheit temperature for a given temperature in degrees centigrade.

59. Show that $f(x) = x^2 - 9,\ x \leq 0$ and $g(x) = -\sqrt{x + 9}$ are inverse functions. You will need to recall that $\sqrt{x^2} = |x|$, $(\sqrt{x})^2 = x$, and the definition of $|x|$ (section 1–1) to do this problem properly.

60. Show that $f(x) = 3 - \sqrt{x + 14}$ and $g(x) = x^2 - 6x - 5$, $x \leq -3$ are inverse functions. See problem 59.

61. Find the inverse of the general linear function $f(x) = ax + b$, where a and b are real-valued constants. Under what conditions would this function not have an inverse?

62. Find the expression for the inverse of the general quadratic function $f(x) = ax^2 + bx + c$, $a \neq 0$, if $x \geq \dfrac{-b}{2a}$.

63. It is possible to find the inverse of certain functions by "undoing" the operations implied in their definition. For example, consider $f(x) = 2x - 3$. To calculate a value when given a value for x, we (1) multiply by 2 and then (2) subtract 3.

To undo this we should (1) add 3 and then (2) divide by 2. This would be $f^{-1}(x) = \frac{x + 3}{2}$, or $f^{-1}(x) = \frac{1}{2}x + \frac{3}{2}$.

Another example is $f(x) = \frac{1}{5}x + 7$. To evaluate for a given x we (1) divide by 5 and then (2) add 7. To undo we (1) subtract 7 and then (2) multiply by 5. This would be $f^{-1}(x) = 5(x - 7)$, or $f^{-1}(x) = 5x - 35$.

This method is useful whenever the variable x appears only once in an expression. Use this method to find the inverse function for problems 25 through 38.

Skill and review

1. Combine $\frac{2}{x + 3} - \frac{3}{x - 2}$.
2. Graph $f(x) = \frac{2}{x + 3}$.
3. Graph $f(x) = \frac{2}{(x + 3)(x - 1)}$.
4. Graph $f(x) = \frac{2x^2}{(x + 3)(x - 1)}$.
5. Solve $\left|\frac{5x - 2}{x + 1}\right| < 2$.
6. Graph $f(x) = x^3 - x^2 - x + 1$.

4–6 Decomposition of rational functions

Find the sum $\frac{1}{1 \cdot 2} + \frac{1}{2 \cdot 3} + \cdots + \frac{1}{99 \cdot 100}$

A programmable calculator could be programmed to compute this sum, but it turns out that a little algebra will do the same job faster and more accurately. In this section we introduce the algebra necessary for this task.

Calculation will show that $\frac{1}{x - 1} + \frac{3}{x - 4}$ can be combined into $\frac{4x - 7}{(x - 1)(x - 4)}$. We sometimes need to be able to decompose a rational expression like $\frac{4x - 7}{(x - 1)(x - 4)}$ back into a sum of two or more fractions. Besides solving the problem posed above, this finds a great deal of use in advanced mathematics, such as the Calculus, and in discrete mathematics.

We first consider the case where we can factor the denominator into a product of linear factors. We do this by assuming the existence of certain values, as shown in the examples. The rational expression is said to be decomposed into **partial fractions.**

Linear factors in the denominator

Example 4–6 A illustrates the procedure for decomposing a rational expression in which

- the degree of the numerator is less than the degree of the denominator and
- the denominator factors into a product of linear factors.

The procedure is to assume a separate rational expression for each linear factor, with some unknown numerator.

■ ***Example 4–6 A***

Decompose the rational expression into partial fractions: $\dfrac{4x - 7}{x^2 - 5x + 4}$

$$\frac{4x - 7}{(x - 1)(x - 4)}$$ Factor the denominator

Note that this is the expression we saw above, so we know the answer—of course, we will assume we don't.

$$\frac{4x - 7}{(x - 1)(x - 4)} = \frac{A}{x - 1} + \frac{B}{x - 4}$$ Assume a separate rational expression for each factor of the denominator

Now solve for A and B by first multiplying each member by the LCD $(x - 1)(x - 4)$.

$$\frac{(x - 1)(x - 4)}{1} \cdot \frac{4x - 7}{(x - 1)(x - 4)} = \frac{(x - 1)(x - 4)}{1} \cdot \frac{A}{x - 1} + \frac{(x - 1)(x - 4)}{1} \cdot \frac{B}{x - 4}$$

[1] $4x - 7 = A(x - 4) + B(x - 1)$

A useful technique for finding A and B is to let $x = 4$ and then $x = 1$ in equation [1]. These are the values for which one of the factors is zero.

[1] $4x - 7 = A(x - 4) + B(x - 1)$

$4(4) - 7 = A(4 - 4) + B(4 - 1)$ Let $x = 4$

$9 = 0A + 3B$

$9 = 3B$

$3 = B$

[1] $4x - 7 = A(x - 4) + B(x - 1)$

$4(1) - 7 = A(1 - 4) + B(1 - 1)$ Let $x = 1$

$-3 = -3A + 0B$

$-3 = -3A$

$1 = A$

Now, $A = 1$ and $B = 3$, so

$$\frac{4x - 7}{(x - 1)(x - 4)} = \frac{A}{x - 1} + \frac{B}{x - 4}$$
$$= \frac{1}{x - 1} + \frac{3}{x - 4}$$

It can be shown that *if one of the linear factors of the denominator is of multiplicity greater than one it must be assumed that it appears once for each positive integer up to and including its multiplicity.* This is illustrated in example 4–6 B.

■ Example 4–6 B

Decompose the rational expression into partial fractions: $\dfrac{3x^2 - 15x + 14}{(x - 1)(x - 2)^2}$

We assume the existence of values A, B, C such that

$$\frac{3x^2 - 15x + 14}{(x - 1)(x - 2)^2} = \frac{A}{x - 1} + \frac{B}{x - 2} + \frac{C}{(x - 2)^2}$$

Multiply both members by the LCD $(x - 1)(x - 2)^2$:

[1] $\quad 3x^2 - 15x + 14 = A(x - 2)^2 + B(x - 1)(x - 2) + C(x - 1)$

Now find A, B, and C.

Let $x = 2$ in equation [1]:

$$3(2)^2 - 15(2) + 14 = A(2 - 2)^2 + B(2 - 1)(2 - 2) + C(2 - 1)$$
$$-4 = C(1)$$
$$-4 = C$$

Let $x = 1$ in equation [1]:

$$3(1)^2 - 15(1) + 14 = A(1 - 2)^2 + B(1 - 1)(1 - 2) + C(1 - 1)$$
$$2 = A(-1)^2$$
$$2 = A$$

To obtain B we can let x be any value other than 1 and 2 and use the known values for A and C; 0 is a logical value to use for x.

Let $x = 0$ in equation [1]:

$$3(0)^2 - 15(0) + 14 = A(0 - 2)^2 + B(0 - 1)(0 - 2) + C(0 - 1)$$
$$14 = 4A + 2B - C$$
$$14 = 4(2) + 2B - (-4) \qquad A = 2, C = -4$$
$$1 = B$$

Thus, the solution is $\dfrac{3x^2 - 15x + 14}{(x - 1)(x - 2)^2} = \dfrac{2}{x - 1} + \dfrac{1}{x - 2} + \dfrac{-4}{(x - 2)^2}$. ■

Another very important point is: *If the degree of the numerator is greater than or equal to that of the denominator we must do long division first.*

■ ***Example 4–6 C***

Decompose $\frac{2x^2 - 3x + 7}{(x - 1)(x + 2)}$ into partial fractions.

$(x - 1)(x + 2) = x^2 + x - 2$ Multiply the denominator

$$\begin{array}{r} 2 \\ x^2 + x - 2\overline{)2x^2 - 3x + 7} \\ \underline{2x^2 + 2x - 4} \\ -5x + 11 \end{array}$$ Divide using long division

Thus, $\frac{2x^2 - 3x + 7}{(x - 1)(x + 2)} = 2 + \frac{-5x + 11}{(x - 1)(x + 2)}$.

$\frac{-5x + 11}{(x - 1)(x + 2)} = \frac{A}{x - 1} + \frac{B}{x + 2}$ Assume the values A, B

Solving produces $A = 2$, $B = -7$, so $\frac{2x^2 - 3x + 7}{(x - 1)(x + 2)} = 2 + \frac{2}{x - 1} + \frac{-7}{x + 2}$. ■

Prime quadratic factors in the denominator

Recall (from section 4–2) that a quadratic expression $ax^2 + bx + c$, $a \neq 0$, is prime (over the real numbers) if the discriminant $b^2 - 4ac < 0$. If the denominator of a rational expression has quadratic factors that are prime, we employ a similar procedure to that shown above.

The difference in procedures is that we must assume the numerators are of the form $ax + b$, and not simply constants as before. Here we restrict ourselves to cases where the prime quadratic factors appears once—this is because the process of finding the values of the assumed variables can become very complicated otherwise.

■ ***Example 4–6 D***

Decompose the rational expression into partial fractions:
$\frac{x^2 + 3x + 2}{(x^2 + x + 1)(x - 1)}$

We assume values A, B, C such that $\frac{x^2 + 3x + 2}{(x^2 + x + 1)(x - 1)} = \frac{Ax + B}{x^2 + x + 1}$ $+ \frac{C}{x - 1}$ and proceed as before.

Multiply by the LCD $(x^2 + x + 1)(x - 1)$:

[1] $x^2 + 3x + 2 = (Ax + B)(x - 1) + C(x^2 + x + 1)$

Let $x = 1$: $1 + 3 + 2 = (A + B)(0) + C(1 + 1 + 1)$
$2 = C$

We now let x take on two other values to obtain two equations with which to find A and B:

Let $x = 0$: $2 = B(-1) + 2(1)$ $C = 2$
$0 = B$

Let $x = -1$: $1 - 3 + 2 = (-A)(-2) + 2(1)$ $B = 0, C = 2$
$-1 = A$

The solution is $\dfrac{x^2 + 3x + 2}{(x^2 + x + 1)(x - 1)} = \dfrac{-x}{x^2 + x + 1} + \dfrac{2}{x - 1}$. ■

Mastery points

Can you

- Decompose rational expressions in which the denominator is a product of linear factors into partial fractions?
- Decompose rational expressions in which the denominator is a product of prime quadratic and linear factors into partial fractions (assuming the factors are limited to exponents of one)?

Exercise 4–6

Decompose the expressions into partial fractions.

1. $\dfrac{x - 10}{x^2 - 5x + 4}$

2. $\dfrac{7x + 5}{x^2 - 2x - 15}$

3. $\dfrac{-6x - 2}{x^2 - 1}$

4. $\dfrac{x^3 - 13x - 30}{x^2 - 9}$

5. $\dfrac{4x^3 - 6x^2 - 1}{2x^2 - 3x + 1}$

6. $\dfrac{4x^3 - 12x^2 + 7x + 14}{2x^2 - 5x - 3}$

7. $\dfrac{18x^3 - 51x^2 + 14x + 28}{6x^2 - 13x - 5}$

8. $\dfrac{11x + 4}{6x^2 - 37x + 6}$

9. $\dfrac{3x^2 - 4x - 1}{(x - 1)^2(x - 2)}$

10. $\dfrac{2x^2 - 5x + 1}{(x - 1)^2(x - 2)}$

11. $\dfrac{13x^2 - 12x + 12}{2x^2(x - 2)}$

12. $\dfrac{4x^2 + 3x - 12}{(x + 4)(x^2 - 16)}$

13. $\dfrac{3x^3 - 11x^2 + x - 17}{(x - 3)^2(x + 1)^2}$

14. $\dfrac{7x^2 - 2x + 23}{(x - 3)^2(x + 1)^2}$

15. $\dfrac{-5x^3 + 32x^2 - 17x - 22}{(x - 3)^2(x + 1)^2}$

16. $\dfrac{x^3 - 4x^2 + 13x + 2}{(x - 3)^2(x + 1)^2}$

17. $\dfrac{x^2 + 4x + 4}{(x - 1)(x^2 + x + 1)}$

18. $\dfrac{x^2 + 6x + 12}{(x + 2)(x^2 + 3x + 3)}$

19. $\dfrac{-x^2 + 3x - 4}{x^3 + 2x^2 + 4x}$

20. $\dfrac{5x^2 + x + 2}{(x + 1)(x^2 + 1)}$

21. $\dfrac{x^2 - 11x - 2}{(x - 3)(x^2 + x + 1)}$

22. $\dfrac{2x^2 - 2x + 8}{(x^2 + 5)(x + 1)}$

23. $\dfrac{x^2 + 6x - 5}{(x + 3)(x^2 + 2x + 4)}$

24. $\dfrac{31x + 12}{(x - 3)(x^2 + 3x + 3)}$

25. In an electronics circuit with two resistances in parallel, the reciprocal of the combined resistances is $\dfrac{2x + 15}{x^2 + 15x + 50}$. Decompose this term using partial fractions.

26. In a situation in which one machine requires x hours to produce 5 items and a second machine requires $x + 3$ hours to do the same thing, the rate at which both machines work is $\dfrac{2x + 3}{x^2 + 3x}$. Decompose this term using partial fractions.

27. In the sum $\dfrac{1}{1 \cdot 2} + \dfrac{1}{2 \cdot 3} + \cdots + \dfrac{1}{99 \cdot 100}$, the nth term is of the form $\dfrac{1}{n(n + 1)}$. Decompose this term using the method of partial fractions and use the result to compute the sum.

28. In the sum $\dfrac{2}{1 \cdot 3} + \dfrac{2}{3 \cdot 5} + \dfrac{2}{5 \cdot 7} + \cdots + \dfrac{2}{99 \cdot 101}$, the nth term is of the form $\dfrac{2}{n(n + 2)}$. Decompose this term using the method of partial fractions and use the result to compute the sum.

29. In the sum $\frac{2}{1 \cdot 3} + \frac{2}{2 \cdot 4} + \frac{2}{3 \cdot 5} + \cdots + \frac{2}{99 \cdot 101}$, the nth term is of the form $\frac{2}{n(n+2)}$. Decompose this term using the method of partial fractions and use the result to compute the sum (see problem 28).

30. In the sum $\frac{3}{1 \cdot 4} + \frac{5}{4 \cdot 9} + \frac{7}{9 \cdot 16} + \cdots + \frac{199}{99^2 \cdot 100^2}$, the nth term is of the form $\frac{2n+1}{n^2(n+1)^2}$. Decompose this term using the method of partial fractions and use the result to compute the sum.

Skill and review

1. Compute **a.** 8^3 **b.** $8^{1/3}$ **c.** 8^{-3} **d.** $8^{-1/3}$.

2. If $2^5 = a^5$, what is a?

3. If $2^a = 2^5$, what is a?

4. Graph $f(x) = 2x^2 - x - 6$.

5. Graph $f(x) = (x - 1)(x + 2)(x - 2)$.

6. Solve $x^3 - x^2 + 1 > x$.

7. Graph $f(x) = \frac{x^2 + 1}{x^2 - 1}$.

Chapter 4 summary

- **Quadratic function** A function of the form $f(x) = ax^2 + bx + c$, $a \neq 0$; its graph is a parabola.
- **Vertex form for a quadratic function** $f(x) = a(x - h)^2 + k$. The vertex is at (h,k); it opens up if $a > 0$, down if $a < 0$.
- **Polynomial function** A function of the form $f(x) = a_nx^n + \cdots + a_2x^2 + a_1x + a_0$, $a_n \neq 0$.
- If $f(c) = 0$ for some function f and a real number c then c is a zero of the function.
- **Rational zero theorem** If $\frac{p}{q}$ is a rational number in lowest terms (p and q are therefore integers) and $\frac{p}{q}$ is a zero of a polynomial function, then p is a factor of the constant term a_0, and q is a factor of the leading coefficient a_n.
- **Remainder theorem** If f is a nonconstant polynomial function and c is a real number, then the remainder when $x - c$ is divided into $f(x)$ is $f(c)$.
- **Multiplicity of zeros** If $(x - c)^n$, $n \in N$, divides a function f, and $(x - c)^{n+1}$ does not divide f, we say that c is a root of multiplicity n.
- If a zero of a polynomial is of **even multiplicity** the graph just touches, but does not cross, the x-axis at that point. If the zero is of **odd multiplicity** the function crosses the axis at the zero.
- Every polynomial of positive degree n over R is the product of a real number and one or more prime linear or quadratic polynomials over R.
- **Bounds theorem for real zeros** Let c be a real number and $f(x)$ be a polynomial function with real coefficients and positive leading coefficient; consider all the coefficients in the last line of the synthetic division algorithm as applied to the value c. Then c is

  an upper bound if $c \geq 0$ and these coefficients are all positive or zero.

  a lower bound if $c \leq 0$ and these coefficients alternate between nonnegative and nonpositive values.
- **Descartes' rule of signs** Let $f(x)$ be a function defined by a polynomial with real coefficients. Then,

  the number of *positive* real zeros is equal to the number of variations in sign in $f(x)$ or is less than this number by a multiple of 2.

  the number of *negative* real zeros is equal to the number of variations in sign in $f(-x)$ or is less than this number by a multiple of 2.
- **Rational function** A function of the form $f(x) = \frac{P(x)}{Q(x)}$, where $P(x)$ and $Q(x)$ are polynomial expressions in the variable x.
- **Vertical asymptote** A vertical line that a function gets closer and closer to as x approaches a certain value.
- **Horizontal asymptote** A horizontal line that a function gets closer and closer to as $|x|$ gets larger and larger.
- **Slant asymptote** A slanted straight line that a function gets closer and closer to as $|x|$ gets larger and larger.
- In general a rational function has a vertical asymptote wherever the denominator takes on the value zero.

- **Graphing rational functions** To graph rational functions we use the following information.

  Horizontal and slant asymptotes

  Vertical asymptotes

  Intercepts

  Plotting points

- **Arithmetic operations for functions** Let f and g be functions and x a value in the domain of f and g. Then

  $(f + g)(x) = f(x) + g(x)$

  $(f - g)(x) = f(x) - g(x)$

  $(f \cdot g)(x) = f(x) \cdot g(x)$

  $(f/g)(x) = \dfrac{f(x)}{g(x)}$, if $g(x) \neq 0$

- **Composition of functions** Let f and g be functions, $x \varepsilon$ domain of g, $g(x) \varepsilon$ domain of f. Then $(f \circ g)(x) = f(g(x))$.

- **Inverse functions** If $(f \circ g)(x) = x$ and $(g \circ f)(x) = x$ for all x in the domains of functions f and g, then f and g are said to be inverse functions.
- A function f has an inverse function f^{-1}, if and only if it is one to one.
- If an ordered pair (a,b) is in a function f, and if f has an inverse function f^{-1}, then (b,a) is an ordered pair in f^{-1}.
- The graph of f^{-1} is a reflection of the graph of f about the line $y = x$.
- **To find $f^{-1}(x)$ for a one-to-one function f:**

  Replace $f(x)$ by y.

  Replace each x by y and y by x.

  Solve for y.

  Replace y by $f^{-1}(x)$.

Chapter 4 review

[4–1] Graph the following parabolas. Compute the intercepts and vertex.

1. $y = x^2 - 3x - 18$ **2.** $y = -3x^2$
3. $y = x^2 - 4x$ **4.** $y = -x^2 + 5x + 6$
5. $y = 9 - x^2$ **6.** $y = 3x^2 + 4x - 4$
7. $y = x^2 - 5x - 1$ **8.** $y = x^2 + 2x + 2$
9. $y = x^2 - x + 5$ **10.** $y = -x^2 - 4x$

Solve problems 11–14 by creating an appropriate second-degree equation and finding the vertex.

11. A homeowner has 200 feet of fencing to fence off a rectangular area behind the home. The home will serve as one boundary (the fence is only necessary for three of the four sides). What are the dimensions of the maximum area that can be fenced off, and what is the area?

12. What is the area of the largest rectangle that can be created with 400 feet of fence? What are the length and width of this rectangle?

13. Suppose that 400 feet of fencing are available to fence in an area. Will a half-circle (perimeter including the straight side) or a rectangle contain a greater area? (*Hint:* For a fixed perimeter the radius of a half-circle is a fixed value, and can be found. See the result of problem 12 also.)

14. If an object is thrown vertically into the air with an initial velocity of v_0 ft/s, then its distance above the ground s, for time t, is given by $s = v_0t - 16t^2$. Suppose an object is thrown upward with initial velocity 512 ft/s; find how high the object will go and when it will return to the ground.

Graph the following functions.

15. $h(x) = \begin{cases} -2x - 1, & x < -1 \\ x + 2, & x \geq -1 \end{cases}$

16. $g(x) = \begin{cases} x + 2, & x < 1 \\ 2x^2 - 4x + 5, & x \geq 1 \end{cases}$

[4–2] List all possible rational zeros for the following polynomials.

17. $2x^4 - 3x^2 + 6$ **18.** $2x^3 - 4x^2 + 2x - 10$
19. $x^5 - x^3 - 4$ **20.** $3x^4 - 8$

Use synthetic division to (a) divide each polynomial by the divisor indicated and (b) to evaluate the function at the value indicated.

21. $f(x) = 2x^4 - 5x^3 + 2x^2 - 1$
a. $x - 3$ **b.** $f(3)$

22. $g(x) = -2x^3 - 5x^2 - 3x + 2$
a. $x + 4$ **b.** $g(-4)$

23. $f(x) = \frac{1}{2}x^3 - 3x^2 + \frac{3}{4}x - 3$
a. $x - 4$ **b.** $f(4)$

For each function:

a. Use Descartes' rule of signs to find the number of possible real zeros.
b. List all possible rational zeros of each polynomial.
c. Find all rational zeros; state the multiplicity when greater than one.
d. Factor each as much as possible over the real numbers.
e. If there are any possible irrational zeros state the least positive integer upper bound and the greatest negative integer lower bound for these zeros.

24. $f(x) = 2x^4 + x^3 - 24x^2 - 9x + 54$
25. $g(x) = 2x^4 - x^3 - 9x^2 + 4x + 4$
26. $h(x) = 2x^4 - 9x^3 - 13x^2 + 4x + 4$
27. $f(x) = 16x^5 - 48x^4 - 40x^3 + 120x^2 + 9x - 27$

[4–3] Graph the following polynomial functions. State all intercepts.

28. $g(x) = \frac{1}{4}(x - 4)(x^2 - 16)$
29. $h(x) = (x^2 - 4)(4x^2 - 9)$
30. $g(x) = (x^2 - 9)(4x^2 - 9)(x - 3)$
31. $f(x) = (2x^2 - 3x - 5)^2$
32. $h(x) = (x^2 - x - 20)(x^2 - 4)(2x + 1)$
33. $f(x) = (x - 3)^2(x + 1)$
34. $h(x) = (x - 2)^3(2x + 3)^2$

[4–4] Graph the following rational functions. State all asymptotes and intercepts.

35. $f(x) = \dfrac{2}{(x - 3)^3}$
36. $h(x) = \dfrac{-3}{(x + 5)^2}$
37. $f(x) = \dfrac{3}{x^2 - 4x - 45}$
38. $h(x) = \dfrac{2x}{2x^2 - 7x + 5}$
39. $g(x) = \dfrac{x^3}{(x - 1)(x^2 - 9)}$
40. $g(x) = \dfrac{x^2 - x - 6}{x^2 + 3}$
41. $f(x) = \dfrac{x^2 - x - 6}{x - 3}$

[4–5] Find expressions for $f + g$, $f - g$, $f \cdot g$, f/g, and $(f \circ g)(x)$ and $(g \circ f)(x)$ for the given functions f and g.

42. $f(x) = 3 - \frac{1}{2}x$; $g(x) = \frac{1}{2}x - 3$
43. $f(x) = x^4 - 1$; $g(x) = \sqrt{8 - x}$
44. $f(x) = \dfrac{x - 3}{2x}$; $g(x) = \dfrac{x}{2x - 1}$
45. $f(x) = -2x$; $g(x) = 3x$
46. $f(x) = x$; $g(x) = -3$

Find the inverse function of the given function. All the functions are one to one.

47. $g(x) = \dfrac{2x - 5}{4}$
48. $h(x) = \dfrac{x - 4}{2x + 1}$
49. $g(x) = x^2 + 8, x \geq 0$
50. $g(x) = 8x^3 - 27$
51. $g(x) = \sqrt[3]{1 - x} - 3$
52. $f(x) = x^2 - 7x + 6, x \geq 3\frac{1}{2}$
53. A solar engineer found that a solar hot water heating installation could heat 40 gallons per day in January and 200 in August. Create a linear function that models this situation of capacity c, as a function of time in months t, and use this function to predict the number of gallons of hot water heating capacity that the system might have in June.

[4–6] Decompose the following rational functions into partial fractions.

54. $\dfrac{8x + 11}{2x^2 - 5x - 3}$
55. $\dfrac{13x^2 - 52x + 32}{(x - 3)^2(2x + 1)}$
56. $\dfrac{2x^2 - 5x + 8}{(x - 2)(x^2 - x + 4)}$
57. $\dfrac{-2x^3 + 18x^2 - 53x + 62}{(x + 2)(x - 2)^3}$

Chapter 4 test

Graph the following parabolas. Compute the intercepts and vertex.

1. $y = x^2 + 5x - 14$
2. $y = -2x^2 + 8$
3. $y = 3x^2 + 5x - 2$
4. $y = -x^2 + 4x$

Solve the following two problems by creating an appropriate second-degree equation and finding the vertex.

5. A homeowner has 50 feet of fencing to fence off a rectangular area behind the home. The home will serve as one boundary (the fence is only necessary for three of the four sides). What are the dimensions of the maximum area which can be fenced off, and what is the area?

6. If an object is thrown vertically into the air with an initial velocity of v_0 ft/s, then its distance above the ground s, for time t, is given by $s = v_0t - 16t^2$. Suppose an object is thrown upward with initial velocity 48 ft/s; find how high the object will go and when it will return to the ground.

7. Graph the function $g(x) = \begin{cases} -\frac{1}{2}x - \frac{5}{2}, & x < -1 \\ x^2 + 2x - 1, & x \geq -1 \end{cases}$.

List all possible rational zeros for the following polynomials.

8. $x^4 - 3x^2 + 8$ **9.** $4x^3 - 4x^2 + 2x - 12$

10. Use synthetic division to (a) divide the polynomial by the divisor indicated and (b) to evaluate the function at the value indicated.

$$f(x) = 3x^4 - 2x^3 - 30x^2 - 20$$

a. $x + 3$ **b.** $f(-3)$

For each function in problems 11–13:

a. Use Descartes' rule of signs to find the number of possible real zeros.
b. List all possible rational zeros of each polynomial.
c. Find all rational zeros; state the multiplicity when greater than one.
d. Factor each as much as possible over the real numbers.
e. If there are any possible irrational zeros state the least positive integer upper bound and the greatest negative integer lower bound for these zeros.

11. $f(x) = 4x^3 - 4x^2 - x + 1$
12. $g(x) = x^4 + 3x^3 + 4x^2 + 3x + 1$
13. $h(x) = 3x^5 - 5x^4 - 23x^3 + 53x^2 - 16x - 12$

Graph the following polynomial functions. Compute all intercepts.

14. $f(x) = x^3 - 6x^2 + 3x + 10$
15. $g(x) = x^3 + 3x^2 - 9x - 27$
16. $f(x) = (x^2 - 1)^2$
17. $h(x) = (x^2 - 3x - 10)(x^2 - 3x + 10)$

Graph the following rational functions. Compute all intercepts and state all asymptotes.

18. $f(x) = \dfrac{2}{(x + 1)^2}$ **19.** $f(x) = \dfrac{-1}{(x - 2)^3}$

20. $f(x) = \dfrac{1}{x^2 + 2x - 24}$ **21.** $f(x) = \dfrac{-x}{x^2 - 4}$

22. $g(x) = \dfrac{x^2 - 7x + 12}{x^2 + 1}$ **23.** $f(x) = \dfrac{x^2 - x - 12}{x - 5}$

Find expressions for $f + g$, $f - g$, $f \cdot g$, f/g, and $(f \circ g)(x)$ and $(g \circ f)(x)$ for the given functions f and g.

24. $f(x) = 2x + 5$; $g(x) = x^2 - 2x - 4$
25. $f(x) = x^4 - 2$; $g(x) = 2\sqrt{x + 1}$
26. $f(x) = \dfrac{x + 2}{x}$; $g(x) = \dfrac{x}{2x - 1}$

Find the inverse function of the given function. All the functions are one to one.

27. $g(x) = 5x + 4$ **28.** $f(x) = \dfrac{5x + 1}{4x}$

29. $g(x) = x^2 - 4$, $x \geq 0$
30. $f(x) = x^2 - x - 6$, $x \leq \frac{1}{2}$
31. A thermocouple is an electronic device that can be used to measure temperature; its voltage output depends on (is a function of) temperature. A technician measured the output of a thermocouple to be 60 millivolts (mV) at 50° F and 80 mV at 100° F. Find a linear function that fits this data and use it to predict what the temperature is when the output of the thermocouple is 65 mV.

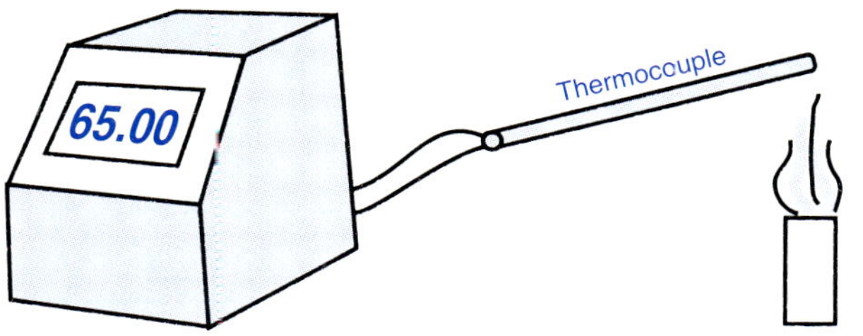

Decompose the following rational functions into partial fractions.

32. $\dfrac{5x^2 + 3x + 1}{(x + 1)^2(x - 2)}$ **33.** $\dfrac{4x^2 + 4x + 10}{(x - 1)(x^2 + x + 4)}$

C H A P T E R 5

The Trigonometric Functions

In this and the following three chapters we will study trigonometry. The word trigonometry means "triangle measurement."[1] Trigonometry was used by the ancient Greeks to study astronomy, and was well developed by the eighteenth century. By then it was also used in surveying and in the physics developed by Sir Isaac Newton and others. In the age of the computer, trigonometry has become more important than ever. Computer robots are programmed using trigonometric descriptions of their motions, computer graphics use trigonometry extensively to generate screen images, and voice recognition by computers uses Fourier transforms, a concept built on the trigonometric functions.

5–1 The trigonometric ratios

Find the speed relative to the ground of a boat heading straight across a river at 16 knots if the current is moving at 4.3 knots.

The mathematics we study in this section can be used to solve this problem. It requires some basic physics and a law that has been known for well over 2,000 years.

Degree measure for angles

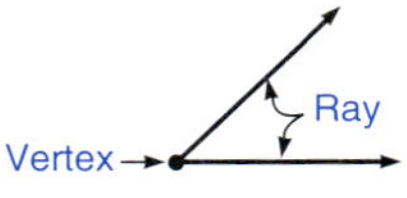

Figure 5–1

An **angle** is composed of two rays, both beginning at what is called the **vertex** of the angle. Figure 5–1 shows a representation of a ray and of an angle.

Angles are often measured in degrees. The notation for degrees is °. A common and useful interpretation of angle measure is "the amount of rotation" of one ray away from the other. In this context 90° corresponds to a

[1]The word *trigonometry* is credited to Bartholomaus Pitiscus (1561–1613).

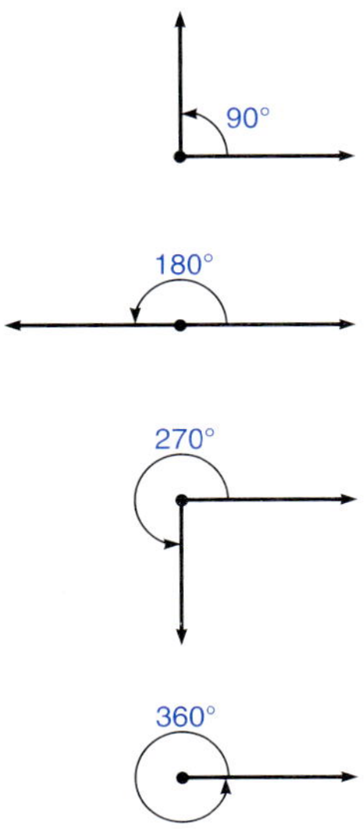

Figure 5–2

quarter-rotation, 180° to a half-rotation, 270° to three-quarters of a rotation, and 360° to a full rotation. See figure 5–2. An angle with measure between 0° and 90° is said to be **acute;** an angle with measure 90° is **right** and an angle with measure between 90° and 180° is **obtuse.**

There are two systems in wide use for measuring angles in degrees: the **degree, minute, second (DMS) system** and the **decimal degree system.**

In the first system, parts of an angle are broken down into 60ths, called minutes, and minutes are broken down into 60ths, called seconds. This means there are $60^2 = 3{,}600$ seconds in one degree. The symbols °, ′, and ″ are used for degrees (°), minutes (′), and seconds (″), respectively. It is much like our system for keeping time. For example, 16°4′22″ means 16 degrees, 4 minutes, 22 seconds. Observe that the 4 minutes is $\frac{4}{60}$ degree and that 22 seconds is $\frac{22}{60^2}$ degree. We use this idea in example 5–1 A. Calculations are easiest in decimal degrees, and so it is important to be able to convert to this form if necessary. Many electronic calculators have special keys to convert angles in degrees, minutes, and seconds to decimal form. This is illustrated in example 5–1 A for two typical calculators, as well as for the TI-81 graphing calculator, which does not have special keys for this purpose.

■ *Example 5–1 A*

Convert 12°20′34″ to decimal form. Round to the nearest 0.001°.

$12 + \frac{20}{60} + \frac{34}{60^2}$ degrees — Rewrite minutes and seconds as fractional parts of a degree

$12.3427777\ldots°$ — Compute the decimal form of the fractions

$12.343°$ — Round to the nearest 0.001°

Typical key sequences:

Calculator 1: 12 [° ′ ″] 20 [° ′ ″] 34 [° ′ ″]

Calculator 2: 12.2034 [→H] "H" stands for Hours.

TI-81: 12 [+] 20 [÷] 60 [+] 34 [÷] 60 [x^2] [ENTER] ■

Several properties of triangles

A **triangle** is a closed figure of three sides and three angles. It is a theorem that *the sum of the measures of the angles of a triangle is 180°.* We often denote angles using Greek letters. The letters most often used are θ (theta), α (alpha), and β (beta). If we know the measure of two of the angles in a triangle we can compute the measure of the third by subtraction.

■ *Example 5–1 B*

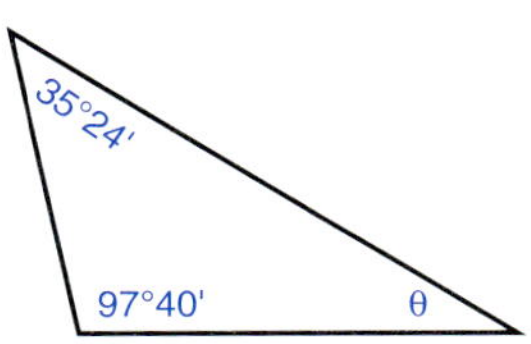

Find the measure of angle θ in the triangle.

$$
\begin{aligned}
35°24' + 97°40' &= 132°64' \\
&= 133°4' \qquad 64' = 1°4' \\
180° - 133°4' &= 179°60' \qquad 1° = 60' \\
&\underline{-\ 133°4'} \\
&\ 46°56'
\end{aligned}
$$

Thus the measure of θ is 46°56′. ■

A **right triangle** is a triangle in which one of the angles is a right (90°) angle. In such a triangle, two of the angles must be acute, since their sum is 90°. The side of a right triangle opposite the right angle is called the **hypotenuse.** This is always the longest side of the triangle because it is a theorem of geometry that *the longest side of any triangle is opposite the largest angle (and that the shortest side is opposite the smallest angle).*

It is important to recognize which side is opposite which acute angle. Figure 5–3 shows a right triangle in which the angles are labeled A, B, and C, and the sides are a, b, and c. Angle A is highlighted. Observe that side a is opposite to angle A and that side b is said to be adjacent to angle A. Side b is opposite angle B, and side a is adjacent to angle B.

One of the most-used facts in mathematics is the Pythagorean theorem.[2]

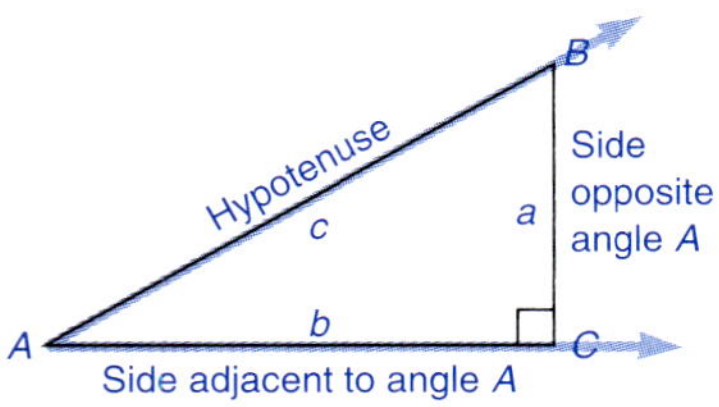

Figure 5–3

The Pythagorean theorem

In a right triangle with hypotenuse of length c and sides of lengths a and b,

$$a^2 + b^2 = c^2$$

This theorem allows us to find the measure of any side of a right triangle when we know the measure of the other two sides.

■ *Example 5–1 C*

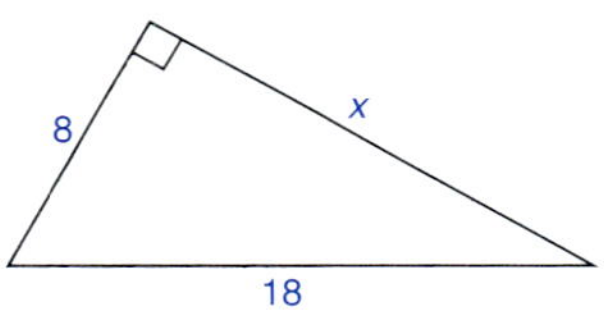

Find the length of side x in the triangle.

Using the Pythagorean theorem we obtain

$$x^2 + 8^2 = 18^2$$
$$x^2 + 64 = 324$$
$$x^2 = 260$$
$$x = \sqrt{260}$$
$$x = 2\sqrt{65} \approx 16.1 \qquad \sqrt{260} = \sqrt{4 \cdot 65} = \sqrt{4} \cdot \sqrt{65}$$

■

A physical law

As we noted earlier the Pythagorean theorem has many applications. One applies to a fact of physics, concerning the flight of an aircraft through the air.[3] The **airspeed** of an aircraft is its speed relative to the air. It is the speed that the instruments in the aircraft actually measure. It is generally *not* the speed of the aircraft as measured from the ground. The **heading** of an aircraft is the direction in which it is pointed. The aircraft does not actually travel in this direction unless it is pointed directly into the wind or is pointed with the wind. Usually the wind causes the aircraft to travel in a direction different than that in which it is pointed. See figure 5–4.

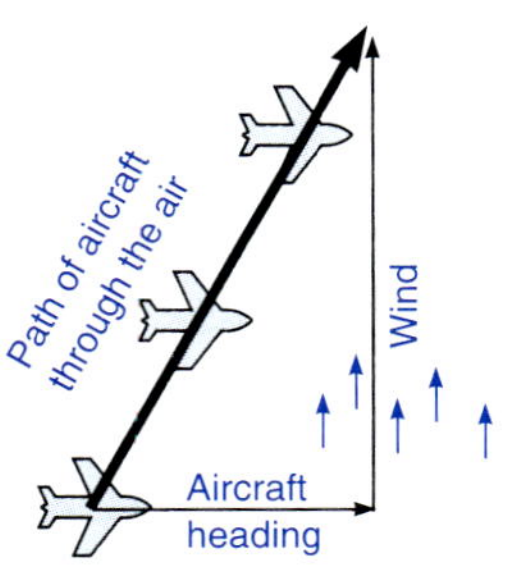

Figure 5–4

[2]The word *theorem* means a statement that has been proved to be true. The proof of this theorem is credited to the Greek mathematician Pythagoras (sixth century B.C.), who is said to have sacrificed an ox as an offering of thanks. In the last two thousand years, literally hundreds of proofs of this theorem have been given.

[3]We are describing a specific application of vectors. The topic is discussed further in chapter 8.

If the direction of the wind and the heading of an aircraft are perpendicular, then they can be described by the sides of a right triangle. The lengths of the sides describe the airspeed of the aircraft and the speed of the wind. We show the sides as arrows, called vectors, to indicate the direction of the wind and the direction in which the aircraft is pointed. The vector (arrow) for the wind begins at the end of the vector for the aircraft.

The length of the hypotenuse represents the speed of the aircraft relative to the ground, called its **ground speed.** The hypotenuse is drawn as a vector also, beginning at the beginning of the vector for the aircraft, and ending at the end of the vector for the wind. The direction of this vector represents the direction in which the aircraft actually moves, relative to the ground.

■ *Example 5–1 D*

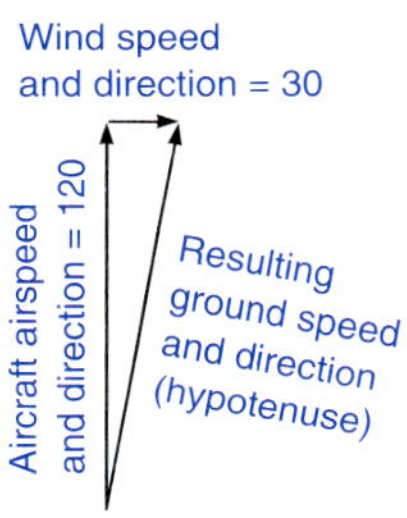

An aircraft is flying with an airspeed of 120 knots[4] and a heading of due north. A 30 knot wind is blowing from the west. What is the aircraft's ground speed v, to the nearest knot?

Represent the speeds and directions given above in the right triangle as shown. The ground speed v is the length of the hypotenuse.

$$v^2 = 120^2 + 30^2$$
$$v^2 = 15{,}300$$
$$v = \sqrt{15{,}300} \approx 123.69$$

Thus, the ground speed is about 124 knots. ■

The trigonometric ratios

In the fifteenth and sixteenth centuries in Germany, trigonometry developed into the form we now present. We begin with a definition of three **trigonometric ratios.** These are, along with their abbreviations, **sine** (sin), **cosine** (cos), and **tangent** (tan). Later we will define three more ratios.

The primary trigonometic ratios

If θ is either of the two acute angles in a right triangle, then

Ratio	Definition
$\sin\theta =$	$\dfrac{\text{length of side opposite } \theta}{\text{length of hypotenuse}}$
$\cos\theta =$	$\dfrac{\text{length of side adjacent to } \theta}{\text{length of hypotenuse}}$
$\tan\theta =$	$\dfrac{\text{length of side opposite to } \theta}{\text{length of side adjacent to } \theta}$

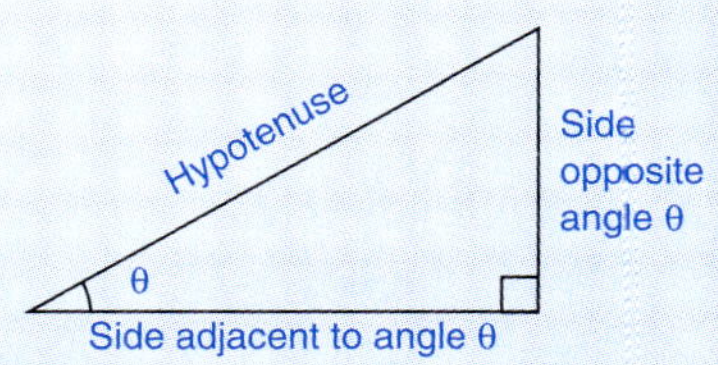

Figure 5–5

The first ratio means "the sine of angle θ" and is usually read "sine theta." The other two ratios are interpreted similarly. These ratios are used in astronomy, surveying, engineering, science, and mathematics—there is in fact virtually no area of science and technology that does not use them somewhere. Example 5–1 E illustrates the application of these definitions to a particular triangle.

[4]One knot means 1 nautical mile (≈ 6,080.27 ft) per hour. Recall that 1 mile is 5,280 feet.

■ Example 5–1 E

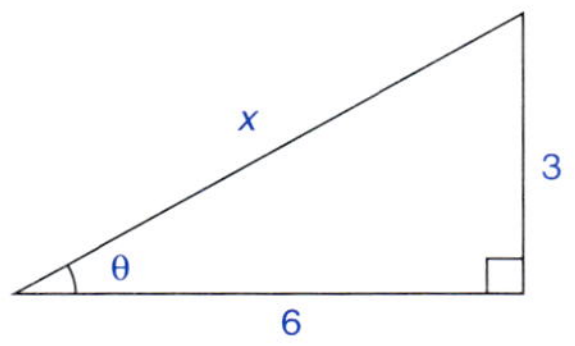

1. Find the sine, cosine, and tangent ratios for angle θ.

$x^2 = 3^2 + 6^2$

Use the Pythagorean theorem to find x

$x^2 = 45$

$x = \sqrt{45} = \sqrt{9 \cdot 5} = 3\sqrt{5}$

$$\sin\theta = \frac{\text{length of side opposite }\theta}{\text{length of hypotenuse}} = \frac{3}{3\sqrt{5}} = \frac{1}{\sqrt{5}} = \frac{\sqrt{5}}{5}$$

$\frac{1}{\sqrt{5}} \cdot \frac{\sqrt{5}}{\sqrt{5}} = \frac{\sqrt{5}}{5}$

$$\cos\theta = \frac{\text{length of side adjacent to }\theta}{\text{length of hypotenuse}} = \frac{6}{3\sqrt{5}} = \frac{2\sqrt{5}}{5}$$

$\frac{\overset{2}{\cancel{6}}}{\underset{1}{\cancel{3}}\sqrt{5}} \cdot \frac{\sqrt{5}}{\sqrt{5}} = \frac{2\sqrt{5}}{5}$

$$\tan\theta = \frac{\text{length of side opposite }\theta}{\text{length of side adjacent to }\theta} = \frac{3}{6} = \frac{1}{2}$$

■

The final three trigonometric ratios are called the **cosecant** (csc), **secant** (sec), and **cotangent** (cot). These ratios can be defined as the reciprocals of the primary ratios.

> **Reciprocal pair ratios**
>
> If θ is either acute angle of a right triangle, then
>
> $$\csc\theta = \frac{1}{\sin\theta}, \quad \sec\theta = \frac{1}{\cos\theta}, \quad \cot\theta = \frac{1}{\tan\theta}$$

It is worth stressing that the following pairs of ratios are reciprocals:

sine and cosecant
cosine and secant
tangent and cotangent

The definitions above also imply the following relations:

$$\sin\theta = \frac{1}{\csc\theta}, \quad \cos\theta = \frac{1}{\sec\theta}, \quad \tan\theta = \frac{1}{\cot\theta}$$

These relations and the definitions mean that if we know one ratio we can invert it to find another ratio. The example 5–1 F illustrates this.

■ Example 5–1 F

Find the value of the other member of the appropriate reciprocal pair.

1. $\sin\alpha = \frac{2}{3}$ $\qquad \csc\alpha = \frac{3}{2}$, since $\frac{1}{\frac{2}{3}} = \frac{3}{2}$

2. $\sec\alpha = 1.6$ $\qquad \cos\alpha = \frac{1}{1.6} = \frac{1}{1.6} \cdot \frac{10}{10} = \frac{10}{16} = \frac{5}{8}$

■

Example 5–1 G applies the idea of reciprocal functions to a particular triangle.

■ Example 5–1 G

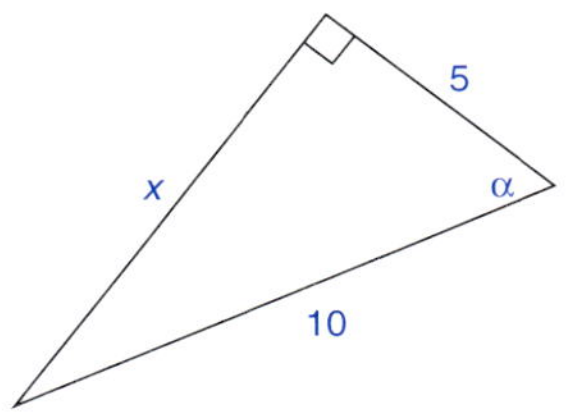

Find the value of the six trigonometric ratios for angle α in the triangle.

$10^2 = x^2 + 5^2$ Use the Pythagorean theorem to find x
$75 = x^2$
$5\sqrt{3} = x$ $\sqrt{75} = \sqrt{25 \cdot 3} = 5\sqrt{3}$

$$\sin \alpha = \frac{x}{10} = \frac{5\sqrt{3}}{10} = \frac{\sqrt{3}}{2}, \csc \alpha = \frac{2}{\sqrt{3}} \cdot \frac{\sqrt{3}}{\sqrt{3}} = \frac{2\sqrt{3}}{3}, \quad \csc \alpha = \frac{1}{\sin \alpha}$$

$$\cos \alpha = \frac{5}{10} = \frac{1}{2}, \quad \sec \alpha = 2, \quad \sec \alpha = \frac{1}{\cos \alpha}$$

$$\tan \alpha = \frac{x}{5} = \frac{5\sqrt{3}}{5} = \sqrt{3}, \cot \alpha = \frac{1}{\sqrt{3}} \cdot \frac{\sqrt{3}}{\sqrt{3}} = \frac{\sqrt{3}}{3}, \quad \cot \alpha = \frac{1}{\tan \alpha}$$ ■

If we know one of the trigonometric ratios of an acute angle, we can construct a right triangle with an angle for which that ratio is true. We can use this triangle to compute the other five trigonometric ratios.

■ Example 5–1 H

Solve each problem.

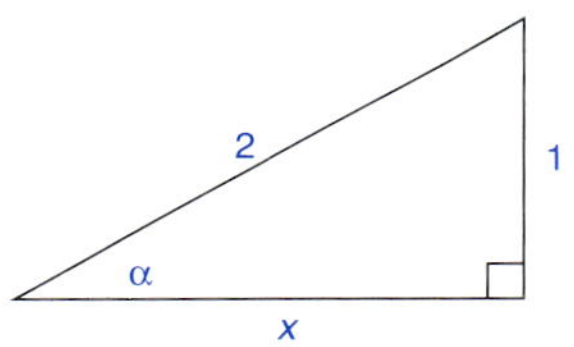

1. In a right triangle with acute angle α, $\csc \alpha = 2$. Construct a right triangle for which this is true, and use the triangle to find the values of the other five trigonometric ratios for angle α.

Since $\csc \alpha = 2$, $\sin \alpha = \frac{1}{2}$. Since $\sin \alpha = \frac{\text{length of side opposite } \alpha}{\text{length of hypotenuse}} = \frac{1}{2}$, we can use a triangle in which the hypotenuse has length 2 and the length of the side opposite angle α is 1, as shown in the figure.

$2^2 = x^2 + 1^2$ Find the value of side x
$3 = x^2$
$\sqrt{3} = x$

$$\cos \alpha = \frac{x}{2} = \frac{\sqrt{3}}{2}, \quad \sec \alpha = \frac{1}{\cos \alpha} = \frac{2}{\sqrt{3}} \cdot \frac{\sqrt{3}}{\sqrt{3}} = \frac{2\sqrt{3}}{3},$$

$$\tan \alpha = \frac{1}{x} = \frac{1}{\sqrt{3}} = \frac{\sqrt{3}}{3}, \cot \alpha = \frac{1}{\tan \alpha} = \frac{\sqrt{3}}{1} = \sqrt{3}$$

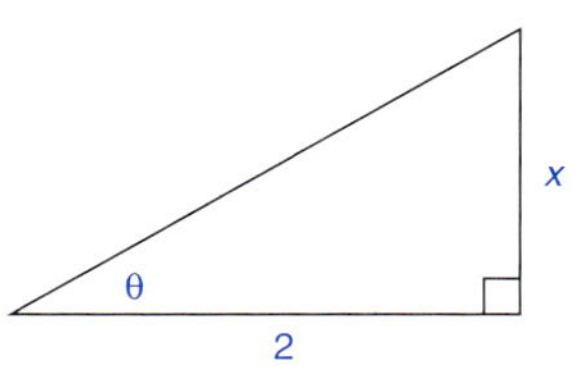

2. Suppose $\tan \theta = \dfrac{x}{2}$ and $x > 0$. Construct a triangle for which this is true, and use it to compute $\cos \theta$.

$\tan \theta = \dfrac{x}{2} = \dfrac{\text{opp}}{\text{adj}}$. The triangle shown meets this description. The length of the hypotenuse is $\sqrt{x^2 + 2^2} = \sqrt{x^2 + 4}$.

$$\cos \theta = \frac{\text{adj}}{\text{hyp}} = \frac{2}{\sqrt{x^2 + 4}}.$$ ■

Note As shown in part 1 of example 5–1 H, *when given the value of one of the reciprocal trigonometric ratios, it is best to immediately convert that value to a primary trigonometric ratio.*

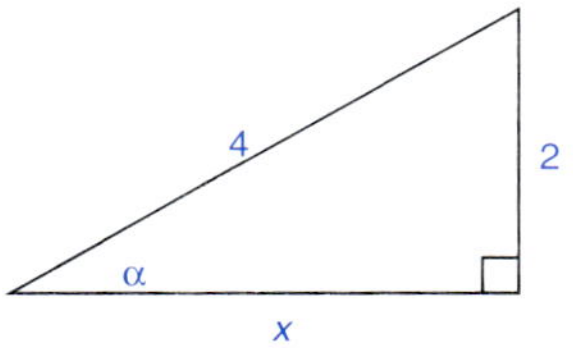

Figure 5–6

A logical question is whether any other triangle could be used in part 1 of example 5–1 H. The answer is yes. For example, since $\frac{2}{4}$ reduces to $\frac{1}{2}$ we could use the triangle shown in figure 5–6. However, our values for the six trigonometric ratios would not change. For example, we would discover that $x = 2\sqrt{3}$, so that $\cos \alpha = \frac{x}{4} = \frac{2\sqrt{3}}{4} = \frac{\sqrt{3}}{2}$, which is the same value we found in the example. In fact, the lengths of the three sides of this second triangle are each double the corresponding length in the previous triangle. Thus, each ratio would reduce to the previous values.

Similarly we could start with any fraction equivalent to $\frac{1}{2}$. This means that we could use an unlimited number of triangles in such problems and always arrive at the same values of the six trigonometric ratios.

Mastery points

Can you

- State whether a given angle is acute, right, or obtuse?
- Convert an angle given in degrees, minutes, and seconds into decimal form?
- Use the fact that the sum of the measures of the angles of any triangle is 180° to determine unknown angles?
- Use the Pythagorean theorem to find the length of the third side in a right triangle when given the lengths of two of the sides?
- State the definitions of the six trigonometric ratios for an acute angle θ in a right triangle?
- Find the values of the six trigonometric ratios for a given acute angle in a right triangle, given the lengths of at least two of the sides?
- Find the values of the five other trigonometric ratios when given the value of one trigonometric ratio for an acute angle of a right triangle?

Exercise 5–1

Convert each angle to its measure in decimal degrees. Round the answer to the nearest 0.001° where necessary. Also, state whether each angle is acute or obtuse.

1. 13°25′ **2.** 111°56′ **3.** 0°12′ **4.** 42°37′
5. 25°33′19″ **6.** 87°2′13″ **7.** 165°47′ **8.** 19°15′
9. 33°5′55″ **10.** 0°19′12″ **11.** 159°59′ **12.** 20°1′

In the following problems find the measure of angle θ.

13.

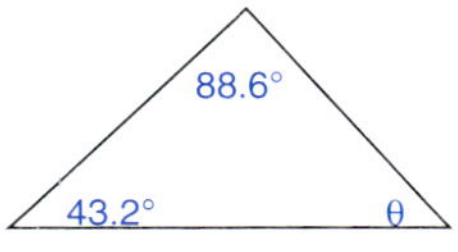

14.

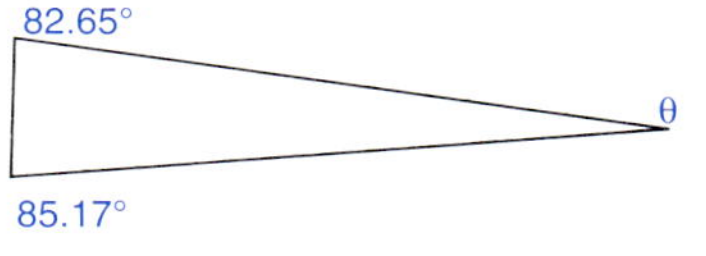

15.

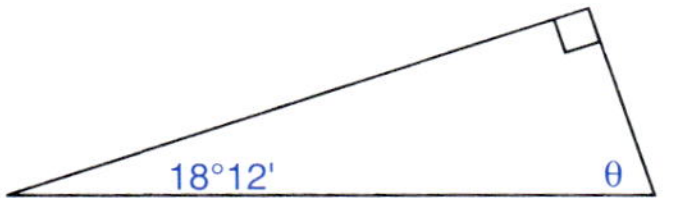

16.

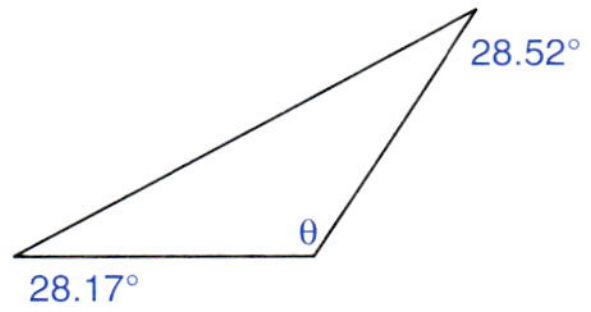

17.

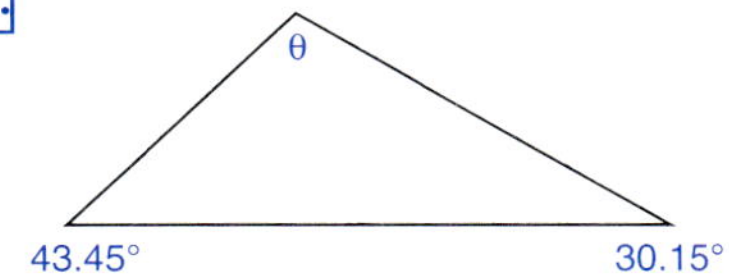

18.

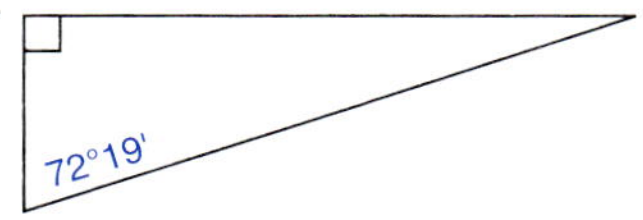

In the following problems find the length of the missing side. The lengths *a*, *b*, and *c* refer to the diagram. Leave your answer in exact form (in terms of rational numbers and radicals) unless the data is given in decimal form. In these cases, round the answer to the same number of decimal places as the data.

	a	*b*	*c*
19.	9	12	?
20.	10	?	26
21.	?	8	10
22.	5	10	?
23.	12	?	18
24.	?	6.8	9.2
25.	$\sqrt{5}$	3	?
26.	13.2	19.6	?
27.	100	150	?
28.	0.66	1.42	?
29.	$\sqrt{7}$	3	?
30.	4	?	$\sqrt{23}$
31.	6.3	?	15.0
32.	2	?	3
33.	$3\sqrt{2}$	$4\sqrt{5}$	?
34.	$3\sqrt{2}$	?	$4\sqrt{5}$
35.	?	19	28
36.	1,002	3,512	?
37.	1	1	?
38.	30	?	50

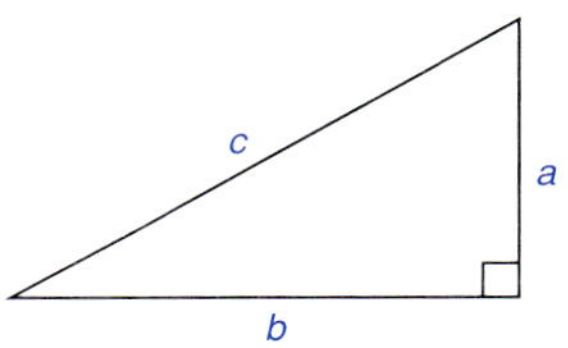

39. A flagpole is 55 feet tall and is supported by a wire attached to the top of the pole and to the ground 26 feet from the base of the pole. How long is the wire, to the nearest foot?

40. A flagpole is 93 feet tall and is supported by a wire that is 157 feet long, attached at the top of the pole and to the ground some distance from the base of the pole. Find the distance of the wire's ground attachment point from the base of the pole, to the nearest foot.

41. A surveyor has made the measurements shown in the diagram in order to compute the width of a pond. How wide is the pond, to the nearest 0.1 foot?

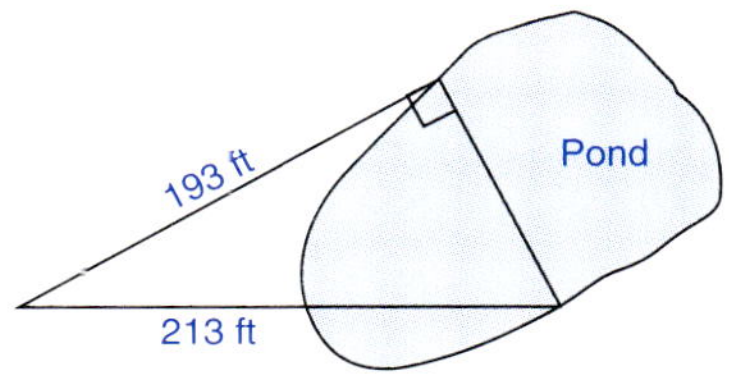

The diagram, called an impedance diagram, is used to compute total impedance Z in a certain electronic circuit. X_L is called the inductive reactance, and R is the resistance. All units are ohms. Use the diagram for problems 42–45.

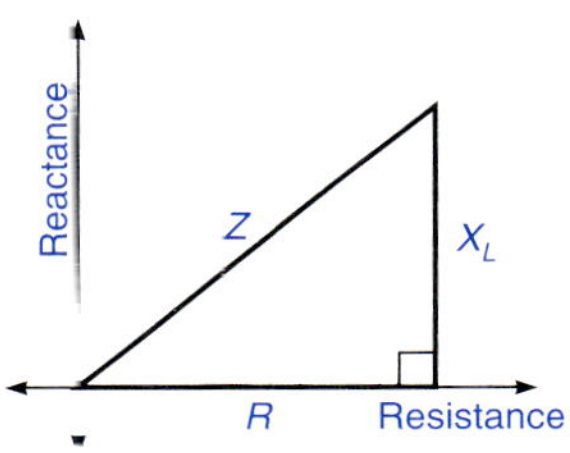

42. If the inductive reactance X_L is 40.0 ohms and the resistance R is 56.6 ohms, calculate the total impedance Z to the nearest 0.1 ohm.

43. If $X_L = 5.68$ ohms and $R = 19.25$ ohms, find Z to the nearest 0.01 ohms.

44. If $Z = 213$ ohms and $R = 183$ ohms, find X_L to the nearest ohm.

45. If $Z = 4{,}340$ ohms and $X_L = 2{,}150$ ohms, find R to the nearest 10 ohms.

46. A machinist has to cut a rectangular piece of steel along its diagonal. The saw that will be used can cut this thickness of steel at the rate of 0.75 inch per minute. If the piece is 13.8 inches long and 9.6 inches wide, calculate how many minutes, to the nearest minute, it will take to cut the piece.

47. Assume the same situation as problem 46, except that the piece is 15.0 centimeters long and 10.5 centimeters wide, and that the saw will cut 0.8 centimeters per minute.

48. The ladder on a fire truck can extend 125 feet. If the truck is 25 feet from a building, how high up the building can the ladder reach, to the nearest tenth of a foot?

49. If the fire truck of problem 48 moves 5 more feet from the building (to 30 feet), does the height up the building that the ladder can reach decrease by 5 feet?

Find the value of the other member of the appropriate reciprocal pair.

50. $\cos\theta = \frac{1}{2}$ **51.** $\cos\alpha = \frac{1}{3}$

52. $\csc\beta = 1.5$ **53.** $\tan\beta = \sqrt{2}$

54. $\sin\beta = \dfrac{\sqrt{3}}{5}$ **55.** $\cot\theta = \sqrt{5}$

56. $\tan\alpha = 2.25$ **57.** $\cos\theta = \dfrac{\sqrt{6}}{8}$

The following problems refer to the figure shown. Observe that side a is opposite angle A, side b is opposite angle B, and side c is the hypotenuse. You are given parts of right triangle ABC. Use this information to compute the six trigonometric ratios for the angle specified.

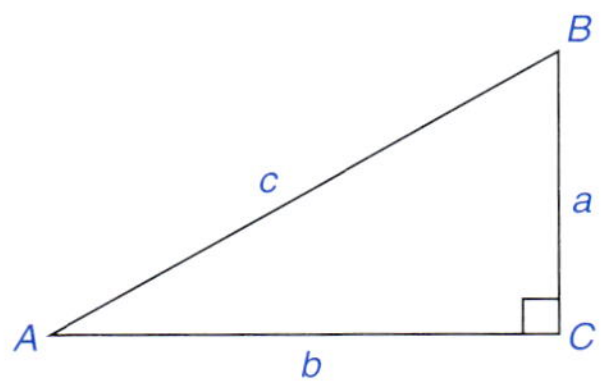

	a	b	c	Find ratios for this angle		a	b	c	Find ratios for this angle
58.	3	4		A	**59.**	5		13	B
60.	1	3		B	**61.**	4	$\sqrt{10}$		B
62.	5	$\sqrt{7}$		B	**63.**		2	$\sqrt{13}$	B
64.	2		$\sqrt{5}$	A	**65.**	4		7	A
66.	12	13		A	**67.**	5	12		B
68.		6	10	B	**69.**	10		15	A
70.	$\sqrt{3}$	4		A	**71.**	1	1		B
72.	1		2	B	**73.**		5	8	A
74.	9	5		B	**75.**	2		8	B
76.	x	y		A	**77.**	x		z	B
78.		y	$2y$	B	**79.**	$\dfrac{z}{3}$		z	A

In the following problems you are given one of the trigonometric ratios for an angle. Use this (a) to sketch a triangle for which the ratio is true, then (b) use this triangle to find the other five trigonometric ratios for that angle.

80. $\sin\alpha = \frac{4}{5}$ **81.** $\cos\alpha = \frac{1}{4}$ **82.** $\cos\alpha = 0.5$ **83.** $\tan\alpha = 4$ **84.** $\sec\alpha = 3$

85. $\cos\alpha = 0.2$ **86.** $\sin\alpha = \dfrac{5}{13}$ **87.** $\csc\alpha = 1.6$ **88.** $\cos\alpha = 0.9$ **89.** $\cos\alpha = \dfrac{\sqrt{3}}{5}$

90. In right triangle ABC, $\sin A = x$ and $x > 0$. Sketch a triangle for which this is true and use it to find $\sin B$ in terms of the variable x. $\left(\textit{Hint: } x = \dfrac{x}{1}.\right)$

91. In right triangle ABC, $\tan A = x$ and $x > 0$. Sketch a triangle for which this is true and use it to find $\sin B$ in terms of the variable x.

92. In right triangle ABC, $\sec A = x$ and $x > 0$. Sketch a triangle for which this is true and use it to find $\sec B$ in terms of the variable x.

93. In right triangle ABC, $\tan B = x$ and $x > 0$. Sketch a triangle for which this is true and use it to find $\tan A$ in terms of the variable x.

Solve the following problems.

94. Find the ground speed, to the nearest knot, of an aircraft flying with a heading due east and an airspeed of 132 knots, if there is a wind blowing from the north at 23 knots.

95. Find the ground speed, to the nearest knot, of an aircraft flying with a heading due west and an airspeed of 105 knots, if there is a wind blowing from the north at 18 knots.

96. Find the ground speed, to the nearest knot, of an aircraft flying with a heading due south and an airspeed of 178 knots, if there is a wind blowing from the east at 25 knots.

97. Find the speed relative to the ground, to the nearest 0.1 knot, of a boat heading straight across a river at 16 knots if the current is moving at 4.3 knots.

98. To the nearest 0.1 inch, find the length of the diagonal of an $8\frac{1}{2}$ inch by 11 inch piece of paper.

99. In the "Mathematics of Warfare" by F. W. Lanchester (from *The World of Mathematics* by James R. Newman) Mr. Lanchester presents the idea that, all other things being equal, the strengths of fighting forces add in a manner proportional to the squares of their numbers. Referring to the diagram, this means that a force of size B is equal to two forces of sizes a and b; that C is equal to the combined strengths of a, b, and c, and so forth. Assuming that forces a, b, c, d, and e are of size 20, 5, 12, 8, and 10, respectively, find the size of force E that is equivalent to the combined strengths of these forces. Find this force to the nearest unit.

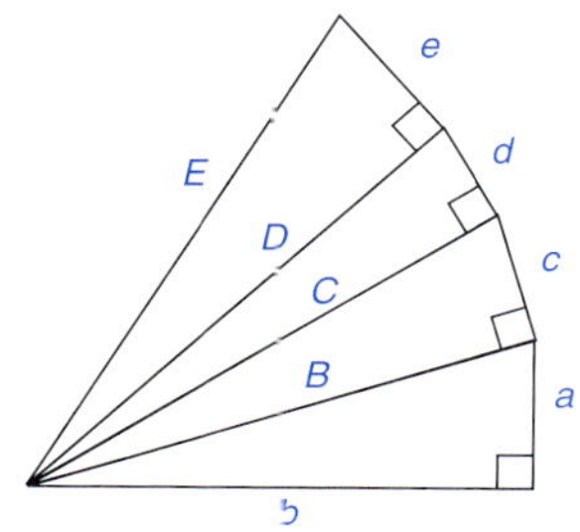

Skill and review

1. Find the equation of the line that passes through the points $(-2,5)$ and $(3,-10)$.
2. Solve the inequality $|2x - 3| < 13$.
3. Graph the parabola $y = x^2 - 3x - 4$.
4. Use the rational zero theorem and synthetic division to find all zeros of the polynomial $2x^4 + 5x^3 - 5x^2 - 5x + 3$.
5. Graph the polynomial function $f(x) = (x + 1)(x - 2)^2(x + 3)$.

5–2 *Angle measure and the values of the trigonometric ratios*

The diagram illustrates a piece of wood that is being mass produced to form the bottom of a planter. Find dimension a in the figure if $b = 8\frac{1}{4}$ inches.

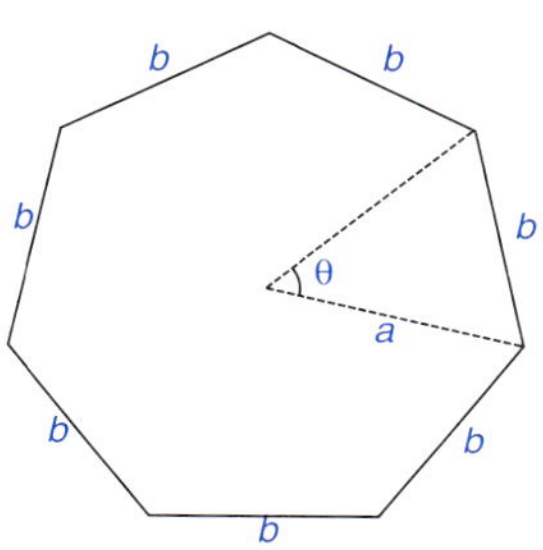

The mathematics necessary to solve this problem involve using the trigonometric ratios. We will see how to solve this problem in this section, after studying some basic principles on which we build our theory.

Trigonometric ratios for equal angles are equal

It is possible to have angles of the same measure in different right triangles. Angles A and A' in figure 5–7 are examples. The trigonometric ratios will have the same value for these angles. This is because these triangles are *similar*—that is, they have the same shape, but perhaps different sizes. It is a theorem of geometry that corresponding ratios in similar figures are equal. Thus, in

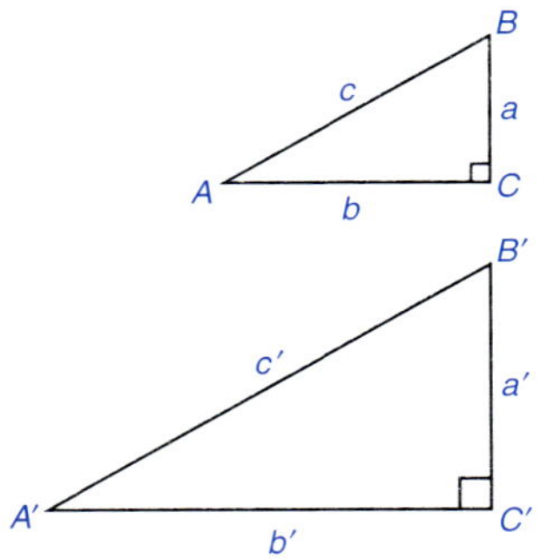

Figure 5–7

figure 5–7, $\frac{a}{c} = \frac{a'}{c'}$, so $\sin A = \sin A'$. For the same reasons, the other five trigonometric ratios are also equal.

Note Read A' as "A-prime," B' as "B-prime," etc.

These facts mean that the values of the trigonometric ratios for an acute angle do not depend on the particular right triangle in which it appears. *For an acute angle with a given measure, the values of the trigonometric ratios will always be the same.*

Values for angles of measure 30°, 45°, and 60°

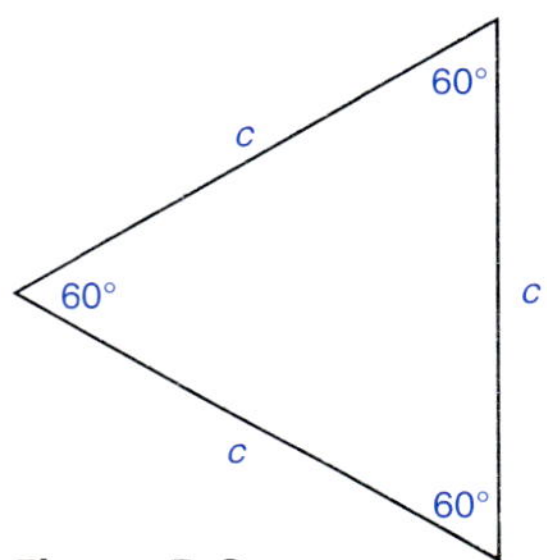

Figure 5–8

We now find the trigonometric ratios for some special angles—30°, 45°, and 60°. Figure 5–8 shows an equilateral triangle, which is a triangle in which all sides have equal length. We label this length c.

We construct the line AC, as shown in figure 5–9, which forms a right triangle with acute angles $A = 30°$ and $B = 60°$. The length a is half of c, so $a = \frac{c}{2}$. We find b next.

$$b^2 = c^2 - a^2 = c^2 - \left(\frac{c}{2}\right)^2, \text{ so } b^2 = \frac{3c^2}{4}, \text{ or } b = \frac{\sqrt{3}}{2}c.$$

Angle A is a 30° angle, so we will write sin 30° to mean the value of the sine ratio associated with an angle of measure 30°. Thus,

$$\sin 30° = \frac{a}{c} = \frac{\frac{c}{2}}{c} = \frac{1}{2}; \cos 30° = \frac{b}{c} = \frac{\frac{\sqrt{3}}{2}c}{c} = \frac{\sqrt{3}}{2}$$

$$\tan 30° = \frac{a}{b} = \frac{\frac{c}{2}}{\frac{\sqrt{3}}{2}c} = \frac{1}{\sqrt{3}} = \frac{\sqrt{3}}{3}$$

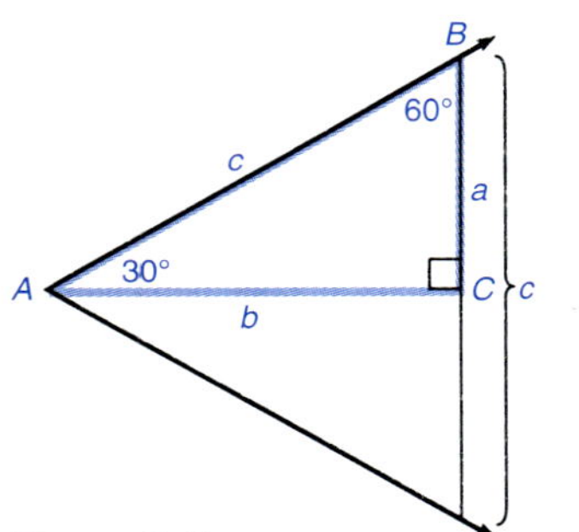

Figure 5–9

The values for a 60° angle can be obtained from angle B in figure 5–9.

In the exercises we will compute the sine, cosine, and tangent ratios for a 45° angle; these are shown in table 5–1, along with the values obtained above.

	Sine	Cosine	Tangent
30°	$\frac{1}{2}$	$\frac{\sqrt{3}}{2}$	$\frac{\sqrt{3}}{3}$
45°	$\frac{\sqrt{2}}{2}$	$\frac{\sqrt{2}}{2}$	1
60°	$\frac{\sqrt{3}}{2}$	$\frac{1}{2}$	$\sqrt{3}$

Table 5–1

General values

It is actually impossible to find the exact values of the trigonometric ratios for most angles. Tables of approximate values were calculated long ago. The earliest known table of trigonometric values, for the equivalent of the sine ratio, was created by Hipparchus of Nicaea about 150 B.C. In the second century A.D., Ptolemy constructed a table of values of the sine ratio for acute angles in increments of one-quarter degree. Today we use calculators to approximate these values. When using a calculator it is important that the calculator be in **degree mode.** This means that the calculator is expecting the measure of the angle in decimal degrees. *Check your calculator's manual to make sure it is in degree mode.* This is usually done with a key marked [DRG] or simply [DEG]. "DRG" means degrees, radians, grads. We discuss radian measure in a later section. Grads, or grades, is the metric measure for an angle. There are 100 grads in a right angle. We will not use this measure in this text.

To select degree mode on the Texas Instruments TI-81 it is necessary to select "Deg" under the [MODE] feature. TO do this, select [MODE], darken in the "Deg" mode indicator (use the four cursor moving "arrow" keys) and select [ENTER].

Example 5–2 A shows typical calculator keystrokes to calculate each value.

■ ***Example 5–2 A***

Find each value rounded to four decimal places.

1. sin 34.51° 34.51 [sin] Display [0.5665500655]

sin 34.51° ≈ 0.5666 TI-81: [SIN] 34.51 [ENTER]

2. sec 33.5°

Since there is no secant key on a calculator we use the fact that sec 33.5° $= \dfrac{1}{\cos 33.5°}$. Compute cos 33.5° and divide it into one; the [1/x] key is designed for this type of situation.

33.5 [cos] [1/x] Display [1.199204943]

TI-81: [(] [COS] 33.5 [)] [x^{-1}]

Note that on the TI-81 the [1/x] key is the [x^{-1}] key.
sec 33.5° ≈ 1.1992

3. cos 13°43′

Recall that angles in the DMS system must be converted to decimal degrees. We show the calculation with and without special calculator keys. (See also example 5–1 A.)

No special keys: 13 [+] 43 [÷] 60 [=] [cos]

Display [0.9714801855]

Calculator A: 13 [° ′ ″] 43 [° ′ ″] [cos]

Calculator B: 13.43 [→H] [cos]

TI-81: [COS] [(] 13 [+] 43 [÷] 60 [)] [ENTER]

cos 13°43′ ≈ 0.9715 ■

Finding an angle from a known trigonometric ratio

It is important to be able to reverse the operations discussed above. For example, if θ is an acute angle and sin θ = $\frac{1}{2}$, what is θ? We can see from table 5–1 that θ must be 30°. The calculator is programmed to solve this problem.

This is done with the *inverse trigonometric ratios* called the inverse sine ($\sin^{-1}$), inverse cosine ($\cos^{-1}$), and inverse tangent ($\tan^{-1}$) ratios. The superscript -1 does *not* indicate a reciprocal value in the way that, say, $2^{-1} = \frac{1}{2}$. We will study these ratios in more detail later. For now we illustrate how to find the acute angle whose sine, cosine, or tangent value is known.

$\sin^{-1}$	$\cos^{-1}$	$\tan^{-1}$
[sin]	[cos]	[tan]

For this, most calculators use the appropriate key (sin, cos, tan), prefixed by another key such as [SHIFT], [2nd], [INV], or [ARC]. The appropriate function is generally shown above the key itself. The *result is always an angle in decimal degrees* (when the calculator is in degree mode). We will show the necessary two keystrokes as one.

■ ***Example 5–2 B***

Find θ in the following problems using the calculator. Assume θ is an acute angle. Round the answer to the nearest 0.01°.

1. $\cos\theta = 0.4602$

We need to calculate $\cos^{-1}0.4602$.

0.4602 [$\cos^{-1}$] Display [62.59998611]

TI-81: [COS^{-1}] .4602 [ENTER]

$\theta \approx 62.60°$

2. $\csc\theta = 1.0551$

We use the fact that if $\csc\theta = 1.0551$, then $\sin\theta = \frac{1}{1.0551}$. Thus we compute $\sin^{-1}(\frac{1}{1.0551})$. Use the [$1/x$] key.

1.0551 [$1/x$] [$\sin^{-1}$] Display [71.40162609]

TI-81: [SIN^{-1}] 1.0551 [x^{-1}] [ENTER]

$\theta \approx 71.40°$ ■

Solving right triangles

One application of trigonometry that occurs in many situations is *solving right triangles*—this means *discovering the lengths of all sides and the measures of all angles* of the triangle. We will round the values we compute to the same number of decimal places as the given data.

These types of situations fall into two categories, ones in which we know one side and one acute angle and others in which we know two sides and no angles. Each category is illustrated in example 5–2 C.

Note In general we assume side *a* is opposite angle *A*, side *b* is opposite angle *B*, and the hypotenuse *c* is opposite right angle *C*.

■ ***Example 5–2 C***

Solve the following right triangles.

1. $A = 35.6°$, $a = 13.6$ (one side, one angle)

To solve this triangle we need to find the lengths of sides b and c and the measure of angle B. Since angle C is always 90°, angles A and B total 90°. Thus, angle B is $90° - 35.6° = 54.4°$.

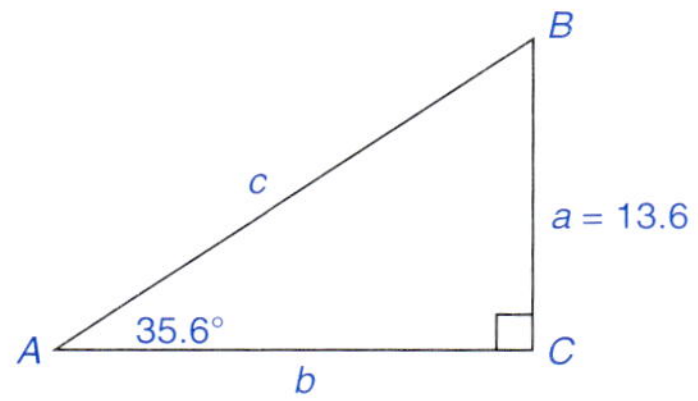

We now note that $\sin A = \dfrac{a}{c}$, so that

$$\sin 35.6^\circ = \frac{13.6}{c}$$

$$c \sin 35.6^\circ = 13.6$$ Multiply each member by c

$$c = \frac{13.6}{\sin 35.6^\circ}$$ Divide each member by sin 35.6°

$$c \approx 23.4$$ 13.6 [÷] 35.6 [sin] [=] Display [23.3627613]

TI-81: 13.6 [÷] [SIN] 35.6 [ENTER]

Now we find b by noting that $\tan A = \dfrac{a}{b}$.

$$\tan 35.6^\circ = \frac{13.6}{b}$$

$$b \tan 35.6^\circ = 13.6$$

$$b = \frac{13.6}{\tan 35.6^\circ}$$

$$b \approx 19.0$$

Since we know the lengths of all sides and all angles we have solved the triangle. To summarize, $a = 13.6$, $b \approx 19.0$, $c \approx 23.4$, $A = 35.6^\circ$, $B = 54.4^\circ$, $C = 90^\circ$.

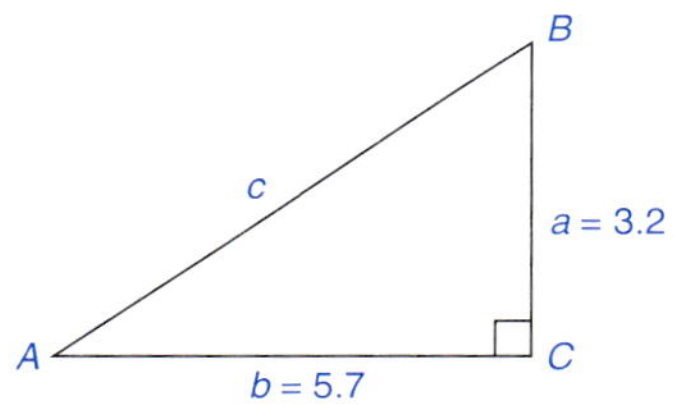

2. $a = 3.2$, $b = 5.7$ (two sides)
We can find the length of side c by the Pythagorean theorem: $c \approx 6.5$.

We can find angle A by noting that $\tan A = \dfrac{a}{b}$.

$$\tan A = \tfrac{3.2}{5.7}$$

$$A = \tan^{-1}(\tfrac{3.2}{5.7})$$ 3.2 [÷] 5.7 [=] [$\tan^{-1}$] Display [29.31000707]

TI-81: [TAN^{-1}] [(] 3.2 [÷] 5.7 [)] [ENTER]

$$A \approx 29.3^\circ$$

$B = 90^\circ - 29.3^\circ = 60.7^\circ$. Thus, $a = 3.2$, $b = 5.7$, $c \approx 6.5$, $A \approx 29.3^\circ$, $B \approx 60.7^\circ$, $C = 90^\circ$.

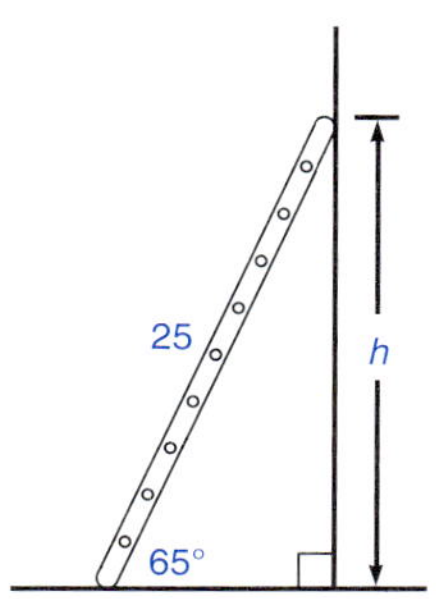

3. A tag on a 25-foot ladder states that, for safety reasons, the angle that the ladder makes with the ground should not exceed 65°. How high can the ladder reach without exceeding this angle, to the nearest 0.1 feet?

We need to find h in the figure. If we observe that h is opposite the known angle and that the length of the hypotenuse of the triangle is known, we see that we can use the sine ratio.

$$\sin 65^\circ = \frac{h}{25}$$

$$25 \sin 65^\circ = h$$ Multiply each member by 25

$$22.7 \approx h$$

The ladder can reach a height of approximately 22.7 feet without exceeding a 65° angle with the ground. ■

Mastery points

Can you

- Compute the value of the trigonometric ratios for a given acute angle, using a calculator?
- Compute the value of an acute angle, given the value of a trigonometric ratio, using a calculator?
- Solve a right triangle when given one side and one acute angle?
- Solve a right triangle when given two sides?

Exercise 5–2

Use a calculator to find four-decimal-place approximations for the following.

1. sin 31.28°	**2.** cos 85.23°	**3.** tan 11.95°	**4.** sec 40.08°
5. cot 28.87°	**6.** csc 5.15°	**7.** sin 40.28°	**8.** tan 76.23°
9. sec 66.47°	**10.** sin 35.56°	**11.** sin 78.33°	**12.** cos 17.45°
13. sin 35°56′	**14.** sin 78°33′	**15.** cos 17°45′	**16.** cos 85°28′
17. tan 40°41′	**18.** tan 35°8′	**19.** cos 23°24′	**20.** cos 56°24′
21. cot 13°3′	**22.** sin 48°8′	**23.** tan 33°38′	**24.** sec 86°22′

25. A surveyor needs to compute R in the following formula as part of finding the area of the segment of a circle: $R = \dfrac{LC}{2 \sin I}$. Find R to one decimal place if $LC = 425.0$ feet and $I = 13.2°$.

26. Compute R using the formula of the previous problem if $LC = 611.1$ meters and $I = 18°20'$. Round the answer to two decimal places.

27. In the mathematical modeling of an aerodynamics problem the following equation arises:

$$y = x \cos A \cos B - x^2 \cos A \sin B - x^3 \sin A.$$

Compute y to two decimal places if $x = 2.5$, $A = 31°$, and $B = 17°$.

28. Compute y to two decimal places using the formula of problem 27 if $x = 1.2$, $A = 10°$, and $B = 15°$.

29. The average power in an AC circuit is given by the formula $P = VI \cos \theta$. Compute P (in watts) if $V = 120$ volts, $I = 2.3$ amperes, and $\theta = 45°$, to the nearest 0.1 watt.

30. Compute P using the formula of the previous problem if $V = 42$ volts, $I = 25$ amperes, and $\theta = 45°$, to the nearest 0.1 watt.

31. A formula that relates the distance across the flats of a piece of hexagonal stock in relation to the distance across the corners is $f = 2r \cos \theta$. A machinist needs to compute f for a piece of stock in which $r = 28$ millimeters and $\theta = 30°$. Compute f to the nearest 0.1 mm.

32. Using the formula of problem 31 find r if $f = 21.4$ inches and $\theta = 25°$.

33. Using the formula of problem 31 find θ to the nearest 0.1 if $f = 36.8$ millimeters and $r = 24.0$.

34. Find the exact values for the sine, cosine, and tangent ratios for an angle of measure 45° by proceeding in the following manner. Draw an isosceles right triangle—a right triangle in which the two legs have the same length. Label this length one. Observe that the two acute angles must be 45°. Now find the length of the hypotenuse and use the definitions of the trigonometric ratios to find the desired values.

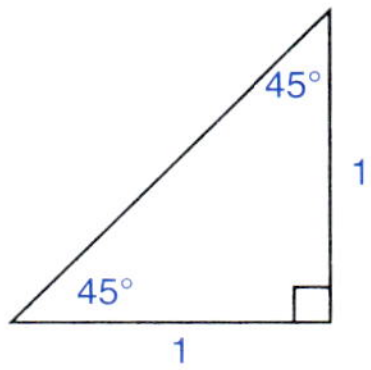

Find the unknown acute angle θ to the nearest 0.01°.

35. $\sin\theta = 0.8007$ **36.** $\cos\theta = 0.1028$ **37.** $\tan\theta = 1.8807$ **38.** $\sin\theta = 0.9484$

39. $\cos\theta = 0.8515$ **40.** $\tan\theta = 1.0014$ **41.** $\sin\theta = \dfrac{35.9}{68.3}$ **42.** $\cos\theta = \dfrac{8.25}{12.5}$

43. $\tan\theta = 2$ **44.** $\csc\theta = 1.1243$ **45.** $\sec\theta = 4.8097$ **46.** $\cot\theta = 2.5$

47. $\sec\theta = \dfrac{6.45}{2.35}$ **48.** $\csc\theta = \sqrt{10.8}$

In the following problems you are given one side and one angle of a right triangle. Solve the triangle. Round all answers to the same number of decimal places as the data.

49. $a = 15.2$, $B = 38.3°$ **50.** $a = 12.6$, $B = 17.9°$ **51.** $a = 11.1$, $A = 13.7°$ **52.** $a = 5.25$, $A = 70.3°$

53. $b = 0.672$, $A = 29.4°$ **54.** $b = 15.2$, $A = 81.3°$ **55.** $b = 21.8$, $B = 78.0°$ **56.** $b = 2.14$, $B = 50.4°$

57. $c = 10.0$, $A = 15.0°$ **58.** $c = 3.45$, $A = 46.2°$ **59.** $c = 122$, $B = 65.5°$ **60.** $c = 31.5$, $B = 62.0°$

In the following problems you are given two sides of a right triangle. Solve the triangle. Round all lengths to the same number of decimal places as the data and all angles to the nearest 0.1°.

61. $a = 13.1$, $b = 15.6$ **62.** $a = 5.67$, $b = 8.91$ **63.** $a = 0.22$, $b = 1.34$ **64.** $a = 2.82$, $b = 1.09$

65. $a = 17.8$, $c = 25.2$ **66.** $a = 311$, $c = 561$ **67.** $b = 51.3$, $c = 111.0$ **68.** $b = 4.55$, $c = 5.66$

69. $a = 12.0$, $c = 13.0$ **70.** $a = 33.1$, $c = 41.0$ **71.** $b = 84.0$, $c = 90.1$ **72.** $b = 0.651$, $c = 0.927$

73. The figure illustrates an impedance diagram used in electronics theory. If Z (impedance) = 10.35 ohms and X_L (inductive reactance) = 4.24 ohms, find θ (phase angle) to the nearest degree and R (resistance) to the nearest 0.01 ohm.

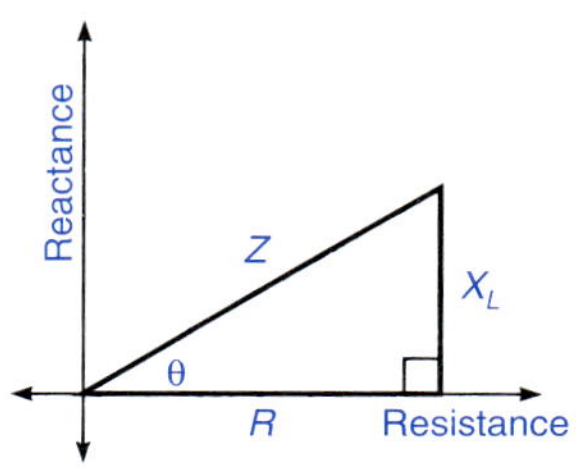

74. Use the impedance diagram of problem 73 to find Z if $\theta = 24.2°$ and $X_L = 22.6$ ohms.

75. The diagram illustrates the measurements a surveyor made to find the width w of a pond; compute the width to the nearest foot.

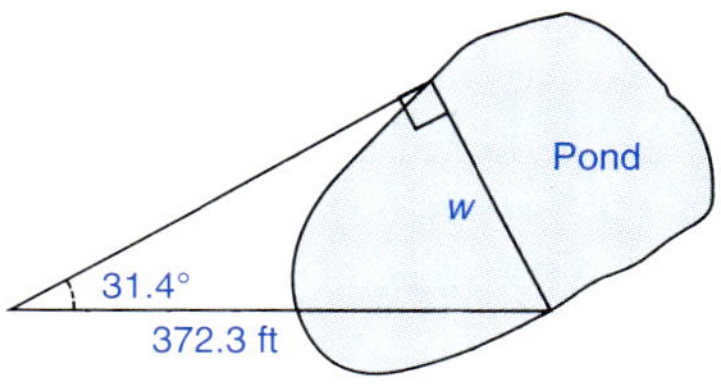

76. The diagram illustrates the tip of a threading tool; find angle θ to the nearest degree.

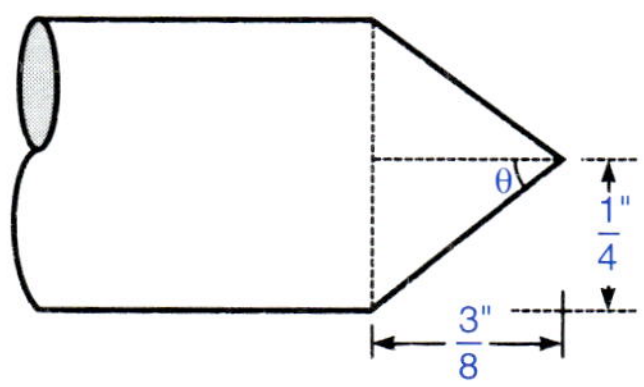

77. The diagram illustrates a piece of wood that is being mass produced to form the bottom of a planter. Find dimension a in the figure to the nearest 0.01 inch if $b = 8\frac{1}{4}$ inches. Note that angle θ is $\frac{360°}{7}$.

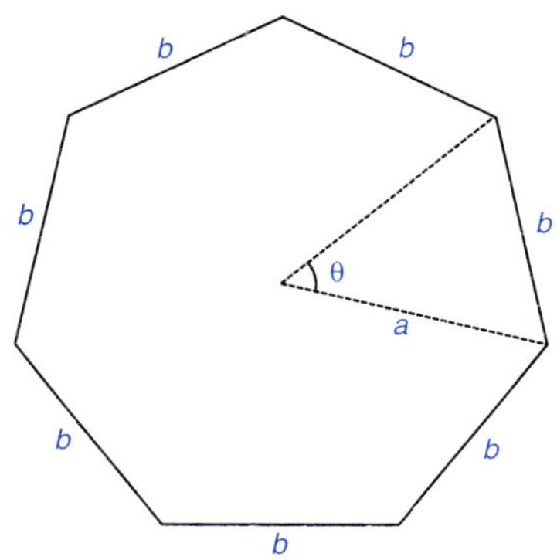

78. A formula found in electronics is $E = \dfrac{P}{I \cos \theta}$, where E is voltage, P is power, I is current, and θ is phase angle. Find E (in volts) if $P = 45.0$ watts, $I = 2.5$ amperes, and $\theta = 15°$. Round the answer to the nearest 0.1 volt.

79. The diagram illustrates the wind triangle problem in air navigation. A plane has an airspeed of 155 mph and heading of due north. It is flying in a wind from the west with a speed of 30 mph. Find the ground speed S and the ground direction θ, each to the nearest unit.

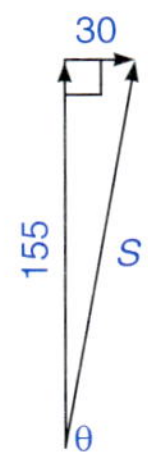

80. An *angle of elevation* is an angle formed by one horizontal ray and another ray that is above the horizontal. Angle θ in the diagram is the angle of elevation to an aircraft that radar shows has a *slant distance* of 12.4 miles from the radar site. If θ is 30.1°, find the elevation h of the aircraft, to the nearest 100 feet. Remember that 1 mile = 5,280 feet.

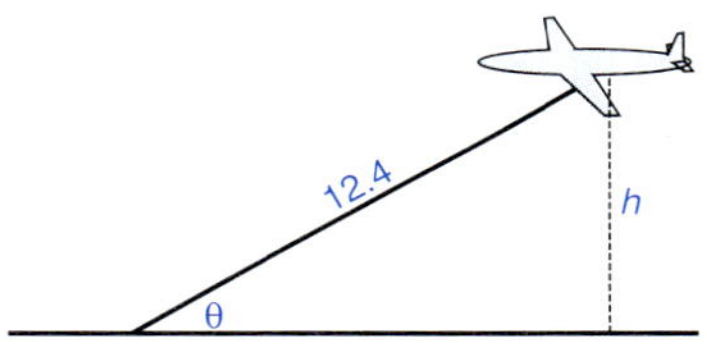

81. If it is known that an aircraft is flying at 28,500 feet and the angle of elevation of a radar beam tracking the aircraft is 8.2°, what is the slant distance d from the radar to the aircraft, to the nearest 100 feet? See problem 80.

82. The diagram illustrates the path of a laser beam on an optics table. Compute the total distance traveled by the beam to the nearest millimeter.

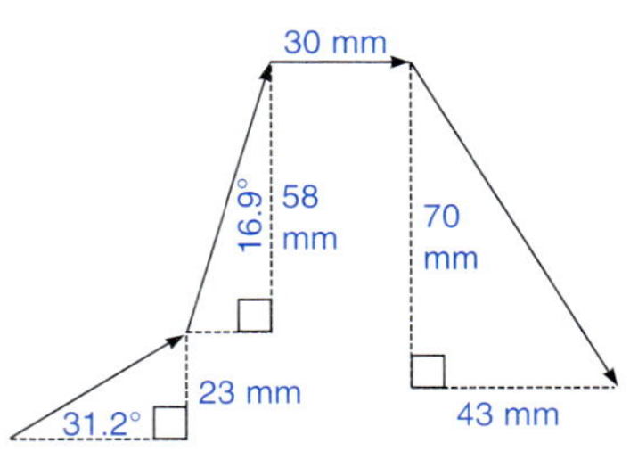

83. An *angle of depression* is an angle formed by one horizontal ray and another ray that is below the horizontal. Angle θ in the figure is the angle of depression formed by the line of sight of an observer in an airport control tower looking at a helicopter on the ground. If θ is 17.2° and the tower is 257 feet high, how far is the aircraft from the base of the tower, to the nearest foot?

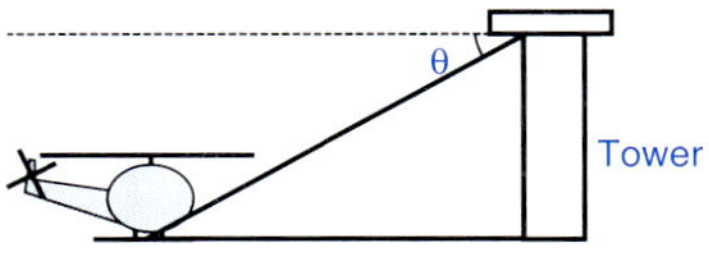

84. If an aircraft is 1.23 miles from the foot of the tower in problem 83, what is θ, to the nearest 0.1°? (Remember, 1 mile = 5,280 ft.)

85. The diagram is a top view of a portion of a spiral staircase that an architect has designed. If $\theta_1 = \theta_2 = \theta_3$, find the length x, to one decimal place. (Caution: Carry out your calculations to as many digits as practical to avoid an accumulation of errors.)

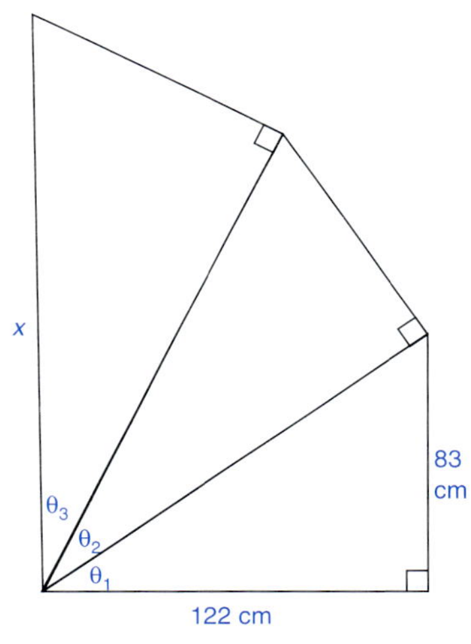

86. If the architect of problem 85 revises the plans so that $\theta_2 = \theta_1 + 5°$ and $\theta_3 = \theta_2 + 5°$, find x to one decimal place.

87. In right triangle ABC, $A = 45°$ and $b = 4$. Solve this triangle using *exact* values only. (Use the exact values for sin 45° etc. and do not approximate radicals as decimals.)

88. In right triangle ABC, $B = 60°$ and $b = 8$. Solve this triangle using *exact* values only. (Use the exact values for sin 60° etc. and do not approximate radicals as decimals.)

Skill and review

1. Graph the rational function $f(x) = \dfrac{2}{x^2 - 9}$.
2. Find the slope of the straight line $3x - 2y = 5$.
3. Solve the nonlinear inequality $x^2 - 2x > 3$.
4. In right triangle ABC, $a = 9.0$ and $b = 16.8$. Find c.

5–3 The trigonometric functions—definitions

In an electronic circuit with an inductive component to the impedance, the current follows the voltage. The difference is often measured in degrees. For example, the current may follow the voltage by 15°. In this case, we could say the phase angle of the current is −15°, relative to the voltage. We could just as easily say that the phase angle of the voltage is 345°, relative to the current. Find the phase angle of the voltage relative to the current if the phase angle of the current relative to the voltage is −88°.

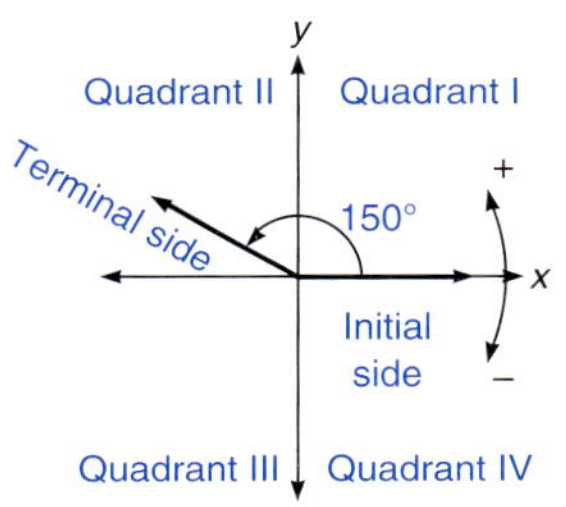

Figure 5–10

The relationship between current and voltage is one phenomenon that can be described with some of the terminology we study in this section.

Angles in standard position

As in the example above, there are many situations where we have to think of angles as being nonacute. This is often a situation in which we wish to describe an amount of rotation. For example, a ship may turn through an angle of 215°, a computed tomography (CT) scanner used in medical diagnosis may move through an angle of 360°, or a surveyor may find the measure of the angle at one corner of a piece of land to be 165°20′. For these situations we often place the angle in a rectangular (x-y) coordinate system.

Angle in standard position

An angle in standard position is formed by two rays, with the vertex at the origin. One ray always lies on the nonnegative portion of the x-axis. This ray is called the **initial side** of the angle. The second ray is called the **terminal side** of the angle. It may be in any quadrant or along any axis.

Figure 5–10 shows an angle in standard position, with measure 150°. We generally use the word ''angle'' instead of the phrase ''angle in standard position.''

In some situations it is convenient to let the measure of an angle be negative. If the measure of the angle is positive, we picture the terminal side as having moved away from the initial side in a counterclockwise direction (↶); if the measure of the angle is negative we picture the terminal side as having moved away from the initial side in a clockwise direction (↷). If an angle's measure is greater than 360° or less than −360° we consider the angle to have ''gone around'' more than once. Several examples of angles in standard position are shown in figure 5–11. In part c of the figure we show the angle as a 360° revolution, followed by an additional 200° turn.

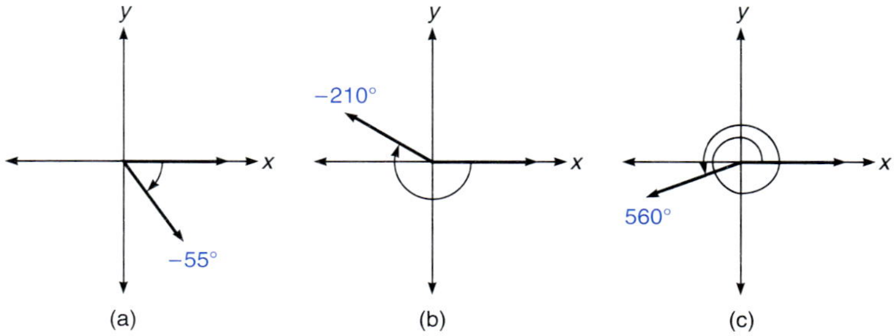

Figure 5–11

Angles that have the same terminal side are said to be **coterminal.** (All angles in standard position have the same initial side.) The 150° angle in figure 5–10 and the −210° angle in figure 5–11(b) are coterminal. We can see this when we realize that in each case the angle formed by the negative side of the x-axis and the terminal side of each angle is 30°. Since ±360° represents one complete revolution, coterminal angles are angles whose degree measures differ by an integer multiple of 360°. This forms the basis for our definition.

> **Coterminal angles**
>
> Two angles α and β are said to be coterminal if
>
> $$\alpha = \beta + n(360°), \quad n \text{ an integer}$$
>
> ***Concept***
>
> Two angles are coterminal if the measure of one can be formed from the measure of the other by adding or subtracting multiples of 360°.

■ ***Example 5–3 A***

In each case find a coterminal angle with measure x such that $0° \le x < 360°$.

1. 875°

$875° - 360° = 515°$ Subtract 360° until x is found

$515° - 360° = 155°$ The required angle is 155°

We could have done this more elegantly by computing $875° - 2(360°)$.

2. −1,000°

$-1{,}000° + 360° = -640°$ Add 360° until x is found

$-640° + 360° = -280°$

$-280° + 360° = 80°$

Or compute as $-1{,}000° + 3(360°) = 80°$. ■

The trigonometric functions

We now define the six trigonometric functions. They have the same names as the six trigonometric ratios, and the same abbreviations.

The trigonometric ratios are actually functions with domain the set of *acute* angles. The trigonometric functions have the set of *all* angles as their domain. For acute angles the trigonometric functions are the same as the trigonometric ratios. The following definition refers to figure 5–12.

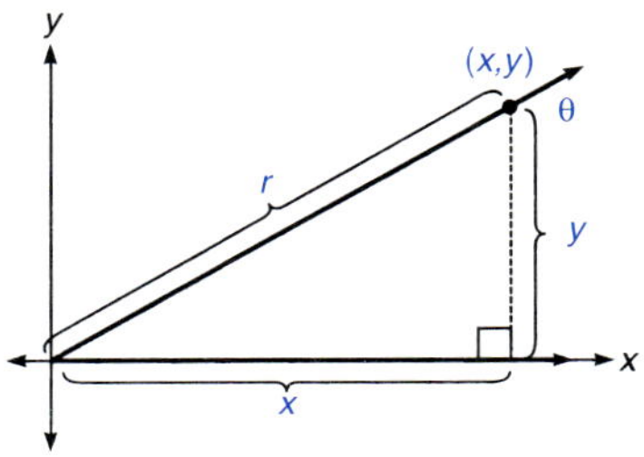

Figure 5–12

The trigonometric functions

Let θ be an angle in standard position, and let (x,y) be any point on the terminal side of the angle, except (0,0). Let $r = \sqrt{x^2 + y^2}$ be the distance from the origin to the point. Then,

$$\sin\theta = \frac{y}{r}, \quad \cos\theta = \frac{x}{r}, \quad \tan\theta = \frac{y}{x}$$

$$\csc\theta = \frac{r}{y}, \quad \sec\theta = \frac{r}{x}, \quad \cot\theta = \frac{x}{y}$$

Note
1. We define r so that $r > 0$.
2. If x or y in the point (x,y) is zero, then those ratios with x or y in the denominator are not defined.
3. Unlike the trigonometric ratios, the trigonometric functions can take on negative values.

It can be proven that for a given angle, *it does not matter what point on the terminal side is chosen; the values of the trigonometric functions will be the same.* This is for the same reasons that the trigonometric ratios do not depend on the size of the right triangle.

It can also be seen that *coterminal angles have the same values for the trigonometric functions.* This is because two coterminal angles have the same terminal side, and the definitions depend solely upon a point on the terminal side.

The definitions of the trigonometric functions imply the following identities for all values of θ for which any denominator is nonzero. These identities look identical to those for the trigonometric ratios. They can be applied to either trigonometric ratios or functions.

Reciprocal function identities

$$\csc\theta = \frac{1}{\sin\theta}, \quad \sin\theta = \frac{1}{\csc\theta}$$

$$\sec\theta = \frac{1}{\cos\theta}, \quad \cos\theta = \frac{1}{\sec\theta}$$

$$\cot\theta = \frac{1}{\tan\theta}, \quad \tan\theta = \frac{1}{\cot\theta}$$

To see that the first reciprocal function identity is true for the trigonometric functions, observe that $\csc\theta = \frac{r}{y} = \frac{1}{\frac{y}{r}} = \frac{1}{\sin\theta}$. It is left as an exercise to show that the rest of these are true. Using the reciprocal function identities, we can usually find the values of the cosecant, secant, and cotangent functions by finding the reciprocal of the sine, cosine, and tangent functions. This will not work when a function has value zero, since the reciprocal of zero is not defined.

Two other identities that can be useful are the following; again, they are true only for those values of θ for which no denominator is 0.

> **Tangent/cotangent identities**
>
> $$\tan\theta = \frac{\sin\theta}{\cos\theta}, \quad \cot\theta = \frac{\cos\theta}{\sin\theta}$$

■ ***Example 5–3 B***

Show that $\tan\theta = \dfrac{\sin\theta}{\cos\theta}$ is an identity for the trigonometric functions.

We show that each side of the equation is equivalent to the same thing.

$\tan\theta$	$\dfrac{\sin\theta}{\cos\theta}$	
$\dfrac{y}{x}$	$\dfrac{\frac{y}{r}}{\frac{x}{r}}$	$\tan\theta = \frac{y}{x}$, $\sin\theta = \frac{y}{r}$, $\cos\theta = \frac{x}{r}$
	$\dfrac{y}{r}\cdot\dfrac{r}{x}$	Algebra of division
	$\dfrac{y}{x}$	Reduce $\frac{ry}{rx}$

Thus $\tan\theta = \dfrac{y}{x}$ and $\dfrac{\sin\theta}{\cos\theta} = \dfrac{y}{x}$, so $\tan\theta = \dfrac{\sin\theta}{\cos\theta}$. ■

Example 5–3 C illustrates finding values of the trigonometric functions for an angle in standard position, given a point on the terminal side of the angle.

■ ***Example 5–3 C***

In each problem a point on the terminal side of an angle θ is given. Use it to find the trigonometric functions for that angle. Also, make a sketch of the angle.

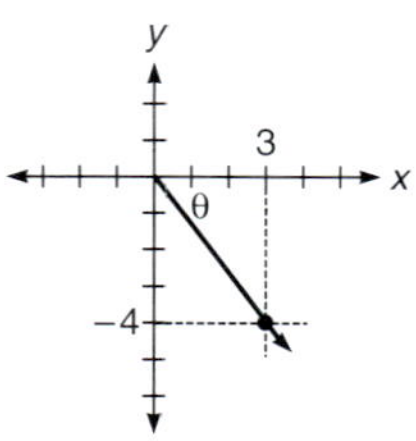

1. (3,−4)

$$\begin{aligned} r &= \sqrt{x^2 + y^2} && \text{Definition} \\ &= \sqrt{3^2 + (-4)^2} && \text{Replace } x, y \\ &= 5 \end{aligned}$$

$$\sin\theta = \frac{y}{r} = -\frac{4}{5}, \ \csc\theta = \frac{1}{\sin\theta} = -\frac{5}{4}$$

$$\cos\theta = \frac{x}{r} = \frac{3}{5}, \ \sec\theta = \frac{1}{\sin\theta} = \frac{5}{3}$$

$$\tan\theta = \frac{y}{x} = -\frac{4}{3}, \ \cot\theta = \frac{1}{\tan\theta} = -\frac{3}{4}$$

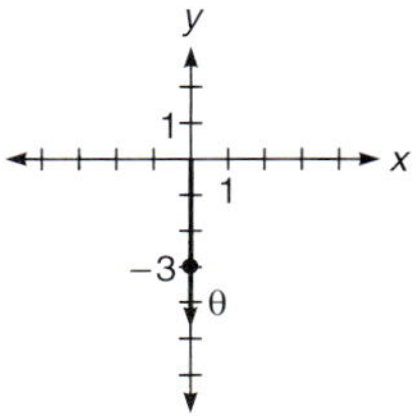

2. $(0, -3)$

$$r = \sqrt{0^2 + (-3)^2} = 3$$

$$\sin\theta = \frac{y}{r} = \frac{-3}{3} = -1,\ \csc\theta = \frac{1}{\sin\theta} = -1$$

$$\cos\theta = \frac{x}{r} = \frac{0}{3} = 0,\ \sec\theta = \frac{1}{\cos\theta} = \frac{1}{0};\ \text{undefined}$$

$$\tan\theta = \frac{y}{x} = \frac{-3}{0},\ \text{undefined};\ \cot\theta = \frac{\cos\theta}{\sin\theta} = \frac{0}{-1} = 0$$

Mastery points

Can you

- When given an angle θ, find a positive coterminal angle with measure x such that $0° \leq x < 360°$?
- Sketch an angle and find the values of the six trigonometric functions when given a point on the terminal side of the angle?

Exercise 5–3

In problems 1–17,

a. draw the initial and terminal side of the given angle.
b. state the measure of the smallest nonnegative angle that is coterminal with the given angle.

1. 420° **2.** −40° **3.** 230° **4.** 1,000° **5.** 800.6° **6.** 1,260°
7. 547.9° **8.** 2,000° **9.** −870° **10.** 625° **11.** 525° **12.** −610°
13. −530.3° **14.** 390° **15.** −720° **16.** −313° **17.** −11.9°

18. An automobile engine is timed to fire the spark plug for cylinder 1 at 8° BTDC (before top dead center), which, for our purposes, is −8°. Assuming this engine rotates in a counterclockwise direction, what is the equivalent amount ATDC (after TDC) (i.e., the least nonnegative angle coterminal with it)?

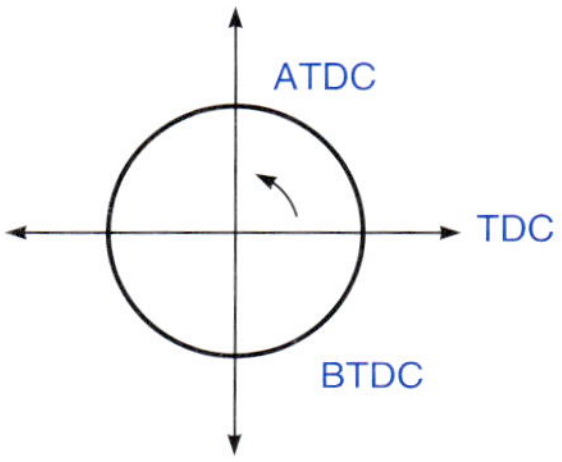

19. If an automobile engine is timed to fire at 13° BTDC, what is the equivalent amount ATDC?

20. If an automobile engine is timed to fire at 8.6° BTDC, what is the equivalent amount ATDC?

21. If an automobile engine is timed to fire at 6.1° BTDC, what is the equivalent amount ATDC?

22. In an electronic circuit with an inductive component to the impedance, the current follows the voltage. For example, the current may follow the voltage by 15°, in which case we could say the phase angle of the current is −15°, relative to the voltage. We could just as easily say that the phase angle of the voltage is 345°, relative to the current. Find the phase angle of the voltage relative to the current if the phase angle of the current relative to the voltage is (a) −88°, (b) −24.33°, (c) −35°56′, (d) −16.56°, (e) −0°14′, and (f) −0.14°. (Find the least nonnegative coterminal angle in each case.)

In the following problems you are given a point that lies on the terminal side of an angle in standard position. In each case, compute the value of all six trigonometric functions for the angle.

23. $(3,6)$ **24.** $(-2,5)$ **25.** $(-5,8)$ **26.** $(-7,-8)$ **27.** $(2,-2)$
28. $(3,0)$ **29.** $(-1,4)$ **30.** $(0,-4)$ **31.** $(-10,-15)$ **32.** $(3,\sqrt{5})$
33. $(-\sqrt{2},6)$ **34.** $(3,-\sqrt{6})$ **35.** $(-\sqrt{3},-\sqrt{2})$ **36.** $(1,-\sqrt{3})$ **37.** $(\sqrt{6},-\sqrt{10})$

Show that each identity is true.

38. $\sec\theta = \dfrac{1}{\cos\theta}$ **39.** $\cot\theta = \dfrac{1}{\tan\theta}$ **40.** $\cot\theta = \dfrac{\cos\theta}{\sin\theta}$ **41.** $\cos\theta = \dfrac{1}{\sec\theta}$ **42.** $\sin\theta = \dfrac{1}{\csc\theta}$

To solve the following two problems we must recall that the equation of a nonvertical straight line can be put in the form $y = mx + b$, where m is the slope and b is the y-intercept. If a straight line passes through the origin then $b = 0$ and the equation becomes $y = mx$.

43. Show that if two different points lie on the terminal side of an angle in standard position, then using either point gives the same value for the sine function. For the sake of simplicity assume the terminal side is not vertical or horizontal. Represent the points as (x_1,y_1) and (x_2,y_2). Note that these points lie on the same line. The equation of any line that passes through the origin is of the form $y = mx$, so we know that for the same value of m, $y_1 = mx_1$ and $y_2 = mx_2$. This means that the points (x_1,y_1) and (x_2,y_2) can be rewritten as (x_1,mx_1) and (x_2,mx_2). Use these versions of the points to compute the length r for each point. Then show that the value of the sine function is the same when computed using either point.

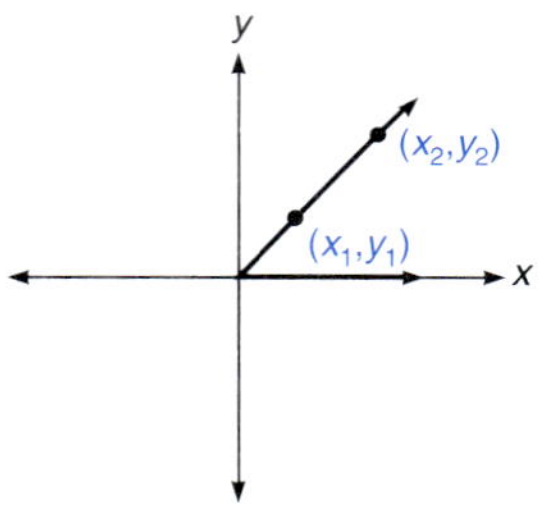

44. Show that if the trigonometric function values are the same for two points, then these points lie on the terminal side of the same angle. Assume for simplicity that these points are not located on either the x-axis or the y-axis. To show that the points lie on the terminal side of the same angle we must show that both points lie on the same line and are in the same quadrant. Let (x_1,y_1) and (x_2,y_2) represent the two points, and consider the value of the tangent function as given by each point. This can be used to show that $y_1 = mx_1$ and $y_2 = mx_2$ (for the same value of m). This means that the two points lie on the same line. Now explain why they must be in the same quadrant.

Skill and review

1. In right triangle ABC, $a = 9.0$, $b = 16.8$. Solve the triangle.
2. Use the graph of $y = \sqrt{x}$ to graph the function $f(x) = \sqrt{x-4} - 2$.
3. Factor $8x^3 - 27$.
4. Rationalize the denominator of $\dfrac{3}{\sqrt{6}}$.
5. Compute $\dfrac{5 - 3^2}{8} - 2$.
6. The circumference C of a circle is the distance around the circle. The radius r is the distance from the center to the edge of the circle. The relation $C = 2\pi r$ has been known for several thousand years. (π is a real number, and $\pi \approx 3.14159$.) Find the circumference of a circle with radius 15 inches, to the nearest tenth of an inch.

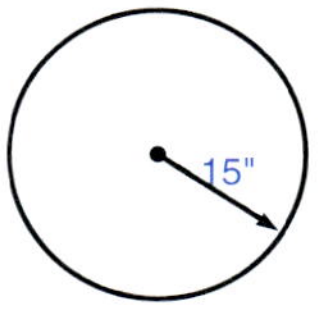

5–4 Values for any angle—the reference angle/ASTC procedure

If a force of 200 pounds is applied to a rope to drag an object, the actual force tending to move the object horizontally is $f(\theta) = 200 \cos \theta$, where θ is the angle the rope makes with the horizontal. Compute the force tending to move the object horizontally if the angle of the rope is 25°.

As in this problem, many physical phenomena can be described using the trigonometric functions. In this section we study more about these functions.

The values of the trigonometric functions for an angle of any measure are related to the values for the acute angles of the first quadrant. These values (for the first quadrant) are the same as those for the trigonometric ratios for acute angles. The values of the trigonometric functions for any angle have a sign and a ''size'' (absolute value). We first discuss the sign of the basic trigonometric functions, then the size.

The ASTC rule—the signs of the trigonometric functions by quadrant

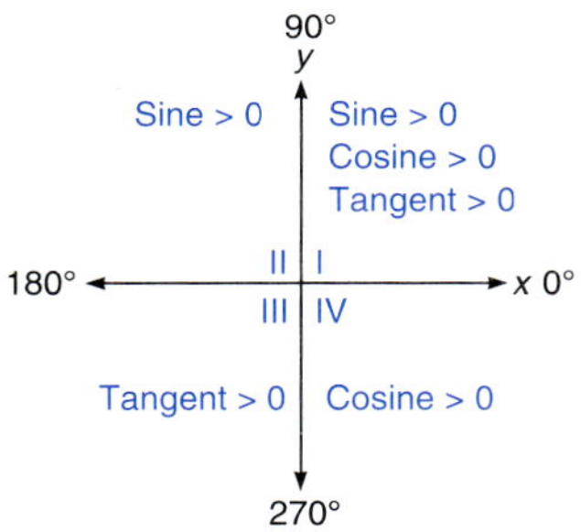

Figure 5–13

The **sign** of the value of a trigonometric function for an angle *depends on the quadrant in which the angle terminates.* Figure 5–13 shows the quadrants in which the sine, cosine, and tangent functions are positive. (They are negative in the other quadrants.) The figure shows that the sine function is positive in quadrants I and II, and therefore negative in quadrants III and IV. This is because the sine function is defined by the ratio $\frac{y}{r}$; since r is always positive this ratio is positive where y is positive, in quadrants I and II. Since the cosine function is $\frac{x}{r}$ and $r > 0$, the cosine is positive where x is positive: quadrants I and IV. The tangent function is defined by $\frac{y}{x}$, so it is positive where x and y are both positive (quadrant I) or both negative (quadrant III).

Figure 5–13 should be memorized. It is reflected in the following four statements, which we call the ASTC rule.

The ASTC rule

In quadrant I,	**A**ll the trigonometric functions are positive.
In quadrant II, the	**S**ine function is positive.
In quadrant III, the	**T**angent function is positive.
In quadrant IV the	**C**osine function is positive.

One memory aid is the sentence ''**A**ll **S**tudents **T**ake **C**alculus.''

Since the sign of the reciprocal of a value is the same as the value, the sign of the cosecant function is the same as the sign of the sine function, that of the secant function is the same as that of the cosine function, and the sign of the cotangent function is the same as that of the tangent function.

The ASTC rule can be used to determine in which quadrant a given angle terminates.

■ Example 5–4 A

Determine in which quadrant the given angle θ terminates.

1. $\sin\theta < 0$, $\tan\theta > 0$

If $\sin\theta < 0$ then θ terminates in quadrants III or IV.
If $\tan\theta > 0$ then θ terminates in quadrants I or III.

Thus, for both conditions to be true, θ must terminate in quadrant III.

2. $\cos\theta < 0$, $\sin\theta > 0$

$\cos\theta < 0$ means θ terminates in quadrant II or III.
$\sin\theta > 0$ means θ terminates in quadrant I or II.

Thus, θ terminates in quadrant II. ■

Reference angles

Angles whose degree measures are integer multiples of 90°, such as 0°, ±90°, ±180°, ±270°, etc. are called **quadrantal angles** because their terminal sides fall between two quadrants. All other angles are nonquadrantal angles. A **reference angle** for a nonquadrantal angle is the acute angle formed by the terminal side of the angle and the *x*-axis. A reference angle is not defined for quadrantal angles. Figure 5–14 shows a reference angle, θ′ (theta-prime) for an angle θ terminating in each quadrant.

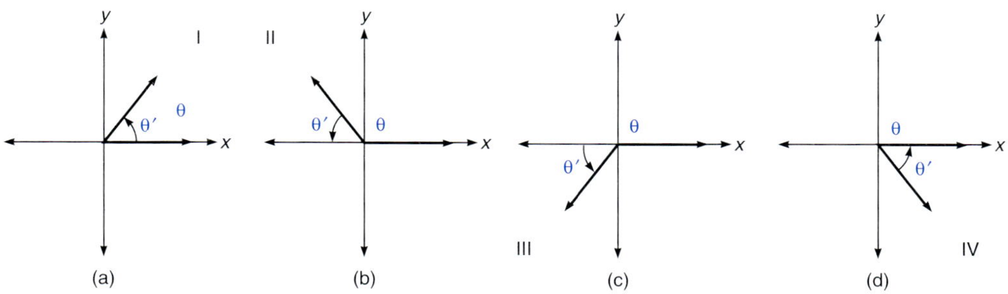

Figure 5–14

A reference angle is always acute (between 0° and 90°) *and is always formed by the terminal side of the angle and the x-axis* (never the *y*-axis). As will be illustrated in example 5–4 B, a good way to find a reference angle is to sketch the angle itself. This should make clear what computation to perform.

■ *Example 5–4 B*

Compute and sketch the reference angle for each angle.

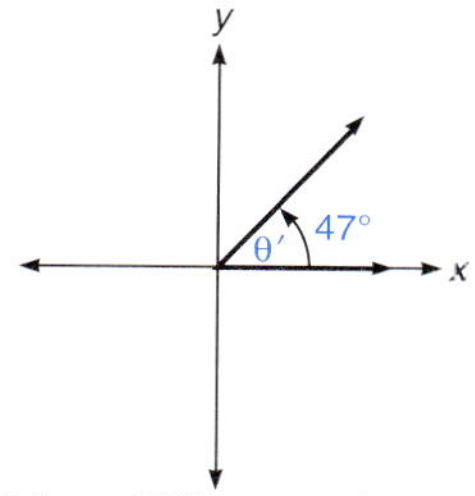

1. 47°

This angle terminates in quadrant I. The reference angle is the same as the angle itself, 47°.

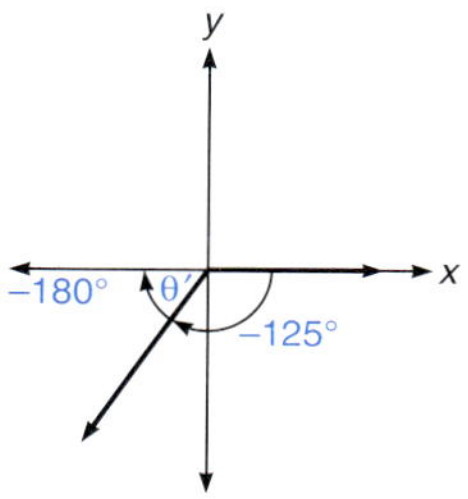

2. −125°

This angle terminates in quadrant III. The positive difference between −125° and −180° is 180° − 125° = 55°, which is the value of the reference angle.

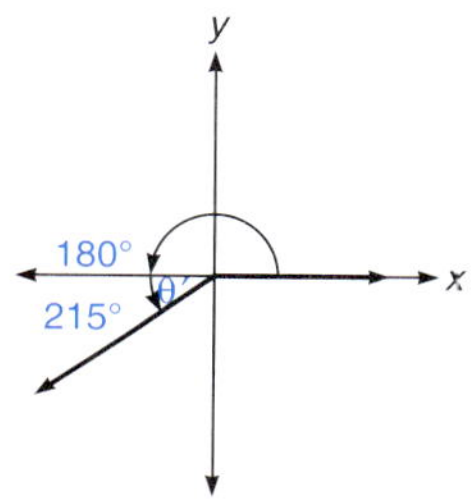

3. 215°

This angle terminates in quadrant III also. Here the reference angle is 215° − 180° = 35°.

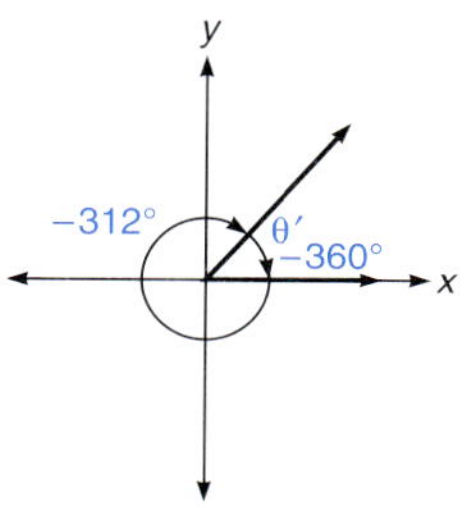

4. −312°

This angle terminates in quadrant I. The value of the reference angle is the positive difference between −312° and −360° = 360° − 312° = 48°. ■

It can be seen that if 0° < θ < 360°, then the reference angle θ′ can be found according to the following formulas:

$$
\begin{aligned}
&\theta \text{ in quadrant I:} && \theta' = \theta \\
&\theta \text{ in quadrant II:} && \theta' = 180^\circ - \theta \\
&\theta \text{ in quadrant III:} && \theta' = \theta - 180^\circ \\
&\theta \text{ in quadrant IV:} && \theta' = 360^\circ - \theta
\end{aligned}
$$

The absolute value of the trigonometric functions for any angle

The absolute value of a trigonometric function for any angle is the same as the trigonometric ratio for the corresponding reference angle. Figure 5–15 illustrates this idea for the angle 150°. If an angle of measure 150° is in standard position, then we find the values of the trigonometric functions by taking a point on its terminal side (the point $B(x,y)$ in the figure) and using the definitions of these functions in terms of x, y, and r.

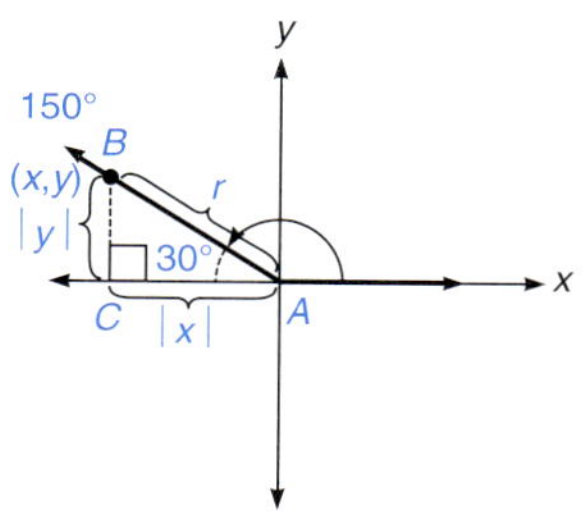

Figure 5–15

As seen in the figure, the absolute value of sin 150° is $\frac{|y|}{r}$. This is also the value of the trigonometric ratio for the reference angle, with measure 30°: $\sin 30^\circ = \frac{\text{length of side opposite } 30^\circ}{\text{length of hypotenuse}}$. We know from section 5–2 that sin 30° = $\frac{1}{2}$. Thus, in absolute value, $|\sin 150^\circ| = \sin 30^\circ = \frac{1}{2}$.

The exact values of the trigonometric functions for certain angles

The facts discussed in the previous paragraphs provide a method for finding the *exact* values of the basic trigonometric functions for any nonquadrantal angle whose reference angle is 30°, 45°, or 60°. We call this the reference angle/ASTC procedure.

> **Reference angle/ASTC procedure**
> To find the value of a trigonometric function for a nonquadrantal angle whose reference angle is 30°, 45°, or 60°:
> 1. Find the value of the reference angle.
> 2. Find the value of the appropriate trigonometric ratio for the reference angle from table 5–1 in section 5–2.
> 3. Determine the sign of this value using the ASTC rule (figure 5–13).

■ ***Example 5–4 C***

Find the exact value of the given trigonometric function for the given angle.

1. cos 210°

$\theta' = 210° - 180° = 30°$	Find the value of the reference angle
$\cos 30° = \dfrac{\sqrt{3}}{2}$	Table 5–1 (memorized value)
$\cos 210° = -\dfrac{\sqrt{3}}{2}$	A 210° angle terminates in quadrant III, where the cosine function is negative

2. tan 840°

$840° - 2(360°) = 120°$	120° and 840° are coterminal
$\theta' = 180° - 120° = 60°$	Reference angle for 840°
$\tan 60° = \sqrt{3}$	Memorized value
$\tan 840° = -\sqrt{3}$	840° terminates in quadrant II, where the tangent function is negative ■

The values of the trigonometric functions for quadrantal angles can be found by selecting any point on the terminal side of the angle and using the definitions.

■ ***Example 5–4 D***

Find the value of the six trigonometric functions for the angle with measure 900°.

$900° - 2(360°) = 180°$ 900° and 180° are coterminal angles

Thus, sin 900° = sin 180°.
The point $(-1,0)$ is on the terminal side of a 180° angle. Use this point to find the values for 900°.

$r = \sqrt{(-1)^2 + 0^2} = 1$ $\quad x = -1, y = 0$

$\sin 900° = \frac{y}{r} = \frac{0}{1} = 0$, $\csc 900° = \frac{1}{\sin 900°} = \frac{1}{0}$; undefined

$\cos 900° = \frac{x}{r} = \frac{-1}{1} = -1$, $\sec 900° = \frac{1}{\cos 900°} = \frac{1}{-1} = -1$

$\tan 900° = \frac{y}{x} = \frac{0}{-1} = 0$, $\cot 900° = \frac{1}{\tan 900°} = \frac{1}{0}$; undefined ■

Approximate values of the trigonometric functions—calculators

As with the trigonometric ratios it is difficult and in most cases impossible to find exact values for the trigonometric functions. The keys marked [sin], [cos], and [tan] calculate approximate values of these three trigonometric functions as necessary. It is not necessary to find a reference angle when using a calculator, since they are programmed to perform this step automatically.[5]

Recall from section 5–2 that the calculator must be in degree mode when entering angle measure in degrees. See example 5–2 A also.

■ ***Example 5–4 E***

Find four decimal place approximations to the following function values.

1. $\sin 133°$ — 133 [sin] Display [0.731353701]

   $\sin 133° \approx 0.7314$ — TI-81: [SIN] 133 [ENTER]

2. $\tan 62°14'$ — 62 [+] 14 [÷] 60 [=] [tan]

   $\tan 62°14' \approx 1.8993$ — TI-81: [TAN] 62 [+] 14 [÷] 60 [ENTER]

   Display [1.899346356]

3. $\sec(-335.6°)$ — 335.6 [±] [cos] [1/x]

   $\sec(-335.6°) \approx 1.0981$ — TI-81: [(] [COS] [(−)] 335.6 [)] [x^{-1}] [ENTER]

   Display [1.098076141] ■

Solutions to trigonometric equations

Recall from section 5–1 that we use the inverse trigonometric functions to find the degree measure of an acute angle when given the value of one of the trigonometric ratios. The same idea can be used to solve trigonometric equations of the form $\sin \theta = k$, $\cos \theta = k$. and $\tan \theta = k$, where k is a known constant. In particular, to find one value of θ in each equation, we use the following facts.

[5]However, some older calculators have limits on how large the measure of an angle may be (for example, 1,000°).

if $\sin \theta = k$, then one solution for θ is $\theta = \sin^{-1}k$
if $\cos \theta = k$, then one solution for θ is $\theta = \cos^{-1}k$
if $\tan \theta = k$, then one solution for θ is $\theta = \tan^{-1}k$

In section 6–4 we will examine this situation in more depth, but for now we will simply rely on these facts, and on the fact that these inverse trigonometric functions are programmed into calculators.

■ Example 5–4 F

Find one solution to each trigonometric equation, to the nearest 0.1°.

1. $\sin \theta = -0.8500$

$\theta = \sin^{-1}(-0.8500) \approx -58.2°$ .85 [+/−] [$\sin^{-1}$]

TI-81: [SIN^{-1}] [(−)] .85 [ENTER]

2. $\cos \theta = -0.2145$

$\theta = \cos^{-1}(-0.2145) \approx 102.4°$ ■

Mastery points

Can you

- Determine in which quadrant an angle terminates when given the signs of two of the trigonometric function values for that angle?
- Compute and sketch the reference angle for a given nonquadrantal angle θ with given degree measure?
- Find the exact value of any trigonometric function for an angle whose reference angle is 30°, 45°, or 60°, using the reference angle/ASTC procedure?
- Find the exact value of any trigonometric function for a quadrantal angle?
- Find the approximate value of any trigonometric function using a calculator?
- Find the approximate value of one solution to an equation of the form $\sin \theta = k$, $\cos \theta = k$, $\tan \theta = k$?

Exercise 5–4

In the following problems you are given the sign of two of the trigonometric functions of an angle in standard position. State in which quadrant the angle terminates.

1. $\sin \theta > 0$, $\cos \theta < 0$
2. $\sec \theta < 0$, $\tan \theta > 0$
3. $\cos \theta > 0$, $\tan \theta > 0$
4. $\cot \theta < 0$, $\csc \theta > 0$
5. $\tan \theta < 0$, $\csc \theta < 0$
6. $\sec \theta > 0$, $\csc \theta < 0$
7. $\csc \theta > 0$, $\cos \theta < 0$
8. $\tan \theta > 0$, $\sin \theta < 0$
9. $\sec \theta > 0$, $\sin \theta < 0$
10. $\cot \theta > 0$, $\sin \theta > 0$
11. $\sin \theta < 0$, $\sec \theta < 0$

For each of the following angles, find the measure of the reference angle θ'.

12. 39.3°
13. 164.2°
14. 213.2°
15. 427.1°
16. −16.8°
17. −255.3°
18. −100.4°
19. 130.7°
20. −671.3°
21. −181.0°
22. 512.8°
23. −279.5°
24. 292.3°
25. −252°
26. 312°

Find the exact trigonometric function value for each angle.

27. sin 135° **28.** cos 120° **29.** sin 210° **30.** cos 330° **31.** tan 300°
32. sin 240° **33.** sin(−120°) **34.** cos(−315°) **35.** cos 660° **36.** csc(−315°)
37. cot 300° **38.** sin 450° **39.** cos(−450°) **40.** tan(−540°) **41.** csc 90°
42. sin 840° **43.** sin(−690°) **44.** cot 215° **45.** sec 150° **46.** tan 330°

Find the trigonometric function value for each angle to four decimal places.

47. sin 113.4° **48.** cos 88.2° **49.** tan 214.6° **50.** csc 345°10′ **51.** cot 412°
52. tan 527.2° **53.** sec(−13°) **54.** sin(−88°) **55.** cos(−355°20′) **56.** tan(−248.6°)
57. csc 285.3° **58.** sec 211° **59.** cos(−133.2°) **60.** sin(−293°50′)

Find one approximate solution to each equation, to the nearest 0.1°.

61. $\sin \theta = 0.25$ **62.** $\sin \theta = \frac{1}{3}$ **63.** $\cos \theta = -0.5$ **64.** $\cos \theta = 0.813$ **65.** $\tan \theta = -\frac{8}{5}$
66. $\tan \theta = 3$ **67.** $\sin \theta = -0.59$ **68.** $\cos \theta = -0.18$

69. In a certain electrical circuit the instantaneous voltage E (in volts) is found by the formula $E = 156 \sin(\theta + 45°)$. Compute E to the nearest 0.01 volt for the following values of θ:
a. 0° **b.** 45° **c.** 100° **d.** −200° **e.** 13.3° **f.** −45°.

70. In a certain electrical circuit the instantaneous current I (in amperes) is found by the formula $I = 1.6 \cos(800t)°$. Find I to the nearest 0.01 ampere for the following values of t:
a. 0 **b.** 0.25 **c.** 0.85 **d.** 1 **e.** −1 **f.** −2.5 **g.** −0.02.

71. If a force of 200 pounds is applied to a rope to drag an object, the actual force tending to move the object horizontally is $f(\theta) = 200 \cos \theta$, where θ is the angle the rope makes with the horizontal. Compute the force tending to move the object horizontally if the angle of the rope is **a.** 0° **b.** 25° **c.** 50°

72. If a rocket is moving through the air at a speed of 1,200 mph, at an angle of $\theta°$ with the horizontal, then the rate at which it is rising is $v(\theta) = 1{,}200 \sin \theta$. Find the rate at which a rocket moving at 1,200 mph is rising if the angle it makes with the horizontal is
a. 50° **b.** 60° **c.** 70° **d.** 80°

Skill and review

1. Use the values 30° and 60° to see if the statement $\sin(2\theta) = 2 \sin \theta$ is true. (Let θ be 30°.)
2. Use the values 30° and 60° to see if the statement $\sin\frac{\theta}{2} = \frac{\sin \theta}{2}$ is true. (Let θ be 60°.)
3. Use the values 30°, 60°, 90° to see if the statement $\sin(\alpha + \beta) = \sin \alpha + \sin \beta$ is true.

5–5 Finding values from other values—reference triangles

A numerically controlled drill is being set up to drill a hole in a piece of steel 6.8 millimeters from the origin at an angle of 135°30′. What are the coordinates of this point?

The advent of numerically controlled machinery has made trigonometry more important than ever. The problem above is one of many situations where this is true.

Finding a general angle from a value and quadrant

In section 5–2 we learned how to find the degree measure of an acute angle if we know the value of one of the trigonometric ratios for that angle. We used the inverse sine, cosine, or tangent function as appropriate. We are now dealing with angles of any measure, but the same procedure can be used to find the value of a reference angle. From this we can find the least nonnegative measure for an angle.

As is illustrated in example 5–4 A we *always find a reference angle* θ' *by finding the inverse sine, cosine, or tangent function value for a positive value of* x. This is because the value of the sine, cosine, and tangent functions are positive for acute angles, and reference angles are acute angles. (The topic of inverse trigonometric functions is explored fully in section 6–4.) We could summarize the procedure as follows.

Finding the least nonnegative measure of an angle from a trigonometric function value and information about a quadrant

1. If necessary use the ASTC rule[6] to determine the quadrant for the terminal side of the angle.
2. Use $\sin^{-1}$, $\cos^{-1}$, or $\tan^{-1}$ to find θ'. Use the absolute value of the given trigonometric function value.
3. Apply θ' to the correct quadrant to determine the value of θ.

Note We find the "least nonnegative value." There are actually an unlimited number of values, since the trigonometric values are the same for all coterminal angles.

In section 5–3 we saw formulas that find θ' if $0° < \theta < 360°$. These formulas can be solved for θ if necessary and thus provide a formula for finding θ given θ'.

Relationship between θ and θ' if $0° < \theta < 360°$

θ in quadrant I:	$\theta' = \theta$	$\theta = \theta'$
θ in quadrant II:	$\theta' = 180° - \theta$	$\theta = 180° - \theta'$
θ in quadrant III:	$\theta' = \theta - 180°$	$\theta = \theta' + 180°$
θ in quadrant IV:	$\theta' = 360° - \theta$	$\theta = 360° - \theta'$

It is interesting to observe that the formulas are the "same" when solved for θ and θ' for every quadrant except quadrant III.

[6] Section 5–4.

■ Example 5–5 A

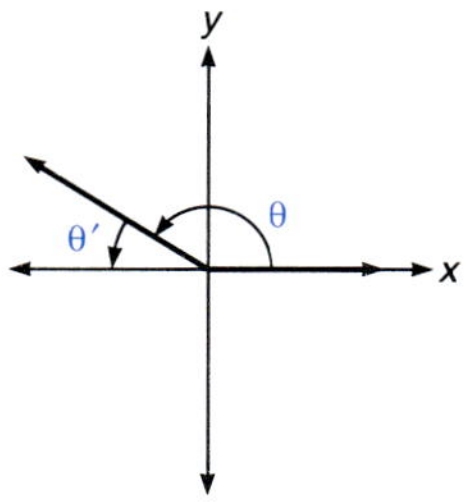

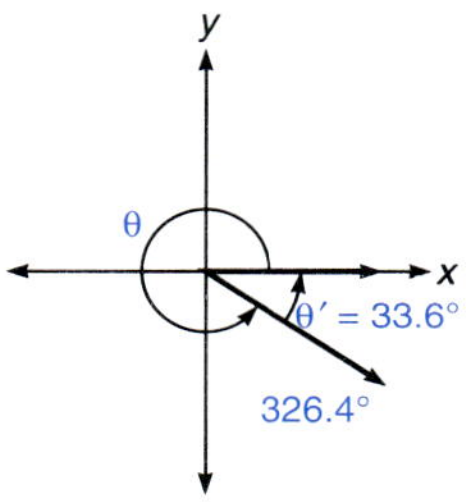

Find the least nonnegative measure of θ to the nearest 0.1°.

1. $\sin \theta = 0.5150$ and $\cos \theta < 0$
Since $\sin \theta > 0$ and $\cos \theta < 0$, θ terminates in quadrant II (see the figure). We find the acute reference angle θ′ just as we did in section 5–2.

$$\theta' = \sin^{-1} 0.5150 \approx 31.0°$$

Thus, $\theta \approx 180° - 31.0° = 149.0°$.

Note The calculator can be used to verify our result by checking that $\sin 149° \approx 0.5150$ and that $\cos 149° < 0$.

2. $\tan \theta = -0.6644$ and $\sin \theta < 0$
Since $\tan \theta < 0$ and $\sin \theta < 0$, θ terminates in quadrant IV.

$$\theta' = \tan^{-1} 0.6644 \approx 33.6°$$ Note we use the positive value 0.6644
$$\theta = 360° - \theta' \approx 326.4°.$$ ■

There are many places in science and technology where we find applications for trigonometric functions. With the advent of numerically controlled, or computer-controlled, machines, these applications are becoming more common.

■ Example 5–5 B

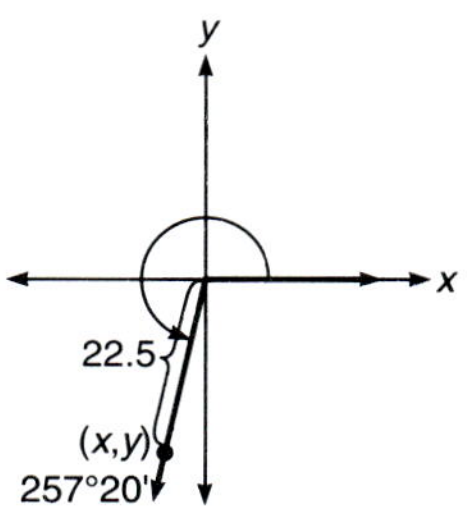

A technician is setting up a numerically controlled grinding wheel. The starting position for the wheel must be at an angle of 257°20′ and must be 22.5 inches from the origin (assuming the machine uses our usual *x*-*y* coordinate system). Find the *x*- and *y*-coordinates of the point at which the grinding wheel must start, to the nearest tenth of an inch.

The figure illustrates the situation. We have $r = 22.5$ inches and $\theta = 257°20'$. By definition, $\sin \theta = \frac{y}{r}$ and $\cos \theta = \frac{x}{r}$, so we find x and y as follows:

$$\sin 257°20' = \frac{y}{22.5}$$
$$y = 22.5 \sin 257°20'$$
$$y \approx -22.0 \text{ inches}$$

Note The calculation is

22.5 [×] [(] 257 [+] 20 [÷] 60 [)] [sin] [=] [−21.9524013]

TI-81: 22.5 [×] [SIN] [(] 257 [+] 20 [÷] 60 [)] [ENTER]

$$\cos 257°20' = \frac{x}{22.5}$$
$$x = 22.5 \cos 257°20'$$
$$x \approx -4.9 \text{ inches}$$

Thus, the starting coordinates, in inches, for the grinder are (−4.9, −22.0). ■

Exact values of the trigonometric functions from a known value—reference triangles

There are many situations in which we know the exact value of one of the trigonometric functions for a given angle and need to find the exact value of one or more of the remaining five trigonometric functions for the same angle. We can do this by using a reference triangle, which is a convenient way of combining the idea of reference angle and right triangle. A **reference triangle** is a right triangle with one leg on the x-axis and one leg parallel to the y-axis. The acute angle on the x-axis is the reference angle for the angle in question. *The lengths of the legs of a reference triangle are treated as directed distances (i.e., positive or negative); the hypotenuse is always positive.* This is illustrated in example 5–5 C. Figure 5–16 shows a reference triangle for each quadrant.

The primary use of a reference triangle is to help find the values of the other five trigonometric functions when the value of one of them is known. A secondary use is to find the value of an angle. This is also illustrated in example 5–5 C.

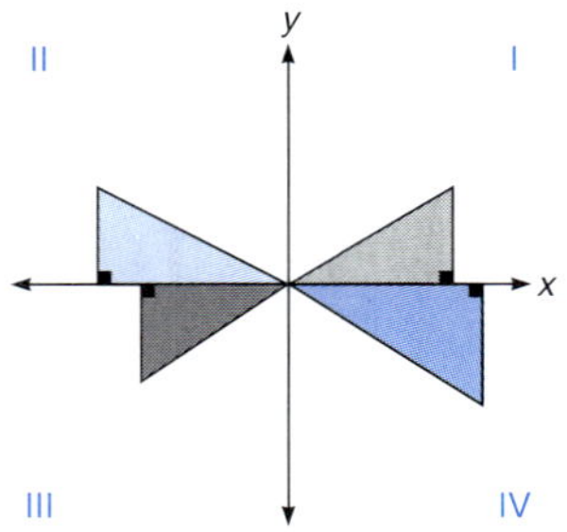

Figure 5–16

■ ***Example 5–5 C***

In each case draw a representation of angle θ and use a reference triangle to help find the values of the other five trigonometric functions. Also, find the least positive value of θ to the nearest 0.1°.

1. $\sin\theta = -\frac{1}{4}$ and $\tan\theta > 0$

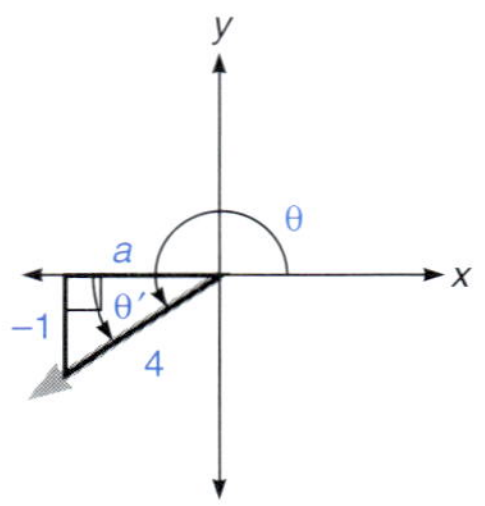

We know θ terminates in quadrant III since $\sin\theta < 0$ and $\tan\theta > 0$. We construct a right triangle in quadrant III in which one acute angle is a reference angle. This is shown in the figure. We label the hypotenuse 4 and the directed side opposite θ′ as −1. Thus, $\sin\theta' = \frac{\text{length of side opposite } \theta'}{\text{length of hypotenuse}} = -\frac{1}{4}$.

Note In a reference triangle the length of the hypotenuse is always positive.

$a^2 + (-1)^2 = 4^2$ Find the value of a using the Pythagorean theorem; since we are squaring values this theorem works for directed distances

$a^2 = 15$

$a = \pm\sqrt{15}$

We choose $a = -\sqrt{15}$ since it is negative as a directed distance.

We can now use the definitions of the trigonometric ratios for θ′ along with the directed distances to find the remaining trigonometric function values for θ.

$$\cos\theta = \frac{\text{adjacent}}{\text{hypotenuse}} = -\frac{\sqrt{15}}{4}, \quad \tan\theta = \frac{\text{opposite}}{\text{adjacent}} = \frac{-1}{-\sqrt{15}} = \frac{\sqrt{15}}{15}$$

$$\csc\theta = \frac{1}{\sin\theta} = -4, \quad \sec\theta = \frac{1}{\cos\theta} = -\frac{4}{\sqrt{15}} = -\frac{4\sqrt{15}}{15}$$

$$\cot\theta = \frac{1}{\tan\theta} = \sqrt{15}$$

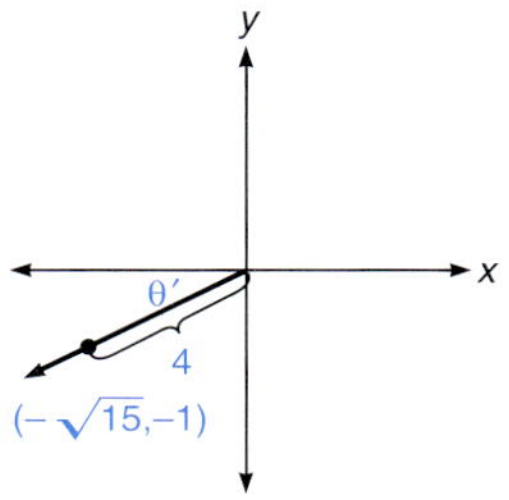

Figure 5–17

We now find an approximation to θ.

$$\sin\theta' = \tfrac{1}{4}\text{, so } \theta' = \sin^{-1}\tfrac{1}{4} \approx 14.5°\text{, so } \theta \approx 180° + 14.5° = 194.5°.$$

Note The reference triangle works because it is equivalent to finding a point on the terminal side of θ and applying the definitions of the trigonometric functions (section 5–3). Finding the reference triangle in part 1 was equivalent to finding the point $(-\sqrt{15},-1)$ to be on the terminal side of angle θ. (See figure 5–17.)

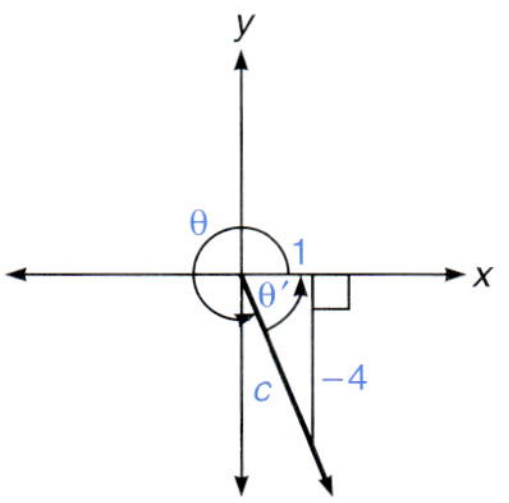

2. $\cot\theta = -\frac{1}{4}$ and $270° < \theta < 360°$

If $\cot\theta = -\dfrac{1}{4}$ then $\tan\theta = -4 = -\dfrac{4}{1} = \dfrac{\text{opposite}}{\text{adjacent}}$. The fact that $270° < \theta < 360°$ means that θ is in quadrant IV. The figure shows a reference triangle for an angle in quadrant IV with tangent $\left(\dfrac{\text{opposite}}{\text{adjacent}}\right)$ of −4.

$$c^2 = 1^2 + (-4)^2$$
$$c = \sqrt{17}$$

$$\sin\theta = \frac{\text{opposite}}{\text{hypotenuse}} = \frac{-4}{\sqrt{17}} = -\frac{4\sqrt{17}}{17},\quad \csc\theta = \frac{1}{\sin\theta} = -\frac{\sqrt{17}}{4}$$

$$\cos\theta = \frac{\text{adjacent}}{\text{hypotenuse}} = \frac{1}{\sqrt{17}} = \frac{\sqrt{17}}{17},\qquad \sec\theta = \frac{1}{\cos\theta} = \sqrt{17}$$

$\tan\theta' = 4$, so $\theta' = \tan^{-1}4 \approx 76.0°$, and $\theta = 360° - \theta' \approx 284°$ ■

Note In section 5–1 we noted it is best to convert values of reciprocal trigonometric ratios to primary trigonometric ratios. This advice applies to the trigonometric functions as well, and it is why we first converted cot θ to tan θ in part 2 of example 5–5 C.

Example 5–5 D shows how to handle a case where the value of a trigonometric function is given in terms of a literal constant (i.e., a letter). This is in fact a situation often encountered in studying the Calculus.

■ *Example 5–5 D*

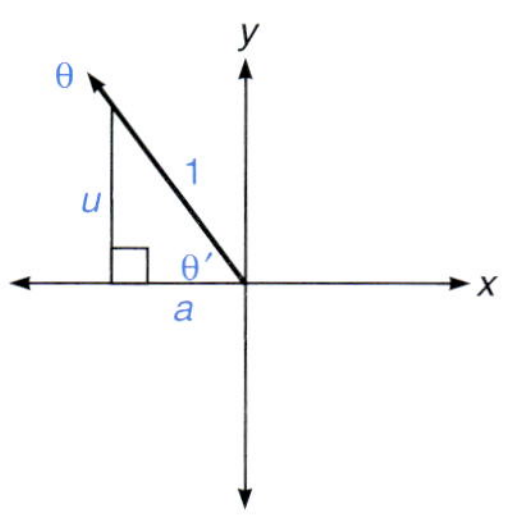

Draw a representation of angle θ and use a reference triangle to help find the values of the other five trigonometric functions if $\sin\theta = u$ and θ terminates in quadrant II.

In this problem the only given value is u. We therefore need to express the values of the trigonometric functions in terms which use only constants and the value u.

The figure shows a reference triangle in quadrant II, where $\sin\theta' = u$. Observe that we use the fact that $\dfrac{u}{1} = u$ to obtain u for the side of the triangle and 1 for the hypotenuse.

We need to find an expression for the length of the third side of the triangle, labeled a in the figure.

$$\begin{aligned} 1 &= u^2 + a^2 && \text{Pythagorean theorem} \\ 1 - u^2 &= a^2 \\ \pm\sqrt{1-u^2} &= a \\ -\sqrt{1-u^2} &= a && \text{Choose } a < 0 \text{ as a directed distance} \end{aligned}$$

$$\cos\theta = \frac{a}{1} = \frac{-\sqrt{1-u^2}}{1} = -\sqrt{1-u^2},\ \sec\theta = -\frac{1}{\sqrt{1-u^2}},$$

$$\tan\theta = -\frac{u}{\sqrt{1-u^2}},\ \cot\theta = -\frac{\sqrt{1-u^2}}{u},\ \csc\theta = \frac{1}{u}$$

Since we do not know the actual value of u we cannot make a determination of an approximate value for angle θ. ■

Mastery points

Can you

- Find an approximation to the least nonnegative measure of an angle, given the value of one of the trigonometric functions, and the sign of a second, for that angle?
- Apply the definitions of the trigonometric functions in appropriate situations?
- Use reference triangles to find the values of the remaining trigonometric functions for a given angle, when given the value of one of the trigonometric functions for that angle?

Exercise 5–5

Find the measure of the least nonnegative angle that meets the conditions given in the following problems, to the nearest 0.1°.

1. $\sin\theta = 0.8251$, $\cos\theta > 0$
2. $\cos\theta = -0.1771$, $\sin\theta < 0$
3. $\tan\theta = 0.6569$, $\sec\theta > 0$
4. $\sin\theta\ -0.6508$, $\tan\theta > 0$
5. $\sec\theta = -1.0642$, $\sin\theta < 0$
6. $\csc\theta = -1.3673$, $\tan\theta > 0$
7. $\tan\theta = -0.0349$, $\csc\theta < 0$
8. $\cos\theta = -0.2222$, $\sin\theta > 0$
9. $\sin\theta = \frac{3}{8}$, $\cos\theta > 0$
10. $\sin\theta = \frac{3}{8}$, $\cos\theta < 0$
11. $\cot\theta = -5$, $\sin\theta > 0$
12. $\tan\theta = -5$, $\sin\theta < 0$
13. $\cos\theta = -\frac{5}{7}$, $\tan\theta > 0$
14. $\cos\theta = -\frac{5}{7}$, $\tan\theta < 0$

In each problem you are given the coordinates of a point on the terminal side of θ. In each case (a) find the exact value of the trigonometric functions for θ and (b) find the least nonnegative measure of θ, to the nearest 0.1°.

15. $(-3,4)$
16. $(-5,-12)$
17. $(12,-5)$
18. $(4,3)$
19. $(-4,-5)$
20. $(5,-4)$
21. $(4,6)$
22. $(-12,16)$

In each case (a) draw a representation of angle θ, (b) use a reference triangle to help find the exact values of the three primary trigonometric functions (sine, cosine, tangent), and (c) find the least positive value of θ to the nearest 0.1°.

23. $\sin\theta = \frac{3}{4}$, $\cos\theta > 0$
24. $\sin\theta = \frac{4}{5}$, $\cos\theta < 0$
25. $\cos\theta = -\frac{1}{2}$, $\tan\theta > 0$
26. $\cos\theta = -\frac{5}{13}$, $\tan\theta < 0$
27. $\sin\theta = 1$
28. $\cos\theta = 1$
29. $\tan\theta = 2$, $\cos\theta < 0$
30. $\tan\theta = 3$, $\cos\theta > 0$
31. $\csc\theta = -5$, $\sec\theta < 0$
32. $\csc\theta = -2$, $\sec\theta > 0$
33. $\csc\theta = -1$
34. $\sec\theta = -1$
35. $\sin\theta = -\frac{3}{4}$, $\tan\theta > 0$
36. $\sin\theta = -\frac{2}{5}$, $\tan\theta < 0$
37. $\sec\theta = 4$, $\csc\theta > 0$
38. $\sec\theta = \sqrt{6}$, $\csc\theta < 0$
39. $\cot\theta = \frac{\sqrt{2}}{3}$, $\sin\theta < 0$
40. $\cot\theta = \frac{1}{3}$, $\sin\theta > 0$
41. $\cos\theta = -\frac{5}{13}$, $\sin\theta > 0$
42. $\cos\theta = -\frac{12}{13}$, $\sin\theta < 0$
43. $\tan\theta = \frac{7}{2}$, $\sec\theta < 0$
44. $\tan\theta = \frac{7}{3}$, $\sec\theta > 0$
45. $\sec\theta = 5$, $\tan\theta > 0$
46. $\sec\theta = 4$, $\tan\theta < 0$
47. $\sin\theta = \frac{1}{\sqrt{5}}$, $\tan\theta < 0$
48. $\sin\theta = \frac{1}{\sqrt{3}}$, $\tan\theta > 0$

In problems 49–54, draw a representation of the angle θ and use a reference triangle to find values of the other five trigonometric functions in terms of u.

49. $\cos\theta = u$ and θ terminates in quadrant I.
50. $\tan\theta = u$ and θ terminates in quadrant I.
51. $\cos\theta = u$ and θ terminates in quadrant III.
52. $\tan\theta = u$ and θ terminates in quadrant III.
53. $\sin\theta = u + 1$ and θ terminates in quadrant I.
54. $\cos\theta = 1 - u$ and θ terminates in quadrant I.

Solve the following problems.

55. A numerically controlled drill is being set up to drill a hole in a piece of steel 6.8 millimeters from the origin at an angle of 135°30′. To the nearest 0.01 millimeter, what are the coordinates of this point?

56. Suppose the hole in problem 55 must be 10.25 inches from the origin at an angle of 13°20′. Find the coordinates of this point to the nearest 0.01 inch.

57. Suppose the hole in problem 55 must be 8.25 centimeters from the origin at an angle of −134.4°. Find the coordinates of this point to the nearest 0.01 centimeter.

58. A numerically controlled drill must drill four holes on a circle whose center is at the origin with radius 17.8 centimeters, as shown in the diagram. The holes must be drilled wherever on this circle the x-coordinate is ±10.0 centimeters. Find the y-coordinate and angle (to the nearest 0.1°) for each of these four holes.

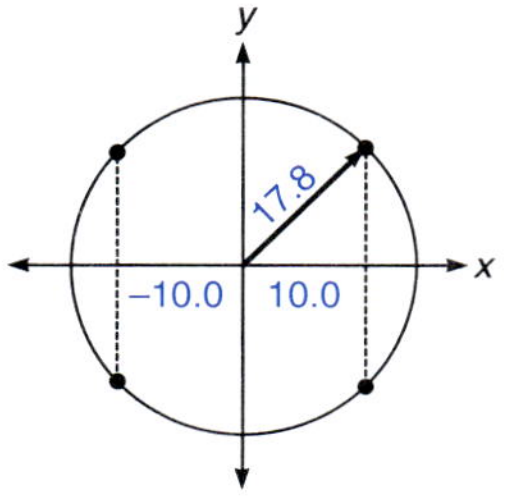

59. Suppose in problem 58 four additional holes must be drilled wherever the y-coordinate is ±15.5 cm. Find the x-coordinate and angle for each of these holes.

60. A technician is aligning a laser device that is used to cut patterns out of cloth. The device is positioned at an angle of 135.20° and at a distance 5.50 feet from the origin. What should the x- and y-coordinates be at this point, to the nearest 0.01 foot?

61. A scanning device used in medical diagnosis has a moving part that moves with great precision in a circle around the patient. Assume the y-axis is perpendicular to the top of the table on which the patient lies and the x-axis is at right angles to the length of the table. The diameter of the machine is 4 feet 3.5 inches. Find the coordinates of the moving part when the angle is 211.5°, to the nearest 0.1 inch.

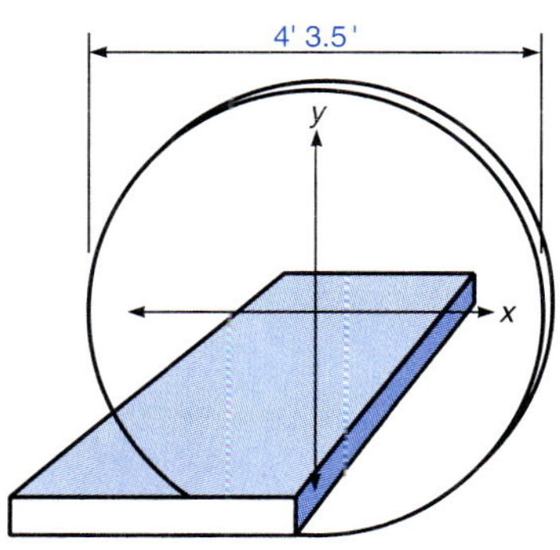

62. In the June 1980 issue of *Popular Science* magazine Mr. R. J. Ransil presented several formulas for calculating saw angles for compound miters. The formulas are:

$$\text{angle } A = 90° - \frac{180°}{\text{number of sides}}$$
$$\tan(\text{arm angle}) = \cot(A) \cdot \sin(\text{slope})$$
$$\sin(\text{tilt angle}) = \cos(A) \cdot \cos(\text{slope})$$

Arm angles and tilt angles are acute.

Calculate the arm angle and tilt angle to the nearest 0.1° for the following numbers of sides and slopes:

Number of sides	Slope (in degrees)
3	5
5	5
7	25
7	30
6	35
8	35

63. A surveying manual describes how to find distance BP in the figure. The distance AP can be found, but trees prevent measuring angle a. Angle b can be measured, but not distance BP. The manual instructs the surveyor to find BP by solving the following sequence of formulas:

$$\sin p = \frac{AB \sin b}{AP}$$
$$a = 180° - (b + p)$$
$$BP = \frac{AP \sin a}{\sin b}$$

Note that the first formula does not give angle p but only $\sin p$. Also, assume p is acute. Solve the sequence of formulas to compute the distance BP to the nearest 0.1 foot if $AB = 512.4$ feet, $AP = 322.6$ feet, and $b = 28.3°$.

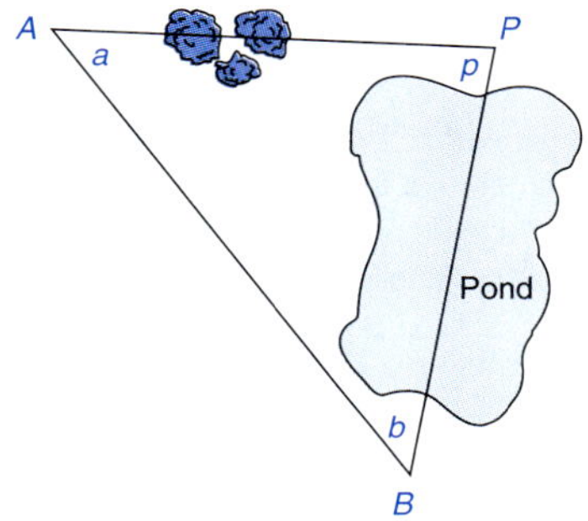

64. Find BP in problem 63 to the nearest 0.1 meter if $AB = 319.2$ meters, $AP = 225.7$ meters, and $b = 31.6°$.

Skill and review

1. What is the reference angle for $-250°$?
2. The point $(3,-4)$ is on the terminal side of an angle in standard position. Find the value of the cosine of this angle.
3. In right triangle ABC, $A = 36.2°$, $c = 10.0$. Solve the triangle.
4. Use the graph of $y = x^3$ as a guide to graph $f(x) = (x + 1)^3 - 1$.
5. Solve the equation $\frac{3x - 5}{12} = 2(x - 3) - 8x$.
6. Solve the inequality $\frac{3x - 9}{x - 2} \le 0$.

5–6 Introduction to trigonometric equations

Solve the equation $2\cos^2\theta - \cos\theta - 1 = 0$.

This section introduces equations involving the trigonometric functions. Recall that equations can be categorized as identities or conditional equations (section 2–1). An *identity* is an equation that is true for every allowed value of its variable (or variables). For example, $2(x + 3) = 2x + 6$ is an identity, since the left member and right member of the equation represent the same value, regardless of the value of x. Similarly, $\frac{3x^2}{3x} = x, \quad x \neq 0$ is an identity.

A *conditional equation* is an equation that is true only for some, but not all, values that may replace the variable. For example, $6x = 12$ is true only if x is replaced by 2, and $x^2 = 9$ is true only if x is replaced by 3 or -3.

Identities

We have seen the following identities

$$[1] \qquad \csc\theta = \frac{1}{\sin\theta}, \quad \sec\theta = \frac{1}{\cos\theta}, \quad \cot\theta = \frac{1}{\tan\theta}$$

$$[2] \qquad \sin\theta = \frac{1}{\csc\theta}, \quad \cos\theta = \frac{1}{\sec\theta}, \quad \tan\theta = \frac{1}{\cot\theta}$$

Knowing these identities permits us to simplify certain trigonometric expressions.

■ *Example 5–6 A*

Simplify each trigonometric expression.

1. $\csc\theta\sin\theta = \frac{1}{\sin\theta}\cdot\sin\theta$ — Replace $\csc\theta$ by $\frac{1}{\sin\theta}$

$= 1$ — $\frac{1}{a}\cdot a = 1$

2. $\frac{1-\csc\theta}{\csc\theta} = \frac{1-\csc\theta}{\csc\theta}$

$= \frac{1}{\csc\theta} - \frac{\csc\theta}{\csc\theta}$ — $\frac{a+b}{c} = \frac{a}{c} + \frac{b}{c}$

$= \sin\theta - 1$ — $\frac{1}{\csc\theta} = \sin\theta$ and $\frac{\csc\theta}{\csc\theta} = 1$ ■

Two more useful identities we have seen are

$$[3] \qquad \tan\theta = \frac{\sin\theta}{\cos\theta}, \quad \cot\theta = \frac{\cos\theta}{\sin\theta}$$

■ *Example 5–6 B*

Simplify the expression.

$\cot\alpha(\sin\alpha - \tan\alpha)$

$\cot\alpha\sin\alpha - \cot\alpha\tan\alpha$ — $a(b+c) = ab + ac$

$\frac{\cos\alpha}{\sin\alpha}\sin\alpha - \frac{1}{\tan\alpha}\tan\alpha$ — $\cot\theta = \frac{\cos\theta}{\sin\theta}$, $\cot\theta = \frac{1}{\tan\theta}$

$\cos\alpha - 1$ — Reduce common numerators and denominators

Note We replaced $\cot\alpha$ by $\frac{\cos\alpha}{\sin\alpha}$ in one term, and by $\frac{1}{\tan\alpha}$ in another term. We use whichever identity better suits the rest of the term. ■

Another very important identity is often called the fundamental identity of trigonometry.

> **Fundamental identity of trigonometry**
> For any angle θ, $\sin^2\theta + \cos^2\theta = 1$.

The term $\sin^2\theta$ means $(\sin\theta)^2$, and $\cos^2\theta$ means $(\cos\theta)^2$. This identity can be shown to be true as follows:

$$\begin{aligned}\sin^2\theta + \cos^2\theta &= (\sin\theta)^2 + (\cos\theta)^2 \\ &= \left(\frac{y}{r}\right)^2 + \left(\frac{x}{r}\right)^2 && \sin\theta = \frac{y}{r},\ \cos\theta = \frac{x}{r} \\ &= \frac{x^2 + y^2}{r^2} \\ &= \frac{r^2}{r^2} = 1 && \text{Recall that } r^2 = x^2 + y^2\end{aligned}$$

Example 5–6 C illustrates applying the fundamental identity of trigonometry.

■ ***Example 5–6 C***

1. Simplify the expression.

$$\left(\frac{1}{\sec\theta}\right)^2 + \left(\frac{1}{\csc\theta}\right)^2$$
$$(\cos\theta)^2 + (\sin\theta)^2$$
$$\cos^2\theta + \sin^2\theta$$
$$1$$

2. Simplify the expression.

$$1 - \sin^2\beta$$
$$(\sin^2\beta + \cos^2\beta) - \sin^2\beta \qquad \text{Replace 1 by } \sin^2\beta + \cos^2\beta$$
$$\cos^2\beta \qquad \sin^2\beta - \sin^2\beta = 0$$

3. Verify the fundamental identity of trigonometry for $\theta = 60°$.

$$\begin{aligned}\sin^2 60° + \cos^2 60° &= \left(\frac{\sqrt{3}}{2}\right)^2 + \left(\frac{1}{2}\right)^2 \\ &= \tfrac{3}{4} + \tfrac{1}{4} = 1\end{aligned}$$

■

Conditional trigonometric equations

We have already seen how to solve simple equations like $\tan\theta = 4$ by using the inverse tangent function to obtain one value of θ ($\tan^{-1}4 \approx 76°$, sections 5–2 and 5–4). There are many other solutions to this equation (in fact an unlimited number), which includes, among others, all angles coterminal to

Quadrant for least nonnegative solution

Function	Function value is positive	Function value is negative
Sine	I	III
Cosine	I	II
Tangent	I	II

Table 5–2

76°. This will be explored in more detail in section 7–4; *here we will restrict ourselves to the least nonnegative solution* to trigonometric equations. Considering the ASTC rule will show that the least nonnegative angle for which a trigonometric function value is positive is in quadrant I. When the function has a negative value ($\sin\theta = -\frac{1}{2}$ for example) the least nonnegative value is in quadrant II for the cosine and tangent function, and quadrant III for the sine function. This is summarized in table 5–2.

As we solve these equations whenever we compute an inverse trigonometric function to solve an equation we use the absolute value of the argument. This gives us the reference angle of the answer (as previously discussed in example 5–5 A). This is illustrated in parts 3 and 4 of example 5–6 D and can be summarized as

$$\text{if } \sin\theta = k, \text{ then } \theta' = \sin^{-1}|k|$$
$$\text{if } \cos\theta = k, \text{ then } \theta' = \cos^{-1}|k|$$
$$\text{if } \tan\theta = k, \text{ then } \theta' = \tan^{-1}|k|$$

■ *Example 5–6 D*

Solve the following conditional equations for the least nonnegative solution to the nearest 0.1°.

1. $2 \sin x = \sqrt{3}$

$2 \sin x = \sqrt{3}$

$\sin x = \dfrac{\sqrt{3}}{2}$ — Divide both members by 2

$x' = \sin^{-1}\dfrac{\sqrt{3}}{2}$ — If $\sin\theta = k$ then $\theta' = \sin^{-1}|k|$

$x' = 60°$

$x = 60°$ — $\sin x > 0$, so x is in quadrant I or II; the least nonnegative solution would be in quadrant I

2. $4 \cos 3\alpha = 3$

$4 \cos 3\alpha = 3$

$\cos 3\alpha = \frac{3}{4}$ — Divide both members by 4

$(3\alpha)' = \cos^{-1}\frac{3}{4}$ — If $\cos\theta = k$, then $k' = \cos^{-1}|k|$

$(3\alpha)' \approx 41.41°$ — .75 [cos⁻¹] Display [41.40962211]

TI-81: [COS⁻¹] .75 [ENTER]

$3\alpha \approx 41.41°$ — Since $\cos 3\alpha > 0$ the least nonnegative value of 3α is in quadrant I; in quadrant I, the reference angle and the angle itself are coterminal

$\alpha \approx 13.8°$ — Divide both members by 3; round the result

3. $\tan \frac{x}{2} = -1.5$

$$\tan \frac{x}{2} = -1.5$$

$$\frac{x'}{2} = \tan^{-1}1.5 \approx 56.31°$$

If $\tan \theta = k$, then $k' = \tan^{-1}|k|$

$$\frac{x}{2} \approx 180° - 56.31°$$
$$\approx 123.69°$$

$\frac{x}{2}$ terminates in quadrant II, since $\tan \frac{x}{2} < 0$ and we want the least nonnegative value of x; thus, $\theta = 180° - \theta'$

$$x \approx 247.4°$$

$2(123.69) \approx 247.4$

4. $2 \sin^2\theta - \sin \theta - 1 = 0$

This equation is quadratic in the variable $\sin \theta$. It can be factored. If this is difficult to see, use substitution for expression (section 1–3) as follows:

$$2 \sin^2\theta - \sin \theta - 1 = 0$$

$$2u^2 - u - 1 = 0$$

Replace $\sin \theta$ by u, so $\sin^2\theta$ is replaced by u^2

$$(2u + 1)(u - 1) = 0$$

$$(2 \sin \theta + 1)(\sin \theta - 1) = 0$$

Replace u by $\sin \theta$

$$2 \sin \theta + 1 = 0 \quad \text{or} \quad \sin \theta - 1 = 0$$

Zero product property

$2 \sin \theta = -1$	$\sin \theta = 1$
$\sin \theta = -\frac{1}{2}$	
θ in quadrant III	
$\theta' = \sin^{-1}\frac{1}{2}$	$\theta' = \sin^{-1}1$
$\theta' = 30°$	$\theta' = 90°$
$\theta = 210°$	$\theta = 90°$

Thus, there are two solutions, 90° and 210°.

Note a. We took the least nonnegative solution for *each* factor.
b. The steps involving the substitution by u can be skipped if you see that $2 \sin^2\theta - \sin \theta - 1$ factors into $(2 \sin \theta + 1)(\sin \theta - 1)$. ■

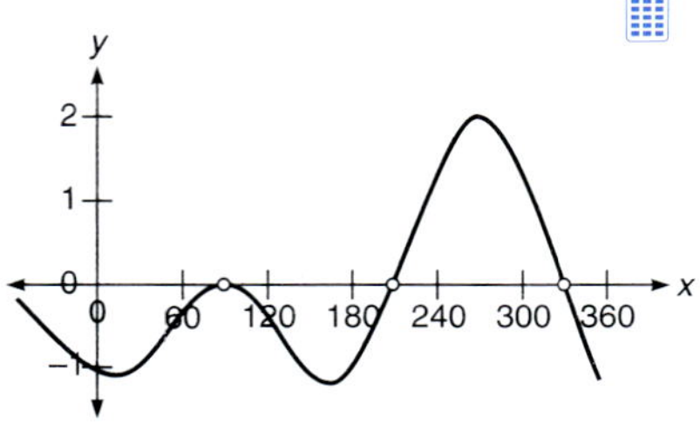

Using a graphing calculator it is possible to visualize the solutions to these equations. The graph below is that of $y = 2 \sin^2\theta - \sin \theta - 1$. Where the graph crosses the x-axis, y is 0, and therefore the points at which the graph crosses the x-axis represent solutions to the equation $2 \sin^2\theta - \sin \theta - 1 = 0$.

The graph shows the two solutions we obtained in part 4 of example 5–6 D (at 90° and at 210°). A third solution occurs at 330°. We did not find this algebraically because we are taking the least nonnegative solution to each part of the equation. Chapter 7 will show how to find all solutions to trigonometric equations. To obtain this graph on the TI-81, make sure the calculator is in degree mode, and use the RANGE values Xmin=−30, Xmax=360, Ymin=−1, Ymax=2, Xscl=60. Enter the equation as [Y=] 2 [(] [SIN] [X|T] [)] [x^2] [−] [SIN] [X|T] [−] 1.

Mastery points

Can you

- Simplify simple trigonometric expressions?
- Solve simple equations involving the trigonometric functions?

Exercise 5–6

Show that each of the expressions in the left member can be transformed into the expression in the right member.

1. $\tan\theta\cot\theta = 1$
2. $\sec\theta\cos\theta = 1$
3. $\cos\theta(1 - \sec\theta) = \cos\theta - 1$
4. $\cot\alpha(\tan\alpha + \sin\alpha) = 1 + \cos\alpha$
5. $\sec\theta(\cot\theta + \cos\theta - 1) = \csc\theta - \sec\theta + 1$
6. $\dfrac{\cos\theta - 1}{\sin\theta} = \cot\theta - \csc\theta$
7. $\dfrac{\cos\alpha - \sin\alpha}{\cos\alpha} = 1 - \tan\alpha$
8. $\dfrac{\sin\theta + \cos\theta - 2}{\cos\theta} = \tan\theta + 1 - 2\sec\theta$
9. $1 - \cos^2\theta = \sin^2\theta$
10. $\cos\theta\cos\theta + \sin^2\theta = 1$
11. $\cos\beta(\sec\beta - \cos\beta) = \sin^2\beta$
12. $-\sin\theta(\sin\theta - \csc\theta) = \cos^2\theta$
13. $(\cos\theta + \sin\theta)(\cos\theta - \sin\theta) + 2\sin^2\theta = 1$
14. $\tan x(\cot x + \csc x) = 1 + \sec x$
15. $\csc\alpha(\cos\alpha - \sin\alpha) = \cot\alpha - 1$
16. $\sin\beta(\cot\beta - \csc\beta + \sin\beta) = \cos\beta - \cos^2\beta$
17. $\dfrac{\sin x - \cos x}{\sin x} = 1 - \cot x$
18. $\cos\alpha(\csc\alpha + \sec\alpha) = \cot\alpha + 1$
19. $\tan\beta(\cot\beta - \cos\beta) = 1 - \sin\beta$
20. Verify by computation that the fundamental identity is true when $\theta = 30°$.
21. Using approximate values check the fundamental identity when
    a. $\theta = 16°50'$ **b.** $\theta = 50°$
22. Use the two identities $\cot\theta = \dfrac{1}{\tan\theta}$ and $\tan\theta = \dfrac{\sin\theta}{\cos\theta}$ to show that $\cot\theta = \dfrac{\cos\theta}{\sin\theta}$.
23. Use approximate values to show that $\tan\theta = \dfrac{\sin\theta}{\cos\theta}$ when $\theta = 32°40'$.
24. Show that $\tan\theta = \dfrac{\sin\theta}{\cos\theta}$ when $\theta = 60°$ (use values from table 5–1).

Solve the following conditional equations for the least nonnegative solution to the nearest 0.1°.

25. $2\cos x = 1$
26. $\sqrt{3}\tan x = 1$
27. $2\sin x = \sqrt{3}$
28. $\sqrt{2}\cos x = 1$
29. $5\sin x = 1$
30. $3\sin x = 2$
31. $2\tan x = 9$
32. $4\cot x = 3$
33. $\dfrac{\sin x}{3} = \dfrac{2}{11}$
34. $\dfrac{\csc x}{3} = 2$
35. $\sin 3x = \frac{1}{2}$
36. $\cos 2x = \frac{1}{2}$
37. $\tan 2x = \sqrt{3}$
38. $\sin 2x = 0.8$
39. $\csc 3x = 3$
40. $4\sin 2x = 3$
41. $2\cos 4x = -1$
42. $-3\sin 2x = 0.75$
43. $-2\tan 3x = 8$
44. $\frac{1}{2}\sin 3x = -\frac{1}{4}$
45. $4\sin^2\theta - 1 = 0$
46. $2\sin^2\theta + \sin\theta - 1 = 0$
47. $2\cos^2\theta - \cos\theta - 1 = 0$
48. $4\sin^2\theta - 4\sin\theta + 1 = 0$
49. $\cos^2\theta - 1 = 0$
50. $2\sin^2\theta - 1 = 0$
51. $4\tan^2\theta + 4\tan\theta + 1 = 0$

Skill and review

1. $\cos\theta = -\frac{1}{3}$ and $\tan\theta < 0$. **a.** Use a reference angle to find the exact value of $\tan\theta$. **b.** Find the value of θ to the nearest 0.1°.
2. A 28 foot ladder is leaning against a building so that its base is 5 feet from the base of the building. Find the measure of the acute angle that the ladder makes with the ground, to the nearest 0.1°.
3. The circumference C of a circle is 28.5 feet. Use $C = 2\pi r$ to find the radius r to the nearest 0.1 feet.
4. Solve the equation $7x^2 + 14x - 10 = 6x^2 + 12x + 5$.

Chapter 5 summary

- $1° = 60'$ (one degree is equivalent to 60 minutes)
  $1' = 60''$ (one minute is equivalent to 60 seconds)
  $1° = 3600''$ (one degree is equivalent to 3,600 seconds)
- The sum of the measures of the angles of a triangle is 180°.
- A **right triangle** is a triangle in which one of the angles is a right (90°) angle. The side of a right triangle opposite the right angle is called the **hypotenuse.** This is always the longest side of the triangle.
- **The Pythagorean theorem** In a right triangle with hypotenuse of length c and sides of lengths a and b, $a^2 + b^2 = c^2$.
- **The primary trigonometric ratios** If θ is either of the two acute angles in a right triangle, then

$$\sin\theta = \frac{\text{length of side opposite } \theta}{\text{length of hypotenuse}}$$

$$\cos\theta = \frac{\text{length of side adjacent to } \theta}{\text{length of hypotenuse}}$$

$$\tan\theta = \frac{\text{length of side opposite to } \theta}{\text{length of side adjacent to } \theta}.$$

- **Reciprocal pair ratios** If θ is either acute angle of a right triangle, $\csc\theta = \dfrac{1}{\sin\theta}$, $\sec\theta = \dfrac{1}{\cos\theta}$, $\cot\theta = \dfrac{1}{\tan\theta}$.
- **Important relations**

$$\sin\theta = \frac{1}{\csc\theta},\ \cos\theta = \frac{1}{\sec\theta},\ \tan\theta = \frac{1}{\cot\theta}.$$

- An **angle in standard position** is formed by two rays with the vertex at the origin. The **initial side** of the angle lies on the nonnegative portion of the x-axis. The **terminal side** of the angle may be in any quadrant or along any axis.
- **Coterminal angles** Two angles are coterminal if the measure of one can be formed from the measure of the other by adding or subtracting multiples of 360°.
- **The trigonometric functions** Let θ be an angle in standard position, and let (x,y) be any point on the terminal side of angle θ, except (0,0). Let $r = \sqrt{x^2 + y^2}$ be the distance from the origin to the point. Then,

$$\sin\theta = \frac{y}{r},\quad \cos\theta = \frac{x}{r},\quad \tan\theta = \frac{y}{x}$$

$$\csc\theta = \frac{r}{y},\quad \sec\theta = \frac{r}{x},\quad \cot\theta = \frac{x}{y}$$

- **Tangent/cotangent identities**

$$\tan\theta = \frac{\sin\theta}{\cos\theta},\quad \cot\theta = \frac{\cos\theta}{\sin\theta}$$

- **Quadrantal angles** have degree measures which are integer multiples of 90°: 0°, ±90°, ±180°, ±270°, etc.
- **The ASTC rule** is illustrated in the figure. This rule gives the sign of the value of a trigonometric ratio for a nonquadrantal angle, based on the quadrant in which the angle terminates.

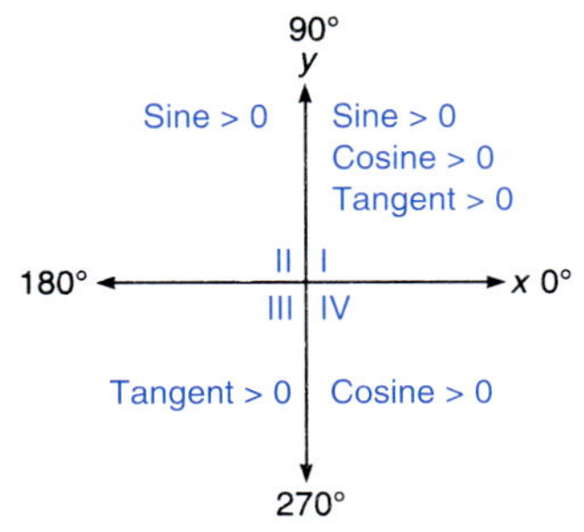

- A **reference angle** for a nonquadrantal angle is the acute angle formed by the terminal side of the angle and the x-axis.
- If $0° < \theta < 360°$ then the reference angle θ' can be found according to the following formulas:

θ in quadrant I: $\theta' = \theta$
θ in quadrant II: $\theta' = 180° - \theta$
θ in quadrant III: $\theta' = \theta - 180°$
θ in quadrant IV: $\theta' = 360° - \theta$

- **Values of sine, cosine, and tangent for special angles**

	Sine	Cosine	Tangent
30°	$\frac{1}{2}$	$\frac{\sqrt{3}}{2}$	$\frac{\sqrt{3}}{3}$
45°	$\frac{\sqrt{2}}{2}$	$\frac{\sqrt{2}}{2}$	1
60°	$\frac{\sqrt{3}}{2}$	$\frac{1}{2}$	$\sqrt{3}$

- **Finding a general angle from a value and quadrant** Find a reference angle θ' by finding the inverse sine, cosine, or tangent function value for a *positive* value of x. Then use the ASTC rule to establish the actual quadrant of the angle θ, according to the following rules:

θ in quadrant I: $\theta = \theta'$
θ in quadrant II: $\theta = 180° - \theta'$
θ in quadrant III: $\theta = 180° + \theta'$
θ in quadrant IV: $\theta = 360° - \theta'$

- **Fundamental identity of trigonometry** For any angle θ,

$$\sin^2\theta + \cos^2\theta = 1$$

- **Solving conditional trigonometric equations**

if $\sin\theta = k$, then $\theta' = \sin^{-1}|k|$
if $\cos\theta = k$, then $\theta' = \cos^{-1}|k|$
if $\tan\theta = k$, then $\theta' = \tan^{-1}|k|$

Chapter 5 review

[5–1] Convert each angle to its measure in decimal degrees. Round the answer to the nearest 0.001° where necessary. Also, state whether each angle is acute or obtuse.

1. 165°47′ **2.** 37°18′

3. Find the measure of angle θ in the figure.

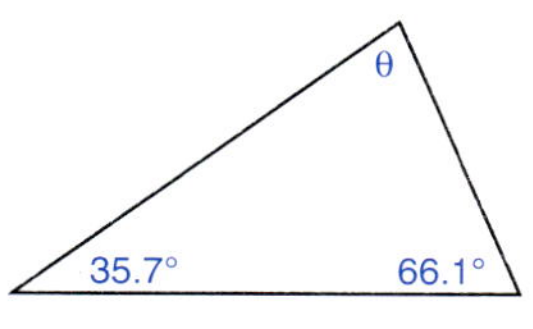

In the following problems find the length of the missing side of right triangle ABC. Leave your answer in both exact form (in terms of rational numbers and radicals) and decimal form, to the nearest tenth.

	a	b	c		a	b	c
4.	3	5	?	**5.**	10	?	20
6.	?	$2\sqrt{2}$	10	**7.**	5	12	?

8. A ladder is 24 feet long, and is leaning against a building so its base is 4.5 feet from the base of the building. How far up the building does the ladder reach?

9. Find the ground speed, to the nearest knot, of an aircraft flying with a heading due west and an airspeed of 120 knots, if there is a wind blowing from the north at 16 knots.

[5–2] Use a calculator to find four-decimal-place approximations for the following.

10. sin 48.3° **11.** cot 58.7°

12. tan 10°20′ **13.** sec 4°38′

14. In the mathematical modeling of an aerodynamics problem the following equation arises:

$$y = x\cos A\cos B - x^2\cos A\sin B - x^3\sin A$$

Compute y to two decimal places if $x = 3.0$, $A = 13.2°$, and $B = 6°$.

Find the unknown acute angle θ to the nearest 0.01°.

15. $\sin\theta = 0.215$ **16.** $\sec\theta = 1.1028$

In the following problems you are given one side and one angle of a right triangle. Solve the triangle. Round all answers to the same number of decimal places as the data.

17. $a = 12.1$, $B = 30.3°$ **18.** $c = 12.6$, $A = 21.9°$

In the following problems you are given two sides of a right triangle. Solve the triangle. Round all lengths to the same number of decimal places as the data and all angles to the nearest 0.1°.

19. $a = 25.1$, $b = 15.0$ **20.** $b = 5.67$, $c = 10.40$

21. The figure illustrates an impedance diagram used in electronics theory. If Z (impedance) = 60.0 ohms and X_L (inductive reactance) = 25.0 ohms, find θ (phase angle) to the nearest degree and R (resistance) to the nearest 0.01 ohm.

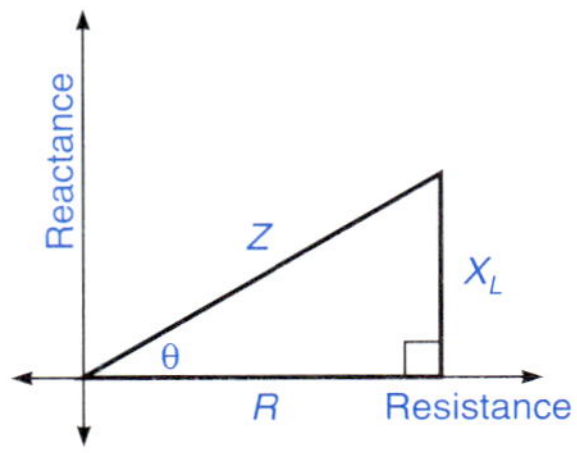

22. An angle of elevation is an angle formed by one horizontal ray and another ray that is above the horizontal. The figure shows measurements made to find the height of a tree. Find the height to the nearest foot, if the angle of elevation is 42.5°, made 86 feet from the foot of the tree.

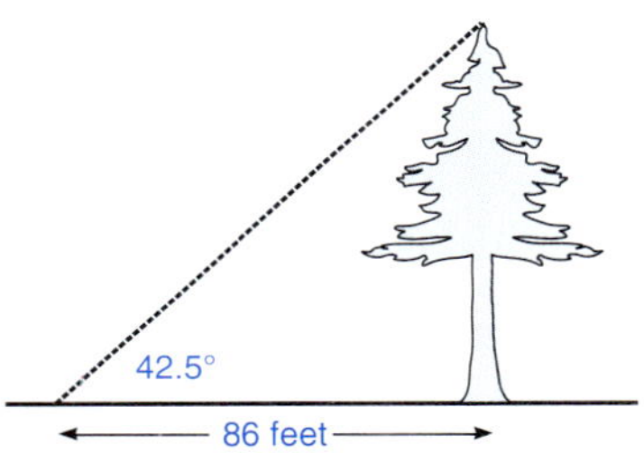

[5–3] In the following exercises state the measure of the smallest nonnegative angle that is coterminal with the given angle.

23. 480° **24.** −140° **25.** 1,256° **26.** −28°45′

[5–4] In the following problems you are given the sign of two of the trigonometric functions of an angle in standard position. State in which quadrant the angle terminates.

27. $\cot \theta > 0$, $\cos \theta < 0$ **28.** $\csc \theta < 0$, $\cot \theta < 0$

For each of the following angles, find the measure of the reference angle θ′.

29. 152.6° **30.** −172.3° **31.** −13.22°
32. 429.40° **33.** −250°

Find the trigonometric function value for each angle. If the reference angle is 30°, 45°, or 60°, find the exact trigonometric function value; otherwise find the required value to four decimal places.

34. sin 135° **35.** cos 48.5° **36.** tan 120°
37. csc 300° **38.** cot(−19°45′)

[5–5]

39. For a surveyor to locate a point by measuring an angle at one station and a distance from another one, the distance BP must be found by solving the following sequence of formulas:

$$\sin p = \frac{AB \sin b}{AP}$$
$$a = 180° - (b + p)$$
$$BP = \frac{AP \sin a}{\sin b}$$

Assume p is acute. Compute the distance BP to the nearest 0.1 foot if $AB = 420$ feet, $AP = 410$ feet, and $b = 20°$.

Find the measure of the least nonnegative angle that meets the conditions given in the following problems, to the nearest 0.1°.

40. $\cos \theta = 0.5252$, $\sin \theta > 0$
41. $\sin \theta = -0.8133$, $\tan \theta > 0$

In each problem you are given the coordinates of a point on the terminal side of θ. In each case (a) find the exact value of the trigonometric functions for θ and (b) find the least nonnegative measure of θ, to the nearest 0.1°.

42. (2,−6) **43.** (−2,4) **44.** $(-5, -\sqrt{2})$

In each case (a) draw a representation of angle θ, (b) use a reference triangle to help find the exact values of the other three primary trigonometric functions (sine, cosine, tangent), and (c) find the least positive value of θ to the nearest 0.1°.

45. $\sin \theta = \frac{4}{7}$, $\cos \theta > 0$ **46.** $\tan \theta = -4$, $\cos \theta > 0$
47. $\csc \theta = \sqrt{3}$, $\sec \theta < 0$

Find values of the three primary trigonometric functions in terms of u.

48. $\cos \theta = u$ and θ terminates in quadrant II.
49. $\tan \theta = u$ and θ terminates in quadrant II.

[5–6] Simplify the following trigonometric expressions.

50. $\sin \theta \csc \theta$ **51.** $\sec \alpha(\cos \alpha - \cot \alpha)$

52. $\dfrac{\sin \theta + 1}{\sin \theta}$ **53.** $\dfrac{\sin \beta - 1}{\cos \beta}$

54. $\cos \theta(\sec \theta - \cos \theta)$

Show that the following are identities.

55. $\cot x\left(\sec x - \tan x + \dfrac{1}{\cot^2 x}\right) = \csc x - 1 + \tan x$

56. $(\sin\alpha - \cos\alpha)(\csc\alpha + \sec\alpha) = \tan\alpha - \cot\alpha$

Solve the following conditional equations for the least nonnegative solution to the nearest 0.1°.

57. $2\sin x = 1$ **58.** $2\cos x = -\sqrt{3}$ **59.** $3\tan x = 5$
60. $\sin 2x = 0.8$ **61.** $\sec 3x = -2$ **62.** $5\sin 2x = 3$
63. $2\cos^2\theta + \cos\theta - 1 = 0$

Chapter 5 test

1. Find the measure of angle θ in the figure.

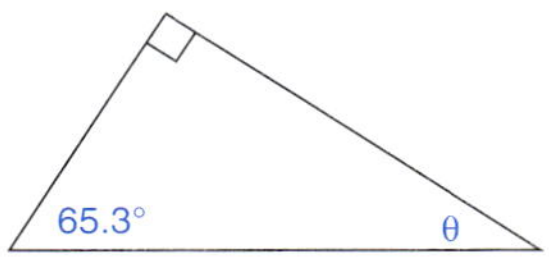

2. Find the length x in the triangle in the figure.

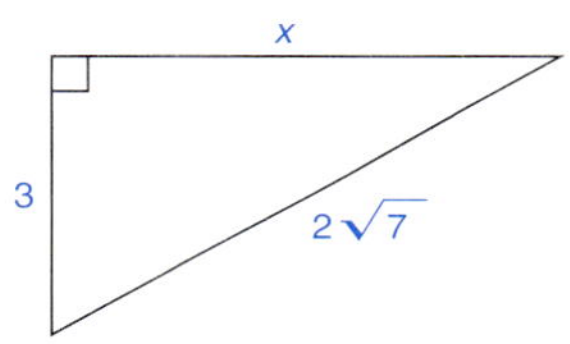

3. A tree that is 46 feet tall has fallen against a building. It reaches 38 feet up a building. How far from the base of the building is the base of the leaning tree, to the nearest foot?

4. Find the ground speed, to the nearest knot, of an aircraft flying with a heading due south and an airspeed of 260 knots, if there is a wind blowing from the west at 30 knots.

Use a calculator to find four-decimal-place approximations for the following.

5. $\sin 27.3°$ **6.** $\tan 11.0°$
7. $\cot 38°10'$ **8.** $\sec 41°38'$

9. In the mathematical modeling of an aerodynamics problem the following equation arises:

$$y = x\cos A\cos B - x^2\cos A\sin B - x^3\sin A$$

Compute y to two decimal places if $x = 1.8$, $A = 6.2°$, and $B = 21°$.

10. If $\cos\theta = 0.8711$, find θ to the nearest 0.1°.

11. In right triangle ABC, $b = 83$ and $B = 19.3°$. Solve the triangle. Leave answers to the nearest 0.1.

12. In right triangle ABC, $b = 83$ and $c = 125$. Solve the triangle. Leave answers to the nearest 0.1.

13. An *angle of depression* is an angle formed by one horizontal ray and another ray that is below the horizontal. The figure shows the situation of an aircraft flying at 30,000 feet, whose radar measures a slant distance of 12 miles to a building. Find the angle of depression of the radar beam (θ). Remember that one mile is 5,280 feet.

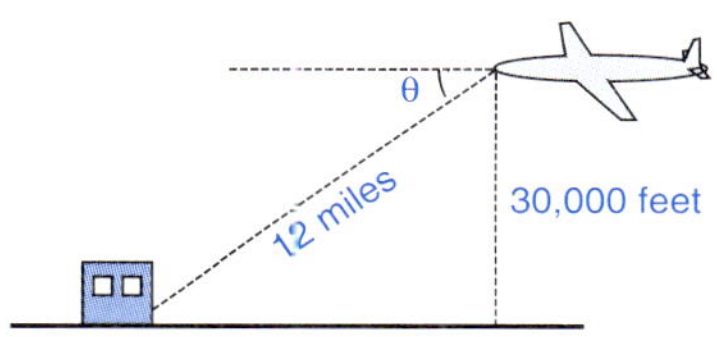

14. Find the least nonnegative angle that is coterminal with an angle of measure 675°.

15. If $\csc\theta > 0$, $\cos\theta < 0$, determine in which quadrant θ terminates.

For each of the following angles, find the measure of the reference angle θ'

16. 211° **17.** −192.1°

Find the trigonometric function value for each angle. If the reference angle is 30°, 45°, or 60°, find the exact trigonometric function value; otherwise find the required value to four decimal places.

18. $\cos 219.8°$ **19.** $\sec 315°$
20. $\cot 128°$ **21.** $\sec 310°$

22. For a surveyor to locate a point by measuring an angle at one station and a distance from another one, the distance BP must be found by solving the following sequence of formulas: $\sin p = \dfrac{AB\sin b}{AP}$; $a = 180° - (b + p)$; $BP = \dfrac{AP\sin a}{\sin b}$. Assume p is acute. Compute the distance BP to the nearest 0.1 meter if $AB = 210$ meters, $AP = 150$ meters, and $b = 15°$.

23. Find the measure of the least nonnegative angle θ for which $\sin\theta = -0.4$ and $\tan\theta < 0$, to the nearest 0.1°.

In each problem you are given the coordinates of a point on the terminal side of θ. In each case (a) find the exact value of the trigonometric functions for θ and (b) find the least nonnegative measure of θ, to the nearest $0.1°$.

24. $(12, -6)$

25. $(-5, \sqrt{5})$

26. $\csc \theta = -\frac{4}{3}$, $\cos \theta > 0$. (a) draw a representation of angle θ, (b) use a reference triangle to help find the exact values of the three primary trigonometric functions (sine, cosine, tangent), and (c) find the least positive value of θ to the nearest $0.1°$.

27. If $\tan \theta = \dfrac{u}{2}$ and θ terminates in quadrant II, find the values of the other five trigonometric functions in terms of u.

Simplify the following trigonometric expressions.

28. $\tan \theta \cot \theta$

29. $\sec \theta(\cos \theta - \cos^3\theta)$

30. Show that $(\sin \theta + \cos \theta)^2 - \sin \theta \cos \theta = 1 + \sin \theta \cos \theta$ is an identity.

Solve the following conditional equations for the least non-negative solution to the nearest $0.1°$.

31. $2 \sin x = \sqrt{2}$

32. $3 \tan x = -5$

33. $\cos 2x = 0.62$

34. $4 \cos^2 x = 3$

C H A P T E R 6

Radian Measure, Properties of the Trigonometric and Inverse Trigonometric Functions

We begin this chapter with a second system of angle measurement, called radians. We then study some additional properties of the trigonometric functions. At the end of the chapter another set of functions are presented—the inverse trigonometric functions.

6–1 Radian measure

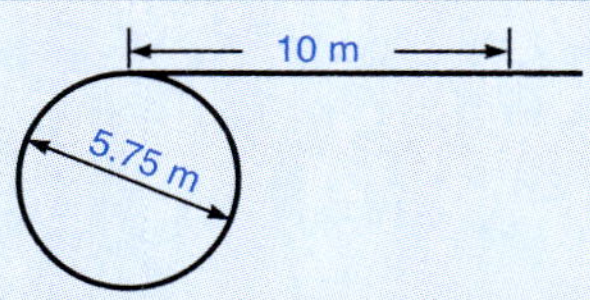

The diameter of a wheel that moves the cable of a ski lift is 5.75 meters, as shown in the diagram. Through what angle, in degrees, does the wheel have to move to advance one of the chairs a distance of 10 meters?

Radian measure is a way to measure angles that is actually used more than degree measure in advanced science and engineering applications. It is useful for solving this problem, as well as many, many other situations.

The unit circle

The circle with radius one and center at the origin is described by the equation

$$x^2 + y^2 = 1$$

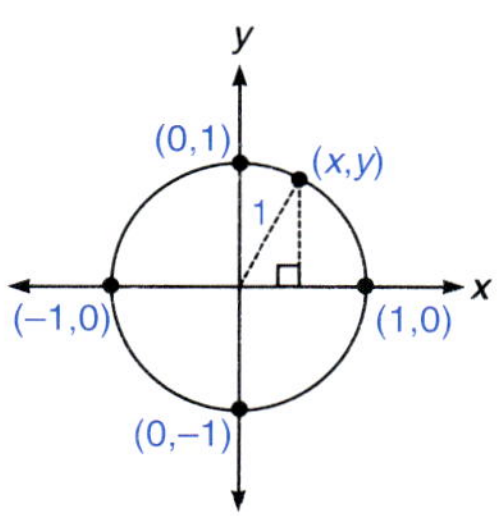

Figure 6–1

It is called the unit circle (see figure 6–1). Observe that the absolute values of the x- and y-coordinates of any point not on an axis describe the lengths of two sides of a right triangle with hypotenuse of length one. The Pythagorean theorem shows that for these points $x^2 + y^2 = 1$. Those points of the circle that are on an axis also satisfy this equation.

The circumference C of a circle with radius r is the distance around the circle. This distance is found using the relation $C = 2\pi r$. Since the radius r for the unit circle is one, its circumference is $C = 2\pi$ (about 6.28 units).

Note The constant π is approximately 3.14159. It is a much-used number, about which entire books have been written. It is an irrational number, and has been approximated to over a billion digits![1]

Radian measure

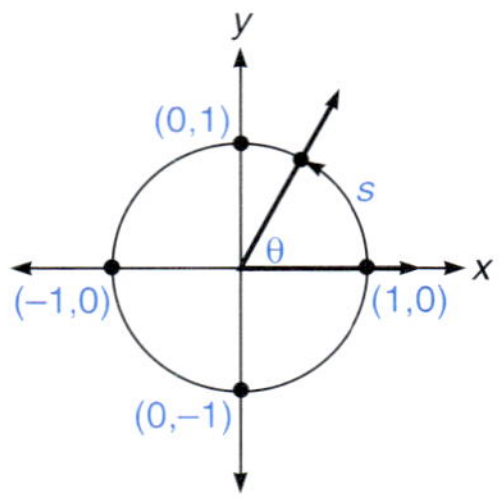

Figure 6–2

A second system of angle measurement uses units called **radians.** This system is used extensively in engineering and scientific applications, as well as in the calculus. We will use it throughout our study of trigonometry. To define this system of angle measurement we use the unit circle.

Let θ be an angle in standard position, and let s represent the distance from the point (1,0) along the circumference of the unit circle to the terminal side of θ. The distance s is called the **arc length** (see figure 6–2). As with degree measure if the distance is measured in a counterclockwise direction we say s is positive, and if in a clockwise direction s is negative.

We define the radian measure of an angle to be this arc length s.

> **Radian measure of an angle in standard position**
> Let θ be an angle in standard position. Let s be the corresponding arc length on the unit circle. Let s be positive if measured in the counterclockwise direction, and negative if measured in the clockwise direction.
> Then s is the radian measure of the angle θ.

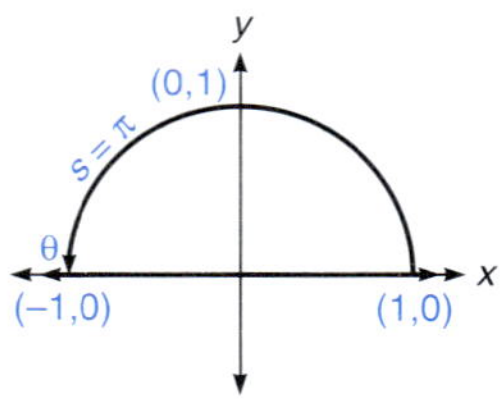

Figure 6–3

For example, an angle of degree measure 180° has an arc length that corresponds to half the circumference of the unit circle. The corresponding radian measure is half of the circumference, or one half of 2π, which is π. Thus, the radian measure of an angle that corresponds to a rotation of one half a circle, in the counterclockwise direction, is π (see figure 6–3).

Conversions between radian and degree measure

Since 360° corresponds to a full revolution, and the circumference of the unit circle (2π) also corresponds to a complete revolution about the unit circle, the following relation is true:

$$\frac{\text{arc length } s}{\text{circumference } (2\pi)} = \frac{\text{measure of angle in degrees}}{360^\circ}$$

If we multiply each member by 2 we obtain the same true statement, but with smaller denominators of π and 180°.

[1]An interesting book on π is *A History of* π by Petr Beckmann, Golem Press, Boulder, Colo., 1977. Gregory V. and David V. Chudnovsky of Columbia University calculated 1,011,196,691 digits of π in 1989.

We use this proportion[2] to convert between degree and radian measure.

Radian/degree proportion

Let θ be an angle in standard position with degree measure $\theta°$ and radian measure s. Then,

$$\frac{s}{\pi} = \frac{\theta°}{180°}$$

The radian measure of an angle is a real number, defined with no units in mind. We often add the word radians after such a measure, but this is not necessary where it is clear that the real number refers to the measure of an angle. Observe that in the radian/degree proportion the ratio of degrees to degrees is unitless also. For example, $\frac{90°}{180°}$ is the same as the unitless ratio $\frac{1}{2}$.

We can describe the measure of an angle in standard position by using degree measure or by stating the arc length to which the angle corresponds on the unit circle. The proportion just described shows the relationship between these two systems.

■ **Example 6–1 A**

Compute the radian or degree measure, given the measure for each angle in degrees or radians.

1. $-210°$

$$\frac{s}{\pi} = \frac{-210°}{180°} \qquad \text{Replace } \theta° \text{ with } -210°$$

$$s = \frac{-210(\pi)}{180} = -\frac{7\pi}{6}$$

Therefore $-210°$ corresponds to $-\frac{7\pi}{6}$.

2. $\frac{7\pi}{5}$

$$\frac{\frac{7\pi}{5}}{\pi} = \frac{\theta°}{180°} \qquad \text{Replace } s \text{ by } \frac{7\pi}{5}$$

$$\frac{7\pi}{5} \cdot \frac{1}{\pi} = \frac{\theta°}{180°} \qquad \text{Division by } \pi \text{ is the same as multiplication by } \frac{1}{\pi}$$

$$\frac{7}{5} \cdot 180° = \theta° \qquad \text{Multiply each member by } 180°$$

$$252° = \theta°$$

$\frac{7\pi}{5}$ (radians) corresponds to $252°$.

[2]A proportion is a statement of equality between two ratios (fractions).

3. 1

Note that this means 1 radian.

$$\frac{1}{\pi} = \frac{\theta°}{180°}$$

$$\frac{180°}{\pi} = \theta° \qquad \text{Multiply each member by } 180°$$

$$57.30° \approx \theta° \qquad \text{Decimal approximation to } \frac{180}{\pi}$$

Thus, one radian corresponds to $\frac{180°}{\pi}$ or about 57.30°. ■

Note It is useful to remember that 1 radian is a little less than 60°, and that 2π radians exactly equals 360°.

Common radian measures

The unit circle can be very helpful in getting a feeling for radian measure. Those values of radian measure that correspond to quadrantal angles (0°, 90°, 180°, etc.) and to angles with reference angles of 30°, 45°, and 60° are common. In particular, the following correspondences are useful: $\frac{\pi}{6}$ and 30°, $\frac{\pi}{4}$ and 45°, and $\frac{\pi}{3}$ and 60°. The unit circle can be conveniently marked in terms of multiples of $\frac{\pi}{6}$ radians (30°) and of multiples of $\frac{\pi}{4}$ radians (45°). This is shown in figure 6–4.

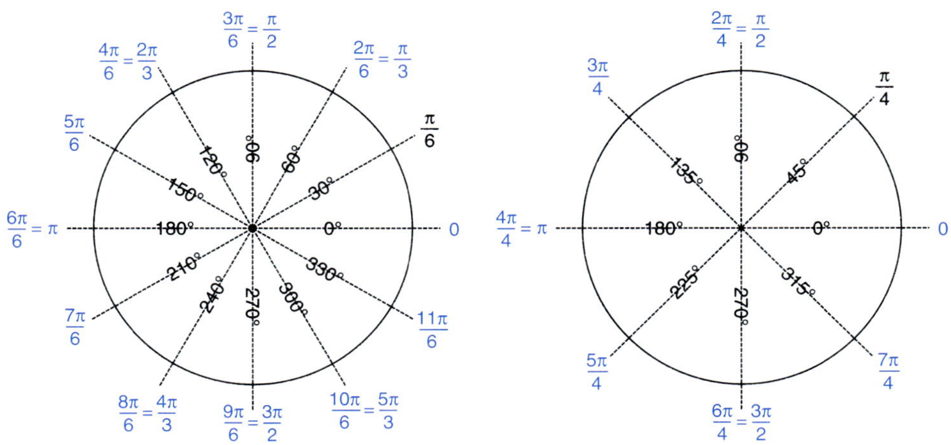

Figure 6–4

Using the unit circle to find values of trigonometric functions

Recall that if (x,y) is a point on the terminal side of an angle θ (in standard position), and $r = \sqrt{x^2 + y^2}$, then $\sin\theta = \frac{y}{r}$ and $\cos\theta = \frac{x}{r}$. On the unit circle, $r = 1$, and therefore if (x,y) is the point on the unit circle that intersects the terminal side of θ, then $\sin\theta = y$, $\cos\theta = x$. Figure 6–5 shows this fact.

Figure 6–5 is very useful to keep handy because it shows the degree and radian measure for many common angles with measure between 0° and 360° (0 and 2π in radian measure). The angles shown are either quadrantal or have reference angles of measure $30°\left(\frac{\pi}{6}\right)$, $45°\left(\frac{\pi}{4}\right)$, or $60°\left(\frac{\pi}{3}\right)$. The figure also shows the point on the terminal side of an angle where it meets the unit circle. As stated, the (x,y) pair is $(\cos\theta, \sin\theta)$. Note that the radian measure is shown for the values 1, 2, 3, . . . , 6 as well. For example, 2 (radians) is near $\frac{2\pi}{3}$ (radians), or 120°, because $\frac{2\pi}{3} \approx 2.1$.

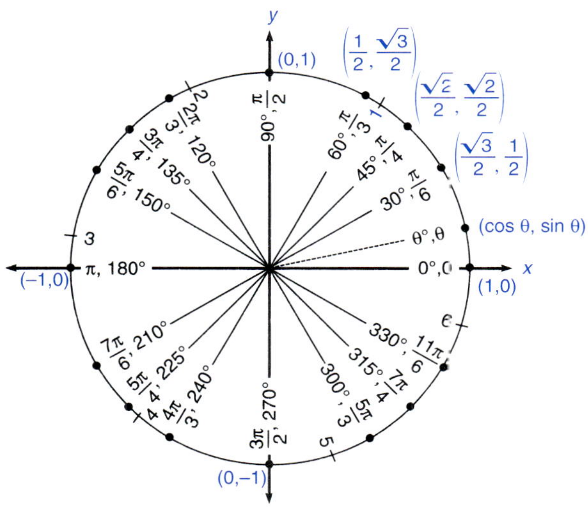

Figure 6–5

Using symmetries in the unit circle

Observe that you can find the sine or cosine value for any of the angles shown by observing the symmetries in figure 6–5. For example, the coordinates at $\frac{4\pi}{3}$ must be $\left(-\frac{1}{2}, -\frac{\sqrt{3}}{2}\right)$. The reason is shown in figure 6–6. Triangle ABO is congruent (same size and shape) to triangle $A'B'O$. This can be shown geometrically because angle AOB is the same size as angle $A'OB'$, and the hypotenuse of each triangle is the same length.

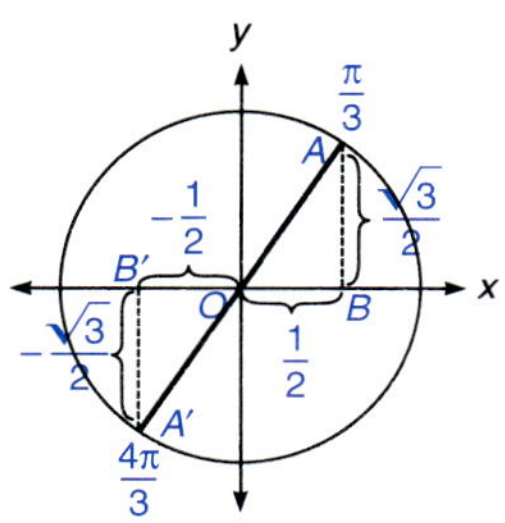

Figure 6–6

Similarly, the coordinate at $\frac{5\pi}{6}$ (figure 6–5) must have the same y-coordinate as at $\frac{\pi}{6}$, but the opposite of the x-coordinate. Thus the coordinates there must be $\left(-\frac{\sqrt{3}}{2},\frac{1}{2}\right)$, and from this point we know that $\sin\frac{5\pi}{6} = \sin 150° = y = \frac{1}{2}$.

Calculators and radian mode

Calculators are programmed to accept angle input in radian measure. All scientific calculators have a key, often marked [DRG] or [MODE], to tell the calculator to accept angles in radian measure. On the TI-81 use the [MODE] key and darken the Rad mode, then use [ENTER] to make the change. For angles that are not coterminal with those in figure 6–5 use the calculator.

■ ***Example 6–1 B***

Find the required value. Use figure 6–5 as an aid when possible; otherwise use a calculator and round the answer to four decimal places.

1. tan 300° The coordinate at 300° is $\left(\frac{1}{2},-\frac{\sqrt{3}}{2}\right)$.

$$\tan 300° = \frac{\sin 300°}{\cos 300°} = \frac{y}{x} = \frac{-\frac{\sqrt{3}}{2}}{\frac{1}{2}} = -\sqrt{3}.$$

2. $\sin\frac{10\pi}{3}$ $\frac{10\pi}{3} = \frac{6\pi}{3} + \frac{4\pi}{3} = 2\pi + \frac{4\pi}{3}$

Thus $\frac{10\pi}{3}$ is coterminal with $\frac{4\pi}{3}$, so $\sin\frac{10\pi}{3} = \sin\frac{4\pi}{3}$.

$$\sin\frac{4\pi}{3} = -\sin\frac{\pi}{3}$$ The points associated with these angles are diagonally opposite on the circle

$$= -\frac{\sqrt{3}}{2}$$

Thus, $\sin\frac{10\pi}{3} = -\frac{\sqrt{3}}{2}$.

3. sin 1.2

Make sure the calculator is in radian mode.

1.2 [sin] Display [0.932039086]

TI-81: [SIN] 1.2 [ENTER]

$\sin 1.2 \approx 0.9320$ ■

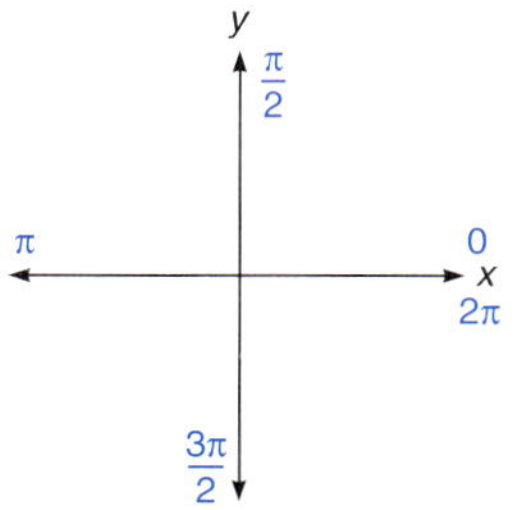

Figure 6–7

Coterminal and reference angles in radian measure

Figure 6–7 shows the smallest positive radian measure of the quadrantal angles, as well as the fact that a full revolution (circle) can be described by 2π radians. Observe that the quadrantal angles 0°, 90°, 180°, 270°, and 360° are, in radians, 0, $\frac{\pi}{2}$, π, $\frac{3\pi}{2}$, and 2π.

In degree measure, all angles that differ in measure by integer multiples of 360° are coterminal. For radian measure the difference is multiples of 2π. If k is an integer (positive, zero, or negative), then integer multiples of 2π are $k \cdot 2\pi$, or $2k\pi$.

> **Coterminal angles, radian measure**
> Two angles α and β are coterminal if $\alpha = \beta + 2k\pi$, k an integer.

Reference angles are found in the same manner as with degree measure (section 5–4) except that 180° becomes π and 360° becomes 2π. If the measure of θ in radians is positive and less than 2π, the following rules give the value of θ', the reference angle.

Quadrant in which θ terminates	Value of θ', the reference angle
I	$\theta' = \theta$
II	$\theta' = \pi - \theta$
III	$\theta' = \theta - \pi$
IV	$\theta = 2\pi - \theta$

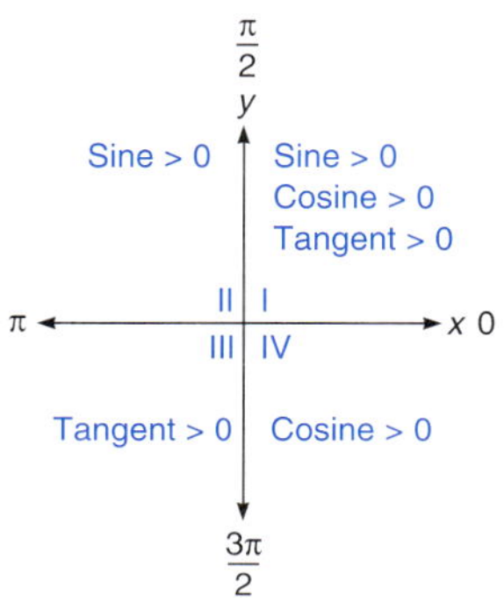

Figure 6–8

The reference angle/ASTC procedure for radian measure

The reference angle/ASTC procedure (section 5–4) can also be used instead of figure 6–5. It is restated here. Table 6–1 is the same as Table 5–1 except that the radian measure of each angle is included. Figure 6–8 is the same as figure 5–13 except that the quadrantal angles less than 2π (360°) are shown in radian measure.

	Sine	Cosine	Tangent
$\frac{\pi}{6}$, 30°	$\frac{1}{2}$	$\frac{\sqrt{3}}{2}$	$\frac{\sqrt{3}}{3}$
$\frac{\pi}{4}$, 45°	$\frac{\sqrt{2}}{2}$	$\frac{\sqrt{2}}{2}$	1
$\frac{\pi}{3}$, 60°	$\frac{\sqrt{3}}{2}$	$\frac{1}{2}$	$\sqrt{3}$

Table 6–1

> **Reference angle/ASTC procedure**
> To find the value of a trigonometric function for a nonquadrantal angle whose reference angle is $\frac{\pi}{6}$, $\frac{\pi}{4}$, or $\frac{\pi}{3}$:
>
> 1. Find the value of the reference angle.
> 2. Find the value of the appropriate trigonometric function for the reference angle from Table 6–1.
> 3. Determine the sign of this value using the ASTC rule (figure 6–8).

In applying this method it is very important to know in which quadrant the angle terminates. We need this information to be able to find the reference angle and to apply the ASTC rule. Example 6–1 C illustrates this process.

■ Example 6–1 C

Use table 6–1 and the ASTC rule to find $\sin \dfrac{11\pi}{3}$

$$\frac{11\pi}{3} = \frac{6\pi}{3} + \frac{5\pi}{3} = 2\pi + \frac{5\pi}{3}$$

Thus $\dfrac{11\pi}{3}$ is coterminal with $\dfrac{5\pi}{3}$, so $\sin \dfrac{11\pi}{3} = \sin \dfrac{5\pi}{3}$. To locate in which quadrant the angle $\dfrac{5\pi}{3}$ terminates, rewrite $\dfrac{5\pi}{3}$ as $\dfrac{10\pi}{6}$ and the quadrantal angles in terms of denominators of 6.

$$0, \quad \frac{\pi}{2}, \quad \pi, \quad \frac{3\pi}{2}, \quad 2\pi$$

$$\frac{0}{6}, \quad \frac{3\pi}{6}, \quad \frac{6\pi}{6}, \quad \frac{9\pi}{6}, \quad \frac{12\pi}{6}$$

Now we see that $\dfrac{9\pi}{6} < \dfrac{10\pi}{6} < \dfrac{12\pi}{6}$ so $\dfrac{10\pi}{6} = \dfrac{5\pi}{3}$ is in quadrant IV.

Step 1: $\theta' = 2\pi - \dfrac{5\pi}{3} = \dfrac{12\pi}{6} - \dfrac{5\pi}{3} = \dfrac{\pi}{3}$ — In quadrant IV, θ' is $360° - \theta$ or $2\pi - \theta$

Step 2: $\sin \dfrac{\pi}{3} = \dfrac{\sqrt{3}}{2}$ — Table 6–1

Step 3: $\sin \dfrac{5\pi}{3} < 0$ — ASTC rule; $\sin \theta < 0$ in quadrant IV

Thus, $\sin \dfrac{5\pi}{3} = -\dfrac{\sqrt{3}}{2}$ and therefore $\sin \dfrac{11\pi}{3} = -\dfrac{\sqrt{3}}{2}$. ■

Arc length and radian measure

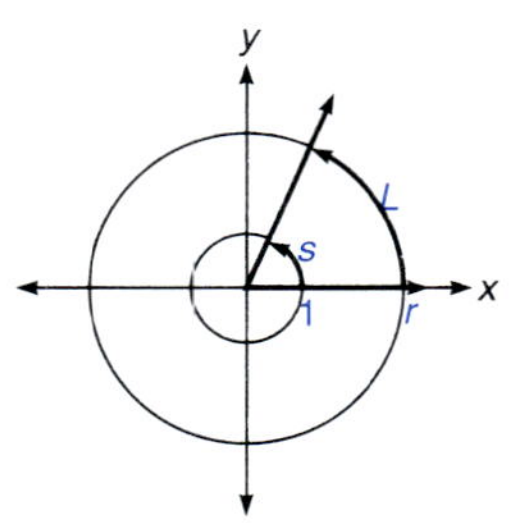

Figure 6–9

A simple relation exists between the radian measure of a central angle s and arc length L determined by that angle on the circumference of any circle (figure 6–9). Geometry tells us that corresponding parts of similar figures form equal ratios. This means, in this case, that $\dfrac{s}{1} = \dfrac{L}{r}$, or $L = rs$. Thus, *if s is the radian measure of a central angle on a circle of radius r, and L is the corresponding arc length,* then

$$L = rs$$

■ Example 6–1 D

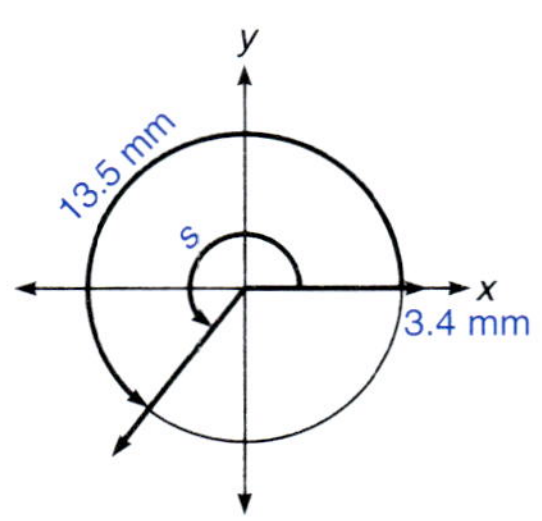

Use the relation $L = rs$ to solve each problem.

1. Find the measure in radians of the central angle corresponding to an arc length of 13.5 mm on a circle of diameter 6.8 mm.

$r = 3.4$ mm — One half the diameter

$L = rs$

$13.5 \text{ mm} = (3.4 \text{ mm})s$ — Substitute known values

$\dfrac{13.5}{3.4} = s$

$3.97 \approx s$ — Rounding to nearest 0.1

Thus, the central angle measures 3.97 radians.

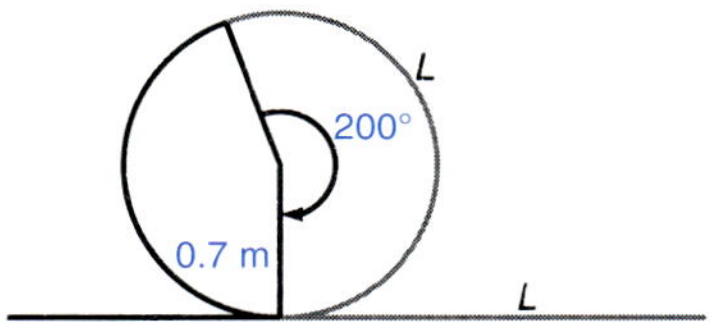

2. A railroad car has wheels with diameter 1.4 m (meters). If the wheels move through an angle of 200°, how far does the train move?

As illustrated, the distance the train will move is the same as the arc length L on the wheel. This length is determined by the central angle of 200°. We will find the measure of the central angle θ in radians, then use the relation $L = rs$.

$$\frac{200°}{180°} = \frac{\theta}{\pi}, \text{ so } \frac{10\pi}{9} = \theta$$

$$L = rs$$

$$L = 0.7\left(\frac{10\pi}{9}\right)$$ The radius r is half the diameter of 1.4 m; $s = \theta$ (in radians)

$$L \approx 2.4 \text{ m}$$

Thus, the train moves 2.4 meters when the wheels move through an angle of 200°. ■

Mastery points

Can you

- Convert between degree and radian measure for angles?
- Find the exact value of a trigonometric function for an angle whose reference angle is $\frac{\pi}{6}$, $\frac{\pi}{4}$, or $\frac{\pi}{3}$ using figure 6–5 or the reference angle/ASTC procedure?
- Use the relation $L = rs$ to solve problems concerning arc length on any circle?

Exercise 6–1

1. State the algebraic relation (equation) that describes the unit circle.

Convert the following degree measures to radian measure. Leave your answers in both exact form and approximated to two decimal places.

2. 30° **3.** 45° **4.** 60°
5. 100° **6.** −200° **7.** −300°
8. −135° **9.** 270° **10.** 750°
11. 127° **12.** −422° **13.** −305°

Convert the following radian measures into degree measure. Leave answers in both exact form and approximated to two decimal places.

14. $\frac{5\pi}{2}$ **15.** $\frac{11\pi}{6}$ **16.** $\frac{2\pi}{7}$
17. $\frac{3\pi}{5}$ **18.** $\frac{10\pi}{9}$ **19.** $\frac{2\pi}{9}$
20. $-\frac{5\pi}{3}$ **21.** $-\frac{17\pi}{6}$ **22.** $-\frac{5\pi}{7}$
23. $\frac{3}{2}$ **24.** $\frac{11}{6}$ **25.** $-\frac{12}{17}$
26. 1.5 **27.** 2 **28.** 3.25
29. −5 **30.** −6

Find the following function values where the angle is given in radian measure. Round your answers to four decimal places.

31. sin 0.9	**32.** cos 1.1	**33.** tan 0.5
34. sec 1.4	**35.** csc 0.7	**36.** cot 1.5
37. sin 2.3	**38.** cos 3.5	**39.** tan 4.1
40. sec 5.2	**41.** csc 2.5	**42.** cot 1.9

Find the exact function values for the following angles.

43. $\sin \frac{2\pi}{3}$	**44.** $\tan \frac{5\pi}{4}$	**45.** $\cos \frac{11\pi}{6}$
46. $\tan \frac{5\pi}{6}$	**47.** $\cos \frac{4\pi}{3}$	**48.** $\sin \frac{5\pi}{6}$
49. $\sin \frac{5\pi}{4}$	**50.** $\cos \frac{7\pi}{4}$	**51.** $\tan \frac{5\pi}{3}$

Solve the following problems.

52. Find the length of the arc determined by a central angle of 2.1 (radians) on a circle of diameter 10 inches.

53. Find the measure, in radians, of a central angle on a circle of radius 4.5 mm (millimeter) determined by an arc length of 12 mm, to the nearest 0.1 radian.

54. Find the length of the arc determined by a central angle of 300° on a circle of diameter 12 mm.

55. Find the length of the arc determined by a central angle of 45° on a circle of radius 8.3 inches, to the nearest 0.1 inch.

56. Find the measure, in both radians and degrees, of the central angle determined by an arc length of 14.5 mm on a circle with diameter 10.3 mm. Round both answers to the nearest tenth.

57. The diameter of a wheel on an automobile is 32.4 inches. If the wheel moves through an angle of 85°, how far will the car move?

58. The diameter of a wheel that moves the cable of a ski lift is 5.75 meters, as shown in the diagram. Through what angle, in degrees, does the wheel have to move to advance one of the chairs a distance of 10 meters?

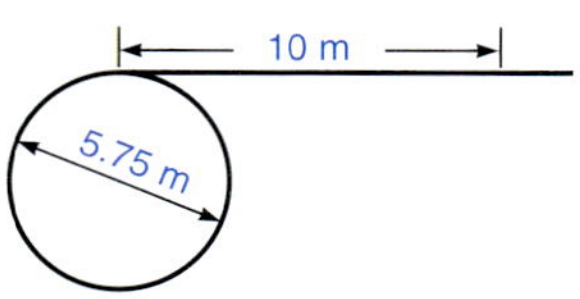

59. The alternator on an automobile engine is attached by a belt to a wheel on the engine. The wheel on the engine has a diameter of 14.88 cm (see the figure), and the wheel on the alternator has a diameter of 9.86 cm. If the wheel driven by the engine moves through an angle of 2.85 radians, through what angle does the alternator move, to the nearest 0.01 radian?

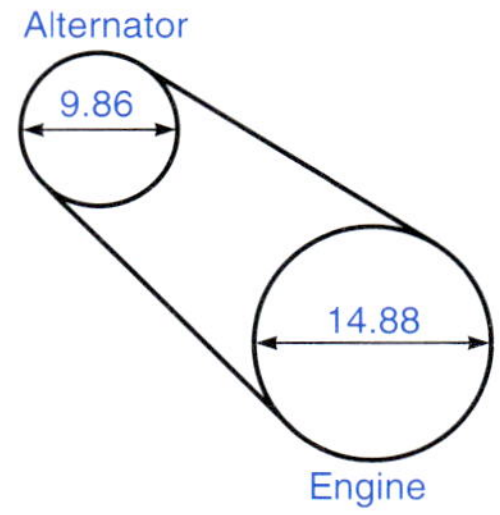

60. A decal is being made to indicate timing marks on a wheel attached to the front of an engine (see the diagram). The radius of the wheel is 86.6 mm. What should the distance be between the −10° and 10° marks, to the nearest millimeter?

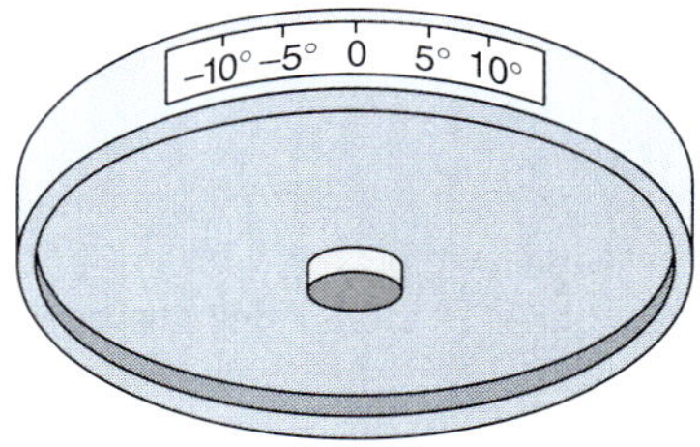

61. In a certain series circuit the applied voltage V, in volts, is determined by the function

$$V = 200 \sin(35t + 1)$$

where t represents time in milliseconds and the expression $35t + 1$ is in radians. Compute V to the nearest 0.1 volt for the following values of t:
a. 0 **b.** 0.1 **c.** 0.8 **d.** 1

62. An equation that arises in finding the trajectory of a rocket is $r = \dfrac{p}{1 + e\cos(s - C)}$. Assume $p = 200$, $e = 1.5$, and $C = 0.5$. Find r if **a.** $s = 1$ **b.** $s = 1.25$

63. The position d at the end of a spring, under certain initial conditions, as a function of time t in seconds, is $d = \frac{1}{3}\cos 8t - \frac{1}{4}\sin 8t$. Compute d to the nearest thousandth for the following values of t: **a.** $\frac{1}{8}$ **b.** $\frac{1}{4}$

64. An equation that can be used to compute $\sin x$, if x is in radians, is called the *Maclaurin series* for the sine function. It is $\sin x = x - \frac{x^3}{3!} + \frac{x^5}{5!} - \frac{x^7}{7!} + \frac{x^9}{9!} - \cdots$, where

$$3! = 1 \cdot 2 \cdot 3 = 6$$
$$5! = 1 \cdot 2 \cdot 3 \cdot 4 \cdot 5 = 120$$
$$7! = 1 \cdot 2 \cdot 3 \cdot 4 \cdot 5 \cdot 6 \cdot 7 = 5{,}040, \text{ etc.}$$

($n!$ is read "n factorial" and is defined as $1 \cdot 2 \cdot 3 \cdot 4 \cdot \cdots \cdot n$.) Although the Maclaurin series goes on forever, good accuracy is obtained by using the first few terms. Use the first four of the five terms shown to compute approximations to

a. $\sin 0.1$ **b.** $\sin 0.5$ **c.** $\sin 1$ **d.** $\sin \frac{\pi}{6}$

Check the results with the sine key of a calculator. (Some computers use a method similar to this for computing the trigonometric functions.)

65. (See problem 64.) The Maclaurin series for the cosine function, if x is in radians, is $\cos x = 1 - \frac{x^2}{2!} + \frac{x^4}{4!} - \frac{x^6}{6!} + \frac{x^8}{8!} - \cdots$. Use the first four of the five terms shown in problem 64 to calculate approximations to

a. $\cos 0.8$ **b.** $\cos 1$ **c.** $\cos 1.3$

d. Approximate $\cos 10°$ by first converting $10°$ to radians.

The following discussion applies to problems 66–75. The area A of a circle is determined by the equation $A = \pi r^2$. The area of a sector of a circle (the shaded portion shown in the diagram) is proportional to the measure of the central angle determining the sector. Thus, the area of a sector determined by an angle θ with measure s (radians) or $\theta°$ is

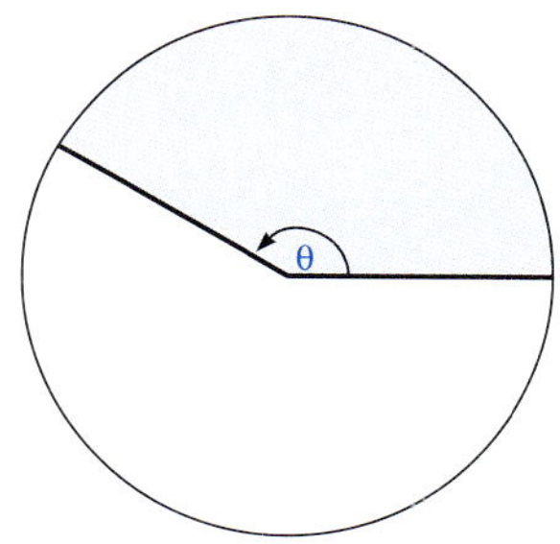

$$\frac{s}{2\pi}A \text{ or } \frac{\theta°}{360°}A$$

Substituting πr^2 for A in each case, and letting A_p mean the area of a sector (p stands for "part"), produces

$$A_p = \frac{s}{2}r^2, \text{ or } A_p = \frac{\theta°}{360°}\pi r^2$$

for the angle measured in radians and degrees, respectively. Note that these expressions only make sense if $0 \le s \le 2\pi$ or $0° \le \theta° \le 360°$. Find the area of the sector determined by each of the following angles and radii. Give both the exact answer and a two-decimal-place approximation.

66. $\theta° = 35°$, $r = 6$ inches

67. $\theta° = 140°$, $r = 7$ mm

68. $s = 4$, $r = 10$ mm

69. $s = 2.4$, $r = 5$ inches

70. $s = \frac{5}{12}$, $r = 6$ inches

71. $s = \frac{2\pi}{5}$, $r = 8$ inches

72. $\theta° = 100°$, $r = 10$ inches

73. $\theta° = 15°$, $r = 9$ mm

74. $s = 3.25$, $r = 44.6$ mm

75. Find the central angle θ necessary to form a sector of area 14.6 cm^2 on a circle of radius 4.85 cm. Find the angle both in radians and degrees, to two decimal places.

76. The figure shows two wheels, with a thin belt wrapped around both wheels. The larger wheel has a radius of 150 cm, and the smaller has a radius of 80 cm. The wheels are 460 cm apart. Ignoring the thickness of the belt, calculate the total length of the belt.

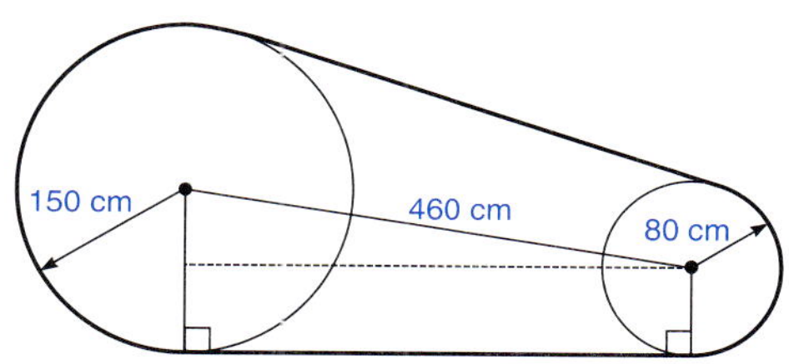

Skill and review

1. Simplify the expression $\frac{\csc x - 1}{\csc x}$ (no fraction should appear in the result).
2. Solve the literal equation $A = (n_1 - 1) + b(n_2 - n_1)$ for n_2.
3. Solve the literal equation $A = (n_1 - 1) + b(n_2 - n_1)$ for n_1.
4. Find the coordinates of the point (a,b) in the figure. Leave the values in exact form (do not put the results in decimal form).
5. Solve the inequality $\left|\frac{1}{2}x + 10\right| > 8$.

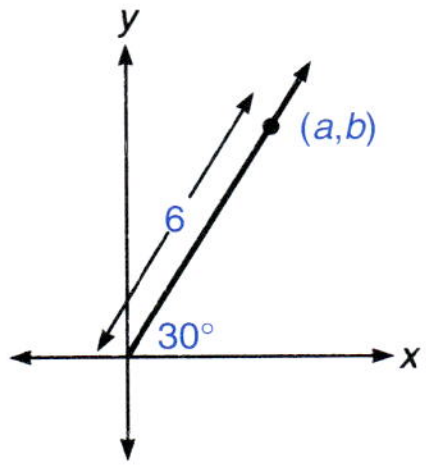

6–2 *Properties of the sine, cosine, and tangent functions*

The activity of sunspots seems to follow an 11-year cycle. Assuming that this activity can be roughly modeled with a sine wave, construct a sine function with period $\frac{360}{11}$°, amplitude 1, phase shift 90°, and vertical translation 2.

To understand many applications of the trigonometric functions, such as the one in this problem, it is necessary to have a good understanding of the properties of these functions. As with any functions we can gain a great deal of knowledge from their graphs.

In more advanced work we use radian measure more than degree measure. Radian measure is stressed for the remainder of this chapter.

Graph and properties of the sine function

Figure 6–10 is the graph of $y = \sin x$ for x (in radians) between 0 and 2π. This graph can be obtained by plotting points for various values of x. The points for the table of values shown are plotted in the figure. When we allow x to take on any value, this graph is repeated over and over. This is because for $x > 2\pi$ or $x < 0$ we have angles that are coterminal with values we have already plotted. Every 2π units (once around the unit circle) we find that the graph repeats itself.

To obtain the graph with the TI-81 calculator, first make sure the calculator is in radian mode. See section 6–1.

x	$\sin x$
0	0
$\frac{\pi}{6} \approx 0.52$	$\frac{1}{2} = 0.5$
$\frac{\pi}{4} \approx 0.79$	$\frac{\sqrt{2}}{2} \approx 0.707$
$\frac{\pi}{3} \approx 1.05$	$\frac{\sqrt{3}}{2} \approx 0.866$
$\frac{\pi}{2} \approx 1.57$	1

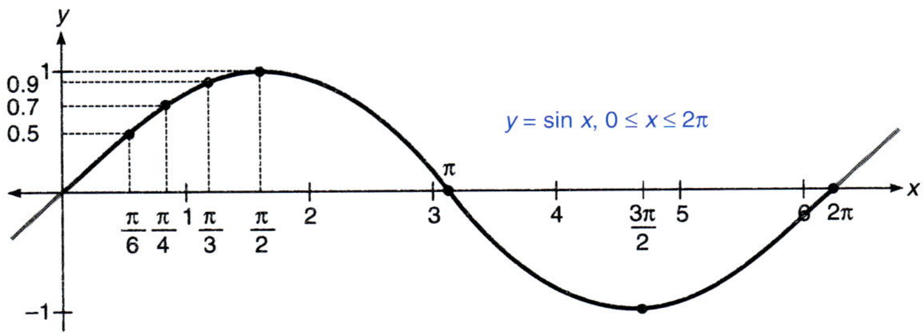

Figure 6–10

To graph the function, do the following:

ZOOM 7	Trig (basic scale settings for many trigonometric functions)
RANGE	Change Xmin to 0, Ymin = −2, Ymax = 2, Yscl = .5
Y= SIN X|T GRAPH	

We can use the graph in figure 6–10 to produce the graph in figure 6–11, which shows the graph of the sine function for all values of x. The graph in figure 6–10 is one *cycle* of the sine function. We refer to figure 6–10 as the **basic sine cycle.**

The repetitious nature of the sine function can be described with the identity

$$\sin x = \sin(x + k \cdot 2\pi), \quad k \text{ any integer}$$

We say that the *sine function is 2π-periodic,* because it repeats every 2π units. We also say that the **amplitude** of the sine function is 1. We define the amplitude of a periodic function like the sine function as half the difference of its greatest and least values.

The graph in figure 6–11 shows that the domain of the sine function is all real numbers, while the range (the y-values) is restricted to the interval from −1 to 1. This and other information is summarized later in table 6–2.

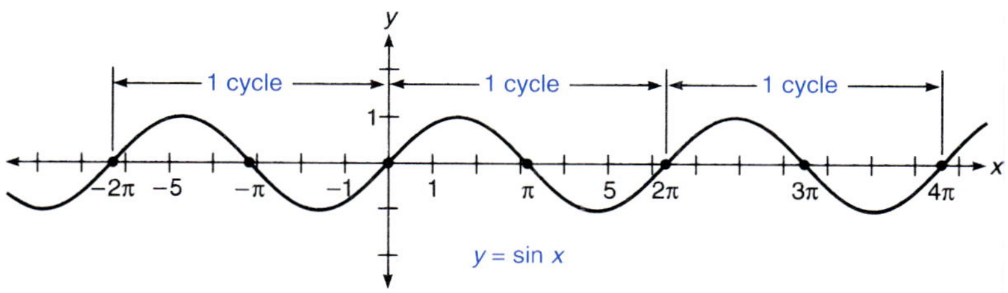

Figure 6–11

Another important point is that $\sin(-x) = -\sin x$ for any value of x. This is illustrated in figure 6–12, where we see that if we go equal distances in the positive and negative directions along the x-axis, the value of the sine function at each place is of the same magnitude (absolute value) but of the opposite sign. *Any function for which $f(-x) = -f(x)$ is true for all x in its domain is called an* ***odd function.*** Thus, *sine is an odd function.* Recall that the even/odd property of functions was introduced in section 3–5. We noted there that the graph of an odd function has symmetry about the origin. Example 6–2 A illustrates one application of the odd function property.

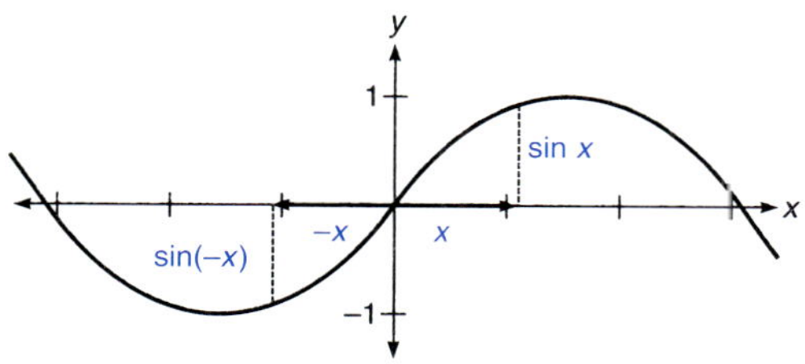

Figure 6–12

Graph and properties of the cosine function

x	$\cos x$
0	1
$\frac{\pi}{6} \approx 0.52$	$\frac{\sqrt{3}}{2} \approx 0.866$
$\frac{\pi}{4} \approx 0.79$	$\frac{\sqrt{2}}{2} \approx 0.707$
$\frac{\pi}{3} \approx 1.05$	$\frac{1}{2} = 0.5$
$\frac{\pi}{2} \approx 1.57$	0
π	-1

The table shows some values of x and $\cos x$. Plotting these and other ordered pairs (x,y) where $y = \cos x$ (x in radians) and then connecting them with a smooth curve produces the graph shown in figure 6–13, for $0 \leq x \leq 2\pi$. Of course a graphing calculator could also be used to obtain this graph. The steps would be practically the same as those previously shown for graphing $y = \sin x$.

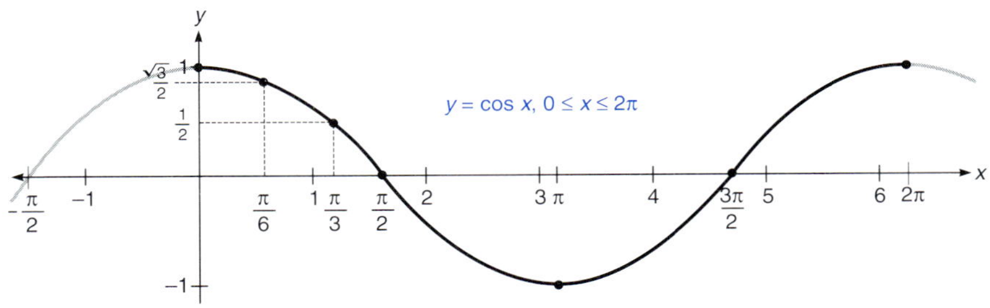

Figure 6–13

Just as with the sine function, *the cosine function is* 2π*-periodic.* Thus for any angle x, $\cos x = \cos(x + k \cdot 2\pi)$, where k is any integer. The graph in figure 6–13 is one *cycle* of the cosine function. We refer to it as the **basic cosine cycle.** If we repeat the basic cosine cycle we obtain the graph of figure 6–14. In figure 6–14, we can see that the domain of the cosine function is all real numbers and that the range is the values from -1 to 1 (as with the sine function). Also, the amplitude of the cosine function is 1.

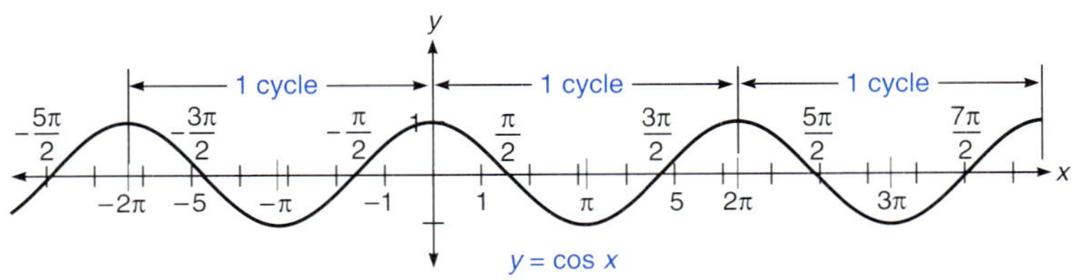

Figure 6–14

The graph in figure 6–14 also shows that $\cos(-x) = \cos x$ for all x. This is further illustrated in figure 6–15. *Any function for which* $f(-x) = f(x)$ *for all* x *is called an* ***even function,*** so the *cosine function is an even function.* Recall from section 3–5 that the graph of a function with the even function property has y-axis symmetry.

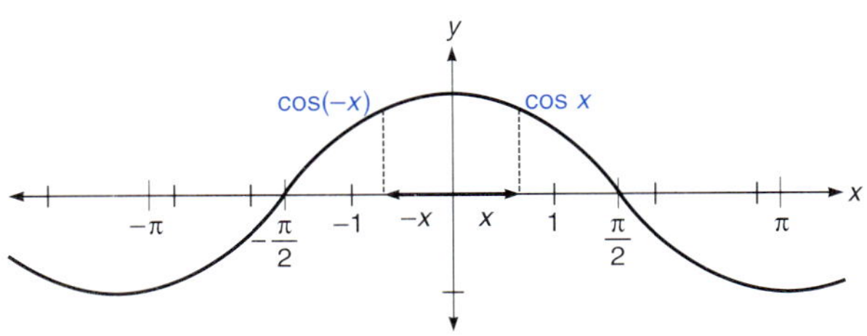

Figure 6–15

Finally, *the graphs of the sine and cosine functions have exactly the same shape*. Either one is identical to the other if it is shifted right or left by a suitable amount. The smallest amount is $\frac{\pi}{2}$, which is described in the statement that $\sin\left(x + \frac{\pi}{2}\right) = \cos x$. This statement can be proved with methods shown in section 7–2.

Graph and properties of the tangent function

To obtain the graph of the tangent function we can compute values and plot points. Some values of $\tan x$ for $0 \leq x < \frac{\pi}{2}$ are shown in the table, and the graph of the tangent function is shown in figure 6–16.

x	$\tan x$
0	0
$\frac{\pi}{6} \approx 0.52$	$\frac{\sqrt{3}}{3} \approx 0.6$
$\frac{\pi}{4} \approx 0.79$	1
$\frac{\pi}{3} \approx 1.05$	$\sqrt{3} \approx 1.7$
1.25	≈ 3.0
1.50	≈ 14.1
1.55	≈ 48.1
$\frac{\pi}{2} \approx 1.57$	undefined

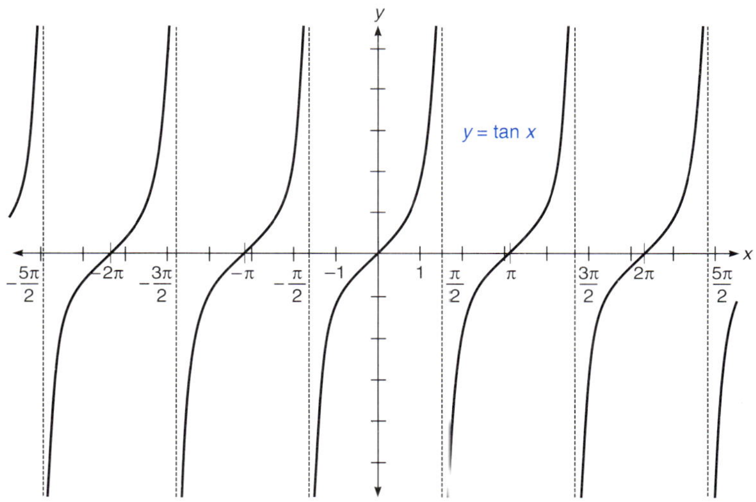

Figure 6–16

Observe that as x gets closer and closer to certain values, such as $\frac{\pi}{2}$, $|\tan x|$ gets larger and larger. At these values $\sin x \approx \pm 1$, so near these values $|\tan x| = \left|\frac{\sin x}{\cos x}\right| \approx \frac{1}{|\cos x|}$. Near these same points $|\cos x|$ is approaching 0. The reciprocal of a very small number is a very large number, so $\frac{1}{|\cos x|}$ gets larger and larger.

To obtain the graph above on the TI-81 use [RANGE −8,8,−5,5], Xscl $= 1.5708 \approx \frac{\pi}{2}$, Yscl = 1. It will not look as nice, and the vertical, dashed lines (discussed below) will be solid. They are actually not part of the graph, and it is only the technical limitations of the calculator that causes them to be drawn at all.

If we recall that $\tan\theta = \dfrac{y}{x}$, where (x,y) is a point on the terminal side of angle θ, and if we examine figure 6–5, we will see that $x = 0$ at $\pm\dfrac{\pi}{2}$, $\pm\dfrac{3\pi}{2}$, $\pm\dfrac{5\pi}{2}$, ... Thus, $\tan\theta$ is not defined for these angles, which can be described as $\dfrac{\pi}{2} + k\pi$, k an integer. Figure 6–16 shows vertical dashed lines at these points. These lines are **vertical asymptotes** of the tangent function. The tangent function is defined for all other values. The vertical asymptotes are lines that the tangent function approaches more and more closely as x approaches those values where the tangent function is undefined.

Because the tangent function gets arbitrarily large or small, the range of the tangent function is all real numbers. We do not define an amplitude for the tangent function.

Examination of figure 6–16 shows that *the tangent function is* π*-periodic.* Thus, for any x, $\tan x = \tan(x + k\pi)$, k any integer. The actual proof of this fact is an exercise in chapter 7. The *tangent function is an odd function,* as was the sine function. This can be shown algebraically as follows:

$$\tan(-x) = \frac{\sin(-x)}{\cos(-x)} \quad \text{Identity}$$

$$= \frac{-\sin x}{\cos x} \quad \text{Sine is odd, cosine is even}$$

$$= -\frac{\sin x}{\cos x}$$

$$= -\tan x.$$

Thus, $\tan(-x) = -\tan x$ for any value x, and therefore tangent is an odd function.

Summary of properties

Table 6–2 summarizes the properties of the three functions we have examined. The unit circle (figure 6–5), the properties found in table 6–2, and the graphs of these three functions, have many applications. Some are illustrated in example 6–2 A.

Function	Domain	Range	Period
$y = \sin x$	R	$-1 \le y \le 1$	2π
$y = \cos x$	R	$-1 \le y \le 1$	2π
$y = \tan x$	$x \ne \dfrac{\pi}{2} + k\pi,\ k \in J$	R	π

$\sin(-x) = -\sin x$ (odd, origin symmetry)
$\cos(-x) = \cos x$ (even, y-axis symmetry)
$\tan(-x) = -\tan x$ (odd, origin symmetry)

Table 6–2

■ ***Example 6–2 A***

Use table 6–2 and the graphs of the sine, cosine, and tangent functions to solve the following problems.

1. Find $\sin\left(-\frac{\pi}{3}\right)$

$$\sin\left(-\frac{\pi}{3}\right) = -\sin\frac{\pi}{3} \qquad \text{Sine is an odd function, so } \sin(-x) = -\sin(x)$$
$$= -\frac{\sqrt{3}}{2}$$

2. Test the function $f(x) = x \sin x$ for the even/odd property. State which type of symmetry the graph of this function would have based on being even, odd, or neither even nor odd.

$$f(x) = x \sin x$$
$$f(-x) = (-x)\sin(-x)$$
$$= (-x)(-\sin x) \qquad \sin(-x) = -\sin x$$
$$= x \sin x$$

Thus, $f(x) = f(-x)$, so this function has the even function property. Its graph would therefore have symmetry about the y-axis.

3. Describe all points for which $\cos x = -\frac{1}{2}$.

Examination of the unit circle (figure 6–5) shows there are two points on $0 \le x < 2\pi$ for which $\cos x = -\frac{1}{2}$. These are $\frac{2\pi}{3}$ and $\frac{4\pi}{3}$. This can also be seen in figure 6–14. Since the cosine function is 2π-periodic, $\cos x = -\frac{1}{2}$ for $x = \frac{2\pi}{3} + 2k\pi$ or $\frac{4\pi}{3} + 2k\pi$, $k \in J$. ■

Linear transformations of the sine and cosine functions

Most applications of the sine, cosine, and tangent functions require that they be transformed in some way to fit measured data or describe theoretical properties. In the remainder of this section, we examine linear transformations of these functions. Graphically these transformations represent moving the graph horizontally or vertically, or squeezing or expanding the graph horizontally or vertically.

Vertical scaling factors and translations

We first discuss *vertical scaling factors and translations.* We know from section 3–4 that, given the graph of $y = f(x)$, then the graph of $y = k \cdot f(x)$ is a vertically scaled version of the graph of $f(x)$.

Thus, the graph of $y = 2 \sin x$ is the same as the graph of $y = \sin x$, but vertically expanded 2 units. This is shown in figure 6–17. Similarly, the graph of $y = -4 \cos x$ is the graph of $y = \cos x$ but expanded vertically by a factor of 4 *and* reversed about the x-axis, because the coefficient -4 is negative.

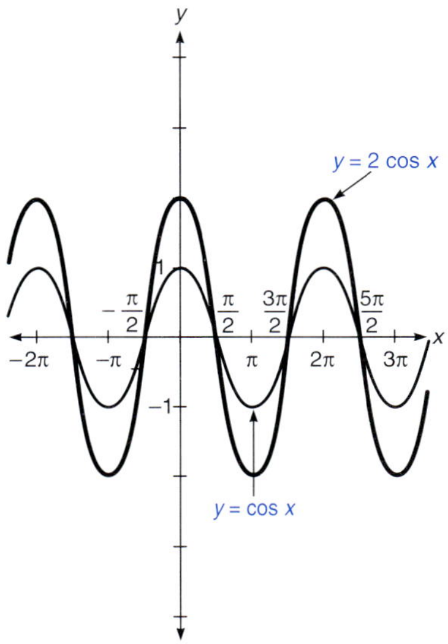

Figure 6–17

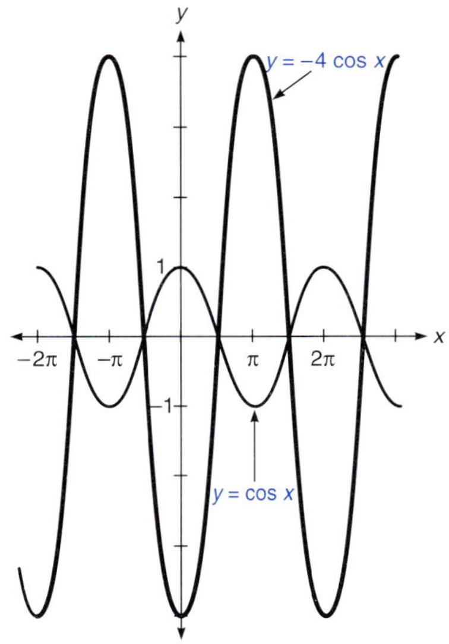

Figure 6–18

This is shown in figure 6–18. In general the graphs of $y = A \sin x$ and $y = A \cos x$, $A \varepsilon R$ are scaled vertically by the factor $|A|$ and the graph is reflected about the horizontal axis if $A < 0$.

We know (section 3–4) that the graph of $y = f(x) + k$ is the graph of $y = f(x)$ but shifted vertically by k units. Thus, for example, the graph of $y = \sin x + 3$ (not to be confused with $y = \sin(x + 3)$) is the graph of $y = \sin x$ but shifted vertically by 3 units. This is illustrated in figure 6–19. Thus, the graphs of $y = \sin x + D$ and $y = \cos x + D$ are the graphs of $y = \sin x$ or $y = \cos x$, shifted vertically D units.

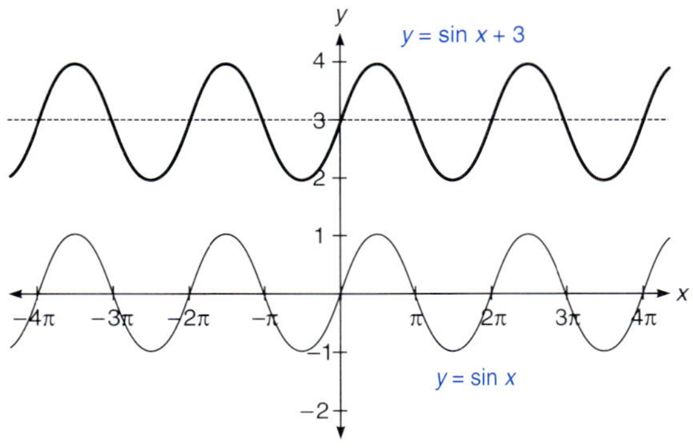

Figure 6–19

■ *Example 6–2 B*

Graph the function $y = 2 \sin x - 3$.

This is the graph of $y = \sin x$ but shifted down 3 units, and with amplitude 2. This is shown in the figure.

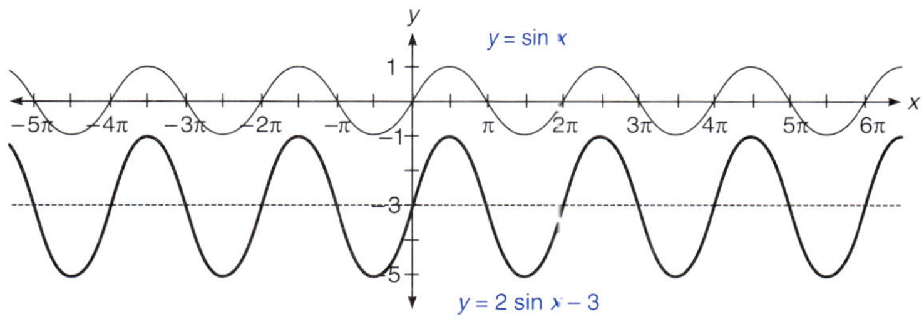

Y=	2	SIN	X|T	−	3	RANGE −16,16,−6,2	Xscl = 1.5708

■

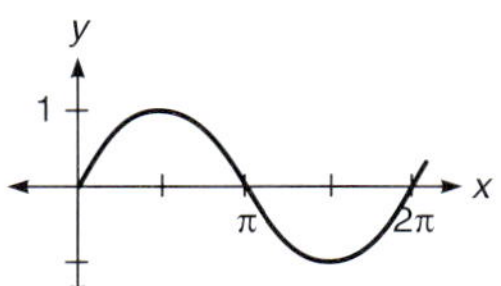

Sine function

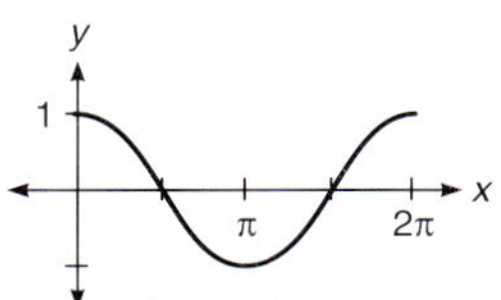

Cosine function

Figure 6–20

Horizontal scaling factors and translations

The **argument** of a function is the expression that represents the domain element. In $y = \sin x$, the argument is x. In $y = \cos 3x$ the argument is $3x$. In $y = \tan\left(x - \frac{\pi}{2}\right)$ the argument is $x - \frac{\pi}{2}$.

Now consider what we know about the sine and cosine functions. *As the argument takes on values between 0 and 2π each of these functions has the graph shown in figure 6–20.* These are the basic sine and cosine cycles discussed earlier. Observe that the basic sine cycle is 0 at its beginning, middle, and end points, while the cosine function starts at 1, ends at 1, and is −1 at its midpoint.

Now consider what the graph of $y = \sin\left(x - \frac{\pi}{4}\right)$ should look like. We know that one basic cycle of the sine function is produced as the argument goes from 0 to 2π. In this case, the argument is $x - \frac{\pi}{4}$. We find the values of x for which the argument goes from 0 to 2π as follows:

$$0 \le x - \frac{\pi}{4} \le 2\pi$$

$$\frac{\pi}{4} \le x \le 2\pi + \frac{\pi}{4} \qquad \text{Add } \frac{\pi}{4} \text{ to each member}$$

$$\frac{\pi}{4} \le x \le \frac{9\pi}{4}$$

Thus, as x takes on the values $\frac{\pi}{4}$ through $\frac{9\pi}{4}$, $x - \frac{\pi}{4}$ takes on the values from 0 through 2π, tracing out the graph of the basic sine function. Thus, the basic sine function starts at $\frac{\pi}{4}$ and ends at $\frac{9\pi}{4}$. See figure 6–21, where we mark the axis in units of $\frac{\pi}{4}$ for convenience.

The distance between the points at $\frac{9\pi}{4}$ and $\frac{\pi}{4}$ is $\frac{9\pi}{4} - \frac{\pi}{4} = \frac{8\pi}{4} = 2\pi$. This is the period of this function. The value $\frac{\pi}{4}$ is called the phase shift of the function. It is the value to which the starting point of the basic cycle shifted.

Y=	SIN	(	X|T
−	π	÷ 4	)

RANGE −5.5,13.5,−2,2

Xscl=0.7854 ≈ $\frac{\pi}{4}$

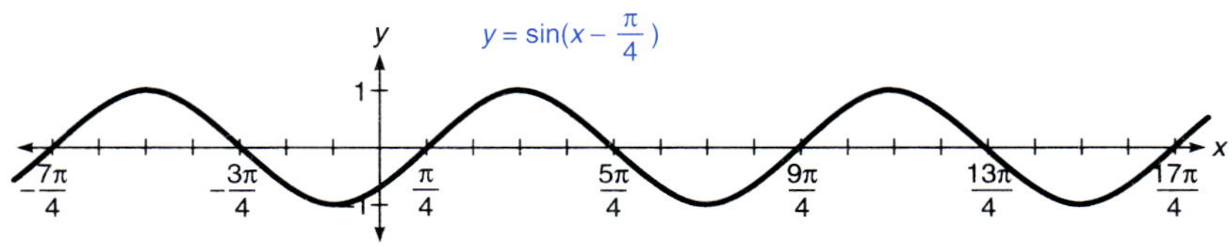

Figure 6–21

This method illustrates how to graph sine and cosine functions where the argument is of the form $Bx + C$. In these cases, we get horizontal scaling (the period) and horizontal shifts (the phase shift).

> **Algebraic methods to graph sine and cosine functions of the form**
>
> $$y = A\sin(Bx + C) + D \quad \text{and}$$
> $$y = A\cos(Bx + C) + D, \quad \text{where } B > 0$$
>
> 1. Solve $0 \le Bx + C \le 2\pi$ so x is the middle member.
>   - This gives the left and right end points for one basic cycle.
>   - The left end point is the phase shift.
>   - The difference between the end points is the period.
> 2. The amplitude is $|A|$.
>   - This is the height of the basic graph above and below the x-axis.
>   - The graph is reflected across the horizontal axis if $A < 0$.
>
>   Draw one basic cycle with the information from steps 1 and 2.
> 3. Repeat the cycle obtained from steps 1, 2 to obtain more of the graph.
> 4. Shift the graph vertically D units.

Period and phase shift—definition and method of computation

We formalize the terms period and phase shift in the following way. We perform step 1: $0 \le Bx + C \le 2\pi$

$$-C \le Bx \le 2\pi - C \qquad \text{Subtract } C \text{ from each member}$$

$$-\frac{C}{B} \le x \le \frac{2\pi - C}{B} \qquad \text{Divide each member by } B$$

The expression $-\frac{C}{B}$ is called the **phase shift,** of the particular sine or cosine function. The difference between the left and right end points of the basic cycle, $\frac{2\pi - C}{B} - \left(-\frac{C}{B}\right) = \frac{2\pi}{B}$ is called the **period** of the function. B is also the number of complete cycles in 2π units.

To compute the phase shift and period, either memorize the definitions $-\frac{C}{B}$ and $\frac{2\pi}{B}$ or solve the inequality $0 \le Bx + C \le 2\pi$ for x. The second method is used in example 6–2 C.

■ *Example 6–2 C*

Graph each function. Show at least two cycles. State the amplitude, period, and phase shift.

1. $y = -2 \cos 4x$

Step 1: $0 \leq 4x \leq 2\pi$

$0 \leq x \leq \frac{\pi}{2}$ Divide each member by 4

Thus one basic cycle of the cosine function starts at 0 and ends at $\frac{\pi}{2}$. The period is $\frac{\pi}{2} - 0 = \frac{\pi}{2}$, and the phase shift is 0.

Step 2: The amplitude is $|-2| = 2$. There is a reflection about the horizontal axis. Use this information to draw one basic cycle; this is shown in the figure.

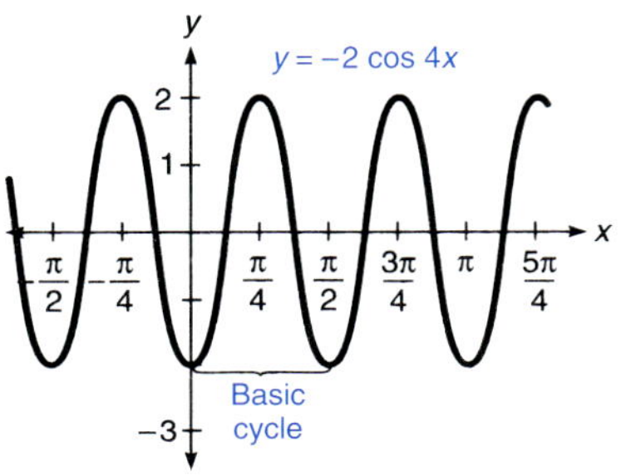

Step 3: Draw additional cycles by repeating this pattern to the right and left of the original basic cycle. Since there is no value for D (it is therefore 0), we do not need to consider step 4.

Y=	(−)	2	COS	(	4	X|T	)	RANGE −2,4,−3,3

Xscl = 0.7854

Note $\frac{\pi}{4} \approx 0.7854.$

2. $y = \sin\left(2x + \frac{\pi}{3}\right)$

Step 1: $0 \leq 2x + \frac{\pi}{3} \leq 2\pi$

$0 \leq 6x + \pi \leq 6\pi$ Multiply each term of each member by 3

$-\pi \leq 6x \leq 5\pi$ Subtract π from each member

$-\frac{\pi}{6} \leq x \leq \frac{5\pi}{6}$ A basic cycle starts at $-\frac{\pi}{6}$ and ends at $\frac{5\pi}{6}$

The phase shift is $-\frac{\pi}{6}$, and period is $\frac{5\pi}{6} - \left(-\frac{\pi}{6}\right) = \pi.$

Step 2: The amplitude is 1. We draw one basic cycle (see the figure). It is convenient to mark the x-axis in increments of $\frac{\pi}{6}$.

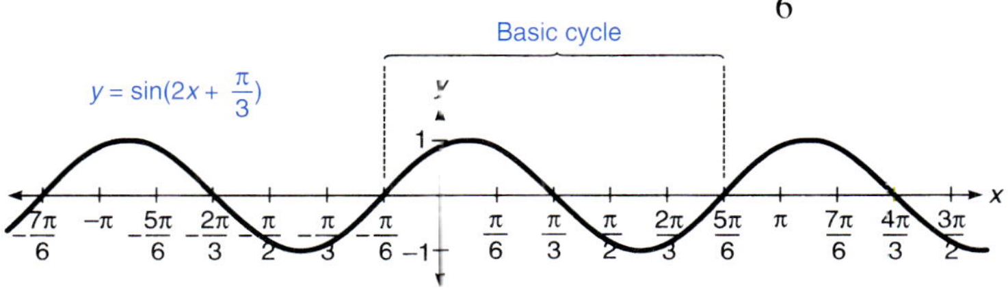

Y=	SIN	(	2	X|T
+	π	÷	3	)

RANGE −3.7,4.7,−2,2

Xscl = .5236 ≈ $\frac{\pi}{6}$

Step 3: We sketch one complete cycle to the left of the basic cycle and a portion of a third cycle to its right.

3. $y = 3\cos\frac{\pi x}{2} + 1$

Step 1: $0 \le \frac{\pi x}{2} \le 2\pi$

$0 \le \pi x \le 4\pi$ Multiply each member by 2

$0 \le x \le 4$ Divide each member by π

Thus one basic cosine cycle begins at 0 and ends at 4. The phase shift is 0, and the period is $4 - 0 = 4$.

Step 2: The amplitude is 3. We draw one basic cosine cycle between 0 and 4, with amplitude 3 (see the figure).

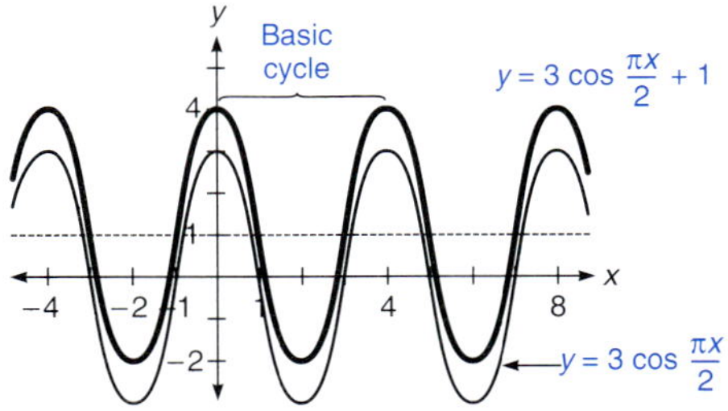

Step 3: We repeat one cycle to the left and one to the right of the basic cycle.

Step 4: We shift the graph up 1 unit. The horizontal line $y = 1$ helps orient the figure.

Y= 3 COS (π X|T ÷ 2) + 1
RANGE −4,8,−3,5

Negative coefficients of *x*

If the coefficient of x, B, is negative in the argument $Bx + C$ we use the odd and even properties to get an equivalent expression with B positive. We can then graph the equivalent expression where the coefficient of x is positive.

■ ***Example 6–2 D***

Rewrite each function so the coefficient of x is positive.

1. $y = 2\cos(-3x)$

$y = 2\cos 3x$ $\cos\theta = \cos(-\theta)$

2. $y = 3\sin(\pi - 2x)$

$y = 3\sin[-(2x - \pi)]$ $\pi - 2x = -(2x - \pi)$

$y = -3\sin(2x - \pi)$ $\sin(-\theta) = -\sin\theta$ ■

Finding an equation from its properties

There are times when we will want to find the equation of a function, given some of its properties.

■ Example 6–2 E

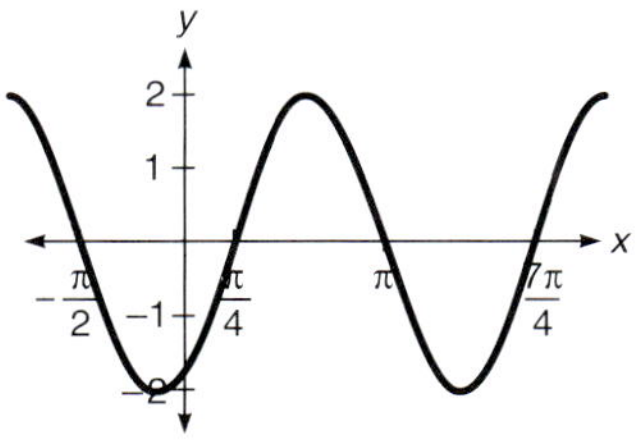

Find the equation of the function in the figure, assuming it is of the form $y = A\sin(Bx + C) + D$.

Since the distance between the high and low points of the graph is 4 we know that A is 2. There is no vertical shift, so D is 0.

One cycle of the sine function begins at $\frac{\pi}{4}$ and ends at $\frac{7\pi}{4}$. We work backward from this information. We want to obtain end points of 0 and 2π:

$\frac{\pi}{4} \leq x \leq \frac{7\pi}{4}$	One basic cycle is between these end points
$\pi \leq 4x \leq 7\pi$	Multiply each term by 4 to clear fractions
$0 \leq 4x - \pi \leq 6\pi$	Subtract π from each member; this makes the left member 0
$\frac{0}{3} \leq \frac{4x - \pi}{3} \leq \frac{6\pi}{3}$	Divide each member by 3; this makes the right member 2π
$0 \leq \frac{4x}{3} - \frac{\pi}{3} \leq 2\pi$	$Bx + C$ is $\frac{4x}{3} - \frac{\pi}{3}$

Thus, the function is $y = 2\sin\left(\frac{4x}{3} - \frac{\pi}{3}\right)$. ■

One of the places trigonometry finds application is in alternating current theory in electronics. Many such applications express x in degrees instead of radians. Our graphing procedures are the same except that our limits for the basic cycle are 0° and 360°.

■ Example 6–2 F

1. An AC (alternating current) signal with peak-to-peak voltage of 170 volts and phase shift of −120°, riding on a DC (direct current) level of 100 volts, could be described by the function $y = 85\sin(x + 120°) + 100$, where y represents volts and x is in degrees. Graph one cycle of this function.

Step 1: $0° \leq x + 120° \leq 360°$
$-120° \leq x \leq 240°$

Phase shift is −120° and period is $240° - (-120°) = 360°$. We have one basic sine cycle between −120° and 240°.

Step 2: The amplitude is 85. The basic cycle is shown in the figure.

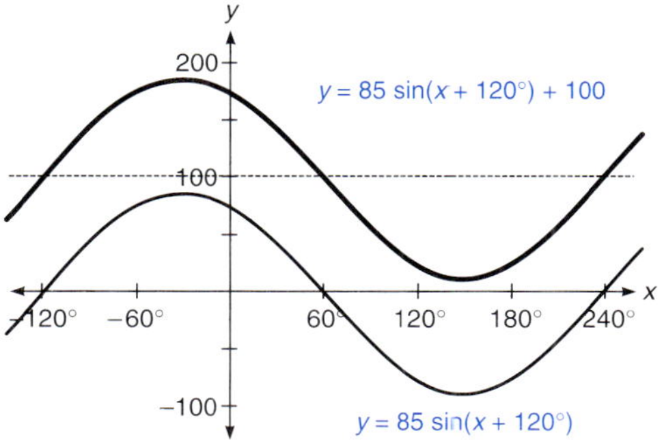

Step 3: We shift the basic cycle upward by 100 units to obtain the final graph.

A graphing calculator must be in degree mode to graph this function. Use the [MODE] key, then use the cursor keys to select Deg, then select this mode with the [ENTER] key.

[Y=] 85 [SIN] [(] [X|T] [+] 120 [)] [+] 100
[RANGE −140,260,−100,200] Xscl = 60, Yscl = 50

2. An electronic signal is to be modeled with the sine function. The peak-to-peak voltage is 340 volts (amplitude is 170 volts). There is a phase shift of 30°, and the period is 120°. The signal is at 200 volts above ground potential (there is a vertical shift of 200). Find the sine function that will model this signal.

The required function is of the form $y = A\sin(Bx + C) + D$. A (amplitude) is 170, and D (vertical shift) is 200.

To find B and C we proceed ''backward.'' We know that we get one basic cycle as x varies between 30° (phase shift) and 30° + 120° (phase shift and period).

$30° \le x \le 30° + 120°$ — We want $0° \le Bx + C \le 360°$

$0° \le x - 30° \le 120°$ — Subtract 30° from each member so the left member is 0°

$0° \le 3x - 90° \le 360°$ — Multiply each member by 3 so the right member is 360°

Thus, the function we want is $y = 170\sin(3x - 90°) + 200$. ■

Mastery points

Can you

- State whether each function, sine, cosine, and tangent is even or odd?
- State the domain and range of the sine, cosine, and tangent functions?
- Use the odd/even properties to compute the values of $\sin x$, $\cos x$, and $\tan x$ for negative values of x?
- Transform equations of the form

$$y = A\sin(Bx + C) + D \text{ or}$$
$$y = A\cos(Bx + C) + D$$

  with $B < 0$ so that $B > 0$, using the odd/even properties?
- Sketch the graph an equation of the form

$$y = A\sin(Bx + C) + D \text{ or}$$
$$y = A\cos(Bx + C) + D$$

  with $B > 0$ and state the amplitude, period, and phase shift?
- Sketch the graph of $y = \tan x$?
- Find an equation of the form $y = A\sin(Bx + C) + D$ or $y = A\cos(Bx + C) + D$, given certain required properties?

Exercise 6–2

1. From memory sketch the graphs of $y = \sin x$, $y = \cos x$, $y = \tan x$.

2. From memory, or using their graphs as an aid, state the domain, range, and period of each of the functions: sine, cosine, and tangent.

3. Using the graph of $y = \sin x$ as a guide, describe all values of x for which $\sin x$ is **a.** 1 **b.** -1 **c.** 0

4. Using the graph of $y = \cos x$ as a guide, describe all values of x for which $\cos x$ is **a.** 1 **b.** -1 **c.** 0

5. Using the graph of $y = \tan x$ as a guide, describe all values of x for which $\tan x$ is 0.

6. Use the graph of $y = \sin x$, as well as the unit circle, to describe all points for which $\sin x$ is
a. $\frac{1}{2}$ **b.** $-\frac{\sqrt{3}}{2}$ **c.** $-\frac{\sqrt{2}}{2}$

7. Use the graph of $y = \cos x$, as well as the unit circle, to describe all points for which $\cos x$ is
a. $\frac{1}{2}$ **b.** $-\frac{\sqrt{3}}{2}$ **c.** $-\frac{\sqrt{2}}{2}$

Use the appropriate property, odd or even, to simplify the computation of the exact value of the (a) sine, (b) cosine, and (c) tangent functions for the following values.

8. $-\frac{\pi}{3}$ **9.** $-\frac{\pi}{6}$ **10.** $-45°$ **11.** $-\frac{2\pi}{3}$

Graph three cycles of the following functions.

12. $y = 5 \sin x$ **13.** $y = 5 \cos x$ **14.** $y = \frac{2}{3} \cos x$ **15.** $y = \frac{1}{5} \sin x$
16. $y = -4 \cos x$ **17.** $y = -2 \sin x$ **18.** $y = -\frac{1}{3} \sin x$ **19.** $y = -\frac{5}{2} \sin x$
20. $y = 2 \sin x + 1$ **21.** $y = 3 \cos x - 2$ **22.** $y = -\frac{3}{4} \cos x - 2$ **23.** $y = -\frac{1}{2} \sin x + 3$

Graph three cycles of the following functions. State the amplitude, period, phase shift, and any vertical shift of each.

24. $y = 2 \sin 4x$ **25.** $y = 3 \cos \frac{x}{2}$ **26.** $y = \cos\left(x - \frac{\pi}{2}\right)$ **27.** $y = 3 \sin(2x + \pi)$
28. $y = \frac{2}{3} \sin(3x + \pi)$ **29.** $y = \frac{5}{8} \cos 5x$ **30.** $y = -\cos 3x$ **31.** $y = -\sin x$
32. $y = -\cos\left(2x + \frac{\pi}{2}\right)$ **33.** $y = -\sin\left(3x - \frac{\pi}{3}\right)$ **34.** $y = \sin(3x + \pi)$ **35.** $y = \cos(2x - \pi)$
36. $y = \cos 2\pi x$ **37.** $y = \sin \pi x$ **38.** $y = 2 \sin 3x + 2$ **39.** $y = 3 \cos 2x - 3$
40. $y = -3 \cos x + 1$ **41.** $y = -\sin 4x + 1$ **42.** $y = 2 \sin(2x - \pi) + 1$ **43.** $y = 3 \sin(3x + \pi) - 3$
44. $y = \sin \pi x + 1$ **45.** $y = 2 \cos \frac{\pi x}{2} - 2$

Use the odd/even properties of the sine and cosine functions to rewrite each of the following functions as an equivalent function in which the coefficient of x is positive.

46. $y = \sin(-2x)$ **47.** $y = \cos(-x)$ **48.** $y = -\cos(-3x)$ **49.** $y = -\sin(-5x)$
50. $y = \sin(-x - 3)$ **51.** $y = \cos(-2x + 4)$ **52.** $y = \sin(-x) - 3$ **53.** $y = \cos(-2x) + 4$
54. $y = -3 \cos\left(-2x + \frac{\pi}{2}\right)$ **55.** $y = 2 \sin\left(-\frac{x}{3} - \pi\right)$

Use the odd/even properties of the sine and cosine functions to rewrite each of the following functions as an equivalent function in which the coefficient of x is positive. Then graph one cycle of the function.

56. $y = \sin(-x)$ **57.** $y = \cos(-2x)$ **58.** $y = \cos\left(-x - \frac{\pi}{3}\right)$ **59.** $y = 2 \sin(-2x + \pi)$
60. $y = -\sin(-2\pi x + \pi)$ **61.** $y = -\cos(-\pi x)$ **62.** $y = \sin(-\pi x + 1)$ **63.** $y = 2 \cos(-3\pi x - 2)$

Assume that each of the following graphs is the graph of a sine function of the form $y = A\sin(Bx + C)$. Find values of A, B, C, and D that would produce each graph, and write the corresponding equation using these values.

64.

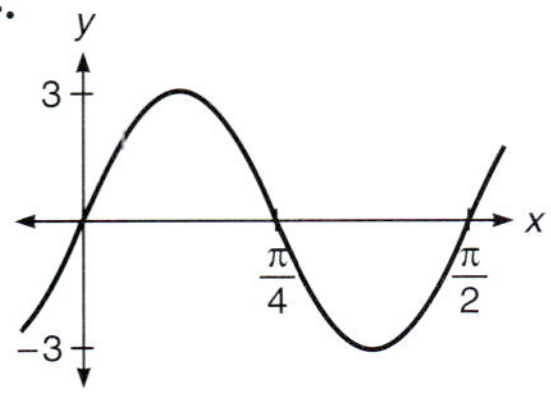

65.

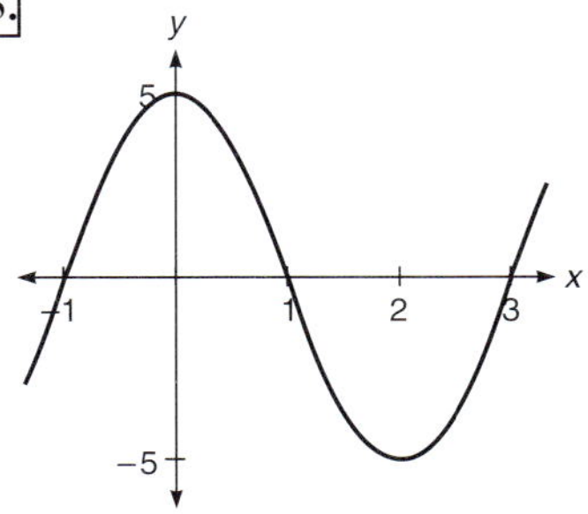

66.

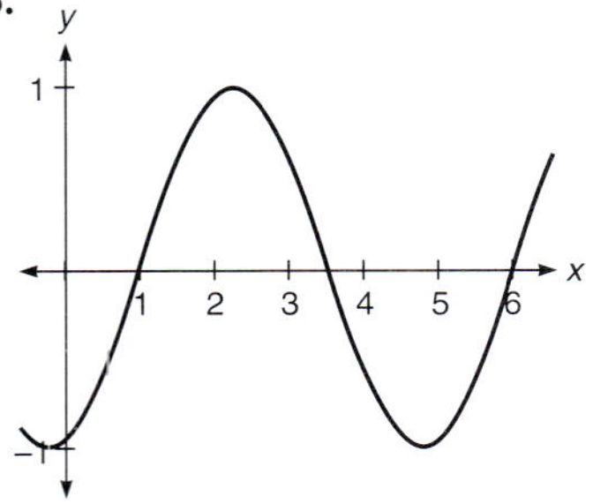

67.

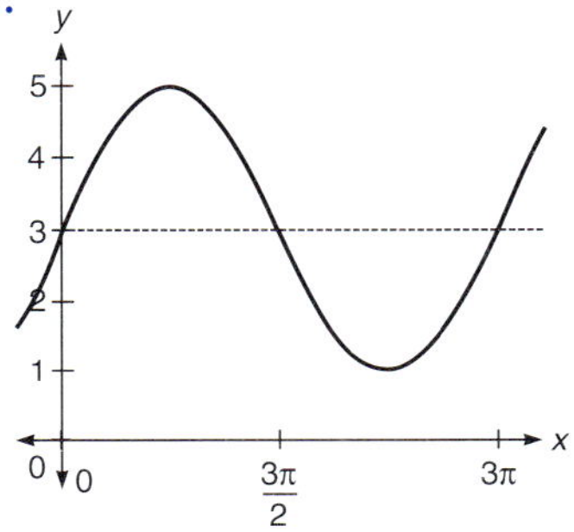

68. Do problem 64, assuming that the graph is a cosine function of the form $y = A\cos(Bx + C) + D$.

69. Do problem 65, assuming that the graph is a cosine function of the form $y = A\cos(Bx + C) + D$.

70. Do problem 66, assuming that the graph is a cosine function of the form $y = A\cos(Bx + C) + D$.

71. Do problem 67, assuming that the graph is a cosine function of the form $y = A\cos(Bx + C) + D$.

Graph one cycle of each of the following functions. Mark the horizontal axis in degrees.

72. $y = 3\sin(x + 60°)$

73. $y = -50\cos(x - 120°)$

74. $y = 25\cos 3x$

75. $y = 10\sin(2x - 180°)$

76. An electronic signal modeled with the sine function has a peak-to-peak voltage of 120 volts (amplitude is 60 volts), phase shift of 90°, and period of 54°. Find an equation of the sine function that will model this signal.

77. An ocean wave is being modeled with the sine function. Its amplitude is 6 feet and its phase shift (with respect to another wave) is −180°. If the period is 720°, find an equation of the sine function that will model this wave.

78. One of the components of a function that could describe the earth's ice ages for the last 500,000 years is described by a sine function with amplitude 0.5, period $\frac{360}{43}°$, 0° phase shift, and vertical translation 23.5. Find an equation for this component.

79. The activity of sunspots seems to follow an 11-year cycle. Assuming that this activity can be roughly modeled with a sine wave, construct a sine function with period $\frac{360}{11}°$, amplitude 1, phase shift 90°, and vertical translation 2.

80. Graph the following functions on the same set of axes.

a. $y = \sin x$ **b.** $y = \sin 3x$ **c.** $y = \sin\frac{x}{3}$

81. Graph the following functions on the same set of axes:

a. $y = \sin x$ **b.** $y = \sin\left(x + \frac{\pi}{2}\right)$

c. $y = \sin x + \frac{\pi}{2}$

82. Graph the following functions on the same set of axes:

a. $y = \cos\left(\frac{\pi}{2} - x\right)$ **b.** $y = \sin x$

Draw a conclusion from the graph. *Hint:* Rewrite part a as $y = \cos\left(-x + \frac{\pi}{2}\right)$.

Test the function for the even/odd property. State which type of symmetry the graph would have based upon being even, odd, or neither even nor odd.

83. $f(x) = x$
84. $f(x) = 3x$
85. $f(x) = 3x^2$
86. $f(x) = -x^2$
87. $f(x) = 2x^4 - 4x^2$
88. $f(x) = 2x^3 - 4x$
89. $f(x) = 3 \sin x$
90. $f(x) = \cos x$
91. $f(x) = \sin^2 x$
92. $f(x) = x^2 \tan x$
93. $f(x) = \dfrac{\cos x}{x}$
94. $f(x) = \sin x + \cos x$
95. $f(x) = (\sin x)(\cos x)$
96. $f(x) = \dfrac{\sin x}{x}$
97. $f(x) = \sin^3 x$

Skill and review

1. Find the degree measure of an angle of $\frac{5\pi}{8}$ radians.
2. Find the radian measure of an angle of measure 345°.
3. Find L in the figure, to the nearest 0.1 millimeter.

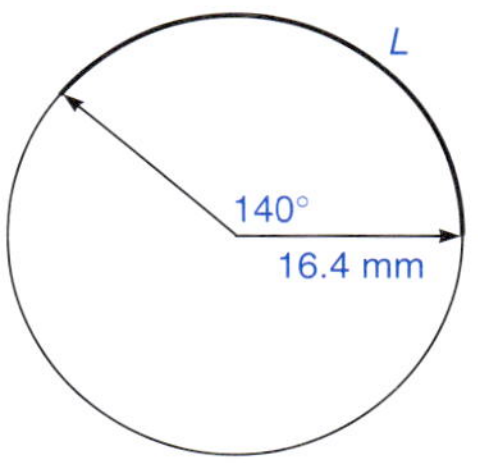

4. θ is an angle in standard position, $\sin \theta = \frac{\sqrt{3}}{5}$, and θ terminates in quadrant II. If the point (a,b) is on the terminal side of θ, and $b = -3$, find a.
5. Solve the right triangle ABC if the measure of angle A is 30° and the length of side c is 12. Leave the answers in exact form.

6–3 The tangent, cotangent, secant, and cosecant functions

The function $y = \cot 2x$ arises in chaos theory, a branch of mathematics that has developed only recently. Chaos theory is used to model things like population growth. Graph three cycles of this function.

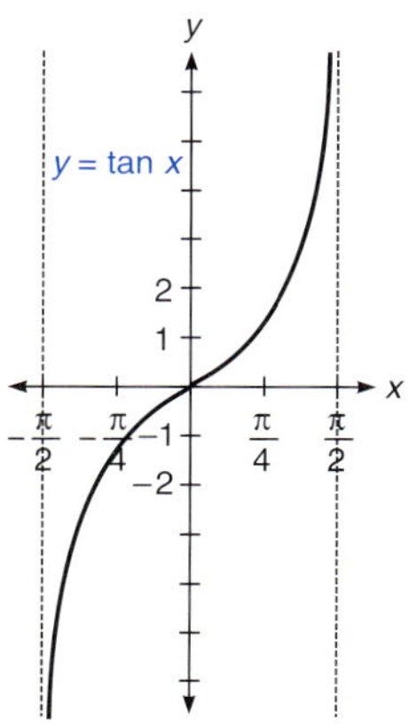

Figure 6–22a

The tangent and cotangent functions

The tangent and cotangent functions are π-periodic, so the basic cycle for each is π units "wide" instead of 2π units, as with the sine and cosine functions. Figure 6–22 shows a basic cycle for the tangent and cotangent functions. The basic graph of the cotangent function can be obtained by plotting points or with a graphics calculator.

The basic tangent cycle starts at $-\frac{\pi}{2}$ and ends at $\frac{\pi}{2}$. The basic cotangent cycle starts at 0 and ends at π.

Graphing functions of the form

$$y = A \tan(Bx + C) \quad \text{and}$$
$$y = A \cot(Bx + C)$$

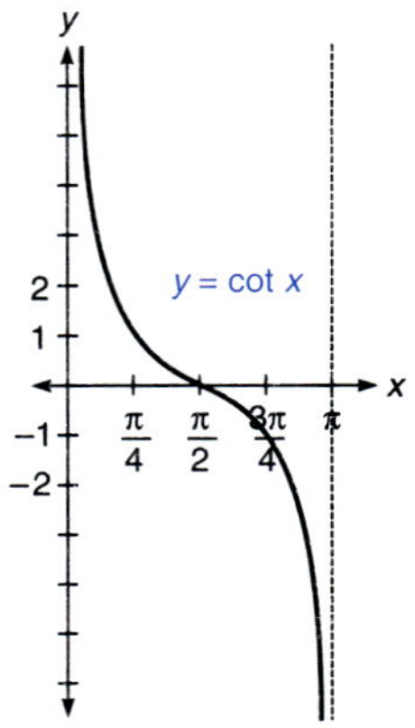

Figure 6–22b

is done in a manner similar to that for the sine and cosine functions. The values of these functions gets arbitrarily large in absolute value, so the concept of amplitude is not especially useful. We consider only cases where $|A| = 1$. We also do not stress phase shift or vertical shift for these functions.

> **To graph functions of the form**
>
> $y = \tan(Bx + C)$ and
>
> $y = \cot(Bx + C)$, $B > 0$
>
> using algebraic methods
>
> - For the tangent function, solve $-\frac{\pi}{2} < Bx + C < \frac{\pi}{2}$ for x; for the cotangent function solve $0 < Bx + C < \pi$ for x.
>   This gives the left and right end points for one basic cycle. The difference between the end points is the period.
> - Use the values from the first step to draw one basic cycle. Repeat this cycle to obtain as much of the graph as desired.

Note If the coefficient of the function is −1, the graph is obtained as stated, but is then reflected across the x-axis (as with any other function).

■ *Example 6–3 A*

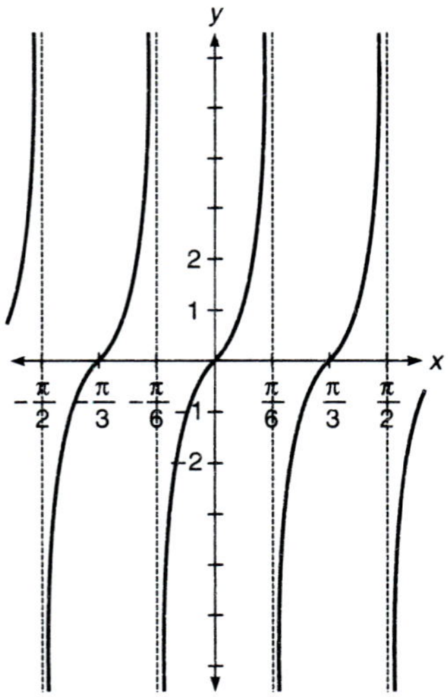

Graph the function. Show at least two basic cycles. State the period.

1. $y = \tan 3x$

Step 1: $-\frac{\pi}{2} < 3x < \frac{\pi}{2}$ Multiply each member by $\frac{1}{3}$

$-\frac{\pi}{6} < x < \frac{\pi}{6}$

The period is $\frac{\pi}{6} - \left(-\frac{\pi}{6}\right) = \frac{\pi}{3}$.

Step 2: We draw a basic cycle between $-\frac{\pi}{6}$ and $\frac{\pi}{6}$, then repeat two cycles, one on each side of the basic cycle.

Y=	TAN	(	3	X|T	)	RANGE −2,2,−4,4

2. $y = \cot\left(2x - \frac{\pi}{3}\right)$

Step 1: $0 < 2x - \frac{\pi}{3} < \pi$ Multiply each member by 3

$0 < 6x - \pi < 3\pi$

$\pi < 6x < 4\pi$

$\frac{\pi}{6} < x < \frac{2\pi}{3}$

The period is $\frac{2\pi}{3} - \frac{\pi}{6} = \frac{\pi}{2}$.

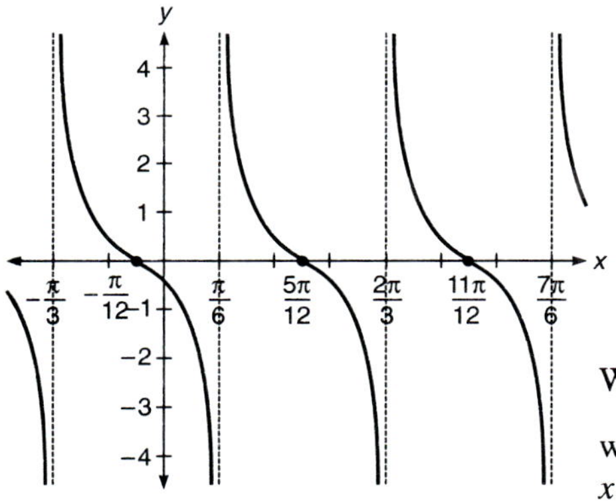

Step 2: Draw a basic cotangent cycle from $\frac{\pi}{6}$ to $\frac{2\pi}{3}$. Repeat this cycle on either side to obtain the graph shown in the figure.

> [Y=] [(] [TAN] [(] 2 [X|T] [−] [π] [÷] 3 [)] [)] [x^{-1}]
> [RANGE −1.5,4,−4,4] Xscl = 0.5236

We already know that the tangent function is odd. Since $\cot x = \frac{1}{\tan x}$, except where $\tan x = 0$, cotangent is also an odd function. Thus, if the coefficient of x is negative we use this fact, as we did for the sine function.

■ ***Example 6–3 B***

1. Rewrite $y = \tan(\pi - x)$ so the coefficient of x is positive.

$y = \tan(\pi - x)$

$y = \tan[-(x - \pi)]$ $\quad$ $\pi - x = -(x - \pi)$

$y = -\tan(x - \pi)$ $\quad$ $\tan(-x) = -\tan x$, since tangent is an odd function.

2. Determine if the function $f(x) = x \cot x$ is even, odd, or neither.

$f(-x) = (-x)\cot(-x)$

$= (-x)(-\cot x)$ $\quad$ $\cot(-x) = -\cot(x)$

$= x \cot x$

$= f(x)$

Since $f(-x) = f(x)$, this is an even function.

The secant and cosecant functions

To graph variations of the secant and cosecant functions we use the fact that they are reciprocals of the cosine and sine functions, respectively. Figures 6–23 and 6–24 show the graphs of the cosecant and secant functions. Observe that the ranges of both functions are $|y| \geq 1$, and that they have vertical asymptotes where their reciprocal function (cosine or sine) is zero. The domain of each function is the set of real numbers excluding the points at which they have vertical asymptotes.

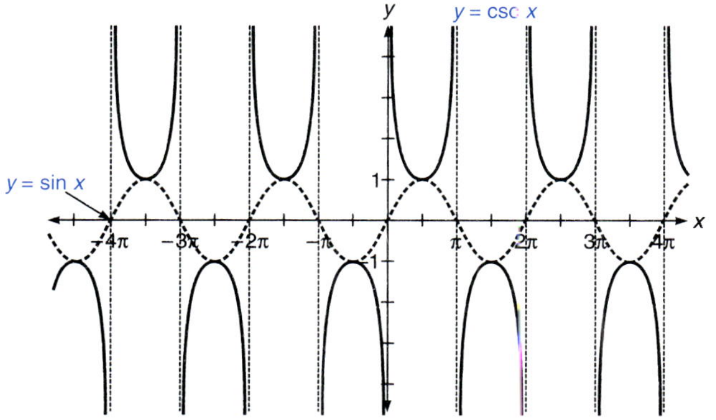

Figure 6–23

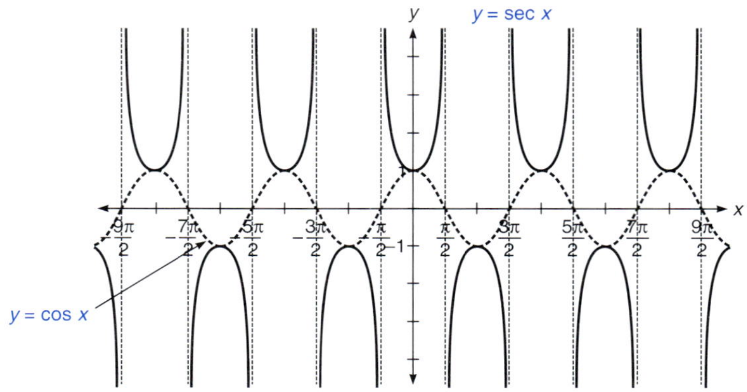

Figure 6–24

To understand the graph of $y = \csc x$, observe that $\csc x = \dfrac{1}{\sin x}$. Where $\sin x = 1$, $\csc x = 1$. Where $\sin x = -1$, $\csc x = -1$. Where $|\sin x| < 1$, $|\csc x| > 1$. As $|\sin x|$ gets closer to zero, $\left|\dfrac{1}{\sin x}\right|$ gets larger and larger. This is where the vertical asymptotes occur. Observe in the figures that the graph of the cosecant and secant functions touch the graphs of the sine and cosine functions, respectively, at their highest points, and that they then increase in absolute value from that point toward their vertical asymptotes. We use the sine and cosine functions to help us graph the cosecant and secant functions.

To graph $y = A\csc(Bx + C)$ or $y = A\sec(Bx + C)$ using algebraic methods

- Graph the appropriate reciprocal function,

$$y = A\sin(Bx + C) \quad \text{or} \quad y = A\cos(Bx + C)$$

- Sketch in vertical asymptotes wherever the sine or cosine function is zero.
- Create the cosecant or secant graph by starting at the highest and lowest points of the sine or cosine graph and sketching values that increase in absolute value from that point as x approaches the vertical asymptotes. Note that these functions are not defined at the asymptotes.

■ *Example 6–3 C*

Graph the function.

1. $y = 3\csc 2x$

This has the same graph as $y = 3\left(\dfrac{1}{\sin 2x}\right)$. We graph $y = 3\sin 2x$, then graphically form the reciprocal function, as shown in the figure.

$$0 \le 2x \le 2\pi$$
$$0 \le x \le \pi$$

There is one basic sine cycle between 0 and π. The resultant graph is shown in the figure.

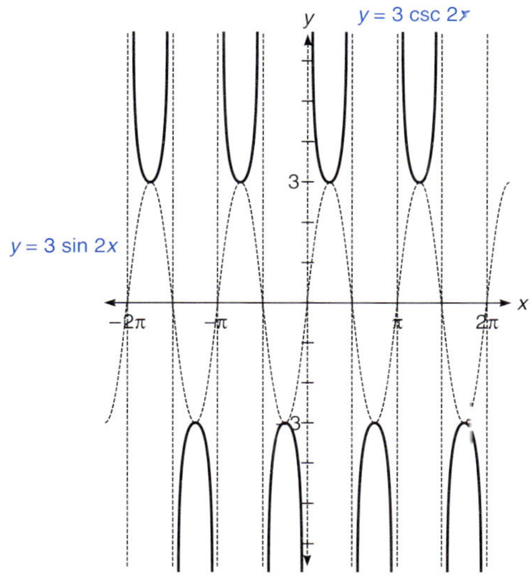

Y=	3	÷	SIN	(	2	X|T	)	RANGE −6.5,6.5,−6,6

Xscl = 1.5708

2. $y = 2\sec(3x - \pi)$
We graph $y = 2\cos(3x - \pi)$.

$$0 \leq 3x - \pi \leq 2\pi$$
$$\pi \leq 3x \leq 3\pi$$
$$\frac{\pi}{3} \leq x \leq \pi$$

There is one basic cosine cycle from $\frac{\pi}{3}$ to π. The resultant graph is shown in the figure.

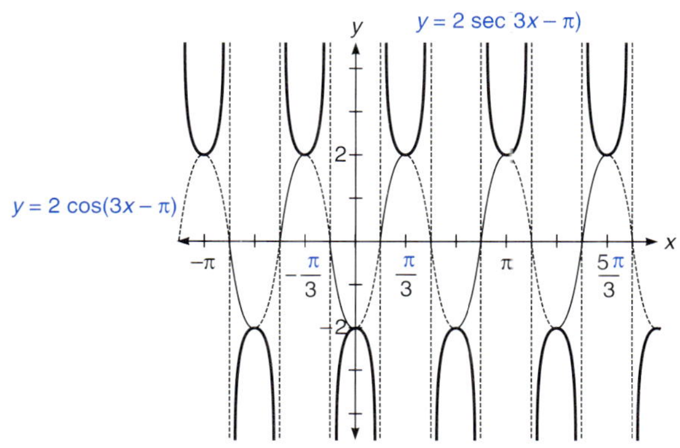

Y=	2	÷	COS	(	3	X|T	−	π	)

RANGE −6,12,−4,4 Xscl = 1.0472

Since $\sec x = \frac{1}{\cos x}$, *secant is an even function.* Csc $x = \frac{1}{\sin x}$, so *cosecant is an odd function* (the proofs are in the exercises).

■ Example 6–3 D

Rewrite $y = \csc\left(\frac{\pi}{2} - 3x\right)$ so that the coefficient of x is positive.

$$y = \csc\left[-\left(3x - \frac{\pi}{2}\right)\right] \qquad \frac{\pi}{2} - 3x = -\left(3x - \frac{\pi}{2}\right)$$

$$y = -\csc\left(3x - \frac{\pi}{2}\right) \qquad \csc(-\theta) = \csc\theta\text{; cosecant is an odd function}$$ ■

Table 6–3 summarizes the properties of the cotangent, secant, and cosecant functions.

Function	Domain	Range	Period
$y = \csc x$	$x \neq k\pi,\ k \in J$	$\lvert y\rvert \geq 1$	2π
$y = \sec x$	$x \neq \frac{\pi}{2} + k\pi,\ k \in J$	$\lvert y\rvert \geq 1$	2π
$y = \cot x$	$x \neq k\pi,\ k \in J$	R	π

$\csc(-x) = -\csc x$ (odd, origin symmetry)
$\sec(-x) = \sec x$ (even, y-axis symmetry)
$\cot(-x) = -\cot x$ (odd, origin symmetry)

Table 6–3

Mastery points

Can you

- Sketch the graphs of the cotangent, cosecant, and secant functions?
- State the domain and range of the cotangent, cosecant, and secant functions?
- Use the appropriate property, even or odd, to transform cotangent, cosecant, and secant functions so the argument is positive?
- Sketch the graphs of functions of the form
  $y = \tan(Bx + C)$ and $y = \cot(Bx + C)$, $B > 0$?
- Sketch the graphs of functions of the form
  $y = A \csc(Bx + c)$ and $y = A \sec(Bx + C)$, $B > 0$?

Exercise 6–3

1. Sketch the graphs of the cotangent, cosecant, and secant functions.
2. State the domain, range, and period for the cotangent, cosecant, and secant functions.
3. Use the identity $\csc x = \frac{1}{\sin x}$ to show that cosecant is an odd function.
4. Use the identity $\sec x = \frac{1}{\cos x}$ to show that secant is an odd function.

5. Use the identity $\cot x = \dfrac{1}{\tan x}$ to show that cotangent is an odd function.

6. Show that the function $f(x) = \sin^2 x \cos x$ is an even function.

7. Show that the function $f(x) = \sin^2 x \tan x$ is an odd function.

8. Show that the function $f(x) = \sin x \tan x$ is an even function.

Graph three cycles of the following functions.

9. $y = 2 \tan x$
10. $y = -\cot x$
11. $y = \tan \dfrac{x}{4}$
12. $y = \cot \dfrac{x}{2}$
13. $y = \cot\left(x - \dfrac{\pi}{2}\right)$
14. $y = 3 \tan(2x + \pi)$
15. $y = -\cot\left(2x + \dfrac{\pi}{2}\right)$
16. $y = -\tan\left(3x - \dfrac{\pi}{3}\right)$
17. $y = \cot 2\pi x$
18. $y = \tan \pi x$

Use the odd/even properties of the tangent and cotangent functions to rewrite each of the following functions as an equivalent function in which the coefficient of x is positive. Then graph one cycle of the function.

19. $y = \tan(-2x)$
20. $y = \cot(-x)$
21. $y = -\cot(-\pi x)$
22. $y = -\tan(-2\pi x)$
23. $y = \tan(-x - \pi)$
24. $y = \cot(-2x + \pi)$

Graph three cycles of the following functions.

25. $y = \frac{2}{3} \csc x$
26. $y = \frac{1}{5} \sec x$
27. $y = -4 \csc x$
28. $y = 2 \sec 4x$
29. $y = 3 \csc \dfrac{x}{2}$
30. $y = \csc\left(x - \dfrac{\pi}{2}\right)$
31. $y = 3 \sec(2x + \pi)$
32. $y = \frac{2}{3} \sec(3x + \pi)$
33. $y = \csc\left(\dfrac{x}{2} - \dfrac{\pi}{3}\right)$
34. $y = \csc 2\pi x$
35. $y = \sec \pi x$

Use the odd/even properties of the sine and cosine functions to graph at least one cycle of each of the following functions.

36. $y = \sec(-2x)$
37. $y = \csc(-x)$
38. $y = 3 \csc\left(-2x + \dfrac{\pi}{2}\right)$
39. $y = 2 \sec\left(-\dfrac{x}{3} - \dfrac{\pi}{2}\right)$

40. Chaos theory is a recent development in mathematics. It finds application in modeling things like population growth. Gilbert Strang of the Massachusetts Institute of Technology has shown[3] that the function $y = \dfrac{1}{2}\left(\cot x - \dfrac{1}{\cot x}\right)$ arises in chaos theory. It can be shown (chapter 7) that this is equivalent to the function $y = \cot 2x$. Graph three cycles of this function.

[3] "A Chaotic Search for *i*," by Gilbert Strang, *The College Mathematics Journal*, Vol. 22, No. 1, January 1991.

Skill and review

1. Graph the function $f(x) = x^4 - x^3 - 7x^2 + x + 6$. Recall that the zeros of the right member are the x-intercepts, and that the rational zero theorem and synthetic divison can be used to help find these zeros.
2. Graph the quadratic function $f(x) = x^2 + 6x - 4$. Find the vertex by completing the square and putting the equation in vertex form $f(x) = a(x - h)^2 + k$.
3. Rationalize the denominator of $\dfrac{\sqrt{3}}{\sqrt{3} - 6}$.
4. Multiply the complex numbers $(3 - 7i)(2 + 3i)$.
5. Test the function $f(x) = x^2 - \cos x$ for the even or odd property. State which type of symmetry the graph of this function would exhibit.
6. Graph the function $f(x) = 3 \sin x$.
7. Graph the function $f(x) = -\cos 2x$.
8. Graph the function $f(x) = 2 \sin\left(x + \dfrac{\pi}{5}\right)$.

6–4 The inverse sine, cosine, and tangent functions

Firefighters use the formula $d = \frac{h}{4} + 2$ to obtain the safe height of a fire ladder resting against a wall, where h is the height up the wall and d is the distance away from the wall at which the ladder should rest. Let θ represent the angle the ladder makes with the ground. Write the value of θ in terms of the variable h.

To solve problems like this we need the functions presented in this section—the inverse trigonometric functions. We have already used them when we solved equations like $\sin \theta = \frac{1}{2}$, where we stated that $\theta' = \sin^{-1}\frac{1}{2}$ and that θ' is 30° or $\frac{\pi}{6}$ (throughout chapter 5).

We know (section 4–5) that the inverse of a function is formed by interchanging the first and second components of all of the ordered pairs in the function, and that a function has an inverse if and only if it is a one-to-one function. We also know (section 3–5) that a function is one to one if and only if it passes the horizontal line test.

An examination of the graphs of any of the trigonometric functions shows that they fail the horizontal line test and are therefore not one to one. These functions are all periodic, and no periodic function can be one to one. Nevertheless, there is a real need for inverses of the trigonometric functions.

We can define these inverse functions by limiting the domain of a given trigonometric function so that this limited part of the function (which is itself a function) is one to one and includes the entire range of the function.

The inverse sine function

Figure 6–25 shows this idea for the sine function. For the sine function, we select that portion of the domain for $-\frac{\pi}{2} \leq x \leq \frac{\pi}{2}$. This new function is one to one and therefore has an inverse. We call this function the **inverse sine function,** sine^{-1}, abbreviated $\sin^{-1}$. The ordered pairs of this function are formed by reversing the coordinates of the ordered pairs in the restricted portion of the sine function previously discussed. The figure shows this idea for several points. For example, $\sin \frac{\pi}{3} = \frac{\sqrt{3}}{2}$, so the ordered pair $\left(\frac{\pi}{3}, \frac{\sqrt{3}}{2}\right)$ is an element of the sine function. Thus, the ordered pair $\left(\frac{\sqrt{3}}{2}, \frac{\pi}{3}\right)$ is an element of the inverse sine function. This is also written $\sin^{-1}\frac{\sqrt{3}}{2} = \frac{\pi}{3}$.

In general, if an ordered pair (x,y) is in the inverse sine function, then the ordered pair (y,x) is in the sine function. Thus, if $y = \sin^{-1}x$, then $x = \sin y$. Using this idea and figure 6–25 we make the following definition.

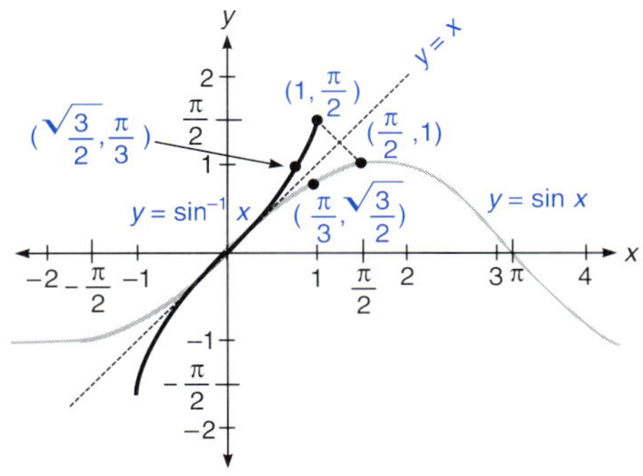

Y=	2nd	SIN	X|T
RANGE −1.5,1.5,−2,2			

Figure 6–25

Inverse sine function

$y = \sin^{-1}x$ means

1. $\sin y = x$
2. $-\frac{\pi}{2} \le y \le \frac{\pi}{2}$
3. $-1 \le x \le 1$

Concept

$\sin^{-1}x$ is the angle in quadrant I or IV whose sine is x.

Note When we refer to quadrant IV we refer to angles with negative measure.

The domain of the sine^{-1} function is $-1 \le x \le 1$, which is the range of the sine function. The range of the sine^{-1} function is $-\frac{\pi}{2} \le \sin^{-1}x \le \frac{\pi}{2}$, which is the domain of the one-to-one portion of the sine function that we selected above (figure 6–25).

Another notation for $\sin^{-1}x$ is **arcsin *x*.** This is because the word arc can refer to an angle, so arcsin x means "the arc (angle) whose sine is x."

It can be seen in its graph that *the inverse sine function is an odd function.* This can help in computations, because it means that $\sin(-x) = -\sin x$. This odd property and table 6–1 can be helpful in solving for exact values of the inverse sine function.

Although we defined the inverse trigonometric functions using radian measure we often want the result in degrees. For the sine^{-1} function this corresponds to a range of $-90° \le \sin^{-1}x \le 90°$.

■ ***Example 6–4 A***

Find the exact value both in radians and degrees.

1. $\sin^{-1}\frac{1}{2}$

Find the angle in table 6–1 whose sine is $\frac{1}{2}$. This is $\frac{\pi}{6}$, or 30°.

2. $\arcsin\left(-\frac{\sqrt{2}}{2}\right)$ — This uses the alternate notation for the inverse sine function

$\arcsin\left(-\frac{\sqrt{2}}{2}\right) = -\arcsin\left(\frac{\sqrt{2}}{2}\right)$ — $\sin^{-1}$ is an odd function

Find the value of $\arcsin\left(\frac{\sqrt{2}}{2}\right)$ first.

Find the angle in table 6–1 whose sine is $\frac{\sqrt{2}}{2}$. This is $\frac{\pi}{4}$ or 45°.

Since $\arcsin\left(\frac{\sqrt{2}}{2}\right) = \frac{\pi}{4}$, $\arcsin\left(-\frac{\sqrt{2}}{2}\right) = -\frac{\pi}{4}$, or −45°. ■

For most values of x, $\sin^{-1}x$ can only be approximated using a calculator. As noted in section 5–2 most calculators use the [sin] key, prefixed by another key such as [SHIFT], [2nd], [INV], or [ARC]. We will assume the key is called the [SHIFT] key.

■ ***Example 6–4 B***

Find arcsin(−0.9249) to the nearest 0.01 radians and 0.1°.

$$y = \arcsin(-0.9249)$$
$$y \approx -1.18 \text{ (radians) or } -67.7°$$

Use .9249 [+/−] [SHIFT] [sin] (TI-81: [2nd] [SIN] [(−)] .9249 [ENTER]) in both radian and degree modes. ■

The inverse cosine function

The inverse cosine function (cosine^{-1}) is defined in a manner similar to the inverse sine function. The one-to-one part chosen is for $0 \leq x \leq \pi$. See figure 6–26.

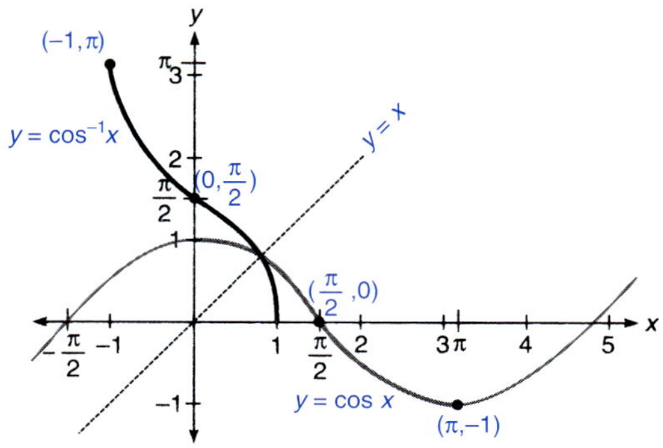

Figure 6–26

[Y=] [2nd] [COS] [X|T] [RANGE −1.5,1.5,−.5,4] Xscl=1.5708

The graph of $y = \cos^{-1}x$ is the reflection of this portion of the cosine function across the line $y = x$. We can see from the graph that the domain of the cosine^{-1} function is $-1 \leq x \leq 1$, and the range is $0 \leq \cos^{-1}x \leq \pi$. The inverse cosine function is defined as follows.

> **Inverse cosine function**
>
> $y = \cos^{-1}x$ means
>
> 1. $\cos y = x$
> 2. $0 \leq y \leq \pi$
> 3. $-1 \leq x \leq 1$
>
> ***Concept***
>
> $\cos^{-1}x$ is the angle in quadrant I or II whose cosine is x.

Arccos x also means $\cos^{-1}x$. Note that in degrees, $0° \leq \cos^{-1}x \leq 180°$.

The inverse cosine function is neither even nor odd. A useful fact to know, however, is the following identity.

$$\cos^{-1}(-x) = \pi - \cos^{-1}x$$

This identity, along with table 6–1, is useful in finding exact values for the inverse cosine function. Observe that in degree measure this can be written

$$\cos^{-1}(-\theta) = 180° - \cos^{-1}\theta$$

Example 6–4 C illustrates using the inverse cosine function.

■ ***Example 6–4 C***

Find the value both in radians and degrees.

1. $\cos^{-1}(-\frac{1}{2})$; find the exact value.

Find $\cos^{-1}\frac{1}{2}$ first. This is $\frac{\pi}{3}$, 60° from table 6–1.

$$\begin{aligned}\cos^{-1}(-\tfrac{1}{2}) &= \pi - \cos^{-1}\tfrac{1}{2} \text{ or } 180° - \cos^{-1}\tfrac{1}{2}\\ &= \pi - \frac{\pi}{3} \quad \text{or } 180° - 60°\\ &= \frac{2\pi}{3} \text{ or } 120°\end{aligned}$$

2. arccos (1)
Since $\cos\theta = 1$ this is 0, or 0°.

3. $\cos^{-1}(-0.5141)$; round radians to 0.01, degrees to 0.1°.

$$y = \cos^{-1}(-0.5141)$$
$$y \approx 2.11 \text{ (radians) or } 120.9°$$

Use .5141 [+/−] [SHIFT] [cos] (TI-81: [2nd] [COS] [(−)] .5141 [ENTER]) in radian mode, then degree mode. ■

The inverse tangent function

The inverse of the tangent function (tangent^{-1}) is based on the basic cycle of the tangent function, from $-\frac{\pi}{2}$ to $\frac{\pi}{2}$. This is one to one. The graph of $y = \tan^{-1}x$ is the reflection of the basic cycle across the line $y = x$. See figure 6–27.

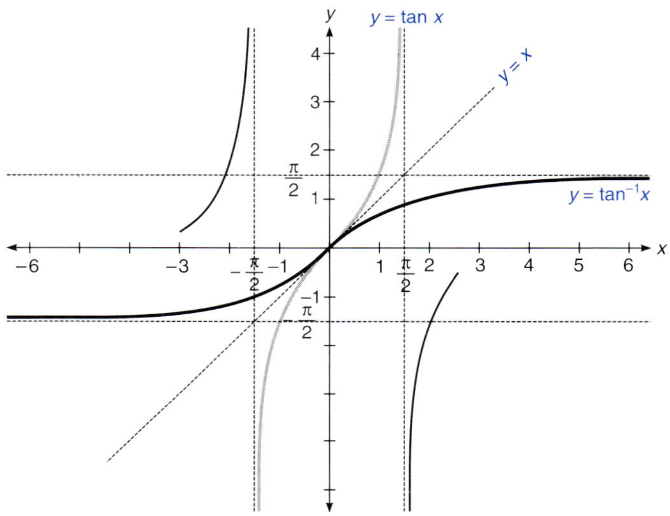

Figure 6–27

Y= | 2nd | TAN | X|T
RANGE −5,5,−2,2
Xscl = 1.5708

The domain of the tangent^{-1} function is all the real numbers, and the range is the values strictly between $-\frac{\pi}{2}$ and $\frac{\pi}{2}$. The range of the tangent^{-1} function is in quadrants I and IV, as is the range of the sine^{-1} function, except that the points $-\frac{\pi}{2}$ and $\frac{\pi}{2}$ are not included. *Arctan x* is another notation for $\tan^{-1}x$.

Inverse tangent function

$y = \tan^{-1}x$ means

1. $\tan y = x$
2. $-\frac{\pi}{2} < y < \frac{\pi}{2}$

Concept

$\tan^{-1}x$ is the angle in quadrant I or IV whose tangent is x.

The inverse tangent function is an odd function. Thus,

$$\tan^{-1}(-x) = -\tan^{-1}x$$

This and table 6–1 can help find exact values of this function, which is illustrated in example 6–4 D.

■ Example 6–4 D

Find the required value both in degrees and radians.

1. $\tan^{-1}(\sqrt{3})$

Table 6–1 shows this is $\frac{\pi}{3}$ (radians) or 60°.

2. arctan 0.9697

$$y = \arctan 0.9697$$
$$y \approx 0.77 \text{ (radians) or } 44.1°.$$

Use .9697 [SHIFT] [tan] (TI-81: [2nd] [TAN] .9697 [ENTER]) both in radian and then degree modes. ■

The domains and ranges of the inverse sine, cosine, and tangent functions are summarized in table 6–4, as well as the property for dealing with $-x$. The quadrants to which the ranges correspond are also indicated.

Function	Domain	Range	Quadrants
$y = \sin^{-1}x$	$-1 \le x \le 1$	$-\frac{\pi}{2} \le y \le \frac{\pi}{2}$	I, IV
$y = \cos^{-1}x$	$-1 \le x \le 1$	$0 \le y \le \pi$	I, II
$y = \tan^{-1}x$	R	$-\frac{\pi}{2} < y < \frac{\pi}{2}$	I, IV

$\sin^{-1}(-x) = -\sin^{-1}x$ (odd)
$\cos^{-1}(-x) = \pi - \cos^{-1}x$ (neither even nor odd)
$\tan^{-1}(-x) = -\tan^{-1}x$ (odd)

Table 6–4

It is often important (especially in the study of the calculus) to simplify expressions that involve composition of the trigonometric and inverse trigonometric functions. This can often be done with the aid of a reference triangle (section 5–4), as illustrated in example 6–4 E. Remember that *any expression of the form $\sin^{-1}x$, $\cos^{-1}x$, and $\tan^{-1}x$ can be interpreted to represent an angle.*

■ Example 6–4 E

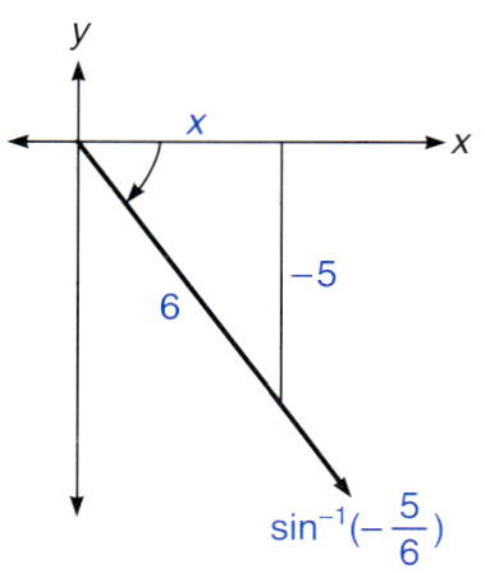

Simplify each expression.

1. $\sec[\sin^{-1}(-\frac{5}{6})]$

$\sin^{-1}(-\frac{5}{6})$ represents an angle in quadrant IV (since $-\frac{5}{6} < 0$). We construct a reference triangle for such an angle as shown. We compute x to be $\sqrt{11}$.

The cosine for this angle is $\frac{\sqrt{11}}{6}$, and the secant is the reciprocal of this value, $\frac{6}{\sqrt{11}}$ or $\frac{6\sqrt{11}}{11}$. Thus, $\sec\left[\sin^{-1}\left(-\frac{5}{6}\right)\right] = \frac{6\sqrt{11}}{11}$.

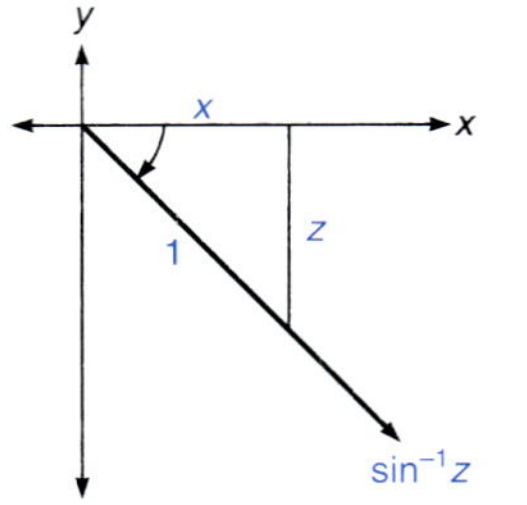

2. $\cos(\sin^{-1}z)$, $z < 0$
Since $z < 0$, $\sin^{-1}z$ is an angle in quadrant IV. The reference triangle is for an angle in quadrant IV whose sine is z, $z < 0$.

We compute $x = \sqrt{1 - z^2}$ (Pythagorean theorem).

Thus the cosine of this angle is $\frac{x}{1} = x = \sqrt{1 - z^2}$.

Thus, $\cos(\sin^{-1}z) = \sqrt{1 - z^2}$ if $z < 0$.

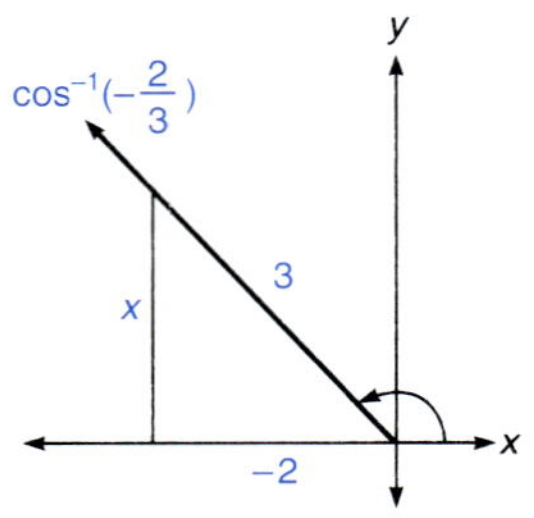

3. $\tan[\cos^{-1}(-\frac{2}{3})]$
$\cos^{-1}(-\frac{2}{3})$ is an angle in quadrant II, since $-\frac{2}{3} < 0$. The figure shows a reference triangle for this angle. We compute $x = \sqrt{5}$, so the tangent of this angle is $\frac{x}{-2} = -\frac{\sqrt{5}}{2}$.

Thus, $\tan[\cos^{-1}(-\frac{2}{3})] = -\frac{\sqrt{5}}{2}$.

4. $\sin^{-1}\left(\sin\frac{7\pi}{6}\right) = \sin^{-1}\left(-\frac{1}{2}\right) = -\frac{\pi}{6}$ ■

One application of the inverse trigonometric functions is to describe an angle in a given situation by using a mathematical expression.

■ Example 6–4 F

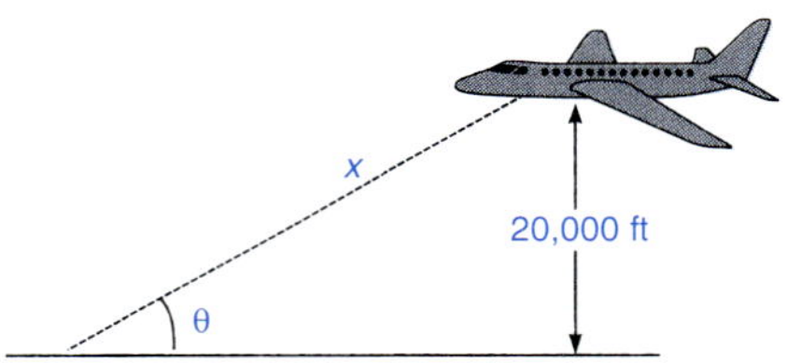

A jet aircraft is flying at an altitude of 20,000 feet. If x represents the distance from a ground observer to the aircraft (also in feet), describe the angle of elevation θ from the observer to the aircraft in terms of an inverse trigonometric function.

We represent the situation as shown in the figure. We can see that $\sin\theta = \frac{20{,}000}{x}$, so $\theta = \sin^{-1}\frac{20{,}000}{x}$. ■

Another way in which the inverse trigonometric functions are used is to describe one of the solutions to a trigonometric equation.

■ Example 6–4 G

Describe one solution, using an inverse trigonometric function.

1. $\cos 2\alpha = 0.55$

$2\alpha = \cos^{-1}0.55$

$\alpha = \frac{1}{2}\cos^{-1}0.55$

2. $3\sin\theta = 0.69$

$\sin\theta = 0.23$

$\theta = \sin^{-1}0.23$

3. $A \sin Bx = C, A \neq 0, B \neq 0$

$$\sin Bx = \frac{C}{A}$$

$$Bx = \sin^{-1}\frac{C}{A}$$

$$x = \frac{1}{B}\sin^{-1}\frac{C}{A}$$

■

The composition of a function with its inverse always produces the same result (section 4–5), x. That is, $f(f^{-1}(x)) = f^{-1}(f(x)) = x$ for all x in the domain of the appropriate function. Example 6–4 E (part 4) showed that $\sin^{-1}(\sin x)$ is not necessarily x. This is similar to saying $\sqrt{x^2}$ is not x for all x. (Think of a value of x for which $\sqrt{x^2} \neq x$.) The problem arises because the sine function does not have an inverse function. Sine^{-1} is the inverse of only a restricted subset of the sine function, for $-\frac{\pi}{2} \leq x \leq \frac{\pi}{2}$. Thus, $\sin^{-1}(\sin x) = x$ only if $-\frac{\pi}{2} \leq x \leq \frac{\pi}{2}$.

Taking the functions in the other direction, $\sin(\sin^{-1}x) = x$ whenever $-1 \leq x \leq 1$, because this is the domain of the sine^{-1} function. Similar reasoning allows us to make the following statements.

$\sin^{-1}(\sin x) = x$ if and only if $-\frac{\pi}{2} \leq x \leq \frac{\pi}{2}$

$\cos^{-1}(\cos x) = x$ if and only if $0 \leq x \leq \pi$

$\tan^{-1}(\tan x) = x$ if and only if $-\frac{\pi}{2} < x < \frac{\pi}{2}$

$\cos(\cos^{-1}x) = x$ if and only if $-1 \leq x \leq 1$

$\sin(\sin^{-1}x) = x$ if and only if $-1 \leq x \leq 1$

$\tan(\tan^{-1}x) = x$ for any x

Mastery points

Can you

- State the domain and range of the inverse sine, cosine, and tangent functions?
- Find exact values, both in radians and degrees, for expressions of the form $\sin^{-1}x$, $\cos^{-1}x$, and $\tan^{-1}x$, for appropriate values of x, using table 6–1?
- Find approximate values, both in radians and degrees, for expressions of the form $\sin^{-1}x$, $\cos^{-1}x$, and $\tan^{-1}x$, using a calculator?
- Simplify expressions that involve combinations of the trigonometric and inverse trigonometric functions, using a reference triangle where appropriate?
- Use the inverse sine, cosine, and tangent functions to describe an angle in a given situation by using a mathematical expression?

Exercise 6–4

1. Sketch the graph of each function.
a. inverse sine **b.** inverse cosine **c.** inverse tangent

2. State the domain and range of each function.
a. inverse sine **b.** inverse cosine **c.** inverse tangent

Find exact values for each of the following expressions in both radians and degrees.

3. $\cos^{-1}\frac{\sqrt{2}}{2}$
4. $\sin^{-1}(-\frac{1}{2})$
5. $\arcsin\frac{\sqrt{3}}{2}$
6. $\arcsin\left(-\frac{\sqrt{2}}{2}\right)$
7. $\sin^{-1}1$
8. $\cos^{-1}(-\frac{1}{2})$
9. $\arccos\frac{\sqrt{3}}{2}$
10. $\cos^{-1}0$
11. $\arccos\left(-\frac{\sqrt{2}}{2}\right)$
12. $\arcsin(-1)$
13. $\tan^{-1}1$
14. $\arctan(-\sqrt{3})$
15. $\arccos(-1)$
16. $\tan^{-1}\left(-\frac{\sqrt{3}}{3}\right)$
17. $\arctan 0$
18. $\cos^{-1}\left(-\frac{\sqrt{3}}{2}\right)$
19. $\sin^{-1}0$
20. $\arcsin\left(-\frac{\sqrt{3}}{2}\right)$

Find approximate values for the following expressions in both radians and degrees. Round the values for radians to two decimal places and for degrees to one decimal place.

21. $\arctan(-0.2553)$
22. $\arccos(-0.9888)$
23. $\tan^{-1}11.08$
24. $\arccos 0.8253$
25. $\tan^{-1}0.9316$
26. $\sin^{-1}0.9323$
27. $\sin^{-1}0.6442$
28. $\arccos(-0.9902)$
29. $\sin^{-1}(-0.9976)$
30. $\arcsin(-0.2955)$
31. $\tan^{-1}(-3.4776)$
32. $\arcsin(-0.2571)$
33. $\arccos(-0.5299)$
34. $\cos^{-1}0.9632$

nplify the following expressions, obtaining exact values.

35. $\cos(\arcsin\frac{3}{5})$
36. $\tan[\arcsin(-0.8)]$
37. $\cos(\arcsin 0.3)$
38. $\tan(\arcsin 0.4)$
39. $\cos(\arctan 0.3)$
40. $\sin(\arctan 0.4)$
41. $\cos(\tan^{-1}(-\frac{3}{5}))$
42. $\sin(\cos^{-1}(-\frac{1}{4}))$
43. $\sin(\arccos\frac{5}{8})$
44. $\cos(\arctan 2)$
45. $\tan[\cos^{-1}(-\frac{2}{3})]$
46. $\sin[\arccos(-0.8)]$
47. $\sin(\tan^{-1}\sqrt{5})$
48. $\csc\left(\cos^{-1}\frac{\sqrt{2}}{6}\right)$
49. $\cot\left(\sin^{-1}\frac{\sqrt{3}}{5}\right)$
50. $\csc\left(\sin^{-1}\frac{\sqrt{2}}{6}\right)$
51. $\sec[\sin^{-1}(-\frac{2}{3})]$
52. $\tan(\arcsin\frac{5}{8})$
53. $\cos(\sin^{-1}z)$, $z > 0$
54. $\cos(\sin^{-1}3z)$, $z < 0$
55. $\tan[\sin^{-1}(1 + z)]$, $1 + z < 0$
56. $\cos(\arcsin\sqrt{z})$
57. $\sec(\arcsin\sqrt{2z})$
58. $\cot(\sin^{-1}\sqrt{z - 1})$
59. $\sin(\cos^{-1}z)$, $z > 0$
60. $\cos(\arccos z)$, $z > 0$
61. $\tan(\arccos z)$, $z > 0$
62. $\sin(\tan^{-1}z)$, $z > 0$
63. $\cos(\arctan z)$, $z < 0$
64. $\tan(\cos^{-1}z)$, $z < 0$
65. $\sin(\cos^{-1}3z)$, $z < 0$
66. $\cos(\arctan 2z)$, $z > 0$
67. $\sec[\tan^{-1}(1 + z)]$, $z > 0$
68. $\sin(\arccos\sqrt{z})$
69. $\cos(\arctan\sqrt{2z})$
70. $\tan(\cos^{-1}\sqrt{z - 1})$
71. $\sin^{-1}\left(\sin\frac{\pi}{6}\right)$
72. $\arcsin\left(\tan\frac{\pi}{4}\right)$
73. $\arcsin\left(\cos\frac{2\pi}{3}\right)$
74. $\sin^{-1}\left(\sin\frac{11\pi}{6}\right)$
75. $\sin^{-1}(\cos 0)$
76. $\arcsin\left(\sin\frac{5\pi}{6}\right)$
77. $\tan^{-1}\left(\sin\frac{\pi}{2}\right)$
78. $\cos^{-1}\left(\sin\frac{7\pi}{6}\right)$
79. $\cos^{-1}\left(\cos\frac{11\pi}{6}\right)$
80. $\tan^{-1}(\cos 0)$
81. $\arccos\left(\tan\frac{5\pi}{4}\right)$
82. $\cos^{-1}\left(\cos\frac{3\pi}{2}\right)$

In the following problems, state the angle θ shown in each diagram in terms of an inverse trigonometric function.

83.
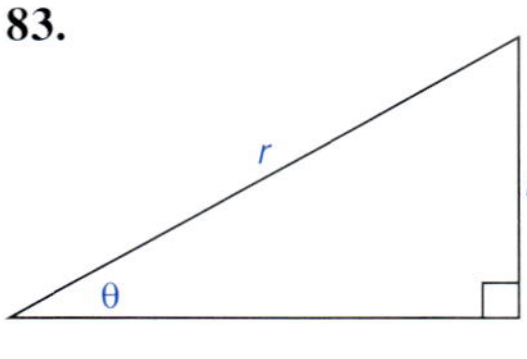

84.
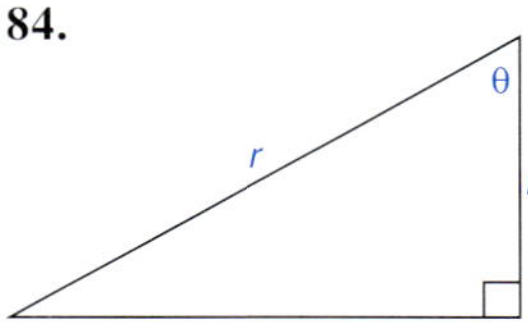

85.
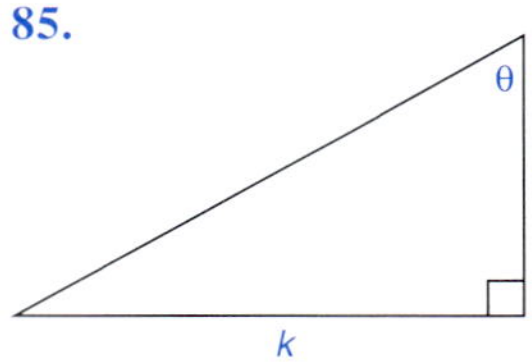

86.
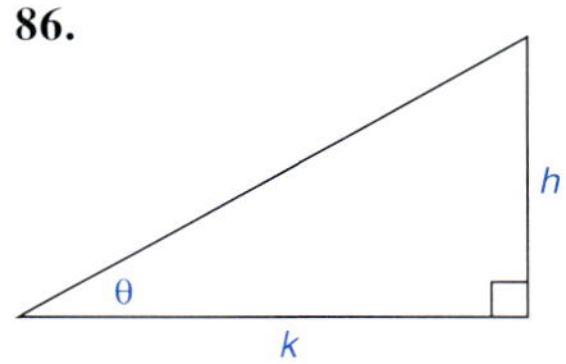

87. A picture hangs on a wall so that the bottom of the picture is 5 feet above the floor. The picture is 2 feet high. Describe the angle determined by the picture at the eye of an observer in terms of an inverse trigonometric function, if the observer's eye is also 5 feet above the floor and the observer is x feet away from the wall.

88. A radar antenna will track the launch of a rocket from a point 12,500 feet from the launch point. Both are at the same ground elevation at launch. If a represents the altitude of the rocket in feet, describe the angle of elevation θ of the rocket at the radar site in terms of an inverse trigonometric function.

89. An aircraft is flying toward an airport at an elevation of 3,500 feet above the airport. Describe the angle of depression θ of the airport at the aircraft in terms of the distance z from the aircraft directly to the airport, using an inverse trigonometric function.

In the following problems describe one value of θ, or x, in exact form, in terms of an inverse trigonometric function.

90. $\sin\theta = 0.75$ **91.** $\cos\theta = -0.8$ **92.** $\tan\theta = 3$ **93.** $\tan\theta = 4.1$

94. $2\sin\theta = 1.6$ **95.** $3\tan\theta = 5$ **96.** $\frac{1}{2}\sin\theta = -0.1$ **97.** $\dfrac{\tan\theta}{5} = 10$

98. $\sin 2\theta = 0.76$ **99.** $\tan 3\theta = 9$ **100.** $\cos\dfrac{\theta}{3} = -0.42$ **101.** $\sin\dfrac{3\theta}{2} = -0.56$

102. $4\cos 3\theta = 3$ **103.** $2\sin 4\theta = 1.5$ **104.** $\dfrac{3\cos 2\theta}{8} = \dfrac{7}{40}$ **105.** $\dfrac{6\sin 5\theta}{5} = \dfrac{10}{13}$

106. $\dfrac{A\cos Bx}{C} = D$ **107.** $A\tan(Bx + C) = D$ **108.** $\cos(x - 2) = 0.2$ **109.** $\sin(2x + 3) = 0.6$

110. Firefighters use the formula $d = \dfrac{h}{4} + 2$ to obtain the safe height of a fire ladder resting against a wall, where h is the height up the wall and d is the distance away from the wall at which the ladder should rest. Let θ represent the angle the ladder makes with the ground. Write the value of θ in terms of the inverse tangent function and the variable h.

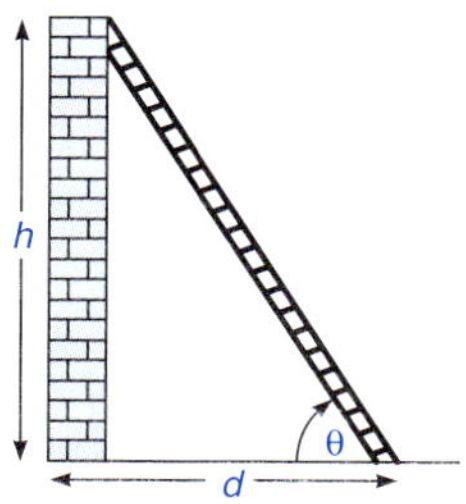

111. Some computer languages provide only an arctangent function. In these situations, we must program our own arcsine function. Use appropriate reference triangles to show that the following is an identity.

$$\arcsin(x) = \begin{cases} -\dfrac{\pi}{2} & \text{if } x = -1 \\ \arctan\left(\dfrac{x}{\sqrt{1-x^2}}\right) & \text{if } |x| < 1 \\ \dfrac{\pi}{2} & \text{if } x = 1 \end{cases}$$

112. Some computer languages provide only an arctangent function. In these situations we must program our own arccosine function. Use appropriate reference triangles to show that the following is an identity.

$$\arccos(x) = \begin{cases} \arctan\left(\dfrac{\sqrt{1-x^2}}{x}\right) & \text{if } 0 < x \le 1 \\ \dfrac{\pi}{2} & \text{if } x = 0 \\ \arctan\left(\dfrac{\sqrt{1-x^2}}{x}\right) + \pi & \text{if } -1 \le x < 0 \end{cases}$$

Skill and review

1. Perform the indicated multiplication, then simplify the result:

$$\cos x(\cos x + \sin x \tan x - \sec x)$$

2. Solve the triangle in the figure.

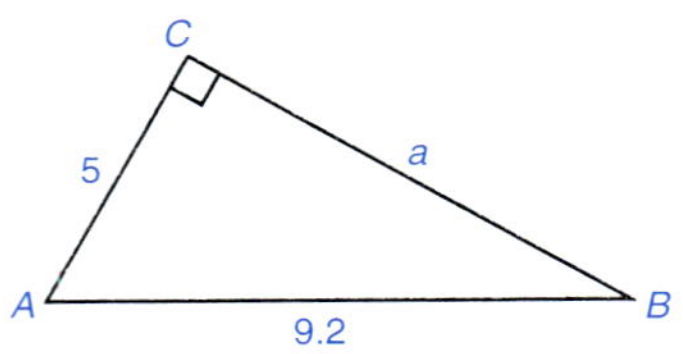

3. Find $\sin \frac{7\pi}{6}$.
4. Find $\tan\left(-\frac{11\pi}{3}\right)$.
5. A circle has radius 12 inches. What is the central angle which corresponds to an arc length of 15 inches? State the result in radians (exact) and degrees (nearest 0.1°).
6. Graph the rational function $f(x) = \frac{2x}{x^2 - 4}$.

6–5 The inverse cotangent, secant, and cosecant functions

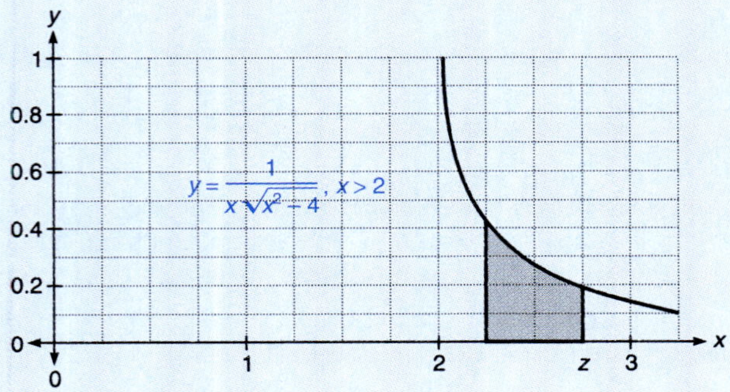

In higher mathematics it can be shown that the area A between the curve $y = \frac{1}{x\sqrt{x^2 - 4}}$ and the x-axis between $x = 2.25$ and $x = z$, $z > 2$, is $A = \frac{1}{2}\left(\sec^{-1}\frac{z}{2} - \sec^{-1}\frac{2.25}{2}\right)$. Calculate A if $z = 3.25$.

The reciprocal trigonometric functions (cotangent, secant, and cosecant) and their inverses were useful for solving triangles before the advent of electronic calculating devices like calculators and computers. Today, they have no practical use in solving triangles. This is why calculators do not have keys for the reciprocal functions, and why programming languages, such as FORTRAN, BASIC, or Pascal, do not support these functions. These functions still have value in simplifying and transforming certain expressions in higher mathematics, however, and that is why we study them here. The introductory problem illustrates one application from higher mathematics.

The inverse cotangent function

We define the inverse cotangent function by reversing the ordered pairs in the basic cotangent cycle; that is, we restrict the domain to $0 < x < \pi$. Figure 6–28 shows the inverse cotangent function, cotangent^{-1}. The domain is R, and the range is $0 < y < \pi$. As we might suspect *arccot x* also means $\cot^{-1}x$.

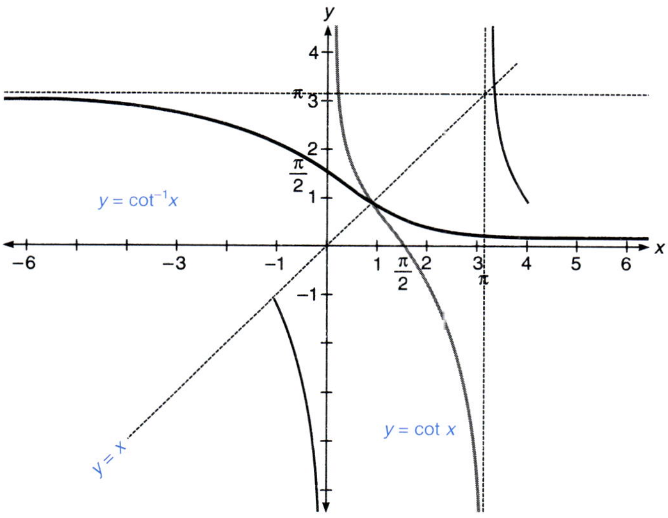

Figure 6–28

> **The inverse cotangent function**
> $y = \cot^{-1}x$ means
> 1. $\cot y = x$
> 2. $0 < y < \pi$

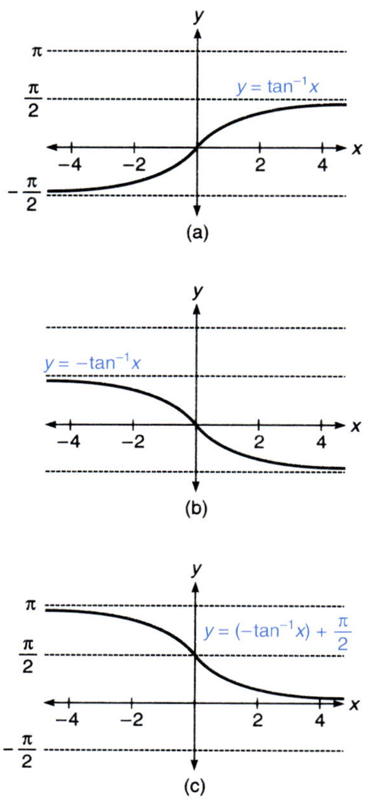

Figure 6–29

There are several ways to compute values of the inverse cotangent function. One way is to use the identity

$$\cot^{-1}x = \frac{\pi}{2} - \tan^{-1}x$$

This identity can be seen by considering the graphs of $y = \cot^{-1}x$ (figure 6–28) and $y = \tan^{-1}x$ (figure 6–27). Figure 6–29 shows the sequence of operations that will transform one graph into the other. Part (a) shows $y = \tan^{-1}x$. Part (b) shows the transformation caused by the scaling factor -1. Part (c) shows the vertical shift caused by adding $\frac{\pi}{2}$. This result is the same as the graph of the inverse cotangent function.

Note that this transformation of graphs does not absolutely guarantee that the identity $\cot^{-1}x = \frac{\pi}{2} - \tan^{-1}x$ is true. It should be proven algebraically. We will not prove it here.

Inverse secant function

Recall the identity $\sec\theta = \dfrac{1}{\cos\theta}$. We will use this to define the inverse of the secant function. Suppose we stated that $y = \sec^{-1}x$. Then $\sec y = x$. We proceed as shown.

$y = \sec^{-1}x$	An expression we want to define
$\sec y = x$	We expect the expression to have this property
$\dfrac{1}{\cos y} = x$	$\sec\theta = \dfrac{1}{\cos\theta}$
$\cos y = \dfrac{1}{x}$	$\dfrac{1}{\cos y} = x$; $1 = x\cos y$; $\dfrac{1}{x} = \cos y$
$y = \cos^{-1}\dfrac{1}{x}$	Since y is the angle whose cosine is $\dfrac{1}{x}$
$\sec^{-1}x = \cos^{-1}\dfrac{1}{x}$	$y = \sec^{-1}x$

We use this sequence of steps to motivate our definition. As expected, *arcsec* x also means $\sec^{-1}x$.

The inverse secant function

$$\sec^{-1}x = \cos^{-1}\frac{1}{x} \quad \text{if } |x| \geq 1$$

We require $|x| \geq 1$ so that $\left|\dfrac{1}{x}\right| \leq 1$, as required by the cosine^{-1} function. The range of the secant function is the range of the cosine^{-1} function since the latter defines the former, *except* that $\dfrac{\pi}{2}$ is not in the range. This is because $\dfrac{\pi}{2} = \cos^{-1}0$, and there is no value of x such that $\dfrac{1}{x} = 0$.

Inverse cosecant function

The identity $\csc\theta = \dfrac{1}{\sin\theta}$, and reasoning similar to that above, leads to the following definition for the inverse cosecant (*arcsecant*) function.

The inverse cosecant function

$$\csc^{-1}x = \sin^{-1}\frac{1}{x} \quad \text{if } |x| \geq 1$$

For reasons similar to those stated for the inverse secant function, the domain of the cosecant^{-1} function is $|x| \geq 1$, and the range is the same as that of the sine^{-1} function, except for 0, which is $\sin^{-1}0$, and $\frac{1}{x}$ can not take on the value 0.

Summary of properties

Table 6–5 summarizes the domains and ranges of the functions introduced above, as well as the identity mentioned above. Note in table 6–5 that quadrant I always corresponds to a nonnegative domain element (a nonnegative value of x), and quadrant II or IV corresponds to negative domain elements.

Example 6–5 A illustrates computations.

Function	Domain	Range	Quadrants
$y = \cot^{-1}x$	R	$0 < y < \pi$	I, II
$y = \sec^{-1}x$	$\|x\| \geq 1$	$0 \leq y \leq \pi, y \neq \frac{\pi}{2}$	I, II
$y = \csc^{-1}x$	$\|x\| \geq 1$	$-\frac{\pi}{2} \leq y \leq \frac{\pi}{2}, y \neq 0$	I, IV

$$\cot^{-1}x = \frac{\pi}{2} - \tan^{-1}x$$

Table 6–5

■ ***Example 6–5 A***

Find the given values in both radians and degrees. Round radians to two decimal places and degrees to one decimal place where necessary.

1. $\cot^{-1}(-4)$

$= \frac{\pi}{2} - \tan^{-1}(-4)$ — $\cot^{-1}x = \frac{\pi}{2} - \tan^{-1}x$

≈ 2.90 (radians) — Calculator in radian mode

$= 90° - \tan^{-1}(-4)$

$\approx 166.0°$ — Calculator in degree mode

Thus, $\cot^{-1}(-4) \approx 166.0°$ or 2.90 (radians).

2. $\sec^{-1}2$

$= \cos^{-1}\frac{1}{2}$ — Definition of $\sec^{-1}\theta$

$= 60°$ or $\frac{\pi}{3}$

Thus, $\sec^{-1}2 = 60°$ or $\frac{\pi}{3}$. ■

As we saw in section 6–4, some expressions that involve both the trigonometric and inverse trigonometric functions can be simplified by using a reference triangle. Example 6–5 B illustrates for the functions covered in this section.

■ Example 6–5 B

Simplify the expression.

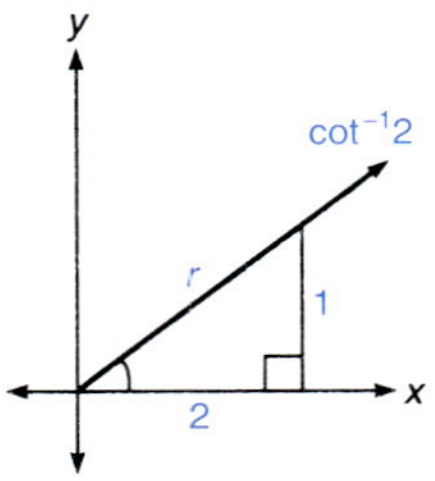

1. $\sec(\cot^{-1}2)$

$\cot^{-1}2$ is an angle in quadrant I. Since its cotangent is 2, its tangent is $\frac{1}{2}$. A reference triangle for a quadrant I angle with tangent $\frac{1}{2}$ is shown in the figure.

$r = \sqrt{5}$ (Pythagorean theorem), so the cosine of the angle is $\dfrac{2}{\sqrt{5}}$, and therefore the secant is $\dfrac{\sqrt{5}}{2}$. Thus, $\sec(\cot^{-1}2) = \dfrac{\sqrt{5}}{2}$.

Note The identity $\cot^{-1}x = \dfrac{\pi}{2} - \tan^{-1}x$ would not be helpful in simplifying this expression. It is useful for computing values of $\cot^{-1}x$.

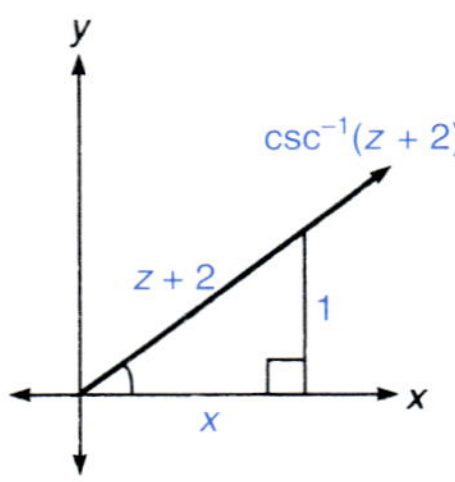

2. $\cos(\csc^{-1}(z + 2)), z > 0$

$\csc^{-1}(z + 2) = \sin^{-1}\dfrac{1}{z + 2}$ by definition. Since $\dfrac{1}{z + 2}$ is positive, $\sin^{-1}\dfrac{1}{z + 2}$ is an angle in quadrant I (see the figure). We find x by the Pythagorean theorem.

$$(z + 2)^2 = 1^2 + x^2$$
$$z^2 + 4z + 4 - 1 = x^2$$
$$\sqrt{z^2 + 4z + 3} = x$$

The cosine of the angle is $\dfrac{\text{adjacent}}{\text{hypotenuse}} = \dfrac{x}{z + 2} = \dfrac{\sqrt{z^2 + 4z + 3}}{z + 2}$. ■

Mastery points

Can you

- State the domain and range of the inverse cotangent, cosecant, and secant functions?
- Find exact values, in both radians and degrees, for expressions of the form $\cot^{-1}x$, $\csc^{-1}x$, and $\sec^{-1}x$, for appropriate values of x, using the definitions of these functions?
- Find approximate values, in both radians and degrees, for expressions of the form $\sin^{-1}x$, $\cos^{-1}x$, and $\tan^{-1}x$, using a calculator and the definitions of these functions?
- Simplify expressions that involve combinations of the trigonometric and inverse trigonometric functions, using a reference triangle where appropriate?

Exercise 6–5

Compute the exact value of the following expressions.

1. $\csc^{-1}2$ **2.** $\text{arcsec}\left(-\frac{2\sqrt{3}}{3}\right)$ **3.** arccot 1 **4.** $\text{arccsc}(-1)$ **5.** $\sec^{-1}(-2)$

6. $\cot^{-1}(-\sqrt{3})$ **7.** $\text{arccsc}\left(\frac{2\sqrt{3}}{3}\right)$ **8.** $\csc^{-1}\sqrt{2}$ **9.** arccot 0 **10.** $\text{arcsec}(-\sqrt{2})$

Compute approximate values of the following expressions in both radians (to hundredths) and degrees (to tenths).

11. $\csc^{-1}3.3534$ **12.** $\cot^{-1}0.5080$ **13.** arcsec(−2.9986) **14.** arccsc(−2.5087) **15.** arccot 5.1997

16. $\sec^{-1}(-2.0126)$ **17.** $\sec^{-1}(-11.1261)$ **18.** $\csc^{-1}(-3.8898)$ **19.** arccsc 3.1790 **20.** arccot(−8.3534)

Simplify the following expressions.

21. $\sin(\csc^{-1}3)$ **22.** $\cos(\cot^{-1}2)$ **23.** cot(arcsec 4) **24.** $\sin(\sec^{-1}1.5)$

25. csc(arccot 5) **26.** $\sec(\text{arccsc } \frac{7}{4})$ **27.** $\cos[\text{arccsc}(-\frac{5}{4})]$ **28.** $\tan[\text{arccsc}(-\frac{5}{3})]$

29. $\tan[\sec^{-1}(-\frac{6}{5})]$ **30.** $\csc[\sec^{-1}(-\frac{8}{7})]$ **31.** $\sin(\csc^{-1}z)$, $z > 0$ **32.** $\cot(\sec^{-1}z)$, $z > 0$

33. $\sec(\cot^{-1}z)$, $z < 0$ **34.** $\tan(\text{arcsec } z)$, $z < 0$ **35.** $\cos(\text{arcsec } 2z)$, $z > 0$ **36.** $\sin(\csc^{-1}3z)$, $z < 0$

37. $\tan[\sec^{-1}(z + 1)]$, $z + 1 > 0$ **38.** $\csc[\sec^{-1}(1 - z)]$, $1 - z > 0$

39. $\csc\left[\text{arccsc } \frac{3}{z}\right]$, $z > 0$ **40.** $\sec\left[\cot^{-1}\left(\frac{2}{z + 1}\right)\right]$, $z + 1 > 0$

41. In higher mathematics it can be shown that the area A between the curve $y = \frac{1}{x\sqrt{x^2 - 4}}$ and the x-axis between $x = 2.25$ and $x = z$, $z > 2$, is (see the figure)

$$A = \frac{1}{2}\left(\sec^{-1}\frac{z}{2} - \sec^{-1}\frac{2.25}{2}\right)$$

Calculate A, to the nearest thousandth, if $z = 3.25$.

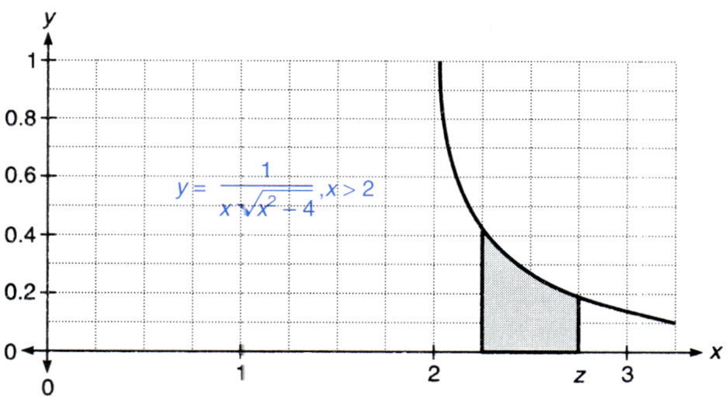

Skill and review

1. Combine: $\frac{2x}{x - 3} - \frac{x}{x + 5}$.

2. If $\sin x = -\frac{1}{2}$ and x terminates in quadrant III, find the measure of x in radians (exact).

3. If $\cos \theta = 1 - u$, and θ terminates in quadrant II, find $\sin \theta$ in terms of u.

4. Find the least nonnegative solution to the equation $3 \sin x = -2$, to the nearest 0.1°.

5. Simplify $\cos(\tan^{-1}5)$.

6. Simplify $\csc(\cos^{-1}(1 - m))$, $1 - m > 0$.

7. Find $\csc\left(\frac{19\pi}{6}\right)$.

8. If the point (−5,8) is on the terminal side of an angle θ in standard position, (a) find $\sin \theta$ and (b) find the least nonnegative measure of θ to the nearest 0.1°.

Chapter 6 summary

- Let θ be an angle in standard position with degree measure $\theta°$ and radian measure s. Then, $\frac{s}{\pi} = \frac{\theta°}{180°}$.
- If (x,y) is the point on the unit circle that intersects the terminal side of θ, then $\sin\theta = y$, $\cos\theta = x$.
- If s is the radian measure of a central angle on a circle of radius r, and L is the corresponding arc length, then $L = rs$.

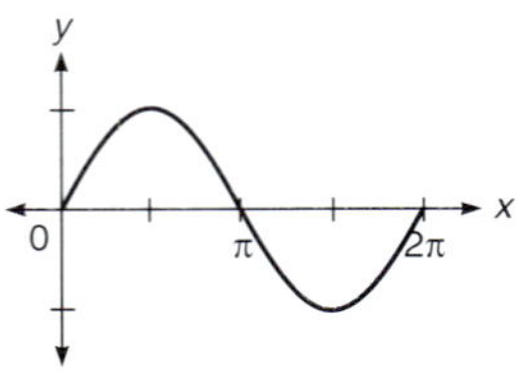

Basic sine cycle

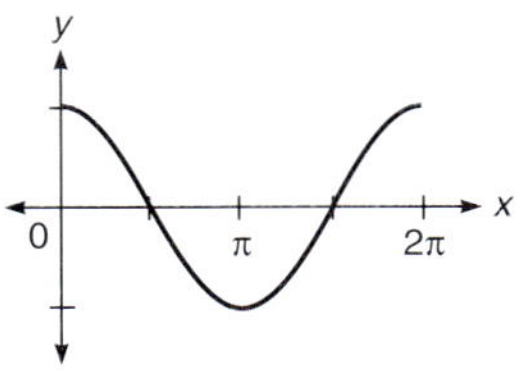

Basic cosine cycle

- The expression $-\frac{C}{B}$ is the phase shift, of the particular sine or cosine function. The value $\frac{2\pi}{B}$ is the period of the function. B is also the number of complete cycles in 2π units.
- Summary of the properties of the sine, cosine, and tangent functions.

Function	Domain	Range	Period
$y = \sin x$	R	$-1 \le y \le 1$	2π
$y = \cos x$	R	$-1 \le y \le 1$	2π
$y = \tan x$	$x \ne \frac{\pi}{2} + k\pi,\ k \in J$	R	π

$\sin(-x) = -\sin x$ (odd, origin symmetry)
$\cos(-x) = \cos x$ (even, y-axis symmetry)
$\tan(-x) = -\tan x$ (odd, origin symmetry)

- Summary of the properties of the cosecant, secant, and cotangent functions.

Function	Domain	Range	Period
$y = \csc x$	$x \ne k\pi,\ k \in J$	$\lvert y \rvert \ge 1$	2π
$y = \sec x$	$x \ne \frac{\pi}{2} + k\pi,\ k \in J$	$\lvert y \rvert \ge 1$	2π
$y = \cot x$	$x \ne k\pi,\ k \in J$	R	π

$\csc(-x) = -\csc x$ (odd, origin symmetry)
$\sec(-x) = \sec x$ (even, y-axis symmetry)
$\cot(-x) = -\cot x$ (odd, origin symmetry)

- **Inverse sine function** $y = \sin^{-1}x$ means
  1. $\sin y = x$
  2. $-\frac{\pi}{2} \le y \le \frac{\pi}{2}$
  3. $-1 \le x \le 1$
- **Inverse cosine function** $y = \cos^{-1}x$ means
  1. $\cos y = x$
  2. $0 \le y \le \pi$
  3. $-1 \le x \le 1$
- **Inverse tangent function** $y = \tan^{-1}x$ means
  1. $\tan y = x$
  2. $-\frac{\pi}{2} < y < \frac{\pi}{2}$
- **Inverse cotangent function** $y = \cot^{-1}x$ means
  1. $\cot y = x$
  2. $0 < y < \pi$
- **Inverse secant function** $\sec^{-1}x = \cos^{-1}\frac{1}{x}$ if $\lvert x \rvert \ge 1$.
- **Inverse cosecant function** $\csc^{-1}x = \sin^{-1}\frac{1}{x}$ if $\lvert x \rvert \ge 1$.
- Summary of the properties of the inverse sine, cosine, and tangent functions.

Function	Domain	Range	Quadrants
$y = \sin^{-1}x$	$-1 \le x \le 1$	$-\frac{\pi}{2} \le y \le \frac{\pi}{2}$	I, IV
$y = \cos^{-1}x$	$-1 \le x \le 1$	$0 \le y \le \pi$	I, II
$y = \tan^{-1}x$	R	$-\frac{\pi}{2} < y < \frac{\pi}{2}$	I, IV

$\sin^{-1}(-x) = -\sin^{-1}x$ (odd)
$\cos^{-1}(-x) = \pi - \cos^{-1}x$
$\tan^{-1}(-x) = -\tan^{-1}x$ (odd)

- Summary of the properties of the inverse cotangent, secant, and cosecant functions.

Function	Domain	Range	Quadrants
$y = \cot^{-1}x$	R	$0 < y < \pi$	I, II
$y = \sec^{-1}x$	$\lvert x \rvert \ge 1$	$0 \le y \le \pi,\ y \ne \frac{\pi}{2}$	I, II
$y = \csc^{-1}x$	$\lvert x \rvert \ge 1$	$-\frac{\pi}{2} \le y \le \frac{\pi}{2},\ y \ne 0$	I, IV

$$\cot^{-1}x = \frac{\pi}{2} - \tan^{-1}x$$

Chapter 6 review

[6–1] Convert the following degree measures to radian measures. Leave answers in both exact form and approximated to two decimal places.

1. 315° **2.** −240° **3.** −148° **4.** 320°

Convert the following radian measures into degree measures. Leave answers both in exact form and approximated to two decimal places.

5. $\frac{7\pi}{9}$ **6.** $\frac{11\pi}{8}$ **7.** −4 **8.** $\frac{3}{8}$

9. Find the length of the arc determined by a central angle of 3.8 radians on a circle of diameter 16 inches.

10. Find the length of the arc determined by a central angle of 230° on a circle of radius 150 millimeters.

11. Find the measure, in both radians and degrees, of the central angle determined by an arc length of 32 inches on a circle of diameter 20 inches.

12. The diagram shows two wheels. Wheel *A* drives wheel *B*. Determine the radian measure of the angle through which wheel *B* will move when wheel *A* moves through an angle of 5 radians.

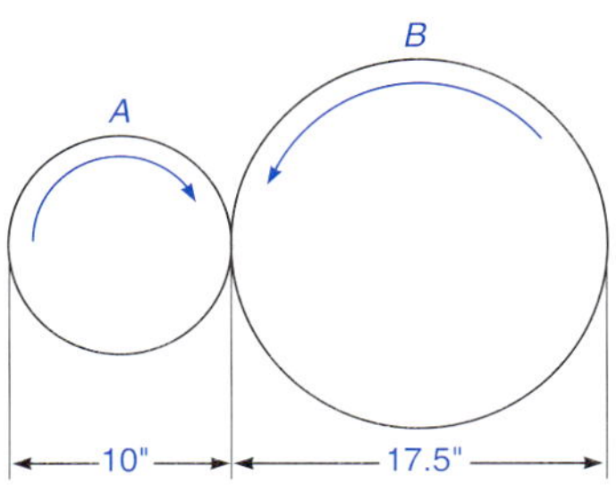

Find the following function values where the angle is given in radian measure. Round your answer to four decimal places.

13. sin 4 **14.** cos 2.52 **15.** sec $\frac{3}{8}$ **16.** cot 1

Find the exact function values for the following angles.

17. $\cos\frac{5\pi}{6}$ **18.** $\cot\frac{5\pi}{3}$

[6–2]

19. Sketch the graph of the sine function; state the domain, range, and period of the sine function.

20. Using the graph of $y = \cos x$ as a guide describe all values of x for which $\cos x$ is 1.

Use the appropriate property, even or odd, to calculate the exact function value.

21. $\sin\left(-\frac{\pi}{3}\right)$ **22.** $\tan\left(-\frac{4\pi}{3}\right)$

Test the function for the even/odd property. State which type of symmetry the graph would have based on being even, odd, or neither even nor odd.

23. $f(x) = x \sin x$ **24.** $f(x) = \tan x \cdot \cos x$

Graph three cycles of the following functions.

25. $y = -3 \sin x$ **26.** $y = 2 \cos x - 1$

27. $y = -\tan x$ **28.** $y = \sin 4x$

29. $y = \cos 3x$ **30.** $y = \tan\frac{x}{3}$

31. $y = \frac{1}{2}\sin\left(x - \frac{\pi}{4}\right)$ **32.** $y = 2\cos\left(x + \frac{\pi}{6}\right)$

33. $y = \cos 2\pi x$ **34.** $y = 3 \sin(\pi - 2x)$

35. $y = 2 \cos(\pi - 3x)$

36. Assume that the graph shown is of the form $y = A \sin(Bx + C) + D$. Find values of *A*, *B*, *C*, and *D* that would produce the graph, and state the equation produced by these values.

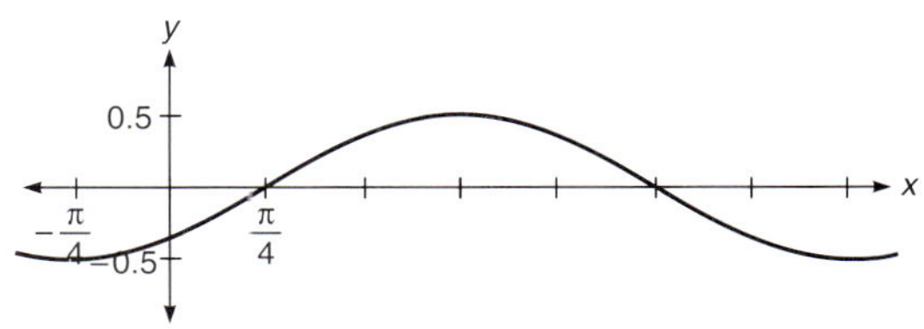

[6–3]

37. Sketch the graph of the cosecant function.

38. Show that the function $f(x) = \sec x \cdot \sin^2 x + x^4$ is an even function.

Graph three cycles of the following functions.

39. $y = \tan 3x$ **40.** $y = \sec\left(x + \frac{\pi}{3}\right)$

41. $y = \tan(-x)$

[6–4]

42. State the domain and range of the inverse cosine function.

Find exact values for each of the following expressions, in both radians and degrees.

43. $\sin^{-1}\frac{1}{2}$ **44.** $\cos^{-1}\frac{\sqrt{2}}{2}$ **45.** $\arctan\sqrt{3}$

46. $\arcsin\left(-\frac{\sqrt{3}}{2}\right)$

Find approximate values for the following expressions in both radians and degrees. Round the values for radians to two decimal places and for degrees to one decimal place.

47. $\arctan(-3.55)$ **48.** $\arccos(-0.4290)$

49. $\tan^{-1}2.5$

Simplify the following expressions. Obtain exact values.

50. $\cos(\arcsin\frac{3}{8})$ **51.** $\tan[\arcsin(-\frac{2}{3})]$

52. $\cos(\arcsin 0.5)$ **53.** $\csc\left(\cos^{-1}\frac{\sqrt{2}}{4}\right)$

54. $\sin[\cos^{-1}(-\frac{1}{3})]$ **55.** $\sin[(\arctan(-5))]$

56. $\cos(\sin^{-1}2z)$, $z > 0$ **57.** $\tan[\sin^{-1}(1 - z)]$, $1 - z < 0$

58. $\sin(\arccos\sqrt{z + 1})$ **59.** $\sec(\tan^{-1}\sqrt{z - 1})$

60. $\sin^{-1}\left(\sin\frac{7\pi}{6}\right)$ **61.** $\sin^{-1}\left(\cos\frac{\pi}{2}\right)$

62. $\arccos\left(\cos\frac{5\pi}{3}\right)$

63. An aircraft is flying toward an airport at an elevation of 6,000 feet above the airport. Describe the angle of depression of the airport at the aircraft in terms of the slant distance z from the aircraft directly to the airport, using an inverse trigonometric function.

In the following problems, describe one value of θ, or x, in exact form, in terms of an inverse trigonometric function.

64. $\cos\frac{\theta}{3} = \frac{1}{3}$ **65.** $\frac{1}{3}\sin 2x = \frac{1}{12}$

66. $a\tan k\theta = b$ (a, b, and k are constants).

[6–5] Compute the exact value of the following expressions.

67. $\csc^{-1}2$ **68.** $\sec^{-1}\sqrt{2}$

Find approximate values for the following expressions in both radians and degrees. Round the values for radians to two decimal places and for degrees to one decimal place.

69. $\cot^{-1}3$ **70.** $\text{arcsec}(-2.66)$

Simplify the following expressions.

71. $\cot(\sec^{-1}3)$ **72.** $\sec\left(\cot^{-1}\frac{1}{z}\right)$, $z > 0$

Chapter 6 test

1. Convert $-250°$ to radian measure. Leave answers in both exact form and approximated to two decimal places.
2. Convert $\frac{7\pi}{5}$ into degree measure. Leave answers both in exact form and approximated to two decimal places.
3. Find the length of the arc, to the nearest tenth, determined by a central angle of 2.5 radians on a circle of diameter 20 inches.
4. Find the measure, in both radians and degrees, to the nearest tenth, of the central angle determined by an arc length of 19 millimeters on a circle of diameter 28 millimeters.
5. The diagram shows a train wheel. The wheel has a diameter of 38 inches. Through what angle, to the nearest degree, will the wheel rotate if the train moves forward a distance of 5 feet?

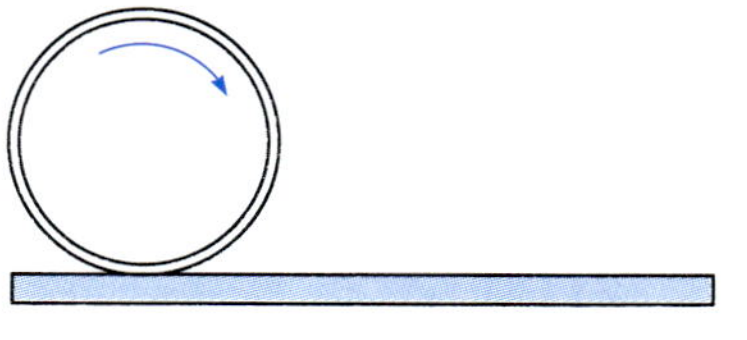

6. Find the value of sec 3.12 to four decimal places.
7. Find the exact function value of $\cos\frac{11\pi}{6}$.
8. Sketch the graph of the cosine function; state the domain, range, and period of the cosine function.

9. Using the graph of $y = \sin x$ as a guide describe all values of x for which $\sin x$ is -1.

10. Use the appropriate property, even or odd, to calculate the exact function value of $\tan\left(-\frac{5\pi}{3}\right)$.

11. Test the function $f(x) = x + \sin x$ for the even/odd property. State which type of symmetry the graph would have based on being even, odd, or neither even nor odd.

Graph three cycles of the following functions.

12. $y = -\frac{1}{2}\sin x$

13. $y = 3\cos x + 1$

14. $y = \sin 3x$

15. $y = \cos\left(x + \frac{\pi}{6}\right)$

16. $y = 3\sin(\pi - 2x)$

17. Assume that the graph is of the form $y = A\sin(Bx + C) + D$. Find values of A, B, C, and D that would produce the graph, and write the corresponding equation.

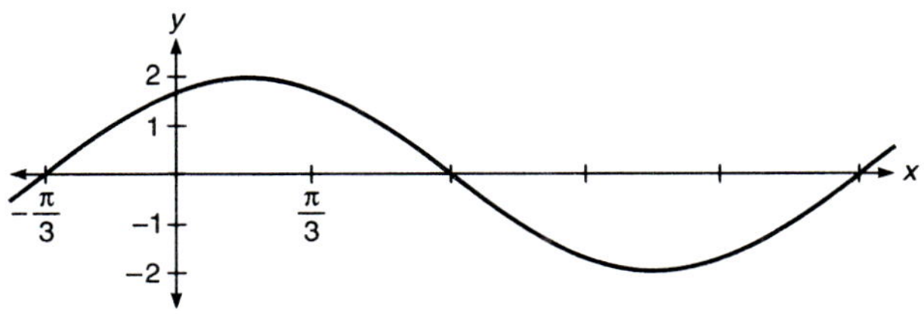

18. Sketch the graph of the secant function.

19. Show that the function $f(x) = \sec x \cdot \sin x + x^3$ is an odd function.

Graph three cycles of the following functions.

20. $y = \tan(-2x)$

21. $y = \csc\left(x - \frac{\pi}{6}\right)$

22. State the domain and range of the inverse cosine function.

23. Find the exact value for $\sin^{-1}\left(-\frac{\sqrt{3}}{2}\right)$ both in radians and degrees.

24. Find an approximate value for $\cos^{-1}(-0.80)$ in both radians and degrees. Round the values for radians to two decimal places and for degrees to one decimal place.

Simplify the following expressions. Obtain exact values.

25. $\tan(\arcsin \frac{3}{4})$

26. $\tan[\arccos(-\frac{1}{3})]$

27. $\csc\left(\sin^{-1}\frac{\sqrt{3}}{4}\right)$

28. $\cos(\tan^{-1}2z)$, $z > 0$

29. $\cos^{-1}\left(\cos\frac{7\pi}{6}\right)$

30. $\csc^{-1}\frac{2}{\sqrt{3}}$

31. Write angle θ in the triangle shown in terms of z and 5, using an inverse trigonometric function.

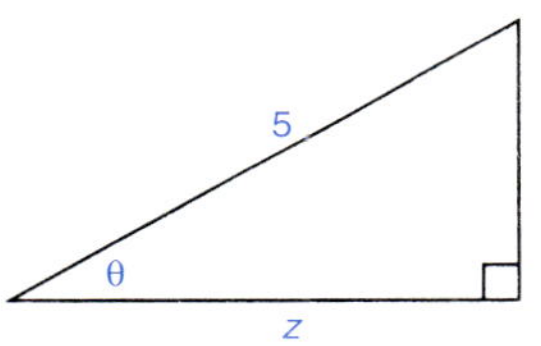

32. Describe one value of x in exact form, in terms of an inverse trigonometric function: $2\cos 3x = \frac{1}{3}$.

33. Find approximate values for the following expression in both radians and degrees. Round the values for radians to two decimal places and for degrees to one decimal place: $\sec^{-1}2.65$.

C H A P T E R

7 Trigonometric Equations

Recall that trigonometric equations were introduced in section 5–6. In this chapter we first study trigonometric identities; these are very important in the study of the calculus and certain engineering applications. We then examine conditional trigonometric equations in more depth than we did previously.

7–1 Basic trigonometric identities

A computer program is being written that must calculate $\frac{1}{\cot\theta + \tan\theta}$ for a varying value θ, which is input into the computer 100 times per second. A more efficient way to compute this expression is desired. Show that calculating the simpler expression $\sin\theta\cos\theta$ will give the same results.

In this section we examine identities. This problem is equivalent to showing that

$$\frac{1}{\cot\theta + \tan\theta} = \sin\theta\cos\theta$$

is an identity. Identities are used to simplify computations in many situations.

Recall from section 5–6 that an **identity** is an equation that is true for every allowed value of its variable (or variables). We have seen the following identities in that and other sections.

Reciprocal identities

$$\csc\theta = \frac{1}{\sin\theta},\quad \sec\theta = \frac{1}{\cos\theta},\quad \cot\theta = \frac{1}{\tan\theta}$$

$$\sin\theta = \frac{1}{\csc\theta},\quad \cos\theta = \frac{1}{\sec\theta},\quad \tan\theta = \frac{1}{\cot\theta}$$

Tangent and cotangent identities

$$\tan\theta = \frac{\sin\theta}{\cos\theta},\quad \cot\theta = \frac{\cos\theta}{\sin\theta}$$

The Pythagorean identities

Recall that $\sin^2\theta + \cos^2\theta = 1$ is the fundamental identity. Two other forms of the fundamental identity are

$$\sin^2\theta = 1 - \cos^2\theta \quad \text{and} \quad \cos^2\theta = 1 - \sin^2\theta.$$

If each term in the fundamental identity is divided by $\cos^2\theta$, we obtain

$$\frac{\sin^2\theta}{\cos^2\theta} + \frac{\cos^2\theta}{\cos^2\theta} = \frac{1}{\cos^2\theta}$$
$$\left(\frac{\sin\theta}{\cos\theta}\right)^2 + 1 = \left(\frac{1}{\cos\theta}\right)^2$$
$$\tan^2\theta + 1 = \sec^2\theta$$

Similarly, if each term of the fundamental identity is divided by $\sin^2\theta$ we obtain the identity $\cot^2\theta + 1 = \csc^2\theta$. These two identities, along with the fundamental identity, are called the **Pythagorean identities.** They are summarized here.

Pythagorean identities	Useful forms	
$\sin^2\theta + \cos^2\theta = 1$	$\sin^2\theta = 1 - \cos^2\theta$	$\cos^2\theta = 1 - \sin^2\theta$
$\sec^2\theta = \tan^2\theta + 1$	$\tan^2\theta = \sec^2\theta - 1$	$\sec^2\theta - \tan^2\theta = 1$
$\csc^2\theta = \cot^2\theta + 1$	$\cot^2\theta = \csc^2\theta - 1$	$\csc^2\theta - \cot^2\theta = 1$

Example 7–1 A illustrates applications of these identities.

■ ***Example 7–1 A***

Simplify each expression into one term.

1. $1 - \cos^2 4\alpha$

$\sin^2 4\alpha$ — $\sin^2 4\alpha + \cos^2 4\alpha = 1$, so $1 - \cos^2 4\alpha = \sin^2 4\alpha$

2. $(1 - \sec x)(1 + \sec x)$

$1 + \sec x - \sec x - \sec^2 x$ — $(a - b)(a + b) = a^2 + ab - ab - b^2$

$1 - \sec^2 x$ — Collect like terms

$-(\sec^2 x - 1)$ — $a - b = -(b - a)$

$-\tan^2 x$ ■

Verifying identities

One use of identities is in simplifying and transforming trigonometric expressions. This is illustrated in example 7–1 B. We proceed by replacing given parts of an expression by equivalent parts from the identities just summarized. *There are many correct sequence of steps!* We proceed by trial and error, guided by past experience. If we can show that one member of an equation can be transformed into the other member by replacing expressions using identities and performing algebraic transformations, then we say we have *verified* the identity.

■ *Example 7–1 B*

Verify that each equation is an identity by showing that the left member of the identity is equivalent to the right member.

1. $\tan \theta \csc \theta = \sec \theta$

$\tan \theta \csc \theta$	
$\dfrac{\sin \theta}{\cos \theta} \cdot \dfrac{1}{\sin \theta}$	$\tan \theta = \dfrac{\sin \theta}{\cos \theta}$, $\csc \theta = \dfrac{1}{\sin \theta}$
$\dfrac{1}{\cos \theta}$	Reduce by a factor of $\sin \theta$
$\sec \theta$	$\sec \theta = \dfrac{1}{\cos \theta}$

2. $\dfrac{1}{\sin \beta - \csc \beta} = -\tan \beta \sec \beta$

$\dfrac{1}{\sin \beta - \csc \beta}$	
$\dfrac{1}{\sin \beta - \dfrac{1}{\sin \beta}}$	$\csc \beta = \dfrac{1}{\sin \beta}$
$\dfrac{1}{\sin \beta - \dfrac{1}{\sin \beta}} \cdot \dfrac{\sin \beta}{\sin \beta}$	Multiply numerator and denominator by $\sin \beta$
$\dfrac{\sin \beta}{\sin^2\beta - 1}$	$\sin \beta\left(\sin \beta - \dfrac{1}{\sin \beta}\right) = \sin^2\beta - 1$
$\dfrac{\sin \beta}{-\cos^2\beta}$	$\cos^2\beta = 1 - \sin^2\beta$, so $-\cos^2\beta = \sin^2\beta - 1$
$-\dfrac{\sin \beta}{\cos \beta} \cdot \dfrac{1}{\cos \beta}$	
$-\tan \beta \sec \beta$	

3. $\dfrac{\cos^2\alpha}{1 + \sin \alpha} = 1 - \sin \alpha$

$\dfrac{\cos^2\alpha}{1 + \sin \alpha}$	
$\dfrac{1 - \sin^2\alpha}{1 + \sin \alpha}$	$\cos^2\theta = 1 - \sin^2\theta$
$\dfrac{(1 - \sin \alpha)(1 + \sin \alpha)}{1 + \sin \alpha}$	$m^2 - n^2 = (m - n)(m + n)$
$1 - \sin \alpha$	■

It is not necessary to transform just one member or the other of an identity. Sometimes it is easier to transform both members, as in example 7–1 C.

■ ***Example 7–1 C***

Verify the identity $\sin^2\theta \tan^2\theta + 1 = \sec^2\theta - \cos^2\theta \sec^2\theta + \cos^2\theta$

$\sin^2\theta \tan^2\theta + 1$	$\sec^2\theta - \cos^2\theta \sec^2\theta + \cos^2\theta$
$\sin^2\theta \tan^2\theta + \sin^2\theta + \cos^2\theta$	$\sec^2\theta(1 - \cos^2\theta) + \cos^2\theta$
$\sin^2\theta (\tan^2\theta + 1) + \cos^2\theta$	$\sec^2\theta \sin^2\theta + \cos^2\theta$
$\sin^2\theta \sec^2\theta + \cos^2\theta$	$\dfrac{1}{\cos^2\theta} \sin^2\theta + \cos^2\theta$
$\sin^2\theta \dfrac{1}{\cos^2\theta} + \cos^2\theta$	$\tan^2\theta + \cos^2\theta$
$\tan^2\theta + \cos^2\theta$	

Since the left member and right member can be transformed into the same expression, they are equivalent. ■

Showing an equation is not an identity

Most equations are not identities. To show that this is the case we need to find a value for the variable for which the statement is not true. This value is called a **counter example;** it shows that the equation is *not* an identity.

■ ***Example 7–1 D***

Show by counter example that $\cos x + \sin x \cot x = 1$ is not an identity.

Choose a value for which each expression is defined; $x = \dfrac{\pi}{4}$ is such a value.

$$\cos x + \sin x \cot x$$
$$\cos \frac{\pi}{4} + \sin \frac{\pi}{4} \cot \frac{\pi}{4} = \frac{\sqrt{2}}{2} + \frac{\sqrt{2}}{2} \cdot 1 = \frac{2\sqrt{2}}{2} = \sqrt{2}$$

Since $\sqrt{2} \neq 1$ we have shown that the given equation is not an identity.

Note There are usually many counter examples for a given equation. In this example almost any other value would have worked as well. (However, $\dfrac{\pi}{3}$ would not work, because the equation is true for this value.) ■

Mastery points

Can you

- State the reciprocal identities, fundamental identity, and the remaining Pythagorean identities from memory?
- Recognize useful forms of the Pythagorean identities?
- Transform forms of the Pythagorean identities into simpler forms?
- Transform one member of an identity into the other member?
- Show that an equation is not an identity by a counter example?

Exercise 7–1

Each of the following expressions can be simplified into the form 1, −1, sin θ, cos θ, tan θ, cot θ, sec θ, csc θ, $\sin^2\theta$, $\cos^2\theta$, $\tan^2\theta$, $\cot^2\theta$, $\sec^2\theta$, or $\csc^2\theta$. Simplify each expression into one of these forms.

1. $\dfrac{\sin\theta}{\tan\theta}$
2. $\dfrac{\cos\theta}{\cot\theta}$
3. $\cot\theta\sec\theta$
4. $\sec\theta\cot\theta$
5. $\cot^2\theta\sin^2\theta$
6. $\sin^2\theta\sec^2\theta$
7. $(\tan^2\theta + 1)(1 - \sin^2\theta)$
8. $(1 - \cos^2\theta)(1 + \cot^2\theta)$
9. $\dfrac{(\sec\theta - 1)(\sec\theta + 1)}{\sin^2\theta}$
10. $\dfrac{(\sec\theta + \tan\theta)(\sec\theta - \tan\theta)}{\cos^2\theta}$
11. $(\csc x + \cot x)(1 - \cos x)$
12. $\cos x + \sin x\tan x$
13. $\sec x - \tan x\sin x$
14. $(\csc x + 1)(\sec x - \tan x)$
15. $\cot x\sec x$
16. $\dfrac{\csc x + \sec x}{\tan x + 1}$
17. $\dfrac{\csc^2\theta - 1}{\csc^2\theta}$
18. $\dfrac{\sec^4 y - \tan^4 y}{\sec^2 y + \tan^2 y}$ (*Hint:* Factor $\sec^4 y - \tan^4 y$.)
19. $\sin x + \cos x\cot x$
20. $\tan x\csc x\cos x$
21. $\dfrac{\csc\theta\sin\theta}{\cot\theta}$
22. $\tan^2\theta(\cot^2\theta + 1)$
23. $\dfrac{\tan\theta\cot\theta}{\sin\theta}$
24. $\cos\theta(\sec\theta - \cos\theta)$
25. $\cos^2\theta(1 + \cot^2\theta)$
26. $\csc^2\theta(1 - \cos^2\theta)$
27. $\sin^2\theta(\csc^2\theta - 1)$
28. $\cot^2\theta - \csc^2\theta$
29. $\tan^2\theta - \sec^2\theta$
30. $\sec^2\theta(\csc^2\theta - 1)$
31. $\dfrac{\cot\theta\sec\theta}{\csc\theta}$
32. $\dfrac{\sec\theta}{\tan\theta\csc\theta}$

Verify the following identities.

33. $\csc\theta + \cot\theta = \dfrac{1 + \cos\theta}{\sin\theta}$
34. $\tan\theta + \sec\theta = \dfrac{1 + \sin\theta}{\cos\theta}$
35. $\dfrac{\csc\theta}{\sec\theta + \tan\theta} = \dfrac{\cos\theta}{\sin\theta + \sin^2\theta}$
36. $\dfrac{\sec\theta}{\csc\theta + \cot\theta} = \dfrac{\tan\theta}{1 + \cos\theta}$
37. $\dfrac{1 + \csc\theta}{1 + \sec\theta} = \cot\theta\left(\dfrac{1 + \sin\theta}{1 + \cos\theta}\right)$
38. $\dfrac{\tan\theta + 1}{\tan\theta - 1} = \dfrac{\sin\theta + \cos\theta}{\sin\theta - \cos\theta}$
39. $\dfrac{\tan^2\theta + \sec^2\theta}{\sec^2\theta} = \sin^2\theta + 1$
40. $\dfrac{\cot^2\theta + \csc^2\theta}{\csc^2\theta} = 1 + \cos^2\theta$
41. $\dfrac{1 + \cot^2\theta}{\tan^2\theta} = \cot^2\theta\csc^2\theta$
42. $\dfrac{1}{\cot\theta + \tan\theta} = \sin\theta\cos\theta$
43. $\dfrac{1}{\sec\theta - \cos\theta} = \cot\theta\csc\theta$
44. $\dfrac{\cot\theta}{\sec\theta} = \csc\theta - \sin\theta$
45. $\dfrac{\cot^2\theta}{\csc\theta + 1} = \csc\theta - 1$
46. $\dfrac{\tan\theta}{\csc\theta} = \sin\theta\tan\theta$
47. $\dfrac{\tan^2\theta}{\sec\theta - 1} = 1 + \sec\theta$
48. $\tan y\sin y = \sec y - \cos y$
49. $2\cos^2 x - 1 = \cos^2 x - \sin^2 x$
50. $\dfrac{\sin x}{1 + \cos x} = \dfrac{1 - \cos x}{\sin x}$
51. $\dfrac{1 + \sin y}{\cos y} = \dfrac{\cos y}{1 - \sin y}$
52. $\dfrac{1}{\sec x - \tan x} = \sec x + \tan x$
53. $\dfrac{\cot x + 1}{\cot x - 1} = \dfrac{\sin x + \cos x}{\cos x - \sin x}$
54. $\dfrac{1 + \sin y}{1 - \sin y} = \dfrac{\csc y + 1}{\csc y - 1}$
55. $\dfrac{\cos x}{\sec x - \tan x} = \dfrac{\cos^2 x}{1 - \sin x}$
56. $\dfrac{\cos x}{\cos x + \sin x} = \dfrac{\cot x}{1 + \cot x}$
57. $\dfrac{1}{1 - \sin x} + \dfrac{1}{1 + \sin x} = 2\sec^2 x$
58. $\dfrac{\cos y}{\csc y + 1} + \dfrac{\cos y}{\csc y - 1} = 2\tan y$
59. $\sin^4 x - \cos^4 x = \sin^2 x - \cos^2 x$
60. $\cot^2 x - \cos^2 x = \cot^2 x\cos^2 x$
61. $\csc^2 y + \sec^2 y = \sec^2 y\csc^2 y$
62. $\dfrac{\tan x - \cot x}{\tan x + \cot x} = \dfrac{\tan^2 x - 1}{\sec^2 x}$
63. $\dfrac{\cot^2 x - 1}{\cot^2 x + 1} = \cos^2 x - \sin^2 x$
64. $\dfrac{1 - \sin x}{1 + \sin x} = (\tan x - \sec x)^2$
65. $\sec^4 x - \sec^2 x = \tan^4 x + \tan^2 x$
66. $\tan x - \cot x = \dfrac{\sin^2 x - \cos^2 x}{\sin x\cos x}$

In problems 67–76 show by counter example that each equation is not an identity.

67. $\sin\theta = 1 - \cos\theta$

68. $\tan^2\theta - \cot^2\theta = 1$

69. $\sec\theta = \dfrac{1}{\csc\theta}$

70. $\sin\theta = \dfrac{1}{\cos\theta}$

71. $\sin^2\theta - 2\cos\theta\sin\theta + \cos^2\theta = 2$

72. $\tan^2\theta - \tan\theta = 0$

73. $\csc\theta + \sec\theta\cot\theta = 2$

74. $\sin\theta + 2\sin\theta\cos\theta = 0$

75. $\dfrac{1-\cos\theta}{1+\cos\theta} = \sin^2\theta$

76. $\dfrac{1}{\tan\theta + \csc\theta} = \sec\theta$

77. a. Verify by calculation that $(\csc^2\theta - 1)(\sec^2\theta - 1) = 1$ for the values $\theta = \dfrac{\pi}{6}$ and $\theta = \dfrac{\pi}{4}$.

b. Is this equation an identity?

78. a. Verify by calculation that $\dfrac{\sin\theta - \cos\theta}{\cos\theta} = \tan\theta - 1$ for the values $\theta = \dfrac{\pi}{3}$ and $\theta = \dfrac{3\pi}{4}$.

b. Is this equation an identity?

79. a. Verify by calculation that $2\sin^2\theta + \sin\theta = 1$ for the values $\theta = \dfrac{\pi}{6}$ and $\theta = \dfrac{3\pi}{2}$.

b. Is this equation an identity?

80. a. Verify by calculation that $\tan^4\theta - \tan^2\theta = 6$ for the values $\theta = \dfrac{\pi}{3}$ and $\theta = \dfrac{4\pi}{3}$.

b. Is this equation an identity?

Skill and review

1. The point $(6,-2)$ is on the terminal side of θ, an angle in standard position. Compute $\sin\theta$, $\cos\theta$, and $\tan\theta$.
2. Find the length of overhang x required to shade 4 feet down the side of a house when the sun is 75° above the horizon, as shown in the figure.

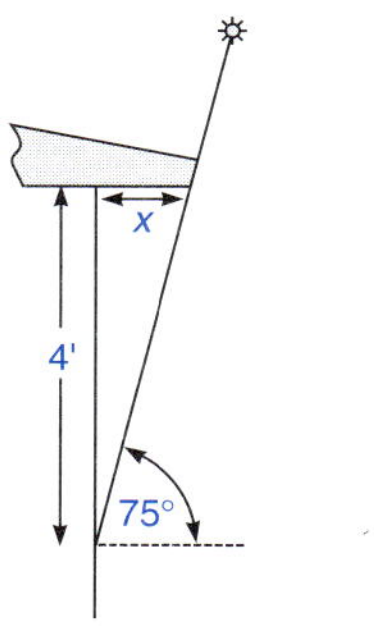

3. Find the degree measure of an angle of radian measure $-\dfrac{6\pi}{5}$.
4. Simplify $\cos(\tan^{-1}(-\frac{5}{2}))$.
5. Use the graph of $y = |x|$ to graph the function $f(x) = |x - 3|$.
6. Graph $y = \cos\left(2x - \dfrac{\pi}{3}\right)$. State the amplitude, period, and phase shift.
7. Solve the equation $4\sin^2 x - 1 = 0$ for $0 \le x < 2\pi$. (This implies answers should be in radian measure.)

7–2 Sum and difference identities

Find the exact value of x in the figure.

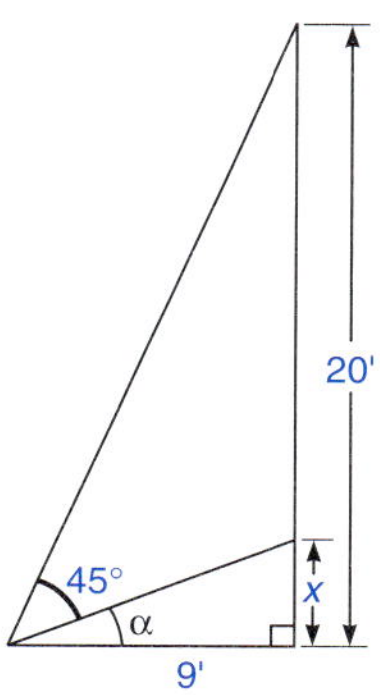

In this section we describe identities that can be used to solve this problem. These identities are also useful in simplifying certain problems in higher mathematics, and they are used to derive other important identities. Some of these derivations are shown in section 7–3.

Four important identities are called the **sum and difference identities.**

Sum and difference identities for sine and cosine

[1] $\cos(\alpha + \beta) = \cos\alpha\cos\beta - \sin\alpha\sin\beta$

[2] $\cos(\alpha - \beta) = \cos\alpha\cos\beta + \sin\alpha\sin\beta$

[3] $\sin(\alpha + \beta) = \sin\alpha\cos\beta + \cos\alpha\sin\beta$

[4] $\sin(\alpha - \beta) = \sin\alpha\cos\beta - \cos\alpha\sin\beta$

The last three of these four identities can be developed using the first. Their verification is left as exercises. A demonstration that identity [1] is true is given in appendix A. These identities have several applications, as illustrated in example 7–2 A.

■ ***Example 7–2 A***

1. Use the fact that $\frac{7\pi}{12} = \frac{\pi}{4} + \frac{\pi}{3}$ to find the exact value of $\cos\frac{7\pi}{12}$.

$$\cos\frac{7\pi}{12} = \cos\left(\frac{\pi}{3} + \frac{\pi}{4}\right)$$

$$= \cos\frac{\pi}{3}\cos\frac{\pi}{4} - \sin\frac{\pi}{3}\sin\frac{\pi}{4} \qquad \text{Identity [1]}$$

$$= \frac{1}{2}\cdot\frac{\sqrt{2}}{2} - \frac{\sqrt{3}}{2}\cdot\frac{\sqrt{2}}{2} = \frac{\sqrt{2}}{4} - \frac{\sqrt{6}}{4}$$

$$= \frac{\sqrt{2} - \sqrt{6}}{4}$$

This answer can and should be checked with a calculator.

2. Show that $\cos(\pi - \theta) = -\cos\theta$ for any angle θ.

$$\cos(\pi - \theta) = \cos\pi\cos\theta + \sin\pi\sin\theta$$
$$= (-1)\cos\theta + 0\sin\theta$$
$$= -\cos\theta$$

■

The cofunction identities

Identity [1] can be used to prove the following identities (the proofs are left for the exercises). These identities are called the **cofunction identities.**

Cofunction identities

[5] $\sin\left(\frac{\pi}{2} - \theta\right) = \cos\theta$ [6] $\cos\left(\frac{\pi}{2} - \theta\right) = \sin\theta$

[7] $\tan\left(\frac{\pi}{2} - \theta\right) = \cot\theta$ [8] $\cot\left(\frac{\pi}{2} - \theta\right) = \tan\theta$

[9] $\sec\left(\frac{\pi}{2} - \theta\right) = \csc\theta$ [10] $\csc\left(\frac{\pi}{2} - \theta\right) = \sec\theta$

The reason for the name of these identities is as follows. When the sum of two angles is 90°, or $\frac{\pi}{2}$ radians, the angles are said to be **complementary.** The angles $\frac{\pi}{2} - \theta$ and θ add up to $\frac{\pi}{2}$, so they are complementary angles. Each is said to be the complement of the other. The cofunction identities say

trig function (angle) = "co" trig function (complement of angle)

Thus, the sine and "co"sine appear in one identity, the tangent and "co"tangent appear in another, and the secant and "co" secant in the third. Whenever the sum of two angles is $\frac{\pi}{2}$ (or 90°) a trigonometric function of one equals the "cotrigonometric" function of the other. Thus for example, the following statements are true:

$$\sin 50^\circ = \cos 40^\circ \qquad 50^\circ + 40^\circ = 90^\circ$$

$$\sec \frac{\pi}{6} = \csc \frac{\pi}{3} \qquad \frac{\pi}{6} + \frac{\pi}{3} = \frac{\pi}{2}$$

$$\cot 130^\circ = \tan(-40^\circ) \qquad 130^\circ + (-40^\circ) = 90^\circ$$

■ ***Example 7–2 B***

1. Rewrite $\csc \frac{2\pi}{5}$ in terms of its cofunction.

$$\csc \frac{2\pi}{5} = \sec\left(\frac{\pi}{2} - \frac{2\pi}{5}\right) = \sec \frac{\pi}{10}$$

2. Simplify the expression: $\frac{\sin 10^\circ}{\cos 80^\circ}$.

$$\frac{\sin 10^\circ}{\cos 80^\circ} = \frac{\cos 80^\circ}{\cos 80^\circ} = 1$$ ■

Sum and difference identities for the tangent function

Two more important identities are sum and difference formulas for the tangent function.

Sum and difference identities for tangent

[11] $$\tan(\alpha + \beta) = \frac{\tan \alpha + \tan \beta}{1 - \tan \alpha \tan \beta}$$

[12] $$\tan(\alpha - \beta) = \frac{\tan \alpha - \tan \beta}{1 + \tan \alpha \tan \beta}$$

The derivation of the first identity is as follows:

$$\tan(\alpha + \beta) = \frac{\sin(\alpha + \beta)}{\cos(\alpha + \beta)}$$

$$= \frac{\sin \alpha \cos \beta + \cos \alpha \sin \beta}{\cos \alpha \cos \beta - \sin \alpha \sin \beta}$$

$$= \frac{\dfrac{\sin \alpha \cos \beta + \cos \alpha \sin \beta}{\cos \alpha \cos \beta}}{\dfrac{\cos \alpha \cos \beta - \sin \alpha \sin \beta}{\cos \alpha \cos \beta}}$$ Divide numerator and denominator by $\cos \alpha \cos \beta$

$$= \frac{\dfrac{\sin \alpha}{\cos \alpha} + \dfrac{\sin \beta}{\cos \beta}}{1 - \dfrac{\sin \alpha}{\cos \alpha} \cdot \dfrac{\sin \beta}{\cos \beta}} = \frac{\tan \alpha + \tan \beta}{1 - \tan \alpha \tan \beta}$$

Example 7–2 C illustrates an application of these identities.

■ Example 7–2 C

Use the fact that $15° = 45° - 30°$ to find the exact value of $\tan 15°$.

$$\tan 15° = \tan(45° - 30°)$$

$$= \frac{\tan 45° - \tan 30°}{1 + \tan 45° \tan 30°} = \frac{1 - \dfrac{\sqrt{3}}{3}}{1 + 1\left(\dfrac{\sqrt{3}}{3}\right)}$$

$$= \frac{1 - \dfrac{\sqrt{3}}{3}}{1 + \dfrac{\sqrt{3}}{3}} \cdot \frac{3}{3} = \frac{3 - \sqrt{3}}{3 + \sqrt{3}} \cdot \frac{3 - \sqrt{3}}{3 - \sqrt{3}} = \frac{12 - 6\sqrt{3}}{6} = 2 - \sqrt{3}$$

Check the result with a calculator. ■

Some problems can be solved by using the identities above and reference triangles (section 5–4). Example 7–2 D illustrates.

■ Example 7–2 D

$\sin \alpha = \frac{2}{3}$, α in quadrant I; $\cos \beta = -\frac{4}{5}$, β in quadrant II. Find the exact value of $\cos(\alpha - \beta)$.

We first create reference triangles for each angle. This allows us to find any necessary trigonometric function values as necessary.

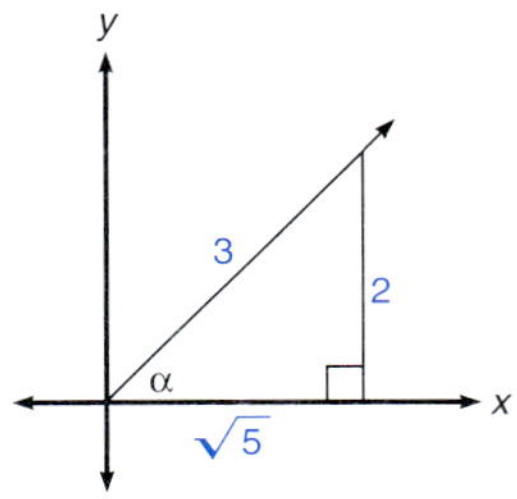

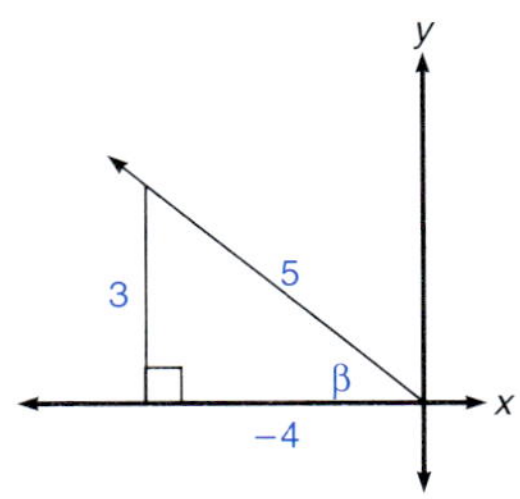

$$\begin{aligned}\cos(\alpha - \beta) &= \cos\alpha\cos\beta + \sin\alpha\sin\beta\\ &= \frac{\sqrt{5}}{3}\cdot\left(\frac{-4}{5}\right) + \frac{2}{3}\ \frac{3}{5}\\ &= -\frac{4\sqrt{5}}{15} + \frac{6}{15}\\ &= \frac{-4\sqrt{5}+6}{15}\end{aligned}$$

We find the necessary values from the reference triangles

Mastery points

Can you

- State the sum and difference identities?
- State and apply the cofunction identities?
- Apply the sum and difference identities to find exact values of sine, cosine, and tangent for certain angles?
- Apply the sum and difference identities to find exact values of sine, cosine, and tangent for $\alpha + \beta$ and $\alpha - \beta$ given information about α and β?
- Verify identities using the sum and difference identities?

Exercise 7–2

Rewrite each function in terms of its cofunction.

1. $\sin 18°$
2. $\cos 42°$
3. $\tan 8°$
4. $\csc 100°$
5. $\sec\frac{\pi}{3}$
6. $\cot\frac{\pi}{6}$
7. $\cos\frac{5\pi}{6}$
8. $\sin\left(-\frac{\pi}{3}\right)$
9. $\sec\left(-\frac{3\pi}{4}\right)$
10. $\csc\left(-\frac{\pi}{4}\right)$

Simplify each expression.

11. $\frac{\cos 65°}{\sin 25°}$
12. $\tan\frac{\pi}{3}\tan\frac{\pi}{6}$
13. $\cos 20° \csc 70°$
14. $\frac{\sin^2 5°}{\cos^2 85°}$
15. $\sin\frac{\pi}{5}\sec\frac{3\pi}{10}$
16. $\cos^2 25° + \cos^2 65°$
17. $\tan^2 8° - \csc^2 82°$
18. $\frac{\cos^2 30°}{1 - \cos^2 60°}$
19. $\tan 40° \tan 50°$
20. $\tan 19° \tan 71°$
21. $\sec\frac{\pi}{6}\sin\frac{\pi}{3}$
22. $\cot\frac{\pi}{5}\cot\frac{3\pi}{10}$
23. $\sin^2 10° + \sin^2 80°$
24. $\tan^2 25° - \csc^2 65°$
25. $\sec^2\frac{\pi}{3} - \cot^2\frac{\pi}{6}$
26. $\cos^2\frac{3\pi}{8} + \cos^2\frac{\pi}{8}$

Use the sum and difference identities to find the exact value of each of the following expressions. Observe that each value is the sum or difference of values chosen from $\frac{\pi}{6}$ (30°), $\frac{\pi}{4}$ (45°), and $\frac{\pi}{3}$ (60°).

27. $\cos\frac{\pi}{12}$
28. $\tan\frac{\pi}{12}$
29. $\sin\frac{5\pi}{12}$
30. $\cos\frac{5\pi}{12}$
31. $\sin\frac{7\pi}{12}$
32. $\tan\frac{7\pi}{12}$
33. $\sin 15°$
34. $\tan 15°$
35. $\cos 105°$
36. $\sin 105°$
37. $\tan 75°$
38. $\cos 15°$

Each of the following problems presents information about two angles, α and β, including the quadrant in which the angle terminates. Use the information to find the required value.

39. $\cos\alpha = \frac{1}{3}$, quadrant I; $\sin\beta = \frac{3}{4}$, quadrant I. Find $\sin(\alpha + \beta)$.

40. $\cos\alpha = -\frac{12}{13}$, quadrant II; $\sin\beta = \frac{1}{2}$, quadrant II. Find $\cos(\alpha - \beta)$.

41. $\sin\alpha = \frac{5}{13}$, quadrant II; $\cos\beta = -\frac{3}{4}$, quadrant III. Find $\tan(\alpha - \beta)$.

42. $\sin\alpha = -\frac{4}{5}$, quadrant IV; $\sin\beta = -\frac{1}{5}$, quadrant IV. Find $\sin(\alpha + \beta)$.

43. $\sin\alpha = -\frac{4}{5}$, quadrant IV; $\cos\beta = \frac{15}{17}$, quadrant IV. Find $\cos(\alpha + \beta)$.

44. $\cos\alpha = -\frac{3}{5}$, quadrant II; $\sin\beta = -\frac{8}{17}$, quadrant III. Find $\sin(\alpha - \beta)$.

45. $\sin\alpha = \frac{2}{3}$, quadrant I; $\cos\beta = -\frac{1}{3}$, quadrant III. Find $\cos(\alpha - \beta)$.

46. $\cos\alpha = \frac{\sqrt{2}}{2}$, quadrant IV; $\sin\beta = -\frac{\sqrt{3}}{2}$, quadrant III. Find $\tan(\alpha + \beta)$.

47. $\sin\alpha = \frac{2}{3}$, quadrant I; $\tan\beta = \frac{1}{4}$, quadrant I. Find $\sin(\alpha + \beta)$.

48. $\tan\alpha = \frac{3}{4}$, quadrant III; $\sin\beta = -\frac{4}{5}$, quadrant III. Find $\cos(\alpha - \beta)$.

49. $\cos\alpha = \frac{5}{13}$, quadrant IV; $\tan\beta = -\frac{5}{12}$, quadrant IV. Find $\tan(\alpha - \beta)$.

50. $\cos\alpha = \frac{1}{2}$, quadrant I; $\cos\beta = \frac{\sqrt{2}}{2}$, quadrant IV. Find $\sin(\alpha + \beta)$.

51. $\tan\alpha = 2$, quadrant III; $\cos\beta = -\frac{3}{5}$, quadrant II. Find $\cos(\alpha - \beta)$.

52. $\sin\alpha = -\frac{15}{17}$, quadrant III; $\tan\beta = -\frac{3}{4}$, quadrant IV. Find $\tan(\alpha + \beta)$.

53. $\sin\alpha = \frac{2}{3}$, quadrant I; $\sin\beta = -\frac{1}{5}$, quadrant III. Find $\sin(\alpha - \beta)$.

54. $\cos\alpha = -\frac{5}{13}$, quadrant III; $\cos\beta = -\frac{8}{17}$, quadrant III. Find $\cos(\alpha + \beta)$.

55. $\cos\alpha = -\frac{\sqrt{2}}{2}$, quadrant II; $\sin\beta = \frac{\sqrt{3}}{2}$, quadrant II. Find $\tan(\alpha + \beta)$.

56. $\cos\alpha = -\frac{3}{5}$, quadrant III; $\sin\beta = \frac{1}{3}$, quadrant II. Find $\tan(\alpha - \beta)$.

Use the sum and difference identities to verify the following identities.

57. $\sin(\pi - \theta) = \sin\theta$
58. $\sin(\pi + \theta) = -\sin\theta$
59. $\cos(\pi - \theta) = -\cos\theta$
60. $\cos(\pi + \theta) = -\cos\theta$
61. $\tan(\pi - \theta) = -\tan\theta$
62. $\tan(\pi + \theta) = \tan\theta$

63. Use the sum formula to show that the sine function is 2π-periodic; that is that $\sin(\theta + 2\pi) = \sin\theta$.

64. Use the sum formula to show that the cosine function is 2π-periodic; that is that $\cos(\theta + 2\pi) = \cos\theta$.

65. Use the sum formula to show that the tangent function is π-periodic; that is that $\tan(\theta + \pi) = \tan\theta$.

The following identities are important because they express a product of factors as a sum of terms. Verify each identity.

66. $\sin\alpha\cos\beta = \frac{1}{2}[\sin(\alpha + \beta) + \sin(\alpha - \beta)]$

67. $\cos\alpha\sin\beta = \frac{1}{2}[\sin(\alpha + \beta) - \sin(\alpha - \beta)]$

68. $\cos\alpha\cos\beta = \frac{1}{2}[\cos(\alpha + \beta) + \cos(\alpha - \beta)]$

69. $\sin\alpha\sin\beta = \frac{1}{2}[\cos(\alpha - \beta) - \sin(\alpha + \beta)]$

70. A picture on a wall is 2 feet tall and 6 feet above eye level; see the diagram. Compute the exact value of $\tan(\alpha - \beta)$.

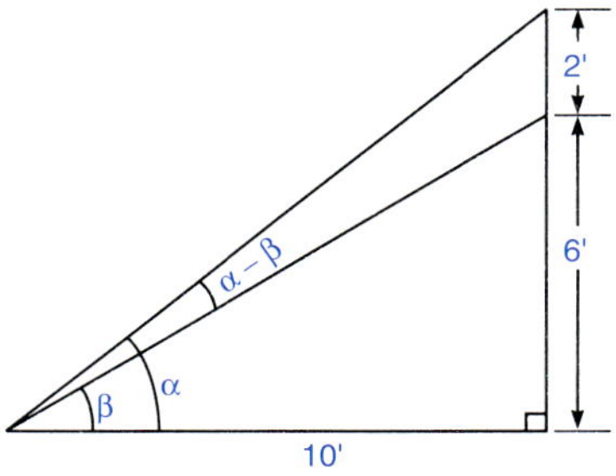

71. Referring to the figure, find the value of $\tan \alpha$ and use this to find the exact value of x. *Hint:* Compute $\tan(\alpha + 45°)$.

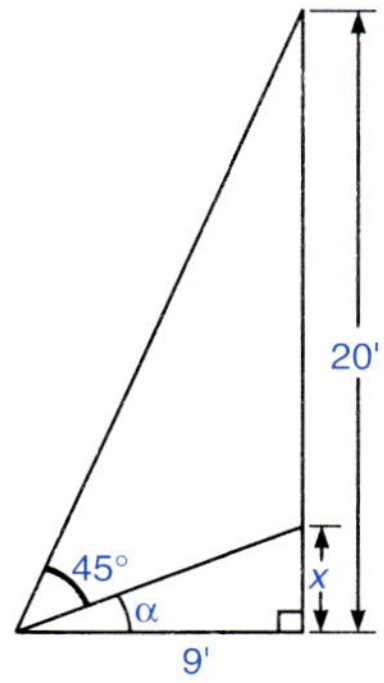

72. Use the identities for $\sin(\alpha + \beta)$ and $\cos(\alpha + \beta)$ to find $\sin \theta$ in the diagram.

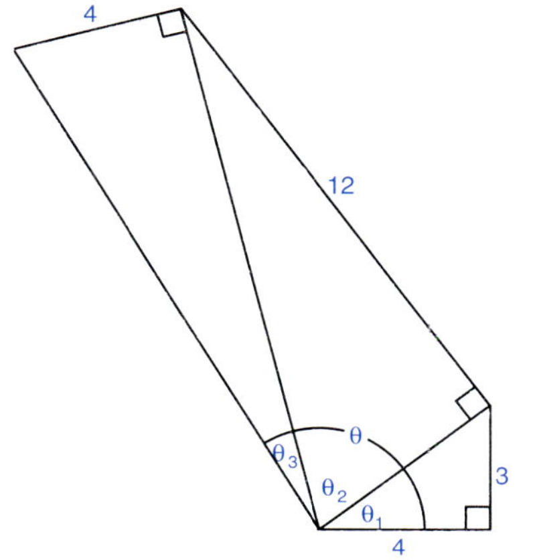

The following problems are designed to show that the sum and difference identities for sine and cosine [2], [3], and [4] and the cofunction identities [5] through [10] are true, using the fact that identity [1] is true. The problems are in the necessary logical order.

73. Use the identity [1] $\cos(\alpha + \beta) = \cos \alpha \cos \beta - \sin \alpha \sin \beta$ to verify the identity [2] $\cos(\alpha - \beta) = \cos \alpha \cos \beta + \sin \alpha \sin \beta$. Do this by replacing β by $(-\beta)$ in the identity for $\cos(\alpha + \beta)$ and simplifying, using the even and odd properties for the sine and cosine functions.

74. Verify the identity [5] $\cos\left(\frac{\pi}{2} - \theta\right) = \sin \theta$ by using identity [2], letting $\alpha = \frac{\pi}{2}$ and $\beta = \theta$.

75. Identity [6] $\sin\left(\frac{\pi}{2} - \theta\right) = \cos \theta$ is really the same as identity [5]. This can be shown as follows. Let $\alpha = \frac{\pi}{2} - \theta$, so that $\theta = \frac{\pi}{2} - \alpha$. Replace $\frac{\pi}{2} - \theta$ by α in the left member of identity [5], then replace θ in the right member by $\frac{\pi}{2} - \alpha$. Then observe that α and θ are arbitrary values, so the result can be rewritten in terms of θ.

76. Verify identity [3] $\sin(\alpha + \beta) = \sin \alpha \cos \beta + \cos \alpha \sin \beta$ as follows.

$$\sin \theta = \cos\left(\frac{\pi}{2} - \theta\right) \qquad \text{This is identity [5], which we know is true}$$

$$\sin(\alpha + \beta) = \cos\left[\frac{\pi}{2} - (\alpha + \beta)\right] \qquad \text{Replace } \theta \text{ by } \alpha + \beta$$

$$\sin(\alpha + \beta) = \cos\left[\left(\frac{\pi}{2} - \alpha\right) - \beta\right] \qquad \text{Regroup } \frac{\pi}{2} - \alpha - \beta$$

Now use identity [2] to expand the right member of this equation, then apply identities [5] and [6] to simplify the result and obtain identity [3].

77. Use identity [3] $\sin(\alpha + \beta) = \sin \alpha \cos \beta + \cos \alpha \sin \beta$ to verify identity [4] $\sin(\alpha - \beta) = \sin \alpha \cos \beta - \cos \alpha \sin \beta$. Do this by replacing β by $(-\beta)$ in identity [3].

78. Verify identity [7] $\tan\left(\frac{\pi}{2} - \theta\right) = \cot \theta$ by using the fact that $\tan x = \frac{\sin x}{\cos x}$.

79. Verify identity [8] $\cot\left(\frac{\pi}{2} - \theta\right) = \tan \theta$. See problem 78 for guidance.

80. Verify identity [9] $\sec\left(\frac{\pi}{2} - \theta\right) = \csc \theta$ by using the fact that $\sec x = \frac{1}{\cos x}$.

81. Verify identity [10] $\csc\left(\frac{\pi}{2} - \theta\right) = \sec \theta$. See problem 80 for guidance.

Skill and review

1. Find the equation of the straight line that passes through the points $(-4,3)$ and $(-8,11)$.

2. In right triangle *ABC*, $a = 3$, $c = 4$. Solve the triangle. Round answers to tenths.

3. The line $y = 3x$ passes through the origin. What angle does this line make with the *x*-axis, to the nearest 0.1°? (*Hint:* Find a point that lies on the line in the first quadrant. This point lies on the terminal side of the angle.)

4. Factor $3x^4 - 5x^3 - 14x^2 + 20x + 8$. Use the rational zero theorem and synthetic division as necessary.

5. If the length of arc *L* in the figure is 14.6, find the length of the arc *T*. (There are many ways to solve this. One way would be to find the value of angle θ, in radians, first. Recall the formula $L = rs$.)

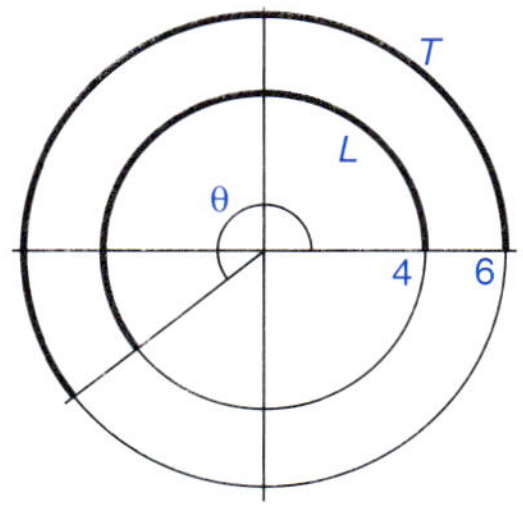

6. Simplify $\sin^{-1}\left(\sin \frac{5\pi}{3}\right)$.

7. Verify the identity $\frac{\csc^2 x - 1}{\sin^2 x} = \cos^2 x \csc^4 x$.

7–3 The double-angle and half-angle identities

In the figure, $\theta_1 = \theta_2$. Find the length of the side marked *x*.

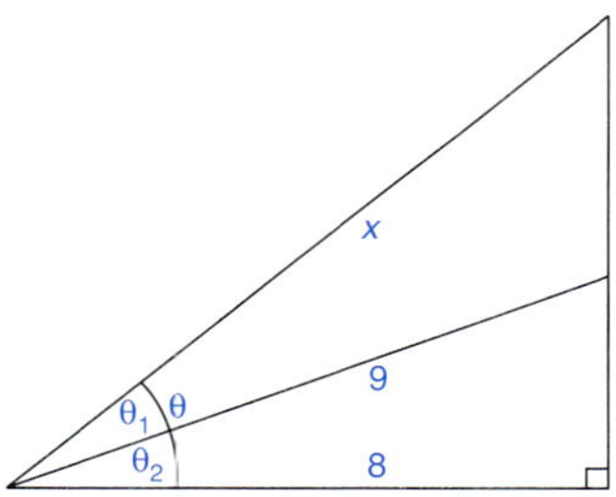

In this section we investigate identities which could be used to solve this problem. They find wider application in advanced mathematics.

Double-angle identities

Some more important identities are the **double-angle identities.** Recall that if we multiply a value by two we say we "double" the value.

Double-angle identities

[1]	$\sin 2\alpha = 2 \sin \alpha \cos \alpha$
[2-a]	$\cos 2\alpha = \cos^2\alpha - \sin^2\alpha$
[2-b]	$\cos 2\alpha = 1 - 2 \sin^2\alpha$
[2-c]	$\cos 2\alpha = 2 \cos^2\alpha - 1$
[3]	$\tan 2\alpha = \frac{2 \tan \alpha}{1 - \tan^2\alpha}$

Observe that we present three identities for $\cos 2\alpha$. This is because identities [2-b] and [2-c] get so much use in the development of other identities.

The proof of identity [1] is as follows.

$$\sin(\alpha + \beta) = \sin\alpha\cos\beta + \cos\alpha\sin\beta \qquad \text{Sum identity from section 7–2}$$
$$\sin(\alpha + \alpha) = \sin\alpha\cos\alpha + \cos\alpha\sin\alpha \qquad \text{Let } \beta = \alpha$$
$$\sin 2\alpha = 2\sin\alpha\cos\alpha \qquad \alpha + \alpha = 2\alpha$$

The verification of the remaining identities is left for the exercises. They are done in a similar way, starting with the identities for $\cos(\alpha + \beta)$ and $\tan(\alpha + \beta)$. Example 7–3 A illustrates applying these identities.

Example 7–3 A

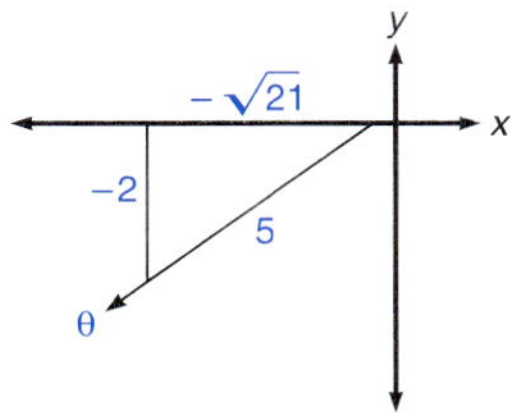

1. If $\sin\theta = -\frac{2}{5}$ and θ terminates in quadrant III, find exact values of $\sin 2\theta$ and $\cos 2\theta$.

We first construct a reference triangle for θ to obtain any required trigonometric function values for that angle.

$$\sin 2\theta = 2\sin\theta\cos\theta$$
$$= 2\left(-\frac{2}{5}\right)\left(-\frac{\sqrt{21}}{5}\right) = \frac{4\sqrt{21}}{25}$$
$$\cos 2\theta = \cos^2\theta - \sin^2\theta$$
$$= \left(-\frac{\sqrt{21}}{5}\right)^2 - \left(-\frac{2}{5}\right)^2 = \frac{21}{25} - \frac{4}{25} = \frac{17}{25}$$

2. Find an identity for $\tan 3\theta$ in terms of $\tan\theta$.

$$\tan 3\theta = \tan(2\theta + \theta) \qquad 3\theta = 2\theta + \theta$$
$$= \frac{\tan 2\theta + \tan\theta}{1 - \tan 2\theta\tan\theta} \qquad \tan(\alpha + \beta) = \frac{\tan\alpha + \tan\beta}{1 - \tan\alpha\tan\beta}$$
$$= \frac{\dfrac{2\tan\theta}{1 - \tan^2\theta} + \tan\theta}{1 - \dfrac{2\tan\theta}{1 - \tan^2\theta}\tan\theta} \qquad \text{Replace } \tan 2\theta \text{ by } \frac{2\tan\theta}{1 - \tan^2\theta}$$
$$= \frac{2\tan\theta + \tan\theta(1 - \tan^2\theta)}{(1 - \tan^2\theta) - 2\tan^2\theta} \qquad \text{Multiply numerator and denominator by } (1 - \tan^2\theta)$$
$$= \frac{3\tan\theta - \tan^3\theta}{1 - 3\tan^2\theta} \qquad \text{Combine}$$

Half-angle identities

A further set of important identities is the half-angle identities.

Half-angle identities

[4] $\sin\frac{\alpha}{2} = \pm\sqrt{\frac{1 - \cos\alpha}{2}}$

[5] $\cos\frac{\alpha}{2} = \pm\sqrt{\frac{1 + \cos\alpha}{2}}$

[6] $\tan\frac{\alpha}{2} = \pm\sqrt{\frac{1 - \cos\alpha}{1 + \cos\alpha}}$

[6-a] $\tan\frac{\alpha}{2} = \frac{1 - \cos\alpha}{\sin\alpha}$

[6-b] $\tan\frac{\alpha}{2} = \frac{\sin\alpha}{1 + \cos\alpha}$

We verify identity [5] as follows:

$$2\cos^2\theta - 1 = \cos 2\theta \quad \text{Identity [2-c]}$$
$$2\cos^2\theta = 1 + \cos 2\theta$$
$$\cos^2\theta = \frac{1 + \cos 2\theta}{2}$$
$$\cos^2 \frac{\alpha}{2} = \frac{1 + \cos \alpha}{2} \quad \text{Replace } \theta \text{ by } \frac{\alpha}{2}$$
$$\cos \frac{\alpha}{2} = \pm\sqrt{\frac{1 + \cos \alpha}{2}} \quad \text{Take the square root of each member}$$

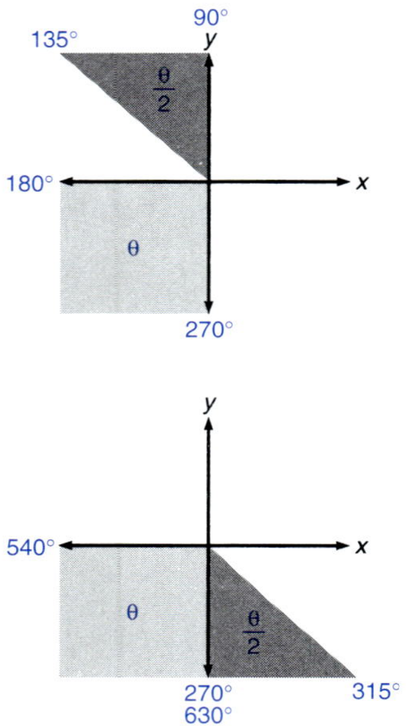

Figure 7–1

The verification of the remaining identities is left for the exercises.

The choice of plus or minus depends on the quadrant in which the angle in question terminates. It is only possible to determine the quadrant if we have information about the measure of the angle. To see why, consider figure 7–1. If $180° < \theta < 270°$ then $90° \le \frac{\theta}{2} \le 135°$; in this case, θ terminates in quadrant III and $\frac{\theta}{2}$ terminates in quadrant II. However, if $540° \le \theta \le 630°$, then $270° \le \frac{\theta}{2} \le 315°$. In this case, θ also terminates in quadrant III, but $\frac{\theta}{2}$ terminates in quadrant IV.

The half-angle identities have applications such as those shown in example 7–3 B.

■ *Example 7–3 B*

1. Use the fact that 22.5° is one half of 45° to find the exact value of sin 22.5°.

$$\sin 22.5° = \sin \frac{45°}{2}$$
$$= \pm\sqrt{\frac{1 - \cos 45°}{2}} \quad \sin\frac{\alpha}{2} = \pm\sqrt{\frac{1-\cos\alpha}{2}};\ \alpha = 45°$$
$$= \sqrt{\frac{1 - \frac{\sqrt{2}}{2}}{2}} \quad \text{We know } \sin 22.5° > 0$$
$$= \sqrt{\frac{2 - \sqrt{2}}{4}} \quad \frac{1 - \frac{\sqrt{2}}{2}}{2} = \frac{1}{2}\left(1 - \frac{\sqrt{2}}{2}\right) = \frac{1}{2}\left(\frac{2 - \sqrt{2}}{2}\right)$$
$$= \frac{\sqrt{2 - \sqrt{2}}}{2}$$

2. $\cos\theta = \frac{3}{5}$ and $\frac{3\pi}{2} < \theta < 2\pi$. Find the exact value for $\cos\frac{\theta}{2}$.

Since $\frac{3\pi}{2} < \theta < 2\pi$, $\frac{3\pi}{4} < \frac{\theta}{2} < \pi$, so $\frac{\theta}{2}$ terminates in quadrant II, where the cosine function is negative.

$$\cos\frac{\theta}{2} = -\sqrt{\frac{1+\cos\theta}{2}}$$ Choose minus since $\frac{\theta}{2}$ terminates in quadrant II where $\cos\frac{\theta}{2} < 0$

$$= -\sqrt{\frac{1+\frac{3}{5}}{2}}$$ Replace $\cos\theta$ with $\frac{3}{5}$

$$= -\sqrt{\frac{1}{2}\cdot\frac{8}{5}} = -\sqrt{\frac{4}{5}} = -\sqrt{\frac{20}{25}} = -\frac{2\sqrt{5}}{5}$$ ■

It is important to understand how to rewrite identities with different forms of the argument. For example the following identities are all the same; the argument of each is shown in different forms.

[1] $\sin 2\alpha = 2\sin\alpha\cos\alpha$ — Identity [1] of the double-angle identities

$\sin 4\alpha = 2\sin 2\alpha\cos 2\alpha$ — Replace α in identity [1] by 2α

$\sin\alpha = 2\sin\frac{\alpha}{2}\cos\frac{\alpha}{2}$ — Replace α in identity [1] by $\frac{\alpha}{2}$

Example 7–3 C illustrates rewriting identities.

■ ***Example 7–3 C***

Rewrite each expression as an expression of the form $a\sin x$, $a\cos x$, or $a\tan x$, for appropriate values of a and x.

1. $\dfrac{4\tan 2\theta}{1-\tan^2 2\theta}$

$$\frac{2\tan\alpha}{1-\tan^2\alpha} = \tan 2\alpha$$ Identity [3]

$$\frac{4\tan\alpha}{1-\tan^4\alpha} = 2\tan 2\alpha$$ Multiply each member by 2

$$\frac{4\tan 2\theta}{1-\tan^2 2\theta} = 2\tan 4\theta$$ Replace α by 2θ

Thus, the answer is $2\tan 4\theta$.

2. $\cos^2 80° - \sin^2 80°$

Compare

$\cos^2\alpha - \sin^2\alpha = \cos 2\alpha$ — Identity [2] of the double-angle identities

$\cos^2 80° - \sin^2 80°$

Since 80° replaces α, we know that $\cos 2\alpha$ becomes $\cos 2(80)° = \cos 160°$. Thus, $\cos^2 80° - \sin^2 80° = \cos 160°$, and the answer is $\cos 160°$. ■

A similar idea is illustrated in example 7–3 D.

■ Example 7–3 D

Find a value of θ for which the statement sin 100° = 2 sin θ cos θ is true, then rewrite the statement replacing θ by this value.

Compare

$$\sin 110° = 2 \sin \theta \cos \theta$$
$$\sin 2\theta = 2 \sin \theta \cos \theta$$

Let $2\theta = 110°$, so $\theta = 55°$.
The statement becomes $\sin 110° = 2 \sin 55° \cos 55°$. ■

The identities of this and the previous sections may be combined to verify new identities.

■ Example 7–3 E

Verify the following identities.

1. $\sin 2\theta = \dfrac{2 \tan \theta}{1 + \tan^2\theta}$

It is best to work with the right member since it is more complicated.

$$\frac{2 \tan \theta}{1 + \tan^2\theta} = 2 \cdot \frac{\tan \theta}{\sec^2\theta} \qquad \tan^2\alpha + 1 = \sec^2\alpha$$
$$= 2 \cdot \tan \theta \cos^2\theta \qquad \cos \alpha = \frac{1}{\sec \alpha}$$
$$= 2 \cdot \frac{\sin \theta}{\cos \theta} \cos^2\theta$$
$$= 2 \sin \theta \cos \theta \qquad \frac{1}{\cos \theta} \cdot \cos^2\theta = \cos \theta$$
$$= \sin 2\theta \qquad \sin 2\alpha = 2 \sin \alpha \cos \alpha$$

2. $\sin 4\theta = 8 \sin \theta \cos^3\theta - 4 \sin \theta \cos \theta$

$$\sin 4\theta = \sin [2(2\theta)]$$
$$= 2 \sin 2\theta \cos 2\theta \qquad \text{Use } \sin 2\alpha = 2 \sin \alpha \cos \alpha, \text{ with } \alpha = 2\theta$$
$$= 2(2 \sin \theta \cos \theta)(2 \cos^2\theta - 1)$$
$$= 4 \sin \theta \cos \theta (2 \cos^2\theta - 1)$$
$$= 8 \sin \theta \cos^3\theta - 4 \sin \theta \cos \theta$$ ■

Mastery points

Can you

- Write the double-angle and half-angle identities?
- Use the double-angle and half-angle identities to find exact values of $\sin 2\theta$, $\cos 2\theta$, $\tan 2\theta$, $\sin \frac{\theta}{2}$, $\cos \frac{\theta}{2}$, and $\tan \frac{\theta}{2}$?
- Use the double-angle and half-angle identities to derive new identities and to verify given identities?
- Rewrite certain identities as a trigonometric function of $k\theta, k \in J$?

Exercise 7–3

Use the identities of this section to rewrite each expression as an expression of the form $a \sin x$, $a \cos x$, or $a \tan x$, for appropriate values of a and x.

1. $2 \sin \frac{\pi}{4} \cos \frac{\pi}{4}$
2. $2 \sin 52° \cos 52°$
3. $\cos^2 3\pi - \sin^2 3\pi$
4. $2 \cos^2 5\pi - 1$
5. $1 - 2 \sin^2 \frac{\pi}{10}$
6. $\dfrac{2 \tan \frac{\pi}{6}}{1 - \tan^2 \frac{\pi}{6}}$
7. $\dfrac{6 \tan 10°}{1 - \tan^2 10°}$
8. $8 \cos^2 \frac{\pi}{2} - 4$
9. $2 \sin 6\theta \cos 6\theta$
10. $4 \sin 2\theta \cos 2\theta$
11. $6 \cos^2 5\theta - 3$
12. $8 \cos^2 3\theta - 4$
13. $\dfrac{10 \tan 3\theta}{1 - \tan^2 3\theta}$
14. $\dfrac{8 \tan \frac{\theta}{2}}{1 - \tan^2 \frac{\theta}{2}}$
15. $2 - 4 \sin^2 7\theta$
16. $\frac{1}{2} - \sin^2 2\theta$
17. $3 \cos^2 3\theta - 3 \sin^2 3\theta$
18. $2 \cos^2 \frac{\theta}{2} - 2 \sin^2 \frac{\theta}{2}$

Find a value of θ for which each statement is true.

19. $\sin 140° = 2 \sin \theta \cos \theta$
20. $\sin \theta \cos \theta = \frac{1}{2} \sin \frac{\pi}{5}$
21. $\cos \frac{5\pi}{6} = \cos^2 \theta - \sin^2 \theta$
22. $2 \tan 86° = \dfrac{4 \tan \theta}{1 - \tan^2 \theta}$
23. $3 \cos 70° = 6 \cos^2 \theta - 3$
24. $\cos 560° = 1 - 2 \sin^2 \theta$
25. $\sin 10° = \sqrt{\dfrac{1 - \cos \theta}{2}}$
26. $\tan \theta = \sqrt{\dfrac{1 - \cos 46°}{1 + \cos 46°}}$
27. $\cos \theta = \sqrt{\frac{1}{2}\left(1 + \cos \frac{\pi}{4}\right)}$
28. $\sin \frac{\pi}{6} = \sqrt{\frac{1}{2}(1 - \cos \theta)}$
29. $\tan \frac{2\pi}{5} = \sqrt{\dfrac{1 - \cos \theta}{1 + \cos \theta}}$
30. $\cos 40° = \sqrt{\dfrac{1 + \cos \theta}{2}}$

Find the exact value of $\sin 2\theta$, $\cos 2\theta$, and $\tan 2\theta$ for each of the following.

31. $\sin \theta = \frac{3}{5}, 0 < \theta < \frac{\pi}{2}$
32. $\sin \theta = -\frac{12}{13}, \pi < \theta < \frac{3\pi}{2}$
33. $\cos \theta = -\frac{4}{5}, \frac{\pi}{2} < \theta < \pi$
34. $\tan \theta = -\frac{3}{4}, \frac{3\pi}{2} < \theta < 2\pi$
35. $\csc \theta = -\frac{8}{5}, \pi < \theta < \frac{3\pi}{2}$
36. $\tan \theta = \frac{5}{12}, \pi < \theta < \frac{3\pi}{2}$

Find the exact value of $\sin \frac{\theta}{2}$, $\cos \frac{\theta}{2}$, and $\tan \frac{\theta}{2}$ for each of the following.

37. $\sec \theta = -\frac{5}{2}, \pi < \theta < \frac{3\pi}{2}$
38. $\tan \theta = -\sqrt{15}, \frac{\pi}{2} < \theta < \pi$
39. $\cot \theta = -2, \frac{3\pi}{2} < \theta < 2\pi$
40. $\cos \theta = \frac{1}{4}, \frac{3\pi}{2} < \theta < 2\pi$

Use the half-angle identities to find the exact value of (a) $\sin \theta$, (b) $\cos \theta$, and (c) $\tan \theta$ for the following values of θ.

41. $15°$, or $\frac{\pi}{12}$
42. $22.5°$, or $\frac{\pi}{8}$

Use the sum/difference identities (from section 7–2) and the results of problems 41 and 42 to compute the exact value of the following. Observe that $75° = 60° + 15°$, and $37.5° = 15° + 22.5°$.

43. $\sin 75°$
44. $\tan 75°$
45. $\cos 37.5°$
46. $\sin 37.5°$
47. Find $\sin 7.5°$; see problem 41.
48. Find $\cos 7.5°$; see problem 41.

Verify the following identities.

49. $\sin 2\theta + 1 = (\sin\theta + \cos\theta)^2$

50. $\cos 2\theta + 2\sin^2\theta = 1$

51. $\cos^4\theta - \sin^4\theta = \cos 2\theta$

52. $\cot\theta = \dfrac{1 + \cos 2\theta}{\sin 2\theta}$

53. $\dfrac{1 + \cos 2\theta}{1 - \cos 2\theta} = \cot^2\theta$

54. $\tan 2\theta = \dfrac{2\tan\theta}{2 - \sec^2\theta}$

55. $\cot\theta - \tan\theta = \dfrac{2\cos 2\theta}{\sin 2\theta}$

56. $2\csc 2\theta = \tan\theta + \cot\theta$

57. $\sin 2\theta - 4\sin^3\theta\cos\theta = \sin 2\theta\cos 2\theta$

58. $\cos 4\theta = 1 - 8\sin^2\theta\cos^2\theta$

59. $\csc^2\theta = \dfrac{2}{1 - \cos 2\theta}$

60. $\dfrac{2\cos^3\theta}{1 - \sin\theta} = 2\cos\theta + \sin 2\theta$

61. $\tan 2\theta = \dfrac{2(\tan\theta + \tan^3\theta)}{1 - \tan^4\theta}$

62. $\cot 4\theta = \dfrac{1 - \tan^2 2\theta}{2\tan 2\theta}$

63. $2\csc 2\theta\sin\theta\cos\theta = 1$

64. $\sec 2\theta = \dfrac{1}{1 - 2\sin^2\theta}$

65. $\cos 2\theta = \dfrac{1 - \tan^2\theta}{1 + \tan^2\theta}$

66. $\csc^2\dfrac{\theta}{2} = \dfrac{2}{1 - \cos\theta}$

67. $\sec^2\dfrac{\theta}{2} = \dfrac{2}{1 + \cos\theta}$

68. $\tan^2\dfrac{\theta}{2} = \dfrac{1 - \cos\theta}{1 + \cos\theta}$

69. $\cos^2\dfrac{\theta}{2} = \dfrac{1 - \cos^2\theta}{2 - 2\cos\theta}$

70. $\dfrac{\csc\theta - \cot\theta}{1 + \cos\theta} = \csc\theta\tan^2\dfrac{\theta}{2}$

71. $2\cos^2\dfrac{\theta}{2} - \cos\theta = 1$

72. $\cos\dfrac{\theta}{2}\sin\dfrac{\theta}{2} = \dfrac{\sin\theta}{2}$

73. $\sin^2\dfrac{\theta}{2} - \cos^2\dfrac{\theta}{2} = -\cos\theta$

74. $\tan^2\dfrac{\theta}{2} + \cos^2\dfrac{\theta}{2} = \dfrac{\cos^2\theta + 3}{2 + 2\cos\theta}$

75. $\tan^2\dfrac{\theta}{2} = \dfrac{2}{1 + \cos\theta} - 1$

76. $\sin^2\dfrac{\theta}{2} = \dfrac{\csc\theta - \cot\theta}{2\csc\theta}$

77. $4\sin^2\dfrac{\theta}{2}\cos^2\dfrac{\theta}{2} = \sin^2\theta$

78. $\dfrac{1 + \sec\theta}{\sec\theta} = 2\cos^2\dfrac{\theta}{2}$

79. $\tan\dfrac{\theta}{2} + \cot\dfrac{\theta}{2} = \dfrac{2}{\sin\theta}$

80. $\cot 2\theta = \dfrac{1}{2}\left(\cot\theta - \dfrac{1}{\cot\theta}\right)$ (See footnote 1.)

81. Show that $\sin 3\theta = 3\sin\theta - 4\sin^3\theta$.

82. Find an identity for $\cos 3\theta$ in terms of $\cos\theta$. See the previous problem.

83. Find identities (a) for $\sin 4\theta$ in terms of $\sin\theta$ and $\cos\theta$ and (b) for $\cos 4\theta$ in terms of $\cos\theta$.

84. Find identities (a) for $\sin 5\theta$ in terms of $\sin\theta$ and (b) for $\cos 5\theta$ in terms of $\cos\theta$.

85. Finding the center of gravity of a certain solid involves the expression $\dfrac{3}{16}a\left(\dfrac{1 - \cos 2\alpha}{1 - \cos\alpha}\right)$. Show that this is equivalent to $\frac{3}{8}a(1 + \cos\alpha)$.

86. Show that $\tan\dfrac{\alpha}{2} = \dfrac{1 - \cos\alpha}{\sin\alpha}$. Do this as follows. Let $\dfrac{\alpha}{2} = \theta$, so that $\alpha = 2\theta$. Replace $\dfrac{\alpha}{2}$ and α in the identity. Then simplify the right member; the most direct route will use $\cos 2\theta = 1 - 2\sin^2\theta$.

87. Show that $\tan\dfrac{\alpha}{2} = \dfrac{\sin\alpha}{1 + \cos\alpha}$. See problem 86.

88. **a.** Use the identity for $\sin\dfrac{\theta}{2}$ with $\theta = 30°$ to find the exact value of $\sin 15°$.
b. Use $\alpha = 45°$, $\beta = 30°$ and $\sin(\alpha - \beta)$ to find the exact value of $\sin 15°$.
c. Show that the values in (a) and (b) are the same. You may find useful the principle that if $a > 0$ and $b > 0$, then $a^2 = b^2$ implies that $a = b$.

89. **a.** Find $\tan 15°$ with the identity for $\tan\dfrac{\alpha}{2}$ (half-angle identity [6]), with $\alpha = 30°$.
b. Rewrite $\tan 15°$ as $\dfrac{\sin 15°}{\cos 15°}$ and use identities [4] and [5], with $\alpha = 30°$ to compute $\tan 15°$.
c. Show that the values in parts a and b are the same.

90. Verify half-angle identities [5] and [6].

91. Verify the double-angle identities [2-a], [2-b], [2-c], and [3].

[1]See the article "A Chaotic Search for *i*," by Gilbert Strang, in *The College Mathematics Journal,* Vol. 22, No. 1, January 1991, for an interesting relationship between this identity and the subject of chaotic behavior in iterative systems.

92. In the figure, $\theta_1 = \theta_2$. Use the identity for $\cos\frac{\theta}{2}$ to find the length of side x.

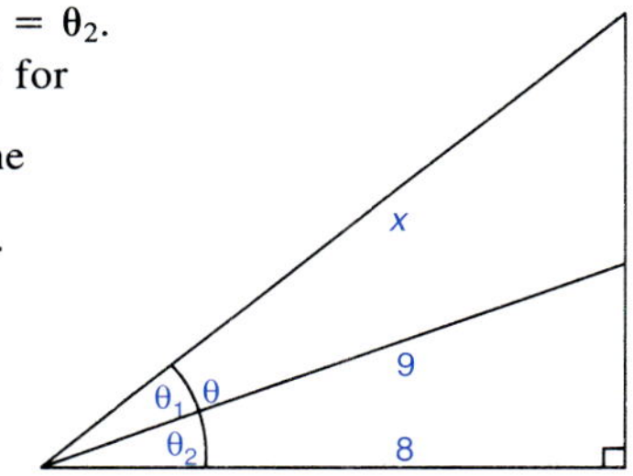

93. In the figure, $\theta_1 = \theta_2$. Use the identity for $\tan\frac{\theta}{2}$ to find the length of side x.

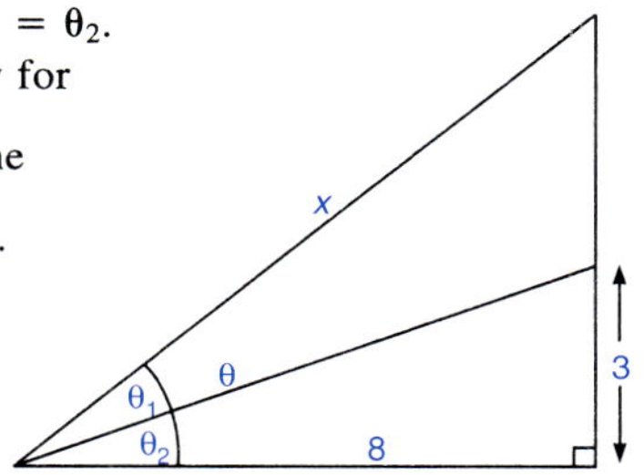

The following identities are important in some situations because they relate the sums and differences of trigonometric expressions to the products of trigonometric expressions. Verify each identity.

94. $\sin 2\alpha + \sin 2\beta = 2\sin(\alpha + \beta)\cdot\cos(\alpha - \beta)$

95. $\sin 2\alpha - \sin 2\beta = 2\sin(\alpha - \beta)\cdot\cos(\alpha + \beta)$

96. $\cos 2\alpha + \cos 2\beta = 2\cos(\alpha + \beta)\cdot\cos(\alpha - \beta)$ (*Hint:* Convert everything to cosine.)

97. $\cos 2\alpha - \cos 2\beta = -2\sin(\alpha + \beta)\cdot\sin(\alpha - \beta)$ (*Hint:* Convert everything to cosine.)

Skill and review

1. Use the fact that $\frac{\pi}{12} = \frac{\pi}{4} - \frac{\pi}{6}$ and the identity for $\cos(\alpha - \beta)$ to compute $\cos\frac{\pi}{12}$.
2. Graph the rational function $f(x) = \frac{2x}{x^2 - x - 6}$.
3. Find the point of intersection of the two straight lines $2x - y = 3$ and $x + 3y = 5$.
4. Rewrite $\frac{1 - \cot\theta}{\csc\theta + 1}$ in terms involving only the sine and cosine functions and simplify the result.
5. Find the least nonnegative solution to the equation $-3\sin x = 1$, to the nearest 0.1°.

7–4 Conditional trigonometric equations

Show that the area A of the shaded rectangle is given by $A = \sin\theta\sqrt{1.44 - \sin^2\theta}$.

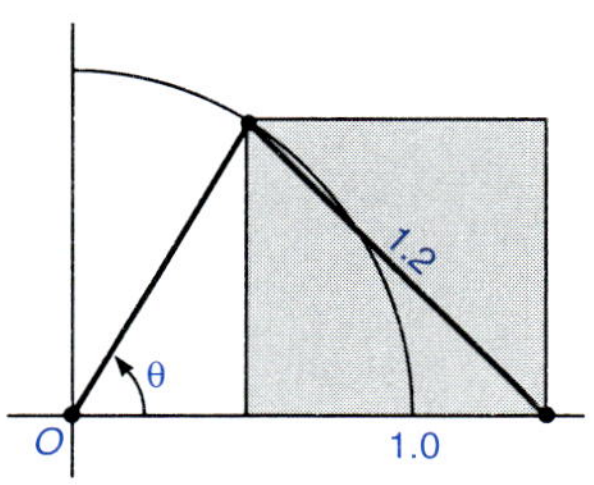

This equation for area is an example of a conditional trigonometric equation. Given a value for θ we can easily compute the area A. Given a value of the area A, it would be more difficult to discover the corresponding value of θ. Conditional trigonometric equations were introduced in section 5–6. In this section we examine these equations in a more general way.

Remember that *whenever we compute an inverse trigonometric function to solve an equation we use the absolute value of the argument, which gives us the reference angle of the answer.*

Primary solutions

In this section we solve for values in both degree and radian measure. We also determine all solutions that fall between $0° \le x < 360°$ or, in radian measure, $0 \le x < 2\pi$. We call such solutions **primary solutions.** Example 7–4 A illustrates.

■ Example 7–4 A

Find all primary solutions for the following trigonometric equations. Find the solutions in degrees (nearest tenth) and radians (four decimal places).

1. $\cos\theta = -\frac{1}{2}$

$\cos\theta = -\frac{1}{2}$ — $\cos\theta < 0$ so all solutions are in quadrants II and III

$\theta' = \cos^{-1}\frac{1}{2}$ — Use the absolute value of $-\frac{1}{2}$

$\theta' = 60°$ or $\dfrac{\pi}{3}$ radians — Exact values, obtained from the unit circle, figure 6–5

$\theta = 180° \pm 60°$ or $\pi \pm \dfrac{\pi}{3}$ — In quadrant II $\theta = 180° - \theta'$; in quadrant III $\theta = 180° + \theta'$

$\theta = 120°$ or $240°$ (degrees) or $\dfrac{2\pi}{3}$ or $\dfrac{4\pi}{3}$ (radians).

2. $5\sin\alpha = -2$

$\sin\alpha = -\frac{2}{5}$

Since $\sin\alpha < 0$ all solutions terminate in quadrants III and IV.

$\alpha' = \sin^{-1}\frac{2}{5}$ — As noted earlier, we use $\left|-\frac{2}{5}\right|$

$\alpha' \approx 23.6°$ or 0.4115 radians — Degree mode, radian mode

Degrees:

$$\alpha \approx 180° + 23.6° \text{ or } 360° - 23.6°$$

Radians:

$$\alpha \approx \pi + 0.4115 \text{ or } 2\pi - 0.4115$$

Thus, in degrees $\alpha \approx 203.6°$ or $336.4°$; in radians $\alpha \approx 3.5531$ or 5.8717.

3. $\tan^2 x + 4\tan x = 1$

$\tan^2 x + 4\tan x - 1 = 0$ — Quadratic, but will not factor

$$\tan x = \frac{-4 \pm \sqrt{4^2 - 4(1)(-1)}}{2(1)}$$

$a = 1$, $b = 4$, and $c = -1$ in $\dfrac{-b \pm \sqrt{b^2 - 4ac}}{2a}$

$$= \frac{-4 \pm \sqrt{20}}{2} = \frac{-4 \pm 2\sqrt{5}}{2}$$

$\sqrt{20} = \sqrt{4 \cdot 5} = 2\sqrt{5}$

$$= -2 \pm \sqrt{5}$$

$\tan x = -2 + \sqrt{5}$	$\tan x = -2 - \sqrt{5}$
$x' = \tan^{-1}(-2 + \sqrt{5})$ $\approx 13.3°$, 0.2318 (radians) x is in quadrants I or III since $-2 + \sqrt{5} > 0$ $x \approx 13.3°$ or $180° + 13.3°$ ≈ 0.2318 or $\pi + 0.2318$	$x' = \tan^{-1}\left|-2 - \sqrt{5}\right|$ $= \tan^{-1}(2 + \sqrt{5})$ $\approx 76.7°$, 1.3390 (radians) x is in quadrants II or IV since $-2 - \sqrt{5} < 0$ $x \approx 180° - 76.7°$ or $360° - 76.7°$ $\approx \pi - 1.3390$ or $2\pi - 1.3390$

$x \approx 13.3°, 193.3°, 103.3°, 283.3°$ or $0.2318, 3.3734, 1.8026, 4.9442$. ■

Using a graphing calculator to solve an equation

In section 4–3, in the subsection "The TI-81 and Newton's method," we presented a way to solve equations using a graphing calculator and a built-in program called NEWTON. Example 7–4 B shows how to apply this method to part 2 of example 7–4 A.

■ Example 7–4 B

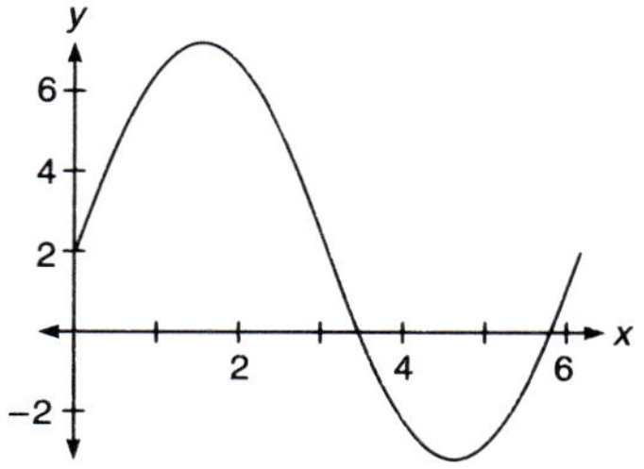

Solve $5 \sin \alpha = -2$ using graphical methods.

This is equivalent to $5 \sin \alpha + 2 = 0$; find the zeros of the function $y = 5 \sin x + 2$.

Graph $y = 5 \sin x + 2$. This is shown in the figure, with Xmin = −.1, Xmax = 6.3, Ymin = −3, Ymax = 7. *The calculator is in radian mode.*

Use the trace feature to position the cursor near the zero between 3 and 4. Now execute the program NEWTON. The value 3.5531095 appears. This is one of the zeros.

Graph the function again, select trace, and position the cursor near the second zero and run the program NEWTON again. The value 5.871668461 appears. This is an approximation to the second zero.

Repeating these steps in degree mode will find approximations to the zeros in degree mode. Of course, Xmin should be something like −10°, and Xmax about 360°. ■

Using identities to help solve an equation

When an expression involves more than one trigonometric function we often use identities to rewrite the equation in terms of a single trigonometric function. This is illustrated in the next example.

■ Example 7–4 C

Find all primary solutions for the following trigonometric equations. Find the solutions in degrees and radians.

1. $\tan \theta - \cot \theta = 0$

$$\tan \theta - \frac{1}{\tan \theta} = 0 \qquad \cot \theta = \frac{1}{\tan \theta} \text{ where } \tan \theta \neq 0$$

$$\tan^2\theta - 1 = 0 \qquad \text{Multiply each term by } \tan \theta$$

$$\tan^2\theta = 1$$

$$\tan \theta = \pm 1$$

When $\tan \theta = 1$, θ' is 45° or $\frac{\pi}{4}$ (see the unit circle, figure 6–4, or table 6–1), so using this fact and the ASTC rule we determine that the primary solutions in degrees are 45°, 135°, 225°, 315° and in radians are $\frac{\pi}{4}, \frac{3\pi}{4}, \frac{5\pi}{4}, \frac{7\pi}{4}$.

2. $2\cos^2 x - 3\sin x - 3 = 0$

$2(1 - \sin^2 x) - 3\sin x - 3 = 0$ $\qquad \cos^2\theta = 1 - \sin^2\theta$

$-2\sin^2 x - 3\sin x - 1 = 0$

$2\sin^2 x + 3\sin x + 1 = 0$

$(2\sin x + 1)(\sin x + 1) = 0$

$2\sin x + 1 = 0$	$\sin x + 1 = 0$
$\sin x = -\frac{1}{2}$	$\sin x = -1$
$x = 210°, 330°$ or $\frac{7\pi}{6}, \frac{11\pi}{6}$	$x = 270°$ or $\frac{3\pi}{2}$

The primary solutions are 210°, 270°, 330° or $\frac{7\pi}{6}, \frac{3\pi}{2}, \frac{11\pi}{6}$. ■

Finding all solutions to a trigonometric equation

We need to take note of the fact that there are an infinite number of solutions to the equations we solved in the preceding problems. Because the trigonometric functions are periodic, the set of all solutions can be found by adding all integral values of the appropriate period (2π or π) to the solution.

This can be illustrated for part 1 of example 7–4 A, where we found that the primary solutions to $\cos\theta = -\frac{1}{2}$ are $\frac{2\pi}{3}$ and $\frac{4\pi}{3}$ (in radians). However, the cosine function is 2π-periodic, which means that $\cos(\theta + 2k\pi) = \cos\theta$ for any value of θ and for integer values of k. Thus, the actual set of all radian-valued solutions for this problem is $\frac{2\pi}{3} + 2k\pi$ and $\frac{4\pi}{3} + 2k\pi$, $k \in J$. This idea is illustrated in figure 7–2 and example 7–4 D.

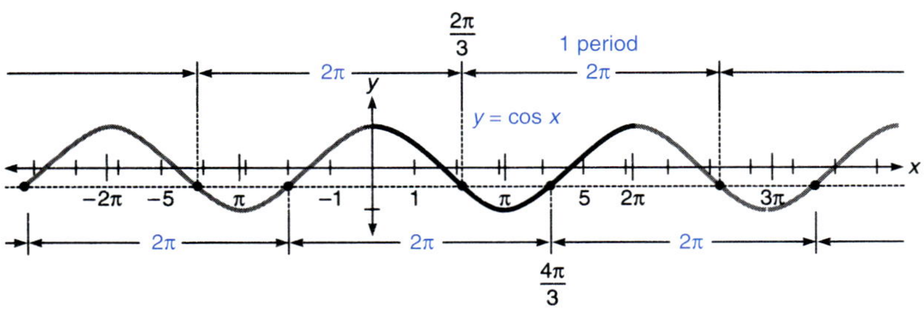

Figure 7–2

■ ***Example 7–4 D***

Find *all* solutions for the following trigonometric equations. Find the solutions in degrees (nearest tenth) and radians (four decimal places).

1. $3 \sin 2x = -2$

$\sin 2x = -\frac{2}{3}$

Since $\sin 2x < 0$, $2x$ is in quadrants III or IV.

$(2x)' = \sin^{-1} \frac{2}{3}$

$(2x)' \approx 41.8°$ or 0.7297 radians

Note 1. Do not divide both members by 2 at this point. It is necessary to find the solutions for $2x$ *before* dividing by 2.

2. Although we show the intermediate values above as 41.8° and 0.7297, it is important to keep the maximum accuracy of the calculator up to the last step of the problem. All calculators have the capability to store at least one value in memory, which should be used to avoid tedious and error-prone reentry of values.

$$2x \approx \begin{cases} 180° + 41.8° \text{ or } 360° - 41.8° \text{ (degrees)} \\ \pi + 0.7297 \text{ or } 2\pi - 0.7297 \text{ (radians)} \end{cases}$$

$2x$ is an angle in quadrants III or IV

$$2x \approx \begin{cases} 221.8°, 318.2° \text{ (degrees)} \\ 3.8713, 5.5535 \text{ (radians)} \end{cases}$$

Primary solutions for $2x$

To describe all solutions we add multiples of the period of the sine function, 360° or 2π.

$$2x \approx \begin{cases} 221.8° + k \cdot 360°, 318.2° + k \cdot 360° \\ 3.8713 + 2k\pi, 5.5535 + 2k\pi \end{cases}$$

We now divide each solution by 2.

$$x \approx \begin{cases} 110.9° + k \cdot 180°, 159.1° + k \cdot 180° \\ 1.9357 + k\pi, 2.7768 + k\pi \end{cases}$$

This describes *all* solutions to the equation. To find primary solutions for x we would compute the values above for $k = 0$ and $k = 1$. If $k = 2$, the solutions are greater than 360° (2π), and if k is negative the solutions are negative.

2. $\tan \frac{x}{2} = \frac{\sqrt{3}}{3}$

$$\tan \frac{x}{2} = \frac{\sqrt{3}}{3}$$

$$\left(\frac{x}{2}\right)' = \tan^{-1} \frac{\sqrt{3}}{3} = 30°, \frac{\pi}{6}$$

$$\frac{x}{2} = 30°, 210°, \text{ or } \frac{\pi}{6}, \frac{7\pi}{6}$$

These are the primary solutions for $\frac{x}{2}$

The tangent function is π-periodic. Thus we add integer multiples of 180° (π) to obtain all solutions.

$$\frac{x}{2} = \begin{cases} 30° + k \cdot 180°, 210° + k \cdot 180° \text{ (degrees)} \\ \frac{\pi}{6} + k\pi, \frac{7\pi}{6} + k\pi \text{ (radians)} \end{cases}$$

This describes all solutions. However, $210° - 30° = 180°$, and similarly $\frac{7\pi}{6} - \frac{\pi}{6} = \pi$, so the solutions can be described more compactly.

$$\frac{x}{2} = 30° + k \cdot 180°, \text{ or } \frac{\pi}{6} + k\pi$$

$$x = 60° + k \cdot 360° \text{ or } \frac{\pi}{3} + 2k\pi \qquad \text{Multiply each member by 2}$$

All solutions for x are $x = 60° + k \cdot 360°$ or $\frac{\pi}{3} + 2k\pi$. ■

Equations involving multiples of the angle

If an equation mixes multiples of values with the values themselves, we can eliminate the multiple value with an appropriate identity.

■ ***Example 7–4 E***

Solve $\sin 2\theta - \sin \theta = 0$; find primary solutions.

$$2 \sin \theta \cos \theta - \sin \theta = 0 \qquad \sin 2\alpha = 2 \sin \alpha \cos \alpha$$

$$\sin \theta (2 \cos \theta - 1) = 0 \qquad \text{Factor}$$

$$\sin \theta = 0 \text{ or } 2 \cos \theta - 1 = 0 \qquad \text{Zero product property}$$

$\sin \theta = 0$	$2 \cos \theta - 1 = 0$
	$\cos \theta = \frac{1}{2}$
$\theta = 0°, 180°$, or $0, \pi$ (radians)	$\theta = 60°, 300°$, or $\frac{\pi}{3}, \frac{5\pi}{3}$ (radians)

The solutions are 0°, 60°, 180°, 300° or $0, \frac{\pi}{3}, \pi, \frac{5\pi}{3}$ (radians). ■

Mastery points

Can you

- Solve linear and quadratic trigonometric equations?
- Solve trigonometric equations involving multiple angles?
- Solve trigonometric equations by applying an appropriate identity?

Exercise 7–4

Find all primary solutions to the following trigonometric equations. Leave answers in both degrees and radians. All answers should be exact.

1. $\tan\theta + 1 = 0$
2. $\sin\theta - 1 = 0$
3. $2\cos\theta - 1 = 0$
4. $2\cos\theta + 1 = 0$
5. $\sqrt{3}\tan\theta - 1 = 0$
6. $\cot\theta + \sqrt{3} = 0$
7. $\csc\theta + 2 = 0$
8. $\sec\theta - 2 = 0$
9. $3\sin^2\theta - 3 = 0$
10. $3\csc^2\theta = 3$
11. $\sec^2\theta = 1$
12. $\tan^2\theta - 1 = 0$
13. $(\cos\theta - 1)(\sin\theta + 1) = 0$
14. $(\sec\theta + 2)(\csc\theta - 2) = 0$
15. $(2\cos^2\theta - 1)(\cot\theta - 1) = 0$
16. $(3\tan^2\theta - 1)(\sqrt{3}\sec\theta - 2) = 0$
17. $\sin^2\theta - \sin\theta = 0$
18. $\cos^2\theta + \cos\theta = 0$
19. $\tan^2\theta - \sqrt{3}\tan\theta = 0$
20. $\cos^2\theta - \frac{1}{2}\cos\theta = 0$
21. $2\sin^2\theta + \sin\theta - 1 = 0$
22. $\cos^2\theta + 2\cos\theta + 1 = 0$
23. $2\sin^3\theta - \sin\theta = 0$
24. $2\cos^2\theta + 3\cos\theta = 2$
25. $2\sin\theta\cos\theta - \sin\theta = 0$
26. $2\sin\theta\cos\theta + \cos\theta = 0$
27. $\sqrt{3}\tan\theta\cot\theta + \cot\theta = 0$
28. $2\tan^2\theta\cos\theta - \tan^2\theta = 0$
29. $\tan x \cot x = 0$
30. $\sin x \cos x = 0$
31. $2\sin x - \csc x + 1 = 0$
32. $2\cos x + \sec x - 3 = 0$
33. $\tan x + \cot x = -2$
34. $2 - \sin x - \csc x = 0$
35. $2\sin^2 x - \cos x = 1$
36. $2\cos^2 x - 3\sin x = 3$
37. $4\tan^2 x = 3\sec^2 x$
38. $4\cot^2 x - 3\csc^2 x = 0$
39. $\sin^2 x - \cos^2 x = 0$
40. $\cot^2 x + \csc^2 x = 0$
41. $2\tan^2 x \sin x = \tan^2 x$
42. $\sin^2 x \cos x - \cos x = 0$

Solve the following equations using the quadratic formula if necessary and a calculator. Find the primary solutions in both radians and degrees. Round radian answers to hundredths and degree answers to tenths.

43. $6\sin^2 x - 2\sin x - 1 = 0$
44. $3\cos^2 x + \cos x - 2 = 0$
45. $\cot^2 x - 3\cot x - 2 = 0$
46. $\tan^2 x + 5\tan x + 2 = 0$
47. $\sec^2 x - 2\sec x - 4 = 0$
48. $2\csc^2 x - \csc x - 5 = 0$
49. $\tan x + 2\sec x = 3$
50. $3\cot x - \csc x - 1 = 0$

Find all solutions to the following trigonometric equations in both degrees and in radians.

51. $\cos x = \frac{1}{2}$
52. $\sin x = 1$
53. $\cot x = -\sqrt{3}$
54. $\cos x = -\dfrac{\sqrt{3}}{2}$
55. $\sin x = -\dfrac{\sqrt{2}}{2}$
56. $\tan x = -1$
57. $\tan x = 1$
58. $\sec x = \dfrac{2}{\sqrt{3}}$
59. $\csc x = 2$
60. $\tan\dfrac{x}{2} = 1$
61. $\sin\dfrac{x}{2} = \dfrac{\sqrt{3}}{2}$
62. $\sin 3x = 0$
63. $\cos 3x = -1$
64. $\sec\dfrac{x}{2} = 1$
65. $3\cot 2x = \sqrt{3}$
66. $2\sin 3x = -1$
67. $2\cos 4x = -1$
68. $-\sqrt{3}\tan 5x = 1$
69. $2\cos 2x + 1 = 0$
70. $\tan 2\theta - 1 = 0$
71. $\cot 2\theta - \sqrt{3} = 0$
72. $2\cos 3\theta = -1$
73. $2\sin 2\theta = 1$
74. $\sin\dfrac{\theta}{3} = \dfrac{\sqrt{3}}{2}$
75. $\sec 3\theta = 2$
76. $\csc 2\theta = -\dfrac{2\sqrt{3}}{3}$
77. $\sqrt{3}\tan\dfrac{\theta}{4} = 1$
78. $\cot\dfrac{\theta}{3} = \dfrac{\sqrt{3}}{3}$

Find the primary solutions to the following trigonometric equations in both radians and degrees. Find solutions in radians to hundredths and in degrees to the nearest tenth of a degree, where necessary.

79. $\cos 2\theta + \sin\theta = 0$
80. $\cos 2\theta - \cos\theta = 0$
81. $\sin 2\theta + \sin\theta = 0$
82. $\cos^2\theta - \sin^2\theta = 1$
83. $\cos 2\theta = 1 - \sin\theta$
84. $\cos 2\theta = \cos\theta - 1$
85. $\sin\dfrac{\theta}{2} = \tan\dfrac{\theta}{2}$
86. $\sin\dfrac{\theta}{2} = \cos\theta$
87. $2\sec\theta = \csc^2\dfrac{\theta}{2}$
88. $\sin^2\dfrac{\theta}{2} = \cos\theta$
89. $\tan\dfrac{\theta}{2} = \cos\theta - 1$
90. $\cot\theta - \tan\dfrac{\theta}{2} = 0$
91. $\sin 2\theta - \cos\theta = \cos^2\theta$

In the mathematical modeling of an aerodynamics problem the following equation arises:

$$y = x \cos A \cos B - x^2 \cos A \sin B - x^3 \sin A$$

Problems 92 and 93 use this equation.

92. If $A = 0.855$, $B = 1.052$, and $y = 0$, solve for x to the nearest 0.01.

93. If $B = 0.7$, $x = 2$, and $y = -8$, find A to the nearest 0.01. Find the least nonnegative solution(s). (*Hint:* $\sin^2 A = 1 - \cos^2 A$.)

A mechanical device is constructed as shown in the diagram. The arm OA moves through angle θ, from 0° to 90°. Two positions are shown. Point A moves along a circle of radius 1.0 meters, and point B moves horizontally only. The distance AB is fixed by arm AB at 1.2 meters. The area of the shaded rectangle is the product of its length and width, $A = \ell w$.

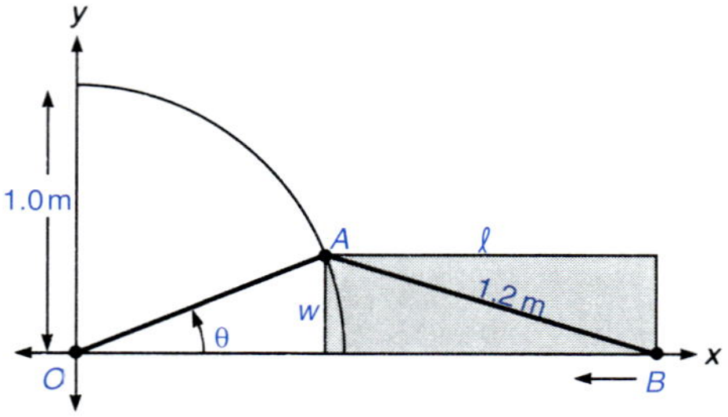

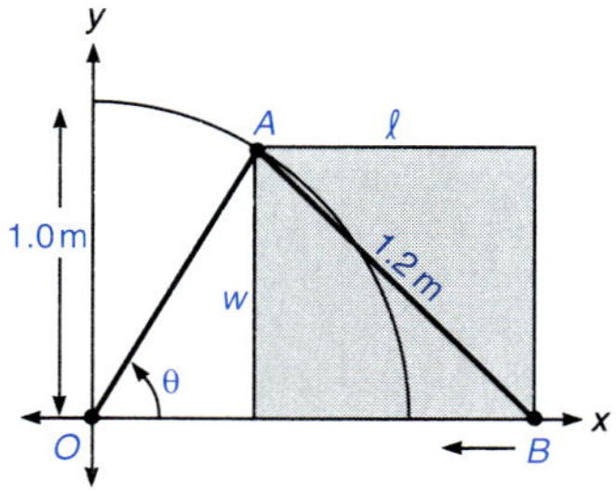

94. Show that $A = \sin\theta\sqrt{1.44 - \sin^2\theta}$. (The units are square meters.)

95. Find A, to the nearest 0.01 m^2, when $\theta = 0°, 30°, 45°, 60°, 90°$.

96. Find θ when $A = 0.5$ m^2. Round the answer to the nearest 0.1°.

Skill and review

1. Graph the function $y = -2 \sin 4x$.
2. Graph the quadratic function $f(x) = x^2 + 4x - 8$ by completing the square and putting it in vertex form.
3. Use a reference triangle to find $\tan\theta$ when $\sin\theta = \dfrac{\sqrt{3}}{6}$ and $\cos\theta < 0$.
4. Factor $-5a^8 + 5a^2x^6$.
5. Find the equation of the straight line that passes through the point (1,3) and has x-intercept at 4.
6. Perform the indicated subtraction to collect $\dfrac{3}{x^2 - 1} - \dfrac{1}{x - 1}$ into one term, in simplest form.

Chapter 7 summary

- **Reciprocal identities**

$$\csc\theta = \frac{1}{\sin\theta}, \quad \sec\theta = \frac{1}{\cos\theta}, \quad \cot\theta = \frac{1}{\tan\theta}$$

$$\sin\theta = \frac{1}{\csc\theta}, \quad \cos\theta = \frac{1}{\sec\theta}, \quad \tan\theta = \frac{1}{\cot\theta}$$

- **Tangent and cotangent identities**

$$\tan\theta = \frac{\sin\theta}{\cos\theta}, \quad \cot\theta = \frac{\cos\theta}{\sin\theta}$$

- **Fundamental identity of trigonometry**

$$\sin^2\theta + \cos^2\theta = 1$$

Pythagorean identities	**Useful forms**
$\sin^2\theta + \cos^2\theta = 1$	$\sin^2\theta = 1 - \cos^2\theta$
	$\cos^2\theta = 1 - \sin^2\theta$
$\sec^2\theta = \tan^2\theta + 1$	$\tan^2\theta = \sec^2\theta - 1$
	$\sec^2\theta - \tan^2\theta = 1$
$\csc^2\theta = \cot^2\theta + 1$	$\cot^2\theta = \csc^2\theta - 1$
	$\csc^2\theta - \cot^2\theta = 1$

- **Sum and difference identities for sine and cosine**

[1] $\cos(\alpha + \beta) = \cos \alpha \cos \beta - \sin \alpha \sin \beta$

[2] $\cos(\alpha - \beta) = \cos \alpha \cos \beta + \sin \alpha \sin \beta$

[3] $\sin(\alpha + \beta) = \sin \alpha \cos \beta + \cos \alpha \sin \beta$

[4] $\sin(\alpha - \beta) = \sin \alpha \cos \beta - \cos \alpha \sin \beta$

- When the sum of two angles is 90°, or $\frac{\pi}{2}$ radians, the angles are said to be **complementary.**

- **Cofunction identities**

[5] $\sin\left(\frac{\pi}{2} - \theta\right) = \cos \theta$ [6] $\cos\left(\frac{\pi}{2} - \theta\right) = \sin \theta$

[7] $\tan\left(\frac{\pi}{2} - \theta\right) = \cot \theta$ [8] $\cot\left(\frac{\pi}{2} - \theta\right) = \tan \theta$

[9] $\sec\left(\frac{\pi}{2} - \theta\right) = \csc \theta$ [10] $\csc\left(\frac{\pi}{2} - \theta\right) = \sec \theta$

- **Sum and difference identities for tangent**

[11] $\tan(\alpha + \beta) = \dfrac{\tan \alpha + \tan \beta}{1 - \tan \alpha \tan \beta}$

[12] $\tan(\alpha - \beta) = \dfrac{\tan \alpha - \tan \beta}{1 + \tan \alpha \tan \beta}$

- **Double-angle identities**

[1] $\sin 2\alpha = 2 \sin \alpha \cos \alpha$

[2-a] $\cos 2\alpha = \cos^2\alpha - \sin^2\alpha$

[2-b] $\cos 2\alpha = 1 - 2 \sin^2\alpha$

[2-c] $\cos 2\alpha = 2 \cos^2\alpha - 1$

[3] $\tan 2\alpha = \dfrac{2 \tan \alpha}{1 - \tan^2\alpha}$

- **Half-angle identities**

[4] $\sin \frac{\alpha}{2} = \pm\sqrt{\dfrac{1 - \cos \alpha}{2}}$

[5] $\cos \frac{\alpha}{2} = \pm\sqrt{\dfrac{1 + \cos \alpha}{2}}$

[6] $\tan \frac{\alpha}{2} = \pm\sqrt{\dfrac{1 - \cos \alpha}{1 + \cos \alpha}}$

[6-a] $\tan \frac{\alpha}{2} = \dfrac{1 - \cos \alpha}{\sin \alpha}$

[6-b] $\tan \frac{\alpha}{2} = \dfrac{\sin \alpha}{1 + \cos \alpha}$

Chapter 7 review

[7–1] Verify the following identities.

1. $\dfrac{\cot \theta}{\cos \theta} = \csc \theta$
2. $\sec \theta \tan \theta = \sec^2\theta \sin \theta$
3. $\dfrac{\tan^2\theta}{\sec^2\theta - 1} = \sin^4\theta \csc^4\theta$
4. $\dfrac{\csc^2\theta - 1}{\sec^2\theta - 1} = \cot^4\theta$
5. $\dfrac{\csc \theta \tan \theta}{\sin \theta} = \csc \theta \sec \theta$
6. $\sin^2\theta - \cos^2\theta = 2 \sin^2\theta - 1$

Obtain an equivalent expression involving only the sine and cosine functions. Simplify the resulting expression.

7. $\csc \theta - \sec \theta$
8. $\tan \theta + \cot \theta$
9. $\dfrac{\sec \theta}{\tan \theta - \cot \theta}$
10. $\dfrac{1 - \cot \theta}{\csc \theta + 1}$
11. $\dfrac{\sec^2\theta - 1}{\sec^2\theta}$
12. $\dfrac{1 - \cot^2\theta}{\csc^2\theta - 1} \cos^2\theta$

Verify that each equation is an identity.

13. $\csc x - \tan x \cot x = \csc x - 1$
14. $\sin^2 x + \sin^2 x \cot^2 x = 1$
15. $\csc^2 x - \sec^2 x = \csc^2 x \sec^2 x(\cos^2 x - \sin^2 x)$
16. $\dfrac{1}{\csc x - \cot x} = \dfrac{\sin x}{1 - \cos x}$
17. $(\tan x - 1)(\csc^2 x - \cot^2 x) = \sec x(\sin x - \cos x)$
18. $\dfrac{\csc^2 x - 1}{\sin^2 x} = \cos^2 x \csc^4 x$
19. $\dfrac{1}{1 + \csc x} + \dfrac{1}{1 - \csc x} = -2 \tan^2 x$
20. $\sin^2 x + \sin^2 x \cos^2 x = 1 - \cos^4 x$
21. $\tan^4 x + \tan^2 x = \dfrac{\sec^2 x}{\cot^2 x}$
22. $\dfrac{1 - \cot x}{1 + \csc x} = \dfrac{\sin x - \cos x}{\sin x + 1}$

[7–2] Use the sum and difference identities to find the exact value of each of the following. Observe that each value is the sum or difference of values chosen from $\frac{\pi}{6}$ (30°), $\frac{\pi}{4}$ (45°), and $\frac{\pi}{3}$ (60°).

23. $\sin \frac{\pi}{12}$ **24.** $\tan\left(-\frac{\pi}{12}\right)$

25. $\sin 75°$ **26.** $\cos(-15°)$

Each of the following problems presents information about two angles, α and β, including the quadrant in which the angle terminates. Use the information to find the required value.

27. $\cos \alpha = \frac{3}{5}$, quadrant I; $\sin \beta = \frac{5}{13}$, quadrant I. Find $\sin(\alpha + \beta)$.

28. $\cos \alpha = -\frac{12}{13}$, quadrant III; $\sin \beta = \frac{2}{3}$, quadrant I. Find $\cos(\alpha - \beta)$.

29. $\tan \alpha = -\frac{5}{12}$, quadrant II; $\sec \beta = -3$, quadrant II. Find $\tan(\alpha - \beta)$.

Use the sum and difference identities to verify the following identities.

30. $\cos\left(\frac{3\pi}{2} + \theta\right) = \sin \theta$

31. $\sin\left(\frac{\pi}{4} - \theta\right) = \frac{\sqrt{2}}{2}(\cos \theta - \sin \theta)$

32. $\frac{\cos(\alpha + \beta)}{\sin \alpha \cos \beta} = \cot \alpha - \tan \beta$

33. Use the sum formula to show that the sine function is 2π-periodic; that is that $\sin(\theta + 2\pi) = \sin \theta$.

34. A picture on a wall is 2 feet tall and 6 feet above eye level (see the diagram). Compute the exact value of $\sin(\alpha - \beta)$.

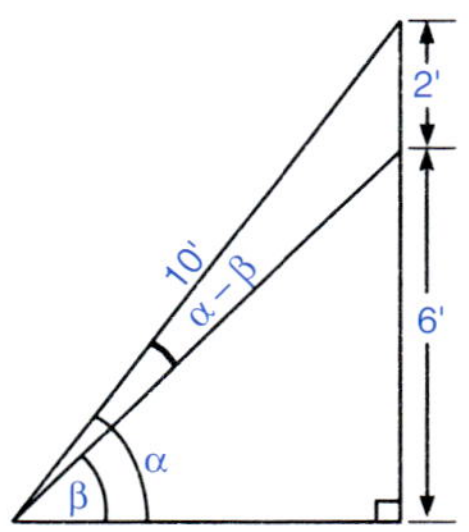

[7–3] Using the double-angle identities, find the angle θ that makes the following statements true.

35. $\cos \theta = \cos^2 62° - \sin^2 62°$

36. $\sin \theta = 2 \sin 5\pi \cos 5\pi$

37. $\tan \theta = \dfrac{2 \tan \frac{7\pi}{12}}{1 - \tan^2 \frac{7\pi}{12}}$

38. $\cos 24° = 1 - 2 \sin^2\theta$

39. $a \sin b\theta = 6 \sin \frac{\theta}{2} \cos \frac{\theta}{2}$; find a and b.

40. Given $\cos \theta = -\frac{5}{12}$, θ terminates in quadrant III. Find the exact value of (a) $\tan 2\theta$; (b) $\sin 2\theta$.

41. Given $\sin \theta = \frac{4}{5}$, θ terminates in quadrant II. Find the exact value of (a) $\cos 2\theta$; (b) $\tan 2\theta$.

42. Verify the identity $\sin 2x - \cos x = \cos x(2 \sin x - 1)$.

43. Verify the identity $1 + \cos 2x = 2 \cos^2 x$.

Using the half-angle identities, find the exact value of the following.

44. $\tan 22.5°$ **45.** $\cos \frac{\pi}{8}$

46. Given $\cos x = \frac{12}{13}$, $0 < x < \frac{\pi}{2}$. Find (a) $\sin \frac{x}{2}$ and (b) $\tan \frac{x}{2}$.

47. Given $\sin x = -\frac{2}{3}$, $3\pi < x < \frac{7\pi}{2}$. Find (a) $\cos \frac{x}{2}$ and (b) $\tan \frac{x}{2}$.

48. Verify the identity $\cot \frac{\theta}{2} = \frac{1 + \cos \theta}{\sin \theta}$.

49. Verify the identity $\sec^2\frac{\theta}{2} - \tan^2\frac{\theta}{2} = 1$.

[7–4] Solve the following conditional equations. Find all primary solutions in both radians and degrees.

50. $2 \sin x - 1 = 0$ **51.** $3 \cot^2 x - 1 = 0$

52. $(\sin x - 1)(2 \cos x - 1) = 0$

53. $(4 \sin^2 x - 1)(\sec x - 2) = 0$

54. $\cot^2\theta - \cot \theta = 0$ **55.** $\sec^2\theta - 4 = 0$

56. $2 \cos^2\theta - \cos \theta - 1 = 0$

57. $2 \sin \theta - \csc \theta + 1 = 0$

58. $2 \cot \theta \cos \theta = \cot^2\theta$ **59.** $2 \sin^2\theta - 3 \cos \theta = 3$

60. $2 \sin 4x = 1$

61. $2 \cos \frac{x}{2} - \sqrt{3} = 0$

62. $\sqrt{3} \tan \frac{x}{2} + 1 = 0$

63. $\sin^2 \frac{x}{4} = \frac{1}{2}$

64. $3 \tan^2 \frac{\theta}{4} = 9$

65. $2 \cos 5\theta = 1$

66. $\cos \theta + \sin 2\theta = 0$

Solve the following conditional equations. Find all primary solutions in radians only. Round solutions to hundredths.

67. $3 \cos^2 x - 1 = 0$

68. $\sin x - 2 \csc x = 5$

69. $\cos \frac{x}{2} - \sin x = 0$

70. $\left(\tan \frac{x}{4} - \frac{\sqrt{3}}{3}\right)\left(\sin 3x + \frac{1}{2}\right) = 0$

71. $3 \sin^2 2x - \sin 2x - 2 = 0$

Chapter 7 test

1. Show that the expression $\csc^2 x \sin x \cos x$ is equivalent to $\cot x$.

2. Write the expression $\frac{\csc x - \sec x}{\tan x + \cot x}$ as an expression in sine and cosine, and simplify.

3. Show by counter example that the equation $\cot x - 2 \tan x = 1$ is not an identity.

4. $a \cos b\theta = 8 \cos^2 2\theta - 8 \sin^2 2\theta$; find a and b.

5. Given $\cos \alpha = -\frac{1}{2}$, α terminates in quadrant III, and $\sin \beta = \frac{15}{17}$, β terminates in quadrant II. Find $\sin(\alpha + \beta)$.

6. Given $\cos x = -\frac{8}{17}$, x terminates in quadrant II. Find $\sin 2x$.

7. Given $\csc x = \frac{5}{3}$, $\frac{5\pi}{2} < x < 3\pi$. Find $\tan \frac{x}{2}$.

8. Use the appropriate half-angle formula to find sec 22.5°.

Verify the following identities.

9. $\frac{1 + \cot \theta}{\csc \theta} = \sin \theta + \cos \theta$

10. $\frac{\cos^2 x - 1}{\sin^2 x} = -1$

11. $\tan^4 x + \tan^2 x = \frac{\sec^2 x}{\cot^2 x}$

12. $\cos\left(\theta - \frac{3\pi}{2}\right) = -\sin \theta$

13. $\cos 2x - \sin 2x = 2 \cos x(\cos x - \sin x) - 1$

14. Find all primary solutions to the equation $4 \cos^2 x - 1 = 0$ (radians and degrees).

15. Find all primary solutions to the equation $(\cot \theta - \sqrt{3})(\sec \theta + 2) = 0$ (radians and degrees).

16. Find all primary solutions to the equation $6 \sin^2 x + 5 \sin x - 1 = 0$ (radians only). Round solutions to the nearest 0.1.

17. Find all primary solutions to the equation $\sin^2 3x = 0.5$ (radians only). Round answers to the nearest 0.1.

CHAPTER 8

Additional Topics in Trigonometry

In the next two sections we learn two methods for solving triangles that are not right triangles. After that we introduce vectors, which are used extensively in physics and engineering. We then introduce complex numbers in polar form, which is also very important in physics and engineering. Finally we see another coordinate system besides the usual rectangular coordinate system—the polar coordinate system.

8–1 The law of sines

Ground-based radar at point A determines that the angle of elevation to an aircraft 42.9 miles away is 13.2°. Radar at point B is on a straight line between a point on the ground directly below the aircraft and the radar at A, and determines that the same aircraft is 13.6 miles away from point B. Find the distance from A to B.

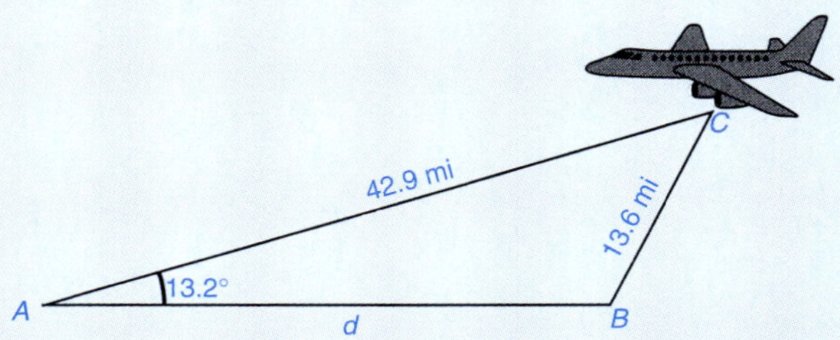

A triangle in which none of the angles is a right angle is called an **oblique** triangle. Triangle ABC in the preceding problem is such a triangle. One method for solving certain oblique triangles is the **law of sines.** The following paragraphs develop this law.

First, we observe that *at least two of the angles in every triangle are acute.* If only one angle were acute (less than 90°) then the other two would be obtuse or right (greater than or equal to 90°). This is impossible since the sum of these two angles would be greater than or equal to 180°, and thus the sum of all three angles would be greater than 180°.

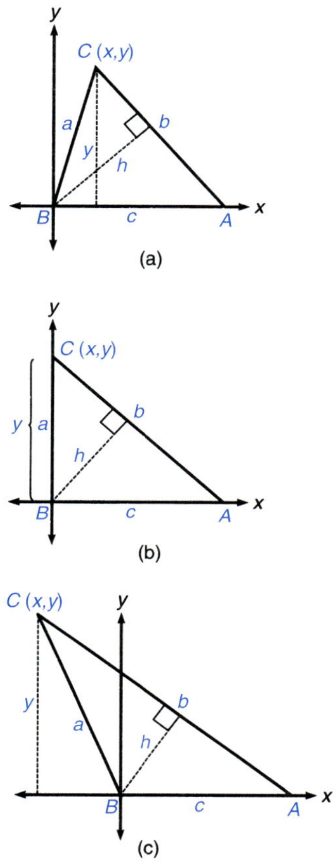

Figure 8–1

Now we consider any triangle ABC, and label two of the acute angles A and C. Angle B may be acute, obtuse, or right. We place the triangle in a coordinate system so that angle B is in standard position and vertex A is on the x-axis. Figure 8–1 shows sketches for the cases where B is (a) acute, (b) right, and (c) obtuse.

From vertex B we construct a line segment perpendicular to side AC and label this line h. From what we know about right triangles we can see that, in all three cases,

$$\sin A = \frac{h}{c} \quad \text{and} \quad \sin C = \frac{h}{a}$$

If we solve for h in each, we obtain

$$h = c \sin A \quad \text{and} \quad h = a \sin C$$

Since $c \sin A$ and $a \sin C$ equal the same quantity (h) they themselves must be equal. Thus,

$$c \sin A = a \sin C$$

$$\frac{c \sin A}{ac} = \frac{a \sin C}{ac} \qquad \text{Divide both members by } ac$$

$$\frac{\sin A}{a} = \frac{\sin C}{c} \qquad \text{Remove common factors}$$

We now derive a similar relation involving angle B and side b. Let (x,y) be the coordinates of the vertex of angle C. From what we know about the trigonometric functions for any angle in standard position (chapter 5) we see, by examining all three cases in the figure, that

$$\sin B = \frac{y}{a} \quad \text{or} \quad y = a \sin B$$

Examining the three triangles, and from what we know about right triangles,

$$\sin A = \frac{y}{b} \quad \text{or} \quad y = b \sin A$$

Thus, $a \sin B = y$ and $y = b \sin A$, so

$$a \sin B = b \sin A$$

$$\frac{\sin A}{a} = \frac{\sin B}{b}$$

Putting these results together we have the law of sines.

The law of sines

In any triangle ABC,

$$\frac{\sin A}{a} = \frac{\sin B}{b} = \frac{\sin C}{c}$$

Concept

The ratio of the sine of an angle to the length of the side opposite that angle is the same for all angles in any triangle.

Note When solving problems only use two of the three ratios at a time.

In the rest of this chapter, we always assume side a is opposite angle A, side b is opposite angle B, and side c is opposite angle C. Example 8–1 A illustrates the law of sines.

■ Example 8–1 A

Solve the triangle ABC if $a = 13.2$, $A = 21.3°$, $B = 61.4°$.

It is a good idea to make a table of values:

a:	13.2	A:	21.3°
b:	?	B:	61.4°
c:	?	C:	?

Find C first.

$$C = 180° - A - B = 180° - 21.3° - 61.4° = 97.3°$$

Now fill in the law of sines.

$$\frac{\sin A}{a} = \frac{\sin B}{b} = \frac{\sin C}{c}$$

$$\frac{\sin 21.3°}{13.2} = \frac{\sin 61.4°}{b} = \frac{\sin 97.3°}{c}$$

To use the law of sines we must always know one of the three ratios completely. In this case, we know the first ratio, so we use it to solve the other two. Using the first and second ratios:

$$\frac{\sin 21.3°}{13.2} = \frac{\sin 61.4°}{b}$$

$$b \sin 21.3° = 13.2 \sin 61.4°$$ Multiply each member by 13.2b (or cross multiply)

$$b = \frac{13.2 \sin 61.4°}{\sin 21.3°} \approx 31.9$$ Divide each member by sin 21.3°

Using the first and third ratios:

$$\frac{\sin 21.3°}{13.2} = \frac{\sin 97.3°}{c}$$

$$c = \frac{13.2 \sin 97.3°}{\sin 21.3°} \approx 36.0$$

Thus we have solved the triangle:

a:	13.2	A:	21.3°
b:	31.9	B:	61.4°
c:	36.0	C:	97.3°

■

The ambiguous case

If we are given only one of the two angles of a triangle it is possible to get two different solutions to the problem. The reason for this is shown in example 8–1B. When we use the law of sines to solve a triangle *for which only one angle is known* we call this the **ambiguous case.**

■ *Example 8–1 B*

Solve triangle ABC if $a = 28.5$, $b = 30.0$, $A = 65°$.

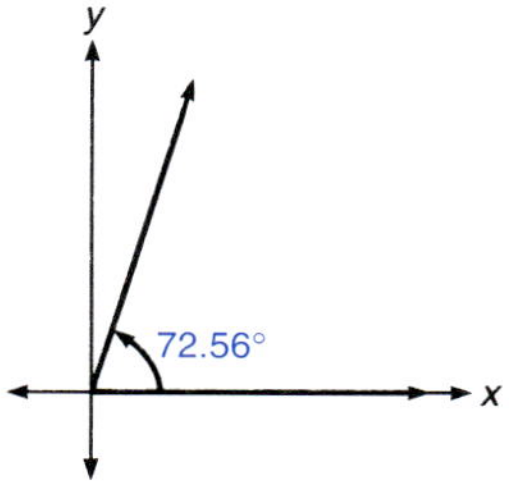

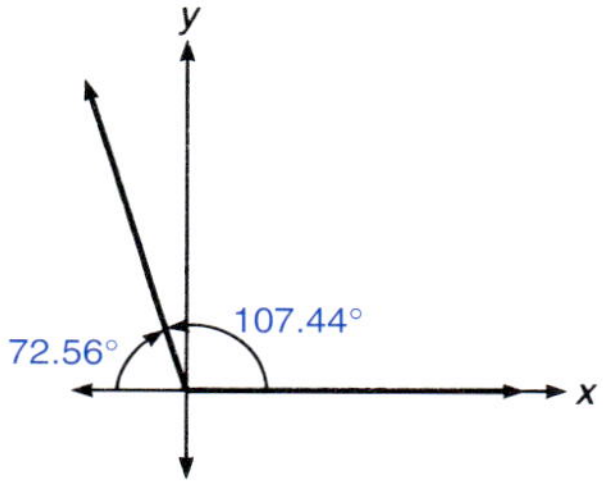

a: 28.5	A: 65°	Make a table of values
b: 30.0	B: ?	
c: ?	C: ?	

$$\frac{\sin 65°}{28.5} = \frac{\sin B}{30.0} = \frac{\sin C}{c}$$ Fill values into the law of sines

$$\frac{\sin 65°}{28.5} = \frac{\sin B}{30.0}$$ Use the first two ratios to find angle B

$$30.0 \sin 65° = 28.5 \sin B$$

$$\frac{30.0 \sin 65°}{28.5} = \sin B$$

A *reference angle* for angle B is found with the inverse sine function.

$$B' = \sin^{-1}\left(\frac{30.0 \sin 65°}{28.5}\right) \approx 72.56°$$

Since angle B is in a triangle, we know its measure is between 0° and 180°. Thus, using B' as a reference, angle B could either be 72.56° or 180° − 72.56° = 107.44°. See the figure.

At this point we must divide the problem into two cases: the case where $B \approx 72.56°$ and the one where $B \approx 107.44°$.

Case 1: $B \approx 72.56°$

a: 28.5	A: 65°
b: 30.0	B: 72.56°
c: ?	C: ?

$$C \approx 180° - 65° - 72.56° \approx 42.44°$$

We can use the value of angle C to find c.

$$\frac{\sin 65°}{28.5} \approx \frac{\sin 42.44°}{c}$$

$$c \approx \frac{28.5 \sin 42.44°}{\sin 65°} \approx 21.2$$

Thus, $C \approx 42.4°$, $c \approx 21.2$.

Case 2: $B \approx 107.44°$

a: 28.5	A: 65°
b: 30.0	B: 107.44°
c: ?	C: ?

$$C \approx 180° - 65° - 107.44° \approx 7.56°$$

$$\frac{\sin 65°}{28.5} \approx \frac{\sin 7.56°}{c}$$

$$c \approx \frac{28.5 \sin 7.56°}{\sin 65°} \approx 4.1$$

Thus, $C \approx 7.6°$, $c \approx 4.1$.

We can summarize these two solutions in two tables.

Case 1		**Case 2**	
a: 28.5	A: 65°	a: 28.5	A: 65°
b: 30.0	B: 72.6°	b: 30.0	B: 107.4°
c: 21.2	C: 42.4°	c: 4.1	C: 7.6°

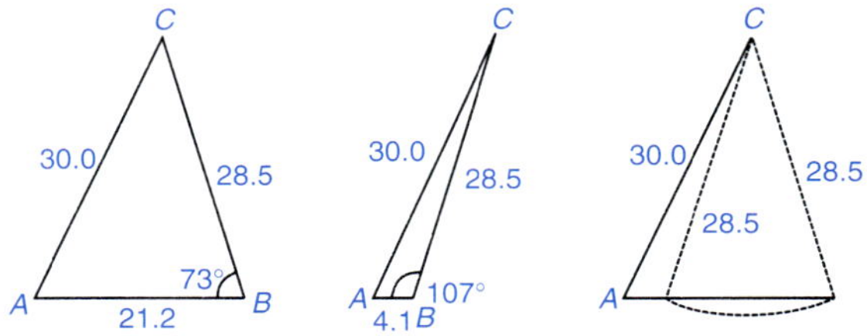

Figure 8–2

Figure 8–2 shows the two triangles and both triangles together, where we can see why the ambiguous case was possible—with the given information side a could be in one of two positions, giving two possible triangles.

As a final check on our work, we observe that the sum of the angles in each case is 180°, and that in each case the longest side is opposite the largest angle and the shortest side is opposite the smallest angle. ■

The ambiguous case does not always produce two triangles. When only one triangle is possible, the measure of the third angle will be a negative number in case 2.

Mastery points

Can you

- State and use the law of sines to solve oblique triangles?
- Recognize and solve the ambiguous case when using the law of sines?

Exercise 8–1

In the following problems, round answers to the same number of decimal places as the data, unless otherwise specified.

Solve the following oblique triangles using the law of sines.

1. $a = 12.5$, $A = 35°$, $B = 49°$
2. $b = 17.1$, $B = 100°$, $C = 10°$
3. $a = 1.25$, $B = 13.6°$, $C = 132°$
4. $c = 9.04$, $A = 51.6°$, $B = 40.0°$
5. $b = 92.5$, $A = 47°$, $B = 100°$
6. $c = 10.2$, $A = 16.7°$, $B = 89.2°$
7. $a = 0.452$, $A = 67.6°$, $C = 91.8°$
8. $b = 0.508$, $B = 13.1°$, $C = 5.2°$
9. $c = 5.00$, $A = 100°$, $B = 45°$
10. $a = 10.9$, $B = 76.9°$, $C = 100°$
11. $a = 12.5$, $b = 13.2$, $B = 49°$
12. $b = 37.1$, $c = 21.3$, $B = 100°$
13. $a = 4.25$, $c = 2.86$, $A = 132°$
14. $c = 9.04$, $a = 21.3$, $C = 10.0°$
15. $b = 92.5$, $c = 98.6$, $B = 43.7°$
16. $c = 10.2$, $a = 16.7$, $A = 89.2°$
17. $a = 4$, $b = 22$, $A = 30°$
18. $a = 0.452$, $c = 0.606$, $C = 91.8°$
19. $b = 6.35$, $c = 4.29$, $C = 42.3°$
20. $b = 0.508$, $c = 1.09$, $C = 5.2°$
21. $c = 5.00$, $b = 8.00$, $B = 45.0°$
22. $a = 10.9$, $c = 16.9$, $C = 100.0°$

23. Ground-based radar at point A determines that the angle of elevation to an aircraft 42.9 miles away is 13.2°. Radar at point B is on a straight line between a point on the ground directly below the aircraft and the radar at A and determines that the same aircraft is 13.6 miles away from point B. To the nearest 0.1 mile, find the distance from A to B (see the diagram). Assume $B > 90°$.

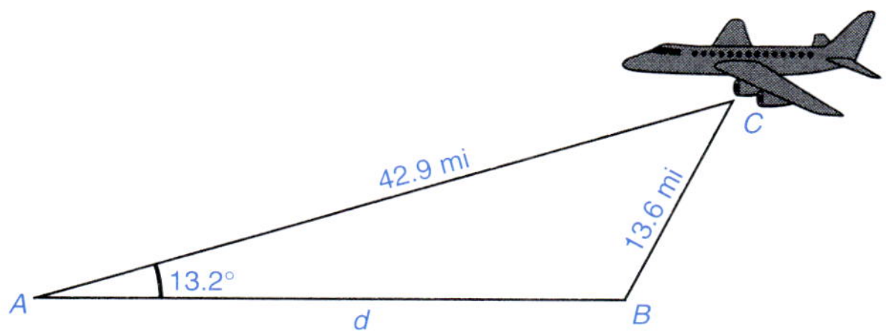

24. Two forest rangers sight a fire. Their reports are plotted on a map and yield the results shown in the diagram. If the locations of the rangers are 8.23 miles apart, how far is the fire from station A, to the nearest 0.1 mile?

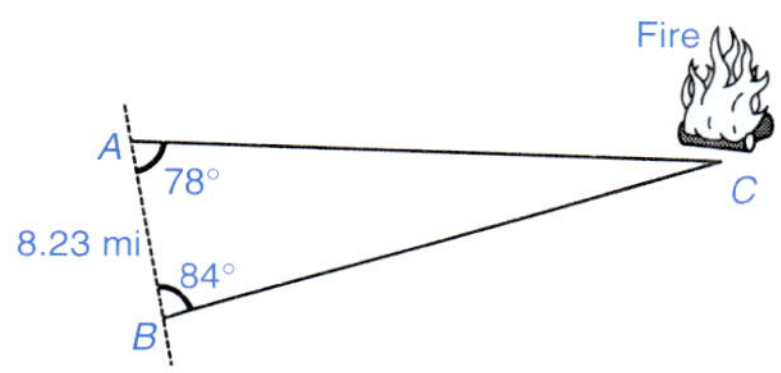

25. The diagram shows a situation in which astronomers made measurements at two locations of a slow-moving asteroid. Using their measurements, find the distance d to the asteroid, to the nearest 100 miles.

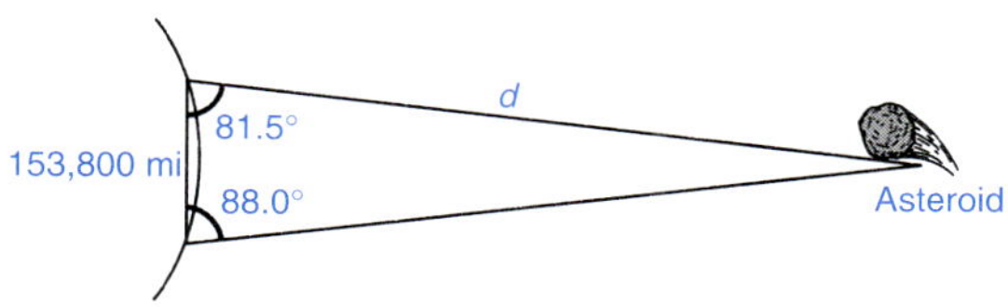

26. A surveyor made the measurements shown in the diagram. Find the distance across the lake, to the nearest foot.

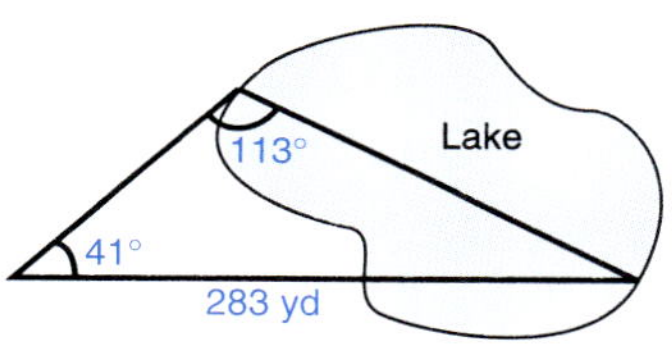

27. The diagram illustrates a situation in which a ship is moving at 12.6 knots heading due east. It is moving through a current moving east of north at 8.4 knots. The result is that the ship is moving at an angle of 58° east of north. Find the true speed S of the ship.

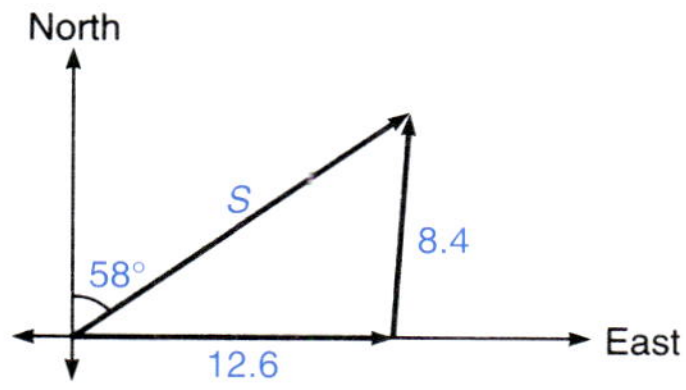

28. A ship travels due east to a point 8,000 meters from its starting point. It then turns toward the south through an angle of 65° and proceeds until it crosses a line of sight from the starting position to itself that is 35° south of east. At this point, how far is the ship from its starting point, to the nearest 10 meters?

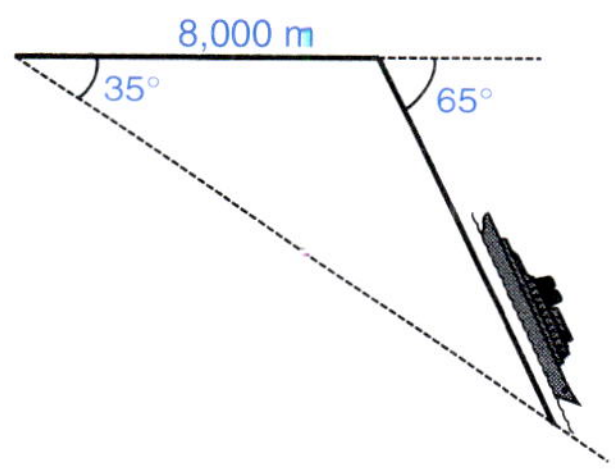

29. Two cities are 75 miles apart. An aircraft that is between the two cities is being tracked from radar in each city. City A's radar shows that the aircraft is at an angle of elevation of 40°; city B's radar shows that the slant distance of the plane to city B is 58 miles and that the angle of elevation there is less than 40°. What is the slant distance d from the plane to city A, to the nearest mile?

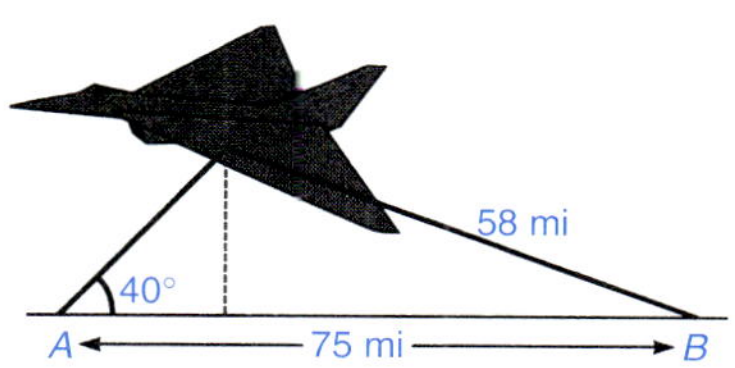

30. Find the height of the aircraft in problem 29, to the nearest 100 feet (1 mile = 5,280 feet). Use the diagram for help.

31. Show that in any oblique triangle ABC, if h is the altitude of the triangle relative to side b, then an expression for h is $h = \dfrac{b \cdot \tan A \cdot \tan C}{\tan A + \tan C}$. (*Hint:* The diagram shows a triangle in which angle A is acute (on the left) and obtuse. Note that in each case $\tan A = \dfrac{y}{x}$ and $\tan C = \dfrac{y}{b - x}$. Also, $h = y$ in each case.)

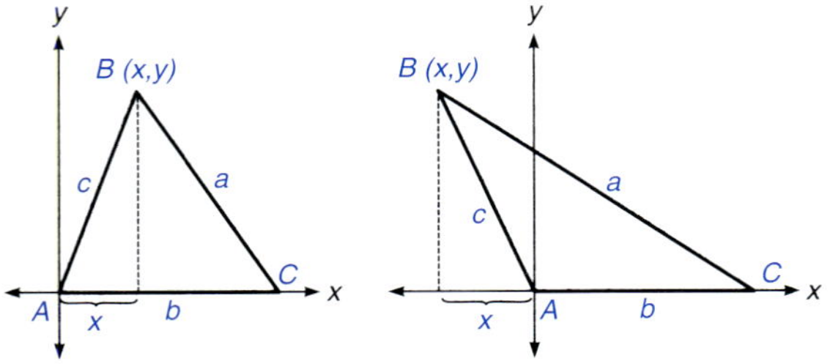

32. Quadrilateral $ABCD$ is shown in the diagram, where $AB = 17.3$, $AD = 18.9$, angle $A = 110°$, angle $ABD = 52°$, angle $BDC = 41°$, and angle $C = 93°$. Find the length of CD to the nearest 0.1. (*Hint:* Draw diagonal BD.)

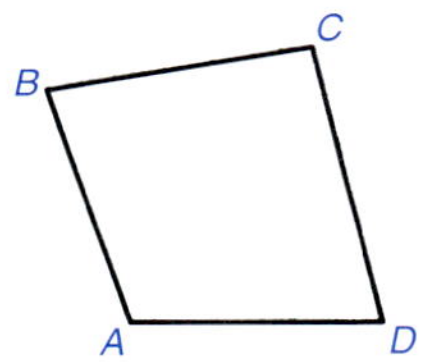

33. Show that in any triangle ABC,

[1] $a = b \cos C + c \cos B$

[2] $b = c \cos A + a \cos C$

[3] $c = a \cos B + b \cos A$

(*Hint:* The diagram shows angle A in standard position when the angle is acute, right, or obtuse. Show why the statements $\cos A = \dfrac{x}{c}$ and $\cos C = \dfrac{b - a}{x}$ are true in each case, and then use them to show that [2] is true.)

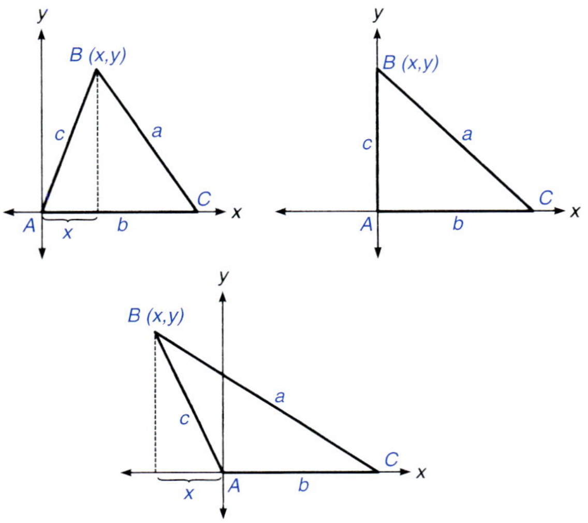

34. An army observation point is 325 yards northeast (i.e., 45° north of east) of a second point. At this point, a tank is sighted on a line of sight 37° south of east. The same tank is sighted at the second point along a sight 18° north of east. To the nearest yard, how far is the tank from the first observation point?

35. Recall that the formula for the area of a triangle is $\frac{1}{2}bh$ (one half the product of the base and height). Use the formula to show that the area of any triangle ABC is $\frac{1}{2}bc \sin A$.

36. The figure shows three wires that are attached from a common point to the side of a building. Find the lengths of b and c, to the nearest inch.

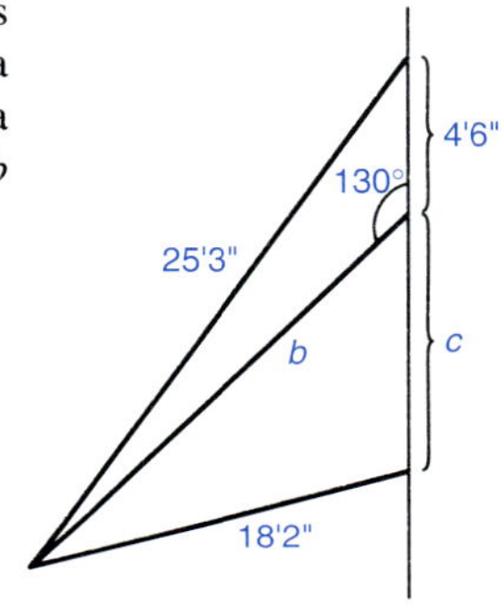

37. In the problems of section 1–2 we presented the ancient Egyptian formula for finding the area of a four-sided figure like $ABCD$ in the figure:

$$\tfrac{1}{4}(ab + ad + bc + cd)$$

and noted that it is inaccurate. Problem 35 shows that the area of any triangle ABC is $\frac{1}{2}bc \sin A$. It is also $\frac{1}{2}ac \sin B$ and $\frac{1}{2}ab \sin C$. Geometrically, this is one half the product of two sides and the sine of the angle between those two sides.

a. Use this result to show that the area of the four-sided figure can be described by $\frac{1}{4}(ab \sin A + ad \sin D + bc \sin B + cd \sin C)$.

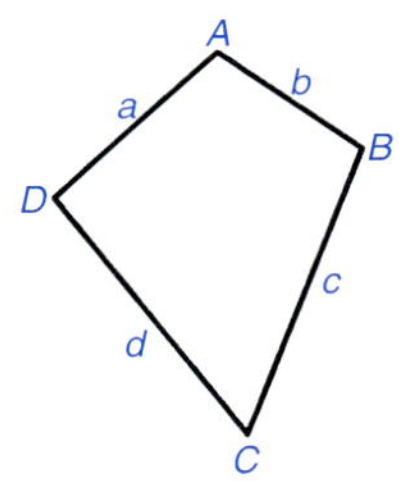

b. Use the result of part a, along with the fact that for $0 < \theta < 180°$, $0 < \sin \theta < 1$, to show that the Eygptian formula is always too large, except for rectangles, when it is exact. (Assume angles A, B, C, and D are all less than 180° in measure.)

Skill and review

1. Solve the right triangle ABC if $a = 6.5$ and $c = 9.0$. Round answers to the nearest 0.1.
2. Solve the trigonometric equation $\sec x = -\frac{2}{3}\sqrt{3}$ for all primary solutions, in radians.
3. Verify the identity $(\sin\theta - \sec\theta)(\csc\theta + \cos\theta) = \dfrac{\sin^2\theta\cos^2\theta - 1}{\sin\theta\cos\theta}$.
4. Factor $2x^5 - x^4 - 10x^3 + 5x^2 + 8x - 4$.
5. What is the length of the arc determined by a central angle of 24° on a circle with radius 18 meters?

8–2 The law of cosines

A surveyor made the measurements shown in the diagram to calculate the distance across a lake. Compute this distance.

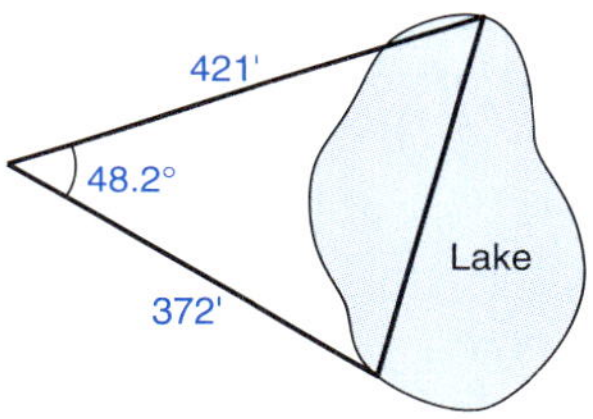

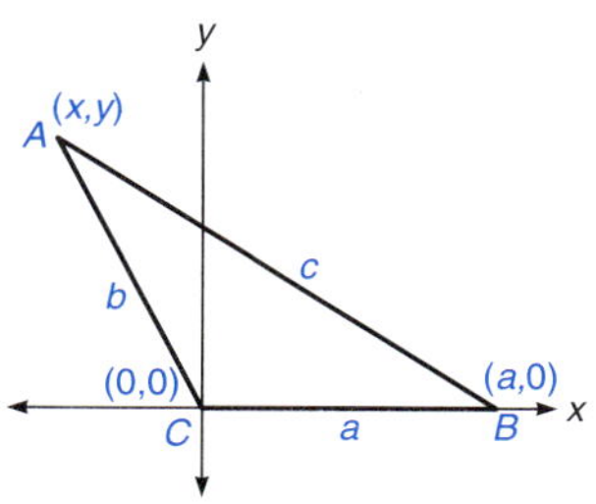

Figure 8–3

The triangle in this problem is an oblique triangle. Unfortunately it cannot be solved by the law of sines. This is because we do not know any of the three ratios completely. In these cases we use the law of cosines, which we develop in this section. We use the distance formula to develop this law.

Let triangle ABC be any triangle. Put angle C in standard position and side a on the positive portion of the x-axis. See figure 8–3. Let (x,y) be the point at vertex A. (The figure shows C as an obtuse angle, but the algebraic statements that follow would also apply if C were acute or right.) Recall the distance formula (section 3–1):

$$d = \sqrt{(x_2 - x_1)^2 + (y_2 - y_1)^2}$$

Apply this to length c. Use $(x_2,y_2) = (a,0)$, and $(x_1,y_1) = (x,y)$.

$$c = \sqrt{(a - x)^2 + (0 - y)^2}$$
$$c^2 = (a - x)^2 + y^2$$
$$c^2 = a^2 - 2ax + x^2 + y^2$$

We know that $x^2 + y^2 = b^2$ (also by the distance formula).

$$c^2 = a^2 - 2ax + b^2 \qquad \text{Replace } x^2 + y^2 \text{ by } b^2$$
$$c^2 = a^2 + b^2 - 2ax$$

We know that $\cos C = \dfrac{x}{b}$ by the definition of the cosine function (section 5–3). Thus, $x = b\cos C$.

$$c^2 = a^2 + b^2 - 2ab\cos C \qquad \text{Replace } x \text{ by } b\cos C$$

This result is called the **law of cosines** for angle C. If we put angle A or angle B in standard position, we arrive at two other versions of this law.

> **The law of cosines**
> For any triangle ABC,
> $$a^2 = b^2 + c^2 - 2bc \cos A$$
> $$b^2 = a^2 + c^2 - 2ac \cos B$$
> $$c^2 = a^2 + b^2 - 2ab \cos C$$

There is a pattern in each of the three equations above. The variable in the left member is the length of the side opposite the angle in the right member. The sides in the right member appear twice.

The law of cosines should be used whenever the law of sines cannot be used. In particular, the law of cosines is used when we know the lengths of two sides and the angle included between those sides, or when we know the lengths of all three sides. Example 8–2 A illustrates both situations.

Example 8–2 A

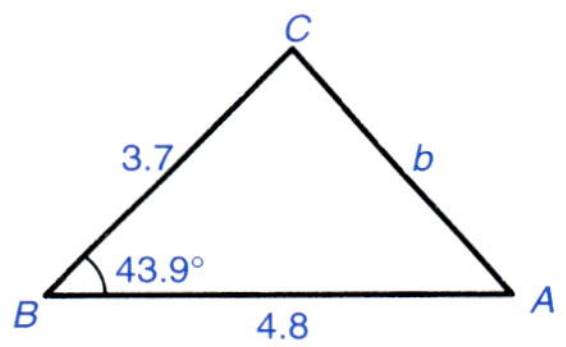

Solve the triangle.

1. $a = 3.7$, $c = 4.8$, $B = 43.9°$

Since we know angle B we use the form of the law of cosines in which angle B appears.

$$b^2 = a^2 + c^2 - 2ac \cos B$$
$$b^2 = 3.7^2 + 4.8^2 - 2(3.7)(4.8) \cos 43.9°$$
$$b = \sqrt{3.7^2 + 4.8^2 - 2(3.7)(4.8) \cos 43.9°}$$
$$b \approx 3.337$$ We will round this to 3.3 in the final answer

We can now use the law of sines to find angle A or C. It is best to find angle A first, because we know it is not the largest angle in the triangle (angle C is because side c is the longest side) and it is therefore acute. This eliminates the ambiguous case of the law of sines.

$$\frac{3.7}{\sin A} = \frac{3.337}{\sin 43.9°}, \text{ so } A \approx 50.2°$$

$$C = 180° - A - B \approx 180° - 50.2° - 43.9° \approx 85.9°$$

$a = 3.7$	$A \approx 50.2°$
$b \approx 3.3$	$B = 43.9°$
$c = 4.8$	$C \approx 85.9°$

Note It is never necessary to use the law of cosines more than once to solve a triangle. Complete the solution with the law of sines.

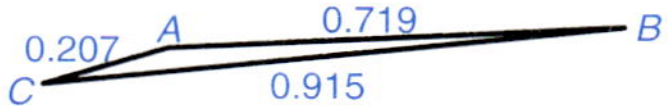

2. $a = 0.915$, $b = 0.207$, $c = 0.719$

We do not know any of the angles, so we cannot use the law of sines. (Without any angles we cannot know the value of any of the three ratios in the law of sines.) Thus, we use the law of cosines to find one of the angles.

It is best to find the largest angle first (see the following note); this will be angle A since it is opposite the longest side a. We must use the form of the law of cosines that includes angle A.

$$a^2 = b^2 + c^2 - 2bc \cos A$$

We solve for $\cos A$ before replacing the variables with known values.

$$2bc \cos A = b^2 + c^2 - a^2$$

$$\cos A = \frac{b^2 - c^2 - a^2}{2bc}$$

Substituting known values, we get

$$\cos A = \frac{0.207^2 + 0.719^2 - 0.915^2}{2(0.207)(0.719)} \approx -0.9320$$

Since $\cos A < 0$ we know A is in quadrant II (obtuse). The inverse cosine function gives results in quadrants I and II, so we use the inverse cosine function directly.

$$A \approx \cos^{-1}\left(\frac{0.207^2 + 0.719^2 - 0.915^2}{2(0.207)(0.719)}\right) \approx 158.74°$$

Now that we know the measure of the largest angle in the triangle, we can use the law of sines or the law of cosines to find another angle. The law of sines is generally easier to calculate. Using the law of sines we find one of the acute angles: $B \approx 4.70°$.

$$C = 180° - 158.74° - 4.70° \approx 16.56°$$

The triangle is solved since we now know the measures of the three sides and three angles.

$a = 0.915$ $\quad A \approx 158.7°$
$b = 0.207$ $\quad B \approx 4.7°$
$c = 0.719$ $\quad C \approx 16.6°$

Note

1. There is no ambiguous case when using the law of cosines because the range of the inverse cosine function is $0° \le y \le 180°$, which includes the range of the measure of the angles in a triangle, $0° < y < 180°$.
2. When solving a triangle remember that only one angle (the largest) can be obtuse (measure more than 90°), and the largest angle is opposite the longest side.

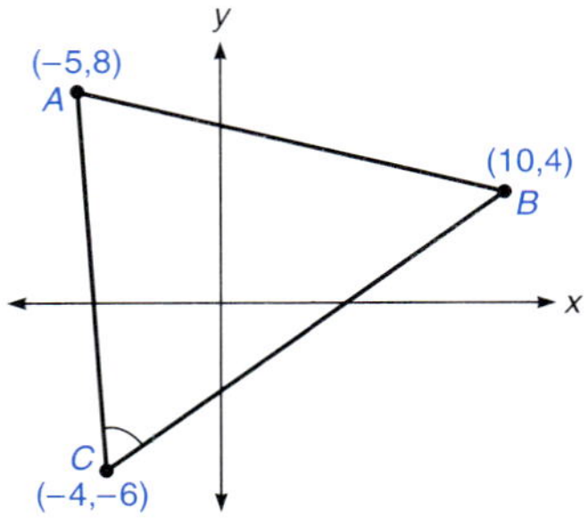

3. Find the measure of angle C in a triangle whose vertices are at the points $A(-5,8)$, $B(10,4)$, and $C(-4,-6)$. See the figure.

We use the distance formula to find the lengths AB, BC, AC.

$$AB = \sqrt{(10 - (-5))^2 + (4 - 8)^2} = \sqrt{241}$$
$$BC = \sqrt{(-4 - 10)^2 + (-6 - 4)^2} = 2\sqrt{74}$$
$$AC = \sqrt{(-4 - (-5))^2 + (-6 - 8)^2} = \sqrt{197}$$

To find angle C, we use the version of the law of cosines with angle C.

$$c^2 = a^2 + b^2 - 2ab\cos C$$
$$\cos C = \frac{a^2 + b^2 - c^2}{2ab}$$
$$= \frac{(2\sqrt{74})^2 + (\sqrt{197})^2 - (\sqrt{241})^2}{2(2\sqrt{74})(\sqrt{197})}$$
$$= \frac{296 + 197 - 241}{4(\sqrt{74})(\sqrt{197})} \approx 0.5218$$
$$C \approx 58.5°$$

■

Mastery points

Can you

- Use the law of cosines to solve triangles when the law of sines cannot be used?

Exercise 8–2

Solve the following oblique triangles. You will have to use the law of cosines as the first step. Round answers to the same number of decimal places as the data.

1. $a = 3.2$, $b = 5.9$, $C = 39.4°$
2. $a = 4.9$, $b = 3.2$, $C = 78.2°$
3. $b = 61.3$, $c = 23.9$, $A = 124.0°$
4. $b = 123.0$, $c = 89.4$, $A = 19.5°$
5. $a = 31.4$, $c = 17.0$, $B = 100.3°$
6. $a = 67.25$, $c = 13.56$, $B = 76.30°$
7. $a = 23.5$, $b = 19.4$, $c = 35.0$
8. $a = 61.7$, $b = 80.0$, $c = 102.0$
9. $a = 0.214$, $b = 0.500$, $c = 0.399$
10. $a = 1.75$, $b = 0.98$, $c = 1.03$
11. $a = 13.2$, $b = 5.9$, $C = 139.4°$
12. $a = 14.9$, $b = 13.2$, $C = 45.0°$
13. $b = 61.3$, $c = 43.9$, $A = 24.5°$
14. $b = 23.9$, $c = 89.4$, $A = 79.5°$
15. $a = 30.0$, $c = 20.0$, $B = 112.0°$
16. $a = 6.72$, $c = 1.55$, $B = 76.35°$
17. $a = 235$, $b = 194$, $c = 354$
18. $a = 61.7$, $b = 80.0$, $c = 42.0$
19. $a = 0.21$, $b = 0.49$, $C = 1.50°$
20. $a = 10.0$, $b = 13.9$, $c = 17.5$

21. A surveyor made the measurements shown in the diagram to calculate the distance across a lake. Compute the distance to the nearest 0.1 foot.

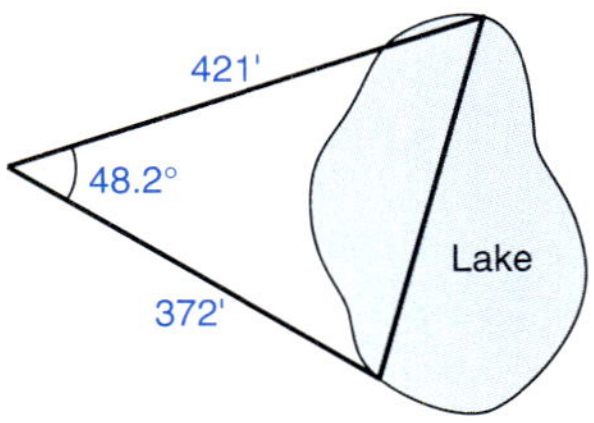

22. Calculate the measure of the smallest angle in a triangle whose sides have measure 22.1 cm, 32.6 cm, and 40.5 cm.

23. Calculate the measure of the largest angle in a triangle whose sides have measure 12.3 in., 16.2 in., and 19.0 in.

24. A numerically controlled laser cloth cutter is being set up to cut a triangular pattern. The vertices of the triangle are at $A(2,6)$, $B(5,8)$, and $C(6,2)$. (a) Find the length of side AB. (b) Find the measure of the smallest angle in the triangle.

25. In the same situation as problem 24 the vertices of another triangular piece of cloth are determined to be at $A(0,5)$, $B(2,3)$, and $C(8,4)$. Determine the measure of the largest of the three angles, A, B, or C, to the nearest 0.1°.

26. Two ships are being tracked by radar. One ship is determined to be at 17.6 miles from the radar, while the second is 22.5 miles from the radar. The lines of sight from the radar to the two ships form an angle of 47.2° (see the diagram). Find the distance between the two ships to the nearest 0.1 mile.

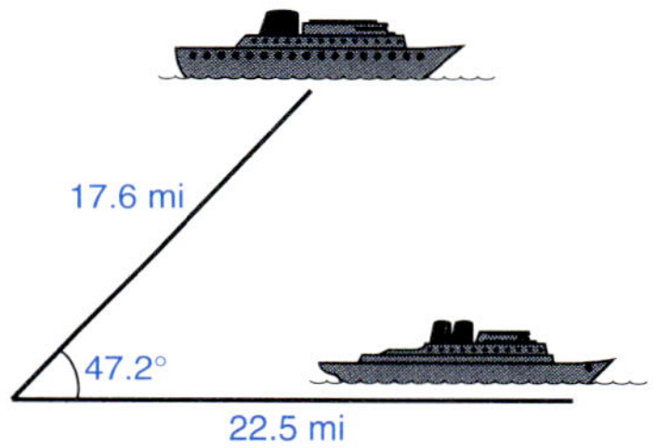

27. In the situation described in problem 26, what would be the angle formed by the two lines of sight to the ships if the ships were 31.5 miles apart?

28. A ship leaves a harbor heading due east and travels 17.3 kilometers (km). It then turns north through a 33° angle and travels for another 22.0 km. How far is the ship from its starting point, to the nearest kilometer?

29. A plane takes off and travels southeast (45° south of east) for 27 miles, then turns due south and travels for 16 miles. How far is it from its starting position, to the nearest mile?

30. The points (5,3), (−2,1), and (1,−4) form a triangle. Find the measure of the smallest angle in this triangle, to the nearest 0.1°.

31. Find the measure of the largest angle in the triangle of problem 30, to the nearest 0.1°.

32. Two observers are 87 meters apart, and a rangefinder shows that a certain building is 111 meters from one observer and 114 meters from the other. What is the angle formed by the two lines of sight from the building to the observers, to the nearest degree?

33. The triangle in the diagram is a right triangle. Can the law of cosines be used to find the length of c? Compare using the law of cosines to using the Pythagorean theorem.

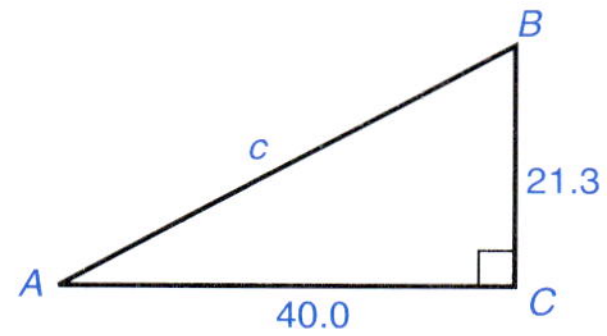

34. The diagram shows three triangles in which angle C is (a) acute, (b) right, and (c) obtuse. If we use the Pythagorean theorem and apply it to the right triangle in (b) and apply the law of cosines to angle C in all three situations, the resulting equations are the two equations

$$c^2 = a^2 + b^2$$
$$c^2 = a^2 + b^2 - 2ab\cos C$$

We can view the term $-2ab \cos C$ as a "correction factor" to the Pythagorean theorem that will give the correct result even when angle C is not 90°.

Solve for side c in cases (a), (b), and (c), using the law of cosines. Discuss how the correction factor increases the length of side c, compared to case (b), in case (c), and how it decreases it compared to case (b), as in case (a).

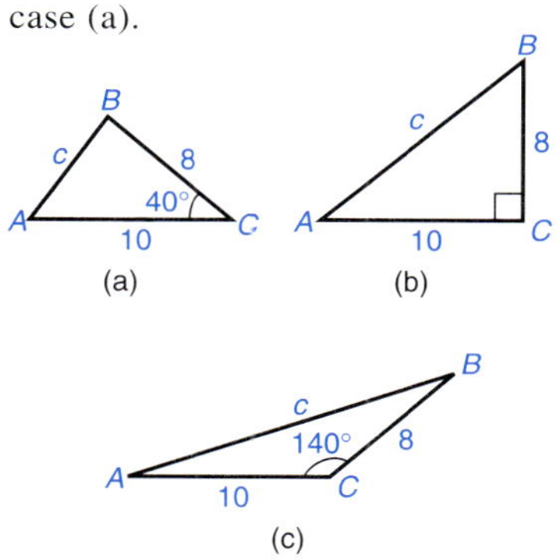

Skill and review

1. Solve the triangle ABC if $a = 6.8$, $b = 12.0$, and $B = 43°$.
2. Solve the right triangle ABC if $a = 6.8$ and $B = 43°$.
3. Graph the function $f(x) = 2x^4 + 5x^3 - 8x^2 - 17x - 6$.
4. Use the graph of $y = x^2$ to graph the function $f(x) = -(x - 3)^2 + 1$.
5. If $\sin\theta = -\frac{2}{3}$ and $\cos\theta < 0$, find the value of $\tan\theta$.

8–3 Vectors

> A sled loaded with lumber will not move until the horizontal component of the applied force is 1,200 pounds or more. If a winch being used to move the sled can apply a maximum force of 1,700 pounds, what is the largest angle of elevation at which the winch can act on the sled and move it?

This problem is one application of vectors, which we study in this section.

If we know that a plane flying over a certain spot is flying at 100 mph, we cannot tell where it will be in 1 hour without also knowing its direction. This combination of speed and direction is called *velocity*. Many other natural phenomena are described by a magnitude and direction: forces of all types, accelerations, alternating voltage in electricity theory are all examples. A conceptual tool used to describe two such pieces of information is the **vector.** It is noteworthy that the concept of vectors extends into an area of mathematics called *linear algebra,* which has applications in every field of knowledge, from physics and economics to medicine and sociology.

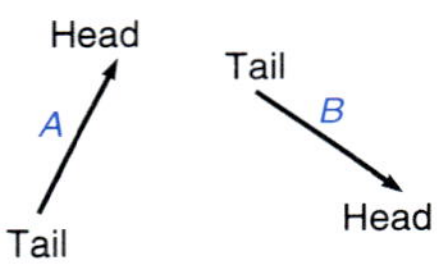

Figure 8–4

We imagine a vector as a directed line segment. That is, a finite portion of a straight line that is considered to be pointing in one direction. Our representation of vectors would commonly be called "arrows." Examples of vectors are shown in figure 8–4. Observe that we use the terms **head** and **tail** to describe the "end" and "beginning" of a vector, respectively, and that we often use capital letters, such as A and B, to name a vector.

One way to describe a vector is by specifying its length and direction. The length is called the **magnitude** of the vector, and for a vector A we denote its magnitude by $|A|$ (the same notation as that for absolute value of a real number). The **direction** of a vector is specified by an angle, θ, usually specified in degrees, measured in the same way as angles in standard position. When we specify a vector this way we say it is in **polar form.**

> **Polar form of a vector**
> A vector A in polar form is the ordered pair $A = (|A|,\theta_A)$.
> $|A|$ is the magnitude of vector A; $|A| \geq 0$.
> θ_A is the direction of vector A.

Note The polar form of a vector is not unique; all coterminal values of θ_A are equivalent.

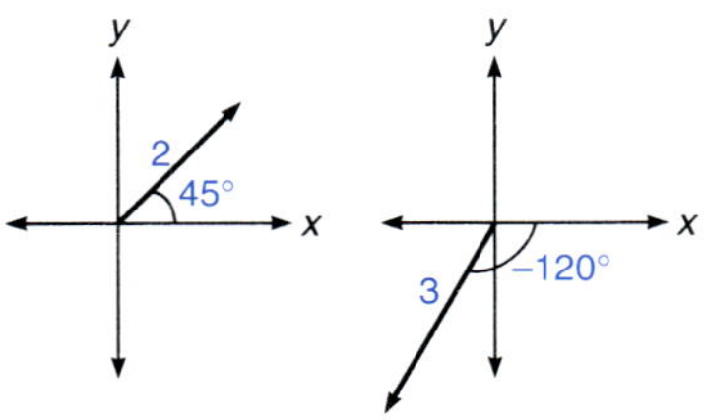

Figure 8–5

Figure 8–5 illustrates the vectors $(2,45°)$ and $(3,-120°)$.

A vector can also be specified by using its *relative rectangular coordinates.* For a vector in standard position, this is equivalent to specifying the coordinates of its head (see figure 8–7). If a vector is not in standard position, its relative rectangular coordinates are relative to the coordinates of the vector's tail. We refer to these coordinates as describing the **rectangular form** of a vector.

Rectangular form of a vector
A vector A in rectangular form is $A = (A_x, A_y)$, where A_x is called the **horizontal component** of the vector and A_y is called the **vertical component** of the vector.

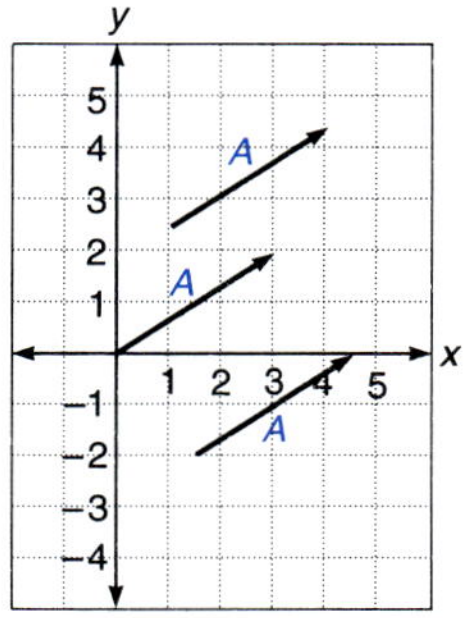

Figure 8–6

Figure 8–6 illustrates the vector $A(3,2)$, shown in three different positions.

Converting polar form to rectangular form

There is a useful relation for converting the polar form of a vector to its rectangular form. If we recall the definitions of section 5–2 for an angle in standard position, we can see the following.

To convert polar form to rectangular form
Given vector $A = (|A|,\theta_A) = (A_x, A_y)$,

$$A_x = |A| \cos \theta_A \quad \text{and} \quad A_y = |A| \sin \theta_A$$

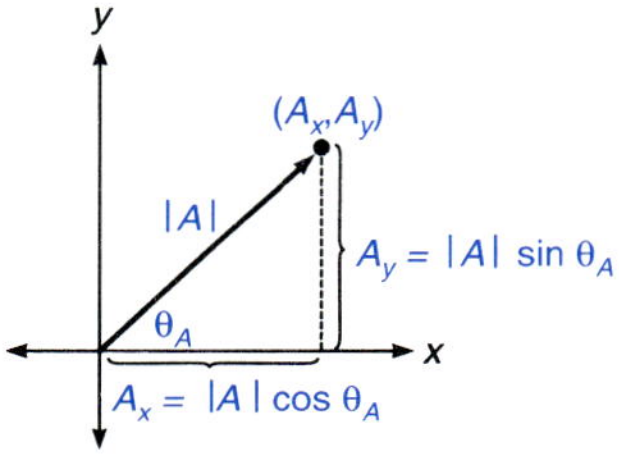

Figure 8–7

This is true because, by the definitions of section 5–2, $\cos \theta_A = \dfrac{A_x}{|A|}$ and $\sin \theta_A = \dfrac{A_y}{|A|}$. Figure 8–7 illustrates this relationship, where we view A_x and A_y as directed horizontal and vertical distances, respectively.

Example 8–3 A illustrates converting a vector in polar to form to its rectangular form.

■ ***Example 8–3 A***

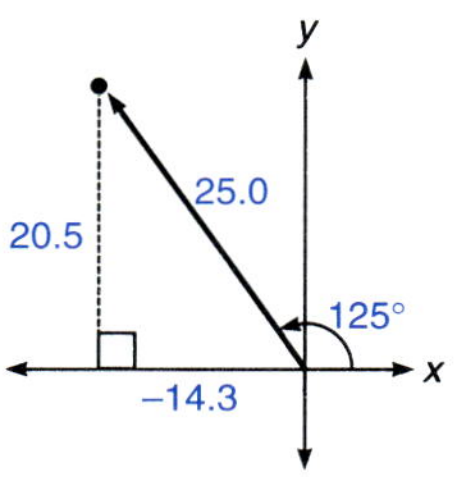

Convert the polar form of the vector to the rectangular form (approximate to the nearest tenth). $A = (25.0,125°)$.

$$A_x = |A| \cos \theta_A = 25.0 \cos 125° \approx -14.3$$
$$A_y = |A| \sin \theta_A = 25.0 \sin 125° \approx 20.5$$

Thus, the rectangular form is $(-14.3,20.5)$. ■

Converting rectangular form to polar form

When we convert from rectangular to polar form we always give the direction of the vector θ_V so that it has the smallest possible absolute value. This means *we will choose* θ_V *so that* $-180° < \theta_V \leq 180°$. One reason for doing this is that this is the result obtained from electronic calculators (see the discussion following example 8–3 B).

Examining figure 8–7 shows that, for a given vector $V = (V_x,V_y) = (|V|,\theta_V)$, $|V| = \sqrt{V_x^2 + V_y^2}$, and $\tan \theta_V' = \dfrac{V_y}{V_x}$ if $V_x \neq 0$. The angle $\theta_V' = \tan^{-1}\dfrac{V_y}{V_x}$ is only the reference angle and θ_V depends on the quadrant in which the vector occurs.

Reference angles obtained with the inverse tangent function fall in the range $-90° < \theta' < 90°$ (quadrant I and quadrant IV). If $V_x > 0$ this is the value of θ_V, since that angle should be in quadrant I or IV.

If $V_x < 0$, θ_V can be obtained by adding or subtracting 180° to or from θ'_V. If $\theta'_V < 0$, add 180°, and if $\theta'_V > 0$, subtract 180°. This can be summarized as follows.

To convert rectangular form to polar form

Given vector $V = (V_x, V_y) = (|V|, \theta_V)$. Then,

$$|V| = \sqrt{V_x^2 + V_y^2},\ \theta'_V = \tan^{-1}\frac{V_y}{V_x} \text{ if } V_x \neq 0, \text{ and}$$

$$\theta_V = \begin{cases} \theta'_V & \text{if } V_x > 0 \\ \theta'_V - 180° \text{ if } \theta'_V > 0 & \text{if } V_x < 0 \\ \theta'_V + 180° \text{ if } \theta'_V < 0 & \end{cases}$$

Note If $V_x = 0$ then θ_V is 90° if $V_Y > 0$, and $-90°$ if $V_Y < 0$. This is clear from a sketch of the vector.

The examples and the exercises will make clear why the rule works when $V_x < 0$. Example 8–3 B illustrates finding the polar form of a vector when its rectangular form is known.

■ ***Example 8–3 B***

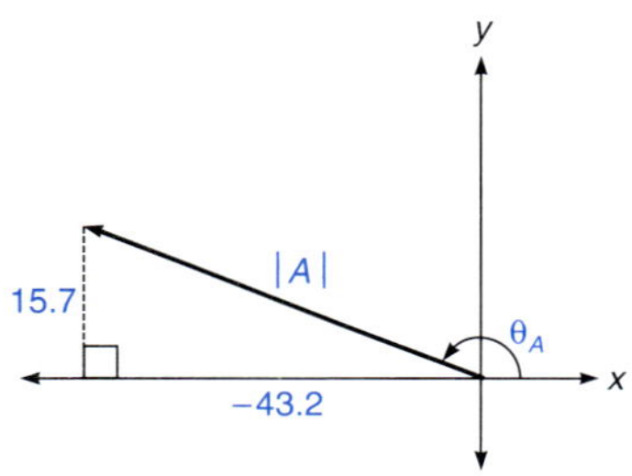

Vector $A = (-43.2, 15.7)$. Find the polar form of A.

$$|A| = \sqrt{A_x^2 + A_y^2}$$
$$= \sqrt{(-43.2)^2 + (-15.7)^2} \approx 46.0 \quad \text{Nearest tenth}$$
$$\theta'_A = \tan^{-1}\left(\frac{15.7}{-43.2}\right) \approx -20.0°$$
$$\theta_A \approx -20° + 180° \approx 160° \quad A_x < 0, \theta'_V < 0$$

Thus, $A \approx (46.0, 160°)$. ■

Using special calculator keys

Most engineering/scientific calculators are programmed to perform the conversions of examples 8–3 A and 8–3 B. These calculators have keys marked ''R→P'' or simply ''→P'' (rectangular to polar conversion) and ''P→R'' or ''→R'' (polar to rectangular conversion), or something equivalent. The results are stored in locations referred to as x and y. Typical keystrokes are illustrated here. Example 8–3 A would be done as follows. (The TI-81 is discussed below.)

$A = (25.0, 125°)$

25 [P → R] 125 [=] Display: [−14.33941091]

[x↔y] Display: [20.47880111]

Thus, $A \approx (-14.3, 20.5)$.

Example 8–3 B would be done in the following way.

$A = (-43.2, 15.7)$

43.2 [+/−] [R→P] 15.7 [=] Display: [45.96444278]

[x↔y] Display: [160.0275433]

Thus, $A \approx (46.0, 160°)$.

The TI-81 uses the values X, Y, R, θ. (Y, R, θ are [ALPHA] 1, [×], and 3, respectively.) It also uses the two MATH functions "R ▸ P(" (Rectangular to polar) and "P ▸ R(" (Polar to rectangular).

Example 8–3 A:

[MODE] Deg [ENTER] Make sure the calculator is in degree mode.

[MATH] 2 25 [ALPHA] [.] 125 [)] [ENTER]

Display: [−14.33941091]

[ALPHA] 1 [ENTER] Display: [20.47880111]

Example 8–3 B:

[MATH] 1 [(−)] 43.2 [ALPHA] [.] 15.7 [)] [ENTER]

Display: [45.96444278]

[ALPHA] 3 [ENTER] Display: [160.0275433]

Addition of vectors

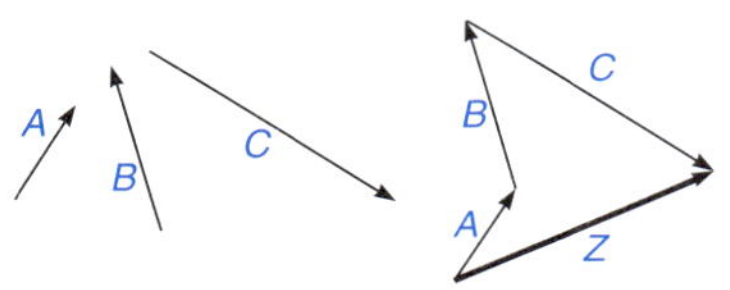

Figure 8–8

It has been shown experimentally that natural phenomena that are described by vectors combine as if they were connected tail to head in a series. The result is a vector with its tail at the tail of the first vector in the series, and its head at the head of the last vector in the series. The resulting vector is called the **resultant vector.** This is illustrated in figure 8–8, where vectors *A, B,* and *C* combine into the resultant vector *Z.* This idea is the basis for the definition of addition of vectors, which we develop here.

Example 8–3 C illustrates that this process is equivalent to summing all the horizontal components and, separately, the vertical components. This is the basis for the definition of the addition of two vectors.

> **Vector sum**
>
> Let Z be the resultant (vector sum) of two vectors $A = (A_x, A_y)$ and $B = (B_x, B_y)$. Then we say $Z = A + B$, where
>
> $$Z_x = A_x + B_x \quad \text{and} \quad Z_y = A_y + B_y$$

Observe that this definition describes how to add two vectors whose rectangular form is known. When vectors are known in polar form they must first be converted to rectangular form. This is illustrated in example 8–3 C.

■ Example 8–3 C

Find the vector sum of the given vectors.

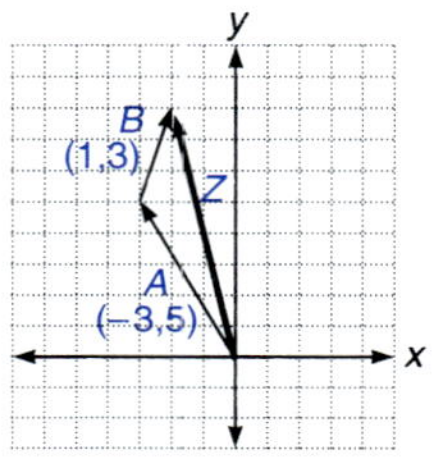

1. $A = (-3,5)$ and $B = (1,3)$

$$Z_x = A_x + B_x = -3 + 1 = -2$$
$$Z_y = A_y + B_y = 5 + 3 = 8$$
$$Z = (-2,8)$$

This is illustrated in the figure.

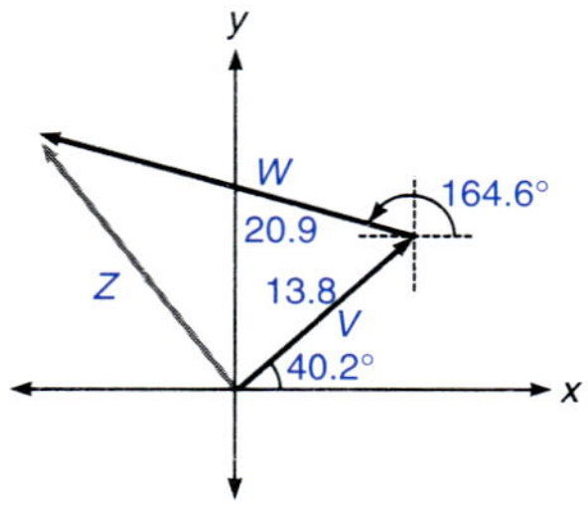

2. $V = (13.8,40.2°)$, $W = (20.9,164.6°)$

Note that these vectors are in polar form. See the figure.

$$\begin{aligned} Z_x &= V_x + W_x \\ &= |V| \cos \theta_V + |W| \cos \theta_W \\ &= 13.8 \cos 40.2° + 20.9 \cos 164.6° \approx -9.6092 \\ Z_y &= V_y + W_y \\ &= |V| \sin \theta_V + |W| \sin \theta_W \\ &= 13.8 \sin 40.2° + 20.9 \sin 164.6° \approx 14.4574 \end{aligned}$$

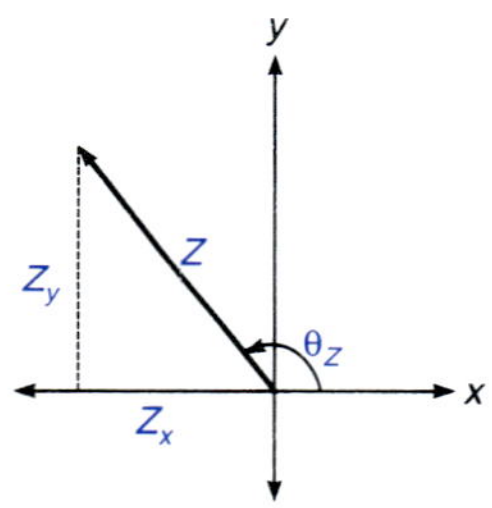

Thus, $Z \approx (-9.6,14.5)$ (to nearest tenth).

We should find Z in polar form, since V and W were given in this form.

$$Z = \sqrt{Z_x^2 + Z_y^2} = \sqrt{(-9.6093)^2 + 14.4589^2} \approx 17.4 \qquad \text{Nearest 0.1}$$

$$\theta_Z' = \tan^{-1}\frac{Z_y}{Z_x} \approx \tan^{-1}\left(\frac{14.4589}{-9.6093}\right) \approx -56.4°$$

$$\theta_Z = \theta_Z' + 180° \approx 123.6° \qquad \text{Since } Z_x < 0, \text{ and } \theta_Z' \text{ is negative, we add } 180°$$

Thus, in polar form, $Z \approx (17.4,123.6°)$. ■

Problem 76 in the exercises presents a program that adds two or more vectors given in polar form.

The next example illustrates a general principle of navigating aircraft (and, analogously, ships at sea). The speed of the aircraft relative to the air is called the *airspeed*, and the direction in which the aircraft is pointed is its *heading*. The airspeed and heading combine into the *heading vector H*. The *wind vector W* is the speed and direction of the wind. If we add these two vectors we get the *true course* and *ground speed* of the plane, $T = H + W$, where T is the speed and direction relative to the earth's surface.

■ *Example 8–3 D*

An aircraft is flying with heading 5° north of east and airspeed 123 knots. The wind is blowing at 32 knots in the direction 10° west of south. Find the true course and ground speed of the aircraft (see the figure).

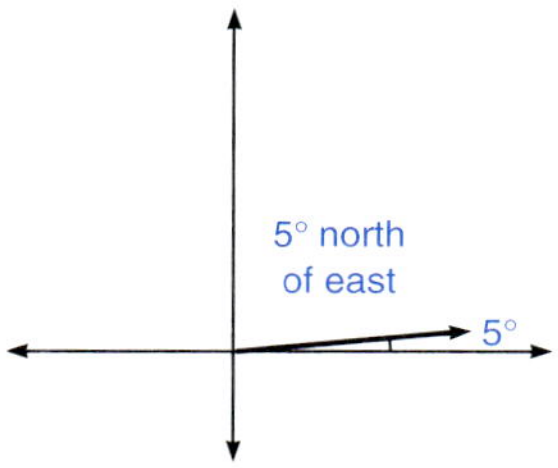

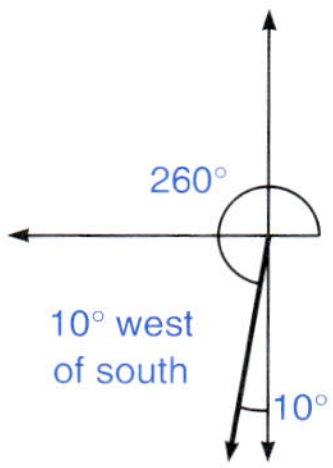

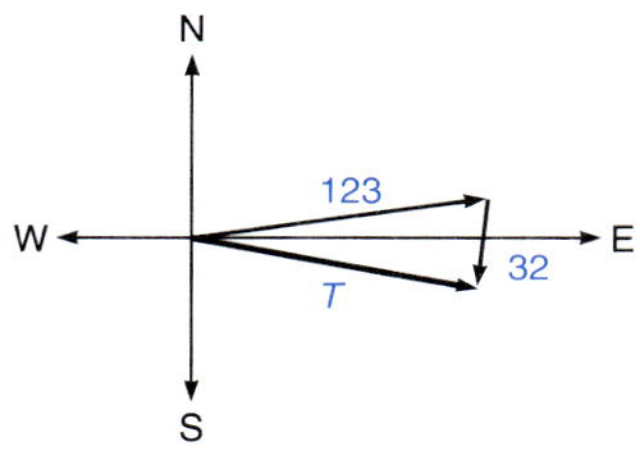

$$H = (123{,}5°);\ W = (32{,}260°)$$
$$T = H + W$$
$$T_x = H_x + W_x$$
$$= |H| \cos \theta_H + |W| \cos \theta_W$$
$$= 123 \cos 5° + 32 \cos 260° \approx 116.98$$
$$T_y = H_y + W_y$$
$$= |H| \sin \theta_H + |W| \sin \epsilon_W$$
$$= 123 \sin 5° + 32 \sin 260° \approx -20.79$$

$$\theta'_T = \tan^{-1}\frac{T_y}{T_x} = \tan^{-1}\left(\frac{-20.79}{116.98}\right) \approx -10°$$ To the nearest degree

$\theta_T = \theta'_T$ since $T_x > 0$, $|T| = \sqrt{T_x^2 + T_y^2} = \sqrt{116.98^2 + (-20.79)^2} \approx 119$

Thus, the ground speed of the aircraft is 119 knots and the true course is 10° south of east. ■

The zero vector and the opposite of a vector

It is useful to define a zero vector and the opposite of a vector. The zero vector is defined so its length is zero. The direction does not matter. The opposite of a vector is defined to have equal length but opposite direction. We do this by adding or subtracting 180°. Both definitions are in terms of polar form.

> **Zero vector**
> The vector $0 = (0,\theta)$, where θ is any angle, is the zero vector.
>
> **Opposite of a vector**
> Given a vector $V = (|V|,\theta_V)$, then $-V$ means its opposite, and $-V = (|V|, \theta_V \pm 180°)$.

In computing the opposite we generally choose whichever value, $\theta_V + 180°$ or $\theta_V - 180°$, has the smallest absolute value. It is easy to show that for any vector V,

$$V + (-V) = 0$$

Example 8–3 E illustrates uses of the zero vector and the opposite of a vector.

■ Example 8–3 E

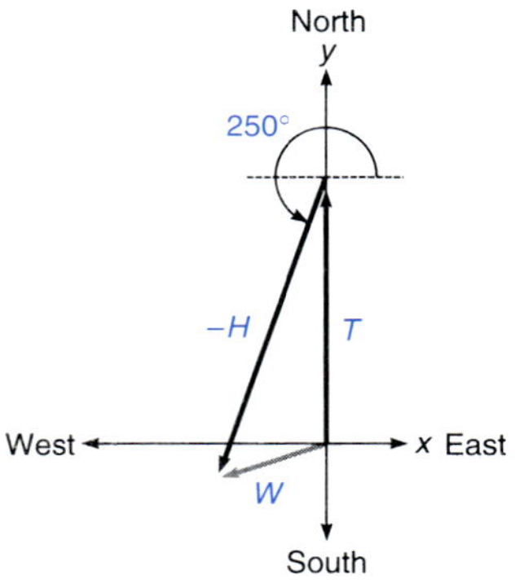

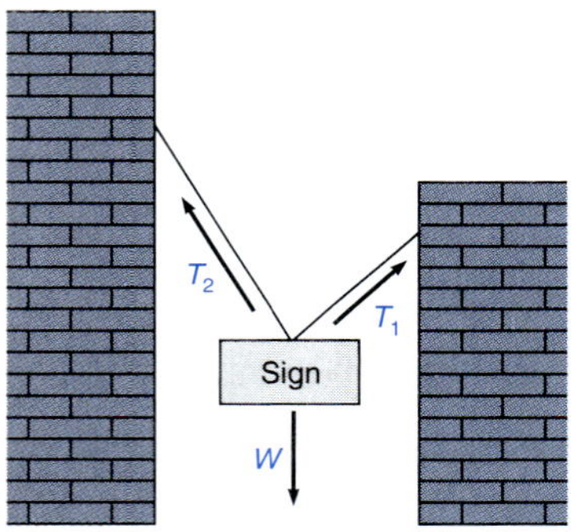

1. An aircraft's on-board inertial navigation computer shows that the aircraft is traveling due north at 200 knots with respect to the ground, and that the aircraft is headed 20° east of north with an airspeed of 225 knots. Find the wind vector *W*. Interpret this vector.

We know that true course and ground speed *T* are due north and 200 knots. Thus, $T = (200,90°)$. Heading and airspeed are 20° east of north and 225 knots. Thus, $H = (225,70°)$.

$H + W = T$ Aircraft heading + wind = true course
$W = T - H$ Solve[1] for *W*
$= T + (-H)$
$W = (200,90°) + (225,250°)$ $-H = (225,70° + 180°)$

Performing the addition shows that $W \approx (77.8,-171.6°)$. This tells us that the wind is blowing in a direction 8.4° south of west at 77.8 knots.

2. A large sign is suspended between two buildings by two wires, as in the figure. One wire acts at an angle of 45° above the horizontal and has a tension (force) T_1, of 400 pounds. If the sign weighs 800 pounds (vector *W*), compute a vector that describes T_2, the tension and direction of the second wire, to the nearest unit.

We use a fact from physics to describe the situation. Since the sign is motionless, all the forces acting on it must be balanced, or add to zero. Thus we proceed as follows.

$T_1 + T_2 + W = 0$ All forces balanced
$T_2 = -T_1 - W$ Solve for T_2
$T_1 = (400,45°)$, so $-T_1 = (400,45° + 180°) = (400,225°)$
$W = (800,270°)$, so $-W = (800,270° - 180°) = (800,90°)$

Computing $-T_1 - W$ shows that to the nearest unit $T_2 \approx (589,119°)$. ■

Mastery points

Can you

- Find the horizontal and vertical components of a vector?
- Find the magnitude and direction of a vector when given the horizontal and vertical components?
- Add or subtract vectors?
- Apply vectors to navigation and force problems?

[1] We are assuming we can solve a vector-valued equation as we solve real-valued equations. In fact, it can be proved that this is valid.

Exercise 8–3

Convert each vector from its polar to its rectangular form. Leave all answers to the nearest tenth unless the reference angle is 30°, 45°, or 60°. In that case find the exact values.

1. $(40,30°)$
2. $(15.2,33.6°)$
3. $(100.0,122.3°)$
4. $(4.2,97.3°)$
5. $(10.0,200.0°)$
6. $(18,120°)$
7. $(25,300°)$
8. $(82.0,341.9°)$
9. $(6,-45°)$
10. $(5.9,59.2°)$
11. $(7.8,-264.3°)$
12. $(20.0,-333.0°)$

Convert each vector to polar form. Round to the nearest tenth unless an exact form is possible (if the reference angle is 30°, 45°, or 60°).

13. $(3.0,4.0)$
14. $(31.2,6.9)$
15. $(-3.0,5.2)$
16. $(-12.5,31.0)$
17. $(\sqrt{3},-2)$
18. $(3,-3)$
19. $(5,-10)$
20. $(-12.5,-20.3)$
21. $(-6.8,3.4)$
22. $(-8,4)$
23. $(-\sqrt{2},-\sqrt{8})$
24. $(-1,0.4)$

Add the following vectors. Leave the resultant in rectangular form. Round the resultant to the nearest tenth.

25. $(-3,8)$, $(2,12)$
26. $(4,3\frac{1}{2})$, $(-1,-2\frac{1}{2})$
27. $(\sqrt{2},5)$, $(\sqrt{8},1)$, $(\sqrt{50},-6)$
28. $(5,-13)$, $(-\frac{1}{2},8)$, $(7\frac{1}{2},-1)$

Add the following vectors. Leave the resultant in polar form. Round the resultant to the nearest tenth.

29. $(30.0,30°)$, $(15.2,33.6°)$
30. $(100,0°)$, $(4.2,97.3°)$
31. $(10.0,200°)$, $(29.3,250°)$
32. $(13.7,300.0°)$, $(82.0,341.9°)$
33. $(3.2,-45.0°)$, $(5.9,-59.2°)$
34. $(37.9,-100.5°)$, $(69.2,-170.1°)$
35. $(41.9,-213.0°)$, $(7.7,-264.3°)$
36. $(20.0,-333.0°)$, $(39.2,-359.0°)$
37. $(3.5,19.2°)$, $(2.7,83.1°)$, $(4.3,145.7°)$
38. $(7.1,13.8°)$, $(6.2,131°)$, $(10.4,215°)$
39. $(15.3,311°)$, $(20.9,117°)$, $(13.2,83°)$
40. $(3.2,19.5°)$, $(5.1,45.0°)$, $(6.0,180°)$
41. $(3.5,-25°)$, $(6.8,25°)$, $(4.2,50°)$
42. $(25,-30°)$, $(25,-60°)$, $(25,-100°)$

43. An aircraft is moving in the direction 25° north of west at 150 knots. Find the east–west and north–south components of its velocity to the nearest knot and interpret the results.

44. An aircraft is moving in the direction 35° east of south at 120 knots. Find the east–west and north–south components of its velocity to the nearest knot and interpret the results.

45. An aircraft is moving in the direction 15° north of east at 200 knots. Find the east–west and north–south components of its velocity to the nearest knot and interpret the results.

46. A rocket is climbing with a speed of 825 knots and an angle of climb of 58.6°. (The angle of climb is the angle measured from the ground to its flight path.) Find the horizontal and vertical components of the rocket's velocity, to the nearest knot.

47. An aircraft is traveling in a direction 30° west of north. Its speed is 456 knots. Find the east–west and north–south components of its velocity, to the nearest knot.

48. At the location of a ship, the Gulf Stream ocean current is moving to the northwest at a speed of 8.2 knots. Find the east–west and north–south components of its velocity, to the nearest knot.

49. A ship has left an east-coast harbor and has been sailing in a direction 32° north of east for 2.5 hours, at a speed of 18 knots. (a) How far north of the harbor has it gone, to the nearest nautical mile? (b) How far east of the harbor has it gone, to the nearest nautical mile?

50. A force is acting on a tree stump at a 40° angle of elevation (40° measured from the horizontal up to the force vector). See the diagram. If the force is 2,500 pounds, find its vertical and horizontal components, to the nearest pound.

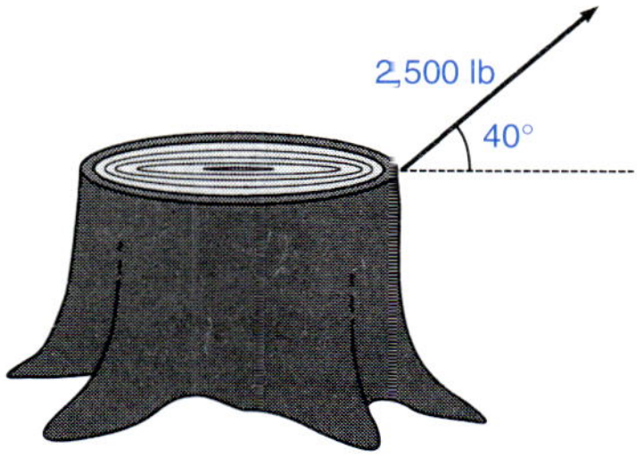

51. A 2,250-pound force is pulling on a sled loaded with lumber at an angle of elevation of 33°. If the sled will not move until the horizontal component of the force exceeds 1,900 pounds, will the sled move?

52. A sled loaded with lumber will not move until the horizontal component of the applied force is 1,200 pounds or more. If a winch being used to move the sled can apply a maximum force of 1,700 pounds, what is the largest angle of elevation at which the winch can act on the sled and move it?

53. Consider a force of 1,000 pounds acting at an angle of elevation of 15° on a point.
 a. Compute the horizontal and vertical components of the force.
 b. Double the force to 2,000 pounds and recompute the horizontal and vertical components. Do they double also?
 c. Double the angle of elevation to 30° (keep the force at 1,000 pounds). Recompute the horizontal and vertical components. Do they double also?

54. A plane is flying over Minneapolis with a ground speed of 200 miles per hour and true course due east. After one hour it turns to a true course of 60° south of east, maintaining the same ground speed. After flying for an additional half-hour the navigator notes its position on a map. How far and in what direction is the plane from Minneapolis, to the nearest unit?

55. A plane is flying over Orlando with ground speed 135 miles per hour, and true course 23° north of east. After 1 hour, it turns to a true course of 40° south of west, maintaining the same ground speed. After flying for an additional hour the navigator notes its position on a map. How far and in what direction is the plane from Orlando, to the nearest unit?

56. Two forces are acting on a point, 12.6 pounds in the direction 123° and 15.8 pounds in the direction 211°. Compute the magnitude and direction of the resultant force to the nearest tenth.

57. Two forces are acting on a point, 2.6 newtons in the direction 18.3° and 15.8 newtons in the direction −86.2°. Compute the magnitude and direction of the resultant force, to the nearest tenth.

58. Three forces are acting on a point: 27.6 newtons in the direction 18.3°, 32.1 newtons at 223.0°, and 46.8 newtons at −30.0°. Find the resultant force acting on the point, to the nearest 0.1 newton.

59. Three forces are acting on a point: 199 pounds at 19.0°, 175 pounds at 131.0°, and 96 pounds at 130.0°. Find the resultant force acting on the point, to the nearest 0.1 pound.

60. A ship leaves its harbor traveling 10° north of east. After 1 hour it turns to the direction 40° south of east. After 2 more hours, it turns to the direction 15° west of south. The ship travels for 1 half-hour more and then stops. The ship has maintained a steady speed of 16 knots (nautical miles per hour) for the entire trip. How many nautical miles, and in what direction, is the ship from its starting position, to the nearest unit?

61. A ship leaves its harbor traveling 15° west of south. After 1 hour it turns to the direction 34° south of west. After 2 more hours, it turns to the direction 10° north of west. The ship travels for 1 half-hour more and then stops. The ship has maintained a steady speed of 20 knots for the entire trip. How many nautical miles, and in what direction, is the ship from its starting position, to the nearest unit?

62. An aircraft is flying with an airspeed of 123 knots and a heading of 30° west of north. The wind is blowing in the direction 15° south of west at 26 knots. Add the heading and wind vectors to find the aircraft's true course and ground speed, to the nearest integer.

63. If the wind in problem 62 now shifts to 35 knots in the direction 10° west of south, find the aircraft's new true course and ground speed, to the nearest integer.

64. A ship is traveling through an ocean current that flows in the direction 5° west of north at 7.2 knots. The ship's heading is 10° north of west, and its speed relative to the water is 19.6 knots. Add these two vectors to find the ship's true course and speed, to the nearest tenth.

65. A ship is traveling through an ocean current that flows in the direction 15° west of north at 7.2 knots. The ship's heading is 10° north of east, and its speed relative to the water is 26.1 knots. Add these two vectors to find the ship's true course and speed, to the nearest tenth.

66. The voltage in an alternating current circuit adds vectorially. If one voltage E_1 is 122 volts with phase angle 30° and a second voltage E_2 is 86 volts with phase angle 21°, find the magnitude and phase angle of the resultant voltage E_T to the nearest unit.

67. (Refer to problem 66.) In an AC circuit E_1 is 240 volts at −45° and E_2 is 115 volts at +45°. Find the resultant E_T to the nearest unit.

68. An aircraft has a ground speed of 135 knots and true course 35° east of north. If its heading is due north and airspeed is 120 knots, find the direction and speed of the wind, to the nearest unit.

69. An aircraft has a ground speed of 80 knots and true course 15° west of north. If the wind is directly from the northeast at 12 knots, find the plane's heading and airspeed, to the nearest unit.

70. A ship is traveling at 14 knots, relative to the water, with heading 20° south of west. If the true speed and direction of the ship is 12 knots due west, find the speed and direction of the ocean current, to the nearest tenth.

71. The ocean current in a certain area is 6.4 knots with direction 8° east of south. A ship in the area is traveling with true direction of 12 knots at 25° east of north. Find the ship's heading and speed relative to the water, to the nearest tenth.

72. Two cables support a one-ton (2,000 pound) sign between two buildings. One of the cables has a tension of 1,500 pounds and acts at an angle of 33° above the horizontal. The other cable is attached to the other building. Find the tension in the other cable as well as its direction relative to the horizontal, to the nearest unit.

73. Two cables support a sign between two buildings. The tension and direction of one cable is 456 pounds at 63° above the horizontal. If the sign weighs 650 pounds, find the tension in the other cable, as well as its direction relative to the horizontal, to the nearest unit.

74. Prove that vector addition is commutative. That is, if A and B are vectors, then $A + B = B + A$. (*Hint:* Real number addition is commutative, and the horizontal and vertical components of a vector are real values.)

75. Prove that vector addition is associative. That is, if A, B, and C are vectors, then $(A + B) + C = A + (B + C)$. (*Hint:* Real number addition is associative, and the horizontal and vertical components of a vector are real values.)

76. Write a program for a computer or graphing calculator that will add two or more vectors when given in polar form.

Skill and review

1. Solve the right triangle ABC if $c = 12.0$ and $A = 23°$.
2. Solve the triangle ABC if $c = 12.0$, $a = 7.5$, and $A = 23°$.
3. Solve the triangle ABC if $c = 12.0$, $a = 7.5$, and $B = 23°$.
4. Graph the rational function $f(x) = \dfrac{5}{x^2 - 4x + 3}$.

8–4 Complex numbers in polar form

$I = \sqrt{\dfrac{P}{Z}}$ is a formula from electronics. Use it to determine I given the complex values $P = 5 + 2i$ and $Z = 1 - 4i$.

Many applications, such as the one above, which require complex numbers are best done with complex numbers in polar form. This is the subject of this section.

Basic definitions

In section 1–7 we made the following definitions.

> **Imaginary unit**
>
> $$i = \sqrt{-1}$$
>
> **Complex number (rectangular form)**
>
> A number of the form $a + bi$, a and b both real numbers.
>
> **Complex conjugate**
>
> The complex conjugate of $a + bi$ is $a - bi$.
>
> **Equality of complex numbers**
>
> $a + bi = c + di$ if and only if $a = c$ and $b = d$.

We also defined addition, subtraction, multiplication, and division for complex numbers.

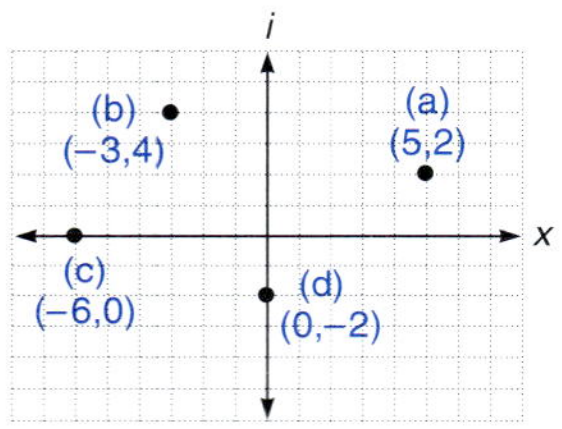

Figure 8–9

Observe that we now refer to the form $a + bi$ as the rectangular form of a complex number. Complex numbers in rectangular form can be graphed as ordered pairs in a rectangular coordinate system by letting horizontal distances represent the real part (a) and vertical distances represent the imaginary part (b) of each complex number. We use a vertical axis marked i and a horizontal axis marked x for this purpose. Figure 8–9 shows the graph of the values (a) $5 + 2i$, (b) $-3 + 4i$, (c) -6, and (d) $-2i$. A graph of complex numbers such as the figure is called an **Argand diagram.** ■

Polar form of a complex number

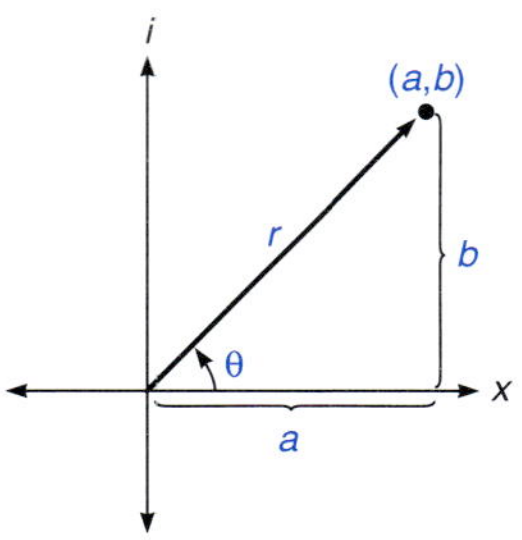

Figure 8–10

We can use the graph of a complex number as a guide to develop another way to represent a complex number. Given the complex number $a + bi$, let r represent the distance from the origin to the point (a,b), and let θ represent the angle in standard position determined by the ray containing the origin and (a,b). See figure 8–10. Using the definitions of section 5–2 we obtain $\cos\theta = \dfrac{a}{r}$ and $\sin\theta = \dfrac{b}{r}$. Thus, $a = r\cos\theta$ and $b = r\sin\theta$. This means we can rewrite $a + bi$ as $r\cos\theta + (r\sin\theta)i$, or $r(\cos\theta + i\sin\theta)$. The expression $\cos\theta + i\sin\theta$ occurs so often it is abbreviated as **cis** θ, so $a + bi = r \text{ cis } \theta$. This form is called the **polar form** of a complex number.[2] Also, note that $r = \sqrt{a^2 + b^2}$ and that $\tan\theta = \dfrac{b}{a}$.

[2]In 1893 Irving Stringham first used the notation $\text{cis } \beta = \cos\beta + i\sin\beta$.

> **Polar form of a complex number**
> If $z = a + bi$ is a complex number that determines an angle θ, then
> $$z = r \text{ cis } \theta$$
> is its polar form, where cis θ means $\cos\theta + i \sin\theta$, and
> $$r = \sqrt{a^2 + b^2}$$
> The value r is called the **modulus** of z, which is also written $|z|$.

Note The value of θ is not unique. All coterminal values produce the same rectangular form. Thus, 2 cis 10° is equivalent to 2 cis 370°.

Polar-rectangular conversions

As with vectors (section 8–3), we generally give a value of θ so that $-180° < \theta \le 180°$. A method for converting from rectangular form to polar form is a paraphrase of the method from converting vectors from rectangular to polar form.

> Given a complex number $z = a + bi = r \text{ cis } \theta$,
> $$r = \sqrt{a^2 + b^2}, \quad \tan\theta = \tan^{-1}\frac{b}{a} \text{ (if } a \ne 0\text{)}, \quad \text{and}$$
> $$\theta = \begin{cases} \theta' & \text{if } a > 0 \\ \theta' - 180° \text{ if } \theta' > 0 & \text{if } a < 0 \\ \theta' + 180° \text{ if } \theta' < 0 & \end{cases}$$

Note If $a = 0$ then θ is 90° if $b > 0$, and $-90°$ if $b < 0$. A sketch will make the choice clear.

■ *Example 8–4 A*

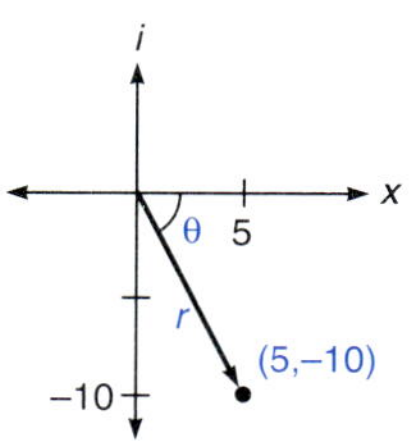

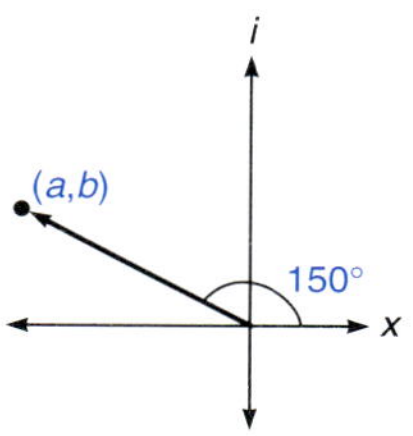

Convert between polar and rectangular form.

1. $5 - 10i$

The graph is shown in the figure.

$$r = |5 - 10i| = \sqrt{5^2 + (-10)^2} = \sqrt{125} = 5\sqrt{5}$$
$$\theta' = \tan^{-1}\frac{b}{a} = \tan^{-1}\left(\frac{-10}{5}\right) = \tan^{-1}(-2) \approx -63.4°$$
$$\theta = \theta' \approx -63.4° \text{ since } a > 0$$

Thus, the polar form of $5 - 10i$ is $5\sqrt{5}\text{ cis}(-63.4°) \approx 11.2 \text{ cis}(-63.4°)$.

2. $5 \text{ cis } 150° = 5(\cos 150° + i \sin 150°)$
$$= 5\left(-\frac{\sqrt{3}}{2} + i \cdot \frac{1}{2}\right) = -\frac{5\sqrt{3}}{2} + \frac{5}{2}i$$

Thus, $5 \text{ cis } 150° = -\frac{5\sqrt{3}}{2} + \frac{5}{2}i$ or about $-4.3 + 2.5i$. ■

Polar-rectangular conversions on a calculator

As stated in section 8–3 most calculators are programmed to perform polar/rectangular conversions. This conversion works equally well for vectors (section 8–3) and for complex numbers. Typical keystrokes are illustrated here. (The TI-81 steps are shown below.)

Example 8–4 A, part 1 would be done as follows:

$z = 5 - 10i$

5 [R → P] 10 [+/−] [=] Display: [11.18033989]

[x↔y] Display: [−63.43494882]

Thus, $z \approx 11.2 \text{ cis}(-63.4°)$.

Example 8–4 A, part 2 would be done in the following way:

$z = 5 \text{ cis } 150°$

5 [P → R] 150 [=] Display: [−4.330127019]

[x↔y] Display: [2.5]

Thus, $z \approx -4.3 + 2.5i$.

TI-81 (degree mode)

Example 8–4 A, part 1:

[MATH] 1 5 [ALPHA] [.] [(−)] 10 [)] [ENTER] Display: [11.18033989]

[ALPHA] 3 [ENTER] Display: [−63.43494882]

Example 8–4 A, part 2:

[MATH] 2 5 [ALPHA] [.] 150 [)] [ENTER] Display: [−4.330127019]

[ALPHA] 1 [ENTER] Display: [2.5]

Multiplication and division of complex numbers in polar form

Multiplication and division of complex numbers in rectangular form is quite complicated. The following theorems show that these procedures are quite simple when the complex numbers are in polar form.

Complex multiplication—polar form

$$(r_1\text{cis } \theta_1)(r_2\text{cis } \theta_2) = r_1r_2\text{cis}(\theta_1 + \theta_2).$$

Complex division—polar form

$$\frac{r_1\text{cis } \theta_1}{r_2\text{cis } \theta_2} = \frac{r_1}{r_2}\text{ cis}(\theta_1 - \theta_2),\ r_2 \neq 0.$$

Concept

To multiply, multiply the moduli and add the angles. To divide, divide the moduli and subtract the angles.

We can see that the first theorem is true by converting the complex numbers into rectangular form and performing the multiplication as defined in section 1–7.

$(r_1 \text{ cis } \theta_1)(r_2 \text{ cis } \theta_2)$
$= (r_1\cos\theta_1 + ir_1\sin\theta_1)(r_2\cos\theta_2 + ir_2\sin\theta_2)$
$= r_1r_2\cos\theta_1\cos\theta_2 + ir_1r_2\cos\theta_1\sin\theta_2 + ir_1r_2\sin\theta_1\cos\theta_2 + i^2r_1r_2\sin\theta_1\sin\theta_2$
$= r_1r_2\cos\theta_1\cos\theta_2 - r_1r_2\sin\theta_1\sin\theta_2 + ir_1r_2\cos\theta_1\sin\theta_2 + ir_1r_2\sin\theta_1\cos\theta_2$

Recall that $i^2 = -1$

$= r_1r_2[(\cos\theta_1\cos\theta_2 - \sin\theta_1\sin\theta_2) + i(\sin\theta_1\cos\theta_2 + \cos\theta_1\sin\theta_2)]$

Now use the identities for $\cos(\alpha + \beta)$ and $\sin(\alpha + \beta)$ from section 7–2

$= r_1r_2[\cos(\theta_1 + \theta_2) + i\sin(\theta_1 + \theta_2)]$
$= r_1r_2\text{cis}(\theta_1 + \theta_2)$

The proof of the process for division in polar form is left as an exercise.

In electronics, Ohm's law states $V = IZ$, where V means voltage, I means current, and Z means impedance. The units are volts, amperes, and ohms, respectively. Often complex numbers are used to describe the values of volts, amperes, and ohms. The fact that the angles add in forming the product models the physical situation in an electronics circuit. This use is illustrated in example 8–4 B.

■ ***Example 8–4 B***

Multiply or divide the complex numbers.

1. In a certain electronic circuit current $I = 4 \text{ cis } 300°$ amperes and impedance $Z = 2 \text{ cis } 150°$. Use Ohm's law, $V = IZ$, to compute voltage V.

$$V = IZ = (4 \text{ cis } 300°)(2 \text{ cis } 150°) = 8 \text{ cis } 450° = 8 \text{ cis } 90° \text{ (volts)}$$

90° coterminal to 450°

2. $\dfrac{15 \text{ cis } 30°}{18 \text{ cis } 80°} = \dfrac{15}{18} \text{ cis } (30° - 80°) = \dfrac{5}{6} \text{ cis } (-50°)$ ■

De Moivre's theorem

Consider the following computations for successive powers of a complex number $r \text{ cis } \theta$.

$$(r \text{ cis } \theta)^1 = r \text{ cis } \theta$$
$$(r \text{ cis } \theta)^2 = (r \text{ cis } \theta)(r \text{ cis } \theta) = r^2\text{cis } 2\theta$$
$$(r \text{ cis } \theta)^3 = (r \text{ cis } \theta)(r \text{ cis } \theta)^2 = (r \text{ cis } \theta)(r^2 \text{ cis } 2\theta) = r^3\text{cis } 3\theta$$
$$(r \text{ cis } \theta)^4 = (r \text{ cis } \theta)(r \text{ cis } \theta)^3 = (r \text{ cis } \theta)(r^3 \text{ cis } 3\theta) = r^4\text{cis } 4\theta$$

It is logical to assume that this pattern continues. This is true, and the result is called **De Moivre's theorem.** It can be proved for positive integers using the method of finite induction in section 12–4. It actually turns out that the exponent can be any real number.

De Moivre's theorem

$(r \text{ cis } \theta)^n = r^n\text{cis } n\theta$ for any real number n.

This theorem is illustrated in example 8–4 C.

■ ***Example 8–4 C***

Use De Moivre's theorem to compute $(5 \text{ cis } 137°)^3$; leave the answer in polar form.

$$= 5^3 \text{ cis } (3 \cdot 137°)$$
$$= 125 \text{ cis } 411°$$
$$= 125 \text{ cis } 51°$$

■

De Moivre's theorem for roots

In the complex number system every number except 0 has n nth roots; that is, two square roots, three cube roots, four fourth roots, etc. These can be expressed by De Moivre's theorem by replacing n by $\frac{1}{n}$; recall that $x^{\frac{1}{2}} = \sqrt{x}$, $x^{\frac{1}{3}} = \sqrt[3]{x}$, $x^{\frac{1}{4}} = \sqrt[4]{x}$, etc.

De Moivre's theorem for roots

The n nth roots of r cis θ are of the form

$$r^{\frac{1}{n}} \text{cis}\left(\frac{\theta}{n} + \frac{k \cdot 360°}{n}\right),\ 0 \le k < n$$

where k and n are positive integers.

We can show that any number of the form above is an nth root of r cis θ by raising it to the nth power.

$$\left[r^{\frac{1}{n}} \text{cis}\left(\frac{\theta}{n} + \frac{k \cdot 360°}{n}\right)\right] = \left(r^{\frac{1}{n}}\right)^n \text{cis}\left[n\left(\frac{\theta}{n} + \frac{k \cdot 360°}{n}\right)\right]$$
$$= r \text{ cis}(\theta + k \cdot 360°)$$
$$= r \text{ cis } \theta \qquad \theta + k \cdot 360° \text{ is coterminal to } \theta$$

If $k \ge n$, we get a repetition of a previous root. The proof of this is left as an exercise. The proof of the fact that the roots are all distinct and that there are no other roots is beyond the scope of this text.

■ ***Example 8–4 D***

Find the roots.

1. Find the three cube roots of 1.

$$1 = 1 \text{ cis } 0°$$

Evaluate $1^{\frac{1}{3}} \text{cis}\left(\frac{0°}{3} + \frac{k \cdot 360°}{3}\right) = \text{cis}(k \cdot 120°)$ for $k = 0, 1, 2$.

$k = 0$: $\text{cis } 0° = \cos 0° + i \sin 0° = 1 + 0i = 1$

$k = 1$: $\text{cis } 120° = \cos 120° + i \sin 120° = -\frac{1}{2} + \frac{\sqrt{3}}{2}i$

$k = 2$: $\text{cis } 240° = \cos 240° + i \sin 240° = -\frac{1}{2} - \frac{\sqrt{3}}{2}i$

Thus the three cube roots of 1 are 1, $-\frac{1}{2} + \frac{\sqrt{3}}{2}i$, and $-\frac{1}{2} - \frac{\sqrt{3}}{2}i$.

2. Find decimal approximations (to tenths) of the four fourth roots of $10 - 2\sqrt{39}i$.

$$10 - 2\sqrt{39}i \approx 16 \text{ cis } 308.68°$$

$$16^{\frac{1}{4}}\text{cis}\left(\frac{308.68°}{4} + \frac{k \cdot 360°}{4}\right) \approx 2 \text{ cis}(77.17 + k \cdot 90°)$$

Evaluate $2 \text{ cis}(77.17 + k \cdot 90°)$ for $k = 0, 1, 2, 3$.

$k = 0$: $2 \text{ cis } 77.17° = 2(\cos 77.17° + i \sin 77.17°) = 0.4 + 2.0i$
$k = 1$: $2 \text{ cis } 167.17° = 2(\cos 167.17° + i \sin 167.17°) = -2.0 + 0.4i$
$k = 2$: $2 \text{ cis } 257.17° = 2(\cos 257.17° + i \sin 257.17°) = -0.4 - 2.0i$
$k = 3$: $2 \text{ cis } 347.17° = 2(\cos 347.17° + i \sin 347.17°) = 2.0 - 0.4i$

Thus, the four fourth roots of $10 - 2\sqrt{39}i$ are approximately $0.4 + 2.0i$, $-2.0 + 0.4i$, $-0.4 - 2.0i$, and $2.0 - 0.4i$. ■

Mastery points

Can you

- Multiply and divide complex numbers in polar form?
- State and use De Moivre's theorem for integral powers?
- State and use De Moivre's theorem for roots?

Exercise 8–4

Write the polar form of each complex number. Round the results to the nearest tenth.

1. $5 - 2i$ **2.** $\sqrt{2} + 3i$ **3.** $-1 + 3i$ **4.** $\sqrt{3} - 2i$
5. $-3 + 4i$ **6.** $13 - 9i$

Write the polar form of each complex number. Leave the result in exact form.

7. $\sqrt{3} + i$ **8.** $1 - \sqrt{3}i$ **9.** $3 + 3i$ **10.** $-1 + \sqrt{3}i$
11. $-1 - i$ **12.** $\sqrt{5} - \sqrt{5}i$ **13.** $5i$ **14.** $-3i$

Write the rectangular form of the following numbers. Round the results to the nearest tenth.

15. $3 \text{ cis } 15°$ **16.** $5 \text{ cis } 20°$ **17.** $4.5 \text{ cis } 35°$ **18.** $10 \text{ cis } 40°$
19. $\sqrt{2} \text{ cis } 315°$ **20.** $200 \text{ cis } 8°$ **21.** $13.6 \text{ cis}(-25°)$ **22.** $12 \text{ cis}(-6°)$

Write the rectangular form of the following numbers. Leave the result in exact form.

23. $\sqrt{3} \text{ cis } 30°$ **24.** $4 \text{ cis } 210°$ **25.** $10 \text{ cis } 300°$ **26.** $6 \text{ cis } 135°$
27. $\sqrt{10} \text{ cis } 180°$ **28.** $2 \text{ cis } 90°$ **29.** $\sqrt{8} \text{ cis } 315°$ **30.** $5 \text{ cis } 240°$

Multiply or divide the following complex numbers. Leave the result in polar form.

31. $(5 \text{ cis } 30°)(3 \text{ cis } 45°)$ **32.** $(2 \text{ cis } 18°)(4.5 \text{ cis } 100°)$ **33.** $(5.4 \text{ cis } 300°)(2 \text{ cis } 300°)$

34. $(0.5 \text{ cis } 230°)(80 \text{ cis } 200°)$ **35.** $\dfrac{20 \text{ cis } 100°}{5 \text{ cis } 20°}$ **36.** $\dfrac{100 \text{ cis } 45°}{200 \text{ cis } 15°}$

37. $\dfrac{40 \text{ cis } 80°}{18 \text{ cis } 160°}$ **38.** $\dfrac{90 \text{ cis } 300°}{50 \text{ cis } 100°}$

Use De Moivre's theorem to compute the power indicated. Leave the answer in the form in which the problem is stated (polar or rectangular).

39. $(8 \text{ cis } 100°)^3$

40. $(5 \text{ cis } 10°)^4$

41. $(3 \text{ cis } 200°)^3$

42. $(2 \text{ cis } 300°)^5$

43. $(0.5 - 1.2i)^8$ (round to nearest tenth)

44. $(0.8 + 0.6i)^{10}$ (round to nearest tenth)

Solve the following problems.

45. Find the 3 cube roots of 8 in exact form.

46. Find the 4 fourth roots of -1 in exact form.

47. Find the 4 fourth roots of 81 in exact form.

48. Find the 6 sixth roots of -64 in exact form.

49. Find the 3 cube roots of $75 - 100i$ to the nearest tenth.

50. Find the 4 fourth roots of $\sqrt{3} + 3i$ to the nearest tenth.

51. In electronics, one version of Ohm's law says that $I = \dfrac{V}{Z}$, where I is current, V is voltage, and Z is impedance. Find I in a circuit in which V is 125 cis 25° and Z is 50 cis 45°.

52. Find I in a circuit in which $V = 200$ cis 40° and $Z = 4$ cis 50°. See problem 51.

53. Find V in a circuit where $I = 10$ cis 15° and $Z = 5$ cis 30°. See problem 51.

54. Find Z in a circuit where $I = 40$ cis 200° and $V = 10$ cis 125°. See problem 51.

55. In a parallel electronic circuit with two legs, total impedance Z_T is $\dfrac{Z_1 Z_2}{Z_1 + Z_2}$. Find Z_T in a circuit in which $Z_1 = 2 + i$ and $Z_2 = 3 - 5i$. Leave the answer in polar form.

56. Find Z_T in a parallel circuit in which $Z_1 = 12 + 3i$ and $Z_2 = 4 - 2i$. Leave the answer in polar form. See problem 55.

57. Use $I = \sqrt{\dfrac{P}{Z}}$ to determine I if $P = 5 + 2i$ and $Z = 1 - 4i$. Leave the result in rectangular form, to the nearest hundredth. Use the first square root ($k = 0$ in De Moivre's theorem).

58. Use $I = \sqrt{\dfrac{P}{Z}}$ to determine I if $P = -2 + 2i$ and $Z = 2 - i$. Leave the result in rectangular form, to the nearest hundredth. Use the first square root ($k = 0$ in De Moivre's theorem).

59. Is the following an identity: $a \text{ cis}(-\theta) = -a \text{ cis } \theta$?

Multiplication by i can be interpreted as a 90° rotation. If z represents the complex number given in each of the following problems, compute and graph z, iz, i^2z, i^3z.

60. $4 + 2i$

61. $-3 + i$

62. $5i$

63. 6

64. $1 - i$

65. $-1 - i$

66. Let $z_1 = -2 + 2i$, $z_2 = 1 - \sqrt{3}i$. Form the product in two ways: (a) by multiplication in rectangular form and (b) by changing each value into polar form (in exact form) and performing the multiplication in polar form. Then (c) convert the answer to (b) back to rectangular form and verify that the answers to (a) and (b) are the same.

67. A numerically controlled machine is programmed to rotate a laser beam according to mathematical rules. The laser initially points to the point $1 + i$.

a. Find the rectangular form of a complex number z such that the angle of the product $z(1 + i)$ is 30° greater than the angle of $1 + i$, without changing the modulus of $1 + i$.

b. Give the rectangular form of the point to which the laser points after this rotation. Round the answer to two decimal places.

c. Give the rectangular form of the point to which the laser points after eight such rotations, starting at the point $1 + i$. Round the answer to two decimal places.

68. In this section, we stated that $\dfrac{r_1 \text{cis } \theta_1}{r_2 \text{cis } \theta_2} = \dfrac{r_1}{r_2} \text{ cis}(\theta_1 - \theta_2)$. Prove that this is true. Use the proof in the text that $(r_1 \text{cis } \theta_1)(r_2 \text{cis } \theta_2) = r_1 r_2 \text{cis}(\theta_1 + \theta_2)$ as a guide.

69. *Warning:* This problem requires a *great* deal of algebraic manipulation. De Moivre's theorem for roots states that the n nth roots of r cis θ are of the form

$$(r \text{ cis } \theta)^{\frac{1}{n}} = r^{\frac{1}{n}} \text{cis}\left(\frac{\theta}{n} + \frac{k \cdot 360°}{n}\right),\ 0 \le k < n,$$

where k and n are positive integers. Show that if $k \ge n$, then the expression is a repetition of another root. That is, it is the same as the expression for some value of $k < n$. Do this in the following manner.

First, if $k \ge n$, then $\frac{k}{n} = a + \frac{b}{n}$, where $b < n$. $\left(\text{Think of the example } 22 \ge 5, \text{ so } \frac{22}{5} = 4 + \frac{2}{5}.\right)$ This means $k = an + b$, $b < n$.

Next, show that the following steps are true:

$$r^{\frac{1}{n}} \text{cis}\left(\frac{\theta}{n} + \frac{k \cdot 360°}{n}\right)$$

$$= r^{\frac{1}{n}} \text{cis}\left[\frac{\theta}{n} + \frac{(an + b) \cdot 360°}{n}\right] \qquad \text{Why?}$$

$$= r^{\frac{1}{n}} \text{cis}\left[\left(\frac{\theta}{n} + \frac{b \cdot 360°}{n}\right) + a \cdot 360°\right].$$

Show the algebra for this step

Finally expand this last expression into terms of sine and cosine (i.e., expand "cis α"), and then apply the identities for the cosine and sine of a sum (section 7–2).

Skill and review

1. Add the vectors (3.8,28°) and (5.1,134°). Leave in polar form, with the magnitude to the nearest 0.1 and the angle to the nearest integer.
2. If vector $A = (190,30°)$ and $C = (150,68°)$, and $A + B = C$, find vector B in polar form. Leave results to the nearest unit.
3. Solve triangle ABC if $a = 12.6$, $b = 19.1$, and $c = 28.0$. Round answers to tenths.
4. Rationalize the denominator of $\frac{1}{\sqrt[3]{4a^4b}}$.
5. If $\tan \theta = 1 + u$ and θ terminates in quadrant I, describe $\sin \theta$ in terms of u.

8–5 Polar coordinates

The pattern of strongest radiation of a certain radio antenna is described by the equation $r = 1 + 2 \sin 2\theta$. This pattern is said to have sidelobes. Convert this equation into rectangular form (rewrite it in terms of x- and y-coordinates).

Some natural phenomena, such as the motion of the planets, the contour of a cam on an automobile camshaft, the path traveled by a client on many of the rides in an amusement park, or the field strength around a radio transmitter, can be described most simply by describing the phenomenon in terms of distance from some point as a line moves in a circle. The polar coordinate system is a coordinate system that is better suited to describing these phenomena than are rectangular coordinates.[3] A polar coordinate system is a series of concentric circles and an angle reference line (see figure 8–11). The common center of the circles is called the **pole.**

[3]James Bernoulli is often credited with the creation of polar coordinates in 1691, although Isaac Newton used them earlier.

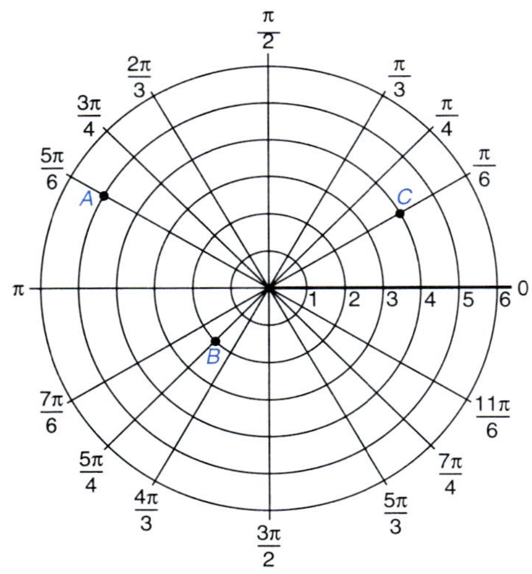

Figure 8–11

Definitions

> **Polar coordinates**
> The polar coordinates of a point are an ordered pair of the form (r,θ), where r is the *radius* and θ is an angle.

A point in polar form is located by finding the radius line corresponding to the angle θ, often stated in radians, and moving r units from the center along this line. Figure 8–11 shows the graphs[4] of the points $\left(5,\frac{5\pi}{6}\right)(A)$, $\left(2,\frac{5\pi}{4}\right)(B)$, and $\left(4,\frac{\pi}{6}\right)(C)$. In each of these cases $r > 0$.

If two points have equal radii and coterminal angles, they will have the same graph. For this reason we define such points to be equivalent. If $r < 0$ we interpret this to mean a change of direction by π radians (the opposite direction).

> **Equivalence of points**
> 1. $(r,\alpha) = (r,\beta)$ if α and β are coterminal angles or $r = 0$.
> 2. $(-r,\theta) = (r,\theta + \pi)$.

Note $(-r,\theta) = (r,\theta - \pi)$ because $\theta + \pi$ and $\theta - \pi$ are coterminal.

[4]Polar coordinate paper is widely available.

This definition means that (1,0), $(1,2\pi)$, $(1,4\pi)$, $(1,-2\pi)$, $(1,-4\pi)$, $(-1,\pi)$, $(-1,3\pi)$, $(-1,-\pi)$, etc. all describe the same point! This can occasionally lead to some confusion.

Example 8–5 A illustrates the basic definitions.

■ ***Example 8–5 A***

1. List 2 other coordinates for the point $\left(2,\frac{3\pi}{4}\right)$, with one so that $r < 0$.

$$\left(2,\frac{3\pi}{4}\right) = \left(2,\frac{3\pi}{4} + 2\pi\right) = \left(2,\frac{11\pi}{4}\right)$$ Adding 2π gives a coterminal angle

$$\left(2,\frac{3\pi}{4}\right) = \left(-2,\frac{3\pi}{4} + \pi\right) = \left(-2,\frac{7\pi}{4}\right).$$ Adding or subtracting π gives a coterminal angle in which r changes sign

2. Plot the point $\left(-5,\frac{11\pi}{6}\right)$.

$$\left(-5,\frac{11\pi}{6}\right) = \left(5,\frac{11\pi}{6} - \pi\right) = \left(5,\frac{5\pi}{6}\right),$$

which is plotted at A in figure 8–11. ■

Polar-rectangular coordinate conversions

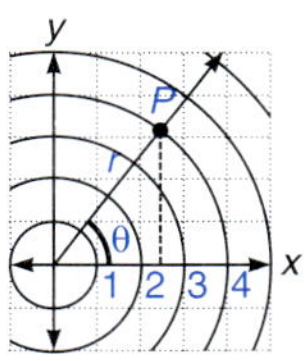

Figure 8–12

There is a way to relate polar and rectangular coordinates. Figure 8–12 shows polar and rectangular coordinates superimposed. From the definitions of section 5–3 we know that if $P = (x,y) = (r,\theta)$, $r > 0$, then $\cos\theta = \frac{x}{r}$ and $\sin\theta = \frac{y}{r}$.

Thus, *to convert from polar to rectangular coordinates* we have only to use

$$x = r\cos\theta$$
$$y = r\sin\theta$$

In fact, it will be an exercise to show that we can use the same relations when $r < 0$. Example 8–5 B illustrates a polar-to-rectangular conversion.

■ ***Example 8–5 B***

Convert the polar coordinates $\left(2,\frac{\pi}{3}\right)$ to rectangular coordinates.

$$x = r\cos\theta = 2\cos\frac{\pi}{3} = 2\cdot\frac{1}{2} = 1$$

$$y = r\sin\theta = 2\sin\frac{\pi}{3} = 2\cdot\frac{\sqrt{3}}{2} = \sqrt{3}$$

Thus, the rectangular coordinates are $(1,\sqrt{3})$. ■

To convert from rectangular to polar coordinates we use the fact that $r^2 = x^2 + y^2$ and $\tan\theta = \frac{y}{x}$ if $x \neq 0$. This also means $\theta' = \tan^{-1}\frac{y}{x}$.

As with vectors and complex numbers (sections 8–3 and 8–4) we always leave the angle θ so that it is the smallest possible absolute value. In this case, using radian measure, we always choose θ so that $-\pi < \theta \leq \pi$.

A rule for finding θ is essentially the same as that for finding θ_V for vectors and θ for complex numbers.

Given polar coordinates for point P, $P = (x,y) = (r,\theta)$,

$$r = \sqrt{x^2 + y^2}, \quad \tan\theta' = \tan^{-1}\frac{y}{x} \text{ if } x \neq 0, \quad \text{and}$$

$$\theta = \begin{cases} \theta' & \text{if } x > 0 \\ \theta' - \pi \text{ if } \theta' > 0 & \text{if } x < 0 \\ \theta' + \pi \text{ if } \theta' < 0 & \end{cases}$$

Note If $x = 0$ then θ is π if $y > 0$, and $-\pi$ if $y < 0$. This is clear from a sketch.

■ ***Example 8–5 C***

Convert the rectangular coordinates into polar coordinates.

1. $(-2\sqrt{3},2)$

$r^2 = (-2\sqrt{3})^2 + 2^2 = 16$, $r = 4$

$$\theta' = \tan^{-1}\frac{2}{-2\sqrt{3}} = \tan^{-1}\left(-\frac{\sqrt{3}}{3}\right) = -\tan^{-1}\frac{\sqrt{3}}{3}$$

$\tan^{-1}$ is an odd function, so $\tan^{-1}(-x) = -\tan^{-1} x$

so $\theta' = -\dfrac{\pi}{6}$. $x < 0$, $\theta' < 0$, so $\theta = \theta' + \pi = -\dfrac{\pi}{6} + \pi = \dfrac{5\pi}{6}$.

Therefore, the required polar coordinates are $\left(4,\dfrac{5\pi}{6}\right)$.

2. $(-2,-5)$

$$r^2 = (-2)^2 + (-5)^2 = 29, \; r = \sqrt{29} \approx 5.39$$
$$\theta' = \tan^{-1}\tfrac{5}{2} \approx 1.19 \text{ (radians)}$$

$x < 0$, $\theta' > 0$, so $\theta = \theta' - \pi = \tan^{-1}\frac{5}{2} - \pi$ (exactly), or approximately $1.19 - \pi \approx -1.95$.

Thus, the polar coordinates are $(\sqrt{29},\tan^{-1}2.5 - \pi)$ exactly, or approximately $(5.39,-1.95)$. ■

Conversions with a calculator

Engineering/scientific calculators are programmed to perform the conversions of example 8–5 C, using the same method and keys shown in section 8–3 for vectors and in section 8–4 for complex numbers. These calculators have keys marked R → P and P → R, or something equivalent. The only difference is that the calculator must be in radian mode. See the sections for keystokes for polar to rectangular conversions (example 8–5 B) and rectangular to polar conversions (example 8–5 C).

Conversion of equations between rectangular and polar form

Recall that a nonvertical line in analytic geometry is an equation of the form $y = mx + b$, and a circle with center at the origin and radius r is an equation of the form $x^2 + y^2 = r^2$. These and other equations can also be written using polar coordinates using the variables r and θ.

It is often useful to discover the polar coordinate version of a rectangular coordinate equation. We say that *a rectangular equation and a polar equation are equivalent if they describe the same set of points,* assuming the appropriate rectangular/polar conversions of the points themselves.

> **Conversion of equations from rectangular to polar form**
>
> To convert an equation in rectangular form into an equivalent equation in polar form, use the relations used to convert a point from rectangular to polar form:
>
> $$x = r\cos\theta, \quad y = r\sin\theta, \quad \text{and } r^2 = x^2 + y^2$$

When possible we customarily write a polar equation in which r is described as a function of θ. That is, to the extent possible, we put all terms with r in one member of the equation, and all other terms in the other member.

Example 8–5 D illustrates conversions from rectangular to polar form.

■ Example 8–5 D

Convert each rectangular equation into polar form.

1. The line $y = 3x - 2$.

$$y = 3x - 2$$

$$r\sin\theta = 3(r\cos\theta) - 2 \qquad x = r\cos\theta,\ y = r\sin\theta$$

$$2 = 3r\cos\theta - r\sin\theta \qquad \text{It is customary to solve a polar equation for } r \text{ if possible.}$$

$$2 = r(3\cos\theta - \sin\theta)$$

$$r = \frac{2}{3\cos\theta - \sin\theta}$$

2. The line $y = -2x$.

$$y = -2x$$
$$r \sin \theta = -2r \cos \theta \qquad x = r \cos \theta,\ y = r \sin \theta$$
$$\sin \theta = -2 \cos \theta \qquad \text{Divide both members by } r\text{; this assumes } r \neq 0 \text{ (see below)}$$
$$\frac{\sin \theta}{\cos \theta} = -2 \qquad \text{Divide both members by } \cos \theta\text{; this assumes } \cos \theta \neq 0 \text{ (see below)}$$
$$\tan \theta = -2$$

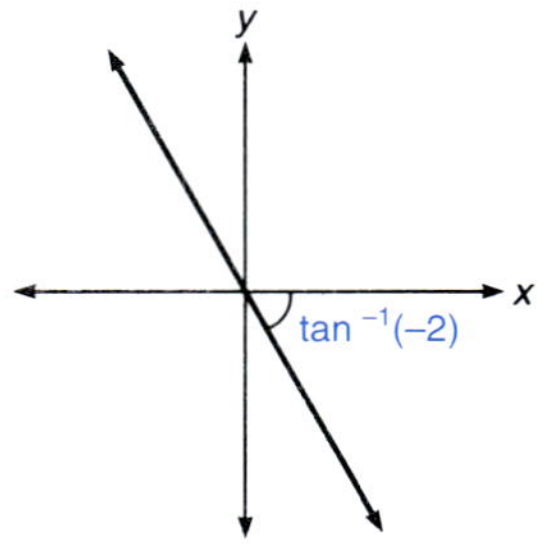

This equation is equivalent to the equation $y = -2x$. To see this, first observe that it does not mention r. This means r can be any value. The values of θ for which $\tan \theta = -2$ are in quadrants II and IV, where the tangent function takes on negative values. All points (r,θ) for which $\tan \theta = -2$ and r takes on any value are shown in the figure. This is the line $y = -2x$.

It was all right to assume $r \neq 0$ here because the resulting solution includes the pole as a solution (this is where $r = 0$).

It was also valid to assume $\cos \theta \neq 0$, for two reasons. One is that we arrive at an equation that satisfies the requirements and thus we do not need to consider the case where $\cos \theta = 0$. Second, when $\cos \theta = 0$, $\sin \theta$ is ± 1. These values do not solve the equation $\sin \theta = -2 \cos \theta$, so that $\cos \theta = 0$ cannot occur in this situation.

3. The hyperbola $x^2 - y^2 = 2$. (Hyperbolas are discussed in section 11–3.)

$$x^2 - y^2 = 2$$
$$(r \cos \theta)^2 - (r \sin \theta)^2 = 2$$
$$r^2\cos^2\theta - r^2\sin^2\theta = 2$$
$$r^2(\cos^2\theta - \sin^2\theta) = 2$$
$$r^2\cos 2\theta = 2 \qquad \cos 2\theta = \cos^2\theta - \sin^2\theta$$
$$r^2 = \frac{2}{\cos 2\theta}$$
$$r^2 = 2 \sec 2\theta$$

■

> **Conversion of equations from polar to rectangular form**
>
> To convert from polar to rectangular coordinates we use the relations
>
> $$\sin \theta = \frac{y}{r}, \quad \cos \theta = \frac{x}{r}, \quad r^2 = x^2 + y^2$$

Example 8–5 E illustrates converting polar equations into rectangular form.

■ Example 8–5 E

Convert the polar equation into rectangular form.

1. $r = 5 \sec \theta$

$$r = \frac{5}{\cos \theta}$$
$$r \cos \theta = 5$$
$$r \cdot \frac{x}{r} = 5$$
$$x = 5$$

2. $r^2 = \cos 2\theta$

We do not have any relation for cos 2θ, so we use an identity (section 7–3) to replace it.

$$r^2 = \cos^2\theta - \sin^2\theta$$
$$r^2 = \left(\frac{x}{r}\right)^2 - \left(\frac{y}{r}\right)^2$$
$$r^2 = \frac{x^2}{r^2} - \frac{y^2}{r^2}$$
$$r^4 = x^2 - y^2$$
$$(x^2 + y^2)^2 = x^2 - y^2$$

3. $r = 1 + \cos \theta$

$$r = 1 + \frac{x}{r}$$
$$r^2 = r + x$$
$$x^2 + y^2 = r + x$$
$$x^2 + y^2 - x = r$$
$$(x^2 + y^2 - x)^2 = r^2$$
$$(x^2 + y^2 - x)^2 = x^2 + y^2$$
$$x^4 - 2x^3 + y^4 + 2x^2y^2 - 2xy^2 - y^2 = 0$$ ■

Note Observe that r can be replaced by squaring both members as necessary to obtain r^2, or any even power of r.

Mastery points

Can you

- Graph points in the polar coordinate system?
- Give alternate polar coordinates for a given point?
- Convert between polar and rectangular coordinates?
- Convert equations between polar and rectangular form?

Exercise 8–5

Graph the following points in polar coordinates.

1. $(3,0)$	**2.** $(4,\pi)$	**3.** $\left(2,\frac{\pi}{6}\right)$	**4.** $\left(1,\frac{\pi}{3}\right)$	**5.** $\left(2,\frac{3\pi}{4}\right)$	**6.** $\left(3,\frac{7\pi}{6}\right)$
7. $\left(6,\frac{11\pi}{6}\right)$	**8.** $\left(5,\frac{\pi}{2}\right)$	**9.** $(-2,\pi)$	**10.** $\left(-4,\frac{5\pi}{3}\right)$	11. $\left(-1,\frac{\pi}{3}\right)$	**12.** $\left(-5,\frac{\pi}{6}\right)$
13. $(4,2)$	**14.** $(3,5)$	15. $(-5,6)$	**16.** $\left(-4,\frac{\pi}{2}\right)$	**17.** $\left(5,\frac{4\pi}{3}\right)$	**18.** $\left(4,\frac{5\pi}{6}\right)$

List three of the many other coordinates for the point, with one point having $r < 0$ and two points having $r > 0$.

19. $\left(2,\frac{\pi}{6}\right)$ **20.** $\left(1,\frac{\pi}{3}\right)$ **21.** $\left(6,\frac{11\pi}{6}\right)$ **22.** $\left(-5,\frac{\pi}{6}\right)$ **23.** $(2,2)$ **24.** $\left(4,\frac{17\pi}{6}\right)$

Convert the following polar coordinates into rectangular coordinates. Leave the result in exact form.

25. $\left(4,\frac{\pi}{2}\right)$ **26.** $\left(2,\frac{\pi}{3}\right)$ **27.** $\left(5,\frac{5\pi}{6}\right)$ **28.** $\left(1,\frac{11\pi}{6}\right)$ **29.** $\left(4,\frac{4\pi}{3}\right)$ **30.** $\left(2,\frac{5\pi}{3}\right)$

Convert the following polar coordinates into rectangular coordinates to two-decimal place accuracy.

31. $(2,1)$ **32.** $(5,1.2)$ **33.** $(3,0.82)$ **34.** $(3,5)$ **35.** $(4,4)$ **36.** $(1,6)$

Convert the following rectangular coordinates into polar coordinates. Leave the result in exact form.

37. $(-2\sqrt{3},-2)$ **38.** $(3,-3)$ **39.** $(-2,0)$ **40.** $(-1,\sqrt{3})$ **41.** $(-4,-4)$ **42.** $(0,1)$

Convert the following rectangular coordinates into polar coordinates to two-decimal place accuracy.

43. $(2,3)$ **44.** $(-5,2)$ **45.** $(1,-4)$ **46.** $(-4,-3)$ **47.** $(5,4)$ **48.** $(-3,5)$

Convert the following rectangular equations into polar equations.

49. $y = 4x$ **50.** $y = -2x$ **51.** $y = -3x + 2$ **52.** $y = 5x - 3$ **53.** $y = mx + b,\ b \neq 0$
54. $y = 2$ **55.** $y^2 - 2x^2 = 5$ **56.** $y^2 - x = 4$ **57.** $3x^2 + 2y^2 = 1$ **58.** $x^2 + y^2 = 3$

Convert the following polar equations into rectangular equations.

59. $r = \sin\theta$ **60.** $2r = \cos\theta$ **61.** $r = 2\sec\theta$ **62.** $r = 3\csc\theta$
63. $r = 3\sin 2\theta$ **64.** $r = 2\cos\theta$ **65.** $r^2 = \sin 2\theta$ **66.** $r = \cos 2\theta$
67. $r^2 = \tan\theta$ **68.** $r\sin\theta = 5$ **69.** $r = \dfrac{3}{1 - 2\sin\theta}$ **70.** $r = \dfrac{5}{4 - \cos\theta}$

71. Show that a polar equation of $2xy = 5$ is $r^2 = 5\csc 2\theta$.

72. In the text we noted that $x = r\cos\theta$ if $r > 0$. Show that this is also true for a point given in polar coordinates where $r < 0$. (*Hint:* Consider a point $P = (r,\theta)$, where $r < 0$. Then $P = (-r,\theta + \pi)$, where $-r > 0$. Therefore, since $-r > 0$, $x = -r\cos(\theta + \pi)$ is true. Proceed from here.)

73. In the text we noted that $y = r\sin\theta$ if $r > 0$. Show that this is also true for a point given in polar coordinates where $r < 0$. See the hint in problem 72.

74. The shape of a cam that drives a certain sewing machine needle is described by the polar equation $r = 3 - 2\cos\theta$. Convert this equation into rectangular form.

75. The path that an industrial robot must follow to paint a pattern on a part being manufactured is described by the curve $r = 1 - 2\sin\theta$. Convert this equation into rectangular form.

76. The pattern of strongest radiation of a certain bidirectional radio antenna is described by the curve $r = 1 + \sin 2\theta$. Convert this equation into rectangular form.

77. The pattern of strongest radiation of a certain radio antenna is described by the equation $r = 1 + 2\sin 2\theta$. This pattern is said to have sidelobes. Convert this equation into rectangular form.

78. In the October 1983 issue of *Scientific American,* Jearl Walker described several rides, the Scrambler and the Calypso, at the Geauga Lake Amusement Park near Cleveland, Ohio. Assume the path taken by the Scrambler is described by the polar equation $r = 2\cos 3\theta$. Convert this equation into rectangular form. It will be necessary to rewrite $\cos 3\theta$ in terms of $\cos\theta$. See problem 82 in section 7–3.

79. Assume the path of the Calypso (problem 78) is described by the equation $r = 1 - 3\cos\theta$. Convert this equation into rectangular form.

Skill and review

1. Multiply the complex numbers 2 cis 30° and 5 cis 45°.
2. Find the exact form of the 3 cube roots of 1,000.
3. Solve the triangle ABC if $a = 125$, $b = 85$, and $C = 50°$. Leave all results to the nearest integer.
4. Put the function $f(x) = 2x^2 + 4x - 5$ into vertex form (by completing the square) and graph.
5. Combine $\frac{2x - 3}{x^2 - 16} - \frac{5}{x - 4}$.
6. Divide $\frac{2 + 3i}{5 - 2i}$.
7. Simplify $\left(\frac{x^2x^{-5}}{x^3}\right)^{-2}$.
8. Multiply $3x^{-2}(\frac{1}{3}x^2 - 2x + 1)$. Rewrite the result so there are no negative exponents, and combine into one term.

Chapter 8 summary

- **The law of sines** In any triangle ABC, $\frac{\sin A}{a} = \frac{\sin B}{b} = \frac{\sin C}{c}$. This assumes side a is opposite angle A, side b is opposite angle B, side c is opposite angle C.
- **The law of cosines** For any triangle ABC,
  $a^2 = b^2 + c^2 - 2bc \cos A$
  $b^2 = a^2 + c^2 - 2ac \cos B$
  $c^2 = a^2 + b^2 - 2ab \cos C$
  This assumes side a is opposite angle A, side b is opposite angle B, and side c is opposite angle C.
- **Polar form of a vector** A vector A in polar form is the ordered pair $A = (|A|,\theta_A)$.
  $|A|$ is the magnitude of vector A; $|A| \geq 0$.
  θ_A is the direction of vector A.
- **Rectangular form of a vector** A vector A in rectangular form is $A = (A_x,A_y)$.
  A_x is the horizontal component of the vector,
  A_y is the vertical component of the vector.
- **Relation between the polar and rectangular form of a vector** Given vector $A = (|A|,\theta_A) = (A_x,A_y)$,
  $A_x = |A| \cos \theta_A$
  $A_y = |A| \sin \theta_A$
- **Vector sum** Let Z be the resultant (vector sum) of two vectors A and B. Then $Z = A + B$, and if
  $A = (A_x,A_y)$ and $B = (B_x,B_y)$, then
  $Z_x = A_x + B_x$ and $Z_y = A_y + B_y$
- **Polar form of a complex number** If $z = a + bi$ is a complex number that determines an angle θ, the r cis θ is its polar form, where cis θ means $\cos \theta + i \sin \theta$, $r = \sqrt{a^2 + b^2}$ and $\tan \theta = \frac{b}{a}$. The value r is called the modulus of z, which is also written $|z|$.
- **Complex multiplication—polar form**
  $(r_1 \text{ cis } \theta_1)$
  $(r_2 \text{ cis } \theta_2) = r_1r_2 \text{ cis}(\theta_1 + \theta_2)$
- **Complex division—polar form**
  $\frac{r_1 \text{ cis } \theta_1}{r_2 \text{ cis } \theta_2} = \frac{r_1}{r_2} \text{ cis}(\theta_1 - \theta_2)$, $r_2 \neq 0$
- **De Moivre's theorem** $(r \text{ cis } \theta)^n = r^n \text{cis } n\theta$ for any real number n.
- **De Moivre's theorem for roots** The n nth roots of r cis θ are of the form $(r \text{ cis } \theta)^{\frac{1}{n}} = r^{\frac{1}{n}}\text{cis}\left(\frac{\theta}{n} + \frac{k \cdot 360°}{n}\right)$, $0 \leq k < n$, where k and n are positive integers.
- The polar coordinates of a point are an ordered pair of the form (r,θ), where r is the radius, or radius vector, and θ is an angle.
- **Equivalence of points in polar coordinates**
  1. $(r,\alpha) = (r,\beta)$, if α and β are coterminal angles.
  2. $(-r,\theta) = (r,\theta + \pi)$.
- **Relation between polar and rectangular coordinates** If $P = (x,y) = (r,\theta)$, $r > 0$, then $x = r \cos \theta$ and $y = r \sin \theta$.
- **To convert a rectangular equation into a polar equation** use the relations $x = r \cos \theta$, $y = r \sin \theta$, and $r^2 = x^2 + y^2$.
- **To convert a polar equation into a rectangular equation** use the relations $\sin \theta = \frac{y}{r}$, $\cos \theta = \frac{x}{r}$, and $r^2 = x^2 + y^2$.

Chapter 8 review

[8–1] Solve the following oblique triangles using the law of sines. Round answers to the nearest tenth.

1. $a = 10.6$, $A = 47.9°$, $B = 10.3°$
2. $b = 3.55$, $B = 23.8°$, $C = 5.2°$
3. $a = 10.0$, $b = 13.0$, $B = 79.0°$
4. $a = 12.6$, $c = 7.0$, $C = 32.7°$
5. A lost pilot is given a position report by triangulation with two radar sites. The situation is shown in the diagram. Find the distance of the aircraft from radar site A, to the nearest mile.

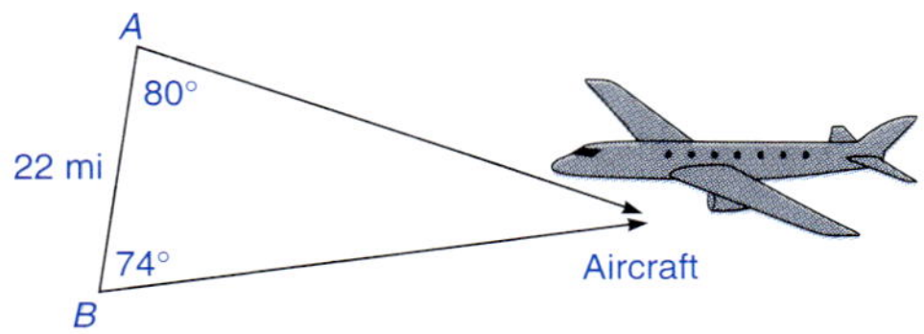

[8–2] Solve the following oblique triangles to the nearest tenth. Use the law of cosines.

6. $a = 4.1$, $b = 6.8$, $C = 29.4°$
7. $b = 60.0$, $c = 20.0$, $A = 92.1°$
8. $a = 21.4$, $c = 27.0$, $B = 112°$
9. $a = 43.5$, $b = 17.8$, $c = 35.0$
10. $a = 31.7$, $b = 80.0$, $c = 105$
11. A technician is setting up a numerically controlled grinding machine. A triangular pattern is to be ground and, therefore, must be coordinatized. The vertices of the triangle are at $A(-2,5)$, $B(4,7)$, and $C(5,2)$. Solve the resulting triangle. Round answers to the nearest 0.1.
12. A ship leaves a harbor heading due west and travels 23.3 km. It then turns north through a 63° angle and travels for another 10.0 km. How far is the ship from its starting point, to the nearest kilometer?

[8–3]

13. Given the vector (27.2,29.0°), find the horizontal and vertical components, to the nearest tenth.
14. The horizontal and vertical components of a vector are 19.6 and 30.5, respectively. Find the magnitude and direction of the vector.
15. A rocket climbs with a speed of 450 knots and an angle of climb of 34.6°. (The angle of climb is the angle measured from the ground to its flight path.) Find the horizontal and vertical components of the rocket's velocity, to the nearest knot.
16. A force is acting on a cart at a 13.5° angle of elevation (13.5° measured from the ground up to the force vector). If the force is 256 pounds, find the horizontal and vertical components, to the nearest pound.

Add the following vectors. Round results to the nearest tenth.

17. (33.0,14.7°), (15.2,33.6°)
18. (10.2,112.3°), (4.2,19.3°)
19. (3.5,29.2°), (1.7,43.1°), (4.3,115.0°)
20. (7.1,13.8°), (6.6,142.0°), (11.9,215.0°)
21. Two forces are acting on a point, 126 pounds in the direction 223° and 158 pounds in the direction 311°. Compute the magnitude and direction of the resultant force, to the nearest unit.

[8–4] Write the polar form of each complex number. Round answers to the nearest tenth when necessary.

22. $3 - 2i$ **23.** $\sqrt{3} + 3i$ **24.** $-1 - 2i$

Write the rectangular form of each complex number. Round answers to the nearest tenth.

25. 3 cis 35° **26.** 5 cis 243°

Write the rectangular form of each complex number. Leave results in exact form.

27. 3 cis 240° **28.** 10 cis 330°

Multiply the complex numbers.

29. (2 cis 25°)(3 cis 45°) **30.** (2 cis 18°)(6.5 cis 122°)

Divide the complex numbers.

31. $\dfrac{40 \text{ cis } 120°}{5 \text{ cis } 20°}$ **32.** $\dfrac{50 \text{ cis } 45°}{100 \text{ cis } 9°}$

33. Compute the cube of 2 cis 130°.
34. Compute the fourth power of 2 cis 150°.
35. Compute an approximation to $(0.8 + 0.6i)^8$; round results to the nearest 0.01.
36. Find the 4 fourth roots of 16 in exact rectangular form.
37. Find the 3 cube roots of $6 - 5i$ to the nearest 0.01. Leave the answer in rectangular form.
38. In electronics, one version of Ohm's law is $I = \dfrac{V}{Z}$, where I is current, V is voltage, and Z is impedance. Find I in a circuit in which $V = 130$ cis 25° and $Z = 30$ cis 75°.

[8–5] Graph the following points in polar coordinates.

39. (2,0) **40.** $(2,\pi)$ **41.** $\left(3,\dfrac{5\pi}{6}\right)$ **42.** $\left(-6,\dfrac{11\pi}{6}\right)$

Convert the following polar coordinates into rectangular coordinates. Leave the result in exact form.

43. $\left(3,\frac{11\pi}{6}\right)$ **44.** $\left(4,\frac{2\pi}{3}\right)$ **45.** $\left(-2,\frac{5\pi}{3}\right)$

Convert the following polar coordinates into rectangular coordinates. Round results to the nearest tenth.

46. $\left(2,\frac{\pi}{5}\right)$ **47.** $(5,2)$ **48.** $(-1,-4)$

Convert the following rectangular coordinates into polar coordinates. Round results to the nearest hundredth.

49. $(2,1)$ **50.** $(-5,3)$ **51.** $(-1,-4)$

Convert the following rectangular equations into polar equations.

52. $y = -3x$ **53.** $y = 4x + 2$ **54.** $2y^2 - x^2 = 5$
55. $y^2 - 3x = 0$ **56.** $x^2 + y^2 = 9$ **57.** $x = 2$

Convert the following polar equations into rectangular equations.

58. $r = \sin\theta$ **59.** $r = 2\sec\theta$ **60.** $r^2 = \sin 2\theta$
61. $r^2 = \tan\theta$ **62.** $r\sin\theta = 2$ **63.** $r = \dfrac{3}{2 - \sin\theta}$

Chapter 8 test

1. In triangle ABC, $b = 22.6$, $A = 13.5°$, and $C = 82.1°$. Solve this triangle. Round answers to the nearest tenth.

2. In triangle ABC, $b = 22.6$, $c = 24.0$, and $C = 62.1°$. Solve this triangle. Round answers to the nearest tenth.

3. In triangle ABC, $a = 25.9$, $c = 16.2$, and $B = 100°$. Solve this triangle. Round answers to the nearest tenth.

4. In triangle ABC, $a = 2.55$, $b = 3.12$, and $c = 4.00$. Solve this triangle. Round answers to the nearest tenth.

5. The distance to a boat on a lake is being found by triangulation from two points on shore. The situation is shown in the diagram. Find the distance to the boat from site B to the nearest yard.

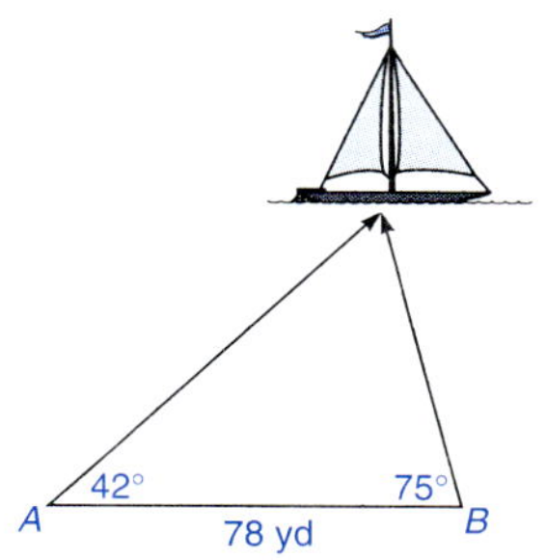

6. Given the three points $A(6,8)$, $B(-3,5)$, and $C(10,-4)$, find the angle formed by line segments AB and BC, to the nearest 0.1°.

7. Find the horizontal and vertical components of the vector $(2,30°)$, in exact form.

8. The horizontal and vertical components of a vector are 4.0 and 5.0. Find the magnitude and direction of the vector, to the nearest tenth.

9. Add the vectors $(5.4,19.0°)$ and $(8.0,123°)$. Round the results to the nearest tenth.

10. A sign is suspended between two buildings. One cable from which the sign is suspended has a tension of 535 pounds and acts at an angle of elevation of 62° (i.e., 62° with respect to the horizontal). If the weight of the sign is 1,000 pounds, find the tension and angle of elevation in the other cable, to the nearest unit.

11. Write the polar form of the complex number $4 - 5i$. Round to hundredths.

12. Write the rectangular form of the complex number 2 cis 120°. Leave the result in exact form.

13. Multiply (2 cis 325°)(7 cis 145°). Leave the result in polar form.

14. Divide $\dfrac{6 \text{ cis } 140°}{2 \text{ cis } 20°}$. Leave the result in polar form.

15. Compute $(3 \text{ cis } 150°)^3$.

16. Find the 4 fourth roots of -16 in exact form.

17. Graph the polar coordinates $\left(-4,\frac{\pi}{6}\right)$.

18. Convert the polar coordinates $\left(3,\frac{\pi}{8}\right)$ to rectangular coordinates, to the nearest tenth.

19. Convert the rectangular coordinates $(-\sqrt{3},-1)$ into polar coordinates, in exact form.

20. Convert the rectangular equation $y = -3x + 5$ into polar form.

21. Convert the rectangular equation $2y^2 - x = 5$ into polar form.

22. Convert the polar equation $r = 2\csc\theta$ into rectangular form.

23. Convert the polar equation $r^2 = \cos 2\theta$ into rectangular form.

C H A P T E R

9 Exponential and Logarithmic Functions

This chapter discusses two general classes of functions, exponential functions and logarithmic functions. Each is the inverse of the other, and both have many applications in banking (compound interest), biology (population growth), physics (chain reactions), electronics (charging a resistance-capacitance network), psychology (learning curves), and computer science (the height of a balanced binary tree). Both functions also have many other applications in these and other areas.

9–1 Exponential functions and their properties

If m represents the visual magnitude of a star, then the ratio R of the brightness of the star to a star of the first magnitude is approximately $R(m) = 2.5^{1-m}$. Graph this function.

This function, called an exponential function, is the topic of this section.

Real number exponents

We previously defined integral and rational exponents. For example,

$2^3 = 2 \cdot 2 \cdot 2 = 8$	Positive integer exponent
$2^0 = 1$	Zero exponent
$2^{-3} = \frac{1}{8}$	Negative integer exponent
$2^{\frac{1}{7}} = \sqrt[7]{2} \approx 1.104$	Rational exponent

We have not defined the meaning of exponents that are irrational, such as π or $\sqrt{2}$. For example, what would 2^π mean? Since $\pi \approx 3\frac{1}{7}$, we know $2^\pi \approx 2^{3\frac{1}{7}}$, which is $2^3 \cdot 2^{\frac{1}{7}} = 8\sqrt[7]{2}$.

This shows how we can approximate values for a base with an irrational exponent by considering a rational number with a value close to the irrational value. In more advanced mathematics it is possible to define exponents with irrational values in a manner similar to this. For the purposes of this text, we assume that this definition has been made.

It can be proved that the properties of exponents that are true for integer and rational exponents hold for any real exponent. Some of these properties are summarized here.

Properties of exponents

If $x, y, z \in R$, then

[1]	$x^y \cdot x^z = x^{y+z}$	[2]	$\dfrac{x^y}{x^z} = x^{y-z}, x \neq 0$	[3]	$(xy)^z = x^z y^z$
[4]	$\left(\dfrac{x}{y}\right)^z = \dfrac{x^z}{y^z}$	[5]	$(x^y)^z = x^{yz}$	[6]	$x^{-y} = \dfrac{1}{x^y}$

Example 9–1 A illustrates the properties of exponents.

■ ***Example 9–1 A***

Use the properties of exponents to simplify each expression.

1. $3^\pi \cdot 3^2$

$= 3^{\pi+2}$

2. $(5^{\sqrt{2}})^{\sqrt{8}}$

$= 5^{(\sqrt{2} \cdot \sqrt{8})}$

$= 5^{\sqrt{16}} = 5^4$

$= 625$

3. $\dfrac{\pi^{\sqrt{8}}}{\pi^{\sqrt{2}}}$

$= \pi^{(\sqrt{8} - \sqrt{2})}$

$= \pi^{(2\sqrt{2} - \sqrt{2})}$

$= \pi^{\sqrt{2}}$ ■

Exponential function—definition

With the knowledge that any real exponent has meaning, we can now define a class of functions in which the domain element is the exponent of a fixed base.

Exponential function

An exponential function is a function of the form

$$f(x) = b^x, \quad b > 0 \quad \text{and} \quad b \neq 1$$

The constant value b is called the **base** of the function. The variable x can represent any real number, and therefore the domain of an exponential function is the set of real numbers.

The function $f(x) = 5^x$ is an example of an exponential function. The value of f for various domain elements is computed.

$$f(2) = 5^2 = 25$$
$$f(0) = 5^0 = 1$$
$$f(-3) = 5^{-3} = \frac{1}{125}$$

Graphs of exponential functions

As further examples of the graphs of exponential functions consider the graph of $f(x) = 2^x$, the exponential function with base 2, and the function $f(x) = 3^x$, the exponential function with base 3. Some of the (x,y) ordered pairs for these functions is shown in table 9–1. These values are plotted and connected by smooth curves in figure 9–1. Information for plotting the graphs on the TI-81 graphing calculator is also shown. Observe that both of these functions have the same y-intercept, (0,1). This is because any base, raised to the zero power, is one. Also, $3^x > 2^x$ for $x > 0$, and $3^x < 2^x$ for $x < 0$.

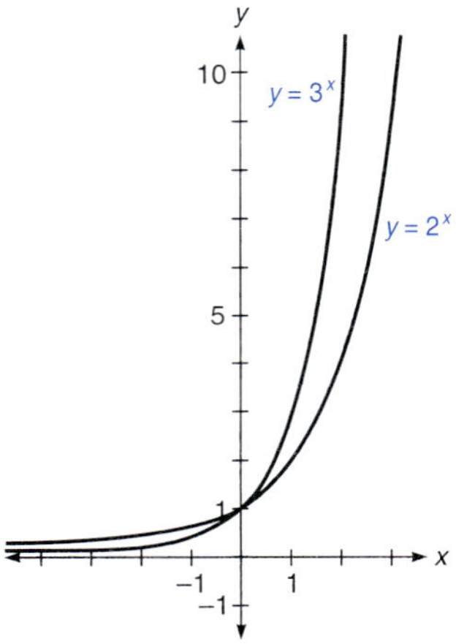

Figure 9–1

x	2^x	3^x
−3	$\frac{1}{8}$	$\frac{1}{27}$
−2	$\frac{1}{4}$	$\frac{1}{9}$
−1	$\frac{1}{2}$	$\frac{1}{3}$
0	1	1
1	2	3
2	4	9
3	8	27

Table 9–1

[Y=] 2 [∧] [X|T] [ENTER] 3 [∧] [X|T]
[RANGE −4,4,−1,10]

The x-axis is a **horizontal asymptote** for both curves. Neither function has an x-intercept since $0 = b^x$ has no solution. Also, as x gets greater, the $y = f(x)$ values for both curves keep getting greater also so the function is increasing (section 3–5).

The graphs in figure 9–1 illustrate the behavior of exponential functions for $b > 1$. When $0 < b < 1$ we get graphs similar to these, but that are decreasing. Table 9–2 and figure 9–2 illustrate the curves for the functions $f(x) = (\frac{1}{2})^x$ and for $f(x) = (\frac{1}{3})^x$.

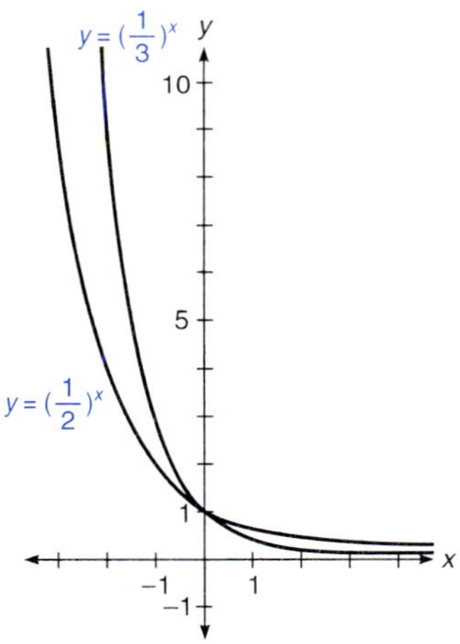

Figure 9–2

x	$(\frac{1}{2})^x$	$(\frac{1}{3})^x$
−3	8	27
−2	4	9
−1	2	3
0	1	1
1	$\frac{1}{2}$	$\frac{1}{3}$
2	$\frac{1}{4}$	$\frac{1}{9}$
3	$\frac{1}{8}$	$\frac{1}{27}$

Table 9–2

[Y=] .5 [∧] [X|T] [ENTER] [(] 1
[÷] 3 [)] [∧] [X|T]
[RANGE −4,4,−1,10]

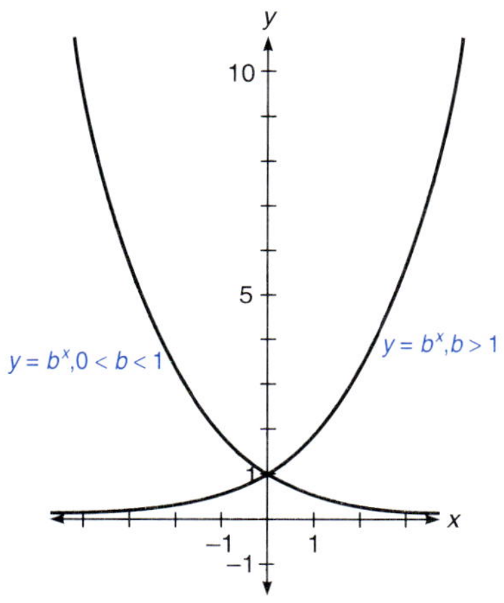

Figure 9–3

Figure 9–3 illustrates the general graphs of $f(x) = b^x$ for $b > 1$ and for $0 < b < 1$. When b is greater than 1 the graph is increasing, and when b is less than 1 the graph is decreasing. From the graphs of figure 9–3 we can see that *an exponential function is one to one*. We can see this because the graphs pass the horizontal line test (section 3–5). We can also observe from the graph that the domain of the exponential function is all the real numbers, and its range is all $y > 0$. This is summarized as follows.

Features of exponential function and its graph

An exponential function is a function of the form

$$f(x) = b^x, \quad b > 0 \quad \text{and} \quad b \neq 1, \text{ with}$$

Domain: $\{x \mid x \in R\}$

Range: $\{y \mid y \in R, y > 0\}$

- Exponential functions are one-to-one functions.
- The y-intercept is $(0, 1)$ (assuming no vertical or horizontal translations).
- The x-axis is a horizontal asymptote.
- If $b > 1$, the function is increasing.
- If $b < 1$, the function is decreasing.

To graph an exponential function using algebraic techniques we use the cases illustrated in figure 9–3 along with the observations we made earlier: translations (section 3–4) and point plotting. Of course the graphing calculator can be used as well. The appropriate information for the TI-81 is shown. This is all illustrated in example 9–1 B.

■ *Example 9–1 B*

Graph the function. State whether the function is increasing or decreasing. Note the value of any intercepts.

1. $f(x) = 4.5^x$

Since $4.5 > 1$ the function is an increasing function.

y-intercept: $f(0) = 4.5^0 = 1$

Additional points:

x	-1	1	1.5
$y = 4.5^x$	0.2	4.5	9.5

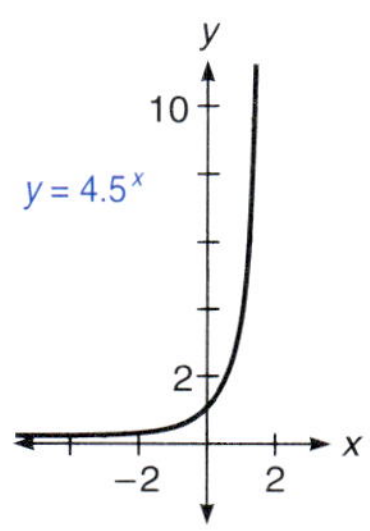

Y= 4.5 ^ X|T

RANGE −2,2,−1,10

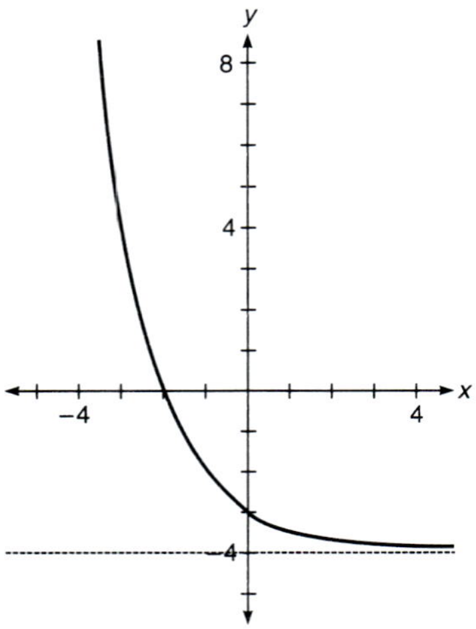

$\boxed{\text{Y=}}$ 2 $\boxed{\wedge}$ $\boxed{(-)}$ $\boxed{\text{X|T}}$ $\boxed{-}$ 4
$\boxed{\text{RANGE } -4,4,-5,8}$

2. $f(x) = 2^{-x} - 4$

This is the graph of $2^{-x} = (\frac{1}{2})^x$ shifted down 4 units. Thus, the usual horizontal asymptote of $y = 0$ (the x-axis) is shifted down to $y = -4$.

y-intercept: $f(0) = 2^0 - 4 = 1 - 4 = -3$

x-intercept:

$0 = 2^{-x} - 4$	Replace $f(x)$ by 0
$4 = 2^{-x}$	
$2^2 = 2^{-x}$	
$2 = -x$	
$-2 = x$	See example 9–1 C for this step

This is a decreasing function.

Additional values:

x	−4	−3	−1	1	2
$y = 2^{-x} - 4$	12	4	−2	−3.5	−3.75

■

Solving exponential equations with the one-to-one property

Recall from section 3–3 that a function is one to one if all of the second components of the ordered pairs are different. The fact that exponential functions are one to one implies the following about the situation where $b^m = b^n$. Suppose two ordered pairs (m,b^m) and (n,b^n) are in the exponential function $f(x) = b^x$, and $b^m = b^n$. Since the function is one to one, all second components are different. Therefore, if $b^m = b^n$ in the ordered pairs (m,b^m) and (n,b^n) (two second components are the same), these ordered pairs must actually be the same ordered pair. Thus, m and n must be the same, so we conclude that $m = n$. This fact implies the following statement.

> **One-to-one property for exponential functions**
>
> If $b^x = b^y$, then $x = y$.

Example 9–1 C illustrates using the one-to-one property to solve equations in which each member can be expressed as an exponential expression. Observe that *to solve these exponential equations we put both sides of the equation in terms of the same base* and then apply the one-to-one property. For now, we examine equations that have integer or rational solutions. Section 9–5 deals with equations whose solutions are not necessarily integer or rational.

■ *Example 9–1 C*

Solve for x in the following exponential equations.

1. $9^x = \frac{1}{27}$

$(3^2)^x = 3^{-3}$	$9 = 3^2$; $\frac{1}{27} = \frac{1}{3^3} = 3^{-3}$
$3^{2x} = 3^{-3}$	$(x^y)^z = x^{yz}$
$2x = -3$	One-to-one property
$x = -\frac{3}{2}$	Divide both members by 2

2. $4^x = \sqrt{32}$

$4^x = \sqrt{2^5}$

$(2^2)^x = 2^{5/2}$ — Rewrite both members as powers of 2

$2^{2x} = 2^{5/2}$ — $(x^y)^z = x^{yz}$

$2x = \frac{5}{2}$ — One-to-one property

$x = \frac{5}{4}$ ■

Mastery points

Can you

- State the features of an exponential function and its graph?
- Solve certain simple exponential equations?
- Use the properties of exponents to simplify expressions involving exponents?
- Graph certain exponential functions?

Exercise 9–1

1. Define the general equation of the exponential function with base b. Make sure you state the restrictions on b.

2. Describe the behavior of the function $f(x) = b^x$ with respect to the words "increasing" and "decreasing" and the value of b.

Use the properties of exponents to perform the indicated operations.

3. $2^\pi \cdot 2$ **4.** $5^{\sqrt{5}} \cdot 5^{\sqrt{20}}$ **5.** $7^\pi \cdot 7^{3\pi}$ **6.** $3^\pi \cdot 9^\pi$

7. $\dfrac{4^{\sqrt{12}}}{4^{\sqrt{3}}}$ **8.** $\dfrac{2^{5\pi}}{2^{2\pi}}$ **9.** $(3^{\sqrt{2}})^{\sqrt{2}}$ **10.** $(3^{\sqrt{2\pi}})^{\sqrt{8\pi}}$

Solve the following exponential equations for x.

11. $3^x = 27$ **12.** $2^x = 512$ **13.** $3^x = \sqrt{27}$ **14.** $4^x = 2^5$

15. $9^3 = 3^x$ **16.** $10^{2x} = 1{,}000$ **17.** $2^x = \frac{1}{8}$ **18.** $4^x = \frac{1}{64}$

19. $4^x = \frac{1}{8}$ **20.** $8^x = \sqrt{128}$ **21.** $3^x = \sqrt[3]{243}$ **22.** $4^{x/2} = 8$

23. $(\sqrt{2})^x = 16$ **24.** $2^{-x} = 0.25$

Graph each function. State whether the function is increasing or decreasing. Label any y-intercepts.

25. $f(x) = 5^x$ **26.** $f(x) = 6^x$ **27.** $f(x) = 4^{-x}$ **28.** $f(x) = 8^{-x}$

29. $f(x) = 0.3^x$ **30.** $f(x) = 0.9^x$ **31.** $f(x) = 0.7^{-x}$ **32.** $f(x) = 0.35^{-x}$

33. $f(x) = 4^{x+1}$ **34.** $f(x) = 3^{x-1}$ **35.** $f(x) = 3^{1-x}$ **36.** $f(x) = 2^{-x+1}$

37. $f(x) = 4^{-x+2} + 1$ **38.** $f(x) = 2^{x/2} + 1$ **39.** $f(x) = 2^{-x}$ **40.** $f(x) = 3 - 3^x$

41. $f(x) = 2^{-x} + 2$ **42.** $f(x) = 2^x - 2$

Solve the following problems.

43. If a bank account paid 5% interest, compounded continuously, on the balance above \$10, then, if the initial amount deposited were \$13, the balance after time x (in years) would be closely described by the function $f(x) = 3(1.05)^x + 10$. Graph this function for $0 \le x \le 15$.

44. If x represents the strength of a sound in bels (one bel equals 10 decibels) then the factor F that represents the ratio of the given noise to a reference noise level is $F(x) = 10^x$. Graph this function.

45. If m represents the visual magnitude of a star, then the ratio R of the brightness of the star to a star of the first magnitude is approximately $R(m) = 2.5^{1-m}$. Graph this function.

46. A certain automobile seems to depreciate about 15% every year. Its value V after n years is given by the function $V = P(0.85^n)$, where P is the purchase price. Assume that P is 1 and graph the resulting function.

47. Assume the automobile of the previous problem cost \$12,000 when new. Approximate its value after 4 years.

48. The function $S = S_0 2.8^{-d}$ could describe the strength of a signal in a telephone cable at a distance d, measured in a suitable unit, from the source, where S_0 is the initial signal strength. Graph this function assuming that $S_0 = 1$.

49. In computer science, a binary tree has many uses. The number of pieces of data that can be stored in a binary tree of height h is $f(h) = 2^{h+1}$. Graph this function.

50. The value of 2^{10} is 1,024; this is close to 1,000, which is often indicated by the prefix "kilo." The amount of memory on a computer is often described in terms of kilobytes (a byte being one basic unit of memory). Thus, a 2 KB memory means a 2 kilobyte memory, or $2 \cdot 1{,}024 = 2{,}048$ bytes of memory. Find the exact number of bytes in the following values. (Note that 1,000 KB is often called 1 MB, for 1 megabyte.)

a. 16 KB **b.** 32 KB **c.** 512 KB **d.** 1,000 KB

51. As noted in problem 50, $2^{10} \approx 10^3$. This provides a way to estimate large powers of 2 in terms of powers of 10. For example, $2^{34} = 2^4 \cdot 2^{30} = 16(2^{10})^3 \approx 16(10^3)^3 = 16 \cdot 10^9$ (16 billion). Thus, 2^{34} is approximately 16 billion. Estimate the following values in terms of powers of 10.

a. 2^{13} **b.** 2^{21} **c.** 2^{43}

52. Exponential functions change rapidly for small changes in the value of the argument. (In $f(x) = 2^x$, x is the "argument.") We can illustrate this by a story. A person once saved the life of a certain very rich monarch. The monarch, wishing to reward the individual, asked how to repay this individual. The person told the monarch, "Take a chessboard; put 1 cent on the first square, 2 cents on the second, 4 cents on the third, and so on, doubling the amount each time. I wish to be paid the amount that you put on the last (64th) square." This sounded good to the monarch, who agreed. Estimate the amount paid to the individual in terms of powers of 10, in dollars. See problem 51.

53. One of the following is an estimate, in terms of a power of 10 (see problems 51 and 52) of how many grains of sand it would take to fill up a sphere the size of the earth, which has a radius slightly less than 4,000 miles. The volume of a sphere is $\frac{4}{3}\pi r^3$, where r is the radius of the sphere. Assume that there are about one million grains of sand in a cubic inch, and determine which of the following is the closest estimate.

a. 10^{20} **b.** 10^{30} **c.** 10^{50}
d. 10^{90} **e.** 10^{200} **f.** 10^{500}

54. Using an estimate of one million grains of sand in a cubic inch, estimate the number of grains of sand which it would take to fill the known universe. Assume the known universe is a sphere which is 20 billion light years in radius; a light year is approximately six trillion miles. Determine which of the values in the previous problem is closest to your estimate.

Skill and review

1. Graph $f(x) = \dfrac{x - 1}{x^2 - 4}$.
2. Graph $f(x) = (x - 1)(x^2 - 4)$.
3. Solve $(x - 1)(x^2 - 4) > 0$.
4. Factor $6x^3 + 5x^2 - 2x - 1$.
5. Graph $f(x) = -x^3 + 1$.
6. Solve $x^{2/3} - x^{1/3} - 6 = 0$.

9–2 Logarithmic functions—introduction

The number of binary digits (bits) that a digital computer requires to represent a positive integer N is the smallest integer i such that $2^i \geq N$. Find the number of bits required to represent the integer 35,312.

In this section we define logarithmic functions. We will see that these functions are inverses of exponential functions, and that they prove to be useful in almost any situation in which we use exponential functions. Specifically, these functions will quickly find the value of the exponent i in the section opening problem. As an introduction to these functions we first introduce logarithms.[1]

Logarithms

Consider the statement $2^x = 8$. We can see by inspection that the solution to this statement is 3. In other words, the exponent of the base 2 that "produces" 8 is 3. We also say that the logarithm, to the base 2, of 8 is 3. We use the word logarithm synonymously with the word exponent: A **logarithm** is an exponent. In fact, we continue to use the word logarithm largely for historical reasons. Example 9–2 A illustrates finding logarithms in certain situations—remember that a logarithm is an exponent.

■ ***Example 9–2 A***

Find the unknown logarithm (exponent) in each case.

1. $4^x = 16$
Since $4^2 = 16$ we know that the logarithm x is 2.

2. What is the logarithm of 64 to the base 2?

We want the exponent of the base 2 that produces 64. This is 6 since $2^6 = 64$, so the logarithm, to the base 2, of 64 is 6. ■

A symbolic method was devised to describe the phrase "the logarithm of x to the base b." We write $\log_b x$. Thus, $\mathbf{\log_b x}$ means "the exponent of base b that produces x." For example, $\log_2 16$ is 4, since 4 is the exponent of 2 that produces 16. Example 9–2 B uses this notation.

■ ***Example 9–2 B***

Find the value of the given logarithmic expression.

1. $\log_3 9$
The base is 3; what exponent (logarithm) of 3 gives 9? Since $3^2 = 9$, $\log_3 9 = 2$.

[1]Logarithms were introduced by a Scottish baron, John Napier (1550–1617), in 1614, as a method of simplifying many complex computations. Napier invented the word "logarithm," which means "ratio number." In 1624, Johannes Kepler used the contraction "log," and Henry Briggs, professor of geometry at Oxford, further developed the idea of logarithms in the same year.

2. $3 \log_2 \frac{1}{8}$

We first evaluate $\log_2 \frac{1}{8}$. What exponent of 2 gives $\frac{1}{8}$? We know that $\frac{1}{8} = \frac{1}{2^3} = 2^{-3}$, so the exponent of 2 that gives $\frac{1}{8}$ is -3, and $\log_2 \frac{1}{8} = -3$. Therefore, $3 \log_2 \frac{1}{8} = 3(-3) = -9$. ■

A general property of logarithms is that since $b^0 = 1$, $\log_b 1 = 0$ for any permissible value of b. Also, since $b^1 = b$, $\log_b b = 1$ for $b > 0$ and $b \neq 1$. This is summarized as follows.

> **Properties of logarithms**
> For any base b, $b > 0$ and $b \neq 1$:
> Logarithm of one: $\log_b 1 = 0$
> Logarithm of the base: $\log_b b = 1$

Equivalence of logarithmic and exponential forms

The statement $4 = \log_2 16$ means 4 is the exponent of 2 that produces 16, or $2^4 = 16$. This shows an equivalence between logarithmic and exponential forms. Let us now formalize a definition of the logarithm to the base b of x using this idea of equivalence of forms.

> **Equivalence of logarithmic and exponential form**
> $y = \log_b x$ if and only if $b^y = x$, where $b > 0$, $b \neq 1$.

Observe that we put the same restrictions on the value of b as we did for exponential functions in section 9–1.

We have defined logarithms in terms of exponents; $b^y = x$ is an exponential equation and $y = \log_b x$ is a logarithmic equation. Thus, we can change any exponential equation into its equivalent logarithmic form, and vice versa.

Notice that *we can rewrite a logarithmic equation as an exponential equation, or vice versa, by moving only the base from one side of the equation to the other.* This rewriting process is illustrated in example 9–2 C.

■ *Example 9–2 C*

1. $3 = \log_{10} 1{,}000$ — The base is 10

$10^3 = 1{,}000$ — Move the base to the other side of the equation; since 10 is a *base* 3 becomes an exponent

2. $\log_m 6 = 18$ — Base is m

$6 = m^{18}$ — Move the base to the other side of the equation

Write each exponential equation as a logarithmic equation.

3. $3^5 = 243$ — The base is 3

$5 = \log_3 243$ — Move the base to the other side; since it must stay a base it must be written $\log_3$

4. $x = 7^2$ Base is 7

$\log_7 x = 2$ Move the base to the other side of the equation ■

Solving logarithmic equations by rewriting as exponential equations

As illustrated in example 9–2 D, logarithmic equations can often be solved by putting the equation in exponential form.

■ ***Example 9–2 D***

Solve the following equations.

1. $\log_5 x = -2$

$x = 5^{-2}$

$x = \dfrac{1}{5^2}$

$x = \dfrac{1}{25}$

2. $\log_4 64 = z$

$64 = 4^z$

$4^3 = 4^z$

$z = 3$ ■

Estimating values of logarithms

Although the values of most logarithms are irrational numbers, it is useful to be able to estimate their values as integers. This uses the following property, which is true when $b > 1$:

$$\text{if } x < y, \text{ then } \log_b x < \log_b y.$$

This is because logarithmic functions with $b > 1$ are increasing functions (as we will see later). In most practical situations involving logarithms, $b > 1$. Example 9–2 E illustrates estimating logarithms using this property.

■ ***Example 9–2 E***

The number of binary digits (bits) that a digital computer requires to represent a positive integer N is the smallest integer greater than or equal to the value $\log_2 N$. Find the number of bits required to represent the integer 5,218.

We need to approximate $\log_2 5{,}218$.

$2^{12} = 4{,}096$ and $2^{13} = 8{,}192$, so $\log_2 4{,}096 = 12$ and $\log_2 8{,}192 = 13$, so $12 < \log_2 5{,}218 < 13$. Therefore, 13 is the number of bits required. ■

Logarithmic functions

We now define the general logarithmic function.

> **Logarithmic function**
>
> A logarithmic function is a function of the form
>
> $$f(x) = \log_b x,\ b > 0 \text{ and } b \neq 1,$$
>
> with
>
> Domain: $\{x \mid x \in R, x > 0\}$
>
> Range: $\{y \mid y \in R\}$.

The logarithmic function to the base b is the inverse function of the exponential function to the base b. Thus, the domain of the logarithmic function is the range of the corresponding exponential function, and the range is the domain of the corresponding exponential function.

The logarithmic and exponential functions are inverses because if an ordered pair (x,y) satisfies the function $f(x) = \log_b x$, its reversal satisfies the function $f(x) = b^x$. For example, the ordered pair (8,3) satisfies the function $f(x) = \log_2 x$, since $3 = \log_2 8$, and the ordered pair (3,8) satisfies the function $f(x) = 2^x$, since $8 = 2^3$. This is a direct result of our definition of logarithms.

The fact that these functions are inverses implies two properties that we now develop. Recall from section 4–5 that if two functions, f and g, are inverses then $f(g(x)) = x$ and $g(f(x)) = x$. Assume that $f(x) = \log_b x$ and $g(x) = b^x$. Then,

$$\begin{aligned} f(g(x)) &= \log_b(g(x)) && f(x) = \log_b x \\ &= \log_b(b^x) && g(x) = b^x \\ &= x && \text{We know } f(g(x)) = x \end{aligned}$$

Also, $g(f(x)) = b^{f(x)} = b^{\log_b x} = x$

Thus, we know that the following two properties are true.

Composition of exponential and logarithmic functions

$$\log_b b^x = x$$
$$b^{\log_b x} = x$$

Example 9–2 F illustrates simplifying certain expressions composed of a logarithmic expression and an exponential expression that are the inverse of each other.

■ ***Example 9–2 F***

Simplify the following, using the properties cited above.

1. $\log_3 3^{10} = 10$ $\quad \log_a a^x = x$

2. $27^{\log_{27} 15} = 15$ $\quad a^{\log_a x} = x$

3. $\log_7 7 = 1$ $\quad \log_b b = 1$

4. $\log_3 1 = 0$ $\quad \log_b 1 = 0$ ■

Mastery points

Can you

- State the definition of the statement $y = \log_b x$?
- Convert between exponential and logarithmic forms of equations?
- Solve simple logarithmic equations?
- Estimate the values of logarithms?
- Use the various properties of logarithms to simplify appropriate expressions?

Exercise 9–2

Find the unknown logarithm in each case.

1. $3^x = 27$ **2.** $2^y = 128$ **3.** $5^z = 125$ **4.** $2^y = 8$ 5. $3^x = \frac{1}{27}$
6. $2^n = \frac{1}{8}$ **7.** $2^w = 0.25$ **8.** $4^z = \frac{1}{64}$ **9.** $10^k = 0.1$ **10.** $10^r = 0.001$

Find the value of the expression.

11. $\log_2 8$ **12.** $\log_5 25$ **13.** $\log_4 256$ **14.** $\log_{10} 0.01$
15. $\log_2 \frac{1}{64}$ **16.** $\log_3 \frac{1}{81}$ **17.** $5 \log_3 27$ **18.** $2 \log_{10} 100$
19. $3 \log_5 \frac{1}{125}$ **20.** $2 \log_3 9 + 5 \log_6 36$ 21. $5(3 \log_2 \frac{1}{8} + 2 \log_{10} 0.1)$ **22.** $3 \log_5 5^2$
23. $\log_4 4^3$ **24.** $(\log_4 4)^3$

Put each logarithmic equation into exponential form.

25. $\log_2 8 = 3$ **26.** $\log_{10} 100 = 2$ **27.** $\log_{10} 0.1 = -1$ **28.** $\log_4 x = 3$
29. $\log_{12}(x + 3) = 2$ **30.** $\log_2 y = 5$ 31. $\log_3 5 = x + 2$ **32.** $\log_m x = 2y + 1$

Put each exponential equation into logarithmic form.

33. $2^4 = 16$ **34.** $3^4 = 81$ **35.** $x^2 = m + 3$ **36.** $4 = y^{2x-1}$
37. $m^y = x + 1$ **38.** $(x - 1)^3 = 5$ 39. $(2x - 3)^{x+y} = y + 2$ **40.** $(3x)^2 = 4y$

Solve the following equations for x.

41. $\log_2 x = 4$ **42.** $\log_3 x = 2$ **43.** $\log_2 4 = x$ **44.** $\log_x 64 = 3$
45. $\log_x 64 = 6$ **46.** $\log_{10} x = -2$ **47.** $\log_{16} x = 0.25$ **48.** $\log_x \frac{1}{8} = -3$
49. $\log_x 0.1 = -1$ **50.** $\log_x k = 1$ **51.** $\log_k x = 2$ **52.** $\log_k k^3 = x$

Estimate the values of the following logarithms by stating two consecutive integers that bracket the value, or the value itself if possible.

53. $\log_2 100$ **54.** $\log_3 100$ **55.** $\log_4 100$ **56.** $\log_5 500$ **57.** $\log_2 0.3$ **58.** $\log_2 0.8$

Simplify the following expressions.

59. $\log_2 2^4$ **60.** $10^{\log_{10} 100}$ **61.** $\log_5 5$ **62.** $\log_6 1$ **63.** $\log_4 4^5$ **64.** $2^{\log_2 19}$
65. $\log_{10} 10^{18}$ **66.** $\log_{10} 10^{-4}$ **67.** $\log_a a$ **68.** $m^{\log_m 7}$ **69.** $5 \log_5 1$ **70.** $3 \log_3 3$
71. $-5 \log_2 2$ **72.** $\frac{1}{2} \log_5 5$ 73. $4^{\log_2 9}$ **74.** $9^{\log_3 4}$ **75.** $6^{\log_6 5^4}$ **76.** $3^{\log_3 1}$

77. The number of binary digits (bits) that a digital computer requires to represent a positive integer N is the smallest integer greater than or equal to the value $\log_2 N$. Find the number of bits required to represent the following integers.
a. 843 **b.** 9,400 **c.** 16,000 **d.** 35,312

78. The cost of a typical \$1 item after a year of inflation at a rate of r percent per year is approximately $V = 2.7^r$. Rewrite this equation in logarithmic form.

79. A relation that relates power I, relative to some fixed power taken as a basic unit, to decibels (d is a measure of sound level) is $\dfrac{d}{10} = \log_{10} I$. Rewrite this as an exponential equation.

80. The population of a certain bacterial culture is found to fit the relation $P = 5(1.25^t)$, where P is the population after time t. Rewrite this in logarithmic form.

81. The following definition is incorrect; fix the exponential portion of the definition so it is correct. *Definition:* $\log_b x = y$ if and only if $x^y = b$, $b > 0$, $b \neq 1$.

82. Several of the following statements are incorrect. Fix them. For any base b, $b > 0$, $b \neq 1$,
a. $\log_b 1 = 1$ **b.** $\log_b b = 0$
c. $\log_b b^x = x$ **d.** $b^{\log_x b} = x$

Skill and review

1. Rewrite 9^{2x} as a power of 3.
2. Find the inverse function f^{-1} of the function $f(x) = 2 - 3x$ and graph both f and f^{-1}.
3. Solve $2x^6 + 15x^3 - 8 = 0$.
4. Graph $f(x) = x^4 - x$.
5. Solve $\frac{2x - 5}{3} - \frac{3x + 12}{2} = 4$.
6. Solve $2xy = \frac{x + y}{3}$ for y.

9–3 Properties of logarithmic functions

The time t necessary for an amount of money P to grow to an amount A at a fixed interest rate i, compounded daily, is approximated by the relation $t = \frac{1}{i}\log_{2.7}\left(\frac{A}{P}\right)$. Rewrite the right member so that the division $\frac{A}{P}$ is avoided.

In this section we study several properties of logarithms that are used extensively in solving logarithmic equations; they would allow us to do what is asked in this problem.

The one-to-one property of logarithmic functions

Logarithmic functions are one to one (since they have inverse functions). The first of the following properties was stated previously (section 9–1); the second property states the same thing about logarithmic functions.

One-to-one property for exponential functions

If $b^x = b^y$, then $x = y$.

One-to-one property for logarithmic functions

If $\log_b x = \log_b y$, then $x = y$.

The one-to-one property for logarithmic functions is used to solve certain logarithmic equations, as illustrated in example 9–3 A.

■ ***Example 9–3 A***

Solve the equation:

$$\log_3 5x = \log_3(3x + 2)$$

$$5x = 3x + 2 \quad \text{One-to-one property}$$

$$x = 1$$

■

Three important properties of logarithmic functions

Logarithmic functions have several important algebraic properties. We introduce these properties here, examine why they are true, and see some examples of their use.

Product-to-sum property of logarithms

$$\log_b(xy) = \log_b x + \log_b y, \text{ if } x > 0 \text{ and } y > 0.$$

Concept

The logarithm of a product is the same as the sum of the logarithms of each factor in the product.

The following shows why this property is true. Let $p = \log_b(xy)$, so $xy = b^p$. Let $q = \log_b x$, so $x = b^q$. Let $r = \log_b y$, so $y = b^r$.

Now we focus on xy:

$xy = x \cdot y$	
$b^p = b^q \cdot b^r$	Replace xy by b^p, x by b^q, y by b^r
$b^p = b^{q+r}$	Property of exponents
$p = q + r$	One-to-one property
$\log_b(xy) = \log_b x + \log_b y$	Replace p by $\log_b(xy)$, q by $\log_b x$, r by $\log_b y$

Example 9–3 B illustrates some ways in which this property is used.

■ *Example 9–3 B*

Apply the product-to-sum property of logarithms in each problem.

1. If $\log_5 2 \approx 0.4307$ and $\log_5 3 \approx 0.6826$, find an approximation for $\log_5 12$.

$\log_5 12 = \log_5(2 \cdot 2 \cdot 3)$	Factor 12 completely
$= \log_5 2 + \log_5 2 + \log_5 3$	Product-to-sum property
$\approx 0.4307 + 0.4307 + 0.6826$	Replace by given values
≈ 1.5440	

2. Solve the logarithmic equation $\log_3 x + \log_3(x + 8) = 2$.

$\log_3 x + \log_3(x + 8) = 2$	
$\log_3(x(x + 8)) = 2$	Product-to-sum property
$x(x + 8) = 3^2$	Rewrite as exponential equation
$x^2 + 8x - 9 = 0$	Multiply in left member
$(x - 1)(x + 9) = 0$	Factor left member
$x - 1 = 0$ or $x + 9 = 0$	Zero product property
$x = 1$ or -9	Solve each linear equation

The domain of any logarithmic function $f(x) = \log_b x$ is the nonnegative real numbers. Therefore, neither $\log_3 x$ nor $\log_3(x + 8)$ is defined for $x = -9$. Either of these undefined expressions means that *we must reject the solution* -9.

Thus, the result is the value 1.

Check for $x = 1$

$\log_3 x + \log_3(x + 8) = 2$	Original equation
$\log_3 1 + \log_3 9 = 2$	Replace x by 1
$0 + 2 = 2$	$\log_3 1 = 0$ and $\log_3 9 = 2$ ■

Just as one adds the logarithms of the factors of a product, one subtracts the logarithms of the factors of a quotient.

Quotient-to-difference property of logarithms

$$\log_b\left(\frac{x}{y}\right) = \log_b x - \log_b y, \text{ if } x > 0 \text{ and } y > 0.$$

Concept

The logarithm of a quotient is the same as the difference of the logarithms of the numerator and denominator.

It is left as an exercise to show why this property is true. Example 9–3 C illustrates this property.

Example 9–3 C

Use the quotient-to-difference property to solve the logarithmic equation $\log_{10}(x + 99) - \log_{10}x = 2$ for x.

$\log_{10}\frac{x + 99}{x} = 2$	Quotient-to-difference property
$\frac{x + 99}{x} = 10^2$	Rewrite as exponential equation
$\frac{x + 99}{x} = 100$	
$x + 99 = 100x$	Multiply each member by x
$99 = 99x$	Add $-x$ to both members
$1 = x$	Divide both members by 99

Since $\log_{10}(x + 99)$ and $\log_{10}x$ are both defined for $x = 1$ this value will check. ■

The following is one more important property of logarithms.

Exponent-to-coefficient property of logarithms

$$\log_b x^r = r \log_b x, \text{ if } x > 0.$$

Concept

The logarithm of an expression with an exponent is equivalent to the product of that exponent and the logarithm of that expression without the exponent.

The following shows why this is true. Let $m = \log_b x^r$, so $x^r = b^m$. Let $n = \log_b x$, so $b^n = x$.

$b^n = x$	We begin with this statement
$(b^n)^r = x^r$	Raise both sides to power r
$b^{nr} = x^r$	$(a^m)^n = a^{mn}$
$b^{nr} = b^m$	$x^r = b^m$ (because $m = \log_b x^r$)
$nr = m$	Exponential functions are one to one
$r \cdot n = m$	Rewrite order of left member
$r \cdot \log_b x = \log_b x^r$	$n = \log_b x$ and $m = \log_b x^r$

As with the previous properties, this one is often applied in a variety of situations, as illustrated in example 9–3 D.

■ *Example 9–3 D*

Apply the exponent-to-coefficient property of logarithms in each problem.

1. Write $\log_3\dfrac{9x^3y^5}{3z}$ in terms of $\log_3 x$, $\log_3 y$, and $\log_3 z$.

$\log_3\dfrac{9x^3y^5}{3z}$

$= \log_3(9x^3y^5) - \log_3(3z)$ — Quotient-to-difference property

$= \log_3 9 + \log_3 x^3 + \log_3 y^5 - (\log_3 3 + \log_3 z)$ — Product-to-sum property

$= 2 + 3\log_3 x + 5\log_3 y - 1 - \log_3 z$ — Exponent-to-coefficient property

$= 1 + 3\log_3 x + 5\log_3 y - \log_3 z$

2. Solve the logarithmic equation $\log_2 x^4 = 40$ for x; assume $x > 0$.

$\log_2 x^4 = 40$

$4 \cdot \log_2 x = 40$ — Exponent-to-coefficient property

$\log_2 x = 10$ — Divide both members by 4

$x = 2^{10}$ — Rewrite as an exponential equation

$x = 1{,}024$ — Evaluate 2^{10} ■

Note We assume $x > 0$ in part 2 of example 9–3 D so that the exponent-to-coefficient property would apply. An alternate solution is required without this assumption:

$$\log_2 x^4 = 40$$
$$x^4 = 2^{40}$$
$$x = \pm\sqrt[4]{2^{40}}$$
$$x = \pm 2^{10} = \pm 1{,}024$$

The properties of logarithms and the one-to-one property of exponential functions are summarized here. They should be memorized.

Summary of properties

If $b^x = b^y$, then $x = y$	One-to-one property of exponential functions
If $\log_b x = \log_b y$, then $x = y$	One-to-one property of logarithmic functions
$\log_b(xy) = \log_b x + \log_b y$	Product-to-sum property
$\log_b\left(\dfrac{x}{y}\right) = \log_b x - \log_b y$	Quotient-to-difference property
$\log_b x^r = r \cdot \log_b x$	Exponent-to-coefficient property

Mastery points

Can you

- Use the algebraic properties of logarithmic functions summarized here to solve certain logarithmic and exponential equations and transform certain logarithmic expressions?

Exercise 9–3

Solve the following logarithmic equations.

1. $\log_2 3x = 3$
2. $\log_4 5x = -2$
3. $\log_2 3x = \log_2 3$
4. $\log_5(1 - 2x) = \log_5 6$
5. $\log_3 5x = \log_3(2x + 1)$
6. $\log_2 3x = \log_2(x - 2)$
7. $\log_3 9 = x$
8. $\log_2 \frac{1}{16} = x$
9. $\log_2(5x - 1) = -4$
10. $\log_2 \dfrac{4}{x} = 3$
11. $\log_2 5x - 1 = -4$
12. $\log_2(-x) = \log_2 \frac{1}{8}$
13. $\log_5(x + 1) = \log_{10} 100$
14. $\log_3(x + 2) = \log_2 \frac{1}{8}$
15. $\log_2 2x + \log_2(x + 1) = 3$
16. $\log_2(x + 1) + \log_2(x - 1) = 4$
17. $\log_2 5 + \log_2(3 - 2x) = \log_2 6$
18. $\log_4 3 = \log_4 x + \log_4(x - 2)$
19. $\log_4 x - \log_4 3 = 2$
20. $\log_4(x + 1) - \log_4 x = 3$
21. $\log_5 x - \log_5 3 = \log_5 2$
22. $\log_5 2 - \log_5 x = \log_5 3$
23. $\log_3 2x = \log_3 2 + \log_3 x$
24. $\log_3 2x = \log_3 2 - \log_3 x$
25. $\log_2 x^4 = 12,\ x > 0$
26. $\log_5 x^3 = \log_5 5^9,\ x > 0$
27. $\log_2(x - 2) + \log_2(x + 3) = \log_2(x^2 - 3x + 2)$
28. $\log_{10}(x - 2) - \log_{10}(x + 3) = \log_{10}(3x + 2)$

Rewrite the following expressions in terms of $\log_a x$, $\log_a y$, and $\log_a z$ for the given value of a.

29. $\log_6(2xy)$
30. $\log_{10}(3xyz)$
31. $\log_4(4xyz) - \log_4 z$
32. $\log_4(2xy) + \log_4(3x)$
33. $\log_3 \dfrac{3xy}{2z}$
34. $\log_2 \dfrac{4x}{y}$
35. $\log_{10} \dfrac{1}{3xyz}$
36. $\log_3 \dfrac{2xyz}{15}$
37. $\log_2 4x^3y^2z^5$
38. $\log_3 9x^2y$
39. $\log_4 \dfrac{8y^4z^3}{x^3}$
40. $\log_{10} \dfrac{3x^{12}y^2z}{1{,}000}$

Assume $\log_a 2 \approx 0.3562$, $\log_a 3 \approx 0.5646$, and $\log_a 5 \approx 0.8271$. Use these values to find approximate values for the following logarithms.

41. $\log_a 6$
42. $\log_a 30$
43. $\log_a 36$
44. $\log_a 10$
45. $\log_a 81$
46. $\log_a 300$
47. $\log_a 0.2$
48. $\log_a 0.5$

49. Referring to the values given above for $\log_a x$, suppose it is also known that $\log_a 14 = 1.3562$. What is a? (Look at the decimal values.)

50. Estimate the value of $\log_2 9 + \log_3 30 + \log_5 20$, to the nearest integer.

51. An equation that occurs when measuring sound levels relative to an initial sound level of 100 is $\alpha = 10 \log_{10}\left(\dfrac{I}{100}\right)$. Rewrite the right member of this equation using the properties of logarithms.

52. The time t necessary for an amount of money P to grow to an amount A at a fixed interest rate i, compounded daily, is approximated by the relation $t = \dfrac{1}{i}\log_{2.7}\left(\dfrac{A}{P}\right)$. Rewrite the right member using the property of the logarithm of a quotient.

53. Use the properties of logarithms to prove that $\log_a \sqrt[n]{x} = \dfrac{\log_a x}{n}$.

54. Prove that $\log_a \dfrac{x}{y} = \log_a x - \log_a y$; use the proof of the fact that $\log_a(xy) = \log_a x + \log_a y$ as a guide.

55. One property of logarithms is $\log_a(xy) = \log_a x + \log_a y$. Show that the following is not a property: $\log_a(x + y) = \log_a x + \log_a y$. Do this by finding values of a, x, and y for which you know all the values and show that the left member of the equation is not equal to the right member for those values.

56. Is the following a property of logarithms (see problem 55)?

$$\log_a(xy) = (\log_a x)(\log_a y)$$

Skill and review

1. If $2^3 < 2^x < 2^4$, what can be said about x?
2. Solve $3^{2x} = 27$ for x.
3. If $\log_x 125 = 3$, what is x?
4. Solve $x^3 + 2x^{3/2} - 3 = 0$.
5. Solve $|2x - 5| = 10$.
6. Graph $f(x) = x^2 + 3x - 5$.

9–4 Values and graphs of logarithmic functions

Probabilistic risk assessment is used to predict the reliability of electronic equipment, aircraft, nuclear power plants, spacecraft, etc. A reliability function R is defined to be $R(t) = e^{-t/\text{MTBF}}$, where t represents time and MTBF means "mean time between failures," the average time it takes for a given piece of equipment to fail. Assuming a MTBF of 1,500 hours for a certain computer, compute the probability that the computer will run for at least 1,000 hours without failure.

In this section we study the mathematical tools we need to answer questions like this one.

Until the 1970s logarithms were used extensively for performing computations involving multiplications, divisions, and extractions of roots. Indeed, this is the very purpose for which Napier created logarithms. Also, the values of logarithms were found using printed tables.

In the 1970s, electronic computing devices made these applications obsolete but made other uses of logarithms more important. For example, we might use a $\boxed{y^x}$ key on a calculator to compute $3.7^{2.1}$. The calculator or computer uses logarithms internally to find the required value. Logarithms are also used extensively in computer science to describe the performance of algorithms. Of course logarithms, like everything else in this text, are also used in more advanced mathematics courses.

The answer to many problems involve the numeric computation of a logarithm. Calculators are programmed to produce the values of logarithms to two bases, 10 and e. The base e is a constant with a value about 2.7. It is discussed later in this section, after we discuss the base 10.

Common logarithms

A scientific electronic calculator is programmed to produce values of **common logarithms.** Common logarithms are logarithms to the base 10. Usually $\log_{10}x$ is abbreviated as simply $\log x$ (the base is assumed to be 10).

Common logarithm

$\log x$ means $\log_{10}x$; it is called the common logarithm of x.

The necessary keystrokes for computing approximations to $\log x$ for a given calculator may differ, but the $\boxed{\log}$ key is practically universal. Some calculators may also require a "function" or "2nd" key.

■ Example 9–4 A

Use a calculator to approximate the values of the following common logarithms. Round the results to 4 decimal places.

	Typical scientific calculator	TI-81
1. log 50 = 1.6990	50 [log]	[LOG] 50 [ENTER]
2. log 0.5 = −0.3010	.5 [log]	[LOG] .5 [ENTER] ■

Note from example 9–4 A that log 0.5 is negative. The common logarithm of x when $0 < x < 1$ must always be negative. This is because $\log 1 = 0$ and the common logarithm function is an increasing function so if $x < 1$, $\log x < \log 1 = 0$. (We will see the graph of $y = \log x$ shortly. It will confirm that this is an increasing function.)

■ Example 9–4 B

Use a calculator to approximate log 1,230,000,000,000,000 to 5 decimal places.

Since this value is too large to directly enter into a calculator, write it in scientific notation first:

$$1{,}230{,}000{,}000{,}000{,}000 = 1.23 \times 10^{15}.$$

Calculators will accept values in this form, but we will illustrate a more general method, which uses the product-to-sum property.

$$\begin{aligned}\log(1.23 \times 10^{15}) &= \log 1.23 + \log 10^{15} && \text{Product-to-sum property}\\ &= 0.0899 + 15 \log 10 && \text{Exponent-to-coefficient property}\\ &= 0.0899 + 15 && \log 10 = 1\\ &= 15.0899\end{aligned}$$

■

Natural logarithms

Earlier we mentioned that calculators will also calculate logarithms to the base e. The symbol e is used, like π, to represent a certain value. Like π, e is an irrational number; it has been calculated to over 100,000 decimal places, and is approximately 2.718 281 828 459 045 235. The symbol e is credited to Leonard Euler, and first appeared in the year 1727. The origin of e is discussed after example 9–4 G.

Logarithms to the base e are called **natural logarithms,** and $\log_e x$ is often abbreviated[2] as $\ln x$.

> **Natural logarithm**
> $\ln x$ means $\log_e x$ and is called the natural logarithm of x.

[2]Irving Stringham used this notation in 1893.

To obtain values of natural logarithms we use a key that is usually marked [ln] on a calculator. This is illustrated in example 9–4 C.

■ ***Example 9–4 C***

Approximate the natural logarithm; round results to 4 decimal places.

1. ln 100 ≈ 4.6052 100 [ln] TI-81: [LN] 100 [ENTER]

2. ln 10 ≈ 2.3026 10 [ln] TI-81: [LN] 10 [ENTER] ■

Graph of the common and natural logarithm functions

The exponential function to the base e and the natural logarithm function are inverses of each other, since they both use the same base, e. Similarly, the exponential function to the base 10 is the inverse function of the common logarithm function.

The graphs of the common and natural logarithm functions are easily found by reflecting the graphs of the functions $f(x) = 10^x$ and $f(x) = e^x$ about the line $y = x$. This is shown in figure 9–4, which allows us to see the following properties of these functions.

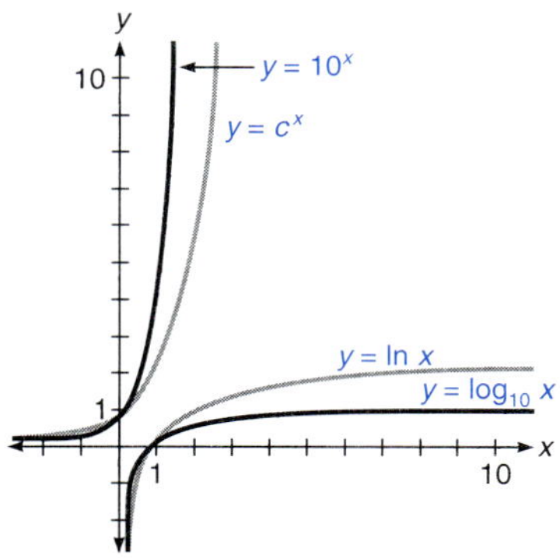

Figure 9–4

[Y=] [10^x] [X|T] [ENTER]
[e^x] [X|T] [ENTER] [LOG]
[X|T] [ENTER] [LN] [X|T]
[RANGE −2,8,−2,8]

The domains and ranges for the natural and common logarithm and exponential functions to base 10 and base *e*

Function	Domain	Range
log x, ln x	$\{x \mid x > 0\}$	R
10^x, e^x	R	$\{y \mid y > 0\}$

Observe also that *the common and natural logarithm functions are increasing functions*. Indeed, it can be shown that $f(x) = \log_b x$ is an increasing function if $b > 1$, and a decreasing function if $0 < b < 1$.

There are many instances where we need to find a decimal approximation of 10 or e raised to some power. This is done with a calculator, as shown in example 9–4 D.

■ ***Example 9–4 D***

1. Approximate $10^{3.28}$ to the nearest integer.

Since $10^3 = 1{,}000$ the value should be somewhat more than 1,000. A typical sequence of key strokes is

3.28 [shift] [log] or 3.28 [10^x] — Typical scientific calculator

[2nd] [LOG] 3.28 [ENTER] — TI-81

$10^{3.28} = 1{,}905$ to the nearest integer.

2. Approximate $e^{4.1}$ to the nearest tenth.

Since $e \approx 3$, $e^{4.1} \approx 3^4 = 81$. Using a sequence like

4.1 [shift] [ln] or 4.1 [e^x] — Typical scientific calculator

[2nd] [LN] 3.28 [ENTER] — TI-81

we obtain 60.3, to the nearest tenth. ■

The properties of composition of exponential with logarithmic function, and composition of logarithmic with exponential function, from section 9–2, when put in terms of common and natural logarithms, state:

> **Composition of exponential/logarithm functions for base 10 and *e***
>
> [1] $\log 10^x = x$ [3] $\ln e^x = x$
>
> [2] $10^{\log x} = x$ [4] $e^{\ln x} = x$

Example 9–4 E illustrates applications of these properties.

■ ***Example 9–4 E***

Simplify each expression.

1. Find log 10,000.

$$\begin{aligned} \log 10{,}000 &= \log 10^4 \\ &= 4 \end{aligned}$$ $\log 10^x = x$

2. Simplify $\ln(2e^{5x})$.

$$\begin{aligned} \ln(2e^{5x}) &= \ln 2 + \ln e^{5x} && \text{Product-to-sum property} \\ &= \ln 2 + 5x && \ln e^x = x \end{aligned}$$

3. Simplify $e^{\ln(3x)^4}$.

$$\begin{aligned} e^{\ln(3x)^4} &= (3x)^4 && e^{\ln x} = x \\ &= 81x^4 \end{aligned}$$ ■

Change-of-base formula

There are many instances where we need to know the value of a logarithm to an arbitrary base. For example, in computer science the bases 2, 8, or 16 are common. In section 9–5 we will also see further instances of the need to compute logarithms to any base.

Suppose then, that we need to compute the quantity $y = \log_b x$, where b is a base for which a calculator is not preprogrammed. Assume that we do have the values to another base, m, available to us. We can develop a formula that will give us the $\log_b x$ in terms using only values of logs to the base m.

Let	
$y = \log_b x$	
$b^y = x$	Rewrite in exponential form
$\log_m b^y = \log_m x$	Take the logarithm to the base m of both members
$y \log_m b = \log_m x$	Exponent-to-coefficient property
$y = \dfrac{\log_m x}{\log_m b}$	Replace y by $\log_b x$

Change-of-base formula

$$\log_b x = \frac{\log_m x}{\log_m b}$$

In practice, we often use $m = 10$, so that the change-of-base property becomes the following.

Change-to-common log formula

$$\log_b x = \frac{\log x}{\log b}$$

This diagram can help us remember the formula: $\log_b x = \dfrac{\log x}{\log b}$. Of course, we can use the base e also; then the property would look like $\log_b x = \dfrac{\ln x}{\ln b}$. Example 9–4 F illustrates the change-of-base formula.

■ ***Example 9–4 F***

Use the change-to-common log formula to compute $\log_7 100$ to four decimal places.

$$\log_7 100 = \frac{\log 100}{\log 7} \approx \frac{2}{0.8451} \approx 2.3666$$

100 [log] [÷] 7 [log] [=]	Typical scientific calculator
[LOG] 100 [÷] [LOG] 7 [ENTER]	TI-81

It is a good idea to check this value in the following way:

Since $7^2 = 49$ and $7^3 = 343$, we know $2 < \log_7 100 < 3$. ■

Graphs of logarithmic functions

In section 4–5 we observed that the graphs of a function and its inverse function are symmetric about the line $y = x$. Since the logarithmic and exponential functions are inverses their graphs are mirror images about this line. This is illustrated in figure 9–5; part a shows the case for $b < 1$, and part b for $b > 1$.

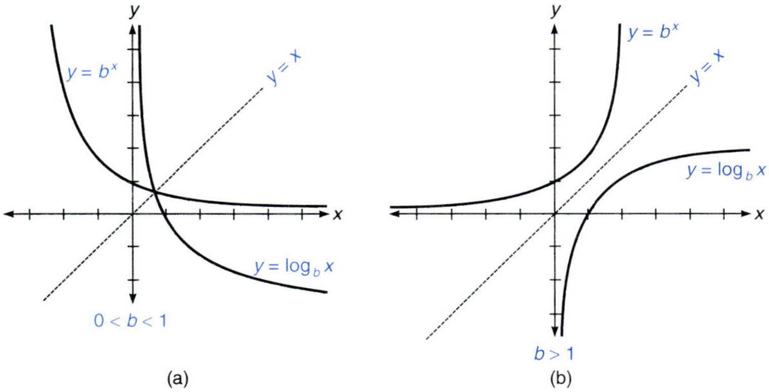

Figure 9–5

Since we already know how to graph exponential functions (from section 9–1), and since a logarithmic function is the inverse of some exponential function, we can use exponential functions to graph logarithmic functions in the following way.

> **To graph a logarithmic function**
> 1. Replace $f(x)$ by y.
> 2. Replace every x by y and every y by x.
> 3. Solve for y. (This is the inverse of the logarithmic function.)
> 4. Graph this equation. (This is the inverse of the desired graph.)
> 5. Reflect the result about the line $y = x$ to obtain the desired graph.

This method of graphing logarithmic functions has the advantage that we do not have to memorize any more basic graphs.

Another method for graphing these functions is to memorize the basic graph of $y = \log_b x$ and use vertical and horizontal translations and vertical scaling, as covered in section 3–4.

Of course the graphing calculator or computer can be used to obtain these graphs also. The essential information for the TI-81 is shown.

■ *Example 9–4 G*

Graph the following logarithmic functions. State the value of any intercepts.

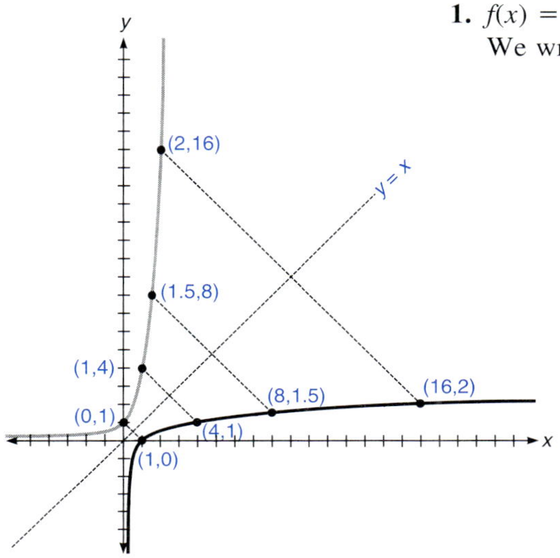

1. $f(x) = \log_4 x$

We write as $y = \log_4 x$, then find the inverse function:

$x = \log_4 y$	Exchange x and y
$y = 4^x$	Solve for y

Graph the function $y = 4^x$ first, then reflect the result about the line $y = x$. This is shown in the figure.

x-intercept:

$0 = \log_4 x$	Replace $f(x)$ by 0
$4^0 = x$	
$1 = x$	

Points for $y = 4^x$.

x	4^x
−1	0.25
1	4
1.5	8
2	16

Use the change-of-base formula to rewrite $f(x) = \log_4 x$ as $\dfrac{\log x}{\log 4}$.

Y=	LOG	X|T	÷	LOG 4	RANGE −1,17,−2,3

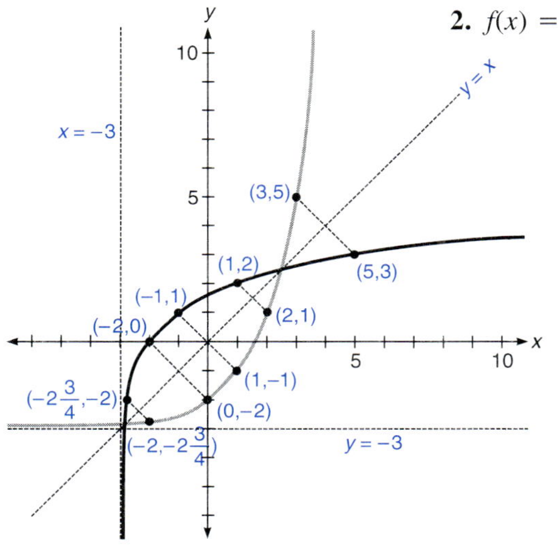

2. $f(x) = \log_2(x + 3)$

We find the inverse:

$y = \log_2(x + 3)$	$y = f(x)$
$x = \log_2(y + 3)$	Exchange x and y
$y + 3 = 2^x$	Write in exponential form
$y = 2^x - 3$	Solve for y

Graph $y = 2^x - 3$, the inverse function, then reflect about the line $y = x$, as shown.

y-intercept:

$y = \log_2(0 + 3)$	Replace x by 0 in $y = \log_2(x + 3)$
$y = \log_2 3$	y-intercept is $(0, \log_2 3)$

x-intercept:

$0 = \log_2(x + 3)$	Replace y by 0 in $y = \log_2(x + 3)$
$2^0 = x + 3$	
$-2 = x$	x-intercept is $(-2,0)$

Points for $y = 2^x - 3$.

x	$2^x - 3$
−2	−2.75
−1	−2.5
1	−1
2	1
3	5

Use the change-of-base formula to rewrite f as $f(x) = \dfrac{\log(x + 3)}{\log 2}$.

Y=	LOG	(	X|T	+	3	)	÷	LOG	2

RANGE −4,10,−4,10

The number *e*

One way to appreciate where the value of e comes from is shown in the following sequence, where we calculate the value of $\left(1 + \dfrac{1}{n}\right)^n$ for increasing values of n.

n	$\left(1 + \dfrac{1}{n}\right)^n$	
1	$\left(1 + \dfrac{1}{1}\right)^1$	$= 2$
10	$\left(1 + \dfrac{1}{10}\right)^{10}$	$\approx 2.5937\ldots$
100	$\left(1 + \dfrac{1}{100}\right)^{100}$	$\approx 2.7048\ldots$
1,000	$\left(1 + \dfrac{1}{1{,}000}\right)^{1{,}000}$	$\approx 2.7169\ldots$
10,000	$\left(1 + \dfrac{1}{10{,}000}\right)^{10{,}000}$	$\approx 2.7181\ldots$

As we can see the values of $\left(1 + \dfrac{1}{n}\right)^n$ do not change much as n gets greater and greater.[3] In fact, these values get closer and closer to the value $e \approx 2.71828$.

Although we cannot prove it here, the expression $\left(1 + \dfrac{x}{n}\right)^n$ gets closer and closer to the value e^x as n gets larger and larger.

Compound interest

It turns out that the expression $\left(1 + \dfrac{x}{n}\right)^n$ occurs in many applied situations. For example, consider compound interest. Suppose a bank pays a simple interest rate of $i = 8\%$, compounded quarterly (four times per year). This means that the bank really pays $2\% \left(\dfrac{8\%}{4}\right)$ per quarter. Suppose a principal P of \$100 is deposited. The table shows how much is in the account after each quarter, using the fact that to compute total amount at the end of a quarter, increase the previous amount by 2%. This is accomplished in one step by multiplying the previous amount by 102% or 1.02.

[3]The TI-81 handles this calculation up to n = one billion (1 [EE] 12), but fails at $n = 10$ billion (1 [EE] 13). Calculators do have limitations.

End of quarter	Computation	Amount	Computations to date
1	\$100(1.02)	\$102	$\$100(1.02)^1$
2	\$102(1.02)	\$104.04	$\$100(1.02)^2$
3	\$104.04(1.02)	\$106.12	$\$100(1.02)^3$
4	\$106.12(1.02)	\$108.24	$\$100(1.02)^4$
5	\$108.24(1.02)	\$110.40	$\$100(1.02)^5$
. . .	. . .	. . .	. . .
m	. . .	. . .	$\$100(1.02)^m$

Thus to compute the present value A in a bank account paying a yearly simple interest rate i compounded n times per year after t years, on an initial deposit, P (the principal) one computes

$$[1] \qquad A = P\left(1 + \frac{i}{n}\right)^{nt}$$

For one year $t = 1$, and the amount is $A = P\left(1 + \frac{i}{n}\right)^n$. As n, the number of compounding periods, increases, this quantity gets closer and closer to the value given by $A = Pe^i$, which defines the amount paid in an account (after one year) in which the interest is said to be compounded continuously. After time t, in years, the amount in an account on which the interest at rate i is compounded continuously is

$$[2] \quad A = Pe^{it}$$

Example 9–4 H illustrates both of these formulas.

■ ***Example 9–4 H***

1. A bank account pays 10% simple interest, compounded quarterly. What will be the value of a deposit of \$1,000 after two years?

Use formula [1] with $P = 1{,}000$, $i = 10\% = 0.1$, $n = 4$, and $t = 2$:

$$\begin{aligned} A &= 1{,}000\left(1 + \frac{0.1}{4}\right)^{4(2)} \\ &= 1{,}000(1.025)^8 \\ &\approx 1{,}000(1.21840) \\ &\approx 1{,}218.40 \end{aligned}$$

Thus, \$1,000 grows to \$1,218.40 after two years if interest is compounded quarterly.

2. Assume the money of part 1 of this example is deposited in an account in which the interest is compounded continuously. What is the amount after two years?

Use formula [2] with $i = 0.1$ and $t = 2$:

$$\begin{aligned} A &= 1{,}000e^{0.1(2)} \\ &= 1{,}000e^{0.2} \\ &\approx 1{,}000(1.22140) \approx 1{,}221.40 \end{aligned}$$

Thus, \$1,000 grows to \$1,221.40 after two years if interest is compounded continuously. ■

Mastery points

Can you

- Use a calculator to compute the common or natural logarithm of any positive real number?
- Use the change-to-common log formula to compute the value of a logarithm to any base?
- Sketch the graphs of the natural and common logarithm functions and state their domains and ranges?
- Graph logarithmic equations?
- Use the formulas $A = P\left(1 + \frac{i}{n}\right)^{nt}$ and $A = Pe^{it}$ to compute compound interest?

Exercise 9–4

Use a calculator to approximate the values of the following logarithms. Round the results to 4 decimal places.

1. log 52 **2.** log 17 **3.** log 2.55 **4.** log 190 **5.** log 10.6 **6.** log 2,500
7. log 0.85 **8.** log 0.003 **9.** log 8,720 **10.** ln 52 **11.** ln 17 **12.** ln 2.55
13. ln 190 **14.** ln 10.6 **15.** ln 2,500 **16.** ln 0.85 **17.** ln 0.003 **18.** ln 8,720
19. log 7,920,000,000,000,000
20. log 2,003,400,000,000,000,000,000,000
21. log 90,000,000,000,000,000,000
22. log 718,420,000,000,000
23. log 0.000 000 000 000 000 2
24. log 0.000 000 000 000 000 009 129
25. log 0.000 000 000 120 004

26. Compute the common logarithm of Avogadro's number (from chemistry), to four decimal places; Avogadro's number is 6.024×10^{23}.

27. Approximate the common logarithm of Planck's constant (from physics), 6.63×10^{-27}, to four decimal places.

28. The speed of light in a vacuum is about 2.99776×10^{10} centimeters per second. Compute the common logarithm of this value to four decimal places.

29. Assume the age of the known universe to be 18 billion years; compute the common logarithm of this value, to four decimal places.

Approximate the given logarithm; round to 4 decimal places.

30. $\log_5 19$ **31.** $\log_{20} 2{,}000$ **32.** $\log_2 3.89$ **33.** $\log_8 0.78$ **34.** $\log_{0.25} 8$ **35.** $\log_3 0.95$

36. Sketch the graph of the common logarithm function.

37. Sketch the graph of the natural logarithm function.

38. State the domain and range of the common and natural logarithm functions.

Graph the following logarithmic functions. State the value of all intercepts.

39. $f(x) = \log_3 x$ **40.** $f(x) = \log_4(x - 2)$ **41.** $f(x) = \log_2(x - 1)$ **42.** $f(x) = \log_3 2x$
43. $f(x) = \log_4 x - 1$ **44.** $f(x) = \log_2 x + 2$ **45.** $f(x) = \log_2 3x$ **46.** $f(x) = \log_3(-x)$

Approximate the following values to 2 decimal places.

47. $10^{2.9}$ **48.** $10^{4.82}$ **49.** $10^{-0.33}$ **50.** $e^{3.1}$ **51.** $e^{1.85}$ **52.** $e^{10.6}$
53. $e^{4.8}$ **54.** $10^{3.02}$ **55.** $10^{2/3}$

Simplify the following expressions.

56. $\log 1{,}000$ **57.** $\ln 2e^{3x}$ **58.** $\ln e^{(2x+1)}$ **59.** $\log 10^{(4-3x)}$ **60.** $10^{\log 100}$ **61.** $e^{\ln 100}$

62. $e^{\ln x^2}$ **63.** $10^{\log (x-1)^2}$ **64.** $e^{\ln 3x} + \ln e^{3x}$ **65.** $\ln 5e^{x}$ **66.** $10^{\log \sqrt{2}}$

67. \$1,800 is deposited in a bank account that computes its interest continuously. The simple interest rate is 8.5% per year. Use the formula $A = Pe^{it}$ to compute the amount in the account after $5\frac{1}{2}$ years.

68. Find the amount of money on deposit in a bank account after 18 months if the initial deposit was \$5,000 and the simple yearly interest rate is 11.5% compounded continuously. See problem 67.

69. Use the formula $A = P\left(1 + \dfrac{i}{n}\right)^{nt}$ to find the value, after 2 years, of an account in which \$1,000 was deposited at an interest rate of 8% per year if the interest is compounded monthly.

70. Use the formula $A = P\left(1 + \dfrac{i}{n}\right)^{nt}$ to find the value after 18 months of an account in which \$1,800 was deposited at an interest rate of 6.5% per year, if the interest is compounded monthly.

71. The logarithms created by John Napier in the seventeenth century were neither common nor natural logarithms. According to Howard Eves in his book *Great Moments in Mathematics Before 1650,* Napier's values could be found by the function $\text{Nap}\log x = 10^7\log_{1/e}\left(\dfrac{x}{10^7}\right)$.
Show that
$\text{Nap}\log x = 10^7(7\ln 10 - \ln x)$.

72. Referring to problem 71, compute the values that Napier would have obtained for his system of logarithms for the following values of x.
a. 1 **b.** 10 **c.** 100 **d.** one million
Leave answers in scientific notation with four-digit accuracy.

73. The Weber-Fechner law, from psychology, states that sound loudness S is given by $S = k\log\left(\dfrac{I}{I_0}\right)$, where I is the intensity of sound compared to an initial reference intensity, I_0. Assuming that $k = 12$, find S if I is 6 times the value of I_0. Round to one decimal place.

74. The Smith chart is used in electronics to study the performance of antennas. It uses the property that $\log n$ and $\log \dfrac{1}{n}$ are equal distances from zero. Show that this is true; that is, show that $|\log n| = \left|\log \dfrac{1}{n}\right|$ for all $n > 0$.

75. The surge impedance Z_0, in ohms, in a two-wire conductor is given by the relation $Z_0 = k\log\dfrac{b}{a}$, where a is the radius of the wires and b is the distance between the centers of the wires. If $k = 276$, find Z_0 for wire of radius $\frac{1}{8}$ inches, with centers separated by a distance of $\frac{3}{4}$ inches. Round to the nearest unit.

76. Probabilistic risk assessment is used to predict the reliability of pieces of electronic equipment, aircraft, nuclear power plants, spacecraft, etc. A reliability function R is defined to be $R(t) = e^{-t/\text{MTBF}}$, where t represents time and MTBF means mean time between failures, the average time it takes for a given piece of equipment to fail. Assuming a MTBF of 1,500 hours for a certain computer, compute R for 1,000 hours (the probability that the computer will run for at least 1,000 hours without failure). Round to two decimal places.

77. In designing a heating/cooling system that depends on water flowing through a pipe buried in the earth, one uses the formula $Q = 0.07L\left(\dfrac{T_{\text{in}} - T_{\text{out}}}{\log\dfrac{T_{\text{earth}} - T_{\text{in}}}{T_{\text{earth}} - T_{\text{out}}}}\right)$, where Q is heat transfer in BTU/hour, T is the temperature at the pipe inlet (in) and outlet (out) and of the earth, and L is the pipe length. Assuming the temperature of the earth to be 54°, the temperature at the inlet to be 30°, and at the outlet to be 42°, with a pipe length of 80 feet, find Q.

78. An oblate spheroid is similar in appearance to an egg; the earth has the shape of an oblate spheroid. The surface area S of such an object is given by $S = 2\pi a^2 + \pi\dfrac{b^2}{\varepsilon}\ln\left(\dfrac{1+\varepsilon}{1-\varepsilon}\right)$, where a, b, and ε (eccentricity) are parameters which describe the spheroid. Find S to the nearest tenth if $a = 14$ inches, $b = 8$ inches, and $\varepsilon = \frac{1}{4}$.

79. In medicine, the formula

$$D_{L_{CO}} = \frac{V_A}{(P_B - 47)(t_2 - t_1)} \cdot \log \frac{F_{A_{CO_{t_1}}}}{F_{A_{CO_{t_2}}}}$$ is part of computing carbon monoxide (CO) diffusing capacity ($D_{L_{CO}}$) across the alveolocapillary membrane, where V_A = alveolar volume, P_B = barometric pressure, $t_2 - t_1$ = time interval of measurement, $F_{A_{CO_{t_1}}}$ = fraction of CO in alveolar gas before diffusion, and $F_{A_{CO_{t_2}}}$ = fraction of CO in alveolar gas at end of diffusion. Find $D_{L_{CO}}$ if $V_A = 40$ centiliters, $P_B = 52$, $t_2 - t_1 = 10$ seconds, $F_{A_{CO_{t_1}}} = 0.01$, and $F_{A_{CO_{t_2}}} = 0.004$. The units will be centiliters per second.

Skill and review

1. Solve $2x^2 - 9x + 4 = 0$.
2. Solve $2x^4 - 9x^2 + 4 = 0$.
3. Solve $2x - 9\sqrt{x} + 4 = 0$.
4. Solve $2(x - 3)^2 - 9(x - 3) + 4 = 0$.
5. Rewrite $\log_a \frac{2x^4}{3y^3z}$ in terms of logarithms to the base a.
6. Solve $\log_2 x = -3$.
7. Graph $f(x) = \log_3(x - 1)$.
8. Solve $\frac{x^2 - 4}{x^2 - 1} > 2$.

9–5 Solving logarithmic and exponential equations/applications

The Richter scale was invented in 1935 by Charles F. Richter to measure the intensity of earthquakes. Each number on the scale represents an earthquake 10 times stronger than one of the next lower magnitude. For example, an earthquake of 5 on the Richter scale is 10 times stronger than one of measure 4 on the Richter scale. Suppose one earthquake measures 4.5 on the Richter scale, and a second measures 6.2. To the nearest unit, how many times stronger is the second earthquake than the first?

In this section we study ways to solve problems like this, which can be described using logarithmic and exponential equations. We first look at the various techniques available to us to solve these types of equations, and then we see a few of the many places where these equations arise.

Equation solving techniques

In this section we introduce more techniques for solving logarithmic and exponential equations.

Some equations can be solved by taking the logarithm of both members of the equation. Although we will use common logarithms, the procedure and result is the same using natural logarithms. Example 9–5 A illustrates this method of solving equations.

■ *Example 9–5 A*

Solve the equation for x; also find a decimal approximation to the answer to the nearest 0.1.

$3^{x+1} = 2^{3x-1}$

$\log 3^{x+1} = \log 2^{3x-1}$ — Take the common logarithm of both members

$(x + 1)\log 3 = (3x - 1)\log 2$ — Exponent-to-coefficient property

$x \log 3 + \log 3 = 3x \log 2 - \log 2$	Perform the indicated multiplications
$x \log 3 - 3x \log 2 = -\log 2 - \log 3$	Put x terms all in one member
$x(\log 3 - 3 \log 2) = -(\log 2 + \log 3)$	Factor x in the left member
$x = -\dfrac{\log 2 + \log 3}{\log 3 - 3 \log 2}$	Divide each member by $\log 3 - 3 \log 2$
$x \approx 1.8$	Approximate value

2 [log] [+] 3 [log] [=] [÷] [(] 3 [log] [−] 3 [×] 2 [log] [)] [=] [+/−]

TI-81: [(−)] [(] [(] [LOG] 2 [+] [LOG] 3 [)] [÷] [(] [LOG] 3 [−] 3 [LOG] 2 [)] [)] [ENTER] ■

An expression of the form $(\log x)^2$ is often written as $\log^2 x$ for convenience. It is important to distinguish between $\log^2 x$ and $\log x^2$. $\log^2 x$ means $(\log x)^2$, whereas $\log x^2$ means $\log(x^2)$.

The expression $\log^2 x$ means to evaluate the logarithm, then square that value. $\log x^2$ means square x, then evaluate the logarithm. For example, if $x = 100$,

$$\log^3 100 = (\log 100)^3 = 2^3 = 8$$
$$\log 100^3 = 3 \log 100 = 3(2) = 6.$$

Example 9–5 B shows that the technique of substitution for expression can help in some cases (section 1–3).

■ ***Example 9–5 B***

1. Under certain conditions the equation $y = \frac{1}{2}(e^x + e^{-x})$ describes the shape of a cable (such as a phone line) hanging between two poles. Solve this equation for x when y has the value 3.

$$y = \tfrac{1}{2}(e^x + e^{-x})$$
$$3 = \tfrac{1}{2}(e^x + e^{-x})$$
$$6 = e^x + e^{-x}$$
$$6 = e^x + \frac{1}{e^x}$$

Let $u = e^x$

$$6 = u + \frac{1}{u}$$
$$6u = u^2 + 1$$
$$u^2 - 6u + 1 = 0$$
$$u = 3 \pm 2\sqrt{2}$$
$$e^x = 3 \pm 2\sqrt{2}$$
$$x = \ln(3 \pm 2\sqrt{2})$$
$$x \approx 1.76 \text{ or } -1.76$$

2. Solve $(\log x)(2 \log x + 1) = 6$

$$2 \log^2 x + \log x - 6 = 0$$
$$2u^2 + u - 6 = 0$$
$$(2u - 3)(u + 2) = 0$$
$$2u - 3 = 0 \text{ or } u + 2 = 0$$
$$u = \tfrac{3}{2} \text{ or } u = -2$$
$$\log x = \tfrac{3}{2} \text{ or } \log x = -2$$
$$10^{3/2} = x \text{ or } 10^{-2} = x$$
$$x = 10\sqrt{10} \text{ or } x = \frac{1}{100}$$
$$x \approx 31.6 \text{ or } 0.01$$

■

Graphing calculators and computers provide a powerful tool for estimating the values of solutions to equations that can be expressed in terms of one variable. Example 9–5 C reviews one way to solve equations with graphing calculators. This was shown in more detail in section 4–3.

■ ***Example 9–5 C***

Solve $3 = \frac{1}{2}(e^x + e^{-x})$ graphically. (Part 1 of example 9–5 B.)

Graph the function $y = \frac{1}{2}(e^x + e^{-x}) - 3$. The solutions to the original problem are the zeros of this graph.

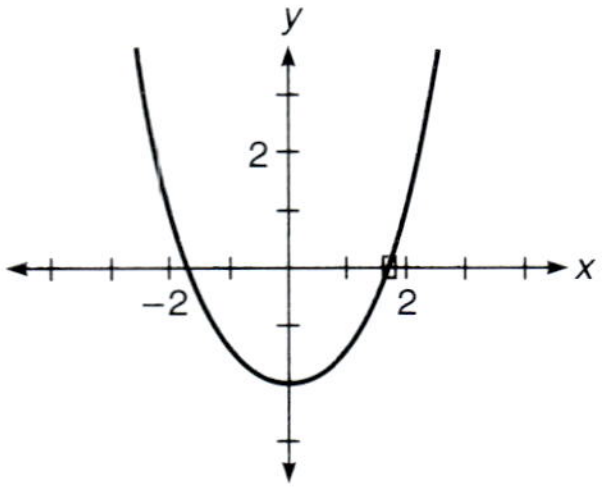

The graph is as shown. Use TRACE to position the cursor as shown, then PRGM NEWTON (section 4–3) to obtain a highly accurate value for the zero. Doing this displays the value 1.762747174. Analysis of the TI-81 program NEWTON will show that it delivers a result that differs from the actual value by less than one part in 10 billion. That is, $\left|\dfrac{\text{actual } x - \text{estimated } x}{\text{actual } x}\right| \le 10^{-10}$.

It can be seen from the graph that the other zero is the negative of the first, so it is not necessary to compute it. (Formally, $y = \frac{1}{2}(e^x + e^{-x}) - 3$ is an even function (section 3–5), which means its graph is symmetric about the y-axis.) ■

It is important to note that *the graphical method shown in example 9–5 C is of limited usefulness for many of the equations in this section.* This is because the graphs are not easy to view in their entirety and because there are mathematical limitations to all calculating devices. For example, the answer to $\log(\log x) = 2$ is 10^{100}, a value that is beyond the capability of most calculators and computers. The answers to $\log\sqrt{x} = \sqrt{\log x}$ are 1 or 10,000. Whatever graph is used will not show both solutions. We must reset the range to discover them both. This is an impractical procedure in these extreme cases.

Applications

Logarithmic and exponential equations appear in many business and scientific applications.

Compound interest applications

The formula $A = Pe^{it}$ relates present amount A, principal P, and time t in years in an account or loan paying (or demanding) simple interest rate i, compounded continuously. (This was introduced in the previous section.) Example 9–5 D illustrates this formula.

■ ***Example 9–5 D***

An account pays 12% interest compounded continuously. How long would it take $1,000 to double in value in this account?

$P = 1{,}000$, $i = 0.12$, and $A = 2{,}000$; we want t.

$A = Pe^{it}$	
$2{,}000 = 1{,}000e^{0.12t}$	Replace variables with the given values
$2 = e^{0.12t}$	Divide each member by 1,000
$\ln 2 = \ln e^{0.12t}$	Take the natural logarithm of both members
$\ln 2 = 0.12t$	$\ln e^x = x$
$\dfrac{\ln 2}{0.12} = t$	Divide both members by 0.12; this is the exact value of t
$5.78 \approx t$	2 [ln] [÷] 0.12 [=]

Thus, it takes about 5.8 years for the $1,000 to attain the value $2,000. ■

Growth and decay applications

A large class of problems are called **growth and decay problems.** In a situation where the rate of growth or decay of something is a constant proportion of the amount present, exponential equations describe the situation.

> **Growth and decay**
>
> If a quantity varies continuously at a certain rate r then the quantity q after time t starting with an original quantity q_0, is given by the growth/decay equation
>
> $$q = q_0e^{rt}$$
>
> When $r > 0$ we have growth, and when $r < 0$ we have decay.

Example 9–5 E illustrates a growth and decay situation.

■ ***Example 9–5 E***

How long will it take the earth's population to double if it grows continuously at the rate of 2.7 percent per year?

If an original population q_0 doubles, then at that time $q = 2q_0$. We are also told that $r = 0.027$.

$q = q_0e^{rt}$	Growth/decay equation
$2q_0 = q_0e^{0.027t}$	Replace variables by given values
$2 = e^{0.027t}$	Divide both members by q_0.
$\ln 2 = \ln e^{0.027t}$	Take the natural logarithm of both members
$\ln 2 = 0.027t$	$\ln e^x = x$
$t = \dfrac{\ln 2}{0.027}$	Divide both members by 0.027
$t \approx 25.7$	

Thus, at 2.7% annual growth the earth's population will double in about 26 years. ■

Two additional applications are illustrated in example 9–5 F.

■ **Example 9–5 F**

Solve the following applications.

Decibel levels

The intensity of sound is described in units called decibels. One decibel = 0.1 bel, named after Alexander Graham Bell. The **intensity level,** α (alpha) of a sound is defined by the equation $\alpha = 10 \log \frac{I}{I_0}$, where the units are decibels, where I is the present intensity (power) of a sound, and I_0 is an initial or baseline intensity.

1. Find the change in power that would result in a 1 decibel increase in the intensity level of a sound; that is, find the value of I for which $\alpha = 1$.

$$\alpha = 10 \log \frac{I}{I_0}$$

$$1 = 10 \log \frac{I}{I_0}$$ Replace α by 1

$$0.1 = \log \frac{I}{I_0}$$ Divide each member by 10

$$10^{0.1} = \frac{I}{I_0}$$ Rewrite in exponential form

$$10^{0.1}I_0 = I$$ Multiply both members by I_0; exact value of I

$$1.259I_0 = I$$ $10^{0.1} \approx 1.259$

Thus, the intensity must be about 1.26 times that of the initial intensity to get a 1 decibel increase.

Resistance-capacitance (RC) time constants

In electronic circuit theory, an **RC time constant** describes how long it will take a capacitor to take on about 63% of its electrical charge. For a given fraction q of full charge, $0 < q < 1$, the time t, in RC time constants, is given by the relation $q = 1 - e^{-t}$. This type of circuit can be used in applications like the delay in automobile windshield wipers, or to turn off an automobile's headlights after a time delay.

2. How many RC time constants are necessary to charge a capacitor to 50% of its full capacity; that is, for what value of t is $q = 0.50$?

$$q = 1 - e^{-t}$$

$$0.5 = 1 - e^{-t}$$ Replace q by 0.5

$$e^{-t} = 0.5$$

$$(e^{-t})^{-1} = 0.5^{-1}$$ Raise both members to the -1 power to change the sign of the exponent of e

$$e^t = 2$$ $(e^{-t})^{-1} = e^t$; $0.5^{-1} = (\frac{1}{2})^{-1} = 2$

$$\ln e^t \approx \ln 2$$ Take the natural logarithm of both members

$$t \approx \ln 2$$ $\ln e^x = x$

$t \approx 0.69$, to the nearest hundredth

Thus, it takes about 0.69 time constants for a capacitor to be charged to 50% of its full capacity. ■

Obtaining growth and decay formulas from measurements (optional)

In the laboratory, we often have several measurements with which to obtain a growth/decay formula that describes the situation. Example 9–5 G illustrates how to obtain the correct values of q_0, r, and t.

■ ***Example 9–5 G***

A population of bacteria is assumed to be growing continuously at a fixed rate. There are initially 28 μg (micrograms) of the bacteria; 3 hours later there are 40 μg.

a. How many micrograms of bacteria will there be after 6 hours?

b. When will there be 100 μg of bacteria?

$$
\begin{aligned}
q_0 &= 28 \text{ and } q = 40 \text{ when } t = 3 \\
q &= q_0e^{rt} && \text{Basic growth/decay formula} \\
40 &= 28e^{3r} \\
\tfrac{10}{7} &= e^{3r} \\
\ln \tfrac{10}{7} &= \ln e^{3r} \\
\ln \tfrac{10}{7} &= 3r \\
r &= \tfrac{1}{3} \ln \tfrac{10}{7} \\
r &\approx 0.1189
\end{aligned}
$$

Thus, the equation that describes the growth of this population is $q = 28e^{0.1189t}$ or $q = 28(1.1262^t)$.

Note Since $e^{\ln(10/7)} = \frac{10}{7}$, the equation is also $q = 28(\frac{10}{7})^{t/3}$.

a. Now find q when $t = 6$: $q = 28(1.1262^6) = 57.1$ μg.

b. We now find when the population will be $q = 100$.

$$
\begin{aligned}
100 &= 28e^{0.1189t} \\
\tfrac{25}{7} &= e^{0.1189t} \\
\ln \tfrac{25}{7} &= 0.1189t \\
t &= \frac{\ln \frac{25}{7}}{0.1189} \approx 10.7
\end{aligned}
$$

Thus, the population will be 100 μg after 10.7 hours of growth (or 7.7 hours after it reaches 40 μg). ■

The TI-81 calculator is preprogrammed to help with problems like that in example 9–5 G. Through a feature called ExpReg (exponential regression) the calculator can find the equations required directly. This is illustrated in example 9–5 H.

■ Example 9–5 H

A population of bacteria is assumed to be growing continuously at a fixed rate. There are initially 28 μg (micrograms) of the bacteria; 3 hours later there are 40 μg.

a. How many micrograms of bacteria will there be after 6 hours?

b. When will there be 100 μg of bacteria?

We have two (x,y) or (time, quantity) pairs: (0,28) and (3,40). We want the value of x in the ordered pair $(x,100)$.

We obtain the exponential equation that "interpolates" (passes through) the first two ordered pairs. Proceed as follows:

[STAT] DATA 2 [ENTER]	Clear out any old data
[STAT] DATA 1	Edit data
0 [ENTER] 28 [ENTER] 3 [ENTER] 40	
[STAT] 4 [ENTER]	Select exponential regression

$a = 28$, $b = 1.12624788$

The equation is $q = 28(1.12624788^t)$.

a. 6 [STO▶] [X|T] [ENTER]	Put 6 in x
[VARS] LR 4 [ENTER]	Evaluate $28(1.12624788^x)$ with $x = 6$

The result is 57.1 μg.

Note We can easily graph $y = 28(1.12624788^x)$ as follows

[Y=] [CLEAR] [VARS] LR 4	Enter the equation obained above
[RANGE −10,10,−1,100] YSCL=5	Use these settings for the graph

b. The best way to solve part b is to realize that a logarithmic equation is the inverse of an exponential equation. Recall that the ordered pairs of a function are reversed in its inverse (section 4–5). We reverse the ordered pairs above to obtain (28,0) and (40,3). Then, we want y in the ordered pair $(100,y)$. Instead of doing exponential regression, we do logarithmic regression. This gives the values of a and b in the equation $y = b \ln x + a$.

[STAT] DATA 2 [ENTER]	Clear out any old data
[STAT] DATA 1	Edit data
28 [ENTER] 0 [ENTER] 40 [ENTER] 3	
[STAT] 3 [ENTER]	Select logarithmic regression

$a = -28.02723797$, $b = 8.411019757$, which corresponds to $y = 8.411 \ln x - 28.03$.

100 [STO▶] [X|T] [ENTER]	Put 100 in x
[VARS] LR 4 [ENTER]	Evaluate $8.411 \ln x - 28.03$ with $x = 100$

The result is 10.7 hours. ■

Mastery points

Can you

- Solve certain exponential and logarithmic equations?
- Solve certain applications problems involving exponential and logarithmic equations?

Exercise 9–5

Solve for x.

1. $8^x = 32^{3-2x}$
2. $5^{3x} = 25^x$
3. $27^{4x} = 9^{x+2}$
4. $(\sqrt{3})^x = 9^{x-2}$
5. $(\sqrt{8})^{2x-2} = 4^{3x}$
6. $8^{x+1} = 4^{2-x}$
7. $\log(x-1) + \log(x+3) = \log 4$
8. $\log(x-1) + \log(2x) = \log 4$
9. $\log(x+1) - \log(x-3) = \log 4$
10. $\log(x+1) - \log(2x) = \log 4$
11. $\log(x-1) + \log(x+3) = 2$
12. $\log(x-1) + \log(2x) = 2$
13. $\log(x-1) - \log(x-3) = 2$
14. $\log(x+1) - \log(2x) = 2$

Solve the following equations for x; also, find a decimal approximation to the answer, to the nearest tenth.

15. $14.2 = 2^x$
16. $100 = 20^x$
17. $25 = x^4$
18. $18 = x^{2.3}$
19. $34 = 17^x$
20. $12 = 2^{x+3}$
21. $(x+2)^4 = 200$
22. $45 = 5^{1-x}$
23. $25 = 8^{1/x}$
24. $146 = (x+3)^{0.6}$
25. $41^{2x-1} = 2^x$
26. $172^{1-x} = 6^x$
27. $57^{x/2} = 5^{x+1}$
28. $0.88^{x+2} = 1.6^x$
29. $5^{x+1} = 3^{x-1}$
30. $2^{-x} = 6^{x+2}$
31. $\log_2 x = 0.33$
32. $\log_5 10 = x$
33. $\log_x 10 = 5$
34. $\log(x-2) = 3$
35. $\log_3 30 = 2x$
36. $\log_{x-1} 18 = 2$
37. $\log_{2x} 14 = 3$
38. $\log_{2x} 8 = 3$

Solve the following equations for x.

39. $\log x^2 = (\log x)^2$
40. $\log 2^x = \log 3^{2x-1}$
41. $\log(\log x) = 3$
42. $\log(\log x^2) = 2$
43. $\log(\log 3x) = \log 2$
44. $\log^2 x - \log x^2 = 3$
45. $\log_2 x + \log_3 x = 5$
46. $\log 2^{x+1} + \log 3^x = 2$
47. $1 = e^x - 2e^{-x}$
48. $3 = \dfrac{e^x + e^{-x}}{e^x - e^{-x}}$
49. $e^{2x} - 3e^x = 4$
50. $10^{x-1} = e^{x+1}$
51. $4 \log(\log x + 4) = 3$
52. $\ln x = \dfrac{8}{\ln x - 2}$

Use the formula $A = Pe^{it}$ in problems 53 to 62.

53. \$850 is deposited in an account paying 7.25% interest, compounded continuously. Find the amount in the account after $2\frac{1}{4}$ years.

54. \$2,000 is deposited in an account paying 5% interest, compounded continuously. Find the amount in the account after $3\frac{1}{2}$ years.

55. Find the amount that is an account after 4 years if the account pays 8.5% compounded continuously and if the initial principal was \$2,500.

56. \$45,000 is invested at a rate of 6% simple interest, compounded continuously. What is the value of the investment after $4\frac{1}{2}$ years?

57. How much money should be invested at 10% interest, compounded continuously, so that the value of the investment will be \$5,000 after 6 years?

58. How much money should be invested at 6% interest, compounded continuously, so that the value of the investment will be \$2,000 after 4 years?

59. At what interest rate does money double in 12 years, if the interest is compounded continuously?

60. At what interest rate does money triple in 12 years, if the interest rate is compounded continuously?

61. How long does it take the value in an account that pays 5% interest compounded continuously to triple in value?

62. How long does it take the value in an account that pays 7% interest compounded continuously to double in value?

Use the growth/decay formula $q = q_0 e^{rt}$ and the following information about carbon 14 in problems 63–68. The amount of carbon 14 in a living organism is constant while the organism is alive. At the death of the organism, the carbon 14 is not replenished and begins to decay. Thus, the percentage of the original amount of carbon 14 that is still present gives an estimate of the time that has passed since the organism died. Radioactive carbon 14 diminishes by radioactive decay according to the equation $q = q_0 e^{-0.000124t}$.

63. Compute the amount of carbon 14 still remaining in a sample that originally contained 100 milligrams (mg), after 8,000 years.

64. Compute the amount of carbon 14 still remaining in a sample that originally contained 10 milligrams (mg), after 3,500 years.

65. Estimate the age of a piece of charcoal (i.e., wood) if 30% of the original amount of carbon 14 is still present. That is, find t for which $q = 0.3q_0$.

66. Estimate the age of a piece of a sample in which 70% of the original amount of carbon 14 is still present.

67. What is the half-life of carbon 14?

68. After 1,200 years 18 μg (micrograms) of carbon 14 remained in a certain sample. How much was originally present?

Use the formula $\alpha = 10 \log \dfrac{I}{I_0}$ for problems 69–72.

69. The power of a sound increases by a factor of 20; that is, $I = 20I_0$. What is the resulting change in the decibel level?

70. The power of a sound decreases by a factor of 6; that is, $I = \frac{1}{6}I_0$. What is the resulting change in the decibel level?

71. How much must the power of a sound change to undergo a 3-decibel increase in intensity level? That is, solve for I if $\alpha = 3$.

72. How much must the power of a sound change to undergo a 5-decibel decrease in intensity level? That is, solve for I if $\alpha = -5$.

Use the formula $q = 1 - e^{-t}$ in problems 73–76.

73. What is the charge q on a capacitor after 3 RC time constants, to the nearest percentage? That is, find q when $t = 3$.

74. What is the charge q on a capacitor after 1.5 RC time constants, to the nearest percentage?

75. How many RC time constants are necessary for the charge on a capacitor to reach 45% of its full charge? That is, compute t for $q = 0.45$.

76. How many RC time constants are necessary for the charge on a capacitor to reach 90% of its full charge? That is, compute t for $q = 0.90$.

77. Solve $q = 1 - e^{-t}$ for t.

78. Allometry is the study of relationships between size and shape of organs of living animals, and the size and shape of organisms themselves. The equation of simple allometry is $y = \alpha x^{\beta}$, where α (alpha) and β (beta) are constants, and x and y describe size. Solve this equation for β.

79. Show that $b^x = e^{x \ln b}$ for $b > 0$.

80. The idea "n-factorial" is defined as $n! = 1 \cdot 2 \cdot 3 \cdot \cdots \cdot n$; for example, "3-factorial" is $3! = 1 \cdot 2 \cdot 3 = 6$, and $6! = 1 \cdot 2 \cdot 3 \cdot 4 \cdot 5 \cdot 6 = 720$. An approximation formula, known as Stirling's formula, for factorials when n is large is $n! \approx \left(\dfrac{n}{e}\right)^n \sqrt{2\pi n}$. Compute the approximation value for $n = 30$, to the nearest unit.

81. A formula that arises in studying normal distributions in probability theory is $y = \dfrac{e^{-x^2/2}}{\sqrt{2\pi}}$. Solve this equation for x.

82. In computer science, an AVL tree is a method of storing information in an efficient manner. It is a theorem of computer science that the height h of an AVL tree with N internal nodes is approximated by $h = 1.5 \log_2(N + 2)$. Solve this equation for N.

83. In testing a computer program in which the probability of a failure on any one test is $\dfrac{1}{h}$, the probability M that there will be no failure in N tests is $M = \left(1 - \dfrac{1}{h}\right)^N$. Solve this for N.

84. Refer to problem 83, with $h = 1{,}000$.

a. What is the probability that there will be no failure in 2,000 tests?

b. If we want the probability of a failure to be at least 0.01, how many tests would we expect to run? (That is, find N so that $M \geq 0.01$.)

85. Exponential functions are said to "grow larger" much faster than polynomial functions. Demonstrate this by computing the values for $f(x) = x^2$, a polynomial function, and for $g(x) = 2^x$, an exponential function for $x = 5$, 10, 20, and 40.

86. A method for computing values of e^x is given by the following equation: $e^x = 1 + x + \frac{x^2}{2!} + \frac{x^3}{3!} + \frac{x^4}{4!} + \cdots$, where 2!, 3!, etc. are defined as shown in problem 80. The expression on the right (also called a power series) goes on forever, but one only needs a few terms for good accuracy for small values of x. Use the first six terms (five are given above) of this equation to compute an approximation for $e^{0.2}$. Check your answer with your calculator's $\boxed{e^x}$ key.

87. A problem from Mesopotamia, thousands of years old, asks how long it takes for money to double at 20% annually. The answer given is $3 + \frac{47}{60} + \frac{13}{60^2} + \frac{20}{60^3}$ (the Mesopotamians used a sexagesimal [base 60] system of numeration). How accurate is this answer? (Calculate how long it takes money to double at 20% annually yourself, then compare your answer to the Mesopotamian answer.)

88. It is estimated that 23% of a certain radioactive substance decays in 30 hours. What is the half-life of this substance?

89. The population of a bacteria was determined to increase by 15% in 15 hours. How long will it take the population to double, assuming it is increasing continuously at a fixed rate?

90. Assume that inflation has acted at a fixed rate for a 10-year period, and that in this time what initially cost one dollar now costs \$1.56. What was the rate of inflation for this period? How much did an item which initially cost one dollar cost after five years?

91. A population of bacteria is assumed to be growing continuously at a fixed rate. There are initially 10 μg (micrograms) of the bacteria; 2 hours later there are 40 μg.

a. How many micrograms of bacteria will there be after 3 hours?

b. When will there be 100 μg of bacteria?

92. The Richter scale was invented in 1935 by Charles F. Richter to measure the intensity of earthquakes. Each number on the scale represents an earthquake 10 times stronger than one of the next lower magnitude. For example, an earthquake of 5 on the Richter scale is 10 times stronger than one of measure 4 on the Richter scale. Suppose one earthquake measures 4.5 on the Richter scale, and a second measures 6.2. To the nearest unit how many times stronger is the second earthquake than the first?

93. In approximately the year 50 B.C., the Roman statesman Cicero was a governor in Asia Minor. He decided a case in which creditors had loaned money to the town of Salamis in Cyprus at 48% interest. Roman law permitted only 12% interest. Cicero decreed that only 12% compounded interest could be charged. On this basis, the deputies of Salamis determined that they owed 106 talents. Use the formula $A = P\left(1 + \frac{i}{n}\right)^n$ (from the previous section) to determine the original amount of the loan, using an interest rate of 12% compounded yearly, and assuming the loan was for 5 years.

94. Calculators often use the following method for computing $\log x$. Certain roots of 10 are stored:

$10^{1/2} = 3.162278$, $10^{1/4} = 1.778279$
$10^{1/8} = 1.333521$, $10^{1/16} = 1.154782$,
$10^{1/32} = 1.074608$, $10^{1/64} = 1.036633$, etc.

Then, to compute, say, log 4, we find, by successive divisions, that

$4 = 10^{1/2} \cdot 10^{1/16} \cdot 10^{1/32} \cdot 10^{1/128} \cdot 10^{1/2048}$, etc., so
$\log 4 \approx 1/2 + 1/16 + 1/32 + 1/128 + 1/2{,}048 \approx 0.6021$
(to four decimal places).

This method is reasonably efficient once enough roots of 10 are calculated and permanently stored into the calculator's memory. (To assure four-decimal place accuracy, 15 roots will suffice; on a typical eight-place calculator 32 roots will suffice.) Describe two ways in which this method could be modified to calculate natural logarithms.

Skill and review

1. Graph $f(x) = 2^{1-x}$.
2. Graph $f(x) = \dfrac{2}{(x-1)(x-5)}$.
3. Solve $|3 - 2x| < 13$.
4. Solve $\left|\dfrac{3-2x}{x}\right| < 13$.
5. Graph $f(x) = x^5 - 4x^4 + 2x^3 + 4x^2 - 3x$.
6. Simplify $\dfrac{2}{x-3} + \dfrac{2}{x+3} - \dfrac{5}{x+1}$.
7. Compute $(\frac{2}{5} - \frac{3}{4}) \div 2$.

Chapter 9 summary

- An exponential function is a function of the form $f(x) = b^x$, $b > 0$ and $b \neq 1$, with domain: $\{x \mid x \in R\}$ and range: $\{y \mid y \in R, y > 0\}$.
- **Logarithm** A logarithm is an exponent.
- **Equivalence of logarithmic and exponential form** $y = \log_b x$ if and only if $b^y = x$.
- **Logarithmic function** A function of the form $f(x) = \log_b x$, $b > 0$ and $b \neq 1$, with

  Domain: $\{x \mid x \in R, x > 0\}$
  Range: $\{y \mid y \in R\}$
- **Properties of exponential and logarithmic functions** (Assume $b > 0$ and $b \neq 1$):

  If $b^x = b^y$, then $x = y$
  If $\log_b x = \log_b y$, then $x = y$
  $y = \log_b x$ if and only if $b^y = x$
  $\log_b(xy) = \log_b x + \log_b y$
  $$\log_b\left(\frac{x}{y}\right) = \log_b x - \log_b y$$
  $\log_b x^r = r \log_b x$
  $\log_b b = 1$
  $\log_b 1 = 0$
  $\log_b(b^x) = x$
  $b^{(\log_b x)} = x$
  $$\log_b x = \frac{\log_m x}{\log_m b}$$
  $$\log_b x = \frac{\log x}{\log b}$$
- **Common logarithms** Logarithms to the base 10 or $\log_{10} x$, which is abbreviated as simply $\log x$.
- **The symbol *e*** An irrational number; $e \approx 2.718$.
- **Natural logarithms** Logarithms to the base e or $\log_e x$, which is abbreviated as simply $\ln x$.

Chapter 9 review

[9–1]

1. Describe the behavior of the general exponential function with respect to the words "increasing" and "decreasing" and the value of b.

Use the properties of exponents to perform the indicated operations.

2. $5^{\sqrt{8}} \cdot 5^{\sqrt{50}}$
3. $\dfrac{4^{\sqrt{50}}}{4^{\sqrt{8}}}$
4. $(3^{\sqrt{2\pi}})^{\sqrt{8}}$

Graph the given function. State whether the function is increasing or decreasing. Label any intercepts.

5. $f(x) = 8^x$
6. $f(x) = 0.4^x$
7. $f(x) = 0.6^{-x}$
8. $f(x) = 4^{x+2}$
9. $f(x) = 3^{-x+1}$
10. $f(x) = 3^{x/2} - 1$

Solve the following exponential equations.

11. $9^x = 27$
12. $9^x = \sqrt{27}$
13. $10^{-2x} = 1{,}000$
14. $16^x = \frac{1}{8}$
15. $3^{x/2} = 9$
16. $(\sqrt{2})^x = 8$
17. $2^{-x} = 0.125$

[9–2] Find the value of the expression.

18. $\log_4 128$

19. $\log_{10} 0.001$

20. $3 \log_{25}\frac{1}{125}$

21. $5(3 \log_4\frac{1}{8} + 2 \log_{10} 100)$

22. The following definition is incorrect; fix the logarithmic portion of the definition so it is correct. Definition:

$$\log_b x = y \text{ if and only if } x^y = b,\ b > 0,\ b \neq 1$$

Put each logarithmic equation into exponential form.

23. $\log_4 0.25 = -1$

24. $\log_5(x - 3) = 2$

25. $\log_2 y = 8$

26. $\log_3 9 = x + 2$

27. $\log_m x = y + 1$

Put each exponential equation into logarithmic form.

28. $x^3 = m - 3$

29. $4 = y^{2x} - 1$

30. $4 = y^{2x-1}$

31. $(x - 1)^y = 5$

32. $(x + 3)^{x+y} = y - 2$

33. $(5x)^2 = 3y$

Solve the following equations for x.

34. $\log_4 8 = x$

35. $\log_x 16 = 4$

36. $\log_{10}(x + 1) = -2$

37. $\log_x\frac{1}{8} = 3$

38. $\log_x k = 3$

Estimate the values of the following logarithms by stating two consecutive integers that bracket the value.

39. $\log_4 100$

40. $\log_{10} 15{,}600$

Simplify the following expressions.

41. $\log_5 5^m$

42. $x^{\log_x 5}$

[9–3] Solve the following logarithmic equations.

43. $\log_2 3x = -3$

44. $\log_5(x^2 - x) = \log_5 6$

45. $\log_3\dfrac{5}{x} = \log_3(2x - 3)$

46. $\log_{0.5}(5x - 1) = -4$

47. $\log_2(-3x) = \log_2\frac{1}{8}$

48. $\log_5(x - \frac{1}{2}) = \log_{10} 100$

49. $\log_3(x - 2) = \log_2\frac{1}{16}$

Solve the following logarithmic equations.

50. $\log_5(x + 3) - \log_5(x - 1) = 2$

51. $\log_2 5 + \log_2(3 - 2x) = \log_2 x$

52. $\log_5 x - \log_5 3 = \log_5 2$

53. $\log_2(x - 4) + \log_2(x + 3) = \log_2(x^2 - 3x + 2)$

54. $\log_2 x^2 = 12,\ x > 0$

55. $\log_4 x^3 = \log_4 16^9,\ x > 0$

Rewrite the following expressions in terms of $\log_a x$, $\log_a y$, and $\log_a z$, for the given value of a.

56. $\log_2 8x^4y^2z$

57. $\log_4\dfrac{8yz^3}{x^2}$

58. $\log_{10}\dfrac{x^5y^3z}{100}$

Assume $\log_a 2 = 0.3562$, $\log_a 3 = 0.5646$, and $\log_a 5 = 0.8271$. Use these values to find approximate values for the following logarithms.

59. $\log_a 30$

60. $\log_a 36$

61. $\log_a 0.2$

[9–4] Use a calculator to find the values of the following common logarithms. Round the results to 4 decimal places.

62. $\log 0.935$

63. $\log 6{,}250$

64. $\log 5{,}021{,}400{,}000{,}000{,}000{,}000{,}000$

65. $\log 0.000\ 000\ 000\ 000\ 000\ 004\ 13$

Compute the given logarithm. Round the results to 4 decimal places.

66. $\log_5 40$

67. $\log_{20} 1{,}000$

68. $\log_8 1.5$

69. $\log_{0.25} 20$

Use a calculator to find the values of the following natural logarithms. Round the results to 4 decimal places.

70. $\ln 12.31$

71. $\ln 0.0035$

Calculate the following values to 4 decimal places.

72. 10^{π}

73. $10^{-2.5}$

74. $e^{4.8}$

Graph the following logarithmic functions.

75. $f(x) = \log_5(x + 3)$

76. $f(x) = \log_2 2x$

77. $f(x) = \log_4 x - 3$

78. $f(x) = \log_2(-x)$

Simplify the following expressions.

79. $\log 10{,}000$

80. $\ln(e^{3x})^2$

81. $\ln e^{\sqrt{3}}$

82. $\log 10^{1-3x}$

83. $10^{\log 30}$

84. $e^{\ln 3x^2}$

85. $e^{\ln 2x} + \ln e^{2x}$

86. $\ln \sqrt{5}e^x$

87. \$2,500 is deposited in a bank account that computes its interest continuously. The simple interest rate is $6\frac{1}{4}\%$ per year. Use the formula $A = A_0e^{it}$ to compute the amount in the account after $3\frac{1}{2}$ years, where A is the amount after t years, A_0 is the initial deposit, and i is the simple yearly interest rate.

[9–5] Solve for x.

88. $8^x = 16^{4+5x}$

89. $(\sqrt{8})^{x-2} = 8^{3x}$

Solve the following equations for x, to the nearest hundredth.

90. $200 = x^{2.3}$

91. $80 = 2^{x+3}$

92. $45 = 9^{1-2x}$

93. $4.8 = 1.6^x$

94. $\log_2 x = 0.6$

95. $\log_5 30 = x$

96. $\log_x 100 = 5$

97. $\log_3 784 = 2x$

98. $\log_{2x-3} 3{,}000 = 6$

Solve the following equations for x.

99. $\log x^2 = \log^2 x$ **100.** $\log 2^{x-1} = \log 3^{2x-1}$
101. $\log(\log x) = 2$ **102.** $\log(\log x^2) = 1$
103. $\log(\log 3x) = \log 2$ **104.** $\log^2 x - \log x^2 = 15$
105. $\log_2 x + \log_3 x = 8$ **106.** $\log 3^{x+1} + \log 2^x = 10$
107. $3 = e^x - 4e^{-x}$ **108.** $5 = \dfrac{e^x + e^{-x}}{e^x - e^{-x}}$
109. $5^{x-1} = e^{x+1}$ **110.** $4 \log x(\log x + 4) = 0$
111. $\ln x = \dfrac{12}{\ln x - 4}$

112. How much money should be invested at 10% interest, compounded continuously, so that the value of the investment will be \$5,000 after 6 years. Use the formula $A = Pe^{it}$.

113. At what interest rate does money double in 9 years, if the interest is compounded continuously?

114. It is estimated that 23% of a certain radioactive substance decays in 30 hours. What is the half-life of this substance?

Chapter 9 test

Use the properties of exponents to perform the indicated operations.

1. $\dfrac{8^{\sqrt{18}}}{8^{\sqrt{32}}}$ **2.** $(2^{\sqrt{2}})^{\sqrt{8}}$

Graph the given function. State whether the function is increasing or decreasing. Label any intercepts.

3. $f(x) = 3^x$ **4.** $f(x) = 0.3^x$
5. $f(x) = 3^{x+2}$ **6.** $f(x) = 3^x - 1$

Solve the following exponential equations.

7. $9^x = 3$ **8.** $27^x = \sqrt{27}$
9. $25^x = \frac{1}{125}$ **10.** $(\sqrt{5})^x = 25$
11. $100^{-x} = 0.001$

Find the value of the expression.

12. $\log_8 128$ **13.** $3 \log_2 \frac{1}{16}$
14. $2(3 \log_8 \frac{1}{8} + \log_{100} 10)$

15. The following definition is incorrect; fix the *exponential* portion of the definition so it is correct.

$$\log_a x = y \text{ if and only if } x^y = a,\ a > 0,\ a \neq 1$$

Put each logarithmic equation into exponential form.

16. $\log_5 0.25 = -x$ **17.** $\log_3(x - 3) = 2$
18. $2 \log_m x = y$

Put each exponential equation into logarithmic form.

19. $2^5 = 32$ **20.** $(x - 1)^3 = m$ **21.** $z = y^{2x-1}$

Solve the following equations for x.

22. $\log_4 \frac{1}{8} = x$ **23.** $\log_{10}(2x - 5) = 3$
24. $\log_{x+1} \frac{1}{8} = 3$

Graph the following logarithmic functions.

25. $f(x) = \log_3 x$ **26.** $f(x) = \log_5(x + 1)$
27. $f(x) = \log_2 x - 1$

Solve the following logarithmic equations.

28. $\log_2 \dfrac{x}{2} = -3$ **29.** $\log_2(x^2 - 14x) = \log_2 32$
30. $\log_3(x + 3) = \log_3(4x - 12)$
31. $\log_3(5x - 1) = -2$
32. $\log_3(3x - 2) = \log_4 \frac{1}{16}$
33. $\log_2(x + 6) - \log_2(x - 1) = 5$
34. $\log_5(x + 1) - \log_5 3 = \log_5 2$
35. $2 \log_{10}(x + 2) = \log_{10}(x + 14)$
36. $\log_2 x^3 = 15,\ x > 0$

Rewrite the following expressions in terms of $\log_a x$, $\log_a y$, and $\log_a z$, for the given value of a.

37. $\log_3 \dfrac{9x^3y}{z^4}$ **38.** $\log_{10} \dfrac{x^{10}y^3z}{1{,}000}$

Assume $\log_a 2 = 0.3562$, $\log_a 3 = 0.5646$, and $\log_a 5 = 0.8271$. Use these values to find approximate values for the following logarithms.

39. $\log_a 12$ **40.** $\log_a 40$ **41.** $\log_a 0.4$

Use a calculator to find the values of the following logarithms. Round the results to 4 decimal places.

42. $\log 31{,}020{,}000{,}000{,}000$
43. $\log 0.000\,000\,001\,03$
44. $\log_5 50$ **45.** $\ln 1{,}000$

Calculate the following values to 4 decimal places.

46. $10^{\pi-2}$ **47.** $3\sqrt{e}$

Simplify the following expressions.

48. $\ln[(e^x)^2]$ **49.** $(\ln e^x)^2$ **50.** $\log 10^{18}$

51. $10^{\log 27}$ **52.** $e^{\ln 5x} + \ln e^{5x}$

53. $2,500 is deposited in a bank account that computes its interest continuously. The simple interest rate is $7\frac{1}{2}\%$ per year. Use the formula $A = A_0e^{it}$ to compute the amount in the account after $3\frac{1}{4}$ years.

Solve the following equations for x, to the nearest hundredth.

54. $27^x = 9^{4+5x}$ **55.** $178 = x^{1.9}$ **56.** $80 = 3^{x-2}$

57. $\log_5 x = 3$ **58.** $\log_2 12 = x$ **59.** $\log_x 34 = 2.5$

Solve the following equations for x.

60. $\log^3 x = \log x^4$ **61.** $4^{x-2} = 3^{6x-1}$

62. $\log_3(\log_2 x) = 2$ **63.** $\log(\log x^2) = 1$

64. $\log^2 x - \log x^2 = 35$ **65.** $\log_2 x + \log_3 x = 3$

66. $9 = \dfrac{e^x + e^{-x}}{e^x - e^{-x}}$ **67.** $3^{x-1} = e^{x+1}$

68. How much money should be invested at 6% interest, compounded continuously, so that the value of the investment will be $2,000 after 4 years. Use the formula $A = Pe^{it}$.

69. At what interest rate does money double in 12 years, if the interest is compounded continuously? Round to the nearest 0.1%. (See the formula from problem 68.)

70. The half-life of a certain radioactive substance is 30 hours. How much of an initial amount of 25 grams of this substance will remain after 15 hours? Use $q = q_0e^{rt}$, and round the result to the nearest 0.1 grams.

C H A P T E R 10

Systems of Linear Equations and Inequalities

In Sections 10–1, 10–2, and 10–3 we study what are called "systems of n linear equations in n variables." These systems of equations are found in practically all areas where mathematics is used, including in electronics technology, engineering, economics, biology, business, etc. The study of these systems is extended and generalized in an area of mathematics called *linear algebra.*

Section 10–4 investigates systems of linear inequalities.[1] These appear widely also, but have their most common application in linear programming, a method of finding the "best" (i.e., cheapest, fastest, etc.) way to solve problems in business and management. Section 10–5 introduces the algebra of matrices, which finds application in science, engineering, business, and economics.

10–1 Solving systems of linear equations—the addition method

On Babylonian clay tablets, 4,000 years old, is found a problem that refers to quantities called a "first silver ring" and a "second silver ring." If these quantities are represented by x and y, the tablet asks for the solution to the following system of equations: $\frac{x}{7} + \frac{y}{11} = 1$ and $\frac{6x}{7} = \frac{10y}{11}$. The answer is given as $\frac{x}{7} = \frac{11}{7 + 11} + \frac{1}{72}$ and $\frac{y}{11} = \frac{7}{7 + 11} - \frac{1}{72}$. Show that this answer is correct.

These two equations are called a system of two equations in two variables (unknowns). Obviously the solution of these systems has interested humankind for a long time. This is the topic of this section.

[1]Systems of nonlinear equations and inequalities are presented in section 11–4.

Systems in two variables

> **System of two linear equations in two variables**
> A system of two linear equations in two variables is a set of equations of the form
>
> $$a_1x + b_1y = c_1$$
> $$a_2x + b_2y = c_2$$
>
> where a_1 and b_1 are not both zero and a_2 and b_2 are not both zero. The values of a_1, a_2, b_1, b_2, c_1, and c_2 are constants.

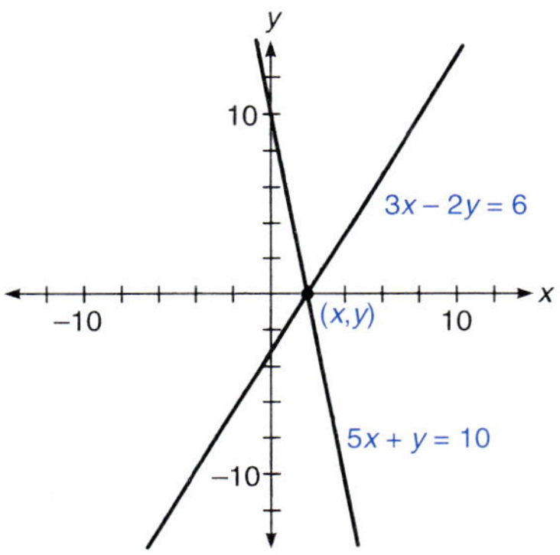

Figure 10–1

Systems of linear equations were described in something like this modern form at least 300 years ago. An example of such a system is

$$\begin{array}{ll} [1] & 3x - 2y = 6 \\ [2] & 5x + y = 10 \end{array}$$

The graph of each of these linear equations is a straight line; see figure 10–1.

In general, two straight lines intersect at a point, like (x,y), in the figure. In section 3–2 we saw that the *method of substitution for expression* can be used to eliminate one variable and find the point of intersection.

In this section we learn how to algebraically compute the value of the point (x,y) by the process called the addition method. The idea is to create a new equation in which one of the variables is not present using addition.

We could do this with these equations (see the following steps) by multiplying both members of equation [2] by the value 2, giving the new system equations [1] and [2a] (below). If we add the two left members of equations [1] and [2a] together, and the two right members together, we obtain equation [3].

$$\begin{array}{llcll} [1] & 3x - 2y = 6 & \quad & [1] & 3x - 2y = 6 \\ [2] & 5x + y = 10 & & [2a] & \underline{10x + 2y = 20} \\ & & & [3] & 13x = 26 \\ & & & & x = 2 \end{array}$$

We solved for x by eliminating y. We now use the same idea to obtain an equation in which x is not present; to do this we can multiply equation [1] by 5 and equation [2] by -3, giving the system [1b] and [2b], and equation [4].

$$\begin{array}{llcll} [1] & 3x - 2y = 6 & \quad & [1b] & 15x - 10y = 30 \\ [2] & 5x + y = 10 & & [2b] & \underline{-15x - 3y = -30} \\ & & & [4] & -13y = 0 \\ & & & & y = 0 \end{array}$$

The point (x,y) is $(2,0)$. We can verify this by checking that $(2,0)$ is a solution to both equations:

$$[1] \quad \begin{aligned} 3x - 2y &= 6 \\ 3(2) - 2(0) &= 6 \\ 6 &= 6 \checkmark \end{aligned} \qquad [2] \quad \begin{aligned} 5x + y &= 10 \\ 5(2) - 0 &= 10 \\ 10 &= 10 \checkmark \end{aligned}$$

An additional verification comes from the graph in figure 10–1, where the point of intersection appears to be at, or close to $(2,0)$. In fact, modern graphing calculators make it easy to graph both straight lines and use the TRACE function to put the cursor over the solution, and thus obtain an approximation to the actual solution.

To generalize, *to solve a system of two linear equations* in two variables $\begin{aligned} a_1x + b_1y &= c_1 \\ a_2x + b_2y &= c_2 \end{aligned}$ means to *find all points (x,y) that satisfy both equations.*

To solve a system of two linear equations by addition

1. Multiply each of the equations by values that produce two equivalent equations with the coefficients of x being additive inverses.
2. Add the left and right members of these equations to eliminate x.
3. Solve the resulting equation for y.
4. Repeat the process with the coefficients of y and solve for x.

Note One can choose to eliminate either the x or the y variable first.

Step 1 above requires multiplying each equation by certain values. These values are chosen to obtain the least common multiple (LCM) of the two coefficients. The **LCM** is the *smallest positive integer into which each of the (integer) coefficients divide.* Also, if the coefficients have the same sign we make one of the multipliers negative. This is summarized as follows.

Choosing a multiplier for each equation

1. Choose multipliers that make the x (or y) coefficients the LCM of the coefficients.
2. If the coefficients have the same sign, make one multiplier negative.

This is illustrated in the following examples.

Notation $(n)\rightarrow$

In the examples, we use the notation $(n)\rightarrow$ to show that an equation was transformed by multiplying each term of both members by the integer n.

■ Example 10–1 A

Solve the system by addition.

$$\begin{aligned} 2x - 3y &= -7 \\ 3x + 5y &= 1 \end{aligned}$$

Eliminate the x terms. The LCM of 2 and 3 is 6 (the smallest integer into which 2 and 3 divide). Since the values 2 and 3 have the same sign, choose one negative multiplier. Multiply the first equation by 3, and the second by -2. This is shown as follows.

$$\begin{aligned} 2x - 3y &= -7 \quad (3)\rightarrow & 6x - 9y &= -21 \\ 3x + 5y &= 1 \quad (-2)\rightarrow & -6x - 10y &= -2 \\ \hline & & -19y &= -23 \\ & & y &= \frac{23}{19} \end{aligned}$$

Eliminate the y terms. The LCM of -3 and 5 is 15. The coefficients have opposite signs, so each multiplier can be positive.

$$\begin{aligned} 2x - 3y &= -7 \quad (5)\rightarrow & 10x - 15y &= -35 \\ 3x + 5y &= 1 \quad (3)\rightarrow & 9x + 15y &= 3 \\ \hline & & 19x &= -32 \\ & & x &= -\tfrac{32}{19} \end{aligned}$$

Thus, the solution is $(x,y) = (-\frac{32}{19},\frac{23}{19})$ ■

In example 10–1 A the two lines intersected at a point. However, two other things can happen. The lines may be parallel, and never meet, or the equations may represent the same line.

If the lines never meet (they are parallel) we say the system of equations is **inconsistent.** Algebraically we say that *a system of equations is inconsistent if there is no solution to that system.*

If the equations describe the same line, we say the system is **dependent.** Algebraically, *a system of equations is dependent if one equation can be derived from the rest of the equations.*

What happens in both cases is illustrated in example 10–1 B.

■ Example 10–1 B

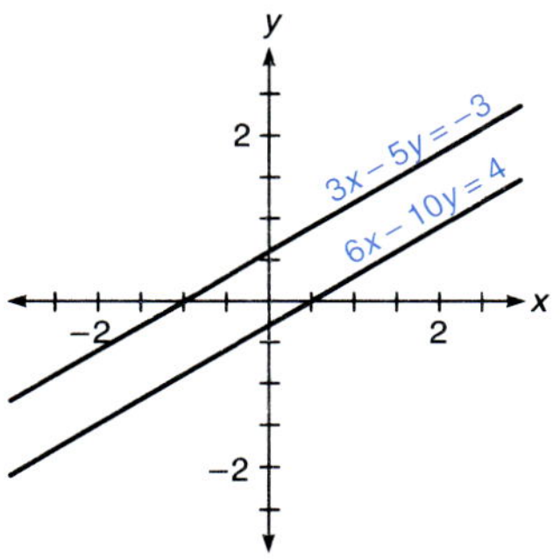

1. Solve the system $\begin{aligned} 6x - 10y &= 4 \\ 3x - 5y &= -3 \end{aligned}$.

To eliminate the x terms, we multiply the second equation by -2:

$$\begin{aligned} 6x - 10y &= 4 \\ -6x + 10y &= 6 \\ \hline 0 &= 10 \end{aligned}$$

The figure shows why we obtained the false statement $0 = 10$—the two lines in the system are parallel, and therefore never meet. The system is inconsistent.

2. Solve the system $\begin{array}{r} 2x - 3y = 6 \\ -4x + 6y = -12 \end{array}$.

Multiplying the first equation by 2:

$$\begin{array}{r} 4x - 6y = 12 \\ -4x + 6y = -12 \\ \hline 0 = 0 \end{array}$$

We obtain the statement $0 = 0$ because the first equation is just a multiple of the second equation (the first equation can be derived from the second equation). Both equations are the same line, and *the system is dependent.* The solution is therefore all points on the line $2x - 3y = 6$ (or $-4x + 6y = -12$). ■

As just illustrated, *an inconsistent system leads to a false statement. A dependent system leads to a statement that is true regardless of the value of x or y.*

Systems in more than two variables

A system of n equations involving n variables is a generalization of a system of two equations in two variables. Our objective is to find the values of all n variables that simultaneously satisfy all n equations. Our answer is written as an "n-tuple" (an ordered pair (x,y) is also called a "2-tuple"). It is possible for these systems to be dependent or inconsistent, just as with the case where $n = 2$ (as in example 10–1 B), although there will not always be a geometric interpretation of the result.

An example of a system of three equations involving three variables is

$$\begin{array}{ll} [1] & x - 2y - z = 2 \\ [2] & x + 4y + 2z = 2 \\ [3] & -2x - 2y + z = -6 \end{array}$$

A system of n linear equations in n variables can be solved using the addition method. The idea is to *reduce the number of variables and equations, one step at a time.* A method that ensures finding a solution, when it exists, uses the idea of a "key equation."

To solve a system of *n* equations in *n* variables by addition

1. Select one equation as the key equation (see guideline below).
2. Use the key equation to eliminate one variable from all other equations.
3. Repeat steps 1 and 2 for this new system of $n - 1$ equations in $n - 1$ variables, until reaching 2 equations in 2 variables. Then go to step 4.
4. Obtain numerical values for two variables.
5. Substitute back into the key equation(s) to obtain the complete solution.

Note step 3 is only necessary for $n \geq 4$. A guideline for making the selection in steps 1 and 2 follows.

> **Selecting the key equation**
> Choose an equation in which the coefficient of one of the variables is 1. Use this key equation to eliminate the same variable from the other equations.

Of course there may be no equation in which any coefficient is one. In this case there may be no obvious choice as the key equation. This method is illustrated in example 10–1 C.

■ ***Example 10–1 C***

Solve the system of n equations in n variables.

1. [1] $x - 2y - z = 2$
[2] $x + 4y + 2z = 2$
[3] $-2x - 2y + z = -6$

Step 1: We select equation [1] as the key equation to eliminate x.

Step 2: Eliminate x from equation [2] using the key equation [1]:

[1] $x - 2y - z = 2$
[2] $x + 4y + 2z = 2$ $(-1)\rightarrow$

$$\begin{array}{rrrrl} & x - 2y - z = & 2 \\ & -x - 4y - 2z = & -2 \\ \hline & -6y - 3z = & 0 \\ [4] & 2y + z = & 0 \end{array}$$

Divide each member by -3

Eliminate x from equation [3] using the key equation [1]:

[1] $x - 2y - z = 2$ $(2)\rightarrow$
[3] $-2x - 2y + z = -6$

$$\begin{array}{rr} & 2x - 4y - 2z = 4 \\ & -2x - 2y + z = -6 \\ \hline [5] & -6y - z = -2 \end{array}$$

Step 4: Equations [4] and [5] are a system of two equations in two variables, so solve as in example 10–1 A.

[4] $2y + z = 0$
[5] $-6y - z = -2$

$y = \frac{1}{2}$, $z = -1$

Step 5: Substitute for y and z in the key equation, [1].

[1] $x - 2y - z = 2$
$x - 2(\frac{1}{2}) - (-1) = 2$
$x = 2$

Thus the solution is the ordered triple $(x,y,z) = (2,\frac{1}{2},-1)$.

2. Solve the system of equations for the solution (x,y,z):

[1] $2x - y + 3z = 0$
[2] $4x + 3y = 2$
[3] $2x + 2y - 3z = -3$

We can shorten the amount of work in this problem if we observe that the variable z is not present in equation [2]. Also the coefficients of z make elimination of z easy using equations [1] and [3]. We therefore choose equation [3] as the key equation to eliminate z from equation [1].

Step 1: Choose [3] as the key equation to eliminate z.

Step 2:
[1] $2x - y + 3z = 0$
[3] $\underline{2x + 2y - 3z = -3}$
[4] $4x + y = -3$

Step 4: Equations [2] and [4] form a system of two equations and two variables.

[2] $4x + 3y = 2$
[4] $4x + y = -3$
$x = -\frac{11}{8}, y = \frac{5}{2}$

Step 5: Use the key equation [3] to find z:

[3] $2x + 2y - 3z = -3$

$2(-\frac{11}{8}) + 2(\frac{5}{2}) - 3z = -3$ — $x = -\frac{11}{8}, y = \frac{5}{2}$

$-\frac{11}{4} + 5 - 3z = -3$ — Perform arithmetic

$-\frac{11}{4} + 8 = 3z$ — Add +3 and $3z$ to each member

$-11 + 32 = 12z$ — Multiply each member by 4 to eliminate denominators

$\frac{7}{4} = z$

The solution is the 3-tuple $(x,y,z) = (-\frac{11}{8}, \frac{5}{2}, \frac{7}{4})$.

> We recognize that a system is dependent by obtaining the equation 0 = 0 at some point; in this case, we will not attempt to describe the solution set, but simply state "dependent." If we arrive at a statement which is false, such as 0 = 2, the system is inconsistent, and there are no points in the solution. We state "inconsistent" in this case.

3. Solve the system of equations.

[1] $x + 6y + 3z = 5$
[2] $x + 2y + z = 3$
[3] $4y + 2z = 2$

Use equation [2] as the key equation to eliminate x.

[1] $x + 6y + 3z = 5$ → $x + 6y + 3z = 5$
[2] $x + 2y + z = 3$ (-1)→ $\underline{-x - 2y - z = -3}$
$4y + 2z = 2$
[4] $2y + z = 1$ — Divide by 2

Since equation [3] does not contain x, we do not modify it. We now use the system of two equations in two variables [3] and [4] to find y and z.

$$\begin{array}{lll} [3] & 4y + 2z = 2 & \\ [4] & 2y + z = 1 & (-2)\rightarrow \end{array} \qquad \begin{array}{r} 4y + 2z = 2 \\ \underline{-4y - 2z = -2} \\ 0 = 0 \end{array}$$

The statement $0 = 0$ tells us that *this system is dependent.* ■

Note It is possible for an inconsistent system to produce a statement like $0 = 0$ also. An example is the system $x + y + z = 1$, $2x + 2y + 2z = 2$, $x + y + z = 2$.

In this text we will not investigate systems of three or more variables that are dependent or inconsistent except to state the following. In an inconsistent system of n equations in n variables there are no solutions. In a dependent system of n equations in n variables there are infinitely many solutions. We will continue to focus our attention on systems that are independent and consistent. These systems have one solution.

There are many applied situations that lead to systems of linear equations.

■ ***Example 10–1 D***

1. The golden ratio is a value that appears in many places throughout nature and art. Among other things, it has been observed that rectangles whose length and width are in this ratio are considered pleasing to the eye. This ratio is approximately 8 to 5.

   Now, suppose that an artist has a piece of stainless steel stock that is 14 feet long, and wishes to use the entire length to form a rectangle in the golden ratio. How long should the length and width be?

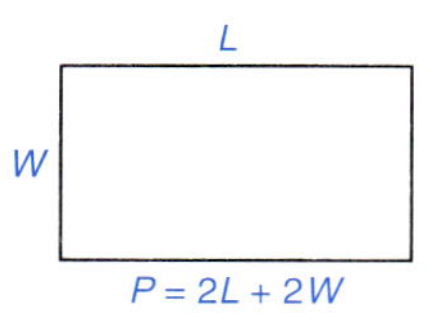

For rectangles we know that perimeter $P = 2L + 2W$, L is length, and W is width (see the figure). For this case we know that $P = 14$, so we can write $14 = 2L + 2W$, or $7 = L + W$ (divide each member by 2). A ratio of 8 to 5 for length and width means

$$\frac{8}{5} = \frac{L}{W}$$
$$5L = 8W$$
$$5L - 8W = 0$$

Thus, we have a system of two equations in two variables: $\begin{array}{r} L + W = 7 \\ 5L - 8W = 0 \end{array}$.

Solving shows that the length should be $4\frac{4}{13}$ feet and the width should be $2\frac{9}{13}$ feet.

2. An investor invested a total of \$10,000 in a two-part mutual fund; one part is more risky than the other, but pays a higher return. The investor has forgotten how much was invested in each part of the fund, but knows that they paid 5% and 8% last year, and that the total income from the investment was \$695. Compute how much must have been invested at each rate.

Let x be the amount invested at 5%, and y the amount at 8%.

[1] $x + y = 10{,}000$ The total of the two amounts is \$10,000
[2] $0.05x + 0.08y = 695$ 5% of x plus 8% of y totals \$695

To eliminate decimals we multiply equation [2] by 100:

[2] $5x + 8y = 69{,}500$

Solving the system [1] $x + y = 10{,}000$, [2] $5x + 8y = 69{,}500$ gives $x = 3{,}500$ and $y = 6{,}500$, so \$3,500 was invested at 5% and \$6,500 was invested at 8%.

3. A parabola is the graph of an equation of the form $y = ax^2 + bx + c$, $a \neq 0$ (section 4–1). Three noncollinear[2] points determine unique values of a, b, and c—that is, there is only one parabola that passes through any three noncollinear points. In computer modeling of geometric shapes, a computer might use a system of three equations in the three variables a, b, and c to find these unique values. A similar process occurs when a computer art program creates what are called Bezier curves.

Find the parabola that passes through the three points $(-2,5)$, $(1,2)$, and $(4,10)$.

Substituting the values for each point into the equation $y = ax^2 + bx + c$, we obtain a system of three equations in three variables.

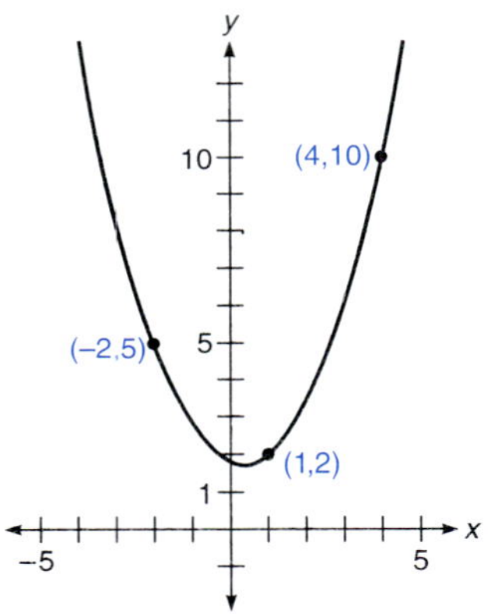

$$y = ax^2 + bx + c$$

$x = -2, y = 5$: $5 = a(-2)^2 + b(-2) + c$, or [1] $5 = 4a - 2b + c$
$x = 1, y = 2$: $2 = a(1)^2 + b(1) + c$, or [2] $2 = a + b + c$
$x = 4, y = 10$: $10 = a(4)^2 + b(4) + c$, or [3] $10 = 16a + 4b + c$

Solving produces $a = \frac{11}{18}$, $b = -\frac{7}{18}$, $c = \frac{16}{9}$, so the equation is $y = \frac{11}{18}x^2 - \frac{7}{18}x + \frac{16}{9}$. For illustration, the graph of this parabola is shown in the figure. ■

Mastery points

Can you

- Solve a system of n linear equations in n variables, giving the solution as an n-tuple or stating dependent or inconsistent as appropriate?
- Solve certain applications using systems of linear equations?

[2]Three points that are not on a straight line are said to be noncollinear.

Exercise 10–1

Solve the following systems of two equations in two variables; if the system is dependent or inconsistent state this.

1. $x + 10y = -7$
$-2x + 5y = 4$

2. $4x + 13y = 5$
$2x - 4y = -\frac{41}{13}$

3. $-3y = -6$
$2x + 7y = 4$

4. $x - 4y = -\frac{73}{6}$
$-6x - y = -2$

5. $\frac{2}{3}x + \frac{3}{5}x = \frac{14}{3}$
$x - \frac{2}{5}y = -6$

6. $-3x + 4y = -17$
$2x - \frac{3}{4}y = -4$

7. $-\frac{1}{2}x + 9y = -1$
$\frac{1}{2}x + \frac{3}{14}y = \frac{57}{14}$

8. $-6x + 12y = -15$
$-6x - 6y = 0$

9. $10x + 2y = 2$
$10y = -5$

10. $3x + 3y = 1$
$-6x - 6y = 2$

11. $6x + 12y = \frac{11}{2}$
$-3x - 6y = -\frac{11}{4}$

12. $-4x - \frac{2}{7}y = 10$
$-x - \frac{5}{7}y = -2$

13. $-2x - 3y = -10$
$x - 6y = 35$

14. $-2x - 8y = \frac{6}{7}$
$4x + 8y = -\frac{5}{7}$

15. $-\frac{2}{5}x + \frac{7}{9}y = 3$
$x + \frac{2}{3}y = 7$

16. $\frac{2}{7}x + 3y = -69$
$-x + \frac{2}{3}y = 7$

17. $4x + 4y = 12$
$9x + 9y = 18$

18. $-4x + 10y = 3$
$12x + 2y = 23$

19. $-2x - 6y = 6$
$4x + 12y = -12$

20. $4x - 6y = -\frac{27}{2}$
$7x + 18y = -\frac{33}{2}$

21. $-8x + 8y = -63$
$x + 4y = \frac{17}{2}$

22. $x + y = -2$
$-\frac{3}{2}x + \frac{13}{4}y = -28$

23. $-2x + 3y = -5$
$10x + 9y = 9$

24. $7x = \frac{1}{6}$
$6x - 5y = \frac{9}{14}$

Solve the following systems of three equations in three variables; if the system is dependent or inconsistent state this.

25. $x + y - 5z = -9$
$-x + y + 2z = 9$
$5x + 2y = -4$

26. $-5x - y + 3z = -14$
$-2x + 2y - 6z = 16$
$x + 7y + 2z = -5$

27. $-x - 6y - z = 22$
$\frac{1}{3}x + y - 3z = -9$
$2x - \frac{2}{5}y + z = 16$

28. $-5x + y + 9z = -37$
$4x - y - z = 11$
$\frac{9}{2}x + 2y - 5z = 24$

29. $-x + 3y - 3z = 29$
$x + \frac{4}{5}y - 5z = 30$
$-3x - 3y + \frac{12}{5}z = -30$

30. $-3x + \frac{8}{5}z = 14$
$5x - \frac{2}{3}y + 3z = 1$
$-x - 4y - 2z = -32$

31. $-3x + 6z = 6$
$\frac{9}{2}x + \frac{9}{2}y + 3z = 6$
$x + 7y - 2z = 40$

32. $-x + 6y + 6z = -11$
$-4x + \frac{2}{3}y + 4z = -4$
$5x + y + 2z = 10$

33. $y + z = 0$
$-3x + 2y + 2z = 0$
$6x + 2y - 6z = -40$

34. $-5x + 5y + 4z = 14$
$3x + y - 4z = -10$
$x - 2y + 2z = 0$

35. $x - y + z = -14$
$\frac{2}{5}x + 3y - 2z = 7$
$\frac{2}{5}x - 4y + 9z = -25$

36. $2x + z = 2$
$x + \frac{3}{10}y - 2z = 9$
$-5x + \frac{1}{5}y + 8z = -24$

Solve the following problems.

37. The perimeter of a certain rectangle is 44 centimeters long; if the ratio of length to width is 8 to 5, find the values of the length and width.

38. The ratio of length to width of a given rectangle is 3 to 2; if the perimeter is 100 inches, find each dimension.

39. The length of a rectangle is 5 inches longer than the width. The perimeter is 36 inches. Find the two dimensions.

40. The length of a certain rectangle is 3 inches longer than twice the width. The perimeter is 120 inches. Find the two dimensions.

41. The width of a rectangle is 10 millimeters less than half the length. If the perimeter is 150 millimeters, find the length and width.

42. The ratio of width to length of a certain rectangle is 1.5 to 4.0. If the perimeter of the rectangle is 25 centimeters, find the length and width.

43. An investor invested a total of \$12,000 into a two-part mutual fund; one part paid 5% and the other part paid 10% last year. If the total income from the investments was \$800, compute how much must have been invested at each rate.

44. A total of \$20,000 was invested, part at 4% and the rest at 12%. The total income from both investments was \$2,000. How much was invested at each rate?

45. \$900 was earned on \$10,000 last year; part of the \$10,000 was invested at 6% and the rest at 12%. How much was invested at each rate?

46. If the total income from a \$40,000 investment last year was \$5,000, and the money was invested in two funds paying 6% and 16%, how much was invested at each rate?

47. A parabola is the graph of an equation of the form $y = ax^2 + bx + c$. Find the values of a, b, and c so that the parabola will pass through the points $(-4,1)$, $(0,3)$, and $(2,6)$.

48. Find the equation of the parabola that passes through the points $(-1,1)$, $(\frac{1}{2},3)$ and $(2,10)$. See problem 47.

49. A nonvertical straight line is the graph of an equation of the form $y = mx + b$. Two points uniquely determine a straight line. Find the equation of the straight line that passes through the points $(5,-1)$ and $(8,6)$.

50. Find the equation of the straight line that passes through the points $(-2,8)$ and $(5,1)$. See problem 49.

51. A straight line passes through the points $(-2,3)$ and $(1,5)$; a second straight line passes through the points $(0,2)$ and $(4,1)$. Find the point at which these two straight lines intersect. See problem 49.

52. Find the equation of the straight line that passes through the points $(2,5)$ and $(6,8)$. See problem 49.

53. Find the equation of the straight line that passes through the points $(-2,-3)$ and $(0,2)$. See problem 49.

54. On Babylonian clay tablets, 4,000 years old, is found a problem that refers to quantities called a ''first silver ring'' and a ''second silver ring.'' If these quantities are represented by x and y, the tablet asks for the solution to the following system of equations: $\frac{x}{7} + \frac{y}{11} = 1$ and $\frac{6x}{7} = \frac{10y}{11}$. The answer is given as $\frac{x}{7} = \frac{11}{7 + 11} + \frac{1}{72}$ and $\frac{y}{11} = \frac{7}{7 + 11} - \frac{1}{72}$. Show that this answer is correct.

Skill and review

1. Simplify $\sqrt{\frac{4x^2}{27y^3z}}$.

2. Solve $\frac{2x - 1}{3} = \frac{5 - 3x}{4}$.

3. Solve $\frac{2x - 1}{3} = \frac{5 - 3x}{x}$.

4. Solve $\left|\frac{2x - 1}{3}\right| > 5$.

5. Solve $\left|\frac{2x - 1}{x}\right| < 5$.

10–2 Systems of linear equations—matrix elimination

A certain mix of animal feed contains 10% protein; a second mix is 45% protein. How many pounds of each must be mixed to obtain 250 pounds of a 30% protein mix?

This section discusses another method of solving systems of linear equations. The stated problem could be solved by such a system.

The choice of symbols used for the variables in a system of equations is not important; it is the coefficients that determine the solution. A matrix is a mathematical tool that allows us to focus exclusively and efficiently on these coefficients.

Matrix

A matrix is a rectangular array of numbers. The numbers that make up the matrix are called its elements.

The words matrix and array are used to mean the same thing. Matrices are shown by enclosing the array in brackets, as illustrated by the following examples.

$$\text{I}\quad \begin{bmatrix} 1 & 2 \\ 3 & 4 \end{bmatrix} \qquad \text{II}\quad \begin{bmatrix} 4 & -7 & 3 \\ 2 & 3 & -4 \\ 3 & 1 & 0 \end{bmatrix} \qquad \text{III}\quad \begin{bmatrix} \sqrt{2} & -1 & 0 \\ 8 & 0 & 5 \end{bmatrix} \qquad \text{IV}\quad \begin{bmatrix} 9 \\ -2 \\ \pi \\ 5 \end{bmatrix}$$

Matrices are classified by stating the numbers of rows and columns they contain, *always stating the number of rows first.* The examples above are:

I 2×2	("two by two")	(two rows by two columns)
II 3×3	("three by three")	(three rows by three columns)
III 2×3	("two by three")	(two rows by three columns)
IV 4×1	("four by one")	(four rows by one column)

When the number of rows equals the number of columns the matrix is said to be **square.** Examples I and II are square matrices; square matrices are also described as being of an **order.** Example I is an order 2 matrix, and example II is an order 3 matrix.

A symbolic definition[3] of a matrix A with m rows and n columns is

$$A = \begin{bmatrix} a_{1,1} & a_{1,2} & a_{1,3} & \cdots & a_{1,n} \\ a_{2,1} & a_{2,2} & a_{2,3} & \cdots & a_{2,n} \\ \vdots & \vdots & \vdots & & \vdots \\ a_{m,1} & a_{m,2} & a_{m,3} & \cdots & a_{m,n} \end{bmatrix}$$

We usually use capital letters to denote a matrix, and the corresponding lower case letter, with subscripts, to denote the elements of that array. Thus an element of A is $a_{i,j}$, $1 \le i \le m$, $1 \le j \le n$. We read this element as "the *i-j*'th element of A" or "*a* sub *ij*." For example $a_{2,3}$ is "*a* sub two three," the element in row 2 and column 3 of matrix *a*.

Historically, matrices developed as a shorthand notation useful in describing and solving systems of equations. In solving systems of linear equations by the addition method (section 10–1) we focus our attention on the coefficients of the variables, not the letter that represents the variable. In other words, we would solve both of the following systems in the same way.

$$\begin{aligned} 2x - 3y &= 5 \\ 3x + y &= 13 \end{aligned} \qquad \text{or} \qquad \begin{aligned} 2a - 3b &= 5 \\ 3a + b &= 13 \end{aligned}$$

The information in both systems is contained in the 2×3 matrix

$$\begin{bmatrix} 2 & -3 & 5 \\ 3 & 1 & 13 \end{bmatrix}$$

Each row of the matrix corresponds to one of the equations. The first two columns of the matrix correspond to the coefficients of the variables x and y, and the third column corresponds to the constants.

[3]This modern notation was perfected by C. E. Cullis in his book *Matrices and Determinoids*, (Cambridge), Vol. I (1913).

The method of matrix elimination parallels the steps we could do to solve this system of equations by addition. To illustrate matrix elimination we will solve this system, using the matrix shown.

First we could multiply the second row by 3 and add it to the first row (this will eliminate the value -3). This is similar to using a key equation in section 10–1. This would produce the matrix

$$\begin{bmatrix} 3(3) + 2 & 3(1) + (-3) & (13) + 5 \\ 3 & 1 & 13 \end{bmatrix} \text{ which is } \begin{bmatrix} 11 & 0 & 44 \\ 3 & 1 & 13 \end{bmatrix}$$

We could divide the first row by 11. This is the same as saying that $11x + 0y = 44$, so $x + 0y = 4$.

$$\begin{bmatrix} 1 & 0 & 4 \\ 3 & 1 & 13 \end{bmatrix}$$

If we now multiply the first row by -3 and add the result to the second row we obtain

$$\begin{bmatrix} 1 & 0 & 4 \\ (-3 + 3) & 0 + 1 & (-12 + 13) \end{bmatrix} = \begin{bmatrix} 1 & 0 & 4 \\ 0 & 1 & 1 \end{bmatrix}$$

The matrix now corresponds to the system of equations $\begin{matrix} x + 0y = 4 \\ 0x + y = 1, \end{matrix}$ or $\begin{matrix} x = 4 \\ y = 1 \end{matrix}$, which has the point (4,1) for its solution. This solution is the rightmost column of the matrix. This is also the solution to the original system of equations.

Before we further discuss solving systems of linear equations by this method we need some more definitions. The order 2 matrix $\begin{bmatrix} 1 & 0 \\ 0 & 1 \end{bmatrix}$ as in the variable columns of $\begin{bmatrix} 1 & 0 & 4 \\ 0 & 1 & 1 \end{bmatrix}$ above is called the order 2 **identity matrix.** The order 3 identity matrix is $\begin{bmatrix} 1 & 0 & 0 \\ 0 & 1 & 0 \\ 0 & 0 & 1 \end{bmatrix}$, and in general the **order n identity matrix** has ones on the **main diagonal** (upper-left to lower-right) and zeros everywhere else.

An **augmented matrix** is a matrix formed as we did above that represents a system of n equations in n variables. It has n rows (one for each equation) and $n + 1$ columns—the first n columns contain the coefficients of the variables in the equations, and the last column contains the constants. This matrix can be used to obtain the solutions to the system of equations.

The elimination method uses the following operations, called **row operations.** It can be proven that *row operations do not change the solution set of a system of equations.*

> **Row operations**
> 1. Multiply or divide each entry of a row by a nonzero value.
> 2. Add a nonzero multiple of one row to a nonzero multiple of another row, and replace either row by the result.
> 3. Rearrange the order of the rows.

The **matrix elimination method** for solving a system of n equations in n variables consists of the following steps.

1. Create the augmented matrix.
2. Use the row operations to obtain the identity matrix in the first n columns.
3. The solution is in the column of constants.

To obtain the identity matrix we will **sweep out** each of the first n columns. This means to use one nonzero value in a column to eliminate the rest of the nonzero elements in that column.

We will use the idea of a *key row* (KR) for sweeping out columns; each row will serve as a key row once (unless we simply rearrange rows). Using key rows helps keep track of what we have done; without this idea the procedure can become confusing. We will see that this is the same concept as "key equations" from section 10–1. Remember that *a row may be a key row only once.*

It is helpful to box in the element in the key row that is also in the column we are sweeping out. This is shown in example 10–2 A.

■ ***Example 10–2 A***

Solve each system of equations.

1. $3x - 2y = 8$
$-2x - 5y = 1$

Step 1: The augmented matrix is $\begin{bmatrix} 3 & -2 & 8 \\ -2 & -5 & 1 \end{bmatrix}$.

Step 2: We now use row operations to sweep out column 1. We let the first row be the key row, so we box in the element in row 1, column 1. Our objective is to make the entry in row 2, column 1, zero.

As shown below it is a good idea to use a "scratch pad" to perform the arithmetic.

$$\begin{bmatrix} 3 & -2 & 8 \\ -2 & -5 & 1 \end{bmatrix}$$

Add twice the first row to three times the second

2R1:	6	−4	16
3R2:	−6	−15	3
	0	−19	19 → Row 2

$$\begin{bmatrix} 3 & -2 & 8 \\ 0 & -19 & 19 \end{bmatrix}$$

The first column is swept out

$$\begin{bmatrix} 3 & -2 & 8 \\ 0 & 1 & -1 \end{bmatrix}$$ Divide the second row by -19

$$\begin{bmatrix} 3 & -2 & 8 \\ 0 & \boxed{1} & -1 \end{bmatrix}$$ Now the second row is the key row, and we want to sweep out the second column; box in the element in row 2 column 2; add twice the second row to the first row

$$\begin{array}{rrrl} \text{R1:} & 3 & -2 & 8 \\ \text{2R2:} & 0 & 2 & -2 \\ & 3 & 0 & 6 \rightarrow \text{Row 1} \end{array}$$

$$\begin{bmatrix} 3 & 0 & 6 \\ 0 & 1 & -1 \end{bmatrix}$$ The second column is now swept out

$$\begin{bmatrix} 1 & 0 & 2 \\ 0 & 1 & -1 \end{bmatrix}$$ Divide the first row by 3

The third column gives the solution $(x,y) = (2,-1)$.

2. $2x - y + z = 10$
$x + 2y - z = -3$
$3x + y + 2x = 11$

The augmented matrix is $\begin{bmatrix} 2 & -1 & 1 & 10 \\ 1 & 2 & -1 & -3 \\ 3 & 1 & 2 & 11 \end{bmatrix}$.

Let us first focus on column 3, the ''z'' column. Since column 3 has a 1 in row 1, we choose this row to be our key row. We will sweep out column 3 using row 1, so we box in the element in row 1 column 3. We will add multiples of row 1 to multiples of the other rows to make the entries in column 3 in those rows zero.

We can make the entry in row 2 zero by adding row 1 (the key row) to row 2; ''scratch pad'' work is shown below the matrices.

$$\begin{bmatrix} 2 & -1 & \boxed{1} & 10 \\ 1 & 2 & -1 & -3 \\ 3 & 1 & 2 & 11 \end{bmatrix} \text{Key row} \quad \text{becomes} \begin{bmatrix} 2 & -1 & \boxed{1} & 10 \\ 3 & 1 & 0 & 7 \\ 3 & 1 & 2 & 11 \end{bmatrix}.$$

$$\begin{array}{rrrrl} \text{R1:} & 2 & -1 & 1 & 10 \\ \text{R2:} & 1 & 2 & -1 & -3 \\ & 3 & 1 & 0 & 7 \rightarrow \text{New row 2} \end{array}$$

Let us indicate the last step by the notation R2 $\leftarrow$ KR + R2, which shows that ''row 2 (R2) is replaced by the sum of the key row (KR) and row 2 (R2).''

We now make the last entry in column 3 zero by adding -2 times row 1 to row 3, and putting the result in row 3:

$$\begin{bmatrix} 2 & -1 & \boxed{1} & 10 \\ 3 & 1 & 0 & 7 \\ 3 & 1 & 2 & 11 \end{bmatrix} \text{Key row} \quad \text{becomes} \begin{bmatrix} 2 & -1 & \boxed{1} & 10 \\ 3 & 1 & 0 & 7 \\ -1 & 3 & 0 & -9 \end{bmatrix}.$$

$$\begin{array}{rrrrl} -2\text{R1:} & -4 & 2 & -2 & -20 \\ \text{R3:} & 3 & 1 & 2 & 11 \\ \hline & -1 & 3 & 0 & -9 \rightarrow \text{R3} \end{array}$$

Our notation for this is R3 ← −2(KR) + R3. *It is important to note that column 3 is swept out.* This means there is only one nonzero entry in that column. *Also, row 1 may not be reused as a key row.* (Assuming we do not change the order of the rows.)

Now sweep out column 2; we want to arrange column 2 so that all entries but one are zero. We first choose the next key row. Row 2 would be a good choice because column 2 has an entry of 1 in row 2, which will make our arithmetic that much easier. Thus, we box in the entry in column 2 and row 2.

$$\begin{bmatrix} 2 & -1 & 1 & 10 \\ 3 & \boxed{1} & 0 & 7 \\ -1 & 3 & 0 & -9 \end{bmatrix} \text{ becomes } \begin{bmatrix} 5 & 0 & 1 & 17 \\ 3 & \boxed{1} & 0 & 7 \\ -10 & 0 & 0 & -30 \end{bmatrix}.$$

The ''scratch pad'' work is:

R1:	2	−1	1	10		−3KR:	−9	−3	0	−21
KR:	3	1	0	7		R3:	−1	3	0	−9
	5	0	1	17 → R1			−10	0	0	−30 → R3

Observe that we can divide each entry in row 3 by −10 to obtain

$$\begin{bmatrix} 5 & 0 & 1 & 17 \\ 3 & 1 & 0 & 7 \\ 1 & 0 & 0 & 3 \end{bmatrix}$$

Now we sweep out column 1; only row 3 is left to serve as a key row.

$$\begin{bmatrix} 5 & 0 & 1 & 17 \\ 3 & 1 & 0 & 7 \\ \boxed{1} & 0 & 0 & 3 \end{bmatrix} \begin{matrix} -5(\text{KR}) + \text{R1} \rightarrow \\ -3(\text{KR}) + \text{R2} \rightarrow \\ \text{Key row} \end{matrix} \quad \begin{bmatrix} 0 & 0 & 1 & 2 \\ 0 & 1 & 0 & -2 \\ 1 & 0 & 0 & 3 \end{bmatrix}$$

Note The notation above states that we added −5 times the key row to row 1, and −3 times the key row to row 2, and the key row is row 3.

If we now rearrange the order of the rows we obtain the 3 × 3 identity matrix and the values of x, y, and z in the last column:

$$\begin{bmatrix} 1 & 0 & 0 & 3 \\ 0 & 1 & 0 & -2 \\ 0 & 0 & 1 & 2 \end{bmatrix}.$$

This matrix corresponds to the set of equations $x = 3$, $y = -2$, and $z = 2$. Thus, our solution is $(x,y,z) = (3,-2,2)$.

3. $2x - y + w = 10$
$x + 2y - z = 3$
$y - z + 3w = -2$
$3x + 4w = 5$
Put in zero coefficients where a variable does not appear in an equation. This gives the augmented matrix

$$\begin{bmatrix} 2 & -1 & 0 & 1 & 10 \\ 1 & 2 & -1 & 0 & 3 \\ 0 & 1 & -1 & 3 & -2 \\ 3 & 0 & 0 & 4 & 5 \end{bmatrix}$$

Since there are only two nonzero entries in column 3 we will sweep it out first.

$$\begin{bmatrix} 2 & -1 & 0 & 1 & 10 \\ 1 & 2 & \boxed{-1} & 0 & 3 \\ 0 & 1 & -1 & 3 & -2 \\ 3 & 0 & 0 & 4 & 5 \end{bmatrix} \begin{matrix} \\ \text{Key row} \\ -\text{KR} + \text{R3} \rightarrow \\ \\ \end{matrix} \begin{bmatrix} 2 & -1 & 0 & 1 & 10 \\ 1 & 2 & -1 & 0 & 3 \\ -1 & -1 & 0 & 3 & -5 \\ 3 & 0 & 0 & 4 & 5 \end{bmatrix}$$

Columns 2 or 4 are good choices to be swept out next. Column 1 is the poorest choice because it has the fewest zeros. We choose column 4 next, using row 1 as the key row.

$$\begin{bmatrix} 2 & -1 & 0 & \boxed{1} & 10 \\ 1 & 2 & -1 & 0 & 3 \\ -1 & -1 & 0 & 3 & -5 \\ 3 & 0 & 0 & 4 & 5 \end{bmatrix} \begin{matrix} \text{Key row} \\ \\ -3(\text{KR}) + \text{R3} \rightarrow \\ -4(\text{KR}) + \text{R4} \rightarrow \end{matrix} \begin{bmatrix} 2 & -1 & 0 & 1 & 10 \\ 1 & 2 & -1 & 0 & 3 \\ -7 & 2 & 0 & 0 & -35 \\ -5 & 4 & 0 & 0 & -35 \end{bmatrix}$$

Only rows 3 and 4 may now be used as key rows.

We will sweep out column 2, using row 3 as the key row.

$$\begin{bmatrix} 2 & -1 & 0 & 1 & 10 \\ 1 & 2 & -1 & 0 & 3 \\ -7 & \boxed{2} & 0 & 0 & -35 \\ -5 & 4 & 0 & 0 & -35 \end{bmatrix} \begin{matrix} \text{KR} + 2(\text{R1}) \rightarrow \\ -\text{KR} + \text{R2} \rightarrow \\ \text{Key row} \\ -2(\text{KR}) + \text{R4} \rightarrow \end{matrix} \begin{bmatrix} -3 & 0 & 0 & 2 & -15 \\ 8 & 0 & -1 & 0 & 38 \\ -7 & 2 & 0 & 0 & -35 \\ 9 & 0 & 0 & 0 & 35 \end{bmatrix}$$

We can now sweep out column 1; only row 4 is left to serve as a key row.

$$\begin{bmatrix} -3 & 0 & 0 & 2 & -15 \\ 8 & 0 & -1 & 0 & 38 \\ -7 & 2 & 0 & 0 & -35 \\ \boxed{9} & 0 & 0 & 0 & 35 \end{bmatrix} \begin{matrix} \text{KR} + 3(\text{R1}) \rightarrow \\ 8(\text{KR}) - 9(\text{R2}) \rightarrow \\ 7(\text{KR}) + 9(\text{R3}) \rightarrow \\ \text{Key row} \end{matrix} \begin{bmatrix} 0 & 0 & 0 & 6 & -10 \\ 0 & 0 & 9 & 0 & -62 \\ 0 & 18 & 0 & 0 & -70 \\ 9 & 0 & 0 & 0 & 35 \end{bmatrix}$$

At this point all of the columns have been swept out. Each has all but one zero entries, and none of these nonzero entries is in the same row.

Rearrange the order of the rows so the nonzero entries are on the main diagonal.

$$\begin{bmatrix} 9 & 0 & 0 & 0 & 35 \\ 0 & 18 & 0 & 0 & -70 \\ 0 & 0 & 9 & 0 & -62 \\ 0 & 0 & 0 & 6 & -10 \end{bmatrix}$$

Divide each row by its nonzero entry in columns 1 through 4.

$$\begin{bmatrix} 1 & 0 & 0 & 0 & \frac{35}{9} \\ 0 & 1 & 0 & 0 & -\frac{35}{9} \\ 0 & 0 & 1 & 0 & -\frac{62}{9} \\ 0 & 0 & 0 & 1 & -\frac{5}{3} \end{bmatrix}$$

The solution is the 4-tuple $(w,x,y,z) = (\frac{35}{9}, -\frac{35}{9}, -\frac{62}{9}, -\frac{5}{3})$. ■

The example 10–2 B illustrates what happens when a system is dependent or inconsistent.

■ ***Example 10–2 B***

Solve each system of equations using matrix elimination.

1. $x + y - z = -2$
$2x - y + 2z = 6$
$3x + z = 3$

Augmented matrix Swept out matrix

$$\begin{bmatrix} 1 & 1 & -1 & -2 \\ 2 & -1 & 2 & 6 \\ 3 & 0 & 1 & 3 \end{bmatrix} \qquad \begin{bmatrix} 3 & 0 & 1 & 4 \\ 4 & 1 & 0 & 2 \\ 0 & 0 & 0 & 1 \end{bmatrix}$$

The last row of the swept out matrix corresponds to the statement $0x + 0y + 0z = 1$, or $0 = 1$. This is a false statement, which means that there is no solution to this system. This system of equations is *inconsistent*. Thus there is no solution.

Note Whenever a system is inconsistent, at least one row will contain a false statement such as $0 = 1$ in the last example.

2. $2x - 3y + z = 4$
$x + y - 2z = -3$
$3x - 2y - z = 1$

Augmented matrix Swept out matrix

$$\begin{bmatrix} 2 & -3 & 1 & 4 \\ 1 & 1 & -2 & -3 \\ 3 & -2 & -1 & 1 \end{bmatrix} \qquad \begin{bmatrix} -1 & 0 & 1 & 1 \\ 1 & -1 & 0 & 1 \\ 0 & 0 & 0 & 0 \end{bmatrix}$$

If there are one or more rows of zeros, when we have swept out as many columns as nonzero rows, and there is no false statement (such as $1 = 0$), the system is dependent but consistent (not inconsistent). In this case we have swept out two columns and there are two nonzero rows. Thus, this system is dependent.

Note There are other methods of determining when a system is dependent or inconsistent than those shown here. Section 10–3 shows a simpler method, for example. Thus we will not dwell on this detail here. ■

Matrix elimination provides another way to solve systems of linear equations when they occur in applications.

■ Example 10–2 C

A chemical company stores a certain herbicide in two concentrations of herbicide and water: 10% solution and 25% solution. It needs to manufacture 1,000 gallons of a 15% solution by mixing the correct amounts of the 10% and 25% solutions. How many gallons of each of the 10% and 25% solutions should be mixed to obtain the required product?

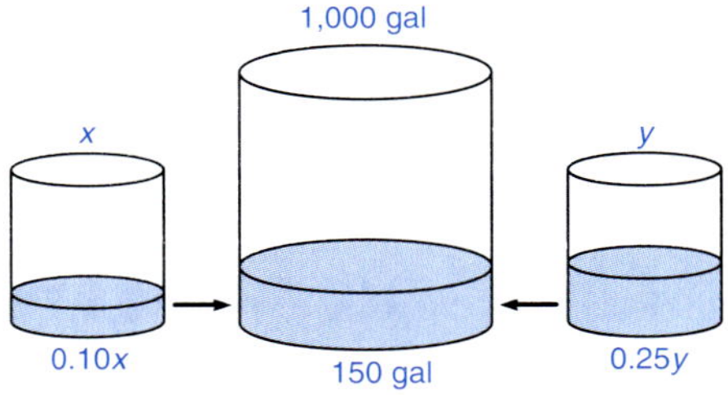

This type of problem can be solved by focusing on two aspects of the problem: the total amounts of the solutions, and the total amounts of the herbicide itself. For example, in the 10% solution we know that 10% of the total is herbicide and the rest water; thus, if x is the amount of the 10% solution then 10% of x $(0.10x)$ is herbicide.

Let x be the amount of the 10% solution, and y be the amount of the 25% solution. Then $x + y = 1{,}000$, since the total amount will be 1,000 gallons.

Now, focus on the herbicide itself. We need 1,000 gallons of a 15% solution; this will contain 15% of 1,000, or 150, gallons of herbicide. This 150 gallons of herbicide comes from 10% of x and 25% of y.

$$\begin{aligned} 0.10x + 0.25y &= 150 \\ 10x + 25y &= 15{,}000 \\ 2x + 5y &= 3{,}000 \end{aligned}$$

We now have two equations in two variables.

$$\begin{aligned} [1] \quad x + y &= 1000 \\ [2] \quad 2x + 5y &= 3{,}000 \end{aligned}$$

Augmented matrix

$$\begin{bmatrix} 1 & 1 & 1{,}000 \\ 2 & 5 & 3{,}000 \end{bmatrix}$$

Solution matrix

$$\begin{bmatrix} 1 & 0 & \frac{2{,}000}{3} \\ 0 & 1 & \frac{1{,}000}{3} \end{bmatrix}$$

The solution is $x = \frac{2{,}000}{3}$, $y = \frac{1{,}000}{3}$. Thus the final product should consist of $666\frac{2}{3}$ gallons of the 10% solution and $333\frac{1}{3}$ gallons of the 25% solution. ■

Mastery points

Can you

- Solve a system of n linear equations in n variables by matrix elimination?
- Describe a dependent solution set?
- Solve applications whose solutions can be found by n linear equations in n variables?

Exercise 10–2

Solve the following systems of equations by matrix elimination; after describing the solution set, state dependent or inconsistent if appropriate.

1. $2x + \frac{2}{3}y = -2$
$-3x + y = 15$

2. $2x + \frac{1}{2}y = 9$
$-3x + y = -10$

3. $-2x + 5y = 9$
$x + 3y = \frac{13}{2}$

4. $2x + 3y = 7$
$-5x + 2y = 11$

5. $-3x + 4y = -1$
$4x + y = 14$

6. $-3x + 2y = -2$
$3x + 2y = 0$

7. $\frac{2}{5}x + \frac{1}{3}y = 1$
$-2x + 2y = -16$

8. $\frac{1}{3}x + \frac{1}{4}y = 2$
$-2x - y = -16$

9. $\frac{3}{4}x + 2y = 0$
$-3x + y = -27$

10. $2x + y = 0$
$3x - 5y = -65$

11. $\frac{1}{2}x + 3y = 21$
$2x - 2y = 0$

12. $-3x + 2y = 3$
$6x + y = 9$

13. $4x + 5y = -17$
$-x + 10y = 38$

14. $4x - 2y = 8$
$\frac{1}{2}x + 3y = 27$

15. $-2x + \frac{1}{3}y = 14$
$4x + \frac{2}{3}y = -20$

16. $6x - 2y = -9$
$8x + 5y = 11$

17. $3x + y = 20$
$-2x + 2y = 0$

18. $2x + 3y = 8$
$-4x - 6y = -16$

19. $-3x + y = -4$
$6x + 2y = 20$

20. $2x + 5y = 1$
$5x - y = -11$

21. $\frac{2}{3}x - 3y = 10$
$-2x + y = -14$

22. $4x - 3y = 0$
$-8x + 3y = -5$

23. $x + y - 5z = -9$
$-x + y + 2z = 9$
$5x + 2y = -4$

24. $-5x - y + 3z = -14$
$-2x + 2y - 6z = 16$
$x + 7y + 2z = -5$

25. $-x - 6y - z = 22$
$x + y - 3z = -5$
$2x - 2y + z = 24$

26. $-5x + y + 2z = -16$
$4x - y - z = 11$
$x + 2y - 5z = 17$

27. $-x + 3y - 3z = 15$
$x + 4y - 3z = 22$
$-3x - 3y + 6z = -20$

28. $-3x + 8z = 14$
$5x - 2y + 3z = -8$
$-x - 4y - 2z = -2$

29. $-3x + z = 3$
$9x + 2y + 3z = -3$
$3x + y - 2z = 4$

30. $-x + 6y + 6z = -11$
$\frac{2}{3}y + 4z = 4$
$5x + 3y + 2z = 4$

31. $y + z = 0$
$-3x + 2y + 2z = 0$
$5x + 2y - 6z = -40$

32. $-5x + 5y + 4z = 14$
$3x + y - 4z = -10$
$x - 2y + 2z = 0$

33. $x - y + z = -14$
$-2x + 3y - 2z = 31$
$4x - 4y + 7z = -59$

34. $2x + z = 2$
$x + 3y - 2z = -3$
$-5x + 2y + 8z = -32$

35. $x - z = 5$
$-2x + 4y + 2w = 4$
$-2x - 2y - 3z - 2w = -6$
$2x - 5z + 5w = 21$

36. $2x - z + w = -2$
$3x + z + 4w = 7$
$4y - 2z + 5w = -9$
$y + 2z + 2w = 6$

37. $6x + 4y + z = 1$
$3x + y - 3w = -11$
$-4x + w = 0$
$-x + 4y = -9$

38. $-3z + 5w = -10$
$-x + z + 3w = -7$
$-3x - 3y + z = -6$
$-5x - y - 5z + 5w = -16$

39. $x + \frac{1}{2}y + 3z - 3w = -5$
$2x - \frac{3}{2}y + 3z + 5w = 16$
$-6x - w = -8$
$x - 6z = -1$

40. $4x + y - 5z = 14$
$x + y - 5z + 3w = \frac{25}{2}$
$-6x + 3y + 2z + 2w = -1$
$z - 3w = -2$

41. $3x + 2y - 3z - 3w = -3$
$-3x - y + 7z + 4w = -5$
$x - 3y - z + w = 7$
$4x + y + 4z = -12$

42. $-5x + y - z - 3w = 10$
$x - 3z = -11$
$x + 3y - z + w = -6$
$-3x - 3y + 5w = 1$

43. $-3x + 6y + 5z - 3w = 1$
$4x + 2z - 3w = 36$
$2x - 5y - 3z = 9$
$x + 2y + 5z - 4w = 29$

44. $-6x + 3y - 5w = -11$
$3x + 2y + 4w = 19$
$-3x - y + 2z + 3w = -4$
$y + 8z + w = -3$

45. $y + 8z + 2w = 6$
$3x + y + 4w = 11$
$x - 6y + 4z + 5w = 23$
$2x + 10y - 4z = -17$

46. $-4x + y + w = 16$
$x + y + 2z + 4w = 8$
$x - 4y + 2z - 4w = -21$
$x + 4y + 2z - 6w = -19$

Solve the following problems.

47. Kirchhoff's law, from circuit theory in electronics, states that the sum of the voltages around any loop of a circuit is 0. In a certain circuit with two loops, with i_1 the current in one loop and i_2 the current in the second loop, the application of Kirchhoff's law gives the system
$60i_1 - 10i_2 = 116$
$10i_1 - 30i_2 = 8$. Solve for the currents i_1 and i_2.

48. For a certain electronics circuit Kirchhoff's law (problem 47) gives the system
$35i_1 + 12i_2 + 5i_3 = 197$
$60i_1 - 20i_2 - 10i_3 = 260$. Find i_1, i_2, and i_3.
$15i_1 + 10i_2 + 5i_3 = 95$

49. A certain scale is known to be very inaccurate for weights about 200 pounds. Three items, I_1, I_2, and I_3 must be weighed, and it is known that their weights are in the 200 pound range. Thus, the items are weighed together, and the following results are noted:
$I_1 + I_2 = 380$
$I_1 + I_3 = 390$. Find the weight of each of the three items.
$I_2 + I_3 = 410$

50. A chemical company stores a certain herbicide in two concentrations of herbicide and water: 6% solution and 15% solution. It needs to manufacture 1,000 gallons of a 10% solution by mixing the correct amounts of the two solutions. How many gallons of each should be mixed to obtain the required product?

51. Certain amounts of a 20% and a 50% solution of alcohol are to be mixed to obtain 500 liters of a 30% solution. How many liters of each solution should be mixed together?

52. A certain mix of animal feed contains 10% protein; a second mix is 45% protein. How many pounds of each must be mixed to obtain 250 pounds of a 30% protein mix?

53. A trucking firm keeps a stock of 25% antifreeze solution and of 90% antifreeze solution on hand. How much of each should it mix to obtain 80 gallons of a 50% solution?

54. A paint manufacturer has two concentrations of its paint base on hand; one contains 4% linseed oil, the other contains 10% linseed oil. How many gallons of each should it mix to obtain 200 gallons of base of which 6.5% is linseed oil?

55. The following system of equations is found in a Chinese book from about 250 B.C.:

$$3x + 2y + z = 39$$
$$2x + 3y + z = 34$$
$$x + 2y + 3z = 26$$

It is solved by using steps similar to those shown in this section. Solve this system of equations.

Skill and review

1. Find the point of intersection of the two lines $y = 2x + 3$ and $2x - 3y = 6$.
2. Find $\log_3 81$.
3. Solve $9^{3x+1} = 27^x$.
4. Solve $\log(x - 1) - \log(x + 1) = 2$.
5. Solve $\log(x - 1) + \log(x + 1) = \log 2$.
6. Solve $x^3 - x^2 - x + 1 < 0$.

10–3 Systems of linear equations—Cramer's rule

Find the area of the triangle with vertices (−2,6), (3,−2), and (6,12).

Finding the area of a polygon such as that in this problem is one of the many types of problems that can be solved using systems of equations. Cramer's rule, studied in this section, provides another way to solve these systems.

Determinants

As defined in section 10–2 a square matrix of order n is an array (or matrix) of n rows and n columns. The word *order* implies the matrix in question is square. The general matrices of order 2 and order 3 are

$$\begin{bmatrix} a_{1,1} & a_{1,2} \\ a_{2,1} & a_{2,2} \end{bmatrix} \quad \text{and} \quad \begin{bmatrix} a_{1,1} & a_{1,2} & a_{1,3} \\ a_{2,1} & a_{2,2} & a_{2,3} \\ a_{3,1} & a_{3,2} & a_{3,3} \end{bmatrix}$$

Associated with every square matrix is a real number called a **determinant.** The determinant[4] of an order 2 matrix is indicated by enclosing the matrix elements in vertical bars (as in the absolute value of a real number), and is defined as follows.

Determinants of an order 2 matrix

The determinant of a 2 × 2 matrix $\begin{bmatrix} a_{1,1} & a_{1,2} \\ a_{2,1} & a_{2,2} \end{bmatrix}$ is written $\begin{vmatrix} a_{1,1} & a_{1,2} \\ a_{2,1} & a_{2,2} \end{vmatrix}$ and has the value $a_{1,1}a_{2,2} - a_{2,1}a_{1,2}$.

Observe that the value is computed as the difference of the products of the elements on the two diagonals, with the diagonal from upper left to lower right used first.

$$\begin{vmatrix} a_{1,1} & a_{1,2} \\ a_{2,1} & a_{2,2} \end{vmatrix}$$

■ **Example 10–3 A**

Compute the determinant of $\begin{vmatrix} 2 & -4 \\ -3 & 5 \end{vmatrix}$

$$(2)(5) - (-3)(-4) = 10 - 12 = -2$$ ■

The definition of the determinant of an order n matrix, $n > 2$, is defined in terms of the determinant of an order $n - 1$ matrix. This definition requires two additional definitions.

[4]The earliest determinant notations go back to the originator of determinants in Europe, Gottfried Wilhelm Leibniz (1693).

> **Minor of an element of a matrix**
> Given an nth order matrix A, the minor of element $a_{i,j}$ is the $(n - 1)$ order matrix formed by deleting the ith row and jth column of matrix A. We refer to this minor as $m_{i,j}$.

■ ***Example 10–3 B***

Find the minor $m_{2,3}$ of the matrix. This minor is the order 3 matrix found by deleting row 2 and column 3.

$$\begin{bmatrix} 4 & 0 & -1 & 3 \\ -2 & 3 & -5 & 6 \\ 0 & 5 & 1 & 2 \\ 0 & 8 & -3 & 7 \end{bmatrix} \text{ becomes } \begin{bmatrix} 4 & 0 & -1 & 3 \\ -2 & 3 & -5 & 6 \\ 0 & 5 & 1 & 2 \\ 0 & 8 & -3 & 7 \end{bmatrix} \text{ so } m_{2,3} = \begin{bmatrix} 4 & 0 & 3 \\ 0 & 5 & 2 \\ 0 & 8 & 7 \end{bmatrix}$$ ■

> **Sign matrix for an order n matrix**
> The sign matrix for an order n matrix is an order n matrix where each element has the value $(-1)^{i+j}$, where i is the row and j is the column. It looks like the following.
>
> n columns, n rows
>
> $$\begin{bmatrix} + & - & + & - & + & - & \cdots \\ - & + & - & + & - & \cdots & \\ + & - & + & - & \cdots & & \\ - & + & - & \cdots & & & \\ + & - & \cdots & & & & \\ - & \cdots & & & & & \\ \cdots & & & & & & \end{bmatrix}$$

In the sign matrix, the upper left (i.e., 1,1) entry is $+$, and the pattern alternates with plus and minus signs through the rows and columns. The entry for a given row and column in the sign matrix can either be found by examination of the matrix itself or by computing $(-1)^{i+j}$, where i is the row and j is the column.

We are now prepared to define the determinant of matrices of orders greater than 2. It may look complicated, but the examples below will make it clear.

> **Determinant of an order n matrix**
> The determinant of an order n matrix, $n > 2$, is the sum of the products of each element of any row or column, $+1$ or -1, depending on the sign matrix and the determinant of its respective minor.

The following procedure reflects this definition.

> **Procedure for computing the determinant of an order n matrix.**
> 1. Choose the row or column with the most zero elements.
> 2. For each nonzero element $a_{i,j}$ in this row or column compute the product of the following three factors:
>   - $a_{i,j}$; the element itself
>   - $(-1)^{i+j}$; the corresponding element in the sign matrix (+1 or −1)
>   - $|m_{i,j}|$; the determinant of its minor
> 3. Add up all the values found in step 2.

Note
- In step 1 we can actually choose any row or column. Choosing the row or column with the most zeros minimizes the amount of work in the following steps.
- We generally perform steps 2 and 3 together. They are shown separately to help us understand them.

It has been proven that we get the same value for the determinant regardless of which row or column we choose for this "expansion."

The TI-81 calculator can compute determinants for matrices up to 6 by 6. This is illustrated in part 3 of the following example.

■ ***Example 10–3 C***

Compute each determinant.

1. $\begin{vmatrix} 0 & 5 & 2 \\ 3 & -2 & -1 \\ -3 & 4 & 6 \end{vmatrix}$

Step 1: No row or column has more than one zero. We choose row 1.

Matrix — Sign Matrix

$$\begin{vmatrix} 0 & \boxed{5} & \boxed{2} \\ 3 & -2 & -1 \\ -3 & 4 & 6 \end{vmatrix} \quad \begin{bmatrix} + & \boxed{-} & \boxed{+} \\ - & + & - \\ + & - & + \end{bmatrix}$$

Steps 2 and 3: For each nonzero element in row 1 we form the product of the three factors shown in the steps.

$$-5\begin{vmatrix} 3 & -1 \\ -3 & 6 \end{vmatrix} + 2\begin{vmatrix} 3 & -2 \\ -3 & 4 \end{vmatrix} = 5(15) + 2(6) = -63$$

2. $\begin{vmatrix} 3 & 6 & 4 \\ 0 & 0 & -2 \\ -3 & -5 & 0 \end{vmatrix}$

Since row 2 has the most zeros we expand about row 2.

$$-(-2)\begin{vmatrix} 3 & 6 \\ -3 & -5 \end{vmatrix} = 2(3) = 6$$

3. $\begin{vmatrix} 2 & -1 & 3 & -4 \\ 4 & 3 & -1 & 2 \\ 1 & -5 & -2 & -3 \\ 1 & 5 & 2 & -6 \end{vmatrix}$

Expand around the first row. The calculation for one of the order 3 matrices is shown below.

$$= 2\begin{vmatrix} 3 & -1 & 2 \\ -5 & -2 & -3 \\ 5 & 2 & -6 \end{vmatrix} - (-1)\begin{vmatrix} 4 & -1 & 2 \\ 1 & -2 & -3 \\ 1 & 2 & -6 \end{vmatrix}$$
$$+ 3\begin{vmatrix} 4 & 3 & 2 \\ 1 & -5 & -3 \\ 1 & 5 & -6 \end{vmatrix} - (-4)\begin{vmatrix} 4 & 3 & -1 \\ 1 & -5 & -2 \\ 1 & 5 & 2 \end{vmatrix}$$
$$= 2(99) + 77 + 3(209) + 4(-22)$$
$$= 814$$

The calculation of one of the order 3 determinants is shown.

$$\begin{vmatrix} 3 & -1 & 2 \\ -5 & -2 & -3 \\ 5 & 2 & -6 \end{vmatrix} = 3\begin{vmatrix} -2 & -3 \\ 2 & -6 \end{vmatrix} - (-1)\begin{vmatrix} -5 & -3 \\ 5 & -6 \end{vmatrix}$$
$$+ 2\begin{vmatrix} -5 & -2 \\ 5 & 2 \end{vmatrix} = 3(18) + 45 + 2(0) = 99$$

The determinant can be computed on the TI-81 as follows:

MATRIX EDIT 1	Enter the values into matrix [A]
4 ENTER 4 ENTER	This is a 4 × 4 matrix

Enter the values by row. Use the (−) key for negative values. Use the ENTER key after each value.

[MATRIX] 5	det
[A]	2nd 1
ENTER	The result is 814

To view the matrix, enter [A] ENTER.

Cramer's rule

It turns out that determinants can, among other things, be used to solve systems of linear equations by Cramer's rule.

> **Cramer's rule**
>
> Assume a given system of n linear equations in n variables. Let D represent the determinant of the coefficient matrix. Let D_x be the determinant of D with the x column replaced by the column of constants, D_y the determinant of D with the y column replaced by the column of constants, etc. Then,
>
> $$x = \frac{D_x}{D}, \quad y = \frac{D_y}{D}, z = \frac{D_z}{D}, \text{ etc.}$$

■ ***Example 10–3 D***

Use Cramer's rule in each problem.

1. Solve the system $\begin{matrix} 3x - 2y = 26 \\ 3y = 5 \end{matrix}$.

$$D = \begin{vmatrix} 3 & -2 \\ 0 & 3 \end{vmatrix} = 9 - 0 = 9$$ The coefficients from $\begin{matrix} 3x - 2y \\ 0x + 3y \end{matrix}$

$$D_x = \begin{vmatrix} 26 & -2 \\ 5 & 3 \end{vmatrix} = 78 - (-10) = 88$$ Replace the x column in D by the coefficients $\begin{matrix} 26 \\ 5 \end{matrix}$

$$D_y = \begin{vmatrix} 3 & 26 \\ 0 & 5 \end{vmatrix} = 15 - 0 = 15$$ Replace the y column in D by the coefficients $\begin{matrix} 26 \\ 5 \end{matrix}$

So by Cramer's rule, $x = \dfrac{D_x}{D} = \dfrac{88}{9}$ and $y = \dfrac{D_y}{D} = \dfrac{15}{9} = \dfrac{5}{3}$.

2. Solve the system.

$2x - y = 5$
$x + 3z = 0$
$y - 2z + w = -2$
$x - w = 3$

$$D = \begin{vmatrix} 2 & -1 & 0 & 0 \\ 1 & 0 & 3 & 0 \\ 0 & 1 & -2 & 1 \\ 1 & 0 & 0 & -1 \end{vmatrix} = 11$$

$$D_x = \begin{vmatrix} 5 & -1 & 0 & 0 \\ 0 & 0 & 3 & 0 \\ -2 & 1 & -2 & 1 \\ 3 & 0 & 0 & -1 \end{vmatrix} = 18; \quad D_y = \begin{vmatrix} 2 & 5 & 0 & 0 \\ 1 & 0 & 3 & 0 \\ 0 & -2 & -2 & 1 \\ 1 & 3 & 0 & -1 \end{vmatrix} = -19$$

$$D_z = \begin{vmatrix} 2 & -1 & 5 & 0 \\ 1 & 0 & 0 & 0 \\ 0 & 1 & -2 & 1 \\ 1 & 0 & 3 & -1 \end{vmatrix} = -6; \quad D_w = \begin{vmatrix} 2 & -1 & 0 & 5 \\ 1 & 0 & 3 & 0 \\ 0 & 1 & -2 & -2 \\ 1 & 0 & 0 & 3 \end{vmatrix} = -15$$

Thus, $x = \frac{18}{11}$, $y = -\frac{19}{11}$, $z = -\frac{6}{11}$, and $w = -\frac{15}{11}$. ■

Neither of the systems illustrated in example 10–3 D are inconsistent or dependent. *When the determinant of the coefficient matrix is zero ($D = 0$), the system is either inconsistent or dependent.* If any of the other determinants D_x, D_y, D_z, are not zero the system is inconsistent; if they are all zero, the system is dependent.

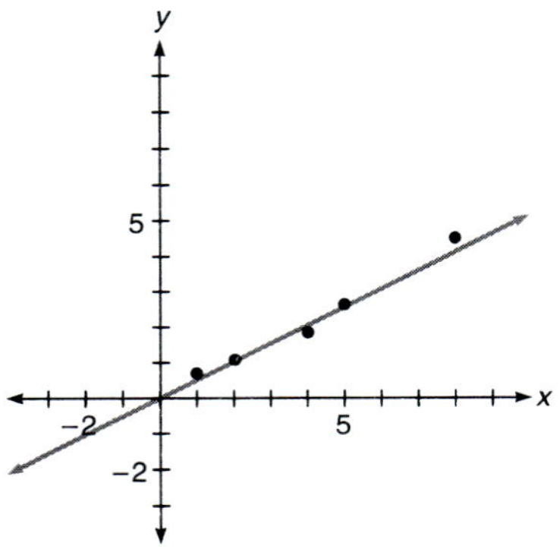

Figure 10–2

Linear regression

In the modeling of many situations it is desirable to find the equation $y = mx + b$ of a straight line that best fits a set of measured data. For example, in figure 10–2 we see the graph of a line that closely fits the points (1,0.7), (2,1.1), (4,1.8), (5,2.6), and (8,4.4).

This line is called the **least-squares line,** and the process of finding the linear is called **linear regression.** We present a method for finding the values m and b for the equation $y = mx + b$ of the least-squares line.

Let

Y equal the sum of the y data.

X equal the sum of the x data.

P equal the sum of the products of the x and y in each observation.

S equal the sum of the squares of the x data.

N equal the number of observations.

Then it can be shown that

$$\begin{aligned} Xm + Nb &= Y \\ Sm + Xb &= P \end{aligned}$$

■ *Example 10–3 E*

Find the least-squares line for the data: (1,0.7), (2,1.1), (4,1.8), (5,2.6), and (8,4.4). The computations for the example data are most easily done using a table (table 10–1).

	x	y	xy	x^2
	1	0.7	0.7	1
	2	1.1	2.2	4
	4	1.8	7.2	16
	5	2.6	13.0	25
	8	4.4	35.2	64
Totals	20	10.6	58.3	110
	X	Y	P	S

Table 10–1

Also, $N = 5$, the number of observations. Using the values from the table our equations become

$$\begin{aligned} Xm + Nb &= Y \qquad [1]\ 20m + 5b = 10.6 \\ Sm + Xb &= P \qquad [2]\ 110m + 20b = 58.3 \end{aligned}$$

Using Cramer's rule we compute m and b:

$$D = \begin{vmatrix} 20 & 5 \\ 110 & 20 \end{vmatrix} = -150$$

$$D_m = \begin{vmatrix} 10.6 & 5 \\ 58.3 & 20 \end{vmatrix} = -79.5$$

$$D_b = \begin{vmatrix} 20 & 10.6 \\ 110 & 58.3 \end{vmatrix} = 0$$

$$m = \frac{D_m}{D} = \frac{-79.5}{-150} = 0.53 \qquad b = \frac{D_b}{D} = \frac{0}{-150} = 0$$

Thus, the least-squares line $y = mx + b$ is $y = 0.53x$. ■

Calculators and computers can be used to find the values m and b for the least-squares line. The procedure is the same as that shown in example 3–2 D (chapter 3) for two points, except that more than two points are being entered.

Mastery points

Can you

- Find the determinant of matrices of order 2 and above?
- Use Cramer's rule to solve systems of linear equations?

Exercise 10–3

Compute the determinant of the following matrices.

1. $\begin{bmatrix} 1 & -4 \\ -3 & 3 \end{bmatrix}$

2. $\begin{bmatrix} 1 & 6 \\ -\frac{1}{2} & 3 \end{bmatrix}$

3. $\begin{bmatrix} -7 & \frac{3}{4} \\ \frac{2}{3} & 6 \end{bmatrix}$

4. $\begin{bmatrix} -3 & -4 \\ 6 & 8 \end{bmatrix}$

5. $\begin{bmatrix} -3\pi & -4\pi \\ 2 & 3 \end{bmatrix}$

6. $\begin{bmatrix} \sqrt{2} & -\sqrt{8} \\ 3 & -2 \end{bmatrix}$

7. $\begin{bmatrix} 4 & 0 & -5 \\ 1 & 5 & -2 \\ 0 & 3 & 7 \end{bmatrix}$

8. $\begin{bmatrix} 2 & -1 & 2 \\ 4 & -3 & 2 \\ -6 & 1 & 3 \end{bmatrix}$

9. $\begin{bmatrix} 2 & \frac{2}{3} & -1 \\ 4 & -1 & \frac{1}{2} \\ -3 & 0 & -2 \end{bmatrix}$

10. $\begin{bmatrix} 3 & -2 & -3 \\ 1 & 5 & -2 \\ 0 & 3 & 0 \end{bmatrix}$

11. $\begin{bmatrix} 4 & 3 & -2 \\ -1 & 0 & 2 \\ -2 & 1 & 4 \end{bmatrix}$

12. $\begin{bmatrix} 2 & -1 & 0 \\ 4 & -1 & 2 \\ -3 & 1 & -2 \end{bmatrix}$

13. $\begin{bmatrix} -1 & 1 & 3 \\ 1 & -2 & 5 \\ 0 & 0 & 7 \end{bmatrix}$

14. $\begin{bmatrix} 2 & -1 & 1 \\ 1 & -3 & 2 \\ 0 & 1 & 3 \end{bmatrix}$

15. $\begin{bmatrix} -3 & 6 & 0 \\ 4 & -1 & 0 \\ -3 & 1 & -5 \end{bmatrix}$

16. $\begin{bmatrix} \frac{1}{4} & 0 & -\frac{1}{3} \\ -4 & 2 & 2 \\ 0 & \frac{3}{4} & -3 \end{bmatrix}$

17. $\begin{bmatrix} \sqrt{2} & 0 & 3 \\ \sqrt{8} & -5 & 2 \\ -\sqrt{2} & -1 & 7 \end{bmatrix}$

18. $\begin{bmatrix} 4 & -8 & 3 \\ 0 & 5 & -12 \\ 0 & 0 & \frac{1}{2} \end{bmatrix}$

19. $\begin{bmatrix} 4 & 0 & -5 & 1 \\ 5 & -2 & 0 & 3 \\ 7 & 0 & 1 & 0 \\ 4 & -2 & 0 & 3 \end{bmatrix}$

20. $\begin{bmatrix} 3 & -2 & -1 & 1 \\ -5 & -2 & 2 & 3 \\ 0 & 0 & -1 & 4 \\ 0 & -6 & 0 & 3 \end{bmatrix}$

21. $\begin{bmatrix} 0 & -3 & 2 & 0 \\ -2 & 5 & 10 & 0 \\ -4 & 3 & 0 & 1 \\ 2 & -3 & 1 & 0 \end{bmatrix}$

22. $\begin{bmatrix} 4 & 0 & -2 & 1 \\ 2 & -2 & 0 & 3 \\ 1 & 0 & 1 & 3 \\ 4 & -2 & 0 & 3 \end{bmatrix}$

23. $\begin{bmatrix} 4 & 5 & 1 & 0 \\ -2 & 1 & 3 & 7 \\ 0 & 1 & 2 & 0 \\ 4 & -2 & 0 & 3 \end{bmatrix}$

24. $\begin{bmatrix} 3 & -2 & -1 & 0 \\ -5 & -1 & 2 & 3 \\ 0 & 0 & -1 & -4 \\ 0 & -2 & 0 & 3 \end{bmatrix}$

25. $\begin{bmatrix} 0 & 2 & -4 & 0 \\ -2 & 5 & 6 & 0 \\ 0 & 3 & 0 & 5 \\ 2 & -3 & 1 & 0 \end{bmatrix}$

26. $\begin{bmatrix} -2 & 1 & -2 & 3 \\ 2 & -4 & 0 & 3 \\ 1 & 2 & 1 & 3 \\ 4 & -2 & 0 & 3 \end{bmatrix}$

27. $\begin{bmatrix} 3 & 0 & 2 & 1 \\ 5 & 0 & 0 & 3 \\ 5 & 0 & 1 & -3 \\ 4 & -2 & 2 & 3 \end{bmatrix}$

28. $\begin{bmatrix} 3 & 0 & -1 & 1 \\ -5 & -2 & 0 & 3 \\ -3 & 2 & -1 & 4 \\ 0 & -6 & 0 & 7 \end{bmatrix}$

Solve the following systems of equations by Cramer's rule; if the system is dependent or inconsistent state that.

29. $-3x - 4y = 0$
 $-x + 9y = 4$

30. $-9y = 1$
 $-4x - 6y = -1$

31. $9y = -5$
 $-8x - 8y = 12$

32. $3x + 2y = 12$
 $-8x - 2y = 2$

33. $-2x - 5y = 9$
 $-7x - 8y = -4$

34. $-7x + 6y = -6$
 $-10x + 3y = 3$

35. $6x + 8y + 3z = -4$
 $x - 3y = -1$
 $5x + 9y + 7z = -6$

36. $-5x - 6y - 3z = 5$
 $-5x + 5y = 7$
 $x + 8y - 3z = 9$

37. $9x - 3y - 3z = -6$
 $x - 4y - 6z = 7$
 $3x + 2y = -3$

38. $7y + 5z = 5$
 $9x - 2y + 9z = 8$
 $8x + 2z = 9$

39. $-x + 2y = -2$
 $x + 5y - 2z = 3$
 $4x + 2y = 6$

40. $-4x + 3y + 3z = 9$
 $5x - 4z = 3$
 $-x - 6y + 2z = -5$

41. $x - y = 4$
 $-4x + 9y - 3z = -3$
 $-3x - y = 7$

42. $8x - 6y + 8z = -5$
 $-5x + 4y + 6z = -2$
 $9x + 3y + 7z = -6$

43. $6x + 8y = -3$
 $4x - 5y - 2z = 0$
 $10x + 3y - 2z = 2$

44. $3x - z = 6$
 $2y + 3z = -8$
 $y = 5$

45. $2x + 2y + 3z - 2w = 2$
$-x + 2y + 5z - w = -3$
$-4x + 4y - 2z - 4w = -2$
$-4x + 4y - 4w = 2$

46. $-4x + 5y - 3z + 5w = 2$
$6x - 2y + 6z = 2$
$4y + 6z - 4w = 2$
$4x - 2y - 2w = -1$

47. $x - y + 4z - 4w = 4$
$3x + 4y + 2z = -1$
$5x - 2z + 6w = 6$
$-2x + 4y - 2z + w = 2$

48. $y - w = -2$
$-y + 4z - 3w = -1$
$-x - 3z - 4w = 1$
$4x + 5y + 2z + 6w = 3$

49. $2x - y + 6z - 4w = 1$
$5x - y + 2z + 4w = 4$
$2x - 3y + 4z = 5$
$-3y + 4z - 4w = 5$

50. $3x + 4y - 2w = 0$
$2y - z - w = 3$
$z + w = -2$
$x + z = 2$

51. Solve for the variable C in the following system:
$2A - B + 3C - D = 5$
$A + B - 2D + E = 0$
$-B - C = 10$
$3A - C + D + E = -4$
$C - E = -20$

52. Solve for E in the system of problem 51.

In problems 53 through 56 use Cramer's rule to compute the least-squares line for each set of data; compute the values of m and b to two decimal places. Use the formulas given in example 10–3 E.

53. (0,0.5), (1,1.8), (2,3.0), (4,5.0), (6,7.6)

54. (1,−2.0), (3,−7.8), (4,−11.5), (5,−14.4)

55. (1.5,−4.8), (2,−4.0), (3,−2.0), (4.5,1.4), (6,4.7), (6.5,5.6)

56. (0,−64), (1,−56), (2,−40), (3,−35), (4,−20), (5,−10)

57. A company is studying failure rates in its line of power steering pumps for automobiles. It has measured the following data, where the first element of each ordered pair is the age of the pump in years, and the second is percentage of failures for pumps of that age:

(1,3), (2,5), (3,6), (4,9)

For example, at 2 years old, 5% of the pumps fail. Find the least-squares line for these data and use it to predict the percentage of pump failures, to the nearest tenth of a percentage, in the fifth and sixth years.

58. Use Cramer's rule to derive a formula, in terms of X, Y, P, N, and S, which will compute m and b directly. That is, solve the system $\begin{matrix} Xm + Nb = Y \\ Sm + Xb = P \end{matrix}$ for m and for b.

59. The data in the table represents the world's record for the 1-mile run in the year shown; use it to (a) find its least-squares line and then (b) predict the year when a 3 minute 35 second mile will be run. To make things easier, rewrite the year 1875 as 0, 1895 as 20, etc., subtracting 1875 from each year. Also, describe the time in seconds, using 4:24.5 as the base. For example, 4:17 − 4:24.5 = −7.5 (seconds). Thus, for example the data pair 1895, 4:17 is the ordered pair (20,−7.5).

Year	Time
1875	4:24.5
1895	4:17
1915	4:12.6
1923	4:10.4
1934	4:06.8
1945	4:01.4
1954	3:59.4
1965	3:53.6
1975	3:49.4

If the three points that form the vertices of a triangle are (x_1,y_1), (x_2,y_2), and (x_3,y_3), then it can be shown that the area of the triangle is the absolute value of $\frac{1}{2}\begin{vmatrix} x_1 & y_1 & 1 \\ x_2 & y_2 & 1 \\ x_3 & y_3 & 1 \end{vmatrix}$. Problems 60 through 63 refer to this fact.

60. Find the area of the triangle with vertices (0,2), (5,3), and (2,8).

61. Find the area of the triangle with vertices (−2,6), (3,−2), and (6,12).

62. Describe the set of all points (x,y) that form the third vertex of a triangle with vertices at (1,3) and (5,1) and area 10.

63. Find the area of the quadrilateral (four-sided figure) with vertices (0,2), (3,0), (5,8), and (1,10).

64. Show that $\begin{vmatrix} x & y & 1 \\ x_1 & y_1 & 1 \\ x_2 & y_2 & 1 \end{vmatrix} = 0$ produces the equation of the straight line through the distinct points (x_1,y_1) and (x_2,y_2).

65. Find the equation of the straight line that passes through the points $(2,-3)$ and $(5,4)$. See problem 64.

66. Find the equation of the straight line with x-intercept -4 and y-intercept 2. See problem 64.

67. A parabola is the graph of an equation of the form $y = ax^2 + bx + c$. State the equation of the parabola that will pass through the points $(-4,2)$, $(1,3)$, and $(2,8)$.

68. Find the equation of the parabola that passes through the points $(-1,0)$, $(1,3)$ and $(3,10)$. See problem 67.

69. Problem 64 showed one way to find the equation of a straight line that passes through two points; another method is to realize that a nonvertical straight line is the graph of an equation of the form $y = mx + b$. Two points uniquely determine a straight line. Find the equation of the straight line that passes through the points $(5,-1)$ and $(8,6)$ by substituting these values into the equation $y = mx + b$ and thereby obtaining a system of two equations in two unknowns.

70. Find the equation of the straight line that passes through the points $(-2,8)$ and $(5,1)$. Refer to problem 69.

71. A straight line passes through the points $(-2,-1)$ and $(3,2)$; a second straight line passes through the points $(-6,2)$ and $(5,-7)$. Find the point at which these two straight lines intersect.

72. Kirchhoff's law, from circuit theory in electronics, states that the sum of the voltages around any loop of a circuit is 0. In a certain circuit with two loops, with i_1 the current in one loop and i_2 the current in the second loop, the application of Kirchhoff's law gives the system $20i_1 - 10i_2 = 40$ and $10i_1 - 4i_2 = 25$. Solve for the currents i_1 and i_2.

73. For a certain electronics circuit Kirchhoff's law (problem 72) gives the following system. Find the currents i_1, i_2, and i_3.

$$\begin{aligned} 35i_1 + 12i_2 + 5i_3 &= 50 \\ 30i_1 - 20i_2 - 10i_3 &= -40 \\ 15i_1 + 10i_2 + 5i_3 &= 60 \end{aligned}$$

74. A certain scale is known to be very inaccurate for weights about 200 pounds. Three items I_1, I_2, and I_3 must be weighed, and it is known that their weights are in the 200 pound range. Thus, the items are weighed together, and the following results are noted. Find the weight of each of the three items.

$$\begin{aligned} I_1 + I_2 &= 370 \\ I_1 + I_3 &= 395 \\ I_2 + I_3 &= 415 \end{aligned}$$

75. An inheritance of \$36,000 is to be given to three charities, x, y, and z, in the ratios 3:4:5. How much will each charity get?

$\left(\textit{Hint: } x + y + z = 36{,}000, \dfrac{x}{3} = \dfrac{y}{4} = \dfrac{z}{5}.\right)$

76. Divide \$84,000 three ways so the ratios of each amount are 5:6:10. See problem 75.

77. A problem with finding approximate solutions to systems of equations is determining when a solution is "correct." For example, consider the system

$$\begin{aligned} 0.12658x + 0.25315y &= 0.37973 \\ 0.88606x + 1.77213y &= 2.65819 \end{aligned}$$

The "solution" $(3,0)$ gives an error of only -0.00001 in the first equation and 0.00001 in the second, yet this solution is actually not too close to the "true" solution. Find the true solution using Cramer's rule.

78. A problem similar to problem 77 occurs when evaluating determinants. Compute the determinant of each of the following matrices, and compare the results:

$$\begin{bmatrix} 11 & 19 & 9 \\ 25 & 48 & 24 \\ -124 & 12 & 65 \end{bmatrix} \begin{bmatrix} 11.01 & 19 & 9 \\ 25 & 48 & 24 \\ -124 & 12 & 65 \end{bmatrix}$$

79. In the text we defined the determinant of a matrix of order greater than 2 in a different fashion than that for an order 2 matrix. Also, we did not define the determinant of an order 1 matrix. Find a definition for the determinant of an order 1 matrix that would then allow us to find the determinant of an order 2 matrix using the definition for matrices of order greater than 2.

80. In problems 60–63 we stated a formula which gives the area of a triangle with vertices at points $P_1(x_1,y_1)$, $P_2(x_2,y_2)$, and $P_3(x_3,y_3)$. Show that this formula is equivalent to the formula $\frac{1}{2}[(x_1y_2 - x_2y_1) + (x_2y_3 - x_3y_2) + (x_3y_1 - x_1y_3)]$.

81. Consider the figure shown. It shows how a polygon can be divided up into triangles. The area of the polygon is the sum of the areas of the triangles.

a. Use the determinant of problem 80 to show that the following is a formula for the area of a four-sided polygon (a quadrilateral):
$\frac{1}{2}[(x_1y_2 - x_2y_1) + (x_2y_3 - x_3y_2) + (x_3y_4 - x_4y_3) + (x_4y_1 - x_1y_4)]$.

b. Similarly, show that the area of a five-sided polygon is given by the formula:
$\frac{1}{2}[(x_1y_2 - x_2y_1) + (x_2y_3 - x_3y_2) + (x_3y_4 - x_4y_3) + (x_4y_5 - x_5y_4) + (x_5y_1 - x_1y_5)]$.

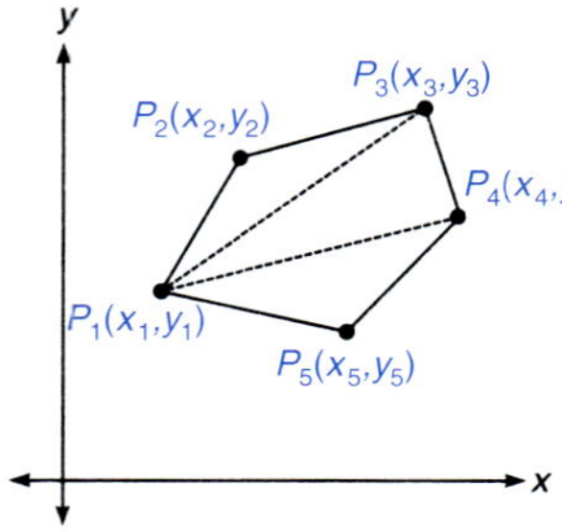

82. The figure shows a piece of land whose area is to be found. A north-south base line and an east-west base line are laid out and a survey made as shown. The distances shown are in meters from the origin. Find the area to the nearest square meter.

Do this as follows. Establish a formula for the area of a polygon of six sides. Use problem 81 as a guide for doing this. Then use the measurements shown to establish coordinates for each vertex of the piece of land and apply the formula.

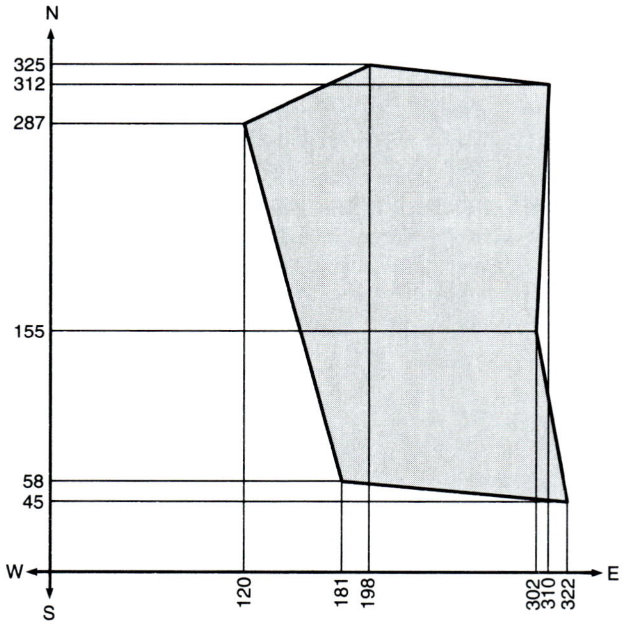

Skill and review

1. Solve $2x - 3 < 8$.
2. Is the point $(2,-1)$ a solution to the statement $2x + y > 2$?
3. Graph the lines $y = 2x - 1$ and $y = -\frac{1}{3}x + 1$ in the same graph. Label the point where the lines intersect.
4. Add $7\frac{1}{2}\%$ of \$1,200 to 5% of \$1,800.
5. Solve $3(2x + 3) - 2(5x + 3) = x$.
6. Solve $\frac{4x - 1}{x} < -5x$.
7. Solve $\log^2 x - \log x = 6$.
8. Graph $f(x) = \log_2(x + 1)$.
9. Graph $f(x) = x^2 + 3x - 4$.

10–4 Systems of linear inequalities

A certain animal food is available from two sources, A and B. A supplies 5 grams per pound (5 gm/lb) of protein and 4 gm/lb of carbohydrates. B supplies 8 gm/lb and 3 gm/lb of protein and carbohydrates. A costs 15 cents per pound, and B is 18 cents. If the *minimum* daily requirement for this animal is 20 gm of protein and 12 gm of carbohydrates, how much of A and B should be used to minimize costs?

This problem can be solved by a method called linear programming, which we will investigate after examining linear inequalities in two variables.

In the previous sections we have focused our attention on systems of linear equations; we now turn our attention to systems of linear inequalities in two variables. The solution sets in these cases are best indicated by their graphs. We will see that a very important application of these systems is linear programming, mentioned above, which is used extensively in the discipline called operations research.

Linear inequalities in two variables

A linear inequality in two variables would be, for example, $2x - 3y < 6$. We are interested in describing all points (x,y) that make this inequality true. We know that the corresponding equality $2x - 3y = 6$ is a straight line.

Table 10–2 shows various ordered pairs, plotted in figure 10–3, along with the statement "true" or "false" depending on the truth value of $2x - 3y < 6$ for that ordered pair (x,y). For example, if $(x,y) = (3,-2)$ (point A in the figure), we calculate

$$2x - 3y < 6$$
$$2(3) - 3(-2) < 6 \qquad \text{Replace } x \text{ with 3 and } y \text{ with } -2$$
$$12 < 6, \text{ which is false.}$$

Point	$2x - 3y < 6$	True/False
A (3,−2)	$12 < 6$	False
B (−1,−3)	$7 < 6$	False
C (5,1)	$7 < 6$	False
D (−1,−1)	$1 < 6$	True
E (−1,2)	$-8 < 6$	True
F (4,3)	$-1 < 6$	True

Table 10–2

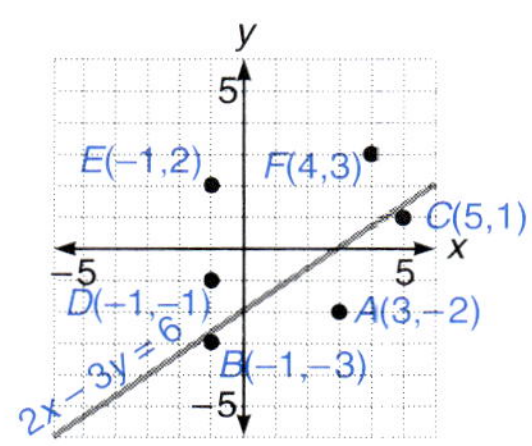

Figure 10–3

The points for which the inequality is true are circled in the figure. All of them are above the line $2x - 3y = 6$ (D,E,F), while those below the line (A, B, and C) all make it false. This might lead us to guess that any point above the line will make the inequality true, and any point below will make it false. It can be proven that this guess is correct. A straight line divides the plane up into two halves, and one of these two half-planes is the solution to a corresponding strict linear inequality in two variables.

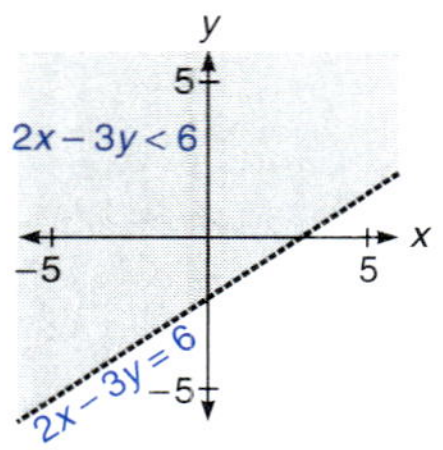

Figure 10–4

We could indicate the solution set to the linear inequality $2x - 3y < 6$ by shading in the half-plane above the line $2x - 3y = 6$. This is illustrated in figure 10–4. Observe also that we draw the line $2x - 3y = 6$ as a dashed line. This is to indicate that it is not part of the solution; any point (x,y) on this line satisfies $2x - 3y = 6$, not $2x - 3y < 6$.

If the inequality were the weak[5] inequality $2x - 3y \leq 6$, we would show the line as a solid line since the line would be part of the solution.

> **To graph a linear inequality in two variables**
> - Graph the corresponding linear equality (the straight line). If the inequality is a weak inequality draw the straight line as a solid line, otherwise a dashed line.
> - Try a test point from one of the half-planes in the inequality. If this point makes the inequality true then that half-plane is the solution set, so shade in that half-plane; otherwise shade in the other half-plane.

Example 10–4 A illustrates this procedure.

■ *Example 10–4 A*

Graph the solution set of each inequality.

1. $3y \geq 6x - 12$

Graph $3y = 6x - 12$ as a solid line since this is a weak inequality. Do this by plotting the intercepts and drawing the line through them.

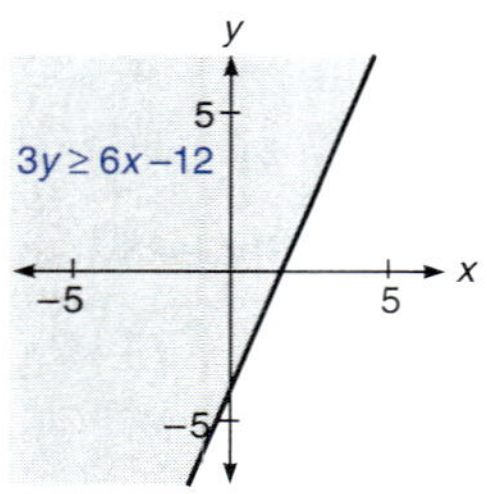

$3y > 6x - 12$	
$3(0) > 6(0) - 12$	Try the test point (0,0) in the inequality
$0 > -12$	True

Since (0,0) satisfies the inequality the half-plane which includes the origin (0,0) is in the solution set. Shade in this half-plane. The solution set is the half-plane along with the line $3y = 6x - 12$.

Note In a weak inequality the line itself is part of the solution set.

2. $x < 2$

Graph $x = 2$ as a (vertical) dashed line.

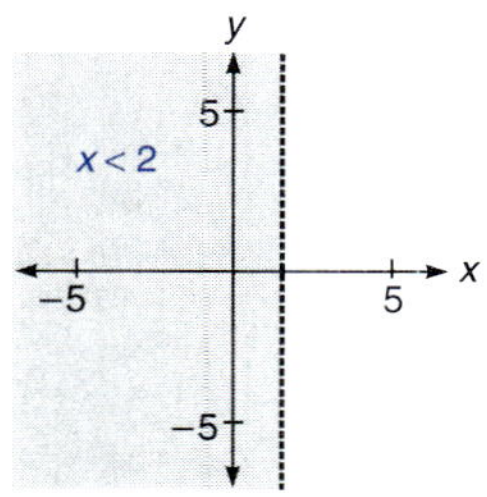

$x < 2$	Now determine which half-plane is the solution
$0 < 2$	Try (0,0) in the inequality; this is true

Thus, the solution set is the half-plane containing the origin. Shade in this half-plane. ■

Systems of linear inequalities in two variables

The graph of a *system* of linear inequalities in two variables consists of the intersection of the solution sets of each inequality; graphically, this means where the graphs of all the inequalities overlap.

[5]Recall that a weak inequality is represented by $\geq$ or $\leq$ (section 1–1).

■ ***Example 10–4 B***

Graph the solution set to each system of inequalities.

1. $x - 3y \geq 3$
$2x + y < 4$
Part (a) of the figure shows the solution to the inequality $x - 3y \geq 3$. Part (b) shows the solution to $2x + y < 4$. Part (c) shows the solution set. Observe the solid line, which is part of the solution set. This is where the line $x - 3y = 3$ intersects the inequality $2x + y < 4$.

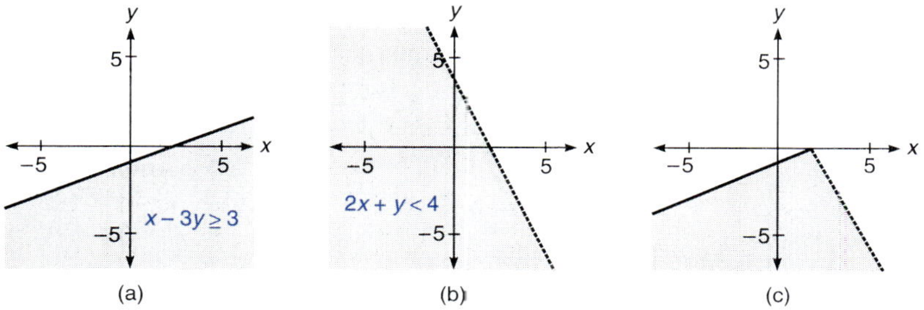

2. $3x + y < 9$
$x \leq 3, y > -6$
The solutions to each of these three inequalities are shown in parts (a), (b), and (c) of the figure, and the final answer in part (d).

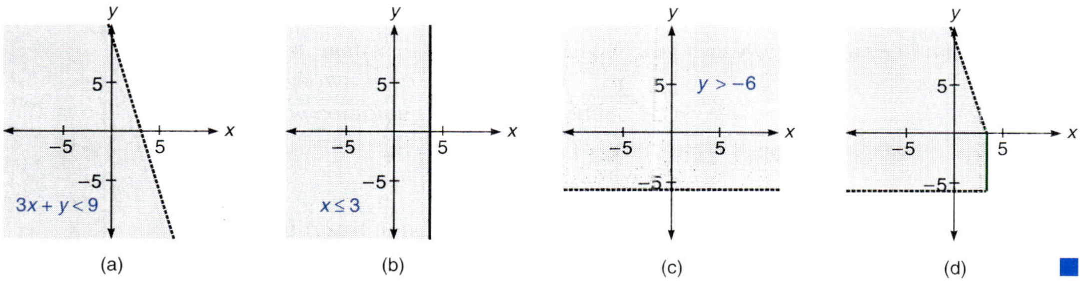

Linear programming

Linear programming is a mathematical tool used by organizations to maximize or minimize chosen values. For example, it might be used to maximize profits or minimize costs in a company. We will show only examples that employ two variables, but note that linear programming, developed in the 1940s for the Air Force by the mathematician George Dantzig, is used in more complicated situations every day in many applications. Students will encounter this subject in more depth in a course on finite mathematics; courses in linear programming and operations research would present the full power of this topic.

A **linear programming problem** in two variables is a problem that can be described by a set of linear inequalities, called **constraints,** and a linear equation in two variables, called the **objective function.** The constraints form a set of **feasible solutions.** The set of linear equalities that correspond to the set of inequalities form a boundary, and it can be proven that *the maximum or minimum value of the objective function can be found at one of the vertices of this boundary.* This is called the **fundamental principle of linear progamming.**

An example will illustrate. Suppose a company makes two products, tables and chairs. The company makes a profit of \$3 on each table it sells and \$2 on each chair. Thus, if x is the number of tables sold per day, and y is the number of chairs sold per day, then the day's profit P could be described as $P = 3x + 2y$. This is our *objective function* for this problem, and naturally the company wishes to maximize this value.

Now, suppose the company consists of two departments, assembly and painting. The assembly department requires 4 hours to assemble a table and 3 hours to assemble a chair. It can only assemble one table or one chair at a time. Now, $4x$ is the time required to assemble x tables at 4 hours per table, and $3y$ is the time required to assemble y chairs at 3 hours per chair. The total time cannot exceed the 24 hours in one day. The department operates 24 hours per day, so it is limited by the *constraint* $4x + 3y \leq 24$.

For example, if two tables and five chairs are made in one day, it would take $4(2) = 8$ hours to make the tables, and $3(5) = 15$ hours to make the chairs; the total of 23 hours is less than the 24 allowed, so this is a possible number of tables and chairs for the assembly department, and is therefore a feasible solution. Profit would be $3(2) + 2(5) = \$16$.

Suppose that the paint department requires 3 hours to paint a table, and 1 hour to paint a chair. However, due to drying times, the department can only operate 12 hours per day. This means that on a daily basis, $3x + y \leq 12$, another *constraint.*

Also note that x and y cannot be negative, since this would mean that a negative number of tables or chairs was produced in a day.

We can now formulate our linear programming problem for this company as follows: given the constraints

$$\begin{aligned} 4x + 3y &\leq 24 \\ 3x + y &\leq 12 \\ x \geq 0&, y \geq 0 \end{aligned}$$

maximize the objective function $P = 3x + 2y$.

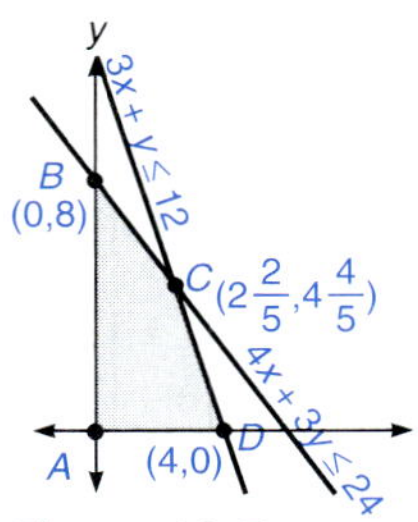

Figure 10–5

The set of constraints is graphed in figure 10–5. The point of intersection C of the first two constraints is found by any of the methods shown in sections 10–1 through 10–3. The line segments connecting the points A, B, C, and D form a polygon, which is shaded in. This shaded area defines what is called the *set of feasible solutions;* our goal is to find the feasible solution that gives the largest value for profit P. The *maximum* value of P can be found *at one of the vertices A, B, C,* or D. Table 10–3 shows the value of P for each of these points.

	Point	$P\ (= 3x + 2y)$
A	(0,0)	0
B	(0,8)	16
C	$(2\frac{2}{5}, 4\frac{4}{5})$	16.8
D	(4,0)	12

Table 10–3

We can see that the maximum value of P is \$16.80 per day, corresponding to a production of 2.4 tables and 4.8 chairs per day. This solution assumes that it is possible to make fractional parts of tables and chairs in a day.

We can see why the solution is one of the vertices if we consider the graph of the objective function $P = 3x + 2y$, or $y = -\frac{3}{2}x + \frac{P}{2}$. For differing values of P, this is a family of parallel lines (figure 10–6). As P increases, the lines move "up." When the lines move past a point, they no longer intersect the set of feasible solutions. It can be seen that this point will be a vertex.

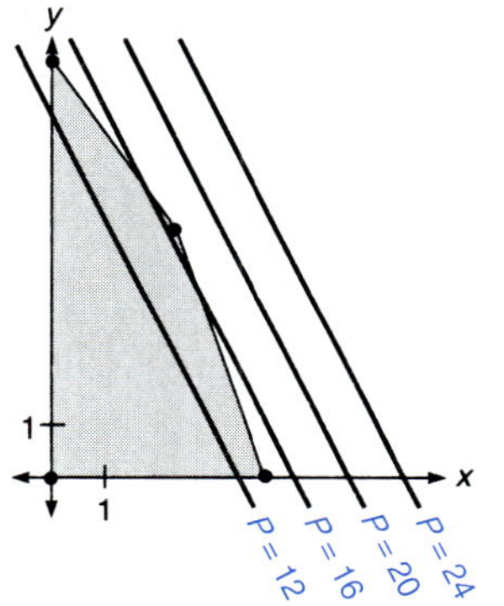

Figure 10–6

Note If the line corresponding to the objective function is parallel to an edge, there may in fact be many solutions, any of which maximizes the objective function.

■ *Example 10–4 C*

1. Maximize the value of the objective function, $Z = x + 3y$, with regard to the constraints indicated.

$$y - \tfrac{1}{2}x \le 1$$
$$4y \ge 2 + x$$
$$2x + 3y \le 6$$
$$x \ge 0,\ y \ge 0$$

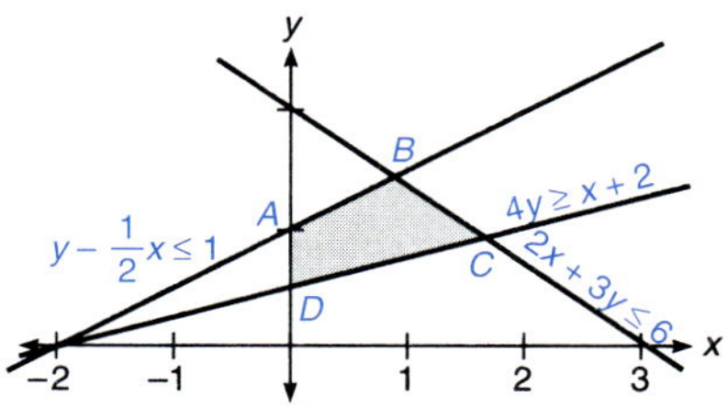

The graph of the feasible solutions is shown in the figure. The points that form the vertices of the polygon that is the set of feasible solutions are labeled A, B, C, and D. Point B is the intersection of the lines $y - \frac{1}{2}x = 1$ and $2x + 3y = 6$. Point C is the intersection of the lines $4y = x + 2$ and $2x + 3y = 6$.

The coordinate of these points and the value of $Z = x + 3y$ at these points is shown in the table where we see that $Z = 5\frac{1}{7}$, at point B, is the maximum value of Z. Thus, the solution is the point $B(\frac{6}{7}, 1\frac{3}{7})$; at this point $Z = 5\frac{1}{7}$.

Point	$Z = x + 3y$
A (0,1)	3
$B = (\frac{6}{7}, \frac{10}{7})$	$5\frac{1}{7}$
C $(\frac{18}{11}, \frac{10}{11})$	$4\frac{4}{11}$
D $(0, \frac{1}{2})$	$1\frac{1}{2}$

2. Two fertilizers are available; one is a 10-5-10 mix and the second is 5-10-25. The first number refers to percentage of nitrogen, the second to percentage of phosphorus, and the third to percentage of potash. The first fertilizer sells for \$0.10 per pound, and the second for \$0.06 per pound. A farmer must put the following minimum amounts of each nutrient on a field: 6 pounds of nitrogen, 5 pounds of phosphorus, and 15 pounds of potash. How much of each fertilizer should the farmer use to minimize cost?

Let x be the number of pounds of the 10-5-10 mix, and let y be the number of pounds of the 5-10-25 mix. Now we consider each nutrient.

Nitrogen: This is supplied by 10% of the first fertilizer ($0.10x$) and 5% of the second ($0.05y$) and this must be at least 6 pounds. Thus,

$$0.10x + 0.05y \geq 6$$
$$10x + 5y \geq 600 \quad \text{Multiply each member by 100}$$
$$2x + y \geq 120 \quad \text{Divide each member by 5}$$

Phosphorus: This comes from 5% of x and 10% of y, and must be at least 5 pounds. Thus,

$$0.05x + 0.10y \geq 5$$
$$5x + 10y \geq 500 \quad \text{Multiply each member by 100}$$
$$x + 2y \geq 100 \quad \text{Divide each member by 5}$$

Potash: The minimum amount of 15 pounds comes from 10% of x and 25% of y, so,

$$0.10x + 0.25y \geq 15$$
$$10x + 25y \geq 1500$$
$$2x + 5y \geq 300$$

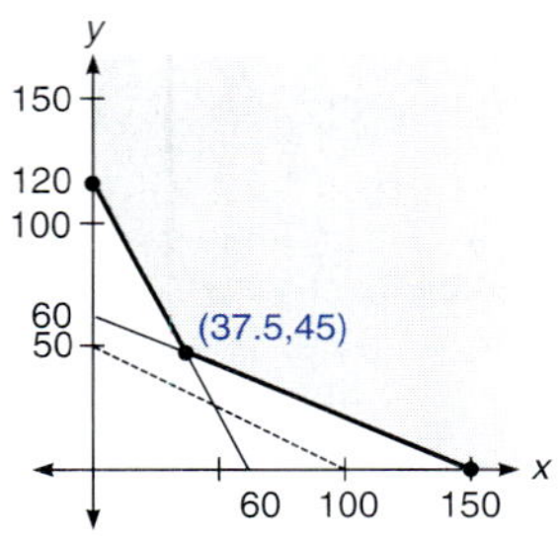

The cost function C is $C = 0.10x + 0.06y$. Thus, we need to solve the following linear programming problem. Minimize, $C = 0.10x + 0.06y$ if

$$2x + y \geq 120$$
$$x + 2y \geq 100$$
$$2x + 5y \geq 300$$

We graph the set of feasible solutions. The constraint $x + 2y \geq 100$ does not provide any feasible solutions. (The line $x + 2y = 100$ is graphed as a dashed line.)

We must check the points at the vertices (0,120), (150,0), and (37.5,45). This is shown in the table. We can see that the cost is minimized by using 37.5 pounds of the first fertilizer and 45 pounds of the second. The cost will be \$6.45. ■

Point	$C = 0.10x + 0.06y$
(0,120)	7.20
(150,0)	15.00
(37.5,45)	6.45

Mastery points

Can you

- Graph the solution to a linear inequality in two variables?
- Graph the solution to a system of linear inequalities in two variables?
- Solve a linear programming problem that is defined in two variables?
- Convert certain problems into a linear programming problem in two variables and solve that problem?

Exercise 10–4

Graph the solution set of the following linear inequalities.

1. $3x - 2y > 6$ **2.** $x + 3y < 9$ **3.** $5x > y - 1$ **4.** $2x + 6y \geq 18$ **5.** $9 - 2x > 3y$
6. $3y - x < 0$ **7.** $x + 2y \geq -4$ **8.** $-x - 4 < y$ **9.** $6 \geq x + 7y$ **10.** $-9x + 3 > y$
11. $15y < 12x + 5$ **12.** $x - 8 \leq \frac{y}{2}$ **13.** $x > 2$ **14.** $y < -1$ **15.** $2x \leq 5$
16. $\frac{5}{3}y - \frac{7}{2} < 3x$ **17.** $\frac{5}{6}x - 3y \geq 9$ **18.** $y - \frac{x}{5} < \frac{3}{10}$ **19.** $0.3x - 1.2y \leq 4$ **20.** $1.2y \leq 2x - 0.5$
21. $1.5 > x + 0.1y$

22. The dimensions in yards of a piece of land are length x and width y. The length is to be increased by 5 yards, while the width is to be decreased by 4 yards. It is desired that the new area be at least equal to the old. The new area is $(x + 5)(y - 4)$, and the old is xy. Thus we want $(x + 5)(y - 4) \geq xy$. Graph the solution to this inequality.

Graph the solution to the following systems of linear inequalities.

23. $2x - y > 5$, $x + y < 6$
24. $x < 3$, $y - x < 3$
25. $y < x + 4$, $x + 2y > -1$
26. $7x + 2y \geq 14$, $2x + 2y < 3$
27. $2x - 3 \leq y$, $3x - 6 \geq y$
28. $y - 2x > 4$, $2x + 5y \leq 10$
29. $x + 10y > -6$, $x - 4y > 3$
30. $4x + 13y < 52$, $\frac{2}{3}x + \frac{2}{5}y \geq -6$
31. $-y < 1$, $3x + 4y < 12$
32. $-6x - y > 0$, $-4x + 13y < 0$
33. $8x - 4y < -4$, $-3x + 3y > 3$
34. $2x - 3y < 11$, $10x + 2y > 13$
35. $\frac{3}{2}x + \frac{5}{4}y < 10$, $2x + 3y > -4$
36. $-4x - 3y > -5$, $6x + 12y > 3$
37. $10y < 6 - x$, $-4x - 2y < 10$
38. $x - 5y > 10$, $5x + y < 10$

39. The distance a vehicle with constant velocity travels is $d = rt$ (distance equals the product of rate and time). A car is traveling along a highway at a speed of 1.5 ± 0.5 kilometers per minute (kpm). Thus the rate r is at least 1 kpm, so the minimum distance traveled after t minutes is $d \geq 1t$. (a) Write a similar inequality that describes the maximum distance traveled, then (b) graph the solution to this system. Assume $t > 0$. The solution represents the set of possible distances traveled by the car after t minutes.

40. A box has dimensions that depend on two variables, x and y. Its length is $x + y - 3$ and its width is $2x - y + 6$. The length and width must have positive values. Thus, we require that $x + y - 3 > 0$ and $2x - y + 6 > 0$. Also, $x > 0$ and $y > 0$ must be true. Graph this system of inequalities.

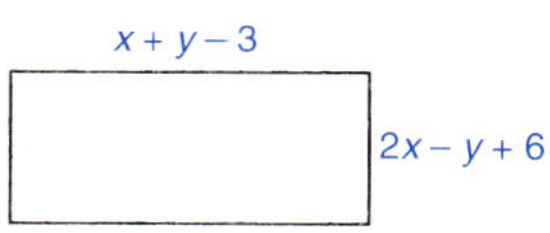

In the following problems, maximize the value of the objective function P with regard to the constraints supplied. In all cases it is assumed that $x \geq 0$ and $y \geq 0$.

41. $-x + y \leq 2$, $x + y \leq 6$, $P = 2x + y$

42. $-x + 2y \leq 4$, $\frac{7}{6}x + y \leq 7$, $P = x + 2y$

43. $x + 4y \leq 18$, $4x + 5y \leq 28$, $P = 2x + 3y$

44. $-11x + 10y \leq 5$, $-6x + y \leq -24$, $P = x + \frac{1}{2}y$

45. $-26x + 21y \leq 14$, $2x + y \leq 12$, $P = 2x + \frac{1}{3}y$

46. $3x + 3y \leq 16$, $7x + 6y \leq 35$, $P = 2x + 3y$

47. $5x + 16y \leq 80$, $-25x + 16y \geq -100$, $P = 2x + 4y$

48. $2x + y \leq 10$, $4x + 3y \leq 24$, $P = x + y$

49. $x + 4y \leq 20$, $x + 2y \leq 12$, $P = \frac{1}{3}x + \frac{1}{2}y$

50. $-x + y \leq 3$, $7x + 4y \leq 56$, $P = 2x - \frac{1}{2}y$

51. $2x + 5y \leq 25$, $3x + 2y \leq 21$, $P = x - \frac{1}{3}y$

52. $x + 6y \leq 18$, $2x + y \leq 14$, $P = -x + 2y$

53. $-3x + 4y \leq 20$, $2x + y \leq 16$, $P = -\frac{1}{2}x + 2y$

54. $-3x + 8y \leq 28$, $2x + y \leq 13$, $P = x - y$

55. $-x + 2y \leq 4$, $x + 3y \leq 11$, $x + y \leq 7$, $P = x + 2y$

56. $-\frac{3}{2}x + y \leq 3$, $x + y \leq 8$, $4x + y \leq 20$, $P = 2x + 3y$

57. $-7x + 4y \leq 2$, $y \leq 4$, $\frac{4}{3}x + y \leq 12$, $P = 2x + y$

58. $-2x + 5y \leq 15$, $2x + 3y \leq 17$, $\frac{3}{2}x + y \leq 9$, $P = -x + 3y$

59. $-x + y \leq 6$, $3x + 2y \leq 22$, $x + y \leq 8$, $P = 3x - y$

60. $\frac{1}{2}x + y \leq 8$, $\frac{2}{3}x + y \leq \frac{26}{3}$, $2x + y \leq 18$, $P = 12 - x - y$

61. $y \leq 3$, $\frac{1}{2}x + y \leq \frac{11}{2}$, $4x + y \leq 30$, $P = \frac{1}{5}x - \frac{1}{8}y$

62. $-4x + 3y \leq 3$, $-x + 3y \leq 12$, $3x + y \leq 24$, $P = \frac{1}{2}x + 2y$

63. $-2x + 3y \leq 15$, $5x + 3y \leq 36$, $x + y \leq 8$, $P = 5x + 3y$

64. $-x + y \leq 3$, $7x + 4y \leq 67$, $x + y \leq 10$, $P = -x + \frac{1}{2}y$

65. $-x + y \leq 8$, $5x + 6y \leq 70$, $5x + 2y \leq 40$, $P = 3x + y$

66. $-8x + 9y \leq 27$, $10x + 7y \leq 94$, $2x + y \leq 18$, $P = 2x + 5y$

In the following problems, minimize the value of the objective function C with regard to the constraints supplied. In all cases it is assumed that $x \geq 0$ and $y \geq 0$.

67. $-x + y \geq 2$, $x + y \geq 6$, $C = 2x + y$

68. $-x + 2y \geq 4$, $\frac{7}{6}x + y \geq 7$, $C = x + 2y$

69. $x + 4y \geq 18$, $4x + 5y \geq 28$, $C = 2x + 3y$

70. $-11x + 10y \geq 5$, $-6x + y \geq -24$, $C = x + \frac{1}{2}y$

71. $-26x + 21y \geq 14$, $2x + y \geq 12$, $C = 2x + \frac{1}{3}y$

72. $3x + 3y \geq 16$, $7x + 6y \geq 35$, $C = 2x + 3y$

73. $4x + y \geq 9$, $x + y \geq 6$, $x + 5y \geq 10$, $C = x + 2y$

74. $x + y \geq 9$, $\frac{7}{2}x + y \geq \frac{23}{2}$, $x + 3y \geq 12$, $C = 2x + 3y$

75. $\frac{7}{2}x + y \geq 8$, $x + 7y \geq 9$, $x + y \geq 4$, $C = 2x + y$

76. $x + y \geq 8$, $2x + 3y \geq 17$, $\frac{3}{2}x + y \geq 9$, $C = 5x + 4y$

77. $-x + y \geq -2$, $3x + 2y \geq 22$, $x + y \geq 8$, $C = 3x - y$

78. $5x + y \geq 6$, $x + 5y \geq 6$, $\frac{3}{4}x + y \geq 3$, $C = 12 - x - y$

79. A furniture company produces two products, tables and chairs. Tables sell for \$29 while chairs sell for \$10. It takes 3 hours to assemble a table and 1 hour to assemble a chair. In a production run there are 300 hours available for assembly. It takes 2 hours to finish a table, and $\frac{5}{6}$ of an hour for a chair. The finishing department has 240 hours available in a production run. Maximize total income from the sales of tables and chairs.

80. In a company there are assembly and paint departments. The company produces two products, A and B. Product A requires 2 hours for assembly and 1 hour for painting; B requires 5 hours for assembly and 2 hours for painting. A sells for \$3, and B for \$10. The assembly department is limited to 200 hours for a production run, and the paint department to 100 hours. How many of each product should be produced to maximize the income from sales?

81. A coal mining company has crews comprised of workers and excavation machines. It has two types of crews, type A and type B. Type A crews have 12 workers and 2 machines, and type B crews have 20 workers and 4 machines. Type A crews produce 13 tons of coal per hour and type B crews produce 25 tons per hour. The company has 260 workers and 50 machines. How many crews of each type should be allocated to maximize coal production?

82. A large logging company has two types of crews of workers and supervisors, called A-crews and B-crews. It has found that A-crews, which have a crew of 1 supervisor and 4 loggers, log 20 trees per day, and that B-crews, which have a crew of 2 supervisors and 6 loggers, log 30 trees per day. The company has 40 supervisors and 150 loggers on its payroll. What mix of crews would produce the most trees per day?

83. A certain animal food is available from two sources, A and B. A supplies 5 grams per pound (5 gm/lb) of protein and 4 gm/lb of carbohydrates. B supplies 8 gm/lb and 3 gm/lb of protein and carbohydrates. A costs 15 cents per pound, and B is 18 cents. If the *minimum* daily requirement for this animal is 20 gm of protein and 12 gm of carbohydrates, how much of A and B should be used to minimize costs?

84. A company builds toy automobiles in two sizes, large and small. The large size sells for \$3, and the small for \$1. The following constraints apply to producing these products. The large car requires 14 ounces of plastic to produce, and the small car requires 6 ounces of plastic to produce. The company is limited to 6,000 ounces of plastic. It takes 2 minutes to produce a small car, and 3.5 minutes for a large. Total labor is limited to 2,000 minutes. The small cars require 5 small decorative decals, while the large cars only require one of the same type. The company has 2,000 of these small decals. How many of each size should be built to maximize the total dollar amount of sales?

Skill and review

1. Write the identity matrix of order 4.

2. Rewrite $\left|3 - \frac{\sqrt{2}}{2}\right|$ without absolute value notation.

3. Solve $|4 - x| < 10$.

4. Solve $2x - 3 = 0$.

5. Solve $2x^2 - 3x = 5$.

6. Solve $|2x^2 - 3x| = 5$.

7. Simplify $\sqrt[3]{16x^5y^2z}$.

8. Graph $f(x) = |x - 2|$.

10–5 Systems of linear equations—matrix algebra

An insect lives in three stages: egg, larval, and adult. In the first stage females have no progeny. In the second they have four daughters. In the third they have three daughters. The survival rate in stage 1 is 22.9%, and in stage 2 it is 12.5%. Assume the number of females in each stage in some initial generation is 1,000. Find the number of females in each stage after three generations.

This is one of many, many types of problems that can be solved using matrix algebra, the topic of this section.

In this section we will also see that systems of linear equations can be described in terms of matrices. This process has developed into an area of mathematics called linear algebra, which finds wide application in the social sciences as well as engineering and mathematics.

Addition and subtraction of matrices

Operations of addition, subtraction, and multiplication are defined for matrices. Addition and subtraction are defined in a natural way. To add or subtract matrices we add or subtract each element.

> **Addition and subtraction of matrices**
> Let A and B be two matrices of the same dimensions, $m \times n$. $C = A + B$ is an $m \times n$ matrix in which $c_{i,j} = a_{i,j} + b_{i,j}$, and $D = A - B$ is an $m \times n$ matrix in which $d_{i,j} = a_{i,j} - b_{i,j}$ for $1 \le i \le m$, $1 \le j \le n$.

Observe that these operations are not defined for matrices whose dimensions differ.

The TI-81, as well as most graphing calculators, can store matrices and do most matrix operations. This will be illustrated in example 10–5 F.

Example 10–5 A illustrates addition and subtraction of matrices.

■ ***Example 10–5 A***

Perform the indicated addition or subtraction.

1. $\begin{bmatrix} 2 & -1 \\ 3 & 5 \end{bmatrix} + \begin{bmatrix} 3 & -4 \\ -3 & 0 \end{bmatrix} = \begin{bmatrix} (2+3) & (-1-4) \\ (3-3) & (5+0) \end{bmatrix} = \begin{bmatrix} 5 & -5 \\ 0 & 5 \end{bmatrix}$

2. $\begin{bmatrix} -4 & \frac{1}{2} \\ 2 & -3 \\ 0 & 1 \end{bmatrix} - \begin{bmatrix} -3 & \frac{3}{2} \\ 2 & 5 \\ -6 & 1 \end{bmatrix} = \begin{bmatrix} (-4+3) & (\frac{1}{2} - \frac{3}{2}) \\ (2-2) & (-3-5) \\ (0+6) & (1-1) \end{bmatrix} = \begin{bmatrix} -1 & -1 \\ 0 & -8 \\ 6 & 0 \end{bmatrix}$

3. $\begin{bmatrix} -3 & 2 \\ 1 & 5 \\ -2 & 0 \end{bmatrix} + \begin{bmatrix} -3 & 2 & 1 \\ 5 & -2 & 0 \end{bmatrix}$ Not defined since the dimensions are not the same. ■

The scalar product

There are two forms of multiplication of matrices. The first is **scalar multiplication.** In the context of matrix algebra, real numbers are also called **scalars.** The product of a scalar and a matrix is a matrix in which each element is the product of the scalar and the element.

> **Scalar product**
> If A is a matrix of dimension $m \times n$ and $k \in R$, then the scalar product of k and A is $C = kA$, where $c_{i,j} = ka_{i,j}$ for $1 \le i \le m$, $1 \le j \le n$.

■ ***Example 10–5 B***

Form the scalar product of -5 and $A = \begin{bmatrix} -6 & 5 & 1 \\ 2 & -1 & 3 \\ 0 & 4 & -2 \end{bmatrix}$.

$$-5A = -5\begin{bmatrix} -6 & 5 & 1 \\ 2 & -1 & 3 \\ 0 & 4 & -2 \end{bmatrix} = \begin{bmatrix} -5(-6) & -5(5) & -5(1) \\ -5(2) & -5(-1) & -5(3) \\ -5(0) & -5(4) & -5(-2) \end{bmatrix}$$

$$= \begin{bmatrix} 30 & -25 & -5 \\ -10 & 5 & -15 \\ 0 & -20 & 10 \end{bmatrix}$$ ■

The dot product

Before we talk about the second form of matrix multiplication we discuss another "product," the dot product.

An ***n*-dimension vector** is a matrix of n values, in which the number of rows or columns is 1. Thus, $[2, -4]$ is a 2-dimensional vector, $[-3, 4, 0, 11]$ is a 4-dimensional vector, and $\begin{bmatrix} 9 \\ 3 \\ -5 \end{bmatrix}$ is a 3-dimensional vector. Since one of the dimensions is one we often simplify our notation for a matrix element. In this case, if V is an n-dimensional vector then v_i, $1 \le i \le n$ describes the n elements in V.

The **dot product** of two n-dimensional vectors is a scalar (real number) that is the sum of the products of the corresponding elements in the vectors. The dot product is indicated by the symbol $\cdot$.

> **Dot product of two *n*-dimensional vectors**
>
> Let U and V be two n-dimensional vectors. Then the dot product k of U and V, $k = U \cdot V$, is a scalar such that
>
> $$k = u_1v_1 + u_2v_2 + \cdots + u_nv_n.$$

Observe that the dot product requires that the vectors have the same number of elements. Example 10–5 C illustrates forming the dot product.

■ ***Example 10–5 C***

Form the dot product of the given vectors.

1. $[-2, 0, 3, 1] \cdot [4, -5, 10, 6] = (-2)(4) + (0)(-5) + (3)(10) + (1)(6)$
$= 28$

2. $[3, 1, -2] \cdot \begin{bmatrix} 4 \\ -4 \\ 5 \end{bmatrix} = (3)(4) + (1)(-4) + (-2)(5) = -2$ ■

The matrix product

The **matrix product** is defined under certain conditions placed on the dimensions.

> **Matrix product**
> If A is an $m \times k$ matrix, and B is a $k \times n$ matrix, then the matrix product $C = AB$ is an $m \times n$ matrix where $c_{i,j}$ is the dot product of the ith row of A and the jth column of B.

For the matrix product to be defined the number of columns of the first factor must equal the number of rows of the second factor. The following illustrates this, as well as how the dimensions of the result are determined.

$$\begin{array}{ccc} A & B & = \; C \\ m \times k & k \times n & m \times n \end{array}$$

(inner dimensions k and k: =; outer dimensions m and n: dimensions of result)

For an example we will form the product of $A = \begin{bmatrix} -2 & 0 \\ 5 & 1 \\ 2 & -4 \end{bmatrix}$ and $B = \begin{bmatrix} 4 & -1 & 2 & 5 \\ 1 & 3 & -1 & 3 \end{bmatrix}$. A is a 3×2 matrix and B is a 2×4 matrix, so $C = AB$ will be a 3×4 matrix.

The element in the second row, third column of C, $c_{2,3}$, is the dot product of the vectors that are the second row of A and the third column of B; thus,

$$c_{2,3} = [5\ 1] \cdot \begin{bmatrix} 2 \\ -1 \end{bmatrix} = 5(2) + 1(-1) = 9$$

$$\begin{bmatrix} -2 & 0 \\ 5 & 1 \\ 2 & -4 \end{bmatrix} \begin{bmatrix} 4 & -1 & 2 & 5 \\ 1 & 3 & -1 & 3 \end{bmatrix} = \begin{bmatrix} -8 & 2 & -4 & -10 \\ 21 & -2 & 9 & 28 \\ 4 & -14 & 8 & -2 \end{bmatrix}$$

As another example:

$$c_{1,4} = [-2,0] \cdot \begin{bmatrix} 5 \\ 3 \end{bmatrix} = -2(5) + 0(3) = -10$$

$$\begin{bmatrix} -2 & 0 \\ 5 & 1 \\ 2 & -4 \end{bmatrix} \begin{bmatrix} 4 & -1 & 2 & 5 \\ 1 & 3 & -1 & 3 \end{bmatrix} = \begin{bmatrix} -8 & 2 & -4 & -10 \\ 21 & -2 & 9 & 28 \\ 4 & -14 & 8 & -2 \end{bmatrix}$$

The complete array is

$$\begin{bmatrix} -2 & 0 \\ 5 & 1 \\ 2 & -4 \end{bmatrix} \begin{bmatrix} 4 & -1 & 2 & 5 \\ 1 & 3 & -1 & 3 \end{bmatrix} = \begin{bmatrix} -8 & 2 & -4 & -10 \\ 21 & -2 & 9 & 28 \\ 4 & -14 & 8 & -2 \end{bmatrix}$$

Example 10–5 D also illustrates matrix multiplication.

■ ***Example 10–5 D***

Compute each product.

1. $\begin{bmatrix} 2 & -3 \\ -1 & 5 \end{bmatrix}\begin{bmatrix} 4 & -2 & 0 \\ 1 & 3 & -4 \end{bmatrix}$

This is the product of a 2×2 and a 2×3 matrix; the result will be a 2×3 matrix.

$$= \begin{bmatrix} (2)(4) + (-3)(1) & (2)(-2) + (-3)(3) & (2)(0) + (-3)(-4) \\ (-1)(4) + (5)(1) & (-1)(-2) + (5)(3) & (-1)(0) + (5)(-4) \end{bmatrix}$$

$$= \begin{bmatrix} 5 & -13 & 12 \\ 1 & 17 & -20 \end{bmatrix}$$

2. $\begin{bmatrix} 6 \\ 2 \\ -2 \end{bmatrix}[-10\ 15]$

$(3 \times 1) \times (1 \times 2)$ tells us that the multiplication is defined, and that the result will be a 3×2 matrix.

$$\begin{bmatrix} (6)(-10) & (6)(15) \\ (2)(-10) & (2)(15) \\ (-2)(-10) & (-2)(15) \end{bmatrix} = \begin{bmatrix} -60 & 90 \\ -20 & 30 \\ 20 & -30 \end{bmatrix}$$

3. $\begin{bmatrix} 5 & 1 \\ 2 & -3 \end{bmatrix}\begin{bmatrix} a & b \\ c & d \end{bmatrix} = \begin{bmatrix} 5a + c & 5b + d \\ 2a - 3c & 2b - 3d \end{bmatrix}$

4. $\begin{bmatrix} -2 & 1 \\ 0 & 3 \\ 2 & 0 \end{bmatrix}\begin{bmatrix} 1 & 0 \\ -2 & 0 \\ 4 & 4 \end{bmatrix}$; this product is not defined.

$$(3 \times \underbrace{2) \times (3}_{\text{not equal}} \times 2)$$

■

After developing several necessary facts we will show how to solve systems of equations by matrix multiplication.

The identity matrix

We defined the order-n identity matrix in section 10–2. It is a square matrix of n rows and n columns, with every element zero except on the main diagonal (upper-left to lower-right), where every element is a 1. Whenever an element is on the main diagonal the row number is the same as the column number. This is used for a more formal definition.

> **Identity matrix of order-*n***
>
> The identity matrix of order n is the $n \times n$ matrix I_n such that
>
> $$i_{a,b} = \begin{cases} 1 \text{ if } a = b \\ 0 \text{ if } a \neq b \end{cases} \quad 1 \leq a, b \leq n.$$

Note Where the value of n is not important or implied we often use I instead of I_n.

The matrix I_n is called the **identity element for matrix multiplication,** because it can be shown that if A is a square matrix then $IA = AI = A$. For example,

$$\begin{bmatrix} 1 & 0 \\ 0 & 1 \end{bmatrix}\begin{bmatrix} -2 & 3 \\ 4 & 5 \end{bmatrix} = \begin{bmatrix} -2 & 3 \\ 4 & 5 \end{bmatrix}$$ Multiplication by I_2

$$\begin{bmatrix} 1 & 0 & 0 \\ 0 & 1 & 0 \\ 0 & 0 & 1 \end{bmatrix}\begin{bmatrix} 4 \\ 2 \\ 7 \end{bmatrix} = \begin{bmatrix} 4 \\ 2 \\ 7 \end{bmatrix}$$ Multiplication by I_3

Inverse of a matrix

For a *square* matrix C, if there is a matrix M such that $CM = MC = I$, we say that M is the inverse of C; we then usually call this matrix C^{-1} (the same notation we use for the inverse of a function). As with functions the superscript "-1" is not an exponent. It simply means "the matrix multiplicative inverse of."

> **Multiplicative inverse of a square matrix**
> For a square matrix C, if there exists a matrix C^{-1} such that
> $$C^{-1}C = I \text{ and } CC^{-1} = I$$
> we call C^{-1} the multiplicative inverse of C.

It can be proven that a square matrix C has a multiplicative inverse if and only if its determinant $|C|$ is nonzero. One method of finding the multiplicative inverse of a square matrix is shown below. In this procedure, we form the appropriate **augmented matrix.** This means to create a new matrix of twice as many columns as the original, where the rightmost columns form the identity matrix.

■ ***Example 10–5 E***

Find the multiplicative inverse of the given matrix.

1. $A = \begin{bmatrix} 2 & -5 \\ 3 & -1 \end{bmatrix}$

$$\begin{bmatrix} 2 & -5 & 1 & 0 \\ 3 & -1 & 0 & 1 \end{bmatrix}$$ Form the augmented matrix

Now transform the left two columns into I_2 using row operations (see section 10–2).

$$\begin{bmatrix} 2 & -5 & 1 & 0 \\ 3 & -1 & 0 & 1 \end{bmatrix}$$ Use -1 to sweep out column 2

$$\begin{bmatrix} -13 & 0 & 1 & -5 \\ 3 & -1 & 0 & 1 \end{bmatrix}$$ $R1 \leftarrow -5(R2) + R1$

This notation means that row 1 is replaced with the sum of -5 times row 2 and row 1. We will now use -13 to sweep out column 1.

$$\begin{bmatrix} -13 & 0 & 1 & -5 \\ 0 & -13 & 3 & -2 \end{bmatrix}$$ $R2 \leftarrow 3(R1) + 13(R2)$

$$\left[\begin{array}{cccc} 1 & 0 & -\frac{1}{13} & \frac{5}{13} \\ 0 & 1 & -\frac{3}{13} & \frac{2}{13} \end{array}\right] \qquad \text{Divide each row by } -13$$

Observe that we have I_2 in the left two columns.

$$A^{-1} = \begin{bmatrix} -\frac{1}{13} & \frac{5}{13} \\ -\frac{3}{13} & \frac{2}{13} \end{bmatrix}$$

This can be verified by computing $A^{-1}A$ and AA^{-1}, and noting that each product produces I_2.

2. $C = \begin{bmatrix} 1 & -3 & 0 \\ 2 & 0 & -1 \\ -2 & 3 & 4 \end{bmatrix}$

$$\left[\begin{array}{cccccc} 1 & -3 & 0 & 1 & 0 & 0 \\ 2 & 0 & -1 & 0 & 1 & 0 \\ -2 & 3 & 4 & 0 & 0 & 1 \end{array}\right] \qquad \text{Form the augmented matrix}$$

Now transform the left three columns into I_3. That is sweep out columns 1, 2, and 3. First sweep out column 3, using row 2.

$$\left[\begin{array}{cccccc} 1 & -3 & 0 & 1 & 0 & 0 \\ 2 & 0 & -1 & 0 & 1 & 0 \\ 6 & 3 & 0 & 0 & 4 & 1 \end{array}\right] \qquad R3 \leftarrow 4R2 + R3$$

Column 3 is now swept out, using row 2. We can now only use rows 1 or 3 to sweep out columns 1 and 2. We use the -3 in column 2.

$$\left[\begin{array}{cccccc} 1 & -3 & 0 & 1 & 0 & 0 \\ 2 & 0 & -1 & 0 & 1 & 0 \\ 7 & 0 & 0 & 1 & 4 & 1 \end{array}\right] \qquad R3 \leftarrow R3 + R1$$

Now columns 2 and 3 are swept out, and only row 3 may be used as a key row.

$$\left[\begin{array}{cccccc} 0 & -21 & 0 & 6 & -4 & -1 \\ 0 & 0 & -7 & -2 & -1 & -2 \\ 7 & 0 & 0 & 1 & 4 & 1 \end{array}\right] \qquad \begin{array}{l} R1 \leftarrow 7(R1) - R3 \\ R2 \leftarrow 7(R2) - 2(R3) \end{array}$$

$$\left[\begin{array}{cccccc} 7 & 0 & 0 & 1 & 4 & 1 \\ 0 & -21 & 0 & 6 & -4 & -1 \\ 0 & 0 & -7 & -2 & -1 & -2 \end{array}\right] \qquad \text{Rearrange the order of the rows}$$

$$\left[\begin{array}{cccccc} 1 & 0 & 0 & \frac{1}{7} & \frac{4}{7} & \frac{1}{7} \\ 0 & 1 & 0 & -\frac{2}{7} & \frac{4}{21} & \frac{1}{21} \\ 0 & 0 & 1 & \frac{2}{7} & \frac{1}{7} & \frac{2}{7} \end{array}\right] \qquad \text{Divide each row by its first nonzero element}$$

$$C^{-1} = \begin{bmatrix} \frac{1}{7} & \frac{4}{7} & \frac{1}{7} \\ -\frac{2}{7} & \frac{4}{21} & \frac{1}{21} \\ \frac{2}{7} & \frac{1}{7} & \frac{2}{7} \end{bmatrix}$$

This can be verified by checking that $C^{-1}C = CC^{-1} = I_3$. ■

Example 10–5 F illustrates how to perform matrix operations on the TI-81 graphing calculator. This calculator can store up to three matrices of size up to 6×6.

■ ***Example 10–5 F***

Let $A = \begin{bmatrix} 2 & -1 & 3 \\ 0 & 4 & -2 \\ -3 & 1 & 5 \end{bmatrix}$ and $B = \begin{bmatrix} 2 & 0.5 & -4 \\ 0 & 1 & 6 \\ -0.5 & 1 & 4 \end{bmatrix}$. Use the TI-81 to solve each problem.

1. Enter each matrix into the TI-81.

[MATRX] EDIT 1 3 [ENTER] 3 [ENTER] 2 [ENTER] [(−)] 1 [ENTER] 3 [ENTER] 0 [ENTER] 4 [ENTER] [(−)] 2 [ENTER] [(−)] 3 [ENTER] 1 [ENTER] 5 [QUIT]

[[A]] [ENTER]   [[A]] is [2nd] 1; The array A appears on the display.

To enter B, start with [MATRX] EDIT 2 3 [ENTER] 3 [ENTER], then enter the array's values as for A.

2. Compute $3A + B$.

3 [[A]] [+] [[B]] [ENTER] $\begin{bmatrix} 8 & -2.5 & 5 \\ 0 & 13 & 0 \\ -9.5 & 4 & 19 \end{bmatrix}$

3. Compute AB.

[[A]] [[B]] [ENTER] $\begin{bmatrix} 2.5 & 3 & -2 \\ 1 & 2 & 16 \\ -8.5 & 4.5 & 38 \end{bmatrix}$

4. Compute $|A|$.

[MATRIX] 5 [[A]] [ENTER] $|A| = 74$

5. Compute A^{-1}; round each element to four decimal places.

[MATH] NUM 1 [[A]] [x^{-1}] [,] 4 [)] [ENTER]   [,] is [ALPHA] [.]

Use the round function to show four decimal places. Use the left and right arrow keys to see the full array.

$$\begin{bmatrix} 0.2973 & 0.1081 & -0.1351 \\ 0.0811 & 0.2568 & 0.0541 \\ 0.1622 & 0.0135 & 0.1081 \end{bmatrix}$$

To show exact values (rational numbers) compute $|A|A^{-1}$. This gives the numerators of the result's elements. The denominators are $|A|$. On the TI-81 you must use the [×] (multiply) key between $|A|$ and A^{-1}. ■

Solving systems of equations by matrix multiplication

We define **equality of matrices** to mean that *two matrices are equal if and only if each element of one matrix is equal to each element of the other.* This implies that two matrices can be equal only if their dimensions are identical. By way of example, if we state that $\begin{bmatrix} 5 & b \\ c & 3 \end{bmatrix} = \begin{bmatrix} x & y \\ z & w \end{bmatrix}$ then we know that $x = 5$, $y = b$, $z = c$, and $w = 3$.

The **multiplication property of equality** can be proved for matrices A, X, and Y:

$$\text{if } X = Y, \text{ then } AX = AY.$$

Associativity states that when multiplying the expression ABC we can multiply AB first or BC first. Specifically, if A, B, and C are matrices then

$$A(BC) = (AB)C$$

This property can also be proved true for matrix multiplication in which all the products are defined (i.e., the dimensions match up properly).

A system of n linear equations in n variables can be described using matrix multiplication and the definition of equality of matrices. For example,

$$\begin{aligned} 3x - 2y &= 5 \\ x + 3y &= -9 \end{aligned}$$

can be described as

$$\begin{bmatrix} 3 & -2 \\ 1 & 3 \end{bmatrix}\begin{bmatrix} x \\ y \end{bmatrix} = \begin{bmatrix} 5 \\ -9 \end{bmatrix}$$

because we can recreate the equations as follows.

$$\begin{bmatrix} 3x - 2y \\ x + 3y \end{bmatrix} = \begin{bmatrix} 5 \\ -9 \end{bmatrix}$$ Multiply the matrices in the left member

$$\begin{aligned} 3x - 2y &= 5 \\ x + 3y &= -9 \end{aligned}$$ By equality of matrices

If we let C represent the coefficient matrix $\begin{bmatrix} 3 & -2 \\ 1 & 3 \end{bmatrix}$, X represents the matrix $\begin{bmatrix} x \\ y \end{bmatrix}$, and K represent the matrix of constants $\begin{bmatrix} 5 \\ -9 \end{bmatrix}$, then the system above can be described by the matrix equation

$$CX = K$$

If a matrix C^{-1} could be found such that $C^{-1}C = I$, we could solve the system by the following process.

$CX = K$	The original system of equations expressed using matrices
$C^{-1}CX = C^{-1}K$	Multiply by C^{-1}
$IX = C^{-1}K$	$C^{-1}C = I$
$X = C^{-1}K$	X is the matrix of variables

Example 10–5 G illustrates how to solve a system of linear equations using matrix multiplication.

■ Example 10–5 G

Solve the system of equations using matrix multiplication.

1. $2x - 5y = 3$
$3x - y = -4$

This system can be described by the matrix equation $CX = K$ as shown.

$$\overset{C}{\begin{bmatrix} 2 & -5 \\ 3 & -1 \end{bmatrix}} \overset{X}{\begin{bmatrix} x \\ y \end{bmatrix}} \overset{=}{=} \overset{K}{\begin{bmatrix} 3 \\ -4 \end{bmatrix}}$$

Thus, as computed above, we need to compute $C^{-1}K$. We found C^{-1} in example 10–5 E, so we can proceed:

$$X = C^{-1}K$$

$$\begin{bmatrix} x \\ y \end{bmatrix} = \begin{bmatrix} -\frac{1}{13} & \frac{5}{13} \\ -\frac{3}{13} & \frac{2}{13} \end{bmatrix} \begin{bmatrix} 3 \\ -4 \end{bmatrix}$$ Replace C^{-1} and K

$$\begin{bmatrix} x \\ y \end{bmatrix} = \begin{bmatrix} -\frac{23}{13} \\ -\frac{17}{13} \end{bmatrix}$$ Multiply the right member

$x = -\frac{23}{13}$ and $y = -\frac{17}{13}$ Use equality of matrices

2. $x - y = 3$
$2y + 3z = -2$
$-2x + 3y = 5$

We write the system as a matrix product, then solve the matrix equation for the variable array X.

$$\begin{bmatrix} 1 & -1 & 0 \\ 0 & 2 & 3 \\ -2 & 3 & 0 \end{bmatrix} \begin{bmatrix} x \\ y \\ z \end{bmatrix} = \begin{bmatrix} 3 \\ -2 \\ 5 \end{bmatrix}$$

$$CX = K$$

so

$$X = C^{-1}K$$

Thus we must find C^{-1} to find X.

$$\left[\begin{array}{rrr|rrr} 1 & -1 & 0 & 1 & 0 & 0 \\ 0 & 2 & 3 & 0 & 1 & 0 \\ -2 & 3 & 0 & 0 & 0 & 1 \end{array}\right]$$

Augmented matrix

$$\left[\begin{array}{rrr|rrr} 1 & 0 & 0 & 3 & 0 & 1 \\ 0 & 1 & 0 & 2 & 0 & 1 \\ 0 & 0 & 1 & -\frac{4}{3} & \frac{1}{3} & -\frac{2}{3} \end{array}\right]$$

Row operations used to form I_3 in the first three columns

$$C^{-1} = \begin{bmatrix} 3 & 0 & 1 \\ 2 & 0 & 1 \\ -\frac{4}{3} & \frac{1}{3} & -\frac{2}{3} \end{bmatrix}$$

Thus we proceed,

$$\begin{array}{ccccc} X & = & C^{-1} & & K \end{array}$$

$$\begin{bmatrix} x \\ y \\ z \end{bmatrix} = \begin{bmatrix} 3 & 0 & 1 \\ 2 & 0 & 1 \\ -\frac{4}{3} & \frac{1}{3} & -\frac{2}{3} \end{bmatrix} \begin{bmatrix} 3 \\ -2 \\ 5 \end{bmatrix} = \begin{bmatrix} 14 \\ 11 \\ -8 \end{bmatrix}$$

$x = 14$, $y = 11$, $z = -8$

To solve this problem on the TI-81 we enter the array C and K. Since this calculator calls its arrays A, B, and C, we use the terminology

$$X = A^{-1}B$$

First, matrices A and B (that is, C and K above) are entered into the TI-81 as shown in example 10–5 F.

Now compute the matrix product $A^{-1}B$:

[A] x^{-1} [B] ENTER

The result appears: $\begin{bmatrix} 14 \\ 11 \\ -8 \end{bmatrix}$. Thus $(x,y,z) = (14,11,-8)$. ■

Mastery points

Can you

- Form the dot product of vectors?
- Compute scalar and matrix products?
- Find the inverse of a matrix by using an augmented matrix?
- Use matrix multiplication and the inverse of a matrix to solve systems of linear equations?

Exercise 10–5

Add or subtract the given matrices as indicated.

1. $\begin{bmatrix} -1 & 3 \\ -2 & 5 \end{bmatrix} + \begin{bmatrix} 3 & -2 \\ 1 & -5 \end{bmatrix}$

2. $\begin{bmatrix} 3 & -6 \\ 1 & 0 \end{bmatrix} - \begin{bmatrix} 3 & -2 \\ 5 & 0 \end{bmatrix}$

3. $\begin{bmatrix} -1 & 3 \\ -2 & 5 \end{bmatrix} - \begin{bmatrix} -1 & 3 \\ -2 & 5 \end{bmatrix}$

4. $\begin{bmatrix} 3 & -6 \\ 1 & 0 \end{bmatrix} + \begin{bmatrix} 3 & -2 \\ 5 & 0 \end{bmatrix}$

5. $\begin{bmatrix} -1 & -2 & -1 \\ 3 & -2 & 5 \end{bmatrix} + \begin{bmatrix} 0 & 3 & -2 \\ 1 & -2 & -3 \end{bmatrix}$

6. $\begin{bmatrix} -4 & 1 & 3 \\ -6 & 1 & 0 \end{bmatrix} - \begin{bmatrix} 1 & -3 & 5 \\ 0 & -6 & 2 \end{bmatrix}$

7. $\begin{bmatrix} 1 & 2 & 3 \\ -2 & -3 & 5 \end{bmatrix} - \begin{bmatrix} 4 & 3 & -1 \\ 2 & 15 & -2 \end{bmatrix}$

8. $\begin{bmatrix} -5 & 2 & 6 \\ 1 & 2 & 1 \end{bmatrix} + \begin{bmatrix} 13 & -12 & 5 \\ 10 & 1 & 0 \end{bmatrix}$

9. $\begin{bmatrix} -1 & -2 \\ -1 & 3 \\ -2 & 5 \end{bmatrix} + \begin{bmatrix} 0 & 3 \\ -2 & 1 \\ -2 & -3 \end{bmatrix}$

10. $\begin{bmatrix} -4 & 1 \\ 3 & -6 \\ 1 & 0 \end{bmatrix} - \begin{bmatrix} 1 & -3 \\ 5 & 0 \\ -6 & 2 \end{bmatrix}$

11. $\begin{bmatrix} 1 & 2 \\ 3 & -2 \\ -3 & 5 \end{bmatrix} - \begin{bmatrix} 4 & 3 \\ -1 & 2 \\ 15 & -2 \end{bmatrix}$

12. $\begin{bmatrix} -5 & 2 \\ 6 & 1 \\ 2 & 1 \end{bmatrix} + \begin{bmatrix} 13 & -12 \\ 5 & 10 \\ 1 & 0 \end{bmatrix}$

Compute the given scalar product.

13. $4\begin{bmatrix} -1 & 2 \\ 4 & 5 \end{bmatrix}$

14. $3\begin{bmatrix} 3 & -2 \\ 1 & 0 \end{bmatrix}$

15. $-5\begin{bmatrix} 0 & 3 \\ -2 & 5 \end{bmatrix}$

16. $-2\begin{bmatrix} 8 & -6 \\ 1 & 10 \end{bmatrix}$

17. $\frac{1}{2}\begin{bmatrix} 4 & 1 & -2 \\ 2 & 6 & -2 \end{bmatrix}$

18. $\frac{2}{3}\begin{bmatrix} -15 & 9 & 3 \\ 1 & 6 & 1 \end{bmatrix}$

19. $-1\begin{bmatrix} 2 & -13 \\ 5 & 0 \\ 3 & 2 \end{bmatrix}$

20. $0\begin{bmatrix} 0 & 2 \\ -3 & 2 \\ -3 & 5 \end{bmatrix}$

Form the dot product of the given vectors.

21. [3, −4], [−2, 5]

22. [11, 0, −3, 2], [4, −2, 2, 6]

23. [3, 1, −2], $\begin{bmatrix} 4 \\ -4 \\ 5 \end{bmatrix}$

24. [−2, 5], [3, 1]

25. $\left[\sqrt{2}, \frac{1}{3}, -5\right], \left[\sqrt{8}, 6, \frac{\pi}{5}\right]$

26. [−4, 0, 3, −5], $\begin{bmatrix} -4 \\ 2 \\ 0 \\ 2 \end{bmatrix}$

27. Find a vector v such that $[-3, 1, -2, 5] \cdot v = 1$.

28. Find a vector v such that $[5, 2, -4, 3] \cdot v = \frac{1}{2}$

Compute the indicated matrix products.

29. $\begin{bmatrix} -1 & 1 \\ -2 & 5 \end{bmatrix}\begin{bmatrix} 0 & -2 \\ 1 & -5 \end{bmatrix}$

30. $\begin{bmatrix} 3 & -2 \\ 1 & 0 \end{bmatrix}\begin{bmatrix} 3 & -2 \\ 5 & 0 \end{bmatrix}$

31. $\begin{bmatrix} -1 & 3 \\ -2 & 4 \end{bmatrix}\begin{bmatrix} 1 & 3 \\ -2 & 6 \end{bmatrix}$

32. $\begin{bmatrix} 3 & -3 \\ 1 & 0 \end{bmatrix}\begin{bmatrix} 3 & -2 \\ 5 & 1 \end{bmatrix}$

33. $\begin{bmatrix} 1 & 2 & -1 \\ 0 & 2 & 3 \\ -2 & 5 & 2 \end{bmatrix}\begin{bmatrix} 0 & 3 & -1 \\ 5 & 2 & 0 \\ 1 & -2 & -3 \end{bmatrix}$

34. $\begin{bmatrix} 4 & 2 & 5 \\ 1 & 3 & -6 \\ 1 & 0 & 4 \end{bmatrix}\begin{bmatrix} 1 & -3 & 5 \\ 0 & -6 & 2 \\ 2 & 0 & -1 \end{bmatrix}$

35. $\begin{bmatrix} 1 & -1 & 0 \\ 2 & 3 & -2 \\ -3 & 5 & 1 \end{bmatrix}\begin{bmatrix} 4 & 3 \\ -1 & 2 \\ 15 & -2 \end{bmatrix}$

36. $\begin{bmatrix} -5 & 2 & 6 \\ 1 & 2 & 1 \end{bmatrix}\begin{bmatrix} 1 & -4 & -12 \\ 5 & 10 & 1 \\ 3 & 4 & 0 \end{bmatrix}$

37. $\begin{bmatrix} -1 & -2 & -1 \\ 3 & -2 & 5 \end{bmatrix}\begin{bmatrix} 0 & 3 \\ -2 & 1 \\ -2 & -3 \end{bmatrix}$

38. $\begin{bmatrix} -4 & 1 & 3 \\ -6 & 1 & 0 \end{bmatrix}\begin{bmatrix} 1 & -3 \\ 5 & 0 \\ -6 & 2 \end{bmatrix}$

39. $\begin{bmatrix} 1 & 2 & 3 \\ -2 & -3 & 5 \end{bmatrix}\begin{bmatrix} 4 & 3 \\ -1 & 2 \\ 15 & -2 \end{bmatrix}$

40. $\begin{bmatrix} -5 & 2 & 6 \\ 1 & 2 & 1 \end{bmatrix}\begin{bmatrix} 13 & -12 \\ 5 & 10 \\ 1 & 0 \end{bmatrix}$

41. $\begin{bmatrix} 1 & -3 \\ -1 & 4 \end{bmatrix}\begin{bmatrix} -4 & -21 & 10 \\ 1 & 3 & -4 \end{bmatrix}$

42. $[-10, 1, -5]\begin{bmatrix} 1 \\ 2 \\ -8 \end{bmatrix}$

43. $\begin{bmatrix} 5x & 1 \\ 4y & -3 \end{bmatrix}\begin{bmatrix} -4x & 3 \\ y & 9 \end{bmatrix}$

44. $\begin{bmatrix} x & 2 \\ 4y+1 & 2 \end{bmatrix}\begin{bmatrix} -4x-1 & 3 \\ y+1 & 2 \end{bmatrix}$

45. $\begin{bmatrix} 2 & 3 & -1 \\ 4 & \frac{1}{2} & 8 \end{bmatrix}\begin{bmatrix} -4 & -6 \\ 10 & 2 \\ 3 & -4 \end{bmatrix}$

46. $\begin{bmatrix} 2 & 1 \\ -1 & 4 \\ 3 & 8 \end{bmatrix}\begin{bmatrix} -1 & -6 & 1 & 3 \\ 1 & 2 & 0 & 2 \end{bmatrix}$

47. $\begin{bmatrix} 2 & 3 & 2 & -1 \\ 4 & -6 & 2 & 8 \end{bmatrix}\begin{bmatrix} -4 & -6 & 5 \\ 11 & 1 & 3 \\ -4 & -2 & -7 \\ 2 & 0 & -3 \end{bmatrix}$

48. $\begin{bmatrix} 1 & -6 & 2 & 3 \\ 2 & -1 & 4 & 8 \end{bmatrix}\begin{bmatrix} -4 & -6 & 3 \\ -4 & -2 & -7 \\ 2 & 5 & 11 \\ 1 & 0 & -3 \end{bmatrix}$

49. Let $A = \begin{bmatrix} 2 & -1 \\ 4 & 6 \\ -4 & 5 \end{bmatrix}$, $B = \begin{bmatrix} 3 & -5 & 2 & 7 \\ -1 & 3 & -6 & 1 \end{bmatrix}$, and $C = \begin{bmatrix} 3 & 0 & -1 \\ 4 & 2 & 0 \\ 2 & 5 & -1 \\ 8 & 3 & -4 \end{bmatrix}$. By computation determine whether $(AB)C = A(BC)$.

50. In the text it was stated that a square matrix has an inverse if and only if its determinant is not zero. The determinant of matrix $A = \begin{bmatrix} 3 & 1 & 5 \\ 2 & 1 & -3 \\ -1 & -1 & 11 \end{bmatrix}$ is zero. (a) Verify this by computation. (b) Attempt to find A^{-1} and observe what happens.

Find the inverse of each matrix. If the matrix does not have an inverse, state this. Also see problem 50.

51. $\begin{bmatrix} 12 & -5 \\ -3 & -1 \end{bmatrix}$

52. $\begin{bmatrix} 5 & -5 \\ -3 & 4 \end{bmatrix}$

53. $\begin{bmatrix} -3 & \frac{1}{2} \\ 2 & 3 \end{bmatrix}$

54. $\begin{bmatrix} 1 & -\frac{2}{3} \\ 6 & -1 \end{bmatrix}$

55. $\begin{bmatrix} 0 & -3 & 0 \\ 2 & 0 & -1 \\ 2 & 3 & 4 \end{bmatrix}$

56. $\begin{bmatrix} 1 & 3 & 2 \\ -2 & 0 & -1 \\ 3 & 0 & 4 \end{bmatrix}$

57. $\begin{bmatrix} 1 & 0 & 3 \\ 2 & 0 & -1 \\ 2 & 3 & 0 \end{bmatrix}$

58. $\begin{bmatrix} 2 & 3 & 1 \\ 0 & 1 & 1 \\ 2 & 0 & 4 \end{bmatrix}$

59. $\begin{bmatrix} 0 & -3 & 0 & 2 \\ 2 & -1 & 0 & 3 \\ -5 & 2 & 0 & -1 \\ 2 & 0 & -6 & 1 \end{bmatrix}$

60. $\begin{bmatrix} 2 & -3 & 0 & 0 \\ 2 & -2 & 1 & 3 \\ -3 & 2 & 0 & -1 \\ 2 & 0 & -2 & 1 \end{bmatrix}$

61. $\begin{bmatrix} 0 & 0 & 4 & 1 \\ 2 & -1 & 0 & 3 \\ 3 & 2 & 0 & 1 \\ 2 & 2 & 6 & 1 \end{bmatrix}$

62. $\begin{bmatrix} 1 & 0 & 3 & -2 \\ 0 & -1 & 0 & 3 \\ 5 & 1 & -2 & -1 \\ 0 & 2 & 1 & 1 \end{bmatrix}$

Solve the system of equations using matrix multiplication. *The matrix of coefficients of each problem corresponds to the matrix from the problem indicated in parentheses, where its inverse was computed.* Thus, there is no need to recompute the inverse of this matrix.

63. $12x - 5y = -6$ (51)
$-3x - y = -3$

64. $5x - 5y = -15$ (52)
$-3x + 4y = 10$

65. $-3x + \frac{1}{2}y = -7$ (53)
$2x + 3y = -2$

66. $x - \frac{2}{3}y = -5$ (54)
$6x - y = -12$

67. $-3y = 3$ (55)
$2x - z = 1$
$2x + 3y + 4z = 13$

68. $x + 3y + 2z = 3$ (56)
$-2x - z = 2$
$3x + 4z = 2$

69. $x + 3z = -6$ (57)
$2x - z = -5$
$2x + 3y = -4$

70. $2x + 3y + z = -4$
$y + z = -1$ (58)
$2x + 4z = 5$

71. $-3y + 2w = 8$ (59)
$2x - y + 3w = 7$
$-5x + 2y - w = -10$
$2x - 6z + w = -9$

72. $2x - 3y = -8$ (60)
$2x - 2y + z + 3w = 2$
$-3x + 2y - w = 6$
$2x - 2z + w = -11$

73. $4z + w = -2$ (61)
$2x - y + 3w = 5$
$3x + 2y + w = 11$
$2x + 2y + 6z + w = 4$

74. $x + 3z - 2w = 21$
$-y + 3w = -7$ (62)
$5x + y - 2z - w = 3$
$2y + z + w = 5$

Solve the system of equations using matrix multiplication.

75. $x + \frac{2}{3}y = -1$
$-3x + y = 12$

76. $2x + \frac{1}{2}y = -1$
$3x - y = -5$

77. $-2x + y = 1$
$4x - y = 0$

78. $-x + 3y = 13$
$2x + y = 2$

79. $x + y - 5z = -9$
$-x + 2z = 6$
$2x + 2y = 2$

80. $-y + 3z = 9$
$-2x + 2y + z = 1$
$x + 3y + 2z = 7$

81. $-x - y - 2z = 1$
$x + y - 4z = -4$
$-\frac{1}{2}y + 2z = 3$

82. $-2x + 2z = -8$
$4x - y - z = 8$
$x + 2y - 2z = 5$

In the following problems use the three scalars $a = 2$, $b = -3$, and $c = 5$, and two matrices, $A = \begin{bmatrix} 2 & -3 \\ -1 & 4 \end{bmatrix}$ and $B = \begin{bmatrix} 0 & 3 \\ 4 & -1 \end{bmatrix}$, to compute the values of the following polynomial matrix expressions.

83. aAB **84.** bA^2 **85.** $aA - bB$ **86.** $bA^2 - aB^2 - cB$ **87.** $(A - B)^2$ **88.** $AB - BA$

Solve the following problems.

89. In archeology the problem of placing sites and artifacts in proper chronological order is called sequence dating.[6] In one situation with five types of pottery and four graves, the computation

$$\begin{bmatrix} 1 & 1 & 0 & 1 & 1 \\ 0 & 0 & 1 & 0 & 1 \\ 0 & 1 & 1 & 0 & 1 \\ 1 & 1 & 1 & 1 & 1 \end{bmatrix} \begin{bmatrix} 1 & 0 & 0 & 1 \\ 1 & 0 & 1 & 1 \\ 0 & 1 & 1 & 1 \\ 1 & 0 & 0 & 1 \\ 1 & 1 & 1 & 1 \end{bmatrix}$$

must be made. The result can be used to put the grave sites in chronological order. Perform this computation.

90. The figure is from graph theory, an area of mathematics that has application in the sciences and business. It shows four nodes labeled 1 through 4 and paths from some nodes to others. This graph can be represented by the matrix $A = \begin{bmatrix} 0 & 1 & 1 & 0 \\ 0 & 0 & 0 & 1 \\ 0 & 1 & 0 & 1 \\ 1 & 0 & 0 & 0 \end{bmatrix}$, where a 1 in row i column j means there is a path from node i to node j. For example, there is a 1 in row 2 column 4, which shows that node 2 has a path to node 4. The rest of the row is zeros because node 2 has a path only to node 4. If we compute $A^2 = A \cdot A$ the matrix would give the number of paths of length 2 from one node to the other. Perform this computation and use the result to determine how many paths of length 2 there are from node 1 to node 4.

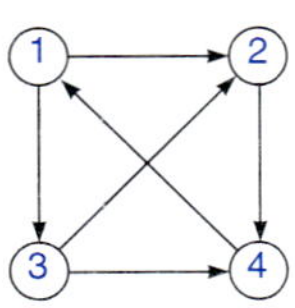

[6]From "Mathematics in Archaeology," by Gareth Williams, Stetson University, from *The Two-Year College Mathematics Journal*, Vol. 13, No. 1, January 1982.

91. Create a matrix A that describes the graph in the figure (see problem 90). Compute A^2 and determine how many paths of length 2 there are from node 1 to node 2.

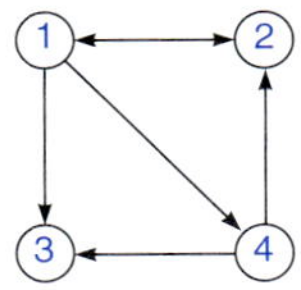

92. Referring to problem 90, compute A^3 and determine how many paths of length 3 there are from node 1 to node 4.

93. Referring to problem 91 compute A^3 and determine how many paths of length 3 there are from node 1 to node 2.

94. The figure could represent many situations in which there are three states. We will imagine it is a maze with three rooms. If a mouse is put in room 1, it has two choices for moving into another room. Assuming it

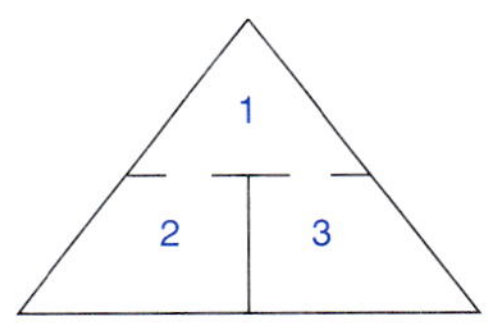

chooses randomly the chance of moving into either room 2 or 3 is $\frac{1}{2}$. In rooms 2 or 3, it has no choices for moving into another room. Thus the chance of moving into room 1 from room 2 is 1. The array $A = \begin{bmatrix} 0 & \frac{1}{2} & \frac{1}{2} \\ 1 & 0 & 0 \\ 1 & 0 & 0 \end{bmatrix}$ can describe this situation, where the entry in row i column j describes the probability that, if the mouse is in room i it will next move to room j. In the array $A^2 = A \cdot A$, $a_{i,j}$ would tell the probability that, if the mouse started in room i, it is in room j after two room changes. Compute A^2 and determine the probability that a mouse which started in room 2 is in room 3 after 2 room changes (i.e., that it went from room 2 to room 1 to room 3).

The basic idea presented here is used to create what is called Markov chains (the powers of the matrix A) to study manufacturing and biological processes, economies, and chemical reactions among other things.

95. The array for the maze in the figure is $A = \begin{bmatrix} 0 & \frac{1}{2} & \frac{1}{2} & 0 \\ \frac{1}{2} & 0 & 0 & \frac{1}{2} \\ 1 & 0 & 0 & 0 \\ 0 & 1 & 0 & 0 \end{bmatrix}$ (see problem 94). Compute A^2 and use it to determine the probability that a mouse which started in room 2 is in room 3 after two moves.

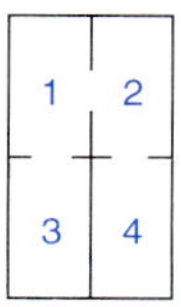

96. Referring to problem 95, compute A^3 and determine the probability that a mouse that started in room 3 is in room 4 after 3 moves.

97. Let $L = \begin{bmatrix} 0 & 4 & 3 \\ 0.229 & 0 & 0 \\ 0 & 0.125 & 0 \end{bmatrix}$. This represents a creature that lives in three stages. An example would be an insect that has an egg, larval, and adult stage. In the first stage, females have no progeny. In the second, they have four daughters. In the third, they have three daughters. The survival rate in stage one is 22.9%, and in stage two it is 12.5%. Let $V = \begin{bmatrix} 1{,}000 \\ 1{,}000 \\ 1{,}000 \end{bmatrix}$ represent the number of females in each stage in some initial generation. (L is called a LESLIE matrix.) $LV = \begin{bmatrix} 7{,}000 \\ 229 \\ 125 \end{bmatrix}$ is the number of females in each stage after one life cycle (generation), and in general L^nV is the number after n cycles. (The elements of each state are rounded to the nearest integer.)

Compute L^3V, and find the number of females in each stage after three cycles (generations).

98. If a figure, such as the triangle in the figure, is to be rotated through an angle θ (theta), every point (x,y) must be transformed to a new point (x',y'). It can be shown that the new point can be found by matrix multiplication of the original point. The matrix is $R = \begin{bmatrix} \cos\theta & \sin\theta \\ -\sin\theta & \cos\theta \end{bmatrix}$. Note that $\cos\theta$ (read cosine of angle theta) and $\sin\theta$ (read sine of angle theta) are functions defined in trigonometry. They give a certain value for any angle θ.

If $P = (x,y)$ is to be rotated, the new point $P' = P \cdot R$. (Note that R is on the *right* of the point P.) For an angle of rotation of 30°, $R = \begin{bmatrix} \cos 30° & \sin 30° \\ -\sin 30° & \cos 30° \end{bmatrix} = \begin{bmatrix} \frac{\sqrt{3}}{2} & \frac{1}{2} \\ -\frac{1}{2} & \frac{\sqrt{3}}{2} \end{bmatrix}$. Assume a triangle is defined by the points (3,2), (−3,2), and (0,−4).

a. Draw this triangle.

b. Rotate each point by 30° by using the matrix R as given above, then draw the rotated triangle determined by these three new points.

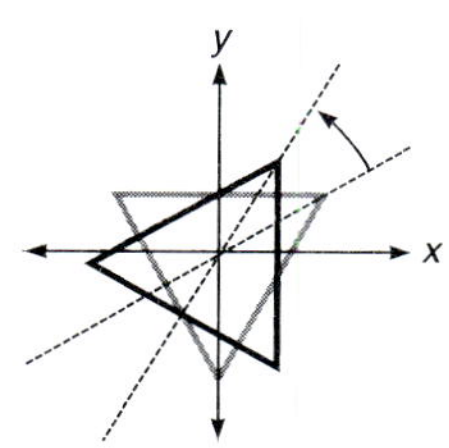

Skill and review

1. Graph the system of inequalities $\begin{array}{l} 2x + y < 6 \\ y \le 3x - 1 \end{array}$.

2. Solve the system of equations
$2x - y - 2z = -7$
$x + y + 4z = 2$
$3x + 2y - 2z = -3$

3. Find the point at which the lines $2x + 3y = -6$ and $x - 4y = 8$ intersect.

4. Find the equation of the line that passes through the points (−2,4) and (3,8).

5. Find the equation of a circle with center at (−2,4) and passes through the point (3,8).

6. Simplify $\sqrt{\frac{3}{8x^5y}}$.

7. Factor $81x^4 - 1$.

8. Combine $\frac{3a}{2b} - \frac{5a}{3c} + \frac{1}{a}$.

Chapter 10 summary

- **Dependent system** A system of n equations in n variables with more than one solution.
- **Inconsistent system** A system of n equations in n variables with no solution.
- **Recognizing dependent and inconsistent systems**
  - If we obtain a statement that is always true, such as $0 = 0$, the system of equations is dependent.
  - If we obtain a statement that is never true, such as $0 = 1$, the system of equations is inconsistent.
- **Row operations on matrices**
  1. If we multiply or divide each entry of a row by any nonzero value we do not change the solution set of the system.
  2. If we add a nonzero multiple of one row to a nonzero multiple of another row, and replace either row by the result, we do not change the solution set of the system of linear equations.
  3. Rearranging the order of the rows does not change the solution set of the system.
- **Matrix elimination** A method of solving systems of n equations in n variables. Each row serves as a key row once.
- **Identity matrix** The identity matrix of order n is the $n \times n$ matrix I such that $i_{a,b} = \begin{cases} 1 \text{ if } a = b \\ 0 \text{ if } a \neq b \end{cases}$, $1 \leq a, b \leq n$.
- **Minor** Given an nth order matrix A, the minor of element $a_{i,j}$ is the determinant of the $n - 1$ order matrix formed by deleting the ith row and jth column of matrix A.
- **Sign matrix of order n**

$$\underbrace{\begin{bmatrix} + & - & + & - & + & - & \cdots \\ - & + & - & + & - & \cdots & \\ + & - & + & - & \cdots & & \\ - & + & - & \cdots & & & \\ + & - & \cdots & & & & \\ - & \cdots & & & & & \\ \cdots & & & & & & \end{bmatrix}}_{}\quad (n \text{ columns}, n \text{ rows})$$

- **Determinant of an order 2 matrix**
$\begin{vmatrix} a_{1,1} & a_{1,2} \\ a_{2,1} & a_{2,2} \end{vmatrix} = a_{1,1}a_{2,2} - a_{2,1}a_{1,2}$.
- **Determinant of an order n matrix, $n > 2$** Given an order n matrix, the determinant is the sum of the products of each element of any row or column, the corresponding element of the sign matrix, and its respective minor.
- The determinant of a system of n equations in n variables is 0 if and only if the system is dependent or inconsistent.
- **Cramer's rule** Given a system of n linear equations in the n variables v_i, $1 \leq i \leq n$, let D represent the determinant of the coefficient matrix, and let D_i be the determinant of the matrix composed of the coefficient matrix D with the ith column of D replaced by the column of constants. Then for each i, $1 \leq i \leq n$, $v_i = \dfrac{D_i}{D}$.
- **Linear programming problem in two variables** A problem that can be described by a nonempty set of linear inequalities, called constraints, and a linear equation in two variables, called the objective function. The constraints form a set of feasible solutions. The set of linear equalities that correspond to the set of inequalities form a boundary to the feasible solutions.
- **Fundamental principle of linear programming** In a linear programming problem the objective function is always maximized or minimized at a vertex of the graph of feasible solutions.
- **Scalar** A real number.
- **Vector** A one-dimensional matrix.
- **Inverse of a matrix** For a square matrix C, if there exists a matrix C^{-1} such that $C^{-1}C = I$, and $CC^{-1} = I$, C^{-1} is the inverse of C and vice versa.
- A square matrix C has an inverse if and only if $|C| \neq 0$.

Chapter 10 review

[10–1] Solve the following systems of equations.

1. $3x + 2y = -6$
$-\frac{3}{2}x + y = 6$

2. $2x + \frac{3}{2}y = 5$
$-3x + \frac{1}{2}y = -2$

3. $-x + 4y = 20$
$2x + y = -22$

4. $0.4x + 0.5y = -2.9$
$-0.1x + y = 1.4$

5. $x - 5z = -7$
$-2x + y + 2z = 9$
$5x + 2y = -4$

6. $-4x - y + 3z = -7$
$-2x + 2y - z = -6$
$x + 2z = -3$

7. $x + y - z = 7$
$-2x + 4y + 2w = 4$
$-2y - 3z - 2w = 0$
$2x - 5z + 5w = 21$

8. $2x - z + 2w = -3$
$3x + z + 4w = 4$
$4y - 2z - w = -15$
$y + 2z + 2w = 6$

9. The perimeter of a certain rectangle is 182 centimeters; if the ratio of length to width is 8 to 5, find the values of the length and width.

10. The length of a certain rectangle is 2 inches less than twice the width. The perimeter is 158 inches. Find the two dimensions.

11. A total of $15,000 was invested, part at 6% and the rest at 12%. The total income from both investments was $1,530. How much was invested at each rate?

12. A parabola is the graph of an equation of the form $y = Ax^2 + Bx + C$. Find the values of A, B, and C so that the parabola will pass through the points $(-4,37)$, $(0,3)$, and $(2,10)$, and write the resulting equation.

[10–2] Solve the following systems of equations by matrix elimination; after describing the solution set, state dependent or inconsistent if appropriate.

13. $-2x + 5y = 7$
$2x + 3y = 9$

14. $2x + 5y = -17$
$-5x + \frac{2}{3}y = 3$

15. $-5x + y + 2z = 4$
$4x - y - z = -5$
$\frac{1}{2}x + 2y - 5z = 14$

16. $-x + 3y - 3z = 15$
$x + y - 3z = 7$
$-3x - y + 6z = -10$

17. $x - 3z + 5w = -9$
$-x + z + 3w = -7$
$3x - 3y + z = 15$
$-5x + y - 5z + 5w = -19$

18. $x + \frac{1}{2}y + 3z - 3w = -6$
$2x - \frac{3}{4}y + 3z + 5w = 16$
$-6x + y - w = -12$
$x - 2z = \frac{1}{3}$

19. For a certain electronics circuit Kirchhoff's law gives the system
$25i_1 + 20i_2 + 5i_3 = 50$
$40i_1 - 20i_2 - 10i_3 = 40$
$5i_1 + 10i_2 + 5i_3 = 45$
Find the currents i_1, i_2, and i_3.

20. A chemical company stores a certain herbicide in two concentrations of herbicide and water: 8% solution and 20% solution. It needs to manufacture 1,000 gallons of a 12% solution. How many gallons of each of the solutions should be mixed to obtain the required product?

[10–3] Solve the following systems of equations by Cramer's rule.

21. $-3x - 4y = 0$
$x + 9y = 4$

22. $5x + \frac{1}{3}y = 12$
$-8x + 2y = \frac{5}{8}$

23. $x + 8y + 3z = -4$
$x - 3y = 5$
$-x + 9y + 7z = -6$

24. $2x + 5y = -2$
$x - 5y - 2z = 1$
$4x + 2y = 0$

25. $2y + 3z - w = 2$
$-x + 2y + 5z = 0$
$-4x - 2z - 4w = -2$
$-4y - 4w = 1$

26. $x - 2y - 4w = 4$
$3x + y + 2z = -2$
$x - 2z + 6w = 0$
$4y - 2z + w = 2$

27. Solve for the variable D in the system:
$2A - B + 3C - D = 5$
$A + B - 2D + E = 0$
$-B - C = 10$
$3A - C + D + E = -4$
$C - E = -20$

28. If the three points that form the vertices of a triangle are (x_1,y_1), (x_2,y_2), (x_3,y_3), then it can be shown that the area of the triangle is the absolute value of $\frac{1}{2}\begin{vmatrix} x_1 & y_1 & 1 \\ x_2 & y_2 & 1 \\ x_3 & y_3 & 1 \end{vmatrix}$.

Find the area of the triangle with vertices $(3\frac{1}{2},2)$, $(5,-3)$, and $(-2,8)$.

[10–4] Graph the solution set of the following linear inequalities.

29. $3x - 4y > 12$

30. $-12x + 6y < 18$

31. $x + 2y \geq -9$

32. $-9x + 3 > 4y$

33. $6y > 2$

34. $\frac{5}{3}y - \frac{7}{6} < 3x$

35. $2.4x - 1.2y \leq 4.8$

Graph the solution set of the following systems of linear inequalities.

36. $2x - y > 5$
$x + 2y < -6$

37. $3x + 2y \geq 12$
$2x + 2y < 9$

38. $\frac{1}{2}x + \frac{5}{4}y > 10$
$4x + 3y > -12$

In the following problems, maximize the value of the objective function P with regard to the constraints supplied. It is assumed that $x \geq 0$ and $y \geq 0$.

39. $-14x + 15y \leq 9$
$2x + 10y \leq 23$
$9x + 8y \leq 48$
$P = 4x + 2y$

40. $x + 6y \leq 18$
$x + 3y \leq 10$
$2x + y \leq 10$
$P = x + 3y$

41. A furniture factory is asked by its parent company to produce two products, tables and chairs. The factory's profit on tables is $3 and on chairs is $1. It takes 4 hours to assemble a table and $1\frac{1}{2}$ hours to assemble a chair. In a production run there are 300 hours available for assembly. It takes 2 hours to finish a table, and $\frac{5}{8}$ hour for a chair. The finishing department has 200 hours available in a production run. The factory may make any mix of tables and chairs it chooses since the parent company has other factories also. Maximize total profit from the production of tables and chairs for this factory.

42. A large logging company has crews of workers and supervisors. It has found that a crew of 1 supervisor and 4 loggers log 14 trees per day, and a crew of 3 supervisors and 18 loggers log 45 trees per day. The company has 40 supervisors and 200 loggers on its payroll. What mix of crews would produce the most trees per day? (For simplicity, assume that fractional parts of crews make sense.)

[10–5] Form the dot product of the given vectors.

43. $[3,-\frac{1}{4}][2,5]$

44. $[-1, 10, -3, 2], [4, -2, -2, \frac{1}{2}]$

45. $[\frac{1}{4}, 1, -2], \begin{bmatrix} 4 \\ -4 \\ 5 \end{bmatrix}$

46. $\left[\sqrt{2}, \frac{1}{3}, -\frac{15}{\sqrt{\pi}}\right], \left[\sqrt{8}, 6, \frac{\pi}{5}\right]$

47. Find a vector v such that $[-2, 1, 6, 5] \cdot v = 3$.

Compute the indicated matrix products.

48. $\begin{bmatrix} 2 & -3 \\ 1 & 4 \end{bmatrix}\begin{bmatrix} -4 & -8 & 10 \\ 1 & 2 & -4 \end{bmatrix}$

49. $\begin{bmatrix} -x & 2 \\ 4y & -3 \end{bmatrix}\begin{bmatrix} 4x & 3 \\ y & 9 \end{bmatrix}$

50. $\begin{bmatrix} 2 & -3 & 1 \\ 4 & \frac{1}{2} & 6 \\ 0 & -2 & 3 \\ 10 & -1 & 2 \end{bmatrix}\begin{bmatrix} -4 & -8 & 10 & 1 \\ 2 & -4 & 3 & \frac{3}{4} \\ -1 & 0 & 4 & 1 \end{bmatrix}$

Find the inverse of each matrix. If the matrix does not have an inverse, state this.

51. $\begin{bmatrix} 2 & -5 \\ -3 & -1 \end{bmatrix}$

52. $\begin{bmatrix} 1 & -3 & 0 \\ 2 & 0 & -1 \\ 3 & 5 & 4 \end{bmatrix}$

Solve the system of equations using matrix multiplication.

53. $2x - 3y = 1$
$x + y = -2$

54. $x - y = 0$
$y + 3z = 2$
$-2x + y = -5$

Chapter 10 test

Solve the following systems of n equations in n variables by elimination.

1. $2x - 3y = -1$
$4x + 9y = 8$

2. $2x + 2y - z = 6$
$3x - 4y + z = 3$
$x - 2y + 3z = -2$

3. The length of a certain rectangle is 8 inches longer than three times the width. The perimeter is 208 inches. Find the two dimensions.

4. A total of \$12,000 was invested, part at 6% and the rest at 10%. The total income from both investments was \$1,020. How much was invested at each rate?

5. A parabola is the graph of an equation of the form $y = ax^2 + bx + c$. Find the equation of the parabola that will pass through the points $(-2,13)$, $(1,4)$, and $(2,9)$.

Solve the following systems of equations by matrix elimination.

6. $2x - y + 2z = 12$
$4x - y - z = 7$
$x - 2y - 5z = -9$

7. $2x - y + 2z = 4$
$x + 2y - w = -3$
$3x - 2y + z + w = 4$
$y - 5w = 3$

8. A company has two antifreeze mixtures on hand. One is a 30% solution (30% is alcohol) and the other is a 70% solution. How much of each should be mixed to obtain 500 gallons of a 45% solution?

Solve the following systems of equations by Cramer's rule.

9. $x + \frac{1}{3}y = 4$
$-x + 2y = \frac{5}{4}$

10. $2x + 5y + z = -2$
$x - 5y - 2z = 1$
$6x - 2z = -2$

11. If the three points that form the vertices of a triangle are (x_1,y_1), (x_2,y_2), (x_3,y_3), then it can be shown that the area of the triangle is the absolute value of $\frac{1}{2}\begin{vmatrix} x_1 & y_1 & 1 \\ x_2 & y_2 & 1 \\ x_3 & y_3 & 1 \end{vmatrix}$.
Find the area of the triangle with vertices $(3,-1)$, $(6,-3)$, and $(-2,8)$.

Graph the solution set of the linear inequalities.

12. $x + 2y \geq -10$

13. $5y - 40 < 10x$

14. Graph the solution to the system of linear inequalities.
$3x + y > 6$
$x - 2y \leq -6$

15. In the following problem, maximize the value of the objective function P with regard to the constraints supplied. Also assume $x \geq 0$ and $y \geq 0$.

$$2x + 6y \leq 18$$
$$x + 3y \leq 10$$
$$2x + y \leq 10$$
$$P = x + 2y$$

16. A company makes two mixtures of dog food nutrient supplement which it calls Regular and Prime. There are three ingredients, A, B, and C, in each mix. The table shows how much of each ingredient (in milligrams, mg) is in each gram of the food and the cost in cents for that mix. It also shows the minimum daily requirement (MDR) for each ingredient, in milligrams.

Mix	A	B	C	Cost per gram
Prime	2	5	8	6
Regular	2	3	1	3
MDR	6	12	8	

A dog owner wants to feed sufficient quantities of each supplement so that the owner's dog gets the MDR for each ingredient at the least expense. How many grams of each supplement should the owner feed the dog per day?

17. A large logging company has crews of workers and supervisors. It has found that a crew of 2 supervisors and 8 loggers log 22 trees per day, and a crew of 3 supervisors and 9 loggers log 31 trees per day. The company has 40 supervisors and 144 loggers on its payroll. What mix of crews would produce the most trees per day? (Assume that fractional parts of crews make sense.)

Form the dot product of the given vectors.

18. $[-1, 10, -3, 2], [4, -2, -2, \frac{1}{2}]$

19. $[\frac{1}{4}, 1, -3], \begin{bmatrix} -8 \\ 4 \\ -5 \end{bmatrix}$

Compute the indicated matrix products.

20. $\begin{bmatrix} 2 & -3 \\ 1 & 5 \end{bmatrix} \begin{bmatrix} 3 & -2 & 4 \\ 1 & 2 & -4 \end{bmatrix}$

21. $\begin{bmatrix} 2 & -1 & 1 \\ 0 & 2 & 6 \end{bmatrix} \begin{bmatrix} -2 & 3 \\ 1 & 1 \\ 2 & -4 \end{bmatrix}$

Find the inverse of each matrix. If the matrix does not have an inverse, state this.

22. $\begin{bmatrix} 2 & 5 \\ -3 & 0 \end{bmatrix}$

23. $\begin{bmatrix} 2 & -2 & 3 \\ 1 & 0 & -1 \\ 1 & -3 & 2 \end{bmatrix}$

24. Solve using matrix multiplication. $\begin{aligned} 3x - 3y &= 2 \\ 2x + y &= -2 \end{aligned}$

25. If $\begin{bmatrix} 2 & 5 \\ -3 & 4 \end{bmatrix} \begin{bmatrix} a & b \\ c & d \end{bmatrix} = \begin{bmatrix} 12 & 11 \\ 5 & 18 \end{bmatrix}$, find a, b, c, and d.

26. The figure shows five points and the paths that connect them. Construct a 5×5 array A that shows the paths, then compute A^2 and use it to determine the number of paths of length 2 from node 1 to node 5.

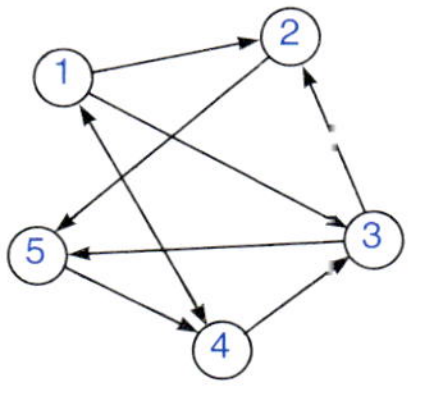

C H A P T E R 11

The Conic Sections

In this chapter we study more of analytic geometry, the area of mathematics that connects algebra and geometry. Recall that in chapter 3 we studied points, lines, and circles. We also studied parabolas in chapter 4.

In this chapter we study the parabola in more detail, and learn about the ellipse and the hyperbola.[1] The point, line, circle, parabola, ellipse, and hyperbola comprise a family of curves called the **conic sections.** This is because each curve can be found by slicing a right circular cone, as shown in figure 11–1. This fact was discovered by the Greek Menaechmus.[2]

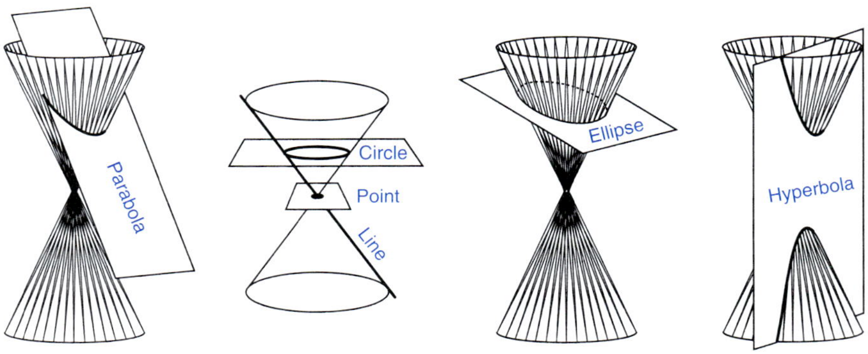

Figure 11–1

[1]The names parabola, ellipse, and hyperbola were created by the Greek Apollonius of Perga, (approximately 262 to 190 B.C.). They mean comparison, deficiency, and excess, respectively. These meanings come from the fact that these geometric figures were used by the Greeks in solving quadratic equations.

[2]Menaechmus was a student of Eudoxus, perhaps the best of the ancient Greek mathematicians. Eudoxus was in turn a student of the Platonic Academy in Athens, founded by Plato, best known for his work in philosophy. Plato has been called the "maker of mathematicians"; legend has it that above the doors of his school was inscribed the motto "Let no one ignorant of geometry enter here." This mathematical activity spanned approximately the years 400 B.C. to 300 B.C.

11–1 The parabola

A stunt person is going to jump off a 40-foot-high building, running at an estimated velocity of 4 feet per second (horizontally). How far from the base of the building will this person land?

The parabola is the mathematical object that models this and many other situations. In section 4–1 we graphed parabolas. We now define a parabola from a geometric viewpoint. *A parabola is the set of all points equidistant*[3] *from a line and a point that is not on that line.* The line is called the **directrix,** and the point is called the **focus.** Figure 11–2 illustrates this.

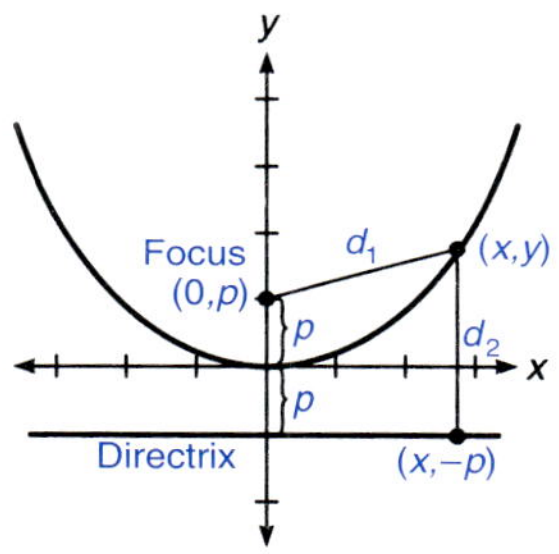

Figure 11–2

In figure 11–2 we have placed the focus at $(0,p)$, and the directrix is the line $y = -p$. Thus, $|p|$ equals half the distance from the focus to the directrix.

Let (x,y) be any point on the parabola (see the figure). Then we can proceed as follows:

$d_1 = \sqrt{x^2 + (y - p)^2}$	Apply the distance formula (section 3–1) to the points (x,y) and $(0,p)$
$d_2 = y + p$	Distance formula or examination of the figure
$d_1 = d_2$	Definition of parabola
$\sqrt{x^2 + (y - p)^2} = y + p$	Replace d_1 and d_2 by the values above
$x^2 + (y - p)^2 = (y + p)^2$	Square both members
$x^2 + y^2 - 2py + p^2 = y^2 + 2py + p^2$	Square the binomials
$x^2 = 4py$	Simplify the equation
$y = \dfrac{1}{4p}x^2$	Divide both members by $4p$, and interchange the members

Thus, we arrive at the analytic (algebraic) description of a parabola.

[3]Equidistant means "equal distance."

Parabola

$$y = \frac{1}{4p}x^2$$

is the equation of a parabola with vertex at (0,0), focus $(0,p)$, and directrix the line $y = -p$. The y-axis is an axis of symmetry.

The parabola opens up if $p > 0$ and down if $p < 0$. Note that the only intercept for an equation of this form is at the vertex, (0,0).

In section 4–1 we graphed equations of the form $y = ax^2 + bx + c$, $a \neq 0$, as a parabola, by completing the square, finding the vertex, and using linear transformations. Transformations can be applied as follows:

$$y = \frac{1}{4p}x^2$$ Basic parabola with vertex at (0,0)

$$y = \frac{1}{4p}(x - h)^2$$ Parabola translated horizontally h units

$$y = \frac{1}{4p}(x - h)^2 + k$$ Parabola translated horizontally h units and vertically k units

We can generalize the description above as follows:

Parabola (general)

$$y = \frac{1}{4p}(x - h)^2 + k$$

is the equation of a parabola with vertex at (h,k).

The focus is p units above or below the vertex, at $(h,k + p)$. The directrix is a horizontal line p units above or below the vertex, at $y = k - p$. The graph is symmetric (forms a mirror image) about the vertical line $x = h$. This line is called the **axis of symmetry.**

To graph a parabola of the form $y = \frac{1}{4p}(x - h)^2 + k$ using algebraic methods:

- The vertex is at (h,k).
- The focus is at $(h,k + p)$.
- The directrix is the line $y = k - p$.
- Compute the x-intercepts: set $y = 0$ and solve for x.
- Compute the y-intercept: set $x = 0$ and solve for y.
- The line $x = h$ is an axis of symmetry.

In general, to graph a parabola, we compute the focus, vertex, x- and y-intercepts, and directrix. We plot these values and the axis of symmetry. If we do not have enough points for a reasonably accurate graph we plot a few additional points.

Of course a modern graphing calculator can be used to draw the graph. (The steps for the TI-81 are shown throughout this chapter.) It is still useful to calculate the focus, vertex, x- and y-intercepts, and directrix.

Section 4–1 illustrated graphing parabolas. Example 11–1 A illustrates graphing a parabola using the information provided by the vertex and intercepts. It also illustrates finding the focus and directrix.

■ ***Example 11–1 A***

Graph the parabola; clearly state the x- and y-intercepts, focus, directrix, and vertex.

$y = x^2 - 6x + 4$

$y = x^2 - 6x + 9 - 9 + 4$ Complete the square (section 4–1)

$y = (x - 3)^2 - 5$

$y = (x - 3)^2 + (-5)$

Vertex: $(3,-5)$

p: $\dfrac{1}{4p} = 1$

$1 = 4p$ Multiply each member by $4p$

$\frac{1}{4} = p$ Divide each member by 4

Focus: $(3,-5 + \frac{1}{4}) = (3,-4\frac{3}{4})$

Directrix: $y = -5 - \frac{1}{4} = -5\frac{1}{4}$

Axis of symmetry: The line $x = 3$.

y-intercept: Let $x = 0$.

$$y = 0^2 - 6(0) - 4 = 4; (0,4)$$

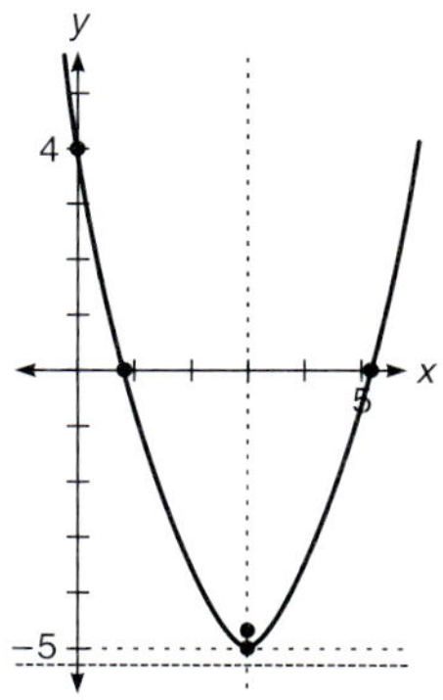

x-intercepts: Let $y = 0$.

$0 = (x - 3)^2 - 5$ Use this form for x

$5 = (x - 3)^2$

$\pm\sqrt{5} = x - 3$ Square root of both members

$3 \pm \sqrt{5} = x$ Add 3 to both members

$x \approx 0.8, 5.2; (0.8,0)(5.2,0)$

Y=	X|T	x^2	−	6	X|T	+	4	RANGE −1,6,−6,5

■

All of the development to this point has focused on equations in which the x variable is squared. A parabola also results when it is the y variable that is squared (and the x variable is not squared). In this situation, the parabola opens to the right or left instead of up or down. Rather than try to learn a new technique for this situation, we can combine an old one with what we have just seen.

If we exchange the variables x and y in an equation, then obtain important points that are part of this new equation, we can obtain important points that relate to the original equation by reversing the values in the ordered pairs.

What we are doing when we do this is creating the inverse of a given relation, studying this inverse relation, and then applying what we discover to the original relation. (See section 4–5.) These ideas are illustrated in example 11–1 B.

■ Example 11–1 B

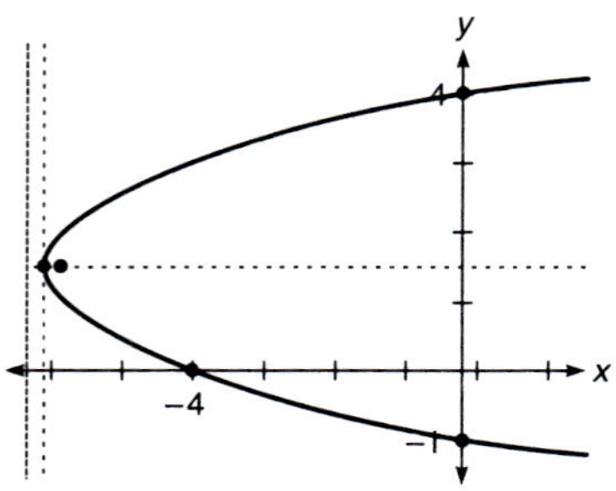

Graph the parabola $x = y^2 - 3y - 4$.

This parabola is quadratic in the variable y, not x. One way to graph it is to compute key points for its inverse, then reverse the points.

$$x = y^2 - 3y - 4$$

$$y = x^2 - 3x - 4 \qquad \text{Replace } x \text{ by } y \text{ and } y \text{ by } x\text{; This is the inverse relation to the original relation}$$

$$y = x^2 - 3x + \tfrac{9}{4} - \tfrac{9}{4} - 4 \qquad \text{Complete the square}$$

$$y = (x - \tfrac{3}{2})^2 - \tfrac{25}{4} \qquad \text{Vertex at } (\tfrac{3}{2}, -\tfrac{25}{4}) = (1\tfrac{1}{2}, -6\tfrac{1}{4})$$

As we compute key points for this inverse relation we keep a record of the reverse of these points—these are the points in the original relation.

$y = x^2 - 3x - 4$	$x = y^2 - 3y - 4$
Vertex: $(1\frac{1}{2}, -6\frac{1}{4})$	$(-6\frac{1}{4}, 1\frac{1}{2})$
p: $\dfrac{1}{4p} = 1$	
$1 = 4p$	
$\frac{1}{4} = p$	
Focus: $(1\frac{1}{2}, -6\frac{1}{4} + \frac{1}{4}) = (1\frac{1}{2}, -6)$	$(-6, 1\frac{1}{2})$
Directrix: $y = -6\frac{1}{4} - \frac{1}{4} = -6\frac{1}{2}$	$x = -6\frac{1}{2}$
Axis of symmetry: $x = 1\frac{1}{2}$	$y = 1\frac{1}{2}$
y-intercept: Let $x = 0$.	x-intercept:
$y = 0^2 - 3(0) - 4 = 4$; (0,4)	(4,0)
x-intercepts: Let $y = 0$.	y-intercepts:
$0 = x^2 - 3x - 4$	
$0 = (x - 4)(x + 1)$	
$x - 4 = 0$ or $x + 1 = 0$	
$x = 4$ or $x = -1$; (4,0) and $(-1,0)$	(0,4) and $(0,-1)$

Plotting the vertex, focus, x-intercept,and y-intercepts, as well as knowing the axis of symmetry, and that a parabola in which the y variable is squared opens horizontally, allows us to draw the graph of the original relation.

A graphing calculator cannot be used directly to graph this parabola. This is because this relation is not a function. (Section 3–5 showed that any graph which fails the ''vertical line test'' is not a function.) Algebraically this means that the equation can not be solved for y without obtaining at least two solutions. The following shows how to graph this equation with a graphing calculator.

$x = y^2 - 3y - 4$ First solve for y by completing the square (section 4–1)
$y^2 - 3y = x + 4$
$y^2 - 3y + (\frac{3}{2})^2 = x + 4 + (\frac{3}{2})^2$
$(y - \frac{3}{2})^2 = x + \frac{25}{4}$
$y - \frac{3}{2} = \pm\sqrt{x + \frac{25}{4}}$
$y = \pm\sqrt{x + \frac{25}{4}} + \frac{3}{2}$

Now graph each function $y = \sqrt{x + \frac{25}{4}} + \frac{3}{2}$ and $y = -\sqrt{x + \frac{25}{4}} + \frac{3}{2}$. Do this by entering:

[Y=] [√] [(] [X|T] [+] 6.25 [)]	$Y_1 = \sqrt{x + 6.25}$
[ENTER] [Y-VARS] [ENTER] [+] 1.5	$Y_2 = Y_1 + 1.5$
[ENTER] [(−)] [Y-VARS] [ENTER] [+] 1.5	$Y_3 = -Y_1 + 1.5$

Now use the cursor arrows to place the blinking cursor on the = in ":Y_1=√(X+6.25)" and hit [ENTER]. This turns off the graphing of Y_1. Now select [Range −7,1,−2,5] [GRAPH]. ■

Example 11–1 C illustrates finding the equation of a parabola given some of its algebraic properties.

■ *Example 11–1 C*

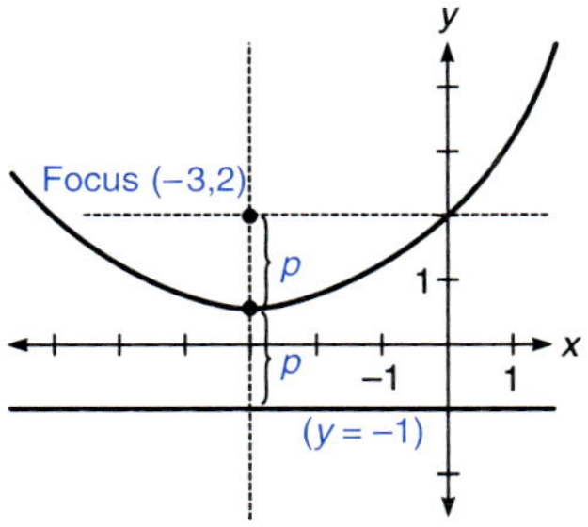

Determine the equation of a parabola with focus at (−3,2), opening vertically, and directrix at $y = -1$.

The x value of the vertex is −3, since it is directly below the focus at (−3,2).

The vertex is half-way between the focus and directrix, so its y value must be half way "between" the y values 2 and −1. This value is their average: $\frac{2 + (-1)}{2} = \frac{1}{2}$.

Thus the vertex is at $(-3,\frac{1}{2})$. The distance between the focus and vertex is $|p|$, so $|p| = 1\frac{1}{2}$. Since the parabola opens upward $p > 0$, so $p = 1\frac{1}{2}$.

$y = \frac{1}{4p}(x - h)^2 + k$ Basic equation of a parabola with vertex at (h,k)

$y = \frac{1}{4(1\frac{1}{2})}(x + 3)^2 + \frac{1}{2}$ $p = 1\frac{1}{2}$, $(h,k) = (-3,\frac{1}{2})$

$y = \frac{1}{6}(x + 3)^2 + \frac{1}{2}$ $\frac{1}{4(1\frac{1}{2})} = \frac{1}{4(\frac{3}{2})} = \frac{1}{6}$

$6y = (x + 3)^2 + 3$ Multiply each member by 6

$6y = x^2 + 6x + 12$ $(x + 3)^2 + 3 = (x^2 + 6x + 9) + 3$

$y = \frac{1}{6}x^2 + x + 2$ Divide each member by 6 ■

An important property of parabolas

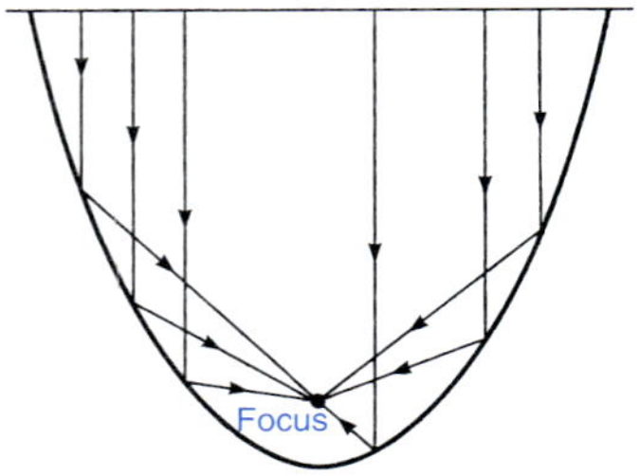

Figure 11–3

A parabola has the important property that if a mirror is shaped in the form of a parabola, then parallel light rays entering the reflector parallel to the axis of symmetry are reflected to the focus (see figure 11–3). This is the principle behind dish antennas used to receive TV signals from space satellites. The same principle, in reverse, is why the reflecting portion of a flashlight or automobile headlight is parabolic.

Example 11–1 D illustrates how to find the equation of a parabola given certain information about its dimensions.

■ *Example 11–1 D*

The width of the parabolic mirror determines how much light or radio waves are collected. Suppose a parabolic mirror is constructed so that the focus is 6″ from the vertex. The mirror is to be 4′ across. Determine the height of the mirror, d in the figure.

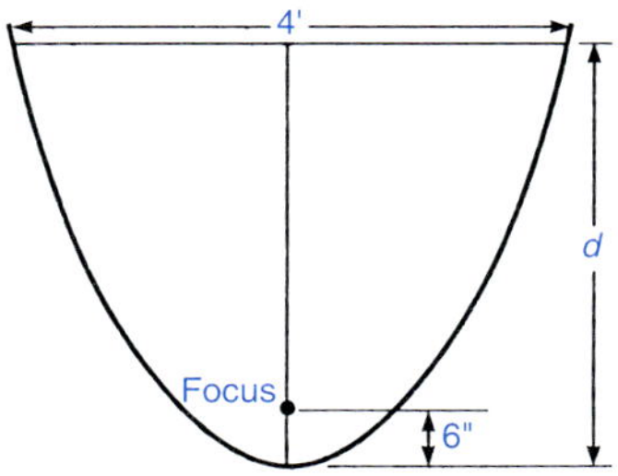

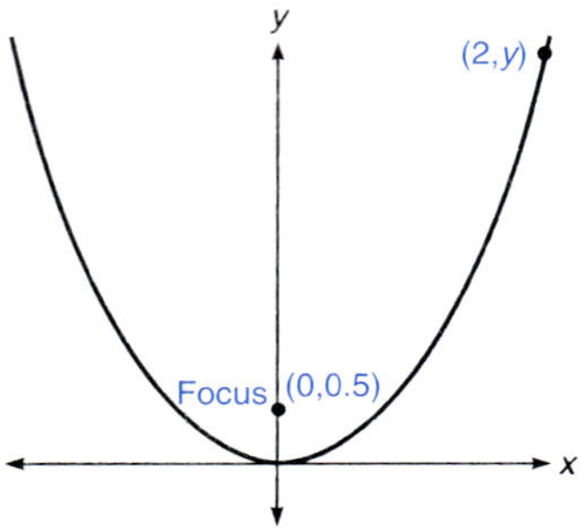

If we put this figure in a coordinate system as shown, we have the vertex at (0,0) and $p = \frac{1}{2}$ ft.

$$y = \frac{1}{4p}x^2 \quad \text{Basic parabola with vertex at the origin}$$

$$y = \frac{1}{4(\frac{1}{2})}x^2 \quad p = \tfrac{1}{2}$$

$$y = \tfrac{1}{2}x^2 \quad \text{The equation of the antenna}$$

If we find the value of y shown in the coordinate $(2,y)$ we will have the value of d we desire. For this we insert the value $x = 2$ into our equation.

$$y = \tfrac{1}{2}(2^2) \quad \text{Find } y \text{ in } (2,y) \text{ by replacing } x \text{ by 2 in the equation}$$

$$y = 2 \quad \text{The distance } d \text{ is 2 feet}$$ ■

The path of a falling object

Near the surface of the earth an object with no initial vertical velocity falls a distance $d = 16t^2$, where t is time in seconds. For instance in $\frac{1}{2}$ second an object will fall $d = 16(\frac{1}{2})^2 = 4$ feet. If the object has a constant horizontal velocity v, its trajectory will be a parabola. This can be seen as follows (see

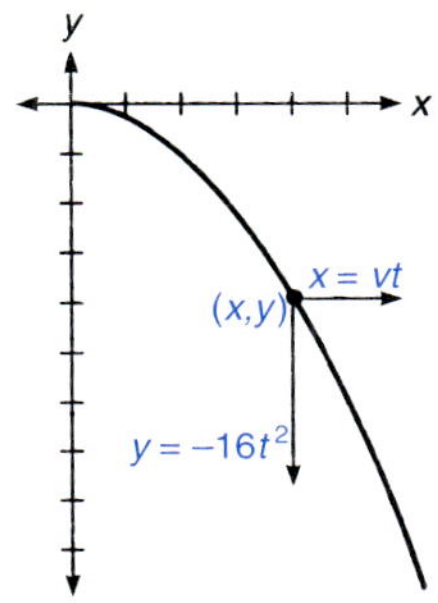

Figure 11–4

figure 11–4). The horizontal distance covered is the product of rate and time, so it is $x = vt$ (v in feet per second and t in seconds), and the vertical distance is $y = -16t^2$ (we use the negative value to indicate the object is falling). Thus,

Horizontal motion	Vertical motion
$x = vt$	
$x^2 = v^2t^2$	$y = -16t^2$
$\dfrac{x^2}{v^2} = t^2$	$-\dfrac{y}{16} = t^2$

If we replace t^2 in the horizontal motion equation by $-\dfrac{y}{16}$ we obtain $\dfrac{x^2}{v^2} = -\dfrac{y}{16}$, so $y = -\dfrac{16}{v^2}x^2$. (This is an example of the method of substitution for expression as illustrated in section 3–2.) For a given value of v this equation is a parabola. Example 11–1 E illustrates a use of this information.

■ ***Example 11–1 E***

An object with horizontal velocity 2 ft/sec begins to fall. How far will it have fallen when it has moved 5 feet horizontally?

$$v = 2 \text{ ft/sec}, x = 5: \quad y = -\frac{16}{v^2}x^2$$

$$y = -\frac{16}{2^2} \cdot 5^2 = -4(25) = -100$$

Thus, the object will have fallen 100 feet when it has moved 5 feet horizontally. ■

Mastery points

Can you

- Graph an equation of the form $y = ax^2 + bx + c$ or $x = y^2 + by + c$ as a parabola?
- Find the equation of a parabola, given certain initial conditions?

Exercise 11–1

Graph each parabola; clearly state the focus and directrix. State the intercepts and vertex where not clear from the graph.

1. $y = -2x^2$ **2.** $y = \frac{1}{2}x^2$ **3.** $y = 3x^2$ **4.** $y = -\frac{1}{3}x^2$
5. $y = x^2 - 4$ **6.** $y = 2x^2 - 4$ **7.** $y = -x^2 + 1$ **8.** $y = x^2 - 9$
9. $y = 2(x - 3)^2$ **10.** $y = 3(x + 1)^2$ **11.** $y = \frac{1}{2}(x + 2)^2$ **12.** $y = -\frac{1}{2}(x - 2)^2$
13. $y = -(x + 1)^2$ **14.** $y = (x + 2)^2 + 3$ **15.** $y = 3(x - \frac{1}{2})^2 - 1$ **16.** $y = 2(x - 1)^2 - 1$
17. $y = -2(x + 2)^2 + 1$ **18.** $y = x^2 - 4x - 3$ **19.** $y = x^2 + 5x + 15$ **20.** $y = -x^2 + 5x + 10$
21. $y = -x^2 + 6x - 7$ **22.** $y = x^2 - 6x + 9$ **23.** $y = x^2 + 2x + 1$ **24.** $y = 2x^2 + 5x - 3$
25. $y = 2x^2 - x - 3$ **26.** $y = -3x^2 - 10x + 8$ **27.** $y = -x^2 + 16$ **28.** $y = 9 - x^2$
29. $y = x^2 - 3x - 5$ **30.** $y = x^2 + 4x + 1$ **31.** $y = 2x^2 - 6x - 1$ **32.** $y = 3x^2 + 9x + 11$
33. $y = -3x^2 - 4x + 7$ **34.** $y = -x^2 + 5x + 10$

Graph each parabola; state the focus and directrix. State the intercepts and vertex where not clear from the graph.

35. $x = y^2 - 7y - 8$ **36.** $x = y^2 + 3y - 18$ **37.** $x = y^2 - 9$
38. $x = y^2 - 4y + 4$ **39.** $x = -y^2 + y + 2$ **40.** $x = -y^2 - 4y + 8$

Find the equation of a parabola with the given properties.

41. focus: $(2,-3)$; directrix: $y = -6$
42. focus: $(-1,4)$; directrix: $y = 0$
43. focus: $(-3,-1)$; directrix: $y = 2$
44. focus: $(0,2)$; directrix: $y = 5$
45. vertex: $(3,-1)$; directrix: $y = -3$
46. vertex: $(-2,\frac{1}{3})$; directrix: $y = 1$
47. focus: $(0,3)$; vertex: $(0,0)$
48. focus: $(-4,2)$; vertex: $(-4,4)$
49. vertex: $(3,-1)$; x-intercepts: 2, 4
50. vertex: $(-2,-3)$; y-intercept: -1; opens vertically

Refer to the figure for problems 51–56. Assume all values are in inches.

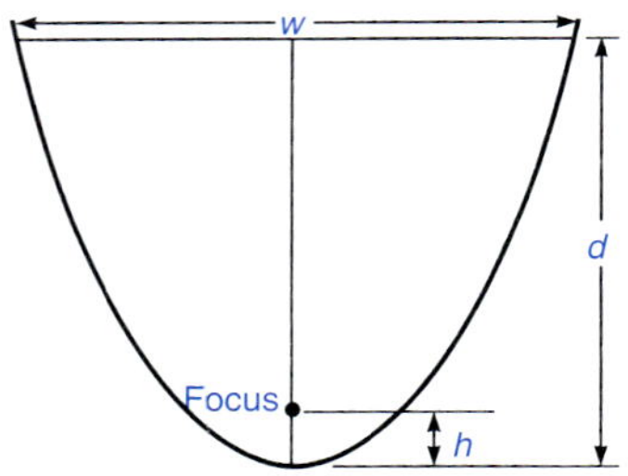

51. $h = 8$, $d = 24$; find w.
52. $h = 3$, $d = 9$; find w.
53. $h = 5$, $w = 12$; find d.
54. $h = 20$, $w = 50$; find d.
55. $w = 24$, $d = 20$; find h.
56. $w = 30$, $d = 34$; find h.

57. The cable on a suspension bridge takes the form of a parabola. The figure shows such a bridge, which is 250 feet long. At a it's height is 30 feet. At b it is 45 feet. Find an equation that would describe this parabola assuming the low point of the cable is at the origin.

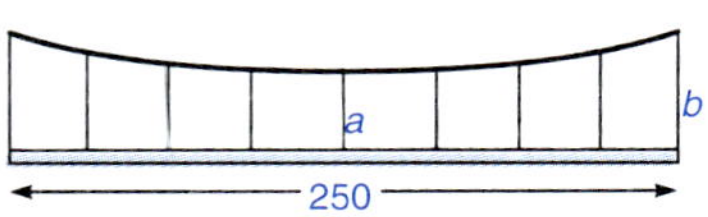

58. The path of an object launched near the surface of the earth at a velocity of $8\sqrt{2}$ feet per second and at an angle of 30° would follow the path described by $y = \frac{1}{\sqrt{3}}x - \frac{1}{6}x^2$, $x \geq 0$. Graph this parabola for those values for which $x \geq 0$ and $y \geq 0$. Label the intercepts and vertex.

59. As in problem 58, if the angle at which the object were launched were 45°, the trajectory of the object would be described by $y = x - \frac{1}{4}x^2$, $x \geq 0$. Graph this parabola for those values for which $x \geq 0$ and $y \geq 0$. Label the intercepts and vertex.

Use the formula $y = -\frac{16}{v^2}x^2$ for problems 60 through 64.

60. A stunt person is going to jump off a 40-foot-high building, running at an estimated horizontal velocity v of 4 feet per second. How far from the base of the building will this person land?

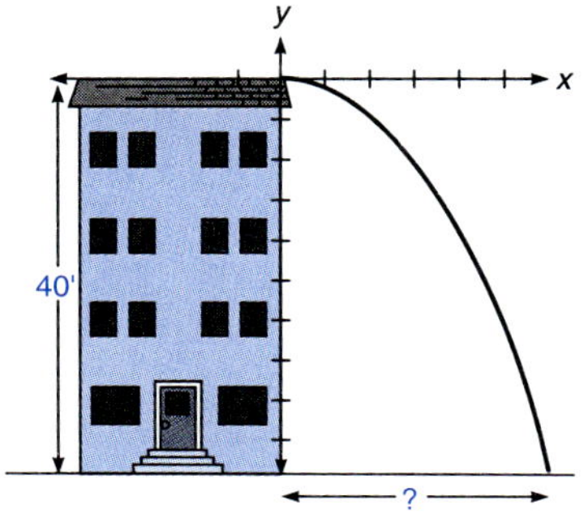

61. If the same individual (problem 60) runs at 8 feet per second how far from the building will this person land? The horizontal velocity doubled (4 ft/s to 8 ft/s); did the horizontal distance traveled double also?

62. What would be the required velocity to land 10 feet from the base of the building of problem 60?

63. What would be the required velocity to land 4 feet from the base of the building of problem 60?

64. For a fixed horizontal velocity, does doubling the height of the fall also double the horizontal distance of the landing point from the base of the building?

Skill and review

1. Solve $\frac{x^2}{4} + 3y^2 = 1$ for y.
2. Multiply the matrices $\begin{bmatrix} -2 & 3 & 0 \\ 1 & 5 & -3 \\ 4 & 2 & 6 \end{bmatrix}\begin{bmatrix} -2 & 3 \\ 1 & 5 \\ -3 & 2 \end{bmatrix}$.
3. Solve the system
$$\begin{aligned} 2x + 3y - z &= 5 \\ 4x - 6y + z &= -4 \\ 2x + 6y - 3z &= 11 \end{aligned}$$
4. Solve the system $\begin{matrix} 2x - y \le 4 \\ x + y \ge 3 \end{matrix}$.
5. Compute $\log_5 10$ to 4 decimal places.
6. Solve $\log(2x - 1) + \log(3x + 1) = \log 4$.
7. Solve $|2x - 5| < 10$.

11–2 The ellipse

An asteroid has been found orbiting the sun with an elliptical orbit such that the sun is at one focus, and the asteroid is 200 million miles from the sun at its farthest point, and 100 million miles from the sun at its closest point. Find an equation that describes the path of the asteroid.

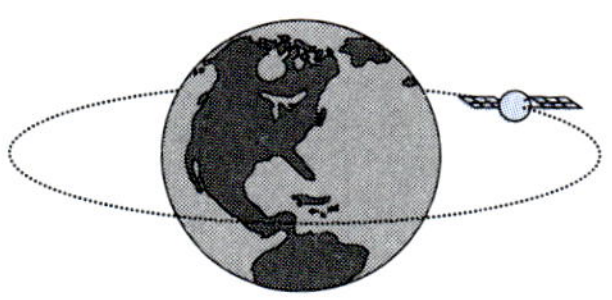

An ellipse is a "flattened circle." It is the path that the earth takes around the sun and that satellites take around the earth. As stated in this problem it is the path that an asteroid takes when it goes around the sun. This fact was first discovered by the astronomer Johann Kepler (1571–1630).

Like the parabola the ellipse can be defined from a geometrical viewpoint. Fix two points, called the foci (c and $-c$ in figure 11–5). Now find all points such that the sum of the two distances from that point to each focus ($d_1 + d_2$ in the figure) is a constant.

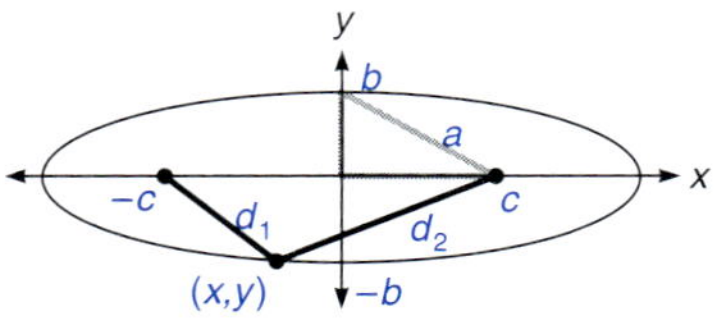

Figure 11–5

Figure 11–5 shows an ellipse placed so its center is at the origin. The foci are placed on the x-axis, equidistant from the origin; they are at $(-c,0)$ and $(c,0)$. The point (x,y) represents any point on the ellipse. The y-intercept is labeled b. We call the distance from the y-intercept to the focus a. The right triangle shown illustrates that $a^2 = b^2 + c^2$. We can develop an analytic description of this ellipse as follows.

The sum of d_1 and d_2 is a constant. If we consider (x,y) to be at $(0,b)$ (one of the y-intercepts) we can see that this constant is $2a$. We thus proceed algebraically from the statement $d_1 + d_2 = 2a$.

$d_1 = \sqrt{(x-(-c))^2 + (y-0)^2}$	Distance formula with $(-c,0)$ and (x,y)
$\quad = \sqrt{(x+c)^2 + y^2}$	
$d_2 = \sqrt{(x-c)^2 + (y-0)^2}$	Distance formula with $(c,0)$ and (x,y)
$\quad = \sqrt{(x-c)^2 + y^2}$	
$d_1 + d_2 = 2a$	Definition of ellipse
$\sqrt{(x+c)^2 + y^2} + \sqrt{(x-c)^2 + y^2} = 2a$	Replace d_1 and d_2 in $d_1 + d_2 = 2a$ by the values above

Appendix A describes the algebra that shows that this equation leads to the relation $\frac{x^2}{a^2} + \frac{y^2}{b^2} = 1$, where we define b so that $b^2 = c^2 - a^2$.

By letting x and then y be zero we find that the y-intercept is at b (which we knew already), and the x-intercept is at a. Also, solving for c in $a^2 + b^2 = c^2$, we find that $c = \sqrt{a^2 - b^2}$. The **foci** in this development are at $x = \pm c$. By our definition of a, we guaranteed that $a > b$. *It can be shown that if $a < b$ the foci are on the y-axis and $c = \sqrt{b^2 - a^2}$.* We thus obtain the following result.

> **Ellipse**
>
> The graph of an equation of the form
>
> $$\frac{x^2}{a^2} + \frac{y^2}{b^2} = 1$$
>
> is an ellipse whose center is at the origin.
>
> - If $a > b$, then $c = \sqrt{a^2 - b^2}$ and the foci are at $(\pm c,0)$.
> - If $a < b$, then $c = \sqrt{b^2 - a^2}$ and the foci are at $(0,\pm c)$.

A **line segment** is a part of a line with a finite length. The line segment along the axis on which the foci lie is called the **major axis;** the other axis is the **minor axis** (see figure 11–6). *The length of the major axis is the greater of 2a and 2b, and the length of the minor axis is the lesser of these two values.*

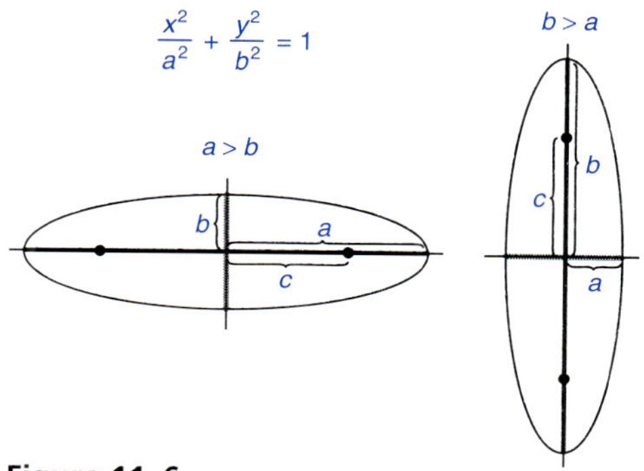

Figure 11–6

> **To graph an ellipse in the form $\dfrac{x^2}{a^2} + \dfrac{y^2}{b^2} = 1$**
>
> - The major axis is along the x-axis if $a > b$ and along the y-axis if $b > a$.
> - Plot the end points of the major and minor axes. These are also the x- and y-intercepts.
> - $c = \sqrt{a^2 - b^2}$ if $a > b$ or $c = \sqrt{b^2 - a^2}$ if $b > a$.
> - The foci are c units from the center along the major axis.

It is not possible to graph an ellipse whose equation is given in the form here by simply entering it into a graphing calculator. A method for using the graphing calculator is discussed at the end of this section.

Example 11–2 A illustrates graphing ellipses with center at the origin.

■ Example 11–2 A

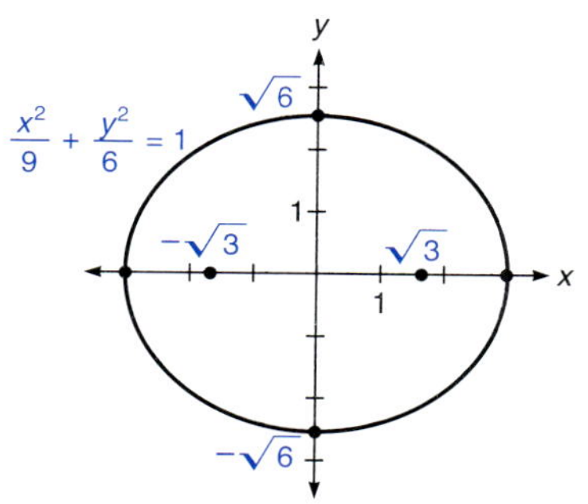

Graph each ellipse.

1. $\dfrac{x^2}{9} + \dfrac{y^2}{6} = 1$

y-intercepts: Let $x = 0$.

$$\frac{0^2}{9} + \frac{y^2}{6} = 1$$
$$\frac{y^2}{6} = 1$$
$$y^2 = 6$$
$$y = \pm\sqrt{6} \approx \pm 2.4$$

x-intercepts: Let $y = 0$.

$$\frac{x^2}{9} + \frac{0^2}{6} = 1$$
$$\frac{x^2}{9} = 1$$
$$x^2 = 9$$
$$x = \pm 3$$

$$c = \sqrt{a^2 - b^2} = \sqrt{9 - 6} = \sqrt{3}$$

Since the x-intercepts are farther apart than the y-intercepts, the x-axis contains the major axis. This is therefore where the foci are.

2. $8x^2 + y^2 = 10$

To determine a, b, and c we require that the equation be in the form $\frac{x^2}{a^2} + \frac{y^2}{b^2} = 1$. *The right member must be 1.*

$$\frac{4x^2}{5} + \frac{y^2}{10} = 1 \quad \text{Divide each term by 10 so the right member becomes 1}$$

$$\frac{x^2}{\frac{5}{4}} + \frac{y^2}{10} = 1 \quad \frac{(4x^2) \div 4}{5 \div 4} = \frac{x^2}{\frac{5}{4}}$$

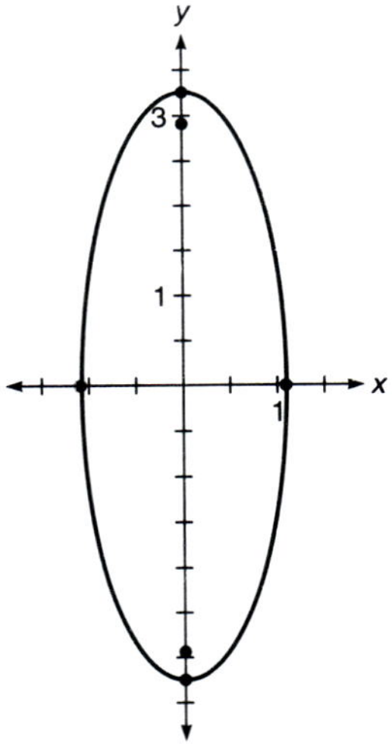

Thus, $a^2 = \frac{5}{4}$ and $b^2 = 10$.

y-intercepts:

$$\frac{0^2}{\frac{5}{4}} + \frac{y^2}{10} = 1$$

$$y^2 = 10$$

$$y = \pm\sqrt{10} \approx \pm 3.2$$

x-intercepts:

$$\frac{x^2}{\frac{5}{4}} + \frac{0^2}{10} = 1$$

$$x^2 = \frac{5}{4}$$

$$x = \pm\sqrt{\frac{5}{4}} = \pm\frac{\sqrt{5}}{2} \approx \pm 1.1$$

$$c = \sqrt{b^2 - a^2}$$

$$c = \sqrt{10 - \frac{5}{4}} = \sqrt{\frac{40}{4} - \frac{5}{4}} = \sqrt{\frac{35}{4}} = \frac{\sqrt{35}}{2} \approx 2.96$$

The y-axis contains the major axis since $b > a$. ■

A more general form of the equation of an ellipse takes horizontal and vertical translations into account.

> **General form of the equation of an ellipse**
>
> $$\frac{(x-h)^2}{a^2} + \frac{(y-k)^2}{b^2} = 1$$

This is an ellipse that is translated h units horizontally and k units vertically. Thus its *center* is at (h,k) instead of $(0,0)$. In this form, the values of a, b, and c are distances from the center (h,k).

> **To graph any ellipse**
>
> - The major axis is parallel to the x-axis if $a > b$ and parallel to the y-axis if $b > a$.
> - Plot the end points of the major axis and minor axis a and b units, as appropriate, from (h,k).
> - $c = \sqrt{a^2 - b^2}$ if $a > b$ or $c = \sqrt{b^2 - a^2}$ if $b > a$.
> - The foci are c units from the center (h,k) along the major axis.

This is illustrated in example 11–2 B.

■ *Example 11–2 B*

Graph the ellipse $\frac{(x+2)^2}{16} + \frac{(y-1)^2}{9} = 1$. Label the points at the end of the major and minor axes, center, and foci.

$$\frac{(x-(-2))^2}{16} + \frac{(y-1)^2}{9} = 1$$ Rewrite in the general form

Center: $(-2,1)$ $(h,k) = (-2,1)$

Since $16 > 9$ the major axis is parallel to the x-axis.

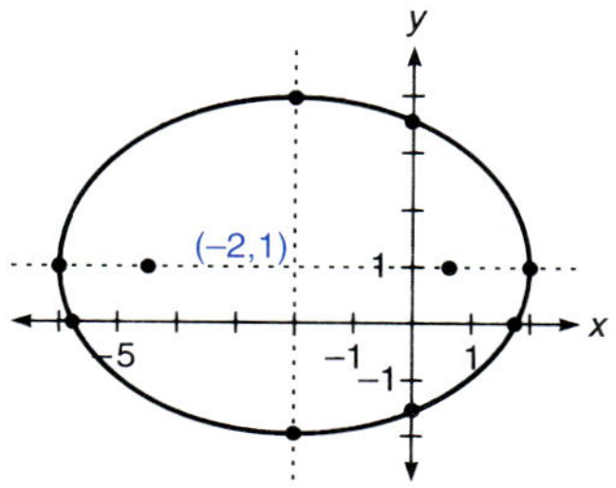

End points of the major axis: $a = \sqrt{16} = 4$, so the end points of the major axis are 4 units to the right and left of the center $(-2,1)$. These are at $(-2 \pm 4,1)$ or $(2,1)$ and $(-6,1)$.

End points of the minor axis: $b = \sqrt{9} = 3$, so there are points 3 units above and below the center. These are $(-2,1 \pm 3)$ or $(-2,4)$ and $(-2,-2)$.

Foci: $c = \sqrt{16-9} = \sqrt{7}$, and the major axis is parallel to the x-axis, so the foci are $\pm\sqrt{7}$ right and left of the center. These are $(-2 \pm \sqrt{7},1)$ or $(-2-\sqrt{7},1)$ and $(-2+\sqrt{7},1)$, or about $(-4.6,1)$ and $(0.6,1)$. ■

By completing the square we can put any equation of the form

$$Ax^2 + Bx + Cy^2 + Dy + F = 0$$

into the general form of an ellipse *if A and B both have the same sign.*[4] If the right side is positive we then have the equation of an ellipse. Example 11–2 C illustrates this procedure.

■ *Example 11–2 C*

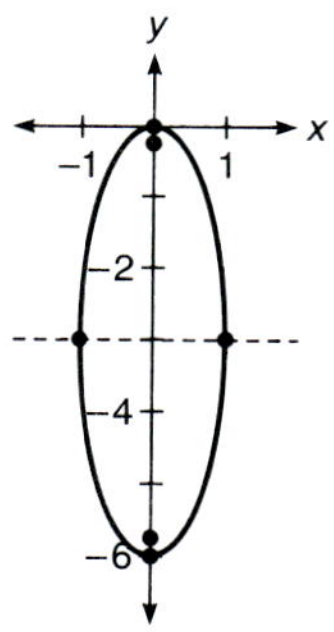

Convert each equation into the standard form of the equation of an ellipse. Identify the points that terminate the major and minor axes and the foci, then graph.

1. $9x^2 + y^2 + 6y = 0$

$9x^2 + y^2 + 6y + 9 = 9$ Complete the square on y

$9x^2 + (y+3)^2 = 9$

$x^2 + \frac{(y+3)^2}{9} = 1$ Divice each term by 9

Center: $(0,-3)$, $a = 1$, $b = \sqrt{9} = 3$

The major axis is parallel to the y-axis since $b > a$.

End points of major axis: $(0,-3 \pm 3)$ or $(0,-6)$ and $(0,0)$

End points of minor axis: $(0 \pm 1,-3)$ or $(-1,-3)$ and $(1,-3)$

Foci: (Along the major (vertical) axis)

$c = \sqrt{9-1} = \sqrt{8} = 2\sqrt{2}$

Foci are $(0,-3 \pm 2\sqrt{2}) \approx (0,-5.8)$ and $(0,-0.2)$.

[4]If A and B have opposite signs, the figure is a hyperbola, covered in section 11–3.

2. $x^2 + 2x + 3y^2 - 12y - 5 = 0$

$$x^2 + 2x + 3(y^2 - 4y) = 5$$
$$x^2 + 2x + 1 + 3(y^2 - 4y + 4) = 5 + 1 + 3(4)$$
$$(x + 1)^2 + 3(y - 2)^2 = 18$$
$$\frac{(x + 1)^2}{18} + \frac{(y - 2)^2}{6} = 1$$

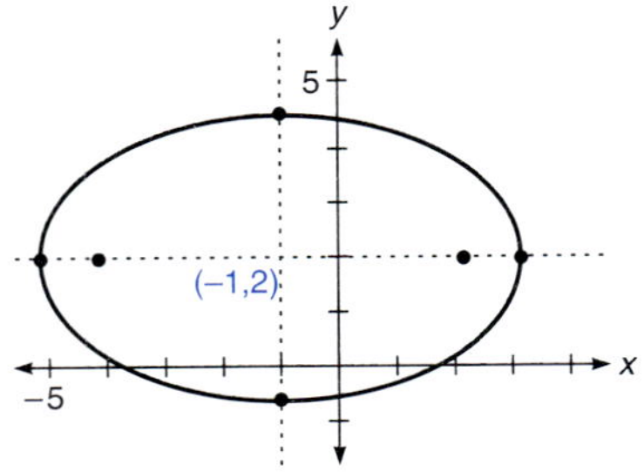

Center: $(-1,2)$, $a = \sqrt{18} = 3\sqrt{2}$, $b = \sqrt{6}$

Major axis: parallel to x-axis because $a > b$.

End points of major axis: $a = 3\sqrt{2}$; $(-1 \pm 3\sqrt{2},2) \approx (-5.2,2)$ and $(3.2,2)$

End points of minor axis: $b = \sqrt{6}$; $(-1,2 \pm \sqrt{6}) \approx (-1,-0.4)$ and $(-1,4.4)$

Foci: $c = \sqrt{18 - 6} = \sqrt{12} = 2\sqrt{3}$; $(-1 \pm 2\sqrt{3},2) \approx (-4.5,2)$ and $(2.5,2)$. ■

To graph any equation that is in terms of rectangular coordinates (x and y) on a graphing calculator, the relation (equation) must be solved for y. This is how we can use a graphing calculator to graph ellipses. Example 11–2 D illustrates.

■ *Example 11–2 D*

Graph the ellipse on a graphing calculator.

$\frac{(x + 2)^2}{16} + \frac{(y - 1)^2}{9} = 1$ (example 11–2 B)

$$\frac{(y - 1)^2}{9} = 1 - \frac{(x + 2)^2}{16}$$
$$(y - 1)^2 = 9\left(1 - \frac{(x + 2)^2}{16}\right)$$
$$y - 1 = \pm 3\sqrt{1 - \frac{(x + 2)^2}{16}}$$
$$y = 1 \pm 3\sqrt{1 - \frac{(x + 2)^2}{16}}$$

We will store the expression $3\sqrt{1 - \frac{(x + 2)^2}{16}}$ in Y_1, then let $Y_2 = 1 + Y_1$, and $Y_3 = 1 - Y_1$. We also must turn off Y_1, so it will not be graphed.

[Y=] 3 [√] [(] 1 [−] [(] [X|T] [+] 2 [)] [x^2] [÷] 16 [)] [ENTER]

1 [+] [Y-VARS] [ENTER] [ENTER]

1 [−] [Y-VARS] [ENTER]

Now use the up and down arrow keys to position the cursor on the "=" in "Y_1=," then select [ENTER]. This turns off Y_1 so it will not be graphed.

[RANGE −7,3,−3,5]

The graph appears to have holes at each end. This is due to the inherent inaccuracies of the calculator.

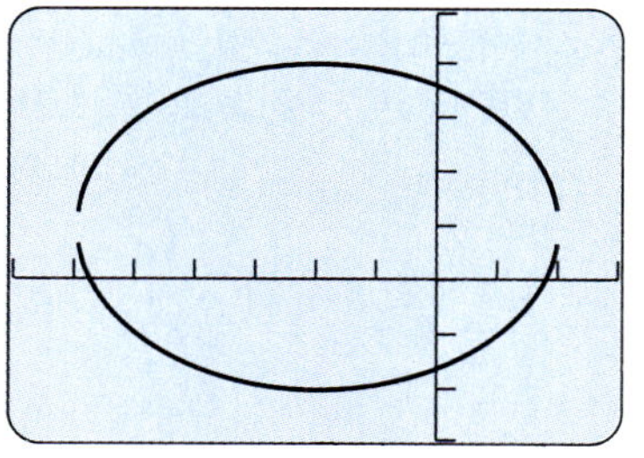

■

There are times when we wish to find the equation of an ellipse, given information such as the foci or intercepts, as illustrated in example 11–2 E.

■ **Example 11–2 E**

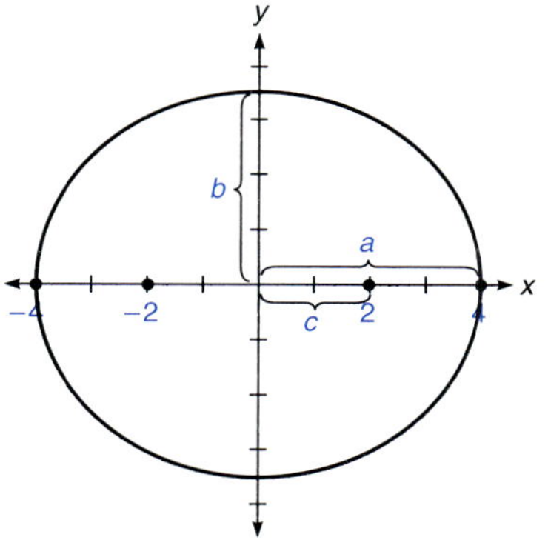

In a certain ellipse the x-intercepts are at ± 4 and the foci are on the x-axis at ± 2. Find the equation of the ellipse.

The major axis is the x-axis since the foci are on this axis. Therefore, $a = 4$ and $c = 2$.

$$\begin{aligned} c &= \sqrt{a^2 - b^2} \\ 2 &= \sqrt{4^2 - b^2} \\ 4 &= 16 - b^2 \\ b^2 &= 12 \end{aligned}$$

We put these values into the general equation $\frac{x^2}{a^2} + \frac{y^2}{b^2} = 1$ to obtain

$$\frac{x^2}{16} + \frac{y^2}{12} = 1.$$

■

One way to actually draw an ellipse is to put two tacks in a drawing board, put a loop of string around the tacks, and draw the figure with the string stretched taught (figure 11–7). The tacks are at the foci, and the constant sum depends on the length of the string and the distance between the foci (tacks). Example 11–2 F illustrates this idea.

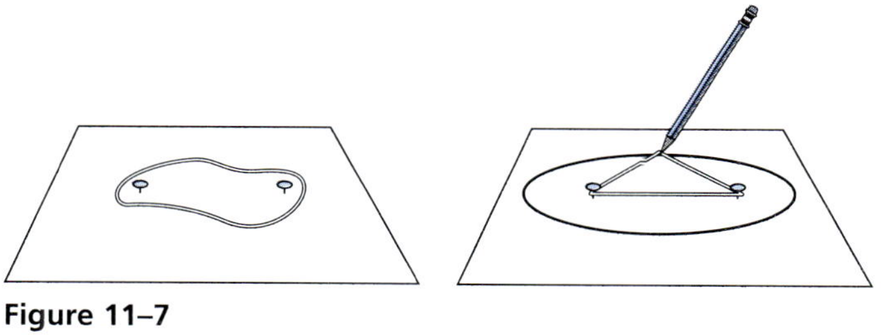

Figure 11–7

■ Example 11–2 F

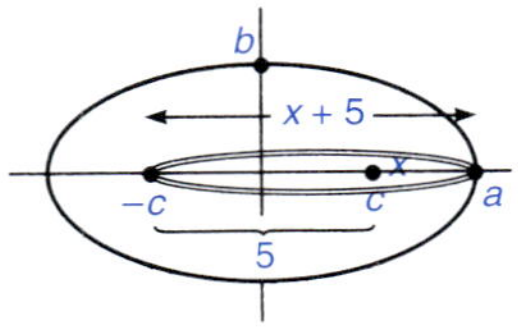

Two tacks are put in a board (as in figure 11–7) 5 inches apart and a loop of string with length 12 inches is looped around the tacks. The ellipse is drawn. Find an equation that describes the ellipse, assuming the center of the ellipse is at the origin.

We can find the value of a, the x-intercept, as follows. Consider the loop of string shown, stretched around the left focus and the point a. We can see that the length of the string is $2(x + 5)$, where x is the distance between a and c. Since the length of the string is 12″, we solve for x:

$$2(x + 5) = 12$$
$$x = 1$$

Since c is $5 \div 2 = 2.5$, then $a = 2.5 + 1 = 3.5$.

$$c = \sqrt{a^2 - b^2}$$
$$2.5 = \sqrt{3.5^2 - b^2}$$
$$b^2 = 6$$

Plugging the values of $a^2 = 12.25$ and $b^2 = 6$ into $\frac{x^2}{a^2} + \frac{y^2}{b^2} = 1$ we obtain

$$\frac{x^2}{12\frac{1}{4}} + \frac{y^2}{6} = 1$$
$$\frac{x^2}{\frac{49}{4}} + \frac{y^2}{6} = 1$$
$$\frac{4x^2}{49} + \frac{y^2}{6} = 1$$ ■

Mastery points

Can you

- Graph an ellipse when given its equation in the form $\frac{x^2}{a^2} + \frac{y^2}{b^2} = 1$ or the general form $\frac{(x-h)^2}{a^2} + \frac{(y-k)^2}{b^2} = 1$?
- Convert certain equations, that are quadratic in both variables x and y into the general form $\frac{(x-h)^2}{a^2} + \frac{(y-k)^2}{b^2} = 1$?
- Find the equation of an ellipse given certain conditions?

Exercise 11–2

Graph the equation. State the intercepts and foci where not clear from the graph.

1. $\frac{x^2}{16} + \frac{y^2}{9} = 1$ **2.** $\frac{x^2}{12} + \frac{y^2}{9} = 1$ **3.** $\frac{x^2}{9} + \frac{y^2}{25} = 1$ **4.** $\frac{x^2}{4} + \frac{y^2}{5} = 1$

5. $\frac{x^2}{9} + \frac{y^2}{4} = 1$ **6.** $\frac{x^2}{16} + \frac{y^2}{25} = 1$ **7.** $\frac{4x^2}{5} + y^2 = 1$ **8.** $x^2 + \frac{25y^2}{4} = 1$

9. $\frac{x^2}{16} + y^2 = 1$ **10.** $x^2 + \frac{y^2}{9} = 1$ **11.** $\frac{x^2}{49} + \frac{y^2}{25} = 1$ **12.** $\frac{x^2}{121} + \frac{y^2}{36} = 1$

Graph the equation. State the points at the end of the major and minor axes, center, and foci where not clear from the graph.

13. $\frac{(x+3)^2}{4} + \frac{(y-1)^2}{16} = 1$ **14.** $\frac{(x+1)^2}{4} + \frac{(y-2)^2}{9} = 1$ **15.** $\frac{(x-2)^2}{25} + \frac{(y+3)^2}{9} = 1$

16. $\frac{(x-2)^2}{4} + (y+1)^2 = 1$ **17.** $\frac{(x-1)^2}{36} + (y+2)^2 = 1$ **18.** $(x-3)^2 + \frac{(y+1)^2}{36} = 1$

19. $\frac{(x+3)^2}{9} + \frac{y^2}{25} = 1$ **20.** $\frac{(x-2)^2}{100} + \frac{(y-3)^2}{49} = 1$

Convert each equation into standard form if necessary. Then graph the equation, stating the points at the end of the major and minor axes, center, and foci where not clear from the graph.

21. $x^2 + 3y^2 = 27$ **22.** $3x^2 + 4y^2 = 12$ **23.** $36x^2 + 9y^2 = 324$ **24.** $16x^2 + y^2 = 64$
25. $9x^2 + 2y^2 = 18$ **26.** $8x^2 + y^2 = 16$ **27.** $6x^2 + 3y^2 = 27$ **28.** $4x^2 + y^2 = 4$
29. $x^2 + 9y^2 = 9$ **30.** $9x^2 + 4y^2 = 36$ **31.** $12x^2 + 6y^2 = 12$ **32.** $5x^2 + 3y^2 = 15$
33. $4x^2 + 5y^2 = 5$ **34.** $3x^2 + y^2 = 3$ **35.** $x^2 + 2y^2 - 8y = 0$ **36.** $2x^2 - 4x + y^2 = 9$
37. $4x^2 - 4x + 8y^2 + 48y = -57$ **38.** $2x^2 - 8x + y^2 = 0$ **39.** $x^2 - 6x + 2y^2 + 20y = 1$
40. $x^2 + 2x + 2y^2 + 12y + 1 = 0$ **41.** $x^2 + 2y^2 + 8 = 0$ **42.** $9x^2 + 4y^2 = 0$
43. $4x^2 - 8x + 9y^2 + 36y + 4 = 0$ **44.** $x^2 + 4x + 4y^2 - 8y + 4 = 0$ **45.** $16x^2 + 25y^2 + 100y = 300$
46. $2x^2 + 4x + 8y^2 = 30$

Find the equation of the ellipse with the required properties in problems 47–52.

47. foci: $(-2,0)$ and $(2,0)$; one y-intercept at 3
48. foci: $(-3,0)$ and $(3,0)$; one x-intercept at 5
49. foci: $(0,4)$ and $(0,-4)$; one y-intercept at 8
50. foci: $(0,1)$ and $(0,-1)$; one x-intercept at 1
51. x-intercepts: $(\pm 3,0)$; y-intercepts: $(\pm 2,0)$
52. x-intercepts: $(\pm 2,0)$; y-intercepts: $(\pm 4,0)$

53. Two tacks are put in a board 4 inches apart and a string tied in a loop with length 10 inches is looped around the tacks. The ellipse is drawn. Find an equation that describes the ellipse.

54. It is desired to construct an ellipse with height (along the minor axis) of 4 feet and length (along the major axis) of 6 feet. (See the figure.) Find out where to place the two tacks for the foci and how long to make the length of string.

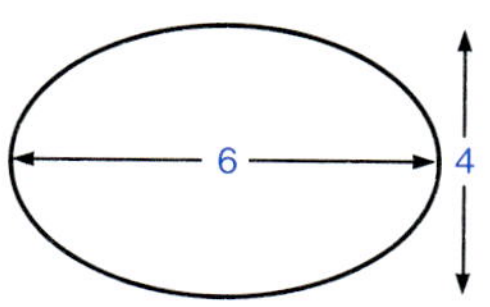

55. An asteroid has been found orbiting the sun with an elliptical orbit such that the sun is at one focus, and the asteroid is 200 million miles from the sun at its farthest point, and 100 million miles from the sun at its closest point. Find an equation that describes this orbit, using the values 200 and 100, as shown in the figure.

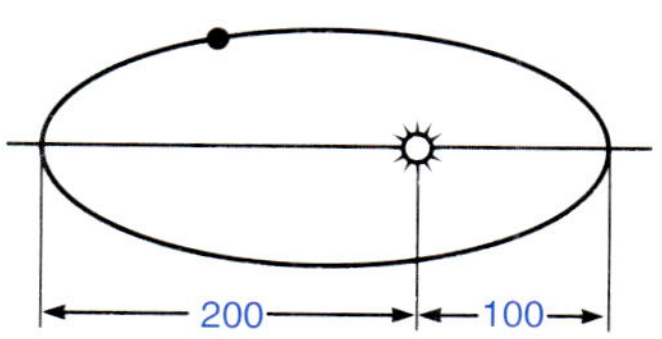

56. (Refer to problem 55.) Find an equation that describes the orbit of the asteroid if the distance from the sun was always the same, 200 million miles. Thus, the nearest and farthest points are both 200 in this case.

57. When water leaves a garden hose nozzle, its path through the air is a parabola (if we neglect air resistance; see the figure, where we picture the hose nozzle at the origin). As the direction of the nozzle is raised from the horizontal, the path traced by the highest point of the water's path is part of an ellipse.[5] If the water's velocity is $8\sqrt{2}$ feet per second the paths will be those shown in the figure. The ellipse shown has center at (0,1), minor axis of length 2 and major axis of length 4. Find it's equation.

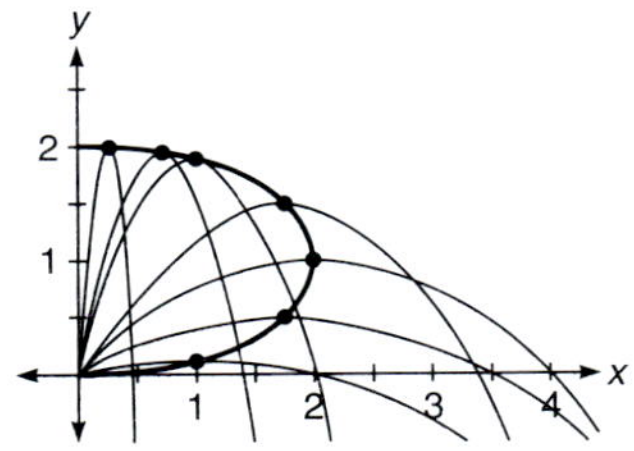

[5]From a problem proposed by Alan Wayne in the May, 1989 issue of *The College Mathematics Journal.*

The *eccentricity e* of an ellipse is defined as the ratio $\frac{c}{a}$ when $a > b$, or $\frac{c}{b}$ when $b > a$. Examining the case where $a > b$, $a^2 = b^2 + c^2$, we know that $c < a$. As $0 \le c < a$, the ratio $e = \frac{c}{a}$ varies from 0 to 1: $0 \le e < 1$. When $e = 0$, the ellipse is a circle; as e approaches 1 the ellipse gets flatter and flatter. Find the eccentricity of the ellipse in each problem.

58. Problem 1. **59.** Problem 2. **60.** Problem 3. **61.** Problem 4. **62.** Problem 5.
63. Problem 6. **64.** Problem 7. **65.** Problem 8. **66.** Problem 9. **67.** Problem 10.
68. Problem 33. **69.** Problem 34. **70.** Problem 35. **71.** Problem 36.

Skill and review

1. Graph $y = 2(x - 1)^2 - 4$.
2. Graph $x^2 - 4x + y^2 + 12y + 12 = 0$.
3. Solve $x^{2/3} + 7x^{1/3} = 8$.
4. Rationalize the denominator: $\frac{2\sqrt{3}}{\sqrt{6} - \sqrt{2}}$.
5. Graph $f(x) = \frac{2x}{x^2 - 9}$.
6. Factor $x^3 - 3x^2 + x + 2$ over R.

11–3 The hyperbola

Two long-range navigation (LORAN) radio navigation stations are 130 miles apart. A ship receiving the signals from these stations determines that the difference in the distances from the ship to each station is 50 miles. Find the equation that describes this situation.

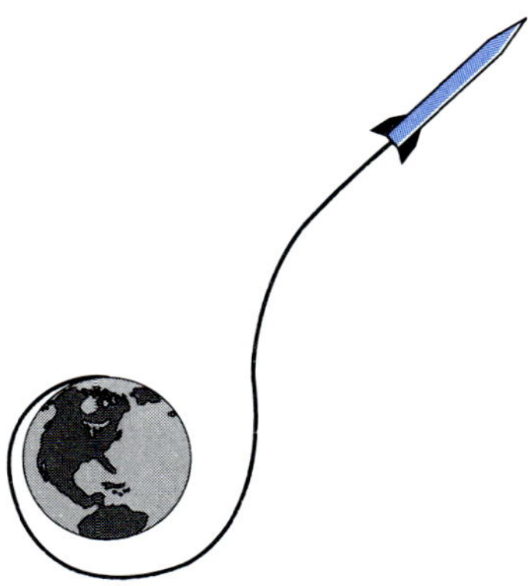

The equation that describes the situation in this problem is a hyperbola. A hyperbola has two parts, each of which resembles a parabola. If a space vehicle is given a velocity in excess of what it needs to escape the earth's gravity, it leaves the earth on a hyperbolic path. As suggested in the opening problem, hyperbolas also form the basis for the marine and aviation system of navigation called LORAN.

To define a hyperbola geometrically, fix two points, called the foci. (At c and $-c$ in figure 11–8.) For every point (x,y) there is a distance d_1 to one focus and a distance d_2 to the other. The **hyperbola** is the set of all points such that the absolute value of the difference between d_1 and d_2 is a constant.

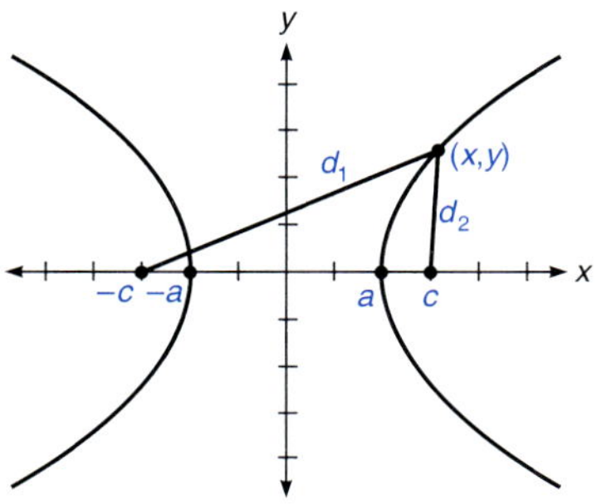

Figure 11–8

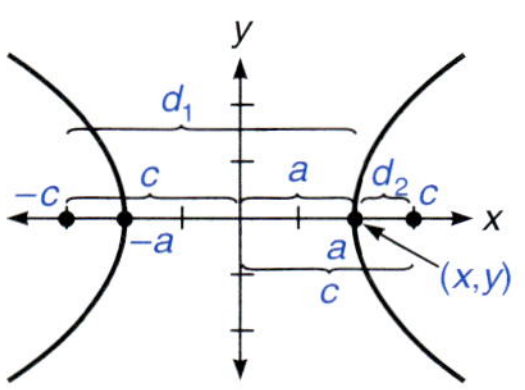

Figure 11–9

To develop an algebraic description of a hyperbola we proceed as follows. Consider figure 11–8. In this case, the foci are placed on the x-axis, at c and $-c$ $(c > 0)$. The x-intercepts are at a and $-a$, and (x,y) represents any point on the hyperbola. We are told that $|d_1 - d_2|$ is some constant. By letting the point (x,y) be at the point $(a,0)$ it can be seen that this constant is $2a$ (see figure 11–9).

$$|d_1 - d_2| = |(a + c) - (c - a)| = |2a|$$

If $a > 0$,

$$|d_1 - d_2| = 2a$$

Now find d_1 and d_2 with the distance formula.

$d_1 = \sqrt{(x - (-c))^2 + (y - 0)^2}$ — Use the distance formula on (x,y) and $(-c,0)$

$d_2 = \sqrt{(x - c)^2 + (y - 0)^2}$ — Use the distance formula on (x,y) and $(c,0)$

$|d_1 - d_2| = 2a$ — Established above

$\left|\sqrt{(x + c)^2 + y^2} - \sqrt{(x - c)^2 + y^2}\right| = 2a$ — Substitute for d_1 and d_2

By carrying out the details of this calculation, and defining a value b so that $b^2 = c^2 - a^2$, the last statement can be transformed into $\frac{x^2}{a^2} - \frac{y^2}{b^2} = 1$. This transformation is left as an exercise. Observe that if $b^2 = c^2 - a^2$, then $c^2 = a^2 + b^2$; we use this below.

By letting x and then y be zero, we find that there is no y-intercept and the x-intercept is at a. The origin, with respect to any translated axes, is called the **center** of the hyperbola. By putting the foci on the y-axis we obtain a similar equation. We thus obtain the following result.

Hyperbola

The graph of an equation of the form $\frac{x^2}{a^2} - \frac{y^2}{b^2} = 1$ is a hyperbola with center at the origin. The hyperbola opens right and left. The x-intercepts are at $(\pm a,0)$, and there are no y-intercepts. The foci are at $(\pm c,0)$, where $c^2 = a^2 + b^2$.

The graph of an equation of the form $\frac{y^2}{a^2} - \frac{x^2}{b^2} = 1$ is a hyperbola with center at the origin. The hyperbola opens up and down. The y-intercepts are at $(0,\pm a)$, and there are no x-intercepts. The foci are at $(0,\pm c)$, where $c^2 = a^2 + b^2$.

A **major axis** and a **minor axis** is also defined. The major axis is the line segment between the foci. The minor axis is perpendicular to the major axis. It is the line segment extending $|b|$ units above and below the center.

As in the case of the ellipse, a more general form of the equation of a hyperbola takes horizontal and vertical translations into account.

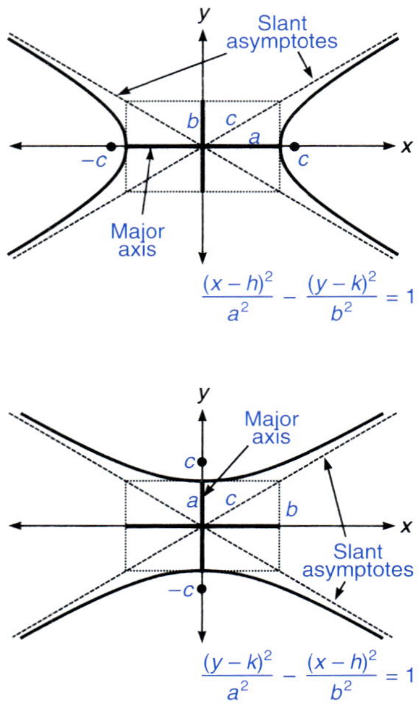

Figure 11–10

> **Standard form of the equation of a hyperbola**
>
> [1] $\dfrac{(x-h)^2}{a^2} - \dfrac{(y-k)^2}{b^2} = 1$ Opens horizontally
>
> [2] $\dfrac{(y-k)^2}{a^2} - \dfrac{(x-h)^2}{b^2} = 1$ Opens vertically

Each of these equations has its *center* at (h,k) instead of (0,0), and its foci must be computed relative to this new center. The values a and b are *distances* from the center (h,k) at which *we can plot the end points of the major and minor axes.*

Hyperbolas also have **slant asymptotes,** which are a great aid in graphing. The graph of the hyperbola gets closer and closer to these lines as the value of $|x|$ gets larger and larger. We do not need to determine the equation of these asymptotes—the following procedure for graphing allows them to be constructed without determining their equations. At the end of this section we discuss why these lines are in fact slant asymptotes. The following procedure for graphing a hyperbola refers to figure 11–10.

> **To graph a hyperbola**
>
> - Find the values of a and b and the center (h,k). Determine whether the major axis is horizontal or vertical.
> - Draw a rectangle with sides a units from the center along the major axis and b units from the center along the minor axis.
> - Construct slant asymptotes. These are the lines that form the diagonals of the rectangle.
> - Compute c using $c = \sqrt{a^2 + b^2}$. The foci are c units from the center, along the major axis. Observe in the figure that c can also be viewed as the hypotenuse of a triangle with sides of length a and b.
> - Sketch the hyperbola using the rectangle and slant asymptotes as a guide.

Example 11–3 A illustrates graphing a hyperbola.

■ *Example 11–3 A*

Graph each hyperbola. State the foci and center. Show the rectangle that is used to draw the slant asymptotes.

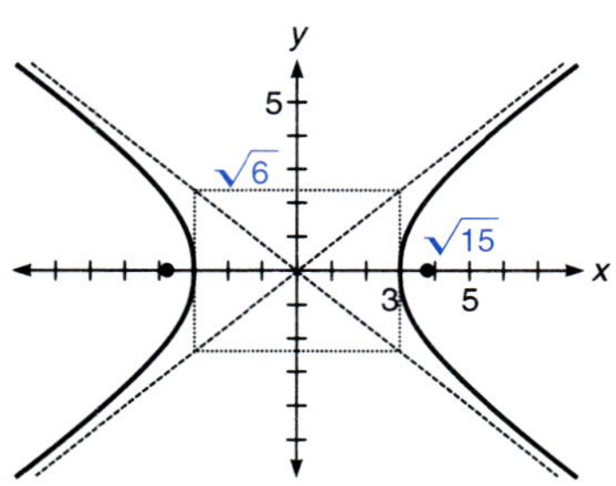

1. $\dfrac{x^2}{9} - \dfrac{y^2}{6} = 1$

The center is at (0,0). The major axis is horizontal, and the minor axis is vertical.

$$a = \sqrt{9} = 3$$
$$b = \sqrt{6} \approx 2.4$$

x-intercepts: $(\pm 3,0)$

Construct a rectangle as shown and draw the slant asymptotes.

$$c = \sqrt{a^2 + b^2} = \sqrt{9 + 6} = \sqrt{15} \approx 3.9$$

Since the major axis is horizontal the foci are along the x-axis at $(-\sqrt{15},0)$ and $(\sqrt{15},0)$. Draw the hyperbola so it touches the end points of the major axis and approaches the slant asymptotes.

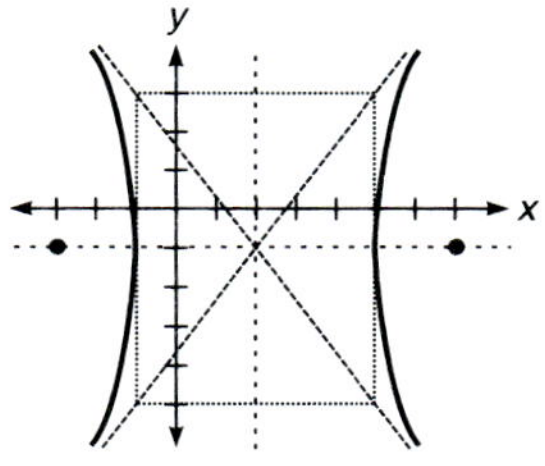

2. $\dfrac{(x-2)^2}{9} - \dfrac{(y+1)^2}{16} = 1$

Center is at $(2,-1)$. The major axis is horizontal.

$$a = \sqrt{9} = 3; \qquad b = \sqrt{16} = 4$$

Construct a rectangle with sides 3 units right and left of the center and 4 units above and below it. Draw in the slant asymptotes.

$$c = \sqrt{a^2 + b^2} = \sqrt{9 + 16} = \sqrt{25} = 5.$$

The foci are 5 units right and left of $(2,-1)$ at $(2 \pm 5,-1)$, or $(-3,-1)$ and $(7,-1)$.

Note By completing the square we can put any equation of the form $Ax^2 + Bx + Cy^2 + Dy + E = 0$ into the standard form of a hyperbola if A and B have opposite signs.

3. $9y^2 + 18y - 4x^2 + 24x = 36$

$9(y^2 + 2y) - 4(x^2 - 6x) = 36$ Factor 9 from y terms, -4 from x terms

$9(y^2 + 2y + 1) - 4(x^2 - 6x + 9) = 9(1) - 4(9) + 36$ Complete the square

$9(y+1)^2 - 4(x-3)^2 = 9$

$\dfrac{9(y+1)^2}{9} - \dfrac{4(x-3)^2}{9} = \dfrac{9}{9}$ Divide each term by 9 to obtain 1 in the right member

$(y+1)^2 - \dfrac{(x-3)^2}{\frac{9}{4}} = 1$

Center is at $(3,-1)$; major axis is vertical.

$$a = \sqrt{1} = 1; \qquad b = \sqrt{\tfrac{9}{4}} = \tfrac{3}{2} = 1.5$$

End points of major axis: $(3,-1 \pm 1)$ or $(3,-2)$ and $(3,0)$
End points of minor axis: $(3 \pm 1.5,-1)$ or $(1.5,-1)$ and $(4.5,-1)$

$$c = \sqrt{a^2 + b^2} = \sqrt{1 + \tfrac{9}{4}} = \sqrt{\tfrac{13}{4}} = \frac{\sqrt{13}}{2} \approx 1.8$$

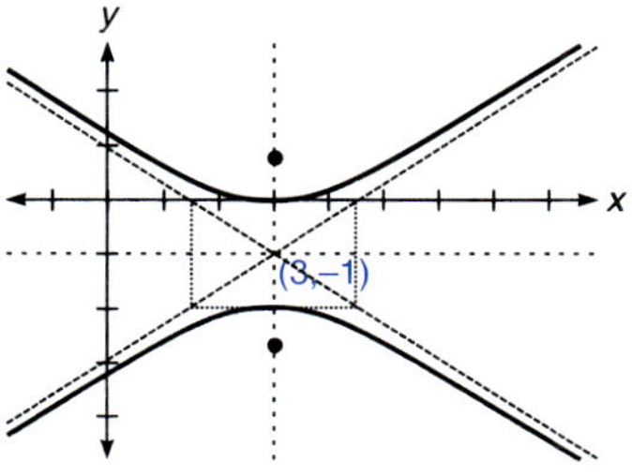

Foci: $\left(3,-1 \pm \dfrac{\sqrt{13}}{2}\right) \approx (3,-2.8)$ and $(3,0.8)$

Draw a rectangle using these end points, and draw the slant asymptotes. Then draw the hyperbola so it touches the end points of the major axis and approaches the slant asymptotes.

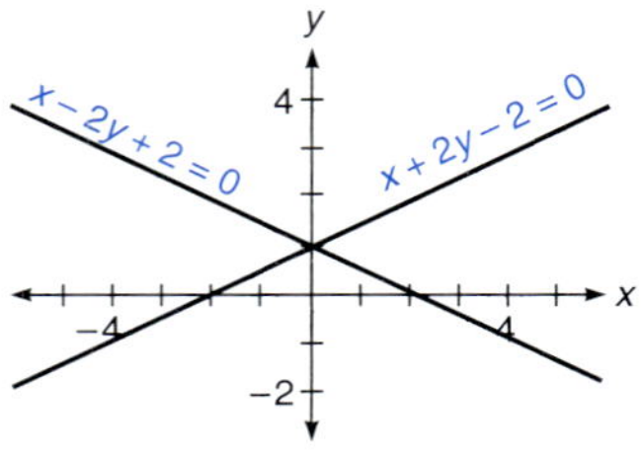

4. $x^2 - 4y^2 + 8y = 4$

$x^2 - 4(y - 1)^2 = 0$ Coefficient of y^2 is 1

This equation cannot be put into the standard form of a hyperbola since the right member will never be 1. However, the equation is the difference of two squares and will factor.

$x^2 - 4(y - 1)^2 = 0$

$[x - 2(y - 1)][x + 2(y - 1)] = 0$ $m^2 - n^2 = (m - n)(m + n)$

$(x - 2y + 2)(x + 2y - 2) = 0$

$x - 2y + 2 = 0$ or $x + 2y - 2 = 0$ Zero product property

Each of the equations $x - 2y + 2 = 0$ and $x + 2y - 2 = 0$ is a straight line. This could be considered a "degenerate" hyperbola. ■

As with the ellipse, we must solve an equation for y to graph it with a graphing calculator.

■ ***Example 11–3 B***

Graph with a graphing calculator.

$(y + 1)^2 - \dfrac{(x - 3)^2}{\frac{9}{4}} = 1$ (part 3, example 11–3 A.)

$(y + 1)^2 = \frac{4}{9}(x - 3)^2 + 1$

$y + 1 = \pm\sqrt{\frac{4}{9}(x - 3)^2 + 1}$

$y = -1 \pm\sqrt{\frac{4}{9}(x - 3)^2 + 1}$

Put the expression $\sqrt{\frac{4}{9}(x - 3)^2 + 1}$ into Y_1, then let $Y_2 = -1 - Y_1$, and $Y_3 = -1 + Y_2$.

[Y=] [√] [(] [(] 4 [÷] 9 [)] [(] [X|T] [−] 3 [)] [x^2]
[+] 1 [)] [ENTER] [(−)] 1 [−] [Y-VARS] 1 [ENTER]
[(−)] 1 [+] [Y-VARS] 1

Now turn off the graphing of Y_1 by positioning the cursor on the "=" part of "Y_1=" and selecting [ENTER].

[RANGE −1,7,−4,2] ■

One of the most common uses of the hyperbola is in navigation. In particular, a system of navigation by radio waves called LORAN (long-range navigation) is based on the hyperbola.[6]

■ ***Example 11–3 C***

In the LORAN radio-aided navigation system, two transmitting stations are the foci of a hyperbola, and the receiving radio is on a hyperbola determined by the foci and the location of a ship or aircraft.

Two LORAN radio navigation stations are 100 miles apart. A ship receiving the signals from these stations determines that the difference in the

[6]Other hyperbolic navigation systems are the Decca Navigation System, Omega, and the satellite-based Global Positioning System.

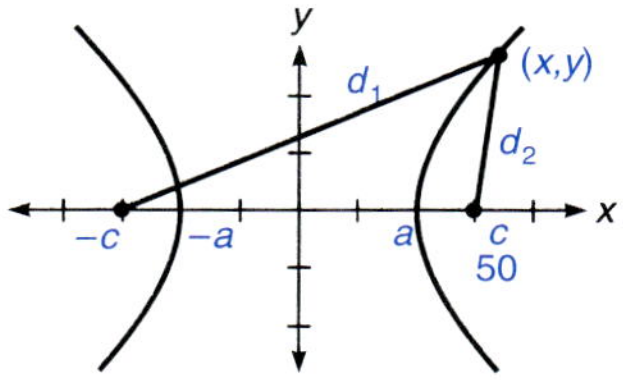

distances from the ship to each station is 80 miles. Find the equation of a hyperbola that describes the position of the ship.

The figure shows the ship at some point (x,y). We know that $|d_1 - d_2| = 80$. The stations are at foci 50 units from the origin (since they are 100 miles apart). Thus $c = 50$.

When we developed the equation of a hyperbola at the beginning of this section we noted that the difference between the distances is $2a$. Thus,

$2a = 80$, so $a = 40$

$c^2 = a^2 + b^2$; $50^2 = 40^2 + b^2$; $30 = b$

$$\frac{x^2}{1{,}600} - \frac{y^2}{900} = 1$$ Replace values in $\frac{x^2}{a^2} - \frac{y^2}{b^2} = 1$ ■

The general quadratic equation in two variables

The general quadratic equation in two variables is $Ax^2 + By^2 + Cx + Dy + E = 0$, where A and B are not both zero. The graph of this equation depends largely upon the values of A and B. We categorize these equations in the following way.

> **General quadratic equation in two variables:**
>
> $$Ax^2 + By^2 + Cx + Dy + E = 0$$
>
> - If A or B (but not both) is zero, the equation is a parabola.
> - If $A = B$ (and neither is zero), then the equation is a circle.
> - If $|A| \neq |B|$ but A and B have the same sign, the equation is an ellipse.
> - If A and B have opposite signs the equation is a hyperbola.

In each case it is possible that no graph exists or that is corresponds to another geometric object. For example, the equation $x^2 + y^2 = -4$ falls in the category of a circle, but does not have a graph. The equation $x^2 + y^2 = 0$ also falls in the category of a circle, but its graph is the single point (0,0). The best way to determine the category of an equation is to put it in one of the following forms, usually by completing the square.

Straight line:	$ax + by + c = 0$, a and b not both 0
Parabola:	$y = \frac{1}{4p}(x - h)^2 + k$ or $x = \frac{1}{4p}(y - k)^2 + h$
Circle:	$(x - h)^2 + (y - k)^2 = r^2$
Ellipse:	$\frac{(x - h)^2}{a^2} + \frac{(y - k)^2}{b^2} = 1$
Hyperbola:	$\frac{(x - h)^2}{a^2} - \frac{(y - k)^2}{b^2} = 1$ or $\frac{(y - k)^2}{a^2} - \frac{(x - h)^2}{b^2} = 1$

Example 11–3 D shows the process of categorizing equations by their geometric significance.

■ ***Example 11–3 D***

Categorize the graph of each equation as a point, line, circle, parabola, ellipse, or hyperbola (or no graph). Put the equation in one of the listed forms.

1. $3x^2 + 2x - 5y = 8$

This is most likely a parabola since it is quadratic in only one variable, x.

$$y = \frac{3}{5}(x + \tfrac{1}{3})^2 - \frac{5}{3}$$ Complete the square on x

The graph is a parabola.

2. $2x^2 - 3y^2 + 9y = 5\frac{3}{4}$

This is most likely a hyperbola since the coefficients of x^2 and y^2 have opposite signs.

$$\frac{(y - \frac{3}{2})^2}{\frac{1}{3}} - \frac{x^2}{\frac{1}{2}} = 1$$ Complete the square on y

This is a hyperbola.

3. $2x^2 + 4x + 2y^2 + 6 = 0$

This is probably a circle because the coefficients of x^2 and y^2 have the same sign and value.

$$(x + 1)^2 + y^2 = -2$$ Complete the square

The left member is nonnegative and the right member is negative; thus the graph of this equation has no points. ■

Equations of slant asymptotes

As already illustrated, slant asymptotes are a great aid in graphing hyperbolas. To see why these asymptotes exist consider the hyperbola $\frac{x^2}{a^2} - \frac{y^2}{b^2} = 1$. If we solve for y^2, we obtain $y^2 = \frac{b^2}{a^2}x^2 - b^2$. As $|x|$ gets greater (graphically, as we move right and left of the origin), the term $\frac{b^2}{a^2}x^2$ gets greater and greater, but the term $-b^2$ is fixed in size. As a percentage of the total value of $\frac{b^2}{a^2}x^2 - b^2$, b^2 tends to become less and less. To see this, let $a = b = 1$, giving $x^2 - 1$ and x^2. The table shows how the percentage of error diminishes as $|x|$ grows. The last column shows the percentage of error between $x^2 - 1$ and x^2.

x	$\lvert x\rvert$	x^2	$x^2 - 1$	$\lvert\text{error}\rvert$	$\frac{\lvert\text{error}\rvert}{x^2 - 1} \times 100\%$
±10	10	100	99	1	1.00%
±20	20	400	399	1	0.25%
±30	30	900	899	1	0.11%
±100	100	10,000	9,999	1	0.01%

Even a 1% error is usually not detectable in a graph. Thus, the graph of $x^2 - 1$ and x^2 are practically the same when $|x|$ gets greater and greater.

Generalizing, we would say that as $|x|$ gets greater and greater, the difference between $\frac{b^2}{a^2}x^2 - b^2$ and $\frac{b^2}{a^2}x^2$ gets smaller and smaller (as a percentage of $\frac{b^2}{a^2}x^2 - b^2$). Continuing the development above,

$$y^2 \approx \frac{b^2}{a^2}x^2 \qquad \text{Ignore } -b^2 \text{ when } |x| \text{ is large}$$

$$y \approx \pm\frac{b}{a}x$$

The lines $y = \frac{b}{a}x$ and $y = -\frac{b}{a}x$ are the slant asymptotes for the hyperbola $\frac{x^2}{a^2} - \frac{y^2}{b^2} = 1$. The graph of the hyperbola gets closer and closer to these lines as the value of $|x|$ gets greater and greater.

Similar reasoning will show that $y - k = \frac{b}{a}(x - h)$ and $y - k = -\frac{b}{a}(x - h)$ are the equations of the slant asymptotes for a hyperbola of the form $\frac{(x-h)^2}{a^2} - \frac{(y-k)^2}{b^2} = 1$. See exercises 70 and 71 also.

Mastery points

Can you

- Graph a hyperbola when given the standard equation

$$\frac{(x-h)^2}{a^2} - \frac{(y-k)^2}{b^2} = 1 \text{ or } \frac{(y-k)^2}{a^2} - \frac{(x-h)^2}{b^2} = 1?$$

- Transform certain equations that are quadratic in two variables into the standard form of the equation of a hyperbola?
- Categorize equations of the form $Ax^2 + By^2 + Cx + Dy + E = 0$ as lines, parabolas, circles, ellipses, and hyperbolas?

Exercise 11–3

Graph each hyperbola. If necessary transform the equation first. State the coordinates of the end points of the major axis and foci where not clear from the graph.

1. $\frac{x^2}{25} - \frac{y^2}{16} = 1$
2. $\frac{y^2}{20} - \frac{x^2}{16} = 1$
3. $\frac{y^2}{25} - \frac{x^2}{16} = 1$
4. $\frac{x^2}{20} - \frac{y^2}{16} = 1$
5. $x^2 - \frac{y^2}{6} = 1$
6. $\frac{y^2}{12} - x^2 = 1$
7. $\frac{x^2}{16} - \frac{y^2}{4} = 1$
8. $\frac{x^2}{4} - \frac{y^2}{25} = 1$
9. $\frac{y^2}{9} - \frac{x^2}{4} = 1$
10. $\frac{y^2}{16} - \frac{x^2}{9} = 1$
11. $\frac{4x^2}{25} - y^2 = 1$
12. $\frac{16y^2}{9} - x^2 = 1$
13. $4y^2 - x^2 = 2$
14. $6x^2 - 18y^2 = 36$
15. $9x^2 - y^2 = 36$
16. $y^2 - 25x^2 = 25$
17. $16y^2 - x^2 = -16$
18. $2y^2 - 3x^2 = -18$
19. $2x^2 - 9y^2 = -36$
20. $6x^2 - 6y^2 = -1$
21. $25y^2 - 16x^2 = 400$
22. $16x^2 - 9y^2 = 144$
23. $8x^2 - 3y^2 = 4$
24. $5y^2 - 4x^2 = 10$

25. $\frac{(x-2)^2}{100} - \frac{(y+3)^2}{25} = 1$
26. $\frac{(y-1)^2}{20} - (x+1)^2 = 1$
27. $\frac{(y+2)^2}{4} - x^2 = 1$
28. $\frac{(x-3)^2}{9} - \frac{(y+1)^2}{4} = 1$
29. $\frac{(x+1)^2}{25} - \frac{(y-1)^2}{36} = 1$
30. $\frac{(y-3)^2}{9} - (x-2)^2 = 1$
31. $(x-3)^2 - \frac{(y+1)^2}{16} = 1$
32. $y^2 - \frac{(x-5)^2}{25} = 1$
33. $(y-2)^2 - \frac{(x-2)^2}{4} = 1$
34. $\frac{4(x-1)^2}{25} - \frac{y^2}{4} = 1$
35. $\frac{16(y-1)^2}{25} - \frac{x^2}{9} = 1$
36. $\frac{25(x-1)^2}{36} - \frac{9(y+1)^2}{4} = 1$
37. $2x^2 - 4x - y^2 - 4y - 10 = 0$
38. $y^2 + 6y - 12x^2 = 3$
39. $4x^2 + 8x - 3y^2 + 24y + 4 = 0$
40. $x^2 - 2y^2 + 4y = 10$
41. $x^2 - 4x + 3y^2 - 24y + 49 = 0$
42. $3x^2 + 18x - 2y^2 - 4y + 19 = 0$
43. $x^2 - 2x - 4y^2 - 8y = 7$
44. $3x^2 + 24x - y^2 + 2y + 50 = 0$
45. $2y^2 - 12y - 4x^2 = 6$
46. $3x^2 - 12x - 2y^2 - 6 = 0$
47. $x^2 - 3x - 3y^2 + 6y = 0$
48. $5y^2 - 20y - 2x^2 + 2x = 5$

Categorize the graph of each equation as a point, line, circle, parabola, ellipse, or hyperbola (or no graph). Put each equation in the standard form for whichever geometric figure it represents. Graph each figure; state centers and foci as appropriate.

49. $x^2 + 2x - y^2 + 8y = 16$
50. $y^2 - 6y - x + 4 = 0$
51. $4x^2 - 8x - y^2 - 4y = 4$
52. $2y - 4x = 1$
53. $9y^2 - 4x^2 + 36 = 0$
54. $x^2 - 4x + y^2 - 8y + 16 = 0$
55. $2x^2 + 12x + y^2 - 8y + 32 = 0$
56. $4x^2 - 24x - 25y^2 - 50y = 14$
57. $4x^2 - 12x + 4y^2 + 20y + 30 = 0$
58. $3x - 2y + 8 = 0$
59. $25x^2 + 16y^2 = 100$
60. $x^2 - 2x + 2y^2 + 12y + 11 = 0$
61. $y = x^2 + 2x + 4$
62. $3x^2 + 4y^2 - 16y + 4 = 0$
63. $4x^2 + 8x + 4y^2 - 4y = 27$
64. $x^2 = 7y + 4$
65. $4y = 4x^2 - 20x + 23$
66. $x^2 - 2x + y^2 + 6y + 10 = 0$

67. Two LORAN radio navigation stations are 130 miles apart. A ship receiving the signals from these stations determines that the difference in the distances from the ship to each station is 50 miles. Find the equation of a hyperbola that describes this situation.

68. Two LORAN radio navigation stations are 80 miles apart. A ship receiving the signals from these stations determines that the difference in the distances from the ship to each station is 60 miles. Find the equation of a hyperbola that describes this situation.

69. Suppose that the cost y of producing t items is $y = \sqrt{t}$, and the profit x on t items is $x = \frac{\sqrt{t-4}-1}{2}$. Then $y^2 = t$, and

$$2x = \sqrt{t-4} - 1$$
$$4x^2 + 4x + 1 = t - 4$$
$$4x^2 + 4x + 5 = t$$

Replacing t by y^2, $y^2 = 4x^2 + 4x + 5$. Transform this relation and graph it.

70. In the text we saw that the lines $y = \pm\frac{b}{a}x$ are the slant asymptotes for the hyperbola $\frac{x^2}{a^2} - \frac{y^2}{b^2} = 1$. Use a similar development to determine the equations of the slant asymptotes of a hyperbola of the form $\frac{y^2}{a^2} - \frac{x^2}{b^2} = 1$.

71. In the text we saw that $y - k = \pm\frac{b}{a}(x - h)$ are the equations of the slant asymptotes for a hyperbola of the form $\frac{(x-h)^2}{a^2} - \frac{(y-k)^2}{b^2} = 1$. Determine the equations of the slant asymptotes for a hyperbola of the form $\frac{(y-k)^2}{a^2} - \frac{(x-h)^2}{b^2} = 1$.

72. In the text we saw that the definition of a hyperbola with foci at $(a,0)$ and $(-a,0)$ led to the statement

$$\left|\sqrt{(x+c)^2 + y^2} - \sqrt{(x-c)^2 + y^2}\right| = 2a$$

and stated that by carrying out the details of this calculation, it can be shown that this statement is equivalent to $\frac{x^2}{a^2} - \frac{y^2}{b^2} = 1$ where $b^2 = c^2 - a^2$. Complete this calculation, but, to make the calculations easier, start from the equivalent statement $\sqrt{(x+c)^2 + y^2} = \sqrt{(x-c)^2 + y^2} \pm 2a$.

Skill and review

Graph each relation.

1. $x^2 + y^2 - 8y = 0$
2. $3x^2 + 4y^2 = 12$
3. $3x - 2y = 6$
4. $2x^2 - 4x + 4y^2 = 2$
5. $y = x^2 - 6x - 8$

11–4 Systems of nonlinear equations and inequalities

If a rock is dropped into a well and a splash is heard after 3 seconds, how deep is the well?

The solution to this problem involves two separate equations—one that describes the rock as it falls into the well and another that describes the sound of the splash as it comes out of the well. The answer to the problem is found by solving the system of these two equations, using methods studied in this section.

Systems of nonlinear equations

Chapter 10 discussed systems of linear equations. In this section we discuss systems of two equations in two variables that may be nonlinear. To solve such a system means to find all the ordered pairs that satisfy each equation in the system. Geometrically this corresponds to the points of intersection of the graphs of each equation.

By way of example, consider the following. Hyperbolas form the theoretical basis for a system of navigation called LORAN (illustrated in section 11–3). This system is based on radio waves and charts. The idea is that a ship or aircraft receives radio signals from two transmitters. The ship does not know the distance to either transmitter, but can determine very accurately when each signal arrives at the ship. This permits computation of the time difference in the arrival of each signal, and this time interval corresponds to the distance the radio waves travel in that time interval. This difference tells the ship's navigator that the ship lies on a certain hyperbola. By doing the same thing with some other radio transmitter, it is known that the ship lies on some other hyperbola. *By determining where these hyperbolas intersect the position of the ship or aircraft is discovered.* Thus, locating the ship is equivalent to determining the point(s) of intersection of two hyperbolas. In practice this is done graphically, on special navigation charts, or by computers.

We use the method of substitution for expression to find these points. This method was first shown in section 3–2 and is restated here.

To solve a system of two equations using substitution for expression

- Solve one of the equations for one of the variables, say y. The other member of this equation is an expression in terms of x.
- In the *other* equation replace each instance of y by this expression in x.
- Solve this new equation for x.
- Use these values of x in either original equation to find y.

The role of x and y in these steps can be reversed when that is more convenient. Also we might solve for y^2 or some other expression in the first step. This method is illustrated in example 11–4 A.

■ Example 11–4 A

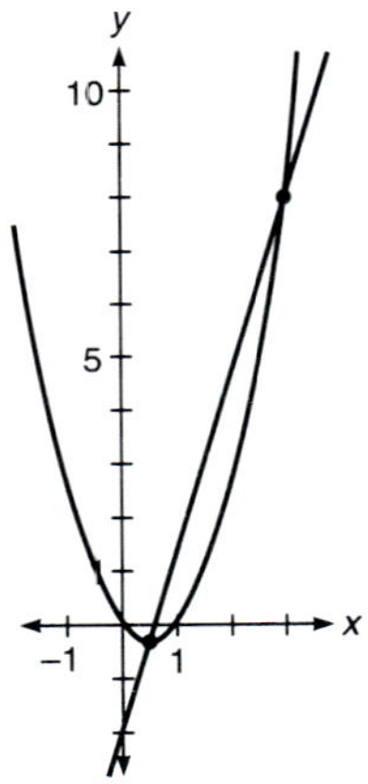

1. Find the point(s) where the line $3x - y = 2$ and the parabola $y = x^2 - x$ meet.

$3x - y = 2$	Equation of the line
$y = 3x - 2$	Solve for y
$y = x^2 - x$	Equation of parabola
$3x - 2 = x^2 - x$	Replace y with $3x - 2$ in $y = x^2 - x$
$0 = x^2 - 4x + 2$	Add $-3x + 2$ to each member
$x = 2 \pm \sqrt{2}$	Using the quadratic formula

This result (two values of x) implies that the line meets the parabola at two places. We still need the y values; these can be obtained from either the equation of the line, $y = 3x - 2$, or the parabola, $y = x^2 - x$. Either will give the same values of y. We use the line since it is easier.

x	$y = 3x - 2$
$2 + \sqrt{2}$	$y = 3(2 + \sqrt{2}) - 2 = 4 + 3\sqrt{2}$
$2 - \sqrt{2}$	$y = 3(2 - \sqrt{2}) - 2 = 4 - 3\sqrt{2}$

Thus, we find that there are two points where the line $y = 3x - 2$ meets the parabola $y = x^2 - x$; at $(2 + \sqrt{2}, 4 + 3\sqrt{2}) \approx (3.4,8.2)$, and at $(2 - \sqrt{2}, 4 - 3\sqrt{2}) \approx (0.6,-0.2)$. This is shown graphically in the figure.

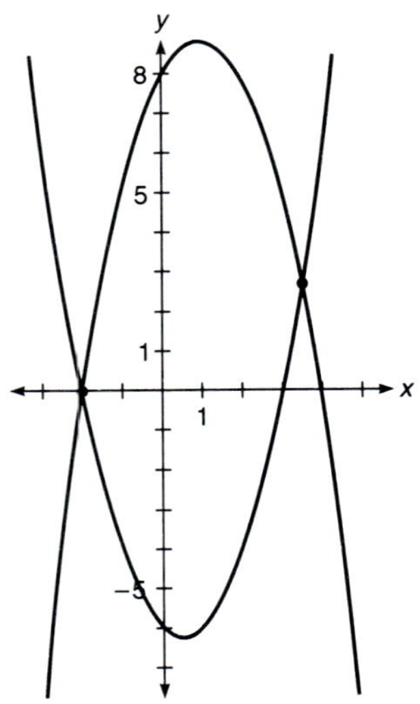

2. Find the point(s) where the parabolas $y = x^2 - x - 6$ and $y = -x^2 + 2x + 8$ meet.

$y = x^2 - x - 6$	First equation; solved for y
$y = -x^2 + 2x + 8$	Second equation
$x^2 - x - 6 = -x^2 + 2x + 8$	Replace y in the second equation
$2x^2 - 3x - 14 = 0$	We want one of the members to be 0
$(2x - 7)(x + 2) = 0$	Factor
$2x - 7 = 0$ or $x + 2 = 0$	Zero product property
$x = 3\frac{1}{2}$ or -2	Solve each linear equation

Now compute the corresponding y values from either equation.

x	$y = x^2 - x - 6$	
$3\frac{1}{2}$	$y = (\frac{7}{2})^2 - \frac{7}{2} - 6 = 2\frac{3}{4}$	Replace x with $3\frac{1}{2}$
-2	$y = (-2)^2 - (-2) - 6 = 0$	Replace xy with -2

Thus, the curves intersect at the points $(3\frac{1}{2},2\frac{3}{4})$ and $(-2,0)$. This is illustrated in the figure.

3. Find the points of intersection of the hyperbolas $x^2 - 2y^2 = 4$ and $8y^2 - 2x^2 = 1$.

$x^2 = 2y^2 + 4$ — Solve the first equation for x^2

$8y^2 - 2(2y^2 + 4) = 1$ — Replace x^2 with $2y^2 + 4$ in the second equation

$y = \pm\frac{3}{2}$ — Solve for y

Now find x.

$x^2 = 2y^2 + 4$ — First equation

$x^2 = 2(\frac{9}{4}) + 4$ — $y^2 = \frac{9}{4}$

$x = \pm\sqrt{\frac{17}{2}} = \pm\frac{\sqrt{34}}{2}$ — $\sqrt{\frac{17}{2}} = \frac{\sqrt{17}}{\sqrt{2}} \cdot \frac{\sqrt{2}}{\sqrt{2}} = \frac{\sqrt{34}}{2}$

Thus, we obtain four solutions: $\left(\frac{\sqrt{34}}{2}, \frac{3}{2}\right)$, $\left(-\frac{\sqrt{34}}{2}, \frac{3}{2}\right)$, $\left(\frac{\sqrt{34}}{2}, -\frac{3}{2}\right)$, $\left(-\frac{\sqrt{34}}{2}, -\frac{3}{2}\right)$. These solutions are shown in the figure.

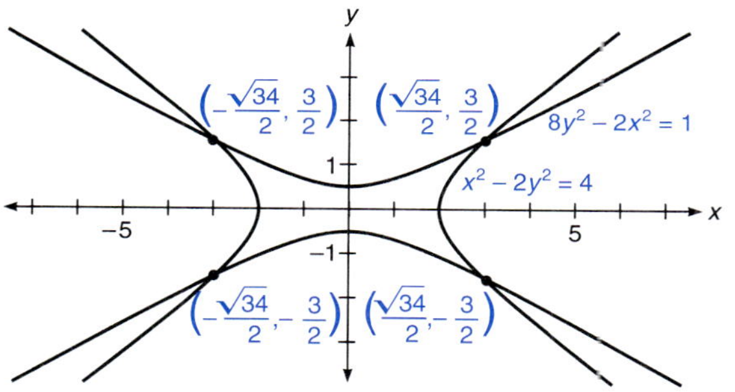

Graphing calculators can be conveniently used to obtain approximate solutions to systems in which both equations represent functions. In this situation, both equations are solved for y, or can be easily solved for y. Thus parts 1 and 2 of example 11–4 A are suitable for a graphing calculator. Example 11–4 B illustrates how to use the calculator in this case.

Part 3 cannot be easily done on a graphing calculator because at least one of the equations represents a relation that is not a function. Graphically, at least one of the graphs fails the vertical line test (see section 3–5). Of course, we saw how to deal with hyperbolas and ellipses in previous sections, so it is possible to attack these problems with a graphing calculator also.

■ ***Example 11–4 B***

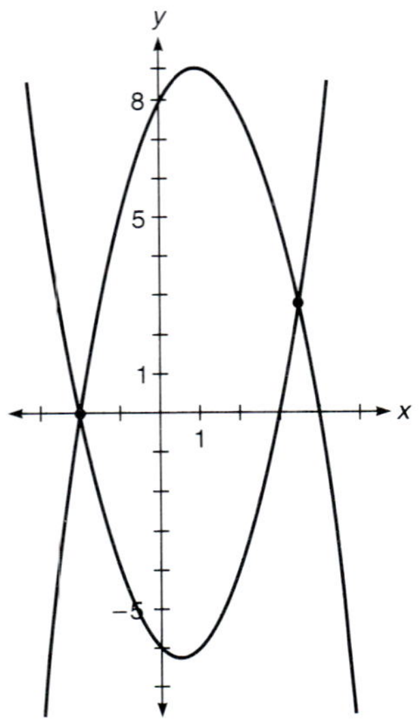

Solve the system of equations $y = x^2 - x - 6$ and $y = -x^2 + 2x + 8$ by obtaining approximate solutions with a graphing calculator.

We will illustrate two methods for solving this problem.

Method 1: Graph both equations in the same coordinate system. The point(s) of intersection of the graphs are the solutions. In the figure, there are two such points. In example 11–4 A, these were shown to be (3.5,2.75) and (−2,0). Use the TRACE and ZOOM features to estimate values near these actual values.

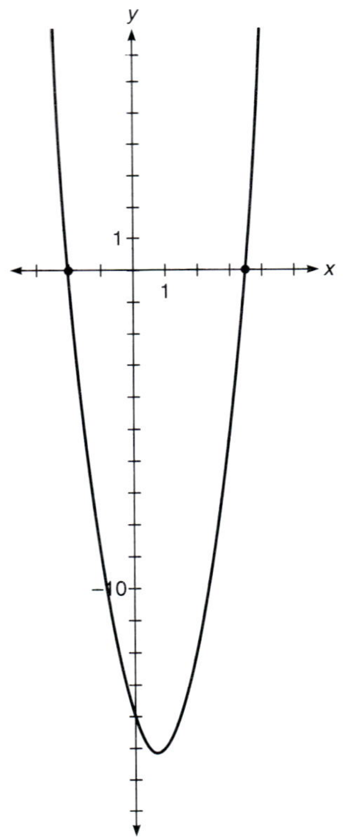

Method 2: A second method, which will obtain just the x value, is to graph the difference of the two x-expressions, in this case,

$$\begin{aligned} y &= (x^2 - x - 6) - (-x^2 + 2x + 8) \\ &= 2x^2 - 3x - 14 \end{aligned}$$

The points in which we are interested are the zeros of this new function. The graph is shown in the figure, where we can accurately estimate the x values to be −2 and 3.5. The y values can be found by evaluating either of the two equations

$$y = x^2 - x - 6 \quad \text{and} \quad y = -x^2 + 2x + 8$$

for y using $x = -2$ and $x = 3.5$.

Note In this case $0 = 2x^2 - 3x - 14$ can be accurately solved by factoring or the quadratic formula. We can also use the program NEWTON presented in section 4–3. ■

Nonlinear inequalities

Section 10–4 discussed linear inequalities. An example is $2x + y < 8$. The solution to such an inequality is a half-plane bounded by the corresponding equality $2x + y = 8$.

The solution to nonlinear inequalities is also a part of the plane, although not a half-plane. The method for finding this part of the plane is similar to that for linear inequalities.

To solve a nonlinear inequality

- Graph the corresponding equality. This divides the plane up into two or more regions.
- Try a test point from each region in the original inequality to determine which regions form the solution set.

Note There can be more than two regions formed by a nonlinear inequality. All regions should be checked.

Example 11–4 C illustrates graphing nonlinear inequalities.

■ ***Example 11–4 C***

Graph the solution to the nonlinear inequality.

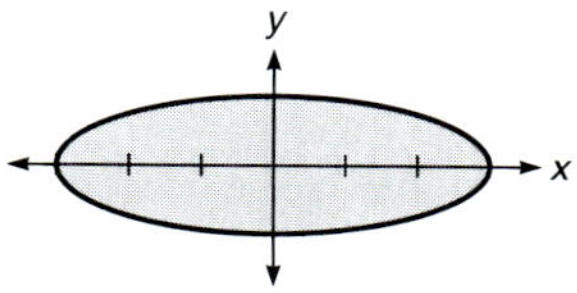

1. $\dfrac{x^2}{9} + y^2 \le 1$

Graph the corresponding equality $\dfrac{x^2}{9} + y^2 = 1$. This is an ellipse centered at the origin. The x-intercepts are $(\pm 3,0)$ and the y-intercepts are $(0,\pm 1)$. The ellipse divides the plane into the region inside and outside of it. The test point of (0,0) is true since $\dfrac{0^2}{9} + 0^2 \le 1$ is true. Thus, the part of the plane that contains the origin is part of the solution set. Also, this is a weak inequality, so the ellipse itself is part of the solution set. Indicate this by drawing the ellipse with a solid line.

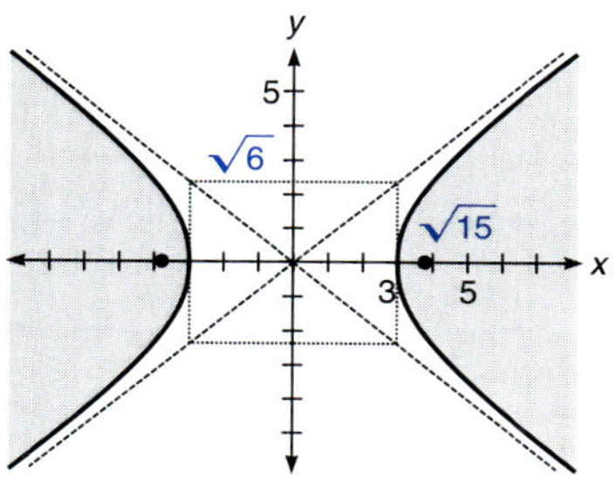

2. $\dfrac{x^2}{9} - \dfrac{y^2}{6} \ge 1$

The graph of $\dfrac{x^2}{9} - \dfrac{y^2}{6} = 1$ is part 1 of example 11–3 A. We take test points in each of the *three* regions formed by the hyperbola: $(-5,0)$, (0,0), and (5,0). The points $(-5,0)$ and (5,0) satisfy the original inequality, but the test point (0,0) does not. The solution is therefore the two regions that do not include the origin. ■

Systems of nonlinear inequalities

A system of nonlinear inequalities is a system of inequalities in which at least one of the inequalities is nonlinear. The solution set to such a system is the intersection of the solution set of each of the inequalities. Graphically this is where the solutions to the individual inequalities overlap. This is the same as for systems of linear inequalities (section 10–4). Example 11–4 D illustrates.

■ **Example 11–4 D**

Graph the solution set to the system of nonlinear inequalities $\frac{x^2}{9} - \frac{y^2}{4} < 1$ and $3y + x \geq -3$.

We first graph the corresponding equalities.

$$\frac{x^2}{9} - \frac{y^2}{4} = 1 \text{ is a hyperbola with } a = 3, b = 2.$$

$3y + x = -3$ is a straight line with y-intercept -1 and x-intercept -3. Now try test points to determine the parts of the plane that solve each individual inequality.

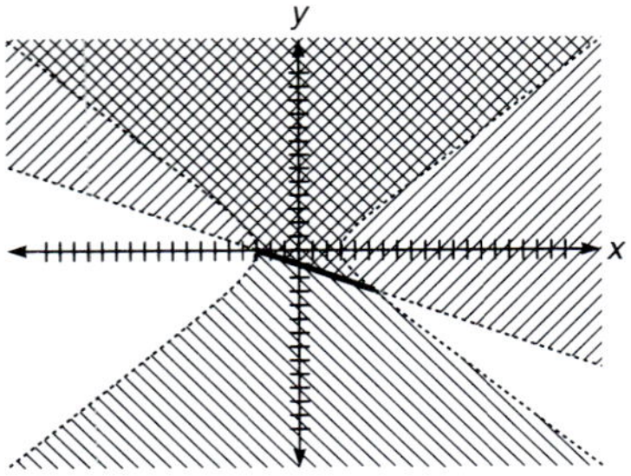

The solution to the first inequality is the part of the plane determined by the hyperbola that contains the point (0,0). The half-plane that contains (0,0) is part of the solution to the second inequality.

The edge of the hyperbola can *not* be part of the solution, since the hyperbola is described by a strong inequality ($<$). The line $3y + x = -3$ can form part of the solution because this is a weak inequality ($\geq$). This is shown by graphing with a solid line.

The solution is where the solution sets overlap. This is the cross-hatched portion of the graph along with the part of the straight line that is solid. ■

Mastery points

Can you

- Solve systems of two equations in which one or both is nonlinear?
- Solve a nonlinear inequality?
- Solve systems of two or more inequalities in which one or more is nonlinear?

Exercise 11–4

Solve the following systems of equations.

1. $y = 2x + 1$
   $y = x^2 + x - 5$
2. $y = -x + 2$
   $y = x^2 - 3x - 1$
3. $y = x^2 - x$
   $y = 4 - x^2$
4. $y = x^2 - 9$
   $y = 4 - x^2$
5. $y = 3x^2 - 2x - 4$
   $y = x^2 + x + 1$
6. $y = x^2 - 3x - 4$
   $y = 2 - x - 3x^2$
7. $y = x^2 - 3x - 4$
   $y = x - 7$
8. $y = x^2 + x - 20$
   $y + 3x + 2 = 0$
9. $y = x^2 + 3x - 8$
   $y - x = 2$
10. $y = x^2 - 4x + 2$
    $y + x = -6$
11. $4x^2 + y^2 = 4$
    $y = 2x - 1$
12. $x^2 + 2y^2 = 2$
    $y + x - 1 = 0$
13. $2x^2 + y^2 = 1$
    $2x + y = 1$
14. $2x^2 + 3y^2 = 6$
    $x = y + 2$
15. $2x^2 + y^2 = 3$
    $2x - y = 1$
16. $\frac{x^2}{4} + y^2 = 1$
    $x + y = 1$
17. $x^2 + \frac{y^2}{3} = 1$
    $y = x - 2$
18. $y^2 + 2x^2 = 4$
    $y = x - 1$
19. $x^2 - y^2 = 1$
    $2y^2 - x^2 = 2$
20. $3y^2 - x^2 = 6$
    $y^2 - x^2 = 1$
21. $4y^2 - x^2 = 8$
    $2x^2 - y^2 = 2$
22. $3y^2 - x^2 = 3$
    $3x^2 - 2y^2 = 6$

23. A circle has center at (1,2) and is tangent to the line $y = \frac{1}{2}x - 3$. Find the equation of the circle. (*Hint:* Construct the radius which touches the line $y = \frac{1}{2}x - 3$ and find its equation. Find the point of intersection of these two lines, etc.) (See problem 25 for an alternate method of solving the problem.)

24. Two circles and a straight line will be engraved on a steel plate by a computer-controlled grinding machine. The circles are as shown in the figure, and the line passes through the two points where the circles intersect. Find the equation of the straight line.

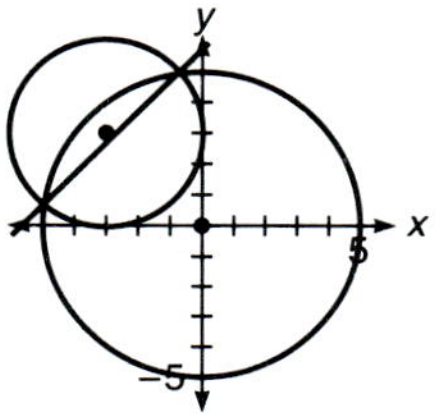

25. A circle with center at (2,5) is tangent to the line $y = -x - 1$. Find the equation of the circle. (*Hint:* Write the equation of the circle, assuming its radius is some number r. Replace y in the equation of the circle by $-x - 1$ [since the circle and line touch at some point]. Solve for x. Because of the constraints in the problem there should only be one value of x in this solution. Find out how this could happen.)

Graph the solution to the following nonlinear inequalities.

26. $y > x^2 - 1$

27. $y < x^2 - 1$

28. $y < x^2 + 1$

29. $y \geq x^2 + 1$

30. $x^2 + y^2 \leq 9$

31. $x^2 + y^2 > 9$

32. $x^2 + y^2 > 16$

33. $x^2 + y^2 < 16$

34. $4x^2 + y^2 < 4$

35. $x^2 + \frac{y^2}{4} \geq 1$

36. $\frac{x^2}{4} + y^2 \geq 1$

37. $\frac{x^2}{4} + y^2 < 1$

38. $x^2 - \frac{y^2}{4} < 1$

39. $x^2 - \frac{y^2}{4} > 1$

40. $y^2 - \frac{x^2}{4} \geq 1$

41. $y^2 - \frac{x^2}{4} \leq 1$

Graph the solutions to the systems of inequalities.

42. $y > x^2 - 3$
$y \leq -x^2 + 4$

43. $y < x^2 - 1$
$y \leq x + 2$

44. $y < x^2 + 1$
$y > -2$

45. $y > x^2 + 2$
$2y < x + 6$

46. $x^2 + y^2 < 9$
$x + y > 0$

47. $x^2 + y^2 > 9$
$x + y > 0$

48. $x^2 + y^2 > 1$
$x^2 + y^2 \leq 16$

49. $x^2 + y^2 < 16$
$x^2 + y^2 < 4$

50. $x^2 + y^2 > 4$
$x^2 + y^2 < 1$

51. $x^2 + y^2 > 4$
$x^2 + y^2 > 9$

52. $x^2 + \frac{y^2}{4} < 1$
$x + \frac{y}{4} < \frac{1}{2}$

53. $x^2 + \frac{y^2}{4} > 1$
$y > x^2 - 4$

54. $\frac{x^2}{4} + y^2 \geq 1$
$y < -x^2 + 4$

55. $\frac{x^2}{4} + y^2 < 1$
$x - y > 2$

56. $x^2 - \frac{y^2}{4} < 1$
$2x \leq y + 2$

57. $x^2 - \frac{y^2}{4} > 1$
$\frac{x^2}{4} + y^2 < 1$

58. $y^2 - \frac{x^2}{4} > 1$
$x^2 + \frac{y^2}{4} < 1$

59. $y^2 - \frac{x^2}{4} < 1$
$x^2 + y^2 > 1$

60. In section 11–1 we saw that a falling object with horizontal velocity v (in ft/s) and no initial vertical velocity will follow the path $y = -\frac{16}{v^2}x^2$, where y is vertical distance in feet and x is horizontal distance in feet.

Is there a point in the fall of such an object where the vertical distance fallen equals the horizontal distance traveled? Note that at such a point the object would be on the path $y = -x$. See the figure.

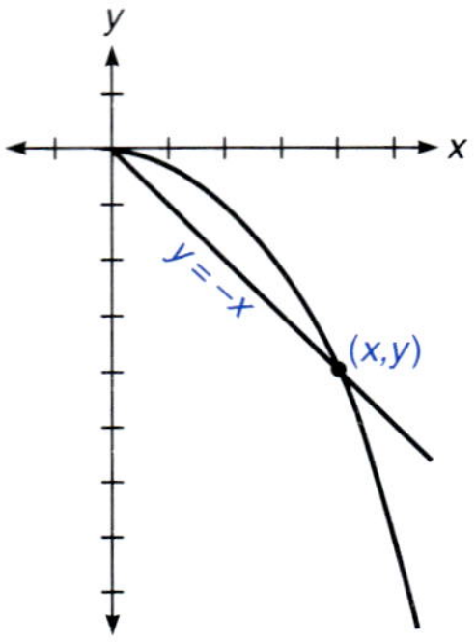

61. If a rock is dropped into a well, it falls so that its distance s in feet is $s = 16t^2$, where time t is in seconds. Sound travels at about 1,100 feet per second. Thus the distance s in feet that sound travels in t seconds is $s = 1{,}100t$. If a rock is dropped into a well and the splash is heard after 3 seconds, how deep is the well? (*Hint:* Let z be the time it takes the rock to fall and hit the water. Let $3 - z$ be the time it takes for the sound to come back.)

62. A series of numbers $u_1, u_2, u_3, \ldots$ is constructed so that $u_{n+1} = \frac{1}{1 + u_n}$, $u_1 = 1$. For example, $u_2 = \frac{1}{1 + u_1} = \frac{1}{1 + 0} = \frac{1}{2}$. For another example, $u_3 = \frac{1}{1 + u_2} = \frac{1}{1 + \frac{1}{2}} = \frac{2}{3}$. Continuing in this fashion gives the series of numbers shown in the table.

u_1	u_2	u_3	u_4	u_5	u_6	u_7
1	$\frac{1}{2}$	$\frac{2}{3}$	$\frac{3}{5}$	$\frac{5}{8}$	$\frac{8}{13}$	$\frac{13}{21}$
1	0.5	0.667	0.60	0.625	0.616	0.619

This series of fractions approaches a certain value, which is equivalent to solving the system of equations

$$y = \frac{1}{1 + x}$$
$$y = x$$

Solve this system and find the real number that the series approaches.

63. A rectangular piece of cloth has length 5 yards (yd) and width 3 yards. The length is to be increased by x yards and the width decreased by y yards. It is desired to have the new area at least as large as the old area. The new area is $(5 + x)(3 - y)$ yd^2, and the old area is $5 \cdot 3 = 15$ yd^2. Thus we want equation [1] $(5 + x)(3 - y) \geq 15$. Under the assumption that $x > 0$ it can be shown that this is equivalent to equation [2] $y \leq \frac{3x}{x + 5}$.

a. Show that equation [1] can be transformed into equation [2].

b. Graph the solution set to equation [2], including the constraints $x, y > 0$. (*Hint:* Graphing rational functions was covered in section 4–4.)

Skill and review

Graph each relation.

1. $y - 3x = -9$
2. $y - 3x^2 = -9$
3. $y^2 - 3x^2 = 9$
4. $y^2 + 3x^2 = 9$
5. $3y^2 + 3x^2 = 9$
6. $y = 2x^5 + x^4 - 10x^3 - 5x^2 + 8x + 4$

Chapter 11 summary

- **A parabola** is the set of all points equidistant from a line and a point not on that line. The line is called the directrix, and the point is called the focus.

$$y = \frac{1}{4p}(x - h)^2 + k$$

is the equation of a parabola with vertex at (h,k), focus at $(h, k + p)$, and directrix the line $y = k - p$.

- **An ellipse** is the set of all points such that the sum of the distances from each point to two other points, the foci, is constant. The line segment along the axis on which the foci lie is called the major axis; the other is the minor axis.

$$\frac{(x - h)^2}{a^2} + \frac{(y - k)^2}{b^2} = 1$$

is the equation of an ellipse with center (h,k).

- **A hyperbola** is the set of all points such that the absolute value of the difference of the distances from each point on the hyperbola to two other points, the foci, is constant. The general equation of a hyperbola takes two forms.

[1] $\frac{(x - h)^2}{a^2} - \frac{(y - k)^2}{b^2} = 1$ Opens horizontally

[2] $\frac{(y - k)^2}{a^2} - \frac{(x - h)^2}{b^2} = 1$ Opens vertically

Chapter 11 review

[11–1] Graph each parabola; state the x- and y-intercepts, focus, directrix, and vertex.

1. $y = -\frac{1}{8}x^2$ **2.** $y = x^2 - 6x + 5$
3. $y = -3x^2 - 4x + 4$ **4.** $y = x^2 + 4x + 6$
5. $x = y^2 - 7y - 8$ **6.** $x = y^2 - 4y + 4$

Determine the equation of a parabola with the given properties.

7. focus: $(1,-3)$, directrix: $y = 2$
8. focus: $(-3,-1)$, directrix: $y = -2$
9. vertex: $(2,-1)$, directrix: $y = -\frac{3}{4}$
10. focus: $(-4,2)$, vertex: $(-4,1)$
11. vertex: $(3,-1)$, x-intercepts: $2\frac{1}{2}, 3\frac{1}{2}$

Refer to the figure for the following problems.

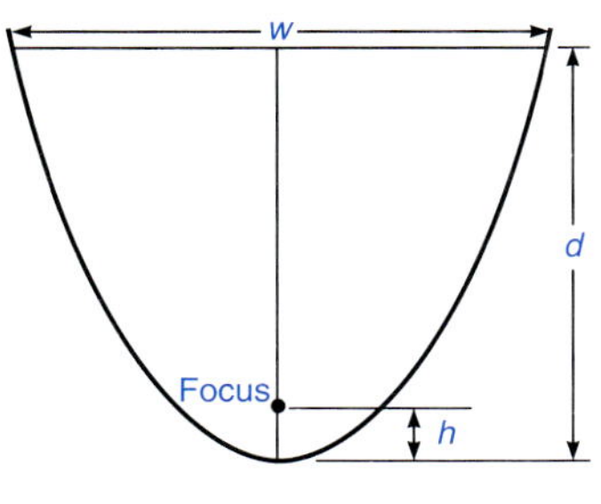

12. $h = 8$, $d = 20$; find w **13.** $h = 6$, $w = 30$; find d
14. $w = 25$, $d = 25$; find h

[11–2] Convert each equation into the standard form for an ellipse if necessary. Then graph the equation. State the coordinates of the foci and end points of the major and minor axes where not clear from the graph.

15. $\frac{x^2}{4} + y^2 = 1$ **16.** $\frac{(x + 3)^2}{16} + \frac{(y + 1)^2}{25} = 1$
17. $12x^2 + 6y^2 = 24$ **18.** $4x^2 + 8y^2 = 4$
19. $x^2 - 6x + 4y^2 + 16y + 9 = 0$

Find the equation of the ellipse with the required properties.

20. foci: $(-3,0)$ and $(3,0)$; one x-intercept at 4
21. x-intercepts: $(\pm 3,0)$; y-intercepts: $(0,\pm 4)$

22. Two tacks are put in a board 6 inches apart and a string tied in a loop with length 8 inches is looped around the tacks. The ellipse is drawn. Find an equation that describes the ellipse.

[11–3] Graph each hyperbola. If necessary transform the equation into one of the two standard forms. Show the slant asymptotes and foci. State the coordinates of the foci and endpoints of the major axis where not clear from the graph.

23. $\frac{y^2}{20} - \frac{x^2}{5} = 1$ **24.** $x^2 - \frac{y^2}{2} = 1$
25. $4y^2 - x^2 = 4$ **26.** $8x^2 - 3y^2 = 8$
27. $\frac{(x - 1)^2}{16} - \frac{(y + 3)^2}{8} = 1$
28. $y^2 + 8y - 3x^2 + 12x + 1 = 0$

Categorize the graph of each equation as a point, line, circle, parabola, ellipse, or hyperbola (or no graph). Put each equation in the standard form for whichever geometric figure it represents.

29. $x^2 + 2x - 2y^2 + 8y = 16$
30. $9y^2 + 4x^2 - 36 = 0$
31. $2x^2 + 12x + y^2 - 6y + 23 = 0$
32. $4x^2 - 12x + 4y^2 - 30 = 0$
33. $y = x^2 + 3x - 4$
34. $x^2 + 8x + 4y^2 - 4y = 19$
35. $4x - 4y^2 + 20y - 23 = 0$

[11–4] Solve the following systems of nonlinear equations.

36. $y = x^2 - 3x + 4$
$y = \frac{2}{3}x + 4$

37. $\dfrac{x^2}{2} + y^2 = 1$
$y = -2x - 1$

38. $x^2 + 3y^2 - 8y - 1 = 0$
$x = y - 2$

39. $\dfrac{(x-1)^2}{12} - \dfrac{y^2}{16} = 1$
$x - 2y = 6$

40. $2y^2 - 6y - x^2 - 6x = 20$
$4x - 3 = y$

41. $x^2 + 5x - y^2 + 6y = 0$
$y = x - 2$

42. $y = -2x - 1$
$y = x^2 + x - 5$

43. $y = 3x^2 - 2x - 5$
$y = x^2 + 2x + 1$

44. A circle has center at $(-2,3)$ and is tangent to the line $y = 3x - 2$. Find the equation of the circle.

Graph the solutions to the nonlinear inequalities.

45. $y > x^2 - 6x - 8$
46. $4x^2 + y^2 \le 16$
47. $4x^2 - y^2 \ge 16$

Graph the solutions to the systems of nonlinear inequalities.

48. $y < -x^2 + 4$
$y > x - 2$

49. $x^2 + y^2 < 16$
$\dfrac{x^2}{4} + \dfrac{y^2}{16} > 1$

50. $\dfrac{y^2}{9} - x^2 > 1$
$\dfrac{x^2}{4} + \dfrac{y^2}{16} < 1$

Chapter 11 test

Graph each parabola; state the x- and y-intercepts, focus, directrix, and vertex.

1. $y = -4x^2$
2. $y = 2x^2 + 3x - 9$
3. $y = -x^2 - 2x + 8$
4. $x = y^2 - 4y - 12$

Determine the equation of a parabola with the given properties.

5. focus: (1,3), directrix: $y = 2$
6. vertex: $(-2,0)$, directrix: $y = 2$

Refer to the figure for problems 7 and 8.

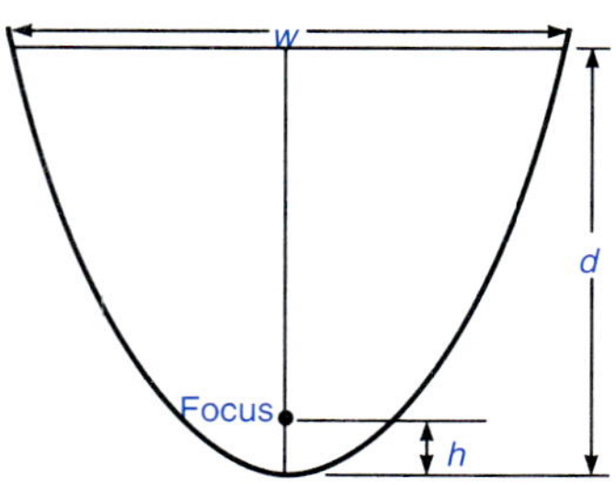

7. $h = 2$, $d = 12$; find w
8. $w = 20$, $d = 10$; find h

Convert each equation into the standard form for an ellipse if necessary. Then graph the equation. State the end points of the major and minor axes and foci where not clear from the graph.

9. $4x^2 + y^2 = 16$
10. $4x^2 + y^2 - 10y = 11$
11. $2x^2 - 8x + 3y^2 + 9y = 15\frac{1}{4}$

12. Find the equation of an ellipse with x-intercepts at $(\pm5,0)$ and y-intercepts at $(0,\pm4)$.

13. Two tacks are put in a board 8 inches apart and a string tied in a loop with length 24 inches is looped around the tacks. The ellipse is drawn. Find an equation that describes the ellipse.

Graph each hyperbola. If necessary transform the equation into one of the two standard forms. State the end points of the major axis and foci where not clear from the graph.

14. $\dfrac{y^2}{9} - \dfrac{x^2}{4} = 1$
15. $4x^2 - y^2 = 8$
16. $\dfrac{(x-2)^2}{25} - \dfrac{(y+1)^2}{9} = 1$
17. $y^2 + 2y - 4x^2 + 16x = 19$

Categorize the graph of each equation as a point, line, circle, parabola, ellipse, or hyperbola (or no graph). Put each equation in the standard form for whichever geometric figure it represents.

18. $4x + 20y - 23 = 0$
19. $9y^2 - 3x^2 - 18 = 0$
20. $2x^2 + y^2 - 6y + 9 = 0$
21. $2x^2 + 2x - 2y^2 + 8y = 5$
22. $4x^2 - 12x + 4y^2 = 0$
23. $x^2 + 8x + y^2 - 4y = 20$
24. $x^2 + 3x - 6 - y = 0$

Solve the following systems of nonlinear equations.

25. $y = x^2 - 2x + 4$
$y = 2x + 1$

26. $\dfrac{x^2}{3} + y^2 = 1$
$y = x - 1$

27. $x^2 + 3y^2 - 8y - 2 = 0$
$y = x^2 - 2$

28. A circle has center at $(-1,3)$ and is tangent to the line $y = 2x - 2$. Find the equation of the circle.

Graph the solutions to the nonlinear inequalities.

29. $x^2 + 3y^2 > 9$

30. $3x^2 - 9y^2 \leq 27$

Graph the solutions to the systems of nonlinear inequalities.

31. $y > x^2 - 9$
$y < x + 2$

32. $x^2 + y^2 > 1$
$\dfrac{x^2}{4} + y^2 \leq 1$

33. An artist wants to construct two ellipses and a circle as shown. Find the equations of all three figures relative to an x-y coordinate system with center at the origin.

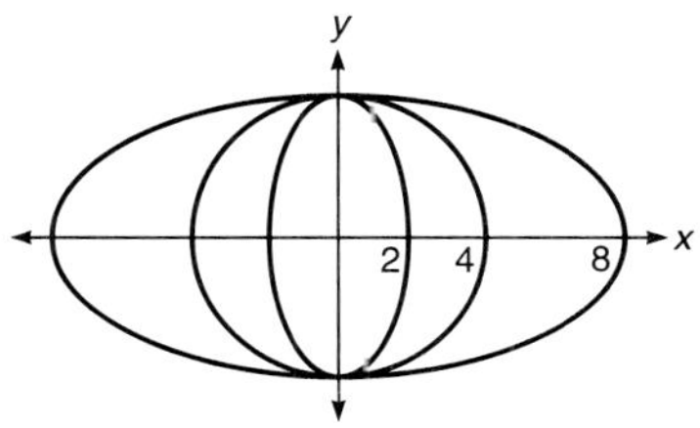

C H A P T E R 12

Topics in Discrete Mathematics

In this chapter we investigate several topics that have wide application in science and business as well as in advanced mathematics. The term **discrete mathematics** refers to the fact that most of these topics can be discussed largely in terms of integers. Discrete is here used in the sense of ''distinct.'' Discrete mathematics has taken on new importance with the advent of electronic digital computers.

12–1 Sequences

A new employee is hired at $26,000 and is told to expect an 8% raise each year for the first six years. What is the employee's pay in each year from the first to the sixth?

This problem describes what is called a series. In this section we study the mathematics that can deal with series, and therefore that can deal with this type of problem.

General principles

A **sequence** is a list of numbers; for example, 1, 3, 5, 7, 9, . . . is a sequence. The ellipsis (. . .) indicates the list goes on indefinitely, making this an infinite sequence. The numbers in the list are called **terms.** Since there is a first, second, third, etc. term in the list, the terms can be paired up with the positive integers.

Sequence

A **finite sequence** is a list of numbers that can be paired up with the set of positive integers 1, 2, 3,. . . , n for some positive integer n.

An **infinite sequence** is a list of numbers that can be paired up with the set of positive integers 1, 2, 3, ...

We refer to the values 1, 2, 3, . . . , n as the **domain** of the sequence. In general, *we define a sequence by a formula for its nth term.* If we call the sequence A, the nth term is called a_n. If the sequence is called B, we call the nth term b_n, etc. Example 12–1 A illustrates this.

■ Example 12–1 A

1. An infinite sequence is defined by the expression $a_n = 3n + 2$. List the first three terms of the sequence.

$$\begin{aligned} a_n &= 3n + 2 \\ a_1 &= 3(1) + 2 = 5 \\ a_2 &= 3(2) + 2 = 8 \\ a_3 &= 3(3) + 2 = 11 \end{aligned}$$

Thus, the sequence looks like 5, 8, 11, . . . $3n + 2$, ...

2. List the first four terms of the sequence in which the nth term is $a_n = (-1)^n(2^n - n)$.

$$\begin{aligned} a_1 &= (-1)^1(2^1 - 1) = -1 \\ a_2 &= (-1)^2(2^2 - 2) = 2 \\ a_3 &= (-1)^3(2^3 - 3) = -5 \\ a_4 &= (-1)^4(2^4 - 4) = 12 \end{aligned}$$

Thus, the sequence looks like -1, 2, -5, 12, . . . $(-1)^n(2^n - n)$. . . . ■

There are times when we wish to find an expression for the general term a_n. This is not always easy or even possible, but in many instances examination of the values of the terms will provide some guidance. This is illustrated in example 12–1 B.

■ Example 12–1 B

Find an expression for the general term of each sequence.

1. 7, 12, 17, 22, 27, . . .
   Subtract 2 from each term, giving the sequence 5, 10, 15, 20, . . . in which the nth term is of the form $5n$. The general term of the original sequence is 2 more than this: $a_n = 5n + 2$.

2. -3, 9, -15, 21, . . .
   The alternating signs can be accounted for by a factor of $(-1)^n$. We thus consider the positive valued sequence 3, 9, 15, 21, ...

   Subtracting 3 from each term yields the sequence 0, 6, 12, 18, . . . in which we have multiples of 6, with general term $6(n - 1)$.

   Thus, the sequence 3, 9, 15, 21, . . . can be expressed by the general term $6(n - 1) + 3 = 6n - 3$, and the general term of the original sequence is $a_n = (-1)^n(6n - 3)$.

3. A manufacturer is testing an electronics device in a high-heat situation. The total number of failed devices is recorded each hour. In the first 5 hours, the total numbers of failed devices recorded each hour are 3, 8, 12, 18, and 23. Create a sequence that approximates this pattern and use this to predict the number of failed devices that will be counted in the seventh hour.

If we make a list of the *increase* in the number of failed devices each hour we obtain the sequence 5, 4, 6, 5. Note that the average of these values is 5. It would seem logical to simulate the actual sequence with the following one: 3, 8, 13, 18, 23, in which the increase is a constant, 5. Using this sequence, we obtain a sixth element of 28 and a seventh element of 33. Thus, it might be reasonable to predict that the number of failed devices that will be counted in the seventh hour will be 33. The general term for this sequence is $a_n = 5n - 2$. ■

Arithmetic sequences

Many sequences encountered in practice are a variety in which each succeeding element can be found by adding a constant to the previous term. For example, if a_1 is 5, and we add 3 to get each next term, then the sequence is 5, 8, 11, 14, ... These sequences are called **arithmetic sequences.**

> **Arithmetic sequence**
> An arithmetic sequence A is a sequence in which $a_{n+1} = a_n + d$ for all terms of the sequence and for some (fixed) real number d.

Observe that the definition implies that $a_{n+1} - a_n = d$ for all terms of the sequence. In other words the difference between successive terms is a constant. We call d the **common difference.**

■ ***Example 12–1 C***

1. 40, 34, 28, 22, 16 is an arithmetic sequence, with $d = -6$.
2. 2, 5, 9, 14, 20 is not an arithmetic sequence since the difference between terms is not constant. ■

The general term of an arithmetic sequence

To find a general expression for any term a_n of an arithmetic sequence consider the following pattern.

$$\begin{aligned} &a_1 \\ a_2 &= a_1 + d \\ a_3 &= a_1 + 2d \\ a_4 &= a_1 + 3d \end{aligned}$$

This suggests the following.

> **General term of an arithmetic sequence**
> If A is an arithmetic sequence with first term a_1 and common difference d, then the general term a_n is
> $$a_n = a_1 + (n - 1)d$$

Example 12–1 D illustrates some uses of this formula for the general term of an arithmetic sequence.

■ ***Example 12–1 D***

1. Given an arithmetic sequence in which $a_1 = 100$ and $a_{21} = 10$. Find a_{50}.
We must find d to apply the formula. The fact $a_{21} = 10$ gives us a value for n and for a_n to substitute into the general formula; we also use the value 100 for a_1.

$a_n = a_1 + (n - 1)d$ General term of an arithmetic sequence

$a_{21} = a_1 + (21 - 1)d$ General term when $n = 21$

$10 = 100 + (21 - 1)d$ Replace a_{21} with 10, a_1 with 100

$-\frac{9}{2} = d$

We can now find a_{50}.

$a_{50} = 100 + 49(-\frac{9}{2})$ $a = 100, d = -\frac{9}{2}, n = 50$

$= -120\frac{1}{2}$

2. Find the number of terms in the arithmetic sequence $-9, -4, 1, 6, \ldots, 111$.
We can determine that $a_1 = -9$ and $d = 5$ from the first few terms. We know that 111 is the nth term; we just don't know n yet.

$$a_n = a_1 + (n - 1)d$$
$$111 = -9 + (n - 1)(5)$$
$$25 = n$$

Thus, 111 is the 25th term, so the sequence has 25 terms. ■

Geometric sequences

Another common type of sequence is called a **geometric sequence.** Geometric sequences are used to describe the growth of everything from populations to bank accounts.

In a geometric sequence, each next element can be found by multiplying the previous term by a constant, called r, for **common ratio.** For example, if a_1 is 5, and the common ratio is 2, then the sequence is

$$\begin{array}{lllll} 5, & 5 \cdot 2, & 5 \cdot 2 \cdot 2, & 5 \cdot 2 \cdot 2 \cdot 2, \ldots & \\ 5, & 5 \cdot 2, & 5 \cdot 2^2, & 5 \cdot 2^3, \ldots & \\ 5, & 10, & 20, & 40, & 80, \ldots \end{array}$$

> **Geometric sequence**
> A geometric sequence is a sequence in which $a_{n+1} = r \cdot a_n$ for all terms of the sequence and for some real number r. We require $r \neq 0$ and $a_1 \neq 0$.

The restrictions on r and a_1 are so that any given geometric sequence has only one description in terms of a_1 and r. See problem 91 in the exercises for an illustration of what happens without these restrictions.

Observe that the definition implies that $\dfrac{a_{n+1}}{a_n} = r$ for all terms of the sequence. This can be used to determine if a sequence is or is not a geometric sequence.

■ *Example 12–1 E*

1. 1, −2, 4, −8, 16 is a geometric sequence, with $r = -2$, because
$$\frac{-2}{1} = \frac{4}{-2} = \frac{-8}{4} = \frac{16}{-8} = -2$$
2. 2, 4, 6, 8 is not a geometric sequence since the ratio of successive terms is not constant. (It is in fact an arithmetic sequence.) Observe that $\frac{4}{2} = 2$ but $\frac{6}{4} = 1.5$. ■

The general term of a geometric sequence

To find a general expression for the nth term a_n of a geometric sequence consider the expressions

$$\begin{aligned} &a_1 \\ &a_2 = a_1 r \\ &a_3 = a_2 r = (a_1 r)r = a_1 r^2 \\ &a_4 = a_3 r = (a_1 r^2)r = a_1 r^3 \end{aligned}$$

which suggest the following.

General term of a geometric sequence

If A is a geometric sequence with first term a_1 and common ratio r, then the general term a_n is

$$a_n = a_1 r^{n-1}$$

The use of the general term of a geometric sequence is shown in example 12–1 F.

■ *Example 12–1 F*

1. Find the fourth term of a geometric sequence in which $a_1 = 5$ and $r = 1\frac{1}{4}$.
$$a_4 = a_1 r^3 = 5\left(\frac{5}{4}\right)^3 = 5\left(\frac{5^3}{4^3}\right) = \frac{5^4}{4^3} = \frac{625}{64}$$
2. Given a geometric sequence in which $a_1 = 160$ and $a_8 = \frac{5}{4}$, find a_4.

$a_n = a_1 r^{n-1}$	General formula
$a_8 = a_1 r^{8-1}$	General formula for $n = 8$
$\frac{5}{4} = 160r^7$	Replace a_8 with $\frac{5}{4}$ and a_1 with 160
$\frac{5}{4} \cdot \frac{1}{160} = \frac{1}{160} \cdot 160r^7$	Multiply each member with $\frac{1}{160}$
$\frac{1}{128} = r^7$	
$(\frac{1}{2})^7 = r^7$	
$\frac{1}{2} = r$	

We can now compute a_4.
$$a_4 = a_1 r^3 = 160(\tfrac{1}{2})^3 = 160(\tfrac{1}{8}) = 20$$
■

Notes on the general expression of a sequence

It needs to be noted that the first few terms of a sequence do not necessarily give enough information to find the general term; consider the sequence 1, 2, 4, ... One expression for a_n is 2^{n-1}. This general term produces 8 for the fourth term. However, the expression $c_n = \frac{1}{2}n^2 - \frac{1}{2}n + 1$ also produces 1, 2, 4 for the first three terms, but 7 for the fourth term. In fact, given the first n terms of a sequence it is possible to derive an unlimited number of expressions for the general term; we will not pursue this further here,[1] however, except to say that to fully specify a sequence we must actually include a rule or expression for the general term.

Further, an expression does not always suffice for the general term. For some sequences, no expression is possible. For example, we could specify a sequence with the *rule* that the nth term is the nth digit in the decimal expansion for π. Thus, the sequence looks like 3, 1, 4, 1, 5, 9, 2, ... It is impossible (even in theory) to find a general expression for the nth term of this sequence.

Mastery points

Can you

- Find the terms of a sequence, given the general term?
- Find any given term of a sequence from the general term?
- Find an expression for the general term of a sequence, given the first few terms of the sequence?
- Identify arithmetic and geometric sequences?

Exercise 12–1

List the first four terms of each sequence.

1. $a_n = \frac{5}{2}n - 3$
2. $a_n = \dfrac{3n - 4}{2}$
3. $b_n = 2^n - n^2$
4. $b_n = (\frac{1}{2})^n$
5. $a_n = 3$
6. $b_n = 30 - n^2$
7. $c_n = n^2 - 4n + 2$
8. $a_n = (n - 5)^3$
9. $a_n = \dfrac{\sqrt{n}}{n + 1}$
10. $c_n = \dfrac{n}{n - 2}$
11. $b_n = (n - 2)(n + 3)$
12. $a_n = 11 + \dfrac{n}{5}$

Find an expression for the general term of each sequence.

13. 2, 5, 8, 11, . . .
14. 3, 6, 9, 12, . . .
15. −20, −16, −12, . . .
16. −8, −3, 2, 7, . . .
17. $-1, \frac{1}{2}, -\frac{1}{3}, \frac{1}{4}$ . . .
18. $\frac{2}{3}, \frac{3}{4}, \frac{4}{5}, \frac{5}{6}, \ldots$
19. $\frac{1}{6}, \frac{2}{11}, \frac{3}{16}, \frac{4}{21}, \ldots$
20. $\frac{3}{2}, \frac{5}{4}, \frac{7}{6}, \frac{9}{8}, \ldots$
21. $\frac{1}{4}, -\frac{4}{9}, \frac{9}{16}, -\frac{16}{25}, \ldots$
22. 2, 6, 12, 20, . . .
23. 1, −1, 1, −1, . . .
24. 1, 0.1, 0.01, 0.001, . . .

[1]See the last two problems in the exercises for further development.

25. A biology researcher measured the population of a certain insect under laboratory conditions every 5 hours and obtained the values 300, 400, 530, 710. Approximately what value might the researcher expect for the next measurement?

26. A manufacturer makes integrated circuits by putting a certain number of circuits on a circular silicon wafer. The manufacturer can vary the number of circuits per wafer by varying the diameter of the wafer. The manufacturer has found that the number of bad circuits per wafer varies in the manner shown in the table. How many bad circuits would the manufacturer expect on a wafer with diameter 8″?

Diameter	Number of bad circuits
2″	4
3″	9
4″	16
5″	25

Characterize the sequence from the problem indicated as arithmetic, geometric, or neither. State the common difference or common ratio as appropriate.

27. Problem 1
28. Problem 2
29. Problem 3
30. Problem 4
31. Problem 5
32. Problem 6
33. Problem 7
34. Problem 8
35. Problem 9
36. Problem 10
37. Problem 11
38. Problem 12
39. Problem 13
40. Problem 14
41. Problem 15
42. Problem 16
43. Problem 17
44. Problem 18
45. Problem 19
46. Problem 20
47. Problem 21
48. Problem 22
49. Problem 23
50. Problem 24

Solve the following problems.

51. Find the 12th term of an arithmetic sequence in which $a_1 = 6$ and $d = 5$.

52. Given an arithmetic sequence in which $a_1 = 10$ and $a_{21} = 220$, find a_{25}.

53. Find the number of terms in the arithmetic sequence 7, 11, 15, . . . , 135.

54. Find the 15th term of an arithmetic sequence in which $a_1 = -200$ and $d = 3$.

55. Given an arithmetic sequence in which $a_1 = -40$ and $a_{15} = 40$, find a_{14}.

56. Find the number of terms in the arithmetic sequence 150, $148\frac{1}{2}$, 147, . . . , 39.

57. Find the 27th term of an arithmetic sequence in which $a_1 = 6$ and $a_{40} = 300$.

58. Given an arithmetic sequence in which $a_1 = -46$ and $a_{21} = 150$, find a_{11}.

59. Find the number of terms in the arithmetic sequence $-42, -39\frac{1}{2}, \ldots, -9\frac{1}{2}$.

60. Find a_{200} of the arithmetic sequence where $a_1 = 19$ and $a_{10} = 28$.

61. In an arithmetic sequence, $a_{15} = 49$ and $a_{28} = 88$. Find a_4.

Find a_n for the following geometric sequences for the given values of a_1, r, and n.

62. $a_1 = 10, r = -2, n = 5$
63. $a_1 = \frac{1}{4}, r = -2, n = 6$
64. $a_1 = \frac{1}{40}, r = 4, n = 4$
65. $a_1 = -12, r = -\frac{1}{2}, n = 5$
66. $a_1 = 400, r = 0.1, n = 3$
67. $a_1 = \frac{2}{3}, r = \frac{3}{2}, n = 3$

68. Find the 4th term of a geometric sequence in which $a_1 = 4$ and $a_2 = 1$.

69. Given a geometric sequence in which $a_1 = 5$ and $a_5 = \frac{5}{81}$, find a_4.

70. Given a geometric sequence in which $a_3 = \frac{1}{3}$ and $a_6 = -\frac{1}{81}$, find a_2.

71. Find the sixth term of a geometric sequence in which $a_1 = \frac{9}{25}$ and $r = 5$.

72. Given a geometric sequence in which $a_1 = \frac{1}{32}$ and $a_4 = -\frac{1}{4}$, find a_5.

73. Given a geometric sequence in which $a_2 = \frac{5}{4}$ and $a_4 = 5$, find a_1.

74. A pendulum swings a distance of 20 inches on its first swing. Each subsequent swing is 95% of the previous distance. How far does the pendulum swing on the (a) fourth swing, and (b) eighth swing?

75. A new employee is hired at \$26,000 and is told to expect an 8% raise each year for the first six years. What is the employee's pay in each year from the first to the sixth?

76. A ball is dropped from a height of 10 feet. If the ball rebounds three-fourths of the height of its previous fall with each bounce, how high does it rebound on the (a) third bounce, (b) sixth bounce, and (c) nth bounce?

77. A machine is to be depreciated by what is called the "constant percentage method." A certain, fixed percentage will be deducted from the machine's value every year. Suppose the machine cost \$15,000 and has a scrap value of \$3,000 after six years. Find the rate r of depreciation. That is, solve $15{,}000\,(1 - r)^6 = 3{,}000$ for r. Find r to the nearest 0.1%.

Problems 78 through 83 are related. In problems 78, 79, and 80, assume two arithmetic sequences a and b such that $a_n = 2n + 5$ and $b_n = 1 - n$.

78. Define a new sequence c such that $c_n = a_n + b_n$.
a. Write the first four terms of the sequences a, b, and c. **b.** Is the sequence c arithmetic?
c. Find an expression for c_n and use it to compute c_{20}.

79. Define a new sequence d such that $d_n = 3a_n$.
a. Write the first four terms of the sequences a and d.
b. Is d an arithmetic sequence?
c. Find an expression for d_n and use it to compute d_{20}.

80. Define a new sequence e such that $e_n = a_n \cdot b_n$.
a. Write the first four terms of the sequences a, b, and e. **b.** Is e an arithmetic sequence?
c. Find an expression for e_n and use it to compute e_{20}.

81. Suppose a_n and b_n are two arithmetic sequences, and a new sequence c is defined such that $c_n = a_n + b_n$. Is the new sequence an arithmetic sequence? Prove or disprove this statement.

82. Suppose a_n is an arithmetic sequence and k is some constant. Is the sequence b defined as $b_n = ka_n$ an arithmetic sequence? Prove or disprove this statement.

83. Suppose a_n and b_n are two arithmetic sequences, and a new sequence c is defined such that $c_n = (a_n)(b_n)$. Is the new sequence an arithmetic sequence? Prove or disprove this statement.

Problems 84 through 89 are related. In problems 84, 85, and 86 assume two geometric sequences a and b such that $a_n = 3^n$ and $b_n = 3(2^n)$.

84. Define a new sequence c such that $c_n = a_n + b_n$.
a. Write the first four terms of the sequences a, b, and c. **b.** Is the sequence c geometric?
c. Find an expression for c_n and use it to compute c_5.

85. Define a new sequence d such that $d_n = \frac{1}{2}a_n$.
a. Write the first four terms of the sequences a and d.
b. Is d a geometric sequence?
c. Find an expression for d_n and use it to compute d_5.

86. Define a new sequence e such that $e_n = a_n \cdot b_n$.
a. Write the first four terms of the sequences a, b, and e. **b.** Is e a geometric sequence?
c. Find an expression for e_n and use it to compute e_5.

87. Suppose a_n and b_n are two geometric sequences, and a new sequence c is defined such that $c_n = a_n + b_n$. Is the new sequence a geometric sequence? Prove or disprove this statement.

88. Suppose a_n is a geometric sequence and k is some constant. Is the sequence b defined as $b_n = ka_n$ a geometric sequence? Prove or disprove this statement.

89. Suppose a_n and b_n are two geometric sequences, and a new sequence c is defined such that $c_n = (a_n)(b_n)$. Is the new sequence a geometric sequence? Prove or disprove this statement.

90. A store charges \$5 to develop a roll of film. With each roll of film the customer gets a coupon. With four coupons the customer gets a roll developed free. Let A be the sequence in which a_n represents the *average* price for developing n rolls of film, one after the other. Find an expression for a_n.

Hint: The greatest integer function may be helpful. This is the function $f(x) = [x]$, in which $[x]$ is the greatest integer less than or equal to x. For $x > 0$ this is equivalent to throwing away the fractional part of a number. For example, $[1.8] = 1$, $[9\frac{1}{2}] = 9$, etc.

91. In the definition of a geometric sequence we required $r \neq 0$ and $a_1 \neq 0$. If we remove these restrictions, which of the following would be a geometric sequence?
a. 5, 0, 0, 0, 0, . . .
b. −3, 0, 0, 0, . . .
c. 0, 0, 0, 0, . . .
d. $a_1 = \sqrt{2}$, $r = 0$
e. $a_1 = 0$, $r = \sqrt{2}$
f. $a_1 = 0$, $r = -25{,}000$

92. The following is sometimes called the "Coxeter-Ulam" algorithm.
First, select any natural number as the first element of the sequence.
Perform the following procedure to get the next element of the sequence.

- If the element is even, divide by 2.
- If the element is odd, multiply by 3 and add 1.

Continue this procedure to derive new elements in the sequence, but stop if the element is one.
This algorithm will generate a sequence of numbers and each sequence seems to have a certain property in common. Generate a few of these sequences and determine this property. (No one has been able to prove that all such sequences have this property.) (*Hint:* Start with a first element of 2, 3, 4, 5, 6. Then try 21, then 7, as the first element. Try other values at random.)

As stated in the text, given the first, say, 3 terms of a sequence (or the first n terms, for that matter), it is possible to find an unlimited number of expressions for the general term. One way to provide some more examples is to take any geometric sequence with terms a_n, and assume there is some other sequence b_n that has the same three first elements. Substitute the pairs of values $(1,b_1)$, $(2,b_2)$, and $(3,b_3)$ into $b_n = An^2 + Bn + C$. (This is called finding a "quadratic interpolation formula.")

For example, consider the geometric sequence 2, 3, $4\frac{1}{2}$, with $a_1 = 2$, $r = \frac{3}{2}$. Consider these as ordered pairs $(n,b_n) = (1,2), (2,3), (3,4\frac{1}{2})$ and substitute.

$$b_n = An^2 + Bn + C$$

$$n = 1:\quad 2 = A(1^2) + B(1) + C;\ 2 = A + B + C$$

$$n = 2:\quad 3 = A(2^2) + B(2) + C;\ 3 = 4A + 2B + C$$

$$n = 3:\quad 4\tfrac{1}{2} = A(3^2) + B(3) + C;\ 4\tfrac{1}{2} = 9A + 3B + C$$

Now, solve this system of three equations in three unknowns (chapter 10) for A, B, and C. We obtain $A = B = \frac{1}{4}$, $C = \frac{3}{2}$ so the expression is $b_n = \frac{1}{4}n^2 + \frac{1}{4}n + \frac{3}{2}$.

In the geometric sequence, the formula is $a_n = 2(\frac{3}{2})^{n-1}$, and $a_4 = 6\frac{3}{4}$, whereas $b_4 = 6\frac{1}{2}$. Thus, given the three terms 2, 3, $4\frac{1}{2}$, we can find at least two different sequences that begin with these terms.

93. Using this example, find a quadratic expression that defines a sequence that begins with the same three terms as the given sequence but is different in the fourth term.
a. geometric sequence with $a_1 = 9$, $r = \frac{2}{3}$
b. geometric sequence with $a_1 = 3$, $r = \frac{2}{3}$
c. 3, 1, 4, 1, . . . , where a_n is the nth term in the decimal expansion of π

94. Ramsey theory, named for Frank Plumpton Ramsey, an English mathematician in the first half of the twentieth century, discusses finding order in disorder. One unexpected implication of this theory is the following.

Take the arithmetic progression 1, 2, 3, 4, 5, 6, 7, 8, 9. Underline some of the values, and leave the rest not underlined. Ramsey theory indicates that either three of the underlined or three of the not-underlined values will form an arithmetic progression.[2] For example, consider the arrangement

$$1\quad \underline{2}\quad 3\quad \underline{4}\quad 5\quad 6\quad \underline{7}\quad 8\quad 9$$

The not-underlined values 1, 3, and 5 form an arithmetic progression. In

$$1\quad \underline{2}\quad 3\quad \underline{4}\quad \underline{5}\quad 6\quad \underline{7}\quad 8\quad 9$$

the values 3, 6, and 9 form an arithmetic progression.

Create two other arrangements by underlining some (or none) of the values in the sequence 1 . . . 9, then find at least one arithmetic progression in each.

[2]For an understandable, short proof of this fact see the excellent article "Ramsey Theory" by Ronald L. Graham and Joel H. Spencer in the July 1990 issue of *Scientific American* magazine.

Skill and review

1. If $3 + 6 + 9 + \cdots + 3n = 231$, what is $1 + 2 + 3 + \cdots + n$?

2. Find the sum $(1 - 5) + (5 - 9) + (9 - 13) + \cdots + (81 - 85)$.

3. If $x_1 + x_2 + x_3 + \cdots + x_n = 420$, what is $3x_1 + 3x_2 + 3x_3 + \cdots + 3x_n$?

4. If $a_1 + a_2 + a_3 + \cdots + a_n = 500$ and $b_1 + b_2 + b_3 + \cdots + b_n = 200$, what is $(a_1 + b_1) + (a_2 + b_2) + (a_3 + b_3) + \cdots + (a_n + b_n)$?

5. Graph $\dfrac{y^2}{16} - \dfrac{x^2}{9} = 1$.

6. Solve $\left|3 - \frac{1}{2}x\right| > 12$.

12–2 Series

A vacuum pump removes one-fifth of the remaining air in a tank with each stroke. (a) How much air remains in the tank after the fifth stroke? (b) How many strokes would be necessary to remove 98% of the air?

The solution to the problem posed here involves the mathematics of series, which is what we study in this section.

An expression that indicates the summation of the terms of a sequence is called a **series.** Such a sum might represent the total distance traveled by an accelerating object, the population of a town after a few years, or the present value of an annuity that will pay for someone's education.

For the finite sequence

$$5, 10, 15, 20$$

the expression

$$5 + 10 + 15 + 20$$

is its series. The resulting value, 50, is the sum of the series. We would also refer to 5, 10, 15, and 20 as the terms of the series.

Sigma notation

Sigma notation[3] is a convenient way to express a series in more compact form. Σ is the capital letter *sigma* in the Greek alphabet. We use it to indicate the word "sum," or "series." For example, $\sum_{i=1}^{4}(3i + 2)$ is read "the sum of the terms $3i + 2$ as i takes on the integer values from 1 through 4." It means that the ith term of a series is $3i + 2$, and that we are interested in terms 1 through 4, giving the series

i	1	2	3	4
term	$(3 \cdot 1 + 2) +$	$(3 \cdot 2 + 2) +$	$(3 \cdot 3 + 2) +$	$(3 \cdot 4 + 2)$
series		$5 + 8 + 11 + 14$		

Note that the letter i was not significant in the notation; we could have used any letter we wished. Example 12–2 A illustrates expanding sigma notation.

■ **Example 12–2 A**

Expand the following sigma expressions.

1. $\sum_{j=3}^{6}(4j - 1)$

$(4 \cdot 3 - 1) + (4 \cdot 4 - 1) + (4 \cdot 5 - 1) + (4 \cdot 6 - 1)$
$11 + 15 + 19 + 23$

[3]The Σ symbol was first used by Leonhard Euler in 1755.

2. $\sum_{k=2}^{5} (-1)^k \left(\frac{k}{k-1}\right)^2$

$(-1)^2 (\frac{2}{1})^2 + (-1)^3 (\frac{3}{2})^2 + (-1)^4 (\frac{4}{3})^2 + (-1)^5 (\frac{5}{4})^2$

$4 - \frac{9}{4} + \frac{16}{9} - \frac{25}{16}$ ■

If the number of terms in the series is infinite, the corresponding series may not have a sum. For example, if the sequence is 1, 3, 5, 7, . . . , the series is 1 + 3 + 5 + 7 +. . . , which cannot be summed. We can, however, add up finite subsequences. For example,

$$1 = 1$$
$$1 + 3 = 4$$
$$1 + 3 + 5 = 9$$
$$1 + 3 + 5 + 7 = 16, \text{ etc.}$$

Sum of the terms of a finite arithmetic sequence

To obtain an expression for the sum of the first n terms (S_n) of an arithmetic sequence we write the expression for the sum forward and also backward, as shown, and add up the terms on each side of the equations.

$$\begin{array}{rcccccccccc} S_n & = & a_1 & + & (a_1 + d) & + & (a_1 + 2d) & + \cdots + & [a_1 + (n-1)d] \\ S_n & = & a_n & + & (a_n - d) & + & (a_n - 2d) & + \cdots + & [a_n - (n-1)d] \\ \hline 2S_n & = & (a_1 + a_n) & + & (a_1 + a_n) & + & (a_1 + a_n) & + \cdots + & (a_1 + a_n) \end{array}$$

n terms of $(a_1 + a_n)$

$$2S_n = n(a_1 + a_n)$$
$$S_n = \frac{n}{2}(a_1 + a_n)$$

Sometimes we do not know the value of a_n; we can derive a formula for this purpose.

$S_n = \frac{n}{2}(a_1 + a_n)$ Sum of first n terms of an arithmetic series

$a_n = a_1 + (n - 1)d$ Expression for a_n

$S_n = \frac{n}{2}[a_1 + (a_1 + (n - 1)d)]$ Replace a_n by $a_1 + (n - 1)d$

$S_n = \frac{n}{2}[2a_1 + (n - 1)d]$ Formula for S_n.

This formula does not require that a_n be calculated to find S_n. This is summarized as follows.

Sum of the first n terms of an arithmetic sequence

The sum S_n of the first n terms of an arithmetic sequence with first term a_1 and nth term a_n is

$$S_n = \frac{n}{2}(a_1 + a_n)$$

or

$$S_n = \frac{n}{2}[2a_1 + (n - 1)d]$$

S_n is called the nth **partial sum.**

Example 12–2 B illustrates summing finite arithmetic sequences.

■ ***Example 12–2 B***

Find the required sum.

1. $a_n = 2n - 1$; sum the first 5 terms.
Compute a_1 and a_5 first.

$a_1 = 1; a_5 = 9$ — Use $a_n = a_1 + (n - 1)d$

$S_5 = \frac{5}{2}(1 + 9) = 25$ — Use $S_n = \frac{n}{2}(a_1 + a_n)$; $n = 5$, $a_1 = 1$, $a_5 = 9$

2. Add up the values 2, 9, 16, 23, . . . , 65.
$a_1 = 2$, $d = 7$. We need to know how many terms there are.

$a_n = a_1 + (n - 1)d$ — General expression for nth term

$65 = 2 + (n - 1)(7)$ — Find n for which $a_n = 65$

$10 = n$

Thus, 65 is a_{10}, and we therefore want to find S_{10}.
$S_{10} = \frac{10}{2}(2 + 65) = 5(67) = 335$ ■

Sum of the terms of a finite geometric sequence

Just as the expression just discussed determines the nth partial sum of an arithmetic sequence, there is an expression that determines the nth partial sum of a geometric sequence. The derivation of this formula is left as an exercise.

Sum of the first n terms of a geometric sequence

The nth partial sum S_n of the first n terms of a geometric sequence with first term a_1 and ratio r, $r \neq 1$, is

$$S_n = \frac{a_1(1 - r^n)}{1 - r}$$

S_n is often called the nth **partial sum.**

■ ***Example 12–2 C***

Find the required nth partial sum.

1. A geometric series with $a_1 = 3$, $r = \frac{1}{2}$, and $n = 6$.

$$S_6 = \frac{3(1 - (\frac{1}{2})^6)}{1 - \frac{1}{2}} = 5\tfrac{29}{32}$$

2. $\sum_{k=1}^{7} 3(2)^k$

The series is $3 \cdot 2 + 3 \cdot 2^2 + 3 \cdot 2^3 + \cdots + 3 \cdot 2^7$. Thus $a_1 = 6$ and $r = 2$. We want S_7.

$$S_7 = \frac{6(1 - 2^7)}{1 - 2} = 762$$

Infinite geometric series

Suppose a sprinter runs a 100-meter race. One view of the race is as follows. First, the runner runs to the 50-meter position, or half the distance to the finish ($\frac{1}{2}$ of the total distance). Then, the runner runs to the 75-meter position, or half the remaining distance ($\frac{1}{4}$ of the total distance). Next, the runner runs half the remaining distance, getting to the 87.5-meter mark ($\frac{1}{8}$ of the total distance). The runner goes on and on in this fashion, always attaining a goal and then running half the distance remaining to the finish. See figure 12–1.

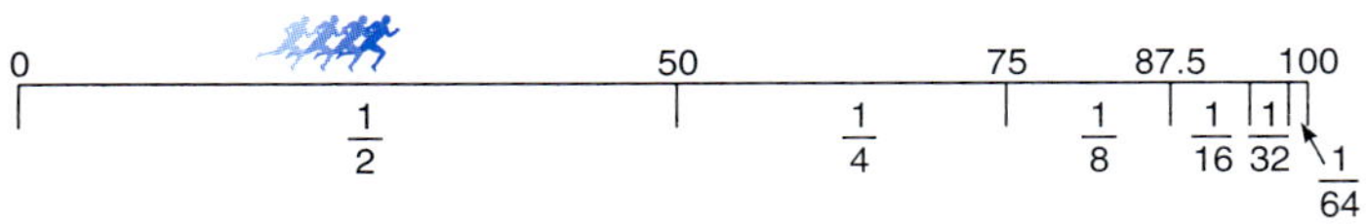

Figure 12–1

Note that the runner covers the following parts of the course: $\frac{1}{2} + \frac{1}{4} + \frac{1}{8} + \frac{1}{16} + \cdots + \frac{1}{2^n} + \cdots$. Since the runner completes one race, it makes sense to say that $\frac{1}{2} + \frac{1}{4} + \frac{1}{8} + \frac{1}{16} + \cdots + \frac{1}{2^n} + \cdots = 1$.

Now the values $\frac{1}{2}, \frac{1}{4}, \frac{1}{8}, \frac{1}{16}, \ldots, \frac{1}{2^n}, \ldots$ determine a geometric series that does not terminate. Such a series is said to be an **infinite geometric series.** The nth partial sum would be $\sum_{i=1}^{n} (\frac{1}{2})^n = \frac{\frac{1}{2}(1 - (\frac{1}{2}))^n}{1 - \frac{1}{2}} = 1 - \left(\frac{1}{2}\right)^n$. As n gets greater and greater, the fraction $(\frac{1}{2})^n$ in the expression $1 - (\frac{1}{2})^n$ gets closer and closer to 0, and so the nth partial sum gets closer and closer to the value 1. This is in accord with what we said about running the race. In like fashion, the value of r^n gets less and less as n gets greater and greater in the expression $\dfrac{a_1(1 - r^n)}{1 - r}$, as long as $|r| < 1$. This means that the expression gets closer and closer to $\dfrac{a_1(1 - 0)}{1 - r} = \dfrac{a_1}{1 - r}$, leading to the following definition.

> **Sum of an infinite geometric series**
> If $|r| < 1$ the sum S of the terms of an infinite geometric series is
> $$S = \frac{a_1}{1 - r}$$

Note If $|r| \geq 1$ the sum is not defined. To see why, consider the series $1 + 3 + 9 + 27 + \cdots + 3^{n-1} + \cdots$. As each term is added, this sum grows larger and larger, by ever increasing amounts.

Summing infinite geometric sequences is illustrated in example 12–2 D.

■ ***Example 12–2 D***

Find the sum of the infinite geometric series.

1. $\sum_{i=1}^{\infty} 3(\frac{2}{3})^i$

The symbol ∞ (infinity) means there is no last term. This is an infinite geometric series with $a_1 = 2$ and $r = \frac{2}{3}$, and, since $|r| < 1$, the infinite sum is defined. $S = \dfrac{2}{1 - \frac{2}{3}} = \dfrac{2}{\frac{1}{3}} = 6$.

2. $\sum_{i=1}^{\infty} (\frac{4}{3})^i$

$a_1 = \frac{4}{3}$ and $r = \frac{4}{3}$. Since $|r| > 1$ this series does not have a sum. ■

Rational form of repeating decimal numbers

Repeating decimal numbers can be viewed as the sums of infinite geometric series. This can be used to find their rational form. For example, we are familiar with the fact that $\frac{2}{3} = 0.66\overline{6}$, but what is $0.77\overline{7}$? Example 12–2 E answers this question.

■ ***Example 12–2 E***

Find the rational number form of the repeating decimal.

1. $0.77\overline{7}$

This is $\dfrac{7}{10} + \dfrac{7}{100} + \dfrac{7}{1{,}000} + \cdots$ or $\dfrac{7}{10} + 7\left(\dfrac{1}{10}\right)^2 + 7\left(\dfrac{1}{10}\right)^3 + \cdots$, which is an infinite geometric series with $a_1 = 0.7$ and $r = 0.1$. Thus, $S = \dfrac{a}{1 - r} = \dfrac{0.7}{1 - 0.1} = \dfrac{0.7}{0.9} = \dfrac{7}{9}$.

2. $0.343434\overline{34}$

This can be written as $0.34 + 0.0034 + 0.000034 + \cdots$, or $34\left(\frac{1}{100}\right) + 34\left(\frac{1}{100}\right)^2 + 34\left(\frac{1}{100}\right)^3 + \cdots$, which is an infinite geometric series with $a_1 = 0.34$ and $r = 0.01$, so $S = \frac{0.34}{1 - 0.01} = \frac{0.34}{0.99} = \frac{34}{99}$.

3. $0.531\overline{31}$

If $x = 0.531\overline{31}$, then $10x = 5.31\overline{31}$. We first find the value of $0.31\overline{31}$, which is an infinite geometric progression in which $a_1 = 0.31$ and $r = 0.01$, so $S = \frac{0.31}{1 - 0.01} = \frac{0.31}{0.99} = \frac{31}{99}$.

Thus, $10x = 5\frac{31}{99} = \frac{526}{99}$, so $x = \frac{1}{10}\left(\frac{526}{99}\right) = \frac{263}{495}$. ■

Infinite geometric series find wide application in solving problems from areas outside of mathematics.

■ *Example 12–2 F*

A ball is dropped from a height of 24 meters. Each time it strikes the floor, the ball rebounds to a height that is three-fourths of the previous height. Find the total vertical distance that the ball travels before it comes to rest on the floor. See the figure.

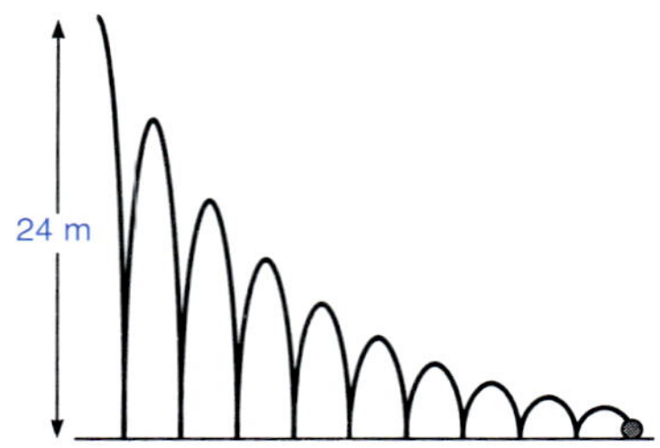

The sequence of distances that the ball travels would be 24, <u>18</u>, 18, <u>13.5</u>, 13.5, <u>10.125</u>, 10.125, etc. The subsequence (underlined) of every other term beginning at 18, which is 18, 13.5, 10.125, etc., is an infinite geometric progression.

We can find its sum and double this value, then add 24. This is an infinite geometric series with $a_1 = 18$ and $r = 0.75$, so $S = \frac{18}{1 - 0.75} = 72$. Thus, the ball travels $24 + 2(72) = 168$ feet before coming to rest on the floor. ■

Mastery points

Can you

- Expand sigma expressions into series?
- Find the sum of finite arithmetic and geometric series?
- Find the sum of certain infinite geometric series?
- Solve applications using arithmetic and geometric series?

Exercise 12–2

Expand the following sigma expressions.

1. $\sum_{j=1}^{4} (4j + 1)$ **2.** $\sum_{j=3}^{5} (3j^2 + 1)$ **3.** $\sum_{j=2}^{5} j(j + 1)$ **4.** $\sum_{j=1}^{3} (j + 3)(j - 1)$ **5.** $\sum_{j=3}^{4} \frac{j}{j + 1}$

6. $\sum_{j=3}^{5} \frac{3j - 2}{j}$ 7. $\sum_{j=1}^{6} (-1)^j\left(\frac{4}{3j}\right)$ **8.** $\sum_{j=1}^{4} (-1)^j(j + 1)^j$ **9.** $\sum_{j=1}^{4} \left(\sum_{k=1}^{j} k^2\right)$ **10.** $\sum_{j=1}^{3} \left(\sum_{k=1}^{j} (k - 1)^2\right)$

Find the sum of the series determined by the given arithmetic sequence.

11. $3, 6, 9, \ldots, 96$

12. $2, 8, 14, \ldots, 62$

13. $-10, -14, -18, \ldots, -66$

14. $\frac{1}{3}, \frac{2}{3}, 1, 1\frac{1}{3}, \ldots, 20\frac{2}{3}$

15. $-8, -7\frac{1}{4}, -6\frac{1}{2}, \ldots, 1$

16. $-20, -5, 10, \ldots, 85$

17. $a_1 = 3, d = -5$; find S_{12}

18. $a_1 = -\frac{3}{4}, d = -\frac{1}{4}$; find S_{30}

19. $a_1 = -2, d = 2$; find S_{22}

20. $a_1 = 3\frac{1}{3}, d = -2\frac{2}{3}$; find S_{12}

21. $4, -2, -8, \ldots$; find S_{14}

22. $-11, -9\frac{1}{2}, -8, \ldots$; find S_{21}

23. $a_1 = 2, a_3 = 7$; find S_{10}

24. $a_3 = 15, a_6 = 30$; find S_6

25. $a_5 = 50, a_8 = 68$; find S_6

Find the required nth partial sum for each geometric sequence.

26. $a_1 = 2, r = 3$; find S_4

27. $a_1 = -1, r = -2$; find S_{12}

28. $a_1 = \frac{2}{3}, r = \frac{1}{3}$; find S_7

29. $a_1 = -\frac{1}{4}, r = 1\frac{1}{4}$; find S_4

30. $-2, \frac{4}{3}, -\frac{8}{9}, \ldots$; find S_4

31. $3, 18, 108, \ldots$; find S_3

32. $-5, 15, -45, \ldots$; find S_6

33. $\frac{8}{27}, \frac{4}{9}, \frac{2}{3}, \ldots$; find S_9

34. $3, 6, 12, \ldots, 96$

35. $2, 6, 18, \ldots, 4{,}374$

36. $\sum_{k=1}^{7} 3^k$

37. $\sum_{k=1}^{6} \frac{1}{9}(3^k)$

38. $\sum_{k=1}^{10} (-2)^k$

39. $\sum_{k=1}^{5} (\frac{2}{5})^k$

40. $\sum_{k=1}^{8} -3(-\frac{2}{3})^k$

41. $\sum_{k=1}^{12} \frac{16}{3}(\frac{1}{2})^k$

Find the sum of the given infinite geometric series. If the series has no sum state that.

42. $\sum_{i=1}^{\infty} (\frac{2}{3})^i$

43. $\sum_{i=1}^{\infty} 3(\frac{1}{3})^i$

44. $\sum_{i=1}^{\infty} -4(\frac{1}{2})^i$

45. $\sum_{i=1}^{\infty} \frac{2}{3}(\frac{1}{3})^i$

46. $\sum_{i=1}^{\infty} \frac{1}{8}(2^i)$

47. $\sum_{i=1}^{\infty} (-\frac{2}{3})^i$

48. $\sum_{i=1}^{\infty} (-\frac{9}{10})^i$

49. $\sum_{i=1}^{\infty} (\frac{4}{3})^i$

50. $14 + 7 + \frac{7}{2} + \cdots$

51. $3 + 2 + \frac{4}{3} + \cdots$

52. $4 + 5 + \frac{25}{4} + \cdots$

53. $1 - \frac{2}{3} + \frac{4}{9} - \cdots$

Find the rational number form of the repeating decimal number.

54. $0.666\overline{6}$

55. $0.222\overline{2}$

56. $0.5151\overline{51}$

57. $0.2828\overline{28}$

58. $0.216216\overline{216}$

59. $0.882882\overline{882}$

60. $0.21342134\overline{2134}$

61. $0.51555155\overline{5155}$

62. $0.23636\overline{36}$

63. $0.343535\overline{35}$

64. $0.216060\overline{60}$

65. $0.4014\overline{14}$

Solve the following problems.

66. A ball is dropped from a height of 10 meters. Each time it strikes the floor, the ball rebounds to a height that is 40% of the previous height. Find the total distance that the ball travels before it comes to rest on the floor.

67. A pendulum swings a distance of 20 inches on its first swing. Each subsequent swing is 95% of the previous distance. How far does the pendulum swing before it stops?

68. Neglecting air resistance, a freely falling body near the surface of the earth falls vertically 16 feet during the first second, 48 feet during the second second, 80 feet during the third second, and so on. Under these conditions how far will a body fall in the eighth second?

69. Referring to problem 68, how far will a body fall in 8 seconds?

70. A freely falling body in a vacuum near the surface of the earth will fall 4.9 m (meters) in the first second, 14.7 m in the second second, 24.5 m in the third second, and in general will fall 9.8 m farther in a given second than in the previous second. How far will such a body fall in 15 seconds?

71. How long will it take a freely falling body in a vacuum near the surface of the earth to fall 250 meters? Find the time to the nearest second. (See problem 70.)

72. An aircraft is flying at 400 mph, and is being followed at a distance of 1 mile (5,280 ft) by another aircraft moving at the same velocity. This second aircraft begins to accelerate so that each second it covers 30 more feet than it covered in the previous second. How long will it take before the second aircraft overtakes the first, from the time the second aircraft begins to accelerate?

73. A biologist in a laboratory estimates that a culture of bacteria is growing by 12% per hour. How long will it be, to the nearest hour, before the population doubles?

74. A certain bacterial culture triples in size each hour. If there were originally 1,000 bacteria how many hours would it take the bacteria to (a) surpass 1 million (b) surpass 10 million?

75. After doing a monarch a big favor one of the monarch's subjects asked to be rewarded in the following manner. Take a chessboard and put one grain of wheat on the first square, two on the second, four on the third, eight on the fourth, and so on, doubling the amount each time. The subject asked for the wheat that would be on the board. How many grains of wheat is this? (A chessboard has 64 squares.)

76. In a grocery store a clerk stacks boxes of cereal in a floor display so that there are 30 boxes of cereal in the first row, 27 in the second, 24 in the third, etc. How many boxes of cereal are in the floor display?

77. The same clerk as in problem 76 has 400 boxes to stack in a similar manner, except that each row is to have two fewer boxes than the one below it. The top row should have three boxes. How many boxes should be put in the first row to begin this display (the clerk may not be able to use all the boxes, but wants the display to be as high as possible)?

78. A parent put $500 into a bank account on the day a child was born. On each birthday the parent put in $100 more than the last deposit. How much money was deposited up to and including the child's eighteenth birthday?

79. It is estimated that over a 10-year period in a certain country the value of money went down 5% per year; that is, each year a dollar (the country's currency is in dollars) bought only 95% of what it bought the year before. By what percentage did the value of the dollar fall in this 10-year period?

80. A company needs $100,000 six years from now. It plans to obtain the money by making six deposits in an account. It will withdraw the interest earned each year, so this can be neglected. However, the company is growing and estimates that it can make each deposit 15% larger than the previous deposit. How large should the first deposit be?

81. A well-drilling company charges for well drilling according to the following schedule. One dollar for the first foot, and an additional $0.25 per foot for each foot after that. What would it charge to drill a well 250 feet deep?

82. A vacuum pump removes one-fifth of the remaining air in a tank with each stroke. **a.** How much air remains in the tank after the fifth stroke? **b.** How many strokes would be necessary to remove 98% of the air?

83. A donor to a college's development fund gave $50,000 and promised to give 80% of the previous year's donation each year, for an indefinite period. Excluding considerations such as inflation and the like, what is the value of this grant?

84. Derive the formula for the sum of a finite geometric series. Do this by writing the sum S as $S = a_1 + a_1r + a_1r^2 + \cdots + a_1r^{n-1}$. Then compute rS, and consider the sum $S - rS$.

85. Add up the integers from 1 to 100.

86. A basic formula in financial mathematics is the *present value formula.* Present value is the amount of money that would need to be invested at some rate of return to achieve some predetermined return in the future. If C_1 represents cash flow for period 1, C_2 for period 2, etc., and r_1 the rate of return (a percentage) that could be received on C_1, r_2 the rate of return that could be received on C_2, etc., then the present value, PV is

$$\text{PV} = \frac{C_1}{1 + r_1} + \frac{C_2}{(1 + r_2)^2} + \frac{C_3}{(1 + r_3)^3} + \cdots$$

The *perpetuity formula* assumes that the cash flows and rates of return are all the same. If these values are C and r, then under these conditions

$$\text{PV} = \frac{C}{1 + r} + \frac{C}{(1 + r)^2} + \frac{C}{(1 + r)^3} + \cdots$$

Note that if $r > 0$, $1 + r > 1$, and therefore $\frac{1}{1 + r} < 1$.

Show that the present value in the perpetuity formula reduces to $\frac{C}{r}$.

87. Two trains are 200 miles apart, headed toward each other on the same track, each traveling at 20 mph. A fly leaves one of the trains and flies directly toward the other at 60 mph (it is a very athletic fly). When it gets to the other train it turns around and repeats the journey to the first train. It repeats this process until the trains crash. How far did the fly fly?

88. On an ancient Babylonian tablet the value $1 + 2 + 2^2 + 2^3 + \cdots + 2^9$ is summed. Find this value.

89. How many record-breaking total yearly snowfalls (relative to their own life) would a person expect to see in a lifetime?[4] We will assume that the amount of snowfall in one year is unrelated to the amount that fell in the previous year. In the first year, the chance of a record is one out of one, or 1; the second year has a fifty-fifty chance of being more or less than the previous year (or one out of two, or $\frac{1}{2}$). In the third year the chance is $\frac{1}{3}$ (one out of three), since the heaviest snowfall for those 3 years could have fallen in any one of them. Similarly, in the nth year the chance of a new record is $\dfrac{1}{n}$. To find the total number of record snowfalls we must add up the sequence of probable record snowfalls for each year: $1 + \frac{1}{2} + \frac{1}{3} + \cdots + \dfrac{1}{n}$. Find the expected number of record snowfalls that a person will see in their first 10 years of life.

[4]Problems 89 and 90 are extracted from "Snowfalls and Elephants, Pop Bottles, and π," by Ralph Boas, from *The Two-Year College Mathematics Journal,* Vol. 11, No. 2, March 1980. Boas attributes problem 90 to R. C. Buck, University of Wisconsin.

90. A mobile is being constructed of straws (see figure). It is desired to arrange things so that the bottom straw projects beyond the point of support. It turns out that, measured in straw-lengths, the successive offsets from the bottom should be $\frac{1}{2}, \frac{1}{4}, \frac{1}{6}, \frac{1}{8}$, etc. Since $\frac{1}{2} + \frac{1}{4} + \frac{1}{6} + \frac{1}{8} \approx 1.042$, it takes just four straws to achieve this effect. How many straws would it take to achieve a $1\frac{1}{2}$ straw offset?

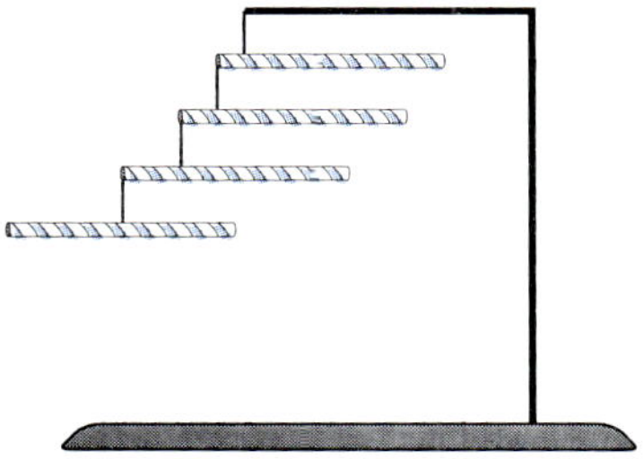

91. The following problem[5] comes from an ancient Hindu manuscript excavated in Pakistan in 1881. A certain person travels 5 yojanas (1 yojana = 8,000 times the height of a person) on the first day, and 3 yojanas more than the previous day on each successive day. Another person has a head start of 5 days, and travels 7 yojanas a day. In how many days will they meet?

92. A familiar nursery rhyme is:

As I was going to St. Ives,
I met a man with seven wives;
Every wife had seven sacks,
Every sack had seven cats,
Every cat had seven kits,
Kits, cats, sacks and wives,
How many were going to St. Ives?

[5]From "Hindu Romance with Quadratic Equations" by Dr. Gurcharan Singh Bhalla, *The Amatyc Review,* Fall 1987.

Skill and review

1. Find the 33rd term of an arithmetic series with $a_1 = 3$ and $d = 5$.
2. Find the fifth term of a geometric series with $a_1 = 3$ and $r = 5$.
3. Solve $\left|\dfrac{x+2}{x}\right| < 4$.
4. Graph $f(x) = x^2 + 5x - 6$.
5. Find the equation of the circle that has center at $(2,-1)$ and passes through the origin.
6. Graph $f(x) = x^3 - 3x^2 + x + 2$.

12–3 The binomial expansion and more on sigma notation

When a certain computer program runs, the number of steps that it requires to process k data elements is given by $\sum_{i=1}^{k} (2i^2 + 5i - 12)$. Find an expression for this quantity in terms of k.

In this section we study the mathematics related to solving this type of problem.

Pascal's triangle

Consider the sequence of indicated products for the expression $(x + y)^n$, for $n = 1, 2, 3, 4, \ldots$.

$$
\begin{aligned}
(x + y)^1 &= x + y \\
(x + y)^2 &= x^2 + 2xy + y^2 \\
(x + y)^3 &= x^3 + 3x^2y + 3xy^2 + y^3 \\
(x + y)^4 &= x^4 + 4x^3y + 6x^2y^2 + 4xy^3 + y^4 \\
(x + y)^5 &= x^5 + 5x^4y + 10x^3y^2 + 10x^2y^3 + 5xy^4 + y^5 \quad \text{etc.}
\end{aligned}
$$

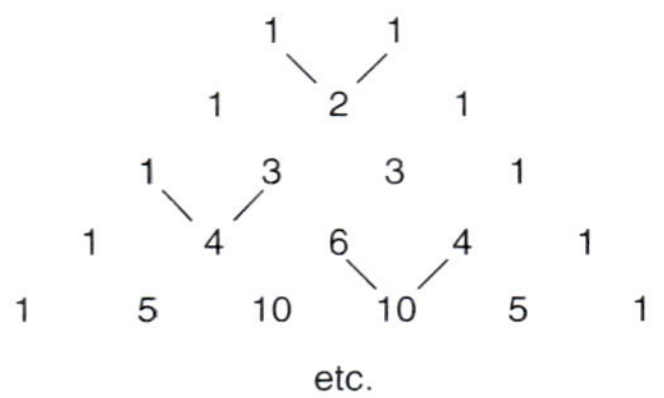

Figure 12–2

The coefficients can be found using **Pascal's triangle**[6] (figure 12–2). It can be seen that the next row in this tableau of numbers is formed by adding the elements of the row above it two at a time. For example, (as shown in figure 12–2) $1 + 1 = 2$, $1 + 3 = 4$, $6 + 4 = 10$.

To expand $(x + y)^k$ we might note the following.

- Row k of Pascal's triangle is the numeric coefficients.
- The exponent for x begins in the leftmost term with k, and the exponent for y is zero.
- As we move from one term to the following term the exponent for x decreases by one and that for y increases by one.
- The sum of the exponents in each term is always k.
- There are $n + 1$ terms.

Thus, we could compute $(x + y)^6$ by forming the next row of Pascal's triangle, which would be 1 6 15 20 15 6 1, and forming the terms

$$1x^6y^0 + 6x^5y^1 + 15x^4y^2 + \cdots \text{etc.}$$

It would be difficult to determine, say, the third term of $(x + y)^{20}$ in this manner. We do know it would be of the form $Kx^{18}y^2$, but finding K would require determining the coefficients for $n = 1, 2, 3, \ldots, 19$ first. There is a formula which can be used to find these coefficients and expand $(x + y)^n$ at the same time. To understand it we need to define **factorials** first.

[6]Used by Blaise Pascal (1623–1662), a French mathematician. It is also depicted at the front of Chu Shih-Chieh's *Ssü-yüan yü-chien (Precious Mirror of the Four Elements)*, which appeared in China in 1303.

Factorials

> ***n*-factorial**
>
> n-factorial, $n!$, is defined as
>
> $$n! = n \cdot (n-1) \cdot (n-2) \cdot \cdots \cdot 3 \cdot 2 \cdot 1 \text{ for } n \geq 1,\ n \in N$$
>
> As a special case, $0! = 1$.

For example, $5! = 5 \cdot 4 \cdot 3 \cdot 2 \cdot 1 = 120$

and

$$\begin{aligned} 8! &= 8 \cdot 7 \cdot 6 \cdot 5 \cdot 4 \cdot 3 \cdot 2 \cdot 1 \\ &= 8 \cdot 7 \cdot 6 \cdot 5! \\ &= 366 \cdot 120 = 40{,}320 \end{aligned}$$

Note Most calculators will calculate $n!$ with a key marked $x!$. On the TI-81 this is [MATH] 5.

Combinations

Factorials are useful in what is called **combinatorics,** an area of mathematics and statistics that concerns itself with counting in complicated situations. For example, combinatorics is used to find the number of ways a person can win a lottery, or the number of trials a computer program may make to solve a given problem.

One thing that is important in combinatorics as well as our development here is called **combinations.** As an example, the number of ways to form a committee of 3 from a group of 8 people is called ''the number of combinations of 3 things taken from 8 available things,'' or ''8 choose 3,'' which, it turns out, is computed as

$$\frac{8!}{(8-3)! \cdot 3!} = \frac{8 \cdot 7 \cdot 6 \cdot 5!}{5! \cdot 3!} = \frac{8 \cdot 7 \cdot 6}{3 \cdot 2 \cdot 1} = 56$$

This computation is also written $\binom{8}{3}$ and is defined in general as follows.

> **Combinations**
>
> The number of k-element combinations of n available things, denoted by $\binom{n}{k}$ (''n choose k'') is
>
> $$\binom{n}{k} = \frac{n!}{(n-k)!\,k!}$$
>
> if $n \geq k \geq 0$, $k, n \in N$

It is useful to know that

[1] $\binom{n}{n} = 1$ [2] $\binom{n}{1} = n$ [3] $\binom{n}{0} = 1$

It is left as an exercise to verify that these statements are true.

Note Most calculators will calculate $\binom{n}{r}$ with a key marked ${}_nC_r$. On the TI-81 this is [MATH] PRB 3.

■ *Example 12–3 A*

Expand and simplify each expression.

1. $\binom{7}{3} = \dfrac{7!}{4!3!}$ Definition of $\binom{n}{k}$

$= \dfrac{7 \cdot 6 \cdot 5 \cdot 4!}{4! \cdot 3!} = \dfrac{7 \cdot 6 \cdot 5}{3 \cdot 2 \cdot 1} = 7 \cdot 5 = 35$

Calculator: 7 [${}_nC_r$] 3 [=]

TI-81: 7 [MATH] PRB 3 3 [ENTER]

2. $\binom{n+2}{n} = \dfrac{(n+2)!}{[(n+2)-n]! \cdot n!}$ Definition of $\binom{n}{k}$

$= \dfrac{n! \cdot (n+1) \cdot (n+2)}{2! \cdot n!}$

$= \dfrac{(n+1)(n+2)}{2} = \dfrac{n^2 + 3n + 2}{2}$ ■

The binomial expansion formula

With this notation at hand[7] we can state the binomial expansion formula.

Binomial expansion formula

$$(x + y)^n = \sum_{i=0}^{n} \binom{n}{i} x^{n-i}y^i,\ n \in N$$

This formula incorporates the information about the coefficients of each term from Pascal's triangle and the observations we made about the exponents of each factor in a given term.[8] Its application is illustrated in example 12–3 B.

[7]The symbol $n!$ was introduced in 1808 by Christian Kramp of Strasbourg, in his *Élémens d' arithmétique universelle.* In 1846, Rev. Harvey Goodwin introduced the notation $\underline{|n}$ for the same thing. This notation may still be found in older books. Euler used the notation $\left(\frac{n}{k}\right)$ in 1778 and nineteenth-century writers shortened this to the current form $\binom{n}{k}$.

[8]A (difficult) proof of this formula uses the method of finite induction (section 12–4).

■ ***Example 12–3 B***

1. Expand $(2a - b)^4$ using the binomial expansion formula.

$$(x + y)^n = \sum_{i=0}^{n} \binom{n}{i} x^{n-i}y^i$$ Binomial expansion formula

$$(2a - b)^4 = \sum_{i=0}^{4} \binom{4}{i} (2a)^{4-i}(-b)^i$$ Replace x with $2a$, y with $-b$, and n with 4

$$= \binom{4}{0}(2a)^{4-0}(-b)^0 + \binom{4}{1}(2a)^{4-1}(-b)^1 + \binom{4}{2}(2a)^{4-2}(-b)^2$$

$$+ \binom{4}{3}(2a)^{4-3}(-b)^3 + \binom{4}{4}(2a)^{4-4}(-b)^4$$

$$= 1(2a)^4(-b)^0 + 4(2a)^3(-b) + 6(2a)^2(-b)^2 + 4(2a)(-b)^3 + 1(2a)^0(-b)^4$$

$$= 16a^4 - 32a^3b + 24a^2b^2 - 8ab^3 + b^4$$

2. Find the fourth term in the expansion of $(x - 3y^2)^{10}$.

The fourth term results when $i = 3$, so the term needed is for $i = 3$, $n = 10$, and y replaced with $-3y^2$:

$$\binom{10}{3} x^{10-3}(-3y^2)^3 = -120x^7(27)y^6$$

$$= -3{,}240x^7y^6$$ ■

Properties of sigma notation

Complicated uses of sigma notation are often encountered in computer science, economics, and statistics, as well as in advanced mathematics. In these cases, it is often useful to know some properties of sigma notation and to be able to manipulate the notation in certain ways.

> **Properties of sigma notation**
>
> Let k be a constant and let $f(i)$ and $g(i)$ represent expressions in the index variable i. Then the following properties are true:
>
> Sum of constants property: $\sum_{i=1}^{n} k = nk$
>
> Constant factor property: $\sum_{i=1}^{n} [k \cdot f(i)] = k \cdot \sum_{i=1}^{n} f(i)$
>
> Sum of terms property: $\sum_{i=1}^{n} [f(i) + g(i)] = \sum_{i=1}^{n} f(i) + \sum_{i=1}^{n} g(i)$

The sum of constants property states that if the expression is a constant, we get the product of the upper index and the constant. For example,

$$\sum_{i=1}^{4} 6 = 6 + 6 + 6 - 6 = 4(6) = 24$$

The constant factor property states that a common factor may be factored out of a sum. For example,

$$\sum_{i=1}^{3} 5i = 5 \cdot 1 + 5 \cdot 2 + 5 \cdot 3 = 5(1 + 2 + 3) = 5\sum_{i=1}^{3} i$$

The sum of terms property states that the summation of an expression in two (or more, actually) terms is equivalent to summing each term separately. For example,

$$\begin{aligned}\sum_{i=1}^{3}(5i + i^2) &= (5 \cdot 1 + 1^2) + (5 \cdot 2 + 2^2) + (5 \cdot 3 + 3^2)\\ &= (5 \cdot 1 + 5 \cdot 2 + 5 \cdot 3) + (1^2 + 2^2 + 3^2)\\ &= \sum_{i=1}^{3} 5i + \sum_{i=1}^{3} i^2\end{aligned}$$

Sums of certain series

An expression for the sum of the series where the general term is i, i^2, and i^3 is known. The sum of integer series (below) is an arithmetic series, so its sum can be derived from the formula for the sum of an arithmetic series; this is assigned in the exercises. The expressions for the sum of squares of integers series and the sum of cubes of integers series (below) are proved in section 12–4.

Sums of certain series

Sum of integer series: $\sum_{i=1}^{n} i = \frac{n(n+1)}{2}$

Sum of squares of integers series: $\sum_{i=1}^{n} i^2 = \frac{n(n+1)(2n+1)}{6}$

Sum of cubes of integers series: $\sum_{i=1}^{n} i^3 = \left(\frac{n(n+1)}{2}\right)^2$

These sums, combined with the properties of sigma notation, can be used to sum certain series. This is shown in example 12–3 C.

■ ***Example 12–3 C***

Simplify the following series.

1. $\sum_{i=1}^{4} [5i + (\frac{4}{5})^i]$

$= \sum_{i=1}^{4} 5i + \sum_{i=1}^{4} (\frac{4}{5})^i$ — Sum of terms property

$= 5\sum_{i=1}^{4} i + \sum_{i=1}^{4} (\frac{4}{5})^i$ — Constant factor property

$\sum_{i=1}^{4} i = \frac{4 \cdot 5}{2} = 10$ — Sum of integer series

$\sum_{i=1}^{4} (\frac{4}{5})^i$ is a geometric series with $a_1 = r = \frac{4}{5} = 0.8$.

$$\text{We want } S_4 = \frac{a_1(1 - r^n)}{1 - r} = \frac{0.8(1 - 0.8^4)}{1 - 0.8} = 2.3616.$$

Thus, we continue:

$$= 5\sum_{i=1}^{4} i + \sum_{i=1}^{4} (\tfrac{4}{5})^i = 5 \cdot 10 + 2.3616 = 52.3616$$

2. $\displaystyle\sum_{i=1}^{6} (2i - 3)^3$

$$= \sum_{i=1}^{6} (8i^3 - 36i^2 + 54i - 27)$$
Expand $(2i - 3)^3$

$$= 8\sum_{i=1}^{6} i^3 - 36\sum_{i=1}^{6} i^2 + 54\sum_{i=1}^{6} i - \sum_{i=1}^{6} 27$$
Sum of terms property and constant factor property

$$= 8\left[\frac{6(7)}{2}\right]^2 - 36\left[\frac{6(7)(13)}{6}\right] + 54\left[\frac{6(7)}{2}\right] - 6(27)$$
Sum of cubes, squares, and integer series, and sum of constants property

$$= 8(21^2) - 36(91) + 54(21) - 6(27) = 1{,}224$$

3. Find an expression for $\displaystyle\sum_{i=1}^{k} (i - 2)^2$, $k > 1$, in terms of the value k.

$$\sum_{i=1}^{k} (i - 2)^2 = \sum_{i=1}^{k} (i^2 - 4i + 4) = \sum_{i=1}^{k} i^2 - 4\sum_{i=1}^{k} i + \sum_{i=1}^{k} 4$$
$$= \frac{k(k + 1)(2k + 1)}{6} - 4\left(\frac{k(k + 1)}{2}\right) + k \cdot 4$$
$$= \frac{2k^3 + 3k^2 + k}{6} - 2k^2 - 2k + 4k$$
$$= \tfrac{1}{3}k^3 + \tfrac{1}{2}k^2 + \tfrac{1}{6}k - 2k^2 - 2k + 4k$$
$$= \tfrac{1}{3}k^3 - \tfrac{3}{2}k^2 + \tfrac{13}{6}k$$

Mastery points

Can you

- State the definition of $\binom{n}{k}$?
- State the binomial expansion theorem?
- Use the three properties given for sigma notation to simplify appropriate expressions?
- Use the expressions given for $\sum_{i=1}^{n} i$, $\sum_{i=1}^{n} i^2$, and $\sum_{i=1}^{n} i^3$ to simplify appropriate series?

Exercise 12–3

Expand and simplify each expression.

1. $\dfrac{7!}{5!}$ **2.** $\dfrac{8!}{5!3!}$ **3.** $\dbinom{8}{5}$ **4.** $\dbinom{10}{3}$

5. $\dbinom{6}{6}$ **6.** $\dbinom{20}{16}$ **7.** $\dbinom{n+3}{n}$ **8.** $\dbinom{k}{k-1}$

Expand and simplify the following expressions using the binomial expansion formula.

9. $(ab - 3)^4$ **10.** $(ab^2 + 2c^3)^5$ **11.** $(2p^4 + q)^6$ **12.** $(a^4 + 2b^3)^5$

13. $(a^3b^2 - 2c)^7$ **14.** $(2p^2 - q)^6$ **15.** $\left(\dfrac{p}{2} + 2\right)^6$ **16.** $\left(3a - \dfrac{2b}{3}\right)^5$

17. Find the fifth term of $(a^3 + 2b^5)^{15}$.

18. Find the fourth term of $(p^4 - q)^{18}$.

19. Find the fourth term of $(p - 3q)^{22}$.

20. Find the third term of $\left(2a - \dfrac{b}{8}\right)^6$.

Compute the sum of the following series.

21. $\sum_{i=1}^{56} 8$ **22.** $\sum_{i=1}^{18} (i + 3)$ **23.** $\sum_{i=1}^{23} (4i - 6)$ **24.** $\sum_{i=1}^{9} (2i^2 - 4)$

25. $\sum_{i=1}^{9} (3 - 4i + i^2)$ **26.** $\sum_{i=1}^{9} (i - 4)^2$ **27.** $\sum_{i=1}^{12} (2i + 1)^2$ **28.** $\sum_{i=1}^{10} (i^3 - 4i + 2)$

29. $\sum_{i=1}^{7} (2i^3 - 3)$ **30.** $\sum_{i=1}^{9} (i - 3)^3$ **31.** $\sum_{i=1}^{8} (i^3 + 6i^2 + 8i - 1)$ **32.** $\sum_{i=1}^{8} (8i^3 - 16i^2)$

33. $\sum_{i=1}^{4} [i^2 - (\frac{1}{4})^i]$ **34.** $\sum_{i=1}^{6} [6i^2 + 3i - (\frac{1}{3})^i]$ **35.** $\sum_{i=1}^{6} [4(\frac{1}{4})^i - 2(\frac{3}{2})^i]$ **36.** $\sum_{i=1}^{5} [16(\frac{1}{2})^i - 3(\frac{3}{4})^i]$

37. Find an expression for $\sum_{i=1}^{k} (6i^2 - 4i + 2)$ in terms of k.

38. Find an expression for $\sum_{i=1}^{k} (12i^2 + 2i - 7)$ in terms of k.

39. When a certain computer program runs, the number of steps that it requires to process k data elements is given by $\sum_{i=1}^{k} (2i^2 + 5i - 12)$. Find an expression for this quantity in terms of k.

40. Find an expression for $\sum_{i=1}^{k} (i^2 - i - 1)$ in terms of k.

41. Create Pascal's triangle down to the eighth row. The first row is "1 1".

42. Add up the values in a few of the rows of Pascal's triangle. Make a conjecture about what the value is in the kth row.

43. Show that $\dbinom{n}{n} = 1$, $\dbinom{n}{1} = n$, and $\dbinom{n}{0} = 1$.

44. State the binomial expansion theorem.

45. Prove that $\sum_{i=1}^{n} i = \dfrac{n(n+1)}{2}$. Note that $1 + 2 + \cdots + n$ is an arithmetic series (section 12–2).

46. Show that $\dbinom{n}{k+1} + \dbinom{n}{k} = \dbinom{n+1}{k+1}$ for $n \ge k \ge 0$.

47. Show that $\dbinom{n}{k} = \dbinom{n}{n-k}$ for $n \ge k \ge 0$.

48. Show that $(1 + k)^n \ge 1 + nk$ for all values of $k \ge 0$ and for all integers $n \ge 0$. (*Hint:* Expand $(1 + k)^n$.)

49. Prove that $\sum_{i=0}^{n} \dbinom{n}{i} = 2^n$. (*Hint:* This is a form of the binomial expansion with $x = y = 1$.)

Skill and review

1. If $1 + 2 + \cdots + n = \frac{n(n+1)}{2}$, what is $1 + 2 + \cdots + n + (n+1)$ equal to?
2. If $\frac{1}{1 \cdot 2} + \frac{1}{2 \cdot 3} + \frac{1}{3 \cdot 4} + \cdots + \frac{1}{n(n+1)} = \frac{n}{n+1}$, what is $\frac{1}{1 \cdot 2} + \frac{1}{2 \cdot 3} + \frac{1}{3 \cdot 4} + \cdots + \frac{1}{n(n+1)} + \frac{1}{(n+1)(n+2)}$ equal to?
3. Find an expression for $\frac{1}{3} + \frac{1}{9} + \frac{1}{27} + \cdots + \frac{1}{3^n}$.
4. Add up the even integers $2 + 4 + \cdots + 240$.
5. Solve $\frac{x-1}{2} - \frac{x-1}{3} = x$.
6. Graph $f(x) = (x-1)^3 - 1$.

12–4 Finite Induction

Prove that the sum of the squares of the first n natural numbers is $\frac{n(n+1)(2n+1)}{6}$. That is, prove that $1^2 + 2^2 + 3^2 + \cdots + n^2 = \frac{n(n+1)(2n+1)}{6}$.

This series is not arithmetic or geometric, so the methods of the previous sections are ineffective. The formula was correctly guessed over 700 years ago, probably by trial and error. In this section we show a method that allows us to prove that the formula really does work for any value of n.

The series $\sum_{i=1}^{n} i = 1 + 2 + 3 + \cdots + n$ is an arithmetic series, and it can be shown that its sum is $\frac{n(n+1)}{2}$, using the methods of section 12–2 for summing arithmetic series. That is,

$$[1] \qquad 1 + 2 + 3 + \cdots + n = \frac{n(n+1)}{2}$$

for any positive integer n.

We will prove this result here using the method of proof mentioned above, called **finite induction.** This is used to show that statements such as equation [1] are true for every positive integer. These types of statements arise often in advanced mathematics and the analysis of algorithms in computer science.

The idea of finite induction is as follows. Suppose we know that some statement is true for the integers 1, 2, and 3. For example:

$$1 + 2 + \cdots + n = \frac{n(n+1)}{2} \text{ is true for } n = 1, 2, 3.$$

Calculation will verify that this statement is true. For $n = 3$ the check is $1 + 2 + 3 = \frac{3(3+1)}{2}$, so $6 = 6$.

Now, suppose that we could prove the following: Whenever the statement above is true for an integer it also works for the next integer. This would mean that the statement must be true for $n = 4$, since we can see that it works for $n = 3$. Now, if it is true for $n = 4$, then the same logic says it must work for $n = 5$. We can proceed along these lines to any value of n we wish. This is the concept of finite induction.

Let us see if we can show that the supposition we used above is true. That is, can we show that, if [1] is true up to some integer k (statement [2]),

$$[2] \qquad 1 + 2 + 3 + \cdots + k = \frac{k(k+1)}{2}$$

then it must also be true for the next integer, $k + 1$? The formula for $k + 1$ would be

$$[3] \qquad 1 + 2 + 3 + \cdots + k + (k+1) \stackrel{?}{=} \frac{(k+1)(k+2)}{2},$$

which we obtain by substituting $k + 1$ into equation [1] instead of k. *What we want to do is show that equation [3] is true, given that equation [2] is true.* (We show that a statement needs to be shown true by the symbol $\stackrel{?}{=}$.)

We proceed as follows. We know that equation [2] is true up to some value of k (in this case, 3). Now, add the next value, $k + 1$, to both members of equation [2]:

$$[4] \qquad 1 + 2 + 3 + \cdots + k + (k+1) = \frac{k(k+1)}{2} + (k+1).$$

Equation [4] must be true, since we have simply added the same quantity to both members of a true equation. Now, the left member of equation [4] is the same as the left member of equation [3]; if we could show that the right member of equation [4] is the same as the right member of equation [3], we would have shown that equation [4] is really equation [3], and, since equation [4] is true, so is equation [3]. Proceeding with the right member of equation [4],

$$\begin{aligned}\frac{k(k+1)}{2} + (k+1) &= \frac{k(k+1) + 2(k+1)}{2} \\ &= \frac{(k+1)(k+2)}{2}\end{aligned}$$

which is the right member of equation [3].

Although this finishes our proof, it is unlikely that a first-time reader would understand completely what we have done, so let us review what happened.

1. We wanted to prove that the formula [1] is true *for any positive integer n.* Imagine picking a value for n.
2. We checked that equation [1] is true for $n = 1$, 2, and 3, by hand. We could check for even more values if we wished.

3. We thus knew that equation [1] was true up to at least a value of 3. Equation [2] is just equation [1] rewritten for $n = k$. Thus, equation [2] was true if $k = 3$ (and $k = 1$ or $k = 2$).
4. We showed that, if equation [2] was true for some arbitrary but fixed k, then so was equation [3].
5. Equation [2] is true for $k = 1, 2$, or 3. Equation [3] states that, if equation [2] is true for 3, it is true for 4 (i.e., $k + 1$). Thus, equation [2] is true for 4.
6. Now, we know that equation [2] is also true for 4, and so equation [3] shows that equation [2] is true for 5 $(4 + 1)$.
7. Repeating the logic of step 6, we can see that equation [2] is true for $k = 6, 7, 8$, etc. In fact, we can clearly repeat the steps to arrive at the conclusion that equation [2] is true for $k = n$, no matter how large n is. When $k = n$, equation [2] becomes equation [1], which is therefore true for n.

Observe that we really did not need to check equation [1] for n up to 3 by hand; just checking it for 1 would have been enough, since our logic would have then shown it must be true for 2 $(1 + 1)$ and 3 $(2 + 1)$.

With this example in mind, we state the principle of finite induction.

Principle of finite induction

If

1. a statement is true for $n = 1$ and
2. it can be shown that if the statement is true for $n = k$ then it must also be true for $n = k + 1$,

then the statement is true for any positive integer.

Example 12–4 A illustrates proofs by finite induction for situations in which we need to show that a certain sum of terms is equal to some expression.

■ ***Example 12–4 A***

1. Prove that $1^2 + 2^2 + 3^2 + \cdots + n^2 = \dfrac{n(n+1)(2n+1)}{6}$

   First, show the statement is true for $n = 1$:

$$1^2 = \frac{1(1+1)(2 \cdot 1 + 1)}{6}, \text{ so } 1 = 1$$

   Next, assume the statement is true for $n = k$ ($k = 1$, for example), and then show that this implies that the statement must be true for $n = k + 1$. Assume that

$$[1]\ 1^2 + 2^2 + 3^2 + \cdots + k^2 = \frac{k(k+1)(2k+1)}{6}$$

is true, and use this to prove that the same statement is true for $k + 1$; for $k + 1$ this is

[2] $$1^2 + 2^2 + 3^2 + \cdots + (k+1)^2 \stackrel{?}{=} \frac{(k+1)[(k+1)+1][2(k+1)+1]}{6}$$

or

$$1^2 + 2^2 + 3^2 + \cdots + (k+1)^2 \stackrel{?}{=} \frac{(k+1)(k+2)(2k+3)}{6}$$

Let us call this statement (equation [2]) our **"goal statement."**

Proceed as follows. Assume equation [1] is true for k. Now, add the next term, $(k + 1)^2$, to both members of equation [1].

$$1^2 + 2^2 + 3^2 + \cdots + k^2 + (k+1)^2 = \frac{k(k+1)(2k+1)}{6} + (k+1)^2$$

The left member is the same as the left member of the goal statement, equation [2]. We need to show that the right member is the same as the right member of the goal statement.

$$\begin{aligned}\frac{k(k+1)(2k+1)}{6} + (k+1)^2 &= \frac{k(k+1)(2k+1) + 6(k+1)^2}{6}\\ &= \frac{(k+1)[k(2k+1) + 6(k+1)]}{6}\\ &= \frac{(k+1)(2k^2+7k+6)}{6}\\ &= \frac{(k+1)(k+2)(2k+3)}{6}\end{aligned}$$

which is the right member of the goal statement.

Thus, we have shown that the original statement is true for $n = 1$, and that, if it is true for $n = k$, it must also be true for $n = k + 1$, thus proving, by the principle of finite induction, that the statement is true for any value of n.

2. Prove by induction that

$$\frac{1}{2\cdot 7} + \frac{1}{7\cdot 12} + \frac{1}{12\cdot 17} + \cdots + \frac{1}{(5n-3)(5n+2)} = \frac{n}{2(5n+2)}$$

First the case where $n = 1$:

$$\tfrac{1}{14} = \tfrac{1}{14}$$

Now assume the statement is true for $n = k$:

[1] $$\frac{1}{2\cdot 7} + \frac{1}{7\cdot 12} + \frac{1}{12\cdot 17} + \cdots + \frac{1}{(5k-3)(5k+2)} = \frac{k}{2(5k+2)}$$

We want to show the statement is true for $n = k + 1$; the goal statement is

$$[2] \qquad \frac{1}{2 \cdot 7} + \frac{1}{7 \cdot 12} + \frac{1}{12 \cdot 17} + \cdots + \frac{1}{(5(k+1) - 3)(5(k+1) + 2)} \stackrel{?}{=} \frac{k+1}{2(5(k+1)+2)} \text{ or}$$

$$\frac{1}{2 \cdot 7} + \frac{1}{7 \cdot 12} + \frac{1}{12 \cdot 17} + \cdots + \frac{1}{(5k+2)(5k+7)} \stackrel{?}{=} \frac{k+1}{2(5k+7)}$$

Add the next term to both members of equation [1].

$$\frac{1}{2 \cdot 7} + \frac{1}{7 \cdot 12} + \frac{1}{12 \cdot 17} + \cdots + \frac{1}{(5k-3)(5k+2)} + \frac{1}{(5(k+1) - 3)(5(k+1) + 2)} = \frac{k}{2(5k+2)} + \frac{1}{(5(k+1) - 3)(5(k+1) - 2)} \text{ or}$$

$$\frac{1}{2 \cdot 7} + \frac{1}{7 \cdot 12} + \frac{1}{12 \cdot 17} + \cdots + \frac{1}{(5k+2)(5k+7)} = \frac{k}{2(5k+2)} + \frac{1}{(5k+2)(5k+7)}.$$

The left member is the same as the left member of the goal statement [2], so if the right members are the same, then equation [2] is true. We simplify the right member:

$$\frac{k}{2(5k+2)} + \frac{1}{(5k+2)(5k+7)} = \frac{k(5k+7)}{2(5k+2)(5k+7)} + \frac{1(2)}{2(5k+2)(5k+7)}$$

$$= \frac{5k^2 + 7k + 2}{2(5k+2)(5k+7)} = \frac{(k+1)(5k+2)}{2(5k+2)(5k+7)}$$

$$= \frac{k+1}{2(5k+7)}$$

This is the right member of the goal statement, equation [2], so equation [2] is correct whenever equation [1] is correct. Thus, the original statement is true. ■

The principle of finite induction can be applied to situations in which we are not simply showing that some summation formula is true. These applications are harder to understand, and may require some facts beyond the algebraic manipulations shown above. For example we will show that $n^2 + n$ is always divisible by 2, when n is a positive integer, in example 12–4 B.

We will use the fact that divisibility by a certain value means that that value is a factor. Thus, we will use statements like:

If an integer j is divisible by 2, then $j = 2m$ for some integer m.

If an integer j is divisible by 3, then $j = 3m$ for some integer m.

We may also find it easier to work from the goal statement.

■ **Example 12–4 B**

Prove by induction that $n^2 + n$ is divisible by 2 for all $n \varepsilon N$.

First, show that the statement is true when $n = 1$: $1^2 + 1 = 2$ which is divisible by 2. Now, assume that the statement is true for $n = k$; that is, assume that $k^2 + k$ is divisible by 2. Our goal is to show that this implies that $(k + 1)^2 + (k + 1)$ is divisible by 2. In this case, it is more convenient to work from the goal statement.

$$\begin{aligned}(k + 1)^2 + (k + 1) &= k^2 + 3k + 2 \\ &= (k^2 + k) + (2k + 2)\end{aligned}$$

Write this last expression this way because we have the expression $k^2 + k$ in mind.

Now, since $k^2 + k$ is divisible by 2, it can be written as $2m$ for some integer m; thus we can proceed:

$$\begin{aligned}&= 2m + 2k + 2 \qquad \text{Replace } k^2 + k \text{ by } 2m \\ &= 2(m + k + 1)\end{aligned}$$

which is clearly divisible by 2.

Thus, assuming that $n^2 + n$ is divisible by 2 for some k, we have shown that it is also divisible by 2 for the next integer, $k + 1$. Hence by induction $n^2 + n$ is divisible by 2 for all $n \varepsilon N$. ■

Example 12–4 B shows that induction can be used for statements other than simple formulas.

A note to the skeptic[9]

After seeing this method of proof for the first time it might seem that almost anything, whether true or not, can be proved this way. This is not the case, however. Two examples will illustrate.

First, consider the statement $1 + 3 + 5 + \cdots + (2n - 1) = \dfrac{n^2 + n}{2}$ for every positive integer. This statement can be shown true for $n = 1$, but if we assume it is true for k we will not be able to. This is because the original statement is false! Now consider the statement $4 + 10 + 16 + \cdots + (6n - 2) = 3n^2 + n - 2$, which is not even true for $n = 1$, but, if assumed true for k can be shown true for $k + 1$! The reader is invited to explore both of these examples in the exercises.

These two examples do not "prove" that proof by induction works only on true statements, but hopefully will convince the reader that this is in fact the case. The principle of finite induction can be proved true in higher mathematics.

[9] "There is no use trying," (Alice) said. "One can't believe impossible things." "I daresay you haven't had much practice," said the Queen. "When I was your age, I always did it for half-an-hour a day. Why, sometimes I've believed as many as six impossible things before breakfast." (From *Alice in Wonderland,* by Lewis Carroll.)

Mastery points

Can you

- Prove statements using finite induction?

Exercise 12–4

Prove that the following statements are true for all $n \in N$ using finite induction.

1. $2 + 4 + 6 + \cdots + 2n = n(n + 1)$

2. $1 + 3 + 5 + \cdots + (2n - 1) = n^2$

3. $4 + 9 + 14 + \cdots + (5n - 1) = \dfrac{n(5n + 3)}{2}$

4. $1 + 4 + 7 + \cdots + (3n - 2) = \dfrac{n(3n - 1)}{2}$

5. $1 + 5 + 9 + \cdots + (4n - 3) = 2n^2 - n$

6. $1^2 + 2^2 + 3^2 + \cdots + n^2 = \dfrac{n(n + 1)(2n + 1)}{6}$

(See footnote 10.)

7. $1 + 8 + 30 + 80 + \cdots + \dfrac{n^2(n + 1)(n + 2)}{6} = \dfrac{n(n + 1)(n + 2)(n + 3)(4n + 1)}{120}$

8. $\dfrac{1}{1 \cdot 2} + \dfrac{1}{2 \cdot 3} + \dfrac{1}{3 \cdot 4} + \cdots + \dfrac{1}{n(n + 1)} = \dfrac{n}{n + 1}$

9. $\dfrac{1}{2} + \dfrac{1}{2^2} + \dfrac{1}{2^3} + \cdots + \dfrac{1}{2^n} = \dfrac{2^n - 1}{2^n}$

10. $1^3 + 2^3 + 3^3 + \cdots + n^3 = \left[\dfrac{n(n + 1)}{2}\right]^2$

11. Show that $n^3 + 2n$ is divisible by 3 for any natural number n.

12. Show that $(1 + a)^n \geq 1 + na$ for any natural number n, assuming $a \geq 0$.

13. $\dfrac{1}{1 \cdot 3} + \dfrac{1}{3 \cdot 5} + \dfrac{1}{5 \cdot 7} + \cdots + \dfrac{1}{(2n - 1)(2n + 1)} = \dfrac{n}{2n + 1}$

14. $1 + 4 + 4^2 + \cdots + 4^{n-1} = \dfrac{4^{n-1}}{3}$

15. $2 + 6 + 18 + \cdots + 2(3^{n-1}) = 3^n - 1$

16. $3 + 12 + 48 - \cdots + 3(4^{n-1}) = 4^n - 1$

17. $8 + 4 + 2 + \cdots + \dfrac{1}{2^{n-4}} = \dfrac{2^n - 1}{2^{n-4}}$

18. $\dfrac{1}{2 \cdot 5} + \dfrac{1}{5 \cdot 8} + \dfrac{1}{8 \cdot 11} + \cdots + \dfrac{1}{(3n - 1)(3n + 2)} = \dfrac{n}{6n + 4}$

19. $\dfrac{1}{1 \cdot 4} + \dfrac{1}{4 \cdot 7} + \dfrac{1}{7 \cdot 10} + \cdots + \dfrac{1}{(3n - 2)(3n + 1)} = \dfrac{n}{3n + 1}$

20. $\dfrac{1}{a(a + b)} + \dfrac{1}{(a + b)(a + 2b)} + \dfrac{1}{(a + 2b)(a + 3b)} + \cdots + \dfrac{1}{[a + (n - 1)b](a + nb)} = \dfrac{n}{a(a + nb)}$

21. $\dfrac{1}{1 \cdot 2 \cdot 3} + \dfrac{1}{2 \cdot 3 \cdot 4} + \dfrac{1}{3 \cdot 4 \cdot 5} + \cdots + \dfrac{1}{n(n + 1)(n + 2)} = \dfrac{n(n + 3)}{4(n + 1)(n + 2)}$

22. $\dfrac{1}{1 \cdot 3 \cdot 5} + \dfrac{1}{3 \cdot 5 \cdot 7} + \dfrac{1}{5 \cdot 7 \cdot 9} + \cdots + \dfrac{1}{(2n - 1)(2n + 1)(2n + 3)} = \dfrac{n(n + 2)}{3(2n + 1)(2n + 3)}$

[10]Carl Boyer notes that the Babylonians may have known this result thousands of years ago. Also, this and the next problem are both found in the *Precious Mirror of the four Elements,* a book that appeared in China in 1303.

23. $\dfrac{1}{1 \cdot 4 \cdot 7} + \dfrac{1}{4 \cdot 7 \cdot 10} + \dfrac{1}{7 \cdot 10 \cdot 13} + \cdots + \dfrac{1}{(3n-2)(3n+1)(3n+4)} = \dfrac{n(3n+5)}{8(3n+1)(3n+4)}$

24. $(1 \cdot 2) + (2 \cdot 3) + \cdots + [n(n+1)] = \dfrac{(n+1)^3 - (n+1)}{3}$

25. In the text we considered the statement $1 + 3 + 5 + \cdots + (2n-1) = \dfrac{n^2+n}{2}$ where we said that this statement can be shown true for $n = 1$, but if we assume it is true for k and try to show it true for $k + 1$ we will not be able to. Try this and see what happens.

26. Also in the text we stated that the statement $4 + 10 + 16 + \cdots + (6n-2) = 3n^2 + n - 2$ is not true even for $n = 1$, but, if assumed true for k can be shown true for $k + 1$. Show that this is indeed the case.

27. Refer to problems 25 and 26. The two series (a) $1 + 3 + 5 + \cdots + (2n-1)$ and (b) $4 + 10 + 16 + \cdots + (6n-2)$ are arithmetic series. Find an expression for the sum of the first n terms of each series. See section 12–2 for arithmetic series, if necessary. (Note that we are "fixing" the previous two problems by finding the correct expression for the right member of each equation.)

28. You are given n coins ($n \geq 2$). They look identical, but one of them is counterfeit and weighs less than all of the others. You have a balancing scale, and need to determine which of the coins is counterfeit.

If you have two coins you can detect the light coin in one weighing. Three coins can be done in one weighing: weigh two coins; if they balance, the third coin is the counterfeit, otherwise the balance shows the light counterfeit coin. Four coins would require two weighings as shown in the diagram. The first weighing shows that the light coin is c or d. The second weighing shows that d is the light coin.

Two weighings would suffice for five coins also. We can show this inductively, since we can group the coins into a group of four coins and one coin. If the light coin is in the group of four, two weighings will be enough. If the four are the same, the fifth coin is the light coin.

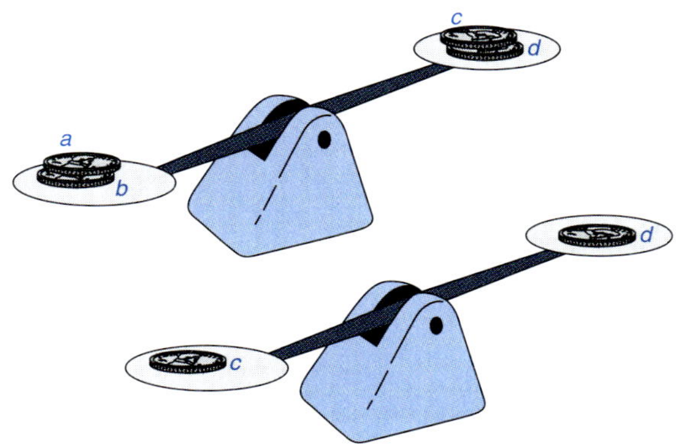

In this manner we can show inductively that two weighings suffice for any number of coins. For n coins, group them into a group of $n - 1$ coins and 1 coin. If the lighter coin is not among the $n - 1$ coins, which takes two or fewer weighings, then it is the single coin.

Unfortunately, this statement is not true. Try to actually apply this idea to six coins to see that it is false. Then explain where the logic we applied was in error.

Skill and review

1. If $1 + 4 + 7 + \cdots + (3n-2) = \dfrac{n(3n-1)}{2}$, what is $1 + 4 + 7 + \cdots + (3n-2) + [3(n+1) - 2]$?

2. Graph $3x - 4y = 12$.

3. Graph $3x^2 - 4y^2 = 12$.

4. Solve $\begin{matrix} 2x - 3y \leq 12 \\ x + 2y \geq 4 \end{matrix}$.

5. Solve $2x - 1 > \dfrac{5}{x+1}$.

12–5 Introduction to combinatorics

> Suppose NASA has 19 astronauts suitable for the next space mission. A crew consists of five astronauts. How many different crews are possible?

The answer to questions like this are found using counting methods that are introduced in this section. These methods are important in the study of probability, which is studied in section 12–6.

The basic concepts of probability depend on our ability to determine the number of possible ways that an experiment (such as rolling a die eight times) can occur. We must know what is possible before we can determine what is probable.

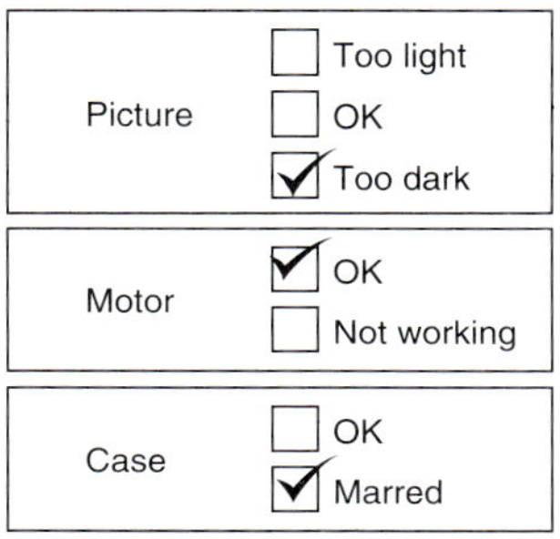

Figure 12–3

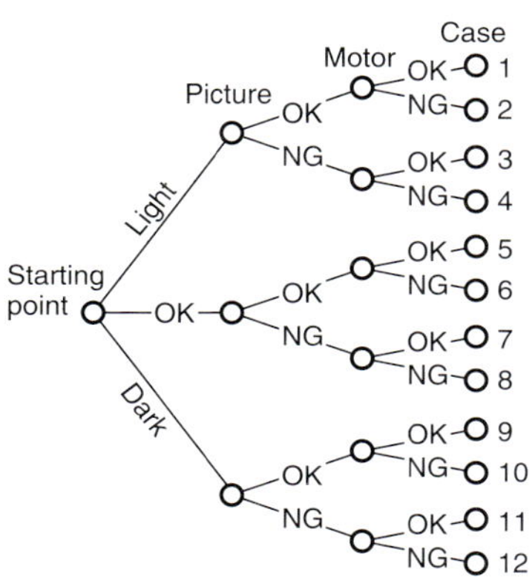

Figure 12–4

Multiplication of choices

To illustrate one of the basic principles of counting, consider an assembly line that produces a camera. A final test is made of each camera. A picture is taken, and it might be too light, alright, or too dark. The motor may or may not work, and the case may be marred or not. Each of these characteristics is marked on a slip. In how many ways can this slip be marked? Figure 12–3 shows one such marking.

The possible markings can be displayed on a tree diagram (figure 12–4); NG stands for no good in the figure (anything that is not "OK"). From a starting point we represent the three qualities of a picture, then for each of these possibilities the two characteristics of a motor, and for each of these (now 6) possibilities the two conditions of the case. This gives 12 possible markings of each slip.

These markings can also be shown with a list. If we list the picture quality first, then the motor, then the case, using L = light, O = OK, D = dark, N = no good, we obtain the list *LOO, LON, LNO, LNN, OOO, OON, ONO, ONN, DOO, DON, DNO, DNN*. If we had chosen to list the markings for, say, the motor first this would still be considered the same list, just as the tree diagram could have started with the first branch indicating the markings for the motor or case.

It is no coincidence that $12 = 3 \quad 2 \cdot 2$. This example illustrates the following property, called the multiplication-of-choices property.

> **Multiplication-of-choices property**
> If a choice consists of k decisions, where the first can be made n_1 ways and for each of these choices the second can be made in n_2 ways, and in general the ith choice can be made in n_i ways, then the complete choice can be made in $n_1 \cdot n_2 \cdot \cdots \cdot n_k$ ways.

Each complete choice is called an **outcome.** Example 12–5 A illustrates using the multiplication-of-choices property.

■ ***Example 12–5 A***

Determine the number of outcomes.

1. This year the Fine Arts Auditorium is hosting four plays and 12 concerts. A student pass entitles a student to see one play and one concert. In how many ways can a student see one play and one concert?

   Using the multiplication-of-choices property we calculate $4 \cdot 12$ or 48 different ways.

2. A newsstand carries five different newspapers, four different sporting magazines, six fashion magazines, and 12 general interest magazines. In how many ways can a shopper buy one of each of these types of magazines?

   In this case the choice of one of each of the magazines involves four choices. Using the number of ways in which each choice can be made, and the multiplication-of-choices property we calculate $5 \cdot 4 \cdot 6 \cdot 12 = 1{,}440$ ways.

3. A race has five individuals in it. In how many ways can the five runners finish the race (neglecting ties)?

   Any one of the five runners can finish in first place; having made this "choice," any of the four remaining runners can finish in second place. There are then three possibilities for third place, two for fourth place, and, finally, one choice for fifth place. Thus, there are $5 \cdot 4 \cdot 3 \cdot 2 \cdot 1 = 120$ different ways for the race to end. ■

Factorial notation

In part 3 of example 12–5 A the answer was obtained by forming a product of all the integers from 1 to 5; this type of product occurs often enough in counting problems that the following notation was defined (see also section 12–3).[11]

> **n-factorial**
>
> n-factorial, denoted by $n!$, is defined as
>
> $$n! = n \cdot (n-1) \cdot (n-2) \cdot \cdots \cdot 3 \cdot 2 \cdot 1 \text{ for } n \geq 1,\ n \in N$$
>
> As a special case, $0! = 1$.

Thus, for example,

$$6! = 6 \cdot 5 \cdot 4 \cdot 3 \cdot 2 \cdot 1 = 720 \quad \text{Six-factorial is 720}$$
$$10! = 10 \cdot 9 \cdot 8 \cdot 7 \cdot 6! = 10 \cdot 9 \cdot 8 \cdot 7 \cdot 720 = 3{,}628{,}800$$
$$1! = 1$$

We define $0!$ to be 1 strictly for convenience in certain formulas that we will see later.

[11]The symbol $n!$ was introduced by Christian Kramp of Strasbourg in 1808.

Observe that 6! is also $6 \cdot 5!$, or $6 \cdot 5 \cdot 4!$, or $6 \cdot 5 \cdot 4 \cdot 3!$, etc. In general

$$\begin{aligned} n! &= n(n-1)! \\ &= n(n-1)(n-2)! \\ &= n(n-1)(n-2)(n-3)!, \text{ etc.} \end{aligned}$$

We can make use of this fact to simplify certain expressions involving factorials, as illustrated in example 12–5 B. We also note that most calculators can calculate $n!$.

■ ***Example 12–5 B***

Simplify the following expressions.

1. $\dfrac{12!}{8!}$

$$\frac{12!}{8!} = \frac{12 \cdot 11 \cdot 10 \cdot 9 \cdot 8!}{8!} = 12 \cdot 11 \cdot 10 \cdot 9 = 11{,}880$$

2. $\dfrac{15!}{12!3!}$

$$\frac{15!}{12!3!} = \frac{15 \cdot 14 \cdot 13 \cdot 12!}{12! \cdot 3 \cdot 2 \cdot 1} = \frac{15 \cdot 14 \cdot 13}{3 \cdot 2} = 5 \cdot 7 \cdot 13 = 455$$ ■

Permutations and combinations

A club with four members wants to choose a president and a secretary. In how many ways can this be done? Using the multiplication-of-choices property, we see that we have four choices to fill the president's position, and then there remain three choices to fill the secretary's position. Thus, there are $4 \cdot 3 = 12$ ways to fill these two positions. If the persons are person A, B, C, and D, we can list the 12 possible ways:

Selection	**1**	**2**	**3**	**4**	**5**	**6**	**7**	**8**	**9**	**10**	**11**	**12**
President	A	A	A	B	B	B	C	C	C	D	D	D
Secretary	B	C	D	A	C	D	A	B	D	A	B	C

Note that *the order of selection was important.* For example, the selection of A then B (selection 1) means that A is president and B is secretary, whereas the selection of B first then A (selection 4) means that B is president and A is secretary.

Suppose the same club wants instead to form a two-person committee. The same analysis shows all the ways in which we can select a two-person committee; however, we now don't care about who is selected first and who is selected second. Thus, selection 1 and selection 4 are equivalent as a two person committee. In fact, we can verify that in this case there are only half as many possible selections, or six different two-person committees. This is because we do not care about the $2 \cdot 1$ different ways each committee could be ordered.

In both of these situations a person could not be selected twice. This is called **selection without repetition (or without replacement).** This is the situation we will continue to consider here. In the situation of selecting a president and secretary, where *order is important,* each different selection is called a **permutation.**[12] In selecting committees, where *order is not important,* each selection is called a **combination.**

In selecting our president and secretary we would say we wanted the number of permutations of two things taken from four available things. This is written symbolically as ${}_4P_2$. Similarly in selecting our committees we wanted ${}_4C_2$, or the number of combinations of two things taken from four available things. We will develop formulas for each of these situations.

Permutations

Consider the following examples of ${}_nP_r$ for different values of n and r:

${}_5P_3 = 5 \cdot 4 \cdot 3$ Out of five people, choose a president, a vice-president and a secretary

${}_7P_4 = 7 \cdot 6 \cdot 5 \cdot 4$ Out of seven books choose four and arrange them on a shelf

We define ${}_nP_r$ in the following way.

Permutations[13]

The number of permutations of r distinct elements selected from n available elements, where $r \leq n$ is

$${}_nP_r = \underbrace{n(n-1)(n-2)\ldots(n-[r-1])}_{r \text{ factors}}$$

Note that this is the first r factors in $n! = n(n-1)(n-2)\cdots 3 \cdot 2 \cdot 1$. Also, note that ${}_nP_n = n!$.

Note Most modern calculators have keys which will compute ${}_nP_r$.

In the exercises we illustrate by example that an alternative formula for ${}_nP_r$ is

$${}_nP_r = \frac{n!}{(n-r)!}$$

This can be useful under certain circumstances, as illustrated in part 2 of example 12–5 C.

[12]The first known enumerations of permutations are from the Hebrew Book of Creation (c. A.D. 100), according to D. E. Knuth, *The Art of Computer Programming,* Vol. 3, Addison-Wesley Publishing Company, Inc., Reading, Massachusetts, 1973.

[13]The notation ${}_nP_r$ was introduced by Rev. Harvey Goodwin of Cambridge University circa 1869.

■ ***Example 12–5 C***

Determine the number of outcomes.

1. During periods of radio silence between two ships messages are sent by means of signal flags. If there are eight different flags, how many messages can be sent by placing three flags, one above the other, on a flagpole?

Since there is a definite order implied by the placement of the flags the problem involves permutations. From eight available things, choose three in order. The number of ways to do this is ${}_8P_3 = 8 \cdot 7 \cdot 6 = 336$ different messages.

Typical keystrokes on a calculator are 8 $\boxed{{}_nP_r}$ 3 $\boxed{=}$. On a TI-81 this function is under the $\boxed{\text{MATH}}$ key, in the PRB menu. Select 8 $\boxed{\text{MATH}}$ PRB 2 3 $\boxed{\text{ENTER}}$.

2. A trucking firm has 20 trucks and 18 drivers. The trucks differ in age and so are considered different by the drivers. In how many ways can the trucks be assigned to the drivers?

From 20 available trucks we need to select 18, and the order matters. Thus we need to compute ${}_{20}P_{18} = 20 \cdot 19 \cdot 18 \cdot \cdots \cdot 3$.

An $\boxed{{}_nP_r}$ calculator key will perform the calculation for us easily, but if a calculator does not have this key, but does have a factorial key $\boxed{x!}$ we could instead use the alternative form (from above):

$$
\begin{aligned}
{}_{20}P_{18} &= \frac{20!}{(20-18)!} = \frac{20!}{2!} = 1.216451004 \times 10^{18} \\
&\approx 1{,}216{,}451{,}004{,}000{,}000{,}000
\end{aligned}
$$

The result is so large we can only expect an approximate result using a calculator. The actual result, found with a sophisticated computer program,[14] is 1,216,451,004,088,320,000. ■

Indistinguishable permutations of *n* elements

Consider the possible arrangements of the letters of the word NOON. Since there are four letters we might conclude that there are ${}_4P_4 = 4! = 24$ ways to arrange the letters. However, the two O's and the two N's are indistinguishable. For example, the permutation in which the two N's are switched produces NOON also, which is indistinguishable from the original permutation.

[14]In this case Theorist®. Other powerful products that would calculate this number are Mathematica® and Maple.

For each arrangement, if the N's are only permuted among themselves, there is no distinguishable change; since there are 2! permutations of the two N's, we must divide the 24 by 2!. We must also divide by a second 2! for the O's. Thus, the number of distinguishable permutations of the letters in the word NOON is $\frac{4!}{2!2!} = 6$. We now make the following generalization.

> **Indistinguishable permutations**
> The number of distinct permutations P of n elements, where n_1 are alike of one kind, n_2 are alike of one kind, . . . , and n_k are alike of one kind, where $n_1 + n_2 + \cdots + n_k = n$, is given by
> $$P = \frac{n!}{n_1!n_2! \ldots n_k!}$$

Note that many of the n_i's may be one, which we can ignore in the computation. Example 12–5 D illustrates this idea.

■ ***Example 12–5 D***

Determine the number of outcomes.

1. Find the number of distinct permutations of the letters in the word MISSISSIPPI.

There are eleven letters of which there is one M, four I's, four S's, and two P's. Therefore the number of permutations is $P = \frac{11!}{1!4!4!2!} = 34{,}650$.

2. If a signal consists of nine flags one above the other on a flagpole, and there are nine flags, with three red, three white, and three blue, how many signals can be created?

$$P = \frac{9!}{3!3!3!} = 1{,}680$$ ■

Combinations

Recall from the discussion of two-person committees that order is not important in combinations. A combination is the same as a subset of some set of distinguishable elements. As illustrated with committees, the number of combinations of n things taken r at a time ${}_nC_r$ is the same as ${}_nP_r$ if we divide by the number of permutations of r things. (For the committees, r was 2.) Thus, ${}_nC_r = \frac{{}_nP_r}{r!}$. It can be shown that this is the same value as $\frac{n!}{r!(n-r)!}$.

Combinations

The number of combinations (subsets) of r distinct elements selected from n available elements, where $r \leq n$ is

$${}_nC_r = \frac{n!}{r!(n-r)!}$$

Another notation[15] for ${}_nC_r$ is $\binom{n}{r}$.

Note Most modern calculators have keys that will compute ${}_nC_r$. The steps are practically the same as the computation for ${}_nP_r$.

Example 12–5 E illustrates counting combinations.

■ Example 12–5 E

Determine the number of outcomes.

1. An individual has 12 (distinguishable) shirts and wants to pack 3 of them for a trip. In how many different ways can this be done?

   We are interested in how many 3-element subsets of 12 elements there are; we do not care about the order in which the shirts are selected. Thus, we want ${}_{12}C_3 = \frac{12!}{3!9!} = \frac{12 \cdot 11 \cdot 10 \cdot 9!}{3!9!} = \frac{12 \cdot 11 \cdot 10}{6} = 220.$

2. On a ten-question test, in how many ways can a student get exactly seven questions correct (assuming that all questions are either right or wrong).

   We want to know: out of ten elements, how many ways can we choose exactly seven of them. This is ${}_{10}C_7 = \frac{10!}{7!3!} = 120.$ ■

In part 2 of example 12–5 E, if we asked instead how many ways can a student get exactly three questions out of ten wrong, which is of course the same as getting seven questions correct, this would be ${}_{10}C_3 = \frac{10!}{3!7!} = 120$. This illustrates the fact that ${}_nC_r = {}_nC_{n-r}$, which can be demonstrated as follows.

■ Example 12–5 F

Show that $\binom{n}{n-r} = \binom{n}{r}$.

$$\binom{n}{n-r} = \frac{n!}{(n-r)![n-(n-r)]!} = \frac{n!}{(n-r)!r!} = \binom{n}{r}$$ ■

[15]Introduced in section 12–3.

Further counting problems

Many counting problems cannot be answered with just one counting property. The example 12–5 G illustrates using more than one of the properties we have examined.

■ ***Example 12–5 G***

1. A class contains 30 students, 18 females and 12 males. How many different committees of 7 students can be formed if there must be 4 females and 3 males on the committee?

   It is surprising how often looking at a problem a certain way can help. In this problem, we first simply ask how many committees of 4 females can be formed, and then how many committees of 3 males. These figures are ${}_{18}C_4 = 3{,}060$ and ${}_{12}C_3 = 220$. Now use the multiplication principle, because for each of the female committees we can choose any one of the male committees. The result is $3{,}060 \cdot 220 = 673{,}200$ committees.

2. How many different 7-card hands from a standard deck of 52 cards[16] are possible if the hand is to contain 3 hearts, 2 diamonds, and 2 cards that are not a heart or a diamond?

   There is a great deal of similarity between this problem and the previous one. The class was divided into 2 categories (male and female), and the deck of cards is divided into 4 categories (13 of each of the four suits—hearts, diamonds, clubs, and spades). A card hand can be viewed as one committee. Thus, we proceed in the same manner, and first ask how many hands there are of 3 hearts (${}_{13}C_3 = 286$), how many hands there are of 2 diamonds (${}_{13}C_2 = 78$), and how many hands there are of black cards (clubs and spades) (${}_{26}C_2 = 325$). Now, since we do not care about the order in which cards are selected we can use the multiplication-of-choices property: $286 \cdot 78 \cdot 325 = 7{,}250{,}100$ such hands.

3. A race is to be run between two stables. The Adams stable will enter three of its seven horses, and the Baker stable will enter three of its eight horses. In how many ways can six horses finish the race?

   The multiplication-of-choices principle tells us that the six horses to enter the race can be selected in ${}_7C_3 \cdot {}_8C_3 = 35 \cdot 56 = 1{,}960$ ways. However, the order in which the horses finish is important! For each of the 1,960 ways in which the race can start, there are $6! = 720$ ways (${}_6P_6$) for it to finish. Thus, there are $1{,}960 \cdot 720 = 1{,}411{,}200$ ways for the horses to be selected and then finish the race. ■

[16]A standard deck of cards has four suits, the two red suits (diamonds and hearts) and the two black suits (clubs and spades). Each suit has 13 cards: an ace, the numbered cards 2, 3, 4, 5, 6, 7, 8, 9, 10, and three face cards, jack, queen, king.

Mastery points

Can you

- Use the multiplication-of-choices property to solve counting problems?
- Compute and use permutations to solve counting problems?
- Compute and use combinations to solve counting problems?
- Combine these counting methods to solve problems?

Exercise 12–5

See figures 12–3 and 12–4 for the following problems.

1. There are 3 people in a race, A, B, and C.
   a. Draw a tree diagram of the ways in which the race can be finished.
   b. List all possible ways in which the race can be finished (for example, CBA).
2. A die is to be thrown twice.
   a. Draw a tree diagram of the ways in which the numbers on the die can appear.
   b. List all possible sequences of numbers that can occur (for example, 5, 2 if 5 appears on the first throw and 2 on the second).
3. A person is flipping a coin and noting heads (H) or tails (T) each time. The person does this three times.
   a. Draw a tree diagram of this experiment.
   b. List all possible outcomes (for example, HHT for head-head-tail).
4. An individual is tossing a die and flipping a coin.
   a. Draw a tree diagram in which the first branch reflects what happens with the coin (head/tail).
   b. Draw a tree diagram in which the first branch reflects what happens with the die (1/2/3/4/5/6).
   c. List all possible outcomes of this experiment. A typical outcome is H4, for heads on the coin and 4 on the die. This could just as well be described as 4H, since we are merely listing outcomes, not ways in which the outcome can be achieved.

Solve the following problems.

5. An individual has four pairs of slacks and six shirts. How many different combinations of shirts and slacks can this person wear?
6. A certain shoe store customer complained that a good shoe store would stock every style in every color and size. If the store has 65 styles and each style comes in 12 sizes, 2 widths, and 3 colors, how many different pairs of shoes would the store have to stock to have one of each? Would it be a good idea for a store to order shoes this way?
7. The Ahmes[17] (or Rhind) Papyrus is an ancient Egyptian papyrus. It contains a variation of the following problem. There are seven buildings for storing grain; each building is guarded by seven cats, each of which eats seven mice. If it were not for the cats, each of these mice would eat seven ears of corn, each of which could produce seven bushels of corn. How many bushels of grain are saved by the cats?
8. A certain building has 12 entrances. In how many different ways can someone go in one entrance and out a different one?
9. A class has 12 females and 15 males. A female and a male are to be selected as student government representatives. In how many different ways can the selection be made?
10. A menu offers a choice of 5 appetizers, 3 salads, 12 entrees, 4 kinds of potatoes and 5 vegetables. A meal consists of one of each. In how many ways can a person select a meal?
11. In how many different ways can a student answer all the questions on a quiz consisting of 8 true or false questions?
12. From a standard deck of playing cards, in how many ways can a person select one heart, one club, one diamond, and one spade?

[17]Ahmes was the Egyptian scribe who copied this papyrus in about 1650 B.C.

Evaluate each expression.

13. $7!$
14. $9!$
15. $\frac{10!}{5!}$
16. $\frac{12!}{6!}$
17. $\frac{12!}{3!9!}$
18. $\frac{20!}{17!3!}$
19. ${}_6P_4$
20. ${}_8P_3$
21. ${}_6P_6$
22. ${}_{15}P_5$
23. ${}_{18}P_6$
24. ${}_{20}P_5$
25. ${}_{20}P_1$
26. ${}_4P_3$
27. Show that ${}_nP_n = n!$.
28. Show that ${}_nP_1 = n$.

Solve the following problems.

29. In a nine-horse race, how many different first-second-third place finishes are possible?

30. A president, vice-president and secretary are to be elected from a club with 25 members. In how many different ways can these offices be filled?

31. A basketball team has 15 players. In how many ways can a captain and cocaptain be chosen from the players?

32. In how many different ways can eight students be seated in a row of eight chairs?

33. In how many ways can seven books be arranged on a shelf?

34. In horse racing, a perfecta bet picks the first place finisher in a race and the second place finisher. If a race has 11 horses, how many perfecta bets are possible?

35. In horse racing, a trifecta bet picks the first, second, and third place finishers in a race. If a race has 11 horses, how many trifecta bets are possible?

36. A contractor wishes to build eight houses, all different in design. In how many ways can these houses be placed if there are five lots on one side of the street and three lots on the other side?

37. There are 15 players on a baseball team. Nine are selected for the starting lineup, and they are given a certain batting order. We will call this the initial batting order. How many different initial batting orders are possible?

38. A football coach checks a player for six performance traits and lists the three strongest on a card; the coach lists these three strongest traits in order of the player's proficiency. How many different evaluations are possible?

39. A neighborhood children's club's secret password is "zyzzybalubah." How many different 12-letter "words" can be formed using all the letters of the secret word?

40. How many different words can be formed using all the letters of the word "mammal"?

41. In how many different ways can the monomial $3a^2b^4c^5$ be written without using exponents? (One way is "abc(3)abbbcccc.")

42. In how many different ways can the monomial $3a^2b^4c^5$ be written without using exponents, if the numerical coefficient must appear first? (See problem 41.)

Evaluate each expression.

43. ${}_{15}C_{10}$
44. ${}_{20}C_6$
45. ${}_8C_5$
46. ${}_{18}C_{16}$
47. ${}_{14}C_2$
48. ${}_{14}C_{12}$
49. Show that ${}_nC_n = 1$.
50. Show that ${}_nC_1 = n$.

51. A child is to be allowed to choose four different candies from a box containing 20 different kinds. How many different selections are possible?

52. How many different five-card hands can be dealt from a standard 52-card deck of playing cards?

53. On an examination consisting of 12 essay questions the student may omit any 4. In how many different ways can the student select the problems to be answered (the order of selection is not important)?

54. Suppose NASA has 19 astronauts suitable for the next space mission. A crew consists of 5 astronauts. How many different crews are possible?

55. A pizza parlor has eight different toppings for its pizzas. A regular pizza has any two of these toppings (they must be different). How many combinations of toppings are there for a regular pizza?

56. Ten individuals want to form two teams of five players each. In how many different ways can this be done?

57. Seven distinct points lie on a circle. How many different inscribed triangles can be drawn such that all of their vertices come from these points?

58. How many different straight lines can be drawn through the six points shown in the figure? (One such line is shown.) No three points are collinear.

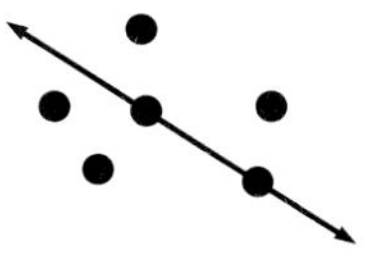

The following problems may require the multiplication-of-choices property, permutations, and/or combinations.

59. Fourteen children are playing a game of musical chairs. If there is a row of ten chairs in which the children can sit when the music stops, how many different groups of four children could be eliminated when the music stops?

60. From a standard deck of playing cards how many ways can a person select an ace, a king, a queen, and a jack (without regard to the order of selection)?

61. Suppose there are 17 players on a baseball team.
- **a.** In how many different ways can a team of 9 be chosen if every player can play every position?
- **b.** In how many different ways can a captain and a cocaptain be chosen (assuming the cocaptain and captain are different positions)?
- **c.** How many different batting orders are possible (considering all possible 9-player teams and all possible batting orders for 9 players)?
- **d.** In how many different ways can the team membership be reduced to 12 players?

62.
- **a.** How many ways can three males and four females sit in a row?
- **b.** How many ways can three males and four females sit in a row if males and females must alternate?
- **c.** How many arrangements of males and females are possible (i.e., permutations in which males are indistinguishable and females are indistinguishable)?

63. In how many different ways can a student answer all the questions on a quiz consisting of ten true/false questions?

64. Selecting from the set of digits {1, 2, 3, 4, 5} how many of the following are possible?
- **a.** Three-digit numbers
- **b.** Three-digit odd numbers
- **c.** Three-digit numbers, where the first and last digit must be even
- **d.** Three-digit numbers using only odd digits

65. Answer problem 64 if repetition of a digit is not allowed.

The following problems may involve several counting properties to solve.

66. In how many ways can a group of eight males and six females be divided into two groups consisting of four males and four females?

67. Ten teams are in a league. If each team is required to play every other team twice during the season, what is the total number of league games that will be played?

68. A shopper is choosing 6 different frozen dinners from a selection of 17 and 4 different fruits from a selection of 11. In how many different ways can the selections be made?

69. If a group consists of 18 men and 12 women, in how many different ways can a committee of 6 be selected if:
- **a.** the committee is to have an equal number of men and women?
- **b.** the committee is to be all women?
- **c.** there are no restrictions on membership on the committee?

70. At the beginning and end of every meeting of a certain club, each member must give the ritual handshake to every other member. If there are 20 members present at the meeting, how many different handshakes will take place?

71. In a certain computer there are 256 CPUs (central processor units); each is connected to each of the others. How many such connections are there?

72. A test contains three groups of questions, A, B, and C, that contain five, four, and three questions, respectively. If a student must select three questions from group A and two from each of the remaining groups, how many different tests are possible?

73. In horse racing, a double trifecta bet picks the first, second, and third place finishers in the first two races. If there are nine horses in the first race and eight horses in the second, how many different bets are possible?

74. Show that $\binom{n}{r} = \frac{n}{r}\binom{n-1}{r-1}$.

75. Show that $\binom{n}{r}\binom{r}{k} = \binom{n}{k}\binom{n-k}{r-k}$.

76. Show that $\binom{n}{r+1} + \binom{n}{r} = \binom{n+1}{r+1}$.

77. We defined ${}_nP_r = n(n-1)(n-2) \cdots (n-[r-1])$. Show that ${}_nP_r = \frac{n!}{(n-r)!}$ gives the same result for $n = 10$ and $r = 3$.

78. We have two ways to compute ${}_nP_r$:

$${}_nP_r = n(n-1)(n-2) \cdots (n-[r-1])$$

$${}_nP_r = \frac{n!}{(n-r)!}$$

Consider a calculator that has only a factorial $\boxed{x!}$ key and consider computing ${}_{300}P_3$ and ${}_{30}P_{28}$ to find an advantage that each method has over the other.

79. Show that $\frac{{}_nP_r}{r!} = \frac{n!}{r!(n-r)!}$ is true for $n = 10$ and $r = 3$.

80. The "party puzzle" states the following. Choose any six people at a party (or anywhere else). Then one of the following statements is true. Either there is a group of three who all know one another, or there is a group of three in which none of the members knows either of the other two.[18]

[18]A proof that one of these two types of groups can always be found is in the article "Ramsey Theory," by Ronald L. Graham and Joel H. Spencer, *Scientific American*, July 1990.

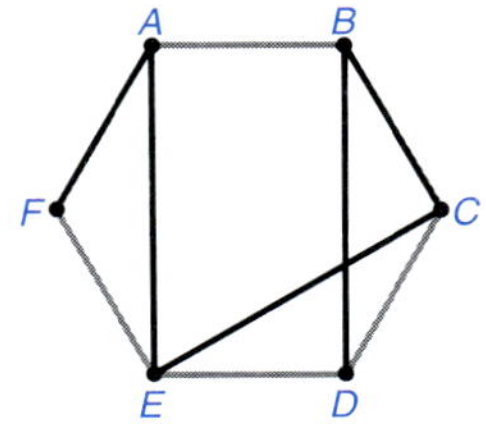

For example, the diagram shows six people, A, B, C, D, E, and F. The solid lines show who knows whom. Thus, person A knows persons F and E, and person D only knows person B. In this case, persons B, E, and F form a group of three, none of whom knows the other two.

a. In each of the three situations shown in the three-part figure, find a group of three people who either mutually know each other or who mutually do not know each other.

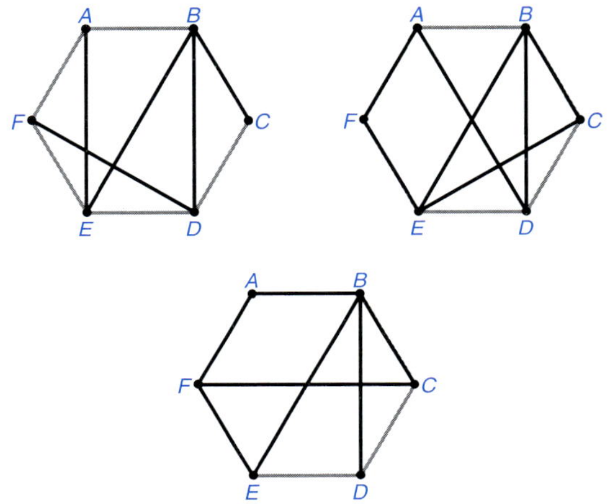

b. Using the diagrams as shown simplifies solving this problem. However, if we were to check each possible group of three people in a given situation the work would be tedious. How many groups of three people are there, given six people?

Skill and review

1. Find $\sum_{i=1}^{\infty} \left(\frac{1}{2}\right)^i$.

2. Find $2 + 7 + 12 + 17 + \cdots + 97$.

3. Solve $\left|\frac{2-3x}{4}\right| \le 10$.

4. Graph $f(x) = \frac{x^2-1}{x^2-4}$.

5. Graph $f(x) = (x-2)^2(x+1)^3$.

6. Solve $x + \frac{1}{x} = 3$.

7. Solve $\frac{x+y}{x} = \frac{x-y}{2}$ for y.

12–6 Introduction to probability

A certain test for determining whether or not an individual has used certain drugs is 90% accurate. That is, it will detect 90% of those drug users who are tested. The test also gives 10% false positives. That is, of those tested who do *not* use drugs the test will falsely report that the individual does use them. Suppose that 100,000 workers are to be tested for drug use, and that 15% of these workers use the drugs in question. If the test reports that an individual uses the drugs, what is the probability that this is *not* true?

This problem is just one of the many places where probability occurs every day in society. Probability is the subject of this section.

Terminology

Probability is a means of measuring uncertainty. The theory of probability is said to have begun in 1654 when a French nobleman and gambler, the Chevalier de Méré, asked his friend, the famous French mathematician-philosopher-writer Blaise Pascal, questions about gambling. Pascal's attempt to answer the Chevalier's questions, along with correspondence between Pascal and the French mathematician Pierre de Fermat, began the modern theory of probability. Today probability is used to explain atomic physics and human behavior; to derive the charge for an insurance policy; to decide the proper medical treatment for a disease; and to predict the chance that it will rain tomorrow, that a traffic light will last at least 1 year without failure, or the probability that an engine on a multiengine jet will fail on an overseas flight.

In the study of probability, any happening whose result is uncertain is called an **experiment.** The different possible results of the experiment are called **outcomes,** and the set of all possible outcomes of an experiment is called the **sample space** of the experiment, denoted by S. In this text all sample spaces are finite—that is, they have a limited number of outcomes.

An **event** is a subset of the sample space. If an event is the empty set it is said to be **impossible.** If an event has only one element it is called a **simple event.** Any nonempty event that is not simple is called a **compound event.**

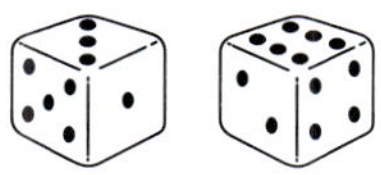

To illustrate, consider the *experiment* of rolling two dice.[19] When the first die is rolled, any value from 1 through 6 can appear, and the same is true for the second die. Let S be all the possible outcomes of rolling the two dice (the *sample space*). Let A be the *event* of rolling a total of 7, and let B be the *event* of rolling a total of 2. Figure 12–5 illustrates S, A, and B, where S is all of the number pairs. For example, (1,3) represents getting a 1 on the first die, and a 3 on the second.

1,1	1,2	1,3	1,4	1,5	1,6
2,1	2,2	2,3	2,4	2,5	2,6
3,1	3,2	3,3	3,4	3,5	3,6
4,1	4,2	4,3	4,4	4,5	4,6
5,1	5,2	5,3	5,4	5,5	5,6
6,1	6,2	6,3	6,4	6,5	6,6

Figure 12–5

[19] A standard die has six sides. Each side has a different number of from one through six dots. Whenever we refer to a die it will mean a standard die.

The event A is the set of two-tuples in which the sum of the elements is 7. The event $A = \{(6,1), (5,2), (4,3), (3,4), (2,5), (1,6)\}$. A is a *compound* event. The event B is the set $B = \{(1,1)\}$. B is a *simple* event.

Example 12–6 A illustrates how to determine a sample space.

■ Example 12–6 A

Determine the sample space for each experiment.

1. Tossing a six-sided die.
The sample space consists of six outcomes, which are represented by the integers from 1 to 6.

$$S = \{1, 2, 3, 4, 5, 6\}$$

2. Flipping a coin twice.
The possible outcomes are two heads, head, then tail, tail, then head, two tails.
If we use H for a head and T for a tail, the sample space would be

$$S = \{\text{HH, HT, TH, TT}\}.$$

In this sample space the event of a head and a tail can occur in two ways, HT and TH, and is a compound event. The event of getting two heads can occur only one way, and is a simple event. ■

Probability of an event

If we wish to determine the probability of an event, we must know how many outcomes make up the event and how many outcomes make up the sample space. The number of outcomes in event A is represented by $n(A)$, and the number of outcomes in the sample space is represented by $n(S)$. If the outcomes of a sample space are *equally likely,* then every outcome in the sample space has the same chance of occurring. We will concern ourselves only with equally likely events.

Probability of an event

If an event A is made up of $n(A)$ equally likely outcomes from a sample space S that has $n(S)$ equally likely outcomes, then the probability of the event A, represented by $P(A)$, is

$$P(A) = \frac{n(A)}{n(S)}$$

We know that $0 \le n(A)$ and $0 \le n(S)$, because each value represents counting a number of events. Also, A represents a subset of S, so $n(A) \le n(S)$. Thus,

$$0 \le n(A) \le n(S)$$

Dividing each member by $n(S)$, when $n(S) > 0$, we obtain

$$\frac{0}{n(S)} \leq \frac{n(A)}{n(S)} \leq \frac{n(S)}{n(S)}$$

or

$$0 \leq P(A) \leq 1$$

The following summarizes these and several other principles of probability.

Basic probability principles

If the event A has $n(A)$ equally likely outcomes and the sample space S has $n(S)$ equally likely outcomes, and $n(S) > 0$, then

1. $0 \leq P(A) \leq 1$
2. $P(A) = \dfrac{n(A)}{n(S)}$
3. $P(A) = 0$ means event A cannot occur, and is called an impossible event.
4. $P(A) = 1$ means event A must occur and is called a certain event.

Example 12–6 B illustrates the terminology and values associated with the probability of an event.

■ ***Example 12–6 B***

When rolling a single six-sided die, name the following events and give their probability.

	Event	Name	Probability
1.	$\{1\}$	Rolling a one	$\frac{1}{6}$
2.	$\{2, 4, 6\}$	Rolling an even number	$\frac{3}{6}$ or $\frac{1}{2}$
3.	$\{7\}$	Rolling a 7 (impossible event)	0
4.	$\{1, 2, 3, 4, 5, 6\}$	Rolling a 1, 2, 3, 4, 5, or 6	1

■

In the examples of rolling a single die or flipping a coin twice, it was easy to list all of the possible outcomes in the sample space. When the problem has a large number of possible outcomes, we will not always try to list all possible outcomes. Instead we will use the counting techniques seen in section 12–5 to determine the number of outcomes that make up an event or the sample space. Example 12–6 C illustrates finding probabilities in which the sample space has a large number of elements.

■ ***Example 12–6 C***

Find the probability of the given event.

1. A coin is flipped 3 times. What is the probability of exactly two heads? The sample space is $S = \{$HHH, HHT, HTH, HTT, THH, THT, TTH, TTT$\}$. The event of getting exactly two heads is $A = \{$HHT, HTH, THH$\}$. Thus, $P(A) = \dfrac{n(A)}{n(S)} = \dfrac{3}{8}$.

2. To win the State of Michigan lottery a person must correctly select the six integers drawn by the lottery commission from the set of integers from 1 to 47. If a person buys one ticket, what is the probability of winning?

The sample space is the set of all collections of 6 different numbers chosen from 1 through 47. By way of example, the figure shows the selection of 3, 8, 9, 21, 33, and 35. The order of selection of the integers is not important, so the size of the sample space is the number of combinations of size six, chosen from 47 available items, ${}_{47}C_6$. Thus,

$$n(S) = {}_{47}C_6 = \frac{47!}{41!\,6!} = 10,737,573.$$

The event A of winning is one of these selections, so $n(A) = 1$.

Thus, the probability of winning the lottery $P(A)$ is $P(A) = \dfrac{n(A)}{n(S)}$

$$= \frac{1}{10{,}737{,}573} \approx 0.000000093.$$

Note In a calculator the display may look like $\boxed{9.313091515^{-08}}$. This is in scientific notation and means approximately $9.3 \times 10^{-8} \approx 0.000000093$. (See section 1–2 to review scientific notation.)

3. Five cards are selected from a standard pack of playing cards.[20] What is the probability that all five cards are hearts?

The sample space S is the set of all possible five-card hands chosen from 52 cards. The event A is the set of all possible five-card hands chosen from the 13 hearts.

$$P(A) = \frac{n(A)}{n(S)} = \frac{{}_{13}C_5}{{}_{52}C_5} = \frac{1{,}287}{2{,}598{,}960} = \frac{33}{66{,}640} \approx 0.00050$$

4. A congressional committee contains eight Democrats and six Republicans. A subcommittee of four people is randomly chosen. What is the probability that all four people will be Republicans?

The sample space S is the set of all four-person committees that can be selected from 14 people. The event R of an all-Republican committee is the number of four-person committees that can be chosen from the 6 Republicans.

$$P(R) = \frac{n(R)}{n(S)} = \frac{{}_6C_4}{{}_{14}C_4} = \frac{15}{1{,}001} \approx 0.015$$

■

Mutually exclusive events

If two events from the same sample space have no outcomes in common, they are called **mutually exclusive events.** For example, if a single die is tossed, the event A of rolling a number less than 3 ($A = \{1,2\}$) and the event B of rolling a number greater than 4 ($B = \{5,6\}$) are mutually exclusive. In this situation the probability of the compound event of A or B is the sum of $P(A)$ and $P(B)$.

[20]See footnote 16.

Probability of mutually exclusive events

If the events A and B are mutually exclusive, then the probability that A or B will occur is

$$P(A \text{ or } B) = P(A) + P(B)$$

Example 12–6 D illustrates finding probabilities of compound events which are composed of mutually exclusive events.

■ ***Example 12–6 D***

Find the probability of the given event.

1. A card is drawn from a standard deck of 52 cards. What is the probability of an ace or a jack?

   The card that is drawn cannot be an ace and a jack at the same time. Therefore, the two events are mutually exclusive, and therefore the probability of one or the other is the sum of the probability of each individually. If $P(A)$ is the probability of an ace and $P(J)$ is the probability of a jack, we have

$$P(A) = \frac{\text{number of aces}}{\text{number of cards in deck}} = \frac{n(A)}{n(S)} = \frac{4}{52} = \frac{1}{13}$$

$$P(J) = \frac{\text{number of jacks}}{\text{number of cards in deck}} = \frac{n(J)}{n(S)} = \frac{4}{52} = \frac{1}{13}$$

$$P(A \text{ or } J) = P(A) + P(J) = \tfrac{1}{13} + \tfrac{1}{13} = \tfrac{2}{13}$$

2. A single die is tossed. What is the probability of a 5 or an even number?

   Since 5 is not even, the two events are mutually exclusive. Let F be the event of a 5: $F = \{5\}$. Then $P(F) = \dfrac{n(F)}{n(S)} = \tfrac{1}{6}$. Let E be the event of an even number: $E = \{2, 4, 6\}$. Then $P(E) = \dfrac{n(E)}{n(S)} = \tfrac{3}{6} = \tfrac{1}{2}$. Therefore,

   $P(F \text{ or } E) = P(F) + P(E) = \tfrac{1}{6} + \tfrac{1}{2} = \tfrac{2}{3}$. ■

Events that are not mutually exclusive

Consider the problem where we are asked to find the probability, when rolling a single die, of an even number or a number greater than 4. These two events could be described as $E = \{2, 4, 6\}$ and $G = \{5,6\}$. These events are not mutually exclusive, because the outcome of a 6 is part of both events.

To count the probability of the event E or G we cannot simply add the probability of each event separately. For example, calculating as we do for mutually exclusive events

$$P(E \text{ or } G) = P(E) + P(G) = \tfrac{3}{6} + \tfrac{2}{6} = \tfrac{5}{6}$$

is incorrect. The event E or G is the set $\{2, 4, 5, 6\}$, and $P(E \text{ or } G) = \tfrac{4}{6}$. Thus $\tfrac{5}{6}$ is too large by $\tfrac{1}{6}$.

The problem is that the simple event of rolling a 6 is in both compound events E and G. Thus, it was counted twice instead of once. We can adjust the method for calculating probabilities of mutually exclusive events by subtracting those probabilities that are counted twice. This leads to the following conclusion.

General addition rule of probability

If the events A and B are from the same sample space, then the probability that A or B will occur is

$$P(A \text{ or } B) = P(A) + P(B) - P(A \text{ and } B)$$

Note If the events A and B are mutually exclusive they cannot occur at the same time, and $P(A \text{ and } B) = 0$.

Now we can calculate the probability of the event E or G (above) correctly. First, note that the events E and G is the event $\{6\}$, since this is the only event both in the set E and G. Thus, $P(E \text{ and } G) = \frac{1}{6}$.

$$P(E \text{ or } G) = P(E) + P(G) - P(E \text{ and } G) = \tfrac{3}{6} + \tfrac{2}{6} - \tfrac{1}{6} = \tfrac{4}{6} = \tfrac{2}{3}$$

These ideas are further illustrated in example 12–6 E.

■ ***Example 12–6 E***

1. A card is drawn from a standard deck of 52 cards. What is the probability of a diamond or a face card?

Let D be the event of a diamond and F be the event of a face card. The events D and F are not mutually exclusive, since they have the three cards king, queen, and jack of diamonds in common. We find the probability of the event D or F as follows:

$$P(D) = \frac{\text{number of diamonds}}{\text{number of cards in deck}} = \frac{n(D)}{n(S)} = \frac{13}{52}$$

$$P(F) = \frac{\text{number of face cards}}{\text{number of cards in deck}} = \frac{n(F)}{n(S)} = \frac{12}{52}$$

4 face cards per suit

$$P(D \text{ and } F) = \frac{\text{number of face cards that are diamonds}}{\text{number of cards in deck}} = \frac{n(D \text{ and } F)}{n(S)} = \frac{3}{52}$$

Therefore,

$$P(D \text{ or } F) = P(D) + P(F) - P(D \text{ and } F) = \tfrac{13}{52} + \tfrac{12}{52} - \tfrac{3}{52} = \tfrac{22}{52} = \tfrac{11}{26}$$ ■

Complement of an event

The **complement of an event** A is the set of all outcomes in the sample space that are not part of the event A. We denote the complement of an event A by A', which we read as "A's complement." By definition, A and A' are mutually exclusive events. The following can be seen to be true.

$n(A \text{ or } A') = n(S)$

$P(A \text{ or } A') = P(S) = 1$

$P(A \text{ and } A') = 0$ (since A and A' cannot happen at the same time).

Using the general addition rule of probability, we obtain

$$P(A \text{ or } A') = P(A) + P(A') - P(A \text{ and } A')$$
$$1 = P(A) + P(A') - 0 \quad \text{Substitute values from above}$$
$$1 - P(A) = P(A')$$

Thus we can state the following.

> **Probability of complementary events**
> If A is any event and A' is its complement, then
> $$P(A') = 1 - P(A)$$

Example 12–6 F shows how to find the probability of an event by first finding the probability of the complementary event.

■ ***Example 12–6 F***

Find the probability of the given event.

1. A card is drawn from a standard deck of 52 cards. What is the probability of not selecting a face card?

The face cards are the kings, queens, and jacks of the four suits. Thus there are 12 face cards. Let F be the event of drawing a face card. Then $P(F) = \frac{12}{52} = \frac{3}{13}$. F' is the event of drawing a card which is not a face card. Then,

$$P(F') = 1 - P(F) = 1 - \tfrac{3}{13} = \tfrac{10}{13}$$

2. Five cards are selected from a standard deck of cards. What is the probability that one or more of the cards is not a heart?

If H is the event of drawing five hearts, then H' is the event of drawing five cards in which at least one is not a heart.

$$P(H') = 1 - P(H) = 1 - \frac{{}_{13}C_5}{{}_{52}C_5} = \frac{66{,}640}{66{,}640} - \frac{33}{66{,}640} = \frac{66{,}607}{66{,}640} \approx 0.9995$$ ■

Mastery points

Can you

- Determine the number of outcomes in an event?
- Determine the number of outcomes in a sample space?
- Determine the probability of an event?

Exercise 12–6

A coin is tossed two times. Find the probability of

1. exactly one head. **2.** one head and one tail. **3.** at least one head **4.** no heads.

A coin is tossed three times. Find the probability of

5. all tails. **6.** no tails. **7.** at least two tails. **8.** at most two tails.

A coin is tossed four times. Find the probability of

9. exactly two heads. **10.** exactly two tails. **11.** less than two heads. **12.** more than two heads.

A card is drawn from a standard deck of playing cards. What is the probability of

13. a ten? **14.** a seven? **15.** a club? **16.** a heart?
17. a card from 4 through 9, inclusive? **18.** a card from 3 through 6, inclusive?
19. a red 7? **20.** a black 8? **21.** a black 4 or 5? **22.** a red king?

A card is drawn from a standard deck of playing cards. Find the probability that the card is

23. a heart or a 7.
24. a diamond or a queen.
25. from 2 through 6, inclusive, or a spade.
26. from 5 through 8, inclusive, or a club.
27. not a 10.
28. not a jack.
29. not from 4 through 10, inclusive.
30. not from 2 through 5, inclusive.
31. not a club.
32. not a heart.
33. not red.
34. not black.

A roulette wheel contains the numbers from 1 through 36. Eighteen of these numbers are red, and the other 18 are black. There are two more numbers, 0 and 00, which are green. The wheel is spun, and a ball allowed to fall on one of these 38 locations at random, as the wheel stops. What is the probability that the ball will land on

35. the 10? **36.** a red number? **37.** a number that is not green?
38. a black or green number? **39.** a white number? **40.** an odd number?
41. an even, nonzero number?

A bowl contains 24 balls. Six are red, 10 are blue, and 8 are white. If one ball is randomly selected, what is the probability that the ball is

42. red? **43.** red or white? **44.** red, white, or blue?
45. black? **46.** not white? **47.** not red or white?

Find the probability of the given event. You may need to use some of the counting techniques of section 12–5.
An automobile parts supplier has 18 alternators on hand of a certain model. Ten of the alternators are new, and 8 are remanufactured. If 6 of the alternators are chosen randomly for a shipment, what is the probability that the shipment

48. contains all remanufactured alternators?
49. contains all new alternators?
50. contains half new and half remanufactured alternators?
51. contains 2 new and 4 remanufactured alternators?

Five cards are drawn from a standard deck of playing cards. Find the probability of each event.

52. All five cards are spades. **53.** All five cards are red. **54.** All five cards are black.
55. None of the cards are clubs. **56.** None of the cards are face cards. **57.** All of the cards are face cards.
58. Three black and two red cards. **59.** One black and four red cards. **60.** Three clubs and two hearts.
61. Four diamonds and one spade.

62. In a certain state lottery, 6 numbers must be chosen correctly from 46 numbers. What is the probability of making this choice correctly?

63. In a minilottery you must choose 2 numbers correctly from the numbers 1 through 15. What is the chance of doing this successfully?

64. A certain state instant-winner game has ten circles. Two of the circles cover the word WIN and eight cover the word SORRY. The game is played by scratching the cover off of two (and only two) of the circles. If both reveal the word WIN the player wins. What is the probability of winning this game?

65. A certain assembly line built 50 television sets today. Of those, 4 are defective.
   a. If 1 of the TVs is chosen at random, what is the probability that it is defective?
   b. If a shipment of 6 TVs was sent out before testing, what is the probability that at least 1 of these was defective?

66. A hospital has 100 units of type A blood on hand. Of those, 3 are infected with hepatitis.
   a. If a patient receives 1 unit of this blood, what is the probability of getting a unit that is infected?
   b. If a patient receives 4 units of this blood, what is the probability of not getting any units that are infected?

A doctor has four patients waiting, patients A, B, C, and D. They do not have appointments, and arrived at the same time. If the doctor chooses the order in which to see the patients at random, what is the probability that

67. B will be chosen first?

68. A will be chosen second?

69. A and B are seen before C or D?

70. C is seen before D?

Apply the following to problems 71–76. The idea of probability goes far beyond that presented in this section. By way of example, consider the formula $P(k,t) = \dfrac{e^{-n} \times n^k}{k!}$, where $n = \dfrac{t}{\text{MTBF}}$, and e is the constant introduced in section 9–4. This formula gives the probability of k failures in time period t, in a situation in which something, say an aircraft engine, fails based on what is called a normal distribution. MTBF stands for mean time between failures, and is the average time between failures. For example, if an aircraft engine is expected to operate 2,000 hours between failures (MTBF = 2,000), then the probability of two failures in 500 hours would be found by using $n = \dfrac{500}{2{,}000} = 0.25$.

$$P(2, 500) = \frac{e^{-0.25} \cdot 0.25^2}{2!} \approx 0.024$$

71. Suppose a computer's MTBF is 4,000 hours. Find the probability of one failure in 3,000 hours of operation.

72. Suppose a computer's MTBF is 1,000 hours of operation. Find the probability of two failures in 3,000 hours of operation.

73. A computer's MTBF is 2,000 hours. Find the probability of at least one failure in 3,000 hours of operation. (This is 1, less the probability of no failures in 3,000 hours of operation.)

74. A computer's MTBF is 1,000 hours of operation. Find the probability of at least two failures in 3,000 hours of operation. (This is 1 less the sum of the probabilities of 0 and 1 failures.)

75. Suppose an automobile's alternator has a MTBF of 1,000 hours. Find the probability of at least 1 failure in 1,500 hours of operation. (1, less the probability of no failures.)

76. Referring to problem 75, find the probability of no failures in 800 hours of operation.

77. A certain test for determining whether or not an individual has used certain drugs is 90% accurate. That is, it will detect 90% of those drug users who are tested. The test also gives 10% false positives. That is, of those tested who do *not* use drugs the test will falsely report that the individual does use them. Suppose that 100,000 workers are to be tested for drug use, and that 15% of these workers use the drugs in question. If the test reports that an individual uses the drugs, what is the probability that this is *not* true?

78. The current test for the HIV (AIDS) virus (called ELISA) has sensitivity 0.983, and specificity 0.998. Sensitivity means the probability that a person with the virus will test positive. Specificity means the probability that a person who does not have the virus will test negative. If 500 people out of 1 million carry the virus, and those million are tested with the ELISA test, what is the probability that a person who tests positive actually has the virus? (In practice, those who test positive are given another test, which is much more accurate but much more expensive. Even the second test is still not perfect, however.)

Skill and review

1. Solve $S = \frac{1}{2}[a - b(a + c)]$ for a.
2. If $f(x) = x^3 - x^2 - x$, compute $f(a - 1)$.
3. Find the inverse function of $f(x) = \dfrac{1 - 2x}{3}$.
4. Find $\log_4 64$.
5. Solve $\log(x + 3) + \log(x - 1) = 1$.
6. Graph $f(x) = \sqrt{x - 3} - 1$.

12–7 Recursive definitions and recurrence relations—optional

The Fibonacci sequence is the sequence 1, 1, 2, 3, 5, 8, . . . , where each element after the first two is the sum of the two previous elements. Find a way to compute the value of any given element without having to compute all previous values.

Recursive definitions

In addition to advanced mathematics, the material in this section is used in computer science.[21] It is not common in most other areas where mathematics is applied—this is why this section is optional.

In section 12–1 we discussed sequences that are defined by an expression or rule for the nth term. For example, $a_n = 2n - 3$ defines a sequence by an expression for the nth term. If we define another sequence as the digits in the decimal expansion of $\sqrt{2}$, we specify a rule for the sequence without specific reference to the nth term.

Another way to define the nth term of a sequence is to give a rule that depends on previously known terms; this is called a **recursive definition.** An example of this would be the rule $a_n = \begin{cases} 3 \text{ if } n = 0 \\ 2a_{n-1} \text{ if } n > 0 \end{cases}$. We are indexing the sequence beginning at 0 instead of 1 because this makes some of the following discussion simpler. This rule says that the first value in the sequence is 3, and each value after that is twice the previous value. Thus, this rule gives the sequence 3, 6, 12, 24, 48, ... The part of the definition that is $2a_{n-1}$ is called its **recursive part,** and the part that is 3 is called the **terminal part.** Any recursive definition needs a terminal part, or there is no place to "start." With this rule, to find, say, a_{50} we would have to first know a_{49}, and to find a_{49} we would have to know a_{48}, etc., until we arrived at the terminal part of the definition. This is a "weakness" of recursive definitions—you cannot just jump in any where you want!

It can be seen that the sequence 3, 6, 12, 24, 48, . . . is a geometric sequence, with $a_0 = 3$ and $r = 2$, so we could formulate the rule[22] $a_n = 3(2^n)$,

[21]This chapter relies heavily on *Mathematics for the Analysis of Algorithms,* by Daniel Green and Donald Knuth, Birkhäuser, Boston, 1982. To see just how mathematical computer science can be see *The Art of Computer Programming,* a three volume tour de force by Donald Knuth. It is published by Addison-Wesley, Reading, Massachusetts.

[22]When we index a series from 0 instead of 1 the resulting formulas will be different from those presented earlier.

$n = 0, 1, 2, \ldots$, from which we could deduce that $a_{50} = 3(2^{50})$. This second definition is sometimes called a **closed form definition.** Thus, the following are two different rules for the same sequence.

$$a_n = \begin{cases} 3 \text{ if } n = 0 \\ 2a_{n-1} \text{ if } n > 0 \end{cases}$$ recursive

$$a_n = 3(2^n),\ n = 0, 1, 2, \ldots$$ closed form

For computational purposes, a closed form rule is more useful than a recursive rule, but there are times when the closed form rule is hard to find or simply does not exist. In computer science, many algorithms (procedures) are most easily described in a recursive manner. For example, the computation of factorials can be described recursively: if $n > 1$, then $n! = n \cdot (n - 1)!$. This can be translated into a one-line computer program in most modern programming languages. Another example is discovering the shortest path that a traveling salesperson could follow to visit every customer in a certain region once.

Example 12–7 A illustrates applying recursive definitions to list the elements of a sequence.

■ ***Example 12–7 A***

Find the first five terms of each recursively defined sequence.

1. $a_n = \begin{cases} 2 \text{ if } n = 0 \\ a_{n-1} + 3 \text{ if } n > 0 \end{cases}$

$$\begin{aligned} a_0 &= 2 \\ a_1 &= a_0 + 3 = 2 + 3 = 5 \\ a_2 &= a_1 + 3 = 5 + 5 = 8 \\ a_3 &= a_2 + 3 = 8 + 3 = 11 \end{aligned}$$ Terminal part

Thus, the sequence is 2, 5, 8, 11, 14, . . . (an arithmetic sequence).

2. $a_n = \begin{cases} 1 \text{ if } n = 0 \\ 3a_{n-1} \text{ if } n > 0 \end{cases}$

$$\begin{aligned} a_0 &= 1 \\ a_1 &= 3a_0 = 3(1) = 3 \\ a_2 &= 3a_1 = 3(3) = 9 \end{aligned}$$ Terminal part

This is 1, 3, 9, 27, 81, . . . (a geometric sequence).

3. $a_n = \begin{cases} 1 \text{ if } n = 0,\ 3 \text{ if } n = 1 \\ 2a_{n-1} + 3a_{n-2} \text{ if } n > 1 \end{cases}$

In this sequence each term after the first two depends on the two previous terms.

$$\begin{aligned} a_0 &= 1,\ a_1 = 3 \\ a_2 &= 2a_1 + 3a_0 = 2(3) + 3(1) = 9 \\ a_3 &= 2a_2 + 3a_1 = 2(9) + 3(3) = 27 \\ a_4 &= 2a_3 + 3a_2 = 2(27) + 3(9) = 81 \end{aligned}$$ Terminal part

Thus, the first five terms of the sequence are 1, 3, 9, 27, 81, which appears to be the beginning of a geometric sequence.

4. $a_n = \begin{cases} 2 \text{ if } n = 0,\ 3 \text{ if } n = 1 \\ 2a_{n-1} + 3a_{n-2} \text{ if } n > 1 \end{cases}$

$$a_0 = 2,\ a_1 = 3$$
$$a_2 = 2a_1 + 3a_0 = 2(3) + 3(2) = 12$$
$$a_3 = 2a_2 + 3a_1 = 2(12) + 3(3) = 33$$
$$a_4 = 2a_3 + 3a_2 = 2(33) + 3(12) = 102$$

Thus, the sequence is 2, 3, 12, 33, 102, ... Observe that the recursive rule is the same for this sequence, but the first term is different, and the resulting sequence is not a geometric sequence.

5. $a_n = \begin{cases} 1 \text{ if } n = 0,\ 1 \text{ if } n = 1 \\ a_{n-1} + a_{n-2} \text{ if } n > 1 \end{cases} = 1, 1, 2, 3, 5, 8, \ldots$ ■

The sequence in part 5 of example 12–7 A, 1, 1, 2, 3, 5, 8, . . . , is called the **Fibonacci sequence.** Leonardo of Pisa, also called Fibonacci, described the sequence in a problem[23] in his book *Liber abaci* (book of the abacus) in 1202. This sequence actually occurs in solving certain problems in computer science.[24]

It is often important to have an expression for a_n in a sequence (that is to have a closed form definition of the sequence). This is not always possible, but it can be done for the sequences presented in example 12–7 A. Example 12–7 B shows how this is done when the sequence fits a pattern we saw in section 12–1.

■ Example 12–7 B

Find an expression for a_n for the sequences in parts 1 and 2 of example 12–7 A.

1. The sequence is an arithmetic sequence: 2, 5, 8, 11, . . . with $a_0 = 2$, $a_1 = 5$, and $d = 3$.

$$a_n = a_1 + (n - 1)d$$
$$a_n = 5 + 3(n - 1) = 3n + 2$$

General formula for *n*th term of an arithmetic sequence

Thus, $a_n = 3n + 2$, $n = 0, 1, 2, \ldots$.

2. 1, 3, 9, 27, 81 is a geometric sequence with $a_0 = 1$, $a_1 = 3$, and $r = 3$.

$$a_n = a_1 r^{n-1}$$
$$a_n = 3(3^{n-1}) = 3^n$$

General formula for *n*th term of a geometric sequence.

Thus, $a_n = 3^n$, $n = 0, 1, 2, \ldots$. ■

[23]Leonardo wrote "How many pairs of rabbits will be produced in a year, beginning with a single pair, if in every month each pair bears a new pair which becomes productive from the second month on?" His answer, under the conditions he gave, is 377.

[24]In fact, this sequence is found in fields as varied as biology and art. It is found in the seed pattern of the sunflower and in pine cones and is related to the dimensions of the chambers of the nautilus seashell. It has produced so much interest that there is a periodical called the *Fibonacci Quarterly,* produced by the Fibonacci Association.

Recurrence relations

To find an expression for a_n in the remaining three sequences of example 12–7 A, we resort to **recurrence relations.** We will illustrate finding recurrence relations for the case where the recursive definition expresses a_n as a linear combination of previous terms. The following procedure is used.

> **Recurrence relations**
>
> A procedure for finding expressions for a_n for recursively defined sequences where the recursive part is of the form
>
> $$a_n = c_{n-1}a_{n-1} + c_{n-2}a_{n-2} + \cdots + c_{n-j}a_{n-j},\ c_i \in R$$
>
> is as follows:
>
> 1. Subtract the right member from both members, producing an equation of the form
> $$a_n - c_{n-1}a_{n-1} - c_{n-2}a_{n-2} - \cdots - c_{n-j}a_{n-j} = 0$$
> Note that the indices of a are in decreasing order.
> 2. Replace each a_i by x^i. This produces a polynomial for the form
> $$x^n - c_{n-1}x^{n-1} - c_{n-2}x^{n-2} - \cdots - c_{n-j}x^{n-j} = 0$$
> Then replace n by j; this produces a polynomial with a constant term $-c_{n-j}$.
> 3. Solve the polynomial, obtaining the roots $\alpha_1, \alpha_2, \ldots, \alpha_j$.
> 4. The solution is of the form $a_n = C_1\alpha_1^n + C_2\alpha_2^n + \cdots + C_j\alpha_j^n$. The values of the C_i are found by using the known values of $a_0, a_1, \ldots$ to obtain a system of equations.

As an example, consider the series in part 3 of example 12–7 A.

$$a_n = \begin{cases} 1 \text{ if } n = 0,\ 3 \text{ if } n = 1 \\ 2a_{n-1} + 3a_{n-2} \text{ if } n > 1 \end{cases} = 1, 3, 9, 27, 81, \ldots$$

The terms of the sequence appear to be a geometric sequence with $a_n = 3^n$. However, we cannot prove this directly from the definition.[25] Instead, we use recurrence relations, as follows.

	$a_n = 2a_{n-1} + 3a_{n-2}$	Recursive part of definition
Step 1:	$a_n - 2a_{n-1} - 3a_{n-2} = 0$	Arrange so all nonzero terms are on one member with subscripts in decreasing order
Step 2:	$x^n - 2x^{n-1} - 3x^{n-2} = 0$	Replace each a_i by x^i
	$x^2 - 2x^{2-1} - 3x^{2-2} = 0$	Replace n by 2
	$x^2 - 2x - 3 = 0$	
Step 3:	$(x - 3)(x + 1) = 0$	Factor
	$x = 3$ or -1	

[25]This could be proved by the method of finite induction (section 12–4). See problems 23 and 24 in the exercises.

Step 4: $a_n = A(3)^n + B(-1)^n$

a_n is a linear combination of the solutions, each to the power n. We can find the values of A and B by using a_0 and a_1.

$$a_n = A(3)^n + B(-1)^n$$
$$n = 0: a_0 = A(3)^0 + B(-1)^0 \text{ so } 1 = A + B$$
$$n = 1: a_1 = A(3)^1 + B(-1)^1 \text{ so } 3 = 3A - B$$

We solve this system of two linear equations in two variables using any of the methods of chapter 10, or substitution (section 3–2). We find that $A = 1$, $B = 0$. Thus, $a_n = (1)(3^n) + 0(-1)^n$, or $a_n = 3^n$.

Example 12–7 C illustrates further.

■ ***Example 12–7 C***

Find an expression for a_n for the sequences in parts 4 and 5 of example 12–7 A.

1. $a_n = \begin{cases} 2 \text{ if } n = 0, 3 \text{ if } n = 1 \\ 2a_{n-1} + 3a_{n-2} \text{ if } n > 1 \end{cases}$

$$a_n = 2a_{n-1} + 3a_{n-2}$$
$$a_n - 2a_{n-1} - 3a_{n-2} = 0$$
$$x^n - 2x^{n-1} - 3x^{n-2} = 0$$
$$x^2 - 2x - 3 = 0$$
$$x = 3 \text{ or } -1$$
$$a_n = A3^n + B(-1)^n$$
$$n = 0: a_0 = 2 = A(3^0) + B(-1)^0 \text{ so } 2 = A + B$$
$$n = 1: a_1 = 3 = A(3^1) + B(-1)^0 \text{ so } 3 = 3A - B$$
$$A = \tfrac{5}{4}, B = \tfrac{3}{4}$$

$$a_n = \tfrac{5}{4}(3^n) + \tfrac{3}{4}(-1)^n, \text{ or } a_n = \tfrac{1}{4}[5(3^n) + 3(-1)^n].$$

2. $a_n = \begin{cases} 1 \text{ if } n = 0, 1 \text{ if } n = 1 \\ a_{n-1} + a_{n-2} \text{ if } n > 1 \end{cases}$

$$a_n = a_{n-1} + a_{n-2}$$
$$a_n - a_{n-1} - a_{n-2} = 0$$
$$x^n - x^{n-1} - x^{n-2} = 0$$
$$x^2 - x - 1 = 0$$
$$x = \frac{1 + \sqrt{5}}{2} \text{ or } \frac{1 - \sqrt{5}}{2}$$
$$a_n = A\left(\frac{1 + \sqrt{5}}{2}\right)^n + B\left(\frac{1 - \sqrt{5}}{2}\right)^n$$

Now find A and B using $a_0 = 1$ and $a_1 = 1$.

$$a_n = A\left(\frac{1 + \sqrt{5}}{2}\right)^n + B\left(\frac{1 - \sqrt{5}}{2}\right)^n$$
$$n = 0: a_0 = 1 = A\left(\frac{1 + \sqrt{5}}{2}\right)^0 + B\left(\frac{1 - \sqrt{5}}{2}\right)^0 \text{ so } 1 = A + B$$
$$n = 1: a_1 = 1 = A\left(\frac{1 + \sqrt{5}}{2}\right)^1 + B\left(\frac{1 - \sqrt{5}}{2}\right)^1$$

We now solve for A and B using substitution (section 3–2).

$1 = A + B$ — First equation from above

$B = 1 - A$ — Solve for B

$1 = A\left(\frac{1+\sqrt{5}}{2}\right) + B\left(\frac{1-\sqrt{5}}{2}\right)$ — Second equation from above

$2 = A(1+\sqrt{5}) + B(1-\sqrt{5})$ — Multiply each term by 2

$2 = A(1+\sqrt{5}) + (1-A)(1-\sqrt{5})$ — Replace B by $1 - A$

$\sqrt{5} + 1 = 2A\sqrt{5}$ — Add $\sqrt{5}$ to both members

$\frac{\sqrt{5}+1}{2\sqrt{5}} = A$ — Divide both members by $2\sqrt{5}$

$$B = 1 - A = \frac{2\sqrt{5}}{2\sqrt{5}} - \frac{\sqrt{5}+1}{2\sqrt{5}} = \frac{\sqrt{5}-1}{2\sqrt{5}}$$

Thus, $a_n = \left(\frac{\sqrt{5}+1}{2\sqrt{5}}\right)\left(\frac{1+\sqrt{5}}{2}\right)^n + \left(\frac{\sqrt{5}-1}{2\sqrt{5}}\right)\left(\frac{1-\sqrt{5}}{2}\right)^n$ is the general term of the Fibonacci sequence. ■

Roots of multiplicity greater than one

When we solve the polynomial equation that is part of the preceding procedure we may find roots that are of multiplicity greater than 1.

> **Roots of multiplicity $m > 1$**
> Roots of multiplicity m, $m > 1$, are treated by using powers of n from 0 to $m - 1$ as additional coefficients with the appropriate roots. For example, if γ (gamma) is a root of multiplicity 3, then the expression for a_n would include
> $$c_i\gamma^n + nc_{i+1}\gamma^n + n^2c_{i+2}\gamma^n$$

This is illustrated in example 12–7 D.

■ ***Example 12–7 D***

Find an expression for the nth term of the sequence.

1. $a_n = \begin{cases} 1 \text{ if } n = 0, 3 \text{ if } n = 1 \\ 4a_{n-1} - 4a_{n-2} \text{ if } n > 1 \end{cases}$

$$a_n - 4a_{n-1} + 4a_{n-2} = 0$$
$$x^2 - 4x + 4 = 0$$
$$(x-2)^2 = 0$$
$x = 2$ is a root of multiplicity 2.
$$a_n = A(2^n) + Bn(2^n)$$
$$n = 0\text{: } a_0 = 1 = A$$
$$n = 1\text{: } a_1 = 3 = 2A + 2B$$
$$A = 1,\ B = \tfrac{1}{2}$$

$a_n = 2^n + \frac{n}{2}(2^n)$, or $a_n = 2^n\left(\frac{n+2}{2}\right)$.

2. $a_n = \begin{cases} 1 \text{ if } n = 0, 3 \text{ if } n = 1, 4 \text{ if } n = 2 \\ a_{n-1} + a_{n-2} - a_{n-3} \text{ if } n > 2 \end{cases}$

$$a_n = a_{n-1} + a_{n-2} - a_{n-3}$$
$$a_n - a_{n-1} - a_{n-2} + a_{n-3} = 0$$
$$x^n - x^{n-1} - x^{n-2} + x^{n-3} = 0$$
$$x^3 - x^2 - x + 1 = 0$$
$$x^2(x - 1) - 1(x - 1) = 0$$
$$(x - 1)(x^2 - 1) = 0$$
$$(x - 1)(x + 1)(x - 1) = 0$$

$x = 1$ and -1 are zeros. 1 has multiplicity 2.

$$a_n = A(1^n) + Bn(1^n) + C(-1)^n$$
$$a_n = A + nB + (-1)^n C \quad \text{General form of solution}$$

$n = 0$: $a_0 = A + 0B + (-1)^0C$, so [1] $1 = A + C$

$n = 1$: $a_1 = A + B + (-1)^1C$, so [2] $3 = A + B - C$

$n = 2$: $a_2 = A + 2B + (-1)^2C$, so [3] $4 = A + 2B + C$

Solving shows that $A = \frac{5}{4}$, $B = \frac{3}{2}$, $C = -\frac{1}{4}$.

$$a_n = \tfrac{5}{4} + n(\tfrac{3}{2}) + (-1)^n(-\tfrac{1}{4}) = \frac{5 + 6n - (-1)^n}{4}$$

■

Mastery points

Can you

- Compute terms in recursively defined sequences?
- Find an expression for a_n in certain recursively defined sequences when that sequence is an arithmetic or geometric sequence?
- Find an expression for a_n in certain recursively defined sequences using recursion relations?

Exercise 12–7

Find the first five terms of each recursively defined sequence. Then find an expression for a_n.

1. $a_n = \begin{cases} 3 \text{ if } n = 0 \\ a_{n-1} + 5 \text{ if } n > 0 \end{cases}$

2. $a_n = \begin{cases} -10 \text{ if } n = 0 \\ a_{n-1} + 3 \text{ if } n > 0 \end{cases}$

3. $a_n = \begin{cases} 5 \text{ if } n = 0 \\ 2a_{n-1} \text{ if } n > 0 \end{cases}$

4. $a_n = \begin{cases} 2 \text{ if } n = 0 \\ -3a_{n-1} \text{ if } n > 0 \end{cases}$

5. $a_n = \begin{cases} -2 \text{ if } n = 0, 3 \text{ if } n = 1 \\ 2a_{n-1} + 3a_{n-2} \text{ if } n > 1 \end{cases}$

6. $a_n = \begin{cases} 2 \text{ if } n = 0, 3 \text{ if } n = 1 \\ 5a_{n-1} - 4a_{n-2} \text{ if } n > 1 \end{cases}$

7. $a_n = \begin{cases} 3 \text{ if } n = 0, 1 \text{ if } n = 1 \\ 3a_{n-1} + 4a_{n-2} \text{ if } n > 1 \end{cases}$

8. $a_n = \begin{cases} -2 \text{ if } n = 0, 4 \text{ if } n = 1 \\ 5a_{n-1} - 6a_{n-2} \text{ if } n > 1 \end{cases}$

9. $a_n = \begin{cases} -2 \text{ if } n = 0, 4 \text{ if } n = 1 \\ 3a_{n-1} + 6a_{n-2} \text{ if } n > 1 \end{cases}$

10. $a_n = \begin{cases} -2 \text{ if } n = 0, 4 \text{ if } n = 1 \\ 5a_{n-1} - a_{n-2} \text{ if } n > 1 \end{cases}$

Find an expression for the nth term of the sequence.

11. $a_n = \begin{cases} 1 \text{ if } n = 0, 3 \text{ if } n = 1 \\ 6a_{n-1} - 9a_{n-2} \text{ if } n > 1 \end{cases}$

12. $a_n = \begin{cases} 2 \text{ if } n = 0, 3 \text{ if } n = 1 \\ -6a_{n-1} - 9a_{n-2} \text{ if } n > 1 \end{cases}$

13. $a_n = \begin{cases} 4 \text{ if } n = 0, 3 \text{ if } n = 1 \\ 2a_{n-1} - a_{n-2} \text{ if } n > 1 \end{cases}$

14. $a_n = \begin{cases} -2 \text{ if } n = 0, 2 \text{ if } n = 1 \\ 8a_{n-1} - 16a_{n-2} \text{ if } n > 1 \end{cases}$

15. $a_n = \begin{cases} 1 \text{ if } n = 0, 1 \text{ if } n = 1, 3 \text{ if } n = 2 \\ a_{n-1} + a_{n-2} - a_{n-3} \text{ if } n > 2 \end{cases}$

16. $a_n = \begin{cases} 1 \text{ if } n = 0, 1 \text{ if } n = 1, 3 \text{ if } n = 2 \\ 6a_{n-1} - 12a_{n-2} + 8a_{n-3} \text{ if } n > 2 \end{cases}$

17. State a weakness of recursive definitions that a closed form expression for the definition does not have.

18. Show that the general expression for the Fibonacci sequence (example 12–7 C) can be transformed into $a_n = \frac{1}{\sqrt{5}}\left(\left(\frac{1+\sqrt{5}}{2}\right)^{n+1} - \left(\frac{1-\sqrt{5}}{2}\right)^{n+1}\right)$.

19. In the text we stated that the sequence $a_n = \begin{cases} 2 \text{ if } n = 0 \\ a_{n-1} + 3 \text{ if } n > 0 \end{cases}$ is an arithmetic sequence. We did not prove this, but only observed this. Using the definition of arithmetic sequence prove that this sequence is indeed an arithmetic sequence.

20. In the text we observed, without proof, that the sequence $a_n = \begin{cases} 1 \text{ if } n = 0 \\ 3a_{n-1} \text{ if } n > 0 \end{cases}$ is a geometric sequence. Using the definition of geometric sequence prove that this sequence is indeed a geometric sequence.

21. In the text the sequence $a_n = \begin{cases} 1 \text{ if } n = 0, 3 \text{ if } n = 1 \\ 2a_{n-1} + 3a_{n-2} \text{ if } n > 1 \end{cases}$ was seen to be a geometric sequence, whereas the sequence $a_n = \begin{cases} 2 \text{ if } n = 0, 3 \text{ if } n = 1 \\ 2a_{n-1} + 3a_{n-2} \text{ if } n > 1 \end{cases}$ was not a geometric sequence, even though in both cases $a_n = 2a_{n-1} + 3a_{n-2}$ for $n > 1$. Find the value A so that $a_n = \begin{cases} 2 \text{ if } n = 0, A \text{ if } n = 1 \\ 2a_{n-1} + 3a_{n-2} \text{ if } n > 1 \end{cases}$ is a geometric sequence. (*Hint:* For a sequence to be geometric $\frac{a_1}{a_0} = r$ and $\frac{a_2}{a_1} = r$ must be true for some $r > 0$, $r \neq 1$.)

22. Given $a_n = \begin{cases} A \text{ if } n = 0, B \text{ if } n = 1 \\ 5a_{n-1} + 6a_{n-2} \text{ if } n > 1 \end{cases}$, find values of A and B so that the sequence is a geometric sequence. (See problem 21 for a hint.)

23. Use finite induction (section 12–4) to prove that $b_n = 3^n$ defines the same sequence as $a_n = \begin{cases} 1 \text{ if } n = 0, 3 \text{ if } n = 1 \\ 2a_{n-1} + 3a_{n-2} \text{ if } n > 1 \end{cases}$. That is, show that for all $n \in N$, each element a_n is of the form 3^n. (*Hint:* Show the cases for $n = 0$ or $n = 1$ by hand. Then, assume that $a_k = 2a_{k-1} + 3a_{k-2}$ is of the form 3^k for all natural numbers up to k, and use this to show that $a_{k+1} = 2a_k + 3a_{k-1}$ is of the form 3^{k+1}. Note that if a_k is of the form 3^k, then $a_{k-1} = 3^{k-1}$ and $a_{k-2} = 3^{k-2}$, since the statement is true for all values $0, 1, \ldots, k$.)

24. Use finite induction to prove that $b_n = 2n + 1$ defines the same sequence as $a_n = \begin{cases} 1 \text{ if } n = 0, 3 \text{ if } n = 1 \\ 2a_{n-1} - a_{n-2} \text{ if } n > 1 \end{cases}$ (See problem 23.)

Skill and review

1. The first term in an arithmetic progression is 4, and the 58th term is 67. Find the 96th term.
2. An artist is going to fill in each of the four sections of the work illustrated here.

   a. If the artist has 4 colors, and wants each section to be a different color, how many ways can the artist color the painting?
   b. If the artist has 6 colors, and wants each section to be a different color, how many ways can the artist color the painting?
   c. If the artist has 5 colors, and if the artist is willing to repeat colors in sections (including adjacent sections), how many ways can the artist color the painting?
3. Solve the inequality $x^2 - 10x - 18 > 6$.
4. Add up the numbers $1 + \frac{1}{3} + \frac{1}{9} + \frac{1}{27} + \frac{1}{81} + \frac{1}{243}$.

Chapter 12 summary

- **Arithmetic sequence** $a_{n+1} = a_n + d$

  $a_n = a_1 + (n - 1)d$

  $S_n = \frac{n}{2}(a_1 + a_n)$

  $S_n = \frac{n}{2}[2a_1 + (n - 1)d]$

- **Geometric sequence** $a_{n+1} = r \cdot a_n$

  $a_n = a_1 r^{n-1}$

  $S_n = \frac{a_1(1 - r^n)}{1 - r}$

  $S_\infty = \frac{a_1}{1 - r}, \; |r| < 1$

- **Pascal's triangle**

  1 1

  1 2 1

  1 3 3 1

  1 4 6 4 1

  1 5 10 10 5 1

  etc.

- $n! = n \cdot (n - 1) \cdot (n - 2) \cdots 2 \cdot 1$
- ${}_nC_k = \frac{n!}{(n - k)!\,k!}$ if $n \geq k \geq 0$, $k, n \in N$.
- [1] $\binom{n}{n} = 1$ [2] $\binom{n}{1} = n$ [3] $\binom{n}{0} = 1$
- ${}_nP_r = n(n - 1)(n - 2) \cdots (n - [r - 1])$
- **Binomial expansion formula** $(x + y)^n = \sum_{i=0}^{n} \binom{n}{i} x^{n-i}y^i$
- Let k be a constant, and let $f(i)$ and $g(i)$ represent expressions in the index variable i. Then the following are true.

  Sum of constants property: $\sum_{i=1}^{n} k = nk$

  Constant factor property: $\sum_{i=1}^{n} [k \cdot f(i)] = k \cdot \sum_{i=1}^{n} f(i)$

  Sum of terms property: $\sum_{i=1}^{n} [f(i) + g(i)] = \sum_{i=1}^{n} f(i) + \sum_{i=1}^{n} g(i)$

  Sum of integer series: $\sum_{i=1}^{n} i = \frac{n(n + 1)}{2}$

  Sum of squares of integers series: $\sum_{i=1}^{n} i^2 = \frac{n(n + 1)(2n + 1)}{6}$

  Sum of cubes of integers series: $\sum_{i=1}^{n} i^3 = \left(\frac{n(n + 1)}{2}\right)^2$

- **Principle of finite induction**
  If (1) a statement is true for $n = 1$, and
  (2) it can be shown that if the statement is true for $n = k$ then it must also be true for $n = k + 1$, then the statement is true for any positive integer.
- **Multiplication-of-choices property** If a choice in which the order in which each choice is made is not important consists of k decisions, where the first can be made n_1 ways and for each of these choices the second can be made in n_2 ways, and in general the ith choice can be made in n_i ways, then the complete choice can be made in $n_1 \cdot n_2 \cdot \cdots \cdot n_k$ ways. Each complete choice is also called an outcome.
- **Indistinguishable permutations** The number of distinct permutations P of n elements, where n_1 are alike of one kind, n_2 are alike of another kind, . . . , and n_k are alike of another kind, where $n_1 + n_2 + \cdots + n_k = n$, is $P = \frac{n!}{n_1!n_2! \cdots n_k!}$.
- **The probability of an event** If an event A is made up of $n(A)$ equally likely outcomes from a sample space S that has $n(S)$ equally likely outcomes, then the probability of the event A, represented by $P(A)$, is $P(A) = \frac{n(A)}{n(S)}$.
- **Basic probability principles** If the event A has $n(A)$ equally likely outcomes and the sample space S has $n(S)$ equally likely outcomes, and $n(S) > 0$, then
  1. $0 \leq P(A) \leq 1$.
  2. $P(A) = \frac{n(A)}{n(S)}$.
  3. $P(A) = 0$ means event A cannot occur, and is called an impossible event.
  4. $P(A) = 1$ means event A must occur and is called a certain event.
- **Probability of mutually exclusive events** If the events A and B are mutually exclusive, then the probability that A or B will occur is

  $$P(A \text{ or } B) = P(A) + P(B)$$

- **General addition rule of probability** If the events A and B are from the same sample space, then the probability that A or B will occur is

  $$P(A \text{ or } B) = P(A) + P(B) - P(A \text{ and } B)$$

- **Probability of complementary events** If A is any event and A' is its complement, then $P(A') = 1 - P(A)$.

Chapter 12 review

[12–1] List the first four terms of each sequence.

1. $a_n = 6n - 2$ **2.** $b_n = n^2 - \dfrac{1}{n}$ **3.** $a_n = (n - 1)^2$

Find an expression for the general term of each sequence.

4. 3,7,11,15, . . . **5.** −200,−160,−120, . . .
6. $2, \frac{3}{2}, \frac{4}{3}, \frac{5}{4}, \ldots$

7. A traffic engineer measured the average number of cars passing through a certain intersection every 15 minutes and obtained the values 300, 350, 425, 525, ... What might the engineer expect for the next measurement?

Characterize the sequence as arithmetic, geometric, or neither. State the common difference or common ratio as appropriate.

8. 2, 8, 32, 128, . . . **9.** 2, 8, 16, 26, . . .
10. 2, 8, 14, 20, . . .

11. Find the number of terms in the arithmetic sequence 150, 148, 146, . . . , 118.

12. Suppose a_n and b_n are two arithmetic sequences, and a new sequence c is defined such that $c_n = a_n + 2b_n$. Is the new sequence an arithmetic sequence? Prove or disprove this statement.

Find a_n for the following geometric sequences for the given values of a_1, r, and n.

13. $a_1 = 1{,}024$, $r = -\frac{1}{2}$, $n = 5$ **14.** $a_1 = \frac{1}{40}$, $r = 2$, $n = 7$
15. $a_1 = 1$, $r = 0.1$, $n = 3$

16. Given a geometric sequence in which $a_1 = 81$ and $a_6 = \frac{1}{3}$. Find a_4.

17. Given a geometric sequence in which $a_3 = \frac{5}{4}$ and $a_5 = 5$ find a_6. Assume $r > 0$.

18. Suppose a_n and b_n are two geometric sequences, and a new sequence c is defined such that $c_n = (a_n)(2b_n)$. Is the new sequence a geometric sequence? Prove or disprove this statement.

19. A ball is dropped from a height of 12 feet. If the ball rebounds two-thirds of the height of its previous fall with each bounce, how high does it rebound on the
a. second bounce? **b.** fourth bounce? **c.** nth bounce?

[12–2] Expand the following sigma expressions.

20. $\sum_{j=1}^{4} (2j + 3)$ **21.** $\sum_{j=1}^{5} \dfrac{2j + 1}{j + 1}$

22. $\sum_{j=1}^{4} (-1)^j(j + 1)^2$ **23.** $\sum_{j=1}^{3} \left(\sum_{k=1}^{j} (k - 1)\right)$

Find the sum of the series determined by the given arithmetic sequence.

24. 3, 9, 15, . . . , 99
25. −10, −14, −18, . . . , −66
26. $-8, -6\frac{1}{2}, -5, \ldots, 2\frac{1}{2}$
27. $a_1 = -\frac{3}{4}$, $d = \frac{1}{4}$; find S_{32}

Find the required nth partial sum for each geometric sequence.

28. $a_1 = \frac{2}{3}$, $r = 3$; find S_5 **29.** $\frac{4}{3}, -\frac{8}{9}, \ldots$; find S_5

30. $\sum_{k=1}^{6} \frac{1}{16}(2^k)$ **31.** $\sum_{k=1}^{10} (-3)^k$

Find the sum of the given infinite geometric series. If the sum of the series is not defined state that.

32. $\sum_{i=1}^{\infty} (\frac{1}{3})^i$ **33.** $\sum_{i=1}^{\infty} 4(\frac{1}{2})^i$

34. $3 - 2 + \frac{4}{3} - \cdots$

Find the rational number form of the repeating decimal number.

35. $0.323\overline{232}$ **36.** $0.312312\overline{312}$ **37.** $0.322\overline{2}$

38. A ball is dropped from a height of 18 meters. Each time it strikes the floor the ball rebounds to a height that is 60% of the previous height. Find the total distance that the ball travels before it comes to rest on the floor.

39. A biologist in a laboratory estimates that a culture of bacteria is growing by 15% per hour. How long will it be, to the nearest hour, before the population doubles?

[12–3] Expand and simplify each expression.

40. $\binom{12}{3}$ **41.** $\binom{21}{18}$ **42.** $\binom{n + 2}{n - 1}$

Expand and simplify the following expressions using the binomial expansion formula.

43. $(2x - y)^4$ **44.** $(a^2b - 3)^5$
45. Find the 14th term of $(5a^2 + b^5)^{16}$.

Compute the sum of the following series.

46. $\sum_{i=1}^{25} (i + 3)$ **47.** $\sum_{i=1}^{10} (4i^2 - 1)$ **48.** $\sum_{i=1}^{5} [i^2 - (\frac{1}{3})^i]$

49. Find a general expression for $\sum_{i=1}^{k} (i^2 - i + 1)$.

50. Show that $\binom{n}{0} = 1$.

[12–4] Prove that the following statements are true for all $n \in N$ using finite induction.

51. $4 + 7 + 10 + \cdots + (3n + 1) = \dfrac{n(3n + 5)}{2}$

52. $1^2 + 2^2 + 3^2 + \cdots + n^2 = \dfrac{n(n + 1)(2n + 1)}{6}$

53. Show that $n^3 - n$ is divisible by 3 for any natural number n.

[12–5]

54. There are three prizes in a box, A, B, and C.
  a. Draw a tree diagram of the ways in which the prizes can be drawn from the box.
  b. List all possible ways in which the prizes can be drawn (for example, CBA).

55. An individual has four horses and six saddles. How many different combinations of horses and saddles can this person ride?

56. A certain school offers ten courses. In how many different ways can someone take two of them, one after the other?

57. In how many different ways can a student answer all the questions on a quiz consisting of eight multiple-choice questions, where each question offers three choices?

58. From a standard deck of playing cards, in how many ways can a person select two hearts and a diamond?

59. Compute ${}_{10}P_2$.

60. In an eight-horse race, how many different first-second-third place finishes are possible?

61. A president and vice-president are to be elected from a club with 12 members. In how many different ways can these offices be filled?

62. Twenty people bought chances on a raffle in which there are first, second, third, and fourth prizes. No person can win two prizes. How many ways can these prizes be awarded?

63. How many different words can be formed using all the letters of the word ''madam''?

64. Compute ${}_{18}C_{16}$.

65. Show that ${}_nC_k = {}_nC_{n-k}$.

66. On an examination consisting of 12 essay questions the student may omit any three. In how many different ways can the student select the problems to be answered (the order is not important)?

67. An artist has six pigments that can be added to a basic white paint to form other colors. How many colors can be formed using equal amounts of two pigments?

68. A shopper wants to make a salad using three fruits. If there are seven fruits available, how many salads are possible?

69. From a standard deck of playing cards how many ways can a person select two aces and three kings, without regard to the order of selection?

70. Suppose there are 15 players on a baseball team.
  a. In how many different ways can 12 players be sent to a charity event?
  b. In how many different ways can a team of 9 be chosen if every player can play every position?
  c. In how many different ways can a captain and a co-captain be chosen (assuming the cocaptain and captain are different positions)?
  d. How many different batting orders are possible (considering all possible 9-player teams and all possible batting orders for 9 players)?

71. **a.** How many ways can eight individuals, who happen to be four females and four males, sit in a row?
  b. How many ways can four females and four males sit in a row if females and males must alternate?
  c. How many arrangements of females and males sitting in a row are possible (i.e., permutations in which females are indistinguishable and males are indistinguishable)?

72. In how many different ways can a student answer eight of the questions on a quiz consisting of ten multiple-choice questions, if each question offers three choices?

73. Selecting from the set of digits {1, 2, 3, 4, 5, 6} (repeat selections are allowed) how many of the following are possible?
  a. Four-digit numbers
  b. Four-digit odd numbers
  c. Three-digit numbers where the first digit must be even
  d. Three-digit numbers using only even digits.

74. Answer problem 73 if repetition of a digit is not allowed.

75. Eight teams are in a bowling league. If each team is required to play every other team twice during the season, what is the total number of league games that will be played?

76. A restaurant has four appetizers, six main dishes, and eight desserts. For a fixed price two diners at the same table can choose two from each of these categories. In how many ways can this choice be made?

77. If a group consists of 10 men and 12 women, in how many different ways can a committee of six be selected if:

a. The committee is to have an equal number of men and women?

b. The committee is to be all the same sex?

c. There are no restrictions on membership on the committee?

78. A test contains three groups of questions, A, B, and C, which contain eight, four, and six questions, respectively. If a student must select five questions from group A and three from each of the remaining groups, how many different tests are possible?

[12–6] A coin is tossed four times. Find the probability of

79. one head and three tails.

80. no tails.

A card is drawn from a standard deck of playing cards. What is the probability of

81. a five.

82. a diamond.

83. a card from four through ten, inclusive.

84. a black king.

85. a diamond or a jack.

86. a diamond or a face card.

87. a card that is not a club.

A bowl contains 14 balls. Four are red, 4 are blue, and 6 are white. If one ball is randomly selected, what is the probability that the ball is

88. red?

89. red or blue?

90. not blue?

In a nursery there are 12 rubber plants. Four of the plants are diseased, although this is not detectable. If 6 of the plants are selected for a delivery, what is the probability that

91. two of the plants are diseased?

92. **a.** all of the diseased plants are in the shipment?

b. none of the diseased plants are in the shipment?

Five cards are drawn from a standard deck of playing cards. Find the probability of each event.

93. All five cards are clubs.

94. All five cards are red.

95. None of the cards are red.

96. Two black and three red cards.

97. In a certain state lottery six numbers must be chosen correctly from 49 numbers. What is the probability of making this choice correctly?

98. In a nursery there are 12 rubber plants. Four of the plants are diseased, although this is not detectable. If a customer buys 1 of the 12 plants, what is the probability that the customer gets a diseased plant?

[12–7] Find the first five terms of each recursively defined sequence. Then find an expression for a_n.

99. $a_n = \begin{cases} 2 \text{ if } n = 0 \\ a_{n-1} + 6 \text{ if } n > 0 \end{cases}$

100. $a_n = \begin{cases} 3 \text{ if } n = 0 \\ 2a_{n-1} \text{ if } n > 0 \end{cases}$

101. $a_n = \begin{cases} 2 \text{ if } n = 0, 3 \text{ if } n = 1 \\ 2a_{n-1} + a_{n-2} \text{ if } n > 1 \end{cases}$

102. $a_n = \begin{cases} 2 \text{ if } n = 0, 3 \text{ if } n = 1 \\ 2a_{n-1} - a_{n-2} \text{ if } n > 1 \end{cases}$

103. $a_n = \begin{cases} 1 \text{ if } n = 0, 3 \text{ if } n = 1 \\ 4a_{n-1} - 4a_{n-2} \text{ if } n > 1 \end{cases}$

Chapter 12 test

List the first four terms of each sequence.

1. $a_n = (-1)^{n+1}(-n + 3)$

2. $b_n = n - \dfrac{1}{n}$

Find an expression for the general term of each sequence.

3. 6, 10, 14, 18, . . .

4. 3, 2, $\frac{5}{3}, \frac{3}{2}, \frac{7}{5}$, . . .

5. Find the number of terms in the arithmetic sequence 103, 106, 109, . . . , 184.

6. Find a_5 for the arithmetic sequence in which $a_7 = 12$ and $a_{13} = 28$.

7. Suppose A is an arithmetic sequence, and a new sequence B is defined such that $b_n = 3a_n$. Is sequence B an arithmetic sequence? Prove or disprove this statement.

8. Find a_4 for the geometric sequence in which $a_1 = 243$ and $r = -\frac{2}{3}$.

9. Given a geometric sequence in which $a_3 = 200$ and $a_5 = 25$, find a_7.

10. Suppose A is a geometric sequence, and a new sequence B is defined such that $b_n = a_n + 1$. Is the new sequence a geometric sequence? Prove or disprove this statement.

11. A ball is dropped from a height of 36 feet. If the ball rebounds two-thirds of the height of its previous fall with each bounce, how high does it rebound on the
a. second bounce? **b.** fourth bounce? **c.** nth bounce?

Expand the following sigma expressions.

12. $\sum_{j=1}^{4} \frac{2j}{j^2 + 1}$ **13.** $\sum_{i=1}^{5} (-1)^i(i - 3)^2$

14. $\sum_{j=1}^{3} \left(\sum_{k=1}^{j} k^2 \right)$

Find the sum of the series determined by the given arithmetic sequence.

15. $-20, -18, -16, \ldots, 18, 20, 22$

16. $a_1 = 80$, $d = \frac{1}{4}$; find S_{32}

Find the required nth partial sum for each geometric sequence.

17. $a_1 = \frac{1}{3}$, $r = 6$; find S_5. **18.** $6, -1, \ldots$; find S_5.

19. $\sum_{k=1}^{8} 512(-\frac{1}{2})^k$ **20.** $\sum_{k=1}^{5} (0.3)^k$

Find the sum of the given infinite geometric series. If the sum of the series is not defined state that.

21. $\sum_{i=1}^{\infty} 3^i$ **22.** $\sum_{i=1}^{\infty} 27(\frac{1}{3})^i$

Find the rational number form of the repeating decimal number.

23. $0.2727\overline{27}$ **24.** $0.433\overline{3}$

25. A ball is dropped from a height of 40 meters. Each time it strikes the floor the ball rebounds to a height that is 75% of the previous height. Find the total distance that the ball travels before it comes to rest on the floor.

26. An investment grows at a rate of 10% per year. How long will it be, to the nearest tenth of a year, before the investment doubles?

Expand and simplify each expression.

27. $\binom{8}{4}$ **28.** $\binom{n}{n-1}$

Expand and simplify the following expressions using the binomial expansion formula.

29. $(x^2 - 3y)^4$

30. Find the 11th term of $(3a^2 + b^5)^{14}$.

Compute the sum of the following series.

31. $\sum_{i=1}^{14} (i + 3)^2$ **32.** $\sum_{i=1}^{5} [i - (\frac{1}{4})^i]$

33. Find a general expression for $\sum_{i=1}^{k} (2i + 1)$.

34. If $\binom{n}{2} = 66$, find n.

Prove that the following statements are true for all $n \in N$ using finite induction.

35. $5 + 9 + 13 + \cdots + (4n + 1) = 2n^2 + 3n$

36. $3 + \frac{3}{2} + \frac{3}{4} + \cdots + \frac{3}{2^{n-1}} = \frac{6(2^n - 1)}{2^n}$

37. Show that $n^2 + 7n + 12$ is divisible by 2 for any natural number n.

38. A traveler wishes to visit France (F), England (E), and Germany (G).
a. Draw a tree diagram of the ways in which the traveler can visit these three countries once, one after the other.
b. List all possible ways in which the countries can be visited (for example, FEG).

39. An individual has five pairs of shoes, four pairs of pants, and six shirts. How many different combinations of shoes, pants, and shirts can this person wear?

40. In how many different ways can a student answer all the questions on a quiz consisting of ten multiple-choice questions, where each question offers four choices?

41. From a standard deck of playing cards, in how many ways can a person select a hand consisting of two hearts and two clubs?

42. Compute ${}_{15}P_3$.

43. A background color and a foreground color are to be chosen from the color menu on a computer art program. There are 15 colors, and the background and foreground colors should be different. In how many different ways can these colors be chosen?

44. How many sentences can be formed using all the words in the sentence "to be or not to be, that is not a question," neglecting punctuation and meaning?

45. A computer has been programmed to print out all the different ways the 26 letters of the alphabet A, B, C, . . . , Z can be written down with no repetition, using all 26 letters.
a. How many ways is this?
b. If the computer prints 10 of these ways per second, how long will it take to print out all of them?

46. Compute ${}_{30}C_{26}$.

47. Simplify ${}_{n}C_{n-2}$; the answer should not use factorial notation.

48. On an examination consisting of ten essay questions the student must answer any eight. In how many different ways can the student select the problems to be answered (the order is not important)?

49. A club of 23 students is to choose 2 to represent the club at a student government meeting. How many ways can these 2 students be chosen?

50. From an ordinary deck of playing cards how many ways can a person select two kings, three queens, and a jack, without regard to the order of selection?

Suppose there are 20 players on a baseball team in problems 51 through 54.

51. In how many different ways can a team of 9 be chosen if every player can play every position?

52. In how many different ways can a captain and a cocaptain be chosen (assuming the cocaptain and captain are different positions)?

53. How many different batting orders are possible (considering all possible 9-player teams and all possible batting orders for 9 players)?

54. How many different ways can a team of 9 be chosen if three of the players are pitchers and each team must have exactly one pitcher?

55. In how many different ways can a student answer six of the questions on a quiz consisting of ten multiple-choice questions, if each question offers four choices?

56. Using the ten digits 0, 1, . . . , 9, how many three-digit numbers can be formed if repetition of a digit is not allowed?

57. Twenty-four players enter a tennis tournament. Each player is required to play one match with every other player. What is the total number of matches that will be played?

58. In a modern European literature course students are offered a choice of what they must read from six countries. The number of reading choices are categorized as post-war P or contemporary C as follows:

Country	P	C	Country	P	C
France	3	2	Russia	3	2
Germany	2	2	Spain	3	1
Italy	1	3	United States	2	3

For example, for the United States, the student is offered two post-war writers and three current writers.

a. If a student must choose any two works from each country, how many ways can this selection be made?

b. If a student must choose one post-war and one current work from each country, how many ways can this be done?

59. A coin is tossed four times. Find the probability of two heads and two tails.

A card is drawn from a standard deck of playing cards. What is the probability of

60. a red five? **61.** a red card or a face card?

62. not getting a 5?

63. A bowl contains 22 balls. Twelve are red, 4 are blue, and 6 are white. If 1 ball is randomly selected, what is the probability that the ball is not white?

64. An employee of a state weights and measures department selects a carton that contains 30 bags of potato chips; 4 of the bags are underweight. The employee selects 5 of the bags at random. What is the probability that at least 1 of the 5 bags selected is underweight? (Note that this event is the complement of the event in which none of the bags is underweight.)

65. In a radio contest, a caller will win if the caller selects two numbers between 1 and 20 inclusive correctly. What is the probability of winning?

66. Show that, for $r \geq 1$, $\dfrac{{}_{n}C_{r}}{{}_{n}C_{r-1}} = \dfrac{n-(r-1)}{r}$.

Find the first five terms of each recursively defined sequence. Then find an expression for a_n.

67. $a_n = \begin{cases} 5 \text{ if } n = 0 \\ a_{n-1} + 2 \text{ if } n > 0 \end{cases}$

68. $a_n = \begin{cases} 2 \text{ if } n = 0 \\ 3a_{n-1} \text{ if } n > 0 \end{cases}$

69. $a_n = \begin{cases} 2 \text{ if } n = 0,\ 4 \text{ if } n = 1 \\ 2a_{n-1} - a_{n-2} \text{ if } n > 1 \end{cases}$

70. $a_n = \begin{cases} 2 \text{ if } n = 0,\ 3 \text{ if } n = 1 \\ 3a_{n-1} - a_{n-2} \text{ if } n > 1 \end{cases}$

Appendix A

Development of Several Formulas

$\cos(\alpha + \beta) = \cos \alpha \cos \beta - \sin \alpha \sin \beta$

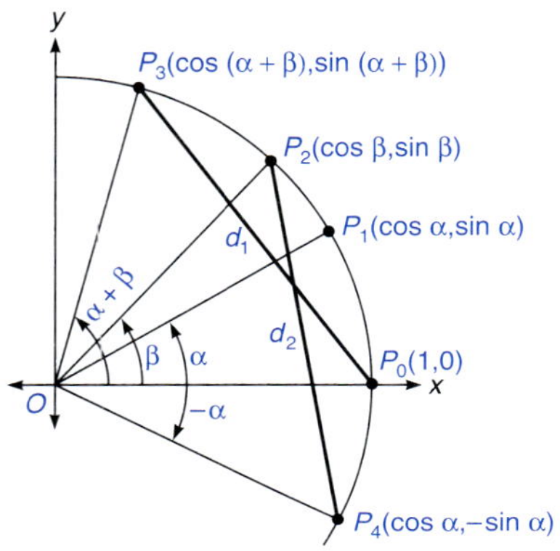

Figure A–1

A proof that this identity, which is discussed in section 7–2, is true is beyond the scope of this text, but an argument for its correctness can be obtained in the following way.

Let α and β be two angles in standard position (see figure A–1). Let P_1 be the point where the terminal side of α intersects the unit circle, and let P_2 be the point where angle β intersects the unit circle. Let P_3 be the point where the angle $\alpha + \beta$ (the sum of the angles α and β) intersects the circle. Let P_0 be the point (1,0). Finally, let P_4 be the point where the terminal side of angle $-\alpha$ intersects the unit circle.

On the unit circle the x- and y-coordinates of a point are the cosine and sine values for the appropriate angle. Thus P_1 has coordinates $(\cos \alpha, \sin \alpha)$. The coordinates for the other points are shown in the figure.

Angle $\alpha + \beta$, or angle P_0OP_3 in standard position, has the same measure as angle P_4OP_2. It is a geometric property that central angles of a circle having equal measure have chords of equal length. Thus, the chords P_3P_0 and P_2P_4 have the same length. The length of a line segment with end points (x_1,y_1) and (x_2,y_2) is given by the distance formula

$$d = \sqrt{(x_2 - x_1)^2 + (y_2 - y_1)^2}$$

We apply this to the chords P_3P_0 and P_2P_4.

Let d_1 = length of P_3P_0, and let d_2 = length of P_2P_4.

$$d_1 = d_2$$

$$\sqrt{(\cos(\alpha + \beta) - 1)^2 + (\sin(\alpha + \beta) - 0)^2}$$
$$= \sqrt{(\cos \beta - \cos \alpha)^2 + (\sin \beta - (-\sin \alpha))^2}$$

Square both sides.

$$(\cos(\alpha + \beta) - 1)^2 + (\sin(\alpha + \beta))^2 = (\cos \beta - \cos \alpha)^2 + (\sin \beta + \sin \alpha)^2$$

Performing the indicated operations we obtain

$$\cos^2(\alpha + \beta) - 2\cos(\alpha + \beta) + 1 + \sin^2(\alpha + \beta) = \cos^2\beta - 2\cos \alpha \cos \beta$$
$$+ \cos^2\alpha + \sin^2\beta + 2\sin \alpha \sin \beta + \sin^2\alpha$$

$$[\cos^2(\alpha + \beta) + \sin^2(\alpha + \beta)] - 2\cos(\alpha + \beta) + 1 = (\cos^2\beta + \sin^2\beta)$$
$$+ (\cos^2\alpha + \sin^2\alpha) + 2\sin \alpha \sin \beta - 2\cos \alpha \cos \beta$$

Using the fundamental identity $\sin^2\theta + \cos^2\theta = 1$, we obtain

$$1 - 2\cos(\alpha + \beta) + 1 = 1 + 1 + 2\sin \alpha \sin \beta - 2\cos \alpha \cos \beta$$
$$-2\cos(\alpha + \beta) = 2\sin \alpha \sin \beta - 2\cos \alpha \cos \beta$$
$$\cos(\alpha + \beta) = \cos \alpha \cos \beta - \sin \alpha \sin \beta$$

Equation of the ellipse

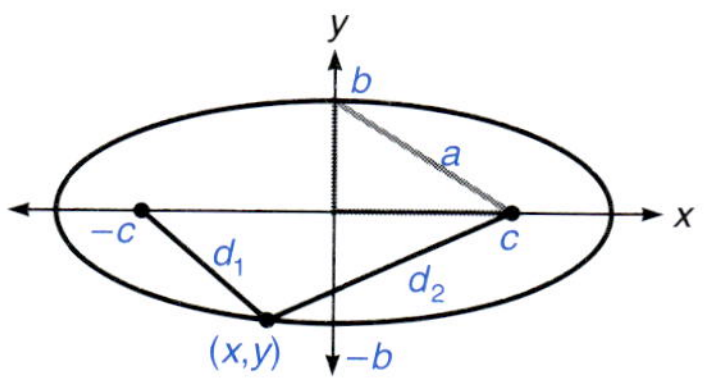

The figure shows an ellipse (section 11–2) placed so its center is at the origin. The foci are placed on the x-axis equidistant for the origin; they are at $(-c,0)$ and $(c,0)$. The point (x,y) represents any point on the ellipse. The y-intercept is labeled b. We call the distance from the y-intercept to the focus a. The right triangle shown illustrates that $a^2 = b^2 + c^2$. We can develop an analytic description of this ellipse as follows.

The sum of d_1 and d_2 is a constant. If we consider (x,y) to be at $(0,b)$ (one of the y-intercepts) we can see that this constant is $2a$. We thus proceed algebraically from the statement $d_1 + d_2 = 2a$.

$$d_1 = \sqrt{(x-(-c))^2 + (y-0)^2}$$
Distance formula with $(-c,0)$ and (x,y)

$$= \sqrt{(x+c)^2 + y^2}$$

$$d_2 = \sqrt{(x-c)^2 + (y-0)^2}$$
Distance formula with $(c,0)$ and (x,y)

$$= \sqrt{(x-c)^2 + y^2}$$

$$d_1 + d_2 = 2a$$
Definition of ellipse

$$\sqrt{(x+c)^2 + y^2} + \sqrt{(x-c)^2 + y^2} = 2a$$
Replace d_1 and d_2 in $d_1 + d_2 = 2a$ by the values above

$$\sqrt{(x+c)^2 + y^2} = 2a - \sqrt{(x-c)^2 + y^2}$$

$$\left[\sqrt{(x+c)^2 + y^2}\right]^2 = \left[2a - \sqrt{(x-c)^2 + y^2}\right]^2$$
Square both members

$$(x+c)^2 + y^2 = 4a^2 - 4a\sqrt{(x-c)^2 + y^2} + (x-c)^2 + y^2$$

$$x^2 + 2cx + c^2 + y^2 = 4a^2 - 4a\sqrt{(x-c)^2 + y^2} + x^2 - 2cx + c^2 + y^2$$

$$4cx = 4a^2 - 4a\sqrt{(x-c)^2 + y^2}$$
Simplify terms

$$cx = a^2 - a\sqrt{(x-c)^2 + y^2}$$
Divide each term by 4

$$a\sqrt{(x-c)^2 + y^2} = a^2 - cx$$
Rearrange terms

$$\left[a\sqrt{(x-c)^2 + y^2}\right]^2 = (a^2 - cx)^2$$
Square both members

$$a^2[(x-c)^2 + y^2] = a^4 - 2a^2cx + c^2x^2$$

$$a^2x^2 - 2a^2cx + a^2c^2 + a^2y^2 = a^4 - 2a^2cx + c^2x^2$$

$$a^2x^2 + a^2c^2 + a^2y^2 = a^4 + c^2x^2$$

$$a^2x^2 - c^2x^2 + a^2y^2 = a^4 - a^2c^2$$

$$(a^2 - c^2)x^2 + a^2y^2 = a^2(a^2 - c^2)$$

$$b^2x^2 + a^2y^2 = a^2b^2$$
$a^2 = b^2 + c^2$, so $a^2 - c^2 = b^2$

$$\frac{b^2x^2}{a^2b^2} + \frac{a^2y^2}{a^2b^2} = \frac{a^2b^2}{a^2b^2}$$
Divide each term by a^2b^2

$$\frac{x^2}{a^2} + \frac{y^2}{b^2} = 1$$

Also, solving for c in $a^2 + b^2 = c^2$ we find that $c = \sqrt{a^2 - b^2}$.

Thus, an analytic description of the ellipse is $\dfrac{x^2}{a^2} + \dfrac{y^2}{b^2} = 1$, where $c = \sqrt{a^2 - b^2}$.

Appendix B

Answers and Solutions

Chapter 1

Exercise 1–1

Answers to odd-numbered problems

1. $\{4, 5, 6, 7, 8, 9, 10, 11\}$
3. $\{1, 3, 5, 7, 9, 11, 13, 15, 17, 19\}$
5. $\{-6, -3, 0, 3, 6, 9, 12\}$
7. 0.4, terminating
9. $0.230769\overline{230769}$, repeating
11. -276 **13.** $\frac{31}{40}$ **15.** $\frac{48}{23}$ **17.** $-\frac{1}{16}$

19. $\dfrac{bx - ay}{ab}$ **21.** $\dfrac{-xy - 3y^2 - 8x^2}{12xy}$

23. $-14x^7$ **25.** $\dfrac{3x^3}{25y^3}$ **27.** $\dfrac{3a^2 + 6b^2}{10a^2}$

29. $(-2, 8)$

31. $[-8,0)$

33. $(-\sqrt{2},\pi]$

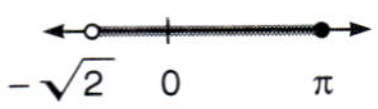

35. $(\infty,4)$

37. $[-2,\infty)$

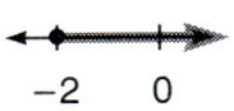

39. $\{x \mid -5 \le x \le -1\}$

41. $\{x \mid x < 1\}$

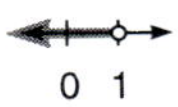

43. $\left\{x \,\middle|\, -\dfrac{\pi}{2} \le x < \dfrac{3\pi}{2}\right\}$

45. $\{x \mid -\frac{1}{2} < x \le 1\frac{1}{2}\}, (-\frac{1}{2},1\frac{1}{2}]$
47. $\{x \mid 5\frac{1}{2} \le x < 7\}, [5\frac{1}{2},7)$
49. $\{x \mid -2 < x < \frac{1}{2}\}, (-2,\frac{1}{2})$
51. $\{x \mid x > 5\}, (5,\infty)$ **53.** 4 **55.** -2
57. $\sqrt{10} + 3$ **59.** $\frac{7}{4}$ **61.** 25
63. $\sqrt{2} - 3$ **65.** $2x^4$

67. $\dfrac{x^2y^6}{z^8}$ **69.** $-5x^2$

71. $\dfrac{5|x|}{2y^2}$ **73.** $(x - 2)^2|x + 1|$

75. if $x > 0$, $\dfrac{x^2}{|x|} = \dfrac{x^2}{x} = x$;

if $x < 0$, $\dfrac{x^2}{|x|} = \dfrac{x^2}{-x} = -x$

77. $-\frac{3}{5}$ or -0.6

Solutions to skill and review problems

1. $2 \cdot 3(x^2 \cdot x^3) = 6x^{2+3} = 6x^5$

2. $2n + (-2n)$
$0n$
0

3. $8 - (-3)$
$8 + 3 = 11$

4. $2 \cdot 1{,}000{,}000{,}000 = 2 \cdot 10^9$, b

5. $0.3 = \dfrac{3}{10}$, $0.03 = \dfrac{3}{100}$,

$0.003 = \dfrac{3}{1000}$, $0.0003 = \dfrac{3}{10{,}000}$,

$0.00003 = \dfrac{3}{100{,}000}$; c

6. $-3[2(4[\frac{1}{2}(2 - 3) + 2] - 1) + 7] + 4$
$-3[2(4[\frac{1}{2}(-1) + 2] - 1) + 7] + 4$
$-3[2(4[-\frac{1}{2} + \frac{4}{2}] - 1) + 7] + 4$
$-3[2(4[\frac{3}{2}] - 1) + 7] + 4$
$-3[2(6 - 1) + 7] + 4$
$-3[10 + 7] + 4$
$-51 + 4$
-47

7. $2a(2a - 2b - ac)$
$2a(2a) - 2a(2b) - 2a(ac)$
$4a^2 - 4ab - 2a^2c$

8. $(2a - c)(3a + 2c)$
$2a(3a + 2c) - c(3a + 2c)$
$6a^2 + 4ac - 3ac - 2c^2$
$6a^2 + ac - 2c^2$

Solutions to trial exercise problems

3. $\{1,2,3,4,5,6,7,8,9,10,11,12,13,14,15,16,17,18,19,20\}$, $x \in N$ and $x < 21$
$\{1,\not{2},3,\not{4},5,\not{6},7,\not{8},9,\not{10},11,\not{12},13,\not{14},15,\not{16},17,\not{18},19,\not{20}\}$, x is odd
$\{1,3,5,7,9,11,13,15,17,19\}$

4. $\dfrac{178}{185} = 0.9\ 621\ 621\ 621\ \ldots$;
$\{1, 2, 6, 9\}$

16. $\left(\dfrac{3}{7} - \dfrac{7}{12}\right) \div \left(\dfrac{3}{7} + \dfrac{7}{12}\right)$

$\left[\dfrac{3(12) - 7(7)}{7(12)}\right] \div \left[\dfrac{3(12) + 7(7)}{7(12)}\right]$

$\left(-\dfrac{13}{84}\right) \div \left(\dfrac{85}{84}\right)$

$\left(-\dfrac{13}{84}\right) \cdot \dfrac{84}{85}$

$-\dfrac{13}{85}$

21. $\dfrac{x - y}{4x} - \dfrac{2x + y}{3y}$

$\dfrac{3y(x - y) - 4x(2x + y)}{4x(3y)}$

$\dfrac{3xy - 3y^2 - 8x^2 - 4xy}{12xy}$

$\dfrac{-xy - 3y^2 - 8x^2}{12xy}$

33. $(-\sqrt{2},\pi]$

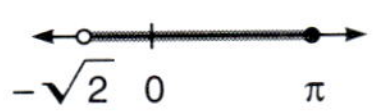

58. $-|\sqrt{10} - 6|$
$-(6 - \sqrt{10})$
$\sqrt{10} - 6$

62. $-|-\sqrt{2}|$
$-(\sqrt{2})$
$-\sqrt{2}$

70. $\left|\dfrac{3x^2}{2y}\right|$

$\dfrac{3|x^2|}{2|y|}$

$\dfrac{3x^2}{2|y|}$

78. a.

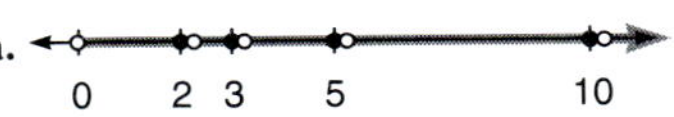

b. $(0, 2], (2, 3], (3, 5], (5, 10], (10, \infty)$

c. $\dfrac{15}{2} = 7.5$¢/oz, $\dfrac{20}{3} = 6.7$¢/oz,

$\dfrac{30}{5} = 6$¢/oz, $\dfrac{40}{10} = 4$¢/oz, 3.5¢/oz

Exercise 1–2

Answers to odd-numbered problems

1. $2x^{11}$ **3.** -32 **5.** $6a^7b^3$ **7.** $128x$

9. $\frac{3x^4}{y^3}$ **11.** $8x^9y^{15}$ **13.** $\frac{81a^4}{b^6}$ **15.** $\frac{y^3}{x^2}$

17. $-\frac{1}{27}$ **19.** $\frac{4}{x^4}$ **21.** $-\frac{6x^7}{y}$ **23.** 1

25. $\frac{3y^5}{x^5}$ **27.** $\frac{8b^{21}}{a^{12}}$ **29.** $\frac{-8x^{15}}{125y^6}$

31. $\frac{b^6c^{16}}{9a^{10}}$ **33.** $\frac{2}{x^4}$ **35.** $\frac{9}{16x^2y^8}$ **37.** x^{4n}

39. x **41.** $\frac{x^{8n}}{y^{8n-16}}$ **43.** 3.65×10^{15}

45. -1.9002×10^{13} **47.** -2.92×10^{-14}

49. 3.502×10^{-12}

51. 25,020,000,000,000

53. $-0.000\ 000\ 000\ 138\ 4$

55. 9,230,000 **57.** 9.1×10^{-28} grams

59. trinomial, degree is 2

61. polynomial, degree is 3

63. trinomial, degree is 6

65. not a polynomial because of the $\sqrt{x}$

67. -557 **69.** $36\frac{1}{3}$ **71.** $\frac{1}{4}$

73. $2x^2 - 4x + 6$ **75.** $6a - 7b + c$

77. $-2x^2y + 2xy$ **79.** $9x - 2y$

81. $-16a + 7b$ **83.** $10x^5 - 4x^4 + 14x^2$

85. $-10a^4b^2 - 6a^4b^3 + 4a^3b^4$

87. $25a^2 - 9$ **89.** $15x^2 + 2xy - y^2$

91. $2a^2 + 2b^2 - 5ab + 2ac - bc$

93. $10x^4 - 19x^3 + 25x^2 - 23x + 7$

95. $5b^4 + 3b^3 - 14b^2 + 9b - 9$

97. $x^3 - 7xy^2 - 6y^3$

99. $9a^3 + 21a^2b + 4ab^2 - 4b^3$

101. $6a^2 - 4ab + 2ac - 9a + 6b - 3c$

103. $x^3 + 2x^2y - 4xy^2 - 8y^3$

105. $8x^3 + 60x^2 + 150x + 125$

107. $\frac{2x^3y}{3}$ **109.** $3a^2 - 4b^2 + 6b^4$

111. $\frac{2}{3}x^2z^2 + xz - \frac{4}{3}y^2$ **113.** $x - 2$

115. $x^3 - 5x^2 + 10x - 20 + \frac{48}{x + 2}$

117. $3x^2 - 4x + 1 + \frac{2}{2x + 3}$

119. $4x^2 + 7x + 17 + \frac{38}{x - 2}$

121. $4x + 3 + \frac{-x + 2}{x^2 - x + 1}$

123. $3x^2 + 7 + \frac{-x + 22}{x^2 - 3}$

125. **a.** $2t_1 - 2t_2 + 3t_3$
b. $3t_1^2 + 7t_1t_2 - 9t_1t_3 + 4t_2^2 - 12t_2t_3$
c. $-24x_1^5x_2^7$

127. First show that $(a^2 + b^2)(c^2 + d^2) = (ac + bd)^2 + (ad - bc)^2$:
$(a^2 + b^2)(c^2 + d^2)$
$= a^2c^2 + a^2d^2 + b^2c^2 + b^2d^2$
and $(ac + bd)^2 + (ad - bc)^2$
$= (a^2c^2 + 2abcd + b^2d^2) + (a^2d^2 - 2abcd + b^2c^2)$
$= a^2c^2 + a^2d^2 + b^2c^2 + b^2d^2$
Now show that
$(a^2 + b^2)(c^2 + d^2) = (ac - bd)^2 + (ad + bc)^2$:
$(a^2 + b^2)(c^2 + d^2)$
$= a^2c^2 + a^2d^2 + b^2c^2 + b^2d^2$
and $(ac - bd)^2 + (ad + bc)^2$
$= (a^2c^2 - 2abcd + b^2d^2) + (a^2d^2 + 2abcd + b^2c^2)$
$= a^2c^2 + a^2d^2 + b^2c^2 + b^2d^2$

129. 77.7

131. Square: $(x - 2y)(x - 2y)$
$= x^2 - 4xy + 4y^2$
Rectangle: $(a + b)(a + 2b)$
$= a^2 + 3ab + 2b^2$

133. Area = Triangle + Rectangle
$= \frac{1}{2}(3x - y)(y) + (3x - y)(x)$
$= \frac{1}{2}(3xy - y^2) + (3x^2 - xy)$
$= \frac{3}{2}xy - \frac{1}{2}y^2 + 3x^2 - xy$
Area $= 3x^2 + \frac{1}{2}xy - \frac{1}{2}y^2$

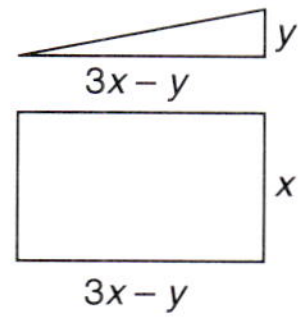

135. $\frac{1}{2}(a + c) \cdot \frac{1}{2}(b + d)$
$= \frac{1}{2} \cdot \frac{1}{2}(a + c)(b + d)$
$= (\frac{1}{2} \cdot \frac{1}{2})[(a + c)(b + d)]$
$= \frac{1}{4}[ab + ad + bc + bd]$

137. **a.** 7, 23 **b.** 22, 27 **c.** 7, 17
d. 8, 35 **e.** 13, 87

139. $(5 \times 10^{12}) \div (2.5 \times 10^{-10})$
2×10^{22}
5 [EXP] 12 [÷] 2.5 [EXP] 10 [+/−] [=]
TI-81: 5 [EE] 12 [÷] 2.5 [EE] [(−)] 10 [ENTER]

141. $\sqrt{4 \times 10^{18}}$
2×10^9
4 [EXP] 18 [$\sqrt{x}$]
TI-81: [2nd] [x^2] 4 [EE] 18 [ENTER]

Solutions to skill and review problems

1. $360 = 10 \cdot 36$
$= 2 \cdot 5 \cdot 6 \cdot 6$
$= 2 \cdot 5 \cdot 2 \cdot 3 \cdot 2 \cdot 3$
$= 2^3 \cdot 3^2 \cdot 5$

2. $3x^2y^3(x^3y - 4x + 2)$

3. $(x - 4)(x + 4)$ **4.** $(x + 2)(x + 4)$

5. $(x - 2)(x + 8)$ **6.** $(2x - 3)(3x + 1)$

7. $x^3 - 1$

Solutions to trial exercise problems

8. $3a^{-1}b^4(3^{-1}a^2b)$
$3^{1-1}a^{-1+2}b^{4+1}$
$3^0a^1b^5$
ab^5

15. $\frac{1}{x^2y^{-3}}$
$\frac{y^3}{x^2}$

21. $(2x^4y)(-3x^3y^{-2})$
$-6x^7y^{-1}$
$-\frac{6x^7}{y}$

26. $\frac{2x^{-2}y^0z^3}{-2^{-3}x^2y^2z^{-1}}$
$-\frac{2(2^3)z^3z^1}{x^2x^2y^2}$
$-\frac{16z^4}{x^4y^2}$

27. $\left(\frac{9a^5b^4}{18ab^{11}}\right)^{-3}$
$\left(\frac{a^4}{2b^7}\right)^{-3}$
$\frac{a^{-12}}{2^{-3}b^{-21}}$
$\frac{2^3b^{21}}{a^{12}}$
$\frac{8b^{21}}{a^{12}}$

35. $\left(\frac{x^2y}{x^2y^2}\right)^2\left(\frac{3x^0y^{-2}}{4xy}\right)^2$
$\left(\frac{1}{y}\right)^2\left(\frac{3}{4xy \cdot y^2}\right)$
$\frac{1}{y^2} \cdot \frac{9}{16x^2y^6}$
$\frac{9}{16x^2y^8}$

36. $\left(\frac{a^5bc^2}{2abc}\right)^3\left(\frac{2^2a^3b^2c}{ab^{-2}c^3}\right)^{-1}$
$\left(\frac{a^4c}{2}\right)^3\left(\frac{4a^2b^4}{c^2}\right)^{-1}$
$\frac{a^{12}c^3}{8} \cdot \frac{4^{-1}a^{-2}b^{-4}}{c^{-2}}$
$\frac{a^{12}c^3}{8} \cdot \frac{c^2}{4a^2b^4}$
$\frac{a^{10}c^5}{32b^4}$

41. $\left(\frac{x^ny^{2-n}}{x^{-n}y^{n-2}}\right)^4$
$\left(\frac{x^nx^n}{y^{n-2}y^{-(2-n)}}\right)^4$
$\left(\frac{x^{2n}}{y^{2n-4}}\right)^4$
$\frac{x^{8n}}{y^{8n-16}}$

47. $-0.000\ 000\ 000\ 000\ 029\ 2$
$\quad 3 \quad 3 \quad 3 \quad 3 \quad 2$
-2.92×10^{-14}

64. $\underline{3(x+1)^1} + \underline{(3x-2)^1} + 9$; trinomial, degree is 1

81. $[-(3a - b) - (2a + 3b)]$
$- [(a - 6b) - (3b - 10a)]$
$[-3a + b - 2a - 3b]$
$- [a - 6b - 3b + 10a]$
$[-5a - 2b] - [11a - 9b]$
$-5a - 2b - 11a + 9b$
$-16a + 7b$

97. $(x + 2y)(x - 3y)(x + y)$
$(x + 2y)(x^2 - 2xy - 3y^2)$
$x^3 - 2x^2y - 3xy^2 + 2x^2y - 4xy^2 - 6y^3$
$x^3 - 7xy^2 - 6y^3$

105. $(2x + 5)^3$
$(2x + 5)(2x + 5)(2x + 5)$
$(2x + 5)(4x^2 + 20x + 25)$
$8x^3 + 40x^2 + 50x + 20x^2 + 100x + 125$
$8x^3 + 60x^2 + 150x + 125$

117. $\dfrac{6x^3 + x^2 - 10x + 5}{2x + 3}$
$= 3x^2 - 4x + 1 + \dfrac{2}{2x + 3}$

$$\begin{array}{r}
3x^2 - 4x + 1 \\
2x + 3 \overline{)\,6x^3 + x^2 - 10x + 5} \\
\underline{6x^3 + 9x^2 \qquad\qquad} \\
-8x^2 - 10x + 5 \\
\underline{-8x^2 - 12x \quad} \\
2x + 5 \\
\underline{2x + 3} \\
2
\end{array}$$

123. $\dfrac{3x^4 - 2x^2 - x + 1}{x^2 - 3}$
$= 3x^2 + 7 + \dfrac{-x + 22}{x^2 - 3}$

$$\begin{array}{r}
3x^2 \qquad + 7 \\
x^2 - 3 \overline{)\,3x^4 + 0x^3 - 2x^2 - x + 1} \\
\underline{3x^4 \qquad - 9x^2 \qquad\qquad} \\
7x^2 - x + 1 \\
\underline{7x^2 \qquad - 21} \\
-x + 22
\end{array}$$

134. a. Two sequences of attacks that total 16 beats.
$(A + B)^2 = A^2 + AB + BA + B^2$
$A = 3, B = 1$: $16 = (3 + 1)^2$
$= 3^2 + 3 \cdot 1 + 1 \cdot 3 + 1^2$
$= 9 + 3 + 3 + 1$
$A = 1, B = 3$: $16 = (1 + 3)^2$
$= 1^2 + 1 \cdot 3 + 3 \cdot 1 + 3^2$
$= 1 + 3 + 3 + 9$

b. Use the expansion of $(A + B + C)^2$ with values for A, B, and C to generate two sequences that total 36 beats.
$(A + B + C)^2 = A^2 + AB + AC + BA + B^2 + BC + CA + CB + C^2$
$A = 1, B = 2, C = 3$:
$36 = (1 + 2 + 3)^2$
$= 1^2 + 1 \cdot 2 + 1 \cdot 3 + 2 \cdot 1 + 2^2 + 2 \cdot 3 + 3 \cdot 1 + 3 \cdot 2 + 3^2$
$= 1 + 2 + 3 + 2 + 4 + 6 + 3 + 6 + 9$
$A = 1, B = 4, C = 1$:
$36 = (1 + 4 + 1)^2$
$= 1^2 + 1 \cdot 4 + 1 \cdot 1 + 4 \cdot 1 + 4^2 + 4 \cdot 1 + 1 \cdot 1 + 1 \cdot 4 + 1^2$
$= 1 + 4 + 1 + 4 + 16 + 4 + 1 + 4 + 1$

142. a. There are 365 days × 24 hours/day = 8,760 hours per year. The amount of energy in joules reaching the surface of the earth per hour is therefore

$$\frac{\text{total energy in joules, per year}}{\text{number of hours in year}} = \frac{3.9 \times 10^6 \times 10^9}{8760} \approx 4.452 \times 10^{11} \text{ joules per hour.}$$

$$\frac{\text{total energy}}{\text{energy per ton}} = \frac{4.452 \times 10^{11} \text{ joules}}{45.5 \text{ joules per ton}} \approx 9.78 \times 10^9 \text{ tons (about 10 billion tons).}$$

b. $\dfrac{350 \times 10^9 \text{ joules}}{45.5 \text{ joules per ton}} \approx 7.7 \times 10^9$ or 7.7 billion tons.

Exercise 1–3

Answers to odd-numbered problems

1. $3(4x^2 - 3xy - 6)$
3. $-4a^2b(5a^2b - 15a + 6b)$
5. $(a - b)(6x + 5y)$
7. $(2x - y)(5a - 1)$
9. $(n + 5)(2m - 1 - p)$
11. $(c + d)(a - 2b)$
13. $(a + 3b)(5x - y)$
15. $(3x + 2)(2x + 3)$
17. $(x + 4y)(x + 3y)$
19. $(2a + 5b)(3a - b)$
21. $(x - 2)(x - 16)$
23. $(3x - 5)(3x + 5)$
25. $(x - 2y)(x + 2y)(x^2 + 4y^2)$
27. $(3x - 1)(9x^2 + 3x + 1)$
29. $(2a + 5)(4a^2 - 10a + 25)$
31. $(a - b)(a^2 + ab + b^2)(a^6 + a^3b^3 + b^6)$
33. $(y + 7)(y - 6)$
35. $4(m - n - 8)(m - n + 1)$
37. $(x - 2y)(3x)$ **39.** $20x(4x + 3)$
41. $2x^3(2x^3 + 1)(4x^6 - 2x^3 + 1)$
43. $(a - b)(a + b)(2x + 3)(x - 1)$
45. $(m - 7)(m + 7)$ **47.** $(x + 5)(x + 1)$
49. $(7a + 1)(a + 5)$
51. $(2a + 3)(a + 6)$
53. $(ab + 4)(ab - 2)$
55. $(3a + b)(9a^2 - 3ab + b^2)$
57. $5x(3x + y)(5x + 1)$
59. $10(x - y)^2$ **61.** $4(m - 2n)(m + 2n)$
63. $(a - b - 2x - y)(a - b + 2x + y)$
65. $3(x^2 - 3y)(x^4 + 3x^2y + 9y^2)$
67. $6xy^2(2x - 3y)(x - y)$
69. $4(x - 3y)(x + 3y)$
71. $(x + 2y)(3a - b)$
73. $(3a^3 - bc)(9a^6 + 3a^3bc + b^2c^2)$
75. $(5a + 3)(a - 7)$
77. $(a - 2)(a + 2)(a - 1)(a + 1)$
79. $(2a - 5b)(2a + 3b)$
81. $(y - 2)(y + 2)(y^2 + 4)$
83. $2(2a + 1)(a + 2)$
85. $(x + y + 1)(x + y - 9)$
87. $(3a + 5)(2a - 1)$
89. $4ab(x + 3y)(1 - 2ab)$ **91.** $(2a - 5b)^2$
93. $5x(2x - 1)(2x + 1)(4x^2 + 1)$
95. $3ab(a^2 - 3b^2)^2$
97. $(3a - x - 5y)(3a + x + 5y)$
99. $(3x - 13)(x + 7)$
101. $3x^2(2xy^3 + 3z^2)(4x^2y^6 - 6xy^3z^2 + 9z^4)$
103. $(3 - x)(3 + x)(a - 3)^2$
105. $\pi(r_2 - r_1)(r_2 + r_1)$
107. $(2x + y)(x + 3y)$
$= 2x^2 + 6xy + xy + 3y^2$
$= 2x^2 + 7xy + 3y^2$

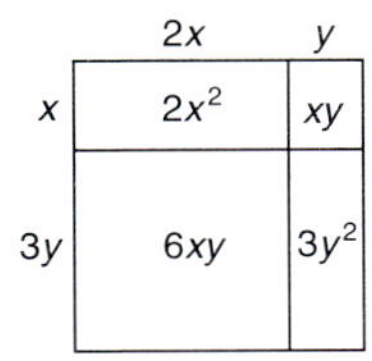

109. As a difference of two squares:
$x^6 - 1 = (x^3 - 1)(x^3 + 1)$
$= (x - 1)(x^2 + x + 1)(x + 1)(x^2 - x + 1)$
$= (x - 1)(x + 1)(x^2 + x + 1)(x^2 - x + 1)$
As a difference of two cubes:
$x^6 - 1 = (x^2 - 1)(x^4 + x^2 + 1)$
$= (x - 1)(x + 1)(x^4 + x^2 + 1)$
Thus, $(x - 1)(x + 1)(x^2 + x + 1)(x^2 - x + 1) = (x - 1)(x + 1)(x^4 + x^2 + 1)$, so $x^4 + x^2 + 1$ must be the same as $(x^2 + x + 1)(x^2 - x + 1)$.

111. The following is a program for a TI-81 programmable calculator which will compute the greatest common factor of two integers. The integers must be in the variables F and S. *Note:* All program lines are terminated with the [ENTER] key, which is not shown.

Display	Keystrokes—use [ENTER] at the end of each line.
Prgm3:GCF	[PRGM] ▶ 3 [TAN] [PRGM] [COS]
:If F>S	[PRGM] 3 [ALPHA] [COS] [2nd]
	[MATH] 3 [ALPHA] [LN]
:Goto 1	[PRGM] 2 1
:F→T	[ALPHA] [COS] [STO▸] 4
:S→F	[ALPHA] [LN] [STO▸] [COS]
:T→S	[ALPHA] 4 [STO▸] [LN]
:Lbl 1	[PRGM] 1 1
:IPart(F/S)→T	[MATH] ▶ 2 [(] [ALPHA] [COS] [÷]
	[ALPHA] [LN] [)] [STO▸] 4
:F − TS→R	[ALPHA] [COS] [−] [ALPHA] 4
	[ALPHA] [LN] [STO▸] [×]
:If R=0	[PRGM] 3 [ALPHA] [×] [2nd]
	[MATH] 1 0
:Goto 2	[PRGM] 2 2
:S→F	[ALPHA] [LN] [STO▸] [COS]
:R→S	[ALPHA] [×] [STO▸] [LN]
:Goto 1	[PRGM] 2 1
:Lbl 2	[PRGM] 1 2
:abs S→S	[2nd] [x^{-1}] [ALPHA] [LN] [STO▸]
	[LN] [2nd] [CLEAR]

To use this program to find the GCF of 140 and 196, do the following:

140 [STO▸] [COS] [ENTER] 196 [STO▸]
[LN] [ENTER] [PRGM] 3 [ENTER] [ALPHA]
[LN] [ENTER]

and the result, 28, appears. This program could be made more user friendly, but it is used as is in problem 112.

Solutions to skill and review problems

1. $2x$ must be 5, so x must be $2\frac{1}{2}$.

2. Never, since 4 is positive and x^2 is positive or zero. Adding positive values gives positive results.

3. $-\dfrac{-1}{2}$ is the same as $\dfrac{1}{2}$. Think of $-(-0.5)$ to help see this.

4. $\dfrac{3x^3}{6x^6} = \dfrac{3}{6} \cdot \dfrac{xxx}{xxxxxx} = \dfrac{1}{2} \cdot \dfrac{\cancel{xxx}}{\cancel{xxx}xxx} = \dfrac{1}{2x^3}$

5. $\dfrac{3}{4} \cdot \dfrac{12}{5} = \dfrac{3}{1} \cdot \dfrac{3}{5} = \dfrac{9}{5}$

6. $\dfrac{2}{3} + \dfrac{1}{4} = \dfrac{2 \cdot 4 + 3 \cdot 1}{3 \cdot 4} = \dfrac{11}{12}$

7. $(2x - 3)(x + 1) - (x - 2)(x - 1)$
$= (2x^2 - x - 3) - (x^2 - 3x + 2)$
$= 2x^2 - x - 3 - x^2 + 3x - 2$
$= x^2 + 2x - 5$

8. $\frac{5}{8} \div 2 = \frac{5}{8} \div \frac{2}{1} = \frac{5}{8} \cdot \frac{1}{2} = \frac{5}{16}$

Solutions to trial exercise problems

9. $2m(n + 5) - 1(n + 5) - p(n + 5)$
$(n + 5)(2m - 1 - p)$

30. $(xy - 3z)[(xy)^2 + 3z(xy) + (3z)^2]$
$(xy - 3z)(x^2y^2 + 3xyz + 9z^2)$

39. $9(3x + 1)^2 - (x - 3)^2$
$9a^2 - b^2$
Replace $3x + 1$ by a, $x - 3$ by b
$(3a - b)(3a + b)$
$[3(3x + 1) - (x - 3)][3(3x + 1) + (x - 3)]$
Replace a by $3x + 1$, b by $x - 3$
$(8x + 6)(10x)$
$2(4x + 3)(10x)$
$20x(4x + 3)$

42. $x^2(x^2 - 9) + 2x(x^2 - 9) - 15(x^2 - 9)$
$(x^2 - 9)(x^2 + 2x - 15)$
$(x - 3)(x + 3)(x + 5)(x - 3)$

43. $2x^2(a^2 - b^2) + x(a^2 - b^2) - 3a^2 + 3b^2$
$2x^2(a^2 - b^2) + x(a^2 - b^2) - 3(a^2 - b^2)$
$(a^2 - b^2)(2x^2 + x - 3)$
$(a - b)(a + b)(2x + 3)(x - 1)$

85. $(x + y)^2 - 8(x + y) - 9$
$z^2 - 8z - 9 \qquad z = x + y$
$(z + 1)(z - 9)$
$(x + y + 1)(x + y - 9)$

103. $a^2(9 - x^2) - 6a(9 - x^2) - 9(x^2 - 9)$
$a^2(9 - x^2) - 6a(9 - x^2) + 9(9 - x^2)$
$(9 - x^2)(a^2 - 6a + 9)$
$(3 - x)(3 + x)(a - 3)(a - 3)$
$(3 - x)(3 + x)(a - 3)^2$

112. (See page 608.)

Exercise 1–4

Answers to odd-numbered problems

1. $\dfrac{4p^4q^4}{3}$ **3.** $\dfrac{4}{3}$ **5.** $\dfrac{a - 3}{4}$

7. $-8 - 7p$ **9.** $\dfrac{6(a^2 + ab + b^2)}{a + b}$

11. $\dfrac{-2}{a^2 + 4a + 16}$ **13.** $-\dfrac{a + 6}{a + 3}$

15. $\dfrac{15x^2 - 4y^2}{10xy}$ **17.** $\dfrac{2x^2 - 3x - 3}{x^2 - 1}$

19. $\dfrac{-7}{x - 4}$ **21.** $\dfrac{45a + 6ab - 20b}{10ab}$

23. $\dfrac{12a^2}{b^2}$ **25.** $\dfrac{2x + 5}{x(x - 3)}$

27. $\dfrac{-a(13a + 9)}{(a + 5)(a + 2)(a - 3)}$ **29.** $\dfrac{3a - 5}{2a - 3}$

112. The following TI-81 program will compute the required values a, b, c, and d. It uses the program GCF of problem 111.

Display	Keystrokes—use [ENTER] at the end of each line.
Prgm5:QTRI	[PRGM] 5 9 4 [×] [x^2]
:Input A	[PRGM] ▶ 2 [ALPHA] [MATH]
:Input B	[PRGM] ▶ 2 [ALPHA] [MATRX]
:Input C	[PRGM] ▶ 2 [ALPHA] [PRGM]
:AC→Z	[ALPHA] [MATH] [ALPHA] [PRGM]
	[STO▶] 2
:Z/abs Z → D	[ALPHA] 2 [÷] [2nd] [x^{-1}] [ALPHA]
	2 [STO▶] [x^{-1}]
:1 → M	1 [STO▶] [÷]
:abs AC → N	[2nd] [x^{-1}] [ALPHA] [MATH]
	[ALPHA] [PRGM] [STO▶] [LOG]
:Lbl 3	[PRGM] 1 3
:If MN ≠ abs Z	[PRGM] 3 [ALPHA] [÷] [ALPHA]
	[LOG] [2nd] [MATH] 2 [2nd] [x^{-1}]
	[ALPHA] 2
:Goto 6	[PRGM] 2 6
:If abs (M + DN)=abs B	[PRGM] 3 [2nd] [x^{-1}] [(] [ALPHA]
	[÷] [+] [ALPHA] [x^{-1}] [ALPHA]
	[LOG] [)] [2nd] [MATH] 1
	[2nd] [x^{-1}] [ALPHA] [MATRX]
:Goto 5	[PRGM] 2 5
:Lbl 6	[PRGM] 1 6
M+1 → M	[ALPHA] [÷] [+] 1 [STO▶] [÷]
:abs Z/M → N	[2nd] [x^{-1}] [ALPHA] 2 [÷] [ALPHA]
	[÷] [STO▶] [LOG]
:If M>N	[PRGM] 3 [ALPHA] [÷] [2nd]
	[MATH] 3 [ALPHA] [LOG]
:Goto 2	[PRGM] 2 2
:Goto 3	[PRGM] 2 3
:Lbl 2	[PRGM] 1 2
:Disp "No"	[PRGM] ▶ 1 [2nd] [ALPHA] [+]
	[LOG] 7 [+]

Display	Keystrokes—use [ENTER] at the end of each line.
:Stop	[PRGM] 8
:Lbl 5	[PRGM] 1 5
:(B/abs B)N → N	[(] [ALPHA] [MATRX] [÷] [2nd]
	[x^{-1}] [ALPHA] [MATRX] [)] [ALPHA]
	[LOG] [STO▶] [LOG]
Z/N → M	[ALPHA] 2 [÷] [ALPHA] [LOG]
	[STO▶] [÷]
:A → F	[ALPHA] [MATH] [STO▶] [COS]
:M → S	[ALPHA] [÷] [STO▶] [LN]
:Prgm3	[PRGM] ▶ ▶ 3
:S → D	[ALPHA] [LN] [STO▶] [x^{-1}]
:N → F	[ALPHA] [LOG] [STO▶] [COS]
:C → S	[ALPHA] [PRGM] [STO▶] [LN]
:Prgm3	[PRGM] ▶ ▶ 3
:(N/abs N)S → E	[(] [ALPHA] [LOG] [÷] [2nd] [x^{-1}]
	[ALPHA] [LOG] [)] [ALPHA] [LN]
	[STO▶] [SIN]
:A/D → F	[ALPHA] [MATH] [÷] [ALPHA] [x^{-1}]
	[STO▶] [COS]
:M/D → G	[ALPHA] [÷] [÷] [ALPHA] [x^{-1}]
	[STO▶] [TAN]
:Disp D	[PRGM] ▶ 1 [ALPHA] [x^{-1}]
:Disp E	[PRGM] ▶ 1 [ALPHA] [SIN]
:Disp F	[PRGM] ▶ 1 [ALPHA] [COS]
:Disp G	[PRGM] ▶ 1 [ALPHA] [TAN] [2nd]
	[CLEAR]

To factor $10x^2 + x - 24$, do: [PRGM] 5 [ENTER] 10 [ENTER] 1 [ENTER] [(−)] 24 [ENTER], and the values 5, 8, 2, −3 appear, which means the expression is $(5x + 8)(2x - 3)$.

31. $\dfrac{4(a-2)}{15(a-4)}$ **33.** $\dfrac{x^3 - x^2 - 5x - 3}{4}$

35. $\dfrac{6y^2 + 107y - 190}{10y^2 - 40}$

37. $\dfrac{3y^2 - y + 15}{y^3 + 3y^2 - 4y - 12}$

39. $\dfrac{3m^2 - 13m + 12}{m + 3}$ **41.** $\dfrac{1}{2x - 10}$

43. $\dfrac{5}{3}$ **45.** $\dfrac{ab + b}{2ab - 3}$ **47.** $\dfrac{15x - 20}{8x + 4}$

49. $\dfrac{1}{7}$ **51.** $m(m + n)$ **53.** $-\dfrac{1}{3}$

55. $\dfrac{-x + 3}{3x - 9}$ **57.** $\dfrac{2a^2}{3}$

59. $\dfrac{R_2R_3V_1 + R_1R_3V_2 + R_1R_2V_3}{R_2R_3 + R_1R_3 + R_1R_2}$

61. $\dfrac{2r_1r_2}{r_1 + r_2}$ **63.** $\dfrac{3}{x^2 + 3x}$

65. $\dfrac{P}{Q} + \dfrac{R}{S} = \dfrac{PS}{QS} + \dfrac{RQ}{SQ} = \dfrac{PS + QR}{QS}$

$\dfrac{P}{Q} - \dfrac{R}{S} = \dfrac{PS}{QS} - \dfrac{RQ}{SQ} = \dfrac{PS - QR}{QS}$

67. 529 hours

69. An even number (integer) greater than two is not prime because all even numbers are divisible by two. A prime number must be divisible by only one and itself. Even integers must have 0, 2, 4, 6, or 8 for their last digit. The following is a programming solution for the TI-81. It also works for even integers.

```
Prgm6:PRIME
:Input N
:2 → D
:If FPart (N/D)=0
:Goto 5
:√N → L
:3 → D
:Lbl 1
:If FPart (N/D)=0
:Goto 5
:D+2 → D
:If D ≤ L
:Goto 1
:Disp "PRIME"
:Stop
:Lbl 5
:Disp D
:N/D → N
:Disp N
```

Solutions to skill and review problems

1. a. $\sqrt{5^2} = 5$ **b.** $\sqrt{10^2} = 10$ **c.** $\sqrt{20^2} = 20$

2. a. $\sqrt[3]{2^3} = 2$ **b.** $\sqrt[3]{4^3} = 4$ **c.** $\sqrt[6]{2^6} = 2$

3. a. $\sqrt{36} = 6$ **b.** $2 \cdot 3 = 6$

4. $2 \cdot 2^3x^2x^2y^3y$
$2^4x^4y^4$
$16x^4y^4$

5. a. 8 **b.** 8 **c.** 16 **d.** 44 **e.** 56

6. Observe that $81 = 3^4$
$3^4a^4b^8 = 3 \cdot 3^xa^2a^yb^5b^z$
$3^4a^4b^8 = 3^{1+x}a^{2+y}b^{5+z}$
$4 = 1 + x$ so $x = 3$
$4 = 2 + y$ so $y = 2$
$8 = 5 + z$ so $z = 3$

Solutions to trial exercise problems

11. $\dfrac{8 - 2a}{a^3 - 64}$
$$\frac{2(4 - a)}{(a - 4)(a^2 + 4a + 16)}$$
$$\frac{-2(a - 4)}{(a - 4)(a^2 + 4a + 16)}$$
$$\frac{-2}{a^2 + 4a + 16}$$

19. $\dfrac{3x - 5}{x - 4} + \dfrac{3x + 2}{-(x - 4)}$
$$\frac{3x - 5}{x - 4} - \frac{3x + 2}{x - 4}$$
$$\frac{(3x - 5) - (3x + 2)}{x - 4}$$
$$\frac{-7}{x - 4}$$

24. $\dfrac{a - 6}{6a + 18}(a + 3)$
$$\frac{a - 6}{6(a + 3)} \cdot \frac{a + 3}{1}$$
$$\frac{a - 6}{6}$$

27. $\dfrac{-6a}{(a - 3)(a + 2)} - \dfrac{7a}{(a + 5)(a + 2)}$
$$\frac{-6a(a + 5)}{(a - 3)(a + 2)(a + 5)}$$
$$- \frac{7a(a - 3)}{(a + 5)(a + 2)(a - 3)}$$
$$\frac{(-6a^2 - 30a) - (7a^2 - 21a)}{(a + 5)(a + 2)(a - 3)}$$
$$\frac{-13a^2 - 9a}{(a + 5)(a + 2)(a - 3)}$$
$$\frac{-a(13a + 9)}{(a + 5)(a + 2)(a - 3)}$$

43. $\dfrac{6\left(\frac{1}{3} + \frac{1}{2}\right)}{6\left(\frac{2}{3} - \frac{1}{6}\right)}$
$$\frac{2 + 3}{2(2) - 1}$$
$$\frac{5}{3}$$

54. $\dfrac{\frac{6}{x(x - 1)} - 2}{\frac{3}{x - 1} + 2}$
$$\frac{x(x - 1)\left(\frac{6}{x(x - 1)} - 2\right)}{x(x - 1)\left(\frac{3}{x - 1} + 2\right)}$$
$$\frac{6 - 2(x)(x - 1)}{3x + 2(x)(x - 1)}$$
$$\frac{-2x^2 + 2x + 6}{2x^2 + x}$$

66. $\text{MTBF}_S = \left(\dfrac{1}{\text{MTBF}_1} + \dfrac{1}{\text{MTBF}_2} + \dfrac{1}{\text{MTBF}_3}\right)^{-1}$
$$= \left(\frac{1}{\text{MTBF}_1} \cdot \frac{\text{MTBF}_2 \cdot \text{MTBF}_3}{\text{MTBF}_2 \cdot \text{MTBF}_3} + \frac{1}{\text{MTBF}_2} \cdot \frac{\text{MTBF}_1 \cdot \text{MTBF}_3}{\text{MTBF}_1 \cdot \text{MTBF}_3} + \frac{1}{\text{MTBF}_3} \cdot \frac{\text{MTBF}_1 \cdot \text{MTBF}_2}{\text{MTBF}_1 \cdot \text{MTBF}_2}\right)^{-1}$$
$$= \left(\frac{\text{MTBF}_2 \cdot \text{MTBF}_3}{\text{MTBF}_1 \cdot \text{MTBF}_2 \cdot \text{MTBF}_3} + \frac{\text{MTBF}_1 \cdot \text{MTBF}_3}{\text{MTBF}_2 \cdot \text{MTBF}_1 \cdot \text{MTBF}_3} + \frac{\text{MTBF}_1 \cdot \text{MTBF}_2}{\text{MTBF}_3 \cdot \text{MTBF}_1 \cdot \text{MTBF}_2}\right)^{-1}$$
$$= \left(\frac{\text{MTBF}_2 \cdot \text{MTBF}_3 + \text{MTBF}_1 \cdot \text{MTBF}_3 + \text{MTBF}_1 \cdot \text{MTBF}_2}{\text{MTBF}_1 \cdot \text{MTBF}_2 \cdot \text{MTBF}_3}\right)^{-1}$$
$$= \frac{\text{MTBF}_1 \cdot \text{MTBF}_2 \cdot \text{MTBF}_3}{\text{MTBF}_2 \cdot \text{MTBF}_3 + \text{MTBF}_1 \cdot \text{MTBF}_3 + \text{MTBF}_1 \cdot \text{MTBF}_2}$$

Exercise 1–5

Answers to odd-numbered problems

1. 17 **3.** 2 **5.** -5 **7.** $2|x|$ **9.** $5y^4|x^3|$ **11.** $\dfrac{4|x^3|}{3|y^5|}$ **13.** $|x^2 - 3|$ **15.** $2\sqrt[3]{5}$ **17.** $10\sqrt{2}$ **19.** $2ab\sqrt[5]{2a^2}$ **21.** $10xy^4z^6\sqrt{2y}$ **23.** $2y\sqrt[4]{x^3y^2}$ **25.** a^4 **27.** $5ab^3c\sqrt[3]{5a}$ **29.** $4\sqrt{2}$ **31.** $\dfrac{6\sqrt{2a}}{5}$ **33.** $\dfrac{2\sqrt{6}}{9}$ **35.** $\dfrac{2\sqrt[3]{3}}{3}$ **37.** $\dfrac{2x^2y^3\sqrt{15xz}}{5z}$ **39.** $\dfrac{2x^2y\sqrt[5]{2x^2w^2z}}{w^2z}$ **41.** $\dfrac{2a\sqrt[3]{2a^2b^2c}}{b^3c}$

43. $\dfrac{2\sqrt[4]{3x^2y^2z^2}}{3z^2}$ **45.** $\dfrac{\sqrt{10xy}}{5y}$

47. $-2\sqrt{2}$ **49.** $14\sqrt{3}$ **51.** $-\sqrt{3}$

53. $2\sqrt[3]{2} + 10\sqrt[3]{3}$ **55.** $8a\sqrt{b}$

57. $-30a$ **59.** $2x + 2\sqrt[3]{2x^2} - x\sqrt[3]{4}$

61. $12 - 22\sqrt{x} + 8x$ **63.** $45 - 18\sqrt{2}$

65. $8x^2 - 8x\sqrt{2x} + 4x$ **67.** $\dfrac{5 + \sqrt{10}}{3}$

69. $\dfrac{a^2 + 2a\sqrt{b} + b}{a^2 - b}$ **71.** $\dfrac{x\sqrt{3} - \sqrt{x}}{3x - 1}$

73. $\dfrac{\sqrt{7} - 4}{15}$ **75.** $-\dfrac{4\sqrt{15}}{15}$

77. $\dfrac{\sqrt{6} - \sqrt{2}}{2}$ **79.** $\dfrac{\sqrt{32} - \sqrt{5}}{4}$

81. $a\sqrt{a}$ **83.** $y\sqrt{2x}$

---

85. Recall, $a^3 - b^3$
$= (a - b)(a^2 + ab + b^2)$ and $a^3 + b^3$
$= (a + b)(a^2 - ab + b^2)$.

a. Using $a^3 - b^3$
$= (a - b)(a^2 + ab + b^2)$,
let $a = \sqrt[3]{x}$ and $b = \sqrt[3]{y}$ so that
$x - y = (\sqrt[3]{x} - \sqrt[3]{y})[(\sqrt[3]{x})^2 + \sqrt[3]{x}\sqrt[3]{y} + (\sqrt[3]{y})^2]$
$= (\sqrt[3]{x} - \sqrt[3]{y})(\sqrt[3]{x^2} + \sqrt[3]{xy} + \sqrt[3]{y^2})$
Thus, $Q(x,y) = \sqrt[3]{x^2} + \sqrt[3]{xy} + \sqrt[3]{y^2}$.

b. $a^3 + b^3 = (a + b)(a^2 - ab + b^2)$;
let $a = \sqrt[3]{x}$ and $b = \sqrt[3]{y}$:
$x + y = (\sqrt[3]{x} + \sqrt[3]{y})(\sqrt[3]{x^2} - \sqrt[3]{xy} + \sqrt[3]{y^2})$

c. $8x - y = (2\sqrt[3]{x})^3 - (\sqrt[3]{y})^3$
$= (2\sqrt[3]{x} - \sqrt[3]{y})[(2\sqrt[3]{x})^2 + 2\sqrt[3]{x}\sqrt[3]{y} + (\sqrt[3]{y})^2]$
$= (2\sqrt[3]{x} - \sqrt[3]{y})(4\sqrt[3]{x^2} + 2\sqrt[3]{xy} + \sqrt[3]{y^2})$
$8x - y = (\sqrt{8x} - \sqrt{y})(\sqrt{8x} + \sqrt{y})$
$= (2\sqrt{2x} - \sqrt{y})(2\sqrt{2x} + \sqrt{y})$

87. $\dfrac{\sqrt[3]{x}}{\sqrt[3]{2x^2} - \sqrt[3]{3x}} \cdot \dfrac{(\sqrt[3]{2x^2})^2 + \sqrt[3]{2x^2}\sqrt[3]{3x} + (\sqrt[3]{3x})^2}{(\sqrt[3]{2x^2})^2 + \sqrt[3]{2x^2}\sqrt[3]{3x} + (\sqrt[3]{3x})^2}$

$\dfrac{\sqrt[3]{x}(\sqrt[3]{4x^4} + \sqrt[3]{6x^3} + \sqrt[3]{9x^2})}{2x^2 - 3x}$

$\dfrac{\sqrt[3]{4x^5} + \sqrt[3]{6x^4} + \sqrt[3]{9x^3}}{2x^2 - 3x}$

$\dfrac{x\sqrt[3]{4x^2} + x\sqrt[3]{6x} + x\sqrt[3]{9}}{2x^2 - 3x}$

$\dfrac{x(\sqrt[3]{4x^2} + \sqrt[3]{6x} + \sqrt[3]{9})}{x(2x - 3)}$

$\dfrac{\sqrt[3]{4x^2} + \sqrt[3]{6x} + \sqrt[3]{9}}{2x - 3}$

89. $\sqrt{\dfrac{2 - \sqrt{3}}{2 + \sqrt{3}}}$ $\qquad \dfrac{1 - \dfrac{1}{\sqrt{3}}}{1 + \dfrac{1}{\sqrt{3}}}$

$\sqrt{\dfrac{2 - \sqrt{3}}{2 + \sqrt{3}} \cdot \dfrac{2 - \sqrt{3}}{2 - \sqrt{3}}}$ $\qquad \dfrac{1 - \dfrac{1}{\sqrt{3}}}{1 + \dfrac{1}{\sqrt{3}}} \cdot \dfrac{\sqrt{3}}{\sqrt{3}}$

$\sqrt{\dfrac{(2 - \sqrt{3})^2}{1}}$ $\qquad \dfrac{\sqrt{3} - 1}{\sqrt{3} + 1} \cdot \dfrac{\sqrt{3} - 1}{\sqrt{3} - 1}$

$\sqrt{(2 - \sqrt{3})^2}$ $\qquad \dfrac{4 - 2\sqrt{3}}{2}$

$2 - \sqrt{3}$ $\qquad 2 - \sqrt{3}$

---

91. a. $\sqrt[3]{\sqrt{5}} = \sqrt[6]{5}$. This seems logical because $(\sqrt[6]{5})^6 = 5$ and $(\sqrt{\sqrt[3]{5}})^6 = (\sqrt[3]{5})^3 = 5$.

b. $\sqrt[m]{\sqrt[n]{x}} = \sqrt[mn]{x}$.

Solutions to skill and review problems

1. $\dfrac{4 + 9}{12} = \dfrac{13}{12}$ **2.** $\dfrac{1}{3} \cdot \dfrac{3}{4} = \dfrac{1}{1} \cdot \dfrac{1}{4} = \dfrac{1}{4}$

3. $3\left(\dfrac{1}{a^2}\right) = \dfrac{3}{a^2}$ **4.** $3a^2b^3$

5. $\dfrac{1}{\sqrt[3]{-8}} = \dfrac{1}{-2} = -\dfrac{1}{2}$ **6.** $\dfrac{8a^8}{2a^2} = 4a^6$

7. $\dfrac{-2a^{-2}}{8a^8} = -\dfrac{1}{4a^{10}}$

Solutions to trial exercise problems

11. $\dfrac{\sqrt{16x^6}}{\sqrt{9y^{10}}}$
$\dfrac{4\,|x^3|}{3\,|y^5|}$

18. $\sqrt[3]{8{,}000}$
$\sqrt[3]{2^6 \cdot 5^3}$
$2^2 \cdot 5$
20

27. $\sqrt[3]{25a^2b^4c}\sqrt[3]{25a^2b^5c^2}$
$\sqrt[3]{5^4a^4b^9c^3}$
$5ab^3c\sqrt[3]{5a}$

36. $\sqrt[3]{\dfrac{3}{20}}$
$\dfrac{\sqrt[3]{3}}{\sqrt[3]{2^2 \cdot 5}} \cdot \dfrac{\sqrt[3]{2 \cdot 5^2}}{\sqrt[3]{2 \cdot 5^2}}$
$\dfrac{\sqrt[3]{2 \cdot 3 \cdot 5^2}}{\sqrt[3]{2^3 \cdot 5^3}}$
$\dfrac{\sqrt[3]{150}}{10}$

41. $\sqrt[3]{\dfrac{16a^5}{b^7c^2}}$
$\sqrt[3]{\dfrac{16a^5}{b^7c^2} \cdot \dfrac{b^2c}{b^2c}}$
$\dfrac{\sqrt[3]{2^4a^5b^2c}}{\sqrt[3]{b^9c^3}}$
$\dfrac{2a\sqrt[3]{2a^2b^2c}}{b^3c}$

45. $\dfrac{\sqrt[4]{8x^5y^7}}{\sqrt[4]{50x^3y^9}}$
$\sqrt[4]{\dfrac{8x^5y^7}{50x^3y^9}}$
$\sqrt[4]{\dfrac{4x^2}{25y^2}}$
$\sqrt[4]{\dfrac{4x^2}{5^2y^2} \cdot \dfrac{5^2y^2}{5^2y^2}}$
$\dfrac{\sqrt[4]{100x^2y^2}}{\sqrt[4]{5^4y^4}}$
$\dfrac{\sqrt[4]{100x^2y^2}}{5y}$
$\dfrac{\sqrt[4]{10^2x^2y^2}}{5y}$
$\dfrac{\sqrt{10xy}}{5y}$

59. $\sqrt[3]{4x}(\sqrt[3]{2x^2} + \sqrt[3]{4x} - \sqrt{x^2})$
$\sqrt[3]{8x^3} + \sqrt[3]{16x^2} - \sqrt[3]{4x^3}$
$2x + 2\sqrt[3]{2x^2} - x\sqrt[3]{4}$

63. $(5\sqrt{3} - 2\sqrt{6})(\sqrt{3} + \sqrt{12})$
$5(3) + 5\sqrt{36} - 2\sqrt{18} - 2\sqrt{72}$
$15 + 30 - 6\sqrt{2} - 12\sqrt{2}$
$45 - 18\sqrt{2}$

71. $\dfrac{\sqrt{2x}}{\sqrt{6x} + \sqrt{2}} \cdot \dfrac{\sqrt{6x} - \sqrt{2}}{\sqrt{6x} - \sqrt{2}}$
$\dfrac{\sqrt{12x^2} - \sqrt{4x}}{6x - \sqrt{12x} + \sqrt{12x} - 2}$
$\dfrac{2x\sqrt{3} - 2\sqrt{x}}{6x - 2}$
$\dfrac{2(x\sqrt{3} - \sqrt{x})}{2(3x - 1)}$
$\dfrac{x\sqrt{3} - \sqrt{x}}{3x - 1}$

76. $-\frac{5}{2\sqrt{2}}\left(-\frac{1}{\sqrt{2}}\right) - \frac{1}{2}\left(\frac{\sqrt{5}}{3}\right)$

$\frac{5}{4} - \frac{\sqrt{5}}{6}$

$\frac{6(5) - 4\sqrt{5}}{4(6)}$

$\frac{30 - 4\sqrt{5}}{24}$

$\frac{2(15 - 2\sqrt{5})}{24}$

$\frac{15 - 2\sqrt{5}}{12}$

77. $\sqrt{\frac{3 - \frac{1}{\sqrt{2}}}{2}}$

$\sqrt{\frac{1}{2}\left(3 - \frac{\sqrt{2}}{2}\right)}$

$\sqrt{\frac{1}{2} \cdot \frac{6 - \sqrt{2}}{2}}$

$\sqrt{\frac{6 - \sqrt{2}}{4}}$

$\frac{\sqrt{6 - \sqrt{2}}}{2}$

84. $\sqrt[8]{a^8b^{12}c^{16}}$

$\sqrt[8\div4]{a^{8\div4}b^{12\div4}c^{16\div4}}$

$\sqrt{a^2b^3c^4}$

$abc^2\sqrt{b}$

86. $\frac{3xy}{\sqrt[3]{x} + \sqrt[3]{y}} \cdot \frac{\sqrt[3]{x^2} - \sqrt[3]{xy} + \sqrt[3]{y^2}}{\sqrt[3]{x^2} - \sqrt[3]{xy} + \sqrt[3]{y^2}}$

$\frac{3xy(\sqrt[3]{x^2} - \sqrt[3]{xy} + \sqrt[3]{y^2})}{x + y}$

Exercise 1–6

Answers to odd-numbered problems

1. $2\sqrt{2}$ **3.** $\frac{1}{4}$ **5.** $\frac{1}{4}$ **7.** $4\sqrt{x}$
9. $3\sqrt[4]{x^3}$ **11.** $2xy\sqrt[4]{2x^2y^3}$ **13.** $2x\sqrt{2x}$
15. 5 **17.** $8a^{\frac{3}{4}}$ **19.** b^2
21. $2x^{\frac{2}{7}}y^{\frac{3}{5}}z^{\frac{1}{2}}$ **23.** $\frac{1}{8}x^{\frac{1}{2}}y^{\frac{3}{4}}$ **25.** $\frac{1}{ab^{\frac{1}{2}}}$
27. $a^{\frac{3}{4}}$ **29.** $3a^{\frac{1}{4}}$ **31.** $x^{\frac{1}{4}}y^{\frac{4}{3}}$
33. $x^{\frac{1}{6}}y^{\frac{1}{5}}z$ **35.** $a^{2m}b^4$ **37.** 2^my^n
39. 75.3760 **41.** −2.8854
43. −9.6549 **45.** 15.1539
47. 1.4238 **49.** 15.8322 **51.** 0.5
53. 849.1202 **55.** 167.6478

57. To find the new value of D_l replace L_b by $4L_b$.
$D_l = c(4L_b)^{1.5} = c(4^{1.5})L_b{}^{1.5}$.
Compare this new value to the original value of $cL_b^{1.5}$:
$\frac{c(4^{1.5})L_b^{1.5}}{cL_b^{1.5}} = 4^{1.5} = 4^{3/2} = (\sqrt{4})^3 = 8$
Thus the new leg diameter must be eight times the original diameter if the body length increases by a factor of 4.

59. \$394.91 **61.** 1 **63.** 1

Solutions to skill and review problems

1. a. not real

2. $10i^2 + 35i - 6i - 21$
$-10 + 29i - 21$
$-31 + 29i$

3. a. −1 **b.** 1 **c.** −1

4. $3(2) + \sqrt{12} - 3\sqrt{6} - \sqrt{18}$
$6 + 2\sqrt{3} - 3\sqrt{6} - 3\sqrt{2}$

5. $\frac{2\sqrt{3}}{\sqrt{6} + \sqrt{2}} \cdot \frac{\sqrt{6} - \sqrt{2}}{\sqrt{6} - \sqrt{2}}$

$\frac{2\sqrt{18} - 2\sqrt{6}}{6 + \sqrt{12} - \sqrt{12} - 2}$

$\frac{6\sqrt{2} - 2\sqrt{6}}{4}$

$\frac{2(3\sqrt{2} - \sqrt{6})}{4}$

$\frac{3\sqrt{2} - \sqrt{6}}{2}$

Solutions to trial exercise problems

5. $(-8)^{-\frac{2}{3}} = \frac{1}{(-8)^{\frac{2}{3}}} = \frac{1}{(\sqrt[3]{-8})^2}$

$= \frac{1}{(-2)^2} = \frac{1}{4}$

23. $\left(\frac{1}{4}x^{\frac{1}{4}}y^{\frac{1}{2}}\right)\left(\frac{1}{2}x^{\frac{1}{4}}y^{\frac{1}{4}}\right)$

$= \frac{1}{4}\left(\frac{1}{2}\right)x^{\frac{1}{4}+\frac{1}{4}}y^{\frac{1}{2}+\frac{1}{4}} = \frac{1}{8}x^{\frac{1}{2}}y^{\frac{3}{4}}$

31. $\frac{x^{\frac{5}{8}}y^{\frac{2}{3}}}{x^{\frac{3}{8}}y^{\frac{-2}{3}}}$

$= x^{\frac{5}{8}-\frac{3}{8}}y^{\frac{2}{3}-\frac{-2}{3}}$

$= x^{\frac{1}{4}}y^{\frac{4}{3}}$

32. $\left(\frac{x^{\frac{-3}{5}}y^{\frac{3}{4}}}{z^{\frac{3}{10}}}\right)^{20}$

$= \frac{x^{-12}y^{15}}{z^6}$

$= \frac{y^{15}}{x^{12}z^6}$

49. $(\sqrt[3]{-500})^{\frac{4}{3}}$
500 [$x^{1/y}$] 3 [=] [y^x] [(] 4 [÷] 3 [)] [=]
TI-81: [(] [MATH] 4 500 [)] [^] [(] 4 [÷] 3 [)] [ENTER]
15.8322

60. $B_n = 45{,}000\left[\frac{\frac{0.10}{12}\cdot\left(1 + \frac{0.10}{12}\right)^0}{\left(1 + \frac{0.10}{12}\right)^{360} - 1}\right]$

$= \frac{45{,}000 \cdot 0.10}{12} \cdot \frac{1}{\left(1 + \frac{0.10}{12}\right)^{360} - 1}$

$= 375 \cdot \frac{1}{\left(1 + \frac{0.10}{12}\right)^{360} - 1} \approx 19.91$

1 [+] 0.10 [÷] 12 [=] [x^y] 360 [−] 1 [=] [$1/x$] [×] 375 [=]
TI-81: 375 [(] 1 [÷] [(] [(] 1 [+] .1 [÷] 12 [)] [^] 360 [−] 1 [)] [)] [ENTER]

$I_n = 394.91\left[1 - \left(1 + \frac{0.10}{12}\right)^{1-1-360}\right]$

$= 394.91\left[1 - \left(1 + \frac{0.10}{12}\right)^{-360}\right]$

≈ 375.00

1 [−] [(] 1 [+] 0.10 [÷] 12 [)] [x^y] 360 [+/−] [=] [×] 394.91 [=]
TI-81: 394.91 [(] 1 [−] [(] 1 [+] .1 [÷] 12 [)] [^] [(−)] 360 [)] [ENTER]

Exercise 1–7

Answers to odd-numbered problems

1. $-5 + 8i$ **3.** $13 + 8i$
5. $5 + 62i$ **7.** 34 **9.** $21 - 20i$
11. $-30 - i$ **13.** $-\frac{14}{29} - \frac{23}{29}i$
15. $\frac{5}{13} + \frac{12}{13}i$ **17.** $-\frac{18}{13} + \frac{12}{13}i$
19. $-6\sqrt{2}$ **21.** $5\sqrt{2}i$
23. $13 - 7\sqrt{2}i$ **25.** $15 - \sqrt{3}i$
27. $6 + 2\sqrt{3} + (-2\sqrt{2} + 3\sqrt{6})i$
29. $\frac{4 - 3\sqrt{3}}{11} - \frac{6\sqrt{2} + \sqrt{6}}{11}i$
31. $-\frac{7\sqrt{2}}{22} - \frac{27}{11}i$ **33.** $-\sqrt{3}i$

35. -1 **37.** $-i$ **39.** $-i$ **41.** i

43. $-7 + 11i$ **45.** $-\frac{125}{58} + \frac{95}{58}i$

47. -5 **49.** $T = \frac{6{,}190}{43{,}709} + \frac{1{,}094}{43{,}709}i$

51. $6 - 2i = X_C$

53. The value is complex for $x - 16 < 0$, so $x < 16$.

55. *Subtraction:*
$(a + bi) - (c + di)$
$= (a - c) + (bi - di)$
$= (a - c) + (b - d)i$
Rule: $(a + bi) - (c + di)$
$= (a - c) + (b - d)i$
Multiplication:
$(a + bi)(c + di)$
$= ac + bdi^2 + adi + bci$
$= (ac - bd) + (ad + bc)i$
Rule: $(a + bi)(c + di)$
$= (ac - bd) + (ad + bc)i$
Division:
$\frac{a + bi}{c + di} = \frac{a + bi}{c + di} \cdot \frac{c - di}{c - di}$
$= \frac{ac - bdi^2 - adi + bci}{c^2 - d^2i^2 - cdi + cdi}$
$= \frac{(ac + bd) + (bc - ad)i}{c^2 + d^2}$
$= \frac{ac + bd}{c^2 + d^2} + \frac{bc - ad}{c^2 + d^2}i$
Rule: $\frac{a + bi}{c + di} = \frac{ac + bd}{c^2 + d^2} + \frac{bc - ad}{c^2 + d^2}i$

57. After approximately 18 iterations, the value of z repeats the value $0.1074991191 + 0.0636941246i$. The following is a program for a TI-81:

```
Prgm2: JULIA
:.5→A
:−.2→B
:Lbl 1
:A²−B²→C
:2AB→D
:C+.1→A
:D+.05→B
:Disp A
:Disp B
:Pause
:Goto 1
```

Solutions to skill and review problems

1. $-5[3x - 2(1 - 4x)]$
$-5[3x - 2 + 8x]$
$-5[11x - 2]$
$-55x + 10$

2. $x + 5 = 12$; 7, since $7 + 5 = 12$

3. $5x = 20$; 4, since $5 \cdot 4 = 20$

4. $\frac{x}{6} = 48$; 288, since $\frac{288}{6} = 48$

5. $3(2 - 3x) = 1 - 10x$
Replace x by -5:
$3[2 - 3(-5)] = 1 - 10(-5)$
$3(2 + 15) = 1 + 50$
$3(17) = 51$
$51 = 51$
yes

6. any value, since $x + x$ combines into $2x$ regardless of considering the value of x

7. $C = \frac{5}{9}(72 - 32)$
$= \frac{5}{9}(40)$
$= \frac{200}{9} = 22\frac{2}{9}°$ centigrade

8. $0.06(1{,}000 - 2x)$
$0.06(1{,}000) - 0.06(2)x$
$60 - 0.12x$

9. $8\% = \frac{8}{100}$, so d. 0.08

10. $0.08(12{,}000) = 960$

11. $0.06(4{,}000) + 0.1(12{,}000)$
$240 + 1{,}200$
$1{,}440$

Solutions to trial exercise problems

11. $i[(5 - 3i)(-2 + 4i) - (2 - i)^2]$
$i[(5 - 3i)(-2 + 4i) - (2 - i)(2 - i)]$
$i[(-10 + 20i + 6i - 12i^2)$
$\quad - (4 - 2i - 2i + i^2)]$
$i[(2 + 26i) - (3 - 4i)]$
$i[-1 + 30i]$
$-i + 30i^2$
$-30 - i$

15. $\frac{6 + 4i}{6 - 4i} = \frac{3 + 2i}{3 - 2i}$
$\frac{3 + 2i}{3 - 2i} \cdot \frac{3 + 2i}{3 + 2i}$
$\frac{9 + 6i + 6i + 4i^2}{9 + 6i - 6i - 4i^2}$
$\frac{5 + 12i}{13}$
$\frac{5}{13} + \frac{12}{13}i$

27. $(2 + \sqrt{-6})(3 - \sqrt{-2})$
$(2 + \sqrt{6}i)(3 - \sqrt{2}i)$
$6 - 2\sqrt{2}i + 3\sqrt{6}i - \sqrt{12}i^2$
$6 + \sqrt{12} - 2\sqrt{2}i + 3\sqrt{6}i$
$6 + 2\sqrt{3} + (-2\sqrt{2} + 3\sqrt{6})i$

33. $\frac{\sqrt{-6} + \sqrt{6}}{\sqrt{-2} - \sqrt{2}}$
$\frac{\sqrt{6}i + \sqrt{6}}{\sqrt{2}i - \sqrt{2}}$
$\frac{\sqrt{6} + \sqrt{6}i}{-\sqrt{2} + \sqrt{2}i}$
$\frac{\sqrt{6} + \sqrt{6}i}{-\sqrt{2} + \sqrt{2}i} \cdot \frac{-\sqrt{2} - \sqrt{2}i}{-\sqrt{2} - \sqrt{2}i}$
$\frac{-\sqrt{12} - \sqrt{12}i - \sqrt{12}i - \sqrt{12}i^2}{2 + 2\sqrt{2}i - 2\sqrt{2}i - 2i^2}$
$\frac{-2\sqrt{12}i}{4}$
$\frac{-4\sqrt{3}i}{4}$
$-\sqrt{3}i$

39. $i^{-5} = \frac{1}{i^5} = \frac{1}{i} = \frac{1}{i} \cdot \frac{i}{i} = \frac{i}{-1} = -i$

47. $(2 - i)^3 - 3(2 - i)^2 + (2 - i)$
$(2 - i)(2 - i)(2 - i)$
$\quad - 3(2 - i)(2 - i) + (2 - i)$
$(2 - i)(3 - 4i) - 3(3 - 4i) + 2 - i$
$6 - 8i - 3i + 4i^2 - 9 + 12i + 2 - i$
-5

Chapter 1 review

1. $\{0, 1, 2, 3, 4, 5, 6, 7\}$

2. $\{1, 2, 3, 4, \ldots\}$

3. $\left\{\frac{1}{2}, \frac{2}{3}, \frac{3}{4}, \ldots, \frac{100}{101}\right\}$ **4.** $0.4166\overline{6}$

5. $0.230769\overline{230769}$ **6.** $-\frac{43}{4}$ **7.** $-\frac{7}{24}$

8. $-\frac{29}{45}$ **9.** $\frac{5}{112}$ **10.** $\frac{12bx - 5ay}{20ab}$

11. $\frac{8a + 13ab + 3b}{4ab}$ **12.** $\frac{3x^3}{25y^3}$

13. $\frac{6a^2 + b^2}{5a^2}$

14. $\left[-2\frac{1}{2}, -\frac{3}{4}\right)$

15. $\left(-\frac{1}{4}, \infty\right)$

16. $\{x \mid x \le -1\}$

17. $\left\{x \mid -\frac{\pi}{3} \le x < \pi\right\}$

18. $\left(-\frac{1}{2}, 3\frac{1}{4}\right]$; $\left\{x \mid -\frac{1}{2} < x \le 3\frac{1}{4}\right\}$

19. $[\pi,\infty)$; $\{x \mid x \geq \pi\}$ 20. $-\dfrac{1}{2}$

21. $\pi + 9$ 22. $\sqrt{2} - 5$ 23. $5x^2$

24. $\dfrac{3x^2}{2|y^3|}$ 25. $5|2x - 1|$

26. $-(x - 2)^2|x + 1|$ 27. $\dfrac{27x^6}{y^3}$

28. $-\dfrac{3x}{y^2}$ 29. $\dfrac{x^3y^2}{25}$ 30. $\dfrac{6x}{y}$

31. $-\dfrac{36x^6}{y^6}$ 32. $\dfrac{3y^5}{x^5}$ 33. $-\dfrac{y^6}{64}$

34. $\dfrac{36}{a^{20}}$ 35. $\dfrac{1}{x^{2n}}$ 36. $\dfrac{x^{8n}}{y^{4n-12}}$

37. 4.2182×10^{16} 38. -4.605×10^{-11}

39. 0.000 000 405 2

40. −340,900,000,000 41. 27 42. 21

43. −515 44. $-15x^6 - 21x^4 + x^3 + 6$

45. $5a^3 - 49a^2 + 115a + 25$

46. $-3x^5 + 13x^4 - 5x^3 - 7x^2 - 15x + 15$

47. $-3x^4 - x^2 + 1$

48. $-10a^4 - 3 - 4b^8$

49. $y^3 + y^2 + y + 1$

50. $2x^3 - 3x^2 - x - 4\frac{1}{2} + \dfrac{-\frac{1}{2}}{2x + 1}$

51. $x^3(5 - x)(5 + x)$

52. $(x + 4)(x + 9)$

53. $a(2a - 1)(4a - 5)$

54. $(3ab - 4)(ab + 2)$

55. $b(2a + 5b)(4a^2 - 10ab + 25b^2)$

56. $5a(x - 1)(x + 1)(3x - 1)(3x + 1)$

57. $(5x - y)(x - 10y)$

58. $(a - b + 2x + y)(a - b - 2x - y)$

59. $2(3x^2 - y)(9x^4 + 3x^2y + y^2)$

60. $12(x - 2y)(x + 2y)$

61. $(x + 2y)(3a - b)$

62. $(2a^3 - bc)(4a^6 + 2a^3bc + b^2c^2)$

63. $7a^2 - 32a - 21$ 64. $4(x^2 - 3a^2)^2$

65. $(3x^3 + 1)(x^3 - 3)$

66. $4b(x + 3y)(1 - 2a)(1 + 2a)$

67. $(a - 2x - 10y)(a + 2x + 10y)$

68. $(3x + 13)(x - 7)$

69. $3x^2(xy^3 + 3z^2)(x^2y^6 - 3xy^3z^2 + 9z^4)$

70. $(x - 1)(x + 1)(x - 2)^2$

71. $\dfrac{a}{3a + 1}$ 72. $\dfrac{-y}{2x + 3}$

73. $\dfrac{2x + 1}{4x^2 + 2x + 1}$ 74. $\dfrac{x - 2}{(x + 2)(2x - 1)}$

75. $\dfrac{15x^2 - 2xy - 6y^2}{15xy}$ 76. $\dfrac{2x - 8}{x - 2}$

77. $\dfrac{-8x^2 + 4x - 5}{4x(4x - 5)}$ 78. $\dfrac{12}{b}$

79. $\dfrac{2x(x - 4)}{4x + 3}$ 80. $\dfrac{1}{(x + 1)(x + 2)}$

81. $\dfrac{-2x^2 + 12x - 10}{(x - 3)(x - 2)}$

82. $\dfrac{-2x^2 - 13x + 13}{(x + 3)(x - 3)(x - 1)}$

83. $\dfrac{-6x + 5y}{2(2x + y)}$ 84. $\dfrac{38a - 12b}{6a + 9b - 10}$

85. $\dfrac{-2a + 2b + 3}{5a - 5b - 3}$ 86. −2 87. 4

88. $6\sqrt[3]{2}$ 89. $3x^2y^3z\sqrt{6y}$

90. $2a^2bc^2\sqrt[3]{6b^2c^2}$ 91. $3a^2b\sqrt[3]{3a^2b}$

92. $5ab^2\sqrt[4]{bc^3}$ 93. $b\sqrt[3]{a^2}$

94. $8a^7b^5\sqrt{2a}$ 95. $\dfrac{\sqrt{3a}}{a}$

96. $\dfrac{2xy^2\sqrt{10wxyz}}{5w^2z}$ 97. $9ab\sqrt{2b}$

98. $-c\sqrt[4]{3b^3c^2}$

99. $3xy^2 - 3xy\sqrt{y} + 3y^2\sqrt{2x}$

100. $3x - 6x\sqrt[4]{27x} + 9x\sqrt[4]{9x^2}$

101. $\dfrac{6x - 5\sqrt{6x} - 2\sqrt{3x} + 5\sqrt{2}}{2(3x - 1)}$

102. $\dfrac{a\sqrt{a} - \sqrt{ab} + a - \sqrt{b}}{a^2 - b}$

103. $\dfrac{3\sqrt{2} + 2\sqrt{6}}{16}$ 104. $\dfrac{\sqrt{9 - \sqrt{2}}}{3}$

105. $5x\sqrt{x}$ 106. $8x^6$ 107. $\dfrac{\sqrt{x}}{3x^2}$

108. $2|x^3|y^4\sqrt{2y}$ 109. $\dfrac{\sqrt{2}|x^3|}{4|y^5|}$

110. $4|x^3(x - 3)|$ 111. $\left(x^{\frac{17}{12}}y^{\frac{1}{4}}\right)$

112. $\dfrac{4x^{\frac{8}{15}}z^{\frac{2}{3}}}{y^{\frac{1}{2}}}$ 113. $\dfrac{x^{\frac{1}{4}}}{2y^{\frac{5}{4}}}$ 114. $\dfrac{x^{\frac{8}{5}}y^2z^2}{256}$

115. $x^{\frac{c}{2}}y^{\frac{b}{3}}$ 116. $64^mx^{3m+2n}y^n$

117. 50.2304 118. 625 119. 3.5342

120. 1.5551 121. $-19\frac{1}{2} + 15i$

122. $5 + 62i$ 123. $7\frac{8}{9} + 11\frac{2}{9}i$

124. $-\frac{24}{29} - \frac{27}{29}i$ 125. $\frac{5}{17} + \frac{14}{17}i$

126. $12 - 6\sqrt{2}i$ 127. $-\dfrac{3}{7} + \dfrac{9\sqrt{3}}{7}i$

128. $-28 - 16\sqrt{2}i$ 129. $-\frac{1}{10} + \frac{3}{10}i$

130. $-16 + 2i$ 131. $\frac{57}{29} - \frac{41}{29}i$

132. $-2 - 10i$

Chapter 1 test

1. $\frac{1}{3}, \frac{1}{2}, \frac{3}{5}, \frac{2}{3}, \frac{5}{7}$ 2. 14 3. $\frac{5}{24}$

4. $\dfrac{a^2 - 5ab - 12b^2}{4ab}$ 5. $\dfrac{6a^2 + b^2}{5a^2}$

6. $(-2\frac{1}{2}, -\frac{3}{4}]$

7. $\{x \mid x \geq -3\}$

8. $\{x \mid -1\frac{1}{2} \leq x < 2\}$, $[-1\frac{1}{2}, 2)$ 9. −4

10. $\pi - 2$ 11. $3x^2|y|$ 12. $6x^9y^3$

13. $-16x^4y^4$ 14. $-3x^3y^5$ 15. $\dfrac{36}{a^{20}}$

16. $\dfrac{1}{x^n}$ 17. 2.05×10^{11} 18. 0.000213

19. 3 20. $-\frac{7}{3}$

21. $-2x^6 - x^4 + 6x^3 + 4$

22. $a^4 - 50a^2 + 625$

23. $2x^2 + 3x + 10 + \dfrac{15}{x - 2}$

24. $4a(a - 2)(a + 2)$

25. $(3x - 2)(3x + 1)$

26. $(x - 2)(x + 2)(x^2 + 4)$

27. $[(2x - 1)(4x^2 + 2x + 1)]$
$[(2x + 1)(4x^2 - 2x + 1)]$

28. $(x + 1)(x - 3)$

29. $(3c + d)(a - 2b)$ 30. $\dfrac{x}{x + 1}$

31. $\dfrac{x + 1}{2x + 1}$ 32. $\dfrac{2x - 3}{x - 2}$

33. $\dfrac{x^2 + x + 1}{3x(2x + 1)}$ 34. $\dfrac{x^2 - 3x + 2}{x^2 + 3x + 2}$

35. $\dfrac{2b - 9a}{2 + 6ab}$ 36. $\dfrac{a - b - 1}{a - b + 1}$ 37. $4\sqrt[3]{2}$

38. $5x^2y\sqrt{2yz}$ 39. $4b^3\sqrt[3]{4a^2b}$

40. $6b^2c\sqrt[4]{a^3b}$ 41. $\sqrt{6x}$ 42. $\dfrac{2x\sqrt[3]{xyz^2}}{yz}$

43. $2a\sqrt{5ab}$

44. $-2xy\sqrt{x} + 4x\sqrt{xy} - 2xy\sqrt{3}$

45. $\dfrac{a\sqrt{6} - \sqrt{3a} - \sqrt{6a} + \sqrt{3}}{2a - 1}$

46. $\dfrac{\sqrt{3} + 3\sqrt{5}}{6}$ 47. $2x\sqrt[3]{2}$ 48. $\dfrac{\sqrt[4]{x}}{2x^3y}$

49. $2\sqrt{5}|x^3|y^4$ 50. $5x^2|x - 3|$

51. $\dfrac{x}{y^{\frac{1}{4}}}$ 52. $\dfrac{81xz^{\frac{4}{3}}}{y^{\frac{1}{2}}}$ 53. $\dfrac{1}{4x^{\frac{1}{6}}y}$

54. $\dfrac{256}{x^4y^2z^2}$ **55.** $a^{\frac{1}{2}}b^{\frac{1}{m}}$ **56.** 0.0494

57. $-226 - 481i$ **58.** $-\frac{24}{29} - \frac{27}{29}i$

59. $2\sqrt{3} - 3i$ **60.** $\dfrac{10}{7} - \dfrac{2\sqrt{3}}{7}i$

61. $\frac{2}{3} + \frac{4}{3}i$

Chapter 2

Exercise 2–1

Answers to odd-numbered problems

1. $\{\frac{5}{16}\}$ **3.** $\{-8\}$ **5.** $\{88\}$ **7.** $\{-\frac{6}{5}\}$

9. $\{\frac{5}{7}\}$ **11.** $\{\frac{2}{5}\}$ **13.** R **15.** $\{-\frac{7}{3}\}$

17. $\{-\frac{59}{16}\}$ **19.** $\{0\}$ **21.** $\{-\frac{1}{9}\}$

23. $\{4\}$ **25.** R **27.** 1.2356

29. 0.3571 **31.** $\dfrac{V - k}{g} = t$

33. $\dfrac{2S + gt^2}{2t} = V$

35. $\dfrac{2S - dn^2 + nd}{2n} = a$

37. $\dfrac{d - d_1 - jd_3}{k - 1 - d_3} = d_2$ **39.** $y = \dfrac{3x}{10}$

41. $x = -\frac{7}{3}y$ **43.** $\dfrac{V + br^2}{r^2} = a$

45. $8y = x$ **47.** $\dfrac{3V + \pi h^3}{3\pi h^2} = R$

49. $\dfrac{by - 4b - 3a}{a} = x$ **51.** $R_0 = \dfrac{RT}{a + T}$

53. $\dfrac{p - s}{s(p - 1)} = f$

55. \$10,000 at 8% and \$5,000 at 6%

57. \$10,000 at 14% gain, \$8,000 at 9% loss

59. \$5,000 at 5% and \$6,000 at 8%

61. \$12,285.71 at 5% and \$5,714.29 at 9%

63. 40 gallons of 35% and 40 gallons of 65% **65.** 1,500 gallons of 4% solution

67. 128.6 liters of 20% and 171.4 liters of 55% **69.** **a.** 1 hr 17 min **b.** 6 hr 3 min **71.** 11 hours 31 minutes **73.** 45 mph for the truck and 60 mph for the car

75. $56\frac{1}{4}$ mph **77.** $\frac{7}{12}$ mile

79. assume $\dfrac{a}{b} = \dfrac{c}{d}$; multiply each member by bd

$$bd\left(\frac{a}{b}\right) = bd\left(\frac{c}{d}\right)$$
$$d(a) = b(c)$$
$$ad = bc$$

Solutions to skill and review problems

1. $(x - 2)(x + 5) = 0$
$x - 2 = 0$ or $x + 5 = 0$
$x = 2$ or $x = -5$

2. $(2x - 3)(x + 2)$
$2x^2 + 4x - 3x - 6$
$2x^2 + x - 6$

3. $4x^2 - 16x$
$4x(x - 4)$

4. $4x^2 - 1$
$(2x - 1)(2x + 1)$

5. $6x^2 - 5x - 4$
$(3x - 4)(2x + 1)$

6. $\sqrt{-20}$
$\sqrt{4 \cdot 5}i$
$2\sqrt{5}i$

7. $\sqrt{8 - 4(3)(-2)}$
$\sqrt{8 + 24}$
$\sqrt{32} = \sqrt{2^5} = 2^2\sqrt{2} = 4\sqrt{2}$

8. $\dfrac{8 - \sqrt{32}}{4}$

$\dfrac{8 - 4\sqrt{2}}{4}$

$\dfrac{4(2 - \sqrt{2})}{4}$

$2 - \sqrt{2}$

9. Area = length × width
= 8(6) = 48 in.^{2}
Perimeter = 2(length) + 2(width)
= 2(8) + 2(6)
= 28 inches

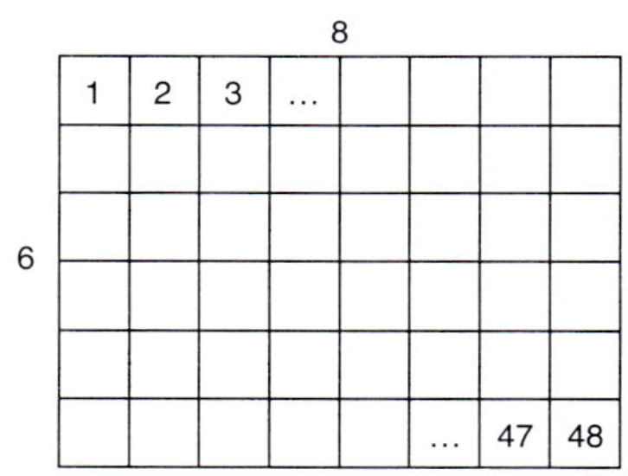

10. distance = rate × time

time $= \dfrac{\text{distance}}{\text{rate}} = \dfrac{135}{45} = 3$ hours

Solutions to trial exercise problems

5. $\frac{1}{4}x + 3 = \frac{3}{8}x - 8$
$8(\frac{1}{4}x + 3) = 8(\frac{3}{8}x - 8)$
$2x + 24 = 3x - 64$
$88 = x$
$\{88\}$

11. $\dfrac{2 - 3x}{4} = \dfrac{x}{2}$
$2(2 - 3x) = 4x$
$4 - 6x = 4x$
$4 = 10x$
$\frac{4}{10} = x$
$\{\frac{2}{5}\}$

29. $150x - 13.8 = 0.04(1{,}500 - 1{,}417x)$
$150x - 13.8 = 60 - 56.68x$
$206.68x = 73.8$
$x = \dfrac{73.8}{206.68} \approx 0.3571$

35. $S = \dfrac{n}{2}[2a + (n - 1)d]$; for a
$2S = n[2a + (n - 1)d]$
$2S = 2an + dn^2 - nd$
$2S - dn^2 + nd = 2an$
$\dfrac{2S - dn^2 + nd}{2n} = a$

39. $\dfrac{x + 2y}{x - 2y} = 4$; for y
$x + 2y = 4x - 8y$
$10y = 3x$
$y = \dfrac{3x}{10}$

51. $T = \dfrac{aR_0}{R - R_0}$; for R_0
$T(R - R_0) = aR_0$
$TR - TR_0 = aR_0$
$TR = aR_0 + TR_0$
$TR = R_0(a + T)$
$\dfrac{TR}{a + T} = R_0$

59. Let x be the smaller amount, which was invested at 5%. Then the larger amount was x + 1,000, and was invested at 8%. The return from the larger investment minus the return from the smaller investment is \$230.
$0.08(x + 1{,}000) - 0.05x = 230$
$0.08x + 80 - 0.05x = 230$
$0.03x = 150$
$x = \dfrac{150}{0.03} = 5{,}000$; thus, \$5,000 was invested at 5% and \$6,000 at 8%.

62. x = additional amount invested at 8%
\$6,000 at 5% earns \$300
x at 8% earns $0.08x$
The total amount earned is 300 + $0.08x$
The total investment is x + 6,000
We want $0.06(x + 6{,}000) = 300 + 0.08x$
$0.06x + 360 = 300 + 0.08x$
$60 = 0.02x$
$\dfrac{60}{0.02} = x$
$3{,}000 = x$
Thus we want to invest \$3,000 at 8%.

65. Let x = amount of 4% pesticide solution. The total amount of solution will be $3,000 + x$. The total amount of pesticide will be 10% of 3,000 (which is 300) and 4% of x, $(0.04x)$, or $300 + 0.04x$. We want this amount of to be 8% of $3,000 + x$:

$0.08(3,000 + x) = 300 + 0.04x$
$240 + 0.08x = 300 + 0.04x$
$0.04x = 60$
$x = \frac{60.0}{0.04} = 1,500$

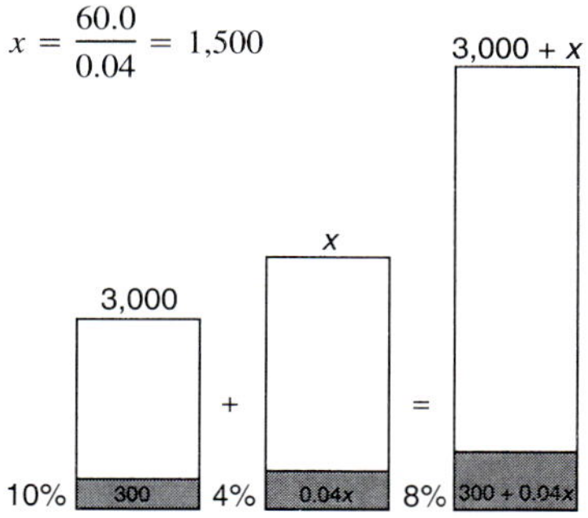

Thus, 1,500 gallons of the 4% solution should be added to the 3,000 gallons of 10% solution.

68. rate × time = work, so rate $= \frac{\text{work}}{\text{time}}$; first rate is $\frac{5,000}{35} = \frac{1,000}{7}$ flyers/minute; second rate is $\frac{5,000}{50} = 100$ flyers/minute. Combined rate is $\frac{1,000}{7} + 100$.

rate × time = work

a. $\left(\frac{1,000}{7} + 100\right)t = 5,000$
$\frac{1,000}{7}t + 100t = 5,000$
$1,000t + 700t = 35,000$;
$t = \frac{35,000}{1,700} = 20.5882$ minutes
= 20 minutes 35 seconds

b. $\left(\frac{1,000}{7} + 100\right)t = 8,000$
$\frac{1,000}{7}t + 100t = 8,000$
$1,000t + 700t = 56,000$
$t = \frac{56,000}{1,700} = 32$ min 56 sec

74. x = speed of current; upstream the boat's rate is $16 - x$, and downstream it is $16 + x$; times are equal, and $t = \frac{d}{r}$, so

$\frac{20}{16 + x} = \frac{14}{16 - x}$
$20(16 - x) = 14(16 + x)$
$320 - 20x = 224 + 14x$
$96 = 34x$
$x = 2\frac{14}{17}$ mph for the speed of the current

Exercise 2–2

Answers to odd-numbered problems

1. $\{-1, 8\}$ **3.** $\{0, 5\}$ **5.** $\{-1, 3\}$
7. $\{-2, \frac{1}{3}\}$ **9.** $\{-8, 1\}$ **11.** $\{-3, 1\}$
13. $\left\{-\frac{3y}{5}, 2y\right\}$ **15.** $\left\{\frac{3a}{2}, -\frac{5a}{6}\right\}$
17. $\{\pm 3\}$ **19.** $\{\pm 2\sqrt{10}\}$
21. $\left\{\pm\frac{2\sqrt{10}}{3}\right\}$ **23.** $\left\{\pm\frac{4\sqrt{10}}{5}\right\}$
25. $\{3 \pm \sqrt{10}\}$ **27.** $\left\{-1 \pm \frac{2\sqrt{6}}{3}\right\}$
29. $\left\{-\frac{c}{b} \pm \frac{\sqrt{ad}}{ab}\right\}$ **31.** $\frac{5 \pm \sqrt{97}}{6}$
33. $\frac{-2 \pm \sqrt{94}}{6}$ **35.** $3 \pm \sqrt{14}$
37. $5\left(x - \frac{4 + 2\sqrt{19}}{5}\right)\left(x - \frac{4 - 2\sqrt{19}}{5}\right)$
39. $2\left(x - \frac{-3 + \sqrt{17}}{2}\right)\left(x - \frac{-3 - \sqrt{17}}{2}\right)$
41. 16 **43.** $w = \frac{\sqrt{3,999} + 3}{2} \approx 33.1$ ft, length $= \frac{3\sqrt{3,999} - 1}{2} \approx 94.4$ ft
45. length = 52 m; w = 23 m **47.** \$3.33
49. $\frac{\sqrt{265} + 13}{2} \approx 14.6$ hours and $\frac{\sqrt{265} + 19}{2} \approx 17.6$ hours
51. $\frac{\sqrt{321} - 1}{4} \approx 4.3$ hours
53. $\{x \mid x \neq 1\frac{1}{2}\}$ **55.** $\{z \mid z \neq -2\}$
57. $\{m \mid m \neq 0, 2\}$ **59.** $\{x \mid x \neq 7, -3\}$
61. R **63.** $\{\pm 2, \pm\sqrt{7}i\}$ **65.** $\{5 \pm \sqrt{13}\}$
67. $\left\{\frac{73 + 3\sqrt{137}}{32}\right\}$ **69.** $\{1, 729\}$
71. $\{\frac{1}{2}, 2\}$
73. $a\left(\frac{-b + \sqrt{b^2 - 4ac}}{2a}\right)^2 + b\left(\frac{-b + \sqrt{b^2 - 4ac}}{2a}\right) + c$
$= a\left(\frac{b^2 - 2b\sqrt{b^2 - 4ac} + b^2 - 4ac}{4a^2}\right) + \frac{-b^2 + b\sqrt{b^2 - 4ac}}{2a} + c$
$= \frac{b^2 - 2b\sqrt{b^2 - 4ac} + b^2 - 4ac}{4a} + \frac{-2b^2 + 2b\sqrt{b^2 - 4ac}}{4a} + \frac{4ac}{4a} = 0$
The case for $\frac{-b - \sqrt{b^2 - 4ac}}{2a}$ is almost the same.
75. $2 - 3i$ or $-2 + 3i$
77. $a = \sqrt{\frac{c + \sqrt{c^2 + d^2}}{2}}$, $b = \frac{d}{\sqrt{2}\left(\sqrt{c + \sqrt{c^2 + d^2}}\right)}$ when c and d are not both 0. When c and d are zero, let $b = 0$.

Solutions to skill and review problems

1. $(3\sqrt{2x})^2 = 3^2(\sqrt{2x})^2 = 9(2x) = 18x$
2. $(3 + \sqrt{2x})(3 + \sqrt{2x}) = 9 + 3\sqrt{2x} + 3\sqrt{2x} + 2x = 9 + 6\sqrt{2x} + 2x$
3. $\sqrt{x} = 4$
$(\sqrt{x})^2 = 4^2$
$x = 16$
4. $\sqrt[3]{x} = 4$
$(\sqrt[3]{x})^3 = 4^3$
$x = 64$
5. $\sqrt{\frac{2}{3} + 2} = \sqrt{\frac{8}{3}} = \frac{\sqrt{8}}{\sqrt{3}} = \frac{2\sqrt{2}}{\sqrt{3}} \cdot \frac{\sqrt{3}}{\sqrt{3}} = \frac{2\sqrt{6}}{3}$; d

Solutions to trial exercise problems

9. $\frac{x}{2} + \frac{7}{2} = \frac{4}{x}$
$2x\left(\frac{x}{2}\right) + 2x\left(\frac{7}{2}\right) = 2x\left(\frac{4}{x}\right)$
$x^2 + 7x = 8$
$x^2 + 7x - 8 = 0$
$(x + 8)(x - 1) = 0$
$\{-8, 1\}$

10. $(p + 4)(p - 6) = -16$
$p^2 - 2p - 24 = -16$
$p^2 - 2p - 8 = 0$
$(p - 4)(p + 2) = 0$
$\{-2, 4\}$

12. $x^2 - 4ax + 3a^2 = 0$
$(x - 3a)(x - a) = 0$
$x - 3a = 0$ or $x - a = 0$
$x = 3a$ or $x = a$
$\{a, 3a\}$

27. $3(x + 1)^2 = 8$
$(x + 1)^2 = \frac{8}{3}$
$x + 1 = \pm\sqrt{\frac{8}{3}}$
$x = -1 \pm \frac{2\sqrt{6}}{3}$
$\left\{-1 \pm \frac{2\sqrt{6}}{3}\right\}$

36. $\frac{1}{x + 2} - x = 5$
$1 - x(x + 2) = 5(x + 2)$
$0 = x^2 + 7x + 9$
$a = 1, b = 7, c = 9;$
$x = \frac{-7 \pm \sqrt{7^2 - 4(1)(9)}}{2(1)}$
$x = \frac{-7 \pm \sqrt{13}}{2}$

39. $2x^2 + 6x - 4$
$2(x^2 + 3x - 2)$
$2\left(x - \frac{-3 + \sqrt{17}}{2}\right)\left(x - \frac{-3 - \sqrt{17}}{2}\right)$

43. w = width, so length $= 3w - 5$;
$100^2 = w^2 + (3w - 5)^2$
$w = \frac{\sqrt{3{,}999} + 3}{2} \approx 33.1$ ft,
length $= 3w - 5$
$= \frac{3\sqrt{3{,}999} - 1}{2} \approx 94.4$ ft

49. x = time for one press, so $x + 3$ is the time for the other press; rate × time = work, so rate $= \frac{\text{work}}{\text{time}}$. One rate is $\frac{10{,}000}{x}$, second is $\frac{10{,}000}{x + 3}$; combined rate is $\frac{10{,}000}{x} + \frac{10{,}000}{x + 3}$, so using rate × time = work we obtain
$\left(\frac{10{,}000}{x} + \frac{10{,}000}{x + 3}\right)8 = 10{,}000$
Multiply each member by $\frac{1}{10{,}000}$
$\left(\frac{1}{x} + \frac{1}{x + 3}\right)8 = 1$
$\frac{(x + 3) + x}{x(x + 3)} \cdot 8 = 1$
$\frac{8(2x + 3)}{x^2 + 3x} = 1$
$16x + 24 = x^2 + 3x$
$0 = x^2 - 13x - 24$
$x = \frac{\sqrt{265} + 13}{2}$
So the rates are $\frac{\sqrt{265} + 13}{2} \approx 14.6$ hours and $\frac{\sqrt{265} + 19}{2} \approx 17.6$ hours.

51. x = time in no wind condition; $x + \frac{1}{2}$ = time into wind. Rate $= \frac{\text{distance}}{\text{time}}$; so rate in no wind $= \frac{600}{x}$, rate in wind is $\frac{600}{x + \frac{1}{2}} = \frac{1{,}200}{2x + 1}$, and the difference of the rates is 15 mph, so
$\frac{600}{x} - \frac{1{,}200}{2x + 1} = 15$
$\frac{600(2x + 1) - 1{,}200x}{x(2x + 1)} = 15$
$600 = 15(2x^2 + x)$
$40 = 2x^2 + x$
$0 = 2x^2 + x - 40, x = \frac{\sqrt{321} - 1}{4}$
≈ 4.3 hours

59. $\frac{2 - 3x}{x^2 - 4x - 21}$
$x^2 - 4x - 21 = 0$
$(x - 7)(x + 3) = 0$
$x - 7 = 0$ or $x + 3 = 0$
$x = 7$ or $x = -3$
$\{x \mid x \neq 7, -3\}$

71. $4y^{-4} + 4 = 17y^{-2}$
$u = y^{-2}$, so $u^2 = y^{-4}$
$4u^2 - 17u + 4 = 0$
$(4u - 1)(u - 4) = 0$
$u = y^{-2} = \frac{1}{4}$ or 4;
$(y^{-2})^{-1/2} = (\frac{1}{4})^{-1/2}$ or
$(y^{-2})^{-1/2} = 4^{-1/2}$
$y = 2$ or $y = \frac{1}{2}$
$\{\frac{1}{2}, 2\}$

75. $(a + bi)^2 = -5 - 12i;\ a^2 - b^2 + 2abi = -5 - 12i$, so $a^2 - b^2 = -5$ and $2ab = -12$, or $ab = -6$, or $b = -\frac{6}{a}$ thus,
$a^2 - \left(-\frac{6}{a}\right)^2 = -5$
$a^4 + 5a^2 - 36 = 0,\ a^2 = 4,\ a = \pm 2;$ choose $a = 2$; then $b = -\frac{6}{2} = -3$; thus one value for $a + bi$ is $2 - 3i$. If $a = -2$, b is 3, giving $-2 + 3i$.

77. $(a + bi)^2 = c + di$
$a^2 - b^2 + 2abi = c + di$
$a^2 - b^2 = c,\ 2ab = d;\ b = \frac{d}{2a}$
$a^2 - \left(\frac{d}{2a}\right)^2 = c$
$4a^4 - 4a^2c - d^2 = 0$; this is quadratic in a^2, so
$a^2 = \frac{-(-4c) \pm \sqrt{(-4c)^2 - 4(4)(-d^2)}}{2(4)}$
$a = \pm\sqrt{\frac{4c \pm \sqrt{16c^2 + 16d^2}}{8}}$
$= \pm\sqrt{\frac{4c \pm 4\sqrt{c^2 + d^2}}{8}}$
$= \pm\sqrt{\frac{c \pm \sqrt{c^2 + d^2}}{2}}$. Because a must be real, we require that $c \pm \sqrt{c^2 + d^2} \geq 0$; since $\sqrt{c^2 + d^2} \geq c$ we choose $c + \sqrt{c^2 + d^2}$, and choose $a \geq 0$, obtaining $a = \sqrt{\frac{c + \sqrt{c^2 + d^2}}{2}}$.
Using $b = \frac{d}{2a}$ it can be shown that
$b = \frac{d}{\sqrt{2}\left(\sqrt{c + \sqrt{c^2 + d^2}}\right)}$ when c and d are not both 0. When c and d are zero, let $b = 0$.

78. a.
$A = h\left(\frac{a + b}{2}\right)$
$= 36p\left(\frac{24p + 30p}{2}\right) = 36p(27p) = 972p^2$

b. $972p^2 = 972 + m$
$p^2 = \frac{972 + m}{972} = 1 + \frac{m}{972}$
$p = \sqrt{1 + \frac{m}{972}}$
If $m = 100$, then $p = \sqrt{1 + \frac{100}{972}}$
≈ 1.050
New dimensions: $a = 24p \approx 25.2$ units, $b = 30p \approx 31.5$ units, and $h = 36p \approx 37.8$ units.

Exercise 2–3

Answers to odd-numbered problems

1. $2x + 3$ **3.** $4x + 3$
5. $x - 10\sqrt{x} + 25$ **7.** $9x + 18$
9. $2x - 4\sqrt{2x} + 4$
11. $x + 3 - 4\sqrt{x - 1}$ **13.** $4x - 4$
15. $x + 12 + 6\sqrt{x + 3}$
17. $-x + 2 - 2\sqrt{1 - x}$ **19.** $\{32\}$

21. $\{54\}$ **23.** Φ **25.** $\{-4\}$
27. $\{-7\frac{2}{3}\}$ **29.** $\{-64\}$ **31.** $\{22\}$
33. $\{-3, 27\}$ **35.** $\{-2, 8\}$ **37.** $\{-2, 7\}$
39. $\{2\}$ **41.** $\{1\}$ **43.** $\{9\}$ **45.** $\{9\}$
47. $\{5\}$ **49.** $\{2, 3\}$ **51.** $\{3, 11\}$
53. $\{0, 8\}$ **55.** $\{2\}$ **57.** Φ
59. $x = \pm\sqrt{-2y^2 + 18y - 27}$
61. $\frac{\pi}{6}D^3 = A$ **63.** $s = t^2 + 7t + 9$
65. 300 feet $= S$ **67.** $\frac{4}{3}\pi r^3 = V$
69. $i2\sqrt{\frac{A}{P}} - 2$

$i + 2 = 2\sqrt{\frac{A}{P}}$

$(i + 2)^2 = 2^2\left(\sqrt{\frac{A}{P}}\right)^2$

$i^2 + 4i + 4 = 4\left(\frac{A}{P}\right)$

$P\left(1 + i + \frac{i^2}{4}\right) = A$

$P\left(1 + \frac{i}{2}\right)^2 = A$

Solutions to skill and review problems

1. $3[2(\frac{1}{2}) + 1] > \frac{1}{2} + 6$

$3[2] > 6\frac{1}{2}$

$6 > 6\frac{1}{2}$; no

2.
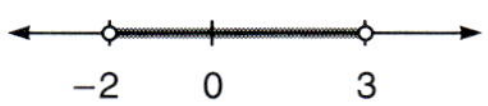

3. no since it is equivalent to $\frac{3}{6} < \frac{2}{6}$
4. yes since $-\frac{3}{6} < -\frac{2}{6}$ is true
5. $\frac{-2}{-2 - 3} > -2$

$\frac{2}{5} > -2$ is true

6. only f, or $-5 > -2$ is false

Solutions to trial exercise problems

12. $(x + \sqrt{x + 1})^2$
$(x + \sqrt{x + 1})(x + \sqrt{x + 1})$
$x^2 + x\sqrt{x + 1} + x\sqrt{x + 1} + (x + 1)$
$x^2 + x + 1 + 2x\sqrt{x + 1}$

33. $\sqrt[4]{x^2 - 24x} = 3$
$(\sqrt[4]{x^2 - 24x})^4 = (3)^4$
$x^2 - 24x = 81$
$x^2 - 24x - 81 = 0$
$x = -3$ or 27
$\{-3, 27\}$

43. $\sqrt{m}\sqrt{m - 8} = 3$
$(\sqrt{m}\sqrt{m - 8})^2 = 3^2$
$m(m - 8) = 9$
$m^2 - 8m - 9 = 0$
$m = -1$ or 9
-1 does not check.
$\{9\}$

51. $\sqrt{2n + 3} - \sqrt{n - 2} = 2$
$(\sqrt{2n + 3})^2 = (\sqrt{n - 2} + 2)^2$
$2n + 3 = (n - 2) + 4\sqrt{n - 2} + 4$
$(n + 1)^2 = (4\sqrt{n - 2})^2$
$n^2 + 2n + 1 = 16n - 32$
$n^2 - 14n + 33 = 0$
$n = 3$ or 11
$\{3, 11\}$

55. $(2y + 3)^{1/2} - (4y - 1)^{1/2} = 0$
$[(2y + 3)^{1/2}]^2 = [(4y - 1)^{1/2}]^2$
$2y + 3 = 4y - 1$
$2 = y$
$\{2\}$

63. $\sqrt{s - t} = t + 3$; for s
$(\sqrt{s - t})^2 = (t + 3)^2$
$s - t = t^2 + 6t + 9$
$s = t^2 + 7t + 9$

Exercise 2–4

Answers to odd-numbered problems

1.
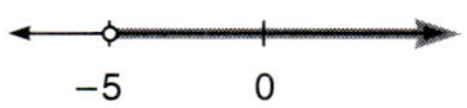

$\{x \mid x > -5\}$

3.
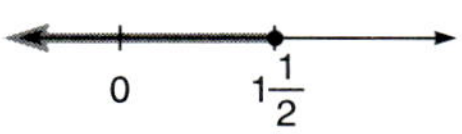

$\{x \mid x \le 1\frac{1}{2}\}$

5.
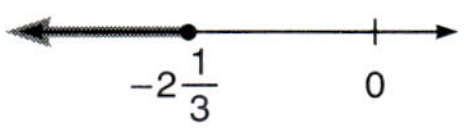

$\{x \mid x \le -2\frac{1}{3}\}$

7.
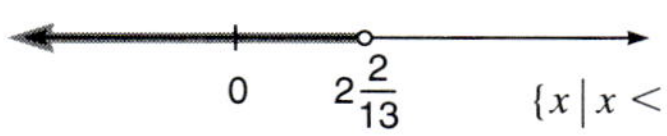

$\{x \mid x < 2\frac{2}{13}\}$

9.
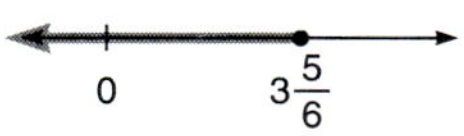

$\{x \mid x \le 3\frac{5}{6}\}$

11.

$\{x \mid x < 2\frac{7}{10}\}$

13.

$\{x \mid x \ge -3\frac{3}{5}\}$

15.

$\{x \mid x \le 0\}$

17.
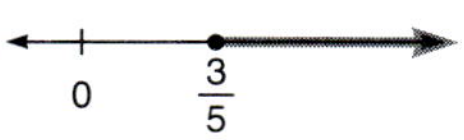

$\{x \mid x \ge \frac{3}{5}\}$

19.

$\{x \mid x \le \frac{3}{4}\}$

21.

$\{x \mid x < -\frac{3}{10}\}$

23.
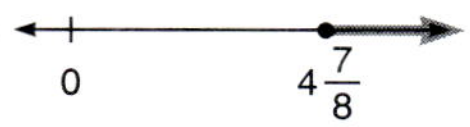

$\{x \mid x \ge 4\frac{7}{8}\}$

25.

$\{x \mid x < 4\frac{1}{5}\}$

27.

$\{x \mid -8 \le x \le 3\}$

29.
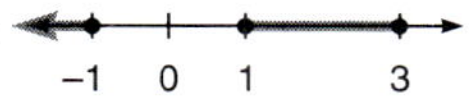

$\{x \mid x \le -1 \text{ or } 1 \le x \le 3\}$

31.
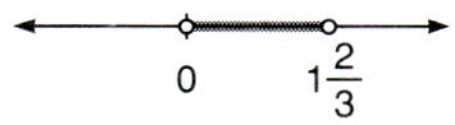

$\{q \mid 0 < q < 1\frac{2}{3}\}$

33.
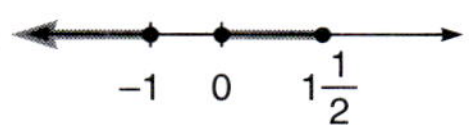

$\{r \mid r \le -1 \text{ or } 0 \le r \le 1\frac{1}{2}\}$

35.
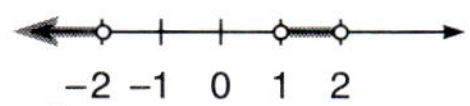

$\{x \mid x < -2 \text{ or } 1 < x < 2\}$

37.
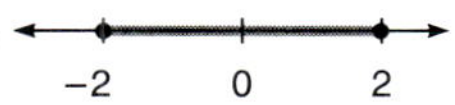

$\{x \mid -2 \le x \le 2\}$

39.
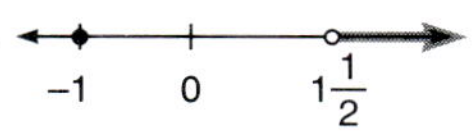

$\{x \mid x = -1 \text{ or } x > 1\frac{1}{2}\}$

41.

$\{x \mid x > 0\}$

43.
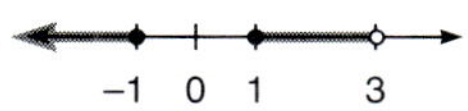

$\{x \mid x \le -1 \text{ or } 1 \le x < 3\}$

45.
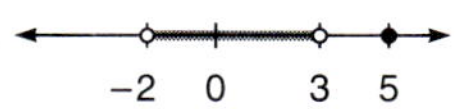

$\{x \mid -2 < x < 3 \text{ or } x = 5\}$

47.
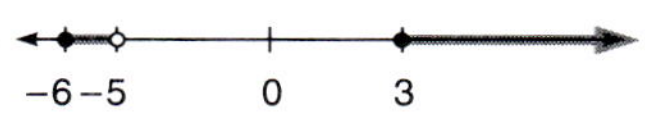

$\{x \mid -6 \le x < -5 \text{ or } x \ge 3\}$

49. 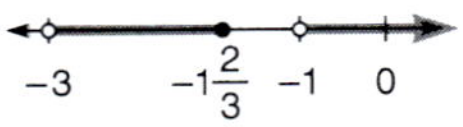

$\{x \mid -3 < x \le -1\frac{2}{3} \text{ or } x > -1\}$

51. $x \ge \frac{5}{2}$ **53.** $x \le \frac{9}{2}$ **55.** $x \le -1$ or $x \ge 6$ **57.** $x \le -\frac{1}{2}$ or $x \ge 1\frac{1}{2}$

59. a. $x \ge 79$ **b.** $80\frac{1}{4}$ = average

61. $0 < W < 20$

63. The side must be between 4 and 21 inches long. **65.** $0 < x \le 35 + 5\sqrt{65}$

67. $3 < t \le \dfrac{4 + \sqrt{10}}{2}$

69. a. conforms **b.** conforms **c.** conforms **d.** does not conform **e.** does not conform **f.** does not conform

71. between 3 hours 49 minutes and 10 hours

Solutions to skill and review problems

1. $\dfrac{2x - 3}{4} = 2$

$2x - 3 = 8$

$2x = 11$

$x = \frac{11}{2}$

2. $\dfrac{2x^2 - 4}{7} = x$

$2x^2 - 4 = 7x$

$2x^2 - 7x - 4 = 0$

$(2x + 1)(x - 4) = 0$

$2x + 1 = 0$ or $x - 4 = 0$

$2x = -1$ or $x = 4$

$x = -\frac{1}{2}$ or $x = 4$

3. If $|x| = 8$, then $x = 8$ or $x = -8$ (c).

4. If $|x| < 8$, then $-8 < x < 8$ (b). (Try some values for x.)

5. If $|x| > 8$, then $x < -8$ or $x > 8$ (c). (Try some values for x.)

6. $\left|\dfrac{1 - x}{2}\right| < x$

$\left|\dfrac{1 - 3}{2}\right| < 3$

$|-1| < 3$

$1 < 3$

True (Yes)

7. R.

Solutions to trial exercise problems

9. $-3(x - 2) + 2(x + 1) \ge 5(x - 3)$

$-3x + 6 + 2x + 2 \ge 5x - 15$

$-6x \ge -23$

$x \le \frac{23}{6}$

$\{x \mid x \le 3\frac{5}{6}\}$

29. $(x - 3)(x + 1)(x - 1) \le 0$

critical points: -1 (true), 1 (true), 3 (true)

test points: -2 (true), 0 (false), 2 (true), 4 (false)

$\{x \mid x \le -1 \text{ or } 1 \le x \le 3\}$

37. $(x^2 - 4x + 4)(x^2 - 4) \le 0$

$(x - 2)^3(x + 2) \le 0$

critical points: -2 (true), 2 (true)

test points: -3 (false), 0 (true), 3 (false)

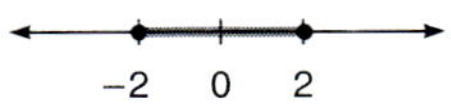

$\{x \mid -2 \le x \le 2\}$

43. $\dfrac{x^2 - 1}{x - 3} \le 0$

critical points: -1 (true), 1 (true), 3

test points: -2 (true), 0 (false), 2 (true), 4 (false)

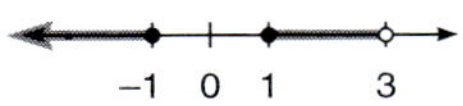

$\{x \mid x \le -1 \text{ or } 1 \le x < 3\}$

49. $\dfrac{x}{x + 1} - \dfrac{2}{x + 3} \le 1$

$\dfrac{x(x + 3) - 2(x + 1)}{(x + 1)(x + 3)}$

$-\dfrac{(x + 1)(x + 3)}{(x + 1)(x + 3)} \le 0$

$\dfrac{-3x - 5}{(x + 1)(x + 3)} \le 0$

critical points: -3, $-\frac{5}{3}$ (true), -1

test points: -4 (false), -2 (true), $-1\frac{1}{10}$ (false), 0 (true)

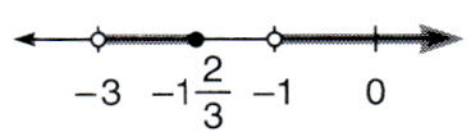

$\{x \mid -3 < x \le -1\frac{2}{3} \text{ or } x > -1\}$

61. $P = 2@ + 2w$ (perimeter is twice the length plus twice the width)

$P = 2(30) + 2w$

$P = 60 + 2w$

We want $60 + 2w < 100$

$2w < 40$

$w < 20$

We also want $w > 0$, so we require $0 < w < 20$

67. $\dfrac{1}{1{,}500} \cdot \dfrac{1{,}500}{t} + \dfrac{1}{1{,}500} \cdot \dfrac{1{,}500}{t - 3}$

$\ge \dfrac{1}{1{,}500} \cdot 3{,}000$

$\dfrac{1}{t} + \dfrac{1}{t - 3} \ge 2$

$\dfrac{2t - 3}{t(t - 3)} - 2 \ge 0$

$\dfrac{2t - 3}{t(t - 3)} - \dfrac{2t(t - 3)}{t(t - 3)} \ge 0$

$\dfrac{-2t^2 + 8t - 3}{t(t - 3)} \ge 0$

critical points are 0, 3, and $\dfrac{4 \pm \sqrt{10}}{2}$ ($\approx 3.6, 0.4$) (true); using test points of -1, 0.2, 1, 3.5, and 4 we find

$0 < t \le \dfrac{4 - \sqrt{10}}{2}$ or

$3 < t \le \dfrac{4 + \sqrt{10}}{2}$. If $t - 3 > 0$ then $t > 3$, so the solution is

$3 < t \le \dfrac{4 + \sqrt{10}}{2}$.

72. critical points: $3x - 5 = 6x$

$-5 = 3x$

$-1\frac{2}{3} = x$

test points: -2, -1

$x = -2$: $3(-2) - 5 \le 6(-2)$

$-11 \le -12$ False

$x = -1$: $3(-1) - 5 \le 6(-1)$

$-8 \le -6$ True

Solution: $x \ge -1\frac{2}{3}$

Exercise 2–5

Answers to odd-numbered problems

1. $\{\pm\frac{8}{5}\}$ **3.** $\{-5\frac{1}{2}, \frac{1}{2}\}$ **5.** $\{-1, 4\}$

7. $\{-4\}$ **9.** $\{\frac{1}{3}, 3\}$

11. $\left\{\dfrac{5 \pm \sqrt{97}}{4}, \dfrac{5 \pm \sqrt{47}i}{4}\right\}$

13. $\{x \mid x < -1\frac{1}{6} \text{ or } x > \frac{1}{6}\}$

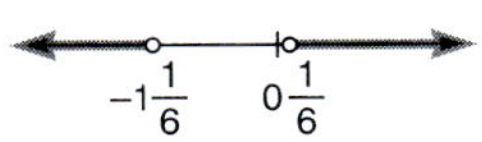

15. $\{x \mid -26 \le x \le 34\}$

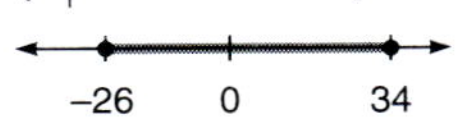

17. Φ

19. $\{x \mid x \le -\frac{23}{10} \text{ or } x \ge -\frac{7}{10}\}$

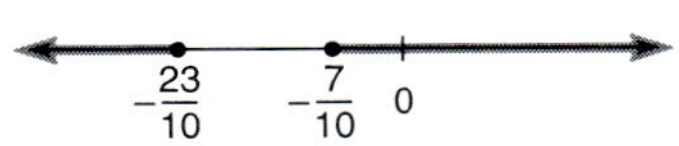

21. $\{x \mid x > \frac{9}{8} \text{ or } x < -\frac{9}{8}\}$

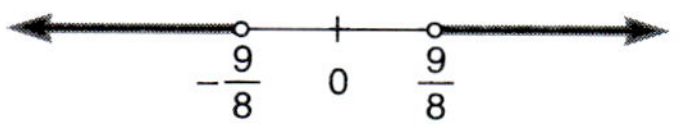

23. R

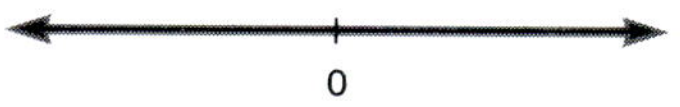

25. $\{x \mid x > 7\frac{1}{3} \text{ or } x < -7\frac{1}{3}\}$
27. $\{x \mid x < -10 \text{ or } x > 15\}$
29. $\{x \mid x < -34 \text{ or } x > 38\}$
31. $\{-7\frac{1}{3}, 7\frac{1}{3}\}$
33. $\{-10, 15\}$ **35.** $\{-34, 38\}$
37. $x \mid -7\frac{1}{3} < x < 7\frac{1}{3}\}$
39. $\{x \mid -10 < x < 15\}$
41. $\{x \mid -34 < x < 38\}$
43. $\{-1\frac{1}{4}, 1\frac{1}{4}\}$
45. $\{x \mid -\frac{1}{3} < x < 1\frac{1}{3}\}$
47. $\{x \mid x \le -2\frac{2}{3} \text{ or } x \ge 2\frac{2}{3}\}$
49. $\{x \mid -32\frac{1}{2} \le x \le 35\frac{1}{2}\}$
51. $\{x \mid x \le -2\frac{23}{36} \text{ or } x \ge -\frac{31}{36}\}$
53. $\{x \mid x < -2 \text{ or } x > 2\}$
55. $\{x \mid x \ge 4 \text{ or } x \le -8\}$
57. The dimension y is $5\frac{5}{8}$ and the dimension x is $7\frac{3}{8}$ inches.
59. $0 \le x \le 18$

27. $25 < |5 - 2x|$
$|5 - 2x| > 25$
$5 - 2x > 25 \text{ or } 5 - 2x < -25$
$-20 > 2x \text{ or } 30 < 2x$
$-10 > x \text{ or } 15 < x$
$\{x \mid x < -10 \text{ or } x > 15\}$

42. $8 > \left|\frac{3 - 2x}{5}\right|$
$8 > \frac{3 - 2x}{5} > -8$
$40 > 3 - 2x > -40$
$37 > -2x > -43$
$-\frac{37}{2} < x < \frac{43}{2}$
$\{x \mid -18\frac{1}{2} < x < 21\frac{1}{2}\}$

56. $\frac{3}{4} \le \left|\frac{3x + 1}{8}\right|$
$\left|\frac{3x + 1}{8}\right| \ge \frac{3}{4}$
$\frac{3x + 1}{8} \ge \frac{3}{4} \text{ or } \frac{3x + 1}{8} \le -\frac{3}{4}$
$3x + 1 \ge 6 \text{ or } 3x + 1 \le -6$
$3x \ge 5 \text{ or } 3x \le -7$
$x \ge \frac{5}{3} \text{ or } x \le -\frac{7}{3}$
$\{x \mid x \ge \frac{5}{3} \text{ or } x \le -\frac{7}{3}\}$

60. $(x - y)^2 \ge 0$
$x^2 - 2xy + y^2 \ge 0$
$x^2 + y^2 \ge 2xy$
$\frac{x^2 + y^2}{2} \ge xy$

Solutions to skill and review problems

	x	y	$2x - y = ?$	$= 5?$
1.	1	−3	$2(1) - (-3) = 5$	True
2.	−3	−11	$2(-3) - (-11) = 5$	True
3.	3	1	$2(3) - 1 = 5$	True
4.	0	−5	$2(0) - (-5) = 5$	True

5. If $x = 2$ and $y = -3$, which of the statements is true?
 a. $3(2) + (-3) = 3$
 b. $-2 + 5(-3) = -17$
 c. $-3 + 9 = 3(2)$
 d. $2 = -3 + 5$ All are true.

6. If $x = -2$ and $y = 4$, which of the statements is true?
 a. $3(-2) + 4 = -2$
 b. $-(-2) + 5(4) \ne 18$
 c. $4 + 10 \ne 3(-2)$
 d. $-2 \ne 4 + 6$ Only a is true.

7. Solve $2x + y = 8$ for y.
$y = -2x + 8$

8. Solve $x - 2y = 4$ for y.
$x - 4 = 2y$
$\frac{x - 4}{2} = y$

Solutions to trial exercise problems

9. $\left|\frac{3x - 5}{4}\right| = 1$
$\frac{3x - 5}{4} = 1 \text{ or } \frac{3x - 5}{4} = -1$
$3x - 5 = 4 \text{ or } 3x - 5 = -4$
$3x = 9 \text{ or } 3x = 1$
$x = 3 \text{ or } x = \frac{1}{3}$
$\{\frac{1}{3}, 3\}$

10. $|x^2 - 2x| = 3$
$x^2 - 2x = 3 \text{ or } x^2 - 2x = -3$
$x^2 - 2x - 3 = 0$ or
$x^2 - 2x + 3 = 0$
$x = -1 \text{ or } 3 \text{ or } x = 1 \pm \sqrt{2}i$
$\{-1, 3, 1 \pm \sqrt{2}i\}$

15. $\left|\frac{x - 4}{3}\right| \le 10$
$-10 \le \frac{x - 4}{3} \le 10$
$-30 \le x - 4 \le 30$
$-26 \le x \le 34$
$\{x \mid -26 \le x \le 34\}$

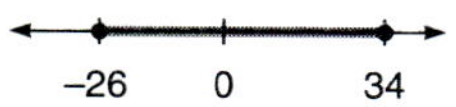

21. $\left|3x - \frac{x}{3}\right| > 3$
$\left|\frac{9x}{3} - \frac{x}{3}\right| > 3$
$\left|\frac{8x}{3}\right| > 3$
$\frac{8x}{3} > 3 \text{ or } \frac{8x}{3} < -3$
$8x > 9 \text{ or } 8x < -9$
$x > \frac{9}{8} \text{ or } x < -\frac{9}{8}$
$\{x \mid x > \frac{9}{8} \text{ or } x < -\frac{9}{8}\}$

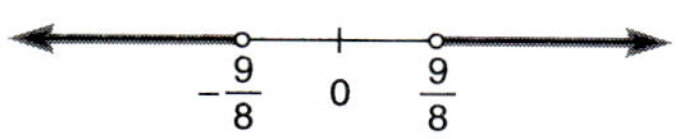

Chapter 2 review

1. $\{4\frac{4}{9}\}$ **2.** $\{-22\frac{1}{3}\}$ **3.** $\{2\frac{8}{11}\}$ **4.** R
5. $\{0\}$ **6.** 0.6595 **7.** −9.9619
8. $Q = \frac{px - m}{p}$ **9.** $b = \frac{W - 2kc - R}{k}$
10. $P_1 = \frac{nP_2 - 5c - 5P}{n}$ **11.** $x = \frac{y}{1 - y}$
12. $x = \frac{y}{y - 1}$ **13.** \$3,550 at 7% and \$4,450 at 5%
14. \$3,545.45 at 12%, \$11,454.55 at 10%
15. \$2,900 at 5% and \$2,100 at 9%
16. 1,333.3 pounds of the 10% mixture, 666.7 pounds of the 25% mixture
17. 7.5 tons of 55% copper
18. $23\frac{1}{3}$ minutes **19.** 3 mph
20. $21\frac{2}{3}$ mph **21.** $-\frac{5}{2}$ or 6
22. $-\frac{1}{2}$ or 3 **23.** $\frac{5}{2}$ or $-\frac{3}{2}$
24. $-\frac{5b}{3a}$ or $\frac{b}{2a}$ **25.** $\pm\frac{2\sqrt{30}}{9}$
26. $\frac{3}{2} \pm \sqrt{3}$ **27.** $-1 \pm \sqrt{2}$

28. $-\frac{c}{b} \pm \frac{\sqrt{ad}}{ab}$ **29.** $\frac{15 \pm \sqrt{345}}{10}$

30. $\frac{1 \pm \sqrt{17}i}{6}$ **31.** -2 or 4

32. $\frac{-37 \pm \sqrt{193}}{12}$

33. $3\left(x - \frac{4 + 2\sqrt{13}}{3}\right)\left(x - \frac{4 - 2\sqrt{13}}{3}\right)$

34. $5\left(x - \frac{-3 + \sqrt{29}}{5}\right)\left(x - \frac{-3 - \sqrt{29}}{5}\right)$

35. 24 **36.** 12 and 7.5 **37.** 10 units
38. 74 hours **39.** 3 hours **40.** $x \neq 3$
41. $x \neq -2$ and $x \neq -6$
42. $x = \pm 3$ or $x = \pm\frac{3}{2}\sqrt{2}$ **43.** 1 or 13

44. $\frac{9}{4}$ **45.** 1 or 16 **46.** $\pm\frac{\sqrt{2}}{4}$ or $\pm\frac{1}{2}i$

47. $\frac{1}{6}$ or -1 **48.** Φ (the empty set)

49. $4\frac{4}{5}$ **50.** 8 **51.** $\frac{19}{2}$

52. $k + 1$ **53.** $\frac{\pi r^2}{1 - \pi R^2} = A$

54. $x > \frac{12}{17}$

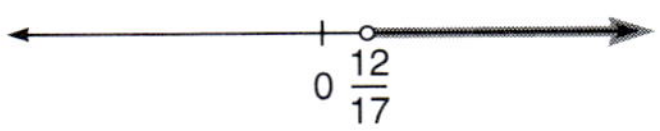

55. $x \leq 3$

56. $x \geq -30$

57. $x < \frac{27}{14}$

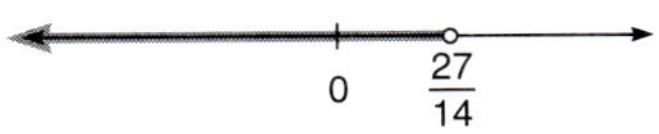

58. $-2 \leq r \leq 0$ or $r \geq 3$

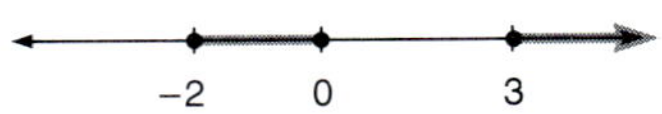

59. $-\frac{3}{4} < w < \frac{4}{3}$

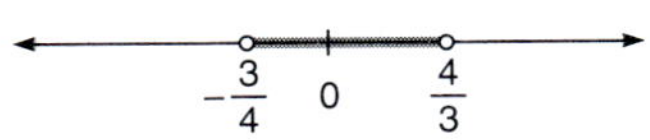

60. $-4 \leq x \leq -2\frac{1}{2}$ or $2\frac{1}{2} \leq x \leq 4$

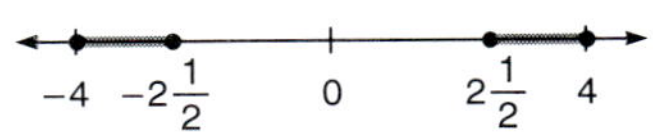

61. $-1 \leq x \leq 1$ or $x = 3$

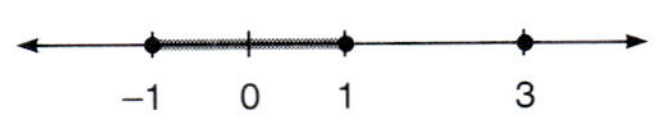

62. $x \neq 4$

63. $x \leq 1$ or $x > 3$

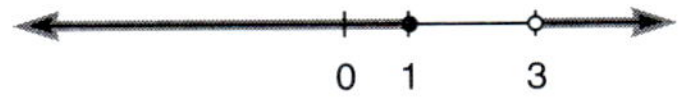

64. $3 < x \leq 5$

65. $x < -3$ or $-2 < x < 3$

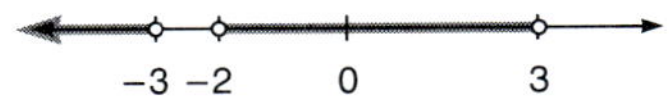

66. $x < -5$ or $-\frac{5}{2} - \frac{1}{2}\sqrt{7} \leq x \leq -\frac{5}{2} + \frac{1}{2}\sqrt{7}$

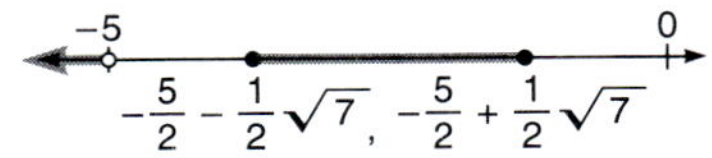

67. $-3 < x < 1$

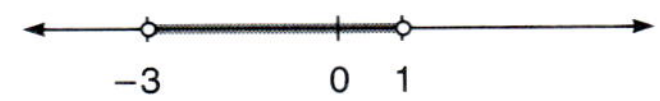

68. $x \leq 3\frac{7}{9}$ or $4 < x < 6$

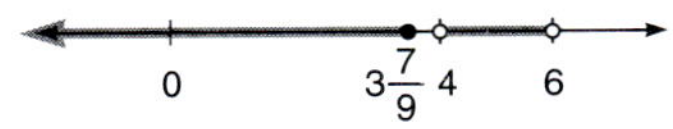

69. $-\frac{3}{16}$ or $\frac{9}{16}$
70. $-5, 10, \frac{5}{2}(1 \pm \sqrt{7}i)$
71. $0, \pm\sqrt{2}i$ **72.** $-16 \leq x \leq 24$
73. $-1, 2, \frac{1}{2}(1 \pm \sqrt{7}i)$
74. $-24\frac{1}{2} < x < 25\frac{1}{2}$
75. $x > 38$ or $x < -34$ **76.** $-34, 38$
77. $-4 < x < 1$ **78.** $x \geq -\frac{13}{3}$ or $x \leq -\frac{29}{3}$ **79.** $x > \frac{14}{15}$ or $x < -\frac{2}{5}$

Chapter 2 test

1. $\{\frac{16}{7}\}$ **2.** $\{\frac{7}{8}\}$ **3.** $\{1\}$ **4.** -2.4407

5. $x = \frac{m + pQ}{p}$ **6.** $P_2 = \frac{5P + nP_1 + 5c}{n}$

7. $y = \frac{x}{x + 1}$ **8.** \$3,000 at 9% and \$9,000 at 5% **9.** 18.67 tons
10. 14.3 minutes **11.** $1\frac{3}{7}$ miles per hour
12. -3 or $\frac{5}{3}$ **13.** $-\frac{3}{2}$ or $\frac{1}{5}$

14. $\pm\frac{5\sqrt{2}}{2}$ **15.** $1 \pm \frac{2}{3}\sqrt{6}$

16. $\frac{3 \pm \sqrt{33}}{2}$ **17.** $-1 \pm \frac{\sqrt{3}}{3}i$

18. $2\left[x - \left(\frac{1}{4} + \frac{\sqrt{97}}{4}\right)\right]\left[x - \left(\frac{1}{4} - \frac{\sqrt{97}}{4}\right)\right]$

19. 24 units **20.** 2 units
21. 23.5 hours **22.** 100 mph

23. $x \neq \pm 9$ **24.** $\pm\frac{1}{2}, \pm 3$ **25.** 4

26. $\frac{1}{2}, -1$ **27.** 4 **28.** $b = \frac{1}{\pi} - \frac{r^2}{A}$

29. $x > 3$

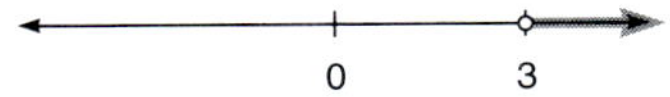

30. $x \geq -\frac{4}{3}$

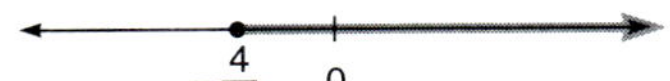

31. $-2 < x < 2$ or $x > 3$

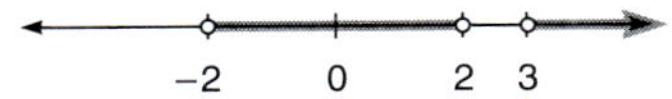

32. $-1 \leq x \leq 1$ or $x = 2$

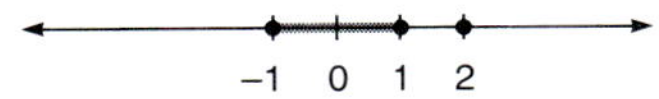

33. $-4 < x < 3$

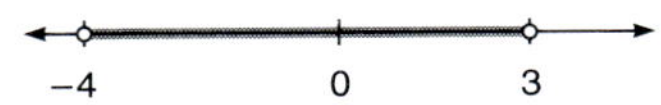

34. $-3 \leq x < -2$ or $x > 4$

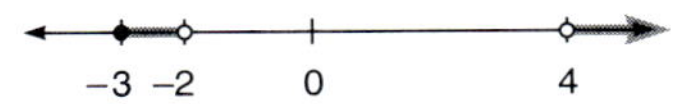

35. $-\frac{3}{2}, \frac{13}{2}$ **36.** $-\frac{8}{3} \leq x \leq 4$
37. $x < -2$ or $x > \frac{10}{3}$
38. $x \geq \frac{7}{5}$ or $x \leq -1$ **39.** $0 < x < 3$
40. $-1 \pm \sqrt{11}, -1 \pm 3i$ **41.** $\frac{1}{3}$

42. $-\frac{3}{2} \pm \frac{\sqrt{7}}{2}i$ **43.** $\frac{5}{2} \pm \frac{\sqrt{41}}{2}$

44. $\pm\sqrt{2}$ **45.** $\frac{17}{3}$ **46.** $\frac{17}{16} \pm \frac{\sqrt{449}}{16}$

47. -11 **48.** $-\frac{2}{3} \pm \frac{\sqrt{11}}{3}i$

Chapter 3

Exercise 3–1

Answers to odd-numbered problems

Answers to problems 1–15 will vary.
1. $(0,-8), (1,-5), (2,-2)$ **3.** $(-4,-6), (0,-3), (4,0)$ **5.** $(0,-2), (1,-1), (2,0)$
7. $(0,-2), (1,1), (2,4)$ **9.** $(1,1), (2,2), (3,3)$

11. $(-1,7)$, $(0,7)$, $(1,7)$
13. $(-2,-6)$, $(0,-3)$, $(2,0)$
15. $(0,-\frac{10}{3})$, $(1,-\frac{8}{3})$, $(2,-2)$

17.

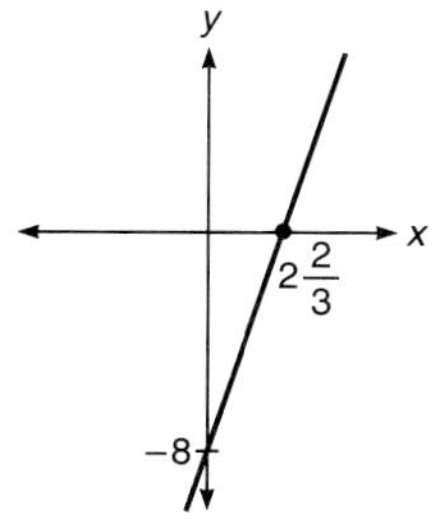

19.

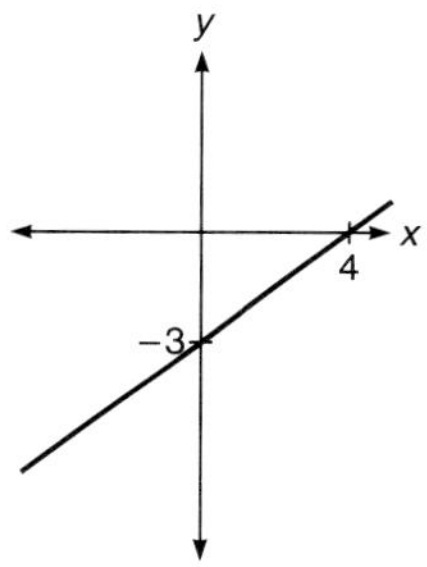

21.

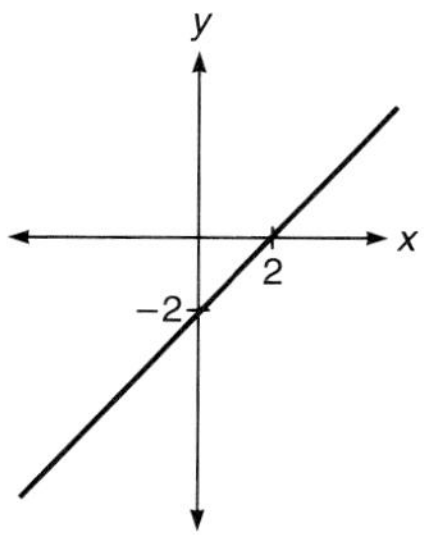

23.

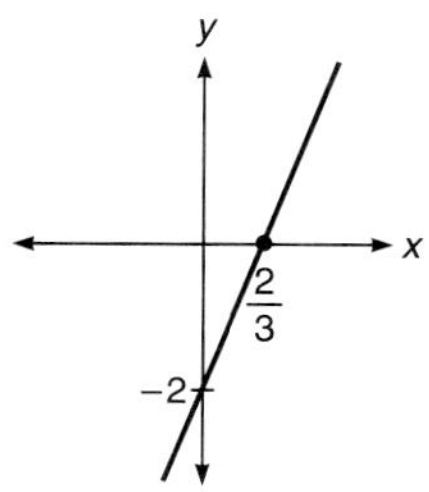

25.

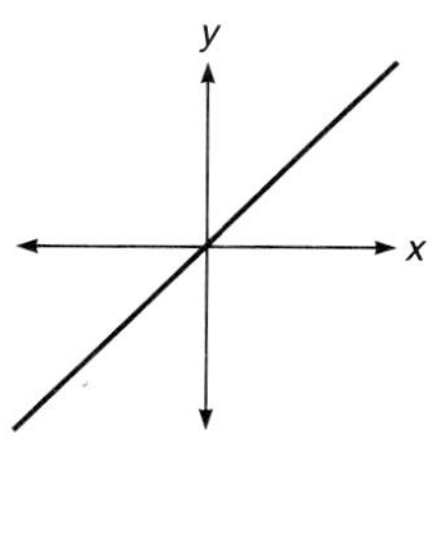

27.

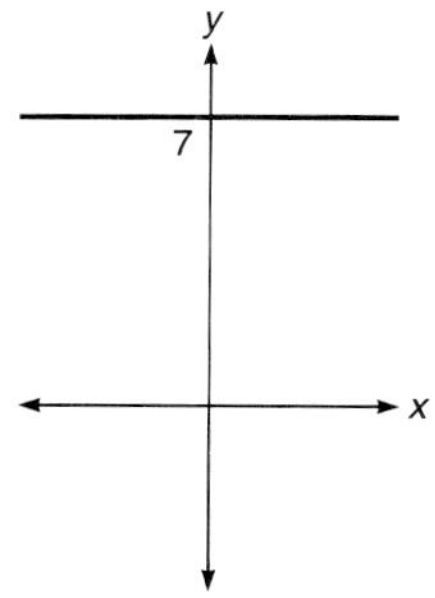

29.

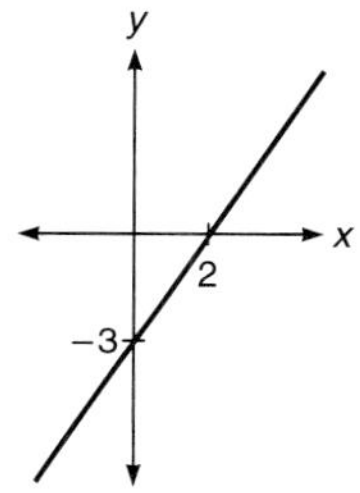

31.

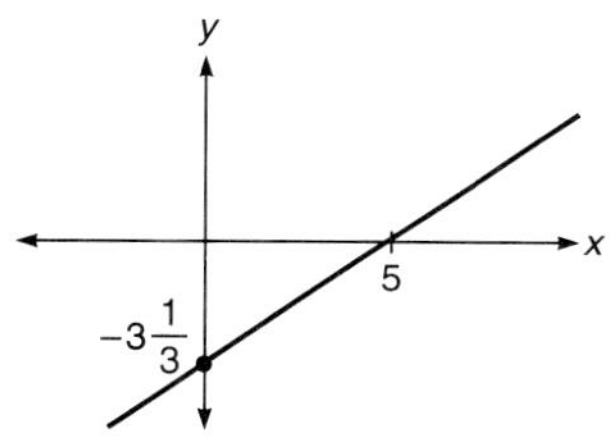

33.

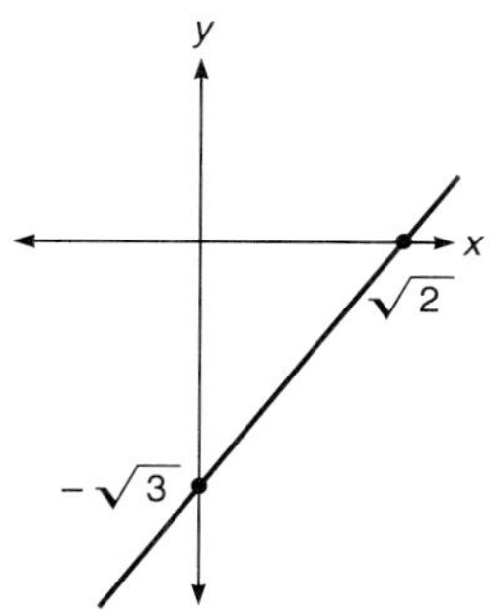

35.

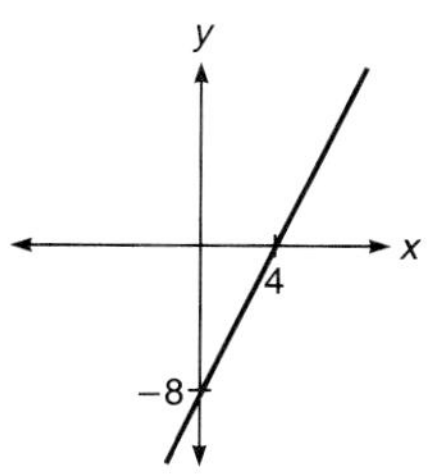

37.

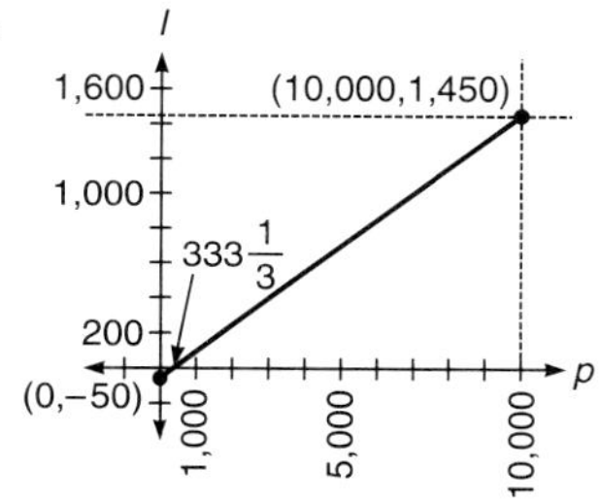

39.

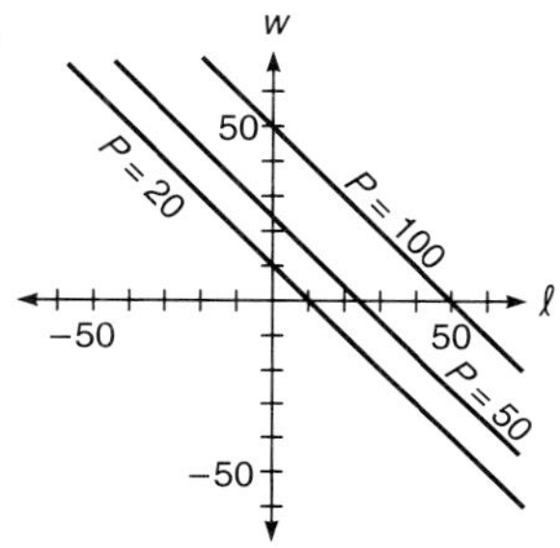

41.

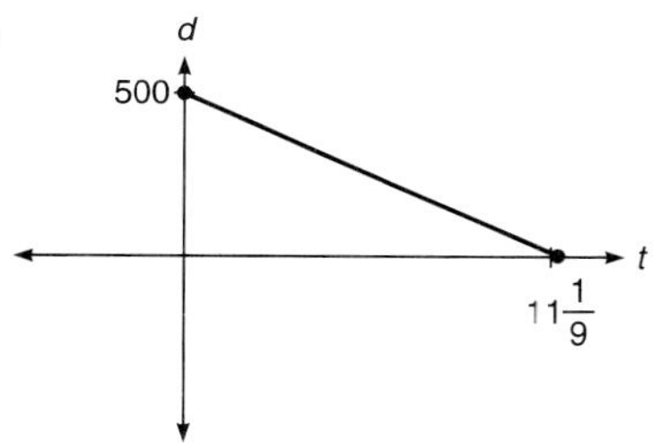

43.

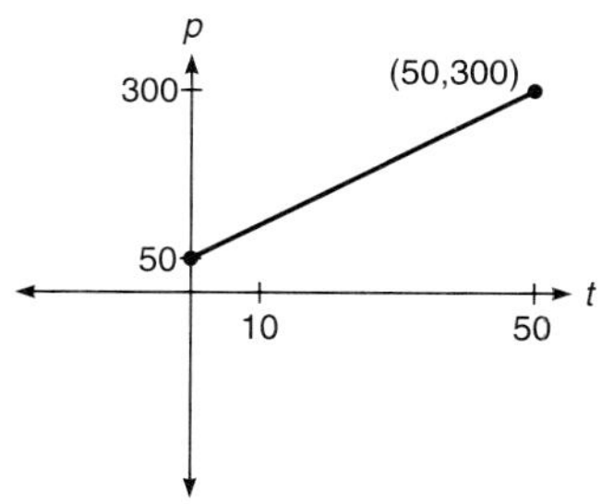

45. $(-1,7)$ **47.** $(4,5)$ **49.** $(2,-1\frac{1}{2})$
51. $(2\frac{7}{12},1\frac{2}{3})$ **53.** $(-2\frac{1}{2},6)$
55. $\left(1,\dfrac{7\sqrt{2}}{2}\right)$ **57.** $5\sqrt{2}$ **59.** $2\sqrt{10}$
61. 1 **63.** $6\sqrt{5}$ **65.** $3\sqrt{2}$ **67.** $2\frac{1}{2}$
69. $\dfrac{2\sqrt{101}}{5}$ **71.** $2\sqrt{73}$
73. $3\sqrt{a^2 + 4b^2}$ **75.** 4
77. $y = -\frac{4}{3}x + \frac{35}{3}$ **79.** $y = \frac{3}{7}x - \frac{10}{7}$
81. $a = \frac{5}{3}$, $b = -\frac{34}{15}$
83. 6

85. **a.** $d = |x_2 - x_1| + |y_2 - y_1|$
b. Taxicab distance is always longer unless the two points lie on the same horizontal or vertical line, in which case they are the same.

87. $a = 10$

89. The second line is $a_2x + b_2y + c_2 = 0$; $a_2 = ka_1$, $b_2 = kb_1$, $c_2 = kc_1$, so the second line is also $ka_1x + kb_1y + kc_1 = 0$, and since $k \neq 0$ we can divide each term by it obtaining $a_1x + b_1y + c_1 = 0$, which is the first line.

91. 44.5 square units

Solutions to skill and review problems

1. Compute $\frac{a-b}{c-d}$ if $a = 9$, $b = -3$, $c = -5$, $d = -1$.
$\frac{9-(-3)}{-5-(-1)} = \frac{12}{-4} = -3$

2. Solve $3y - 2x = 5$ for y.
$3y = 2x + 5$
$y = \frac{2x+5}{3}$

3. Solve $ax + by + c = 0$ for y.
$by = -ax - c$
$y = \frac{-(ax+c)}{b}$
$y = -\frac{ax+c}{b}$

4. If $y = 3x - b$ contains the point $(-2, 4)$, find b.
$4 = 3(-2) + b$
$4 = -6 + b$
$10 = b$

5.

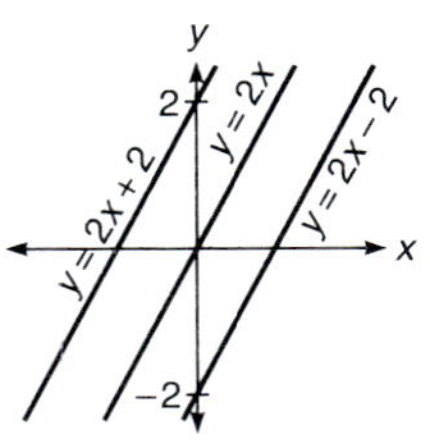

6. Solve the equation $2x^2 - x = 3$.
$2x^2 - x - 3 = 0$
$(2x - 3)(x + 1) = 0$
$2x - 3 = 0$ or $x + 1 = 0$
$2x = 3$ or $x = -1$
$x = \frac{3}{2}$ or -1; $\{-1, 1\frac{1}{2}\}$

7. Solve the equation $|2x - 3| = 5$.
$2x - 3 = 5$ or $2x - 3 = -5$
$2x = 8$ or $2x = -2$
$x = 4$ or $x = -1$
$\{-1, 4\}$

8. Simplify $\sqrt{48x^4}$.
$\sqrt{2^4 \cdot 3 \cdot x^4} = 2^2x^2\sqrt{3} = 4x^2\sqrt{3}$

9. Calculate $\frac{5}{8} - \frac{1}{4} + \frac{2}{3}$.
$\frac{5}{8} - \frac{2}{8} + \frac{2}{3}$
$\frac{3}{8} + \frac{2}{3} = \frac{9+16}{24} = \frac{25}{24}$

Solutions to trial exercise problems

13. Answers to problem 13 will vary. We solve for y and select values of x that produce integer values of y (for convenience).
$\frac{1}{2}x - \frac{1}{3}y = 1$
$-\frac{1}{3}y = -\frac{1}{2}x + 1$
$y = \frac{3}{2}x - 3$
$(-2,-6)$, $(0,-3)$, $(2,0)$

29. $\frac{1}{2}x - \frac{1}{3}y = 1$
$y = \frac{3x-6}{2}$
$(1,-1\frac{1}{2})$, $(2,0)$, $(3,1\frac{1}{2})$
x-intercept ($y = 0$):
$\frac{1}{2}x - 0 = 1$
$x = 2$; $(2,0)$
y-intercept ($x = 0$):
$y = \frac{0-6}{2} = -3$; $(0,-3)$

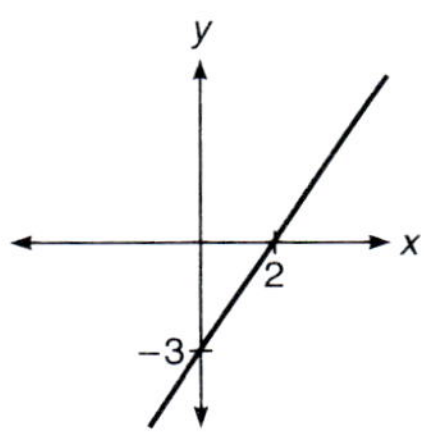

33. $\sqrt{3}x - \sqrt{2}y = \sqrt{6}$
$y = \frac{\sqrt{6}x - 2\sqrt{3}}{2}$
$\left(1, \frac{\sqrt{6} - 2\sqrt{3}}{2}\right)$,
$(2, \sqrt{6} - \sqrt{3})$,
$\left(3, \frac{3\sqrt{6} - 2\sqrt{3}}{2}\right)$
x-intercept ($y = 0$):
$\sqrt{3}x - 0 = \sqrt{6}$
$x = \frac{\sqrt{6}}{\sqrt{3}} = \sqrt{2}$; $(\sqrt{2}, 0)$
y-intercept ($x = 0$):
$y = \frac{0 - 2\sqrt{3}}{2}$; $(0, -\sqrt{3})$

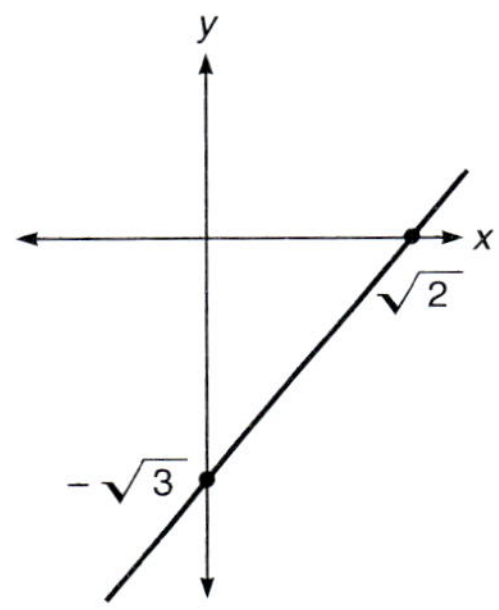

---

38. $I = 0.10p - 20$
$0 \leq p \leq 10{,}000$
I-intercept ($p = 0$):
$I = 0 - 20$
$(0, -20)$
p-intercept ($I = 0$):
$0 = 0.10p - 20$
$p = \frac{20}{0.10} = 200$
$(200,0)$
At $p = 0$, plot $(0,-20)$.
At $p = 10{,}000$ plot $(10{,}000, 980)$.

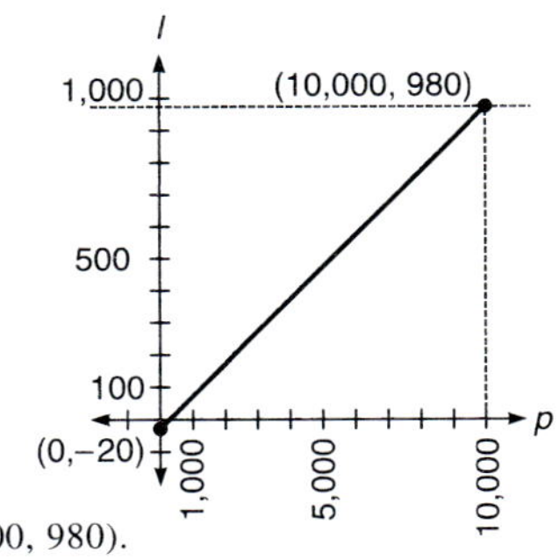

51. $(\frac{2}{3}, 3)$, $(4\frac{1}{2}, \frac{1}{3})$
$$\left(\frac{\frac{2}{3} + 4\frac{1}{2}}{2}, \frac{3 + \frac{1}{3}}{2}\right) = (2\tfrac{7}{12}, 1\tfrac{2}{3})$$

69. $(3, \frac{1}{5})$, $(-1, \frac{3}{5})$
$$d = \sqrt{(3-(-1))^2 + (\tfrac{1}{5} - \tfrac{3}{5})^2}$$
$$= \sqrt{4^2 + (\tfrac{2}{5})^2} = \sqrt{16 + \tfrac{4}{25}}$$
$$= \sqrt{\frac{16(25)}{25} + \frac{4}{25}} = \sqrt{\frac{404}{25}}$$
$$= \frac{\sqrt{4(101)}}{\sqrt{25}} = \frac{2\sqrt{101}}{5}$$

86. Let $M\left(\frac{x_1 + x_2}{2}, \frac{y_1 + y_2}{2}\right)$ be the midpoint; we need to show that the distance from M to P_1 equals the distance from M to P_2.

$$\sqrt{\left(x_1 - \frac{x_1 + x_2}{2}\right)^2 + \left(y_1 - \frac{y_1 + y_2}{2}\right)^2}$$
$$= \sqrt{\left(x_2 - \frac{x_1 + x_2}{2}\right)^2 + \left(y_2 - \frac{y_1 + y_2}{2}\right)^2}$$

Squaring both sides:

$$\left(x_1 - \frac{x_1 + x_2}{2}\right)^2 + \left(y_1 - \frac{y_1 + y_2}{2}\right)^2$$
$$= \left(x_2 - \frac{x_1 + x_2}{2}\right)^2 + \left(y_2 - \frac{y_1 + y_2}{2}\right)^2$$
$$\left(\frac{2x_1}{2} - \frac{x_1 + x_2}{2}\right)^2 + \left(\frac{2y_1}{2} - \frac{y_1 + y_2}{2}\right)^2$$
$$= \left(\frac{2x_2}{2} - \frac{x_1 + x_2}{2}\right)^2 + \left(\frac{2y_2}{2} - \frac{y_1 + y_2}{2}\right)^2$$
$$\left(\frac{x_1 - x_2}{2}\right)^2 + \left(\frac{y_1 - y_2}{2}\right)^2 = \left(\frac{x_2 - x_1}{2}\right)^2 + \left(\frac{y_2 - y_1}{2}\right)^2$$
$$\tfrac{1}{4}(x_1 - x_2)^2 + \tfrac{1}{4}(y_1 - y_2)^2$$
$$= \tfrac{1}{4}(x_2 - x_1)^2 + \tfrac{1}{4}(y_2 - y_1)^2$$
$$(x_1 - x_2)^2 + (y_1 - y_2)^2$$
$$= (x_2 - x_1)^2 + (y_2 - y_1)^2.$$

It is not too difficult to show that each side is the same by performing the indicated squaring operations.

Exercise 3–2

Answers to odd-numbered problems

1. $-\frac{1}{8}$ **3.** $\frac{13}{3}$ **5.** $\frac{5}{4}$ **7.** $\frac{43}{144}$
9. m is not defined **11.** 0 **13.** 10
15. $-\frac{5\sqrt{2}}{2}$ **17.** $-2\sqrt{3}$ **19.** $-\frac{p-q}{2q}$
21. Use (1,1) and (2,−1):

$$m = \frac{-1-1}{2-1} = \frac{-2}{1} = -2$$

23. $m = \frac{3}{5}$

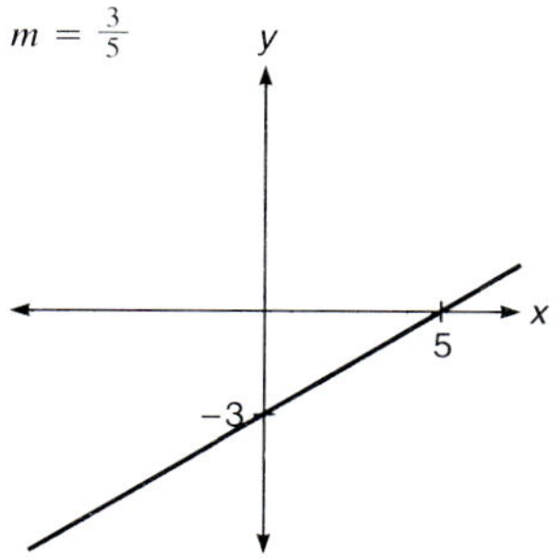

25. $m = -4$

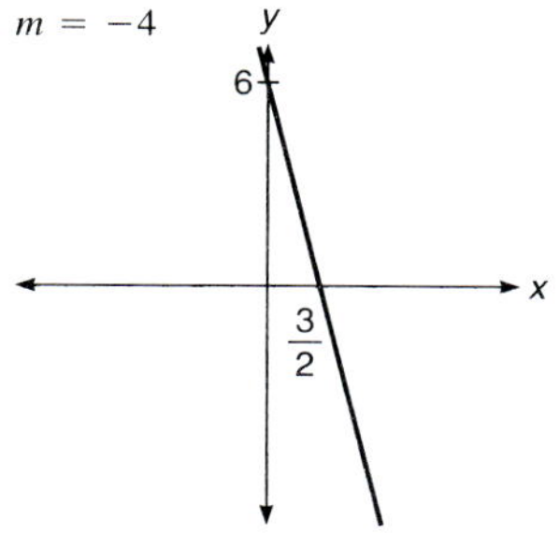

27. $m = \frac{1}{3}$

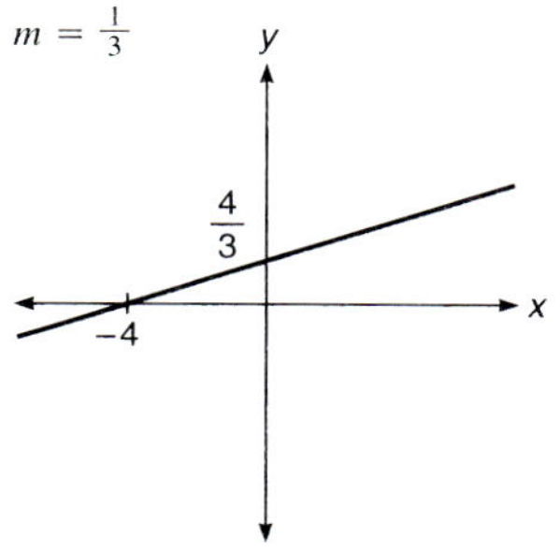

29. $m = \frac{1}{12}$

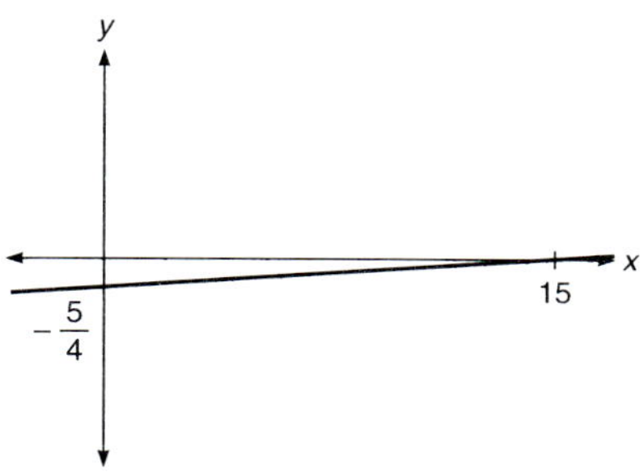

31. $m = 0$

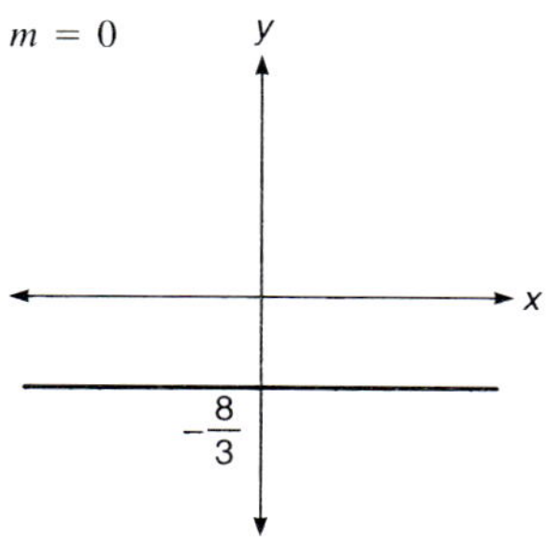

33. $m = \frac{2}{5}$

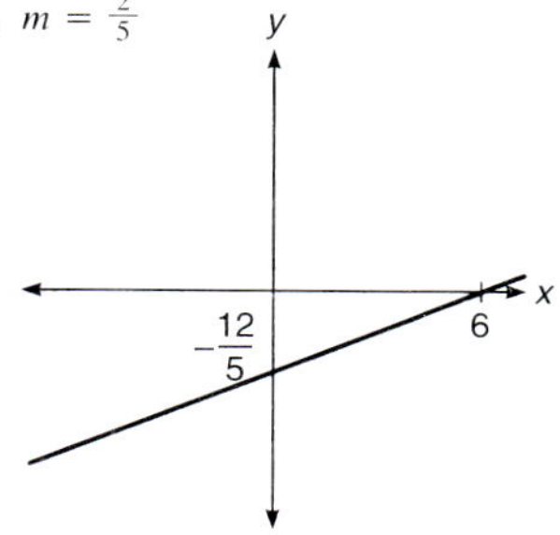

35. m is undefined

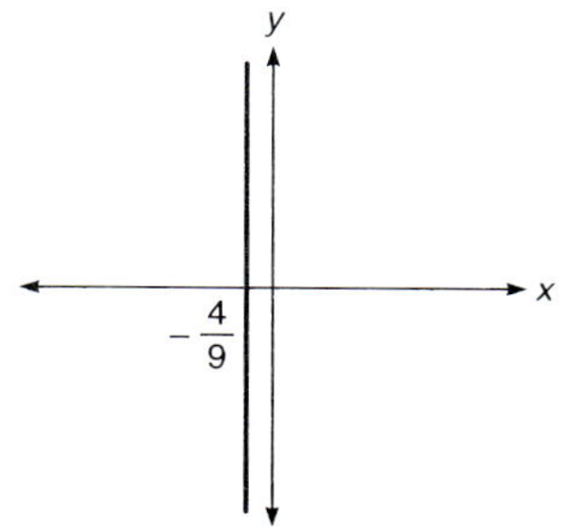

37. $m = 0$

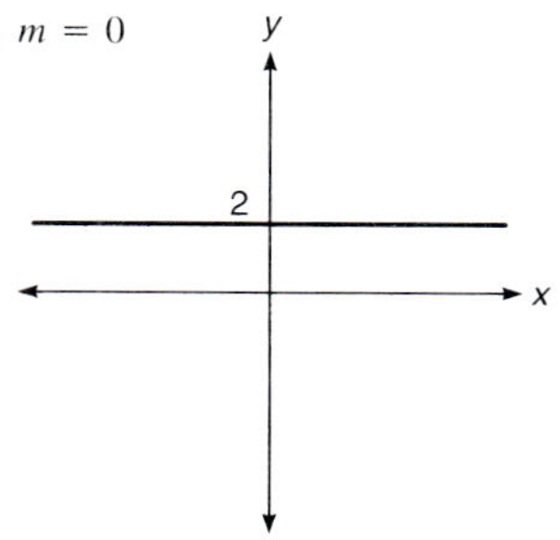

39. $y = -2x - 1$ **41.** $y = -4x + 9\frac{3}{4}$
43. $y = \frac{1}{a}x + (b - 1)$ **45.** $y = \frac{3}{8}x + \frac{17}{8}$
47. $y = \frac{7}{8}x - \frac{1}{4}$ **49.** $y = \frac{22}{3}x - 120$
51. $y = -4x - \frac{6}{5}$ **53.** $y = -\frac{5}{64}x + \frac{11}{16}$
55. $y = -\frac{5\sqrt{2}}{2}x$
57. $y = \frac{4n}{m-n}x + \frac{m^2 - 3mn - 2n^2}{m-n}$
59. $y = 5x - 3$ **61.** $y = -2x - 2$
63. $y = -5$ **65.** $y = \frac{3}{5}x + 5$
67. $y = -\frac{4}{3}x + 1\frac{4}{5}$ **69.** $y = 5x + 15$
71. $y = -x$ **73.** $x = -1$
75. $y = -2x - 2$ **77.** $x = 2$
79. **a.** −11.4°; **b.** −35.8° **81.** −52.1°
83. **a.** \$16,751; **b.** \$21,112
85. $(1\frac{3}{5}, -3\frac{1}{5})$

87. Let $P_1(x_1,y_1)$ and $P_2(x_2,y_2)$ be two different points on the line $y = 3x - 4$. Then $y_1 = 3x_1 - 4$, and $y_2 = 3x_2 - 4$.

$$m = \frac{y_2 - y_1}{x_2 - x_1} = \frac{(3x_2 - 4) - (3x_1 - 4)}{x_2 - x_1} = \frac{3x_2 - 3x_1}{x_2 - x_1} = \frac{3(x_2 - x_1)}{x_2 - x_1} = 3$$

89. Let $P_1(x_1,y_1)$ and $P_2(x_2,y_2)$ be two different points on the line $y = \frac{1}{3}x + 2$. Then $y_1 = \frac{1}{3}x_1 + 2$, and $y_2 = \frac{1}{3}x_2 + 2$.

$$m = \frac{y_2 - y_1}{x_2 - x_1} = \frac{(\frac{1}{3}x_2 + 2) - (\frac{1}{3}x_1 + 2)}{x_2 - x_1} = \frac{\frac{1}{3}x_2 - \frac{1}{3}x_1}{x_2 - x_1} = \frac{\frac{1}{3}(x_2 - x_1)}{x_2 - x_1} = \frac{1}{3}$$

91. $y - y_1 = m(x - x_1)$ is $y - y_1 = \frac{y_2 - y_1}{x_2 - x_1}(x - x_1)$, or $(y - y_1)(x_2 - x_1) = (x - x_1)(y_2 - y_1)$. Now plug the point $P_1 = (x_1,y_1)$ in for (x,y), obtaining $(y_1 - y_1)(x_2 - x_1) = (x_1 - x_1)(y_2 - y_1)$, or $0 = 0$, which makes P_1 a solution. Now plug in $P_2 = (x_2,y_2)$ and observe that both sides are the same, making this point also a solution.

93. $(3\frac{1}{3},-2\frac{2}{3})$ **95.** $(-3\frac{1}{4},2\frac{3}{4})$

97. $(-4\frac{3}{4},-1\frac{1}{8})$

99. The point of intersection is found by substitution:

$y = -3x + 15 \quad y = \frac{1}{3}x + 2$

$$\begin{aligned} -3x + 15 &= \tfrac{1}{3}x + 2 \\ -9x + 45 &= x + 6 \\ 39 &= 10x \\ x &= \tfrac{39}{10} \end{aligned}$$

$y = -3(\frac{39}{10}) + 15$

$y = \frac{33}{10}$

Thus the point $(h,k) = (\frac{39}{10},\frac{33}{10})$.

Using the distance formula we find a and b. a is the distance from the point (h,k) to $(0,15)$:

$$a = \sqrt{\left(\frac{39}{10} - 0\right)^2 + \left(\frac{33}{10} - 15\right)^2}$$

$$a = \sqrt{\left(\frac{39}{10}\right)^2 + \left(-\frac{117}{10}\right)^2}$$

$$a = \sqrt{\frac{15{,}210}{100}}, \text{ so } a^2 = \frac{1{,}521}{10}$$

b is the distance from the point (h,k) to $(0,2)$:

$$b = \sqrt{\left(\frac{39}{10} - 0\right)^2 + \left(\frac{33}{10} - 2\right)^2}$$

$$b = \sqrt{\left(\frac{39}{10}\right)^2 + \left(\frac{13}{10}\right)^2}$$

$$b = \sqrt{\frac{1{,}690}{100}}, \text{ so } b^2 = \frac{169}{10}$$

Thus, $a^2 + b^2 = \frac{1{,}521}{10} + \frac{169}{10} = \frac{1{,}690}{10} = 169$. $c = 13$, so $c^2 = 169$.

Thus, $a^2 + b^2 = c^2$.

101. $(-3,7)$ **103.** $(1\frac{7}{8},-\frac{5}{8})$

105. $(3,33)$, $(7,73)$ **107.** $(\frac{1}{2},0)$, $(3,2\frac{1}{2})$

109. $(-1,-1)$, $(6,13)$

111. $(-4,-9)$, $(2,3)$ **113.** $(17,270)$

115. $(-1,-8)$, $(2\frac{1}{2},-2\frac{3}{4})$

Solutions to skill and review problems

1. Evaluate $3x^2 + 2x - 10$ for $x = -5$.

$3(-5)^2 + 2(-5) - 10$

$75 - 10 - 10$

55

2. Evaluate $3x^2 + 2x - 10$ for $x = c + 1$.

$3(c + 1)^2 + 2(c + 1) - 10$

$3(c^2 + 2c + 1) + 2c + 2 - 10$

$3c^2 + 8c - 5$

3. Solve $2x^2 - 2x - 5 = 0$.

$$x = \frac{-(-2) \pm \sqrt{(-2)^2 - 4(2)(-5)}}{2(2)} = \frac{2 \pm \sqrt{44}}{4} = \frac{2 \pm 2\sqrt{11}}{4} = \frac{2(1 \pm \sqrt{11})}{4} = \frac{1 \pm \sqrt{11}}{2}$$

4. Simplify $\sqrt{\frac{2x}{5y^3}}$.

$$\frac{\sqrt{2x}}{\sqrt{5y^3}} = \frac{\sqrt{2x}}{y\sqrt{5y}} \cdot \frac{\sqrt{5y}}{\sqrt{5y}} = \frac{\sqrt{10xy}}{y(5y)} = \frac{\sqrt{10xy}}{5y^2}$$

5. Solve $|2x - 6| < 8$.

$-8 < 2x - 6 < 8$

$-2 < 2x < 14$

$-1 < x < 7$

6. Solve $\frac{x - 2}{4} = \frac{2x + 1}{3}$.

$3(x - 2) = 4(2x + 1)$

$3x - 6 = 8x + 4$

$-10 = 5x$

$-2 = x$

7. Compute $(\frac{2}{3} - \frac{1}{4}) \div 5$.

$$\frac{2(4) - 1(3)}{3(4)} \div 5 = \frac{5}{12} \cdot \frac{1}{5} = \frac{1}{12}$$

Solutions to trial exercise problems

7. $(7,\frac{7}{12})$, $(-5,-3)$

$(x_1,y_1) = (7,\frac{7}{12})$, $(x_2,y_2) = (-5,-3)$

$$m = \frac{y_2 - y_1}{x_2 - x_1} = \frac{-3 - \frac{7}{12}}{-5 - 7} = \frac{-\frac{36}{12} - \frac{7}{12}}{-12} = \frac{-\frac{43}{12}}{-12} = \frac{43}{12} \cdot \frac{1}{12} = \frac{43}{144}$$

17. $(\sqrt{27},3)$, $(\sqrt{12},9)$

$$m = \frac{9 - 3}{\sqrt{12} - \sqrt{27}} = \frac{6}{2\sqrt{3} - 3\sqrt{3}} = \frac{6}{-\sqrt{3}} \cdot \frac{\sqrt{3}}{\sqrt{3}} = -\frac{6\sqrt{3}}{3} = -2\sqrt{3}$$

33. $\frac{1}{3}x - \frac{5}{6}y = 2$

$y = \frac{2}{5}x - \frac{12}{5}$; $m = \frac{2}{5}$

intercepts: $(0,-2\frac{2}{5})$, $(6,0)$

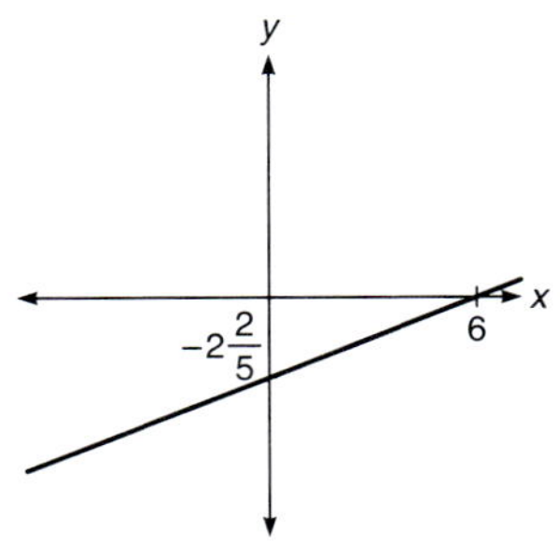

43. (a,b), $m = \frac{1}{a}$

$y - b = \frac{1}{a}(x - a)$

$y - b = \frac{1}{a}x - 1$

$y = \frac{1}{a}x - 1 + b$

$y = \frac{1}{a}x + (b - 1)$

53. $(4,\frac{3}{8})$, $(12,-\frac{1}{4})$

$m = \frac{-\frac{1}{4} - \frac{3}{8}}{12 - 4} = \frac{-\frac{5}{8}}{8} = -\frac{5}{8}\cdot\frac{1}{8} = -\frac{5}{64}$

$y - (-\frac{1}{4}) = -\frac{5}{64}(x - 12)$

$y + \frac{1}{4} = -\frac{5}{64}x + \frac{15}{16}$

$y = -\frac{5}{64}x + \frac{11}{16}$

67. A line that is perpendicular to the line $4y + 5 = 3x$ and passes through the point $(\frac{3}{2},-\frac{1}{5})$.

$4y + 5 = 3x$

$4y = 3x - 5$

$y = \frac{3}{4}x - \frac{5}{4}$ Solved for y; $m = \frac{3}{4}$.

Use $m = -\frac{4}{3}$, the negative reciprocal of $\frac{3}{4}$, since we want a line perpendicular to the original line.

$y - (-\frac{1}{5}) = -\frac{4}{3}(x - \frac{3}{2})$

$y + \frac{1}{5} = -\frac{4}{3}x + 2$

$y = -\frac{4}{3}x + 1\frac{4}{5}$

81. First find the wind chill factor, wcf, for a $-11.5°$ temperature for 15 mph and 20 mph winds. At 15 mph we use the (temperature, wcf) points $(-10,-45)$ and $(-15,-51)$. We compute y in the ordered pair $(-11.5,y)$. The wcf is $-46.8°$. At 20 mph we use the (temperature, wcf) points $(-10,-52)$ and $(-15,-60)$. We compute y in the ordered pair $(-11.5,y)$ and obtain the wcf $-54.4°$.

Now we have the ordered pairs (mph, wcf) of $(15,-46.8°)$ and $(20,-54.4°)$. We use these to compute y in the ordered pair $(18.5,y)$. The value of y is -52.12, so the required wind chill factor for $-11.5°$ and 18.5 mph is $-52.1°$.

86. Let $P_1(a, b)$ and $P_2(c, d)$ be any two points on the line $y = 5x - 2$. Then $b = 5a - 2$, and $d = 5c - 2$. Now we put P_1 and P_2 into the definition of m:

$m = \frac{d - b}{c - a}$

$= \frac{(5c - 2) - (5a - 2)}{c - a}$ Replace d by $5c - 2$ and b by $5a - 2$

$= \frac{5c - 5a}{c - a}$ Remove parentheses and combine like terms

$= \frac{5(c - a)}{c - a}$ Factor 5 from the numerator

$= 5$ Reduce by $c - a$

Thus, no matter what two points we choose on this line we will obtain the slope 5.

95. $y = x + 6$; $3y + x = 5$

$3(x + 6) + x = 5$ Replace y in second equation by $x + 6$

$x = -\frac{13}{4}$ Solve for x

$y = x + 6$ First equation

$y = -\frac{13}{4} + \frac{24}{4}$ Replace x by $-\frac{13}{4}$

$= \frac{11}{4}$ Solve for y

The point is $(-3\frac{1}{4},2\frac{3}{4})$.

108. $y = 4x^2 + 6x - 1$; $y = -2x + 4$

$-2x + 4 = 4x^2 + 6x - 1$

$0 = 4x^2 + 8x - 5$

$0 = (2x + 5)(2x - 1)$

$x = -\frac{5}{2}, \frac{1}{2}$

$y = -2(-\frac{5}{2}) + 4 = 9$

$y = -2(\frac{1}{2}) + 4 = 3$

$(-2\frac{1}{2},9)$, $(\frac{1}{2},3)$

116. $y = x^2 + 3x + 13$; $y = -x^2 - 6x + 9$

$-x^2 - 6x + 9 = x^2 + 3x + 13$

$0 = 2x^2 + 9x + 4$

$0 = (2x + 1)(x + 4)$

$x = -4, -\frac{1}{2}$

$y = -(-4)^2 - 6(-4) + 9 = 17$

$y = -(-\frac{1}{2})^2 - 6(-\frac{1}{2}) + 9 = \frac{47}{4}$

$(-4,17)$, $(-\frac{1}{2},11\frac{3}{4})$

Exercise 3–3

Answers to odd-numbered problems

1. A function is a relation in which no first element repeats. **3.** function, one to one; domain $\{-3, 1, 4, 5\}$, range $\{1, 2, 5, 8\}$

5. not a function; the first element 2 repeats; domain $\{-10, 2, 4\}$, range $\{-5, 9, 12, 13\}$

7. $\{(-2,10), (3,5), (5,3), (\frac{3}{4},7\frac{1}{4}), (7,1)\}$; function, one to one; domain: $\{-2, 3, 5, \frac{3}{4}, 7\}$, range: $\{10, 5, 3, 7\frac{1}{4}, 1\}$

9. $\{(1,1), (8,2), (27,3), (-1,-1), (-8,-2), (-27, -3)\}$; function, one to one; domain: $\{\pm1, \pm8, \pm27\}$, range: $\{\pm1, \pm2, \pm3\}$

11. $D = R$; $f(-4) = -23$; $f(0) = -3$; $f(\frac{1}{2}) = -\frac{1}{2}$; $f(7) = 32$; $f(3\sqrt{2}) = 15\sqrt{2} - 3$; $f(c - 1) = 5c - 8$

13. $D = \{x \mid x \geq \frac{1}{2}\}$; $g(-4)$, $g(0)$ are not defined since -4, 0 are not in the domain of g; $g(\frac{1}{2}) = 0$; $g(7) = \sqrt{13}$; $g(3\sqrt{2}) = \sqrt{6\sqrt{2} - 1}$; $g(c - 1) = \sqrt{2c - 3}$

15. $D = \{x \mid x \neq -3\}$; $f(-4) = 9$; $f(0) = -\frac{1}{3}$; $f(\frac{1}{2}) = 0$; $f(7) = \frac{13}{10}$; $f(3\sqrt{2}) = \frac{13 - 7\sqrt{2}}{3}$; $f(c - 1) = \frac{2c - 3}{c + 2}$

17. $D = R$; $m(-4) = 41$; $m(0) = -11$; $m(\frac{1}{2}) = -\frac{43}{4}$; $m(7) = 129$; $m(3\sqrt{2}) = 43 - 3\sqrt{2}$; $m(c - 1) = 3c^2 - 7c - 7$

19. $D = \{x \mid x \geq 1\}$; $f(-4)$, $f(0)$, $f(\frac{1}{2})$ are not defined since -4, 0, $\frac{1}{2}$ are not in the implied domain; $f(7) = \frac{\sqrt{3}}{2}$; $f(3\sqrt{2}) = \frac{\sqrt{3\sqrt{2} - 1}}{\sqrt{3\sqrt{2} + 1}}$ or $\frac{(3\sqrt{2} - 1)\sqrt{17}}{17}$; $f(c - 1) = \frac{\sqrt{c - 2}}{\sqrt{c}}$

21. $D = R$; $h(-4) = -68$; $h(0) = -4$; $h(\frac{1}{2}) = -3\frac{7}{8}$; $h(7) = 339$; $h(3\sqrt{2}) = 54\sqrt{2} - 4$; $f(c - 1) = c^3 - 3c^2 + 3c - 5$ **23.** domain: $x \neq -3$; $g(-4) = 4$; $g(0) = 0$; $g(\frac{1}{2}) = \frac{1}{7}$; $g(7) = \frac{7}{10}$; $g(3\sqrt{2}) = 2 - \sqrt{2}$; $g(c - 1) = \frac{c - 1}{c + 2}$

25. $2x - 3 + h$ **27. a.** 105 **b.** 0

29. a. -3 **b.** -2 **c.** $\frac{5}{2}$ **d.** $\frac{8}{3}$

31. -3 **33.** $-\frac{1}{5}$ **35.** 208

37. $5x + 2$ **39.** $2x - 2$

41.

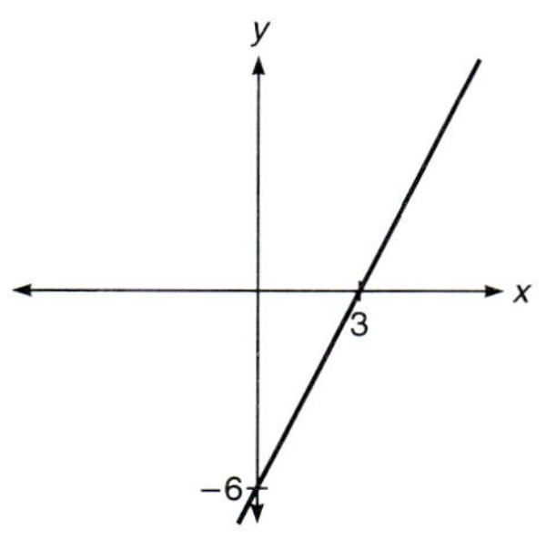

43.

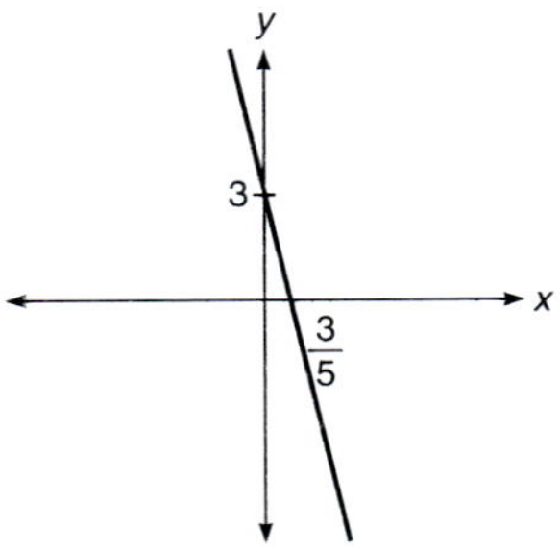

45.

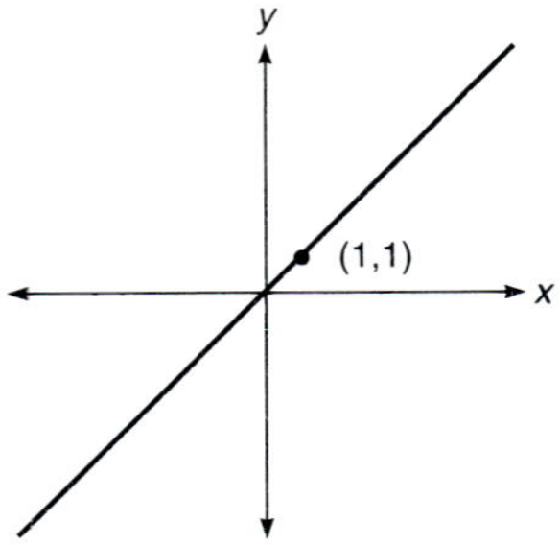

47. $y = 0$ is the x-axis

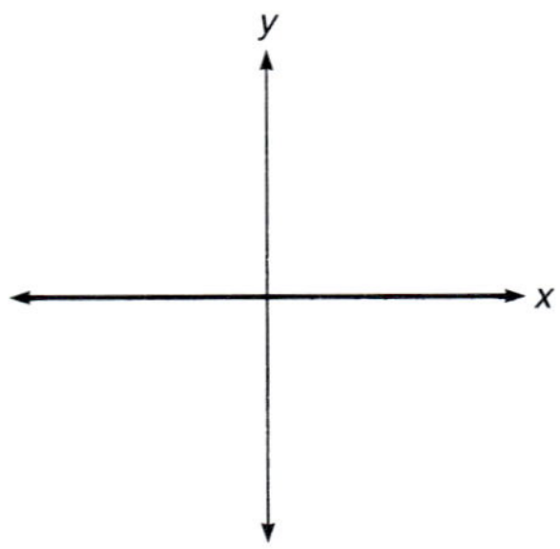

49. $C(m) = 0.34m + 500$

51. $8{,}333\frac{1}{3}$ miles **53.** $v = \frac{1}{20}h + 8$

55. $W = \frac{44}{25}a + \frac{574}{5}$; using $a = 40$ we predict $185\frac{1}{5}$ pounds

57. $A = 800\pi - 40\pi h$

59. $V = 4x^3 - 140x^2 + 1{,}200x$

Solutions to skill and review problems

1. $m = \dfrac{4-3}{-2-1} = -\dfrac{1}{3}$

$y - 3 = -\frac{1}{3}(x - 1)$

$y - 3 = -\frac{1}{3}x + \frac{1}{3}$

$y = -\frac{1}{3}x + 3\frac{1}{3}$

2. $2y - x = 4$

$2y = x + 4$

$y = \frac{1}{2}x + 2;\ m = \frac{1}{2}$

The y-intercept of 3 is the point (0,3):

$y - 3 = \frac{1}{2}(x - 0)$

$y = \frac{1}{2}x + 3$

3. $8x^3 - 1$

$(2x - 1)((2x)^2 + (1)(2x) + 1^2)$

$(2x - 1)(4x^2 + 2x + 1)$

4. $(3x - 2)^3$

$(3x - 2)(3x - 2)(3x - 2)$

$(9x^2 - 12x + 4)(3x - 2)$

$27x^3 - 36x^2 + 12x - 18x^2 + 24x - 8$

$27x^3 - 54x^2 + 36x - 8$

5. $\dfrac{2x-1}{3} - \dfrac{x-1}{2} = 6$

$6\left(\dfrac{2x-1}{3}\right) - 6\left(\dfrac{x-1}{2}\right) = 6(6)$

$2(2x - 1) - 3(x - 1) = 36$

$4x - 2 - 3x + 3 = 36$

$x + 1 = 36$

$x = 35$

6. $|x - 3| > 1$

$x - 3 > 1$ or $x - 3 < -1$

$x > 4$ or $x < 2$

7. $\dfrac{x+3}{x-1} < 1$

Critical points:

a. Solve corresponding equality

$\dfrac{x+3}{x-1} = 1$

$x + 3 = x - 1$

$0 = -4$ (no solution)

b. Zeros of denominators

$x - 1 = 0$

$x = 1$

CP: 1

TP: 0, 2

$x = 0$: $\dfrac{0+3}{0-1} < 1$

$-3 < 1$, True

$x = 2$: $\dfrac{2+3}{2-1} < 1$

$5 < 1$, False

solution: $x < 1$

Solutions to trial exercise problems

19. $f(x) = \dfrac{\sqrt{x-1}}{\sqrt{x+1}}$; domain: $x + 1 > 0$ and $x - 1 \geq 0$, so, $x > -1$ and $x \geq 1$. Both conditions are satisfied if $x \geq 1$. $D = \{x \mid x \geq 1\}$

$f(-4), f(0), f(\frac{1}{2})$ are not defined since $-4, 0, \frac{1}{2}$ are not in the implied domain.

$f(7) = \dfrac{\sqrt{7-1}}{\sqrt{7+1}} = \dfrac{\sqrt{6}}{\sqrt{8}} = \sqrt{\dfrac{3}{4}} = \dfrac{\sqrt{3}}{2}$

$f(3\sqrt{2}) = \dfrac{\sqrt{3\sqrt{2}-1}}{\sqrt{3\sqrt{2}+1}} \cdot \dfrac{\sqrt{3\sqrt{2}-1}}{\sqrt{3\sqrt{2}-1}}$

$= \dfrac{3\sqrt{2}-1}{\sqrt{18-1}} = \dfrac{3\sqrt{2}-1}{\sqrt{17}} \cdot \dfrac{\sqrt{17}}{\sqrt{17}}$

$= \dfrac{(3\sqrt{2}-1)\sqrt{17}}{17}$

$f(c-1) = \dfrac{\sqrt{(c-1)-1}}{\sqrt{(c-1)+1}} = \dfrac{\sqrt{c-2}}{\sqrt{c}}$

27. a. $g(3) = 2\sqrt{2}$ so $f(g(3)) = f(2\sqrt{2})$

$= 2(2\sqrt{2})^4 - 3(2\sqrt{2})^2 + 1$

$= 2(2^4)(\sqrt{2})^4 - 3(2^2)(\sqrt{2})^2 + 1$

$= 2(16)(4) - 3(8) + 1 = 105$

b. $g(\frac{2}{3}) = 1$, so $f(g(\frac{2}{3})) = f(1) = 0$

35. $(f(-3))^2 - 3(g(1))^2$

$[(5(-3) - 1)]^2 - 3[(2(1) + 2)]^2$

$(-16)^2 - 3(4)^2$

208

56. We have two points (temperature, time) (t,T): (74°,3:05) and (625°,4:15). If we put the times in minutes these points are (74,0) and (625,70). If $T = mt + b$, then $74 = m(0) + b$, so $b = 74$. Using the second point $625 = m(70) + 74$, $551 = 70m$, $m = \dfrac{551}{70}$, so $T = \dfrac{551}{70}t + 74$, where t is in minutes after 3:05 P.M. 4:00 corresponds to $t = 55$, so at this time $T = \dfrac{551}{70}(55) + 74 \approx 507°$.

Exercise 3–4

Answers to odd-numbered problems

1. $y = x^2 - 4$; graph of $y = x^2$ shifted down 4 units

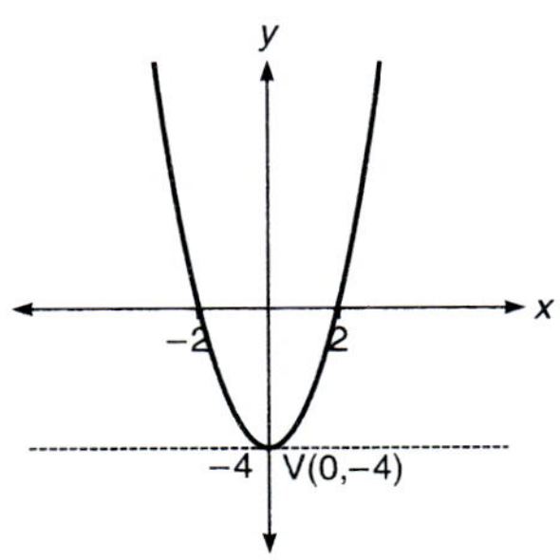

3. $y = x^2 + 3$; graph of $y = x^2$ shifted up 3 units

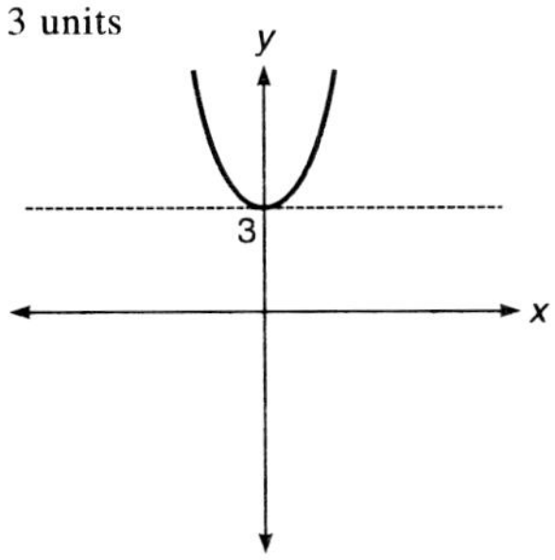

5. $y = (x - 1)^2$; graph of $y = x^2$ shifted right 1 unit

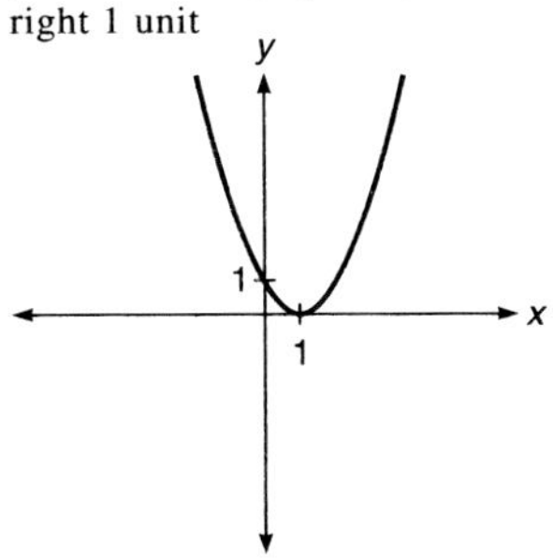

7. $y = (x + 3)^2$; $y = (x - (-3))^2$; graph of $y = x^2$ shifted left 3 units

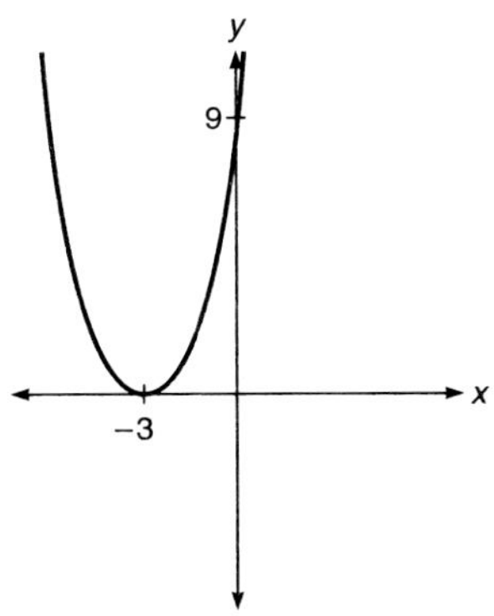

9. $y = (x + 3)^2 - 3$; $y = (x - (-3))^2 - 3$; graph of $y = x^2$ shifted down 3 units, left 3 units

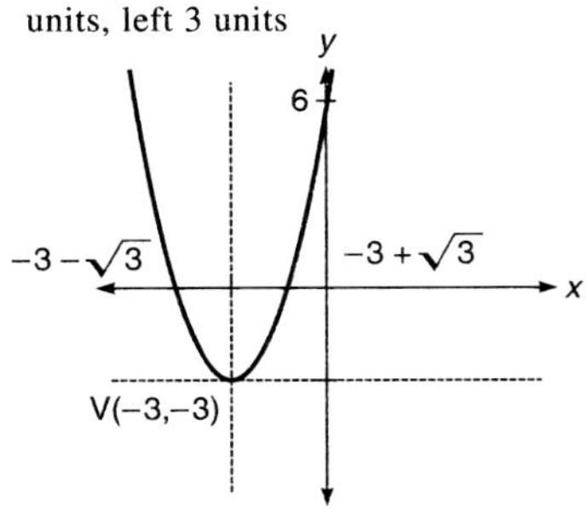

11. $y = (x + 2)^2 + 1$; $y = (x - (-2))^2 + 1$; graph of $y = x^2$ shifted up 1 unit, left 2 units

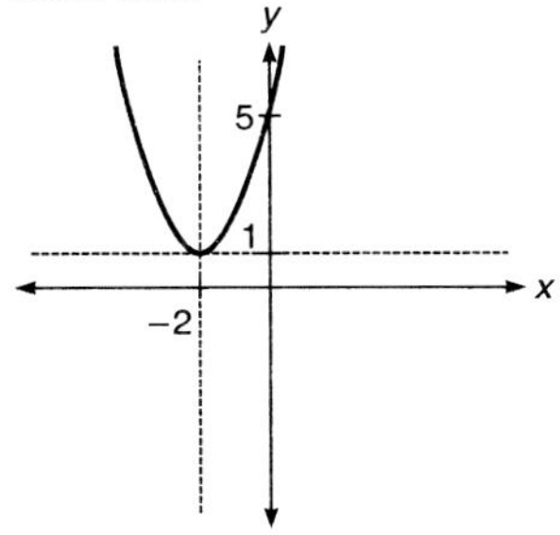

13. $y = \sqrt{x} - 2$; graph of $y = \sqrt{x}$ shifted down 2 units

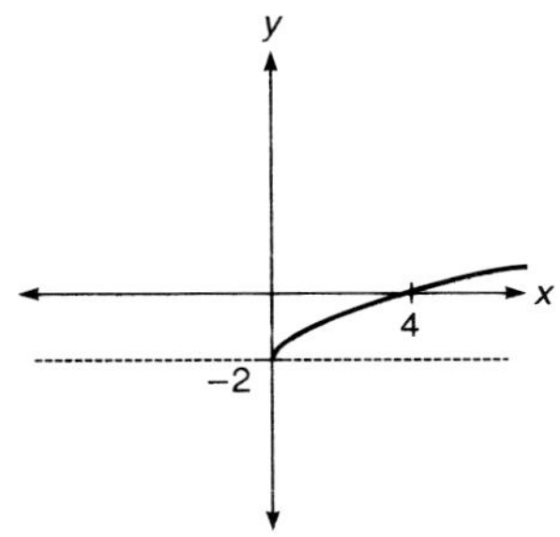

15. $y = \sqrt{x} + 2$; graph of $y = \sqrt{x}$ shifted up 2 units

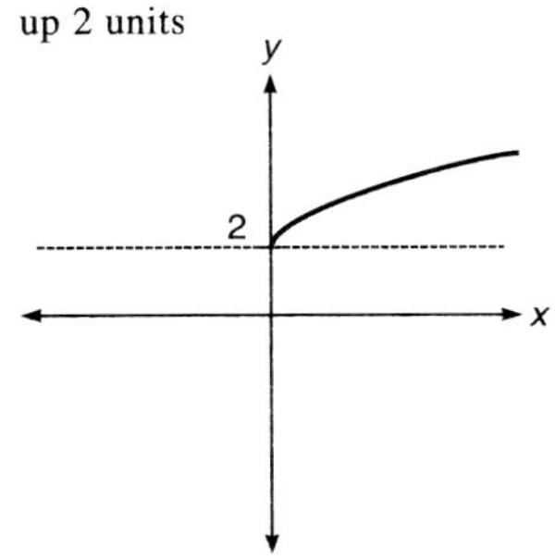

17. $y = \sqrt{x + 1}$; $y = \sqrt{x - (-1)}$; graph of $y = \sqrt{x}$ shifted left 1 unit

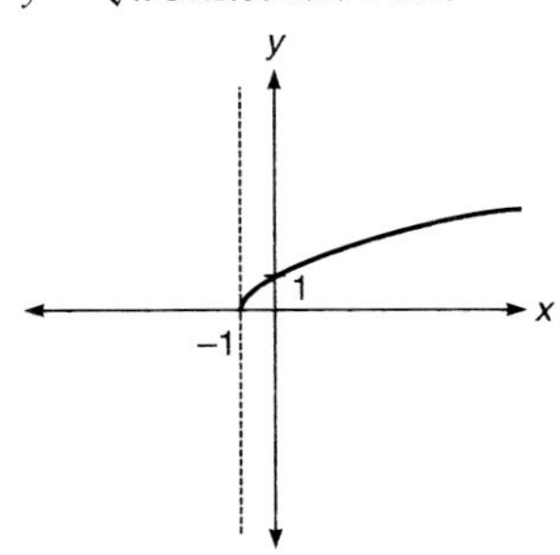

19. $y = \sqrt{x - 3}$; graph of $y = \sqrt{x}$ shifted right 3 units

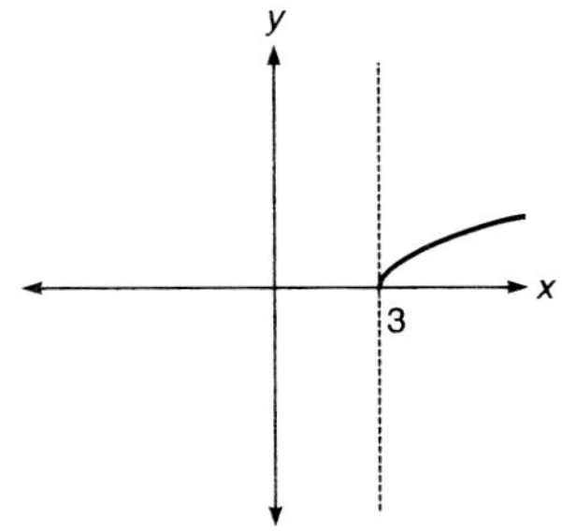

21. $y = \sqrt{x - 3} + 2$; graph of $y = \sqrt{x}$ shifted right 3 units, up 2 units

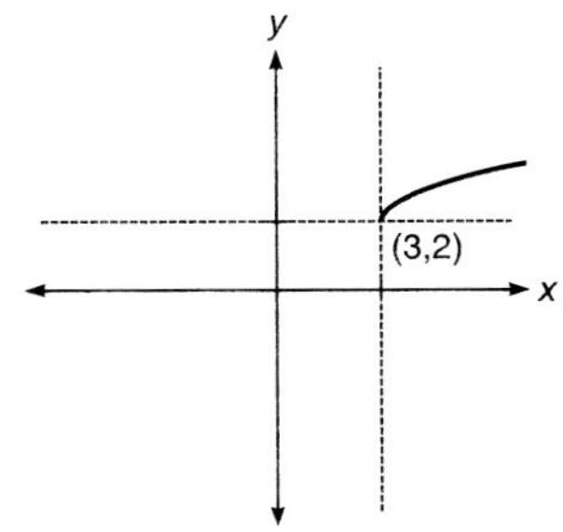

23. $y = \sqrt{x + 5} + 5$; $y = \sqrt{x - (-5)} + 5$; graph of $y = \sqrt{x}$ shifted left 5 units, up 5 units

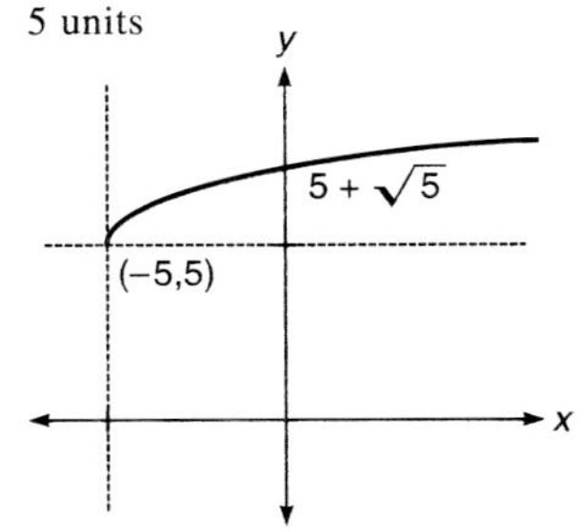

25. $y = (x - 2)^3$; graph of $y = x^3$ shifted right 2 units

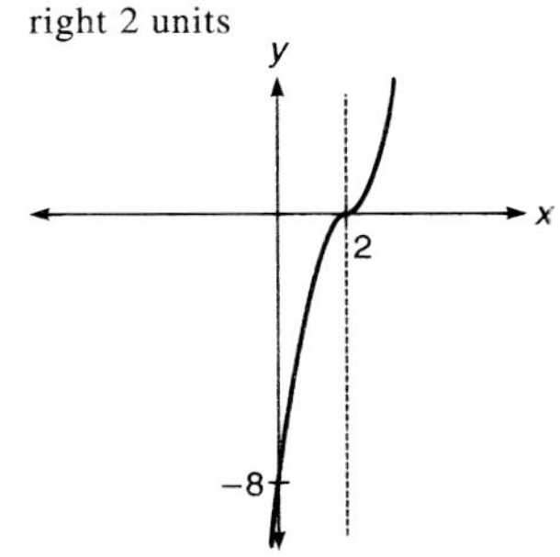

27. $y = x^3 - 8$; graph of $y = x^3$ shifted down 8 units

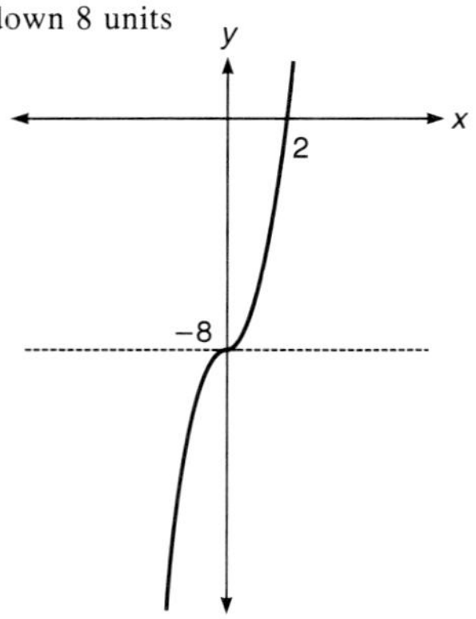

29. $y = (x + 1)^3 - 2$; $y = (x - (-1))^3 - 2$; graph of $y = x^3$ shifted left 1 unit, down 2 units

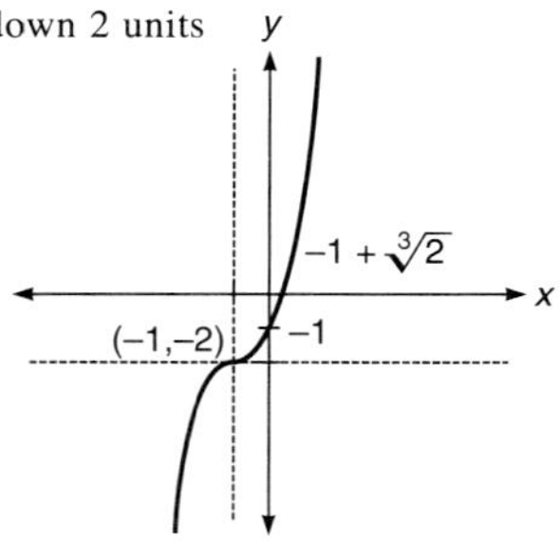

31. $y = (x + 2)^3 + 2$; $y = (x - (-2))^3 + 2$; graph of $y = x^3$ shifted up 2 units, left 2 units

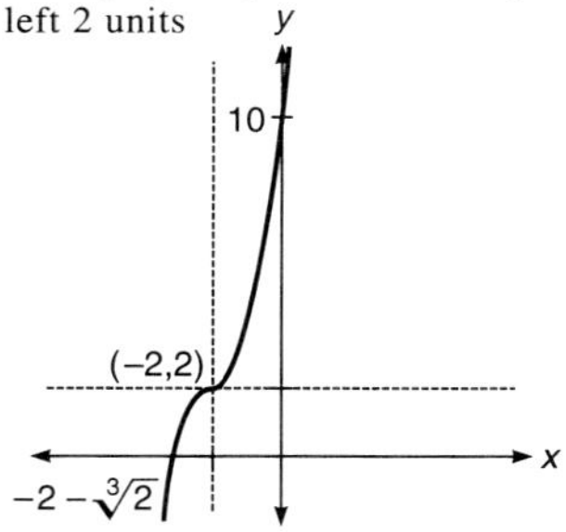

33. $y = \dfrac{1}{x} + 2$; graph of $y = \dfrac{1}{x}$ shifted up 2 units

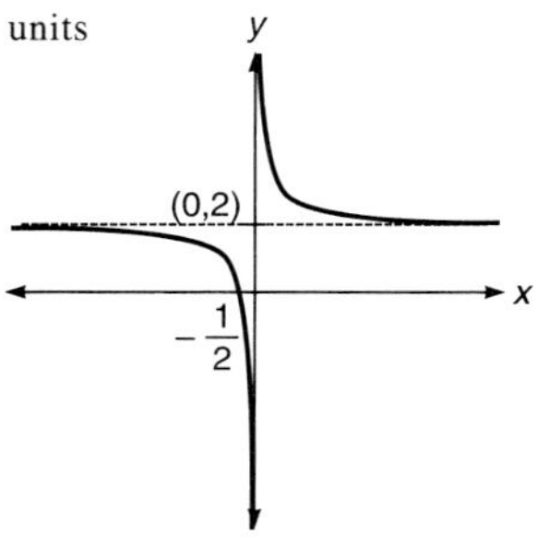

35. $y = \dfrac{1}{x - 6}$; graph of $y = \dfrac{1}{x}$ shifted right 6 units

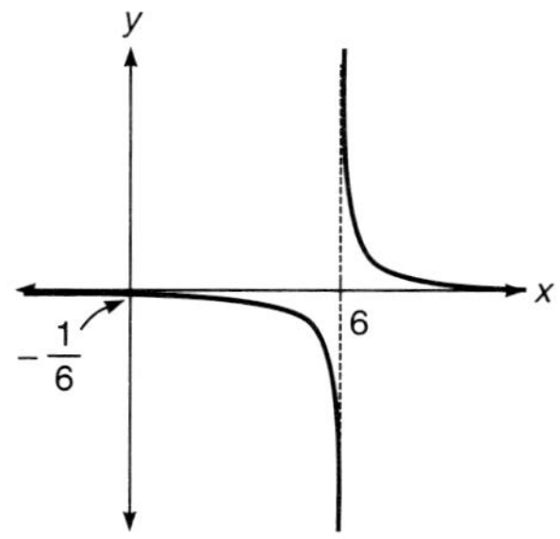

37. $y = \dfrac{1}{x - 3} - 5$; graph of $y = \dfrac{1}{x}$ shifted right 3 units, down 5 units

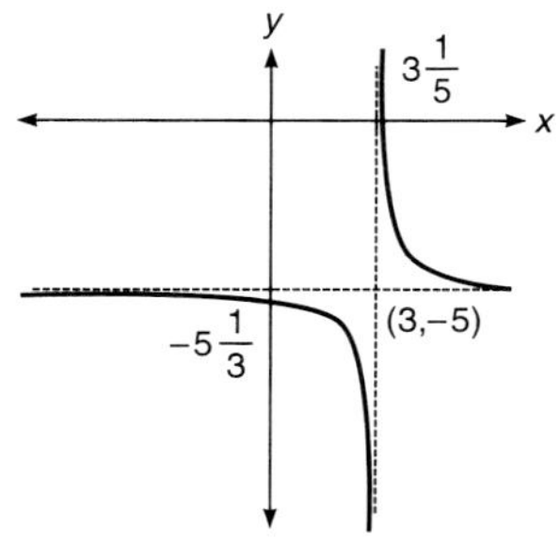

39. $y = \dfrac{1}{x - 1} + 1$; graph of $y = \dfrac{1}{x}$ shifted up 1 unit, right 1 unit

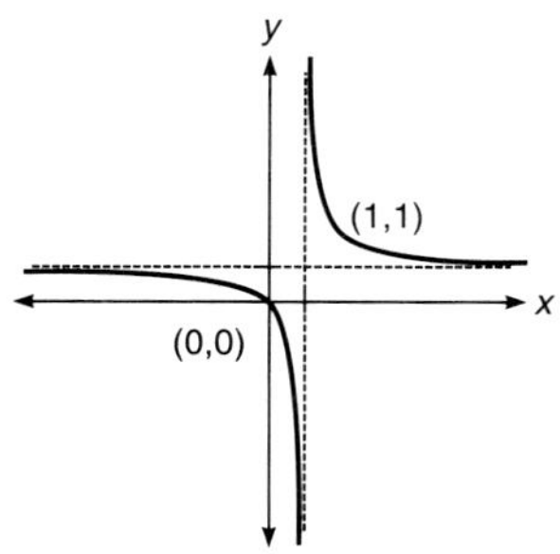

41. $y = |x + 2|$; $y = |x - (-2)|$; graph of $y = |x|$ shifted left 2 units

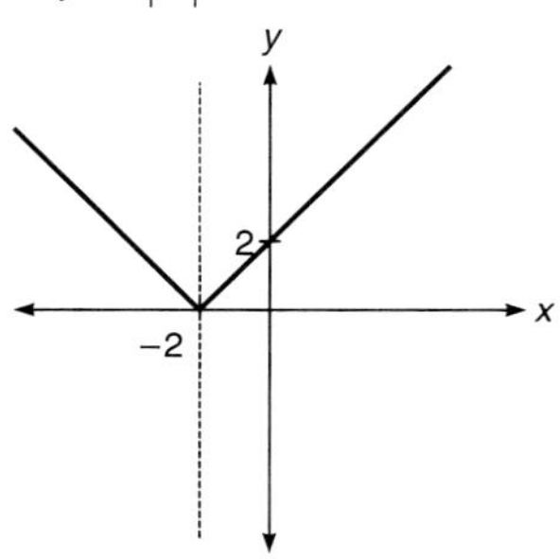

43. $y = |x| + 2$; graph of $y = |x|$ shifted up 2 units

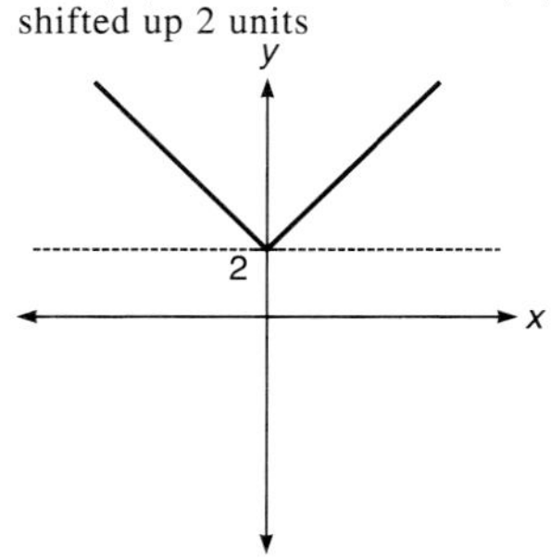

45. $y = |x - 5| - 4$; graph of $y = |x|$ shifted right 5 units, down 4 units

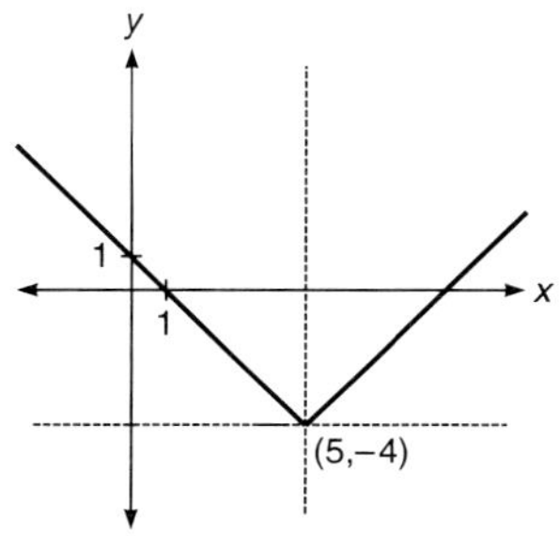

47. $y = |x + 3| + 3$; $y = |x - (-3)| + 3$; graph of $y = |x|$ shifted left 3 units, up 3 units

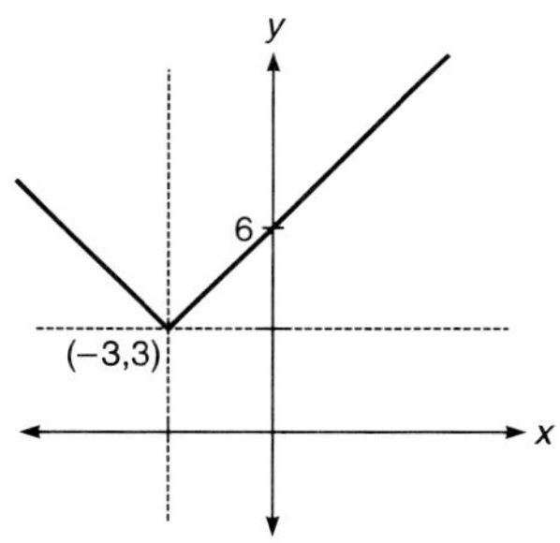

49. $y = 3(x - 1)^2 + 2$; graph of $y = x^2$ shifted up 2 units, right 1 unit, vertically scaled 3 units

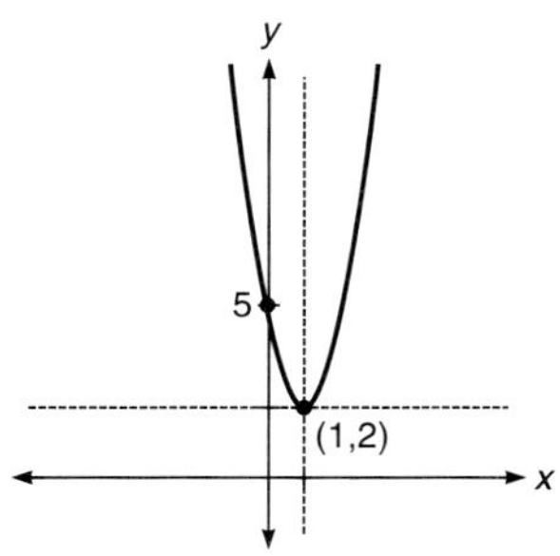

51. $y = \frac{1}{2}(x - 2)^3$; graph of $y = x^3$ shifted right 2 units, vertically scaled $\frac{1}{2}$ units

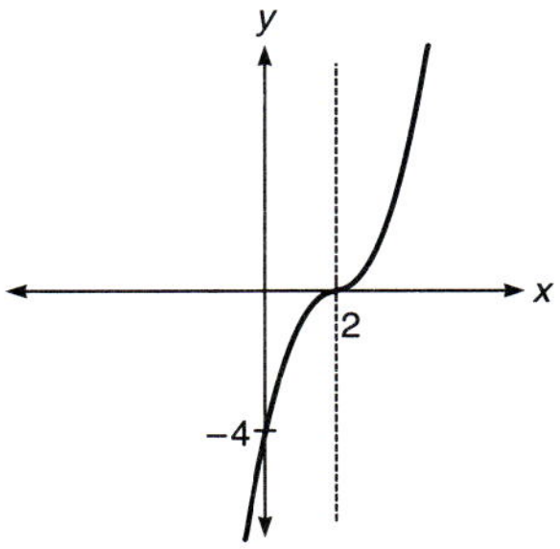

53. $y = |3x - 6| - 2$; $y = 3|x - 2| - 2$; graph of $y = |x|$ shifted down 2 units, right 2 units, vertically scaled 3 units

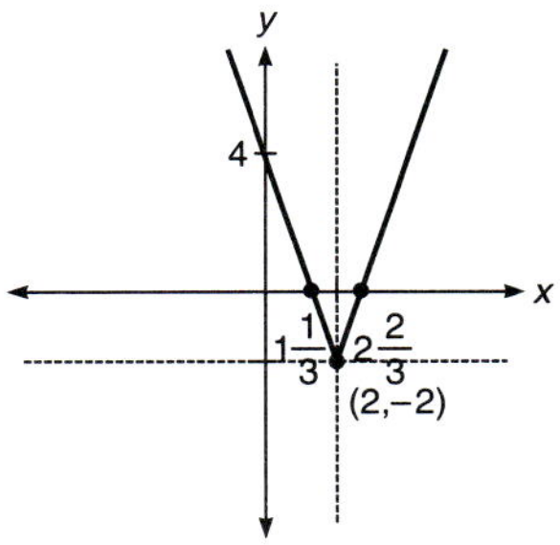

55. $y = -4\sqrt{x + 3} + 2$;
$y = -4\sqrt{x - (-3)} + 2$; graph of $y = \sqrt{x}$ shifted up 2 units, left 3 units, vertically scaled −4 units

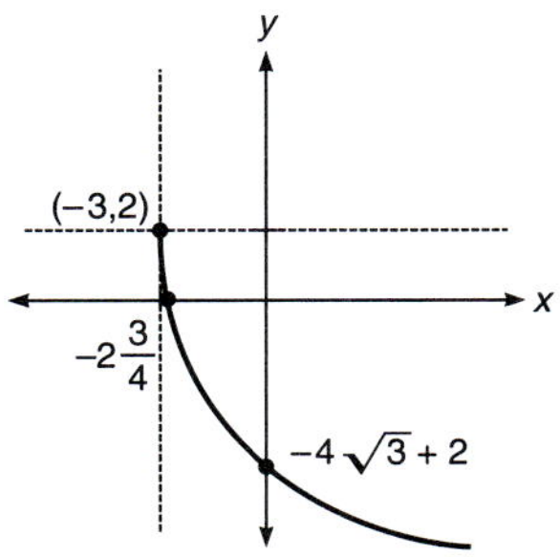

57. $y = \dfrac{-2}{x + 3} - 4$; $y = \dfrac{-2}{x - (-3)} - 4$;

graph of $y = \dfrac{1}{x}$ shifted down 4 units, left 3 units, vertically scaled −2 units

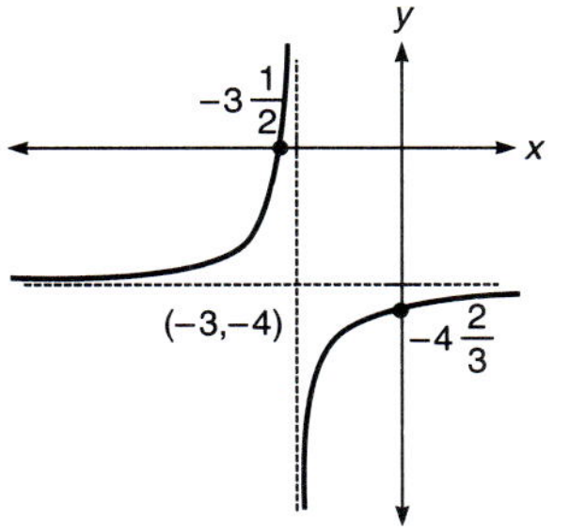

59. $y = -2(x + 1)^2 + 3$;
$y = -2(x - (-1))^2 + 3$; graph of $y = x^2$ shifted up 3 units, left 1 unit, vertically scaled −2 units

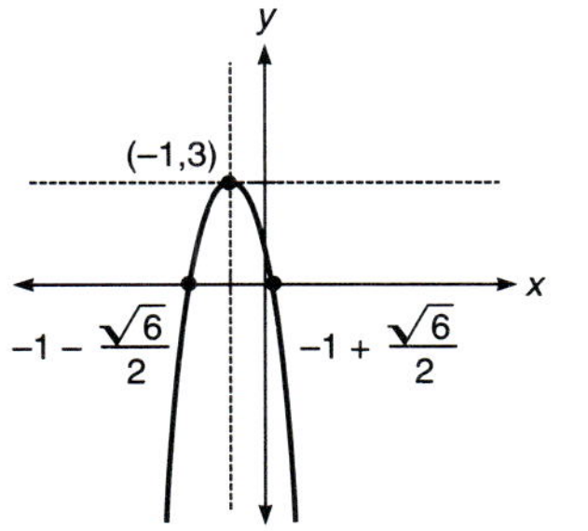

61. $y = -\frac{3}{2}(x + 1)^3$; $y = -\frac{3}{2}(x - (-1))^3$; graph of $y = x^3$ shifted left 1 unit, vertically scaled $-1\frac{1}{2}$ units

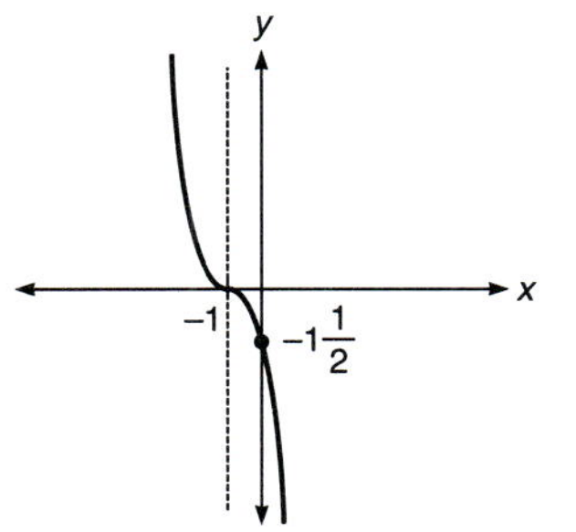

63. $y = \left|\dfrac{x}{3}\right| + 1$; graph of $y = |x|$ shifted up 1 unit, vertically scaled $\frac{1}{3}$ unit

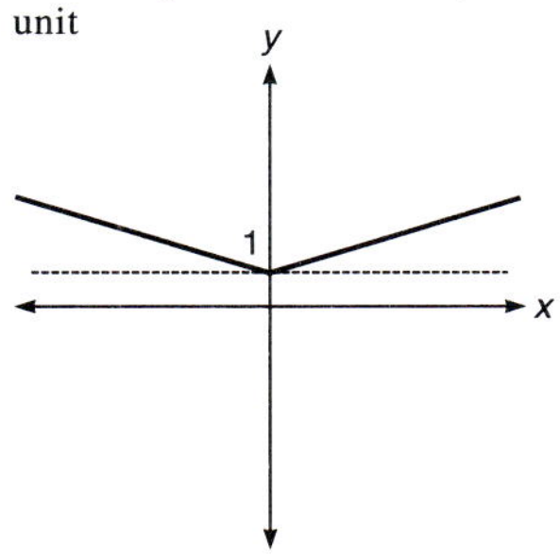

65.

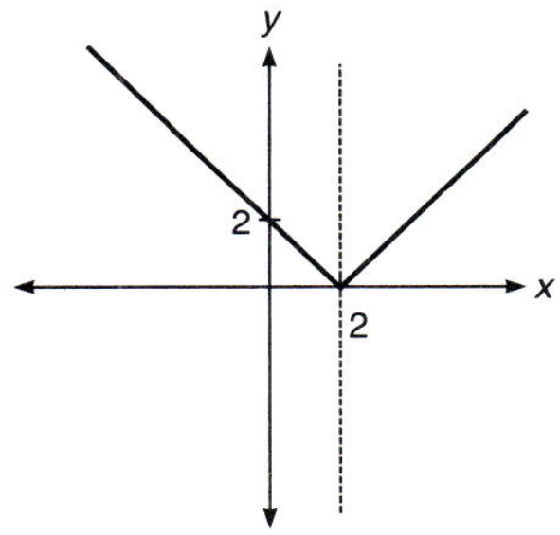

67.

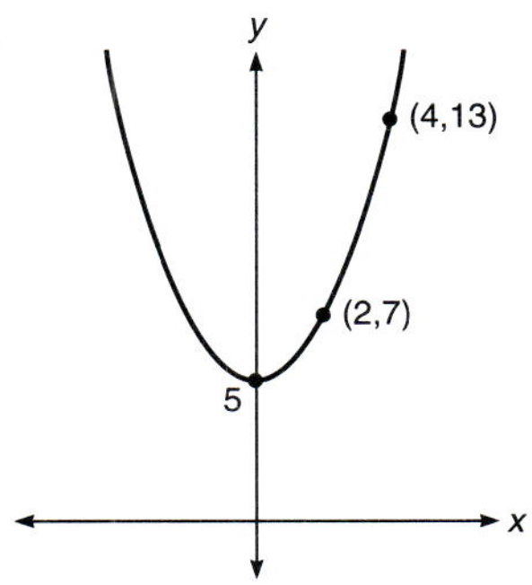

Solutions to skill and review problems

1. Find the distance between the points (1,2) and (6,8).

$(x_1,y_1) = (1,2)$; $(x_2,y_2) = (6,8)$

$$\begin{aligned} d &= \sqrt{(x_2 - x_1)^2 + (y_2 - y_1)^2} \\ &= \sqrt{(6 - 1)^2 + (8 - 2)^2} \\ &= \sqrt{5^2 + 6^2} \\ &= \sqrt{61} \end{aligned}$$

2. Find the midpoint of the line segment which joins the points (1,2) and (6,8).
$(x_1,y_1) = (1,2)$; $(x_2,y_2) = (6,8)$

$$\text{midpoint} = \left(\frac{x_1 + x_2}{2}, \frac{y_1 + y_2}{2}\right)$$
$$= \left(\frac{1 + 6}{2}, \frac{2 + 8}{2}\right)$$
$$= (3\tfrac{1}{2}, 5).$$

3. Find the equation that describes all points equidistant from the two points (1,2) and (6,8). Let (x,y) be a point which is equidistant from these two points. Then, using the distance formula (answer 1 above):

$\sqrt{(x - 1)^2 + (y - 2)^2}$
$= \sqrt{(x - 6)^2 + (y - 8)^2}$
$(x^2 - 2x + 1) + (y^2 - 4y + 4)$
$= (x^2 - 12x + 36) + (y^2 - 16y + 64)$
$-2x + 1 - 4y + 4$
$= -12x + 36 - 16y + 64$
$10x + 12y - 95 = 0$

4. Find where the lines [1] $2y - 3x = 5$ and [2] $x + y = 3$ intersect.
$y = -x + 3$
Solve [2] for y
$2(-x + 3) - 3x = 5$
Replace y by $-x + 3$ in [1]
$-2x + 6 - 3x = 5$
$1 = 5x$
$\frac{1}{5} = x$
$y = -\frac{1}{5} + 3$
Replace x by $\frac{1}{5}$ in $y = -x + 3$
$y = 2\frac{4}{5}$
The point is $(\frac{1}{5}, 2\frac{4}{5})$.

5. Find the equation of a line that is perpendicular to the line $y = -2x + 3$ and passes through the point $(1,-2)$. The slope of $y = -2x + 3$ is -2. We want a slope of $m = \frac{1}{2}$, since the slopes of perpendicular lines are negative reciprocals of each other.
$y - y_1 = m(x - x_1)$
Point-slope formula
$y - (-2) = \frac{1}{2}(x - 1)$
$y + 2 = \frac{1}{2}x - \frac{1}{2}$
$y = \frac{1}{2}x - 2\frac{1}{2}$

6. Solve $x^2 - 4x = 32$.
$x^2 - 4x - 32 = 0$
$(x - 8)(x + 4) = 0$
$x - 8 = 0$ or $x + 4 = 0$
$x = 8$ or $x = -4$

7. Solve $\frac{x}{x + y} = 3$ for x.
$x = 3(x + y)$
$x = 3x + 3y$
$-3y = 2x$
$-\frac{3}{2}y = x$

Solutions to trial exercise problems

9. $y = (x + 3)^2 - 3$
$y = (x - (-3))^2 - 3$
Graph of $y = x^2$ shifted down 3 units, left 3 units.
Vertex at $(-3,-3)$.
Intercepts:
$x = 0$: $y + 3 = (0 + 3)^2$
$y = 6$
$y = 0$: $3 = (x + 3)^2$
$\pm\sqrt{3} = x + 3$
$-3 \pm \sqrt{3} = x$

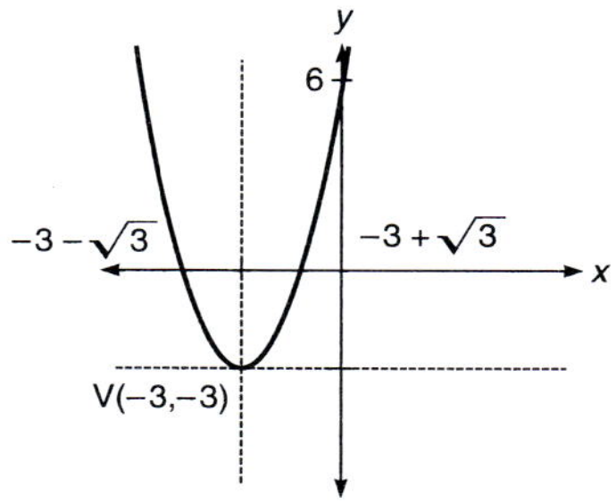

21. $y = \sqrt{x - 3} + 2$
Graph of $y = \sqrt{x}$ shifted left 3 units, up 2 units.
Vertex at (3,2).
Intercepts:
$x = 0$: $y = \sqrt{-3} + 2$, not real so no y-intercept
$y = 0$: $0 = \sqrt{x - 3} + 2$
$-2 = \sqrt{x - 3}$; a square root is nonnegative, so no x-intercept.

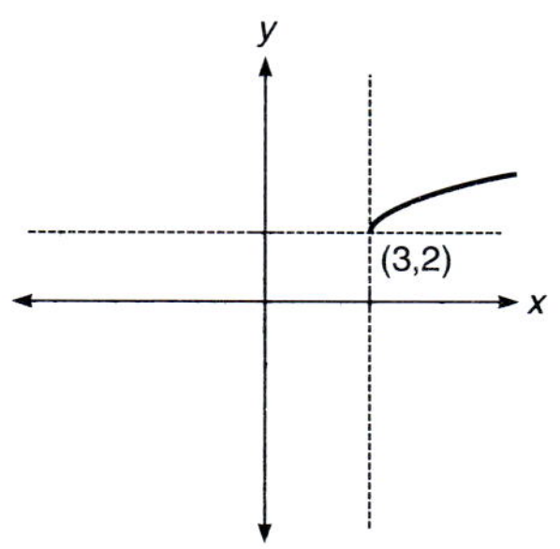

29. $y = (x + 1)^3 - 2$
$y = (x - (-1))^3 - 2$
Graph of $y = x^3$ shifted left 1 unit, down 2 units; origin at $(-1,-2)$.
Intercepts:
$x = 0$: $y + 2 = 1^3$, $y = -1$
$y = 0$: $2 = (x + 1)^3$
$\sqrt[3]{2} = x + 1$
$-1 + \sqrt[3]{2} = x \approx 0.3$

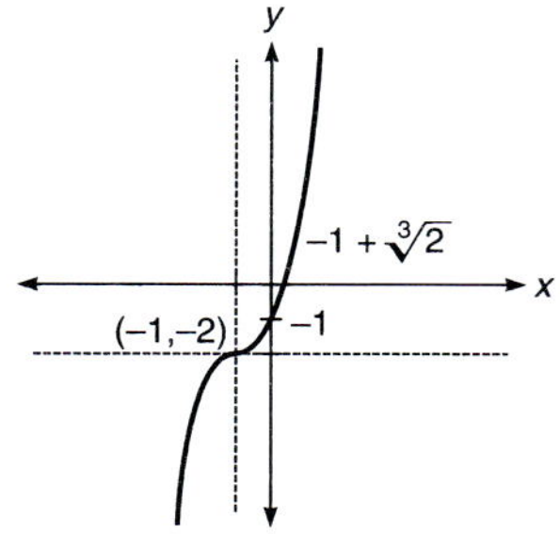

37. $y = \frac{1}{x - 3} - 5$

Graph of $y = \frac{1}{x}$ shifted right 3 units, down 5 units; origin at $(3,-5)$.
Intercepts:
$x = 0$: $y = -\frac{1}{3} - 5$
$y = -5\frac{1}{3}$
$y = 0$: $5 = \frac{1}{x - 3}$
$5(x - 3) = 1$
$x - 3 = \frac{1}{5}$
$x = 3\frac{1}{5}$

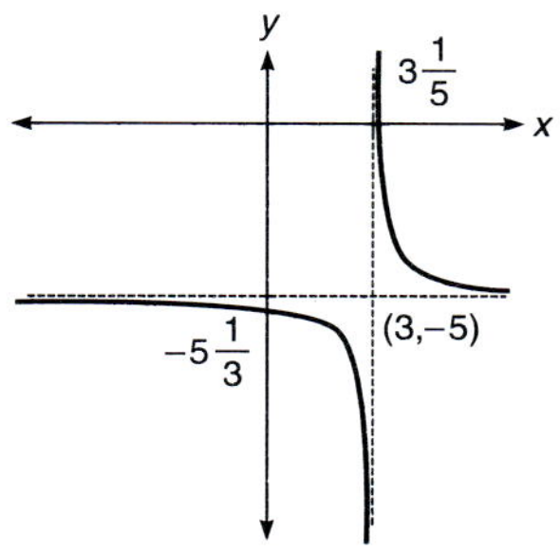

45. $y = |x - 5| - 4$
Graph of $y = |x|$ shifted right 5 units, down 4 units; origin at $(5, -4)$.
Intercepts:
$x = 0$: $y = |-5| - 4$
$= 1$
$y = 0$: $4 = |x - 5|$
$x - 5 = 4$ or $x - 5 = -4$
$x = 9$ or $x = 1$

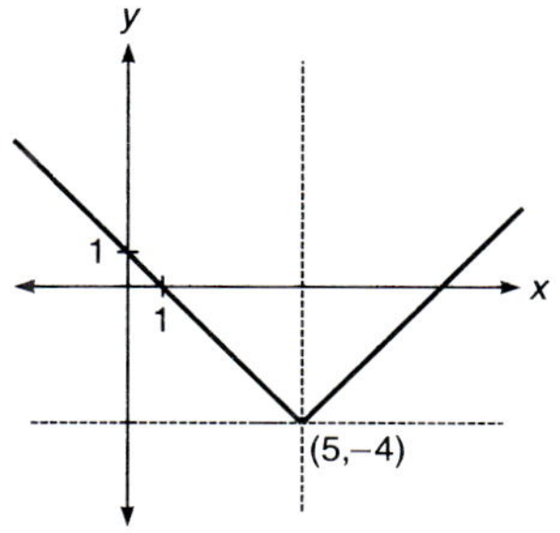

53. $y = |3x - 6| - 2$
$y = 3|x - 2| - 2$
Graph of $y = |x|$ shifted down 2 units, right 2 units, vertically scaled 3 units; origin at $(2, -2)$.
Intercepts:
$x = 0$: $y = |-6| - 2 = 4$
$y = 0$: $2 = 3|x - 2|$
$\frac{2}{3} = |x - 2|$, so
$x - 2 = \frac{2}{3}$ or $x - 2 = -\frac{2}{3}$
$x = 2\frac{2}{3}$ or $x = 1\frac{1}{3}$

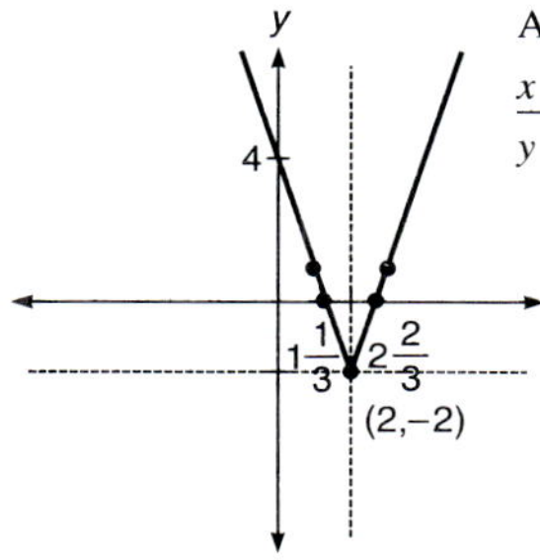

Additional points:

x	1	3
y	1	1

57. $y = \dfrac{-2}{x + 3} - 4$
$y = \dfrac{-2}{x - (-3)} - 4$
Graph of $y = \dfrac{1}{x}$ shifted down 4 units, left 3 units, vertically scaled -2 units; origin at $(-3, -4)$.
Intercepts:
$x = 0$: $y = -\frac{2}{3} - 4 = -4\frac{2}{3}$
$y = 0$: $0 = \dfrac{-2}{x + 3} - 4$
$4 = \dfrac{-2}{x + 3}$
$4x + 12 = -2$
$4x = -14$
$x = -3\frac{1}{2}$

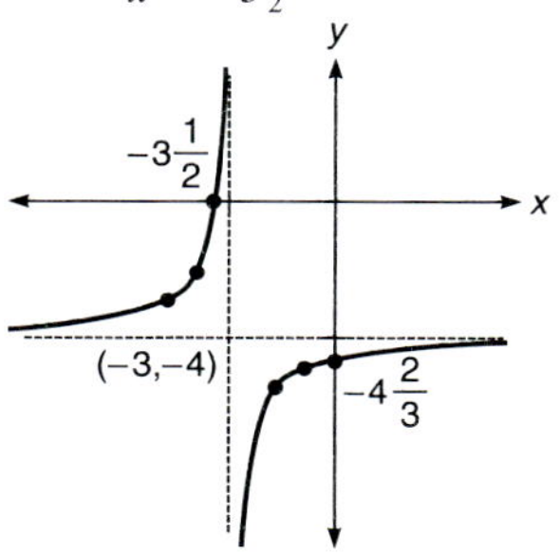

Additional points:

x	-5	-4	-2	-1
y	-3	-2	-6	-5

60. $y = \sqrt{4x - 8} - 3$
$y = \sqrt{4(x - 2)} - 3$
$y = 2\sqrt{x - 2} - 3$
Graph of $y = \sqrt{x}$ shifted down 3 units, right 2 units, vertically scaled 2 units; origin at $(2, -3)$.
Intercepts:
$x = 0$: $y = \sqrt{-8} - 3$, so no y-intercept
$y = 0$: $3 = 2\sqrt{x - 2}$
$\frac{3}{2} = \sqrt{x - 2}$
$\frac{9}{4} = x - 2$
$\frac{17}{4} = x$

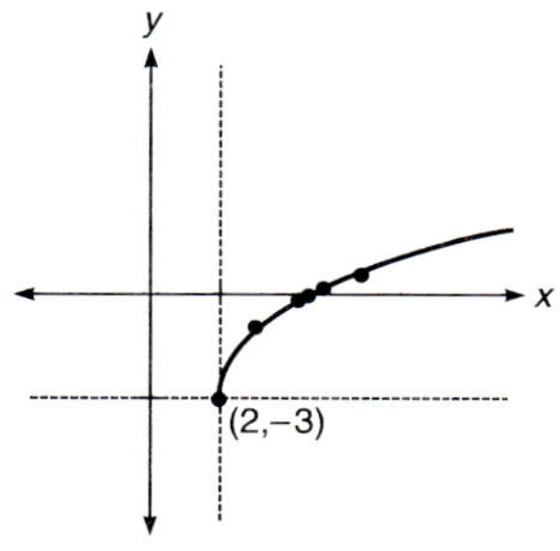

Additional points:

x	3	4	5	6
y	-1	-0.2	0.46	1

Exercise 3–5

Answers to odd-numbered problems

1. $C(0,0)$, $r = \sqrt{16} = 4$

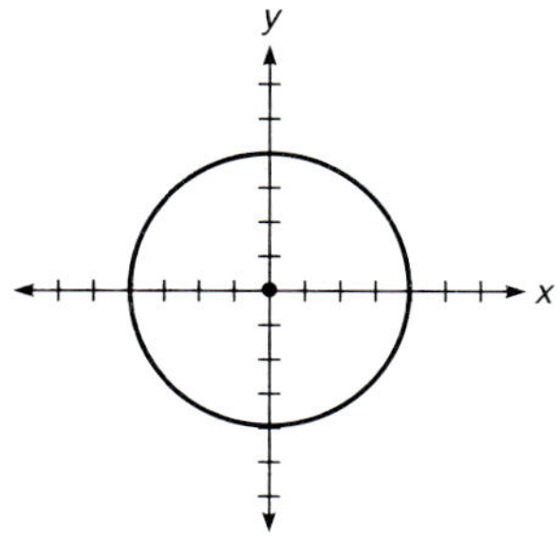

3. $C(-2,0)$; $r = \sqrt{20} = 2\sqrt{5} \approx 4.47$

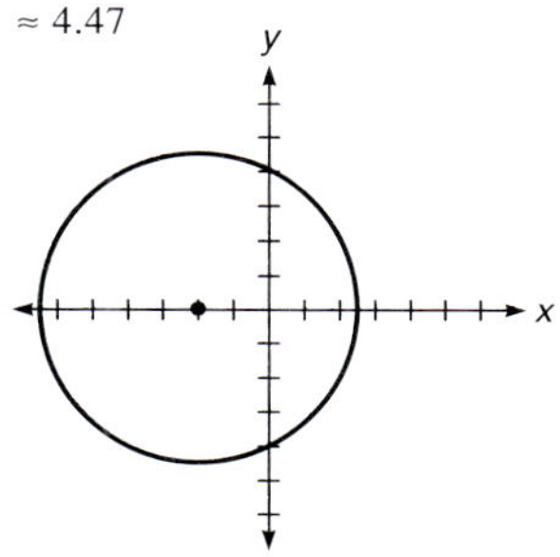

5. $C(0,4)$, $r = 3$

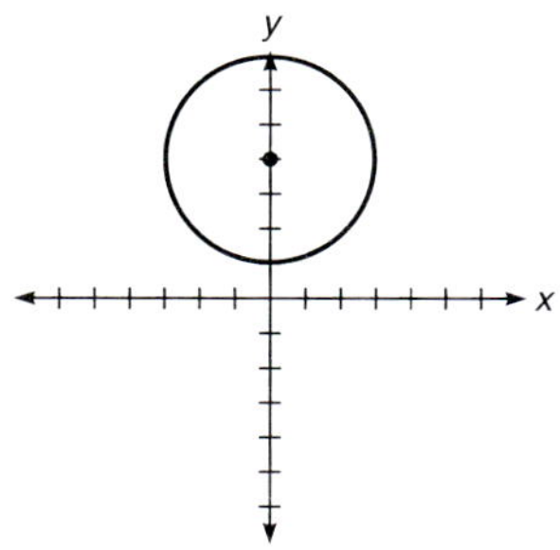

7. $C(1,4)$, $r = \sqrt{8} = 2\sqrt{2} \approx 2.8$

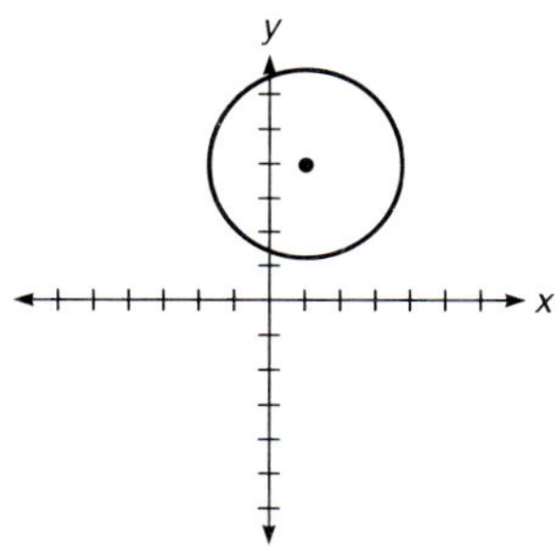

9. $C(-3,2)$, $r = \sqrt{20} = 2\sqrt{5} \approx 4.47$

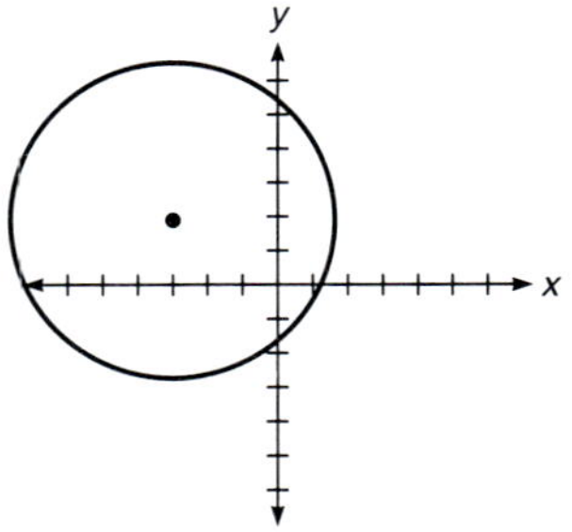

11. $x^2 + (y - 3)^2 = 15$
$C(0,3)$, $r = \sqrt{15} \approx 3.9$

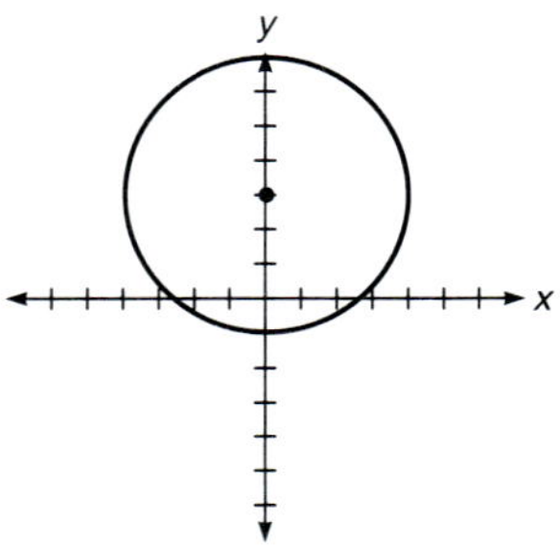

13. $(x + \frac{5}{2})^2 + y^2 = \frac{41}{4}$
$C(-\frac{5}{2},0)$, $r = \sqrt{\frac{41}{4}} = \frac{\sqrt{41}}{2} \approx 3.2$

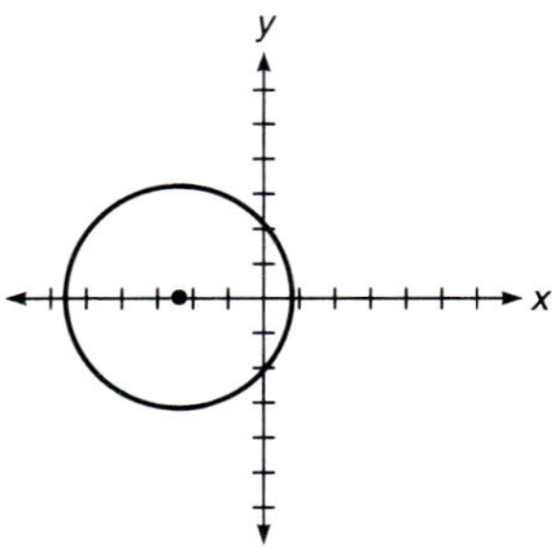

15. $(x - \frac{1}{2})^2 + (y - 2)^2 = \frac{53}{4}$
$C(\frac{1}{2},2)$, $r = \frac{\sqrt{53}}{2} \approx 3.6$

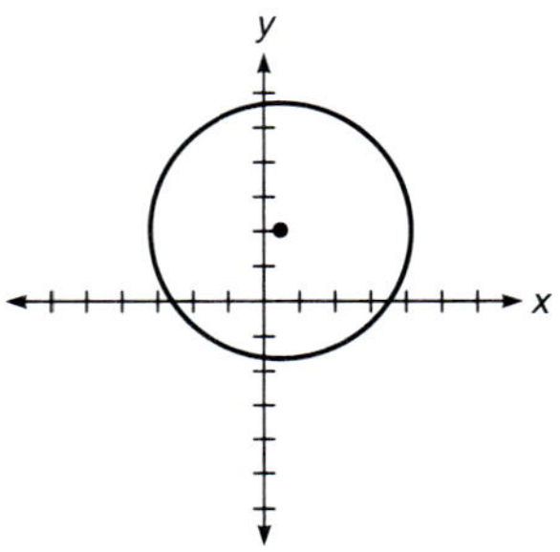

17. $(x - 1)^2 + (y + 2)^2 = 0$
$C(1,-2)$, $r = 0$
With $r = 0$ this "circle" is just the point $(1,-2)$.

19. $(x + 2)^2 + y^2 = -2$ Since the left side of the equation is nonnegative there are no real solutions to the equation, and so there is no graph.

21. $x^2 + (y - \frac{1}{6})^2 = \frac{121}{36}$
$C(0,\frac{1}{6})$, $r = \frac{11}{6}$

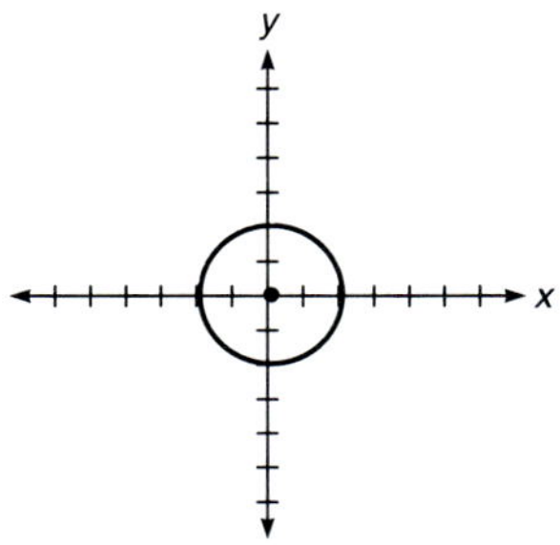

23. $(x - \frac{1}{5})^2 + (y + \frac{1}{2})^2 = \frac{89}{100}$
$C(\frac{1}{5},-\frac{1}{2})$, $r = \frac{\sqrt{89}}{10} \approx 0.9$

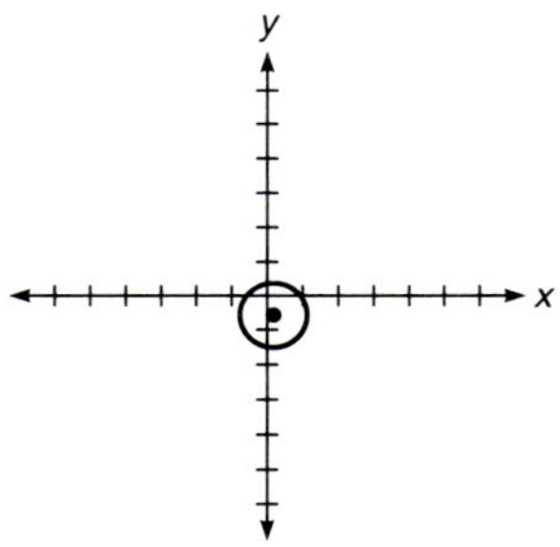

25. $(x - \frac{3}{2})^2 + (y + \frac{1}{2})^2 = \frac{1}{2}$
$C(\frac{3}{2},-\frac{1}{2})$, $r = \sqrt{\frac{1}{2}} = \frac{\sqrt{2}}{2} \approx 0.7$

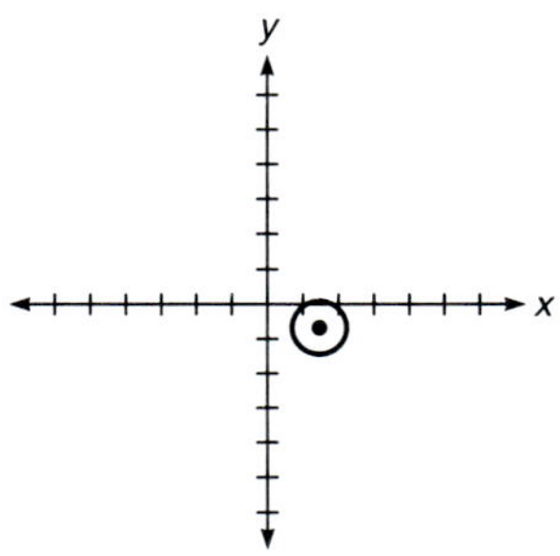

27. $x^2 + 6x + y^2 - 4y + 9 = 0$
29. $x^2 - 4x + y^2 - (6 - 2\sqrt{2})y + 10 - 6\sqrt{2} = 0$ 31. $x^2 + y^2 - 6y - 55 = 0$
33. $x^2 - 2x + y^2 + 6y - 63 = 0$
35. $x^2 - 6x + y^2 - 10y - 24 = 0$
37. function (passes vertical line test); not one to one (fails horizontal line test)
39. not a function (fails vertical line test)
41. 40% 43. 0.35 45. a. 0.7; b. -0.6 47. -2.2, -0.4, 1.3, 1.8
49. -1.25 to 0.5, 1.5 to 2.7
51. even, y-axis symmetry
53. odd, origin symmetry
55. even, y-axis symmetry
57. odd, origin symmetry
59. neither even nor odd
61. neither even nor odd
63. odd, origin symmetry
65. even, y-axis symmetry
67. odd, origin symmetry
69. $y = \frac{2}{3}x - 7$
71. $(x - 3)^2 + (y - 2)^2 = \frac{9}{16}$

Solutions to skill and review problems

1. Graph $f(x) = x^2 - 4$.
$y = x^2 - 4$
Graph of $y = x^2$ shifted down 4 units.
Vertex at $(0,-4)$.
Intercepts:
$x = 0$: $y = 0^2 - 4 = -4$
$y = 0$: $0 = x^2 - 4$
$4 = x^2$; $\pm 2 = x$

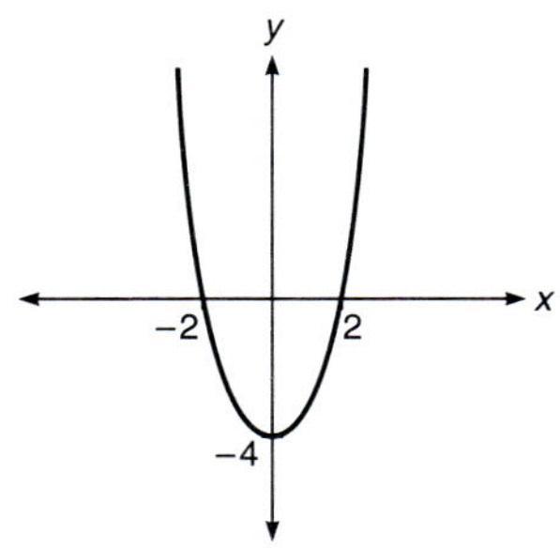

2. Graph $f(x) = (x - 4)^2$.
$y = (x - 4)^2$
Graph of $y = x^2$ shifted right 4 units.
Vertex at (4,0).
Intercepts:
$x = 0$: $y = (0 - 4)^2 = 16$
$y = 0$: $0 = (x - 4)^2$
$0 = x - 4$; $4 = x$

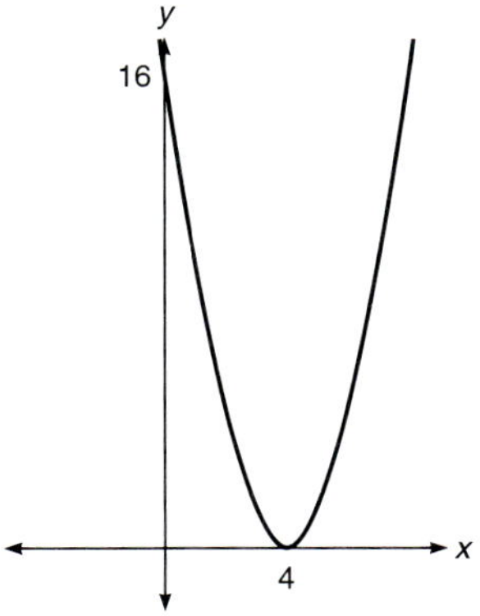

3. Graph $f(x) = (x - 4)^2 - 4$.
$y = (x - 4)^2 - 4$
Graph of $y = x^2$ shifted right 4 units, down 4 units.
Vertex at $(4,-4)$.
Intercepts:
$x = 0$: $y = (0 - 4)^2 - 4$
$= 12$
$y = 0$: $0 = (x - 4)^2 - 4$
$4 = (x - 4)^2$
$\pm 2 = x - 4$
$4 \pm 2 = x$; $x = 2$ or 6

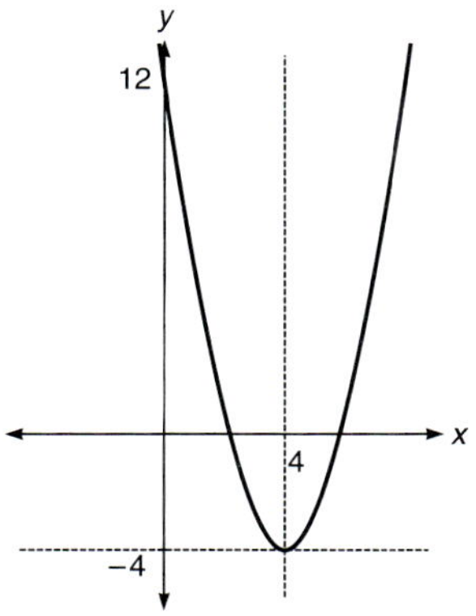

4. Solve $|2x - 3| = 8$.
$2x - 3 = 8$ or $2x - 3 = -8$
$2x = 11$ or $2x = -5$
$x = 5\frac{1}{2}$ or $x = -2\frac{1}{2}$
$\{-2\frac{1}{2}, 5\frac{1}{2}\}$

5. Factor $x^6 - 64$.
$(x^3 - 8)(x^3 + 8)$
Difference of two squares
$(x - 2)(x^2 + 2x + 4)(x + 2)(x^2 - 2x + 4)$
Difference of two cubes

6. Find the equation of the line that passes through the points $(-4,1)$ and $(3,-5)$.
$(x_1,y_1) = (-4,1)$; $(x_2,y_2) = (3,-5)$
$$m = \frac{y_2 - y_1}{x_2 - x_1} = \frac{-5 - 1}{3 - (-4)} = \frac{-6}{7}$$
$$= -\frac{6}{7}$$
$y - y_1 = m(x - x_1)$
$y - 1 = -\frac{6}{7}(x - (-4))$
$y - 1 = -\frac{6}{7}x - \frac{24}{7}$
$y = -\frac{6}{7}x - \frac{24}{7} + \frac{7}{7}$
$y = -\frac{6}{7}x - \frac{17}{7}$

Solutions to trial exercise problems

9. $(x + 3)^2 + (y - 2)^2 = 20$
$C(-3,2)$, $r = \sqrt{20} = 2\sqrt{5} \approx 4.47$

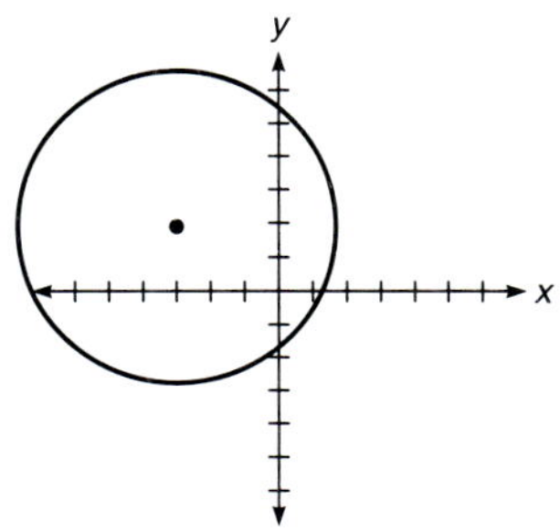

21. $3x^2 + 3y^2 - y - 10 = 0$
$x^2 + y^2 - \frac{1}{3}y = \frac{10}{3}$
$\frac{1}{2}(-\frac{1}{3}) = -\frac{1}{6}$, $(-\frac{1}{6})^2 = \frac{1}{36}$,
$x^2 + y^2 - \frac{1}{3}y + \frac{1}{36} = \frac{10}{3} + \frac{1}{36}$
$x^2 + (y - \frac{1}{6})^2 = \frac{121}{36}$
$C(0,\frac{1}{6})$, $r = \frac{11}{6}$

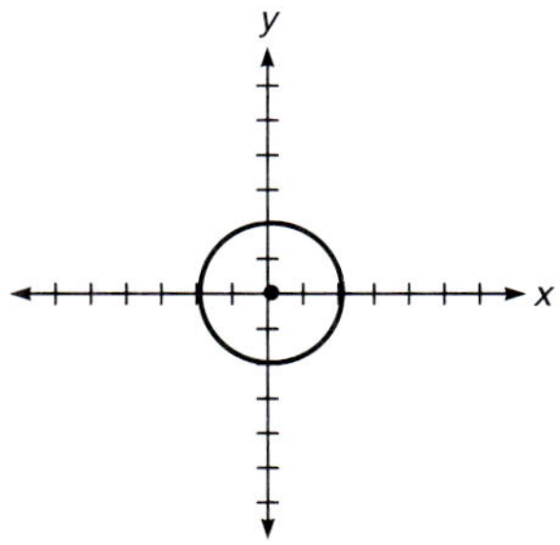

29. $(h,k) = (2,3 - \sqrt{2})$, $r = \sqrt{5}$
$(x - 2)^2 + (y - (3 - \sqrt{2}))^2 = (\sqrt{5})^2$
$(x - 2)^2 + (y - (3 - \sqrt{2}))^2 = 5$
$x^2 - 4x + 4 + y^2 - 2(3 - \sqrt{2})y + (3 - \sqrt{2})^2 = 5$
$x^2 - 4x + 4 + y^2 - 2(3 - \sqrt{2})y + 9 - 6\sqrt{2} + 2 = 5$
$x^2 - 4x + y^2 - (6 - 2\sqrt{2})y + 10 - 6\sqrt{2} = 0$

---

50. For what values of x is $f(x) = x$?
The graph shows the line $y = x$, superimposed on the graph of f. Where it meets the graph of f, $f(x) = x$. These are approximately -0.75, 0, 0.65.

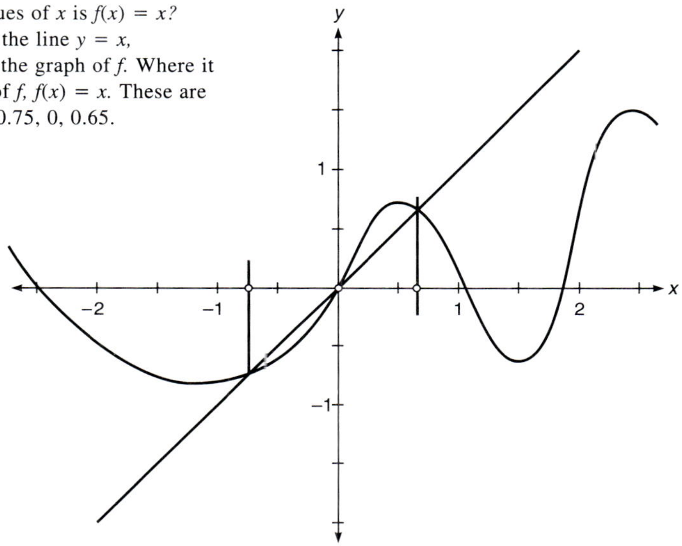

54. $f(x) = x^5 - 4x^3 - x$
$f(-x) = (-x)^5 - 4(-x)^3 - (-x)$
$= -x^5 + x^4 + x - f(x)$
$= -(x^5 - 4x^3 - x) = -x^5 + 4x^3 + x$
$f(-x) = -f(x)$: odd, origin symmetry

58. $f(x) = x^4 - x - 2$
$f(-x) = x^4 + x - 2$
$-f(x) = -x^4 + x + 2$
thus $f(-x) \neq f(x)$ and $f(-x) \neq -f(x)$, so f is neither even nor odd

70.

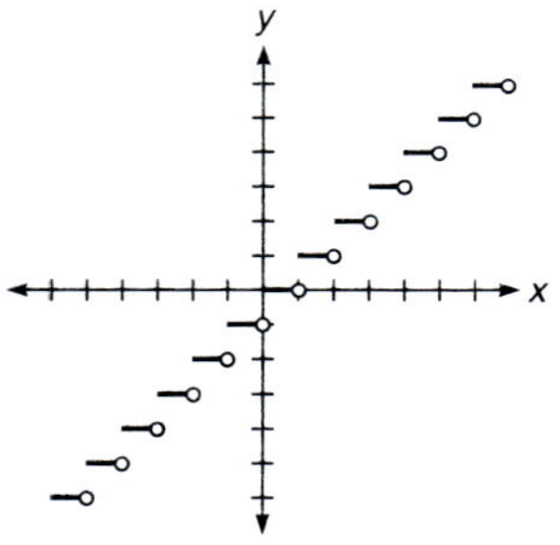

Chapter 3 review

In problems 1 through 6 the three points you use may differ from those shown, but the x- and y-intercepts, and the graph, should be the same.

1. $(0,-4)$, $(1,-\frac{13}{2})$, $(2,-9)$, $(-1\frac{3}{8},0)$

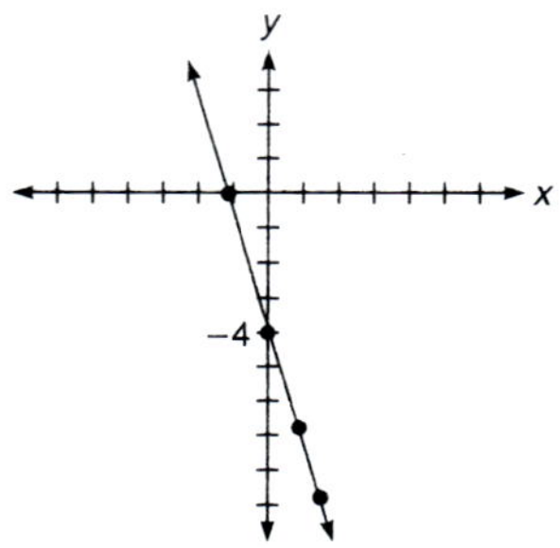

2. $(-3,-1)$, $(0,-\frac{3}{4})$, $(3,-\frac{1}{2})$, $(9,0)$

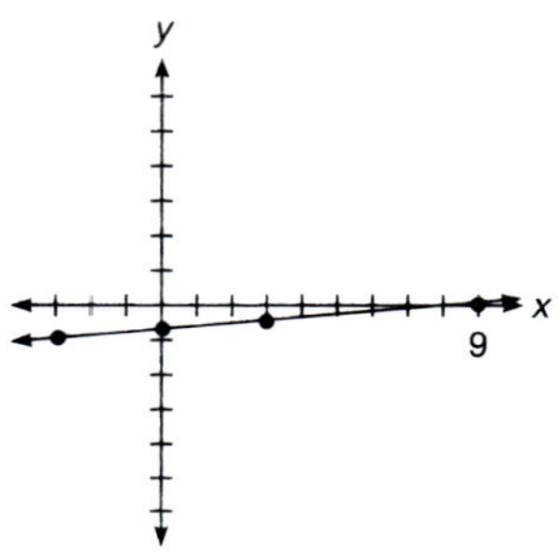

3. $(-2,1)$, $(0,4)$, $(2,7)$, $(-2\frac{2}{3},0)$

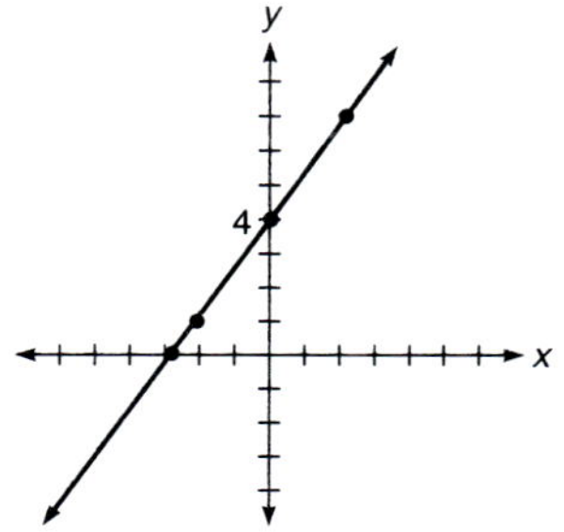

4. $(-4,-2)$, $(-4,0)$, $(-4,2)$

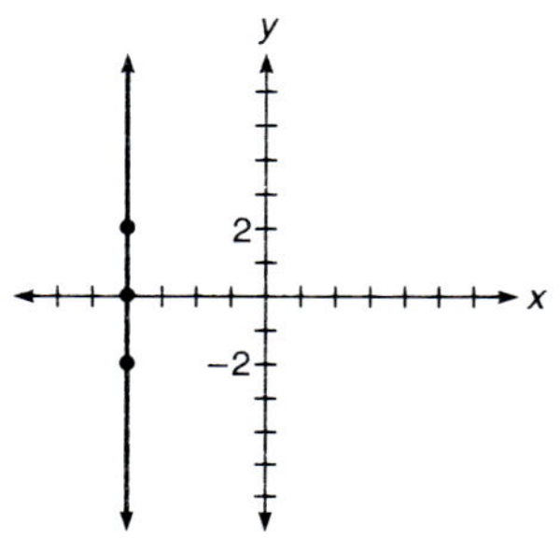

5. $(-4,-12)$, $(0,-3)$, $(4,6)$, $(1\frac{1}{3},0)$

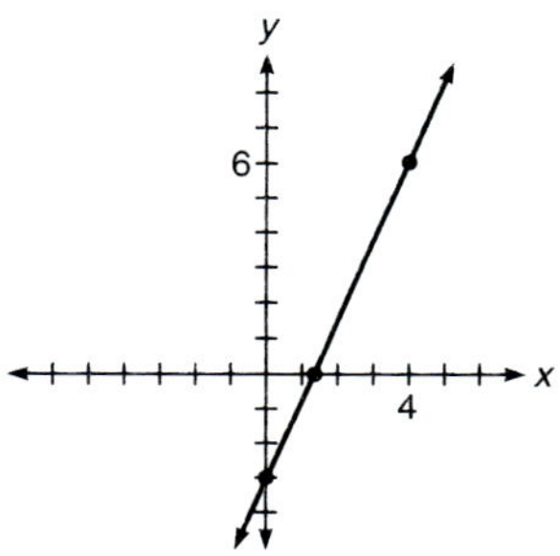

6. $(-2,-9)$, $(0,-6)$, $(4,0)$

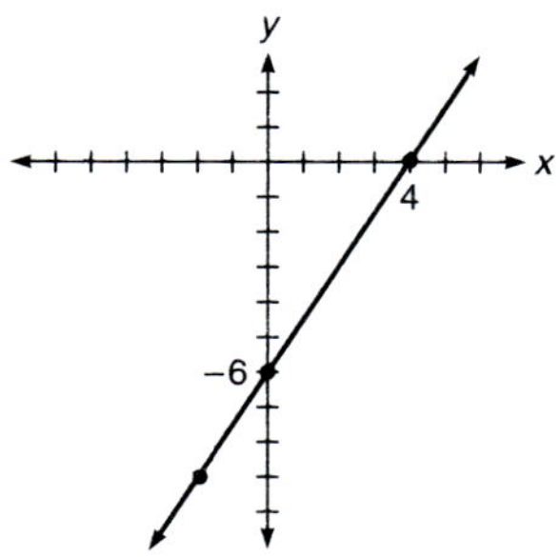

7. $(0,-100)$ I-intercept
$(1{,}111\frac{1}{9}, 0)$ p-intercept

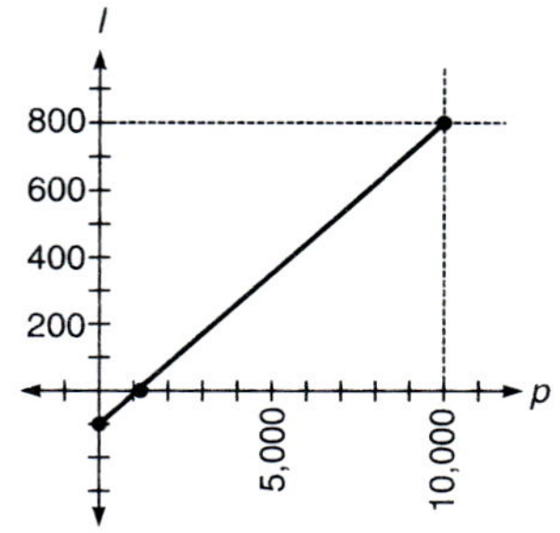

8. $(2,-2\frac{1}{2})$ **9.** $(-4,\frac{3}{2}\sqrt{2})$

10. $\sqrt{65}$ **11.** $3\sqrt{3}$

12. $(3,0)$, $(0,-\frac{5}{3})$
$m = \frac{5}{9}$

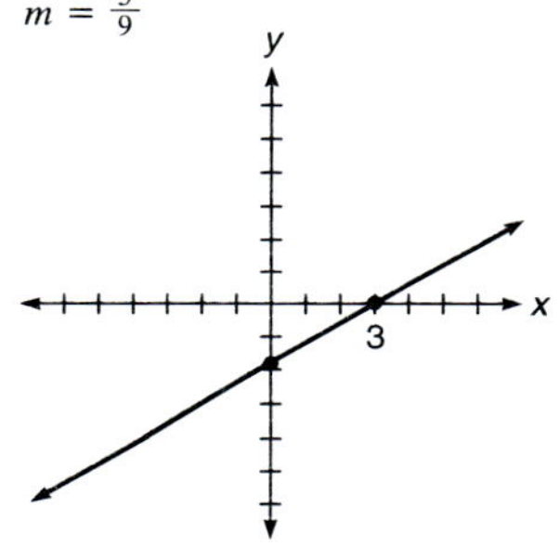

13. $(2,0)$, $(0,3)$
$m = -1\frac{1}{2}$

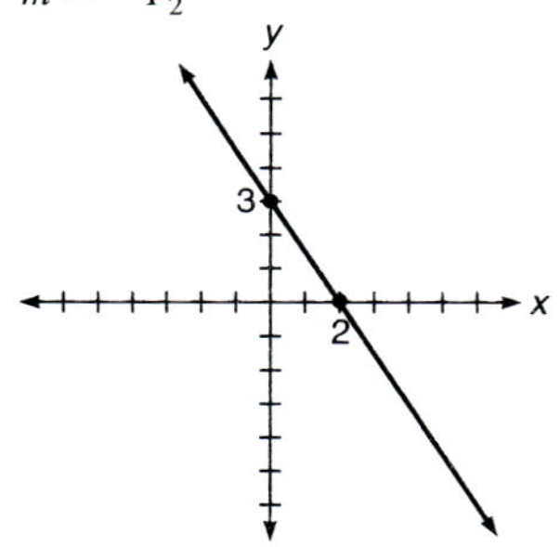

14. $(-2,0)$, $(0,\frac{4}{3})$
$m = \frac{2}{3}$

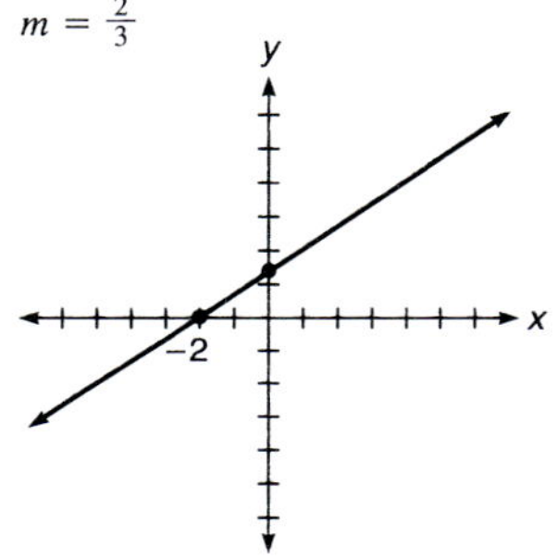

15. $(0,\frac{4}{9})$
$m = 0$

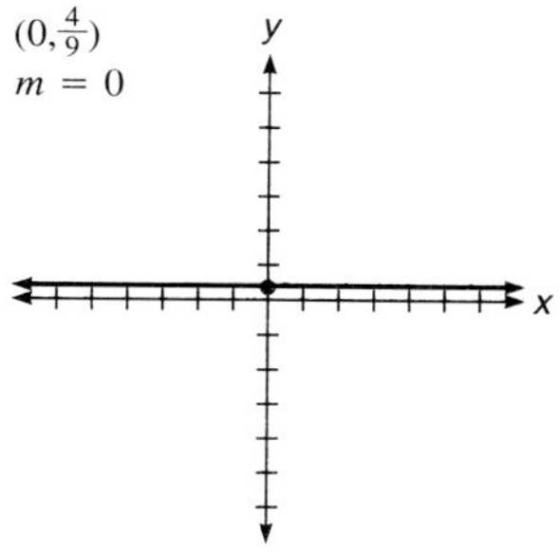

16. $(8.5,0)$
m is undefined

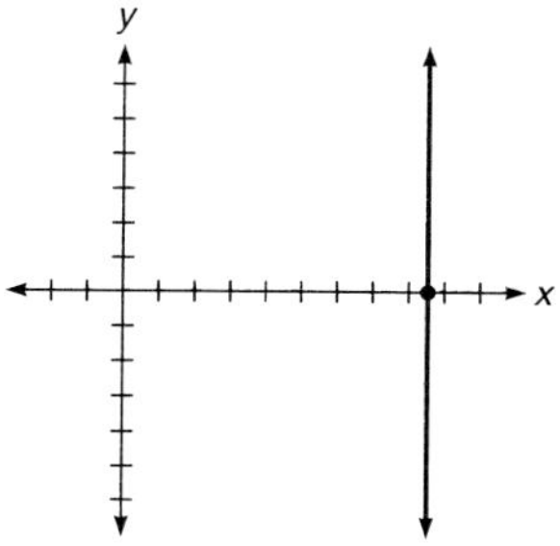

17. $\frac{3}{8}$ **18.** $-\frac{4}{3}$ **19.** $\frac{5\sqrt{3}}{3}$
20. $y = -\frac{4}{3}x + \frac{11}{3}$ **21.** $y = 4x - 11$
22. $y = -\frac{2}{3}x - 3$ **23.** $y = -6x - 1$
24. $y = \frac{3}{2}x - \frac{5}{2}$ **25.** $y = \frac{2}{5}x - \frac{9}{5}$
26. $(2\frac{2}{7},\frac{6}{7})$ **27.** $(1\frac{1}{2},6\frac{1}{2})$
28. $16x + 6y - 27 = 0$
29. Let (a,b) and (c,d) be two points on the line $y = 3x - 4$ so that $a \neq c$. Then $b = 3a - 4$ and $d = 3c - 4$.

$$m = \frac{y_2 - y_1}{x_2 - x_1}$$
$$m = \frac{d - b}{c - a} \text{ if } c \neq a$$
$$m = \frac{(3c - 4) - (3a - 4)}{c - a}$$
$$m = \frac{3c - 3a}{c - a}$$
$$m = \frac{3(c - a)}{c - a}$$
$$m = 3$$

30. 125 seconds **31.** It is not a function since it is not true that all the first elements are different. Domain $\{-3, 1, 4\}$; range $\{2, 4, 5, 8\}$. **32.** Function since all first elements are different. Not one to one since it is not true that all the second elements are different. Domain $\{-3, -2, -1, 0, 1\}$; range $\{1, 3, 4, 5\}$.

33. Function since all first elements are different. One to one since all the second elements are different. Domain $\{-10, 2, 3, 4\}$; range $\{-5, 2, 12, 13\}$. **34.** Not a function since it is not true that all the first elements are different. Domain $\{3, \pi, 17\}$; range $\{-\sqrt{2}, \frac{8}{13}, \sqrt{2}, \pi\}$. **35.** $\{(-3,3), (9,-1), (\sqrt{18}, 2 - \sqrt{2}), (\frac{3}{4}, 1\frac{3}{4}), (\pi, 2 - \frac{\pi}{3})\}$; one-to-one function **36.** $\{(-3,\sqrt{13}), (-3,-\sqrt{13}), (-2,\sqrt{10}), (-2,-\sqrt{10}), (-1,\sqrt{7}), (-1,-\sqrt{7}), (0,2), (0,-2), (1,1), (1,-1)\}$; not a function **37.** Implied domain: $x \neq \frac{3}{4}$; $f(-4) = \frac{15}{19}$, $f(0) = -\frac{1}{3}$, $f(\frac{1}{2}) = -\frac{3}{4}$,
$$f(3\sqrt{5}) = \frac{-44}{12\sqrt{5} - 3},$$
$$f(c - 2) = \frac{-c^2 + 4c - 3}{4c - 11}$$
38. Implied domain: $x \leq \frac{1}{2}$; $g(-4) = 6\sqrt{3}$, $g(0) = 2\sqrt{3}$, $g(\frac{1}{2}) = 0$, $3\sqrt{5} > \frac{1}{2}$, so it is not in the domain of g, $g(c - 2) = 2\sqrt{15 - 6c}$
39. Implied domain: R; $v(-4) = -5$, $v(0) = 3$, $v(\frac{1}{2}) = 1\frac{3}{4}$, $v(3\sqrt{5}) = -42 - 6\sqrt{5}$, $v(c - 2) = -c^2 + 2c + 3$
40. Implied domain: $x < -3$ or $x > 2$; $g(-4) = -\frac{2}{3}\sqrt{6}$. 0 is not in the domain of g, $\frac{1}{2}$ is not in the domain of g, $g(3\sqrt{5})$
$$= \frac{-4}{\sqrt{3\sqrt{5} + 39}},\ g(c - 2) = \frac{-4}{\sqrt{c^2 - 3c - 4}}$$
41. $2x - 5 + h$
42. **a.** 89, **b.** -1, **c.** $a^6 - 3a^3 + 1$
43. **a.** 29, **b.** 1, **c.** $\frac{-19a - 11b}{a + b}$
44. **a.** $\frac{9}{10}$, **b.** $-\frac{27}{34}$
45. $(\frac{3}{5},0), (0,-3)$

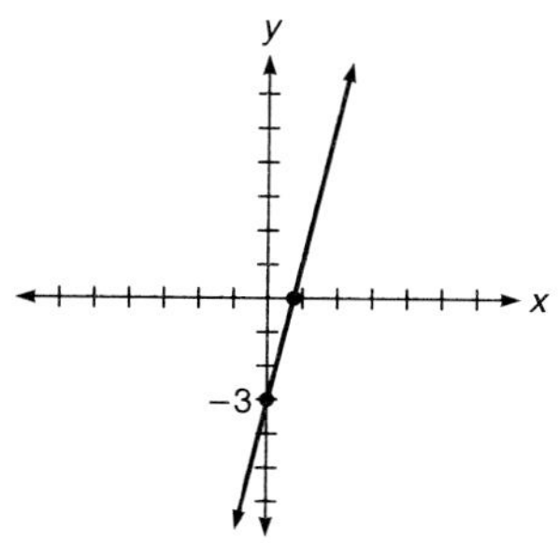

46. $f(x) = 1.25x - 21.5$, $-4°$

47. vertex: $(3\frac{1}{2},5)$; y-intercept: $(0,17.25)$

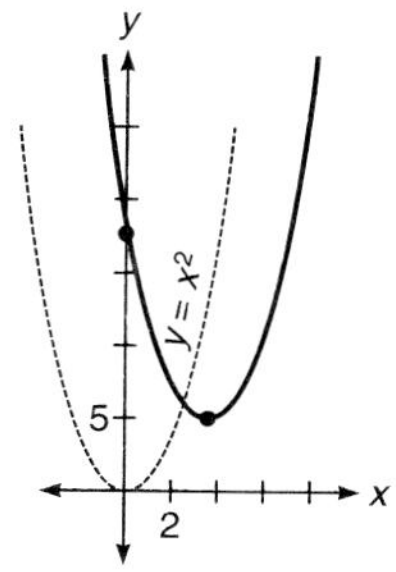

48. vertex: $(-1,-2)$; x-intercept: $(-1 + \sqrt{2},0)$, $(-1 - \sqrt{2},0)$; y-intercept: $(0,-1)$

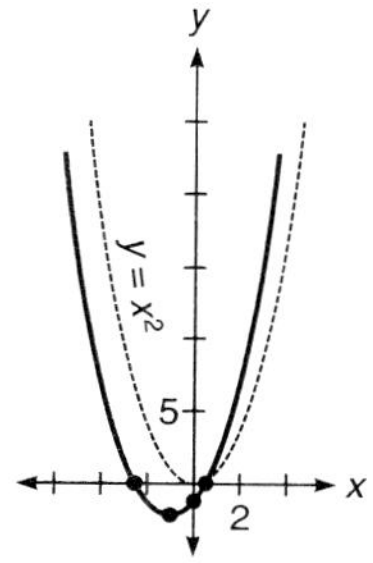

49. origin shifted to $(-2\frac{1}{2},-5)$; x-intercept: $(22.5,0)$; y-intercept: $\frac{\sqrt{10} - 10}{2} \approx -3.4$

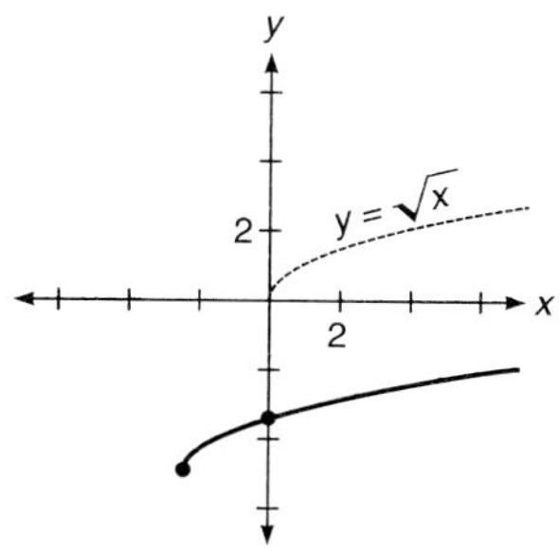

50. origin: $(-4,4)$; y-intercept $(0,6)$

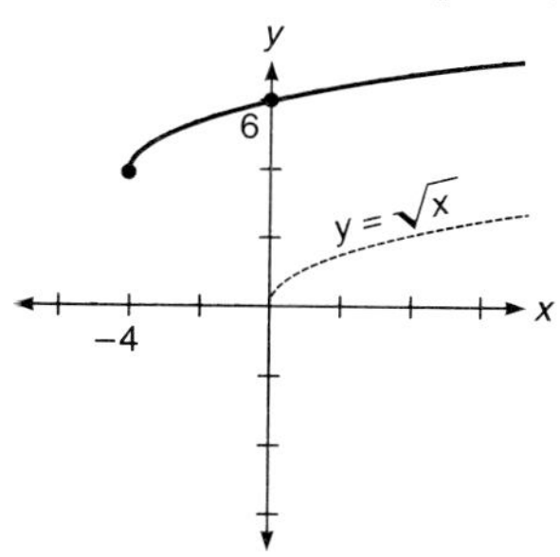

51. origin: (2,8); y-intercept: (0,0); x-intercept: (0,0)

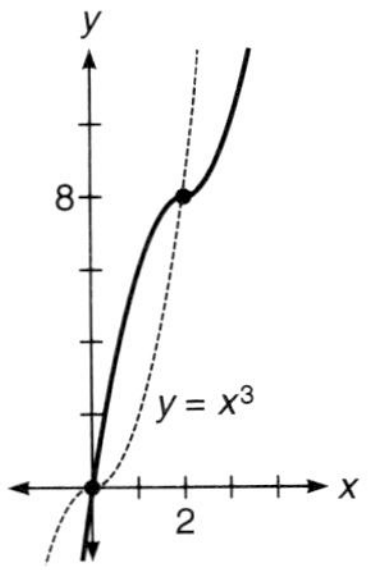

52. x-intercept: (2,0); y-intercept: (0,−8)

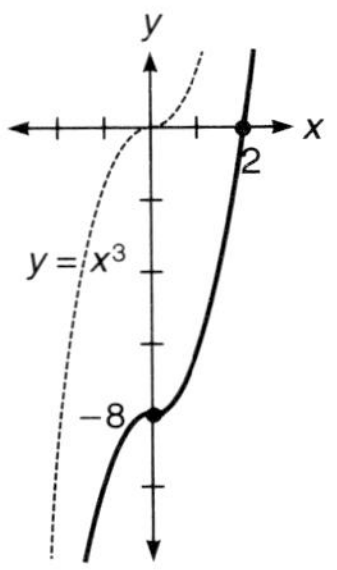

53. origin: (3,−5); x-intercept: $(3\frac{1}{5},0)$; y-intercept: $(0,-5\frac{1}{3})$

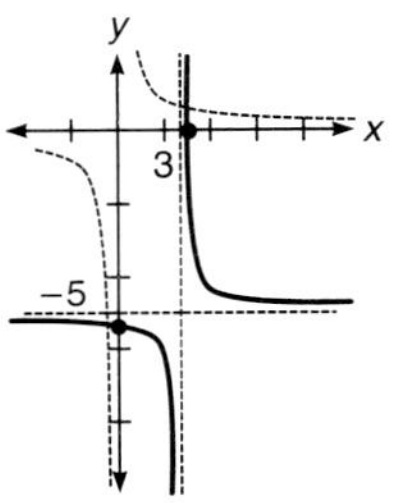

54. graph of $y = \dfrac{1}{x}$ raised up 2 units; x-intercept: $(-\frac{1}{2},0)$

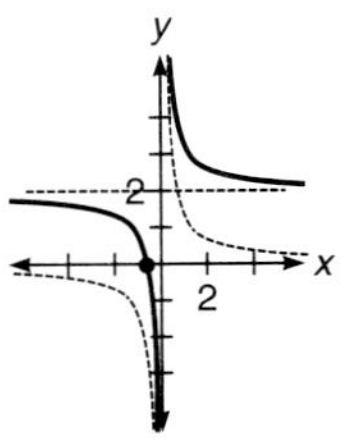

55. origin: (−5,−5); x-intercept: (0,0), (−10,0); y-intercept: (0,0)

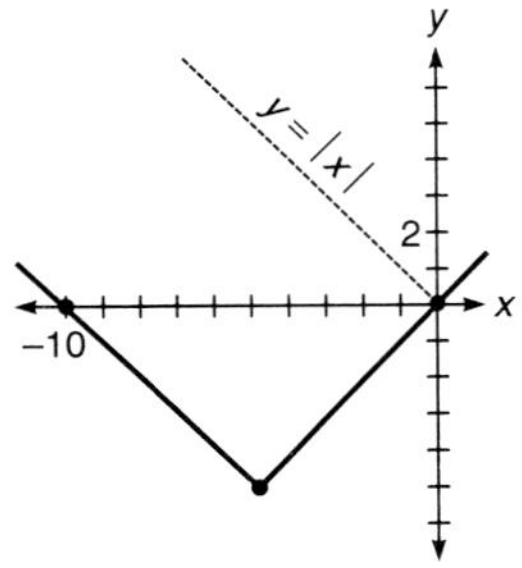

56. graph of $y = |x|$ shifted left 2 units, up 3 units; y-intercept: (0,5)

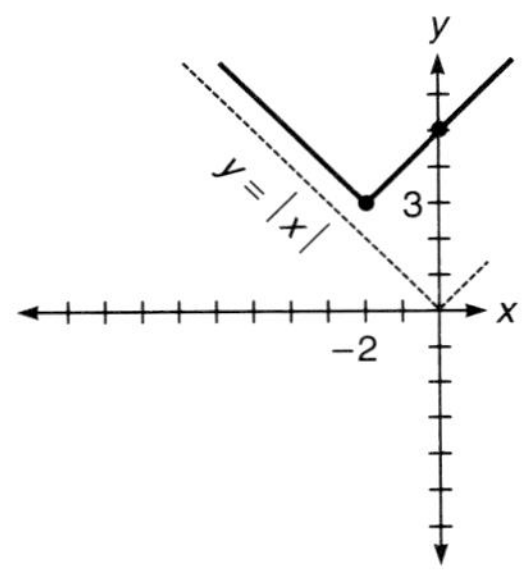

57. graph of $y = x^2$ but flipped over, vertically scaled by 3, and origin shifted to (1,4); x-intercept: (−0.2,0), (2.2,0); y-intercept: (0,1)

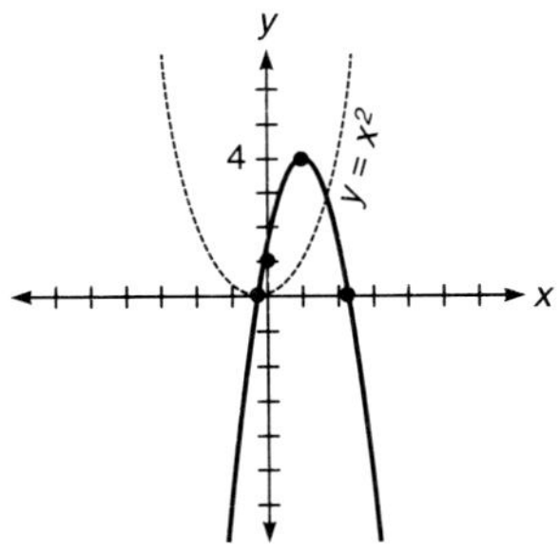

58. graph of $y = \sqrt{x}$ shifted to a new origin of (8,−5), vertically scaled by 3 units; x-intercept: $(10\frac{7}{9},0)$

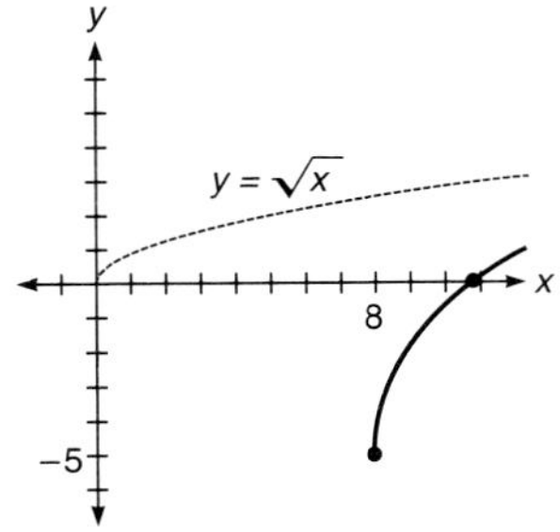

59. graph of $y = x^3$ shifted to a new origin of (2,1), vertically scaled by 2 units; x-intercept: $\left(2 - \dfrac{\sqrt[3]{4}}{2},0\right)$; y-intercept: (0,−15)

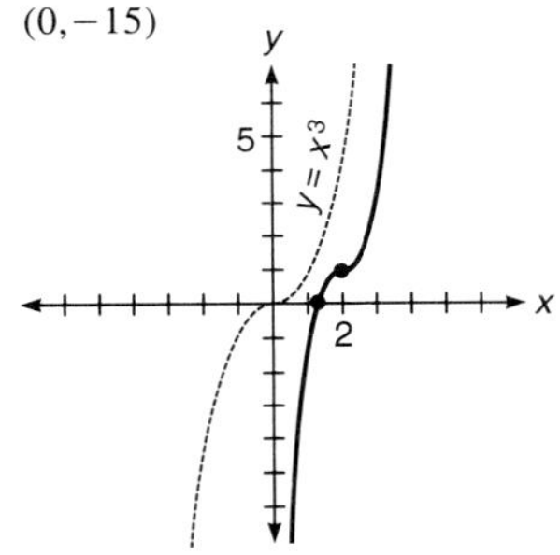

60. graph of $y = \dfrac{1}{x}$ shifted to a new origin of (−4,−1), vertically scaled by 4; all intercepts at the origin

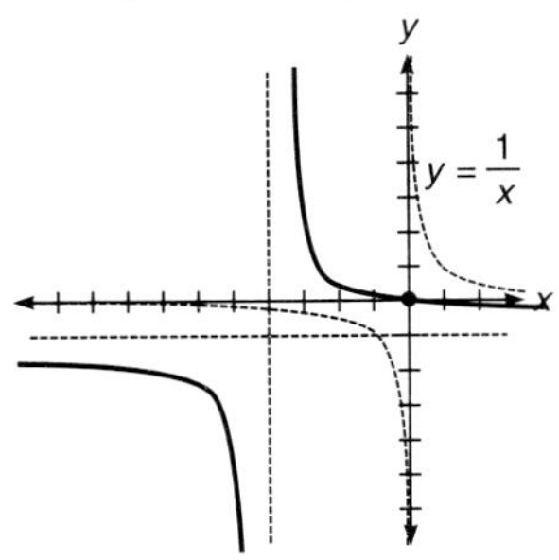

61. center: $(0,-4)$; $r = \sqrt{12} = 2\sqrt{3} \approx 3.46$

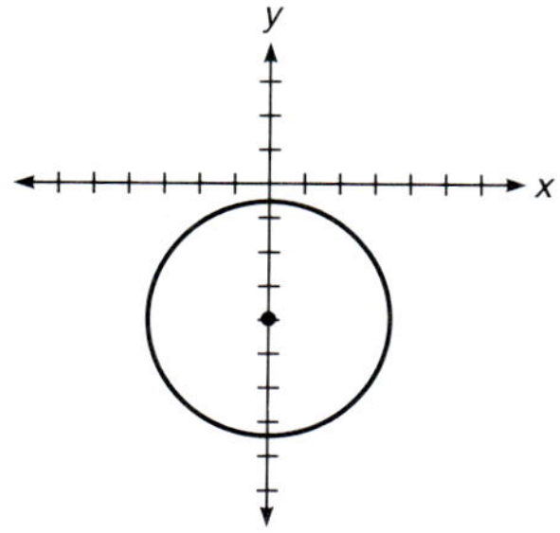

62. $(x - 3)^2 + (y - 0)^2 = 25$; center: $(3,0)$; $r = 5$

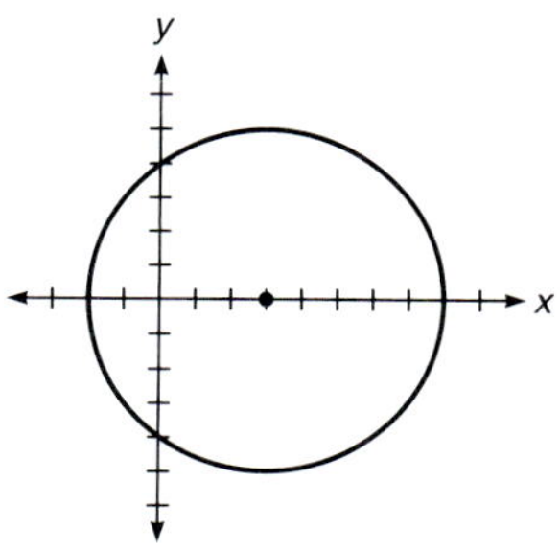

63. $(x - 1)^2 + (y - 2)^2 = 9$; center: $(1,2)$; $r = 3$

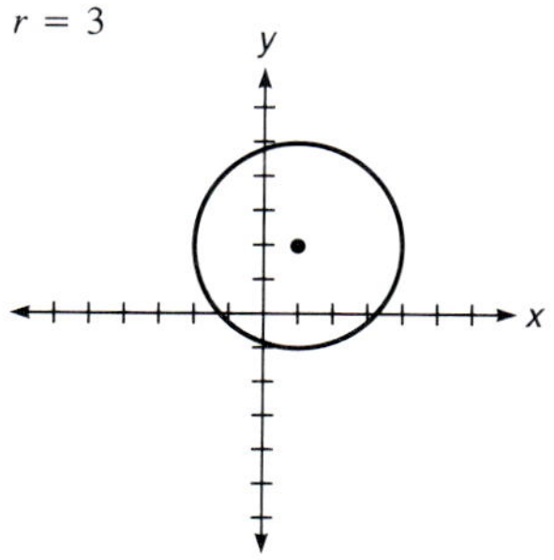

64. $(x - \frac{1}{2})^2 + (y - (-\frac{3}{2}))^2 = \frac{15}{2}$; center: $(\frac{1}{2}, -1\frac{1}{2})$; $r = \dfrac{\sqrt{30}}{2} \approx 2.7$

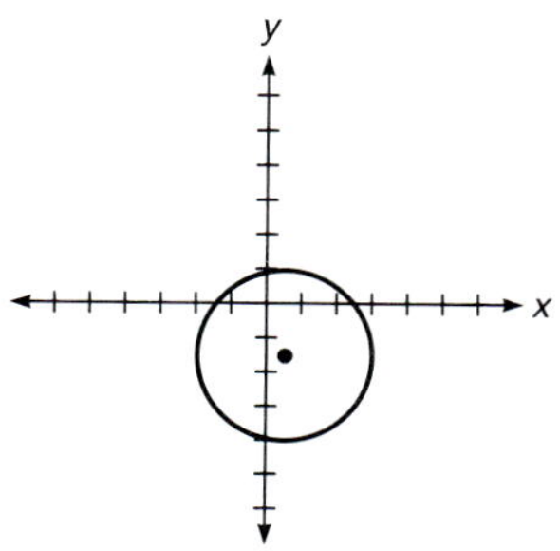

65. $(x - \frac{5}{2})^2 + (y - (-\frac{3}{2}))^2 = 2$; center: $(2\frac{1}{2}, -1\frac{1}{2})$; $r = \sqrt{2} \approx 1.4$

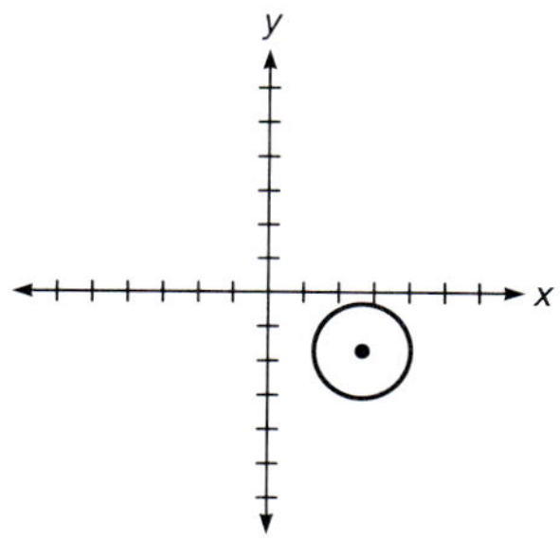

66. $(x + \frac{7}{2})^2 + (y - 2)^2 = 80$
67. $(x - 1)^2 + (y + 3)^2 = 9$
68. $(x - 1)^2 + (y - 3)^2 = 73$
69. function, not one to one
70. function, one to one
71. function, one to one
72. even; y-axis symmetry
73. odd; origin symmetry
74. odd; origin symmetry
75. odd; origin symmetry
76. neither odd nor even
77. neither odd nor even
78. $y = \frac{5}{3}x - \frac{35}{3}$ **79.** 40 and 55
80. 9% **81.** 70
82. approximately generations 15–20, 30–45, 55–60, and 70–75

Chapter 3 test

1. $(-3,10)$, $(0,5)$, $(3,0)$

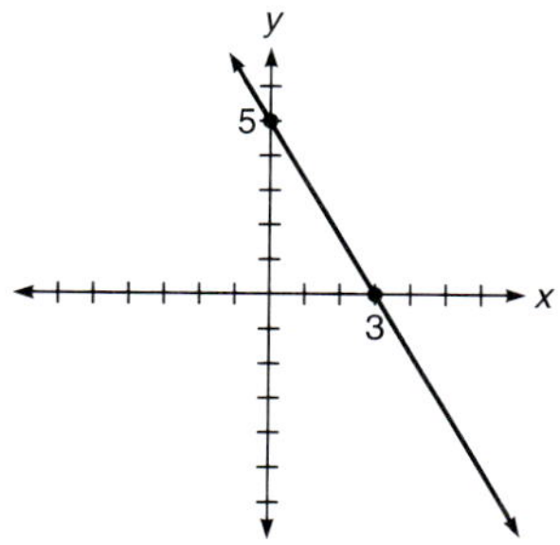

2. $(-1,-8)$, $(0,-6)$, $(3,0)$

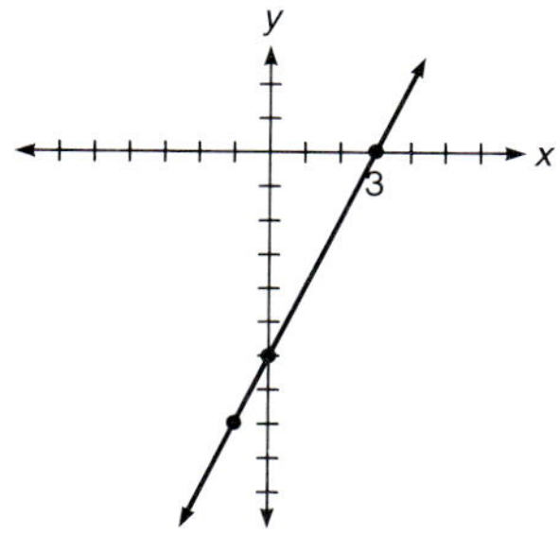

3. $(5,0)$, $(5,1)$, $(5,2)$

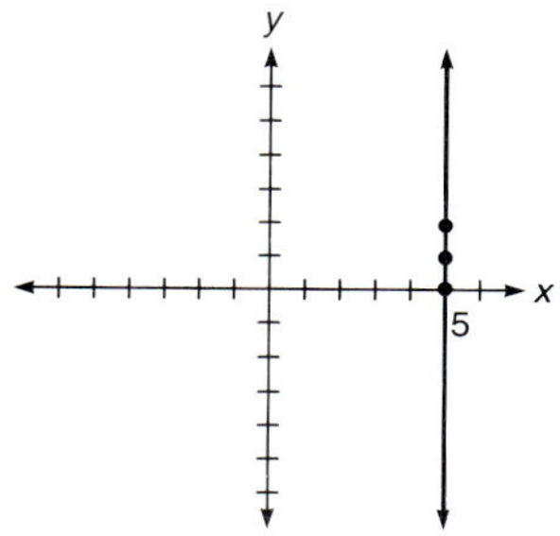

4. $(-5,-8)$, $(0,-4)$, $(5,0)$

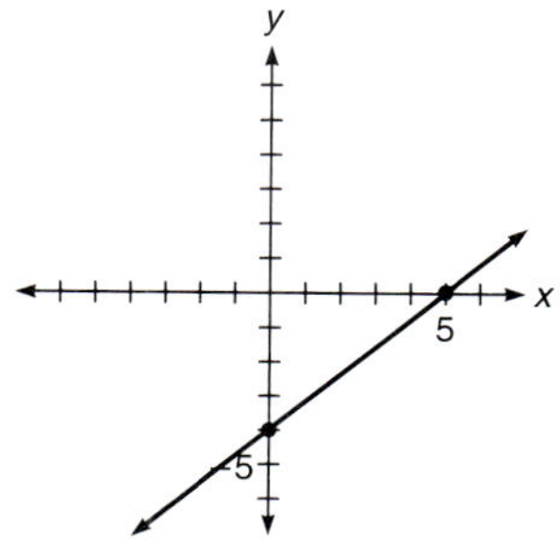

5. $(\frac{1}{2}, 4)$
6. $\sqrt{41}$
7. $m = 2$; $(3\frac{1}{2}, 0)$; $(0,-7)$

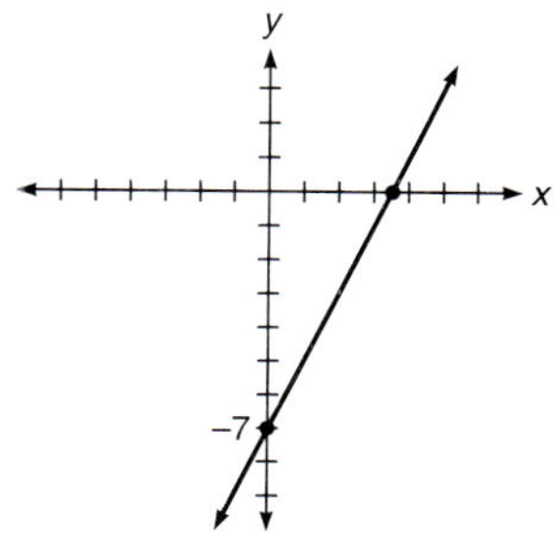

8. $m = -\frac{5}{4}$; $(-4,0)$, $(0,-5)$

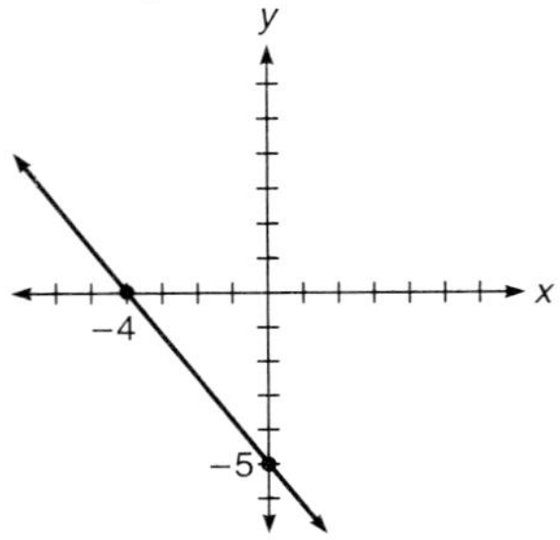

9. y-intercept is -1; no x-intercept; $m = 0$

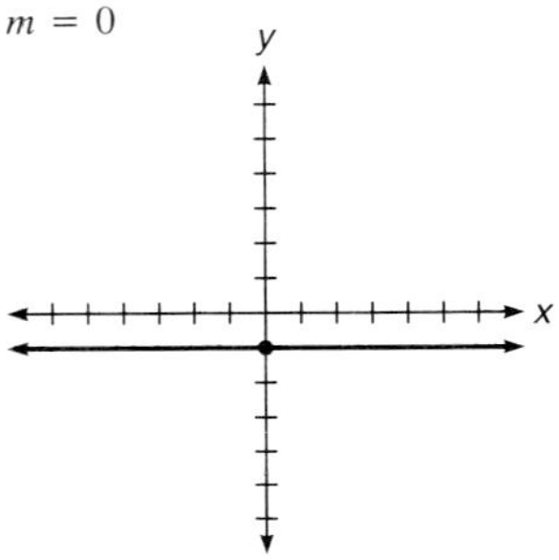

10. $-\frac{1}{4}$ **11.** $y = -\frac{1}{2}x + 2$

12. $y = -5x - 15$ **13.** $y = -5x + 7$

14. $(x,y) = (2,0)$ **15.** $4x + 6y - 39 = 0$

16. $\{(1,1), (2,\sqrt{3}), (3,\sqrt{5}), (4,\sqrt{7})\}$; a one-to-one function

17. domain: $x \neq 3$; $f(-2) = \frac{2}{5}$; $f(0) = 0$; 3 is not in the domain of f.

$$f(c - 3) = \frac{c - 3}{c - 6}$$

18. $g(x) = \sqrt{6 - 2x}$; domain: $x \leq 3$; $g(-2) = \sqrt{10}$; $g(0) = \sqrt{6}$ $g(3) = 0$; $g(c - 3) = \sqrt{12 - 2c}$

19. domain: R; $v(-2) = 3$; $v(0) = 3$; $v(3) = -12$; $v(c - 3) = 4c - c^2$

20. $4x + 2h$

21. vertex: $(-1,-2)$; x-intercept: $(-1 \pm \sqrt{2},0)$; y-intercept: $(0,-1)$

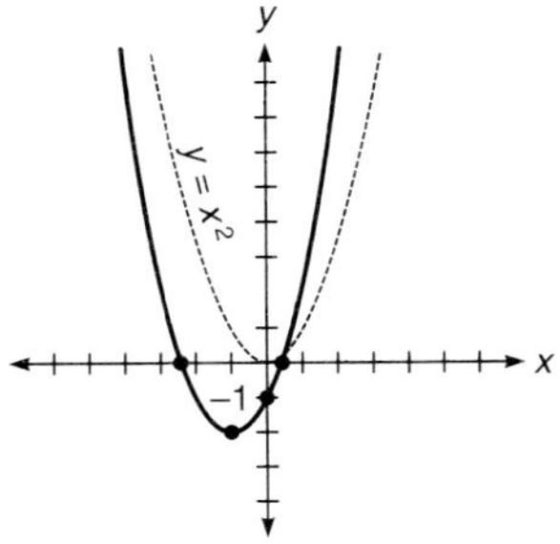

22. graph of $y = \sqrt{x}$ but shifted so new origin is at $(-4,-2)$; intercepts at $(0,0)$

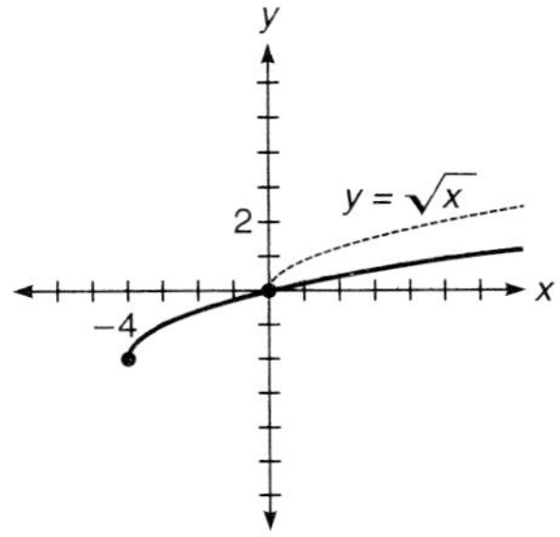

23. graph of $y = x^3$ with origin shifted to $(2,3)$; x-intercept: $(2+\sqrt[3]{-3},0)$; y-intercept: $(0,-5)$

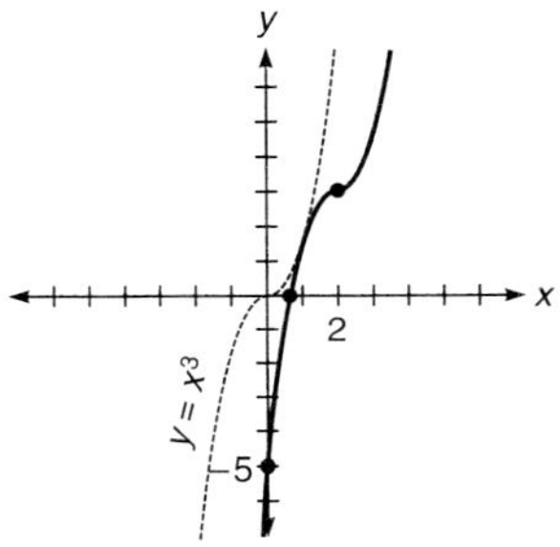

24. graph of $y = \dfrac{1}{x}$ with origin shifted to $(3,-2)$; x-intercept: $(3\frac{1}{2},0)$; y-intercept: $(0,-2\frac{1}{3})$

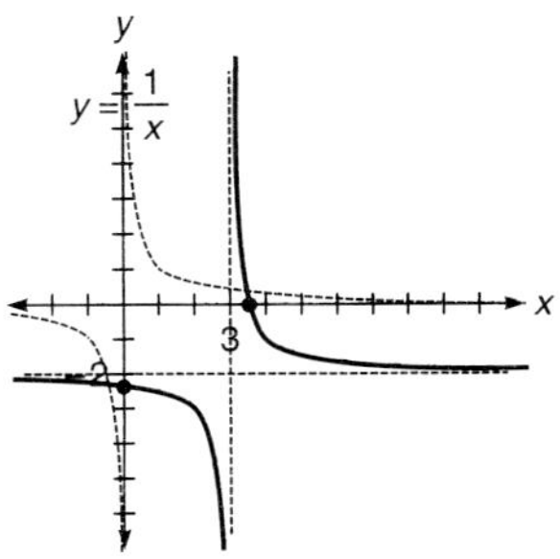

25. graph of $y = |x|$ with origin shifted to $(-2,3)$; y-intercept: $(0,5)$

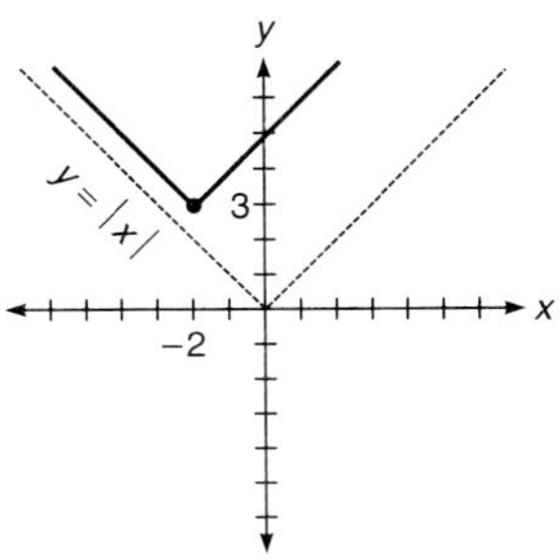

26. graph of $y = x^3$ but with origin moved to $(2,1)$, vertically scaled by $\frac{1}{2}$; x-intercept: $(2 + \sqrt[3]{-2},0)$; y-intercept: $(0,-3)$

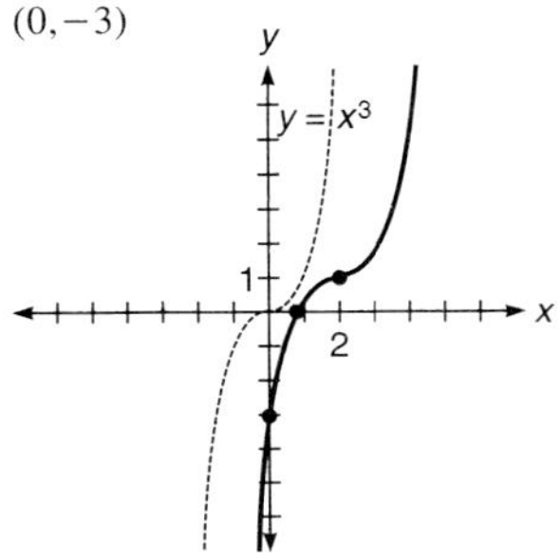

27. center: $(2,-4)$; $r = \sqrt{16} = 4$

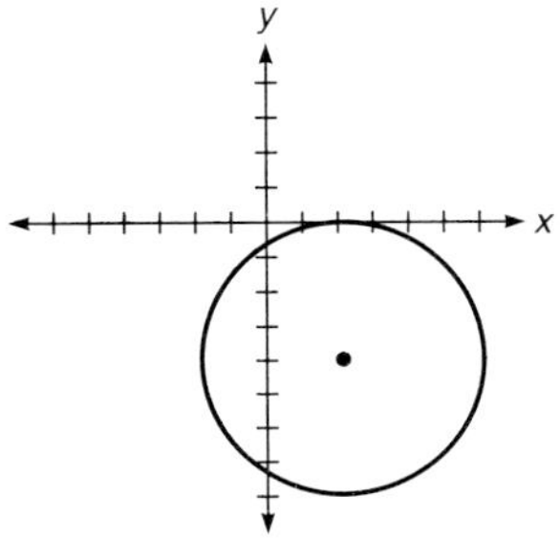

28. $(x - 3)^2 + (y - (-1\frac{1}{2}))^2 = \frac{65}{4}$; center: $(3,-1\frac{1}{2})$, $r = \dfrac{\sqrt{65}}{2} \approx 4$

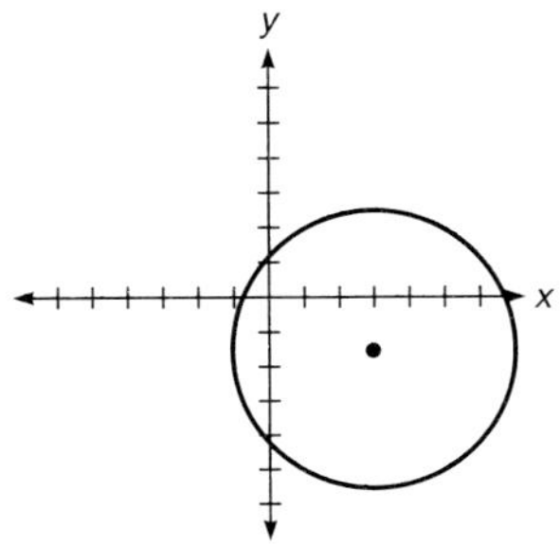

29. $(x + \frac{7}{2})^2 + (y - 2)^2 = 80$

30. $(x - 1)^2 + (y - 3)^2 = 73$

31. **a.** -2; **b.** 4; **c.** -3; **d.** -3

32. $-9, -7, -2, 5, 7.5$

33. $-8, -1, 4, 9$

34. -8 to -4, 1.5 to 6

35. -9 to -8, -4 to 1.5, 6 to 10

36. -9 to 10 **37.** -3.5 to 4

38. even; y-axis symmetry

39. neither even nor odd

40. even; y-axis symmetry

41. 94.7° Celsius; 88.2° Celsius

Chapter 4

Exercise 4–1

Answers to odd-numbered problems

1. vertex: (1,3); intercepts: (0,4)

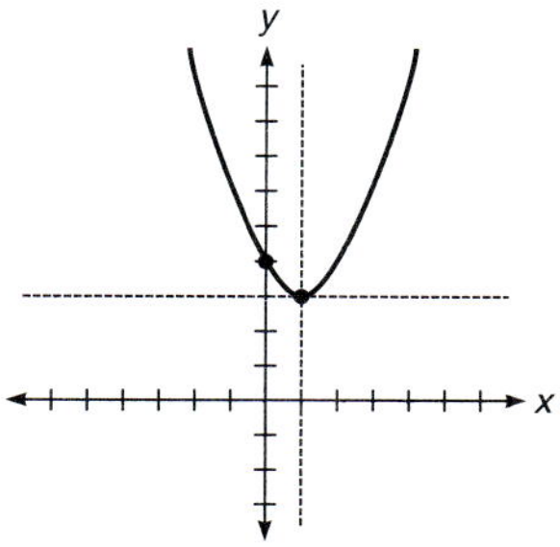

3. vertex: (−3,−4); intercepts: (0,14), $(-3-\sqrt{2},0)$, $(-3+\sqrt{2},0)$

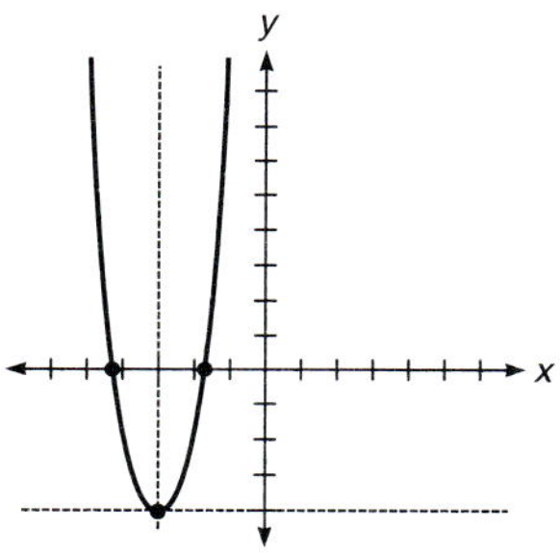

5. vertex: (5,−1); intercepts: (0,−26)

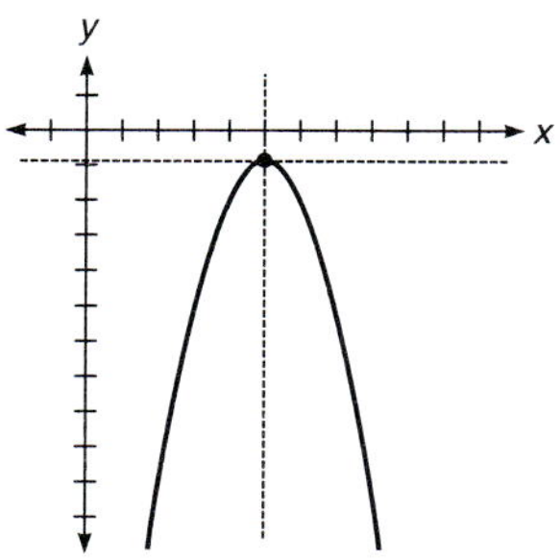

7. $y = (x - \frac{1}{2})^2 - 6\frac{1}{4}$
vertex: $(\frac{1}{2}, -6\frac{1}{4})$; intercepts: (0,−6), (−2,0), (3,0)

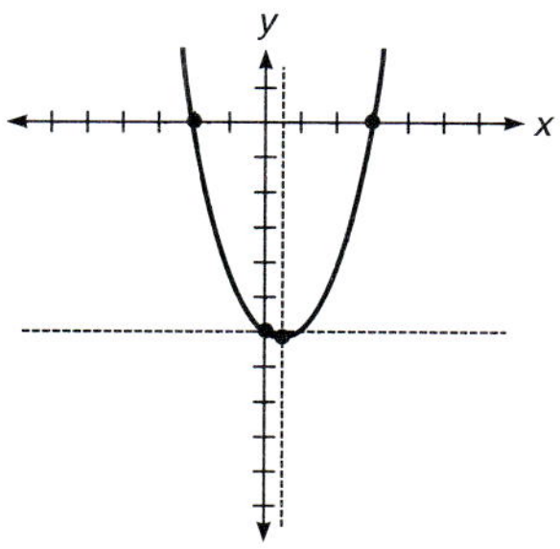

9. vertex: (0,0); intercepts: (0,0)

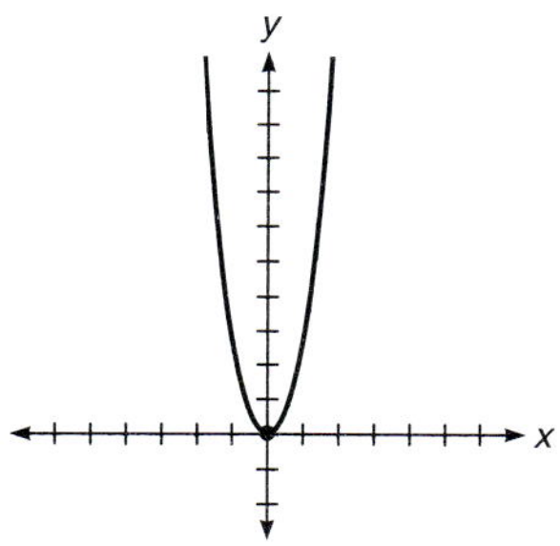

11. $y = (x + \frac{3}{2})^2 - \frac{9}{4}$
vertex: $(-1\frac{1}{2}, -2\frac{1}{4})$; intercepts: (0,0), (−3,0)

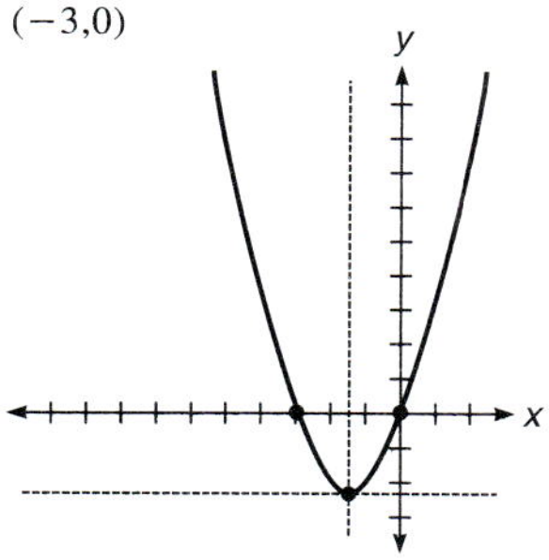

13. $y = -(x - \frac{3}{2})^2 + 42\frac{1}{4}$
vertex: $(1\frac{1}{2}, 42\frac{1}{4})$; intercepts: (0,40), (−5,0), (8,0)

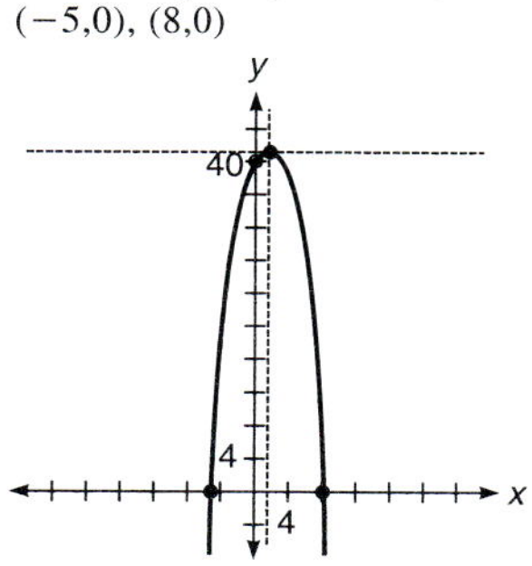

15. $y = x^2 - 4$
vertex: (0,−4); intercepts: (0,−4), (−2,0), (2,0)

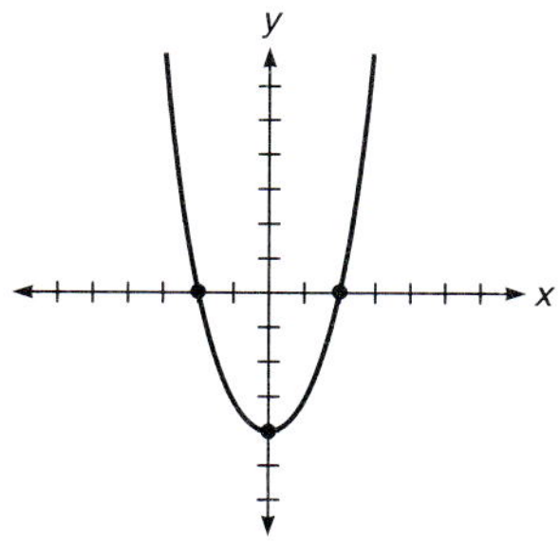

17. $y = 3(x + 1)^2 - 5$
vertex: (−1,−5); intercepts: (0,−2), $\left(-1 \pm \dfrac{\sqrt{15}}{3}, 0\right) \approx (-2.3,0), (0.3,0)$

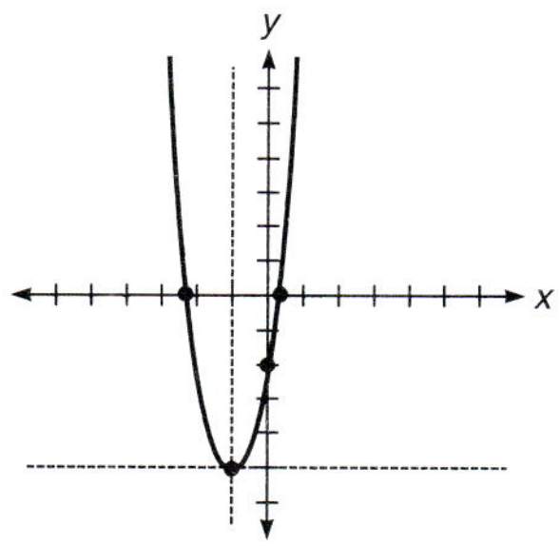

19. $y = (x - \frac{5}{2})^2 - 14\frac{1}{4}$
vertex: $(2\frac{1}{2}, -14\frac{1}{4})$; intercepts: (0,−8), $\left(\dfrac{5 \pm \sqrt{57}}{2}, 0\right) \approx (-1.3,0), (6.3,0)$

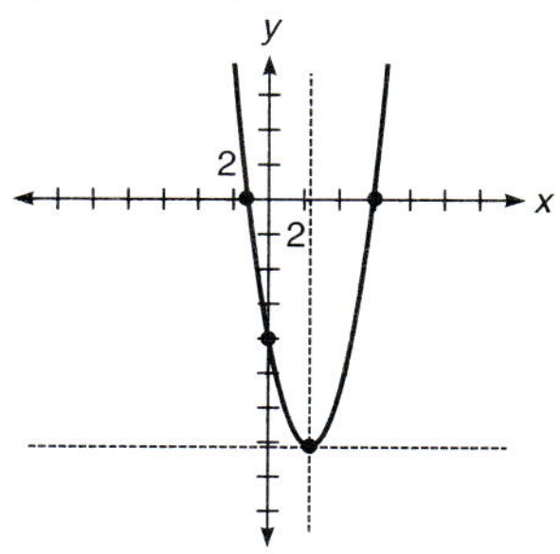

21. $y = 2(x - 1)^2 - 6$
vertex: $(1,-6)$; intercepts: $(0,-4)$, $(1\pm\sqrt{3},0) \approx (-0.7,0)$, $(2.7,0)$

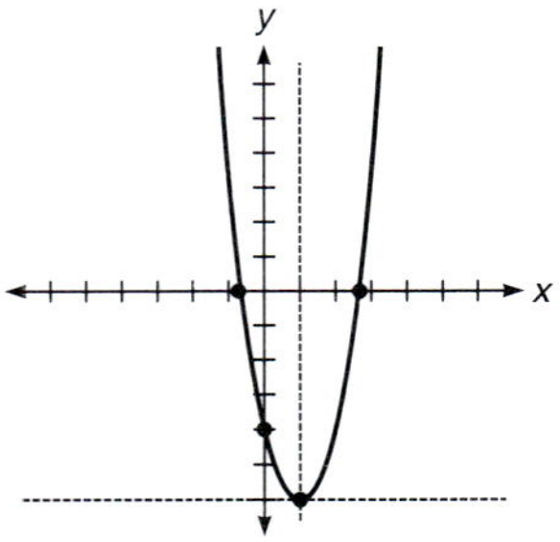

23. vertex: $(0,4)$; intercepts: $(0,4)$

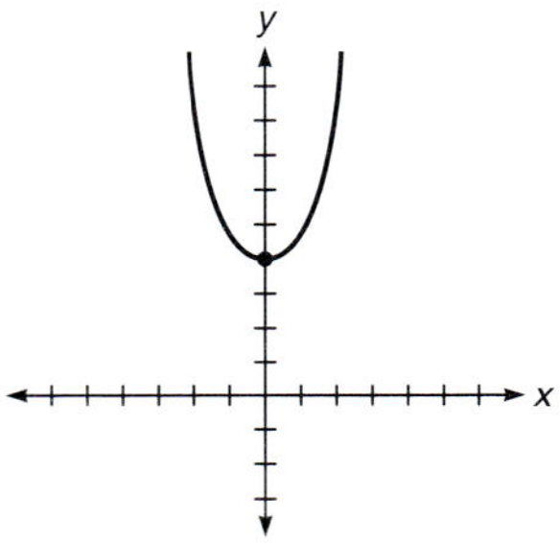

25. $y = (x - \frac{1}{2})^2 + 4\frac{3}{4}$
vertex: $(\frac{1}{2}, 4\frac{3}{4})$; intercepts: $(0,5)$

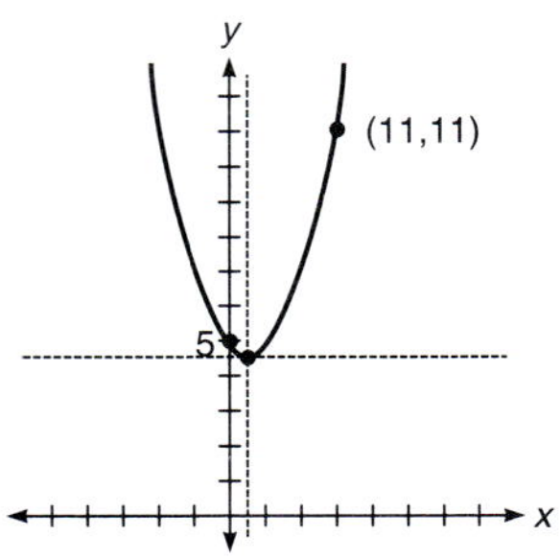

27. $y = -(x + \frac{1}{2})^2 + \frac{1}{4}$
vertex: $(-\frac{1}{2}, \frac{1}{4})$; intercepts: $(0,0)$, $(-1,0)$

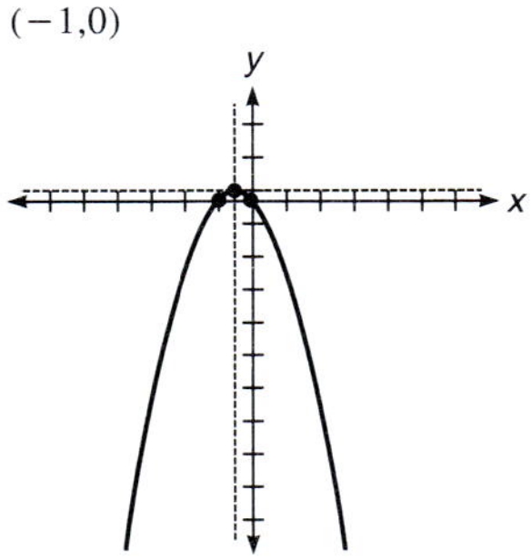

29. 65 ft, 130 ft, 8,450 ft^2 **31.** a square with dimension 65 ft; area is 4,225 sq. ft **33.** The maximum height of $s = 64$ feet is reached after $t = 2$ seconds. The object is thrown, and it returns to earth after 4 seconds. **35.** The maximum velocity is 9 m/s, 3 meters from the inside wall. **37.** A production of 50 units will produce the maximum profit of \$1,500. **39.** The numbers are 4 and -4 and the product is -16. **41.** circle

43. $4a(ax^2 + bx + c) = 4a(0)$

$$4a^2x^2 + 4abx + 4ac = 0$$
$$4a^2x^2 + 4abx + 4ac + b^2 = b^2$$
$$4a^2x^2 + 4abx + b^2 = b^2 - 4ac$$
$$(2ax + b)^2 = b^2 - 4ac$$
$$2ax + b = \pm\sqrt{b^2 - 4ac}$$
$$2ax = -b \pm \sqrt{b^2 - 4ac}$$
$$x = \frac{-b \pm \sqrt{b^2 - 4ac}}{2a}$$

45.

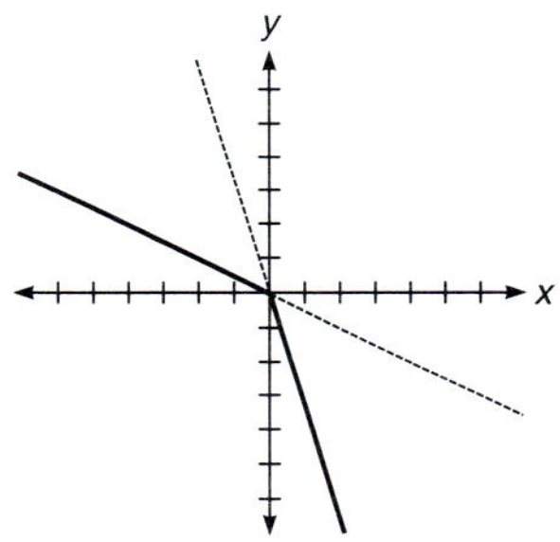

47.

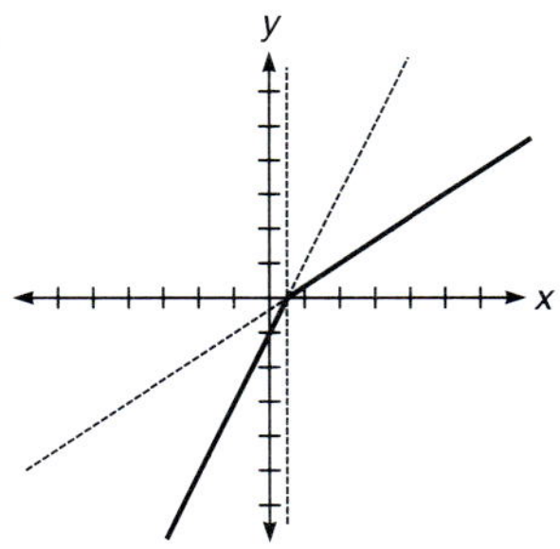

49.

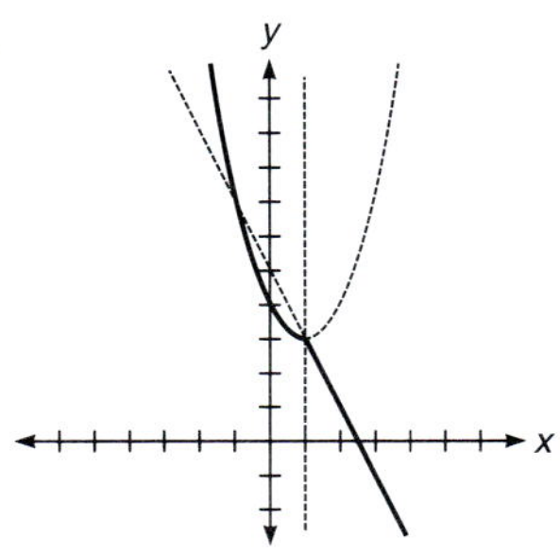

51.

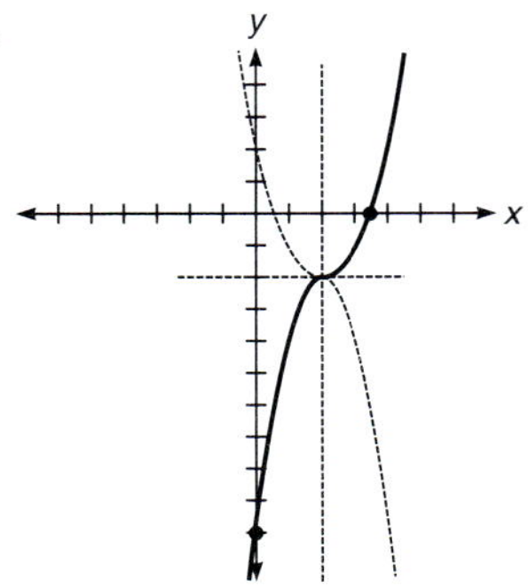

53.

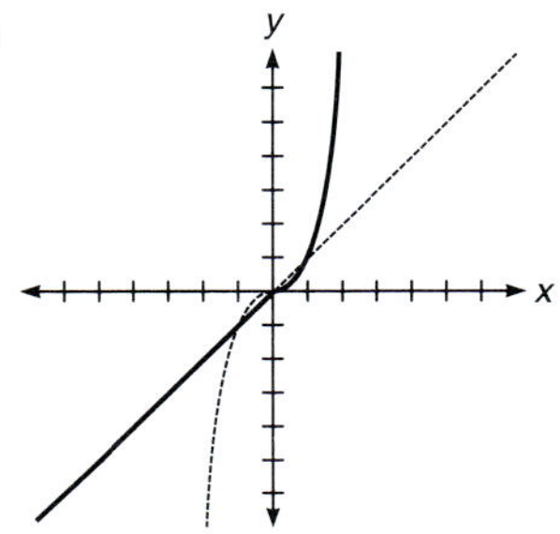

55.

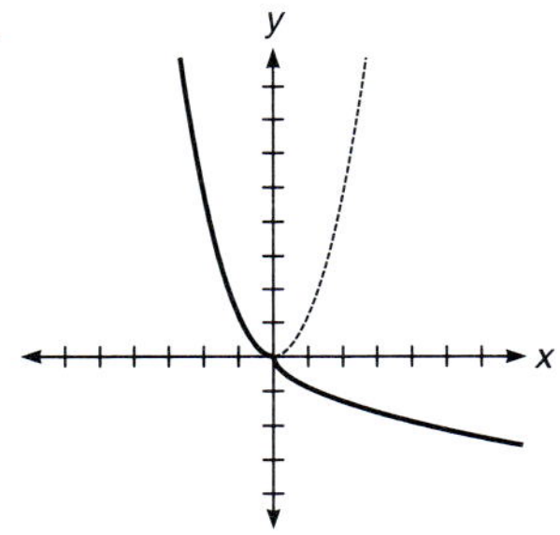

57.

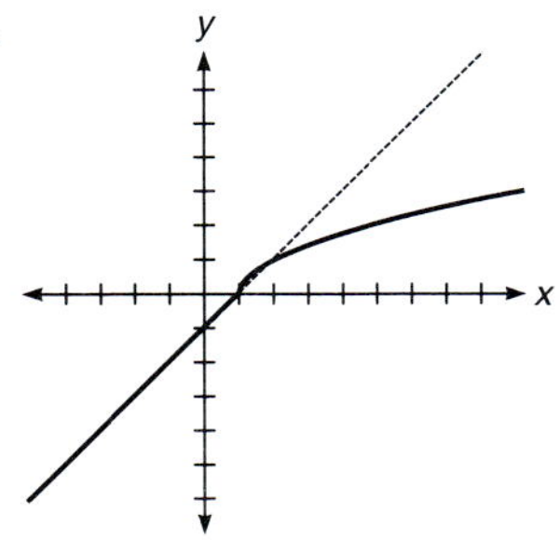

Solutions to skill and review problems

1. Factor: $3x^2 + x - 10$.
   $(3x - 5)(x + 2)$
2. Factor: $3x^2 + 13x - 10$.
   $(3x - 2)(x + 5)$

3. Factor: $x^4 - 16$.
$(x^2 - 4)(x^2 + 4)$
$(x - 2)(x + 2)(x^2 + 4)$

4. List all the prime divisors of 96.
$96 = 6 \cdot 16$
$= 2 \cdot 3 \cdot 2^4$
$= 2^5 \cdot 3$
2 and 3 are the only prime divisors.

5. List all the positive integer divisors of 96.
2, 3, 4, 6, 8, 12, 16, 24, 32, 48, 96

6. If $f(x) = 2x^3 - x^2 - 6x + 20$, find $f(-2)$.
$f(-2) = 2(-2)^3 - (-2)^2 - 6(-2) + 20$
$= 2(-8) - 4 + 12 + 20$
$= 12$

7. Use long division to divide $2x^3 - x^2 - 6x + 20$ by $x^2 + 2$.

$$\begin{array}{r} 2x - 1 \\ x^2 + 2 \overline{)\, 2x^3 - x^2 - 6x + 20} \\ \underline{2x^3 + 4x } \\ -x^2 - 10x + 20 \\ \underline{-x^2 - 2} \\ -10x + 22 \end{array}$$

Quotient is $2x - 1$, remainder is $-10x + 22$.

8. Graph $f(x) = \sqrt{x - 2} - 3$.
The graph of $y = \sqrt{x - 2} - 3$ is the graph of $y = \sqrt{x}$ but with the "origin" shifted to $(2,-3)$.

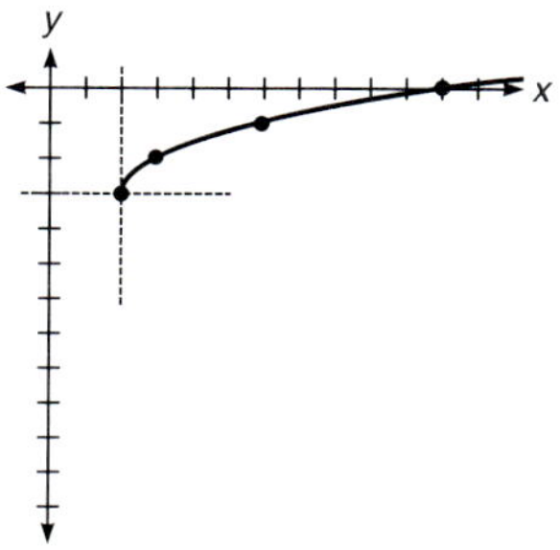

Intercepts:
$x = 0$: $y = \sqrt{-2} - 3$; no real solution so no y-intercept
$y = 0$: $0 = \sqrt{x - 2} - 3$
$3 = \sqrt{x - 2}$
$9 = x - 2$
$11 = x$
Additional points:

x	3	6
y	−2	−1

9. Compute $f \circ g(x)$ and $g \circ f(x)$ if $f(x) = x^4 - 6x^2 + 8$ and $g(x) = \sqrt{x + 1}$.
$f \circ g(x) = f(g(x)) = [g(x)]^4 - 6[g(x)]^2 + 8$
$= [\sqrt{x + 1}]^4 - 6[\sqrt{x + 1}]^2 + 8$
$= [(\sqrt{x + 1})^2]^2 - 6[\sqrt{x + 1}]^2 + 8$
$= [x + 1]^2 - 6(x + 1) + 8$
$= x^2 + 2x + 1 - 6x - 6 + 8$
$= x^2 - 4x + 3$
$g \circ f(x) = g(f(x)) = \sqrt{(x^4 - 6x^2 + 8) + 1}$
$= \sqrt{x^4 - 6x^2 + 9}$
$= \sqrt{(x^2 - 3)^2}$
$= |x^2 - 3|$

Solutions to trial exercise problems

5. $y = -(x - 5)^2 - 1$
Vertex: $(5,-1)$
Intercepts:
$x = 0$: $y = -(-5)^2 - 1 = -26$, $(0,-26)$
$y = 0$: $0 = -(x - 5)^2 - 1$
$1 = -(x - 5)^2$; no real solution since the left side is positive and the right side is negative. Thus, no x-intercepts.
Additional points:

x	3	4	6	7
y	−5	−2	−2	−5

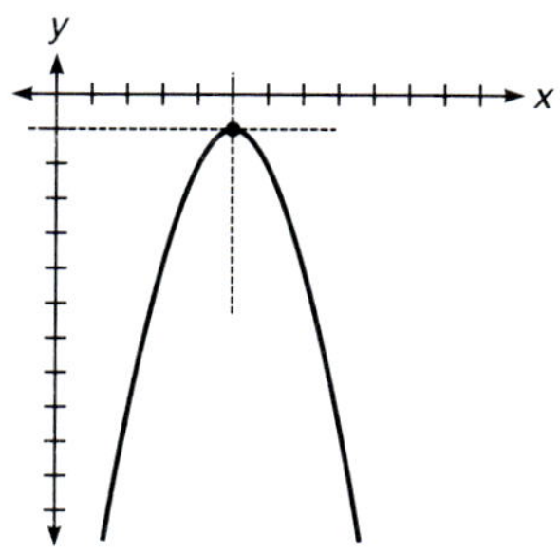

13. $y = -x^2 + 3x + 40$
$y = -(x^2 - 3x) + 40$
$\frac{1}{2} \cdot (-3) = -\frac{3}{2}$; $(-\frac{3}{2})^2 = \frac{9}{4}$
$y = -(x^2 - 3x + \frac{9}{4}) + 40 + 1(\frac{9}{4})$
$y = -(x - \frac{3}{2})^2 + 42\frac{1}{4}$
Vertex: $(1\frac{1}{2}, 42\frac{1}{4})$
Intercepts:
$x = 0$: $y = -0^2 + 0 + 40 = 40$; $(0,40)$
$y = 0$: $0 = -x^2 + 3x + 40$
$0 = x^2 - 3x - 40$
$0 = (x - 8)(x + 5)$
$x = -5$ or 8; $(-5,0)$, $(8,0)$

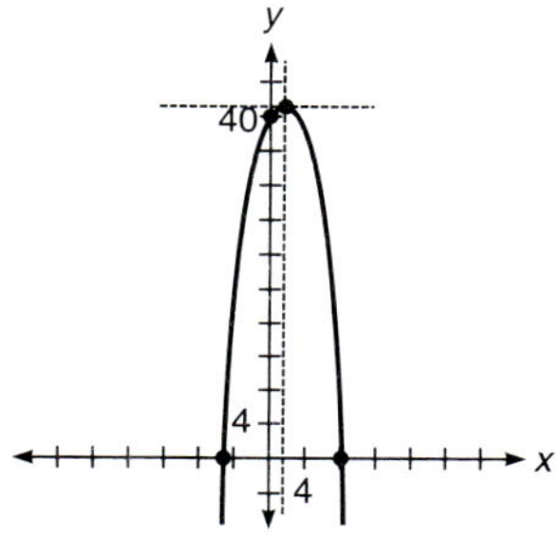

21. $y = 2x^2 - 4x - 4$
$y = 2(x^2 - 2x) - 4$
$\frac{1}{2} \cdot (-2) = -1$; $(-1)^2 = 1$
$y = 2(x^2 - 2x + 1) - 4 - 2(1)$
$y = 2(x - 1)^2 - 6$
Vertex: $(1,-6)$
Intercepts:
$x = 0$: $y = 0 - 0 - 4 = -4$; $(0,-4)$
$y = 0$: $0 = 2x^2 - 4x - 4$
$0 = x^2 - 2x - 2$
$x = 1 \pm \sqrt{3}$; $(1 \pm \sqrt{3},0) \approx (-0.7,0)$, $(2.7,0)$

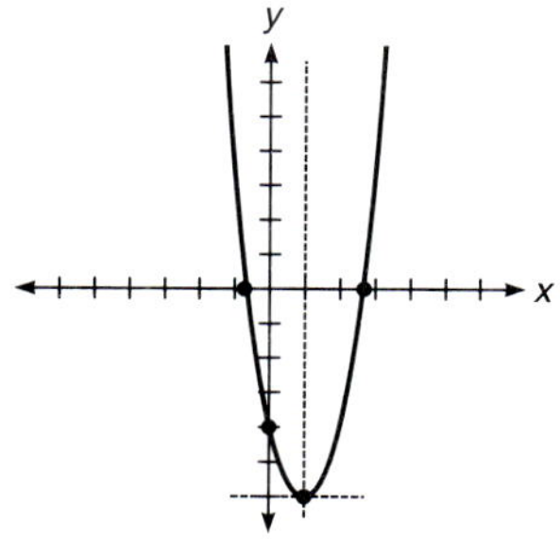

26. $y = -2x^2 - 2x + 1$
$y = -2(x^2 + x) + 1$
$\frac{1}{2} \cdot 1 = \frac{1}{2}; (\frac{1}{2})^2 = \frac{1}{4}$
$y = -2(x^2 + x + \frac{1}{4}) + 1 + 2(\frac{1}{4})$
$y = -2(x + \frac{1}{2})^2 + \frac{3}{2}$
Vertex: $(-\frac{1}{2}, 1\frac{1}{2})$
Intercepts:
$x = 0$: $y = 0 - 0 + 1 = 1$; $(0,1)$
$y = 0$: $0 = -2x^2 - 2x + 1$

$$x = \frac{-1 \pm \sqrt{3}}{2}$$

$$\left(\frac{-1 - \sqrt{3}}{2}, 0\right), \left(\frac{-1 + \sqrt{3}}{2}, 0\right)$$

$\approx (-1.4,0), (0.4,0)$

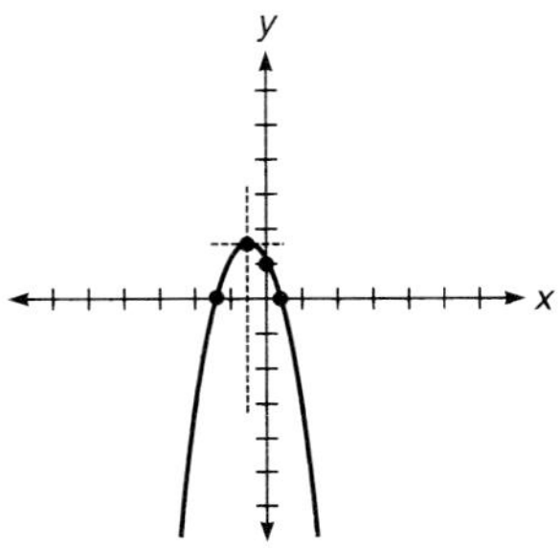

32. The radius of the semicircle is x. The circumference of a circle is $C = 2\pi r$, so the circumference of the semicircle is half this:

$$\frac{C}{2} = \frac{2\pi r}{2} = \pi r = \pi x$$

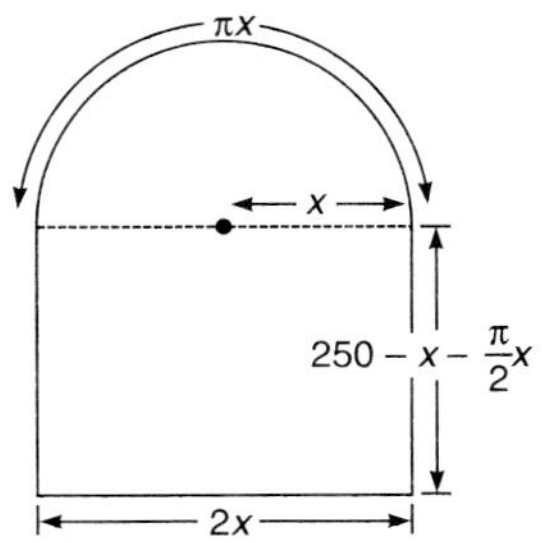

The base of the figure has length $2x$. Since the total length of chain is 500 ft the other dimension of the rectangle is $\frac{1}{2}(500 - 2x - \pi x) = 250 - x - \frac{\pi}{2}x$. The area of a circle is $A = \pi r^2$, so the area of the semicircle is half this, $\frac{1}{2}\pi r^2 = \frac{\pi}{2}x^2$. The area of the rectangular part is $2x(250 - x - \frac{\pi}{2}x)$.

Total area is:

$$\begin{aligned}A &= 2x(250 - x - \tfrac{\pi}{2}x) + \tfrac{\pi}{2}x^2 \\ &= 500x - 2x^2 - \pi x^2 + \tfrac{\pi}{2}x^2 \\ &= 500x - x^2(2 + \tfrac{2\pi}{2} - \tfrac{\pi}{2}) \\ &= 500x - (2 + \tfrac{\pi}{2})x^2 \\ &\approx -3.571x^2 + 500x \\ &\approx -3.571(x^2 - \tfrac{500}{3.571}x) \\ &\approx -3.571(x^2 - 140.0x) \\ &\approx -3.571(x^2 - 140.0x + 4{,}900) + 3.571(4{,}900) \\ &\approx -3.571(x - 70.0)^2 + 17{,}496.9\end{aligned}$$

Vertex: $(70.0, 17{,}496.9) = (x,A)$
The vertex is the maximum point: $x = 70.0$, and $A = 17{,}496.9$.
Thus $x = 70$ ft and area $= 17{,}497$ ft^2.

36. $P = 14I - 0.20I^2$

$$\begin{aligned}&= -0.2(I^2 - \tfrac{14}{0.2}I) \\ &= -0.2(I^2 - 70I) \\ &= -0.2(I^2 - 70I + 1{,}225) + 0.2(1{,}225) \\ &= -0.2(I - 35)^2 + 245\end{aligned}$$

Vertex: $(35,245) = (I,P)$
Thus a current of 35 amperes produces a maximum power of 245 watts.

39. Let the numbers be x and $x - 8$. Then their product is $P = x(x - 8)$.
$P = x^2 - 8x$
$P = x^2 - 8x + 16 - 16$
$P = (x - 4)^2 - 16$
Vertex: $(4,-16) = (x,P)$. Thus, $x = 4$, $x - 8 = -4$, $P = -16$. Since the parabola opens upward this is a minimum. Thus, the numbers are 4 and -4 and the product is -16.

42. $ax^2 + bx + c = 0$
$ax^2 + bx = -c$

$$x^2 + \frac{b}{a}x = -\frac{c}{a}$$

$$\frac{1}{2}\left(\frac{b}{a}\right) = \frac{b}{2a}, \left(\frac{b}{2a}\right)^2 = \frac{b^2}{4a^2}, \text{ so}$$

$$x^2 + \frac{b}{a}x + \frac{b^2}{4a^2} = -\frac{c}{a} + \frac{b^2}{4a^2}$$

$$\left(x + \frac{b}{2a}\right)^2 = \frac{b^2}{4a^2} - \frac{c}{a}\left(\frac{4a}{4a}\right)$$

$$\left(x + \frac{b}{2a}\right)^2 = \frac{b^2}{4a^2} - \frac{4ac}{4a^2}$$

$$\left(x + \frac{b}{2a}\right)^2 = \frac{b^2 - 4ac}{4a^2}$$

$$x + \frac{b}{2a} = \pm\sqrt{\frac{b^2 - 4ac}{4a^2}}$$

$$x + \frac{b}{2a} = \pm\frac{\sqrt{b^2 - 4ac}}{2|a|}$$

. Since $\pm |a| = \pm a$ (this can be believed with a few examples or proven using the definition of $|a|$ (section 1–2)), we rewrite this as

$$x + \frac{b}{2a} = \pm\frac{\sqrt{b^2 - 4ac}}{2a}$$

$$x = -\frac{b}{2a} \pm \frac{\sqrt{b^2 - 4ac}}{2a}$$

$$x = \frac{-b \pm \sqrt{b^2 - 4ac}}{2a}$$

50. $f(x) = \begin{cases} x^2 + 2, & x < 0 \\ -2x^2 + 2, & x \geq 0 \end{cases}$

Graph $y = x^2 + 2$:
Parabola with vertex at $(0,2)$. Plot additional points, say $(-2,6)$ and $(2,6)$.
Graph $y = -2x^2 + 2$:
Parabola with vertex at $(0,2)$. Opens downward.
x-intercepts: $(y = 0)$: $0 = -2x^2 + 2$
$2x^2 = 2$
$x^2 = 1$
$x = \pm 1$
Darken in the first line for $x < 0$; darken in the second line for $x \geq 0$.

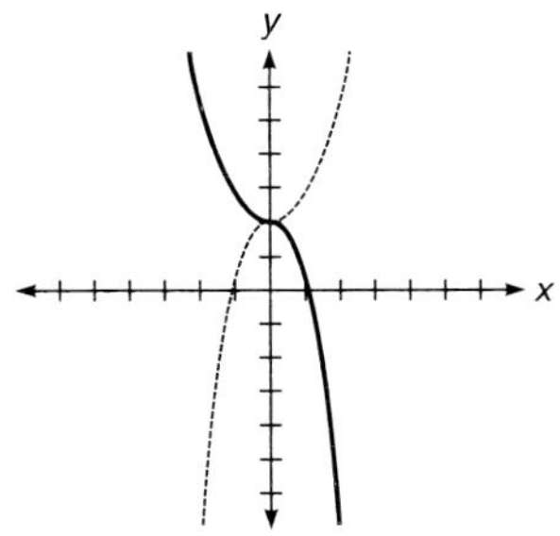

58. $f(x) = ax^2 + bx + c$

$$\begin{aligned}&= a\left(x^2 + \frac{b}{a}x\right) + c \\ &= a\left(x^2 + \frac{b}{a}x + \left(\frac{b}{2a}\right)^2\right) + c - a\left(\frac{b}{2a}\right)^2 \\ &= a\left(x + \frac{b}{2a}\right)^2 + c - \frac{b^2}{4a} \\ &= a\left(x - \left(-\frac{b}{2a}\right)\right)^2 + \frac{4ac}{4a} - \frac{b^2}{4a} \\ &= a\left(x - \left(-\frac{b}{2a}\right)\right)^2 + \frac{4ac - b^2}{4a}\end{aligned}$$

Thus the vertex is at the point $\left(-\frac{b}{2a}, \frac{4ac - b^2}{4a}\right)$.

Exercise 4–2

Answers to odd-numbered problems

1. no zeros **3.** $\frac{2}{3}$ **5.** -11

7. $-2, 5$ **9.** $\pm6, \pm3, \pm2, \pm1$

11. $\pm\frac{4}{3}, \pm\frac{2}{3}, \pm\frac{1}{3}, \pm4, \pm2, \pm1$

13. $\pm\frac{5}{2}, \pm\frac{1}{2}, \pm5, \pm1$

15. $\pm\frac{1}{3}, \pm9, \pm3, \pm1$

17. $\pm10, \pm5, \pm2, \pm1, \pm\frac{1}{5}, \pm\frac{2}{5}$

19. $\pm\frac{1}{2}, \pm\frac{1}{4}, \pm\frac{1}{8}, \pm4, \pm2, \pm1$

21. $\pm\frac{1}{2}, \pm\frac{1}{4}, \pm2, \pm1$

23. $\pm\frac{4}{5}, \pm\frac{1}{5}, \pm\frac{2}{5}, \pm\frac{1}{10}, \pm\frac{1}{2}, \pm4, \pm2, \pm1$

25. $\pm2, \pm1$

27. a. $3x^3 + 10x^2 + 41x + 164 + \dfrac{651}{x-4}$
b. $f(4) = 651$

29. a. $x^2 - x + 2 + \dfrac{3}{x-1}$
b. $f(1) = 3$

31. a. $x^4 - 3x^3 + 6x^2 - 19x + 57 + \dfrac{-166}{x+3}$
b. $h(-2) = -166$

33. a. $\frac{1}{2}x^2 + \frac{3}{4} + \dfrac{\frac{3}{2}}{x-6}$
b. $f(6) = \frac{3}{2}$

35. a. 0 or 2 positive zeros; 0 or 2 negative zeros
b. $\pm6, \pm3, \pm2, \pm1$
c. $-2, -1, 1, 3$
d. $f(x) = (x-3)(x+2)(x-1)(x+1)$

37. a. 1 or 3 positive zeros; no negative zeros
b. $\frac{3}{4}, \frac{3}{2}, \frac{1}{4}, \frac{1}{2}, 3, 1$
c. $\frac{3}{2}, \frac{1}{2}, 1$
d. $f(x) = 4(x - \frac{3}{2})(x - \frac{1}{2})(x - 1)$

39. a. 0, 2 or 4 positive zeros; no negative zeros
b. 81, 27, 9, 3, 1
c. 3, with multiplicity 2
d. $f(x) = (x-3)^2(x^2 - 2x + 9)$

41. a. one positive zero; one negative zero
b. $\pm1, \pm5$ **c.** -1
d. $f(x) = (x+1)(x^2 - x + 1)$
$(x - \sqrt[3]{5})(x^2 + \sqrt[3]{5}x + (\sqrt[3]{5})^2)$
e. irrational zero: $\sqrt[3]{5}$

43. a. 0 or 2 positive zeros; 0 or 2 negative zeros
b. $\pm\frac{2}{3}, \pm\frac{1}{3}, \pm6, \pm3, \pm2, \pm1$
c. $\frac{1}{3}$ and -2
d. $f(x) = 3(x - \frac{1}{3})(x + 2)$
$(x - \sqrt{3})(x + \sqrt{3})$
e. $\pm\sqrt{3}$

45. a. 0 or 2 positive zeros; one negative zero
b. $\pm1, \pm2$
e. The function f has one negative irrational zero between -1 and 0. It has 0 or 2 positive irrational zeros between 0 and 2.

47. a. one positive zero; 1 or 3 negative zeros
b. $\pm1, \pm3$ **c.** $-3, 1$
d. $f(x) = 2(x+3)(x-1)(x^2 + x + 1)$

49. a. 0 or 2 positive zeros; 0 or 2 negative zeros
b. $\pm\frac{1}{3}, \pm1$
c. $1, -1$ (mult 2), $\frac{1}{3}$
d. $f(x) = 3(x-1)(x+1)^2(x - \frac{1}{3})$

51. a. one positive zero; 0 or 2 negative zeros
b. $\pm1, \pm2, \pm5, \pm10$ **c.** -2
d. $f(x) = (x+2)\left(x - \dfrac{1-\sqrt{21}}{2}\right)\left(x - \dfrac{1+\sqrt{21}}{2}\right)$
e. irrational zeros: $\dfrac{1 \pm \sqrt{21}}{2}$

53. a. 0 or 2 positive zeros; 1 or 3 negative zeros
b. $\pm1, \pm2, \pm4$ **c.** -2
d. $f(x) = (x+2)(x^4 - 2x^3 + x + 2)$
e. There are 0 or 2 positive irrational zeros between the values 0 and 2. There are 0 or 2 negative irrational zeros between the values 0 and -1.

55. a. 0 or 2 positive zeros; 0 or 2 negative zeros
b. $\pm\frac{1}{3}, \pm\frac{1}{9}, \pm1, \pm3, \pm9$
c. $\pm\frac{1}{3}, \pm3$
d. $f(x) = 9(x - \frac{1}{3})(x + \frac{1}{3})(x-3)(x+3)$

57. a. no positive zeros; 1, 3 or 5 negative zeros
b. $-\frac{1}{4}, -\frac{1}{2}, -1, -2, -4$
c. $-\frac{1}{2}$(mult 2), -1
d. $f(x) = 4(x+1)(x + \frac{1}{2})^2(x^2 + 2x + 4)$

59.

	1	0	0	0	-3
		$\sqrt[4]{3}$	$\sqrt[4]{3^2}$	$\sqrt[4]{3^3}$	3
$\sqrt[4]{3}$	1	$\sqrt[4]{3}$	$\sqrt[4]{3^2}$	$\sqrt[4]{3^3}$	0

Note: $\sqrt[4]{3^4} = 3$

$x^4 - 3 = (x - \sqrt[4]{3})(x^3 + \sqrt[4]{3}x^2 + \sqrt[4]{3^2}x + \sqrt[4]{3^3})$
$x^4 - 3 = (x - \sqrt[4]{3})(x^3 + \sqrt[4]{3}x^2 + \sqrt[4]{9}x + \sqrt[4]{27})$

61. $\pm\dfrac{b}{a}, \pm\dfrac{d}{a}, \pm\dfrac{e}{a}, \pm\dfrac{bd}{a}, \pm\dfrac{be}{a}, \pm\dfrac{de}{a}, \pm\dfrac{bde}{a}$,
$\pm\dfrac{b}{c}, \pm\dfrac{d}{c}, \pm\dfrac{e}{c}, \pm\dfrac{bd}{c}, \pm\dfrac{be}{c}, \pm\dfrac{de}{c}, \pm\dfrac{bde}{c}$,
$\pm\dfrac{b}{ac}, \pm\dfrac{d}{ac}, \pm\dfrac{e}{ac}, \pm\dfrac{bd}{ac}, \pm\dfrac{be}{ac}, \pm\dfrac{de}{ac}, \pm\dfrac{bde}{ac}$,
$\pm b, \pm d, \pm e, \pm bd, \pm be, \pm de, \pm bde$,
$\pm\dfrac{1}{a}, \pm\dfrac{1}{c}, \pm\dfrac{1}{ac}$

63. $f(x) = a_4x^4 + a_3x^3 + a_2x^2 + a_1x + a_0$, $a_4 \neq 0$

Solutions to skill and review problems

1. $f(x) = (x-2)^2 - 1$
Graph of $y = x^2$ shifted right 2 units and down 1 unit.
Vertex: $(2,-1)$
Intercepts:
$x = 0$: $y = (-2)^2 - 1 = 3$; $(0,3)$
$y = 0$: $0 = (x-2)^2 - 1$
$1 = (x-2)^2$
$\pm 1 = x - 2$
$2 \pm 1 = x$
$1, 3 = x$; $(1,0)$, $(3,0)$

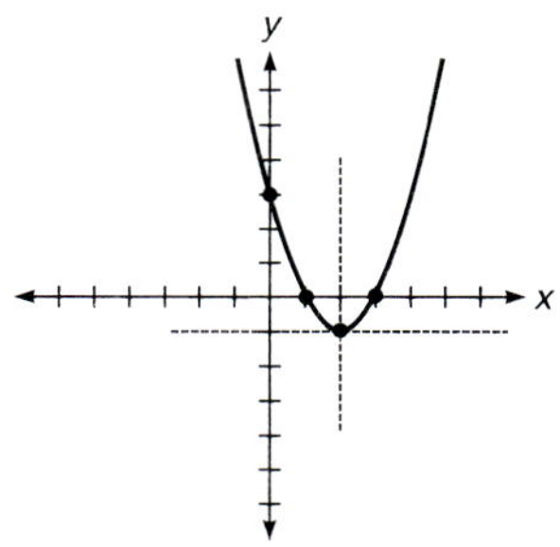

2. $f(x) = x^2 + x - 4$
$f(x) = x^2 + x + \frac{1}{4} - 4 - \frac{1}{4}$
$f(x) = (x + \frac{1}{2})^2 - 4\frac{1}{4}$
Graph of $y = x^2$ shifted left $\frac{1}{2}$ unit, down $4\frac{1}{4}$ units.
Vertex: $(-\frac{1}{2}, -4\frac{1}{4})$
Intercepts:
$x = 0$: $y = 0^2 + 0 - 4 = -4$; $(0,-4)$
$y = 0$: $0 = (x + \frac{1}{2})^2 - 4\frac{1}{4}$
$\frac{17}{4} = (x + \frac{1}{2})^2$
$\pm\sqrt{\frac{17}{4}} = x + \frac{1}{2}$
$-\frac{1}{2} \pm \frac{\sqrt{17}}{2} = x$
$-2.6, 1.6 \approx x$; $(-2.6,0)$, $(1.6,0)$

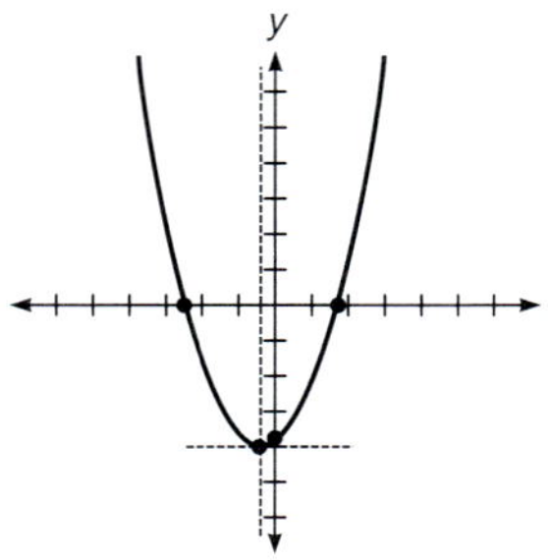

3. $f(x) = |x - 2| - 3$
Graph of $y = |x|$ translated; origin $(2,-3)$.
Intercepts:
$x = 0$: $y = |-2| - 3 = -1$; $(0,-1)$
$y = 0$: $0 = |x - 2| - 3$
$3 = |x - 2|$
$x - 2 = 3$ or $x - 2 = -3$
$x = 5$ or $x = -1$; $(-1,0)$, $(5,0)$

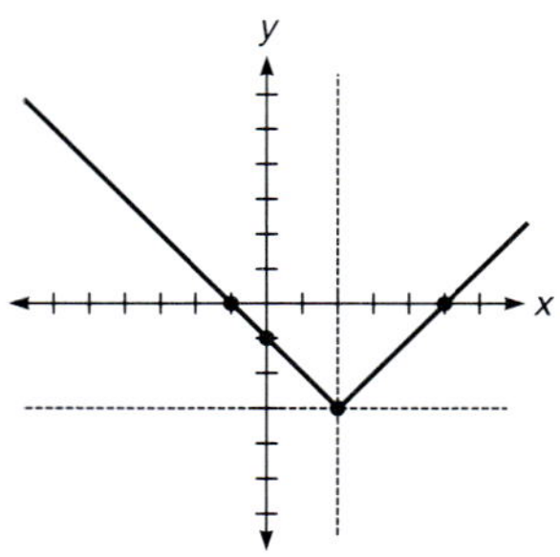

4. $f(x) = x^3 - 1$
Graph of $y = x^3$ shifted down 1 unit; origin $(0,-1)$.
Intercepts:
$x = 0$: $y = 0^3 - 1 = -1$; $(0,-1)$
$y = 0$: $0 = x^3 - 1$
$1 = x^3$
$1 = x$; $(1,0)$
Additional point: $(-1,-2)$

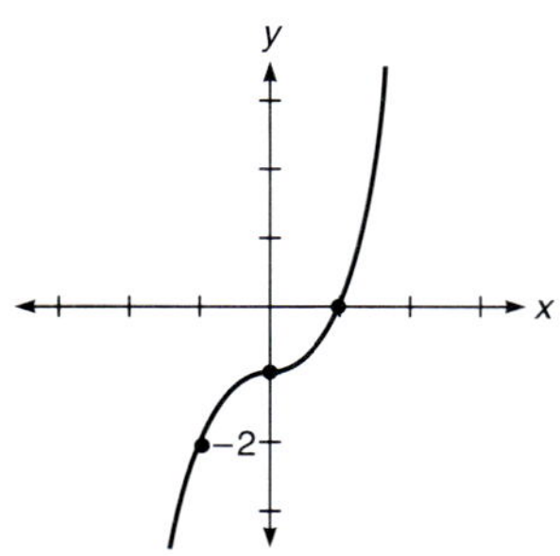

5. Find all zeros of $f(x) = 2x^5 + 7x^4 + 2x^3 - 11x^2 - 4x + 4$

If $\frac{p}{q}$ is a rational zero, p divides 4 and q divides 2:
$\pm\frac{4}{2}, \pm\frac{4}{1}, \pm\frac{2}{2}, \pm\frac{2}{1}, \pm\frac{1}{2}, \pm\frac{1}{1}$ or $\pm1, \pm2, \pm4, \pm\frac{1}{2}$.

	2	7	2	−11	−4	4
		2	9	11	0	−4
1	2	9	11	0	−4	0

$(x - 1)$ is a factor of $f(x)$
$f(x) = (x - 1)(2x^4 + 9x^3 + 11x^2 - 4)$

	2	9	11	0	−4
		−2	−7	−4	4
−1	2	7	4	−4	0

$(x + 1)$ is a factor of $f(x)$
$f(x) = (x - 1)(x + 1)(2x^3 + 7x^2 + 4x - 4)$

	2	7	4	−4
		−4	−6	4
−2	2	3	−2	0

$(x + 2)$ is a factor of $f(x)$
$f(x) = (x - 1)(x + 1)(x + 2)(2x^2 + 3x - 2)$
$f(x) = (x - 1)(x + 1)(x + 2)(2x - 1)(x + 2)$
$f(x) = 2(x - 1)(x + 1)(x - \frac{1}{2})^2(x + 2)^2$
Rational zeros are $\pm1, \frac{1}{2}$ (multiplicity 2), -2 (multiplicity 2)

6. Solve $|2x - 3| < 9$
$-9 < 2x - 3 < 9$
$-6 < 2x < 12$
$-3 < x < 6$

Solutions to trial exercise problems

21. $2 - 3x^2 + 4x^3$; rewrite as $4x^3 - 3x^2 + 2$: In $\frac{p}{q}$ p divides 2 and q divides 4, so we have $\pm\frac{2}{4}, \pm\frac{2}{2}, \pm\frac{2}{1}, \pm\frac{1}{4}, \pm\frac{1}{2}, \pm\frac{1}{1}$ or $\pm\frac{1}{2}, \pm\frac{1}{4}, \pm2, \pm1$.

25. $8x^3 - 8x + 16$; rewrite as $8(x^3 - x + 2)$ and focus on $x^3 - x + 2$. In $\frac{p}{q}$ p divides 2 and q divides 1, so we have $\pm\frac{2}{1}, \pm\frac{1}{1}$ or $\pm2, \pm1$.

33. $f(x) = \frac{1}{2}x^3 - 3x^2 + \frac{3}{4}x - 3$;
a. $x - 6$; **b.** $f(6)$
a.

	$\frac{1}{2}$	−3	$\frac{3}{4}$	−3
		3	0	$\frac{9}{2}$
6	$\frac{1}{2}$	0	$\frac{3}{4}$	$\frac{3}{2}$

So $\dfrac{\frac{1}{2}x^3 - 3x^2 + \frac{3}{4}x - 3}{x - 6}$

$= \frac{1}{2}x^2 + \frac{3}{4} + \dfrac{\frac{3}{2}}{x - 6}$.

b. $f(6) = \frac{3}{2}$

35. a. $f(x) = x^4 - x^3 - 7x^2 + x + 6$
There are 2 sign changes in $f(x)$; the number of positive roots is 0 or 2.
$f(-x) = x^4 + x^3 - 7x^2 - x + 6$
There are 2 sign changes in $f(-x)$; the number of negative roots is 0 or 2.

b. Possible rational zeros are $\pm6, \pm3, \pm2, \pm1$.
Test the rational zeros and factor.

	1	−1	−7	1	6
		3	6	−3	−6
3	1	2	−1	−2	0

$(x - 3)$ is a factor of $f(x)$
$f(x) = (x - 3)(x^3 + 2x^2 - x - 2)$

	1	2	−1	−2
		−2	0	2
−2	1	0	−1	0

$(x + 2)$ is a factor of $f(x)$
$f(x) = (x - 3)(x + 2)(x^2 - 1)$
$f(x) = (x - 3)(x + 2)(x - 1)(x + 1)$
Factor $x^2 - 1$

c. The rational zeros are $-2, -1, 1, 3$ (From the factors of part d).

d. $f(x) = (x - 3)(x + 2)(x - 1)(x + 1)$

41. a. $f(x) = x^6 - 4x^3 - 5$
There is one sign change so there is one positive root.
$f(-x) = x^6 + 4x^3 - 5$
There is one sign change so there is one negative root.

b. possible rational zeros: $\pm 1, \pm 5$. We can factor the expression for $f(x)$.
$x^6 - 4x^3 - 5 = (x^3 - 5)(x^3 + 1)$
$= (x^3 - 5)(x + 1)(x^2 - x + 1)$
$x^3 - 5$ has the **irrational zero** $\sqrt[3]{5}$, and $x^2 - x + 1$ has only complex zeros, so this expression cannot be factored further using rational zeros.

c. rational zeros: -1; irrational zero: $\sqrt[3]{5}$

d. $f(x) = (x^3 - 5)(x + 1)(x^2 - x + 1)$

---

57. a. $f(x) = 4x^5 + 16x^4 + 37x^3 + 43x^2 + 22x + 4$
no sign changes; no positive roots.
$f(-x) = -4x^5 + 16x^4 - 37x^3 + 43x^2 - 22x + 4$
five sign changes; 1, 3, or 5 negative roots.

b. possible rational zeros: $-\frac{1}{4}, -\frac{1}{2}$, $-1, -2, -4$; test rational zeros and factor

	4	16	37	43	22	4
		−4	−12	−25	−18	−4
−1	4	12	25	18	4	0

$(x + 1)$ is a factor of $f(x)$
$f(x) = (x + 1)(4x^4 + 12x^3 + 25x^2 + 18x + 4)$

	4	12	25	18	4
		−2	−5	−10	−4
$-\frac{1}{2}$	4	10	20	8	0

$(x + \frac{1}{2})$ is a factor of $f(x)$

$f(x) = (x + 1)(x + \frac{1}{2})(4x^3 + 10x^2 + 20x + 8)$

$f(x) = (2)(x + 1)(x + \frac{1}{2})(2x^3 + 5x^2 + 10x + 4)$
Common factor of 2

	2	5	10	4
		−1	−2	−4
$-\frac{1}{2}$	2	4	8	0

$(x + \frac{1}{2})$ is a factor for the second time
$(x + \frac{1}{2})^2$ is a factor of $f(x)$

$f(x) = 2(x + 1)(x + \frac{1}{2})^2(2x^2 + 4x + 8)$

$f(x) = 2(2)(x + 1)(x + \frac{1}{2})^2(x^2 + 2x + 4)$
Common factor of 2
$x^2 + 2x + 4$ is prime on R

c. rational zeros: $-\frac{1}{2}$ (mult 2), -1

d. $f(x) = 4(x + 1)(x + \frac{1}{2})^2(x^2 + 2x + 4)$

Exercise 4–3

Answers to odd-numbered problems

1. $y = (x - 2)(x + 1)(x + 3)$
intercepts: $(0,-6)$, $(-3,0)$, $(-1,0)$, $(2,0)$

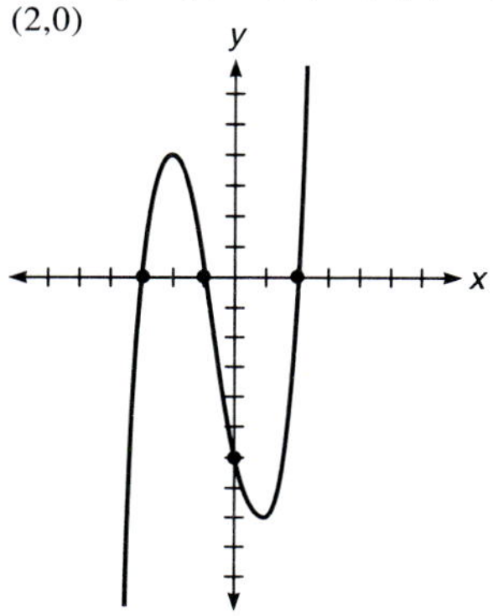

3. $y = (x - 1)(x + 1)(x - 3)(x + 3)$
intercepts: $(0,9)$, $(-3,0)$, $(-1,0)$, $(1,0)$, $(3,0)$

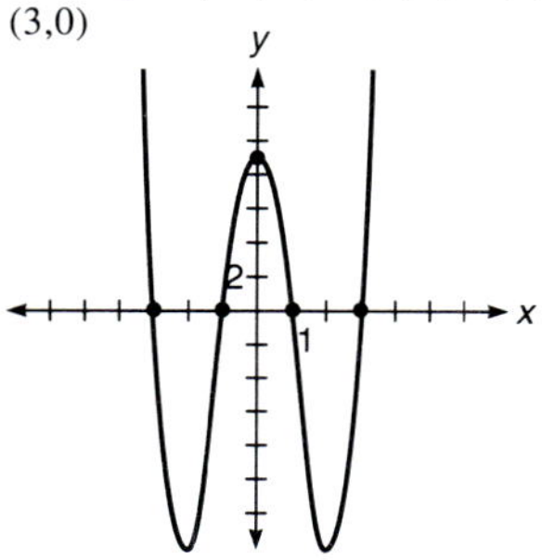

5. $y = (x - 2)(x + 2)(2x - 5)(2x + 5)(x + 3)$
intercepts: $(300,0)$, $(-3,0)$, $(-2.5,0)$, $(-2,0)$, $(2,0)$, $(2.5,0)$

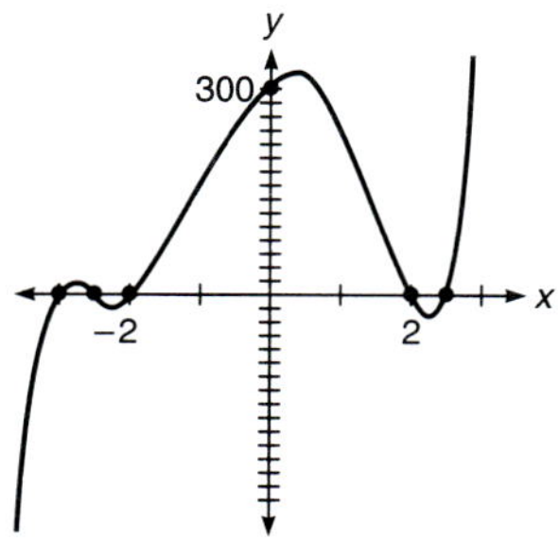

7. $y = (x - 1)^2(x + 1)$
intercepts: $(0,1)$, $(-1,0)$, $(1,0)$

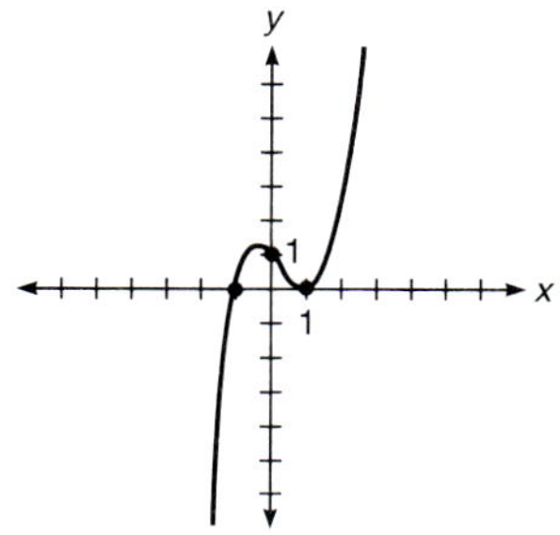

9. $y = (x + 2)^3(2x - 3)^2$
intercepts: $(0,72)$, $(-2,0)$, $(1\frac{1}{2},0)$

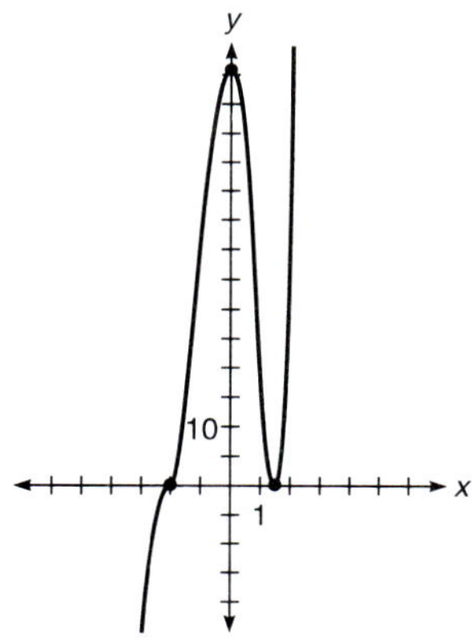

11. $y = (x - 2)^2(x^2 + 3x + 6)$
intercepts: $(0,24)$, $(2,0)$

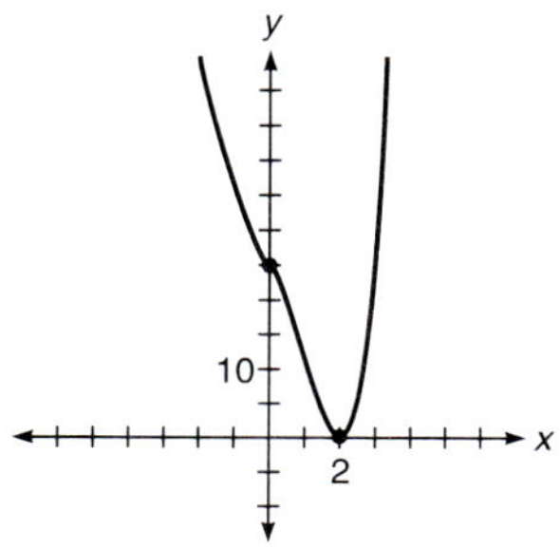

13. $y = (x - 3)(x - 1)(x + 1)(x + 2)$
intercepts: $(0,6)$, $(-2,0)$, $(-1,0)$, $(1,0)$, $(3,0)$

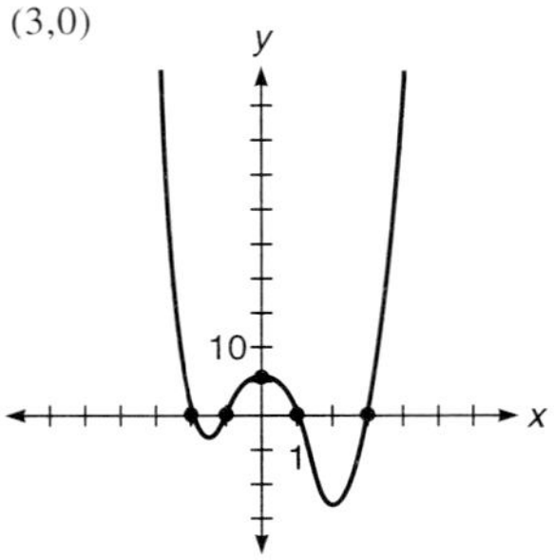

15. $y = (x - 1)(2x - 1)(2x - 3)$
intercepts: $(0,-3)$, $(1,0)$, $(\frac{1}{2},0)$, $(1\frac{1}{2},0)$

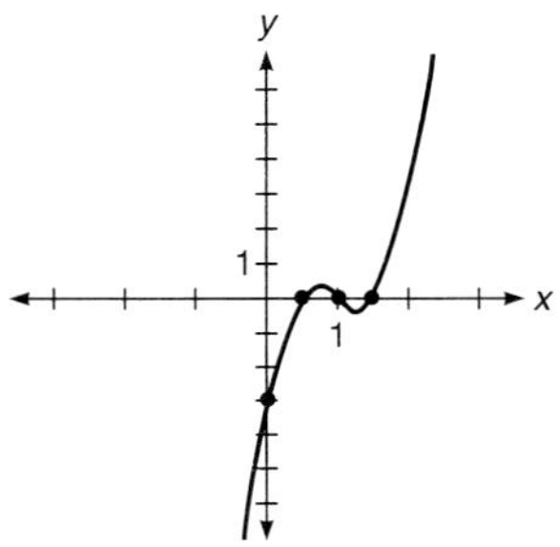

17. $y = (x - 3)^2(x^2 - 2x + 9)$
intercepts: $(0,81)$, $(3,0)$

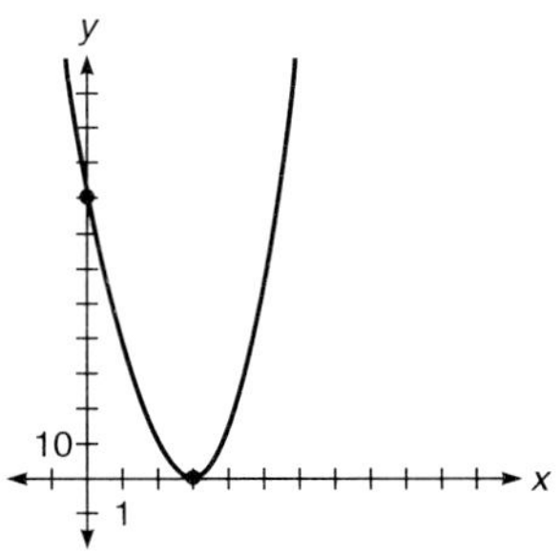

19. $y = (x + 1)(3x + 2)(2x - 1)$
intercepts: $(0,-2)$, $(-1,0)$, $(-\frac{2}{3},0)$, $(\frac{1}{2},0)$

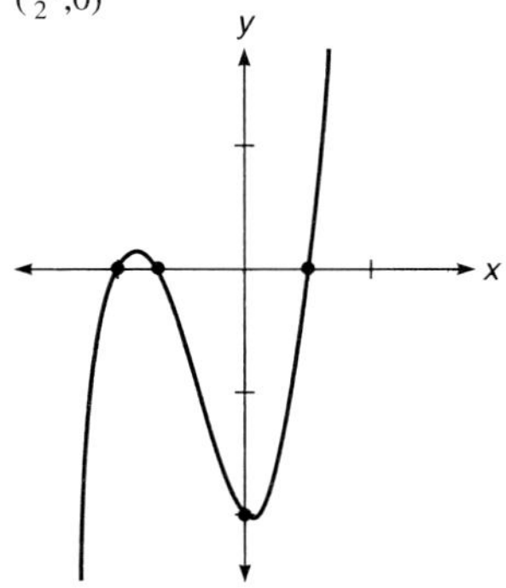

21. $y = (2x - 1)(2x + 1)(x - 2)(x^2 + 2x + 4)$
intercepts: $(0,8)$, $(-\frac{1}{2},0)$, $(\frac{1}{2},0)$, $(2,0)$

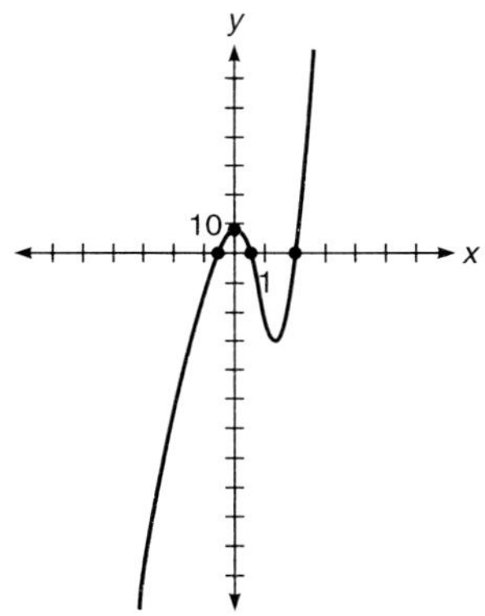

23. a.

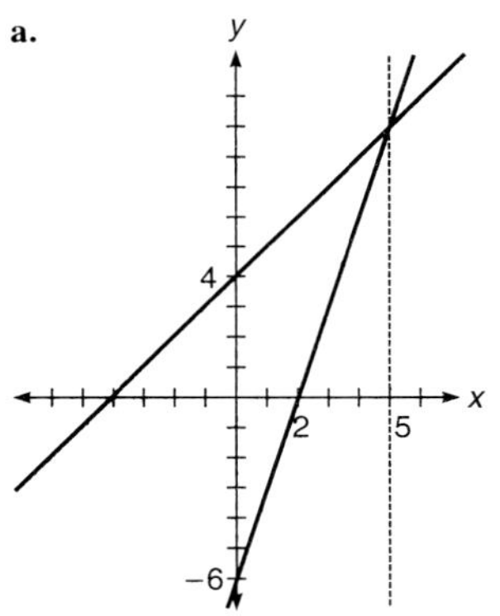

b. At least five items must be produced to break even or make a profit.

25. a.

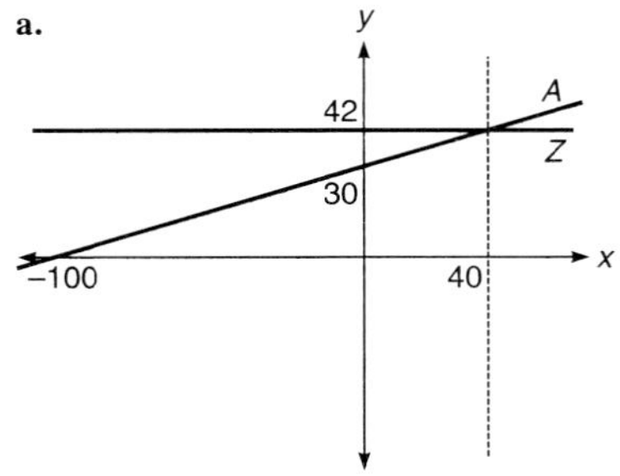

b. When $x < 40$ A is cheaper; when $x > 40$ Z is cheaper.

27. $A(x) = x(x + 3)$, or $A(x) = x^2 + 3x$ describes area A as a function of width x.

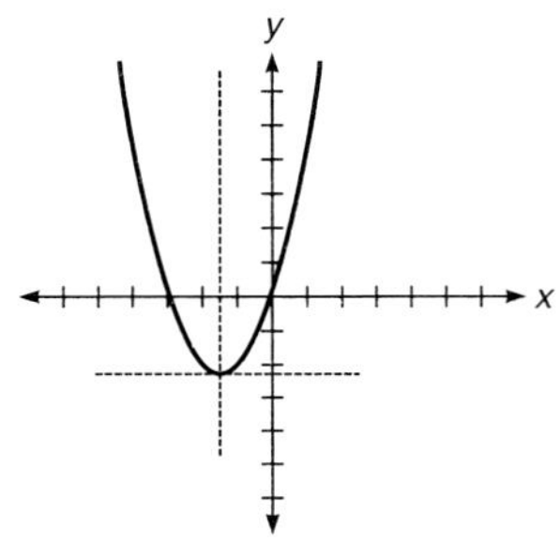

29. $A(x) = 6x^2 + 16x - 24$ in^2
Graph $y = 6x^2 + 16x - 24$.

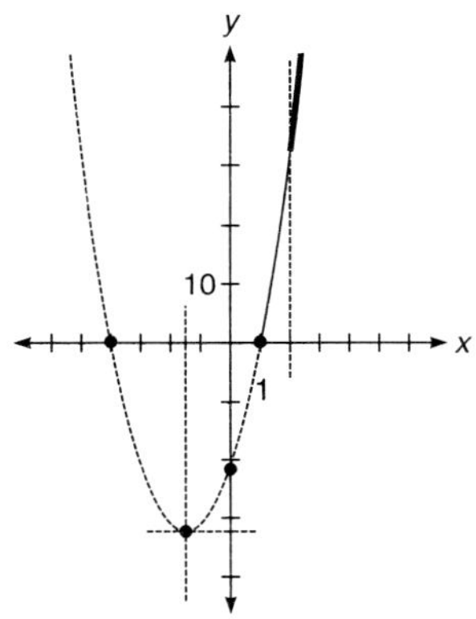

The value of x must be greater than 2 because width is $x - 2$, and must be a positive quantity.

31. a. $f(x) = 5x$

$2f(x) = 2(5x) = 10x$

$f(2x) = 5(2x) = 10x$

Thus, $2f(x) = f(2x)$ for this function.

b. $f(x) = -3x$

$kf(x) = k(-3x) = -3kx$

$f(kx) = -3(kx) = -3kx$

Thus, $kf(x) = f(kx)$.

c. $f(x) = x^2 - 2x - 8$; assume f is 3-scalable.

Then, $3f(x) = 3(x^2 - 2x - 8) = 3x^2 - 6x - 24$

Also, $f(3x) = (3x)^2 - 2(3x) - 8 = 9x^2 - 6x - 8$.

If $3f(x) = f(3x)$, then

$3x^2 - 6x - 24 = 9x^2 - 6x - 8$

$6x^2 + 16 = 0$

$6x^2 = -16$

This has no real solutions for x, and in any case even if there were solutions the solution set would have to be all real numbers if $3f(x) = f(3x)$ is to be true for all real numbers. Thus, f is not 3-scalable.

33. 0.46682

35. 0.69996, −1.72043

37. 0.69091, −0.72720, −2.65081

39. ±0.61803, ±1, ±1.30278, ±1.61803, ±2.30278

Solutions to skill and review problems

1. Graph $f(x) = x^2 + 2x - 1$.

$y = x^2 + 2x - 1$

$y = x^2 + 2x + 1 - 1 - 1$

$y = (x + 1)^2 - 2$

parabola; vertex: $(-1,-2)$; intercepts:

$x = 0$: $y = 0^2 + 2(0) - 1 = -1$; $(0,-1)$

$y = 0$: $0 = (x + 1)^2 - 2$

$(x + 1)^2 = 2$

$x + 1 = \pm\sqrt{2}$

$x = -1 \pm\sqrt{2}$; $\approx (-2.4,0)$, $(0.4,0)$

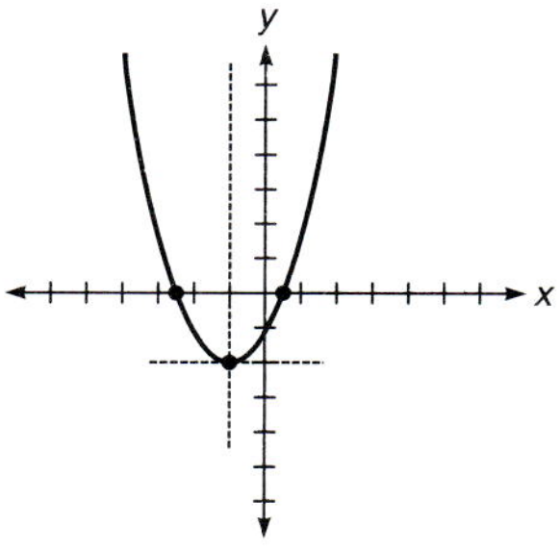

2. Solve $\left|\frac{2x - 3}{4}\right| \geq \frac{1}{2}$.

$\frac{2x - 3}{4} \geq \frac{1}{2}$ or $\frac{2x - 3}{4} \leq -\frac{1}{2}$

$4\left(\frac{2x - 3}{4}\right) \geq 4\left(\frac{1}{2}\right)$ or $4\left(\frac{2x - 3}{4}\right) \leq 4\left(-\frac{1}{2}\right)$

$2x - 3 \geq 2$ or $2x - 3 \leq -2$

$2x \geq 5$ or $2x \leq 1$

$x \geq \frac{5}{2}$ or $x \leq \frac{1}{2}$

3. Solve $x^{-2} - x^{-1} - 12 = 0$.

$x^2(x^{-2} - x^{-1} - 12) = x^2(0)$

$x^0 - x^1 - 12x^2 = 0$

$12x^2 + x - 1 = 0$

$(4x - 1)(3x + 1) = 0$

$4x - 1 = 0$ or $3x + 1 = 0$

$4x = 1$ or $3x = -1$

$x = \frac{1}{4}$ or $x = -\frac{1}{3}$

4. Solve $\sqrt{2x - 2} = x - 5$.

$(\sqrt{2x - 2})^2 = (x - 5)^2$

$2x - 2 = x^2 - 10x + 25$

$0 = x^2 - 12x + 27$

$0 = (x - 3)(x - 9)$

$x - 3 = 0$ or $x - 9 = 0$

$x = 3$ or $x = 9$

The value 3 does not check, so the answer is 9.

5. Combine $\frac{3}{x - 1} - \frac{2}{x + 1} + \frac{1}{x}$.

$\frac{3(x + 1) - 2(x - 1)}{(x - 1)(x + 1)} + \frac{1}{x}$

$\frac{x + 5}{x^2 - 1} + \frac{1}{x}$

$\frac{x(x + 5) + (x^2 - 1)}{x(x^2 - 1)}$

$\frac{2x^2 + 5x - 1}{x^3 - x}$

6. Simplify $\sqrt[3]{\frac{4x^2y^7}{3z^8}}$

$= \frac{y^2\sqrt[3]{4x^2y}}{z^2\sqrt[3]{3z^2}} \cdot \frac{\sqrt[3]{3^2z}}{\sqrt[3]{3^2z}} =$

$\frac{y^2\sqrt[3]{4(3^2)x^2yz}}{z^2\sqrt[3]{3^3z^3}} =$

$\frac{y^2\sqrt[3]{36x^2yz}}{z^2(3z)} = \frac{y^2\sqrt[3]{36x^2yz}}{3z^3}$

7. Rewrite $|5 - 2\pi|$ without absolute value symbols.

$5 - 2\pi < 0$ so $|5 - 2\pi|$

$= -(5 - 2\pi) = 2\pi - 5$

Solutions to trial exercise problems

15. $g(x) = 4x^3 - 12x^2 + 11x - 3$

Using possible rational zeros and synthetic division we find that

$y = (x - 1)(2x - 1)(2x - 3)$.

Intercepts:

$x = 0$: $y = 0 - 0 + 0 - 3 = -3$; $(0,-3)$

$y = 0$: $0 = (x - 1)(2x - 1)(2x - 3)$

$x = 1, \frac{1}{2}, \frac{3}{2}$; $(1,0)$, $(\frac{1}{2},0)$, $(1\frac{1}{2},0)$

Additional points: (0.75,0.19), (1.25,−0.19), (2,3)

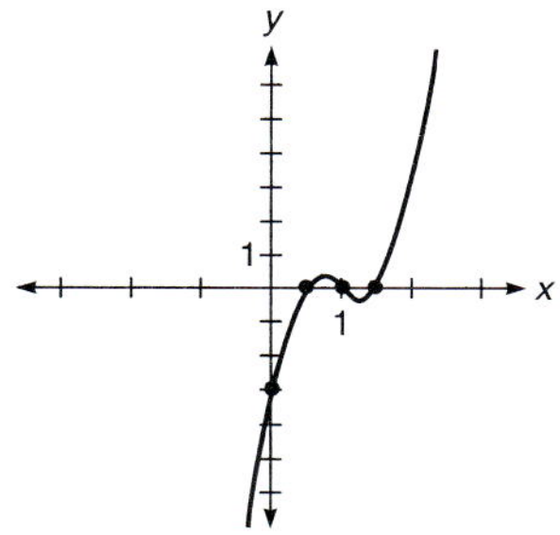

27. Let x = width; then length is $x + 3$. The area A is the product of length and width. Thus, $A(x) = x(x + 3)$, or $A(x) = x^2 + 3x$ describes area A as a function of width x.

Graph: $y = x^2 + 3x + \frac{9}{4} - \frac{9}{4}$

$y = (x + 1\frac{1}{2})^2 - 2\frac{1}{4}$

Parabola; vertex at $(-1\frac{1}{2},-2\frac{1}{4})$;

intercepts:

$x = 0$: $y = x^2 + 3x$

$y = 0^2 + 0 = 0$; $(0,0)$

$y = 0$: $0 = x(x + 3)$

$x = 0$ or -3; $(0,0)$, $(-3,0)$

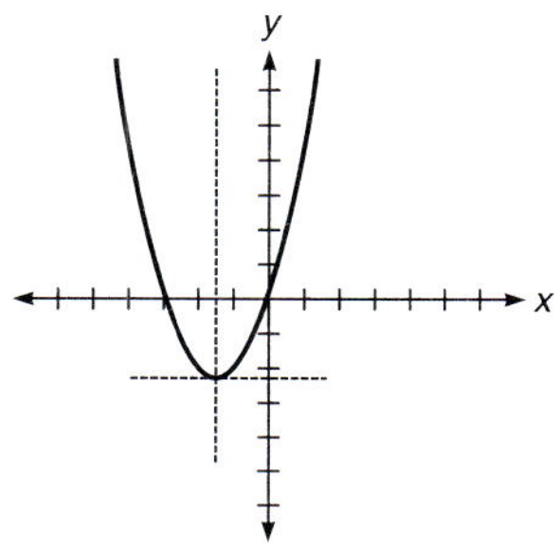

32. **a.** $f(a) = 5a$, and $f(b) = 5b$

$f(a) + f(b) = 5a + 5b$

$f(a + b) = 5(a + b) = 5a + 5b$

Thus, $f(a) + f(b) = f(a) + f(b)$

b. $f(a) = -3a + 1, f(b) = -3b + 1$

$f(a) + f(b) = -3a + 1 - 3b + 1$

$= -3a - 3b + 2$

$f(a + b) = -3(a + b) + 1 = -3a - 3b + 1$

Thus $f(a + b) \neq f(a) + f(b)$

c. Show that the function $f(x) = x^2 - 2x - 8$ is not additive.

$f(a) = a^2 - 2a - 8,$

$f(b) = b^2 - 2b - 8$

$f(a) + f(b)$

$= (a^2 - 2a - 8) + (b^2 - 2b - 8)$

$= a^2 - 2a + b^2 - 2b - 16$

$f(a + b)$

$= (a + b)^2 - 2(a + b) - 8$

$= a^2 + 2ab + b^2 - 2a - 2b - 8,$

which is not equal to $f(a) + f(b)$.

Exercise 4–4

Answers to odd-numbered problems

1. graph of $y = \frac{1}{x}$ shifted right 2 units, vertically scaled 3 units; intercepts: $(0,-1\frac{1}{2})$; asymptotes: $x = 2$, $y = 0$

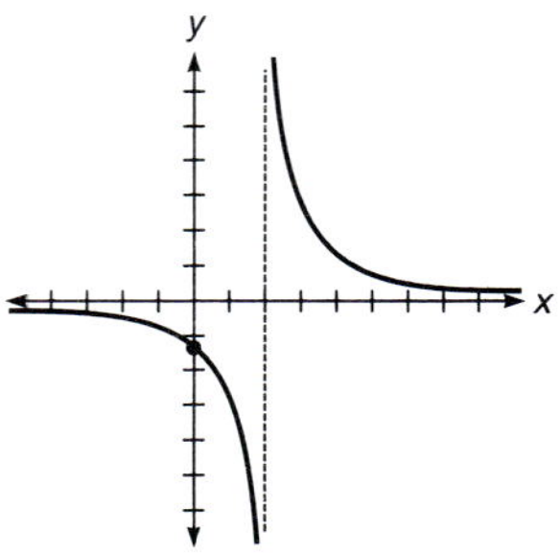

3. graph of $y = \frac{1}{x}$ shifted left 4 units, vertically scaled -2 units; intercepts: $(0,-\frac{1}{2})$; asymptotes: $x = -4$, $y = 0$

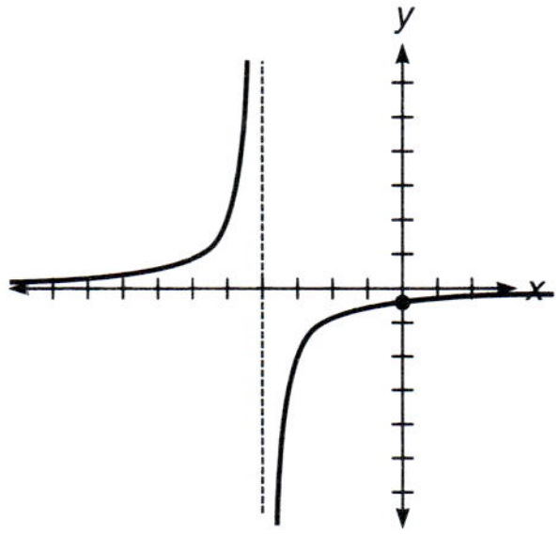

5. graph of $y = \frac{1}{x^2}$ shifted right 2 units, vertically scaled 3 units; intercepts: $(0,\frac{3}{4})$; asymptotes: $x = 2$, $y = 0$

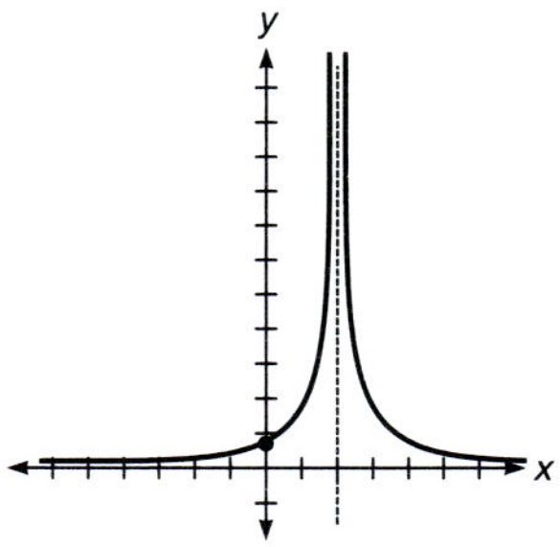

7. graph of $y = \frac{1}{x^4}$ shifted left $\frac{1}{2}$ unit; intercepts: $(0,16)$; asymptotes: $x = -\frac{1}{2}$, $y = 0$

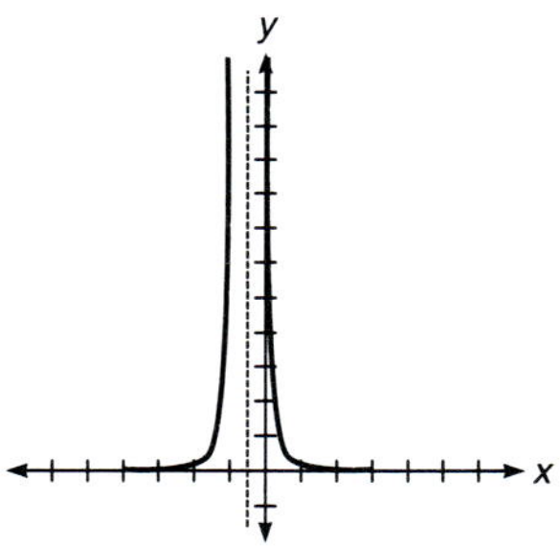

9. graph of $y = \frac{1}{x^3}$ shifted right 2 units, vertically scaled 3 units; intercepts: $(0,-\frac{3}{8})$; asymptotes: $x = 2$, $y = 0$

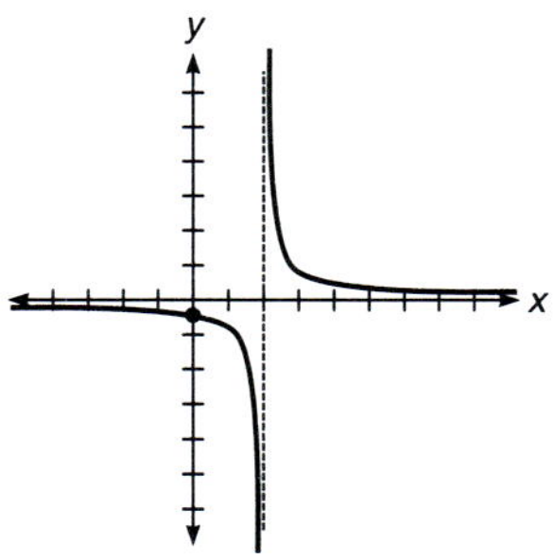

11. graph of $y = \frac{1}{x^2}$ shifted right 2 units, vertically scaled -4 units; intercepts: $(0,-1)$; asymptotes: $x = 2$, $y = 0$

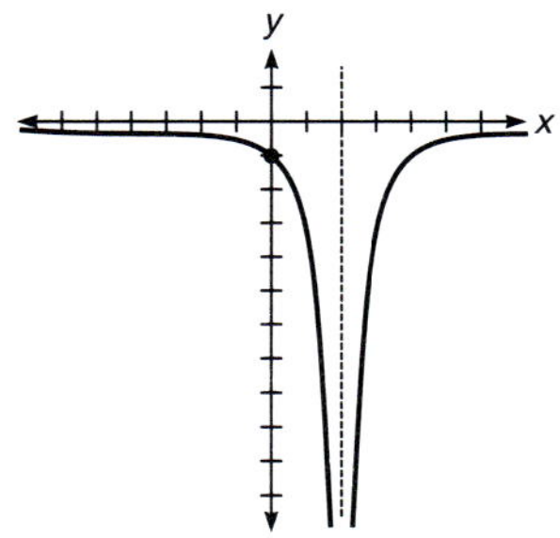

13. graph of $y = \frac{1}{x}$ shifted right 1 unit, up 2 units; intercepts: $(0,1)$, $(\frac{1}{2},0)$; asymptotes: $x = 1$, $y = 2$

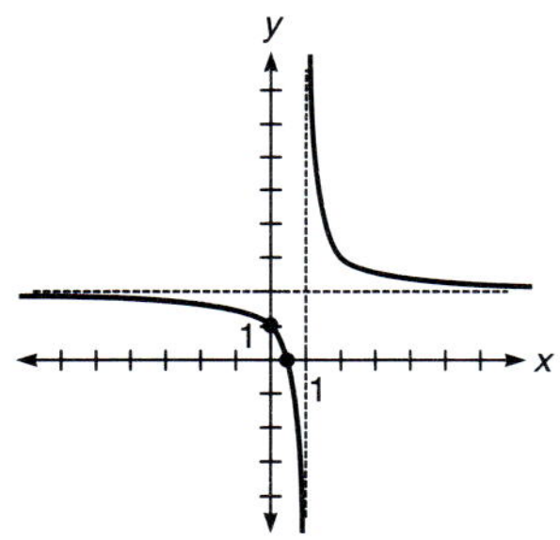

15. graph of $y = \frac{1}{x^2}$ shifted right 3 units, up 2 units; intercepts: $(0,2\frac{1}{9})$; asymptotes: $x = 3$, $y = 2$

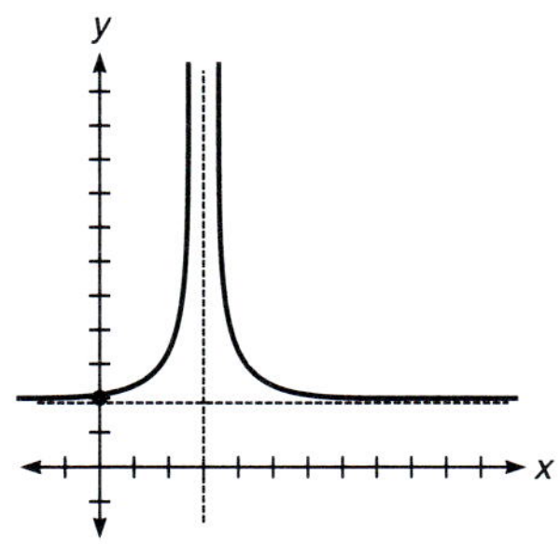

17. intercepts: $(0,-\frac{1}{6})$; asymptotes: $x = -3$, $x = 6$, $y = 0$

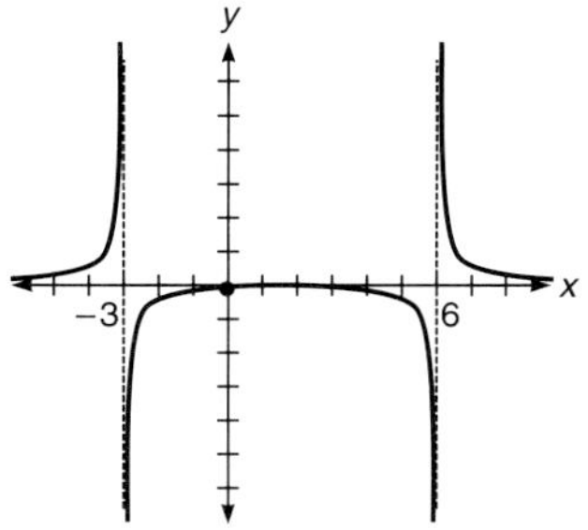

19. intercepts: $(0,\frac{1}{2})$; asymptotes: $x = -4$, $x = 2$, $y = 0$

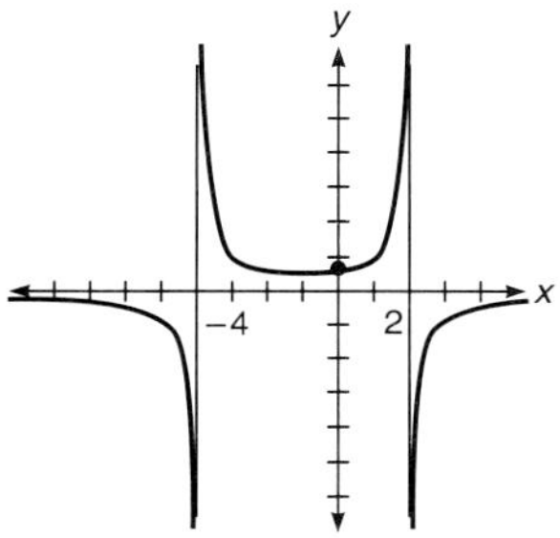

21. intercepts: $(0,\frac{1}{4})$, $(\frac{1}{2},0)$; asymptotes: $x = \pm 2$, $y = 0$

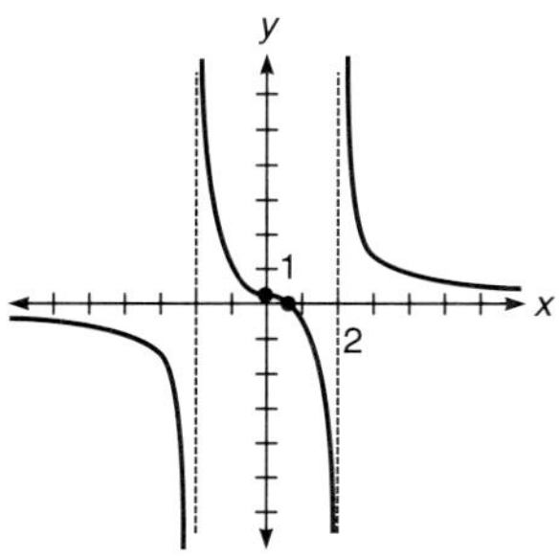

23. intercepts: $(0,0)$; asymptotes: $x = -1$, $x = 5$, $y = 0$

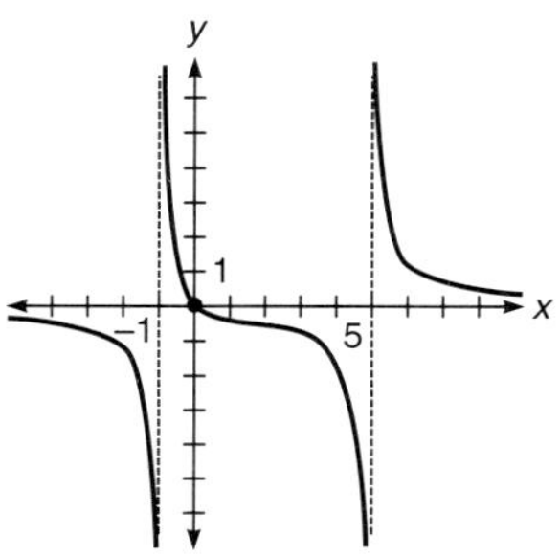

25. intercepts: $(-1\frac{1}{2},0)$; asymptotes: $x = 0$, $x = 4$, $y = 0$

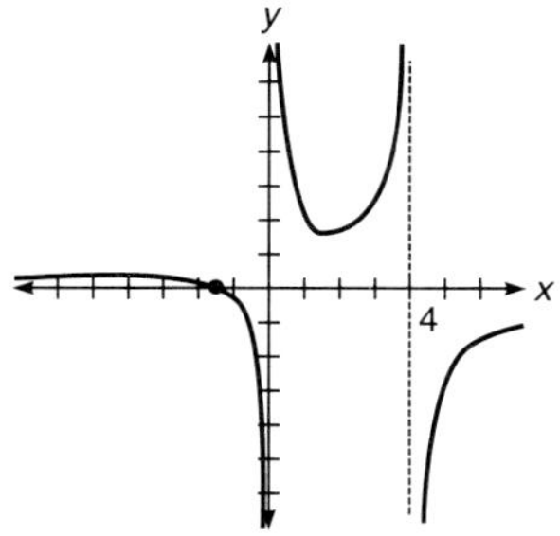

27. intercepts: $(0,-\frac{1}{18})$, $(\pm 0.6,0)$; asymptotes: $x = -3$, $x = 2$, $x = 3$, $y = 0$

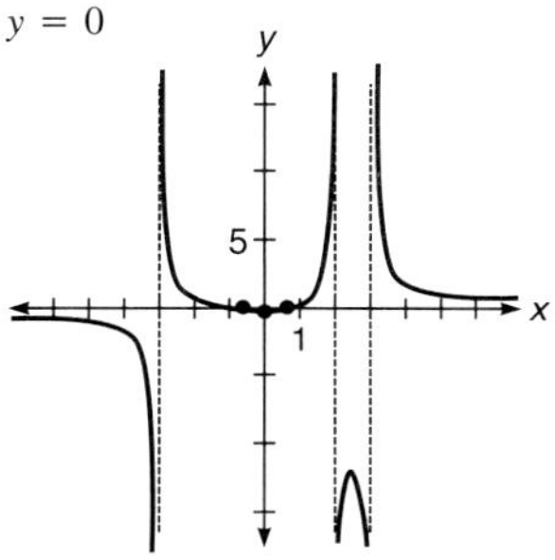

29. intercepts: $(\frac{1}{3},0)$; asymptotes: $x = -1$, $x = 0$, $x = 2$, $y = 0$

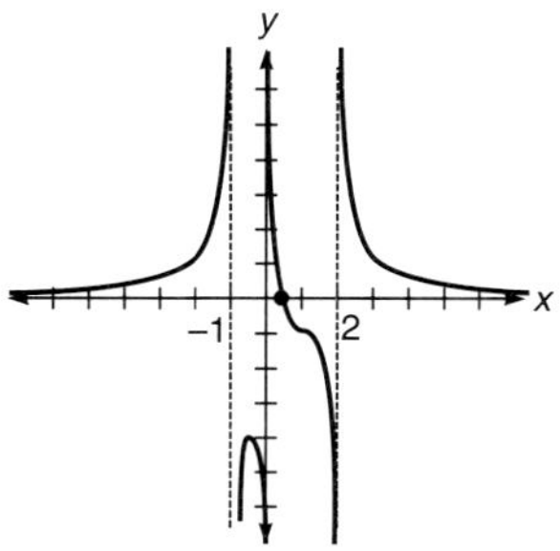

31. intercepts: $(0,1\frac{1}{3})$, $(-1 \pm \sqrt{3},0)$; asymptotes: $x = -3$, $x = 1$, $y = 2$

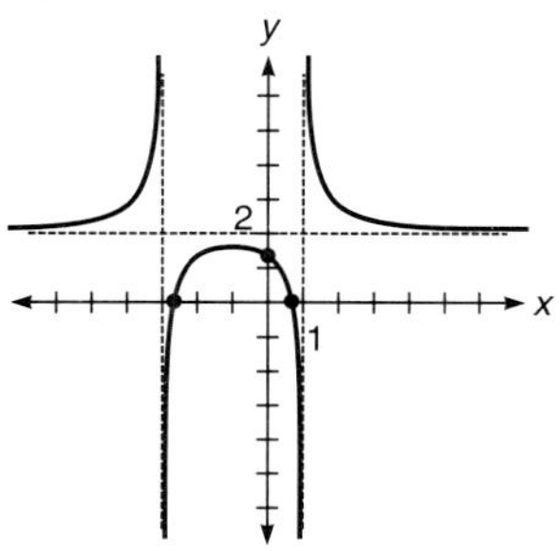

33. intercepts: $(0,-4)$, $\left(\frac{9 \pm \sqrt{1{,}041}}{8},0\right)$; asymptotes: $x = -3$, $x = 5$, $y = -4$

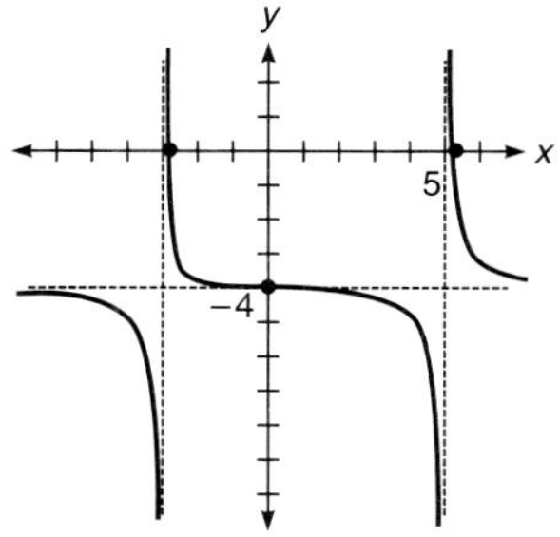

35. intercepts: $(0,0)$; asymptotes: $x = -1$, $y = 1$

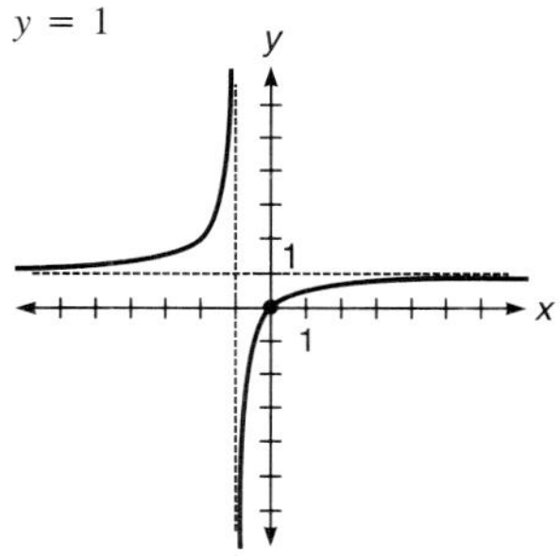

37. intercepts: $(0,0)$; asymptotes: $x = 3$, $y = -2$

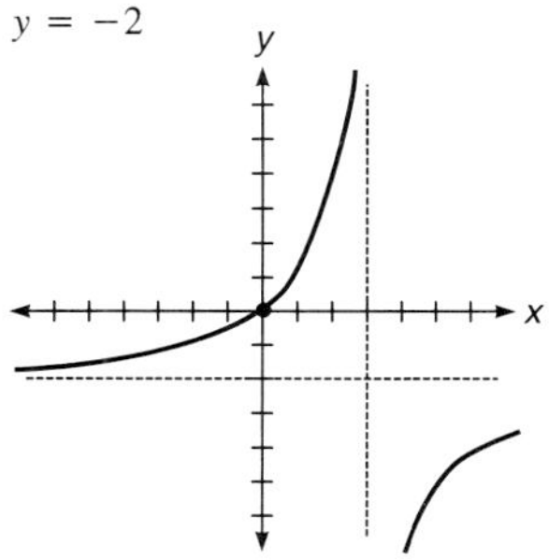

39. intercepts: $(0,\frac{3}{5})$, $(\pm\sqrt{3},0)$; asymptotes: $x = -1$, $x = 5$, $y = 1$

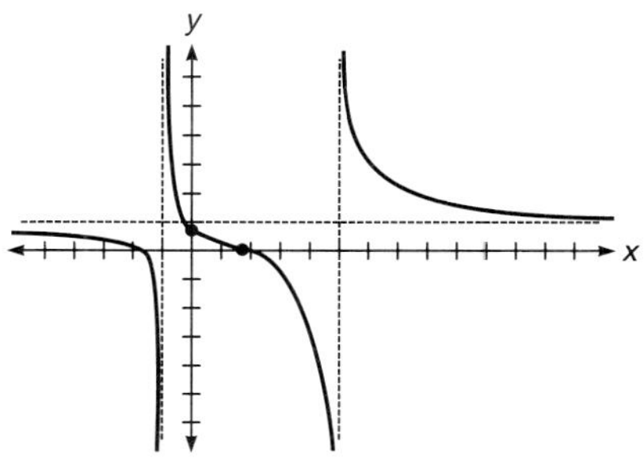

41. intercepts: (0,0.25); asymptotes: $x = \pm 2$, $y = -3$

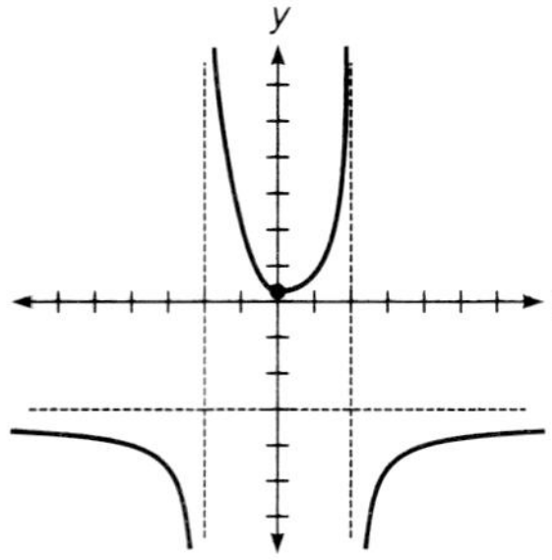

43. intercepts: (0,0); asymptotes: $x = 1$, $y = \frac{1}{2}x + \frac{1}{2}$

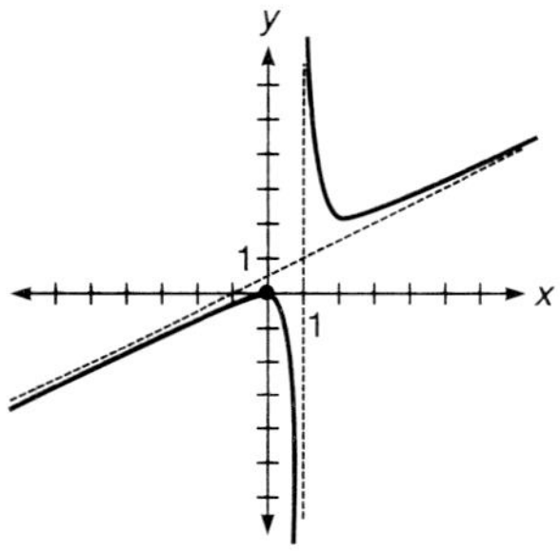

45. intercepts: (0,0); asymptotes: $x = 1$, $y = x + 2$

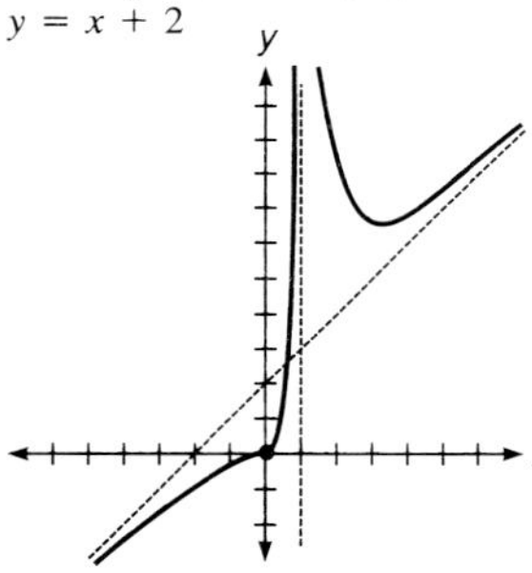

47. intercepts: (0,0), ($\pm$1,0), (0,0); asymptotes: $x = \pm 2$, $y = x$

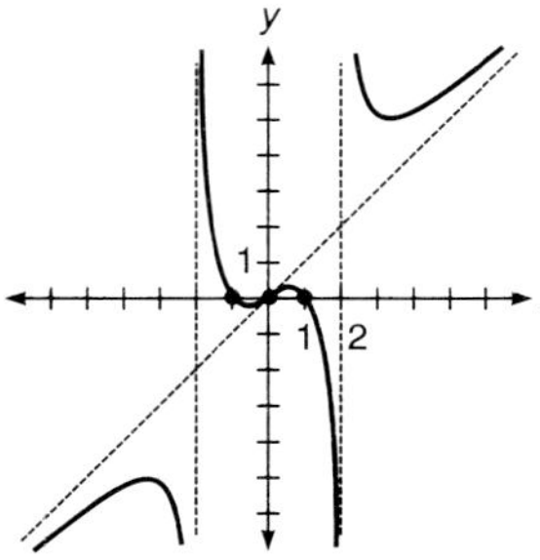

49. intercepts: (0,−2), (−2,0); asymptotes: $x = -4$, $x = 1$, $y = x - 3$

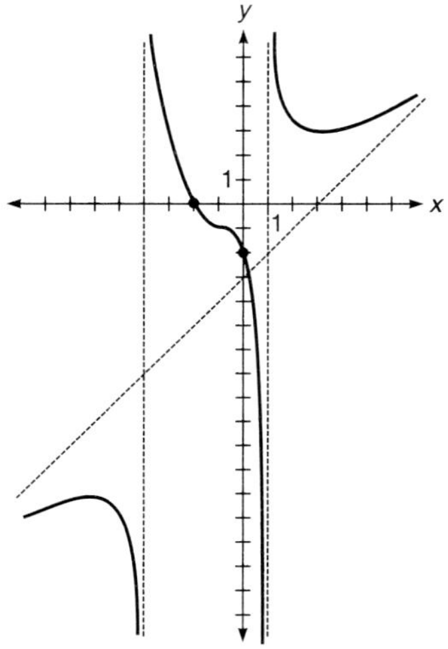

51. intercepts: (0,0), $\left(\pm\frac{\sqrt{2}}{2},0\right)$; asymptotes: $x = \pm 1$, $x = -2$, $y = 2x - 4$

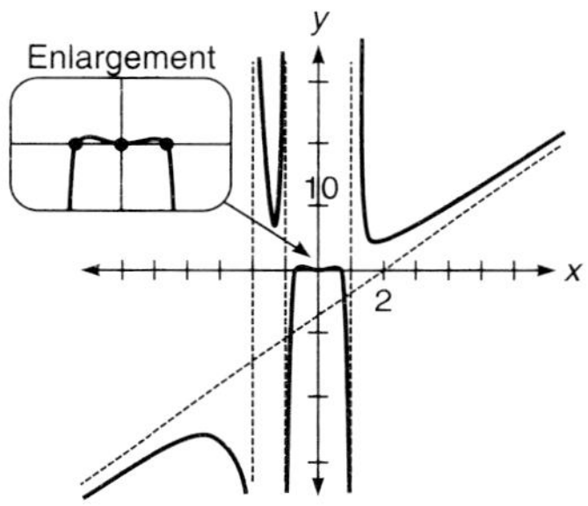

53. intercepts: $(0,\frac{1}{3})$, (−1,0); asymptotes: $x = -3$, $y = 1$

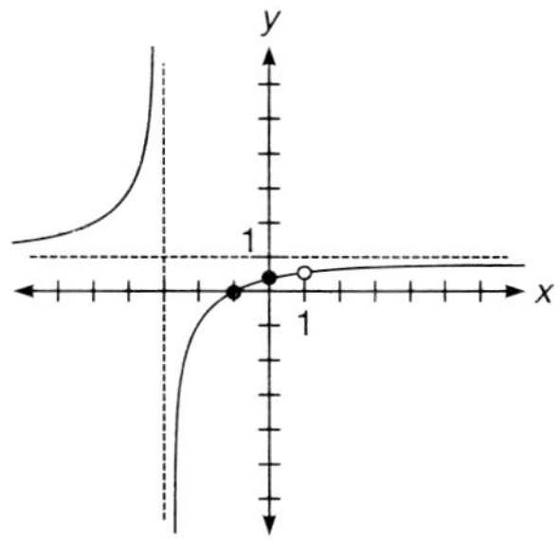

55. intercepts: $(0,-\frac{1}{2})$; asymptotes: $x = -2$, $x = 1$, $y = 0$

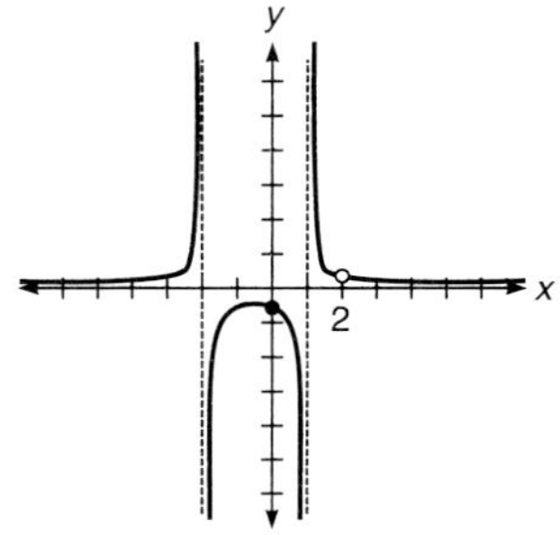

57. $y = x + 4$
intercepts: (−4,0), (0,4)

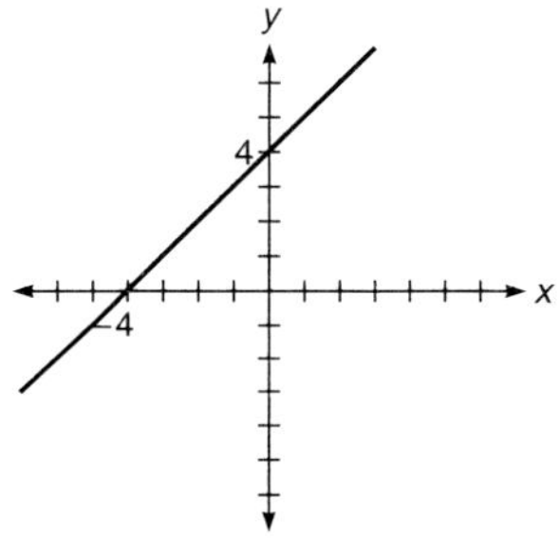

59. intercepts: (0,−2), $(\frac{1}{4},0)$

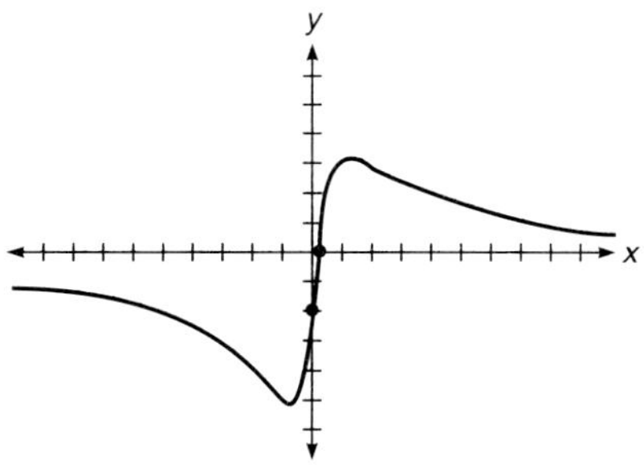

61. intercepts: (0,−1), ($\pm$2,0)

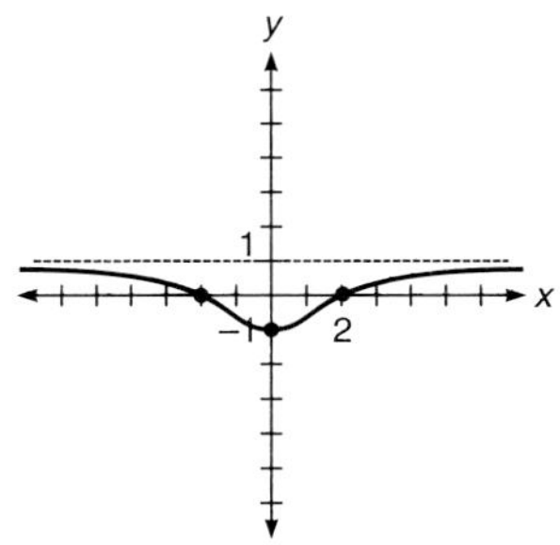

63.

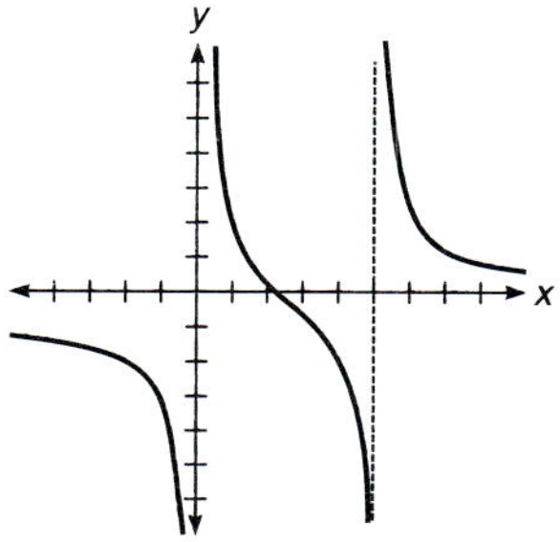

65. a. $y = \dfrac{20x}{x - 20}$

b.

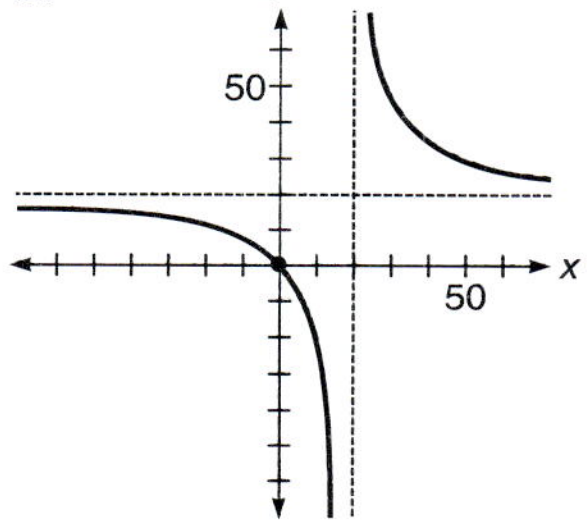

67.

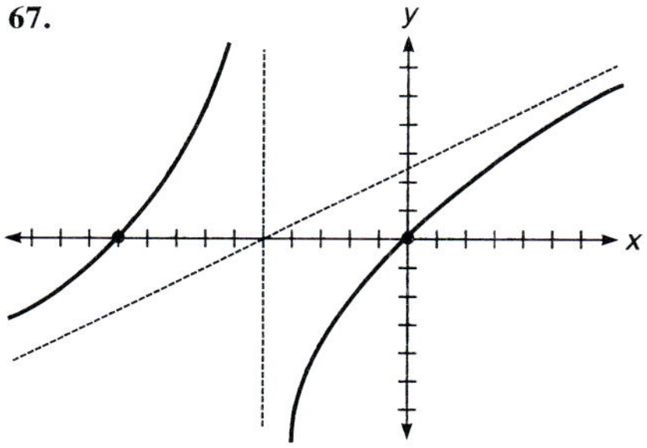

69. a. \$750 **b.** \$250 **c.** \$125

Solutions to skill and review problems

1. $f(5) = 2(5) - 3 = 7$

2. $g(-4) = (-4)^2 + 2(-4) + 3 = 11$

3. $g(2) = 2^2 + 2(2) + 3 = 11$
$f(g(2)) = f(11) = 2(11) - 3 = 19$

4. $f(-1) = 2(-1) - 3 = -5$
$g(f(-1)) = g(-5) = (-5)^2 + 2(-5) + 3 = 18$

5. Solve $x = 2y + 7$ for y.
$x = 2y + 7$
$2y = x - 7$
$y = \dfrac{x - 7}{2}$

6. Solve $x = \dfrac{1}{y - 2}$ for y.
$x = \dfrac{1}{y - 2}$
$x(y - 2) = 1$
$xy - 2x = 1$
$xy = 2x + 1$
$y = \dfrac{2x + 1}{x}$

7. Graph $f(x) = 2x^4 - x^3 - 14x^2 + 19x - 6$
$y = 2(x - 1)(x - 2)(x + 3)(x - \frac{1}{2})$
(using possible rational zeros and synthetic division)
Intercepts:
$x = 0$: $y = -6$; $(0,-6)$
$y = 0$:
$0 = 2(x - 1)(x - 2)(x + 3)(x - \frac{1}{2})$
$x = -3, \frac{1}{2}, 1$; $(-3,0)$, $(\frac{1}{2},0)$, $(1,0)$, $(2,0)$
Additional points: $(-2,-60)$, $(-1,-36)$, $(1.5,-2.25)$, $(2.5,16.5)$

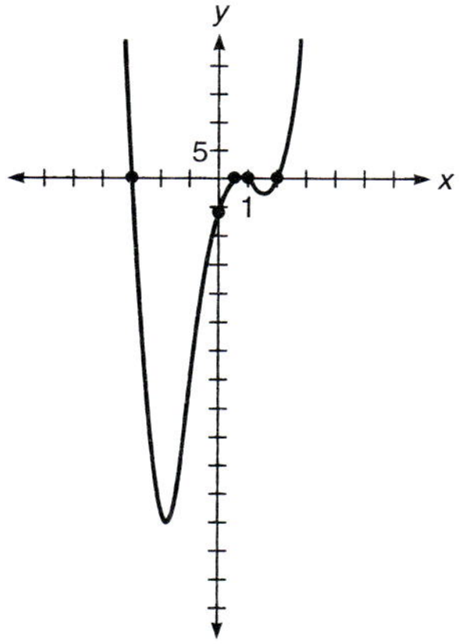

8. Solve $|4 - 3x| = 16$
$4 - 3x = 16$ or $4 - 3x = -16$
$-12 = 3x$ or $20 = 3x$
$-4 = x$ or $6\frac{2}{3} = x$
$\{-4, 6\frac{2}{3}\}$

Solutions to trial exercise problems

11. $y = \dfrac{-4}{(x - 2)^2}$

Same as $y = \dfrac{1}{x^2}$ translated right 2 units and scaled vertically by -4. Vertical asymptote: $x = 2$. Horizontal asymptote: $y = 0$ (x-axis).
Intercepts:
$x = 0$: $y = \dfrac{-4}{(-2)^2} = -1$; $(0,-1)$
$y = 0$: $0 = \dfrac{-4}{(x - 2)^2}$ has no solutions.
Additional points: $(1,-4)$, $(3,-4)$, $(4,-1)$

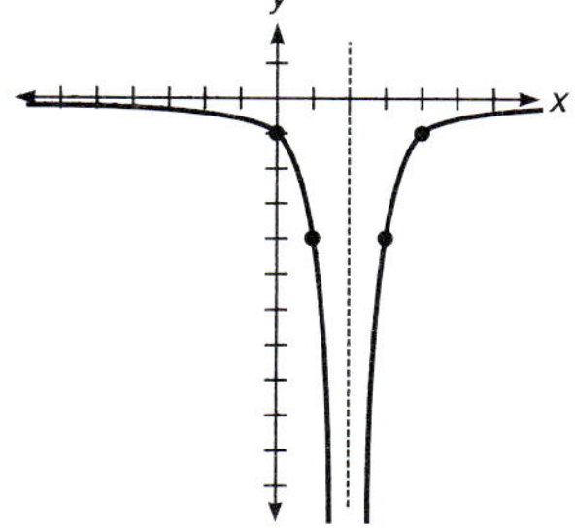

15. $y = \dfrac{1}{(x - 3)^2} + 2$

Same as $y = \dfrac{1}{x^2}$, translated. Vertical asymptote at $x = 3$; horizontal asymptote at $y = 2$.
Origin: $(3,2)$
Intercepts:
$x = 0$: $y = \dfrac{1}{(-3)^2} + 2 = 2\frac{1}{9}$; $(0,2\frac{1}{9})$
$y = 0$: $0 = \dfrac{1}{(x - 3)^2} + 2$
$-2 = \dfrac{1}{(x - 3)^2}$
$-2(x^2 - 6x + 9) = 1$
$-2x^2 + 12x - 19 = 0$
$2x^2 - 12x + 19 = 0$; no real solutions
Additional points: $(2,3)$, $(4,3)$, $(6,2\frac{1}{9})$

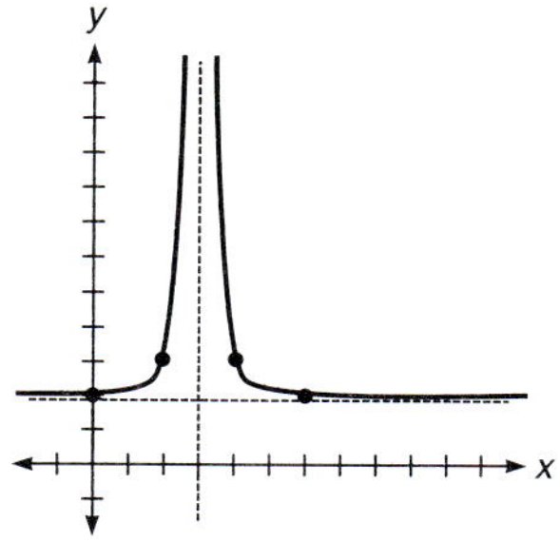

27. $y = \dfrac{3x^2 - 1}{(x - 2)(x^2 - 9)}$

$= \dfrac{3x^2 - 1}{(x - 2)(x - 3)(x + 3)}$

Vertical asymptotes: $x = 2, \pm 3$; horizontal asymptote: $y = 0$ (x-axis).

Intercepts:

$x = 0$: $y = \dfrac{-1}{(-2)(-9)} = -\dfrac{1}{18}$; $\left(0, -\dfrac{1}{18}\right)$

$y = 0$: $0 = \dfrac{3x^2 - 1}{(x - 2)(x^2 - 9)}$

$0 = 3x^2 - 1$

$\dfrac{1}{3} = x^2$

$\pm\dfrac{\sqrt{3}}{3} = x$; $(\pm 0.6, 0)$

Additional points: $(-4, -1.1)$, $(-2, 0.55)$, $(1.5, 1.7)$, $(2.25, -14.4)$, $(2.5, -12.9)$, $(2.75, -20.1)$, $(4, 3.4)$, $(5, 1.5)$

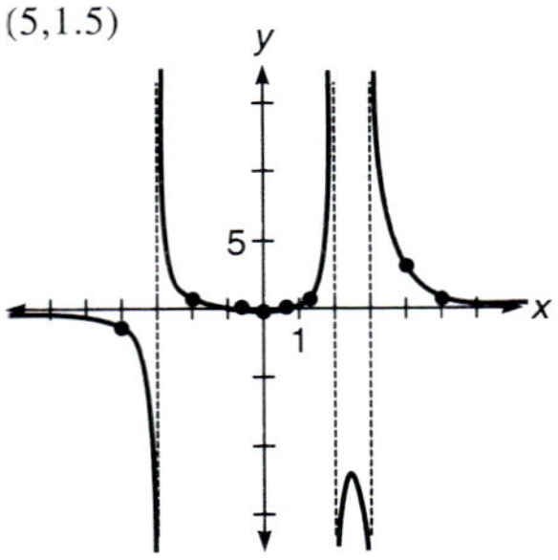

41. $y = \dfrac{-3x^2 + 2x - 1}{x^2 - 4}$

$= -3 + \dfrac{2x - 13}{(x - 2)(x + 2)}$

Horizontal asymptote: $y = -3$; vertical asymptotes: $x = \pm 2$

Intercepts:

$x = 0$: $y = \dfrac{-1}{-4} = \dfrac{1}{4}$; $(0, 0.25)$

$y = 0$: $0 = \dfrac{-3x^2 + 2x - 1}{x^2 - 4}$

$0 = -3x^2 + 2x - 1$

$0 = 3x^2 - 2x + 1$

No real solutions.

Additional points: $(-5, -4.1)$, $(-3, -6.8)$, $(-1, 2)$, $(3, -4.4)$, $(5, -3.1)$, $(9, -2.94)$, $(11, -2.92)$

The value of y at $x = 5$ is less than -3 and at $x = 9$ is more than -3. The coordinate where $y = -3$ can be found by replacing y by -3 and solving.

$-3 = \dfrac{-3x^2 + 2x - 1}{x^2 - 4}$

$-3x^2 + 12 = -3x^2 + 2x - 1$

$13 = 2x$

$6.5 = x$

The point $(6.5, -3)$ is plotted.

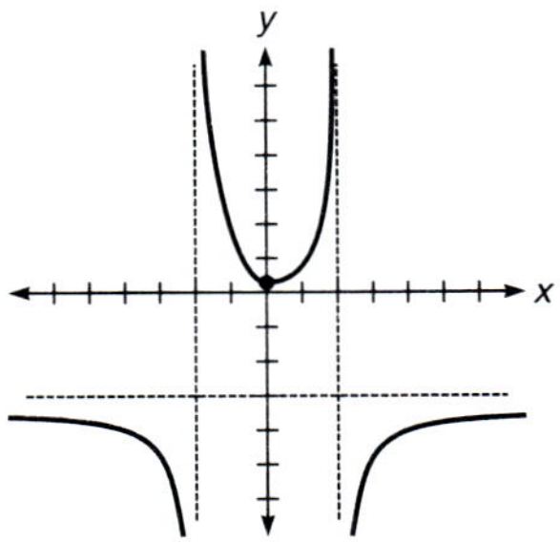

51. $y = \dfrac{2x^4 - x^2}{(x + 2)(x^2 - 1)}$

$= \dfrac{2x^4 - x^2}{x^3 + 2x^2 - x - 2}$

$= 2x - 4 + \dfrac{9x^2 - 8}{(x + 2)(x - 1)(x + 1)}$

Slant asymptote: $y = 2x - 4$; vertical asymptotes: $x = -2, \pm 1$.

Intercepts:

$x = 0$: $y = \dfrac{0}{-2} = 0$; $(0,0)$

$y = 0$: $0 = \dfrac{2x^4 - x^2}{(x + 2)(x^2 - 1)}$

$0 = 2x^4 - x^2$

$0 = x^2(2x^2 - 1)$

$x^2 = 0$ or $2x^2 - 1 = 0$

$x = 0$ or $x^2 = \frac{1}{2}$

$x = 0$ or $x = \pm\dfrac{\sqrt{2}}{2}$; $(0,0)$, $(\pm 0.7, 0)$

Additional points: $(-5, -17.0)$, $(-4, -16.5)$, $(-3, -19.1)$, $(-1.75, 30.4)$, $(-1.5, 12.6)$, $(-1.25, 7.9)$, $(-0.5, 0.1)$, $(0.5, 0.07)$, $(1.5, 1.8)$, $(2, 2.3)$, $(3, 3.8)$

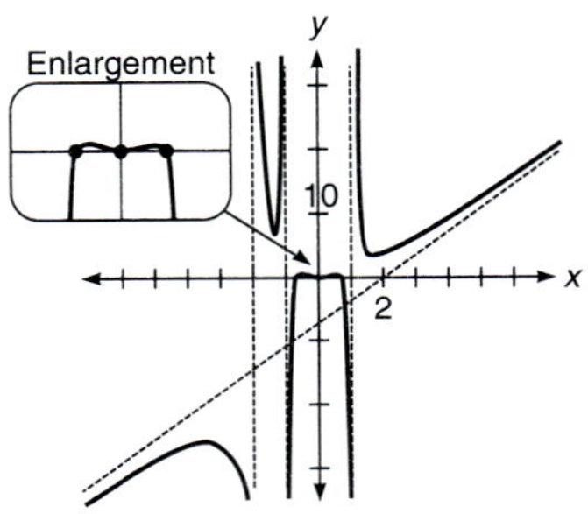

57. $y = \dfrac{x^3 + 4x^2 + 3x + 12}{x^2 + 3}$

$= \dfrac{x^2(x + 4) + 3(x + 4)}{x^2 + 3}$

$= \dfrac{(x + 4)(x^2 + 3)}{x^2 + 3}$

$= x + 4$

This is a straight line with intercepts at $(0,4)$ and $(-4,0)$. Since $x^2 + 3 \neq 0$ for all real values of x there are no restrictions on the domain.

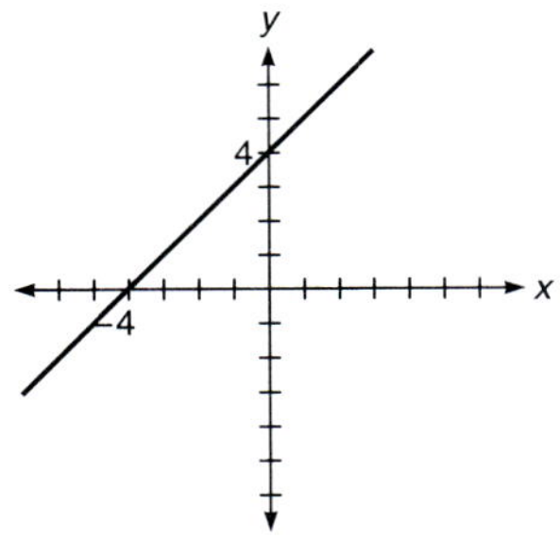

Exercise 4–5

Answers to odd-numbered problems

1. $x + 3$; $5x - 13$; $-6x^2 + 34x - 40$; $\dfrac{3x - 5}{-2x + 8}$; $-6x + 19$; $-6x + 18$

3. $x + 4 + \sqrt{x - 4}$; $x + 4 - \sqrt{x - 4}$; $(x + 4)\sqrt{x - 4}$; $\dfrac{(x + 4)\sqrt{x - 4}}{x - 4}$; $\sqrt{x - 4} + 4$; $\sqrt{x}$

5. $\dfrac{3x^2 - 4x + 3}{2x^2 - 2x}$; $\dfrac{-x^2 - 4x + 3}{2x^2 - 2x}$; $\dfrac{x - 3}{2x - 2}$; $\dfrac{x^2 - 4x + 3}{2x^2}$; $\dfrac{-2x + 3}{2x}$; $-\dfrac{x - 3}{x + 3}$

7. $x^4 - x^2 + 3 + \sqrt{\dfrac{x}{x + 1}}$; $x^4 - x^2 + 3 - \sqrt{\dfrac{x}{x + 1}}$; $(x^4 - x^2 + 3)\left(\sqrt{\dfrac{x}{x + 1}}\right)$; $\dfrac{(x + 1)(x^4 - x^2 + 3)\sqrt{\dfrac{x}{x + 1}}}{x}$; $\dfrac{3x^2 + 5x + 3}{x^2 + 2x + 1}$; $\sqrt{\dfrac{x^4 - x^2 + 3}{x^4 - x^2 + 4}}$

9. $x + 3$; $x - 3$; $3x$; $\dfrac{x}{3}$; 3; 3

11. $\sqrt[3]{x-5}+(x^3+5)$; $\sqrt[3]{x-5}-x^3-5$; $(x^3+5)\sqrt[3]{x-5}$; $\dfrac{\sqrt[3]{x-5}}{x^3+5}$; x; x

13. a. $f(x) = 2x - 7$; $g(x) = \frac{1}{2}x + 3\frac{1}{2}$

$$\begin{aligned} f(g(x)) &= 2(g(x)) - 7 \\ &= 2(\tfrac{1}{2}x + 3\tfrac{1}{2}) - 7 \\ &= x + 7 - 7 = x \end{aligned}$$

$$g(f(x)) = \tfrac{1}{2}(2x - 7) + 3\tfrac{1}{2} = x$$

b.

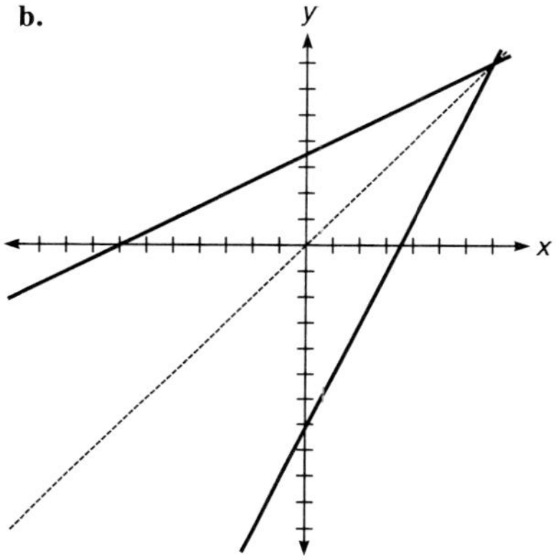

15. a. $f(x) = \frac{1}{3}x + \frac{8}{3}$; $g(x) = 3x - 8$

$$\begin{aligned} f(g(x)) &= \tfrac{1}{3}(3x - 8) + \tfrac{8}{3} \\ &= x - \tfrac{8}{3} + \tfrac{8}{3} = x \\ g(f(x)) &= 3(\tfrac{1}{3}x + \tfrac{8}{3}) - 8 \\ &= x + 8 - 8 = x \end{aligned}$$

b.

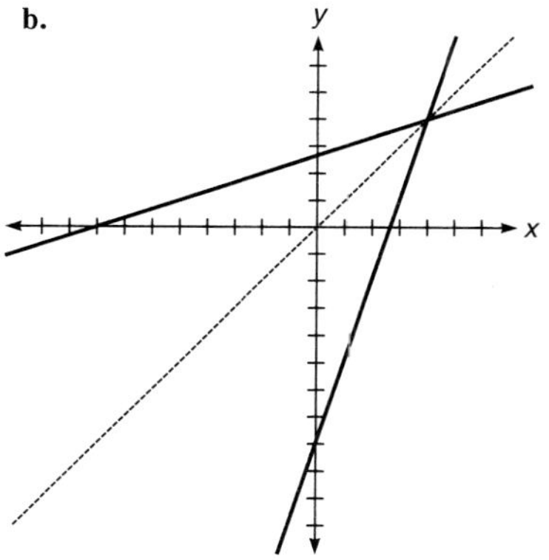

17. a. $f(x) = x^2 - 9, x \ge 0$;

$$\begin{aligned} g(x) &= \sqrt{x+9} \\ f(g(x)) &= (\sqrt{x+9})^2 - 9 \\ &= x + 9 - 9 = x \\ g(f(x)) &= \sqrt{(x^2-9)+9} = \sqrt{x^2} = x \end{aligned}$$

b.

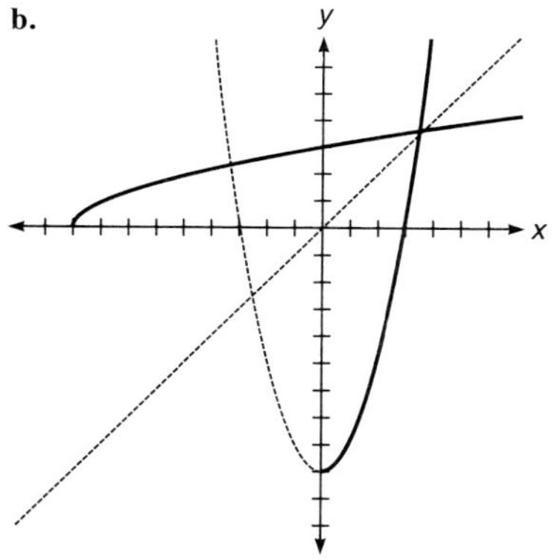

19. a. $f(x) = x^3$; $g(x) = \sqrt[3]{x}$

$$\begin{aligned} f(g(x)) &= (\sqrt[3]{x})^3 = x \\ g(f(x)) &= \sqrt[3]{x^3} = x \end{aligned}$$

b.

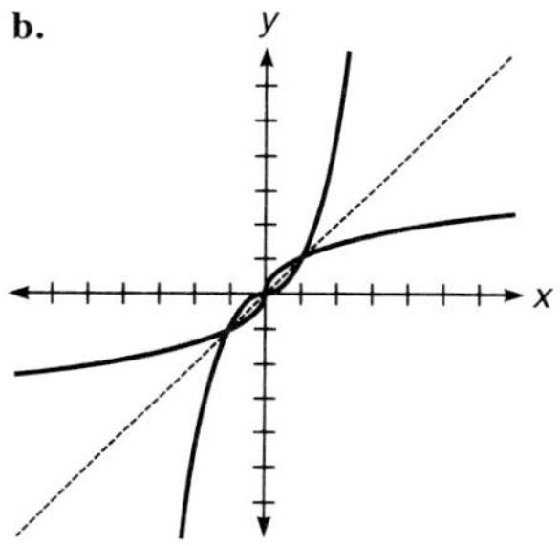

21. a. $f(x) = x^2 - 2x + 3, x \ge 1$

$$\begin{aligned} g(x) &= \sqrt{x-2} + 1 \\ f(g(x)) &= (\sqrt{x-2} + 1)^2 - 2(\sqrt{x-2} + 1) + 3 \\ &= ((x-2) + 2\sqrt{x-2} + 1) - 2\sqrt{x-2} - 2 + 3 \\ &= x \\ g(f(x)) &= \sqrt{(x^2 - 2x + 3) - 2} + 1 \\ &= \sqrt{x^2 - 2x + 1} + 1 \\ &= \sqrt{(x-1)^2} + 1 \\ &= (x - 1) + 1 = x \end{aligned}$$

b.

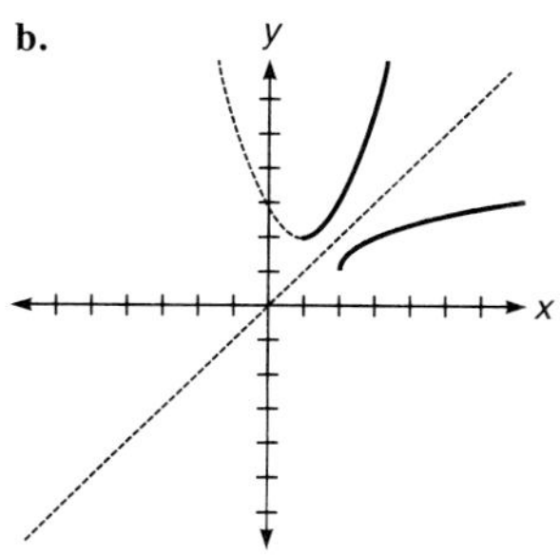

23. a. $f(x) = \dfrac{2x}{x-3}$; $g(x) = \dfrac{3x}{x-2}$

$$\begin{aligned} f(g(x)) &= \frac{2\left(\dfrac{3x}{x-2}\right)}{\dfrac{3x}{x-2} - 3} \\ &= \frac{\dfrac{6x}{x-2}}{\dfrac{3x - 3(x-2)}{x-2}} \cdot \frac{x-2}{x-2} \\ &= \frac{6x}{6} = x \\ g(f(x)) &= \frac{3\left(\dfrac{2x}{x-3}\right)}{\dfrac{2x}{x-3} - 2} \\ &= \frac{\dfrac{6x}{x-3}}{\dfrac{2x - 2(x-3)}{x-3}} \cdot \frac{x-3}{x-3} \\ &= \frac{6x}{6} = x \end{aligned}$$

b.

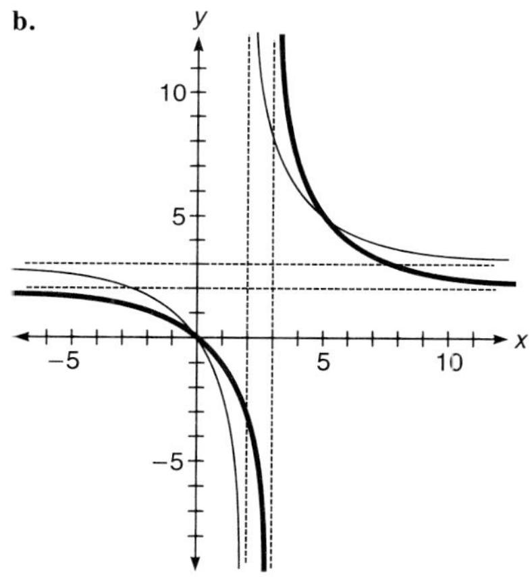

25. $f^{-1}(x) = \frac{1}{4}x + \frac{5}{4}$

27. $h^{-1}(x) = -\frac{2}{5}x + \frac{24}{5}$

29. $g^{-1}(x) = \sqrt{x+9}$

31. $f^{-1}(x) = \sqrt{9 - x^2}$

33. $h^{-1}(x) = x^2 + 4, x \ge 0$

35. $g^{-1}(x) = \dfrac{\sqrt[3]{4(x+9)}}{2}$

37. $f^{-1}(x) = \dfrac{x^3+5}{4}$ **39.** $f^{-1}(x) = \dfrac{3}{4x+5}$

41. $g^{-1}(x) = \dfrac{-x}{x-1}$ **43.** $h^{-1}(x) = \dfrac{-x-1}{x-1}$

45. $h^{-1}(x) = 1 + \sqrt{x+10}$

47. $g^{-1}(x) = \dfrac{-3 + \sqrt{8x+25}}{4}$

49. $C(x) = \frac{x^3}{2}$ **51.** $V_e(t) = \frac{1}{4}t - 2$

53. $A(t) = 80t^3$ **55.** $A^{-1}(x) = \frac{1}{4}x - 4$

57. $R^{-1}(x) = \frac{20x}{20 - x}$

59. $f(g(x)) = (-\sqrt{x + 9})^2 - 9$
$= (x + 9) - 9 = x$
$g(f(x)) = -\sqrt{(x^2 - 9) + 9} = -\sqrt{x^2}$
$= -|x|$; since $x \leq 0$, $|x|$
$= -x$, so $-|x| = -(-x) = x$

61. $f^{-1}(x) = \frac{1}{a}x - \frac{b}{a}$; the inverse does not exist if $a = 0$.

Solutions to skill and review problems

1. Combine $\frac{2}{x + 3} - \frac{3}{x - 2}$.

$\frac{2}{x + 3} - \frac{3}{x - 2}$

$\frac{2(x - 2) - 3(x + 3)}{(x + 3)(x - 2)}$

$\frac{-x - 13}{x^2 + x - 6}$

2. Graph $f(x) = \frac{2}{x + 3}$.

Vertical asymptote: $x = -3$; horizontal asymptote: $y = 0$ (x-axis); intercepts:
$x = 0$: $y = \frac{2}{3}$; $(0,\frac{2}{3})$
$y = 0$: $0 = \frac{2}{x + 3}$; no solution
Additional points: $(-6,-0.67)$, $(-4,-2)$, $(-2,2)$, $(1,0.5)$

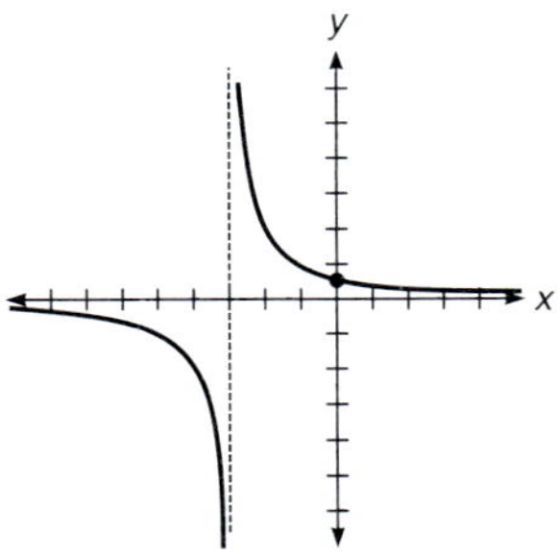

3. Graph $f(x) = \frac{2}{(x + 3)(x - 1)}$.

Vertical asymptotes: $x = -3, 1$; horizontal asymptote: $y = 0$ (x-axis); intercepts:

$x = 0$: $y = \frac{2}{3(-1)} = -\frac{2}{3}$; $(0,-\frac{2}{3})$

$y = 0$: $0 = \frac{2}{(x + 3)(x - 1)}$; no solution

Additional points: $(-4,0.4)$, $(-2,-0.67)$, $(-1,-0.5)$, $(2,0.4)$

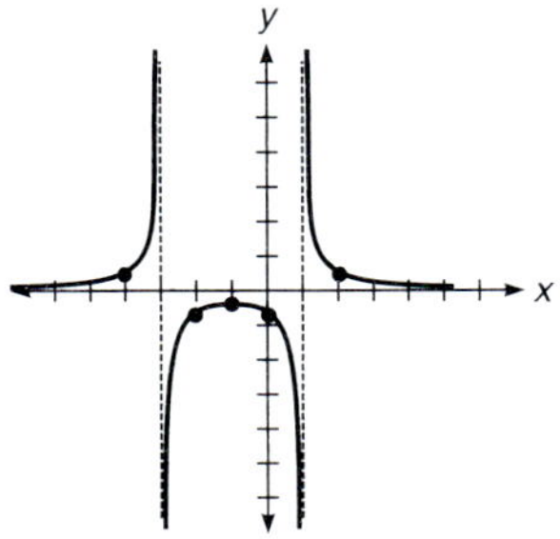

4. Graph $f(x) = \frac{2x^2}{(x + 3)(x - 1)}$.

$= \frac{2x^2}{x^2 + 2x - 3}$

$= 2 + \frac{-4x + 6}{x^2 + 2x - 3}$

$= 2 + \frac{-4x + 6}{(x + 3)(x - 1)}$

Vertical asymptotes: $x = -3, 1$; horizontal asymptote: $y = 2$; intercepts:

$x = 0$: $y = \frac{0}{-3} = 0$; $(0,0)$

$y = 0$: $0 = \frac{2x^2}{(x + 3)(x - 1)}$
$0 = x$; $(0,0)$

Additional points: $(-6,3.4)$, $(-4,6.4)$, $(-2,-2.7)$, $(-1,-0.5)$, $(2,1.6)$, $(4,1.5)$, $(6,1.6)$

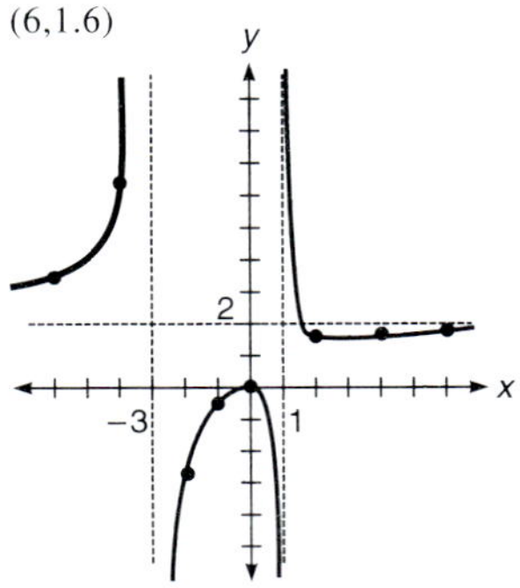

5. Solve $\left|\frac{5x - 2}{x + 1}\right| < 2$.

This inequality is nonlinear so we must use the critical point/test point method. Find critical points:

a. Solve the corresponding equality.

$\left|\frac{5x - 2}{x + 1}\right| = 2$

$\frac{5x - 2}{x + 1} = 2$ or $\frac{5x - 2}{x + 1} = -2$

$5x - 2 = 2x + 2$ or $5x - 2 = -2x - 2$

$3x = 4$ or $7x = 0$

$x = \frac{4}{3} = 1\frac{1}{3}$ or $x = 0$

These are critical points.

b. Find zeros of denominators.

$x + 1 = 0$
$x = -1$

Critical points are $-1, 0, 1\frac{1}{3}$. They form the 4 intervals shown.

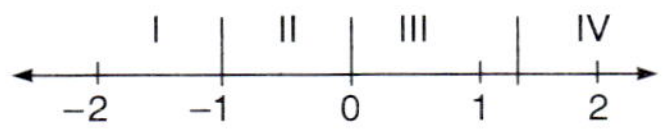

Choose a test point from each interval, such as $-2, -\frac{1}{2}, 1, 2$. Try these in the original inequality.

$x = -2$: $\left|\frac{5(-2) - 2}{-2 + 1}\right| < 2$; $12 < 2$; false

$x = -\frac{1}{2}$: $\left|\frac{5(-\frac{1}{2}) - 2}{-\frac{1}{2} + 1}\right| < 2$; $9 < 2$; false

$x = 1$: $\left|\frac{5(1) - 2}{1 + 1}\right| < 2$; $1\frac{1}{2} < 2$; true

$x = 2$: $\left|\frac{5(2) - 2}{2 + 1}\right| < 2$; $2\frac{2}{3} < 2$; false

Only interval III forms the solution: $\{x \mid 0 < x < 1\frac{1}{3}\}$.

6. Graph $f(x) = x^3 - x^2 - x + 1$.

$$\begin{aligned} y &= x^3 - x^2 - x + 1 \\ &= x^2(x - 1) - 1(x - 1) \\ &= (x - 1)(x^2 - 1) \\ &= (x - 1)(x - 1)(x + 1) \\ y &= (x - 1)^2(x + 1) \end{aligned}$$

Intercepts:

$x = 0$: $y = 1$; $(0,1)$

$y = 0$: $0 = (x - 1)^2(x + 1)$

$x = -1$ or 1; $(-1,0)(1,0)$

The zero 1 has multiplicity 2 (even multiplicity) so the graph does not cross the x-axis at 1. Additional points: $(-1.5,-3.1)$, $(-0.5,1.1)$, $(2,3)$

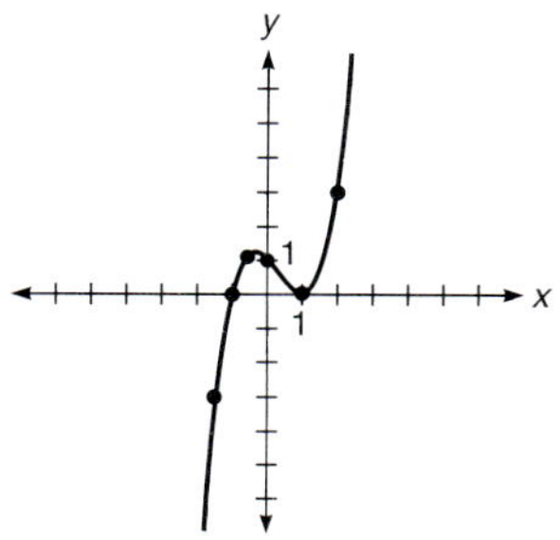

22. a. $f(x) = \sqrt{x + 9} - 2$; $g(x) = x^2 + 4x - 5$, $x \geq -2$

$$\begin{aligned} f(g(x)) &= \sqrt{(x^2 + 4x - 5) + 9} - 2 \\ &= \sqrt{x^2 + 4x + 4} - 2 \\ &= \sqrt{(x + 2)^2} - 2 = (x + 2) - 2 = x \\ g(f(x)) &= (\sqrt{x + 9} - 2)^2 + 4(\sqrt{x + 9} - 2) - 5 \\ &= ((x + 9) - 4\sqrt{x + 9} + 4) + 4\sqrt{x + 9} - 8 - 5 = x \end{aligned}$$

b.

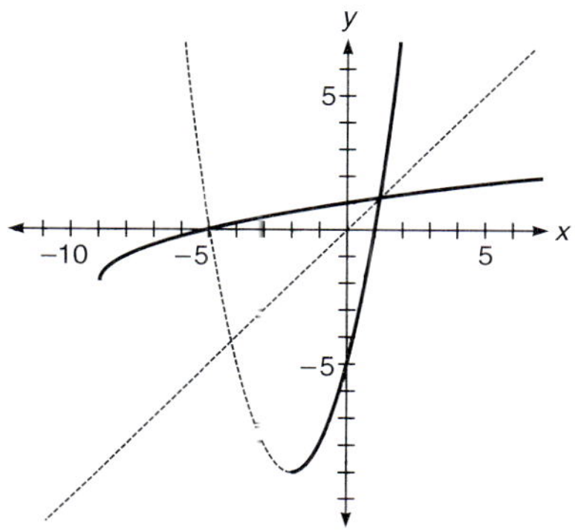

24. a. $f(x) = \dfrac{x - 3}{x - 2}$; $g(x) = 2 - \dfrac{1}{x - 1}$

$$\begin{aligned} f(g(x)) &= \frac{\left(2 - \dfrac{1}{x - 1}\right) - 3}{\left(2 - \dfrac{1}{x - 1}\right) - 2} \\ &= \frac{-1 - \dfrac{1}{x - 1}}{-\dfrac{1}{x - 1}} \cdot \frac{x - 1}{x - 1} \\ &= \frac{(-x + 1) - 1}{-1} = x \\ g(f(x)) &= 2 - \frac{1}{\dfrac{x - 3}{x - 2} - 1} \\ &= 2 - \frac{1}{\dfrac{x - 3}{x - 2} - 1} \cdot \frac{x - 2}{x - 2} \\ &= 2 - \frac{x - 2}{(x - 3) - (x - 2)} = 2 - \frac{x - 2}{-1} \\ &= 2 + (x - 2) = x \end{aligned}$$

b.

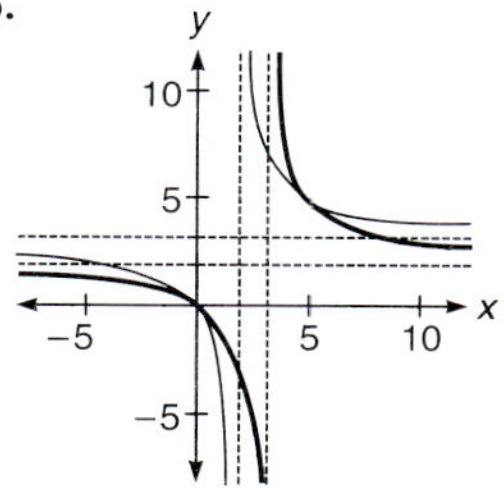

Solutions to trial exercise problems

5. $f(x) = \dfrac{x - 3}{2x}$; $g(x) = \dfrac{x}{x - 1}$

$$\frac{x - 3}{2x} + \frac{x}{x - 1} = \frac{(x - 3)(x - 1) + x(2x)}{2x(x - 1)} = \frac{x^2 - 4x + 3 + 2x^2}{2x^2 - 2x} = \frac{3x^2 - 4x + 3}{2x^2 - 2x}$$

$$\frac{x - 3}{2x} - \frac{x}{x - 1} = \frac{(x - 3)(x - 1) - x(2x)}{2x(x - 1)} = \frac{x^2 - 4x + 3 - 2x^2}{2x^2 - 2x} = \frac{-x^2 - 4x + 3}{2x^2 - 2x}$$

$$\frac{x - 3}{2x} \cdot \frac{x}{x - 1} = \frac{x - 3}{2} \cdot \frac{1}{x - 1} = \frac{x - 3}{2x - 2}$$

$$\frac{x - 3}{2x} / \frac{x}{x - 1} = \frac{x - 3}{2x} \cdot \frac{x - 1}{x} = \frac{x^2 - 4x + 3}{2x^2}$$

$$f[g(x)] = f\left[\left(\frac{x}{x - 1}\right)\right] = \frac{\dfrac{x}{x - 1} - 3}{2 \cdot \dfrac{x}{x - 1}} \cdot \frac{x - 1}{x - 1} = \frac{x - 3(x - 1)}{2x} = \frac{-2x + 3}{2x}$$

$$g[f(x)] = g\left[\left(\frac{x - 3}{2x}\right)\right] = \frac{\dfrac{x - 3}{2x}}{\dfrac{x - 3}{2x} - 1} \cdot \frac{2x}{2x} = \frac{x - 3}{(x - 3) - 2x} = \frac{x - 3}{-x - 3} = -\frac{x - 3}{x + 3}$$

39. $f(x) = \dfrac{3 - 5x}{4x}$

$$\begin{aligned} y &= \frac{3 - 5x}{4x} \\ x &= \frac{3 - 5y}{4y} \\ 4xy &= 3 - 5y \\ 4xy + 5y &= 3 \\ y(4x + 5) &= 3 \\ y &= \frac{3}{4x + 5} \\ f^{-1}(x) &= \frac{3}{4x + 5} \end{aligned}$$

47. $g(x) = 2x^2 + 3x - 2$, $x \geq -\frac{3}{4}$

$$\begin{aligned} y &= 2x^2 + 3x - 2,\ x \geq -\tfrac{3}{4} \\ x &= 2y^2 + 3y - 2,\ y \geq -\tfrac{3}{4} \\ 0 &= 2y^2 + 3y + (-x - 2) \\ y &= \frac{-3 \pm \sqrt{8x + 25}}{4},\ y \geq -\tfrac{3}{4} \\ y &= -\tfrac{3}{4} \pm \frac{\sqrt{8x + 25}}{4} \end{aligned}$$

Since $y \geq -\frac{3}{4}$,

$$g^{-1}(x) = \frac{-3 + \sqrt{8x + 25}}{4}$$

62. $f(x) = ax^2 + bx + c$
$y = ax^2 + bx + c$
$x = ay^2 + by + c$
$0 = ay^2 + by + c - x$
$$y = \frac{-b \pm \sqrt{b^2 - 4a(c - x)}}{2a}$$
$$y = \frac{-b \pm \sqrt{b^2 - 4ac + 4acx}}{2a}$$
If $x \geq \frac{-b}{2a}$ in the domain we want
$y \geq \frac{-b}{2a}$ so we choose
$$f^{-1}(x) = \frac{-b + \sqrt{b^2 - 4ac + 4acx}}{2a}$$

Exercise 4–6

Answers to odd-numbered problems

1. $\frac{-2}{x-4} + \frac{3}{x-1}$

3. $\frac{-4}{x-1} + \frac{-2}{x+1}$

5. $2x + \frac{4}{2x-1} + \frac{-3}{x-1}$

7. $3x - 2 + \frac{-3}{3x+1} + \frac{3}{2x-5}$

9. $\frac{2}{(x-1)^2} + \frac{3}{x-2}$

11. $\frac{3}{2x} + \frac{-3}{x^2} + \frac{5}{x-2}$

13. $\frac{2}{x-3} + \frac{-2}{(x-3)^2} + \frac{1}{x+1} + \frac{-2}{(x+1)^2}$

15. $\frac{5}{(x-3)^2} + \frac{-5}{x+1} + \frac{2}{(x+1)^2}$

17. $\frac{3}{x-1} + \frac{-2x-1}{x^2+x+1}$

19. $\frac{-1}{x} + \frac{5}{x^2+2x+4}$

21. $\frac{3x}{x^2+x+1} - \frac{2}{x-3}$

23. $\frac{3x+1}{x^2+2x+4} - \frac{2}{x+3}$

25. $\frac{1}{x+5} + \frac{1}{x+10}$ **27.** $\frac{99}{100}$

29. $\frac{14{,}949}{10{,}100} \approx 1.4801$

Solutions to skill and review problems

1. Compute **a.** 8^3 **b.** $8^{1/3}$ **c.** 8^{-3} **d.** $8^{-1/3}$
a. $8^3 = 8 \cdot 8 \cdot 8 = 512$
b. $8^{1/3} = \sqrt[3]{8} = 2$
c. $8^{-3} = \frac{1}{8^3} = \frac{1}{512}$
d. $8^{-1/3} = \frac{1}{8^{1/3}} = \frac{1}{2}$

2. If $2^5 = a^5$, what is a?
$a = 2$

3. If $2^a = 2^5$, what is a?
$a = 5$

4. Graph $f(x) = 2x^2 - x - 6$.
This is a parabola; we complete the square.
$y = 2(x^2 - \frac{1}{2}x) - 6$
$y = 2(x^2 - \frac{1}{2}x + \frac{1}{16}) - 6 - 2(\frac{1}{16})$
$\frac{1}{2}(-\frac{1}{2}) = -\frac{1}{4}$; $(-\frac{1}{4})^2 = \frac{1}{16}$
$y = 2(x - \frac{1}{4})^2 - 6\frac{1}{8}$
Vertex: $(\frac{1}{4}, -6\frac{1}{8})$; intercepts:
$x = 0$: $y = -6$; $(0,-6)$
$y = 0$: $0 = 2x^2 - x - 6$
$0 = (2x + 3)(x - 2)$
$x = -\frac{3}{2}, 2$; $(-1\frac{1}{2},0)$, $(2,0)$

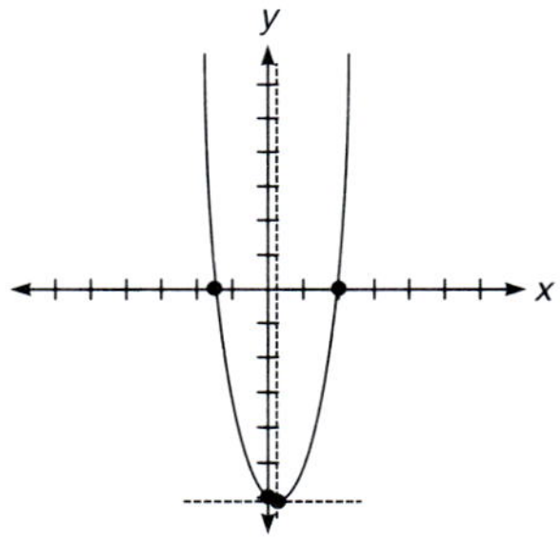

5. Graph $f(x) = (x - 1)(x + 2)(x - 2)$.
Intercepts:
$x = 0$: $y = (-1)(2)(-2) = 4$; $(0,4)$
$y = 0$: $0 = (x - 1)(x + 2)(x - 2)$
$x = -2, 1, 2$; $(-2,0)$, $(1,0)$, $(2,0)$
Additional points: $(-2.25,-3.5)$, $(-1,6)$, $(1.5,-0.9)$, $(2.5,3.4)$

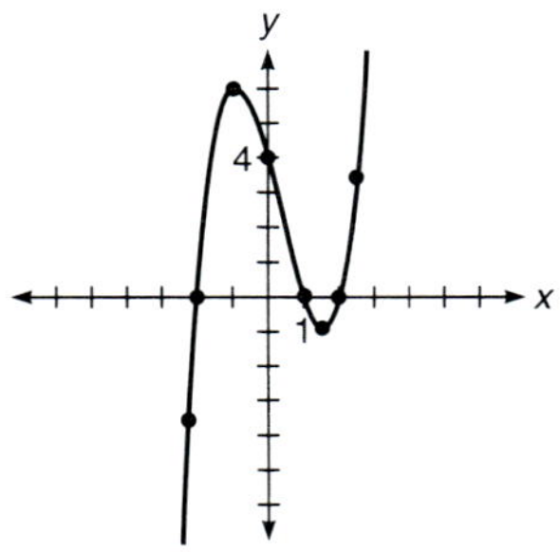

6. Solve $x^3 - x^2 + 1 > x$.
This is a nonlinear inequality; it must be solved using the critical point, test point method. Find critical points from (a) the corresponding equality and (b) zeros of denominators. Solve the corresponding equality:
$x^3 - x^2 + 1 = x$
$x^3 - x^2 - x + 1 = 0$
$x^2(x - 1) - 1(x - 1) = 0$
$(x - 1)(x^2 - 1) = 0$
$(x - 1)(x + 1)(x - 1) = 0$
$x = \pm 1$
Critical points: Find test points in each interval and test in the original inequality. We will use ± 2, 0.
$x^3 - x^2 + 1 > x$
$x = -2$: $-11 > -2$; false
$x = 0$: $1 > 0$; true
$x = 2$: $5 > 2$; true
Thus the solution set is intervals II and III.
$\{x \mid -1 < x < 1 \text{ or } x > 1\}$

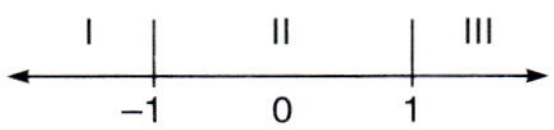

7. Graph $f(x) = \frac{x^2 + 1}{x^2 - 1}$.
$$y = \frac{x^2 + 1}{x^2 - 1} = 1 + \frac{2}{(x - 1)(x + 1)}$$
Horizontal asymptote: $y = 2$; vertical asymptotes: $x = \pm 1$; intercepts:
$x = 0$: $y = \frac{1}{-1} = -1$; $(0,-1)$
$y = 0$: $0 = \frac{x^2 + 1}{x^2 - 1}$
$0 = x^2 + 1$; no real solutions so no x-intercepts
Additional points: $(\pm 3,1.25)$, $(\pm 2,1.7)$, $(\pm 0.5,-1.7)$

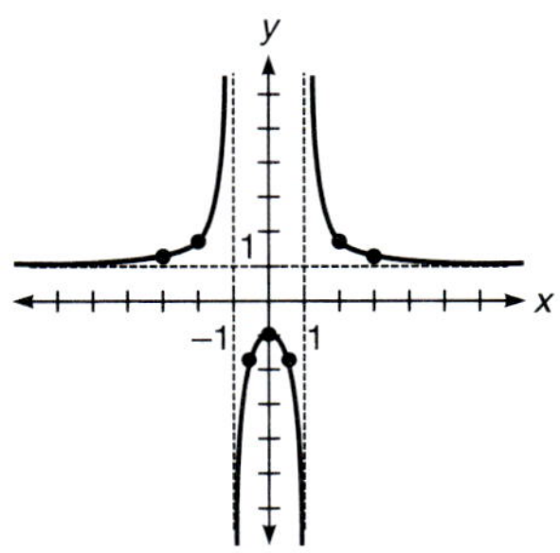

Solutions to trial exercise problems

13. $\dfrac{3x^3 - 11x^2 + x - 17}{(x-3)^2(x+1)^2} = \dfrac{A}{x-3} + \dfrac{B}{(x-3)^2} + \dfrac{C}{x+1} + \dfrac{D}{(x+1)^2}$

$\dfrac{3x^3 - 11x^2 + x - 17}{(x-3)^2(x+1)^2} \cdot (x-3)^2(x+1)^2$

$= \dfrac{A}{x-3} \cdot (x-3)^2(x+1)^2 + \dfrac{B}{(x-3)^2} \cdot (x-3)^2(x+1)^2$

$+ \dfrac{C}{x+1} \cdot (x-3)^2(x-1)^2 + \dfrac{D}{(x+1)^2} \cdot (x-3)^2(x+1)^2$

$3x^3 - 11x^2 + x - 17 = A(x-3)(x+1)^2 + B(x+1)^2$

$+ C(x-3)^2(x+1) + D(x-3)^2$

Let $x = 3$: $-32 = A(0) + B(16) + C(0) + D(0)$

$-2 = B$

Let $x = -1$: $-32 = A(0) + B(0) + C(0) + D(16)$

$-2 = D$

We now make any other two choices for x.

Let $x = 0$: $-17 = -3A + (-2) + 9C + 9(-2)$

$B = -2,\ D = -2$

$-17 = -3A - 2 + 9C - 18$

$3 = -3A + 9C$

[1] $1 = -A + 3C$

Let $x = 1$: $-24 = -8A + 4(-2) + 8C + 4(-2)$

$-24 = -8A - 8 + 8C - 8$

$B = -2,\ D = -2$

$-8 = -8A + 8C$

[2] $1 = A - C$

By equation [1], $A = 3C - 1$; plugging this into equation [2] we obtain

$1 = (3C - 1) - C$

$1 = 2C - 1$

$2 = 2C$

$C = 1$

Since $A = 3C - 1$, $A = 3 - 1 = 2$.

Thus, $\dfrac{3x^3 - 11x^2 + x - 17}{(x-3)^2(x+1)^2} =$

$\dfrac{2}{x-3} + \dfrac{-2}{(x-3)^2} + \dfrac{1}{x+1} + \dfrac{-2}{(x+1)^2}$.

21. $\dfrac{x^2 - 11x - 2}{(x-3)(x^2+x+1)} = \dfrac{A}{x-3} + \dfrac{Bx + C}{x^2+x+1}$

$\dfrac{x^2 - 11x - 2}{(x-3)(x^2+x+1)} \cdot (x-3)(x^2+x+1)$

$= \dfrac{A}{x-3} \cdot (x-3)(x^2+x+1)$

$+ \dfrac{Bx+C}{x^2+x+1} \cdot (x-3)(x^2+x+1)$

$x^2 - 11x - 2 = A(x^2+x+1) + (Bx+C)(x-3)$

Let $x = 3$: $-26 = A(13)$

$-2 = A$

Let $x = 0$: $-2 = -2(1) + C(-3)$

$A = -2$

$C = 0$

Let $x = 1$: $-12 = -2(3) + B(-2)$

$A = -2,\ C = 0$

$B = 3$

$\dfrac{x^2 - 11x - 2}{(x-3)(x^2+x+1)} = \dfrac{3x}{x^2+x+1} - \dfrac{2}{x-3}$

27. $\dfrac{1}{n(n+1)} = \dfrac{A}{n} + \dfrac{B}{n-1}$

$\dfrac{1}{n(n+1)} \cdot n(n+1) = \dfrac{A}{n} \cdot n(n+1) + \dfrac{B}{n+1} \cdot n(n+1)$

$1 = A(n+1) + Bn$

Let $n = 0$: $1 = A$

Let $n = -1$: $1 = -B$; $B = -1$

$\dfrac{1}{n(n+1)} = \dfrac{1}{n} + \dfrac{-1}{n+1}$

$\dfrac{1}{n(n+1)} = \dfrac{1}{n} - \dfrac{1}{n+1}$

Thus, for example, $\dfrac{1}{1 \cdot 2} = \dfrac{1}{1} - \dfrac{1}{2}$ and $\dfrac{1}{99 \cdot 100} = \dfrac{1}{99} - \dfrac{1}{100}$.

Thus, $\dfrac{1}{1 \cdot 2} + \dfrac{1}{2 \cdot 3} + \dfrac{1}{3 \cdot 4} + \cdots + \dfrac{1}{98 \cdot 99} + \dfrac{1}{99 \cdot 100} =$

$\left(\dfrac{1}{1} - \dfrac{1}{2}\right) + \left(\dfrac{1}{2} - \dfrac{1}{3}\right) + \left(\dfrac{1}{3} - \dfrac{1}{4}\right) + \cdots + \left(\dfrac{1}{98} - \dfrac{1}{99}\right)$

$+ \left(\dfrac{1}{99} - \dfrac{1}{100}\right) =$

$\dfrac{1}{1} - \dfrac{1}{2} + \dfrac{1}{2} - \dfrac{1}{3} + \dfrac{1}{3} - \dfrac{1}{4} + \cdots + \dfrac{1}{98} - \dfrac{1}{99} + \dfrac{1}{99}$

$- \dfrac{1}{100} = \dfrac{1}{1} - \dfrac{1}{100} = \dfrac{99}{100}$

Chapter 4 review

1. vertex: $(1\frac{1}{2}, -20\frac{1}{4})$; x-intercept: $(-3,0)$, $(6,0)$; y-intercept: $(0,-18)$

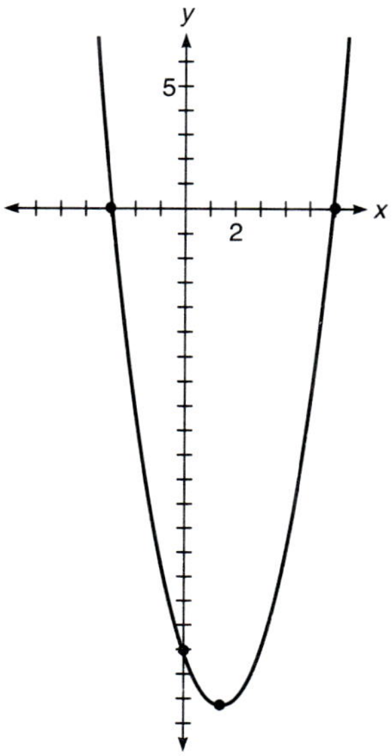

2. vertex and all intercepts at $(0,0)$

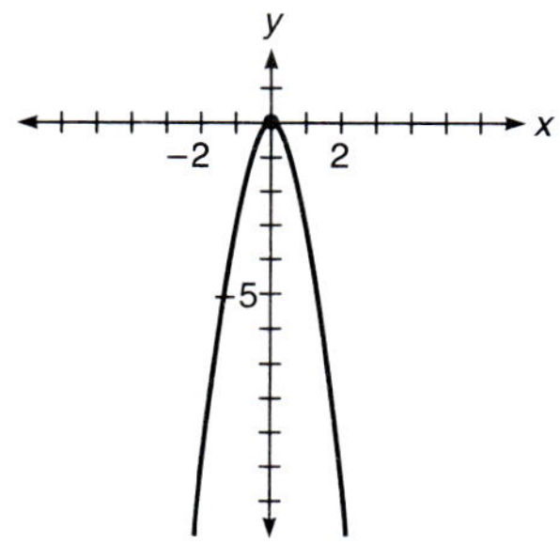

3. vertex: $(2,-4)$; x-intercept: $(0,0)$, $(4,0)$; $(0,0)$ is also the y-intercept

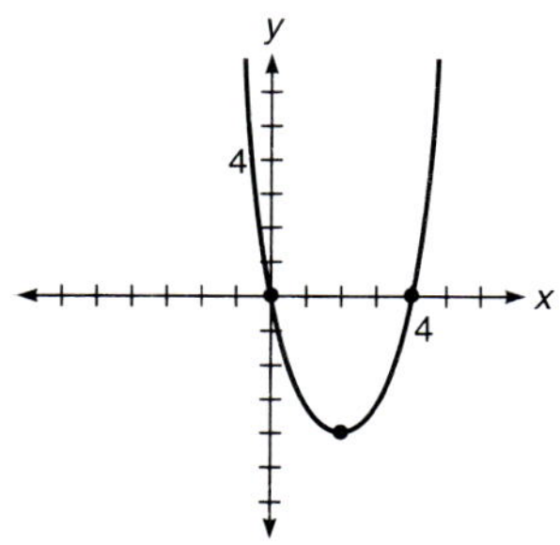

4. vertex: $(2\frac{1}{2}, 12\frac{1}{4})$; x-intercept: $(-1,0)$, $(6,0)$; y-intercept: $(0,6)$

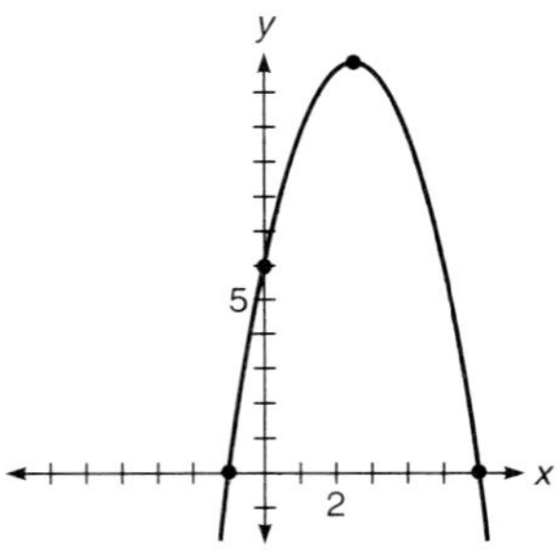

5. vertex: $(0,9)$; x-intercept: $(-3,0)$, $(3,0)$; y-intercept: $(0,9)$

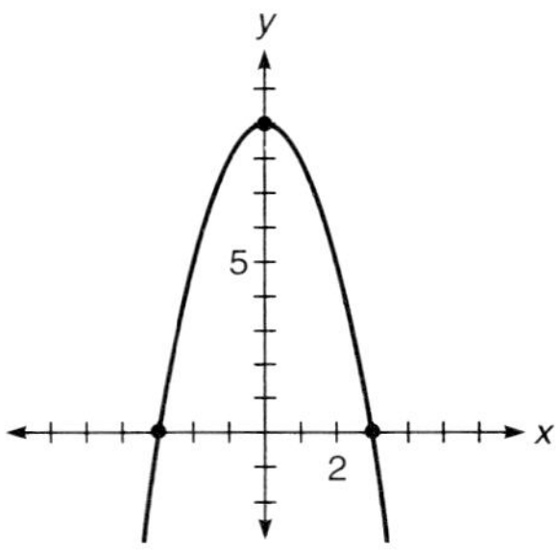

6. vertex: $(-\frac{2}{3}, -5\frac{1}{3})$; x-intercept: $(-2,0)$, $(\frac{2}{3},0)$; y-intercept: $(0,-4)$

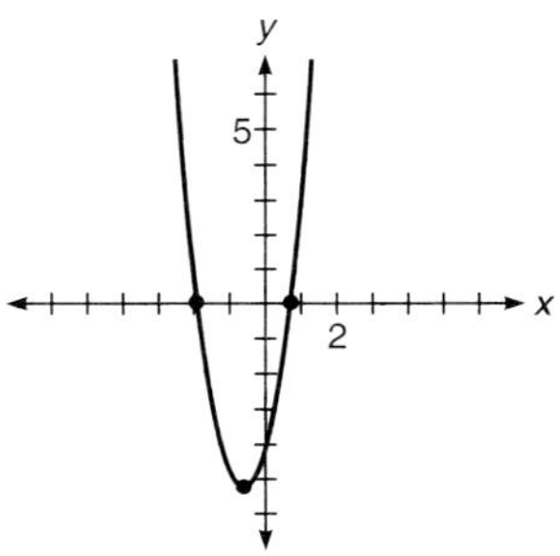

7. vertex: $(2\frac{1}{2}, -7\frac{1}{4})$; x-intercept: $(-0.2,0)$, $(5.2,0)$; y-intercept $(0,-1)$

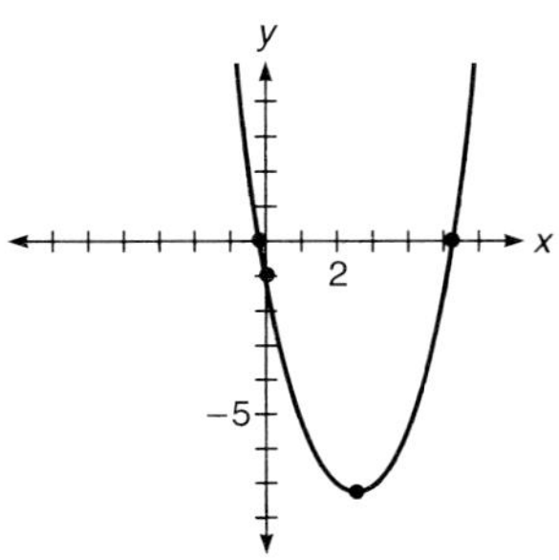

8. vertex: $(-1,1)$; y-intercept: $(0,2)$; additional points: $(-3,5)$, $(-2,2)$, $(1,5)$

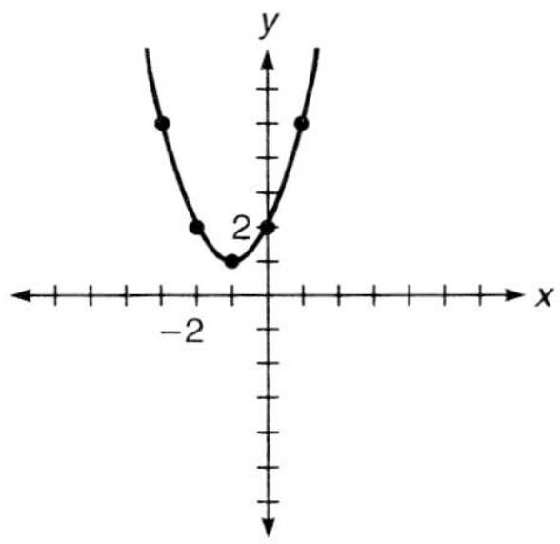

9. vertex: $(\frac{1}{2}, 4\frac{3}{4})$; y-intercept: $(0,5)$

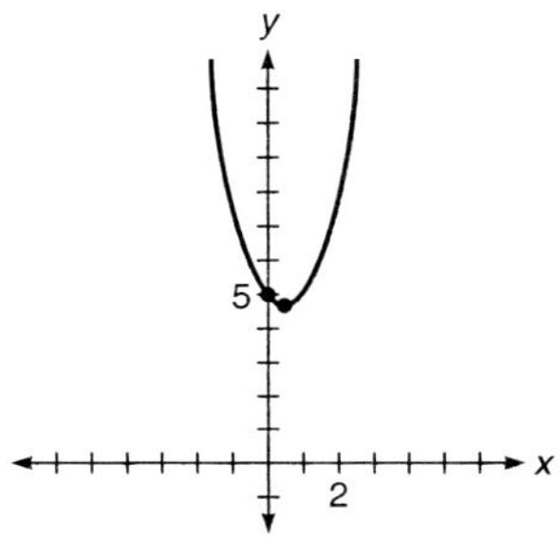

10. vertex: $(-2,4)$; x-intercept: $(0,0)$, $(-4,0)$; y-intercept: $(0,0)$

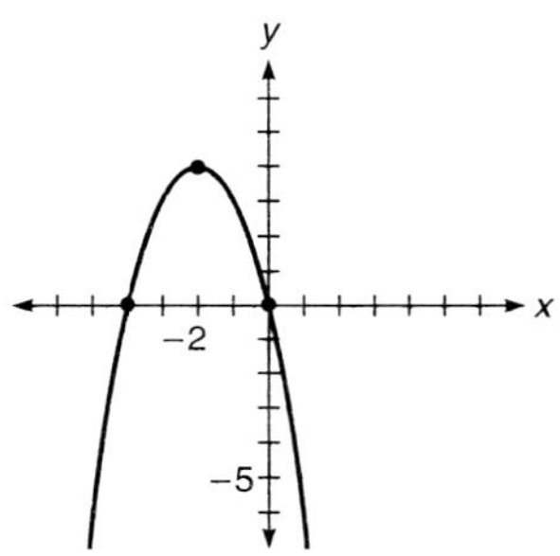

11. The dimensions are 50 and 100; in this case the area is 5,000 sq. ft.

12. The dimensions are 100 feet on a side, and the area is 10,000 sq. ft.

13. The rectangle (a square) will give a larger area for a given perimeter.

14. The object will rise to a maximum height of 4,096 ft after 16 seconds; the object returns to the ground after 32 seconds.

15.

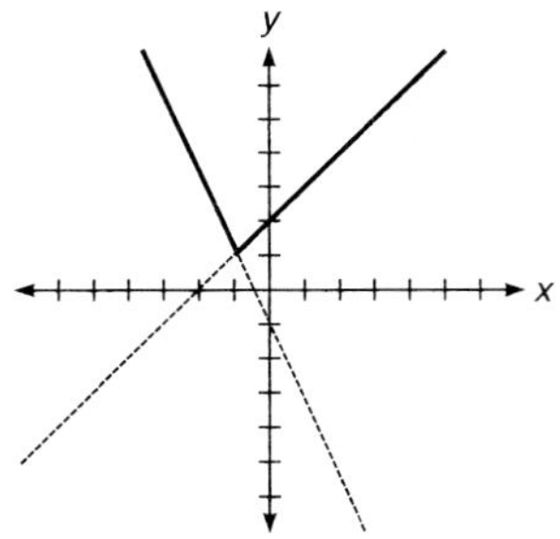

16.

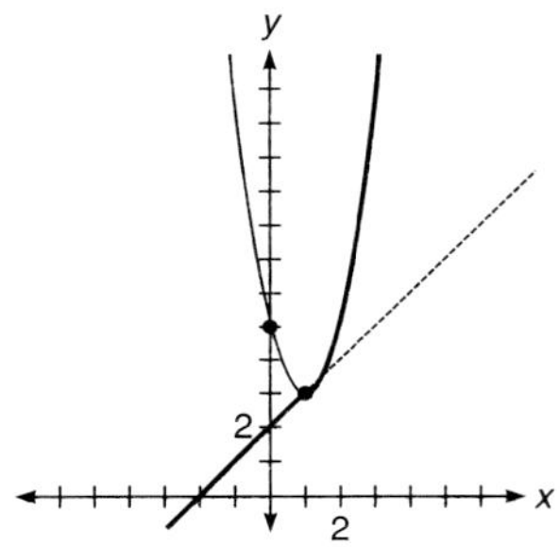

17. $\pm1, \pm2, \pm3, \pm6, \pm\frac{1}{2}, \pm\frac{3}{2}$

18. $\pm1, \pm2, \pm5, \pm10, \pm\frac{1}{2}, \pm\frac{5}{2}$

19. $\pm1, \pm2, \pm4$

20. $\pm1, \pm2, \pm4, \pm8, \pm\frac{1}{3}, \pm\frac{2}{3}, \pm\frac{4}{3}, \pm\frac{8}{3}$

21. $f(x) \div (x-3) = 2x^3 + x^2 + 5x + 15 + \dfrac{44}{x-3}$; $f(3) = 44$

22. $g(x) \div (x+4) = -2x^2 + 5x - 23 + \dfrac{94}{x+4}$; $g(-4) = 94$

23. $f(x) \div (x-4) = \frac{1}{2}x^2 - x - \frac{13}{4} - \dfrac{16}{x-4}$; $f(4) = -16$

---

24. **a.** 0 or 2 positive real zeros; 0 or 2 negative real zeros

b. $\pm(1, 2, 3, 6, 9, 18, 27, 54, \frac{1}{2}, \frac{3}{2}, \frac{9}{2}, \frac{27}{2})$

c,d. $f(x) = (x+2)(x-3)(2x^2+3x-9)$

$= (x+2)(x-3)(x+3)(2x-3)$

All zeros are $-3, -2, \frac{3}{2}, 3$.

25. **a.** 0 or 2 positive real zeros; 0 or 2 negative real zeros

b. $\pm(1, 2, 4, \frac{1}{2})$

c,d. $f(x) = (x-1)(x-2)(2x^2+5x+2)$

$= (x-1)(x-2)(x+2)(2x+1)$,

All zeros are $1, 2, -2, -\frac{1}{2}$

26. **a.** 0 or 2 positive real zeros; 0 or 2 negative real zeros
b. $\pm(1, 2, 4, \frac{1}{2})$
c,d. $h(x) = 2(x + \frac{1}{2})(x^3 - 5x^2 - 4x + 4)$; $-\frac{1}{2}$ is the only rational zero
e. -2 is the greatest negative integer lower bound; 6 is the least positive integer upper bound.

27. **a.** 1 or 3 positive real zeros; 0 or 2 negative real zeros
b. $\pm(1, 3, 9, 27, \frac{1}{2}, \frac{3}{2}, \frac{9}{2}, \frac{27}{2}, \frac{1}{4}, \frac{3}{4}, \frac{9}{4}, \frac{27}{4}, \frac{1}{8}, \frac{3}{8}, \frac{9}{8}, \frac{27}{8}, \frac{1}{16}, \frac{3}{16}, \frac{9}{16}, \frac{27}{16})$
c,d. $f(x) = (x - 3)(2x - 3)(2x + 3)(2x - 1)(2x + 1)$
All the zeros for f are $3, \pm\frac{3}{2}, \pm\frac{1}{2}$.

28. $(g(x) = \frac{1}{4}(x - 4)^2(x + 4)$
y-intercept: 16; x-intercept: ± 4

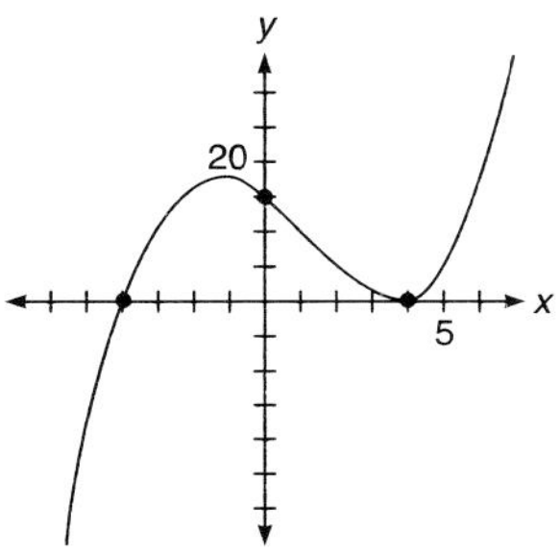

29. $h(x) = (x - 2)(x + 2)(2x - 3)(2x + 3)$
x-intercepts: $\pm 2, \pm 1\frac{1}{2}$; y-intercept: 36

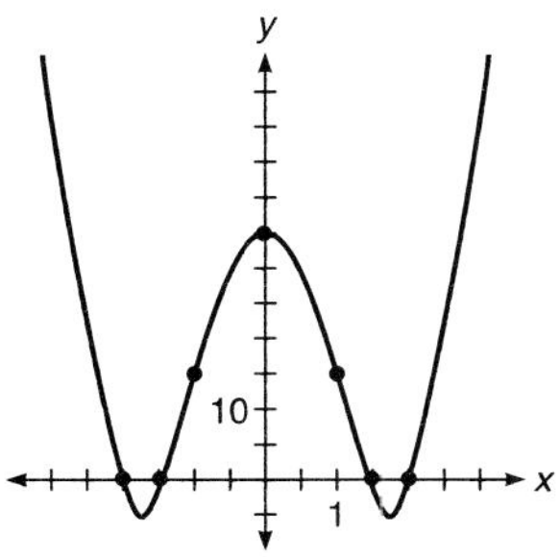

30. $g(x) = (x - 3)(x + 3)(2x - 3)$
$(2x + 3)(x - 3)$
x-intercepts: $\pm 3, \pm 1\frac{1}{2}$; y-intercept: -243

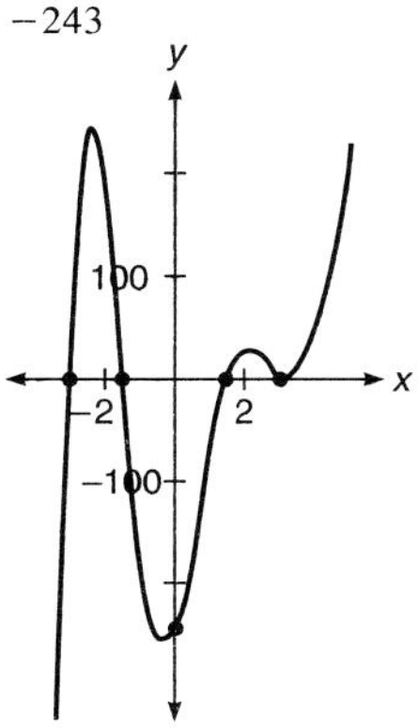

31. $f(x) = (2x - 5)^2(x + 1)^2$
x-intercepts: $2\frac{1}{2}, -1$; y-intercept: 25

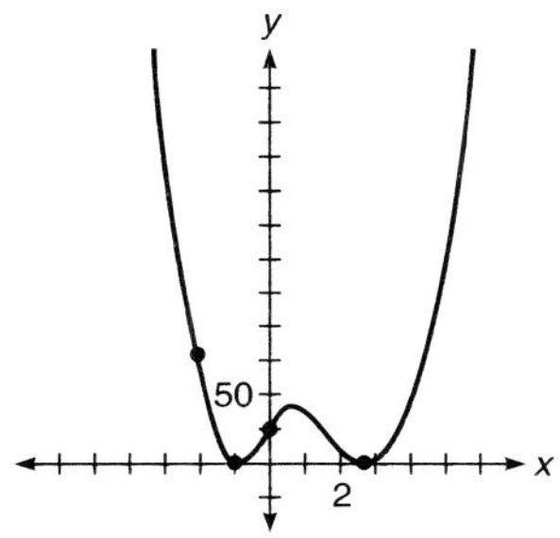

32. $h(x) = (x - 5)(x + 4)(x - 2)$
$(x + 2)(2x + 1)$
x-intercepts: $-4, -2, -\frac{1}{2}, 2, 5$;
y-intercept: 80

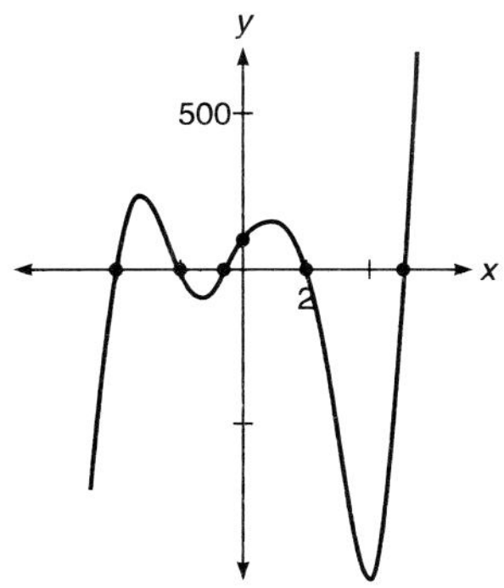

33. x-intercepts: $-1, 3$; y-intercept: 9

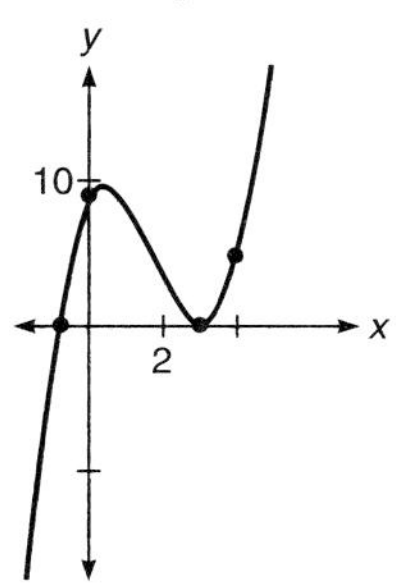

34. x-intercepts: $2, -1\frac{1}{2}$; y-intercept: -72

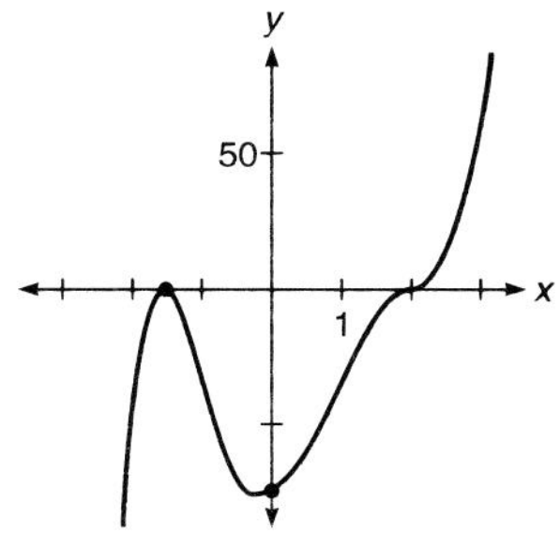

35. asymptotes: $x = 3$, $y = 0$; y-intercept: $-\frac{2}{27}$

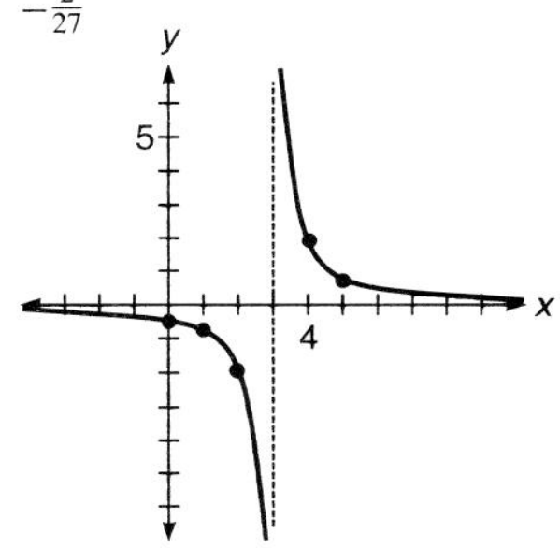

36. asymptotes: $x = -5$, $y = 0$;
y-intercept: $-\frac{3}{25}$

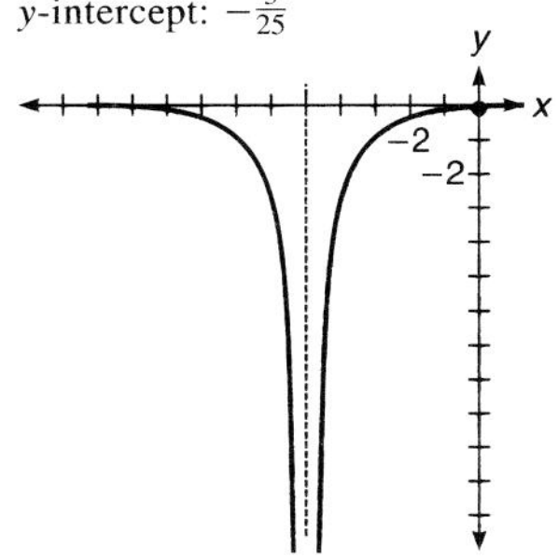

37. asymptotes: $x = -5$, $x = 9$, $y = 0$;
y-intercept: $-\frac{1}{15}$

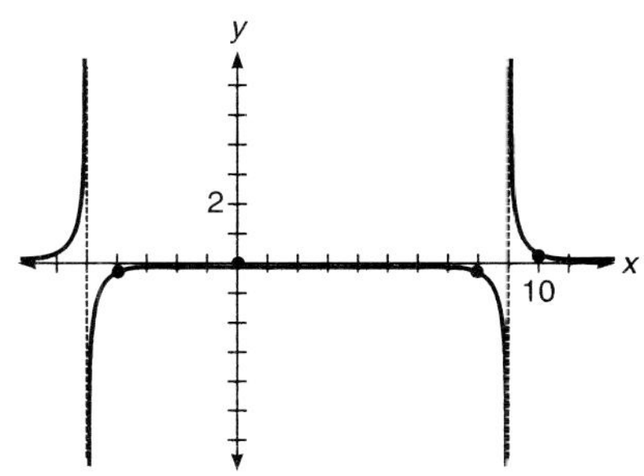

38. asymptotes: $x = 1$, $x = 2\frac{1}{2}$, $y = 0$; intercepts at origin

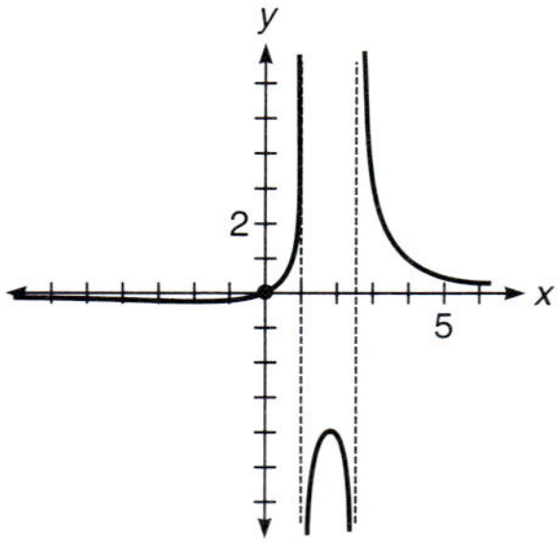

39. asymptotes: $x = -3$, $x = 1$, $x = 3$, $y = 1$; y-intercept: 0, x-intercept: 0

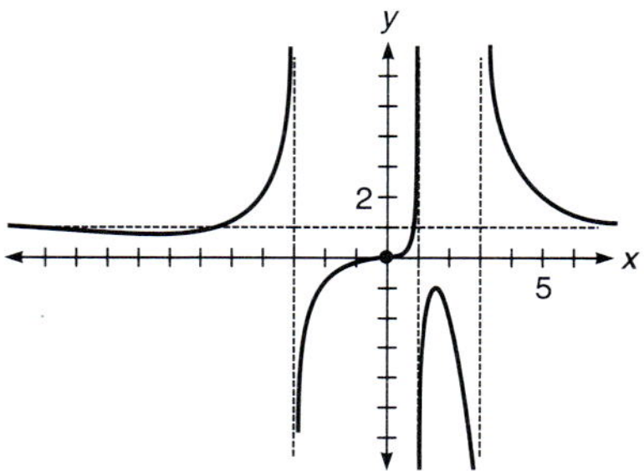

40. asymptote: $y = 1$; x-intercepts: -2, 3, y-intercept: -2

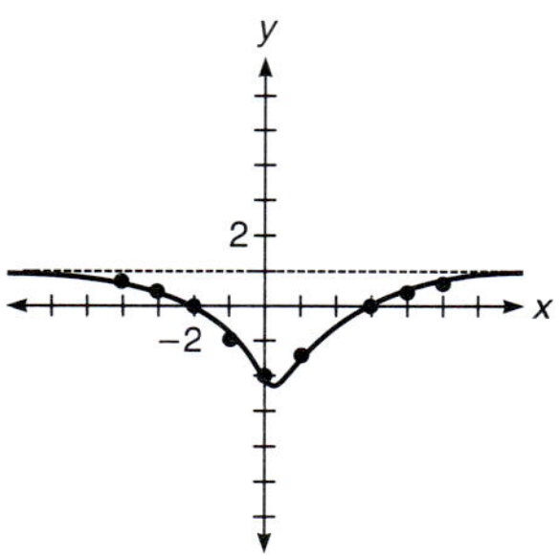

41. $f(x) = \dfrac{x^2 - x - 6}{x - 3} = x + 2$ when $x \neq 3$; x-intercept: -2 y-intercept: $f(0) = 2$

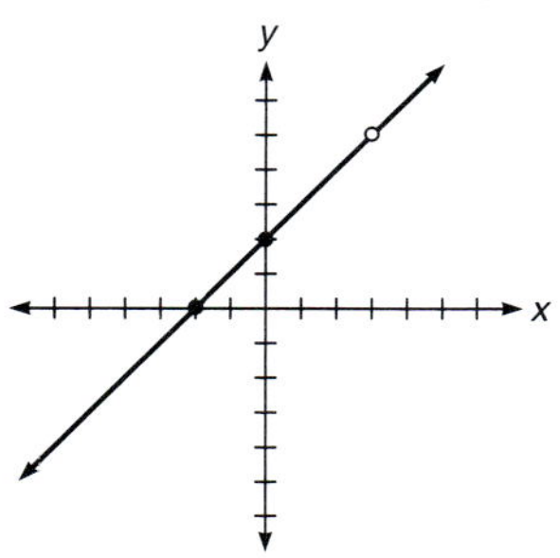

42. 0; $-x + 6$; $-\frac{1}{4}x^2 + 3x - 9$; $\dfrac{-(x - 6)}{x - 6} = -1$ if $x \neq 6$; $-\frac{1}{4}x + 4\frac{1}{2}$; $-\frac{1}{4}x - 1\frac{1}{2}$

43. $x^4 - 1 + \sqrt{8 - x}$; $x^4 - 1 - \sqrt{8 - x}$; $(x^4 - 1)(\sqrt{8 - x})$; $\dfrac{x^4 - 1}{\sqrt{8 - x}}$; $x^2 - 16x + 63$; $\sqrt{9 - x^4}$

44. $\dfrac{4x^2 - 7x + 3}{2x(2x - 1)}$; $\dfrac{-7x + 3}{2x(2x - 1)}$; $\dfrac{x - 3}{2(2x - 1)}$; $\dfrac{2x^2 - 7x + 3}{2x^2}$; $\dfrac{-5x + 3}{2x}$; $\dfrac{-x + 3}{6}$

45. x; $-5x$; $-6x^2$; $-\frac{2}{3}$; $-6x$; $-6x$

46. $x - 3$; $x + 3$; $-3x$; $-\frac{x}{3}$; -3; -3

47. $g^{-1}(x) = \dfrac{4x + 5}{2}$

48. $h^{-1}(x) = -\dfrac{x + 4}{2x - 1}$

49. $g^{-1}(x) = \sqrt{x - 8}$

50. $g^{-1}(x) = \dfrac{\sqrt[3]{x + 27}}{2}$

51. $g^{-1}(x) = -x^3 - 9x^2 - 27x - 26$

52. $f^{-1}(x) = \dfrac{7 + \sqrt{4x + 25}}{2}$

53. $f(x) = \dfrac{160}{7}x + \dfrac{120}{7}$; 154 gallons

54. $\dfrac{5}{x - 3} + \dfrac{-2}{2x + 1}$

55. $\dfrac{4}{x - 3} + \dfrac{-1}{(x - 3)^2} + \dfrac{5}{2x + 1}$

56. $\dfrac{1}{x - 2} + \dfrac{x - 2}{x^2 - x + 4}$

57. $\dfrac{-4}{x + 2} + \dfrac{2}{x - 2} + \dfrac{-2}{(x - 2)^2} + \dfrac{3}{(x - 2)^3}$

Chapter 4 test

1. vertex: $(-2\frac{1}{2}, -20\frac{1}{4})$; x-intercept: $(-7,0)$, $(2,0)$; y-intercept: $(0,-14)$

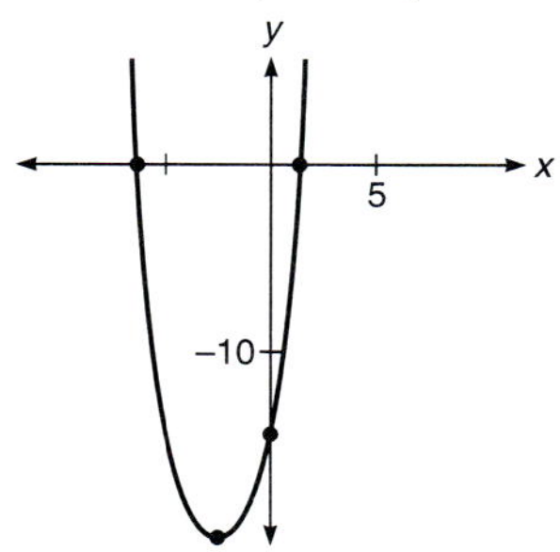

2. vertex: $(0,8)$; x-intercept: $(-2,0)$, $(2,0)$; y-intercept: $(0,8)$

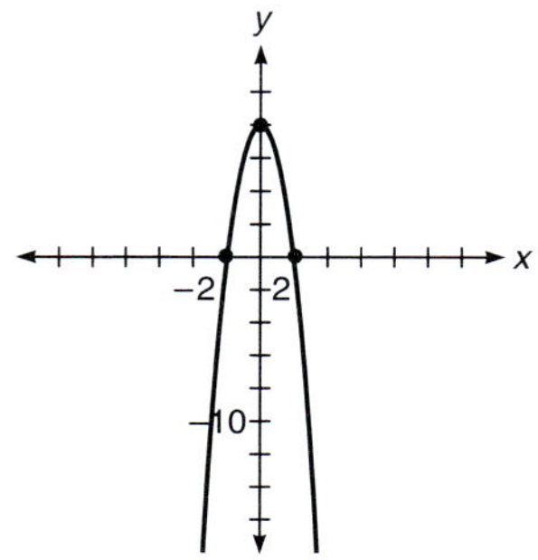

3. vertex: $(-\frac{5}{6}, -4\frac{1}{12})$; x-intercept: $(-2,0)$, $(\frac{1}{3},0)$; y-intercept: $(0,-2)$

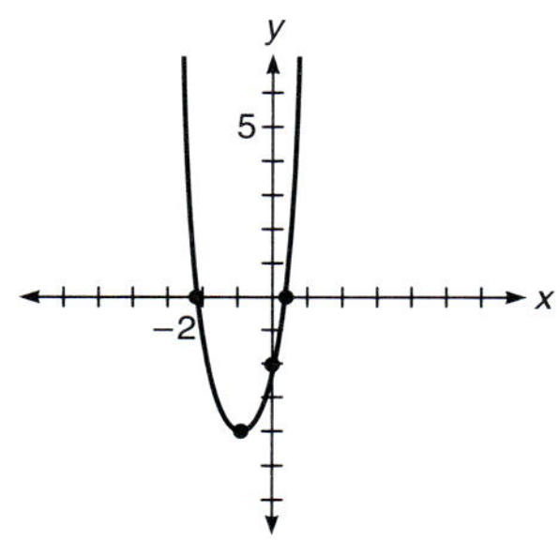

4. vertex: $(2,4)$; x-intercept: $(0,0)$, $(4,0)$; y-intercept is the origin also

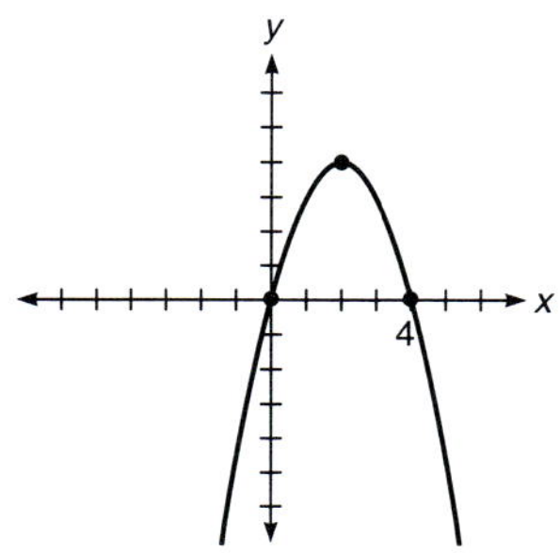

5. The dimensions should be 12.5 ft by 25 ft, and the area will be 312.5 ft^2.

6. The object will be at its highest point after 1.5 seconds, and it will be 36 feet high at that time; it returns to its starting point after $t = 3$ seconds.

7.

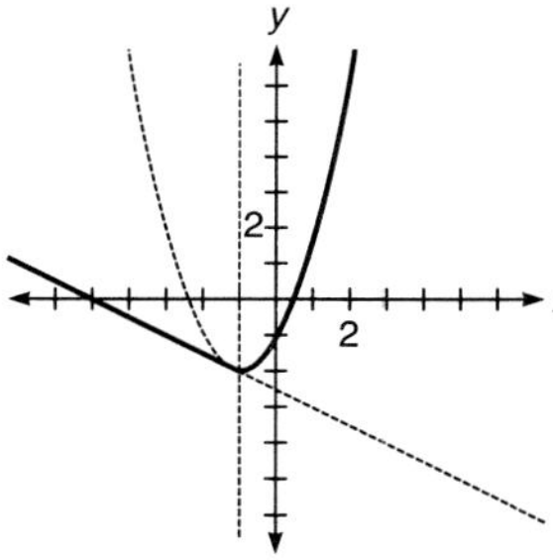

8. $\pm(1, 2, 4, 8)$

9. $\pm(1, 2, 3, 4, 6, 12, \frac{1}{2}, \frac{3}{2}, \frac{1}{4}, \frac{3}{4})$

10. $f(x) \div (x + 3) = 3x^3 - 11x^2 + 3x - 9 + \dfrac{7}{x + 3}$, and $f(-3) = 7$

11. **a.** 0 or 2 positive real zeros; one negative real zero
  b. $\pm(1, \frac{1}{2}, \frac{1}{4})$
  c,d. $f(x) = (x - 1)(2x - 1)(2x + 1)$
  Real zeros are $1, \pm\frac{1}{2}$.

12. **a.** no positive real zeros; 0, 2, or 4 real negative roots
  b. possible rational zeros: ± 1
  c,d. $f(x) = (x + 1)^2(x^2 + x + 1)$
  rational zeros are -1, multiplicity 2
  e. There are no possible irrational zeros.

13. **a.** 1 or 3 positive real zeros; 0 or 2 negative real zeros
  b. $\pm(1, 2, 3, 4, 6, 12, \frac{1}{3}, \frac{2}{3}, \frac{4}{3})$
  c,d. $f(x) = (x - 1)(x - 2)^2(x + 3)(3x + 1)$
  Real zeros are $1, 2, -3, -\frac{1}{3}$.
  The zero 2 has multiplicity 2.

14. $f(x) = (x - 2)(x + 1)(x - 5)$
  x-intercepts: $-1, 2, 5$; y-intercept: 10

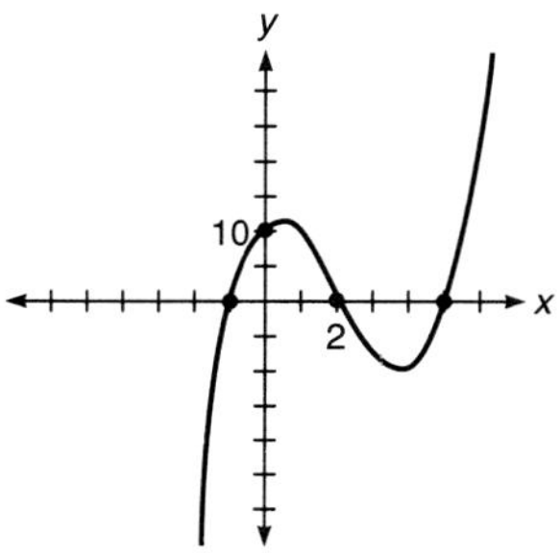

15. $g(x) = (x + 3)^2(x - 3)$
  x-intercepts: $3, -3$; y-intercept: -27

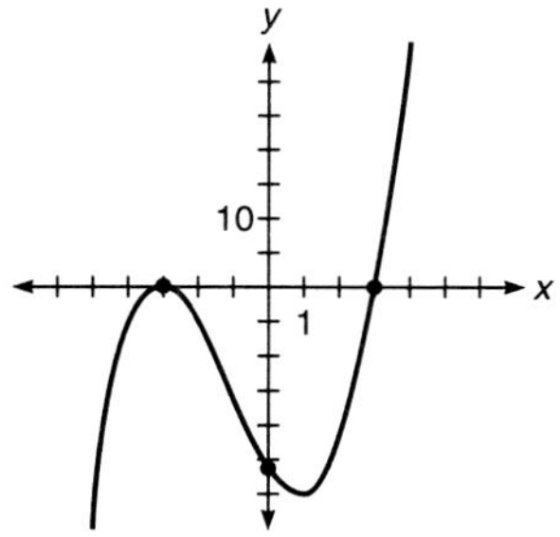

16. $f(x) = (x - 1)^2(x + 1)^2$
  x-intercepts at ± 1; y-intercept at 1

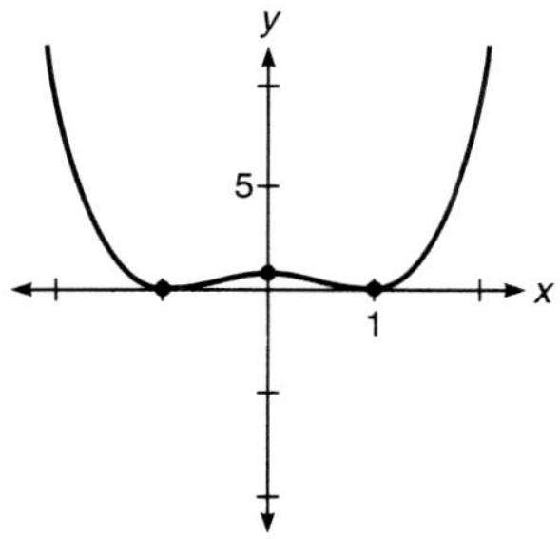

17. $h(x) = (x - 5)(x + 2)(x^2 - 3x + 10)$
  x-intercepts at $-2, 5$; y-intercept at -100

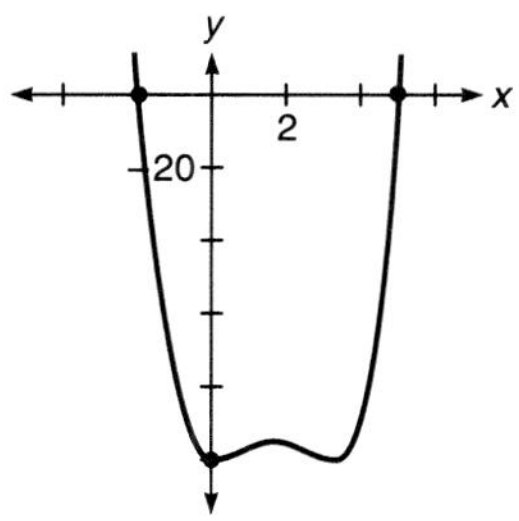

18. y-intercept: 2; asymptotes: $x = -1$, $y = 0$

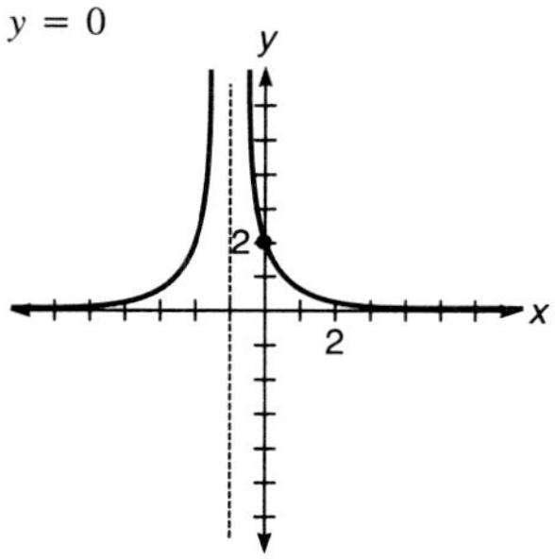

19. y-intercept: $\frac{1}{8}$; asymptotes: $x = 2$, $y = 0$

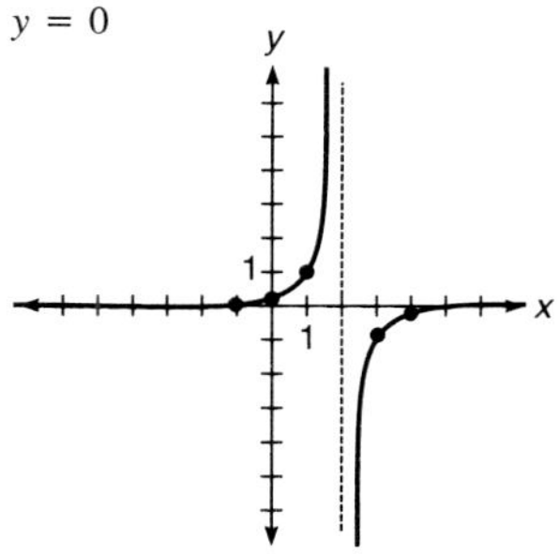

20. $f(x) = \dfrac{1}{(x - 4)(x + 6)}$
  y-intercept: $-\frac{1}{24}$; asymptotes: $x = -6$, $x = 4$, $y = 0$

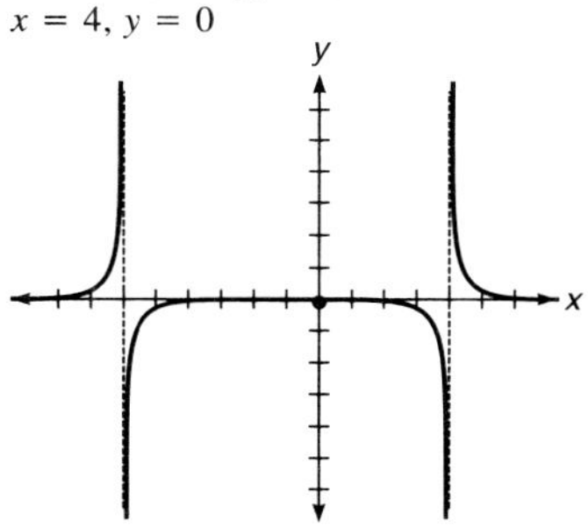

21. $f(x) = \dfrac{-x}{(x - 2)(x + 2)}$
  asymptotes: $x = \pm 2$, $y = 0$; intercepts at (0,0)

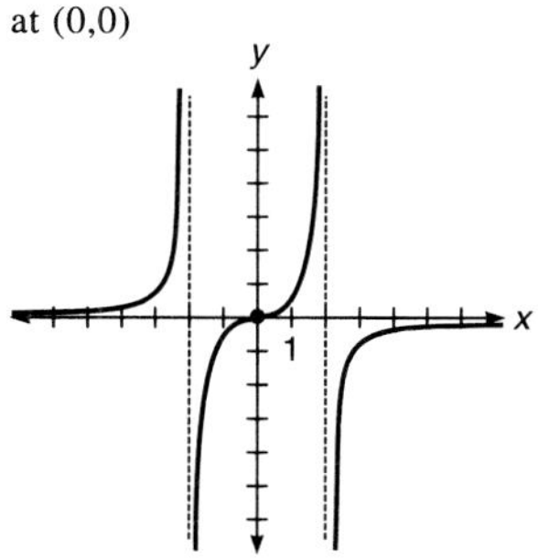

22. $g(x) = 1 + \dfrac{-7x + 11}{x^2 + 1}$

asymptote: $y = 1$; x-intercept: 3 or 4, y-intercept: 12

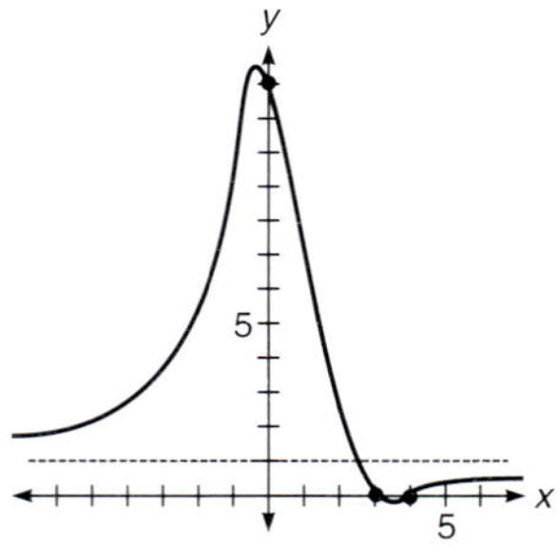

23. $f(x) = x + 4 + \dfrac{8}{x - 5}$ the line $x + 4$ is a slant asymptote; vertical asymptote at $x = 5$; x-intercepts are at -3 and 4; y-intercept at $f(0) = 2\frac{2}{5}$

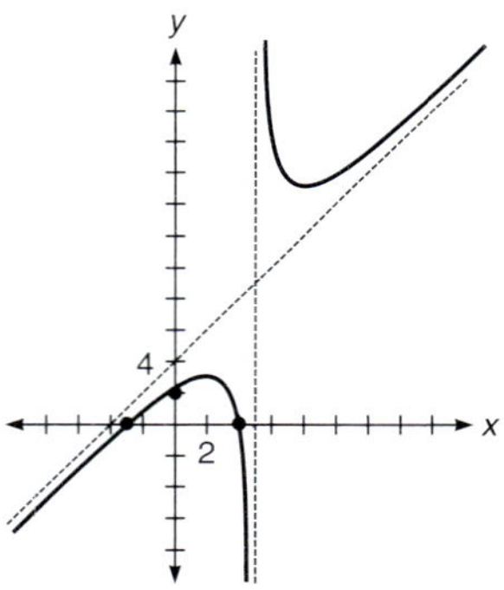

24. $x^2 + 1$; $-x^2 + 4x + 9$; $2x^3 + x^2 - 18x - 20$; $\dfrac{2x + 5}{x^2 - 2x - 4}$; $2x^2 - 4x - 3$; $4x^2 + 16x + 11$

25. $x^4 - 2 + 2\sqrt{x + 1}$; $x^4 - 2 - 2\sqrt{x + 1}$; $2(x^4 - 2)\sqrt{x + 1}$; $\dfrac{x^4 - 2}{2\sqrt{x + 1}}$; $16x^2 + 32x + 14$; $2\sqrt{x^4 - 1}$

26. $\dfrac{3x^2 + 3x - 2}{x(2x - 1)}$; $\dfrac{x^2 + 3x - 2}{x(2x - 1)}$; $\dfrac{x + 2}{2x - 1}$; $\dfrac{2x^2 + 3x - 2}{x^2}$; $\dfrac{5x - 2}{x}$; $\dfrac{x + 2}{x + 4}$

27. $g^{-1}(x) = \dfrac{x - 4}{5}$ **28.** $f^{-1}(x) = \dfrac{1}{4x - 5}$

29. $g^{-1}(x) = \sqrt{x + 4}$

30. $f^{-1}(x) = \dfrac{1 - \sqrt{4x + 25}}{2}$

31. $f(x) = 2.5x - 100$; 62.5° F

32. $\dfrac{2}{x + 1} + \dfrac{-1}{(x + 1)^2} + \dfrac{3}{x - 2}$

33. $\dfrac{3}{x - 1} + \dfrac{x + 2}{x^2 + x + 4}$

Chapter 5

Exercise 5–1

Answers to odd-numbered problems

1. 13.417° **3.** 0.2° **5.** 25.555° **7.** 165.783° **9.** 33.099° **11.** 159.983° **13.** 48.2° **15.** 71°48′ **17.** 106.40° **19.** 15 **21.** 6 **23.** $6\sqrt{5}$ **25.** $\sqrt{14}$ **27.** $50\sqrt{13}$ **29.** 4 **31.** $\sqrt{185.31} \approx 13.6$ **33.** $7\sqrt{2}$ **35.** $3\sqrt{47}$ **37.** $\sqrt{2}$ **39.** 60.8 feet **41.** 90.1 feet **43.** 20.07 ohms **45.** 3,770 ohms **47.** 22.9 minutes **49.** The reach of the ladder decreases by about 1 foot, not 5 feet.

51. $\sec \alpha = 3$ **53.** $\cot \beta = \dfrac{\sqrt{2}}{2}$

55. $\tan \theta = \dfrac{\sqrt{5}}{5}$ **57.** $\sec \theta = \frac{4}{3}\sqrt{6}$

In problems 59 through 79 values are given in the order sin, csc, cos, sec, tan, cot.

59. $\dfrac{b}{c} = \frac{12}{13}, \frac{13}{12}, \dfrac{a}{c} = \frac{5}{13}, \frac{13}{5}, \dfrac{b}{a} = \frac{12}{5}, \frac{5}{12}$

61. $\dfrac{\sqrt{65}}{13}, \dfrac{\sqrt{65}}{5}, \frac{2}{13}\sqrt{26}, \dfrac{\sqrt{26}}{4}, \dfrac{\sqrt{10}}{4}, \frac{2}{5}\sqrt{10}$

63. $\frac{2}{13}\sqrt{13}, \dfrac{\sqrt{13}}{2}, \frac{3}{13}\sqrt{13}, \dfrac{\sqrt{13}}{3}, \frac{2}{3}, \frac{3}{2}$

65. $\frac{4}{7}, \frac{7}{4}, \dfrac{\sqrt{33}}{7}, \frac{7}{33}\sqrt{33}\ \frac{4}{33}\sqrt{33}, \dfrac{\sqrt{33}}{4}$

67. $\dfrac{b}{c} = \frac{12}{13}, \frac{13}{12}, \dfrac{a}{c} = \frac{5}{13}, \frac{13}{5}, \dfrac{b}{a} = \frac{12}{5}, \frac{5}{12}$

69. $\frac{2}{3}, \frac{3}{2}, \dfrac{\sqrt{5}}{3}, \frac{3}{5}\sqrt{5}, \frac{2}{5}\sqrt{5}, \dfrac{\sqrt{5}}{2}$

71. $\dfrac{\sqrt{2}}{2}, \sqrt{2}, \dfrac{\sqrt{2}}{2}, \sqrt{2}, 1, 1$

73. $\dfrac{\sqrt{39}}{8}, \frac{8}{39}\sqrt{39}, \frac{5}{8}, \frac{8}{5}, \dfrac{\sqrt{39}}{5}, \frac{5}{39}\sqrt{39}$

75. $\frac{1}{4}\sqrt{15}, \frac{4}{15}\sqrt{15}, \frac{1}{4}, 4, \sqrt{15}, \dfrac{\sqrt{15}}{15}$

77. $\dfrac{\sqrt{z^2 - x^2}}{z}, \dfrac{z}{\sqrt{z^2 - x^2}}, \dfrac{x}{z}, \dfrac{z}{x}, \dfrac{\sqrt{z^2 - x^2}}{x}, \dfrac{x}{\sqrt{z^2 - x^2}}$

79. $\frac{1}{3}, 3, \dfrac{2\sqrt{2}}{3}, \frac{3}{4}\sqrt{2}, \dfrac{\sqrt{2}}{4}, 2\sqrt{2}$

81.

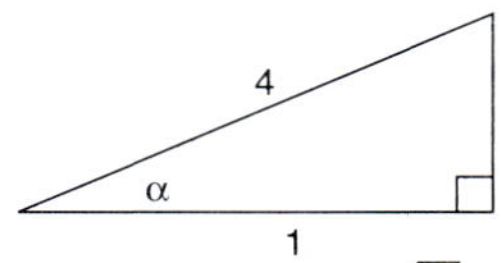

$\sec \alpha = 4$, $\sin \alpha = \dfrac{\sqrt{15}}{4}$, $\csc \alpha = \dfrac{4}{15}\sqrt{15}$, $\tan \alpha = \sqrt{15}$, $\cot \alpha = \dfrac{\sqrt{15}}{15}$

83.

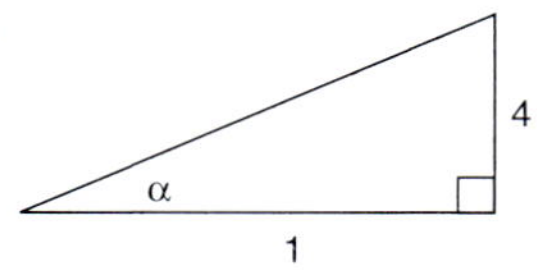

$\cot \alpha = \frac{1}{4}$, $\sin \alpha = \dfrac{4}{17}\sqrt{17}$, $\csc \alpha = \dfrac{\sqrt{17}}{4}$, $\cos \alpha = \dfrac{\sqrt{17}}{17}$, $\sec \alpha = \sqrt{17}$

85.

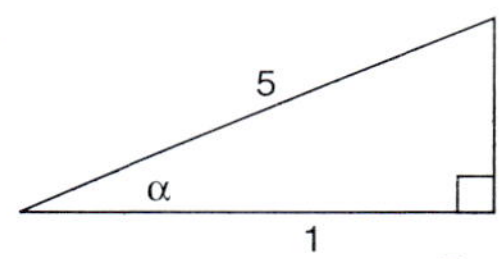

$\sec \alpha = 5$, $\sin \alpha = \dfrac{2\sqrt{6}}{5}$, $\csc \alpha = \frac{5}{12}\sqrt{6}$, $\tan \alpha = 2\sqrt{6}$, $\cot \alpha = \dfrac{\sqrt{6}}{12}$

87.

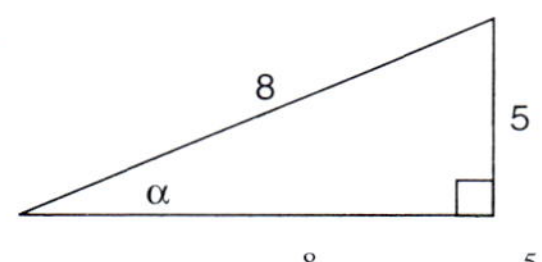

$\csc \alpha = 1.6 = \frac{8}{5}$, $\sin \alpha = \frac{5}{8}$, $\cos \alpha = \dfrac{\sqrt{39}}{8}$, $\sec \alpha = \dfrac{8}{39}\sqrt{39}$, $\tan \alpha = \dfrac{5}{39}\sqrt{39}$, $\cot \alpha = \dfrac{\sqrt{39}}{5}$

89.

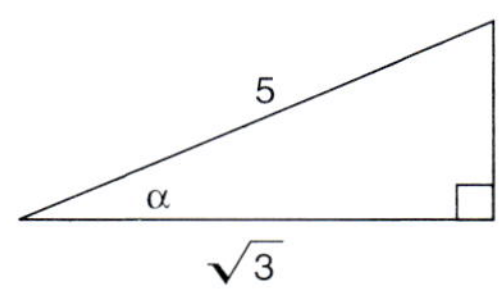

$\sec \alpha = \frac{5}{3}\sqrt{3}$, $\sin \alpha = \dfrac{\sqrt{22}}{5}$, $\csc \alpha = \frac{5}{22}\sqrt{22}$, $\tan \alpha = \dfrac{\sqrt{66}}{3}$, $\cot \alpha = \dfrac{\sqrt{66}}{22}$

91.

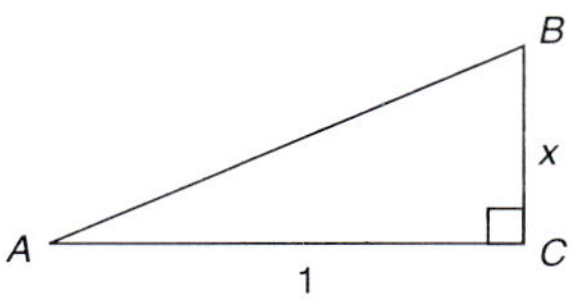

$\dfrac{x}{\sqrt{1+x^2}}$

93.

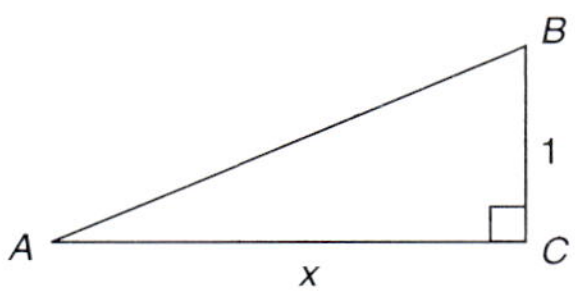

$\dfrac{1}{x}$

95. 107 knots **97.** 16.6 knots

99. 27

Solutions to skill and review problems

1. Find the equation of the line that passes through the points $(-2,5)$ and $(3,-10)$.

$m = \dfrac{y_2 - y_1}{x_2 - x_1} = \dfrac{-10-5}{3-(-2)} = \dfrac{-15}{5} = -3$

$y - y_1 = m(x - x_1)$

$y - 5 = -3(x-(-2))$

$y = -3x - 1$

2. Solve the inequality $|2x - 3| < 13$.

$-13 < 2x - 3 < 13$

$-10 < 2x < 16$

$-5 < x < 8$

3. Graph the parabola $y = x^2 - 3x - 4$.

$y = x^2 - 3x + \frac{9}{4} - 4 - \frac{9}{4}$

$y = (x - \frac{3}{2})^2 - \frac{25}{4}$; vertex at $(1\frac{1}{2}, -6\frac{1}{4})$

x-intercept:

$0 = x^2 - 3x - 4$

$0 = (x-4)(x+1)$

$x = -1$ or 4

y-intercept: $y = 0^2 - 3(0) - 4 = -4$

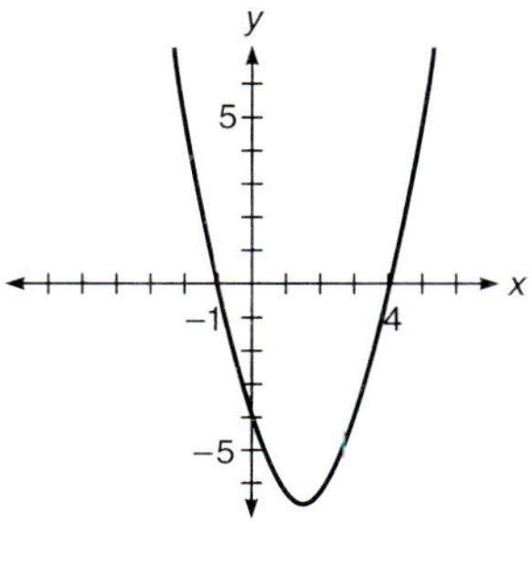

4. Use the rational zero theorem and synthetic division to find all zeros of the polynomial $2x^4 + 5x^3 - 5x^2 - 5x + 3$. Synthetic division produces the zeros $-3, -1, \frac{1}{2}, 1$.

5. Graph the polynomial $f(x) = (x+1)(x-2)^2(x+3)$.

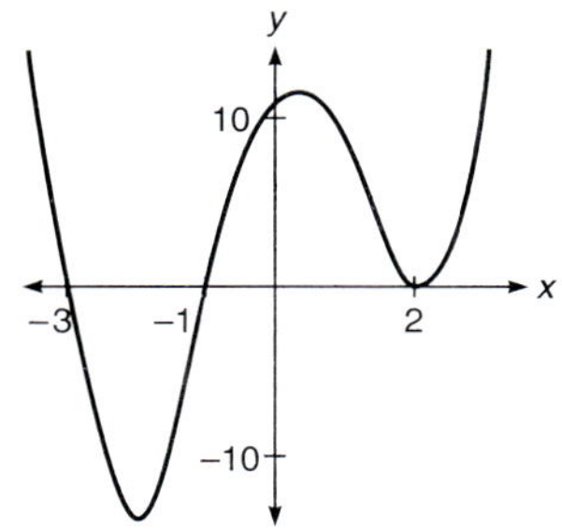

Solutions to trial exercise problems

6. $87°2'13'' = \left(87 + \dfrac{2}{60} + \dfrac{13}{3{,}600}\right)°$

$\approx 87.037°$

17. $(180 - 43.45 - 30.15)° = 106.40°$

25. $c^2 = a^2 + b^2$

$c^2 = (\sqrt{5})^2 + 3^2$

$c^2 = 5 + 9$

$c^2 = 14$

$c = \sqrt{14}$

34. $c^2 = a^2 + b^2$

$(4\sqrt{5})^2 = (3\sqrt{2})^2 + b^2$

$16 \cdot 5 = 9 \cdot 2 + b^2$

$62 = b^2$

$\sqrt{62} = b$

47. $d^2 = 15^2 + 10.5^2$, so $d \approx 18.3$ cm, and $\dfrac{18.3 \text{ cm}}{0.8 \text{ cm/min}} = 22.9$ minutes to make the cut.

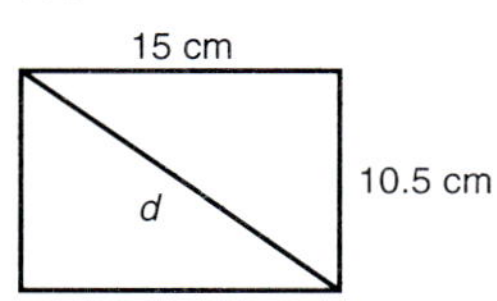

56. $\tan\alpha = 2.25 = 2\frac{1}{4}$;

$\cot\alpha = \dfrac{1}{\tan\alpha} = \dfrac{1}{2\frac{1}{4}} = \dfrac{1}{\frac{9}{4}} = \dfrac{4}{9}$

63. $a^2 + 2^2 = (\sqrt{13})^2$

$a^2 = 9$

$a = 3$

$\sin B = \dfrac{b}{c} = \dfrac{2}{\sqrt{13}} = \dfrac{2}{13}\sqrt{13}$, $\csc B = \dfrac{\sqrt{13}}{2}$

$\cos B = \dfrac{a}{c} = \dfrac{3}{\sqrt{13}} = \dfrac{3}{13}\sqrt{13}$, $\sec B = \dfrac{\sqrt{13}}{3}$

$\tan B = \dfrac{b}{a} = \dfrac{2}{3}$, $\cot B = \dfrac{3}{2}$

79. $\left(\dfrac{z}{3}\right)^2 + b^2 = z^2$

$b^2 = z^2 - \dfrac{z^2}{9} = \dfrac{8}{9}z^2$

$b = \dfrac{2\sqrt{2}}{3}z$

$\sin A = \dfrac{a}{c} = \dfrac{\frac{z}{3}}{z} = \dfrac{z}{3} \cdot \dfrac{1}{z} = \dfrac{1}{3}$, $\csc A = 3$

$\cos A = \dfrac{b}{c} = \dfrac{\frac{2\sqrt{2}}{3}z}{z} = \dfrac{2\sqrt{2}}{3}$

$\sec A = \dfrac{3}{2\sqrt{2}} \cdot \dfrac{\sqrt{2}}{\sqrt{2}} = \dfrac{3}{4}\sqrt{2}$

$\tan A = \dfrac{a}{b} = \dfrac{\frac{z}{3}}{\frac{2\sqrt{2}}{3}z} = \dfrac{z}{3} \cdot \dfrac{3}{2\sqrt{2}z}$

$= \dfrac{1}{2\sqrt{2}} \cdot \dfrac{\sqrt{2}}{\sqrt{2}} = \dfrac{\sqrt{2}}{4}$

$\cot A = \dfrac{1}{\frac{1}{2\sqrt{2}}} = 2\sqrt{2}$

87. $\csc\alpha = 1.6 = \dfrac{1.6}{1} = \dfrac{16}{10} = \dfrac{8}{5}$, so

$\sin\alpha = \frac{5}{8}$, $\cos\alpha = \dfrac{\sqrt{39}}{8}$,

$\sec\alpha = \dfrac{8}{\sqrt{39}} = \dfrac{8}{39}\sqrt{39}$,

$\tan\alpha = \dfrac{5}{\sqrt{39}} = \dfrac{5}{39}\sqrt{39}$, $\cot\alpha = \dfrac{\sqrt{39}}{5}$

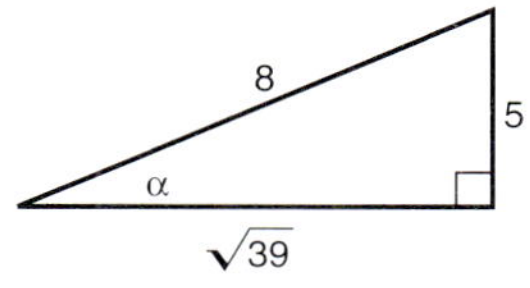

93. $\tan B = x = \frac{x}{1} = \frac{\text{opp}}{\text{adj}}$;

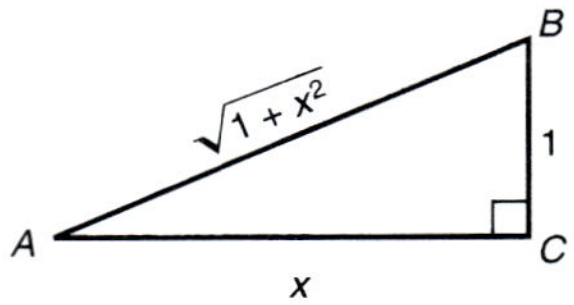

$\tan A = \frac{\text{opp}}{\text{adj}} = \frac{1}{x}$

97. $v^2 = 16^2 + 4.3^2$

$v \approx 16.6$ knots

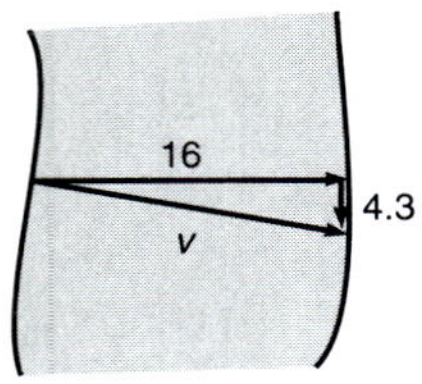

Exercise 5–2

Answers to odd-numbered problems

1. 0.5192 **3.** 0.2116 **5.** 1.8137
7. 0.6465 **9.** 2.5048 **11.** 0.9793
13. 0.5868 **15.** 0.9524 **17.** 0.8596
19. 0.9178 **21.** 4.3143 **23.** 0.6652
25. 930.6 sq. ft **27.** −7.56
29. 195.2 watts **31.** 48.5 mm
33. 39.9° **35.** 53.2° **37.** 62.0°
39. 31.6° **41.** 31.7° **43.** 63.4°
45. 78.0° **47.** 68.6°
49. $A = 51.7°$, $c \approx 19.4$, $b \approx 12.0$
51. $B = 76.3°$, $c \approx 46.9$, $b \approx 45.5$
53. $B = 60.6°$, $a \approx 0.379$, $c \approx 0.771$
55. $A = 12.0°$, $c \approx 22.3$, $a \approx 4.6$
57. $B = 75.0°$, $b \approx 9.7$, $a \approx 2.6$
59. $A = 24.5°$, $a \approx 50.6$, $b \approx 111.0$
61. $c \approx 20.4$, $A \approx 40.0°$, $B \approx 50.0°$
63. $c \approx 1.36$, $A \approx 9.3°$, $B \approx 80.7°$
65. $b \approx 17.8$, $A \approx 44.9°$, $B \approx 45.1°$
67. $a \approx 98.4$, $A \approx 62.5°$, $B \approx 27.5°$
69. $b \approx 5.0$, $A \approx 67.4°$, $B \approx 22.6°$
71. $a \approx 32.6$, $A \approx 21.2°$, $B \approx 68.8°$
73. 9.44 ohms, 24.2° **75.** $w \approx 194$ feet
77. 9.51 inches
79. $S \approx 158$ mph, $\theta \approx 11°$
81. 199,800 feet (37.8 miles)
83. 830 feet **85.** 215.9 cm
87. $a = 4$, $c = 4\sqrt{2}$, $B = 45°$

Solutions to skill and review problems

1. Graph the rational function

$f(x) = \frac{2}{x^2 - 9}$.

$f(x) = \frac{2}{(x-3)(x+3)}$.

Vertical asymptotes: $x = 3$ and $x = -3$

y-intercept: $f(0) = \frac{2}{0-9} = -\frac{2}{9}$

x-intercepts: $0 = \frac{2}{x^2 - 9}$

No solution, so no x-intercepts.

Additional points:

x	−4	−3.5	−2.5	−2	2	2.5	3.5	4
y	0.3	0.6	−0.7	−0.4	−0.4	−0.7	0.6	0.3

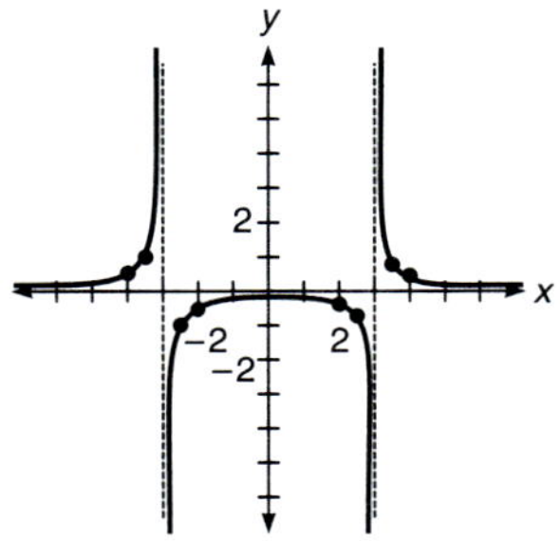

2. Solve for y: $3x - 2y = 5$

$-2y = -3x + 5$

$2y = 3x - 5$

$y = \frac{3}{2}x - \frac{5}{2}$

m = coefficient of $x = 1\frac{1}{2}$.

3. Solve equality to find critical points:

$x^2 - 2x = 3$

$x^2 - 2x - 3 = 0$

$(x-3)(x+1) = 0$

$x - 3 = 0$ or $x + 1 = 0$

$x = 3$ or $x = -1$

Choose test points on each interval.

Use −2, 0, 4: $x^2 - x > 3$

$x = -2$: $(-2)^2 - (-2) > 3$

$6 > 3$; True

$x = 0$: $0 - 0 > 3$; False

$x = 4$: $4^2 - 4 > 3$

$12 > 3$; True

The result is those intervals where the test points make $x^2 - 2x > 3$ true.

Set-builder notation

$\{x \mid x < -1 \text{ or } x > 3\}$

$(-\infty, -1)$ or $(3, \infty)$

4. $c^2 = 16.8^2 + 9.0^2$

$c = \sqrt{16.8^2 + 9.0^2} \approx 19.1$

Solutions to trial exercise problems

6. $\csc 5.15° \approx 11.1404$

5.15 [sin] [1/x]

TI–81: [(] [SIN] 5.15 [)] [x^{-1}]

[ENTER]

21. $\cot 13°3' \approx 4.3143$

13 [+] 3 [÷] 60 [=]

[tan] [1/x]

TI–81: [(] [TAN] [(] 13 [+] 3

[÷] 60 [)] [)] [x^{-1}]

[ENTER]

42. $\cos\theta = \frac{8.25}{12.5}$, $\theta \approx 48.7°$

8.25 [÷] 12.5 [=] [shift] [cos]

TI–81: [2nd] [COS] [(] 8.25 [÷]

12.5 [)] [ENTER]

44. $\csc\theta = 1.1243$, so $\sin\theta = \frac{1}{1.1243}$

$\theta \approx 62.8°$: 1.1243

[1/x] [shift] [sin]

TI–81: [2nd] [SIN] 1.1243 [x^{-1}]

[ENTER]

49. $a = 15.2$, $B = 38.3°$

$A = 90° - 38.3° = 51.7°$

$\cos 38.3° = \frac{15.2}{c}$;

$c = \frac{15.2}{\cos 38.3°} \approx 19.4$

$\tan 38.3° = \frac{b}{15.2}$;

$b = 15.2 \tan 38.3° \approx 12.0$

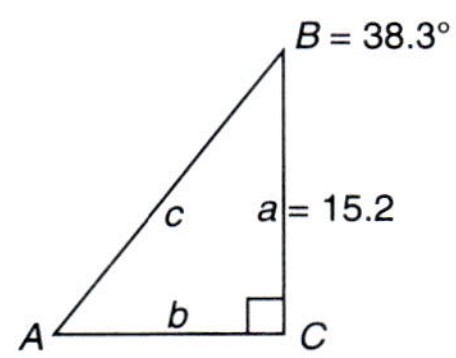

59. $c = 122,\ B = 65.5°$

$A = 90° - 65.5° = 24.5°$

$\cos 65.5° = \dfrac{a}{122};$

$a = 122 \cos 65.5° \approx 50.6$

$\sin 65.5° = \dfrac{b}{122};$

$b = 122 \sin 65.5° \approx 111.0$

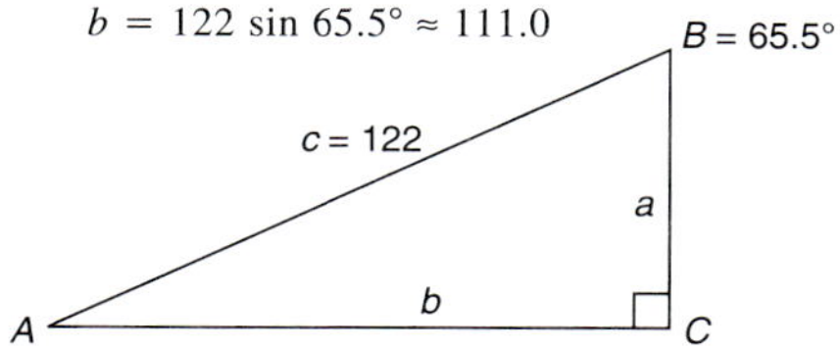

67. $b = 51.3,\ c = 111.0$

$a^2 + 51.3^2 = 111^2$

$a = \sqrt{111^2 - 51.3^2} \approx 98.4$

$\cos A = \dfrac{51.3}{111},\ A \approx 62.5°$

$\sin B = \dfrac{51.3}{111},\ B \approx 27.5°$

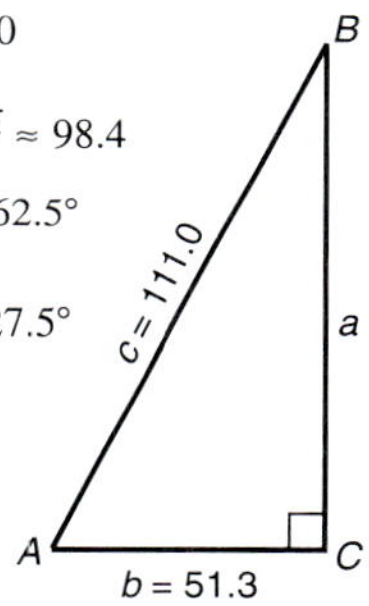

85. $a = \sqrt{122^2 + 83^2} \approx 147.5567687$ cm

$\tan\theta = \dfrac{83}{122};\ \theta \approx 34.22854584°$

$\cos\theta = \dfrac{a}{c};$

$c = \dfrac{a}{\cos\theta} = \dfrac{147.5567687}{\cos 34.22854584°}$

≈ 178.4672131 cm

$\cos\theta = \dfrac{c}{x};$

$x = \dfrac{c}{\cos\theta} = \dfrac{178.4672131}{\cos 34.22854584°}$

≈ 215.8528302 cm

Thus $x \approx 215.9$ cm.

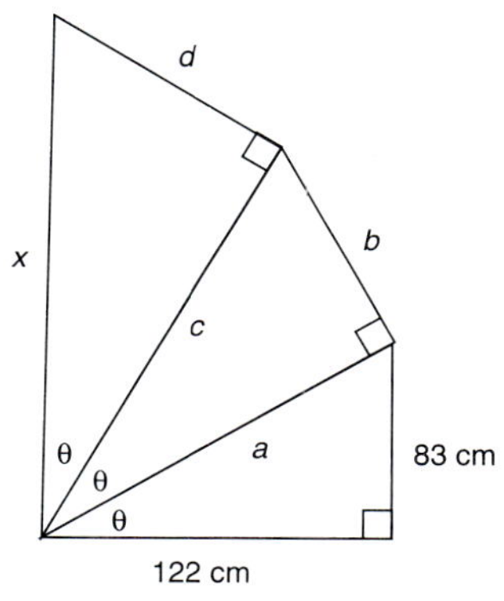

88. $A = 90° - 60° = 30°$

$\sin 60° = \dfrac{8}{c},\ \dfrac{\sqrt{3}}{2} = \dfrac{8}{c},\ \sqrt{3}\,c = 16,$

$c = \dfrac{16}{\sqrt{3}} \cdot \dfrac{\sqrt{3}}{\sqrt{3}} = \dfrac{16\sqrt{3}}{3};$

$\tan 60° = \dfrac{8}{a},\ \dfrac{\sin 60°}{\cos 60°} = \dfrac{8}{a},\ \dfrac{\frac{\sqrt{3}}{2}}{\frac{1}{2}} = \dfrac{8}{a},$

$\sqrt{3} = \dfrac{8}{a},\ a = \dfrac{8}{\sqrt{3}} \cdot \dfrac{\sqrt{3}}{\sqrt{3}} = \dfrac{8\sqrt{3}}{3}.$

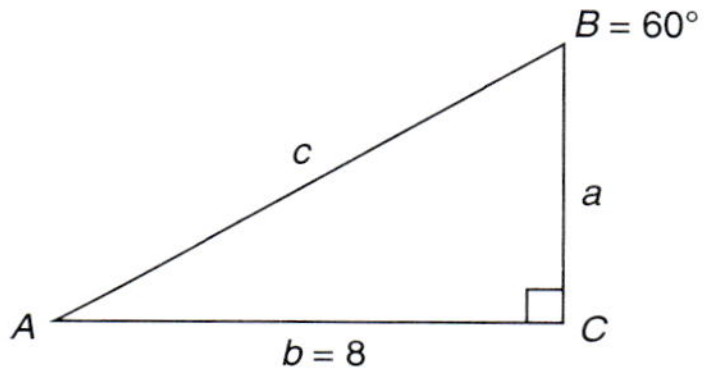

Exercise 5–3

Answers to odd-numbered problems

1. 60°

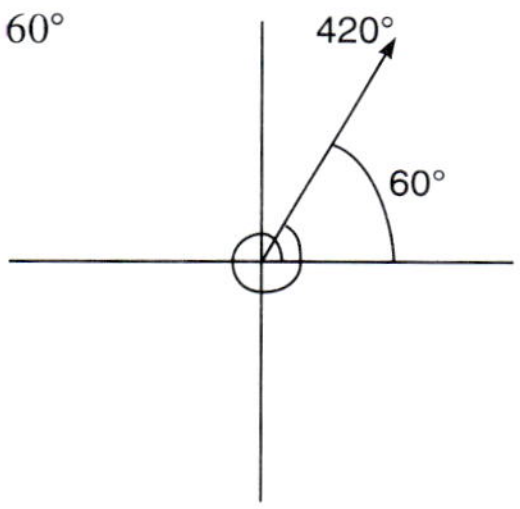

3. 230°

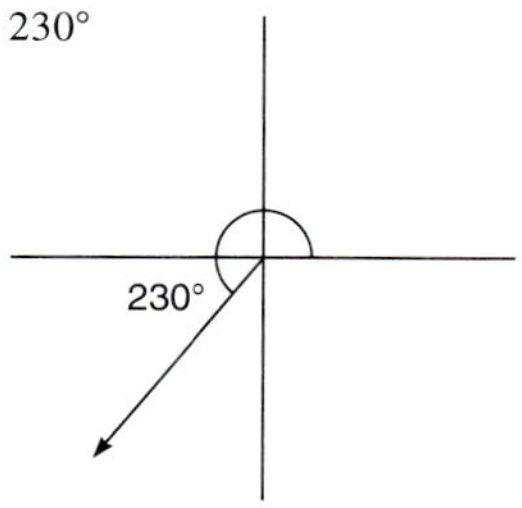

5. 80.6°

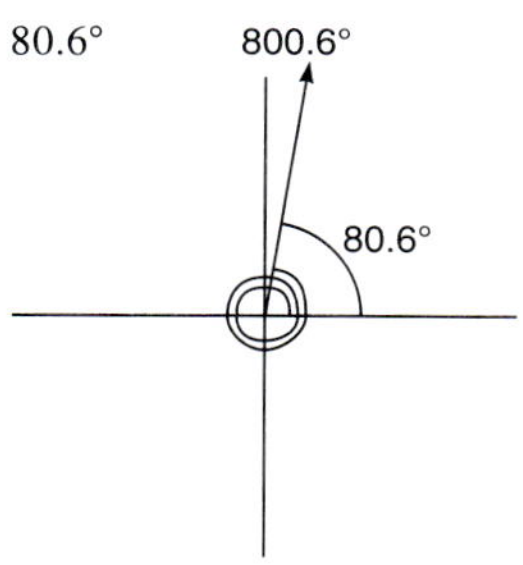

7. 187.9°

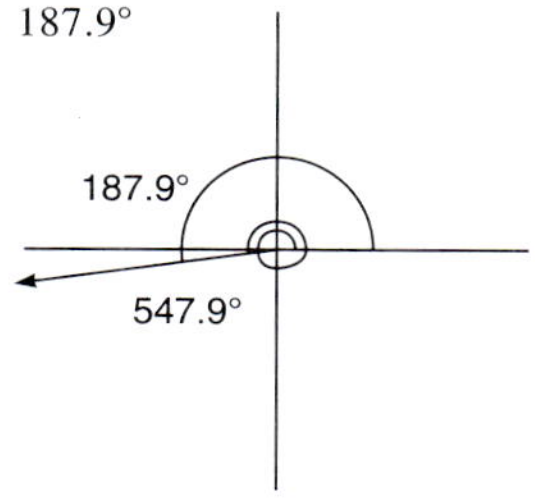

9. 210°

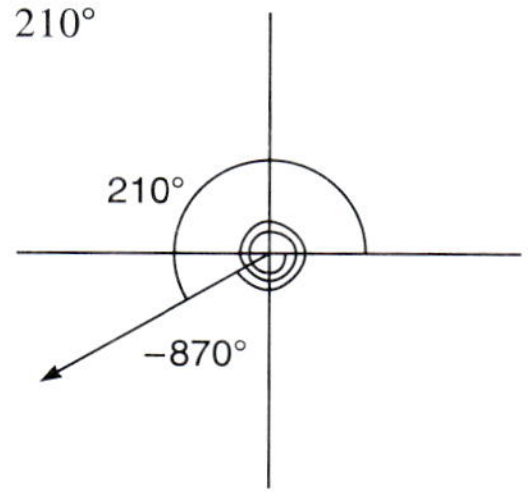

11. 165°

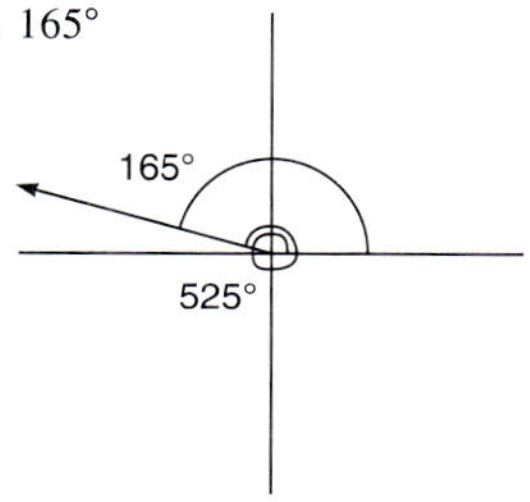

13. 189.7°

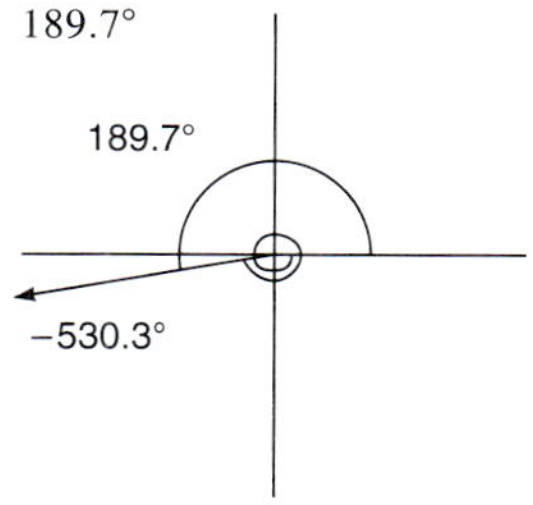

15. 0°

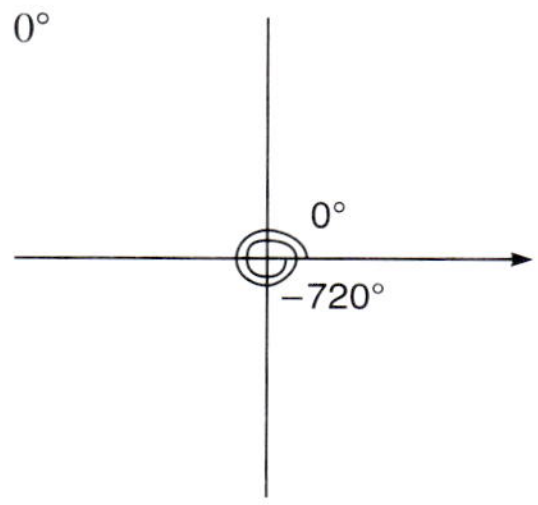

17. 348.1°

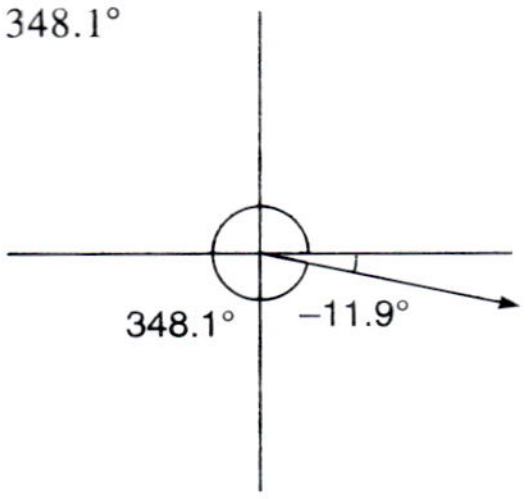

19. 347° **21.** 353.9°

In problems 23 through 37 answers are in the order sin, csc, cos, sec, tan, cot.

23. $\frac{2\sqrt{5}}{5}, \frac{\sqrt{5}}{2}, \frac{\sqrt{5}}{5}, \sqrt{5}, \frac{6}{3} = 2, \frac{1}{2}$

25. $\frac{8\sqrt{89}}{89}, \frac{\sqrt{89}}{8}, \frac{-5\sqrt{89}}{89}, -\frac{\sqrt{89}}{5}, -\frac{8}{5}, -\frac{5}{8}$

27. $-\frac{\sqrt{2}}{2}, -\sqrt{2}, \frac{\sqrt{2}}{2}, \sqrt{2}, -1, -1$

29. $\frac{4\sqrt{17}}{17}, \frac{\sqrt{17}}{4}, -\frac{\sqrt{17}}{17}, -\sqrt{17}, -4, -\frac{1}{4}$

31. $-\frac{3\sqrt{13}}{13}, -\frac{\sqrt{13}}{3}, -\frac{2\sqrt{13}}{13}, -\frac{\sqrt{13}}{2}, 1\frac{1}{2}, \frac{2}{3}$

33. $\frac{3\sqrt{38}}{19}, \frac{\sqrt{38}}{6}, -\frac{\sqrt{19}}{19}, -\sqrt{19}, -3\sqrt{2}, -\frac{\sqrt{2}}{6}$

35. $-\frac{\sqrt{10}}{5}, -\frac{\sqrt{10}}{2}, -\frac{\sqrt{15}}{5}, -\frac{\sqrt{15}}{3}, \frac{\sqrt{6}}{3}, \frac{\sqrt{6}}{2}$

37. $-\frac{\sqrt{10}}{4}, -\frac{2\sqrt{10}}{5}, \frac{\sqrt{6}}{4}, \frac{2\sqrt{6}}{3}, -\frac{\sqrt{15}}{3}, -\frac{\sqrt{15}}{5}$

39. $\cot\theta = \frac{x}{y}$ Definition of $\cot\theta$

$= \frac{1}{\frac{y}{x}}$

$= \frac{1}{\tan\theta}$

41. $\cos\theta = \frac{x}{r}$

$= \frac{1}{\frac{r}{x}}$

$= \frac{1}{\sec\theta}$

43. Using (x_1, mx_1) we obtain

$r_1 = \sqrt{x_1^2 + m^2x_1^2} = \sqrt{x_1^2(1 + m^2)}$

$= |x_1|\sqrt{1 + m^2}$

Using (x_2, mx_2) we obtain

$r_2 = \sqrt{x_2^2 + m^2x_2^2} = \sqrt{x_2^2(1 + m^2)}$

$= |x_2|\sqrt{1 + m^2}$

For the first point:

$\sin\theta = \frac{y}{r} = \frac{mx_1}{|x_1|\sqrt{1 + m^2}}$

$= \pm\frac{m}{\sqrt{1 + m^2}}$

For the second point:

$\sin\theta = \frac{y}{r} = \frac{mx_2}{|x_2|\sqrt{1 + m^2}}$

$= \pm\frac{m}{\sqrt{1 + m^2}}$

Thus $\sin\theta$ has the same absolute value, $\frac{m}{\sqrt{1 + m^2}}$ in either case. The sign of x_1 is the same as the sign of x_2, so both values of $\sin\theta$ will have the same sign also.

Solutions to skill and review problems

1. Solve the triangle.

$c^2 = 9^2 + 16.8^2$

$c = \sqrt{9^2 + 16.8^2} \approx 19.1$

$\tan A = \frac{9}{16.8}, A \approx 28.2°$

$\tan B = \frac{16.8}{9}, B \approx 61.8°$

2. Use the graph of $y = \sqrt{x}$ to graph the function $f(x) = \sqrt{x - 4} - 2$. The graph of $f(x)$ is the graph of $y = \sqrt{x}$ but shifted to the right 4 units and down 2 units. Thus the graph of $y = \sqrt{x}$ at the origin shifts to a new origin at $(4, -2)$.

x-intercept (let $y = 0$):

$0 = \sqrt{x - 4} - 2$

$2 = \sqrt{x - 4}$

Square each side

$4 = x - 4$

$8 = x$

y-intercept (let $x = 0$):

$y = \sqrt{0 - 4} - 2$

Since $\sqrt{-4}$ is imaginary there is no y-intercept.

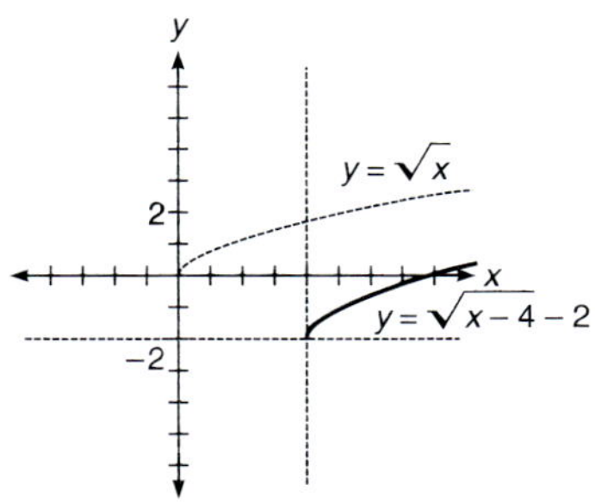

3. $8x^3 - 27 = ((2x)^3 - 3^3)$

$= (2x - 3)((2x)^2 + (2x)(3) + 3^2)$

$= (2x - 3)(4x^2 + 6x + 9)$

4. $\frac{3}{\sqrt{6}} \cdot \frac{\sqrt{6}}{\sqrt{6}} = \frac{3\sqrt{6}}{6} = \frac{\sqrt{6}}{2}$

5. $\frac{5 - 3^2}{8} - 2 = \frac{5 - 9}{8} - 2 = \frac{-4}{8} - 2$

$= -\frac{1}{2} - \frac{4}{2} = \frac{-1 - 4}{2} = -\frac{5}{2}$

or $-2\frac{1}{2}$

6. $C = 2\pi(15) \approx 2(3.14)(15) \approx 94.2$ inches

Solutions to trial exercise problems

35. $(-\sqrt{3}, -\sqrt{2})$

$r = \sqrt{(-\sqrt{3})^2 + (-\sqrt{2})^2} = \sqrt{5}$

$\sin\theta = \frac{-\sqrt{2}}{\sqrt{5}} = -\frac{\sqrt{10}}{5}$

$\csc\theta = \frac{1}{\sin\theta} = -\frac{\sqrt{5}}{\sqrt{2}} = -\frac{\sqrt{10}}{2}$

$\cos\theta = \frac{-\sqrt{3}}{\sqrt{5}} = -\frac{\sqrt{15}}{5}$

$\sec\theta = \frac{1}{\cos\theta} = -\frac{\sqrt{5}}{\sqrt{3}} = -\frac{\sqrt{15}}{3}$

$\tan\theta = \frac{-\sqrt{2}}{-\sqrt{3}} = \frac{\sqrt{6}}{3}$

$\cot\theta = \frac{1}{\tan\theta} = \frac{\sqrt{3}}{\sqrt{2}} = \frac{\sqrt{6}}{2}$

40. $\cot\theta = \frac{x}{y}$ if $y \neq 0$, definition of $\cot\theta$

$= \frac{\frac{x}{r}}{\frac{y}{r}}$ Ok since $r \neq 0$

$= \frac{\cos\theta}{\sin\theta}$ definition of $\cos\theta$, $\sin\theta$

If $y = 0$, then $\sin\theta = 0$ $\left(\text{since it is } \frac{y}{r}\right)$. Thus if we restrict $\sin\theta$ so $\sin\theta \neq 0$, then $y \neq 0$, and the fraction $\frac{x}{y}$ is defined.

Exercise 5–4

Answers to odd-numbered problems

1. II **3.** I **5.** IV **7.** II **9.** IV
11. III **13.** 15.8° **15.** 67.1°
17. 75.3° **19.** 49.3° **21.** 1.0°
23. 80.5° **25.** 72° **27.** $\frac{\sqrt{2}}{2}$
29. $-\frac{1}{2}$ **31.** $-\sqrt{3}$ **33.** $-\frac{\sqrt{3}}{2}$

35. $\frac{1}{2}$ **37.** $-\frac{\sqrt{3}}{3}$ **39.** 0

41. 1 **43.** $\frac{1}{2}$ **45.** $-\frac{2\sqrt{3}}{3}$

47. 0.9178 **49.** 0.6899 **51.** 0.7813

53. 1.0263 **55.** 0.9967 **57.** −1.0367

59. −0.6845 **61.** 14.5° **63.** 120°

65. −58.0° **67.** −36.2°

69. a. 110.31 **b.** 156 **c.** 89.48

d. −65.93 **e.** 132.73 **f.** 0

71. a. 200 lb **b.** 181.3 lb **c.** 128.6 lb

Solutions to skill and review problems

1. Let θ be 30°.
$\sin(2\theta) = 2\sin\theta$
$\sin 60° = 2\sin 30°$
$\frac{\sqrt{3}}{2} = 1$; Since these values are not equal, the statement $\sin(2\theta) = 2\sin\theta$ is not necessarily true.

2. Let θ be 60°.
$\sin\frac{\theta}{2} = \frac{\sin\theta}{2}$
$\sin 30° = \frac{\sin 60°}{2}$
$\frac{1}{2} = \frac{\sqrt{3}}{4}$; Since these values are not equal, the statement $\sin\frac{\theta}{2} = \frac{\sin\theta}{2}$ is not necessarily true.

3. Let $\alpha = 30°$, $\beta = 60°$.
$\sin(\alpha + \beta) = \sin\alpha + \sin\beta$
$\sin(30° + 60°) = \sin 30° + \sin 60°$
$\sin 90° = \sin 30° + \sin 60°$
$1 = \frac{1 + \sqrt{3}}{2}$; Since these values are not equal the statement $\sin(\alpha + \beta) = \sin\alpha + \sin\beta$ is not necessarily true.

Solutions to trial exercise problems

6. $\sec\theta > 0$, $\csc\theta < 0$
$\cos\theta > 0$, $\sin\theta < 0$
I, IV; III, IV
IV

19. 130.7°; θ = 130.7° in quadrant II, so $\theta' = 180° - \theta = 180° - 130.7° = 49.3°$

31. tan 300°; θ′ = 60°; $\tan 60° = \sqrt{3}$. In quadrant IV so tan 300° < 0: $\tan 300° = -\sqrt{3}$.

43. sin(−690°); −690° coterminal with 30°, $\sin(-690°) = \sin 30° = \frac{1}{2}$.

52. tan 527.2°, −0.2272
527.2 [tan]
TI–81: [TAN] 527.2 [ENTER]

65. $\tan\theta = -\frac{8}{5}$; $\theta = \tan^{-1}(-\frac{8}{5})$
$\approx -57.995° \approx -58.0°$
8 [÷] 5 [=] [±] [SHIFT] [tan]
TI–81: [2nd] [TAN] [(] [(−)] 8 [÷] 5 [)] [ENTER]

Exercise 5–5

Answers to odd-numbered problems

1. 55.6° **3.** 33.3° **5.** 200.0° **7.** 358°

9. 22.0° **11.** 168.7° **13.** 224.4°

In problems 15 through 21 part a answers are in the order sin, csc, cos, sec, tan, cot.

15. a. $\frac{4}{5}, \frac{5}{4}, -\frac{3}{5}, -\frac{5}{3}, -\frac{4}{3}, -\frac{3}{4}$
b. $\theta \approx 126.9°$

17. a. $-\frac{5}{13}, -\frac{13}{5}, \frac{12}{13}, \frac{13}{12}, -\frac{5}{12}, -\frac{12}{5}$
b. $\theta \approx 337.4°$

19. a. $-\frac{5\sqrt{41}}{41}, -\frac{\sqrt{41}}{5}, -\frac{4\sqrt{41}}{41}, -\frac{\sqrt{41}}{4}, \frac{5}{4}, \frac{4}{5}$
b. $\theta \approx 231.3°$

21. a. $\frac{3\sqrt{13}}{13}, \frac{\sqrt{13}}{3}, \frac{2\sqrt{13}}{13}, \frac{\sqrt{13}}{2}, \frac{3}{2}, \frac{2}{3}$
b. $\theta \approx 56.3°$

23. a.

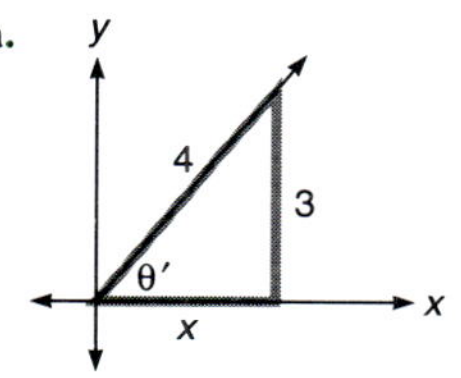

b. $\sin\theta = \frac{\sqrt{7}}{4}$, $\tan\theta = \frac{3\sqrt{7}}{7}$
c. $\theta = \theta' \approx 48.6°$

25. a.

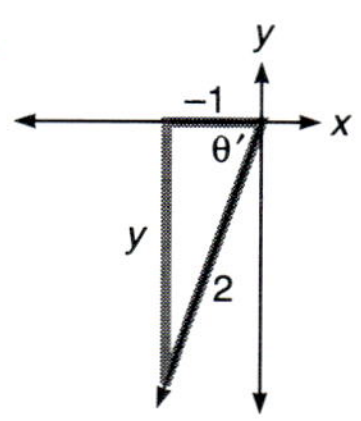

b. $\sin\theta = -\frac{\sqrt{3}}{2}$, $\tan\theta = \sqrt{3}$
c. $\theta' = 60°$, $\theta = 240°$

27. a.

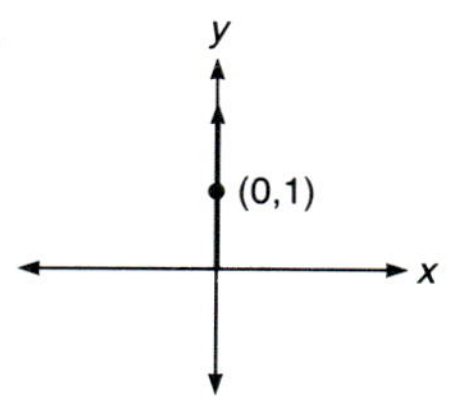

b. $\cos\theta = 0$, tan θ (undefined)
c. $\theta = 90°$

29. a.

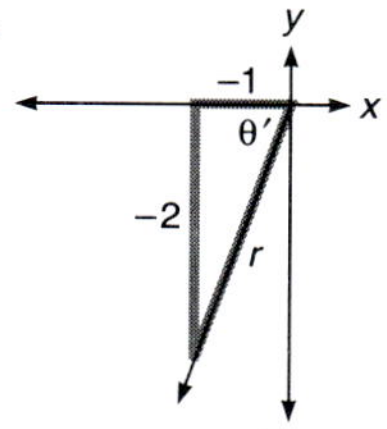

b. $\sin\theta = -\frac{2\sqrt{5}}{5}$, $\cos\theta = -\frac{\sqrt{5}}{5}$
c. $\theta' \approx 63.4°$, $\theta \approx 243.4°$

31. a.

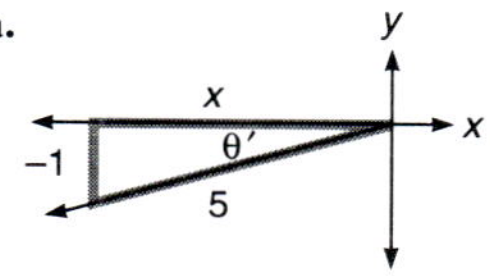

b. $\cos\theta = -\frac{2\sqrt{6}}{5}$,
$\tan\theta = \frac{\sqrt{6}}{12}$, $\sin\theta = -\frac{1}{5}$
c. $\theta' \approx 11.5°$, $\theta \approx 191.5°$

33. a.

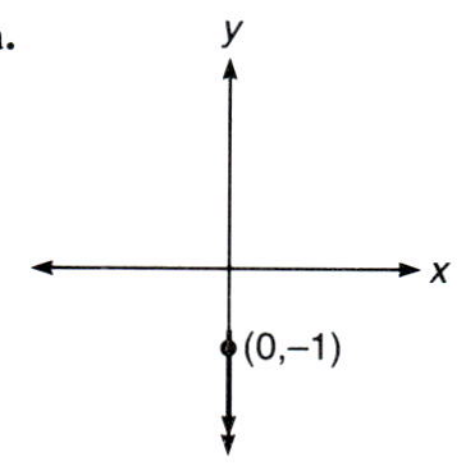

b. $\cos\theta = 0$, $\sin\theta = -1$, tan θ (undefined)
c. $\theta = 270°$

35. a.

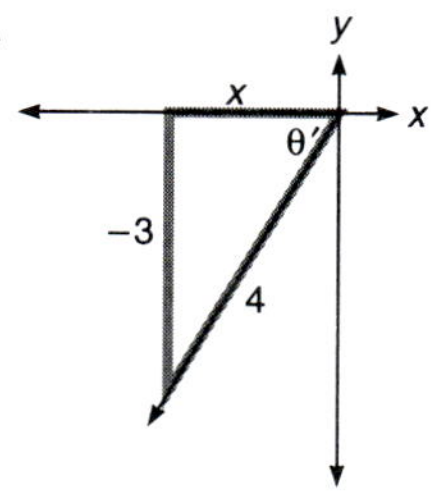

b. $\cos\theta = -\frac{\sqrt{7}}{4}$, $\tan\theta = \frac{3\sqrt{7}}{7}$
c. $\theta' \approx 48.6°$, $\theta \approx 228.6°$

37. a.

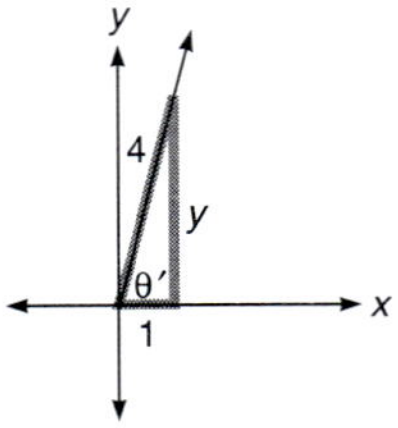

b. $\sin\theta = \dfrac{\sqrt{15}}{4}$,

$\tan\theta = \sqrt{15}$, $\cos\theta = \frac{1}{4}$

c. $\theta = \theta' \approx 75.5°$

39. a.

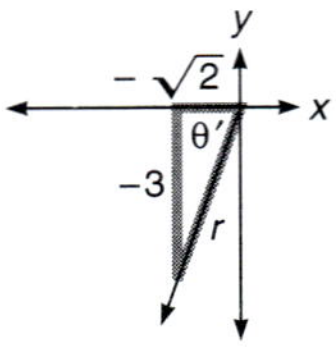

b. $\sin\theta = -\dfrac{3\sqrt{11}}{11}$,

$\cos\theta = -\dfrac{\sqrt{22}}{11}$, $\tan\theta = \dfrac{3\sqrt{2}}{2}$

c. $\theta' \approx 64.8°$, $\theta \approx 244.8°$

41. a.

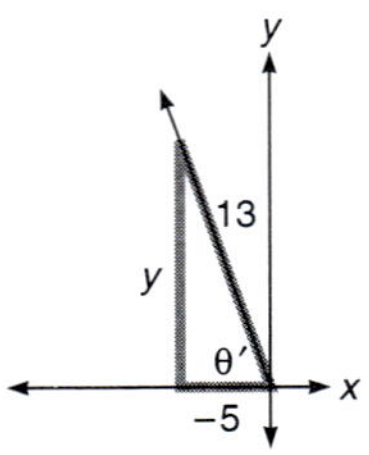

b. $\sin\theta = \frac{12}{13}$, $\tan\theta = -\frac{12}{5}$

c. $\theta' \approx 67.4°$, $\theta \approx 112.6°$

43. a.

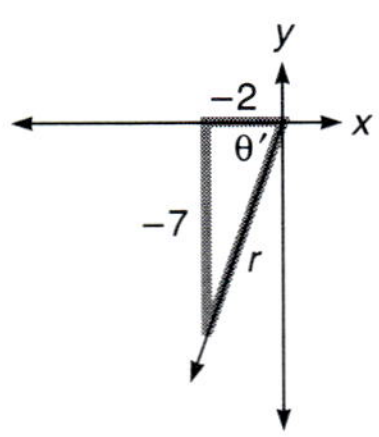

b. $\sin\theta = -\dfrac{7\sqrt{53}}{53}$, $\cos\theta = -\dfrac{2\sqrt{53}}{53}$

c. $\theta' \approx 74.1°$, $\theta \approx 254.1°$

45. a.

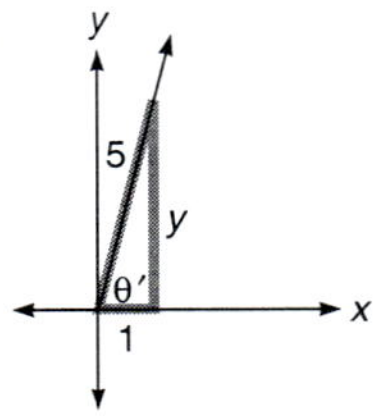

b. $\sin\theta = \dfrac{2\sqrt{6}}{5}$, $\tan\theta = 2\sqrt{6}$,

$\cos\theta = \frac{1}{5}$

c. $\theta = \theta' \approx 78.5°$

47. a.

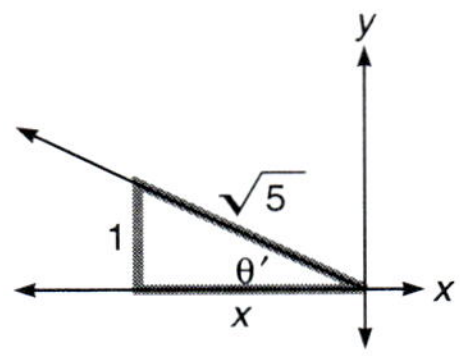

b. $\cos\theta = -\dfrac{2\sqrt{5}}{5}$, $\tan\theta = -\frac{1}{2}$

c. $\theta' \approx 26.6°$, $\theta \approx 153.4°$

49. $\sec\theta = \dfrac{1}{u}$, $\sin\theta = \sqrt{1-u^2}$,

$\csc\theta = \dfrac{1}{\sqrt{1-u^2}}$, $\tan\theta = \dfrac{\sqrt{1-u^2}}{u}$,

$\cot\theta = \dfrac{u}{\sqrt{1-u^2}}$

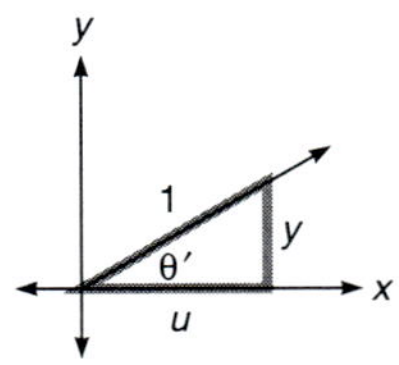

51. $\sec\theta = \dfrac{1}{u}$, $\sin\theta = -\sqrt{1-u^2}$,

$\csc\theta = -\dfrac{1}{\sqrt{1-u^2}}$,

$\tan\theta = \dfrac{-\sqrt{1-u^2}}{u}$,

$\cot\theta = \dfrac{-u}{\sqrt{1-u^2}}$

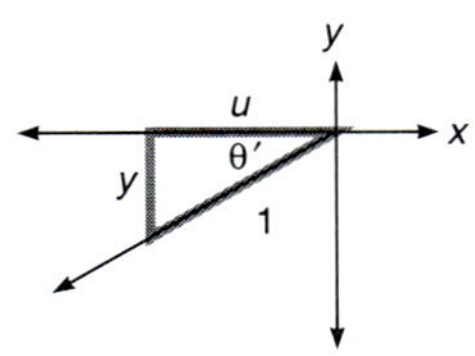

53. $\csc\theta = \dfrac{1}{u+1}$,

$\cos\theta = \sqrt{-u^2-2u}$,

$\sec\theta = \dfrac{1}{\sqrt{-u^2-2u}}$,

$\tan\theta = \dfrac{u+1}{\sqrt{-u^2-2u}}$,

$\cot\theta = \dfrac{\sqrt{-u^2-2u}}{u+1}$

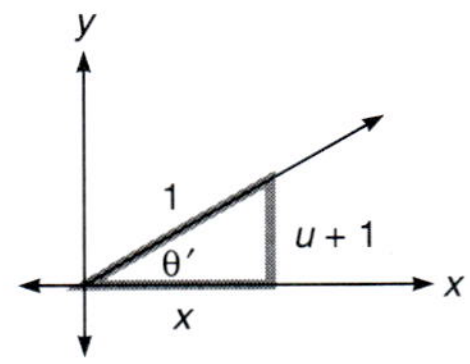

55. $y \approx 4.77$ mm, $x \approx -4.85$ mm

57. $y \approx -5.89$ cm, $x \approx -5.77$ cm

59. The x-coordinates are ±8.8.
The angles are 60.6°, 119.4°, 240.6°, 299.4°.

61. $x \approx -1'10.0''$, $y \approx -1'1.5''$

63. 663.4 ft

Solutions to skill and review problems

1. −250° coterminal with −250° + 360° = 110°. 110° is in quadrant II. Thus, $\theta' = 180° - \theta = 180° - 110° = 70°$.

2. $r = \sqrt{x^2+y^2} = \sqrt{3^2+(-4)^2} = 5$

$\cos\theta = \dfrac{x}{r} = \dfrac{3}{5}$

3. $B = 90° - 36.2° = 53.8°$

$\sin 36.2° = \dfrac{a}{10}$, $a = 10\sin 36.2° \approx 5.9$

$\cos 36.2° = \dfrac{b}{10}$, $b = 10\cos 36.2°$

≈ 8.1

4. $y = (x + 1)^3 - 1$ is the graph of $y = x^3$ shifted down 1 unit and to the left 1 unit.

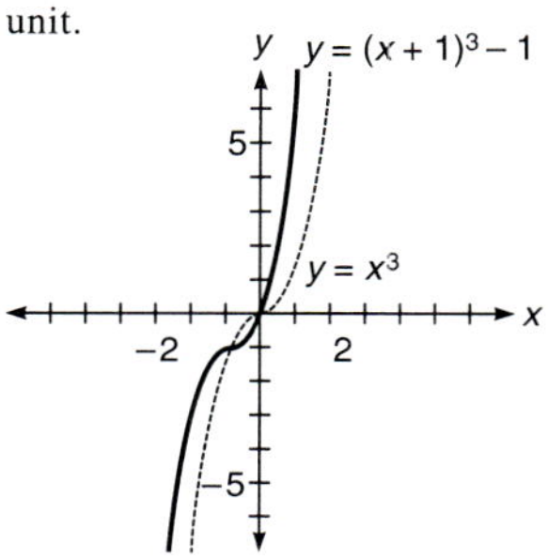

x-intercept:

(set $y = 0$) $0 = (x + 1)^3 - 1$

$1 = (x + 1)^3$

$1 = x + 1$

$0 = x$

y-intercept:

(set $x = 0$) $y = (0 + 1)^3 - 1$

$y = 0$

Additional points:

x	-2	-1	0	1
y	-2	-1	0	7

5. $\dfrac{3x - 5}{12} = 2(x - 3) - 8x$

$\dfrac{3x - 5}{12} = -6x - 6$

$\dfrac{3x - 5}{12} \cdot 12 = 12(-6x - 6)$

$3x - 5 = -72x - 72$

$75x - 5 = -72$

$75x = -67$

$x = -\dfrac{67}{75}$

6. $\dfrac{3x - 9}{x - 2} \leq 0$

This is a nonlinear inequality. We use the critical point/test point method.

Critical points:

Solve the equality: $\dfrac{3x - 9}{x - 2} = 0$

$3x - 9 = 0$

$3x = 9$

$x = 3$

Find zeros of denominators: $x - 2 = 0$

$x = 2$

Critical points are 2 and 3.

Use test points 0, $2\frac{1}{2}$, 4.

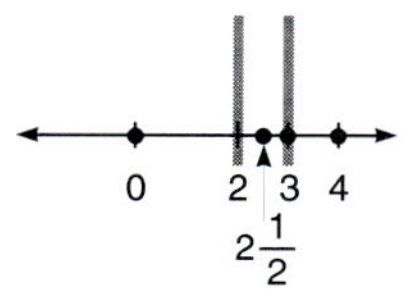

$\dfrac{3x - 9}{x - 2} \leq 0$

$x = 0$: $\frac{9}{2} \leq 0$, false

$x = 2.5$: $-3 \leq 0$, true

$x = 4$: $\frac{3}{2} \leq 0$, false

The solution is the interval between 2 and 3, along with the point $x = 3$.

$\{x \mid -2 < x \leq 3\}$ (set-builder notation)

(2,3] (interval notation)

33. a.

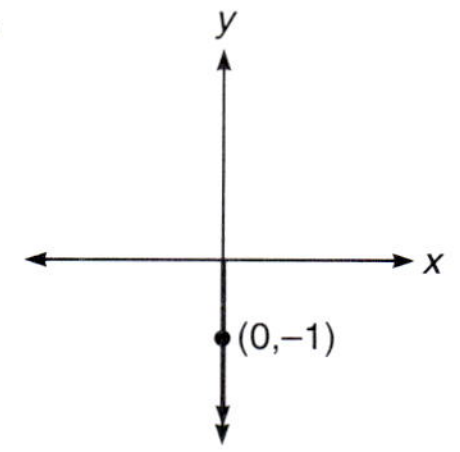

b,c. $\csc\theta = -1$, $\sin\theta = \dfrac{1}{\csc\theta} = \dfrac{1}{-1} = -1$

θ is 270°; pick a point, say $(0, -1)$ on the terminal side of the angle.

$r = \sqrt{x^2 + y^2} = \sqrt{0^2 + (-1)^2} = 1.$

$\cos\theta = \dfrac{x}{r} = \dfrac{0}{1} = 0;$

$\tan\theta = \dfrac{y}{x} = \dfrac{-1}{0}$ (undefined)

51. $\cos\theta = u$ and θ terminates in quadrant III.

$y = -\sqrt{1^2 - u^2} = -\sqrt{1 - u^2}$

$\sec\theta = \dfrac{1}{\cos\theta} = \dfrac{1}{u}$

$\sin\theta = \dfrac{\text{opp}}{\text{hyp}} = \dfrac{y}{1} = y = -\sqrt{1 - u^2}$

$\csc\theta = \dfrac{1}{\sin\theta} = -\dfrac{1}{\sqrt{1 - u^2}}$

$\tan\theta = \dfrac{\text{opp}}{\text{adj}} = \dfrac{y}{u} = \dfrac{-\sqrt{1 - u^2}}{u}$

$\cot\theta = \dfrac{-u}{\sqrt{1 - u^2}}$

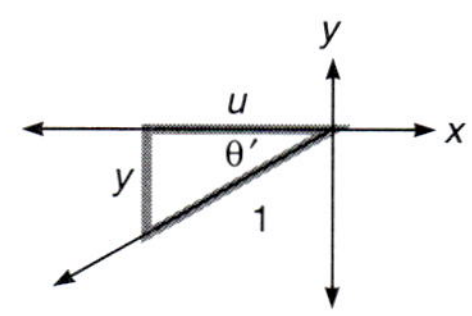

59. $x^2 + 15.5^2 = 17.8^2$, so $x \approx 8.8$.

Thus the x-coordinates are ± 8.8.

$\sin\theta_1 = \dfrac{15.5}{17.8}$, so $\theta_1 \approx 60.6°$.

Thus $\theta_1 \approx 60.6°$

$\theta_2 \approx 180° - 60.6° \approx 119.4°$

$\theta_3 \approx 180° + 60.6° \approx 240.6°$

$\theta_4 \approx 360° - 60.6° \approx 299.4°$

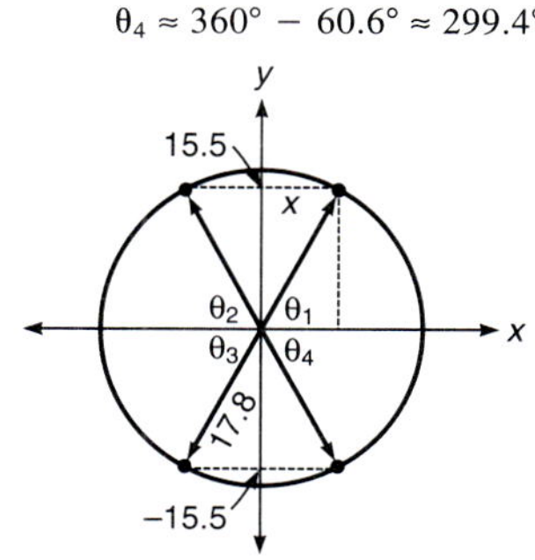

Solutions to trial exercise problems

13. $\cos\theta = -\frac{5}{7}$, $\tan\theta > 0$

$\cos\theta' = \frac{5}{7}$, so $\theta' \approx 44.4°$.

$\cos\theta < 0$, $\tan\theta > 0$ so θ is in quadrant III

$\theta = 180° + \theta' \approx 180° + 44.4° \approx 224.4°$

16. $(-5,-12)$; $r = \sqrt{(-5)^2 + (-12)^2} = 13$

a. $\sin\theta = \dfrac{y}{r} = -\dfrac{12}{13}$, $\qquad \csc\theta = \dfrac{1}{\sin\theta} = -\dfrac{13}{12}$

$\cos\theta = \dfrac{x}{r} = -\dfrac{5}{13} \qquad \sec\theta = \dfrac{1}{\cos\theta} = -\dfrac{13}{5}$

$\tan\theta = \dfrac{y}{x} = -\dfrac{-12}{-5} = \dfrac{12}{5} \qquad \cot\theta = \dfrac{1}{\tan\theta} = \dfrac{5}{12}$

$\sin\theta' = \dfrac{12}{13}$ so $\theta' \approx 67.4°$

b. $(-5,-12)$ is in quadrant III, so θ terminates in quadrant III. Thus $\theta = 180° + \theta' \approx 180° + 67.4° \approx 247.4°$.

61. $4'3.5'' = 4\frac{3.5}{12}'' \approx 4.292'$; $r \approx \frac{4.292}{2}$ ft ≈ 2.146 ft; $\theta = 211.5°$.

$\sin\theta = \frac{y}{r}$; $y = r\sin\theta$; $y \approx 2.146 \sin 211.5° \approx -1.121$ ft.

0.121 ft $\times$ $12''$/ft $\approx 1.5''$, so $y \approx -1'1.5''$

$\cos\theta = \frac{x}{r}$, $x = r\cos\theta$, $x \approx 2.146\cos 211.5° \approx -1.830$ ft

0.830 ft $\times$ $12''$/ft $\approx 10.0''$, so $x \approx -1'10.0''$

63. $\sin p = \frac{AB \sin b}{AP} = \frac{512.4 \cdot \sin 28.3°}{322.6} \approx 0.75302$, $p \approx 48.852°$

$a = 180° - (b + p) \approx 102.848°$

$BP = \frac{AP \sin a}{\sin b} \approx \frac{322.6 \cdot \sin 102.848°}{\sin 28.3°} \approx 663.4$ ft

Exercise 5–6

Answers to odd-numbered problems

1. $\tan\theta \cot\theta$

$\tan\theta \cdot \frac{1}{\tan\theta}$

1

3. $\cos\theta(1 - \sec\theta)$

$\cos\theta - \cos\theta \cdot \sec\theta$

$\cos\theta - \cos\theta \cdot \frac{1}{\cos\theta}$

$\cos\theta - 1$

5. $\sec\theta(\cot\theta + \cos\theta - 1)$

$\sec\theta \cdot \cot\theta + \sec\theta \cdot \cos\theta - \sec\theta$

$\frac{1}{\cos\theta} \cdot \frac{\cos\theta}{\sin\theta} + \frac{1}{\cos\theta} \cdot \cos\theta - \sec\theta$

$\frac{1}{\sin\theta} + 1 - \sec\theta$

$\csc\theta - \sec\theta + 1$

7. $\frac{\cos\alpha - \sin\alpha}{\cos\alpha}$

$\frac{\cos\alpha}{\cos\alpha} - \frac{\sin\alpha}{\cos\alpha}$

$1 - \tan\alpha$

9. $1 - \cos^2\theta$

$(\sin^2\theta + \cos^2\theta) - \cos^2\theta$

$\sin^2\theta$

11. $\cos\beta(\sec\beta - \cos\beta)$

$\cos\beta \cdot \sec\beta - \cos^2\beta$

$\cos\beta \cdot \frac{1}{\cos\beta} - \cos^2\beta$

$1 - \cos^2\beta$

$\sin^2\beta$

(See problem 9.)

13. $(\cos\theta + \sin\theta)(\cos\theta - \sin\theta) + 2\sin^2\theta$

$\cos^2\theta - \cos\theta\sin\theta + \sin\theta\cos\theta - \sin^2\theta + 2\sin^2\theta$

$\cos^2\theta + \sin^2\theta$

1

15. $\csc\alpha(\cos\alpha - \sin\alpha)$

$\csc\alpha \cdot \cos\alpha - \csc\alpha \cdot \sin\alpha$

$\frac{1}{\sin\alpha} \cdot \cos\alpha - \frac{1}{\sin\alpha} \cdot \sin\alpha$

$\frac{\cos\alpha}{\sin\alpha} - 1$

$\cot\alpha - 1$

17. $\frac{\sin x - \cos x}{\sin x}$

$\frac{\sin x}{\sin x} - \frac{\cos x}{\sin x}$

$1 - \cot x$

19. $\tan\beta(\cot\beta - \cos\beta)$

$\tan\beta \cdot \cot\beta - \tan\beta \cdot \cos\beta$

$\tan\beta \cdot \frac{1}{\tan\beta} - \frac{\sin\beta}{\cos\beta} \cdot \cos\beta$

$1 - \sin\beta$

21. a. $\sin^2\theta + \cos^2\theta = 1$

$\sin^2 16°50' + \cos^2 16°50' = 1$

$(0.28959)^2 + (0.95715)^2 = 1$

$1 = 1$

b. $\sin^2\theta + \cos^2\theta = 1$

$\sin^2 50° + \cos^2 50° = 1$

$(0.76604)^2 + (0.64279)^2 = 1$

$1 = 1$

The accuracy of these results depends on how much accuracy is used on a calculator.

23.

$\tan 32°40'$	$\frac{\sin 32°40'}{\cos 32°40'}$
0.64117	$\frac{0.53975}{0.84182} \approx 0.64117$

25. 60° **27.** 60° **29.** 11.5°
31. 77.5° **33.** 33.1° **35.** 10°
37. 30° **39.** 6.5° **41.** 30° **43.** 34.7°
45. 30° or 210° **47.** 0° or 120°
49. 0° or 180° **51.** 153.4°

Solutions to skill and review problems

1. a. $\cos\theta < 0$ and $\tan\theta < 0$, so θ is in quadrant II.

$y = +\sqrt{3^2 - (-1)^2} = \sqrt{8} = 2\sqrt{2}$.

$\tan\theta = \frac{\text{opp}}{\text{adj}} = \frac{y}{-1} = -2\sqrt{2}$

b. $\cos\theta' = \frac{1}{3}$, so $\theta' \approx 70.5°$

$\theta = 180° - \theta' \approx 180° - 70.5° \approx 109.5°$.

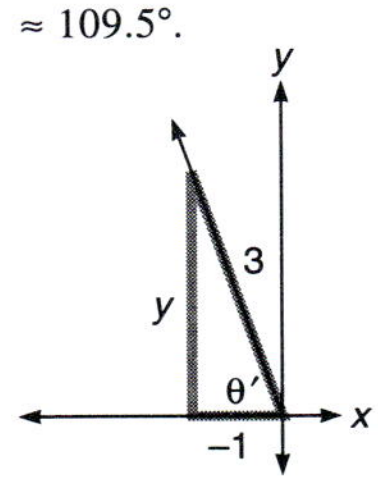

2. $\cos\theta = \frac{5}{28}$

$\theta = \cos^{-1}\frac{5}{28} \approx 79.7°$

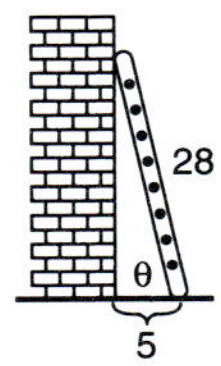

3. $C = 2\pi r$

$28.5 = 2\pi r$

$r = \frac{28.5}{2\pi} \approx 4.5$ ft

4. $7x^2 + 14x - 10 = 6x^2 + 12x + 5$

$x^2 + 2x - 15 = 0$

$(x + 5)(x - 3) = 0$

$x + 5 = 0$ or $x - 3 = 0$

$x = -5$ or $x = 3$

$x = -5$ or 3

Solutions to trial exercise problems

42. $-3\sin 2x = 0.75$

$\sin 2x = -0.25$

$(2x)' = \sin^{-1}(0.25)$

$(2x)' \approx 14.18°$

$2x \approx 194.48°$

(Least positive solution in quadrant II.)

$x \approx 97.2°$

46. $2\sin^2\theta + \sin\theta - 1 = 0$

$(2\sin\theta - 1)(\sin\theta + 1) = 0$

$2\sin\theta - 1 = 0$ or $\sin\theta + 1 = 0$

$\sin\theta = \frac{1}{2}$ or $\sin\theta = -1$

$\theta = 30°$ or $270°$

Chapter 5 review

1. 165.783° **2.** 37.3° **3.** 78.2°
4. $\sqrt{34} \approx 5.8$ **5.** $10\sqrt{3} \approx 17.3$
6. $2\sqrt{23} \approx 9.6$ **7.** 13 **8.** 23.6 feet
9. 121.0 knots **10.** 0.7466
11. 0.6080 **12.** 0.1823 **13.** 1.0033
14. −4.18 **15.** 12.42° **16.** 24.94°
17. $A = 59.7°$, $c \approx 14.0$, $b \approx 7.1$
18. $B = 68.1°$, $a \approx 4.7$, $b \approx 11.7$
19. $c \approx 29.2$, $A \approx 59.1°$, $B \approx 30.9°$
20. $a \approx 8.72$, $A \approx 57.0°$, $B \approx 33.0°$
21. $R \approx 54.54$ ohms, $\theta \approx 24.6°$
22. 79 feet **23.** 120° **24.** 220°
25. 176° **26.** 331°15′ **27.** III
28. IV **29.** 27.4° **30.** 7.7°
31. 13.22° **32.** 69.40° **33.** 70°
34. $\frac{\sqrt{2}}{2}$ **35.** 0.6626 **36.** $-\sqrt{3}$
37. $-\frac{2\sqrt{3}}{3}$ **38.** −2.7852 **39.** 778.7 ft
40. 58.3° **41.** 234.4°

Part a of 42, 43, and 44 is in this order: sin, csc, cos, sec, tan, cot.

42. a. $-\frac{3\sqrt{10}}{10}, -\frac{\sqrt{10}}{3}, \frac{\sqrt{10}}{10}, \sqrt{10}, -3, -\frac{1}{3}$
b. $\theta \approx 288.4°$

43. a. $\frac{2\sqrt{5}}{5}, \frac{\sqrt{5}}{2}, -\frac{\sqrt{5}}{5}, -\sqrt{5}, -2, -\frac{1}{2}$
b. $\theta \approx 116.6°$

44. a. $-\frac{\sqrt{6}}{9}, -\frac{3\sqrt{6}}{2}, -\frac{5\sqrt{3}}{9}, -\frac{3\sqrt{3}}{5}, \frac{\sqrt{2}}{5}, \frac{5\sqrt{2}}{2}$
b. $\theta \approx 195.8°$

45. a.

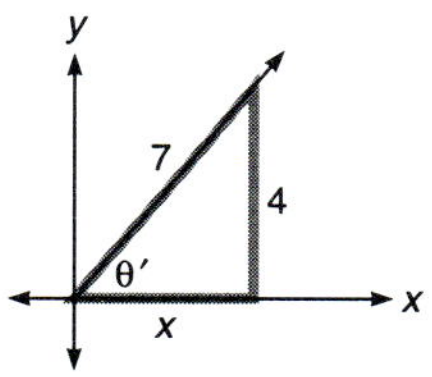

b. $\cos\theta = \frac{\sqrt{33}}{7}$, $\tan\theta = \frac{4\sqrt{33}}{33}$
c. $\theta \approx 34.8°$

46. a.

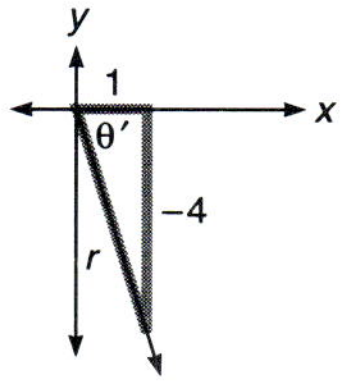

b. $\sin\theta = -\frac{4\sqrt{17}}{17}$, $\cos\theta = \frac{\sqrt{17}}{17}$
c. $\theta \approx 284.0°$

47. a.

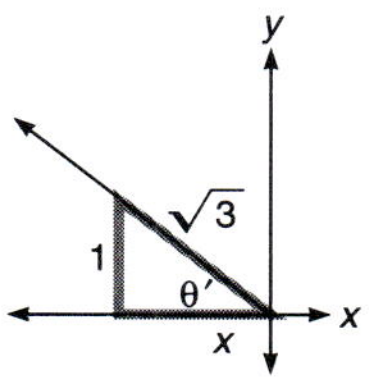

b. $\sin\theta = \frac{1}{\sqrt{3}}$, $\cos\theta = \frac{\sqrt{6}}{3}$, $\tan\theta = \frac{\sqrt{2}}{2}$
c. $\theta \approx 144.7°$

48. $\sin\theta = \sqrt{1-u^2}$, $\tan\theta = \frac{\sqrt{1-u^2}}{u}$

49. $\sin\theta = \frac{-u}{\sqrt{u^2+1}}$, $\cos\theta = \frac{-1}{\sqrt{u^2+1}}$

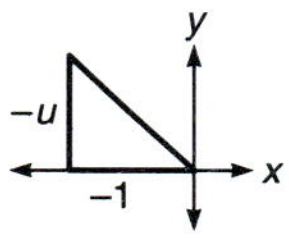

50. $\sin\theta\csc\theta$
$\sin\theta \cdot \frac{1}{\sin\theta}$
1

51. $\sec\alpha(\cos\alpha - \cot\alpha)$
$\sec\alpha \cdot \cos\alpha - \sec\alpha \cdot \cot\alpha$
$\frac{1}{\cos\alpha} \cdot \cos\alpha - \frac{1}{\cos\alpha} \cdot \frac{\cos\alpha}{\sin\alpha}$
$1 - \frac{1}{\sin\alpha}$
$1 - \csc\alpha$

52. $\frac{\sin\theta + 1}{\sin\theta}$
$\frac{\sin\theta}{\sin\theta} + \frac{1}{\sin\theta}$
$1 + \csc\theta$

53. $\frac{\sin\beta - 1}{\cos\beta}$
$\frac{\sin\beta}{\cos\beta} - \frac{1}{\cos\beta}$
$\tan\beta - \sec\beta$

54. $\cos\theta(\sec\theta - \cos\theta)$
$\cos\theta \cdot \sec\theta - \cos^2\theta$
$\cos\theta \cdot \frac{1}{\cos\theta} - \cos^2\theta$
$1 - \cos^2\theta$
$\sin^2\theta + \cos^2\theta - \cos^2\theta$
$\sin^2\theta$

55. $\cot x\left(\sec x - \tan x + \frac{1}{\cot^2 x}\right)$
$\cot x \cdot \sec x - \cot x \cdot \tan x + \frac{\cot x}{\cot^2 x}$
$\frac{\cos x}{\sin x} \cdot \frac{1}{\cos x} - 1 + \frac{1}{\cot x}$
$\frac{1}{\sin x} - 1 + \tan x$
$\csc x - 1 + \tan x$

56. $(\sin\alpha - \cos\alpha)(\csc\alpha + \sec\alpha)$
$\sin\alpha \cdot \csc\alpha + \sin\alpha \cdot \sec\alpha - \cos\alpha \cdot \csc\alpha - \cos\alpha \cdot \sec\alpha$
$\sin\alpha \cdot \frac{1}{\sin\alpha} + \sin\alpha \cdot \frac{1}{\cos\alpha} - \cos\alpha \cdot \frac{1}{\sin\alpha} - \cos\alpha \cdot \frac{1}{\cos\alpha}$
$1 + \frac{\sin\alpha}{\cos\alpha} - \frac{\cos\alpha}{\sin\alpha} - 1$
$\tan\alpha - \cot\alpha$

57. 30° **58.** 30° **59.** 59.0°
60. 26.6° **61.** 40° **62.** 18.4°
63. $\theta = 60°$ or 180°

Chapter 5 test

1. 24.7° **2.** $\sqrt{19}$ **3.** 26 feet
4. 262 knots **5.** 0.4586 **6.** 0.1944
7. 1.2723 **8.** 1.3380 **9.** −0.11
10. 29.4° **11.** $A = 70.7°$, $c \approx 251.1$, $a \approx 237.0$ **12.** $a \approx 93.5$, $A \approx 48.4°$, $B \approx 41.6°$ **13.** 28.3° **14.** 315°
15. II **16.** 31° **17.** 12.1°
18. −0.7683 **19.** $\sqrt{2}$ **20.** −0.7813
21. 1.5557 **22.** 342.7 meters
23. 336.4°

24. a. $\sin\theta = -\frac{\sqrt{5}}{5}$, $\csc\theta = -\sqrt{5}$,
$\cos\theta = \frac{2\sqrt{5}}{5}$, $\sec\theta = \frac{\sqrt{5}}{2}$,
$\tan\theta = -\frac{1}{2}$, $\cot\theta = -2$
b. $\theta \approx 333.4°$

25. a. $\sin\theta = \frac{\sqrt{6}}{6}$, $\csc\theta = \sqrt{6}$,

$\cos\theta = -\frac{\sqrt{30}}{6}$, $\sec\theta = -\frac{\sqrt{30}}{5}$,

$\tan\theta = -\frac{\sqrt{5}}{5}$, $\cot\theta = -\sqrt{5}$

b. $\theta \approx 155.9°$

26. a.

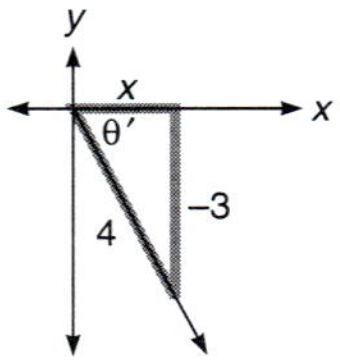

b. $\sin\theta = -\frac{3}{4}$, $\cos\theta = \frac{\sqrt{7}}{4}$,

$\tan\theta = -\frac{3\sqrt{7}}{7}$

c. $\theta \approx 311.4°$

27. $\sin\theta = -\frac{u}{\sqrt{u^2+4}}$, $\csc\theta = -\frac{\sqrt{u^2+4}}{u}$,

$\cos\theta = -\frac{2}{\sqrt{u^2+4}}$, $\sec\theta = -\frac{\sqrt{u^2+4}}{2}$

$\tan\theta = \frac{u}{2}$, $\cot\theta = \frac{2}{u}$

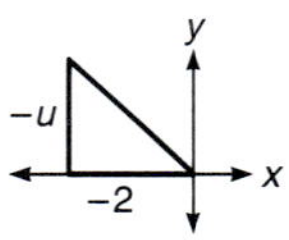

28. $\tan\theta\cot\theta$

$\tan\theta \cdot \frac{1}{\tan\theta}$

1

29. $\sec\theta(\cos\theta - \cos^3\theta)$

$\sec\theta \cdot \cos\theta - \sec\theta \cdot \cos^3\theta$

$\frac{1}{\cos\theta} \cdot \cos\theta - \frac{1}{\cos\theta} \cdot \cos^3\theta$

$1 - \cos^2\theta$

$\sin^2\theta + \cos^2\theta - \cos^2\theta$

$\sin^2\theta$

30. $(\sin\theta + \cos\theta)^2 - \sin\theta\cos\theta$

$\sin^2\theta + 2\sin\theta\cos\theta + \cos^2\theta - \sin\theta\cos\theta$

$\sin^2\theta + \cos^2\theta + 2\sin\theta\cos\theta - \sin\theta\cos\theta$

$1 + \sin\theta\cos\theta$

31. 45° **32.** 121.0° **33.** 25.8°

34. 30° or 150°

Chapter 6

Exercise 6–1

Answers to odd-numbered problems

1. $x^2 + y^2 = 1$ **3.** $\frac{\pi}{4} \approx 0.79$

5. $\frac{5\pi}{9} \approx 1.75$ **7.** $-\frac{5\pi}{3} \approx -5.24$

9. $\frac{3\pi}{2} \approx 4.71$ **11.** $\frac{127\pi}{180} \approx 2.22$

13. $-\frac{61\pi}{36} \approx -5.32$ **15.** 330° **17.** 108°

19. 40° **21.** −510° **23.** $\frac{270°}{\pi} \approx 85.94°$

25. $-\frac{2{,}160°}{17\pi} \approx -40.4°$ **27.** $\frac{360°}{\pi} \approx 114.6°$

29. $-\frac{900°}{\pi} \approx -286.5°$ **31.** 0.7833

33. 0.5463 **35.** 1.5523 **37.** 0.7457

39. 1.4235 **41.** 1.6709 **43.** $\frac{\sqrt{3}}{2}$

45. $\frac{\sqrt{3}}{2}$ **47.** $-\frac{1}{2}$ **49.** $-\frac{\sqrt{2}}{2}$

51. $-\sqrt{3}$ **53.** 2.7 radians **55.** 6.5 inches **57.** 24.0 in. **59.** 4.30

61. a. 168.3 volts **b.** −195.5 volts **c.** −132.7 volts **d.** −198.4 volts

63. a. −0.030 **b.** −0.366 **65.** Values obtained with TI-81 calculator.

a. 0.6967025778 **b.** 0.5402777778

c. 0.2673002653 **d.** 0.984807753

67. $\frac{343}{18}\pi \approx 59.86$ mm² **69.** 30 in.²

71. $\frac{64}{5}\pi \approx 40.21$ in.² **73.** $\frac{27}{8}\pi \approx 10.60$ mm²

75. 1.24 radians, 71.12°

Solutions to skill and review problems

1. $\frac{\csc x - 1}{\csc x}$

$\frac{\csc x}{\csc x} - \frac{1}{\csc x}$

$1 - \sin x$

2. $A = (n_1 - 1) + b(n_2 - n_1)$

$A = n_1 - 1 + bn_2 - bn_1$

$A - n_1 + 1 + bn_1 = bn_2$

$\frac{A + n_1(b-1) + 1}{b} = n_2$

3. $A = (n_1 - 1) + b(n_2 - n_1)$

$A = n_1 - 1 + bn_2 - bn_1$

$A + 1 - bn_2 = n_1 - bn_1$

$A + 1 - bn_2 = n_1(1 - b)$

$\frac{A + 1 - bn_2}{1 - b} = n_1$

4. $\sin\theta = \frac{y}{r}$, so $\sin 30° = \frac{b}{6}$;

$b = 6\sin 30° = 6 \cdot \frac{1}{2} = 3$

$\cos\theta = \frac{x}{r}$, so $\cos 30° = \frac{a}{6}$;

$a = 6\cos 30° = 6 \cdot \frac{\sqrt{3}}{2} = 3\sqrt{3}$.

5. Solve the inequality $\left|\frac{1}{2}x + 10\right| > 8$.

$\frac{1}{2}x + 10 > 8$ or $\frac{1}{2}x + 10 < -8$

$\frac{1}{2}x > -2$ or $\frac{1}{2}x < -18$

$x > -4$ or $x < -36$

$\{x \mid x > -4 \text{ or } x < -36\}$

Solutions to trial exercise problems

12. $\frac{s}{\pi} = \frac{-422°}{180°}$; $s = \frac{-422° \cdot \pi}{180°} = -\frac{211\pi}{90} \approx -7.37$

21. $\theta° = \frac{180°}{\pi} \cdot \left(-\frac{17\pi}{6}\right) = -510°$

29. $\theta° = \frac{180°}{\pi} \cdot (-5) = -\frac{900}{\pi} \approx -286.5°$

40. $\sec 5.2 = \frac{1}{\cos 5.2} \approx 2.1344$

44. $\frac{5\pi}{4}$ is in quadrant III, so $\tan\frac{5\pi}{4} > 0$ and θ' is $\frac{5\pi}{4} - \pi = \frac{\pi}{4}$.

$\tan\frac{\pi}{4} = 1$, so $\tan\frac{5\pi}{4} = 1$.

45. $\frac{11\pi}{6}$ is in quadrant IV, so $\cos\frac{11\pi}{6} > 0$, and $\theta' = 2\pi - \frac{11\pi}{6} = \frac{\pi}{6}$.

$\cos\frac{\pi}{6} = \frac{\sqrt{3}}{2}$, so $\cos\frac{11\pi}{6} = +\frac{\sqrt{3}}{2}$.

48. $\frac{5\pi}{6}$ is in quadrant II so $\sin\frac{5\pi}{6} > 0$, and $\theta' = \pi - \frac{5\pi}{6} = \frac{\pi}{6}$.

$\sin\frac{\pi}{6} = \frac{1}{2}$, so $\sin\frac{5\pi}{6} = +\frac{1}{2}$.

56. $L = 14.5$ mm, $r = \frac{10.3}{2} = 5.15$ mm

$L = rs$; $14.5 = 5.15s$; $s \approx 2.816$ radians

$\frac{\theta°}{180°} = \frac{s}{\pi}$; $\frac{\theta°}{180°} = \frac{2.816}{\pi}$;

$\theta° = \frac{2.816 \cdot 180°}{\pi} \approx 161.3°$

57. Find L where $r = \dfrac{32.4}{2} = 16.2$ inches and $\theta° = 85°$.

$85° = \dfrac{17}{36}\pi$

$L = rs$

$L = 16.2 \cdot \dfrac{17\pi}{36} \approx 24.0$ in.

73. $A_p = \dfrac{\theta°(\pi r^2)}{360°} = \dfrac{15° \cdot \pi \cdot 9^2}{360°} = \dfrac{27}{8}\pi$
$\approx 10.60 \text{ mm}^2$

Exercise 6–2

Answers to odd-numbered problems

1. See figures 6–10, 6–13, and 6–16.

3. a. $\dfrac{\pi}{2} + 2k\pi$ **b.** $\dfrac{3\pi}{2} + 2k\pi$

c. $k\pi$ **5.** $k\pi$

7. a. $\dfrac{\pi}{3} + 2k\pi, \dfrac{5\pi}{3} + 2k\pi$

b. $\dfrac{5\pi}{6} + 2k\pi, \dfrac{7\pi}{6} + 2k\pi$

c. $\dfrac{3\pi}{4} + 2k\pi, \dfrac{5\pi}{4} + 2k\pi$

9. a. $-\dfrac{1}{2}$ **b.** $\dfrac{\sqrt{3}}{2}$ **c.** $-\dfrac{\sqrt{3}}{3}$

11. a. $-\dfrac{\sqrt{3}}{2}$ **b.** $-\dfrac{1}{2}$ **c.** $\sqrt{3}$

13. Amplitude is 5.

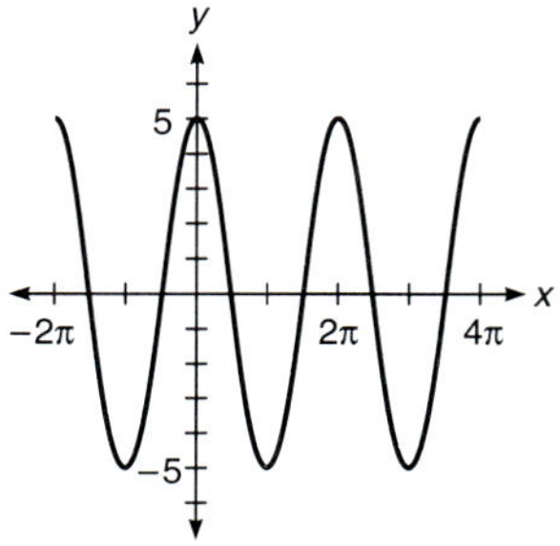

15.

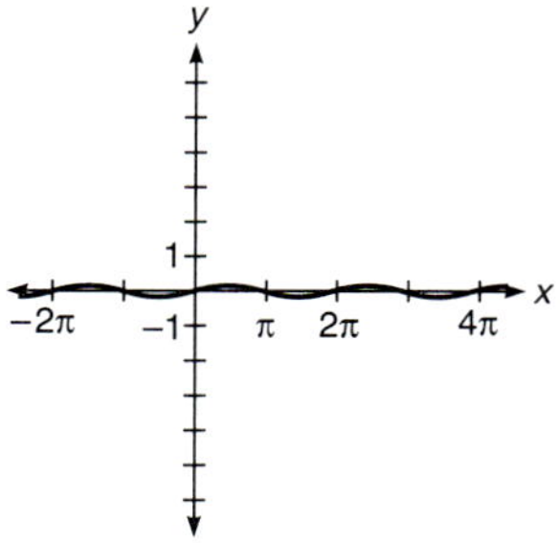

17.

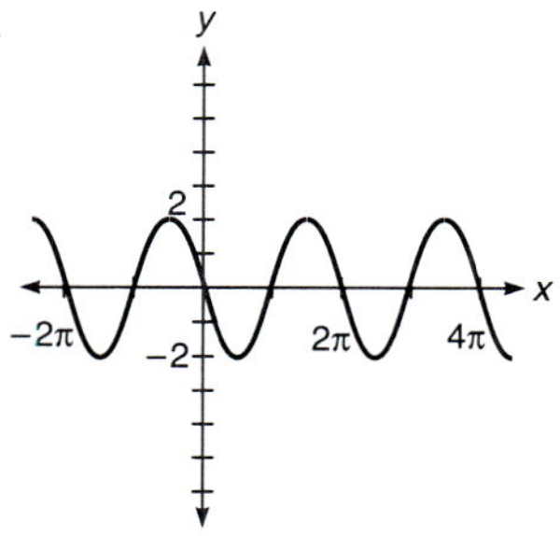

19.

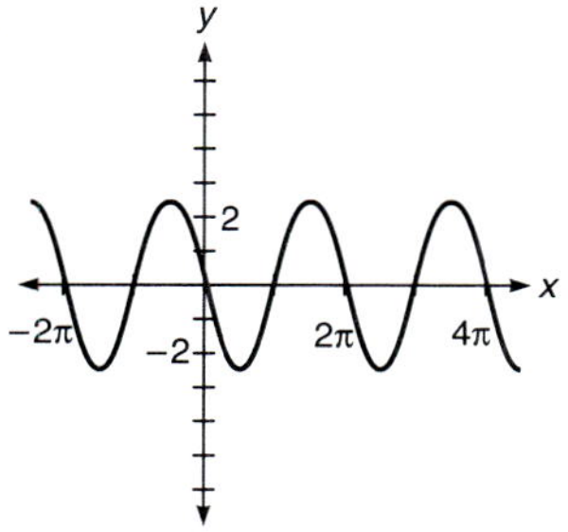

21.

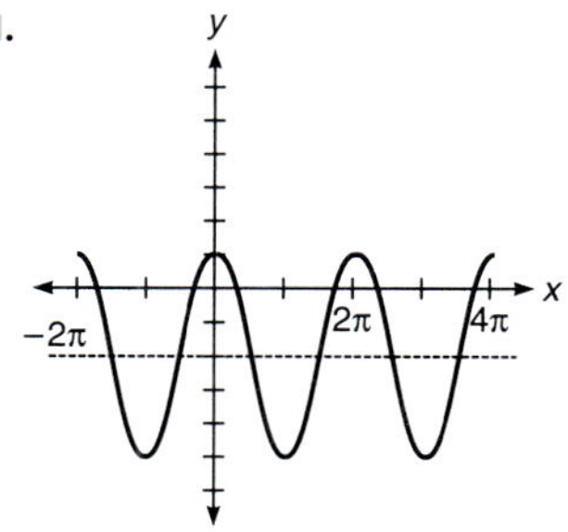

23.

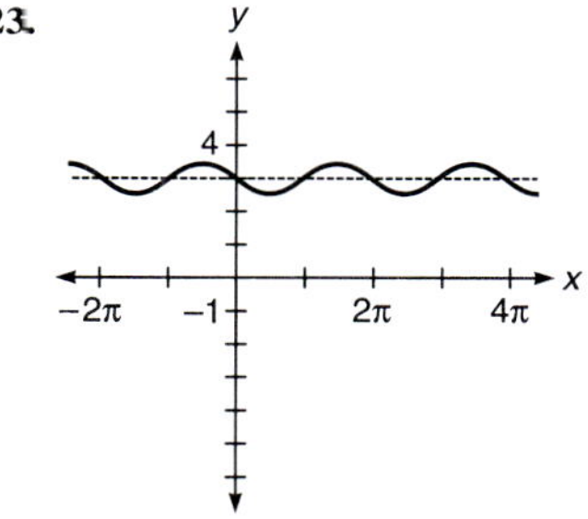

25. Amplitude is 3, phase shift is 0, period is 4π.

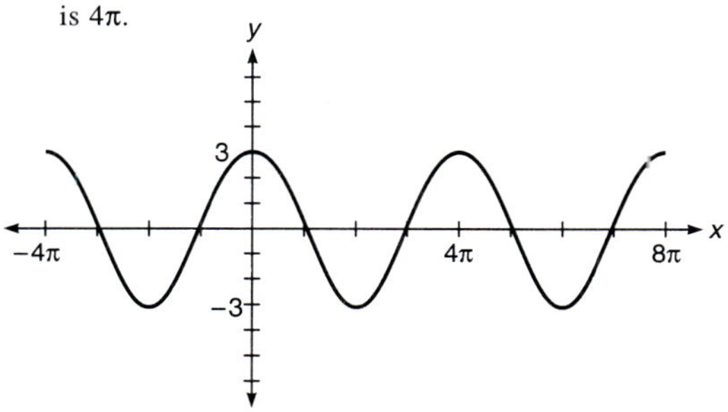

27. Amplitude is 3, phase shift is $-\dfrac{\pi}{2}$, period is π.

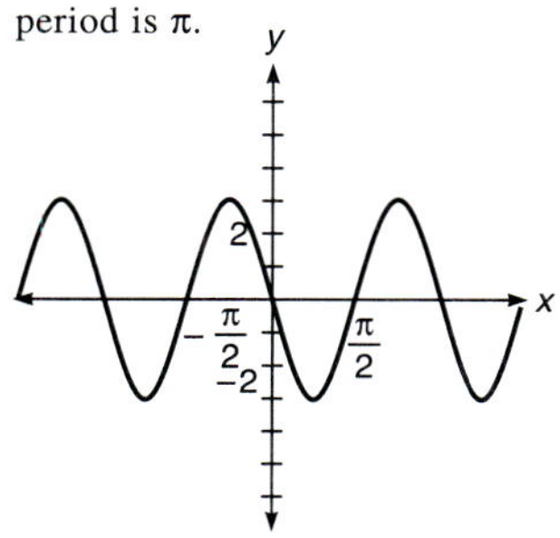

29. Amplitude is $\dfrac{5}{8}$, phase shift is 0, period is $\dfrac{2\pi}{5}$.

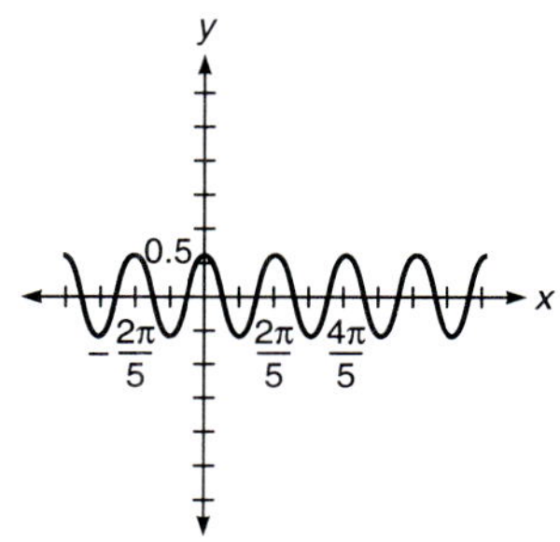

31. Amplitude is 1, phase shift is 0, period is 2π.

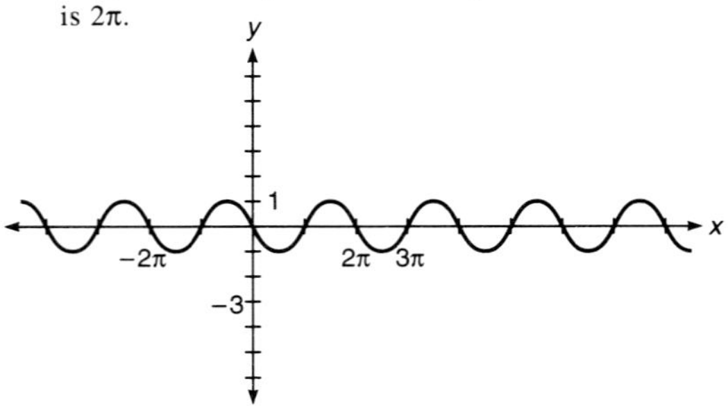

33. Amplitude is 1, phase shift is $\frac{\pi}{9}$, period is $\frac{2}{3}\pi$.

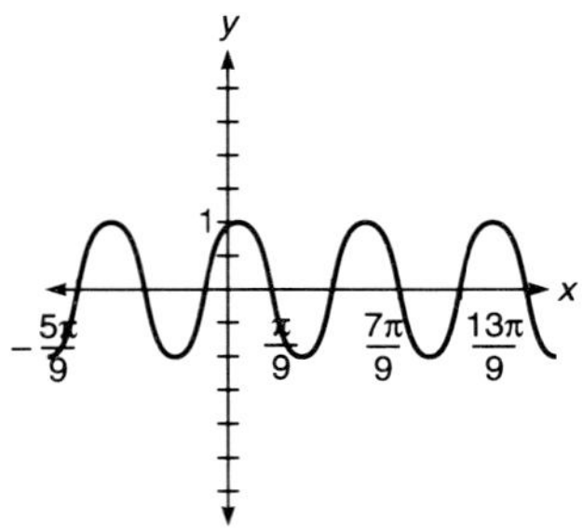

35. Amplitude is 1, phase shift is $\frac{\pi}{2}$, period is π.

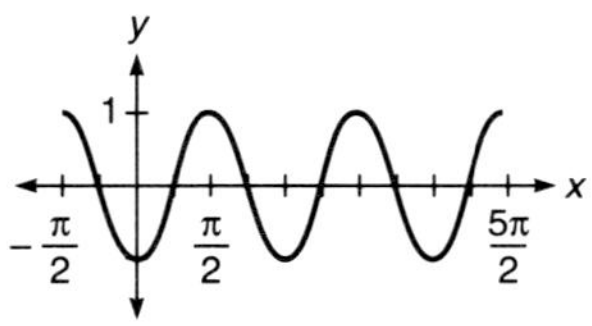

37. Amplitude is 1, phase shift is 0, period is 2.

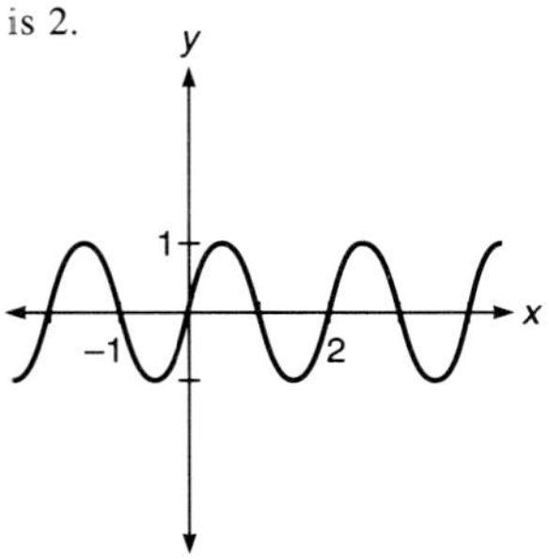

39. Amplitude is 3, phase shift is 0, period is π; vertical shift 3 units downward.

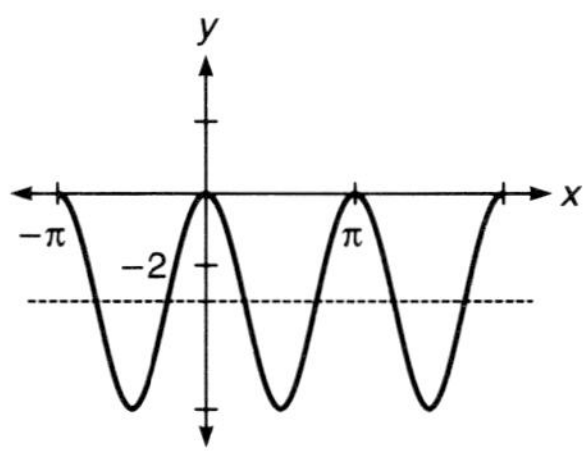

41. Amplitude is 1, phase shift is 0, period is $\frac{\pi}{2}$; vertical shift up 1 unit.

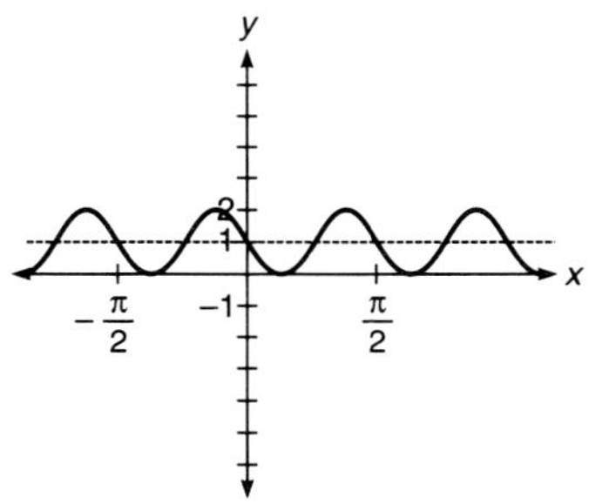

43. Amplitude is 3, phase shift is $-\frac{\pi}{3}$, period is $\frac{2\pi}{3}$; vertical shift downward 3 units.

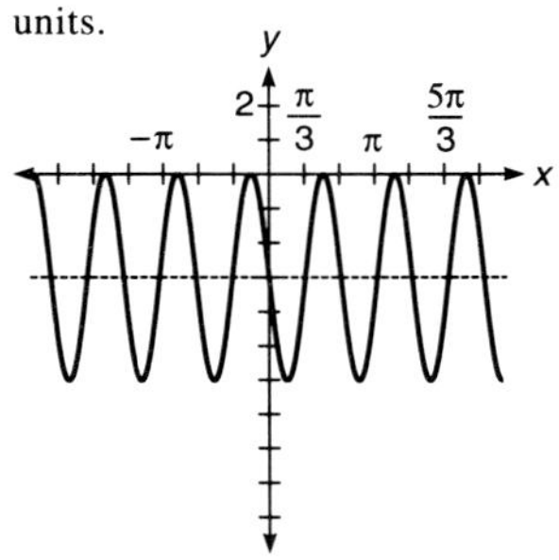

45. Amplitude is 2; phase shift is 0, period is 4; vertical shift is 2 units downward.

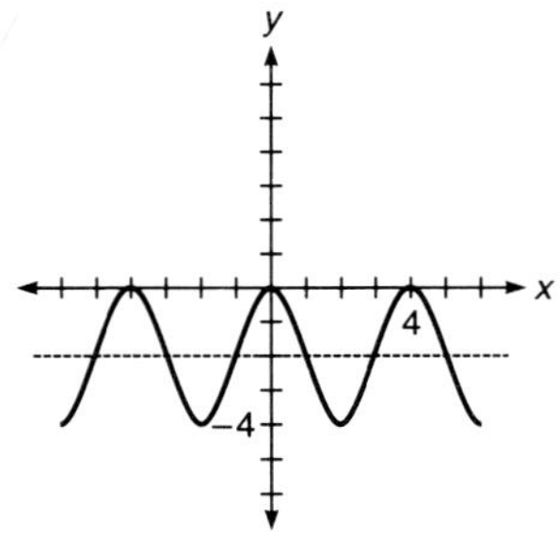

47. $y = \cos x$ **49.** $y = \sin 5x$

51. $y = \cos(2x - 4)$

53. $y = \cos 2x + 4$

55. $y = -2\sin\left(\frac{x}{3} + \pi\right)$

57. $y = \cos 2x$

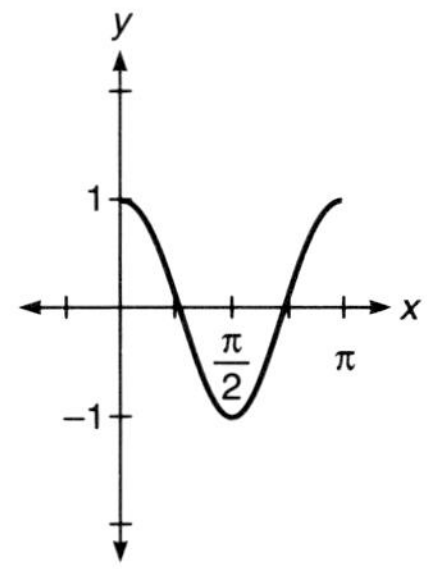

59. $y = -2\sin(2x - \pi)$

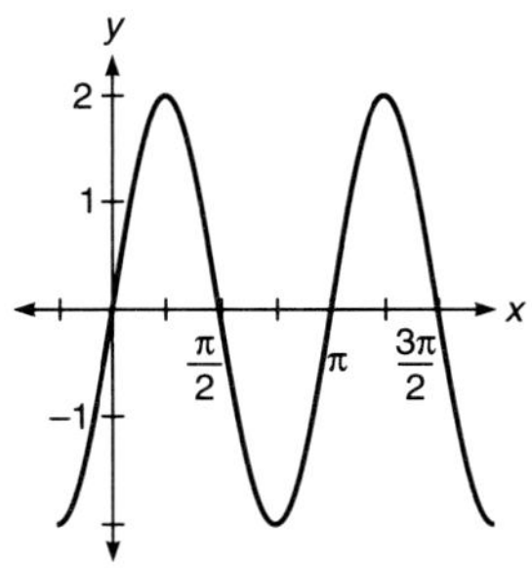

61. $y = -\cos \pi x$

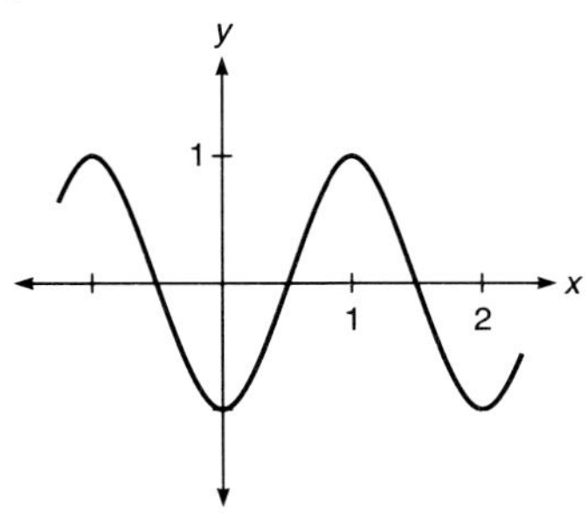

63. $y = 2\cos(3\pi x + 2)$

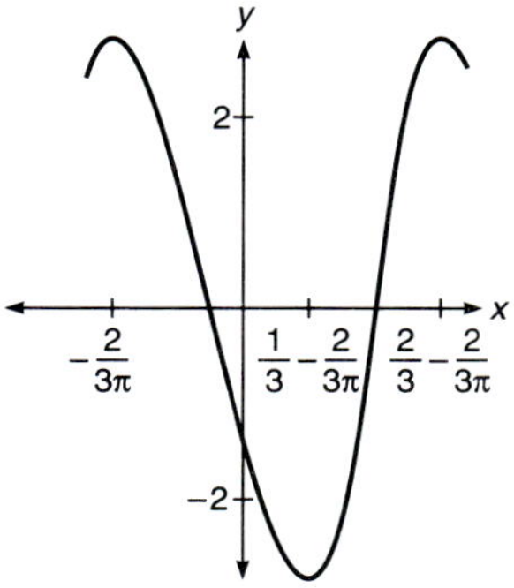

65. $y = 5\sin\left(\frac{\pi}{2}x + \frac{\pi}{2}\right)$

67. $y = 2\sin\frac{2x}{3} + 3$ **69.** $y = 5\cos\frac{\pi x}{2}$

71. $y = 2\cos\left(\frac{2}{3}x - \frac{\pi}{2}\right) + 3$

73.

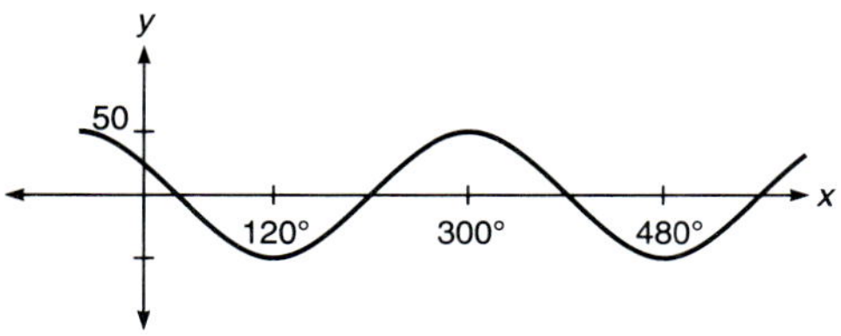

75.

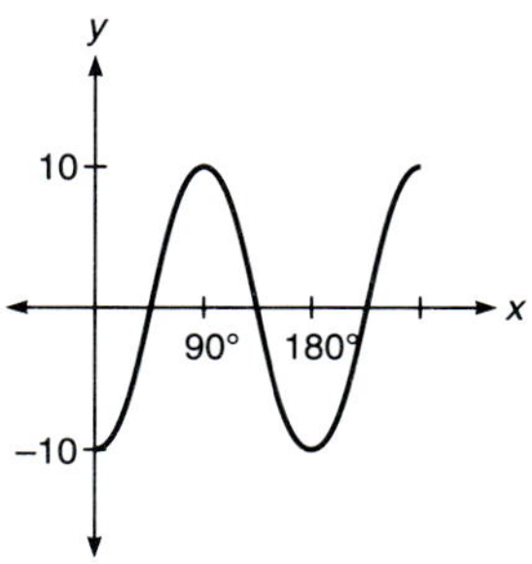

77. $y = 6\sin\left(\frac{x}{2} + 90°\right)$

79. $y = \sin(11x - 990°) + 2$

81.

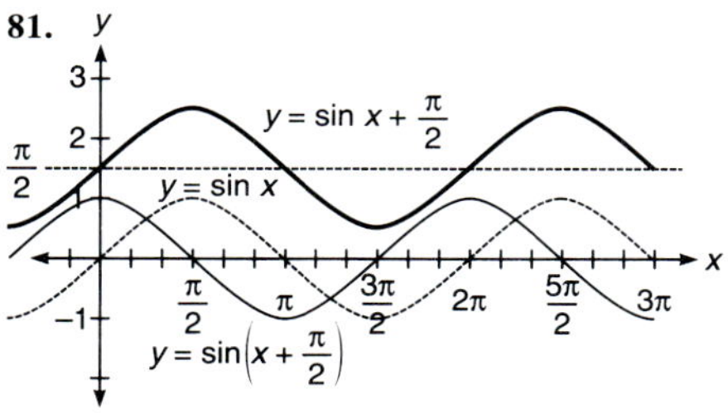

83. odd; symmetry across the origin
85. even; symmetry about the y-axis
87. even; symmetry about the y-axis
89. odd; symmetry across the origin
91. even; symmetry about the y-axis
93. odd; symmetry across the origin
95. odd; symmetry across the origin
97. odd; symmetry across the origin

Solutions to skill and review problems

1. $\frac{\theta°}{180°} = \frac{s}{\pi}$

$\theta° = \frac{180°}{\pi}s$

$\theta° = \frac{180°}{\pi}\cdot\frac{5\pi}{8}$

$\theta° = 112.5°$

2. $\frac{23}{12}\pi \approx 6.02$ radians

3. $140° = \frac{7}{9}\pi$

$L = rs$

$L = 16.4\cdot\frac{7}{9}\pi \approx 40.1$ mm

4. Using a reference triangle we learn that $\tan\theta' = -\frac{\sqrt{3}}{\sqrt{22}}$. Also, for any point (x,y) on the terminal side of an angle α, $\tan\alpha = \frac{y}{x}$. For $\theta' = \frac{b}{a} = \frac{-3}{a}$.

Thus, $\frac{-3}{a} = -\frac{\sqrt{3}}{\sqrt{22}}$, so $a\sqrt{3} = 3\sqrt{22}$,

so $a = \frac{3\sqrt{22}}{\sqrt{3}}\cdot\frac{\sqrt{3}}{\sqrt{3}} = \frac{3\sqrt{66}}{3} = \sqrt{66}$.

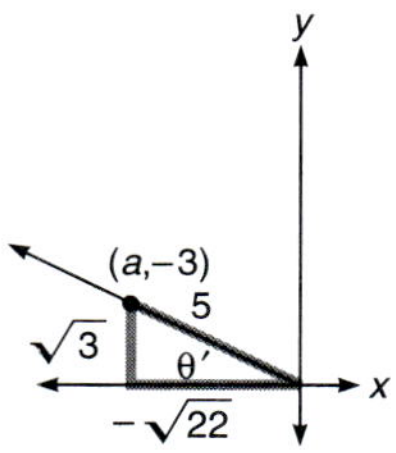

5. $\sin 30° = \frac{a}{12}, \frac{1}{2} = \frac{a}{12}, a = 6$

$\cos 30° = \frac{b}{12}, \frac{\sqrt{3}}{2} = \frac{b}{12}, b = 6\sqrt{3}$

$B = 90° - 30° = 60°$

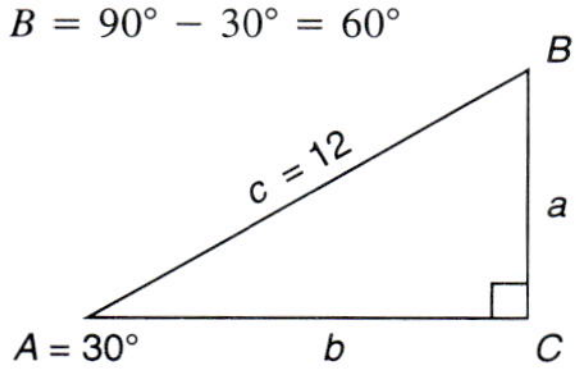

Solutions to trial exercise problems

11. $\sin\left(-\frac{2\pi}{3}\right) = -\sin\frac{2\pi}{3} = -\frac{\sqrt{3}}{2}$

$\cos\left(-\frac{2\pi}{3}\right) = \cos\frac{2\pi}{3} = -\frac{1}{2}$

(θ' is $\frac{\pi}{3}$ and $\frac{2\pi}{3}$ is in quadrant II, where $\cos\theta < 0$).

$\tan\left(-\frac{2\pi}{3}\right) = -\tan\frac{2\pi}{3}$

$= -(-\sqrt{3}) = \sqrt{3}$

$\tan\frac{2\pi}{3} < 0$ because $\frac{2\pi}{3}$ is in quadrant II, where $\tan\theta < 0$.

33. $y = -\sin\left(3x - \frac{\pi}{3}\right)$ Amplitude is 1.

Graph is reflected about the x-axis with respect to the graph of $y = \sin x$.

$0 \le 3x - \frac{\pi}{3} \le 2\pi$

$\frac{\pi}{3} \le 3x \le 2\pi + \frac{\pi}{3}$

$\frac{\pi}{3} \le 3x \le \frac{7\pi}{3}$

$\frac{1}{3}\cdot\frac{\pi}{3} \le \frac{1}{3}\cdot 3x \le \frac{1}{3}\cdot\frac{7\pi}{3}$

$\frac{\pi}{9} \le x \le \frac{7\pi}{9}$; one basic sine cycle between $\frac{\pi}{9}$ and $\frac{7\pi}{9}$.

Phase shift is $\frac{\pi}{9}$; period is $\frac{7\pi}{9} - \frac{\pi}{9} = \frac{2}{3}\pi$.

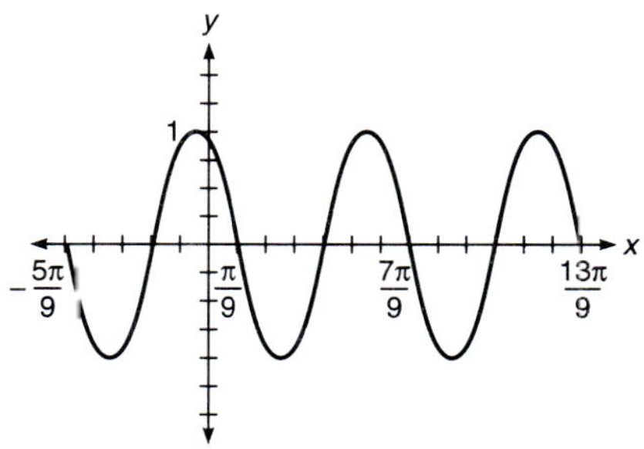

55. $y = 2\sin\left(-\frac{x}{3} - \pi\right)$

$y = 2\sin\left[-\left(\frac{x}{3} + \pi\right)\right]$

$y = -2\sin\left(\frac{x}{3} + \pi\right)$

65. Amplitude $= |A| = 5$; $A = 5$; $D = 0$, since there is no vertical translation.
A basic sine cycle runs from -1 to 3.
To find B and C:
$-1 \le x \le 3$ Basic cycle
Convert left member to 0.
$0 \le x + 1 \le 4$
Convert right member to 2π.
$0 \le \dfrac{x+1}{2} \le 2$ Divide by 2.
$0 \le \dfrac{x+1}{2}\pi \le 2\pi$ Multiply by π.
$0 \le \dfrac{\pi}{2}x + \dfrac{\pi}{2} \le 2\pi$
Thus $Bx + C = \dfrac{\pi}{2}x + \dfrac{\pi}{2}$,
so $B = C = \dfrac{\pi}{2}$.
The equation is $y = 5\sin\left(\dfrac{\pi}{2}x + \dfrac{\pi}{2}\right)$.

75. Amplitude is 10.
$0 \le 2x - 180° \le 360°$
$180° \le 2x \le 540°$
Add 180° to each member.
$90° \le x \le 270°$; one basic sine cycle.

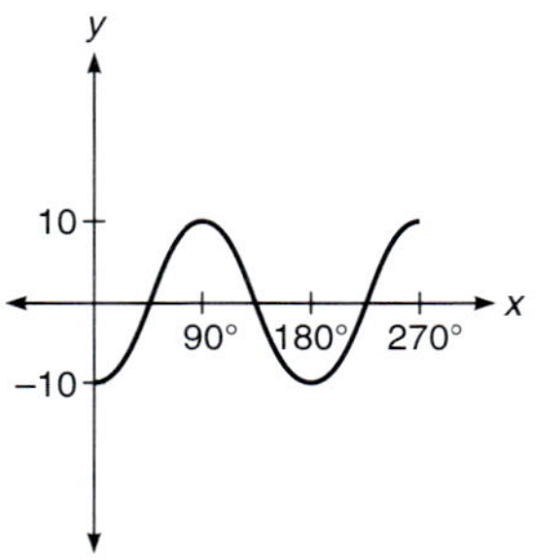

93. $f(x) = \dfrac{\cos x}{x}$

$f(-x) = \dfrac{\cos(-x)}{-x} = \dfrac{\cos x}{-x} = -\dfrac{\cos x}{x}$

$-f(x) = -\dfrac{\cos x}{x}$

Thus $f(-x) = -f(x)$, so the function is odd. The symmetry would be across the origin.

Exercise 6–3

Answers to odd-numbered problems

1. Using figure 6–22 we obtain the following graph of the cotangent function.

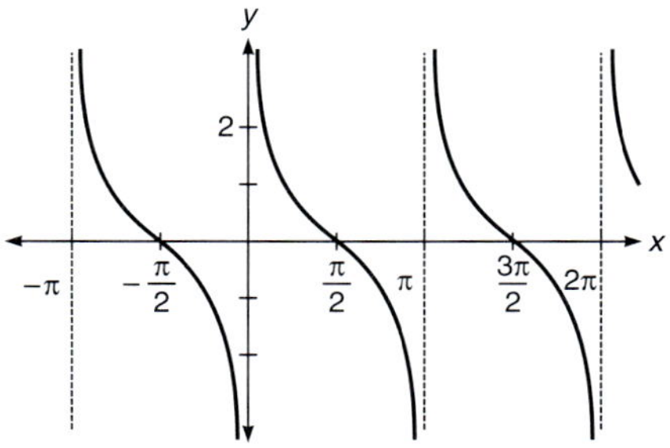

Figures 6–23 and 6–24 are the graphs of the cosecant and secant functions.

3. $\csc(-x) = \dfrac{1}{\sin(-x)} = \dfrac{1}{-\sin x}$

$= -\dfrac{1}{\sin x} = -\csc x$

Since $\csc(-x) = -\csc x$, cosecant is an odd function.

5. $\cot(-x) = \dfrac{1}{\tan(-x)} = \dfrac{1}{-\tan x}$

$= -\dfrac{1}{\tan x} = -\cot x$

7. $f(-x) = \sin^2(-x)\tan(-x)$
$= (-\sin x)^2(-\tan x)$
$= \sin^2 x(-\tan x)$
$= -\sin^2 x \tan x$
$= -f(x)$
Thus f is an odd function.

9.

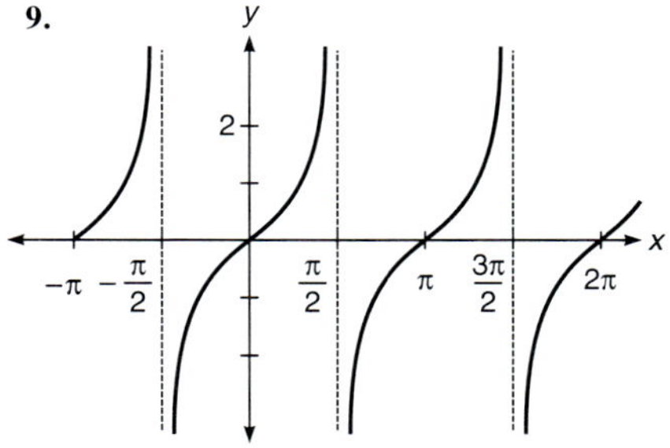

11.

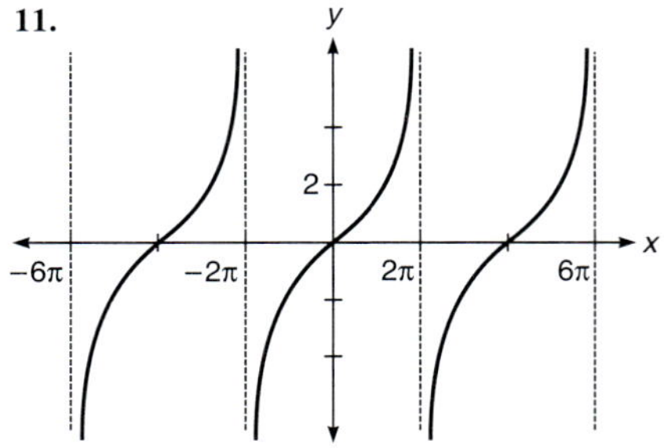

13.

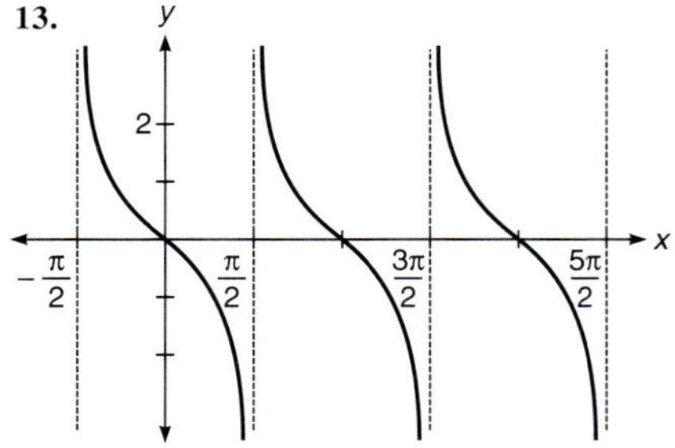

15.

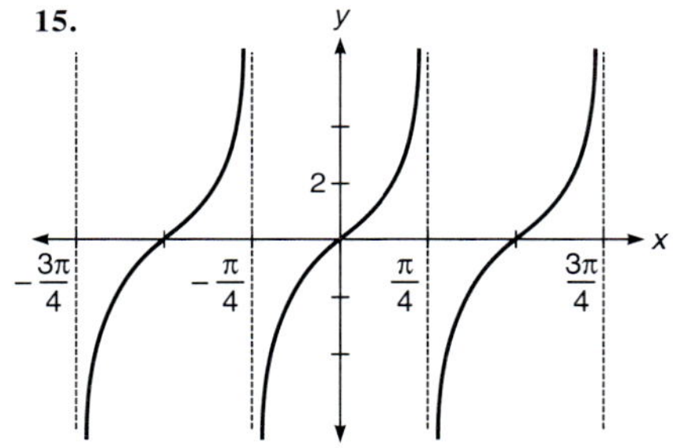

17.

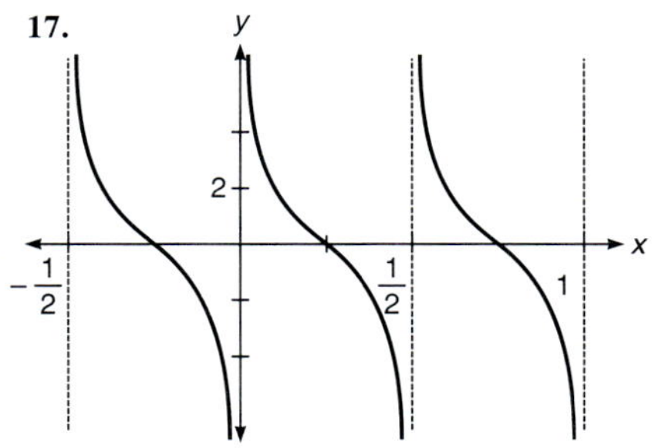

19. $-y = -\tan 2x$

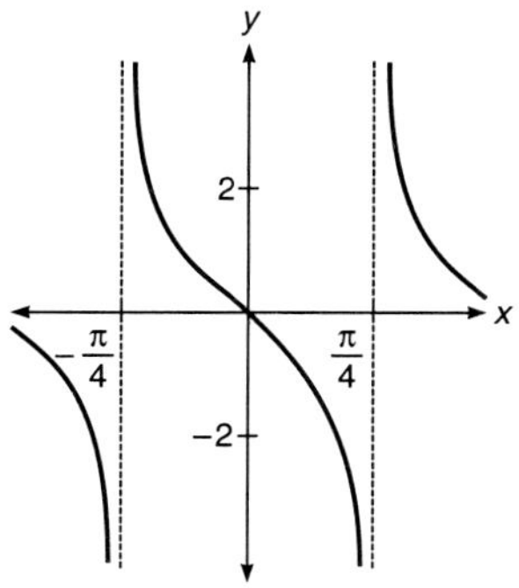

21. $y = \cot \pi x$

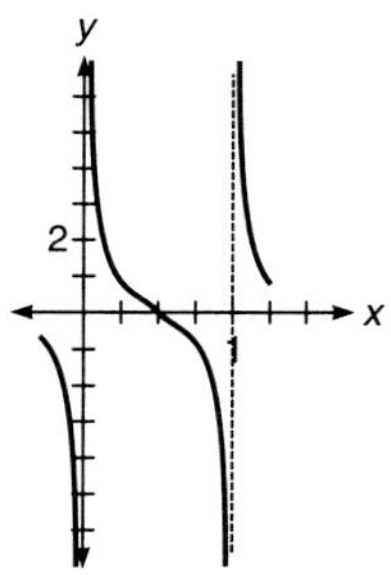

23. $y = -\tan(x + \pi)$

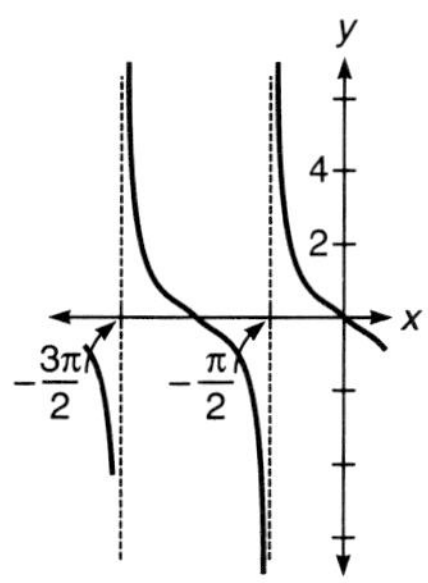

25.

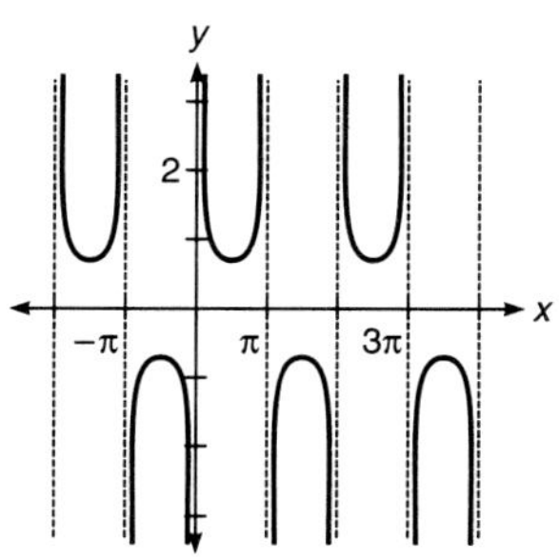

27.

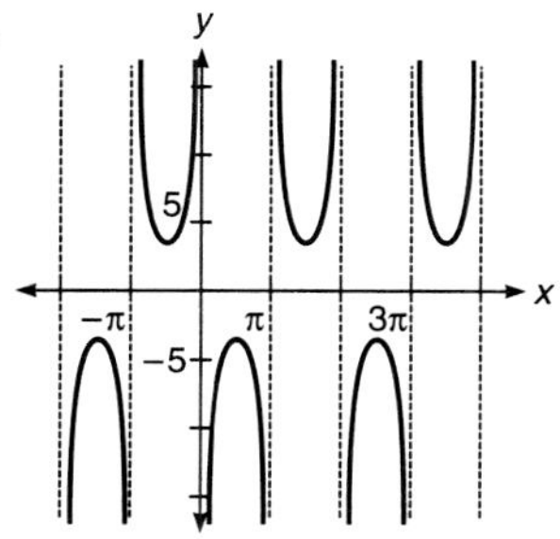

29.

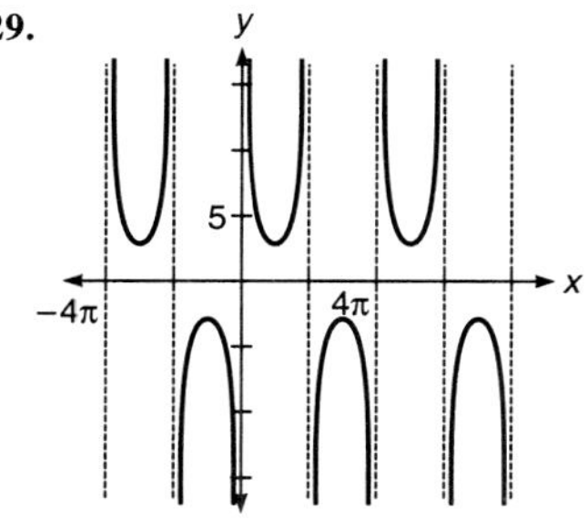

31.

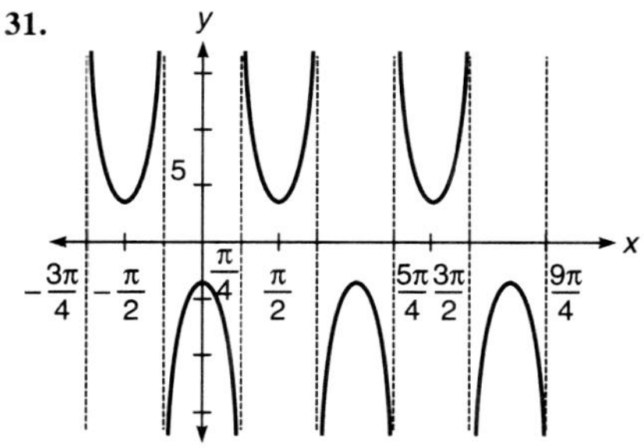

33.

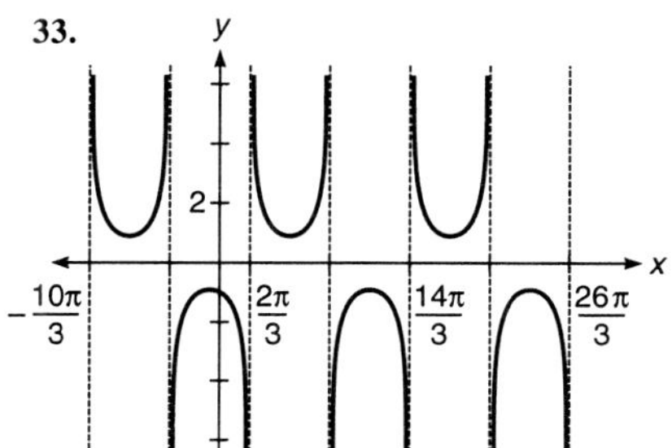

35.

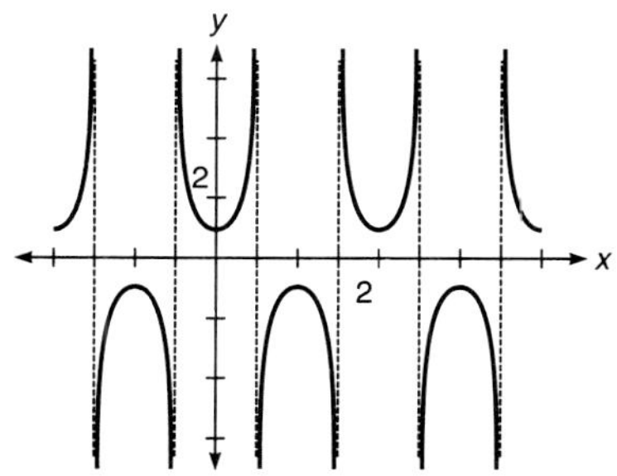

37. $y = -\csc x$

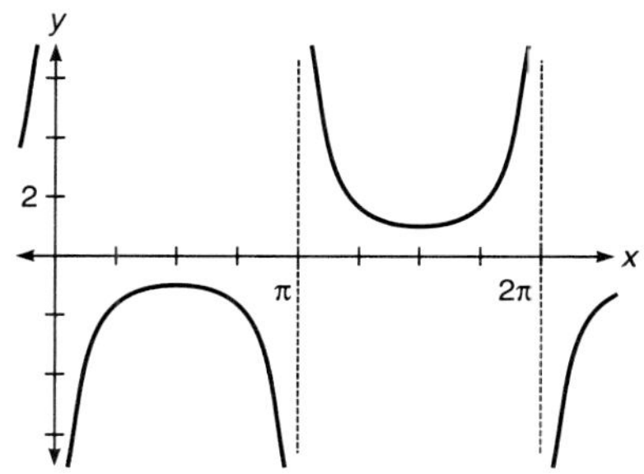

39. $y = 2 \sec\left(\frac{x}{3} + \frac{\pi}{2}\right)$

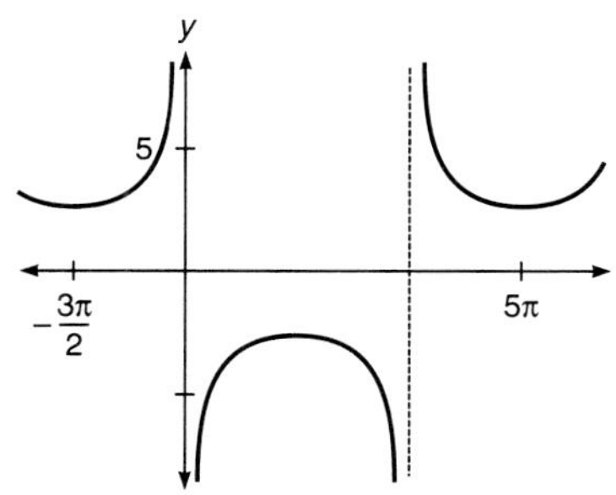

Solutions to skill and review problems

1. Graph the function $f(x) = x^4 - x^3 - 7x^2 + x + 6$.

Recall that the zeros of the right member are the x-intercepts, and that the rational zero theorem and synthetic division can be used to help find these zeros. Possible rational zeros are the factors of $\frac{6}{1} = 6$. These are $\pm 1, \pm 2, \pm 3, \pm 6$.

Synthetic division with 1 shows the remainder is 0, so $x - 1$ is a factor of $f(x)$.

	1	−1	−7	1	6
		1	0	−7	−6
1	1	0	−7	−6	0

Thus, $f(x) = (x - 1)(x^3 - 7x - 6)$.

Synthetic division with -1:

	1	0	−7	−6
		−1	1	6
−1	1	−1	−6	0

Thus, $f(x) = (x - 1)(x + 1)(x^2 - x - 6)$
$= (x - 1)(x + 1)(x - 3)(x + 2)$

The x-intercepts of f are the zeros above, which are $-2, -1, 1, 3$.

The y-intercept is at $f(0) = 6$.

Additional points:

x	−2.5	−1.5	−0.5	0.5	1.5	2	2.5	3.3
y	14.4	−2.8	3.9	4.7	−6.6	−12	−11.8	15.7

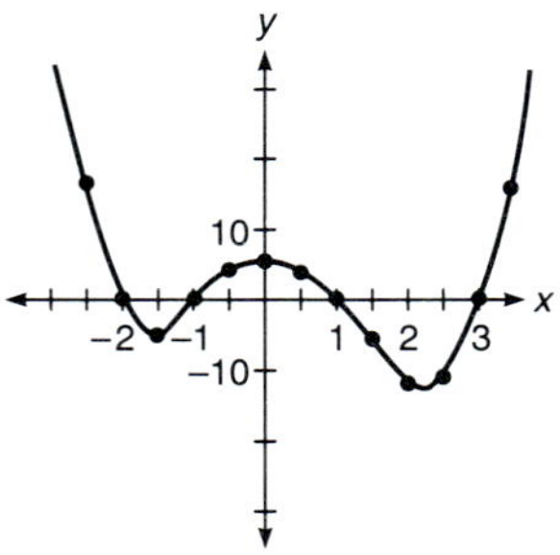

2. $f(x) = x^2 + 6x - 4$
$= x^2 + 6x + 3^2 - 4 - 3^2$
$= (x + 3)^2 - 13$
$= (x - (-3))^2 - 13$

Vertex at $(h,k) = (-3,-13)$.

y-intercept: $f(0) = -4$; $(0,-4)$

x-intercepts: $0 = x^2 + 6x - 4$

$$x = \frac{\pm\sqrt{52} - 6}{2} \approx -6.6,\ 0.6;\ (-6.6,0),$$
$(0.6,0)$

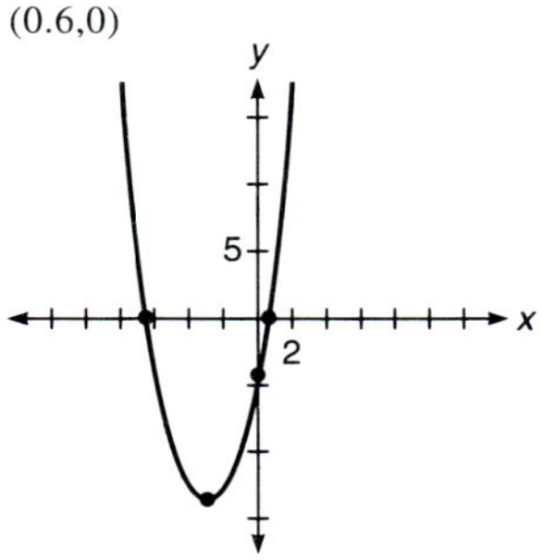

3. $$\frac{\sqrt{3}}{\sqrt{3} - 6} \cdot \frac{\sqrt{3} + 6}{\sqrt{3} + 6} = \frac{3 + 6\sqrt{3}}{3 - 36}$$
$$= \frac{3(1 + 2\sqrt{3})}{-33} = -\frac{1 + 2\sqrt{3}}{11}$$

4. $(3 - 7i)(2 + 3i)$
$6 + 9i - 14i - 21i^2$
$6 - 5i + 21$; $i^2 = -1$
$27 - 5i$

5. $f(x) = x^2 - \cos x$
$f(-x) = (-x)^2 - \cos(-x)$
$= x^2 - \cos x = f(x)$
$(-x^2) = x^2$; $\cos(-x) = \cos x$

Since $f(-x) = f(x)$ the function is even. It would have y-axis symmetry.

6. The graph of $f(x) = 3 \sin x$ is the graph of $y = \sin x$, but with an amplitude of 3.

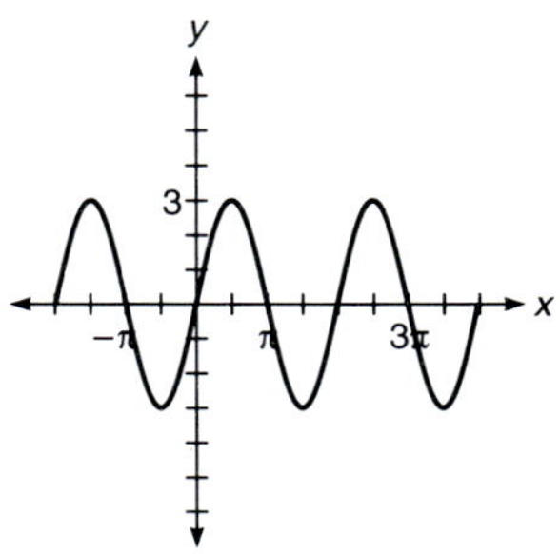

7. $0 \le 2x \le 2\pi$
$0 \le x \le \pi$

The graph of $f(x) = -\cos 2x$ is the graph of $y = \cos x$ but with one complete cycle from 0 to π, and flipped over about the x-axis.

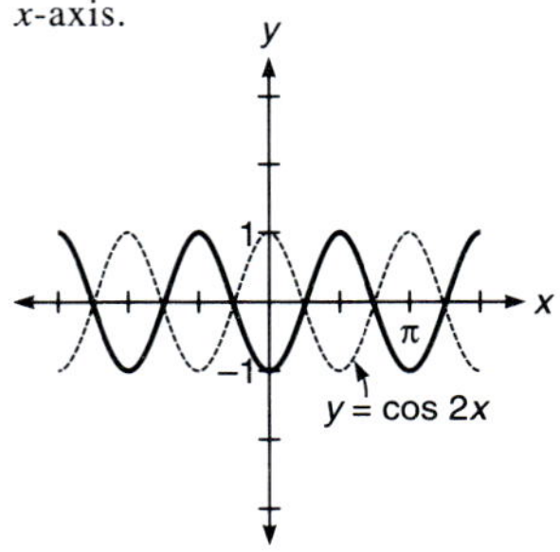

8. $0 \le x + \frac{\pi}{5} \le 2\pi$
$-\frac{\pi}{5} \le x \le 2\pi - \frac{\pi}{5}$
$-\frac{\pi}{5} \le x \le \frac{9\pi}{5}$

One complete sine cycle.

Period is $\frac{9\pi}{5} - \left(-\frac{\pi}{5}\right) = 2\pi$.

Other cycles start at $\frac{9\pi}{5}$ and at $\frac{9\pi}{5} + 2\pi = \frac{19\pi}{5}$.

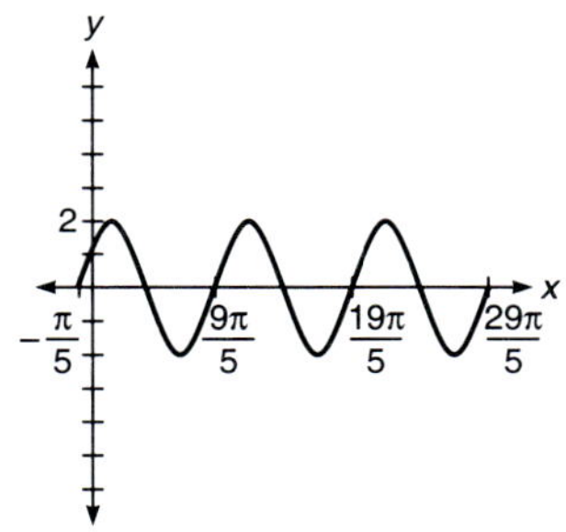

Solutions to trial exercise problems

6. $f(-x) = \sin^2(-x)\cos(-x)$
$= [\sin(-x)]^2 \cos(-x)$
$= [-\sin x]^2 \cos x$
$= [\sin x]^2 \cos x$
$= \sin^2 x \cos x$
$= f(x)$

Thus f is an even function.

15. $y = -\cot\left(2x + \dfrac{\pi}{2}\right)$

$0 \le 2x + \dfrac{\pi}{2} \le \pi$

$-\dfrac{\pi}{2} \le 2x \le \dfrac{\pi}{2}$

$-\dfrac{\pi}{4} \le x \le \dfrac{\pi}{4}$

One basic cotangent cycle between $-\dfrac{\pi}{4}$ and $\dfrac{\pi}{4}$.

See the answer to problem 15 for the graph.

23. $y = \tan(-x - \pi)$

$= \tan[-(x + \pi)]$

$= -\tan(x + \pi)$

This is the graph of $y = \tan(x + \pi)$, flipped about the x-axis.

$-\dfrac{\pi}{2} \le x + \pi \le \dfrac{\pi}{2}$

$-\dfrac{3\pi}{2} \le x \le -\dfrac{\pi}{2}$

One basic tangent cycle between $-\dfrac{3\pi}{2}$ and $-\dfrac{\pi}{2}$.

See the answer to problem 23 for the graph.

31. $y = 3\sec(2x + \pi)$

Graph three cycles of $y = 3\cos(2x + \pi)$.

$0 \le 2x + \pi < 2\pi$

$-\pi \le 2x \le \pi$

$-\dfrac{\pi}{2} \le x \le \dfrac{\pi}{2}$

One basic secant cycle between $-\dfrac{\pi}{2}$ and $\dfrac{\pi}{2}$.

Graph one basic cosine cycle between $-\dfrac{\pi}{2}$ and $\dfrac{\pi}{2}$, then use it to sketch the related secant function. See the answer to problem 31 for the graph.

39. $y = 2\sec\left(-\dfrac{x}{3} - \dfrac{\pi}{2}\right)$

$= 2\sec\left[-\left(\dfrac{x}{3} + \dfrac{\pi}{2}\right)\right]$

$= 2\sec\left(\dfrac{x}{3} + \dfrac{\pi}{2}\right)$ $\quad \sec(-\theta) = \sec\theta$

$0 \le \dfrac{x}{3} + \dfrac{\pi}{2} \le 2\pi$

$-\dfrac{\pi}{2} \le \dfrac{x}{3} \le \dfrac{3\pi}{2}$

$-\dfrac{3\pi}{2} \le x \le \dfrac{9\pi}{2}$

Graph a cosine cycle from $-\dfrac{3\pi}{2}$ to $\dfrac{9\pi}{2}$, then "flip over" the graph. See the answer to problem 39 for the graph.

40. $0 < 2x < \pi$

$0 < x < \dfrac{\pi}{2}$

Period is $\dfrac{\pi}{2}$.

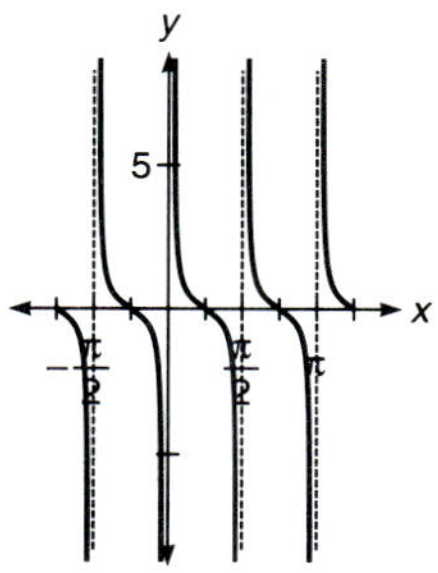

Exercise 6–4

Answers to odd-numbered problems

1. a.

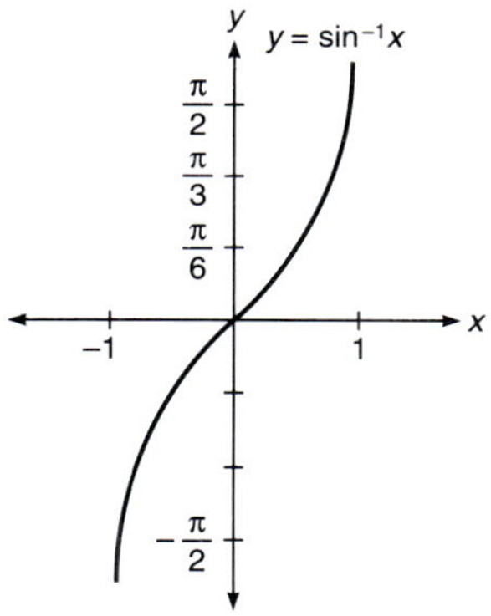

b.

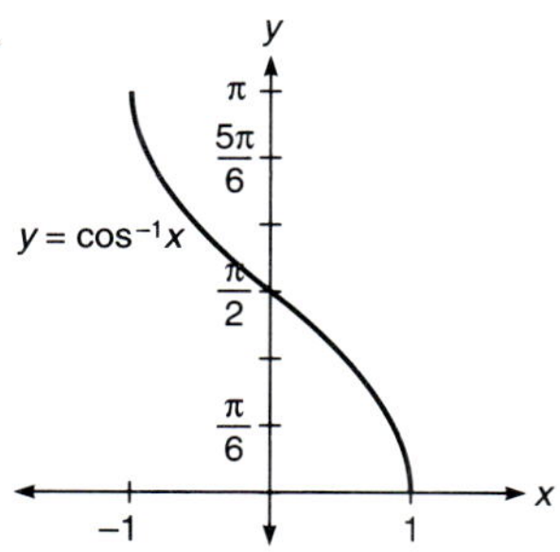

c.

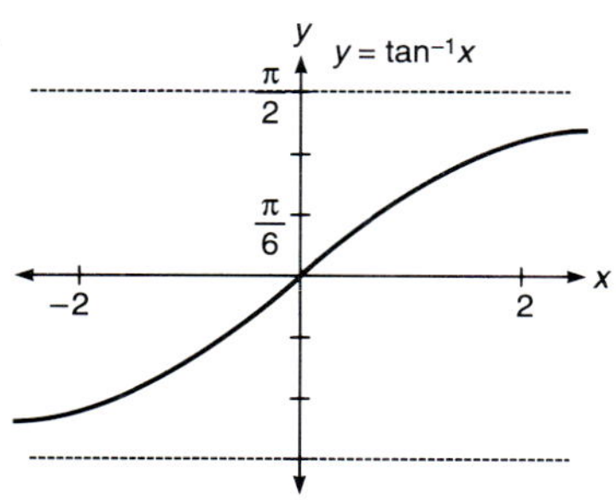

3. $\dfrac{\pi}{4}$, 45° **5.** $\dfrac{\pi}{3}$, 60° **7.** $\dfrac{\pi}{2}$, 90°

9. $\dfrac{\pi}{6}$, 30° **11.** $\dfrac{3\pi}{4}$, 135° **13.** $\dfrac{\pi}{4}$, 45°

15. π, 180° **17.** 0, 0° **19.** 0, 0°

21. -0.25, $-14.3°$ **23.** 1.48, 84.8°

25. 0.75, 43.0° **27.** 0.70, 40.1°

29. -1.50, $-86.0°$ **31.** -1.29, $-74.0°$

33. 2.13, 122.0° **35.** $\dfrac{4}{5}$ **37.** $\dfrac{\sqrt{91}}{10}$

39. $\dfrac{10\sqrt{109}}{109}$ **41.** $\dfrac{5\sqrt{34}}{34}$ **43.** $\dfrac{\sqrt{39}}{8}$

45. $-\dfrac{\sqrt{5}}{2}$ **47.** $\dfrac{\sqrt{30}}{6}$ **49.** $\dfrac{\sqrt{66}}{3}$

51. $\dfrac{3\sqrt{5}}{5}$ **53.** $\sqrt{1 - z^2}$

55. $\dfrac{1 + z}{\sqrt{-2z - z^2}}$ **57.** $\dfrac{1}{\sqrt{1 - 2z}}$

59. $\sqrt{1 - z^2}$ **61.** $\dfrac{\sqrt{1 - z^2}}{z}$

63. $\dfrac{1}{\sqrt{1 + z^2}}$ **65.** $\sqrt{1 - 9z^2}$

67. $\sqrt{z^2 + 2z + 2}$ **69.** $\dfrac{1}{\sqrt{1 + 2z}}$

71. $\dfrac{\pi}{6}$ **73.** $-\dfrac{\pi}{6}$ **75.** $\dfrac{\pi}{2}$ **77.** $\dfrac{\pi}{4}$

79. $\dfrac{\pi}{6}$ **81.** 0 **83.** $\sin^{-1}\dfrac{m}{r}$

85. $\tan^{-1}\dfrac{k}{h}$ **87.** $\tan^{-1}\dfrac{2}{x}$ **89.** $\sin^{-1}\dfrac{3{,}500}{z}$

91. $\cos^{-1}(-0.8)$ **93.** $\tan^{-1}4.1$

95. $\tan^{-1}\frac{5}{3}$ **97.** $\tan^{-1}50$ **99.** $\frac{1}{3}\tan^{-1}9$

101. $\frac{2}{3}\sin^{-1}(-0.56)$ **103.** $\frac{1}{4}\sin^{-1}0.75$

105. $\frac{1}{5}\sin^{-1}\frac{25}{39}$ **107.** $\dfrac{1}{B}\left(\tan^{-1}\dfrac{D}{A} - C\right)$

109. $\frac{1}{2}(\sin^{-1}0.6 - 3)$

111. $\arcsin 1 = \dfrac{\pi}{2}$

$\arcsin(-1) = -\arcsin 1 = -\dfrac{\pi}{2}$

Thus, the first and third parts of the identity are correct.

If $x = 0$, the formula $\arcsin(x) = \arctan\left(\dfrac{x}{\sqrt{1 - x^2}}\right)$ is correct, since

$\arcsin 0 = \arctan\left(\dfrac{0}{\sqrt{1 - 0^2}}\right) = 0$

The figures show the two remaining cases for $\theta = \arctan\left(\dfrac{x}{\sqrt{1 - x^2}}\right)$, when $1 > x > 0$ and when $-1 < x < 0$. In each case $r = 1$.

Examining the reference triangles shows that in each case $\sin\theta = \frac{x}{1} = x$, so that in each case $\theta = \arcsin(x)$.

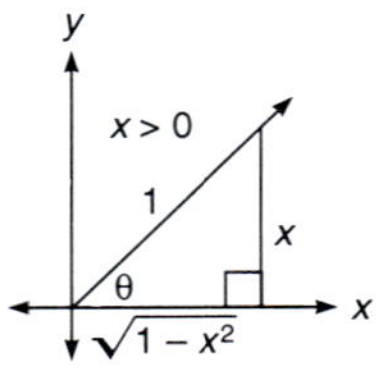

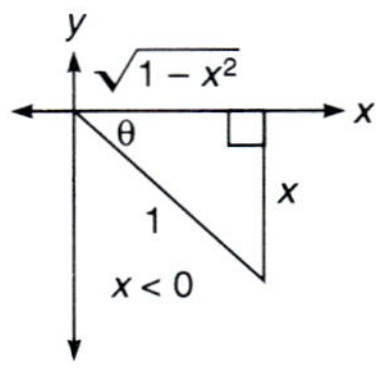

Solutions to skill and review problems

1. $\cos x(\cos x + \sin x \tan x - \sec x)$

$\cos x \cdot \cos x + \cos x \cdot \sin x \cdot \tan x - \cos x \cdot \sec x$

$\cos^2 x + \cos x \cdot \sin x \cdot \dfrac{\sin x}{\cos x} - \cos x \cdot \dfrac{1}{\cos x}$

$\cos^2 x + \sin^2 x - 1$

$1 - 1 = 0$

2. $a = \sqrt{9.2^2 - 5^2} \approx 7.7$

$\sin B = \dfrac{5}{9.2}$; $B = \sin^{-1}\dfrac{5}{9.2} \approx 32.9°$

$\cos A = \dfrac{5}{9.2}$; $A = \cos^{-1}\dfrac{5}{9.2} \approx 57.1°$

3. $\dfrac{7\pi}{6}$ terminates in quadrant III,

so $\sin\dfrac{7\pi}{6}$ is negative, and

$\theta' = \theta - \pi = \dfrac{7\pi}{6} - \dfrac{6\pi}{6} = \dfrac{\pi}{6}$

$\sin\dfrac{\pi}{6} = \dfrac{1}{2}$, so $\sin\dfrac{7\pi}{6} = -\dfrac{1}{2}$

4. Find a positive angle coterminal with $-\dfrac{11\pi}{3}$. Add 4π to get $-\dfrac{11\pi}{3} + 4\pi = -\dfrac{11\pi}{3} + \dfrac{12\pi}{3} = \dfrac{\pi}{3}$. Thus, $\tan\left(-\dfrac{11\pi}{3}\right) = \tan\dfrac{\pi}{3} = \sqrt{3}$.

5. $L = rs$

$15 = 12s$

$L = 15,\ r = 12$

$s = \frac{5}{4}$ (radians)

$\dfrac{\theta°}{180°} = \dfrac{s}{\pi}$

$\theta° = \dfrac{180}{\pi}°\left(\dfrac{5}{4}\right) \approx 71.6°$

6. $f(x) = \dfrac{2x}{x^2 - 4} = \dfrac{2x}{(x-2)(x+2)}$

Vertical asymptotes at -2 and 2.

x-intercepts: (solve $f(x) = 0$)

$0 = \dfrac{2x}{x^2 - 4}$, so $x = 0$.

Intercept at (0,0).

y-intercept: (compute $f(0)$)

$f(0) = \dfrac{2(0)}{0 - 4} = 0.$

Intercept at (0,0).

Additional points:

x	-4	-3	-1.5	-1	1	1.5	3	4
y	-0.7	-1.2	1.7	0.7	-0.7	-1.7	1.2	0.7

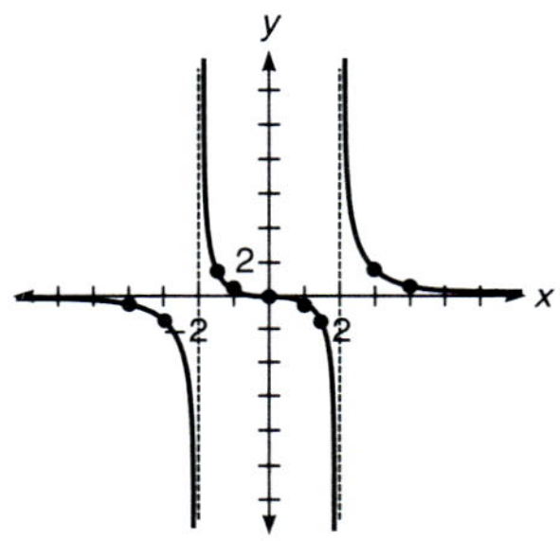

Solutions to trial exercise problems

20. $\arcsin\left(-\dfrac{\sqrt{3}}{2}\right) = -\arcsin\dfrac{\sqrt{3}}{2}$; $-\dfrac{\pi}{3}$, $-60°$

31. $\tan^{-1}(-3.4776)$

3.4776 [±]

[INV] [TAN]

TI-81: [2nd] [TAN] [(−)] 3.4776

[ENTER]

-1.29, $-74.0°$

45. $\tan[\cos^{-1}(-\frac{2}{3})]$

The figure shows an angle θ whose cosine is $-\frac{2}{3}$.

$y = \sqrt{5}$; $\tan\theta = \dfrac{y}{-2} = \dfrac{\sqrt{5}}{-2} = -\dfrac{\sqrt{5}}{2}$.

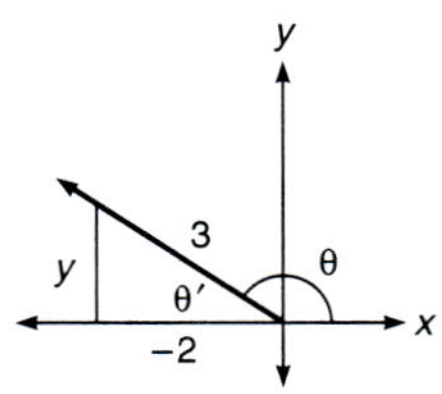

67. $\sec[\tan^{-1}(1 + z)],\ z > 0$

$r = \sqrt{1^2 + (1+z)^2} = \sqrt{z^2 + 2z + 2}$

$\cos\theta = \dfrac{1}{r}$; $\sec\theta = \dfrac{1}{\cos\theta}$

$= r = \sqrt{z^2 + 2z + 2}$

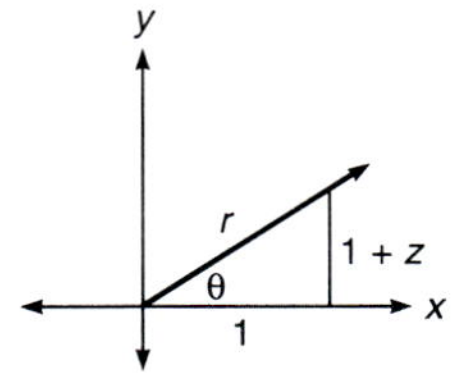

79. $\cos^{-1}\left(\cos\dfrac{11\pi}{6}\right) = \cos^{-1}\dfrac{\sqrt{3}}{2} = \dfrac{\pi}{6}$

87. $\tan\theta = \dfrac{2}{x}$, so $\theta = \tan^{-1}\dfrac{2}{x}$.

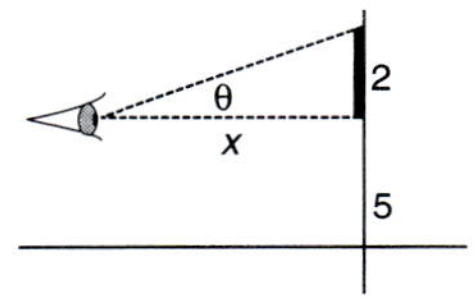

105. $\dfrac{6\sin 5\theta}{5} = \dfrac{10}{13}$

$6\sin 5\theta = \frac{50}{13}$

$\sin 5\theta = \frac{1}{6} \cdot \frac{50}{13}$

$\sin 5\theta = \frac{25}{39}$

$5\theta = \sin^{-1}\frac{25}{39}$

$\theta = \frac{1}{5}\sin^{-1}\frac{25}{39}$

109. $\sin(2x + 3) = 0.6$

$2x + 3 = \sin^{-1}0.6$

$2x = \sin^{-1}0.6 - 3$

$x = \frac{1}{2}(\sin^{-1}0.6 - 3)$

Exercise 6–5

Answers to odd-numbered problems

1. $\dfrac{\pi}{6}$ **3.** $\dfrac{\pi}{4}$ **5.** $\dfrac{2\pi}{3}$ **7.** $\dfrac{\pi}{3}$ **9.** $\dfrac{\pi}{2}$

11. 0.30, 17.3° **13.** 1.91, 109.5°

15. 0.19, 10.9° **17.** 1.66, 95.2°

19. 0.32, 18.3° **21.** $\sin(\csc^{-1}3) = \sin(\sin^{-1}\frac{1}{3}) = \frac{1}{3}$ **23.** $\dfrac{\sqrt{15}}{15}$

25. $\sqrt{26}$ **27.** $\frac{3}{5}$ **29.** $-\frac{\sqrt{11}}{5}$ **31.** $\frac{1}{z}$

33. $\frac{\sqrt{z^2+1}}{z}$ **35.** $\frac{1}{2z}$ **37.** $\sqrt{z^2+2z}$

39. $\frac{3}{z}$ **41.** 0.216

Solutions to skill and review problems

1. $\frac{2x}{x-3} - \frac{x}{x+5}$

$\frac{2x(x+5) - x(x-3)}{(x-3)(x+5)}$

$\frac{x^2+13x}{x^2+2x-15}$

2. $\sin x' = \frac{1}{2}$, so $x' = \frac{\pi}{6}$. Since x terminates in quadrant III, $x = \pi + x'$ $= \pi + \frac{\pi}{6} = \frac{7\pi}{6}$.

3. $y = \sqrt{1^2 - (1-u)^2} = \sqrt{2u - u^2}$;

$\sin\theta = \frac{y}{1} = y = \sqrt{2u-u^2}$.

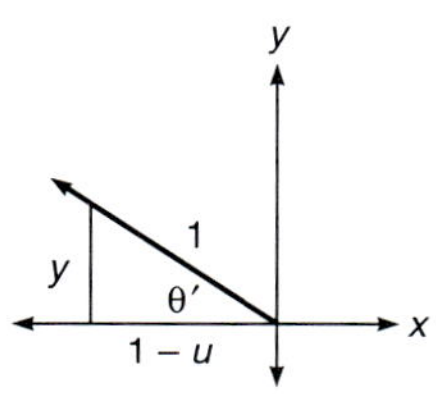

4. $3\sin x = -2$

$\sin x = -\frac{2}{3}$

$\sin x' = \frac{2}{3}$, so $x' \approx 41.8°$.

Since $\sin x < 0$, x terminates in quadrants III or IV. The least nonnegative value is in quadrant III. Thus, $x = 180° + x' \approx 221.8°$.

5. $\cos(\tan^{-1}5)$

$r = \sqrt{26}$; $\cos\theta = \frac{1}{r} = \frac{1}{\sqrt{26}} = \frac{\sqrt{26}}{26}$

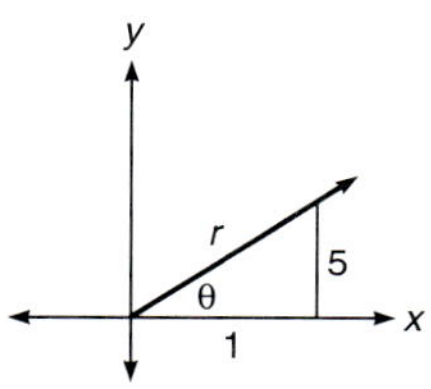

6. $y = \sqrt{1^2 - (1-m)^2} = \sqrt{2m - m^2}$,

$\csc\theta = \frac{1}{\sin\theta} = \frac{1}{\frac{y}{1}} = \frac{1}{y} = \frac{1}{\sqrt{2m-m^2}}$

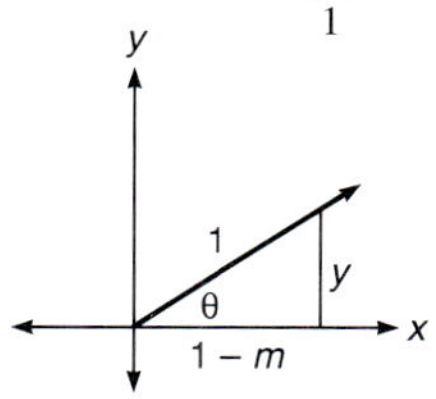

7. $\csc\left(\frac{19\pi}{6}\right)$; $\frac{19\pi}{6} - 2\pi = \frac{19\pi}{6}$ $- \frac{12\pi}{6} = \frac{7\pi}{6}$. Thus, $\csc\left(\frac{19\pi}{6}\right) =$ $\csc\left(\frac{7\pi}{6}\right) = \frac{1}{\sin\left(\frac{7\pi}{6}\right)}$.

$\frac{7\pi}{6}$ terminates in quadrant III, so $\sin\frac{7\pi}{6}$ < 0 and $\theta' = \frac{7\pi}{6} - \pi = \frac{\pi}{6}$.

$\sin\frac{\pi}{6} = \frac{1}{2}$, so $\sin\frac{7\pi}{6} = -\frac{1}{2}$.

Thus, $\csc\left(\frac{19\pi}{6}\right) = \frac{1}{-\frac{1}{2}} = -2$.

8. a. $r = \sqrt{89}$; $\sin\theta = \frac{y}{r} = \frac{8}{\sqrt{89}}$

b. $\theta' = \tan^{-1}\left|\frac{y}{x}\right| = \tan^{-1}\frac{8}{5} \approx 58.0°$

Since θ terminates in quadrant II, $\theta = 180° - \theta' \approx 180° - 58.0° \approx$ 122.0°.

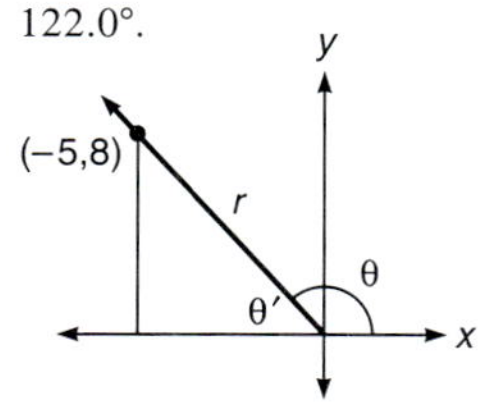

Solutions to trial exercise problems

7. $\operatorname{arccsc}\left(\frac{2\sqrt{3}}{3}\right) = \sin^{-1}\left(\frac{1}{\frac{2\sqrt{3}}{3}}\right)$

$= \sin^{-1}\frac{\sqrt{3}}{2} = \frac{\pi}{3}$

17. $\sec^{-1}(-11.1261) = \cos^{-1}\left(-\frac{1}{11.1261}\right)$

≈ 1.66 (calculator in radian mode)

$\approx 95.2°$ (calculator in degree mode)

11.1261 [$\frac{1}{x}$] [±] [INV] [COS]

TI-81: [2nd] [COS] [(−)]

11.1261 [x^{-1}] [ENTER]

27. $\cos[\operatorname{arccsc}(-\frac{5}{4})] = \cos[\sin^{-1}(-\frac{4}{5})]$

$x = 3$; $\cos\theta = \frac{x}{5} = \frac{3}{5}$

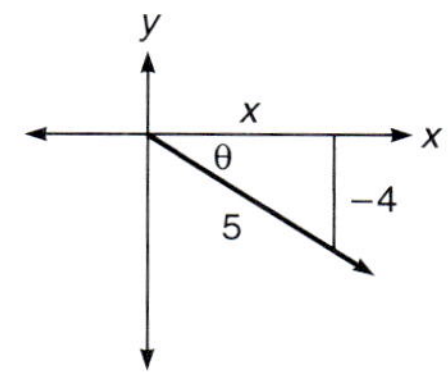

40. $\sec\left(\cot^{-1}\left(\frac{2}{z+1}\right)\right)$, $z + 1 > 0$

Since $\cot\theta = \frac{2}{z+1}$, $\tan\theta = \frac{z+1}{2}$

$r = \sqrt{(z+1)^2 + 2^2} = \sqrt{z^2+2z+5}$;

$\sec\theta = \frac{1}{\cos\theta} = \frac{1}{\frac{2}{r}} = \frac{r}{2}$

$= \frac{1}{2}\sqrt{z^2+2z+5}$

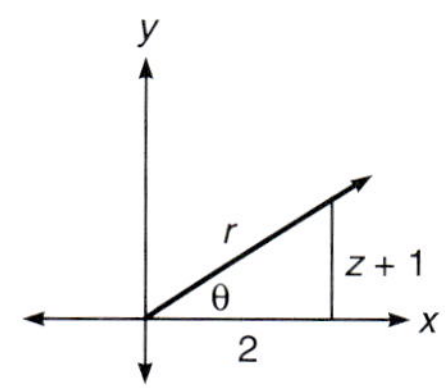

Chapter 6 review

1. $\frac{7}{4}\pi \approx 5.50$ **2.** $-\frac{4}{3}\pi \approx -4.19$

3. $-\frac{37}{45}\pi \approx -2.58$ **4.** $\frac{16}{9}\pi \approx 5.59$

5. 140° **6.** 247.5° **7.** $-\frac{720}{\pi}°$ $\approx -229.18°$ **8.** $\frac{135}{2\pi}° \approx 21.49°$

9. 30.4 inches **10.** 602.1 millimeters

11. $3\frac{1}{5}$ (radians) $\approx 183.3°$ **12.** $2\frac{6}{7}$ radians

13. -0.7568 **14.** -0.8130 **15.** 1.0747

16. 0.6421 **17.** $-\frac{\sqrt{3}}{2}$ **18.** $-\frac{\sqrt{3}}{3}$

19. domain: R; range: $-1 \le y \le 1$; period: 2π

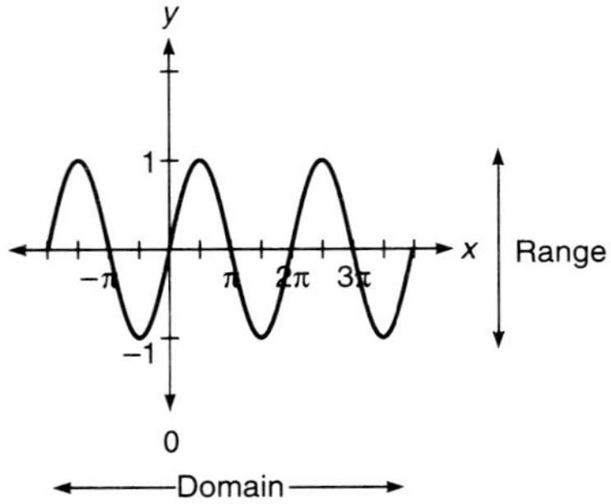

20. Even multiples of π: . . . , -2π, 0, 2π, 4π, ... Thus $x = 2k\pi$, k an integer.

21. $-\dfrac{\sqrt{3}}{2}$ **22.** $-\sqrt{3}$

23. even function, since $f(-x) = -f(x)$; symmetry about the y-axis **24.** odd function; symmetry about the origin

25.

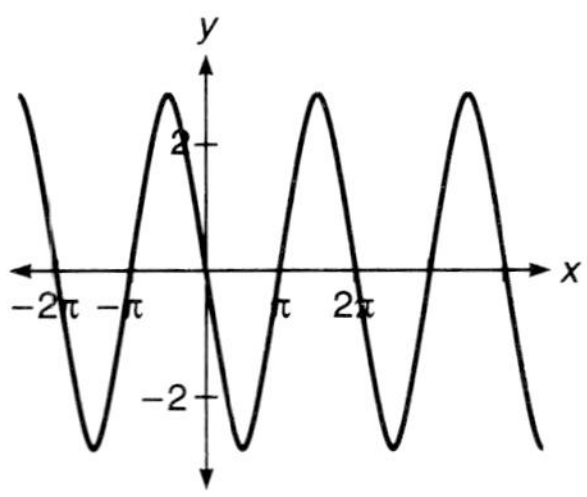

26.

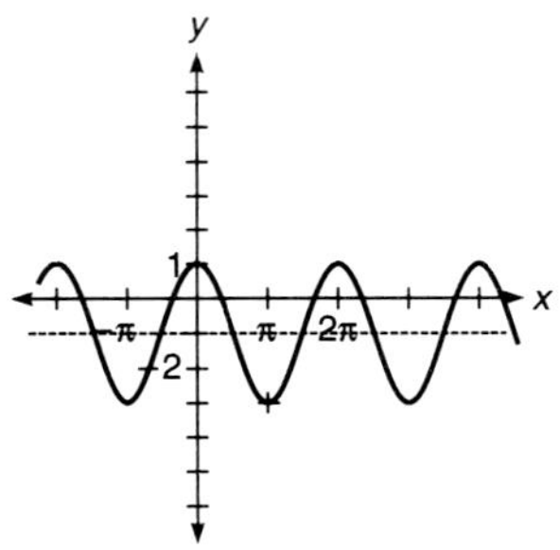

27.

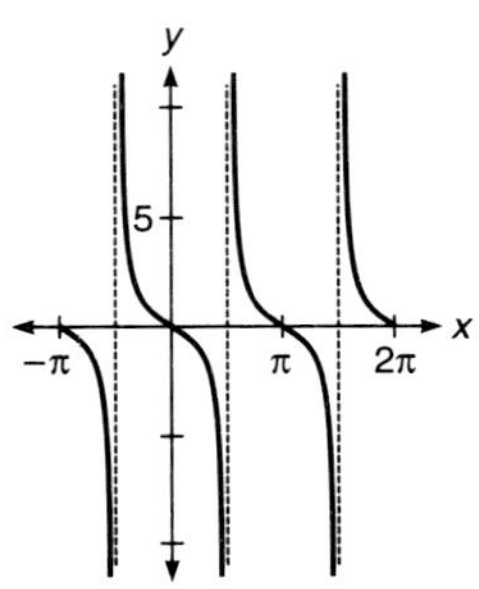

28.

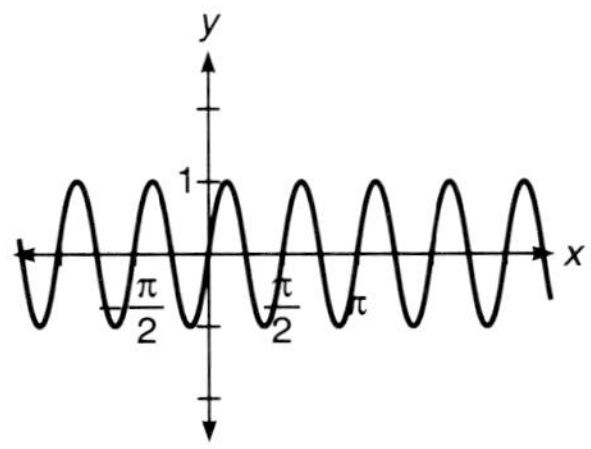

29.

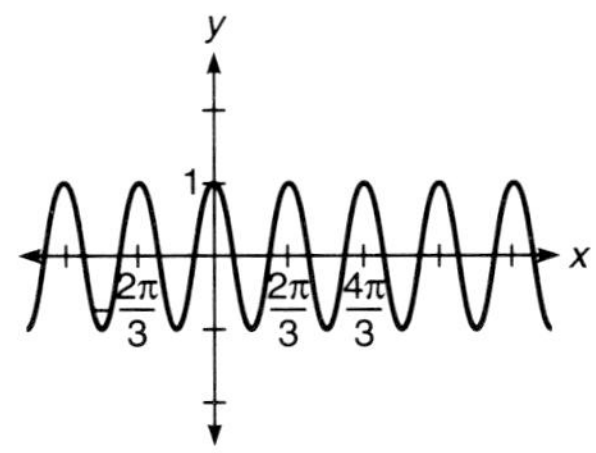

30.

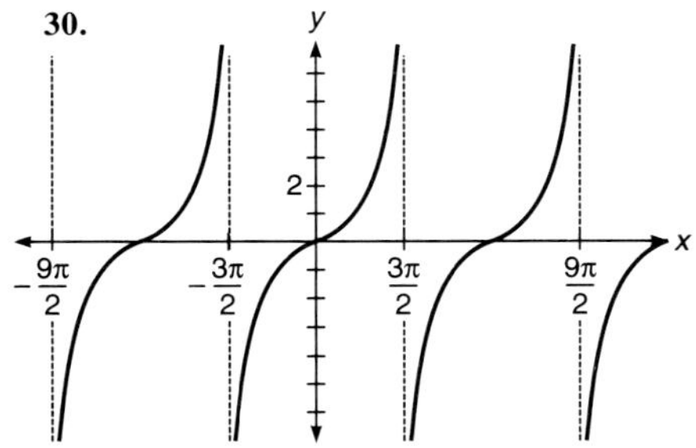

31.

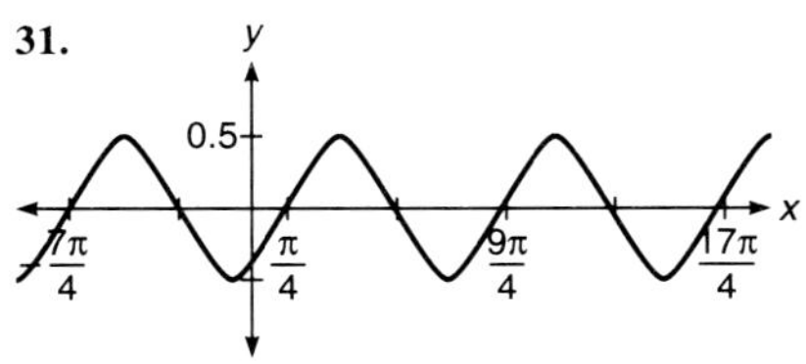

32.

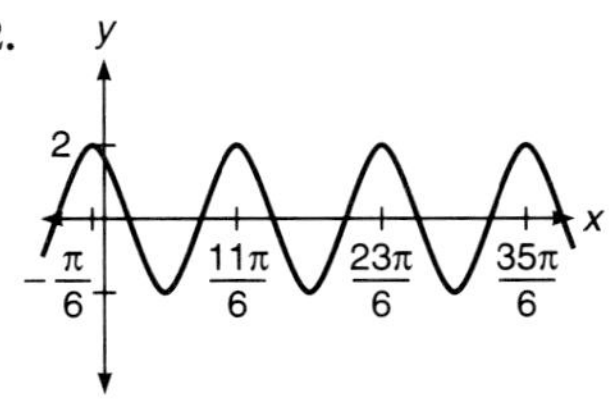

33.

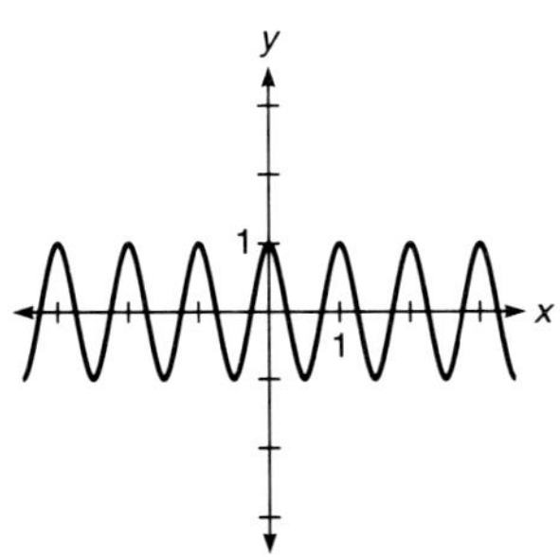

34. $y = 3\sin(\pi - 2x) = -3\sin(2x - \pi)$

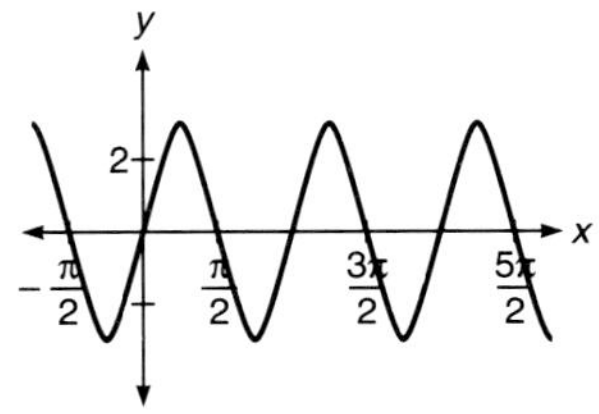

35. $y = 2\cos(\pi - 3x) = 2\cos(3x - \pi)$

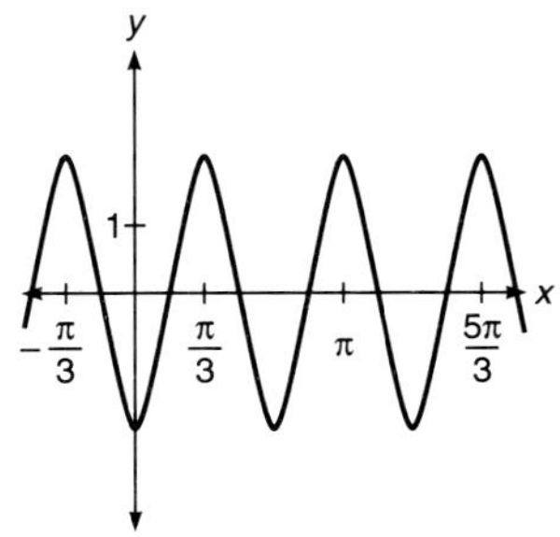

36. $A = \dfrac{1}{2}$, $B = 1$, $C = -\dfrac{\pi}{4}$, $D = 0$, $y = \dfrac{1}{2}\sin\left(x - \dfrac{\pi}{4}\right)$

37.

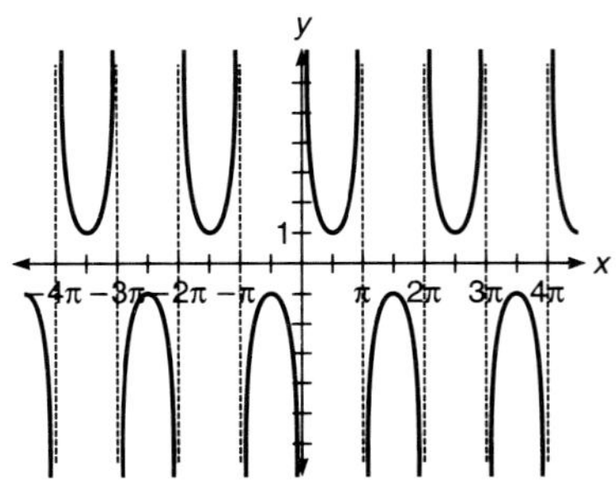

38.
$$
\begin{aligned}
f(x) &= \sec x \cdot \sin^2 x + x^4 \\
f(-x) &= \sec(-x)[\sin(-x)]^2 + (-x)^4 \\
&= \sec x[-\sin x]^2 + x^4 \\
&= \sec x[\sin x]^2 + x^4 \\
&= \sec x \cdot \sin^2 x + x^4 \\
&= f(x)
\end{aligned}
$$
Since $f(-x) = f(x)$, f is an even function.

39.

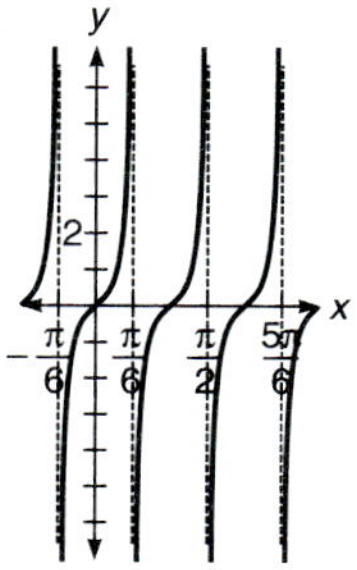

40.

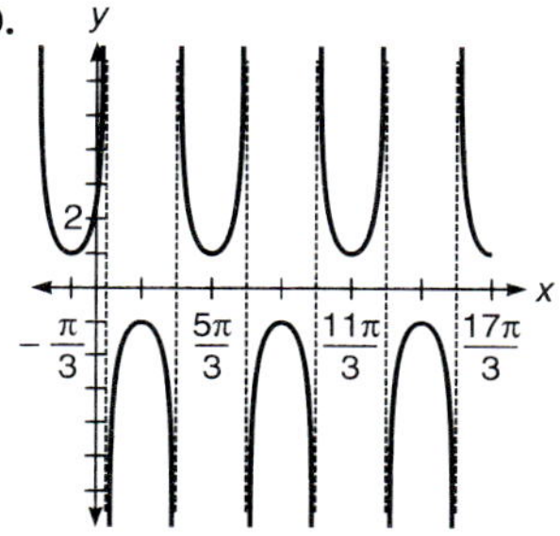

41. $y = \tan(-x) = -\tan x$
The graph of $y = -\tan x$ is given in problem 27. **42.** domain: $|x| \leq 1$; range: $0 \leq y \leq \pi$.

43. $\frac{\pi}{6}$, 30° **44.** $\frac{\pi}{4}$, 45° **45.** $\frac{\pi}{3}$, 60°

46. $-\frac{\pi}{3}$, $-60°$ **47.** -1.30, $-74.3°$

48. 2.01, 115.4° **49.** 1.19, 68.2°

50. $\frac{\sqrt{55}}{8}$ **51.** $-\frac{2\sqrt{5}}{5}$ **52.** $\frac{\sqrt{3}}{2}$

53. $\frac{2\sqrt{14}}{7}$ **54.** $\frac{2\sqrt{2}}{3}$ **55.** $-\frac{5\sqrt{26}}{26}$

56. $\sqrt{1-4z^2}$ **57.** $\frac{1-z}{\sqrt{2z-z^2}}$

58. $\sqrt{-z}$ **59.** $\sqrt{z}$ **60.** $-\frac{\pi}{6}$

61. 0 **62.** $\frac{\pi}{3}$ **63.** $\sin^{-1}\frac{6{,}000}{z}$

64. $3\cos^{-1}\frac{1}{3}$ **65.** $\frac{1}{2}\sin^{-1}\frac{1}{4}$

66. $\frac{1}{k}\tan^{-1}\frac{b}{a}$ **67.** $\frac{\pi}{6}$ **68.** $\frac{\pi}{4}$

69. 0.32, 18.4° **70.** 1.96, 112.1°

71. $\frac{\sqrt{2}}{4}$ **72.** $\sqrt{1+z^2}$

Chapter 6 test

1. $-\frac{25}{18}\pi$, -4.36 **2.** 252° **3.** 25 inches

4. $\frac{19}{14}$, 77.8° **5.** 181° **6.** -1.0002

7. $\frac{\sqrt{3}}{2}$ **8.** domain: all reals (R); range: $-1 \leq y \leq 1$; period: 2π; figure 6–14 shows the sketch **9.** $x = \frac{3\pi}{2} + 2k\pi$, k an integer

10. $\sqrt{3}$

11. $f(x) = x + \sin x$

$$\begin{aligned} f(-x) &= (-x) + \sin(-x) \\ &= (-x) + (-\sin x) \\ &= -(x + \sin x) \\ &= -f(x) \end{aligned}$$

Since $f(-x) = -f(x)$, f is an odd function. Its graph would have symmetry about the origin.

12.

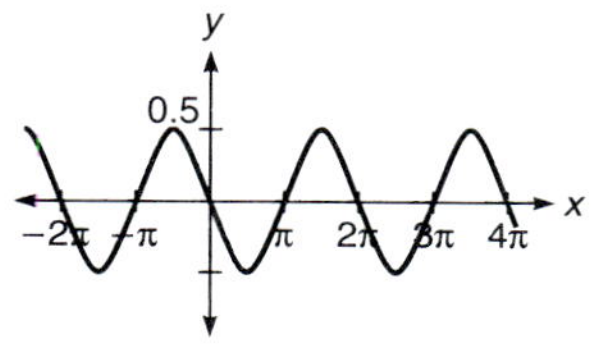

13.

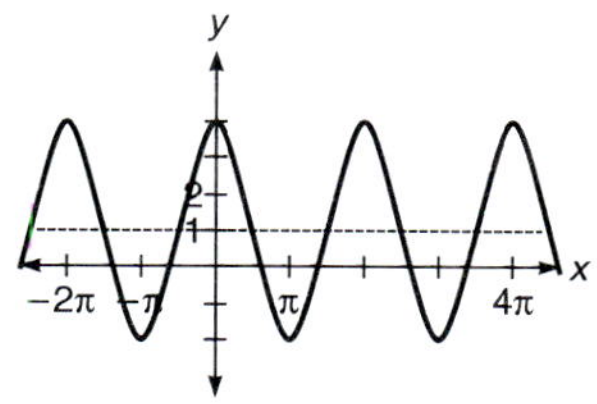

14.

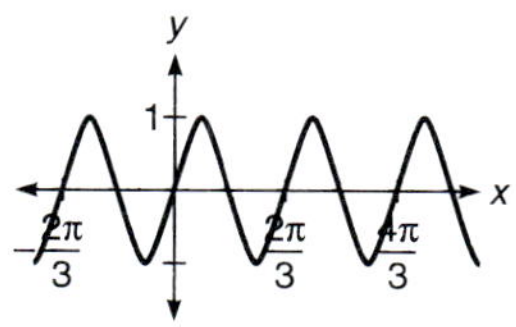

15.

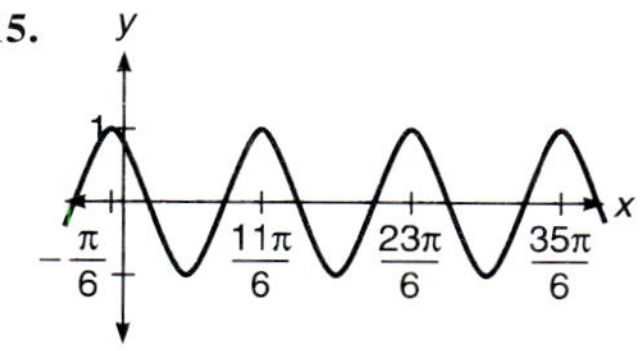

16. $y = 3\sin(\pi - 2x) = -3\sin(2x - \pi)$

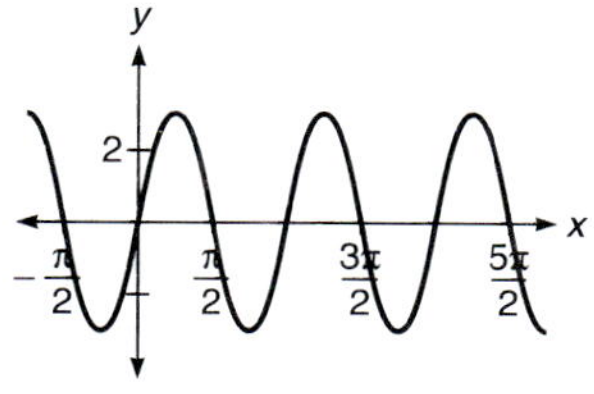

17. $A = +2$, $D = 0$, $B = 1$, $C = \frac{\pi}{3}$;

$$y = 2\sin\left(x + \frac{\pi}{3}\right)$$

18.

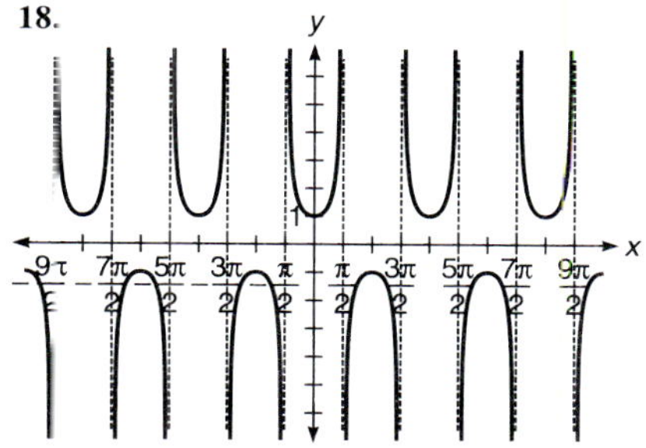

19. $f(x) = \sec x \cdot \sin x + x^3$

$$\begin{aligned} f(-x) &= \sec(-x) \cdot \sin(-x) + (-x)^3 \\ &= \sec x \cdot (-\sin x) + (-x^3) \\ &= -\sec x \cdot \sin x - x^3 \\ &= -(\sec x \cdot \sin x + x^3) \\ &= -f(x) \end{aligned}$$

Thus, $f(x)$ is an odd function.

20. $y = \tan(-2x) = -\tan 2x$

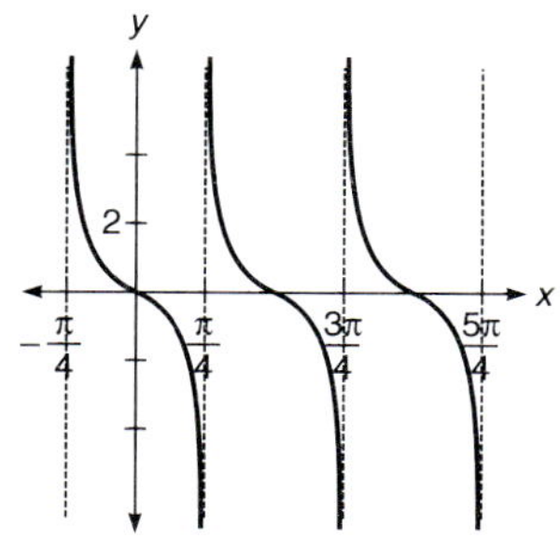

21.

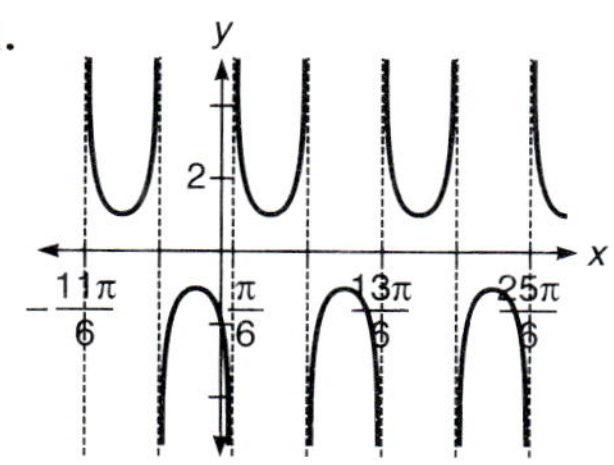

22. domain: $-1 \leq x \leq 1$; range: $0 \leq y \leq \pi$ **23.** $-\frac{\pi}{3}$, $-60°$

24. 2.50, 143.1° **25.** $\frac{3\sqrt{7}}{7}$ **26.** $-2\sqrt{2}$

27. $\frac{4\sqrt{3}}{3}$ **28.** $\frac{1}{\sqrt{1+4z^2}}$ **29.** $\frac{5\pi}{6}$

30. $\frac{\pi}{3}$ **31.** $\cos^{-1}\frac{z}{5}$ **32.** $\frac{1}{3}\cos^{-1}\frac{1}{6}$

33. 1.18, 67.8°

Chapter 7

Exercise 7–1

Answers to odd-numbered problems

1. $\dfrac{\sin\theta}{\tan\theta}$
$\dfrac{\sin\theta}{\dfrac{\sin\theta}{\cos\theta}}$
$\sin\theta \cdot \dfrac{\cos\theta}{\sin\theta}$
$\cos\theta$

3. $\cot\theta \sec\theta$
$\dfrac{\cos\theta}{\sin\theta} \cdot \dfrac{1}{\cos\theta}$
$\dfrac{1}{\sin\theta}$
$\csc\theta$

5. $\cot^2\theta \sin^2\theta$
$\dfrac{\cos^2\theta}{\sin^2\theta} \cdot \sin^2\theta$
$\cos^2\theta$

7. $(\tan^2\theta + 1)(1 - \sin^2\theta)$
$\sec^2\theta(\sin^2\theta + \cos^2\theta - \sin^2\theta)$
$\sec^2\theta \cos^2\theta$
$\dfrac{1}{\cos^2\theta}\cos^2\theta$
1

9. $\dfrac{(\sec\theta - 1)(\sec\theta + 1)}{\sin^2\theta}$
$\dfrac{\sec^2\theta - 1}{\sin^2\theta}$
$\dfrac{\tan^2\theta}{\sin^2\theta}$
$\tan^2\theta \dfrac{1}{\sin^2\theta}$
$\dfrac{\sin^2\theta}{\cos^2\theta} \cdot \dfrac{1}{\sin^2\theta}$
$\dfrac{1}{\cos^2\theta}$
$\sec^2\theta$

11. $(\csc x + \cot x)(1 - \cos x)$
$\csc x - \csc x \cos x + \cot x - \cot x \cos x$
$\dfrac{1}{\sin x} - \dfrac{1}{\sin x}\cos x + \cot x - \dfrac{\cos x}{\sin x}\cos x$
$\dfrac{1}{\sin x} - \cot x + \cot x - \dfrac{\cos^2 x}{\sin x}$
$\dfrac{1}{\sin x} - \dfrac{\cos^2 x}{\sin x}$
$\dfrac{1 - \cos^2 x}{\sin x}$
$\dfrac{\sin^2 x}{\sin x}$
$\sin x$

13. $\sec x - \tan x \sin x$
$\sec x - \dfrac{\sin x}{\cos x} \cdot \sin x$
$\dfrac{1}{\cos x} - \dfrac{\sin^2 x}{\cos x}$
$\dfrac{1 - \sin^2 x}{\cos x}$
$\dfrac{\cos^2 x}{\cos x}$
$\cos x$

15. $\cot x \sec x$
$\dfrac{\cos x}{\sin x} \cdot \dfrac{1}{\cos x}$
$\dfrac{1}{\sin x}$
$\csc x$

17. $\dfrac{\csc^2\theta - 1}{\csc^2\theta}$
$\dfrac{\cot^2\theta}{\csc^2\theta}$
$\dfrac{\cos^2\theta}{\sin^2\theta} \cdot \sin^2\theta$
$\cos^2\theta$

19. $\sin x + \cos x \cot x$
$\sin x + \cos x \dfrac{\cos x}{\sin x}$
$\dfrac{\sin^2 x}{\sin x} + \dfrac{\cos^2 x}{\sin x}$
$\dfrac{\sin^2 x + \cos^2 x}{\sin x}$
$\dfrac{1}{\sin x}$
$\csc x$

21. $\dfrac{\csc\theta \sin\theta}{\cot\theta}$
$\csc\theta \sin\theta \tan\theta$
$\dfrac{1}{\sin\theta} \cdot \sin\theta \cdot \tan\theta$
$\tan\theta$

23. $\dfrac{\tan\theta \cot\theta}{\sin\theta}$
$\tan\theta \cot\theta \csc\theta$
$\tan\theta \cdot \dfrac{1}{\tan\theta} \cdot \csc\theta$
$\csc\theta$

25. $\cos^2\theta(1 + \cot^2\theta)$
$\cos^2\theta \csc^2\theta$
$\dfrac{\cos^2\theta}{\sin^2\theta}$
$\cot^2\theta$

27. $\sin^2\theta(\csc^2\theta - 1)$
$\sin^2\theta \csc^2\theta - \sin^2\theta$
$1 - \sin^2\theta$
$\cos^2\theta$

29. $\tan^2\theta - \sec^2\theta$
$\tan^2\theta - (\tan^2\theta + 1)$
-1

31. $\dfrac{\cot\theta \sec\theta}{\csc\theta}$
$\cot\theta \sec\theta \sin\theta$
$\dfrac{\cos\theta}{\sin\theta} \cdot \dfrac{1}{\cos\theta} \sin\theta$
1

33. $\csc\theta + \cot\theta$
$\dfrac{1}{\sin\theta} + \dfrac{\cos\theta}{\sin\theta}$
$\dfrac{1 + \cos\theta}{\sin\theta}$

35. $\dfrac{\csc\theta}{\sec\theta + \tan\theta}$
$\dfrac{\dfrac{1}{\sin\theta}}{\dfrac{1}{\cos\theta} + \dfrac{\sin\theta}{\cos\theta}}$
$\dfrac{\dfrac{1}{\sin\theta}}{\dfrac{1 + \sin\theta}{\cos\theta}}$
$\dfrac{1}{\sin\theta} \cdot \dfrac{\cos\theta}{1 + \sin\theta}$
$\dfrac{\cos\theta}{\sin\theta + \sin^2\theta}$

37. $\dfrac{1 + \csc\theta}{1 + \sec\theta}$
$\dfrac{1 + \dfrac{1}{\sin\theta}}{1 + \dfrac{1}{\cos\theta}}$
$\dfrac{\dfrac{\sin\theta + 1}{\sin\theta}}{\dfrac{\cos\theta + 1}{\cos\theta}}$
$\dfrac{\sin\theta + 1}{\sin\theta} \cdot \dfrac{\cos\theta}{\cos\theta + 1}$
$\dfrac{\cos\theta}{\sin\theta} \cdot \dfrac{\sin\theta + 1}{\cos\theta + 1}$
$\cot\theta\left(\dfrac{1 + \sin\theta}{1 + \cos\theta}\right)$

39. $\dfrac{\tan^2\theta + \sec^2\theta}{\sec^2\theta}$
$\dfrac{\tan^2\theta}{\sec^2\theta} + \dfrac{\sec^2\theta}{\sec^2\theta}$
$\tan^2\theta \cos^2\theta + 1$
$\dfrac{\sin^2\theta}{\cos^2\theta}\cos^2\theta + 1$
$\sin^2\theta + 1$

41. $\dfrac{1 + \cot^2\theta}{\tan^2\theta}$
$\dfrac{1}{\tan^2\theta} + \dfrac{\cot^2\theta}{\tan^2\theta}$
$\cot^2\theta + \cot^2\theta \cot^2\theta$
$\cot^2\theta(1 + \cot^2\theta)$
$\cot^2\theta \csc^2\theta$

43. $\dfrac{1}{\sec - \cos\theta}$
$\dfrac{1}{\dfrac{1}{\cos\theta} - \cos\theta}$
$\dfrac{1}{\dfrac{1 - \cos^2\theta}{\cos\theta}}$
$\dfrac{1}{\dfrac{\sin^2\theta}{\cos\theta}}$
$\dfrac{\cos\theta}{\sin^2\theta}$
$\dfrac{\cos\theta}{\sin\theta} \cdot \dfrac{1}{\sin\theta}$
$\cot\theta \csc\theta$

45. $\dfrac{\cot^2\theta}{\csc\theta + 1}$
$\dfrac{\csc^2\theta - 1}{\csc\theta + 1}$
$\dfrac{(\csc\theta - 1)(\csc\theta + 1)}{\csc\theta + 1}$
$\csc\theta - 1$

47. $\dfrac{\tan^2\theta}{\sec\theta - 1}$
$\dfrac{\sec^2\theta - 1}{\sec\theta - 1}$
$\dfrac{(\sec\theta - 1)(\sec\theta + 1)}{\sec\theta - 1}$
$\sec\theta + 1$

49. $2\cos^2 x - 1$
$2\cos^2 x - (\sin^2 x + \cos^2 x)$
$\cos^2 x - \sin^2 x$

51. $\dfrac{1+\sin y}{\cos y}$

$\dfrac{1+\sin y}{\cos y}\cdot\dfrac{1-\sin y}{1-\sin y}$

$\dfrac{1-\sin^2 y}{\cos y(1-\sin y)}$

$\dfrac{\cos^2 y}{\cos y(1-\sin y)}$

$\dfrac{\cos y}{1-\sin y}$

53. $\dfrac{\cot x+1}{\cot x-1}$

$\dfrac{\dfrac{\cos x}{\sin x}+1}{\dfrac{\cos x}{\sin x}-1}$

$\dfrac{\dfrac{\cos x}{\sin x}+\dfrac{\sin x}{\sin x}}{\dfrac{\cos x}{\sin x}-\dfrac{\sin x}{\sin x}}\cdot\dfrac{\sin x}{\sin x}$

$\dfrac{\cos x+\sin x}{\cos x-\sin x}$

55. $\dfrac{\cos x}{\sec x-\tan x}$

$\dfrac{\cos x}{\dfrac{1}{\cos x}-\dfrac{\sin x}{\cos x}}$

$\dfrac{\cos x}{\dfrac{1-\sin x}{\cos x}}$

$\cos x\cdot\dfrac{\cos x}{1-\sin x}$

$\dfrac{\cos^2 x}{1-\sin x}$

57. $\dfrac{1}{1-\sin x}+\dfrac{1}{1+\sin x}$

$\dfrac{(1+\sin x)+(1-\sin x)}{1-\sin^2 x}$

$\dfrac{2}{\cos^2 x}$

$2\sec^2 x$

59. $\sin^4 x-\cos^4 x$

$(\sin^2 x-\cos^2 x)(\sin^2 x+\cos^2 x)$

$(\sin^2 x-\cos^2 x)(1)$

$\sin^2 x-\cos^2 x$

61. $\csc^2 y+\sec^2 y$

$\dfrac{1}{\sin^2 y}+\dfrac{1}{\cos^2 y}$

$\dfrac{\cos^2 y+\sin^2 y}{\sin^2 y\cos^2 y}$

$\dfrac{1}{\sin^2 y\cos^2 y}$

$\csc^2 y\sec^2 y$

63. $\dfrac{\cot^2 x-1}{\cot^2 x+1}$

$\dfrac{\dfrac{\cos^2 x}{\sin^2 x}-\dfrac{\sin^2 x}{\sin^2 x}}{\dfrac{\cos^2 x}{\sin^2 x}+\dfrac{\sin^2 x}{\sin^2 x}}\cdot\dfrac{\sin^2 x}{\sin^2 x}$

$\dfrac{\cos^2 x-\sin^2 x}{\cos^2 x+\sin^2 x}$

$\cos^2 x-\sin^2 x$

65. $\sec^4 x-\sec^2 x$

$\sec^2 x(\sec^2 x-1)$

$(\tan^2 x+1)[(\tan^2 x+1)-1]$

$(\tan^2 x+1)(\tan^2 x)$

$\tan^4 x+\tan^2 x$

In each of problems 67–75, let $\theta=\dfrac{\pi}{3}$;

$\sin\dfrac{\pi}{3}=\dfrac{\sqrt{3}}{2}$; $\cos\dfrac{\pi}{3}=\dfrac{1}{2}$; $\tan\dfrac{\pi}{3}=\sqrt{3}$.

67. $\sin\theta=1-\cos\theta$

$\sin\dfrac{\pi}{6}=1-\cos\dfrac{\pi}{6}$

$\dfrac{1}{2}\neq 1-\dfrac{\sqrt{3}}{2}$

69. $\sec\theta=\dfrac{1}{\csc\theta}$

$2=\dfrac{1}{\dfrac{2}{\sqrt{3}}}$

$2\neq\dfrac{\sqrt{3}}{2}$

71. $\sin^2\theta-2\cos\theta\sin\theta+\cos^2\theta=2$

$\left(\dfrac{\sqrt{3}}{2}\right)^2-2\left(\dfrac{1}{2}\right)\left(\dfrac{\sqrt{3}}{2}\right)+\left(\dfrac{1}{2}\right)^2=2$

$\dfrac{3}{4}-\dfrac{\sqrt{3}}{2}+\dfrac{1}{4}=2$

$1-\dfrac{\sqrt{3}}{2}\neq 2$

73. $\csc\theta+\sec\theta\cot\theta=2$

$\dfrac{2}{\sqrt{3}}+2\cdot\dfrac{1}{\sqrt{3}}=2$

$\dfrac{4}{\sqrt{3}}\neq 2$

75. $\dfrac{1-\cos\theta}{1+\cos\theta}=\sin^2\theta$

$\dfrac{1-\frac{1}{2}}{1+\frac{1}{2}}=\left(\dfrac{\sqrt{3}}{2}\right)^2$

$\frac{1}{3}\neq\frac{3}{4}$

77. **a.** $\theta=\dfrac{\pi}{6}$:

$(\csc^2\theta-1)(\sec^2\theta-1)=1$

$\left(\csc^2\dfrac{\pi}{6}-1\right)\left(\sec^2\dfrac{\pi}{6}-1\right)=1$

$(2^2-1)\left(\left(\dfrac{2}{\sqrt{3}}\right)^2-1\right)$

$(3)(\frac{4}{3}-1)$

1

$\theta=\dfrac{\pi}{4}$:

$\left(\csc^2\dfrac{\pi}{4}-1\right)\left(\sec^2\dfrac{\pi}{4}-1\right)=1$

$((\sqrt{2})^2-1)((\sqrt{2})^2-1)$

$(2-1)(2-1)$

1

b. Yes:

$(\csc^2\theta-1)(\sec^2\theta-1)=1$

$\cot^2\theta\tan^2\theta$

$\dfrac{1}{\tan^2\theta}\tan^2\theta$

1

79. **a.** $2\sin^2\theta+\sin\theta=1$

$\theta=\dfrac{\pi}{6}$:

$2(\frac{1}{2})^2+\frac{1}{2}=1$

$1=1$

$\theta=\dfrac{3\pi}{2}$:

$2(-1)^2+(-1)=1$

$1=1$

b. No;

let $\theta=0$:

$2(0^2)+0=1$

$0\neq 1$

Solutions to skill and review problems

1. $r = \sqrt{x^2 + y^2} = \sqrt{40} = 2\sqrt{10}$

$\sin\theta = \frac{y}{r} = \frac{-2}{2\sqrt{10}} = -\frac{\sqrt{10}}{10}$;

$\cos\theta = \frac{x}{r} = \frac{6}{2\sqrt{10}} = \frac{3\sqrt{10}}{10}$;

$\tan\theta = \frac{y}{x} = -\frac{2}{6} = -\frac{1}{3}$

2. $\tan 15° = \frac{x}{4}$; $x = 4\tan 15° \approx 1.07$ feet

3. $\frac{\theta°}{180°} = \frac{s}{\pi}$; $\theta° = \frac{180°}{\pi}s$

$= \frac{180°}{\pi}\left(-\frac{6\pi}{5}\right) = -216°$

4. $r = \sqrt{29}$; $\cos\theta = \frac{2}{r} = \frac{2}{\sqrt{29}} = \frac{2\sqrt{29}}{29}$

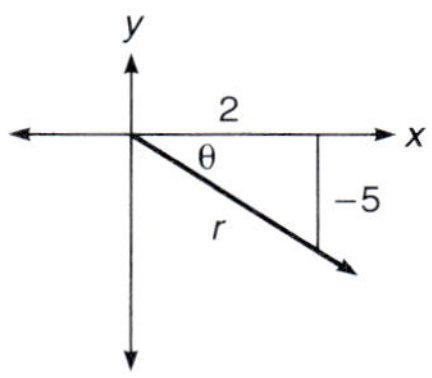

5. The graph of $y = |x - 3|$ is the graph of $y = |x|$, shifted three units to the right.

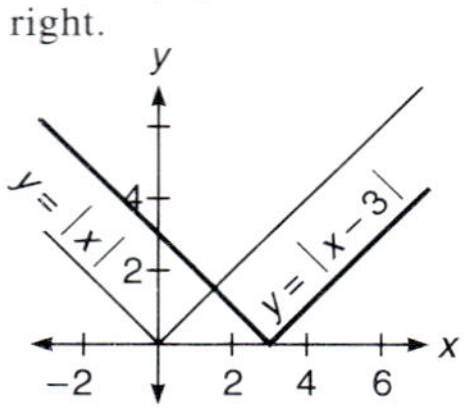

6. $0 \le 2x - \frac{\pi}{3} \le 2\pi$

$\frac{\pi}{3} \le 2x \le \frac{7\pi}{3}$

$\frac{\pi}{6} \le x \le \frac{7\pi}{6}$; one complete cycle from $\frac{\pi}{6}$ to $\frac{7\pi}{6}$; period is $\frac{7\pi}{6} - \frac{\pi}{6} = \pi$; amplitude is 1

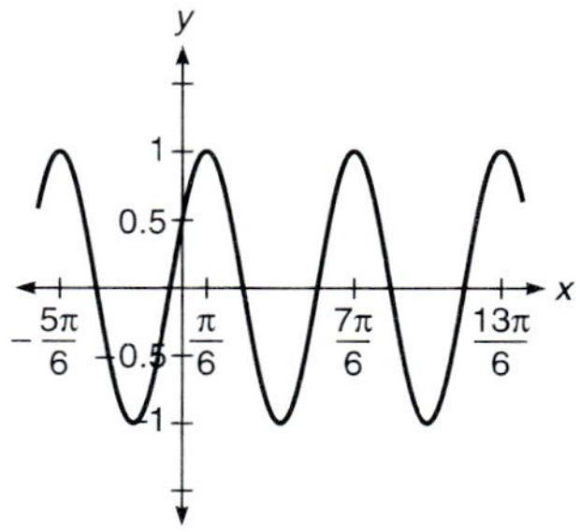

7. $4\sin^2 x - 1 = 0$, $0 \le x < 2\pi$ (This implies answers should be in radians measure.)

$(2\sin x - 1)(2\sin x + 1) = 0$

$2\sin x - 1 = 0$ or $2\sin x + 1 = 0$

$2\sin x = 1$ or $2\sin x = -1$

$\sin x = \frac{1}{2}$ or $\sin x = -\frac{1}{2}$; The reference angle for both cases is $x' = \frac{\pi}{6}$. $\sin x$ is positive in quadrants I and II.

I: $x = x' = \frac{\pi}{6}$

II: $x = \pi - x' = \pi - \frac{\pi}{6} = \frac{5\pi}{6}$

$\sin x$ negative in quadrants III and IV.

III: $x = \pi + x' = \pi + \frac{\pi}{6} = \frac{7\pi}{6}$

IV: $x = 2\pi - x' = 2\pi - \frac{\pi}{6} = \frac{11\pi}{6}$

x is any one of $\frac{\pi}{6}, \frac{5\pi}{6}, \frac{7\pi}{6}$, or $\frac{11\pi}{6}$.

Exercise 7–2

Answers to odd-numbered problems

1. $\cos 72°$ **3.** $\cot 82°$ **5.** $\csc\frac{\pi}{6}$

7. $\sin\left(-\frac{\pi}{3}\right)$ **9.** $\csc\frac{5\pi}{4}$ **11.** 1

13. 1 **15.** 1 **17.** −1 **19.** 1 **21.** 1

23. 1 **25.** 1 **27.** $\frac{\sqrt{6}+\sqrt{2}}{4}$

29. $\frac{\sqrt{6}+\sqrt{2}}{4}$ **31.** $\frac{\sqrt{6}+\sqrt{2}}{4}$

33. $\frac{\sqrt{6}-\sqrt{2}}{4}$ **35.** $\frac{\sqrt{2}-\sqrt{6}}{4}$

37. $2+\sqrt{3}$ **39.** $\frac{2\sqrt{14}+3}{12}$

41. $\frac{-15-12\sqrt{7}}{36-5\sqrt{7}}$ **43.** $\frac{77}{85}$

45. $\frac{-\sqrt{5}-4\sqrt{2}}{9}$ **47.** $\frac{8+\sqrt{5}}{51}\sqrt{17}$

49. $-\frac{119}{120}$ **51.** $-\frac{\sqrt{5}}{5}$ **53.** $\frac{-4\sqrt{6}+\sqrt{5}}{15}$

55. $2+\sqrt{3}$

57. $\sin(\pi - \theta) = \sin\pi\cos\theta - \cos\pi\sin\theta$
$= 0\cos\theta - (-1)\sin\theta$
$= \sin\theta$

59. $\cos(\pi - \theta) = \cos\pi\cos\theta + \sin\pi\sin\theta$
$= (-1)\cos\theta + 0\sin\theta$
$= -\cos\theta$

61. $\tan(\pi - \theta) = \frac{\tan\pi - \tan\theta}{1 + \tan\pi\tan\theta}$
$= \frac{0 - \tan\theta}{1 + 0\tan\theta}$
$= -\tan\theta$

63. $\sin(\theta + 2\pi) = \sin\theta\cos 2\pi + \cos\theta\sin 2\pi$
$= \sin\theta(1) + \cos\theta(0)$
$= \sin\theta$

65. $\tan(\theta + \pi) = \frac{\tan\theta + \tan\pi}{1 - \tan\theta\tan\pi}$
$= \frac{\tan\theta + 0}{1 - \tan\theta(0)}$
$= \tan\theta$

67. $\frac{1}{2}[\sin(\alpha+\beta) - \sin(\alpha-\beta)]$
$\frac{1}{2}[\sin\alpha\cos\beta + \cos\alpha\sin\beta - (\sin\alpha\cos\beta - \cos\alpha\sin\beta)]$
$\frac{1}{2}[2\cos\alpha\sin\beta]$
$\cos\alpha\sin\beta$

69. $\frac{1}{2}[\cos(\alpha-\beta) - \cos(\alpha+\beta)]$
$\frac{1}{2}[\cos\alpha\cos\beta + \sin\alpha\sin\beta - (\cos\alpha\cos\beta - \sin\alpha\sin\beta)]$
$\frac{1}{2}[2\sin\alpha\sin\beta]$
$\sin\alpha\sin\beta$

71. $3\frac{12}{29}$

73. $\cos(\alpha+\beta) = \cos\alpha\cos\beta - \sin\alpha\sin\beta$
This was shown true in the text.
$\cos(\alpha + (-\beta)) = \cos\alpha\cos(-\beta) - \sin\alpha\sin(-\beta)$
Replace β by $-\beta$. This is valid since the identity is true for all angles and α.
$\cos(\alpha-\beta) = \cos\alpha\cos\beta - \sin\alpha[-\sin\beta]$
$\alpha + (-\beta) = \alpha - \beta$; $\cos(-\theta) = \cos\theta$; $\sin(-\theta) = -\sin\theta$.
$\cos(\alpha-\beta) = \cos\alpha\cos\beta + \sin\alpha\sin\beta$
This statement is true since the preceding statements are true.

75. $\cos\left(\frac{\pi}{2} - \theta\right) = \sin\theta$ Proved true above.

Let $\alpha = \frac{\pi}{2} - \theta$, then $\theta = \frac{\pi}{2} - \alpha$.

$\cos\alpha = \sin\left(\frac{\pi}{2} - \alpha\right)$

Substitution of expression (section 1–3).

$\sin\left(\frac{\pi}{2} - \theta\right) = \cos\theta$

The variable name α or θ is unimportant.

77. $\sin(\alpha + \beta) = \sin\alpha\cos\beta + \cos\alpha\sin\beta$

$\sin(\alpha + (-\beta)) = \sin\alpha\cos(-\beta) + \cos\alpha\sin(-\beta)$

$\sin(\alpha - \beta) = \sin\alpha\cos\beta + \cos\alpha[-\sin\beta]$

Cosine is an even function, sine is odd.

$= \sin\alpha\cos\beta - \cos\alpha\sin\beta$

79. $\cot\left(\frac{\pi}{2} - \theta\right) = \frac{1}{\tan\left(\frac{\pi}{2} - \theta\right)} = \frac{1}{\cot\theta} = \tan\theta$

81. $\csc\left(\frac{\pi}{2} - \theta\right) = \frac{1}{\sin\left(\frac{\pi}{2} - \theta\right)} = \frac{1}{\cos\theta} = \sec\theta$

Solutions to skill and review problems

1. $(x_1,y_1) = (-4,3)$; $(x_2,y_2) = (-8,11)$

$m = \frac{y_2 - y_1}{x_2 - x_1} = \frac{11 - 3}{-8 - (-4)} = -2$

$y - y_1 = m(x - x_1)$

$y - 3 = -2(x - (-4))$

$y - 3 = -2x - 8$

$y = -2x - 5$

2. $b = \sqrt{4^2 - 3^2} = \sqrt{7}$

$\cos B = \frac{3}{4}$, so $B = \cos^{-1}\frac{3}{4} \approx 41.4°$

$\sin A = \frac{3}{4}$, so $A = \sin^{-1}\frac{3}{4} \approx 48.6°$

3. The point $(x,y) = (2,6)$ lies on the terminal side of the angle. $\tan\theta = \frac{y}{x}$

$= \frac{6}{2} = 3$, so $\theta = \tan^{-1}3 \approx 71.6°$

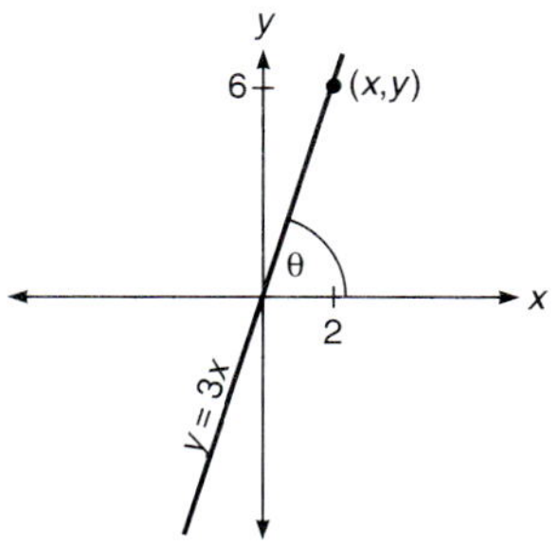

4. Possible rational zeros of $3x^4 - 5x^3 - 14x^2 + 20x + 8$ have numerators that are factors of 8, and denominators that are factors of 3. Thus, the possible rational zeros are $\pm1, \pm2, \pm4, \pm8, \pm\frac{1}{3}, \pm\frac{2}{3}, \pm\frac{4}{3}$, and $\pm\frac{8}{3}$.

	3	−5	−14	20	8
		−6	22	−16	−8
−2	3	−11	8	4	0

−2 is a zero,
so $3x^4 - 5x^3 - 14x^2 + 20x + 8$
$= (x + 2)(3x^3 - 11x^2 + 8x + 4)$

	3	−11	8	4
		6	−10	−4
2	3	−5	−2	0

2 is a zero of $3x^3 - 11x^2 + 8x + 4$, so
$3x^4 - 5x^3 - 14x^2 + 20x + 8$
$= (x + 2)(x - 2)(3x^2 - 5x - 2)$
$= (x + 2)(x - 2)(3x + 1)(x - 2)$
$= (x + 2)(x - 2)^2(3x + 1)$.

Note that if the zero $-\frac{1}{3}$ were used in the synthetic division, the result would be $3(x + 2)(x - 2)^2(x + \frac{1}{3})$.

5. Using the circle with arc length L:
$L = rs$; $14.6 = 4s$; $3.65 = s$
Using the circle with arc length T:
$T = rs$; $T = 6(3.65) = 21.9$

6. $\sin^{-1}\left(\sin\frac{5\pi}{3}\right) = \sin^{-1}\left(-\frac{\sqrt{3}}{2}\right)$

$= -\sin^{-1}\frac{\sqrt{3}}{2} = -\frac{\pi}{3}$

7. $\frac{\csc^2 x - 1}{\sin^2 x}$

$\frac{\cot^2\theta}{\sin^2\theta}$

$\cot^2\theta + 1 = \csc^2\theta$, so $\csc^2\theta - 1 = \cot^2\theta$

$\cot^2\theta \cdot \frac{1}{\sin^2\theta}$

$\frac{\cos^2\theta}{\sin^2\theta} \cdot \csc^2\theta$

$\cos^2\theta \cdot \frac{1}{\sin^2\theta} \cdot \csc^2\theta$

$\cos^2\theta\csc^2\theta\csc^2\theta$

$\cos^2\theta\csc^4\theta$

Solutions to trial exercise problems

8. $\sin\left(-\frac{\pi}{3}\right) = \cos\left(\frac{\pi}{2} - \left(-\frac{\pi}{3}\right)\right) = \cos\frac{5\pi}{6}$

17. $\tan^2 8° - \csc^2 82° = \tan^2 8° - \sec^2 8° = -1$
(Since $\tan^2\theta + 1 = \sec^2\theta$, $\tan^2\theta - \sec^2\theta = -1$.)

28. $\tan\frac{\pi}{12} = \tan\left(\frac{\pi}{4} - \frac{\pi}{6}\right) = \frac{\tan\frac{\pi}{4} - \tan\frac{\pi}{6}}{1 + \tan\frac{\pi}{4}\tan\frac{\pi}{6}}$

$= \frac{1 - \frac{\sqrt{3}}{3}}{1 + 1\left(\frac{\sqrt{3}}{3}\right)} \cdot \frac{3}{3} = \frac{3 - \sqrt{3}}{3 + \sqrt{3}} \cdot \frac{3 - \sqrt{3}}{3 - \sqrt{3}}$

$= \frac{12 - 6\sqrt{3}}{6} = 2 - \sqrt{3}$

40. $\cos(\alpha - \beta) = \cos\alpha\cos\beta + \sin\alpha\sin\beta$

$$= -\frac{12}{13}\cdot\left(-\frac{\sqrt{3}}{2}\right) + \frac{5}{13}\cdot\frac{1}{2} = \frac{12\sqrt{3}+5}{26}$$

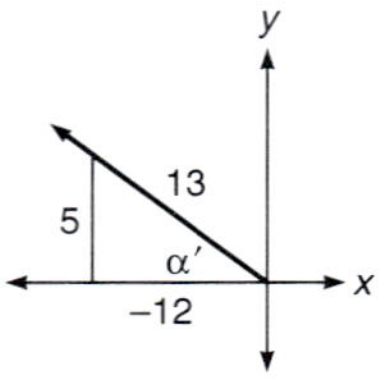

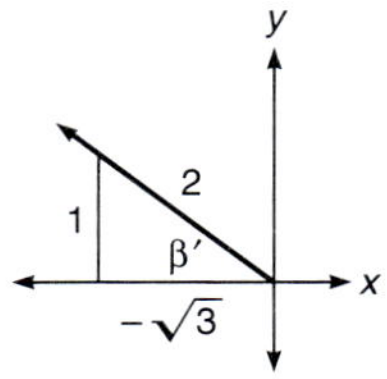

48. $\cos(\alpha - \beta) = \cos\alpha\cos\beta + \sin\alpha\sin\beta$

$$= -\tfrac{4}{5}(-\tfrac{3}{5}) + (-\tfrac{3}{5})(-\tfrac{4}{5}) = \tfrac{24}{25}$$

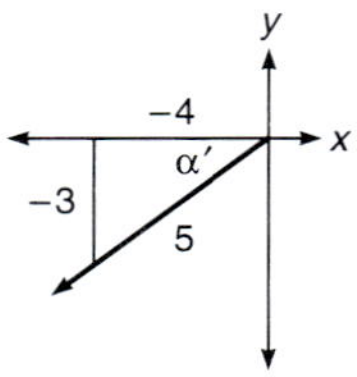

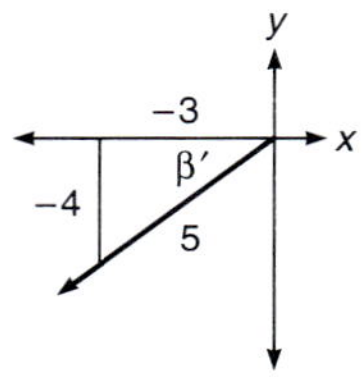

70. $\tan\alpha = \dfrac{2+6}{10} = \dfrac{4}{5}$; $\tan\beta = \dfrac{6}{10} = \dfrac{3}{5}$

$$\tan(\alpha - \beta) = \frac{\tan\alpha - \tan\beta}{1 + \tan\alpha\tan\beta} = \frac{\frac{4}{5} - \frac{3}{5}}{1 + \frac{4}{5}\cdot\frac{3}{5}} = \frac{\frac{1}{5}}{1 + \frac{12}{25}}\cdot\frac{25}{25} = \frac{5}{25+12} = \tfrac{5}{37}$$

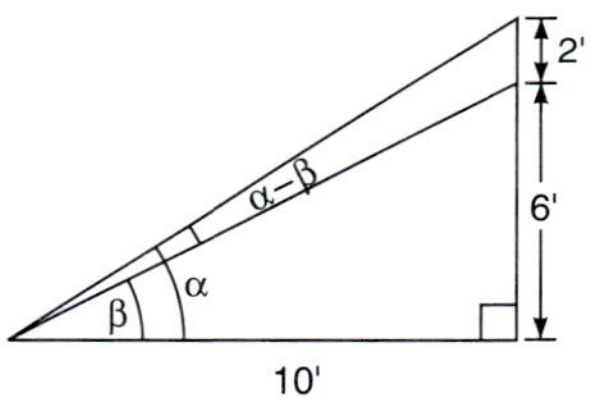

76. $\sin\theta = \cos\left(\dfrac{\pi}{2} - \theta\right)$

This is identity [6], which we know is true.

$\sin(\alpha+\beta) = \cos\left[\dfrac{\pi}{2} - (\alpha+\beta)\right]$

Replace θ by $\alpha + \beta$. This is substitution of expression (section 1–3).

$\sin(\alpha+\beta) = \cos\left[\left(\dfrac{\pi}{2} - \alpha\right) - \beta\right]$

Regroup $\dfrac{\pi}{2} - \alpha - \beta$.

$= \cos\left(\dfrac{\pi}{2} - \alpha\right)\cos\beta + \sin\left(\dfrac{\pi}{2} - \alpha\right)\sin\beta$

Using identity [2].

$= \sin\alpha\cos\beta + \cos\alpha\sin\beta$

Using identities [6] and [5].

Thus, identity [3] is true.

Exercise 7–3

Answers to odd-numbered problems

1. $\sin\dfrac{\pi}{2}$ **3.** $\cos 6\pi$ **5.** $\cos\dfrac{\pi}{5}$

7. $3\tan 20°$ **9.** $\sin 12\theta$ **11.** $3\cos 10\theta$

13. $5\tan 6\theta$ **15.** $2\cos 14\theta$

17. $3\cos 6\theta$ **19.** $70°$ **21.** $\dfrac{5\pi}{12}$

23. $35°$ **25.** $20°$ **27.** $\dfrac{\pi}{8}$ **29.** $\dfrac{4\pi}{5}$

31. $\frac{24}{25}, \frac{7}{25}, \frac{24}{7}$ **33.** $-\frac{24}{25}, \frac{7}{25}, -\frac{24}{7}$

35. $\dfrac{10\sqrt{39}}{64}, \dfrac{7}{32}, \dfrac{5\sqrt{39}}{7}$

37. $\dfrac{\sqrt{70}}{10}, -\dfrac{\sqrt{30}}{10}, -\dfrac{\sqrt{21}}{3}$

39. $\dfrac{\sqrt{50 - 20\sqrt{5}}}{10}, -\dfrac{\sqrt{50 + 20\sqrt{5}}}{10}, 2 - \sqrt{5}$

41. **a.** $\dfrac{\sqrt{2-\sqrt{3}}}{2}$ **b.** $\dfrac{\sqrt{2+\sqrt{3}}}{2}$ **c.** $2 - \sqrt{3}$

43. $\dfrac{\sqrt{6+3\sqrt{3}} + \sqrt{2-\sqrt{3}}}{4}$

45. $\dfrac{\sqrt{4+\sqrt{6}+2\sqrt{3}+2\sqrt{2}} - \sqrt{4+\sqrt{6}-2\sqrt{3}-2\sqrt{2}}}{4}$

47. $\dfrac{\sqrt{2-\sqrt{2+\sqrt{3}}}}{2}$

49. $\sin 2\theta + 1$

$2\sin\theta\cos\theta + 1$

$2\sin\theta\cos\theta + \sin^2\theta + \cos^2\theta$

$\sin^2\theta + 2\sin\theta\cos\theta + \cos^2\theta$

$(\sin\theta + \cos\theta)^2$

51. $\cos^4\theta - \sin^4\theta$

$(\cos^2\theta - \sin^2\theta)(\cos^2\theta + \sin^2\theta)$

$\cos 2\theta(1)$

$\cos 2\theta$

53. $\dfrac{1+\cos 2\theta}{1-\cos 2\theta}$

$\dfrac{1 + (2\cos^2\theta - 1)}{1 - (1 - 2\sin^2\theta)}$

$\dfrac{2\cos^2\theta}{2\sin^2\theta}$

$\dfrac{\cos^2\theta}{\sin^2\theta}$

$\cot^2\theta$

55. $\dfrac{2\cos 2\theta}{\sin 2\theta}$

$\dfrac{2(\cos^2\theta - \sin^2\theta)}{2\sin\theta\cos\theta}$

$\dfrac{\cos^2\theta - \sin^2\theta}{\sin\theta\cos\theta}$

$\dfrac{\cos^2\theta}{\sin\theta\cos\theta} - \dfrac{\sin^2\theta}{\sin\theta\cos\theta}$

$\dfrac{\cos\theta}{\sin\theta} - \dfrac{\sin\theta}{\cos\theta}$

$\cot\theta - \tan\theta$

57. $\sin 2\theta - 4\sin^3\theta\cos\theta$

$2\sin\theta\cos\theta - 4\sin^3\theta\cos\theta$

$2\sin\theta\cos\theta(1 - 2\sin^2\theta)$

$\sin 2\theta\cos 2\theta$

59. $\dfrac{2}{1-\cos 2\theta}$

$\dfrac{2}{1-(2\cos^2\theta-1)}$

$\dfrac{2}{2-2\cos^2\theta}$

$\dfrac{2}{2(1-\cos^2\theta)}$

$\dfrac{1}{\sin^2\theta}$

$\csc^2\theta$

61. Left side:

$\tan 2\theta$

$\dfrac{2\tan\theta}{1-\tan^2\theta}$

Right side:

$\dfrac{2(\tan\theta+\tan^3\theta)}{1-\tan^4\theta}$

$\dfrac{2\tan\theta(1+\tan^2\theta)}{(1-\tan^2\theta)(1+\tan^2\theta)}$

$\dfrac{2\tan\theta}{1-\tan^2\theta}$

63. $2\csc 2\theta\sin\theta\cos\theta$

$2\sin\theta\cos\theta\csc 2\theta$

$\sin 2\theta\cdot\dfrac{1}{\sin 2\theta}=1$

65. $\dfrac{1-\tan^2\theta}{1+\tan^2\theta}$

$\dfrac{1-\dfrac{\sin^2\theta}{\cos^2\theta}}{1+\dfrac{\sin^2\theta}{\cos^2\theta}}$

$\dfrac{\dfrac{\cos^2\theta-\sin^2\theta}{\cos^2\theta}}{\dfrac{\cos^2\theta+\sin^2\theta}{\cos^2\theta}}$

$\dfrac{\cos^2\theta-\sin^2\theta}{\cos^2\theta+\sin^2\theta}$

$\dfrac{\cos 2\theta}{1}$

$\cos 2\theta$

67. $\sec^2\dfrac{\theta}{2}$

$\dfrac{1}{\cos^2\dfrac{\theta}{2}}$

$\dfrac{1}{\left(\pm\sqrt{\dfrac{1+\cos\theta}{2}}\right)^2}$

$\dfrac{1}{\dfrac{1+\cos\theta}{2}}$

$\dfrac{2}{1+\cos\theta}$

69. $\cos^2\dfrac{\theta}{2}$

$\left(\pm\sqrt{\dfrac{1+\cos\theta}{2}}\right)^2$

$\dfrac{1+\cos\theta}{2}$

$\dfrac{1+\cos\theta}{2}\cdot\dfrac{1-\cos\theta}{1-\cos\theta}$

$\dfrac{1-\cos^2\theta}{2-2\cos\theta}$

71. $2\cos^2\dfrac{\theta}{2}-\cos\theta$

$2\left(\pm\sqrt{\dfrac{1+\cos\theta}{2}}\right)^2-\cos\theta$

$2\dfrac{1+\cos\theta}{2}-\cos\theta$

$1+\cos\theta-\cos\theta$

1

73. $\sin^2\dfrac{\theta}{2}-\cos^2\dfrac{\theta}{2}$

$\dfrac{1-\cos\theta}{2}-\dfrac{1+\cos\theta}{2}$

$\dfrac{-2\cos\theta}{2}$

$-\cos\theta$

75. Left side:

$\tan^2\dfrac{\theta}{2}$

$\left(\sqrt{\dfrac{1-\cos\theta}{1+\cos\theta}}\right)^2$

$\dfrac{1-\cos\theta}{1+\cos\theta}$

Right side:

$\dfrac{2}{1+\cos\theta}-1$

$\dfrac{2}{1+\cos\theta}-\dfrac{1+\cos\theta}{1+\cos\theta}$

$\dfrac{2-(1+\cos\theta)}{1+\cos\theta}$

$\dfrac{1-\cos\theta}{1+\cos\theta}$

77. $4\sin^2\dfrac{\theta}{2}\cos^2\dfrac{\theta}{2}$

$4\dfrac{1-\cos\theta}{2}\cdot\dfrac{1+\cos\theta}{2}$

$4\dfrac{1-\cos^2\theta}{4}$

$\sin^2\theta$

---

79. $\tan\dfrac{\theta}{2}+\cot\dfrac{\theta}{2}$

$\dfrac{\sin\theta}{1+\cos\theta}+\dfrac{1}{\dfrac{1-\cos\theta}{\sin\theta}}$

$\dfrac{\sin\theta}{1+\cos\theta}+\dfrac{\sin\theta}{1-\cos\theta}$

$\dfrac{\sin\theta(1-\cos\theta)+\sin\theta(1+\cos\theta)}{1-\cos^2\theta}$

$\dfrac{2\sin\theta}{\sin^2\theta}$

$\dfrac{2}{\sin\theta}$

81. $\sin 3\theta$

$\sin(2\theta+\theta)$

$\sin 2\theta\cos\theta+\cos 2\theta\sin\theta$

$(2\sin\theta\cos\theta)(\cos\theta)+(1-2\sin^2\theta)(\sin\theta)$

$2\sin\theta\cos^2\theta+\sin\theta-2\sin^3\theta$

$2\sin\theta(1-\sin^2\theta)+\sin\theta-2\sin^3\theta$

$2\sin\theta-2\sin^3\theta-\sin\theta-2\sin^3\theta$

$3\sin\theta-4\sin^3\theta$

83. a. $\sin 4\theta$

$\sin 2(2\theta)$

$2\sin 2\theta\cos 2\theta$

$2(2\sin\theta\cos\theta)(1-2\sin^2\theta)$

$4\sin\theta\cos\theta-8\sin^3\theta\cos\theta$

Depending on how $\cos 2\theta$ is expanded, other possible answers are

$\sin 4\theta=8\cos^3\theta\sin\theta-4\cos\theta\sin\theta$

$\sin 4\theta=4\cos^3\theta\sin\theta-4\sin^3\theta\cos\theta$

b. $\cos 4\theta=\cos 2(2\theta)$

$=2\cos^2(2\theta)-1$

$=2[\cos 2\theta]^2-1$

$=2[2\cos^2\theta-1]^2-1$

$=2[4\cos^4\theta-4\cos^2\theta+1]-1$

$=8\cos^4\theta-8\cos^2\theta+1$

85. $\frac{3}{16}a\frac{1-(\cos^2\alpha-\sin^2\alpha)}{1-\cos\alpha}$

$\frac{3}{16}a\frac{(\sin^2\alpha+\cos^2\alpha)-(\cos^2\alpha-\sin^2\alpha)}{1-\cos\alpha}$

$\frac{3}{16}a\frac{2\sin^2\alpha}{1-\cos\alpha}=\frac{3}{8}a\frac{1-\cos^2\alpha}{1-\cos\alpha}$

$\frac{3}{8}a\frac{(1-\cos\alpha)(1+\cos\alpha)}{1-\cos\alpha}$

$\frac{3}{8}a(1+\cos\alpha)$

87. $\tan\frac{\alpha}{2}=\frac{\sin\alpha}{1+\cos\alpha}$; let $\frac{\alpha}{2}=\theta$, so $\alpha=2\theta$.

Use substitution of expression:

$$\tan\theta=\frac{\sin 2\theta}{1+\cos 2\theta}$$
$$=\frac{2\sin\theta\cos\theta}{1+(2\cos^2\theta-1)}$$
$$=\frac{2\sin\theta\cos\theta}{2\cos^2\theta}$$
$$=\frac{\sin\theta}{\cos\theta}=\tan\theta$$

89. a. Problem 41 shows that $\tan 15° = 2-\sqrt{3}$.

b. Problem 41 shows that $\sin 15° = \frac{\sqrt{2-\sqrt{3}}}{2}$, and that

$\cos 15° = \frac{\sqrt{2+\sqrt{3}}}{2}$, so $\tan 15° = \frac{\frac{\sqrt{2-\sqrt{3}}}{2}}{\frac{\sqrt{2+\sqrt{3}}}{2}}$.

c. $$\frac{\frac{\sqrt{2-\sqrt{3}}}{2}}{\frac{\sqrt{2+\sqrt{3}}}{2}}=\frac{\sqrt{2-\sqrt{3}}}{2}\cdot\frac{2}{\sqrt{2+\sqrt{3}}}=\frac{\sqrt{2-\sqrt{3}}}{\sqrt{2+\sqrt{3}}}$$
$$=\frac{\sqrt{2-\sqrt{3}}}{\sqrt{2+\sqrt{3}}}\cdot\frac{\sqrt{2+\sqrt{3}}}{\sqrt{2+\sqrt{3}}}=\frac{\sqrt{(2-\sqrt{3})(2+\sqrt{3})}}{2+\sqrt{3}}$$
$$=\frac{1}{2+\sqrt{3}}=\frac{1}{2+\sqrt{3}}\cdot\frac{2-\sqrt{3}}{2-\sqrt{3}}=\frac{2-\sqrt{3}}{1}=2-\sqrt{3}.$$

91. $\cos(\alpha+\beta)=\cos\alpha\cos\beta-\sin\alpha\sin\beta$

$\cos(\alpha+\alpha)=\cos\alpha\cos\alpha-\sin\alpha\sin\alpha$

$\cos 2\alpha=\cos^2\alpha-\sin^2\alpha$

$=(1-\sin^2\alpha)-\sin^2\alpha$

$\sin^2\theta+\cos^2\theta=1$, so $\cos^2\theta=1-\sin^2\theta$

$=1-2\sin^2\alpha$

$\cos 2\alpha=\cos^2\alpha-\sin^2\alpha$

$=\cos^2\alpha-(1-\cos^2\alpha)$

$\sin^2\theta+\cos^2\theta=1$, so $\sin^2\theta=1-\cos^2\theta$

$=2\cos^2\alpha-1$

$$\tan(\alpha+\beta)=\frac{\tan\alpha+\tan\beta}{1-\tan\alpha\tan\beta}$$
$$\tan(\alpha+\alpha)=\frac{\tan\alpha+\tan\alpha}{1-\tan\alpha\tan\alpha}$$
$$\tan 2\alpha=\frac{2\tan\alpha}{1-\tan^2\alpha}$$

93. $10\frac{34}{55}$

95. Left side:

$\sin 2\alpha-\sin 2\beta=2\sin\alpha\cos\alpha-2\sin\beta\cos\beta$

Right side:

$2\sin(\alpha-\beta)\cdot\cos(\alpha+\beta)$

$2(\sin\alpha\cos\beta-\cos\alpha\sin\beta)(\cos\alpha\cos\beta-\sin\alpha\sin\beta)$

$2(\sin\alpha\cos^2\beta\cos\alpha-\sin^2\alpha\cos\beta\sin\beta-\cos^2\alpha\sin\beta\cos\beta$
$+\cos\alpha\sin^2\beta\sin\alpha)$

$2(\sin\alpha\cos^2\beta\cos\alpha+\cos\alpha\sin^2\beta\sin\alpha-\sin^2\alpha\cos\beta\sin\beta$
$-\cos^2\alpha\sin\beta\cos\beta)$

$2[\sin\alpha\cos\alpha(\cos^2\beta+\sin^2\beta)-\sin\beta\cos\beta(\sin^2\alpha+\cos^2\alpha)]$

$2[\sin\alpha\cos\alpha(1)-\sin\beta\cos\beta(1)]$

$2\sin\alpha\cos\alpha-2\sin\beta\cos\beta$

97. Left side:

$\cos 2\alpha-\cos 2\beta=(2\cos^2\alpha-1)-(2\cos^2\beta-1)=2\cos^2\alpha$
$-2\cos^2\beta$

Right side:

$-2\sin(\alpha+\beta)\cdot\sin(\alpha-\beta)$

$-2(\sin\alpha\cos\beta+\cos\alpha\sin\beta)(\sin\alpha\cos\beta-\cos\alpha\sin\beta)$

$-2(\sin^2\alpha\cos^2\beta-\cos^2\alpha\sin^2\beta)$

$-2[(1-\cos^2\alpha)(\cos^2\beta)-\cos^2\alpha(1-\cos^2\beta)]$

$-2[\cos^2\beta-\cos^2\alpha\cos^2\beta-\cos^2\alpha+\cos^2\alpha\cos^2\beta]$

$-2(\cos^2\beta-\cos^2\alpha)$

$2\cos^2\alpha-2\cos^2\beta$

Solutions to skill and review problems

1. $$\cos\frac{\pi}{12}=\cos\left(\frac{\pi}{4}-\frac{\pi}{6}\right)=\cos\frac{\pi}{4}\cos\frac{\pi}{6}+\sin\frac{\pi}{4}\sin\frac{\pi}{6}$$
$$=\frac{\sqrt{2}}{2}\cdot\frac{\sqrt{3}}{2}+\frac{\sqrt{2}}{2}\cdot\frac{1}{2}$$
$$=\frac{\sqrt{6}+\sqrt{2}}{2}.$$

2. $f(x) = \dfrac{2x}{x^2 - x - 6} = \dfrac{2x}{(x-3)(x+2)}$

Vertical asymptotes at $x = -2$ and $x = 3$.

Intercepts are at the origin.

Additional points:

x	-4	-3	-1	1	2	4	5
y	-0.6	-1	0.5	-0.3	-1	1.3	0.7

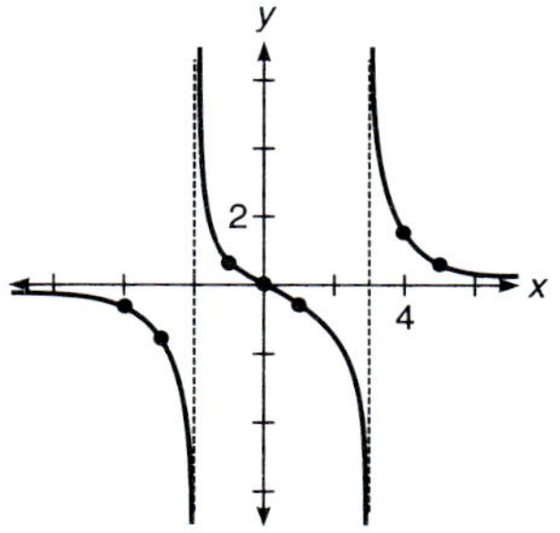

3. $2x - y = 3$

$-y = -2x + 3$

$y = 2x - 3$

The first equation is solved for y.

$x + 3y = 5$

$x + 3(2x - 3) = 5$

Substitute $2x - 3$ for y.

$7x = 14$

$x = 2$

Solve for y.

$y = 2x - 3$

$y = 2(2) - 3 = 1$

The point of intersection is $(x,y) = (2,1)$.

4. $\dfrac{1 - \dfrac{\cos\theta}{\sin\theta}}{\dfrac{1}{\sin\theta} + 1} \cdot \dfrac{\sin\theta}{\sin\theta} = \dfrac{\sin\theta - \cos\theta}{1 + \sin\theta}$

5. $-3 \sin x = 1$

$\sin x = -\frac{1}{3}$

$x' = \sin^{-1}\frac{1}{3} \approx 19.5°$

$\sin x < 0$ in quadrants III and IV. The least nonnegative solution is in quadrant III.

Therefore, $x = 180° + x' = 199.5°$.

Solutions to trial exercise problems

6. $\dfrac{2\tan\dfrac{\pi}{6}}{1 - \tan^2\dfrac{\pi}{6}}$

$\dfrac{2\tan\alpha}{1 - \tan^2\alpha} = \tan 2\alpha$; Let $\alpha = \dfrac{\pi}{6}$, so $2\alpha = \dfrac{\pi}{3}$,

and $\tan 2\alpha = \tan\dfrac{\pi}{3}$

8. $8\cos^2\dfrac{\pi}{2} - 4$

$2\cos^2\alpha - 1 = \cos 2\alpha$

$8\cos^2\alpha - 4 = 4\cos 2\alpha$

Multiply each member by 4.

Let $\alpha = \dfrac{\pi}{2}$, so $2\alpha = \pi$, and $4\cos 2\alpha = 4\cos\pi$.

11. $6\cos^2 5\theta - 3$

$2\cos^2\alpha - 1 = \cos 2\alpha$

$6\cos^2\alpha - 3 = 3\cos 2\alpha$

Multiply each member by 3. Let $\alpha = 5\theta$, so 2α is 10θ, and $3\cos 2\alpha$ represents $3\cos 10\theta$.

25. $\sin 10° = \sqrt{\dfrac{1 - \cos\theta}{2}}$

$\sin\dfrac{\alpha}{2} = \sqrt{\dfrac{1 - \cos\alpha}{2}}$

$\dfrac{\alpha}{2} = 10°$, so $\alpha =$ [illegible] $= 20°$

33. $\cos\theta = -\dfrac{4}{5}, \dfrac{\pi}{2} < \theta < \pi$

$\sin 2\theta = 2\sin\theta\cos\theta$

$= 2 \cdot \frac{3}{5}(-\frac{4}{5}) = -\frac{24}{25}$

$\cos 2\theta = \cos^2\theta - \sin^2\theta$

$= (-\frac{4}{5})^2 - (\frac{3}{5})^2 = \frac{7}{25}$

$\tan 2\theta = \dfrac{\sin 2\theta}{\cos 2\theta} = -\dfrac{24}{7}$

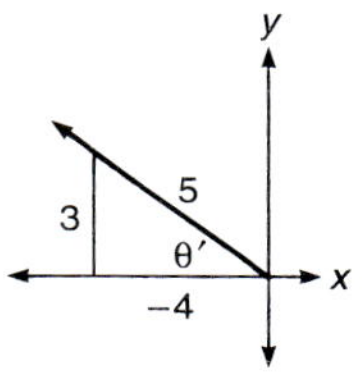

39. $\cot\theta = -2, \dfrac{3\pi}{2} < \theta < 2\pi$; $\cos\theta = \dfrac{2\sqrt{5}}{5}$.

$\dfrac{3\pi}{4} \le \dfrac{\theta}{2} \le \pi$, $\left(\dfrac{\theta}{2} \text{ in quadrant II}\right)$ so $\sin\dfrac{\theta}{2} > 0$,

$\cos\dfrac{\theta}{2} < 0$, $\tan\dfrac{\theta}{2} < 0$.

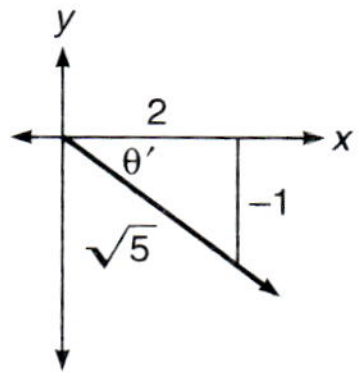

$\sin\dfrac{\theta}{2} = \sqrt{\dfrac{1 - \cos\theta}{2}} = \sqrt{\dfrac{1 - \dfrac{2\sqrt{5}}{5}}{2}} = \sqrt{\dfrac{1}{2}\left(\dfrac{5 - 2\sqrt{5}}{5}\right)}$

$= \sqrt{\dfrac{5 - 2\sqrt{5}}{10}} = \dfrac{\sqrt{5 - 2\sqrt{5}}}{\sqrt{10}} = \dfrac{\sqrt{50 - 20\sqrt{5}}}{10}$

$\cos\dfrac{\theta}{2} = -\sqrt{\dfrac{1 + \cos\theta}{2}} = -\dfrac{\sqrt{50 + 20\sqrt{5}}}{10}$

$\tan\dfrac{\theta}{2} = \dfrac{1 - \cos}{\sin\theta} = \dfrac{1 - \dfrac{2}{\sqrt{5}}}{-\dfrac{1}{\sqrt{5}}} = -\sqrt{5}\left(1 - \dfrac{2}{\sqrt{5}}\right)$

$= -\sqrt{5} + 2 = 2 - \sqrt{5}$

64. $\sec 2\theta$

$\dfrac{1}{\cos 2\theta}$

$\dfrac{1}{1 - 2\sin^2\theta}$

70. Left side:

$\dfrac{\csc\theta - \cot\theta}{1 + \cos\theta}$

$\dfrac{\dfrac{1}{\sin\theta} - \dfrac{\cos\theta}{\sin\theta}}{1 + \cos\theta}$

$\dfrac{\dfrac{1 - \cos\theta}{\sin\theta}}{1 + \cos\theta}$

$\dfrac{1 - \cos\theta}{\sin\theta} \cdot \dfrac{1}{1 + \cos\theta}$

$\dfrac{1}{\sin\theta} \cdot \dfrac{1 - \cos\theta}{1 + \cos\theta}$

$\csc\theta\, \dfrac{1 - \cos\theta}{1 + \cos\theta}$

Right side:

$\csc\theta \tan^2\dfrac{\theta}{2}$

$\csc\theta\left(\pm\sqrt{\dfrac{1 - \cos\theta}{1 + \cos\theta}}\right)^2$

$\csc\theta\, \dfrac{1 - \cos\theta}{1 + \cos\theta}$

82. $\cos 3\theta = \cos(2\theta + \theta)$

$= \cos 2\theta \cos\theta - \sin 2\theta \sin\theta$

$= (2\cos^2\theta - 1)(\cos\theta) - (2\sin\theta\cos\theta)(\sin\theta)$

$= 2\cos^3\theta - \cos\theta - 2\sin^2\theta\cos\theta$

$= 2\cos^3\theta - \cos\theta - 2(1 - \cos^2\theta)(\cos\theta)$

$= 2\cos^3\theta - \cos\theta - 2\cos\theta + 2\cos^3\theta$

$= 4\cos^3\theta - 3\cos\theta$

84. This problem uses the results of problem 83.

a. $\sin 5\theta$

$\sin(4\theta + \theta)$

$\sin 4\theta\cos\theta + \cos 4\theta\sin\theta$

$\cos\theta(4\sin\theta\cos\theta - 8\sin^3\theta\cos\theta)$
$+ \sin\theta(8\cos^4\theta - 8\cos^2\theta + 1)$

$4\sin\theta\cos^2\theta - 8\sin^3\theta\cos^2\theta + 8\sin\theta\cos^4\theta$
$- 8\sin\theta\cos^2\theta + \sin\theta$

We know $\cos^2\theta = 1 - \sin^2\theta$, so
$\cos^4\theta = (1 - \sin^2\theta)^2 = 1 - 2\sin^2\theta + \sin^4\theta$

Replace these in the equation.

$4\sin\theta(1 - \sin^2\theta) - 8\sin^3\theta(1 - \sin^2\theta) + 8\sin\theta(1 - 2\sin^2\theta + \sin^4\theta) - 8\sin\theta(1 - \sin^2\theta) + \sin\theta$

$16\sin^5\theta - 20\sin^3\theta + 5\sin\theta$

b. $\cos 5\theta$

$\cos(4\theta + \theta)$

$\cos 4\theta\cos\theta - \sin 4\theta\sin\theta$

$\cos\theta(8\cos^4\theta - 8\cos^2\theta + 1) - \sin\theta(4\cos^3\theta\sin\theta - 4\sin^3\theta\cos\theta)$

$8\cos^5\theta - 8\cos^3\theta + \cos\theta - 4\cos^3\theta\sin^2\theta + 4\sin^4\theta\cos\theta$

$8\cos^5\theta - 8\cos^3\theta + \cos\theta - 4\cos^3\theta(1 - \cos^2\theta)$
$+ 4(1 - \cos^2\theta)^2\cos\theta$

$16\cos^5\theta - 20\cos^3\theta + 5\cos\theta$

93. $\tan\theta_2 = \tan\dfrac{\theta}{2} = \dfrac{3}{8}$; $\cos\theta = \dfrac{8}{x}$

$\tan\dfrac{\theta}{2} = \sqrt{\dfrac{1 - \cos\theta}{1 + \cos\theta}}$, so $\dfrac{3}{8} = \sqrt{\dfrac{1 - \dfrac{8}{x}}{1 + \dfrac{8}{x}}}$

Square both members.

$\dfrac{9}{64} = \dfrac{1 - \dfrac{8}{x}}{1 + \dfrac{8}{x}}$

If $\dfrac{a}{b} = \dfrac{c}{d}$, then $ad = bc$.

$9 + \dfrac{72}{x} = 64 - \dfrac{512}{x}$

Multiply each member by x.

$9x + 72 = 64x - 512$

$584 = 55x$

$x = \frac{584}{55} = 10\frac{34}{55}$

Exercise 7–4

Answers to odd-numbered problems

1. $\dfrac{3\pi}{4}$ (135°), $\dfrac{7\pi}{4}$ (315°)

3. $\dfrac{\pi}{3}$ (60°), $\dfrac{5\pi}{3}$ (300°)

5. $\dfrac{\pi}{6}$ (30°), $\dfrac{7\pi}{6}$ (210°)

7. $\dfrac{7\pi}{6}$ (210°), $\dfrac{11\pi}{6}$ (330°)

9. $\dfrac{\pi}{2}$ (90°), $\dfrac{3\pi}{2}$ (270°)

11. 0 (0°), π (180°) **13.** 0 (0°), $\dfrac{3\pi}{2}$ (270°)

15. $\dfrac{\pi}{4}$ (45°), $\dfrac{3\pi}{4}$ (135°), $\dfrac{5\pi}{4}$ (225°), $\dfrac{7\pi}{4}$ (315°)

17. 0 (0°), π (180°), $\dfrac{\pi}{2}$ (90°)

19. 0 (0°), π (180°), $\dfrac{\pi}{3}$ (60°), $\dfrac{4\pi}{3}$ (240°)

21. $\dfrac{3\pi}{2}$ (270°), $\dfrac{\pi}{6}$ (30°), $\dfrac{5\pi}{6}$ (150°)

23. 0 (0°), π (180°), $\dfrac{\pi}{4}$ (45°), $\dfrac{3\pi}{4}$ (135°), $\dfrac{5\pi}{4}$ (225°), $\dfrac{7\pi}{4}$ (315°)

25. 0 (0°), π (180°), $\dfrac{\pi}{3}$ (60°), $\dfrac{5\pi}{3}$ (300°)

27. $\dfrac{5\pi}{6}$ (150°), $\dfrac{11\pi}{6}$ (330°), $\dfrac{\pi}{2}$ (90°), $\dfrac{3\pi}{2}$ (270°)

29. 0 (0°), π (180°), $\dfrac{\pi}{2}$ (90°), $\dfrac{3\pi}{2}$ (270°)

31. $\dfrac{\pi}{6}$ (30°), $\dfrac{5\pi}{6}$ (150°), $\dfrac{3\pi}{2}$ (270°)

33. $\dfrac{3\pi}{4}$ (135°), $\dfrac{7\pi}{4}$ (315°)

35. $\dfrac{\pi}{3}$ (60°), $\dfrac{5\pi}{3}$ (300°), π (180°)

37. $\dfrac{\pi}{3}$ (60°), $\dfrac{2\pi}{3}$ (120°), $\dfrac{4\pi}{3}$ (240°), $\dfrac{5\pi}{3}$ (300°)

39. $\dfrac{\pi}{4}$ (45°), $\dfrac{3\pi}{4}$ (135°), $\dfrac{5\pi}{4}$ (225°), $\dfrac{7\pi}{4}$ (315°)

41. 0 (0°), π (180°), $\dfrac{\pi}{6}$ (30°), $\dfrac{5\pi}{6}$ (150°)

43. 0.65, 2.49, 3.42, 6.01, 37.4°, 142.6°, 195.9°, 344.1° **45.** 0.27, 2.08, 3.42, 5.22, 15.7°, 119.3°, 195.7°, 299.3° **47.** 1.26, 5.03, 2.51, 3.77, 72°, 288°, 144°, 216°

49. 5.08 (290.8°), 0.56 (32.3°)

51. $\dfrac{\pi}{3} + 2k\pi$ $(60° + k \cdot 360°)$, $\dfrac{5\pi}{3} + 2k\pi$ $(300° + k \cdot 360°)$

53. $\frac{5\pi}{6} + k\pi$ (150° + $k \cdot$ 180°)

55. $\frac{5\pi}{4} + 2k\pi$ (225° + $k \cdot$ 360°),

$\frac{7\pi}{4} + 2k\pi$ (315° + $k \cdot$ 360°)

57. $\frac{\pi}{4} + k\pi$ (45° + $k \cdot$ 180°)

59. $\frac{\pi}{6} + 2k\pi$ (30° + $k \cdot$ 360°),

$\frac{5\pi}{6} + 2k\pi$ (150° + $k \cdot$ 360°)

61. $\frac{2\pi}{3} + 4k\pi$ (120° + $k \cdot$ 720°),

$\frac{4\pi}{3} + 4k\pi$ (240° + $k \cdot$ 720°)

63. $\frac{\pi}{3} + \frac{2k\pi}{3}$ (60° + $k \cdot$ 120°)

65. $\frac{\pi}{6} + k\frac{\pi}{2}$ (30° + $k \cdot$ 90°)

67. $\frac{\pi}{6} + k\frac{\pi}{2}$ (30° + $k \cdot$ 90°),

$\frac{\pi}{3} + k\frac{\pi}{2}$ (60° + $k \cdot$ 90°)

69. $\frac{\pi}{3} + k\pi$ (60° + $k \cdot$ 180°),

$\frac{2\pi}{3} + k\pi$ (120° + $k \cdot$ 180°)

71. $\frac{\pi}{12} + k\frac{\pi}{2}$ (15° + $k \cdot$ 90°)

73. $\frac{\pi}{12} + k\pi$ (15° + $k \cdot$ 180°),

$\frac{5\pi}{12} + k\pi$ (75° + $k \cdot$ 180°)

75. $\frac{\pi}{9} + \frac{2k\pi}{3}$ (20° + $k \cdot$ 120°),

$\frac{5\pi}{9} + \frac{2k\pi}{3}$ (100° + $k \cdot$ 120°)

77. $\frac{2\pi}{3} + 4k\pi$ (120° + $k \cdot$ 720°)

79. $\frac{7\pi}{6}$ (210°), $\frac{11\pi}{6}$ (330°), $\frac{\pi}{2}$ (90°)

81. 0 (0°), π(180°), $\frac{2\pi}{3}$ (120°), $\frac{4\pi}{3}$ (240°)

83. 0 (0°), π(180°), $\frac{\pi}{6}$ (30°), $\frac{5\pi}{6}$ (150°)

85. 0 (0°) **87.** $\frac{\pi}{3}$ (60°) or $\frac{5\pi}{3}$ (300°)

89. 0 (0°), $\frac{3\pi}{2}$ (270°) **91.** $\frac{\pi}{2}$ (90°), $\frac{3\pi}{2}$ (270°), π(180°), 0.93 (53.1°) **93.** $\frac{\pi}{2}$, 1.31 **95.** 0, 0.55, 0.69, 0.72, 0.66

Solutions to skill and review problems

1. The basic graph, $y = \sin x$, is reflected about the x-axis because of the coefficient -2.

$0 \le 4x \le 2\pi$

$0 \le x \le \frac{\pi}{2}$

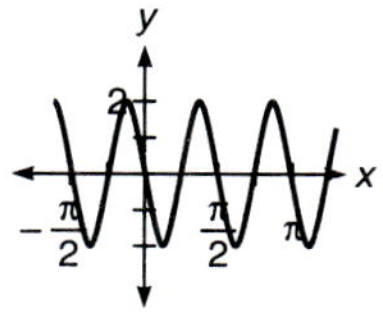

2. $f(x) = x^2 + 4x - 8$

$f(x) = x^2 + 4x + 4 - 4 - 8$

$f(x) = x^2 + 4x + 4 - 12$

$f(x) = (x + 2)^2 - 12$

Vertex: $(-2, -12)$

Intercepts:

$f(0) = -8$: $(0, -8)$

$0 = x^2 + 4x - 8$

$x = -2 \pm 2\sqrt{3} \approx (1.5, 0), (-5.5, 0)$

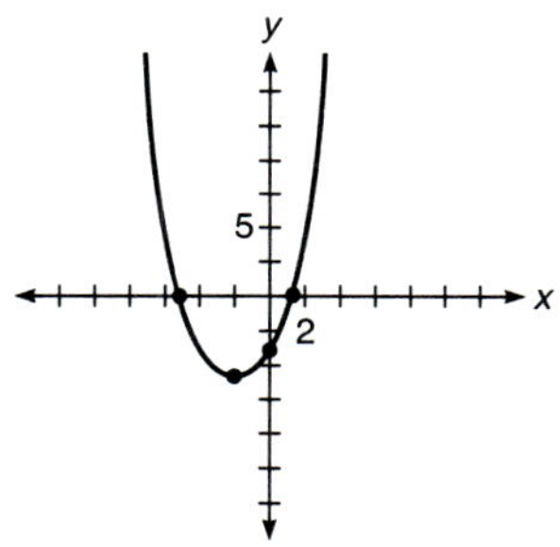

3. $x = -\sqrt{6^2 - (\sqrt{3})^2} = -\sqrt{33}$;

$\tan\theta = \frac{\sqrt{3}}{x} = -\frac{\sqrt{3}}{\sqrt{33}} = -\sqrt{\frac{3}{33}}$

$= -\sqrt{\frac{1}{11}} = -\frac{\sqrt{11}}{11}$

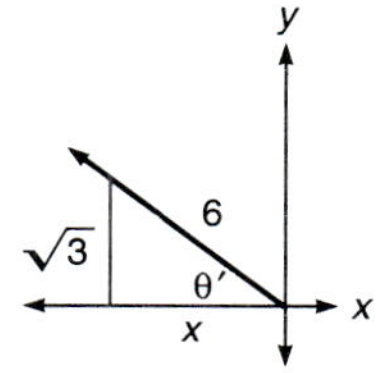

4. $-5a^8 + 5a^2x^6$

$-5a^2(a^6 - x^6)$

$-5a^2(a^3 - x^3)(a^3 + x^3)$

$-5a^2(a - x)(a^2 + ax + x^2)(a + x)$

$(a^2 - ax + x^2)$

5. The x-intercept is the point (4,0).

Let $(x_1,y_1) = (1,3)$, and $(x_2,y_2) = (4,0)$.

$m = \frac{y_2 - y_1}{x_2 - x_1} = \frac{0 - 3}{4 - 1} = -1$

$y - y_1 = m(x - x_1)$

$y - 3 = -1(x - 1)$

$y = -x + 4$

6. $\frac{3}{x^2 - 1} - \frac{1}{x - 1}$

$\frac{3}{(x-1)(x+1)} - \frac{1}{x-1} \cdot \frac{x+1}{x+1}$

$\frac{3 - (x+1)}{(x-1)(x+1)}$

$\frac{-x + 2}{x^2 - 1}$

Solutions to trial exercise problems

4. $2\cos\theta + 1 = 0$

$2\cos\theta = -1$

$\cos\theta = -\frac{1}{2}$

$\theta' = \cos^{-1}\frac{1}{2} = \frac{\pi}{3}$ (60°)

$\cos\theta < 0$ in quadrants II and III, so $\theta = \pi - \theta'$

$= \pi - \frac{\pi}{3} = \frac{2\pi}{3}$ (180° − 60° = 120°) and $\theta = \pi + \theta'$

$= \pi + \frac{\pi}{3} = \frac{4\pi}{3}$ (180° + 60° = 240°)

20. $\cos^2\theta - \frac{1}{2}\cos\theta = 0$

$\cos\theta\,(\cos\theta - \frac{1}{2}) = 0$

$\cos\theta = 0$ or $\cos\theta - \frac{1}{2} = 0$

$\frac{\pi}{2}$ (90°), $\frac{3\pi}{2}$ (270°) $\cos\theta = \frac{1}{2}$

$\frac{\pi}{3}$ (60°), $2\pi - \theta' = 2\pi - \frac{\pi}{3} = \frac{5\pi}{3}$ (300°)

27. $\sqrt{3}\tan\theta\cot\theta + \cot\theta = 0$

$\cot\theta(\sqrt{3}\tan\theta + 1) = 0$

$\cot\theta = 0$ or $\sqrt{3}\tan\theta + 1 = 0$

$\dfrac{\cos\theta}{\sin\theta} = 0$ or $\tan\theta = -\dfrac{\sqrt{3}}{3}$

$\cos\theta = 0$ $\quad \dfrac{5\pi}{6}\ (150°)$ and $\dfrac{11\pi}{6}\ (330°)$

$\dfrac{\pi}{2}\ (90°), \dfrac{3\pi}{2}\ (270°)$

34. $2 - \sin x - \csc x = 0$

$2 - \sin x - \dfrac{1}{\sin x} = 0$

$2\sin x - \sin^2 x - 1 = 0$

$\sin^2 x - 2\sin x + 1 = 0$

$(\sin x - 1)^2 = 0$

$\sin x - 1 = 0$

$\sin x = 1$

$x = \dfrac{\pi}{2}\ (90°)$

41. $2\tan^2 x\sin x = \tan^2 x$

$2\tan^2 x\sin x - \tan^2 x = 0$

$\tan^2 x(2\sin x - 1) = 0$

$\tan^2 x = 0$ or $2\sin x - 1 = 0$

$\tan x = 0$ or $\sin x = \frac{1}{2}$

$0\ (0°), \pi\ (180°), \dfrac{\pi}{6}\ (30°), \dfrac{5\pi}{6}\ (150°)$

46. $\tan^2 x + 5\tan x + 2 = 0$

$a = 1, b = 5, c = 2$: $\tan x = \dfrac{-5 \pm \sqrt{5^2 - 4(2)}}{2} = \dfrac{-5 \pm \sqrt{17}}{2}$

$\tan x = \dfrac{-5 + \sqrt{17}}{2} \approx -0.4384$

$x = \tan^{-1}\left|\dfrac{-5 + \sqrt{17}}{2}\right|$

$\approx 0.413\ (23.7°)$

$\tan x < 0$ in quadrants II and IV.

$x = \pi - x' \approx \pi - 0.413 \approx 2.74$

$= 2\pi - x' \approx 2\pi - 0.413 \approx 5.87$

$x = 180° - x' \approx 180° - 23.7° \approx 156.3°$

$= 360° - x' \approx 360° - 23.7° \approx 336.3°$

$\tan x = \dfrac{-5 - \sqrt{17}}{2} \approx -4.5616$

$x = \tan^{-1}\left|\dfrac{-5 - \sqrt{17}}{2}\right|$

$\approx 1.355\ (77.6°)$

$\tan x < 0$ in quadrants II and IV.

$x = \pi - x' \approx \pi - 1.355 \approx 1.79$

$= 2\pi - x' \approx 2\pi - 1.355 \approx 4.93$

$x = 180° - x' \approx 180° - 77.6° \approx 102.4°$

$= 360° - x' \approx 360° - 77.6° \approx 282.4°$

57. $\tan x = 1$

$x = \tan^{-1}1 = \dfrac{\pi}{4}\ (45°)$

Primary solutions are in quadrants I and III: $\dfrac{\pi}{4}\ (45°)$ and $\dfrac{5\pi}{4}$ $(225°)$. These differ by π (180°), so we can write all solutions with one of them: $\dfrac{\pi}{4} + k\pi\ (45° + k \cdot 180°)$.

64. $\sec\dfrac{x}{2} = 1$; $\cos\dfrac{x}{2} = 1$

Primary solutions: $\dfrac{x}{2} = \cos^{-1}1 = 0\ (0°)$

All solutions: $\dfrac{x}{2} = 0 + 2k\pi\ (0° + k \cdot 360°)$;

$x = 4k\pi\ (k \cdot 720°)$

74. $\sin\dfrac{\theta}{3} = \dfrac{\sqrt{3}}{2}$

$\left(\dfrac{\theta}{3}\right)' = \sin^{-1}\dfrac{\sqrt{3}}{2} = \dfrac{\pi}{3}\ (60°)$

Primary solutions: $\dfrac{\theta}{3} = \dfrac{\pi}{3}\ (60°), \dfrac{2\pi}{3}\ (120°)$

All solutions: $\dfrac{\theta}{3} = \dfrac{\pi}{3} + 2k\pi\ (60° + k \cdot 360°), \dfrac{2\pi}{3} + 2k\pi$ $(120° + k \cdot 360°)$; $\theta = \pi + 6k\pi\ (180° + k \cdot 1080°)$, $2\pi + 6k\pi$ $(360° + k \cdot 1080°)$.

81. $\sin 2\theta + \sin\theta = 0$

$2\sin\theta\cos\theta + \sin\theta = 0$

$\sin\theta(2\cos\theta + 1) = 0$

$\sin\theta = 0$

$0\ (0°), \pi\ (180°)$

$2\cos\theta + 1 = 0$

$\cos\theta = -\frac{1}{2}$

$\dfrac{2\pi}{3}\ (120°), \dfrac{4\pi}{3}\ (240°)$

88. $\sin^2\dfrac{\theta}{2} = \cos\theta$

$\left(\pm\sqrt{\dfrac{1 - \cos\theta}{2}}\right)^2 = \cos\theta$

$\dfrac{1 - \cos\theta}{2} = \cos\theta$

$1 - \cos\theta = 2\cos\theta$

$1 = 3\cos\theta$

$\frac{1}{3} = \cos\theta$

$\theta' = \cos^{-1}\frac{1}{3}$

$\theta = \cos^{-1}\frac{1}{3}$ or $2\pi - \cos^{-1}\frac{1}{3}$

$\theta = 1.23\ (70.5°)$ or $5.05\ (289.5°)$

92. $0 = x\cos 0.855\cos 1.052 - x^2\cos 0.855\sin 1.052 - x^3\sin 0.855$

$0 = 0.32538x - 0.56987x^2 - 0.75457x^3$

$x(0.75457x^2 + 0.56987x - 0.32538) = 0$

$x = 0$ or $0.75457x^2 + 0.56987x - 0.32538 = 0$

Solve the quadratic equation with the quadratic formula.

$x \approx 0, -1.14, 0.38$

93. $-8 = 2\cos A\cos 0.7 - 4\cos A\sin 0.7 - 8\sin A$
$-8 = (2\cos 0.7)\cos A - (4\sin 0.7)\cos A - 8\sin A$
$-8 = 1.5297\cos A - 2.5769\cos A - 8\sin A$
$-8 = -1.0472\cos A - 8\sin A$
$8\sin A - 8 = -1.0472\cos A$
Divide each member by 8.
$\sin A - 1 = -0.1309\cos A$
$\sin A = 1 - 0.1309\cos A$
$(\sin A)^2 = (1 - 0.1309\cos A)^2$
$\sin^2 A = 1 - 0.2618\cos A + 0.017134\cos^2 A$
$1 - \cos^2 A = 1 - 0.2618\cos A + 0.017134\cos^2 A$
$0 = 1.0171\cos^2 A - 0.2618\cos A$
$0 = \cos A(1.0171\cos A - 0.2618)$
$\cos A = 0$ or $1.0171\cos A - 0.2618 = 0$
$A = \cos^{-1}0$ or $1.0171\cos A = 0.2618$
$A = \frac{\pi}{2}$ or $\cos A = 0.25739$
$A \approx 1.310476103$
Thus, A is $\frac{\pi}{2}$ or 1.31

96. $\sin\theta\sqrt{1.44 - \sin^2\theta} = 0.5$
$(\sin\theta\sqrt{1.44 - \sin^2\theta})^2 = (0.5)^2$
$\sin^2\theta(1.44 - \sin^2\theta) = 0.25$
$1.44\sin^2\theta - \sin^4\theta = 0.25$
$\sin^4\theta - 1.44\sin^2\theta + 0.25 = 0$
Let $u = \sin^2\theta$.
$u^2 - 1.44u + 0.25 = 0$
$u \approx 1.2381$ or $u \approx 0.20193$
$\sin^2\theta \approx 1.2381$ or $\sin^2\theta \approx 0.20193$
$\sin\theta \approx \pm\sqrt{1.2381}$ or $\sin\theta \approx \pm\sqrt{0.20193}$
$\sin\theta \approx \pm 1.1127$ or $\sin\theta \approx \pm 0.44936$
No solution $\quad \theta \approx \pm 26.7°$
Because $0 \le \theta \le 90°$, we choose the positive value for θ.
$\theta \approx 26.7°$

Chapter 7 review

1. $\frac{\cot\theta}{\cos\theta}$
$\frac{\cos\theta}{\sin\theta}\cdot\frac{1}{\cos\theta}$
$\frac{1}{\sin\theta}$
$\csc\theta$

2. $\sec\theta\tan\theta$
$\frac{1}{\cos\theta}\cdot\frac{\sin\theta}{\cos\theta}$
$\frac{1}{\cos^2\theta}\sin\theta$
$\sec^2\theta\sin\theta$

3. $\frac{\tan^2\theta}{\sec^2\theta - 1} = \sin^4\theta\csc^4\theta$
$\frac{\tan^2\theta}{\tan^2\theta} \qquad \sin^4\theta\frac{1}{\sin^4\theta}$
$1 \qquad 1$

4. $\frac{\csc^2\theta - 1}{\sec^2\theta - 1}$
$\frac{\cot^2\theta}{\tan^2\theta}$
$\cot^2\theta\frac{1}{\tan^2\theta}$
$\cot^2\theta\cot^2\theta$
$\cot^4\theta$

5. $\frac{\csc\theta\tan\theta}{\sin\theta}$
$\csc\theta\tan\theta\frac{1}{\sin\theta}$
$\frac{1}{\sin\theta}\cdot\frac{\sin\theta}{\cos\theta}\cdot\frac{1}{\sin\theta}$
$\frac{1}{\cos\theta}\cdot\frac{1}{\sin\theta}$
$\csc\theta\sec\theta$

6. $\sin^2\theta - \cos^2\theta$
$\sin^2\theta - (1 - \sin^2\theta)$
$2\sin^2\theta - 1$

7. $\frac{\cos\theta - \sin\theta}{\sin\theta\cos\theta}$

8. $\frac{1}{\sin\theta\cos\theta}$

9. $\frac{\sin\theta}{\sin^2\theta - \cos^2\theta}$

10. $\frac{\sin\theta - \cos\theta}{\sin\theta + 1}$

11. $\sin^2\theta$

12. $\sin^2\theta - \cos^2\theta$

13. $\csc x - \tan x\cot x$
$\frac{1}{\sin x} - \tan x\frac{1}{\tan x}$
$\csc x - 1$

14. $\sin^2 x + \sin^2 x\cot^2 x = 1$
$\sin^2 x + \sin^2 x\frac{\cos^2 x}{\sin^2 x}$
$\sin^2 x + \cos^2 x$
1

---

15. $\csc^2 x\sec^2 x(\cos^2 x - \sin^2 x)$
$\csc^2 x\sec^2 x\cos^2 x - \csc^2 x\sec^2 x\sin^2 x$
$\csc^2 x\frac{1}{\cos^2 x}\cos^2 x - \frac{1}{\sin^2 x}\sec^2 x\sin^2 x$
$\csc^2 x - \sec^2 x$

16. $\frac{1}{\csc x - \cot x}$
$\frac{1}{\frac{1}{\sin x} - \frac{\cos x}{\sin x}}$
$\frac{1}{\frac{1 - \cos x}{\sin x}}$
$\frac{\sin x}{1 - \cos x}$

17. $(\tan x - 1)(\csc^2 x - \cot^2 x) \quad = \sec x(\sin x - \cos x)$
$(\tan x - 1)[(\cot^2 x + 1) - \cot^2 x] \quad \sec x\sin x - \sec x\cos x$
$(\tan x - 1)(1) \quad \frac{1}{\cos x}\sin x - \frac{1}{\cos x}\cos x$
$\tan x - 1 \quad \tan x - 1$

18. $\frac{\csc^2 x - 1}{\sin^2 x}$
$\frac{\cot^2 x}{\sin^2 x}$
$\cot^2 x\frac{1}{\sin^2 x}$
$\frac{\cos^2 x}{\sin^2 x}\cdot\frac{1}{\sin^2 x}$
$\cos^2 x\cdot\frac{1}{\sin^4 x}$
$\cos^2 x\csc^4 x$

19. $\frac{1}{1 + \csc x} + \frac{1}{1 - \csc x}$
$\frac{(1 - \csc x) + (1 + \csc x)}{(1 + \csc x)(1 - \csc x)}$
$\frac{2}{1 - \csc^2 x}$
$\frac{2}{-(\csc^2 x - 1)}$
$-\frac{2}{\cot^2 x}$
$-2\tan^2 x$

20. $\sin^2 x + \sin^2 x\cos^2 x$
$\sin^2 x(1 + \cos^2 x)$
$(1 - \cos^2 x)(1 + \cos^2 x)$
$1 - \cos^4 x$

21. $\tan^4 x + \tan^2 x$
$\tan^2 x(\tan^2 x + 1)$
$\tan^2 x\cdot\sec^2 x$
$\frac{1}{\cot^2 x}\cdot\sec^2 x$
$\frac{\sec^2 x}{\cot^2 x}$

22. $\frac{1 - \cot x}{1 + \csc x}$
$\frac{1 - \frac{\cos x}{\sin x}}{1 + \frac{1}{\sin x}}\cdot\frac{\sin x}{\sin x}$
$\frac{\sin x - \cos x}{\sin x + 1}$

23. $\frac{\sqrt{6}-\sqrt{2}}{4}$ **24.** $-2+\sqrt{3}$

25. $\frac{\sqrt{6}+\sqrt{2}}{4}$ **26.** $\frac{\sqrt{6}+\sqrt{2}}{4}$

27. $\frac{63}{65}$ **28.** $-\frac{12\sqrt{5}+10}{39}$

29. $\frac{270-169\sqrt{2}}{28}$

30. $\cos\left(\frac{3\pi}{2}+\theta\right)=\cos\frac{3\pi}{2}\cos\theta-\sin\frac{3\pi}{2}\sin\theta$
$=0\cos\theta-(-1)\sin\theta=\sin\theta$

31. $\sin\left(\frac{\pi}{4}-\theta\right)=\sin\frac{\pi}{4}\cos\theta-\cos\frac{\pi}{4}\sin\theta$
$=\frac{\sqrt{2}}{2}\cos\theta-\frac{\sqrt{2}}{2}\sin\theta=\frac{\sqrt{2}}{2}(\cos\theta-\sin\theta)$

32. $\frac{\cos(\alpha+\beta)}{\sin\alpha\cos\beta}=\frac{\cos\alpha\cos\beta-\sin\alpha\sin\beta}{\sin\alpha\cos\beta}$
$=\frac{\cos\alpha\cos\beta}{\sin\alpha\cos\beta}-\frac{\sin\alpha\sin\beta}{\sin\alpha\cos\beta}=\frac{\cos\alpha}{\sin\alpha}-\frac{\sin\beta}{\cos\beta}$
$=\cot\alpha-\tan\beta$

33. $\sin(\theta+2\pi)=\sin\theta\cos 2\pi+\cos\theta\sin 2\pi$
$=\sin\theta(1)+\cos\theta(0)$
$=\sin\theta$

34. $\frac{\sqrt{2}}{10}$ **35.** 124° **36.** 10π

37. $\frac{7\pi}{6}$ **38.** 12° **39.** $a=3,\ b=1$

40. a. $\frac{-5\sqrt{119}}{47}$ **b.** $\frac{5\sqrt{119}}{72}$

41. a. $-\frac{7}{25}$ **b.** $\frac{24}{7}$

42. $\sin 2x-\cos x$
$2\sin x\cos x-\cos x$
$\cos x(2\sin x-1)$

43. $1+\cos 2x$
$1+(2\cos^2 x-1)$
$2\cos^2 x$

44. $\frac{\sqrt{12-8\sqrt{2}}}{2}$ **45.** $\frac{\sqrt{2+\sqrt{2}}}{2}$

46. a. $\frac{\sqrt{26}}{26}$ **b.** $\frac{1}{5}$

47. a. $\frac{\sqrt{18-6\sqrt{5}}}{6}$ **b.** $-\frac{3+\sqrt{5}}{2}$

48. $\cot\frac{\theta}{2}=\frac{1}{\tan\frac{\theta}{2}}=\frac{1}{\frac{\sin\theta}{1+\cos\theta}}=\frac{1+\cos\theta}{\sin\theta}$

49. $\sec^2\frac{\theta}{2}-\tan^2\frac{\theta}{2}$
$\left(\tan^2\frac{\theta}{2}+1\right)-\tan^2\frac{\theta}{2}$
1

50. $\frac{\pi}{6}$ (30°), $\frac{5\pi}{6}$ (150°)

51. $\frac{\pi}{3}$ (60°), $\frac{2\pi}{3}$ (120°), $\frac{4\pi}{3}$ (240°), $\frac{5\pi}{3}$ (300°)

52. $\frac{\pi}{2}$ (90°), $\frac{\pi}{3}$ (60°), $\frac{5\pi}{3}$ (300°)

53. $\frac{\pi}{6}$ (30°), $\frac{5\pi}{6}$ (150°), $\frac{7\pi}{6}$ (210°), $\frac{11\pi}{6}$ (330°), $\frac{\pi}{3}$ (60°), $\frac{5\pi}{3}$ (300°)

54. $\frac{\pi}{2}$ (90°), $\frac{3\pi}{2}$ (270°), $\frac{\pi}{4}$ (45°), $\frac{5\pi}{4}$ (225°)

55. $\frac{\pi}{3}$ (60°), $\frac{2\pi}{3}$ (120°), $\frac{4\pi}{3}$ (240°), $\frac{5\pi}{3}$ (300°)

56. $\frac{2\pi}{3}$ (120°), $\frac{4\pi}{3}$ (240°), 0 (0°)

57. $\frac{\pi}{6}$ (30°), $\frac{5\pi}{6}$ (150°), $\frac{3\pi}{2}$ (270°)

58. $\frac{\pi}{2}$ (90°), $\frac{3\pi}{2}$ (270°), $\frac{\pi}{6}$ (30°), $\frac{5\pi}{6}$ (150°)

59. $\frac{2\pi}{3}$ (120°), $\frac{4\pi}{3}$ (240°), π (180°)

60. $\frac{\pi}{24}, \frac{13\pi}{24}, \frac{25\pi}{24}, \frac{37\pi}{24}, \frac{5\pi}{24}, \frac{17\pi}{24}, \frac{29\pi}{24}, \frac{41\pi}{24}$ **61.** $\frac{\pi}{3}$ (60°) **62.** $\frac{5\pi}{3}$ (300°)

63. π (180°) **64.** $\frac{4\pi}{3}$ (240°) **65.** $\frac{\pi}{15}, \frac{7\pi}{15}, \frac{13\pi}{15}, \frac{19\pi}{15}, \frac{5\pi}{3}, \frac{\pi}{3}, \frac{11\pi}{15}, \frac{17\pi}{15}, \frac{23\pi}{15}, \frac{29\pi}{15}$
or 12°, 84°, 156°, 228°, 300°, 60°, 132°, 204°, 276°, 348°

66. $\frac{\pi}{2}$ (90°), $\frac{3\pi}{2}$ (270°), $\frac{7\pi}{6}$ (210°), $\frac{11\pi}{6}$ (330°) **67.** 0.96, 2.19, 4.10, 5.33

68. 3.52, 5.90 **69.** $\frac{\pi}{3}, \frac{5\pi}{3}, \pi$

70. $\frac{2\pi}{3}, \frac{11\pi}{18}, \frac{23\pi}{18}, \frac{35\pi}{18}, \frac{7\pi}{18}, \frac{19\pi}{18}, \frac{31\pi}{18}$

71. $\frac{\pi}{4}, \frac{5\pi}{4}$, 1.94, 2.78, 5.08, 5.92

Chapter 7 test

1. $\csc^2 x\sin x\cos x$
$\frac{1}{\sin^2 x}\sin x\cos x$
$\frac{\cos x}{\sin x}$
$\cot x$

2. $\frac{\csc x-\sec x}{\tan x+\cot x}$
$\frac{\frac{1}{\sin\theta}-\frac{1}{\cos\theta}}{\frac{\sin\theta}{\cos\theta}+\frac{\cos\theta}{\sin\theta}}$
$\frac{\frac{1}{\sin\theta}-\frac{1}{\cos\theta}}{\frac{\sin\theta}{\cos\theta}+\frac{\cos\theta}{\sin\theta}}\cdot\frac{\sin\theta\cos\theta}{\sin\theta\cos\theta}$
$\frac{\cos\theta-\sin\theta}{\sin^2\theta+\cos^2\theta}$
$\frac{\cos\theta-\sin\theta}{1}$
$\cos\theta-\sin\theta$

3. $\cot x-2\tan x=1$
$\cot\frac{\pi}{4}-2\tan\frac{\pi}{4}=1$
$1-2(1)=1$
$-1\neq 1$

4. $a=8,\ b=4$ **5.** $\frac{8\sqrt{3}-15}{34}$

6. $-\frac{240}{289}$ **7.** 3 **8.** $\sqrt{4-2\sqrt{2}}$

9. $\frac{1+\cot\theta}{\csc\theta}$
$\frac{1}{\csc\theta}+\frac{\cot\theta}{\csc\theta}$
$\sin\theta+\cot\theta\,\frac{1}{\csc\theta}$
$\sin\theta+\frac{\cos\theta}{\sin\theta}\cdot\sin\theta$
$\sin\theta+\cos\theta$

10. $\frac{\cos^2 x-1}{\sin^2 x}$
$\frac{-(1-\cos^2 x)}{\sin^2 x}$
$\frac{-\sin^2 x}{\sin^2 x}$
-1

11. $\tan^4 x+\tan^2 x$
$\tan^2 x(\tan^2 x+1)$
$\tan^2 x\sec^2 x$
$\frac{1}{\cot^2 x}\sec^2 x$
$\frac{\sec^2 x}{\cot^2 x}$

12. $\cos\left(\theta-\frac{3\pi}{2}\right)$
$\cos\theta\cos\frac{3\pi}{2}+\sin\theta\sin\frac{3\pi}{2}$
$\cos\theta\,(0)+\sin\theta\,(-1)$
$-\sin\theta$

13. $\cos 2x - \sin 2x$
$(2\cos^2 x - 1) - (2 \sin x \cos x)$
$2\cos^2 x - 2 \sin x \cos x - 1$
$2 \cos x(\cos x - \sin x) - 1$

14. $\frac{\pi}{3}$ (60°), $\frac{2\pi}{3}$ (120°), $\frac{4\pi}{3}$ (240°), $\frac{5\pi}{3}$ (300°)

15. $\frac{\pi}{6}$ (30°), $\frac{7\pi}{6}$ (210°), $\frac{2\pi}{3}$ (120°), $\frac{4\pi}{3}$ (240°)

16. 0.16, 2.97, $\frac{3\pi}{2}$ **17.** $\frac{\pi}{12}, \frac{3\pi}{4}, \frac{17\pi}{12}, \frac{\pi}{4}, \frac{11\pi}{12}, \frac{19\pi}{12}, \frac{5\pi}{12}, \frac{13\pi}{12}, \frac{21\pi}{12}, \frac{7\pi}{12}, \frac{5\pi}{4}, \frac{23\pi}{12}$

Chapter 8

Exercise 8–1

Answers to odd-numbered problems

1. $C = 96°$, $b \approx 16.4$, $c \approx 21.7$
3. $A = 34.4°$, $b \approx 0.52$, $c \approx 1.64$
5. $C = 33°$, $c \approx 51.2°$, $a \approx 68.7$
7. $B = 20.6°$, $b \approx 0.172$, $c \approx 0.489$
9. $C = 35°$, $a \approx 8.58$, $b \approx 6.16$
11. $A \approx 45.6°$, $C \approx 85.4°$, $c \approx 17.4$
13. $B \approx 18.0°$, $C \approx 30.0°$, $b \approx 1.77$
15. $C \approx 47.4°$, $A \approx 88.9°$, $a \approx 133.9$;
$C \approx 132.6°$, $A \approx 3.7°$, $a \approx 8.7$
17. no solution
19. $B \approx 85.0°$, $A \approx 52.7°$, $a \approx 5.07$;
$B \approx 95.01°$, $A \approx 42.7°$, $a \approx 4.32$
21. $C \approx 26.23°$, $A \approx 108.77°$, $a \approx 10.71$
23. 32.3 miles
25. 843,400 miles
27. 15.8 knots
29. 25 miles
31. By the definitions of section 5–3, tan $A = \frac{y}{x}$ in each figure. By the trigonometric ratios (for a right triangle) it can be seen in each case that $\tan C = \frac{y}{b - x}$. (Note that x is negative in the right figure, so that $b - x$ is larger than b itself.) Also, as noted in the problem, $y = h$. Putting these values in the expression for h we obtain

$$\frac{b \cdot \tan A \cdot \tan C}{\tan A + \tan C}$$

$$\frac{b \cdot \frac{y}{x} \cdot \frac{y}{b - x}}{\frac{y}{x} + \frac{y}{b - x}} = \frac{\frac{by^2}{x(b - x)}}{\frac{yb}{x(b - x)}} = y = h$$

33. Let (x,y) be the point at B. It is on the terminal side of angle A. Then $\cos A = \frac{x}{r}$, where r is the length of AB. But then $r = c$, so $\cos A = \frac{x}{c}$. Next, using right triangles we see that in each figure $\cos C = \frac{b - x}{a}$. Note that when A is obtuse (the right-hand figure) x is negative, so $b - x$ is the length of $|b| + |x|$.

$\cos C = \frac{b - x}{a}$ $\quad\quad$ $\cos A = \frac{x}{c}$
$a \cos C = b - x$ $\quad\quad$ $c \cos A = x$
$b - a \cos C = x$
$b - a \cos C = c \cos A$
$b = c \cos A + a \cos C$

Thus, [2] is true.

[1] and [3] can be shown true by putting angles B and C in standard position and proceeding in the same manner. In fact this is not really necessary, since the labeling in a triangle is arbitrary, and thus, for example, we could obtain [1] by changing the label B to A, C to B, and A to C, and labeling the sides appropriately.

35. Consider any triangle ABC; place it as shown in the figure for problem 33, so angle A is in standard position. The figure covers the cases where A is acute, right, or obtuse. Then it can be seen that if h is the height of the triangle then $h = y$. We know that $\sin A = \frac{y}{c} = \frac{h}{c}$, so $h = c \sin A$. The area is $\frac{1}{2}bh = \frac{1}{2}b(c \sin A) = \frac{1}{2}bc \sin A$.

37. a. It can be seen that the sum of the area of the four triangles shown in the figure is $\frac{1}{2}ab \sin A + \frac{1}{2}cd \sin C + \frac{1}{2}ad \sin D + \frac{1}{2}bc \sin B$. This total is twice as large as the total area of the four-sided figure, so the area of the four-sided figure is $\frac{1}{2}$ this sum, or $\frac{1}{4}(ab \sin A + ad \sin D + bc \sin B + cd \sin C)$.

b. The difference between the Egyptian formula and the correct formula is the factors sin A, sin B, sin C, and sin D. The value of the sine of each angle is between 0 and 1. Thus, $ab \geq ab \sin A$, $ad \geq ad \sin D$, $bc \geq bc \sin B$, $cd \geq cd \sin C$, so $ab + ad + bc + cd \geq ab \sin A + ad \sin D + bc \sin B + cd \sin C$.

$\frac{1}{4}(ab + ad - bc + cd) \geq$

$\frac{1}{4}(ab \sin A - ad \sin D + bc \sin B + cd \sin C)$

If the figure is a rectangle, $A = B = C = D = 90°$, and $\sin A = \sin B = \sin C = \sin D = 1$, so both expressions give the same value.

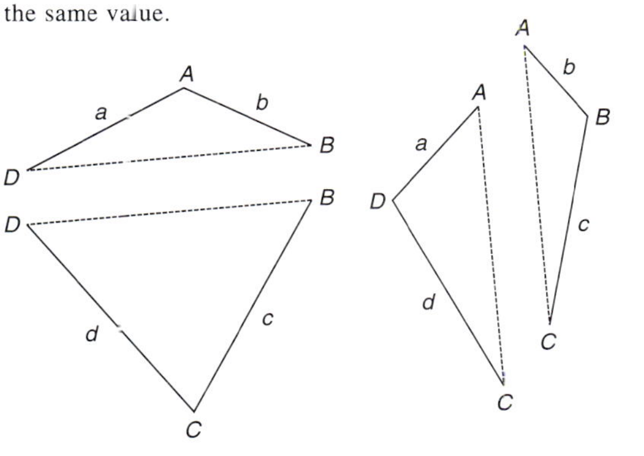

Solutions to skill and review problems

1. $b^2 = \sqrt{9^2 - 6.5^2} \approx 6.2$

$\sin A = \frac{6.5}{9}$ so $A = \sin^{-1}\frac{6.5}{9} \approx 46.2°$;

$\cos B = \frac{6.5}{9}$, $B = \cos^{-1}\frac{6.5}{9} \approx 43.8°$

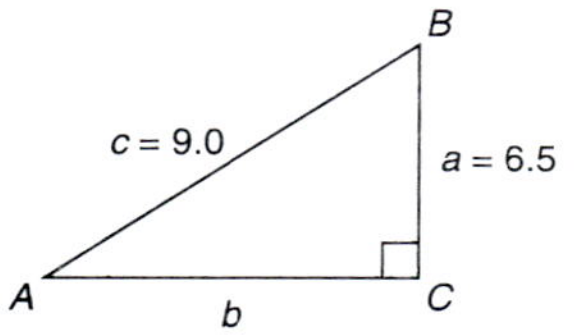

2. $\sec x = -\frac{2}{3}\sqrt{3}$

$\cos x = -\frac{3}{2\sqrt{3}} \cdot \frac{\sqrt{3}}{\sqrt{3}} = -\frac{\sqrt{3}}{2}$

$x' = \cos^{-1}\frac{\sqrt{3}}{2} = \frac{\pi}{6}$; $\cos x < 0$ so x terminates in quadrant II or III. $x = \pi - x'$ or $\pi + x' = \frac{5\pi}{6}$ or $\frac{7\pi}{6}$

3. $(\sin\theta - \sec\theta)(\csc\theta + \cos\theta)$

$\sin\theta\csc\theta + \sin\theta\cos\theta - \sec\theta\csc\theta - \sec\theta\cos\theta$

$1 + \sin\theta\cos\theta - \frac{1}{\cos\theta}\cdot\frac{1}{\sin\theta} - 1$

$\sin\theta\cos\theta - \frac{1}{\cos\theta}\cdot\frac{1}{\sin\theta}$

$\frac{\sin^2\theta\cos^2\theta}{\sin\theta\cos\theta} - \frac{1}{\sin\theta\cos\theta}$

$\frac{\sin^2\theta\cos^2\theta - 1}{\sin\theta\cos\theta}$

---

4. Possible rational zeros of $2x^5 - x^4 - 10x^3 + 5x^2 + 8x - 4$ are ± 1, ± 2, ± 4, and $\pm\frac{1}{2}$.

	2	−1	−10	5	8	−4
		2	1	−9	−4	4
1	2	1	−9	−4	4	0

1 is a zero

$(x - 1)(2x^4 + x^3 - 9x^2 - 4x + 4)$.

	2	1	−9	−4	4
		−2	1	8	−4
−1	2	−1	−8	4	0

−1 is a zero

$2x^5 - x^4 - 10x^3 + 5x^2 + 8x - 4$
$= (x - 1)(x + 1)(2x^3 - x^2 - 8x + 4)$
$2x^3 - x^2 - 8x + 4$
$= x^2(2x - 1) - 4(2x - 1)$
$= (2x - 1)(x^2 - 4)$
$= (2x - 1)(x - 2)(x + 2)$, so
$2x^5 - x^4 - 10x^3 + 5x^2 + 8x - 4$
$= (x - 1)(x + 1)(2x - 1)(x - 2)(x + 2)$

5. $\frac{\theta°}{180°} = \frac{s}{\pi}$, so $s = \frac{\pi}{180°}\theta°$,

so $s = \frac{\pi}{180°}\cdot 24° = \frac{2\pi}{15}$.

$L = rs$

$L = 18\left(\frac{2\pi}{15}\right) = \frac{12\pi}{5} \approx 7.5$ meters

Solutions to trial exercise problems

7. $a = 0.452$, $A = 67.6°$, $C = 91.8°$

$B = 180° - 67.6° - 91.8° = 20.6°$

$\frac{\sin 67.6°}{0.452} = \frac{\sin 20.6°}{b} = \frac{\sin 91.8°}{c}$

$\frac{\sin 67.6°}{0.452} = \frac{\sin 20.6°}{b}$ $\quad \frac{\sin 67.6°}{0.452} = \frac{\sin 91.8°}{c}$

$b \approx 0.172$ $\quad c \approx 0.489$

13. $a = 4.25$, $c = 2.86$, $A = 132°$

$\frac{\sin 132°}{4.25} = \frac{\sin B}{b} = \frac{\sin C}{2.86}$

$\frac{\sin 132°}{4.25} = \frac{\sin C}{2.86}$

so $\sin C = \frac{2.86 \sin 132°}{4.25}$

$C' = \sin^{-1}\frac{2.86\sin 132°}{4.25} \approx 30.01°$

$C \approx 30.01°$ or $180° - 30.01° \approx 149.99°$

Case 1: $C \approx 30.01°$

$B \approx 180° - 132° - 30.01° \approx 17.99°$

$\frac{\sin 132°}{4.25} \approx \frac{\sin 17.99°}{b}$

$b \approx 1.77$

Case 2: $C \approx 149.99°$

$B = 180° - 132° - 149.99°$

≈ -101.99 (no solution)

Thus, the only solution is $B \approx 18.0°$, $C \approx 30.0°$, $b \approx 1.77$.

15. $b = 92.5$, $c = 98.6$, $B = 43.7°$

$\frac{\sin A}{a} = \frac{\sin 43.7°}{92.5} = \frac{\sin C}{98.6}$

$\frac{\sin 43.7°}{92.5} = \frac{\sin C}{98.6}$;

$\sin C = \frac{98.6 \sin 43.7°}{92.5}$

$C' = \sin^{-1}\frac{98.6\sin 43.7°}{92.5} \approx 47.429°$

$C \approx 47.429°$ or $180° - 47.429°$
$\approx 132.571°$

Case 1: $C \approx 47.429°$

$A = 180° - 43.7° - 47.429° \approx 88.871°$

$\frac{\sin 88.871°}{a} = \frac{\sin 43.7°}{92.5}$

$a \approx 133.86$

Solution 1: $C \approx 47.4°$, $A \approx 88.9°$, $a \approx 133.9$

Case 2: $C \approx 132.571°$

$A = 180° - 43.7° - 132.571° \approx 3.729°$

$\frac{\sin 3.729°}{a} = \frac{\sin 43.7°}{92.5}$

$a \approx 8.71$

Solution 2: $C \approx 132.6°$, $A \approx 3.7°$, $a \approx 8.7$

27. Angle $A = 90° - 58° = 32°$;

$\frac{\sin 32°}{8.4} = \frac{\sin B}{12.6}$;

$\sin B = \frac{12.6 \sin 32°}{8.4}$, so $B' \approx 52.64°$.

$C \approx 180° - 32° - 52.64° \approx 95.36°$.

$\frac{\sin 32°}{8.4} = \frac{\sin 95.36°}{S}$; $S \approx 15.78$. Thus, $S \approx 15.8$ knots.

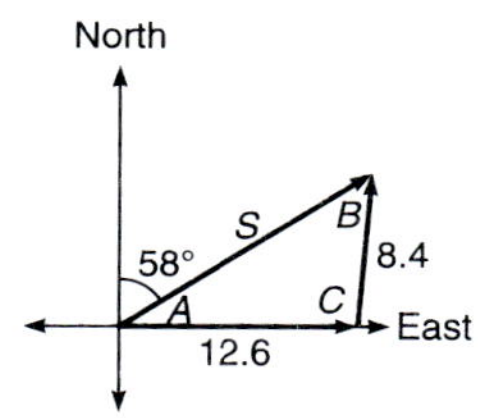

Exercise 8–2

Answers to odd-numbered problems

1. $c \approx 4.0$, $A \approx 30.7°$, $B \approx 109.9°$
3. $a \approx 77.2$, $B \approx 41.1°$, $C \approx 14.9°$
5. $b \approx 38.3$, $A \approx 53.8°$, $C \approx 25.9°$
7. $C \approx 109.0°$, $A \approx 39.4°$, $B \approx 31.6°$
9. $B \approx 105.3°$, $A \approx 24.4°$, $C \approx 50.3°$
11. $c \approx 18.1$, $A \approx 28.3°$, $B \approx 12.3°$
13. $a \approx 28.1$, $C \approx 40.5°$, $B \approx 115.0°$
15. $b \approx 41.8$, $C \approx 26.3°$, $A \approx 41.7°$
17. $C \approx 110.9°$, $A \approx 38.3°$, $B \approx 30.8°$
19. $c \approx 0.28$, $A \approx 1.12°$, $B \approx 177.38°$
21. 326.9 ft **23.** $C \approx 82.4°$
25. $B \approx 125.5°$ **27.** 102.9°
29. 39.9 miles **31.** $A \approx 75.0°$

33. Yes, the law of cosines can be used:
$c^2 = a^2 + b^2 - 2ab\cos C$
$c^2 = 21.3^2 + 40^2 - 2(21.3)(40)\cos 90°$
$c^2 = 21.3^2 + 40^2 - 2(21.3)(40)(0)$
$c^2 = 21.3^2 + 40^2$
$c \approx 45.3$
Since $\cos 90° = 0$, the law of cosines is the same as the Pythagorean theorem when the angle used is 90°.

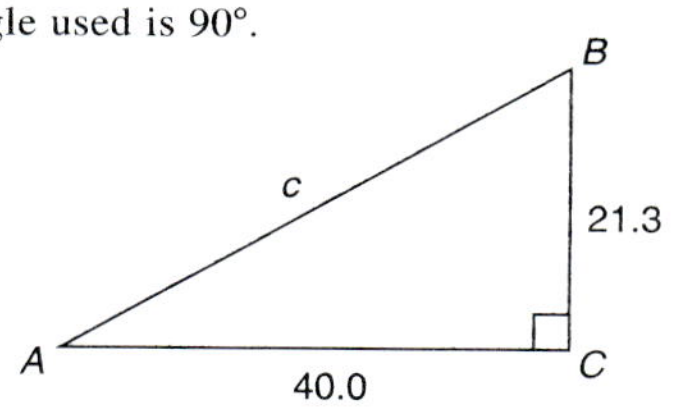

Solutions to skill and review problems

1. $\dfrac{\sin A}{a} = \dfrac{\sin B}{b} = \dfrac{\sin C}{c}$

$\dfrac{\sin A}{6.8} = \dfrac{\sin 43°}{12} = \dfrac{\sin C}{c}$

$\sin A = \dfrac{\sin 43°}{12}(6.8) \approx 0.3865$, so

$A' \approx 22.73°$

$A \approx 22.73°$ or $180° - 22.73° \approx 157.27°$

Case 1: $A \approx 22.73°$

$C \approx 180° - 22.73° - 43°$
$\approx 114.27°$

$c = \dfrac{12\sin 114.27°}{\sin 43°} \approx 16.0$

$c \approx 16.0$, $A \approx 22.7°$, $C \approx 114.3°$

Case 2: $A \approx 157.27°$

$C \approx 180° - 157.27° - 43°$
$\approx -20.27°$

No solution.

2. $\tan 43° = \dfrac{b}{6.8}$; $b = 6.8\tan 43° \approx 6.3$

$A = 90° - 43° = 47°$

$\cos 43° = \dfrac{6.8}{c}$; $c = \dfrac{6.8}{\cos 43°} \approx 9.3$

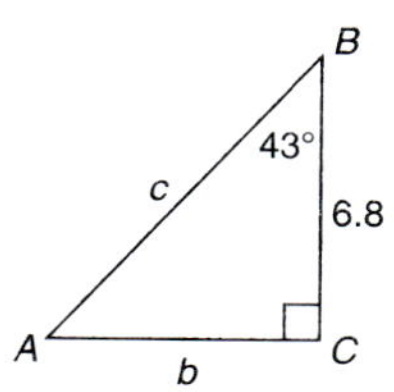

3. $f(x) = 2x^4 + 5x^3 - 8x^2 - 17x - 6$
We look for zeros because these are x-intercepts. We factor the expression on the right at the same time. Possible rational zeros are $\pm1, \pm2, \pm3, \pm6, \pm\frac{1}{2}$, and $\pm\frac{3}{2}$. By synthetic division we find -1 is a zero, so $f(x) = (x + 1)(2x^3 + 3x^2 - 11x - 6)$.
The value 2 is a zero, so $f(x) = (x + 1)(x - 2)(2x^2 + 7x + 3)$.
Thus, $f(x) = (x + 1)(x - 2)(2x + 1)(x + 3)$, and the x-intercepts are -1, 2, $-\frac{1}{2}$, -3.
The y-intercept is $f(0) = -6$. We plot additional points between the x-intercepts:

x	-4	-2	$-\frac{3}{4}$	1	1.5	3
y	126	-12	0.8	-24	-22.5	168

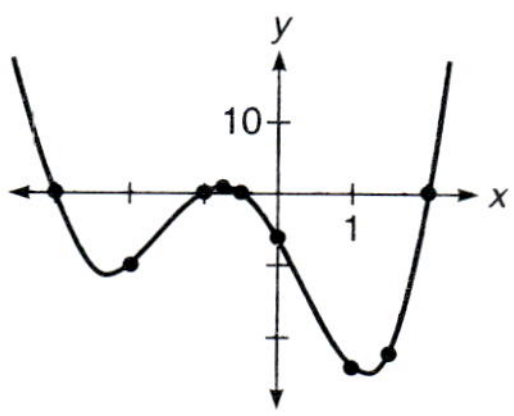

4. $f(x) = -(x - 3)^2 + 1$
The graph of $f(x)$ is the graph of x^2 but flipped over and with vertex at (3,1).
Intercepts: $f(0) = -8$ $(0,-8)$

$0 = -(x - 3)^2 + 1$
$(x - 3)^2 = 1$
$x - 3 = \pm1$
$x = 3 \pm 1$; (2,0) and (4,0).

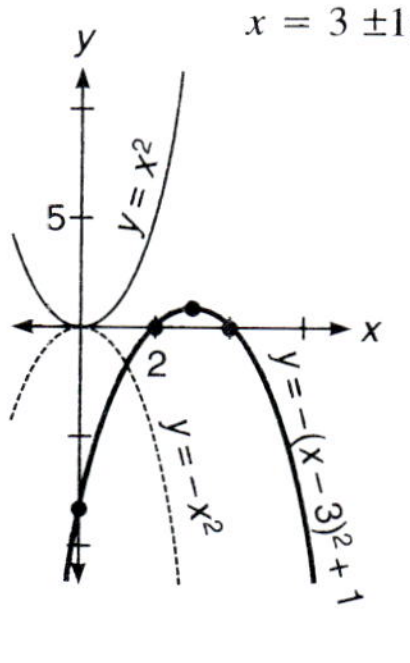

5. $\sin\theta = -\frac{2}{3}$ and $\cos\theta < 0$

$\tan\theta = \dfrac{-2}{-\sqrt{5}} = \dfrac{2\sqrt{5}}{5}$.

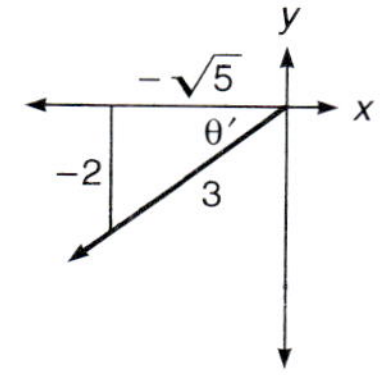

Solutions to trial exercise problems

3. $b = 61.3$, $c = 23.9$, $A = 124.0°$
$a^2 = b^2 + c^2 - 2bc\cos A$
$a^2 = 61.3^2 + 23.9^2 - 2(61.3)(23.9)\cos 124° \approx 5967.4$
$a \approx 77.249$

$\dfrac{\sin A}{a} = \dfrac{\sin B}{b} = \dfrac{\sin C}{c}$

$\dfrac{\sin 124°}{77.249} = \dfrac{\sin B}{61.3} = \dfrac{\sin C}{23.9}$

Find angle C first; it is the smallest and therefore acute.

$\sin C \approx \dfrac{23.9\sin 124°}{77.249}$; $C \approx 14.9°$

$B \approx 180° - 14.9° - 124° \approx 41.1°$

Thus, $a \approx 77.2$, $B \approx 41.1°$, $C \approx 14.9°$.

7. $a = 23.5$, $b = 19.4$, $c = 35.0$
$c^2 = a^2 + b^2 - 2ab\cos C$,

so $\cos C = \dfrac{a^2 + b^2 - c^2}{2ab}$

$= \dfrac{23.5^2 + 19.4^2 - 35^2}{2(23.5)(19.4)}$

so $C \approx 108.97° \approx 109.0°$

23.5 [x^2] [+] 19.4 [x^2] [−] 35 [x^2] [=] [÷] 2 [÷] 23.5 [÷] 19.4 [=] [SHIFT] [cos]

TI-81: [2nd] [COS] [(] [(] 23.5 [x^2] [+] 19.4 [x^2] [−] 35 [x^2] [)] [÷] [(] 2 [×] 23.5 [×] 19.4 [)] [)] [ENTER]

Since C is the largest angle in the triangle, we know A and B are acute.

$\dfrac{\sin A}{a} = \dfrac{\sin B}{b} = \dfrac{\sin C}{c}$

$\dfrac{\sin A}{23.5} = \dfrac{\sin B}{19.4} = \dfrac{\sin 108.97°}{35}$

$\sin A = \dfrac{\sin 108.97°}{35}(23.5)$; $A \approx 39.4°$

$B \approx 180° - 39.4° - 109.0° \approx 31.6°$

25. Use the distance formula, which states that for two points (x_1,y_1), (x_2,y_2), the distance d between them is $d = \sqrt{(x_2 - x_1)^2 + (y_2 - y_1)^2}$. We find that $a = \sqrt{37}$, $b = \sqrt{65}$, and $c = \sqrt{8}$. The largest angle is opposite the longest side, b. Thus, $b^2 = a^2 + c^2 - 2ac \cos B$, so $\cos B = \dfrac{a^2 + c^2 - b^2}{2ac} = \dfrac{37 + 8 - 65}{2(\sqrt{37})(\sqrt{8})} \approx -0.5812$, $B \approx 125.5°$.

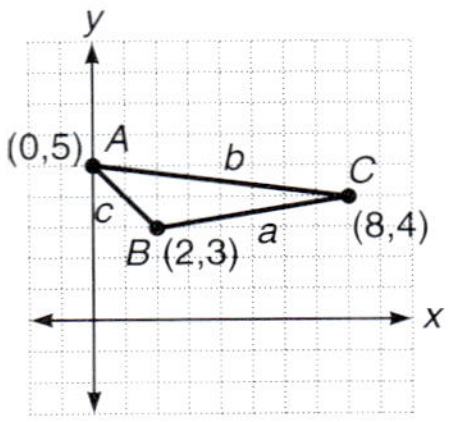

26. $d^2 = 17.6^2 + 22.5^2 - 2(17.6)(22.5) \cos 47.2°$; $d \approx 16.7$ miles

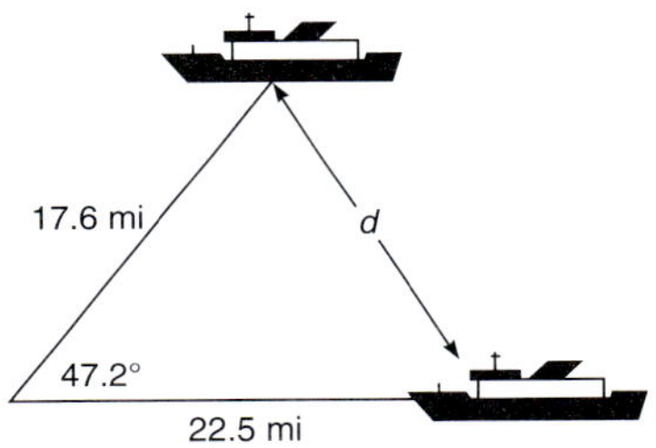

Exercise 8–3

Answers to odd-numbered problems

1. $(20\sqrt{3},20)$ **3.** $(-53.4,84.5)$ **5.** $(-9.4,-3.4)$ **7.** $\left(12\frac{1}{2},-\frac{25\sqrt{3}}{2}\right)$ **9.** $(3\sqrt{2},-3\sqrt{2})$ **11.** $(-0.8,7.8)$ **13.** $(5,53.1°)$ **15.** $(6.0,120.0°)$ **17.** $(2.6,-49.1°)$ **19.** $(11.2,-63.4°)$ **21.** $(7.6,153.4°)$ **23.** $(3.16,-116.6°)$ **25.** $(-1,20)$ **27.** $(8\sqrt{2},0)$ **29.** $(45.2,31.2°)$ **31.** $(36.5,-122.1°)$ **33.** $(9.0,-54.2°)$ **35.** $(47.1,139.7°)$ **37.** $(6.3,89.3°)$ **39.** $(20.3,83.9°)$ **41.** $(12.9,21.0°)$ **43.** west at 136 knots and north at 63 knots **45.** east at 193 knots and north at 52 knots **47.** west: 228 knots; north: 395 knots **49. a.** 24 nm **b.** 38 nm **51.** No. The horizontal component of the force is 1,887 pounds. This is not enough to move the sled. **53. a.** 966, 259 **b.** 1932, 518; yes **c.** 866, 500; no **55.** 40 miles in a direction 59° south of east **57.** $(15.4,-76.8°)$ **59.** $(270.7,87.5°)$ **61.** 63 nm, 40° south of west **63.** 99 knots, 47° north of west **65.** 26.5 knots, 25.7° north of east **67.** 266 volts at 341° **69.** 8° west of north, 87 knots **71.** 76.3°, 17.7 knots **73.** 320 pounds, 50° with the horizontal **75.** Let $A = (a_1,a_2)$, $B = (b_1,b_2)$, and $C = (c_1,c_2)$ be three vectors in rectangular form. Then proceed as follows:

$(A + B) + C$

$[(a_1,a_2) + (b_1,b_2)] + (c_1,c_2)$

The parentheses indicate we add A and B first.

$(a_1 + b_1, a_2 + b_2) + (c_1,c_2)$

Two vectors: one is $A + B$, and the other is C.

$((a_1 + b_1) + c_1, (a_2 + b_2) + c_2)$

One vector: $(A + B) + C$.

$(a_1 + (b_1 + c_1), a_2 + (b_2 + c_2))$

We can rearrange because real numbers are associative, and the components are real numbers.

$A + (B + C)$

Solutions to skill and review problems

1. $c = 12.0$ and $A = 23°$

$B = 90° - 23° = 67°$

$\sin 23° = \dfrac{a}{12}$, so $a = 12 \sin 23° \approx 4.7$

$\cos 23° = \dfrac{b}{12}$, so $b = 12 \cos 23° \approx 11.0$

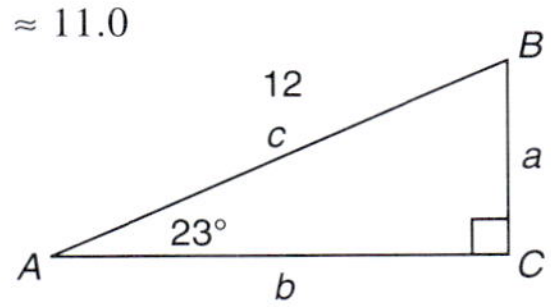

2. $c = 12.0$, $a = 7.5$, and $A = 23°$. We are not told that this is a right triangle, so we must use either the law of sines or the law of cosines. Since we have an angle (A) and the side opposite that angle (a), we use the law of sines.

$\dfrac{\sin 23°}{7.5} = \dfrac{\sin B}{b} = \dfrac{\sin C}{12}$

$\sin C = \dfrac{\sin 23°}{7.5}(12)$; $C' \approx 38.69°$ or $180° - 38.69° \approx 141.31°$

Case 1: $C \approx 38.69°$

$B = 180° - 38.69° - 23° \approx 118.31°$

$\dfrac{\sin 23°}{7.5} = \dfrac{\sin 118.31°}{b}$; $b \approx 16.90$

Case 2: $C \approx 141.31°$

$B = 180° - 141.31° - 23° \approx 15.69°$

$\dfrac{\sin 23°}{7.5} = \dfrac{\sin 15.69°}{b}$; $b \approx 5.19$

Thus, $b \approx 16.9$, $C \approx 38.7°$, $B \approx 118.3°$, or $b \approx 5.2$, $C \approx 141.3°$, $B \approx 15.7°$.

3. $c = 12.0$, $a = 7.5$, and $B = 23°$. We are not told this is a right triangle. We do not know any angle and the length of the side opposite, so we cannot use the law of sines. Thus, we use the law of cosines.

$b^2 = a^2 + c^2 - 2ac \cos B$

$b^2 = 7.5^2 + 12^2 - 2(7.5)(12) \cos 23°$

$b \approx 5.879$

$\dfrac{\sin A}{7.5} = \dfrac{\sin 23°}{5.879} = \dfrac{\sin C}{12}$

Angle C may be obtuse, since it is the largest angle in the triangle (because it is opposite the largest side). Thus, it is better to find angle A next, since it must be acute. $\sin A = \dfrac{\sin 23°}{5.879}(7.5)$, so $A \approx 29.90°$. Thus, $C \approx 180° - 23° - 29.9° \approx 127.1°$. Therefore, $b \approx 5.9$, $A \approx 29.9°$, and $C \approx 127.1°$.

4. $f(x) = \dfrac{5}{x^2 - 4x + 3} = \dfrac{5}{(x - 3)(x - 1)}$

Vertical asymptotes at 1 and 3.

No x-intercepts, since $0 = \dfrac{5}{x^2 - 4x + 3}$ has no solution.

y-intercept is $f(0) = \frac{5}{3} = 1\frac{2}{3}$.

Additional points:

x	−2	−1	0.5	1.5	2	2.5	4	5
y	0.3	0.6	4	−6.7	−5	−6.7	1.7	0.6

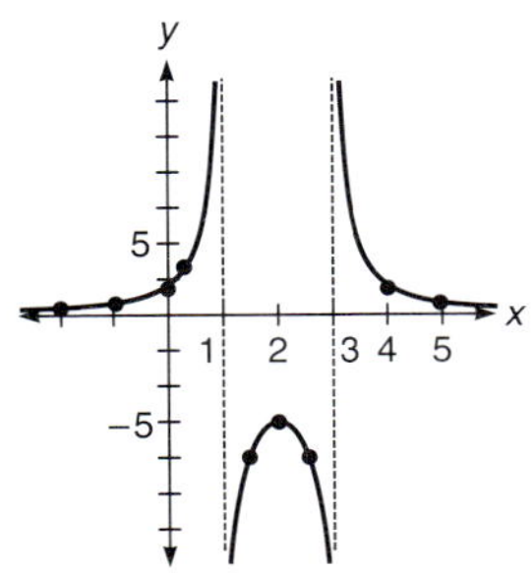

Solutions to trial exercise problems

3. $(100.0,122.3°) = (100 \cos 122.3°, 100 \sin 122.3°) \approx (-53.4,84.5)$

23. $(-\sqrt{2},-\sqrt{8})$; $|A| = \sqrt{(-\sqrt{2})^2 + (-\sqrt{8})^2} \approx \sqrt{10} \approx 3.16$, $\theta' \approx 63.43°$; $\theta \approx 63.43° - 180° \approx -116.57°$; $(3.16,-116.6°)$

39. $(15.3, 311°)$
$= (15.3 \cos 311°, 15.3 \sin 311°)$
$\approx (10.038, -11.547)$ [1]
$(20.9, 117°)$
$= (20.9 \cos 117°, 20.9 \sin 117°)$
$\approx (-9.488, 18.622)$ [2]
$(13.2, 83°)$
$= (13.2 \cos 83°, 13.2 \sin 83°)$
$\approx (1.609, 13.102)$ [3]
Adding [1] + [2] + [3] gives
$(2.158, 20.177) \approx (20.3, 83.9°)$

43. $V = (150, 155°)$;
$V_x = 150 \cos 155° \approx -136$ (136 knots due west)
$V_y = 150 \sin 155° \approx 63$ (63 knots due north)
The aircraft is moving west at 136 knots and north at 63 knots.

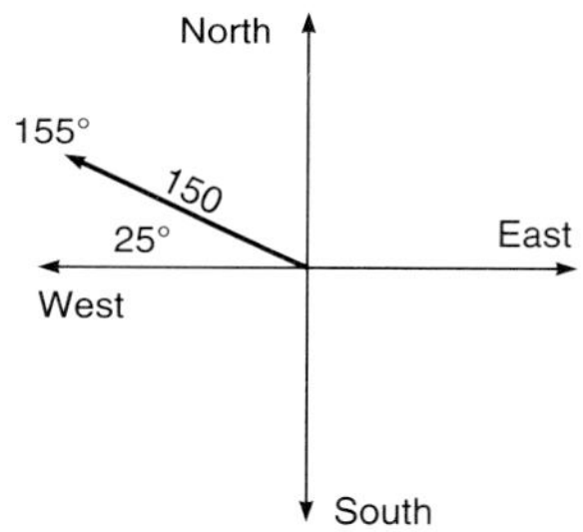

49. 18 knots (18 nautical miles per hour) $\times$ 2.5 hours = 45 nm (nautical miles).
$V = (45, 32°)$;
$V_x = 45 \cos 32° \approx 38$ nm; distance east of the harbor (part b)
$V_y = 45 \sin 32° \approx 24$ nm; distance north of the harbor (part a).

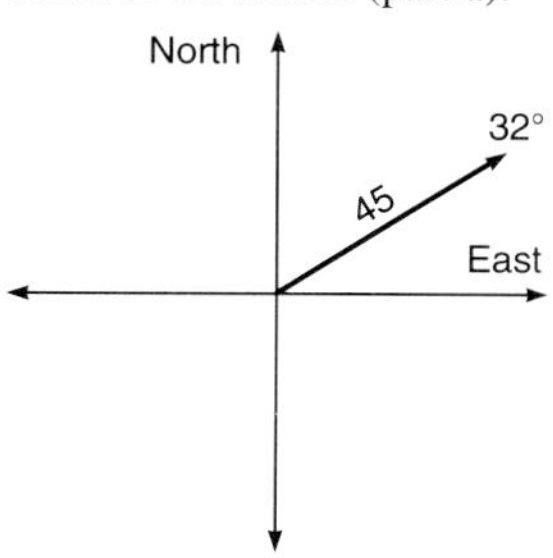

52. $f = 1{,}700 \cos\theta$. We require $f \geq 1{,}200$, so $1{,}200 \geq 1{,}700 \cos\theta$, or $\frac{12}{17} \geq \cos\theta$. $\cos^{-1}\frac{12}{17} \approx 45.1°$. Thus, $\theta \leq 45.1°$ will move the sled. Note that $\theta \leq 45.1°$ is correct, and not $\theta \geq 45.1°$. This can be seen in the figure. If θ increases, f clearly decreases. Mathematically, the value of $\cos\theta$ increases as θ decreases (for acute angles).

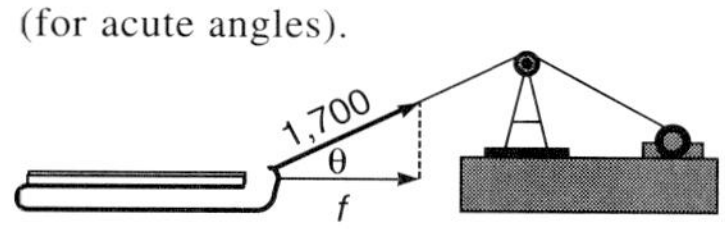

54. At 200 mph, the first leg of the trip is 200 miles. The second is 200 mph $\times$ 0.5 hour = 100 miles. To find d and θ, we add the vectors $(200, 0°)$ and $(100, -60°)$.
$(200, 0°) = (200 \cos 0°, 200 \sin 0°)$
$= (200, 0)$
$(100, -60°) = (100 \cos(-60°), 100 \sin(-60°)) \approx (50, -86.60)$
$(250, -86.60) \approx (265, 341°)$
Thus, $d \approx 265$ miles, and $\theta \approx 341°$. The aircraft is 265 miles from Minneapolis. $360° - 341° = 19°$, so the aircraft is in a direction 19° south of east, relative to the city.

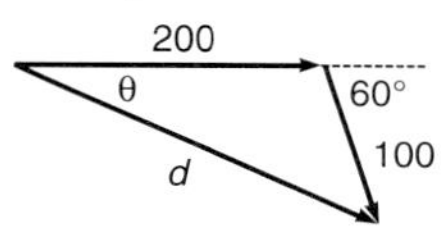

59. $(199, 19.0°) = (199 \cos 19°, 199 \sin 19°) \approx (188.16, 64.79)$
$(175, 131°) = (175 \cos 131°, 175 \sin 131°) \approx (-114.81, 132.07)$
$(96, 130°) = (96 \cos 130°, 96 \sin 130°) \approx (-61.71, 73.54)$
Adding the rectangular form gives
$(11.64, 270.40) \approx (270.7, 87.5°)$

64. $(19.6, 170°) = (19.6 \cos 170°, 19.6 \sin 170°) \approx (-19.3, 3.4)$
$(7.2, 95°) = (7.2 \cos 95°, 7.2 \sin 95°) \approx (0.63, 7.17)$
$(-19.93, 10.58) \approx (22.6, 152.0°)$
Thus, its true course is $180° - 152° = 28°$ north of west, at a speed of 22.6 knots.

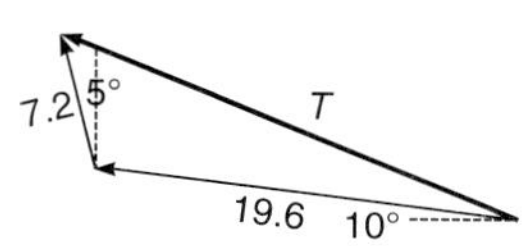

66. $(122, 30°) = (122 \cos 30°, 122 \sin 30°) \approx (105.66, 61.00)$
$(86, 21°) = (86 \cos 21°, 86 \sin 21°) \approx (80.29, 30.82)$
Adding the rectangular forms gives
$(185.94, 91.82) \approx (207, 26°)$
Magnitude is 207 volts, phase angle is 26°.

71. We let W represent the water current vector.
$H + W = T$
$H = T - W$
$= (12, 65°) - (6.4, -82°)$
$= (12, 65°) + (6.4, -82° + 180°)$
$= (12, 65°) + (6.4, 98°)$
$= (5.07, 10.88) + (-0.89, 6.34)$
$= (4.18, 17.21) \approx (17.7, 76.3°)$
Thus the ship's heading is 76.3°, and its speed is 17.7 knots.

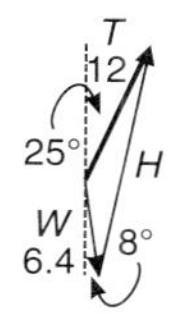

73. The sign is stationary, so the forces acting on it are balanced (they add to zero).
$T_1 + T_2 + W = 0$
$T_1 = -T_2 - W$
$= -(456, 63°) - (650, 270°)$
$= (456, 63° + 180°) + (650, 270° - 180°)$
To negate a vector, add or subtract 180° from its direction angle.
$= (456, 243°) + (650, 90°)$
$\approx (-207.02, -406.3) + (0, 650)$
Convert to rectangular form
$\approx (-207.02, 243.7)$
$\approx (320, 130°)$
Convert back to polar form
Thus, the tension in the second cable is 320 pounds, and it makes an angle θ of 50° (180° − 130°) with the horizontal.

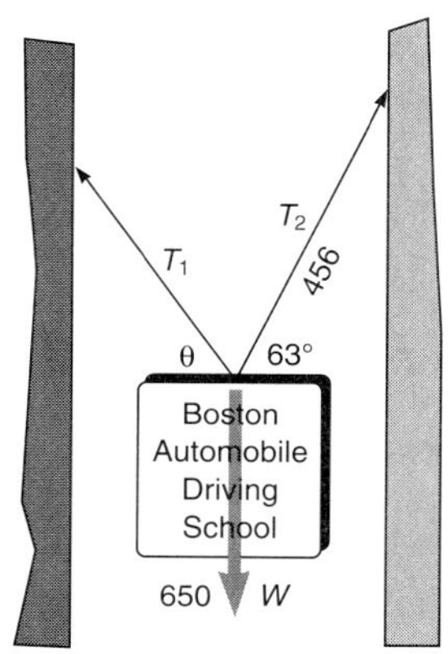

74. Let $A = (a_1,a_2)$ and $B = (b_1,b_2)$ be two vectors in rectangular form. Then proceed as follows:

$A + B$

$(a_1,a_2) + (b_1,b_2)$

$(a_1 + b_1, a_2 + b_2)$

Definition of vector addition

$(b_1 + a_1, b_2 + a_2)$

a_1, a_2, b_1, b_2 are real numbers, so their indicated sum commutes

$B + A$

Definition of vector addition

76. The following is for the TI-81:

PRGM EDIT 2

Choose a free location to enter the program. Say 2 by way of example.

A D D V C T R S Enter these characters as the name of the program

```
:0→A
:0→B
:Lbl 1
:Input R
:If R=0
:Goto 2
:Input θ
:P▸R(R,θ)
:A+X→A
:B+Y→B
:Goto 1
:Lbl 2
:A→X
:B→Y
:R▸P(X,Y)
:Disp "X,Y"
:Disp X
:Disp Y
:Disp "R,θ"
:Disp R
:Disp θ
```

To run the program input R, then θ, for each vector in polar form. When all vectors have been entered, enter zero (0) for R. The program converts each vector into rectangular form as it is entered, and accumulates the values (x,y) in variables A and B. When all vectors are entered, the accumulated values in A and B are converted to polar form.

Exercise 8–4

Answers to odd-numbered problems

1. 5.4 cis(−21.8°) **3.** 3.2 cis 108.4° **5.** 5 cis 126.9° **7.** 2 cis 30° **9.** $3\sqrt{2}$ cis 45° **11.** $\sqrt{2}$ cis(−135°) **13.** 5 cis 90° **15.** $2.9 + 0.8i$ **17.** $3.7 + 2.6i$ **19.** $1 - i$ **21.** $12.3 - 5.7i$ **23.** $\frac{3}{2} + \frac{\sqrt{3}}{2}i$ **25.** $5 - 5\sqrt{3}i$ **27.** $-\sqrt{10}$ **29.** $2 - 2i$ **31.** 15 cis 75° **33.** 10.8 cis(−120°) **35.** 4 cis 80° **37.** $\frac{20}{9}$cis(−80°) **39.** 512 cis(−60°) **41.** 27 cis(−120°) **43.** $-8.2 - 0.1i$ **45.** $2, -1 + \sqrt{3}i, -1 - \sqrt{3}i$ **47.** $3, 3i, -3, -3i$ **49.** $-1.1 + 4.9i, -3.7 - 3.4i, 4.8 - 1.5i$ **51.** 2.5 cis(−20°) **53.** 50 cis 45° **55.** 2.04 cis 6.19° **57.** $0.75 + 0.86i$ **59.** no **61.** $-3 + i, -1 - 3i, 3 - i, 1 + 3i$

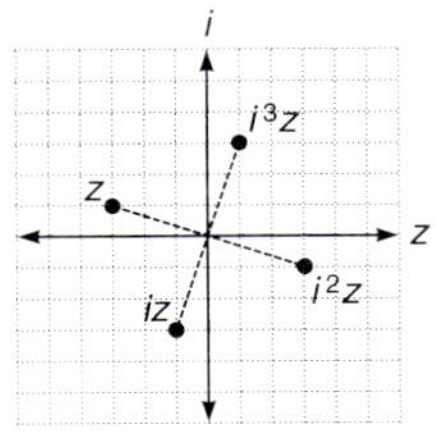

63. $6, 6i, -6, -6i$

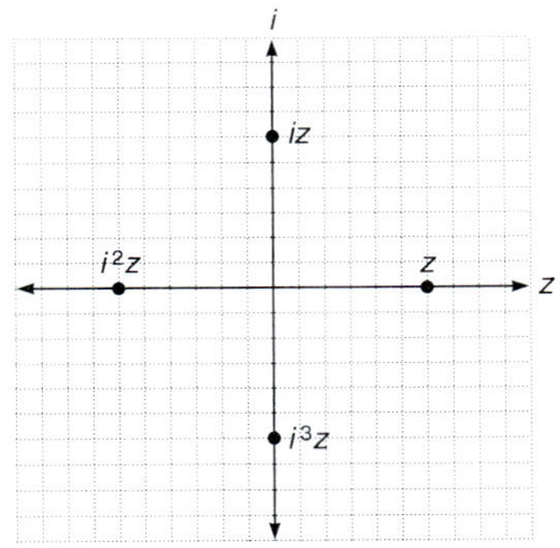

65. $-1 - i, 1 - i, 1 + i, -1 + i$

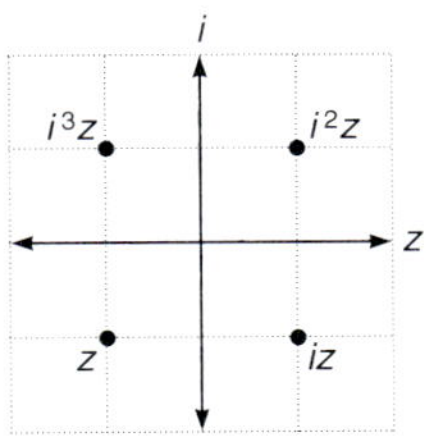

67. **a.** 1 cis 30° or $\frac{\sqrt{3}}{2} + \frac{1}{2}i$ **b.** $0.37 + 1.37i$ **c.** $0.37 - 1.37i$

69. $r^{\frac{1}{n}}\text{cis}\left(\frac{\theta}{n} + \frac{k \cdot 360°}{n}\right)$

Replace k by $an + b$, $b < n$.

$$r^{\frac{1}{n}}\text{cis}\left[\frac{\theta}{n} + \frac{(an + b) \cdot 360°}{n}\right]$$

$$r^{\frac{1}{n}}\text{cis}\left(\frac{\theta}{n} + \frac{an \cdot 360°}{n} + \frac{b \cdot 360°}{n}\right)$$

$$r^{\frac{1}{n}}\text{cis}\left(\frac{\theta}{n} + a \cdot 360° + \frac{b \cdot 360°}{n}\right)$$

$$r^{\frac{1}{n}}\text{cis}\left[\left(\frac{\theta}{n} + \frac{b \cdot 360°}{n}\right) + a \cdot 360°\right]$$

$$r^{\frac{1}{n}}\cos\left[\left(\frac{\theta}{n} + \frac{b \cdot 360°}{n}\right) + a \cdot 360°\right]$$

$$+ ir^{\frac{1}{n}}\sin\left[\left(\frac{\theta}{n} + \frac{b \cdot 360°}{n}\right) + a \cdot 360°\right]$$

$$\cos\left[\left(\frac{\theta}{n} + \frac{b \cdot 360°}{n}\right) + a \cdot 360°\right]$$

$$= \cos\left(\frac{\theta}{n} + \frac{b \cdot 360°}{n}\right)\cos(a \cdot 360°)$$

$$- \sin\left(\frac{\theta}{n} + \frac{b \cdot 360°}{n}\right)\sin(a \cdot 360°)$$

$= \cos\left(\frac{\theta}{n} + \frac{b \cdot 360°}{n}\right)$, because $\cos(a \cdot 360°) = 1$ and $\sin(a \cdot 360°) = 0$, when a is an integer.

$$\sin\left[\left(\frac{\theta}{n} + \frac{b \cdot 360°}{n}\right) + a \cdot 360°\right]$$

$$= \sin\left(\frac{\theta}{n} + \frac{b \cdot 360°}{n}\right)\cos(a \cdot 360°)$$

$$+ \cos\left(\frac{\theta}{n} + \frac{b \cdot 360°}{n}\right)\sin(a \cdot 360°)$$

$= \sin\left(\frac{\theta}{n} + \frac{b \cdot 360°}{n}\right)$, because $\cos(a \cdot 360°) = 1$ and $\sin(a \cdot 360°) = 0$, when a is an integer.

Thus,

$$r^{\frac{1}{n}}\cos\left[\left(\frac{\theta}{n} + \frac{b \cdot 360°}{n}\right) + a \cdot 360°\right]$$

$$+ ir^{\frac{1}{n}}\sin\left[\left(\frac{\theta}{n} + \frac{b \cdot 360°}{n}\right) + a \cdot 360°\right]$$

$$= r^{\frac{1}{n}}\cos\left(\frac{\theta}{n} + \frac{b \cdot 360°}{n}\right)$$

$$+ ir^{\frac{1}{n}}\sin\left(\frac{\theta}{n} + \frac{b \cdot 360°}{n}\right)$$

$= r^{\frac{1}{n}}\text{cis}\left(\frac{\theta}{n} + \frac{b \cdot 360°}{n}\right)$, where $b < n$.

This last expression is one of the previous roots.

Solutions to skill and review problems

1. $(3.8,28°) = (3.8 \cos 28°, 3.8 \sin 28°) \approx (3.36,1.78)$
$(5.1,134°) = (5.1 \cos 134°, 5.1 \sin 134°) \approx (-3.54,3.67)$; adding the rectangular forms gives $(-0.19,5.45)$
$r = \sqrt{0.19^2 + 5.45^2} \approx 5.5$;
$$\theta' = \tan^{-1}\frac{5.45}{-0.19} \approx -88.00°$$
The x-component, -0.19, is negative, and $\theta' < 0$, so $\theta = \theta' + 180° \approx -88° + 180° \approx 92°$. Thus the resultant vector is $(5.5,92°)$.

2. $B = C - A = C + (-A)$; $-A = (190,30° + 180°) = (190,210°)$
C: $(150,68°) = (150 \cos 68°, 150 \sin 68°) \approx (56.19,139.08)$
$-A$: $(190,210°) = (190 \cos 210°, 190 \sin 210°) \approx (-164.54,-95)$; adding gives $(-108.35,44.08) \approx (117,158°)$

3. $a = 12.6$, $b = 19.1$, and $c = 28.0$; this may not be a right triangle, so we must use the law of sines or the law of cosines. We cannot use the law of sines since we do not have any side and the angle opposite that side. Therefore we use the law of cosines to find one of the angles.

The law of cosines does not have an ambiguous case, so we use it to find the largest angle, which may be acute or obtuse. The largest angle is opposite the longest side, which is c in this case.
$c^2 = a^2 + b^2 - 2ab \cos C$, so $\cos C$
$$= \frac{a^2 + b^2 - c^2}{2ab} = \frac{12.6^2 + 19.1^2 - 28^2}{2(12.6)(19.1)}$$
≈ -0.5411, so $C \approx 122.757°$
$$\frac{\sin A}{12.6} = \frac{\sin B}{19.1} = \frac{\sin 122.757°}{28};$$ A and B are both acute, so we can find either next.
$$\sin A = \frac{\sin 122.757°}{28}(12.6) \approx 0.3784, \text{ so}$$
$A \approx 22.2°$. $B \approx 180° - 22.2° - 122.8° \approx 35.0°$. Thus $A \approx 22.2°$, $B \approx 35.0°$, $C \approx 122.8°$.

4. $$\frac{1}{\sqrt[3]{4a^4b}} = \frac{1}{\sqrt[3]{2^2a^3ab}} = \frac{1}{a\sqrt[3]{2^2ab}}$$
$$= \frac{1}{a\sqrt[3]{2^2ab}} \cdot \frac{\sqrt[3]{2a^2b^2}}{\sqrt[3]{2a^2b^2}} = \frac{\sqrt[3]{2a^2b^2}}{a\sqrt[3]{2^3a^3b^3}}$$
$$= \frac{\sqrt[3]{2a^2b^2}}{a(2ab)} = \frac{\sqrt[3]{2a^2b^2}}{2a^2b}$$

5. $\tan \theta = 1 + u$ and θ terminates in quadrant I
$r^2 = (1 + u)^2 + 1^2 = u^2 + 2u + 2$, so $r = \sqrt{u^2 + 2u + 2}$
$$\sin \theta = \frac{1 + u}{r} = \frac{1 + u}{\sqrt{u^2 + 2u + 2}}$$

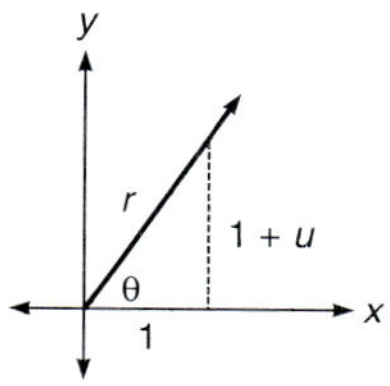

Solutions to trial exercise problems

4. $\sqrt{3} - 2i$
$r = \sqrt{(\sqrt{3})^2 + (-2)^2} = \sqrt{7} \approx 2.6$
$$\theta' = \tan^{-1}\frac{-2}{\sqrt{3}} \approx -49.1°;$$
$a > 0$ so $\theta = \theta'$
The point is $2.6 \text{ cis}(-49.1°)$.

7. $\sqrt{3} + i$
$r = \sqrt{(\sqrt{3})^2 + 1^2} = 2$
$$\theta' = \tan^{-1}\frac{1}{\sqrt{3}} = \tan^{-1}\frac{\sqrt{3}}{3} = 30°$$
The point is in quadrant I, so it is $2 \text{ cis } 30°$.

15. $3 \text{ cis } 15°$
$3 \cos 15° + (3 \sin 15°)i = 2.9 + 0.8i$

26. $6 \text{ cis } 135°$
$6 \cos 135° + (6 \sin 135°)i$
$$= 6 \cdot \left(-\frac{\sqrt{2}}{2}\right) + 6 \cdot \frac{\sqrt{2}}{2}i =$$
$-3\sqrt{2} + 3\sqrt{2}i$

41. $(3 \text{ cis } 200°)^3 = 3^3 \text{ cis}(3 \cdot 200°) = 27 \text{ cis } 600° = 27 \text{ cis}(600° - 2 \cdot 360°) = 27 \text{ cis}(-120°)$

44. $(0.8 + 0.6i)^{10}$
Transform into polar form to use De Moivre's theorem.
$r = \sqrt{0.8^2 + 0.6^2} = 1$
$$\theta = \theta' = \tan^{-1}\frac{6}{8} \approx 36.87°$$
$(0.8 + 0.6i)^{10} \approx (1 \text{ cis } 36.87°)^{10} \approx 1^{10} \text{ cis } 368.7° \approx \text{cis } 8.7° \approx 1.0 + 0.2i$

55. $Z_1 = 2 + i \approx \sqrt{5} \text{ cis } 26.565°$
$Z_2 = 3 - 5i \approx \sqrt{34} \text{ cis } 300.964°$
$Z_1 + Z_2 = 5 - 4i \approx \sqrt{41} \text{ cis } 321.340°$
$$\frac{Z_1Z_2}{Z_1 + Z_2} \approx \frac{(\sqrt{5} \text{ cis } 26.565°)(\sqrt{34} \text{ cis } 300.964°)}{\sqrt{41} \text{ cis } 321.340°}$$
$$\approx \frac{\sqrt{170} \text{ cis } 327.529°}{\sqrt{41} \text{ cis } 321.340°} \approx 2.04 \text{ cis } 6.19°$$

---

68. $$\frac{r_1 \text{cis } \theta_1}{r_2 \text{cis } \theta_2} = \frac{r_1}{r_2} \cdot \frac{\text{cis } \theta_1}{\text{cis } \theta_2} = \frac{r_1}{r_2} \cdot \frac{\cos \theta_1 + i \sin \theta_1}{\cos \theta_2 + i \sin \theta_2}$$
$$\frac{r_1}{r_2} \cdot \frac{\cos \theta_1 + i \sin \theta_1}{\cos \theta_2 + i \sin \theta_2} \cdot \frac{\cos \theta_2 - i \sin \theta_2}{\cos \theta_2 - i \sin \theta_2} = \frac{r_1}{r_2} \cdot$$
$$\frac{\cos \theta_1 \cos \theta_2 - i \cos \theta_1 \sin \theta_2 + i \sin \theta_1 \cos \theta_2 - i^2 \sin \theta_1 \sin \theta_2}{\cos^2\theta_2 - i^2 \sin^2\theta_2}$$
$$\frac{r_1}{r_2} \cdot \frac{\cos \theta_1 \cos \theta_2 + \sin \theta_1 \sin \theta_2 + i(\sin \theta_1 \cos \theta_2 - \cos \theta_1 \sin \theta_2)}{\cos^2\theta_2 + \sin^2\theta_2}$$
$$\frac{r_1}{r_2} \cdot \frac{\cos(\theta_1 - \theta_2) + i \sin(\theta_1 - \theta_2)}{1} = \frac{r_1}{r_2}\text{cis}(\theta_1 - \theta_2)$$

Exercise 8–5

Answers to odd-numbered problems

The figure shows the answers to odd-numbered problems 1 through 17.

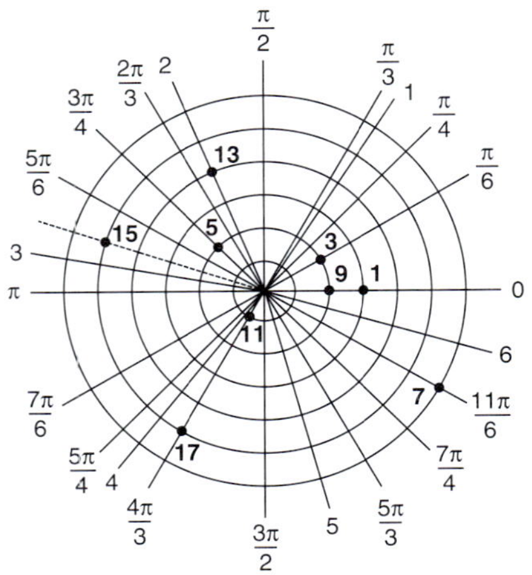

Many answers are possible in problems 19–23.

19. $\left(-2,\frac{7\pi}{6}\right), \left(2,\frac{13\pi}{6}\right), \left(2,\frac{25\pi}{6}\right)$

21. $\left(-6,\frac{5\pi}{6}\right), \left(6,\frac{23\pi}{6}\right), \left(6,\frac{35\pi}{6}\right)$

23. $(-2,2+\pi), (2,2+2\pi), (2,2+4\pi)$

25. $(0,4)$ **27.** $\left(-\frac{5\sqrt{3}}{2},\frac{5}{2}\right)$

29. $(-2,-2\sqrt{3})$ **31.** $(1.08,1.68)$

33. $(2.05,2.19)$ **35.** $(-2.61,-3.03)$

37. $\left(4,-\frac{5\pi}{6}\right)$ **39.** $(2,\pi)$

41. $\left(4\sqrt{2},-\frac{3\pi}{4}\right)$ **43.** $(3.61,0.98)$

45. $(4.12,-1.33)$ **47.** $(6.40,0.67)$

49. $\tan\theta = 4$ **51.** $r = \dfrac{2}{\sin\theta + 3\cos\theta}$

53. $r = \dfrac{b}{\sin\theta - m\cos\theta}$

55. $r^2 = \dfrac{5}{\sin^2\theta - 2\cos^2\theta}$

57. $r^2 = \dfrac{1}{3\cos^2\theta + 2\sin^2\theta}$

59. $x^2 + y^2 - y = 0$ **61.** $x = 2$

63. $x^6 + 3x^4y^2 - 36x^2y^2 + 3x^2y^4 + y^6 = 0$

65. $x^4 + 2x^2y^2 - 2xy + y^4 = 0$

67. $x^3 + xy^2 - y = 0$

69. $3y^2 + 12y - x^2 + 9 = 0$

71. $2xy = 5$

$2(r\cos\theta)(r\sin\theta) = 5$

$2r^2\cos\theta\sin\theta = 5$

$r^2(2\sin\theta\cos\theta) = 5$

$\sin 2\alpha = 2\sin\alpha\cos\alpha$

$r^2\sin 2\theta = 5$

$r^2 = \dfrac{5}{\sin 2\theta}$

$r^2 = 5\csc 2\theta$

73. Consider a point $P = (r,\theta)$, where $r < 0$. Then $P = (-r,\theta+\pi)$, where $-r > 0$. Therefore, since $-r > 0$, $y = -r\sin(\theta+\pi)$ is true.

$y = -r\sin(\theta+\pi)$

$= -r(\sin\theta\cos\pi + \cos\theta\sin\pi)$

$= -r(\sin\theta(-1) + \cos\theta(0))$

$= -r(-\sin\theta)$

$= r\sin\theta$

Thus, $y = r\sin\theta$, even if $r < 0$.

75. $(x^2 + y^2 + 2y)^2 = x^2 + y^2$

77. $(x^2 + y^2)^3 = (x^2 + y^2 + 4xy)^2$

79. $(x^2 + y^2 + 3x)^2 = x^2 + y^2$

Solutions to skill and review problems

1. $2\text{ cis }30° \cdot 5\text{ cis }45° = 2(5)$

$\text{cis}(30° + 45°) = 10\text{ cis }75°$

2. $1{,}000 = 1{,}000\text{ cis }0°$.

Evaluate: $1{,}000^{1/3}\text{cis}\left(\frac{0°}{3} + \frac{k\cdot 360°}{3}\right)$

for $k = 0, 1, 2$. $1{,}000^{1/3} = 10$.

$10\text{ cis}(k\cdot 120°)$ for $k = 0, 1, 2$.

$k = 0$: $10(\cos 0° + i\sin 0°) = 10$

$k = 1$: $10(\cos 120° + i\sin 120°) =$

$10\left(-\frac{1}{2} + \frac{\sqrt{3}}{2}\right) = -5 + 5\sqrt{3}$

$k = 2$: $10(\cos 240° + i\sin 240°) =$

$10\left(-\frac{1}{2} - \frac{\sqrt{3}}{2}\right) = -5 - 5\sqrt{3}$

3. $a = 125$, $b = 85$, and $C = 50°$

$c^2 = a^2 + b^2 - 2ab\cos C$

$c^2 = 125^2 + 85^2 - 2(125)(85)\cos 50°$

$c^2 \approx 9190.76$; $c \approx 95.868$

$\dfrac{\sin A}{125} = \dfrac{\sin B}{85} \approx \dfrac{\sin 50°}{95.868}$

Find angle B next; it must be acute.

$\sin B \approx \dfrac{\sin 50°}{95.868}(85)$; $B \approx 42.8°$

$A \approx 180° - 42.8° - 50° \approx 87.2°$

Thus, $c \approx 96$, $A \approx 87°$, $B \approx 43°$.

4. $f(x) = 2x^2 + 4x - 5$

$= 2(x^2 + 2x) - 5$

$= 2(x^2 + 2x + 1) - 5 - 2(1)$

$= 2(x + 1)^2 - 7$

Vertex: $(-1,-7)$

Intercepts: $f(0) = -5$; $(0,-5)$ is the y-intercept

$0 = 2(x+1)^2 - 7$

$2(x+1)^2 = 7$

$(x+1)^2 = \dfrac{7}{2}$

$x + 1 = \pm\sqrt{\dfrac{7}{2}} = \pm\dfrac{\sqrt{14}}{2}$

$x = -1 \pm \dfrac{\sqrt{14}}{2} \approx -2.87, 0.87$

$(-2.9,0)$ and $(0.9,0)$ are the x-intercepts.

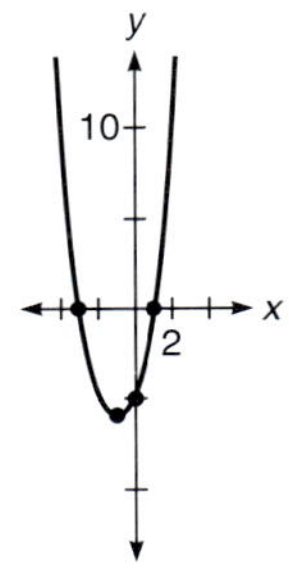

5. $\dfrac{2x-3}{x^2-16} - \dfrac{5}{x-4}$

$\dfrac{2x-3}{(x-4)(x+4)} - \dfrac{5(x+4)}{(x-4)(x+4)}$

$\dfrac{(2x-3) - 5(x+4)}{(x-4)(x+4)}$

$\dfrac{-3x-23}{x^2-16}$

6. $\dfrac{2+3i}{5-2i}\cdot\dfrac{5+2i}{5+2i} = \dfrac{10 + 4i + 15i + 6i^2}{25 - 4i^2}$

$= \dfrac{10 + 19i - 6}{25 + 4} = \dfrac{4 + 19i}{29} = \dfrac{4}{29} + \dfrac{19}{29}i$

7. $\left(\dfrac{x^2x^{-5}}{x^3}\right)^{-2} = \left(\dfrac{x^{-3}}{x^3}\right)^{-2} = \dfrac{x^6}{x^{-6}} = x^{12}$

8. $3x^{-2}(\frac{1}{3}x^2 - 2x + 1)$

$3(\frac{1}{3})x^{-2}x^2 - 3(2)x^{-2}x + 3x^{-2}$

$x^0 - 6x^{-1} + 3x^{-2}$

$1 - \dfrac{6}{x} + \dfrac{3}{x^2}$

$\dfrac{x^2}{x^2} - \dfrac{6x}{x^2} + \dfrac{3}{x^2}$

$\dfrac{x^2 - 6x + 3}{x^2}$

Solutions to trial exercise problems

15. $(-5,6) = (5,6 - \pi) \approx (5,2.86)$; $6 - \pi = (6 - \pi)\cdot\dfrac{180°}{\pi} \approx 164°$. To plot $(-5,6)$, plot a point 5 units from the center, at an angle about 164°. The graph is shown in the answer to the odd problems for this section.

21. Many answers are possible. To change the sign of r add an odd multiple of π to θ; for the rest add an even multiple of π.

$\left(-6,\dfrac{11\pi}{6} - \pi\right) = \left(-6,\dfrac{5\pi}{6}\right)$,

$\left(6,\dfrac{11\pi}{6} + 2\pi\right) = \left(6,\dfrac{23\pi}{6}\right)$,

$\left(6,\dfrac{11\pi}{6} + 4\pi\right) = \left(6,\dfrac{35\pi}{6}\right)$

31. $(2,1) = (2\cos 1, 2\sin 1) \approx (1.08,1.68)$

39. $(-2,0)$; this point is on the x-axis. It is easiest to solve by examination; $r = 2$, and $\theta = \pi$. Thus, the point is $(2,\pi)$.

45. $(1,-4)$

$r = \sqrt{1^2 + 4^2} = \sqrt{17} \approx 4.12$

$\theta' = \tan^{-1}(-4) \approx -1.326$

$x > 0$ so $\theta = \theta'$; $(4.12,-1.33)$

53. $y = mx + b,\ b \neq 0$

$r\sin\theta = mr\cos\theta + b$

$r\sin\theta - mr\cos\theta = b$

$r(\sin\theta - m\cos\theta) = b$

$r = \dfrac{b}{\sin\theta - m\cos\theta}$

55. $y^2 - 2x^2 = 5$

$(r\sin\theta)^2 - 2(r\cos\theta)^2 = 5$

$r^2\sin^2\theta - 2r^2\cos^2\theta = 5$

$r^2(\sin^2\theta - 2\cos^2\theta) = 5$

$r^2 = \dfrac{5}{\sin^2\theta - 2\cos^2\theta}$

63. $r = 3\sin 2\theta$

$r = 3(2\sin\theta\cos\theta)$

$r = 6\dfrac{y}{r}\cdot\dfrac{x}{r}$

$r^3 = 6xy$

Square both members so that we can express the left side in terms of r^2.

$r^6 = 36x^2y^2$

$(r^2)^3 = 36x^2y^2$

$(x^2 + y^2)^3 = 36x^2y^2$

$x^6 + 3x^4y^2 + 3x^2y^4 + y^6 = 36x^2y^2$

$x^6 + 3x^4y^2 - 36x^2y^2 + 3x^2y^4 + y^6 = 0$

69. $r = \dfrac{3}{1 - 2\sin\theta}$

$r(1 - 2\sin\theta) = 3$

$r\left(1 - \dfrac{2y}{r}\right) = 3$

$r - 2y = 3$

$r = 2y + 3$

$r^2 = (2y + 3)^2$

$x^2 + y^2 = 4y^2 + 12y + 9$

$0 = 3y^2 + 12y - x^2 + 9$

78. Assume the path taken by the Scrambler is described by the polar equation $r = 2\cos 3\theta$. Convert this equation into rectangular form. It will be necessary to rewrite $\cos 3\theta$ in terms of $\cos\theta$. Problem 82 in section 7–3 shows that $\cos 3\theta = 4\cos^3\theta - 3\cos\theta$.

$r = 2\cos 3\theta$

$r = 2(4\cos^3\theta - 3\cos\theta)$

$r = 8\cos^3\theta - 6\cos\theta$

$r = 8\left(\dfrac{x}{r}\right)^3 - 6\dfrac{x}{r}$

$r = \dfrac{8x^3}{r^3} - \dfrac{6x}{r}$

Multiply each member by r^3.

$r^4 = 8x^3 - 6xr^2$

$(x^2 + y^2)^2 = 8x^3 - 6x(x^2 + y^2)$

$x^4 + 2x^2y^2 + y^4 = 8x^3 - 6x^3 - 6xy^2$

$x^4 - 2x^3 + 2x^2y^2 + 6xy^2 + y^4 = 0$

Chapter 8 review

1. $C = 121.8°$, $b \approx 2.6$, $c \approx 12.1$

2. $A = 151.0°$, $a \approx 4.3$, $c \approx 0.8$

3. $A \approx 49.0°$, $C \approx 52.0°$, $c \approx 10.4$

4. Case 1: $b \approx 12.2$, $A \approx 76.5°$, $B \approx 70.8°$;
Case 2: $b \approx 9.0$, $A \approx 103.5°$, $B \approx 43.8°$

5. 48 miles

6. $c \approx 3.8$, $A \approx 31.9°$, $B \approx 118.7°$

7. $a \approx 63.9$, $B \approx 69.7°$, $C \approx 18.2°$

8. $b \approx 40.2$, $A \approx 29.5°$, $C \approx 38.5°$

9. $A \approx 106.3°$, $B \approx 23.1°$, $C \approx 50.5°$

10. $C \approx 135.5°$, $A \approx 12.2°$, $B \approx 32.3°$

11. $a = \sqrt{26}$, $b = \sqrt{58}$, $c = \sqrt{40}$; $B \approx 82.9°$, $A \approx 41.6°$, $C \approx 55.5°$

12. 29 km

13. $V_x \approx 23.8$, $V_y \approx 13.2$

14. $|V| \approx 36.3$, $\theta_v \approx 57.3°$

15. horizontal: 370 knots; vertical: 256 knots

16. horizontal: 249 pounds; vertical: 60 pounds

17. $(47.6,20.6°)$ **18.** $(10.8,89.5°)$

19. $(7.2,69.9°)$ **20.** $(8.1,-172.4°)$

21. magnitude $\approx$ 205 pounds; direction $\approx -87°$

22. $\sqrt{13}\text{ cis}(-33.7°)$ or $3.6\text{ cis}(-33.7°)$

23. $2\sqrt{3}\text{ cis }60°$ **24.** $2.2\text{ cis}(-116.6°)$

25. $2.5 + 1.7i$ **26.** $-2.3 - 4.5i$

27. $-1.5 - 1.5\sqrt{3}i$ **28.** $5\sqrt{3} - 5i$

29. $6\text{ cis }70°$ **30.** $13\text{ cis }140°$

31. $8\text{ cis }100°$ **32.** $0.5\text{ cis }36°$

33. $8\text{ cis }30°$ **34.** $16\text{ cis }240°$

35. $0.42 - 0.91i$ **36.** $2, 2i, -2, -2i$

37. $1.93 - 0.46i$, $-0.57 + 1.90i$, $-1.36 - 1.44i$

38. $4\frac{1}{3}\text{cis}(-50°)$ **39.** see figure

40. see figure **41.** see figure

42. $\left(-6,\dfrac{11\pi}{6}\right) = \left(6,\dfrac{11\pi}{6} - \pi\right) = \left(6,\dfrac{5\pi}{6}\right)$; see figure

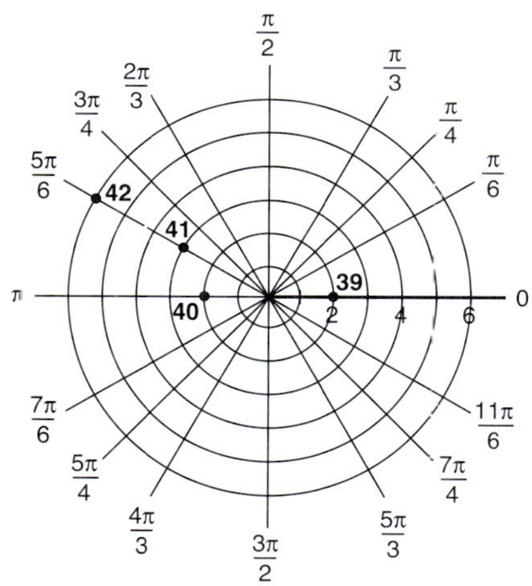

43. $\left(\frac{3}{2}\sqrt{3},-\frac{3}{2}\right)$ **44.** $(-2,2\sqrt{3})$

45. $(-1,\sqrt{3})$ **46.** $(1.6,1.2)$

47. $(-2.1,4.5)$ **48.** $(0.7,-0.8)$

49. $(2.24,0.46)$ **50.** $(5.83,2.60)$

51. $(4.12,-1.82)$ **52.** $\tan\theta = -3$

53. $r = \dfrac{2}{\sin\theta - 4\cos\theta}$

54. $r^2 = \dfrac{5}{2\sin^2\theta - \cos^2\theta}$

55. $r = \dfrac{3\cos\theta}{\sin^2\theta}$; $r = 3\cot\theta\csc\theta$
(alternate form of answer)

56. $r = 3$ **57.** $r = \dfrac{2}{\cos\theta}$; $r = 2\sec\theta$
(alternate form of answer)

58. $x^2 + y^2 - y = 0$ **59.** $x = 2$
60. $(x^2 + y^2)^2 - 2xy = 0$
61. $x^3 + xy^2 - y = 0$ **62.** $y = 2$
63. $4x^2 + 3y^2 - 6y - 9 = 0$

Chapter 8 test

1. $B = 84.4°$, $a \approx 5.3$, $c \approx 22.5$
2. $B \approx 56.3°$, $A \approx 61.6°$, $a \approx 23.9$
3. $b \approx 32.8$, $A \approx 50.9°$, $C \approx 29.1°$
4. $C \approx 89.1°$, $A \approx 39.6°$, $B \approx 51.3°$
5. 59 yards **6.** $B \approx 53.1°$
7. $(\sqrt{3},1)$
8. magnitude: 6.4; direction: 51.3°
9. (8.5,84.9°)
10. tension: 584 pounds; angle above the horizontal: 65°
11. 6.40 cis(−51.3) **12.** $-1 + \sqrt{3}i$
13. 14 cis 110° **14.** 3 cis 120°
15. 27 cis 90°
16. $\sqrt{2} + \sqrt{2}i$, $-\sqrt{2} + \sqrt{2}i$, $-\sqrt{2} - \sqrt{2}i$, $\sqrt{2} - \sqrt{2}i$
17.

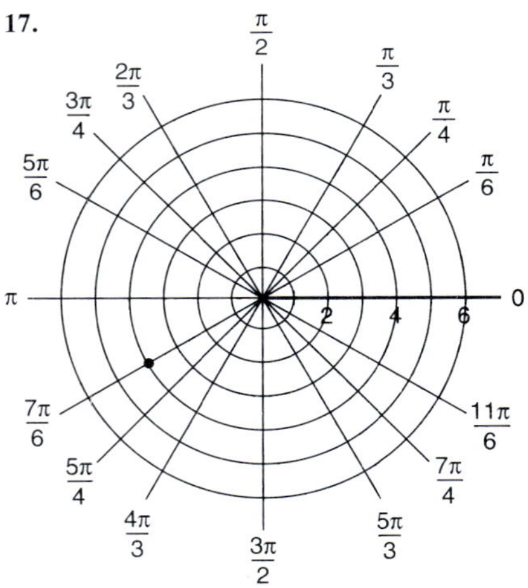

18. (2.8,1.1)
19. $\left(2,-\dfrac{5\pi}{6}\right)$ **20.** $r = \dfrac{5}{\sin\theta + 3\cos\theta}$
21. $2r^2\sin^2\theta - r\cos\theta - 5 = 0$
22. $y = 2$
23. $(x^2 + y^2)^2 - x^2 + y^2 = 0$

Chapter 9

Exercise 9–1

Answers to odd-numbered problems

1. $f(x) = b^x$, $b > 0$ and $b \neq 1$ **3.** $2^{\pi+1}$
5. $2{,}401^{\pi}$ **7.** $4^{\sqrt{3}}$ **9.** 9 **11.** 3
13. $\frac{3}{2}$ **15.** 6 **17.** −3 **19.** $-\frac{3}{2}$ **21.** $\frac{5}{3}$
23. 8 **25.** increasing

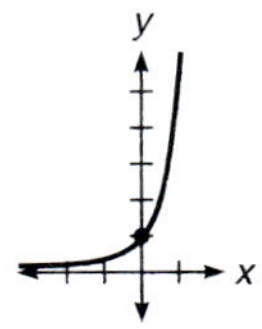

27. decreasing

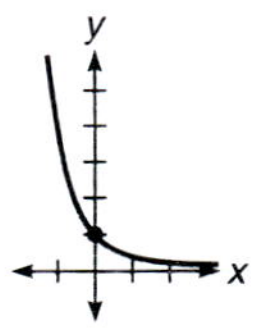

29. decreasing

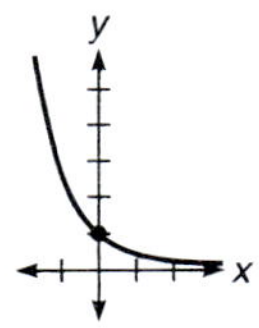

31. increasing

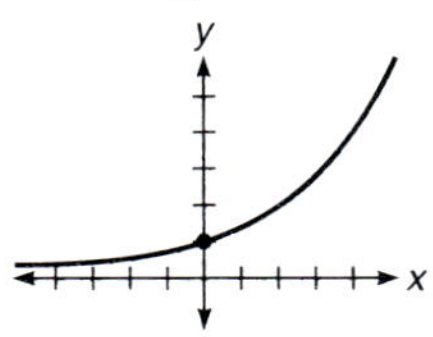

33. increasing

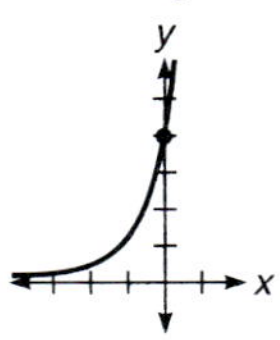

35. decreasing

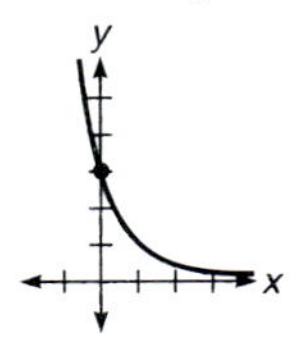

37. decreasing

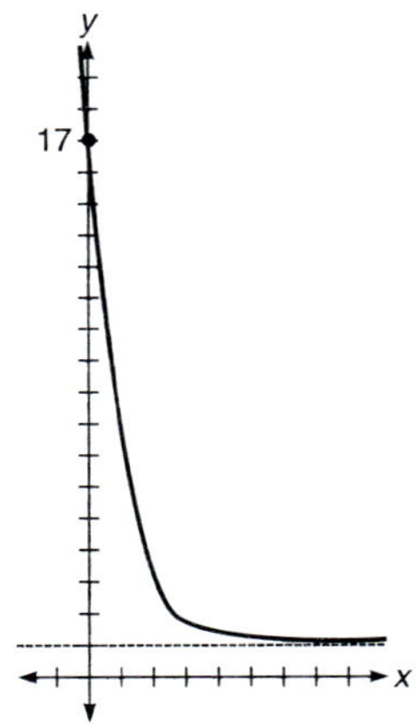

39. increasing

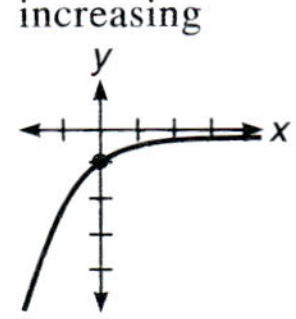

41. decreasing

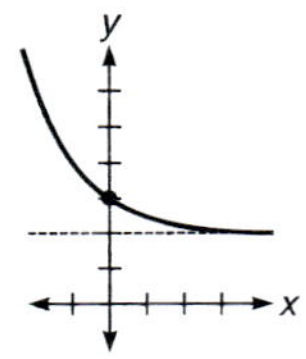

43.

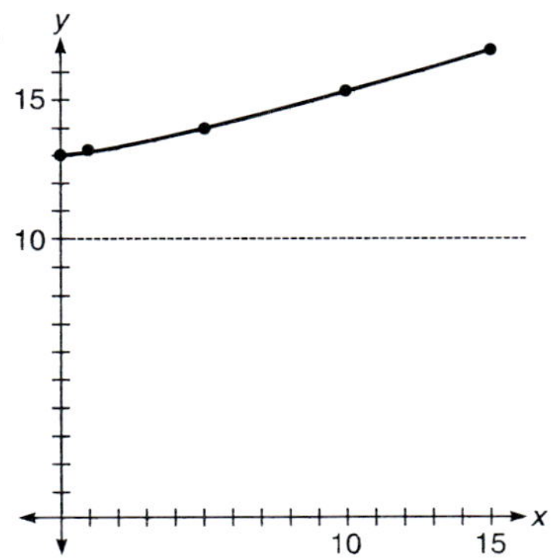

45.

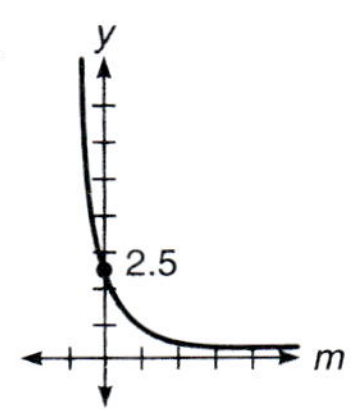

47. \$6,264.08

49.

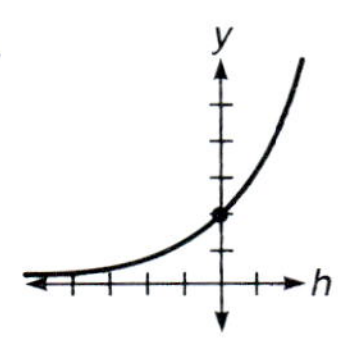

51. **a.** 8,000 **b.** 2,000,000
c. 8,000,000,000,000 (8 trillion)

53. 10^{30}; b

Solutions to skill and review problems

1. $y = \dfrac{x-1}{(x-2)(x+2)}$

Vertical asymptotes at ±2; horizontal asymptote is $y = 0$ (the x-axis).

Intercepts:

$x = 0$: $y = \dfrac{-1}{-4} = \dfrac{1}{4}$; (0,0.25)

$y = 0$: $0 = \dfrac{x-1}{x^2-4}$

$0 = x - 1$

$1 = x$; (1,0)

Additional points: (−3,−0.8), (−1,0.67), (3,0.4)

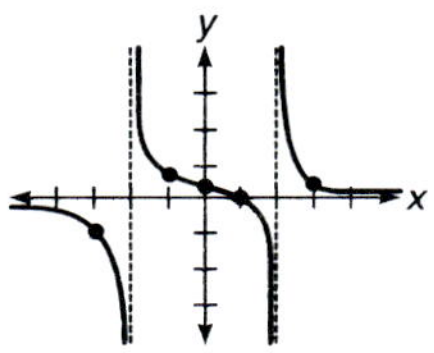

2. $y = (x-1)(x-2)(x+2)$

Intercepts:

$x = 0$: $y = (-1)(-4) = 4$; (0,4)

$y = 0$: $0 = (x-1)(x-2)(x+2)$

$x = -2, 1, 2$; (−2,0), (1,0), (2,0)

Additional points: (−2.25,−3.45), (−1,6), (2.5,3.38)

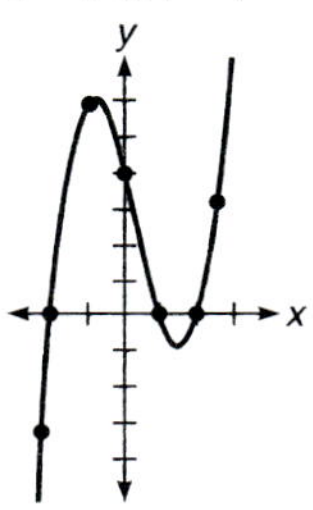

3. $(x-1)(x^2-4) > 0$

To find critical points solve the corresponding equality and find zeros of denominators.

$$(x-1)(x-2)(x+2) = 0$$
$$x = -2, 1, 2$$

Select test points in the intervals determined by the critical points.

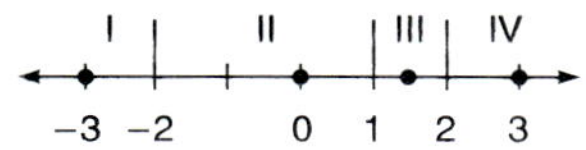

Try these values in the original inequality.

$x = -3$: $(-4)(5) > 0$; false

$x = 0$: $(-1)(-4) > 0$; true

$x = 1.5$: $(0.5)(-1.75) > 0$; false

$x = 3$: $(2)(5) > 0$; true

Solution: $\{x \mid -2 < x < 1 \text{ or } x > 2\}$

4. $6x^3 + 5x^2 - 2x - 1$

Possible zeros are $\pm\frac{1}{6}, \pm\frac{1}{3}, \pm\frac{1}{2}, \pm 1$.

Synthetic division will show that −1 is a zero.

	6	5	−2	−1
		−6	1	1
−1	6	−1	−1	0

Thus, $6x^3 + 5x^2 - 2x - 1$
$= (x+1)(6x^2 - x - 1)$
$= (x+1)(3x+1)(2x-1)$

5. $y = -x^3 + 1$

This is the graph of $y = x^3$ but "flipped over" and shifted up one unit.

Intercepts:

$x = 0$: $y = 0^3 + 1 = 1$; (0,1)

$y = 0$: $0 = -x^3 + 1$

$x^3 = 1$

$x = 1$; (1,0)

Additional points: (−1.5,4.4), (−1,2), (1.5,−2.4)

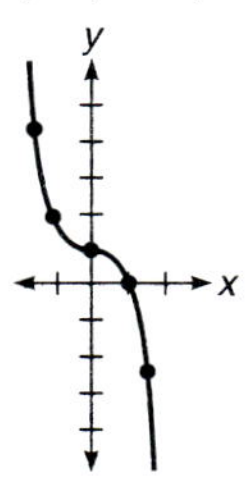

6. $x^{2/3} - x^{1/3} - 6 = 0$

Let $u = x^{1/3}$; then $u^2 = (x^{1/3})^2 = x^{2/3}$.

$u^2 - u - 6 = 0$

$(u-3)(u+2) = 0$

$u = 3$ or $u = -2$

$x^{1/3} = 3$ or $x^{1/3} = -2$

Replace u by $x^{1/3}$.

$x = 27$ or $x = -8$

Cube each member.

Solution set: {−8,27}

Solutions to trial exercise problems

7. $\dfrac{4\sqrt{12}}{4\sqrt{3}}$

$\dfrac{4^2\sqrt{3} - \sqrt{3}}{4\sqrt{3}}$

23. $(\sqrt{2})^x = 16$

$(2^{1/2})^x = 2^4$

$2^{x/2} = 2^4$

$\dfrac{x}{2} = 4$

$x = 8$

37. $f(x) = 4^{-x+2} + 1$
$= 4^2 \cdot 4^{-x} + 1$
$= 16 \cdot (4^{-1})^x + 1$
$= 16(\frac{1}{4})^x + 1$; $b = \frac{1}{4}$

Decreasing since $b < 1$.

This is the graph of $y = (\frac{1}{4})^x$ with a vertical scaling factor of 16 and shifted up one unit.

y-intercept: $f(0) = 2^2 + 1 = 17$; (0,17)

x-intercept: $0 = 16(\frac{1}{4})^x + 1$

$-1 = 16(\frac{1}{4})^x$; no solution as the left member is negative and the right is nonnegative.

Additional points: (1,5), (2,2), (3,1.25)

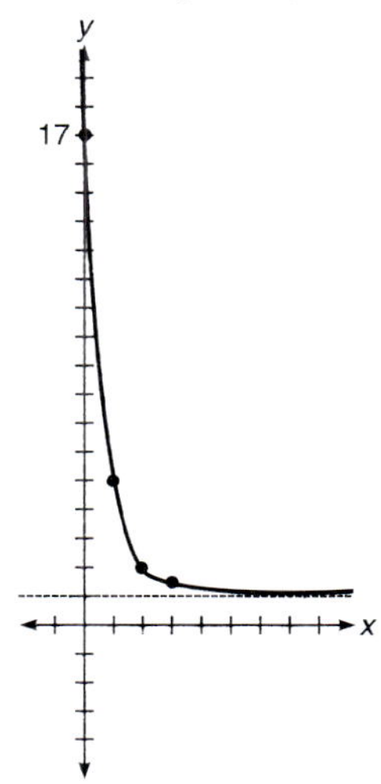

45. $R(m) = 2.5^{1-m}$
$= 2.5^1(2.5^{-m})$
$= 2.5\left(\frac{1}{2.5}\right)^m$
$= 2.5(0.4)^m$; $b = 0.4$
Additional points: $(-1,6.25)$, $(0,2.5)$, $(1,1)$

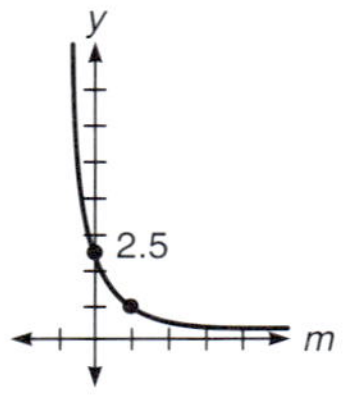

Exercise 9–2

Answers to odd-numbered problems

1. 3 **3.** 3 **5.** -3 **7.** -2 **9.** -1 **11.** 3 **13.** 4 **15.** -6 **17.** 15 **19.** -9 **21.** -55 **23.** 3 **25.** $2^3 = 8$ **27.** $10^{-1} = 0.1$ **29.** $12^2 = x + 3$ **31.** $3^{x+2} = 5$ **33.** $4 = \log_2 16$ **35.** $2 = \log_x(m + 3)$ **37.** $y = \log_m(x + 1)$ **39.** $x + y = \log_{2x-3}(y + 2)$ **41.** 16 **43.** 2 **45.** 2 **47.** 2 **49.** 10 **51.** k^2 **53.** $6 < \log_2 100 < 7$ **55.** $3 < \log_4 100 < 4$ **57.** $-2 < \log_2 0.3 < -1$ **59.** 4 **61.** 1 **63.** 5 **65.** 18 **67.** 1 **69.** 0 **71.** -5 **73.** 81 **75.** 625 **77. a.** 10 bits **b.** 14 bits **c.** 14 bits **d.** 16 bits **79.** $10^{d/10} = I$ **81.** $\log_b x = y$ if and only if $b^y = x$, $b > 0$ and $b \neq 1$.

Solutions to skill and review problems

1. 9^{2x}
$(3^2)^{2x}$
3^{4x}

2. $f(x) = 2 - 3x$
$y = 2 - 3x$
$x = 2 - 3y$
$3y = -x + 2$
$y = -\frac{1}{3}x + \frac{2}{3}$
$f^{-1}(x) = -\frac{1}{3}x + \frac{2}{3}$

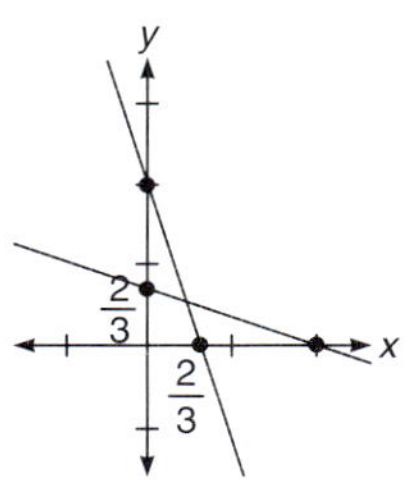

3. $2x^6 + 15x^3 - 8 = 0$
Let $u = x^3$, then $u^2 = x^6$.
$2u^2 + 15u - 8 = 0$
$(2u - 1)(u + 8) = 0$
$2u - 1 = 0$ or $u + 8 = 0$
$2u = 1$ or $u = -8$
$u = \frac{1}{2}$ or $u = -8$
$x^3 = \frac{1}{2}$ or $x^3 = -8$
Replace u by x^3.
$x = \sqrt[3]{\frac{1}{2}}$ or $x = \sqrt[3]{-8}$
$x = \frac{\sqrt[3]{4}}{2}$ or $x = -2$
Note: $\sqrt[3]{\frac{1}{2}} = \sqrt[3]{\frac{1}{2}\cdot\frac{4}{4}} = \frac{\sqrt[3]{4}}{\sqrt[3]{8}} = \frac{\sqrt[3]{4}}{2}$
Solution set: $\left\{-2, \frac{\sqrt[3]{4}}{2}\right\}$

4. $y = x^4 - x$
$= x(x^3 - 1)$
$= x(x - 1)(x^2 + x + 1)$
$x^2 + x + 1$ is prime on R.
Intercepts:
$x = 0$: $y = 0^4 - 0$; $(0,0)$
$y = 0$: $0 = x(x - 1)(x^2 + x + 1)$
$x = 0$ or 1; $(0,0)$, $(1,0)$
Additional points: $(-1.5,6.6)$, $(-1,2)$, $(-0.5,0.6)$, $(0.5,-0.4)$, $(1.5,3.6)$

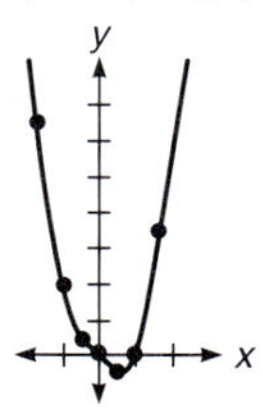

5. $\frac{2x - 5}{3} - \frac{3x + 12}{2} = 4$
$6 \cdot \frac{2x - 5}{3} - 6 \cdot \frac{3x + 12}{2} = 6(4)$
$2(2x - 5) - 3(3x + 12) = 24$
$4x - 10 - 9x - 36 = 24$
$-70 = 5x$
$-14 = x$
Solution set: $\{-14\}$

6. $2xy = \frac{x + y}{3}$
$3(2xy) = 3 \cdot \frac{x + y}{3}$
$6xy = x + y$
$6xy - y = x$
$y(6x - 1) = x$
$y = \frac{x}{6x - 1}$

Solutions to trial exercise problems

21. $5(3 \log_2 \frac{1}{8} + 2 \log_{10} 0.1) = 5[3(-3) + 2(-1)] = -55$ **31.** $3^{x+2} = 5$
39. move the base $2x - 3$ to the other side.
$x + y = \log_{2x-3}(y + 2)$
51. $\log_k x = 2$
$k^2 = x$
57. $\log_2 0.3$
$0.25 < 0.3 < 0.5$
$\frac{1}{4} < 0.3 < \frac{1}{2}$
$2^{-2} < 0.3 < 2^{-1}$
so, $-2 < \log_2 0.3 < -1$
73. $4^{\log_2 9}$
$(2^2)^{\log_2 9}$
$(2^{\log_2 9})^2$, since $(a^m)^n = (a^n)^m$
9^2
81
80. $\frac{P}{5} = 1.25^t$; the base is 1.25, and t is the exponent: $\log_{1.25}\frac{P}{5} = t$

Exercise 9–3

Answers to odd-numbered problems

1. $\frac{8}{3}$ **3.** 1 **5.** $\frac{1}{3}$ **7.** 2 **9.** $\frac{17}{80}$ **11.** $\frac{1}{40}$ **13.** 24 **15.** $\frac{-1 + \sqrt{17}}{2}$ **17.** $\frac{9}{10}$ **19.** 48 **21.** 6 **23.** $x > 0$ **25.** 8 **27.** no solution; solution set is the null set **29.** $\log_6 2 + \log_6 x + \log_6 y$
31. $1 + \log_4 x + \log_4 y$
33. $1 + \log_3 x + \log_3 y - \log_3 2 - \log_3 z$
35. $-\log_{10} 3 - \log_{10} x - \log_{10} y - \log_{10} z$
37. $2 + 3 \log_2 x + 2 \log_2 y + 5 \log_2 z$
39. $\frac{3}{2} + 4 \log_4 y + 3 \log_4 z - 3 \log_4 x$
41. 0.9208 **43.** 1.8416 **45.** 2.2584
47. -0.8271 **49.** 7
51. $\alpha = 10 \log_{10} I - 20$
53. $\log_a \sqrt[n]{x} = \log_a x^{1/n} = \frac{1}{n}\log_a x$
55. Let $a = 2$, $x = y = \frac{1}{2}$.
$\log_a(x + y) = \log_a x + \log_a y$
Assume this is true.
$\log_2(\frac{1}{2} + \frac{1}{2}) = \log_2 \frac{1}{2} + \log_2 \frac{1}{2}$
Replace a by 2, x and y by $\frac{1}{2}$.
$\log_2(1) = \log_2 \frac{1}{2} + \log_2 \frac{1}{2}$
$0 = (-1) + (-1)$
$\log_2 1 = 0$, $\log_2 \frac{1}{2} = -1$.
$0 = -2$
A false statement.
The original ''identity'' does not work for the selected values of a, x, and y, so it is not an identity.

Solutions to skill and review problems

1. x must be between 3 and 4.

2. $3^{2x} = 3^3$; $2x = 3$; $x = 1\frac{1}{2}$

3. Since $5^3 = 125$, the base x must be 5.

4. $x^3 + 2x^{3/2} - 3 = 0$

Let $u = x^{3/2}$; then $u^2 = (x^{3/2})^2 = x^3$.

$u^2 + 2u - 3 = 0$

$(u - 1)(u + 3) = 0$

$u = 1$ or $u = -3$

$x^{3/2} = 1$ or $x^{3/2} = -3$

$(x^{3/2})^2 = 1^2$ or $(x^{3/2})^2 = (-3)^2$

$x^3 = 1$ or $x^3 = 9$

$x = 1$ or $x = \sqrt[3]{9}$

However, $\sqrt[3]{9}$ cannot check in $x^{3/2} = -3$ since $\sqrt[3]{9} > 0$ for any exponent. Thus the solution is $x = 1$.

5. $|2x - 5| = 10$

$2x - 5 = 10$ or $2x - 5 = -10$

$2x = 15$ or $2x = -5$

$x = 7\frac{1}{2}$ or $x = -2\frac{1}{2}$

6. $f(x) = x^2 + 3x - 5$

This is a parabola. We complete the square.

$y = x^2 + 3x + \frac{9}{4} - 5 - \frac{9}{4}$

$\frac{1}{2} \cdot 3 = \frac{3}{2}$; $(\frac{3}{2})^2 = \frac{9}{4}$

$y = (x + \frac{3}{2})^2 - \frac{29}{4}$

Vertex: $(-1\frac{1}{2}, -7\frac{1}{4})$.

Intercepts:

$x = 0$: $y = 0^2 + 3(0) - 5 = -5$; $(0,-5)$

$y = 0$: $0 = x^2 + 3x - 5$

$x = -\dfrac{3}{2} \pm \dfrac{\sqrt{29}}{2} \approx -4.2, 1.2$;

$(-4.2,0)$, $(1.2,0)$

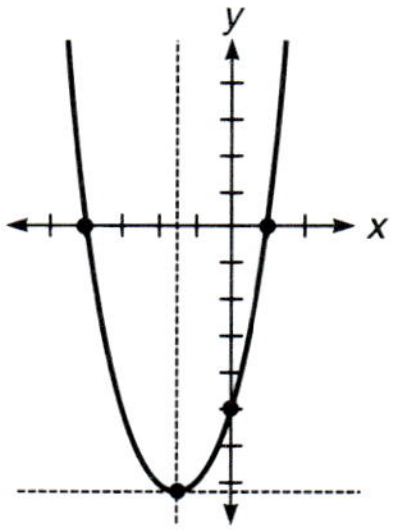

Solutions to trial exercise problems

13. $\log_5(x + 1) = \log_{10}100$

$\log_5(x + 1) = 2$

$x + 1 = 5^2$

$x = 24$

23. $\log_3 2x = \log_3 2 + \log_3 x$

$\log_3 2x = \log_3 2x$

This last equation is an identity and is true for any value for which $\log_3 2x$ is defined. Thus, the solution is all x for which $\log_3 2x$ is defined, which is $\{x \mid x > 0\}$.

27. $\log_2(x - 2) + \log_2(x + 3)$

$= \log_2(x^2 - 3x + 2)$

$\log_2[(x - 2)(x + 3)]$

$= \log_2(x^2 - 3x + 2)$

$x^2 + x - 6 = x^2 - 3x + 2$

$4x = 8$

$x = 2$

However, the solution 2 is not in the domain of the term $\log_2(x - 2)$ so *there is no solution* (the solution set is the null set).

39. $\log_4 \dfrac{8y^4z^3}{x^3}$

$\log_4(8y^4z^3) - \log_4 x^3$

$\log_4 8 + \log_4 y^4 + \log_4 z^3 - 3 \log_4 x$

$\frac{3}{2} + 4 \log_4 y + 3 \log_4 z - 3 \log_4 x$

47. $\log_a 0.2$

$\log_a \frac{1}{5}$

$\log_a 1 - \log_a 5$

$0 - 0.8271$

-0.8271

49. $\log_a 14 \quad = 1.3562$

$-\log_a 2 \quad = 0.3562$

$\log_a 14 - \log_a 2 = 1$

$\log_a \dfrac{14}{2} = 1$

$\log_a 7 = 1$

$a^1 = 7$

$a = 7$

Exercise 9–4

Answers to odd-numbered problems

1. 1.7160 **3.** 0.4065 **5.** 1.0253
7. −0.0706 **9.** 3.9405 **11.** 2.8332
13. 5.2470 **15.** 7.8240 **17.** −5.8091
19. 15.8987 **21.** 19.9542
23. −15.6990 **25.** −9.9208
27. −26.1785 **29.** 10.2553
31. 2.5372 **33.** −0.1195
35. −0.0467

37.

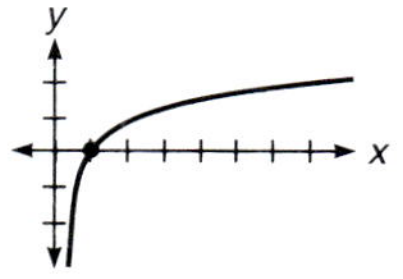

39.

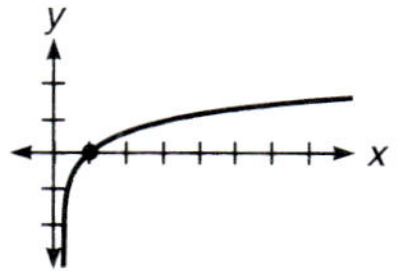

41.

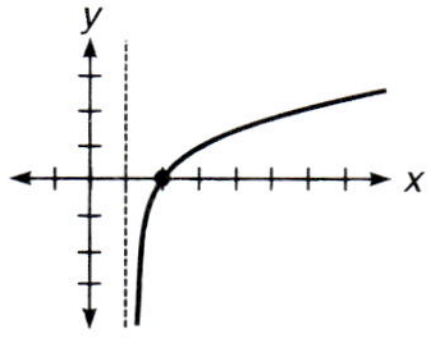

43.

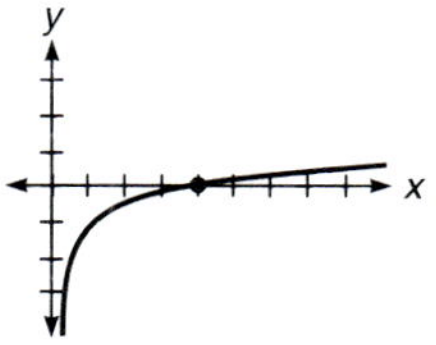

45.

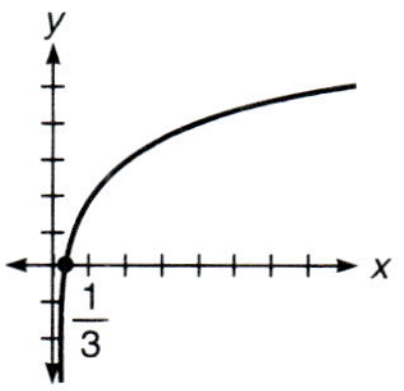

47. 794.33 **49.** 0.47 **51.** 6.36
53. 121.51 **55.** 4.64 **57.** $3x + \ln 2$
59. $4 - 3x$ **61.** 100 **63.** $(x - 1)^2$
65. $x + \ln 5$ **67.** \$2,872.80
69. \$1,172.89

71. Nap log $x = 10^7 \log_{1/e}\left(\frac{x}{10^7}\right)$

Let y = Nap log x.

$y = 10^7 \log_{1/e}\left(\frac{x}{10^7}\right)$

$\frac{y}{10^7} = \log_{1/e}\left(\frac{x}{10^7}\right)$

Divide both members by 10^7.

$\left(\frac{1}{e}\right)^{y/10^7} = \frac{x}{10^7}$

Rewrite as an exponential equation

$(e^{-1})^{y/10^7} = \frac{x}{10^7}$

$e^{-y/10^7} = \frac{x}{10^7}$

$\log_e \frac{x}{10^7} = -\frac{y}{10^7}$

Rewrite as a logarithmic equation with base e and exponent $\frac{-y}{10^7}$.

$\ln \frac{x}{10^7} = -\frac{y}{10^7}$

$\log_e z$ is $\ln z$.

$-10^7 \ln \frac{x}{10^7} = y$

Multiply each member by -10^7.

$y = -10^7(\ln x - \ln 10^7)$

$\ln \frac{x}{10^7} = \ln x - \ln 10^7$.

$y = 10^7(-\ln x + 7 \ln 10)$

$y = 10^7(7 \ln 10 - \ln x)$

Then, Nap log $x = 10^7(7 \ln 10 - \ln x)$.

73. 9.3 **75.** 215 ohms **77.** -223 BTU/hour **79.** 0.32 centiliters per second

Solutions to skill and review problems

1. $2x^2 - 9x + 4 = 0$

$(2x - 1)(x - 4) = 0$

$2x - 1 = 0$ or $x - 4 = 0$

$x = \frac{1}{2}$ or $x = 4$

2. $2x^4 - 9x^2 + 4 = 0$

Let $u = x^2$; then $u^2 = x^4$.

$2u^2 - 9u + 4 = 0$

$u = \frac{1}{2}$ or $u = 4$

Solve as in the previous problem.

$x^2 = \frac{1}{2}$ or $x^2 = 4$

$u = x^2$

$x = \pm\frac{\sqrt{2}}{2}$ or $x = \pm 2$

Extract square root of both sides.

3. $2x - 9\sqrt{x} + 4 = 0$

Let $u = \sqrt{x}$; then $u^2 = x$.

$2u^2 - 9u + 4 = 0$

$u = \frac{1}{2}$ or $u = 4$

See previous two problems.

$\sqrt{x} = \frac{1}{2}$ or $\sqrt{x} = 4$

$u = \sqrt{x}$

$x = \frac{1}{4}$ or $x = 16$

Square both sides.

4. $2(x - 3)^2 - 9(x - 3) + 4 = 0$

Let $u = x - 3$.

$2u^2 - 9u + 4 = 0$

$u = \frac{1}{2}$ or $u = 4$

See previous three problems.

$x - 3 = \frac{1}{2}$ or $x - 3 = 4$

$u = x - 3$.

$x = 3\frac{1}{2}$ or $x = 7$

5. $\log_a \frac{2x^4}{3y^3z}$

$\log_a 2x^4 - \log_a 3y^3z$

$\log_a 2 + \log_a x^4 - (\log_a 3 + \log_a y^3 + \log_a z)$

$\log_a 2 + 4 \log_a x - \log_a 3 - 3 \log_a y - \log_a z$

6. $\log_2 x = -3$

$2^{-3} = x$

$\frac{1}{8} = x$

7. $f(x) = \log_3(x - 1)$

We compute points for the inverse function and reverse them.

Find f^{-1}.

$y = \log_3(x - 1)$

$x = \log_3(y - 1)$

$y - 1 = 3^x$

$y = 3^x + 1$

Computed points:

$y = 3^x + 1$	$y = \log_3(x - 1)$
$(-1, 1\frac{1}{3})$	$(1\frac{1}{3}, -1)$
(0,2)	(2,0)
(1,4)	(4,1)
(2,10)	(10,2)

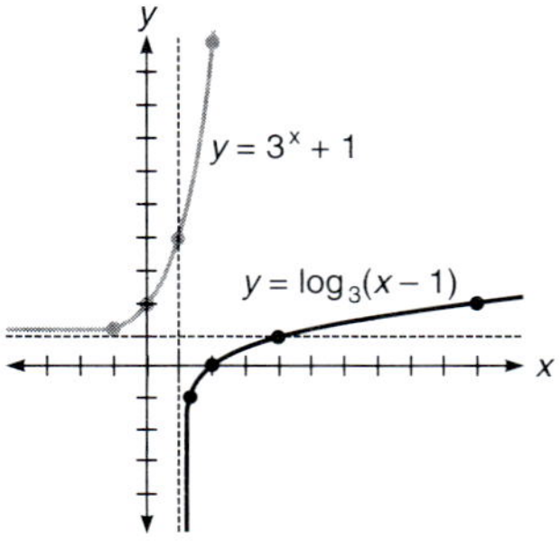

8. $\frac{x^2 - 4}{x^2 - 1} > 2$

Find critical points by (a) solving the corresponding equality and (b) finding zeros of denominators.

Solve the corresponding equality.

$\frac{x^2 - 4}{x^2 - 1} = 2$

$x^2 - 4 = 2x^2 - 2$

$-2 = x^2$

No real solutions.

Find zeros of denominators.

$x^2 - 1 = 0$

$x^2 = 1$

$x = \pm 1$

Critical points are ± 1.

I II III

-3 -2 -1 0 1 2 3

Select trial points from each interval I, II, and III. We will test 0, ± 2.

$\frac{x^2 - 4}{x^2 - 1} > 2$

$x = -2$: $\frac{(-2)^2 - 4}{(-2)^2 - 1} > 2$; $\frac{0}{3} > 2$; false

$x = 0$: $\frac{0 - 4}{0 - 1} > 2$; $4 > 2$; true

$x = 2$: $\frac{2^2 - 4}{2^2 - 1} > 2$; $\frac{0}{3} > 2$; false

Thus, interval II is the solution set: $\{x \mid -1 < x < 1\}$

Solutions to trial exercise problems

25. log 0.000 000 000 120 004

$\log(1.20004 \times 10^{-10})$

$\log 1.20004 + \log 10^{-10}$

$0.0792 + (-10)$

-9.9208

31. $\log_{20} 2{,}000 = \frac{\log 2{,}000}{\log 20} \approx 2.5372$

2,000 [log] [÷] 20 [log] [=]

TI-81: [log] 2000 [÷] [log] 20 [ENTER]

45. $f(x) = \log_2 3x$

Calculate inverse function.

$y = \log_2 3x$

$x = \log_2 3y$

$3y = 2^x$

$y = \frac{1}{3}(2^x)$

Calculated points:

$y = \frac{1}{3}(2^x)$	$y = \log_2 3x$
$(-1,\frac{1}{6})$	$(\frac{1}{6},-1)$
$(0,\frac{1}{3})$	$(\frac{1}{3},0)$
$(1,\frac{2}{3})$	$(\frac{2}{3},1)$
$(2,1\frac{1}{3})$	$(1\frac{1}{3},2)$
$(3,2\frac{2}{3})$	$(2\frac{2}{3},3)$
$(4,5\frac{1}{3})$	$(5\frac{1}{3},4)$

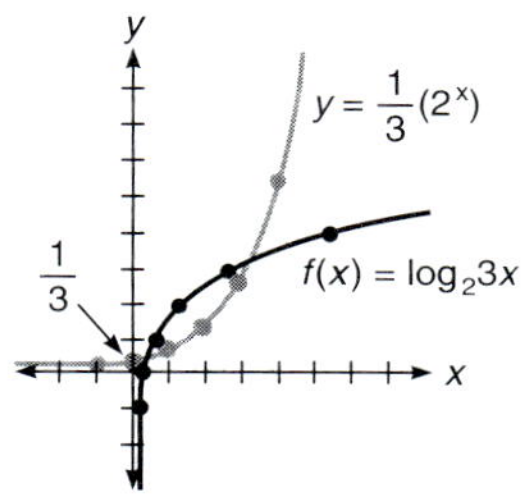

49. $10^{-0.33} \approx 0.47$

63. $10^{\log(x-1)^2} = (x-1)^2$, since $10^{\log x} = x$

73. $k = 12$, and $I = 6I_0$.

$$S = 12 \log\left(\frac{6I_0}{I_0}\right) = 12 \log 6 \approx 9.3$$

77. $L = 80$, $T_{in} = 30$, $T_{out} = 42$, $T_{earth} = 54$

$$Q = 0.07(80)\frac{30-42}{\log\dfrac{54-30}{54-42}} = 5.6\frac{-12}{\log\dfrac{24}{12}} = \frac{-67.2}{\log 2} \approx -223 \text{ BTU/hour}$$

Exercise 9–5

Answers to odd-numbered problems

1. $\frac{15}{13}$ **3.** $\frac{2}{5}$ **5.** -1 **7.** $2\sqrt{2}-1$ **9.** $\frac{13}{3}$ **11.** $2\sqrt{26}-1$ **13.** $\frac{299}{99}$ **15.** $\dfrac{\log 14.2}{\log 2} \approx 3.8$ **17.** $\pm\sqrt[4]{25} \approx \pm 2.2$ **19.** $\dfrac{\log 34}{\log 17} \approx 1.2$ **21.** $-2 \pm \sqrt[4]{200} \approx -5.8$ or 1.8 **23.** $\dfrac{\log 8}{\log 25} \approx 0.6$ **25.** $\dfrac{\log 41}{2\log 41 - \log 2} \approx 0.6$ **27.** $\dfrac{2\log 5}{\log 57 - 2\log 5} \approx 3.9$ **29.** $\dfrac{\log 3 + \log 5}{\log 3 - \log 5} \approx -5.3$ **31.** $2^{0.33} \approx 1.26$ **33.** $\sqrt[5]{10} = 10^{1/5} \approx 1.58$ **35.** $\dfrac{\log 30}{2\log 3} \approx 1.55$ **37.** $\dfrac{\sqrt[3]{14}}{2} \approx 1.21$ **39.** 1 or 100 **41.** $10^{1,000}$ **43.** $\frac{100}{3}$ **45.** $10^{\frac{5(\log 2)(\log 3)}{\log 3 + \log 2}}$ **47.** $\ln 2$ **49.** $\ln 4$ **51.** $10^{10^{3/4}-4}$ **53.** \$1,000.61 **55.** \$3,512.37 **57.** \$2,744.06 **59.** 5.78% **61.** 21.97 years **63.** 37.08 mg **65.** 9,709; the charcoal is about 10,000 years old. **67.** 5,589.9 or about 5,600 years **69.** $10 \log 20 \approx 13$ **71.** $I = 10^{0.3}I_0$. Thus, the power of a sound must change by a factor of $10^{0.3} \approx 2$ for a 3-decibel change in intensity. **73.** 95% **75.** 0.60 time constants **77.** $t = -\ln(1-q)$ **79.** Let $y = b^x$.

$\ln y = \ln b^x$

$\ln y = x \ln b$

$e^{\ln y} = e^{x \ln b}$

$y = e^{x \ln b}$

$b^x = e^{x \ln b}$

81. $\pm\sqrt{-2\ln(y\sqrt{2\pi})}$ **83.** $\dfrac{\log M}{\log\left(1-\dfrac{1}{h}\right)} = N$

85.

x	x^2	2^x
5	25	32
10	100	1,024
20	400	1,048,576
40	1,600	1.09951×10^{12}

87. $5 \ln 2 \approx 3.4657359\ldots$ Thus, it takes about $3\frac{1}{2}$ years. The Mesopotamian value is $3 + \dfrac{47}{60} + \dfrac{13}{60^2} + \dfrac{20}{60^3} \approx 3.787\ldots$ **89.** 74.4 hours **91. a.** 80.0 µg **b.** after 3.3 hours of growth (or 1.3 hours after it reaches 40 µg) **93.** 60.1 talents

Solutions to skill and review problems

1. $y = 2^{1-x}$

$= 2(2^{-x})$

$= 2(\frac{1}{2})^x$; $b = \frac{1}{2}$

y-intercept: $f(0) = 2$; $(0,2)$

Additional points: $(-2,8)$, $(-1,4)$, $(1,1)$, $(2,\frac{1}{2})$

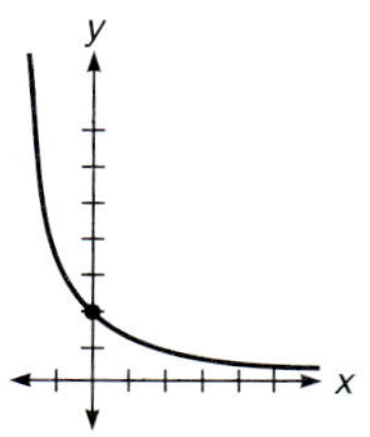

2. $f(x) = \dfrac{2}{(x-1)(x-5)}$.

Vertical asymptotes: 1 and 5.

Horizontal asymptote: x-axis.

y-intercept: $f(0) = \frac{2}{5}$; $(0,\frac{2}{5})$

Additional points: $(-1,0.2)$, $(0.5,0.9)$, $(2,-0.67)$, $(3,-0.5)$, $(4,-0.67)$, $(5.5,0.9)$, $(6,0.4)$

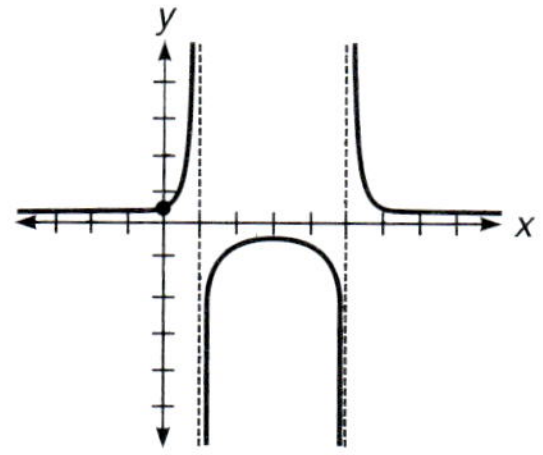

3. $|3-2x| < 13$

$-13 < 3 - 2x < 13$

$-16 < -2x < 10$

$8 > x > -5$

$\{x \mid -5 < x < 8\}$

4. $\left|\frac{3-2x}{x}\right| < 13$

This is nonlinear so it is solved by the critical point/test point method.

Critical points:

Solve the corresponding equality.

$\left|\frac{3-2x}{x}\right| = 13$

$\frac{3-2x}{x} = 13$ or $\frac{3-2x}{x} = -13$

$3 - 2x = 13x$ or $3 - 2x = -13x$

$3 = 15x$ or $11x = -3$

$x = \frac{1}{5}$ or $x = -\frac{3}{11}$

Find zeros of denominators.

$x = 0$

Critical points are $-\frac{3}{11}, 0, \frac{1}{5}$.

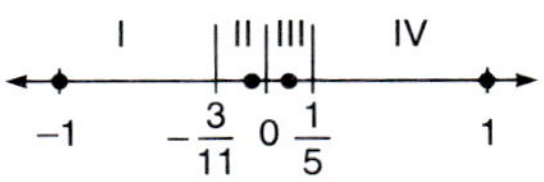

Test points are $-1, -0.1, 0.1, 1$.

$\left|\frac{3-2x}{x}\right| < 13$

$x = -1$: $|-5| < 13$; true

$x = -0.1$: $|-32| < 13$; false

$x = 0.1$: $|28| < 13$; false

$x = 1$: $|1| < 13$; true

The solution set is intervals I and IV: $\{x \mid x < -\frac{3}{11}$ or $x > \frac{1}{5}\}$.

5. $y = x^5 - 4x^4 + 2x^3 + 4x^2 - 3x$

$= x(x^4 - 4x^3 + 2x^2 + 4x - 3)$

Possible zeros of $x^4 - 4x^3 + 2x^2 + 4x - 3$ are $\pm 1, \pm 3$. Using synthetic division produces the following factorization.

$y = x(x-1)^2(x+1)(x-3)$

Since 1 is a root of even multiplicity the function does not cross the x-axis there.

Intercepts:

$x = 0$: $y = 0$; (0,0)

$y = 0$: $0 = x(x-1)^2(x+1)(x-3)$; (0,0)

(1,0), (−1,0), (3,0)

Additional points:

(−1.25,−6.7),

(−0.5,1.97),

(0.5,−0.47),

(2,−6), (2.5,−9.8)

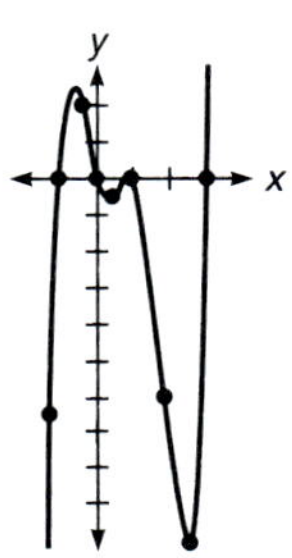

6. $\frac{2}{x-3} + \frac{2}{x+3} - \frac{5}{x+1}$

$\frac{2(x+3) + 2(x-3)}{(x+3)(x-3)} - \frac{5}{x+1}$

$\frac{4x}{x^2-9} - \frac{5}{x+1}$

$\frac{4x(x+1) - 5(x^2-9)}{(x^2-9)(x+1)}$

$\frac{-x^2 + 4x + 45}{x^3 + x^2 - 9x - 9}$

7. $(\frac{2}{5} - \frac{3}{4}) \div 2$

$\left(\frac{2(4) - 3(5)}{5(4)}\right) \cdot \frac{1}{2}$

$\frac{-7}{20} \cdot \frac{1}{2} = -\frac{7}{40}$

Solutions to trial exercise problems

5. $(\sqrt{8})^{2x-2} = 4^{3x}$

$[(2^3)^{1/2}]^{2x-2} = (2^2)^{3x}$

$2^{3x-3} = 2^{6x}$

$3x - 3 = 6x$

$x = -1$

7. $\log(x-1) + \log(x+3) = \log 4$

$\log[(x-1)(x+3)] = \log 4$

$(x-1)(x+3) = 4$

$x^2 + 2x - 3 = 4$

$x^2 + 2x - 7 = 0$

$x = -1 \pm 2\sqrt{2}$

We require $x - 1 > 0$ or $x > 1$ so we choose $x = 2\sqrt{2} - 1$.

11. $\log(x-1) + \log(x+3) = 2$

$\log[(x-1)(x+3)] = 2$

$10^2 = x^2 + 2x - 3$

$x^2 + 2x - 103 = 0$

$x = -1 \pm 2\sqrt{26}$. Because we require $x - 1 > 0$, or $x > 1$, we choose the solution $x = 2\sqrt{26} - 1$.

23. $25 = 8^{1/x}$

$\log 25 = \log 8^{1/x}$

$\log 25 = \frac{1}{x}\log 8$

$\frac{\log 25}{\log 8} = \frac{1}{x}$

$x = \frac{\log 8}{\log 25} \approx 0.6$

27. $57^{x/2} = 5^{x+1}$

$\log 57^{x/2} = \log 5^{x+1}$

$\frac{x}{2}\log 57 = (x+1)\log 5$

$x \log 57 = 2(x+1)\log 5$

$x \log 57 = 2x \log 5 + 2 \log 5$

$x \log 57 - 2x \log 5 = 2 \log 5$

$x(\log 57 - 2 \log 5) = 2 \log 5$

$x = \frac{2 \log 5}{\log 57 - 2 \log 5} \approx 3.9$

37. $\log_{2x} 14 = 3$

$(2x)^3 = 14$

$2x = \sqrt[3]{14}$

$x = \frac{\sqrt[3]{14}}{2} \approx 1.21$

45. $\log_2 x + \log_3 x = 5$

Use the change-to-common log formula.

$\frac{\log x}{\log 2} + \frac{\log x}{\log 3} = 5$

$\frac{\log x}{\log 2}(\log 2)(\log 3) + \frac{\log x}{\log 3}(\log 2)(\log 3)$

$= 5(\log 2)(\log 3)$

$(\log 3)(\log x) + (\log 2)(\log x) = 5(\log 2)(\log 3)$

$\log x(\log 3 + \log 2) = 5(\log 2)(\log 3)$

$\log x = \frac{5(\log 2)(\log 3)}{\log 3 + \log 2}$

$x = 10^{\frac{5(\log 2)(\log 3)}{\log 3 + \log 2}}$

49. $e^{2x} - 3e^x = 4$

Let $u = e^x$ so that $u^2 = e^{2x}$.

$u^2 - 3u - 4 = 0$

$(u-4)(u+1) = 0$

$u = -1$ or 4

$e^x = -1$ or 4

-1 is not in the range of e^x so we proceed with $e^x = 4$.

$x = \ln 4$ (by the property that if $b^x = y$ then $\log_b y = x$, with base e).

52. $\ln x = \frac{8}{\ln x - 2}$

$(\ln x)^2 - 2 \ln x = 8$

$(\ln x)^2 - 2 \ln x - 8 = 0$

Let $u = \ln x$.

$u^2 - 2u - 8 = 0$

$(u-4)(u+2) = 0$

$(\ln x - 4)(\ln x + 2) = 0$

$\ln x = 4$ or $\ln x = -2$

so $x = e^4$ or e^{-2}

57. $A = Pe^{it}$, $A = 5{,}000$, $i = 0.1$, $t = 6$

$5{,}000 = Pe^{(0.1)(6)}$

$P = \frac{5{,}000}{e^{0.6}} \approx \$2{,}744.06$

61. $A = Pe^{it}$, $A = 3P$, $i = 0.05$; find t.

$3P = Pe^{0.05t}$

$3 = e^{0.05t}$

$\ln 3 = 0.05t$; $\ln e^x = x$

$t = \frac{\ln 3}{0.05} = 20 \ln 3 \approx 21.97$ years

63. Use $q = q_0 e^{-0.000124t}$, $q_0 = 100$, $t = 8{,}000$; $q = 100e^{-0.000124(8{,}000)} \approx$ 37.08 mg

68. Use $q = q_0e^{-0.000124t}$, $t = 1{,}200$, $q = 18$; find q_0.
$18 = q_0e^{-0.000124(1{,}200)}$
$$q_0 = \frac{18}{e^{-0.000124(1{,}200)}} \approx 20.89\ \mu g$$

75. $q = 1 - e^{-t}$, $q = 0.45$
$0.45 = 1 - e^{-t}$
$e^{-t} = 1 - 0.45$
$e^{-t} = 0.55$
$\ln e^{-t} = \ln 0.55$
$-t = \ln 0.55$
$t = -\ln 0.55 \approx 0.60$ time constants

89. $q = q_0e^{rt}$, $q = 1.15q_0$ (increase of 15% puts the population at 115%), $t = 15$.
$1.15q_0 = q_0e^{15r}$
$1.15 = e^{15r}$
$\ln 1.15 = \ln e^{15r}$
$\ln 1.15 = 15r$
$r = \dfrac{\ln 1.15}{15} \approx 0.009317$. Thus, for this bacteria $q = q_0e^{0.009317t}$. Now find t for which $q = 2q_0$.
$2q_0 = q_0e^{0.009317t}$
$2 = e^{0.009317t}$
$\ln 2 = \ln e^{0.009317t}$
$\ln 2 = 0.009317t$
$$t = \frac{\ln 2}{0.0093} \approx 74.4 \text{ hours}$$

91. $q_0 = 10$ and $q = 40$ when $t = 2$.
Basic growth/decay formula:
$q = q_0e^{rt}$
$40 = 10e^{2r}$
$4 = e^{2r}$
$\ln 4 = \ln e^{2r}$
$\ln 4 = 2r$
$r = \frac{1}{2}\ln 4 \approx 0.6931$
Thus, the equation is $q = 10e^{0.6931t}$.
a. $q(3) = 10e^{0.6931(3)} \approx 80.0\ \mu g$
b. Find t for $q = 100$.
$100 = 10e^{0.6931t}$
$10 = e^{0.6931t}$
$\ln 10 = 0.6931t$
$$t = \frac{\ln 10}{0.6931} \approx 3.32$$
Thus, the population will be 100 μg after 3.3 hours of growth (or 1.3 hours after it reaches 40 μg).

92. $10^{6.2-4.5} = 10^{1.7} \approx 50.1$, so the second is about 50 times stronger than the first.

94. Method 1: $\ln x = \log_e x = \dfrac{\log x}{\log e}$. The value of $\log e$ (the common logarithm of the value e) could be stored in the calculator. Then to compute $\ln x$, compute $\log x$ as described in the problem and divide by $\log e$.
Method 2: Store roots of e instead of 10:
$e^{1/2} = 1.6487213$
$e^{1/4} = 1.2840254$
$e^{1/8} = 1.1331484$
$e^{1/16} = 1.0644945$
$e^{1/32} = 1.0317434$
$e^{1/64} = 1.0157477$
etc.
Then to compute say $\ln 6$ we find by successive divisions that
$6 = e \cdot e^{1/2} \cdot e^{1/4} \cdot e^{1/32} \cdot e^{1/128} \cdot e^{1/512} \cdot e^{1/2{,}048} \cdot e^{1/4{,}096} \cdots$
so
$$\begin{aligned}\ln 6 &\approx \ln(e \cdot e^{1/2} \cdot e^{1/4} \cdot e^{1/32} \cdot e^{1/128} \cdot e^{1/512} \cdot e^{1/2{,}048} \cdot e^{1/4{,}096})\\ &= \ln e + \ln e^{1/2} + \ln e^{1/4} + \ln e^{1/32} + \ln e^{1/128} + \ln e^{1/512} + \ln e^{1/2{,}048} + \ln e^{1/4{,}096}\\ &= 1 + \frac{1}{2} + \frac{1}{4} + \frac{1}{32} + \frac{1}{128} + \frac{1}{512} + \frac{1}{2{,}048} + \frac{1}{4{,}096}\\ &\approx 1.7917\end{aligned}$$
which is correct to four decimal places.

Chapter 9 review

1. If $b > 1$ the exponential function is increasing; if $0 < b < 1$ the function is decreasing. **2.** $5^{7\sqrt{2}}$ **3.** $4^{3\sqrt{2}}$
4. $3^{4\pi}$
5. increasing

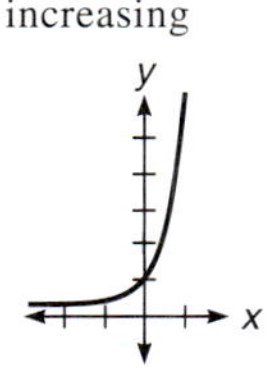

6. decreasing

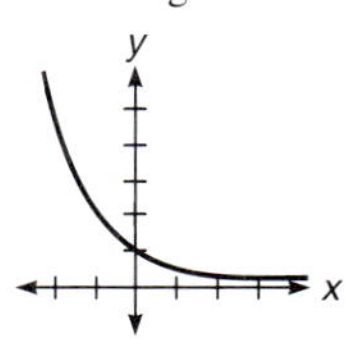

7. increasing

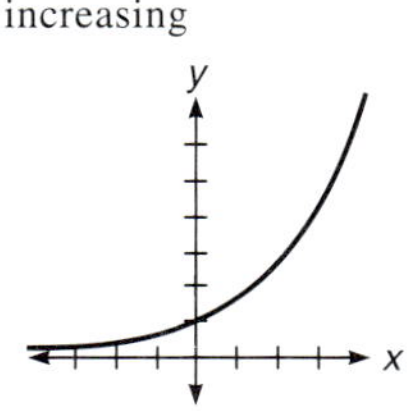

8. increasing; y-intercept: (0,16)

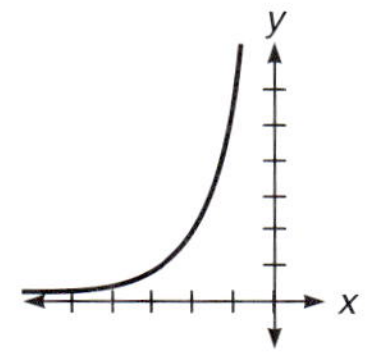

9. decreasing

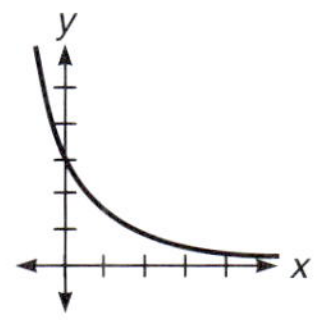

10. increasing

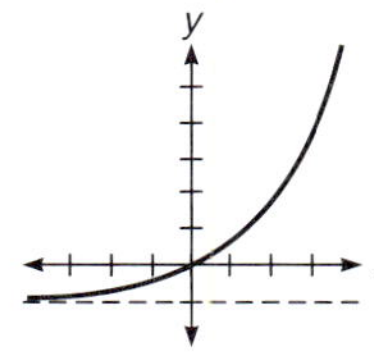

11. $\frac{3}{2}$ **12.** $\frac{3}{4}$ **13.** $-\frac{3}{2}$ **14.** $-\frac{3}{4}$
15. 4 **16.** 6 **17.** 3 **18.** $\frac{7}{2}$ **19.** -3
20. $-\frac{9}{2}$ **21.** $-\frac{5}{2}$ **22.** Definition: $\log_x b = y$ if and only if $x^y = b$, $b > 0$, $b \neq 1$.
23. $4^{-1} = 0.25$ **24.** $5^2 = x - 3$
25. $2^8 = y$ **26.** $3^{x+2} = 9$ **27.** $m^{y+1} = x$
28. $\log_x(m - 3) = 3$ **29.** $\log_y 5 = 2x$
30. $\log_y 4 = 2x - 1$ **31.** $\log_{x-1} 5 = y$
32. $\log_{x+3}(y - 2) = x + y$
33. $\log_{5x} 3y = 2$ **34.** $\frac{3}{2}$
35. 2 **36.** $-\frac{99}{100}$ **37.** $\frac{1}{2}$ **38.** $\sqrt[3]{k}$
39. $3 < \log_4 100 < 4$ **40.** $4 < \log_{10} 15{,}600 < 5$ **41.** m **42.** 5 **43.** $\frac{1}{24}$
44. -2 or 3 **45.** $\frac{5}{2}$ **46.** $\frac{17}{5}$ **47.** $-\frac{1}{24}$
48. $25\frac{1}{2}$ **49.** $2\frac{1}{81}$ **50.** $\frac{7}{6}$ **51.** $\frac{15}{11}$
52. 6 **53.** 7 **54.** 64 **55.** 4,096
56. $3 + 4\log_2 x + 2\log_2 y + \log_2 z$
57. $\frac{3}{2} + \log_4 y + 3\log_4 z - 2\log_4 x$
58. $5\log_{10} x + 3\log_{10} y + \log_{10} z - 2$
59. 1.7479 **60.** 1.8416 **61.** -0.8271
62. -0.0292 **63.** 3.7959 **64.** 21.7008
65. -17.3840 **66.** 2.2920 **67.** 2.3059
68. 0.1950 **69.** -2.1610 **70.** 2.5104
71. -5.6550 **72.** 1,385.4557
73. 0.0032 **74.** 121.5104

75.

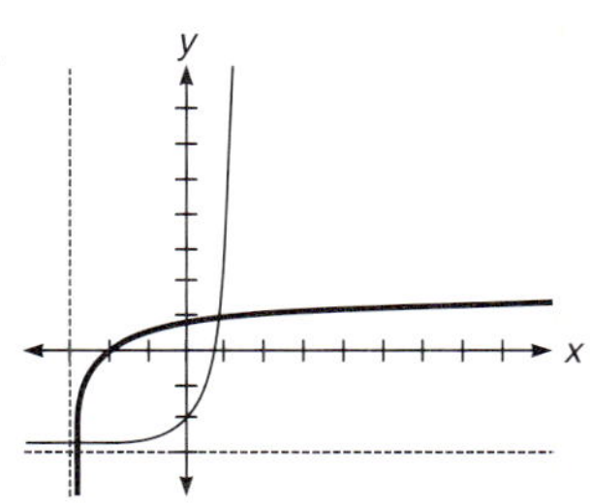

76.

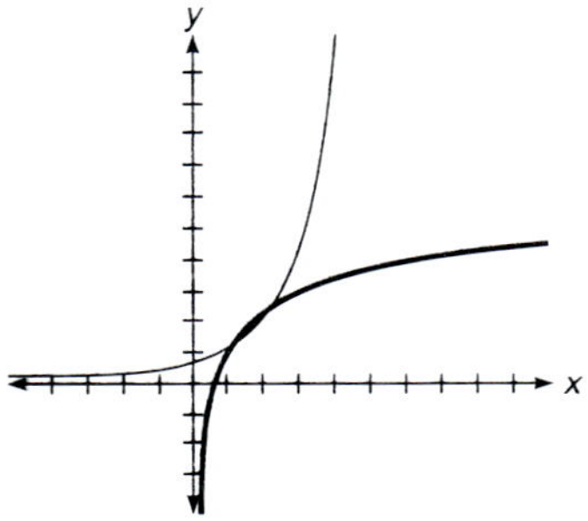

77.

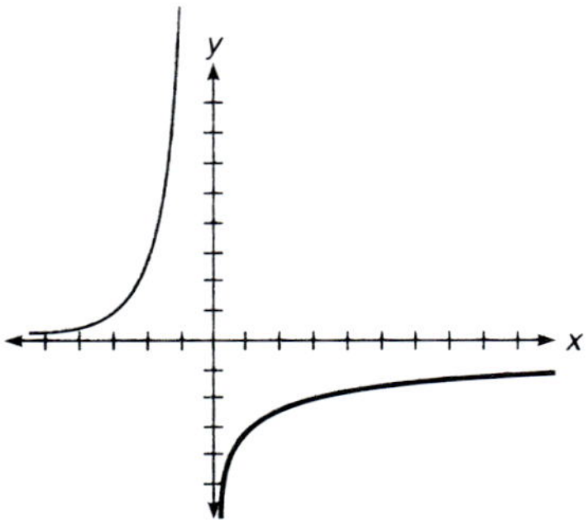

78.

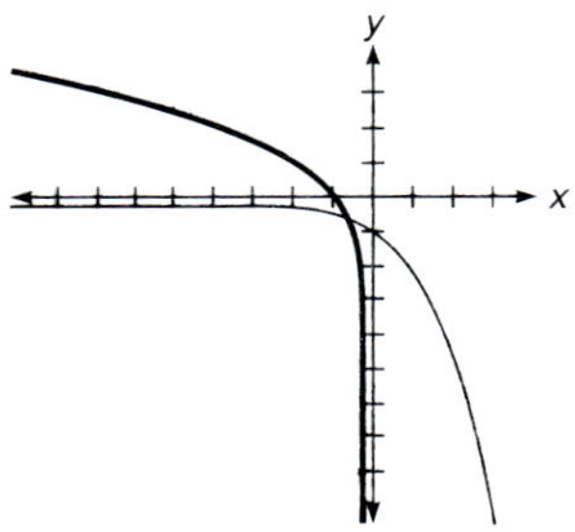

79. 4 **80.** $6x$ **81.** $\sqrt{3}$ **82.** $1 - 3x$
83. 30 **84.** $3x^2$ **85.** $4x$
86. $\frac{1}{2} \ln 5 + x$ **87.** $3,111.30
88. $-\frac{16}{17}$ **89.** $-\frac{2}{5}$ **90.** 10.01
91. 3.32 **92.** −0.37 **93.** 3.34
94. 1.52 **95.** 2.11 **96.** 2.51
97. 3.03 **98.** 3.40 **99.** 1 or 100
100. $\dfrac{\log 3 - \log 2}{2 \log 3 - \log 2}$ **101.** 10^{100}
102. $\pm 10^5$ **103.** $\frac{100}{3}$ **104.** 10^{-3} or 10^5
105. $10^{\frac{8(\log 3)(\log 2)}{\log 2 + \log 3}}$ **106.** $\dfrac{10 - \log 3}{\log 6}$
107. ln 4 **108.** $\frac{1}{2} \ln \frac{3}{2}$ **109.** $\dfrac{\ln 5 + 1}{\ln 5 - 1}$
110. 0.0001 or 1 **111.** e^{-2} or e^6
112. $2,744.06 **113.** 7.7%
114. 79.6 hours

Chapter 9 test

1. $\dfrac{1}{8^{\sqrt{2}}}$ **2.** 16
3. increasing

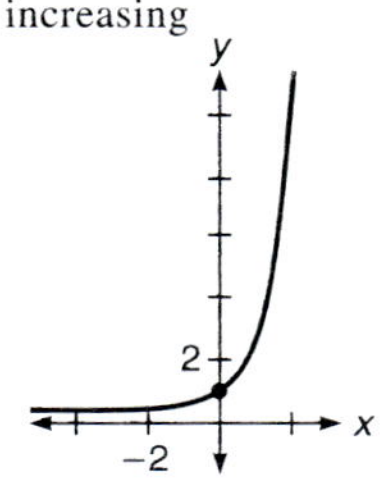

4. decreasing

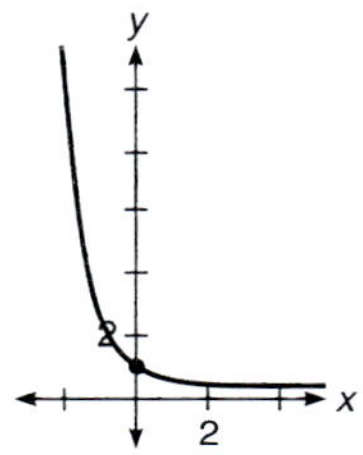

5. increasing

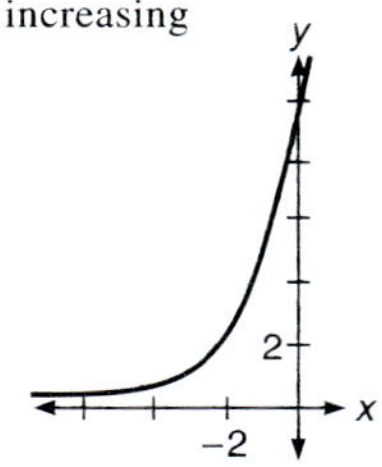

6. increasing

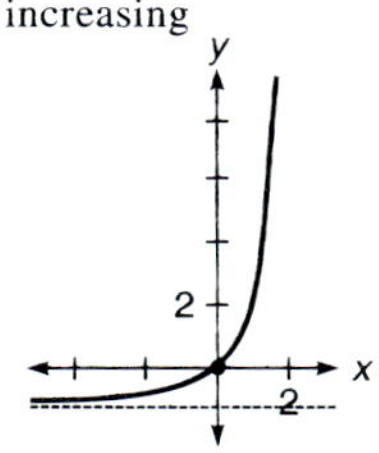

7. $\frac{1}{2}$ **8.** $\frac{1}{2}$ **9.** $-\frac{3}{2}$ **10.** 4
11. $\frac{3}{2}$ **12.** $\frac{7}{3}$ **13.** −12 **14.** −5
15. $\log_a x = y$ if and only if $a^y = x$, $a > 0$, $a \neq 1$ **16.** $5^{-x} = 0.25$ **17.** $3^2 = x - 3$
18. $m^{y/2} = x$ **19.** $\log_2 32 = 5$
20. $\log_{x-1} m = 3$ **21.** $\log_y z = 2x - 1$
22. $-\frac{3}{2}$ **23.** $502\frac{1}{2}$ **24.** $-\frac{1}{2}$

25.

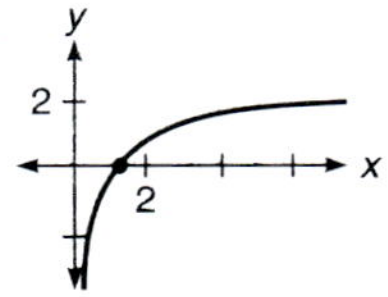

26.

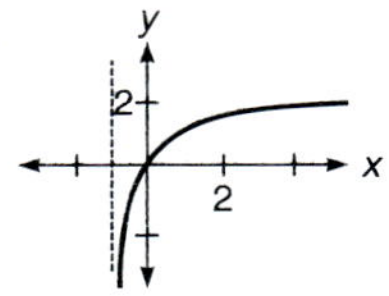

27.

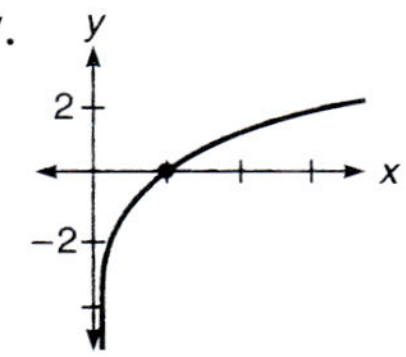

28. $\frac{1}{4}$ **29.** −2 or 16 **30.** 5 **31.** $\frac{2}{9}$
32. $\frac{19}{27}$ **33.** $\frac{38}{31}$ **34.** 5 **35.** 2 **36.** 32
37. $2 + 3 \log_3 x + \log_3 y - 4 \log_3 z$
38. $10 \log_{10} x + 3 \log_{10} y + \log_{10} z - 3$
39. 1.2770 **40.** 1.8957 **41.** −0.4709
42. 13.4916 **43.** −8.9872
44. 2.4307 **45.** 6.9078 **46.** 13.8546
47. 4.9462 **48.** $2x$ **49.** x^2 **50.** 18
51. 27 **52.** $10x$ **53.** $3,190.06
54. −1.14 **55.** 15.29 **56.** 5.99
57. 125 **58.** 3.59 **59.** 4.10
60. 0.01, 1, or 100
61. $\dfrac{2 \log 4 - \log 3}{\log 4 - 6 \log 3} \approx -0.3216$
62. 512 **63.** $\pm 10^5$
64. 10,000,000 or 0.00001
65. $10^{3\left(\frac{1}{\log 2} + \frac{1}{\log 3}\right)^{-1}} \approx 3.5787$
66. $\frac{1}{2} \ln 1.25 \approx 0.1116$
67. $\dfrac{\ln 3 + 1}{\ln 3 - 1} \approx 21.2814$
68. $1,573.26 **69.** 5.8%
70. 17.7 grams

Chapter 10

Exercise 10–1

Answers to odd-numbered problems

1. $(-3, -\frac{2}{5})$ **3.** $(-5, 2)$ **5.** $(-2, 10)$
7. $(8, \frac{1}{3})$ **9.** $(\frac{3}{10}, -\frac{1}{2})$ **11.** dependent
13. $(11, -4)$ **15.** $(\frac{155}{47}, \frac{261}{47})$
17. inconsistent **19.** dependent
21. $(8, \frac{1}{8})$ **23.** $(\frac{3}{2}, -\frac{2}{3})$ **25.** $(-2, 3, 2)$

27. (6,−5,2) **29.** (1,5,−5) **31.** (−4,6,−1)
33. (0,−5,5) **35.** (−10,3,−1)
37. $L = 13\frac{7}{13}$ cm, $W = 8\frac{6}{13}$ cm
39. $L = 11\frac{1}{2}$ in, $W = 6\frac{1}{2}$ in
41. $L = 56\frac{2}{3}$ mm, $W = 18\frac{1}{3}$ mm
43. $8,000 at 5%, $4,000 at 10%
45. $5,000 for each investment
47. $y = \frac{1}{6}x^2 + \frac{7}{6}x + 3$ **49.** $y = \frac{7}{3}x - \frac{38}{3}$
51. $(-\frac{28}{11},\frac{29}{11})$ **53.** $y = \frac{5}{2}x + 2$

Solutions to skill and review problems

1. $\sqrt{\dfrac{4x^2}{27y^3z}}$

$\dfrac{\sqrt{4x^2}}{\sqrt{27y^3z}} = \dfrac{2x}{3y\sqrt{3yz}} \cdot \dfrac{\sqrt{3yz}}{\sqrt{3yz}}$

$= \dfrac{2x\sqrt{3yz}}{3y(3yz)} = \dfrac{2x\sqrt{3yz}}{9y^2z}$

2. $\dfrac{2x-1}{3} = \dfrac{5-3x}{4}$

$4(2x - 1) = 3(5 - 3x)$
Cross multiply.
$8x - 4 = 15 - 9x$
$17x = 19$
$x = \frac{19}{17}$

3. $\dfrac{2x-1}{3} = \dfrac{5-3x}{x}$

$x(2x - 1) = 3(5 - 3x)$
Cross multiply.
$2x^2 - x = 15 - 9x$
$2x^2 + 8x - 15 = 0$
$x = \dfrac{-4 \pm \sqrt{46}}{2}$
Quadratic formula.

4. $\left|\dfrac{2x-1}{3}\right| > 5$

$\dfrac{2x-1}{3} > 5$ or $\dfrac{2x-1}{3} < -5$
If $|x| > a$ then $x > a$ or $x < -a$.
$2x - 1 > 15$ or $2x - 1 < -15$
$2x > 16$ or $2x < -14$
$x > 8$ or $x < -7$

5. $\left|\dfrac{2x-1}{x}\right| < 5$

This inequality is nonlinear. It must be solved using the critical point/test point method. Critical points:
Solve the corresponding equality.

$\left|\dfrac{2x-1}{x}\right| = 5$

$\dfrac{2x-1}{x} = 5$ or $\dfrac{2x-1}{x} = -5$

$2x - 1 = 5x$ or $2x - 1 = -5x$
$-1 = 3x$ or $7x = 1$
$-\frac{1}{3} = x$ or $x = \frac{1}{7}$

Find zeros of denominators.
$x = 0.$
Critical points: $-\frac{1}{3}$, 0, $\frac{1}{7}$

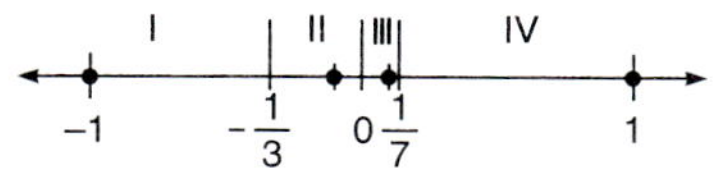

We choose test points from each interval: −1, −0.1, 0.1, 1.

$\left|\dfrac{2x-1}{x}\right| < 5$

$x = -1$: $|3| < 5$; true
$x = -0.1$: $|12| < 5$; false
$x = 0.1$: $|-8| < 5$; false
$x = 1$: $|1| < 5$; true
The soluton set is intervals I and IV:
$\{x \mid x < -\frac{1}{3} \text{ or } x > \frac{1}{7}\}$.

Solutions to trial exercise problems

7. [1] $-1 = -\frac{1}{2}x + 9y$
[2] $\frac{57}{14} = \frac{1}{2}x + \frac{3}{14}y$
Multiply [1] by 2 and multiply [2] by 14.
[1] $-x + 18y = -2$
[2] $7x + 3y = 57$
Add 7 times [1] to [2].
[3] $129y = 43$
$y = \frac{1}{3}$
Add −6 times [2] to [1].
[4] $-43x = -344$
$x = 8$
$(8,\frac{1}{3})$

25. [1] $x + y - 5z = -9$
[2] $-x + y + 2z = 9$
[3] $5x + 2y = -4$
Add [1] to [2].
[4] $2y - 3z = 0$
Add 5 times [2] to [3].
[5] $7y + 10z = 41$
Add 7 times [4] to −2 times [5].
[6] $-41z = -82$
$z = 2$
[7] $2y - 6 = 0$
Insert value of z into [4].
$y = 3$
Insert value of y and z into [1].
[8] $x + 2 = 0$
$x = -2$
(−2,3,2)

41. $W = \frac{1}{2}L - 10$
$P = 150 = 2L + 2W$
$75 = L + W$
Solve $\begin{matrix} L - 2W = 20 \\ L + W = 75 \end{matrix}$ to find $L = 56\frac{2}{3}$ mm, $W = 18\frac{1}{3}$ mm.

45. If the two investments are x and y then $x + y = 10{,}000$ and $0.06x + 0.12y = 900$, or $x + 2y = 15{,}000$, so we solve the system $\begin{matrix} x + y = 10{,}000 \\ x + 2y = 15{,}000 \end{matrix}$ to find $x = y = \$5{,}000$.

49. Since the points satisfy $y = mx + b$, we know $\begin{matrix} -1 = 5m + b \\ 6 = 8m + b \end{matrix}$, which we solve to find m and b: $m = \frac{7}{3}$, $b = -\frac{38}{3}$, so the equation is $y = \frac{7}{3}x - \frac{38}{3}$.

Exercise 10–2

Answers to odd-numbered problems

1. (−3,6) **3.** $(\frac{1}{2},2)$ **5.** (3,2)
7. (5,−3) **9.** (8,−3) **11.** (6,6)
13. (−8,3) **15.** (−6,6) **17.** (5,5)
19. $(\frac{7}{3},3)$ **21.** (6,−2)
23. (−2,3,2) **25.** (6,−5,2)
27. $(1,5,-\frac{1}{3})$ **29.** $(-\frac{4}{3},6,-1)$
31. (0,−5,5) **33.** (−10,3,−1)
35. (3,2,−2,1) **37.** (1,−2,3,4)
39. $(1,-2,\frac{1}{3},2)$ **41.** inconsistent
43. (5,−1,2,−4) **45.** $(3,-2,\frac{3}{4},1)$
47. $i_1 = 2$, $i_2 = \frac{2}{5}$
49. $I_1 = 180$, $I_2 = 200$, $I_3 = 210$
51. $333\frac{1}{3}$ liters of the 20% solution and $166\frac{2}{3}$ liters of the 50% solution
53. $49\frac{3}{13}$ gallons of 25% solution, and $30\frac{10}{13}$ gallons of 90% solution.
55. $(\frac{37}{4},\frac{17}{4},\frac{11}{4})$

Solutions to skill and review problems

1. Solve the system
[1] $2x - y = -3$
[2] $2x - 3y = 6$
$2y = -9 \leftarrow [1] - [2]$
$y = -\frac{9}{2}$
[1] $2x - (-\frac{9}{2}) = -3$
Substitute y into [1].
$2x = -\frac{15}{2}$
$x = -\frac{15}{4}$
Multiply each member by $\frac{1}{2}$.
$(-3\frac{3}{4},-4\frac{1}{2})$

2. $3^4 = 81$, so $\log_3 81 = 4$

3. $9^{3x+1} = 27^x$

$(3^2)^{3x+1} = (3^3)^x$

$3^{6x+2} = 3^{3x}$

$6x + 2 = 3x$

$x = -\frac{2}{3}$

4. $\log(x-1) - \log(x+1) = 2$

$\log \dfrac{x-1}{x+1} = 2$ since $\log \dfrac{m}{n}$

$= \log m - \log n$

$\dfrac{x-1}{x+1} = 10^2$ because if $\log x = y$, then

$x = 10^y$.

$x - 1 = 100x + 100$

$-\dfrac{101}{99} = x$

This solution makes both expressions $\log(x-1)$ or $\log(x+1)$ undefined, since $\log m$ is only defined if $m > 0$. Thus, there is no solution.

5. $\log(x-1) + \log(x+1) = \log 2$

$\log(x-1)(x+1) = \log 2$

$\log mn = \log m + \log n$

$(x-1)(x+1) = 2$

If $\log m = \log n$ then $m = n$.

$x^2 - 1 = 2$

$x^2 = 3$

$x = \pm\sqrt{3}$

The value $-\sqrt{3}$ makes $\log(x-1)$ and $\log(x+1)$ undefined, so the solution is $\sqrt{3}$.

6. $x^3 - x^2 - x + 1 < 0$

This is a nonlinear inequality. It must be solved by the critical point/test point method. Critical points:

Solve the corresponding equality.

$x^3 - x^2 - x + 1 = 0$

$x^2(x-1) - 1(x-1) = 0$

$(x-1)(x^2-1) = 0$

Factor by grouping.

$(x-1)(x-1)(x+1) = 0$

$x = \pm 1$.

Find zeros of denominators; in this case there are none.

Critical points are ± 1.

I	II	III

−2 −1 0 1 2

Choose test points in each interval and try in the original inequality.

$x^3 - x^2 - x + 1 < 0$

$x = -2$: $-9 < 0$; true

$x = 0$: $1 < 0$; false

$x = 2$: $3 < 0$; false

Thus the solution is interval I: $x < -1$.

Solutions to trial exercise problems

7. $\begin{bmatrix} \frac{2}{5} & \frac{1}{3} & 1 \\ -2 & 2 & -16 \end{bmatrix}$

Multiply [1] by 15. $\begin{bmatrix} 6 & 5 & 15 \\ -2 & 2 & -16 \end{bmatrix}$

Divide [2] by 2.

$\begin{bmatrix} 6 & 5 & 15 \\ -1 & 1 & -8 \end{bmatrix}$ $[1] \leftarrow 6[2] + [1]$

Note: This notation means to add 6 times row 2 to row 1 and replace 1 with this result.

$\begin{bmatrix} 0 & 11 & -33 \\ -1 & 1 & -8 \end{bmatrix}$ Divide [1] by 11.

$\begin{bmatrix} 0 & 1 & -3 \\ -1 & 1 & -8 \end{bmatrix}$ $[2] \leftarrow [1] - [2]$

$\begin{bmatrix} 0 & 1 & -3 \\ 1 & 0 & 5 \end{bmatrix}$

Rearrange rows and set coefficients to 1.

$\begin{bmatrix} 1 & 0 & 5 \\ 0 & 1 & -3 \end{bmatrix}$ Solution: $(5, -3)$

29. $\begin{bmatrix} -3 & 0 & 1 & 3 \\ 9 & 2 & 3 & -3 \\ 3 & 1 & -2 & 4 \end{bmatrix}$

$[2] \leftarrow 2[3] - [2]$

$\begin{bmatrix} -3 & 0 & 1 & 3 \\ -3 & 0 & -7 & 11 \\ 3 & 1 & -2 & 4 \end{bmatrix}$

$[2] \leftarrow 7[1] + [2]$

$[3] \leftarrow 2[1] + [3]$

$\begin{bmatrix} -3 & 0 & 1 & 3 \\ -24 & 0 & 0 & 32 \\ -3 & 1 & 0 & 10 \end{bmatrix}$

Divide [2] by 8.

$\begin{bmatrix} -3 & 0 & 1 & 3 \\ -3 & 0 & 0 & 4 \\ -3 & 1 & 0 & 10 \end{bmatrix}$

$[1] \leftarrow -[2] + [1]$

$[3] \leftarrow -[2] + [3]$

$\begin{bmatrix} 0 & 0 & 1 & -1 \\ -3 & 0 & 0 & 4 \\ 0 & 1 & 0 & 6 \end{bmatrix}$

Rearrange rows and set coefficients to 1.

$\begin{bmatrix} 1 & 0 & 0 & -\frac{4}{3} \\ 0 & 1 & 0 & 6 \\ 0 & 0 & 1 & -1 \end{bmatrix}$

Solution: $(-\frac{4}{3}, 6, -1)$

39. $\begin{bmatrix} 1 & \frac{1}{2} & 3 & -3 & -5 \\ 2 & -\frac{3}{2} & 3 & 5 & 16 \\ -6 & 0 & 0 & -1 & -8 \\ 1 & 0 & -6 & 0 & -1 \end{bmatrix}$

Multiply [1] by 2, and [2] by 2.

$\begin{bmatrix} 2 & 1 & 6 & -6 & -10 \\ 4 & -3 & 6 & 10 & 32 \\ -6 & 0 & 0 & -1 & -8 \\ 1 & 0 & -6 & 0 & -1 \end{bmatrix}$

$[2] \leftarrow 3[1] + [2]$

$\begin{bmatrix} 2 & 1 & 6 & -6 & -10 \\ 10 & 0 & 24 & -8 & 2 \\ -6 & 0 & 0 & -1 & -8 \\ 1 & 0 & -6 & 0 & -1 \end{bmatrix}$

Divide [2] by 2.

$\begin{bmatrix} 2 & 1 & 6 & -6 & -10 \\ 5 & 0 & 12 & -4 & 1 \\ -6 & 0 & 0 & -1 & -8 \\ 1 & 0 & -6 & 0 & -1 \end{bmatrix}$

$[1] \leftarrow 2[4] - [1]$

$[2] \leftarrow 5[4] - [2]$

$[3] \leftarrow 6[4] + [3]$

$\begin{bmatrix} 0 & -1 & -18 & 6 & 8 \\ 0 & 0 & -42 & 4 & -6 \\ 0 & 0 & -36 & -1 & -14 \\ 1 & 0 & -6 & 0 & -1 \end{bmatrix}$

Divide [2] by 2.

$\begin{bmatrix} 0 & -1 & -18 & 6 & 8 \\ 0 & 0 & -21 & 2 & -3 \\ 0 & 0 & -36 & -1 & -14 \\ 1 & 0 & -6 & 0 & -1 \end{bmatrix}$

$[1] \leftarrow 6[3] + [1]$

$[2] \leftarrow 2[3] + [2]$

$\begin{bmatrix} 0 & -1 & -234 & 0 & -76 \\ 0 & 0 & -93 & 0 & -31 \\ 0 & 0 & -36 & -1 & -14 \\ 1 & 0 & -6 & 0 & -1 \end{bmatrix}$

Divide [2] by 31.

$\begin{bmatrix} 0 & -1 & -234 & 0 & -76 \\ 0 & 0 & -3 & 0 & -1 \\ 0 & 0 & -36 & -1 & -14 \\ 1 & 0 & -6 & 0 & -1 \end{bmatrix}$

$[1] \leftarrow -78[2] + [1]$

$[3] \leftarrow -12[2] + [3]$

$[4] \leftarrow -2[2] + [4]$

$\begin{bmatrix} 0 & -1 & 0 & 0 & 2 \\ 0 & 0 & -3 & 0 & -1 \\ 0 & 0 & 0 & -1 & -2 \\ 1 & 0 & 0 & 0 & 1 \end{bmatrix}$

Rearrange rows and set coefficients to 1.

$\begin{bmatrix} 1 & 0 & 0 & 0 & 1 \\ 0 & 1 & 0 & 0 & -2 \\ 0 & 0 & 1 & 0 & \frac{1}{3} \\ 0 & 0 & 0 & 1 & 2 \end{bmatrix}$

Solution: $(1, -2, \frac{1}{3}, 2)$

51. Let t = required amount of 20% solution and f = required amount of 50% solution. Then $t + f = 500$. Now, 30% of the 500 liters is to be alcohol, or 150 liters. This alcohol comes from 20% of t and 50% of f, so that we also have the equation $0.20t + 0.50f = 150$, or $2t + 5f = 1{,}500$. Thus, we solve $\begin{matrix} t + f = 500 \\ 2t + 5f = 1{,}500 \end{matrix}$ for t and f. The solution is $\left(\frac{1{,}000}{3}, \frac{500}{3}\right)$, so we need $333\frac{1}{3}$ liters of the 20% solution and $166\frac{2}{3}$ liters of the 50% solution.

Exercise 10–3

Answers to odd-numbered problems

1. -9 **3.** $-42\frac{1}{2}$ **5.** $-\pi$ **7.** 149 **9.** $\frac{34}{3}$ **11.** -6 **13.** 7 **15.** 105 **17.** $-54\sqrt{2}$ **19.** -2 **21.** 74 **23.** 406 **25.** -220 **27.** -112 **29.** $x = -\frac{16}{31}, y = \frac{12}{31}$ **31.** $x = -\frac{17}{18}, y = -\frac{5}{9}$ **33.** $x = \frac{92}{19}, y = -\frac{71}{19}$ **35.** $x = -\frac{59}{110}, y = \frac{17}{110}, z = -\frac{37}{55}$ **37.** $x = -\frac{11}{10}, y = \frac{3}{20}, z = -\frac{29}{20}$ **39.** $x = \frac{8}{5}, y = -\frac{1}{5}, z = -\frac{6}{5}$ **41.** $x = -\frac{3}{4}, y = -\frac{19}{4}, z = -\frac{49}{4}$ **43.** inconsistent **45.** $x = -\frac{5}{4}, y = -\frac{27}{2}, z = 2, w = -\frac{51}{4}$ **47.** $x = \dfrac{417}{2}, y = \dfrac{19}{2}, z = -\dfrac{1{,}329}{4}, w = -\dfrac{567}{2}$ **49.** $x = \frac{17}{13}, y = -\frac{29}{13}, z = -\frac{14}{13}, w = -\frac{17}{26}$ **51.** $\frac{87}{28}$ **53.** $y = 1.16x + 0.57$ **55.** $y = 2.13x - 8.18$ **57.** The line is $y = 1.9x + 1$. For the fifth year the line predicts $y = 1.9(5) + 1 = 10.5\%$ failures, and for the sixth year it predicts $y = 1.9(6) + 1 = 12.4\%$ failures. **59. a.** $y = -0.3426 + 0.8851$. This assumes x is the year less 1,875, and y is the time in seconds less 4:24.5 (in seconds, or 264.5). **b.** 2,022 **61.** 47 **63.** 31 **65.** $-7x + 3y + 23 = 0$ **67.** $y = 0.8x^2 + 2.6x - 0.4$ **69.** $y = \frac{7}{3}x - \frac{38}{3}$ **71.** $(-\frac{57}{26}, -\frac{29}{26})$ **73.** $i_1 = \frac{4}{3}, i_2 = -\frac{55}{3}, i_3 = \frac{134}{3}$ **75.** $x = 9{,}000$, $y = 12{,}000$, $z = 15{,}000$ **77.** (1,1) is the correct solution, which can be verified by substitution into the system itself. **79.** Define the determinant of an order 1 matrix as the value of the single element. This permits order 2 determinants to be evaluated by expansion about rows and columns, just like the determinants of order greater than 2.

81. The area of the five-sided polygon is the sum of the areas marked 1, 2, and 3 in the figure. Each of these is a triangle.

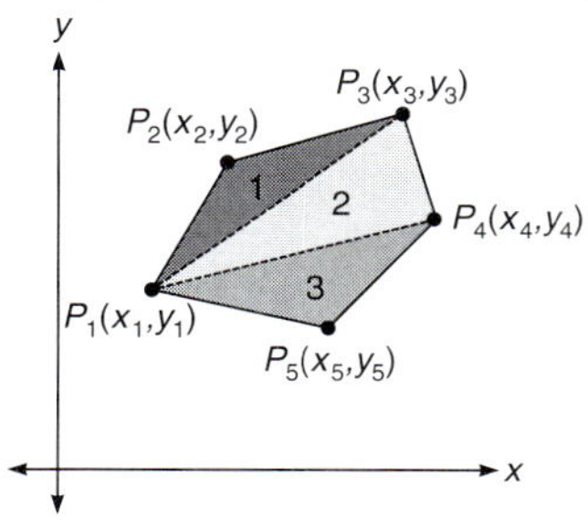

$$\begin{aligned}
\text{Area}_{\text{total}} &= \text{Area}_1 + \text{Area}_2 + \text{Area}_3 \\
&= \frac{1}{2}\begin{vmatrix} x_1 & y_1 & 1 \\ x_2 & y_2 & 1 \\ x_3 & y_3 & 1 \end{vmatrix} + \frac{1}{2}\begin{vmatrix} x_1 & y_1 & 1 \\ x_3 & y_3 & 1 \\ x_4 & y_4 & 1 \end{vmatrix} + \frac{1}{2}\begin{vmatrix} x_1 & y_1 & 1 \\ x_4 & y_4 & 1 \\ x_5 & y_5 & 1 \end{vmatrix} \\
&= \tfrac{1}{2}[(x_2y_3 - x_3y_2) - (x_1y_3 - x_3y_1) + (x_1y_2 - x_2y_1)] + \\
&\quad \tfrac{1}{2}[(x_3y_4 - x_4y_3) - (x_1y_4 - x_4y_1) + (x_1y_3 - x_3y_1)] + \\
&\quad \tfrac{1}{2}[(x_4y_5 - x_5y_4) - (x_1y_5 - x_5y_1) + (x_1y_4 - x_4y_1)] \\
&= \tfrac{1}{2}[(x_1y_2 - x_2y_1) + (x_2y_3 - x_3y_2) + (x_3y_4 - x_4y_3) + (x_4y_5 - x_5y_4) + (x_5y_1 - x_1y_5)].
\end{aligned}$$

The solution for the four-sided figure is similar.

Solutions to skill and review problems

1. $2x - 3 < 8$
$2x < 11$
$x < 5\frac{1}{2}$

2. $2x + y > 2$, $x = 2$, $y = -1$
$2(2) + (-1) > 2$
$3 > 2$; true
Thus, $(2,-1)$ is a solution to the statement $2x + y > 2$.

3. $y = 2x - 1$ and
$y = -\frac{1}{3}x + 1$ so
$2x - 1 = -\frac{1}{3}x + 1$
$6x - 3 = -x + 3$
$7x = 6$
$x = \frac{6}{7}$
$y = 2x - 1 = 2(\frac{6}{7}) - 1 = \frac{5}{7}$.

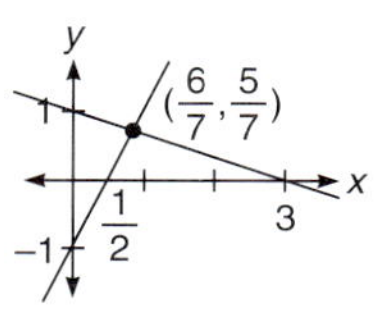

4. $0.075(1{,}200) + 0.05(1{,}800) = \$180$

5. $3(2x + 3) - 2(5x + 3) = x$
$6x + 9 - 10x - 6 = x$
$3 = 5x$
$\frac{3}{5} = x$

6. $\dfrac{4x-1}{x} < -5x$

This is nonlinear. Use the critical point/test point method.

Critical points:

Solve corresponding equality:

$\dfrac{4x-1}{x} = -5x$

$4x - 1 = -5x^2$

$5x^2 + 4x - 1 = 0$

$(5x - 1)(x + 1) = 0$

$5x - 1 = 0$ or $x + 1 = 0$

$x = \frac{1}{5}$ or $x = -1$

Find zeros of denominators: $x = 0$

Critical points: $-1, 0, \frac{1}{5}$.

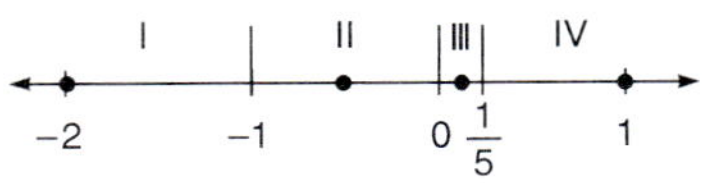

Test points (one from each interval):

$-2, -0.5, 0.1, 1$

$\dfrac{4x-1}{x} < -5x$

$x = -2$: $4.5 < 10$; true

$x = -0.5$: $6 < 2.5$; false

$x = 0.1$: $-6 < -0.5$; true

$x = 1$: $3 < -5$; false

The solution is intervals I and III:

$x < -1$ or $0 < x < \frac{1}{5}$.

7. $\log^2 x - \log x = 6$

Let $u = \log x$:

$u^2 - u = 6$

$u^2 - u - 6 = 0$

$(u - 3)(u + 2) = 0$

$u = 3$ or $u = -2$

$\log x = 3$ or $\log x = -2$

$x = 10^3$ or $x = 10^{-2}$

8. $y = \log_2(x + 1)$

Find inverse function.

$x = \log_2(y + 1)$

$y + 1 = 2^x$

Inverse function:

$y = 2^x - 1$

Computed values:

$y = 2^x - 1$	$y = \log_2(x + 1)$
$(-1, -\frac{1}{2})$	$(-\frac{1}{2}, -1)$
(0,0)	(0,0)
(1,1)	(1,1)
(2,3)	(3,2)
(3,7)	(7,3)

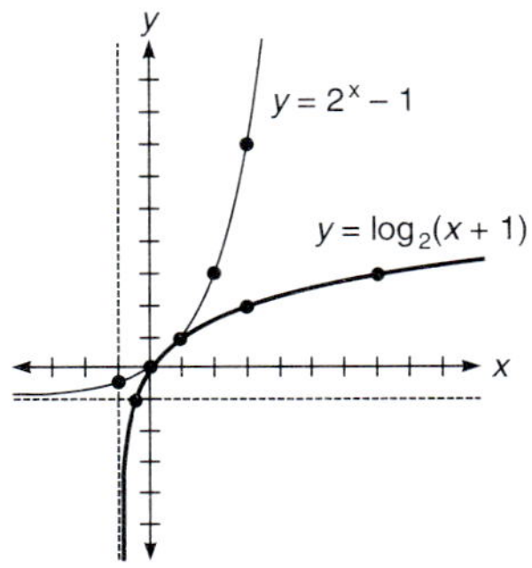

9. $y = x^2 + 3x - 4$

Parabola

$y = x^2 + 3x - 4$

Complete the square.

$y = x^2 + 3x + \frac{9}{4} - 4 - \frac{9}{4}$

$\frac{1}{2} \cdot 3 = \frac{3}{2}$; $(\frac{3}{2})^2 = \frac{9}{4}$

$y = (x + \frac{3}{2})^2 - \frac{25}{4}$

Vertex: $(-1\frac{1}{2}, -6\frac{1}{4})$

Intercepts:

$x = 0$: $y = 0 + 0 - 4 = -4$; $(0, -4)$

$y = 0$: $0 = x^2 + 3x - 4$

$0 = (x + 4)(x - 1)$

$x = -4$ or 1; $(-4,0)$, $(1,0)$

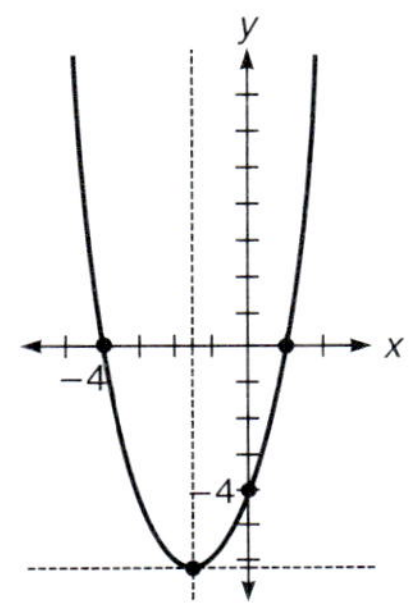

Solutions to trial exercise problems

9. $\begin{vmatrix} 2 & \frac{2}{3} & -1 \\ 4 & -1 & \frac{1}{2} \\ -3 & 0 & -2 \end{vmatrix} = -3\begin{vmatrix} \frac{2}{3} & -1 \\ -1 & \frac{1}{2} \end{vmatrix} + (-2)\begin{vmatrix} 2 & \frac{2}{3} \\ 4 & -1 \end{vmatrix}$

$= -3(-\frac{2}{3}) - 2(-\frac{14}{3}) = \frac{34}{3}$

23. $\begin{vmatrix} 4 & 5 & 1 & 0 \\ -2 & 1 & 3 & 7 \\ 0 & 1 & 2 & 0 \\ 4 & -2 & 0 & 3 \end{vmatrix} = -1\begin{vmatrix} 4 & 1 & 0 \\ -2 & 3 & 7 \\ 4 & 0 & 3 \end{vmatrix} + 2\begin{vmatrix} 4 & 5 & 0 \\ -2 & 1 & 7 \\ 4 & -2 & 3 \end{vmatrix}$

$= -\left\{4\begin{vmatrix} 3 & 7 \\ 0 & 3 \end{vmatrix} - \begin{vmatrix} -2 & 7 \\ 4 & 3 \end{vmatrix}\right\} + 2\left\{4\begin{vmatrix} 1 & 7 \\ -2 & 3 \end{vmatrix} - 5\begin{vmatrix} -2 & 7 \\ 4 & 3 \end{vmatrix}\right\}$

$= -[4(9) - (-34)] + 2[4(17) - 5(-34)]$

$= -[70] + 2[238] = 406$

37. $D_x = \begin{vmatrix} -6 & -3 & -3 \\ 7 & -4 & -6 \\ -3 & 2 & 0 \end{vmatrix} = -132$; $D_y = \begin{vmatrix} 9 & -6 & -3 \\ 1 & 7 & -6 \\ 3 & -3 & 0 \end{vmatrix} = 18$

$D_z = \begin{vmatrix} 9 & -3 & -6 \\ 1 & -4 & 7 \\ 3 & 2 & -3 \end{vmatrix} = -174$; $D = \begin{vmatrix} 9 & -3 & -3 \\ 1 & -4 & -6 \\ 3 & 2 & 0 \end{vmatrix} = 120$

$x = \frac{D_x}{D} = -\frac{11}{10}$, $y = \frac{D_y}{D} = \frac{3}{20}$, $z = \frac{D_z}{D} = -\frac{29}{20}$

45. $D_x = \begin{vmatrix} 2 & 2 & 3 & -2 \\ -3 & 2 & 5 & -1 \\ -2 & 4 & -2 & -4 \\ 2 & 4 & 0 & -4 \end{vmatrix} = -40,\ D_y = \begin{vmatrix} 2 & 2 & 3 & -2 \\ -1 & -3 & 5 & -1 \\ -4 & -2 & -2 & -4 \\ -4 & 2 & 0 & -4 \end{vmatrix} = -432$

$D_z = \begin{vmatrix} 2 & 2 & 2 & -2 \\ -1 & 2 & -3 & -1 \\ -4 & 4 & -2 & -4 \\ -4 & 4 & 2 & -4 \end{vmatrix} = 64,\ D_w = \begin{vmatrix} 2 & 2 & 3 & 2 \\ -1 & 2 & 5 & -3 \\ -4 & 4 & -2 & -2 \\ -4 & 4 & 0 & 2 \end{vmatrix} = -408$

$D = \begin{vmatrix} 2 & 2 & 3 & -2 \\ -1 & 2 & 5 & -1 \\ -4 & 4 & -2 & -4 \\ -4 & 4 & 0 & -4 \end{vmatrix} = 32,\ x = -\frac{5}{4},\ y = -\frac{27}{2},\ z = 2,\ w = -\frac{51}{4}$

52. $D = \begin{vmatrix} 2 & -1 & 3 & -1 & 0 \\ 1 & 1 & 0 & -2 & 1 \\ 0 & -1 & -1 & 0 & 0 \\ 3 & 0 & -1 & 1 & 1 \\ 0 & 0 & 1 & 0 & -1 \end{vmatrix} = 1\begin{vmatrix} 2 & -1 & -1 & 0 \\ 1 & 1 & -2 & 1 \\ 0 & -1 & 0 & 0 \\ 3 & 0 & 1 & 1 \end{vmatrix} + (-1)\begin{vmatrix} 2 & -1 & 3 & -1 \\ 1 & 1 & 0 & -2 \\ 0 & -1 & -1 & 0 \\ 3 & 0 & -1 & 1 \end{vmatrix}$

$\begin{vmatrix} 2 & -1 & -1 & 0 \\ 1 & 1 & -2 & 1 \\ 0 & -1 & 0 & 0 \\ 3 & 0 & 1 & 1 \end{vmatrix} = -(-1)\begin{vmatrix} 2 & -1 & 0 \\ 1 & -2 & 1 \\ 3 & 1 & 1 \end{vmatrix} = (1)\left[2\begin{vmatrix} -2 & 1 \\ 1 & 1 \end{vmatrix} - (-1)\begin{vmatrix} 1 & 1 \\ 3 & 1 \end{vmatrix}\right] = 2(-3) + (-2) = -8,$

$\begin{vmatrix} 2 & -1 & 3 & -1 \\ 1 & 1 & 0 & -2 \\ 0 & -1 & -1 & 0 \\ 3 & 0 & -1 & 1 \end{vmatrix} = -(3)\begin{vmatrix} -1 & 3 & -1 \\ 1 & 0 & -2 \\ -1 & -1 & 0 \end{vmatrix} - (-1)\begin{vmatrix} 2 & -1 & -1 \\ 1 & 1 & -2 \\ 0 & -1 & 0 \end{vmatrix} + \begin{vmatrix} 2 & -1 & 3 \\ 1 & 1 & 0 \\ 0 & -1 & -1 \end{vmatrix}$

$= -3\left[(-1)\begin{vmatrix} 3 & -1 \\ 0 & -2 \end{vmatrix} - (-1)\begin{vmatrix} -1 & -1 \\ 1 & -2 \end{vmatrix}\right] + \left[-(-1)\begin{vmatrix} 2 & -1 \\ 1 & -2 \end{vmatrix}\right] + \left[-(-1)\begin{vmatrix} 2 & 3 \\ 1 & 0 \end{vmatrix} + (-1)\begin{vmatrix} 2 & -1 \\ 1 & 1 \end{vmatrix}\right]$

$= -3[-1(-6) - (-1)(3)] + [-(-1)(-3)] + [(-3) + (-1)(3)] = -3[9] + [-3] + [-6] = -36,$

so $D = 1(-8) - 1(-36) = 28$

$D_E = \begin{vmatrix} 2 & -1 & 3 & -1 & 5 \\ 1 & 1 & 0 & -2 & 0 \\ 0 & -1 & -1 & 0 & 10 \\ 3 & 0 & -1 & 1 & -4 \\ 0 & 0 & 1 & 0 & -20 \end{vmatrix} = 1\begin{vmatrix} 2 & -1 & -1 & 5 \\ 1 & 1 & -2 & 0 \\ 0 & -1 & 0 & 10 \\ 3 & 0 & 1 & -4 \end{vmatrix} + (-20)\begin{vmatrix} 2 & -1 & 3 & -1 \\ 1 & 1 & 0 & -2 \\ 0 & -1 & -1 & 0 \\ 3 & 0 & -1 & 1 \end{vmatrix}$

$= (-73) - 20(-36) = 647$

Therefore, $E = \frac{D_E}{D} = \frac{647}{28}$

55. $(1.5,-4.8), (2,-4.0), (3,-2.0), (4.5,1.4), (6,4.7), (6.5,5.6)$

$X = 1.5 + 2 + 3 + 4.5 + 6 + 6.5 = 23.5$

$Y = -4.8 - 4 - 2 + 1.4 + 4.7 + 5.6 = 0.9$

$P = (1.5)(-4.8) + (2)(-4.0) + (3)(-2.0) + (4.5)(1.4) + (6)(4.7) + (6.5)(5.6) = 49.7$

$S = 1.5^2 + 2^2 + 3^2 + 4.5^2 + 6^2 + 6.5^2 = 113.75$

$N = 6$

Solve $\begin{matrix} 23.5m + 6b = 0.9 \\ 113.75m + 23.5b = 49.7 \end{matrix}$

$D_m = \begin{vmatrix} 0.9 & 6 \\ 49.7 & 23.5 \end{vmatrix} = -277.05$

$D_b = \begin{vmatrix} 23.5 & 0.9 \\ 113.75 & 49.7 \end{vmatrix} = 1065.575$

$D = \begin{vmatrix} 23.5 & 6 \\ 113.75 & 23.5 \end{vmatrix} = -130.25$

$m = \frac{D_m}{D} \approx 2.13,\ b = \frac{D_b}{D} \approx -8.18$, so the line is $y = 2.13x - 8.18$.

64. We know that one equation for a straight line is $y - y_1 = \dfrac{y_2 - y_1}{x_2 - x_1}(x - x_1)$ if $x_2 \neq x_1$.
We can transform this to $(y - y_1)(x_2 - x_1)$
$= (y_2 - y_1)(x - x_1)$, which is true even when $x_2 = x_1$.
$x_2y - x_1y - x_2y_1 + x_1y_1 = xy_2 - x_1y_2 - xy_1 + x_1y_1$
$x_2y - x_1y - x_2y_1 = xy_2 - x_1y_2 - xy_1$
We put the terms on the same side of the equation, in descending order of variable names and subscripts, to make comparison with other equations easier:
$-x_2y_1 + x_2y + x_1y_2 - x_1y - xy_2 + xy_1 = 0$
Expanding $\begin{vmatrix} x & y & 1 \\ x_1 & y_1 & 1 \\ x_2 & y_2 & 1 \end{vmatrix}$ will give the left member of this last equation.

71. First, find the equation of the line passing through $(-2,-1)$ and $(3,2)$:
Solve $\begin{matrix} -1 = -2m + b \\ 2 = 3m + b \end{matrix}$: $m = \frac{3}{5}$, $b = \frac{1}{5}$, so the first line is $y = \frac{3}{5}x + \frac{1}{5}$.
Now, the second line, through $(-6,2)$ and $(5,-7)$: Solve $\begin{matrix} 2 = -6m + b \\ -7 = 5m + b \end{matrix}$:
$m = -\frac{9}{11}$, $b = -\frac{32}{11}$, so the second line is $y = -\frac{9}{11}x - \frac{32}{11}$.
To find the point of intersection we solve the system $\begin{matrix} y = \frac{3}{5}x + \frac{1}{5} \\ y = -\frac{9}{11}x - \frac{32}{11} \end{matrix}$.
To use Cramer's rule it is easier to rewrite this system as $\begin{matrix} 3x - 5y = -1 \\ 9x + 11y = -32 \end{matrix}$.
$D = \begin{vmatrix} 3 & -5 \\ 9 & 11 \end{vmatrix} = 78$, $D_x = \begin{vmatrix} -1 & -5 \\ -32 & 11 \end{vmatrix} = -171$,
$D_y = \begin{vmatrix} 3 & -1 \\ 9 & -32 \end{vmatrix} = -87$, $x = -\frac{57}{26}$, $y = -\frac{29}{26}$, so the point is $(-\frac{57}{26}, -\frac{29}{26})$.

Exercise 10–4

Answers to odd-numbered problems

1.
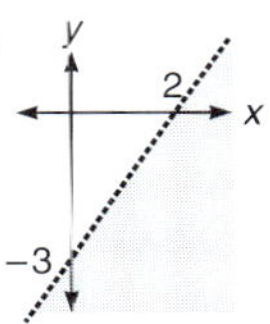

3.
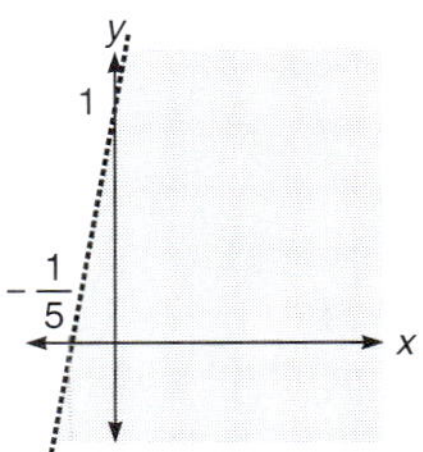

5.
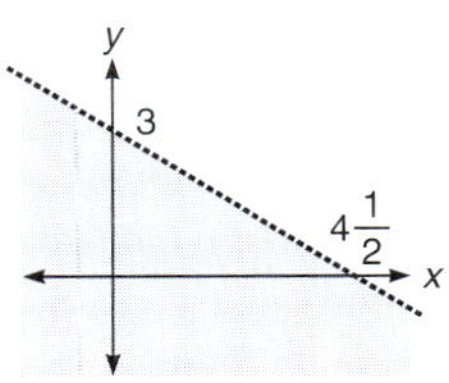

7.
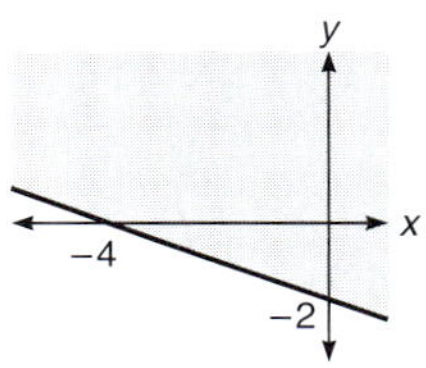

9.
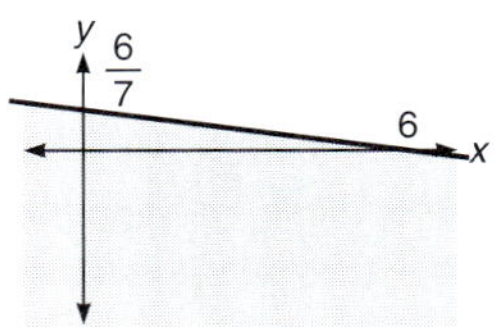

11.
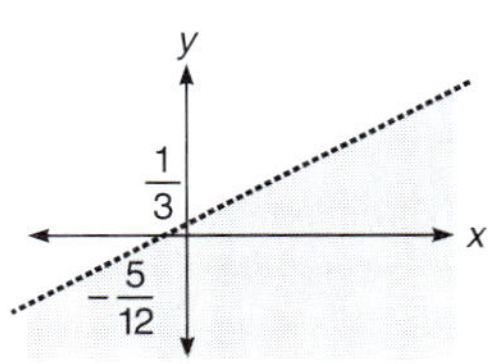

13.
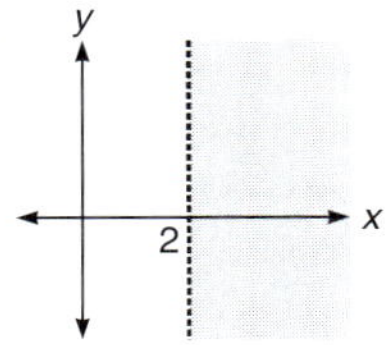

15.
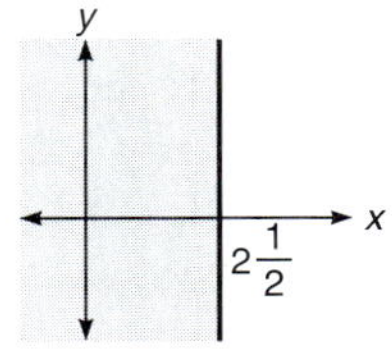

17.
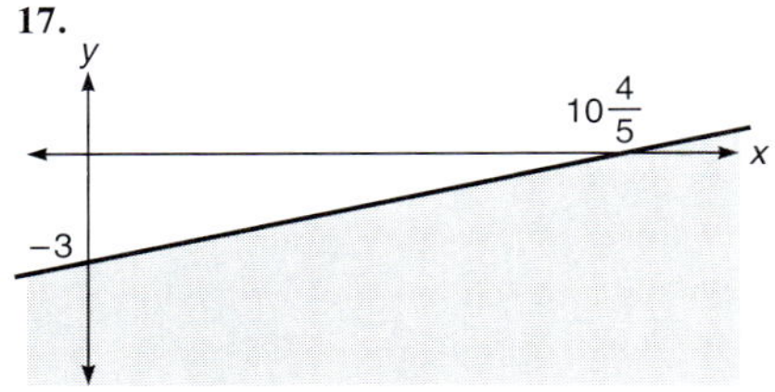

19.
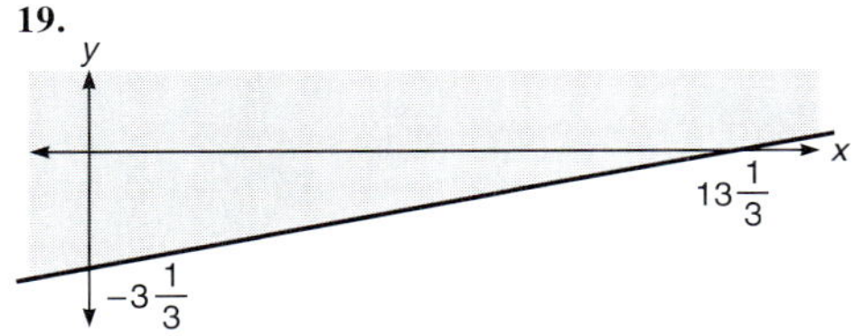

21.
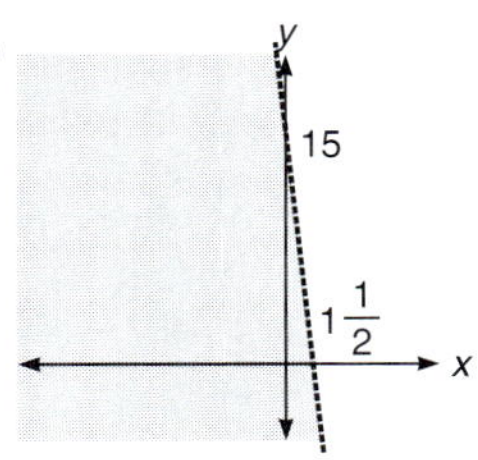

23.
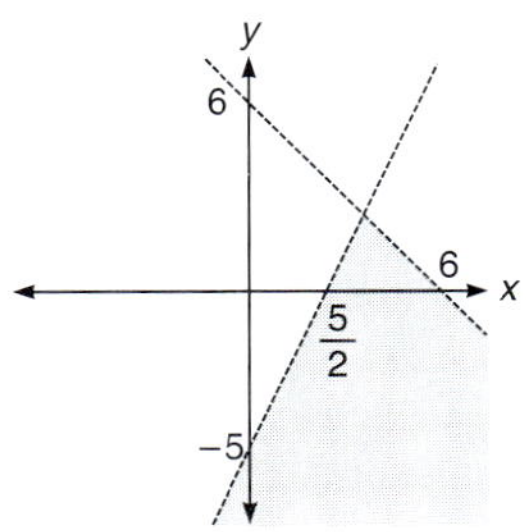

25.

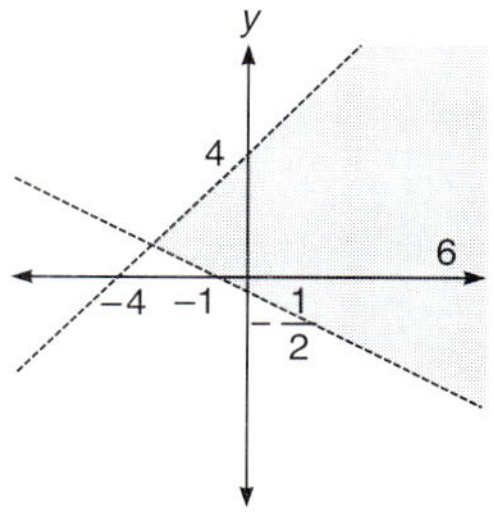

27.

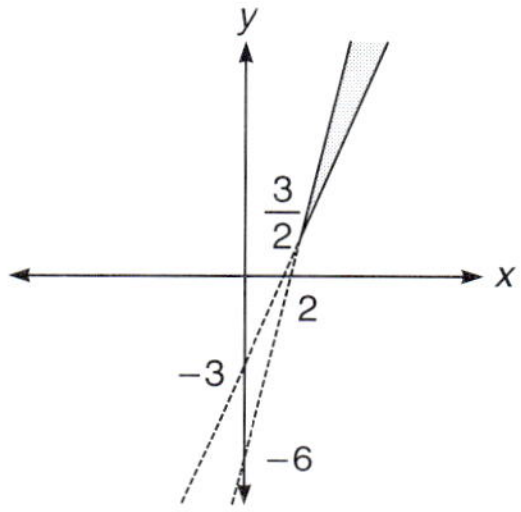

29.

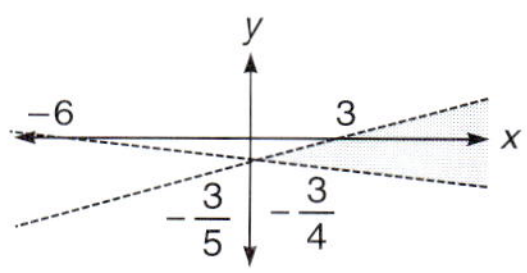

31.

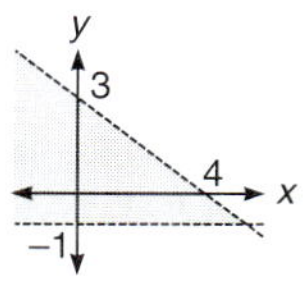

33.

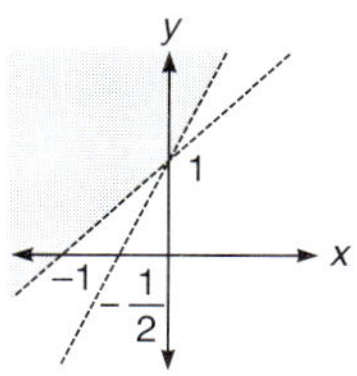

35.

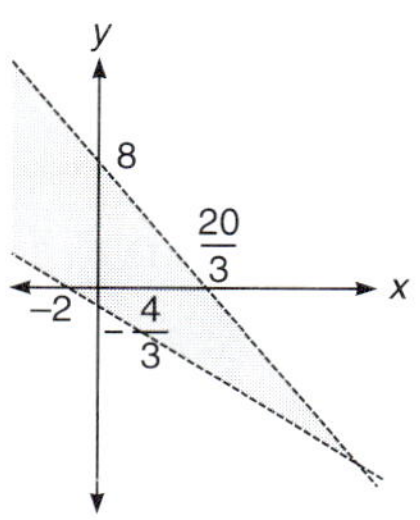

37.

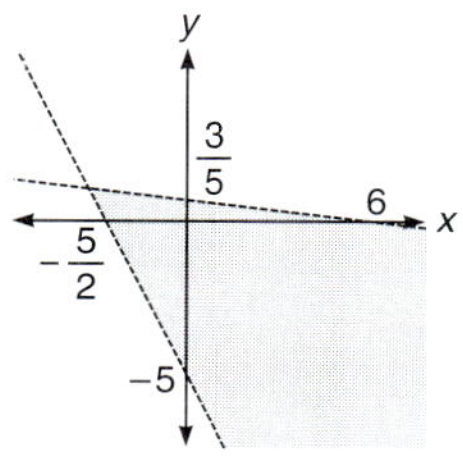

39. a. $r \le 1.5 + 0.5 = 2$, so $d \le 2t$.

b. Graph the system $d \ge t$ and $d \le 2t$.

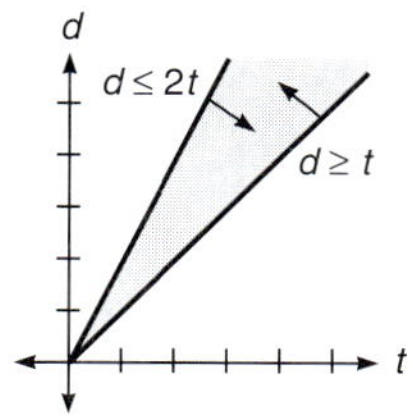

41. $P = 12$ at $(6,0)$ **43.** $P = 16$ at $(2,4)$
45. $P = 12$ at $(6,0)$ **47.** $P = 24\frac{1}{2}$ at $(6,\frac{25}{8})$
49. $P = 4$ at $(12,0)$ **51.** $P = 7$ at $(7,0)$
53. $P = 14$ at $(4,8)$ **55.** $P = 9$ at $(5,2)$
57. $P = 18$ at $(9,0)$ **59.** $P = 22$ at $(7\frac{1}{3},0)$
61. $P = 1\frac{1}{2}$ at $(7\frac{1}{2},0)$ **63.** $P = 36$ at $(6,2)$ or $(7\frac{1}{5},0)$ **65.** $P = 24$ at $(8,0)$
67. $C = 6$ at $(0,6)$ **69.** $C = 16$ at $(2,4)$
71. $C = 8\frac{2}{3}$ at $(3\frac{1}{2},5)$ **73.** $C = 7$ at $(5,1)$
75. $C = 5\frac{3}{5}$ at $(1\frac{3}{5},2\frac{2}{5})$
77. $C = 12\frac{2}{5}$ at $(5\frac{1}{5},3\frac{1}{5})$
79. The maximum income is \$2,980 and comes from producing 20 tables and 240 chairs.
81. Production is maximized at 315 tons with 5 type–A crews and 10 type–B crews.
83. We minimize cost at $52\frac{16}{17}$ cents by using $\frac{36}{17}$ lb of A and $\frac{20}{17}$ lb of B.

Solutions to skill and review problems

1. $\begin{bmatrix} 1 & 0 & 0 & 0 \\ 0 & 1 & 0 & 0 \\ 0 & 0 & 1 & 0 \\ 0 & 0 & 0 & 1 \end{bmatrix}$

2. $\left|3 - \frac{\sqrt{2}}{2}\right| = 3 - \frac{\sqrt{2}}{2}$, since $|x| = x$ if $x \ge 0$

3. $|4 - x| < 10$
$-10 < 4 - x < 10$
$-14 < -x < 6$
$14 > x > -6$
$-6 < x < 14$

4. $2x - 3 = 0$
$2x = 3$
$x = \frac{3}{2}$
$1\frac{1}{2}$

5. $2x^2 - 3x = 5$
$2x^2 - 3x - 5 = 0$
$(2x - 5)(x + 1) = 0$
$2x - 5 = 0$ or $x + 1 = 0$
$2x = 5$ or $x = -1$
$x = \frac{5}{2}$ or $x = -1$

6. $|2x^2 - 3x| = 5$
$2x^2 - 3x = 5$
$x = -1$ or $2\frac{1}{2}$ (problem 5)
$2x^2 - 3x = -5$
$2x^2 - 3x + 5 = 0$
$x = \frac{3}{4} \pm \frac{\sqrt{31}}{4}i$, quadratic formula
$x = -1, 2\frac{1}{2}, \frac{3}{4} \pm \frac{\sqrt{31}}{4}i$

7. $\sqrt[3]{16x^5y^2z}$
$\sqrt[3]{2^4x^5y^2z}$
$\sqrt[3]{2^3x^3}\sqrt[3]{2x^2y^2z}$
$2x\sqrt[3]{2x^2y^2z}$

8. $f(x) = |x - 2|$
This is the graph of $y = |x|$ shifted two units to the right.
Intercepts:
$x = 0$: $y = |-2| = 2$; $(0,2)$
$y = 0$: $0 = |x - 2|$
$0 = x - 2$
$2 = x$; $(2,0)$

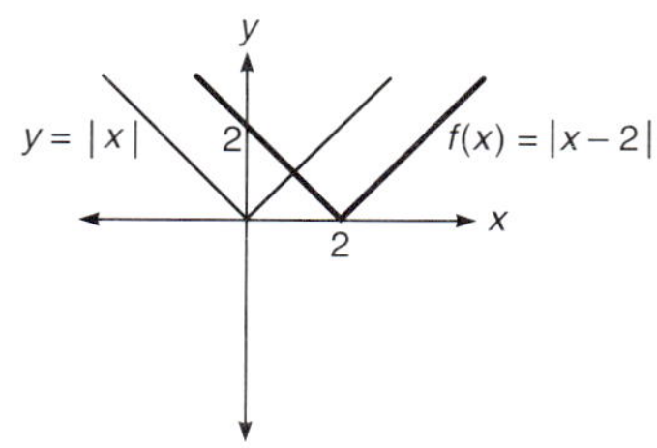

Solutions to trial exercise problems

22. $(x + 5)(y - 4) \geq xy$

$xy - 4x + 5y - 20 \geq xy$

$-4x + 5y - 20 \geq 0$

Thus the solution is also described by the statement $-4x + 5y - 20 \geq 0$.

Graph $-4x + 5y = 20$.

Test point (0,0): $-20 \geq 0$; false

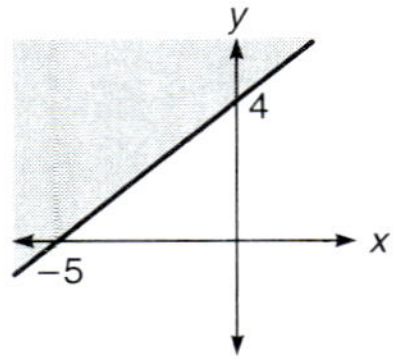

30.

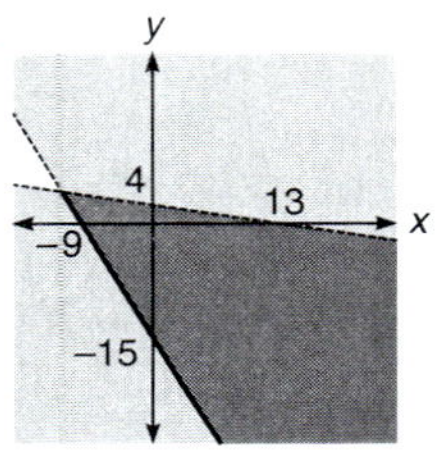

45. $-26x + 21y \leq 14$

$2x + y \leq 12$

$P = 2x + \frac{1}{3}y$

x	y	P
0	0	0
0	$\frac{2}{3}$	$\frac{2}{9}$
$\frac{7}{2}$	5	$8\frac{2}{3}$
6	0	12

Solution:

$P = 12$ at (6,0)

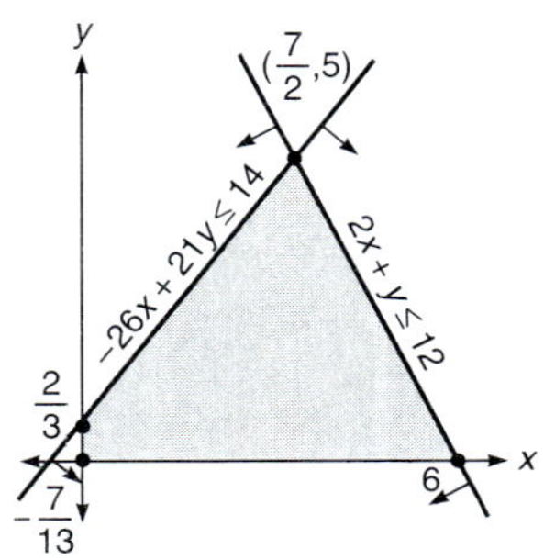

55. $-x + 2y \leq 4$

$x + 3y \leq 11$

$x + y \leq 7$

$P = x + 2y$

$x + 2y = P$

x	y	P
0	0	0
0	2	4
2	3	8
5	2	9
7	0	7

Solution:

$P = 9$ at (5,2)

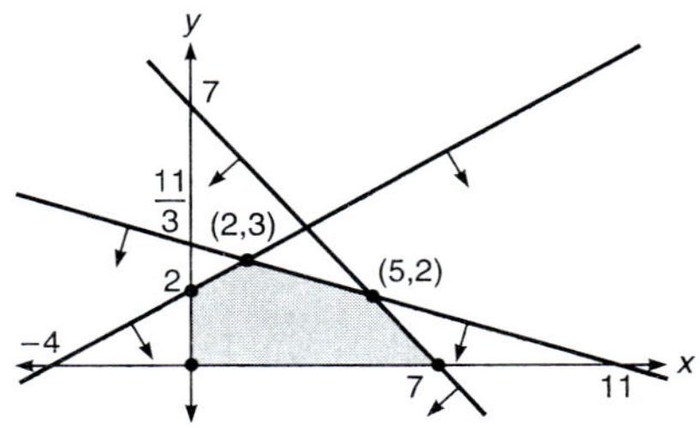

74. $x + y \geq 9$

$\frac{7}{2}x + y \geq \frac{23}{2}$

$x + 3y \geq 12$

$C = 2x + 3y$

	x	y	C
a	0	11.5	34.5
b	1	8	26
c	7.5	1.5	19.5
d	12	0	24

C is a minimum of 19.5 at (7.5,1.5).

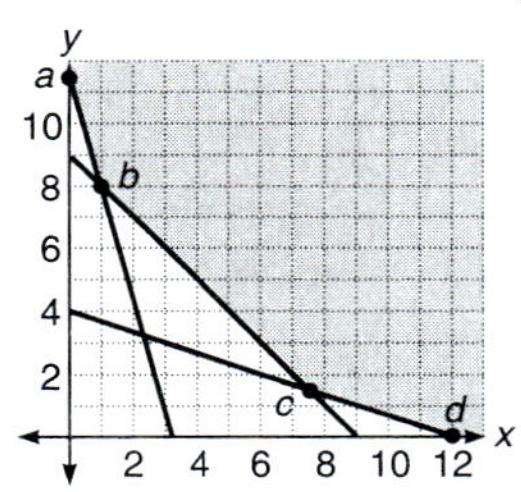

76. $x + y \geq 8$

$2x + 3y \geq 17$

$\frac{3}{2}x + y \geq 9$

$C = 5x + 4y$

	x	y	C
a	0	9	36
b	2	6	34
c	7	1	39
d	8.5	0	42.5

C is a minimum of 34 at (2,6).

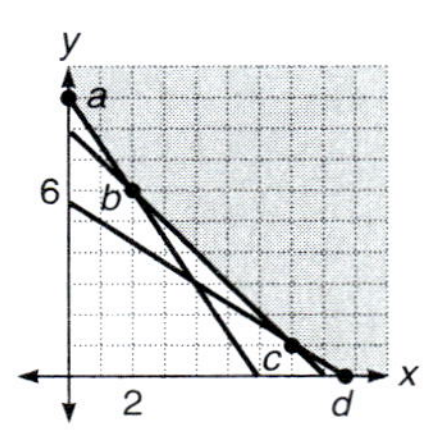

79. Let x = number of tables to produce per production run and y = number of chairs, and C = income per production run. Then $C = 29x + 10y$. The number of hours required to produce x tables is $3x$, and for chairs it is y hours. Therefore, $3x + y \leq 300$. The restrictions on finishing are $2x + \frac{5}{6}y \leq 200$.

Thus, we have the system

$3x + y \leq 300$

$2x + \frac{5}{6}y \leq 240$

$C = 29x + 10y$

$C(0,0) = 0$, $C(0,288) = 2{,}880$, $C(20,240) = 2{,}980$, $C(100,0) = 2{,}900$. The maximum value for C is 2,980 at (20,240). Thus, the maximum income is \$2,980 and comes from producing 20 tables and 240 chairs.

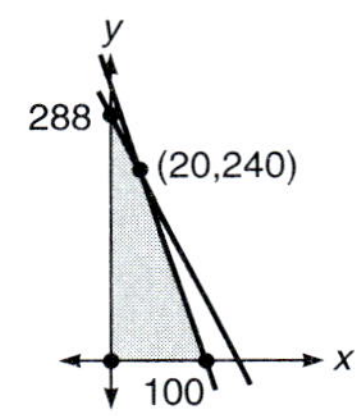

83. x = amount of A, y = amount of B. Based on protein we need $5x + 8y \geq 20$, and based on carbohydrates we need $4x + 3y \geq 12$. Total cost C is $C = 15x + 18y$. Minimize C for the system

$5x + 8y \geq 20$

$4x + 3y \geq 12$

$C = 15x + 18y$

$C(0,4) = 72$, $C(\frac{36}{17},\frac{20}{17}) = 52\frac{16}{17}$, $C(4,0) = 60$. Thus, we minimize cost at $52\frac{16}{17}$ cents by using $\frac{36}{17}$ lb of A and $\frac{20}{17}$ lb of B.

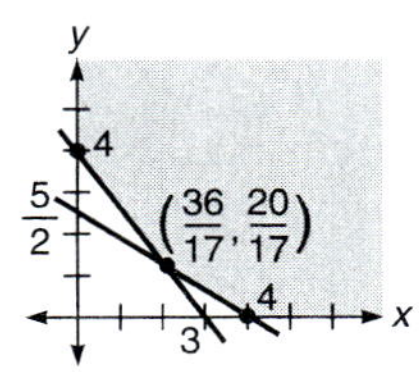

Exercise 10–5

Answers to odd-numbered problems

1. $\begin{bmatrix} 2 & 1 \\ -1 & 0 \end{bmatrix}$ **3.** $\begin{bmatrix} 0 & 0 \\ 0 & 0 \end{bmatrix}$

5. $\begin{bmatrix} -1 & 1 & -3 \\ 4 & -4 & 2 \end{bmatrix}$ **7.** $\begin{bmatrix} -3 & -1 & 4 \\ -4 & -18 & 7 \end{bmatrix}$

9. $\begin{bmatrix} -1 & 1 \\ -3 & 4 \\ -4 & 2 \end{bmatrix}$ **11.** $\begin{bmatrix} -3 & -1 \\ 4 & -4 \\ -18 & 7 \end{bmatrix}$

13. $\begin{bmatrix} -4 & 8 \\ 16 & 20 \end{bmatrix}$ **15.** $\begin{bmatrix} 0 & -15 \\ 10 & -25 \end{bmatrix}$

17. $\begin{bmatrix} 2 & \frac{1}{2} & -1 \\ 1 & 3 & -1 \end{bmatrix}$ **19.** $\begin{bmatrix} -2 & 13 \\ -5 & 0 \\ -3 & -2 \end{bmatrix}$

21. -26 **23.** -2 **25.** $6 - \pi$

27. There are an unlimited number of solutions; an obvious one is [0, 1, 0, 0].

29. $\begin{bmatrix} 1 & -3 \\ 5 & -21 \end{bmatrix}$ **31.** $\begin{bmatrix} -7 & 15 \\ -10 & 18 \end{bmatrix}$

33. $\begin{bmatrix} 9 & 9 & 2 \\ 13 & -2 & -9 \\ 27 & 0 & -4 \end{bmatrix}$ **35.** $\begin{bmatrix} 5 & 1 \\ -25 & 16 \\ -2 & -1 \end{bmatrix}$

37. $\begin{bmatrix} 6 & -2 \\ -6 & -8 \end{bmatrix}$ **39.** $\begin{bmatrix} 47 & 1 \\ 70 & -22 \end{bmatrix}$

41. $\begin{bmatrix} -7 & -30 & 22 \\ 8 & 33 & -26 \end{bmatrix}$

43. $\begin{bmatrix} -20x^2 + y & 15x + 9 \\ -16xy - 3y & 12y - 27 \end{bmatrix}$

45. $\begin{bmatrix} 19 & -2 \\ 13 & -55 \end{bmatrix}$ **47.** $\begin{bmatrix} 15 & -13 & 8 \\ -74 & -34 & -36 \end{bmatrix}$

49. $AB = \begin{bmatrix} 7 & -13 & 10 & 13 \\ 6 & -2 & -28 & 34 \\ -17 & 35 & -38 & -23 \end{bmatrix}$

$(AB)C = \begin{bmatrix} 93 & 63 & -69 \\ 226 & -42 & -114 \\ -171 & -189 & 147 \end{bmatrix}$

$BC = \begin{bmatrix} 49 & 21 & -33 \\ 5 & -21 & 3 \end{bmatrix}$

$A(BC) = \begin{bmatrix} 93 & 63 & -69 \\ 226 & -42 & -114 \\ -171 & -189 & 147 \end{bmatrix}$, so $(AB)C = A(BC)$.

51. $\begin{bmatrix} \frac{1}{27} & -\frac{5}{27} \\ -\frac{1}{9} & -\frac{4}{9} \end{bmatrix}$ **53.** $\begin{bmatrix} -\frac{3}{10} & \frac{1}{20} \\ \frac{1}{5} & \frac{3}{10} \end{bmatrix}$

55. $\begin{bmatrix} \frac{1}{10} & \frac{2}{5} & \frac{1}{10} \\ -\frac{1}{3} & 0 & 0 \\ \frac{1}{5} & -\frac{1}{5} & \frac{1}{5} \end{bmatrix}$

57. $\begin{bmatrix} \frac{1}{7} & \frac{3}{7} & 0 \\ -\frac{2}{21} & -\frac{2}{7} & \frac{1}{3} \\ \frac{2}{7} & -\frac{1}{7} & 0 \end{bmatrix}$

59. $\begin{bmatrix} -\frac{5}{37} & \frac{1}{37} & -\frac{7}{37} & 0 \\ -\frac{13}{37} & \frac{10}{37} & \frac{4}{37} & 0 \\ -\frac{11}{222} & \frac{17}{222} & -\frac{4}{111} & -\frac{1}{6} \\ -\frac{1}{37} & \frac{15}{37} & \frac{6}{37} & 0 \end{bmatrix}$

61. $\begin{bmatrix} -3 & \frac{6}{7} & -\frac{11}{7} & 2 \\ 3 & -1 & 2 & -2 \\ -\frac{1}{2} & \frac{1}{7} & -\frac{3}{7} & \frac{1}{2} \\ 3 & -\frac{4}{7} & \frac{12}{7} & -2 \end{bmatrix}$

63. $x = \frac{1}{3}, y = 2$ **65.** $x = 2, y = -2$

67. $x = 2, y = -1, z = 3$

69. $x = -3, y = \frac{2}{3}, z = -1$

71. $x = 1, y = -2, z = 2, w = 1$

73. $x = 1, y = 3, z = -1, w = 2$

75. $x = -3, y = 3$ **77.** $x = \frac{1}{2}, y = 2$

79. $x = -2, y = 3, z = 2$

81. $x = 2, y = -4, z = \frac{1}{2}$

83. $\begin{bmatrix} -24 & 18 \\ 32 & -14 \end{bmatrix}$; **85.** $\begin{bmatrix} 4 & 3 \\ 10 & 5 \end{bmatrix}$

87. $\begin{bmatrix} 34 & -42 \\ -35 & 55 \end{bmatrix}$ **89.** $\begin{bmatrix} 4 & 1 & 2 & 4 \\ 1 & 2 & 2 & 2 \\ 2 & 2 & 3 & 3 \\ 4 & 2 & 3 & 5 \end{bmatrix}$

91. $\begin{bmatrix} 1 & 1 & 1 & 0 \\ 0 & 1 & 1 & 1 \\ 0 & 0 & 0 & 0 \\ 1 & 0 & 0 & 0 \end{bmatrix}$; The row 1 column 2 entry is 1 so there is one path of length 2 from node 1 to node 2.

93. $\begin{bmatrix} 1 & \boxed{1} & 1 & 1 \\ 1 & 1 & 1 & 0 \\ 0 & 0 & 0 & 0 \\ 0 & 1 & 1 & 1 \end{bmatrix}$; 1 path of length 3 from node 1 to node 2.

95. $\begin{bmatrix} \frac{3}{4} & 0 & 0 & \frac{1}{4} \\ 0 & \frac{3}{4} & \frac{1}{4} & 0 \\ 0 & \frac{1}{2} & \frac{1}{2} & 0 \\ \frac{1}{2} & 0 & 0 & \frac{1}{2} \end{bmatrix}$

The probability that a mouse which started in room 2 is in room 3 after two moves is $\frac{1}{4}$, the entry in row 2 column 3.

97. $L^3V = \begin{bmatrix} 6{,}498 \\ 296 \\ 200 \end{bmatrix}$

Thus, there are 6,498, 296, and 200 females in each stage after three life cycles.

Solutions to skill and review problems

1. Graph the lines $2x + y = 6$ and $y = 3x - 1$. Use a test point for each half-plane. The origin (0,0) is the easiest. Since $y = 3x - 1$ is part of the solution set where it overlaps the solution to $2x + y < 6$, this part of the line is made darker.

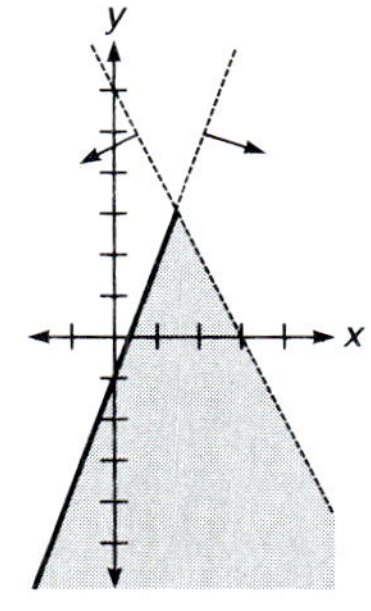

2. Use equation [1] to remove the variable z from equations [2] and [3]:

[1] $2x - y - 2z = -7$

[2] $x + y + 4z = 2$

[3] $3x + 2y - 2z = -3$

Add twice [1] to [2].

[4] $5x - y = -12$

Subtract [1] from [3].

[5] $x + 3y = 4$

Remove y from equations [4] and [5]:

Add 3 times [4] to [5].

$16x = -32$

$x = -2$

Substitute x into equation [4]:

$5(-2) - y = -12$

$2 = y$

Substitute x and y into equation [1]:

$2(-2) - 2 - 2z = -7$

$1 = 2z$

$\frac{1}{2} = z$

Thus, $x = -2, y = 2, z = \frac{1}{2}$.

3. Solve

[1] $2x + 3y = -6$

[2] $x - 4y = 8$

Subtract twice [2] from [1].

$11y = -22$

$y = -2$

Substitute y into equation [2]:

$x - 4(-2) = 8$

$x = 0$

Thus $x = 0$, $y = -2$, and the point is $(0, -2)$.

4. Let $P_1 = (x_1,y_1) = (-2,4)$; $P_2 = (x_2,y_2) = (3,8)$.

$$m = \frac{y_2 - y_1}{x_2 - x_1} = \frac{8-4}{3-(-2)} = \frac{4}{5}$$

$$y - y_1 = m(x - x_1)$$
$$y - 4 = \tfrac{4}{5}(x - (-2))$$
$$5y - 20 = 4(x + 2)$$
$$5y - 4x - 28 = 0$$

5. $(x-h)^2 + (y-k)^2 = r^2$, where (h,k) is the center and r = radius.
r = distance from center $(-2,4)$ to $(3,8)$

$$= \sqrt{(3-(-2))^2 + (8-4)^2}$$
$$= \sqrt{41}$$
$$(x-(-2))^2 + (y-4)^2 = (\sqrt{41})^2$$
$$(x+2)^2 + (y-4)^2 = 41$$

6. $$\sqrt{\frac{3}{8x^5y}} = \frac{\sqrt{3}}{\sqrt{8x^5y}} = \frac{\sqrt{3}}{2x^2\sqrt{2xy}} \cdot \frac{\sqrt{2xy}}{\sqrt{2xy}} = \frac{\sqrt{6xy}}{2x^2(2xy)} = \frac{\sqrt{6xy}}{4x^3y}$$

7. $$81x^4 - 1$$
$$(9x^2 - 1)(9x^2 + 1)$$
$$(3x - 1)(3x + 1)(9x^2 + 1)$$

8. $$\frac{3a}{2b} - \frac{5a}{3c} + \frac{1}{a}$$
$$\frac{3a(3c) - 5a(2b)}{2b(3c)} + \frac{1}{a}$$
$$\frac{9ac - 10ab}{6bc} + \frac{1}{a}$$
$$\frac{a(9ac - 10ab) + 1(6bc)}{a(6bc)}$$
$$\frac{9a^2c - 10a^2b + 6bc}{6abc}$$

Solutions to trial exercise problems

8. $$\begin{bmatrix} -5 & 2 & 6 \\ 1 & 2 & 1 \end{bmatrix} + \begin{bmatrix} 13 & -12 & 5 \\ 10 & 1 & 0 \end{bmatrix} = \begin{bmatrix} -5+13 & 2-12 & 6+5 \\ 1+10 & 2+1 & 1+0 \end{bmatrix} = \begin{bmatrix} 8 & -10 & 11 \\ 11 & 3 & 1 \end{bmatrix}$$

18. $$\frac{2}{3}\begin{bmatrix} -15 & 9 & 3 \\ 1 & 6 & 1 \end{bmatrix} = \begin{bmatrix} \frac{2}{3}(-15) & \frac{2}{3}(9) & \frac{2}{3}(3) \\ \frac{2}{3}(1) & \frac{2}{3}(6) & \frac{2}{3}(1) \end{bmatrix} = \begin{bmatrix} -10 & 6 & 2 \\ \frac{2}{3} & 4 & \frac{2}{3} \end{bmatrix}$$

28. We want a vector $[a, b, c, d]$ such that $[5, 2, -4, 3][a, b, c, d] = \frac{1}{2}$. Of the unlimited number of possibilities, an obvious choice is $[0, \frac{1}{4}, 0, 0]$. Another is $[0, 0, -\frac{1}{8}, 0]$.

35. $$\begin{bmatrix} 1 & -1 & 0 \\ 2 & 3 & -2 \\ -3 & 5 & 1 \end{bmatrix}\begin{bmatrix} 4 & 3 \\ -1 & 2 \\ 15 & -2 \end{bmatrix} = \begin{bmatrix} (1)(4) + (-1)(-1) + (0)(15) & (1)(3) + (-1)(2) + (0)(-2) \\ (2)(4) + (3)(-1) + (-2)(15) & (2)(3) + (3)(2) + (-2)(-2) \\ (-3)(4) + (5)(-1) + (1)(15) & (-3)(3) + (5)(2) + (1)(-2) \end{bmatrix} = \begin{bmatrix} 5 & 1 \\ -25 & 16 \\ -2 & -1 \end{bmatrix}$$

61. $$\begin{bmatrix} 0 & 0 & 4 & 1 & 1 & 0 & 0 & 0 \\ 2 & -1 & 0 & 3 & 0 & 1 & 0 & 0 \\ 3 & 2 & 0 & 1 & 0 & 0 & 1 & 0 \\ 2 & 2 & 6 & 1 & 0 & 0 & 0 & 1 \end{bmatrix} \begin{matrix} \\ \\ \\ [4] \leftarrow 3[1] + -2[4] \end{matrix}$$

$$\begin{bmatrix} 0 & 0 & 4 & 1 & 1 & 0 & 0 & 0 \\ 2 & -1 & 0 & 3 & 0 & 1 & 0 & 0 \\ 3 & 2 & 0 & 1 & 0 & 0 & 1 & 0 \\ -4 & -4 & 0 & 1 & 3 & 0 & 0 & -2 \end{bmatrix} \begin{matrix} \\ \\ [3] \leftarrow 2[2] + 1[3] \\ [4] \leftarrow -4[2] + 1[4] \end{matrix}$$

$$\begin{bmatrix} 0 & 0 & 4 & 1 & 1 & 0 & 0 & 0 \\ 2 & -1 & 0 & 3 & 0 & 1 & 0 & 0 \\ 7 & 0 & 0 & 7 & 0 & 2 & 1 & 0 \\ -12 & 0 & 0 & -11 & 3 & -4 & 0 & -2 \end{bmatrix} \begin{matrix} \\ [2] \leftarrow 2[3] + -7[2] \\ \\ [4] \leftarrow 12[3] + 7[4] \end{matrix}$$

$$\begin{bmatrix} 0 & 0 & 4 & 1 & 1 & 0 & 0 & 0 \\ 0 & 7 & 0 & -7 & 0 & -3 & 2 & 0 \\ 7 & 0 & 0 & 7 & 0 & 2 & 1 & 0 \\ 0 & 0 & 0 & 7 & 21 & -4 & 12 & -14 \end{bmatrix} \begin{matrix} [1] \leftarrow 1[4] + -7[1] \\ [2] \leftarrow 1[4] + 1[2] \\ [3] \leftarrow 1[4] + -1[3] \\ \\ \end{matrix}$$

$$\begin{bmatrix} 0 & 0 & -28 & 0 & 14 & -4 & 12 & -14 \\ 0 & 7 & 0 & 0 & 21 & -7 & 14 & -14 \\ -7 & 0 & 0 & 0 & 21 & -6 & 11 & -14 \\ 0 & 0 & 0 & 7 & 21 & -4 & 12 & -14 \end{bmatrix}$$ Rearrange rows and set coefficients to 1

$$\begin{bmatrix} 1 & 0 & 0 & 0 & -3 & \frac{6}{7} & -\frac{11}{7} & 2 \\ 0 & 1 & 0 & 0 & 3 & -1 & 2 & -2 \\ 0 & 0 & 1 & 0 & -\frac{1}{2} & \frac{1}{7} & -\frac{3}{7} & \frac{1}{2} \\ 0 & 0 & 0 & 1 & 3 & -\frac{4}{7} & \frac{12}{7} & -2 \end{bmatrix}$$

The inverse is $$\begin{bmatrix} -3 & \frac{6}{7} & -\frac{11}{7} & 2 \\ 3 & -1 & 2 & -2 \\ -\frac{1}{2} & \frac{1}{7} & -\frac{3}{7} & \frac{1}{2} \\ 3 & -\frac{4}{7} & \frac{12}{7} & -2 \end{bmatrix}$$

71. The inverse matrix is the answer to problem 59.

$$\begin{bmatrix} -\frac{5}{37} & \frac{1}{37} & -\frac{7}{37} & 0 \\ -\frac{13}{37} & \frac{10}{37} & \frac{4}{37} & 0 \\ -\frac{11}{222} & \frac{17}{222} & -\frac{4}{111} & -\frac{1}{6} \\ -\frac{1}{37} & \frac{15}{37} & \frac{6}{37} & 0 \end{bmatrix}\begin{bmatrix} 8 \\ 7 \\ -10 \\ -9 \end{bmatrix} = \begin{bmatrix} 1 \\ -2 \\ 2 \\ 1 \end{bmatrix};$$

81. The inverse of $\begin{bmatrix} -1 & -1 & -2 \\ 1 & 1 & -4 \\ 0 & -\frac{1}{2} & 2 \end{bmatrix}$ is $\begin{bmatrix} 0 & 1 & 2 \\ -\frac{2}{3} & -\frac{2}{3} & -2 \\ -\frac{1}{6} & -\frac{1}{6} & 0 \end{bmatrix}$.

$$\begin{bmatrix} 0 & 1 & 2 \\ -\frac{2}{3} & -\frac{2}{3} & -2 \\ -\frac{1}{6} & -\frac{1}{6} & 0 \end{bmatrix}\begin{bmatrix} 1 \\ -4 \\ 3 \end{bmatrix} = \begin{bmatrix} 2 \\ -4 \\ \frac{1}{2} \end{bmatrix}; x = 2, y = -4, z = \frac{1}{2}$$

86. $-3\begin{bmatrix}2 & -3\\ -1 & 4\end{bmatrix}^2 - 2\begin{bmatrix}0 & 3\\ 4 & -1\end{bmatrix}^2 - 5\begin{bmatrix}0 & 3\\ 4 & -1\end{bmatrix}$

$= -3\begin{bmatrix}7 & -18\\ -6 & 19\end{bmatrix} - 2\begin{bmatrix}12 & -3\\ -4 & 13\end{bmatrix} - \begin{bmatrix}0 & 15\\ 20 & -5\end{bmatrix}$

$= \begin{bmatrix}-21 & 54\\ 18 & -57\end{bmatrix} - \begin{bmatrix}24 & -6\\ -8 & 26\end{bmatrix} - \begin{bmatrix}0 & 15\\ 20 & -5\end{bmatrix} = \begin{bmatrix}-45 & 45\\ 6 & -78\end{bmatrix}$

91. $\begin{bmatrix}0&1&1&1\\1&0&0&0\\0&0&0&0\\0&1&1&0\end{bmatrix}\begin{bmatrix}0&1&1&1\\1&0&0&0\\0&0&0&0\\0&1&1&0\end{bmatrix} = \begin{bmatrix}1&1&1&0\\0&1&1&1\\0&0&0&0\\1&0&0&0\end{bmatrix}$; The row 1 column 2 entry is 1 so there is one path of length 2 from node 1 to node 2.

95. $A^2 = AA = \begin{bmatrix}0&\frac{1}{2}&\frac{1}{2}&0\\ \frac{1}{2}&0&0&\frac{1}{2}\\ 1&0&0&0\\ 0&1&0&0\end{bmatrix}\begin{bmatrix}0&\frac{1}{2}&\frac{1}{2}&0\\ \frac{1}{2}&0&0&\frac{1}{2}\\ 1&0&0&0\\ 0&1&0&0\end{bmatrix} = \begin{bmatrix}\frac{3}{4}&0&0&\frac{1}{4}\\ 0&\frac{3}{4}&\frac{1}{4}&0\\ 0&\frac{1}{2}&\frac{1}{2}&0\\ \frac{1}{2}&0&0&\frac{1}{2}\end{bmatrix}$ The probability that a mouse that started in room 2 is in room 3 after two moves is $\frac{1}{4}$, the entry in row 2 column 3.

Chapter 10 review

1. $(-3,\frac{3}{2})$ **2.** (1,2) **3.** (−12,2)
4. $(-8,\frac{3}{5})$ **5.** (−2,3,1) **6.** (1,−3,−2)
7. (3,2,−2,1) **8.** (−1,−2,3,1)
9. $L = 56$ cm, $W = 35$ cm
10. $L = 52$ in., $W = 27$ in.
11. \$4,500 at 6%, \$10,500 at 12%
12. $y = 2x^2 - \frac{1}{2}x + 3$ **13.** $(\frac{3}{2},2)$
14. (−1,−3) **15.** (−2,0,−3)
16. dependent **17.** (1,−4,0,−2)
18. $(1,-4,\frac{1}{3},2)$ **19.** $i_1 = 2\frac{3}{5}$, $i_2 = -4\frac{7}{10}$, $i_3 = 15\frac{4}{5}$ **20.** $666\frac{2}{3}$ gallons of 8% solution and $333\frac{1}{3}$ gallons of 20% solution **21.** $x = -\frac{16}{23}$, $y = \frac{12}{23}$
22. $x = \dfrac{571}{304}$, $y = \dfrac{2{,}379}{304}$
23. $x = \dfrac{115}{59}$, $y = -\dfrac{60}{59}$, $z = \dfrac{43}{59}$
24. $x = \frac{1}{4}$, $y = -\frac{1}{2}$, $z = \frac{7}{8}$
25. $x = \frac{23}{18}$, $y = \frac{59}{108}$, $z = \frac{1}{27}$, $w = -\frac{43}{54}$
26. $x = \frac{16}{25}$, $y = -\frac{6}{25}$, $z = -\frac{46}{25}$, $w = -\frac{18}{25}$
27. $D = \frac{6}{7}$; complete solution: $(-\frac{58}{7},-\frac{367}{28},\frac{87}{28},\frac{6}{7},\frac{647}{28})$ **28.** $9\frac{1}{4}$

29.
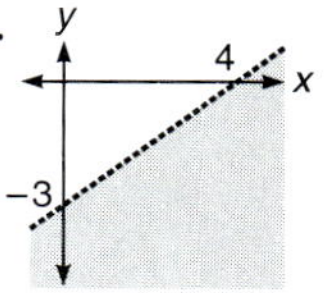

30.
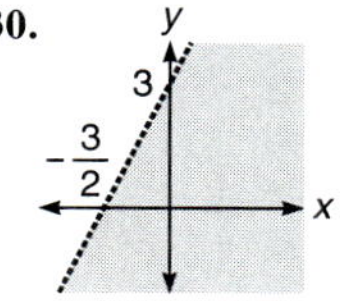

31.
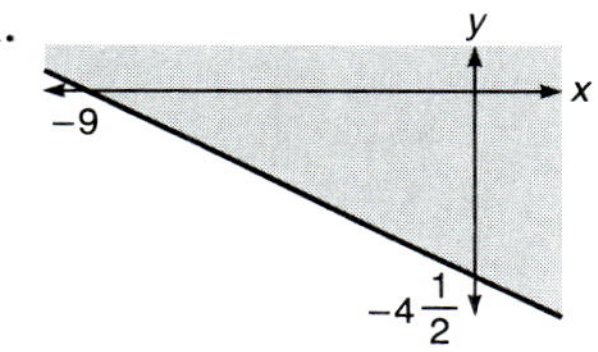

32.
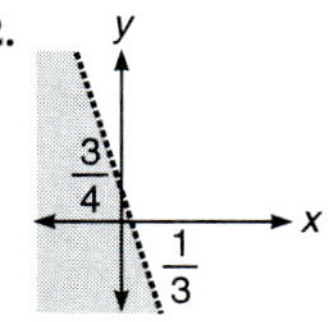

33.
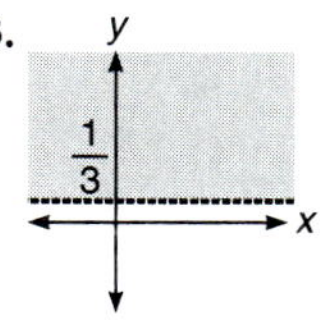

34.
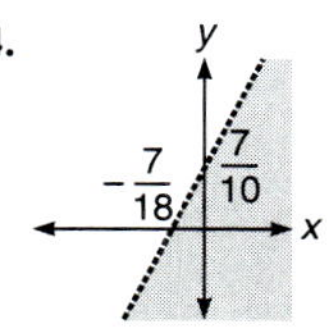

35.
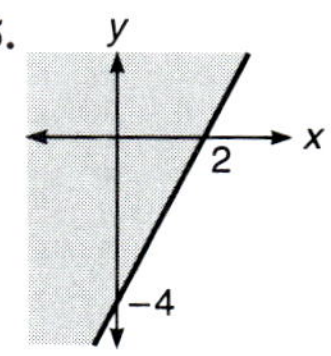

36.
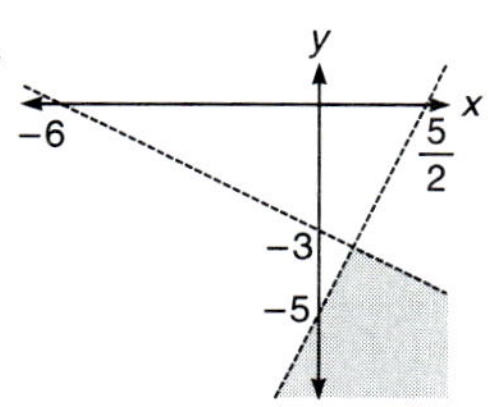

37.
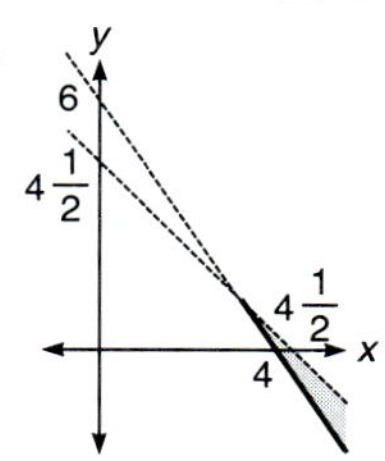

38.
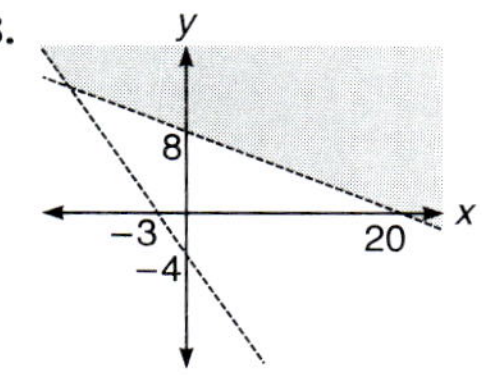

39. P is maximized for $x = 5\frac{1}{3}$, $y = 0$; Its value is $21\frac{1}{3}$.
40. P is maximized at either of the points $(2,2\frac{2}{3})$ or (4,2); its value is 10.
41. Income is maximized at \$225 by producing 75 tables and no chairs per run.
42. Production is maximized at 580 trees by using 20 one-supervisor crews and $6\frac{2}{3}$ three-supervisor crews.
43. $4\frac{3}{4}$ **44.** −17 **45.** −13
46. $6 - 3\sqrt{\pi}$
47. There are an unlimited number of solutions; one is $(0,0,\frac{1}{2},0)$.

48. $\begin{bmatrix}-11 & -22 & 32\\ 0 & 0 & -6\end{bmatrix}$

49. $\begin{bmatrix}-4x^2 + 2y & -3x + 18\\ 16xy - 3y & 12y - 27\end{bmatrix}$

50. $\begin{bmatrix}-15 & -4 & 15 & \frac{3}{4}\\ -21 & -34 & \frac{131}{2} & \frac{83}{8}\\ -7 & 8 & 6 & \frac{3}{2}\\ -44 & -76 & 105 & \frac{45}{4}\end{bmatrix}$

51. $\begin{bmatrix}\frac{1}{17} & -\frac{5}{17}\\ -\frac{3}{17} & -\frac{2}{17}\end{bmatrix}$.

52. $\begin{bmatrix} \frac{5}{38} & \frac{6}{19} & \frac{3}{38} \\ -\frac{11}{38} & \frac{2}{19} & \frac{1}{38} \\ \frac{5}{19} & -\frac{7}{19} & \frac{3}{19} \end{bmatrix}$

53. $x = -1, y = -1$
54. $x = 5, y = 5, z = -1$

Chapter 10 test

1. $(\frac{1}{2}, \frac{2}{3})$ **2.** $(2, \frac{1}{2}, -1)$ **3.** L is length, W is width: $(L, W) = (80'', 24'')$ **4.** S is amount invested at 6%, T = amount invested at 10%, (S, T) = ($4,500, $7,500). **5.** $y = 2x^2 - x + 3$ **6.** $(2, -2, 3)$ **7.** $(0, -2, 1, -1)$ **8.** Let T = amount of 30% solution to use, and S = amount of 70% solution; (T, S) = (312.5 gallons, 187.5 gallons) **9.** $x = \frac{13}{4}, y = \frac{9}{4}$ **10.** $D = 0, D_x = 0, D_y = 0, D_z = 0$. Since all the determinants are 0, the system is dependent. **11.** $8\frac{1}{2}$

12.

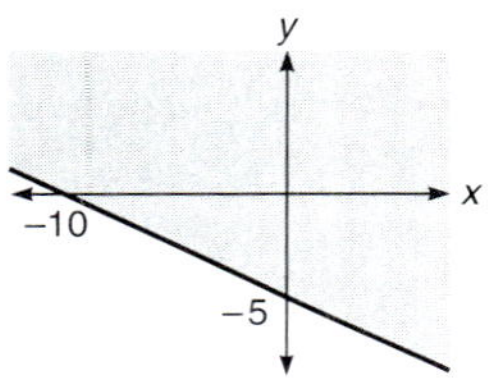

13.

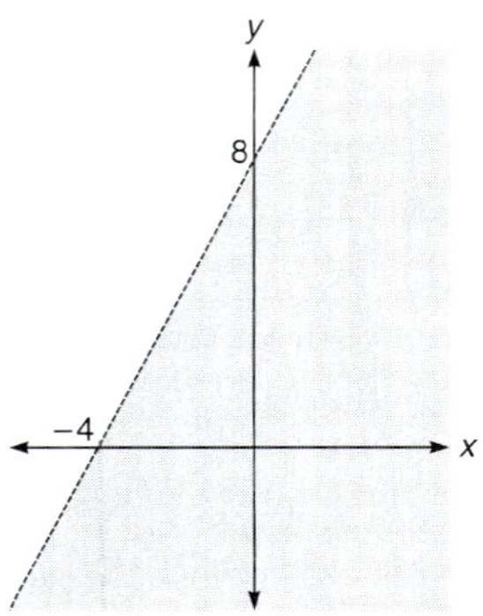

14.

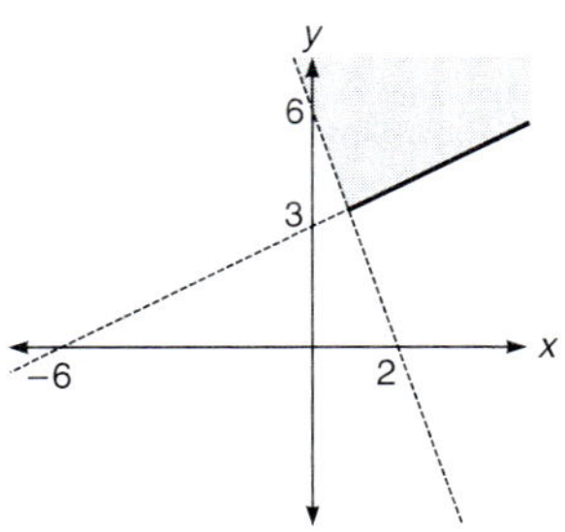

15. C is maximized at $x = 4\frac{1}{5}, y = 1\frac{3}{5}$, with a value of $7\frac{2}{5}$
16. $\frac{12}{19}$ gm of Prime, $2\frac{18}{19}$ gm of Regular
17. 12 of the first type of crews and $5\frac{1}{3}$ of the second type of crews. $429\frac{1}{3}$ trees logged per day.

18. −17 **19.** 17 **20.** $\begin{bmatrix} 3 & -10 & 20 \\ 8 & 8 & -16 \end{bmatrix}$

21. $\begin{bmatrix} -3 & 1 \\ 14 & -22 \end{bmatrix}$ **22.** $\begin{bmatrix} 0 & -\frac{1}{3} \\ \frac{1}{5} & \frac{2}{15} \end{bmatrix}$

23. $\begin{bmatrix} \frac{1}{3} & \frac{5}{9} & -\frac{2}{9} \\ \frac{1}{3} & -\frac{1}{9} & -\frac{5}{9} \\ \frac{1}{3} & -\frac{4}{9} & -\frac{2}{9} \end{bmatrix}$

24. $x = -\frac{4}{9}, y = -\frac{10}{9}$
25. $a = 1, b = -2, c = 2, d = 3$

26. $A^2 = \begin{bmatrix} 1 & 1 & 1 & 0 & 2 \\ 0 & 0 & 0 & 1 & 0 \\ 0 & 0 & 0 & 1 & 1 \\ 0 & 2 & 1 & 1 & 1 \\ 1 & 0 & 1 & 0 & 0 \end{bmatrix}$.

The entry in $A^2_{1,5}$ is 2, so there are 2 paths of length 2 from node 1 to node 5.

Chapter 11

Exercise 11–1

Answers to odd-numbered problems

1. focus: $(0, -\frac{1}{8})$; directrix: $y = \frac{1}{8}$

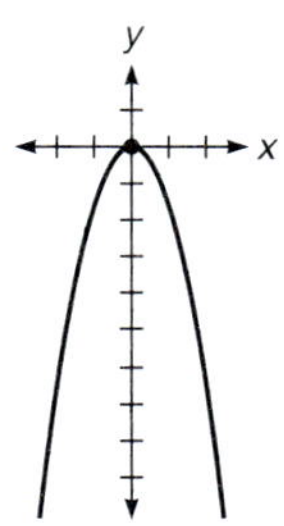

3. focus: $(0, \frac{1}{12})$; directrix: $y = -\frac{1}{12}$

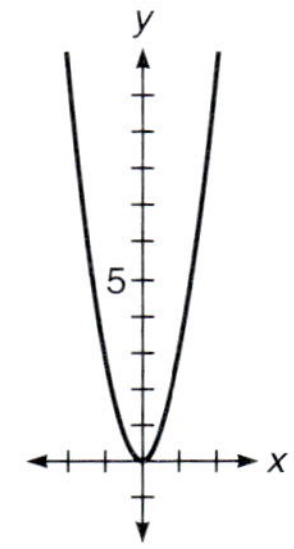

5. focus: $(0, -3\frac{3}{4})$; directrix: $y = -4\frac{1}{4}$

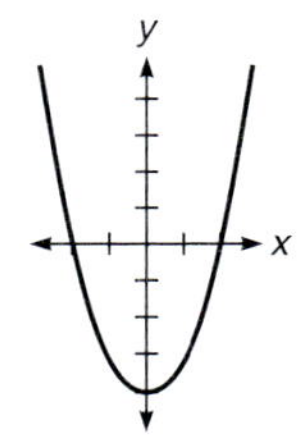

7. focus: $(0, \frac{3}{4})$; directrix: $y = 1\frac{1}{4}$

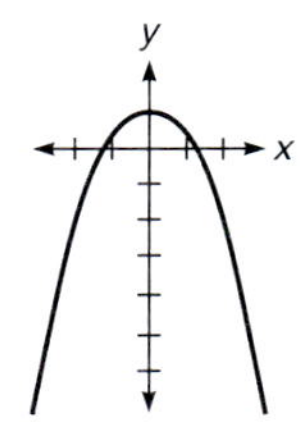

9. focus: $(3, \frac{1}{8})$; directrix: $y = -\frac{1}{8}$

11. focus: $(-2,\frac{1}{2})$; directrix: $y = -\frac{1}{2}$

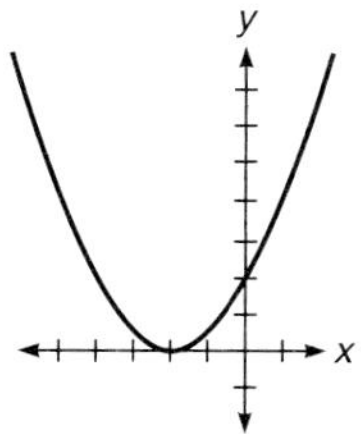

13. focus: $(-1,-\frac{1}{4})$; directrix: $y = \frac{1}{4}$

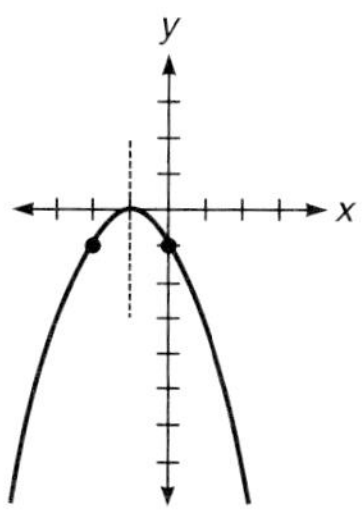

15. focus: $(\frac{1}{2},-\frac{11}{12})$; directrix: $y = -\frac{13}{12}$;
intercepts: $(0,-\frac{1}{4})$, $(\frac{1}{2} \pm \frac{\sqrt{3}}{3},0)$;
vertex: $(\frac{1}{2},-1)$

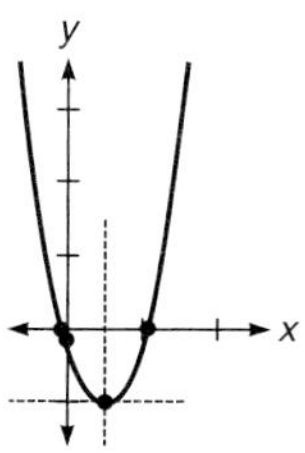

17. focus: $(-2,\frac{7}{8})$; directrix: $y = 1\frac{1}{8}$;
intercepts: $(0,-7)$, $(-2 \pm \sqrt{\frac{1}{2}},0)$

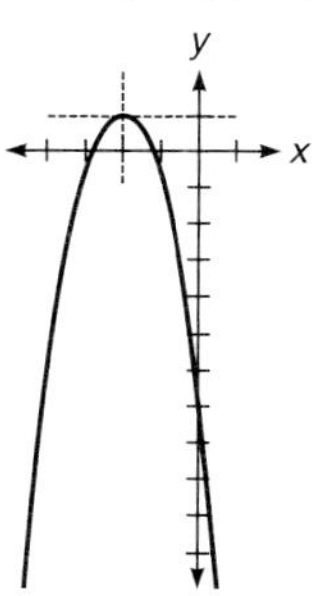

19. focus: $(-2\frac{1}{2},9)$; directrix: $y = 8\frac{1}{2}$;
vertex: $(-2\frac{1}{2},8\frac{3}{4})$

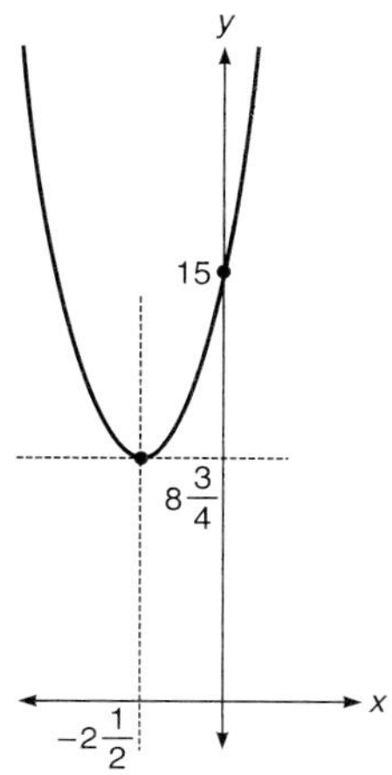

21. focus: $(3,1\frac{3}{4})$; directrix: $y = 2\frac{1}{4}$;
intercepts: $(0,-7)$, $(3 - \sqrt{2},0)$,
$(3 + \sqrt{2},0)$; vertex: $(3,2)$

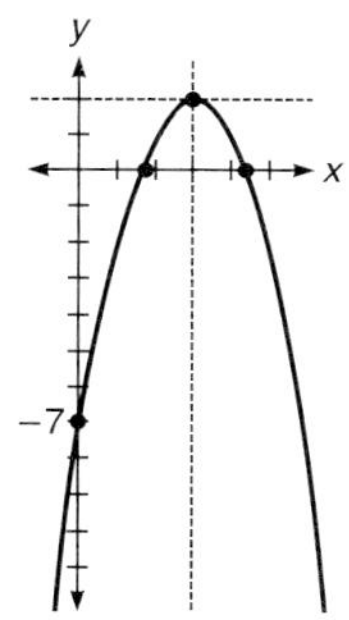

23. focus: $(-1,\frac{1}{4})$; directrix: $y = -\frac{1}{4}$

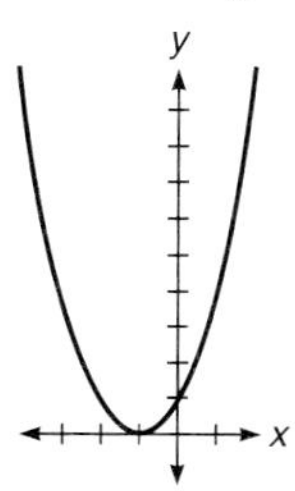

25. focus: $(\frac{1}{4},-3)$; directrix: $y = -3\frac{1}{4}$;
intercepts: $(0,-3)$, $(-1,0)$, $(1\frac{1}{2},0)$;
vertex: $(\frac{1}{4},-3\frac{1}{8})$

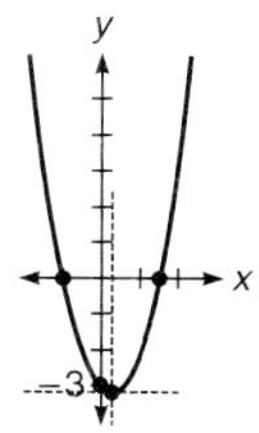

27. focus: $(0,15\frac{3}{4})$; directrix: $y = 16\frac{1}{4}$

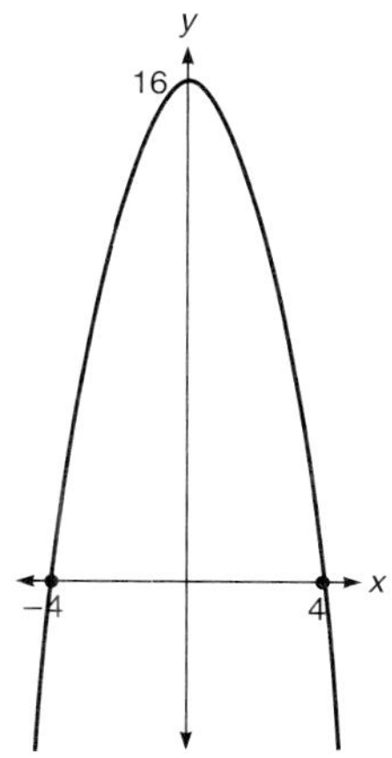

29. vertex: $(1\frac{1}{2},-7\frac{1}{4})$; focus: $(1\frac{1}{2},-7)$;
directrix: $y = -7\frac{1}{2}$; intercepts: $(0,-5)$,
$\left(\frac{3 - \sqrt{29}}{2},0\right)$, $\left(\frac{3 + \sqrt{29}}{2},0\right)$

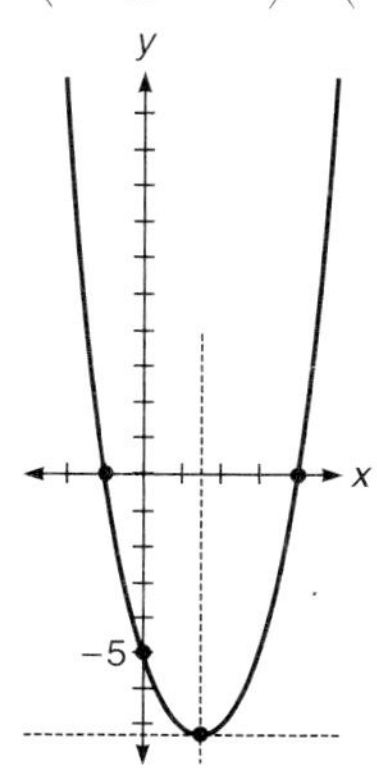

31. vertex: $(1\frac{1}{2},-5\frac{1}{2})$; focus: $(1\frac{1}{2},-5\frac{3}{8})$; directrix: $y = -5\frac{5}{8}$; intercepts: $(0,-1)$, $\left(\frac{3}{2} - \frac{\sqrt{11}}{2},0\right)$, $\left(\frac{3}{2} + \frac{\sqrt{11}}{2},0\right)$

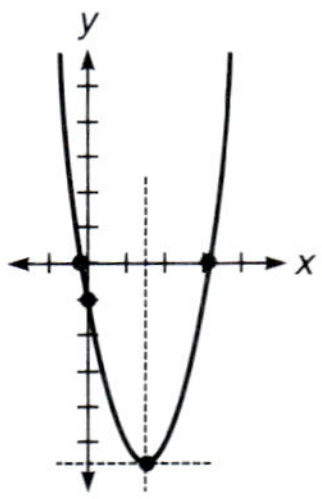

33. vertex: $(-\frac{2}{3},8\frac{1}{3})$; focus: $(-\frac{2}{3},8\frac{1}{4})$; directrix: $y = 8\frac{5}{12}$; intercepts: $(0,7)$, $(-2\frac{1}{3},0)$, $(1,0)$

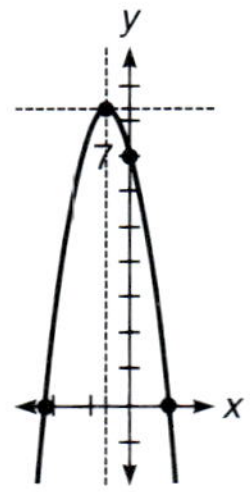

35. vertex: $(-20\frac{1}{4},3\frac{1}{2})$; focus: $(-20,3\frac{1}{2})$; directrix: $x = -20\frac{1}{2}$; intercepts: $(-8,0)$, $(0,8)$, $(0,-1)$

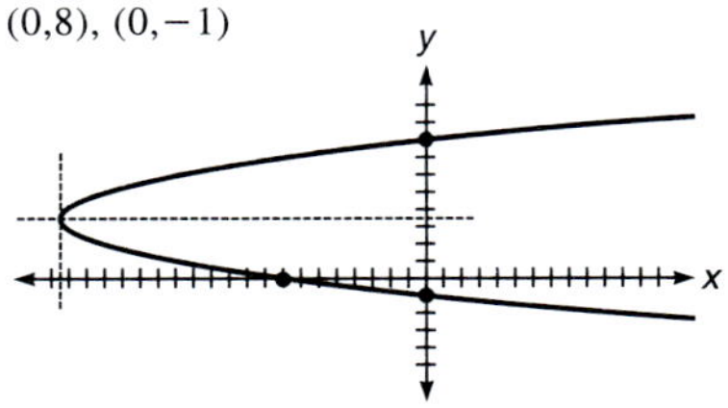

37. vertex: $(-9,0)$; focus: $(-8\frac{3}{4},0)$; directrix: $x = -9\frac{1}{4}$; intercepts: $(-9,0)$, $(0,\pm 3)$

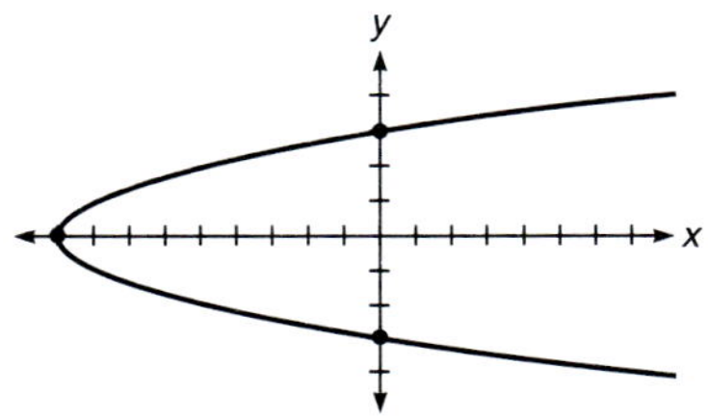

39. vertex: $(2\frac{1}{4},\frac{1}{2})$; focus: $(2,\frac{1}{2})$; directrix: $x = 2\frac{1}{2}$; intercepts: $(2,0)$, $(0,2)$, $(0,-1)$

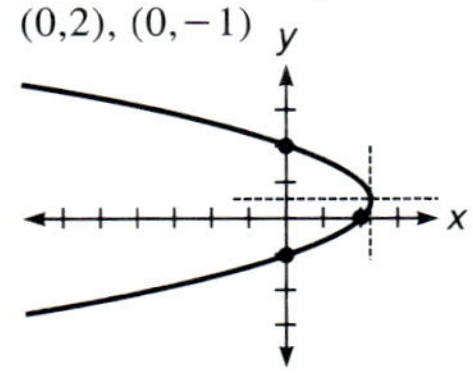

41. $y = \frac{1}{6}x^2 - \frac{2}{3}x - \frac{23}{6}$
43. $y = \frac{1}{6}x^2 - \frac{2}{3}x - \frac{23}{6}$
45. $y = \frac{1}{8}x^2 - \frac{3}{4}x + \frac{1}{8}$
47. $y = \frac{1}{12}x^2$ **49.** $y = x^2 - 6x + 8$
51. $32\sqrt{3}$ **53.** $\frac{9}{5}$ **55.** $\frac{9}{5}$
57. $y = \frac{3}{3{,}125}x^2$
59.

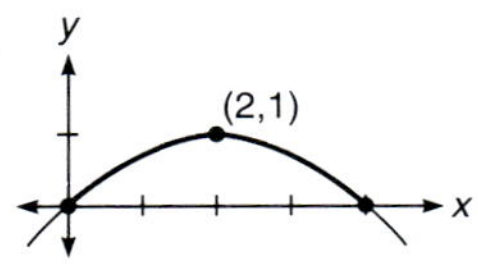

61. $4\sqrt{10} \approx 12.6$ feet; the horizontal distance traveled did double also
63. $\frac{4}{5}\sqrt{10} \approx 2.5$ ft/s

Solutions to skill and review problems

1. $\frac{x^2}{4} + 3y^2 = 1$

$3y^2 = 1 - \frac{x^2}{4}$

$3y^2 = \frac{4 - x^2}{4}$

$y^2 = \frac{4 - x^2}{12}$

$y = \pm\sqrt{\frac{4 - x^2}{12}} = \pm\frac{\sqrt{4 - x^2}}{2\sqrt{3}} \cdot \frac{\sqrt{3}}{\sqrt{3}} = \pm\frac{\sqrt{3}\sqrt{4 - x^2}}{6}$

2. $\begin{bmatrix} -2(-2) + 3(1) + 0(-3) & -2(3) + 3(5) + 0(2) \\ 1(-2) + 5(1) - 3(-3) & 1(3) + 5(5) - 3(2) \\ 4(-2) + 2(1) + 6(-3) & 4(3) + 2(5) + 6(2) \end{bmatrix}$

$= \begin{bmatrix} 7 & 9 \\ 12 & 22 \\ -24 & 34 \end{bmatrix}$

3. Use equation [1] to remove z from equations [2] and [3]:
[1] $2x + 3y - z = 5$
[2] $4x - 6y + z = -4$
[3] $2x + 6y - 3z = 11$
[4] $6x - 3y = 1$ ([1] + [2])
[5] $-4x - 3y = -4$ (−3[1] + 3)
$10x = 5$ ([4] − [5])
$x = \frac{1}{2}$
$-30y = -20$ (4[4] + 6[5])
$y = \frac{2}{3}$

Use equation [1] to find z:
$2x + 3y - z = 5$
$2x + 3y - 5 = z$
$2(\frac{1}{2}) + 3(\frac{2}{3}) - 5 = z$
$-2 = z$
Thus, the solution is $(\frac{1}{2},\frac{2}{3},-2)$.

4. $2x - y \le 4$
$x + y \ge 3$
Graph the straight lines $2x - y = 4$ and $x + y = 3$. Use a test point such as (0,0) to determine which half-plane is applicable to each inequality. Darken in the area in which these two half-planes overlap.

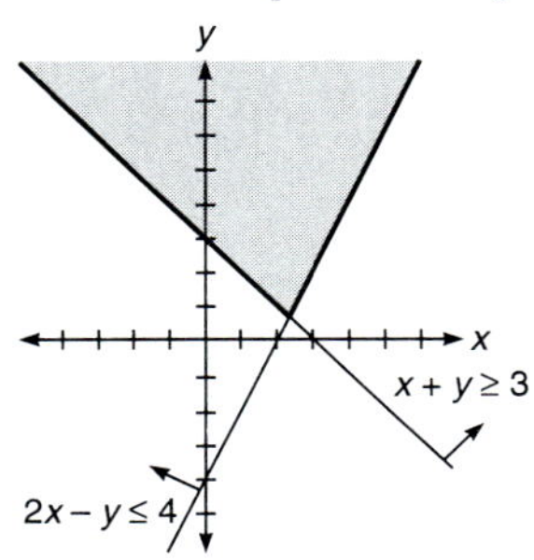

5. $\log_5 10 = \dfrac{\log 10}{\log 5} = \dfrac{1}{\log 5} \approx 1.4307$

6. $\log(2x - 1) + \log(3x + 1) = \log 4$

$\log[(2x - 1)(3x + 1)] = \log 4$

$(2x - 1)(3x + 1) = 4$

$6x^2 - x - 5 = 0$

$(6x + 5)(x - 1) = 0$

$6x + 5 = 0$ or $x - 1 = 0$

$x = -\frac{5}{6}$ or $x = 1$

The negative value is not in the domain of $\log(2x - 1)$ or $\log(3x + 1)$, so the solution is 1.

7. Solve $|2x - 5| < 10$.

$-10 < 2x - 5 < 10$

$-5 < 2x < 15$

$-\frac{5}{2} < x < \frac{15}{2}$

$-2\frac{1}{2} < x < 7\frac{1}{2}$

Solutions to trial exercise problems

21. $y = -x^2 + 6x - 7$

$y = -(x^2 - 6x) - 7$

$y = -(x^2 - 6x + 9) - 7 + 9$

$y = -(x - 3)^2 + 2$; V(3,2)

$\frac{1}{4p} = -1$; $p = -\frac{1}{4}$

focus: $(2, 2 - \frac{1}{4}) = (2, 1\frac{3}{4})$;

directrix: $y = 2 - (-\frac{1}{4}) = 2\frac{1}{4}$; $y = 2\frac{1}{4}$;

intercepts:

$x = 0$: $y = -7$; $(0,-7)$

$y = 0$: $0 = -(x - 3)^2 + 2$

$(x - 3)^2 = 2$

$x - 3 = \pm\sqrt{2}$

$x = 3 \pm \sqrt{2} \approx 4.4, 1.6$

(1.6,0), (4.4,0)

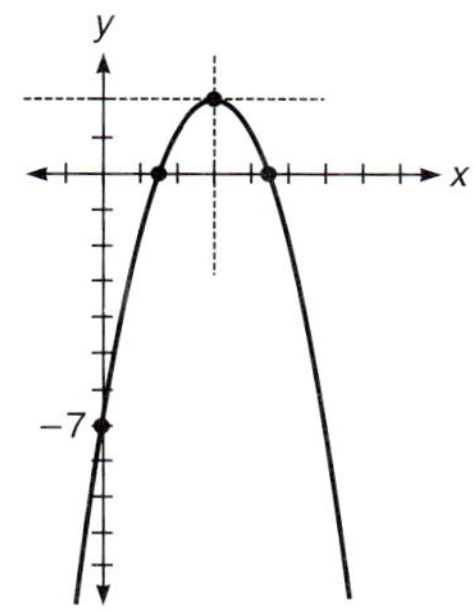

26. $y = -3x^2 - 10x + 8$

$y = -3(x^2 + \frac{10}{3}x) + 8$

$y = -3(x^2 + \frac{10}{3}x + \frac{25}{9}) + 8 + 3(\frac{25}{9})$

$y = -3(x + 1\frac{2}{3})^2 + 16\frac{1}{3}$; $V(-1\frac{2}{3}, 16\frac{1}{3})$

$\frac{1}{4p} = -3$; $p = -\frac{1}{12}$;

focus: $(-\frac{5}{3}, \frac{49}{3} - \frac{1}{12})$, $(-1\frac{2}{3}, 16\frac{1}{4})$;

directrix: $y = \frac{49}{3} - (-\frac{1}{12}) = 16\frac{5}{12}$

$y = 16\frac{5}{12}$

intercepts:

$x = 0$: $y = 8$; $(0,8)$

$y = 0$: $0 = -3x^2 - 10x + 8$

$0 = 3x^2 + 10x - 8$

$0 = (3x - 2)(x + 4)$

$x = \frac{2}{3}$ or -4; $(-4,0)$, $(\frac{2}{3},0)$

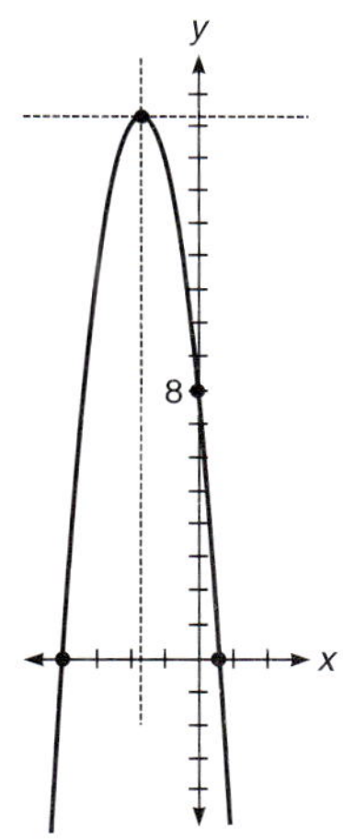

40. $x = -y^2 - 4y + 8$; since this relation expresses x as a function of y we first graph its inverse relation then reflect the graph about the line $y = x$.

$y = -x^2 - 4x + 8$	$x = -y^2 - 4y + 8$
$y = -1(x^2 + 4x) + 8$	
$y = -1(x^2 + 4x + 4) + 8 + 4$	
$y = -1(x + 2)^2 + 12$; V(−2,12)	V(12,−2)
$\frac{1}{4p} = -1$; $p = -\frac{1}{4}$	
focus: $(-2, 12 - \frac{1}{4})$; $(-2, 11\frac{3}{4})$	$(11\frac{3}{4}, -2)$
directrix: $y = 12 - (-\frac{1}{4}) = 12\frac{1}{4}$	$x = 12\frac{1}{4}$
intercepts:	
$x = 0$: $y = 8$; $(0,8)$	(8,0)
$y = 0$: $0 = -1(x + 2)^2 + 12$	
$(x + 2)^2 = 12$	
$x + 2 = \pm\sqrt{12}$	
$x = -2 \pm 2\sqrt{3}$	
$x \approx -5.5, 1.5$; $(-5.5,0)$, $(1.5,0)$	(0,−5.5), (0,1.5)

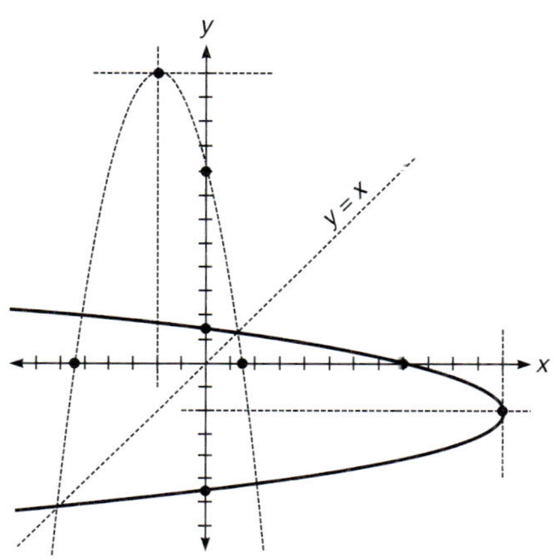

48. focus: $(-4,2)$; vertex: $(-4,4) = (h,k)$
$|p| = 2$, and $p < 0$ since the parabola opens downward (toward the focus).
Thus $p = -2$, and the equation is

$y = \frac{1}{4p}(x-h)^2 + k$

$y = \frac{1}{4(-2)}(x-(-4))^2 + 4$

$y = -\frac{1}{8}(x^2 + 8x + 16) + 4$

$y = -\frac{1}{8}x^2 - x + 2$

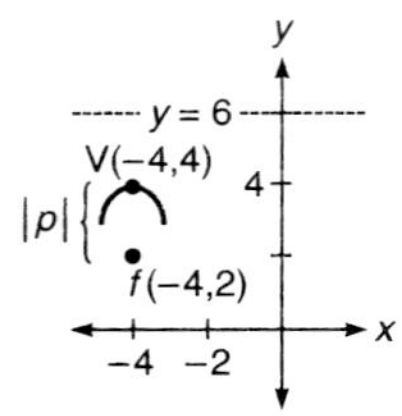

49. vertex: $(3,-1)$; x-intercepts: 2,4

$y = \frac{1}{4p}(x-h)^2 + k$

Replace h by 3, k by -1.

[1] $y = \frac{1}{4p}(x-3)^2 - 1$

To find the value of p we can use the fact that we know the point $(x,y) = (2,0)$ satisfies the equation (since it is one of the equation's x-intercepts). Thus we know that

$0 = \frac{1}{4p}(2-3)^2 - 1$

Replace x by 2, y by 0 in [1].

$1 = \frac{1}{4p}(2-3)^2$

$1 = \frac{1}{4p}$

No need to find p itself.

$y = 1(x-3)^2 - 1$

Replace $\frac{1}{4p}$ by 1 in [1].

$y = x^2 - 6x + 8$

55. $w = 24$, $d = 20$; find h.

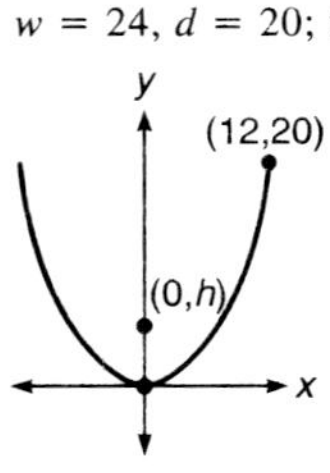

Referring to the figure we see that the vertex is $(0,0)$, so the equation is of the form $y = \frac{1}{4p}x^2$; the point $(12,20)$ satisfies the equation so

$y = \frac{1}{4p}x^2$

$20 = \frac{1}{4p}(144)$

$p = \frac{144}{80} = \frac{9}{5} = h$

58. $y = -\frac{1}{6}x^2 + \frac{1}{\sqrt{3}}x$

$y = -\frac{1}{6}\left(x^2 - \frac{6}{\sqrt{3}}x\right)$, since

$\left(-\frac{1}{6}\right)\left(-\frac{6}{\sqrt{3}}\right) = \frac{1}{\sqrt{3}}$

$y = -\frac{1}{6}(x^2 - 2\sqrt{3}x)$, since

$\frac{6}{\sqrt{3}} \cdot \frac{\sqrt{3}}{\sqrt{3}} = \frac{6\sqrt{3}}{3} = 2\sqrt{3}$

$y = -\frac{1}{6}(x^2 - 2\sqrt{3}x + 3) + \frac{1}{6}(3)$,

since $\frac{1}{2}(-2\sqrt{3}) = -\sqrt{3}$; $(-\sqrt{3})^2 = 3$

$y = -\frac{1}{6}(x - \sqrt{3})^2 + \frac{1}{2}$

vertex: $(\sqrt{3},\frac{1}{2}) \approx (1.7,0.5)$

intercepts: $x = 0$: $y = \frac{1}{\sqrt{3}} \cdot 0 - \frac{1}{6}$

$0^2 = 0$; $(0,0)$

$y = 0$: $0 = \frac{1}{\sqrt{3}}x - \frac{1}{6}x^2$

Multiply each term by $6\sqrt{3}$.

$(6\sqrt{3})(0) = (6\sqrt{3})\left(\frac{1}{\sqrt{3}}x\right) - (6\sqrt{3})\left(\frac{1}{6}x^2\right)$

$0 = 6x - \sqrt{3}x^2$

$0 = x(6 - \sqrt{3}x)$

$x = 0$ or $6 - \sqrt{3}x = 0$

$6 = \sqrt{3}x$

$x = \frac{6}{\sqrt{3}} \cdot \frac{\sqrt{3}}{\sqrt{3}} = 2\sqrt{3}$; $(0,0)$

and $(2\sqrt{3},0) \approx (3.5,0)$

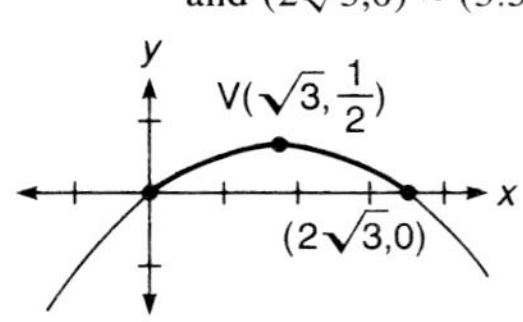

60. $y = -\frac{16}{v^2}x^2$

$v = 4$, $y = -40$.

$-40 = -\frac{16}{4^2}x^2$

$40 = x^2$

$\pm\sqrt{40} = x$

Assume $x > 0$.

$x = 2\sqrt{10} \approx 6.3$

Thus, the stuntperson will land about 6.3 feet from the base of the building.

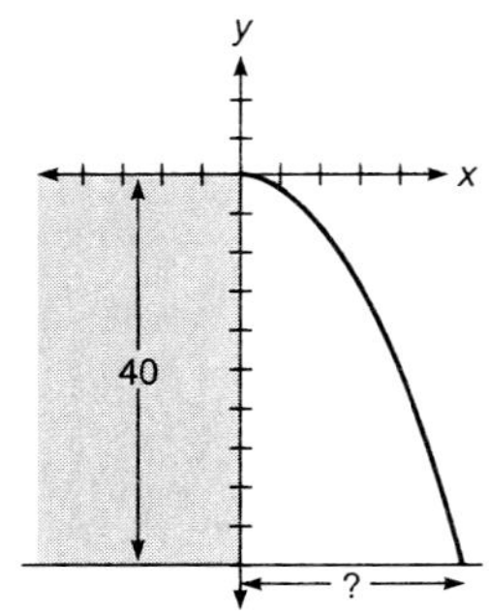

64. Let $-h$ be an initial height, and r a fixed velocity.

$y = -\frac{16}{v^2}x^2$

$y = -h$, $v = r$

$-h = -\frac{16}{r^2}x^2$

$hr^2 = 16x^2$

$x^2 = \frac{hr^2}{16}$

Assume $x > 0$

$x = \sqrt{\frac{hr^2}{16}} = \frac{1}{4}r\sqrt{h}$

Now double the height to $-2h$.

$y = -\frac{16}{v^2}x^2$

$y = -h$, $v = r$

$-2h = -\frac{16}{r^2}x^2$

$2hr^2 = 16x^2$

$x^2 = \frac{hr^2}{8}$

Assume $x > 0$

$x = \sqrt{\frac{hr^2}{8}}$

$x = r\sqrt{\frac{h}{8} \cdot \frac{2}{2}} = r\frac{\sqrt{2h}}{4} = \frac{\sqrt{2}}{4}r\sqrt{h}$

Dividing $\dfrac{\frac{\sqrt{2}}{4}r\sqrt{h}}{\frac{1}{4}r\sqrt{h}} = \sqrt{2}$ shows that the horizontal distance did *not* double when the height doubled. It increased by a factor of $\sqrt{2} \approx 1.4$.

Exercise 11–2

Answers to odd-numbered problems

1. foci: $(\pm\sqrt{7},0)$

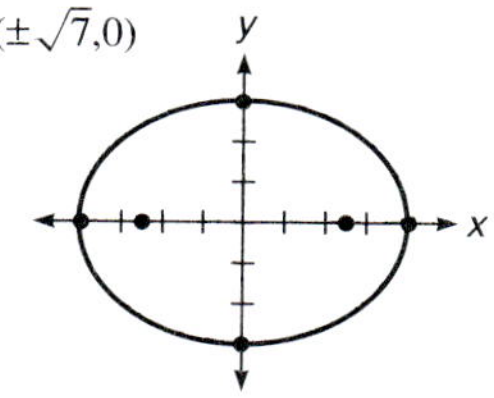

3.

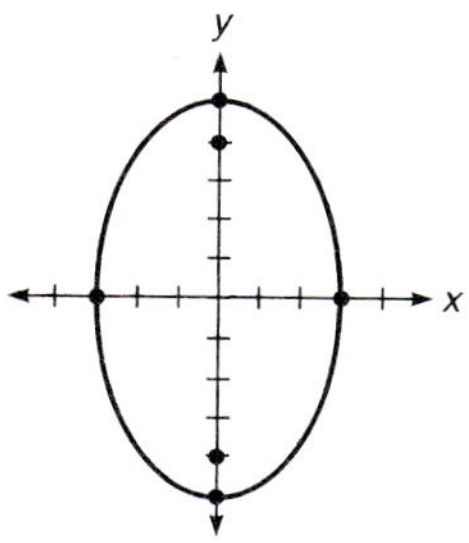

5. foci: $(\pm\sqrt{5},0)$

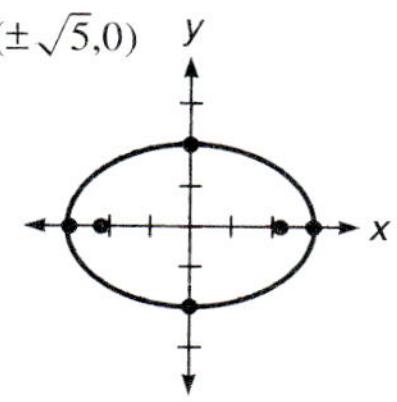

7. foci: $(\pm\frac{1}{2},0)$; intercepts: $\left(\pm\dfrac{\sqrt{5}}{2},0\right)$, $(0,\pm1)$

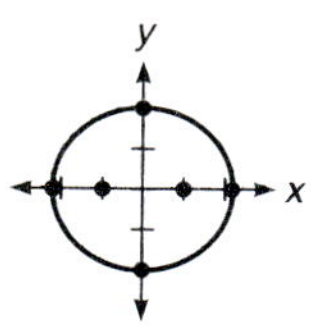

9. foci $(\pm\sqrt{15},0)$; intercepts: $(\pm4,0)$, $(0,\pm1)$

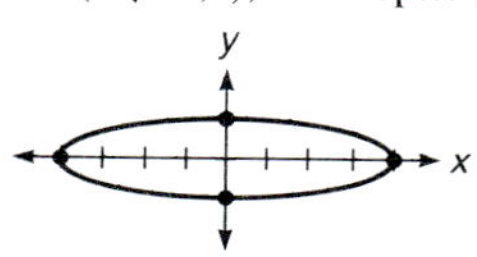

11. foci: $(\pm2\sqrt{6},0)$; intercepts: $(\pm7,0)$, $(0,\pm5)$

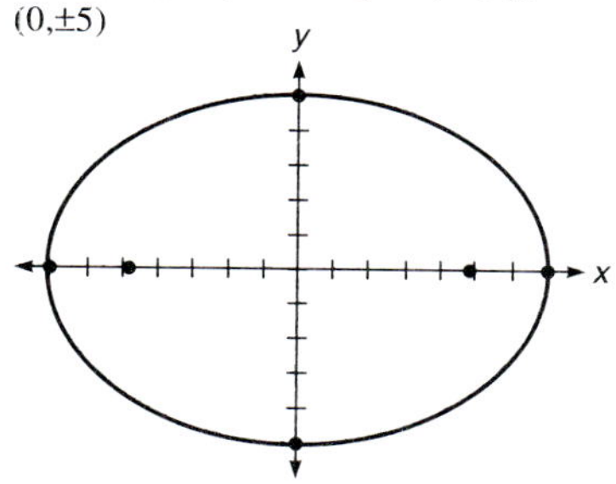

13. foci: $(-3,1\pm2\sqrt{3})$

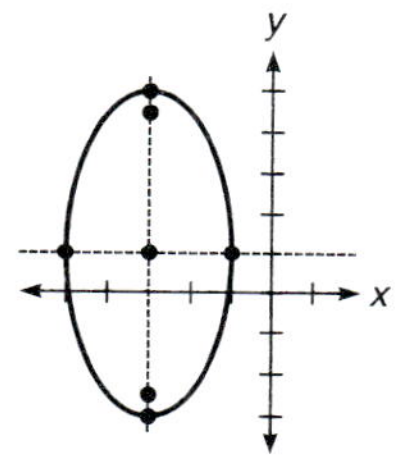

15.

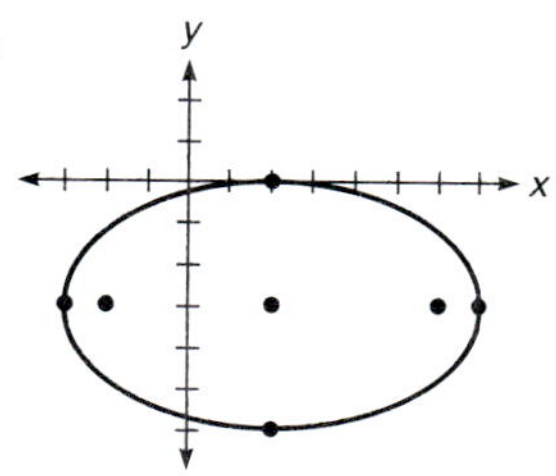

17.

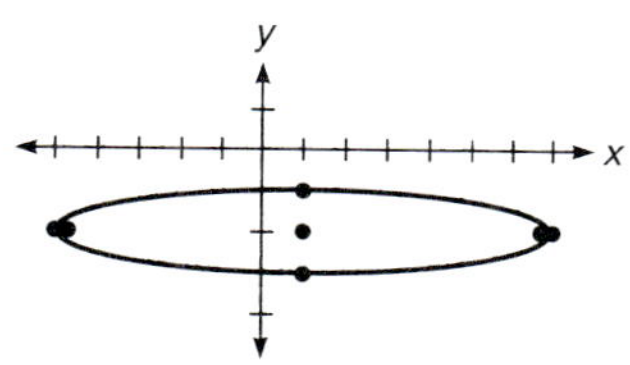

19.

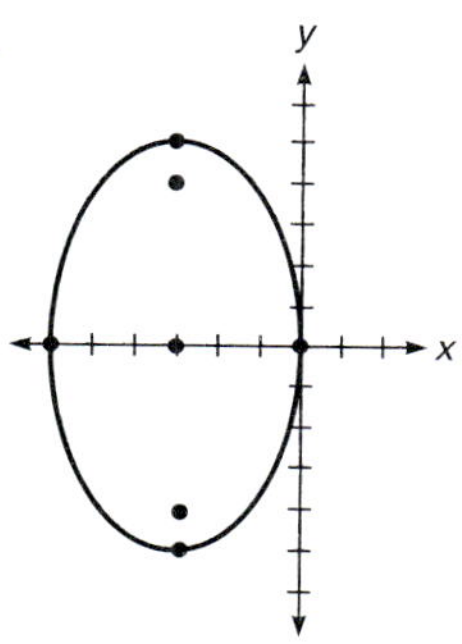

21. $\dfrac{x^2}{27}+\dfrac{y^2}{9}=1$; foci: $(\pm3\sqrt{2},0)$; intercepts: $(\pm3\sqrt{3},0)$, $(0,\pm3)$

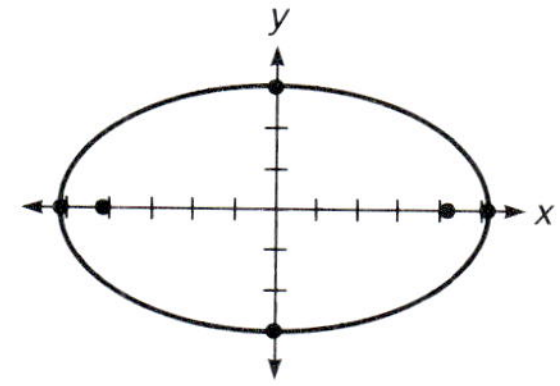

23. $\dfrac{x^2}{9}+\dfrac{y^2}{36}=1$; foci: $(0,\pm3\sqrt{3})$; intercepts: $(\pm3,0)$, $(0,\pm6)$

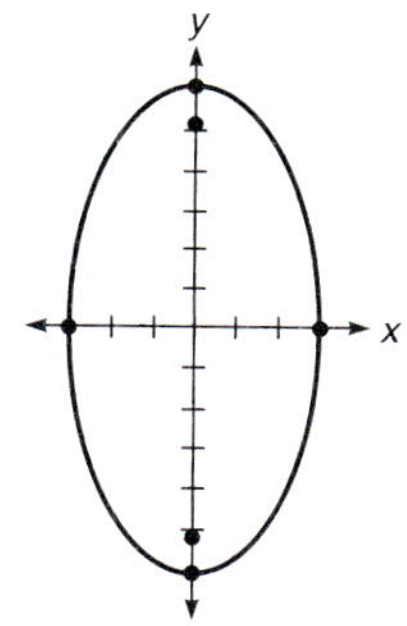

25. $\dfrac{x^2}{2}+\dfrac{y^2}{9}=1$; foci: $(0,\pm\sqrt{7})$; intercepts: $(\pm\sqrt{2},0)$, $(0,\pm3)$

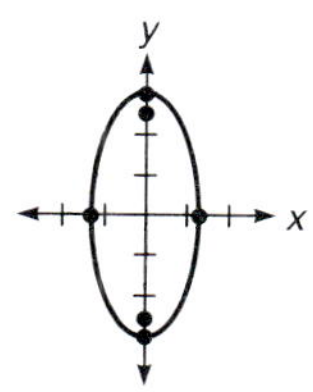

27. $\dfrac{x^2}{\frac{9}{2}}+\dfrac{y^2}{9}=1$; foci: $\left(0,\pm\dfrac{3\sqrt{2}}{2}\right)$; intercepts: $\left(\pm\dfrac{3\sqrt{2}}{2},0\right)$, $(0,\pm3)$

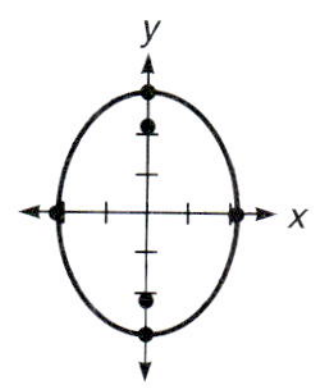

29. $\frac{x^2}{9} + y^2 = 1$; foci: $(\pm 2\sqrt{2},0)$;
intercepts: $(\pm 3,0)$, $(0,\pm 1)$

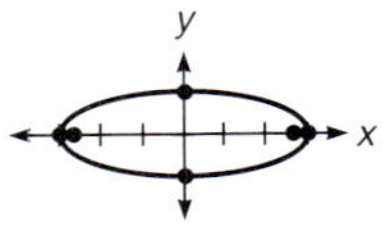

31. $x^2 + \frac{y^2}{2} = 1$; foci: $(0,-1)$ and $(0,1)$;
intercepts: $(0,\pm\sqrt{2})$, $(\pm 1,0)$

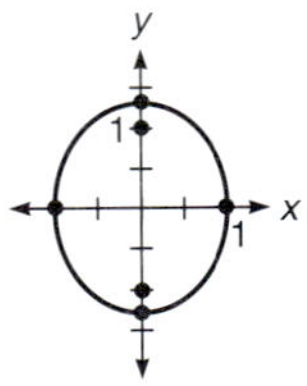

33. $\frac{x^2}{\frac{5}{4}} + y^2 = 1$; foci: $(\pm\frac{1}{2},0)$;
intercepts: $(0,\pm 1)$, $\left(\pm\frac{\sqrt{5}}{2},0\right)$

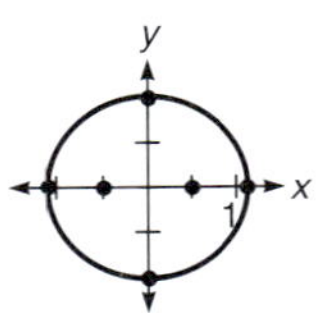

35. $\frac{x^2}{8} + \frac{(y-2)^2}{4} = 1$; center: $(0,2)$; foci: $(\pm 2,2)$; end points of major/minor axes: $(0 \pm 2\sqrt{2},2)$, $(0,0)$, and $(0,4)$

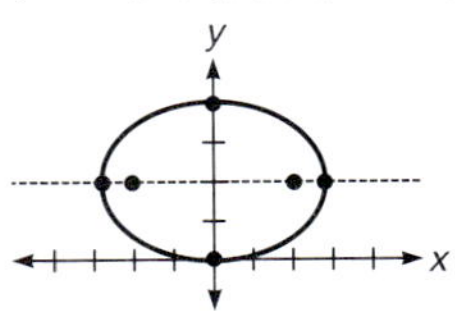

37. $\frac{(x-\frac{1}{2})^2}{4} + \frac{(y+3)^2}{2} = 1$;
center: $(\frac{1}{2},-3)$; foci: $(\frac{1}{2} \pm \sqrt{2},0)$;
end points of major/minor axes:
$(-1\frac{1}{2},-3)$, $(2\frac{1}{2},-3)$, $(\frac{1}{2},-3 \pm \sqrt{2})$

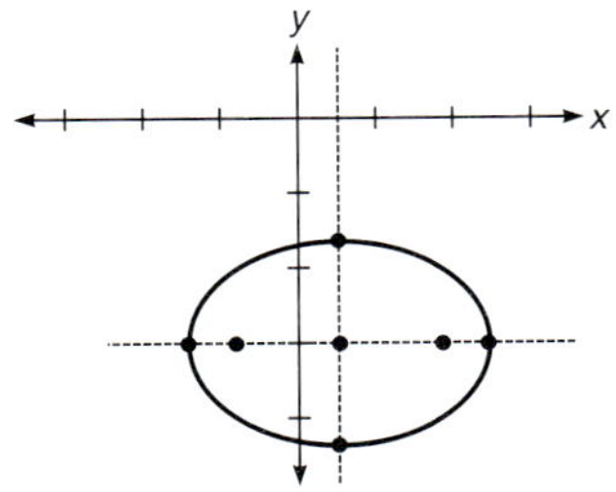

39. $\frac{(x-3)^2}{60} + \frac{(y+5)^2}{30} = 1$; center: $(3,-5)$; foci: $(3 \pm \sqrt{30},-5)$; end points of major/minor axes: $(3 \pm 2\sqrt{15},-5)$, $(3,-5 \pm \sqrt{30})$

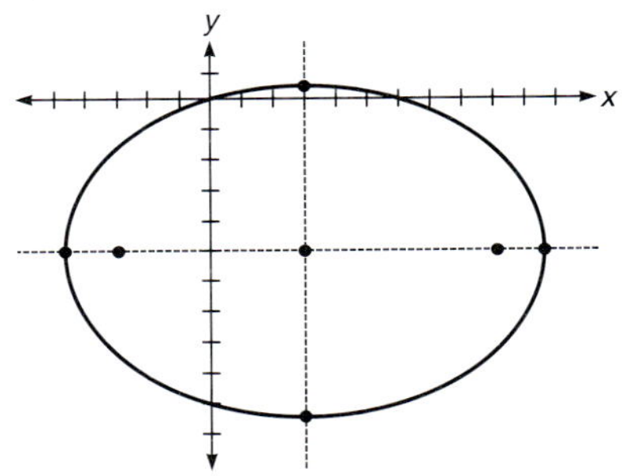

41. $x^2 + 2y^2 + 8 = 0$
$x^2 + 2y^2 = -8$
There is no real solution to this equation, so there is no graph for this relation.

43. $\frac{(x-1)^2}{9} + \frac{(y+2)^2}{4} = 1$; center: $(1,-2)$; foci: $(1 \pm \sqrt{5},-2)$; end points of major/minor axes: $(-2,-2)$, $(4,-2)$, $(1,-4)$, $(1,0)$

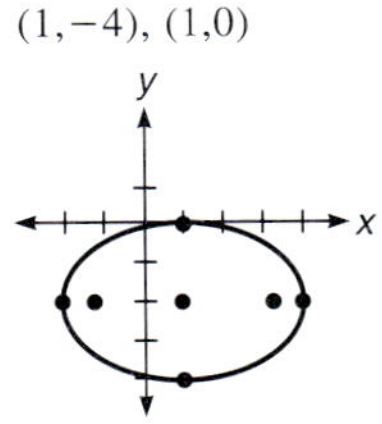

45. $\frac{x^2}{25} + \frac{(y+2)^2}{16} = 1$; center: $(0,-2)$;
foci: $(\pm 3,-2)$; end points of major/minor axes: $(\pm 5,-2)$, $(0,-6)$, $(0,2)$

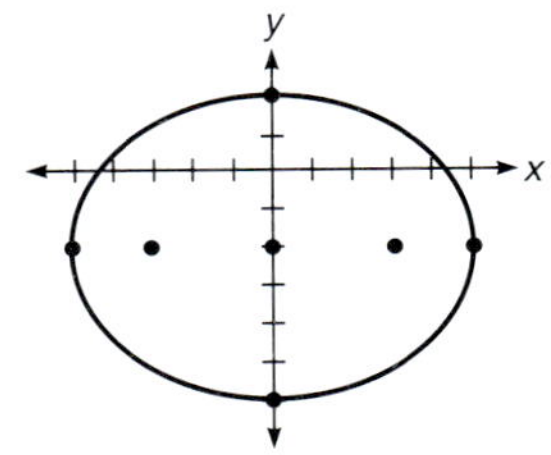

47. $\frac{x^2}{13} + \frac{y^2}{9} = 1$ **49.** $\frac{x^2}{48} + \frac{y^2}{64} = 1$

51. $\frac{x^2}{9} + \frac{y^2}{4} = 1$ **53.** $\frac{x^2}{9} + \frac{y^2}{5} = 1$

55. $\frac{x^2}{22{,}500} + \frac{y^2}{20{,}000} = 1$

57. $\frac{x^2}{4} + (y-1)^2 = 1$ **59.** $\frac{1}{2}$ **61.** $\frac{\sqrt{5}}{5}$

63. $\frac{3}{5}$ **65.** $\frac{\sqrt{3}}{2}$ **67.** $\frac{2\sqrt{2}}{3}$

69. $\frac{\sqrt{6}}{3}$ **71.** $\frac{\sqrt{2}}{2}$

Solutions to skill and review problems

1. $y = 2(x-1)^2 - 4$
Using $y = a(x-h)^2 + k$ we see that this is a parabola with vertex at $(1,-4)$.
intercepts:
$y = 0$: $0 = 2(x-1)^2 - 4$
$4 = 2(x-1)^2$
$2 = (x-1)^2$
$\pm\sqrt{2} = x - 1$
$1 \pm \sqrt{2} = x \approx (2.4,0), (-0.4,0)$
$x = 0$: $y = 2(-1)^2 - 4 = -2$; $(0,-2)$

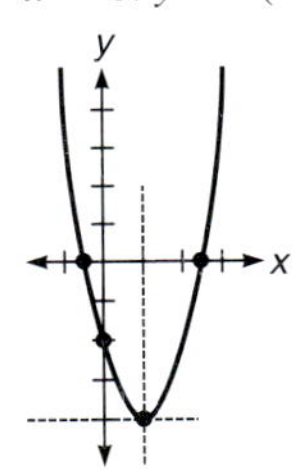

2. $x^2 - 4x + y^2 + 12y + 12 = 0$
$x^2 - 4x + 4 + y^2 + 12y + 36 = -12 + 4 + 36$
$(x - 2)^2 + (y + 6)^2 = 28$
circle; center: $(2,-6)$; radius $= \sqrt{28} = 2\sqrt{7} \approx 5.3$

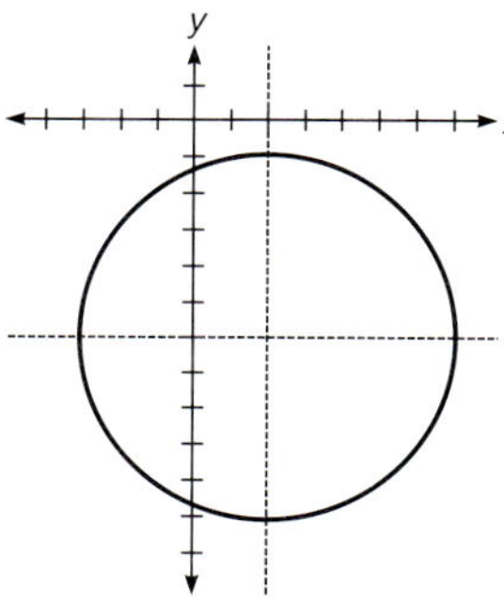

3. $x^{\frac{2}{3}} + 7x^{\frac{1}{3}} = 8$
Let $u = x^{\frac{1}{3}}$; then $u^2 = x^{\frac{2}{3}}$.
$u^2 + 7u = 8$
$u^2 + 7u - 8 = 0$
$(u + 8)(u - 1) = 0$
$u = -8$ or $u = 1$
$x^{\frac{1}{3}} = -8$ or $x^{\frac{1}{3}} = 1$
Cube each member.
$x = -512$ or $x = 1$

4. $\dfrac{2\sqrt{3}}{\sqrt{6} - \sqrt{2}} \cdot \dfrac{\sqrt{6} + \sqrt{2}}{\sqrt{6} + \sqrt{2}}$
$= \dfrac{2\sqrt{18} + 2\sqrt{6}}{6 + \sqrt{12} - \sqrt{12} - 2}$
$= \dfrac{2(3\sqrt{2}) + 2\sqrt{6}}{4}$
$= \dfrac{2(3\sqrt{2} + \sqrt{6})}{4} = \dfrac{3\sqrt{2} + \sqrt{6}}{2}$

5. $y = \dfrac{2x}{(x - 3)(x + 3)}$
vertical asymptotes: $x = \pm 3$
horizontal asymptote: $y = 0$ (x-axis)
intercepts:
$x = 0$: $y = 0$; $(0,0)$
$y = 0$: $0 = \dfrac{2x}{(x - 3)(x + 3)}$
$0 = 2x$
$0 = x$; $(0,0)$
additional points:

x	-5	-4	-2	2	4	5
y	-0.6	-1.1	0.8	-0.8	1.1	0.6

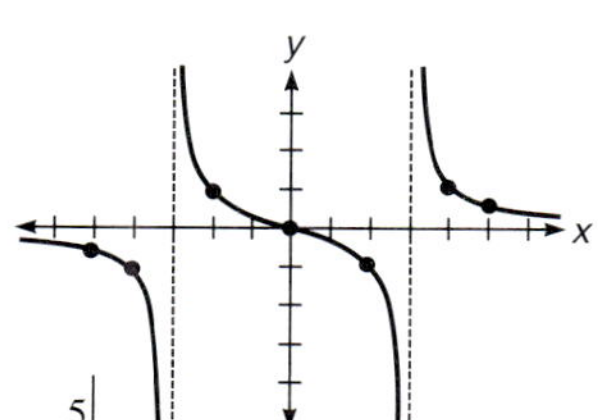

6. $x^3 - 3x^2 + x + 2$
Possible rational zeros are $\pm 1, \pm 2$. Using synthetic division with $x = 2$:

	1	−3	1	2
		2	−2	−2
2	1	−1	−1	0

Thus $x^3 - 3x^2 + x + 2 = (x - 2)(x^2 - x - 1)$. The zeros of $x^2 - x - 1$ are not real. Thus the factorization above is complete over R.

Solutions to trial exercise problems

7. $\dfrac{4x^2}{5} + y^2 = 1$
$\dfrac{x^2}{\frac{5}{4}} + y^2 = 1$
center: $(0,0)$
$a = \dfrac{\sqrt{5}}{2} \approx 1.1$, $b = 1$, $c = \sqrt{\frac{5}{4} - 1} = \frac{1}{2}$
major axis: x-axis
foci: $(\pm\frac{1}{2},0)$
intercepts: $\left(\pm\dfrac{\sqrt{5}}{2},0\right)$, $(0,\pm 1)$

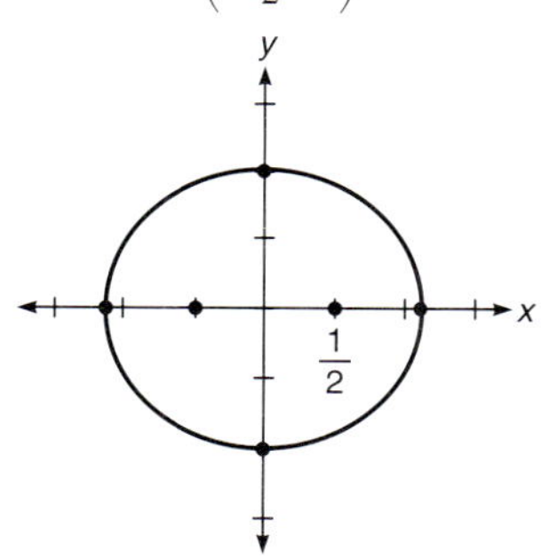

15. $\dfrac{(x - 2)^2}{25} + \dfrac{(y + 3)^2}{9} = 1$
$a = 5$, $b = 3$, $c = 4$; major axis parallel to x-axis.
center: $(h,k) = (2,-3)$
foci: $(h \pm c,k) = (2 \pm 4,-3)$; $(6,-3)$, $(-2,-3)$
end points of major/minor axes:
$(h \pm a,k)$, $(h,k \pm b)$
$(2 \pm 5,-3)$; $(-3,-3)$, $(7,-3)$
$(2,-3 \pm 3)$; $(2,-6)$, $(2,0)$

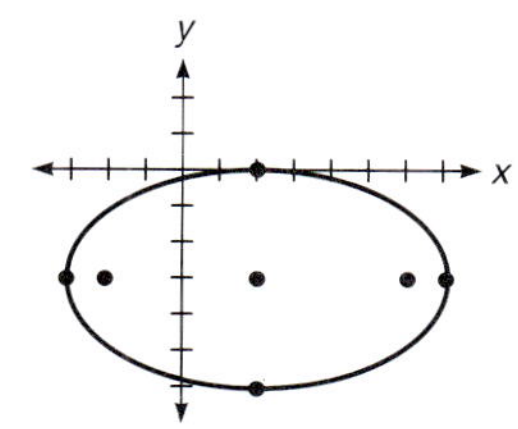

37. $4x^2 - 4x + 8y^2 + 48y = -57$
$4(x^2 - x) + 8(y^2 + 6y) = -57$
$4(x^2 - x + \frac{1}{4}) + 8(y^2 + 6y + 9) = -57 + 4(\frac{1}{4}) + 72$
$4(x - \frac{1}{2})^2 + 8(y + 3)^2 = 16$
$\dfrac{(x - \frac{1}{2})^2}{4} + \dfrac{(y + 3)^2}{2} = 1$
center: $(h,k) = (\frac{1}{2},-3)$
$a = \sqrt{4} = 2$; $b = \sqrt{2}$;
$c = \sqrt{4 - 2} = \sqrt{2}$
major axis parallel to x-axis
foci: $(h \pm c,k) = (\frac{1}{2} \pm \sqrt{2},-3)$
end points of major/minor axes:
$(h \pm a,k)$
$(\frac{1}{2} \pm 2,-3) = (-1\frac{1}{2},-3)$ and $(2\frac{1}{2},-3)$
$(h,k \pm b)$
$(\frac{1}{2},-3 \pm \sqrt{2}) \approx (0.5,-4.4)$, $(0.5,-1.6)$

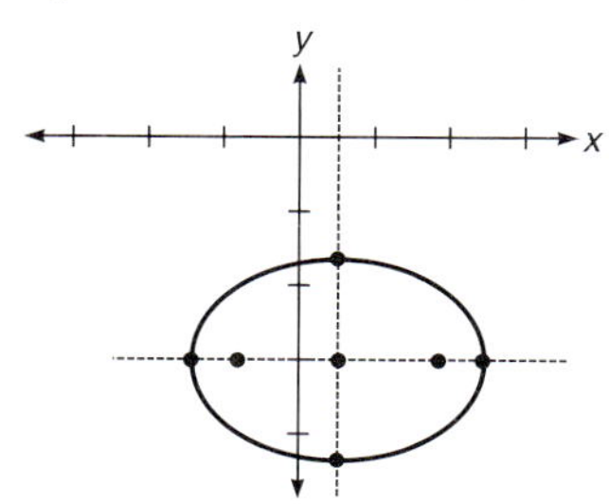

47. foci: $(-2,0)$ and $(2,0)$; one y-intercept at 3

The equation is of this form.

$$[1]\ \frac{x^2}{a^2} + \frac{y^2}{b^2} = 1$$

$\pm b$ are the y-intercepts: $b = 3$

distance to the foci: $c = 2$

The foci are on the major axis, which is parallel to the x-axis, so $a > b$.

$c = \sqrt{a^2 - b^2}$

$2 = \sqrt{a^2 - 3^2}$

$4 = a^2 - 9$

$13 = a^2$

Replace $a^2 = 13$, $b^2 = 9$ in equation [1].

$$\frac{x^2}{13} + \frac{y^2}{9} = 1$$

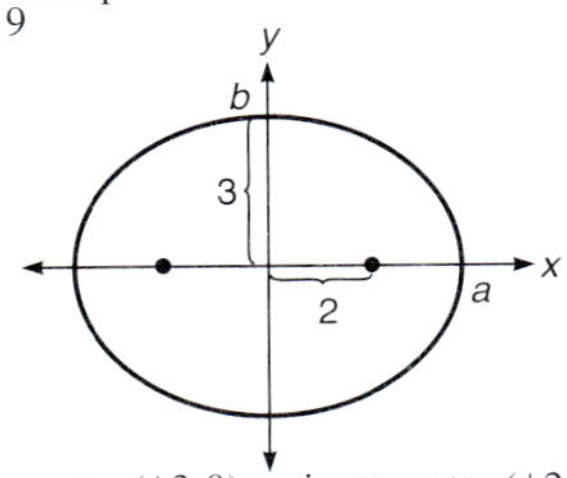

51. x-intercepts: $(\pm 3,0)$; y-intercepts: $(\pm 2,0)$

The equation is of this form.

$$[1]\ \frac{x^2}{a^2} + \frac{y^2}{b^2} = 1$$

x-intercept: $a = 3$

y-intercept: $b = 2$

Replace a and b in [1].

$$\frac{x^2}{9} + \frac{y^2}{4} = 1$$

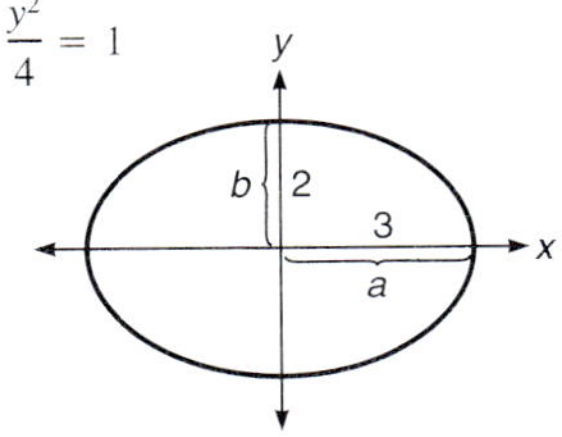

54. We can see that $a = 3$, $b = 2$, so $c = \sqrt{a^2 - b^2} = \sqrt{5}$. Thus, the equation is $\frac{x^2}{9} + \frac{y^2}{4} = 1$, and the foci should be at $(\pm\sqrt{5},0)$. The length of the string l is $2(3 + \sqrt{2}) = 6 + 2\sqrt{2} \approx 8$ ft 10 in.

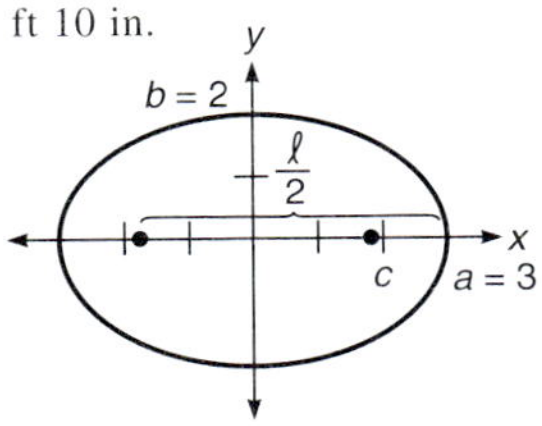

60. $a = 3$, $b = 5$, $c = 4$;

$b > a$ so $e = \frac{c}{b} = \frac{4}{5}$

62. $a = 3$, $b = 2$, $c = \sqrt{5}$;

$a > b$ so $e = \frac{c}{a} = \frac{\sqrt{5}}{3}$

Exercise 11–3

Answers to odd-numbered problems

1. foci: $(\pm\sqrt{41},0)$

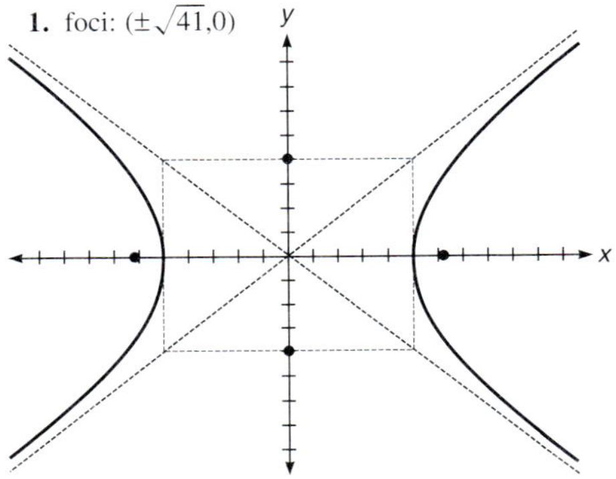

3. foci: $(0,\pm\sqrt{41})$

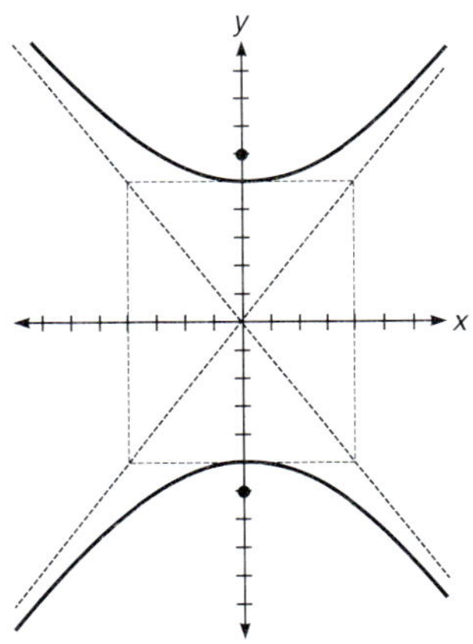

5. foci: $(\pm\sqrt{7},0)$

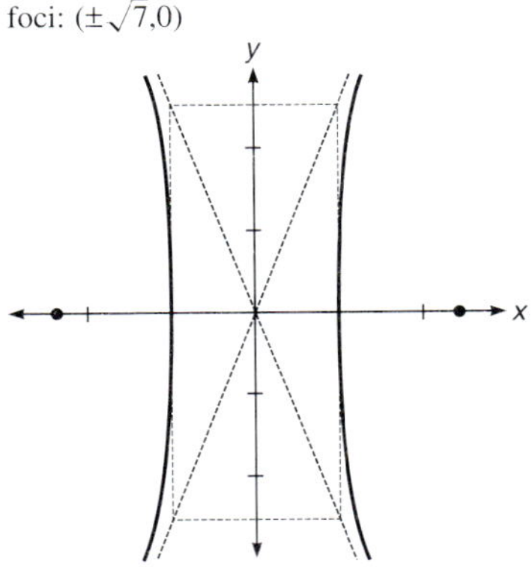

7. foci: $(\pm 2\sqrt{5},0)$

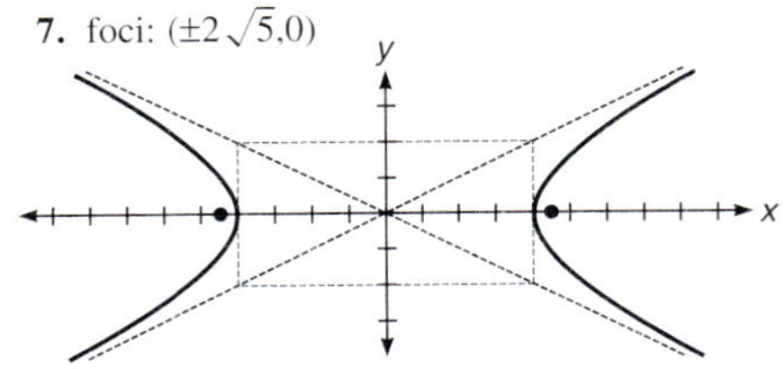

9. foci: $(0,\pm\sqrt{13})$

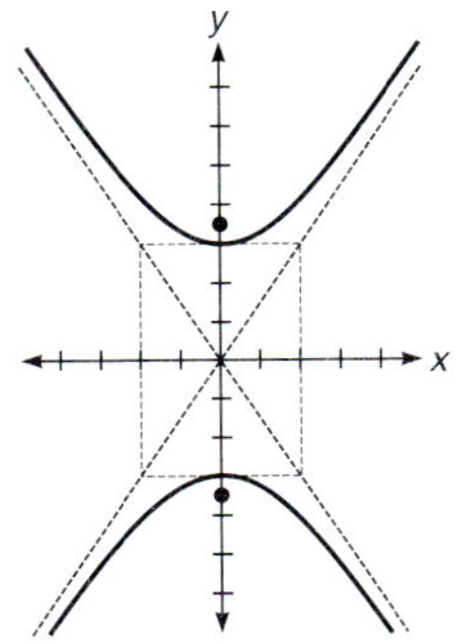

11. $\frac{x^2}{\frac{25}{4}} - y^2 = 1$; foci: $\left(\pm\frac{\sqrt{29}}{2},0\right)$;

end points of major axis: $(\pm 2\frac{1}{2},0)$

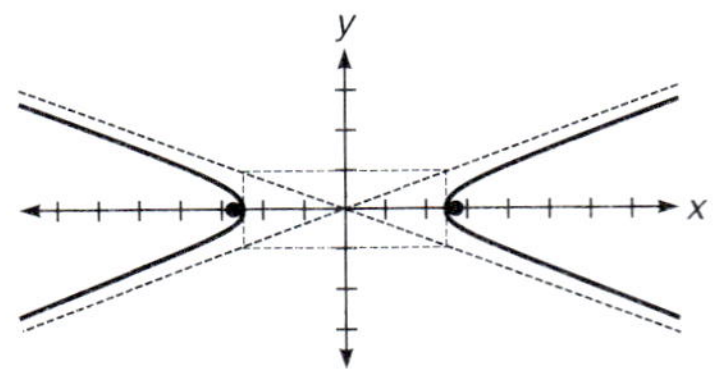

13. $\frac{y^2}{\frac{1}{2}} - \frac{x^2}{2} = 1$; foci: $\left(0,\pm\frac{\sqrt{10}}{2}\right)$;

end points of major axis: $\left(0,\pm\frac{\sqrt{2}}{2}\right)$

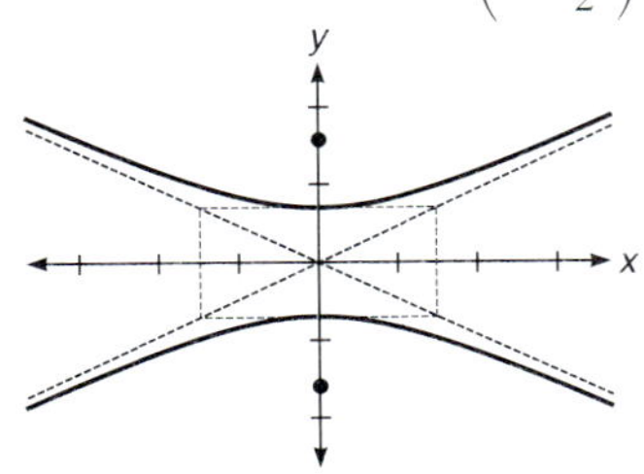

15. $\frac{x^2}{4} - \frac{y^2}{36} = 1$; foci: $(\pm 2\sqrt{10}, 0)$

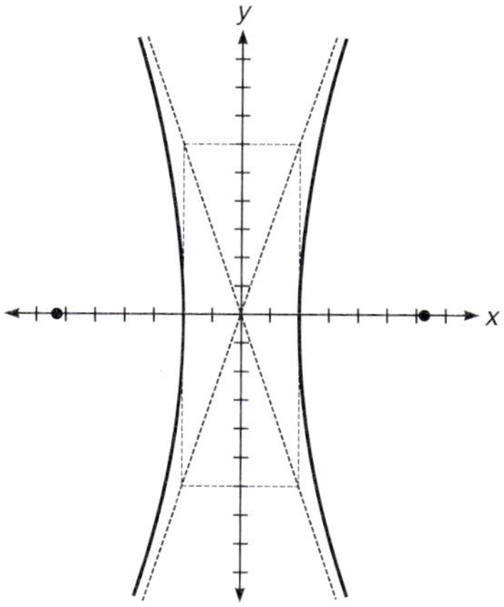

17. $\frac{x^2}{16} - y^2 = 1$; foci: $(\pm\sqrt{17}, 0)$

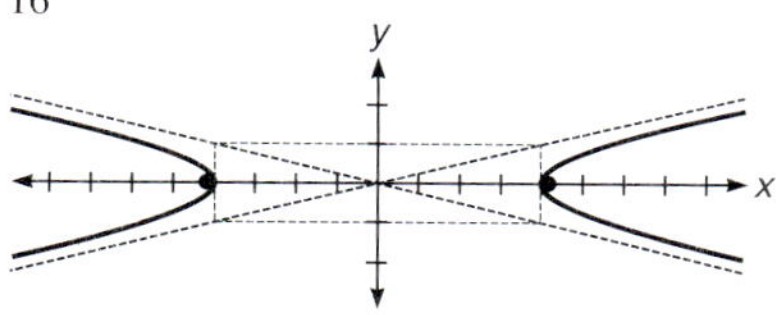

19. $\frac{y^2}{4} - \frac{x^2}{18} = 1$; foci: $(0, \pm\sqrt{22})$

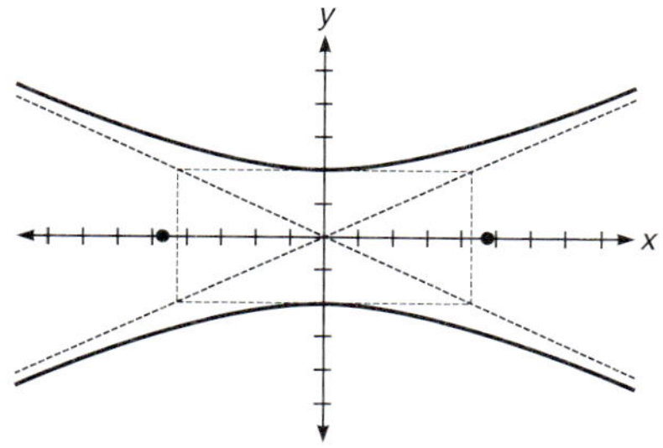

21. $\frac{y^2}{16} - \frac{x^2}{25} = 1$; foci: $(0, \pm\sqrt{41})$

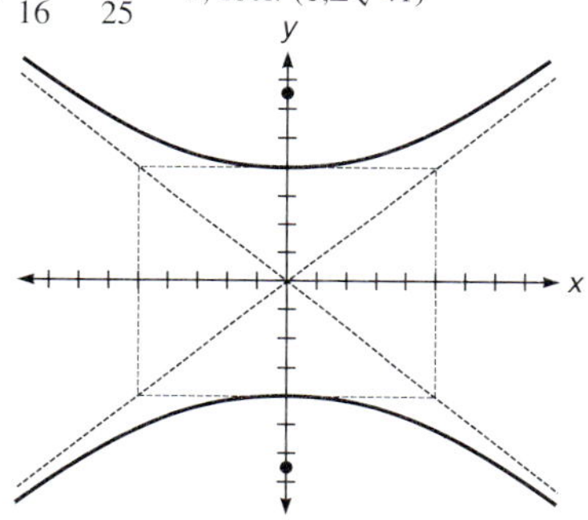

23. $\frac{x^2}{\frac{1}{2}} - \frac{y^2}{\frac{4}{3}} = 1$; foci: $\left(\pm\frac{\sqrt{66}}{6}, 0\right)$;

end points of major axis: $\left(\pm\frac{\sqrt{2}}{2}, 0\right)$

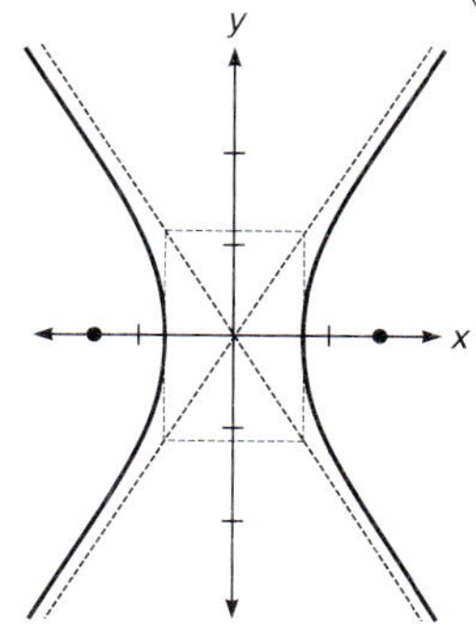

25. foci: $(2 \pm 5\sqrt{5}, -3)$

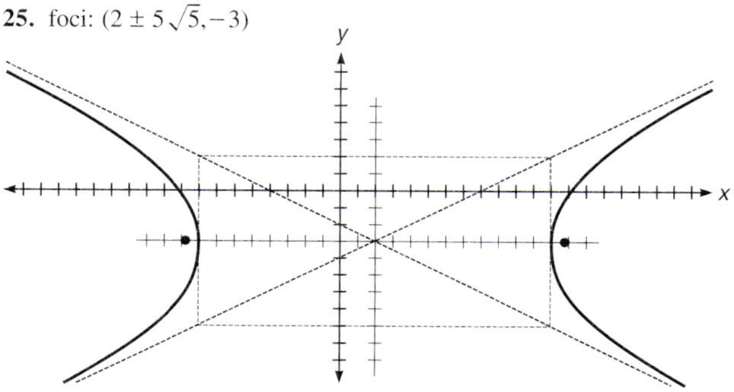

27. foci: $(0, -2 \pm \sqrt{5})$

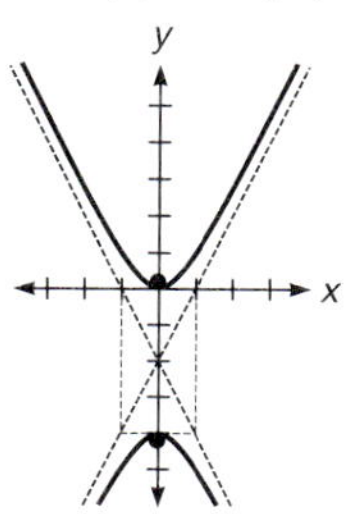

29. foci: $(-1 \pm \sqrt{61}, 0)$

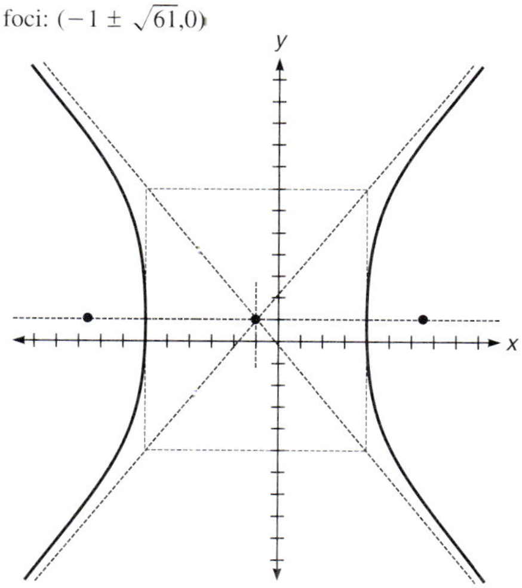

31. foci: $(3 \pm \sqrt{17}, -1)$

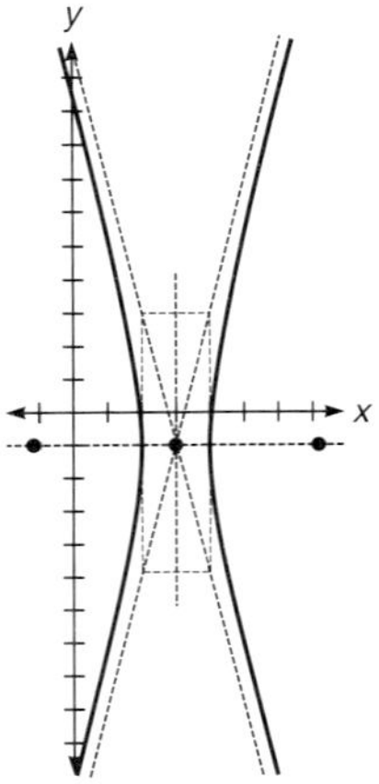

33. foci: $(2, 2 \pm \sqrt{5})$

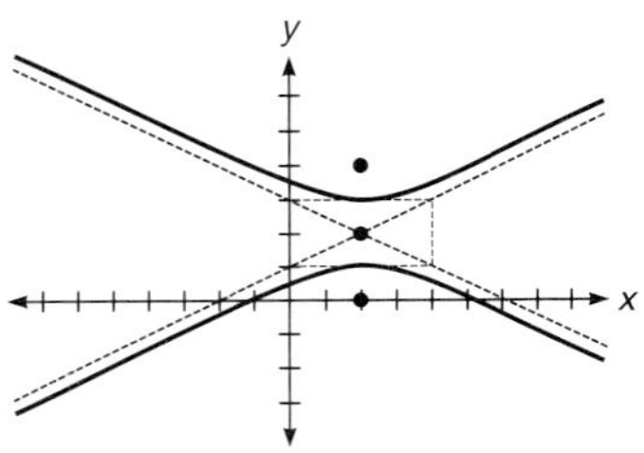

35. $\dfrac{(y-1)^2}{\frac{25}{16}} - \dfrac{x^2}{9} = 1$;
foci: $(1, -2\frac{1}{4})$, $(1, 4\frac{1}{4})$; end points of major axis: $(0, 2\frac{1}{4})$, $(0, -\frac{1}{4})$

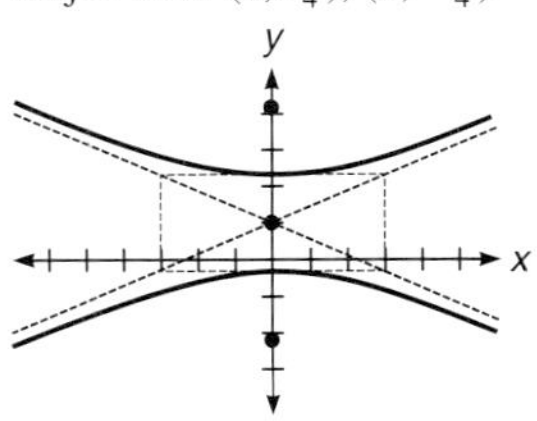

37. $\dfrac{(x-1)^2}{4} - \dfrac{(y+2)^2}{8} = 1$;
foci: $(1 \pm 2\sqrt{3}, -2)$

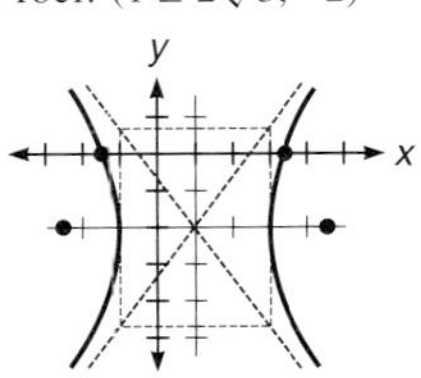

39. $\dfrac{(y-4)^2}{16} - \dfrac{(x+1)^2}{12} = 1$;
foci: $(-1, 4 \pm 2\sqrt{7})$

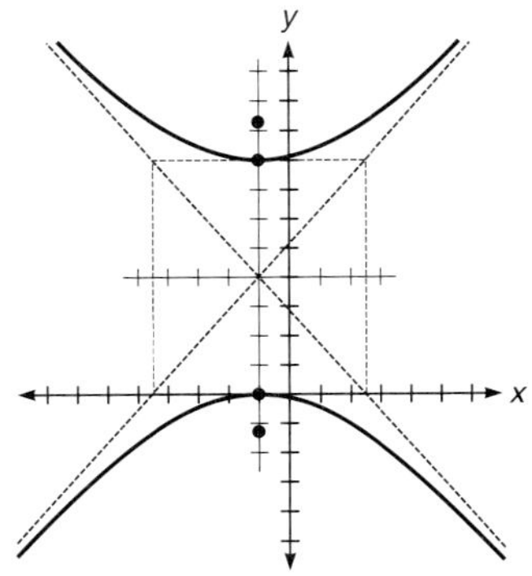

41. $\dfrac{(x-2)^2}{3} - (y-4)^2 = 1$; end points of major axis: $(2 \pm \sqrt{3}, -1)$

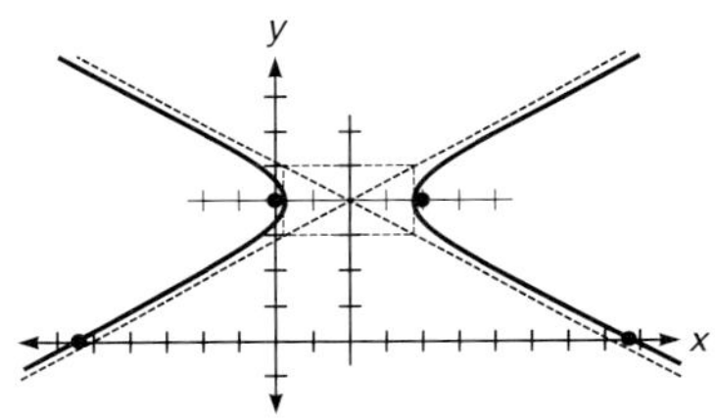

43. $\dfrac{(x-1)^2}{4} - (y+1)^2 = 1$;
foci: $(1 \pm \sqrt{5}, -1)$

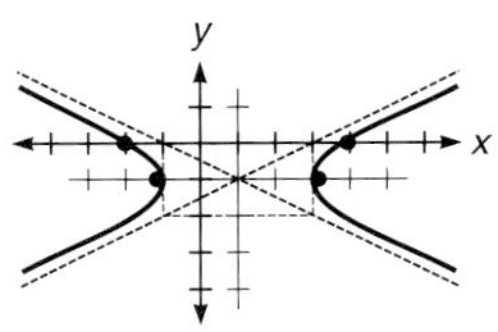

45. $\dfrac{(y-3)^2}{12} - \dfrac{x^2}{6} = 1$; foci: $(0, 3 \pm 3\sqrt{2})$;
end points of major axis: $(0, 3 \pm 2\sqrt{3})$

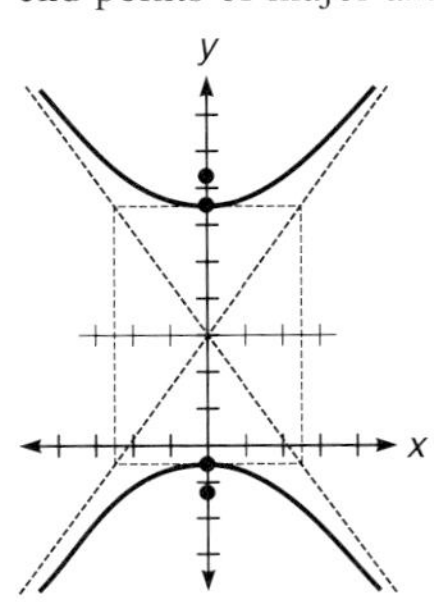

47. $\dfrac{(y-1)^2}{\frac{1}{4}} - \dfrac{(x-\frac{3}{2})^2}{\frac{3}{4}} = 1$; foci: $(1\frac{1}{2}, 2)$, $(1\frac{1}{2}, 0)$; end points of major axis: $(\frac{1}{2}, \frac{1}{2})$, $(\frac{1}{2}, 1)$

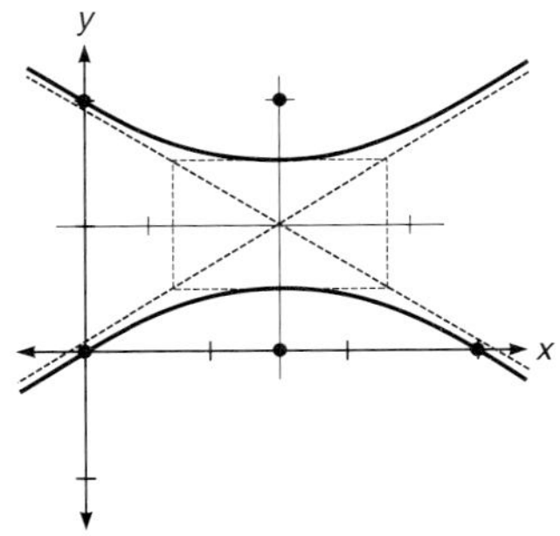

49. $(x+1)^2 - (y-4)^2 = 1$; hyperbola;
foci: $(-1 \pm \sqrt{2}, 4)$

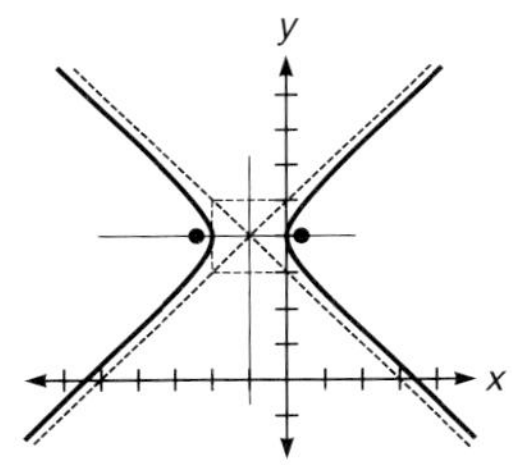

51. $(x-1)^2 - \dfrac{(y+2)^2}{4} = 1$; hyperbola;
foci: $(1, -2 \pm \sqrt{5})$

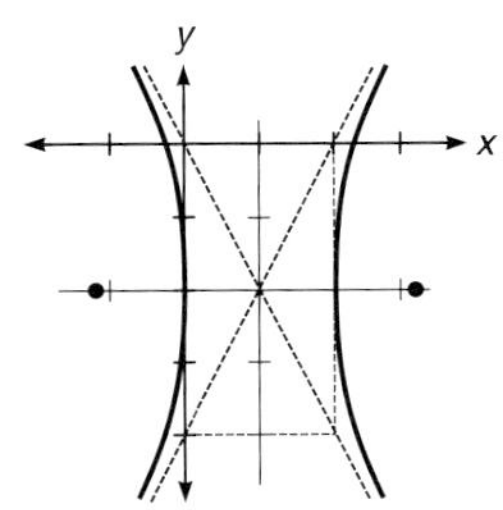

53. $\dfrac{x^2}{9} - \dfrac{y^2}{4} = 1$; foci: $(\pm\sqrt{13}, 0)$; hyperbola

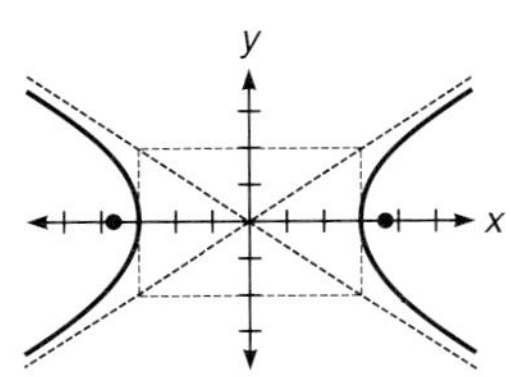

55. $(x+3)^2 + \dfrac{(y-4)^2}{2} = 1$; ellipse

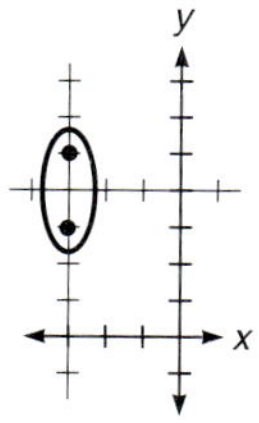

57. $(x - \frac{3}{2})^2 + (y + \frac{5}{2})^2 = 1$; circle;

center: $(\frac{3}{2}, -\frac{5}{2})$

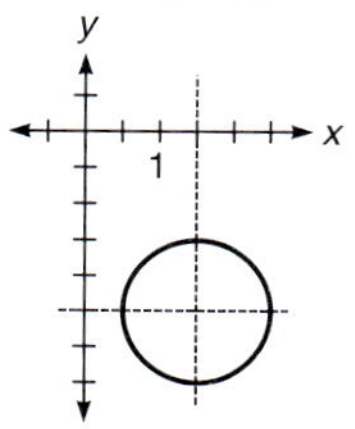

59. $\dfrac{x^2}{4} + \dfrac{y^2}{\frac{25}{4}} = 1$

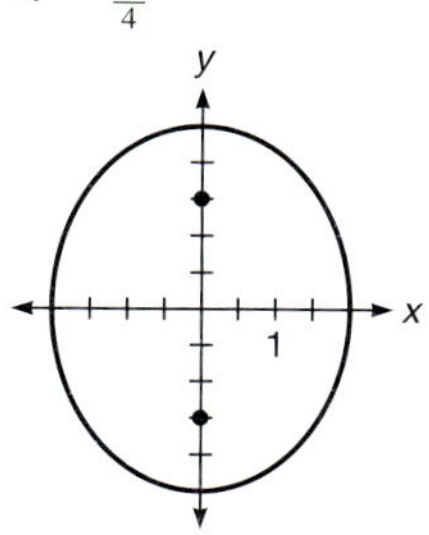

61. $y = (x+1)^2 + 3$; parabola;
focus: $(-1, 3\frac{1}{4})$

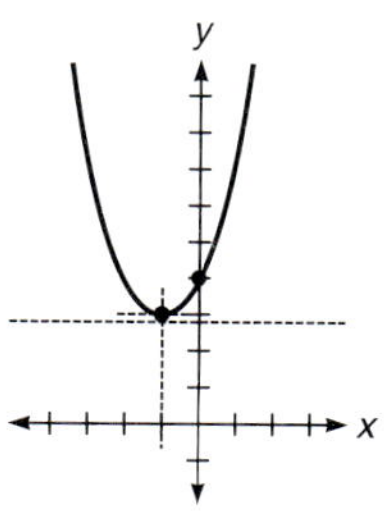

63. $(x+1)^2 + (y - \frac{1}{2})^2 = 8$; circle

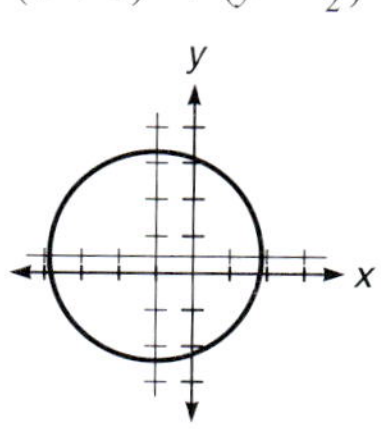

65. $y = (x - \frac{5}{2})^2 + \frac{1}{2}$; focus: $(2\frac{1}{2}, -\frac{1}{4})$

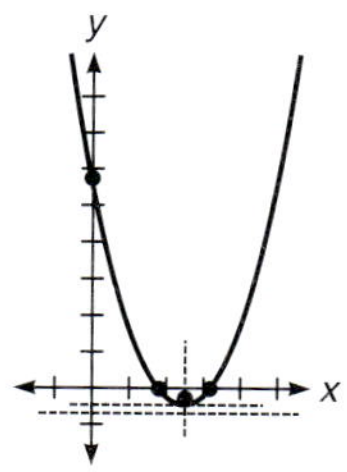

67. $\dfrac{x^2}{625} - \dfrac{y^2}{3{,}600} = 1$

69. $\dfrac{y^2}{4} - (x + \frac{1}{2})^2 = 1$

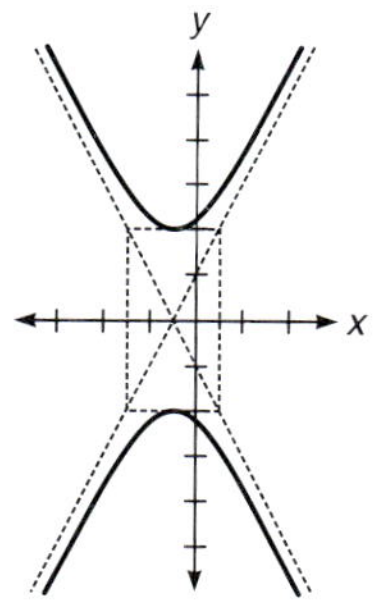

71. $(\sqrt{(x+c)^2 + y^2})^2 = (\sqrt{(x-c)^2 + y^2} \pm 2a)^2$

$(x+c)^2 + y^2 = [(x-c)^2 + y^2] \pm 4a\sqrt{(x-c)^2 + y^2} + 4a^2$

$x^2 + 2cx + c^2 + y^2 = x^2 - 2cx + c^2 + y^2 \pm 4a\sqrt{(x-c)^2 + y^2} + 4a^2$

$4cx - 4a^2 = \pm 4a\sqrt{(x-c)^2 + y^2}$

$cx - a^2 = \pm a\sqrt{(x-c)^2 + y^2}$

$c^2x^2 - 2a^2cx + a^4 = a^2[(x-c)^2 + y^2]$

$c^2x^2 - 2a^2cx + a^4 = a^2x^2 - 2a^2cx + a^2c^2 + a^2y^2$

$c^2x^2 - a^2x^2 - a^2y^2 = a^2c^2 - a^4$

$(c^2 - a^2)x^2 - a^2y^2 = a^2(c^2 - a^2)$

Let $b^2 = c^2 - a^2$; this is valid since $|c| > |a|$, so $c^2 > a^2$, and $c^2 - a^2 > 0$.

$b^2x^2 - a^2y^2 = a^2b^2$

$\dfrac{b^2x^2}{a^2b^2} - \dfrac{a^2y^2}{a^2b^2} = \dfrac{a^2b^2}{a^2b^2}$

$\dfrac{x^2}{a^2} - \dfrac{y^2}{b^2} = 1$

Solutions to skill and review problems

1. $x^2 + y^2 - 8y = 0$
$x^2 + y^2 - 8y + 16 = 16$
$x^2 + (y-4)^2 = 16$
Circle; center is (0,4), radius is 4.

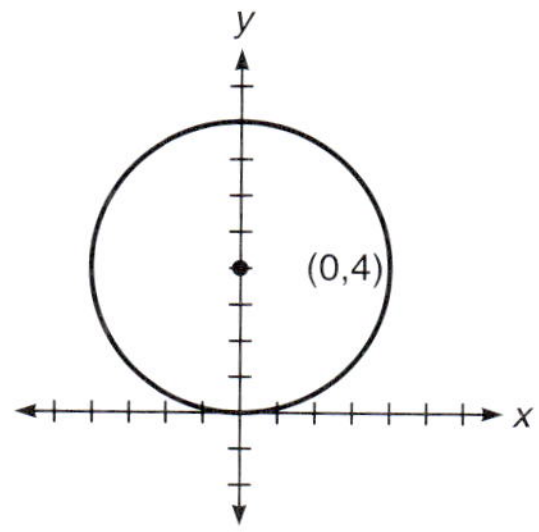

2. $3x^2 + 4y^2 = 12$
$\dfrac{x^2}{4} + \dfrac{y^2}{3} = 1$
Ellipse; $a = 2$, $b = \sqrt{3}$, $c = 1$; foci: $(\pm 1, 0)$

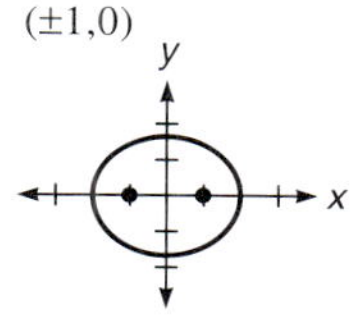

3. $3x - 2y = 6$
Straight line; x-intercept is (2,0); y-intercept is (0,−3).

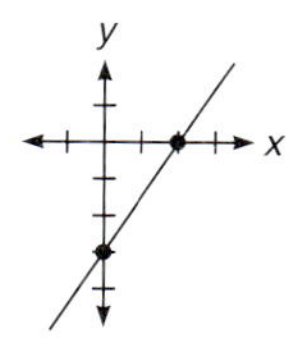

4. $2x^2 - 4x + 4y^2 = 2$
$x^2 - 2x + 2y^2 = 1$
$x^2 - 2x + 1 + 2y^2 = 1 + 1$
$(x-1)^2 + 2y^2 = 2$
$\dfrac{(x-1)^2}{2} + y^2 = 1$
Ellipse; center (1,0); $a = \sqrt{2}$, $b = 1$, $c = 1$; foci: $(1 \pm 1, 0) = (0,0), (2,0)$

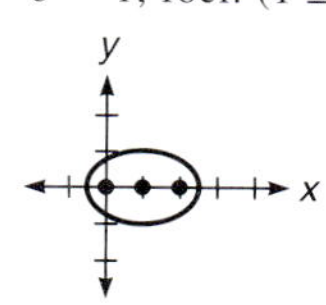

5. $y = x^2 - 6x - 8$
$y = x^2 - 6x + 9 - 8 - 9$
$y = (x - 3)^2 - 17$
Parabola; vertex: $(3,-17)$; intercepts:
$x = 0$: $y = -8$; $(0,-8)$
$y = 0$: $0 = (x - 3)^2 - 17$
$(x - 3)^2 = 17$
$x - 3 = \pm\sqrt{17}$
$x = 3 \pm\sqrt{17}$; $(-1.1,0)$, $(7.1,0)$

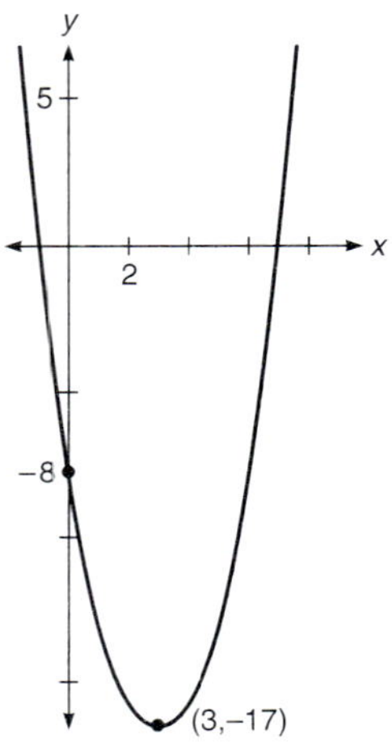

36. $\dfrac{25(x - 1)^2}{36} - \dfrac{9(y + 1)^2}{4} = 1$

$\dfrac{(x - 1)^2}{\frac{36}{25}} - \dfrac{(y + 1)^2}{\frac{4}{9}} = 1$

$a = \frac{6}{5}$, $b = \frac{2}{3}$,

$c = \sqrt{\dfrac{36}{25} + \dfrac{4}{9}} = \sqrt{\dfrac{424}{225}} = \dfrac{2\sqrt{106}}{15}$

center: $(1,-1)$

foci: $\left(1 \pm \dfrac{2\sqrt{106}}{15},-1\right) \approx (-0.4,-1)$, $(2.4,-1)$

end points of major axis: $(-\frac{1}{5},-1)$, $(\frac{11}{5},-1)$

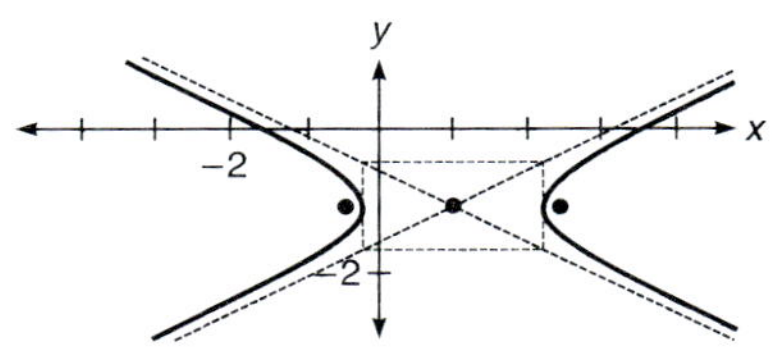

39. $4x^2 + 8x - 3y^2 + 24y + 4 = 0$
$4(x^2 + 2x) - 3(y^2 - 8y) = -4$
$4(x^2 + 2x + 1) - 3(y^2 - 8y + 16) = -4 + 4(1) - 3(16)$
$4(x + 1)^2 - 3(y - 4)^2 = -48$
$3(y - 4)^2 - 4(x + 1)^2 = 48$

$\dfrac{(y - 4)^2}{16} - \dfrac{(x + 1)^2}{12} = 1$

center $(-1,4)$
$c = \sqrt{16 + 12} = 2\sqrt{7}$; foci $(-1,4 \pm 2\sqrt{7})$
$a = 4$, $b = 2\sqrt{3}$; end points of major axis: $(-1,0)$, $(-1,8)$

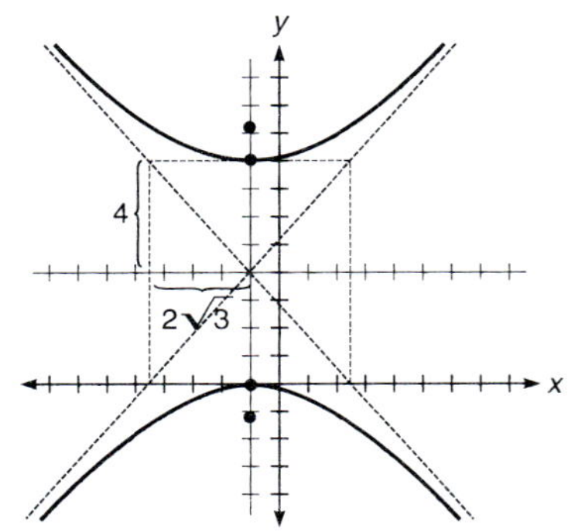

Solutions to trial exercise problems

23. $8x^2 - 3y^2 = 4$

$2x^2 - \dfrac{3y^2}{4} = 1$

$\dfrac{x^2}{\frac{1}{2}} - \dfrac{y^2}{\frac{4}{3}} = 1$

$c = \sqrt{\frac{4}{3} + \frac{1}{2}} = \sqrt{\frac{11}{6}}$

$= \sqrt{\frac{66}{36}} = \frac{\sqrt{66}}{6}$

foci: $\left(\pm\dfrac{\sqrt{66}}{6},0\right)$

$a = \sqrt{\dfrac{1}{2}} = \dfrac{\sqrt{2}}{2}$

$b = \sqrt{\dfrac{4}{3}} = \dfrac{2\sqrt{3}}{3}$

end points of major axis: $\left(\pm\dfrac{\sqrt{2}}{2},0\right)$

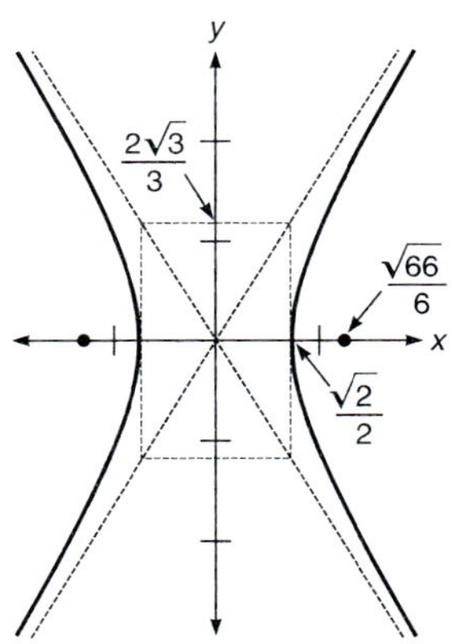

65. $4y = 4x^2 - 20x + 23$; parabola since the equation is quadratic in only one variable.

$y - \frac{23}{4} = x^2 - 5x$

$y - \frac{23}{4} + \frac{25}{4} = x^2 - 5x + \frac{25}{4}$

$y + \frac{1}{2} = (x - \frac{5}{2})^2$

center: $(2\frac{1}{2},-\frac{1}{2})$

$\dfrac{1}{4p} = 1$, $p = \dfrac{1}{4}$; focus:

$(2\frac{1}{2},-\frac{1}{2} + \frac{1}{4}) = (2\frac{1}{2},-\frac{1}{4})$

intercepts:

$x = 0$: $4y = 23$, $y = \frac{23}{4}$

$y = 0$: $\frac{1}{2} = (x - \frac{5}{2})^2$

$\pm\dfrac{\sqrt{2}}{2} = x - \dfrac{5}{2}$

$x = \dfrac{5}{2} \pm \dfrac{\sqrt{2}}{2}$

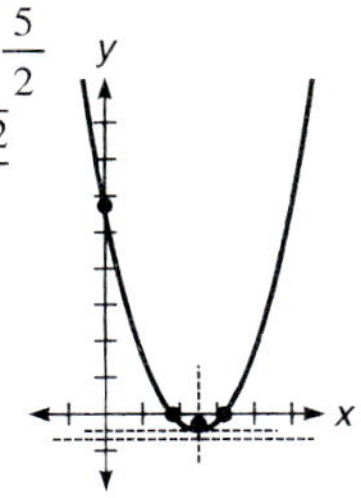

68. We know that $c = \frac{80}{2} = 40$, and $2a = 60$, so $a = 30$.

$c^2 = a^2 + b^2$

$40^2 = 30^2 + b^2$

$700 = b^2$

$\frac{x^2}{a^2} - \frac{y^2}{b^2} = 1$

$\frac{x^2}{900} - \frac{y^2}{700} = 1$

70. $\frac{y^2}{a^2} - \frac{x^2}{b^2} = 1$

$\frac{y^2}{a^2} = \frac{x^2}{b^2} + 1$

$y^2 = \frac{a^2}{b^2}x^2 + a^2$

$y^2 \approx \frac{a^2}{b^2}x^2$, as x gets larger and larger.

$y \approx \pm\frac{a}{b}x$

Exercise 11–4

Answers to odd-numbered problems

1. $(-2,-3)$, $(3,7)$

3. $\left(\frac{1+\sqrt{33}}{4},\frac{15-\sqrt{33}}{8}\right)$, $\left(\frac{1-\sqrt{33}}{4},\frac{15+\sqrt{33}}{8}\right)$

5. $(-1,1)$, $(2\frac{1}{2},9\frac{3}{4})$

7. $(1,-6)$, $(3,-4)$

9. $(-1+\sqrt{11},1+\sqrt{11})$, $(-1-\sqrt{11},1-\sqrt{11})$

11. $\left(\frac{1-\sqrt{7}}{4},\frac{-1-\sqrt{7}}{2}\right)$, $\left(\frac{1+\sqrt{7}}{4},\frac{-1+\sqrt{7}}{2}\right)$

13. $(0,1)$, $(\frac{2}{3},-\frac{1}{3})$

15. $(-\frac{1}{3},-1\frac{2}{3})$, $(1,1)$

17. $(\frac{1}{2},-1\frac{1}{2})$

19. $(2,\sqrt{3})$, $(2,-\sqrt{3})$, $(-2,\sqrt{3})$, $(-2,-\sqrt{3})$

21. $\left(\frac{4\sqrt{7}}{7},\frac{3\sqrt{14}}{7}\right)$, $\left(\frac{4\sqrt{7}}{7},-\frac{3\sqrt{14}}{7}\right)$, $\left(-\frac{4\sqrt{7}}{7},\frac{3\sqrt{14}}{7}\right)$, $\left(-\frac{4\sqrt{7}}{7},-\frac{3\sqrt{14}}{7}\right)$

23. $5x^2 - 10x + 5y^2 - 20y - 56 = 0$

25. $(x-2)^2 + (y-5)^2 = 32$

27.
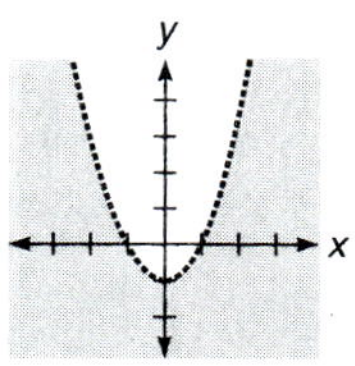

29.
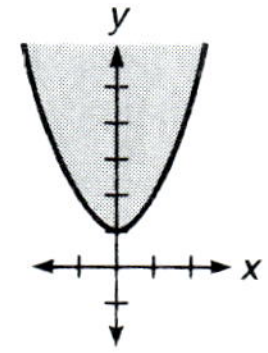

31.
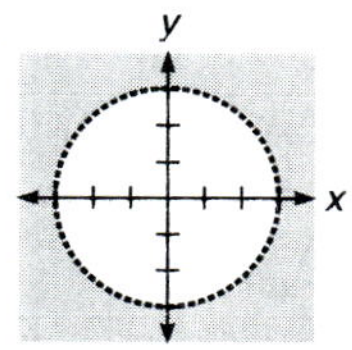

33.
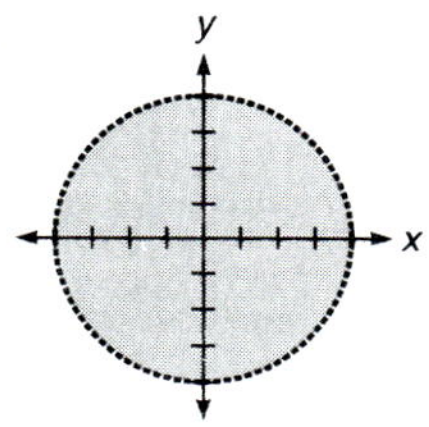

35.
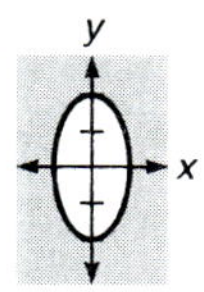

37.
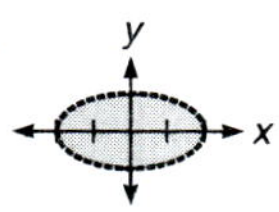

39.
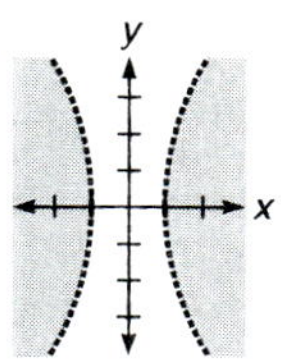

41.
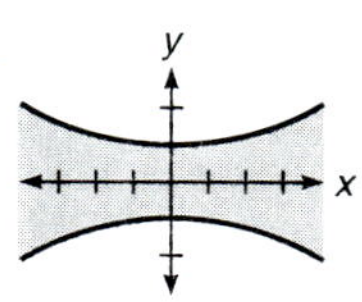

43.
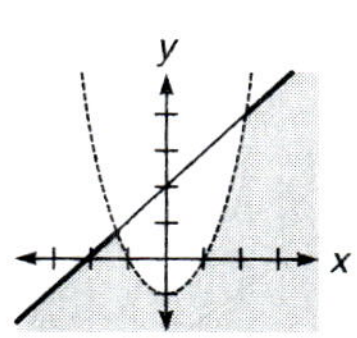

45.
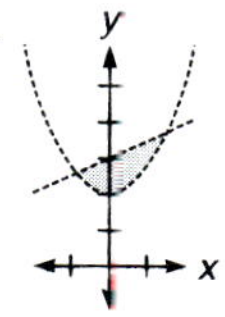

47.
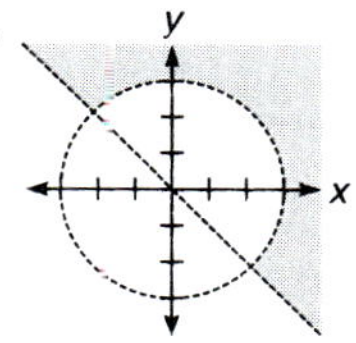

49.
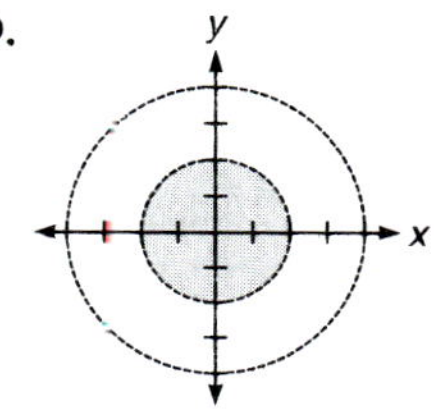

51.
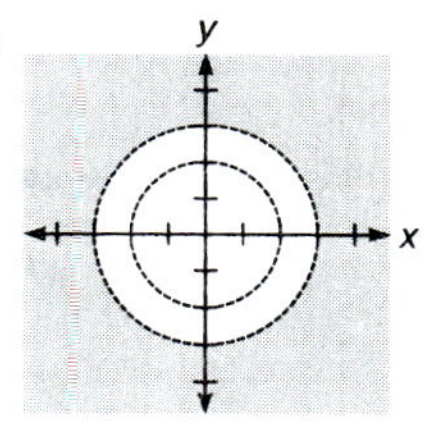

53.
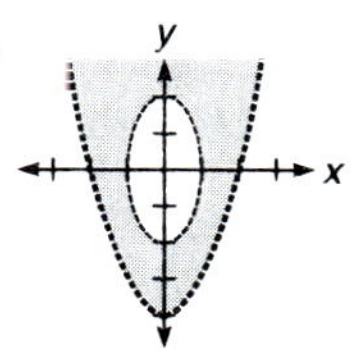

55.
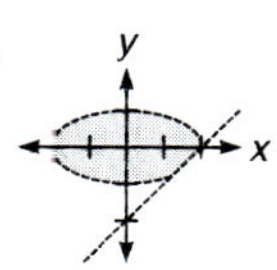

57.
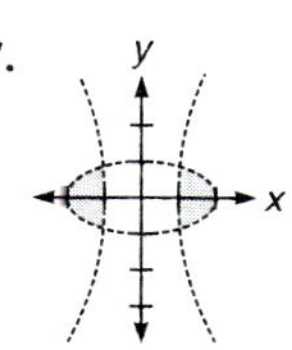

59.
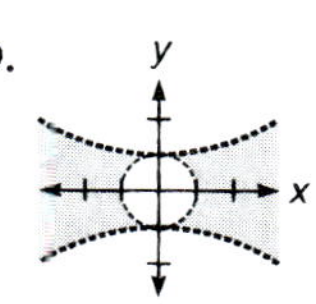

61. 133 feet

63. **a.** $(5 + x)(3 - y) \geq 15$

$15 - 5y + 3x - xy \geq 15$

$-5y - xy \geq -3x$

$5y + xy \leq 3x$

$y(5 + x) \leq 3x$

$y \leq \dfrac{3x}{x + 5}$

Note that we divided an inequality by $x + 5$; we know that $x > 0$ so that $x + 5 > 0$ also. Thus, the direction of the inequality is not affected.

b. We graph the equality

$y = \dfrac{3x}{x + 5} = 3 - \dfrac{15}{x + 5}$.

Horizontal asymptote: $y = 3$

Vertical asymptote: $x = -5$

Intercepts: Origin

Additional points: $(-12,5.1)$, $(-7,10.5)$, $(-3,-4.5)$, $(3,1.1)$, $(10,2)$

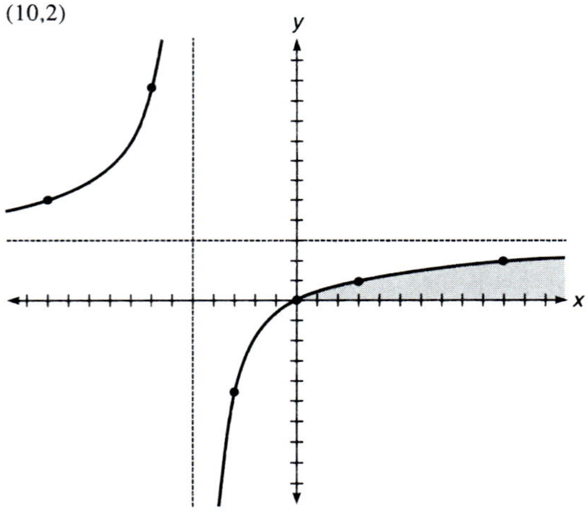

Solutions to skill and review problems

1. $y - 3x = -9$

Straight line

Intercepts:

$x = 0$: $y - 0 = -9$; $(0,-9)$

$y = 0$: $0 - 3x = -9$

$x = 3$; $(3,0)$

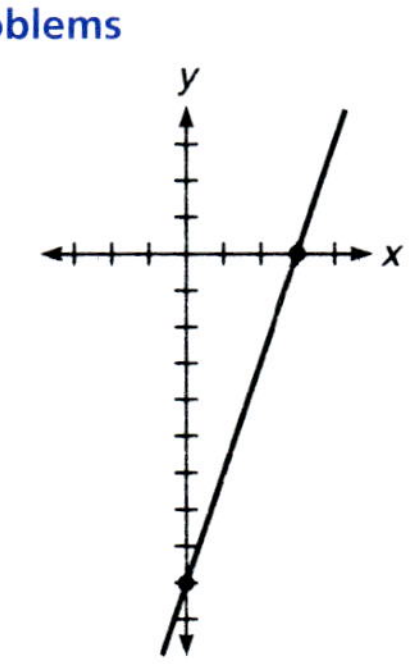

2. $y - 3x^2 = -9$

$y = 3x^2 - 9$

Parabola

Vertex: $(0,-9)$

Intercepts:

$x = 0$: $y = 0 - 9 = -9$; $(0,-9)$

$y = 0$: $0 = 3x^2 - 9$

$3x^2 = 9$

$x^2 = 3$

$x = \pm\sqrt{3}$; $(\pm\sqrt{3},0)$

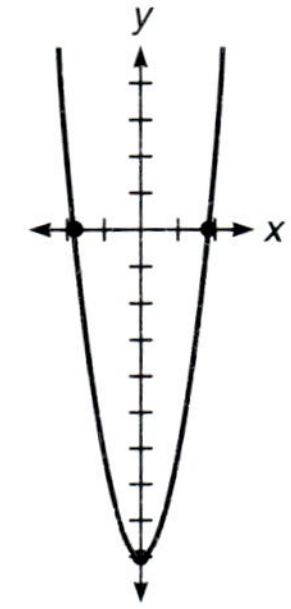

3. $y^2 - 3x^2 = 9$

Hyperbola

Intercepts:

$x = 0$: $y^2 = 9$

$y = \pm 3$; $(0,\pm 3)$

$y = 0$: $-3x^2 = 9$; No real solution.

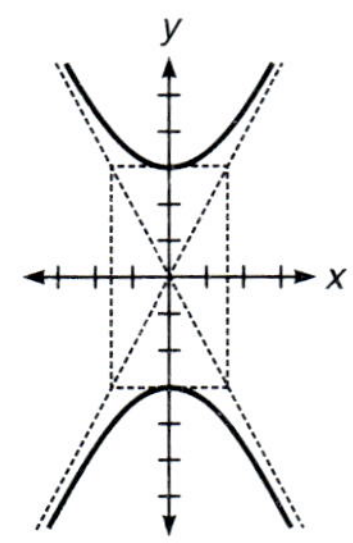

4. $y^2 + 3x^2 = 9$

Ellipse

Intercepts:

$x = 0$: $y^2 = 9$

$y = \pm 3$; $(0,\pm 3)$

$y = 0$: $3x^2 = 9$

$x^2 = 3$

$x = \pm\sqrt{3}$; $(\pm\sqrt{3},0)$

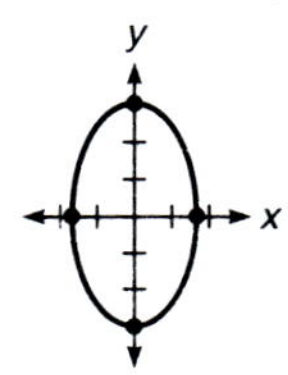

5. $3y^2 + 3x^2 = 9$

$y^2 + x^2 = 3$

Divide each member by 3.

Circle; center at origin, radius $= \sqrt{3}$.

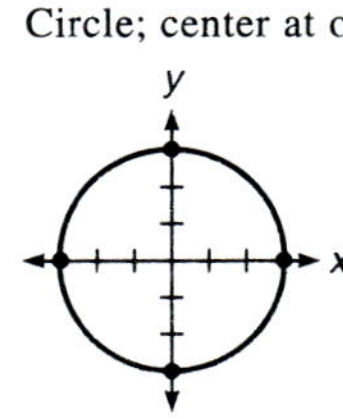

6. $y = 2x^5 + x^4 - 10x^3 - 5x^2 + 8x + 4$

Possible rational zeros: $\pm 1, \pm 2, \pm 4, \pm\frac{1}{2}$.

We use synthetic division to find rational zeros.

	2	1	−10	−5	8	4
		2	3	−7	−12	−4
1	2	3	−7	−12	−4	0

	2	3	−7	−12	−4
		4	14	14	4
2	2	7	7	2	0

	2	7	7	2
		−4	−6	−2
−2	2	3	1	0

$y = (x - 1)(x - 2)(x + 2)(2x^2 + 3x + 1)$

$y = (x - 1)(x - 2)(x + 2)(2x + 1)(x + 1)$

Intercepts:

$x = 0$: $y = 4$; $(0,4)$

$y = 0$: $0 = (x - 1)(x - 2)(x + 2)(2x + 1)(x + 1)$

$x = 1, 2, -2, -\frac{1}{2}, -1$

$(1,0), (2,0), (-2,0), (-\frac{1}{2},0), (-1,0)$

Additional points: $(-1.5,4.4)$, $(-0.75,-0.75)$, $(0.5,5.6)$, $(1.5,-8.8)$

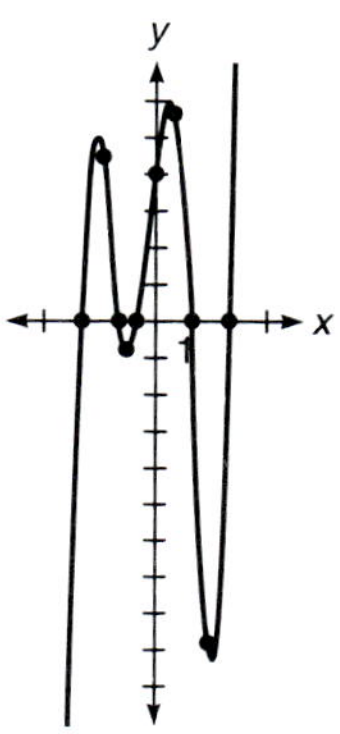

Solutions to trial exercise problems

5. $y = 3x^2 - 2x - 4$

$y = x^2 + x + 1$

$3x^2 - 2x - 4 = x^2 + x + 1$

$2x^2 - 3x - 5 = 0$

$(2x - 5)(x + 1) = 0$

$x = \frac{5}{2}$ or -1

$y = x^2 + x + 1$

$x = -1$: $y = (-1)^2 + (-1) + 1 = 1$

$x = \frac{5}{2}$: $y = (\frac{5}{2})^2 + \frac{5}{2} + 1 = \frac{25}{4} + \frac{10}{4} + \frac{4}{4} = \frac{39}{4}$

The points are $(-1,1)$ and $(2\frac{1}{2},9\frac{3}{4})$.

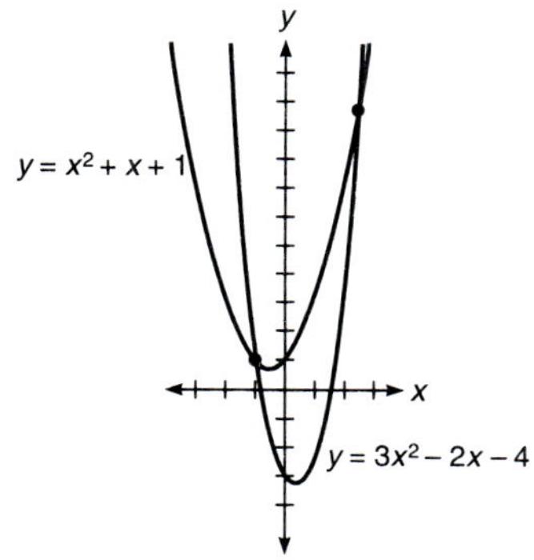

19. $x^2 - y^2 = 1$, $2y^2 - x^2 = 2$, so $x^2 = y^2 + 1$, $x^2 = 2y^2 - 2$

$2y^2 - 2 = y^2 + 1$

$y^2 = 3$

$y = \pm\sqrt{3}$

$x^2 = y^2 + 1$

$x^2 = 3 + 1$

$x = \pm 2$, so the points of intersection are $(2,\sqrt{3})$, $(2,-\sqrt{3})$, $(-2,\sqrt{3})$, $(-2,-\sqrt{3})$, or about $(2,\pm 1.7)$ and $(-2,\pm 1.7)$.

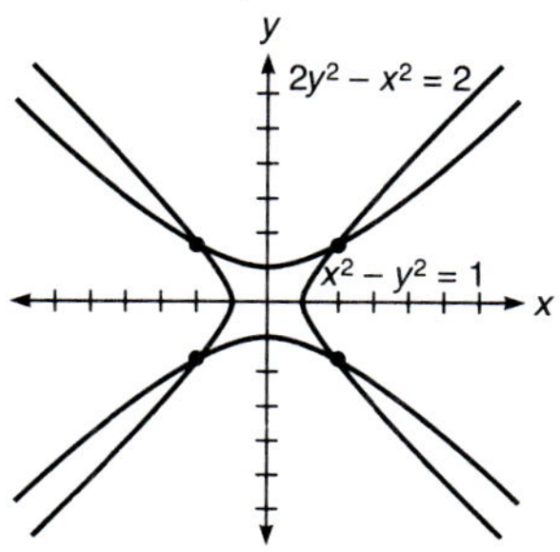

23. Let L represent the line $y = \frac{1}{2}x - 3$. Let (a,b) be the point where the circle is tangent to the line L. Let L' be the line which passes through $(1,2)$ and the point (a,b).

Since the slope of L is $\frac{1}{2}$, the slope of L' is -2 (section 3–2). Using $m = -2$ and the point $(1,2)$ we can find the equation of L' to be $y = -2x + 4$.

We can now find the point (a,b) by solving the system of equations

$y = \frac{1}{2}x - 3$

$y = -2x + 4$

$\frac{1}{2}x - 3 = -2x + 4$

$x - 6 = -4x + 8$

$5x = 14$

$x = \frac{14}{5}$

$y = -2x + 4 = -2(\frac{14}{5}) + 4 = -\frac{8}{5}$

Thus, (a,b) is $(\frac{14}{5},-\frac{8}{5})$.

Now find the distance between $(1,2)$ and $(\frac{14}{5},-\frac{8}{5})$:

$d = \sqrt{(x_2 - x_1)^2 + (y_2 - y_1)^2}$

$= \sqrt{(\frac{14}{5} - 1)^2 + (-\frac{8}{5} - 2)^2}$

$= \sqrt{(\frac{9}{5})^2 + (-\frac{18}{5})^2}$

$= \sqrt{\frac{405}{25}} = \frac{9}{5}\sqrt{5}$

This is the radius of the circle. Thus, the center of the circle is $(h,k) = (1,2)$ and $r = \frac{9}{5}\sqrt{5}$.

Circle: $(x - h)^2 + (y - k)^2 = r^2$

$(x - 1)^2 + (y - 2)^2 = (\frac{9}{5}\sqrt{5})^2$

$x^2 - 2x + 1 + y^2 - 4x + 4 = \frac{81}{5}$

$5x^2 - 10x + 5y^2 - 20y - 56 = 0$

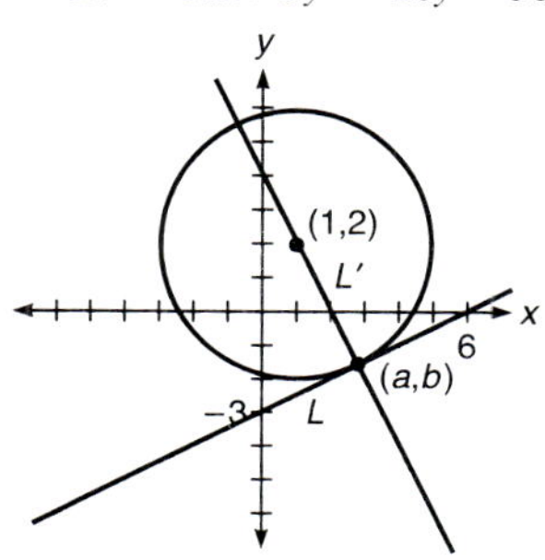

24. The circles have equations $(x + 3)^2 + (y - 3)^2 = 9$ or $x^2 + 6x + y^2 - 6y + 9 = 0$ and $x^2 + y^2 = 25$. To find where they intersect we need to solve one of them for y and substitute into the other equation. Using $x^2 + y^2 = 25$ we obtain $y^2 = 25 - x^2$ and $y = \pm\sqrt{25 - x^2}$

We can see that $y > 0$ for the points that interest us, so we will use $y = \sqrt{25 - x^2}$. Substituting these values for y into the first equation we obtain

$x^2 + 6x + (25 - x^2) - 6(\sqrt{25 - x^2}) + 9 = 0$

$6x + 34 = 6\sqrt{25 - x^2}$

$3x + 17 = 3\sqrt{25 - x^2}$

Now square both sides.

$9x^2 + 102x + 289 = 9(25 - x^2)$

$18x^2 + 102x + 64 = 0$

$9x^2 + 51x + 32 = 0$

$$x = \frac{-17 \pm \sqrt{161}}{6} \approx -0.71857, -4.9481$$

Note that we are keeping the first five nonzero digits in each value for now so that our final answer will have two-place accuracy.

To find the y-values we use $y = \sqrt{25 - x^2}$. From the last result we compute y and find that it is 0.71857 and 4.9481. These are just the absolute values of the x-values, which is not surprising given the symmetry of the points, as seen in the graph. Thus the two points of intersection are $(-0.71857, 4.9481)$, $(-4.9481, 0.71857)$.

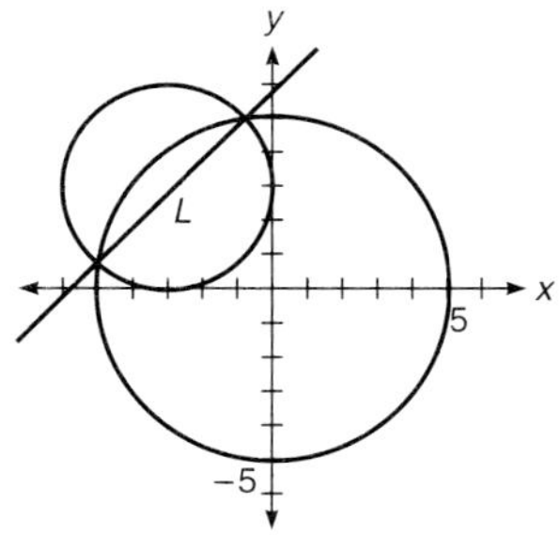

We can see from the graph or by computation that the slope of the line we want is 1, so using $y = x + b$ we compute b from either of the points we have. Using the first we obtain $4.9481 = -0.71857 + b$, so $b \approx 5.67$, to two decimal places. We can also see by some retracing of our steps that b is exactly

$$\frac{17 + \sqrt{161}}{6} - \frac{-17 + \sqrt{161}}{6} = \frac{17}{3}.$$

Thus, an approximate equation of the line is $y = x + 5.67$, and an exact solution is $y = x + 5\frac{2}{3}$.

25. The circle has equation

$$(x - 2)^2 + (y - 5)^2 = r^2$$

The circle touches the line $y = -x - 1$ at one point (since it is tangent to it); at this point, y may be replaced by $-x - 1$:

$(x - 2)^2 + (y - 5)^2 = r^2$

$(x - 2)^2 + ((-x - 1) - 5)^2 = r^2$

$2x^2 + 8x + 40 - r^2 = 0$

Now apply the quadratic formula with $a = 2$, $b = 8$, and $c = 40 - r^2$:

$$x = \frac{-8 \pm \sqrt{64 - 4(2)(40 - r^2)}}{4}$$

$$= \frac{-8}{4} \pm \frac{\sqrt{8r^2 - 256}}{4}$$

$$= -2 \pm \tfrac{1}{4}\sqrt{4(2r^2 - 64)}$$

$$= -2 \pm \tfrac{1}{2}\sqrt{2r^2 - 64}$$

We know that where the line touches the circle there is only one point, and therefore one value of x. This happens only if $2r^2 - 64$ is zero.

$2r^2 - 64 = 0$

$2r^2 = 64$

$r^2 = 32$

Thus, we learn the value of r^2, and so the equation of the circle is

$(x - 2)^2 + (y - 5)^2 = 32$

---

34. $4x^2 + y^2 < 4$

Graph the ellipse $4x^2 + y^2 = 4$; use (0,0) as a test point.

$4x^2 + y^2 < 4$

$4(0) + 0 < 4$

$0 < 4$

True, so the solution is the part of the plane that contains the origin.

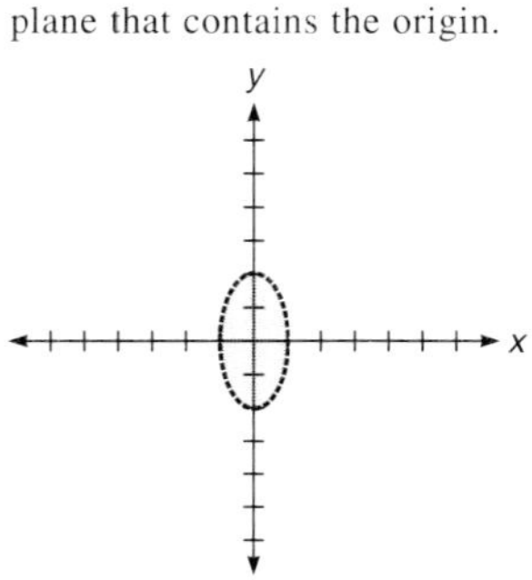

53.

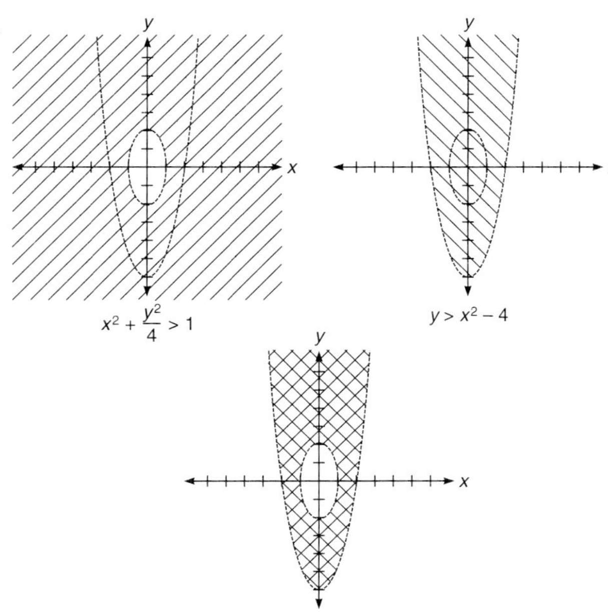

61. Since z is the time it takes to fall to the bottom of the well we know that $s = 16z^2$. Since the time to come back up is $3 - z$ seconds we know that $s = 1{,}100(3 - z)$. Thus,

$s = 16z^2$
$s = 1{,}100(3 - z)$
so
$16z^2 = 1{,}100(3 - z)$
$16z^2 + 1{,}100z - 3{,}300 = 0$
$4z^2 + 275z - 825 = 0$

$$z = \frac{-275 \pm \sqrt{(-275)^2 - 4(4)(-825)}}{2(4)}$$

$$z = \frac{-275 \pm \sqrt{88{,}825}}{8} \approx -71.6,\ 2.879$$

Ignoring the negative value for time, we find that it takes 2.879 seconds for the rock to fall. At this point s is computed as $s = 16(2.879^2) \approx 132.6$ feet. It takes the remaining $3 - 2.879$ or 0.121 seconds for the sound to travel back up the well, so $s = 1{,}100(0.121) = 133.1$ feet. Thus, both calculations show a depth of the well of 133 feet, to the nearest foot. (The results will be the same if more decimal places are used in the approximation of z).

Chapter 11 review

1. vertex: (0,0); focus: (0,−2); directrix: $y = 2$; all intercepts at the origin

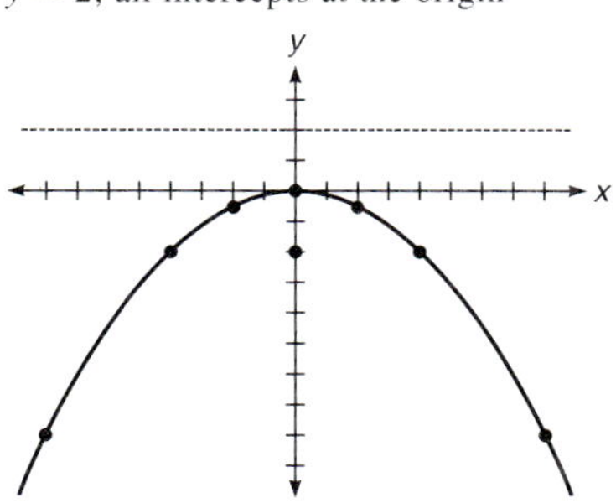

2. vertex: (3,−4); focus is at $(3,-3\frac{3}{4})$; directrix is $y = -4\frac{1}{4}$; intercepts: (1,0), (5,0), (0,5)

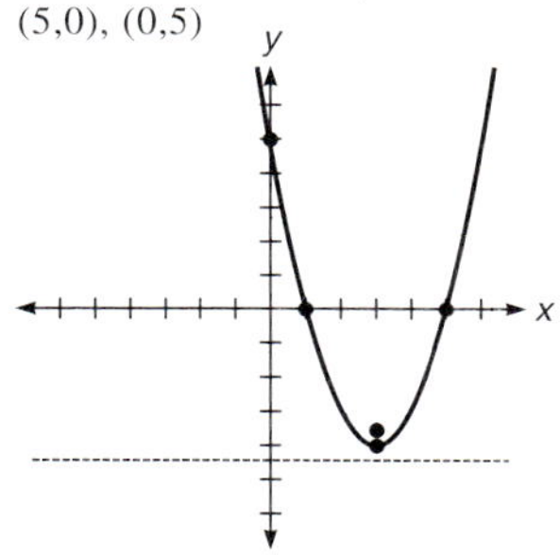

3. vertex: $(-\frac{2}{3},5\frac{1}{3})$;
focus: $(-\frac{2}{3},5\frac{1}{4})$;
directrix: $y = 5\frac{5}{12}$;
intercepts: $(\frac{2}{3},0)$, (−2,0), (0,4)

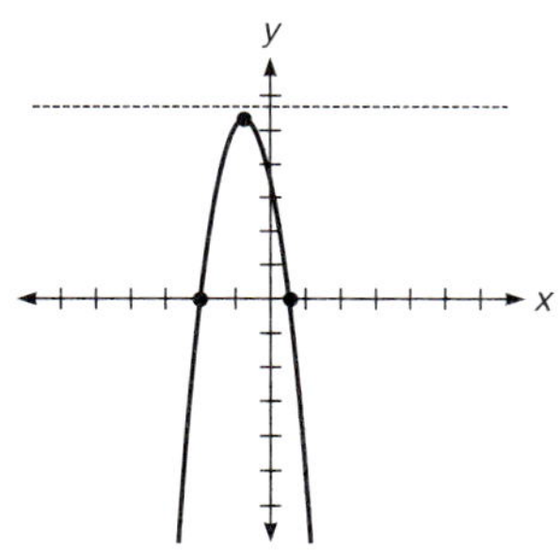

4. vertex: (−2,2);
focus: $(-2,2\frac{1}{4})$;
directrix: $y = 1\frac{3}{4}$;
intercepts: (0,6)

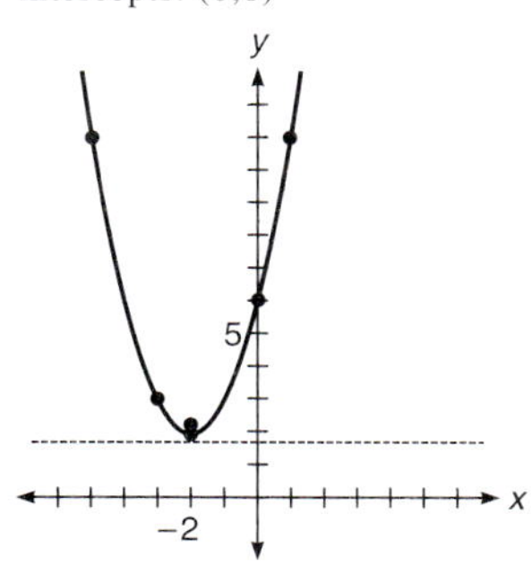

5. vertex: $(-20\frac{1}{4},3\frac{1}{2})$;
focus: $(-20,3\frac{1}{2})$;
directrix: $x = -20\frac{1}{2}$;
intercept: (0,−1), (0,8), (−8,0)

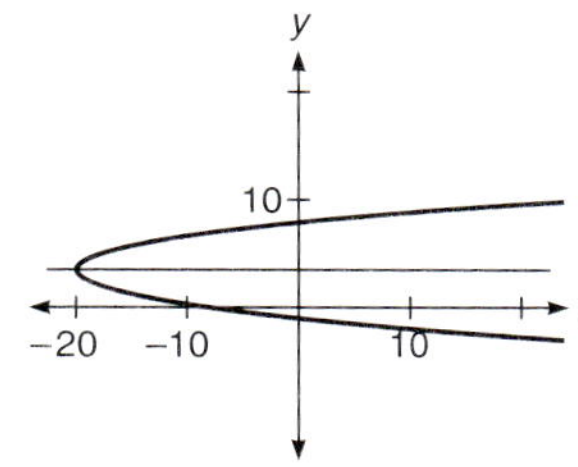

6. vertex: (0,2);
focus: $(\frac{1}{4},2)$
directrix: $x = -\frac{1}{4}$
intercept: (0,2), (4,0)

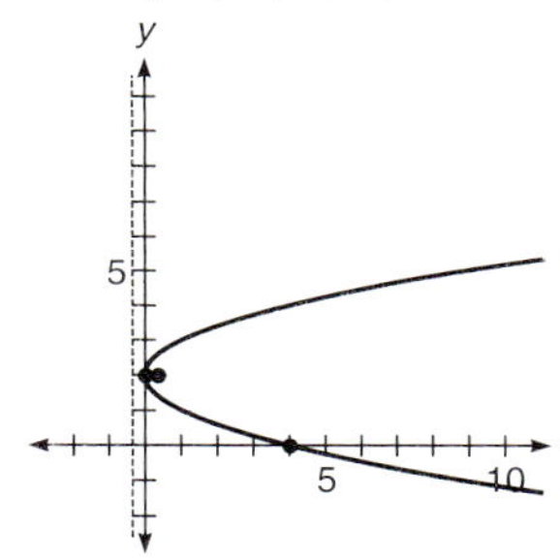

7. $y = -\frac{1}{10}(x - 1)^2 - \frac{1}{2}$
8. $y = \frac{1}{2}(x + 3)^2 - \frac{3}{2}$
9. $y = -(x - 2)^2 - 1$
10. $y = \frac{1}{4}(x + 4)^2 + 1$
11. $y = 4(x - 3)^2 - 1$
12. $w = 16\sqrt{10}$
13. $d = 9\frac{3}{8}$
14. $h = \frac{25}{16}$
15. foci: $(-\sqrt{3},0)$, $(\sqrt{3},0)$

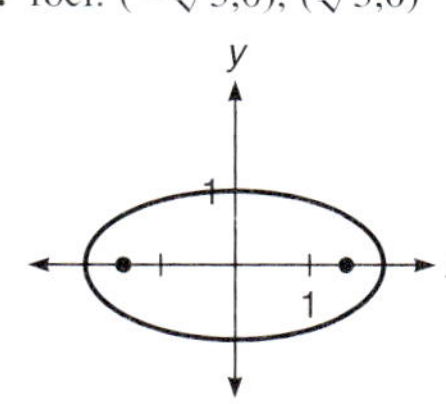

16.

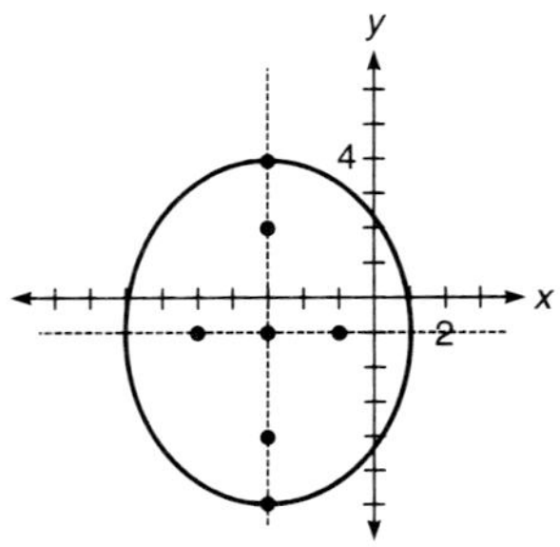

17. foci: $(0,\pm\sqrt{2})$; minor axis: $(\pm\sqrt{2},0)$

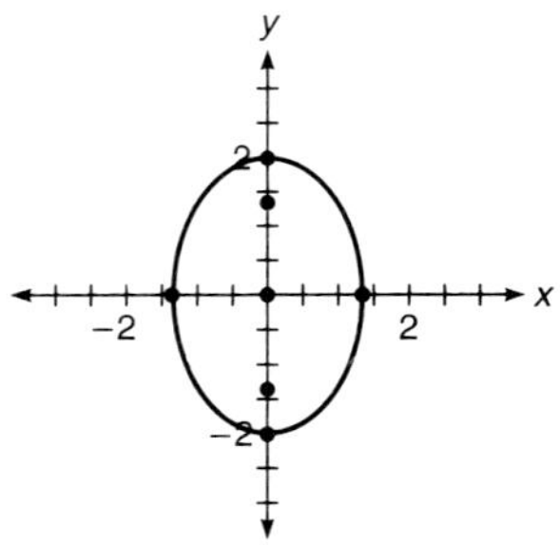

18. foci: $\left(\pm\frac{\sqrt{2}}{2},0\right)$; minor axis: $\left(0,\pm\frac{\sqrt{2}}{2}\right)$

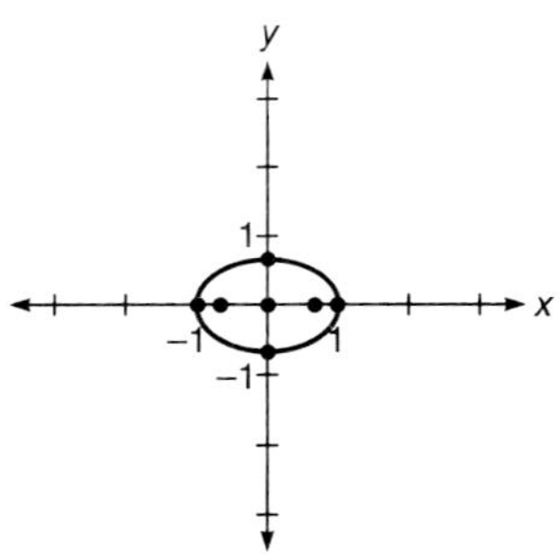

19. foci: $(3\pm2\sqrt{3},-2)$

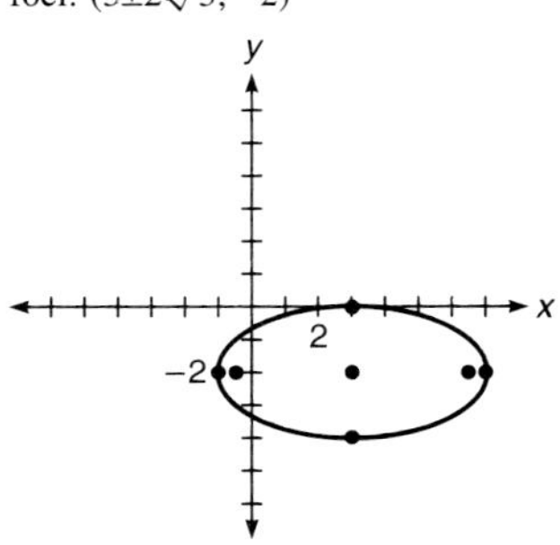

20. $\frac{x^2}{16}+\frac{y^2}{7}=1$ **21.** $\frac{x^2}{9}+\frac{y^2}{16}=1$

22. $\frac{x^2}{16}+\frac{y^2}{7}=1$

23. foci: $(0,\pm5)$

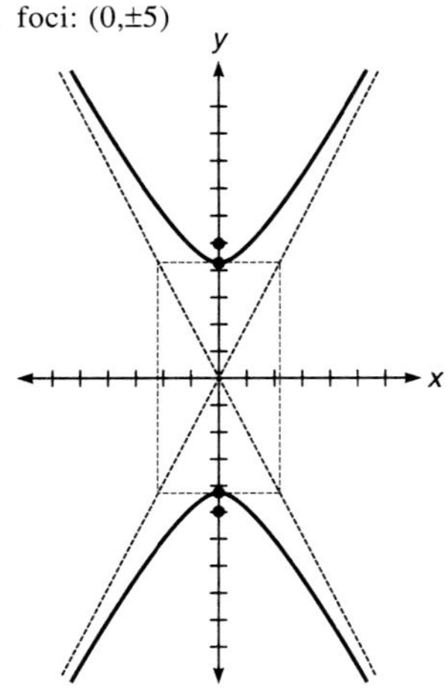

24. foci: $(\pm\sqrt{3},0)$

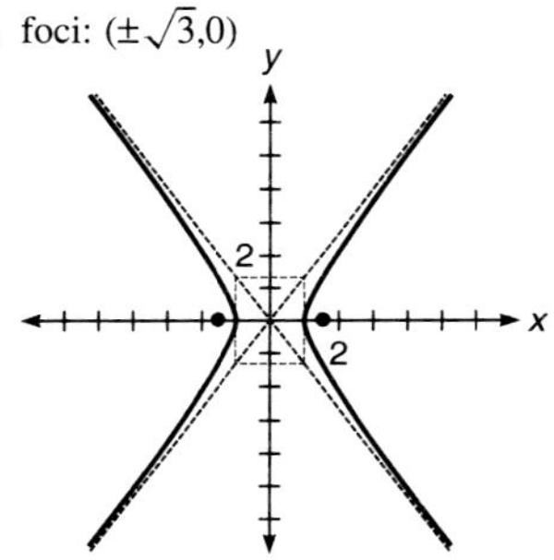

25. $y^2-\frac{x^2}{4}=4$; foci at $(0,\pm\sqrt{5})$

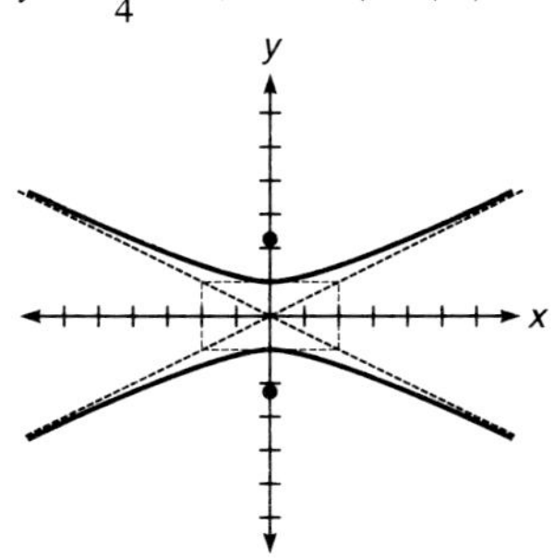

26. $x^2-\frac{y^2}{\frac{8}{3}}=1$; foci: $\left(\pm\frac{\sqrt{33}}{3},0\right)$

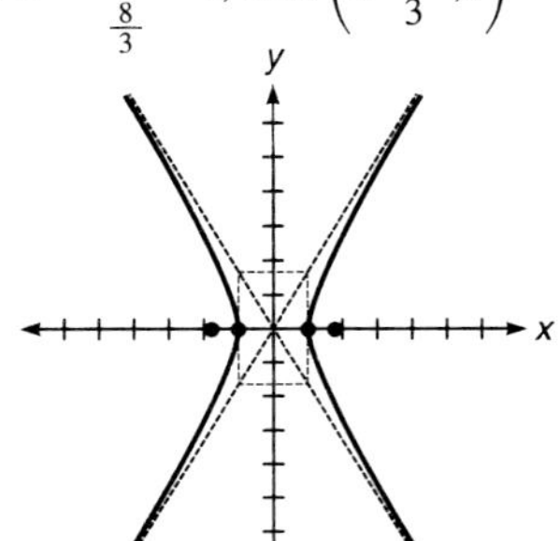

27. foci: $(1\pm2\sqrt{6},-3)$

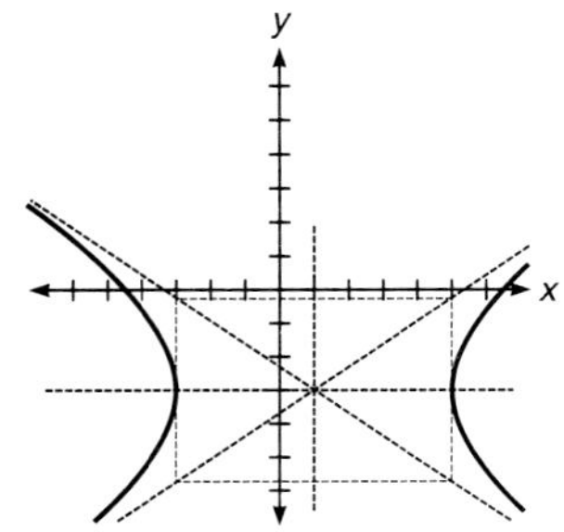

28. $\frac{(y+4)^2}{3}-(x-2)^2=1$

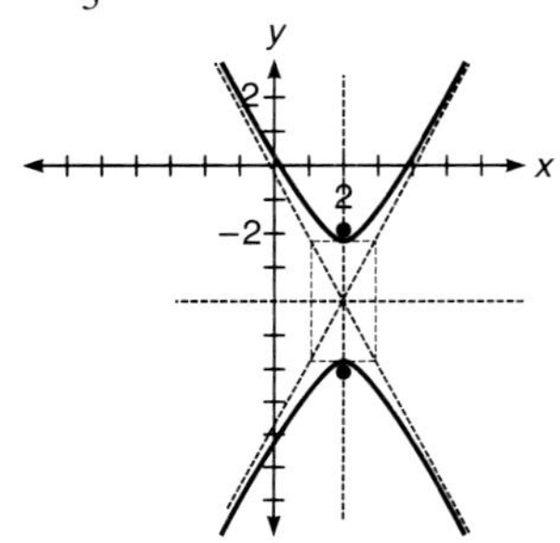

29. hyperbola; $\frac{(x+1)^2}{9}-\frac{(y-2)^2}{\frac{9}{2}}=1$

30. ellipse; $\frac{y^2}{4}+\frac{x^2}{9}=1$

31. ellipse; $\frac{(x+3)^2}{2}+\frac{(y-3)^2}{4}=1$

32. circle; $(x-\frac{3}{2})^2+y^2=\frac{39}{4}$

33. parabola; $y=(x+\frac{3}{2})^2-\frac{25}{4}$

34. ellipse; $\frac{(x+4)^2}{36}+\frac{(y-\frac{1}{2})^2}{9}=1$

35. parabola; $x=(y-\frac{5}{2})^2-\frac{1}{2}$

36. $(0,4)$ and $(3\frac{2}{3},6\frac{4}{9})$

37. $(0,-1)$ and $(-\frac{8}{9},\frac{7}{9})$

38. $(-\frac{1}{2}+\frac{1}{2}\sqrt{6},\frac{3}{2}+\frac{1}{2}\sqrt{6})$ and $(-\frac{1}{2}-\frac{1}{2}\sqrt{6},\frac{3}{2}-\frac{1}{2}\sqrt{6})$

39. $(-\frac{2}{13}+\frac{4}{13}\sqrt{231},-\frac{40}{13}+\frac{2}{13}\sqrt{231})$ and $(-\frac{2}{13}-\frac{4}{13}\sqrt{231},-\frac{40}{13}-\frac{2}{13}\sqrt{231})$

40. $(\frac{39}{31}+\frac{5}{31}\sqrt{41},\frac{63}{31}+\frac{20}{31}\sqrt{41})$ and $(\frac{39}{31}-\frac{5}{31}\sqrt{41},\frac{63}{31}-\frac{20}{31}\sqrt{41})$

41. $(1\frac{1}{15},-\frac{14}{15})$

42. $(-4,7)$ and $(1,-3)$

43. $(3,16)$ and $(-1,0)$

44. $(x+2)^2+(y-3)^2=\frac{121}{10}$

45.

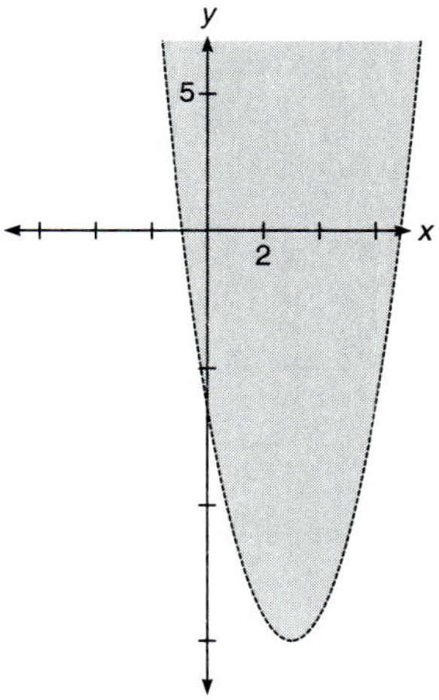

46.

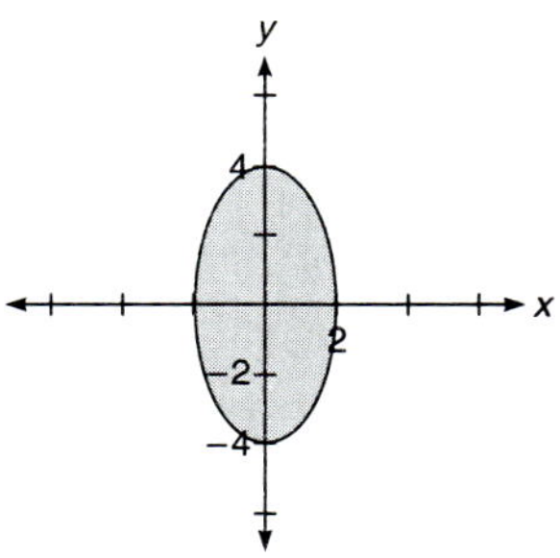

47.

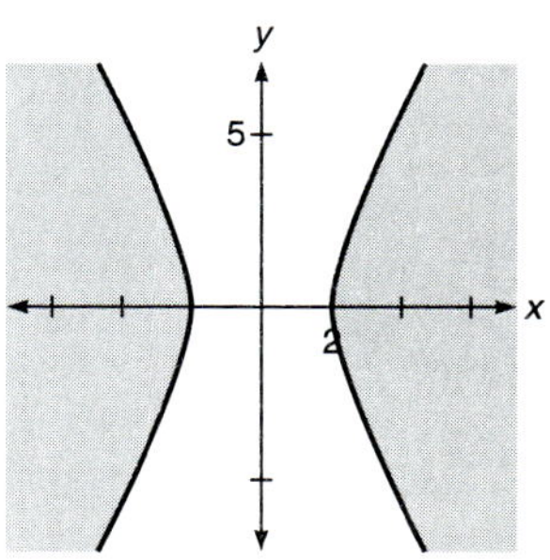

48.

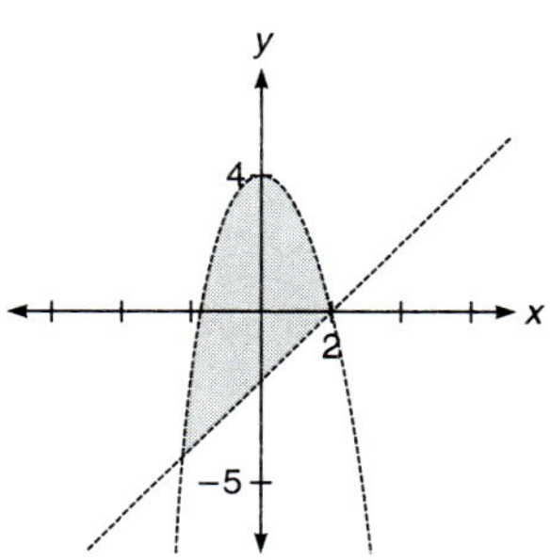

49.

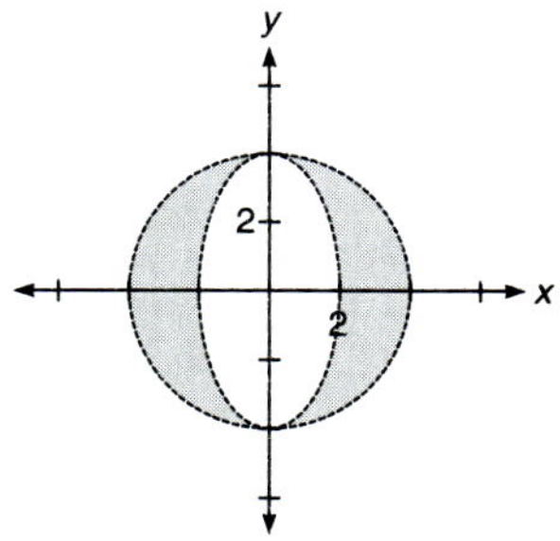

50.

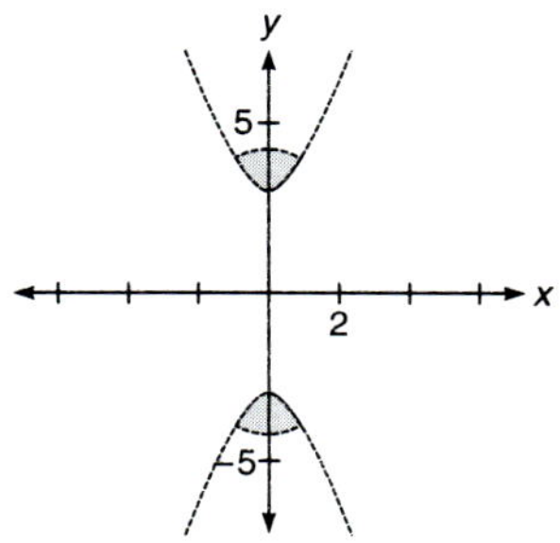

Chapter 11 test

1. all intercepts and vertex at the origin; focus: $(0,-\frac{1}{16})$; directrix: $y = \frac{1}{16}$

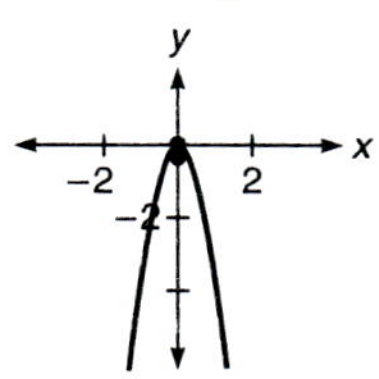

2. intercepts: $(-3,0)$, $(1\frac{1}{2},0)$, $(0,-9)$; focus: $(-\frac{3}{4},-10)$; directrix: $y = -10\frac{1}{4}$; vertex: $(-\frac{3}{4},-10\frac{1}{8})$

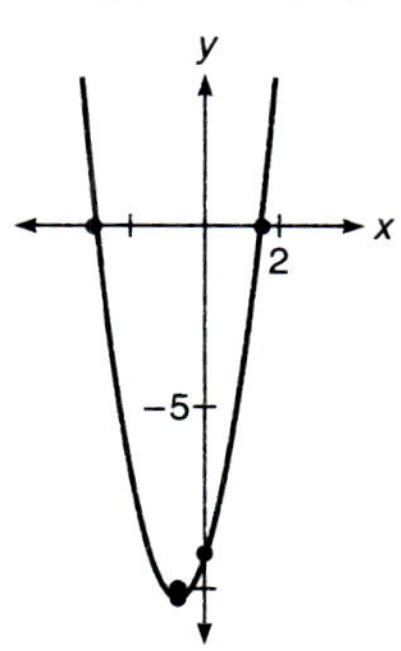

3. intercepts: $(-4,0)$, $(2,0)$, $(0,8)$; focus: $(-1,8\frac{3}{4})$; directrix: $y = 9\frac{1}{4}$; vertex: $(-1,9)$

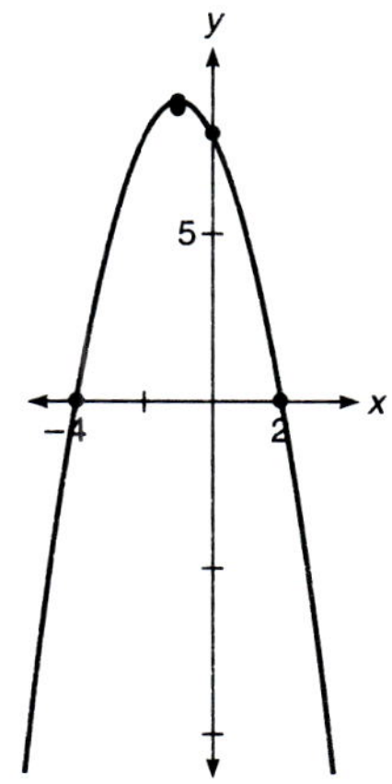

4. intercepts: $(0,-2)$, $(0,6)$, $(-12,0)$; focus: $(-15\frac{3}{4},2)$; directrix: $x = -16\frac{1}{4}$; vertex: $(-16,2)$

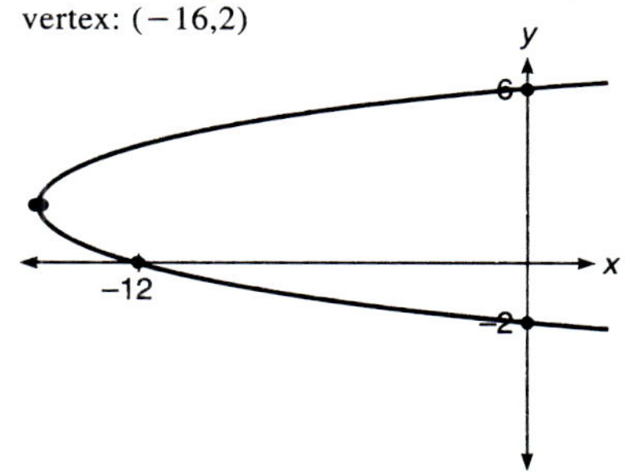

5. $y = \frac{1}{2}(x - 1)^2 + 2\frac{1}{2}$
6. $y = -\frac{1}{8}(x + 2)^2$
7. $w = 8\sqrt{6}$
8. $h = 2\frac{1}{2}$
9. $\dfrac{x^2}{4} + \dfrac{y^2}{16} = 1$; foci: $(0,2 \pm \sqrt{3})$

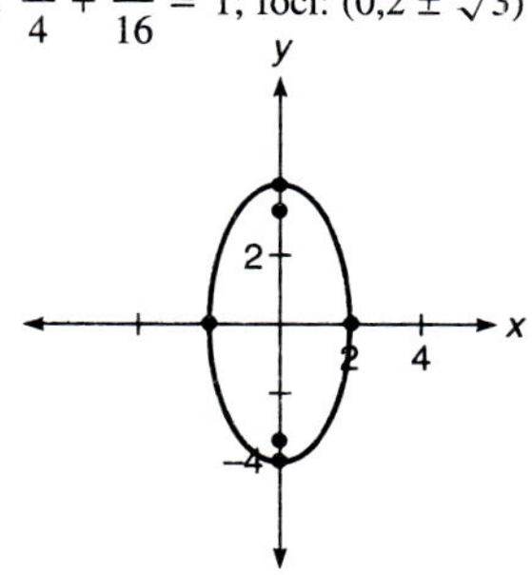

10. $\dfrac{x^2}{9} + \dfrac{(y-5)^2}{36} = 1$; foci: $(0,5 \pm 3\sqrt{3})$

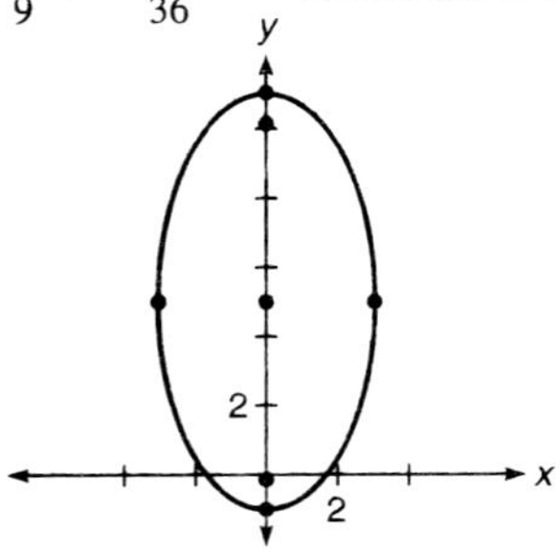

11. $\dfrac{(x-2)^2}{15} + \dfrac{(y+\frac{3}{2})^2}{10} = 1$; ends of minor axis: $(2, -1\frac{1}{2} \pm \sqrt{10})$; ends of major axis: $(2 \pm \sqrt{15}, -1\frac{1}{2})$; foci: $(2 \pm \sqrt{5}, -1\frac{1}{2})$

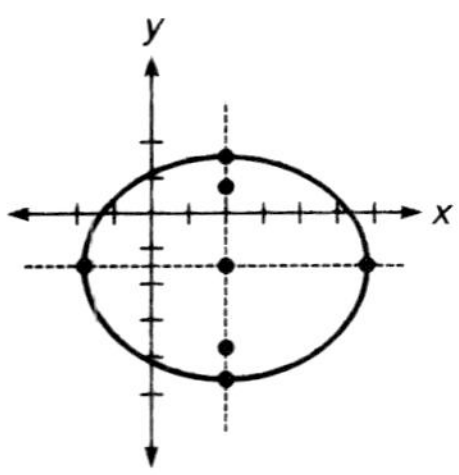

12. $\dfrac{x^2}{25} + \dfrac{y^2}{16} = 1$

13. $\dfrac{x^2}{64} + \dfrac{y^2}{48} = 1$

14. foci: $(0, \pm\sqrt{13})$

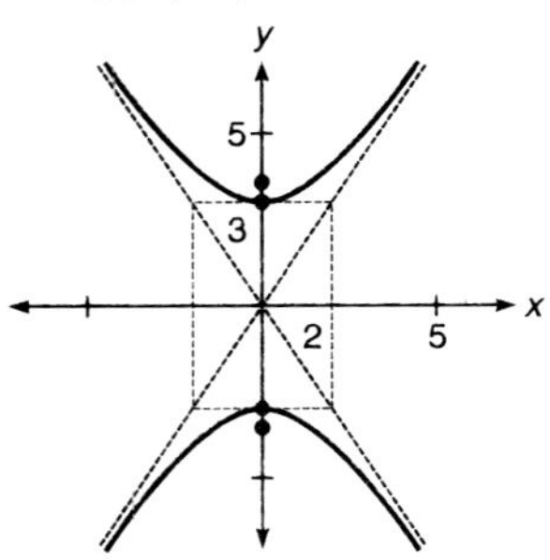

15. $\dfrac{x^2}{2} - \dfrac{y^2}{8} = 1$; ends of major axis: $(\pm\sqrt{2},0)$; foci: $(\pm\sqrt{10},0)$

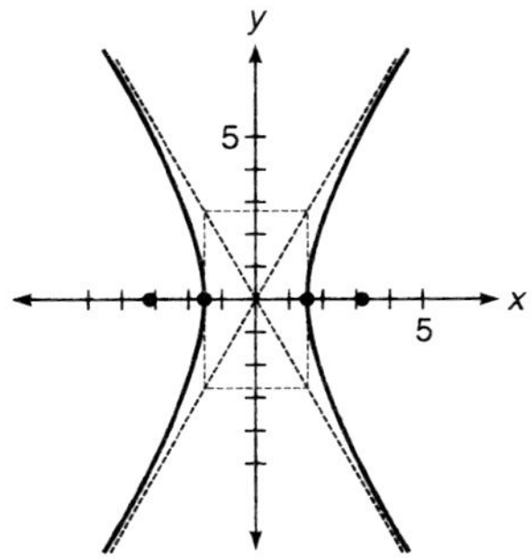

16. $\dfrac{(x-2)^2}{25} - \dfrac{(y+1)^2}{9} = 1$; foci: $(2 \pm \sqrt{34}, -1)$

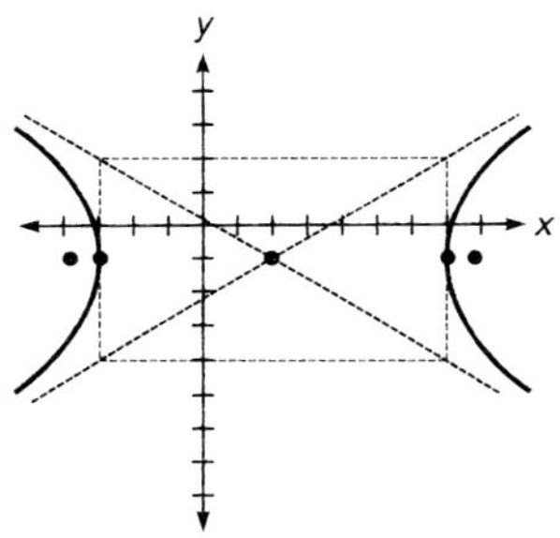

17. $\dfrac{(y+1)^2}{4} - (x-2)^2 = 1$; foci: $(2, -1 \pm \sqrt{5})$

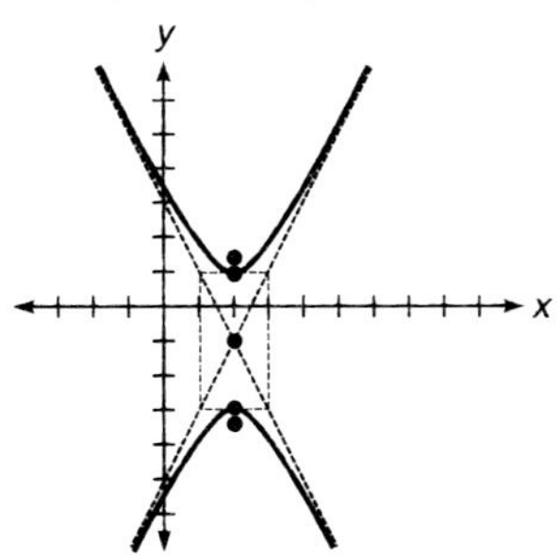

18. straight line; $4x + 20y - 23 = 0$

19. hyperbola; $\dfrac{y^2}{2} - \dfrac{x^2}{6} = 1$

20. degenerate ellipse; actually just the point $(0,3)$; $2x^2 + (y-3)^2 = 0$

21. hyperbola; $\dfrac{(y-2)^2}{\frac{5}{4}} - \dfrac{(x+\frac{1}{2})^2}{\frac{5}{4}} = 1$

22. circle; $\dfrac{(x-\frac{3}{2})^2}{\frac{9}{4}} + \dfrac{y^2}{\frac{9}{4}} = 1$

23. circle; $(x+4)^2 + (y-2)^2 = 40$

24. parabola; $y = (x+\frac{3}{2})^2 - \frac{33}{4}$

25. $(1,3)$ and $(3,7)$

26. $(0,-1)$ and $(1\frac{1}{2}, \frac{1}{2})$

27. $(\sqrt{2},0)$, $(-\sqrt{2},0)$, $\left(\dfrac{\sqrt{39}}{3}, \dfrac{7}{3}\right)$, $\left(-\dfrac{\sqrt{39}}{3}, \dfrac{7}{3}\right)$

28. $(x+1)^2 + (y-3)^2 = \frac{49}{5}$

29.

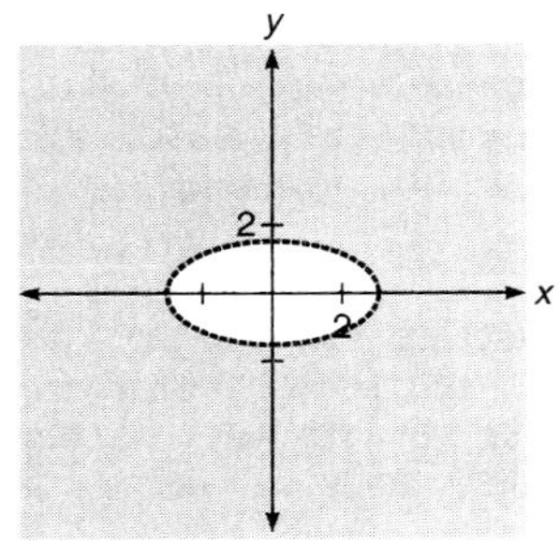

30.

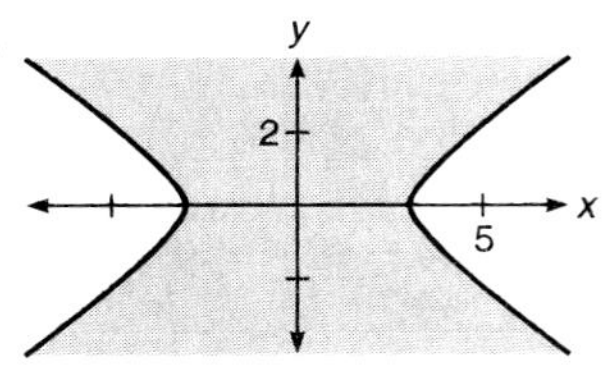

31.

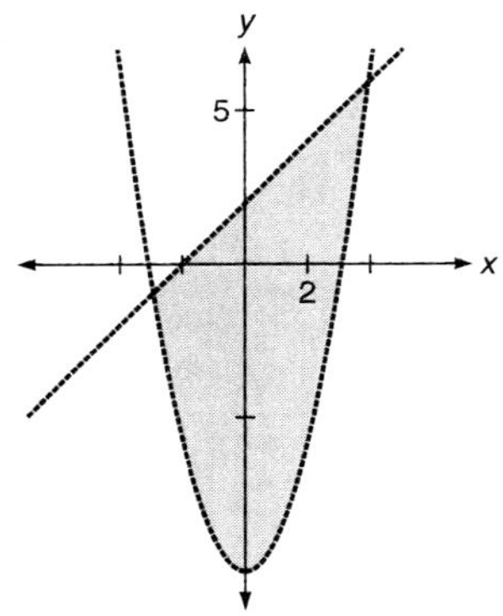

32.

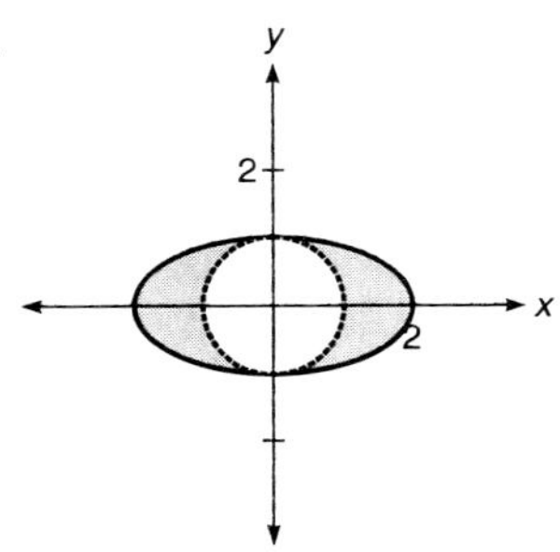

33. outer ellipse: $\frac{x^2}{64} + \frac{y^2}{16} = 1$
inner ellipse: $\frac{x^2}{4} + \frac{y^2}{16} = 1$
circle: $x^2 + y^2 = 16$

Chapter 12

Exercise 12–1

Answers to odd-numbered problems

1. $-\frac{1}{2}, 2, \frac{9}{2}, 7, \ldots$
3. $1, 0, -1, 0, \ldots$
5. $3, 3, 3, 3, \ldots$
7. $-1, -2, -1, 2, \ldots$
9. $\frac{1}{2}, \frac{\sqrt{2}}{3}, \frac{\sqrt{3}}{4}, \frac{2}{5}, \ldots$
11. $-4, 0, 6, 14, \ldots$
13. $3n - 1$ **15.** $4n - 24$
17. $(-1)^n\left(\frac{1}{n}\right)$ **19.** $\frac{n}{5n + 1}$
21. $(-1)^{n+1}\left(\frac{n^2}{(n + 1)^2}\right)$
23. $(-1)^{n+1}$ **25.** approximately 950
27. arithmetic sequence; $d = 2\frac{1}{2}$
29. neither **31.** arithmetic sequence, $d = 0$, and geometric sequence, $r = 1$
33. neither **35.** neither **37.** neither
39. arithmetic; $d = 3$ **41.** arithmetic; $d = 4$ **43.** neither **45.** neither
47. neither **49.** geometric; $r = -1$
51. 61 **53.** 33 **55.** $34\frac{2}{7}$
57. 202 **59.** 14 **61.** 16
63. -8 **65.** $-\frac{3}{4}$ **67.** $\frac{3}{2}$
69. $\frac{5}{27}$ **71.** 1,125 **73.** $\frac{5}{8}$
75. \$26,000; \$28,080; \$30,326.40; \$32,752.51; \$35,372.71; \$38,202.53
77. 23.5% **79. a.** 21, 27, 33, 39
b. yes **c.** $d_n = 6n + 15$; 135
81. Yes. We know that $a_n = a_1 + (n - 1)d_a$ for some constant d_a, and that $b_n = b_1 + (n - 1)d_b$ for some constant d_b.
Thus,
$c_n = a_n + b_n$
$= a_1 + (n - 1)d_a + b_1 + (n - 1)d_b$
$= a_1 + b_1 + (n - 1)(d_a + d_b)$
By definition $a_1 + b_1 = c_1$, and let $d_c = d_a + d_b$, a constant, so that $c_n = c_1 + (n - 1)d_c$, which is an arithmetic sequence.
83. No. Let a be the arithmetic sequence 1, 2, 3, 4, . . . and b the arithmetic sequence 2, 4, 6, 8, ... Then c is the sequence 2, 8, 18, 32, . . . , which is not an arithmetic sequence.
85. a. $\frac{3}{2}, \frac{9}{2}, \frac{27}{2}, \frac{81}{2}$
b. yes **c.** $d_n = \frac{1}{2}(3^n)$; $d_5 = \frac{243}{2}$
87. No. Let a be the geometric sequence $1, \frac{1}{2}, \frac{1}{4}, \ldots$ and b be the geometric sequence 1, 2, 4, ... Then c is the sequence $2, 2\frac{1}{2}, 4\frac{1}{4}, \ldots$, and
$\frac{c_2}{c_1} = \frac{5}{4}$, while $\frac{c_3}{c_2} = \frac{17}{10}$, so there is no constant ratio.
89. Yes. Let
$c_n = (a_n)(b_n)$
$= [a_1(r_a)^{n-1}][b_1(r_b)^{n-1}]$
$= (a_1b_1)(r_ar_b)^{n-1}$
since $a_1 \neq 0$, $b_1 \neq 0$, $r_a \neq 0$, $r_b \neq 0$, then $a_1b_1 \neq 0$, and $r_ar_b \neq 0$, so c_n is a geometric sequence.
91. All of them. Observe that the same sequence of numbers can be generated by different values of a_1 and r.
93. a. $b_n = \frac{1}{2}n^2 - \frac{9}{2}n + 13$, $b_4 = 3$
b. $b_n = -\frac{1}{6}n^2 - \frac{1}{2}n + \frac{11}{3}$, $b_4 = -1$
c. $b_n = \frac{5}{2}n^2 - \frac{19}{2}n + 10$, $b_4 = 12$

Solutions to skill and review problems

1. $3 + 6 + 9 + \cdots + 3n = 231$
$3(1 + 2 + 3 + \cdots + n) = 3(77)$
$1 + 2 + 3 + \cdots + n = 77.$
Thus the sum is 77.
2. $(1 - 5) + (5 - 9) + (9 - 13) + \cdots + (81 - 85)$
$1 - 5 + 5 - 9 + 9 - 13 + \cdots + 77 - 81 + 81 - 85$
$1 - 85$
-84
3. $x_1 + x_2 + x_3 + \cdots + x_n = 420$
$3(x_1 + x_2 + x_3 + \cdots + x_n) = 3(420)$
$3x_1 + 3x_2 + 3x_3 + \cdots + 3x_n = 1260$
4. $(a_1 + a_2 + a_3 + \cdots + a_n) + (b_1 + b_2 + b_3 + \cdots + b_n) = 500 + 200$
$(a_1 + b_1) + (a_2 + b_2) + (a_3 + b_3) + \cdots + (a_n + b_n) = 700$
5. $\frac{y^2}{16} - \frac{x^2}{9} = 1$

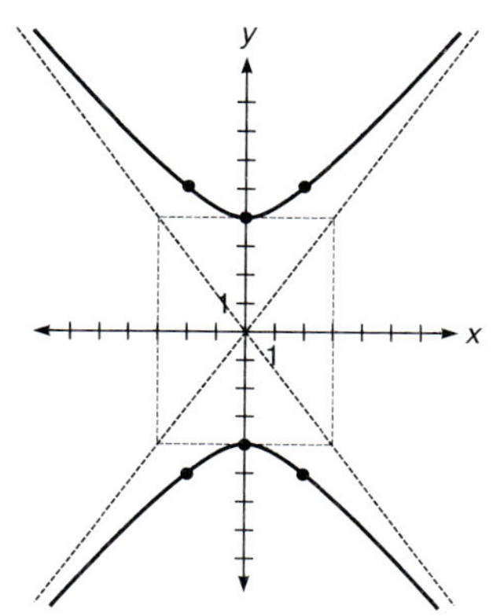

6. $\left|3 - \frac{1}{2}x\right| > 12$
$3 - \frac{1}{2}x > 12$ or $3 - \frac{1}{2}x < -12$
$-\frac{1}{2}x > 9$ or $-\frac{1}{2}x < -15$
$-2(-\frac{1}{2}x) < -2(9)$ or $-2(-\frac{1}{2}x) > -2(-15)$
$x < -18$ or $x > 30$
$x < -18$ or $x > 30$

Solutions to trial exercise problems

15. $-20, -16, -12, \ldots$
$-20 + 0(4), -20 + 1(4), -20 + 2(4), \ldots$
$-20 + (n - 1)(4)$
$-20 + 4n - 4$
$4n - 24$
19. $\frac{1}{6}, \frac{2}{11}, \frac{3}{16}, \frac{4}{21}, \ldots$
$\frac{1}{1(5) + 1}, \frac{2}{2(5) + 1}, \frac{3}{3(5) + 1}, \frac{4}{4(5) + 1}, \ldots, \frac{n}{5n + 1}$

25. The sequence 300, 400, 530, 710, . . . is definitely not an arithmetic sequence since the difference between terms is increasing. We therefore guess that it is a geometric sequence. The ratios of successive terms is $\frac{400}{300} = 1\frac{1}{3} \approx 1.33$, $\frac{530}{400} = 1\frac{13}{40} \approx 1.325$, $\frac{710}{530} = 1\frac{18}{53} \approx 1.34$. It seems reasonable to assume a constant ratio of $1\frac{1}{3}$, and therefore to estimate the next measurement as $710(\frac{4}{3}) = 947$, or about 950.

30. Geometric with ratio $\frac{1}{2}$ since each term is the previous term multiplied by $\frac{1}{2}$.

35. $\frac{1}{2}, \frac{\sqrt{2}}{3}, \frac{\sqrt{3}}{4}, \frac{2}{5}, \ldots$

$\frac{\sqrt{2}}{3} - \frac{1}{2} = \frac{2\sqrt{2} - 3}{6}$, and

$\frac{\sqrt{3}}{4} - \frac{\sqrt{2}}{3} = \frac{3\sqrt{3} - 4\sqrt{2}}{12}$, so there is no constant difference. $\dfrac{\frac{\sqrt{2}}{3}}{\frac{1}{2}} = \frac{2\sqrt{2}}{3}$,

and $\dfrac{\frac{\sqrt{3}}{4}}{\frac{\sqrt{2}}{3}} = \frac{3\sqrt{3}}{4\sqrt{2}} = \frac{3\sqrt{6}}{8}$, so there is no constant ratio. Thus this sequence is neither arithmetic nor geometric.

41. $-20, -16, -12, \ldots$
arithmetic; $d = 4$

55. $a_{15} = a_1 + (15 - 1)d$
$40 = -40 + 14d$
$d = \frac{40}{7}$
so $a_{14} = -40 + 13(\frac{40}{7}) = 34\frac{2}{7}$

61. $a_{15} = a_1 + 14d$
$49 = a_1 + 14d$
$a_{28} = a_1 + 27d$
$88 = a_1 + 27d$
Thus, $a_1 = 49 - 14d$
and $a_1 = 88 - 27d$,
so $49 - 14d = 88 - 27d$
$13d = 39$
$d = 3$
$a_1 = 88 - 27d$
so $a_1 = 88 - 81 = 7$.
Thus, $a_4 = a_1 + 3d = 7 + 3(3) = 16$.

70. $a_3 = \frac{1}{3} = a_1r^2$, $a_6 = -\frac{1}{81} = a_1r^5$,

so $a_1 = \frac{1}{3r^2}$ and

$a_1 = -\frac{1}{81r^5}$, so $\frac{1}{3r^2} = -\frac{1}{81r^5}$,

so $3r^2 = -81r^5$ (divide by r^2), $3 = -81r^3$, $-\frac{1}{27} = r^3$, $r = -\frac{1}{3}$.

$a_1 = \frac{1}{3r^2} = \frac{1}{3(-\frac{1}{3})^2} = 3$.

Thus, we know a_1 and r. $a_2 = a_1r = 3(-\frac{1}{3}) = -1$.

74. The length of each swing forms a geometric sequence with $a_1 = 20$ and $r = 0.95$.

a. $a_4 = 20(0.95)^3 \approx 17.1$ inches

b. $a_8 = 20(0.95)^7 \approx 14.0$ inches

---

90. Every fifth roll of film is developed free. Let n = number of rolls developed, then a_n, the average cost for n rolls, is the ratio of total cost for n rolls to n: $\frac{\text{total cost to develop } n \text{ rolls}}{\text{number of rolls}}$. This value is indicated in the following table.

Number of rolls n	Cost for roll	Total cost	a_n	Form of a_n	Value of $\left[\frac{n}{5}\right]$
1	5	5	$\frac{5}{1}$	$\frac{5(n-0)}{n}$	0
2	5	10	$\frac{10}{2}$	$\frac{5(n-0)}{n}$	0
3	5	15	$\frac{15}{3}$	$\frac{5(n-0)}{n}$	0
4	5	20	$\frac{20}{4}$	$\frac{5(n-0)}{n}$	0
5	0	20	$\frac{20}{5}$	$\frac{5(n-1)}{n}$	1
6	5	25	$\frac{25}{6}$	$\frac{5(n-1)}{n}$	1
7	5	30	$\frac{30}{7}$	$\frac{5(n-1)}{n}$	1
8	5	35	$\frac{35}{8}$	$\frac{5(n-1)}{n}$	1
9	5	40	$\frac{40}{9}$	$\frac{5(n-1)}{n}$	1
10	0	40	$\frac{40}{10}$	$\frac{5(n-2)}{n}$	2
11	5	45	$\frac{45}{11}$	$\frac{5(n-2)}{n}$	2

The numerators in the a_n column are of the form $\frac{5(n-i)}{n}$, where i is the quotient, without the remainder, of $n \div 5$. This value is $\left[\frac{n}{5}\right]$. Thus, $a_n = \dfrac{5\left(n - \left[\frac{n}{5}\right]\right)}{n}$.

Exercise 12–2

Answers to odd-numbered problems

1. $5 + 9 + 13 + 17$
3. $6 + 12 + 20 + 30$ **5.** $\frac{3}{4} + \frac{4}{5}$
7. $-\frac{4}{3} + \frac{2}{3} - \frac{4}{9} + \frac{1}{3} - \frac{4}{15} + \frac{2}{9}$
9. $1 + (1 + 4) + (1 + 4 + 9) + (1 + 4 + 9 + 16)$ **11.** 1,584
13. -570 **15.** $-45\frac{1}{2}$ **17.** -294
19. 418 **21.** -490 **23.** $132\frac{1}{2}$
25. 246 **27.** 1,365 **29.** $-\frac{369}{256}$
31. 129 **33.** $22\frac{163}{864}$ **35.** 6,560
37. $121\frac{1}{3}$ **39.** $\frac{2,062}{3,125}$ **41.** $5\frac{85}{256}$
43. $1\frac{1}{2}$ **45.** $\frac{1}{3}$ **47.** $-\frac{2}{5}$
49. not defined **51.** 9 **53.** $\frac{3}{5}$ **55.** $\frac{2}{9}$
57. $\frac{28}{99}$ **59.** $\frac{98}{111}$ **61.** $\frac{5,155}{9,999}$
63. $\frac{3,401}{9,900}$ **65.** $\frac{1,987}{4,950}$
67. 400 inches **69.** 1,024 feet
71. 7 seconds **73.** 6 hours
75. $2^{64} - 1 \approx 1.8 \times 10^{19}$ grains of wheat. (This is more wheat than has ever existed.)
77. 39 boxes **79.** 37% **81.** \$8,031.25
83. \$250,000 **85.** 5,050 **87.** 300 miles **89.** 3 **91.** About 6.1 days

Solutions to skill and review problems

1. $a_n = a_1 + (n - 1)d$; $a_1 = 3$, $d = 5$, $n = 33$
$a_{33} = 3 + 32(5) = 163$

2. $a_n = a_1r^{n-1}$; $a_1 = 3$, $r = 5$, $n = 5$
$a_5 = 3 \cdot 5^4 = 1,875$

3. $\left|\frac{x+2}{x}\right| < 4$
Nonlinear inequality; use the critical point/test point method.
Critical points:
a. Solve the corresponding equality.
$\left|\frac{x+2}{x}\right| = 4$
$\frac{x+2}{x} = 4$ or $\frac{x+2}{x} = -4$
$x + 2 = 4x \quad x + 2 = -4x$
$2 = 3x \quad 5x = -2$
$\frac{2}{3} = x \quad x = -\frac{2}{5}$
b. Find zeros of denominators. $x = 0$
Critical points are $-\frac{2}{5}$, 0, $\frac{2}{3}$.

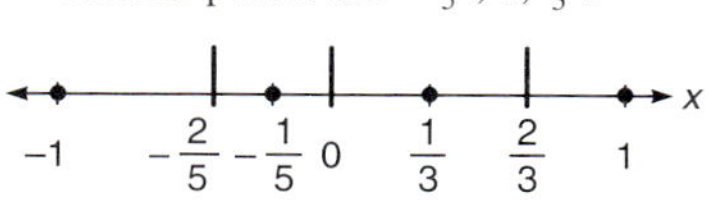

We will use test points of -1, $-\frac{1}{5}$, $\frac{1}{3}$, 1.
$$\left|\frac{x+2}{x}\right| < 4$$
$x = -1$: $|-1| < 4$; true
$x = -\frac{1}{5}$: $|-9| < 4$; false
$x = \frac{1}{3}$: $|7| < 4$; false
$x = 1$: $|3| < 4$; true
$x < -\frac{2}{5}$ or $x > \frac{2}{3}$

4. $f(x) = x^2 + 5x - 6$
Parabola; complete the square.
$y = x^2 + 5x + \frac{25}{4} - 6 - \frac{25}{4}$,
since $\frac{1}{2} \cdot 5 = \frac{5}{2}$ and $(\frac{5}{2})^2 = \frac{25}{4}$
$y = (x + \frac{5}{2})^2 - \frac{49}{4}$
vertex: $(-2\frac{1}{2}, -12\frac{1}{4})$
intercepts:
$x = 0$: $f(0) = -6$; $(0,-6)$
$y = 0$: $0 = x^2 + 5x - 6$
$0 = (x + 6)(x - 1)$
$x = -6$ or 1; $(-6,0)$, $(1,0)$

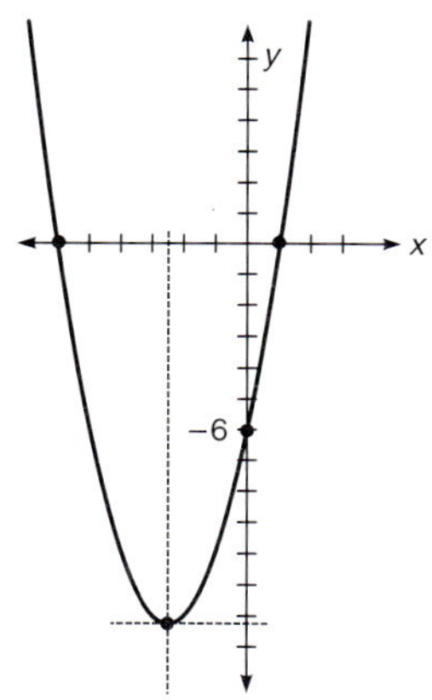

5. Use $(x - h)^2 + (y - k)^2 = r^2$, where (h,k) is the center and r is the radius. To find r, find the distance from the origin $(0,0)$ to $(2,-1)$. This can be done by the distance formula or a simple sketch (see the figure) where we see that $r^2 = 2^2 + 1^2 = 5$.
$(x - 2)^2 + (y - (-1))^2 = 5$
$(x - 2)^2 + (y + 1)^2 = 5$

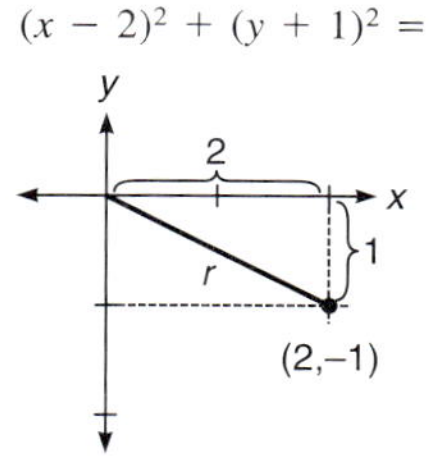

5. $f(x) = x^3 - 3x^2 + x + 2$
Possible zeros are ± 1, ± 2. Synthetic division shows that 2 is a zero, so $x - 2$ is a factor.
$y = (x - 2)(x^2 - x - 1)$
The zeros of $x^2 - x - 1$ are $\frac{1 \pm \sqrt{5}}{2} \approx -0.6, 1.6$ (from the quadratic formula).
Thus, x-intercepts are $(2,0)$, $(-0.6,0)$, $(1.6,0)$.
y-intercept: $f(0) = 2$; $(0,2)$
Additional points: $(-1,-3)$, $(1,1)$, $(1.8,-0.09)$, $(3,5)$

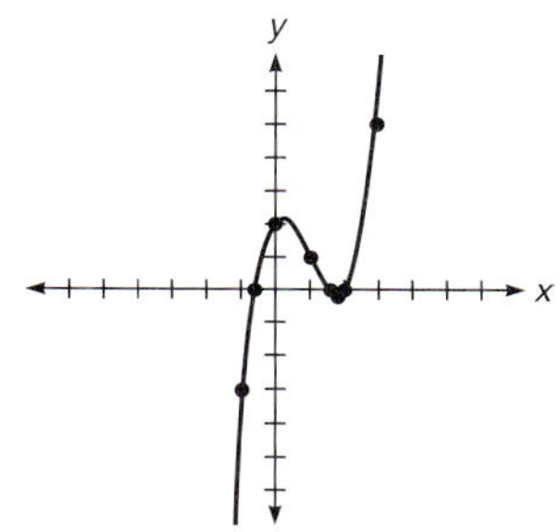

Solutions to trial exercise problems

9. $\sum_{k=1}^{1} k^2 + \sum_{k=1}^{2} k^2 + \sum_{k=1}^{3} k^2 + \sum_{k=1}^{4} k^2$
$1 + (1 + 4) + (1 + 4 + 9) + (1 + 4 + 9 + 16)$

15. $-8, -7\frac{1}{4}, -6\frac{1}{2}, \ldots, 1$
This is an arithmetic sequence with $a_1 = -8$ and $d = \frac{3}{4}$. Using $a_n = a_1 + (n - 1)d$ we obtain
$1 = -8 + (n - 1)(\frac{3}{4})$
$9 = \frac{3}{4}(n - 1)$
$\frac{4}{3}(9) = n - 1$
$n = 13$
Thus, there are 13 terms.
$S_{13} = \frac{13}{2}(-8 + 1) = \frac{13}{2}(-7)$
$= -\frac{91}{2} = -45\frac{1}{2}$.

30. $S_4 = \frac{-2(1 - (-\frac{2}{3})^4)}{1 - (-\frac{2}{3})} = \frac{-2(1 - \frac{16}{81})}{\frac{5}{3}}$
$= \frac{3}{5}(-2)(\frac{65}{81}) = -\frac{26}{27}$

40. $\sum_{k=1}^{8} -3(-\frac{2}{3})^k = 2 - \frac{4}{3} + \frac{8}{27} - \cdots - \frac{28}{37}$;
$a_1 = 2, r = -\frac{2}{3}, n = 8$:
$$S_n = \frac{2(1-(-\frac{2}{3})^8)}{1-(-\frac{2}{3})} = \frac{2(1-\frac{256}{6{,}561})}{\frac{5}{3}} = \frac{3}{5}(2)\left(\frac{6{,}305}{6{,}561}\right) = \frac{2{,}522}{2{,}187} = 1\frac{335}{2{,}187}$$

46. $\sum_{i=1}^{\infty} \frac{1}{8}(2)^i$
$a_1 = \frac{1}{4}, r = 2$
$|r| \geq 1$ so the sum is not defined.

64. $x = 0.21606\overline{060}$
$$\begin{aligned} 10{,}000x &= 2160.606\overline{060} \\ 100x &= 21.606\overline{060} \\ \hline 9{,}900x &= 2{,}139 \\ x &= \frac{2{,}139}{9{,}900} = \frac{713}{3{,}300} \end{aligned}$$

68. We have a sequence 16, 48, 80, . . . , which is an arithmetic progression with $a_1 = 16$ and $d = 32$. Then $a_8 = 16 + (8-1)(32) = 240$ feet.

71. We are given $S_n = 250$, $a_1 = 4.9$, and $d = 9.8$, and need to find n.
$$S_n = \frac{n}{2}[2a_1 + (n-1)d]$$
$$250 = \frac{n}{2}[9.8 + (n-1)(9.8)]$$
$$500 = n(9.8 + 9.8n - 9.8)$$
$$500 = 9.8n^2$$
$$n \approx \pm 7.1$$
Thus, after about 7 seconds a body will have fallen 250 meters.

78. This is an arithmetic series with $a_1 = 500$ and $d = 100$, and we want S_{19} (18 birthdays plus the day of birth).
$S_{19} = \frac{19}{2}[2(500) + 18(100)] = \$26{,}600$.

80. The six deposits are a geometric series with a_1 unknown and $r = 1.15$ (since each deposit is 115% of the previous one). We want $S_6 = 100{,}000$, and $S_6 = a_1\left(\frac{1-r^6}{1-r}\right)$, so
$$100{,}000 = a_1\left(\frac{1-1.15^6}{1-1.15}\right),$$
$100{,}000 \approx 8.753738a_1$, $a_1 \approx 11{,}423.69$.
Thus the first deposit should be about \$11,423.69.

84. Let a be a finite geometric series with n terms and ratio r, and S_n the sum of these n terms. The sum from a_1 to a_n (which is a_1r^{n-1}) is $S_n = a_1 + a_1r + a_1r^2 + \cdots a_1r^{n-1}$. We subtract rS_n from this as follows.
$$\begin{aligned} S_n &= a_1 + a_1r + a_1r^2 + \cdots + a_1r^{n-1} \\ rS_n &= a_1r + a_1r^2 + a_1r^3 + \cdots + a_1r^{n-1} + a_1r^n \\ \hline S_n - rS_n &= a_1 - a_1r^n \\ S_n(1-r) &= a_1(1-r^n) \\ S_n &= \frac{a_1(1-r^n)}{1-r} \end{aligned}$$

87. The easiest analysis of this problem is as follows. The trains are closing the 200-mile distance at 40 mph, so they crash in 5 hours. The fly flew at 60 mph for 5 hours for a total distance of 300 miles. An attempt to analyze the problem as a series is much more complicated.

90. We need to find n such that $\sum_{i=1}^{n} \frac{1}{2i} \geq \frac{3}{2}$. This series is neither arithmetic nor geometric, so we simply proceed by trial and error, and find out that for $n = 11$ the sum is about 1.51.

91. The distance traveled by the first person is the sum of an arithmetic progression with first term 5 and $d = 3$. If x is the number of days in which the persons meet, then the distance traveled by the first person is $\frac{x}{2}[2 \cdot 5 + (x-1)3] = \frac{1}{2}(3x^2 + 7x)$.
The second person travels $5 \cdot 7 = 35$ yojanas in the 5-day head start, and $7x$ yojanas after that. Thus, the second person travels $7x + 35$ yojanas. They meet when the distances are equal:
$$\frac{1}{2}(3x^2 + 7x) = 7x + 35$$
$$3x^2 - 7x - 70 = 0$$
$$x \approx 6.1 \text{ days}$$

Exercise 12–3

Answers to odd-numbered problems

1. 42 **3.** 56 **5.** 1
7. $\frac{(n+3)(n+2)(n+1)}{6}$
9. $a^4b^4 - 12a^3b^3 + 54a^2b^2 - 108ab + 81$
11. $64p^{24} + 192p^{20}q^1 + 240p^{16}q^2 + 160p^{12}q^3 + 60p^8q^4 + 12p^4q^5 + q^6$
13. $a^{21}b^{14} - 14a^{18}b^{12}c + 84a^{15}b^{10}c^2 - 280a^{12}b^8c^3 + 560a^9b^6c^4 - 672a^6b^4c^5 + 448a^3b^2c^6 - 128c^7$
15. $\frac{1}{64}p^6 + \frac{3}{8}p^5 + \frac{15}{4}p^4 + 20p^3 + 60p^2 + 96p + 64$
17. $21{,}840a^{33}b^{20}$ **19.** $-41{,}580p^{19}q^3$
21. 448 **23.** 966 **25.** 132
27. 2,924 **29.** 1,547 **31.** 2,800
33. $29\frac{171}{256}$ **35.** $-61\frac{11}{1{,}024}$

37. $2k^3 + k^2 + k$ **39.** $\frac{2}{3}k^3 + \frac{7}{2}k^2 - \frac{55}{6}k$

41.
```
                    1   1
                  1   2   1
                1   3   3   1
              1   4   6   4   1
            1   5  10  10   5   1
          1   6  15  20  15   6   1
        1   7  21  35  35  21   7   1
      1   8  28  56  70  56  28   8   1
```

43.
$$\binom{n}{n} = \frac{n!}{n!(n-n)!} = \frac{1}{0!} = \frac{1}{1} = 1$$
$$\binom{n}{1} = \frac{n!}{(n-1)!1!} = \frac{n(n-1)!}{(n-1)!} = n$$
$$\binom{n}{0} = \frac{n!}{(n-n)!n!} = \frac{1}{0!} = \frac{1}{1} = 1$$

45. $1 + 2 + 3 + \cdots + n$ is an arithmetic sequence with $a_1 = 1$, $a_n = n$, and $n = n$. $S_n = \frac{n}{2}(a_1 + a_n)$

$$S_n = \frac{n}{2}(1 + n) = \frac{n(n+1)}{2}.$$

47. $\binom{n}{k} = \frac{n!}{k!(n-k)!}$, and

$$\binom{n}{n-k} = \frac{n!}{(n-k)![n-(n-k)]!}$$

$$= \frac{n!}{(n-k)!k!}, \text{ so } \binom{n}{k} = \binom{n}{n-k}$$

49. Using $(x + y)^n = \sum_{i=0}^{n} \binom{n}{i} x^{n-i}y^i$ with $x = y = 1$ we obtain

$$2^n = (1 + 1)^n = \sum_{i=0}^{n} \binom{n}{i} 1^{n-i}1^i = \sum_{i=0}^{n} \binom{n}{i}$$

Solutions to skill and review problems

1. $1 + 2 + \cdots + n = \frac{n(n+1)}{2}$

$$1 + 2 + \cdots + n + (n+1) = \frac{n(n+1)}{2} + (n+1)$$

$$= \frac{(n+1)(n+2)}{2}$$

2. $\frac{1}{1 \cdot 2} + \frac{1}{2 \cdot 3} + \frac{1}{3 \cdot 4} + \cdots + \frac{1}{n(n+1)} = \frac{n}{n+1}$

$$\frac{1}{1 \cdot 2} + \frac{1}{2 \cdot 3} + \frac{1}{3 \cdot 4} + \cdots + \frac{1}{n(n+1)} + \frac{1}{(n+1)(n+2)}$$

$$= \frac{n}{n+1} + \frac{1}{(n+1)(n+2)} = \frac{n+1}{n+2}$$

3. $\frac{1}{3} + \frac{1}{9} + \frac{1}{27} + \cdots + \frac{1}{3^n}$

This is a geometric series; $a_1 = r = \frac{1}{3}$. We want S_n.

$$S_n = a_n\left(\frac{1 - r^n}{1 - r}\right) = \frac{1}{3}\left(\frac{1 - (\frac{1}{3})^n}{1 - \frac{1}{3}}\right) = \frac{1}{3}\left(\frac{1 - \frac{1}{3^n}}{\frac{2}{3}}\right) =$$

$$\frac{1}{3} \cdot \frac{3}{2}\left(\frac{1}{1} - \frac{1}{3^n}\right)$$

$$= \frac{1}{2}\left(\frac{3^n - 1}{3^n}\right) = \frac{3^n - 1}{2(3^n)}$$

4. $2 + 4 + \cdots + 240$. Divice by 2;

$1 + 2 + \cdots + 120$ shows that $n = 120$.

This is an arithmetic series with $a_1 = 2$, $d = 2$, $n = 120$.

$$S_n = \frac{n}{2}(a_1 + a_n)$$

$$S_{120} = \frac{120}{2}(2 + 240) = 14\,520.$$

5. $\frac{x-1}{2} - \frac{x-1}{3} = x$

$$6\left(\frac{x-1}{2}\right) - 6\left(\frac{x-1}{3}\right) = 6x$$

$$3(x - 1) - 2(x - 1) = 6x$$

$$-\frac{1}{5} = x$$

6. $f(x) = (x - 1)^3 - 1$

This is the graph of $y = x^3$, shifted right 1 unit and down 1 unit. Thus the "origin" is shifted to $(1,-1)$.

Intercepts:

$x = 0$: $f(0) = (-1)^3 - 1 = -2$; $(0,-2)$

$y = 0$: $0 = (x - 1)^3 - 1$

$1 = (x - 1)^3$

$\sqrt[3]{1} = x - 1$

$1 = x - 1$

$2 = x$; $(2,0)$

Additional points: $(1,-1)$, $(2.5,2.4)$

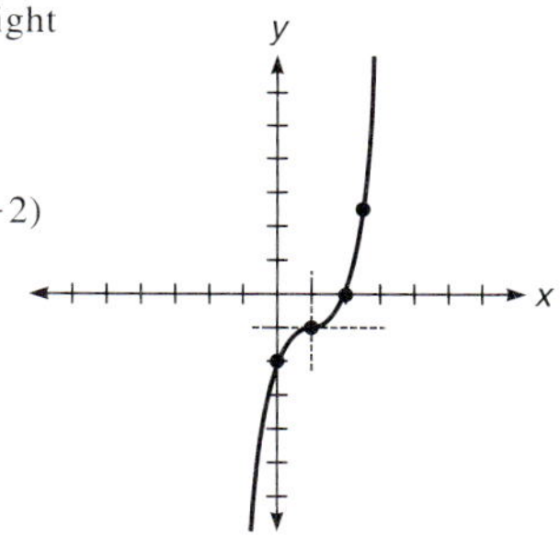

Solutions to trial exercise problems

13. $(a^3b^2 - 2c)^7 =$

$$\binom{7}{0}(a^3b^2)^7(-2c)^0 + \binom{7}{1}(a^3b^2)^6(-2c)^1 + \binom{7}{2}(a^3b^2)^5(-2c)^2 + \binom{7}{3}(a^3b^2)^4(-2c)^3$$

$$+ \binom{7}{4}(a^3b^2)^3(-2c)^4 + \binom{7}{5}(a^3b^2)^2(-2c)^5 + \binom{7}{6}(a^3b^2)^1(-2c)^6 + \binom{7}{7}(a^3b^2)^0(-2c)^7 =$$

$$a^{21}b^{14} + 7(a^{18}b^{12})(-2c) + 21(a^{15}b^{10})(4c^2) + 35(a^{12}b^8)(-8c^3) + 35(a^9b^6)(16c^4)$$

$$+ 21(a^6b^4)(-32c^5) + 7(a^3b^2)(64c^6) + (-128c^7) =$$

$$a^{21}b^{14} - 14a^{18}b^{12}c + 84a^{15}b^{10}c^2 - 280a^{12}b^8c^3 + 560a^9b^6c^4 - 672a^6b^4c^5 + 448a^3b^2c^6 - 128c^7$$

19. Let $i = 3$:

$$\binom{22}{3}p^{22-3}(-3q)^3 = 1{,}540p^{19}(-27)q^3 = -41{,}580p^{19}q^3$$

25. $\sum_{i=1}^{9}(3 - 4i + i^2)$

$$= \sum_{i=1}^{9} 3 - 4\sum_{i=1}^{9} i + \sum_{i=1}^{9} i^2$$

$$= 9(3) - 4\left(\frac{9(10)}{2}\right) + \frac{9(10)(19)}{6} = 132$$

33. $\sum_{i=1}^{4}[i^2 - (\frac{1}{4})^i] = \sum_{i=1}^{4} i^2 - \sum_{i=1}^{4}(\frac{1}{4})^i$

The second expression is a geometric series, $a_1 = r = \frac{1}{4}$;

$S_n = a_n\left(\frac{1 - r^n}{1 - r}\right).$

$= \frac{4(5)(9)}{6} - \frac{1}{4}\left(\frac{1 - (\frac{1}{4})^4}{1 - \frac{1}{4}}\right)$

$= 30 - \frac{85}{256} = 29\frac{171}{256}$

39. $\sum_{i=1}^{k}(2i^2 + 5i - 12)$

$= 2\sum_{i=1}^{k} i^2 + 5\sum_{i=1}^{k} i - \sum_{i=1}^{k} 12 = 2 \cdot \frac{k(k + 1)(2k + 1)}{6} + 5 \cdot \frac{k(k + 1)}{2} - 12 \cdot k$

$= \frac{2}{3}k^3 + \frac{7}{2}k^2 - \frac{55}{6}k$

Exercise set 12–4

Answers to odd-numbered problems

1. Show true for $n = 1$: $2(1) = 1(1 + 1)$; $2 = 2$ √
Find goal statement:
$2 + 4 + 6 + \cdots + 2(k + 1) = (k + 1)[(k + 1) + 1] = (k + 1)(k + 2)$
Assume true for $n = k$:
$2 + 4 + 6 + \cdots + 2k = k(k + 1)$
$2 + 4 + 6 + \cdots + 2k + 2(k + 1) = k(k + 1) + 2(k + 1)$
$= (k + 1)(k + 2)$ √

3. Show true for $n = 1$: $(5(1) - 1) = \frac{1(5(1) + 3)}{2}$; $4 = 4$ √

Find goal statement:

$4 + 9 + 14 + \cdots + (5(k + 1) - 1) = \frac{(k + 1)(5(k + 1) + 3)}{2} = \frac{(k + 1)(5k + 8)}{2}$

Assume true for $n = k$:

$4 + 9 + 14 + \cdots + (5k - 1) = \frac{k(5k + 3)}{2}$

$4 + 9 + 14 + \cdots + (5k - 1) + (5(k + 1) - 1) = \frac{k(5k + 3)}{2} + (5(k + 1) - 1)$

$= \frac{5k^2 + 13k + 8}{2} = \frac{(k + 1)(5k + 8)}{2}$ √

5. Show true for $n = 1$: $(4(1) - 3) = 2(1)^2 - 1$; $1 = 1$ √
Find the goal statement:
$1 + 5 + 9 + \cdots + (4(k + 1) - 3) = 2(k + 1)^2 - (k + 1) = 2k^2 + 3k + 1$
Assume true for $n = k$:
$1 + 5 + 9 + \cdots + (4k - 3) = 2k^2 - k$
$1 + 5 + 9 + \cdots + (4k - 3) + (4(k + 1) - 3) = 2k^2 - k + (4(k + 1) - 3)$
$= 2k^2 + 3k + 1$ √

7. Show true for $n = 1$:

$$\frac{1^2(1+1)(1+2)}{6} = \frac{1(1+1)(1+2)(1+3)(4(1)+1)}{120};\ 1 = 1\ \surd$$

Find goal statement:

$$1 + 8 + 30 + 80 + \cdots + \frac{(k+1)^2((k+1)+1)((k+1)+2)}{6}$$
$$= \frac{(k+1)((k+1)+1)((k+1)+2)((k+1)+3)(4(k+1)+1)}{120} = \frac{(k+1)(k+2)(k+3)(k+4)(4k+5)}{120}$$

Assume true for $n = k$:

$$1 + 8 + 30 + 80 + \cdots + \frac{k^2(k+1)(k+2)}{6} = \frac{k(k+1)(k+2)(k+3)(4k+1)}{120}$$
$$1 + 8 + 30 + 80 + \cdots + \frac{k^2(k+1)(k+2)}{6} + \frac{(k+1)^2((k+1)+1)((k+1)+2)}{6}$$
$$= \frac{k(k+1)(k+2)(k+3)(4k+1)}{120} + \frac{(k+1)^2((k+1)+1)((k+1)+2)}{6}$$
$$= \frac{k(k+1)(k+2)(k+3)(4k+1) + 20(k+1)(k+1)(k+2)(k+3)}{120}$$
$$= \frac{(k+1)(k+2)(k+3)[k(4k+1) + 20(k+1)]}{120} = \frac{(k+1)(k+2)(k+3)[4k^2 + 21k + 20]}{120} = \frac{(k+1)(k+2)(k+3)(k+4)(4k+5)}{120}\ \surd$$

9. Show true for $n = 1$: $\frac{1}{2^1} = \frac{2^1 - 1}{2^1};\ \frac{1}{2} = \frac{1}{2}\ \surd$

Find goal statement: $\frac{1}{2} + \frac{1}{2^2} + \frac{1}{2^3} + \cdots + \frac{1}{2^{k+1}} = \frac{2^{k+1} - 1}{2^{k+1}}$

Assume true for $n = k$:

$$\frac{1}{2} + \frac{1}{2^2} + \frac{1}{2^3} + \cdots + \frac{1}{2^k} = \frac{2^k - 1}{2^k}$$
$$\frac{1}{2} + \frac{1}{2^2} + \frac{1}{2^3} + \cdots + \frac{1}{2^k} + \frac{1}{2^{k+1}} = \frac{2^k - 1}{2^k} + \frac{1}{2^{k+1}} = \frac{2(2^k - 1)}{2(2^k)} + \frac{1}{2^{k+1}} = \frac{2^{k+1} - 1}{2^{k+1}}\ \surd$$

11. Show true for $n = 1$: $1^3 + 2 = 3$, which is divisible by 3. $\surd$
Find goal statement: $(k+1)^3 + 2(k+1)$ is divisible by 3.
Assume true for $n = k$: $k^3 + 2k$ is divisible by 3.
Examine goal statement:
$(k+1)^3 + 2(k+1) = k^3 + 3k^2 + 5k + 3 = k^3 + 2k + 3k^2 + 3k + 3 = [k^3 + 2k] + [3(k^2 + k + 1)]$
We know 3 divides $k^3 + 2k$.
We can see that 3 divides $3(k^2 + k + 1)$.
Therefore, 3, divides their sum $[k^3 + 2k] + [3(k^2 + k + 1)]$.
As shown above, $[k^3 + 2k] + [3(k^2 + k + 1)] = (k+1)^3 + 2(k+1)$.
Thus, 3 divides $(k+1)^3 + 2(k+1)$. $\surd$

13. Show true for $n = 1$:

$$\frac{1}{(2(1) - 1)(2(1) + 1)} = \frac{1}{2(1) + 1};\ \frac{1}{3} = \frac{1}{3}\ \surd$$

Find the goal statement:

$$\frac{1}{1 \cdot 3} + \frac{1}{3 \cdot 5} + \frac{1}{5 \cdot 7} + \cdots + \frac{1}{(2(k+1) - 1)(2(k+1) + 1)} = \frac{k+1}{2(k+1) + 1} = \frac{k+1}{2k+3}$$

Assume true for $n = k$:

$$\frac{1}{1 \cdot 3} + \frac{1}{3 \cdot 5} + \frac{1}{5 \cdot 7} + \cdots + \frac{1}{(2k-1)(2k+1)} = \frac{k}{2k+1}$$
$$\frac{1}{1 \cdot 3} + \frac{1}{3 \cdot 5} + \frac{1}{5 \cdot 7} + \cdots + \frac{1}{(2k-1)(2k+1)} + \frac{1}{(2(k+1) - 1)(2(k+1) + 1)}$$
$$= \frac{k}{2k+1} + \frac{1}{(2(k+1) - 1)(2(k+1) + 1)} = \frac{2k^2 + 3k + 1}{(2k+1)(2k+3)} = \frac{k+1}{2k+3}\ \surd$$

15. Show true for $n = 1$: $2(3^0) = 3^1 - 1$; $2 = 2$ ✓

Find goal statement:

$2 + 6 + 18 + \cdots + 2(3^{(k+1)-1}) = 3^{k+1} - 1$

Assume true for $n = k$:

$2 + 6 + 18 + \cdots + 2(3^{k-1}) = 3^k - 1$

$$\begin{aligned} 2 + 6 + 18 + \cdots + 2(3^{k-1}) + 2(3^{(k+1)-1}) &= 3^k - 1 + 2(3^{(k+1)-1}) \\ &= 3^k - 1 + 2(3^k) \\ &= 3(3^k) - 1 \\ &= 3^{k+1} - 1 \checkmark \end{aligned}$$

17. Show true for $n = 1$: $\frac{1}{2^{-3}} = \frac{2^1 - 1}{2^{-3}}$; $8 = 8$ ✓

Find goal statement:

$$8 + 4 + 2 + \cdots + \frac{1}{2^{(k+1)-4}} = \frac{2^{k+1} - 1}{2^{(k+1)-4}}$$

Assume true for $n = k$:

$$8 + 4 + 2 + \cdots + \frac{1}{2^{k-4}} = \frac{2^k - 1}{2^{k-4}}$$

$$8 + 4 + 2 + \cdots + \frac{1}{2^{k-4}} + \frac{1}{2^{(k+1)-4}} = \frac{2^k - 1}{2^{k-4}} + \frac{1}{2^{(k+1)-4}}$$

$$= \frac{2}{2} \cdot \frac{2^k - 1}{2^{k-4}} + \frac{1}{2^{k-3}} = \frac{2^{k+1} - 1}{2^{k-3}} = \frac{2^{k+1} - 1}{2^{(k-4)+1}} \checkmark$$

19. Show true for $n = 1$: $\frac{1}{(3(1) - 2)(3(1) + 1)} = \frac{1}{3(1) + 1}$; $\frac{1}{4} = \frac{1}{4}$ ✓

Find goal statement:

$$\frac{1}{1 \cdot 4} + \frac{1}{4 \cdot 7} + \frac{1}{7 \cdot 10} + \cdots + \frac{1}{(3(k + 1) - 2)(3(k + 1) + 1)} = \frac{k + 1}{3(k + 1) + 1} = \frac{k + 1}{3k + 4}$$

Assume true for $n = k$:

$$\frac{1}{1 \cdot 4} + \frac{1}{4 \cdot 7} + \frac{1}{7 \cdot 10} + \cdots + \frac{1}{(3k - 2)(3k + 1)} = \frac{k}{3k + 1}$$

$$\frac{1}{1 \cdot 4} + \frac{1}{4 \cdot 7} + \frac{1}{7 \cdot 10} + \cdots + \frac{1}{(3k - 2)(3k + 1)} + \frac{1}{(3(k + 1) - 2)(3(k + 1) + 1)}$$

$$= \frac{k}{3k + 1} + \frac{1}{(3(k + 1) - 2)(3(k + 1) + 1)} = \frac{k(3k + 4)}{(3k + 1)(3k + 4)} + \frac{1}{(3k + 1)(3k + 4)} = \frac{(3k + 1)(k + 1)}{(3k + 1)(3k + 4)} = \frac{k + 1}{3k + 4} \checkmark$$

21. Show true for $n = 1$: $\frac{1}{1 \cdot 2 \cdot 3} = \frac{1(4)}{4(2)(3)}$; $\frac{1}{6} = \frac{1}{6}$ ✓

Find goal statement:

$$\frac{1}{1 \cdot 2 \cdot 3} + \frac{1}{2 \cdot 3 \cdot 4} + \frac{1}{3 \cdot 4 \cdot 5} + \cdots + \frac{1}{(k + 1)((k + 1) + 1)((k + 1) + 2)} = \frac{(k + 1)((k + 1) + 3)}{4((k + 1) + 1)((k + 1) + 2)} = \frac{(k + 1)(k + 4)}{4(k + 2)(k + 3)}$$

Assume true for $n = k$:

$$\frac{1}{1 \cdot 2 \cdot 3} + \frac{1}{2 \cdot 3 \cdot 4} + \frac{1}{3 \cdot 4 \cdot 5} + \cdots + \frac{1}{k(k + 1)(k + 2)} = \frac{k(k + 3)}{4(k + 1)(k + 2)}$$

$$\frac{1}{1 \cdot 2 \cdot 3} + \frac{1}{2 \cdot 3 \cdot 4} + \frac{1}{3 \cdot 4 \cdot 5} + \cdots + \frac{1}{k(k + 1)(k + 2)} + \frac{1}{(k + 1)(k + 2)(k + 3)}$$

$$= \frac{k(k + 3)}{4(k + 1)(k + 2)} + \frac{1}{(k + 1)(k + 2)(k + 3)} = \frac{k(k + 3)^2}{4(k + 1)(k + 2)(k + 3)} + \frac{4}{4(k + 1)(k + 2)(k + 3)} = \frac{k^3 + 6k^2 + 9k + 4}{4(k + 1)(k + 2)(k + 3)}$$

Using both the goal statement and the rational zero theorem as a guide we factor the numerator.

$$= \frac{(k + 1)^2(k + 4)}{4(k + 1)(k + 2)(k + 3)} = \frac{(k + 1)(k + 4)}{4(k + 2)(k + 3)} \checkmark$$

23. Show true for $n = 1$: $\frac{1}{1\cdot 4\cdot 7} = \frac{1(8)}{8(4)(7)}$; $\frac{1}{28} = \frac{1}{28}$ √

Find goal statement:

$$\frac{1}{1\cdot 4\cdot 7} + \frac{1}{4\cdot 7\cdot 10} + \frac{1}{7\cdot 10\cdot 13} + \cdots + \frac{1}{(3(k+1)-2)(3(k+1)+1)(3(k+1)+4)}$$
$$= \frac{(k+1)(3(k+1)+5)}{8(3(k+1)+1)(3(k+1)+4)} = \frac{(k+1)(3k+8)}{8(3k+4)(3k+7)}$$

Assume true for $n = k$:

$$\frac{1}{1\cdot 4\cdot 7} + \frac{1}{4\cdot 7\cdot 10} + \frac{1}{7\cdot 10\cdot 13} + \cdots + \frac{1}{(3k-2)(3k+1)(3k+4)} = \frac{k(3k+5)}{8(3k+1)(3k+4)}$$

$$\frac{1}{1\cdot 4\cdot 7} + \frac{1}{4\cdot 7\cdot 10} + \frac{1}{7\cdot 10\cdot 13} + \cdots + \frac{1}{(3k-2)(3k+1)(3k+4)} + \frac{1}{(3k+1)(3k+4)(3k+7)}$$
$$= \frac{k(3k+5)}{8(3k+1)(3k+4)} + \frac{1}{(3k+1)(3k+4)(3k+7)} = \frac{9k^3+36k^2+35k+8}{8(3k+1)(3k+4)(3k+7)}$$

Use the goal statement to help us factor the numerator.

$$= \frac{(k+1)(3k+8)(3k+1)}{8(3k+1)(3k+4)(3k+7)} = \frac{(k+1)(3k+8)}{8(3k+4)(3k+7)}\ \surd$$

25. Goal statement:

$$1 + 3 + 5 + \cdots + (2(k+1)-1) = \frac{(k+1)^2+(k+1)}{2} = \frac{k^2+3k+2}{2}$$

Assume true for $n = k$, then add the next term to both members.

$$1 + 3 + 5 + \cdots + (2k-1) = \frac{k^2+k}{2}$$

$$1 + 3 + 5 + \cdots + (2k-1) + (2(k+1)-1) = \frac{k^2+k}{2} + (2(k+1)-1)$$

The left side is now the left side of the goal statement; we must show that the right side is the same as the right side of the goal statement.

$$= \frac{k^2+k}{2} + \frac{2(2k+1)}{2} = \frac{k^2+5k+2}{2}$$

This expression is clearly not the same as the goal expression.

27. Given the sum of the first n terms of an arithmetic series

$S_n = a_1 + a_2 + a_3 + \cdots + a_n$, $S_n = \frac{n}{2}(a_1 + a_n)$.

a. $a_1 = 1$, $a_n = (2n-1)$, so $S_n = \frac{n}{2}[1 + (2n-1)] = \frac{n}{2}(2n) = n^2$

b. $a_1 = 4$, $a_n = (6n-2)$, so $S_n = \frac{n}{2}[4 + (6n-2)] = \frac{n}{2}(6n+2) = n(3n+1) = 3n^2 + n$

Solutions to skill and review problems

1. $1 + 4 + 7 + \cdots + (3n-2) + [3(n+1)-2]$

$$= \frac{n(3n-1)}{2} + [3(n+1)-2]$$
$$= \frac{n(3n-1)}{2} + (3n+1)$$
$$= \frac{(n+1)(3n+2)}{2}$$

2. $3x - 4y = 12$

Straight line.

Intercepts:

$x = 0$: $y = -3$; $(0,-3)$

$y = 0$: $x = 4$; $(4,0)$

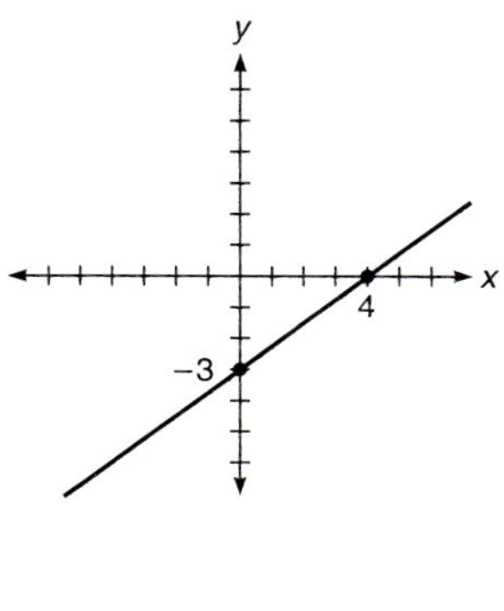

3. $3x^2 - 4y^2 = 12$

$$\frac{x^2}{4} - \frac{y^2}{3} = 1$$

$a = 2$, $b = \sqrt{3}$, $c = \sqrt{7}$

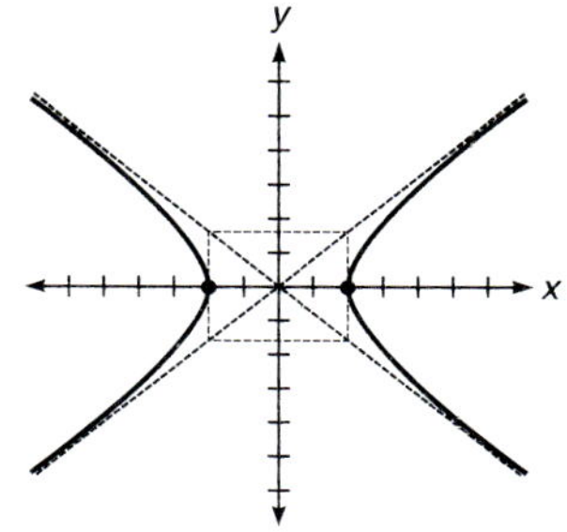

4. $2x - 3y \le 12$
$x + 2y \ge 4$
Graph the straight lines $2x - 3y = 12$ and $x + 2y = 4$. Use (0,0) as a test point to find the appropriate half-planes.

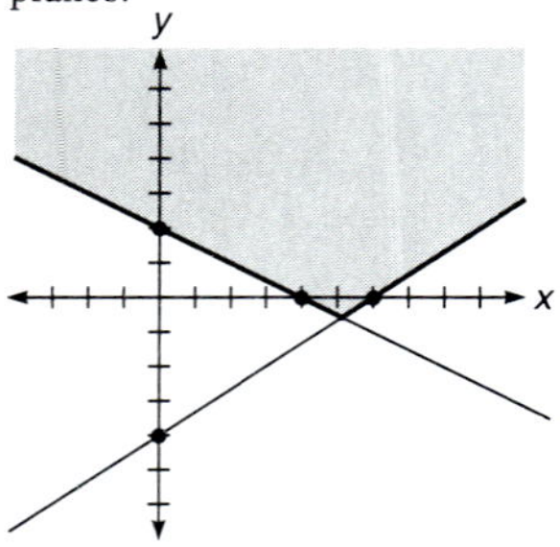

5. $2x - 1 > \dfrac{5}{x + 1}$
This is a nonlinear inequality. Use the critical point/test point method.
Critical points
Solve the corresponding equality:
$2x - 1 = \dfrac{5}{x + 1}$
$(2x - 1)(x + 1) = 5$
$2x^2 + x - 6 = 0$
$(2x - 3)(x + 2) = 0$
$x = \frac{3}{2}$ or -2
Find zeros of denominators:
$x + 1 = 0$; $x = -1$
Critical points are $-2, -1, 1\frac{1}{2}$.

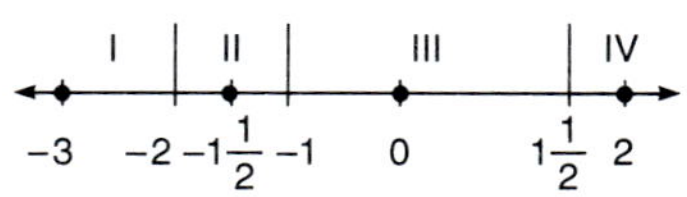

Use $-3, -1.5, 0, 2$ for test points.
$2x - 1 > \dfrac{5}{x + 1}$
$x = -3$: $-7 > -2\frac{1}{2}$; false
$x = -1.5$: $-4 > -10$; true
$x = 0$: $-1 > 5$; false
$x = 2$: $3 > 1\frac{2}{3}$; true
Thus the solution is intervals II and IV: $-2 < x < -1$ or $x > 1\frac{1}{2}$.

Solutions to trial exercise problems

12. Show that $(1 + a)^n \ge 1 + na$ for any natural number n, assuming $a \ge 0$.
Show true for $n = 1$:
$1 + a \ge 1 + a$ √
Find the goal statement:
Replace n by $k + 1$
[2] $(1 + a)^{k+1} \ge 1 + (k + 1)a$
Assume true for $n = k$:
(Replace n by k)
[1] $(1 + a)^k \ge 1 + ka$
We can achieve the left side of statement [2] by multiplying both members of statement [1] by $(1 + a)$. We know $1 + a$ is nonnegative, which is important when multiplying the members of an inequality.
We start with statement [1], which we know to be true.
[1] $(1 + a)^k \ge 1 + ka$
$(1 + a)(1 + a)^k \ge (1 + a)(1 + ka)$
$(1 + a)^{k+1} \ge 1 + ka + a + ka^2$
Now, $1 + ka + a + ka^2 \ge 1 + ka + a$ (since $k > 0$, $a^2 \ge 0$), so $(1 + a)^{k+1} \ge 1 + ka + a + ka^2 \ge 1 + ka + a$, so
[2] $(1 + a)^{k+1} \ge 1 + ka + a$ is true. √

28. Observe that the statement about finding the light coin among five coins includes the assumption that the light coin is actually among the five coins. This is the only basis for selecting the fifth coin when we reject four of the coins.

In this light consider six coins. We group the six coins into two groups: 5 coins and 1 coin. On the basis of our hypothesis we can actually find the light coin among the five coins in two weighings only if we already know that the light coin is among these five coins. Unfortunately the light coin could be in the group which contains one coin. Thus, in the case of six coins we do not meet the hypothesis we required for five coins.

Ironically, two weighings will suffice to find the light coin among six coins, or even seven coins, but not necessarily using the method suggested above for six coins (try to see how). However, two weighings will not suffice for eight coins, which could be shown by trying all possible combinations of weighings of up to eight coins, remembering that no purpose is served by weighing unequal numbers of coins.

Exercise 12–5

Answers to odd-numbered problems

1. a.

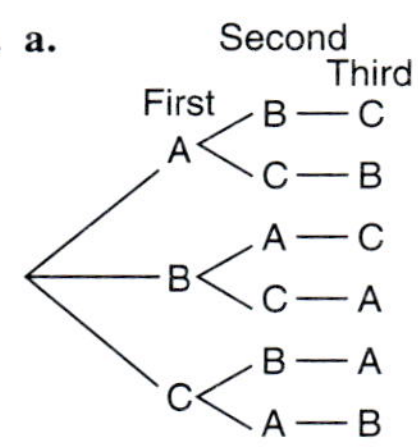

b. ABC, ACB, BAC, BCA, CBA, CAB

3. a.

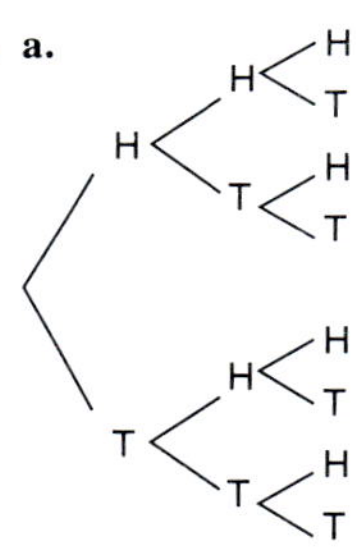

b. HHH, HHT, HTH, HTT, THH, THT, TTH, TTT

5. 24 **7.** 16,807 **9.** 180 **11.** 256 **13.** 5,040 **15.** 30,240 **17.** 220 **19.** 360 **21.** 720 **23.** 13,366,080 **25.** 20 **27.** By definition, ${}_nP_r = n(n - 1)(n - 2) \cdot \cdots \cdot (n - (r - 1))$ so ${}_nP_n = n(n - 1)(n - 2) \cdots (n - (n - 1)) = n(n - 1)(n - 2) \cdots (1) = n!$ **29.** 504 **31.** 210 **33.** 5,040 **35.** 990 **37.** 1,816,214,400 **39.** 9,979,200 **41.** 83,160 **43.** 3,003 **45.** 56 **47.** 91 **49.** By definition, ${}_nC_n = \dfrac{n!}{n!(n - n)!} = \dfrac{n!}{n!0!} = \dfrac{n!}{n!} = 1$ **51.** 4,845 **53.** 495 **55.** 28 **57.** 35

59. 1,001 **61. a.** 24,310 **b.** 272 **c.** 8,821,612,800 **d.** 6,188 **63.** 1,024 **65. a.** 60 **b.** 36 **c.** 6 **d.** 6 **67.** 90 **69. a.** 179,520 **b.** 924 **c.** 593,775 **71.** 32,640 **73.** 169,344

75. $\binom{n}{r}\binom{r}{k}$

$= \dfrac{n!}{r!(n-r)!} \cdot \dfrac{r!}{k!(r-k)!}$

$= \dfrac{n!}{(n-r)!k!(r-k)!}$

$\binom{n}{k}\binom{n-k}{r-k}$

$= \dfrac{n!}{k!(n-k)!} \cdot$

$\dfrac{(n-k)!}{(r-k)![(n-k)-(r-k)]!}$

$= \dfrac{n!}{k!(r-k)!(n-r)!}$

77. Let $n = 10$, $r = 3$.

${}_nP_r = n(n-1)(n-2)\cdots$
$(n-[r-1])$
$n-(r-1) = 8$, so ${}_nP_r$ becomes ${}_{10}P_3$
$= 10 \cdot 9 \cdot 8$.

$\dfrac{n!}{(n-r)!}$ becomes $\dfrac{10!}{(10-3)!}$

$= \dfrac{10 \cdot 9 \cdot 8 \cdot 7!}{7!} = 10 \cdot 9 \cdot 8.$

79. $\dfrac{{}_nP_r}{r!}$

$= \dfrac{n(n-1)(n-2)\cdots(n-[r-1])}{r!}$,

so

$\dfrac{{}_{10}P_3}{3!} = \dfrac{10 \cdot 9 \cdot 8}{3 \cdot 2 \cdot 1}$

Next, $\dfrac{n!}{r!(n-r)!}$

$\dfrac{10!}{3! \cdot 7!} = \dfrac{10 \cdot 9 \cdot 8 \cdot 7!}{3 \cdot 2 \cdot 1 \cdot 7!}$

$= \dfrac{10 \cdot 9 \cdot 8}{3 \cdot 2 \cdot 1}$

Solutions to skill and review problems

1. Infinite geometric series; $a = r = \frac{1}{2}$. Since

$|r| < 1, S_\infty = \dfrac{a}{1-r} = \dfrac{\frac{1}{2}}{1-\frac{1}{2}} = 1.$

2. Arithmetic series; $a_1 = 2$, $d = 5$. To find out how many terms proceed as follows:

$2 + 7 + 12 + 17 + \cdots + 97$

Subtract 2 from each term.

$0 + 5 + 10 + 15 + \cdots + 95$

Divide each term by 5.

$0 + 1 + 2 + 3 + \cdots + 19$

There are thus 20 terms.

$S_n = \dfrac{n}{2}(a_1 + a_n)$

$S_{20} = \dfrac{20}{2}(2 + 97) = 10(99) = 990$

3. $\left|\dfrac{2-3x}{4}\right| \le 10$

$-10 \le \dfrac{2-3x}{4} \le 10$

$-40 \le 2 - 3x \le 40$

$-42 \le -3x \le 38$

$\dfrac{-42}{-3} \ge \dfrac{-3x}{-3} \ge \dfrac{38}{-3}$

$14 \ge x \ge -12\frac{2}{3}$

$-12\frac{2}{3} \le x \le 14$

4. $f(x) = \dfrac{x^2-1}{x^2-4}$

$y = \dfrac{x^2-1}{x^2-4} = \dfrac{x^2-1}{(x-2)(x+2)}$

$= 1 + \dfrac{3}{x^2-4}$ (Long division.)

Horizontal asymptote: $y = 1$

Vertical asymptotes at $x = \pm 2$

Intercepts:

$x = 0$: $y = \dfrac{-1}{-4} = \dfrac{1}{4}$; $\left(0,\dfrac{1}{4}\right)$

$y = 0$: $0 = \dfrac{x^2-1}{x^2-4}$

$0 = x^2 - 1$

$1 = x^2$

$\pm 1 = x$; $(\pm 1,0)$

Additional points: $\left(\pm 3,1\dfrac{3}{5}\right)$

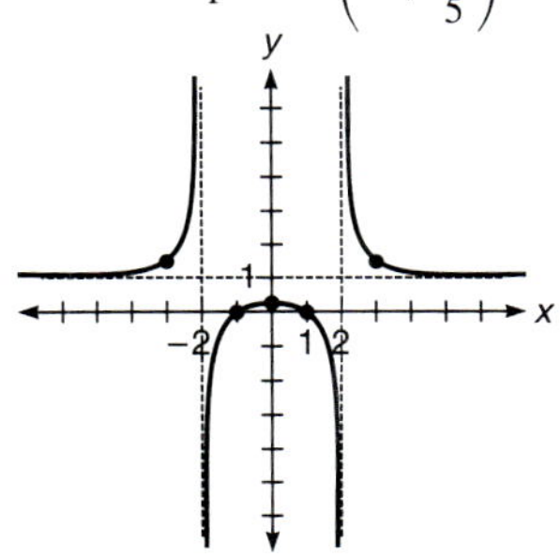

5. $f(x) = (x-2)^2(x+1)^3$

Intercepts:

$x = 0$: $y = (-2)^2(1)^3 = 4$; $(0,4)$

$y = 0$: $0 = (x-2)^2(x+1)^3$

$x = -1$ or 2; $(-1,0)$, $(2,0)$

Additional points: $(-1.5,-1.53)$, $(-0.5,0.8)$, $(1,8)$, $(2.25,2.15)$

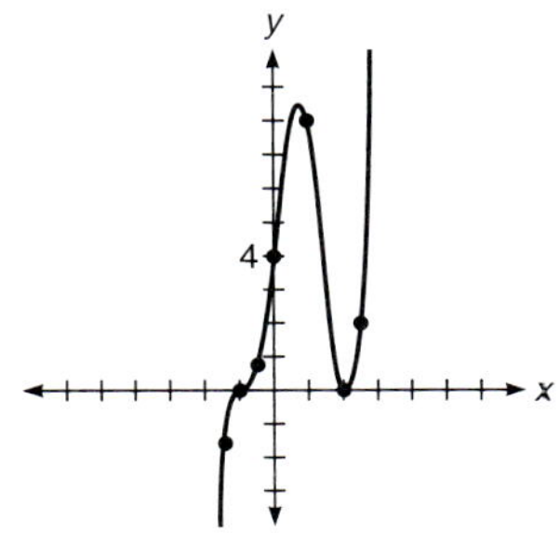

6. $x + \dfrac{1}{x} = 3$

$x^2 + 1 = 3x$

$x^2 - 3x + 1 = 0$

$x = \dfrac{3 \pm \sqrt{5}}{2}$

7. $\dfrac{x+y}{x} = \dfrac{x-y}{2}$

$2(x + y) = x(x - y)$

$2x + 2y = x^2 - xy$

$xy + 2y = x^2 - 2x$

$y(x + 2) = x^2 - 2x$

$y = \dfrac{x^2 - 2x}{x + 2}$

Solutions to trial exercise problems

8. 12 choices for the entrance. After this choice is made there are 11 choices left for the exit: $12 \cdot 11 = 132$

29. Order is important, so we are counting permutations: ${}_9P_3 = 504$

36. There are eight lots and eight houses, so it is ${}_8P_8 = 8! = 40{,}320$. (The side of the street they are on is irrelevant. To see this, try a smaller example, say three houses on three lots—put the lots anywhere you want!)

42. We are interested only in the number of ways to list aabbbbcccccc, which is $\dfrac{11!}{2!4!5!} = 6{,}930$.

56. Once five players are picked for one team, the remaining five are on the other team. The order of selection is not important, so there are ${}_{10}C_5 = 252$ ways.

62. a. Three males and four females are seven people. Thus they can sit in $7! = 5{,}040$ different orders.
b. A female must sit first and last to have alternation. There are $4!$ ways to sit the females and $3!$ ways to sit the males. These orderings of females and males can be selected in $4! \cdot 3! = 144$ ways.
c. We have seven people, of which a group of four and a group of three are indistinguishable, so there are $\frac{7!}{4!3!} =$ 35 distinguishable ways to order them.

64. a. We are not forbidden to repeat digits, so for each of the three digits there are five choices: $5 \cdot 5 \cdot 5 = 125$.
b. A three-digit odd number, with digits selected from this set of digits, ends in 1, 3, or 5. Thus there are only 3 choices for the last digit: $5 \cdot 5 \cdot 3 = 75$. **c.** $2 \cdot 5 \cdot 2 = 20$ **d.** $3 \cdot 3 \cdot 3 = 27$

66. There are ${}_8C_4$ ways to choose the males and ${}_6C_4$ ways to choose the females. For each of the male groups we can choose any of the female groups. Thus there are ${}_8C_4 \cdot {}_6C_4 = 1{,}050$ ways to select the groups.

72. ${}_5C_3 \cdot {}_4C_2 \cdot {}_3C_2 = 180$

76. $\binom{n}{r+1} + \binom{n}{r}$

$$= \frac{n!}{(r+1)![n-(r+1)]!} + \frac{n!}{r!(n-r)!}$$

$$= \frac{n!}{(r+1)![(n-r)-1]!} + \frac{n!}{r!(n-r)!}$$

$$= \frac{(n-r)n!}{(r+1)!(n-r)[(n-r)-1]!} + \frac{(r+1)n!}{(r+1)r!(n-r)!}$$

$$= \frac{(n-r)n!}{(r+1)!(n-r)!} + \frac{(r+1)n!}{(r+1)!(n-r)!}$$

$$= \frac{(n-r)n! + (r+1)n!}{(r+1)!(n-r)!}$$

$$= \frac{n![(n-r)+(r+1)]}{(r+1)!(n-r)!}$$

$$= \frac{n!(n+1)}{(r+1)!(n-r)!} = \frac{(n+1)!}{(r+1)!(n-r)!}$$

and $\binom{n+1}{r+1} =$

$$\frac{(n+1)!}{(r+1)![(n+1)-(r+1)]!}$$

$$= \frac{(n+1)!}{(r+1)!(n-r)!}$$

80. a. In each of the three situations shown find a group of three people who either mutually know each other or who mutually do not know each other.

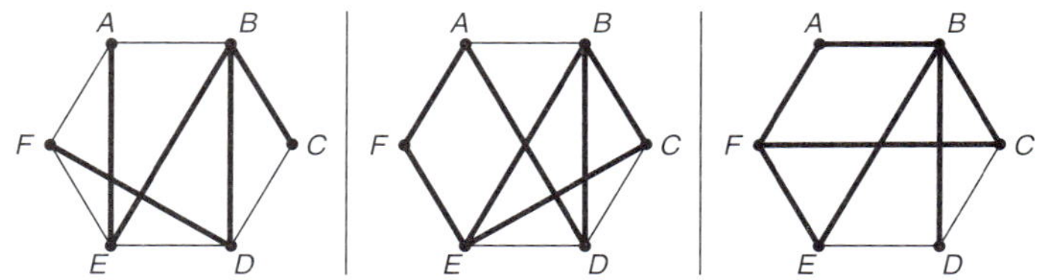

A, C, and D are mutual strangers in the leftmost figure; B, C, and E are mutual acquaintances in the central figure; and A, D, and E are mutual strangers in the rightmost figure.
b. How many groups of three are there, given six people?
There are ${}_6C_3 = 20$ groups, each of which would have to be checked to see if all knew each other or if all were mutual strangers.

Exercise 12–6

Answers to odd-numbered problems

1. $\frac{1}{2}$ **3.** $\frac{3}{4}$ **5.** $\frac{1}{8}$
7. $\frac{1}{2}$ **9.** $\frac{3}{8}$ **11.** $\frac{5}{16}$
13. $\frac{1}{13}$ **15.** $\frac{1}{4}$ **17.** $\frac{6}{13}$
19. $\frac{1}{26}$ **21.** $\frac{1}{13}$ **23.** $\frac{4}{13}$
25. $\frac{7}{13}$ **27.** $\frac{12}{13}$ **29.** $\frac{6}{13}$
31. $\frac{3}{4}$ **33.** $\frac{1}{2}$ **35.** $\frac{1}{38}$
37. $\frac{18}{19}$ **39.** 0 **41.** $\frac{9}{19}$
43. $\frac{7}{12}$ **45.** 0 **47.** $\frac{5}{12}$
49. 0.0113 **51.** 0.1697 **53.** 0.0253
55. 0.2215 **57.** 0.0003 **59.** 0.1496
61. 0.0036 **63.** 0.0095 **65. a.** $\frac{2}{25}$
b. 0.4105 **67.** $\frac{1}{4}$ **69.** $\frac{1}{6}$
71. 0.3543 **73.** 0.777 **75.** 0.777
77. 0.39

Solutions to skill and review problems

1. $S = \frac{1}{2}[a - b(a + c)]$
$2S = a - b(a + c)$
$2S = a - ab - bc$
$2S + bc = a - ab$
$2S + bc = a(1 - b)$
$\frac{2S + bc}{1 - b} = a$

2. $f(x) = x^3 - x^2 - x$
$f(a - 1)$
$= (a - 1)^3 - (a - 1)^2 - (a - 1)$
$= (a^3 - 3a^2 + 3a - 1) - (a^2 - 2a + 1) - (a - 1)$
$= a^3 - 4a^2 + 4a - 1$

3. $y = \frac{1 - 2x}{3}$
$x = \frac{1 - 2y}{3}$
$3x = 1 - 2y$
$2y = 1 - 3x$
$y = \frac{1 - 3x}{2}$
$f^{-1}(x) = \frac{1 - 3x}{2}$

4. $y = \log_4 64$
$4^y = 64$
$y = 3$

5. $\log(x + 3) + \log(x - 1) = 1$
$\log[(x + 3)(x - 1)] = 1$
$\log(x^2 + 2x - 3) = 1$
$x^2 + 2x - 3 = 10^1$
$x^2 + 2x - 13 = 0$
By quadratic formula:
$x = -1 \pm \sqrt{14} \approx -4.7, 2.7$
We select the positive solution since $\log(x - 1)$ is not defined for $x < 1$, since then $x - 1$ is negative.
$x = -1 + \sqrt{14}$

6. $f(x) = \sqrt{x-3} - 1$
This is $y = \sqrt{x}$ shifted right 3 units and down 1 unit. Thus the "origin" shifts to $(3,-1)$.
Intercepts:
$x = 0$: $y = \sqrt{-3} - 1$; No real solution so no y-intercept.
$y = 0$: $0 = \sqrt{x-3} - 1$
$1 = \sqrt{x-3}$
Square both sides.
$1 = x - 3$
$4 = x$; $(4,0)$
Additional points: $(3,-1)$, $(7,1)$, $(12,2)$

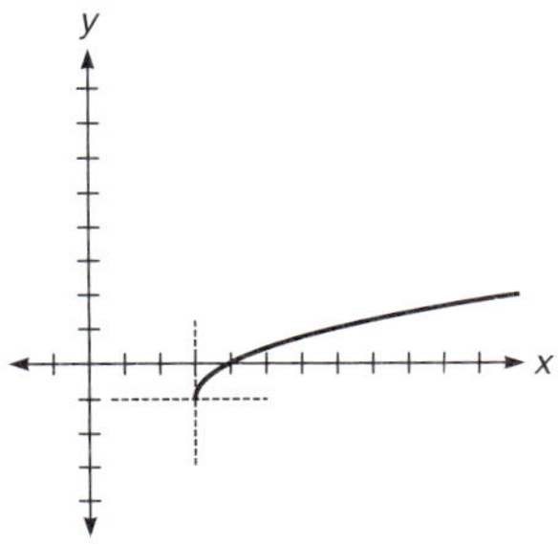

Solutions to trial exercise problems

7. $A = \{HTT, THT, TTH, TTT\}$;
$$P(A) = \frac{n(A)}{n(S)} = \frac{4}{8} = \frac{1}{2}$$

19. There are two red sevens. $P(\text{red seven})$
$$= \frac{\text{number of red sevens}}{\text{number of cards}} = \frac{2}{52} = \frac{1}{26}$$

26. $P(\text{from 5 through 8, inclusive, or a club}) = P(5, 6, 7, 8) + P(\text{club}) - P(5, 6, 7, \text{or 8 of clubs}) =$
$$\frac{16}{52} + \frac{13}{52} - \frac{4}{52} = \frac{25}{52}$$

32. $P(\text{not a heart}) = 1 - P(\text{heart})$
$= 1 - \frac{13}{53} = 1 - \frac{1}{4} = \frac{3}{4}$

38. $P(\text{a black or green number}) = P(\text{black}) + P(\text{green}) = \frac{9}{19} + \frac{1}{19} = \frac{10}{19}$

46. $P(\text{not white})\ 1 - P(\text{white})$
$= 1 - \frac{8}{24} = 1 - \frac{1}{3} = \frac{2}{3}$

50. ${}_{10}C_3 \cdot {}_8C_3 = 6{,}720$ so out of the ${}_{18}C_6$ possible shipments, exactly 6,720 contain 3 new and 3 remanufactured alternators. $P(\text{3 new and 3 manufactured}) = \frac{6{,}720}{{}_{18}C_6} = \frac{6{,}720}{18{,}564}$
≈ 0.3620

56. $P(\text{none of the cards are face cards})$
$$= \frac{\text{number of ways to choose five non face cards}}{\text{number of ways to choose five cards}}$$
$$= \frac{{}_{40}C_5}{{}_{52}C_5} \approx 0.2532$$

60. $P(\text{three clubs and two hearts})$
$$= \frac{{}_{13}C_3 \cdot {}_{13}C_2}{2{,}598{,}960} = \frac{22{,}308}{2{,}598{,}960} \approx 0.0086$$

66. a. $\dfrac{\text{number infected units}}{\text{total number of units}} = \dfrac{3}{100}$

b. $\dfrac{\text{number of ways to choose 4 units out of 97 uninfected units}}{\text{number of ways to choose 4 units out of all 100 units}}$
$$= \frac{{}_{97}C_4}{{}_{100}C_4}$$
$$= \frac{3{,}464{,}840}{3{,}921{,}225} \approx 0.8836$$

68. There are $4! = 24$ possible orders in which to see the four patients. This is the sample space. We need to determine in how many permutations A is in position 2. There are $3!$ ways to position patients B, C, and D in the three remaining slots. Thus the probability of one of these orders of patient selection is $\frac{3!}{4!} = \frac{1}{4}$.

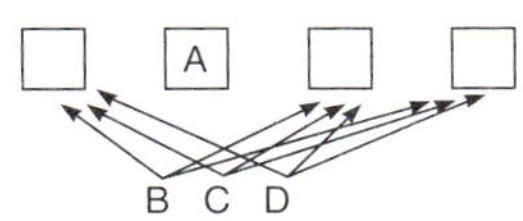

74. This is 1 less the sum of the probabilities of 0 and 1 failures.
$$n = \frac{t}{\text{MTBF}} = \frac{3{,}000}{1{,}000} = 3$$
$$P(\geq 2, 3{,}000) = 1 - P(0, 3{,}000) - P(1, 3{,}000)$$
$$= 1 - \frac{e^{-3} \cdot 3^0}{0!} - \frac{e^{-3} \cdot 3^1}{1!}$$
$$\approx 1 - 0.0498 - 0.1494 \approx 0.801$$

78. Of the 999,500 virus-free people, $0.998 \cdot 999{,}500 = 997{,}501$ will test negative. This means that $999{,}500 - 997{,}501 = 1{,}999$ virus-free individuals will falsely test positive. Of the 500 with the virus, $0.983 \cdot 500 = 492$ will test positive. Thus there are 2,492 positives. Of these the probability of having the virus is $\frac{492}{2{,}491} \approx 0.198$. In other words, if a person tests positive there is about a 20% chance of having the virus.

Exercise 12–7

Answers to odd-numbered problems

1. 3, 8, 13, 18, 23; $a_n = 5n + 3$

3. 5, 10, 20, 40, 80; $a_n = 5 \cdot 2^n$

5. −2, 3, 0, 9, 18; $a_n = \frac{1}{4}(3^n) - \frac{9}{4}(-1)^n$

7. 3, 1, 15, 49, 207; $a_n = \frac{4}{5}(4^n) - \frac{11}{5}(-1)^n$

9. −2, 4, 0, 24, 72; a_n
$$= \left(\frac{14 - 2\sqrt{33}}{2\sqrt{33}}\right)\left(\frac{3 + \sqrt{33}}{2}\right)^n + \left(\frac{-14 - 2\sqrt{33}}{2\sqrt{33}}\right)\left(\frac{3 - \sqrt{33}}{2}\right)^n$$

11. $a_n = 3^n$

13. $a_n = 4 - n$

15. $a_n = \frac{1}{2}(-1)^n + n + \frac{1}{2}$

17. With a recursive definition, to compute a_n we need to first find some or all of the previous terms, $a_0, a_1, \ldots, a_{n-1}$.

19. An arithmetic sequence is a sequence in which $a_{n+1} - a_n = d$ for all n in the domain of the sequence and for some real number d. For the sequence given here we have $a_n = a_{n-1} + 3$ if $n > 0$, so that $a_n - a_{n-1} = 3$ for $n > 0$, or $a_{n+1} - a_n = 3$ for $n > 1$. It is easy to verify that $a_{n+1} - a_n = 3$ for $n = 1$ and $n = 0$, also, so that it is true that $a_{n+1} - a_n = 3$ for all n in the domain of the sequence.

21. -2 or 6

23. The statement we wish to prove is that $a_n = 3^n$ for all $n \in N$,

where $a_n = \begin{cases} 1 \text{ if } n = 0, 3 \text{ if } n = 1 \\ 2a_{n-1} + 3a_{n-2} \text{ if } n > 1 \end{cases}$

Show true for $n = 0, 1$: $a_0 = 1 = 3^0$, so $a_1 = 3 = 3^1$ ✓

Find goal statement: (Replace n by $k + 1$): $a_{k+1} = 3^{k+1}$

Assume true for $n = k$, $k > 1$:

$a_k = 3^k$ for all $n \leq k$, where $k > 1$.

Assume this statement is true.

$a_{k+1} = 2a_k + 3a_{k-1}$

Definition of a_{k+1} for $k > 1$.

$a_k = 3^k$

Assumed true above.

$a_{k-1} = 3^{k-1}$

True because $k - 1 < k$

Replace a_k, a_{k-1} by 3^k, 3^{k-1}:

$a_{k+1} = 2(3^k) + 3(3^{k-1})$
$= 2(3^k) + 3^k$, since $3(3^{k-1}) = 3^k$
$= 3^k[2 + 1]$
$= 3^k(3) = 3^{k+1}$ ✓

Solutions to skill and review problems

1. $a_n = a_1 + (n - 1)d$

$a_1 = 4$: $a_n = 4 + (n - 1)d$

$a_{58} = 67$: $67 = 4 + 57d$, so

$d = \frac{63}{57} = \frac{21}{19}$

a_{96}: $a_{96} = 4 + 95(\frac{21}{19}) = 109$

2. a. There are four choices for the first part, then three choices remain for the second, two choices remain for the third, and there is then one choice for the fourth part: $4 \cdot 3 \cdot 2 \cdot 1 = 24$.

b. ${}_6P_4 = 360$ ways **c.** $5 \cdot 5 \cdot 5 \cdot 5 = 5^4 = 625$

3. $x^2 - 10x - 18 > 6$

Critical points:

$x^2 - 10x - 18 = 6$

$x^2 - 10x - 24 = 0$

$(x - 12)(x + 2) = 0$

$x = -2$ or 12 (Not part of solution set.)

Test points: -3, 0, 13,

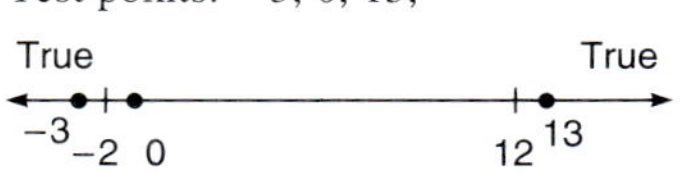

$x^2 - 10x - 18 > 6$

-3: $21 > 6$ (true)

0: $-18 > 6$ (false)

13: $21 > 6$ (true)

Solution: $x < -2$ or $x > 12$

4. This is a geometric series with $a_1 = 1$, $r = \frac{1}{3}$, $n = 6$: $S_n = a_1\frac{1 - r^n}{1 - r}$

$$S_6 = 1\frac{1 - \left(\frac{1}{3}\right)^6}{1 - \frac{1}{3}} = \frac{364}{243} \approx 1.50$$

Solutions to trial exercise problems

5. $a_n = \begin{cases} -2 \text{ if } n = 0, 3 \text{ if } n = 1 \\ 2a_{n-1} + 3a_{n-2} \text{ if } n > 1 \end{cases}$

-2, 3, $2(3) + 3(-2) = 0$, $2(0) + 3(3) = 9$, $2(9) + 3(0) = 18$, . . . or -2, 3, 0, 9, 18, . . .

This sequence is neither geometric nor arithmetic, so we try a recurrence relation.

$a_n = 2a_{n-1} + 3a_{n-2}$

$a_n - 2a_{n-1} - 3a_{n-2} = 0$

$x^n - 2x^{n-1} - 3x^{n-2} = 0$

Replace n by 2.

$x^2 - 2x - 3 = 0$, so $x = 3$ or -1.

Then $a_n = A(3^n) + B(-1)^n$. We find A and B from a_0 and a_1.

$n = 0$: $a_0 = -2 = A + B$

$n = 1$: $a_1 = 3 = 3A - B$

Solving (for example, by adding the two equations we find $1 = 4A$) we find $A = \frac{1}{4}$, $B = -\frac{9}{4}$, so a general term is $a_n = \frac{1}{4}(3^n) - \frac{9}{4}(-1)^n$.

11. $a_n = \begin{cases} 1 \text{ if } n = 0, 3 \text{ if } n = 1 \\ 6a_{n-1} - 9a_{n-2} \text{ if } n > 1 \end{cases}$

$a_n = 6a_{n-1} - 9a_{n-2}$

$a_n - 6a_{n-1} + 9a_{n-2} = 0$

$x^n - 6x^{n-1} + 9x^{n-2} = 0$

$x^2 - 6n + 9 = 0$

$x = 3$ (multiplicity 2)

$a_n = A(3^n) + Bn(3^n)$

$n = 0$: $a_0 = 1 = A$

$n = 1$: $a_1 = 3 = 3A + 3B$, so $B = 0$.

Thus, $a_n = 3^n$.

16. $a_n = \begin{cases} 1 \text{ if } n = 0, 1 \text{ if } n = 1, 3 \text{ if } n = 2 \\ 6a_{n-1} - 12a_{n-2} + 8a_{n-3} \text{ if } n > 2 \end{cases}$

$a_n = 6a_{n-1} - 12a_{n-2} + 8a_{n-3}$

$a_n - 6a_{n-1} + 12a_{n-2} - 8a_{n-3} = 0$

$x^n - 6x^{n-1} + 12x^{n-2} - 8x^{n-3} = 0$

let $n = 3$

$x^3 - 6x^2 + 12x - 8 = 0$

The rational zero theorem (section 4–2) tells us that the only rational zeros are ±1, ±2, ±4, and ±8. Synthetic division or trial and error tells us that 2 is a zero, so the equation is $(x - 2)(x^2 - 4x + 4) = 0$, or $(x - 2)^3 = 0$, so 2 is the only zero, and has multiplicity 3.

$a_n = A(2^n) + Bn(2^n) + Cn^2(2^n)$

$n = 0$: $a_0 = 1 = A$

$n = 1$: $a_1 = 1 = 2A + 2B + 2C$

$n = 2$: $a_2 = 3 = 4A + 8B + 16C$

$A = 1$, $B = -\frac{7}{8}$, $C = \frac{3}{8}$, so

$a_n = 2^n - \frac{7}{8}n(2^n) + \frac{3}{8}n^2(2^n)$ or

$2^n(\frac{3}{8}n^2 - \frac{7}{8}n + 1)$

20. A geometric sequence is a sequence in which $\frac{a_{n+1}}{a_n} = r$ for all n in the domain of the sequence and for some real number r. For this sequence, $a_n = 3a_{n-1}$ for $n > 0$, so $\frac{a_n}{a_{n-1}} = 3$ for $n > 0$, or $\frac{a_{n+1}}{a_n} = 3$ for $n > 1$. It can be verified that this is also true for $n = 0$ and $n = 1$.

21. For the given sequence, $a_2 = 2a_1 + 3a_0 = 2A + 6$. We thus know that $\frac{a_1}{a_0} = \frac{A}{2}$, and $\frac{a_2}{a_1} = \frac{2A + 6}{A}$.

We want these ratios to be equal, so we solve

$\frac{A}{2} = \frac{2A + 6}{A}$

$A^2 = 4A + 12$

$A^2 - 4A - 12 = 0$

$A = -2$ or 6. Both of these values do produce geometric sequences.

Chapter 12 review

1. 4, 10, 16, 22 **2.** 0, $3\frac{1}{2}$, $8\frac{2}{3}$, $15\frac{3}{4}$

3. 0, 1, 4, 9 **4.** $a_n = 3 + (n - 1)(4)$

5. $a_n = -200 + (n - 1)(40)$

6. $a_n = \frac{n + 1}{n}$ **7.** 650 cars

8. geometric sequence; $r = 4$

9. neither **10.** arithmetic sequence; $d = 6$ **11.** 17

12. $c_n = a_n + 2b_n$
$= [a_1 + (n - 1)d_a] + 2[b_1 + (n - 1)d_b]$
$= a_1 + 2b_1 + (n - 1)d_a + 2(n - 1)d_b$
$= (a_1 + 2b_1) + (n - 1)(d_a + 2d_b)$
$= c_1 + (n - 1)d_c$

Thus, c is an arithmetic sequence, and $c_1 = a_1 + 2b_1$ and $d_c = d_a + 2d_b$.

13. 64 **14.** $1\frac{3}{5}$ **15.** 0.01 **16.** 3

17. 10

18. $c_n = a_n(2b_n)$
$= (a_1 r_a^{n-1})(2b_1 r_b^{n-1})$
$= 2a_1b_1(r_ar_b)^{n-1}$
$= c_1 r_c^{n-1}$
(Replace a_1b_1 by c_1, r_ar_b by r_c.)
Thus, c is a geometric sequence in which c_1 is a_1b_1 and r is r_ar_b.

19. a. $\frac{16}{3}$ **b.** $\frac{64}{27}$ **c.** $12 \cdot (\frac{2}{3})^n$

20. $5 + 7 + 9 + 11$ **21.** $\frac{3}{2} + \frac{5}{3} + \frac{7}{4} + \frac{9}{5} + \frac{11}{6}$ **22.** $-4 + 9 - 16 + 25$

23. $0 + (0 + 1) + (0 + 1 + 2)$

24. 867 **25.** -570 **26.** -22

27. 100 **28.** $80\frac{2}{3}$ **29.** $\frac{220}{243}$ **30.** $7\frac{7}{8}$

31. 44,286 **32.** $\frac{1}{2}$ **33.** 4 **34.** $1\frac{4}{5}$

35. $\frac{32}{99}$ **36.** $\frac{104}{333}$ **37.** $\frac{29}{90}$ **38.** 72 meters

39. about 5 hours **40.** 220 **41.** 1,330

42. $\frac{1}{6}(n^3 + 3n^2 + 2n)$

43. $16x^4 - 32x^3y + 24x^2y^2 - 8xy^3 + y^4$

44. $a^{10}b^5 - 15a^8b^4 + 90a^6b^3 - 270a^4b^2 + 405a^2b - 243$ **45.** $70{,}000a^6b^{65}$

46. 400 **47.** 1,530 **48.** $54\frac{122}{243}$

49. $\frac{1}{3}(k^3 + 2k)$

50. $\binom{n}{0} = \dfrac{n!}{(n-0)!0!} = \dfrac{n!}{n!(1)} = 1$

---

51. Case $n = 1$: $3(1) + 1 = \dfrac{1(3(1) + 5)}{2}$

$4 = 4$ ✓

Case $n = k$: $4 + 7 + 10 + \cdots + (3k + 1) = \dfrac{k(3k + 5)}{2}$
(Assume true up to some k).

Case $n = k + 1$: $4 + 7 + 10 + \cdots + (3(k + 1) + 1)$
$= \dfrac{(k + 1)(3(k + 1) + 5)}{2}$
$= \dfrac{(k + 1)(3k + 8)}{2}$ (Goal statement).

Proof for $n = k$:
$4 + 7 + 10 + \cdots + (3k + 1) + (3(k + 1) + 1)$
$= \dfrac{k(3k + 5)}{2} + (3(k + 1) + 1)$
$= \dfrac{k(3k + 5)}{2} + \dfrac{2(3k + 4)}{2}$
$= \dfrac{3k^2 + 11k + 8}{2}$
$= \dfrac{(k + 1)(3k + 8)}{2}$ Right side of goal statement. ✓

52. Case $n = 1$: $1^2 = \dfrac{1(1 + 1)(2(1) + 1)}{6}$

$1 = 1$ ✓

Case $n = k$: $1^2 + 2^2 + 3^2 + \cdots + k^2 = \dfrac{k(k + 1)(2k + 1)}{6}$
(Assume true.)

Case $n = k + 1$: $1^2 + 2^2 + 3^2 + \cdots + (k + 1)^2$
$= \dfrac{(k + 1)((k + 1) + 1)(2(k + 1) + 1)}{6}$
$= \dfrac{(k + 1)(k + 2)(2k + 3)}{6}$ (Goal statement.)

Proof for $n = k$:
$1^2 + 2^2 + 3^2 + \cdots + k^2 + (k + 1)^2$
$= \dfrac{k(k + 1)(2k + 1)}{6} + (k + 1)^2$
$= \dfrac{k(k + 1)(2k + 1) + 6(k + 1)^2}{6}$
$= \dfrac{(k + 1)[k(2k + 1) + 6(k + 1)]}{6}$
$= \dfrac{(k + 1)(2k^2 + 7k + 6)}{6}$
$= \dfrac{(k + 1)(2k + 3)(k + 2)}{6}$ ✓

53. Case $n = 1$: $1^3 - 1 = 0$, which is divisible by 3: $0 = 3 \cdot 0$.
Case $n = k$: Assume $k^3 - k = 3n$ for some natural number n.
Case $n = k + 1$: $(k + 1)^3 - (k + 1)$
$= (k + 1)[(k + 1)^2 - 1]$
$= k^3 + 3k^2 + 2k$
$= k^3 - k + 3k^2 + 3k$
$= 3n + 3k^2 + 3k$
$= 3(n + k^2 + k)$ ✓

54. a.

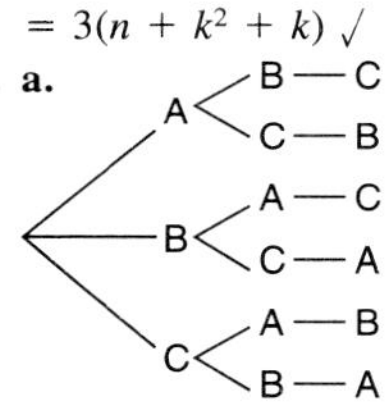

b. ABC, ACB, BAC, BCA, CAB, CBA

55. 24 **56.** 90 **57.** 6,561

58. 1,014 **59.** 90 **60.** 336

61. 132 **62.** 116,280

63. 30 **64.** 153

65. ${}_nC_k = \dfrac{n!}{k!(n - k)!}$

${}_nC_{n-k} = \dfrac{n!}{(n - k)![n - (n - k)]!} = \dfrac{n!}{(n - k)!k!}$

66. 220 **67.** 15 **68.** 35

69. 24 **70. a.** 455 **b.** 5,005 **c.** 210 **d.** 1,816,214,400

71. a. 40,320 **b.** 1,152 **c.** 70

72. 295,245 **73. a.** 1,296 **b.** 648 **c.** 108 **d.** 27 **74. a.** 360 **b.** 180 **c.** 60 **d.** 6

75. 56 **76.** 2,520 **77. a.** 26,400 **b.** 1,134 **c.** 74,613 **78.** 4,480

79. $\frac{1}{4}$ **80.** $\frac{1}{16}$ **81.** $\frac{1}{13}$

82. $\frac{1}{4}$ **83.** $\frac{7}{13}$ **84.** $\frac{1}{26}$

85. $\frac{4}{13}$ **86.** $\frac{11}{26}$ **87.** $\frac{3}{4}$

88. $\frac{2}{7}$ **89.** $\frac{4}{7}$ **90.** $\frac{5}{7}$

91. $\frac{5}{11}$ **92. a.** $\frac{1}{33}$ **b.** $\frac{1}{33}$

93. 0.00050 **94.** 0.0253

95. 0.0253 **96.** 0.3251

97. 0.00000007 **98.** $\frac{1}{3}$

99. 2, 8, 14, 20, 26; $a_n = 6n + 2$
100. 3, 6, 12, 24, 48; $a_n = 3 \cdot 2^n$
101. 2, 3, 8, 19, 46; $a_n = \dfrac{1 + 2\sqrt{2}}{2\sqrt{2}}(1 + \sqrt{2})^n - \dfrac{1 - 2\sqrt{2}}{2\sqrt{2}}(1 - \sqrt{2})^n$
102. 2, 3, 4, 5, 6; $a_n = n + 2$
103. 1, 3, 8, 20, 48; $a_n = 2^n + \dfrac{n}{2}(2^n)$ or $\left(\dfrac{n}{2} + 1\right)2^n$

Chapter 12 test

1. 2, −1, 0, 1 **2.** 0, $1\frac{1}{2}$, $2\frac{2}{3}$, $3\frac{3}{4}$
3. $4n + 2$ **4.** $\dfrac{n+2}{n}$ **5.** 28 **6.** $6\frac{2}{3}$
7. Let $A = 2, 6, 10, 14, \ldots$ be an arithmetic sequence. Then, if $b_n = 3a_n$, $B = 6, 18, 30, 42, \ldots$, which seems to be an arithmetic sequence. Thus, we shall try to show that B is always arithmetic.
$b_n = 3a_n$
$= 3(a_1 + (n - 1)d_a)$
$= 3a_1 + (n - 1)(3d_a)$
$= b_1 + (n - 1)d_b$
Thus B is an arithmetic sequence, where $b_1 = 3a_1$ and $d_b = 3d_a$.
8. −72 **9.** $6\frac{1}{4}$
10. Consider the geometric sequence $A = 1, 2, 4, 8, \ldots$. Then $B = 2, 3, 5, 9, \ldots$, and since the ratio of successive elements is not constant, this is not a geometric sequence. Thus, we cannot conclude that B, where $b_n = a_n + 1$, is necessarily geometric.
11. a. 16 ft **b.** $\frac{64}{9}$ ft **c.** $24(\frac{2}{3})^{n-1}$ ft
12. $1 + \frac{4}{5} + \frac{3}{5} + \frac{8}{17}$
13. $-4 + 1 + 0 + 1 - 4$
14. $1 + (1 + 4) + (1 + 4 + 9)$
15. 22 **16.** 2,684 **17.** $518\frac{1}{3}$
18. $5\frac{31}{216}$ **19.** −170
20. $\dfrac{42{,}753}{100{,}000} \approx 0.42753$ **21.** not defined
22. $13\frac{1}{2}$ **23.** $\frac{3}{11}$ **24.** $\frac{13}{30}$
25. 280 meters **26.** 8 years **27.** 70
28. n
29. $x^8 - 12x^6y + 54x^4y^2 - 108x^2y^3 + 81y^4$
30. $81{,}081a^8b^{50}$ **31.** 1,771
32. $14\dfrac{683}{1{,}024}$ **33.** $k^2 + 2k$ **34.** 12
35. Case $n = 1$: $(4(1) + 1) = 2(1^2) + 3(1)$
$5 = 5$ ✓
Case $n = k$: $5 + 9 + 13 + \cdots + (4k + 1) = 2k^2 + 3k$
Case $n = k + 1$: $5 + 9 + 13 + \cdots + (4(k + 1) + 1)$
$= 2(k + 1)^2 + 3(k + 1)$ (Goal)
$= 2k^2 + 7k + 5$ (Right member expanded.)
Proof for $n = k + 1$:
$5 + 9 + 13 + \cdots + (4k + 1) + (4(k + 1) + 1)$
$= 2k^2 + 3k + (4(k + 1) + 1)$
$= 2k^2 + 7k + 5$ ✓
36. Case $n = 1$: $\dfrac{3}{2^0} = \dfrac{6(2^1 - 1)}{2^1}$; $3 = 3$ ✓
Case $n = k$: $3 + \dfrac{3}{2} + \dfrac{3}{4} + \cdots + \dfrac{3}{2^{k-1}} = \dfrac{6(2^k - 1)}{2^k}$
Case $n = k + 1$: $3 + \dfrac{3}{2} + \dfrac{3}{4} + \cdots + \dfrac{3}{2^{(k+1)-1}} = \dfrac{6(2^{k+1} - 1)}{2^{k+1}}$ (Goal)
Proof for $n = k + 1$:
$3 + \dfrac{3}{2} + \dfrac{3}{4} + \cdots + \dfrac{3}{2^{k-1}} + \dfrac{3}{2^k} = \dfrac{6(2^k - 1)}{2^k} + \dfrac{3}{2^k}$
$= \dfrac{6(2^k) - 3}{2^k} = \dfrac{6(2^k) - 3}{2^k} \cdot \dfrac{2}{2} = \dfrac{6(2)(2^k) - 6}{2^{k+1}} = \dfrac{6(2^{k+1}) - 6}{2^{k+1}} = \dfrac{6(2^{k+1} - 1)}{2^{k+1}}$ ✓
37. $n = 1$: $1^2 + 7 + 12 = 20$, which is divisible by 2. ✓
Assume true for $n = k$; that is, $k^2 + 7k + 12 = 2m$ for some integer m.
For $n = k + 1$: $(k + 1)^2 + 7(k + 1) + 12 = k^2 + 9k + 20 = k^2 + 7k + 12 + 2k + 8 = 2m + 2k + 8 = 2(m + k + 8)$ ✓
38. a.

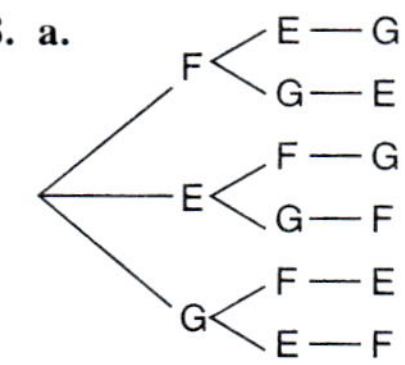

b. FEG, FGE, EFG, EGF, GFE, GEF
39. 120 **40.** 1,048,576 **41.** 24,336
42. 2,730 **43.** 210 **44.** 4,989,600
45. a. $26! \approx 4.0329 \times 10^{26}$
b. 1.28×10^{18} years. (It is thought the universe is less than 20 billion years old; this is 20×10^9 years.) **46.** 27,405
47. $\dfrac{n(n-1)}{2}$ **48.** 45 **49.** 253
50. 96 **51.** 167,960 **52.** 380
53. 6.09×10^{10} **54.** 72,930
55. 860,160 **56.** 720 **57.** 276
58. a. 216,000 **b.** 7,776 **59.** $\frac{3}{8}$
60. $\frac{1}{26}$ **61.** $\frac{8}{13}$ **62.** $\frac{12}{13}$
63. $\frac{8}{11}$ **64.** 0.54 **65.** $\frac{1}{190}$
66. $\dfrac{{}_nC_r}{{}_nC_{r-1}} = \dfrac{\dfrac{n!}{r!(n-r)!}}{\dfrac{n!}{(r-1)![n-(r-1)]!}}$
$= \dfrac{n!}{r!(n-r)!} \cdot \dfrac{(r-1)![n-(r-1)]!}{n!}$
Note that $[n - (r - 1)]! = [n - (r - 1)][n - (r - 1) - 1]! = [n - (r - 1)][n - r]!$
$\dfrac{{}_nC_r}{{}_nC_{r-1}} = \dfrac{(r-1)![n-(r-1)][n-r]!}{r(r-1)!(n-r)!}$
$= \dfrac{n-(r-1)}{r}$
67. 5, 7, 9, 11, 13; $a_n = 5 + 2n$
68. 2, 6, 18, 54, 162; $a_n = 2(3^n)$
69. 2, 4, 6, 8, 10; $a_n = 2(n + 1)$
70. 2, 3, 7, 18, 47; $a_n = \left(\dfrac{3}{2} + \dfrac{\sqrt{5}}{2}\right)^n + \left(\dfrac{3}{2} - \dfrac{\sqrt{5}}{2}\right)^n$

Appendix C

Useful Templates

This appendix includes items that might prove useful to reproduce and have handy. Note that rectangular coordinate graph paper and polar coordinate graph paper are both widely available in book stores.

The Unit Circle

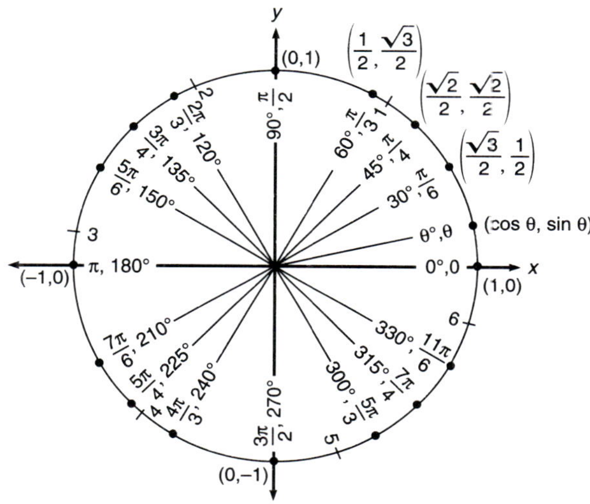

Reference Angles

$0 < \theta < 360°$

$0 < \theta < 2\pi$

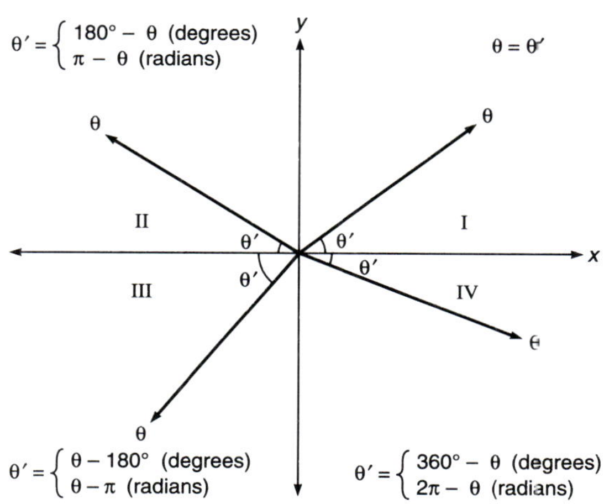

Templates Useful for Graphing Many Trigonometric Functions

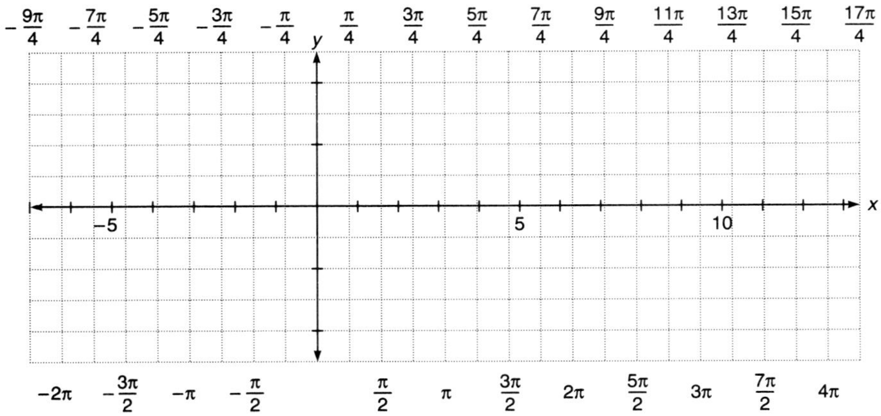

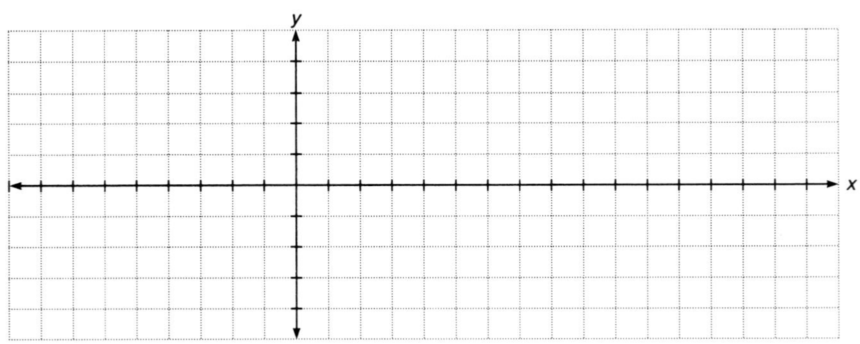

Rectangular Coordinates

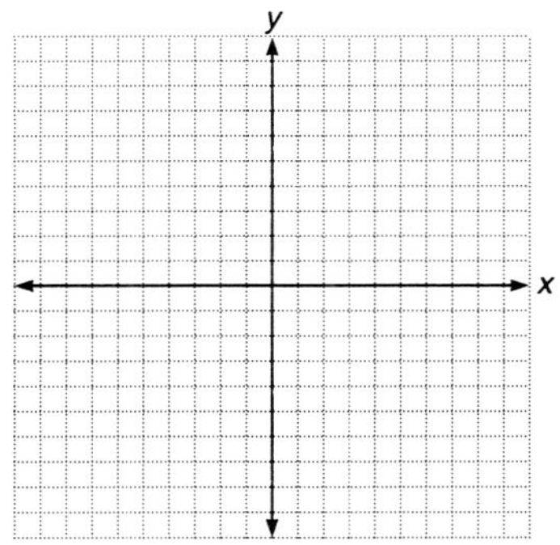

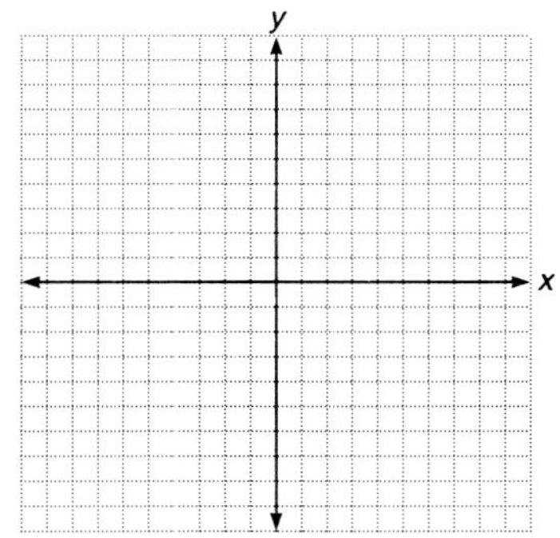

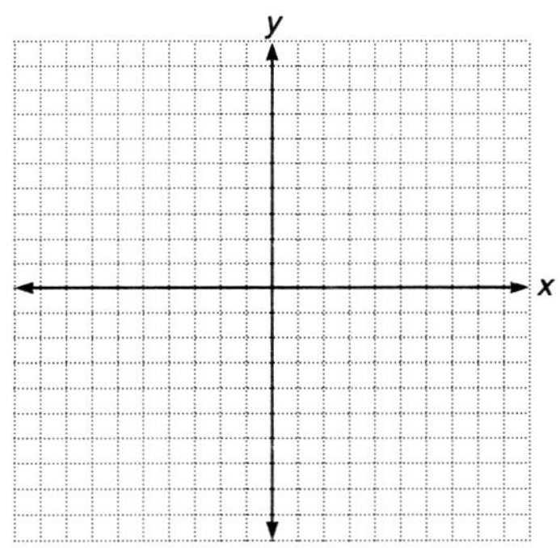

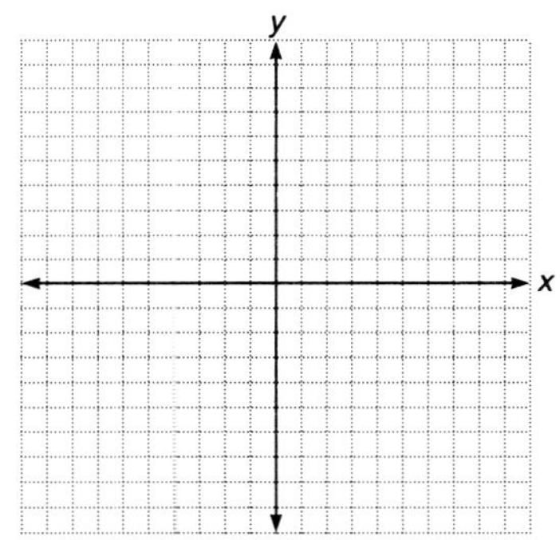

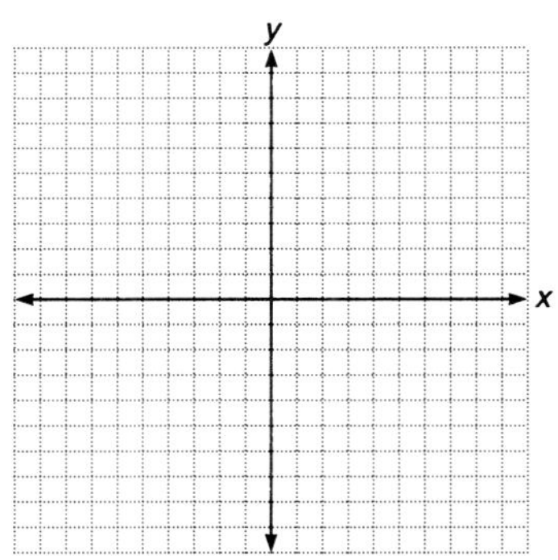

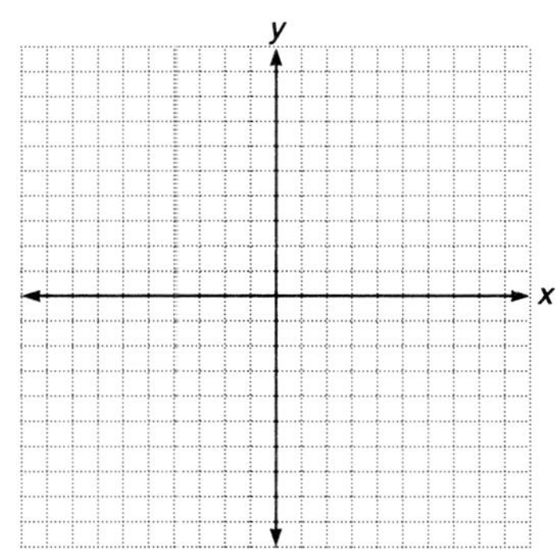

Polar Coordinates

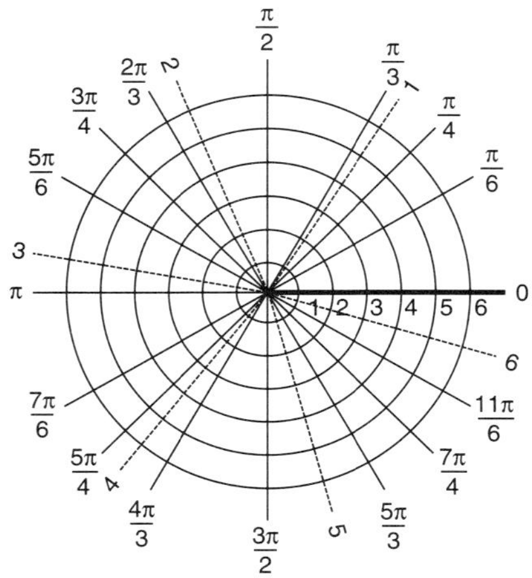

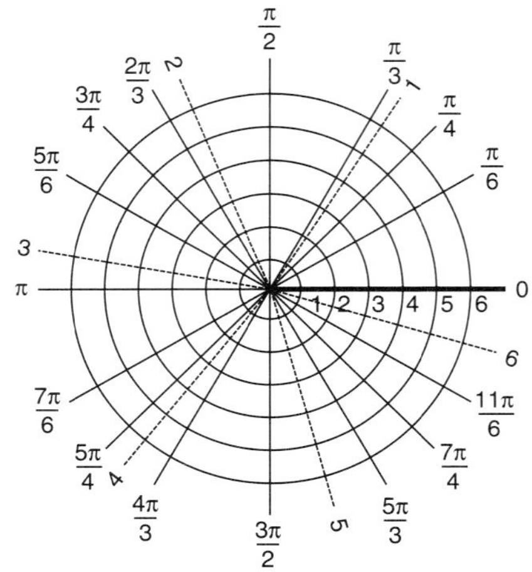

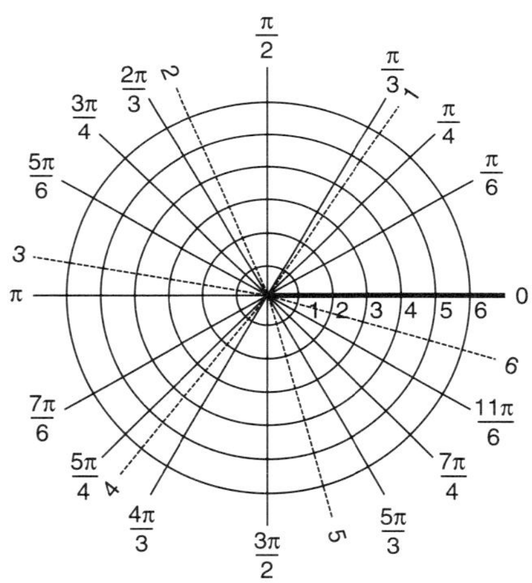

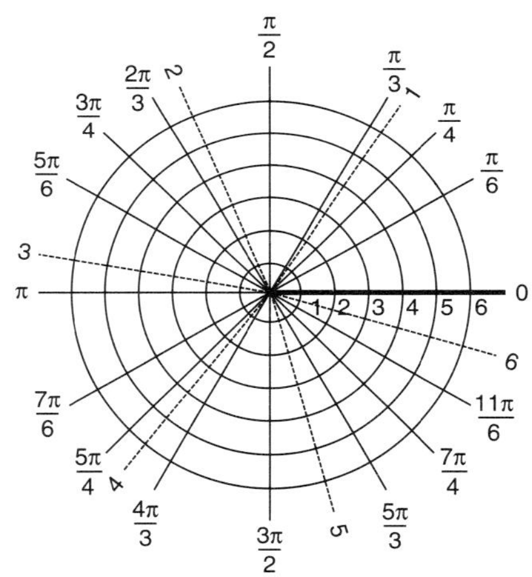

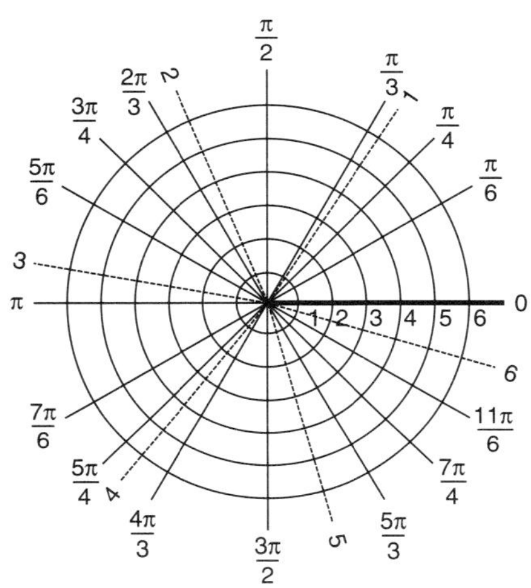

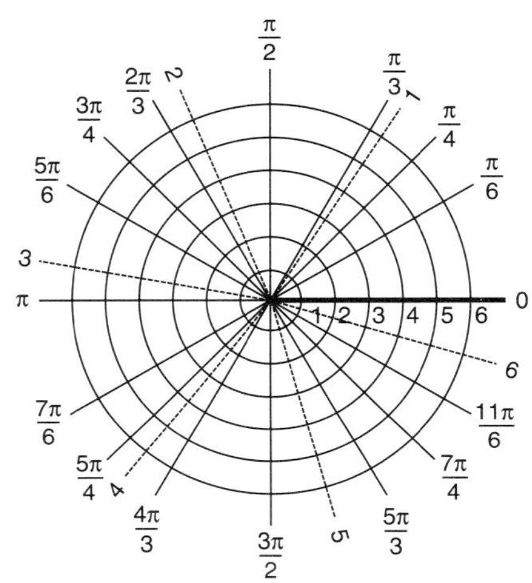

Index of Applications

Engineering

General

Geometry

Government

History

Machining

Mathematics

Medicine

Military Science

Music

Navigation

Physics

Police Science

Psychology

Surveying

Technology

Index

T

U

V

W

Z

The Primary Trigonometric Ratios

If θ is either of the two acute angles in a right triangle, then

$$\sin\theta = \frac{\text{length of side opposite } \theta}{\text{length of hypotenuse}},$$

$$\cos\theta = \frac{\text{length of side adjacent to } \theta}{\text{length of hypotenuse}},$$

$$\tan\theta = \frac{\text{length of side opposite to } \theta}{\text{length of side adjacent to } \theta}.$$

The Trigonometric Functions

Let θ be an angle in standard position, and (x,y) be any point on the terminal side of angle θ, except (0,0). Let $r = \sqrt{x^2 + y^2}$ be the distance from the origin to the point. Then

$$\sin\theta = \frac{y}{r},\ \cos\theta = \frac{x}{r},\ \tan\theta = \frac{y}{x},$$

$$\csc\theta = \frac{r}{y},\ \sec\theta = \frac{r}{x},\ \cot\theta = \frac{x}{y}.$$

Radian Measure

Let θ be an angle in standard position with degree measure $\theta°$ and radian measure s. Then $\dfrac{s}{\pi} = \dfrac{\theta°}{180°}$.

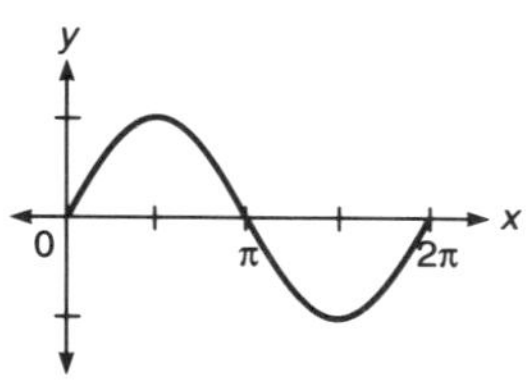

Basic sine cycle

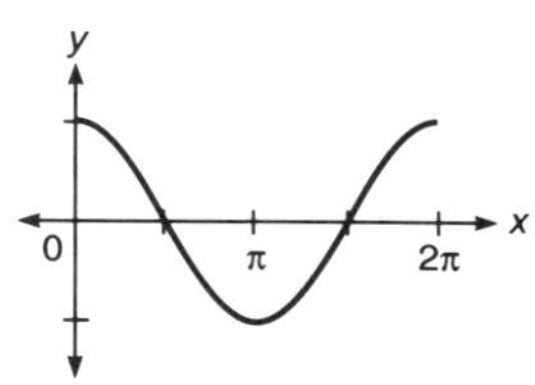

Basic cosine cycle

Inverse Trigonometric Functions

$y = \sin^{-1}x$ means (1) $\sin y = x$

(2) $-\dfrac{\pi}{2} \le y \le \dfrac{\pi}{2}$

(3) $-1 \le x \le 1$

$y = \cos^{-1}x$ means (1) $\cos y = x$

(2) $0 \le y \le \pi$

(3) $-1 \le x \le 1$

$y = \tan^{-1}x$ means (1) $\tan y = x$

(2) $-\dfrac{\pi}{2} < y < \dfrac{\pi}{2}$

$y = \cot^{-1}x$ means (1) $\cot y = x$

(2) $0 < y < \pi$

$$\sec^{-1}x = \cos^{-1}\frac{1}{x} \text{ if } |x| \ge 1.$$

$$\csc^{-1}x = \sin^{-1}\frac{1}{x} \text{ if } |x| \ge 1.$$

$$\cos^{-1}(-x) = \pi - \cos^{-1}x$$

$$\cot^{-1}x = \frac{\pi}{2} - \tan^{-1}x$$

Reciprocal Identities

$$\csc\theta = \frac{1}{\sin\theta},\ \sec\theta = \frac{1}{\cos\theta},\ \cot\theta = \frac{1}{\tan\theta}$$

Tangent, Cotangent Identities

$$\tan\theta = \frac{\sin\theta}{\cos\theta},\ \cot\theta = \frac{\cos\theta}{\sin\theta}$$

Pythagorean Identities

$\sin^2\theta + \cos^2\theta = 1$
$\sec^2\theta = \tan^2\theta + 1$
$\csc^2\theta = \cot^2\theta + 1$

Sum and Difference Identities

$\sin(\alpha + \beta) = \sin\alpha\cos\beta + \cos\alpha\sin\beta$
$\sin(\alpha - \beta) = \sin\alpha\cos\beta - \cos\alpha\sin\beta$
$\cos(\alpha + \beta) = \cos\alpha\cos\beta - \sin\alpha\sin\beta$
$\cos(\alpha - \beta) = \cos\alpha\cos\beta + \sin\alpha\sin\beta$

$$\tan(\alpha + \beta) = \frac{\tan\alpha + \tan\beta}{1 - \tan\alpha\tan\beta}$$

$$\tan(\alpha - \beta) = \frac{\tan\alpha - \tan\beta}{1 + \tan\alpha\tan\beta}$$

Cofunction Identities

$$\sin\left(\frac{\pi}{2} - \theta\right) = \cos\theta,\ \tan\left(\frac{\pi}{2} - \theta\right) = \cot\theta,$$

$$\sec\left(\frac{\pi}{2} - \theta\right) = \csc\theta$$

Double-Angle Identities

$$\sin 2\alpha = 2\sin\alpha\cos\alpha$$

$$\begin{aligned}\cos 2\alpha &= \cos^2\alpha - \sin^2\alpha \\ &= 1 - 2\sin^2\alpha \\ &= 2\cos^2\alpha - 1\end{aligned}$$

$$\tan 2\alpha = \frac{2\tan\alpha}{1 - \tan^2\alpha}$$